Handbook of Validation in Pharmaceutical Processes

Handbook of Validation in Pharmaceutical Processes

Fourth Edition

Edited by

James Agalloco, Phil DeSantis,
Anthony Grilli, and Anthony Pavell

CRC Press is an imprint of the
Taylor & Francis Group, an **informa** business

Fourth edition published 2022
by CRC Press
6000 Broken Sound Parkway NW, Suite 300, Boca Raton, FL 33487–2742

and by CRC Press
2 Park Square, Milton Park, Abingdon, Oxon, OX14 4RN

Third edition published by CRC Press 2008

CRC Press is an imprint of Taylor & Francis Group, LLC

ISBN: 978-0-367-75429-7 (hbk)
ISBN: 978-0-367-75606-2 (pbk)
ISBN: 978-1-003-16313-8 (ebk)

DOI: 10.1201/9781003163138

Contents

About the Editors

James Agalloco is President of Agalloco & Associates, a technical service firm to the pharmaceutical and biotechnology industry. He was previously Director, Worldwide Validation and Automated Technology for Bristol-Myers Squibb. He earned his BS in Chemical Engineering from Pratt Institute in 1968 and his MS also in Chemical Engineering from Polytechnic Institute of New York in 1979. He earned his MBA in Pharmaceutical Studies from Fairleigh Dickinson University in 1983. He is a past president of the Parenteral Drug Association and served as an officer or director from 1982 to 1993. He is a member of USP's Microbiology and Sterility Assurance Expert Committee for 2006–2025. He is a frequent author and lecturer on sterilization, aseptic processing, and process validation.

Phil DeSantis is a pharmaceutical consultant with more than 50 years of pharmaceutical industry experience. He specializes in Pharmaceutical Engineering and Compliance. Phil is a chemical engineer, with a BSChE from the University of Pennsylvania and an MSChE from New Jersey Institute of Technology. He has served as Chair of the PDA Science Advisory Board, is active in ISPE, and has been a frequent lecturer for both organizations. Phil has published or contributed to numerous articles, technical reports, and books in the area of validation and pharmaceutical engineering. In addition, he has lectured on "Steam and Dry Heat Sterilization" as part of the FDA's field investigator training program and for Health Canada on sterilization and critical utilities.

Anthony Grilli is President and Owner of FOCUS Laboratories, a group of contract laboratories serving pharmaceutical, medical device, cosmetic, and other life science industries with cGMP compliant microbiology and analytical chemistry support. Previous to starting his own laboratory network, he managed several other laboratory operations including four SGS North America sites and two Celsis Laboratory Group sites. With more than 25 years of contract laboratory experience, Tony has accumulated a unique perspective from helping hundreds of companies through various microbial validation and remediation concerns. He has successfully hosted dozens of FDA and hundreds of customer audits. Tony has a special interest in validating rapid microbiological methods and has successfully implemented rapid sterility test methodology for the compounding pharmacy industry. He has a Master's degree in Microbiology from Rutgers University and is active in PDA and ASTM committees.

Anthony Pavell is Plant Manager at Fresenius Kabi USA Grand Island New York facility. He has more than 25 years of experience directing production operations, technology transfer of products, and processes into new facilities. He has held an Operations position with Fresenius Kabi, Catalent, Technical Services, and Validation positions with IMA Life (formerly MKCS Inc.), Cardinal Health, Lloyd's Register Serentec, Novo Nordisk, and Burroughs Wellcome. Tony earned a Master of Science in Molecular Biology and Biotechnology from East Carolina University in 1992. He has served on PDA task forces, published articles related to validation, and is a member of the International Society of Pharmaceutical Engineers and the Parenteral Drug Association.

Contributors

James Akers
Akers Kennedy & Associates

Sidney Backstrom
G-CON Manufacturing Inc.

Suraj B. Baloda
Sarmicon LLC

Michael C. Beckloff
Azurity Pharmaceuticals

Francis E. Beideman
Self-employed

Thomas J. Berger
Abbott Laboratories (retired)

Robert Bottome
BioMarin Pharmaceuticals

Rebecca Brewer
Quality Executive Partners

Göran Bringert
Kaye Instruments (retired)

Joseph P. Brower
Aseptic Process Solutions

Allen Burgenson
Lonza Inc.

Timothy S. Charlebois
Pfizer Inc.

Warren Charlton
WHC Bio Pharma Inc.

William V. Collentro
Pharmaceutical Water Specialties LLC

Mark A. Czarneski
Clordisys Solutions Inc.

Don DeCou
West Pharmaceutical Services

Trevor Deeks
Deeks Pharmaceutical Consulting Services LLC

Franco De Vecchi
VPCI Inc.

Denis Drapeau
Pelham Process Sciences Institute

Norman Elder
EQC Consulting

Merrick Endejann
Cellectis, Inc.

Jeff Eshelman
Fresenius-Kabi USA LLC

Karlin Gardner
Mesa Laboratories

Mark Hallworth
Particle Measuring Systems

Crystal Hostler
Mesa Laboratories

Ajaz Hussain
Long Island University

Takuji Ikeda
NITTA Corporation

Yasuhito Ikematsu
Osaka University

Thomas Ingallinera
Pharmaceutics International Inc.

Geoffrey P. Jacobs
Dr. Geoffrey P. Jacobs Associates

Günter Jagschies
Gemini BioProcessing

Kevin M. Jenkins
Quality Excellence Consulting

Maik Jornitz
G-CON Manufacturing Inc.

Charles Levine
Levine Pharmaceutical Consulting

Howard L. Levine
BDO USA Inc.

William G. Lindboe Jr.
Self-employed

Paul Lorchiem
Clordisys Solutions Inc.

Yusuke Matsuda
Taiho Pharmaceutical Co. Ltd

Steven I. Max
The Janssen Company

David Maynard
Maynard & Associates LLC

Kurt McCauley
Mesa Laboratories

Karen Zink McCullough
MMI Associates LLC

Roger D. McDowall
R.D. McDowall Ltd

Robert Mello
Mello PharmAssociates. LLC

Kiyoshi Mochizuki
Xpro Associates, LLC

Jeanne Moldenhauer
Excellent Pharmaceutical Consulting

Mitsuo Mori
Kyowa Kirin Co., Ltd

Hiroaki Nakamura
Earth Environmental Service Co., Ltd

Keisuke Nishikawa
Bayer Yakuhin, Ltd

Steven Ostrove
Ostrove Associates Inc.

William Parker
West Pharmaceutical Services

Anthony Pavell
Fresenius-Kabi USA LLC

Derek Pendlebury
Colder Products Corporation

Frank Riske
BDO USA Inc.

Nicole Robichaud
Mesa Laboratories

George Sheaffer
GES Consulting LLC

John Shirtz
Baxter Healthcare (retired)

Chris Smalley
ValSource Inc.

Satoshi Sugimoto
Takeda Pharmaceutical Company Ltd

Saeed Tafreshi
Intelitec Inc.

Steven C. Tarallo
Lyne Laboratories

Anthony Thatcher
Cellectis, Inc.

Anne B. Tolstrup
AbtBioConsult

A. Mark Trotter
Trotter Biotech Solutions Inc.

Kevin D. Trupp
Abbott Laboratories (retired)

Arthur Vellutato, Jr.
Veltek Asssociates Inc.

Kishore Warrier
Stantec Consulting

Preface to the Fourth Edition

The science of validation has made great strides since the publication of the Third Edition in 2008. Although that volume has remained an important reference for more than a decade, the world of pharmaceutical manufacturing has changed. Advanced therapies, new equipment and process technologies, novel biological pathways and even new ways to build facilities have emerged. Pharmaceutical manufacturing has expanded and validation along with it. Perhaps most significant has been the elevation of validation from a regulatory expectation to a foundational quality system, even more important than the age-old pillars such as documentation, training and change control. Validation has become the essence of the manufacturer's goal, that is to consistently and reliably deliver safe and effective drugs and biologics to the patient population. As it is practiced today, validation embodies so much more than "proof that a process does what it purports to do." From a life-cycle perspective, validation requires an understanding of process and its relation to product quality. Further, validation defines the controls necessary to ensure a steady output of product of the highest level of quality, meeting the expectations of the regulatory bodies, healthcare practitioners and, most importantly, patients.

In this Edition, the editors have carried on the tradition established in the First Edition by bringing together widely recognized experts in the field to address the many topics that contribute to reliable manufacturing processes. Some of the authors are veterans of the book and have provided new and expanded perspectives. Some of the chapters are presented for the first time, keeping pace with important developments in the industry. The basics remain, including metrology, sterilization, aseptic processing, cleaning, critical utilities, nonsterile dosage forms, as well as many others. New subjects include leading edge developments in cell and gene therapy and single-use technologies.

The book is not focused on regulation or colored by the specter of regulatory expectation. It is a practical guide to what is necessary to develop, understand and control pharmaceutical processes with the goal of product quality. In doing so it provides sound advice to satisfy the prevailing regulations as well as the needs of the patients we serve. Its pages are full of the science upon which our industry is based, so necessary to be able to establish and maintain the state of validation we all seek.

Some four decades since the original publication, we hope this version will become both a handbook and a textbook, not only for those of us who identify as "validation professionals" but for all industry professionals who will find it both interesting and informative. For those of us who entered into the practice of validation more than 40 years ago not knowing exactly what was expected and where it would lead, we think that this book provides many of the answers.

Senior Editor's Note – A History of the Fourth Edition

There was never a timetable for the development of this Edition of *Handbook of Validation in Pharmaceutical Processes.* I had always thought it would be something I would work on during retirement. I started actively working on it in early-2018, thinking I had ample time. I enlisted three experienced colleagues to assist me, Phil DeSantis, Anthony Grilli and Anthony Pavell to ease the workload and broaden the perspective. In August 2018, emergency triple bypass surgery was a wake-up call. Seeing the healthcare system from the patient's perspective brought a new urgency to the effort. Despite the excellent prognosis I've received from my physicians I knew it was time to get serious and complete the revision process. The COVID-19 pandemic reenforced the message. Our efforts in the manufacture of healthcare products are vital for global health, and the development of this revised text may help.

Each edition of this text has expanded on the subject. This reflects the increasing reliance on validation as a means for project implementation within the global healthcare industry. The original list of potential chapters for this book considered almost 100 subjects. Triage was necessary, and some of that was easy when we were unable to find suitable authors, or when the authors we identified declined to contribute. Other topics were dropped as too limited in scope or better addressed as a part of something else. What remains is a comprehensive treatment of the field, including the most relevant subjects. We also endeavored to include the latest thinking in what has been an ever-changing subject.

Having operated a single person consulting firm for many years the process of retirement is rather different. There will be no farewell tour or visits with colleagues, no farewell dinner, no funny gifts, and certainly no parting address. I have chosen to use this opportunity as a means to acknowledge those who both shaped my career and who in large part made this book possible. So in lieu of a retirement speech that I will never give, I offer the following:

I have to begin with Fred Carleton, co-editor of the first three editions. It was his initial vision in the early 1980s, that 'Validation' was an important activity, and it would be beneficial to develop a summary of practices that would aid others in executing it with the appropriate scientific rigor. The earlier editions of 1986, 1995 and 2007 and the many citations to these in the literature are a testament to the accuracy of Fred's initial vision. Fred was a serial influencer, and he introduced me, and others. to the Parenteral Drug Association (PDA) which led to countless friendships and numerous collaborations (more on this later). Within PDA, Fred pressed for the establishment of an Education Committee which would provide a means for disseminating the collective knowledge of the industry. Unsurprisingly, many of the initial course offerings related to validation. This was a near immediate success as there seemed to be a near insatiable appetite across the parenteral industry for education in this emerging area. I plunged in willingly, as a committee member, course instructor, and later as Fred's replacement chairing the Education Committee. The various editions are an extension of the educational activity we both supported. Fred remains an inspiration to me and others to this day.

PDA offered peer-to-peer interaction on an unparalleled level. In a field that was expanding in scope and importance almost daily it was extremely challenging to identify 'best practices.' It could rightly be said, "We're making it up as we go along!" As a young manager responsible for validation at a major manufacturer, there was no where else to go to learn how others were dealing with the same validation questions that I was. PDA leaders and members, Michael Anisfeld, Ken Avis, Joyce Aydlett, Hal Baseman, Tom Berger, Ken Chapman, Jack Cole, Doris Conrad, Franco DeVecchi, Jim Fernandez, Ed Fry, John Gillis, Anthony Grilli, Gabriele Gori, Klaus Haberer, Lara Iovino, Richard Johnson, Maik Jornitz, Robert Keiffer, Clarence Kemper, Michael Korczynski, Carol Lampe, Richard Levy, John Lindsay, Nate Manco, Ted Meltzer, Jeanne Moldenhauer, Ted Odlaug, Gordon Personeus, Sol Pflag, Irving Pflug, Ed Smith, Suzanne Stone, Ed Tidswell and Ed Trappler. Art Vellutato, along with many others, shared their insights and knowledge freely to the membership and the industry. We are all better for their contributions.

Another source of support was the Validation Discussion Group. This ad hoc organization started in 1980 assembling validation managers and team leaders to discuss common concerns. The membership shared their experience and learnings in efforts to enhance general practices, identify vulnerabilities and avoid re-inventing the same wheel. The discussions were always engaging, and several lifelong friendships resulted from our semi-monthly interaction. Individuals whose valuable insights helped define present day validation concepts included: Greg Bassett, Wil Brame, Warren Charlton, Norm Elder, Bo Ferenc, Barry Garfinkle, Regina McCairns, Ray Pocoroba, George Sheaffer, Ray Shaw, Ron Simko, and Peg Szymczak.

My career spanned more than 20 years of work in large pharmaceutical firms. I was fortunate to have the opportunity to work with outstanding scientists and engineers whose insights have proved invaluable in shaping my perspectives. Stu Andrews, Domenic Attanasio, Hank Bauer, Mario Bernkopf, Bruce Friedman, George Gassman, Jerry Jackson, John Kimmins, Jan Leschley, Tom Molnar, Stan Nusim, Costa Papastephanou, Roland Pitsch, Ron Pomerantz, Klaus Reininger, Gerry Trotter, and Dick Wood standout as key contributors.

In nearly 30 years of consulting, I must also acknowledge the contributions my clients, vendors and fellow consultants have made. They challenged my ideas, raised concerns, and refined concepts that resulted in continued improvements in approaches and solutions. Notable in this regard are Claude Anger, Larry Beiter, Dave Bekus, Glen Berdela, James Bernstein, Dick Bonucelli, Dave Bradshaw, Larry Callan, Mary Joan Carlin, Alex D'Addio, Ted Egan, Marco Falciani, Giuseppe Fedegari, Lindon Fellows, Michael Ford, Doug Hamilton, David Hussong, Neera Jain, Michelle Jones, Walter Jump, Chris Knutsen, John Kowalski, Nallagounder Kuppusamy, Frank Manella, Dave Matsuhiro, Ken Maves, Nikki Mehringher, Bob Mello, Len Mestrandrea, Anthony Pavell, Art Pelletiere, Clint Pepper, Nora Poliakoff, Chris Prochyshyn, Daniel Py, Gregg Richmond, Doug Rufino, Gail Sofer, Steve Tarallo, Connie Taylor, Andreas Toba, Mike Ultee, Maryellen Usarzewicz and Jeff Werner. There were many others of course, but these were the most influential.

The individuals you work closest with have the greatest impact especially when the assigned projects are only loosely defined and changing over time. The years I spent leading first a site, and eventually a corporate validation effort on a global scale shaped much of what is found in this volume. We approached much of what we did as a joint learning activity, one in which none of us had the answers, but collectively we were able to find them. I am grateful for the opportunity to work with such great people – Rick Beasley, Bernice Bobroski, Becky Brewer, Jackie Castro, Laurie Colman, Tom Cuddy, Jeanine Derbedrosian, Gerry Finken, Cliff Fontaine, Liz Garcia, Bobbi Gordon, Gayle Heffernan, Dennis Kochanski, Steve Kovary, Frank Kuchma, Liam Laffan, Bill Lindboe, Russ Luhrs, Joe Malone, Kim Matthews, Dave Maynard, Maureen Ocleppo, Kenny Paterson, Vincent Rudo, Steve Steinberger, Ron Thiboutot, Al Trapani, and Cyndi Ulrich. We taught each other what we needed to know.

The last contributors to this effort are among my closest friends. I've worked together so closely with each of them it is almost impossible for me to separate their ideas from my own. The newest relationship of these is almost 40 years old, while the longest just passed 50 years. I've argued, agreed, written, re-written, collaborated, taught, learned, and listened with each of them for hundreds of hours. We shared many meals, various alcoholic libations, considerable travel and the occasional round of golf. Each is a validation expert in their own right, and I feel honored to call each of them my friend. Jim Akers, Phil DeSantis and Russ Madsen. They are the brothers I never had. I hope I have held up my share of the conversation over all these years and given to each of them as much as they have given me.

Last, and not least, I must acknowledge my loving wife Linda. She tolerated the endless evenings when I have been immersed in 'The Book', reading, editing and writing rather than spending time with her. Four editions' worth is more than enough. I could not have managed even one without her sacrifice. She is as much a contributor to this edition as she was to the editions that preceded it. I will always love you.

James Agalloco

ADDENDUM

When Jim asked me to be a co-editor of the Fourth Edition I could not have been more honored. I have been a contributor since the beginning and welcome the opportunity to leave something more to the industry that has given so much meaning to my life. I am even more deeply honored to carry on the work of my friend and mentor Fred Carleton, whose kindness and counsel was so important to my career.

I won't list all the people I could and should thank for their support over the years, except to say that my list is pretty much the same as Jim's. Jim and I have been friends and colleagues for more than 50 years, from our mutual beginnings at Merck in Rahway to Squibb and our PDA work. We have worked together, taught together, written together, played golf together (poorly, but joyfully) and spoken nearly daily over that whole time. I have learned more from Jim than from anyone. I hope that I have taught him a few things along the way. He is my best friend, professionally and personally. Thanks, Jim.

Lastly, I want to thank my wife of 50 years, Ellie. She has been the driving force behind everything that I have tried to accomplish in life. I dedicate my part in this book to her patience and support and most of all for always giving me something to come home to.

Phil DeSantis

1 Why Validation?

James Agalloco

CONTENTS

1.1 INTRODUCTION

The origins of validation in the global health care industry can be traced to terminal sterilization process failures in the early 1970s. Individuals in the United States point to the Large Volume Parenterals (LVPs) sterilization problems of Abbott and Baxter, whereas those in the United Kingdom cite the Davenport Incident.[1] Each incident was the result of a nonobvious fault with the sterilization that was not detected because of the inherent limitations of the end product sterility test. As a consequence of these events, non-sterile materials were released to the market, deaths occurred and regulatory investigations were launched. The outcome was the introduction by the regulators of the concept of "Validation."

> Documented evidence which provides a high degree of assurance that a specific process will consistently produce a product meeting its predetermined specifications and quality attributes.[2]

The initial reaction to this regulatory initiative was one of some puzzlement; after all, only a limited number of firms had encountered difficulties, and all of the problems were seemingly associated with the sterilization of LVP containers. It took several years for firms across the industry to understand that the concerns related to process effectiveness were not limited to LVP solutions, and even longer to recognize that those concerns were not restricted to sterile products. Perhaps most unfortunate of all was the lack of enthusiasm on the part of industry. From its earliest days, validation was identified as a new regulatory requirement, to be added onto the list of things that firms must do, with little consideration of its real implications. Some firms believed validation to be little more than a regulatory fad, or a one-time activity that, once completed, could be filed away for use with inspectors. Fortunately, it was considered more objectively by those who initially attempted to perform a "validation". The first efforts reflected what can be termed the "scientific method" of observation of an activity: (1) hypothesis/prediction of cause/effect relationship, (2) experimentation followed by (3) assembly of the results in the form of the experimental report. In the pharmaceutical validation model, this has evolved into the validation protocol (hypothesis and prediction), field execution (experimentation), and summary report preparation (documented observations).

By 1980, when it was becoming evident that validation was here to stay, pharmaceutical firms began to organize their activities more formally. Ad hoc teams and task forces that had started the efforts were gradually replaced by permanent validation departments whose reporting relationship, responsibilities, and scope varied with the organization, but whose purpose was to provide the necessary validation for a firm's products and processes. The individuals in these departments were the first to grapple with validation as their primary responsibility and their methods, concepts and practices have served to define validation ever since.

> Validation—Establishing documented evidence which provides a high degree of assurance that a specific process will consistently produce a product meeting its pre-determined specifications and quality attributes.[3]

The first efforts at validation were simplistic and limited in their understanding of the implications that validation actually entailed. As an example, the first sterilization validations at most firms were performed without prior qualification of the equipment. Once validation had been established as a useful discipline and something more than a passing fad, methods for its execution became substantially more formalized and rigorous.

The validation community made significant strides in clarifying the various components of a sound validation program. Perhaps most important of all was the separation of activities into two major categories: Equipment Qualification and Process Qualification. The former (sometimes divided into Installation and Operational Qualification) focused on the facilities, equipment, and systems needed for the product being processed. This is predominantly a documentation exercise, in which details of the physical components of the system are recorded as a definition of the equipment. Equipment operational capabilities are also established. This activity provides the basis for change control that supports the utility of the validation effort over time. Process Performance Qualification (also known as Process Validation or Performance Qualification) confirms the acceptability of the product manufactured in the equipment and relies heavily on the results of physical, chemical, and microbial tests of samples.

DOI: 10.1201/9781003163138-1

TABLE 1.1
Prerequisites for Validation

- Process Development [21 CFR 820.30—Design Control]—The activities performed to define the process, product or system to be evaluated.
- Process Documentation [21 CFR 211 Subparts F—Production and Process Controls and J—Records and Reports]—The documentation (batch records, procedures, test methods, sampling plans) (software) that define the operation of the equipment to attain the desired result.
- Equipment Qualification [21 CFR 211 Subparts C—Buildings and Facilities and D—Equipment]—The specifications, drawings, checklists and other data that supports the physical equipment (hardware) utilized for the process.
- Calibration—The methods and controls that establish the accuracy of the data.
- Analytical Methods [21 CFR 211 Subpart I—Laboratory Controls]—The means to evaluate the outcome of the process on the materials
- Cleaning—[21 CFR 211.67a Equipment Cleaning and Maintenance]—A specialized process whose intent is to remove traces of the prior product from the equipment.
- Change Control—[21 CFR 211.67c Equipment Cleaning and Maintenance]—A formalized process control scheme that evaluates the changes to documentation, materials, and equipment.

It was soon apparent that validation had to be more closely integrated into the mainstream of Current Good Manufacturing Practice (CGMP) operations in order to maximize its effectiveness. A number of areas can be identified as prerequisites for process or system validation. The origins of these elements can be identified in the CGMP requirements for drugs and devices (Table 1.1).[4]

With this understanding of its dependencies, validation is more easily assimilated into the overall CGMP environment rather than being something apart from it. Although firms will likely have a validation department, it must be supported by activities in other parts of the organization. For example, a poorly developed process performed using uncalibrated equipment making a product that has no standard test methods could never be considered validated. The supportive elements must be properly operated in order to result in a compliant product, and one that can be successfully validated. A later definition that addresses the larger scope of validation within the overall organization appears following:

> Validation is a defined program which, in combination with routine production methods and quality control techniques, provides documented assurance that a system is performing as intended and/or that a product conforms to its predetermined specifications.[5]

1.2 APPLICATION OF VALIDATION

After its first use with LVPs in the early 1970s, the application of validation spread quickly to other sterilization processes. It was also applied for the validation of other pharmaceutical processes, albeit with mixed success. In sterilization validation and to a slightly lesser extent in processes supporting the production of sterile products using aseptic processing, there is little difficulty applying validation concepts. The apparent reasons for this are the common and predominantly quantitative criteria for acceptance of the quality attributes for sterile products. Building consensus on validation of sterile products has largely been achieved across the entire industry. There are numerous regulatory and industry guidance documents outlining validation expectations on the various sterilization processes, as well as numerous publications from individuals and suppliers.

Validation of non-sterile products and their related processes is less certain. Despite the obvious importance of cleaning procedures, cleaning validation was not publicly discussed until the early 1990s. To this day, there is lingering confusion regarding the requirements for validation of this important process. The difficulties with validation are even more complicated for pharmaceutical dosage forms. There are no widely accepted validation requirements for the important quality attributes of drug products. Although the key elements are known, e.g., dissolution, content uniformity, and potency, there are no objective standards upon which to define a validation program. The compendial standards of the various pharmacopeias are poorly suited to validation. The small sample size and absolute nature of the acceptance criteria remain problematic for direct application to large-scale commercial production. After more than 40 years, the absence of universal criteria for dosage forms is distressing.

Applying validation requirements to water and other utility systems is somewhat easier than for the pharmaceutical products themselves. Equipment qualification of utility systems is relatively easy to perform, and samples of the supplied utility (water, steam, environmentally controlled air, compressed gas, solvent, etc.) taken across the system can directly support the acceptability of the preparation, storage (where present), and delivery. Classified and other controlled environments have also proven comparatively easy to validate. Their physical elements lend themselves readily to equipment qualification, and sampling affords direct confirmation of their operational capabilities.

Biotechnology first came of age in the late 1980s into a regulatory environment that expected validation of important processes. As the first biotech products were injectable drugs, it was quite natural for firms to validate their entire production process from the onset. As a consequence, cell culture and purification processes of all types have always been subject to validation expectations. There is a substantial body of validation knowledge on these processes available. In marked contrast, the bulk pharmaceutical chemical segment of the industry was comparatively slow to embrace validation concepts. Although certainly the rigorous environmental expectations associated with many dosage forms and virtually all biotechnology processes weren't present, the important considerations of impurity levels, by-product levels, racemic mixtures, crystal morphology, and trace solvents all suggest that there are important quality attributes to be controlled (and thus validated) as well.

Computerized systems became subject to validation requirements when they were first applied to CGMP functions in the 1980s. For ease of understanding, parallels between computerized systems and physical systems were utilized. The computer hardware can be qualified like the process equipment to which it is connected, whereas the computer software has some similarities to the operating procedures utilized to operate the equipment. This approach may be an oversimplification of the required activities for the software, but it provides some clarity to the uninitiated. Computerized system validation continues to be a subject of substantial interest, but it is no longer the misunderstood behemoth task it appeared to be when first encountered. The early efforts of the Pharmaceutical Manufacturers Association's (PMA's) Computerized Systems Validation Committee and the later development of Good Automated Manufacturing Practices (GAMP) have reduced the uncertainty associated with the use of computerized systems substantially.[6] Present-day concerns with computerized systems focus on data security, electronic signature, electronic records, and data integrity.

One useful concept taken from the validation of computerized systems as it evolved was the "life cycle model."[7] Originally utilized for computer software, it was later applied to the entire computerized system. It suggests that considerations of system qualification, maintenance, and improvement be incorporated at the onset of the design process. Its utility for computerized systems is substantial; however, it may have even greater functionality for pharmaceutical processes. In the early 1990s, the United States Food and Drug Administration (FDA) launched an initiative related to the demonstration of the consistency of processes and data from clinical lots through to commercial manufacture.[8] They mandated the conduct of Preapproval Inspections to affirm that commercial materials had their basis in the pivotal clinical trial materials. The utility of the "life cycle model" in this context is clear. Its application to pharmaceutical development, scale-up, and commercial production allows for a coordination of supportive information in the same manner as software and computerized systems validation. A landmark publication in this area was Kenneth Chapman's paper entitled "The PAR Approach to Process Validation".[9] It addressed the developmental influence on the ability to successfully validate commercial operations, a message that had been somewhat forgotten until more recently. Ajaz Hussain, then of the FDA, voiced concerns relative to the lack of process knowledge on the part of many pharmaceutical firms.[10] That the FDA believed that such a missive was necessary supports the lack of appreciation for Mr. Chapman's earlier effort.

> The goal of development is to identify the process variables necessary to ensure the consistent production of a product or intermediate.[11]

Application of the "life cycle model" to pharmaceutical operations addresses the compliance and quality expectations of the industry in an appropriate manner and should be a near universal goal. This was formally acknowledged by the FDA in its 2011 revision of its Guidelines on Process Validation in which it belatedly embraced the validation life cycle.[12]

Another regulatory development of some importance is that of Process Analytical Technologies (PAT).[13] The concept was well articulated by Dr. Hussain while he was with the FDA. To many in the industry, PAT seems like an advance of some magnitude that could seemingly replace validation. To those well versed in automation, PAT is nothing more than the extension of long-standing process control practices into pharmaceutical batch production. Engineers familiar with process control will recognize PAT as the installation of feedback control relying on sensors in the process equipment. This is by no means startling, except to those unfamiliar with control loops. PAT has its utility and will improve the quality of products produced by it, of this there can be little doubt. It will not, however, replace validation. In order to use a PAT system, the designer must assure that the installed sensor accurately reflects the process conditions throughout the batch, otherwise it will afford no benefit. The need for that assurance means that the PAT system, rather than replacing validation, will actually have to be validated itself!

1.3 WHY VALIDATION?

First, and certainly foremost, among the reasons for validation is that it is a regulatory requirement for virtually every process in the global health care industry, whether for pharmaceuticals, biologics, or medical devices. Regulatory agencies across the world expect firms to validate their processes. The continuing trend towards harmonization of requirements will eventually result in a common level of expectation for validation worldwide.

Utility for validation beyond compliance is certainly available. The emphasis placed on compliance as a rationale has largely reduced the visibility of the other advantages a firm gleans from having a sound validation program. Some years ago, this author identified a number of tangible and intangible benefits of validation realized at his employer (see Table 1.2).[14] In the intervening years, there has been repeated affirmation of those expectations at other firms large and small. Regrettably, there has been little quantification of these benefits. The predominance of compliance-based validation initiatives generally restricts objective discussion of cost implications for any initiative. But once a process/product is properly validated, it

TABLE 1.2 BENEFITS OF VALIDATION

Increased throughput
Reduction in rejections and reworks
Reduction in utility costs
Avoidance of capital expenditures
Fewer complaints about process-related failures
Reduced testing—in process and finished goods
More rapid/accurate investigations into process deviations
More rapid and reliable start-up of new equipment
Easier scale-up from development work
Easier maintenance of the equipment
Improved employee awareness of processes
More rapid automation

would seem that reduced sample size and increased sampling intervals could be easily justified and thus provide a measurable return on the validation effort. Aside from utility systems, this is hardly ever realized and represents one of the major failings relative to the implementation of validation in our industry.

Validation and validation-like activities are found in a number of industries, regulated and unregulated. Banking, aviation, software, microelectronics, nuclear power, and others all incorporate practices closely resembling the validation of health care production. That verification activities for products, processes, and systems have utility in other areas should not be surprising. The health care industries' fixation on compliance has perhaps blinded us to the real value of validation practices.

1.4 CONCLUSION

Validation is here to stay; it has become an integral part of regulatory requirements and everyday life in the global health care environment. There are millions of pages of validation documentation across the world, but the presence of mountains of paper is not justification for its continued existence. Its presence affords a level of confidence in the quality of products for human health. The extent that the risk to the patient is reduced by a validation effort (or any other activity impacting product quality) will ultimately determine its continued utility. If risk-based thinking is adopted across the industry, as it appears it might be, then certain validations will be become more rigorous, others less so, and others unchanged. If the considerations associated with the implementation of validation for a process become financially driven, there may be additional opportunities. Validation is here to stay, at least for the foreseeable future.

NOTES

1. Chapman K, A History of Validation in the United States—Part I. *Pharmaceutical Technology* 1991;15(10).
2. Fry E, PDA Annual Meeting, 1980.
3. FDA, Guideline on General Principles of Process Validation, May 1987.
4 Agalloco J, Course Notes on Principles of Validation, 1985.
5.. Agalloco J, Validation: Yesterday, Today and Tomorrow. Proceedings of Parenteral Drug Association International Symposium, Basel, Switzerland, Parenteral Drug Association, 1993: 1–11.
6. Chapman K, A History of Validation in the United States—Part II: *Pharmaceutical Technology. Validation of Computer-Related Systems* 1991;15(11).
7. Agalloco J, The Validation Life Cycle. *Journal of Parenteral Science and Technology* 1993;47(3): 142–147.
8. FDA, Process Validation Requirements for Drug Products and Active Pharmaceutical Ingredients Subject to Pre-Market Approval (CPG 7132c.08, Sec 490.100), 2004.
9. Chapman K, The PAR Approach to Process Validation. *Pharmaceutical Technology* 1984;8(12): 24–36.
10. FDA, Pharmaceutical CGMPS for the 21st Century—A Risk-Based Approach, 2004.
11. Agalloco J, Course Notes on Principles of Validation, 2014.
12. FDA, Guidelines on Process Validation, 2011.
13. FDA, Guidance for Industry PAT—A Framework for Innovative Pharmaceutical Development, Manufacturing, and Quality Assurance, 2004.
14. Agalloco J, The Other Side of Process Validation. *Journal of Parenteral Science and Technology* 1986;40(6): 251–252.

2 Facility Design for Validation

Phil DeSantis

CONTENTS

DOI: 10.1201/9781003163138-2

2.1 INTRODUCTION

The design, construction, and start-up of pharmaceutical and biopharmaceutical facilities governed by Current Good Manufacturing Practices (CGMP) regulations requires a different approach than might be applied for any other manufacturing facility projects. This is because such facilities must comply with specific requirements of the applicable regulations, as fully described by Subpart C of 21CFR211, the United States CGMP regulations, which begins:

> §211.42 Design and Construction Features
>
> (a) *Any building or buildings used in the manufacture, processing, packing, or holding of a drug product shall be of suitable size, construction and location to facilitate cleaning, maintenance, and proper operations.*

These regulations go on to provide more guidance, much of which will be covered by this chapter. Based on the author's considerable experience, it is of little value to develop a safe and efficacious product with a reliable process and then expect to produce it in a poorly designed or substandard facility.

Further, in order to meet the ultimate business goal of the manufacturer, which is to deliver the product to the patient, validation of the included processes must be successfully completed. Of course, this requires that the facility and manufacturing equipment that is determined to directly affect product quality must be qualified. This chapter will emphasize the importance of both good design and good project management practices to achieve the ultimate goals of a CGMP project.

Before proceeding, however, it is wise to define and understand what the goals of a project are. A project is a series of interrelated activities of limited scope and duration with a defined outcome. In general, project success is defined by return on investment. A capital project, such as those discussed in this chapter, results in some durable (i.e., long-lasting) assets and may be judged by the financial return on those assets. Capital project success is most often driven by budget and schedule, i.e., what is the level of investment and how long will it be before that investment returns value. Engineers and builders are trained to drive projects to completion in the least amount of time and with the lowest cost, while still reaching the defined outcome. It is important to note that a third driver is necessary to achieve the outcome. The assets must perform as expected. In short, the third driver may be described as "quality".

There is an old adage that has appeared frequently in the literature and is repeated periodically online: "Budget, schedule, quality: pick two." If interpreted figuratively, this may be taken to mean that a project must be adequately funded and given enough time to be planned and executed in order to achieve its desired outcomes. This is true enough. On the other hand, biopharm users have often taken this adage to mean that a tight budget and aggressive schedule always result in poor quality. This, of course, is nonsense. Many CGMP projects are undertaken with the all-important "time to market" consideration as a schedule driver. Responsible firms understand that aggressive scheduling may cost more, but that rigorous budgets may still be applied. In the end, research has shown that when best project management practices are applied, all three goals are accomplished: on time, under budget, and expected quality of the facility and its output [7]. This chapter will look at how these seemingly competing objectives can actually work in synergy. Section 2.2 looks at projects in general and Section 2.3 discusses the relationship between project activities, deliverables, and validation, where validation is an essential element of quality.

2.2 PROJECT MANAGEMENT

We should be clear in our discussion of facilities projects that although we speak of success as measured by return on assets, that real return is dependent upon the performance of the facility and equipment in support of product manufacture. This means an acceptable product delivered to the patient without recalls, rejects, reworks, and with a minimum number of deviations, investigations, delays, and interruptions. Certainly, facilities and equipment do not cause all of these problems; however, poor design can contribute to a significant proportion of them.

One of the most important factors contributing to overall project success is the application of a rigorous "stage-gated" project management system. Stage-gating is the application of defined review stages over the course of project development in which scope, budgets, and schedules are reviewed and authorization for further work is given. The final budget and schedule are not firmly established until the final approval of some defined stage. [6]

In building a project management structure, it is valuable to clearly define the various necessary project activities, milestones, and deliverables as well as the project team members and other stakeholders. Of course, this may be done using a series of documents, plans, procedures, schedules, spreadsheets, and so on. In fact, each of these articles plays an important part in a project management structure. It is valuable to organize these into a user-friendly graphic interface similar to the GMP Project Roadmap shown following [2]. **Although this chapter is not a primer on project management, it is important to understand the stages of a project and how each supports and leads to validation.**

2.2.1 A GMP Project Roadmap

The GMP Project Roadmap is a proprietary configurable representation of an effective capital project management system for GMP projects. Shaded nodes indicate activities most closely related to validation.

The GMP Project Roadmap shown is a generic version provided for illustration and guidance of further discussion on project management and validation. (It is an example and not intended to be critiqued in detail. Project management guidelines are configurable and may be unique to each company and even to individual projects). Figure 2.1 indicates the "front-end" or planning end of a project, whereas Figure 2.2 relates to project execution. Of course, in the real world, there is sometimes overlap among activities, but this chapter will emphasize the importance of front-end loading to ensure project success, especially as it relates to CGMP compliance and validation. This principle is strongly supported by many expert

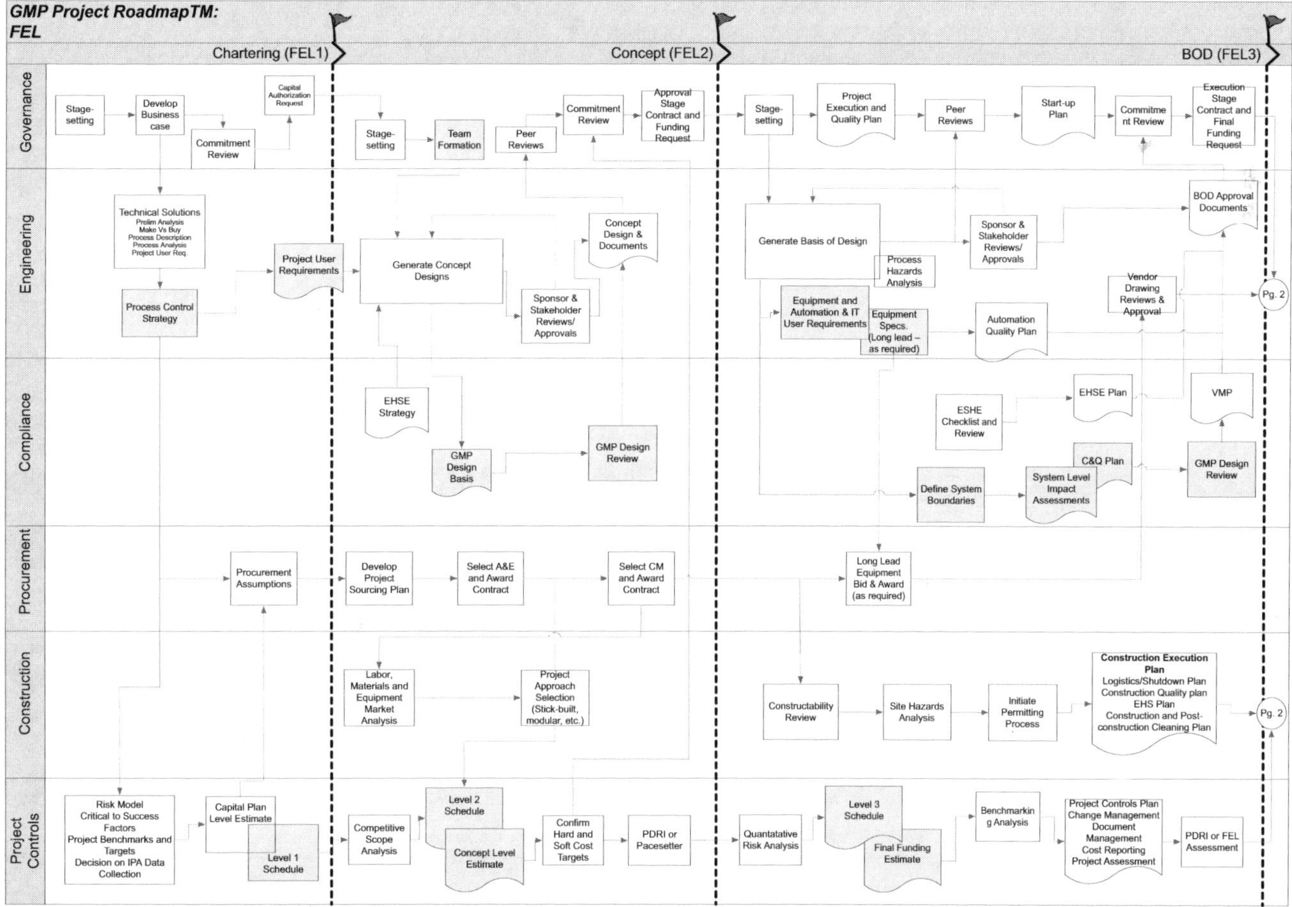

FIGURE 2.1 GMP Project Roadmap—front-end loading.

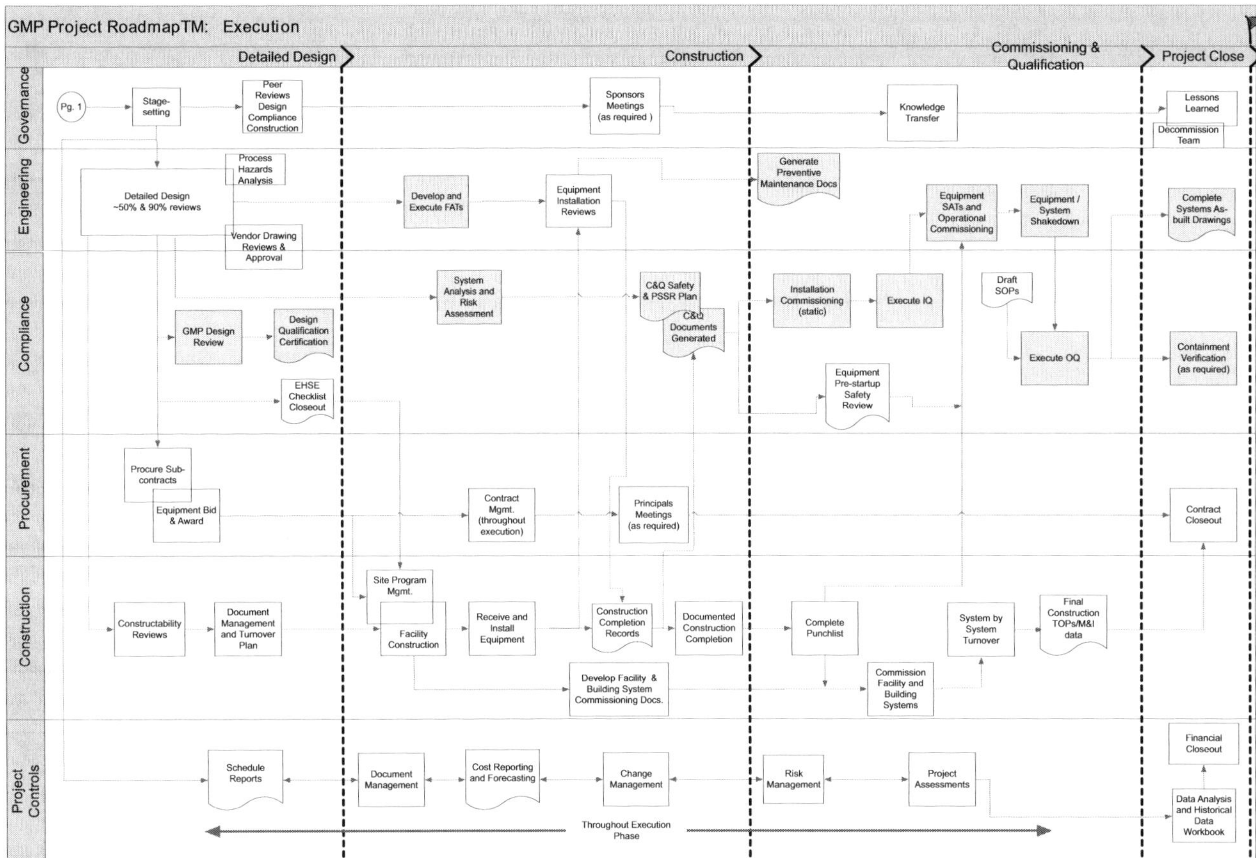

FIGURE 2.2 GMP Project Roadmap—execution.

organizations studying project management [6]. A graphic interface like the Roadmap may be supported by active links to various guidelines and templates to assist the team in accomplishing the tasks described in each node of the diagram. Some of this guidance will be briefly described in the chapter.

2.2.2 Chartering

This phase initiates the project and establishes the business case that the project is meant to satisfy. The business case may be commercial, such as new product introduction or capacity increase. The business case may also be driven by regulatory compliance, for example to improve environmental controls within a facility or even to replace a substandard facility with one that is more compliant. The business case should also define the other major project drivers, such as time to market, cost restrictions, or desired return on investment. Projects without a clear business plan are usually on a rough road toward successful completion.

2.2.3 Concept

Once a business case is established and possible technical solutions are determined, the next phase of design results in a Concept, often summarized in a Conceptual Design Report. This usually includes the following (this is neither an exhaustive nor a required list and varies based on project scope and complexity):

- Site Surveys and Site Master Plan
- Soil Samples
- Process Control Strategy
- Project User Requirements
- Conceptual Layouts
- Conceptual Equipment Arrangement Drawings
- Conceptual Major Equipment List
- Process Flow Diagrams (PFDs)
- Utility Flow Diagrams (UFDs)
- Heat and Material Balances
- Initial Process Hazards Analysis
- People, materials, product, and waste flow diagrams
- Area Classification and Pressurization Scheme
- Heating, ventilation, and air conditioning (HVAC) Concept
- Containment Strategy
- Identification of Critical Utilities and support systems
- Validation Strategy

- Maintenance Strategy
- Design Safety Philosophy Document (DSPD)
- Environmental Impact Review
- Mechanical/Equipment Room Conceptual Configuration
- Material Handling Strategy
- Gowning Requirements
- Conceptual GMP Design Review Report (Optional)
- Conceptual Project Schedule (Level 1)
- Conceptual Project Cost Estimate

2.2.4 Basis of Design

The Basis of Design (BOD) phase is sometimes called Basic Design, Preliminary Engineering, or Design Development. This phase develops the Concept into more detailed drawings and documents that become the basis for the construction drawings and specifications needed to actually build the facility. The BOD drawings are schematic, i.e., they define system components and their relationship and connectivity, but they are not spatially precise.

The following lists typical BOD content sometimes compiled into a Basis of Design Report and collection of drawings. There may be some overlap with Conceptual Design, and the list varies depending on the project's scope and complexity. For small-to-modest project scope, it may be decided to combine the Concept and BOD phases.

- General Arrangement (GA) Drawings
- Preliminary Discipline Specific Drawings
- Area Classification drawing
- Piping and Instrumentation Diagrams (P and IDs)
- Utility Schematics
- HVAC Schematics
- Equipment List
- Motor Load List
- Electrical Single Lines/Schematics
- Line List
- Tie-in List
- Instrument Index
- Room URSs/Room Cards
- Automation Philosophy and Control System Architectural Diagram
- Building Automation Functional Specification
- BMS Architecture Diagram
- Process Hazards Analysis
- Safety Risk Analysis: Health Hazard Criteria
- Regulations, Codes, and Standards Review
- Constructability Review
- Design Safety Plan
- Maintenance Strategy
- Procurement Strategy
- Risk Management Analysis—Cost and Schedule
- Environmental Review
- Document Control and Filing Index
- GMP Design Review
- Project Cost Estimate—Control Budget
- Project Schedule (Level 2)
- System-level Impact Assessments
- Commissioning and Qualification Plan
- Validation Plan

2.2.5 Project Execution—Detailed Design

The documents developed in the Detailed Design phase are specific as to dimension, detail, and orientation and are used by contractors to build the facility, its incorporated systems, and manufacturing equipment. These documents are developed by architects, engineers, and equipment specialists, most often third-party service providers and equipment vendors. This phase is also when detailed commissioning and qualification documents are prepared based on the design information. Detailed drawings are issued for construction (IFC) and are among the documents typically developed during detailed design, as follows:

- GMP Design Review—Critical Systems
- Design Qualification Certification
- Project Schedule—E/P/C/V (Level 3)
- Architectural Finishes, Details, and Specifications
- Mechanical/Equipment Room Layouts
- Civil Details and Installation Specification
- Reflective Ceiling Plan
- Planning/Construction Permits
- HVAC Duct Routing Plans—IFC
- Equipment Specifications
- Construction and Installation Specifications
- Cable Tray Routing Plans—IFC
- Isometrics—IFC
- Pipework Special Design Details
- Pipework Specification
- Instrumentation Specification—IFC
- Loop Drawings—IFC
- Process Control System (PCS) Design Specifications
- Building Automation Design Specification
- Construction Safety Plan
- Maintenance Plans
- Value Engineering Options/Report
- Submittal Log

2.2.6 Project Execution—Construction

During this phase of execution, the facilities are built, equipment installed, and construction turnover dossiers finalized. Turnover dossiers may include:

- Shop drawings
- Construction quality inspection reports
- Pressure test reports
- Slope checks
- Motor inspection and test reports
- Grounding checks
- Weld inspection reports
- Concrete test reports
- Equipment manuals and vendor drawings

- Factory Acceptance Test (FAT) reports
- Construction completion reports/acceptance

2.2.7 Project Execution—Commissioning and Qualification

The final phase of the capital project (remember that PPQ runs are subsequent to the capital portion of the project) is commissioning and qualification. These are explained in detail in the chapter "Commissioning and Qualification" elsewhere in this volume. They are discussed further in the subsequent sections.

2.3 MANAGING FOR VALIDATION

Each of the nodes indicated on the Roadmap is important to project delivery. Rather than review each node and activity, we will focus on those that have the most direct effect on and/or relationship with validation. It is true, however, that the three major factors leading to a qualified facility and validated processes are the same ones that contribute most to overall project success [6]:

- Team organization
- Front-end loading
- Project controls

We shall review each of these in light of how they relate to validation. The discussion will take a very broad view of validation to include its necessary components

2.3.1 Team Organization

When building a project team for any manufacturing project it is necessary to start with Project Management and a Project Owner, usually representing a User group (e.g., Production). Core team additions are technical/process experts, engineers/designers, and construction managers. Some of these may be qualified third-parties (contractors). Part-time stakeholders who may not be fully devoted to the project include regulatory professionals, maintenance personnel, and R and D. GMP projects are unique in that core project teams must include a commissioning/qualification lead, a validation lead, and a Quality unit representative.

It is essential that these last three are necessary parts of the core project team. For any project of significant size, these roles may require full-time focus or at least be considered a main responsibility. The bigger or more complex the project, the more focused these roles must become. They play an essential role in reviewing and approving all project decisions that have an effect on product quality.

2.3.1.1 Commissioning and Qualification Lead

As the Roadmap indicates, the close of a GMP capital project runs through facility and equipment qualification. This typically includes all commissioning and qualification activities that are funded by and managed under the capital project. All manufacturing equipment is qualified along with the facility and utility systems that have a direct effect upon product quality, such as process water purification and distribution. Early planning for commissioning and qualification supported by an expert in these disciplines is necessary.

2.3.1.2 Validation Lead

Process validation is not often managed under the core project management team. Process Performance Qualification (PPQ) lots are often designated for commercial sale and, therefore, are not part of the capital cost. Additionally, process validation requires special expertise often from outside the User or Engineering organizations. Nonetheless, a GMP project is not actually complete until commercial (or clinical) distribution can begin. It is important that the organization plan for the date when actual return on the capital investment can begin in order to determine product success. In addition, process validation expertise is important in reviewing and approving designs and key GMP documents included in the Roadmap.

2.3.1.3 Quality Representative

Often overlooked in GMP projects is the importance of a dedicated Quality unit team member. Experience on many projects has been that designs and plans have been developed without Quality involvement and then had to be changed when Quality entered the picture. Quality must be a core team member, even full-time for larger projects, from team formation through PPQ. It is also important that the designated Quality representative have approval authority for GMP activities and documents. It is not effective in most cases for a junior level team member to participate in reviews and discussions yet defer approvals to a more senior member who is not committed to the project. This is so important that some companies have actually written this Quality project role into an SOP or Quality Standard, making it a GMP violation to structure a GMP project team without Quality on board.

2.3.2 Front-End Loading

The "planning" stages of a project are often referred to as the "front end" and have been proven to be critical to the ultimate project success. All projects should begin with a business case, followed by one or two stages of design that gets a project to a point.

This holds true for validation as well as other measures of success. Team building is one of the initial points of front-end loading (FEL) but is emphasized separately previously for its critical importance. The team ensures that all the required workflows and documents described following are performed, reviewed, and approved as required. Here we focus on those project elements shaded in the diagrams to indicate the influence on GMP compliance and validation.

2.3.2.1 Process Control Strategy

The FDA Guidance for Industry *Process Validation: General Principles and Practices* emphasizes the importance of process knowledge and understanding as the first phase of process validation [15]. The result of that understanding should be a process control strategy that outlines everything that is important to ensure a reliable and repeatable process. Included should be material considerations, equipment, and process considerations including sequences and parametric ranges. This should be available in the initiating phase of the project (FEL 1) or as early as possible in Conceptual Design. This strategy is a necessary element of process validation.

2.3.2.2 Project User Requirements

Project User Requirements are driven first by the Process Control Strategy and then expanded to include all related process, compliance, and business drivers for the facility and project. As the name implies, the User group is the primary source of requirements, but in the capital project sense the term "user" may be expanded to include all project stakeholders. Note that this chapter does not use the common term User Requirements *Specification (URS)*. The author prefers to reserve the term "specification" for its literal use, avoiding possible confusion with subsequent engineering documents.

It really does not matter who drafts the User Requirements, but two important factors need to be considered. First is input from core stakeholders, including the user (after all, these are *user* requirements), technical/process experts, and, because this is a GMP facility requiring validation, Quality. The second important note is that these "requirements" are not technical solutions. They should be limited to what is needed rather than how to accomplish it. For example, if a clean room is required, its classification may be stated (e.g., Grade C or ISO 7), but technical details such as filter grades and number of air changes should be left to the engineers providing the solutions in the subsequent design.

Besides providing a foundation for the initiation of design, another valuable use of User Requirements is to provide a foundation for facility and equipment qualification (henceforth EQ). Each requirement listed may be assessed to determine why it is included. A recommended approach is to designate all requirements that have a direct effect on product quality, for example as QCR (quality-critical requirement or some equivalent acronym). Other requirements may be designated as business, safety, environmental, or whatever other category the team determines to be valuable. By identifying quality-critical requirements, the team provides a source to trace design specification as well as qualification acceptance criteria. For these reasons, User Requirements should be approved by the core project team members including Quality.

More detail on User Requirements and how they are used in validation are found in the chapter "Commissioning and Qualification".

2.3.2.3 GMP Design Basis

It is wise to define what regulations and compliance issues are applicable early in the project. This may be to identify which GMP regulations must be followed as well as regulatory guidelines (e.g., FDA Guidance for Industry). Companies may also have internal Quality standards, design guidelines, and/or procedures (e.g., SOPs) that impact the project. These may be summarized in a document before or at the initiation of Conceptual design. Alternatively, these requirements may be built into the project charter or a similar document.

2.3.2.4 GMP Design Review (Concept)

For mid-to-large size projects, the Conceptual Design will be completed as described previously and a "stage-gate" will be reached. In many companies, this stage-gate is a financial one wherein the project requests funding authorization to proceed. On the GMP side, this is an ideal time to perform a first GMP Design Review. This review is a cooperative effort among core team members and other invitees that reviews the Concept against a predefined set of criteria, often guided by a checklist. Questions may be raised to be answered and followed up by the design team. They should be guided by the User Requirements, the GMP regulations, and other documents specified in the GMP Design Basis. A checklist format has proven to be very effective (an example is provided in the Appendix). A roundtable format has also proven to be most effective but is not the only way this may be done. Once all issues have been resolved (some of which may be deferred to the next design stage), a summary report may be issued for approval by the team. This can be considered a step toward Design Qualification.

2.3.2.5 Equipment User Requirements and Specifications

As FEL proceeds into BOD, individual equipment systems may require dedicated User Requirements. These can be driven by the Process Control Strategy, company or international standards (i.e., ISO) or other needs. These may be developed and formatted similarly to the Project User Requirements and will be subject to the same approvals. Technical or engineering specifications follow and become sources of information for commissioning and EQ protocols.

2.3.2.6 System Boundaries and Impact Assessments

As BOD proceeds through development of system schematics (P and IDs) it is a good practice to set system boundaries. Because many systems are interconnected with other systems (e.g., utilities or other process steps), it is important to know to which system each component shown on a P and ID belongs. This can be done by marking up or color-coding the drawing. This allows for the system to be looked at as a unique entity and a system-level assessment of quality risk completed. The criteria used for this assessment are fully explained in the chapter "Commissioning and Qualification". Determination of significant quality impact will mean that a system requires EQ and not just commissioning.

2.3.2.7 GMP Design Review (Basis of Design)

The GMP Design Review at the completion of BOD (FEL 3) is probably the most important and one that should not be overlooked. This is because most GMP compliance issues on a project level have been addressed in the design at this point; additionally, because it is early enough in the project to correct errors or make changes. There may be more items to review as Detailed Design proceeds, but this is the one that covers the greatest scope. Again, all issues and questions need to be resolved before a GMP Review report can be approved.

2.3.2.8 Validation Planning

Every GMP project needs to plan validation. Depending on the scope of the project, this may include a Commissioning and Qualification Plan, a Process Validation Plan, a Computer Validation Plan, or any combination of these along with other related plans. Commissioning and Qualification are usually considered part of the capital project (because their costs are capitalized), and the total project cost and schedule are not closed out until at least EQ is completed. Studies have shown that the success of a capital project (judged by the triple criteria of cost, schedule and quality) are best assured when C and Q planning is completed as part of FEL, i.e., by the completion of BOD [8].

Because the facility does not begin paying back until PPQ is completed and commercial product is released, it is also important to have a validation plan in place that covers all areas whether or not they are managed under the facility capital project. The earlier these plans are in place, the greater the probability that the work will be completed on time and as expected.

2.3.2.9 Project Controls (Schedule and Estimate)

Although Project Controls are not usually considered GMP activities, they are shaded on the FEL diagram. All activities take time and require resources. Project Controls must consider that validation activities extend the project schedule and also have a cost. More often than not, the capital portion of the project ends with EQ. However, a well-planned project takes into account the costs in material and manpower, as well as the time required to perform and approve PPQ lots that carry the project to the point of financial return. For PPQ lots that will ultimately be distributed for commercial or clinical use, costs are expensed rather than capitalized. Nevertheless, it is wise to consider PPQ and its ancillary activities like analytical and cleaning validation in the overall program schedule. This should include interim milestones (e.g., protocol preparation) as is done for EQ.

2.3.3 Project Execution

The work of validation continues in parallel with the work of the facility project through the detailed design and construction phases. For the capital portion of the project, validation culminates in commissioning and qualification, but for the overall project where specific processes have been identified, validation must proceed through PPQ. For clinical manufacturing facilities that are designed for process development, there may be no specifically identified process, and so the project schedule will end with EQ.

Some firms, if not most, when considering PPQ will account for cost and schedule separately from the capital portion of the project. This does not mean they are ignored or managed less rigorously (although in the author's experience this is sometimes the case). To avoid this, the team organization discussed in Section 2.3.1 should identify an overall program manager, separate from the facility project manager, who oversees the PPQ related activities. These include technology transfer (e.g., from Process Development to Operations), analytical validation, and material logistics for PPQ lots

2.3.3.1 Final Design Review and Design Qualification

As detailed design proceeds specific process and critical building systems are specified and may require targeted GMP design reviews. This is the case when these systems are unique or custom designed. Detailed formal GMP Design Review is usually not required for standard commercial process equipment such as autoclaves and blenders. These designs have been proven through extensive industrial use and the specificity of each system will be confirmed through normal engineering review as shown in the Execution diagram. In addition, these designs are further confirmed through the EQ process.

If a system is entirely unique to the project, e.g., Purified Water or Water for Injection (WFI) distribution, a formal system GMP design review may be instituted. However, even in these cases, the BOD documentation is often complete enough to allow the GMP Design Review to be completed at that stage.

Once the GMP Design Review has been completed and approved by key stakeholders, including Quality, the team may issue a certification that the design has been qualified to meet User Requirements and compliance to applicable regulations. This may be designated as Design Qualification (DQ) as defined in the European Union and PIC/s GMP Guidelines [5,10]

2.3.3.2 Factory Acceptance Tests

The Project Execution Plan or Procurement Plan may define if and where FAT will be required. FAT is strictly a matter of contract between the project and equipment vendors and is not a regulatory requirement. It may be a valuable exercise, especially if automation is included with the system (see the chapter "Control Systems Validation" for additional detail). FAT is highlighted in the diagram because it is allowable to document some of the FAT test results to be leveraged into EQ protocols. This must be predetermined, i.e., tests that will be leveraged predefined, and documentation must follow the same standards as set for EQ. The chapter "Commissioning and Qualification" discusses leveraging in more detail.

2.3.3.3 Preventative Maintenance

In order for systems to be qualified, maintenance procedures must be in place. For individual systems, this includes preventative maintenance procedures, including schedules for periodic, run-time, or predictive maintenance. Maintenance is an essential element of validation and facilitates the continuing process verification described in the FDA Guidance [15].

2.3.3.4 Commissioning and Qualification

Commissioning plans and EQ protocols may be developed during the Detailed Design phase of the project as specific system information and vendor documentation is received and approved by Engineering. It is important that detailed and approved design documents be used as reference. These need to be compared with User Requirements to identify the specific system characteristics and functionality that satisfy the User Requirements identified as essential to product quality. Again, details of the sequence of EQ protocol preparation and execution is found in the chapter "Commissioning and Qualification".

2.3.3.5 Containment Verification

For facilities encompassing highly active or toxic materials an additional step to verify containment may be required. Depending on the containment design, this may be included in the qualification of environmental controls, but for more sophisticated containment devices (e.g., glove boxes/isolators), this may require a separate protocol. Although this may be primarily a safety or health concern, it may also prevent cross-contamination of products and therefore is a validation consideration.

2.3.3.6 As-Built Drawings

For qualified process equipment and facility systems, usually P and IDs, must be verified and maintained as part of the EQ records. This is usually done by marking up or "red-lining" these drawings according to a specific approved procedure. The red-lined drawings are retained in the validation files and are used as a reference to develop certified as-built schematics that are used by the facility for future operations and as a basis for future changes.

2.3.4 Process Validation/Process Performance Qualification

The continuation of the project through PPQ runs has been discussed in the sections prior. It is not shown on the diagrams because they are not usually considered under the management structure of the facility capital project. Planning and structuring a process validation program is covered in great detail elsewhere in this volume. It is included here for emphasis and completeness.

2.4 FACILITY DESIGN

The foregoing sections emphasize the importance of planning, organization, and communication throughout a facility project to facilitate and ensure validation manifested as qualified equipment and facility systems and validated manufacturing processes. Clearly, these are a necessary element of validation. Experience on many projects over the course of the author's long career has shown that when projects fail, it is often the lack of attention paid to validation that is the root cause. However, even when validation receives the appropriate attention, it is insufficient when placed into substandard facilities and equipment with poorly designed utility systems. Good facility design is necessary in order to ensure that processes can be validated and that they will continue to perform reliably and consistently. This includes facility layout, equipment, environment, and critical utilities.

The GMP regulations provide only general guidance on facility design as in the excerpt provided in the Introduction and the following sections of the regulations. The possible exception to this general approach is EU GMP Annex 1 for sterile products, which has several more specific guidelines. None of these regulations, however, are definitive with regard to design criteria. Fortunately, after several decades of facility design under GMP, manufacturers and their supporting contractor engineers and architects have become expert in understanding proven design concepts. There are numerous publications compiling these concepts, among which are a series of ISPE Baseline Guides for facility and equipment design [9]. These and other publications provide significant details and alternative approaches to comply with the general requirements of the regulations and, in doing so, ensure continuously validated processes [10].

The following are a digest of points to be considered when designing a biopharmaceutical facility for GMP compliance and validation. These points were originally compiled as part of a GMP Facilities Design Guide developed by Schering-Plough Global Engineering Services around 2009. These are not requirements but are intended to satisfy the GMP regulations.

2.4.1 General Design Considerations

2.4.1.1 The production facility will be designed, constructed, and maintained in a manner to minimize the impact of the external environment.

2.4.1.2 The premises will be designed and constructed such that they prevent entry of insects and other animals and the migration of extraneous material from the outside into the manufacturing areas and from one area to another.

2.4.1.3 Buildings and facilities will have adequate space for the orderly placement of equipment and materials to prevent cross-contamination and/or mix-ups between different components, drug product containers, closures, labeling, in-process materials, and drug products. The layout should also provide for appropriate access for routine operation, maintenance, and repairs.

2.4.1.4 Facility design will permit appropriate flow of materials, equipment, and personnel such that the risk of mix-ups or contamination is minimized. Unidirectional flow is recommended. If this is not possible, specific procedural controls are necessary. Special consideration shall be given to multi-product facilities, as extra precautions are necessary to prevent cross-contamination. The facility should provide for separate material and equipment flows until the material or product is contained in a closed system. Personnel, material, and product flows will be reviewed during GMP Design Reviews.

2.4.1.5 Corridor dimensions and door openings shall take into account the movement of large equipment and materials. Appropriate wall protection (e.g., bumpers, railings) should be utilized where necessary to protect against damage to equipment, walls, and doors. Mobile equipment should also be designed to minimize damage to the facilities.

2.4.1.6 Operations will be located and performed within specifically defined areas of adequate size. There shall be separate or defined areas as necessary to prevent contamination or mix-ups during the course of the following procedures:

- Receipt and holding of materials, components and labeling pending approval for use.
- Holding rejected materials, components, and labeling before disposition.
- Storage of released components, in process materials, and drug products after release and labeling.
- Manufacturing, packaging and labeling operations.
- Quarantine storage prior to release of drug products or active pharmaceutical ingredients (APIs).
- Control and laboratory operations.
- Sampling of intermediates, APIs, and finished or unpackaged drug products.
- Personnel dress/gowning, hygiene, and food consumption.

2.4.1.7 Controlled and Classified areas are to be segregated from, and kept at a positive pressure to, General non-production areas.

2.4.1.8 As a general rule, the pressure cascade should be from the highest to lowest classification, but special consideration may be necessary (for example in air locks) to contain or limit exposure to toxic or high-potency materials, drug substances, and drug products. (see HVAC Design Requirements).

2.4.1.9 In areas where potent or toxic materials are processed, consideration should be given to enhanced cleanability. These areas should have a separate HVAC system.

2.4.1.10 Room finishes and materials selection should be appropriate to the activities performed in the respective areas. Personnel safety is an important consideration. Floors and surfaces should be nonslip even in wet conditions. Consideration will also be given to the following characteristics: cleanability, durability, functionality (e.g., low humidity, antistatic), and maintenance (ease of repair or renovation)

2.4.1.11 In Controlled and Classified areas, the walls, floors, ceilings, etc. should be constructed so that the primary surfaces are accessible for cleaning. Exposed horizontal surfaces in Controlled Areas are generally acceptable if they are accessible for cleaning. They should, however, be sloped to allow free draining.

Horizontal surfaces should be avoided as much as possible in Classified areas. Windows, doors, and fixtures should be flush with the primary surfaces.

2.4.1.12 As much as possible, process drainage should be through the wall into a noncontrolled area (mechanical support area).

Floor drains are allowed in Controlled and Classified areas only when required (e.g., in a washroom). They must be of adequate size and should be designed for clean service (i.e., food and pharmaceutical) with a removable trap that may be easily cleaned and sanitized. In all cases, process drains must be provided with an air break or a suitable device to prevent backflow from the drain to the process vessel.

Drains in Controlled and Classified Areas (Grades C and D) should be capped when not in use. Drains in Grade A and B areas are not allowed unless specifically approved.

2.4.1.13 The Premises shall be designed in a manner to prevent unauthorized access to production and support areas. Additional controls may be required for limited access areas such as sterile manufacturing, label control, controlled drug storage, and quarantine areas.

2.4.1.14 In most circumstances, air locks with interlocking doors should be installed when accessing Classified Areas or between Classified areas of different grades. A notable exception is when there is a local Grade A operating environment established within a room that is established and maintained at Grade B.

2.4.1.15 Another allowable exception shall be a locally controlled environment operating at Grade C (e.g., a weigh booth) within a Grade D environment.

2.4.1.16 In Gowning Areas and air locks, architectural finishes and materials should comply with the requirements of the higher classification into which the area opens.

2.4.1.17 Doors, windows, walls, ceilings, and floors will be intact with no holes or cracks evident.

2.4.1.18 Doors giving direct access to the exterior from production areas are to be used for emergency purposes only and will be alarmed. These doors will be adequately sealed.

2.4.1.19 In-process storage areas are General areas if the product stored in these areas is not exposed. The facility should generally comply with the requirements specified for General areas. There should be suitable electronic systems or other suitable means to identify (e.g., labeling, signs, bar coding, etc.) and segregate materials.

2.4.1.20 Sampling and weighing rooms should provide a level of protection equivalent to, or greater than, the level of protection required by the product at that stage of its processing. (e.g., materials that will be processed in a Grade C Classified area must be sampled and weighed in a Grade C Classified area or equivalent locally controlled environment—a weigh booth).

2.4.1.21 When providing utilities or instrumentation service to Controlled or Classified Areas, there should be mechanical rooms independent of the processing room. These areas should be designed to minimize the presence of piping and instrumentation in the processing areas. These areas should also be designed to minimize the need for service personnel to enter the Controlled or Classified Area for routine maintenance/calibration and minimize the need for processing personnel to enter the mechanical room or chase during routine processing operations.

2.4.1.22 Adequate, clean locker rooms shall be provided for personnel entering the manufacturing facility from outside. These shall be physically separated from Controlled or Classified manufacturing areas by at least one intermediate area (e.g., an anteroom or a corridor). Personnel should enter from the "street" or public side and exit into the operations area with the appropriate dedicated plant garments.

Locker rooms shall be designed to maintain adequate separation between street clothes and plant uniforms. This may include separate lockers or sections for storage of plant uniforms. Also, locker rooms should be suitably marked or separated to indicate the limit where street clothes and shoes may be worn. Stepover benches provide suitable barriers but are not required.

2.4.1.23 Break rooms, bathrooms, and mechanical support areas should meet the operating needs and building code requirements. These areas should be separated physically from the manufacturing areas by at least one intermediate area (room, vestibule, corridor, etc.) and when accessing Controlled or Classified manufacturing areas, by at least two intermediate areas (room, corridor, gowning room, air lock, etc.). (Note: The intervening zone may or may not be the same classification as the manufacturing zone.) There should be no break rooms or rest rooms within Classified or Controlled (Nonclassified) areas.

2.4.1.24 Adequate maintenance workshop space and spare parts storage facilities should be available or provided, which should have direct access to utility areas, including those requiring gowning.

2.4.1.25 Systems for sewage, trash, and refuse disposal will be designed and maintained in a sanitary manner. Waste flow should not contaminate the process it is derived from or serve as a source for cross-contamination to other areas. Drains shall be of adequate size and shall be provided with an air break or other mechanical device to prevent back-siphonage. Where pass-throughs are utilized, separate waste and material pass-throughs should be provided.

2.4.1.26 Any permanent equipment parts in contact with the floor, walls, or ceiling should be sealed or installed so that access for cleaning is assured.

2.4.1.27 Wiring leading to systems and fixtures in Controlled and Classified areas should be enclosed in appropriate conduit to enhance cleanability. Grounding wires should be designed using an insulated wire instead of bare wire.

2.4.2 Warehousing/Shipping Areas

2.4.2.1 Finishes and materials used in the warehouse area should emphasize durability to traffic and stress. They should be clean and dry, as appropriate, and maintained within temperature and humidity limits that are consistent with material requirements. Storage areas should be designed to ensure safe storage conditions. All areas should be signed and labeled.

2.4.2.2 Shipping and receiving areas should be designed with ample space to promote a logical and orderly movement of materials and to ensure that forklift traffic does not impede upon personnel traffic during normal operation. Shipping and receiving areas should be separated to minimize opportunities for improper staging or materials/product mix-up

2.4.2.3 Shipping and receiving bays should protect materials and product from the weather. Considerations should be made for minimizing the impact of outside environmental conditions, such as the use of automated doors, docking systems, or other barriers.

2.4.2.4 A transition point from the warehouse environment to manufacturing areas should be clearly marked and may require a vestibule or airlock to maintain the integrity of a controlled area. Shipping and receiving areas should not be designed with direct access to Controlled or Classified areas, whereas access to General area may require only a marked transition zone.

2.4.2.5 A separate area shall be provided for sampling and incoming inspection activities. The physical and environmental conditions of this area should be consistent with the area in which the materials will be later processed.

2.4.3 Packaging Areas

2.4.3.1 Primary packaging areas, where exposed product and primary packaging components are handled (including product-contact dosing aids), should be designed to the same level of control as the associated manufacturing areas.

2.4.3.2 Primary packaging areas should be physically separated from secondary areas as well as from adjacent primary packaging (filling) areas via floor-to-ceiling partitions or walls

2.4.3.3 Physical or spatial separation must exist between secondary packaging areas/lines to prevent mix-ups and cross-contamination. Partial-height partitioning is acceptable. Unidirectional design layout is recommended.

2.4.3.4 Adequate space must be provided for storage and staging of components, in-process materials, and finished goods. Limited access storage areas need to exist, as required, for labeling and coding

2.4.3.5 Packaging areas and packaging lines shall be adequately and clearly marked.

2.4.4 Laboratory/Quality Control Areas

2.4.4.1 Laboratory design must ensure sample integrity and the validity of the test results. Most analytical laboratories may be designated as Support areas. However, where environmental conditions may affect analyses (e.g., humidity, viable particulate), laboratories may be Controlled or Classified.

2.4.4.2 Laboratory areas should normally be separated from production areas. However, those used for in-process controls can be located within production areas provided the process operations do not adversely affect the accuracy of the laboratory results and the laboratory activities do not adversely affect the production process or materials. Laboratories within the operating area and not separated by an intermediate space (e.g., anteroom, corridor) should have the same level of architectural design and area classification as the areas they support. For in-process laboratory areas supporting multiple products, separation from the production area is required.

2.4.4.3 Laboratories for biological testing will be separated from the production areas. Additional requirements may exist in such laboratories for handling these particular substances. These laboratories are Classified and, depending upon materials handled, may require additional hazard classification (e.g., NIH Biocontainment guidelines)

2.4.4.4 Quality control laboratories should be designed to suit the operations to be carried out in them. Sufficient space should be given to avoid mix-up of samples and cross-contamination.

2.4.4.5 There should be adequate storage space for the receipt, handling, and storage of samples, standards, and records.

2.4.5 Ancillary Areas

2..4.5.1 Whenever possible, support staff and general office space should be located outside of the manufacturing areas. If necessary, offices within the production area should have cleanable finishes consistent with the adjacent surroundings and/or provide a barrier to the operating area.

2.4.5.2 Maintenance workshops shall have ample space and be physically separated from the production areas. Space and suitable storage facilities should be allocated for equipment change parts and spare parts. General personnel corridors should be routed outside of the production areas. Adequate space should also be provided in the manufacturing areas for the storage of clean room maintenance toolboxes, etc.

2.4.6 API and Biotechnology Manufacturing Areas

2.4.6.1 There should be facilities to maintain master cell banks in segregated areas under storage conditions designed to maintain viability and prevent contamination.

2.4.6.2 Primary biological manufacturing areas (bioreaction) should be designed to prevent contamination of the culture and reaction products. Although largely closed operations, these areas may involve multiple manipulations (e.g., sampling). Therefore, they should be Classified, typically Grade C or better.

2.4.6.3 Appropriate precautions should be taken to prevent potential viral contamination from pre-viral to post-viral removal/inactivation steps. Therefore, open operations prior to the viral removal/inactivation steps should be performed in areas that are separate from other operating activities and have separate air handling units.

2.4.6.4 Vents and drains from areas containing live biological organisms, including viruses, shall be contained and subject to treatment in order to protect the surrounding environment.

2.4.6.5 If open systems are used for biological purification operations, there should be Classified conditions to preserve product quality. The level of classification should be established based on specific process needs. For example, there should be a Grade B environment where manipulations and connections are performed in and to open vessels. If the process requires the maintenance of sterility, a Grade A environment is required for making aseptic connections for transfers.

2.4.6.6 APIs manufacturing by traditional (e.g., organic synthesis) methods may take place in a General area up to and including the final isolation and purification steps. Exposed product subsequent to final purification should be in a Controlled area. (Closed operations where there is no product exposure may be in a General area.) Note that intermediate steps prior to API synthesis are sometimes outside of the building in some climates. This is more common with large scale operations such as distillation.

2.4.7 Non-sterile Drug Product Production Areas

2.4.7.1 All non-sterile drug product manufacturing operations in which raw or in-process materials, product, and product contact components are exposed should be in Controlled areas. Entirely enclosed operations may be in General areas. Corridors within non-sterile manufacturing operations may be Controlled or General depending upon whether product will be exposed.

2.4.7.2 The facility shall provide for specific gowning locations, dependent upon the area classification, where personnel can properly dress prior to entering the area. The size of the gowning area should be appropriate for the number of users. Intermediate changing areas may be needed within the facility to enter into cleaner areas. Finishes should match those of the area being serviced. Clean garment storage and used garment collection bins must be provided at each gowning area. Alternatively, a laundry pass-through system for disposal may be utilized.

2.4.7.3 A suitable transition zone or airlock shall be used to separate areas of differing air classification at the points where personnel, materials, or equipment enter and exit such areas. Transition zones from Support to General areas may be a simple door (e.g., from Lab to corridor). Transition zones from Warehouse to General corridor may be a normally closed door with a clearly marked floor area for lay down of pallets or containers. Transition zones into Controlled environments should comprise a vestibule with opposing doors. Airlocks should be employed when potential airborne cross-contamination is of particular concern (e.g., classified environments, processing of powders). Airlocks shall have interlocked doors to prevent more than one door from being open at the same time (except in emergency).

2.4.7.4 Process systems should be designed to minimize drug product exposure to the environment. Closed or contained systems are preferred. Where possible, maintenance access to the equipment should be from noncontrolled areas.

2.4.8 Sterile Manufacturing Areas

2.4.8.1 The sterile facility design shall utilize a core concept whereby the aseptic environments are isolated from other less-controlled areas. Critical operating areas (e.g., aseptic filling, lyophilizer loading) within the sterile core should be visible from the less-controlled areas to permit observation by authorized personnel.

2.4.8.2 Access control features should be incorporated into the design of Classified areas. There should be separate areas within the clean area for various operations of component preparation, product compounding/preparation, and filling. Technical space (e.g., for autoclaves and lyophilizers) should not be directly accessible from Classified areas.

2.4.8.3 Compounding of product prior to filter sterilization and aseptic processing shall be in a Grade C area. Compounding of product prior to filling and terminal sterilization shall be in at least a Grade D area.

2.4.8.4 Preparation and holding of filling and packaging components and other materials prior to sterilization shall be in a Grade D area.

2.4.8.5 Aseptic operations, including filling, sampling, and connections, or in which sterile product or components are exposed, shall be performed under Grade A conditions, with a surrounding Grade B area. (see exception in Section 2.4.8.7 following). Filling of non-sterile product prior to terminal sterilization shall also be in a Grade A area, with at least a Grade C background.

Note: Filling of inhalation products (this does not include aqueous nasal sprays) shall be under Grade B.

2.4.8.6 Where aseptic operations are performed, preferred design will utilize full barrier isolators (BIs) or Restricted Access Barrier Systems (RABS) suitable for "advanced aseptic processing". Design details and preferences should be discussed with the project team.

2.4.8.7 Where BIs are used for aseptic operations the background environment shall be at least Grade C. RABS require a Grade B background.

2.4.8.8 Capping of filled and stoppered vials may be as an aseptic process carried out within the sterile core (A/B classification or full BI) using sterilized caps. In this case, this should be segregated from the filling operation to minimize exposure of the product to particulate from the capping operation. Segregation within the sterile core may be a separate room, a full BI or spatial separation incorporating a RABS to limit particulate generation in the surrounding room.

2.4.8.9 Alternatively, capping may be as a clean operation outside the sterile core. In the latter case, the stoppered operation shall be supplied with Grade A quality air within at least a Grade D background until the cap has been applied and crimped. In this case, a RABS should be employed to limit product exposure to the surrounding environment. Also, in this configuration, there should be a means to detect and reject un-stoppered or incompletely stoppered vials prior to capping.

2.4.8.10 Unless specific environmental conditions are required for product stability or integrity, inspection and secondary packaging may be within General areas.

2.4.8.11 Access corridors to gown rooms for Grades A/B and C should be Grade D or better. Gown rooms should be designed as airlocks and used to provide physical separation of the different stages of gowning (pre-gowning, sterile gowning). They should be supplied with HEPA-filtered air. The final stage of the changing room should be the same grade as the area into which it leads.

2.4.8.12 Entry to the sterile manufacturing suite (Grades A/B and C) shall be through separate airlocks for personnel versus materials/equipment. The airlocks shall have mechanically interlocked doors to prevent more than one door from being open at the same time. Sliding doors are prohibited.

2.4.8.13 Personnel, equipment, and production waste materials may exit through a common airlock. Where there is no material airlock, exit from Grade C areas may be through the gown room. Grade A/B areas require a separate de-gown room.

2.4.8.14 Grade A/B gown rooms require a two-stage (pre-gowning, gowning) arrangement utilizing adjacent but separate rooms. The initial room may be Grade C or D and should include a handwashing facility. The second (gowning) room must be Grade B or better. and incorporate a clear barrier (i.e., stepover bench) to demarcate the ungowned from the gowned side.

2.4.8.15 Sinks and drains are prohibited in the aseptic areas (Grade A/B). Floor or wall drains (where necessary) in other Classified areas should be capped.

2.4.8.16 Equipment and component sterilizers (autoclaves, ovens, etc.) and lyophilizers shall be built into room architectural wall, floor, and ceiling assemblies in order to maintain continuously smooth, clean, and sanitizable conditions. Where possible, maintenance access to this equipment should be from nonclassified areas.

2.4.8.17 Operating and control panels serving the aseptic area should be located in adjacent mechanical rooms. Alternatively, they may be constructed integral to process rooms such that operator access is on the clean side and maintenance access is from the nonclassified mechanical rooms.

2.4.8.18 Individual manufacturing processes (i.e., multiple products or similar product, different batches) within a common room should have adequate physical separation (e.g., isolators, barriers, mass air flow devices) to prevent cross contamination of product.

2.4.8.19 Classified areas should include provisions for continuous remote particulate monitoring.

2.5 HVAC DESIGN

Environments or areas are defined as classified and nonclassified. Applications and design criteria for each of these environments are outlined following. The directional airflow cascade scheme should be from "more clean" to "less clean" area. Special design for containment may be necessary where highly active or toxic materials are handled or where biocontainment guidelines apply.

The tables following are guidelines. These may be modified based upon specific product or operational requirements.

2.5.1 CLASSIFIED AREAS

2.5.1.1 Classified areas include processing operations for sterile API, biotech manufacturing, high potency products, inhalation drugs

TABLE 2.1
HVAC Requirements—Classified Areas

Class	Grade A (unidirectional)	Grade B	Grade C	Grade D
Fed. Std. 209e (retired) (operational)	Class 100	Class 10,000	Class 100,000	Not specified
ISO 14644	Class 4.8	Class 5	Class 7	Class 8
at rest operational	Class 4.8	Class 7	Class 8	not defined
Particle limits	See Table 2.3 following.			
Application	Local zone of high-risk operations, i.e., aseptic filling zone, stopper bowls, open ampoules or vials, aseptic connections, loading lyophilizer, primary closure process.	Background to Grade A aseptic operations, filling of inhalation products, purification of biotech products (open processes)	Less critical stages of aseptic processing, preparation of solution prior to filter sterilization, filling of terminally sterilized products, biotech purification (closed processes)	Less critical stages of aseptic processing, include preparation/holding of components prior to sterilization, compounding of terminally sterilized products, access corridors for aseptic and biotech operations.
Filtration	Unidirectional air flow with 99.97% efficiency HEPA or better	Ceiling-mounted HEPA filtration, low-wall return-air; plenum design preferred.	Ceiling-mounted HEPA filtration, low level return-air.	Central or ceiling-mounted HEPA filters. Low level returns preferred.
Directional airflow	Highest positive pressure.	Positive to Grade C, 0.04 to 0.06" WC (10–15 Pa) between areas	Negative to Grade A/B (Class 100), Positive to Grade D, 0.04 to 0.06" WC (10–15 Pa) between areas	Positive to nonclassified area, 0.04 to 0.06" WC (10–15 Pa) between areas
Design temperature	68°F ± 2°F (20°C ± 1°C)	68°F ± 2°F (20°C ± 1°C)	70°F ± 2°F (21°C ± 1°C)	70°F ± 2°F (21°C ± 1°C)
design RH %	50% ± 5%	50% ± 5%	50% ± 5%	55% ± 5%
Changes per hour and recovery guide	Linear velocity adequate to maintain unidirectional flow in the critical zone. Typical 90 ± 10 ft/min at 1 foot from the filter face	The number of air changes must be adequate to guarantee the grade (classification) performance requirements. The design basis should consider the room size, equipment arrangement, personnel required, air flow patterns and process operations, with a 15-to-20-minute recovery time.	The number of air changes must be adequate to guarantee the grade (classification) performance requirements, and should be related to the size of the room, the equipment, air flow patterns, process operations, and personnel present in the room, with a 15 to 20 min. recovery time	The number of air changes must be adequate to guarantee the grade (classification) performance requirements and should be related to the room size, operations, air flow patterns, and personnel flows, with a 15-to-20-minute recovery time under at-rest conditions. HEPA filtration is required.

and all aseptic operations. Table 2.2 following is derived from the Worldwide Quality Standards (modified for clarity) and provides the design philosophy and basic requirements for Classified Environments:

2.5.1.2 Gown room and air locks between areas of different classification should be designed to the higher class. Area classifications are designated according to EU Annex 1 for more universal application. Firms may choose their own nomenclature.

TABLE 2.2
Total Particles per Cubic Meter

Class	At rest		In operation	
	0.5 μM	5 μM	0.5 μM	5 μM
A	3520	20	3520	20
B	3520	29	352,000	2900
C	352,000	2900	3,520,000	29,000
D	3,520,000	29,000	not defined	

2.5.2 Nonclassified Areas

2.5.2.1 Nonclassified areas are identified as Controlled, General, or Support. Controlled Operations are used where non-sterile drug products or drug substances are produced and where product or product contact surfaces are exposed to the environment. Controlled areas are designed to achieve Grade D conditions for the purpose of maintaining control and ease of cleaning. Operational requirements for these areas shall be designated by the manufacturing organization. The actual process operations may be open or closed. General Operations are those areas where manufacturing activities that do not expose the product or product contact surfaces to the environment are carried out. (Exposure of vessel interiors during cleaning only may occur in a General area). Support is a broad category for non-production areas. These may be for storage, staging, analytical laboratories, amenities, and other non-production activities. All spaces need to be designed such that the environment does not impact the quality, purity, and integrity of the products or of the test results associated with product manufacture or quality release. (Note that certain laboratory activity, e.g., microbiological, may require a Classified environment.)

2.5.2.2 Note that the Controlled designation is typically not used for sterile manufacturing or biotech operations where microbial control is critical. These areas should be Classified.

2.5.2.3 Table 2.4 following is based on typical industrial quality standards.

Gown rooms, transition zones and air locks between areas of different classification should be designed to the higher class.

TABLE 2.4
HVAC Requirements—Nonclassified Areas

Class	Controlled	General	Support (offices, labs)	Support (warehouse, mechanical)
Fed. Std. 209e (operational)	Not applicable			
ISO 14644	Class 8	Not applicable		
Particle limits	0.5 μM at rest: 3,520,000/m^3	Not applicable		
Application	Non-sterile drug product manufacture, primary filling and packaging, final API isolation steps	API and intermediate manufacture, closed biotech reactions, closed drug product manufacturing processes	Offices, chemical/ analytical labs, locker rooms, amenities	Warehousing, day staging, shipping/receiving, mechanical rooms, shops, utility areas
Filtration	95% ASHRAE or HEPA, ceiling mount or central bank	Standard industrial or commercial practice to provide comfort conditions.		Not applicable; space heaters may be employed
Directional Airflow	Positive to adjacent General or support areas, sufficient to maintain directional flow of air under normal operating conditions.	May be positive or negative to adjacent non-manufacturing areas depending upon subject operations	Neutral or negative to surrounding Support areas, depending upon operations supported.	Neutral or positive to outside environment.
Design temperature	72°F ± 2°F	72°F ± 4°F Conditioned for human comfort.	72°F ± 4°F Conditioned for human comfort.	72°F ± 4°F Heated and ventilated only to maintain conditions below design temp.
Design RH %[1]	55% ± 5%	55% ± 5%	55%–60% Winter humidification optional.	No RH control.
Air changes per hour	Varies depending upon room size, flow factors; should be adequate to achieve Grade D/ ISO 8	Based upon calculated heat load.		As required to maintain temperature uniformity.

2.6 PROCESS WATER AND PURE STEAM SYSTEMS

2.6.1 General Requirements

2.6.1.1 Process water and pure steam (used directly in manufacturing process or sterilization) are critical systems, as they directly impact product quality. These systems must be designed, constructed, commissioned, and qualified to provide water and steam that meets defined User Requirements.

2.6.1.2 Process water systems and user points must be clearly marked. Exposed permanent piping should be avoided as much as possible in production areas, with only use points exposed in the room, where possible. If water lines must be routed in the process room, they should not be run over areas where product may be exposed.

2.6.2 Water Pretreatment

2.6.2.1 Water pretreatment is a series of unit operations required to purify the feed water to a level that ensures that the final treatment step will consistently produce water or steam meeting the chemical and microbiological specifications.

2.6.2.2 Pretreatment system design must be based upon sufficient data on incoming source water quality to account for seasonal variations (typically at least 1 year of data, reported quarterly)

2.6.2.3 When more than one source is used, equivalent data is required for each source of feed water.

2.6.2.4 The feed water data should provide evidence that the feed water consistently complies with World Health Organization (WHO) and local drinking water standards. If it does not, additional treatment steps (e.g., chlorination) may be prescribed.

2.6.2.5 Supplemental testing should be conducted beyond what may be reported by normal source water testing. As a minimum, this should include tests for

- Total Hardness
- Total Solids
- Alkalinity
- Silica (total and colloidal)
- Total Organic Content

2.6.2.6 The use of ultraviolet (UV) lights is allowable in pretreatment systems to help limit microbial growth. However, UV lights are not a recommended practice as the primary means to account for high microbial counts in source water. In these cases, chlorine or other effective chemical pretreatment is preferred.

2.6.2.7 In-line filters may be used when particle removal is of concern at any unit operation within the pretreatment system. Filter efficiency should be selected based on the expected size particle to be removed. The use of microbially retentive filters is discouraged and is not allowed in recirculating lines.

2.6.2.8 Sample points should be provided to evaluate the performance of each unit operation. When parallel units are specified, (e.g., dual multimedia filters, softeners or carbon beds) sample points should be located after each unit.

2.6.2.9 The location of sample points should be accessible for sampling at working level and should also be safe and secure. Drain funnels for sample points should be provided.

2.6.2.10 The materials of construction for pretreatment stages may be copper, PVC, CPVC (or other suitable polymer). The valves and fittings used should be compatible with piping materials used (sanitary fittings are not required). Sample valves should be of sanitary design.

2.6.3 Water Systems Final Purification

2.6.3.1 For non-compendial (e.g., deionized) process waters, final treatment may be reverse osmosis (RO), deionization, or a combination of the two.

2.6.3.2 The preferred final purification methodology for Purified Water USP/EP is RO followed by continuous deionization (CDI).

2.6.3.3 The RO membranes and CDI unit should be specified to be hot water sanitizable.

2.6.3.4 When using ion exchange technology, CDI is the preferred technology.

2.6.3.5 Material of construction for each unit operation should be designed based on the requirements for sanitization. When a system is designed for hot water sanitization, the internal components should be constructed of stainless steel. No PVC internal components are allowed.

2.6.3.6 Water for Injection (WFI) must be generated using multi-effect or vapor compression distillation technology. Other methods may be acceptable (e.g., RO in combination with other technologies) and must be specifically approved.

2.6.3.7 WFI distillation units should be constructed of 316/316L stainless steel and materials suitable for the high temperature and pressure conditions. Note that other materials suitable for sanitary service may be considered (e.g., for vapor compression stills) upon approval. Construction shall be of sanitary design, as appropriate, with minimum dead-space-diaphragm valves at sample locations.

2.6.4 Water Systems Storage

2.6.4.1 The water storage capacity should be defined to provide adequate available water during periods of use, with allowance made for adequate supply during sanitization of the water treatment system (minimum sanitization allowance shall be 4 hours per day; usually done on an off shift). Water use and capacity analysis should be included as part of the BOD.

2.6.4.2 The storage tank shall be constructed of 316/316L Stainless Steel (316L for interior welded surfaces), ASME rated for pressure (min. 3.5 bar/50 psia) and full vacuum. All welds shall be polished smooth and the interior finish shall be Ra 0.5 μm (Ra 20 μ-inch) or better. The vendor shall provide stainless steel mill certification, surface profilometer report, and vessel ASME (or equivalent) stamp.

2.6.4.3 The tanks should be insulated for heat retention. The exterior skin over insulation may be 304 SS (i.e., 316SS is not required). It shall be finished to a polished appearance approximately 150 grit (US).

The tank shall be heated via an external heat exchanger, either on the main circulating line, on an independent heating loop or both. Steam jackets are not preferred.

2.6.4.4 The tank should include:

- 45 cm (18") man-way with sight-glass and light.
- Sanitary clamp-type fittings, to be located at the top or bottom only.
- Temperature probe
- Pressure relief device
- Optional: microbially retentive, hydrophobic vent filter, to be traced and insulated or steam- jacketed
- Spray ball at the water return line, designed to wet all exposed surfaces

2.6.5 Water System Distribution

2.6.5.1 The distribution tubing shall be 316L stainless steel welded sanitary tubing, with sanitary clamp-type fittings. O-rings to be Kalrez, EPDM, or steam-resistant Viton, all designed to withstand sanitization temperatures. The tubing and fittings interior finish to be Ra 0.5 μm (Ra 20 μ-inches) or better. The exterior surfaces shall be finished to a polished appearance approximately 150 grit (US). The vendors to provide mill certification or other suitable certification of materials (including O-rings) and an isometric of the complete system with the identification of all weld points.

2.6.5.2 The fittings to be limited only as required for instrumentation and other required connections to system components. All clamps in the field to be accessible for maintenance. The direct connections to process systems or equipment should be designed to prevent backflow into the system.

2.6.5.3 The flow in distribution loops should be designed for a minimum velocity of 1.5 m/s (5 ft/s) and/or to maintain turbulent flow at normal full flow conditions as calculated by a minimum Reynolds number of 20,000 with all user points closed. The system design should be to maintain a return flow of approximately 0.5 m/s (1.5 ft/s) when user points (typical diversity factor) are open.

2.6.5.4 The distribution systems should be designed to minimize the number and length of dead legs. In no case should any dead leg be greater that three pipe diameters from the interior wall of the adjoining main (dead legs of this length are not recommended). All pipe runs to be self-draining with a minimum slope of 1:100 to low points.

2.6.5.5 The preferred design temperature for Purified Water distribution loops is ambient (20 °C –22°C). Depending on the specified use, they may also be designed as hot (>70°C) or cold (4°C–8 °C). Trim heat exchangers should be employed to maintain ambient circulating loops below 22 °C. All Purified Water loops should be provided with means (e.g., heat exchanger) to provide for hot water sanitization at 80°C or higher. Alternatively, ozone sanitized systems will be considered.

2.6.5.6 WFI loops should be hot loops designed to be self-sanitizing. Hot loops typically operate at 80 °C, with 70 °C minimum temperature. Ambient or cold WFI loops are allowed if justified by process demands. Any loops operating at a temperature of <70 °C should have a daily scheduled sanitization procedure at a temperature of at least 80 °C.

2.6.5.7 There shall be no microbially-retentive filters in recirculating lines. Where sterile water is required, these may be installed at point of use.

2.6.6 Stainless Steel Tubing Fabrication

2.6.6.1 The specification for high purity water and steam systems should include the requirements for system fabrication. The fabrication specification should address weld quality requirements, accepted welding techniques, welder's qualification, welding procedures, welding process, test coupons, inspection requirements and documentation.

2.6.6.2 The fabrication procedure should include the limits established to accept or reject welds and other installation tests.

2.6.6.3 The documentation requirements should include welding logs, isometric drawings, and installation reports (cleaning, pressure testing, and passivation).

2.6.6.4 The fabrication procedure should address the inspection requirements and methodology. Third-party (independent from installer) inspection is preferred. Boroscope will be considered the required inspection method. X-ray will be used for pipes that are not accessible.

2.6.6.5 Use of chemical cleaning or passivation of all product contact stainless steel welded parts is required. If a passivation procedure is conducted, a detailed report must be provided. The report must include information such as, acid/base type and concentration used, circulation times and temperature, final rinse water quality used, and end point determination.

2.6.7 Valves

2.6.7.1 The valves should be diaphragm type, sanitary design (ITT, Saunders or equivalent); diaphragms to be encapsulated (Teflon, EPDM, Teflon. O-Rings shall be Kalrez, EPDM, steam-resistant Viton) and designed to withstand sanitization temperatures.

2.6.7.2 Zero dead leg valves to be used where possible.

2.6.7.3 The sample valves should be sanitary design; diaphragm or Millipore ESP valves are preferred.

2.6.7.4 All automatic valve user points to be fail closed, unless fail open is the design requirement necessary.

2.6.8 Heat Exchangers

2.6.8.1 The heat exchangers must be designed to be fully drainable and suitable for sanitary operations. The preferred heat exchangers are double tube sheet types. Plate and frame heat exchangers may be allowed provided that they are of a fully welded construction. Differential pressure between the heating /cooling fluid and the water should be provided, with higher pressure on the product side. Differential pressure monitoring should be provided on all heat exchangers as well as pressure relief.

2.6.8.2 Spool pieces should be provided to enable bypassing of heat exchangers for degreasing and passivation.

2.6.9 Pumps

2.6.9.1 The pumps should be centrifugal and single seal sanitary design. Pump construction should be of 316/316L stainless steel (water contact side), with suitable impeller construction and EPDM or encapsulated Teflon gaskets. The pump impeller should be one size smaller than the maximum specified for the casing.

2.6.9.2 The pumps should be connected using sanitary clamp connections for ease of service.

2.6.9.3 The pumps should be specified for use at no greater than 70% of capacity. All pumps should be capable of sanitization by steam with temperature monitoring to be installed at all pumps. Pump curves are required for all pumps.

2.6.9.4 The stand by pumps should not be piped (connected) into the distribution system to avoid the potential of dead legs

2.6.10 Pure Steam Systems

2.6.10.1 Pure Steam is defined as steam that is produced by a steam generator that, when condensed, meets requirements for USP WFI. At sites that manufacture product sold in the European Union, pure steam used to sterilize porous loads should also meet the requirements of European EN285 for non-condensable gases, dryness and superheat [6].

2.6.10.2 The term Pure Steam is preferred. Use of the term "Clean Steam" is considered to be equivalent and subject to the requirements of this Guide.

2.6.10.3 Chemical Free Steam (CFS) is steam produced from pretreated water with no volatile boiler additives, which may be used for humidification or selected process operations not involving sterilization or processing of sterile products. Systems generating Chemical-Free Steam are not required to comply with the other requirements of this section.

2.6.10.4 Pure Steam shall be generated by distillation in a Pure Steam generator unit specifically designed for the purpose. Pure Steam may also be obtained from the first effect of a multi-effect still.

2.6.10.5 Feed water to the generation unit shall meet the requirements of the manufacturer to ensure that condensed Pure Steam will meet the USP/EU WFI monograph.

2.6.10.6 Pure Steam (PS) distribution systems shall be constructed of 316L stainless steel pipe or tubing. Lines shall be sloped to thermostatic type condensate traps. The traps shall allow the system to gravity drain. The distribution system shall be of sanitary welded construction. All fittings shall be sanitary type. Diaphragms valves are not recommended. Sanitary ball valves may be used.

2.6.10.7 Pure Steam or its condensate should not be returned for reuse. Condensate from Pure Steam should be quenched and discharged to drain.

2.6.10.8 Pure Steam sample points shall be provided with a cooling water location nearby for connection to the steam-sampling device. Each use point shall be provided with a valve for sampling of steam using a portable condenser.

2.6.11 Water and Pure Steam Instrumentation and Controls

2.6.11.1 Water and pure steam systems shall be provided with automated control systems adequate to ensure repeatable and reliable operation. Where appropriate, these systems may be interfaced with a higher-level control or monitoring system or linked back to a central control system.

2.6.11.2 At a minimum, instrumentation and control systems should include monitoring and/or testing of the following:

- Temperature, pressure, and conductivity monitoring for each component of the pretreatment system, as applicable.
- Regeneration/backwash of pretreatment unit operations.
- Sanitization sequence for unit operations after primary treatment.
- Alarms and interlocks for the integration of the pretreatment, final treatment and storage and distribution systems, as required.
- "Dump on start" function for WFI stills, based on conductivity of the distillate.
- Storage tank—"dump on high conductivity" as a quality function.
- Temperature control and recording for the storage and distribution system.
- Monitoring of conductivity (on-line) at the supply to the storage tank and on the return from the distribution loop.
- Monitoring of total organic carbon (TOC) at the supply to storage tank and on the return line from the distribution loop. Monitoring may be done on-line, with the application of a suitable SOP.

2.7 PROCESS EQUIPMENT, PIPING, AND INSTRUMENTATION

2.7.1 Equipment Requirements

2.7.1.1 Equipment used in processing of pharmaceutical drug products and APIs should be of appropriate design, adequate size, and suitably located for its intended use, qualification, cleaning, sanitization/sterilization (where appropriate) and maintenance.

2.7.1.2 The requirements for process equipment, piping and instrumentation for a manufacturing system will be based upon a User Requirements document. This document will be specific to the system being designed and will include quality, business and other (e.g., safety) requirements from the user perspective. It is not intended to be an engineering or technical specification and will normally not include engineering details unless they are truly required by the user (e.g., to match the process as described in a regulatory filing).

2.7.1.3 An Engineering Specification will be prepared by an engineer or firm designated by the Project Execution Plan. This will describe the technical details in order to meet the User Requirements and is the main document used to communicate with vendors, designers and/or contractors. This may also be called a technical specification.

2.7.1.4 For systems involving complex automation, especially where a high degree of customization is required, a Functional Requirements Specification (FRS) may be required. In these cases, it is permissible (and even recommended) to separate the mechanical (engineering) specification from the automation (FRS). In most cases of skid-mounted equipment with integral standard automation packages, a single engineering specification will suffice.

2.7.1.5 Design details (e.g., datasheets, isometrics, P and IDs, electrical diagrams, General Arrangement Drawings, etc.) are developed by an engineer or firm designated by the Project Execution Plan. For specialized production equipment, these details are usually supplied by the vendor.

2.7.1.6 Purchase agreements should affirm the user's right to audit and inspect the vendor's fabrication facilities, including quality systems. Vendors shall prepare/present a project quality plan or applicable quality manual for the equipment to be delivered.

2.7.1.7 The equipment must be compatible with the product and process for which the equipment is intended. Product contact materials must be resistant to corrosion and must not contribute to product contamination, either by chemical interaction or by promoting microbial growth. Drug product and biotech manufacture require sanitary design. API processed by organic synthesis do not require sanitary design, except for the final isolation of the drug substances subject to microbial limits.

2.7.1.8 The equipment and control panels associated with the manufacturing system in Controlled areas should be designed with sloped surfaces when feasible to provide ease of cleaning. Panels in Classified areas should be flush mounted to the walls.

2.7.1.9 Equipment access should be provided to allow for ease of cleaning, qualification, inspection, validation, processing, sampling, maintenance, and calibration activities. Special access devices, such as lifters and manways, should be identified in the specification, when appropriate.

2.7.1.10 The lubricants used must be compatible with the product and process. Food grade lubricants and/or GRAS (Generally Recognized As Safe) substances must be used for drug product, bioprocessing, and final purification and isolation for APIs. Refer to 21CFR178.3570 on *Lubricants with Incidental Food Contact* for a listing of FDA-approved lubricants [14].

2.7.1.11 Seals with potential product contact should be suitable for the application and designed to withstand process conditions (e.g., sterilization). The seals should be of minimal wearing/shedding material. Seal lubricating fluid (that could potentially have product contact) must be identified and treated as a potential product contact fluid or material.

2.7.1.12 When equipment is designed for steam sterilization, it should be ASME rated for pressure (min. 3.5 bar/50 psia) and full vacuum.

2.7.2 Materials of Construction

2.7.2.1 The product contact materials of construction and surface finishes must be suitable for the product, the process, and with other materials used in cleaning, sanitization or sterilization requirements. The product contact surfaces should have a smooth surface, that is, free of pits, cracks, crevices, folds, projections, and other imperfections.

7.2.2 The external materials of construction and surface finishes should be compatible with the surrounding area and/or room classification. They should present a clean and/or polished appearance, be smooth, cleanable, and not subject to deterioration or corrosion. Exterior painted surfaces are not considered suitable for Classified areas and are not preferred in Controlled areas.

2.7.2.3 If painting is necessary on non-product contact surfaces, it should be shielded (as applicable) to separate the product from the painted surface. Chemically resistant, non-chipping and non-wearing exterior paint systems must be documented (e.g., baked on powder coat finish or two-part epoxy paint are examples of acceptable painting systems).

2.7.2.4 Vendors are required to provide project-specific certification of product contact materials, including mill certificates where appropriate. In no case shall reference to a model number be considered adequate certification. Where mill certificates are not available, the vendor must provide written certification with specific reference to the purchased equipment identified by serial number, if available, or to the bill of materials for a commodity item (e.g., valves).

2.7.2.5 CIP (Clean in Place) equipment should be considered to facilitate cleaning of the product contact areas.

2.7.2.6 Product contact materials should not require a separate finish (e.g., no paint on product contact surfaces). If a coating is required, the coating for product contact surfaces must be resistant to cracking or chipping. When required for corrosion resistance or other surface properties, surface treatments must be specified (e.g., anodizing, or passivation).

2.7.2.7 The preferred vessel/equipment material for drug product processes is 316/316L Stainless Steel (316L for interior welded surfaces). All welds shall be polished smooth and the interior finish shall be Ra 0.5 μm (Ra 20 μ-inch) or better. Where hard to clean materials (e.g., protein residues) are processed electropolished interior finishes may be required. The vendor shall provide a stainless steel mill certification, surface profilometer report and vessel ASME (or equivalent) stamp.

Note: Vendor standard finishes up to Ra 0.8 μm (32 μ-inch) may be accepted when particularly difficult cleaning issues are not expected (e.g., dedicated equipment, soluble actives).

2.7.2.8 Jackets and exterior skin may be 304 SS and shall be finished to a polished appearance, approximately 150 grit (US).

2.7.2.9 The product contact valves design should be appropriate for the intended use and for cleaning (e.g., sanitary valves for drug product processing and bioprocessing). The product contact valve surface finish should be consistent with the surface finish of the attached process equipment. All product contact surfaces of the valve should be cleanable.

2.7.3 Process Piping

2.7.3.1 The process piping materials and internal contact surface finish should be consistent with the associated process equipment.

2.7.3.2 The product contact piping, tubing, and fittings used in drug product and bioprocessing process must be sanitary in design in compliance with the ASME BPE Guide (latest revision) [1]. Acceptable designs include sanitary clamp (e.g., Tri-Clamp) and European Hygienic fittings. Threaded fittings (e.g., IPT, NPT) and flat flanged fittings must not be used in these applications. In the event that equipment that performs a critical process step is available only with threaded fittings (e.g., compression-type or sanitary thread), the system must obtain the approval of GES management.

2.7.3.3 The process equipment and piping systems should be designed to be self-draining. Process piping systems sterilized in place should be designed to allow for removal of trapped air and condensate

2.7.3.4 All process piping points-of-use, sampling ports, and access ports used (e.g., electrical, material handling) should be appropriately identified.

2.7.3.5 All product contact gasket and O-ring materials should be suitably resistant to processes (e.g., able to withstand chemical reaction), cleaning solutions, and sterilization. The gasket inspection or replacement intervals should be specified based on the life expectancy of the material and extended use, as appropriate.

2.7.3.6 The preferred O-ring material for sanitary service are encapsulated Teflon, EPDM, steam-resistant Viton, and Kalrez 6221/6230. All elastomers in product contact must comply with 21CFR177, sections 2400(d) and 2600.

2.7.4 Process Instrumentation

2.7.4.1 The instrumentation should be chosen for the accuracy and reliability over the entire process range. Factors that affect reliability or accuracy of devices include process conditions, ambient conditions, and location of the instruments.

2.7.4.2 The instrument sensors should be installed in locations that provide ease of removal for calibration and validation purposes and where they are not easily snagged. If fitted in areas where they are prone to damage, instrument sensors should be fitted with protective caps.

2.7.4.3 The product contact materials shall be compatible with the process as well as cleaning and sanitizing materials and conditions. The exterior materials of construction should be similar to the associated process equipment.

2.7.5 Process Automation

2.7.5.1 Identification of all measured or controlled parameters, their range and the degree of control required must be specified. The critical process and environmental parameters that directly affect product quality will also be identified. Measuring instruments must be accurate enough to reliably measure critical parameters. Preferred accuracy is one order of magnitude better than the required operating range (i.e., for temperature control within ±20°C, the sensor should be accurate to ±0.20°C)

2.7.5.2 Various control methodologies are acceptable, depending on the project or site automation architecture and the application. These include stand-alone PLC, PLC-SCADA, DCS, single-loop control, DDC, and BMS. As a minimum, the control system should:

- Measure, control, and record critical process parameters
- Measure and record critical quality attributes, as defined in the User Requirements
- Provide specified interlocks, permissives, and alarms. Provide a record of alarms
- Be capable of generating accurate and complete copies of GMP records in electronic form suitable for inspection and review
- Control sequences of operations; allow for selection of multiple recipes or processes as required
- Provide fail-safe shutdown, with memory protection
- Provide logical and physical security to prevent unauthorized access to the system and its data
- Have controls to ensure electronic record integrity and security
- Have an audit trail to provide a record of user actions

2.7.5.3 All automated systems are subject to a life-cycle development approach, beginning with the User Requirements.

2.7.5.4 For further information on development and validation of automated control systems refer to the chapter "Control Systems Validation".

2.8 COMPRESSED AIR AND PROCESS GASSES

2.8.1 The system materials of construction shall be consistent with a quality specification and include all of the site distribution and storage systems leading to the use point. They should prevent accumulation of moisture/oil and particle generation. Materials such as copper (with brazed connections) and stainless steel have been proven suitable for the operation. Stainless steel should be used downstream from any final filter into the production area. Valves and fittings should be specified to match the piping material finish and rated to operating pressure requirements. Sample points should be provided.

2.8.2 Where technically feasible, the use of non-oil lubricated air compressors is preferred for product-contact applications.

2.8.3 All compressed gases that are used in contact with the product require the use of equipment designed to remove particles as well as to remove oil or water that may be present in the gas. Additionally, point-of-use filtration is recommended for sterile gas requirements.

2.9 MICROENVIRONMENTS

The microenvironment equipment includes closed units such as stability chambers (environmental chambers), incubators, isolation units, biosafety cabinets, glove boxes, and cold boxes (e.g., refrigerators, freezers), as well as partially open equipment such as laminar flow hoods, fume hoods, and down flow hoods. All microenvironments are designed to provide a specified set of environmental conditions different from or more tightly controlled than the surrounding room conditions.

2.9.1 General Design Requirements

2.9.1.1 The operating unit must specify the environmental operating set points (or target values) and ranges for the microenvironment. These parameters are to be documented as part of the User Requirements or other suitable document. The environmental parameters that may be specified include:

- Temperature, relative humidity
- Level of light or oxygen
- Airborne particle count (total and/or viable)

2.9.1.2 Appropriate equipment and instrumentation must be in place to ensure the specified environmental conditions are reliably met and monitored. A disaster recovery plan must be in place for the equipment on site and the associated microenvironment. This will include (where appropriate) emergency/backup power and/or redundancy of equipment in specified circumstances.

2.9.1.3 The design of all new microenvironment equipment should take into account the ergonomics of all operator interfaces. This should consider working height, reach, visibility, lighting, ventilation, cleanability, and change out of filters.

2.9.2 Partially Open Equipment Design

2.9.2.1 Air ingress (direction, "laminarity", velocity or volume and intake vs. exhaust) must be specified, as appropriate.

2.9.2.2 The interaction with the room air in which the unit is located must be considered and appropriate air balancing must be provided.

2.9.2.3 The air filtration requirements must be specified including appropriate filter change-out provisions.

2.9.2.4 The material of construction must provide for ease of cleaning/sanitization/decontamination. There should be appropriate vision panels to facilitate the visibility of the operations. It is preferred that non-vision surfaces that are exposed to the microenvironment be constructed of stainless steel. However, other materials may be used if they are suitably compatible with the use. Exposed painted surfaces are generally not permitted.

2.9.2.5 The instrumentation should include pressure differential and airflow. As appropriate, sash height interlocks should be installed that alarm in the event of an opening that exceeds the qualified height.

2.9.3 Cold Boxes, Incubators, and Stability Chambers

2.9.3.1 The temperature, relative humidity, light, and oxygen concentration set point(s) and operating range(s) must be specified, as appropriate.

2.9.3.2 The water/steam quality used for humidification must be specified. Potable water and/or plant steam may be suitable if the product is not contacted.

2.9.3.3 Interaction with the room/external environment in which it is located must be considered. As appropriate, temperature and relative humidity differentials must be accounted for.

2.9.3.4 The Material of Construction must be cleanable and suitable for the environment. Viewing panels may be installed in the door or wall to facilitate viewing of the contents. The viewing panel must be designed such that condensation on the viewing surface is minimized.

2.9.3.5 The equipment must be designed to facilitate ease of maintenance. Noninvasive preventative maintenance is preferred (e.g., to change fans, filters, or lighting). Door gaskets should be designed to permit expedient changing.

9.3.6 The instrumentation should include an appropriate number of monitoring probes that is justified based on the size of the equipment. The monitoring probe(s) should be separate from the controlling probe., There must be appropriate parameter recording capabilities, preferably continuous. There must be appropriate parameter alarming capabilities (facilities and/or procedural) including audible and visual alarms in the local area.

2.9.3.7 Security and access control must be considered. This is particularly important when controlled drugs are contained in the equipment.

2.9.4 Isolators and Glove Boxes

2.9.4.1 Particle concentration (total and /or nonviable), temperature, relative humidity and oxygen concentration target values and operating ranges as well as working light levels must be specified, as appropriate to the use of the unit.

2.9.4.2 The air filtration requirements, including recirculation vs. fresh air makeup/exhaust, must be specified. The ventilation/exhaust system must serve only one isolator/glove box (multiple connected chambers may constitute a single unit) and must include independent inlet and exhaust air systems.

2.9.4.3 Where required (i.e., aseptic isolator), HEPA filtered and/or unidirectional air should be provided. In these circumstances, a HEPA pre-filter should be provided for the supply air mainly to provide backup in the event of a failure of one of the filters. When the isolator will be aseptic, the air-handling unit (AHU) and filtration system must be capable of providing Grade A conditions.

2.9.4.4 There should be HEPA exhaust filters to prevent contamination of the microenvironment in the event of a backflow or to ensure containment when processing toxic or hazardous materials. All HEPA filter systems should be designed to permit integrity testing in place. HEPA filters should be resistant to the cleaning solvents used to clean the isolator.

2.9.4.5 Filter maintenance is critical in these units; the "bag in–bag out" design for filter change-out has proven suitable and is recommended when toxic or high potency materials are handled.

2.9.4.6 In aseptic isolators, a minimum positive pressure differential of 10 Pa (0.04" WC) under all operating conditions and a minimum differential of 25 Pa (0.10" WC) during product processing is recommended. If the requirements justify a negative pressure (e.g., because of operator safety when handling sterile but toxic materials), then consideration should be given to enclosing the unit in a positive pressure envelope (the room itself can constitute the envelope).

2.9.4.7 Isolators with an active pass-through for product flow (e.g., a "mouse hole") must maintain an air velocity of 0.5 to 0.7 meters per second (1.6 to 2.3 feet per second) from the isolator to the outside environment.

2.9.4.8 Isolator gloves/sleeves and half-suits must be rugged, provide for appropriate dexterity, and be designed to facilitate ease of changing and frequent integrity testing with minimal impact on the contained environment. Both one piece as well as two-piece glove/sleeve construction have proven adequate. The material of construction must be carefully considered and justified based on process and material contact needs. (e.g., Neoprene or Hypalon). Gloves must also be compatible with cleaning agents and not allow pass-through of solvent. It should be possible to integrity test the sleeves and gloves.

2.9.4.9 Half suits (when used) should be double walled and provided individual air supply (preferably controlled by the operator) to permit operator comfort. It is preferred that this air be HEPA filtered, however, the specific quality should be established based on process needs. The material of construction should also be evaluated based on process needs (e.g., nylon-lined PVC).

2.9.4.10 The Material Transfer apparatus (e.g., air lock, or rapid transfer port—RTP) must be sized considering the material being transferred and the port must be designed to transfer material without integrity loss of the microenvironment. Examples of RTP designs that have been successfully used include the Double Porte De Transfert Entanche (DPTE), the CQ Trans Plus, Automatic Transfer Valve, The Passport and High Containment Port.

2.9.4.11 The gaskets must be resistant to sporicidal agents (for aseptic isolators) and designed for ease of change out. Consideration must be given to passage of air between the room and the microenvironment (or, as appropriate, between the microenvironment and the room).

2.9.4.12 If an air lock style transfer port is installed, there must be an interlock (mechanical or automated) to prevent opening of both doors simultaneously. There must be facilities or operational procedures that ensure that operations cease if the interlock fails. Isolation unit design must provide for minimal use of "conventional" doors during non-product exposure operations (these doors are a potential source of contamination) and no use of "conventional" doors during product exposure operations.

2.9.4.13 The materials of construction must be resistant to the operating environment (e.g., heavy solvent load or corrosive conditions) as well as the cleaning/sterilization process (e.g., peroxide used for chemical sterilization). The material of construction must also be resistant to adsorption of chemical sterilants (e.g., polycarbonate has a tendency to adsorb hydrogen peroxide). These considerations also apply to the supporting AHU, the transfer ports, and any other apparatus installed within the microenvironment.

2.9.4.14 If the microenvironment will be operated at a negative pressure, then a flexible material of construction is not recommended. The surface should be smooth with minimal crevices and, as appropriate, rounded corners. The chamber should be equipped with "low point drains" to facilitate proper cleaning. There must be minimal sharp edges inside the equipment to prevent glove tears.

2.9.4.15 The isolators/glove boxes should be located in rooms/areas that facilitate suitably clean and orderly execution of operations. They should be located in a room or area that restricts access to only personnel related to isolator operations. The unit should be located to prevent condensation on the internal walls (e.g., room supply air grille should not impinge directly on the walls of the unit, temperature differentials between room and operating environment, etc.) The quality of air in the surrounding area must be established based on operational needs and worker comfort. Aseptic isolators used to manufacture sterile products must be located in an area designed to at least a Grade C environment. (Note: Area may be designated Grade D in actual use.)

2.9.4.16 If an aseptic isolator will be decontaminated using a chemical agent, then the agent must be sporicidal. Peracetic acid, hydrogen peroxide and chlorine dioxide are frequently used and may be used without qualification of the material itself (use of the material in the equipment, however, must be qualified). Other agents may also be used if suitably qualified. If the agent is purchased, then it must be received, tested, released and stored in a manner similar to other GMP materials. If the agent will be generated on site, then the generation equipment must be qualified in a manner similar to that of other GMP equipment.

2.9.4.17 The equipment should be designed to enable multiple modes of operation (e.g., routine operation, sterilization, cleaning, drying, flushing, etc.) There should be ample provision for compressed gas, steam, chemical decontamination agents, electrical, etc. to match process needs and the connections should match the operational requirements (e.g., sanitary fittings). There should be minimal horizontal surfaces to prevent accumulation of material. The internal design should enable access to all surfaces for cleaning without major dismantling. When temperature or humidity sensitive sanitization/sterilization processes are used, the equipment must be equipped with appropriate baffling, assist fans, and control systems to ensure uniform distribution as such.

2.9.4.18 Special consideration must be given to exhaust filter design. The potential for chemical contamination during the chemical decontamination cycle and subsequent blow back into the operating environment must be considered. Filters should guard all vacuum ports. In the case of an aseptic isolator, a microbial retentive filter must be used.

2.9.4.19 CIP systems are preferred and must consider the type of product as well as complexity of the unit. CIP-assist (e.g., difficult to clean certain areas) has also been used successfully. Under all circumstances, the cleaning cycle must consider disposal of the cleaning agent, thus considering use of cleaning agents at high volume at low pressure against low volume at high pressure. The equipment should also provide for an appropriate drying cycle and the use of a drying gas (air or nitrogen) under appropriate circumstances. All surfaces inside the isolator should be free and the isolator should have a drain to remove the CIP agent.

2.9.4.20 Instrumentation should include an appropriate number of monitoring and controlling probes that is justified based on the planned processes (e.g., operational, decontamination, drying, etc.) as well as the size of the equipment. The monitoring probe(s) must be separate from the controlling probe. There should be appropriate parameter recording capabilities (preferably continuous). It is preferred that aseptic isolators are equipped with continuous nonviable particle monitoring instrumentation and it is preferred that the probe be located at the supply air discharge to the microenvironment.

2.10 CONCLUSION

Validation is dependent on facility and equipment design that will ensure product quality by providing for the proper environments, avoiding mix-ups, preventing cross-contamination, and ensuring process control. Understanding and applying good design is not enough. Beyond that, the management of a biopharmaceutical facility project must take into account the importance of validation as a necessary and critical milestone. The project team must include key members who are focused on validation and all other aspects of product quality. These members must be involved in all relevant reviews and decisions. This integrated approach has proven to be a significant driver toward overall project success.

NOTE

1. Humidity may be allowed below the stated range, e.g., in winter, if process operations allow.

BIBLIOGRAPHY

1. American Society of Mechanical Engineers. *Bioprocessing Equipment*, 2019.
2. DeSantis P. *The GMP Project RoadmapTM*, 2021. http://desantisassociates.com/services/project-management/
3. DeSantis P. *A Framework for Quality Risk Management of Facilities and Equipment*, Pharmaceutical Online, January 2018. www.pharmaceuticalonline.com/doc/a-framework-for-quality-risk-management-of-facilities-and-equipment-0001
4. DeSantis P. *Facilities and Equipment Risk Management: A Quality Systems Approach*, Pharmaceutical Online, December 2017. www.pharmaceuticalonline.com/doc/facilities-and-equipment-risk-management-a-quality-systems-approach-0001
5. Eudralex, Volume 4. *Good Manufacturing Practices (GMP Guidelines).*
6. European Standard EN285. *Sterilization-Steam Sterilizers-Large Sterilizers*, 2015.
7. Independent Project Analysis. *Capital Excellence for Pharmaceutical Projects*, Presentation to Schering-Plough, November 2005.
8. Independent Project Analysis. *Commissioning and Qualification Best Practices*, Conference Notes, November 2003.
9. International Society for Pharmaceutical Engineering.
   a) ISPE Baseline® Guide Volume 1: *Active Pharmaceutical Ingredients*, 2007.
   b) ISPE Baseline® Guide Volume 2: *Oral Solid Dosage Forms*, 3rd Ed., 2016.
   c) ISPE Baseline® Guide Volume 3: *Sterile Dosage Forms*, 3rd Ed., 2018.
   d) ISPE Baseline® Guide Volume 4: *Water and Steam System*, 3rd Ed., 2019.
   e) ISPE Baseline® Guide Volume 5: *Commissioning and Qualification*, 2nd Ed., 2019.
   f) ISPE Baseline® Guide Volume 6: *Biopharmaceutical Manufacturing Facilities*, 2nd Ed., 2013.
10. International Standards Organization, ISO 14644. *Cleanrooms and Associated Controlled Environments.*
11. PIC/s. *GMP Guide for Good Manufacturing Practice for Medicinal Products.*
12. Schering-Plough Corporation. *Global Engineering Services Internal Document, GMP Facilities Design Guide*, 2009.
13. US Code of Federal Regulations, 21 CFR Part 211. *Current Good Manufacturing Practice for Finished Pharmaceuticals.*
14. US Code of Federal Regulations, 21 CFR Part 178.3570. *Lubricants with Incidental Food Contact.*
15. US Food and Drug Administration, Guidance for Industry. *Process Validation: General Principles and Practices*, 2011.

Appendix
GMP Design Review Template

The following template is provided for GMP Design Review as described in Section 2.3.2. The body of the document may be derived from the design guidance provided in Sections 2.4–2.9, with initial examples provided following. The overall format of the document may be constructed like a validation protocol or report following the user's conventional format. It should include:

- Title page
- Approvals
- List of GMP Review participants, including affiliation (may include third-parties)
- List of documents reviewed
- Matrix/questionnaire (example following)

	Requirement	Observation	Notes/actions
1	**Facility Design Requirements**		
1.1	**General Considerations**		
1.	Corridor dimensions and door openings shall take into account the movement of large equipment and materials. Appropriate wall protection (e.g., bumpers, railings) should be utilized where necessary to protect against damage to equipment, walls, and doors. Mobile equipment should also be designed to minimize damage to the facilities.		
2.	Operations will be located and performed within specifically defined areas of adequate size. There shall be separate or defined areas as are necessary to prevent contamination or mix-ups during the course of the following procedures: • Receipt and holding of materials, components, and labeling pending approval for use • Holding of rejected materials, components, and labeling before disposition. • Storage of released components, in-process materials, and drug products after release and labeling. • Manufacturing, packaging, and labeling operations. • Quarantine storage prior to release of drug products or APIs. • Control and laboratory operations. • Sampling of intermediates, APIs and finished or unpackaged drug products. • Personnel dress/gowning, hygiene and food consumption.		
3.	In areas where potent or toxic materials are processed, consideration should be given to enhanced cleanability. These areas should have a separate HVAC system.		
4.	In Controlled and Classified areas, the walls, floors, ceilings, etc. should be constructed so that the primary surfaces are accessible for cleaning. Exposed horizontal surfaces in Controlled areas are generally acceptable if they are accessible for cleaning. They should, however, be sloped to allow free draining.		
5.	Horizontal surface should be avoided as much as possible in Classified areas. Windows, door and fixtures should be flush with primary surfaces.		

Abbreviations: APIs = active pharmaceutical ingredients; HVAC = heating, ventilation, and air-conditioning

3 Modular Facilities—Meeting the Need for Flexibility

Maik Jornitz and Sidney Backstrom

CONTENTS

3.1 INTRODUCTION

In the life cycle of pharmaceutical and biopharmaceutical commercial manufacturing, the one constant has been change. With the advent of biotherapeutics, large-scale commercial manufacturing became common place. Typically, such manufacturing created large-scale purpose-built facilities designed to satisfy "blockbuster" drug demand. Because the cell expression rates and protein yield were low it was not uncommon to utilize large-volume 20,000 L bioreactor systems. And often these manufacturing sites utilized a "six-pack" of the large bioreactor volumes and equally large purification systems. Improvements in the expression levels in mammalian cell culture processes allowed the reactor volumes to decrease substantially, enabling the continued and growing implementation of single-use process technologies. Smaller mass balances and tremendous reduction of cleaning solutions because of single-use process technology resulted in process intensification, and therefore smaller footprint cleanroom infrastructures in facility design projects. Demand for smaller, more flexible, manufacturing facility designs has also been accelerated by the need to produce in-country/for country and by advances in biosimilar technology, which requires multiproduct processing capabilities. Furthermore, smaller environments can provide the containment required in the production of highly potent components and purely aseptic processes in cell therapy production. Smaller volume filling created the opportunity for new aseptic processing technologies, for example presterilized container systems and robotic filling in isolators. These systems are so compact that drop-shipping the needed filling capacity within a prefabricated cleanroom environment is now a real possibility. Some of these smaller infrastructures have the further benefit of being able to be rapidly deployed thus reducing the time to first production runs with the resulting monetary benefits.

Once thought of as a "moon shot" future technology, these modular/podular facilities are now replacing traditionally built and designed methods to create the benefits of budget and delivery time robustness. The industry is also beginning to see architecture and engineering (A and E) firms embrace this cleanroom offering in favor of or in addition to the traditional design build approach. It has been realized that traditional design and on-site built is not just disruptive

DOI: 10.1201/9781003163138-3

but often generates lower productivity, lengthy build times and change orders. Additional factors like "repurposability", containment, and avoidance of contaminants like mold are also considered.

The current state of the pharmaceutical and biopharmaceutical industry largely dictates what is required regarding a cleanroom. Cleanrooms made for a "blockbuster" product would conceivably require more cleanroom space than the production of an "orphan" drug, because the volumes are much larger, especially if the expression rate is at the lower end experienced in many applications. Other factors also have an effect on the size and number of cleanrooms, such as increasing expression rates in mammalian cell culture systems, the advent of single-use technology, advances in personalized and orphan therapies, etc. These factors, either by themselves or in combination with each other, have made smaller scale manufacturing of pharmaceuticals and biopharmaceuticals a growing trend.

This trend has led to an increased interest in modular or podular manufacturing platforms, podular being prefabricated cleanroom spaces. The promise of modular and podular manufacturing areas has been speed and flexibility. Speed and flexibility in this context are relative, the question generally being, is it faster and more flexible than traditional brick and mortar facilities? These two low barriers are easily overcome by most modular and especially podular options. When they are, the next question is, "What's the cost per square foot?" And often times, this becomes the primary factor in the decision as to which cleanroom material and design to choose. The counter question to cost per square foot is "How often was the cost per square foot quote achieved when the project was finished?" This begs the review of which of the options of cleanroom designs and infrastructure is most robust in its cost and delivery prediction.

There are a number of key drivers to consider in deciding which facility design option is the best for a particular need. Although this chapter advances the notion that cost per square foot is not the appropriate center point of that discussion, it won't be disregarded as a factor. Speed and flexibility will also be considered and not just in the context of a comparison with brick and mortar. Other cost factors and not to be forgotten, quality, will be considered as well. Only after considering all of these factors can an informed decision be made on the cleanroom option for a project.

3.2 THE OPTIONS IN CLEANROOM INFRASTRUCTURES

Currently there are four primary cleanroom options. They are: stick-built construction, modular construction on-site, modular construction off-site, and podular prefabricated cleanrooms.

Stick-built construction is usually regarded as building internal infrastructures, cleanrooms, and ancillary spaces in a traditional way at the production site. Wood or metal framing arrives in bulk and is cut and erected where needed. Sheetrock or gypsum wall board is affixed to the framing. Electrical, plumbing, finish work, and other crafts are all completed on-site. This type of material and construction has been widely used in the past, as it was once the only option. The benefits of these material choices are experience and low costs. The weaknesses are the robustness of the wall structure and possible mold contamination when the epoxy coating has been breached (1). Such a process, depending on the scope, requires a large amount of labor at the site and typically takes months, and for large projects, even years to complete. This is considered a major disruption of the site with building materials being shipped and stored on-site, additional costs of insurance, required security, additional safety concerns, and the need for additional parking for construction crews.

Modular construction on-site is generally considered as building processing space using modular panels at the ultimate production location. These wall panels are typically an aluminum honeycomb that has been covered with polyvinyl chloride (PVC). Modular wall panels are much more robust than epoxy-coated drywall. But that additional quality comes at a much higher price. There are many companies that provide such panels. Some companies provide the installation of the panels whereas others simply sell to installers who conduct the on-site work. There are many cleanroom contractors who can assemble such panels on site. The panels are also provided in varying sizes so less on-site cutting is required. Less finish work is also required as panel providers provide finish pieces such as coving at the wall-to-ceiling and wall-to-floor connection points. Local mechanical, electrical, and plumbing crafts are required at the site for this method just as they are in the stick-built approach. Figure 3.1 provides an example of the modular on-site construction. The implementation of a modular wall panel cleanroom infrastructure is faster than stick-built, but it still requires a considerable headcount on-site, including the additional cost factors often not considered in the planning stages.

FIGURE 3.1

Source: AES Clean Technology, Inc.

FIGURE 3.2

Source: General Electric Co.

FIGURE 3.3

Source: Pharmadule.

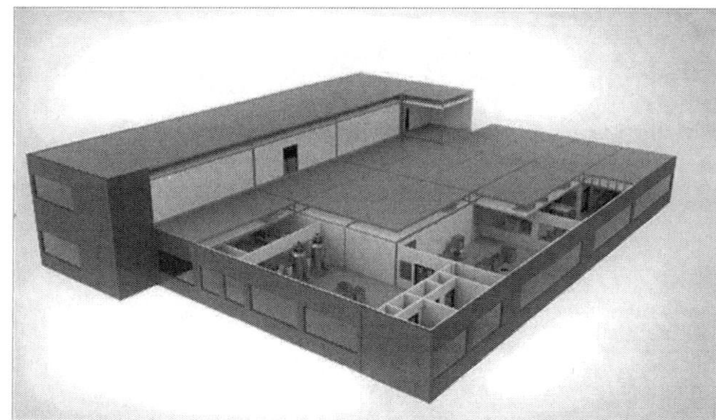

FIGURE 3.4

Source: General Electric Co.

Modular construction off-site is generally considered to be building an entire facility at a manufacturing location module by module, then erecting a structure entirely of those modules and connecting the modules, also at the manufacturing location. These modular facilities can contain 60 to 150 modules or containers, which are typically welded together at the final location. The facility is then tested for functionality and disassembled, whereupon the individual modules are shipped and re-erected at the ultimate production site. Figures 3.2, 3.3, and 3.4 provide an example of this process. At a high level, the advantage to this approach is being able to build facilities where little to no infrastructure exists. Also, being able to receive a full turnkey solution from one provider can be attractive to prospective facility owners. However, assembling the 60 to 150 units means additional construction time needs and for a certain time period and a larger headcount to assemble and weld. Once the facility is built, it will stay where it is, and it serves the application it was designed for, which makes the option inflexible.

Prefabricated, podular cleanrooms are off-site built units that provide clean space including floors, walls, ceilings, windows, and doors in appropriate finishes. In some instances, such modules include their own mechanical space where automation PLCs, HVAC, fire suppression systems, utility connections, etc. are housed, making the cleanroom unit autonomous from the host facility. The cleanrooms are built entirely at a supplier's manufacturing site with much higher productivity levels than an on-site built. At the close of the manufacturing process, they are factory acceptance tested (FAT). After such prequalification testing, they are shipped and moved into the ultimate host facility. Because of the utility infrastructure within the cleanrooms, connections to the host facility are marginal, and infrastructure within the host facility is also minimal. High-level advantages include being able to build the modules rapidly without permitting, the reduced requirement of utility infrastructure at the ultimate destination, the ability to move and repurpose the cleanrooms, and depreciating the cleanrooms in an accelerated manner as process equipment as opposed to the traditional facility time frame. Because these systems are built off-site the delivery time of these units is highly predictable. Any time slippage could be compensated for by working overtime hours. Some examples of the prefabricated cleanroom approach follow in Figures 3.5, 3.6, and 3.7. These cleanroom units are not assembled to become an entire facility but require a shell building. It is of importance to review the shell building space to see whether such units can be introduced into the shell building. If the space is not able to accommodate the cleanroom units, modular wall panel options may be preferred.

FIGURE 3.5

Source: Just Biotherapeutics, Inc.

FIGURE 3.6

Source: Pfizer, Inc.

FIGURE 3.7

Source: Pfizer, Inc.

3.3 CHANGES IN FACILITY SIZE, USE, AND LOCATION AND COSTS INFLUENCE THE DESIGN CHOICE

Historically, the pharmaceutical industry has employed large centralized product-dedicated manufacturing sites to meet global demands for their products as opposed to in-country/for country manufacturing centers to meet demand in the area. However, a multitude of events has changed this paradigm. Recent patent expiries, the resulting losses of market share and lack of production capacity needs has led to the closing of some of these large-scale sites, which now represent more of a burden than an asset. (It is now fairly common to hear of "abandoned assets" in facility discussions with pharma companies.) Costs for insurance, basic utilities, security, and the like persist well after the facility is decommissioned. In addition, rising operation and transportation costs as well as supply chain concerns attendant to distributing drug products to multiple locations have become more important in this analysis. Furthermore, and perhaps most importantly, import taxes by recipient countries who seek to build more industry and infrastructure in their regions has generated more need for "in country/for country" production across the industry. Moreover, increasing success in cell expression rates and advances in continuous processing allow the reduction of required bioprocessing volumes, which can be served by single-use bioreactors. Such bioreactors have a much smaller footprint and height, which allow them to be used in smaller cleanrooms. Costs, both direct and indirect, have also influenced this shift. Direct costs include operating expenses of these large aging facilities that only increase with each year of service because of increasing maintenance needs. Indirect costs include having to manufacture products to forecast in order to be able to meet demand. Resulting product shortages and/or overproduction and expiration of products also has significant effects on profitability. Classical examples of product waste are expiring vaccine doses, which were stockpiled to supply an estimated patient base, instead of being able to supply vaccines on demand with smaller volume sites, which have a lower demand in raw materials and therefore would be able to assure supplies or by converting traditional egg-based processes to cell culture- based processes. Avoiding losses because of cold chain shipments could be accomplished by creating in country/for country small volume sites to supply local markets with minimized needs of distribution distances.

Costs of switching to an autonomous modular approach (versus the "stick-built" approach) have been a stumbling block for some who focus on one aspect only, cost per square foot of cleanroom space. Such a myopic focus though is not an appropriate methodology. Costs have to be considered in light of all of the benefits provided by such facilities. Such benefits include:

- Design costs (conceptual, basic, detailed) are significantly lower than in a traditional approach as the same basic architecture is employed in all of such structures. Such costs are even further reduced when modular sites are cloned.
- Material and transportation costs have to be evaluated for all construction design instances and not just for prefabricated systems. Complexities exist every time building materials are shipped to sites, especially determining when these should be shipped.
- Personnel, engineering, and supervision costs of on-site built approaches are not present in an off-site modular built approach. In an on-site build approach, these costs are incurred for no less than 6–8 months. In an off-site approach, these costs are incurred over 2–3 weeks.
- Lay-down area at the host facility/construction site is not required in an off-site, prefabricated modular built approach. The lay-down area to build a cleanroom on-site is often as large as the cleanroom area itself
- Insurance costs and safety concerns of lengthy on-site construction activities are minimized in an off-site prefabricated modular built approach.
- Detailed and complex infrastructure needs on top of the actual "stick-built" cleanroom space are not required in an off-site modular built approach. Rather, all ductwork and piping are run within the structure of the module making such runs much more compact, efficient and not exposed to potential hazards.
- Operating expenses of "stick-built" are higher because of energy losses inherent in long pipe runs and leaks in the runs
- Repurposability, i.e., is the cleanroom structure built for one product only? Can it be reused? Off-site built modular systems can be readily disconnected from the host facility and moved in a matter of hours.
- Depreciation schedule—autonomous moveable cleanrooms can be depreciated rapidly as equipment versus in-place construction, which is depreciated on the same terms as real estate (8 years vs. 30 years).
- Scalability. For example, is it necessary to shut down existing cleanroom infrastructures when new space is added to the existing space? With off-site built modular units, additional units can be added without affecting the existing operation.
- Ease of cleaning and sanitization. If an excursion occurs, would the production floor be shut down for an extended period of time? With off-site built modular, all surfaces are suitable for vapor-phase hydrogen peroxide (VHP) cleaning systems.
- Time to first product run. How fast can the facility be deployed?
- Qualification and validation costs. With off-site built modular, such costs are reduced as each unit will have the same basic characteristics, architecture, and bill of materials.
- Off-site built modular units will also not require any additional costs or add-ons. Stick-built approaches will generally require the owner to contract for a cleanroom floor and the design and construction of an automated HVAC system.

These are only some of the relevant costs that should be considered in determining which cleanroom system should be employed. But from this it is clear that comparing the price of an off-site modular built system with a stick-built system on a cost per square foot basis is certainly not the appropriate analysis.

3.4 DOES REGULATORY OPINION MATTER IN THE FACILITY DESIGN DECISION?

Although the industry has begun to see the need for change in its manufacturing paradigm, it is also clear that such recognition may not have the same motivation within regulatory authorities. In the case of the U.S. Food and Drug Administration (FDA), the necessity for change is seen as well, as the FDA's 21st Century Initiative supports such change. The initiative declares the need for "*A maximally efficient, agile, flexible pharmaceutical manufacturing sector that reliably produces high quality drugs without extensive regulatory oversight.*" Certainly, there is some idealism in that statement. But it also highlights that failing to change to a more flexible and agile system may lead to being seen as old, aging or obsolete, which is something that industry certainly would not welcome. The inclusion of the terms "maximally efficient", "agile", and "flexible" also suggests that the facilities of yesterday will not fare well going forward. Single product behemoths built in the latter part of the 20th century do not meet any of those terms. As regulatory pressure increases on these older facilities, it can only be expected that decision makers will become more willing to embrace the new facility approaches. Having said this, new paradigms of manufacturing flexibility and rapid deployment need to be supported by regulators. Preapproval inspections must be performed in a timely manner or should be abbreviated when a facility has been cloned and the originator facility has already been inspected.

3.5 ANOTHER KEY DRIVER—FLEXIBILITY

Flexibility is another key factor in the facility approach decision. But what does flexibility mean? Merriam-Webster defines flexible or flexibility as "characterized by a ready capability to adapt to new, different, or changing requirements." The "flexible facility" description is often used to describe single-use processes within bioprocesses. That however does not actually speak to the facility itself. For a facility to be flexible, being capable of adapting to new, different or changing requirements, some appropriate questions include:

1. Can more than one product be produced, simultaneously and concurrently?
2. Can the amount of production change (increased or decreased) to meet demand without interruption in the ongoing process?
3. Can the facility be moved to meet changing global or regional demand or to address regional instability?
4. Can the process be changed (either adding or taking away steps) without interrupting operations?
5. Can parts of the cleanroom structures be sanitized without interrupting other processes?
6. Can multiple tenants work in the same shell building within different cleanroom units?
7. Can the facility methodology be deployed in foreign jurisdictions?[1]

Considering the main options, stick built, modular panels, and prefabricated cleanrooms, it seems obvious that all are more flexible in these areas than the legacy model, i.e., brick and mortar facilities built and dedicated to a single product (2, 3, 4). Such was the approach of choice in the early pharma days when every product was assumed to be a future blockbuster drug. If the demand was never realized or fizzled to less than what was forecasted, either because the patent expired or a competitor or generic emerged, the facility was typically shut down and mothballed. There was no ability to repurpose or move the facility. As such, an asset with a depreciable life of 35–40 years became obsolete well before that time frame. With the increasing amount of mothballed "assets", which really weren't assets at all, flexibility of facilities and processes has become a key criterion in facility decisions.[2]

The failure of brick-and-mortar facilities, combined with stick-built cleanroom structures, to provide flexibility spawned the modular panel and prefabricated cleanroom solutions. Each sought to provide more flexibility. But have they done so? The answers to the following questions shed light on that topic.

3.5.1 Can the Facility Option Support a Multiproduct and Multipurpose Facility to Meet a Developer's Changing Pipeline?

Today, drug developers often produce multiple drugs at one time. Being able to produce the same in one facility allows for the sharing of administrative costs between several drug and drug candidates. Functions such as quality, shipping and receiving, maintenance, facilities support, etc., can all be shared by the drug products and candidates being produced which will lower the cost of each product.

Drug developers also desire the ability to produce multiple drugs at one site in either sequential or concurrent fashion. In the first instance, once production of one drug is complete, another drug can be produced in the same facility. This need can arise when the drug reaches the end of its life cycle either because of patent expiration and/or competition. The need can also arise if health care requirements or even the stability of the local government changes in an emerging or other market.

Multiproduct, multipurpose facilities where several drugs are produced concurrently can be provided using prefabricated autonomous cleanroom modules that are not interconnected to each other via a central air handling system. The lack of common HVAC between the units assures that separate operations can occur within each module cluster. (Figure 3.8). If there is an excursion in one unit or group of units, that can be contained in the affected area and not interfere with other operations. In extreme cases, those units could even be removed from the facility for decontamination and other units brought in for replacements. The ability to clean the prefabricated facilities also allows for concurrent production, as the facilities can be sanitized via vaporized hydrogen peroxide after one form of campaign and before transitioning to another. With no elaborate and interconnected ductwork as exists in the other options, this type of cleaning is more readily achievable in this approach than in others.

With stick-built infrastructures, such an autonomous approach is not likely given the expansive HVAC infrastructure typically required. If multiple products are envisioned, the facility design will have to account for those products with separate zoning at the outset and no changes will be able to be made once in production. Having a facility with one HVAC system only will not allow for multiproduct or multipurpose facilities. As noted previously, cleaning the ductwork during campaign transitions would be difficult to achieve even in the separate HVAC zone regime given the expansive nature of the equipment needed. Examples of the past have shown that some facility and cleanroom infrastructure contaminations are persistent and will affect the company greatly, either via regulatory warnings or production downtime (5, 6).

Modular wall panels can generate a similar outcome, especially when one HVAC system will serve the entire infrastructure. This requires that changes to the product and resulting process(es) cannot be accommodated without redesigning, rebuilding, and retesting the entire system. A similar outcome would be needed if the cleanroom infrastructure requires to be scaled-out. Capacity extensions would disrupt existing processes extensively.

3.5.2 Can the Production Capacity Be Changed to Meet Demand Without Interruption in the Ongoing Process?

Generally, facilities are planned, engineered, and built before drugs are produced. Whether it be a new drug entity or new market, drug makers need to begin these processes well before the drug is approved for sale by the regulatory body with jurisdiction. This reality results in facilities being built pursuant to sales forecast as opposed to actual demand. Producing based on forecast is of course not as accurate as producing to known demand. Hence, it is likely in these situations that the amount of planned production will have to change. Demand

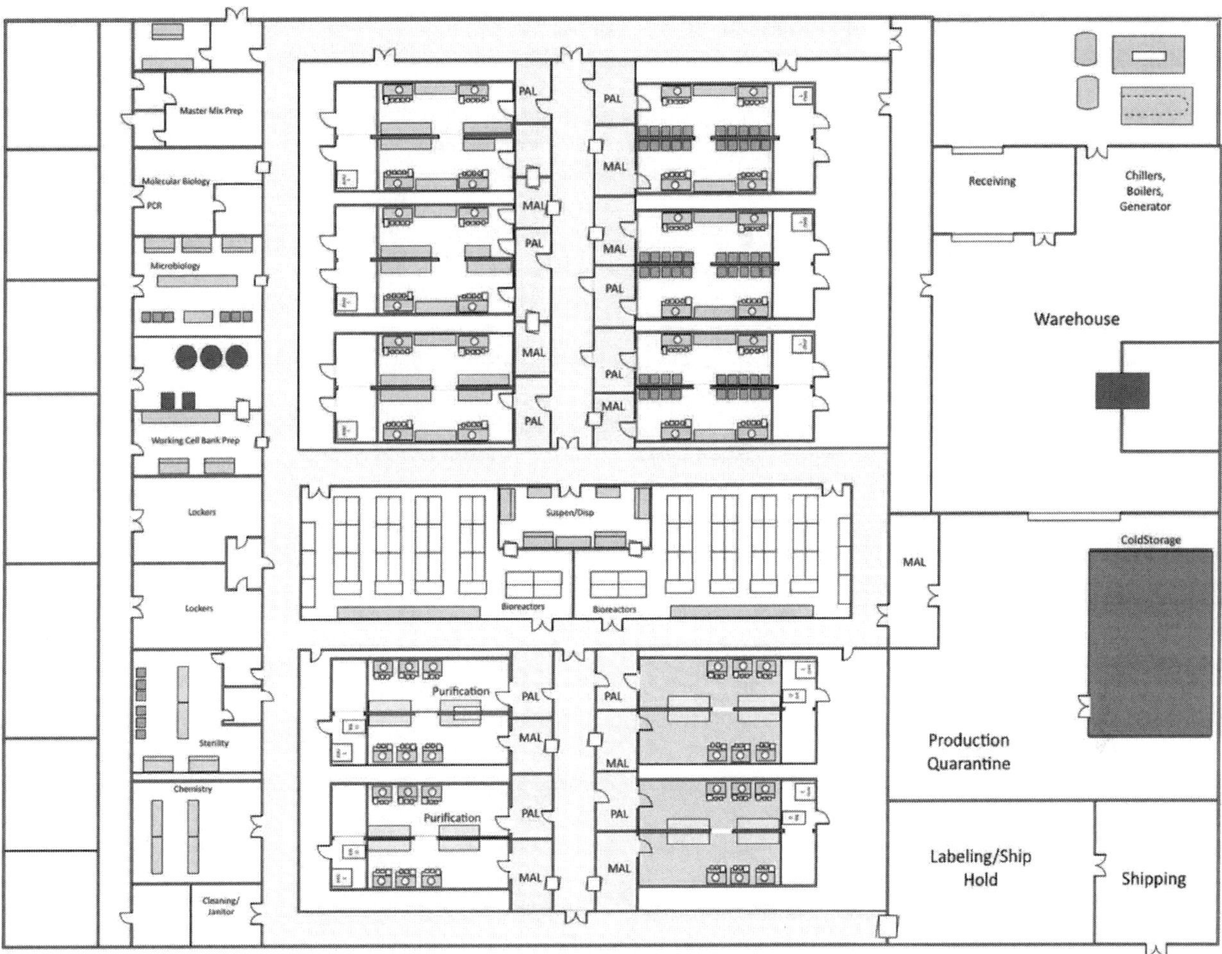

FIGURE 3.8 Example of a multitenant or multipurpose site.

may be less than forecasted because of product effectiveness, competitive landscape, price, etc. Conversely, demand could be greater than expected if the product is more effective than planned, competitors are removed from the market, price is lowered to enhance demand, etc. In either case, the original parameters of production may no longer apply. As an example, the product may require a 2,000 L bioreactor versus a 20,000 L bioreactor. Utilities (that may or may not have been built into the facility infrastructure) supporting that bioreactor will be different. So, having a facility platform that can respond to such changes is a real value to the drug maker.

Modular panel construction on-site, although initially flexible in terms of size and shape, is not very malleable once it is built. Wall panels, doors, ceilings, and the like are permanently placed in position. Ductwork and utilities are provided in intricate detail above the cleanroom structures. Adding to or reducing the footprint of the fixed structures requires on-site construction or demolition, which will interrupt ongoing production. Once construction and/or demolition are complete, rebalancing of the HVAC system and revalidation of the cleanroom space will be required.

With prefabricated cleanrooms, although it may seem that changes in capacity would not be possible with such discrete units, the contrary is true. Because of HVAC segregation, new units can be added when needed, or taken away when not needed, without interrupting the existing process. They can be added in a linear fashion or added to the existing cluster via the use of a corridor or previously placed "knock-out" panel. A reduction in capacity in this context does not mean that previously placed units are shut down. Rather, they can be disconnected and moved out of the facility for deployment elsewhere. If new equipment and process utilities of a different size are required, new modules can replace the old modules, utilizing the same footprint and their own internal HVAC, and being built while the original modules are still in operation. As such, there is little to no facility downtime between campaigns.

The stick-built approach is like modular construction on-site but is even less flexible than the paneled approach. In the paneled approach, walls can be moved more easily than a stick-built wall, which is most often gypsum board covered with epoxy. Moving a wall in this option means demolition. Centralized HVAC distribution requires adding or taking away capacity and rebalancing if rooms are added or taken out of commission. With such being the case, there is no utility in attempting to reconfigure production space using this approach if the user desires to maximize time in production.

3.5.3 Can the Facility Be Moved to Meet Changing Global or Regional Demand or to Address Regional Instability?

Up until recently, asking if a facility or facility infrastructure could be relocated would at least get you an odd stare and at most get you a trip to your friendly neighborhood behavioral health clinic. Now, moving cleanroom infrastructures is readily considered as an advantage to certain approaches (15). Flexibility in this regard could be moving capacity to address softening demand or to meet higher demand in another country and even for production of another product. Also, there have been many examples of today's emerging markets becoming unstable economically or politically. In the past, such changes meant abandoning the asset. Now, drug makers desire to be able to respond to these changes.

With stick-built infrastructures, this flexibility simply is not present. Once built, the infrastructure cannot be disassembled and moved without complete disassembly. For modular panels, although the panels themselves can be disassembled and moved, the non-modular mezzanine structure housing the HVAC equipment cannot be. So, moving this type of facility is not possible either. In the prefabricated approach, the units are mobile. Some even employ air bearings for easy movement into, around and out of facilities. Their minimal connection to the facility and onboard HVAC equipment makes them fairly autonomous within the larger facility as well. As such, these units can be relocated with ease. Disconnecting the units from each other and the host facility takes a matter of days. Then the units can be individually moved on a standard eighteen-wheeler to the next site or port of choice.

With the mobility of prefabricated units, one can even imagine a centralized storage location of the units, which would serve as the supply center for cleanroom capacity when needed. When the cleanroom infrastructure is no longer needed at a particular location, it can be shipped back to the central site for cleaning, sanitization, service, maintenance, and recalibration. This centralized area could also be used if cleanroom assets from disparate regions require retrieval because of political unrest or unfavorable changes in tax structure/economic incentives. With the increasing need to supply the patient base on a local level, the ability to move cleanrooms may go from a "nice to have" to a "need to have" in very short order. As such, cleanrooms may take on even more mobility in the future. (Figure 3.9)

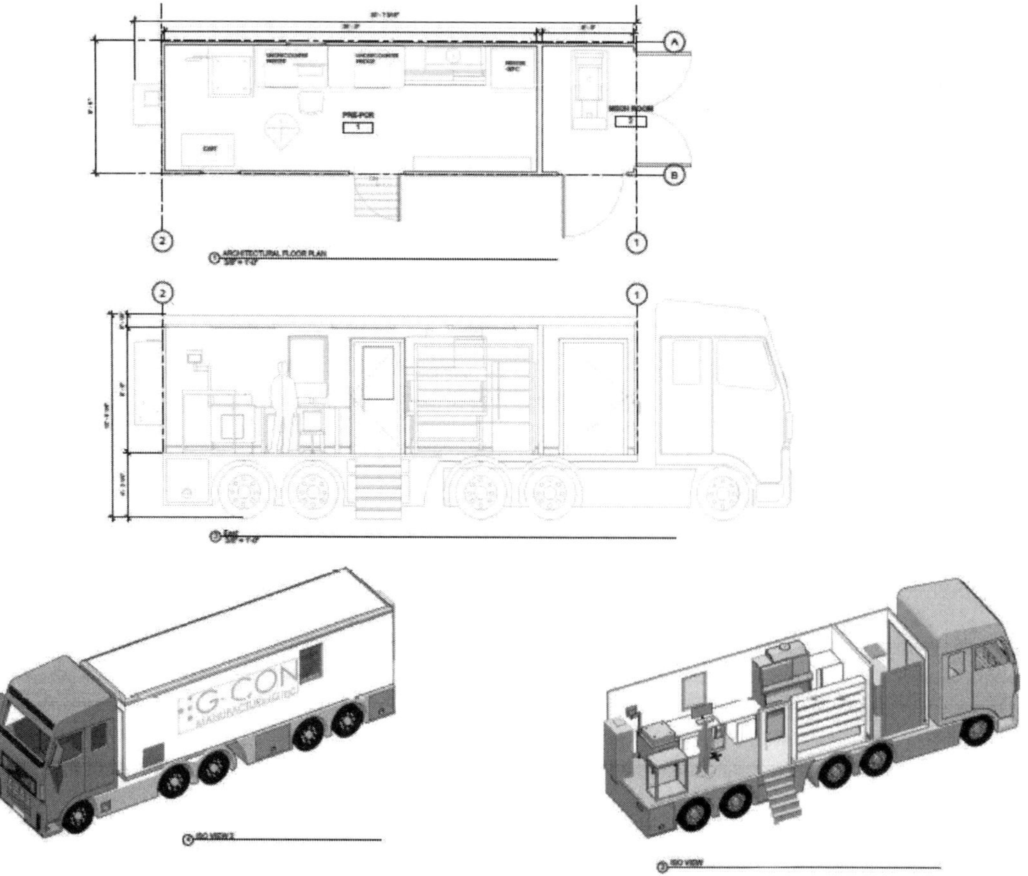

FIGURE 3.9 Example of a mobile processing system, truck or trailer-based unit.

3.5.4 Can a Change in Process Be Accommodated Without Interrupting Operations?[3]

Generally, the regulatory agencies approve the process by which a drug is made. As such, it is often said that "the process is the product." With ever-improving technology though, it is likely that processes will be and/or need to be improved over the life of the product.[4] So having a facility option that can readily handle and adapt to a change in process is of importance.

With prefabricated units, if there is a change in process, units can be added, taken away or switched to account for the same, all without any or significant delays to production. Because of the autonomous nature of each unit, an overall HVAC rebalance is not required. On-site construction is not required. Shutdowns to accommodate the on-site construction are not required. As such, adjusting the process to add or take away steps is not an issue.

For the modular paneled approach and the stick-built approach, a change in process would result in on-site construction, a shutdown to accommodate that construction, additional or different HVAC design and a rebalancing of the HVAC system upon completion. As such, processes cannot be readily changed with either approach unless the process change is simple, such as removing a step and not adjusting the environment in that area. The limitation can be detrimental, especially with new cell and gene therapies, which require scaling of capacities on a rapid basis or possible process turnovers.

3.5.5 Can the Facility Methodology Be Deployed in Foreign Jurisdictions?

The ability to produce internationally has become an area of more interest in pharma and biotech recently. Multinational production provides developers with numerous benefits: (1) avoiding import duties and transportation costs; (2) being able to produce to actual demand instead of forecast; (3) not relying on a single site for worldwide production; and (4) opening markets where importing was not allowed. Furthermore, some countries demand that the therapeutic product is produced in-country/for country, there is no other choice than establishing a facility within the country. So, the question of whether any of the three options being considered can be reliably and rapidly deployed into numerous foreign jurisdictions is very relevant today.

With prefabricated cleanrooms, the infrastructures are built at a central site by an experienced workforce and shipped to the location. A small crew from the manufacturer arrives with the structures and installs and connects the modules to each other. This process can be completed in a matter of days. The modules are then commissioned by the manufacturer's quality team over a two-to-three-week period and the process is complete. A simple warehouse-type structure is required. Other than that, very little infrastructure is needed in the foreign jurisdiction. On-site issues such as permits, union and labor issues, safety trainings, and lay down space and material staging areas are avoided with this approach.

With a stick-built approach, the drug manufacturer must identify a new contractor and engineer who will design and build a custom facility each time for each project. Costing, design, permitting and building will be specific to the site for each project. There will be little to no institutional knowledge in the design and build from one project to the next. Some areas do not have contracting resources and these workforces are required to be brought in. Such a venture can be costly, as these foreign workforces will stay at a site for 6–12 months. In the use of a paneled approach, some efficiency is gained in the use of standard materials, but the exercise of choosing engineers, contractors, designing, costing, etc. is largely the same and is subject to the same issues and delays.

3.5.6 A Real-Life Example

The perspective of cell and gene therapy is a good example in considering flexibility of the options being reviewed. In cell and gene therapy production, rigorous containment is a must. Typically, a cell therapy application requires a relatively small processing space, for patient-by-patient drug production. Patient samples must be handled with the utmost care. The cleanroom structure must be sanitized frequently, generally with vaporized or ionized hydrogen peroxide. Finally, cross contamination from other process spaces must be avoided.

Reviewing the options with this example in mind, prefabricated segregated cleanroom units are first considered. Each unit having its own HVAC and providing unidirectional flow achieves the results required (Figure 3.10). The autonomy of the HVAC units allows for cleaning runs to occur in each unit when needed, e.g., after every patient sample is processed. Unidirectional flow is possible. Although the other systems can be designed to achieve unidirectional flow, neither of them typically provides autonomy of HVAC for each process space. As such, cleaning will have to occur during shutdowns of larger areas comprising multiple rooms. Containment between process spaces will not be feasible either given the shared HVAC design.

3.6 CURRENT VIEWS ON FLEXIBILITY

Recent presenters and authors have described a flexible facility as a ballroom design at a lower cleanroom classification utilizing single-use processing equipment and relying on the single-use components to be the primary barrier. The reason for this approach is to reduce operating costs and therefore the cost of goods sold. But is this lower operating cost approach worth the risk of an entire area being compromised by contamination? Ultimately, this is the end user's decision, but the reliance on single-use technology as primary barrier seems fairly haphazard, at least for costly biologics. Ballroom designs were adopted in the semiconductor industry, but these industry requirements cannot be simply mirrored to the

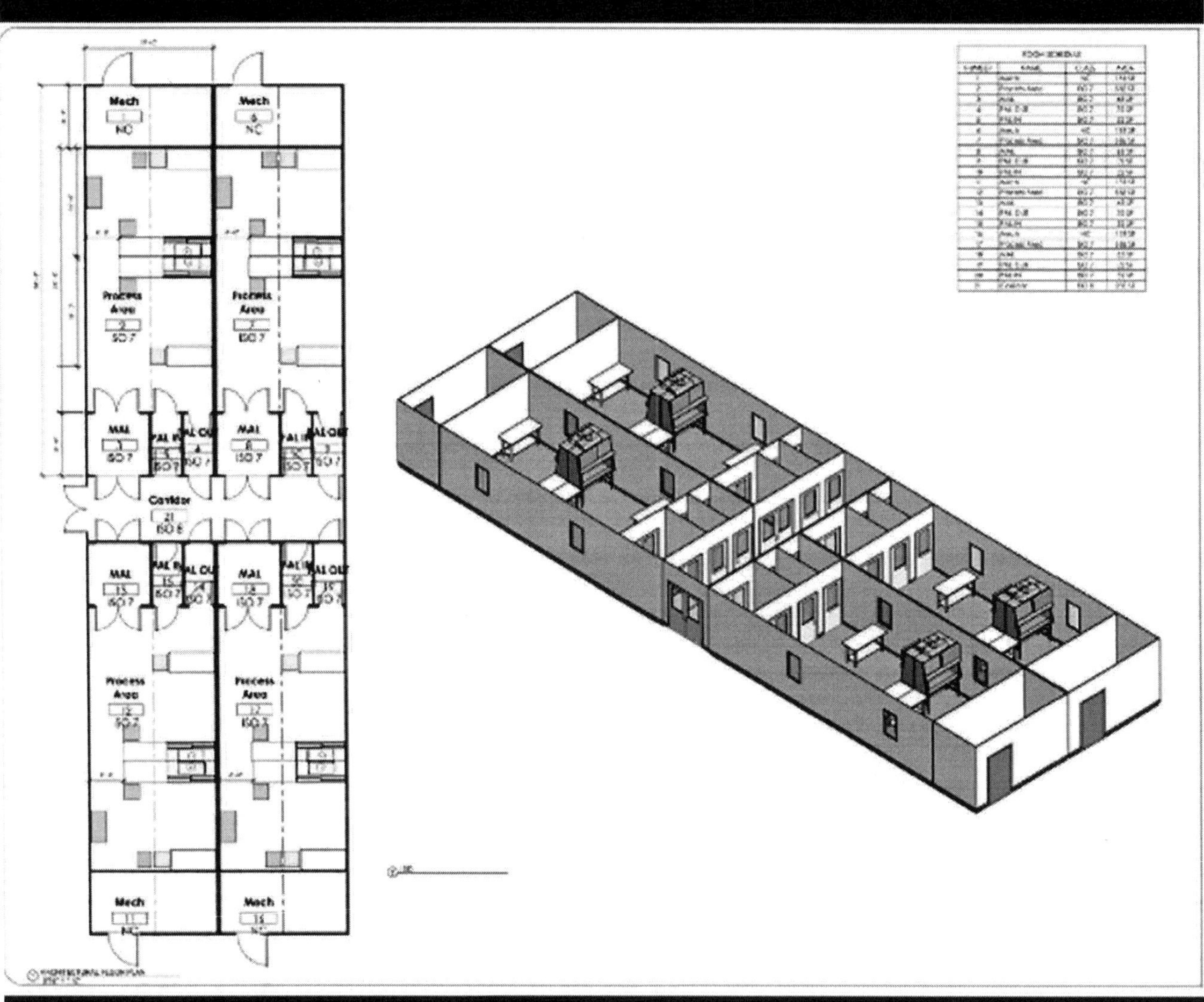

FIGURE 3.10 Cell therapy example—prefabricated units can be scaled without interrupting other units.

biopharmaceutical industry. Ballroom areas being segregated via isolator based or POD based designs seems to make more sense, as a large area can rapidly be outfitted with cleanroom modules, with speed, flexibility, and vital containment.

Another facility description that is often used interchangeably with "flexible" is "modular". Modular facilities can be constructed in a relatively short period of time when compared with that for traditional facility types but ultimately are very similar to traditional facility designs once complete. As previously discussed, the two types of modular facility designs are (1) off-site container built, which are interconnected to form an entire building, and (2) modular panel build, which is generally a cleanroom cluster within a shell building. Both designs lose their modularity as soon as these are completed. Both designs employ a typical HVAC superstructure consistent with single product dedication. Proponents of these designs promote their flexibility, especially in the planning phase, but once constructed, all flexibility is lost. A recent paper offers the same view, stating "Until now, modular facilities have reproduced traditional architecture with regard to embedding utilities piping and HVAC ducts in the interspace between the physical module limits and the suspended ceiling making refurbishment, if required, extremely complicated." (A. Pralong (2013) Single-use technologies and facility layout—a paradigm shift, Biopharma Asia Magazine, Vol 2, Issue 1)

For the biopharmaceutical industry flexibility means:

- Capacity scalability (up and down)
- Multiproduct production
- Short time-to-run or rapid deployment
- Rapid changeover or changes in the layout
- Repurposability
- Mobility

The latter two are new. The industry cannot accept mothballed assets any longer. Assets have to have more than one use. When a cleanroom asset can be repurposed, it has lower investment risk. Mobility is required because of the need for relocation. The ability to move the cleanroom will not be a day-to-day benefit but will be an important benefit if a lease is limited in time or the product being made has volatile demand from region to region.

Scalability of production is a key point in cleanroom decision-making. It is valuable to the manufacturer if the process can be scaled up or down within the cleanroom environment to meet changes in demand. Scalability also refers to the ability to produce multiple products within one facility. For this to be possible, the combination of single-use technology and sanitizable containment options, like podular systems, are required

Sanitization includes the HVAC system, which can most readily be achieved when the cleanroom space has separate air handling and HVAC units.

Mobility is now an important consideration given the advent of single-use equipment. A manufacturer may desire to move production from one facility to another to meet changing demand from region to region or to accommodate for growing or shrinking demand in general. Real estate considerations, such as a short-term lease may require the moving of a site. Changes in process flow may also dictate moving cleanroom spaces, e.g., moving a purification process into a different position in the process or to a decontamination area. In all of these scenarios, mobility of the cleanroom system is required. Moving a fixed built cleanroom system, whether it is of modular panels or stick built, is not possible. With isolator-based or prefabricated POD based systems (Figure 3.11) such is possible. Such systems run independently and are not tied into a centralized facility air handling system. The POD-based system is a cleanroom "box", with its own air handling system and quick connects for the utilities.

3.7 THE CROSSROADS BETWEEN PROCESS AND FACILITY

Modern biopharmaceutical processes have converted from large-volume, rigid stainless-steel designs to small-to-medium volume, flexible single-use unit operations; from reusable to single-use. Past facilities, though, were product-dedicated and designed to accommodate one product for its entire life cycle, i.e., they were most often single-use. Afterwards, the facility was mothballed or had to be completely renovated and redesigned. This capital-intensive approach is now being challenged by innovative ways to design and construct facilities, which will be used for multiple product life cycles or even multiproduct purposes. In contrast to processes, facilities are now moving from being single-use to being reusable. Therefore, processes and facilities cross each other's path to the different utilization approach. An often-erroneous statement is made about a flexible facility describing a single-use process. Facilities and processes are distinctly different and are designed and constructed in different ways. A facility is not necessarily flexible just because the process is single-use. The opposite is often true as traditional facility layouts void the flexibility of single-use processes. If the layout of the facility does not allow easy access or movement, the benefits of flexible process equipment is thwarted. To overcome the frequently stated facility constraints, the process **and** the facility, meaning the cleanroom infrastructures, need to be flexible. Cleanroom spaces that are built in place with interconnected ductwork and that are product dedicated are not flexible. New modular facility designs that are designed as autonomous

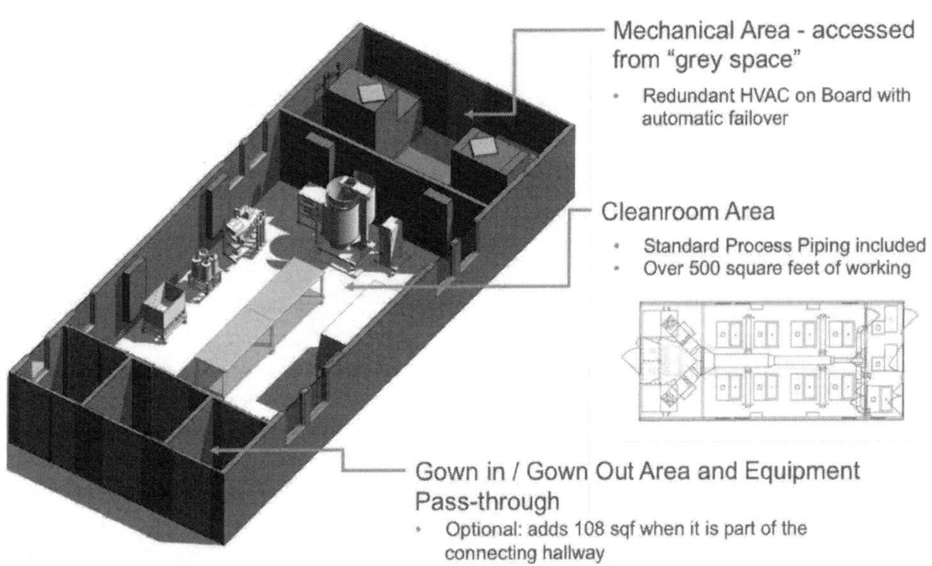
Mechanical Area - accessed from "grey space"
• Redundant HVAC on Board with automatic failover
Cleanroom Area
• Standard Process Piping included
• Over 500 square feet of working
Gown in / Gown Out Area and Equipment Pass-through
• Optional: adds 108 sqf when it is part of the connecting hallway

FIGURE 3.11 Cleanroom POD layout.

Source: G-CON Manufacturing Inc.

units do have the required flexibility. Changes in the cleanroom layout, for example an area extension, can be done without interrupting current processes in the prefabricated approach. The new cleanroom unit would simply be docked against the existing structure, but not interconnected to a centralized HVAC system. Ultimately, the requirement for flexibility is not just an industry "nice to have". The requirement has also been expressed in the FDA's 21st Century Initiative. The vision declares the need for "*A maximally efficient, agile, flexible pharmaceutical manufacturing sector that reliably produces high quality drugs without extensive regulatory oversight*". This vision may be satisfied with new innovative modular or podular cleanroom systems.

3.8 OTHER FACTORS

Up to this point, we have addressed what most would agree are the major factors to consider in making a cleanroom decision. Two other factors that may be of importance are: how turnkey is the option and whether the total financial commitment is well-understood prior to beginning the project. These items are both potential execution risks.

Turnkey refers to something that is ready for immediate use, generally used in the sale or supply of goods or services. In the cleanroom context, the ability to provide a turnkey cleanroom means providing not only the walls, floor, and ceiling but also providing the infrastructure to make it work—such as the HVAC system, the fire suppression system, the control and automation system, door interlocks, utilities, fluid management, filtration, Installation Qualification/Operational Qualification (IQ/OQ) package, etc. With a prefabricated cleanroom approach, a turnkey cleanroom can be provided with all mentioned items provided. Hence, the risk of execution rests upon one service provider who is performing all except for the IQ/OQ at their facility. In the panel approach, items such as HVAC, HVAC controls and automation and fire suppression are more often provided by separate contractors. Although those contractors may be quite diligent, the risk of multiple providers all executing their scope may increase the odds of failure. In the stick-built approach, it is common that a general contractor will be retained, and all specialized functions will be subcontracted by that general contractor. Given the general contractor's contracting with the subcontractors, the client may have only one point of the contact for the job and be able to hold one entity responsible, but there are still multiple points of potential failure.

Generally, all projects will begin with a budget in mind either based on prior similar projects, having received many budgetary quotes, or a combination of both. As noted earlier, cost oftentimes may be the ultimate driver. As such, the total financial commitment of a project is important to understand, and it is critical to have that understanding at the beginning of a project when the budget is approved. In a stick-built approach, the typical construction contract model is in place. Achieving a firm quote and sticking to that throughout the project is often not realistic. With modular panels providers, although some of the components of the job may be well understood at the outset, such as the doors, walls, and windows, the other components that are generally not provided directly by the panel provider are not. With the separate contracts in place and contingencies present in each of those, it is not hard to imagine cost overruns for this type of job. With a prefabricated approach, the cleanroom is sold as a piece of equipment with each option that is provided well defined. Much like choosing the options available in a vehicle or computer purchase, the end user will see the costs for each item chosen at the outset and be able to budget for that. If an experienced cleanroom provider is retained, that provider will have a well-defined basis of design, design specifications, functional specifications, and expected requirements for the host facility. Such will ensure that the installation of and "handshake" between the cleanroom and the facility is seamless.

3.9 THE FUTURE OF FACILITIES

Facilities as we know them from the past will still exist in the future, however, not with the prevalence as seen before. Current drug manufacturers and engineering firms alike will not just favor a particular option but will make choices from the increasing portfolio of facility components and designs. As with every aspect in manufacturing processes, there is no "one-size-fits-all" system for production sites. Future processes and facilities will be driven by the product to be produced within the site and by the economic as well as regional parameters, but there is little doubt that a key consideration will be flexibility.

3.9.1 Supply Chain Benefits

Facilities will be standardized and cloned in a platform approach for in-country/for-country purposes. On the production side, multiple facilities enable the manufacturer to produce to demand instead of based on forecasting, which will lower product inventory and reduce the risk of product expiration. Additional unit operations such as coating and encapsulation can be added with little to no interruption to existing processes.

These modular facilities can be placed into modest shell buildings around the globe. These shell facilities can then hold additional modular facilities for one or several companies. In that scenario, manufacturers can share administrative resources such as using the same quality control, purchasing, operations support, etc. In either case, cleanrooms are deployed faster for new product production or product scale up. Resources are more efficiently used. Both will lower the typical operating cost burden, because the clean space is built around the process.

Although the advances in modular technology are apparent, what is also clear is that these modules will become the building blocks to standardized platform approaches for well-defined processes. A good example of this is the downstream process for a typical monoclonal antibody (mAb) (7). Each mAb typically undergoes multiple chromatography steps, viral inactivation and filtration before finally being formulated and filled. The process itself is well-defined and well understood, which has been taken advantage of by single-use process equipment suppliers. These suppliers have created single-use process unit operations that can be interconnected to a larger process stream. Such unit operations can be placed into cleanroom containment systems and interconnected in an entire facility layout.

In the past, it would be left up to the customer and its architectural and engineering (A and E) partner to design the environment around that process step. Such an approach would result in significant manhours (engineering and construction) and expense in connection with the custom designed and built structure. This approach begs the question: If the step, the scale, and the equipment are the same, can't the facility be the same? And if the facility is the same every time both in terms of size, equipment and materials of construction, wouldn't that lead to a shorter time to validation?

With prefabricated modules designed for the specific process steps, soon process equipment vendors may develop and sell enclosure options for each process step that can be customized to fit the customer need. Thus, in addition to providing the turnkey solution for a particular process step, process equipment manufacturers will also provide the enclosure around the process equipment and thus provide a true integrated turnkey solution, which will greatly abbreviate the current lengthy design phases and lower the cost of facilities. Facilities, similar to the process equipment inside, will become reusable commodities. Conceptual design costs would also decrease substantially.

One modular equipment supplier has already partnered with an A and E firm to generate a standardized 50,000-egg-per-day vaccine facility. That same vendor has worked with another A and E firm to design a standardized 2000-L mAb site as well. Bioprocesses of multiple types and volumes can potentially become a catalogue item, instead of being reinvented over and over.

To further the planning robustness, predesigned, turnkey facility solutions are now being established to shorten the build time and reduce the cost impact. As an example, a 4 × 2,000 L monoclonal antibody turnkey site can be built in 12 months versus 24–36 months. From a standardized, prefabricated cleanroom unit to a predesigned turnkey facility is not a quantum leap, but a mindset change to a standardized, off-the-shelf approach. Instead of spending significant hours on new designs or reinventing the wheel, thinking simply will create speed, cost reductions and minimal execution risk.

3.9.2 Meeting the Aseptic Filling Challenge

Smaller volumes within bioprocesses, the need for more robust containment, and new therapies have led to new, compact fill line designs. These systems utilize automation, robotic fill arms, and pre-sterilized container systems, which avoid human interventions.

Because these systems are compactly designed within an isolator, they can be prequalified within the supplier site and gain the final qualification at the end user site. The next step in the evolution of flexible aseptic processing and filling will be the partnering of the filling line manufacturers with modular companies to provide turnkey filling and enclosure options that can be delivered together. This approach will ease the integration burden and shorten the time frame for delivery and operation of filling equipment (Figure 3.12).

With the first barrier, the isolator, around the fill line, and the prefabricated cleanroom as the second barrier, compact

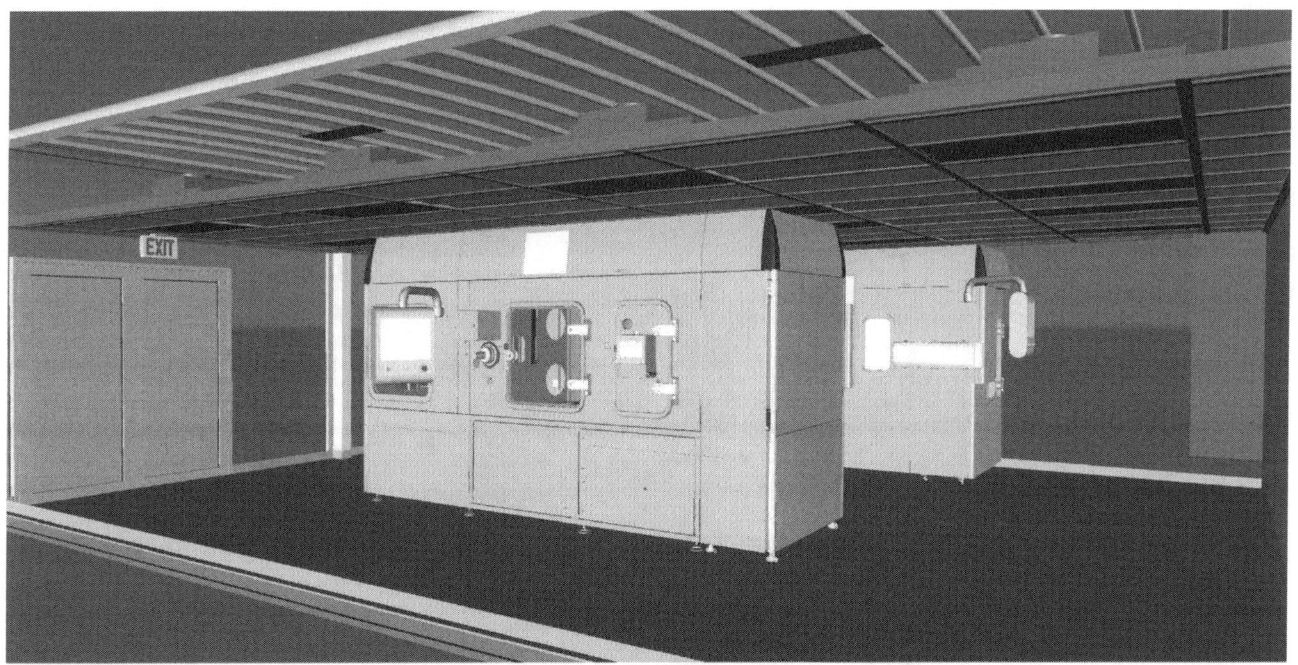

FIGURE 3.12 VanRX filling system within G-CON cleanroom POD.

designs for both, and the possibility of sanitization with vaporized hydrogen peroxide, filling can now stop being the most critical step in the manufacturing process and become a robust and reproducible process step assuring a high level of product quality.

Collaborations between modular companies that, in the past, might have seen themselves as competitors, are now being seen. Modular equipment vendors are starting to carve out their own niches (e.g., modular built off-site versus on-site built panels), and some are starting to work together. At least one major supplier of off-site built cleanrooms has included modular panels in some of its latest designs. Those developments promise to drive further innovation and cost competitiveness in this space.

3.10 CONCLUSION

Facility design requirements are evolving. Bioprocess technologies have been transformed from stainless steel to flexible single-use processes. These innovative technologies have created new opportunities, not just in process designs, but also interconnecting processes, making processes more efficient. Flexibility in facilities has lagged. As such, these flexible processes have been forced into uncompromising, inflexible facility and cleanroom infrastructures.

There is now a movement to fuse flexible processes into flexible facilities as flexible facility platforms become more available. The future of facilities requires modularity, either as panel or prefabricated systems, progressing to potentially clonable platform designs.

There will be headwinds to be sure, just as there were in the early days of single-use equipment. Critics focused on high costs as a reason to maintain the status quo. Not until studies were published showing increased capacity utilization in addition to other benefits did the adoption truly begin. In the same ways, critics of flexible facilities focus on cost per square foot without consideration for other factors such as quality, flexibility, scalability, reduced execution risk, and others. Once the mindset changes because of the experiences of the early adopters, single-use processes and facility platforms will become the optimal flexible solution and will complement each other.

Beyond that, even more innovation is possible, perhaps creating facility platform designs, which will be listed within product catalogues and chosen as process equipment. These platforms could be redlined and modified, but overall lengthy design times would be greatly reduced in such an approach. Each unit operation could have its own platform, which could then be combined into an entire process facility of the future.

NOTES

1. For additional thoughts on what constitutes flexibility in facilities, see Backstrom, Analyzing the Flexibility of Pharmaceutical and Biopharmaceutical Options. *International Journal of Pharmaceutical Science Invention* 2017.
2. It is important to note that even "mothballed" facilities generate costs. These facilities still require basic utilities, insurance, and in some instances, maintenance and security. They are taxed and generally getting the local and state property tax assessor to reduce the value is not readily attainable. Demolition may not be possible because of (1) hazardous chemicals like asbestos or (2) not being able to receive permits from the local or state authority to do so given the adverse tax ramifications to the jurisdiction for having the tax base reduced.
3. Another question that would produce similar results to this one is whether the facility can respond to a change in regulatory requirements.
4. See www.fda.gov/Drugs/DevelopmentApprovalProcess/Manufacturing/ucm169105.htm; Facts About the Current Good Manufacturing Practices (CGMPs) "Accordingly, the 'C' in CGMP stands for 'current,' requiring companies to use technologies and systems that are up-to-date in order to comply with the regulations. Systems and equipment that may have been 'top-of-the-line' to prevent contamination, mix-ups, and errors 10 or 20 years ago may be less than adequate by today's standards."

REFERENCES

1. How Mold Gets into a Cleanroom and How to Deal with Cleanroom Contamination. *PharmEng*, 2015.
2. Sutton S. Podifying Cleanroom Processes. *The Medicine Maker*, 40–43, 2017.
3. Jornitz M, Backstrom S. Evaluating the Benefits of Prefabricated Cleanroom Infrastructure Designs and Costs. *Pharmaceutical Engineering*, 77–81, 2016.
4. Jornitz M. Defining Flexible Facilities. *Pharmaceutical Processing*, 22–23, 2013.
5. Timmerman L. Genzyme Halts Production at Allston Drug Plant after Virus Appears. *Xconomy Boston*, 2009.
6. Reardon S. Contamination Shuts Down NIH Pharmacy Center. *Nature*, 2015.
7. Makowenskyj P, Powers D, Jornitz M. Total Cost of Ownership Considerations for Biopharmaceutical Manufacturing Facilities. *Pharmaceutical Online*, 2017.

ABOUT THE AUTHORS

Maik Jornitz is President and Chief Executive Officer of G-CON Manufacturing, Inc. He has worked for over 30 years in the pharmaceutical and biopharmaceutical industries. He has in-depth experience in filtration, purification, and single-use and flexible facility technologies, including the related regulatory and validation requirements. He has authored over 100 scientific papers, 10 books, 15 book chapters, and a variety of blogs. He also holds over 30 patents related to pharmaceutical processing equipment. Maik has built several successful and sustainable business entities on a local and international level. He is a former Parenteral Drug Association (PDA) Board member (Chair of the Board of Directors) and currently Chair of the PDA Science Advisory Board (SAB), Audit Committee and task force member of the Aging Facility and Post Approval Change Task Force. He is also a member of ISPE, ASTM, and ASME and member of the Science Advisory Board of ICAV and the Biotechnology

Industry Council as well as a multitude of editorial advisory boards. As a faculty member of various training organizations, especially PDA TRI, he trains industry and regulators on a frequent basis. He received his M.Eng. in Bioengineering at the University of Applied Sciences in Hamburg, Germany, and accomplished the PED program at IMD Business School in Lausanne, Switzerland.

Sidney Backstrom is Vice President, Business Management for G-CON Manufacturing, Inc. He functions in multiple areas for G-CON overseeing Quality, Human Resources, IT, Marketing, and general business concerns such as contract negotiations, partnerships, risk and insurance, regulatory, etc. Backstrom has also provided consulting services to Gradalis, Inc., Strike Bio, Inc., the Mary Crowley Cancer Research Center, and several other related entities. He is Secretary and Board member for the Texas Chapter of the Parenteral Drug Association and has sat on the Business Advisory board to Path4 venture capital firm based in Austin, TX, a firm that specializes in the Life Sciences musculoskeletal sector with a focus on early-intervention orthopedic solutions. He received his B.S. in finance from Louisiana State University and his J.D. from Louisiana State University as well. He has authored a book chapter and several articles on topics in the pharmaceutical and biopharmaceutical cleanroom space.

4 Commissioning and Qualification

Phil DeSantis and Steven Ostrove

CONTENTS

4.1 INTRODUCTION

International regulatory authorities require that equipment used for the manufacture of pharmaceutical products for use in humans and animals be "qualified". This requirement is expressed in the EU-GMP regulations (Eudralex Volume 4, Annex 15) as well as in the US FDA Guidance for Industry Process Validation: General Principles and Practices (statutory references contained within that document.[1] Historically the pharmaceutical industry has employed a multitude of definitions for equipment qualification. Early definitions were very broad, such as the following from the ISPE website Glossary from around 2008 (since replaced):

> Qualification—Action of proving and documenting that equipment or ancillary systems are properly installed, work correctly, and actually lead to the expected results. Qualification is part of validation, but the individual qualification steps alone do not constitute process validation.

A more focused definition (discussed in Section 4.4) is:

> Qualification—documented evidence that provides assurance that equipment and systems meet the critical installation and operational requirements necessary to ensure consistent product quality.

Obviously, qualification and validation are closely related but are not identical. Validation is specific to an individual process and the means to execute it. The defined parameters and ranges used relate to a single product and process combination. Qualification supports the equipment's performance across all processes performed on/with it. Qualification establishes the full capabilities of the system for all applications and may not be product or process specific.

In both cases, the intent is similar, that both the process (validation) and the equipment (qualification) performance must be reproducible, meet predetermined attributes and be

DOI: 10.1201/9781003163138-4

well documented as to their state and conditions. The term "qualification" is used for the documentation of the acceptability of equipment or equipment systems. The term "validation" is used for the documentation that proves that the process is reproducible and robust. The former is a necessary prerequisite to the latter.

It is appropriate to take a slightly broader view. Just as the current accepted concept of process validation is a life cycle spanning the development, testing and continuing control of a process, equipment qualification follows a similar life cycle. Initially, a user company qualifies the equipment by specifying, inspecting and documenting in a rigorous manner to establish that the equipment is qualified to perform its designed function reliably. Then, the equipment must be monitored and maintained to ensure that it remains qualified. In a sense, qualification is not something that is performed, but like process validation, it is something that is achieved. Borrowing from thermodynamics, "Qualification (or Validation) is a state function."

This chapter discusses the planning, execution and documentation of qualification as it may be effectively performed throughout the pharmaceutical industry.

Following the ISPE definition, qualification requires that both the physical and operating characteristics of the equipment are verified and documented. It may loosely follow the sequence:

1. What equipment has been specified for the process?
2. Does the received/installed equipment comply with the specification?
3. Does the equipment operate as required for the specific process or use?
4. Is the equipment capable of performing its required function reliably?
5. What procedures and/or practices are in place to ensure the reliable continued performance of the equipment?

Most guidelines for qualification answer the questions posed previously by addressing them in sequence, roughly aligned with questions 2–4. It is common practice to define the steps in the qualification sequence as installation qualification (IQ), operational qualification (OQ) and the performance qualification (PQ—not to be confused with PPQ or process performance qualification, an important step in process validation).

The documented activities may be combined in a single document or kept as separate documents, depending on the corporate philosophy and the complexity of the equipment or system being qualified. All three, whether combined or separate, comprise qualification (often referred to as equipment qualification or EQ).

In many cases, especially where a complex system is custom designed or multiple equipment systems are integrated into a production facility, a fourth element, design qualification (DQ), is included and precedes the other elements. Initially intended to refine the design of medical devices, DQ ensures the system design to be well-documented and suitable for its stated purpose. DQ of pharmaceutical equipment is a requirement under the EU Good Manufacturing Practice (GMP) regulations and is now being used globally. It addresses question #1 prior. A discussion of how DQ may be performed follows next.

The purpose of any qualification program is to document the installation and establish that the equipment is suitable for the intended task(s). This is often (not always) executed according to a predefined protocol or set of protocols. Alternative approaches are discussed following.

Why do we need to qualify manufacturing and associated equipment? It is customarily said that it is required by the regulatory agencies throughout the world. However, this should not be the sole reason. The real reason to perform the qualification is that it makes good business sense. Performed correctly, it will save the company money and time. Yes, there is an upfront cost to the qualification program; but if executed correctly it will save more than that cost during the life of the equipment. This chapter will concentrate primarily on the United States Food and Drug Administration (US FDA) requirements, but the European Union (EU), Japan, and other jurisdictions throughout the world all have similar requirements.

The sustaining parts of any qualification program must answer question #5 prior and include equipment maintenance, calibration, and change control. These are instituted so that a system is maintained in a qualified state during its useful lifetime.

This chapter will discuss the scope of qualification, the elements of qualification and the associated qualification documents. Cost, although not a concern of the regulatory agencies, is certainly a concern for each manufacturing site. The ideas expressed in this chapter should help reduce the cost and time allocated to complete these necessary functions. The examples presented are intended only for illustration and should not be used as "absolute" answers to any qualification program. They are not meant to be all inclusive, but to serve as a guide to the development of a cost effective, fully compliant GMP qualification program that will satisfy current regulatory requirements as set forward by the international regulatory authorities. This chapter will not discuss Process Validation (PV) as the common successor of the qualification program. This will be left to other areas of this book.

4.2 THE REGULATORY BASIS FOR QUALIFICATION

The term "*Qualification*" occurs twice in Title 21 Part 211 Current Good Manufacturing Practices (CGMP) of the Code of Federal Regulations (CFR).[2] These are:

21 CFR 211.25—Personnel Qualifications
21 CFR 211.34—Consultants

These sections deal with the qualification of the personnel and apply to those implementing both process validation and qualification studies. This does not mean that it does not apply to the facility or equipment nor that it does not apply to other sections in the CGMP regulations. It is fully expected and understood that it is necessary to assure the company and the regulatory agencies that the equipment will function as expected, is suitable for its intended use and will not alter or adulterate the product in any way.

In the US, the aforementioned Guidance for Industry provides extensive statutory detail derived from the US CGMP regulations that support the requirements for process validation and by extension, equipment qualification. The Guidance goes on to specify "*Design of Facility and Qualification of Utilities and Equipment*" as part of Phase 2 (*Process Qualification*) for Process Validation.

The EU goes one step beyond the FDA by including Annex 15 *Qualification and Validation* in its regulations. The Annexes to the EU GMP regulations are integral to Eudralex Volume 4 and are therefore considered to be regulations rather than guidelines. Thus, at least in Europe, qualification is required rather than implied. In practice, however, qualification is a requirement everywhere.

4.3 THE ROLE OF COMMISSIONING

All equipment used in the production of a pharmaceutical product may be considered "GMP equipment" for the sake of this discussion. GMP equipment includes not only the process equipment but also the process and product test instruments and supporting utilities. The extent to which this equipment affects product quality, however, may vary and therefore the extent to which it is subject to qualification may also vary. The authors suggest that using the term to exclusively describe equipment that is qualified is an error.

It is important to demonstrate that all equipment that is used in the production of a pharmaceutical product or medical device for use in humans or animals should be documented as to its design as well as demonstrated to perform as required. This concept is consistent across all industries and is generally called "commissioning". A simple working definition of commissioning is:

> Commissioning—the inspection and testing of equipment and systems to ensure that they have been installed according to specifications, perform as expected and are ready for operation (or qualification when it is required)

This set of activities makes sense in any new facility or factory and is considered good engineering practice. Some engineers describe commissioning as the "setting to work" of a system. In many industries, it is the final step prior to occupancy and/or operation. The level of documentation varies from industry to industry and from company to company. One common thread is that all aspects of a system that are considered essential for safe and reliable operation are inspected, started up and tested, or some combination of the three. Without commissioning, we could not have any sense of assurance that our systems would even work, let alone produce quality product.

In the early days of validation, because our industry did not embrace effective commissioning, qualification protocols were expanded to fill in the gaps that engineers may have left behind. Consequently, many errors and omissions were discovered, and protocol deviations were voluminous. Applying the same rules to investigate and resolve all deviations led to lengthy delays, often of questionable value. Commissioning, on the other hand, follows the rules of engineering practice and allows for errors and omissions discovered during inspection and testing to be documented and corrected. Commissioning is performed by subject matter experts (SMEs) familiar with the equipment system requirements and designs. The results are reviewed by independent SMEs with the same expertise. The Quality unit need not participate.

Commissioning may take place in several steps, depending on how a system is specified, designed and built. Systems that are built by specialized vendors, either in their own factories or on the user site, may undergo Factory and/or Site Acceptance testing—FAT and SAT. Depending upon how these tests are planned, executed, observed and documented, some results may be considered part or all of the system commissioning. It is important to recognize, however, that an FAT or SAT planned by a vendor is often purposefully designed for the vendor to complete their contractual obligation and receive final payment. Therefore, we strongly recommend that commissioning of vendor-supplied equipment be planned, documented, approved by and under the oversight of the user firm.

Facility systems and utilities are usually built and commissioned on the site. Although vendors and contractors may assist and perform some independent preliminary testing, it is important that the user firm take control of the commissioning effort.

Proper commissioning ensures that a system is ready to operate safely and effectively according to specifications. If executed and documented by GMP-trained SMEs and the documentation is maintained according to the site Quality Systems, then it may be possible to utilize some of the commissioning results to satisfy qualification protocol requirements, a practice commonly known as "leveraging". The user should be cautious, however, as to which results to leverage. In the example risk analysis template in Table 4.3, the recommendation is to not leverage tests for high-risk functions (e.g., autoclave temperature control or uniformity). This is not a hard and fast rule, but only an example of how risk-based thinking may be applied.

An example of what tests, inspections and documents constitute a commission program is detailed in Section 4.6.3

4.4 WHAT NEEDS TO BE QUALIFIED?

This section serves to highlight the difference between commissioning and qualification. All industries commission facilities and equipment. Commissioning is an engineering activity aimed at the full design and function of a system. Qualification is a GMP activity and focuses on equipment elements that have a direct effect on product quality, follows a more rigorous documentation approach than commissioning and is approved by the Quality unit.

The requirement to formally qualify equipment extends to all products for use in humans, including clinical supplies manufacturing. It also applies to animal health products but does not apply to preclinical animal studies (although many firms have chosen to qualify equipment for preclinical use). The requirement, however, has been interpreted in many ways. Many pharmaceutical manufacturers have decided upon a risk-based approach wherein only equipment that has a "direct" impact on product quality need follow the more rigorous procedures. Systems that do not are commissioned and documented according to engineering practice. Therefore, the additional step (or steps) required for qualification may be defined as in Section 4.2 previously:

> Qualification—documented evidence that provides assurance that equipment and systems meet the critical installation and operational requirements necessary to ensure consistent product quality.

A common checklist that has been used to determine the requirements for equipment qualification is derived and modified from the ISPE Baseline Guide Vol. 5 *Commissioning and Qualification* (Table 4.1):

From the definition, each system may have "critical" requirements that directly affect product quality. Those requirements may relate to its physical structure (e.g., material in product contact) or its operational performance (e.g., ability to control temperature within a specified range). Application of a risk-based approach suggests that any system with at least one such critical requirement must be qualified. A simple approach to by-pass the checklist would be to employ a User Requirements Specification (URS) as described in Section 4.5.1 following. The URS may be used to identify which of the equipment or system requirements has a direct effect upon product quality. Systems without a critical requirement, even those used to support production, may be commissioned-only.

The reasoning behind this approach is simple. As a quality-related program, qualification must fall under the auspices of the firm's Quality unit. This means that Quality must approve the plans as well as the results of the inspections and tests that document the acceptability of the equipment. By limiting this oversight to systems that directly affect product quality, the focus of the Quality unit is not diluted attending to review and approval of activities that do not.

4.4.1 Process Equipment

Clearly, equipment that has the primary function of performing one or more process steps and any other equipment that comes in direct contact with the product is subject to qualification. An example of the latter may be a transfer pump or a holding vessel. Each of these fulfills a requirement that directly affects product quality, even if it is a simple as material in product contact.

What about equipment that has been repurposed to perform a different process step or function? In general, the answer

TABLE 4.1
System Level Evaluation and Risk Assessment[3]

	Yes	No	Note
1. Is the system used for the manufacturing, testing, storage, packaging and labeling of pharmaceutical products and API for commercial or clinical use?			If Yes, proceed to #2. If No, the system is Risk category 0.[4] EQ is not required.
2. Does the system come into direct contact with production materials (e.g., product, excipients), or sterilized primary packaging components?			If Yes, EQ protocol is required.[5] If NO, proceed to #3.
3. Does the system constitute one or more unit operations of a manufacturing, cleaning, final rinse, sanitization or sterilization process?			If Yes, EQ protocol is required. If No, proceed to #4.
4. Is the system used for primary packaging, labeling, or to satisfy a regulatory requirement (e.g., tamper-evidence, child-resistance)?			If Yes, EQ protocol is required. If No, proceed to #5.
5. Does the system control, monitor, record, test or verify a Critical Process Parameter, Critical In-Process Control or Critical Quality Attribute?			If Yes, EQ protocol is required. If No, proceed to #6.
6. Does the system provide an environment or material (e.g., steam, water) that is in direct contact with the product or is otherwise critical to product quality?			If Yes, EQ protocol is required. If No, proceed to #7.
7. Does the system provide a service (e.g., heat/cool, power) necessary for operation of qualified systems, but does not require qualification protocol per #2–6 above?			If Yes, system requires commissioning. Interface to qualified systems is verified in qualification protocols for the qualified systems.

Abbreviations: API = active pharmaceutical ingredient; EQ = equipment qualification.

again is "YES", it needs to be either qualified or re-qualified, at least in part. If the equipment was not used in a similar operation, or if the product or process was different from the original intended use, the equipment needs to be qualified for its new use. Therefore, it is usually a good idea to initially qualify a piece of equipment over its full operating range and functionality. This will allow its use in a variety of potential future operations, not just the one use for which it is initially qualified.

How about equipment that has been transferred from another facility or moved within a facility? Usually, some level of re-qualification is required, depending upon the design and use of the equipment. In the former case, the equipment usually needs installation qualification to define its new location and to ensure that appropriate technical and maintenance information is recorded. In the case of movement within a facility, re-qualification would depend upon the newly defined use of the equipment and the change in utility connections. Typically, firms do not require re-qualification for equipment that is designed to be portable.

Support equipment, such as pumps, agitators and heat exchangers also need to be fully qualified. These units are typically qualified along with and as part of the major unit with which they are associated. When these units are portable and/or may be used to support more than one equipment system (e.g., a portable pump), they may be qualified independently (or within a group of similar units, such as all sanitary lobe pumps).

4.4.2 Utilities

All process utilities that have direct product contact or have a direct impact on the product quality need to be qualified. An example of this may be a process water (e.g., Purified Water) or gas (e.g., compressed air). These are often considered utilities, even though as in the case of water, they might also be considered raw materials. Other utilities considered to have a direct impact on product quality would be a clean-in-place system or a clean steam system used for product or filling component sterilization. In each case, a system risk analysis will determine if qualification is required. If there is no direct impact on the product, the utility may be commissioned only.

Some systems fall into a "gray" area, wherein they are used in the process (i.e., they are "GMP equipment"), but their quality impact is indirect. In this case, they would not have critical requirements but would fall into the category described in Question #7 of the checklist. In these cases, the utility would provide a service to a qualified system without actually having a direct quality impact. Examples of utilities that have indirect impact on the product include the electrical power, boiler steam, or instrument air and cooling water. Again, the thought must always be towards their impact on the product.

4.4.3 Laboratory Equipment

Laboratory equipment used for the in-process or final product testing of pharmaceutical products needs to be qualified prior to its use. This equipment is used to determine the status or release of the product (or intermediates) either to the next process step or for release for sale. Although test equipment often stands alone, it may also be used online, that is, as the product passes through, or offline for those samples collected during the production and taken to the lab for analysis. In addition to the laboratory test equipment qualification, the test methods used must also be fully validated, but validation of analytical methods is not covered in this chapter.

4.4.4 HVAC/Environmental Control

Heating, ventilation, and air conditioning (HVAC) and Controlled Environments present somewhat of a dilemma. Clearly, the environment has the direct effect on product quality, although the air handling equipment carries the burden of providing that environment. Controlled and classified environments require qualification. It is up to each firm to decide what environmental requirements have a direct and significant effect on quality. Usually, these requirements are limited to measurable parameters in the controlled or classified (clean room) area, such as temperature, relative humidity, airborne particulate levels and relative pressures. The air flows, heating and cooling functions (coils), and pre-filtration of the air handling system are usually considered to have an indirect effect. There are so many variables involved in an HVAC system that acceptable environmental results can be achieved with multiple design and operating factors applied. Most firms have used a risk-based approach (see Section 4.5) to conclude that the qualification of these systems should focus on the environment itself rather than the mechanical system, handling the latter by commissioning. It should be noted that, besides the instrumentation to monitor the environmental conditions, final filtration (usually HEPA) is also qualified.

4.4.5 Duplicate Equipment

The question of qualification of identical systems is answered simply. Each system must meet the requirements specified to ensure product quality, therefore, each system must be inspected and tested independently. Although it is permissible to plan the qualification of a family of identical systems (e.g., one protocol covering the qualification of a bank of tablet coaters), qualification for each member of the family must be executed and documented. This is the only way that equivalence of systems is conclusively demonstrated.

The documentation of system equivalence is important to process validation because it may obviate the need for an equal number of replicate process performance qualification (PPQ) runs in identical equipment. This can save considerable time and effort while effectively qualifying the process.

4.5 RISK-BASED QUALIFICATION

The checklist in Table 4.1 prior introduces the concept of risk assessment in its title and the footnotes. Reduced to its core concept, any system with requirements that have a direct impact on product quality is subject to qualification. We may call these "critical quality attributes" or CQAs. The CQAs may be used to define the focus and extent of the qualification exercise. It is clear, however, that every component and function of a system may not have a quality impact, i.e., it may not support a CQA. This is not to say that these noncritical attributes may be ignored. After all, they are probably necessary to make the system work at all, or to satisfy some other requirement, such as safety or business goals. These may be effectively considered in commissioning.

A route to risk-based qualification follows.

4.5.1 User Requirements

Any equipment system must meet the requirements of the user, so it is almost imperative that these be documented. The best approach to a URS is to list requirements without attempting to define technical solutions (An exception to this is when the technical solution is defined in the process development or is part of a regulatory filing.) The URS may be organized in various forms, a segment of one is shown in Table 4.2 following:

Note that these requirements are examples for illustration and not a real exercise. The importance of the URS to qualification is evident in the last two columns. These identify the requirements according to the following definitions and provide the requirement for verification as follows (note that alternative definitions and acronyms may be used, but the concepts themselves are broadly accepted):

Critical Quality Attributes—Critical Quality Attributes (CQA) are measurable properties of the product that are considered critical for establishing the intended purity, efficacy and safety of the product in conjunction with the process validation. A quality attribute is a regulatory or compliance related characteristic that can be measured/tested and will form the basis for process qualification testing. Quality Critical Attributes should be defined, justified, and documented for validation of all products.

Critical Process Parameters—The critical process parameters (CPP) are process variables that must be maintained within the specified range in order to ensure that the product will be within the CQA limits. These parameters are a regulatory or compliance related characteristic that can be measured/tested and will form the basis for equipment operational qualification testing. CPPs will be defined, justified and documented for validation of all products.

Business Essential Attributes- Business Essential Attributes (BEA) are attributes that have been identified by site operations, the business unit or the company as being essential strictly from a business perspective. They define

TABLE 4.2
User Requirements Specification (URS)

Designation	Description	Type[6]	Verification
URS-1	System must be of GMP design to prevent contamination of product, with appropriate materials of construction. Equipment must not be corroded by, or reactive with, materials being processed.	CQA	DQ/IQ
URS-2	Design must facilitate cleaning of the product contact surfaces.	CQA	PQ
URS-3	The system must be capable of mixing powders and liquid into a uniform blend. *[critical process parameter]*	CQA	OQ
URS-4	Tank must have temperature control over a range of 10°C to 50°C, with a control tolerance of ± 5°C or less. *[critical process parameter]*	CPP	OQ
URS-5	System must be able to pump binder solution (250 cps) at a flow rate of 10 to 25 kg/min with a control tolerance ±1 kg/min or less and control total volume pumped to within ±2 kgs or less. *[critical process parameter]*	CPP	OQ
URS-6	The vendor must supply appropriate Operations and Maintenance manuals, schematics, spare parts lists and other documentation required for equipment qualification.	CQA	IQ
URS-7	The control system must have logical and physical security to prevent unauthorized access to the system and its data.	CQA	OQ
URS-8	The control system must be capable of generating accurate and complete copies of GxP records in electronic form suitable for inspection and review.	CQA	OQ
URS-9	The control system must have controls to ensure electronic record integrity and security.	CQA	OQ
URS-10	Equipment must conform to Company XYZ, Inc. equipment or design standards, if applicable.	BEA	Design review
URS-11	The tank must have a working capacity of 75 liters to 225 liters. *[Throughput / size requirement]*	BEA	Commission
URS-12	System must be designed to minimize aeration or foaming. *[Process efficiency requirement]*	BEA	Commission
URS-13	Tank must be capable of installation on the existing tank skid in room 415. The tank must fit all the existing mounts and process piping connections. *[Site limitation]*	BEA	Commission
URS-14	Electrical components must be wash-down and correctly rated for the room's environment. *[site limitation]*	BEA	Commission
URS-15	Spare agitator should be provided	OA	Design Review

capacity parameters necessary to meet the production plan and will include non-FDA regulatory requirements such as personnel health and safety or environmental protection. BEA attributes will be tested during equipment commissioning but will not be tested again during equipment qualification. Business essential attributes must be included in the project.

Optional Attributes—Optional Attributes (OA) are attributes that have been identified as being desirable, but not essential. Optional attributes may increase the parameter range of the equipment; increase the life expectancy of capital equipment or reduce the manpower required to operate a system. These are usually line items quoted separately by the vendor in their bid. Those optional attributes that remain in the project will be qualified if they could be CQAs in the future or commissioned only if they will not. Whether or not an OA requirement remains in the project will be determined by the project team while developing the final equipment specifications, based on a cost/benefit analysis.

4.5.2 Determining Quality Impact

The right-most column of Table 4.2 refers to the verification required for each attribute and is discussed in the following sections. By focusing on the quality-related CQAs and CPPs, the qualification effort may be simplified and allowed to focus on the elements of a system that affect quality and are necessary to ensure effective process validation.

When evaluating impact on product quality, it is important to apply a broad definition of quality attributes. It is not enough to define quality solely as the effect a product may have on a patient. In its broader sense, quality relates to a range of "customers", including physicians, pharmacists, distributors and regulators. Therefore, when determining quality impact, each of the following may apply:

- Drug Safety
  - Efficacy
  - Purity
  - Strength
  - Identity
  - Regulations
- Stability
  - Package integrity
  - Distribution
  - Disposition
  - Documentation
  - Regulatory filings

Having defined quality-related attributes and requirements, the URS may then be used as a guide to the next level of system risk assessment, which looks deeper into the system and its qualification.

4.5.3 Value of Component and Functional Level Risk Assessment

Once critical requirements have been determined (thus, the requirement for qualification), further evaluation of the system's major components and functionality should be carried out in order to address specific qualification requirements. In addition, the information obtained may be used to categorize specific equipment and instrument items for the purpose of more effectively managing operation and maintenance.

Aside from the need to qualify systems having critical requirements, the benefits of system evaluation include:

- Understanding of system structure and function
- Identification of quality-related risks associated with the system, its components and its functions
- Enhanced attention to elements having more risk to/impact on product quality
- Development of a robust Commissioning and Equipment Qualification approach
- Defined methodology and documentation for classifying systems and instruments in Master Equipment and the Master Instrument Lists

In addition, long-term benefits include more effective application of Quality Systems related to facilities and equipment, including:

- Documentation
- Training
- Maintenance
- Calibration
- Change Management
- Deviation Management

This is to say that elements of the Quality Systems applicable to functions and components assessed to have a low or negligible risk to product quality may be handled differently (i.e., different work flows, reviews and approvals) than those having a significant risk.

4.5.4 System Structure and Risk Evaluation

Each system may be made up of multiple components, some mechanical and some described as instrumentation. Each of these has a function or functions and each has more or less of an impact on quality. By evaluating each component and determining its function in relation to Critical Quality Attributes and Critical Process Parameter, significant useful knowledge is attained, and the qualification approach is determined.

The system may be broken down into major components and instruments, usually described on a system schematic diagram. A P and ID (Piping and Instrumentation Diagram) is suitable, but other drawings, sketches or documents that indicate all major components are acceptable.

Most equipment functionality can be linked directly to a physical component. Some software functions, however, may stand alone. Examples include recipes, sequences and calculations. It is important that these functions are also considered in the evaluation of automated systems, whether or not the mechanical and automation qualification protocols will be combined. For the sake of simplicity, the reference to "components" in the text shall also include these software functions.

Component evaluation should consider component function and risk to quality (see Section 4.5.2) and pay particular attention to CPPs, CQAs, critical in-process controls (CPICs) and other critical requirements. The series of questions following may guide the evaluation with answers weighted to arrive at a risk categorization of each component. Any component or function determined to have a significant (high or moderate) risk to quality must be verified in the qualification.

For new systems, it is important to link the component to a requirement in the URS as a form of traceability. This may not be possible for existing systems where a URS has not been developed. In addition, the evaluation should be annotated with a brief description of the type of inspection or test that will be required in commissioning or qualification.

An example of a component/functional level risk evaluation for a simplified process unit follows:

4.5.5 Master Equipment and Instrument Lists

In addition to verification requirements, the evaluation may be used to categorize or classify equipment in Master Equipment Lists (MELs) and Master Instrument Lists (MILs), respectively. These lists become an important element for maintaining equipment in the qualified state.

Each site or section of a site should maintain a single Master List for equipment and one for instruments (MEL and MIL). Lists should be inclusive of *all* items in the maintenance and calibration programs, without regard to whether they are used in manufacturing or their risk assessment category. Separate lists of GMP or qualified equipment/instruments are not recommended unless they are derived directly (i.e., electronically) from the Master List (e.g., from a Computerized

TABLE 4.3
Component and Functional Risk Categories

Cat	Level	Description	Examples	EQ	Operations
0	None	Non-GMP; function not used for or shared with GMP operations. No risk to product.	offices, cafeterias, waste treatment, process development (non-clinical)	Commission as required for safe start-up and to facilitate maintenance and non-GMP regulatory requirements (e.g., environmental). Not subject to GMP-SOPs.	Routine maintenance.
1	Low	GMP functionality, but use does not constitute a primary manufacturing function. No product contact. Defects do not affect patient nor do they result in recall. Malfunction nearly always detectable. May cause system downtime or loss of production, but usually does not result in significant scrap or rework.	GMP non-critical instruments Boilers, chillers, cooling towers, palletizers, accumulator tables, conveyors, air handler fans, coils, pre-filters, water pretreatment steps	Include in system commissioning. Do not require EQ verification. Interface to qualified system documented in EQ protocol; e.g., steam supply to a reactor is documented in reactor protocol, conveyors and turntables are documented in PQ of packaging lines, water pretreatment documented in water purification PQ.	Included in Maintenance and Calibration programs. May not require application of formal Change Management or Investigation systems if parent system is not qualified (as specifically allowed by site SOPs). However, changes should be commissioned to ensure consistent delivery of service (e.g., steam pressure, chilled water temperature).
2	Moderate	Critical requirement, affects quality. Undetected failure may lead to rejected product, major scrap or re-work. Rarely results in recall or danger to patient. Usually detectable or subject to some level of redundancy (e.g., alarm, check weigher), which limits losses.	blender, dryer, tablet coater, capsule fillers, process temperature control, final steps water purification, water distribution, temperature/ RH monitor in Controlled areas	Critical requirements must be verified in EQ. Some requirements may utilize commissioning documentation as evidence. Recommend that actual operation (sequence, control, etc.) be subject to repeat testing per EQ protocol.	Subject to full requirements of all applicable Quality Systems. Changes and investigations should consider the function and its associated risks. Re-commissioning/re-qualification as needed.
3	High	Critical requirements, usually associated with high-risk products (e.g., sterile, high-potency). Undetected failure may lead to total loss of batch or may result in a recall. Malfunction may be difficult to detect (e.g., autoclave temperature uniformity)	autoclave temperature control and distribution, T/RH/dP in Classified areas, fill volume for sterile vials	Critical requirements should be inspected and/or tested in EQ. Not recommended to leverage commissioning results (although preliminary testing in commissioning is standard practice).	Subject to full requirements of all applicable Quality Systems. Changes and investigations should consider high-risk nature of the function. Changes and corrective actions most often require qualification testing.

Abbreviations: dP = pressure differential; EQ = Equipment Qualification; GMP = Good Manufacturing Practice; PQ = Performance Qualification; RH = relative humidity; SOPs = Standard Operating Procedures; T = temperature.

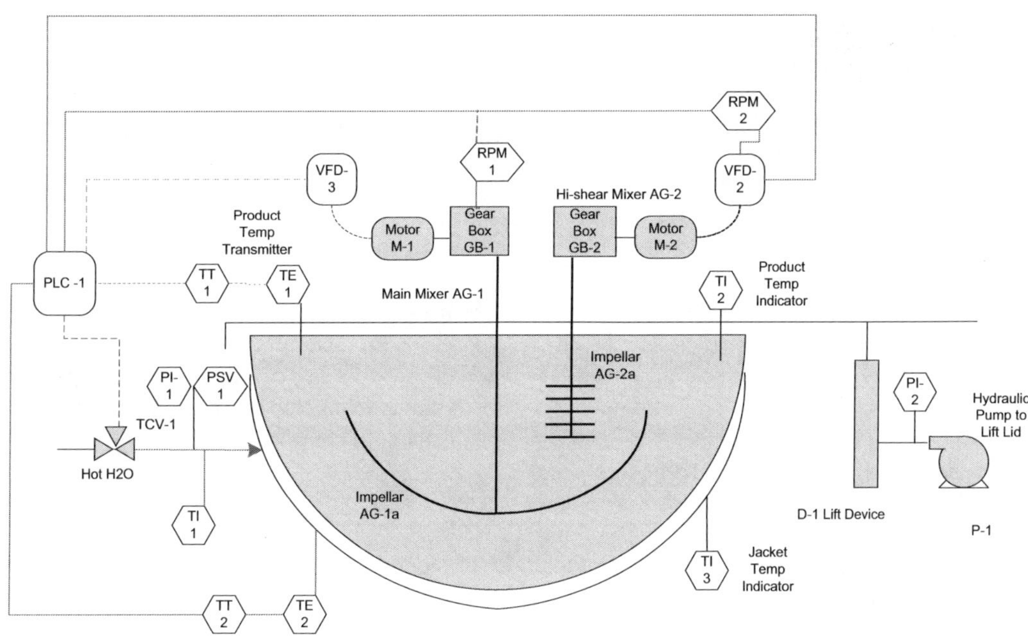

FIGURE 4.1 Liquids-ointments-creams (LOC) processor.

TABLE 4.4
LOC Processor Risk Assessment[1]

Tag no.	Description	Function	Risk category[2]	C and Q activity[3]
K-1	Vessel (kettle)	Contains process	2	Commission all functions. Qualify critical requirements.
AG-1	Main mixer	Bulk mixing; scrapes sidewalls	2 The mixing function requires precise control of power input and is dependent upon impeller type, size and speed. All components of this mixer as listed perform critical functions	Commission and qualify mixing function employing all sub-components listed below. Check range and precision of speed control; check operation under load (e.g., placebo mix) over desired rpm range.
M-1	Motor	Supplies power	2	see note[4]
GB-1	Gearbox	Moderates speed	2	
VFD-1	Variable frequency drive	Controls mixer speed	2	
RPM-1	Speed monitor/ transmitter	Measures speed	2	
AG-1a	Impeller	Mixing	2	
TCV-1	Temperature control valve	Controls batch temperature	2	Commission and qualify
PLC-1	Programmable logic controller	Multi-function control device	2	Subject to developmental testing and validation per corporate standards. Documentation may be combined with EQ documents.
TI-1, TI-3	Product and jacket temperature indictors	Reference gauges	1	Commission only
Utilities Electrical power Hot water	Utilities serving processor	Supply energy	1 These do not contact the product and are subject to monitoring by other devices. Their failure is readily detectable and worst case is that a batch is lost.	Interface of each system to the processor is checked in commissioning and verified in EQ. Distribution of each of the utilities is commissioned in a separate exercise.

Abbreviations: LOC = liquids-ointments-creams; EQ = Equipment Qualification.

[1] Note that this diagram has been simplified for clarity. An actual LOC Processor may include different components than listed here.

[2] This is an example only. It does not represent any specific system and is not the result of a cross-functional analysis.

[3] Activities listed are for illustration only.

[4] Sub-elements are commissioned and qualified with the main functional component.

Maintenance Management System or a validated spreadsheet). The reason for this is because it obviates the possibility of different lists containing disparate information and also better ensures that items are not overlooked.

Both MEL and MIL should be organized in a hierarchal fashion so that each entry is identified as a component of its parent system and, for qualified systems, is traceable back to qualification documentation.

Further, qualified systems require special emphasis when applying the other Quality Systems described in Section 4.5.3. Because they represent a greater risk to product quality, they are subject to more rigorous control under each of these systems, including greater oversight by the Quality unit.

Qualified systems should be subject to site change management and investigational management procedures. The manner in which these systems are applied may be determined by what changes or events are being considered and what component of the system is involved. Those components determined to be within a higher risk category should command more scrutiny. Where the component has been previously evaluated to have less risk to quality, the change request or investigation may be dealt with accordingly, i.e., expedited.

For calibration, a recommended approach is that instruments be categorized as GMP-critical, GMP noncritical or Other. These categories influence the treatment of these instruments under Quality Systems.

Maintenance standards do not require that major mechanical components be categorized. However, risk assessment of components will allow assignment of risk categories that will aid in the effective application of Quality Systems.

Sites have the option of developing SOPs that allow certain changes or events to occur without applying formal change management or investigation (i.e., substitution of functionally equivalent parts when there is no product contact). These procedures should be well-defined and allowable for only low-risk components or functions.

4.5.6 Cross-Functional Analysis

Although it may be acceptable for an assessment at either the system or component level to be carried out by an individual and later reviewed and approved by other Stakeholders, a cross-functional approach is recommended. The cross-functional approach ensures that opinions and decisions are challenged, has the synergistic advantage of multiple inputs and facilitates consensus.

A cross-functional team should consist of the User Group, Technical Group (Maintenance or Engineering), Qualification Group, Quality unit and others, as required. Each of these is responsible to participate in and approve the system and component evaluation.

Sites may provide a record of assessment in any document that is approved by the requisite groups. Examples include Equipment Qualification Plan, Equipment Qualification Protocol and/or stand-alone documents. The format represented in the example prior (a drawing or P and ID that identifies major components plus an evaluation matrix) is a suitable means to document the evaluation.

4.6 THE QUALIFICATION PROGRAM

Prior to recommending a sequence of activities for qualifying equipment, it should be emphasized that there is no regulatory requirement for how this is to be performed and documented. Several of the terms used in the approach recommended do appear in regulatory guidance and even in the EU Annex 15, which constitutes a regulation. Regulatory authorities, however, allow a fair degree of flexibility as to how qualification is performed, documented, reviewed and approved. Alternative approaches are presented here.

4.6.1 ASTM E2500

Published in 2007, ASTM E2500 is entitled *Standard Guide for Specification, Design, and Verification of Pharmaceutical and Biopharmaceutical Manufacturing Systems and Equipment.* Note the absence of the term "qualification" and the substitution of "verification". The authors have no problem with this approach, accepting verification as something that is performed and qualification something that is achieved. Nonetheless, at the time of publication, equipment qualification was perceived by some to have become an exercise in regulatory compliance rather than one in establishing evidence that a system reliably performs as required. There was great emphasis on protocols, acceptance criteria, deviations, investigations and final reports, all under the auspices of the Quality unit. ASTM E2500 shifts the oversight of verification to SMEs who are selected based upon process and equipment knowledge. These SMEs would plan the verification exercise, including defining the system requirements and technical characteristics to be verified. The quality-relevant aspects of the plan would be approved by the Quality unit. Upon completion of the appropriate inspections and tests, the SMEs would ensure that the system met its requirements and issue a report certifying that. The Guide makes no mention of deviations from acceptance criteria, field corrections, adjustments or any similar occurrences. Presumably, any issues found by the SMEs in the verification process may be corrected and confirmed to meet requirements. This approach is very similar to (and may even be considered equivalent to) commissioning as described following.

Several firms have adopted the ASTM approach, at least in part. This approach, however, does not negate the content recommendations following in what may be considered a more traditional approach.

4.6.2 Protocol or Not?

There is no argument from any credible source that equipment qualification (or verification) does not require planning. Equipment Qualification Plans have been used for decades to plan complex projects comprising many process equipment and facility systems. These are usually general in nature and describe the requirement and basic approach to qualify systems, but without the specific inspections and tests described.

The traditional approach to planning qualification activities for equipment systems has been the qualification protocol. The concept of "protocol" was initially derived from process validation protocols, which were in turn derived from the methodology used to prove (or disprove) a scientific theory. In process validation, a process is described that is intended to produce a product with predetermined characteristics (we call these Critical Quality Attributes—CQAs). With critical process parameters (CPPs) defined, a set of acceptance criteria related to the CPPs and resulting CQAs would determine if the process would perform reliably. We know today that process validation relies on so much more that satisfying a protocol and the activity that was once called "validation" has become only a small part of an effective process validation. This is described elsewhere in this text.

Initial validation activities emphasized process, starting with sterilization, and expanding to formulation, API synthesis, cleaning and other processes. The first process validation efforts were supported by very simple and straightforward engineering documentation of equipment. Only after several years of validation experience were equipment qualification protocols introduced. It seemed a natural outgrowth of validation at the time and has persisted until today, ASTM E2500 notwithstanding.

The argument against protocols in the first decade of the 21st century was that the protocol had become more important than what it was intended to prove. More effort seemed to be expended in preparing qualification protocols and then investigating deviations from acceptance criteria than in understanding the relative importance of the various systems. The argument for protocols, and against the ASTM E2500 concept, is that protocols retain the essential governance provided by the Quality unit.

By applying the risk-based approach described in Section 4.5 combined with an effective commissioning effort, much of the ritual around equipment qualification has been reduced, while retaining the Quality unit oversight and final approval. Today, we strive for "skinny" protocols, focused on the system requirements that impact product quality.

For the purpose of further discussion, we shall assume the use of equipment qualification protocols. The sequence and content of the activities is independent of whether or not a protocol or other means of documentation is employed. The necessary activities remain essentially the same regardless of the documentation approach.

4.6.3 Commissioning as a Prerequisite to Qualification

Commissioning is a standard engineering practice and is strongly recommended for all newly installed systems. It is not necessary to complete all system commissioning activities and to obtain approval prior to beginning qualification. It is, however, important that commissioning be sufficiently complete to allow the safe and effective execution of qualification tasks.

As an example, this approach allows initiation of installation qualification if it can be safely performed while the operational aspects of commissioning are being completed. This presumes that there will be no physical changes to the equipment during this exercise. Another example is the initiation of qualification pending the compilation of vendor or contractor documents, provided that none of these are necessary for proper inspection or testing of the system.

It is recommended, however, that commissioning be complete and approved prior to approval of the executed Installation Qualification protocol.

For uncomplicated systems or for minor changes requiring EQ, the site may choose not to Commission. In these cases, tasks listed herein as Commissioning should be included in the qualification protocol.

For existing systems requiring qualification, commissioning functional testing may take the form of an engineering study, wherein system capability is studied and documented.

The Table 4.5 indicates some of the common requirements of a commissioning program.

4.6.4 Principles of Equipment Qualification

Equipment Qualification is the final series of inspections and tests to ensure that critical requirements necessary for product quality are satisfied, and that documents and procedures necessary to properly operate and maintain the system are in place.

Examples of critical installation requirements may include product contact material and instrument accuracy. Examples of critical operating requirements may include temperature and fill weight accuracy. Systems having at least one critical requirement are subject to equipment qualification as described in this section.

Engineering details that do not have a direct effect on product quality and that have been confirmed in Commissioning need not be rechecked in qualification.

Qualification is most often described in terms of installation qualification, operational qualification and performance qualification, or IQ, OQ and PQ. The extent of qualification activities will vary depending on the specific function of the equipment being qualified. Equipment may require IQ only, IQ and OQ, or IQ, OQ and PQ. The number and type of specific inspections and tests will be dependent upon the complexity of the system and its impact upon product quality. The following examples are provided for illustration only:

- Product holding bin without temperature control or agitation: IQ only
- Mixing vessel with agitation and temperature control: IQ, OQ
- Purified water system: IQ, OQ, PQ

Assessment of systems is described in Section 4.5. It includes an evaluation of system components and functions to determine the impact on product quality. These should incorporate User Requirements (when available) as well as Critical

TABLE 4.5
Commissioning

Documents	Notes
Weld Records/Inspection Reports (sanitary tubing only)	As available for existing equipment.
Piping Flushing/Cleaning Reports (sanitary tubing only)	As available for existing equipment.
Vendor Manuals	As available for existing equipment.
Passivation Procedures and Records	As available for existing equipment.
Identification and initial test of Critical Operating Requirement Values	Required. Some or all of this may be repeated in the OQ.
Loop Check Reports	Required for automated systems.
Input/Output lists	Required for automated systems. Verified by loop checks.
Instrument Calibration Reports	Required for instruments that can be calibrated. May be factory or field certified by contractor.
HVAC Test and Balance Reports	Required (Contractor procedures must be approved by Engineering).
HEPA Filter Certifications	Required (Contractor procedures must be approved by Engineering). This may be considered a critical requirement for some areas (e.g., aseptic fill) and will be referred to in the EQ protocol.
System Schematic Walkdown/Markup	Required. Use latest version of drawing. Mark "red-line" to indicate correct installation as well as apparent discrepancies.
Equipment/instrument confirmation vs. specifications	Required for major equipment and instruments identified by a tag no. Include manufacturer, model no., size/range/capacity, material of construction. Information may be derived from nameplate. Specifications may be derived from design data sheets. Also indicate unique identification number assigned by site. For existing systems, information may be recorded as observed.
Functional Test Report	For existing equipment, test requirements may be derived from operational documents in place of an Engineering Specification or from a documented engineering study of the system.

Abbreviations: HEPA = high-efficiency particulate air; HVAC = heating, ventilation and air conditioning; OQ = Operational Qualification.

Process Parameters (CPPs) and Critical Quality Attributes (CQAs) related to system use. The intent of these evaluations is to ensure that appropriate inspections and tests are prescribed for each system element. Records of assessments become part of the validation files. They may be included in a Validation Plan, a Project Quality Plan, Equipment Qualification protocols or other controlled document.

Qualification protocols may combine or separate IQ, OQ and PQ. The choice will depend upon the qualification requirements and the complexity of the system. It is not necessarily required to complete one phase of qualification prior to beginning the next phase. It is, however, necessary that the prior phase be sufficiently complete to allow for accurate and meaningful data to be gathered in the subsequent phase and with no significant deviations outstanding. Also, before approval of a qualification report, the precedent qualification phase must have been approved.

4.6.5 Design Qualification

DQ is an element of equipment qualification that had its inception in the medical device field. In that context, it referred to the design of a specific device or prototype that performed a new and custom function. This DQ qualified the design of the device so that it could be copied in commercial production and distribution.

In EU-GMP Annex 15, DQ is listed as a component of equipment and facility qualification. Rather than take a dogmatic approach to DQ and try to execute a DQ protocol for every system, most firms have adopted a more flexible approach. These varied DQ methods appear to be acceptable to the regulators and serve the purpose of assessing a design prior to the system actually being built.

It makes little sense to attempt to qualify the design of a proven vendor-supplied equipment. What would be the value of attempting to independently qualify the design of a Fedegari autoclave or a GEA high-shear granulator? System designs from highly experienced equipment manufacturers routinely exceed any DQ requirements that could be imposed by the end user. On the other hand, large-scale projects involving newly designed facilities, utilities and process systems entail many GMP considerations that are dependent upon unique designs. In these cases, firms can choose to qualify overall designs in the form of a detailed and documented design review at specified stages of the design (see the typical project sequence in the Appendix). This needs to be done relatively early in the project so that issues may be identified and adjusted or corrected. For larger projects this can be in two stages: at Conceptual Design (layouts, utility lists, areas classifications, process/material/people/product flows, etc.) and Basic Design (or Basis of Design—BOD—system schematics, P and IDs, equipment and instrument lists, etc.). Waiting too far into Detailed Design to perform the design review will often result in changes/corrections that may involve costly reengineering. It is valuable, however, to perform some additional DQ of custom designed systems as the details of the design emerge. An example might be a compendial water distribution system. This, however, can also be accomplished as part of protocol preparation or qualification planning.

The focus of DQ is to ascertain whether the design satisfies User Requirements. For commercial equipment, this is fairly easy and can be accomplished by merely developing the engineering or technical specification to comply with the URS. Then the vendor submittal is reviewed and determined to comply with the specification. In these instances, most firms treat this sequence as an engineering practice, with each review carried out by the technical experts (SMEs). It is not usually required to complete a formal traceability matrix linking the design back to the URS. The circle is closed when the qualification protocol execution provides the evidence that the system is installed and operates as required.

For more overall plant design and specific customized systems with significant risk to quality (e.g., WFI distribution), a detailed design review might be performed and documented. In the case of the overall plant design, as described previously, the firm may decide to provide a detailed report that would be reviewed and approved by a cross-functional team, including Quality. For custom individual systems (including complex computer control systems), the design review is more often handled by SMEs, with Quality sign off only at the qualification stage. Each firm may determine its own approach, but these examples have proven to work well and be acceptable to regulatory authorities.

4.6.6 Installation Qualification

IQ ensures that the equipment as installed meets the User Requirements and that critical installation requirements have been verified. Engineering details that do not directly impact product quality may be documented in Commissioning and need not be repeated in IQ (e.g., circuit breakers, I/P transducers).

IQ requirements are listed as follows. These are not listed in a required order of completion. It is permissible to address some IQ elements in OQ or to combine IQ and OQ into one protocol.

The following are typical critical system requirements, and include only those that have been identified as such for the equipment/system being qualified.

TABLE 4.6
Installation Qualification[1]

Documents	Notes
Objective, scope and background	Should define system use and boundaries.
System description and requirements	May be derived from User Requirements Specifications. Include critical installation requirements.
Commissioning complete	Verify that Commissioning (if applicable) is complete and approved prior to approval of the equipment qualification summary report.
As-built system schematic(s)	Typically P and IDs. These drawings are designated as GMP drawings and must be field verified to confirm all critical equipment and instruments as part of the IQ process. Red-lined drawings may be used for verification purposes. Drawings listed "for reference only" need not be field verified. The protocol should designate which drawing(s) will be field verified, indicating "latest revision" (as revisions may be updated).
Equipment list	Include all system components that are identified by a tag number. Indicate manufacturer, model no., size/capacity, materials in product contact. Serial numbers are not recommended. Information may be derived from commissioning nameplate verification. Critical installation requirements should be re-verified against expected values (may be done during IQ drawing verification).
Instrument list	Include all system instruments that are identified by a tag number. Classify as GMP—critical or noncritical according to system's intended use and operation. Indicate manufacturer, model no., size/range, materials in product contact. Information may be derived from Commissioning nameplate verification. Critical installation requirements should be re-verified against expected values during IQ drawing verification.
Lubricants list	List lubricants with potential product contact. May refer to a master site lubricants list.
Software title, software version, and installation verification	Record for automated systems. If control system is validated under a separate protocol, provide reference link.
Utility connections	Verify utility connections by observation at system boundary. May be verified by line size and physical requirement (e.g., pressure), as required. Electrical service may be verified by observation. Engineering details determined in commissioning (voltage, circuit breaker size, fusing, grounding, chilled water temperature, pump pressures) need not be repeated.
Instrument calibration records	Copies of instrument calibration records for GMP-critical instruments must be included in the executed protocol. At IQ stage, these may be factory certified or contractor calibrated.

(*Continued*)

TABLE 4.6
(Continued)

Documents	Notes
Parts list	Verify the location of parts list(s) required to maintain the equipment. May be located in the commissioning report or in specified manuals.
Maintenance and calibration procedures	List associated maintenance and calibration procedures and confirm that they have been drafted for comment. This list need not be pre-approved in the protocol.
Reference documents	Confirm the list of reference documents available. May include manuals, vendor prints, specifications, material compatibility studies, etc. List may reside in commissioning report. Extent of available documents will vary for existing equipment (may be none).

Abbreviations: GMP = Good Manufacturing Practice; IQ = Installation Qualification; P and ID = piping and instrumentation diagram.
[1]Items listed in IQ should be designated as acceptance criteria if they are critical to the acceptance of the system. Otherwise, they are expected values.

4.6.7 Operational Qualification

OQ ensures that the equipment as installed meets the User Requirements and that critical operational requirements have been verified. Engineering details that do not directly impact product quality may be documented in Commissioning and need not be repeated in OQ (e.g., pump outlet pressures, heating or cooling fluid requirements).

The OQ generally will consist of a test or series of tests designed to challenge critical operating requirements and alarms such as speed, pressure, temperature, flow rate, etc. against predetermined acceptance criteria. A single run of each test may be sufficient. The duration of the test(s) should be adequate to ensure system capability to achieve the desired performance while experiencing variations that might be expected to occur in normal use (e.g., an HVAC system in a Classified Area should be monitored over a period of at least three days in order to respond to normal ambient changes).

Simulated product may be used to conduct the OQ for manufacturing equipment. The operational ranges of testing should include a set of conditions encompassing upper and lower operating limits. Such conditions need not necessarily induce product or process failure.

The completion of a successful OQ allows operating procedures to be finalized and approved. This information should be the basis for training the operators in the requirements for satisfactory operation of the equipment. It is a site decision as to whether these procedures must be approved prior to approval of OQ execution.

The following are typical critical system requirements, including only those that have been identified as such for the equipment/system being qualified.

Operational Qualification

Document	Notes
Objective, scope and background	May refer to or be combined with IQ.
System description and requirements	May be derived from User Requirements and Functional Requirements or, for existing systems, from existing operational and maintenance documents. May refer to or be combined with IQ. Must include critical operating requirements.
IQ complete	Verify that the IQ activities are complete, deviations resolved and executed protocol approved (if required) prior to the approval of OQ.
Sequence of operations[2]	Tests to ensure that the system may be operated according to a sequence of operations representative of normal operations.
Instrument calibration records[2]	Critical instrument calibration records complying with site SOPs must be obtained prior to beginning any OQ testing employing these instruments.
Security, alarm and interlock checks[2]	These may be done in a separate computer system protocol or they may be included in the EQ protocol. Critical alarms and interlock checks completed in Commissioning may be repeated as determined by risk analysis.
Critical requirement test[2]	Tests ensure that critical operating requirements are demonstrated to be achievable within expected ranges. Include failure mode if applicable.
Functional tests[2]	Tests ensure that other functions are achieved as expected. May be employed to prove an extended operating range (beyond required critical requirements) if desired.
Sampling strategy, if applicable	Define sample locations, frequency and analysis to be performed. Note that a minimum number of acceptable samples or measurements should be specified. The sampling plan may include additional samples to allow for errors or omissions.
SOPs	List directly associated operating procedures. This list need not be pre-approved in the protocol. SOPs may be in final draft form at this stage, depending on site procedures.
Acceptance criteria	Required for critical operating requirements.

Abbreviations: EQ = Equipment Qualification; IQ = Installation Qualification; OQ = Operational Qualification; SOPs = Standard Operating Procedures.
[2] Items marked constitute a User Acceptance Test for the system automation when included in the EQ protocol.

4.6.8 Performance Qualification

Not to be confused with PPQ (process performance qualification), PQ is an element of equipment qualification. There is often confusion as to what is meant by PQ and why it is different from OQ. In many cases, it is not. The tests that many firms perform in PQ are the same as those performed in OQ except that often actual process materials are substituted for surrogate materials. This may or may not be valuable, but some firms have decided it reduces the risk prior to committing the system to production or to process validation. We neither recommend for or against this practice. It is strictly a matter of risk assessment and mitigation by the user firm.

One important purpose of PQ that is nearly universally accepted is to qualify the operation of a system of multiple units that may have undergone separate IQ/OQs. Examples may be: equipment operating in unison (critical utility generation and distribution) or sequentially (packaging line) over periods of time representative of expected run times and conditions.

Some helpful recommendations regarding PQ are as follows:

- PQ requirements are analogous to OQ requirements in the table prior. Tests, sampling plans and acceptance criteria need not repeat those employed for OQ but should focus on integrated system performance.
- The PQ of a packaging line may be performed concurrently with packaging validation studies.
- The PQ of critical utilities should involve an extended monitoring and sampling program in order to demonstrate system performance.
- Operational SOPs must be approved prior to beginning PQ studies.
- OQ must be completed and approved before PQ completion may be approved.

4.6.9 Qualification Deviations

If the user follows the advice provided in this chapter, deviations should not be experienced. Where they have been experienced in programs aligned to this guidance, the great majority of deviations are protocol errors rather than system failures. Some advice with regard to protocol preparation, besides employing an effective commissioning, is to "write with your feet". This means, do not attempt to write a protocol and its acceptance criteria from a desk or computer screen. Look at the design documentation, the drawings, the specification and the vendor manuals. Even more importantly, get out into the field during commissioning (or even participate) and look at the equipment to make sure that what you are requiring is realistic.

In the rare event that deviations occur, other than protocol errors, they usually fall within two major categories. Deviations that are failures to meet acceptance criteria require an investigation in compliance with site deviations/investigations procedures. If the investigation concludes that the deviation is caused by an event or condition unrelated to the normal performance of the system (e.g., power failure, lab error, human error), it is permissible to void the test and repeat. Care must be taken in analyzing this type of situation to ensure that problems are not systemic but are unusual occurrences. All deviations found to be system-related require a probable cause, corrective and preventative action (CAPA) and may require a protocol amendment to repeat the test or inspection.

It is never acceptable to write off the failure to meet an acceptance criterion as "It's OK because . . .". Remedial Actions must be completed prior to approval of the executed protocol. Longer term CAPA may be complete or tracked in a separate system until completion.

Failure to follow an approved protocol test or sampling plan (e.g., missed sample, change to an operating parameter that does not change acceptance criteria) or failure to obtain an expected result is a deviation. These deviations may or may not require CAPAs or repetition of the test. Investigation and analysis of the deviation will determine the required action.

Where CAPA is deemed inadequate and the system is determined to be unsuitable for use, the protocol may be closed with a negative summary report stating that the system is not qualified and may not be used in GMP applications until corrections have been made.

All deviations must be documented, resolved, approved and incorporated in the executed protocol. Simple discrepancies, such as typographical or entry errors, may be annotated in the protocol without an investigation.

4.6.10 Other Requirements

- The executed EQ protocol and/or summary report, including all requirements, should be submitted for review and approval in a timely manner.
- Executed EQ protocols or summary reports should not be approved with open deviations.
- Completed protocols may or may not include a covering summary report depending upon site procedures.
- Completed protocols and reports should be approved by the same groups that
- Qualification documentation should be treated as GMP records and filed in a secure location in compliance with site documentation procedures.
- Test Instrumentation used for Equipment Qualification must be calibrated or standardized, as appropriate, and traceable to national standards.
- System instrumentation used to collect data for Equipment Qualification must be calibrated or standardized according to site procedures.
- The completion of satisfactory Equipment Qualification permits progression to the next stage of validation (e.g., Process Validation).
- Draft operational procedures may be used to guide protocol test development. Procedures will be finalized and approved following the OQ activities. The verification of procedures follows the completion of

EQ activities and a list of procedures is not a requisite expected result or acceptance criterion in the protocol.

- Protocol executors should review the protocol with the protocol writer, reviewer or other qualified SME to ensure understanding of the system and protocol requirements. This should be documented by incorporation of a signature sheet within the protocol.
- Documentation such as protocols and reports generated by third party contractors/manufacturers or by company employees for studies to be executed at a third-party location must be assigned to a governing company site. The reason for this is all protocols and reports must be developed, approved, executed, and controlled in a manner consistent with the documentation procedures of the governing site in order to maintain control.

NOTES

1 Note that within this chapter, the terms "equipment" and "system" are used interchangeably.

2 [[Footnote Data Missing]]

3 In general, any system with critical requirements will require an EQ protocol (see Section 4.5.2).

4 Risk category 0 is non-GMP; commission according to good engineering and maintenance practices. See risk categories explained in Section 4.4.4 following.

5 If EQ is required and if the system contains multiple major components, proceed with component-level evaluation and risk assessment (see Section 4).

6 See following for definitions.

REFERENCES

1. 21 CFR Part 210. *Current Good Manufacturing Practice in Manufacturing, Processing, Packing, or Holding of Drugs; General.*
2. 21 CFR Part 211. *Current Good Manufacturing Practice for Finished Pharmaceuticals.*
3. DeSantis P. A Framework for Quality Risk Management of Facilities and Equipment. *Pharmaceutical Online*, January 2018. www.pharmaceuticalonline.com/doc/a-framework-for-quality-risk-management-of-facilities-and-equipment-0001
4. DeSantis P. Facilities and Equipment Risk Management: A Quality Systems Approach. *Pharmaceutical Online*, January 2018.
5. ISPE Baseline® Guide Volume 5: *Commissioning and Qualification*, 2nd Ed., 2019.
6. ISO 14644. *Cleanrooms and Associated Controlled Environments.*
7. PIC/s GMP Guide for Good Manufacturing Practice for Medicinal Products.
8. Eudralex, Volume 4. Good Manufacturing Practices (GMP Guidelines).

Appendix I
Typical Project Sequence

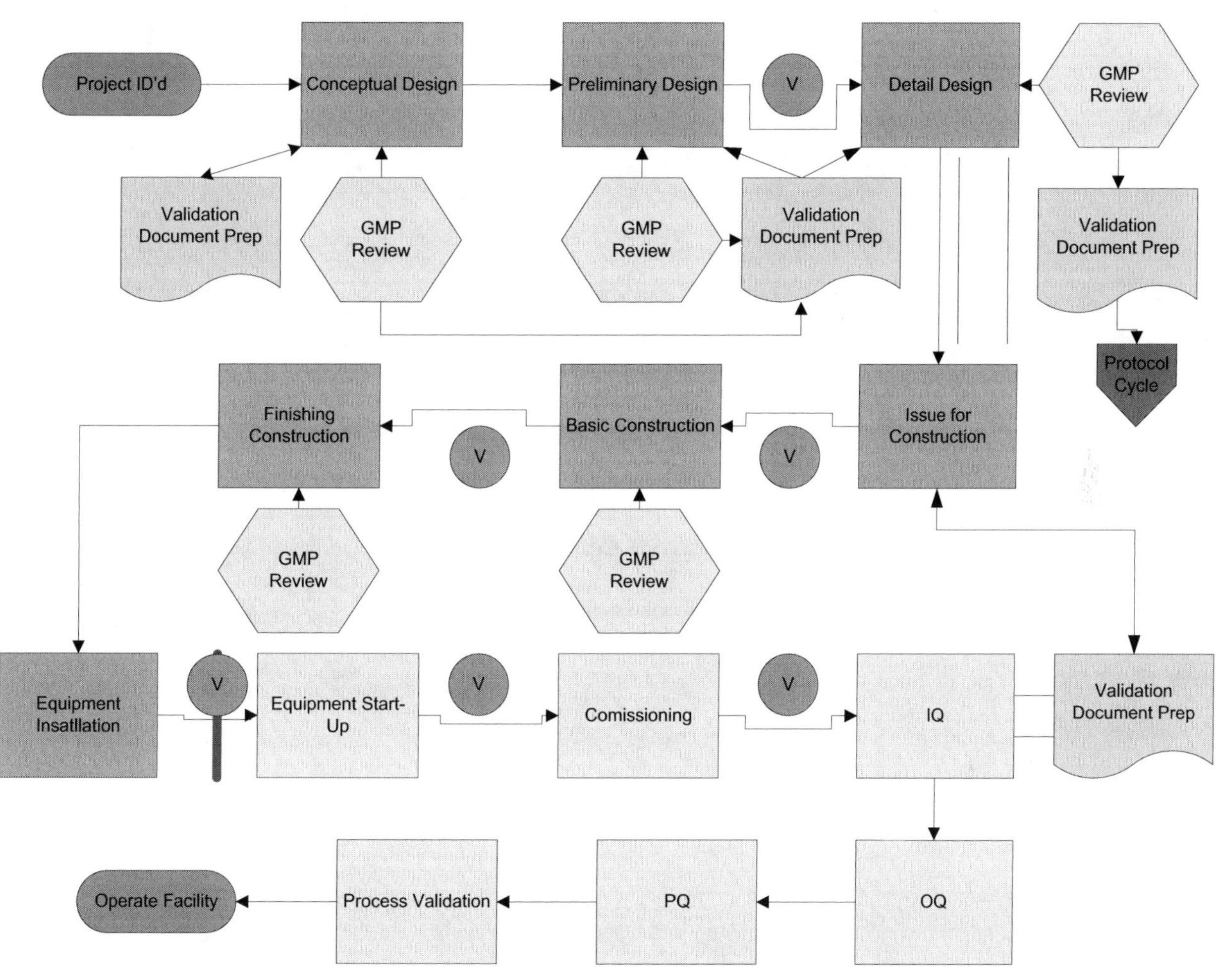

FIGURE 4.2

5 Design and Qualification of Controlled Environments

Franco De Vecchi and Phil DeSantis

CONTENTS

DOI: 10.1201/9781003163138-5

5.1 INTRODUCTION

Environmental control plays an important role in preventing contamination of pharmaceutical products, whether from the air, personnel, or from other products and materials. The level of risk from the environment is tied directly to the risk of environmental contamination reaching the patient. The environment becomes particularly important when manufacturing products intended to be sterile. An aseptic processing environment where products are not terminally sterilized represents probably the highest risk. This risk is greatest while the sterile product or components are directly exposed to the environment and where personnel are close to or intervene directly in the process.

The design of a comprehensive environmental control systems for facilities used for the formulation and packaging of pharmaceutical and biological products (heretofore biopharmaceuticals) requires an understanding of the requirements and must be based upon sound engineering principles. A careful selection of methodologies and structural components is necessary for controlling and in some cases eliminating contamination risk. The controlled environmental validation program must provide assurance that the system meets the approved environmental performance specifications consistently and reproducibly. In one word, it should demonstrate the robustness of the system.

This chapter will discuss the requirements for Clean Rooms and other operating environments and the heating, ventilation, and air conditioning (HVAC) systems that are designed to ensure control of these environments. It will also outline the requirements to qualify controlled environments.

5.2 ENVIRONMENTAL REQUIREMENTS

Although important, microbial contamination is not the only environmental parameter affecting product strength, safety, purity, and effectiveness. Nevertheless, sterile products are the only ones for which specific regulatory requirements or guidelines for the operating environment have been established [7, 18]. For other products, the Current Good Manufacturing Practices (CGMP) regulations are nonspecific, as represented by the US CGMP regulations text [1]

§211.46 Ventilation, Air Filtration, Air Heating and Cooling

(a) *Adequate ventilation shall be provided.*

(b) *Equipment for adequate control over air pressure, micro-organisms, dust, humidity, and temperature shall be provided when appropriate for the manufacture, processing, packing, or holding of a drug product.*

(c) *Air filtration systems, including prefilters and particulate matter air filters, shall be used when appropriate on air supplies to production areas. If air is recirculated to production areas, measures shall be taken to control recirculation of dust from production. In areas where air contamination occurs during production, there shall be adequate exhaust systems or other systems adequate to control contaminants.*

(d) *Air-handling systems for the manufacture, processing, and packing of penicillin shall be completely separate from those for other drug products for human use.*

For sterile products, the regulations and guidelines issued by health authorities provide for specific limits for airborne microbial and particulate contamination. Regarding nonsterile products for which the regulations are not so specific, environmental control must still be considered. Besides airborne particulate and viables, environmental factors that might affect quality are temperature, relative humidity, and cross-contamination from other products or materials. Less frequently cited are vibration, radiation, lighting levels, etc. that may have an effect on the quality of the product. Special attention must be paid to biohazards inherent to processes. These may include containment requirements related to the infectious microorganisms used in processing (e.g., viruses in vaccine production) or materials that are toxic, teratogenic, or carcinogenic.

Therefore, for every product or group of similar products to be manufactured in a biopharmaceutical facility, a definition of environmental requirements is required, considering that these may vary with each phase of the process. These requirements should be documented in a User Requirements document and become the basis for the engineering design of the HVAC system. More complete discussion on the sequence of a design is found in the chapter on "Facility Design for Validation" and the chapter on "Commissioning and Qualification".

Among the regulations and guidelines issued by health authorities that are specific for sterile product manufacturing, the most definitive of these are found in the CGMP regulations published by the European Union health authority (EMA) [7]. Guidelines from PIC/S (the Pharmaceutical Inspection Co-Operation Scheme), to which the US FDA (Food and Drug Administration) belongs, are essentially identical [18]. The US FDA still has an active Guidance for Industry for sterile products published in 2004 that defines environments somewhat differently but is very similar. Other guidelines are published by professional organizations such as the International Society for Pharmaceutical Engineering (ISPE) [10, 11] and the Parenteral Drug Association (PDA) [17]. Many companies have their own internal standards. In general, all of these guidelines are aligned with the regulations.

Throughout the design process, environmental requirements are developed into technical or engineering specifications for areas and equipment and eventually into detailed designs for construction. Ultimately, it will be the User Requirements that dictate what must be verified when qualifying systems.

Because of its importance, an effort should be made by designers, users, constructors, and validators to clearly identify and understand the environmental requirements. The qualification program should be designed to demonstrate that these requirements are met by the design, construction, operation, and performance of the facility and the environmental control systems.

5.3 ENVIRONMENTAL CLASSIFICATION

5.3.1 ISO 14644

Because of regulations and industry guidelines, biopharmaceutical environments that present an environmental risk to product quality must be controlled. Note the distinction regarding product quality. Some manufacturing operations that will be described next may be in environments that are "controlled" in a general sense but not classified according to a regulation or recognized consensus international standard. The more rigorously controlled of these environments are described as "Clean Rooms" or "Classified". The recognized international standard for clean room classification is International Organization for Standardization (ISO) Standard 14644–1 [12]. This Standard classifies by total airborne particle count according to Table 5.1:

The ISO 14644 standard provides specific guidance on number of samples per area and location and method of sampling. Any person or firm responsible for testing for clean room classification or operating clean rooms should become familiar with it. It is important to note that ISO 14644 classifies clean rooms by total airborne particle count only and allows the user to determine the target particle size to use in that effort. The biopharmaceutical industries have universally chosen 0.5 μm for this. ISO 14644 also recommends that if classification is to be extended to a second particle size that it be at least a factor of 10 different from the first size. In biopharmaceuticals this is 5 μm, although this is more often used for confirmation and/or monitoring, while relying on the smaller particle size for classification. This chapter shall refer to areas complying with the ISO 14644–1 as Classified.

To emphasize, ISO 14644 classification is by total airborne particle count only. There are no airborne viable (microorganism) limits. Further, ISO 14644 leaves the choice of clean room application to the user, i.e., it does not designate how each classification shall be used. Finally, ISO 14644 does not indicate the state of the Classified area during testing, defining several optional states including "as built, "at rest", and "operational", again leaving the choice to the user.

Some international bodies, including the PIC/S, EMA, World Health Organization (WHO), and Japan CGMP regulations have gone further in defining clean room requirements for biopharmaceutical manufacturing by specifying both "at rest" and "in operation". The limits published as of this writing are in Table 5.2:

Table 5.2a revises the PIC/S limits in the proposed revision to PIC/S GMP Guidelines Annex 1. These are likely to be published in final form sometime near the publication of this volume. It is expected that regulators will accept these limits.

These limits are for classification of areas. Routine monitoring limits are discussed in the Chapter on "Environmental Monitoring". "At rest" is defined as having environmental control systems operating, all equipment in place, but no

TABLE 5.1
Clean Room Classification

ISO classification number (*N*)	Maximum allowable concentrations (particles/m³) for particles equal to and greater than the considered sizes shown below[a]					
	0.1 μm	0.2 μm	0.3 μm	0.5 μm	1 μm	5 μm
ISO Class 1	*10*[b]	[d]	[d]	[d]	[d]	[e]
ISO Class 2	100	*24*[b]	*10*[b]	[d]	[d]	[e]
ISO Class 3	1,000	237	102	*35*[b]	[d]	[e]
ISO Class 4	10,000	2,370	1,020	352	*83*[b]	[e]
ISO Class 5	100,000	23,700	10,200	3,520	832	[e]
ISO Class 6	1,000,000	237,000	102,000	35,200	8,320	293
ISO Class 7	[c]	[c]	[c]	352,000	83,200	2,930
ISO Class 8	[c]	[c]	[c]	3,520,000	832,000	29,300
ISO Class 9	[c]	[c]	[c]	35,200,000	8,320,000	293,000

Source: Adapted from ISO 14644–1 [12].

[a] All concentrations in the table are cumulative, e.g., for ISO Class 5, the 10,200 particles shown at 0.3 μm includes all particles equal to and greater than this size.

[b] These concentrations will lead to large air sample volumes for classification. Sequential sampling procedures may be applied.

[c] Concentration limits are not applicable in this region of the table because of very high particle concentration.

[d] Sampling and statistical limitations for particles in low concentrations make classification inappropriate.

[e] Sample collection limitations for both particles in low concentrations and sizes >1 μm make classification at this particle size inappropriate because of potential particle losses in the sampling system.

TABLE 5.2
International Clean Room Classifications (Current)

PIC/S Classification (also EMA, WHO, Japan, China)	Particle limit per cubic meter: At rest		In operation	
Grade	0.5 µm	5 µm	0.5 µm	5 µm
A[1]	3520	20	3520	20
B	3520	29	352,000	2900
C	352,000	2900	3,520,000	29,000
D	3,520,000	29,000	Not defined	

Abbreviations: EMA = European Medicines Agency; PIC/S = Pharmaceutical Inspection Co-Operation Scheme; WHO = World health Organization.
[1] Unidirectional flow.

TABLE 5.2A
International Clean Room Classifications (Proposed)

PIC/S classification (also EMA)	Particle limit per cubic meter: At rest		In operation	
Grade	0.5 µm	5 µm	0.5 µm	5 µm
A[1]	3520	N/A	3520	N/A
B	3520	N/A	352,000	2900
C	352,000	2900	3,520,000	29,000
D	3,520,000	29,000	Not defined	

Abbreviations: EMA = European Medicines Agency; PIC/S = Pharmaceutical Inspection Co-Operation Scheme.
[1] Unidirectional flow.

personnel present (except those performing the testing, where necessary). "In operation" adds equipment in normal operation and the normal maximum personnel contingent.

The regulators have never provided rationale for the large difference between the "at rest" and "in operation" limits except that it provides a significant margin of safety when transitioning to operation. Considering the tight controls applied by users regarding personnel practices, cleaning, and sanitization, many clean room qualification and environmental monitoring results typically demonstrate that the "at rest" limits are arbitrary and highly conservative.

Note that US limits are not included in Tables 5.2 or 5.2a. The prevailing US Guidance for Industry *Sterile Drug Products Produced by Aseptic Processing* is aging (2004). It still uses the old Fed. Std. 209 (see following) descriptions and their ISO equivalents (e.g., Class 100/ISO 5). Currently, however, most firms with international interests have adopted something similar to Tables 5.2 and 5.2a. As a member of PIC/S, the US FDA fully accepts these definitions in practice. On the other hand, the FDA continues to focus on operating limits and typically pays little attention to systems at rest.

5.3.2 Airborne Microbial Limits

The PIC/S and EMA limits for Classified Environments are found in the latest proposed revision of Annex 1 (which are essentially identical). The proposed limits published as of this writing are likely to be approved and represent common industry practice. These are in Table 5.3:

TABLE 5.3
PIC/S Airborne Viables Limits (proposed revision) [8]

Grade	Air sample cfu/meter³	Settle plates 90 mm cfu/4 hours	Contact plates 55 mm cfu/plate
A	No growth detected		
B	10	5	5
C	100	50	25
D	200	100	50

Abbreviation: PIC/S = Pharmaceutical Inspection Co-Operation Scheme.

The expectation is that no viables are detected in the Grade A air stream or on surfaces. Settle plates are expected during the duration of the study and should be changed regularly to avoid desiccation. Other notes included in the proposed revision are:

Note 1: All methods indicated for a specific Grade in the table should be used for qualifying the area of that specific Grade. If one of the methods is not used, or alternative methods are used, the approach taken should be appropriately justified.

Note 2: Limits are applied using cfu throughout the document. If different or new technologies are used that present results in a manner different from cfu, the manufacturer should scientifically justify the limits applied and where possible correlate them to cfu.

Note 3: <Note deleted. Refers to qualification of personnel, which is not within the scope of this chapter. See Chapters on "Validation of Training" and "Environmental Monitoring".>

Note 4: Sampling methods should not pose a risk of contamination to the manufacturing operations.

An alternative perspective on airborne viables is provided by USP <1116> by defining the rate of contamination in samples taken from various areas. These recommendations relate to the percentage of samples that exhibit any microbial contamination regardless of the count as follows [20]:

5.3.3 Classified Areas in Biopharmaceutical Manufacturing′

Traditionally, the pharmaceutical industry has applied ISO 14644 Classes 5, 7, and 8 (sometimes Class 6) to its sterile manufacturing facilities. From a regulatory perspective, this custom dates back to the publication of a proposed US FDA addition to

TABLE 5.4
USP <1116> Recommended Initial Viable Contamination Rate

Grade (In operation)	% of Contaminated samples
Isolator (ISO 5)	>0.1
A (ISO 5)	<1
B (ISO 7)	5
C (ISO 8)	10
D[1] (ISO 8)	10

Abbreviation: ISO = International Organization for Standardization; USP = United States Pharmacopeia.

[1] USP defines contamination rates according to ISO Classes. Grade D ISO Class in operation is not defined by PIC/S (the Pharmaceutical Inspection Co-Operation Scheme). In general, a Grade D zone is expected to routinely operate at ISO 8 and many firms have specified contamination rate limits as such.

TABLE 5.5
Application of Classified Areas (Sterile Manufacturing)

Grade (in operation)	Application
A (ISO 5) Unidirectional air flow	• Aseptic fill of sterile products; filling of products to be terminally sterilized when unusually at risk • Handling of exposed sterile components • Transport of partially stoppered sterile vials (e.g., to lyophilizer)
B (ISO 7)	• Background to Grade A (except for isolators) • Autoclave cooldown zone (w/unidirectional flow) • Biotech "aseptic" connections (w/unidirectional flow) • Filling inhalation products
C (ISO 8)	• Sampling, weighing, formulation of materials prior to sterilization by filtration and aseptic fill • Biotech isolation and purification (bioburden control) • Filling of vials to be terminally sterilized • Preparation of solutions to be terminally sterilized when unusually at risk (filled in Grade A)
D (ISO 8)	• Parts and glassware preparation ("kitchen") • Circulating corridors/buffer zone • Background to aseptic isolator • Bioreactors • Non-sterile product processing (China only)

the CGMP regulations in 1976, 21 CFR 212 for Large Volume Parenterals. The proposed regulation was never promulgated, and that section of the CFR was reassigned. In that document, the FDA proposed to make mandatory certain environmental classes commonly employed in sterile manufacturing at the time. These classes were defined in the accepted US standard of the day for clean rooms and controlled environments, Federal Standard 209. These classes were named for the upper limit of airborne 0.5 μm particles in one standard cubic foot of air. The designated classes were Class 100, Class 10,000, and Class 100,000, corresponding to current ISO Classes 5, 7, and 8, respectively. The Fed. Std. 209 designations have all but disappeared and been replaced by the PIC/S Grades or with the ISO Classes. The regulators and industry have designated these Classified areas for use in sterile manufacturing as follows:

5.3.4 Other Environmental Classes

Firms use a variety of descriptions for manufacturing areas other than for sterile products. Some of these are Class 1/ Class 2, Grade E, General, Support, Office, and Lab. Each of these is designated by the user for specific purposes, and each is assigned its own operating specification. Those that are assessed to have a direct effect on the quality of the product require qualification (see Section 5.5 following).

One of the more common and widely used environmental descriptions is "Controlled", more often expanded to "Controlled (non-Classified)" or CnC to distinguish it from the Classified zones described previously. This designation is assigned to areas that have rigorous environmental controls, require controlled or limited access, and have garb or gowning requirements beyond street clothes. Environmental controls for CnC do not usually include airborne particle count or microbial limits. As such, CnC is employed in non-sterile manufacturing, usually in process areas where product is exposed, including primary filling/packaging. It may also be used as a perimeter or first stage of access where microbial control of non-sterile product is required, such as bulk biotech products. The authors do not recommend CnC in sterile manufacturing, where Grade D should be specified for initial access into the sterile manufacturing core. This recommendation is made because Grade D comes with more extensive control over and therefore monitoring of environmental parameters, e.g., airborne particulate and viables. Many designers and users choose to design CnC zones to the same criteria as Grade D, sometimes with somewhat less safety factor or overdesign built in. These areas may be commissioned to ISO Class 8 at rest, but because there are no operating limits, they are not qualified or monitored to ensure these conditions prevail. (See Section 5.5 following).

5.4 DESIGN OF CONTROLLED ENVIRONMENTS

The controlled environments discussed in this section shall be limited to those described as Classified (Grades A—D, ISO 5–8) and CnC (i.e., controlled environments). Other areas will usually be determined by risk assessment to not have a direct effect on product quality (with exceptions noted in Section 5.5). These include manufacturing areas where product is not exposed, such as secondary packaging.

All manufacturing environments are designed using a combination of architectural and mechanical features to protect the product from contamination. Product risk is determined by a combination of factors discussed in Section 5.5 following. For facility design, the type of product (e.g., sterile injectable, oral solid, etc.) and the step in the process (e.g., formulation, primary filling, etc.) will dictate the environmental control system design.

5.4.1 Architecture—Materials

Clean rooms of all classes share common architectural characteristics, with a few modifications as product risk increases. Walls should be smooth, cleanable, and of durable material to be able to withstand disinfecting agents. Modern clean rooms are most often constructed of modular panels constructed of a smooth exterior surface and an interior honeycomb core. Many surface materials are available including painted aluminum, stainless steel, and melamine. Transparent clean room walls may be made of glass or polycarbonate. The modular panels have the advantage of being replaceable, and some clean room systems are even moveable. One of the newer innovations, albeit well established, is modular clean rooms known as "pods" that are assembled off-site and relocated to the user site. Some clean rooms are still "stick-built" with gypsum board over steel framework, coated with smooth, durable material, such as epoxy paint. This continues to become the less preferred approach.

Ceilings are monolithic materials (e.g., painted gypsum board) or more frequently sealed panels in a grid configuration. These may be of the same surface material as the walls or of another smooth, cleanable material.

Floors are monolithic, smooth, cleanable and able to withstand sanitization. Materials range from epoxy terrazzo to self-leveling epoxies and other polymeric systems, to seamless sheet vinyl (heat-sealed joints). Floors are coved to the walls. For higher grade rooms, wall joints and wall to ceiling are often also coved. Drain penetrations in clean room floors should be avoided, with drain lines running through walls into non-clean space. If drains are necessary in formulation areas (Grade C or CnC), these must be of a sanitizable design with a removeable trap. Drains should not be installed in Grades A or B.

Doors and windows may be made of metal, plastic, or other suitable material and should present a neat appearance. Unless constructed of stainless steel (preferred), they should be coated (paint, epoxy, etc.) with a finish of sufficient durability so as not to degrade or shed particles during use or cleaning. The doors may have appropriate vision panels or windows. Doors should be designed to maintain room pressurization (e.g., appropriate clearance, closure system, and gasket material). Interlocking doors to be provided where necessary. Windows should be constructed so that they do not open. Windows should be flush mounted to minimize horizontal surfaces and facilitate cleaning.

5.4.2 Architectural Layout

The architectural layout of clean rooms is as important as their materials in order to prevent contamination. The accepted practice is to configure facilities so that the area of greatest risk is at the core of the facility, surrounded by a clean space of decreasing classification. A simple diagram indicating this for a sterile manufacturing is shown in Figure 5.1.

Personnel and material traffic between Grades is through air locks, designed to maintain the required pressure differentials described in Section 5.6 on HVAC design. Higher classifications operate at higher static pressures to ensure air flow

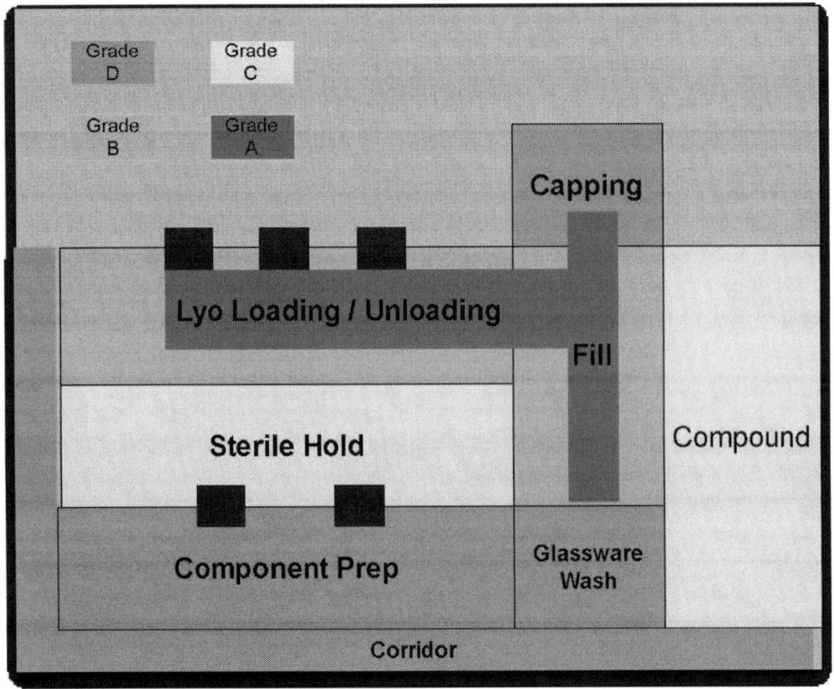

FIGURE 5.1 Clean room layout—personnel/material flows. (Room details are intentionally omitted).

Note: The area surrounding the classified zones is "gray space"; i.e., not controlled (except for human comfort, as required).

from clean to less clean areas. Air locks may be designed as gown rooms for personnel and there may be separate air locks for material. The usual practice is to have separate air locks for personnel and materials as well as separate air locks for entry and exit. Air locks should be the same Grade as the higher of the two adjacent areas. Air locks into Grade B should be two-stage, the outer air lock is Grade C and the inner Grade B for final gowning.

Figures 5.2–5.4 show three different air lock configurations applied to three different Classified areas. Air flow characteristics for air locks are also discussed in more detail in Section 5.4.4 following.

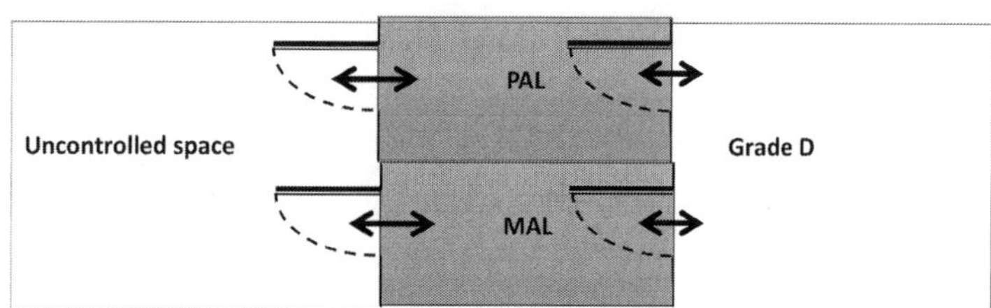

FIGURE 5.2 Grade D air lock—personnel/material flows.

Note: MAL = materials air lock; PAL = personnel air lock.

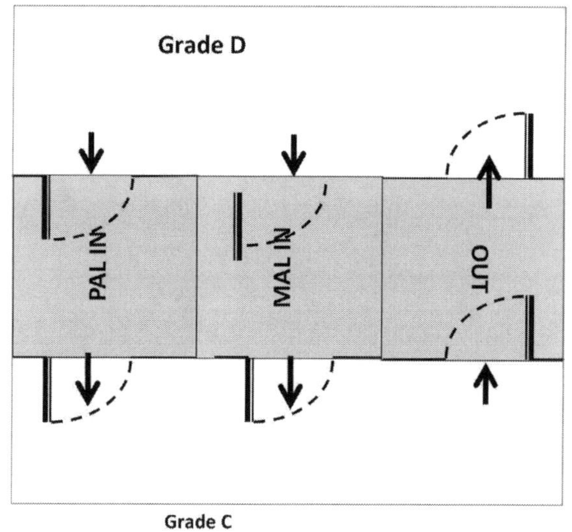

FIGURE 5.3 Grade C air lock—personnel/material flows.

Note: MAL = materials air lock; PAL = personnel air lock.

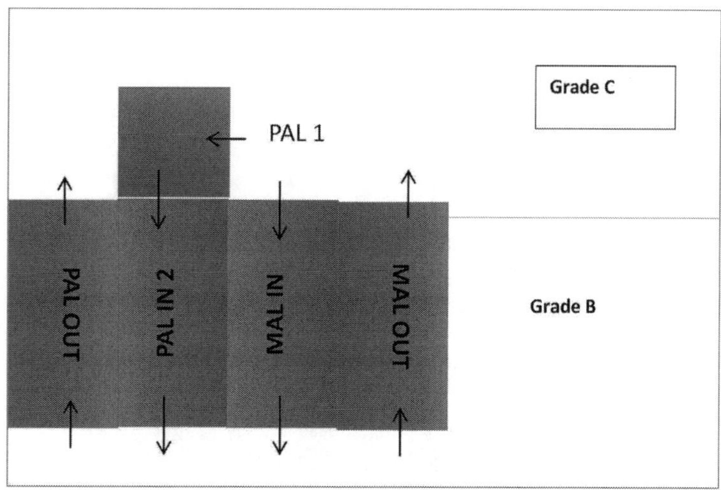

FIGURE 5.4 Grade B air lock (two-stage air lock in)—personnel/material flows.

Note: MAL = materials air lock; PAL = personnel air lock.

5.4.3 HVAC System Design

All heating, ventilation and air conditioning (HVAC) systems are designed to provide some degree of environmental control. The most common reason to control the environment is for human comfort. This generally involves temperature and to some extent relative humidity. Also, a portion of the air circulated needs to be "fresh" in order to displace exhaled carbon dioxide with oxygen. Even in pharmaceutical manufacturing, human comfort is the major driver behind HVAC design.

Human comfort conditions range from 23.5°C–25.5°C (74.3°F–77.9°F) in the Summer to 21°C –23°C (69.8°F –73.4°F) in the Winter. (Why the difference? Energy conservation and mode of dress. People wear warmer clothing in Winter). Relative humidity of 30%–65% is comfortable for most people year-round.

Many, if not most, pharmaceutical products and materials require Controlled Room Temperature, nominally defined as 20°C–25°C, with a mean kinetic temperature (MKT) below 25°C [21]. Spikes of under 24-hour duration up to 40°C are allowed. Interestingly, many products are actually more tolerant of temperature variation than people, and so much of pharmaceutical manufacturing is designed for greater human comfort.

5.4.3.1 Mechanical Design

A typical simple HVAC system is shown in Figures 5.5 and 5.6. The system brings in fresh outside air, conditions it for temperature, cools it to remove humidity (as required), then re-heats it for comfort. The system may have additional components to add or remove humidity as required by the area. The inlet fan supplies the air necessary to account for human occupancy and for particulate removal. In clean rooms, airborne particulate removal dictates the volume of air flow because that is significantly greater than the volume necessary for human comfort.

Figure 5.7 shows a possible HVAC system configuration in a larger plant. Incoming air is conditioned in a primary air handler that distributes the air to secondary air handlers serving various zones of the facility. The primary air handler does the bulk of the heating and cooling, whereas the secondary air handlers provide adjustments, pressurization, and enhanced filtration.

Air may be recirculated back to the rooms for energy efficiency (it requires less additional heating and/or cooling) and also for more effective filtration. Single-pass air flow is sometimes required when exposure to solvents, toxic or otherwise dangerous materials, or extremely dusty conditions are expected. Otherwise, the exhaust fan removes enough air to provide for a steady flow of fresh incoming air for personnel comfort and safety. The exhaust flow is adjusted to maintain a static pressure in the area, which in turn is adjusted relative to adjacent areas (see Section 5.4.3.5)

5.4.3.2 Temperature

The temperature and relative humidity in clean rooms need to be more tightly controlled than in typical office or laboratory environments. The reason for this is more often than not the same reason that other areas need to be controlled and that is human comfort. This should not be confusing when one considers the clean room garb that is required to control the most prevalent source of contamination, that generated by the people themselves. It is critical that people working in clean rooms are kept as comfortable as possible. This reduces sweating and shedding and also helps to avoid errors and unwanted movement that cause contamination. Considering the increasingly rigorous gowning that is required as higher environmental classes are entered, the design temperature shown in Figures 5.8 and 5.9 are decreased accordingly.

5.4.3.3 Relative Humidity

One exception to the "human comfort" rule is that clean rooms usually need to be kept at lower relative humidity than non-clean environments. Because of heavy gowning and also to avoid proliferation of environmental molds, relative humidity

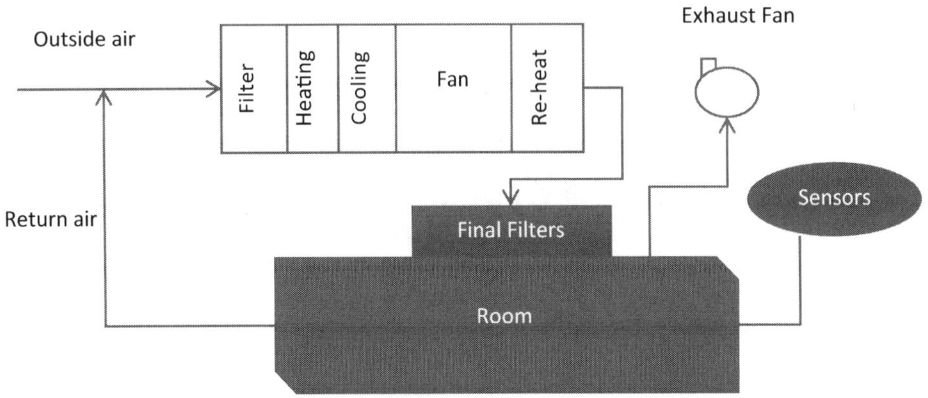

FIGURE 5.5 (Very) simple HVAC System.

Note: HVAC = heating, ventilation, and air conditioning.

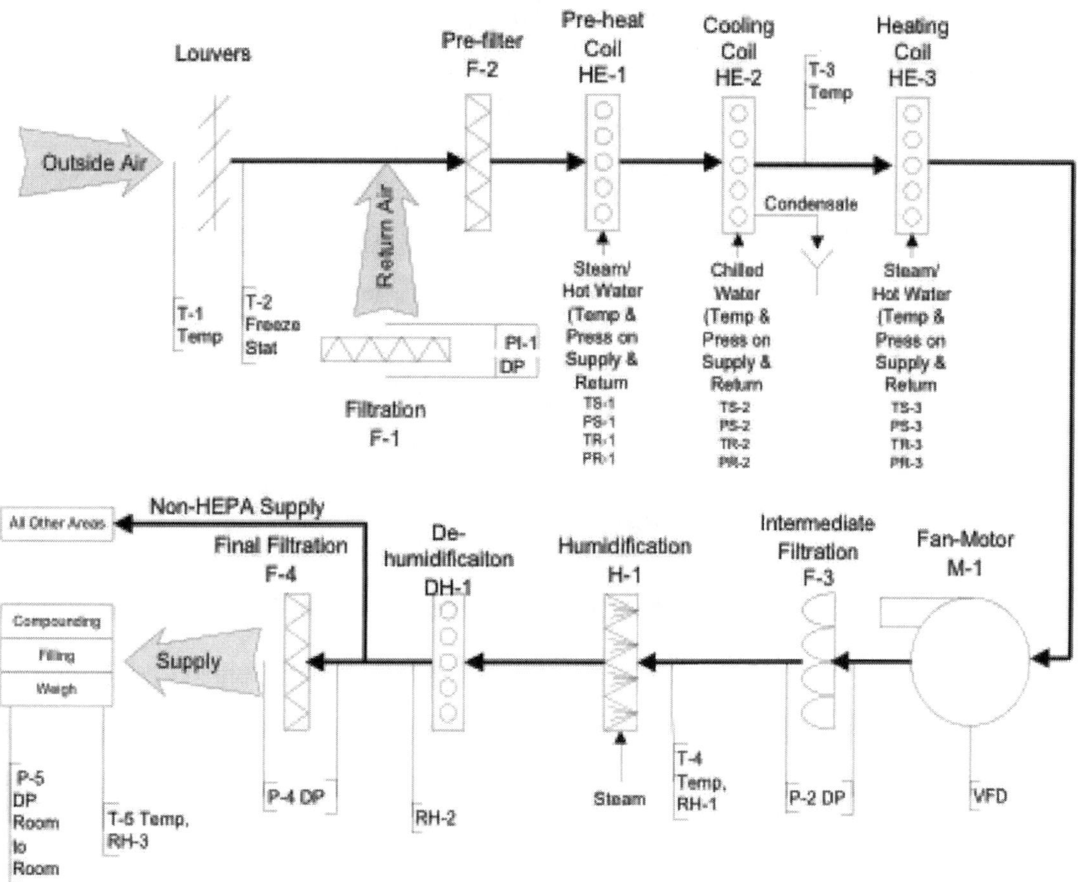

FIGURE 5.6 HVAC system components.

Note: HVAC = heating, ventilation, and air conditioning.

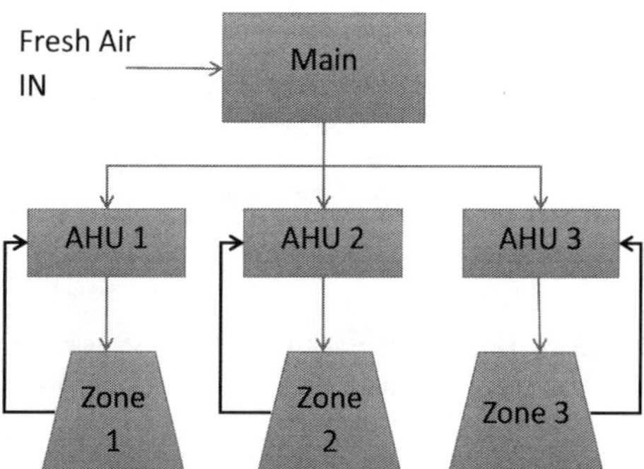

FIGURE 5.7 Air handler network.

Note: AHU = air handling unit; HVAC = heating, ventilation, and air conditioning.

should not exceed 60% and should be designed with a safety factor, as indicated in Figures 5.8 and 5.9.

Humidity is reduced by cooling, much of which is provided in the air handler. Re-heat coils return the cooled air to comfort levels. When more rigorous de-humidification is required, chemical dehumidifiers are employed. These may use solid or liquid desiccant (spray-type) to remove additional moisture. This is a common practice where the higher classes of clean room are installed.

In some locations, winter conditions lead to very dry air that may have to be humidified for human comfort and/or static elimination. When necessary, steam to humidify clean rooms should be free of volatile boiler additives (e.g., amines, hydrazine). Chemical-free steam and Clean Steam are preferred. for

Grades C, D, CnC

Design Temperature:

21±2°C

Operating Temperature:

21±3°C

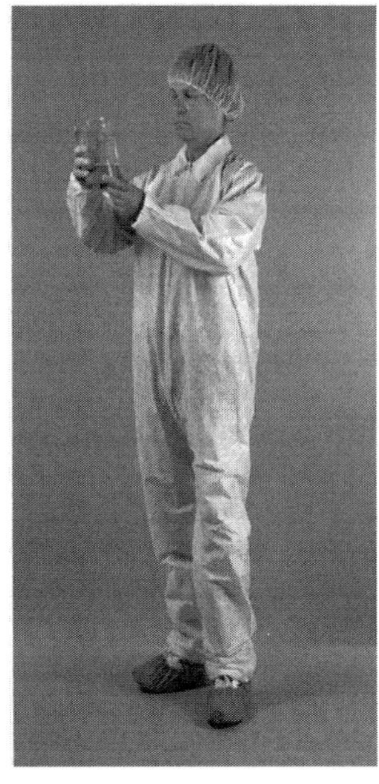

Relative Humidity Design:

50±5%

Operating RH:

30 – 60%

FIGURE 5.8 Non-sterile manufacturing environmental conditions.

Grades A, B

Design Temperature:

19±1°C

Operating Temperature:

19±2°C

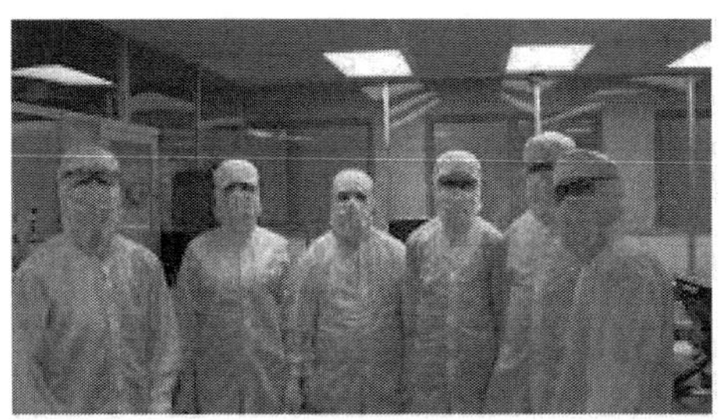

Relative Humidity Design:

45±5%

Operating RH:

35 – 60%

FIGURE 5.9 Sterile manufacturing environmental conditions.

Note: RH = relative humidity.

sterile manufacturing. For non-sterile manufacturing (usually CnC), steam with food-grade additives would be acceptable.

5.4.3.4 Filtration

Filters are an important component of clean room environmental control systems. These are the functional particulate removal devices necessary to remove airborne particulate. Sources of particulate include people, product, and equipment. A walking person in complete aseptic clean room garb sheds an average of 10^5 0.5-μm particles per minute. This includes microorganisms along for the ride. This number increases by more than 100-fold in street clothes [22]. The importance of filtration in assuring control over these contaminants cannot be overemphasized.

Filters are usually employed in a series of increasing efficiency. For manufacturing areas that are not clean rooms, filtration is often two-stage. For clean rooms, a three-stage design is more common. Filter efficiencies are defined by various US and international standards, all describing similar performance levels with different nomenclature. The same filter can have two or more labels if intended for use in different markets. Table 5.6 describes classes used in the US (ASHRAE 52.2) [3] and Europe (EN 779). Filters are rated by the dust spot test, which relates to the efficiency of particle removal down to about 0.4 μm. The dispersed oil particulate (DOP) test indicated employs a uniform aerosol of 0.3 μm particle size.

A typical pre-filtration scheme for a clean room would be Minimum Efficiency Reporting Value (MERV) 10 (F5) followed by MERV 13 (F7). This will vary depending upon the particulate levels of the incoming air and room particulate generation. MERV 16 filters are sometimes used as final filters in CnC areas.

TABLE 5.6
Air Filter Ratings

ASHRAE 52.2 MERV	% Efficiency (dust spot)	EN 779
6	<20%	G4
7	25%–30%	
8	30%–35%	
9	40%-45%	
10	50%–55%	M5(F5)
11	60%–65%	
12	70%–75%	M6(F6)
13	80%–90%	F7
14	90%–95%	F8
15	>95%	F9
16	95% DOP	E11

Abbreviations: ASHRAE = American Society of Heating, Refrigerating and Air-Conditioning Engineers; DOP = dispersed oil particulate; MERV = Minimum Efficiency Reporting Value.

Final filtration in most clean rooms employs High Efficiency Particulate Air (HEPA filters). Biopharmaceutical facilities rarely, if ever, use higher grades designated as ULPA. Table 5.7 shows HEPA (and ULPA) filters as categorized by three different standards, US, EU, and ISO. HEPA filters are graded by their ability to filter 0.3 µm particles, considered the most penetrating particle size (MPPS). Note that the European Norm (EN 1822) and ISO retention ratings are the same. For practical purposes the filters listed in Table 5.7 are the same.

Typical biopharmaceutical applications use H13 or H14 filters. For CnC and Grade D, these may be centrally mounted at the air handler. Because this may raise questions regarding the cleanliness of the supply ducts, most designers choose to mount HEPAs at the ceiling of the room, denoted "terminal" HEPAs. Grades C and better always employ terminal HEPAs.

TABLE 5.7
HEPA Filter Ratings

ASHRAE 52.2 MERV[1]	Retention 0.3 µm[2]	ISO 29463	EN 1822	Retention 0.3 µm
16	99.97%	35H	H13	99.95%
17	99.99%	45H	H14	99.995%
18	99.999%	55H	U15	99.9995%

Abbreviations: ASHRAE American Society of Heating, Refrigerating and Air-Conditioning Engineers; HEPA = High Efficiency Particulate Air; MERV = Minimum Efficiency Reporting Value.

[1] "Unofficial" MERV ratings by the Environmental Protection Agency. ASHRAE 52.2 does not define HEPA filters

[2] 0.3 µm is the "most penetrating particle size"—MPPS

5.4.3.5 Air Flow Patterns

Air flow is another important factor in clean rooms. It is important that the entire volume of the room be subject to effective filtration in order to maintain the desired classification. Unlike normal air conditioning, specific flow patterns are designed. Office and other traditional environments are designed for human comfort and energy efficiency. Air flow in these areas is not focused on particulate removal. Air volume is determined by the need to replenish oxygen and to control temperature.

Clean rooms employ various designs, including filtration, to achieve airborne particulate limits. Among these are increased air flow (air changes per hour—ACH), specialized air diffusers, ceiling HEPA coverage, and type and location of air return registers. All of these have an influence on air cleanliness. There are no specific requirements for any of these, only guidelines and good engineering practices that designers use to satisfy requirements. It is a myth that any specific number of air changes (e.g., 20 ACH) is a *requirement* for any clean room. Even when these are referred to in the various guidelines [19], this is a *guidance number* that may or may not be acceptable based on the other important design criteria cited.

Table 5.8 provides some guideline ranges for commonly used design criteria used for various clean room classifications.

All of the values in Table 5.8 are engineering criteria. None is a regulatory requirement and none of them singularly guarantees or is a direct measure of satisfactory performance. On the other hand, properly assembled into a comprehensive design, along with appropriate architectural and personnel practices, they work well.

TABLE 5.8
Clean Room Design Criteria[1]

Grade	ISO Class Operational	ACH	HEPA coverage	Return
A	5	N/A	100%	Plenum or exhaust to surroundings
B	7	30–90	30%–40%	Plenum or low wall
C	8	25–50	10%–15%	Low wall
D	N/A	12–20	5%–10%	Low wall
CnC[2]	N/A	12–20	5%–10%	Low wall

Abbreviations: ACH = air changes per hour; CnC = Controlled (non-Classified); HEPA = High Efficiency Particulate Air; ISO = International Organization for Standardization.

[1] These ranges are guideline values gathered from multiple projects and various design engineers. They do not represent any specific clean room design.

[2] Designers will allow more flexibility with CnC, often choosing the lower ends of the recommended ranges

5.4.4 Differential Pressure and Air Locks

An important concept of clean room design is that (except for some special situations described following) air flow between adjacent areas should be from the cleaner zone to the less clean. Although clean room walls, doors, and ceilings are designed to seal tightly, there is always a small amount of leakage and, of course, doors open for passage of personnel and materials.

The directionality of air flow is determined by the relative static air pressure in the adjacent space, more commonly differential pressure or dP. Historical practice in clean room design, more recently adopted by regulators as guidance [7, 18], is that the differential between different classes of controlled environments should be 10–15 Pa (pascal—equivalent to .04"—.06" water column) *with doors closed.*

Air locks, as described in Section 5.4.2 previously, are used to preserve the differential between zones of different classification. Air locks should be the same class as the cleaner adjacent zone. The differential between zones should be measured *across* the air lock. The user should not misinterpret the guidelines and attempt to provide 10–15 Pa across each door of the air lock. This has been improperly attempted, leading to a series of pressurization steps that have raised static pressure in the innermost zone to uncomfortable levels.

The following Figures 5.10–5.13 indicate appropriate pressure relationships across various air lock configurations. Figure 5.10 is the most common cascade type air lock between two classes of clean room. Figure 5.11 is a "bubble" air lock, used when the material handled in the cleaner zone is toxic, highly active, or otherwise requires containment. Figure 5.12 is a "sink" air lock used similarly to the "bubble" except

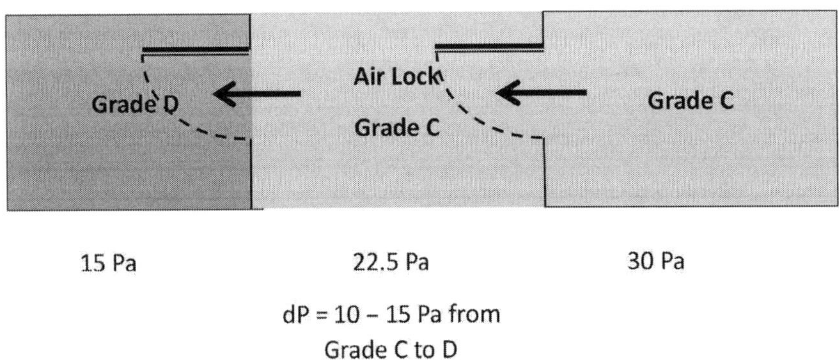

FIGURE 5.10 Typical cascade air lock—air flow direction.

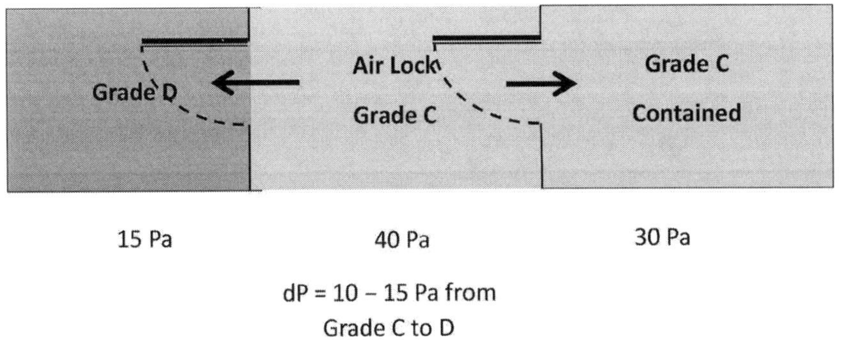

FIGURE 5.11 Bubble air lock—air flow direction.

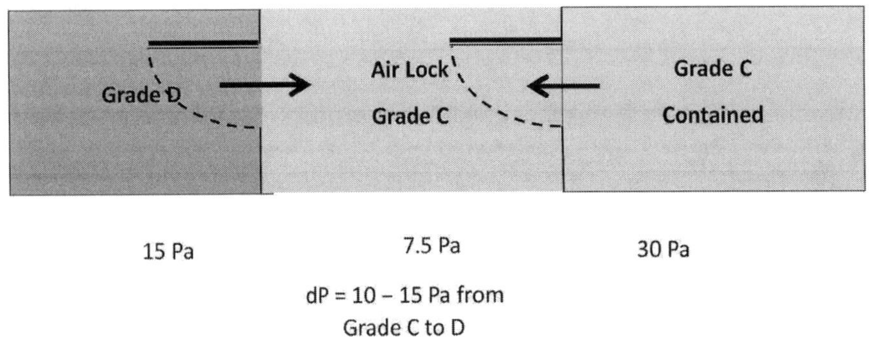

FIGURE 5.12 Sink air lock (exit)—air flow direction.

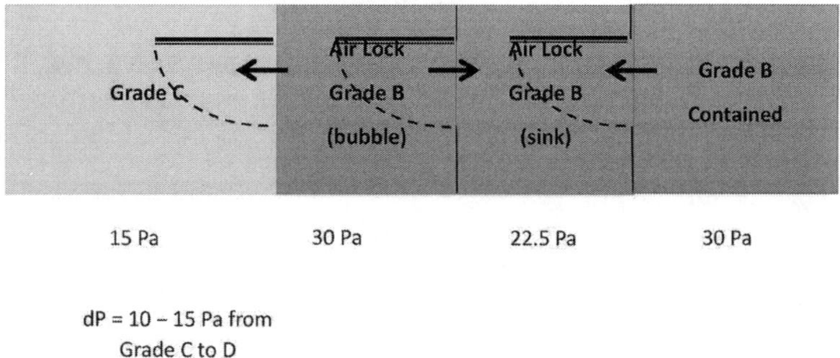

FIGURE 5.13 Combination air lock (sink/bubble)—air flow direction.

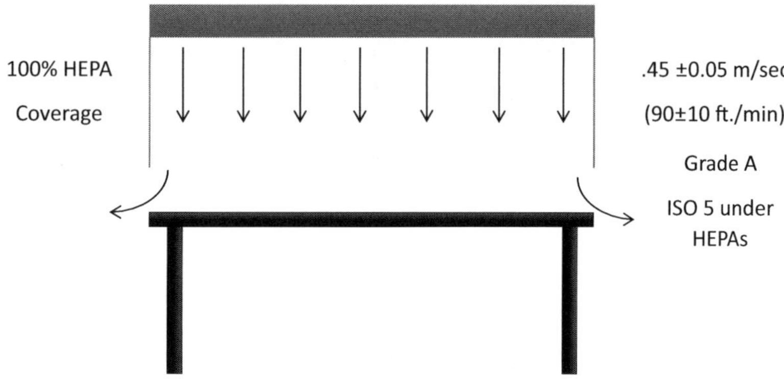

FIGURE 5.14 Unidirectional air flow.

Note: HEPA = High Efficiency Particulate Air.

that it is more often installed at the exit of the cleaner zone. Figure 5.13 represent a very conservative design used when extreme containment of the active material in the cleaner zone is required. This has been employed for the manufacture of sterile live virus vaccines.

Note that in all cases, the cleaner zone is designed to operate at 10–15 Pa higher static pressure than the adjacent less clean zone.

5.4.5 Unidirectional Flow

Unidirectional air flow, sometimes inaccurately referred to as laminar air flow, is used to provide a high degree of protection to a localized area, particularly where sterile product or components that will contact the product are exposed. A simple depiction of unidirectional air flow is shown in Figure 5.14. Air flow is in streamlines that flow around obstructions and away from the work surface. Exit may be through a perforated work surface.

Unidirectional flow is applied in order to avoid turbulence and eddy currents that may bring in contamination from outside the exposure zone. It is often aided by some type of barrier between the work zone and the exposed material, such as a curtain or rigid wall. This is especially the case during aseptic fill of product. Some turbulence is to be expected when the air flow meets horizontal surfaces. This is acceptable as long as eddy currents do not draw in contamination from the surrounding environment.

According to the PIC/S guidelines and widespread industry practice, Grade A is defined as requiring unidirectional flow. Guidance from ISO and other professional organizations (e.g., Institute for Environmental Sciences—IES) [14] have defined the acceptable range of velocity for unidirectional flow as .45 ±0.05 m/sec (90 ± 10 ft./min) measured 150–300 mm from the filter face. Note that some regulations and guidelines mention measurement "at the work surface" [7, 18]. This, however, is an extraordinary expectation and is not a true requirement to maintain unidirectional flow, which naturally slows down as it encounters obstacles. Some installations have found that achieving conforming filter face and work height velocities within the same range is impossible. Testing described following is necessary to confirm flow.

5.4.6 Isolators

Isolators are discussed in detail in the chapter "Qualification and Validation of Advanced Aseptic Processing Technologies". They are discussed briefly here because they have become an important advance in environmental control for biopharmaceuticals. The primary objective of an environmental control system is to assure the quality of the product by preventing contamination that could alter its purity, strength, and safety. The use of isolation technology with built-in or attached decontamination systems provides a higher degree of sterility assurance by eliminating the ingress of contamination from the surrounding areas, especially from the primary source of microbial contamination: personnel. These units may be integrated with component preparation devices (e.g., washers, sterilizing tunnels), transfer devices for product and containers that retain isolation/sterility, cleaning systems, and monitoring equipment. Personnel access may be provided via decontaminated glove ports (half-suits being less utilized) or even eliminated altogether by employing robotics.

In general, isolators are prefabricated, self-contained chambers, or glove boxes, that are capable of being reproducibly sterilized/decontaminated in situ and are designed to prevent the direct exposure of the sterile products to personnel and the surrounding environments. They may exchange air with the surroundings, but only through microbially retentive HEPA filters or well-controlled continuous material ingress/egress ports.

Construction materials should be compatible with the process, cleaning, and decontamination/sterilization procedures employed. The surrounding environment should not directly impinge on the efficiency of the isolator in maintaining the desired internal environmental conditions.

Three points about isolators should be emphasized. The first point is that restricted access barrier systems (RABS) *are not isolators*. RABS are barriers of many different configurations and efficiencies, some of which approach isolation, but none of these is truly isolated from the ambient environment. RABS may be opened to the surroundings, even during operation. Although they may be decontaminated, their structure obviates continued integrity post-decontamination. RABS must be treated as if they are part of the surrounding clean room and are subject to the same design criteria described previously for conventional manned Classified areas.

The second point regarding isolators is that *they are not small clean rooms*. Although they share some of the requirements for Classified areas (e.g., ISO 5 conditions for aseptic operation), they do not share them all. A prime characteristic of isolators is that they can be decontaminated effectively, to the point of a validated sterilization. The second characteristic is that they have no inherent source of microbial contamination (e.g., people). Therefore, it is recommended that the reader rely on the aforementioned chapter "Qualification and Validation of Advanced Aseptic Processing Technologies" for guidance on qualification and validation.

Thirdly, there is debate as to whether unidirectional air flow is required in isolators, although as of this writing PIC/S (and EMA) GMP Annex 1 guidelines still include it. The strong argument made by experts is that unidirectional flow is specified to obviate the incursion of contamination from outside the critical zone. An isolator provides that protection with an impervious physical barrier. [2]. As it stands, most aseptic isolators are provided with unidirectional air flow and firms continue to apply this test.

5.4.7 Warehouses and Temperature-Controlled Units

Warehouses and other environments where product and production materials are stored present a much simpler environmental control solution. For the most part, materials stored or held in these areas are fully enclosed in impervious packaging and require only temperature control. (Note that rarely humidity may need to be controlled for some exposed packing materials, but this can be ensured in the same manner that humidity is controlled in any environment.)

With temperature-controlled areas, including warehouses, refrigerators, and freezers, qualification may be limited to temperature mapping and monitoring only. Refrigerators and freezers have very limited ranges specifically determined by the product or material stored. Refrigerator temperature is usually specified as 2°C–8°C based on criteria established in USP <659> *Packaging and Storage Requirements*. According to USP, freezers operate at −10°C to −25°C, although some products must be stored for long periods at colder temperatures (e.g., Pfizer COVID-19 vaccine −70°C).

Many products and materials designate storage at Controlled Room Temperature, defined by USP <659> as:

- Nominally 20°C–25°C
- MKT ≤25°C (77°F)
- Excursions 15°C–30°C
- Spike (<24 hours) to 40°C

MKT is expressed as:

$$\frac{\Delta H / R}{-\ln\left(\frac{e^{-\Delta H/RT1} + e^{-\Delta H/RT2} + \ldots + e^{-\Delta H/RTn}}{n}\right)}$$

where:

ΔH = activation energy (typically from 60 to 100 kJ/mol for solids and liquids)
R = 8.314472 J/mol-K (universal gas constant)
T = temperature in degrees K
n = the number of sample periods over which data is collected

Note: ln is the natural log and e_x is the natural log base.

5.4.8 Control Systems

For structure and validation of HVAC control systems refer to the chapter "Control Systems Validation". Most HVAC systems are controlled by complex distributed control systems (DCS) usually designated as a Building Management Systems (BMS). These systems have a great many functions, not all of which involve control or monitoring of controlled environments. Examples of these functions are boiler control, chillers, cooling towers, and control of offices and other areas not subject to Current Good Manufacturing Practices (CGMP) regulation.

Building Management Systems monitor all building parameters and control desired settings, record selected parameters, control start-up and shutdown, provide for alarms, alerts, and notifications, etc. Not all of these directly affect product quality.

Good engineering practices (GEP) and good information technology (IT) practice dictates that all system functionality should be verified in commissioning and functional testing. Based on risk assessment described in the chapter "Control System Validation", those functions that directly affect product quality are subject to the full requirements of computer system validation (CSV). Many firms have determined that the measured room parameters are the only quality-critical functions of the control system. To avoid the conflict of what functions of the BMS need to be validated (an approach that is permissible and, if done correctly, compliant with regulations), they have decided to separate these functions into an independent environmental monitoring system (EMS). Figure 5.15 provides an overview of this type of system architecture. Qualification of these functions is discussed in Section 5.6.

5.4.9 Particle Monitoring Systems

Many clean room designs include an installed continuous particle monitoring system. This is most common in the aseptic core, particularly in Grade A. The value of a continuous particle monitoring systems is that these systems can detect particles of sub-micron size level (<1 micron) and can detect immediate changes in the concentration of particles of a specific size within the controlled zone at specified locations.

The usually selected particle size is 0.5 μm and larger as specified in regulatory guidelines. Many firms also monitor at 5 μm, as recommended by the regulations [7]. This is not required for room classification and is specifically recommended against in the ISO Standard (Table 5.1). Nonetheless, there is a supposition that an increase in the 5 μm particle count may be a forewarning of a breakthrough of the smaller particles. The authors are aware of no published data to support this assumption and have discovered only a few anecdotal references (unpublished).

Continuous particle counting provides important evidence of environmental control but does not constitute proof of absence of microbial contamination. There is no scientific correlation between the particle concentration and the presence of microorganisms. Therefore, airborne particle monitoring, which measures total particulate, must be supplemented by airborne monitoring for viable organisms as required by the regulations.

Particle counting technologies are constantly being updated and several different technologies are employed. The user should be aware of the differences and provide for proper qualification of devices prior to using them either for testing during classification or for routine monitoring. Continuous monitoring systems may consist of dedicated or remote probes to gather samples and a network of tubing, whereby a sample is transported to a remote location. Other technologies employ "smart" sensors that measure particulate at the source and an electronic signal is sent to a to a remote monitoring/recording unit. Both types of data gathering and transmission present advantages and disadvantages. Transport of air samples could be difficult in tubing if the transport path and the

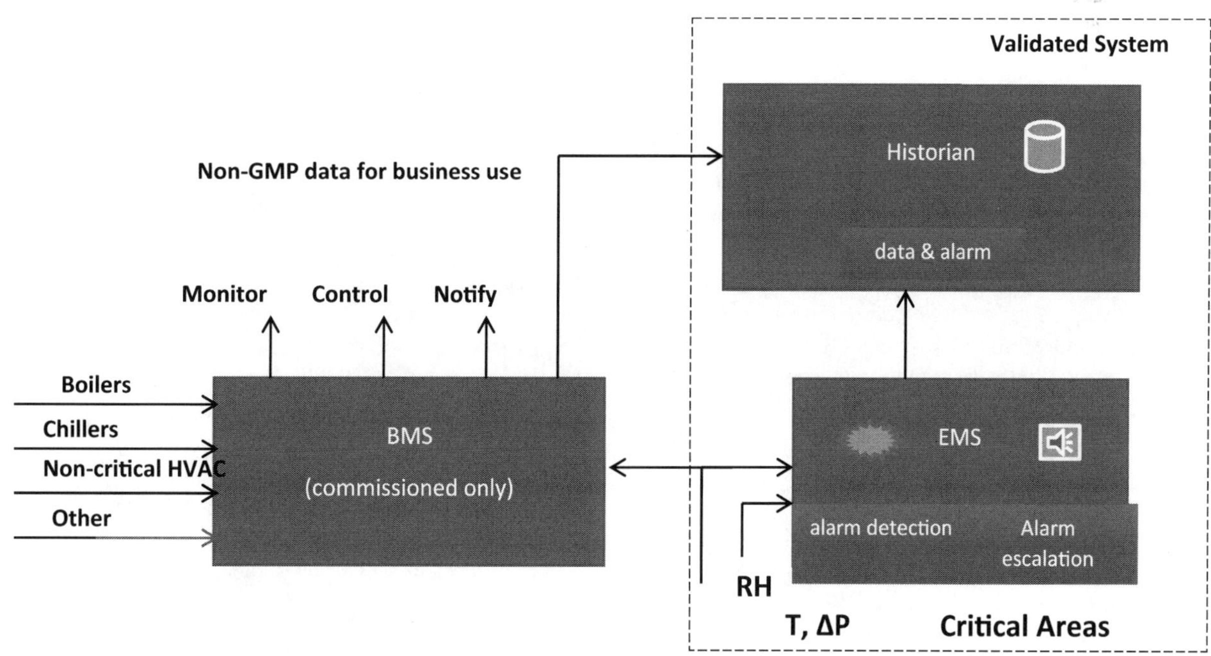

FIGURE 5.15 BMS/EMS architecture.

Note: BMS = building management system; EMS = environmental monitoring system; GMP = Good Manufacturing Practice; HVAC = heating, ventilation and air conditioning; RH = relative humidity.

tubing installation are not well designed. Potential losses or particle accumulation can occur creating data inconsistencies. Locating particle counter sensors in the proximity of critical locations could be difficult because of the device's size and production equipment interaction, although the transmission and integrity of the data can be of better quality. Both types of systems are normally set up to report predefined particle concentrations as alert and action alarms. They may be interfaced to higher level systems, like an EMS, to continuously track and record room status. Such systems can even report adverse trends before alert limits are triggered.

5.5 OTHER CLEAN ROOM FACTORS

Some factors not specific to the design and qualification of controlled environments that influence the ongoing performance of controlled environments must also be considered. Personnel are the major source of contamination and are discussed in the chapters "Validation of Training" and "Environmental Monitoring". Sanitization and disinfection are discussed in the chapters on *Disinfection* and *Cleaning and Disinfection of Laminar Flow Work Stations.*

5.6 COMMISSIONING AND QUALIFICATION

5.6.1 Risk Management

For any system the risk of mechanical failure may affect product quality, safety, the surrounding environment, and business continuity. Appropriate maintenance procedures for the HVAC system must be established as necessary elements of the qualification program. The most effective of these are risk-based in that they evaluate risk from three points of view:

- Severity/criticality
- Probability/frequency of malfunction
- Detectability/predictability

There are several functions available to assess risks using these factors as well as other approaches. Failure Modes and Effects Analysis (FMEA) and its extension Failure Modes Effects and Criticality Analysis (FMECA) are particularly applicable. These are rather detailed analyses that work well for mechanical and electrical systems. By performing these or other risk assessments, users may focus their maintenance programs on preventing potential failures or detecting them before they occur. Considering limited resources, a risk-based maintenance program is usually more effective than a randomly structured one.

The chapter "Commissioning and Qualification" talks extensively about quality risk management (QRM) as it applies to qualification of all facilities and equipment systems. That chapter guides the user to determine which components or functions of a controlled environment system have a direct effect on product quality. These components and functions become subject to equipment qualification procedures. Systems as a whole, including components and functions that indirectly affect quality or have no quality impact, should be commissioned.

For controlled environments, a quality risk assessment performed by some firms and to which the authors subscribe has determined that the environment proper has a direct effect on product quality, whereas the mechanical components that condition and deliver the air do not. The rationale behind this is that primary environmental parameters such as temperature, relative humidity, differential pressure, and airborne particle count can be monitored directly. These are considered to have significant risk to quality whereas mechanical factors like fan speed, pre-filter pressure drop, or coil temperature rise may only have an indirect effect. Therefore, based on the guidance provided by the chapter "Commissioning and Qualification", the qualification of controlled environments described following focuses on the environment itself rather than the mechanical system that produces it.

5.6.2 Commissioning/Functional Testing

Commissioning is an engineering function that entails inspecting a system for proper installation of specified components, sets the system to work, makes adjustments within operating specifications, and performs functional testing. Commissioning is actually more comprehensive than qualification because it covers entire systems rather than quality-critical requirements. Preliminary testing in advance of qualification is often performed. Commissioning is documented according to GEP [23] and allows for the adjustment of parameters, as well as corrections, as long as they are documented. Commissioning is a necessary prerequisite to qualification. More detail on commissioning is found in the cited chapter. For HVAC/Controlled Environment systems the following are included in commissioning:

- Vendor manuals
- Ductwork leak tests
- Ductwork material certification
- Fan motor checks/fan performance
- Coil performance
- Pre-filter pressure drop
- Control system loop checks
- Input/Output lists
- Instrument Calibration Reports
- HVAC Test and Balance Reports
- Room air change calculation (ACH)
- HEPA Filter Certifications
- System Schematic Walkdown/Markup
- Equipment/instrument confirmation vs. specifications
- Identification and initial test of Critical Operating Requirement Values
- Functional Test Report

5.6.3 Overview of Qualification

The scope of qualification for controlled environments is limited to those activities necessary and sufficient to provide a high degree of assurance that clean rooms meet the requirements necessary to ensure product quality. Provided that appropriate environmental monitoring and/or alarm systems are in place, HVAC mechanical and control systems are determined to have an indirect impact on product quality. These HVAC systems should be commissioned to ensure that they are suitable for their intended purpose and have been installed according to applicable engineering standards and specifications. Formal qualification activities (i.e., following standards and practices dictated by CGMP) should focus on the final HEPA filtration (including any ductwork beyond the final filter) and monitoring of area parameters (tests described in Section 5.6.5) to ensure that the clean room requirements are met.

Isolators and unidirectional ("laminar") air flow units that provide Grade A air to local areas within a larger clean room are included. However, this section excludes other locally controlled environments, such as freezers, refrigerators, stability chambers, etc., that are self-contained and where the environment is provided by dedicated specialized equipment and not by traditional HVAC systems. Although sterilization tunnels or other devices that utilize unidirectional and/or HEPA-filtered air are not covered by this chapter, the test methods described for the air flow, filter testing, and air classification may be directly applied.

The details of planning and documenting equipment qualification are found in the "Commissioning and Qualification" chapter. They are not repeated here. The following inspections and tests are to be considered and applied as determined by risk to product quality. In general, these are recommended for all clean rooms.

5.6.4 Installation Qualification

Installation Qualification (IQ) verifies that critical system components meet user requirements and have been properly identified, installed, and calibrated (instruments). Critical components are those judged to have a direct effect on product quality and are identified by component risk analysis. Requirements of Installation Qualification may be confirmed through inspection or by referencing commissioning or factory acceptance test (FAT) documentation. These requirements include:

- Description, location and use of the Classified Area including the defined boundaries and all included Clean rooms.
- Verification of critical system components, including terminal air filters, instruments, and alarms (note that in cases where an EMS is employed, the monitoring sensors, recording, and alarm functionality may be included in that system and documented in a separate protocol). Identification should include manufacturer, model no., size/range/efficiency (as appropriate), materials (product contact only—not typical for HVAC). Where appropriate, this will include material certification of components.
- Verification of Instrument calibration in compliance with site calibration procedures.
- Location of final system filters where the environmental classification has airborne viable and/or non-viable particulate limitations (e.g., Grades A—D).
- Verification of material and proper installation of ductwork downstream of final HEPA filter.
- Completion of a commissioning plan or document with approval by the system user and responsible Engineering group. This shall include a test and balance report.
- Software or firmware title/identification, version and installation verification (may be in separate RMS protocol—see the chapter "Control Systems Validation").
- Transfer of equipment, instrument, and MROM (Maintenance, Repair and Operating Materials) information to the appropriate site maintenance records (e.g., computerized maintenance management system—CMMS).
- Confirmation that maintenance procedures and/or manuals are in place.

5.6.5 Operational Qualification

Operational Qualification (OQ) verifies operational requirements that are critical to quality. These are limited to environmental parameters to which product may be exposed (e.g., temperature) and the operation of final filters required to control airborne contamination. Operational characteristics having indirect impact on the environment, including air handler operation (e.g., fan speed, pre-filters, coil performance) and control elements (e.g., variable air volume controls) should be included in commissioning. Tests to be included are:

- Final air filter pressure differential
- Filter leak test (for H13 and H14 filters or better only)
- Air flow velocity (if unidirectional flow is required)
- Air flow visualization
- Room pressure differential conditions
- Temperature conditions
- Relative humidity conditions
- Alarm and interlock checks
- Room air classification (nonviable particle count)
- Recovery time

OQ is to be performed "at rest" with the HVAC equipment installed, commissioned, and balanced. Other equipment in the room must be installed and functioning without product or material to the extent that is safe. Operating staff is not present in the room (except for test personnel, if required). The operation of equipment verifies the ability of the system to control airborne particulate generated by the equipment. This, however, may be waived if judged to be unsafe.

Testing must be conducted according to written internal or third-party procedures/protocols approved by the site.

Instruments used to gather OQ data may be installed permanently in the system or may be test instruments. They must be calibrated according to the site calibration SOP and must be identified in the OQ documents. Requirements of operational qualification may be confirmed by referencing commissioning or factory acceptance test (FAT) documentation if the site practices for GMP documentation have been adhered to during those stages and the appropriate test methodology has been employed.

Prior to starting OQ, all relevant procedures such as gowning, equipment operation, etc., should be drafted and subject to interim approval for purposes applicable to OQ testing. (These may be modified prior to final approval for PQ).

Unless otherwise noted herein, all testing in support of OQ should comply with the latest published revision of ISO Standard 14644 Cleanrooms and Associated Controlled Environments.

ISO 14644–3 outlines the methodology to be followed for each of the following tests and provides acceptance criteria. These are readily available to users at a minimal cost and are not included here.

The following test matrix (Table 5.9) describes the requirements for qualification in the various clean zones. Table 5.9 does not require that tests be performed in any prescribed order, except as described in the following text. It is common practice to perform certain tests in parallel (e.g., room pressure differential, temperature, and humidity).

The tests listed in Table 5.9 are required for initial qualification. Operational Qualification following a change is determined by a risk analysis depending on the type of change, including area and equipment configuration and change of conditions. For example, HEPA filter change does not impact temperature, humidity, room differential pressure, or air flow pattern (smoke studies). Therefore, for filter changes the required tests are limited to air velocity (Zone A), air flow distribution (zones B, C, D), filter differential pressure (all Grades), filter leak test (H13/H14 or better filters), and room classification (for Grades A and B only). With the exception of the filter change described previously, changes to the equipment in critical areas (e.g., modifications/changes of the geometrical shape on an aseptic filling line in Grade A) require air flow visualization (smoke) studies.

5.6.5.1 Filter Pressure Differential

This qualification test applies only to the final supply filters installed in an air handling system. These may be centrally located at the air handler (filter bank) or at each room air diffuser (terminal filters). Differential pressure across an air filter or filter bank is an indication of particulate loading and an early indicator of potential loss of air flow. The instruments installed in the system to measure pressure differential may be used. One measurement is adequate to establish the differential pressure across newly installed clean final air filters or filter banks (installed devices monitor dP across a bank of filters, not each individual filter in the bank). This should be done and recorded each time a new filter or set of filters is installed. The acceptable clean range is set by the filter manufacturer and is dependent upon the filter rating and air flow velocity.

5.6.5.2 Filter Leak Test

This test is required for installed H13 and H14 HEPA (or better) filters only. It may be performed by site personnel or accredited third-party specialists. Methodology must comply with ISO 14644–3 Cleanrooms and Associated Controlled Environments, Part 3: Test Methods.

TABLE 5.9
OQ Testing Requirements (At Rest)

Test	Grade A[3] (ISO 5 -unidirectional)	Grade B (ISO 5)	Zone C (ISO 7)	Zone D (ISO 8)	CnC (ISO 8—nominal)
Filter pressure differential	Yes	Yes	Yes	Yes	Yes
Filter leak test[1]	Yes	Yes	Yes	Yes	N/A
Air flow distribution	N/A	Yes	Yes	Yes	Yes
Unidirectional air flow velocity	Yes	N/A	N/A	N/A	N/A
Air flow visualization (smoke) study[2]	Yes	Yes	See 5.4.3.5	See 5.4.3.5	Yes
Differential room pressure	N/A	Yes	Yes	Yes	Yes
Temperature	Yes	Yes	Yes	Yes	Yes
Relative humidity	Yes	Yes	Yes	Yes	Yes
Alarms and interlocks	Yes	Yes	Yes	Yes	Yes
Room air classification (nonviable particle count)	Yes	Yes	Yes	Yes	N/A
Recovery time	N/A	Yes	Yes	N/A	N/A

Abbreviations: CnC = Controlled (non-Classified); OQ = Operational Qualification.

1 Required for H13/H14 or better filters only. Most firms certify HEPAs in CnC areas when installed.

2 Air flow visualization studies in non-unidirectional areas are optional based on risk analysis. See Section 5.6.5.3 following

3 Operational Qualification testing for Grade A air supplied for crimping of aseptically filled vials or for local assurance of particulate control (i.e., local laminar air flow devices) is the same as for Grade A, except for Unidirectional air flow and Air flow visualization

5.6.5.3 Air Flow Distribution

The Air Flow Distribution test is applicable to rooms/areas of non-unidirectional (i.e., turbulent) flow. For unidirectional flow devices see 5.6.5.4 following.

The purpose of this test is to determine that the air supply to the room is delivered uniformly enough to ensure that physical parameters throughout the workspace are within specified limits. A recommended acceptance criterion for this test is that all air supply diffusers must provide an air flow within 20% of the specified flow for each diffuser. Total air volume may be calculated in order to compare with the design specifications and to determine the air exchange rate for the room (air changes/hour—ACH). ACH is an indicator only and one of many factors that determine room performance. Therefore, it is not itself a critical parameter and may be tested in commissioning. It should be recorded in the Operational Qualification report in order to provide a baseline for future comparison.

5.6.5.4 Unidirectional Air Flow Velocity

This test is required only for unidirectional air flow devices. Average velocity at 30 cm (app. 1 ft.) from the filter face must be 0.45 m/sec ±.20% (0.36–0.54 m/s). The maximum relative standard deviation across all readings should be no greater than 20%.

5.6.5.5 Air Flow Visualization (Smoke) Studies

This test is required for unidirectional flow areas and devices. It involves the injection of visible particles (usually a neutral density fog, but commonly referred to as "smoke") into the unidirectional air stream. A video recording is recommended to provide evidence of streamlined flow with no eddies, turbulence or backflow that could cause external contamination to be drawn into the clean work area. Evaluation of these results in an isolator, where some configurations may introduce some localized turbulence, must be interpreted carefully with the risk to exposed product being of primary concern.

A second type of Air Flow Visualization test may be required for non-unidirectional zones. This is necessary if areas are not designed with an air lock between clean rooms of differing classification. This includes adjacent zones separated by a door, a curtain or a high-velocity air curtain. This test is not required for clean rooms where product is not exposed. The test uses a smoke generator or smoke stick to indicate the direction of flow under normal (steady-state) and changing (e.g., door opens) conditions.

5.6.5.6 Differential Room Pressure

The room pressure differential must be checked using a calibrated electronic micromanometer or an inclined manometer. Calibrated installed differential pressure measuring devices may also be employed. In many installations, pressure differential is continuously recorded. In cases where it is not, it may be recorded manually at least three times (once per shift) over 24 hours. The differential pressure between the most clean and adjacent less clean zones must be recorded. Continue through less clean zones until the external environment (uncontrolled space) is reached. The acceptance criteria are derived from the engineering design. Excursions from the specified differential are acceptable between rooms of like classification upon opening of doors (see Directional Air Flow Test).

5.6.5.7 Temperature and Relative Humidity

5.6.5.7.1 Manufacturing Areas

These tests should be performed following the Air Flow Uniformity test in 5.6.5.3. previously and may be done in parallel with room pressure differential. OQ studies are performed "at rest". In manufacturing areas, temperature and humidity studies should proceed for a minimum of 24 hours to account for daily changes in external conditions. These tests primarily ensure system capability and correlation to the room temperature as controlled by the HVAC and BMS systems. It is not necessary to perform seasonal testing for manufacturing areas as long as the monitoring system continues to report results within specifications.

Where room conditions are designed for general manufacturing and restrictive product-related conditions are not specified, it is permissible to employ only the installed room monitoring sensors provided that the room is designed with >5 ACH. Where this condition is not met or where restrictive product-related conditions are specified (e.g., temperature or humidity sensitive product) data logging systems independent of the installed monitoring system should be employed, with sensors of the appropriate type placed within the work area. Sensors should be located based upon a risk analysis and placed as close as possible to where product will be exposed. (This may be in one or more places within the room. The rest of the room may be adequately monitored by the installed system sensors.)

5.6.5.7.2 Holding and Storage Areas

In areas where product will be stored for more than 48 hours more extensive temperature mapping is required. Temperature sensors must be dispersed in a geometric pattern or other configuration based on a risk assessment. Sensors should provide representative data for all areas where product may be stored. OQ testing in these areas should proceed for a minimum of 24 hours. For OQ testing of storage areas, mapping may be performed in an empty or unloaded state. PQ testing should be performed with the area containing a representative load and proceed for at least 72 hours. (Note that maximum loading may be neither practical nor necessary. Typically, loaded storage areas provide a more uniform temperature than unloaded areas because of air flow disbursement and heat capacity of the load.)

Where geographic location dictates, mapping should be repeated in opposing seasons (i.e., Summer/Winter) to ensure control. This is recommended if at least one (summer) monthly average temperature for the location is greater than 25°C AND at least one (winter) monthly average daily minimum is less than 15°C.

5.6.5.8 Alarms and Interlocks

System alarms and interlocks associated with environmental parameters affecting product quality must be shown to function properly. Typically, alarms may include temperature, humidity, room pressure differential, and airborne particle count (critical zones). Also included may be interlocks, such as those preventing the concurrent opening of opposing air lock doors. If alarms or interlocks are a function of a Room Monitoring System (RMS), this test may be covered in a separate protocol.

5.6.5.9 Room Classification—Total Particle Count

Total room air classification is required for Zones A through D. Testing is performed according to the methodology outlined in ISO 14644–1 and 14644–3.

5.6.5.10 Recovery Time

This test is applicable to non-unidirectional flow rooms/areas that have an airborne particulate limit. The purpose of the test is to determine that the HVAC system is capable of returning the area to its specified cleanliness level within a reasonable time after a shutdown or upset. The test utilizes a particulate generating device (usually a fogger) to attain a particle count significantly above the specified limit. Re-testing is then carried out at time intervals during the specified cleanup period to establish recovery time. PIC/S Annex 1 provides a guidance value of 20 minutes for recovery. Deviations significantly in excess of this value should be investigated.

5.6.6 Performance Qualification

Performance Qualification (PQ) is carried out in rooms/areas subject to defined particulate and microbial limits (Zones A through D) as well as in CnC areas, which do not. Microbial Monitoring should follow company quality standards and the requirements of the site SOP for environmental monitoring. Microbial Monitoring is carried out "at rest" and "in operation".

For these rooms (Zones A through D, CnC) the following must be in place:

- HVAC equipment installed and commissioned and balanced.
- OQ testing has been completed and any deviations resolved.
- Procedures for access control, gowning, and room sanitization as well as all SOPs defined in the protocol as being necessary for area operation should be approved and implemented.
- Other equipment in the room has been completely installed and are functioning with normal production materials and/or product in process. (Note: Growth medium may be substituted for product in aseptic operations.)
- For all "in operation" tests the operating staff is present in the room and performing normal tasks.

Sample locations are determined based on risk to product exposure according to the principles of QRM. Acceptance criteria for new rooms should be set at or below the alert levels established by the site.

Table 5.10 describes testing to be performed for PQ:

TABLE 5.10
PQ Requirements

Tests (at rest)	Grade A/B	Grade C	Grade D	CnC
Microbial monitoring	Yes	Yes	Yes	N/A
Tests (in operation)	**Grade A/B**	**Grade C**	**Grade D**	**CnC**
Room pressure differential	Yes	Yes	Yes	Yes
Temperature	Yes	Yes	Yes	Yes
Humidity	Yes	Yes	Yes	Yes
Room classification (airborne particle count)	Yes	Yes	Yes	N/A
Air flow visualization (smoke study)	Yes	N/A	N/A	N/A
Microbial monitoring	Yes	Yes	Yes	Optional[1]

Abbreviations: CnC = Controlled (non-Classified); PQ = Performance Qualification.

[1] CnC micro limits are only required when micro limits are applied. Some firms choose to perform this testing for information.

PQ is done "in operation" with the exception of microbial monitoring, which is done both "at rest" and "in operation".

"At rest" microbial monitoring shall be done once after room cleaning and/or sanitization in order to establish a baseline microbial count.

"In operation" microbial monitoring and monitoring of temperature, humidity, and room pressure differential should encompass at least three days of normal operation, with microbial monitoring conducted according to the normal site procedures. Operations should include setup/preparation, cleaning and sanitization, and batch or manufacturing operations. The days need not be consecutive but should coincide with the site's schedule of operations.

For Grades A and B, unidirectional air flows in operation should first be verified by smoke studies with operating personnel performing simulated duties. These should include normal operation plus routine and nonroutine interventions that could alter air flow.

PQ of other physical parameters including temperature, humidity, room pressure differential, and room classification may coincide with aseptic fill (media) challenge studies.

It is not required to include monitoring during cleaning and sanitization as these may introduce environmental conditions (e.g., high humidity) that are outside of the normal operating conditions. Areas must return to within designated operating specifications before beginning operations.

5.6.7 Routine Maintenance

All systems are subject to routine monitoring and preventive maintenance according to site procedures. Refer to the chapter "Environmental Monitoring". The parameters and functions covered by routine plant operation need not be subject to re-qualification or recertification. These include temperature, humidity, room pressure differential, and filter pressure differential. Results and trends for these should, however, be reviewed regularly and in conjunction with the area recertification described in Section 5.6.8.

5.6.8 Routine Recertification

Regular recertification of clean rooms and long-term holding areas (e.g., warehouse) is required in specific cases. Re-qualification and/or recertification of manufacturing areas may be driven by change and/or scheduled periodically. Where changes have not driven recertification of an area, it is necessary to provide evidence that the acceptance criteria indicated in the initial qualification are still met. This should be performed according to the schedule in Table 5.11 and as described in the notes following. It is important to regularly review the results and trends of parameters that are continuously monitored or regularly recorded (e.g., temperature, humidity, room pressure differential, others based on risk analysis).

TABLE 5.11
Routine Recertification[1]

Test[2]	Grade A	Grade B	Grade C	Grade D	CnC
Filter leak test[3]	6 mos.	6 mos.	6 mos.	Annual	2 yrs. (optional)
Air flow distribution	N/A	6 mos.	Annual	Annual	N/A
Unidirectional air flow	6 mos.	N/A	N/A	N/A	N/A
Room classification	Annual [4]	Annual	Annual	Annual	N/A
Recovery[5]	N/A	N/A	N/A	N/A	N/A

Abbreviation: CnC = Controlled (non-Classified).

[1] Frequencies based on proposed draft of EU/PIC/S Annex 1. Many experts have commented that ISO 14644 allows the extension of frequencies based upon consistent satisfactory results over time. The frequencies in Table 5.11 are likely to remain in the regulations, but the reader is advised to stay abreast of final publication and developments in consensus international standards (e.g., ISO 14644)

[2] All recertification testing is performed "at rest"

[3] Filter leak test is required only for H13/H14 filters or better.

[4] For Zone A where continuous fixed particulate monitoring is in place, recertification of the area classification is not required

[5] Recovery testing for routine recertification is not required except in cases where the Air velocity test has shown a significant reduction (>20%) in the number of air changes per hour (ACH). For changes in HVAC design, room layout, or equipment configuration, Recovery testing must be considered within the risk analysis associated with the change.

5.7 CONCLUSION

As with all manufacturing systems and equipment, controlled environments are founded upon proven engineering principles. Engineers and designers have developed and improved these over decades of application in the biopharmaceutical industry. Mechanical equipment used to control the environment in clean rooms is reliable and efficient. When commissioned according to GEP, qualification of the quality-critical parameters of the environment is effectively achieved.

REFERENCES

1. 21 CFR Part 211. *Current Good Manufacturing Practice for Finished Pharmaceuticals*, 2020.
2. Agalloco J. Paradise lost: Misdirection in the implementation of isolation technology. *Pharmaceutical Manufacturing* 2016, 15(4), 34.
3. American Society of Heating, Ventilation and Air Conditioning Engineers (ASHRAE), Standard 52.2. *Method of Testing General Ventilation Air Cleaning Devices for Removal Efficiency by Particle Size*, 1999.
4. DeSantis P. A Framework for Quality Risk Management of Facilities and Equipment. *Pharmaceutical Online*, January 2018. www.pharmaceuticalonline.com/doc/a-framework-for-quality-risk-management-of-facilities-and-equipment-0001
5. DeSantis P. Facilities and Equipment Risk Management: A Quality Systems Approach. *Pharmaceutical Online*, December 2017. www.pharmaceuticalonline.com/doc/facilities-and-equipment-risk-management-a-quality-systems-approach-0001
6. DeSantis P. *HVAC Course Notes*, 2015–2020.
7. Eudralex, Volume 4. *Good Manufacturing Practices (GMP Guidelines)*, Annex 1, *Manufacture of Sterile Medicinal Products*, 2008.
8. European Medicines Agency (EMA). *Annex 1 Proposed Revisions* (draft), February 2020.
9. European Standard EN-1822-1. *High Efficiency Air Filters (HEPA and ULPA): Classification Performance Testing.*
10. International Society for Pharmaceutical Engineering, ISPE Baseline® Guide, Volume 3. *Sterile Products Manufacturing Facilities*, 2018.
11. International Society for Pharmaceutical Engineering, ISPE Good Practice Guide. *Heating, Ventilation and Air Conditioning Systems*, 2009.
12. International Standards Organization, ISO 14644-1. *Cleanrooms and Associated Controlled Environments: Environments Part 1: Classification of Air Cleanliness by Particle Concentration*, 2015.
13. International Standards Organization, ISO 14644-2. *Cleanrooms and Associated Controlled Environment Part 2: Monitoring to Provide Evidence of Cleanroom Performance Related to Air Cleanliness by Particle Concentration*, 2015.
14. International Standards Organization, ISO 14644-3. *Cleanrooms and Associated Controlled Environments Part 3: Test Methods*, 2005.

15. International Standards Organization, ISO 29463-1. *High Efficiency Filters and Filter Media for Removing Particles from Air, Part 1: Classification, Performance, Testing and Marking*, 2017.
16. Parenteral Drug Association. *Points to Consider for Aseptic Processing*, Parts 1 and 2, January 2015 and May 2016.
17. Parenteral Drug Association, Technical Report 13. *Environmental Monitoring*, 2013.
18. PIC/s GMP Guide for Good Manufacturing Practice for Medicinal Products, Annex 1. *Manufacture of Sterile Medicinal Products*, 2018.
19. US Food and Drug Administration, Guidance for Industry. *Sterile Drug Products Produced by Aseptic Processing—Current Good Manufacturing Practice*, 2004.
20. United States Pharmacopeia, USP 43 <1116>, *Microbiological Control and Monitoring of Aseptic Processing Environments*, 2020.
21. United States Pharmacopeia, USP 43 <1150>, *Pharmaceutical Stability.*
22. Cleanroom Technology (online). *Study into Human Particle Shedding*, July 2011. www.cleanroomtechnology.com/news/article_page/Study_into_human_particle_shedding/62768

6 Validation of Pharmaceutical Water Systems

William V. Collentro

CONTENTS

6.1 INTRODUCTION

A pharmaceutical water system consists of multiple purification unit operations designed to remove chemical and microbial contaminants in feed water. The multicomponent nature of a pharmaceutical water system impacts the nature of the validation process. Pharmaceutical water system design, validation, operation, and maintenance require knowledge of both the technical aspects of the entire system as well as the documentation requirements and testing to verify a continuous state of control. Several years ago, the Pharmaceutical Manufacturer's Association Water Committee arranged a conference to discuss the state of pharmaceutical water systems (1). Attendance and participation at the conference was limited to individuals working in pharmaceutical and related industries. The number of participants and quality of material presented in the individual sessions was outstanding, clearly demonstrating the depth of pharmaceutical water system knowledge throughout the industry. The tedious effort of many individuals to develop and present material associated with chemical testing resulted in the USP 23 supplemental changes to pass/fail testing, significantly improving the ability to continuously monitor system performance (2). Further, changes in water purification technologies such as thin-film composite reverse osmosis membranes, continuous electrodeionization, and electrolytic ozone generation advanced the "tool box" for production of compendial waters. From a validation perspective, incorporation of "risk analysis" has provided a useful guide for system design.

DOI: 10.1201/9781003163138-6

With advancements in water purification technology and publication of numerous reference "standards", the evolution of *process validation* of pharmaceutical water systems has become increasingly popular. While addressing a question regarding the purpose of each pharmaceutical water system component from an individual "validating" a pharmaceutical water system at a facility, it was indicated that "they didn't need to know why things were there or what their function was, but just validate the system". Further, system sampling and analysis focus primarily on point-of-use results. USP *General Notices Section* <1231>, "Water for Pharmaceutical Purposes", presents a realistic discussion of the importance of system component online monitoring, "grab" sampling/analysis, and data trending for a pharmaceutical water system (3). Additional recommendations for unit operation product water monitoring are presented in the literature (4,5).

This chapter presents a summary of the documentation path for pharmaceutical water systems. It includes details and examples of specific documents. An overview of the types of compendial waters is presented. Because sterile/packaged waters are produced from bulk Purified Water or Water for Injection, discussions of water purification unit operation design/specification, storage, distribution, and delivery focus on bulk compendial water systems are included. Finally, the operation, monitoring, sampling, analysis, data collection, data trending, and, most importantly, preventative maintenance program throughout the life cycle of a pharmaceutical water system are presented.

6.2 TYPES OF COMPENDIAL WATER

6.2.1 Note

A brief summary of the current United States Pharmacopeia (USP) attributes, test reference sections, and labelling criteria are included. The current edition of USP or other applicable pharmacopeia should be used to verify the attributes (6).

6.2.1.1 Bulk Compendial Waters

- Purified Water
  - Feed water meets the U.S. EPA National Primary Drinking Water Regulations (EPA-NPDWR) or equivalent
  - Contains "No Added Substance"
  - Not for injection
  - Conductivity limit per USP *Physical Tests* Section <645>, "Water Conductivity"
  - Total Organic Carbon (TOC) limit per USP *Physical Tests* Section <643>, "Total Organic Carbon"
  - "Microbial guidance per USP *General Information* Section <1231>, "Water for Pharmaceutical Purposes"
- Water for Injection
  - Feed water meets the U.S. EPA National Primary Drinking Water Regulations (EPA-NPDWR) or equivalent
  - Contains "No Added Substance"
  - Used for preparation of parenteral solutions
  - Conductivity limit per USP *Physical Tests* Section <645>
  - TOC limit per USP *Physical Tests* Section <643>
  - Bacterial endotoxins <0.25 EU/mL per USP *Biological Tests and Assays* Section <85>, "Bacterial Endotoxins Test"
- Microbial guidance per USP *General Information* Section <1231>, "Water for Pharmaceutical Purposes"
- Water for Hemodialysis
  - Feed water meets the U.S. EPA National Primary Drinking Water Regulations (EPA-NPDWR) or equivalent
  - Contains "No Added Antimicrobial Agent""
  - Produced and used "on-site" under the direction of qualified personnel
  - Not intended for injection
  - Conductivity limit per USP *Physical Tests* Section <645>
  - TOC limit per USP *Physical Tests* Section <643>
  - Bacterial endotoxins <1 EU/mL per USP *Biological Tests and Assays* Section <85>, "Bacterial Endotoxins Test"
  - Total Aerobic Count ≤ 100 cfu/mL determined per USP Microbiological Enumeration Test Section <62>. "Examination of Nonsterile Products: Microbial Enumeration Tests"
  - Absence of *Pseudomonas aeruginosa* determined per USP *Microbiological Test* Section <62>, "Microbiological Examination of Nonsterile Products. Tests for Specific Organisms"
  - Packaged and stored in unreactive containers designed to prevent bacterial entry
  - Stored at room temperature

6.2.1.2 Sterile Waters ("Packaged")

- Sterile Purified Water
  - Purified Water (bulk) that is sterilized and suitably packaged. It contains no antimicrobial agent.
  - Not for parenteral use
  - Meets TOC limits per USP Physical Tests Section <643>
  - Conductivity limit per USP *Physical Tests* Section <645>
  - Sterile per USP *Microbiological Tests* Section <71>, "Sterility Testing"
  - Packaged and stored in suitable tight container
  - Labeled to indicate the method of production and to indicate that it is not for parenteral administration
- Sterile Water for Injection
  - Water for Injection (bulk) that is sterilized and and suitable packaged
  - Contains no antimicrobial agent
  - Contains no added substances
  - Meets TOC limits per USP Physical Tests Section <643>
  - Conductivity limit per USP *Physical Tests* Section <645>
  - Sterile per USP *Microbiological Tests* Section <71>, "Sterility Testing"

    - Bacterial Endotoxins <0.25 EU/mL per USP *Biological Tests and Assays* Section <85>, "Bacterial Endotoxins Test"
    - Meets the requirements for Particulate Matter in Injections, Physical Tests and Determinations Section <788>
    - Packaged and stored in single-dose (Type I or Type II) or plastic containers ≤ 1 L
    - Label: "No antimicrobial agent or other substance has been added, Not suitable for intravascular injection without first having been made approximately isotonic by the addition of a suitable solvent"
- Sterile Water for Inhalation
    - Water for Injection (bulk) that is sterilized and suitably packaged
    - Contains no antimicrobial agent
    - Not for parenteral use or for other sterile compendial dose form
    - Meets the requirements for TOC limits per USP Physical Tests Section <643>
    - Conductivity limit per USP *Physical Tests* Section <645>
    - Sterile per USP *Microbiological Tests* Section <71>, "Sterility Testing"
    - Bacterial endotoxins <0.5 EU/mL per USP *Biological Tests and Assays* Section <85>, "Bacterial Endotoxins Test"
    - Preserve in glass (Type I or Type II) or plastic containers
    - Label: "For inhalation therapy only, not for parenteral administration"
- Sterile Water for Irrigation
    - Water for Injection (bulk) that is sterilized and suitably packaged
    - Contains no antimicrobial agent
    - Contains no added substances
    - Meets the requirements for TOC limits per USP Physical Tests section <643>
    - Conductivity limit per USP *Physical Tests* Section <645>
    - Sterile per USP *Microbiological Tests* Section <71>, "Sterility Testing"
    - Bacterial endotoxins <0.25 EU/mL per USP *Biological Tests and Assays* Section <85>, "Bacterial Endotoxins Test"
    - Packaged in single-dose glass (Type I or Type II) or plastic containers that may contain >1 L
    - Label: "No antimicrobial agent has been added"," For Irrigation Only", "Not for Injection"
- Bacteriostatic Water for Injection
    - Water for Injection (bulk) that is sterilized, packaged, and contains one or more suitable antimicrobial agents. Antimicrobial agent(s) to meet requirements of USP *Microbial Tests* Section <51>, "Antimicrobial Effectiveness Testing" with label claims per USP *Chemical Test and Assays* Section <341>, "Antimicrobial Agents—Content"
    - Passes USP *Monograph* chemical tests for calcium, carbon dioxide, and sulfate.
    - Meets pH requirements of USP *Physical Tests* Section <791>, "pH", acceptable criteria 4.5–7.0
    - Sterile per USP *Microbiological Tests* Section <71>, "Sterility Testing"
    - Bacterial endotoxins <0.5 EU/mL per USP *Biological Tests and Assays* Section <85>, "Bacterial Endotoxins Test"
    - Particulate Matter meets USP *Physical Tests and Determination* Section <788>, "Particulate Matter in Injectables"
    - Microbial guidance per USP *General Information* Section <1231>, "Water for Pharmaceutical Purposes"
    - Packaged in single or multiple dose glass (Type I or Type II) or plastic containers with volume ≤ 30 mL size.
    - Label: "NOT FOR USE IN NEWBORNS"

6.3 VALIDATION APPROACH

Validation of a pharmaceutical water system should initiate with generation of a Validation Master Plan. This document provides a summary of the sequence of steps in the validation process as well as a summary of the required documentation. Generally, the Validation Master Plan will be structured in a format established by the "owner" (company). Subsequently, material in the Validation Master Plan will vary but should include the following:

- An "approval page" containing the names and titles of individuals responsible for the generation, review, and approval of the Validation Master Plan. As a minimum, this should include the individuals responsible for regulatory affairs, quality, engineering, and manufacturing/process.
- Revision history for the Validation Master Plan
- Type of pharmaceutical water required including bacteria and bacterial endotoxin limits as well as any "special" chemical or microbial attributes. As indicated previously, the USP Monographs for pharmaceutical water present the minimum criteria. Further, the bacterial Action and Alert Levels are unique for the specific application using compendial water at the facility. As an example, the bacterial Alert and Action Levels will be different for Purified Water employed for ophthalmic solutions or certain topical solutions when compared with Purified Water used for most solid-dose products. Purified Water used for rinsing of components used in production of parenteral solutions, prior to final rinse with Water for Injection, may include an "internal" bacterial endotoxin specification consistent with the bulk Water for Injection limit of 0.25 EU/mL. Water for Injection employed for biotechnology applications may require a very low bacterial endotoxin level

such as <0.01 or 0.001 EU/mL. Company generated "Standards" may require that Purified Water and/or Water for Injection TOC and/or conductivity limits be less than those specified in *Official Monographs*. Finally, the presence of certain gram-negative microorganisms in delivered Purified Water or Water for Injection may be undesirable and restricted even if the bacterial Alert and Action Levels have not been reached.

- The purpose, scope, and objectives of the validation process for the pharmaceutical water system should be defined. This should include the documentation required to provide a fully controlled system capable of delivering water meeting the specified attributes to all points-of-use throughout the life cycle of the system.
- The Validation Master Plan should state the individual disciplines and preferably the individuals from each discipline required to provide input and information for development of a responsive conceptual design for the system. Data gathered by this multidiscipline group will be used to generate the Basis of Design and User Requirements Specification (URS) for the system.
- A summary of the documents required for validation shall be provided. Documentation may include details such as notes, calculations, project meeting minutes, correspondence, preliminary drawings, etc. It is suggested that, as a minimum, documentation include the following:
  - Data used to develop the URS as a Basis of Design.
  - Conceptual Design with Process Flow Diagrams
  - User Requirements Specification (URS)
  - Design Documents with Piping and Instrumentation Diagrams (P and IDs)
  - Functional Requirement Specification (FRS)
  - System Detailed Specifications
  - Risk Assessment—Failure Mode and Effect Analysis
  - Equipment Supplier(s) and Subcontractor(s)—Submittal Package(s) and/or Proposals(s)
  - Equipment Supplier's Turnover Package
  - Instrument and Alarm List
  - Factory Acceptance Test (FAT) results (if applicable)
  - Site Acceptance Test (if applicable)
  - Installation Qualification (IQ)
  - Standard Operating Procedures (SOPs)
  - Maintenance Manual
  - Operating Logs—Master Sheets
  - Training Manual
  - Operation Qualification (OQ)
  - Performance Qualification (PQ)
  - Executed IQ with any Deviations and Resolution
  - Executed OQ with any Deviations and Resolution
  - PQ raw data (chemical, physical, and microbial)
  - PQ tabulated results
  - Validation Summary Report

6.4 VALIDATION STEPS—BASIS OF DESIGN

- Engineering, Regulatory, Quality (QA and QC), Laboratory, Manufacturing, Management, and Production Input
  - List of all points-of-use with the following:
    - Description
    - Physical location
    - Identification (name and/or tag number)
    - Type (manual or automatic)
    - Maximum instantaneous demand (flow rate)
    - Maximum daily volumetric demand
    - 24-hour profile for demand (use versus time of day)
    - Initial and future number of operating shifts each day
    - Initial and future number of operating days per week
    - Type of Compendial Water required (Purified Water, Water for Injection, or both)
    - Bacterial Endotoxin "special" requirements
    - Bacteria Alert and Action Levels
    - Undesirable gram-negative organisms
    - TOC or conductivity levels less than *Monograph* limits
  - Detailed description of each point-of-use
    - "Hard piped"
    - Laboratory sink (with or without hose) and hose management program including cleaning/sanitization program
    - Autoclave with feedwater valve/tubing details
    - Equipment, parts, and/or labware washer with details
    - Pressurized feedwater applications such as Purified Water or Water for Injection feed to a Pure Steam Generator
    - Manual or automatic valves with "hoses" to one or more process tanks and hose management program including cleaning/sanitization program
  - "Delivered Water" Accessories
    - Terminal filtration
    - Flow totalizing meters and any requirements for final flow reduction
    - Flow "orifices"
  - Clean-in-Place (CIP) applications
    - Projected back pressure from cleaning accessories such as spray balls
    - Cleaning "wands"
    - "Drum" or other specialty washers
  - Transfer panels
    - List of "spool pieces" in contact with other process fluids/ingredients and cleaning program
    - Cleaning/sanitization SOPs

- Automatic valves
  - Adjustable opening provisions for flow control in lieu of automatic valves and throttling manual valve in series
  - Control of automatic valve actuator by compendial water system panel or local "process" control panel(s)
  - Provisions for lockout of any local valve control panels during system maintenance, excursion, and/or troubleshooting
  - Simultaneous large volumetric demand for applications such as batch filling with local valve control panels
- Delivered water temperature
  - Impact of compendial water temperature on the stability of other process ingredients (both elevated and decreased temperature compared with ambient temperature)
  - Requirements for point-of-use heat exchanger(s)
  - Requirement for ambient/decreased temperature recirculating sub loops particularly for hot recirculating Water for Injection systems
- Delivered water pressure
  - Shutoff pressure value for automatic valves in support equipment such as glassware washers
  - Minimum pressure for point-of-use accessories such as flow totalizers
- List of countries where product from process will be used
- Sampling limitations at point-of-use such as limitation on access to clean rooms because of the environmental control of the area and/or biological/chemical hazard, requiring "remote sampling" of delivered water
- Physical locations where there are no or limited environmental controls, which could result in microbial and/or chemical contamination of the "delivery" side of a valve. This includes a warm, moist environments that could result in mold in samples.
- Any special provisions for personnel protection or access to the point-of-use valve
- Suggested bacteria enumeration method for samples collected throughout the system.

TABLE 6.1
Example of Point-of-Use Summary Chart

Point-of-use summary chart						
Point-of-use valve number and description	**Description of operation**	**Maximum instantaneous demand**	**Diversity consideration for use**	**Type of compendial water and maximum daily demand**	**Microbial considerations**	**Other considerations**
AV-411, Automatic Zero Dead Leg Sanitary Diaphragm Valve	Purified water feed to process tank, TK-1013	35 gpm	2 batches per day, one on first shift and one on second shift, 7 days per week	Purified Water, 1,500 gals/batch, 3,000 gals/day total.	Topical solution used on open wounds. Absence of pathogens	Ambient temperature, hard piped, requires sampling upstream of AV-411 diaphragm.
HV-412, Manual Zero Dead Leg Sanitary Diaphragm Valve	Purified Water for glassware washer in Chemistry Lab, Room 728	3 gpm	4 washer cycles per day, one each shift and one random, seven days per week	Purified Water, 30 gals. per cycle, 120 gals/day total	Glassware for use in chemical analysis including compendial analysis	Ambient temperature, hard piped, non-pressurized, hose internal to glassware washer may not fully drain.
AV-413, Automatic Zero Dead Leg Sanitary Diaphragm Valve	Purified Water feed to process tank, TK-1244	60 gpm	One batch per day, first shift, 5 days per week Monday through Friday	Purified Water, 3,500 gals/batch, 3,500 gals/day total	Ophthalmic solution. Absence of pathogens.	Ambient temperature, 2" hose 12' long from AV-413 to 2" sanitary ferrule on tank. Hose cleaning/sanitization program needed.
MV-414, Manual Zero Dead Leg Sanitary Diaphragm Valve	Purified Water for CIP of small vessels and support equipment in contact with product	10 gpm	Cleaning and rinsing. Performed three shifts per day generally towards the end of each shift, 7 days per week.	Purified Water, estimated demand 300 gals/shift, 1,200 gals/day total	No special requirements but chemical and microbial attributes similar to Purified Water for production	Elevated temperature desired (40°C–45°C). Point-of-use heat exchanger required
TOTAL	—	108 gpm	—	7,820 gals/day	Purified Water with absence of pathogens	Four ambient points-of-use, one point-of-use heat exchanger.

- Engineering, Maintenance, Manufacturing, Production, Safety, Management, and Facility Personnel Input
  - Feed water to compendial water system
    - Private
      - Source
      - Available chemical and microbial analysis
      - Treatment
      - Disinfecting agent used
    - Public/Municipality
      - Source—ground water, surface water, ground water under the direct influence of surface water
      - Available chemical and microbial results
      - Type of treatment prior to distribution
      - Primary (treatment facility) disinfecting agent(s)
      - Secondary (distribution) disinfecting agent
      - Latest yearly water quality report
      - Contact at municipality
    - Maximum available flow rate
    - Maximum daily volumetric demand limitation if applicable
    - Compliance with National Primary Drinking Water Regulations (40CFR141) or equivalent
  - Wastewater
    - Restrictions and regulations
    - pH
    - Total daily, weekly, monthly, yearly, and or other volumetric limitations
    - Maximum instantaneous discharge rate
    - Maximum temperature
    - Internal drain locations and capacity
  - Chill/Cooling Water
    - Water or water with chemical coolant
    - Maximum flow rate available
    - Maximum ΔT
    - Temperature
  - Electrical
    - Single phase (115-volt, 230-volt)
    - Three phase (208, 230, 460/480 volt)
    - 60 cycle
    - Maximum available current for each supply (amperes)
    - Location of Motor Control Center (MCC) if existing
    - Emergency generator provisions
    - History of electrical "brown outs"
  - Facility Steam
    - Pressure
    - Thermodynamic quality
    - Volatile chemical additives to the facility boiler such as amines
    - Maximum available steam flow
  - Plumbing, electrical, pressure vessel, and other applicable codes
    - Applicable country, state, local, and providence codes
    - ASME (American Society of Mechanical Engineers) Materials of Construction, Section II
    - ASME Welding and Brazing Qualification, Section VIII
    - ASME Code for Unfired Pressure Vessels, Section VIII, Division 1
    - ASME BPE (Bioprocessing Equipment)
    - ASTM (American Society for Testing and Materials)
    - ANSI (American National Standards Institute)
    - NEMA (National Electrical Manufacturer's Association)
    - Fiberglass Vessels—NSF Compliant
    - UL (Underwriters Laboratories)
  - Compressed Air
    - Pressure
    - Oil content
    - Water content
    - Filtered (0.2 μm or HEPA)
  - Projected physical locations of water purification system, storage system, and accessories
    - Indoors
    - Outdoors
    - Environmental conditions of the area including the presence of volatile organic chemicals, moisture/humidity, elevated ambient temperature, and mechanical equipment
    - Ability to provide limited/restricted personnel access to the equipment
  - Floor loading limitations
  - Seismic zone

6.5 CONCEPTUAL DESIGN WITH PROCESS FLOW DIAGRAMS

The information gathered during discussions associated with the Basis of Design is used to develop a conceptual design. It is suggested that this step in the validation process can provide valuable additional feedback from individuals of the project "team" (various disciplines providing input for the Basis of Design). An effective method of presenting and developing the conceptual design for the compendial water system may require multiple meetings using Process Flow Diagrams (PFDs). The initial meeting may use a basic block flow diagram with some initial design information as shown in Figure 6.1 for initial components in the system. This material can be included as part of a Power Point presentation with meeting dialog for feedback. The conceptual design is further developed to include detailed descriptions for individual compendial water system components as shown in Figure 6.2. This step enhances the transition from accumulation of Basis of Design data to preparation and generation of the URS with project "team" input in a highly effective manner.

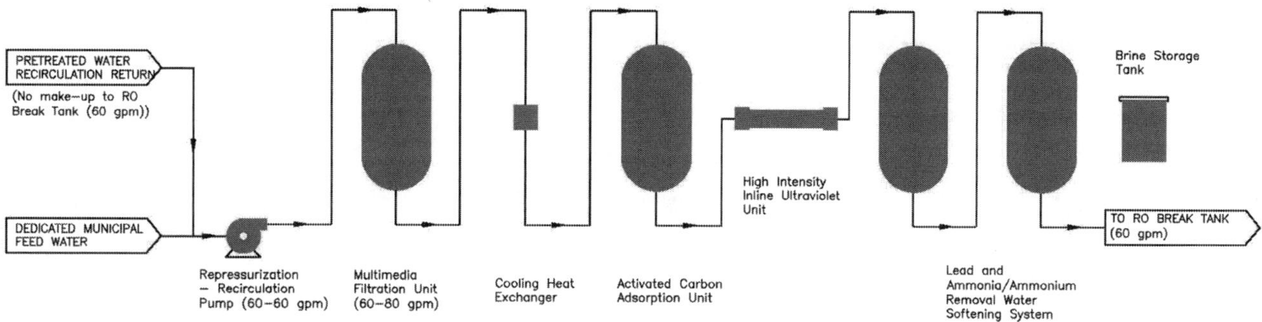

FIGURE 6.1 Process flow diagram—Pretreatment section.

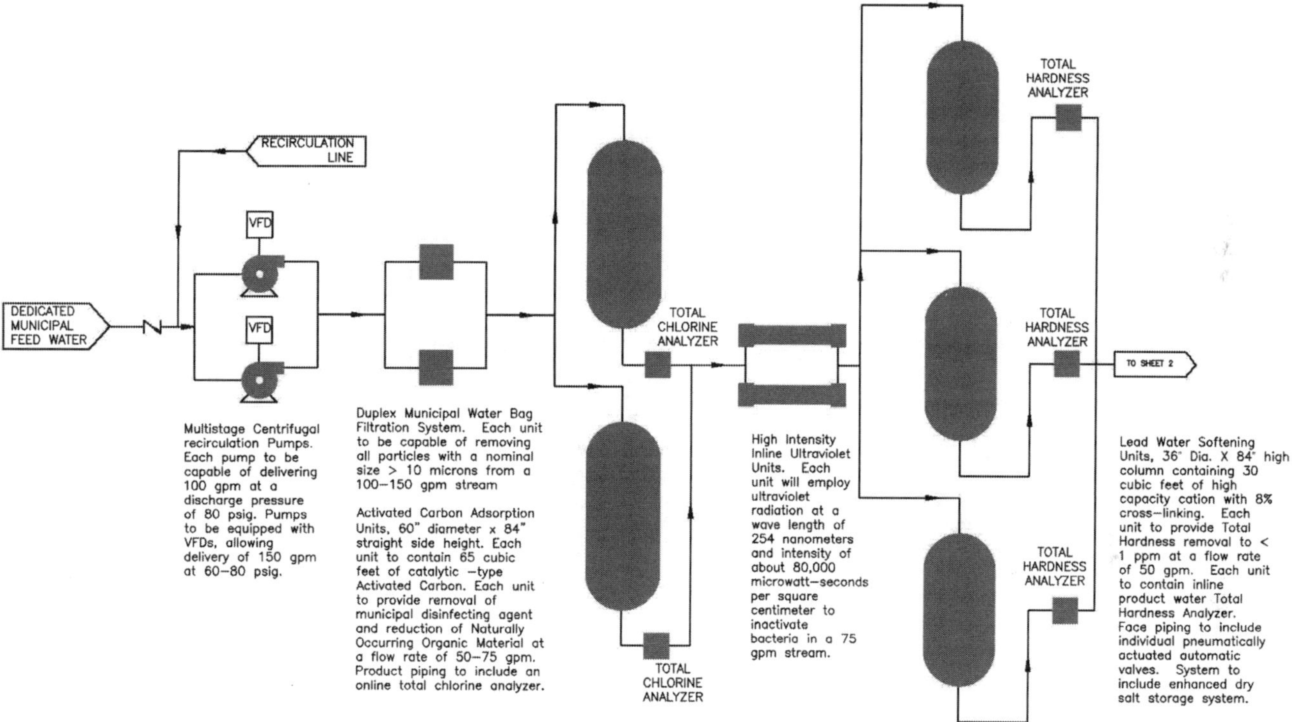

FIGURE 6.2 Process flow diagram with component descriptions—Pretreatment section.

6.6 URS

Information from the Basis of Design as well as the conceptual design phase indicated previously are employed to develop the URS for the pharmaceutical water system. The specific format of the URS is determined by the system "owner". However, as implied by the title of the document, it defines criteria that must be achieved by the system. The URS should definitively set forth system performance and documentation requirements without restrictive details that may change during system procurement and installation. As an example, the URS may state that the system will employ a 20-gpm single pass hot water sanitizable (80°C) reverse osmosis unit using thin film-composite pharmaceutical grade "full-fit" reverse osmosis membranes. However, details such as RO membrane system normal operating pressure, membrane array, percent feed water recovery, and percent ion rejection should not be included. Such items will be set forth in the detailed equipment specification for the project. The intent is to develop the URS in a manner that will not require revision or change throughout the project, including execution of the IQ, OQ, and PQ.

A format for an URS may be structured as follows:

- Document review section with appropriate individual review/approval (name, title, and date)
- Brief purpose of the URS
- General scope of the project
- References
- Responsibilities
- General requirements
- Quality Assurance and Regulatory requirements
- Process Overview

A description of details included in each step is presented following. To provide a concise summary of the content in each section, "extracted" paragraphs from an URS prepared for a USP Purified Water System (also capable of delivering USP Water for Injection), are included to demonstrate the indicated suggested level of detail that should be considered.

- Document Review/Approval
  - Clearly states the type of compendial water system and physical location of the facility. For a facility with multiple systems, it provides descriptive information adequate to describe the location of the system at the facility.
  - Provides "approval" forms for signatures for individuals from regulatory, quality, management, engineering, and/or facility as appropriate.
- Brief Purpose of URS
  - States the compendial water designation, physical location, and references the Validation Master Plan
- General Scope of the Project (typical content following)
  - *Procurement and installation of a 36 gpm Pretreatment System for processing Municipal Water to the facility for feed to a new hot water sanitizable 27 gpm Reverse Osmosis/Continuous Electrodeionization System.*
  - *Procurement and installation of a new hot water sanitizable Reverse Osmosis/Continuous Electrodeionization System with accessories for production of USP/EP Purified Water meeting the chemical criteria set forth in the USP/EP Monographs for Purified Water and internal total viable bacteria Alert and Action Levels including absence of specific gram-negative organisms. Delivered water shall also meet the chemical and microbial attributes for USP/EP Water for Injection.*
  - *Procurement and installation of support accessories, instrumentation, and controls to support the USP/EP Purified Water System.*
  - *Procurement and installation of a USP/EP Purified Water Storage Tank, including accessories, with a capacity of approximately 4,000 gallons.*
  - *Installation of a 2" diameter 316L stainless-steel orbitally welded sanitary distribution system with accessories including a sanitary stainless streel distribution pumping system, and sanitary 316L stainless-steel Heat Exchanger, capable of delivering 70 gpm (maximum) at 50 psig (minimum) and 80°C ±.2°C. Interior to be mechanically polished to Ra 15–20 and electropolished.*
  - *Installation of three point-of-use shell and tube sanitary 316L stainless-steel Cooling Heat Exchangers. Two of the heat exchangers, used for production, shall be capable of cooling 20 gpm of USP/EP Purified Water from 80°C ± 2°C to 25°C ± 2°C using Facility Chill Water at 7°C. The third heat cooling exchangers, used for a laboratory application, shall be capable of cooling 2 gpm of USP/EP Purified Water from 80 ± 2°C to 25°C ± 2°C using Facility Chill Water at 7°C.*
  - *Installation of two points-of-use with manual sanitary diaphragm valves for hot water cleaning of equipment used in production, each capable of delivering 5 gpm at 80°C ± 2°C.*
  - *Complete system documentation including but not limited to Validation Master Plan, Basis of Design, Failure Mode and Effect Analysis (FMEA), Functional Requirement Specification (FRS), Installation, Operation, and Performance Qualification (IQ, OQ, PQ) protocols, Standard Operating Procedures (SOPs), Maintenance Procedures, Spare Parts List, Operating Logs, Training Manual, and Summary Report.*
- References
  - Internal Pharmaceutical Water System Documents
  - Validation Master Plan and Basis of Design with internal document reference number
  - Failure Mode and Effect Analysis (FMEA), Functional Requirement Specification (FRS), Installation Qualification, Operation Qualification, Performance Qualification (IQ, OQ, PQ) protocols, Standard Operating Procedures (SOPs), Maintenance Procedures, Spare Parts List, Operating Logs, Training Manual and Summary Report with "assigned/designated internal document references numbers.
  - USP Pharmaceutical Water References—Current edition of USP including Issues, as appropriate
    - Purified Water Official Monograph
    - *Physical Tests* Section <645>, "Water Conductivity"
    - *Physical Tests* Section <643>, "Total Organic Carbon"
    - *General Information* Section <1231>, "Water for Pharmaceutical Purposes"
  - Codes and Standards (examples)
    - United States Environmental Protection Agency—National Primary Drinking Water Standards (NPDWR), 40CFR141
    - Current Good Manufacturing Practices for Finished Pharmaceuticals, 21CFR210/211
    - NIST, National Institute of Standards and Technology
    - ASTM, American Society for Testing and Materials
    - ASME, American Society of Mechanical Engineers
    - ANSI, American National Standards Institute
    - AWS, American Welding Society

    - NEC, National Electric Code
    - NEMA, National Electrical Manufacturer's Association
    - UL, Underwriter's Laboratory
- Responsibilities—Specific list of all disciplines having responsibility for a definitive list of project items required throughout the life cycle of the pharmaceutical water system
- General Requirements (Example)
    - *The daily demand for USP Purified Water varies from about 6,000 gallons to about 16,200 gallons. System design is based on a Purified Water demand for simultaneous batching operations, once per eight-hour shift, three shifts per day, in a two-hour time period as follows:*
        - *Compounding = 2,400 gallons/hour*
        - *Wash/Rinse = 600 gallons/hour*
        - *Total = 3.000 gallons/hour*
    - *System design parameters consider both the maximum daily demand and maximum instantaneous demand. The maximum instantaneous demand is projected at 3,000 gallons/hour or 50 gpm used for batching operations in two areas simultaneously. Each of the two processing areas would be capable of using 25 gpm of Purified Water for about 2 hours in producing batches and cleaning with a maximum Purified Water volumetric requirement of 1,500 gallons each. A maximum of six batches per day is projected for current and future operations. Further, current and future batching will occur over a seven-day week, three shifts per day. The maximum daily design demand is projected as 18,000 gallons. From these design parameters, the required make-up flow rate from the water purification system, Purified Water Storage Tank and flow rate to the 2" Purified Water Distribution loop can be determined. Based on system design, at maximum Purified Water Distribution Loop draw-off (50 gpm), over a 2-hour time period, the demand is 6,000 gallons. During this two-hour time period, the make-up to the 4,200-gallon Purified Water Storage Tank will be 3,240 gallons (27 gpm for 120 minutes). The volume of Purified Water in the Purified Water Storage Tank will decrease by 2,760 gallons during this "design" two-hour time period. Considering a classical reverse osmosis recovery rate of 75% feed water through the Reverse Osmosis Unit, the pretreated water flow rate of the Purified Water Generation System to the RO Break Tank is 36 gpm.*
    - *The Purified Water Generation System pretreatment section will be designed to deliver 36 gpm to a Reverse Osmosis Break Tank, upstream of the Reverse Osmosis System. Pretreatment design is for Municipal Water meeting the NPDWR with a chloramine concentration range of 1.9 to 3.2 mg/L and average concentration of 2.56 mg/L. Pretreatment components shall be selected to allow continuous Reverse Osmosis Unit operation (36 gpm flow to the Reverse Osmosis Break Tank) when a pretreatment system component is in a backwash or regeneration mode. Multiple components for certain unit operation, configured in parallel, shall be employed to meet this objective. The pretreatment section would operate continuously. When make-up flow is not required to the RO Break Tank with pretreated water, it is recirculated.*
    - *A hot water sanitizable Reverse Osmosis Unit and Continuous Electrodeionization Unit with support components will operate in a continuous manner. Pretreated water will flow to a Reverse Osmosis Break Tank. Water from the Reverse Osmosis Break Tank will flow through the Reverse Osmosis System (RO), Continuous Electrodeionization Unit (CEDI) and support components. Product water from the "RO/CEDI" System will flow to the 4,200-gallon Purified Water Storage Tank when needed for make-up to maintain water level. When the water level in the Purified Water Storage Tank reaches the "High" level set point, RO/CEDI product water will be diverted back to the RO Break Tank. This method of operation provides continuous recirculation, minimizing the bacteria proliferation associated with stagnant conditions and maximizing effective operation of both the RO and CEDI Units.*
    - *The 4,200-gallon USP Purified Water Storage Tank will be of sanitary design constructed of 316L stainless steel, mechanically polished to Ra 15–20 and electropolished. The temperature of the water in the tank will be maintained at 80°C, using a sanitary 316L stainless-steel shell and tube type sanitary heat exchanger installed in the Purified Water Distribution Loop. The tank will be equipped with an outer insulation jacket (side wall and base) surrounded by a 304L stainless-steel shroud.*
    - *Purified Water from the Purified Water Storage Tank will flow to a sanitary centrifugal stainless-steel pump (with "installed spare") with a Totally Enclosed Fan Cooled (TEFC) motor. The power supply to the pump motor will be through a Variable Frequency Drive (VFD), which will maintain a flow rate of 40–50 gpm during non-batching time periods and 70 gpm during batching periods, resulting in turbulent flow in all distribution tubing sections at all times.*
    - *The Purified Water Distribution Loop will consist of 2" orbitally welded 316L stainless-steel tubing. Purified Water will be delivered, as required, to the two Processing Area points-of-use through 316L stainless-steel sanitary shell and tube heat*

exchangers (one in each Processing Area) and a "cleaning" point-of-use in each processing area. Delivery tubing, positioned downstream of each heat exchanger, will direct flow to compounding tanks in processing area. Delivered Purified Water in each compounding tank will be cooled to about 25°C. Rinse/cleaning water in each Processing Area will not be cooled but used at 80°C. Prior to use, the cooling heat exchanger and downstream distribution tubing will be flushed with hot Purified Water for a validated period of time adequate to inhibit microbial contamination of the delivered water. Each of the two additional Purified Water Distribution System points-of-use will be dedicated for washing/rinsing of components used for compounding. Facility Chill Water will be used for cooling media in each heat exchanger. The Purified Water Distribution loop will be serpentine-type, directing return water back to the Purified Water Storage Tank through a spray ball system.

- Quality Assurance and Regulatory Requirements (Example)
  - *The components and accessories in the USP Purified Water Generation System should be designed for operation at a maximum pressure of 100 psig (with the exception of the Reverse Osmosis System). Piping, tubing, instrumentation, pressure vessels, membranes, valves, etc. exposed to high pressure in the Reverse Osmosis System should be selected for operation at a maximum pressure of 450 psig.*
  - *All welding procedures and weld operators shall be in accordance with ASME Code, Section IX and applicable section of "ASME Bioprocessing Equipment 2019" (ASME-BPE2019).*
  - *All piping materials, fittings, and connections shall be in accordance with ANSI B31.1 piping code.*
  - *Material certifications or documentation shall be provided for all 316L stainless-steel vessels and all surfaces in contact with RO/CEDI "Loop" water.*
  - *All sanitary 316L stainless-steel contact surfaces shall be passivated prior to being placed in service per the appropriate Sections of ASTM.*
  - *The USP Purified Water Generation System, Storage System, and Distribution System shall be installed, controlled, and monitored by a single microprocessor-based control panel with display screen located in a designated area adjacent to the USP Purified Water System.*
  - *The USP Purified Water Generation System Pump Motors (Pretreatment Feed Water Pump, RO Break Tank Re-pressurization Pump, and RO High Pressure Pump) shall be equipped with NEMA 4 locally mounted Hands-Off-Auto (HOA) hand selector switch, an indicating light, and auxiliary contacts for motor running indication and remote start/stop. The HOA switch shall be positioned in an area within 3 feet of each pump motor.*
  - *All instruments shall be capable of being calibrated using NIST traceable equipment. Calibration procedures for instruments shall be provided. Where possible, instruments will be provided with factory calibration certificates.*
  - *All Purified Water Generation System instrumentation upstream of the RO Break Tank shall be industrial grade type suitable for the application. Materials of construction for analytical elements shall be 316L stainless steel unless otherwise specified.*
  - *All components and accessories in contact with water within the RO/CEDI "Loop", including the RO Break Tank, should be capable of operation at a periodic hot water sanitizing temperature of 80°C. Documentation shall be provided to verify compatibility with the sanitization temperature. Further instrumentation shall be sanitary type with surfaces in contact with water 316L stainless steel.*
  - *316 L Stainless-steel components, accessories, and instrumentation in the RO/CEDI "loop" shall be provided with Material Certificates with reference to Heat Numbers. A copy of the physical parameters for each Heat Number (Mill Reports) shall be provided.*
  - *Fiberglass-Reinforced Polyester vessels used as part of the Purified Water Generation System Pretreatment Section to the RO Break Tank shall be NSF Certified with indicating label.*
  - *The USP Purified Water Storage Tank shall be designed, constructed, and tested in accordance with this specification and the requirements of the ASME Boiler and Pressure Vessel Code, Section VIII, Division 1, and stamped with the ASME "U" or "UM" symbol and registered with the National Board Number unless otherwise noted.*
  - *All Purified Water Distribution System welding procedures and weld operator qualification shall be in accordance with ASME Code, Section IX and applicable section of "ASME Bioprocessing Equipment 2019" (ASME-BPE2019).*
  - *Visual inspection of orbital welds in the USP Purified Water Distribution System is required. The results of each orbital weld inspection shall be retained and included as part of system documentation. As a minimum, daily orbital welding test coupons shall be generated in accordance with the American Welding Society (AWS) approved welding procedures and ASME-BPE2016. The test coupons shall be retained for project records. Orbital weld machine "print*

outs" for each weld shall also be retained. Qualification certificates for each individual performing orbital welding shall be provided. Each orbital weld in the Purified Water Storage and Distribution System should be identified with an "etched" unique number. Documentation for each orbital weld "number" should include the date, time of day, initial of the certified welder, drawing demonstrating physical location, drawing indicating elevation, drawing indicating 316L stainless-steel "Heat Number" reference, and drawing indicating weld number.

- *The USP Purified Water Distribution Pump(s) shall be equipped with a NEMA 4 rated variable frequency drive (VFD) and disconnect panel, a panel mounted Hands-Off-Auto (HOA) hand selector switch, an indicating light, and auxiliary contacts for motor running indication and remote start/stop. The HOA switch shall be positioned in an area within 3 feet of the pump motor.*
- *All instrument sensors in contact with Purified Water shall be sanitary 316L stainless steel, capable of being calibrated using NIST traceable equipment. Calibration procedures for instruments shall be provided. Where possible, instruments will be provided with factory calibration certificates. Materials of construction for sensing elements shall be 316L stainless steel unless otherwise specified.*
- *An in-line Purified Water Distribution System conductivity measurement system shall be included. The conductivity measurement system shall meet all criteria set forth in USP Physical Tests Sections <645>. Conductivity cells shall be positioned in the supply tubing to the Purified Water Distribution Loop and return tubing from the Loop.*
- *An in-line Purified Water Distribution System Total Organic Carbon (TOC) measurement system shall be included. The TOC measurement system shall meet all criteria set forth in USP Physical Tests Sections <643> and monitor return from the USP Purified Water Distribution Loop.*

- Process Overview
  - The Process Overview should describe the reason for selecting individual components in the pharmaceutical water system. Detailed equipment specification for each component may explain the performance characteristics but may not indicate factors considered when selecting, sizing, and specifying the component. Sections in the Process Overview are provided for each component. A process overview example for an activated carbon adsorption system is presented following.
  - *Tempered filtered municipal water from the Pretreatment Section Cooling Heat Exchanger will flow to an Activated Carbon Adsorption System. Municipal feed water to the system contains monochloramine, a mixture of chlorine and ammonia, as the disinfecting agent. A recent annual report from the municipal water provider indicates that the concentration of monochloramine approaches 4 mg/L, the maximum suggested guideline concentration in the United States Environmental Protection Agency's "National Primary Drinking Water Regulations". Disinfecting agent cannot be present in USP/EP Purified Water. Further, both the downstream reverse osmosis membranes (RO Unit) and ion exchange membranes/ion exchange resin in the Continuous Electrodeionization System will be oxidized by disinfecting agent with resulting loss in membrane integrity and/or ion exchange capacity.*
  - *Subsequently, monochloramine must be removed prior to the RO/CEDI Section of the USP Purified Generation System. Activated carbon will remove monochloramine by a chemical reaction/adsorption process. Monochloramine is physically adsorbed to the surface of the activated carbon media where it chemically reacts with activated carbon producing the chloride ion and ammonia gas. Ammonia gas readily dissolves in water and subsequently reacts with water producing the ammonium ion and the hydroxide ion. Both the capacity and efficiency of activated carbon for removing monochloramine are significantly lower than chlorine (hypochlorous acid and hypochlorite ion). Subsequently, proper system design is critical for successful operation. Further, removal of monochloramine must occur upstream of pretreatment section water softeners because ammonia removal provisions, as the ammonium ion, are necessary prior to the RO Unit. If not removed by the water softening units, ammonia (a gas) will pass through the RO membranes and react with RO product water increasing conductivity and decreasing the measured percent rejection of ions.*
  - *Activated carbon will also remove downstream RO Unit membrane fouling material such as Naturally Occurring Organic Material (NOM) also present in the municipal feed water. The referenced municipal water supplier annual report indicates that the Total Organic Carbon concentration in feed water is 1.88–2.71 mg/L with an average value of 2.17 mg/L. TOC is directly related to the NOM level in municipal water. The indicated TOC/NOM concentration reflects the "surface" or groundwater influenced by "surface" source of municipal water. NOM is reduced by activated carbon through physical attraction and adsorption on the surface of the activated*

carbon media. Reducing the TOC/NOM level in feed water to the downstream RO Unit not only reduces organic fouling of the RO membranes but also reduces the rate of membrane microbial fouling by removing nutrient that will accelerate microbial proliferation on the surface of the RO membranes.

- *The Activated Carbon System will consist of two vertical cylindrical vessels, 36 inches in diameter × minimum 84-inch straight side height arranged for operation in parallel. Each vessel shall contain 25 cubic feet of catalytic-type activated carbon selected specifically for monochloramine removal. The bed depth of activated carbon in the unit will be about 3.5 feet, resulting in 100% "freeboard" in each vessel. At the 36 gpm pretreatment section flow rate, the volumetric flow rate through the Activated Carbon Adsorption System will be 0.72 gpm per cubic foot of activated carbon media, the design value meeting criteria for effective removal of monochloramine. Activated carbon media shall be replaced every six months or when the measured product water Total Chlorine level exceeds 0.1 to 0.2 mg/L. The activated carbon adsorption system shall be designed for periodic replacement of activated carbon media. Each activated carbon unit will be equipped with piping and manual isolation valves for operation, and periodic backwash with final rinse-to-drain.*
- *The feed water piping to each activated carbon vessels shall be equipped with manual isolation valves and liquid-filled pressure gauges. The product water piping from each activated carbon vessel shall be equipped with manual isolation valve, liquid-filled pressure gauge, and sample valve. Sampling for Total Chlorine by "test kit" shall be performed at least once per week. As an alternative, an activated carbon adsorption system Total Chlorine Analyzer may be provided for continuous monitoring of product water.*

6.6.1 System Design With P and IDs

System conceptual design, PFDs, and URS are used to generate system design with P and IDs. This step establishes the capacity and dimensions of all system components. Further, required support accessories such as isolation valves, automatic valves, sample valves, piping/tubing reducers, fittings, regulating valves, modulating valves, etc. are selected for specification. The P and IDs provide the location and type of all instruments, range, alarm indication, and control function. Although revision of the P and IDs may occur during the project, the initial P and IDs provide an excellent tool for development of both the FRS and System Detailed Specification. The P and ID revision cycle shall terminate in an "As Built" set of drawings prior to IQ preparation. Ideally, P and ID changes should not be necessary but items such as fitting sizes, part of the supplier's standard products, may require minor changes from initial to "As Built" condition.

The use of component descriptions and "Notes" on P and IDs can be very helpful. As an example, reverse osmosis units operated in a continuous mode (make-up to a storage tank after polishing ion removal such as continuous electrodeionization or recirculation) contain automatic valves to the tank or recirculation return. If one automatic valve closes without the other automatic valve open, the product side of the RO membranes would experience a rapid pressure increase that could result in loss of membrane integrity. A "Note" on the P and ID (with support statement in the detailed specification) would clearly indicate that one automatic valve must be fully open before the other automatic valve begins to close. Time delays for certain alarm conditions may be required. Finally, certain alarms apply during normal operating conditions but are "defeated" during conditions such as hot water sanitization of the RO/CEDI "Loop". Other alarms are only active during the RO/CEDI "Loop" hot water sanitization cycle. Monitored parameters with dual (normal operation and hot water sanitization) control/alarm set points can also be indicated on P and IDs.

System P and IDs are the primary source of information for initial system evaluation, reference, and evaluation. It is important to understand common symbols and "tag numbers" used on P and IDs. This should include the following:

- Tag Descriptors
  - Figure 6.3 provides an example of a tag with descriptor and number.
  - The upper half of the tag indicates the measured parameter and function of the device.
  - The lower half of the tag contains a unique numeric or alphanumeric designation for the device. A prefix to a sequential numeric code may be used such as the designation (PWS for Purified Water System as an example).
- General Identification Letters and Succeeding Letters
  - The first letter in the tab descriptor indicates the measured parameter such as temperature (T), pressure (P), level (L), flow rate (F), and conductivity or other analysis (A)
  - The succeeding letters indicate the function of the instrument such as I (Indicator), C (Controller), T (Transmitter), or IT (Indicating Transmitter).
- Symbols for Instrument Type and Location
  - Figure 6.4 provides a summary of geometric shapes with inserts employed to designate the type and location of devices.
  - Common symbols that frequently appear on pharmaceutical water system P and IDs include locally mounted pressure gauges ("PI" in a circle), temperature probes/elements ("TE" in a circle), conductivity cell ("AE" in a circle with text at the lower right outside of the circle indicating "cond." or "cond., T").
  - PLC functions are also common on a main/central control panel indicating items such as

temperature indicating controller ("TIC") and pressure indicating controller ("PIC").

- Piping/Tubing and Connection Symbols
  - Lines are used to indicate piping/tubing on the P and ID. Lines do not intersect unless they are physically connected.
  - Electrical signals are designated by a dashed line connecting devices.
  - Pneumatic signals are indicated with diagonal "hash marks".

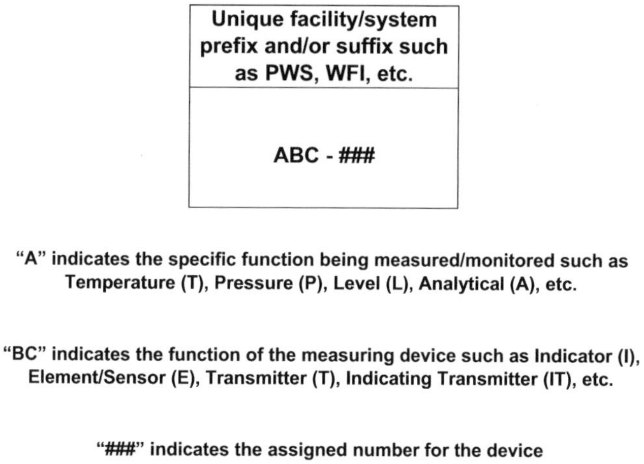

Unique facility/system prefix and/or suffix such as PWS, WFI, etc.
ABC - ###

"A" indicates the specific function being measured/monitored such as Temperature (T), Pressure (P), Level (L), Analytical (A), etc.

"BC" indicates the function of the measuring device such as Indicator (I), Element/Sensor (E), Transmitter (T), Indicating Transmitter (IT), etc.

"###" indicates the assigned number for the device

FIGURE 6.3 Instrument tag number identification convention.

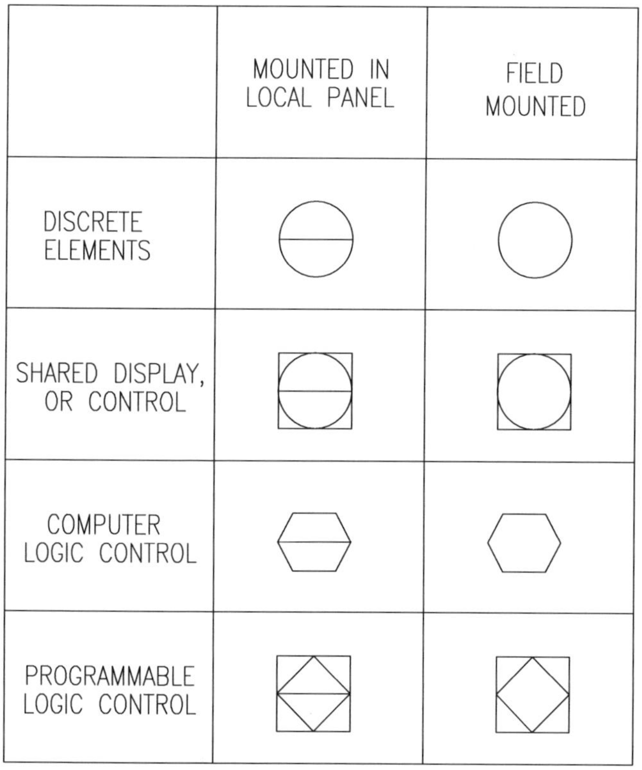

	MOUNTED IN LOCAL PANEL	FIELD MOUNTED
DISCRETE ELEMENTS	⊖	○
SHARED DISPLAY, OR CONTROL	⊖ in square	○ in square
COMPUTER LOGIC CONTROL	⬡ with line	⬡
PROGRAMMABLE LOGIC CONTROL	◇ in square with line	◇ in square

FIGURE 6.4 Drawing instrument identification symbol standard shapes for function and location.

Symbols for valves (ball, butterfly, globe, gate, needle, relief, and diaphragm) may be described in a "Lead Sheet" to the P and IDs. An excellent reference for instrumentation, symbols, and identification, is available (7). Figure 6.5 presents an example of a P and ID for a hot water sanitizable reverse osmosis unit.

6.6.2 Functional Requirement Specification

The FRS should provide a description of the desired control sequence for the system. Although system control requires valves, instruments, and other components from multiple water system components, the FRS can be prepared using a description of individual unit operations and presenting the interface with other components. The P and ID for the system provides an excellent reference during preparation of the FRS.

An example of an FRS section for an activated carbon adsorption system is presented following to demonstrate the discussion and format within a FRS.

The Activated Carbon System consists of three units, configured for parallel operation. The System is capable of providing the desired 120 gpm pretreated water flow with two units in operation. However, during normal operation, all three units will operate in parallel. Each Unit is equipped with a feedwater flow rate meter with 4–20 analog output signal. Each Unit is also equipped with a temperature sensor with analog output signal. A common feedwater pressure sensor with analog output signal and common product water pressure sensor with analog output signal are included. Alarms shall occur for low common feedwater pressure, high common product pressure, low feedwater flow rate to any unit, high feedwater flow rate to any unit, and high common differential pressure through the system. All alarms shall contain a time delay, displayed, and adjustable, on a dedicated screen at the central control panel. Alarms may be defeated (using the highest "access passcode" level for the central control panel) when activities such as maintenance operations are being performed. Further, feedwater pressure, product water pressure, and feedwater flow rate alarms should not occur when a unit is in an ambient backwash cycle (including backwash, settle, and rinse) or hot sanitizing cycle.

Automatic operation of seven individual pneumatically actuated diaphragm valves is controlled for operation, ambient backwash, and hot water sanitization. A dedicated screen on the central control panel shall be provided for adjusting ambient backwash, settle, and rinse elapsed times. A dedicated screen should also be provided for hot water sanitization and cooldown (displacement with cool feedwater). Further, each screen should allow changes to the time between periodic ambient backwash cycles and "lock out" automatic backwash of an upstream multimedia filtration unit, ambient backwash, or hot water sanitization of another activated carbon unit and/or regeneration of a downstream water softening unit when ambient backwash or hot water sanitization of an activated carbon unit is being performed. The upstream municipal water "standby" Repressurization/ Recirculation Pump shall automatically be energized during an ambient backwash cycle. A programmed anticipatory step must be included prior to the post ambient backwash "settle"

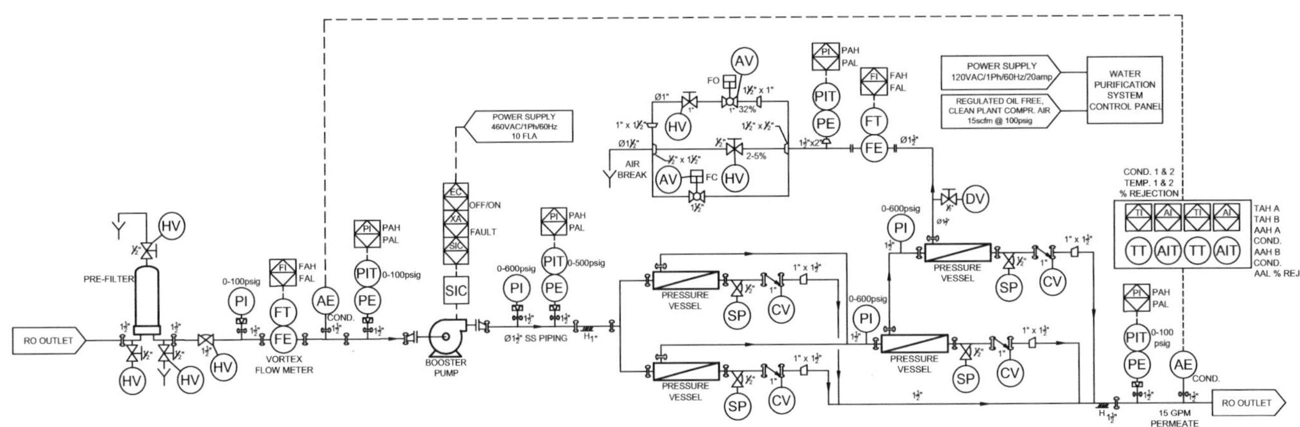

FIGURE 6.5 Example of hot water sanitizable reverse osmosis system components and instrumentation.

step to avoid pumping against a dead head (waste valve closing quickly and VFD trimmed to decrease speed slowly). Controls shall include manual initiation of the ambient backwash cycle for each unit (protected for more than one unit at a time) along with a manual backwash cycle step advance. Manual initiation of the hot water sanitization cycle for each unit shall be included (protected for more than one unit at a time) along with a manual cycle step advance. Although the hot water sanitization cycle will proceed automatically, manual initiation shall be available. The hot water sanitization cycle duration (and cooldown) should be based on the individual temperature sensors on each unit. Hot water sanitization shall be for a time period of 120 minutes once a temperature of 85°C is reached and displacement cooldown for a time period adequate to reduce the column temperature (using the individual temperature sensors) to a value <30°C for a minimum time period of 5 minutes.

An alarm should be energized and display screen exhibited if the automatic hot water sanitization cycle is not completed in the programmed manner indicating "Sanitization Failure—Activated Carbon Unit—". The hot water sanitization screen should include provisions for inserting a time interval between hot water sanitization cycles for each unit, as indicated. If the time interval is exceeded, without manual initiation of a hot water sanitization cycle (for each unit), a banner shall appear on all display screens for the activated carbon units indicating "Sanitization Required—Activated Carbon Unit—". The automated Alarm set points shall be adjustable on a dedicated display screen. Every automatic valve should include manual open, manual close, and automatic provisions on a dedicated screen. Each screen shall include a flashing "Alarm" banner in the upper corner when any alarm is active and a flashing "Output Override" banner in the other upper corner when anything is not in an automatic mode.

6.6.3 System Detailed Specification

Equipment and accessory specifications are a critical tool for defining the details required to meet the expectation set forth in the URS and developed with system design and generation of the system P and IDs. Some specifications are developed "around" standard equipment provided by one or more equipment suppliers. Other specifications provide a minimum description of components by stating the performance requirements of the system. It is important to include adequate details in the specifications to ensure that the components and accessories, including instrumentation and controls, meet the URS objectives while avoiding "over-specification" requiring complete customization of components and support accessories. An example of a suggested level of detail for a Purified Water Storage Tank with accessories is presented following (Table 6.2).

6.6.4 Risk Assessment/Failure Mode and Effect Analysis

A Risk Assessment of the pharmaceutical water system identifies conditions/events that may impact performance of the system as designed and specified. There are several techniques for performing a risk assessment. It is suggested that a "process" approach to risk assessment may not consider the conditions/events that present a degree of risk for successful operation of individual components in a pharmaceutical water system. It is further suggested that a risk assessment can be expanded to explain the consequence and results of potential conditions/events as well as the corrective action. Throughout this chapter, emphasis has been placed on the merits of project "team" input for generation of validation documents, particularly with regard to understanding both regulatory and engineering items. Project "team" preparation/generation of a risk assessment can increase awareness of the multiple items that could impact long-term successful system operation.

An example of sections from a risk assessment for an inline ultraviolet sanitization unit positioned downstream of an activated carbon adsorption unit and upstream of a water softening system is presented following. The general format considers three factors for each condition/event; potential,

TABLE 6.2
Example of Detailed Specification

	USP Purified Water Storage Tank and accessories	
Category	**Specific attribute**	**Additional comments**
Tag No.	PWS-TK-421	USP Purified Water Storage Tank
Model No.	TBD	From tank manufacturer
Quantity	One (1)	—
Sanitary design criteria	Sanitary design	Includes all fittings and manway
Head type	Domed	Flanged and dished
Orientation	Vertical cylindrical	With adjustable support legs
Top type	Domed	Per ASME code
Bottom type	Domed (inverted)	Per ASME code
Reference codes and standards	a. ASME Section VIII, Division 1, 2010 Edition through A11 Addendum (General, Materials of Construction, Welding, Vessel Design and Construction) b. AWS—American Welding Society c. ASME/ANSI F2.1—Food, Drug, and Beverage Equipment d. United States Food and Drug Administration, Department of Health and Human Services, 21CFR Parts 210 and 211 e. ASME Bioprocessing Equipment 2019	
Code stamp data	a. Serial No. TBE b. Design pressure (30 psig and full vacuum) c. Design temperature (350°F) d. National board number—TBE	
Required vessel certificates	a. Completed U-1 Form b. Model No. and Serial No. c. Wetted material of construction: 316L stainless steel d. Jacket material of construction: 316L stainless steel e. Outer shroud material of construction: 304L stainless steel f. Interior surface finish: Ra 15-20 and electropolished g. Outer shroud finish: Ra 30	
Welder qualification	ASME	Certificates required
Drawings	Submittal and As Built	a. Detailed drawings with elevation, dimensions, plan view, and ferrule weldment details. b. Provide "Notes" and complete bill of materials with code reference to drawing
Tank operating parameters	Temperature and pressure range	a. 80°C Normal Operation b. Slight Vacuum to 2 psig
Insulation	2" Thick, Chloride-Free—Shell and Bottom of Vessel	Surrounded by 304L stainless-steel shroud
Inside diameter	84"	Tank without shroud
Outside diameter	TBD	With outer shroud
Straight side height	144"	-
Overall height	TBD	Without accessories
Capacity	3,400 gallons	
Bottom discharge height above floor level	18" minimum	Not to exceed 24"
Sanitary ferrules—top height	a. 1 each 4" for compound rupture disc holder, outer radius b. 5 each 2" sanitary ferrules (Makeup from RO/CEDI loop, upper-level pressure sensor, three spares c. 1 each 4" sanitary ferrule for purified water distribution loop return spray ball assembly	
Sanitary ferrules—bottom head	1 each 3" center mount for feed to purified water distribution loop pump(s)	
Lower sidewall fittings in "alcove"	a. 2" for lower-level pressure sensor b. ½" FNPT with thermowell—0.26" diameter × 9" insertion length c. 18" Circular 316L stainless-steel sanitary manway with hinge, gasket, and multiple closure assemblies	
Accessories—hydrophobic vent filtration system	a. 316L stainless-steel "dome-type" housing for containing a 20" long hydrophobic vent filter, single open end (SOE), double O-ring seal with locking tabs, Ra 15-20 interior finish and electropolished b. Hydrophobic ven filter element—Teflon Membrane, 0.2 μ, polypropylene "cage" c. Thermostatically controlled electrical heating blanket positioned around housing with set point <225°F	

(*Continued*)

TABLE 6.2
(Continued)

	USP Purified Water Storage Tank and accessories	
Category	**Specific attribute**	**Additional comments**
Accessories—compound rupture disc assembly	a. 3" Compound disc assembly mounted in 316L stainless-steel holder with 4" sanitary ferrule "inlet/outlet" connections b. Discs to rupture at about 7 psig pressure and about 35" of water vacuum c. Disc to be equipped with integrity sensor d. Discharge of assembly to be connected to 4" 316L stainless-steel tubing with double elbow assembly to "safe area" at the base of the tank	
Accessories—temperature probe/transmitter	Temperature element with integral transmitter	Positioned in thermowell on lower straight side of tank with analog signal to the central control panel
Accessories—level pressure probes/transmitters	2 each 2" sanitary ferrule 316L stainless-steel pressure sensors with pressure transmitters	Positioned to 2" sanitary ferrules at top and lower straight side of tank providing analog signal to the central control panel for level control (differential pressure)
Accessories—spray ball assembly	Positioned in top center mounted 4" sanitary ferrule	Designed for maximum return flow rate of 60 gpm, 180°coverage, removable with locking "pin"
Minimum required documentation	a. Complete Tank Drawings with front, back, left side, right side, top, and bottom views b. Complete Bill of Materials for all tank components c. Welder Certificates for tank d. U-1 Data Sheet for tank e. Material Certificates for all tank components with heat numbers and corresponding mill analysis f. Material Certificates for all accessories with heat numbers and corresponding mill analysis g. Instrument List with manufacture, part number, and serial number h. Certificate of Compliance and Accuracy—Temperature transmitter i. Certificate of Test—Pressure transmitters j. Complete spare parts list with OEM, model numbers k. Operating and maintenance manual l. Installation procedure, preparation, and integrity test guide—hydrophobic vent filter m. Test Certificate—Rupture Disc	

detection, and impact. The factors are assigned a numeric value from 1–10 with 1 the lowest risk and 10 the greatest. The individual potential, detection, and impact values can be multiplied to determine a total risk factor for the stated conditions/event. During risk assessment assignment of a numeric value for each risk factor, the consequence results on system operations should be considered. Further, the corrective action required to resolve the conditions/event should be discussed. This format presents a responsive Failure Mode and Risk Analysis. The example presented following includes both consequences/results and corrective actions for each condition/event presented (Table 6.3).

6.6.5 Review of Equipment Suppliers and Subcontractors Submittal Packages or Proposals

In response to requests for quotation based on equipment specifications, P and IDs, and general layout drawing(s), water purification equipment manufacturers, tank manufacturers, distribution loop tubing installers, instrumentation and controls provider, and subcontractors (electrical, mechanical, etc.) should provide either submittal packages or detailed proposals. The submittals and proposals shall be reviewed to verify compliance with the specifications. It may be necessary to prepare comments and questions, asking the vendors and contractors to either revise or clarify the scope of supply.

6.6.6 Equipment Supplier and Contractor "Turnover Package"

The "turnover packages" received from equipment suppliers and contractors for the system are a critical part of not only the validation process but also long-term successful system operation. There can never be too much data with regard to details of component, accessory, and installation. Material contained in the turnover packages should be stored in a secure location with limited access. If an item in the system fails during the life cycle of a system, information in the turnover package provides valuable information for replacement of the item with the same model number or a like-to-like equivalent. This significantly expedites preparation of "Change Control" for item replacement. Further, information in the turnover packages may also provide a valuable

TABLE 6.3
Risk Assessment/FMEA Line Item Examples

Condition/event	Consequence/result	Corrective action	Potential	Detection	Impact
Ultraviolet lamp failure	*Ultraviolet lamp failure will result in decreased UV radiation intensity within the sanitizing chamber. This will decrease the ability to inactivate bacteria resulting in higher pretreatment system and RO break tank make-up water total viable bacteria levels.*	*Shutdown unit and replace lamp(s) as soon as possible. Verify that lamp operation is correct after replacement.*	*2*	*1*	*4*
Power supply ballast failure	*Two ultraviolet lamps receive electrical power from a single transformer. Because the unit only contains two lamps, failure of the ballast will inhibit ultraviolet radiation and the ability to inactivate bacteria. Pretreatment system and RO break tank make-up bacteria levels will increase. A "lamp out" alarm will sound.*	*Acknowledge the "Lamp Out" alarm. Shutdown UV Unit. Replace ballast. Restart UV Unit and verify proper operation. Conduct sampling of the pretreatment system to verify that bacteria levels have not significantly increased.*	*3*	*1*	*4*
Quartz sleeve failure	*Quartz sleeve failure results in water contact with the ultraviolet lamps and electrical circuit. An electrical "short" should occur. However, the potential for mercury introduction from the lamp to flowing water through the sanitizing chamber into the pretreatment system feed water is a concern.*	*Shutdown UV Unit and remove broken sleeve(s). Note that sections of the broken sleeve may not be quickly removed from the sanitizing chamber. The sections may remain until the next facility extended shutdown when they should be removed. Verify that "water hammer" was not the cause of sleeve failure. If water hammer is noted, thoroughly investigate upstream components, identify, and correct.*	*1*	*1*	*8*
Ultraviolet lamp O-ring failure	*The O-ring provides a sealing mechanism between the quartz sleeve and sanitizing chamber connection. O-ring failure can result in introduction of contamination and water leakage to the electrical supply for the lamps. Contamination may result in product water chemical and/or microbial impurity level increase. Water leakage to the electrical supply may result in loss of electrical power to the Unit with potential damage to the Unit.*	*Shutdown UV Unit. Drain sanitizing chamber, remove compression nut, sleeve, and O-ring. Inspect O-ring in an attempt to identify the reason for failure. Replace sleeve, O-ring, and stainless-steel compression nut. Replace drain, fill and vent sanitizing chamber, and check for leaks. If no leaks are detected, replace UV lamp and restart UV Unit.*	*2*	*2*	*4*

reference used in response to questions from individuals auditing the system.

A list of suggested items in the turnover packages should include the following:

- Equipment Manufacturer
  - Catalog information
  - Individual component descriptive information
  - Manufacturer, location, contact information
  - Model number
  - Serial number
  - OEM part numbers
  - Factory Calibration Certificates
  - Factory QA/QC Certificates
  - Mill certificates with chemical and physical test results
  - Component Operating Manuals
  - System Operating Manual
  - System Maintenance Manual
  - Component Maintenance Manual
  - Spare Parts List
  - Equipment Manufacturer's Drawings
  - Component Piping Drawings
  - Component Control Drawings
  - Component Electrical Drawings
  - General Arrangement Drawing
- Mechanical, Electrical, Piping Installers
  - Support component information
  - Information, catalog sheets, manufacturer, model number, and serial number for accessories such as steam traps, pressure regulators, backflow preventers, motor starters, electrical disconnects, and valves

- Power wiring drawings
- Control wiring drawings
- Mechanical interface drawings
- Drain piping drawings
- Utility interface drawings

- Stainless-Steel Tubing Installer
  - Material Certificates with Mill Heat Report including chemical and physical test results
  - Individual welder ASME qualification certificate (current) for welding
  - Orbital weld test coupons
  - Boroscopic test results for orbital welds
  - Orbital welding machine printout for each orbital weld
  - Isometric Drawings
    - Weld numbers traceable to orbital weld machine printout
    - Elevation of tubing verifying slope and ability to drain the system
    - Individual Heat Numbers for all tubing and fittings
- Instrumentation and Controls
  - Calibration certificates
  - Mill certificates
  - Instrument and control manufacturer, model number, and serial number
  - Processor information
  - Spare parts list
  - Complete hard copies of "Screen Shots"
  - Point-to-point electrical drawing with individual wire numbers
  - Ladder logic diagrams with full annotation
  - Alarm list
  - Bill of Materials

6.6.7 Instrumentation and Alarm List—Details

As indicated, an Alarm List should be provided from the Instrumentation Controls supplier(s). This list should include the alarm description, any alarm prerequisites, alarm "trigger", and any related interlocks. An example of items that should be included in the alarm list is presented following (Table 6.4).

The Instrument List shall provide System Tag Number, manufacturer, model number, serial number, description, range of measurement, materials of construction, fitting size, and calibration certificate reference.

6.6.8 Factory Acceptance Test (Optional)

A FAT may be used to verify system components and performance as outlined in the Equipment Specifications and Functional Requirement Specification. The FAT may also be used to verify line items in the Installation Qualification and Operation Qualification. Considering that components in the system will be disassembled, shipped, and installed at a facility, the system owner may wish to verify items but perform entire IQ and OQ execution of the installed system. Further, exposing water to items such as activated carbon, reverse osmosis membranes, and membrane filters may result in elevated bacterial levels subsequent to installation.

TABLE 6.4
Instrument and Alarm Listing

PLC reference no.	Alarm description	Alarm prerequisite	Alarm "trigger"	Interlocks
26	(LALL-622) Low-Low Tank, TK-401, Level	Distribution Pump, PU-402, Operational	Set point, Low-Low TK-401 Level	Tank Automatic Fill Valve, AV-313 and Distribution Pump, PU-402
27	(LAHH-622) High-High Tank, TK-401, Level	Distribution Pump, PU-402, Operational	Set point, High-High TK-401 Level	Tank Automatic Fill Valve, AV-313
28	DI-623 Tank, TK-401, Rupture Disc Failure	Distribution Pump, PU-402, Operational	Rupture Disc Fault	Tank Automatic Fill Valve, AV-313

6.6.9 Site Acceptance Testing (Optional)

Site Acceptance Testing (SAT) may be performed as part of system start-up and commissioning. A formal SAT document may be developed to outline the scope of the testing. The primary purpose of an SAT is to verify operation and installation. This can provide excellent input for preparation of the IQ, OQ, SOPs, Maintenance Manual, and Operating Logs.

6.6.10 Installation Qualification—Completion of Preparation

Preparation of the IQ may have commenced prior to system arrival at the facility. As components and accessories arrive at the facility and are installed, preparation of the IQ can be completed. In fact, visual verification of items such as model numbers and serial numbers can minimize deviations observed during IQ execution. It is beneficial to include as much descriptive and support information in the IQ with inclusion of Turnover Package information. As an alternative, items such as accessory operating and maintenance manuals, catalog sheets, and catalog information may be included as an attachment to the IQ. The IQ shall verify that all components and support accessories specified for the system are present. Further, the IQ confirms that specified documentation for components, accessories, and installed items such as orbitally welded stainless-steel tubing are as specified. An example of an IQ section for a hot water sanitizable reverse osmosis unit is presented following for reference (Table 6.5).

TABLE 6.5
Installation Qualification Example Page—Hot Water Sanitizable Reverse Osmosis Unit

Installation verification item —description	Test method/ reference	Verification	Initials/date
Verify, using manufacturer's data, that Reverse Osmosis Unit, PWS-RO-207, feedwater Sanitary Conductivity/Temperature Sensor, PWS-AE-326A, has a Cell Constant of 0.1 cm^{-1}.	*Manufacturer's data—attachment 127*		
Verify, using manufacturer's data, that Reverse Osmosis Unit, PWS-RO-207, feedwater Sanitary Conductivity/Temperature Sensor, PWS-AE-326A, electrode material is 316L stainless steel.	*Manufacturer's data—attachment 127*		
Verify, using manufacturer's data, that Reverse Osmosis Unit, PWS-RO-207, feedwater Sanitary Conductivity/Temperature Sensor, PWS-AE-326A, maximum pressure/temperature rating is 150 psig at 311°F and 450 psig at 77°F	*Manufacturer's data—attachment 127*		
Verify, by direct observation and manufacturer's data, that Reverse Osmosis Unit, PWS- RO-207, feedwater tubing downstream of Sanitary Conductivity/Temperature Sensor, PWS-AE-326A, contains a 2" sanitary ferule short outlet 316L stainless-steel tee with 2" sanitary ferrule fitting (branch side of tee), using a 2" sanitary ferrule gasket and clamp.	*Visual*		
Verify, by direct observation and manufacturer's data, that Reverse Osmosis Unit, PWS-RO-207, feed water 2" Sanitary Automatic Divert-to-Drain Diaphragm Valve, PWS- AV-329, is Manufacturer Model No. XXXXX, Serial No. YYYYYYYY.	*Visual*		
Verify, using manufacturer's literature, that feedwater 2" Sanitary Automatic Divert-to-Drain Diaphragm Valve, PWS-AV-329, is Manufacturer's 2" forged diaphragm-type, with sanitary ferrule end connections, pneumatic actuator, FDA/USP MPTFE/EPDM diaphragm, 316L stainless-steel wetted surfaces, and 15 Ra electropolished internal finish, Heat No. 411672.	*Manufacturer's data—attachment 128*		

6.6.11 Preparation of Standard Operating Procedures

System SOPs must be prepared, reviewed, and approved by the project "team" prior to execution of the OQ. The SOPs establish the precise method of system operation. When preparing SOPs, it is important to remember that the document will require verbatim compliance. In other words, excursions related to system operation using the SOPs should not occur throughout the life cycle of the system.

To avoid inclusion of material not required for normal operation, it is suggested that SOPs not include maintenance procedures. These can be included in a separate document or in a clearly designated separate section of the SOPs indicated as maintenance SOPs. The SOPs should be brief, concise, and highly specific. Lengthy SOPs containing excess material or commentary make it more difficult for operating personnel to focus on required normal operation steps and the sequence of the steps.

Many pharmaceutical water systems employ a central system controller with programmable logic controller (PLC) and display screen. The SOPs should include the detailed screen navigation required for normal operation. This shall include the ability to navigate to a "system overview" screen as well as display screens providing detailed information for portions of the system that physically cannot be shown on the overview screen.

SOPs should provide a sequential list of manual operations such as opening and closing valves, and, where appropriate, PLC operations such as starting or stopping pumps, for system start-up, normal operation, and shutdown. Critical "support" items necessary for successful operation of the system, such as housekeeping practices at points-of-use to avoid back contamination of the distribution system at points-of-use, should be included. Water purification equipment manufacturer's or individual component supplier's operating procedures should be incorporated into the SOPs as appropriate. This avoids the use of multiple documents for normal operation and incorporates procedures into a single integrated document. Further, it eliminates "generic" material that is often present in manufacturer's procedures that unnecessarily increases the complexity for operating personnel.

6.6.12 Preparation of a Maintenance Manual

The Maintenance Manual should contain both short- and long-term required maintenance items. The short-term maintenance items should be physically located at the front of the Maintenance Manual because they will be used more frequently. Short-term maintenance items should include tasks such as filter cartridge replacement and/or hot water sanitization of a reverse osmosis system with support components. Intermediate maintenance items should include tasks such as activated carbon replacement and reverse osmosis membrane cleaning or replacement. Long-term maintenance items should include tasks such as replacement of media in a multimedia filtration unit. The beginning of the Maintenance Manual should contain a detailed Periodic Preventative Maintenance Schedule exhibiting the specific maintenance items that must be performed and the maximum frequency of performance (8, 9). Periodic calibration of instrumentation, generally performed annually, should be included as part of the periodic preventative maintenance. The Preventative Maintenance Schedule may be used to generate facility "work orders" to insure execution. For many facilities, "shutdowns" are generally programmed during the mid-summer and end of year time periods. Facility shutdown "windows" should be considered when establishing the execution period for biannual, annual, or long-term maintenance items.

Maintenance frequency for components and accessories should not exceed the suggested/established time interval stated in the manufacturer's recommendations. The maintenance program, as indicated, should be proactive, eliminating system shutdown or unacceptable variations/excursions of the system, components, or accessories. The frequency may be changed subsequent to operation based on trending data. If, for example, data trending indicates the activated carbon media is required every 13–14 months, annual replacement may be considered to avoid potential breakthrough of municipal disinfecting agent and increase in the concentration of naturally occurring organic material in feedwater to downstream components such as reverse osmosis.

6.6.13 Operating Logs—Master Sheets

Operating Logs should include the display from both field instruments and the central control panel display screen(s) as indicated in the examples following (Tables 6.6 and 6.7). It is critical that the "Acceptable Range" be included. Values outside the "Acceptable Range" must be reported to supervisory personnel immediately.

6.6.14 Training Manual and Training Sessions

System start-up has occurred. The system may be operating continuously or periodically in preparation for execution of the IQ and OQ. The Training Manual can be prepared at this time with training sessions conducted subsequent to OQ preparation and execution. Multiple training sessions may be appropriate with project "team" members. It may be beneficial for individuals participating in OQ preparation to understand/review the function and design of the system components and technology. This can be achieved using material from the Training Manual for presentation of a "preliminary" training session.

As discussed on several occasions in this chapter, the system may consist of multiple water purification unit operations, accessories, storage system(s), and distribution system(s). A Training Manual prepared in a Power Point presentation sequentially discussing each unit operation can be highly effective, because it focuses on the function of each component and ultimately demonstrates the reason for system monitoring and control.

Subsequent to IQ and OQ execution, the Training Manual should be used to conduct training sessions for the project

TABLE 6.6
Example of Log Sheet Section for Field Instruments

Description (direct reading from field instruments)	Tag no.	Entry	Acceptable range	Initials/ date
RO Break Tank Repressurization Pump, PU-412, feed water pressure	PI-501	psig	>2 mm Hg	
RO Break Tank Repressurization Pump, PU-412, discharge pressure	PI-502	psig	30–60 psig	
Oxidizing Ultraviolet Unit, UV-414, radiation intensity	AI-503	%	>80 %	
Oxidizing Ultraviolet Unit, UV-414, lamp run time	AI-504	hours	N/A	
Oxidizing Ultraviolet Unit, UV-414, "lamp out" status	AI-505	On/Off	Off	
Oxidizing Ultraviolet Unit, UV-414, product water pressure	PI-506	psig	30–60 psig	
One Micron Cartridge Filtration System, CF-415, product water pressure	PI-507	psig	30–60 psig	
RO System, RO-417, High Pressure Feed Water Pump, PU-416, discharge pressure	PI-509	psig	125–225 psig	
RO System, RO-417, first array feed water pressure	PI-520	psig	125–225 psig	
RO System, RO-417, second array feed water pressure	PI-521	psig	100–200 psig	
RO System, RO-417, third array feed water pressure	PI-522	psig	70–150 psig	
Post RO In-line Ultraviolet Unit, UV-420, radiation intensity	AI-530	%	> 80%	
Post RO In-line Ultraviolet Unit, UV-420, Run Time	AI-531	hours	N/A	
Post RO In-line Ultraviolet Unit, UV-420, "lamp out" status	XA-532	On/Off	Off	
CEDI-422 dilute feed water pressure	PI-540	psig	40–60 psig	
CEDI-422 concentrate feed water pressure	PI-541	psig	20–40 psig	
CEDI-422 product water pressure	PI-542	psig	10–30 psig	
CEDI-422 waste pressure	PI-543	psig	10–40 psig	
CEDI-422 waste flow rate	MTR-544	gpm	1–2 gpm	
0.1 Micron Final Filtration Unit, MF-423, feed water pressure	PI-550	psig	10–30 psig	
0.1 Micron Final Filtration Unit, MF-423, product water pressure	PI-551	psig	10–30 psig	

TABLE 6.7
Example Log Sheet—Display Screen Input

Description (display screen reading)	Tag no.	Entry	Acceptable range	Initials/ date
RO Break Tank, TK-410, level	LE-515	%	60%–90%	
RO Break Tank, TK-410, water temperature	TE-516	°C	20°C–30°C	
Plant Steam Supply Modulating Valve, position	PCV-613	%	0%	
RO Break Tank Repressurization Pump, PU-412, speed	VFD	%	60%–80%	
RO System, RO-417, feed water flow rate	FE-508	gpm	90–100 gpm	
RO System feed water conductivity and temperature	AE-510A	μS/cm	2–350 μS/cm	
	TE-510A	°C	25-30°C	
RO System, RO-417, feed water pressure	PE-511	psig	30–60 psig	
RO System, RO-417, High Pressure Pump, PU-208, speed	VFD	%	70%–90%	
RO System, RO-417, High Pressure Pump, PU-416, discharge pressure	PE-509	psig	125–225 psig	
RO System, RO-417, wastewater flow rate	FE-512	gpm	20–25 gpm	
RO System, RO-417, wastewater pressure	PE-513	psig	70–150 psig	
RO System, RO-417, product water pressure	PE-514	psig	40–60 psig	
RO System Product water, conductivity and temperature	AE-510B	μS/cm	2–12 μS/cm	
	TE-510B	°C	20°C–30°C	
RO System, RO-417, percent ion rejection	AE-510 A/B	%	80%–90%	
CEDI-422 Product Water, conductivity and temperature	AE-545	μS/cm	≤1.5 μS/cm	
	TE-545	°C	20°C–30°C	
0.1 Micron final Filtration System, MF-423, product water conductivity/temperature	AE-546	μS/cm	<0.25 μS/cm	
	TE-546	°C	20°C–30°C	
"Override" displayed	Yes/No	Red banner	No	
"Alarm" displayed	Yes/No	Red banner	No	
"RO Sanitization Required" displayed	Yes/No	Red banner	No	
NOTES:				

team and any other individuals interfacing with the system. The training sessions should be documented with a "sign-off" sheet containing the name, date of presentation, signatures of attendees, and signature of individual presenting the material. Further, the session(s) should be recorded for future retraining or new employee training.

6.6.15 Operation Qualification—Completion of Preparation

The purpose of the OQ is to verify normal operating conditions for the system including items such as periodic backwash of particulate removal filters and regeneration of water softening units. Further, the OQ should create, using adequate description, all potential transient or excursion conditions, verifying that the system, as designed and installed, responds to the conditions by generating an alarm and, if appropriate, terminating operation. Unlike the IQ, where inspection can focus on an individual component, the OQ must address the interaction of various unit operations within the system. Evaluation of intracomponent control and functionality must be verified for the entire system. Finally, indicated transient conditions should not be produced by introducing chemical and/or microbial contaminants into the system. As an example, a "high conductivity" alarm should not be generated by intentionally adding high conductivity water to the system but by changing the high conductivity set point.

The Turnover Package from the system instrumentation and control supplier shall contain a detailed table indicating the method of simulating transient/alarm conditions with the anticipated system response. This table shall be used during generation and subsequent execution of the OQ. An example of a partial OQ section is presented following (Table 6.8).

Once prepared, reviewed, and approved, the OQ can be executed subsequent to execution and approval of the IQ and SOPs.

6.6.16 Performance Qualification—Preparation and Execution

The PQ for a pharmaceutical water system consists of an intense sampling and analysis program conducted after execution of the IQ and OQ for the system. Generally, it is suggested that this program be conducted for a minimum period of four to six weeks. During this time period, samples should be obtained from each point-of-use within the system at least

TABLE 6.8
Operation Qualification—Example Section

High Reverse Osmosis Unit Feed Water Conductivity (AAH-721A)

Initial Conditions: The Purified Water System shall be in "normal operating mode".

Initial Status Verification

- Audible alarms are silent
- No active alarms indicated on the system display screen
- Purified Water Generation System operational

Operational Evaluation

Observation of displayed value	Observed results	Initial / date
Navigate to the RO System Overview Screen. Record the **displayed** RO Feed Water Conductivity value (μS/cm), AE-721A.		

Using the "Alarm Summary" Screen, navigate the up/down control arrows in the lower left portion of the screen, highlight the (AE-721A) High Reverse Osmosis System Feed Water Conductivity.

Observation of set point value	Observed results	Initial / date
Note the **set point** value (μS/cm) for (AAH-721A) High Reverse Osmosis System Feed Water Conductivity.		

Depress the Set point Entry display on the screen. A pop-up screen will appear with numeric display. Enter a conductivity value (μS/cm) for (AAH-721A) less than the conductivity value **displayed and recorded** for the operating value for AE-721 A above and press the "return" key in the lower left-hand corner.

Verification of alarm	Actual results	Initial / date
Verify that a red-colored Active Alarm Screen appears indicating: (AAH-721A) High Reverse Osmosis System Feed Water Conductivity. Verify that an audible alarm is energized.		

On the Active Alarm Screen, depress the "Acknowledge Alarm" display. Navigate to the Alarm History Screen.

Verification of Alarm Silence but displayed	Actual results	Initial / date
Verify that the audible alarm is silenced. Verify that the Time, Date, and (AAH-721A) "High Reverse Osmosis System Feed Water Conductivity" appears at the top line on the display screen.		

Navigate to the "Alarm Summary" Screen and use the up/down control arrows in the lower left portion of the screen to highlight the (AAH-721A) High Reverse Osmosis System Feed Water Conductivity line. Depress the Set point Entry display on the screen. A pop-up screen will appear with numeric display. Enter the conductivity (μS/cm) value for (AAH-721A) High Reverse Osmosis System Feed Water Conductivity Set point recorded previously (above) and depress the return key.

Verification of Alarm Silence and not displayed	Actual results	Initial / date
Verify that there is no display on the Active Alarm Screen Verify that there is no display for (AAH-721A) High Reverse Osmosis System Feed Water Conductivity.		

once per working day. Further, samples obtained from the feed water to the system, as well as the product of all components within the system, should be obtained daily to at *least* once per week. There are important issues that should be addressed regarding proper PQ preparation and execution.

- Although not discussed previously, validation of the pharmaceutical water system can be conducted in a retrospective, concurrent, or prospective manner. It is suggested that retrospective validation, performed subsequent to use of water from the system, is similar to "new" system validation with regard to documentation and document preparation and IQ, OQ, and PQ execution. Concurrent validation implies that product may be manufactured during the PQ execution, perhaps after receipt of two weeks of acceptable data. Any product manufactured during concurrent validation should not be released for distribution (quarantined) until all analytical results for samples collected during PQ execution have been received, reviewed, and accepted. Finally, during prospective validation, water is not used for product until all analytical results for samples collected during PQ execution have been received, reviewed, and accepted. It is important to note that prospective validation requires water at each point-of-use that simulates the actual demand conditions during normal operation. Further, delivery practices at each point-of-use should mimic normal operating practices such as the use of delivery hoses and flow totalizing meters.
- Delivered water from all points-of-use in the system should be sampled at least once per working day. The purpose of the PQ is to verify, through appropriate analytical techniques, that the chemical, bacteria, and bacterial endotoxin (for Water for Injection Systems) levels meet the Purified Water (or Water for Injection) *Official Monograph* Specifications *and* that the established total viable bacteria Alert and Action Limits are not exceeded. As discussed earlier, the total viable bacteria Alert and Action Limits for Purified Water System are "product and process dependent." Subsequently, the Alert and Action Limits *will* vary for each system.
- Once the sampling schedule has been established, it is appropriate to consider the nature of the required analysis. The first item that will be addressed relates to the chemical parameters stated in the individual monographs for USP Purified Water and Water for Injection. For USP, the chemical specifications include conductivity and TOC. For EP, the chemical specifications include conductivity, TOC, and nitrates. It is important that pharmaceutical manufacturers clearly recognize the requirement to perform analysis in accordance with the criteria for the part of the world where the product will be distributed.
- It is suggested that point-of-use monitoring for conductivity and TOC employ online monitors/analyzers. Systems should be provided with online conductivity measurement of feed water to a distribution loop and return water from the distribution loop as well as an online TOC analyzer to monitor the distribution loop. During execution of the PQ, "grab" samples of delivered water should be collected for conductivity and TOC on a daily basis. Although it is suggested that online conductivity and TOC results should be the measurement used after successful PQ execution, the intense PQ monitoring program provides an excellent method of comparing "grab" delivered water results with online analytical data. Regulatory investigators focus on "delivered" water quality at points-of-use. The water quality from accessories such as mass totalizing meters with transfer hoses may not be the same as the water quality in the distribution loop.
- During PQ execution, samples should be collected for bacteria. As indicated, each point-of-use should be sampled daily and evaluated using an appropriate microbial monitoring technique. The USP *General Information* Section <1231> provides *suggested* methodology for determining bacteria in Drinking Water, Purified Water, and Water for Injection. USP *General Information* Section <1231> also provides an excellent discussion of factors that should be considered when selecting a bacteria enumeration method (culture media, incubation time period, and incubation temperature). Regulatory inspectors may comment about the volume of water used during bacteria enumeration (Purified Water) if results continually demonstrate a less than detectable value for a sample size <100 mL because trending of data is not possible.
- When bacteria samples are obtained, the conditions should simulate the conditions encountered during routine use of the delivered water. Specifically, a sample point should not be rinsed to drain for a preestablished period unless water will be rinsed to drain for the same time period every time water is used from the point-of-use. Rinsing prior to sampling for analysis must replicate the conditions used during manufacturing. For bacteria samples collected in non-environmentally controlled physical locations, particularly humid locations, the use of sample valve treatment with an isopropyl alcohol solution may be required. While discouraged, point-of-use treatment with isopropyl alcohol prior to sampling may be employed. However, total viable bacteria samples must be free of any residual isopropyl alcohol to produce a representative result. "Grab" TOC sample collection immediately prior to collection of bacteria samples can be employed to verify the absence of isopropyl alcohol.
- Many pharmaceutical companies, particularly with smaller water systems and limited laboratory capability, will use the services of a contract laboratory for bacteria determination. If a contract laboratory is used, it is critical that the proper chain of custody be established. The chain of custody is achieved by assigning a unique number to each sample provided

to the contract laboratory. Samples for bacteria determination provided to a contract laboratory should be stored and shipped at ~2°C–4°C. The bacteria enumeration technique *must* be initiated within 24 hours of sample collection (10).

- The PQ should discuss identification of observed bacteria colonies confirmed as gram-negative. As discussed earlier in this chapter, the URS should identify undesirable gram-negative pathogens for the application based on the product and process. If "*undesirable*" organisms are detected for critical products (topical solutions, inhalants, antacids, ophthalmic solutions, etc.), system evaluation should be initiated to determine their source. Corrective action, such as an increase in the sanitization frequency of the storage and distribution system, may be required. Further, reinitiation of the PQ may be appropriate once the source of contamination has been identified and/or operating procedures, such as sanitization frequency, revised.
- Bacterial endotoxin monitoring should be conducted for all samples collected from delivered water to points-of-use (with the same frequency as bacteria monitoring) for USP Water for Injection Systems and USP Purified Water Systems with "internal" bacterial endotoxin specification. A quantitative bacterial endotoxin measurement technique is suggested.
- As indicated earlier, during PQ execution, samples should be obtained from raw feed water (to the water purification system) and product water from each unit operation in the water purification system. Consistent with the requirements stated in the individual monographs for USP Purified Water and Water for Injection, feed water must meet the National Primary Drinking Water Regulations (NPDWR) defined by the Environmental Protection Agency—EPA (or applicable equivalent). There are a number of regulated organic compounds defined in the NPDWR. Municipal water treatment facilities will generally have an established monitoring program for regulated organic compounds. For private water supplies to a pharmaceutical facility, it may be necessary to use the services of a contract laboratory that is familiar with the NPDWR. This is particularly true if the feed water source is from a groundwater supply where contamination from numerous sources, such as fertilizer, pesticides, or industrial pollutants, may be possible.
- Analysis for inorganic impurities in raw water should include items specifically contained in the NPDWR, as well as other items that could affect the performance of downstream water purification unit operations.
- Microbial monitoring should include Total coliform and total aerobic count by heterotrophic plate count (11). In addition to raw water samples, intra component samples should be obtained and appropriately analyzed. Bacteria levels in intra component samples prior to a membrane process such as reverse osmosis should be determined by heterotrophic plate count. Bacteria levels in samples obtained after bacteria "specific" components, such as reverse osmosis or ultrafiltration, may, more appropriately, be determined by Membrane Filtration of a 100 mL sample. Samples for bacteria determination containing a disinfecting agent should be collected in a bottle containing a sodium thiosulfate tablet (10).
- The nature of chemical analysis conducted in the feed water and product water to and from each unit operation within the water purification system is established by the function of the component. For example, feed water to an activated carbon unit (assuming that the raw feed water is treated with a disinfecting agent) should be analyzed for residual disinfectant concentration, TOC, and Total Suspended Solids (TSS). Product water from the activated carbon unit should be sampled and analyzed for the same components. Feed water and product water chemical values and total viable bacteria levels determined by heterotrophic plate count for the activated carbon unit should be plotted as a function of time. This monitoring program will establish the normal operating conditions and maintenance requirements for the unit, such as required hot water sanitization and/or backwash frequency (based on product water bacteria levels) and activated carbon media replacement (based on an increase in product water TOC levels beyond a preestablished value or disinfecting agent "breakthrough" based on Total Chlorine concentration). During PQ execution, it is suggested that intra component monitoring be performed daily to weekly. Subsequent to PQ execution, it is suggested that intra component monitoring be performed every 2–6 months depending upon the "stability" of the feed water supply. For certain components, monitoring may be required more frequently based on the potential effects of excursions in unit operations on point-of-use water quality with reference to the FMEA.
- When the intense PQ sampling monitoring portion has been successfully completed, a routine sampling and monitoring program should be established. This program, collecting samples from points-of-use and water system intra component locations, should be structured around the characteristics of the distribution loop, required microbial control for the application, and the nature of the manufacturing process. It is suggested, as a minimum, that delivered water from each point-of-use be sampled for bacteria enumeration once per week. Although regulatory guidelines (12) indicate that the PQ last for a one-year time period to capture seasonal and

climatic changes in feed water, the time period must consider parameters associated with feed water, the impact of seasonal changes, and/or significant climatic events.

- The PQ not only verifies that a system operates in a manner consistent with the design parameters but also provides two other highly valuable objectives. It verifies that component selection and design will, throughout the life cycle of the system, function in an integrated manner producing water, delivered from points-of-use, meeting specified requirements (chemical, bacteria, and, where applicable, bacterial endotoxin). The PQ also serves as a valuable tool for establishing/verifying the requirements for routine monitoring, specifically between components in the water purification system, as well as guidelines for the concentration of specific impurities (including bacteria) from individual components, which are necessary to insure proper system operation, maintenance, and periodic sanitization.

6.6.17 PQ Summary Report

Data obtained during execution of the PQ can be summarized in a PQ Summary Report or included as part of a validation summary report. The PQ Summary Report should tabulate individual monitored parameters for each water purification unit operation. Actual results for each parameter should be evaluated considering the anticipated results. Delivered point-of-use results for the system should include online TOC, conductivity, and temperature results obtained while "grab" samples were being collected. An example of a section of tabulated data for delivered Purified Water is presented as follows:

Sample point valve designation	Analysis/type	PQ Day 1	PQ Day 2	PQ Day 3	PQ Day 4	PQ Day 5
PWS-HV-128	TVB-MF	<1 cfu/ 100 mL	<1 cfu/ 100 mL	<1 cfu/ 100 mL	<1 cfu/ 100 mL	<1 cfu/ 100 mL
	Conductivity/ temperature (grab)	0.84 @ 24.7°C	0.88 @ 23.8°C	0.90 @ 23.9°C	0.81 @ 25.1°C	0.85 @ 24.5°C
	Conductivity/ temperature loop supply (online)	0.77 @ 23.8°C	0.72 @ 24.0°C	0.75 @ 24.1°C	0.79 @ 23.6°C	0.75 @ 23.9°C
	Conductivity/ temperature loop return (online)	0.79 @ 24.0°C	0.71 @ 24.4°C	0.80 @ 24.6°C	0.72 @ 23.9°C	0.72 @ 24.0°C
	TOC (grab)	2.77 ppb	3.14 ppb	3.03 ppb	2.99 ppb	3.62 ppb
	TOC loop return (online)	2.13 ppb	2.55 ppb	2.73 ppb	2.03 ppb	2.38 ppb

6.7 CONCLUSION

Successful design, validation, operation, and maintenance of a pharmaceutical water system are critical to the successful operation of pharmaceutical processes. Validation of a compendial water system should demonstrate, using documentation and test results, that the system is capable of consistently producing, distributing, and delivering product water that, as a minimum, meets the chemical and microbial attributes set forth in the applicable monograph.

REFERENCES

(1) Pharmaceutical Manufacturers Association—Water Quality Committee. *Pharmaceutical Grades of Water Update & a Look to the Future*, Orlando, FL, February 5–8, 1989.

(2) United States Pharmacopeia USPNF 2021 Issue 1 *General Information Section <1231>*, Water for Pharmaceutical Purposes, Section 7.1, Chemical Test for Bulk Waters, the United States Pharmacopeial Convention, Rockville, MD, 2021.

(3) United States Pharmacopeia USPNF 2021 Issue 1. *General Information Section <1231>*, Water for Pharmaceutical Purposes, Section 7.1, Chemical Test for Bulk Waters, the United States Pharmacopeial Convention, Rockville, MD, 2021.

(4) Collentro W. Analytical monitoring of compendial water purification systems. *Journal of Validation Technology* 2014 Fall, 20(3), Institute of Validation Technology, Duluth, Minnesota.

(5) Collentro W. Raw water regulations, impurities, and pretreatment monitoring UltraPure water. *Tall Oaks Publishing Company, Inc.* 2011 March, 3(28), 14–22, Littleton, Colorado.

(6) United States Pharmacopeia USPNF 2021 Issue 1. *General Information Section <1231>*, Water for Pharmaceutical Purposes, Sections 3.1, Bulk Monographed Waters and Steam and Section 3.2, Sterile Monographed Waters, the United States Pharmacopeial Convention, Rockville, MD, 2021.

(7) American National Standards, Instrument Symbols and Identification, ANSI/ISA-5.1-2009, approved 18 September 2009, USBN: 978-1-936007-29-5, Research Triangle Park, NC.

(8) Collentro W. What are maintenance considerations for WFI systems without distillation? Ultrapure water. *Media Analytics Ltd.* 2016 April, 33(4), Littleton, Colorado.

(9) Collentro W. Maintenance and monitoring considerations for compendial water systems. *Pharmaceutical Processing* 2013 November/December, 30(10), Reed Business Information, Highlands Ranch, Colorado.

(10) Rice E, Baird R, Eaton A, eds. *Standard Methods for the Examination of Water and Wastewater 23rd Edition*, ISBN: 9780875532875, American Public Health Association, American Water Works Association, Water Environmental Foundation, 2017, Section 9060A, B.

(11) Rice E, Baird R, Eaton A, eds. *Standard Methods for the Examination of Water and Wastewater 23rd Edition*, ISBN: 9780875532875, American Public Health Association, American Water Works Association, Water Environmental Foundation, 2017, Section 9215.

(12) *FDA Guide to Inspection of High Purity Water Systems*, Rockville, MD, Food and Drug Administration, Office of Regulatory Affairs, Office of Regional Operations, Division of Field Investigations, 1993.

(13) MacKenzie W, Hoxie N, Proctor M, Gradus M, Blair K, Petterson D, Kazmierczak J, Addiss D, Fox, K, Rose J, Davis J. A massive outbreak in milwaukee of cryptosporidium infection transmitted through public drinking water. *New England Journal of Medicine* 1994, 331(161–167).
(14) Atherholt T, LeChevallier M, Norton W, Rosen J. Effect of Rainfall on giardia and cryptosporidium. *Journal of the American Water Works Association, Denver CO* 1998, 90, 66–80.
(15) United States Pharmacopeia USPNF 2021 Issue 1. *General Information Section <1231>*, Water for Pharmaceutical Purposes, Section 9.4.5, Source Water Control, the United States Pharmacopeial Convention, Rockville, MD, 2021.
(16) Yoder J, Roberts V, Craun G, Hill V, Hicks L, Alexander N, Radke V, Calderon R, Hlavsa M, Beach M, Centers R. Surveillance for Waterborne Disease and Outbreaks Associated with Drinking Water and Water Not Intended for Drinking—United States, 2005–2006, US Center for Disease Control and Prevention, 2008.
(17) Burlingame G, Rose J, Xagoraraki I, Couliette A, Aslan-Yilmaz A. Water Borne Pathogens, How Do We Test for Viruses, Opflow, American Water Works Association, Dever, CO, May 2009.
(18) Kunin R, ed. *The Role of Silica in Water Treatment: Part 1, Amber-HI-Lites, No. 164*. Philadelphia, PA: Rohm and Haas Company, 1980.
(19) Cleasby JL, Hilmore DL, Dimitracopoulos J. Slow sand and direct inline filtration of surface water. *Journal of the American Water Works Association* 1984, 76(12), 44–56.
(20) Bolton JR, Colton, CA. The ultraviolet disinfection handbook. *Journal of the American Water Works Association*, Denver, CO, 2008, 1–2. ISBN 1-58321-584-0.
(21) Munson TE. FDA views of water system validation. Proceedings of the Pharm Tech Conference' 95, Cherry Hill, NJ, Aster Publishing Corporation, 1985:287:289.
(22) FDA Guide to Inspection of High Purity Water Systems. Rockville, MD, Food and Drug Administration, Office of Regulatory Affairs, Office of Regional Operations, Division of Field Investigations, 1993.
(23) Collentro W. A novel approach to control microbial fouling of reverse osmosis membranes. Presented at the International Water Conference, the Engineer's Society of Western Pennsylvania, Presentation IWC 12–46, San Antonio, Texas, November 4–8, 2012.
(24) Liu Y, Ogden K. Benefits of high energy UV185 nm light to inactivate bacteria. *Water Science & Technology* 2010, 62(12), 2776–2782.
(25) Collentro W. *Pharmaceutical Water—System Design, Operation, and Validation*. 2nd ed. London, UK: Informa Healthcare, 2011: 114–116, ISPN-13:978142007827.
(26) Krpan N, Wu L. Chlorine species passage through polyamide reverse osmosis membranes. Presented at UltraPure Water Pharma, New Brunswick, NJ, 2010.
(27) ASTM F838-15. Standard Test Method for Determining Bacterial Retention of Membrane Filters Utilized for Liquid Filtration, ASTM International, West Conshohocken, PA, 2015.
(28) Meltzer T, Jornitz M. Assuring sterility with ASTM F 838–83. *Pharmaceutical Technology Europe*, Multimedia UK, LLC, Cheshire, UK 2018 December, 20(12).
(29) Thomas A, Durrheim H, Alport M, et al. Detection of L-forms of pseudomonas aeruginosa during microbiological validation of filters. *Pharmaceutical Technology* 1991, 15(10), 74–80.
(30) Howard G, Duberstein R. A case of penetration of 0.2 micron rated membrane filters by bacteria. *Journal of the Parenteral Drug Asociation* 1980, 34(2), 95–102.
(31) Collentro W. Practical examples of UF use in pharmaceutical compendial water systems. *Ultrapure Water, Media Analytics Ltd.* 2017 May, 34(5), Littleton, Colorado.
(32) Dvorin R, Zahn J. Organic and inorganic removal by ultrafiltration. *Ultrapure Water* 1987, 4(9), 44–46.
(33) British Standards Institute, BS EN 285:2015, Sterilization: Steam Sterilizers, Large Sterilizers—British Standard, ISBN 978 0 580 820335, January 2016.
(34) ASME BPE-2019. *Bioprocessing Equipment*. New York, NY: American Society of Mechanical Engineers, 2019: 77–78.
(35) Compressed Gas Association. *Safe Handling of Ozone Containing Mixtures Including the Installation and Operation of Ozone Generating Equipment*, Product No. CGA P-34, Chantilly, VA, 2001.
(36) ASME BPE-2019. *Bioprocessing Equipment*. New York, NY: American Society of Mechanical Engineers, 2019: 23–28.

Appendix

Overview of Bulk Compendial Water System Considerations, Unit Operations, and Accessories

A1 IMPURITIES IN RAW WATER SUPPLIES

Regulated impurities in raw water are presented in the U.S. EPA National Primary Drinking Water Regulations (or local governing body regulations). Impurities in raw water that are relevant to compendial water purifications systems are presented following with brief descriptive summary.

- **Particulate Matter**
  - Material that is not dissolved, but exists in a solid phase.
  - May be dirt, debris, corrosion products from municipal water piping, or internal piping at a facility, etc.
  - Visible to the human eye if size is greater than about 40 microns.
  - If not removed, will impact long-term operation of downstream water purification components such as valves, water softener eductors, reverse osmosis prefilter cartridges, and analytical sensors.
- **Ionic Impurities**
  - Material such as common table salt that dissolves in water to the extent calculated by its solubility constant (K_{sp}) and subsequently dissolves in water established by its ionization constant (K_i), a function of water temperature.
  - Dissolved and ionized material product positive (cations) and negative (anions) ions. Ions may be monovalent or multivalent depending upon the material; however, the total positive ion change equals the total negative ion charge.
  - Ionic material allows an electrical current to pass through the water solution resulting in an increase in conductivity.
  - The ability to conduct an electrical charge through municipal feed water is about the same for most cations with the exception of the hydroxide ion, which exhibits a higher ion mobility. The same is true for anions with the exception of the hydronium radical, often called the hydrogen ion, which has a much higher mobility than other anions.
  - For municipal feed water the amount of total dissolved material in mg/L (ppm) can be estimated by Total Dissolved (and ionized) Solids = 0.5 X (conductivity in µS/cm @ 25°C).
  - The suggested maximum Total Dissolved Solid (TDS) concentration in potable water is 500 mg/L.
  - The ionic concentration and profile must be considered as part of compendial water system design, operation, and maintenance. Further, seasonal and climatic changes in both ion concentration and profile must be considered.
- **Dissolved Gases—Nonreactive**
  - The predominant gases present in air, oxygen and nitrogen, readily dissolve in water with concentration ranging from 5 to 20 mg/L.
  - The solubility of gases in water generally decreases with increasing temperature.
  - Nonreactive gases will not chemically react with water.
  - Nonreactive gases such as oxygen may react with metallic surfaces producing undesirable metallic oxides.
  - Nonreactive gases are generally not a concern during most water purification operations but may contribute to corrosion of stainless-steel surfaces in distillation units, storage systems, and distribution systems. Corrosion of stainless steel will be enhanced by oxidative conditions such as exposure to elevated temperature or dissolved ozone.
- **Dissolved Gases—Reactive**
  - Some gases dissolve in water and chemically react generally in an equilibrium in which the majority of the reactive gas remains as a "reservoir" to produce ionic products. The concentration of the ionic products can be determined using the reactive gas/water equilibrium constant (K_{eq}).
  - The two gases of concern for compendial water systems are ammonia and carbon dioxide demonstrated by the following equations.
    - NH_3 (ammonia gas) + H_2O ↔ NH_4^+ (ammonium ion) + OH− (hydroxide ion)
    - CO_2 (carbon dioxide gas) + $2H_2O$ ↔ HCO_3^- (bicarbonate ion) + H_3O^+ (hydronium ion)
  - Carbon dioxide is present in air and readily dissolves in water.
  - Ammonia may be naturally present at a low concentration in feed water and will be produced during removal of the alternate municipal secondary disinfecting agent monochloramine.

 - Gases will pass through membrane processes such as reverse osmosis. Although a small fraction of ammonium ion and/or bicarbonate ion may be removed by RO membranes, the reactive gas will reequilibrate with product water, increasing conductivity.
 - Unlike reverse osmosis, reactive gases are removed by ion exchange as equilibrium is continuously "forced" from reactants to ion products by the exchange process.
 - Ammonia or carbon dioxide gas present in the feed water to a distillation unit will be present in distillate product with resulting increase in conductivity. The observed increase in conductivity is enhanced by the mobility (equivalent conductance) of the hydroxide ion and, of greater impact, the hydronium ion.
- **Microorganisms**
 - Regulated microorganisms in municipal feed water include cysts, total viable bacteria, *Legionella*, *E. coli*, and viruses.
 - Cysts such as *Cryptosporidium* and *Giardia lamblia* from human or animal waste (13) may be present in surface source water to a municipal treatment facility. Within the United States, specific municipal monitoring and treatment criteria have been established for control of both cysts, which have a size of about 10 microns (14). Cysts can be removed by filtration and generally are not a compendial water purification concern.
 - Municipal total viable bacteria level is determined by a Heterotrophic Plate Count. Only a small fraction of bacteria present in feed water are detected, although the technique provides a good indication of municipal bacteria control (treatment facility, distribution, and disinfecting agent). The suggested "Action Level" for bacteria in feed water to a compendial water system is 500 cfu/mL (15). Feed water sampling (in sampling containers with sodium thiosulfate tablets to remove disinfecting agent) and analysis should be routinely performed with the frequency increased during seasonal and climatic changes. If values approach or exceed the recommended 500 cfu/mL level, it may be necessary to introduce a disinfecting agent such as sodium hypochlorite. Total chlorine levels should not exceed 4 mg/L. Further, it is suggested that a storage system be provided downstream of introduction to provide necessary contact time to inactivate bacteria. High feed water bacteria levels will impact bacteria levels throughout the water purification system. The system design should reflect elevated feed water bacteria levels. Further, operation and maintenance procedures may require increased compendial water system bacteria sampling, monitoring, and sanitization cycles for adequate bacteria control.
 - *Legionella* is a regulated pathogenic gram-negative bacterium that causes respiratory infection. Although generally associated with mist and/or vapor from evaporative cooling water systems, it may be present in source water but should be removed by "primary disinfection" (16) within the treatment facility and subsequently not present in feed water to a facility.
 - Total Coliform bacteria in a 100 mL feed water sample are regulated. Coliform bacteria are easy to culture and indicate the presence of gram-negative pathogenic organisms. The presence of several different organisms may result in a "positive" Total Coliform indication. Positive results should be verified as fecal coliform or *E. coli* in accordance with Drinking Water regulations. Compendial water systems should not contain *E. coli* at any sample location.
 - Enteric viruses may be introduced to municipal treatment facilities from several sources. In general, viruses exhibit a size of 0.01 to 0.1 microns, much smaller than bacteria. Viruses are not part of the "normal" human intestinal flora and are excreted only by infected individuals (17). The U.S. EPA NPDWR require 99.99% removal/inactivation of viruses from both ground and surface water sources.
- **Bacterial Endotoxins**
 - Bacterial endotoxins in feed water are not directly regulated.
 - Are associated with gram-negative bacteria, which have an inner membrane and second outer membrane. The outer membrane is composed of phospholipid and, most importantly, lipopolysaccharide (LPS) and protein. The LPS is very stable and is considered an endotoxin because the toxin is synthesized as part of, or endogenous to, the bacteria cell structure. As a result, as gram-negative bacteria are destroyed, bacterial endotoxins are released.
 - The level of bacterial endotoxin in feed water from a surface source, particularly rivers in warm climates or during hot seasons, may be >5–10 EU/mL.
 - In compendial water purification systems, bacterial endotoxins are associated with biofilm formation and related bacteria control. Further, bacterial endotoxins in USP Water for Injection (Bulk) must be <0.25 EU/mL.
- **Organic Material**
 - For source water from a surface supply or ground water supply influenced by a surface source, naturally occurring organic material (NOM) will be present. In general, minimal NOM is present in water from a ground water source. NOM consists of rotting vegetation and material in the run-off area.
 - The concentration of NOM in raw water is not directly regulated but indirectly regulated by control of disinfection by-products. Feed water concentration of NOM from a surface source will generally range from 1.5 to 10.0 mg/L.

- In water purification systems, NOM will foul ion exchange resin and reverse osmosis membranes ultimately resulting in resin or membrane cleaning/replacement.
- NOM will also provide a nutrient for bacteria proliferation.
- Non-naturally occurring organic compounds are regulated by U.S. NPDWR. Currently about >50 organic contaminants/pollutants are regulated by the U.S. NPDWR.

- **Colloids**
  - Colloids in feed water are not regulated.
  - Colloids are physically larger in size than an ion but smaller than particulate matter (10 microns) and generally exhibit a slight negative charge.
  - In feed water, colloids may be in a complex with NOM and include inorganic elements such as aluminum, silica, and iron (18). Colloids may also be produced during municipal facility treatment with additives such as polyphosphates or orthophosphates used to inhibit staining of domestic water fixtures such as sinks.
  - In water purification systems, colloids can contribute to fouling of reverse osmosis membranes and are not effectively removed by gellular ion exchange anion resin.
- **Municipal Disinfecting Agents**
  - Municipal Treatment facilities introduce a primary disinfecting agent during treatment within the facility. A secondary disinfecting agent is employed for bacteria destruction/inactivation and control of delivered Drinking Water. The type of secondary disinfecting agents with associated health effects, are as follows:
  - Chlorine as Cl_2 may be employed as a secondary disinfecting agent for control of microorganisms in the municipal water distribution system. When introduced to water, chlorine produces a mixture of hypochlorous acid and hypochlorite ion as follows:
    - $Cl_2 + 2H_2O \leftrightarrow$ HOCl (hypochlorous acid) + $H_3O^+ + Cl^-$
    - $HOCl + H_2O \leftrightarrow H_3O^+ + OCl^-$ (hypochlorite ion)
  - The U.S. EPA NPDWR maximum recommended concentration for chlorine as Cl_2 is 4.0 mg/L. Chlorine concentrations exceeding this value may result in eye/nose irritation and stomach discomfort. It is strongly suggested that compendial water system free and total chlorine concentration be measured periodically, particularly with seasonal and climatic change.
  - Chlorine Dioxide as ClO_2 may be used as a secondary disinfecting agent. The U.S. EPA NPDWR maximum concentration is 0.8 mg/L as ClO_2. Chlorine dioxide concentration in excess of the regulated value can result in anemia in infants and younger children. Human nervous system effects may also be noted.
  - Chloramines as Cl_2 are obtained by introduction of ammonia into a chlorinated stream at the beginning of a distribution system. Three chloramine species exist; trichloramine, dichloramine, and monochloramine, However, at a pH of 7.0 or greater, monochloramine, the most powerful chloramine oxidant, is prevalent (95%–100% of total chloramines), produced by the following reaction:
    - NH_3 (ammonia) + HOCl (hypochlorous acid) $\rightarrow$ $H_2O + NH_2Cl$ (monochloramine)

    The U.S. EPA NPDWR maximum recommended concentration for chloramines as Cl_2 is 4.0 mg/L. Chloramine concentrations exceeding this value may result in eye/nose irritation, stomach discomfort, and anemia.
  - For water purification systems, removal of chlorine by activated carbon, high-intensity ultraviolet radiation at a wavelength of 185 nanometers, or introduction of a reducing agent is effective. However, removal of chloramines by the indicated unit operations is more difficult and produces undesirable ammonia.
- **Disinfection By-Products**
  - Disinfecting agents (primary and secondary) will react with NOM present in surface water and ground water influenced by surface water to produce disinfection by-products. Regulated disinfection by-products and health effects are presented as follows:
    - Bromate is a disinfection by-product produced by the reaction of "primary" disinfecting agent ozone with NOM. The maximum U.S. EPA NPDWR concentration for bromate is 0.010 mg/L. Higher concentrations of bromate are associated with increased risk of cancer.
    - Chlorite is a disinfection by-product of "primary and secondary" disinfection by chlorine dioxide as it reacts with NOM. The maximum U.S. EPA NPDWR concentration for chlorite is 1.0 mg/L. Higher concentrations of chlorite are associated with anemia in infants and young children. Human nervous system effects may also occur.
    - Haloacetic Acids (HAA5s) are produced by the reaction of NOM with chlorine as a primary and/or secondary disinfecting agent and, to a lesser extent, chloramines as a secondary disinfecting agent. HAA5 compounds include dichloroacetic acid, trichloroacetic acid, monochloroacetic acid, bromoacetic acid, and dibromoacetic acid. The maximum U.S. EPA NPDWR concentration for HAA5s (total) is 0.060 mg/L. Higher concentrations of HAA5s (total) are associated with increased risk of cancer.
    - Trihalomethanes, the most prevalent disinfection by-products, are produced by the reaction of NOM with chlorine as a primary and/or secondary disinfecting agent and, to a lesser extent, chloramines as a secondary disinfecting

agent. Trihalomethane compounds include chloroform, bromodichloromethane, dibromochloromethane, and bromoform. The maximum U.S. EPA NPDWR concentration for Total Trihalomethanes (TTHM) is 0.080 mg/L (average). Higher concentrations of TTHM are associated with increased risk of cancer (particularly liver and kidney) and central nervous system problems. In compendial water systems, removal of THM compounds, particularly chloroform, by ion exchange, activated carbon, reverse osmosis, or oxidizing ultraviolet radiation is challenging. Further, the low molecular weight volatile halogenated compounds may be present in distillate product from a distillation unit. Generally, although they will slightly increase TOC levels on compendial water, they are often undetected.

A2 OVERVIEW OF COMMON COMPENDIAL WATER SYSTEM UNIT OPERATIONS

- **Pretreatment Techniques**
  - Chemical Injection
    - Chemical injection may be employed for introduction of one or more compounds in the compendial water purification system. The chemical content of all injected substances should be established and verified by a Certificate of Analysis. A suggested guideline for any chemical introduced into a compendial water system is that the downstream water purification unit operations should provide removal such that the injected chemical is not a "Foreign Substance or Impurity". Chemical compounds should not be introduced after production of bulk compendial water with the exception of ozone, which must be removed prior to points-of-use in the distribution system. Injected chemicals may include the following:
      - Disinfecting agents such as sodium hydroxide may be introduced to system feed water in which the bacteria levels are elevated or exceed 500 cfu/mL or to normal feed water with minimal concentration of residual disinfecting agent in raw water, elevated raw water microorganism levels, and/or the presence of "undesirable" species of bacteria such as *E. coli* are noted.
      - Reducing agents such as sodium sulfite or sodium bisulfite may be introduced to remove municipal disinfecting agent prior to primary ion removal techniques such as deionization or reverse osmosis units.
      - Sodium hydroxide may be introduced prior to a reverse osmosis system in an attempt to "force" the equilibrium reaction of carbon dioxide and water to products, principally the bicarbonate ion which will be removed by the RO membrane in lieu of gaseous carbon dioxide, which passes through the RO membrane.
      - An antiscalant may be used prior to reverse osmosis to reduce potential scaling on RO membranes.
  - Particulate Removal Filters
    - Particulate removal is generally the initial purification unit operation in a compendial water purification system. Although particulate removal may be accomplished by disposable cartridge or bag filters, back washable filters can exhibit technically attractive benefits. Back washable particulate removable filters generally consist of one or more vertical cylindrical filter vessels with filter media, distribution systems, and "face" piping. Units employ surface filtration or depth filtration, discussed following.
    - Sand filters provide filtration of particles at the top surface of a sand bed generally at least 20 inches deep. The sand is supported over graduated levels of supporting gravel. Flow is downward through the filter bed with water distribution achieved over the cross section of the filter bed created by backpressure from the lower (outlet) distributor. Filtration of particulate matter with a size ≥10 microns is achieved at the top of the sand media when the face velocity through the filter bed is about 6 gpm/square foot of cross-sectional bed area. Periodic backwash of the filter should be considered when the pressure drop through the sand filter is 7–10 psig greater than the post backwash pressure drop. Backwash is generally performed for about 20 minutes at a flow rate of 12–15 gpm/square foot of cross-sectional bed area. Particulate removal capacity between backwash cycles is limited by the mass of particulate matter that can be removed on the surface of the sand media before exceeding the pressure drop. A sight glass on backwash piping is used to establish the duration and effectiveness of the backwash operation.
    - Depth filtration of particulate matter is achieved using two or more types/sizes of filter media supported over graduated levels of supporting gravel. Multimedia filtration can employ an upper filter media level of anthracite supported over sand, supported over very fine gravel. As water passes downward through the filter, larger particles are removed at the top of the bed with smaller particles removed by the denser and finer media such as sand. Subsequently, filtration is not limited to the top of the filter media but throughout the "depth" of the filter bed. A greater amount of particulate matter can be removed between backwash cycles when compared with sand filtrations. In addition, depth

filtration can provide agglomeration of particles that enhance the ability to remove smaller size material (through a process called "ripening") including high molecular weight NOM and colloids (19). Operation and backwash flow rates are similar to those of sand filters.

- Activated Carbon Adsorption Units
  - Properly designed, operated, and maintained activated carbon adsorption units will completely remove municipal disinfecting agent and a portion of the NOM from feed water. Removal of disinfecting agent occurs by a "chemical reaction-adsorptive" process in which the disinfecting agent is attracted to the surface of the activated carbon media (adsorption) where a chemical reaction occurs. Secondary municipal disinfecting agent chlorine removal is represented by the following chemical equations:
    - $C^* + 2Cl_2$ (Chlorine) + $2H_2O \rightarrow 4HCl + CO_2$

    $C^* + H_2O + HOCl$ (Hypochlorous Acid) $\rightarrow CO^* + H_3O^+ + Cl^-$

    Where C* represents the activated carbon surface and CO* represents a surface oxide on the activated carbon surface.

    The reaction for removal of the municipal secondary disinfecting agent monochloramine is:
    - NH_2Cl (Monochloramine) + $C^* + 2H_2O \rightarrow CO^* + H_3O^+ + Cl^- + NH_3$(Ammonia)
  - Chlorine removal by activated carbon is quite effective. One gram of activated carbon will remove about one gram of chlorine. Conservative design, operating, and maintenance procedures for chlorine removal are 3 gpm/square foot of cross-sectional bed area, 1.0 to 1.5 gpm per cubic feet of activated carbon media, 3–4 foot bed depth, and 3 year minimum bed life. Conversely, removal of monochloramine by activated carbon is much more challenging. Ammonia gas in product water will react with water and is not removed by downstream reverse osmosis, although it will be removed by ion exchange resin. A greater number of municipal treatment facilities are switching from chlorine to monochloramine as a secondary disinfecting agent. Design, operating, and maintenance parameters for monochloramine removal require a minimum 4-foot bed depth, 3 gpm/square foot of cross-sectional bed area, 0.5 gpm/cubic foot of media, consideration of catalytic activated carbon media, and 6–12 month replacement frequency.
  - Depending upon the chemical profile of the NOM removed by physical adsorption to the surface of the activated carbon, reduction may range from 30%–70%.
  - As disinfecting agent is removed by activated carbon, microbial proliferation significantly increases. The highest microbial levels in a compendial water system are generally noted in activated carbon unit product water. Backwash of activated carbon units will provide some reduction in product microbial levels. However, frequent backwash and excessive backwash flow rate are undesirable. Excessive backwash flow rates (>10 gpm/square foot of cross-sectional bed area) can result in excessive expansion of the activated carbon bed with resulting impingement of media on the walls and top of the activated carbon unit column. This results in production of undesirable activated carbon "fines" that will contaminate downstream unit operation, increasing microbial proliferation. Further, as activated carbon removes disinfecting agent and NOM, the density of the media increases. Excessive backwash flow rate and frequency may result in accelerated movement of media from the top to the bottom of the bed with potential premature breakthrough.
  - Microbial control may be achieved by periodic hot water sanitization suggested for a time period of 2 hours at a temperature of 80°C. Sanitization with steam is much less effective than with hot water even with an extended sanitization period. Significant reduction in activated carbon unit microbial levels can be achieved using in-line ultraviolet sanitization units at a wavelength of 254 nanometers and intensity of 30,000 to 70,000 microwatt-seconds/square centimeter.
- Organic Scavenging Units
  - Infrequently used but may be employed when municipal raw water contains very high NOM concentration (TOC >5–10 mg/L). Macroporous styrenic anion resin or acrylic anion resin is used to remove organic material that may foul downstream RO membranes or anion resin using ion exchange as a primary ion removal technique. Periodic multistep removal of organic material is required. Caustic brine soak is used with long rinse time periods. Although good organic reduction is achieved during regeneration, chemical handling, rinse time, and waste disposal limit application of this technology.
- Water Softening Units
  - Units are provided to remove multivalent cations prior to RO systems. Multivalent cations such as magnesium, calcium, barium, aluminum, etc. can result in scaling of RO membranes. Water softeners use cation exchange resin in the sodium form. Multivalent and high molecular weight ions have a greater attraction for ion exchange sites on resin than sodium. The ion exchange process is a reversible reaction. During normal operation, water flows

through the cation resin bed where multivalent cations are removed and replaced with sodium ion. However, eventually the cation sodium exchange sites are reduced and multivalent cations present in product water. Regeneration is performed with a concentrated brine solution (sodium chloride) forcing the ion exchange equilibrium in a direction that displaces multivalent cations and restores the sodium exchange sites.

- Representative equations:
  - Operation: $Ca^{++} + R\text{-}Na^{+} \leftrightarrow R\text{-}Ca^{++} + Na^{+}$
  - Regeneration: $Na^{+} + R\text{-}Ca^{++} \leftrightarrow R\text{-}Na^{+} + Ca^{++}$
- Water softening of feed water to a system using ion exchange as a primary ion removal technique should not be considered. During the deionization process, the monovalent, low molecular weight sodium ions are more difficult to remove than multivalent cations resulting in "sodium leakage".
- Design parameters include 7–10 gpm/square foot of cross-sectional bed area flow rate, cation bed depth ≥3 feet, 50% column "freeboard" (2/3 of column containing resin), and provisions for product water Total Hardness monitoring (grab and/or continuous). Operating product water Total Hardness should be <1 mg/L (expressed as Calcium Carbonate).
- Periodic regeneration with concentrated sodium chloride solution should provide about 15 pounds of NaCl over a 20–30-minute time period. The regeneration cycle includes backwash, NaCl injection, displacement (slow) rinse, and final rinse. Although units are generally regenerated in the same flow direction as operation (downward) through the column, units regenerated in a countercurrent direction require lower salt volumes and less total water per regeneration cycle.
- Maintenance considerations should include periodic sanitization of the brine storage tank to suppress microbial introduction during regeneration, resin bed "core" sampling and analysis to check the cation resin condition (including degree of iron fouling), and verification of regenerant salt volume and concentration during regeneration.
- For systems with feed water containing the secondary disinfecting agent monochloramine, the water softening system should be positioned downstream of activated carbon adsorption (or introduction of reducing agent). Ammonia, produced during removal of monochloramine, can be removed as the ammonium ion. However, it is common to employ series positioned softening systems, the "lead" system for hardness removal and "polishing" system for ammonium ion removal because the multivalent cations with less attracted ammonium ion for ion exchange sites.

- In-Line Ultraviolet Sanitization Units
  - Historically, ultraviolet units employ low or medium pressure mercury vapor lamps with 253.7 (~254) nanometer wavelength and intensity >30,000 μW seconds/square centimeter to inactivate bacteria. Units consist of lamps inserted inside quartz sleeves positioned in stainless-steel sanitizing chambers. "Sanitization" UV radiation at a wavelength of 253.7 nanometers does not destroy bacteria but inhibits replication by modifying deoxyribonucleic acid (DNA) and ribonucleic acid (RNA) (20). The bacteria inactivation capacity, expressed in log reduction, has been discussed in regulatory presentations (21) and guidelines (22).
  - Proper operation of an in-line ultraviolet sanitization unit requires continuous flow of water, inhibiting lamp operation at elevated temperature, and avoiding repeated electrical power cycling. Ultraviolet radiation will degrade plastic material. Subsequent, ultraviolet "light traps" should be considered when feed/product water piping/tubing is not stainless steel (or UV resistant non-metallic material) and for quartz sleeve sealing devices such as compression nuts.
  - Maintenance consideration include periodic lamp, O-ring, and sleeve replacement. Further, a method of verifying the proper calibration of the UV intensity meter is suggested.
- In-Line Ultraviolet Oxidization Units
  - Oxidizing UV radiation at a wavelength of 184.9 nanometers destroys bacteria by production of ozone and the hydroxyl radical, both extremely powerful oxidizing agents (23). Further, the highly oxidative environment results in the oxidation of RO membrane fouling organic material. Finally, effective removal of disinfecting chlorine (hypochlorous acid and/or hypochlorite ion) and monochloramine may be achieved. The in-line ultraviolet units are similar to sanitizing UV units but use 184.9 (~185) nanometer radiation at an intensity greater than approximately 600,000 μW seconds/square centimeter. Water passing through this radiation generates both the hydroxyl radical and ozone, both highly oxidizing substances (24). The hydroxyl radical, although an extremely powerful oxidizing agent with significant absorption rate to target compounds and bacteria, has a very short life, less than a second. Subsequently, oxidation associated with the hydroxyl radical is

essentially limited to the stainless-steel chamber of the in-line UV unit. On the other hand, dissolved ozone will be present in UV system product water and continue to oxidize material including bacteria. Ozone decomposes to oxygen with a half-life as long as 20–165 minutes in ultrahigh purity water but only a few minutes in pretreated water containing inorganic, organic, colloidal, and microorganisms.

- Proper operation of an in-line ultraviolet oxidation unit requires continuous flow of water, inhibiting lamp operation at elevated temperature, and avoiding repeated electrical power cycling. Oxidizing ultraviolet radiation will degrade plastic material, resulting in the need for stainless-steel feed/product water piping/tubing. Quartz sleeve sealing devices such as compression nuts must also be stainless-steel.
- Maintenance considerations include periodic lamp and sleeve and O-ring replacement. Further, a method of verifying the proper calibration of the UV intensity meter is suggested.
- Finally, although oxidizing ultraviolet radiation provides excellent removal of the municipal secondary disinfecting agent chlorine, complete removal of monochloramine requires enhanced sizing. This may be accomplished by operation of two units in series.

- **Primary Ion Removal Techniques**
  - Two Bed Deionization Units
    - Some compendial water purification systems employ two bed deionization as a primary ion removal technique in lieu of reverse osmosis. Ion exchange, unlike RO, will remove reactive gases such as carbon dioxide and ammonia as product bicarbonate and ammonium ions. However, unlike RO, deionization will not effectively remove bacteria, bacterial endotoxins, colloids, disinfecting agent, and disinfection by-products. A marginal reduction in organic material may be noted, generally associated with anion fouling for conventional resin.
    - Two bed deionization systems consist of separate lined steel or fiberglass reinforced plastic vertical cylindrical vessels. The initial vessel contains cation resin and the second vessel, in series, anion resin. As a primary ion removal technique, the two separate ion exchange column configurations versus a single bed with mixed cation and anion is preferred. This configuration not only provides less complex regeneration with acid and caustic but also about 40% greater water volume throughput per cubic foot of anion resin (lower capacity per unit volume than cation).
    - The deionization process consists of equilibrium reactions based on ion attraction to an ion exchange site, and subsequent regeneration with acid (cation) and caustic (anion) demonstrated by the following equations:
      - Cation-Operation
        $Na^+ + R\text{-}H_30^+ \rightarrow R\text{-}Na^+ + H_3O^+$
        $Ca^{+2} + 2R\text{-}H_30^+ \rightarrow R\text{-}Ca^{+2} + 2H_3O^+$
      - Anion—Operation
        $Cl^- + R\text{-}OH^- \rightarrow R\text{-}Cl^- + OH^-$
        $SO_4^{-2} + 2R\text{-}OH^- \rightarrow R\text{-}SO_4^{-2} + 2OH^-$
      - Anion Product—Operation
        - $H_3O^+ + OH^- \rightarrow 2\,H_2O$
      - Cation—Regeneration with Acid
        - $R\text{-}Na^+ + \mathbf{H_3O^+} \rightarrow R\text{-}H_30^+ + Na^+$ (To Waste)
        - $R\text{-}Ca^{+2} + \mathbf{2H_3O^+} \rightarrow 2R\text{-}H_30^+ + Ca^{+2}$ (To Waste)
      - Anion—Regeneration with Caustic
        - $R\text{-}Cl^- + \mathbf{OH^-} \rightarrow R\text{-}OH^- + Cl^-$ (To Waste)
        - $R\text{-}SO_4^{-2} + \mathbf{2OH^-} \rightarrow 2R\text{-}OH^- + SO_4^{-2}$ (To Waste)
    - For classical co-current regenerated units, the flow rate through each ion exchange column should be about 7 gpm/square foot of cross section bed area. About 2/3 of the column should contain resin (50% freeboard). Resin bed depth should be ≥3 feet. As a result of the less than complete removal of all monovalent low molecular weight sodium ion, product water quality is generally 5–10 µS/cm @ 25°C.
    - Periodic regeneration steps include cation backwash, acid injection, acid displacement (slow rinse), cation fast rinse, anion backwash, anion regeneration with caustic, caustic displacement (slow rinse), and final rinse.
    - Annual "core" sampling of both resin beds should be performed with subsequent analysis to determine any degradation such as anion fouling.
  - Reverse Osmosis Systems
    - The majority of compendial water purification systems utilize reverse osmosis as a primary ion removal technique. The reverse osmosis process employs pressurized water flow through a semipermeable membrane that removes impurities. RO systems have feed water, waste, and product water streams. Feed water must be pretreated to remove particulate matter (back washable filters and cartridge prefilters), multivalent cations that will precipitate in the "ion concentrating" process as product water is removed (water softening), and disinfecting agent (activated carbon, reducing agent injection, or 185 nanometer UV). Further, the concentration of NOM in feed water must be controlled to reduce fouling of the membranes (activated carbon or 185 nanometer UV), and bacteria levels controlled by techniques such as 254 nanometer UV sanitization units.

Chemical injection to control scaling of the membranes or enhance removal of reactive gasses may also be employed.

- In general, RO membranes may remove >99% of the dissolved ionic material in feed water. However, product water conductivity is a function of reactive gasses in feed water, such as carbon dioxide, which re-equilibrate with product water producing the bicarbonate ion and the highly mobile hydronium ion. As a result, RO product water quality is generally 4–10 µS/cm @ 25°C. Reverse Osmosis membranes also remove organic material with a molecular weight >150–250 Daltons, bacterial endotoxins, colloids, silica ("reactive and non-reactive"), and microorganisms. However, bacteria are generally noted in product water from a properly designed, operated, and maintained single pass reverse osmosis unit at a level <100 cfu/100 mL.
- General design and characteristics for compendial primary ion removal RO system are presented as follows:
    - Recovery of feed water as product water is about 75%.
    - Membranes may be hollow fiber or spiral wound configuration, although spiral wound membranes are employed in the vast majority of systems.
    - Membrane material of construction is generally a thin layer of polyamide supported over polysulfone or similar composite materials.
    - RO membranes are generally about 4" or 8" diameter and 40" long (or longer).
    - Membranes are positioned in pressure vessels designed to contain one to five (or more) RO membrane elements. Membrane-to-pressure vessel (or other membrane) seal mechanism is by adapters with single or double O-rings.
- Tap wrapped, "brackish water" and loose-wrapped (full fit) RO membranes are available. However, in an attempt to to eliminate stagnant water between the outside of the RO membrane and the inner wall of the pressure vessel (associated with "brine seals") to reduce microbial proliferation, "pharmaceutical" full fit membranes are often used.
- Pressure vessels are arranged in an "array" as shown in Figure 6.6. The membrane array attempts to maintain water velocity through membranes as product water is removed, increasing the concentration of impurities including ions. Scaling of membranes can occur as a result of precipitation of compounds particularly in the "tail" (final membranes) in an array.
- An RO system manufacturer's computerized projection should be used to establish system array, operating parameters, and number of membranes using a pretreated feed water analysis of the system. The analysis provides a "scaling warning" if solubility limits are exceeded that would result in scaling and associated decrease in product water flow rate.
- In an attempt to minimize potential scaling of membranes, many systems divert more than 25% of feed water to waste, and recycle a portion of the wastewater back to the RO system feed water. Although the

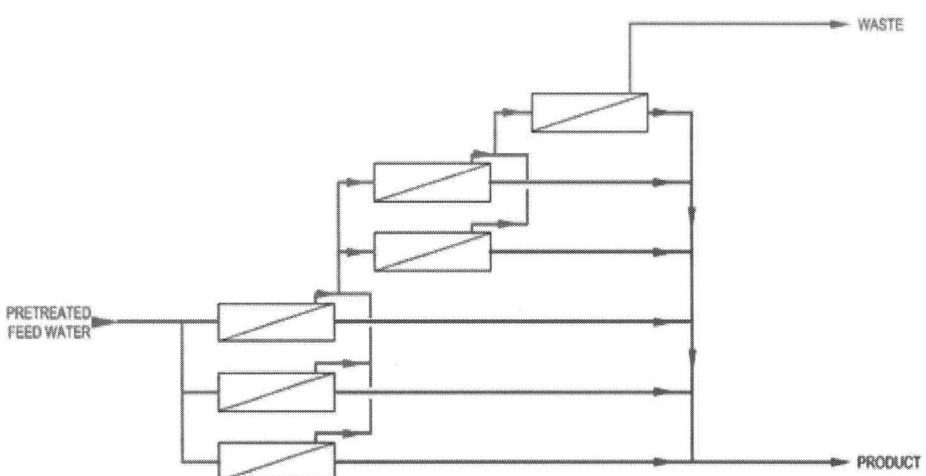

FIGURE 6.6 Reverse osmosis system—3:2:1 pressure vessel array.

amount of water flowing directly to drain remains at 25% of the pretreated feed water, the actual feed water flow rate to the RO system increases by the waste recycled water flow rate. Unfortunately, this process increases the bacteria level of feed water, resulting in increased microbial fouling of RO membranes.

- RO system design should consider the compendial water flow rate demand profile. Specifically, RO product water flow rate requirements projected over a typical operating day and week should be considered. RO systems may deliver product water for a few hours each day and a limited number of days per week or a majority of each day and week. Cyclic operation of a RO system with associated pressure/flow cycling to membranes and sealing mechanisms such as O-rings is undesirable. Although a cycled RO system can be periodically flushed to limit stagnant conditions associated with microbial growth, performance can be enhanced by continuous operation. Continuous operation of the RO system generally requires an upstream feed water tank and may include downstream ion polishing and microbial control provisions. Further, the dynamic nature of this design also provides enhanced microbial control in downstream polishing and support components associated with continuous recirculation. Recirculation can include provisions for periodic hot water sanitization, generally at a temperature of 80°C, and include downstream polishing components and accessories if compatible with the sanitizing temperature. During the recirculation mode, RO System feed water quality approaches, or exceeds, RO product water quality. During recirculation, the actual RO system waste flow to drain can be decreased to a value as low as 5% of the feed water flow rate (25).
- Additional design considerations include provisions for monitoring feed water, product water, waste-to-drain, and waste cycle (if used) flow rates, feed water, waste, and product water pressures, intra-array pressure, feed water and product water conductivity, and feed water temperature.
- Sampling valves with check valves should be considered between each array for troubleshooting.
- A variable frequency drive (VFD) should be considered for the RO feed water pump motor to allow automatic adjustment of feed water pressure because RO product water flow rate decreases with decreasing feed water temperature (higher viscosity).
- Proper maintenance of an RO system is critical to successful compendial water purification system operation, because it removes (or reduces) the concentration of multiple contaminates. Monitoring and maintenance items include the following:
  - Verification of the absence of municipal disinfecting agent in feed water.
  - Periodic determination of RO system feed water and product water bacteria and TOC levels.
  - Feed water contaminants accumulate on RO membrane surfaces in distinct "layers". The layer furthest from the RO membrane contains scalants such as compounds of calcium and carbonate ion. The next layer towards the RO membrane surface contains foulants such as colloids and organic material. The layer on the RO membrane surface contains bacteria and bacterial endotoxins. Periodic chemical cleaning of RO membranes should be performed. The cleaning sequence is extremely important. Scalants are removed with a low pH cleaning agent. Subsequent to rinse, foulants are removed with a high pH cleaning agent. Finally, bacteria and bacterial endotoxins are removed with a sanitizing agent such as Peracetic Acid and Hydrogen Peroxide. The suggested cleaning frequency is every 6–12 months depending upon pretreatment system performance, raw water quality, waste recycle, and feed water bacteria levels.
  - In lieu of periodic in-place cleaning, membranes can be removed, replaced with a second set of membranes, and the used membranes cleaned off-site by a contract service organization. This not only eliminates handling of chemicals and waste disposal but provides technically superior individual membrane cleaning in parallel versus series.
  - Another alternative used for some compendial water systems is to just replace the entire set of RO membranes periodically.
- **"Polishing" Ion Removal Techniques and Accessories**
  - Mixed Bed Deionization Units
    - Some compendial water purification systems employ two bed deionization as a primary ion removal technique with downstream mixed bed deionization as an ion polishing technique. Other systems may employ reverse osmosis with mixed bed deionization polishing because it effectively removes reactive gases as product water ions.
    - The mixed bed deionization system may consist of one or more separate lined steel or fiberglass reinforced plastic vertical cylindrical vessels.

The vessels contain a mixture of cation and anion exchange resin. Frequently, 60% of the total resin volume is anion resin and 40% cation resin in an attempt to balance anion/cation exchange capacity. Regenerative mixed bed deionization units or rechargeable mixed bed deionization "canisters" may be used.

- The deionization process consists of equilibrium reactions based on ion attraction to an ion exchange site and subsequent regeneration with acid (cation) and caustic (anion) as discussed for two-bed deionization units. However, the multistep mixed bed regeneration process requirement for concentrated acid and caustic storage and handling coupled with potential waste neutralization present challenges. The use of mixed bed rechargeable canisters may be attractive in lieu of regenerative units.
- For regenerative units, the flow rate through the mixed ion exchange column should be about 7–10 gpm/square foot of cross section bed area. About 1/2 of the column should contain resin (100% freeboard). Resin bed depth is generally about 4 feet. As a polishing ion removal technique, mixed resin ion exchange unit product water conductivity will be <0.10 μS/cm @ 25°C.
- Typical periodic in-place regeneration steps generally include backwash to separate more dense cation from less dense anion, acid introduction through cation resin from bottom up to interface distributor, cation slow (displacement) rinse, cation fast rinse, anion regeneration with caustic from the top of column to interface distributor, anion slow (displacement) rinse, anion fast rinse, decrease in water level to top of anion resin, air mix of entire resin bed, refill of entire column with water, and final rinse.
- Operating and maintenance concerns for the deionization system include anion regeneration for silica removal (requiring regeneration with "warm" caustic). Silica will not be detected by conductivity increase and will result in scaling on heated surfaces downstream such as multiple effect distillation unit and Pure Steam Generator. Bacteria control between regeneration cycles may be a concern because mixing cation and anion exchange resin provides a neutral bed pH with carbonaceous material.
- Annual "core" sampling of the regenerative mixed bed subsequent to regeneration should be performed with analysis to determine any degradation such as anion fouling, resin "fine" generation, and other critical parameters.

- Continuous Electrodeionization Systems
  - Continuous electrodeionization (CEDI) may be positioned downstream of primary ion removal reverse osmosis systems to remove ions present in RO product water in lieu of classical regenerative or rechargeable mixed bed units. CEDI employs ion exchange membranes configured in ion depleting and ion concentrating flow chambers positioned between an anode and cathode operating at high voltage. The chambers are generally filled with mixed ion exchange resin.
  - Within a CEDI unit, anions in RO product water flowing through ion depleting chambers are "pulled" through anion exchange membranes toward the positively charged anode into adjacent ion concentration flow chambers. Simultaneously, cations in RO product water flowing through the ion depleting chambers are "pulled" through cation exchange membranes toward the negatively charged cathode into adjacent ion concentration flow chambers. Assembled units may contain multiple ion depleting and ion concentrating flow chambers. Most of the RO product water flow is directed toward the ion depleting chambers with a small portion of flow directed to the ion concentrating chambers. Deionized product water is collected from the ion depleting chambers whereas ion concentrating chamber flow is directed to waste. RO product water anions and cations attracted to ion exchange sites are continuously regenerated by hydroxyl ion and hydronium ion produced from electronic "splitting" of water in the high electronic field between the anode and cathode.
  - CEDI system design considerations are presented as follows:
    - Generally, about 95% of RO product feed water is recovered as CEDI system product water. Because the waste stream contains a low concentration of ionic impurities (compared with upstream RO system feed water) it is often recovered and fed to a depressurized location (tank) upstream of the RO system. RO product water carbon dioxide concentration and associated bicarbonate ion concentration may limit the recovery rate of CEDI wastewater.
    - Hot water sanitizable CEDI systems are available. These units can be positioned downstream of a hot water sanitizable RO system in a continuously recirculated manner such that the entire RO/CEDI "loop" with accessories is capable of being periodically hot water sanitized at 80°C.
    - CEDI systems with feed water from a well-maintained RO system with conductivity <5–10 μS/cm @ 25°C will have a product water conductivity ≤0.1 μS @ 25°C.
    - Unlike conventional mixed bed deionization units, bacteria levels in CEDI system

product water are generally low minimizing the need for control techniques such as in-line UV sanitization.

- CEDI assembly operating life is a function of RO product water quality (inorganic, organic, and microbial). With feed water from a well-maintained RO system, stack life is generally 3–5 years or greater.
- CEDI units are highly sensitive to the presence of trace concentration of disinfecting agent (26). Oxidative attack will result in an increase in product water conductivity, increase in pressure drop (feed water to product water), and decrease in product water flow rate.
- CEDI units should be equipped with a flow switch to detect loss of adequate water flow. It is suggested that the pressure switch be installed in the waste stream because it has a lower flow rate than the feed water and product water steams. Further, a waste water pressure switch can be non-sanitary. The switch should automatically inhibit electrical power to the unit (anode and cathode) on low pressure.
- Sanitary diaphragm valves should be installed in the ion depleting chamber feed water tubing, ion concentrating chamber feedwater tubing, and common product water tubing. Further, a flow regulating valve with flow rate monitoring provisions should be installed in the common waste line. Pressure indication must be provided in the tubing/piping physically between the valves and the CEDI unit. Pressure control is required to avoid reversal of ion flow between the ion depleting and ion concentrating chambers.
- Product water in-line conductivity monitoring provisions should be considered to verify CEDI system operation.
- Conductivity monitoring of CEDI unit recycled waste should also be considered. Conductivity wastewater monitoring can automatically divert waste to drain if carbon dioxide/bicarbonate level exceeds a preset level.
- CEDI tubing/piping connections, particularly for hot water sanitizable systems, are often stainless steel. Although catastrophic CEDI unit failure is highly unlikely, provisions shall be included for electrical isolation of the unit from the piping/tubing. This is achieved by the use of non-metallic adapters in the feed water, product water, and waste connections fittings on the unit. Further, to avoid electrical hazards by conduction through the water inside the piping/tubing, conductive stainless-steel rods are positioned into feed water, product water, and waste water piping/tubing and appropriately electrically connected to a "ground".
- Maintenance considerations for CEDI systems are presented as follows:
    - Where appropriate, periodic hot water sanitization should be performed for bacteria control.
    - Routine monitoring of CEDI system operating parameters is suggested. This should include verification of the electrical supply parameters to the unit. Continuous conductivity monitoring may not detect interruption/loss in electrical power immediately because of the ion exchange capacity of resin in the units. Delays in restoring electrical power can result in extended recovery time to achieve normal operating product water conductivity value.
    - Although chemical sanitization of non-hot water sanitizable units can be performed, thermal sanitization is extremely effective.
    - Non-metallic transition adapters from the units to feed water, product, and waste piping/tubing on hot water sanitizable systems frequently employ with sanitary ferrules, gaskets, and clamps. Minor deformation of the adapters may occur after several hot water sanitization cycles resulting in small but undesirable water leaks. Periodic inspection for leaks should be performed. Further, an inventory of spare adapters for each connection should be considered.

- Membrane Filtration
    - Some compendial water systems, primarily Purified Water Systems, use membrane filters for bacteria removal and control. Generally, 0.1 or 0.2 µm filters are employed. For 0.2 µm rated filters, the membrane filter is designated 0.2 µm rating if the following challenge test is met:
        - Membrane material is challenged with *Brevundimonas diminuta*, American Type Culture Collection (ATCC)-19146, at a concentration of 1.0×10^7 organisms per square centimeter and a differential pressure of 30 psig and a flow rate of 0.5 to 1.0 gpm/square foot of effective area. (ASTM Test Designation F 838–05). If the filtrate from test indicates the absence of bacteria, the membrane material is designated as 0.2 micron rating (27).
    - The literature discusses the basis for the challenge test as well as suggested integrity testing to verify the pore size rating (28).

- Other bacteria species may represent a "worse-case" test in terms of ability to a penetrate a 0.2 µm membrane filter particularly in a low nutrient environment (29,30). It is beyond the scope of this chapter to discuss membrane filter pore size designation further.
- Membrane filters may be used as final filters in bulk Purified Water generation systems. They are also used as the final filter installed in the polishing system for Purified Water, downstream of storage but prior to distribution. For a Purified Water system, the use of point-of-use 0.2 µm membrane filtration is discouraged.
- Historical concerns associated with the use of 0.2 µm membrane filters include "grow-through", "blow-through", and the accumulation of bacteria associated with the "retentive" nature of the membrane (lack of a waste stream). However, when coupled with a sound system design with membrane unit operations such as RO, periodic system sanitization, integrity testing, and periodic membrane filter replacement, the desired bacteria control can be achieved. Diffusive flow integrity system can be performed "in-place" and is non-intrusive.
- System design considerations include the following:
  - Selection of a sanitary filter housing to contain the membrane filters. T-style and L-style are generally considered. Housing selection is based on the number of membrane filters ("round") and the approximate 10" lengths of each membrane filter ("high"). Although flat gasket seal mechanism of the membrane filter element to the housing is available, the preferred seal mechanism is a double O-ring with single open end ("SOE"). The membrane filter element can be retained in the base of the filter housing by a "hold down" plate or by integral locking top with top "guide plate". The housing selection and seal mechanism should eliminate the potential for feed water bypass of the membrane filter.
  - The filter housing feed water and product water tubing should be equipped with manual isolation valves, filter housing vent valve, and filter housing drain valve. Some housings contain product water drain valves to minimize contamination during filter removal and replacement. A product water sample valve, feed water pressure indicator, and product water indicator should be included.
- Operation and maintenance considerations include the following:
  - Membrane filter removal and replacement frequency is critical to successful bacteria control. Although feed water and product water pressure monitoring are suggested, membrane filter change based on observed pressure drop indicates significant bacteria retention. Membrane filter change should be based on a routine product water sampling and bacteria monitoring program. When bacteria levels start to increase, by "trending" data, membrane filter replacement should be performed. The product trending program may be used to establish a membrane filter change frequency where replacement is performed prior to historical increase in product water bacteria level.
  - The use of single-use gloves to avoid handling of new membrane filters is suggested. Generally, membrane filters are shipped in air-tight plastic sleeves that can also be used to avoid handling the outside of the membrane element although cleaning of surfaces with isopropyl alcohol may be appropriate.
  - It is possible to "roll-over" O-rings during membrane filter installation particularly for L-Style housings where the seal end of the membrane filter may not be visible. Careful pre-rinsing of the O-rings with compendial water may facilitate proper O-ring engagement to the housing.
  - Because membrane filter removal and replacement is an intrusive operation, sanitization (thermal or chemical) should be considered subsequent to membrane filter replacement. Integrity testing of the installed filters is recommended. Diffusive flow integrity testing (non-intrusive) is preferred.
- Ultrafiltration
  - Ultrafiltration is a non-bacteria retentive membrane process with feed water, product water, and wastewater streams. Membrane pore size is larger than reverse osmosis but smaller than 0.1 or 0.2 µm membrane filtration. Ultrafiltration will remove bacteria, bacterial endotoxins, high molecular weight organic material, and colloids but not ions. Various pore size ultrafiltration membranes are available, generally from 0.001 to 0.05 µm. Ultrafiltration membranes may also be classified by molecular weight cutoff, generally about 10,000 to 100,000 Daltons. Ultrafiltration maximum pore size is verified by a latex bead challenge.
  - The ultrafiltration system may be used in a compendial water purification system, positioned downstream of RO with CEDI systems,

as final filtration to produce WFI quality water (31). Ultrafiltration may also be used in a Purified Water system employing ion exchange for reduction of organic material (TOC level), bacteria, bacterial endotoxins, and colloidal matter (32). Unlike RO membranes, which are generally spiral wound configuration, ultrafiltration systems often employ a hollow fiber configuration, although spiral wound membranes are also available. Hollow fiber ultrafiltration membranes often have low trans membrane pressure drop limitation but provide a unique "fast-flush" or "recycle" internal cleaning capability that is both highly effective and non-intrusive.

- Design, operation, and maintenance considerations include the following:
  - Much like 0.2 μm membrane filters, ultrafiltration membrane material will vary. Polysulfone is frequently used, exhibiting excellent chemical and physical stability (it is as "support" material for polyamide in thin-film composite RO membranes).
  - Feed water and product water pressure monitoring/control is required to verify limited transmembrane pressure drop.
  - One or more individual ultrafiltration membrane elements may be configured in parallel "banks" for operation. Periodic internal ("fast flush") cleaning can be performed to effectively remove material on the ultrafiltration membrane surface. During this operation, a valve in the waste piping fully opens, the product water valve from the bank of membranes slowly closes, and the feed water flow rate is increased. Product water is forced back through the ultrafiltration membrane in a "backwash-type" flow with accumulated impurities directed to waste. Effective cleaning can be achieved in a short time period (2–5 minutes). During normal operation, the waste stream from ultrafiltration membranes, about 2–5% of the feed water flow rate, contains most feed water impurities removed by the membrane. However, periodic internal "backwash" not only enhances system performance but increases ultrafiltration membrane life.
  - Terminal ultrafiltration system contaminants are minimal particularly in systems using reverse osmosis. Subsequently, membrane life is longer and chemical cleaning requirements less frequent than those of reverse osmosis membranes.
  - Ultrafiltration systems with 0.001 μm membranes are capable of proving a 3–5 log reduction in bacterial endotoxin levels when properly designed, operated, and maintained. For a Purified Water system, this can enhance biofilm control in downstream storage and distribution systems.
  - With ultrafiltration employ of parallel "banks" of membranes, bank-to-bank cleaning with freshly ultrafiltration bank product water is essential for bacterial endotoxin removal.
  - Ultrafiltration system design should include provisions for individual membrane module integrity testing.
  - Ultrafiltration system automatic valves must be open/closed slowly to avoid water hammer. Further, two-phase flow associated with gases should be avoided.
- **Distillation and Pure Steam Generation**
  - Single Effect Distillation Units
    - Distillation is the primary employed method for production of bulk compendial Water for Injection. Distillation heats pretreated feed water (liquid) to the boiling point, producing water in the gaseous state, steam. During the phase change, feed water impurities remain in the evaporator section of the distillation unit and are continuously or periodically removed by regulated "blowdown" from the evaporator section to waste. Feed water bacterial endotoxins are not present in pure steam from the evaporator and are directed to waste with other concentrated impurities during blowdown. The boiling process in the evaporator occurs at a minimum temperature of 100°C ("sea level" atmospheric pressure). A well designed, operated, and maintained distillation unit produces distillate product water with ≥ 3 log reduction in bacterial endotoxins.
    - Distillation units consist of three sections, described following.
      - Pretreated feed water is heated, generally by facility steam, in an evaporator section. The heating process consists of two distinct steps; heating water from the feed water temperature to the boiling temperature (energy referenced as "sensible heat") and generation of steam at the boiling water temperature (energy referenced as "latent heat"). The latent heat requirement per unit volume of Pure Steam generated from the evaporator section is about 80%–85% of the total energy requirement. This thermodynamic observation is important when discussing both multiple effect distillation and vapor compression distillation. In a single effect distillation unit or Pure Steam generator, facility steam (or other energy source such as electric heaters) is the only

energy source of both sensible and latent heat for producing pure steam. The concentration of feed water impurities in the evaporator section must be controlled by flow regulated blowdown. High evaporator impurity level increases the surface tension of water resulting in entrainment of water in Pure Steam ("carryover") with bacterial endotoxins.

- Pure Steam from the evaporator section flows to a vapor-liquid disengaging section. This section uses either simple mechanical entrapment-type devices of more elaborate centrifugal-type devices to inhibit the flow of any water vapor, directing liquid water in Pure Steam back to the evaporator section. It should be noted that any volatile feed water inorganic impurities such as ammonia gas or volatile organic impurities such as disinfection byproduct trihalomethanes will not be removed by the vapor-liquid disengaging section and will be present in Pure Steam and subsequent distillate product water.
- Pure Steam from the vapor-liquid disengaging section flows to a condensing section where latent heat from the Pure Steam is transferred to either cooling water or, for vapor compression distillation units, to water in the evaporator.
- Single effect distillation units may be used when small quantities of distilled water or bulk compendial Water for Injection are required. The evaporator section operates at, or slightly above, atmospheric pressure. The condensing section is generally elevated with gravity flow of distillate to a storage tank. Energy consumption per unit volume of distillate product is high when compared with that of multiple effect or vapor compression distillation units. Evaporator heating may use steam or electrical heating elements. Pretreatment of feed water is common. Energy and space physical requirements limit application of the technology to less than about 50–100 gallons per hour of distillate product.

- Multiple Effect Distillation Units
  - Single effect distillation is not capable of effectively providing the quantities of distillate product required for the majority of bulk compendial Water for Injection applications. Multiple effect distillation systems provide an effective alternative to single effect distillation by repeated use of latent heat in multiple pressure vessels configured in series with decreasing evaporator pressure from the first "effect" to the final "effect".
  - The boiling temperature of water increases with increasing pressure, demonstrated by the following data.

Pressure (psig)	**Temperature (°C)**
0 = Atmospheric	100
30	134.4
50	147.8
100	170.0
150	185.6

Facility steam is used to generate pure steam at elevated temperature and pressure from the first effect. Pure Steam from the first effect flows through a vapor-liquid disengaging section and subsequently flows to a second effect operating at a lower pressure than the first effect. Pure Steam from the first effect, in a manner similar to facility steam in the first effect, transfers its latent heat to water in the second effect producing Pure Steam from the second effect. The condensed Pure Steam from the first effect flows to a common condensing section. In a similar manner, Pure Steam from the second effect passes through the second effect vapor-liquid disengaging section and then provides latent heat to a third effect operating at a lower pressure than the second effect. This process can occur multiple times limited by the pressure rating of the stainless-steel pressure vessels and the cost for the vessels. Generally, a multiple effect distillation unit will have 2–6 effects. Both condensed Pure Steam supplying latent heat to the second through the final effect and lower pressure Pure Steam generated from the final effect flow to the common condensing unit for the system.

- By limiting facility steam heating requirement to the first effect, multiple use of latent heat at decreasing pressure in subsequent effects reduces energy cost per unit volume of distillate product. The first effect should be equipped with double tube sheet.
- Pressure vessel material of construction is generally 304, 304L, 316, or 316L stainless steel. The pressure vessels should be designed, fabricated, and tested in accordance with the ASME Code for Unfired Pressure Vessels, Section VIII, Division 1 (or equivalent).
- The higher vessel operating temperature requires higher quality of feed water to eliminate highly undesirable corrosion such as chloride stress and chloride pitting, as well as deposition of contaminants such as silica on heat transfer surfaces.
- Although multiple effect distillation units have few "moving parts", high operating

temperature/pressure requires annual preventative maintenance particularly for the vapor-liquid disengaging section on each effect are a concern.

    - Distillate product water is at atmospheric pressure. Frequently, a condensate collection tank with sanitary pump is used to transfer distillate to the top of a Water for Injection storage tank. Generally, Water for Injection is recirculated back to the collection tank if the Water for Injection tank is full (multiple effect distillation unit in standby mode). The condensing unit should be equipped with a hydrophobic vent filtration system. The distillate tubing from the multiple effect distillation unit should be equipped with sampling provisions, conductivity sensor, and an automatic divert-to-drain valve. When the multiple effect distillation unit begins to produce distillate product, flow should be directed to drain (with air break) for a preset time period. Distillate flow may be directed to a Water for Injection storage after the preset time period if conductivity is below a preset value.
    - A pretreated feed water pump may be required because of the operating pressure in the first effect.
    - Periodic preventative maintenance shall include verification of feed water flow rate to each effect and level control for each effect.
- Vapor Compression Distillation Units
    - A vapor compression distillation system may also be used for production of bulk compendial Water for Injection. Units are capable of delivering small to large volumes of distillate product. The type of distillation unit employed is based on pressure vessel limits (size, cost, and pressure rating) when comparing selection of vapor compression distillation versus multiple effect distillation. Vapor compression units consist of a lower pressure rated single evaporator section, vapor-liquid disinfecting section, and condensing technique with to the second through the final effect and compressor. Finally, because of the lower operating pressure, temperature pretreatment may be limited to activated carbon adsorption unit(s) and water softening unit(s).
    - Vapor compression distillation employs multiple heat exchangers and pumps to maximize thermal recovery of sensible heat. As an example, evaporator blowdown sensible heat may be used to preheat pretreated feed water to the evaporator section. The tempered treated feed water to a single evaporator is heated to produce Pure Steam, which is directed to a vapor-liquid disengaging section. Pure Steam from the vapor-liquid disengaging section is fed to a to the second through the final effect and compressor with stainless-steel internal surfaces where the pressure of to the second through the final effect and subsequently the temperature of to the second through the final effect and, are both increased. The higher-pressure to the second through the final effect and flows to a heat exchanger in the evaporator section where its latent heat is transferred, producing to the second through the final effect and in the evaporator section. Pure Steam condensate, distillate, is re-pressurized and transferred to the Water for Injection storage tank. Because compressed to the second through the final effect and is required for production of evaporator to the second through the final effect and, a "supplemental" facility steam heat exchanger is also positioned in the evaporator section, primarily required during start-up of the vapor compression distillation unit.
    - As indicated, the design of a vapor compression unit includes multiple "regenerative" heat exchangers and pumps to recover sensible heat.
    - When the evaporator pretreatment system is limited to activated carbon adsorption to remove disinfecting agent and water softening to remove multivalent cations, continuous evaporator blowdown is required. The dissolved solids in evaporator water will concentrate, increasing the surface tension of water with resulting carryover of water with bacterial endotoxins. Although to the second through the final effect and velocity from the evaporator section is lower for vapor compression distillation units than multiple effect distillation units, carryover is still a concern, dictating the blowdown rate. Concentrating impurities in the evaporator section such as metallic oxides, insoluble carbonate compounds, and siliceous compounds from both total silica (reactive and colloidal) may precipitate on heat transfer surfaces. Periodic chemical cleaning of the interior of the evaporator section may be required to remove the precipitates.
    - Pretreatment of municipal water using chlorine and ammonia (producing chloramines) as a secondary disinfecting agent often require conservatively designed activated carbon units followed by water softening units for removal of multivalent cations, followed by ammonia removal water softeners.
    - Pretreatment system feed water supplies with high concentration of dissolved ionic impurities, high Total Alkalinity, high Total Silica concentration, and/or high TOC level may require additional water purification techniques such as reverse osmosis.
    - If Pure Steam is required in addition to bulk Water for Injection and the vapor compression unit pretreatment system is limited to activated carbon and water softening, recirculating

Water for Injection is generally used as feed water to the Pure Steam Generator.

- Pure Steam Generators
  - Pure Steam may be used for various applications at a facility including steam-in-place (SIP) operations, autoclave sterilization, and sanitization of components. The chemical, physical, and microbial attributes for Pure Steam are presented in a USP *Monograph.* Condensed Pure Steam should have attributes similar to those of compendial Water for Injection. Pure Steam may be produced by a dedicated Pure Steam Generator or from the first effect of a multiple effect distillation unit. Essentially, a Pure Steam Generator is similar to a single effect distillation unit without a condensing section. Some autoclave units are equipped with integral Pure Steam Generators that may use electrical heating for production of Pure Steam.
  - Pure Steam Generator design, operation, and maintenance consideration are presented as follows:
    - Pure Steam Generator feed water quality should be similar to the feed water quality to a multiple effect distillation unit. The single column stainless-steel pressure vessel operates at elevated temperature and pressure with the long-term corrosion concerns discussed for multiple effect distillation units.
    - Unit capacity (steam volumetric flow rate and pressure) should be considered for all applications at a facility.
    - Periodic or continuous blowdown should be provided to control the concentration of feed water contaminants, minimizing "carryover" of water with bacterial endotoxins.
    - A feed water pump or tank with pump may be required to increase the pressure of feed water. Because the Pure Steam Generator is pressurized, potential backflow of feed water should be considered. A feed water tank with air break eliminates this concern.
    - Pure Steam Generator outlet sampling provisions should be provided, including a small cooling heat exchanger for condensing the Pure Steam to WFI.
    - For product used in the European Union, EN285:2015 (33) may be applicable for autoclaves using Pure Steam for sterilization of certain items such as a "porous load" or medical devices with a chamber size >60 liters. This document presents important Pure Steam criteria for autoclave operation. These criteria include non-condensable gases ≤ 3.5%, superheat ≤ 25°C, and dryness ≥ 0.95.

- **Storage and Accessories**
  - General Discussion
    - Compendial water storage tanks provide a physical separation between a compendial water generation system and distribution system. Some small capacity Purified Water system may employ a "tankless" design with a single delivery point. However, the vast majority of compendial water systems are equipped with storage tanks that provide a physical air break between the compendial water generation system and stored water, minimizing the risk of microbial contamination from biofilm in the generation system particularly for Purified Water systems. Further, storage tanks can provide a reservoir of compendial water for batching operations, a location for feed to and return from a distribution system, and a location for microbial control or introduction of chemical sanitizing agent for periodic distribution loop sanitization, if appropriate. Tank material of construction is dictated by the application but is generally 316L stainless steel. Some examples of design parameters impacting storage tank material of construction selection with operating/sanitization technique include, but are not limited to, the following:
      - Stainless-steel storage tank compendial water temperature maintained at a temperature ≥65°C (often 80°C) for elimination of bacteria. Periodic hot water sanitization may also be executed for the storage tank and distribution system.
      - Stainless-steel storage tank compendial water temperature maintained at "ambient" temperature with bacteria and bacterial endotoxin removal by continuous dissolved ozone with adequate contact time.
      - Stainless-steel or select polymer storage tank compendial water temperature at ambient conditions with periodic chemical sanitization of the storage and downstream distribution system.
    - Volume and tank construction should consider critical system parameters set in the URS such as:
      - Maximum instantaneous demand from the distribution system.
      - Makeup flow rate from the compendial water generation system.
      - Total volumetric demand for large compendial water applications such as batching.
      - Hours per day and days per week of operation.
      - Sanitization method (Ozone, heat, liquid sanitizing agent, etc.)
      - Purified Water or Water for Injection application.

- The stored compendial water sanitization method should consider the following:
    - Periodic hot water sanitization
    - Continuous or cyclic
    - Frequency
    - Temperature
    - Duration
    - Supplemental chemical sanitization (frequency)
    - Dead legs
    - Heating/cooling provisions
    - Distribution loop temperature; "cold", ambient, or hot
- Design Considerations
    - General compendial water storage tank design considerations include the following:
    - Tank material of construction
    - Physical area for installation and access to the area
    - Floor loading
    - Orientation—vertical cylindrical is preferred but horizontal cylindrical may be considered for hot applications but is discouraged for ozone applications.
    - Pressure rating—Note that the tank is generally vented to the atmosphere through a hydrophobic vent filtration system. Although most stainless-steel tanks are designed for an internal pressure of 30–50 psig, the actual operating internal tank pressure should not exceed a suggested 1–2 psig unless an unacceptable transient has occurred.
    - Vacuum rating—The hydrophobic vent filtration system should allow filtered air to freely flow to the gaseous space above compendial water in the tank. A slight vacuum may occur during tank "drawdown" or during cooling of hot compendial water as water vapor condenses. Although it is strongly suggested that all stainless-steel tanks be rated for full vacuum, a vacuum condition in the tank indicates that an unacceptable transient condition has occurred.
    - Stainless-steel tanks should be designed, fabricated, and tested in accordance with the ASME Code for Unfired Pressure Vessels, Section VIII, Division1, and so stamped. This not only verifies design condition but demonstrates required criteria for many United States jurisdictions, other jurisdictions, and facility insurance inspector requirements.
    - Stainless-steel tanks should be equipped with a top or side mounted sanitary manway for access to the interior of the tank.
    - Appropriate top, side, and bottom fittings (sanitary ferrule for stainless-steel tanks) may include, but are not limited to, top makeup water inlet, bottom center outlet, top hydrophobic vent filtration system, top mounted rupture disc, top mounted distribution loop return, level control sensors (top and bottom for differential pressure), and side or bottom mounted thermowell for temperature sensor.
    - Heated stainless-steel tanks generally include 2" thick chloride-free insultation encased in a 304L stainless-steel shroud around the sidewall and tank bottom.
    - Stainless-steel tanks may also be equipped with a single or multiple zone heat transfer jacket around the exterior and base to provide heating or cooling. The jacket is surrounded by insulation and an outer shroud. If a jacket and/or insulation are used, the bottom of the shroud should contain multiple "weep holes" for detection of leaks from the jacket or tank to insulation.
    - Diameter and straight side height of the tank should be selected based on area of installation, volumetric criteria, and the application. In general, taller tanks can provide several benefits such as better mixing for "turnover", decreased "outgassing" for ozone applications, greater contact time for ozone applications, and high NPSH to the downstream distribution pumping system.
    - For stainless-steel tanks, "Mill Certificates" for all 316L stainless-steel material is required. Further, the U1 Sheet containing details of construction with Mill Certificate reference shall be provided.
    - The interior mechanical polish for stainless-steel tanks as well as electropolishing requirements should be defined. For non-ozonated, non-heated Purified Water Storage Tanks, the mechanical polish is < 30 Ra and may be electropolished. For ozonated or heated Purified Water Storage Tanks and all WFI Storage Tanks, the suggested finish is ≤ 15 Ra and electropolished.
    - Compendial water distribution loop return to the storage tank is often directed through one or more spray balls that direct water, generally in a 180–360 degree "patten", to the upper sidewall and top of the vessel. Riboflavin testing is used to verify upper tank interior surface "coverage" by the spray ball system. For compendial water storage tanks using hot water, or even ambient water, the intent is to continuously rinse the upper vessel surfaces with water, reducing stagnant conditions. It should be noted that the use of spray balls is not applicable to

ozonated system because they accelerate outgassing of ozone gas during distribution loop circulation back to the tank. Further, it should be noted that spray balls are constructed of thin wall stainless steel. Holes in spray balls can erode over time, increasing the size of the holes with resulting impact on "coverage". Rouging of spray balls and impingement of storage and distribution system rouge on tank upper and top head surfaces (in the spray pattern of the spray ball) can occur. Subsequently, it is strongly recommended that removable spray balls (using a pin-type system with locking provisions) be considered.

- Nonintrusive tank level control systems should be considered. For stainless-steel tanks both load cells, or more commonly employed, differential pressure are used. The analog signal from either of these systems provides actual level display but also allows provisions for multiple set points with alarm provisions for low-low and high-high tank level and control of makeup flow to the tank based on an upper and lower "operating band". An additional benefit of differential pressure-type level control is the ability to detect and energize an alarm if upper vessel pressure or vacuum conditions, associated with an unacceptable excursion, occur.
- A hydrophobic vent filtration system should be positioned on the top of the compendial water storage tank. The system allows displaced air to flow from the tank during makeup flow with increasing tank level and removes bacteria from inlet air during tank drawdown. The hydrophobic vent filter material is Teflon or other hydrophobic material capable of removing bacteria with a size ≤0.2 μm. The vent filter total area should be adequate to allow removal of bacteria during maximum tank drawdown without allowing appreciable vacuum conditions to occur. For stainless-steel tanks with hot compendial water, a heating jacket must be provided around the hydrophobic vent filter housing. Although steam jackets have historically been used, electric heating blankets with thermostatic control and temperature monitoring should be considered to eliminate condensation on the hydrophobic vent filter media with associated "blinding" of the media and resulting tank pressure or vacuum excursion. The hydrophobic filter media is generally encased in an unpigmented polypropylene "support cage" that limits the heating temperature. Another issue of concern relates to water introduction from the spray ball(s). Hydrophobic vent filter housing arrangement shown in ASME BPE 2019 (34) should be considered with housing drain valve and electric heat tracing to minimize the water impact from spray ball(s). Finally, integrity testing of the installed hydrophobic vent filter membrane element should be performed prior to installation and subsequent to removal.
- For stainless-steel compendial water storage tanks, a compound-type rupture disc with integrity strip sensor should be considered. The compound disc will rupture on pressure or vacuum. As indicated earlier, neither of these conditions should occur for a properly designed and operated storage system and accessories. The discharge from the rupture disc assembly should be directed to a "safe area" at the base of the tank. For an ozone system using dissolved ozone for microbial control in the tank, discharge should be through unobstructed tubing to a safe, unoccupied area outside the facility. The integrity strip should provide an alarm if disc failure occurs.
- The use of unpigmented polyethylene or unpigmented polypropylene storage tanks for certain Purified Water applications may be considered. The tanks should be of the vertical cylindrical type with conical bottom. Interior tank access will be required and is often provided by a tight-fitting gasketed top cover. It should be noted that the tank interior may exhibit a green colored slime-algae appearance from exposure to light after operation for a time period as short as two months. Periodic chemical sanitization should be performed. The microbial growth inside the tank may require access with manual wiping to effectively remove.

- Operation and Maintenance Considerations
    - General compendial water storage tank operation and maintenance considerations include the following:
    - Periodic replacement of the hydrophobic vent filter with integrity testing of new filter and removed filter.
    - Periodic inspection of the interior of the tank. If rouging is detected, derouging and repassivation of the storage and distribution system should be performed. It is recommended that a proactive preventative

scheduled derouging and repassivation program be established. Generally, the frequency for derouging and repassivation will be greater for both ozonated storage systems and hot storage systems.

- The rupture disc assembly should be periodically inspected. Further, it is suggested that rupture disc proactive periodic replacement be considered because discs occasionally rupture without pressure and/or vacuum conditions, resulting in intrusive conditions during a non-shutdown time period.
- Periodic calibration of support accessories such as tank level control and temperature monitoring system should be performed generally annually.
- Periodic inspection of tank spray ball(s) should be performed. Further, periodic replacement of spray balls should be considered, particularly if erosion or rouging is noted.
- It is strongly suggested that tank manway gaskets be available prior to opening the manway on a stainless-steel tank. Manway gaskets, generally O-ring design, are critical for sealing the tank.
- Periodic replacement of sanitary ferrule gaskets should be considered as part of a proactive maintenance program.

- **Ozonation**
  - General Discussion
    - Ozone is a highly oxidative gas. When dissolved in compendial water, it can destroy bacteria and bacterial endotoxins. To achieve complete destruction of bacteria, adequate dissolved ozone concentration and contact time are required for compendial water with known conductivity, TOC level, and bacteria level. Residual dissolved ozone concentration in the compendial water storage tank is removed by high intensity 254 nanometer in-line ultraviolet radiation installed downstream of the distribution pumping system, prior to points-of-use. To enhance the control of bacteria and bacterial endotoxins in a compendial water storage and distribution system, periodic dissolved ozone sanitization of the distribution system can be performed by inhibiting withdrawal of compendial water at points-of-use and terminating electrical power to the dissolved ozone destruct in-line ultraviolet unit for a preestablished time period.

      Historically, dissolved ozone has been used as an effective bacteria control method in Purified Water storage and distribution systems. Gaseous ozone must be dissolved in water for bacteria and bacterial endotoxin destruction. However, dissolved ozone could theoretically be considered for use in Water for Injection applications, particularly when points-of-use require ambient temperature delivered WFI. This may also apply to systems producing WFI without the use of distillation.

      Dissolved ozone for compendial water storage systems can be generated by two different methods; production of gaseous ozone by electric discharge and production of dissolved ozone by electrolytic technique.
  - General Design Considerations

    General design conditions for gaseous ozone and electrolytic ozone systems are presented following.
    - As a gas, ozone is an unstable powerful oxidant and must be dissolved in water for effective microbial control. It is more soluble in water than oxygen. Further, as a gas, the maximum 8-hour average human exposure limit is very low, 0.1 mg of O_3/liter (35). To maintain an adequate concentration of dissolved ozone in a compendial water storage tank, continuous introduction is required.
    - Dissolved ozone will not only destroy bacteria but also oxidize trace inorganic contaminants and organic contaminants, including municipal disinfecting agent by-products, present in makeup water from the compendial water generation system. Oxidation of organic material, particularly halogenated low molecular weight disinfection by-products, must be carefully considered because product carbon dioxide reacts with water producing the highly conductive hydronium ion. This may increase measured conductivity to a value exceeding the Stage 1 compendial water conductivity limit.
    - Although ozone is soluble in water, "outgassing" of dissolved ozone in the compendial water storage tank will occur resulting in gaseous ozone in the physical area above the stored (and recirculated) water. When makeup water flows to the tank and level is increasing, the gas above the compendial water will be displaced. Subsequently, ozonated storage tanks require a technique to remove gaseous ozone. Chemical adsorption techniques are discouraged. Thermal gaseous ozone is suggested. Elevated temperature employed in thermal gaseous ozone destruct units results in decomposition of ozone to oxygen. The discharge from the gaseous thermal ozone destruct unit is generally directed to a "safe" unoccupied area outside the facility.
    - As compendial water level in the storage tank decreases, air will be drawn into the tank

through a hydrophobic vent filter physically position on the domed top of the tank. Some systems employ a check valve-type system to eliminate gaseous ozone flow from the tank through the hydrophobic vent filtration system. Hydrophobic vent filter cartridge materials of construction (membrane material and "cage/support" material) must be carefully selected for use in the gaseous ozone environment. A suggested alternative is to physically position the hydrophobic vent filter system between the thermal gaseous ozone destruct unit and discharge to a "controlled" outside discharge area. A heating jacket for the vent filter housing is suggested.

- During normal operation, dissolved ozone in the compendial water storage tank is removed prior to distribution by an in-line ultraviolet unit using a wavelength of 254 nanometers and intensity of 80,000 to 100,000 microwatt-seconds per.cm^2. The ultraviolet unit product water dissolved ozone concentration should be <0.01 mg/L.
- During periodic distribution loop sanitization, point-of-use withdrawal of compendial water is terminated and electrical power to the dissolved ozone destruct ultraviolet unit is inhibited. The concentration of dissolved ozone is automatically increased and allowed to flow through the distribution loop, returning to the tank. The frequency of this operation can vary from once every 8 hours to weekly or greater depending upon specific system operating factors such as the bacteria level in the makeup water.
- To verify the concentration of dissolved ozone during normal operation and periodic distribution loop sanitization, online sensors and "grab" sampling provisions should be provided as follows:
    - Downstream of distribution pump(s) for establishing the dissolved ozone concentration in the compendial water storage tank.
    - Downstream of the dissolved ozone destruct in-line ultraviolet system to verify the absence of dissolved ozone during normal operation.
    - In the distribution loop return tubing to very that a preset dissolved ozone concentration has been reached during periodic distribution loop dissolved ozone sanitization. It is suggested that the sanitization operation time period begin when the preset concentration is reached at this location.
- A compound-type rupture disc should be positioned at the domed top of the tank. To avoid potential gaseous ozone introduction at the physical location of the tank, the rupture disc should be vented to a "safe" area outside the facility.
- Ambient gaseous monitor(s) should be considered near the base of the tank and any other location where gaseous ozone may be present. Considering that ozone is heavier than air, it is suggested that the ozone sensor be positioned about 12–24" above floor level.
- Because the dissolved ozone distribution loop sanitization cycle includes recirculation of ozonated water back to the tank, the use of spray balls with associated ozone "outgassing" is discouraged. Any stagnant water vapor above the stored compendial water will be continuously exposed to gaseous ozone during normal operation. As an alternative, distribution loop return through a section of tubing inside the tank with discharge below the compendial water level is suggested.
- Materials of construction should include 316L stainless steel (tank) with ≤15 Ra mechanical polish and electropolish, and Teflon envelope or Viton sanitary ferrule gaskets and Viton or silicone manway O-ring. Any gasket material shall be compatible with the dissolved and gaseous ozone concentrations.
- The level control system should be nonintrusive type, preferably differential pressure.

- Additional Design Considerations—Gaseous Ozone Systems

  Additional design considerations for gaseous ozone generation systems include the following.

    - A gaseous ozone generation system requires clean, dry, oil-free air without particulate matter. The air is fed to an oxygen generator, which generally consists of two columns containing a molecular sieve that removes nitrogen by adsorption. Only one nitrogen removal column is operating, pressurized, while the second column is vented to the atmosphere. When the molecular sieve is saturated with nitrogen, pressurized air flow is diverted to the other column. The nitrogen saturated column is vented/depressurized with adsorbed nitrogen released. This "Pressure Swing Adsorption" (PSA) provides product feed to the gaseous ozone generator containing 90%–93% oxygen by volume. Although clean, dry, oil-free air can be used to produce ozone, the resulting oxides of nitrogen will increase the conductivity of water, which is undesirable for compendial water applications.
    - In a gaseous ozone generator, oxygen-enriched air passes between an electrode with a ceramic dielectric and a grounded electrode. A high-current electric field established between a narrow gap between the plates results in a "corona" discharge producing the oxygen atom (nascent oxygen) that reacts with molecular

oxygen to produce gaseous ozone. The gaseous product ozone concentration can be adjusted by increasing/decreasing the applied voltage.

- Gaseous ozone is fed through stainless-steel tubing, mounted vertically such that the elevation is above the height of the compendial water storage tank to eliminate the possibility of water "backflow" to the ozone generator. For most systems, the gaseous ozone is introduced to an ejector/eductor that mixes/delivers the gaseous ozone to the compendial water storage tank. A dedicated pump recirculating compendial water "around" the tank may be used or a side stream loop from the actual compendial water distribution loop. Occasionally, systems employ one or more "sparging devices" inside the compendial water storage tank for delivery of gaseous ozone.
- A critical factor in the physical discharge of gaseous ozone within the tank is the vertical elevation above the base of the tank. Because ozone destruction of bacteria is a function of concentration and contact time, a uniform dissolved ozone concentration in the storage tank is highly desirable. If discharge is too low, water with dissolved ozone may be disproportionally "pulled" to the suction of the distribution pump. This not only results in lower dissolved ozone concentration in water above the injection point but also provides an inaccurate tank dissolved ozone concentration because the tank ozone concentration is measured downstream of the distribution pump. If gaseous ozone is injected too high in the tank, greater gaseous ozone outgassing can occur because the water pressure (head) at the injection point is lower. The suggested injection point should be not only at a point that is always submersed in water but about a height equivalent to 25%–30% of the tank volume.
- Although a function of feed water makeup quality, a suggested operating dissolved ozone concentration for gaseous ozone systems is 0.2 to 0.5 mg/L. The suggested dissolved ozone sanitization concentration for gaseous ozone systems is 0.8 to 1.0 mg/L.

- Additional Design Considerations—Electrolytic Ozone System

  Addition design considerations for dissolved (electrolytic) ozone generation systems include the following.

  - Electrolytic ozone generation produces dissolved ozone within a "cell" containing an anode, cathode, and semipermeable membrane. A strong electrical current passes through compendial water flowing through the cell, splitting water into hydrogen gas and the oxygen atom. The oxygen atom reacts with oxygen at the anode, producing ozone dissolved in water. A "cell" requires a compendial water feed stream and high regulated electrical current. A "cell" produces a compendial water stream with dissolved ozone, wastewater stream, and hydrogen gas stream.
  - Individual "cells" are configured in a sandwich-like arrangement with stainless-steel anode and cathode ends and inert nonconducting, ozone tolerant "middle section. Inlet, outlet, and waste tubing is also nonconductive, ozone tolerant material. Electric isolation is critical.
  - Feed water flows through one or more "cells" that are operating in parallel, generally from a side stream of the distribution loop return water, upstream of the loop back pressure regulating valve. Product compendial water is directed to the depressurized return tubing downstream of the back pressure regulating valve. A static mixer in the tubing prior to the compendial water storage tank enhances mixing of dissolved ozone in compendial water.
  - A constant flow of compendial water through the "cells", constant regulated electrical supply, and feed water temperature ≤30°C are suggested.
  - It is suggested that compendial water distribution loop return (with dissolved ozone) be directed to one or more sections of tubing to a vertical height within the tank representing 25%–30% of the tank total volume. The section(s) of 316L stainless-steel tubing may contain drilled holes to assist in producing a uniform concentration of dissolved ozone in compendial water.
  - It is strongly suggested that the hydrogen waste steam be vented to a location outside the facility, free of any restriction that will allow flow to the atmosphere.
  - System accessories such as thermal ozone destruct unit, dissolved ozone destruct unit, hydrophobic vent filter, ambient and dissolved ozone monitoring, etc. are similar to those required for systems using gaseous ozone.
  - Electrolytic ozone systems do not require an ejector for mixing gaseous ozone with compendial water, do not have a section of stainless-steel tubing with concentrated gaseous ozone, exhibit less outgassing of ozone in the storage tank, and do not generate oxides of nitrogen in the storage tank.
  - The suggested normal operating dissolved ozone concentration in the storage tank for electrolytic ozone systems is 0.03 to 0.05

mg/L. The suggested dissolved ozone concentration during distribution loop sanitization is 0.1 to 0.2 mg/L.

- Operating and Maintenance Considerations
  - Proper operation of the ozone generation system should be verified. For gaseous ozone generation systems, maintenance of the oxygen generator is critical. For electrolytic ozone generators, the dissolved ozone concentration from the system prior to the distribution loop return tubing should be determined and anode, cathode, and semipermeable membranes proactively replaced.
  - Hydrophobic vent filter elements should be replaced every six months.
  - It is suggested that proactive dissolved ozone destruct ultraviolet unit lamps, quartz sleeves, and O-rings be replaced every six months but no longer than annually to ensure absence of dissolved ozone in delivered compendial water.
  - Inspection of the interior of the compendial water storage tank should be performed at least annually. The tank manway gasket should be replaced after inspection.
  - Annual proactive compendial water storage tank and distribution loop derouging and repassivation is suggested.
  - Proactive distribution pump seal replacement should be considered annually.
  - Tank and distribution loop valve diaphragms should be replaced annually.
  - Proactive tank and distribution loop sanitary ferrule gasket replacement should be considered every two years.
  - Dissolved ozone sensing elements should be inspected, cleaned, and calibrated annually.
  - Conductivity cells should be calibrated annually to verify the cell constant.

- **Distribution Systems**
  - Distribution Loops
    - The classical compendial water distribution system consists of a serpentine-type loop to deliver water to points-of-use. The specific application dictates loop configuration. As an example, a hot recirculating Water for Injection loop may have one or more sub loops with a pump that recirculates through a cooling heat exchanger to deliver ambient Water for Injection, returning to the suction side of the sub loop pump. Other hot recirculating Water for Injection distribution systems may employ point-of-use heat exchangers. Loop material of construction for hot or ozone applications are generally 316L stainless steel. The use of a "ladder-type" loop with recirculation and appropriate flow/pressure "balancing" for certain applications is not common. The use of non-stainless-steel tubing/piping such as unpigmented polypropylene or polyvinylidene fluoride (PVDF) may be considered for certain applications where appropriate.
    - The velocity of compendial water through a distribution system under all operating conditions should result in full turbulent flow in the return section of the loop at maximum loop draw-off. The Reynold's number, a function of water velocity, inside tubing diameter, density, and viscosity, should be >4,000. Values lower than 4,000 result in laminar or transitional flow with a resulting boundary layer that contributes to biofilm formation.
  - Distribution Pumps
    - Sanitary centrifugal 316L stainless-steel distribution pumps are used for the majority of compendial water distribution systems. Pump flow rate and required pressure determine sizing including the impeller diameter and motor horsepower size using information from the URS. The use of distribution pump motor VFD is increasing, providing flexibility for various combinations of water flow rate demand at multiple points-of-use, particularly batching applications.
    - Many compendial water distribution pumping systems employ a "hard piped" spare pump as backup to an operating pump. In some of these systems, operation of each pump is "rotated" as frequently as weekly. To avoid a dead leg of the backup pump, feed water and discharge valves are open and a hole cut in a check valve in the discharge tubing. This condition produces reverse flow through the standby pump. System design, operation, and maintenance of this arrangement is important.
    - Some Water for Injection distribution pumps employ a double mechanical seal with WFI flush. It is suggested that the number of applications using this seal arrangement is decreasing with use of a single mechanical seal increasing.
    - Perhaps the biggest concern associated with compendial water distribution pumps is proactive seal maintenance (inspection and, if required, replacement). The clearance between a pump impeller and pump casing is intentionally small. With increasing hours of operation, pump seals (both single and double) may wear without exhibiting a leak. Often, pumps with motors are physically located in areas with background noise. If the pump shaft is not completely at right angle to the impeller it is possible that the impeller may "wobble". This may result in slight contact of the pump

impeller, generally rotating at a speed of about 1750 rpm to 3500 rpm, and the pump casing. The sound of stainless steel to stainless steel contact may not be heard. The consequences can be catastrophic. Production of small stainless-steel particles into the compendial water distribution system is unacceptable.

- Heat Exchangers
  - Compendial water heat exchangers are frequently used, particularly for hot Water for Injection systems. The vast majority of applications employ sanitary shell and tube exchangers with flow through the inside of 316L stainless-steel tubes. Heat exchangers may be used for in-line heating or cooling. Heat exchangers may also be used for cooling and/or heating in distribution system sub loops. Finally, point-of-use heat exchangers may be used for cooling.
  - Heat exchangers should be designed, constructed, and tested in accordance with the ASME Code for Unfired Pressure Vessels, Section VIII, Division 1. Interior tube mechanical finish and electropolish, if applicable, should be the same as that of the compendial water distribution loop tubing.
  - Compendial water heat exchangers should be equipped with a double tube sheet.
- Distribution Tubing and Valves
  - Dead legs are a concern because they provide a location for "hideout" of contaminants, particularly bacteria and bacterial endotoxins. Historically, a dead leg was determined by the number of "pipe diameters" (3–6) of a tubing branch from the centerline of the main branch. A much more appropriate method of determining a dead leg uses the length of the smaller branch from the inside wall of the larger (main) tubing divided by the inside diameter of the smaller branch, $L/D < 2$ as a guideline (36).
  - In an attempt to eliminate dead legs, diaphragm valves are used with diaphragm weir at the wall of the distribution tubing. These zero dead leg valves are commonly used in compendial water systems. Also, short outlet tubing tees provide an excellent method of reducing the L/D value for instrumentation such as sanitary pressure gauges or sensors.
  - Orbital welding is employed to join sections of stainless-steel tubing and fittings. A detailed specification is required for proper documentation of orbital welding including, but not limited to the following.
    - Heat numbers and mill certificates for all stainless-steel tubing, fittings, instrumentation, valves, etc. in contact with compendial water.
    - Welder certificate of qualification to accepted standard such as ASME.
    - Welder "stencil" or other identification number on each orbital weld
    - Organization Quality Control Manual.
    - Dimensioned isometric drawing showing all orbital welds with weld number.
    - Dimensioned isometric drawing showing all welds with heat numbers.
    - Orbital welding machine printout for each weld.
    - Weld test coupons for each day, individual welder, and orbital welding machine.
    - Boroscopic inspection reports for orbital welds.
    - Identification of tubing slope with low point drains.
    - Pressure test results (pneumatic preferred).
    - Compliance with specified orbital weld acceptance criteria.
    - Passivation procedure and report with test results.
- Point-of-Use Valves, Fittings, and Accessories
  - Point-of-use valves should be zero dead leg diaphragm-type. If equipped with automatic actuator. valve should fail in a "safe" position (open/closed).
  - Many points-of-use are "hard piped" directly to the point-of-use valve. A sample valve is required to verify "delivered" compendial water quality. The sample valve can be positioned in the main distribution loop directly upstream or downstream of the point-of-use valve. As an alternative an "access port" type valve may be used, zero dead leg type, an integral part (or connected by sanitary ferrule) to the point-of-use valve.
  - Any point-of-use that could be pressurized with potential backflow to the compendial water distribution system should employ a tank with air break or a double block, bleed, and vent arrangement. Pressurized applications could include direct feed to a Pure Steam Generator, glassware washer, or autoclave.
  - Many points-of-use employ hoses to deliver water for compounding, cleaning, laboratory applications, or other applications. A hose management program is critical. Back contamination of hoses or contamination (chemical or microbial) of delivered compendial water from the hose can occur.

7 Validation of Critical Utilities

David Maynard

CONTENTS

7.1 INTRODUCTION

This chapter will review the utilities used in clinical trial and production facilities, whether these facilities are finished drug, APIs (active pharmaceutical ingredients), whether the products are biologicals, solids, liquids, creams, ointment or sterile products. It is important to understand the methods needed to first determine what types of utilities are critical to the processes and second to determine those on the fringe or outside the domain of qualification. This effort will focus on the critical utilities but provide guidance on the non-critical utilities.

This chapter will assist the pharmaceutical engineer, clinical trial scientist, quality assurance professional, qualification engineer, etc. and the management team at the facility in determining which system needs to be addressed and how these systems will be commissioned, qualified and in some cases validated. It will suggest a team approach that is established to advance the project to its swift and successful conclusion.

Further, the decisions that need to be made in first ascertaining which utilities need to be commissioned or qualified and how this commissioning/qualification of utilities is to be accomplished are best made with a cross-functional team that will be assembled to address the project. This is sometimes termed a risk assessment. This team may be augmented by outside contractors or consultants and will be discussed later in the chapter.

The chapter will first establish some definitions that are important to the process and then use these definitions throughout the document to explain the why and how of the qualification of utilities.

7.2 DEFINITIONS USED IN THIS CHAPTER

Non-Critical—(No Impact)—Utility that has no impact on process or product quality (i.e. water—source for cleaning/sanitizing of non-product contact surfaces; steam—used for heating of vessels).

Non-Critical Point of Use—No direct impact on the quality of the product/process for which it is being used.

Support—(Indirect Impact)—Utility that supports a Direct Impact utility, but does not have a direct impact on the quality of a product.

Critical—(Direct Impact)—utility that is in direct contact with the product or that could have a direct impact on the quality of the product.

DOI: 10.1201/9781003163138-7

Critical Point of Use—Direct impact on the quality of the product/process for which it is being used (i.e. water—used for cleaning of surfaces with direct contact to product, used in formulation processes, used in supply to pure steam generator; steam—used in sterilization processes).

Critical Process Parameter (CPP)—A process parameter that is controlled within a predetermined range to ensure that the product meets its critical quality attributes.

Critical Quality Attributes (CQAs)—A set of measured characteristics inherent in the product that describes the product's acceptability for use.

Commissioning—A well-planned, documented and managed engineering approach of inspection and testing of equipment and systems to ensure that they are installed according to specifications and are ready for operation in a safe and functional environment that meets established design requirements or qualification when required. [1]

Installation Qualification (IQ)—Documented evidence that the equipment, system or utility meets all critical installation requirements. [2,3]

Operational Qualification (OQ)—Documented evidence that the equipment, system or utility operates as intended throughout all required ranges.

Performance Qualification (PQ)—Documented evidence that the equipment, system or utility perform as intended and meets all pre-established acceptance criteria

Sampling Plan—Written procedure describing the physical location of sample points, the frequency of samples taken to ensure the system is in control and the equipment to be used in taking the sample

7.3 PLANNING ACTIVITIES FOR THE CRITICAL UTILITY

The first step is to list all of the utilities at the facility or site and determine the criticality of the system. This can be accomplished by performing impact assessment that presents the risks to product posed by the utility. It can also be determined by following a series of questions that continue to refine the analysis until it is clear which path needs to be followed—no commissioning, commissioning only or commissioning and qualification required.

An example of the questions that can be used to determine if a system needs commissioning and/or qualification is as follows:

> Is the utility supporting Good Manufacturing Practice (GMP) activity? If 'No', then there is probably no need to even Commission the system, but if 'Yes', it needs more clarification as in the next question.
>
> Does the utility or direct output come in direct contact or primary packaging contact? If 'Yes', then the system must be Commissioned and Qualified. If 'No', then it needs more clarification as in the next question.
>
> Is the direct output of the utility used in the environment surrounding an exposed product? If 'Yes', then the system must be Commissioned and Qualified. If 'No', it needs more clarification as in the next question.
>
> Is the utility or output used in final cleaning steps (equipment with direct product contact or the primary packaging components)? If 'Yes', then the system must be Commissioned and Qualified. If 'No', it needs more clarification as in the next question.
>
> Are the utility and its direct output used within a sterilization or sanitation process? [4] If 'Yes', then the system must be Commissioned and Qualified. If 'No', it needs more clarification as in the next question.
>
> Does the operation or control of the utility have a direct impact on the critical quality attributes of the product or the critical process parameter of the production systems? If 'Yes', then the system must be Commissioned and Qualified. If 'No', the system needs only to be commissioned.

The second step in any qualification or validation process is to develop a plan of what is to be accomplished. This can be a complex qualification plan that addresses many different utilities in multiple areas of the site or could be a specific qualification strategy or plan that addresses only one specific utility. Each of the plans will include how the system will be commissioned or qualified, who will perform the effort, what type of protocol is to be used and what approval signatures are required. If the commissioning documents are to be leveraged into the qualification, the copies of those documents must be integrated into the qualification documents prior to the approval of the qualification protocol.

The plans should be developed by a cross-functional team consisting of Engineering, Operations, Quality Control, Quality Assurance and the Commissioning or Qualification personnel. Each of these participants will have their own roles and responsibilities that are important to the outcome of the project. This group allows all GMP functions to participate early in the project and help ensure a satisfactory outcome to the testing.

The protocols will include the system limits, the physical parameters and attributes to be tested, the acceptance criteria to be met and the signatories required to approve or certify the qualification and validation actions. The reports will include the synopsis of the testing and verify the acceptance criteria have been met.

The third step is the actual commissioning, qualification and validation of the individual utility. These activities will verify the design, installation and operation of the equipment or systems. As part of the qualification and validation activities, the critical process parameters that have been established for the utility will be verified to ensure the ongoing control and certification of the systems in the daily activities of the facility. Prior to performing any validation testing, the sampling routines for the utility must be established in order to ensure the validation activities will be the same as those used in routine operation. Note: one of the most frequent comments by regulatory agencies occurs when the validation sampling does not accurately reflect the routine use of that particular point of use, including flushing times and methods of use (i.e.

if a hose is used between the point of use, then the sampling should be from the hose and not directly from the point of use). The validation will test the Critical points of use of the system and set in place the routine monitoring of these points.

7.4 SYSTEMS TO BE DISCUSSED IN THIS CHAPTER

The utilities in typical facilities include Gases—compressed air, nitrogen, oxygen, carbon dioxide; Liquids—process water, solvents; Steam—process, clean; House Vacuum; Electrical; Drains—process and waste. There may be other utilities encountered within the facility and the same or similar validation processes can be adapted to the other systems.

7.4.1 Gases

The most common gases used in the pharmaceutical industry are compressed air used for instruments or product contact, and nitrogen used for providing an inert gas in the vial, ampoule or Water for Injection tanks and also used for creating an inert pressure pad in processes where solvents are present.

The validation of each of these is similar in that the equipment used to generate, store and distribute the gas must be first Commissioned and then Qualified. Once the Installation and Operational Qualification Summary Reports for the equipment have been approved, then the distribution system can undergo Qualification to ensure the delivery of the gas to the acceptance criteria established that verifies the specifications for the gas. It is noted that instrument air need only be tested through the Operational Qualification, as it does not come into product contact.

Some of the instruments used to gather compressed gas samples include SAS Microbiological Air Sampler, Mattson Garvin Compressed Gas Sampler—Model P-320, and SMA Compressed Air Sampler.

7.4.2 Liquids

The most prevalent liquid utilities used are Process Water (e.g. soft, deionized, USP Purified, WFI) that are used in cleaning operations and product batching, and water used in the heating and cooling processes that are not in product contact and thus does not require anything more than commissioning of the system.

The validation of each of these is similar in that the equipment used to generate and store the water must be first Commissioned and then Qualified [5]. Once the Installation and Operational Qualification Summary Reports for the equipment have been approved, then the distribution system can undergo Qualification to ensure the delivery of the water to the acceptance criteria established that verifies the specifications for the water. USP Purified Water and the Water for Injection needs to have an extensive testing of all points of use over a one-month period to ensure the quality of the water is delivered routinely. The points of use tested during the performance qualification will typically be those used for routine sampling. In addition, an extended Qualification is executed wherein the distribution system is monitored over a one-year period (including the 30-day period included in the initial performance qualification) to ascertain any seasonal differences that impact the quality of the water.

7.4.3 Steam

Typically there are three types of steam used in our industry—Plant Steam that typically has boiler chemicals entrained, Chemical-Free Steam without boiler additives and Pure Steam that when condensed meets the water requirements of USP Purified Water. Plant steam and chemical-free steam distribution system validation is typically complete at the end of the Operational Qualification. Each of the systems must be of appropriate design (including steam trap location), be properly maintained and be operated using approved procedures.

Pure steam is typically produced by specially built steam generators or from the first effect of a multiple effect still. The feed water is typically Purified Water or Water for Injection or from other sources of known chemical quality as specified by the vendor of the generator. The steam contact surfaces, including the generator and distribution system, must be corrosion-resistant material (316 L stainless steel is the most common material used).

A properly designed and constructed steam generation and distribution system that is operated and maintained correctly will negate the concern expressed by some regulatory agencies. The three physical attributes non-condensable gasses, dryness fraction and superheat appear to have been concerns in older installations, including hospitals, wherein the systems may not have been properly designed, installed or maintained. However, those pharmaceutical facilities that use pure steam for product sterilization purposes should consider the performing of these tests as they may be required by regulatory agencies. These steam quality tests are detailed in UK Department of Health and Social Security document, Health and Technical Memorandum 2010 part 3 [6]and in ISO 11134, "Sterilization of Healthcare Products—Requirements for Validation and Routine Control—Industrial Moist Heat Sterilization" [7,8]. The steam quality limits are also included in the European standard EN 285 [9].

The qualification of each of these is similar in that the equipment used to generate the steam must be first Commissioned and then Qualified (note: Plant Steam does not require more than an Installation Qualification) to ensure the equipment operates properly. Pure Steam will undergo extensive Performance Qualification testing beyond the Operational Qualification to show the quality of steam is maintained over an extended period of time, typically one month.

7.4.4 House Vacuum

House vacuum systems are used for many services, but the distribution systems that come into contact with the product or the primary container require attention. These systems will require the equipment used to generate the vacuum to be first Commissioned and then Qualified. Once the Installation and Operational Qualification Summary Reports for the equipment have been approved, then the reservoir tank and distribution system can undergo Qualification.

7.4.5 Electrical

The electrical systems are often overlooked in the verification/certification of the facility and utilities. These systems require that similar commissioning and qualification activities are performed to ensure the continued deliverance of the power to operate the other utilities and the facility processes. Each of the electrical systems in the facility has specific requirements that will include both quality and quantity attributes that are established by the requirements of the facility. These attributes will include frequency, phase and voltage requirements as well as sufficient capacity to enable the full load required to operate the facility. Other systems may include battery backup, standby generator capacity, clean lines for computer operation, and voltage surge protection.

The electrical system to be tested requires complete documentation including monitoring system identification, electrical schematics that include all pertinent information including wire size, circuit identification, switching equipment and backup systems. There should be written instructions on operation and maintenance of the systems as well as emergency procedures that will come into effect in case of natural or man-made disasters.

The qualification will consist of monitoring the systems to ensure voltage, phase and load conditions can be maintained while the plant is being started and maintained during production usage. The portions of the facility that do not have backup or emergency power generation systems will need to be tested after the loss of supplied electrical power to ensure the facility can come back on line safely, both from a temporary loss of power and from a sustained loss of power. Procedures should be put in place to cover these situations, as typically there will be a need to have a sequential restart to a facility to ensure the safe operation of the plant and systems.

In areas that are subject to 'brown out' or reduced voltage conditions, the electrical supplies to primary process equipment will need to be verified that the equipment can operate in this reduced energy level and continue to meet all performance attributes established.

All alarm monitoring and display systems as well as where there is emergency power equipment in the production facility should be tested using both the primary and secondary power systems. The switchover capability and operation of the equipment must be tested to ensure the smooth transition between the power sources and to verify the operation of the equipment on the backup power system. Computer equipment must be fully tested to ensure no loss of data during a transition from one power source to another.

All protection equipment (overload, safety switches, voltage stabilizers and line suppressors) should be tested for both normal operation and peak-load or worst-case conditions.

7.4.6 Drains

The drains in facilities are often overlooked and sometimes are the source of unanticipated problems (i.e. contamination, backflow and means of causing flooding from external storm sewer systems) and as such require full understanding of their design and connections. For this discussion, consider only the drains to process and sanitary. The sanitary drains remove various wastes from the areas, whereas process drains remove process-specific fluids. The design and construction of the facility will need to be verified to ensure that there is not an interconnection between the two systems. A simple dye test in which dye is placed into the process drain and shown that it does not appear in the sanitary waste exit from the facility will verify the systems are not interconnected. There is a need for complete installation and operational qualification of drains in process areas to ensure that all drain points are interconnected and drain to process waste.

The specific qualification will be similar to other distribution systems and the drawings, material of construction, pipe size, valves, leak testing, safety features, etc. must be verified. Hard connections between water/steam systems must be avoided and an air break must be verified to ensure there is no back siphoning. The use of check valves between process systems should be tested to ensure the systems remain separated. Where solvents are used, the drainage system will be verified to be explosion proof and properly vented.

7.5 TESTING

7.5.1 Common Steps in Commissioning, Qualification and Validation of Utilities

All of the utility systems during the various stages will have a common series of evaluations and tests that ensure the installed equipment and systems meet the required specifications and design elements that assist in ensuring the long-term operation of the utility. These will include the verification of the materials of construction to the design specifications and engineering/construction drawings; the verification of the welding and of the welder performing the welding of the generation and storage systems by the use of video borescope to verify the individual welds and the use of test coupons that certify the welder as being competent of performing the welding operations; the cleaning and, if required, the passivation of the generation, storage and distribution systems to remove the residual fragments of material, welding material, oils and greases used in the construction of the metallic systems and also the removal of other materials in the system; pressure testing or pressure hold tests to ensure the systems are integral and meet the various code requirements including American Society of Mechanical Engineers (ASME), etc.; various safety tests to ensure the systems are protected from over pressure or over temperature conditions. Each of these individual verifications will depend on specific procedures and test equipment that are described in other chapters.

In order to have a traceable inventory of completion of these tasks, each of the preceding will require written and approved documents that certify the veracity and accuracy of the tests that were performed. These tests can be performed in-house or by qualified engineering, consulting or testing organizations that specialize in these activities. The required testing activities will be compiled in pre-approved protocols that will list all of the tests along with their acceptable results or acceptance criteria. The completed protocol will be executed and checked by the performing organization and then approved or certified by the Quality organization.

The documentation will be organized in files that can be retained in a site library or depository of information along with the Commissioning, Installation, or Operational Qualification documents.

7.5.2 Design

Discussion of design to meet need of facility. The design of the critical utility systems is included in the chapter dedicated to the design of the facility to meet the requirements for the manufacturing of the product.

7.5.3 General Tests for All Utilities

- Installation Qualification (IQ).
  a. Verification of Qualification Prerequisites including successful completion of all Commissioning activities.
  b. Verification of System Documentation that could include technical data sheets, functional specification requirements, material of construction, welding documentation, piping insulation documentation, cleaning and passivation reports and pressure testing reports. If these were included in the Commissioning activities, reference copies may be attached or the section of the commissioning report referenced.
  c. When Sanitary piping is used, a separate welding documentation will be required to indicate the weld number, weld log and welder certification; piping isometric verification will be verified.
  d. Verification of Preventive maintenance documentation and spare parts lists including verification of proper entry into a computer-controlled preventive maintenance management system if used at the facility.
  e. Verification that the Process and Instrumentation Drawings (P and IDs), wiring and cabling drawings are accurate and reflect the installed system. These drawings are commonly referred to as the As-Built drawings.
  f. Verify that all generation and distribution components and piping are properly identified.
  g. Verification that all components in the system are in compliance with the specifications and design.
  h. Verification that all instruments that are critical to the operation of the system are calibrated and are properly entered into the computer-controlled calibration management system if used at the facility.
  i. Verification that all supporting utilities are properly installed and are in compliance with requirements of the system.
  j. Verification that all of the unit operations are operating properly and meet all specification and testing requirements.
  k. Verification that proper software version electronic copies are available for backup capability, programmable logic controller (PLC) source logic is complete and clear, there is no dead code in the system and that hard copies of the control logic are available and made part of the protocol.
  l. Input/output (I/O) verification is performed to ensure that all I/O points were addressed and properly connected to field devices per specification
- Operational Qualification (OQ).
  a. Verify that all Operational Qualification prerequisites are complete in that all IQ sections are completed or that any un-executed sections would not impact the execution activities of the Operational Qualification.
  b. Verify that all generation and distribution components and piping are properly identified.
  c. Standard operating procedures (SOPs) are in-place and approved and all required individuals training is documented and available.
  d. Prepare a list of all SOPs, including reference number, title and effective date related to the generation or distribution systems.
  e. Obtain copies of all current SOPs.
  f. Challenge all Operation SOPs to verify the document is suitable and allows the system to operate properly.
  g. Challenge all Preventive Maintenance SOPs to verify the document is suitable and allows the system to be properly maintained.
  h. Obtain copies of all current Preventive Maintenance documentation including work orders or evidence of maintenance having been performed in the proper frequency.
  i. Verify that all instruments that are critical to the operation of the system are in current calibration and calibration certification is available.
  j. Verify that all testing instruments are in current calibration and the calibration certificates are available.
  k. Verify that environmental conditions surrounding the system do not adversely affect the operation of the generation and distribution systems.
  l. Verify any environmental conditions surrounding a PLC-driven system to not impact the operation of the controller.

m. Radiofrequency and electromagnetic interference tests are performed close to the control panels to verify the system is not adversely affected by these disturbances.
n. Power failure and recovery tests are performed to document the effects of these events on the control of the system.
o. Alarm and interlocks are tested to verify the proper operation of the system.
p. Software security access levels are verified to ensure the system cannot be modified without specific authorization.
q. The sequence of operations are verified to challenge the operational sequence of the control system to ensure that the system's functions are followed as described in specifications.

- Performance Qualification (PQ).
  a. General description of the system to be tested including specific information in sufficient detail to create a verbal picture of the system and its component parts as well as its location in the facility and any special requirements of the system.
  b. Develop the specific critical process parameters to be reviewed or certified as part of this phase of qualification.
  c. Verify that all performance qualification prerequisites are complete in that all OQ sections are completed and reviewed, documents are completed and that any un-executed sections would not impact the execution activities of the Performance Qualification.
  d. Verify that all sampling procedures are current and reflect the testing that is to be accomplished in the performance qualification.
  e. Verify that all personnel that will be executing the performance qualification, or involved with the testing of samples, have been trained in the expectations of the protocol and the need to accomplish the testing in the proper sequence and in the correct time period.
  f. Verify that all generation and distribution sampling points are properly identified and are listed in the current sampling procedure.
- Extended Qualification.
  a. The high purity water systems and pure steam systems will require testing at the sampling ports for one full year to indicate the system's ability to maintain proper quality throughout one complete cycle of the seasons.

7.5.4 Specific Tests—Gases

In addition to the general tests, all gas systems critical process parameters, including pressure, flow rate, and capacity must be verified during the Operational qualification. Specific tests must be established to ensure the systems are delivering the gas at the required conditions.

7.5.4.1 Compressed Air

The typical Compressed Air system consists of an oil-free air compressor that compresses the ambient air and an air dryer connected to an air receiver surge tank that supplies air throughout the distribution system. Where a non-oil-free air compressor is used, the Compressed Air passes through a coalescing filter and oil vapor absorber filter where the hydrocarbons are removed and then passes into the surge tank and distribution system.

1. All use points including the compressed air exit point from the air compressor will be sampled for a minimum of seven consecutive days.
2. Testing will be in accordance with documented procedures.
3. Environmental tests (microbiological and nonviable particulate) will be taken for information only for non-sterile use.
4. Environmental tests for sterile use must meet the criteria for the area of use.
5. Suggested acceptance criteria for the Compressed Air generation system include:

Parameter	Parameter specifications
Compressor oil temperature	<123°F reference only (based on specific system)
Cooling tower differential pressure	<7psi (based on specific system)
Maximum discharge Pressure	Reference only
Maximum dew point	Reference only

6. Suggested acceptance criteria for the Compressed Air distribution system at product point of use (using a 0.22 micron in-line filter in controlled areas) include:

Parameter	Parameter specifications [10,11]
Water and oil	None detected [10]
Oil	0.1 mg/m^3 [11]
Odor	No smell [10,11]
Atmospheric dew point	< 50 °F (20mg/m^3)
Microbiology for total count	Reference only (no growth for aerobic organisms for sterile)
Particulate (nonviable) (total count of particles <0.5 microns per cubic foot)	Reference only (meet area quality standards)
Carbon monoxide	NMT 10 ppm; 5 ppm [11]
Carbon dioxide	NMT 50 ppm [10]; 500 ppm [11]
Nitric oxide /Nitrogen dioxide	NMT 2.5 ppm [10]; 2 ppm [11]
Sulfur dioxide	NMT 5 ppm [10]; 1 ppm [11]

7.5.4.2 Nitrogen

The typical Nitrogen system consists of a generation plant, liquid nitrogen backup tank and distribution header. The basis of operation is for ambient air to be treated and separated creating NF/EP quality Nitrogen that is delivered to the distribution header. In an event of high-demand peaks or when the generation plant is shut down for maintenance activities, the liquid nitrogen backup tank will provide gaseous nitrogen to the distribution header.

1. All use points including the nitrogen exit point from the generation plant will be sampled for a minimum of seven consecutive days.
2. Flush sample lines through the stainless-steel sampling collector to ensure all ambient air is removed from the container prior to collecting the test sample.
3. Testing will be in accordance with documented procedures.
4. Environmental tests (microbiological and nonviable particulate) will be taken for information only for non-sterile use.
5. Environmental tests for sterile use must meet the criteria for the area of use.
6. Suggested acceptance criteria include:

Parameter	Parameter specifications [10,11]
Identification match test	Extinguish in absence of oxygen
Purity (nitrogen assay)	>99.00% [10]; >99.50% [11]
Carbon monoxide	≤0.001% [10]; ≤5 ppm [11]
Carbon dioxide [11]	≤300 ppm
Oxygen	<1% [10]; ≤50 ppm [11]
Water [11]	≤67 ppm [11]
Oil	Not discussed [10,11]
Microbiology for total count	Reference only (no growth for aerobic and anaerobic for sterile)
Particulate (nonviable)	Reference only (meet area standards)

7.5.5 Specific Tests—Liquids

In addition to the general tests, all of the water systems' critical process parameters, including temperature, pressure, flow rate and capacity must be verified during the Operational Qualification. Specific tests must be established to ensure the systems are delivering the water at the specified rates to minimize the buildup of contamination in the piping systems.

7.5.5.1 Water Pretreatment

The water pretreatment system is designed based on the local water conditions and the intended use of the water. Typically the system consists of units dedicated to specific functions including removal of suspended solids, chlorination removal, hardness and metal removal, removal of microbiological impurities and control of microbial growth. Each of these has their specific tests based on the end use of the water.

Suggested tests for Deionized water with the acceptance criteria include:

Parameter	Parameter specifications [10]
Microbiology for total count	<500 CFU/ml
Microbiology	Absence of Coliforms
Conductivity	≤1.3 μSiemens/cm @ 25°C

7.5.5.2 Purified Water

1. All use points including the water exit point from the generation plant will be sampled for a minimum of seven consecutive days.
2. Testing will be in accordance with documented procedures.
3. Suggested acceptance criteria include:

Parameter	Parameter specifications [10]
Microbiology for total count	≤100 CFU/ml
Microbiology	Absence of Coliforms
Microbiology	Absence of *P. aeruginosa*
Microbiology	Absence of *B. cepacia*
Endotoxin	≤0.25 EU/ml
Conductivity	≤1.3 μSiemens/cm @ 25°C, USP <645>
Total organic carbon	0.5 ppm (500 ppb) maximum, USP <643>

7.5.5.3 Water for Injection

1. All use points including the water exit point from the generation plant will be sampled for a minimum of thirty consecutive days to release the system for use. The testing will continue for an additional eleven months to verify the system is capable of maintaining specification conditions over seasonal changes.
2. Testing will be in accordance with documented procedures.
3. Suggested acceptance criteria include:

Parameter	Parameter specifications [10]
Microbiology for total count	≤10 CFU / ml
Microbiology	Absence of gram-negative rods
Endotoxin	≤0.25EU/ml
Conductivity	≤1.3 μSiemens/cm @ 25°C, USP <645>
Total organic carbon	0.5 ppm (500ppb) maximum, USP <643>

7.5.6 Specific Tests—Pure Steam

1. All use points including the water exit point from the generation plant will be sampled for a minimum of seven consecutive days to release the system for use.
2. Testing will be in accordance with documented procedures.
3. Acceptance criteria for pure steam condensate must meet the WFI quality and include:

Parameter	Parameter specifications [10]
Microbiology for total count	≤10 CFU/ml
Microbiology	Absence of gram-negative rods
Endotoxin	≤0.25 EU/ml
Conductivity	≤1.3 μSiemens/cm @ 25°C, USP <645>
Total organic carbon	0.5 ppm (500ppb) maximum, USP <643>
Non-condensable gasses	NMT 3.5% expressed in terms of millilitres of gas per 100 ml of condensate, EN 285
Dryness fraction	NLT 0.90, EN 285
Superheat	NMT 25°C, EN 285

7.5.7 Specific Tests—Vacuum

1. Testing will be in accordance with documented procedures.
2. Suggested acceptance criteria include:

Parameter	Parameter specifications
Pressure hold test	≤2 in. Hg in a 30-minute period for a tightly sealed system with minimum use points
Maximum vacuum pressure	Reference only

7.6 CONCLUSION

This chapter has outlined methods to ensure that critical utilities, equipment and systems meet specifications. The specifications shown in the tables preceding are combinations of industry standards in conjunction with compendia (USP, ISO) specifications and guidance. The tables give a starting point for discussions within the individual corporate organizations. Each corporation must establish their acceptance criteria for the utilities.

Maintaining the qualified and validated state requires ongoing awareness of changes to the equipment/systems including cumulative effects of small changes made to enhance performance. All changes to qualified and validated systems must be reviewed to ensure the changes do not compromise the qualification and/or validation.

Periodic verification should include a documented review of history of equipment/systems since the qualification or validation exercise. This review should include changes made by preventive maintenance, calibration and equipment/system modifications. The history review should be discussed with and approved by the same functional areas that were involved with the qualification/validation, including the quality organization.

REFERENCES

1. *Pharmaceutical Engineering Guides for New and Renovated Facilities, Volume 5 Commissioning and Qualification*, ISPE Baseline Pharmaceutical Engineering Guide, International Society for Pharmaceutical Engineering, First Edition, March 2001.
2. *United States Code of Federal Regulations, Food and Drug Administration, 21 CFR Parts 210 and 211, Current Good Manufacturing Practice in Manufacturing, Processing, Packing, or Holding of Drugs*, 1978.
3. *United States Code of Federal Regulations, Food and Drug Administration, 21 CFR Part 820, Good Manufacturing Practice Regulations for Medical Devices*, 1996.
4. Parenteral Drug Association, Technical Report Number 1.
5. *Pharmaceutical Engineering Guides for New and Renovated Facilities, Volume 4 Water and Steam Systems*, ISPE Baseline Pharmaceutical Engineering Guide, International Society for Pharmaceutical Engineering, First Edition, January 2001.
6. *Health Technical Memorandum 2010 (HTM 2010) Part 3 Validation and Verification Sterilization.*
7. *International Standard ISO 11134, Sterilization of Healthcare Products: Requirements for Validation and Routine Control-Industrial Moist Heat Sterilization.*
8. *Draft International Standard ISO 17665, Sterilization of Health Care Products: Moist Heat-Development, Validation and Routine Control of a Sterilization Process for Medical Devices.*
9. *European Standard-Sterilization-Steam Sterilizers-Large Sterilizers (En 285).*
10. *United States Pharmacopeia (USP 29)-National Formulary (NF 24)*, 2006.
11. *European Pharmacopoeia*, January 2005.

8 Calibration and Metrology

Göran Bringert

CONTENTS

8.1 INTRODUCTION

The regulatory authorities in the United States and Europe have expanded the scope of regulations by reference to international standards for Quality and Risk Management. The Good Manufacturing Practice (GMP) regulations are still in effect; however, the enforcement focuses on the critical risk factors determined by risk assessments. In September 2004, the United States Food and Drug Administration (FDA) issued three documents describing the FDA's current thinking:

- Draft Guidance for Industry—Quality Systems Approach to Pharmaceutical Current Good Manufacturing Practice Regulations [1]
- Guidance for Industry—Sterile Drug Products Produced by Aseptic Processing—Current Good Manufacturing Practice [2]
- Pharmaceutical CGMPs for the 21st Century—A Risk-Based Approach Final Report [3]

And Medical Device-specific international standards:

- ANSI/AAMI/ISO 13485:2003 Medical Devices—Quality management systems—Requirements for regulatory purposes [4]
- ANSI/AAAMI/ISO 14971:2000/A1: 2003 Medical devices—Application of risk management to medical devices [5]

All of the preceding documents refer either directly or indirectly to the following documents as standard requirements for calibration:

- ISO 9001:2000 Quality management systems—Requirements [6]
- ISO 10012: 2003 Measurement management systems—Requirements for measurement processes and measuring equipment [7]
- ISO/IEC 17025:2005 General requirements for the competence of testing and calibration laboratories [8]
- GUM Guide to the Expression of Uncertainty in Measurement, (corrected and reprinted 1995) issued by BIPM, IEC, IFCC, ISO, IUPAC, IUPAP and OIML [9]
- ANSI/NCSL Z540–2–1997 U.S. Guide to the Expression of Uncertainty in Measurement [10]

The GMP's in Europe and the United States have common objectives but differ to some degree in their approaches and have specific regulatory requirements for human and veterinary drugs and medical devices.

DOI: 10.1201/9781003163138-8

Validation verifies that processes perform to specifications. Specifications serve to define what performance is needed for a consistent quality output from the process. Measurements of critical parameters are needed to judge the performance of the process. The measurements have to be accurate and repeatable. Accurate and repeatable measurements require adequately calibrated good quality measurement equipment. Current regulations and standards do not specify accuracy requirements as processes and accuracy specifications vary widely across the industry. The regulations hold the organizations of the regulated industries responsible to set specifications and tolerances for calibrations and to verify that calibration laboratories/providers have the competency required for compliance. The reader should keep in mind that the standards and regulations cover implementation of Quality Management Systems and that calibration is only one of several components of the Quality Management System.

Validation of Pharmaceutical Processes is not possible without reliable and repeatable measurements. The "Predicate Rules" (GxPs) require that critical measurements be performed with adequately calibrated measuring devices.

Heat Penetration Studies are performed to calculate the accumulated lethality, F_0, in the load. The accumulated F_0 is the time integral of the lethality function:

$$L = 10^{\left(\frac{T - T_b}{z}\right)}$$

At a base temperature T_b = 121°C and z = 10°C, the effect of 1°C error in measured temperature at 121°C results in approximately 26% error in the lethality calculation.

This is why in section IX sub clause C 2 *Equipment Controls and Instrument Calibration*, of the 2004 Guidance for Industry—Sterile Drug Products Produced by Aseptic Processing—Current Good Manufacturing Practice Validation of Aseptic Processing and Sterilization [2], FDA states:

> For both validation and routine process control, **the reliability of the data generated by sterilization cycle monitoring devices should be considered to be of the utmost importance**. Devices that measure cycle parameters should be routinely calibrated. Written procedures should be established to ensure that these devices are maintained in a calibrated state.
>
> - Temperature and pressure monitoring devices for heat sterilization should be calibrated at suitable intervals. **The sensing devices used for validation studies should be calibrated before and after validation runs**. (Note: The text has been bolded for added emphasis)

ANSI/ISO/DIS 17665:2004 Sterilization of health care products—Moist heat—Development, validation and routine control of a sterilization process for medical devices [11], requires that calibration of the validation system, "measuring chain", shall be verified before and after each stage of validation.

These statements are a clear indication that regulatory authorities consider the integrity of temperature measurements a critical part of the validation of thermal sterilization processes. It is important that the validation Standard Operating Procedure (SOP) reflects the theoretical and practical aspects of how to achieve and maintain high accuracy temperature measurements in conjunction with thermal validation.

Because of the large number of topics covered by the current standards and regulations, this chapter will focus on the practical issues of calibration based on the assumption that the organization has a quality management system that is compliant with applicable requirements. The reader should refer to the complete regulatory documents, referenced in this chapter, to fully assess what has to be in place within the reader's organization for a competent and compliant calibration function for the organization's specific application area. Depending on origin, some of these documents are available free from the Internet whereas all International Organization for Standardization (ISO) standards and regulations are copyright protected and can be purchased as hard copy or downloaded from any national ISO affiliated organization.

8.2 APPLICATION SPECIFIC REGULATION ("PREDICATE RULES")

8.2.1 U.S. FDA GMP

21 CFR Part 58 Good Laboratory Practice for Nonclinical Laboratory Studies [12]

21 CFR Part 211 Current Good Manufacturing Practice for Finished Pharmaceuticals [13]

21 CFR Part 606 Current Good Manufacturing Practice for Blood and Blood Components [14]

21 CFR Part 820 FDA, Subchapter H—Medical Devices, Quality System Regulation [15]

For full text, go to www.fda.gov and under Reference Room select Code of Federal Regulations

8.2.2 Europe GMP

The European GMP is published in EUDRALEX Volume 4—Medicinal Products for Human and Veterinary Use: Good Manufacturing Practice [16] available from http://ec.europa.eu/enterprise/pharmaceuticals/eudralex/homev4.htm

8.2.3 Medical Devices

There are three current standards specific to medical devices recognized by the FDA and EU:

ANSI/AAMI/ISO 13485:2003 Quality System Requirements for Medical Devices [4]

ANSI/AAMI/ISO 14971:2000/A1:2003 amendment, Medical Devices—Application of risk management to medical devices [5]

ANSI/ISO/DIS 17665:2004 Sterilization of health care products—Moist Heat—Development, validation and routine control of a sterilization process for medical devices [11]; a draft standard expected to revise and replace ISO 11134:1994, ISO 13683:1997 and CEN 554:1994 to keep current with technology and moist-heat sterilization practices.

ISO 17025 has two main sections, Management requirements and Technical requirements. Section 4, Management requirements, has 15 sub-clauses covering 8 pages and Section 5, Technical requirements, has 10 sub-clauses covering 14 pages. It is beyond the scope of this chapter to review the detailed requirements, which will be left to the reader to review as needed.

8.2.4 CAPA (Corrective Action, Preventive Action)

ISO 9001 [6], ISO 10012 [7], ISO 17025 [8] and GMPs have mandatory requirements for Corrective and Preventive Action.

8.3 GMPS (PREDICATE RULES)

8.3.1 Regulatory Requirements for Calibration in Europe and the USA

In "Guidance for Industry—Sterile Products Produced by Aseptic Processing—Current Good Manufacturing Practice, September 2004" [2], the FDA has included text boxes with quotes from the Code of Federal Regulations (CFR) and stated:

> The quotes included in the text boxes are not intended to be exhaustive. Readers of this document should reference to complete CFR to ensure that they have complied, in full, with all relevant sections of the regulations.

8.4 GMP CALIBRATION REQUIREMENTS

This section will identify calibration requirements, as of April 2004, defined by CGMP regulations, "the Predicate Rules" including the 21CFR 820 Quality System Regulations and EU regulations.

21 CFR part 58—Good Laboratory Practice for Nonclinical Laboratory Studies [12]
21 CFR part 211—Current Good Manufacturing Practices for Finished Pharmaceuticals [13]
Sec. 211.68 Automatic, mechanical, and electronic equipment.
Sec. 211.160 Laboratory Controls—General requirements.
Sec. 211.165 Testing and release for distribution.
Sec. 211.194 Laboratory records.
21 CFR part 606—Current Good Manufacturing Practice for Blood and Blood Components [14]
Sec. 606.60 Equipment
Sec. 606.100 Standard Operating Procedures
Sec. 606.160 Records
21 CFR part 820 Quality System Regulation [15]
Sec. 820.72 Inspection, measuring and test equipment
ISO/DIS 17665:2004—Sterilization of health care products—Moist heat—Development, validation and routine control of a sterilization process for medical devices [11]

8.5 SUMMARY, STANDARDS AND REGULATIONS

The standards and regulations, in the preceding summary, require that organizations involved in the validation of processes have personnel with a thorough knowledge and understanding of what is needed to achieve and maintain compliance. Calibration is critical, as "reliability of the data generated by sterilization cycle monitoring devices should be considered to be of the utmost importance". (FDA 2004 Guidance for Industry—Sterile Drug Products Produced by Aseptic Processing—Current Good Manufacturing Practice Validation of Aseptic Processing and Sterilization) [2]

ISO 17025–2005 [8] provides General requirements for the competence of testing and calibration laboratories and the Guide to the Expression of Uncertainty in Measurement [9] clause 1.1 states:

> This *Guide* establishes general rules for evaluating and expressing uncertainty in measurement that can be followed at various levels of accuracy and in many fields—from the shop floor to fundamental research. Therefore, the principles of this *Guide* are intended to be applicable to a broad spectrum of measurements.

An organization that intends to comply has to define and document, in standard operating procedures, how the organization achieves and maintains compliance.

8.6 ESTIMATION AND EXPRESSION OF UNCERTAINTY (AN OVERVIEW)

The *Guide* [9] is a complex document that covers far-reaching requirements, and it should be the responsibility of the metrology function within the organization to define the criteria for uncertainties in measurement that are needed for specific compliance requirements. Calibration is used to maintain measurement errors within the acceptable limits needed to ensure that product quality consistently meets predefined product quality specifications. Historically this was done by comparison of how much a measured value differed from a "true" value. The difference was defined as the measurement "error". The international metrology community has agreed that "true" values are by nature indeterminate and that is why the term "true value" is not used in the ISO Guide to the Expression of Uncertainty in Measurement (GUM) [9]. In other words, there will always be a degree of uncertainty in any measurement.

GUM 2.3.1 defines standard uncertainty as:

> "uncertainty of the result of a measurement expressed as a standard deviation."

Figure 8.1 illustrates standard deviation in a normal distribution, based on definitions in the U.S. Guide to the Expression of Uncertainty in Measurement ANSI/NCSL Z540–2–1997 [10]

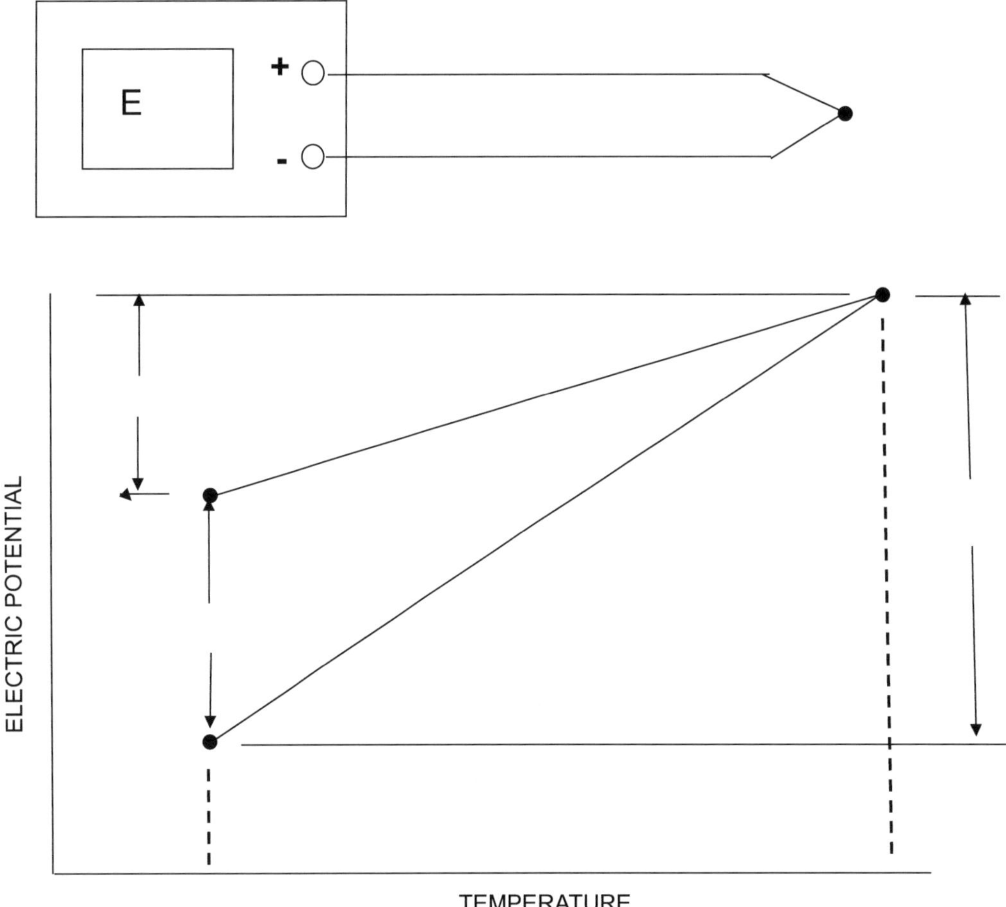

FIGURE 8.1 Standard deviation in a normal distribution.

ANSI/ISO/DIS 17665:2004 [11] requires, and FDA Guidance for Industry—Sterile Drug Products Produced by Aseptic Processing 2004 [2] recommends that the measurement system used for validation should be calibrated before and after validation runs. Calibration requires use of a temperature standard and a stable thermal source. Each of these items contributes some degree of uncertainty to the calibration result in addition to the uncertainties that are inherent in each of the components of the measuring chain (the measurement system used for validation). The *Guide* [10] defines how to combine these uncertainties into a Combined Standard Uncertainty for the measurement system after calibration. A coverage factor is then used as a multiplier of the combined standard uncertainty in order to obtain an expanded uncertainty. If Figure 8.2 represents the combined standard uncertainty of a measurement system with normal distribution, the three confidence intervals shown in the graph represent coverage factors 1, 2 and 2.57 for 68%, 95% and 99% confidence intervals, respectively.

European Standard EN 554:1994 required a minimum Test Uncertainty Ratio (TUR) of 3:1 for validation of thermal sterilization processes. This means that the validation system should have an expanded uncertainty that is at least three times less than the process specification limits. This is illustrated in Figure 8.3. The process under validation has specification limits ±0.5°C which, with an assumed normal distribution, can be represented by the standard uncertainty $1\sigma = 0.1667$°C. The validation system has to have a standard uncertainty at least three times smaller or $1\sigma = 0.055$°C. This is an example of an adequate TUR for validation of a thermal sterilization process.

A larger expanded uncertainty of the validation standard increases the probability for an erroneous validation result that could have severe consequences. This is illustrated by Figure 8.4.

The U.S. version of GUM, ANSI/NCSL Z540–2–1997, U.S. Guide to the Expression of Uncertainty in Measurement [10], is an adaptation of the ISO Guide to the Expression of Uncertainty in Measurement [9] to promote consistent international methods in the expression of measurement uncertainty within the U.S. standardization, calibration, laboratory accreditation and metrology services. The U.S. Guide is identical to the ISO Guide (corrected and printed, 1995) with the exception of minor editorial changes to facilitate its use in the United States.

ISO 17025:2005 [8] makes reference to GUM "for further information on estimation of uncertainty in measurement."

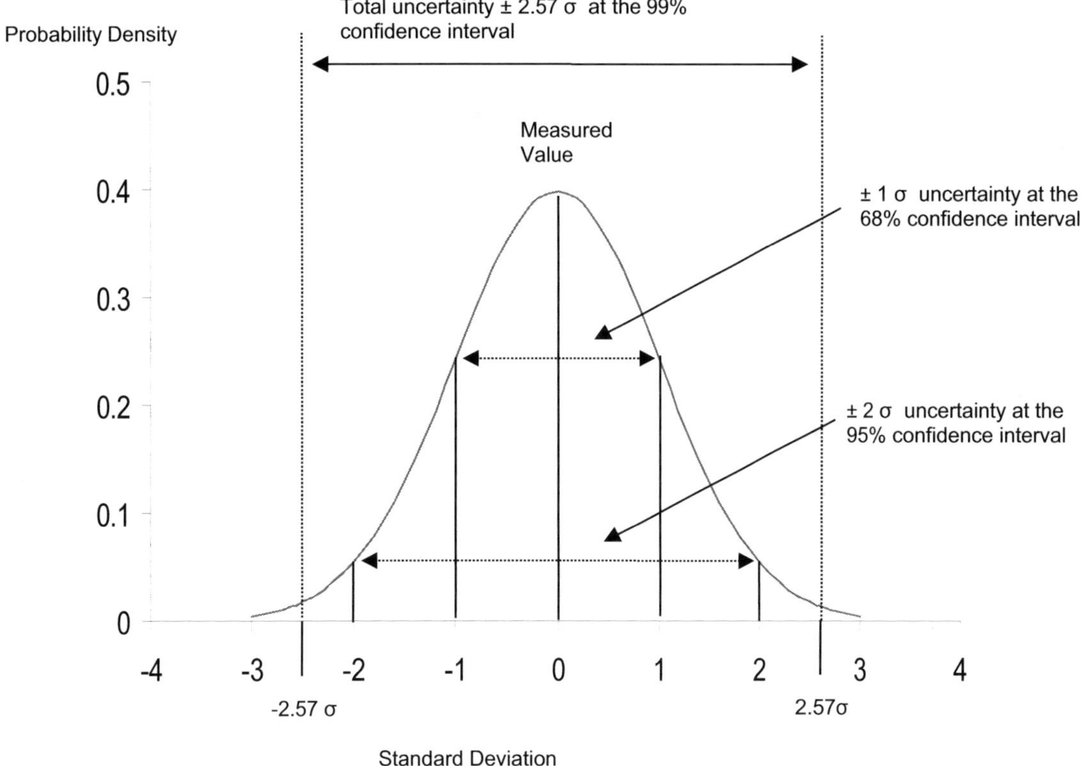

FIGURE 8.2 Adequate test uncertainty ratio for validation.

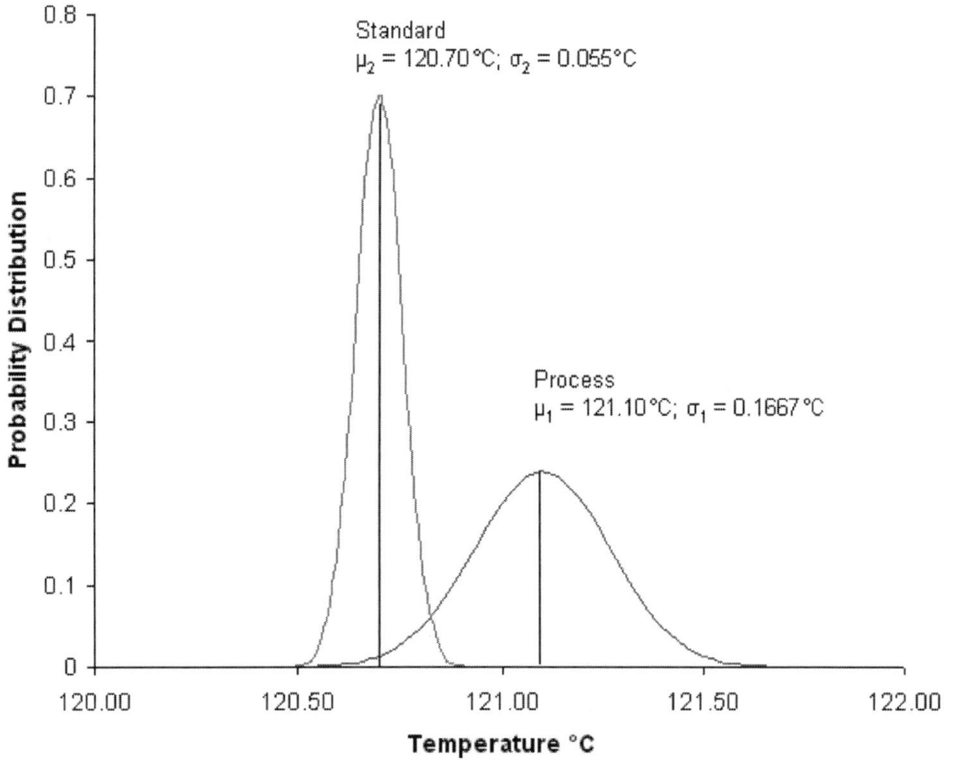

FIGURE 8.3 Adequate TUR for validation.

Note: TUR = test uncertainty ratio.

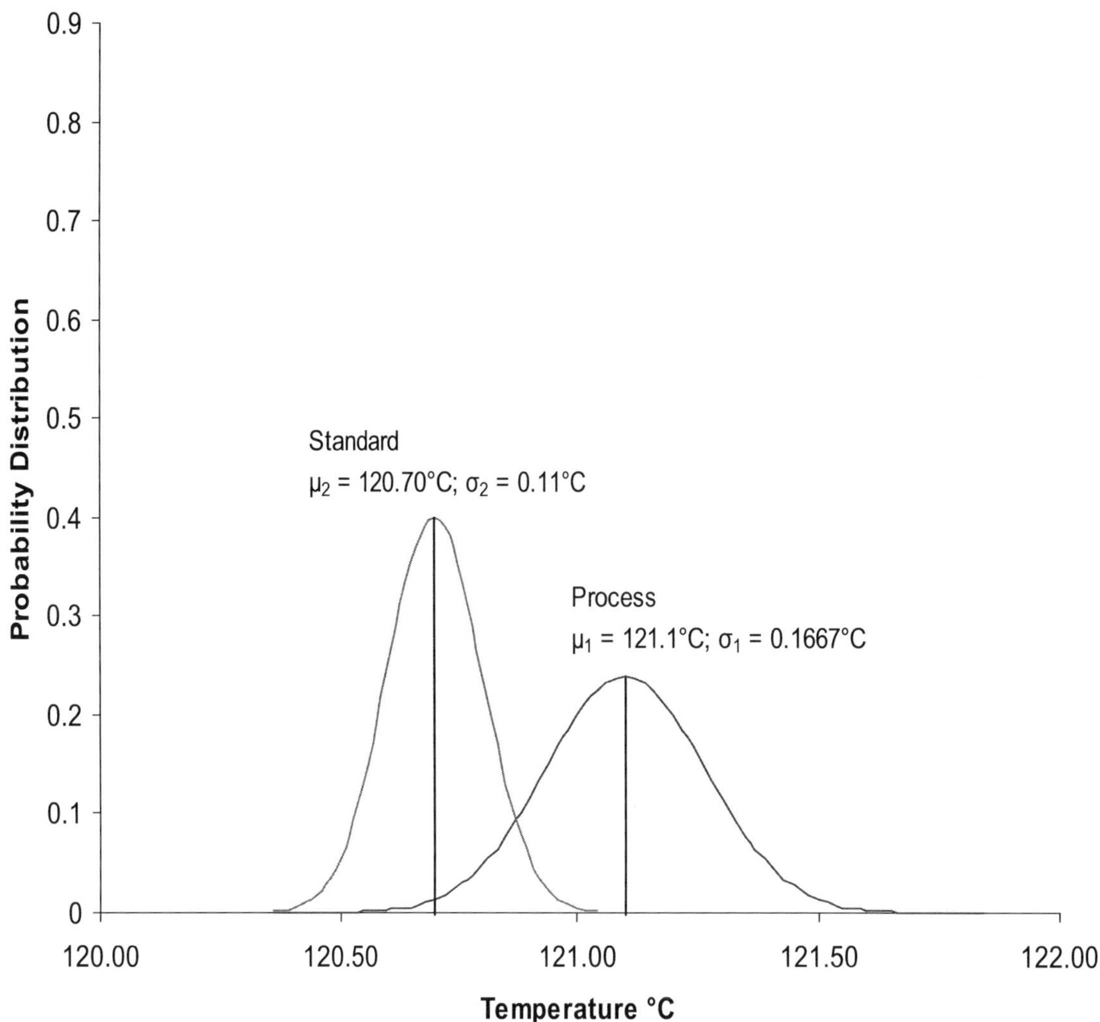

FIGURE 8.4 Inadequate TUR for validation.

Note: TUR = test uncertainty ratio.

ISO 10012:2003 [7] Guidance to clause 7.3.1 Measurement uncertainty refers to the "Guide to the expression of uncertainty in measurement" (GUM) for methods that can be used to combine uncertainties and present the results. It also states that other documented and accepted methods may be used.

To fully understand and master the methods referred to previously, it is necessary to study the documentation and integrate the information and requirements into the user's Quality Management and Risk Management Systems, a topic beyond the scope of this chapter.

8.7 PRACTICAL DISCUSSION OF HOW TO CALIBRATE THE MEASURING CHAIN

The integrity of temperature measurements is a critical part of validation of thermal sterilization processes. It is important that the validation SOP reflects the theoretical and practical aspects of how to achieve and maintain high accuracy temperature measurements in conjunction with thermal validation.

Heat Penetration Studies are performed to calculate the accumulated lethality, F_0, within the load items. The accumulated F_0 is the time integral of the lethality function:

$$L = 10^{\left(\frac{T - T_b}{z}\right)}$$

At a base temperature $T_b = 121°C$ and $z = 10°C$, the effect of 1°C error in measured temperature at 121°C results in approximately 25% error in the lethality calculation.

FDA Definition of Process Validation

"Establishing by objective evidence that a process consistently produces a result or product meeting its predetermined specifications."

(*GMP/Medical Device Quality System Manual, 4. Process Validation*, *1997*)

The required temperature uniformity in the chamber, according to contemporary industry standards for terminal sterilization,

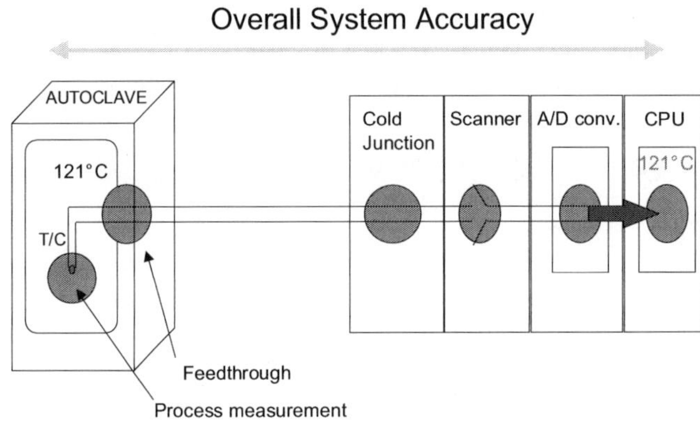

FIGURE 8.5 The measuring chain is composed of all components involved in the measurement system.

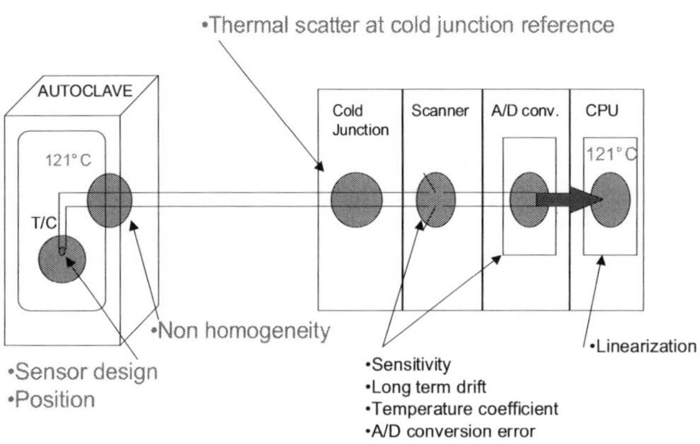

FIGURE 8.6 Random and systematic error sources in the measuring chain.

should be better than or equal to 1°C or 0.5°C depending on the application. The combined standard uncertainty for the instrument used for validation measurements, including temperature sensors, should be at least three times less than the specified range for the process variable. This means that the Overall System Combined Standard Uncertainty should be better than or equal to ± 0.33°C or ± 0.17°C, respectively. All components involved with the measurement (from the tip of each sensor, via the connecting wires, cold junction reference, signal interface, analog to digital conversion, conversion from mill volts to temperature, to display and printout of the measured values) are referred to as the Measuring Chain. Figure 8.5 shows a Measuring Chain using thermocouples.

8.7.1 Error Sources

Several variable error sources can affect the temperature measurement accuracy in validation. Control and management of these error sources should be recognized as the responsibility of the people who perform the validation. Individuals responsible for validation should have the competency to adequately perform the validation studies.

It is important to distinguish between systematic and random errors. Systematic errors are eliminated by calibration, whereas random errors are not eliminated by calibration and can only be minimized through application of knowledge and proper procedures. The operator has to understand how to minimize the influence of random temperature measurement errors to consistently achieve the accuracy required for thermal validation of steam sterilization processes. Procedures documented in the validation SOP and individual training of validation personnel are necessary to maintain the competency of the validation team.

Electronic temperature measurements for validation are acquired using temperature sensors connected to an electronic datalogger or recorder. The components of the Measuring Chain contribute errors, systematic or random, that contribute to the Overall System Accuracy. Figure 8.6 identifies the most significant Random Error Sources (in red) in the Measuring Chain.

Significant random errors can occur in the following areas:

Sensor design
Sensor location
Sensor wire non-homogeneity
Thermal scatter at the cold junction reference

8.7.2 Temperature Sensors

Thermocouple type T (copper/constantan) is the most commonly used thermocouple for temperature measurements in validation applications because of its high accuracy and low cost. Temperature measurement is affected by several ambient conditions, which is why the ASTM Manual on The Use of Thermocouples in Temperature Measurement, Series MNL 12, 1993 [17] has the following statement on its first page:

> Regardless of how many facts are presented herein and regardless of the percentage retained, all will be for naught unless one simple important fact is kept firmly in mind. The thermocouple reports only what it "senses." This may or may not be the temperature of interest. Its entire environment influences the thermocouple and it will tend to attain thermal equilibrium with this environment, not merely part of it. Thus, the environment of each thermocouple installation should be considered unique until proven otherwise. Unless this is done, the designer will likely overlook some unusual, unexpected, influence.

8.7.2.1 Calibration

The Measuring Chain shall be calibrated prior to and after calibration runs. Adequate calibration equipment shall be used when in calibration of the Measuring Chain. A temperature transfer standard, traceable to a primary standard, and a stable thermal source are required to perform calibration of the Measuring Chain. The combined expanded uncertainty of all components of the Measuring Chain, temperature transfer standard and stable thermal source shall be determined by a competent calibration facility and documented to be adequate for the validation of the process equipment.

Measurement Standards are classified based on metrological qualities.

> The primary standard has the highest metrological qualities and is accepted without reference to other standards of the same quantity. Primary standards are normally kept in national measurement laboratories and designated as national standards.
>
> The secondary standard has its value assigned by comparison with a primary standard of the same quantity.
>
> The transfer standard is used for comparison of standards of the same quantity.

This means that the validation study has to have documented evidence that the Measuring Chain was in a calibrated state before and after the validation study was performed. The documented procedures shall describe the criteria chosen for the calibration procedures. This should be backed by historical documentation that gives the rationale for the procedure. The regulations require three consecutive successful validation runs for a successful validation. Based on the company's risk assessment and risk management, it seems realistic to calibrate before the validation study begins and to verify that the Measuring Chain remains in calibration at the end of the third successful run.

Be patient. A frequent mistake in calibrating instrumentation is to take measurements and make adjustments before conditions have stabilized. It may take much longer than expected for a system to become completely stable, because thermal equilibration takes place exponentially and the output may seem to be stable even though it is still changing slowly. Automatic detection of stability to pre-set stability criteria, offered by modern calibration equipment and software, is the most reliable and repeatable method for stability determination.

8.7.3 Temperature Transfer Standard

The accuracy of the transfer standard must be better than that of the instrument being calibrated. This would seem obvious, but it is amazing how often a voltage calibrator is used that has a greater error than the system being calibrated. It is important to recognize that the accuracy of the calibration can be no better than the standard used, and it is a mistake to change the adjustment of a measuring system if it is already more accurate than the standard.

The characteristics of the transfer standard must have been determined by a procedure that is traceable to accepted primary standards. In the United States, the National Institute of Standards and Technology (NIST) is the accepted source of primary standards. The transfer standards need to be calibrated by NIST relative to their primary standards or by a qualified Standards laboratory relative to standards calibrated by NIST. In either case, the test results and test numbers should be known so the calibration procedure can be traced to the primary standards.

The transfer standard must be independent of the measuring system. Because the output of a thermocouple depends on the entire circuit, it is not a desirable transfer standard. A resistance temperature detector (RTD) is a device that indicates changes of temperature by a change of resistance. Because the resistance of an RTD is only a function of its temperature, and the resistance can be measured independently of the system being calibrated, RTDs are ideal temperature transfer standards.

The transfer standard must be stable in shipment and tolerate other handling. As its name implies, the purpose of the transfer standard is to transfer a measured characteristic from one laboratory to another. The characteristics of the standard must be the same when received from NIST as when it was calibrated relative to their standards. Liquid in glass thermometers may be damaged or develop small voids in the liquid during shipment and therefore are not reliable temperature transfer standards. RTDs are sensitive devices that maintain their characteristics only with careful handling and shipment.

8.7.3.1 Stable Thermal Source

Significant random calibration errors that can occur in a dry block temperature reference:

- Transfer calibration error
- Stem conduction error
- Uneven heat transfer
- Immersion depth inadequate
- Well inserts not used
- Instability
- Time needed for stabilization insufficient

8.7.4 Reference Error Using Thermocouples

When calibrating a thermocouple T_1 against an RTD transfer standard T_2, a key contribution to error is the difference in temperature between these devices when placed in the reference (Figure 8.7). This difference is called a *transfer calibration error* and is potentially the largest contribution to calibration errors in dry block references.

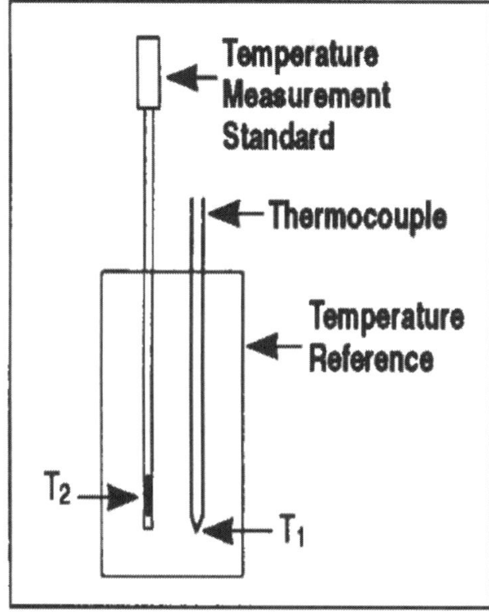

FIGURE 8.7 Transfer calibration error is the temperature difference between the thermocouple tip (T_1) and the measurement standard (T_2).

Transfer calibration error contains two components:

- Stem conduction error, which cools the thermocouple tip (Figure 8.8).
- Uniformity of the reference wells relative to the standard well.

A dry block with one common large diameter well is not suitable for calibration of a measuring chain using multiple thermocouples. The stem conduction will cause heat losses that are greater than the heat radiated from the walls of the single well. Figure 8.9 illustrates this inadequacy.

A dry block with smaller diameter wells and inserts provides closer thermal coupling between the well walls and sensors under calibration, minimizing the effect of stem conduction and transfer calibration error. A dry block designed for maximum transfer calibration accuracy has small diameter wells with inserts that fit the size of the sensors under calibration. (Figure 8.10)

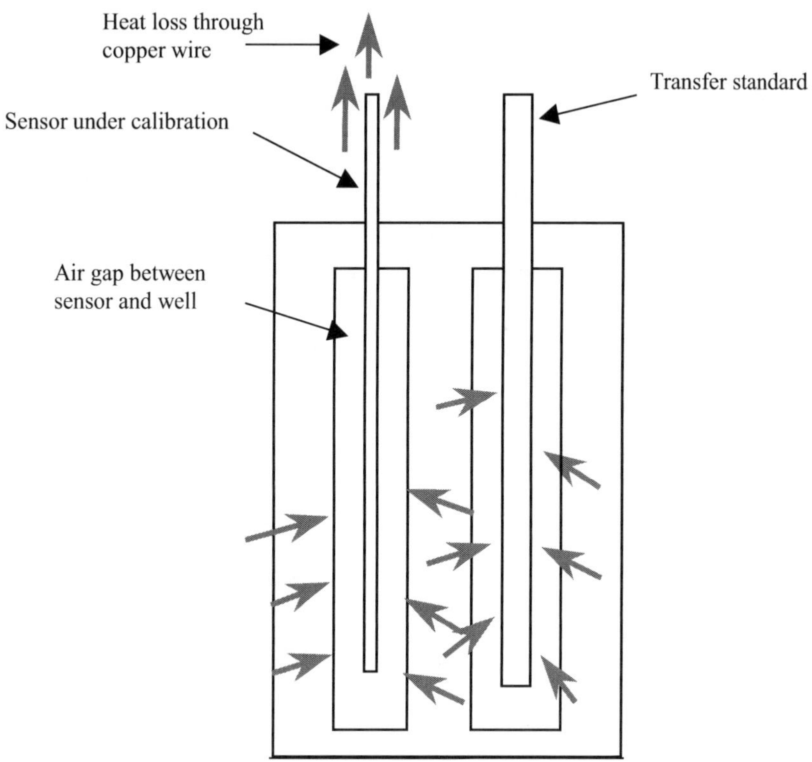

FIGURE 8.8 Stem conduction causes heat loss and generates calibration error.

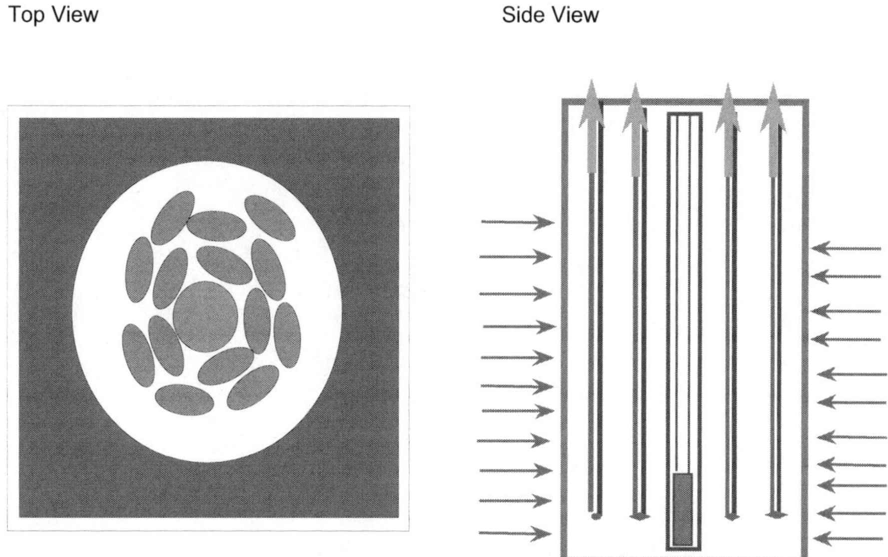

FIGURE 8.9 View of a dry block with one large well and no insert.

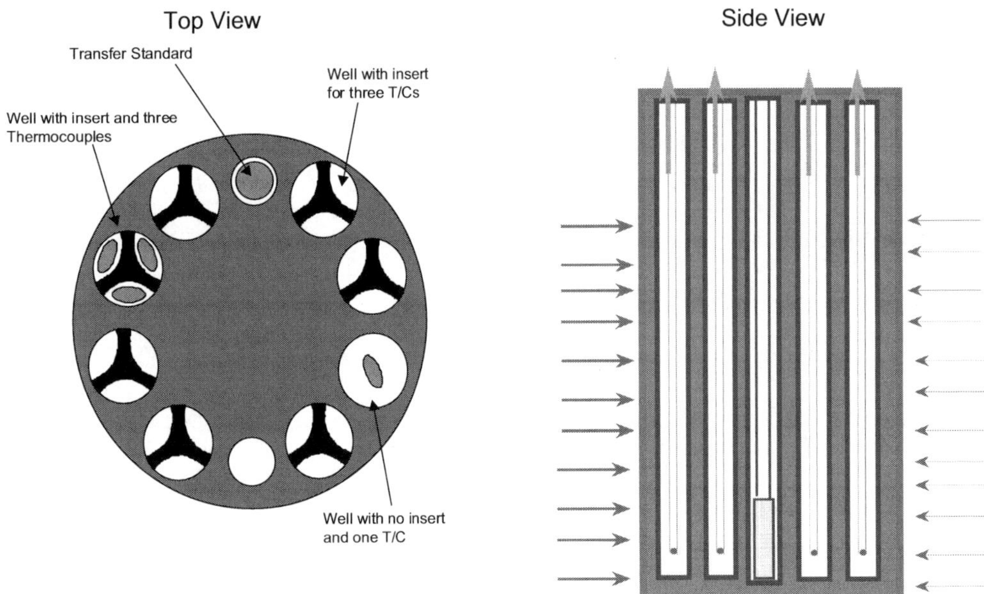

FIGURE 8.10 View of a dry block with inserts to minimize the air space around the sensors.

Note: T/C = thermocouple.

8.7.4.1 Calibration Procedure

Calibration of the Measuring Chain (see Figure 8.5) should be done with the sensors connected to the data logger/recorder and installed in the sterilizer via the feedthrough and out through the open sterilizer door. The sensors should be inserted into the temperature reference bath or dry block, just outside the open sterilizer. Calibration should be performed prior to a validation study and a calibration check should be performed at the conclusion of the validation.

8.7.4.2 Pre-Study Calibration

A two-point calibration should be used with calibration points bracketing the sterilization temperature for the process under validation, e.g. 100°C and 130°C, and the calibration checkpoint should be between the two calibration points to verify the calibration, e.g. 121°C.

8.7.4.3 Post-Study Verification

A two-point comparison between the temperature standard and the temperature sensors should be performed to verify that the calibration of the Measuring Chain is intact.

The calibration should be documented, to provide evidence that the temperature of the reference, the transfer standard and the sensors were stable before the determination of *calibration correction values*. The calibration documentation should include data on the deviation between the temperature standard and each temperature sensor before and after calibration. To ensure traceability, the documentation must list the calibration parameters and equipment including serial numbers and last calibration dates.

8.8 CONCLUSION

Validation of thermal sterilization processes requires accurate temperature measurements to provide reliable results. In order to ensure measurement integrity, it is necessary that validation personnel have adequate training and well-defined processes to follow.

Risk assessment and risk management are now mandatory for processes that are used for manufacture or production of products with critical tolerances. The result of the risk assessment should serve as a basis for the definition of process tolerances and the corresponding measurement tolerances for the process control system. Appropriate measurement tolerances vary by application; although a moist heat sterilization process needs to be measured to ±0.5°C, a dry heat depyrogenation process would be adequately measured with a tolerance of ±1.0°C.

Internationally accepted good metrological standards and process control engineering practice call for application of TUR between the equipment under calibration/validation and the calibration equipment itself. For most applications, the minimum TUR should be 3:1 and preferably higher (Figures 8.2, 8.3 and 8.4). The higher the TUR the more expensive calibration equipment has to be used. This is an area where risk assessment and risk management is used to determine the level of compromise needed to balance the level of acceptable risk versus the cost of more sophisticated calibration equipment. According to ISO 17025 and ISO 10012, management is responsible for the determination and justification of the balance of risk versus cost.

Calibration/verification of the sensors monitoring and controlling the process is as critical as the calibration of the sensors used for validation of the process. Each individual Measuring Chain has to be calibrated/verified against a transfer standard, i.e. sensor, wiring and measurement system, to at least the same expanded uncertainty as required for the calibration of the validation system.

The pressure sensor in the autoclave must be calibrated in place under standard operating conditions. A two-point comparison between the installed pressure sensor and a temporarily connected pressure transfer standard, traceable to a national standard, should be performed. Based on the comparison, zero and span adjustments are done on the installed pressure sensor.

REFERENCES

1. FDA Draft Guidance for Industry: Quality Systems Approach to Pharmaceutical Current Good Manufacturing Practice Regulation, September 2004, U.S. Department of Health and Human Services Food and Drug Administration, Center for Drug Evaluation and Research (CDER) Center for Biologics Evaluation and Research (CBER), Center for Veterinary Medicine (CVM), Office of Regulatory Affairs (ORA).
2. FDA Guidance for Industry: Sterile Drug Products Produced by Aseptic Processing: Current Good Manufacturing Practice, September 2004, U.S. Department of Health and Human Services Food and Drug Administration Center for Drug Evaluation and Research (CDER) Center for Biologics Evaluation and Research (CBER) Office of Regulatory Affairs (ORA).
3. FDA Pharmaceutical CGMPs for the 21st Century: A Risk-Based Approach: Final Report, September 2004, Department of Health and Human Services, U.S. Food and Drug Administration.
4. ANSI/AAMI/ISO 13485:2003 Medical Devices: Quality Management Systems: Requirements for Regulatory Purposes. Association for the Advancement of Medical Instrumentation, Arlington, Virginia, USA.
5. ANSI/AAMI/ISO 14971:2002/A1:2003: Application of Risk Management to Medical Devices. Association for the Advancement of Medical Instrumentation, Arlington, Virginia, USA.
6. ISO 9001:2000 Quality Management Systems: Requirements. International Organization for Standardization, Central Secretariat, Geneva, Switzerland.
7. ISO 10012:2003 Measurement Management Systems: Requirements for Measurement Processes and Measuring Equipment. International Organization for Standardization, Central Secretariat, Geneva, Switzerland.
8. ISO/IEC 17025:2005 General Requirements for the Competence of Testing and Calibration Laboratories. International Organization for Standardization (ISO) and International Electrotechnical Commission, Geneva, Switzerland.
9. ISO Guide to the Expression of Uncertainty in Measurement (GUM), Issued by BIPM, IEC, IFCC, ISO, IUPAC, IUPAP, 1995. International Organization for Standardization, Central Secretariat, Geneva, Switzerland.
10. ANSI/NCSL Z540-2-1997 AMERICAN National Standard for Expressing Uncertainty: U.S. Guide to the Expression of Uncertainty in Measurement. National Conference of Standards Laboratories, Boulder, Colorado, USA.
11. ANSI/ISO/DIS 17665:2004 Sterilization of Health Care Products: Moist Heat: Development, Validation and Routine Control of a Sterilization Process for Medical Devices. International Organization for Standardization, 2004, Geneva, Switzerland.
12. FDA, Good Laboratory Practice for Nonclinical Laboratory Studies, Title 21, Part 58.
13. FDA, Current Good Manufacturing Practice for Finished Pharmaceuticals, Title 21, Part 211, Code of Federal Regulations (USA).
14. FDA, Current Good Manufacturing Practice for Blood and Blood Components, Title 21, Part 606, Code of Federal Regulations (USA).
15. FDA, Subchapter H: Medical Devices, Title 21, Part 820, Quality System Regulation, Code of Federal Regulations (USA).
16. European Good Manufacturing Practice, EDURALEX Volume 4: Medicinal Products for Human and Veterinary Use, August 2004 (update of 1998 edition), European Commission, Enterprise Directorate: General Pharmaceuticals: Regulatory Framework and Market Authorizations. http://ec.europa.eu/enterprise/pharmaceuticals/eudralex/homev4.htm
17. ASTM Manual on the Use of Thermocouples in Temperature Measurement, Manual Series MNL 12, American Society for Testing and Materials 1916 Race Street, Philadelphia, PA 19103.

9 Temperature Measurements

Clarence A. Kemper and Göran S. Bringert

CONTENTS

9.1 INTRODUCTION

Temperature is the most common of all industrial process measurements, and in thermal sterilization processes, it is the most critical. Verification that all temperature measurements are accurate and reliable is one of the most important requirements in the validation of these processes.

> The reliability of the data generated by sterilization cycle monitoring should be considered to be of the utmost importance.
>
> *(FDA 2004 Guidance for Industry—Sterile Drug Products Produced by Aseptic Processing) [1]*

The output of a temperature-measuring system is a result of its entire thermal environment, and the indicated temperature value may change as the system attains thermal equilibrium with this environment. There are several factors that can affect the accuracy of temperature measurements, and all such factors must be considered in calibrating the system. Each temperature measurement installation should be considered unique until proven otherwise. If this is not done, some unusual or unexpected factor may be overlooked.

Prior to calibrating a measuring chain (refer to the earlier chapter in this text on "Calibration and Metrology"), it is important to verify that the entire measuring chain, sensor, wiring, and measuring equipment, and its installation is designed for the application at hand. Calibration and validation Standard Operating Procedures (SOPs) should define application and installation-specific requirements.

The two most commonly used temperature sensors in pharmaceutical processes are the thermocouple (T/C) and

DOI: 10.1201/9781003163138-9

resistance temperature detector (RTD). It is necessary to be aware of their different physical properties as the error sources, which affect the use of either may differ or have similarities. The discussion in this chapter will describe the physical properties of thermocouples and RTDs and address the broader issue of assuring that the indicated value of temperature is an accurate representation of the value being measured.

Thermocouples are the most satisfactory sensors for conducting heat penetration and temperature distribution studies in validation, whereas RTDs are the most satisfactory transfer standards for temperature calibration, and Pt 100 RTD sensors are commonly installed in processing equipment for control and monitoring.

9.2 THERMOCOUPLES

A thermocouple is a simple, versatile temperature sensor constructed by joining two wires of different composition to form a "thermocouple junction". When a thermocouple is connected to a well-designed reference and measuring system, the indicated output is a unique function of the junction temperature. It will be shown that the total output of a thermocouple circuit is not a sensor characteristic, however, so the entire measuring system must be considered in a proper calibration procedure.

The primary reasons for choosing thermocouples for validation of heat penetration and heat distribution are that thermocouples are small, flexible, easily interchangeable, and mechanically resistant and are more convenient to place in difficult-to-reach locations in the load than RTDs.

The inaccuracies in most thermocouple systems do not occur in the sensors; they occur in the instrumentation used to measure the outputs and the circuitry connecting the thermocouple sensors to the measuring system. Additional errors may be caused by sensor designs that are not fit for the specific application and by a location of sensors in an area that is not representative of the temperature of interest. A simplified explanation of thermoelectric theory is included in this chapter as a guide to proper installation of thermocouple circuits. By understanding the source of thermoelectric output, it is easier to avoid the mistakes most often encountered in the use of thermocouples, thereby assuring better measurement accuracy.

9.3 THERMOELECTRIC THEORY

During the 180 years since T. J. Seebeck discovered that current flows in a circuit of two dissimilar conductors whenever the junctions of the conductors are at different temperatures, many investigators have developed theories to explain thermoelectric phenomena. Some, such as Thomson and Bridgman, have based their explanations on thermodynamic considerations; others, such as Mott and Jones, have employed the electron theory of solids [2, 3]. The following explanation of thermoelectric phenomena might be objectionable to both to experts in thermodynamics and atomic physicists but is a concept that can be understood easily and employed to avoid many of the errors encountered in thermocouple circuits.

1. The energy level of an electron in any conductor increases as the temperature of the conductor increases.
2. The amount of energy change for a given temperature change depends on the composition and molecular structure of the conductor.

The material property that expresses the amount of energy increase for a given temperature increase is called the thermoelectric power. The value of thermoelectric power is given in units of microvolts of energy increase for each degree Celsius of temperature increase in the material.

Figure 9.1 depicts a simple thermocouple circuit consisting of two external conductors, A and B, which are connected at a junction where the temperature is T_2. For simplicity, it is assumed that the entire circuit of the voltage-measuring device is at a uniform temperature, T_1. It will be shown that under this assumption the net thermoelectric potential difference is generated only in the external circuit. If the temperature of the voltage-measuring device is equal to T_1 throughout, the two terminals to which the conductors A and B are connected must also be at the same temperature T_1.

The lower portion of Figure 9.1 gives a graphic representation of the thermoelectric potential of the circuit shown in the upper position. The horizontal ordinate is the temperature at a given location in the circuit, and the vertical ordinate is the corresponding electrical potential at that location. Because the thermoelectric power is the amount of energy increase for a given temperature increase, the slope of each line is equal to the thermoelectric power of that conductor. If the thermoelectric powers are different, the slopes of the two lines will be different.

Because the potential at the junction where the two conductors are joined together must have a singular value, Figure 9.1 shows that there will be a net potential difference between the two terminals of the voltage-measuring device.

Consider the energy levels of the electrons in the conductors as the circuit is traversed in a clockwise direction, starting at the terminal where material A is connected to the voltage-measuring device.

Assuming that the temperature T_2 is greater than the temperature T_1, the energy level of the electrons in material A

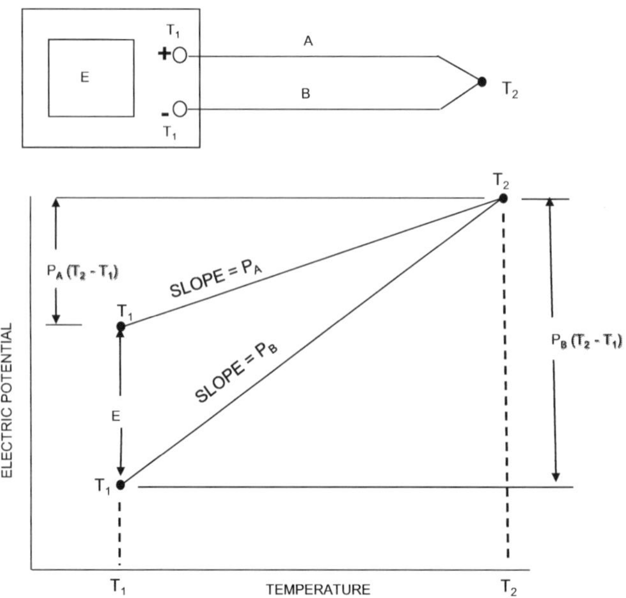

FIGURE 9.1 Basic thermocouple

will increase as the junction with material B is approached because the temperature of the material is increasing. Assuming that the thermoelectric power of the conductor is constant, the amount by which the energy increases is equal to the thermoelectric power of material A, P_A, multiplied by the change in temperature, $(T_2 - T_1)$. As the circuit is traversed from the junction of the two conductors to the terminal where material B is connected to the voltage-measuring device, the energy level of the electrons will decrease by an amount equal to the thermoelectric power of material B, P_B, multiplied by the change in temperature, $(T_1 - T_2)$.

In this example, the thermoelectric power of material B will be greater than that of material A, so the change in energy level in B will be greater than that in A for the same change in temperature. This simple circuit illustrates a characteristic of thermocouple materials that causes some confusion. When the temperature of the external junction is greater than that of the junctions at the voltage-measuring device, the material having the greater thermoelectric power will be the negative lead. Because material B is assumed to have the greater thermoelectric power in the following examples, the terminal to which it is connected will be the negative terminal. The circuit summations are expressed starting at the terminal to which material B is connected and traversing the circuit in a counter-clockwise direction to yield a positive potential difference.

If P is the thermoelectric power and T is the temperature at any location in the circuit, the *gradient* explanation of thermoelectric output states that the net potential generated by the circuit is equal to the cyclic integral of the product of the thermoelectric power and the differential change if temperature [4,5]. That statement is expressed mathematically by Eq. (1), in which E is the net electric potential difference generated by the circuit:

$$E = \int P dT \tag{1}$$

Although the thermoelectric powers of all conductors change slightly with a change in temperature, for the purpose of this discussion, the thermoelectric powers are assumed to be constant within any length of a homogeneous conductor. In reality, the lines in Figure 9.1 would be curved slightly, rather than being straight as they are when it is assumed that each homogeneous conductor has a constant thermoelectric power. If the internal circuits of the measuring instrument are uniform in temperature and the thermoelectric power of each conductor is constant, the integral of Eq. (1) can be evaluated by Eqs. (2) and (3), where P_A is the thermoelectric power of conductor A, P_B is that of conductor B, and P_I is that of the internal circuit.

Equation (2) represents the *conductor* explanation of thermocouple output, which states that the net electrical potential difference generated by each conductor in the circuit is equal to the thermoelectric power of the conductor multiplied by the temperature difference between the ends of the conductor. The net electrical potential difference generated by the total circuit is equal to the sum of the differences of each conductor. The conductor explanation is a simplified form of the gradient explanation.

$$E = P_B(T_2 - T_1) + P_A(T_1 - T_2) + P_I(T_1 - T_1) \tag{2}$$

It is obvious that the last term in Eq. (2) is zero and the contribution of the internal circuit is zero if the temperature is uniform. The net output of the entire circuit under these assumptions is given by Eq. (3):

$$E = P_B(T_2 - T_1) + P_A(T_1 - T_2) \tag{3}$$

An alternative to the gradient or conductor explanation of thermoelectric output is the *junction* explanation [2,3]. It states that the electrical output of each junction of two conductors is equal to the product of the temperature of the junction and the difference between the thermoelectric powers of the two conductors. The net electrical potential difference generated by the total circuit is equal to the sum of the outputs of all junctions in the circuit. Equation (4) is the mathematical expression of the junction explanation for the circuit of Figure 9.1:

$$E = T_1(P_I - P_A) + T_2(P_B - P_A) + T_1(P_A - P_I) \tag{4}$$

It is not quite as obvious in Eq. (4) that the contribution of the internal circuit is zero, but the two terms containing P_I do cancel and the other terms may be rearranged to yield Eq. (5). It is also not obvious that the subtraction of the thermoelectric powers at each junction must be performed in a direction consistent with cyclic integration. Thus, the difference of thermoelectric powers in the second term in Eq. (5) is the negative of that in the first term. This requirement is often confusing to the inexperienced investigator applying the junction explanation.

$$E = T_2(P_B - P_A) + T_1(P_A - P_I) \tag{5}$$

Both Eq. (3) and Eq. (5) may be rewritten to yield Eq. (6), showing that, for this simple circuit of homogeneous conductors having constant thermoelectric powers, the output predicted by either explanation is the same. It is equal to the difference of thermoelectric powers of the two conductors multiplied by the difference between the temperatures at their junctions.

$$E = (P_B - P_A)(T_2 - T_I) \tag{6}$$

Many persons focus on the junctions in evaluating thermocouple circuits, so they often fail to recognize phenomena such as regions of stress within a conductor. When wires are flexed repetitively at one location, the resulting cold-working can create regions of nonhomogeneous thermoelectric power, thereby changing the net electric output of the circuit. By using the gradient or conductor explanation to evaluate a circuit it will be seen that the electrical output is generated where temperature gradients exist in the conductors, and that the thermoelectric power of the conductors in those regions must be known [5]. This is particularly important when thermocouples are used to measure the temperatures of elements within a chamber in which the temperature is different from the surrounding ambient temperature.

Figure 9.2 depicts a slightly more complex thermocouple circuit that adds a third conductor C between the voltage-measuring instrument and each of the other conductors. Because most thermocouple circuits are constructed of duplex wire with the two conductors in close physical proximity, the error introduced by assuming that both conductors are at the same temperature, at any location in the circuit, will be negligible. Thus, in the circuit of Figure 9.2, it is assumed that both of the junctions to the C conductors are at a reference temperature T_r. Because the internal circuit of the voltage-measuring device will have no net output if the temperature is equal to T_1 throughout, the output of the circuit of Figure 9.2 is given by Eq. (7):

$$E = P_C(T_r - T_1) + P_B(T_m - T_r) + P_A(T_r - T_m) + P_C(T_1 - T_r) \quad (7)$$

The terms containing P_C cancel, illustrating an important characteristic of thermocouple circuits: if at all points along the length of a duplex thermocouple pair the temperature is the same in both conductors, the output of that portion of the circuit will be zero if the thermoelectric powers of the two conductors are the same. Duplex copper leads do not contribute to the net output of a circuit if they are made of pure, instrument-grade copper and have not been stressed to create cold-worked regions. Equation (7) can be rewritten to yield Eq. (8):

$$E = (P_B - P_A)(T_m - T_r) \quad (8)$$

That the portion of the circuit makes no net contribution to the output is also shown graphically in the lower portion of Figure 9.2. Because the slope of each line is equal to the thermoelectric power of the conductor, the output curves of the two C conductors are parallel. Therefore, even though the temperature changes from T_r to T_1 in that portion of the circuit, there is no change in the net potential difference.

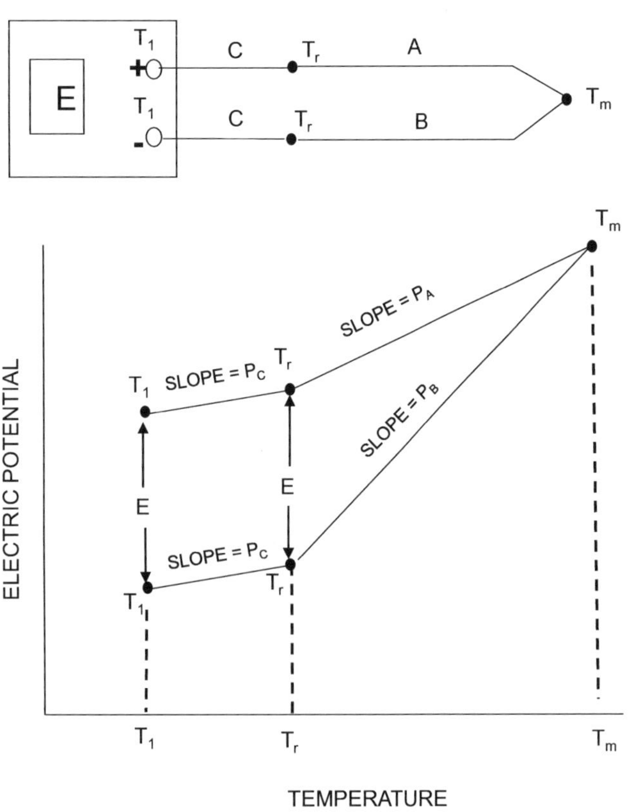

FIGURE 9.2 Simple thermocouple circuit

From Eq. (8), it may be observed that if the temperature T_r is maintained at some known reference value, and the thermoelectric powers of conductors A and B are known, the electrical output of the circuit is proportional to the temperature of the measuring junction T_m. It should be remembered that the thermoelectric powers of conductors vary slightly with temperature, thereby giving all thermocouples a nonlinear output versus temperature rather than the linear output of this simplified explanation. For the purpose of understanding how to avoid circuit errors, however, the linear assumption is adequate.

The fact that the duplex conductors, C, make no net contribution to the electrical output is important, because this allows duplex copper leads to be used in connecting the voltage-measuring device to the junctions at the reference temperature, T_r. Then the temperature of the terminals on the voltage-measuring device can be any value without changing the net electrical output of the circuit. It is relatively easy to maintain the junction of two conductors at a constant, known temperature, but it would be extremely difficult to do so at the terminals of a voltage-measuring device.

The value that has been chosen universally as the standard reference temperature for thermocouple circuits is the equilibrium temperature between ice and air-saturated water, or 0.00°C. A few instruments are sold with oven-controlled reference temperatures of higher values, but all standard tables give the output of thermocouples as a function of the measured temperature, T_m, on the assumption that the reference temperature, T_r, is at 0.00°C [6].

Figure 9.3 depicts a thermocouple circuit composed of several lengths of duplex thermocouple wire. This is the type of circuit that might be used in a typical validation study. Section 1 could be the length of thermocouple wire that goes from

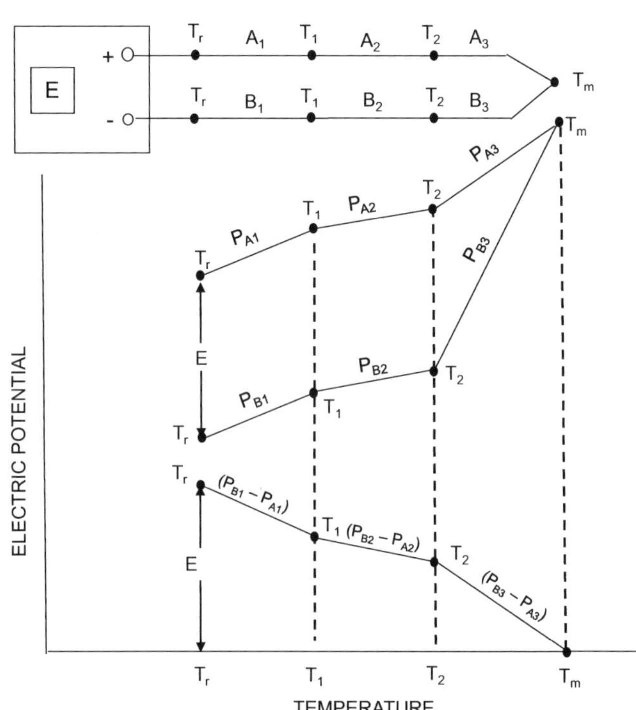

FIGURE 9.3 Typical thermocouple circuit.

the measuring system to a connector outside the autoclave; Section 2 could be the connector; Section 3 could be the length of thermocouple wire that goes from the external connector, through the wall of the autoclave, to the junction inside the autoclave. Because the circuit from the reference junctions to the voltage-measuring device makes no net contribution, the electrical output of the circuit in Figure 9.3 is given by Eq. (9):

$$E = P_{B1}(T_1 - T_r) + P_{B2}(T_2 - T_1) + P_{B3}(T_m - T_1) + P_{A3}(T_2 - T_m) + P_{A2}(T_2 - T_2) + P_{A1}(T_r - T_1) \quad (9)$$

Equation (9) may be rewritten to yield Eq. (10):

$$E = (P_{B1} - P_{A1})(T_1 - T_r) + (P_{B2} - P_{A2})(T_2 - T_1) + (P_{B3} - P_{A3})(T_m - T_2) \quad (10)$$

It may be observed from Eq. (10) and the graphic representation of the output in Figure 9.3 that when at all points along the length of a thermocouple the temperature is the same in both conductors, the output of that portion of the circuit depends only on the difference between the thermoelectric powers of the two conductors and the temperature change along their length. The difference between the thermoelectric powers of two conductors is known as the Seebeck coefficient of the pair [5,6].

The Seebeck coefficient for a single material is always given relative to some reference material. The early evaluations by Peltier, Seebeck and others were done relative to lead, and recent evaluations are relative to platinum-67. Table 9.1 gives the approximate Seebeck coefficients of the most common thermocouple materials relative to ^{67}Pt, and Table 9.2 gives the corresponding values of the most frequently used thermocouple pairs at temperatures near the ice point. The Seebeck coefficient for any pair of conductors is equal to the difference of the Seebeck coefficients of each conductor relative to a standard material. If the Seebeck coefficient of copper relative to ^{67}Pt is +5.9 μV/°C and that of constantan is −32.9 μV/°C, the Seebeck coefficient of a copper-constantan duplex pair (type T thermocouple) is 38.8 μV/°C. The material having the most positive Seebeck coefficient relative to ^{67}Pt will be the positive lead, which is the copper lead in the previous example.

TABLE 9.1
Appropriate Seebeck Coefficients of Common Thermocouple Materials Relative to Platinum-67 at 0.0°C

Material Name	ASTM E-20 letter code	Approximate composition	Seebeck coefficient (μV/°C)
Chromel	EP and KP	90% NI, 10% Cr	25.8
Iron	JP	99.5% Fe	17.9
Copper	TP	100% Cu	5.9
90Pt10Rh	SP	90% Pt, 10% Rh	5.4
87Pr13Rh	RP	87% Pt, 13% Rh	5.3
Alumel	KN	95% NI, 2% Al, 2% Mn, 1% Si	−13.6
Constantan	JN[a]	55% Cu, 45% Ni	−32.5
Constantan	EN and TN	55% Cu, 45% Ni	−32.9

[a] JN is similar to EN and TN, but will generally have a slightly different output

TABLE 9.2
Approximate Seebeck Coefficients of Common Thermocouple Pairs

Thermocouple Name	ASTM E-20 letter code	Seebeck coefficient (μV/°C)
Chromel-constantan	E	58.7
Iron-constantan	J	50.4
Chromel-alumel	K	39.4
Copper-constantan	T	38.8
Platinum 90Pt/10Rh	S	5.4
Platinum 87Pt/13Rh	R	5.3

If the Seebeck coefficient S is substituted for the difference in thermoelectric power, $P_A - P_B$, throughout the circuit in Figure 9.3, Eq. (10) can be rewritten to yield Eq. (11):

$$E = S_1(T_1 - T_r) + S_2(T_2 - T_1) + S_3(T_m - T_2) \quad (11)$$

Figure 9.4 depicts a typical thermocouple circuit of duplex leads for which the output is expressed in terms of the Seebeck coefficients of the conductor pairs. If the thermocouple wire in every section of the circuit is obtained from a single homogeneous length of wire, the Seebeck coefficient in every section will be the same. That condition is expressed by Eq. (12):

$$E = S_1 = S_2 = S_3 = S \quad (12)$$

Substituting the condition of Eq. (12) into Eq. (11) yields Eq. (13):

$$E = S(T_m - T_r) \quad (13)$$

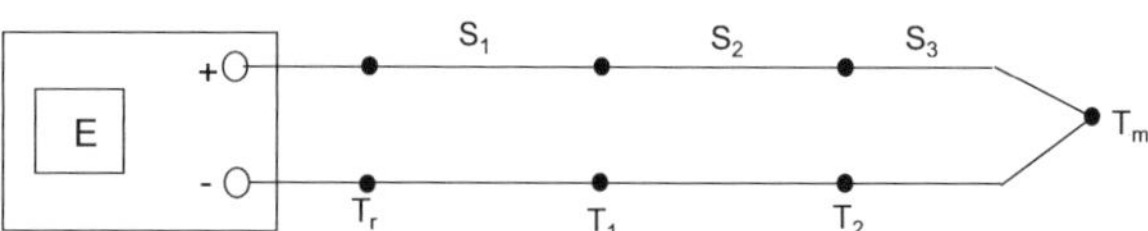

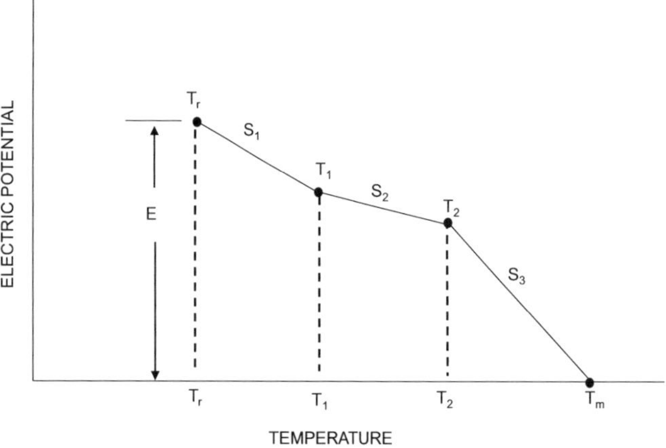

FIGURE 9.4 Typical duplex-lead thermocouple circuit.

If the temperature is expressed in degrees Celsius and the temperature of the reference junctions is maintained at 0.00°C, in Eq. (13) is equal to zero, then the net output of the circuit is given by Eq. (14):

$$E = ST_m \quad (14)$$

Equation (14) is valid only when the Seebeck coefficient is constant. The Seebeck coefficient of any pair of thermocouple wires changes slightly with temperature, but in a homogeneous length of wire, it is a unique function of temperature. Therefore, when the temperature is expressed in degrees Celsius and the reference junctions are at 0.00°C, the true output of a homogeneous thermocouple is a unique function of the measured temperature as given by Eq. (15):

$$E = \int_0^{T_m} S(T)dT \quad (15)$$

Standard values of voltage output as a function of measuring junction temperature, with reference junctions at 0.00°C, have been developed for all commonly used thermocouple pairs. Those values are given by NIST Monograph 175 in the United States and DIN standards in Europe.

9.4 THERMOCOUPLE REFERENCE TEMPERATURE

Equation (13) shows the importance of establishing an accurate reference junction temperature. Any difference between the actual reference temperature and the standard value produces an error that is equal to the temperature difference multiplied by the Seebeck coefficient at the reference temperature. The ice-point has been chosen as the "standard" thermocouple reference temperature because it is a known value of temperature that can be established quite accurately with relatively little effort.

9.4.1 Ice Bath References

Figure 9.5 depicts an ideal thermocouple circuit in which a pair of continuous, homogeneous conductors extend from the measuring junction to their junction with copper. The junctions to copper are immersed in an ice bath and are called the reference junctions of the circuit.

The reference junctions must be inserted to a sufficient depth in the bath to avoid conduction errors [7].

The temperature of the ice bath will be 0.00°C ± 0.01°C if the following procedures are followed:

1. Use a Dewar flask that is at least 10 in. (25.4 cm) deep and 4 in. (10.1 cm) in diameter.
2. Make ice using distilled water and crush it finely.
3. Fill the Dewar flask completely with the crushed ice and fill the voids between the ice particles with distilled water.
4. Insert the thermocouple leads into the central portion of the bath to a depth of at least 4–8 in. (10.1–20.3 cm) depending on the size of the wire.
5. Allow approximately 30 minutes for the ice and water to reach thermal equilibrium.
6. Pack the ice down into the Dewar flask, removing the excess water and adding additional crushed ice to maintain a solidly packed bed of ice with the voids filled by water.
7. Repeat step 6 as required.

9.4.2 Automatic References

Although the mixture of ice and water in a Dewar flask is an ideal reference, it is not very practical outside of the standards laboratory. An excellent alternative to the Dewar flask is an automatic ice bath that maintains a mixture of ice and water in a sealed chamber by means of thermoelectric cooling. Immersion wells extend into the chamber and the reference junctions of the thermocouple circuit are inserted to the bottom of the wells. When used in accordance with the manufacturer's operating instructions, the reference temperature provided by the automatic ice bath is typically 0.00°C ± 0.30°C [8]. The output of the circuit depicted in Figure 9.5 is given by Eq. (14).

In many applications it is not convenient to construct reference junctions on each lead of a thermocouple. Automatic ice baths are available with built-in reference thermocouples attached to terminals to which the external thermocouple leads are connected [8]. Figure 9.6 depicts a circuit using this type of reference.

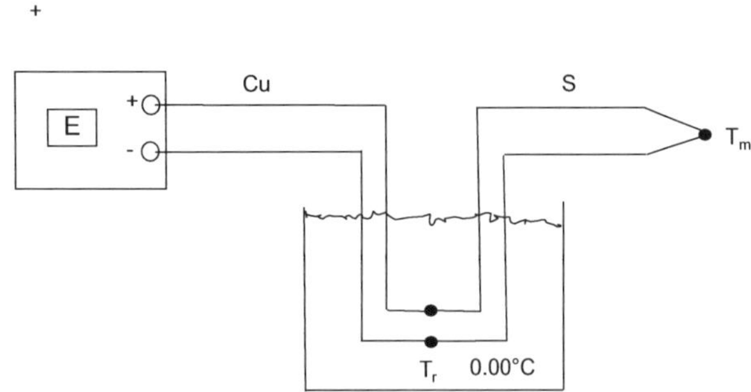

FIGURE 9.5 Thermocouple circuit with ideal reference.

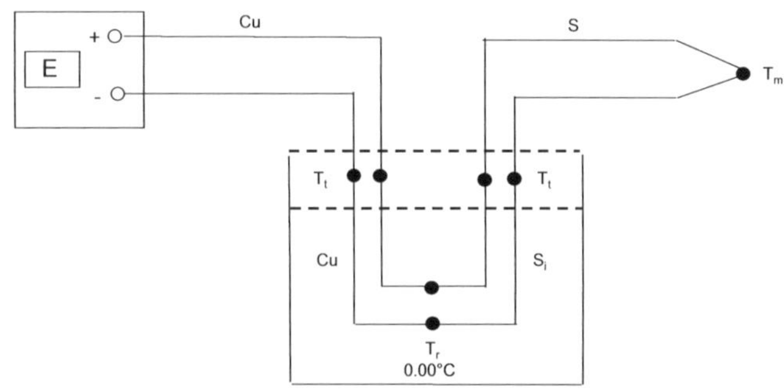

FIGURE 9.6 Thermocouple circuit with reference having internal thermocouples.

Each pair of "input" terminals is for a specific thermocouple type. Internal wires of the same type form reference junctions to copper that are maintained at 0.0°C in the ice bath. The copper leads from the reference junctions are connected to the "output" terminals of the reference.

The output of the circuit depicted in Figure 9.6 is given by Eq. (16). T_t is the temperature of the terminals, S is the Seebeck coefficient of the external thermocouple, and S_i is the Seebeck coefficient of the internal thermocouple. Even though the internal thermocouple is of the same type of material as that in the external portion of the circuit, its Seebeck coefficient may be slightly different because it may be from a different production lot of wire.

$$E = S(T_m - T_1) + S_1(T_t - T_r) \tag{16}$$

The external thermocouple produces a voltage equal to $S(T_m - T_t) + S_i(T_t - T_r)$ and the internal thermocouple produces a voltage equal to $S_i(T_t - T_r)$. The total voltage is the sum of the two. Because T_r is equal to 0.0°C, Eq. (16) can be rewritten to yield Eq. (17):

$$E = ST_m + T_t(S_i - S) \tag{17}$$

The second term of Eq. (17) may be considered an error term. If the Seebeck coefficient of the internal thermocouple is exactly equal to that of the external thermocouple, the second term of Eq. (17) is zero and the output is the same as that of the ideal circuit of Figure 9.5 and Eq. (14).

9.4.3 Thermocouple Compensators

In many industrial temperature-measuring applications, even an automatic ice bath with built-in reference thermocouples may not be practical. Automatic ice baths are expensive and do not operate reliably in ambient temperatures below 0.0°C or above 40.0°C. All instruments and systems being sold today for thermocouple temperature measurement provide an electronic circuit for determining the temperature of the terminals to which the thermocouples are attached. An appropriate reference voltage is added by the system to that produced by the external thermocouple. Early versions of such circuits were called compensators because they compensated for the fact that the terminals to which the thermocouples were connected were not at the ice-point temperature. Figure 9.7 depicts such a circuit.

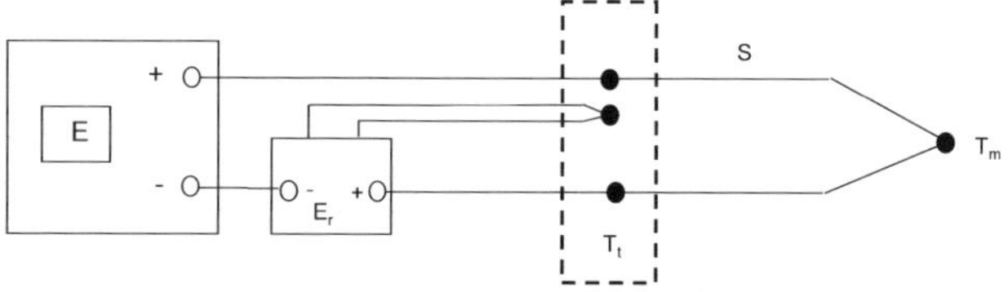

FIGURE 9.7 Thermocouple circuit with compensator.

The compensator produces a voltage that is a function of the terminal temperature. A typical compensator is a resistance bridge with the temperature-sensitive resistor installed near the thermocouple terminals. The bridge is adjusted to have a zero output when the temperature of the resistor is 0.0°C and to produce the proper voltage for the specified thermocouple at a normal ambient temperature [9]. The compensation voltage is added to the voltage produced by the thermocouple, and the total voltage is measured by the voltage-measuring device. Equation (18) gives the output of the circuit depicted in Figure 9.7:

$$E = S(T_m - T_1) + E_r \tag{18}$$

The perfect compensator would have an output equal to that which would be produced by the external thermocouple when its reference junctions were at ice-point and its measuring junction was at the terminal temperature. That characteristic is expressed by Eq. (19):

$$E_r = S(T_t - T_r) \tag{19}$$

Substituting this ideal compensator output into Eq. (18) and setting T_r equal to zero yields the same total output as that given by Eq. (14) for the ideal circuit.

9.4.4 Multichannel Thermocouple Systems

In older multichannel systems, the compensation voltage is added electrically, as depicted in Figure 9.8. When compensation voltage is added electrically to the output of a multipoint scanner, all of the thermocouples in the group must be of the same type. The compensator must be designed to produce the output required for a given thermocouple type, and its output should be adjustable in calibration to match the Seebeck coefficient of the external thermocouples.

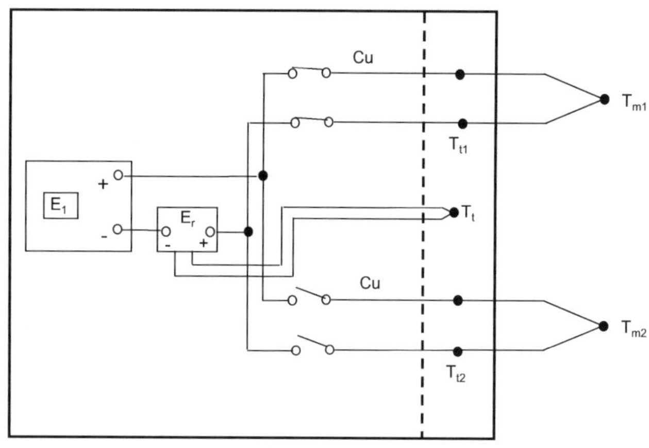

FIGURE 9.8 Multichannel thermocouple system with internal compensator.

Modern multichannel thermocouple-measuring systems use microprocessor capability to add the proper compensation voltage to the measured thermocouple output (Figure 9.9). Rather than adding a compensation voltage electrically, the temperature of the terminals is measured by the system and the value stored in memory. When a channel is programmed to be a thermocouple input, the system automatically computes the appropriate compensation voltage for that type of thermocouple, adds it to the measured voltage, and converts the total voltage to the corresponding temperature.

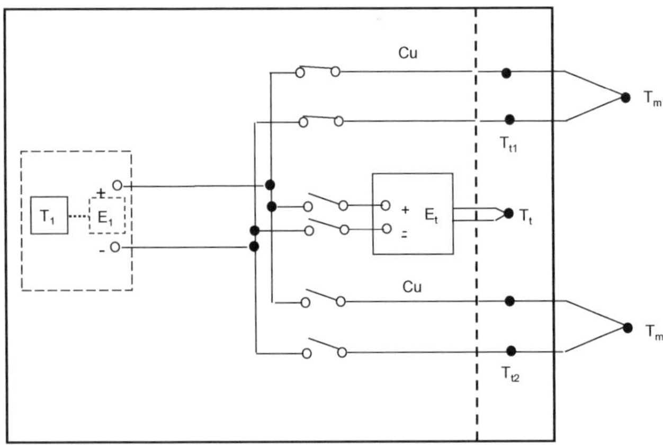

FIGURE 9.9 Multichannel thermocouple system with computer and internal reference.

Whether the system reference voltage is added electrically as in Figure 9.8, or mathematically as in Figure 9.9, it is based on a single measurement of the terminal temperature, which may be different from the temperature of each individual pair of terminals in the group. The total voltage output of the first thermocouple in Figure 9.9 is given by Eq. (20):

$$E = S(T_m - T_1) + S(T_1) \tag{20}$$

The first term of Eq. (20) is the measured voltage produced by the external thermocouple and the second term is computed by the system based on the measured value of the terminal temperature. Equation (20) may be rewritten to yield Eq. (21):

$$E = S(T_{m1}) + S(T_t - T_{t1}) \tag{21}$$

The second term of Eq. (21) is an error term. If the actual temperature of the terminals to which a thermocouple is attached is different from the terminal temperature measured by the system, an error is introduced that is equal to the temperature difference multiplied by the Seebeck coefficient at the terminal temperature.

In some installations, it is not possible, or desirable, to run thermocouple wire from the measuring junction to the recording system. Figure 9.10 depicts one solution that is similar to the multichannel computer system of Figure 9.9. In this case, however, the conversion to copper is at the terminals

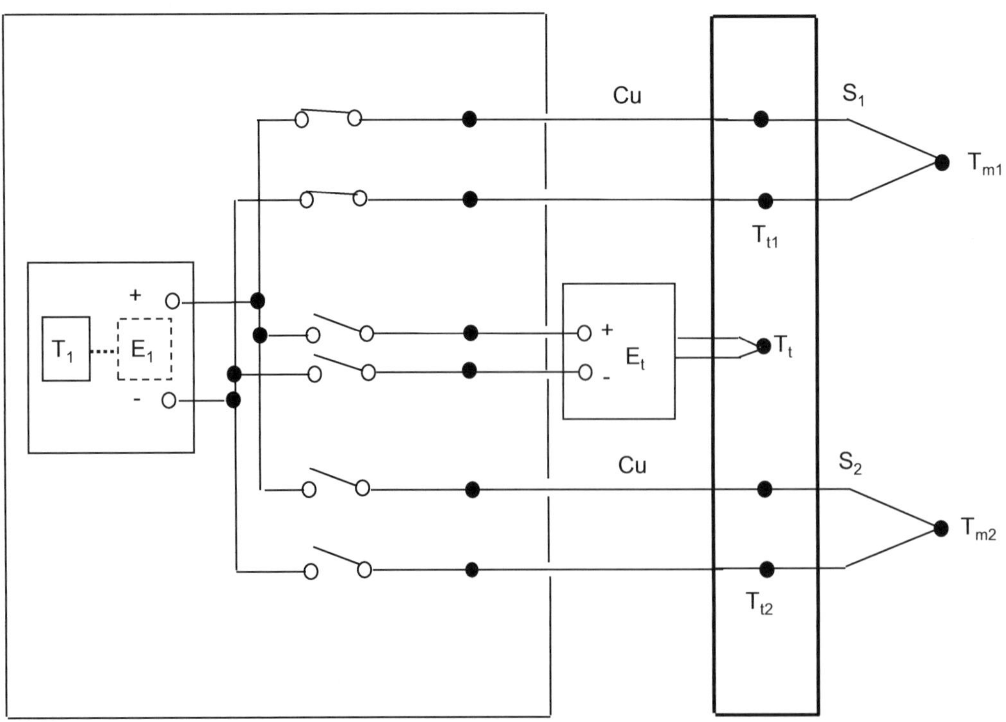

FIGURE 9.10 Multichannel thermocouple system with computer and external UTR reference.

of a remote uniform temperature reference, the temperature of which is measured by some independent means [8]. If the logic of the multichannel data system is designed to operate with a remote reference, the output of each thermocouple is computed in the same fashion as when an internal reference is provided. Equations (20) and (21) apply to this type of installation as well.

9.5 SOURCES AND TYPES OF ERROR

The dictionary defines accuracy as the absence of error, but accuracy is a term that has many different meanings. Any discussion of temperature measurement accuracy must focus on the various sources and types of errors. In a typical thermocouple installation, the three primary sources of error are the thermocouple sensors, the circuit that connects the thermocouple sensors to the measuring system, and the measuring system [10,11].

In discussing errors and accuracy, it is important to distinguish between relative accuracy and absolute accuracy. *Relative accuracy* is the degree to which temperature measurements at different locations can be compared or the degree to which the measurement of a single temperature is repeated. *Absolute accuracy* is the degree to which a measurement gives the absolute thermodynamic value of temperature. In many processes, relative accuracy is sufficient, but in thermal sterilization processes absolute accuracy is essential.

The rate at which microorganisms are destroyed is a strong function of the temperature, so the time required to produce a sterile product depends on the temperature of the product. If the true value of temperature is less than the indicated value, improper sterilization may result.

Another important distinction to make is that of systematic errors and random errors. Systematic errors can be eliminated from the final results by calibration, but random errors can be minimized only by proper selection and installation of the measuring instrumentation. The lack of interchangeability, conformity, and uniformity produce systematic error; nonhomogeneous regions in the circuit and the lack of repeatability produce random errors.

9.5.1 Sensor and Circuit Errors

In thermocouple systems, it is difficult to draw a clear distinction between sensor errors and circuit errors, because a thermocouple is a total integrator of the temperature change from the measuring junction to the reference junction. Conformity and interchangeability are characteristics generally attributed to the sensors; nonhomogeneous effects are attributed to the circuit.

9.5.1.1 Conformity to Standard

Conformity error is the difference between the actual voltage produced by a thermocouple and the standard output voltage for that thermocouple type at the same measured temperature. The reference junctions of the thermocouple circuit are assumed to be at 0.00°C. One specification that is often quoted for thermocouples is the maximum conformity error that thermocouples can have and still meet accepted industrial standards. For standard grade type T (copper constantan) thermocouples, that error is the greater of ±1.0°C or ±0.75%. For special grade type T thermocouples, it is the greater of ±0.5°C or ±0.4% [12]. Selected grade thermocouples supplied by GE Sensing, former Kaye Instruments, have a maximum conformity error of ±0.25°C or ±0.2% at 120°C [13].

It must be emphasized that the conformity error is not indicative of the total measurement error in any particular installation. Conformity errors can be eliminated by calibration at a number of temperatures over the operating range, and there are many other system errors that may be larger than the conformity error.

9.5.1.2 Interchangeability

The degree to which a number of thermocouples all have the same output at the same measured temperature is known as the interchangeability of the thermocouples. Interchangeability is important when comparing two temperatures in an uncalibrated system. When a number of thermocouples are made from the same production lot of wire, the maximum interchangeability error is typically the greater of ±0.1°C or ±0.1%. As with conformity errors, interchangeability errors can be eliminated by calibration. In both cases, it is often sufficient to calibrate the sensors at the two extreme temperatures of the operating range and apply a linear correction to the measurements. If the measuring system does not provide the capability of applying individual calibration corrections to each input, interchangeability error becomes an important consideration, and all thermocouples used at one time should be made from the same production lot of wire.

9.5.1.3 Nonhomogeneous Regions

The thermoelectric power of a conductor is a function of the composition and structure of the material. Most thermocouple conductors are alloys of several elements. Among the commonly used thermocouple materials, only copper and platinum are essentially pure elements; even copper wire must be checked to be sure that it has the proper characteristics. The Seebeck coefficients of thermocouples will vary slightly between production lots of wire because of variations in composition and annealing. Annealing affects the thermoelectric power because it alters the grain structure of the conductor. Similarly, the thermoelectric power of a conductor can be changed slightly if it is stressed to the point of permanent distortion. The phenomenon known as cold working changes the thermoelectric power as well as the physical characteristics of a metal [14]. When a thermocouple circuit is constructed of continuous, homogeneous wire from the measuring junction to the terminals of the measuring system, calibration can eliminate most errors associated with the sensor and the circuit. Tests have shown conclusively that the output of a homogeneous length of thermocouple wire depends only on the total change in temperature from one end to the other;

the location of the change within the wire does not matter. This characteristic is extremely important in calibrated systems because the location of the gradient in the wire during operation will generally be different from the location of the gradient during calibration.

Connectors introduce a section of nonhomogeneous conductors in a thermocouple circuit. When they must be used, connectors should be made of the same materials as the wire and located away from regions of large temperature gradients. Although the materials of thermocouple connectors are essentially the same as the wire, the annealing process used to make a rigid connector pin must be different from that used to make flexible wire. The resulting Seebeck coefficient is usually slightly different.

Repetitive flexing of thermocouple wire at one location can also cause a nonhomogeneous region because of cold working. In the validation process, thermocouples are normally installed through fittings in the walls of sterilizers where they are clamped rigidly. In placing the thermocouples at different locations within the sterilizer, some amount of flexing at the fitting is unavoidable. Because solid wire is much more susceptible to cold working than stranded wire of the same size, only stranded wire should be used in this application, and great care should be exercised to avoid flexing the wire more than necessary. The sterilizer wall is the region of maximum temperature gradient during operation, so even a small change in Seebeck coefficient in that region can cause a significant error.

The effect of nonhomogeneous regions in a circuit is illustrated in Figure 9.11 and the following example. All of the wire in the circuit has a Seebeck coefficient S, but the connector in the circuit has a Seebeck coefficient S_c. The temperatures at the ends of the connector are T_1 and T_2. The output of the circuit shown in Figure 9.11 is given by Eq. (22):

$$E = S(T_m - T_2) + S_c(T_2 - T_a) + S(T_1 - T_r) \tag{22}$$

Equation (22) can be rewritten to yield Eq. (23):

$$E = S(T_m - T_r) + (S_c - S)(T_2 - T_1) \tag{23}$$

The second term of Eq. (23) is the error caused by having a connector in the circuit. The error will be zero if the Seebeck coefficient of the connector is equal to that of the wire or if there is no temperature difference across the connector. It is unlikely that the Seebeck coefficient of a connector will match that of the wire exactly, so it is important to avoid using connectors where they will have large temperature differences imposed on them.

To illustrate the error that would be caused by installing a connector in the wall of a sterilizer, assume the following values for the circuit of Figure 9.11 and Eq. (23):

$T_m = 120.0°C$ $T_2 = 100.0°C$ $T_1 = 50.0°C$ $T_r = 0.00°C$
$S = 40.0\ \mu V/°C$ $S_c = 42.0\ \mu V/°C$

The output according to Eq. (23) is

$$E = 40.0(120.0 - 0.0) + (42.0 - 40.0)(100.0 - 50.0)$$
$$= 4800\ \mu V + 100\ \mu V = 4900\ \mu V$$
$$\text{Error} = 100\ \mu V \text{ or } 2.5°C$$

The values employed in this example are typical of those that would be experienced if a connector in a type T (copper-constantan) thermocouple circuit were installed in the wall of a steam autoclave. The error is 100 μV, or about 2.5°C. A similar error could be caused by a cold-worked region at the wall, but the magnitude of the error would be less. For a connector to have a Seebeck coefficient 5% greater than the wire it is designed to match is typical, but the change because of cold working will be much less.

A second type of nonhomogeneous circuit is illustrated by Figure 9.12. Many thermocouple probes are constructed using lengths of thermocouple wire swaged into stainless-steel tubes. This type of material may be purchased in long sections and cut to form stainless-steel thermocouple probes of the desired length. One end is welded to form the measuring junction and the two wires at the other end are attached to extension leads of matching thermocouple wire. The wire in the stainless-steel tip has a Seebeck coefficient S and the extension wire has a Seebeck coefficient S_c.

The output of this circuit is given by Eq. (24):

$$E = S(T_m - T_1) + S_c(T_1 - T_r) \tag{24}$$

Adding and subtracting the term ST_r and rewriting Eq. (24) yields Eq. (25):

$$E = S(T_m - T_r) + (S_c - S)(T_1 - T_r) \tag{25}$$

The second term of Eq. (25) is an error term. The error will be zero if the Seebeck coefficient of the extension wire is equal to that of the wire in the tip or if the junction between the two is maintained at the reference temperature.

This type of probe should be avoided, unless it can be calibrated under the same conditions encountered in normal operation. Specifically, the value of T_1 must be the same during

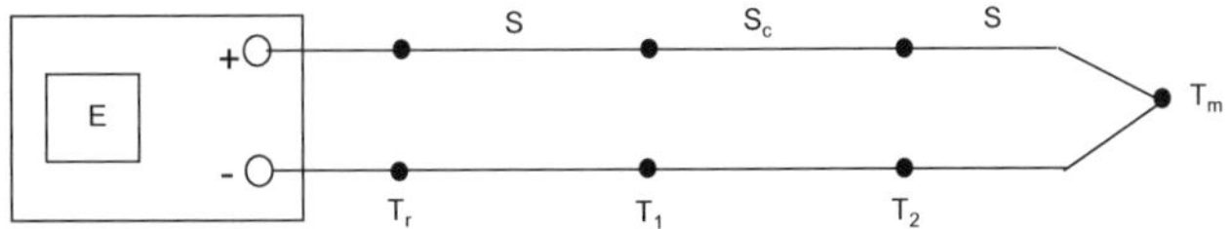

FIGURE 9.11 Thermocouple circuit with connector

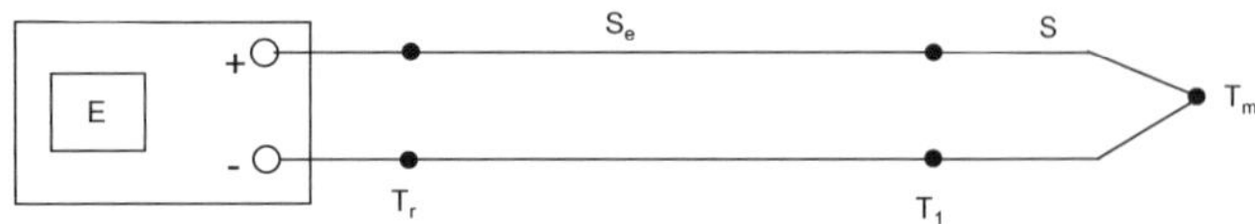

FIGURE 9.12 Thermocouple circuit with two sections

both calibration and operation, or an unrecognized error will be introduced. Assume the following values for the circuit of Figure 9.12 and Eq. (25) when the probe is calibrated in a laboratory:

$$T_m = 120.0°C \; T_l = 30.0°C \; T_r = 0.0°C$$
$$S = 40.0\ \mu V/°C \; Se = 40.2\ \mu V/°C$$

The output of the circuit according to Eq. (25) is

$$\begin{aligned} E &= 40.0(120.0 - 0.0) + (40.2 - 40.0)(30.0 - 0.0) \\ &= 4800 + 6 = 4806\ \mu V \end{aligned}$$

Assuming that the standard Seebeck coefficient is 40.0 μV/°C, the calibration correction is 6 μV or 0.15°C when the probe is measuring 120.0°C.

When this probe is used inside a steam autoclave, the temperature of the junction between the tip and the extension wire will be at autoclave temperature. Assume that all other values are the same as in the calibration example, but the value of T_l is 120.0°C. The output will be

$$\begin{aligned} E &= 40.0(120.0 - 0.0) + (40.2 - 40.0)(120.0 - 0.0) \\ &= 4800 + 24 = 4824\ \mu V \end{aligned}$$

Applying the calibration correction of 6 μV still leaves an error of 18 μV, or almost 0.5°C. This example, even more than the previous one, shows the importance of using thermocouples that have continuous length of homogeneous wire from the measuring junction to the region outside the autoclave.

9.5.1.4 Diffusion of Steam

All insulating materials are permeable to steam after extended exposure.

When stranded wire is installed through the wall of an autoclave, steam will eventually diffuse through the insulation, flow to the lower pressure outside the autoclave through the small passages formed between the strands of wire and condense to form drops of moisture where the insulation ends. This diffusion of moisture along the wire will not cause an error in the output of a thermocouple, but it should be prevented from collecting on terminals or connectors where corrosion could cause problems. Diffusion of moisture along the wire to the outside of the autoclave will not occur if solid wire is used instead of stranded wire, but solid wire is more susceptible to cold working. The flexing of solid wire at the wall of an autoclave may introduce a serious error, whereas moisture dripping from stranded wire is only an inconvenience.

Some thermocouple assemblies are constructed using flexible hose to protect the thermocouple wire inside the autoclave. One end of the flexible hose fits over a length of stainless-steel tubing that forms the thermocouple probe and the other end of the flexible hose connects to a stainless-steel tube that provides a pressure seal at the wall of the autoclave. This design guarantees that there will be no cold working of the homogeneous wire that runs continuously from the measuring junction to a connector outside the autoclave.

Unfortunately, the steam that diffuses through the flexible hose will condense inside when the assembly is cooled down. If some of the moisture collects in the stainless-steel probe near the measuring junction, it will cause an error if the probe is used subsequently to measure temperatures above 100°C. Because the passage from the inside of the probe through the hose is open to the atmosphere, any moisture in the probe will boil at 100°C, absorbing energy from the surrounding material and reducing the temperature of the probe tip. Depending on the amount of moisture, the distance of the moisture from the measuring junction, and the rate of heating at the outer surface of the probe tip, the magnitude of the error caused by moisture in this type of probe can vary from a few tenths of a degree to several degrees.

The presence of moisture in a probe tip is detected readily in calibration, so it should never cause an error in a validation run if the probes are calibrated before each run. If a large amount of moisture is present, it will prevent the tip from ever reaching the calibration temperature, and the steam condensing inside will make the portion of the probe that extends above the calibration bath extremely hot. If a small amount of moisture is present it will boil away, permitting the tip to achieve the proper temperature, but it will retard the rate of heating during the time it is boiling. A probe with moisture will take a noticeably longer time to reach the calibration temperature than a dry one. If a maximum acceptable time to reach calibration temperature is specified, the presence of moisture in a probe will be detected.

Attempts to fill the probe tip with a solid material to prevent moisture from collecting near the measuring junction can cause cold working of the wire because of differential thermal expansion of the wire and the filling material. The resulting errors are more serious than the presence of moisture. Recent tests with a new filling technique indicated that moisture errors may be eliminated in future probes without causing other problems [13].

9.5.1.5 Circuit Resistance

The resistance of a thermocouple circuit has no effect on the voltage generated. The indicated outputs of early industrial thermocouple meters were inversely proportional to the resistance of the external circuits, because galvanometers were used to measure the current flowing in the circuits rather than the voltage potential. Null balance potentiometers generate a balancing potential so no current flows in the circuit, and modern thermocouple meters have measuring circuits with extremely large input impedance compared with that of the circuit. When using either of the latter types of instrumentation, normal levels of thermocouple circuit resistance will not affect the indicated temperature.

Cracked wire or poor electrical contact at connectors can introduce extremely high resistance in a circuit, affecting the accuracy of the voltage measuring device and giving erratic values of the indicated temperatures. The wire in a circuit can be broken but held together by the insulation. When the wire is stretched, the ends come apart and cause an open circuit; when it is relaxed, the ends of the wire may touch, again completing the circuit but with a high resistance at the point of contact. The surface of the copper contacts in a copper-constantan connector can become oxidized, thereby creating a high resistance. If an ohmmeter is used to measure the contact resistance, it may indicate a fraction of an ohm because the excitation voltage of the ohmmeter can break through the oxide film. With only the small potential generated by a thermocouple imposed, however, the resistance may be thousands of ohms. If erratic readings are experienced in a thermocouple circuit having a connector, cleaning the connector contacts may solve the problem. Oxidation of copper contacts can be prevented by plating them with gold.[13]

9.5.2 Measuring System Errors

Different manufacturers state the accuracies of thermocouple measuring systems in different fashions. Some give detailed breakdowns of the error sources; some simply state total error when operating within a limited range of ambient temperatures. Changes in ambient temperatures are the most significant sources of error in thermocouple measuring systems, particularly in multichannel systems with internal references.

9.5.2.1 Resolution

The resolution of a measuring system is the ability to read the output. In analog chart recorders, resolution is determined by the width of the chart paper and the temperature range corresponding to the total width. Because the width of the chart is fixed, a smaller temperature range must be set if better resolution is required. This type of recorder can be purchased with plug-in cards to set the temperature range.

In digital systems, resolution is the value of the least significant digit. The resolution of temperature measurements may be 0.01°, 0.1°, or 1.0°, Fahrenheit or Celsius. Some meters even give a resolution of 0.001°. Measurement accuracy can be no better than the resolution, but it should never be assumed that the accuracy of a measuring instrument is as good as its resolution.

9.5.2.2 Conformity to Standard

All modern thermocouple measuring systems use microcomputers to add compensation voltage to the measured voltage generated by the external thermocouple and to convert the resulting total voltage to the corresponding temperature for that type of thermocouple. The conversion from voltage to temperature typically utilizes a series of straight lines or polynomial functions that approximates the standard tables. The difference between the calculated temperature and the standard temperature at a given voltage is the conformity error at that temperature. At any given measured temperature, the conformity error will always be the same. Maximum conformity error ranges from ±0.02°C in high accuracy systems to as large as ±1.0°C in some systems.

9.5.2.3 Uniformity

Uniformity is the degree to which the measuring system indicates the same value when exactly the same input is applied to different channels of a multichannel system. The largest error in most multichannel thermocouple systems is the uniformity error caused by differences in the temperatures of the terminals to which the thermocouples are attached. It is not unusual to have terminal temperatures differ by 1.0°C. A difference in terminal temperature causes an error equal to the temperature difference multiplied by the ratio of the Seebeck coefficient at the terminal temperature to that at the measured temperature. Even when the terminals are insulated to protect them from external heating and cooling effects, they will be heated nonuniformly by the internal electronics of the measuring system. Once a system has warmed up completely in a steady ambient temperature, the terminal temperatures will be stable. If each thermocouple is calibrated at the ice point (0.00°C), the uniformity error because of the terminal temperature difference will be included in the calibration correction. If the ambient temperature subsequently changes, the terminal temperature difference may also change. Although the systematic uniformity error was eliminated by calibration, an additional random uniformity error may be introduced by a subsequent ambient temperature change.

9.5.2.4 Repeatability

Repeatability is the degree to which the measuring system will indicate the same output over a period of time when exactly the same input is being measured. Repeatability errors can be classified as short term (seconds), medium term (minutes), and long term (weeks). Short-term errors in the indicated output are caused by electrical phenomena. Continuous fluctuations in the output are usually caused by instabilities in the measuring circuit of the system. Sudden jumps of brief duration in the output are usually caused by common-mode voltage differences. The common-mode voltage difference is the potential difference between the sensor and the ground of the measuring system. In steam autoclaves, large static potential differences can be created between ungrounded probes and the ground of the measuring system, particularly when the probes are installed in plastic containers. Proper grounding of the probes can minimize the error caused by this phenomenon. The ratio of the maximum measurement error to the common-mode voltage difference is called the common-mode rejection of the system and is expressed in decibels. A decibel is a measure of voltage ratio or current ratio equal to 20 times the common logarithm of the ratio. The common mode rejection varies from better than 140 db (10 million to 1) in high accuracy systems to less than 100 db (100,000 to 1) in some systems. Medium-term errors in the indicated output are caused by thermal phenomena. Temperature changes in the measuring circuit, in the thermocouple reference, and in the input terminals all cause errors in the indicated output. The magnitude of the measurement error caused by a change in ambient temperature is given by the temperature coefficient of the system. All manufacturers specify the temperature coefficient based on the system being stable before and after the change in ambient temperature; transient errors that occur during the temperature change may be much larger. Temperature coefficients vary from 0.01°C/°C for high accuracy systems to 0.1°C/°C in some systems.

Long-term errors in the indicated output result from component aging. Invalidation studies of this type of error are not important, because the system is calibrated with sufficient frequency to account for any long-term variations.

9.6 CALIBRATION PROCEDURE

Thermocouple systems used to measure temperatures in the validation process should be calibrated before and after each use. Typically, neither the measuring system nor the thermocouples will change their characteristics between calibrations, but the calibration process assures proper operation of the entire system. Because corrections applied to each thermocouple also include the uniformity error of the measuring system, each thermocouple should be connected to the same channel in calibration as in operation. To the extent possible, the entire system should be calibrated under the same ambient temperature and other conditions as it will experience during operation.

9.6.1 Calibration Basics

There are a few basic rules that should be followed in any calibration procedure.

1. Challenge all results. No single measurement should be accepted as being correct unless it is verified by other results. The transfer standard used to determine the temperature of the calibration bath could have an error. If two standards agree, the probability that they both have the same error is extremely low.
2. Be patient. A frequent mistake in calibrating instrumentation is to take measurements and make adjustments before conditions have stabilized. It may take much longer than expected for a system to become completely stable, because thermal errors decay exponentially, and the output may seem to be stable even though it is still changing slowly. Computer-based validation systems are available that provides automatic two-point calibration including automatic stability determination. This eliminates calibration errors caused by operator inconsistencies.
3. The accuracy of the transfer standard must be better than that of the instrument being calibrated. This would seem obvious, but it is amazing how often a voltage calibrator is used that has a greater error than the system being calibrated. Rules such as being 10 times as accurate or even twice as accurate are not absolute: It is only important to recognize that the accuracy of the calibration can be no better than the standard used, and that it is a mistake to change the adjustment of a measuring system if it is already more accurate than the standard.
4. The characteristics of the transfer standard must have been determined by a procedure that is traceable to accepted primary standards. [15] In the United States, the National Bureau of Standards (NIST) is the accepted source of primary standards. The transfer standards used should have been calibrated by the NIST relative to their primary standards or by a qualified Standards laboratory relative to standards that they have had calibrated by the NIST. In either case, the test results and test numbers should be known so the calibration procedure can be traced back to the primary standards.
5. The transfer standard must be independent of the measuring system. Because the output of a thermocouple depends on the entire circuit, it is not a desirable temperature transfer standard. An RTD is a device that indicates changes of temperature by a change of resistance. Because the resistance of an RTD is only a function of its temperature, and the resistance can be measured independently of the system being calibrated, RTDs are ideal temperature transfer standards.
6. The characteristics of the transfer standard must be stable in shipment and other handling. As its name implies, the purpose of the transfer standard is to transfer a measured characteristic from one laboratory to another. The characteristics of the standard must be the same when received from the NIST as when it was calibrated relative to their standards. Liquid in glass thermometers may be damaged or develop small voids in the liquid during shipment and therefore are not reliable temperature transfer standards. RTDs are fairly rugged devices that maintain their characteristics in normal handling and shipment.

9.6.2 Measuring System Calibration

The first step in calibrating a thermocouple system is to check the operation of the measuring system in the voltage mode and adjust it if necessary. Each manufacturer has a recommended procedure and calibration interval, which should be followed.

A precision low level de voltage source having accuracy better than ±1.0 μV ± 0.01% in the range of 0.0 μV to 20,000 μV should be employed in the voltage calibration. The measuring system should be turned on several hours before starting the calibration process to be sure that it has become completely stable. If the system is to be used for important voltage measurements, a second voltage source should be used to check the results of the adjustments. If the only important measurements are thermocouple temperature measurements, the calibration of the sensors will correct for any small voltage errors.

Once the voltage measuring circuits have been adjusted, the thermocouple reference of the system should be checked by connecting thermocouples to the proper input terminals and placing several of their measuring junctions in an ice bath. If a crushed ice bath is used, it should be made and maintained as described in Section 9.4.1. If an automatic ice bath is used, the measuring junctions should be inserted to the bottom of the wells. In either case, allow 10 or 15 minutes for the temperature to stabilize before making any adjustments.

The operation of the thermocouple reference in a multichannel, computer-based system is discussed in Section 9.4.4. When the input terminal temperature of the system is above 0.0°C, a thermocouple with its measuring junction in an ice bath will generate a negative voltage. If the internal reference is adjusted until the indicated temperature is 0.0°C (32.00°F), the output of the internal reference is adjusted to equal the output that is generated by the external thermocouple when its reference junction is at 0.0°C and its measuring junction is at the temperature of the input terminals. The external thermocouple is generating a negative voltage of the same magnitude. As discussed in Section 9.4.3 and shown by Eqs. (18) and (19), this procedure provides the perfect internal reference or compensation voltage for that external thermocouple.

Because the input terminal temperatures and the Seebeck coefficients of each thermocouple in a multichannel system may be slightly different, other thermocouples connected to the measuring system may not indicate exactly 0.0°C when the internal reference is adjusted as described in the previous paragraph. For best overall accuracy, the internal reference should be adjusted until the average of the indicated temperatures of all thermocouples in the ice bath is 0.0°C. If the measuring system can be programmed to compute the average of the outputs of a group of thermocouples, that value can be used directly in the calibration procedure. It should be emphasized that calibration of the internal reference is a measuring system calibration and not a calibration of the external thermocouples.

9.6.3 Thermocouple Calibration

In order to assure absolute accuracy of every temperature measurement, each thermocouple must be calibrated by determining its output when its measuring junction is at two or more known temperatures. Electronic thermocouple calibrators are quite useful in checking systems for proper operation, but they do not provide temperature calibration of the thermocouples being used with the systems.

All temperature sensors should be calibrated at the ice point if 0.0°C is within their normal range of operation. As was discussed in Section 9.4.1, the ice point is a known temperature that can be established quite accurately with relatively little effort. Measuring the ice point temperature is an ideal check for any temperature indicator. It is also important to calibrate a temperature sensor at, or near, the maximum and minimum temperatures to be measured. Some thermocouple measuring systems provide a feature that permits the automatic application of a two-point correction on each thermocouple. These software-controlled validation systems provide fully preprogrammed calibrations, including selective set point control of dry block temperature reference and automatic stability determination.

In steam autoclave measurements, the recommended minimum calibration temperature is 90°C and the recommended maximum calibration temperature is 130°C, with a post calibration verification at 121°C. When selected grade thermocouple wire [13] is calibrated at 90.0°C and 130.0°C, and a linear correction is applied between those temperatures, the maximum conformity error relative to the NIST standard output [5] will be less than +0.1°C. This result has been verified by thousands of calibrations of the selected grade wire. [13]

Typical operating temperatures in hot air ovens are in the vicinity of 200.0°C, and depyrogenation tunnels may be operated at temperature >300.0°C. In validating those processes, the thermocouples should be calibrated at a temperature near the maximum expected operating temperature of the process. If the ice point is used as the second temperature of a two-point calibration of selected grade thermocouple wire and a linear correction is applied, the maximum conformity error relative to the NIST standard output may be as large as ±0.30°C between 0.0°C and 200.0°C and as large as +0.50°C between 0.0°C and 300.0°C. This level of error is normally acceptable in these higher temperature processes, and the error becomes much smaller near the maximum calibration temperature, which is also the normal operating temperature.

If better accuracy is required at higher temperatures, the thermocouples must be calibrated at intermediate points. The maximum expected error in any temperature measurement increases at higher temperatures. When a thermocouple is calibrated at two temperatures and a linear correction is applied between the two temperatures, the maximum expected error because of the thermocouple's characteristics is less than ±0.05°C between 100.0°C and 150.0°C, approximately ±0.10°C between 150.0°C and 200.0°C, and approximately ±0.20°Cbetween 250.0°C and 300.0°C.

The type of equipment and instrumentation that must be used in a temperature calibration facility, and the amount of personnel training required to operate it, depend on the level of accuracy desired. To achieve calibration accuracies of ±0.01°C requires very expensive, elaborate instrumentation and highly trained personnel. Calibration accuracies of better than ±0.1°C can be achieved with relatively inexpensive instrumentation and simple procedures [16]. The less elaborate calibration facility is actually preferred in most validation processes because the level of accuracy is better than required and it is less likely that an error will be introduced by faulty procedure.

The following equipment and instrumentation are required in a basic temperature calibration facility to achieve total calibration accuracy of better than ±0.1°C at temperatures up to 150.0°C and ±0.20°C at temperatures between150°C and 300.0°C:

1. An automatic ice bath [7] or a Dewar flask filled with crushed ice and distilled water as described in Section 9.4.1.
2. A high-temperature reference block [15] or a stirred oil bath with temperature uniformity better than ±0.03°C in the working region.
3. At least three RTDs that have been calibrated traceable to NIST standards to an accuracy of ±0.03°C at the minimum and maximum temperatures in the calibration range and at intervals no larger than 500°C if ±0.1°C accuracy is required or 100.0°C if ±0.2°C accuracy is required.
4. An independent instrument to measure the resistance of the RTDs to an accuracy corresponding to ±0.03°C.
5. A precision resistor with calibration traceable to NIST standards to calibrate the resistance-measuring instrument.

The RTDs should be of a four-wire design, which provides independent leads for the excitation current and for measuring the voltage difference across the resistor. The same excitation current must be used in transfer calibrations as was used in the original calibration of the RTD, because the self-heating error of an RTD is a function of the current. The most common excitation current for a Pt 100 RTD is 1 mA. At least three RTD transfer standards should be available, because two standards must agree at each calibration temperature, and the third is required to determine which of the first two is correct if they do not agree.

A 25 Ω platinum RTD is the primary standard temperature sensor used by all primary calibration laboratories. It is quite expensive and quite delicate. An industrial grade, 100 Ω, platinum RTD is quite acceptable as a transfer standard, and its resistance can be measured to an accuracy of ±0.01 Ω with relatively inexpensive instrumentation [16]. A resistance change of 0.01 Ω corresponds to a temperature change of approximately 0.025°C. The resistance-measuring instrument must be calibrated at two values in the range to be measured. One of the values can be zero resistance, or a shorted input, and the second value should be approximately equal to the maximum RTD resistance to be measured. When 100-Ω RTDs are used to measure temperatures between 0.0°C and 300.0°C, a 150-Ω precision resistor is recommended as the second point. The resistor calibrations should be independently traceable to NIST standards and accurate to ±0.005 Ω.

The resistance- measuring instrument should be capable of measuring the resistance of up to three RTDs and the precision resistor at the same time. The current leads of the precision resistor and the RTDs should be connected in series, so that the same excitation current passes through the precision resistor and the RTD whose resistance is being measured. Adjusting the current to make the instrument indicate the proper value of the precision resistor automatically calibrates it for the RTD reading. In effect, the instrument compares the resistance of the RTD with that of the precision resistor.

The following detailed procedure is recommended for calibrating thermocouples to be used with multichannel measuring system in a validation procedure.

1. Connect all thermocouples to the channels of the measuring system to which they will be connected in the validation run. Each thermocouple must be labeled clearly and a record made of the channel to which each is connected.
2. Turn on the measuring system and the resistance-measuring instrument at least 2 hours before taking any measurements. If an automatic ice bath and a high-temperature reference block are to be used, they should be turned on at the same time. If a crushed ice and distilled water bath is to be used, it should be prepared at least 1/2 hour before being used. Most stirred oil baths require about 15 to 20 minutes to stabilize.
3. Once the measuring system has stabilized, it should be calibrated according to the procedures of Section 9.6.2.
4. Place two RTD transfer standards in the wells of an automatic ice bath or in a crushed ice bath. If an automatic ice bath is used, the RTDs should be inserted to the bottom of the wells and the wells filled with water. At least one manufacturer of automatic ice baths recommends filling the wells with silicone oil having a specific gravity greater than unity [7]. Oil is recommended to prevent the possibility of ice forming in the wells, but water is much more convenient and the formation of ice in such units is an extremely rare occurrence. The temperature accuracy is the same in either case. If a crushed ice bath is used, the RTDs should be inserted to a depth of approximately 30 probe diameters. A 3/16 in. diameter RTD should be inserted to a depth of 6 in. and a 1/4in. diameter RTD to a depth of 8 in. After the probes have been inserted for a few minutes, all excess water should be removed and additional crushed ice added to create a solidly packed bed of ice with the voids filled by water.
5. After the RTDs have reached equilibrium, check the calibration of the resistance measuring instrument by measuring the value of the precision resistor and make an adjustment if necessary. Then measure the resistance of each RTD and compare the measured value with the calibrated value of resistance

at 0.00°C. The measured resistance of a 100-Ω RTD should agree with the calibrated value to within ±0.01 Ω at 0.00°C. If the RTDs indicate the same temperature but both indicate that the ice bath is not 0.00°C ± 0.03°C, check the ice bath. If one of the RTDs has a resistance >0.01 Ω different from the calibrated value, it should be removed from service or recalibrated by a Standards laboratory.

6. Place both RTD transfer standards in the high-temperature reference block, or oil bath, and adjust the temperature to the desired value. Allow at least 10–15 minutes to stabilize if a reference block is used and about 5 minutes when using an oil bath. Measure the resistance of each RTD and determine the corresponding temperature of each from the appropriate calibration tables or equations. The RTDs should indicate the same temperature to within ±0.05°C if the temperature is <150°C and to within ± 0.1°C if the temperature is between 150°C and 300°C. If they do not, a third RTD should be used to determine which of the other RTDs is in error, and the faulty RTD should be removed from service or recalibrated by a Standards laboratory. When proper operation of both transfer standards has been verified, continue to monitor the high-temperature reference with one of the standards.
7. Place the thermocouples in the ice bath and allow at least 10 minutes for them to stabilize. This part of the procedure can be done at the same time as step 6. Once step 6 is complete and the thermocouples have become stable at the ice point temperature, their values at 0.00°C should be recorded for future correction. If the measuring system provides the capability to incorporate calibration corrections in the indicated output, the correction at the first point should be entered. In some systems this can be done automatically by pressing the appropriate keys on the operator's panel.
8. Place the thermocouples in the high-temperature reference and allow sufficient time for them to stabilize at the new temperature. The stabilization time will be approximately 10 minutes if a reference block is used and about 5 minutes in an oil bath. Once the indicated temperatures have become stable, the difference of each from the temperature indicated by the standard should be recorded for future corrections. If the measuring system provides the capability of incorporating calibration corrections in the indicated output, the correction at the second point should be entered.
9. If more than a two-point calibration is to be employed, steps 6 and 8 should be repeated for each calibration temperature.

There are several complete validation systems available that have fully automatic programmable multipoint calibration, improving the repeatability and reliability of the calibrations and providing significant time savings.

Documentation is an important aspect of any calibration procedure. A record must be made of the probe number attached to each channel and the location of each probe in the autoclave or oven during the validation test. The calibration corrections for each thermocouple must be recorded even when they are applied automatically by the measuring system. The calibration certificates of each RTD transfer standard and the precision resistor must include the actual data values obtained. If the calibrations were performed by the NIST, the certificates will contain a test number. If the calibrations were performed by another Standards laboratory, the certificates must contain the NIST test numbers of the instrumentation used by that laboratory provide traceable calibrations of the transfer standards. Every transfer calibration must be documented in order to provide traceability to the primary standard and proof of the accuracy of the final measurement.

9.7 THERMOCOUPLE SUMMARY

One of the most important steps in obtaining accurate temperature measurements with thermocouples is the proper design and installation of the thermocouple circuit. If possible, a continuous length of stranded homogeneous wire should be used from the measuring junction to the terminals of the measuring system. When two or more sections of wire are required by operational considerations, the connections between the sections must be in locations where the temperature in the circuit does not change significantly along its length. Ideally, each section of wire should be from the same production lot. If that is not practical, the wire should be selected to have the best interchangeability possible.

The measuring system must be designed specifically for high accuracy thermocouple measurements. The input terminal section should provide a uniform temperature of all terminals and a means of measuring that temperature accurately. The system's voltage measuring accuracy must be ±1.0 μV or better, and the computation of temperature from the measured voltage should deviate from the standard value by no more than ±0.06°C over the entire measurement range. Most importantly, the thermocouple reference must track changes in ambient temperature accurately, and the voltage measurement must not be affected by such changes, so that the calibration factors determined in the laboratory will still be valid on the production floor.

Finally, the entire system must be calibrated before each use. Although it is not necessary to do a full calibration after each use, it is good practice to verify proper operation by calibrating the system at the process temperature after the validation run. When a properly designed and installed thermocouple system is calibrated by the procedures described in this chapter, the total measurement accuracy should be better than ±0.1°C at 120°C, ±0.2°C at 200°C, and ±0.4°C at 300.0°C.

9.8 RESISTANCE TEMPERATURE DETECTORS

Most temperature sensors, permanently installed by the manufacturer of the sterilization equipment, are RTD sensors used for process control and monitoring of production runs. These RTD sensors are components in measuring chains and therefore the sensors shall be part of the calibration and verification of the measuring chain with calibration intervals defined in the calibration and validation SOPs.

RTDs are also used in wireless loggers (no real-time display) and in radiofrequency (RF) measurement transducers for remote real-time sensing. These are battery-powered, self-contained measuring systems. The wireless logger simplifies access to hostile, remote, and hard-to- reach environments by eliminating the need to hard-wire sensors, greatly reducing study setup time and associated cost. Loggers are available for measuring temperature, humidity, and pressure and come in a wide range of standard configurations to simplify data acquisition. On the other hand, the wireless logger has some minor disadvantages: 1. It is battery operated, the battery life is a function of sampling rate, study duration, and operating temperature. 2. There is no real time indication, only historical data obtained after completion of the study.

Wireless logger systems are made up of three components: wireless Loggers that measure and record process conditions, a Reader station for communicating with the Loggers, and Proprietary Software through which process studies, calculations, and reports are generated. Many manufacturers' software is 21 CFR Part 11 compliant. The Loggers are precision measurement and recording devices, designed for validating and monitoring the most severe temperature, humidity, and pressure applications:

- Steam sterilizers
- Depyrogenation tunnels
- EtO sterilizers
- Retorts
- Freeze dryers
- Dry heat ovens
- Washers
- Incubators
- Stability chambers
- Warehouses

RF measurement transducers are currently used in storage and stability applications. In most cases, the calibration interval recommended by the manufacturer is 6 to 12 months. The relatively long recommended calibration interval puts the responsibility for risk analysis and preventive action on the user. It is imperative that the annual/ semiannual calibration procedure includes an As-found report to verify that the unit has met its calibration tolerances during the period preceding the calibration. In many cases, interim verification of the unit's calibration status is made an integral part of the SOP.

The use of RTDs requires measurement circuits that are different from the thermocouple circuits. The thermocouple electrical output is a direct function of temperature as discussed earlier in this chapter. RTDs, on the other hand, produce a resistance change as a function of temperature variations. The RTD requires a current source to generate a voltage drop across the RTD. The most frequently used resistance value for platinum RTD is 100 ohms at 0°C and its temperature coefficient α is +0.385 ohms/°C at 0°C according to DIN standard. As both the slope and the nominal resistance is small, lead wire resistance can contribute significant

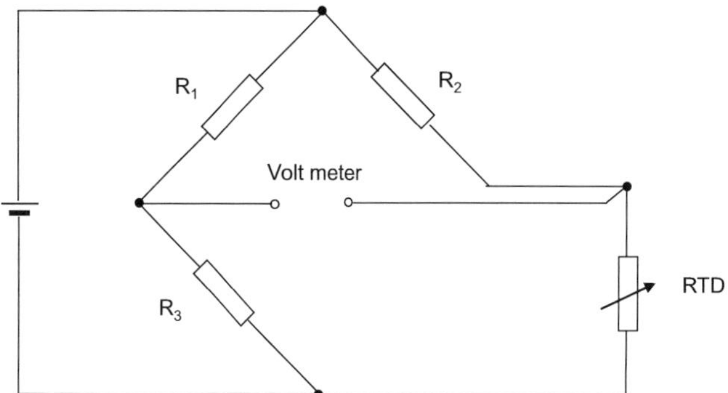

FIGURE 9.13 Three-wire Wheatstone bridge

measurement errors. As an example, 2 ohms lead resistance will cause a $2/0.385 \cong 5.2$°C measurement error and the temperature coefficient of the lead wires can contribute measurable error. An early method to compensate for these errors was the use of a Wheatstone bridge in a three-wire configuration that minimized the errors generated by the lead wires.

If wires A and B are perfectly matched in length, the effect of the lead resistance will cancel as the two leads are in opposite legs of the bridge. Wire C carries no current and acts as a sense lead only for the bridge's output voltage measuring device. The bridge shown in Figure 9.13 has nonlinear characteristics between resistance change and bridge output voltage change. This means that a second equation is needed to convert the bridge output voltage to equivalent RTD resistance that is then converted to temperature.

To meet the uncertainty requirements for critical temperature measurements, it is necessary to use the RTD in a four-wire configuration (Figure 9.14).

A current source drives a current through the RTD via two wires and a high-impedance digital voltmeter (DVM) senses the voltage dropped over the RTD via a second pair of wires. The voltage registered by the DVM is directly proportional to the RTD resistance, therefore only one conversion equation is needed to generate the temperature data. A precision reference resistor is connected in series with the RTD to provide the actual current value needed to calculate the momentary resistance of the RTD. The DVM is insensitive to the resistance of the lead wires as no current flows through them.

This solution requires a fourth extension wire, but that is a small inconvenience compared with the improved accuracy of the measurement.

9.9 SENSOR DESIGN

The temperature sensor should be designed for the application. A sensor designed for measuring the temperature in a LVP bag cannot be used for measuring the temperature in a 1 ml ampoule. Several factors have to be considered when specifying the design of a temperature sensor for a particular application.

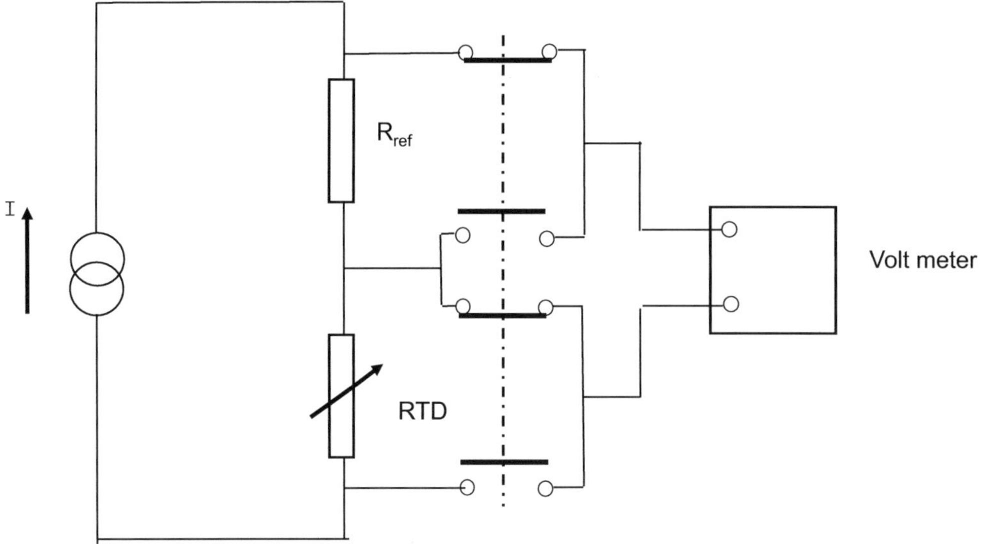

FIGURE 9.14 Four-wire resistance temperature detector measurement circuit.

> Regardless of how many facts are presented herein and regardless of the percentage retained, all will be for naught unless one simple important fact is kept firmly in mind. The thermocouple reports only what it "feels." This may or may not be the temperature of interest. Its entire environment influences the thermocouple and it will tend to attain thermal equilibrium with this environment, not merely part of it. Thus, the environment of each thermocouple installation should be considered unique until proven otherwise. Unless this is done, the designer will likely overlook some unusual, unexpected, influence. Copyright ASTM INTERNATIONAL, Reprinted with permission. [17]

The statement preceding is valid for all types of temperature measurements. The examples following are in some cases relevant for both thermocouples and RTDs; a few are specific to the type of sensor used.

9.9.1 Thermocouple Specific Examples

Twisting bare wires together increases the contact between the leads over the length of the twisted portion. The instrument measures the temperature at the first point of contact, i.e., the furthest point from the tip (Figure 9.15).

Using a twisted thermocouple to measure air temperature in a steam sterilizer would not significantly affect accuracy, because the difference in air temperature between the tip and the last point of contact is negligible. Twisted conductors could produce incorrect data, however, when the thermocouple is used to measure the temperature of liquid in a vial. Inserting this thermocouple (Figure 9.16) causes the instrument to indicate a temperature somewhere between the air and liquid temperature.

Avoid this problem by reducing the junction to the smallest practical size. Use an argon welder to create a thermocouple junction, resulting in a small bead that joins the wires at the tip (Figure 9.17). Strip the wires no more than necessary to create a weld. The insulation that is left on each wire separates the unwelded bare lengths of wire.

Heat Conduction—The copper wires in type T thermocouples can conduct heat into or out of the temperature sensor depending on the cross-sectional area of the copper wire and the temperature difference between the tip and the environment [19].

9.9.2 RTD Specific Examples

Self-Heating—RTDs and Thermistors are subject to self-heating from the current used for excitation. A current of 1 mA through a 100-ohm resistor generates 0.1 mW, which does not create a significant self-heating error.

Common Issues

- Size—A long or large sensing element will report an average temperature over the length of the element. In penetration studies, a small sensor will give a truer reading of the cold spot.
- Shape—A sensor for measuring surface temperature needs to be flat and adhere to the surface.
- Thermal Shunting—The size of the temperature sensor should be small relative to the object being measured in order to minimize the influence on the thermodynamic properties of the object of measurement.
- Response Time—The response time of the sensor is size and mass dependent. The response time should be at least five times shorter than the fastest rate of change in the process to be recorded in order to give a true representation of the process dynamics [18]. This is especially important for determination of D and z values using ampoules in BIER vessels.

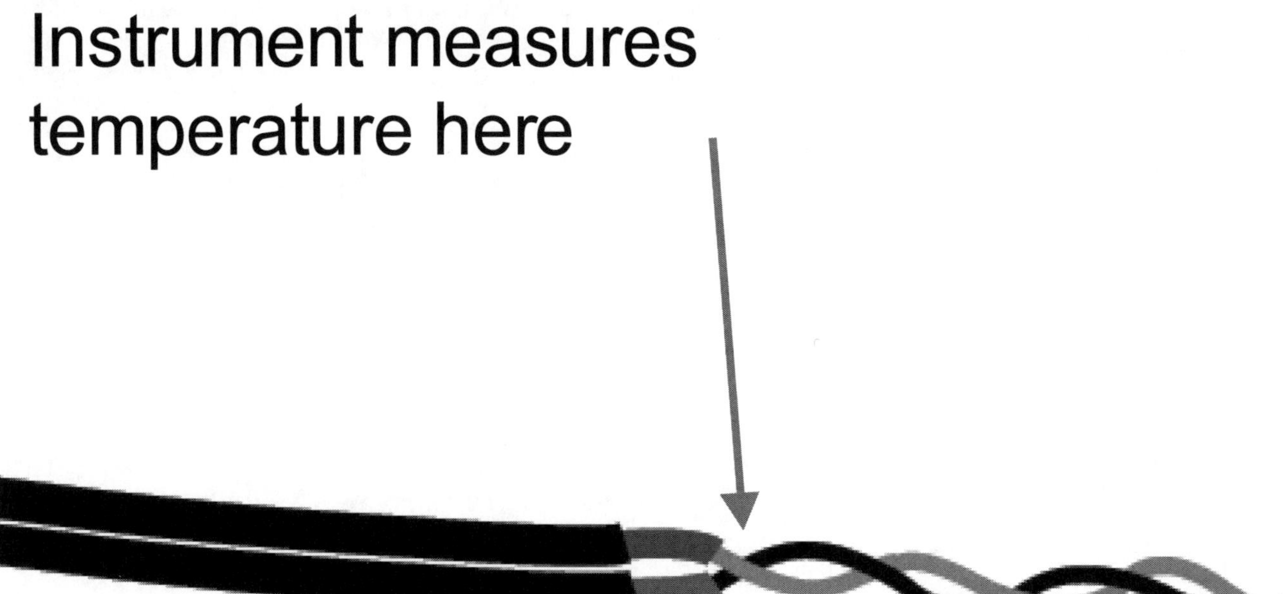

FIGURE 9.15 Twisted thermocouple.

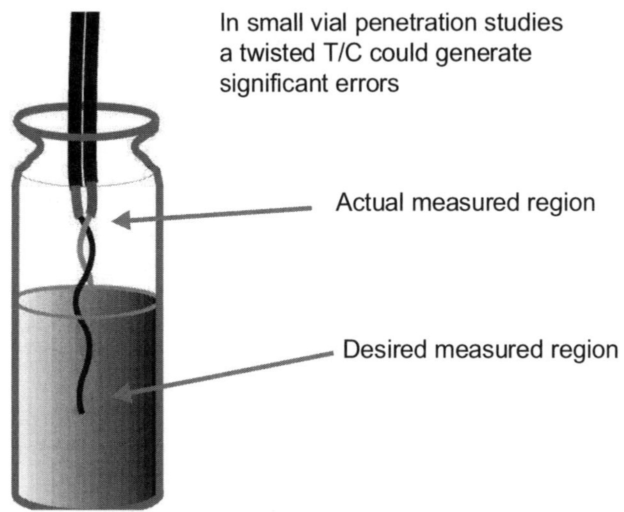

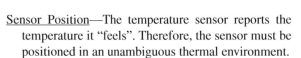

FIGURE 9.16 In a small volume for penetration studies, a twisted thermocouple could generate significant errors, indicating temperature somewhere between the two regions.

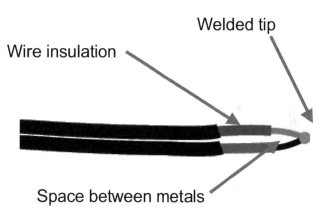

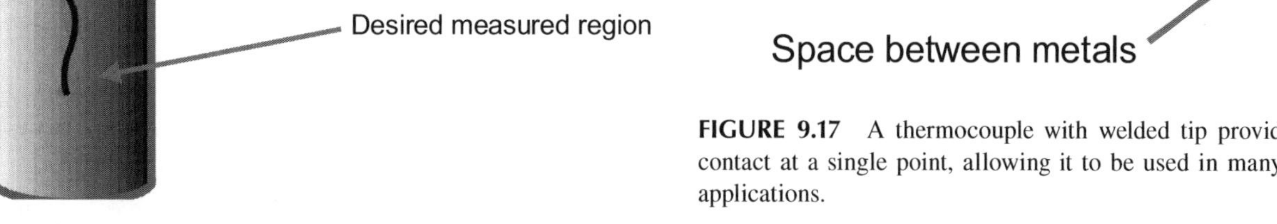

FIGURE 9.17 A thermocouple with welded tip provides secure contact at a single point, allowing it to be used in many different applications.

Sensor Position—The temperature sensor reports the temperature it "feels". Therefore, the sensor must be positioned in an unambiguous thermal environment.

- A sensor measuring temperature distribution in a sterilizer must be freely suspended in the chamber. If the sensor touches the chamber wall, it will report some temperature that lies between the actual chamber temperature and the temperature of the chamber wall.
- A sensor measuring heat penetration must be fixed in position relative to the walls and content of the container.

9.10 CONCLUSION

Temperature measurement is critical to the successful operation and validation of thermal sterilization processes as well as in other processes where temperature plays a vital role. The process and validation rely on measurement taken by either thermocouples or RTDs, and the information provided herein allows for a greater understanding of the system design and the factors influencing its accuracy.

REFERENCES

1. FDA Guidance for Industry: Sterile Drug Products Produced by Aseptic Processing: Current Good Manufacturing Practice, September 2004. U.S. Department of Health and Human Services Food and Drug Administration Center for Drug Evaluation and Research (CDER) Center for Biologics Evaluation and Research (CBER) Office of Regulatory Affairs (ORA).
2. Finch DI. *General Principles of Thermoelectric Thermometry Publication DI.1000.* North Wales, PA: Leeds & Northrup Company, 1969.
3. Roeser WF. Thermoelectric thermometry. *J Appl Phys* 1940, 2(6).
4. Moffat RJ. The gradient approach to thermocouple circuitry. In *Temperature, Its Measurement and Control in Science and Industry*. Vol. 3, Part 2. New York: Reinhold, 1962.
5. Bentley E. The distributed nature of EMF in thermocouples and its consequences. *Aust J Inst Control* 1982 December.
6. Powell, RL et al. Thermocouple Reference Tables Based on the IPTS-68, *NBS Monograph 125.* Washington, DC: US Department of Commerce, 1974.
7. Caldwell R. Temperatures of thermocouple reference junctions in an ice bath. *J RES Natl Bur Stdt.* 1965, 69C (2).
8. Kaye Instruments, *Thermocouple Reference Systems.* Bedford, MA: Kaye Instruments, Inc.
9. Muth Jr. Reference junctions. In *Instruments and Control Systems.* Vol. 40, Part 5. Philadelphia, PA: Reinbach Publications Division of Chilton Company, 1967.
10. Gray WT, Finch DI. Accuracy of temperature measurement. In *Temperature, Its Measurement and Control in Science and Industry*. Vol. 4, Part 2. Pittsburg, PA: Instrument Society of America, 1972.
11. Howard JL. Error accumulation in thermocouple thermometry. In *Temperature, Its Measurement and Control in Science and Industry.* Vol. 4, Part 3. Pittsburg, PA: Instrument Society of America, 1972.
12. *American National Standard for Temperature Measurement Thermocouples,* ANSI-MC96.1. Pittsburg, PA: Instrument Society of America, 1975.
13. *Copper-Constantan Thermocouple Wire, Probes and Accessories.* Bedford, MA: Kaye Instruments, Inc., 1983.
14. Fenton AW. The traveling gradient approach to thermocouple research. In *Temperature, Its Measurement and Control in Science and Industry.* Vol. 4, Part 3. Pittsburg, PA: Instrument Society of America, 1972.
15. Cooper MH Jr, Johnston WW Jr. Traceability, what and how, relating temperature measurements at ORNL. In *Temperature, Its Measurement and Control in Science and Industry.* Vol. 4, Part 2. Pittsburgh, PA: Instrument Society of America, 1972.
16. *Practical Temperature Calibration Standards.* Bedford, MA: Kaye Instruments, Inc., 1983.
17. *Manual on the Use of Thermocouples in Temperature Measurements*, 4th ed., ASTM Manual Series: MNL 12. Philadelphia, PA, April 1993.
18. Murrill PW. *Fundamentals of Process Control Theory.* Instrument Society of America, 1981.
19. Kemper CA, Harper GF. Probe-induced errors in temperature measurement. In *Pharmaceutical Technology*, November 1978.

10 Change Control

Phil DeSantis and Steven Ostrove

CONTENTS

10.1 INTRODUCTION

Within an effective quality system, change control is one of the most important foundational standards. It is particularly important when considering process validation and its related programs: equipment/system qualification, computer/control system validation and analytical methods validation. (For our purposes cleaning, sterilization and aseptic processing shall be considered processes.) Of course, change control is important for all quality systems that do not relate to process validation, but for the purpose of this chapter we will focus on change control related to processes and ultimately to product. We will not, however, consider changes that alter product in such a manner that clinical studies may be required.

In the discussion, we will distinguish between change control at two tiers. The first tier we shall term Change Control and define this as a quality system established to consider, assess, implement and evaluate changes that are considered to have a direct effect on product quality. Change Control at this level is governed in many firms by a global or corporate standard and at virtually every site by a Standard Operating Procedure (SOP).

The second level of change control we will refer to as change management. Change management is aimed at changes that are not deemed to affect product quality. It is an important adjunct to Change Control in a risk-based operating environment, and its various forms and value will be discussed further on in the chapter. The non-capitalized format is purposeful and is meant to convey the relative rigor and formality of the tiers. Change Control is a substantially more rigorous activity as it is directly related to product quality, whereas change management, although a useful practice, does not have the same impact. Change management may be comprised of several different approaches or methods, depending on the type of system wherein the change takes place.

In either case, all processes and systems operating in a Good Manufacturing Practice (GMP) manufacturing facility require changes to be managed and documented. The degree and form of oversight and documentation, particularly with regard to the role of the Quality unit, will be different depending upon the perceived impact of the change to product quality.

Except for the Regulatory implications discussed in Section 10.3, international Current Good Manufacturing Practice (CGMP) regulations and guidelines are consistent as to what is expected when managing change. This chapter will present a risk-based approach wherein cross-functional assessment and evaluation will ensure continued control of processes and product quality assurance.

10.2 TYPES OF CHANGE

Changes affecting a process may be divided into a few major categories. Although this is certainly subject to debate, we might approximate these as follows:

- Material changes: These may include changes to active and excipient materials, their approved sources/vendors, specifications, relative amounts in a formulation, etc. Most, if not all of these changes fall into the Change Control category.
- Equipment changes: These changes may apply to any equipment used in the GMP manufacturing facility. They may be exact replacements, functional equivalents, or even totally new configurations or technologies. Site and intra-site location changes fall into this category (provided there are no accompanying

DOI: 10.1201/9781003163138-10

material or procedural changes). Some of these, but not all, fall under Change Control. Which ones and how to decide are discussed in a later section.

- Procedural changes: These are changes to written procedures, including SOPs, master batch records, data forms, analytical methods and other documents that are used to support GMP manufacturing. Documentation changes may be required because of material or equipment change, but may also be required for other reasons, such as change of responsibility or location of an operation. Scale-up overlaps procedural, equipment and sometimes even material change.
- Software changes: Although computer hardware changes can be considered similar to equipment changes, software changes may be considered to fall into the broader category of procedural change. This seems reasonable because certain elements of software, but not all, may be defined as "batch records" or other records relevant to the manufacturing, testing and release of product. Software as documentation is different enough from traditional documentation to be identified separately.

10.2.1 Risk-Based Approach to Change Control

Each type of change embodies a spectrum of risk from significant risk to product quality (treated under the Change Control program) to risk that is low enough or insignificant enough to be handled by other means of change management. As stated previously, low-risk changes may be handled in several different ways, some of which are discussed in this chapter. Even within the formal and rigorous Change Control program, there may be more than one "work stream" based upon the assessed risk of the change. These work streams may involve differences in responsibility, oversight and approval, as well as differences in implementation and confirmation of the change. Such nuances are discussed further in the Chapter.

10.3 REGULATORY IMPACT OF CHANGE

Before describing general change, particular attention needs to be paid to changes that impact the regulatory filing for the drug product in question. The filing may be a New Drug Application (NDA), Abbreviated New Drug Application, or one of many other forms required by the various international authorities for drug and biologics registration. Sources of regulatory guidance may be found at the following links for the US FDA and European Medicines Agency (EMA). These both contain additional links to an important draft document from the International Council for Harmonisation, ICH Q12, *Technical and Regulatory Considerations for Pharmaceutical Product Lifecycle Management.*

www.fda.gov/drugs/guidance-compliance-regulatory-information/guidances-drugs

www.ema.europa.eu/en/human-medicines-regulatory-information

The US FDA guidance is presented in substantial detail in a series of documents labelled SUPAC (for *Scale-Up and Post-Approval Changes*). This series is focused on non-sterile dosage forms (immediate release, modified release and semi-solids), with an addendum for equipment. They provide concepts that are applicable to all dosage forms, however. They provide a framework for firms to decide what level of notification to the FDA is expected. The SUPAC Guidances describe change categories slightly differently than in Section 10.2 previously, but this is because they are focused on regulatory reporting requirements, whereas Section 10.2 covers many changes that are not subject to reporting.

SUPAC describes changes in the various change categories (a. components and composition, b. site, c. batch size and d. manufacturing, including equipment) on three levels, with each level in each category requiring one of the following:

- Annual report
- Notification of change being effected (CBE)
- Preapproval supplement

Note that the same level in two different categories of change may requires a different notification. For example, a level 2 site change requires a CBE, whereas a level 2 formulation change requires a preapproval supplement.

The varying notifications required by the different international drug agencies make it important for proposed changes to be reviewed by subject matter experts (SMEs) familiar with the regulatory filing, often the Regulatory Affairs group. Some filings describe the process in restrictive terms that make regulatory reporting essential. Others provide for more flexibility, relaxing the reporting requirements.

All proposed changes, however, should be subject to either Change Control or another form of change management, whether or not the authorities require notification.

10.4 RISK-BASED CHANGE CONTROL

Not every change has the same potential impact on product quality. Therefore, to provide for appropriate review and oversight while maintaining an effective utilization of available resources, the concept of quality risk management (QRM) should be applied.

QRM is not limited to identifying high risks and mitigating them. An effective QRM approach also identifies lesser or insignificant risks to product quality that may be dealt with through different applications of review and oversight. A QRM system avoids the trap of grouping all risks within the GMP environment within the same risk category, recognizing that "if everything is critical, nothing is critical". The statement may appear trite; however, it represents the very real experience wherein unimportant issues are given the same attention as important ones, often leading to backlogs and delays and sometimes even important effects on product quality being overlooked by overly burdened staff. This description is particularly descriptive of many Change Control programs. By treating all changes within the GMP manufacturing site the same, i.e., according to the same workflow, some sites defeat the entire purpose of Change Control by applying a universal and strict adherence to the most rigorous procedural requirements while losing focus on the real reason to control change.

10.4.1 Change Control vs. Change Management

The first step in a risk-based system is to define which processes, systems and activities do not need to be included under Change Control. Most formulation (material) and other changes to a process (e.g., sequence of operation, critical process parameters) are dealt with through the Change Control procedure. Most written procedures designated as SOPs or Analytical Methods also require Change Control. These are deemed to directly affect the production and release of product and can be presumed to present a significant risk to product quality. As such, they require the oversight of the Quality unit and all related changes need to be approved by Quality.

On the other hand, there are systems and procedures that may be assessed in their entirety to have a low or insignificant risk to product quality or for some other reason not to require the oversight, review and approval of the Quality unit. An example of this is provided in the chapter on "Commissioning and Qualification", wherein certain facility and equipment systems do not require formal qualification but may be commissioned only. Among these are secondary utilities such as plant steam and cooling water. It is allowable, and often valuable from a risk management perspective, to waive the application of Change Control on these systems and treat them under some other form of change management. Normally, changes to commissioned-only systems are handled with the site's maintenance work order system. Change Control requests need not be completed, and the Quality unit need not be involved. Best practice in the industry is to maintain these records within a validated Computerized Maintenance Management System (CMMS). It should be noted, however, that although these systems may not be qualified, their interface to qualified process units is. Therefore, when these systems are changed, appropriate notification to the User groups and their technical support team will ensure that qualified equipment is not affected (i.e., the utility supply to these units remains equivalent).

The Maintenance system as a whole provides another application, at least partially, for change management. Although each site usually requires SOPs describing the general requirements for preventive maintenance and corrective maintenance, including scheduling and reporting, some of the details of Maintenance (and Calibration as a subset of Maintenance) may be exempted from Change Control. Some sites have taken the option of designating specific maintenance and calibration technical work instructions separately from the SOP system. These technical details are left to the SMEs charged with maintaining and calibrating equipment. They can be changed by this group without the oversight, review and approval of Quality.

Other procedures that might be treated under change management include computer hardware and operating systems development and maintenance, left to the appropriate Information Technology (IT) SMEs. Engineering design, up to equipment qualification, represents another area where change management applies rather than Change Control.

Any activity or system that is entirely devoted to research and/or development may be exempted from Change Control. These functions usually apply their own means of documenting variations in process, procedure and results. This exemption disappears, however, when product is destined for human use, even for clinical trials. Because clinicals are often run parallel to process development, there needs to be a degree of flexibility with regard to changes that are proposed or implemented in clinical manufacturing facilities. The workflows associated with these changes may be different from that for commercial manufacturing, but the Quality unit still maintains oversight.

Each site must apply risk management principles to determine which systems and activities may be dealt with under a change management approach. This is not to say that the SMEs responsible for these systems do not have written procedures or do not keep records. They should and they do. In fact, some firms include the option for Quality audit of these procedures and records (as opposed to formal oversight, review and approval) in their governing SOPs or company Quality Standards.

A risk-based Change Control system that takes into account all or some of the examples cited here allows the Quality unit to focus on the changes that have a real potential impact on product quality. Little or no effort is wasted on change requests that Quality has no expertise to input, nor even a vested interest in reviewing. Change Control based on these principles tends to run more smoothly, with less backlog and better decision-making.

10.4.2 Project Change Management

Projects involving the design and construction of new facilities and equipment require special consideration. As pointed out previously, engineering design up to the point of equipment qualification need not involve formal Change Control. However, because key design decisions have an ultimate effect on product quality, change management within a design project needs to be enhanced.

The best designs are guided by User Requirements (see the chapter on *Commissioning and Qualification*) and the best User Requirements are categorized by defining which requirements are critical to quality. Design requirements like clean room classification, material compatibility and control of critical process parameters are examples. Further, specific risk assessments at the system and component level serve to define the quality impact of potential changes.

Project changes may occur at several different stages of the project, as follows:

- Changes to Project and System User Requirements
- Design changes
- Field changes
- Changes to purchased equipment or systems
- Changes to documents

The following guidelines are offered with regard to managing change within a facilities/equipment project. First, each project should have a change management procedure. This may be a stand-alone procedure, be built into the project execution plan or be a general Engineering department procedure.

Within this procedure, changes should be judged by their nature, rather than by the subject of the change. By this we mean that an engineering change to a system that requires qualification (a so-called "critical system") may be treated as an engineering change rather than a quality-critical one. The initial determination as to the nature of the change (quality-critical or engineering) may be made by the project management team (Engineering). This is allowed because (1) Engineering has management responsibility for projects and (2) the decisions are made before the system or equipment is qualified and placed into GMP operation.

A well-managed design project will perform formal reviews at specified points along the project timeline to ensure compliance with CGMP regulations and company quality standards (see the chapter on "GMP Project Management"). Sometimes called Design Qualification, these reviews include all project stakeholders and require a Quality unit approval at their conclusion. GMP design reviews are guided by the project User Requirements. These reviews provide Engineering with adequate information to discern whether a change affects product quality.

Examples of different types of change are:

- Engineering only
- e.g., location of diffusers, number of air changes
- Approved by a qualified engineer
- Operational changes
- e.g., location of processor within a room
- Approved by engineering and operations
- GMP changes
- e.g., area classification change
- Also approved by Quality unit

To reemphasize, all of these changes take place within project change management and do not involve Change Control. The usual inception of Change Control is after the completion of equipment qualification.

10.4.3 Risk Management Within Change Control

When a change is proposed that has a potential impact on product quality, Change Control is a must. Examples already cited include changes to materials, changes to process parameters or sequence and changes to qualified equipment. As a general rule, changing any procedure that is designated as an SOP, Analytical Method or Master Batch Record requires Change Control.

There are some changes to qualified systems, however, that may be designated in the Change Control procedure as not requiring a formal change request nor requiring review and/or approval of the Quality unit. One example regarding materials may be that changing the vendor's name or specification number of a material may not require a Change Request if those changes are merely in nomenclature and not in the actual analysis of the material. To clarify, the vendor is merged with another company and the material is rebranded. On the other hand, if the vendor's name and specification number are included in the Master Batch Record, that usually requires Change Control.

Equipment provides a major focus area of Change Control. One broadly accepted principle of equipment management is that changing out a component for an exact replacement (same manufacturer, model no.) does not require Change Control and is documented within normal maintenance records.

The question of functional equivalency, however, often enters the discussion when replacement parts are considered. Note that we strongly recommend against the use of the term "like for like" when discussing equipment replacement. This is because different companies define the term differently. Either a component or part is exactly the same or it is not. If it is not, it may or may not be considered functionally equivalent. Whether a functionally equivalent component or part requires Change Control is subject to further discussion.

As explained in the chapter on "Commissioning and Qualification", when qualified equipment systems are broken down into their major components, not all of these may be determined to have a direct effect on product quality. Those that do should be so identified, either in a risk assessment document, in a qualification document and/or in maintenance procedures (e.g., in a CMMS). Inclusion of equipment risk categorization in a CMMS is highly recommended because it provides a ready-reference for assessing changes to systems and equipment down to the component level.

For reference, by "component level" we mean major system components that perform a specific function within an equipment system. By this definition, components are composed of "parts" that have functions, but which only contribute to the function of the component. A simpler way to determine components is to define these as identified by a tag number (sometimes called a functional location) on a system schematic or Piping and Instrumentation Diagram (P and ID).

As discussed in Section 10.4.1, commissioned-only systems do not require Change Control but are usually managed within the maintenance system. Qualified systems, however, require Change Control. When considering these changes, it is important to consider the risk category of the component being changed. Components that are assessed to have a significant risk to quality (we prefer this terminology to "quality critical") must follow the designated Change Control workflow described in Section 10.5 following. Those that are low or insignificant risk may be directed to a change management workflow.

It is beneficial to have a global Quality Standard or site SOP that predefines changes that may be directly referred to change management. For replacement of equipment where the change has no resultant impact on the product or critical operating parameters, the change may be documented according to local maintenance procedures (e.g., work order). The specific items following are examples that may be replaced without Change Control. These must be *non-product contact* and must perform an equivalent function to the component or part they replace:

- Motors (same frame size, rpm, hp)
- Couplings

- Conveyor belts
- Drive belts, v-belts, drive chains, drive shafts
- Standard hardware (nuts, bolts, clamps, etc.)
- I/P transducers
- Electrical/pneumatic switches (manual or auto)
- Reference gauges
- GMP Non-Critical Instruments (do not monitor or control a critical process parameter)
- HVAC filters (non HEPA)
- Indicator lights
- Solenoid valves
- Fuses, circuit breakers, relays
- Fabricated noncontact replacement parts
- Bearings, bushings
- Steam traps (same principle of operation)

In developing the relevant SOP, the site should take care to identify those components or parts that might cause a change to an operating parameter. These usually are determined in the initial risk assessment and qualification documents. Changes to these components require Change Control.

10.4.4 The "Like for Like" Trap

Some firms allow replacement without Change Control of components that fall under the general description of "like for like". All too often these same firms fail to adequately describe the term. "Like for like" then becomes license to substitute any part or component that works, without regard to its potential adverse impact on product quality.

As discussed, exact replacements from the same manufacturer and with the same model number are allowed as maintenance items. This is because the vendor/model information is virtually certain to ensure the same materials, size, range and functionality as the original. Changing either manufacturer or model number presents the possibility that quality will be affected. Although the new item may be "like for like" in the sense that it is deemed functionally equivalent, it still requires Change Control. Subtle differences between the original and the replacement may be difficult to ascertain without more detailed assessment.

Some rules to apply to other than exact replacements are that Change Control is required for:

- Any component or part that contacts the product or primary packaging components
- Any component that controls, monitors or records a critical process parameter or critical quality attribute
- Any component that directly transfers energy (mechanical, thermal or electromagnetic) to perform a process function (e.g., mixing, heating, radiation; note "directly controls" does not apply to the energy source, e.g., boiler steam, cooling water, electrical power).

These rules only require that the Change Control workflow(s) be followed. They do not define what needs to be done in order to challenge, evaluate and accept the change. That is determined by the Change Control process described in Section 10.5.

In summary, lose "like for like". Deal with apparently functionally equivalent changes with a rigorous approach within the Change Control System.

10.5 CHANGE CONTROL

10.5.1 Planned Changes

An effective Change Control program is based on cross-functional input whereby operational, technical and compliance expertise are all brought to bear. No one group is ordinarily informed enough to assess, review and implement a Change Control decision. This approach often results in the overload of the Change Control office, backlogs and occasionally poor decision-making. For every change, the effect on process efficiency, operability and product quality should be considered.

Within Change Control, all changes should begin with a formal change request. Rather than route every change through the same workflow, it is valuable to apply an initial assessment or triage by a small Change Control management team consisting of representatives of Quality, Operations and Technical Support (the latter may be Technical Services or Validation). This team may vary from department to department within a site, but they should be sufficiently familiar with the process and product to steer the change request in the right direction.

This steering group may decide the basic workflow to be applied to the proposal by first determining the broad impact of the change. Some firms use descriptions similar to the following:

- Tier 1—Changes that do not affect the regulatory filing, validation/qualification status, multiple factories or sites or global documents (e.g., Quality Standards)
- Tier 2—Changes that do not fit into Tier 1. Table 10.1 next provides suggested details:

The steering group may also determine the priority of the change and may assign a priority designation and time limit for the change to be moved forward to the next phase of the designated workflow. Examples may be priorities assigned as "Normal" or "High". Normal priority may apply to routine changes, such as those made daily that are required to continually run the business efficiently. High priority may apply to changes initiated to meet a compliance requirement requiring rapid resolution or to resolve an urgent business need. Different firms use different categories; however, the assignment of priority serves to avoid backlogs and prevents important requests from "falling through cracks". Prioritization may also be used to alert site management of important changes that might require higher level attention or approval.

Having determined a category for the change ("tier" or other designation), the steering group may then define the workflow, including the operating, technical and Quality reviews and approvals required. Often, this team may assign a change

TABLE 10.1
List of Suggested Tier 1 Changes

Changes	Criteria/examples
A. Documents (cannot have direct product quality impact):	
(i) Creation/updates/periodic review:	
1. General procedures (non-product specific):	Analytical Instrument Operating Procedures, Equipment Operating Procedures, Equipment Set-Up Instructions, Calibration Administration Procedures, Facility Cleaning Instructions, etc.
2. Administrative procedures:	GMP-driven procedures, such as job skills training, calibration program umbrella document.
3. Process-related documents:	Changes to describe existing practices or to add more details to specific steps for clarity without affecting the validated/qualified process
(ii) Retirement: Documents confirmed as obsolete by Quality can be retired.	
B. Vendor name change (cannot have product quality impact):	
Changes to the manufacturer's name only as well as changes to the name only of distributor, agent, broker and/or similar function. This type of change does not include a change in site, process and/or replacement of one manufacturer with another.	
These changes do not require the approval of the impacted sites. The Supplier list can be updated upon Change Proposal approval.	
C. Secondary packaging (cannot have product quality impact):	
Changes to packaging processes, components and materials that do not impact the product and *do not*:	
• provide additional product integrity protection (e.g., a moisture barrier)	
• come in contact with the drug product	
• impact the regulatory filing	
• impact the regulatory labeling how supplied description (e.g., addition of a new presentation, a change in tablet branding, change in content volume), or ingredients	
• change the storage statement or conditions	
• change the compendial designation (e.g., Ph. Eur, JP, USP, NF)	
NOTE: Changes to label adhesive, ink, or varnish on semipermeable (i.e., plastic) containers that come in direct contact with liquid, ointment, cream, emulsions, suspension, or gel products cannot be considered secondary packaging changes.	
D. Compendial:	
A change should only be classified as a Compendial Change (i.e., changes to USP, Ph Eur., JPE) when the change is solely the result of a Compendium being revised. Examples of changes that need not be designated as a compendial change include:	
• Adding/changing/deleting a test (e.g., particle size) for a compendial item that is not a compendial requirement.	
• A change to the way a company or site complies with the Compendia should not be classified as a Compendial change.	
E. Computer and automated systems (cannot have impact on product quality):	
• Changes limited to computer system hardware/software and do not impact quality processes.	
F. Noncommercial batches (cannot be a part of a commercial registration or have commercial product impact):	
• Changes to noncommercial batches (e.g., development batches and clinical batches) that are managed by R and D and executed at a commercial site prior to the approval of the process validation protocol	
• Associated changes to commercial site documents or equipment are handled according to the commercial change management system requirement.	
G. Engineering (cannot be part of a registration or have product impact):	
1. Equipment upgrade/update:	
• Changes to critical equipment/components where they are evaluated to not affect the validated/qualified state of the process. They cannot have product impact.	
• Equipment/instrument qualification must be performed as required by risk assessment	
2. Maintenance:	
• Changes to maintenance or calibration programs that do not affect the validated/qualified state of equipment and/or process,	

Abbreviations: GMP = Good Manufacturing Practice; JP = Japan Pharmacopoeia; JPE = Japanese Pharmaceutical Excipients; NF = National Formulary; Ph. Eur = European Pharmacopoeia; R and D = Research and Development; USP = United States Pharmacopeia.

owner (or team for an extensive change), SME reviewers for operating and technical groups and Quality oversight. They will also determine if the change requires Regulatory review to determine if notification to regulatory authorities is required.

In general, more extensive and complicated changes (higher "tier") require higher organizational approvals. Levels of review and approval may vary from firm to firm and sometimes from site to site.

The actual challenges to a proposed change should be determined by the designated technical and operational reviewers and approved by Quality. Depending upon the type change (material, equipment, procedural, process, software) the recommended challenge may vary from a simple documentation change, to partial or complete equipment requalification (IQ, OQ, PQ or some combination), to various extent of process validation (e.g., observe and sample, demo batch, full validation sequence). Each change is assessed and planned individually.

A possible Change Control workflow follows. Differences occur within the assigned teams and tests determined:

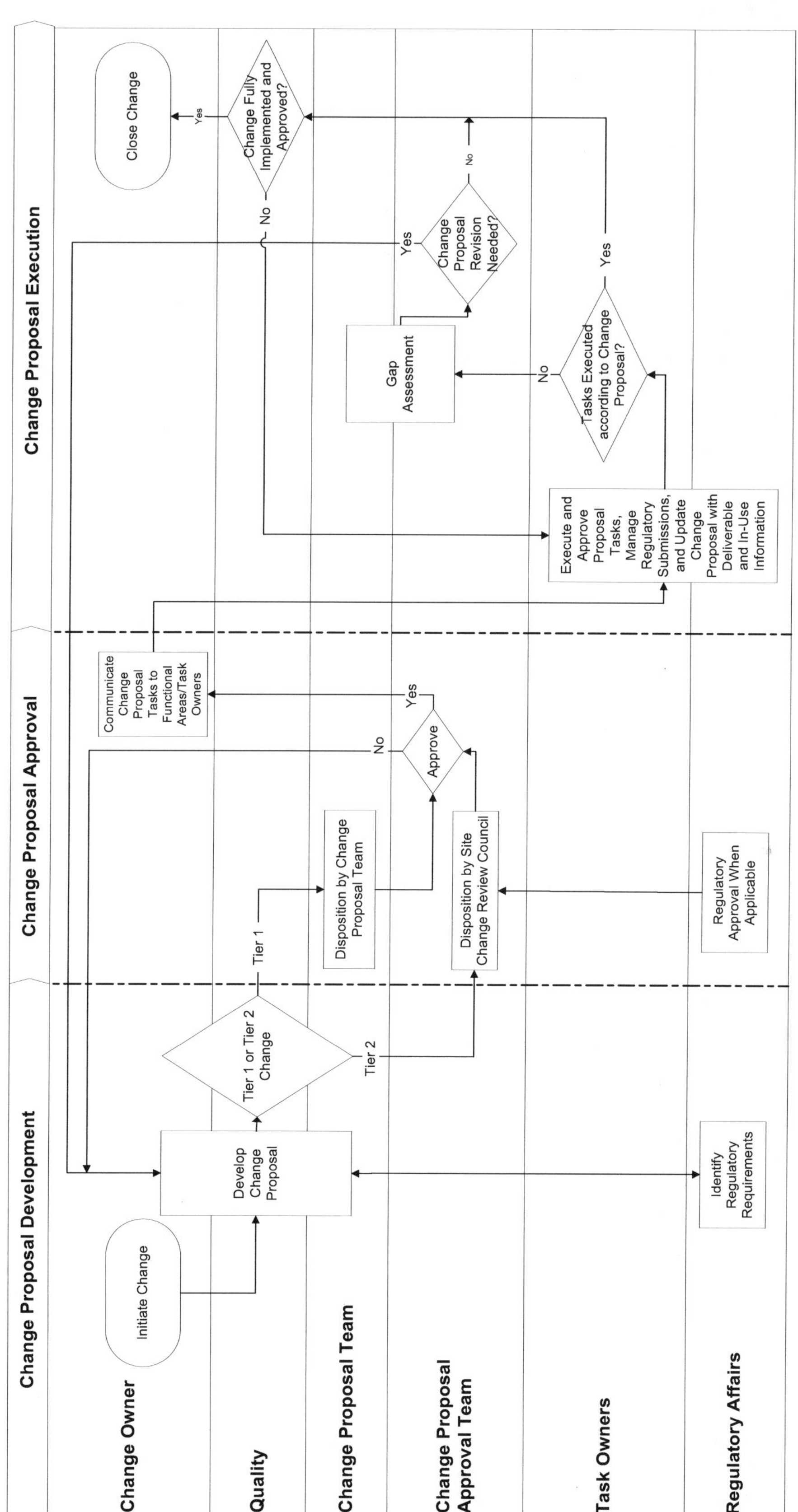

Change Proposal Development
Change Proposal Approval
Change Proposal Execution
Change Owner
Quality
Change Proposal Team
Change Proposal Approval Team
Task Owners
Regulatory Affairs
Initiate Change
Develop Change Proposal
Identify Regulatory Requirements
Tier 1 or Tier 2 Change
Tier 1
Tier 2
Disposition by Change Proposal Team
Disposition by Site Change Review Council
Regulatory Approval When Applicable
Approve
No
Yes
Communicate Change Proposal Tasks to Functional Areas/Task Owners
Execute and Approve Proposal Tasks, Manage Regulatory Submissions, and Update Change Proposal with Deliverable and In-Use Information
Tasks Executed according to Change Proposal?
No
Yes
Gap Assessment
Change Proposal Revision Needed?
Yes
No
Change Fully Implemented and Approved?
No
Yes
Close Change

FIGURE 10.1

10.5.2 Temporary Changes

For changes that are intended to be temporary, the change request should define the tasks and criteria needed to effectively terminate the use of the temporary practice and revert back to the pre-change state and the planned disposition of the impacted batches.

- Temporary changes are used on an exception basis only and cannot be used to circumvent GMPs.
- Examples of situations in which a temporary change may be utilized include, but are not limited to:
  a. Nonroutine inspection(s) of a component to remediate a quality deficiency.
  b. The collection of data to support a regulatory change.
  c. Changes or additions to acceptance criteria for an analytical method or other defined procedure.

10.5.3 Emergency Changes

Emergency changes may occur in instances in which a change may need to be implemented prior to consideration of a change request. This should be a rare occurrence and only implemented in case of an unplanned even that threatens the safety of personnel or the environment or is required to preserve a batch in progress.

Any material produced under an emergency change should be quarantined until a formal change request is routed through the Change Control system, reviewed, tested as required, approved and closed out.

10.6 CONCLUSION

Change Control is a foundational quality system in that it is applicable across all other quality systems (e.g., facilities and equipment, documentation, laboratory operations, material control, packaging and labeling). It is necessary to ensure the continued quality of product when anything that effects that quality is changed. Effective Change Control requires evaluation and planning prior to implementing the change (except in the case of emergencies) and the oversight and approval of the Quality unit.

Change Control is made more effective by employing a risk-based approach, wherein changes are critically assessed to determine their effect on product quality and managed accordingly. Changes to systems or procedures pre-determined to have a low or insignificant risk to quality may be managed with an alternative work flow and may not need approval of the Quality unit.

11 Microbiology of Sterilization Processes

John Shirtz

CONTENTS

11.1 INTRODUCTION

The goal of this chapter is to give a general understanding of pharmaceutical microbiology, its role in the industry, and its impact on the quality of the finished products. Sterility is defined by the total absence of microorganisms. This is a simple concept but difficult to establish in absolute terms. It is often expressed in terms of the Probability of $\leq 10^{-6}$ of a Nonsterile Unit (PNSU), i.e., less than or equal to one chance in one million that viable bioburden microorganisms are present.

Sterility can be attained by the use of validated sterilization processes under appropriate current good manufacturing practices (1, 52) and cannot be demonstrated by reliance on sterility testing (47, 50). There are two basic processes used to assure sterility. Terminal processes wherein the formulated product is sterilized in its final container. Aseptic processes rely on separate sterilization for the individual components followed by assembly and sealing in a pristine environment.

The basic principles for control of sterilization processes, including method development, validation, and ongoing assurance, are as follows (41):

1. Sterilization process development that includes evaluation of the stability and compatibility of materials, container integrity, expected presterilization bioburden, equipment, method, control parameters, etc.
2. Identification of sterilization process parameters that preserve the inherent properties of the materials yet reliably destroy or remove microorganisms.
3. Demonstration that the sterilization process and equipment are capable of operating within the prescribed parameters and corresponding to independent measurements of the critical parameters.
4. Performance of replicate studies that represent the operational range of the equipment and employ actual or simulated product. The use of biological indicators (BIs) for correlation between the measured

DOI: 10.1201/9781003163138-11

physical parameters and the expected lethality is recommended wherever possible (55).
5. Maintenance and monitoring of the validated process during routine operation.
6. Assurance that the bioburden (number and type) of the materials is maintained within predetermined limits during routine operation.

11.2 MICROBIOLOGY

Microorganisms are ubiquitous in nature and as such they will be present everywhere—in the air, on floors, ceilings, personnel, raw materials, excipients, water, instruments, and equipment, i.e., virtually everything involved with the pharmaceutical manufacturing process. Microorganisms are an essential part of some processes such as fermentation for the manufacture of antibiotics or when they are genetically modified through DNA recombinant processes to yield proteins of therapeutic value (11). Microorganisms are objectionable when their presence or the presence of their by-products, such as toxins or pyrogens, might result in deleterious effects for patients. Pharmaceutical products that must be sterile are injections, most ophthalmic products, and even some oral products that are to be used by immunocompromised patients.

Microbiology is the science of microscopic forms of life, with sizes ranging from 0.5 to 50 µm. These life forms are composed of protoplasm bounded by a cell membrane and include water, proteins, lipids, and nucleic acids organized structurally in organelles (64). The DNA within microorganisms is essential because it controls all the biochemical processes of the microorganism; if removed or destroyed, the microorganism will die. That characteristic of DNA is exploited in sterilization processes such as radiation sterilization that destroy DNA (14). In order to grow and replicate, microorganisms require a source of energy to drive the biochemical processes, as well as water and a source of carbon for the production of biomass. The depriving microorganisms of either a source of energy, carbon, or water can be used as control mechanisms to inhibit their growth and proliferation. Microorganisms can be either aerobic which utilize oxygen, or anaerobic which proliferate in the absence of oxygen. The growth of aerobes can be controlled by replacing air with a nitrogen blanket. Complicating that is the fact that some facultative anaerobes have the ability to grow anaerobically if sufficient oxygen is not available. There are >3,000 species of bacteria that reproduce primarily asexually, with genetic variations within species done via exchange of small strands of DNA between two bacterial cells. Depending on the processes used to obtain energy, most microorganisms are chemoheterotrophs meaning they oxidize reduced organic or inorganic molecules to obtain energy, and they also utilize complex organic compounds pre-synthesized by other microorganisms for biomass production. Some are photoautotrophs meaning they use light energy, CO_2, and simple salts for biomass whereas others are photoheterotrophs meaning their energy is obtained from light, and their biomass comes from organic compounds formed by other microorganisms.

The three basic shapes of bacteria, coccus, bacillus, and spiral, can often be used to microscopically determine the type of microbiological contaminant detected. The presence or absence of flagella or a cell wall, or the results of the Gram stain reaction (which can be either negative or positive) can be used to microscopically determine the type of contaminant encountered. Taxonomy can also be assisted by the determination of the presence or absence of spores. Spores are a single-cell state that some microorganisms can attain that supports their reproduction and survival. Bacterial spores are significantly more resistant than vegetative cells to most sterilizing agents; therefore, it is helpful to ascertain prior to the sterilization process if the spore content of the product is within the allowable spore load that the validated sterilization cycle is capable of destroying to an acceptable level (65).[1]

The cultivation of microorganisms involves differing requirements for their optimal growth; however, it is not possible to develop a medium with <u>all</u> of the varied components for <u>all</u> microorganisms. Laboratory analysts who wish to differentiate microorganisms for the purpose of isolation and identification can selectively remove or add components to the media to inhibit the growth of undesirable microorganisms and/or enhance the growth of desired microorganisms. Microorganisms also need water in order to grow. By restricting the amount of water, their proliferation can be altered, hence, if in an aseptic processing suite, you restrict the presence of moisture, it will reduce the growth of microorganisms. The detection and cultivation of microorganisms may also depend on other factors such as the pH of the medium and the temperature of incubation. For bacteria, the United States Pharmacopeia (USP) requires a temperature of incubation of 30°C to 35°C, whereas for molds and yeasts it requires a temperature of 20°C to 25°C (48). Typically, *Geobacillus stearothermophilus* spores are resistant to many sterilizing agents, so the sterilization process should determine their presence/population prior to sterilization in order to make sure that the validated sterilization cycle is adequate to ensure their elimination.

11.3 MICROORGANISMS IN THE PHARMACEUTICAL MANUFACTURING PROCESS

The diversity of microorganisms and their relative resistance to sterilizing agents or processes is wide. The microbiological world is divided into viruses, bacteria, and yeasts and molds (fungi). Viruses, in general, are not thermal, ethylene oxide (ETO), or radiation resistant (7, 8, 18, 23) and consequently they do not present a challenge to the sterilization process. However, classical filtration sterilization using filters of 0.2 mm may be inadequate if one has to rely solely on filtration for sterilization. There are, however, specialized filters that can be used to retain viruses smaller than 0.2 mm. Bacteria are ubiquitous, can live and grow under extreme conditions including temperature extremes, presence of simple inorganic and complex organic compounds, at a variety of pH's, and exhibit growth in a logarithmic fashion. Some diseases are caused by pathogenic bacteria that when present in the body of a patient can produce septicemia, and if not treated can result in death. The presence of pathogenic bacteria in raw materials and excipients is usually an indication of the unsanitary manufacture of these products or of human-derived contamination introduced during their manufacturing process.

As noted previously, certain bacteria have the ability to convert their vegetative form into spores. This sporulation can occur at the end of exponential growth of cells in an appropriate medium or when cells are transferred from a nutrient-rich medium to a nutrient-deficient medium (66). Several factors influence spore formation:

1. Temperature: In general, the optimum temperature for sporulation is equivalent to that for growth, but the range is narrower (37).
2. pH: In general, the optimum pH for sporulation is equivalent to that for growth, but the range is narrower (37), and sporogenes is more fastidious than vegetative cells with respect to pH (18, 34)
3. Oxygen requirements: As cells sporulate, they have an increasing demand in order to sustain the endogenous changes leading to the formation of a spore (24, 37). Oxygen inhibits the growth and sporulation of anaerobic spore formers (26).
4. Manganese: Various researchers have reported that manganese stimulates sporulation and that considerably higher concentrations of manganese were required for spore formation than for normal vegetative growth (26, 40, 64).
5. Carbon and nitrogenous compounds: The presence of certain carbon compounds may increase sporulation, and an interrelationship may occur between the induction of sporulation, spore yield, and the level and nature of carbon and nitrogen sources (26, 63).
6. Composition of sporulation media: Cooked meat medium based on Hartley's digest broth has been shown to support better sporulation of various species than other media investigated (32).

The resistance of spores to sterilizing agents can present unique challenges to destructive sterilization processes, thus the potential presence of spores is of utmost importance in the development of a sterilization cycle. For filtration sterilization, because bacteria can range from fractions of a micrometer to several micrometers, the use of specialized filters <0.2 µm will be necessary for the solution to be filter sterilized. Fungi include yeasts and molds that are generally larger in size than most bacteria. Some yeasts and molds can also form spores similar to bacteria, but their resistance to sterilizing agents is lower and generally not a problem or a concern in sterilization processes.

Microorganisms in pharmaceutical processes originate from varied sources including raw materials, excipients, and ancillary materials. They can also be derived from the water used in manufacturing, formulation, and cleaning of equipment or the environment. They are also present on personnel and the processing environment, so the more stringent the personnel environmental controls for the manufacturing process, the less issues these present. The presence of gram-negative microorganisms in water can give rise to bacterial endotoxins that when injected intravenously can cause an increase in temperature in patients. The Quality Control verification of a batch of product being both sterile and endotoxin-free is accomplished by means of the compendial USP Sterility Test (47) and USP Bacterial Endotoxins Test (48). As stated previously, sterility assurance cannot be demonstrated by sole reliance on sterility testing, but instead by the comprehensive validation of various sterilization cycles and other controls. The USP has monographs, general chapters, and general information chapters that address these various microbiological procedures (42).

11.4 MICROBIOLOGY OF RAW MATERIALS AND EXCIPIENTS

When used for the manufacture of pharmaceutical/biotechnological products, most raw materials are not sterile when received by the pharmaceutical manufacturer. Regardless, the materials should be processed under CGMPs wherein their microbiological bioburden can be maintained within predetermined limits. When this is not possible, especially for products of plant origin or other natural products, the USP recommends that treatment to reduce the microbial bioburden to an acceptable level should be employed (49, 55). Regardless of the method used to reduce the bioburden (ETO, radiation, steam, or dry heat), the potential for generation of undesirable effects and/or toxic microorganisms must be investigated and determined. If the finished product is intended to be sterile, the bioburden population and resistance impacts the design of the sterilization cycle. The endotoxin content of materials must be controlled as well because the finished product must be both sterile and endotoxin-free.

The selection of an appropriate sterilization cycle may be based on the bioburden of the product prior to sterilization, so the characteristics of the microbial flora of the various components need to be determined and controlled. If the materials prior to sterilization have a large population of resistant bacterial spores that can be critical to establishment of a validated sterilization process. Routine testing may need to be conducted to assess the spore bioburden of each batch if the sterilization cycle is to be successful. The minimum Probability of a Non-Sterile Unit (PNSU) should be attained while balancing the method effect on the materials and the destruction of the bioburden (Figure 11.1).

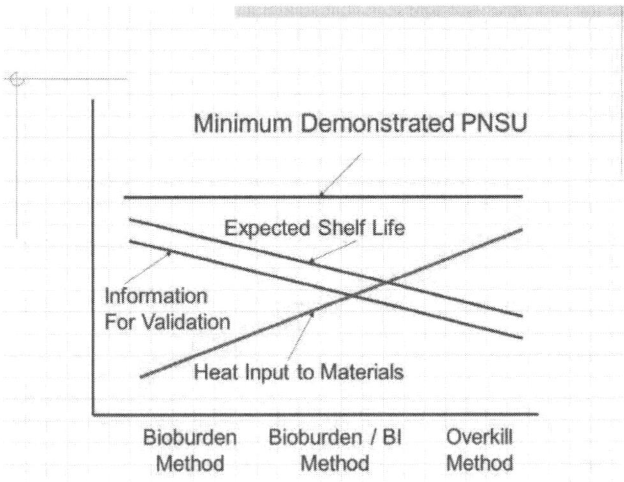

FIGURE 11.1 Balancing the stress on the material.

11.5 WATER MICROBIOLOGY

Water is the most widely used excipient in the pharmaceutical/biotechnological industry. Water is also used in the cleaning and sanitization of the equipment and facilities. Water used as an excipient in formulations intended to be sterile must be USP Water for Injection (WFI). The quality of WFI is governed by the USP monograph on WFI and includes specifications for conductivity and total organic carbon (62). Microbiological relevant guidelines for water are discussed in the USP (62). The USP chapter includes guidance on microbial levels which are *de facto* specifications. From the sterilization point of view, water borne contaminants are most commonly gram-negative microorganisms that have little resistance to sterilization. However, the production of endotoxins by these gram-negative microorganisms can present problems.

There are several different grades of water used for pharmaceutical purposes. These are described in USP monographs that specify use, acceptable methods of preparation, and quality attributes (62). The USP describes the selection process for the water to be used for various pharmaceutical purposes and dosage forms. The USP chapter also includes guidance on the operation, maintenance, and monitoring of such systems.

11.6 MICROBIOLOGY OF THE MANUFACTURING ENVIRONMENT

The manufacturing environment can contribute to the bioburden of materials prior to sterilization. The microbial contamination from the environment should be minimized to ensure that the validated sterilization cycles are not overburdened. Sterilization processes are not intended to correct microbiologically unsanitary conditions that can result in excessive bioburden population. In essence, if one controls potential sources of contamination prior to sterilization, the finished sterile product should be of appropriate microbiological quality. Environmental control is critical for the manufacture of aseptically produced products. However, even if the final product is to be terminally sterilized (51), it is imperative to limit bioburden ingress from the environment.

11.7 STERILITY TESTING AND STERILITY ASSURANCE

The classical affirmation of the sterility of a product is through the USP (47). Unfortunately, this concept is impossible to defend from a statistical point of view. Destructive testing of 20 units from a batch that typically numbers in the thousands cannot reasonably predict the sterility of each and every unit of the batch given that microbial contamination is discrete to an individual unit and can never be uniformly distributed. However, the compendial sterility test, with all its statistical limitations, is a means relied upon by regulatory agencies to assess compliance with the sterility requirement. Although regulators expect the selection of test units to be conducted on a periodic or event basis, the sample size has little or no statistical validity, so the projection of sterility from a limited number of units to a large batch is not warranted (27). Final product sterility is assured through validation of sterilization cycles and control of the microbial bioburden of both the formulation and the environment within predetermined limits prior to sterilization. Sterility assurance is expressed in terms of probability as indicated in Title 21 of the Code of Federal Regulation (CFR) (1) and the USP (50) as follows: "It is generally accepted that terminally sterilized injectable articles purporting to be sterile attain a $\leq 10^{-6}$ PNSU, i.e., assurance of less than or equal to one chance in a million that viable microorganisms are present in the sterilized dosage form." This probability is expressed in a PNSU of 10^{-6}. The compendial sterility test, because of the poor statistical validity and limited media types used and the temperature of incubation, is of limited overall utility in establishing the sterility of the materials tested.

The selection of a sterilization method depends primarily on the nature of the product. If the product is heat labile, it could be sterilized by filtration or a modified heat sterilization cycle based on the control of the microbial bioburden that will provide the appropriate sterility assurance without affecting the stability, integrity, and effectiveness of the product. The economics of sterilization can play an important role in the selection of one mode of sterilization over another. In addition, the nature of the containers used could also dictate the selection process, as well as the interaction of the container with the product as they are affected by the sterilizing mode.

Sterilization processes of many different kinds are available for the preparation of sterile products. These include steam or moist heat, various sterilizing gasses (59), different forms of ionizing radiation, dry heat, sterilization by filtration, and others (30, 43, 53, 54, 60). Many of the present-day sterilization methods and their validation are extensively described elsewhere in this text. All but sterilizing filtration (56) rely on information or assumptions on the relative resistance of the presterilization bioburden and BIs (29, 46, 49, 55, 67). It is essential that the validation practitioner have a sound working knowledge of each.

11.8 BIOBURDEN EVALUATION

In the majority of sterilization processes, it must be proven that the inherent bioburden present does not have greater resistance than the BI. Characterizing the bioburden involves the quantitation, identity, and resistance of the bioburden (53). The USP (44) provides additional methods that are useful for this assessment. The resistance of the bioburden cannot be adequately evaluated by quantitation; it must be determined if the products to be processed might contain microorganisms that are more resistant than the challenge microorganism. After the microbial identifications and quantitation of the bioburden have been completed and analyzed, resistance determinations of the most resistant bioburden component or product must be performed. A literature review is also required.

When microbial identifications of the bioburden are not performed, at least one sublethal cycle should be run to compare the relative inactivation rates of the bioburden with that of the challenge microorganism. Product sterility testing, after exposure to at least one sublethal cycle under appropriate experimental conditions, can ensure that the product's bioburden is not more resistant.

If the quantity, identity, and resistance of the product's bioburden are known, it might be possible to validate and routinely monitor the sterilization process by combining BI and bioburden methods. It must be demonstrated that the BI's degree of challenge to the sterilization process is adequate to ensure that the process will attain the desired PNSU for the bioburden. Combining the BI and bioburden methods to determine the appropriateness of the BI can be time-consuming and result in additional testing costs. However, the required sterilization parameters can be more accurately determined, which can result in reduced processing time and reduced exposure to the sterilant.

11.9 BIOLOGICAL INDICATORS

A BI is a preparation of a specific microorganism that provides a defined and stable resistance to a specific sterilization process. The USP includes a general information chapter that contains information regarding the proper use of BIs, responsibilities of BI manufacturers and users, responsibilities for user-prepared BIs, the types of BIs that can be used, and guidance on the selection of BIs for various sterilization processes (57). The USP also includes a general chapter (45) that provides information regarding how to determine total viable spore count, D-value, recovery, calculation, and Survival and Kill times. A separate chapter devoted to BIs is provided in this book.

It is crucial to ensure that the type of BI used to validate or routinely monitor a given sterilization process is the most appropriate indicator for that process. In addition to identity, quantitation, resistance, storage, general directions for use, and disposal conditions, the manufacturers of BIs are required to provide information regarding the optimal culturing conditions, such as temperature and type of growth media. Irrespective of which BI is chosen, the methods used to recover the challenge microorganism must be validated. This recovery is expressed in terms of the percent recovery of the original inoculum. Recovery studies can be especially challenging when using liquid spore suspensions because of potential interaction between the suspension and the material onto which it is inoculated. The material substrate can alter the resistance characteristics of the inoculum because of such anomalies as spore clumping or the physical sheltering of spores in certain sites within the product (46).

The goal is to kill the indicator microorganisms, which means disabling their ability to reproduce even in their most favorable growth conditions. The user must validate that the incubation time under the prescribed conditions is sufficient to recover delayed growth of the microorganisms after exposure to a given sterilization process. For routine processing, this time period is typically 7 days unless validated for a shorter time period in accordance with current national requirements. In such cases, periodic checks should be run to confirm that the shorter time period yields equivalent recoveries to those obtained from the longer incubation period. It is also important to ensure that the incubation time is sufficient to recover growth from injured microorganisms exposed to sublethal cycles. In some cases, this may mean using a 14-day incubation period. This incubation period is also required by the USP for product sterility testing.

There are different BI configurations suited for different materials possible:

Inoculated Product. The actual product, configured and packaged as it is intended to be sold, can be inoculated with spores of a microorganism such as *Bacillus subtilis* var. *niger*. Direct inoculation usually uses spores suspended in liquid and then placed on the product and dried. The product's surface characteristics will affect the distribution of the spores and may lead to a difference in resistance behavior compared with other challenge systems. It is, therefore, important to achieve an even distribution of spores on the product's surface. After exposure to the cycle, the inoculated unit is immersed in media to check for growth. Control units tested for population are used to confirm the suitability of the challenge units. When inoculation is directly into a solution, it is filtered and tested using methods similar to those utilized for sterility testing. As inoculated units are often prepared locally, the user is responsible for determination of the population and resistance in the inoculated configuration

Inoculated Carrier. A carrier, such as a filter paper strip or stainless-steel coupon, can be inoculated with a population of a resistant microorganism, such as *Bacillus subtilis* var. *niger*, that has been extensively characterized and certified by the manufacturer. The population resistance of this inoculated carrier is established by the BI manufacturer and need not be reconfirmed by the end user when used as described (55).

Inoculated Simulated Product. A simulated product that closely resembles the product can also be directly or indirectly inoculated. This simulated product must present a comparable challenge to the process as the product would in order to be considered an adequate microbial challenge. Each unit must contain a certified inoculum either in liquid form or on a carrier. These are used where the product is potentially inhibitory to the BI precluding direct inoculation.

11.10 THE ROLE OF THE BIOLOGICAL INDICATOR

The biological indicators play a central role in the validation of sterilization processes. They provide *prima facie* evidence that the process has delivered appropriate lethality to the intended locations. BIs by virtue of the size of the microorganisms can be placed in locations where physical measurements are impossible. Physical measurements despite their ease in tabulation, analysis, and comparison must be understood as tools of convenience rather than absolute confirmation of lethality (2). The validation effort intends to demonstrate that the chosen sterilization process will destroy the microorganisms in an orderly and reproducible manner. To that end, it is critical to understand how the microorganisms will respond to the sterilization processes. When homogeneous populations of microorganisms are exposed to a lethal process, they lose their viability in a consistent manner. The rate of this inactivation is directly proportional to the number of microorganisms present at any given time, and thus a constant portion of the surviving population is inactivated for each increment of exposure to the lethal agent. Mathematically, the inactivation process can be described in the same way as a first-order chemical reaction (27):

$$N_t = N_0 e^{-kt}$$

where N_t is the number of surviving microorganisms after time t; N_0 is the number of microorganisms at time zero; k is the microbial inactivation rate constant; and t is the exposure time. If the logarithm of the fraction of survivors (N/N_0) is plotted against exposure time, the resulting survivor curve will be linear, with a negative slope (Figure 11.2). The slope of the line is k/2.303 from which the microbial inactivation rate can be calculated.

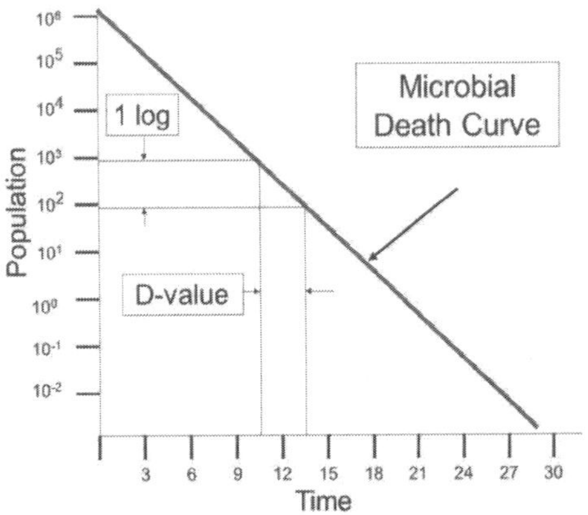

FIGURE 11.2 Death curve and D-value.

Fortunately, there have been extensive studies on the death of microorganisms when exposed to moist heat (30, 31, 33, 34). The most significant observation was that the destruction of microorganisms follows a generally straight line when the log number of viable microorganisms is plotted against time (13). The line is also called the "death curve" and it has proven useful with other sterilization methods as well.

11.11 D-VALUES

The determination of the death curve requires that the conditions of kill be held essentially constant over the duration of the test. The inverse slope of the death curve is called the D-value and is the time (usually in minutes, but seconds and hours could be used for alternative processes) required to reduce the population by 90% or one-logarithm. The D-value of a microorganism is determined using the survivor curve, which is the linear curve obtained by plotting survivor numbers on a logarithmic scale versus exposure times with the sterilizing agent (31, 32, 33, 34, 35). The number of survivor microorganisms is plotted on semilogarithmic graph paper (*y*-axis) and results in a linear or near-linear curve. A process that starts with a BI with 10^6 microorganisms, subjects the BI to fractions of the overall sterilization process, and counts the population of survivors at each time interval can be used to plot the survivor curve and calculate the D-value. Specific instructions for the determination of D-values are detailed in the current USP (43) under chapter <55> *Biological Indicators—Resistance Performance Tests* and in the general information chapter <1229> *Sterilization of Compendial Articles*.[2] The D-value for a microorganism decreases with increasing temperature in thermal processes (10, 36). Its response in nonthermal processes is more complex, as it is affected by the concentration of the sterilizing agent, relative humidity, and temperature.

The Rahn semilogarithmic survivor curve represents the exponential relationship that exists between spore survivorship and time at lethal temperature (36). Thermal degradation of microorganisms by means of steam sterilization has been experimentally shown to obey the laws of mass action and chemical kinetics (19, 24, 34, 36, 38, 39). The primary interest is in killing microorganisms and the number that die, but what is more interesting is the number of microorganisms that actually survive. Using N to indicate the number of viable (surviving) microorganisms present in the system at any given time, the change in the number of viable spores with time is a function of the number of viable spores present and can be represented in mathematical terms as follows:

$$dN = -KN$$

where K is a constant that is typical of the species and conditions of the chosen microorganism. The degradation (sterilization) reaction develops similar to a first-order chemical reaction in which the reaction rate is proportional, at each moment in time, regardless of the number of microorganisms remaining to be degraded or decomposed. Therefore, there is a constant percentage reduction of viable microorganisms for each arbitrary multiple of time t. The time required, then, to reduce the microorganism challenge population to any preset value is a function of the initial concentration of that microorganism. Having expressed the number of survivors in equation form, the equation can then be rearranged into differential equation form and integration of the differential equation is as follows:

$$dN = -K - dt$$

and by converting to base 10 logarithm:

$$\log N = -Kt + C$$

At time zero, $t = 0$ and $N = N_0$; therefore, $\log N_0 = C$. The final equation can then be derived:

$$\frac{\bar{N}_t}{N_0} = 10^{-Kt}$$

where N_t is the number of microorganisms at time t; N_0 is the initial number of microorganisms; K is the reaction time constant (which depends on the species and condition of the microorganism); and t is the reaction time of the steam sterilization, which simply says that the number of survivors of a steam sterilization process will decrease in an exponential (geometric) manner. As shall be seen later, exponential decrease is not the same as exponential growth. If something is known about the resistance of a microorganism, more specifically a BI, the D can be defined as the time it takes to kill 1 log of this BI, then using the foregoing model, 1/D (or –1/D) can be substituted for K. A more resistant BI (D>>1) will have more survivors; therefore, the relation between K and D is inverse. Furthermore, if the formula is standardized by making N0 = 1, then Nt = 10–1/D. Example: By using this formula of $N_t = 10^{-1/D}$ for a range of D-values from 10 to 0.1 minutes and $N_0 = 1$, a probability of survival may be created as shown in Table 11.1. Graphically, this probability of survival (N_t) can be displayed as shown in Figure 11.3.

TABLE 11.1
Process Lethality vs. Time and D-value

Process time	N_t			
Time (min)	**D = 100**	**D = 10**	**D = 1**	**D = 0.1**
0.01	0.999	0.997	0.977	0.794
0.1	0.997	0.077	0.794	0.100
0.2	0.995	0.954	0.630	0.010
0.4	0.990	0.912	0.398	0.0001
0.6	0.986	0.870	0.251	0.000001
0.8	0.981	0.831	0.158	0.00000001
1	0.977	0.794	0.100	1.00×10^{-10}
2	0.954	0.630	0.010	1.00×10^{-20}
3	0.933	0.501	0.001	1.00×10^{-30}
4	0.912	0.398	0.0001	1.00×10^{-40}
5	0.891	0.316	0.00001	1.00×10^{-50}
6	0.870	0.251	0.000001	1.00×10^{-60}
7	0.851	0.199	0.0000001	1.00×10^{-70}
8	0.831	0.158	0.00000001	1.00×10^{-80}
9	0.812	0.125	1×10^{-9}	1.00×10^{-90}
10	0.794	0.100	1×10^{-10}	1.00×10^{-100}

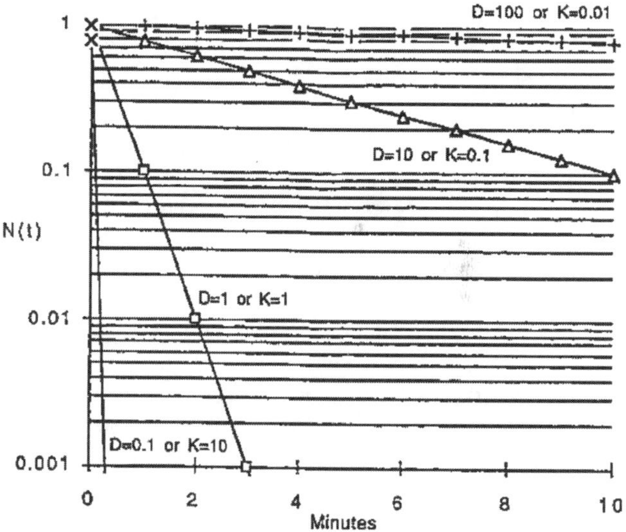

FIGURE 11.3 Death curves (D-value) at different temperatures.

Extensive research on the subjects of disinfection and sterilization has provided a great deal of information on the death of microorganisms (7, 8, 12, 19, 21, 22). Microbiologists differ in their views on the most essential properties of living microorganisms, but the criterion almost universally used to define the death of a microorganism is the failure to reproduce when suitable conditions for reproduction are provided (33, 36).

Before detailed discussion of the various aspects of the D-value, it is essential to examine the fundamental principles involved with the death of a microorganism. Extensive research on the subjects of disinfection and sterilization has provided a great deal of information on the death of microorganism. Microbiologists differ in their views on the most essential properties of living microorganisms, but the criterion almost universally used to define the death of a microorganism is *the failure to reproduce when suitable conditions for reproduction are provided.*

The challenge microorganism chosen for this task is usually a bacterial spore because they are among the microorganisms most resistant to sterilization. Non-sporulating bacteria are more sensitive to inactivation, along with the vegetative forms of yeasts and molds. The larger viruses show resistance similar to that of vegetative bacteria. Spores formed by molds and yeasts are generally more sensitive to inactivation than bacterial spores (23, 66). Bacterial spores are also more stable than vegetative cells, an important consideration for commercial development in that the spore crop may be cultivated, counted, evaluated for resistance, packaged, and distributed without concern for variation within the spore lot. Bacterial spores are also representative of the environmental bioburden, for they are widely distributed in the soil and are found at times in air samples.

Most of the known resistant microbiological spores are found in three genera: *Geobacillus* and *Bacillus*, aerobes that may be facultatively anaerobic, and *Clostridium*, which is usually a strict anaerobe. The most commonly used heat-resistant species are *B. subtilis, G. stearothermophilus, Bacillus coagulans,* and *C. sporogenes. C. thermosaccharolyticum* var. *niger* is commonly used to monitor ethylene oxide and *B. atrophaeus* to monitor dry-heat sterilization owing to their greater resistance to those methods (5, 6, 12, 19, 20, 21, 22, 27, 28, 31, 32, 33, 34, 43, 55).

The D-value is a term used to describe the relative resistance of a particular microorganism to a sterilization process. DT (or D(T)) is defined as the time required at temperature T to reduce a specific microbial population by 90%, or, as the time required for the number of survivors to be reduced by a factor of 10. The letter "D" stands for the fact that the D-value is also referred to as the decimal reduction time, the word decimal being defined by most dictionaries as "pertaining to or founded on the number 10". The preferred method uses semilogarithmic paper on which the *y*-axis is in logarithmic format and the *x*-axis is in arithmetic format. This type of graph paper enables the user to envision a reaction that causes a substantial change (at logarithmic proportions) over a constant rate of time (Figure 11.4).

As depicted in Figure 11.4, this semilogarithmic arrangement allows for the use of two data factors in relative proportion. The D-value can also be determined from a straight line on semi-log paper as the negative reciprocal of the slope of the line fitted to the graph of the logarithm of the number of survivors versus time (33). One cycle on the logarithmic scale represents a 10-fold change in the number of survivors; therefore, the D-value is the time for the straight line to traverse 1 logarithmic cycle (Figure 11.2). This D-value time element is a critical parameter used in both the validation of a process as well as in the routine monitoring of validated processes. Selection of a microorganism with the appropriate D-value for the intended application should be performed only by a qualified microbiologist (2, 12, 27)

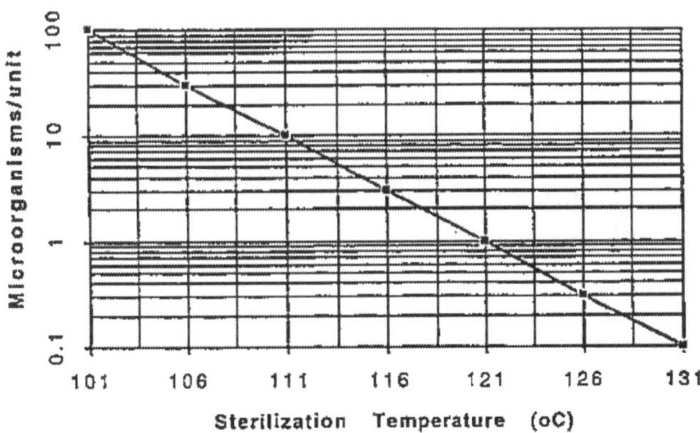

FIGURE 11.4 Typical thermal death survivor curve on semi-log graph paper.

Nonthermal Sterilization Methods

For nonthermal processes such as gas or liquid sterilization (58), there are no widely used means to convert physical measurements of concentration and temperature to lethality; processes for these often use dwell periods determined by multiples (usually 12) of the D-value. Thus, if a D-value is estimated at 6 minutes, a 12D process would have a dwell period of 72 minutes. Vapor processes in which multiple phases are present in the delivery and/or dwell period (63) do not allow for accurate D-value estimation. The inability to establish the phase present during either the D-value assessment or the process execution preclude accurate D-values (24, 31, 50). Radiation sterilization validation has largely abandoned the use of BIs, and so although D-values could be obtained for sterilization by radiation, it is no longer common practice (8, 20, 59, 60). The sterilization chapters elsewhere in this book provide expanded discussion on how BIs are used with the various processes.

Moist and Dry Heat Processes

Thermal process lethality is primarily related to the temperature at which the process is conducted (16, 27). The realities of thermal processing of foods and sealed containers entail cycles including heating, dwell at constant temperature, and cooling. With larger containers, the time spent during heating and cooling can make a substantial contribution to the delivered lethality and refinements to D-value determination were necessary to evaluate actual process performance. D-value estimations at multiple temperatures can be used to estimate the z-value, which can be used to refine the lethality determination when the temperature is not constant during the process (32, 33).

The z-value can be estimated by determination of D-values at different temperatures and plotting of the D-values versus temperature and is expressed in degrees centigrade. The z-value is the number of degrees of temperature change to effect a 10-fold (or one-logarithm) change in the D-value This is essential as thermal sterilization processes are not performed at a constant temperature.

Using moist heat for sterilization in commercial settings requires a means to assess process lethality based upon the temperature to which the materials are exposed. The concept of a lethal rate was developed as an arbitrary but standardized measurement of lethality that corrects for variations in temperature. The basis for the lethal rate of a moist heat process is 1 minute at 121.1°C and the lethal rate is the equivalent time at for any temperature relative to that standard. Example: the lethal rate for 117.0°C relative to 121.1°C (assuming a z-value of 10) is 0.398. This means that for every minute of time at a temperature of 117°C, the process is "credited with" the equivalent of only 0.4 minute at 121.1°C.

This leads directly to the use of F_0 in steam sterilization validation. The term F_0 is defined as the number of equivalent minutes of steam sterilization at temperature 121.1°C delivered to a container or unit of product calculated using a z-value of 10°C (31).[3] The total F_0 of a process incorporates the heating and cooling phases of the cycle and can be calculated by integration of lethal rates with respect to time at discrete intervals. The use of D-values, z-values, and F_0 that are determined based upon the BIs resistance to the sterilization agent is central to the validation of steam sterilization processes (35).

An identically structured approach can be used for dry heat sterilization with the exception that the base temperature is adjusted to 170C.and the delivered lethality is termed F_H (24, 28, 38, 41, 58, 65). The term F_H is similar to F_0 and is used to describe the number of equivalent minutes of dry-heat sterilization at temperature 170°C delivered to a container or unit of product calculated using a z—value of 20°C. Although dry-heat z—values ranging from 13°C to 28°C have been reported in the literature (28, 38, 41, 58, 65), most of these have been in the range of 17°C to 23°C, so 20°C is usually considered an acceptable assumption. Dry heat is also used extensively for the destruction of endotoxin on washed glass containers, but it is also used for sterilizing hospital supplies, such as powders, oils, petroleum jellies, glassware, and stainless-steel equipment that cannot be sterilized with saturated steam (11, 28). Dry-heat sterilization processes are generally less complicated than steam processes, although higher temperatures or longer exposure times are required because microbial lethality associated with dry heat is much lower than that for saturated steam at the same temperature. Calculations for F value do not correlate well with depyrogenation processes because dry heat depyrogenation involves different mechanisms of destruction for endotoxin.

11.12 APPLICATION OF D-, Z- AND F VALUES IN MOIST HEAT STERILIZATION

The F_0 value is used as a measurement of sterilization effectiveness and is the equivalent time at temperature 121.1°C using a z-value of 10°C (19, 35). Therefore, wherever a value is stated in terms of F_0, it is referring to the equivalent time at 121.1°C. If, for example, there is a stated F_0 value of 9, that indicates the process is equivalent to 9 minutes of sterilization at 121.1°C regardless of the actual process temperatures and times observed and used in the cycle.

For all thermal methods of sterilization, the F value is used along with the number of challenge microorganisms, the type of suspending medium, the z-value, and the D-value to determine the microbiological effectiveness of the sterilization cycle. In a perfect world, the F calculation would be fairly simple (37). In that perfect world, the sterilizer chamber would come up to the set point temperature immediately, the cycle would run precisely at that temperature for the desired time period, following which the temperature would immediately drop to ambient. The F value could then simply be the number of minutes the cycle ran at the process temperature. Unfortunately, that perfect world does not exist. There are, however, specialized sterilizers designed to minimize the come-up and come-down times associated with larger sterilizers. These units are known as mini-retorts, or Biological Indicator Evaluation Resistometers (BIER) vessels, so although the thermodynamics of microbial destruction are comparable, the load capacities of these units are extremely limited (31, 32, 33, 34, 35, 43, 44). As a consequence, BIER units have minimal come-up and come-down times and can be referred to as square-wave units from the shape of the graphic image of the chamber temperature during use (Figure 11.5). BIER vessels are commonly used in D- and z- value determinations.

Laboratory and production sterilizers are commonly of a size in which the heating (come-up) and cooling (come-down) times for temperature are relatively slow because of the necessary size of the chamber and mass of the load. If the load consists of a large number of liquid-filled containers, the transfer of the heat from the steam into the containers can be a lengthy process. A primary feature of the F value is the "compensation" it provides for the lower temperatures experienced during both the come-up and come-down periods. The *come-up* time is the phase in the process when the sterilizer cycle is underway, but the load has not yet reached the designated temperature. The *come-down* time is that period of time when the exposure portion of the cycle has been completed and the temperature is descending from the designated operating temperature (Figure 11.6). The time required for the load or chamber to reach operating temperature and to cool off to ambient is usually a function of the chamber size and load.

The F calculation takes into consideration the additional thermal destruction derived from these periods; temperatures during which the load items are not at the desired processing temperature, but nevertheless their effects contribute a detrimental effect on the microbial bioburden. These detrimental effects are not really significant below the temperature of 100°C (as will be seen later in the lethal rate calculation) so no "credit" is accumulated unless the product temperature is at least 100°C. As the process temperature of the sterilizer continues to increase with time during the come-up phase, the F_0 value of each incremental time period increases and

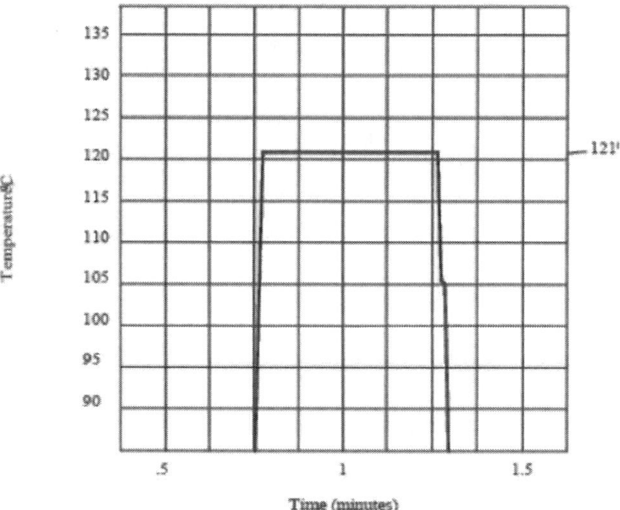

FIGURE 11.5 Graphic image of the cycle produced in a BIER vessel showing the square-wave cycle.

Note: BIER = biological indicator evaluation resistometer

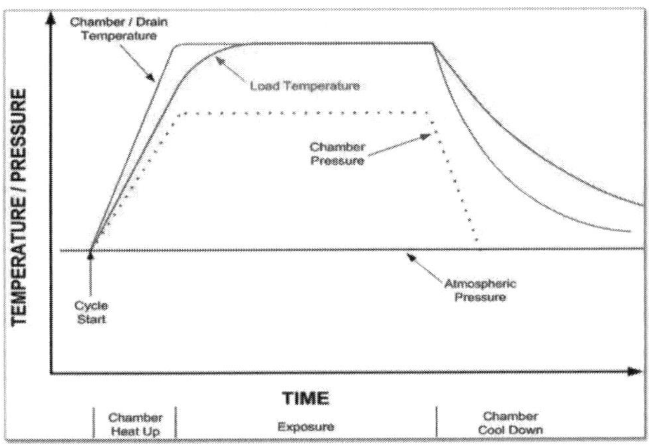

FIGURE 11.6 Graphic image of the cycle produced by a typical terminal sterilizer.

is added to a running total for the process up to that point in time. A similar lethality contribution is made during the cooling until the temperature in the load falls below 100°C.

11.13 LETHAL RATE CALCULATIONS

This sounds reasonable in theory, but how does one determine the lethal rate for every temperature likely to be encountered in the course of process development and validation? Actually, it is quite simple. The lethal rate value can be calculated quite easily by the following formula, or could simply be referenced from one of the numerous publications in which the lethal rates have been calculated [38, 47]:

$$L = \log^{-1}\frac{T_o - T_b}{z} = 10^{\frac{T_o - T_b}{z}}$$

where:

T_o = process temperature
T_b = reference temperature
121.1°C for steam sterilization
170°C for dry heat sterilization
z = 10°C (assumed) for steam sterilization
20°C (assumed) for dry heat sterilization

As an example, assume a process that shows a consistent minute-by-minute temperature of 118°C in a product vial is being processed in a terminal sterilization cycle in which the sterilizer is maintained at 121.1°C. To examine the microbial lethality within that particular unit, it is necessary to know the lethal rate for 118°C:

$$L = \log^{-1}\frac{T_o - T_b}{z} = 10^{\frac{T_o - T_b}{z}}$$

$$L = 10^{118-121.1/10}$$

$$L = 10^{-3.1/10}$$

$$L = 10^{-0.31}$$

$$L = 0.489$$

Therefore, for every minute of the process in which the contents of the subject vial remain at 118°C, 0.489 minute of equivalent time at 121.1°C is added to the total accumulated process. What is the lowest temperature that can be used for lethal rate? Temperatures <100°C generally add an insignificant amount of lethality to the overall sterilization assessment and are typically ignored for processes operating near 121.1°C. For example, the lethal rate calculation for 100°C:

$$L = \log^{-1}\frac{T_o - T_b}{z} = 10^{\frac{T_o - T_b}{z}}$$

$$L = 10^{100-121.1/10}$$

$$L = 10^{-21.1/10}$$

$$L = 10^{-2.11}$$

$$L = 0.008$$

The formula can also be used to calculate the lethal rate in terms of degrees Fahrenheit with an identical result. Substituting the values in the previous example for their equivalent in °F results in the same calculated lethal rate value of 0.008:

$$L = \log^{-1}\frac{T_o - T_b}{z} = 10^{\frac{T_o - T_b}{z}}$$

$$L = 10^{212-250/18}$$

$$L = 10^{-38/18}$$

$$L = 10^{-2.11}$$

$$L = 0.008$$

There is no upper limit for the lethal rate calculation. An F value can be calculated for any temperature above 100°C. For example, a number such as 150.0°C, which is nearly 30° hotter than the reference temperature of 121.1°C, shows an extremely high lethal rate:

$$L = \log^{-1}\frac{T_o - T_b}{z} = 10^{\frac{T_o - T_b}{z}}$$

$$L = 10^{150.0-121.1/10}$$

$$L = 10^{28.9/10}$$

$$L = 10^{28.9}$$

$$L = 776.2$$

What does that figure of 776.2 actually mean? It says that every minute of process time at 150.0°C is equivalent to 776.2 minutes (or 12.9 hours) at 121.1°C. Seems kind of far-fetched, but sterilizers are not usually designed to operate at this temperature, so although it's mathematically correct, there's no useful application of the result. The same formula can be used for calculating the lethal rate of a dry-heat sterilization process, the only difference being the reference temperature and assumed z-value which, for dry heat, are 170°C and 20°C, respectively. For example, a dry-heat sterilization process that operates at 155°C:

$$L = \log^{-1}\frac{T_o - T_b}{z} = 10^{\frac{T_o - T_b}{z}}$$

$$L = 10^{155-170/20}$$

$$L = 10^{-15/20}$$

$$L = 10^{-0.75}$$

$$L = 0.18$$

Therefore, every minute at the process temperature of 155°C is equivalent to 0.18 minute at 170°C.

11.13.1 Mathematical F_0

In mathematical terms, F_0 is expressed as follows:

$$F_0 = \Delta t \times 10^{T-121.1/Z}$$

where:
Δt = time interval between measurements of T
T = temperature of the sterilized product at time t
z = z-value

What this says is that the F_0 value is equal to the number of minutes of exposure time at the observed temperature (Δt) times the log of the value calculated by subtracting the reference temperature from the observed (process) temperature and dividing this quotient by the z-value. The best way to understand this calculation is to run through a few examples. For the sake of simplicity, assume these cycles were conducted in a BIER vessel, so that come-up and come-down times are negligible:

For a process that ran for 12 minutes at exactly 121.1°C:

$$F_0 = \Delta t \times 10^{T-121.1/Z}$$

$$F_0 = 12 \times 10^{121.1-121.1/10}$$

$$F_0 = 12 \times 10^{0}$$

$$F_0 = 12 \times 1 = 12\,\text{mins}$$

F_0 = 12 × 1 = 12 minutes

For a process that ran for 12 minutes at exactly 120.1°C:

$$F_0 = \Delta t \times 10^{T-121.1/Z}$$

$$F_0 = 12 \times 10^{120.1-121.1/10}$$

$$F_0 = 12 \times 10^{-1.0/10}$$

$$F_0 = 12 \times 10^{-1.0}$$

$$F_0 = 12 \times 0.79$$

F_0 = 9.48 minutes

For a process that ran for 12 minutes at exactly 119.1°C:

$$F_0 = \Delta t \times 10^{T-121.1/Z}$$

$$F_0 = 12 \times 10^{119.1-121.1/z}$$

$$F_0 = 12 \times 10^{-2.0/10}$$

$$F_0 = 12 \times 10^{-0.20}$$

$$F_0 = 12 \times 0.63$$

F_0 = 7.56 minutes

It should be fairly evident that a pattern is emerging: the lower the process temperature, the lower the F_0 value of the overall process. The same calculation applies if the process temperature is above the reference temperature:

For a process that ran for 12 minutes at exactly 122.1°C:

$$F_0 = \Delta t \times 10^{T-121.1/Z}$$

$$F_0 = 12 \times 10^{122.1-121.1/z}$$

$$F_0 = 12 \times 10^{0.10}$$

$$F_0 = 12 \times 1.26$$

F_0 = 15.12 minutes

This same F value calculation can be used to express equivalent time relative to any temperature. For example, if the requirement is to determine the $F_{115.0°C}$ value for a cycle in which the temperature ran a slightly cooler 114.8°C. In this case, the formula for lethal rate would be the same as before, the only exception being the use of 115.0 instead of 121.1 as the reference temperature:

$$L = \log^{-1}\frac{T_o - T_b}{z} = 10^{\frac{T_o - T_b}{z}}$$

$$L = 10^{114.8-115/10}$$

$$L = 10^{-0.2/10}$$

$$L = 10^{-0.02}$$

$$L = 0.954$$

In this case, it is easy to see that a process that has run slightly less than the reference temperature would "acquire" nearly the full 1.0 F value. If the temperature runs slightly hotter at 115.6°C, what then is the $F_{115.0}$ value?

$$L = \log^{-1}\frac{T_o - T_b}{z} = 10^{\frac{T_o - T_b}{z}}$$

$$L = 10^{115.6-115.0/10}$$

$$L = 10^{0.6/10}$$

$$L = 10^{0.06}$$

$$L = 1.148$$

Again, the 1.148 minutes equivalent time for a process 0.6°C hotter than the reference temperature is reasonable. The total $F_{115.0}$ for this process would simply be the ΔT times the lethal rate. If using the "near-perfect world" BIER unit and the process ran for 30 minutes, the $F_{115.0}$ in the latter example would be:

30 (minutes) × 1.148 (lethal rate) = 34.44

or somewhat more than that which is necessary to "claim" a minimum $F_{115.0}$ value of 30. (In the example with the temperature of 114.8°C, the calculation would be 30 × 0.954 = 28.62, or slightly less than the $F_{115\cdot0}$ of 30.)

11.13.2 Use of Calculated F_0

The F_0 calculation is used to describe sterilization processes because there are unavoidable variations in even the most precisely controlled sterilizers. For sterilization processes, there is a need to precisely characterize the capability of the process to enable the process engineer to compare the resultant effects of one cycle with those of other cycles with the same intended outcome, or to compare the results with a predetermined level that has been defined as "adequate sterilization".

Modern sterilizers are equipped with microprocessors that control the time, temperature, vacuum, and pressure but can also integrate the lethal rate calculations into real-time F_0 determinations and either report this value along with the other parameters or use this information to control the termination of the cycle. By insertion of a resistance temperature detector (RTD) or thermocouple into a product (or simulated product) container, the microprocessor can determine the F_0 accumulation and can terminate the process accordingly. The monitored location and its required F_0 must of course be correlated through the validation effort to conditions throughout the chamber. By doing so, the variabilities of temperature, load pattern, load size, container size, container penetration, etc. are all negated because the system needs only to monitor the primary objective of this process—delivery of sufficient microbial lethality to the representative product unit (2, 12, 27, 28, 29, 33, 35).

More recently, firms have been evaluating and using terminal sterilization for products that may have been considered "heat sensitive". This is in response to the global regulatory preference for terminal sterilization over aseptic processing. To satisfy this, manufacturers must now evaluate the use of terminal sterilization for all products, not just those products that have traditionally been considered insensitive to the rigorous thermal stresses of an autoclave. The terminal sterilization cycles for heat-sensitive products must be precisely engineered to avoid overprocessing (as well as under-processing) the product. Specialized sterilizer designs are available that have greater ability to control product temperatures during come-up and come-down to minimize the accumulation of excessive heat and lethality. Traditional steam autoclaves can be used for these products, but they ordinarily do not have the capability to control the cycle based upon product temperature. Regardless, a process must be designed that will allow for the imprecision of the sterilization equipment such that it delivers, with a high degree of assurance, an appropriate amount of lethality to every unit within the load. In these situations, the process designer must first determine the level of heat to which the product can be safely exposed. This is usually done through stability studies wherein groups of representative units of the product are exposed to a range of F_0 levels, then tested for quality attributes normally impacted by excessive heat, such as potency, development of degradants, particulates, color, etc. Once this has been determined, a decision on the routine amount of exposure must be made. The process should be designed such that it remains within the validated limits yet allows for unforeseen variations in the cycle to avoid both under- and overprocessed product.

11.14 DETERMINATION OF MINIMUM REQUIRED F_0

The process designer must know precisely how much F_0 to provide for in a new sterilization cycle to meet a desired sterility assurance objective. The target F_0 amount can be determined by evaluating the desired level of sterility assurance required, together with the bioburden of the product being sterilized, and the resistance of indigenous microorganisms in the bioburden using the formula:

$$F_0 = D_{121.1}(\log_{10} A—\log_{10} B)$$

where $D_{121.1}$ is the D-value (at 121.1°C) of the bioburden; A is the bioburden per container; and B is the maximum acceptable sterility assurance level (SAL).

Example: The product being sterilized has a bioburden of 100 spores per container, the D-value of the spore is 3.3 minutes, and the desired SAL is 10^{-6}, i.e., assurance that no more than 1 unit in 1 million units will be non-sterile:

$$F_0 = D_{121.1}(\log_{10} A - \log_{10} B)$$
$$F_0 = (3.3)\ (2-(-6)) = 26.4 \text{ minutes}$$

Therefore, for this sterilization process to achieve the desired destruction of all 100 spores to the extent of a 6-log SAL, all units in the batch must receive at least 26.4 F_0, i.e., the equivalent of 26.4 minutes at 121.1°C.

11.14.1 Calculation of Delivered Process Lethality

When the lethal rate can be calculated, this value can be used to determine the amount of lethality delivered in the process on the assumption that the lethal effect obtained at different temperatures is additive:

$$F_0 = \sum_{T=1}^{x} L\,dt$$

where T = 1 is the first-time increment (>100°C) with an F_0 value, and x is the last time increment (>100°C) with an F_0 value.

Integration of the lethal rates can be calculated using the Trapezoidal Rule that calculates the area under the curve by dividing it into equally spaced parallel cords. The lengths of the cords are y1; y2; y3; .; yn and the distance between the cords d is the time between successive temperature measurements (30, 31, 32, 33, 34, 35). The calculations do not have to be performed manually. Modern data loggers can perform the calculations instantaneously and report the incremental F value and the cumulative F value for the process up to that specific time point. The tabulation is simply an expression of the cumulative F value—in this example, in Table 11.2, the sterilizer started with a cool chamber and was programmed to run a 5-minute cycle at 121.1°C. The individual lethal rates

TABLE 11.2
Example of the Minute-by-Minute Temperature Observations of a Sterilizer Cycle with a 5-Minute Hold Time Along with the Incremental and Cumulative F (Lethal Rate) Values

Time (min)	Temperature (°C)	Lethal rate	Total lethal rate
1	25.0	0.000	0.000
2	55.0	0.000	0.000
3	85.0	0.000	0.000
4	100.0	0.008	0.008
5	110.0	0.079	0.087
6	115.5	0.282	0.369
7	119.0	0.631	1.000
8	121.1	1.000	2.000
9	121.1	1.000	3.000
10	121.1	1.000	4.000
11	121.1	1.000	5.000
12	121.1	1.000	6.000
13	116.0	0.316	6.316
14	108.0	0.200	6.516
15	96.0	0.000	6.516
16	80.0	0.000	6.516
17	65.0	0.000	6.516
18	42.0	0.000	6.516
19	30.0	0.000	6.516
20	25.0	0.000	6.516

allow the system to accumulate total lethality. Nearly all contemporary data loggers and sterilizers have this capability.

11.14.2 Cultivation of Bacterial Spores

Cultivation of bacterial spores differs from that of vegetative cells. The bacterial spore originates from a vegetative cell, wherein a unique process takes place to produce a profound biochemical change that gives rise to the structure known as a spore or endospore. This process is not a part of the reproductive cycle, and it has been characterized as the ultimate example of adaptation of bacteria to starvation. The resultant bacterial endospore is resistant to adverse environments, such as lack of moisture or essential nutrients, toxic chemicals, radiation, and high temperatures.

Although sporulation has been studied for decades, it has only been in the last few years that significant progress has been made in isolating the developmental genes of spores and in unraveling their functions and interactions. The process of sporulation is generally induced by starvation. In a good growth medium, the vegetative rod-shaped cells proliferate and multiply by doubling in length followed by a central division to produce two identical daughter cells; sporulation begins with an asymmetric division producing these two daughter cells that differ markedly in size. The smaller of the daughter cells is called the pre-spore and is engulfed by the mother cell. When this engulfment has just been completed, the membranes surrounding the cytoplasm of the pre-spore (now called the forespore) take on a very amorphous appearance and gradually start to develop the oval shape of the mature spore. A modified form of a cell wall known as the cortex begins to develop at the same time that a proteinaceous spore coat begins to be deposited on the outside surface of the spore. In the final development step, the characteristics of resistance, dormancy, and germinability begin to appear in sequence, and the spore is released by lysis of the mother cell.

11.15 BIOLOGICAL INDICATORS

BIs are used extensively as challenge microorganisms in the validation and monitoring of sterilization processes to verify the actual destruction of viable microorganisms (55). Guidance on their use can be found in the USP (43, 55). The primary advantage of using vendor-supplied materials is that the resistance profile of the microorganism is well characterized and, if used as intended, will provide a reproducible representation of the survivability of a highly resistant microorganism under conditions designed to stress and incapacitate it. Users of purchased BIs should establish a program designed to verify the manufacturers' label claims of population and resistance on each batch of BIs purchased. It is important to understand that the population verification count must be performed in a manner similar to that of the vendor to achieve similar results—some vendors use a blender to break up the fibrous paper of a paper strip, whereas others use sterile glass beads in a sterile test tube, and there are other methods as well. It is incumbent on the user to verify the level of challenge population by using the same method as the vendor. Internal limits of acceptance should be established by the user in concert with the vendor; these limits should provide for a fairly wide margin of variability inherent in this type of testing.

The name of the challenge microorganism is usually not as important as its resistance. Traditionally, *G. stearothermophilus* is used in processes that rely on saturated steam, and *B. atrophaeus* is used to monitor processes that use ethylene oxide and dry heat (39, 40). In all applications, the spore is selected based on its resistance to the process sterilant and the assumption that bioburden microorganisms are less resistant. Other spores or other microorganisms are acceptable for use as process monitors provided their resistance characteristics to the process sterilant have been thoroughly examined.

Another item to consider is the application of the challenge spore. D-values for spore strips are determined by exposure of the paper strip (still contained within the glassine envelope) to the sterilization process in perhaps a "best-case" situation, i.e., the spore strip is directly exposed to the sterilant. As a routine process monitor then, spore strips should not be used in a manner in which the sterilant can more easily kill the spore, e.g., without the glassine envelope, to avoid false-negative results. Most users will place the spore strip in a predetermined worst-case location, on the premise that destruction of the spore under these extreme circumstances can assume destruction of spores in all other less-challenging circumstances. D-values for most commercial spore suspensions are determined by applying the spore to a paper strip that is then tested for resistance just like all the other spore strip lots. How many people are going to use spore suspensions in this manner? The answer is probably none. Most applications using spore suspensions involve the inoculation of something, e.g., a liquid product, an empty vial, a stopper, etc. Unfortunately, spore suspension manufacturers cannot predict or control this use so they provide the resistance information in an unbiased means by inoculating the same type of fibrous paper as used for spore strips and performing the resistance studies as though the object was the spore strip itself. To verify the vendor's reported D-value, the user would need to inoculate samples of the same paper used by the vendor and perform D-value studies of the inoculated strips.

The user must develop resistance data for the spore suspension in the configuration in which it is intended to ultimately be used. For example, if the spore will be used to inoculate liquid product containers, resistance studies should be conducted to examine the thermal resistance of the challenge spores in representative units of this product. There should also be studies conducted to measure the microbiocidal effects of the product itself on the challenge spore to verify minimal spore reduction from simply the product–spore interaction. In the event the product exhibits a sporicidal effect on the spore, an alternative non-sporicidal medium with characteristics similar to those of the product should be chosen.

If the process is one in which only a narrow range of resistance can be tolerated, adjustments to the resistance to precisely target the desired range may be achieved through modification of the suspending medium (26, 28, 42). For example, one study (34) reported resistance studies using *B. stearothermophilus* at approximately 120.9°C in dextrose 5% in saline, dextrose 5% in lactated Ringer's solution, dextrose 5% in water, WFI, Sorensen's buffer, and Butterfield's buffer and showed the following widely divergent D-values (Table 11.3):

More than a threefold difference in resistance was shown by simply changing the suspension solution. In the same study, *C. sporogenes*, at approximately 105°C in the same solutions as the foregoing showed D-values of 2.68, 1.14, 1.34, 13.7, 42.6, and 21.2 minutes, respectively, nearly an eightfold difference! These datasets demonstrate the wide range of resistance characteristics available by using different suspending solutions as well as the lack of correlation among challenge microorganisms, reinforcing the necessity to verify the challenge microorganism under identical conditions to those in which it is to be used.

TABLE 11.3
D-Values in Various Products

Solution	D-value (minutes)
Dextrose 5% in saline	1.30
Dextrose 5% in lactated Ringer's solution	2.12
Dextrose 5% in water	2.42
WFI	2.98
Sorensen's buffer	3.36
Butterfield's buffer	4.70

Abbreviation: WFI = water for injection.

11.16 PHYSICOCHEMICAL INDICATORS AND INTEGRATORS

Device manufacturers often include an additional requirement in their validation effort plan that involves the use of external monitoring systems. Often referred to as process challenge devices (PCDs), they assess the lethality of the process after the cycle has been designed. The PCDs are geometrically distributed around the load rather than in internal locations. Comparisons can then be made between the data obtained from these external PCDs and the BIs placed in internal locations. A PCD must be shown through comparative resistance studies to provide more of a challenge to the process when it is placed in external locations in the load than do the BIs placed in internal locations. They may bear no resemblance to the product. Examples of external PCDs are spore strips double-packaged in plastic bags, in sealed plastic tubing, or in syringes. There are also commercially available PCDs that are sold as ready-to-use packaged systems. It is advisable during the validation studies to evaluate different PCD configurations to determine the best candidate.[61]

Title 21 CFR Part 211 on Good Manufacturing Practices for Finished Pharmaceuticals in section 211.165 (1) reads as follows: "There shall be appropriate laboratory testing, as necessary, of each batch of drug product required to be free of objectionable microorganisms." This statement allows for the use of alternative tests for the sterility test performed on each batch of sterilized product. These indicators/integrators can be used as part of the testing of batches of sterilized products that are to be released under a parametric release program (31, 51).

1. The performance of these physicochemical indicators/integrators must be consistent from lot to lot.
2. The composition of these indicators/integrators must not interact physically or chemically with the products being sterilized.
3. The safety of the operators and laboratory personnel in contact with these items should be ensured at all times.
4. These indicators/integrators are considered as Class II devices and it is necessary to obtain a 510 K approval prior to commercial use.
5. Physicochemical indicators are devices that respond in a measured fashion to one or more critical sterilization parameters. They are used to monitor a physical parameter that indicates that the load has been exposed to that factor, e.g., temperature in a measurable quantifiable manner that can be correlated to microbial lethality. It integrates, for example, the temperature of the sterilization process with the

time of exposure and concentration of the sterilizing agent. A physicochemical indicator is not a substitute for a BI but is useful in indicating if the sterilization cycle is too long and if over- or underexposure by the sterilizing agent has occurred.

6. Physicochemical integrators for steam sterilization react in a predictable fashion to a specified combination of physical parameters of sterilization such as temperature, steam pressure, and time of exposure. Deviations of the preset parameters will be captured by the physicochemical integrator.
7. Physicochemical integrators for ETO sterilization react to a preset combination of parameters such as temperature, humidity, gas concentration, and time of exposure.

11.17 COMPARATIVE RESISTANCE

There are two types of resistance determination studies useful with PCDs: Comparative Resistance and Bioburden Resistance. Their primary purpose is to assist the development of suitable PCDs to monitor routine and validation sterilization cycles. Typically, both comparative and bioburden resistance tests are required for development of PCDs. When these studies are performed together, it is considered a relative resistance study. Resistance studies can also be performed to confirm the appropriateness of already validated PCDs when adopting a new or modified product into and existing sterilization cycle.

Comparative Resistance testing is used to determine the most difficult to sterilize location within a product. It helps identify an appropriate PCD to be used for validation and routine cycle monitoring so that product is not required for microbiological monitoring. A comparative resistance test can also be used to determine an appropriate sterilization cycle thanks to its ability to show lethality in the most difficult to sterilize locations. It compares the resistance of different types of BIs to determine an internal and external BI. Additionally, it can be used to adopt newly developed product or devices that have undergone changes into a current cycle. The study includes at least one fractional cycle that compares samples of the candidate product with the existing product and internal process challenge device (IPCD). It is recommended to run this fractional cycle using the same parameters validated originally.

Bioburden Resistance testing evaluates the resistance of the naturally occurring bioburden on the devices that have undergone manufacturing steps prior to sterilization. This test also compares the resistance of the bioburden to an already categorized PCD. This testing is required and performed according to the recommendations outlined in ANSI/AAMI/ISO 11135–1:2014, AAMI TIR16, and AAMI TIR28 (5).

11.18 STERILIZATION INDICATORS AND INTEGRATORS

The effective execution of sterilization processes can be supported by physical and chemical indicators and integrators that provide an indication that processing has completed. Sterilization indicators respond to sterilization process parameters in a non-qualitative fashion in that they show either simply passing or failing results. They are useful in an operating environment simply as a means to identify whether an item has been exposed to a sterilization process and should not be used to assess the process as having achieved adequate sterilization. They are of minimal use in directly establishing process effectiveness. Sterilization integrators are more sophisticated devices that react quantitatively in response to one or more of the critical sterilization parameters and yield a result that can be correlated to lethality. The most sophisticated of these are radiation dosimeters that are so accurate and robust that their use has essentially displaced the use of BIs for the validation of radiation sterilization. Refer to the USP for further information on these types of devices (59, 62).

Each sterilization mode will have a particular approach to validation of cycles for achieving sterilization of products. A good example is filtration sterilization, in which the concept is to remove a microorganism from the product, not to destroy it as for the other methods of sterilization. Because filtration is dependent on the effectiveness of the filters in retaining the microorganisms, the physical characteristics of the filter to accomplish that purpose must be tested. In addition, challenging the filter with microorganisms will support the capability of the filters used to produce sterile product. This cannot be done in situ and must be accomplished outside of the manufacturing area, preferably in a laboratory, because the test is destructive. Surrogate to the microbiologic testing are physical measurements that can be done in situ such as a bubble point test or diffusive air flow test.

11.19 PARAMETRIC RELEASE OF STERILIZED PRODUCTS

The use of parametric release for sterilized pharmaceutical products requires prior approval by the regulators (59). The principle of parametric release of sterilized products is simple, but the practical application is more difficult. Few organizations in the pharmaceutical industry other than Large Volume Parenteral manufacturers have used parametric release because it involves some significant measures:

1. The mode of sterilization is very well understood and predictable.
2. The lethality of the cycle to be used has been microbiologically determined and the cycle validated using BIs for steam sterilization and ETO sterilization; for radiation sterilization, validation is done using precise dosimeters.
3. When a validated sterilization cycle operates consistently, a combination of critical parameters provides accurate and repeatable data that ensure that the preset lethality has been achieved.
4. A parametric sterilized product will be released without the need to perform a sterility test.
5. Because parametric release is based on preset parameters for a validated cycle, the manufacturers should

ensure that the sterilizers used function as intended and within the preset parameters. This requires the qualification of all production sterilizers to ensure that the critical parameters are always controllable and controlled. Changes in production sterilizers must be assessed in relation to their performance of a validated sterilization cycle. It is necessary to establish a well-planned program of change control. No changes can be made to sterilizer physical parameters unless there is an assessment of the impact of the change on the validated cycle. Not all changes are significant in their impacts on the validated cycle, but it is a good policy to ask that all changes be communicated to a change control system for assessment.

6. Development and validation of cycles are based either on bioburden, or on an overkill approach, or on a combination of bioburden and BIs. When the cycles are based on bioburden, it is mandatory to have a comprehensive microbiological program to assess the various stages of manufacture and of components prior to the sterilization process. Bioburden might change with the season for a given supplier and with different suppliers. This must be taken into consideration when parametric release is contemplated, as the cycles are always linked to the presterilization bioburden. When sterilization cycles are based on the overkill principle, the importance of the bioburden and of microbiologic control is less of a consideration.

11.20 CONCLUSION

The concepts and calculations used for F, D-, and z-value determinations are best understood if they are used on a regular basis. It is recommended that the essential details of these calculations are clearly described within the format of an SOP in order to ensure that the calculations are consistently used in the same manner for confirmation of each new batch of BIs and in every new application of the sterilization process.

Professor Irving J. Pflug, Ph.D., Emeritus Professor of Food Science and Nutrition of the University of Minnesota, has over the past several decades been instrumental in the training of literally thousands of upcoming scientists on the various aspects of this subject. Although these courses are no longer being made available, his books on the subject are available through his website “drpflug.com”.

NOTES

1 It is generally accepted that fungi and viruses have low resistance to wet heating and are inactivated by temperatures as low as 70°C, far less than steam sterilization at 121°C or dry heat sterilization at 170°C. Most studies of fungal and virus resistance to sterilization have been initiated by the food and agricultural industries because fungal or viral contamination of foods and croplands can be a serious problem. Fungi and viruses are generally more susceptible to the rigors of sterilization methods employed by the pharmaceutical industry, so these studies have typically involved temperatures no higher than 60°C. Studies at these temperatures have shown fungi and viruses to be inherently nonresistant compared with bacterial spores.

2 The D-value is not an inherent attribute of the microorganism but can be established at the specific conditions of the test condition. The influence of other factors such as the substrate, matrix, recovery media, and test methodology must be considered in determination of the D-value (32), hence the resistance of a BI should be understood as a consequence of the factors that influence it. Accurate assessment of the D-value should be determined at constant conditions comparable to those used for the sterilization cycle. This information is usually provided by the BI vendor on the specific Certificate of Analysis (C of A) for that lot of BI.

3 For the purpose of process calculations involving the moist heat resistance of indigenous microorganisms, it is appropriate and acceptable to assume a *z*-value of 10°C (18°F) unless an alternative value has been determined from resistance studies (27).

REFERENCES

1. 21 CFR, Part 211 on Good Manufacturing Practices for Finished Pharmaceuticals.
2. Agalloco J. The bugs don’t lie. *PDA Journal of Pharmaceutical Science and Technology* 2019, 73(6), 615–621.
3. Alters MJ, Attia IA, Davis KE. Understanding and utilizing F values. *Pharm Technol* 1978 May.
4. Angelotti R, Maryanski JH, Butler IT, Peeler JT, Campbell JE. Influence of spore moisture content on the dry-heat resistance of *Bacillus subtilis var. niger. Appl Microbiol* 1968, 16.
5. ANSI/AAMI/ISO 11135–1:2014, AAMI TIR16, and AAMI TIR28, in Association for the Advancement of Medical Instrumentation. *Guideline for Industrial Ethylene Oxide Sterilization of Medical Devices*. Arlington, VA: AAMI, 1980.
6. Association for the Advancement of Medical Instrumentation. *Process Control Guidelines for Radiation Sterilization of Medical Devices*. Arlington, VA: AAMI, 1980.
7. Ball CO. Thermal process time for canned foods. *Natl Res Counc Bull* 1923, 7(Part 1), 37.
8. Ball CO. Mathematical solution of problems on thermal processing of canned foods. *Univ Calif. Publ Public Health* 1928, 1(2).
9. Bigelow WD, Bohart GS, Richardson AC, Ball CO. *Heat Penetration in Processing Canned Foods*. Washington, DC: National Canners’ Association, Bulletin 16L, 1920.
10. Bigelow WD. Logarithmic nature of thermal death time curves. *J Infect Dis* 1921, 29.
11. Block SS. *Disinfection, Sterilization, and Preservation*. Philadelphia, PA: Lea & Febiger, 1983.
12. Carleton FJ, Agalloco JP. *Validation of Aseptic Pharmaceutical Processes*. New York: Marcel Dekker, 1986.
13. Chick H, Martin CJ. On the “heat coagulation” of proteins. *J Physiol* 1910, 40(5), 404–430.
14. Christensen EA. Radiation resistance of bacteria and the microbiological control of irradiated medical products. Sterilization and Preservation of Biological Tissues by Ionizing Radiation. Report of a panel, Budapest, IEAE, Vienna, 1970 STOPUB/247, June 1969.

15. Foster JW. Morphogenesis in bacteria: Some aspects of spore formation. *Quart Rev Biol* 1956, 31, 102–118.
16. Fox K, Pflug IL. Effect of temperature and gas velocity on the dry-heat distribution rate of bacterial spores. *Appl Microbiol* 1968, 16.
17. Halverson HO. *J Appl Bacteriol* 1957, 20, 305–314.
18. Halverson HO. *The Bacteria*. Vol. 4. Gunsalus IC, Stanier RY, eds. London and New York: Academic Press: 223–264.
19. Health Industry Manufacturers Association. *Sterilization Cycle Development*. Washington, DC: HIMA Rep 78, 1978.
20. ISO 11135. *Sterilization of Health-Care Products: Ethylene oxide: Requirements for the Development, Validation and Routine Control of a Sterilization Process for Medical Devices*, 2014.
21. ISO18472. *Sterilization of Health Care Products*, 2018.
22. Klapes NA, Vesley D. Vapor-phase hydrogen peroxide as a surface decontaminant and sterilant. *Appl Environ Microbiol* 1990, 56.
23. Larkin EP. Thermal Inactivation of Viruses, United States Army Natick Research and Development Command, Natick, Mass, October 1977.
24. Molin G, Ostlund K. Dry heat inactivation of *Bacillus subtilis var. niger* spores with special reference to spore density. *Can J Microbiol* 1976, 22.
25. Murrell WG. Eleventh symposium of the Society for General Microbiology held at the Royal Institution, London. *Soc Gen Microbiol* 1961, 11, 100–150.
26. Nakata HM. Effect of pH on intermediates produced during growth and sporulation of *Bacillus cereus*. *J Bacteriol* 1963, 86, 577–581.
27. Okinczyc T. *Quality Assurance Statistician*. Greenville, NC: Burroughs Wellcome Co., Personal Communication.
28. Ordal ZJ. *In "Spores II"*. Halvorson HO, ed. Ann Arbor, Minneapolis, USA: Burgess Publishing Co., 1961.
29. Parenteral Drug Association. *Validation of Steam Sterilization Cycles*. Technical Monograph No. 1. Philadelphia, PA: PDA, 1978.
30. Parenteral Drug Association. *Validation of Dry Heat Processes Used for Sterilization and Depyrogenation*. Technical Report No. 3. Philadelphia, PA: PDA, 1981.
31. Parenteral Drug Association. *Parametric Release of Parenteral Solutions Sterilized by Moist Heat Sterilization*. Technical Report No. 8. Philadelphia, PA: PDA, 1987.
32. Patashnik M. A simplified procedure for thermal process evaluation. *Food Technol* 1953, 7.
33. Pflug IJ. Heat sterilization. In Phillips GB, Miller WS, eds. *Industrial Sterilization*. Durham, NC: Duke University Press, 1973.
34. Pflug U, Smith GM. Survivor curves of bacterial spores heated in parenteral solutions. In Barker AN, Wolf J, Ellar DJ, Dring GJ, Gould GW, eds. *Spore Research*. Vol. 2, 1976. Report for Project Year, June 30, 1975 through June 29, 1976 for FDA Contract No. 223-75-3028 entitled "Development of Resistance Parameters for Biological Indicators Used in Sterilization Processes".
35. Pflug IJ. *Syllabus for an Introductory Course in the Microbiology and Engineering of Sterilization Processes*. 4th ed. St. Paul, MN: Environmental Sterilization Services, 1980.
36. Pflug D, Holcomb RG. Principles of thermal destruction of microorganisms. In Block SS, ed. *Disinfection, Sterilization and Preservation*. Philadelphia, PA: Lea & Febiger, 1983: 751–810.
37. Pistolesi D, Mascherpa VF. $F_{0:}$ What It Means, How to Calculate It, How to Use It for Adjustment, Control and Validation of Moist-Heat Sterilization Processes, Fedegari Autoclavi SPA, 1989.
38. Rahn O. Injury and death of bacteria by chemical agents. In Luyet BJ, ed. *Biodynamica Monograph No. 3*. Normandy, MO: Biodynamica, 1945: 9–41.
39. Russell AD. *The Destruction of Bacterial Spores*. Cardiff, Wales: Welsh School of Pharmacy, University of Wales Institute of Science and Technology, 1982.
40. Silverman GJ. *The Resistivity of Microorganisms to Inactivation by Dry Heat*. Contract Ns G-691, Cambridge, MA: Massachusetts Institute of Technology, 1968.
41. Soper CJ, Davies DJG. Principles of sterilization. In Denyer SP, Baird RM, eds. *Guide to Microbiological Control in Pharmaceuticals*. London: Ellis Horwood, 1990.
42. Thompson PJ, Thames OA. Sporulation of *Bacillus stearothermophilus*. *Appl Microbiol* 1967, 15, 975–979.
43. Tsugi K, Harrison S. Dry heat destruction of lipopolysaccharide dry-heat destruction kinetics. *Appl Environ Microbiol* 1978, 35.
44. USP Chapter <1115> Bioburden control of nonsterile drug substances and products and USP Chapter <1229.3> Monitoring of bioburden. In *United Stated Pharmacopeia/National Formulary*. Easton, PA: Mack Publishing Company.
45. USP Chapter <55> *Biological Indicators—Resistance Performance Tests*.
46. USP Chapter <56> *Methods for Determination of Resistance of Microorganisms to Sterilization Processes*.
47. USP Chapter <71> *Sterility Tests*.
48. USP Chapter <85> *Bacterial Endotoxin Test*.
49. USP Chapter <1111> *Microbiological Examination of Nonsterile Products: Acceptance Criteria for Pharmaceutical Preparations and Substances for Pharmaceutical Use*.
50. USP Chapter <1211> *Sterility Assurance*.
51. USPChapter <1222> *Terminally Sterilized Pharmaceutical Products—Parametric Release*.
52. USP Chapter <1229> *Sterilization of Compendial Articles*.
53. USP Chapter <1229.1> *Steam Sterilization by Direct Contact*.
54. USP Chapter <1229.2> *Moist Heat Sterilization of Aqueous Liquids*.
55. USP Chapter <1229.3> *Monitoring of Bioburden*.
56. USP Chapter <1229.4> *Sterilizing Filtration of Liquids*.
57. USP Chapter <1229.5> *Biological Indicators for Sterilization*.
58. USP Chapter <1229.6> *Liquid Phase Sterilization*.
59. USP Chapter <1229.7> *Gas Phase Sterilization*.
60. USP Chapter <1229.8> *Dry Heat Sterilization*.
61. USP Chapter <1229.9> *Physicochemical Integrators and Indicators for Sterilization*.
62. USP Chapter <1229.10> *Radiation Sterilization*.
63. USP Chapter <1229.11> *Vapor Phase Sterilization*.
64. USP Chapter <1231> *Water for Pharmaceutical Purposes*.
65. Vinter V. *In "The Bacterial Spore"*. Gould GW, Hurst A, eds. London and New York: Academic Press, 1969: 73–123.
66. Weinberg ED. Manganese requirement for sporulation and other secondary biosynthetic processes of *Bacillus*. *Appl Microbiol* 1964, 12, 436–441.
67. Winans L, Pflug IJ, Foster TL. Dry heat resistance of selected psychrophiles. *Appl Environ Microbiol* 1977, 34.

ABBREVIATIONS

Abbreviations used in this chapter:

BIs, biologic indicators
BIER, biological indicator evaluator resistometer
CGMP, current good manufacturing practice
ETO, ethylene oxide
IPCD, internal process challenge device
RTD, resistance temperature detector
SAL, sterility assurance level
SOP, standard operating procedure
USP, United States Pharmacopeia
WFI, water for injection.

12 Biological Indicators for Sterilization

Kurt McCauley, Nicole Robichaud, Karlin Gardner and Crystal Hostler

CONTENTS

12.1 INTRODUCTION

A Biological Indicator (BI) is a "test system containing viable microorganisms providing a defined resistance to a specified sterilization process".[1] BIs are a type of lethality indicator that utilizes a biological component, typically bacterial endospores, as the sensing element. Other types of lethality indicators include Chemical Indicators (CIs) and physical measurements (equipment instrumentation). Each type has its advantages and disadvantages, which will be discussed in the next section. This chapter is dedicated to BIs, their uses, the controls utilized in their preparation, and the various formats available for the common sterilization modalities.

12.1.1 Lethality Indicators

Lethality indicators are systems that monitor sterilization processes using a well-characterized technology. Monitoring the calibrated instrumentation on the sterilizer (to verify the predefined conditions have been met) is a lethality indication activity. CIs utilize ink formulations that undergo a color change upon exposure to predefined sterilization conditions. There are six types of CIs described in detail in the ISO 11140 series. Indicator tape on the outside of a wrapped load is a simple form of a CI generally identified as a process indicator (Type 1). It provides no information about the sterility of the pack's contents, but a color change indicates that the pack has been exposed to the sterilant. Single variable, multivariable, integrating, and emulating indicators (Types 3, 4, 5, and 6, respectively) provide more information about a sterilization process and are designed "to be placed inside individual load items and to assess attainment of the critical process variables at the point of placement".[2] Integrating CIs (Type 5) are designed to mimic the performance of BIs by responding to known variables of a sterilization process. Table 12.1 provides additional details on CIs as described in ISO 11140–1.

BIs are the gold standard of lethality indicators as they integrate all sterilization parameters involved, both known and unknown. BIs provide direct evidence of sterility assurance as they challenge the sterilizer's ability to kill calibrated strains of highly resistant spore-forming organisms. BIs are often used in conjunction with the other indicators to monitor sterilization processes. Physical measurements and CIs provide real-time or immediate post-processing results, which is their primary advantage over BIs, which require an incubation period in order to achieve the results.

DOI: 10.1201/9781003163138-12

TABLE 12.1
Description of Chemical Indicator Types and Their Intended Use

Type	ISO 11140-1:2014 definition*	Intended use
Type 1: Process indicators	"Process indicators shall be designed for use with individual items (e.g. packs, containers) to show that the unit has been directly exposed to the sterilization process and to distinguish between processed and unprocessed items."	• Intended to be placed on the outside of individual items. • Shows the item has been exposed to a sterilization process. • Differentiates between processed and unprocessed items. • Example: Indicator tape or indicator labels
Type 2: Indicators for specific tests	"Type 2 indicators are intended for use in specific test procedures as defined in relevant sterilizer/sterilization standards. The requirements for specific test indicators and indicator systems (Type 2 indicators) are provided in ISO 11140-3, ISO 11140-4, and ISO 11140-5."	• Specialty indicator. • Designed for use in specific test procedures outlined in relevant ISO standards. • Example: Bowie–Dick air removal/steam penetration test
Type 3: Single variable indicators	"A single critical process variable indicator shall be designed to react to one of the critical process variables and is intended to indicate exposure to a sterilization process at a stated value (SV) of the chosen critical process variable."	• Intended to be placed inside of items to be sterilized. • Reacts to one critical process variable. • Example: Chemical pellet that melts at a specific temperature
Type 4: Multivariable indicators	"A multicritical process variable indicator shall be designed to react to two or more of the critical process variables and is intended to indicate exposure to a sterilization process at SVs of the chosen critical process variables."	• Intended to be placed inside of items to be sterilized. • Reacts to two or more critical process variables. • Example: Indicator that measures temperature held for a specific amount of time
Type 5: Integrating indicators	"An integrating indicator shall be designed to react to all critical process variables. The SVs are generated to be equivalent to, or exceed, the performance requirements given in the ISO 11138 series for BIs. The minimum SV shall be related to the minimum values required to achieve sterilization as specified in International Standards ISO 11135, ISO 11137 (all parts), ISO 17665 (all parts), or by local regulatory agencies."	• Intended to be placed inside of items to be sterilized. • Reacts to all critical process variables over a specified range of sterilization cycles. • Stated values meet or exceed biological indicator performance requirements found in the ISO 11138 series. • Example: Indicator that can be used in a variety of sterilization cycles
Type 6: Emulating indicators	"An emulating indicator shall be designed to react to all critical process variables for specified sterilization processes. The SVs are generated from process variables of sterilization processes as specified in International Standards ISO 11135, ISO 11137 (all parts), ISO 17665 (all parts), or by local regulatory agencies."	• Intended to be placed inside of items to be sterilized. • Reacts to all critical process variables for specific sterilization cycles. • Stated values are generated from the critical variables of the specified sterilization process. • Response does not necessarily correlate to a biological indicator • Example: Indicator that monitors a specific sterilization cycle (e.g. 132°C, 10 minutes)

*©ISO. This material is adapted from ISO 11140–1:2014, with the permission of the American National Standards Institute (ANSI) on behalf of the International Organization for Standardization. All rights reserved.

12.2 A BRIEF HISTORY OF BIOLOGICAL INDICATORS

The discovery of microorganisms and their role in spreading disease and infection and causing food spoilage is relatively new. Tyndall, Pasteur, Koch, and Lister were among the early pioneers in developing tools to study and combat problematic organisms.

Koch was perhaps the first person to recognize the value of the spore as a measurement tool. In the early 1880s, he successfully isolated spore-forming bacteria from soil samples and experimented with *Bacillus anthracis* spores. He recognized that if a sterilization process can inactivate hard-to-kill spores, then it will likely kill other forms of microorganisms. Koch used soil samples (containing an unknown number of spores with an unknown resistance to the process) to evaluate the effectiveness of the sterilization process. When no growth occurred upon culturing the soil sample, the sterilization process was considered adequate.

In the early 1890s, Kilmer began using sealed reference organisms to judge the success of steam sterilization processes used in the manufacture of sterile dressing, "thus pioneering the use of the biological indicator in industrial sterilization".[3]

Various methods of food preservation (smoking, salting, drying, etc.) have been in use for thousands of years. The process of canning and heat treating to preserve foods became industrialized in the early 1800s, in part to support the soldiers and sailors in various European military conflicts. *Geobacillus stearothermophilus* proved to be a challenge in producing shelf-stable canned foods and was first characterized by Donk[4] in 1917 after he isolated it from canned corn treated at 118°C for 75 minutes. *G. stearothermophilus* spores are now the most widely used microorganism in BIs for moist heat and some chemical sterilization/decontamination processes. *G. stearothermophilus* and other spore-forming species are used worldwide as BIs and are recognized by numerous standard-setting organizations.

12.3 BACTERIAL ENDOSPORES

Like other bacteria, species of *Bacillus, Geobacillus*, and *Clostridium* grow vegetatively; however, when environmental conditions become unfavorable, organisms from these genera have the ability to sporulate leading to the formation of dormant spores. This process is a survival mechanism rather than a reproductive function. When the environment remains favorable, the organisms will continue to grow and reproduce in the vegetative state.

Spores are one of the most stable forms of life known and can remain viable for millions of years in this "dormant" state (Figure 12.1). Although the spore is considered dormant, it does contain active enzymes that trigger the spore to germinate in a short period of time when environmental conditions once again become favorable. In the vegetative state, the cells are susceptible to environmental stresses; however, while in the spore state, the organism gains significant resistance to these stresses.

In short, the characteristics that make bacterial spores ideal for use in BIs include:

1. They can withstand environmental stresses, including sterilization conditions, well beyond that of most other microorganisms.
2. They integrate known and unknown process parameters of a sterilization process.
3. They have a gradual, largely predictable inactivation phase (kill curve/death curve).
4. They can be "calibrated" (manufactured to contain a known spore count with a defined resistance to a sterilization process).
5. They have extreme longevity, which is desirable for maintaining a long shelf-life.
6. They can be grown in sufficient quantities to produce large batches of uniform indicators.
7. They allow results to be binary (a BI positive for growth indicates non-sterility; a BI showing no-growth indicates sterility).

12.4 BI USES THROUGHOUT THE STERILIZATION LIFE CYCLE

A sterilization life cycle generally consists of three elements; cycle development, cycle validation, and routine monitoring. BIs and/or inoculated product are used in each of these three elements.

12.4.1 Cycle Development

Developing a proper sterilization cycle is key to producing sterile and functional product. Increasingly, pharmaceutical and medical device products are becoming heat labile, and moist heat sterilization by the overkill method is not an option in these cases. **It is important to understand the nature of the bioburden (total quantity, identity, resistance, presence/absence of spore formers) as these are the true target of the sterilization process**. It is equally important to know how the product influences the resistance of the organisms. Product components (active ingredient, excipients, preservative, etc.) may protect the organisms by coating them or causing them to clump, increasing their resistance to the sterilization process, or the product components may sensitize the spores, making the organism more vulnerable to the sterilization process.[5]

Performing D-value studies on inoculated product provides this information. A common practice in these studies is to use *G. stearothermophilus* spores (or other recognized organisms referenced in the standards) as a substitute for the bioburden. This is a conservative approach, as these spores are typically substantially more resistant than the bioburden organisms and by design, significantly outnumber the bioburden organisms.

D-value studies are not restricted to product terminal sterilization processes but should also be performed on enclosure components used in aseptic fill. Studies performed on

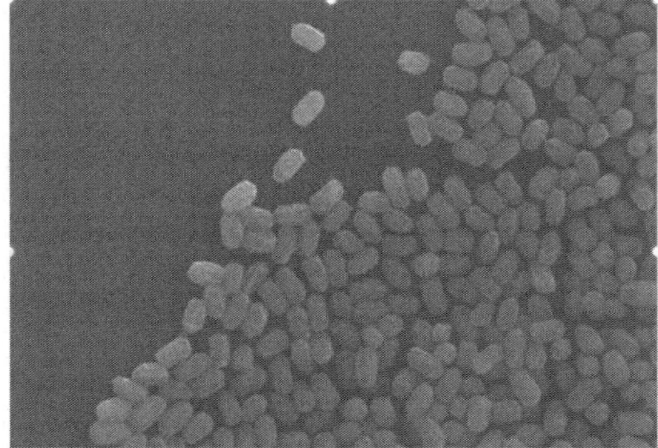
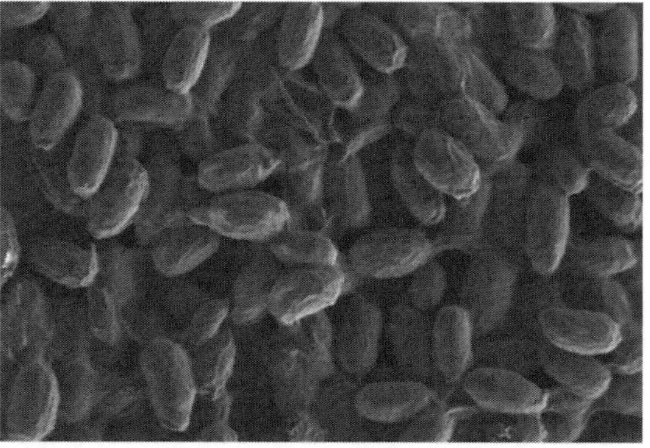

FIGURE 12.1 Micrograph of spores of *B. subtilis* (5k magnification) and *G. stearothermophilus* (7.5k magnification).

stoppers have shown a wide range of D-values depending on their composition and coatings.[5]

D-value studies are key to proper cycle development, and regulatory bodies expect to see these data to support the design of the final process. These studies can be performed by the firm, but commonly they are provided as a service by BI manufacturers. BI manufacturers not only supply spore suspensions, but they also possess resistometers[6] (test equipment required to perform D-value assessments).

12.4.2 Cycle Validation

The purpose of validating a sterilization cycle is to ensure that the process will consistently render the load sterile. The role of the BI in this process is to demonstrate that sterilization conditions are met by monitoring areas in the load that have been identified as having lower delivered lethality. In the case of thermal sterilization processes, these areas are often referred to as "cold spots" or "slow-to-heat areas" and are identified through thermal mapping of the load. In the case of chemical sterilization, CIs may be a useful tool in identifying the areas of lower lethality, which can then be monitored by BIs during the actual cycle validation. The practice of mapping with CIs and then monitoring with BIs is commonly used when validating isolators and associated equipment used in aseptic filling operations.

This book is dedicated to the validation of pharmaceutical processes, and this subject is discussed in greater detail in other chapters.

12.4.3 Routine Monitoring

The purpose of using BIs for routine monitoring of sterilization processes is to ensure that the system has not drifted outside of the validated state. The monitoring frequency will depend on the firm's practices and regulatory requirements. These can range from monitoring each cycle to daily or weekly monitoring.

BIs should be placed in the most difficult to sterilize (worst-case) location in the load and/or chamber, which should have been identified in the process development and validation studies. Retrieving BIs embedded in sterilizer loads can be problematic, as the packaging and/or load will become compromised. An alternative to the embedding of BIs in product is the use of Process Challenge Devices (PCDs).

12.5 PROCESS CHALLENGE DEVICES

PCDs contain an embedded BI and are designed to imitate BI-imbedded product. They can be constructed of actual or simulated product or other configurations (Figure 12.2), provided they simulate the challenge the product presents during a sterilization process. PCDs should be placed in the sterilizer location(s) that have been identified as providing the greatest challenge to sterilant penetration. Upon completion of the cycle, the BIs are retrieved from the PCDs and processed while the actual load remains undisturbed.

Per a recognized standard: "The appropriateness of the PCD used for process definition, validation, or routine monitoring and control shall be determined. The PCD shall present a challenge to the sterilization process that is equivalent or greater than the challenge presented by the natural bioburden at the most difficult to sterilize location within the product."[7]

12.6 BI CONFIGURATIONS (FORMATS)

There are many different BI configurations ranging from simple spore strips to more sophisticated self-contained biological indicators (SCBIs). At a minimum, all BIs contain spores and a substrate onto which the spores are inoculated (i.e. the spore carrier). The majority of BIs also include primary packaging used to contain the inoculated carrier. Additionally, SCBIs contain culture medium within the BI unit that yields certain advantages over those BIs that do not, as discussed following. A particular configuration may be appropriate for numerous sterilization modalities whereas others are specific to defined conditions (Table 12.2).

12.6.1 Strips

Spore strips are the simplest BI format and typically consist of a paper strip packaged into an envelope (primary package) permeable to the sterilant. Spore carriers such as glass fiber or stainless steel are used in processes where cellulose is incompatible with the sterilant (e.g. vapor hydrogen peroxide, nitrogen dioxide).

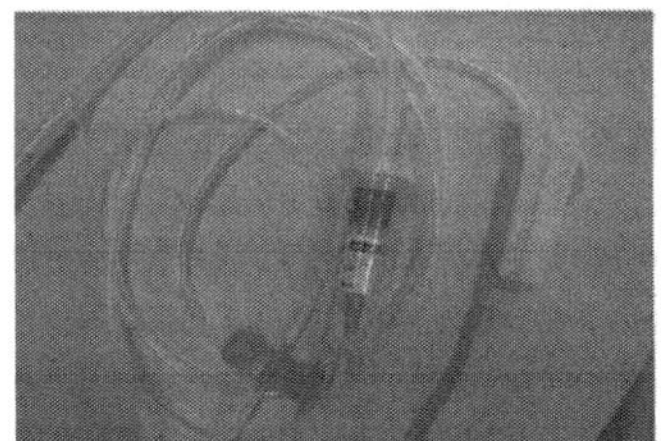

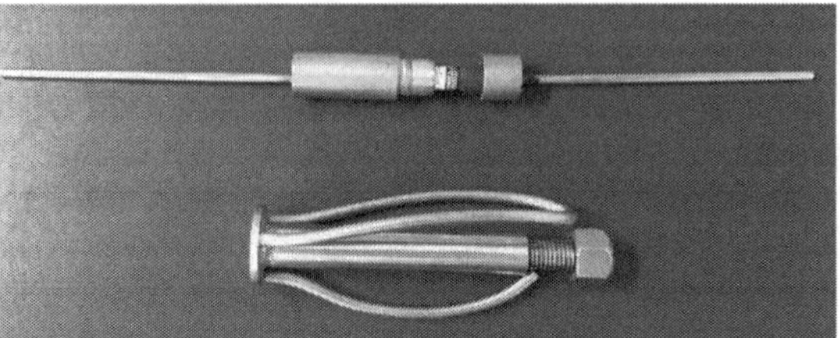

FIGURE 12.2 Examples of PCDs constructed from actual product and other configurations that simulate product during sterilization. Image on the left depicts a BI embedded into a drip chamber. The device on the top of the image on the right depicts BI insertion into a lumen device (pictured open with BI visible). The device on the bottom of the image on the right contains an embedded BI. Sterilant must penetrate the threaded area to gain access to the BI.

Note: BI = biological indicator; PCDs = process challenge devices.

TABLE 12.2
Typical Characteristics and Uses of Commercially Available Biological Indicators

Modality	Process	Indicator organism	BI type*	Population
Moist heat	Sterilization of porous or solid items	*Geobacillus stearothermophilus, Bacillus subtilis "5230"*	Strip or SCBI	10^5 to 10^6
	Sterilization of liquids	*Geobacillus stearothermophilus, Bacillus subtilis "5230"*	LSBI	10^5 to 10^6
Dry heat	Sterilization of solid items	*Bacillus atrophaeus*	Strip	10^6
	Sterilization of nonaqueous liquids	*Bacillus atrophaeus*	LSBI**	10^6
Ethylene oxide	Sterilization of solid items	*Bacillus atrophaeus*	Strip or SCBI	10^6
Vapor-phase hydrogen	Sterilization of solid items	*Geobacillus stearothermophilus*	Strip (non-cellulose carriers)	10^5 to 10^6
peroxide	Decontamination of enclosures	*Geobacillus stearothermophilus*	Strip (non-cellulose carrier)	10^3 to 10^6

Abbreviations: BI = biological indicator; LSBI = liquid-submersible biological indicator; SCBI = self-contained biological indicator.

*Other BI configurations are available as custom items. It is recommended the customer contact the BI manufacturer to discuss the appropriate BI attributes.

**This type of LSBI is unique in that the spore carrier is not an aqueous-based culture medium, which would improperly impart moist heat sterilization conditions on the spores when monitoring a dry heat process. Rather the spore carrier is a solid material, often a silica sand, that is hermetically sealed into a snap-neck glass ampoule. The ampoule is positioned in the liquid in the same manner as described in the LSBI section. Post exposure, the spore carrier is removed from the ampoule and transferred into tubed culture media.

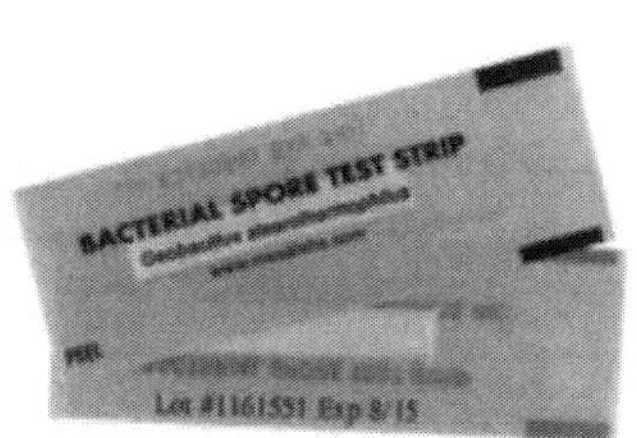

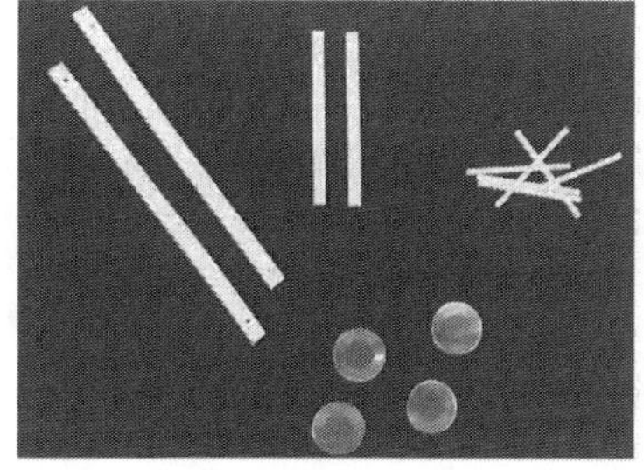
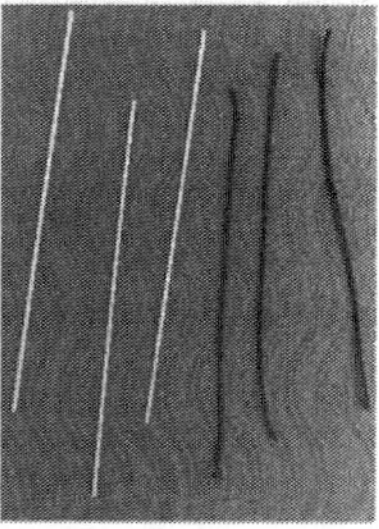

FIGURE 12.3 Standard spore strip packaged into glassine envelope (left) and unpackaged biological indicators (BIs) of various configurations and materials-of-construction (center and right).

The primary packaging serves to protect the spore carrier from outside contamination. It must be a robust design such that it can withstand planned transport as well as the conditions encountered during a sterilization process (e.g. high temperature, rapid rates of pressure change, etc.). The packaging must be permeable to but not absorb the sterilant (in the case of a chemical agent) or release inhibitory substances. Glassine pouches are common primary packaging material for traditional spore strip paper carriers. Tyvek/Tyvek or Tyvek/Mylar pouches are used for many chemical sterilants, especially those that are incompatible with cellulose. The primary packaging is a key component in preventing post-exposure contamination to the spore carrier, which can lead to false-positive BI results (spore strips require a clean bench for aseptic transfer into a culture medium). The packaging can have a significant effect on the resistance of the BIs, which the user should consider if the planned use is to expose the spore carrier naked (unpackaged).

Although most spore strips are designed to be used in their primary packaging, other BIs of various shapes, sizes, and materials-of-construction (Figure 12.3) are commercially available and are designed to be inserted into tight spaces (e.g. lumen, narrow channels, dead legs, etc.). To accommodate placement in the load, these BIs are significantly smaller in their physical size and are designed to be used without primary packaging.

Strip BIs are appropriate for use in monitoring all traditional sterilizing processes of solid loads (moist heat, dry heat, ethylene oxide), as well as many nontraditional processes (Table 12.2).

The incubation period of spore strip BIs is generally seven days unless a reduced incubation time has been validated.

12.6.2 Self-Contained Biological Indicators

SCBIs are considered a second-generation BI and consist of a spore carrier placed into a plastic vial along with a glass ampoule of media, sealed with a plastic cap. The cap contains ports protected with filter paper (Figure 12.4). This design allows sterilant penetration into the vial while preventing microbial contamination.

SCBIs are a "self-contained" system, meaning that they contain both the spores and the culture medium (developed by the manufacturer for defined outgrowth). They are tightly

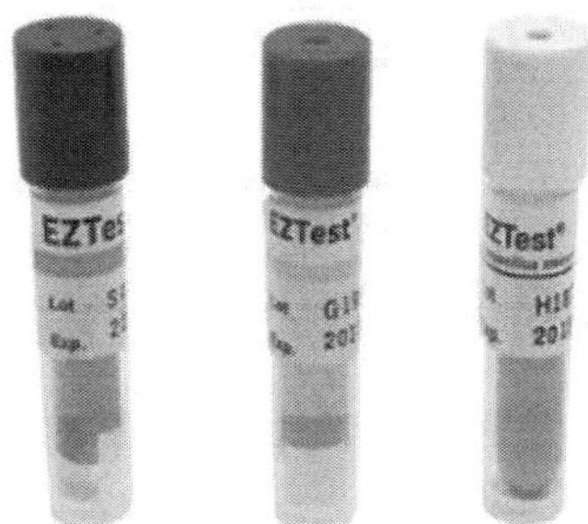

FIGURE 12.4 Self-contained biological indicators (SCBIs) contain both the spore carrier and the culture medium.

controlled and well characterized, and thus have narrowly defined performance boundaries. This high level of control is made possible primarily because the spores can only be used in the supplied culture medium (users of spore strip BIs typically supply their own culture media, which can introduce considerable variation to the resistance). This is a critical factor for the reproducibility of performance characteristics such as the D-value, z-value, and a reduced incubation read-out time.

A clean bench is unnecessary when culturing the SCBI, as the media ampoule is contained within the vial. To culture the spore carrier, the media ampoule is ruptured by compressing the sides of the plastic vial. Spore viability in the BI is easy to interpret because of the design and instruction-for-use provided by the manufacturer. The characteristics of a negative unit, as well as a positive unit (generally a color change) are well defined, leaving no doubts about the viability of the spores upon completion of the test.

Similar to spore strip BIs, SCBIs are appropriate for use in monitoring the traditional sterilizing process of solid loads (e.g. moist heat and ethylene oxide) and may be appropriate with certain nontraditional processes (Table 12.2).

One of the more attractive characteristics of the SCBI is that they are an ideal test system for making reduced incubation time (RIT) claims. The RIT claim is simply the boundary placed on the test system, letting the users know when the test can be considered complete. The manufacturer determines the read-out time after performing testing (using hundreds of BI units) according to a recognized protocol[8]. The critical parameter that must be controlled by both the manufacturer and the end user is the incubation temperature. All other variables are controlled because of the inherent characteristics of the SCBI.

12.6.3 Liquid-Submersible Biological Indicators

Liquid submersible biological indicators (LSBIs), a type of SCBI, have spores inoculated directly into the growth medium, which is hermetically sealed in glass ampoules. LSBIs come in a variety of shapes and sizes to accommodate the sizes of the liquid-filled containers being sterilized. As is the case with the SCBI, LSBIs are tightly controlled, well characterized, and have narrowly defined performance boundaries. This again is made possible because the spores can only be used in the supplied culture medium.

LSBIs are used to monitor moist heat sterilization of liquid products by placing them directly into product-filled containers prior to sterilization (Table 12.2). Smaller ampoules are designed for placement into small containers, whereas larger snap-neck type ampoules (designed to sink in most liquids) are for placement into larger containers. The large ampoules can be held in a defined location (generally centered in the container and slightly off the bottom) using a thin wire attached to the neck of the ampoule (Figure 12.5). The LSBIs can be placed directly into product-filled containers or used in a PCD containing a surrogate liquid with similar heat-transference properties and volume as the actual product.

The spores in these BIs are always in the cultured state (i.e. in direct contact with the growth medium). To test for spore viability, the LSBI simply needs to be incubated at the appropriate temperature and evaluated for color change (Figure 12.5). During storage, care must be taken to keep these units well below their incubation temperature to prevent spoilage (i.e. germination and outgrowth of the test organism). These types of BIs are typically kept under refrigeration until the time of use.

As with all SCBIs, the RIT of LSBIs is determined in the same manner as described previously.

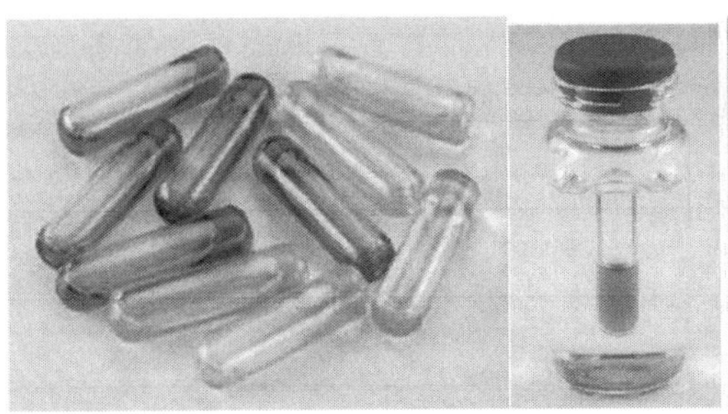

FIGURE 12.5 Binary results of liquid submersible biological indicators (LSBIs) (yellow indicates positive for growth, purple indicates no growth) and LSBIs placed into product-filled containers (center and right).

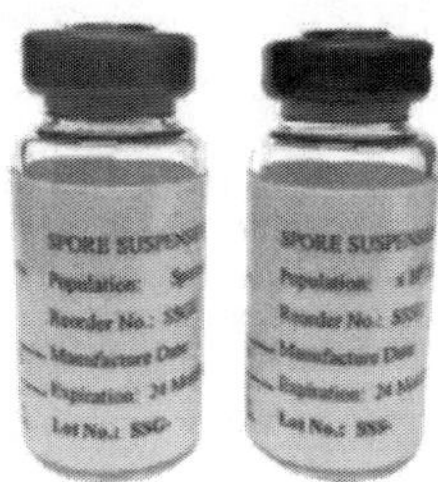

FIGURE 12.6 Example of spore suspensions bulk containers.

12.6.4 Spore Suspensions

Spore suspensions (Figure 12.6) are not BIs unto themselves but can be used to manufacture custom BIs or for the direct inoculation of product. In commercially prepared spore suspensions, the spores are typically suspended in either sterile water or an ethanol/water solution. An aqueous suspension is often used for the inoculation of water-based pharmaceutical products whereas an ethanol suspension is preferred for direct inoculation of solid goods and equipment, because it will dry quickly. Spore suspensions are available in different concentrations typically ranging from 10^4 to 10^7 spores/0.1 ml. When preparing inoculated product, it is important to utilize a suspension with the appropriate spore concentration. Prior to obtaining the suspension, the following should be defined:

1. The targeted spore count for the inoculated product.
2. The volume the inoculated product will hold.
   a. For solid items, consider the size of the area onto which the spores are to be inoculated. Smaller areas will have increased spore concentrations (piling) that may be problematic depending on the sterilization/decontamination modality.
   b. For liquids, especially for small volumes, care should be taken not to over-dilute the product with the suspension (e.g. delivering a 10 μl aliquot from a 10^7/0.1 ml suspension will result in a 10^6 spore load in the product without excessive dilution of the product).

When using a concentrated bulk spore suspension, it may be necessary to dilute the suspension down to an appropriate working dilution. The D-value of spore suspensions (typically assessed on a glassine-packaged paper strip) provided by the manufacturer is of limited value, as it will not reflect the actual use conditions by the customer. However, the D-value is useful when the customer would like to obtain spores that will perform in a similar manner to those used in a previous study, as it is believed that the ratio of resistance between a spore suspension and an inoculated substrate is roughly constant for the same species.

12.7 BI MANUFACTURING CONTROLS

A number of standards exist that offer guidance for BI manufacturers (Table 12.3). ISO 13485 provides guidance for the quality management systems for medical devices. The ISO 11138 series, United States Pharmacopeia (USP), and European Pharmacopoeia (EP) provide details specifically for BI manufacturing controls and performance criteria. Compliance with these standards by the manufacturers is not mandatory; however, in complying, the BI user gains a level of confidence in the quality of the product.

12.8 PREPARATION OF BIs AND MATERIALS OF CONSTRUCTION

BI manufacturers claiming compliance with relevant standards are required to use spores that are traceable to a recognized culture collection. The strains must not be pathogenic or require special handling practices or containment equipment. The majority of BIs produced have 10^5 to 10^6 spores per unit. Hundreds of BIs are consumed by BI manufacturers in order to qualify a lot (population, D/z-value determination). Likewise, BI users may consume hundreds of BIs during process development and validation studies. It is important that all BIs used in these studies are uniform, (contain spores from a single crop) and not from mixed lots. As such, BI manufacturers must produce uniform spore crops of sufficient size in order to produce large BI lots.

The production of a spore crop is initiated by isolating the organism from the seed stock and proving its identity. Once done, the organisms are cultured and allowed to proliferate such that high quantities of vegetative cells are obtained. Eventually the organisms will become stressed because of nutrient depletion, crowding, and the buildup of waste byproducts. These conditions trigger sporulation. Ideally the vast majority of vegetative organisms will sporulate resulting in a high-yield crop. The crop is then harvested by physically removing the spores from the culture medium. Cleaning all residual medium and cellular debris from the spores is critical to the uniform performance of the spores (Figure 12.7). BI manufacturers employ different growing, harvesting, and cleaning techniques to achieve high quantities of clean spores.

12.8.1 BI "Calibration"

Calibration is defined as a "set of operations that establish, under specified conditions, the relationship between values of a quantity . . . and the corresponding values realized by standards."[9]

True calibration of BIs is not practicable as that implies there is a standard organism to which a comparison can be made. As Dr. Pflug so eloquently stated, "There are no standard entities in nature be they man, monkey or microorganism. There are no standard bacteria, not even bacterial spores."[10] How then are BIs calibrated or better stated, characterized?

As discussed earlier, guidance on the manufacture and performance of BIs are described in several standards (Table 12.3). To comply with these standards, BI manufacturers employ various controls during the manufacturing process and then evaluate the performance of the BIs against these standards. Spore counts and resistance data (D/z-values) are

TABLE 12.3
Standards Relevant to BI Manufacturers and Users of BIs

Standard	Title	Overview
ISO 13485	*Medical device—Quality management systems—Requirements for regulatory purposes*	Guidance for medical device industry quality management systems.
ISO 11138-1	*Sterilization of health care products—Biological indicators—Part 1: General requirements*	General requirements for BI production, labeling, test methods, and performance characteristics.
ISO 11138-2	*Sterilization of health care products—Biological indicators—Part 2: Biological indicators for ethylene oxide sterilization processes*	Specific requirements for BIs used for ethylene oxide sterilization, including test organism and performance criteria.
ISO 11138-3	*Sterilization of health care products—Biological indicators—Part 3: Biological indicators for moist heat sterilization processes*	Specific requirements for BIs used for moist heat (steam) sterilization, including test organism and performance criteria.
ISO 11138-4	*Sterilization of health care products—Biological indicators—Part 4: Biological indicators for dry heat sterilization processes*	Specific requirements for BIs used for dry heat sterilization, including test organism and performance criteria.
ISO 11138-5	*Sterilization of health care products—Biological indicators—Part 5: Biological indicators for low-temperature steam and formaldehyde sterilization processes*	Specific requirements for BIs used for low-temperature steam and formaldehyde sterilization, including test organism and performance criteria.
ISO 11138-7	*Sterilization of health care products—Biological indicators—Part 7: Guidance for the selection, use and interpretation of results*	End-user guidance for the selection, use, and interpretation of the results of BIs used in the development, validation, and routine monitoring of sterilization processes.
ISO 18472	*Sterilization of health care products—Biological and chemical indicators—Test equipment*	Specific requirements for the test equipment used to evaluate the response of BIs and CIs to critical process parameters.
EP 5.1.2	*Biological indicators and related microbial preparations*	Guidance for the sterilization of finished products and items that come into direct contact with the final sterilized product.
USP General Chapter <55>	*Biological Indicators—Resistance Performance Tests*	Method and acceptance criteria for viable spore count determination and verification and method for resistance determination.
USP General Chapter <1229.5	*Biological Indicators for Sterilization*	Guidance for use, characterization, and selection of biological indicators.

Abbreviations: BI = biological indicator; CI = chemical indicator; EP = European Pharmacopoeia; ISO = International Organization for Standardization; USP = United States Pharmacopeia.

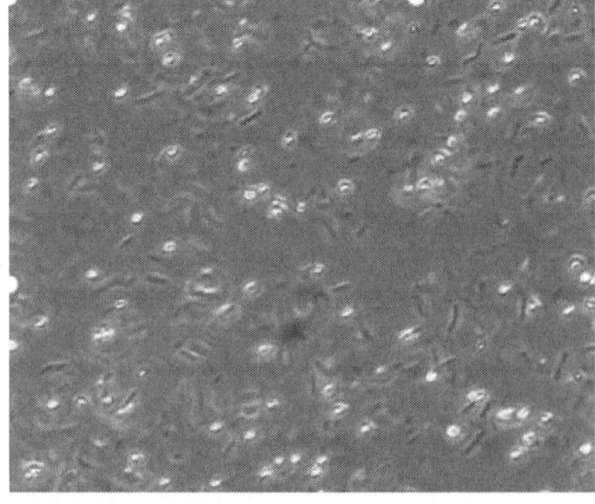

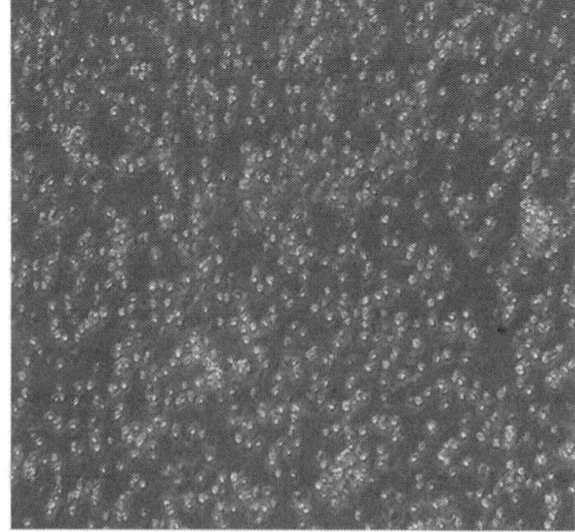

FIGURE 12.7 Phase contrast micrograph of uncleaned (left) and cleaned (right) spore crops. The phase bright objects are the spore form of the organism whereas the dark rods are the vegetative form.

two key characteristics of a BI lot and are briefly discussed next. Unlike instrument calibration, if the BI lot is found to be out of compliance, it cannot be "recalibrated".

The population (spore count) for each BI lot is clearly stated on the manufacturer's Certificate-of-Analysis. Methods for recovering spores from the carriers are detailed in the appropriate standards. The standards may also give guidance on minimum populations allowed for the various sterilization modalities.

BI resistance is determined using a resistometer, formerly known as a BIER (Biological Indicator Evaluator Resistometer) vessel (Figure 12.8). Resistometer performance

FIGURE 12.8 Resistometers for moist heat (left), dry heat (center), and ethylene oxide (right).

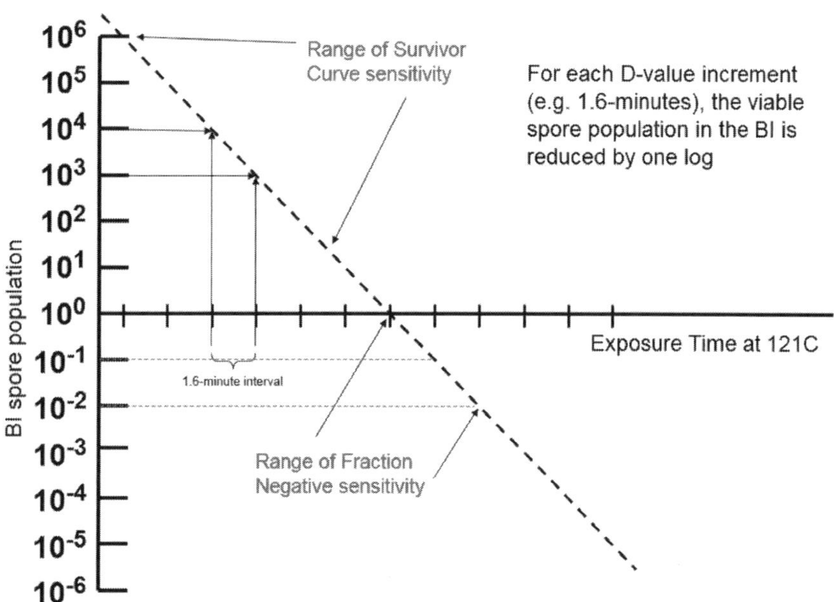

FIGURE 12.9 Graphical representation of the relationship between the Survivor Curve and Fraction Negative methods of D-value determination.

Note: BI = biological indicator.

requirements are outlined in ISO 18472[6]. From the introduction: "Resistometers constitute test equipment designed to create precise and repeatable sterilizing environments, allowing the evaluation of their effect on biological inactivation kinetics, chemical reactions, material degradation and product bioburden. Resistometers allow precise variation of the environmental conditions and cycle sequences in order to produce controlled physical studies."

BI resistance is generally expressed as a D-value. The D-value is defined as the "time or dose required to achieve inactivation of 90 % of a population of the test microorganisms under stated conditions"[1]. Simply stated, the D-value is a killing rate. A D-value can only be calculated when the kill curve exhibits first-order log-linear kinetics (Figure 12.9). BIs exposed to thermal, radiant energy and some chemical modalities exhibit this pattern. Other modalities, including chemical vapors, may exhibit a biphasic curve (Figure 12.10). Although the data from biphasic systems can be forced into a D-value calculation formula, it is not appropriate to do so as these values are meaningless and misleading. D-values are only useful when they are predictable (linear) for the duration of the process. Using a "D-value" that was generated from a multiphasic kill curve will result in a miscalculation of the Sterility Assurance Level.

There are two methods commonly used for D-value determination, the Survivor Curve (SC) and the Fraction Negative

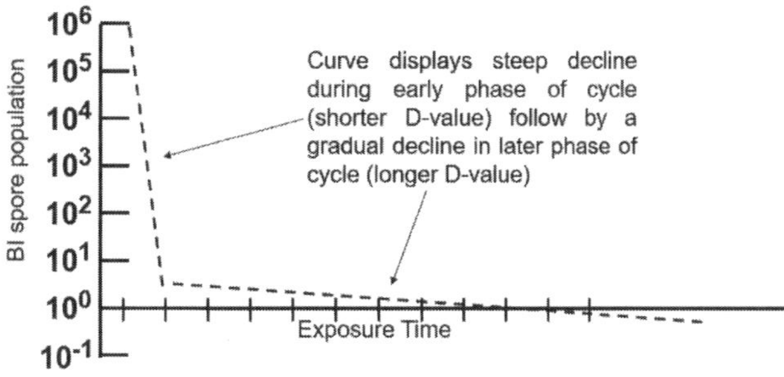

FIGURE 12.10 Graphical representation of a kill curve that exhibits biphasic kill kinetics.

Note: BI = biological indicator.

(FN), both of which are discussed in great detail in another chapter.

In brief, the SC method assesses the number of surviving spores at each exposure test interval by direct enumeration. This is accomplished by performing population assays on BI sets subjected to increasing exposure to the sterilant. The SC method results in an easy-to-understand visual representation when the data are presented graphically (Figure 12.9). This method of D-value determination establishes the resistance for surviving populations >50 spores per indicator.

The FN method focuses on the last surviving spores and gives binary results at appropriate test intervals. At longer test intervals, all BI units will be negative for growth, and at shorter intervals, all BI units will be positive for growth. These data are used for the indirect determination of the D-value using calculations recognized in the standards. The FN method establishes the resistance for surviving populations <10 spores per indicator. Figure 12.9 graphically displays the relationship between the FN and the SC methods.

12.8.2 Read-Out Time

For established sterilization processes, the standard incubation period for BIs (using well-characterized organisms) is seven days at an appropriate temperature for the specific organism. In the mid-1980s, the United States Food and Drug Administration (FDA) Center for Devices and Radiological Health (CDRH) published a protocol for use by BI manufacturers to reduce the incubation of BIs to less than seven days.[8] This reduced incubation time (RIT) protocol requires 100 BIs from three lots be exposed to separate sterilization processes that provide FN results. Of these fractional results, the number of positive BIs from each group of 100 must be within the range of 30%–80% of the total number of BIs. The results are recorded frequently, out to seven days of incubation. The maximum incubation time of the three lots at which at least 97% of the BIs are positive as compared with the number of BIs positive at seven days is considered the minimum incubation time or RIT for that specific BI, exposed to that specific sterilization process, cultured in that specific growth media, and incubated at that specific temperature.

This method is not universally recognized. However, Working Group 4 of ISO TC 198—Sterilization of health care products, is currently developing a standard to describe an internationally agreed approach for the validation of the RIT of BIs.

12.9 BI USER CONTROLS

Many of the standards referenced in Table 12.2 provide guidance for BI users, however, ISO 11138–7[11] is dedicated entirely to providing "guidance for the selection, use and interpretation of results" and is briefly discussed next.

From the introduction: "The selection of an appropriate biological indicator for the particular process used is critical. There is a wide variety of sterilization processes in common use, and biological indicator manufacturers are not able to foresee all possible uses of their product. Manufacturers, therefore, label biological indicators according to their intended use. It is the responsibility of the users of biological indicators to select, use, recover and interpret the results as appropriate for the particular sterilization process used."

Additionally, it is important for the user to establish confidence that the manufacturer is providing a quality product. This can be achieved by assuring that the manufacturer has a robust quality management system in place and is manufacturing product per the relevant standards. This is accomplished through performing quality system audit(s) of the BI manufacturer or by testing critical BI characteristics (e.g. population and resistance) on incoming BI lots.

When performing population verification, the user should be aware of the manufacturer's test methods and materials as these can have a great impact on the results. Likewise, for resistance verification testing, users need to test according to the manufacturers methods to ensure successful confirmation of the label claims. Resistance testing by the user is further complicated if the firm does not possess a resistometer, which is required test equipment for D-value and/or survival-kill testing. Per ISO 11138–7[11], when the manufacturer produces the BI per recognized standards (i.e. the ISO 11138 series), "and the user uses the biological indicator as intended by the biological indicator manufacturer, testing of the resistance characteristics by the user is considered unnecessary".

Third party labs are equipped and available for resistance and population verification in the event that testing is deemed necessary. BI manufacturers also have technical support staff to assist with technical details of the BIs and the methods used for their characterization.

Population assays are a popular incoming inspection activity. Subsequent assays performed on a BI lot should not be considered the new population. The purpose of these assays is to verify the manufacturers label claim and not to recertify the lot. From ISO 11138–1[1]: "Population verification shall be achieved when results fall within 50% to 300% of the manufacturer's stated nominal population. Confirmation test results of the population determined by end users or manufacturers during stated shelf life may meet the 50% to 300% range, but could fall below the minimum population specification as defined in this document. In these cases, the original population is considered to be verified if the confirmation test results are within the 50% to 300% range."

Factors that can affect the population assay results include:

1. Laboratory control
   a. Are test methods sufficiently detailed and in alignment with the BI manufacturer's methods?
   b. Are analysts properly trained and proficient with the test methods?
2. Culture media type/manufacturer
3. Materials and equipment
   a. Size/type of test tubes
   b. Diluent
   c. Frequency of ultrasonic bath (where applicable)
   d. Instruments (e.g. thermometers, timers) in current calibration
   e. Incubator capable of maintaining the correct temperature range
4. Technique for spore removal from carrier
   a. When macerating paper carriers, "vortexing" flat-bottomed tubes containing glass beads has proven effective. Blender cups can also be effective, but some types will result in reduced spore counts.
   b. For solid carriers, the use of surfactants such as USP Fluid D have proven effective. Ultrasonic energy at defined frequencies/times can also be effective.
5. Heat shock temperature/times

Resistance verification by second- or third-party labs is a challenge, and there are no acceptance criteria stated in the standards for these types of tests. An often-used value is ±20% of the labeled D-value. The chances of meeting these criteria are greater when testing SCBIs, as variables such as growth media and culturing proficiency are minimized. With spore strip BIs, these factors are uncontrolled, which can lead to greater variability. ISO 11138–1[1] states "The *D* value shall be within ±20 % of the manufacturer's stated value *when determined by the manufacturer*" (emphasis added). Users should be cautious about using the ±20 % value as their acceptance criteria for resistance verification testing.

Oxborrow et.al.[12] identified sources of variability during round-robin D-value studies. Pflug[13] further analyzed the data and provided additional insight into the variability. Factors that can affect the D-value results include:

1. Laboratory control.
   a. Are test methods sufficiently detailed and in alignment with the BI manufacturer's methods?
   b. Are analysts properly trained and proficient with the test methods?
2. Resistometer technology (make and model).
   a. Is the resistometer compliant with ISO 18472?
   b. Is the resistometer current in maintenance and calibration?
   c. Resistometer settings.
3. Culture media type/manufacturer (not a factor for SCBIs).
4. Post cycle BI handling.
   a. This is not a factor for SCBIs as aseptic transfer is not required.
   b. Quenching of LSBIs in an ice bath or not.
   c. Poor aseptic technique prior to and during BI culturing (not a factor for SCBIs).

Once the user has established confidence in the quality of the BIs, testing on incoming product can be reduced or eliminated depending on the firm's practices.

12.10 CONCLUSION

It is interesting to note that some of the very same microorganisms that have historically proven problematic for humankind are now being utilized for its benefit. Bacterial endospores are an amazing adaptation of nature and they have proven themselves to be well suited for use as lethality indicators.

Finally, from ISO 11138–7, "It is important to note that biological indicators are not intended to indicate that the products in the load being sterilized are sterile. Biological indicators are utilized to test the effectiveness of a given sterilization process and the equipment used, by assessing microbial lethality according to the concept of sterility assurance level." Simply put, BIs are one of several tools required to produce sterile product, and they are still considered the gold standard of lethality process indicators.

REFERENCES

1. ANSI/AAMI/ISO 11138-1:2017 Sterilization of Health Care Products—Biological Indicators—Part 1: General Requirements.
2. ANSI/AAMI/ISO 11140–1:2014 Sterilization of Health Care Products—Chemical Indicators—Part 1: General Requirements.
3. www.kilmerhouse.com/2011/02/fred-kilmer/
4. Donk PJ. A highly resistant thermophilic organism. *Journal of Bacteriology* 1920, 5, 373–337.
5. McCauley KJ, Gillis JR. *The Effect of Carrier Material on the Measured Resistance of Spores.* Pharmaceutical Technology, 2007.
6. ANSI/AAMI/ISO 18472:2018 Sterilization of Health Care Products—Biological and Chemical Indicators—Test Equipment.
7. ANSI/AAMI/ISO 11135:2014 Sterilization of Health Care Products—Ethylene Oxide, Validation, and Routine Control of a Sterilization Process for Medical Devices.
8. The Center for Devices and Radiological Health, FDA Guide for Validation of Biological Indicator Incubation Time, Document Control Number 98984.
9. ANSI/AAMI/ISO 11139:2006, Sterilization of Health Care Products—Vocabulary.

10. Pflug IJ. *Microbiology and Engineering of Sterilization Processes*, Chapter 1 Introduction 1.2.
11. ANSI/AAMI/ISO 11138–7:2019 Sterilization of Health Care Products—Biological Indicators—Part 7: Guidance for the Selection, Use and Interpretation of Results.
12. Oxborrow GS, Twohy CW, Demitrius C. Determining the variability of BIER vessels for EtO and steam. *Medical Device and Diagnostic Industry* 1990 May, 12.
13. Pflug IJ. *Variability in the Data Generated by Laboratories Measuring D-Values of Bacterial Spores.*

13 Steam Sterilization in Autoclaves

Phil DeSantis

CONTENTS

13.1 INTRODUCTION

The validation of steam sterilization in autoclaves constitutes perhaps the most-studied validation problem faced by the pharmaceutical industry. Indeed, it was failure to sterilize certain large-volume parenteral solutions that resulted in several patient deaths in the early 1970s, prompting the United States Food and Drug Administration (FDA) to call for the "validation" of sterilization processes. Because of this, autoclave sterilization was the first validation program undertaken by the industry. This requirement soon spread to other pharmaceutical processes. Sterilization in autoclaves remains a universal issue in nearly all facilities where sterile operations occur and continues to be of paramount concern to both the industry and the various international drug regulatory agencies.

The initial and previously time-honored definition of *validation* is roughly abbreviated to "proof that a process does what it purports to do". This definition tends to foster an emphasis on the testing of the process as required to provide that "proof". For many years, the validation of autoclave sterilization was focused on testing the sterilization cycle, with the goal of achieving repetitive successful results. Little attention was paid to the sterilization mechanism or process, the equipment, or the controls applied.

The modern Validation discipline recognizes the need for an integrated program of development, design, testing, operation, and maintenance. This program must be based on the established relationship among function, structure, appropriate tests and acceptance criteria, as well as ongoing operation. For sterilization, among all processes common to pharmaceutical manufacturing, it is easy to bypass the integrated or "life-cycle" approach and concentrate on the testing aspects only. This may be because the tests generally applied to this process are rigorous and, in themselves, provide substantial assurance of reliability. Still, the ultimate achievement of validation is dependent upon control of the process. This control, in turn, is dependent upon an understanding of the process and the equipment that facilitates it. Note the characterization of validation as a state to be achieved, not a task to be carried out. Failure to follow this approach reduces our assurance and presents the risk of unexpected failure.

The objective of this chapter is to provide a basis for understanding sterilization using moist heat and the sterilizers employed. It will focus on parts, components, and other items that need to be sterilized rather than the terminal sterilization of products in their final containers. That field of endeavor is treated in another chapter in this volume. The approach to validation presented is geared to a practical application of a large body of experience. It is meant to give the user the means to understand the principles of microbial death and the meaning of sterilization. This understanding will be extended to:

- The design and operation of reliable autoclaves
- The characteristics of the loads to be sterilized

DOI: 10.1201/9781003163138-13

- The design of effective sterilization processes or cycles
- The testing of these cycles to provide a high degree of assurance that they will be reliable
- The maintenance of the state of control necessary to ensure the quality of all sterilized materials and products

13.1.1 Mechanism of Sterilization

The microbiology and mathematical modeling of sterilization are described in detail in many references as well as another chapter of this book. It is presented in brief here for completeness.

Steam sterilization under pressure is the most effective sterilant [12]. It is the method of choice when heat and moisture damage is not a problem. The temperature range for the growth of most living organisms is –20°C to 45°C (the range normally experienced on the planet's surface). Exposure to temperatures above this range usually results in the death of the organism, except for some heat-resistant spore formers.

The mechanism responsible for the death of microorganisms is not clearly understood. To date, the most commonly employed criterion for describing microbial death remains the loss of the cell's ability to reproduce. If a sample incubated at suitable conditions in a suitable medium did not exhibit growth within a specified time, it was assumed to be sterile. Unfortunately, this test destroys the sample. Thus, this simplified view is not effective in predicting the sterility of lots or batches of medical products. Traditionally, the sterility of a batch or lot of products was certified by tests such as that described in the *United States Pharmacopeia (USP)* 29. These tests use a small sampling of the sterilizer load to determine the presence of viable (reproducing) organisms in the entire lot. Frequently, this may have been done without consideration for the mechanism of microbial death or the conditions required to facilitate that mechanism. The finished product test in *USP* 29, which requires 20 samples, can detect a contamination level of only 15% with 95% confidence. This corresponds to an approximate probability of survivors of 10^{-1} (10%) [15]. Sample size may be increased to improve confidence of detection. This can be costly and wasteful. Obviously, for this method to ensure absolute sterility, all samples would have to be tested.

An alternative approach to predicting sterilization is the definition of sterility as a probability of survival. This probability is related to knowledge of the mechanism of microbial death and the conditions causing it. The most prevalent description of sterility used today is the reduction of anticipated levels of contamination in a load to the point at which the probability of a non-sterile unit (PNSU) is $<10^{-6}$ (1 in one million) [15].

This transition to thinking of sterilization as a probability function is one that is now firmly accepted throughout the industry and among its regulators. It is closely linked to the concept of validation. Once the levels of microbial contamination and resistance to the sterilizing process are known, the probability of survival can be calculated. Validation involves the measurement of the sterilizing conditions and challenge of the sterilizing cycle to ensure that these conditions have been met.

It is generally believed that microbial death can be linked to the denaturation of critical proteins and nucleic acids within the cell, although clear proof of the theory has not been attained. This denaturization is a result of the disruption of the intramolecular hydrogen bonds that are partially responsible for the spatial orientation of the molecule. Proteins are specifically ordered chains of amino acids, linked by polypeptide bonds. Nucleic acids are polycondensations of ribose sugars joined by phosphate linkages. Each is dependent on a specific spatial orientation to perform its function. As the hydrogen bonds are broken, the structure, and thus the function, is lost. However, the denaturation may be reversible or irreversible. The functional structure of the molecule is lost in stages. If halted before a critical number of hydrogen bonds are cleaved, it is possible for the molecule to return to its original state. For example, DNA gradually changes from a helix to a random coil.

Significant research data support the theory that microbial death may be described as a first-order chemical reaction. This leads to the conclusion that death is essentially a single-molecule reaction. We are probably dealing with the denaturation of a critical molecule within each cell [12].

Bacterial spores are the forms most resistant to thermal death. The spore is the normal resting state in the life of certain groups of organisms, namely, bacilli and clostridia. During this stage, the processes of the cell are carried out at a minimal level. Spores are among the most resistant of all organisms in their ability to withstand hostile environments. Their thermal resistance has been linked to the relative absence of water in their dense central core. There is considerable disagreement on the subject. Some investigators attribute this heat resistance to the existence of the spore core as an insoluble gel or the presence of lipid material. The dry-heat resistance of spores is greatly influenced by the history of the spore relative to water as well as the water content of the spore during the heat treatment. All of the foregoing highlights the importance of moisture in thermal death. Bacterial spores are more rapidly destroyed in the presence of saturated steam than by dry heat. It is possible that the water causes the hydration of a stabilizing polymer (calcium dipicolinate) within the spore. Furthermore, water is linked directly to the denaturizing of proteins and nucleic acids by hydration [12].

13.1.2 Mathematical Modeling

The consideration of microbial death, and more specifically spore inactivation, as a monomolecular reaction with water is consistent with first-order reaction kinetics. That is, the rate of reaction is governed by the concentration of the reactant (spores). Mathematically this is expressed as:

$$\frac{dN_a}{dt} = kC_a$$

where t is time, N_a is number of spores, C_a is the spore concentration, and k is a reaction rate constant at constant temperature. In integral form:

$$\log\left(\frac{C_{a0}}{C_a}\right) = k(t - t_0)$$

$$\log\left(\frac{C_{a0}}{C_a}\right) = k(t - t_0)$$

where the superscript 0 indicates initial conditions. Also, $(t-t_0)$ is usually simplified arbitrarily by setting $(t_0 = 0)$. Thus, a semilogarithmic plot of concentration versus time will yield a straight line of slope k, as shown in Figure 13.1. The rate constant k is expressed as reciprocal minutes (min^{-1}). In turn, the negative reciprocal of the rate constant is equivalent to the number of minutes required at a given temperature to destroy 90% of the organisms present (i.e., a one log reduction). The reciprocal of the rate constant is referred to as the D-value and is expressed in minutes. D is the measure of the relative heat resistance of an organism at a constant temperature.

As shown in Figure 13.1, the simple logarithmic model yields a straight-line survivor curve. Although it does not fit all experimental data, its use is recommended because of its wide applicability and simplicity.

In general, sterilization takes place over a range of temperatures. Therefore, the sterilizing effect must be integrated over a range of temperatures and requires a temperature-dependent model.

A common measure of the temperature dependence of an ordinary chemical reaction is the Q value. Q is defined as the change in the reaction-rate constant k for a change of 10°C. This can be written as:

$$Q = \frac{k_{(T+10^0 C)}}{k_T}$$

The Q value for many first-order chemical reactions is close to 2, whereas for spore destruction with saturated steam, Q is much higher, between 10 and 18.

Another common temperature coefficient model for chemical reactions is the Arrhenius equation. This is written as:

$$k = A\exp\frac{E_A}{RT}$$

where

k = the rate constant
A = a constant
E_A = activation energy
R = universal gas constant

A plot of ln k (determined experimentally) versus 1/T will give a straight-line slope.

This model is consistent with empirical data gathered on the temperature dependence of spore D-values. A typical plot of the effect of temperature on D, the thermal resistance, is shown in Figure 13.2. The negative slope of this thermal resistance curve is called the z-value. z is defined as the temperature

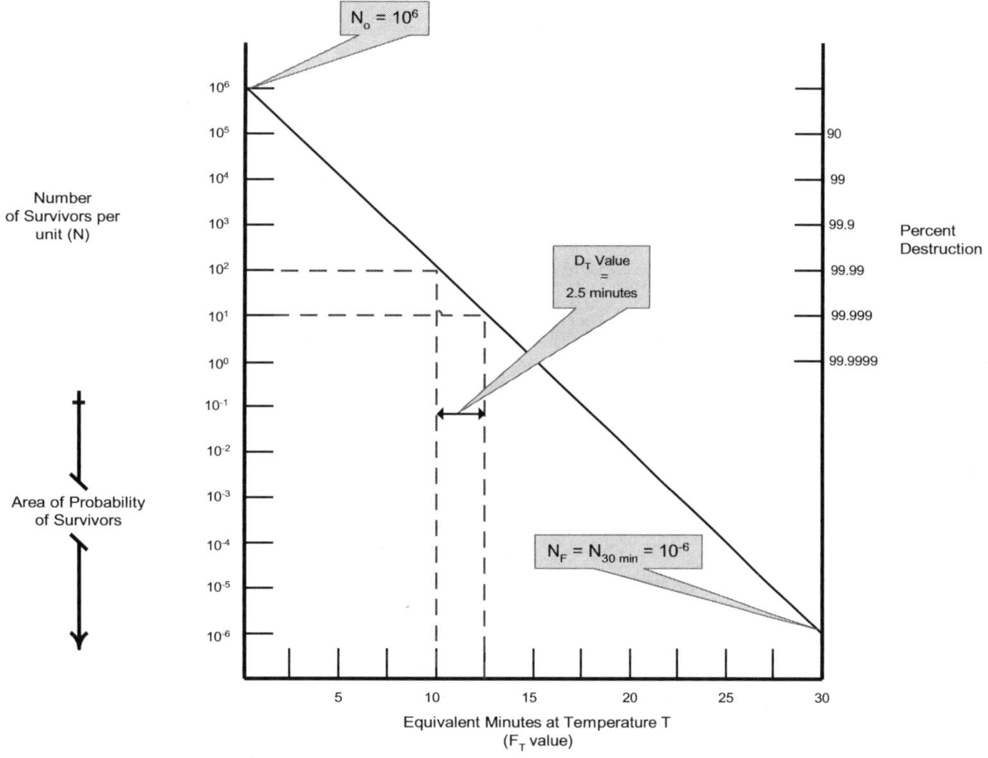

FIGURE 13.1 Microbial Death (Survivor) Curve

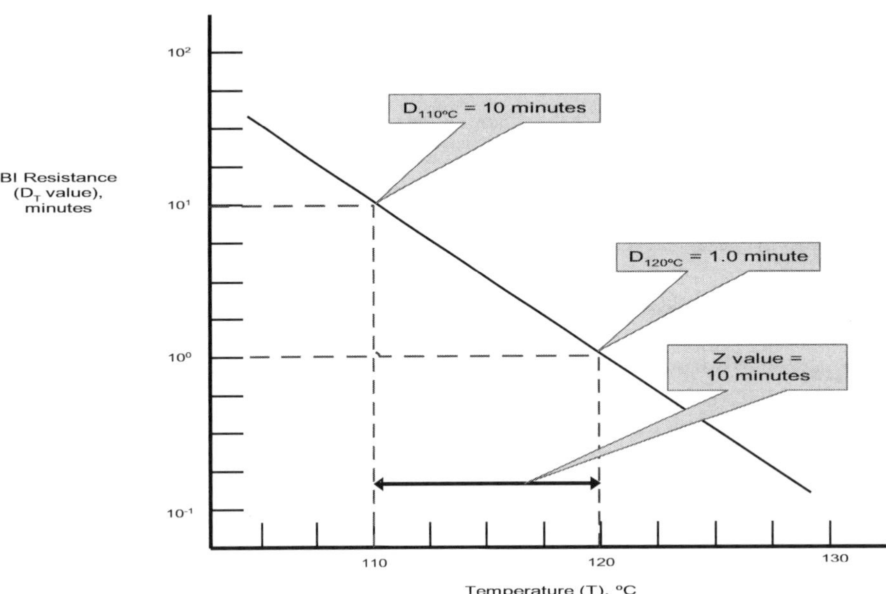

FIGURE 13.2 Thermal Resistance Curve

change required to cause a 1-log decrease in the D-value and is expressed in degrees Celsius. Remembering that D is the reciprocal of the rate constant k, z can be related to Q as follows:

$$z = \log\frac{Q}{10}$$

Similarly, it can be related to E_A in the Arrhenius model:

$$z = \frac{2.33RT^2}{E_A}$$

The use of the D- and z-values to predict microbial death over time and temperature should be approached with caution. The straight-line relations predicted by these models will hold over a limited temperature span and then only for a homogeneous culture of a single species of microorganism. Mixed populations of several levels of heat resistance will produce a curve determined by the relative populations and D-values of the organisms. The usefulness of D and z, though, is that in nature one subpopulation, by virtue of its high thermal resistance (D) and initial concentration, is usually controlling to reach sterility. This subpopulation will follow the model [13].

13.1.3 The Thermal Death Time Curve

The usefulness of the temperature-dependent model in the steam autoclave is to calculate the lethality of the cycle over a range of temperatures (including heat up and cooldown). To do this, a new variable, closely related to D, is introduced. This is called F, the thermal death time. F_{121} is defined as the time in minutes to produce a sterilization effect equivalent to that of the reference temperature of 121°C (250°F). This reference temperature is chosen as a base because it is an economical and effective one for saturated steam sterilization. It may be considered a benchmark, similar to a .300 batting average in baseball. It should not be assumed that 121°C is required to achieve effective sterilization.

Both the thermal resistance curve in Figure 13.2 (log D vs. T) and the thermal death time curve (log F vs. T) are dependent on z as follows:

$$\frac{D_T}{D_{121^0C}} = \frac{F_T}{F_{121^0C}} = 10^{(T-121)/z}$$

where T is the measured temperature. The curves are parallel, both with a slope of z.

The most commonly used value of z for the destruction of microbial spores is 10°C (18°F). This is based on experimental observations for *Geobacillus stearothermophilus* and *Clostridium botulinum*, both highly heat-resistant organisms. These organisms are chosen for divergent reasons. *C. botulinum* was the subject of the pioneering experiments by food scientists attempting to destroy this deadly cause of botulism in canned foods. *G. stearothermophilus* is a readily available and safe indicator organism for use in sterilization studies and has similar resistance.

When the assumption of z = 10°C is made, F may be written as F_0. This is the most commonly used measure of the lethality of a sterilization process spanning a range of temperatures. (Note that, out of respect for the scientists who developed this concept, this is read "F sub-zero" not "F sub-Oh").

F_0 is a summation over time of the instantaneous lethal rates at a series of temperatures. In integral form this is:

$$F_0 = \int 10^{(T-121)/10} \Delta t$$

which approximates to:

$$F_0 = \sum 10^{(T-121)/10} \Delta t$$

where Δt is the chosen time interval and T is the average temperature over that interval. The smaller the interval chosen, the more accurate the calculation will be.

Use of the important value F_0 enables one to simply measure the relative effectiveness of any steam sterilization process. This value, along with an understanding of the number and thermal resistance of the microbial population to be sterilized (bioburden), allows us to estimate the level of sterility. Although sterility is in fact an absolute condition (it is not suitable to refer to an item as being "almost" sterile), it is beyond our means to ascertain absolutely, short of total destructive testing. Therefore, it is common practice to use the F_0 equation to determine the probability of sterility, or sterility assurance level (SAL). Of more common use, the PNSU may be found. As has been previously discussed, the most widely accepted target by the industry and international regulatory agencies for sterilization in autoclaves is a PNSU of 10^{-6} (1 in 1 million) as determined from the presterilization bioburden population and resistance.

The PNSU at a reference temperature of 121°C may be calculated as follows:

$$Log\ N = Log\ N_0 - \frac{F_0}{D_{121}}$$

where:

N = the PNSU

N_0 = the initial bioburden

F_0 = the process lethality equivalent in minutes at 121°C (z = 10°C)

D_{121} = the slope of the thermal resistance curve at 121°C (minutes)

13.2 STERILIZER DESIGN FOR PARTS AND HARD GOODS

The validation of a steam sterilization cycle is dependent on the equipment chosen. The sterilizer and its support systems must be designed and constructed to deliver the effective cycles repeatedly and consistently. Qualification of the sterilizer consists of proper design, installation according to design, operational testing to ensure that design criteria and operational requirements are met, and finally, performance qualification to confirm that the product or materials and equipment are sterilized per specification.

The usefulness of saturated steam for parts/hard goods sterilization has been well documented [7, 12]. The sterilizing effect is accomplished by the heat transfer from the steam to the load and by the hydrating effect of the resultant condensate. The condensate is formed because of the return of the steam to the lower-energy liquid state. This phase change requires the transfer of the latent heat of the steam (that which was required to change it from liquid to vapor: 970 BTU/lb or 1 kcal/kg) to the surroundings, thus heating the sterilizer and its load. The heat transferred by the condensation of saturated steam is many times greater than that which would be transferred from steam above its boiling point, called superheated steam. Heat transferred by superheated steam amounts to only 1 BTU/lb-°F (1 kcal/kg-°C). Also, superheated steam is sometimes known as "dry steam," as it does not form condensate as it cools. Thus, the important hydrating effect is not present.

Sterilization with superheated steam is a dry-heat phenomenon, less efficient than a saturated steam process. Superheat may be avoided by maintaining steam in equilibrium with water at the boiler or steam generator. Also, supplementary heat sources, such as jacket heat, must be controlled so as not to heat the system above the vapor-liquid equilibrium line. Condensation to water causes a volume decrease in excess of 99%. This would result in a pressure decrease if the condensed steam were not immediately replenished, as it is in the sterilizer. It is the condensation-replenishment cycle that allows the steam to penetrate to the surfaces to be heated until they reach an effective sterilization temperature. Sterilizers and sterilization cycles are designed to ensure that saturated steam reaches all of these surfaces.

There are several characteristics common to all modern steam sterilizers in use in the pharmaceutical industry. These include:

- A pressure vessel constructed according to a recognized national or international code (e.g., American Society of Mechanical Engineers—ASME). This must withstand at least 50% in excess of the required internal steam pressures. It may be rectangular or cylindrical in cross section.
- A steam jacket and insulation: These are energy-conserving features designed primarily to heat the metal mass of the vessel and to limit heat loss from within the vessel. Some laboratory and small special-use sterilizers are unjacketed. Where jackets are employed, they should be operated at lower pressure than the chamber to avoid superheat.
- A safety door mechanism to prevent opening while the unit is under pressure (the term "autoclave" means self-closing.) The locking device may be actuated directly by internal pressure or indirectly through an automatic switch. The door itself may be of the swing-out or sliding type.
- A thermostatic steam trap to efficiently remove air and/or condensate from the chamber: This is open when cool and closed when in contact with steam. As air or condensate collects, the trap opens owing to the slight temperature reduction and the air/condensate is discharged. There is also a trap to remove condensate from the steam jacket.
- Process control system (typically a Programmable Logic Controller or PLC for controlling and monitoring the process).
- A Process Data Recorder or Data Collection system.
- A microbial retentive vent filter.
- A chamber pressure indicator
- Pressure relief valves for both the chamber and jacket.
- A vacuum pump or eductor to remove air from the chamber and load

13.3 STERILIZER CONTROL SYSTEMS

A key to effective sterilizer operation lies in the automated process control system. By eliminating the dependence on operator intervention and data recording, automatic temperature and sequence control provides assurance that the "validated" sterilization cycle is consistently and repeatedly delivered.

A typical control system for a new sterilizer includes the following hardware components:

- PLC (Programmable Logic Controller)
- Operator Interface Panel(s)
- Data Recorder/Data Collection System
- Process Variable Sensors
- I/O (Input/Output) Devices

The PLC is the most commonly used primary component of the automated process control system, as it typically completes the sequential control of the process, provides PID control of all proportional valves, controls all devices, receives operator input via the Operator Interface Panels, and provides process information (such as process variable information and alarms) to the operator via displays and/or the Operator Interface Panel. The PLC also typically contains specific recipe information for the various cycles to be utilized. In some cases, the PLC can be used for data collection, but it is much more common to use a separate Data Recorder/Data Collection system.

The Operator Interface Panel can be as simple as switches and displays to as complex as a stand-alone PC running a SCADA HMI software package. These devices are typically used to select the recipe, start the cycle, and display process information during the cycle. The higher-level PC-based SCADA- type Operator Interface Panels can provide detailed cycle reports and trending information.

The Data Recorder/Data Collection system can be as simple as a strip chart recorder to a full-blown MES-type data collection system. In many cases, the PLC can also provide batch data logging functionality. The minimum variables to record for steam sterilization processes are typically time, temperature, and pressure.

Typical sensors include temperature measurement devices (resistance thermal devices [RTDs] or thermocouples) and pressure measurement devices.

The temperature sensor used to control the process temperature shall not be used to provide the batch record process data. A secondary temperature sensor for batch reporting provides a high degree of assurance that the cycle actually ran within its defined limits. Heavy wall thermowells should not be used, as this will affect the time response of the measurement. Thin walled thermowells or temperature elements with stainless-steel sheaths should be used for temperature measurement.

The pressure sensor should be equipped with a sanitary-type diaphragm and connect to the sterilizer using a sanitary-type connection. A sanitary diaphragm introduces additional errors to the pressure measurement because of the stiffness of the diaphragm. This stiffness is related to the size of the diaphragm. This effect is negligible for diaphragms above 3" diameter. This should be considered when sizing the connection to the sterilizer.

For I/O devices, there are both analog types and discrete types. The analog inputs are typically from process sensors and the analog outputs are typically for control of proportional valves. The discrete inputs are typically from switch type (operator and process) devices and the discrete outputs are typically for activating hardware such as valves, pumps, lights, etc. I/O devices using buss connections (Profibus, Controlnet, Hart, Fieldbus, Lonworks, Ethernet, etc.) are not considered a true I/O point. These devices constitute more of a network and have a host of communications capabilities, diagnostics, and maintenance functions available. Typically, these devices provide much more than just the measured variable. Many sensors that have these buss-type connections are available.

The design and development of the software should follow the principals of GAMP 5 for automated process control systems. GAMP details a software life cycle from conception through development to maintenance of the software in a validated state.

13.4 STERILIZATION CYCLES FOR PARTS AND HARD GOODS

Removal of air is a common problem for parts/hard goods sterilization. Air entrained within the load depresses the temperature and prevents the penetration of steam to all the required surfaces. The efficiency of heat transfer (heat transfer coefficient) from the steam to the load is thereby reduced. The first steam sterilizers relied on gravity displacement of the air by steam. Air removal cycles were originally designed for loads of porous materials like hospital surgical packs. All contemporary sterilization processes, or cycles, are designed to remove air mechanically. It has been shown that vacuum as low as 15–20 mmHg (less than 0.4 psia), applied for 8–10 minutes, is required to remove air from some porous loads [12]. This level of vacuum is difficult to achieve and even if applied as described may be inefficient in removing entrapped air from more complex parts and hard good loads. Because of the difficulty in obtaining the high-vacuum conditions needed for efficient air removal, manufacturers developed pulsing systems. These employ a series of alternating steam pulses followed by vacuum excursions. The maximum and minimum pressures are variable. In general, the pulsing system removes air effectively without achieving the level of vacuum required in a simple pre-vacuum cycle. The steam provided serves to rapidly fill the voids attained by the vacuum pulses, forcing out residual air.

The pulsing cycle is among the most prevalent in use throughout the pharmaceutical industry. Many of the loads requiring sterilization contain items for which air removal is difficult. These include coiled hoses, filter housings, and densely packed containers of stoppers. It is important to recognize this difficulty and to specify cycle parameters effective in overcoming it. Figure 13.3 is an example of a multiple, or pulsed, pre-vacuum cycle.

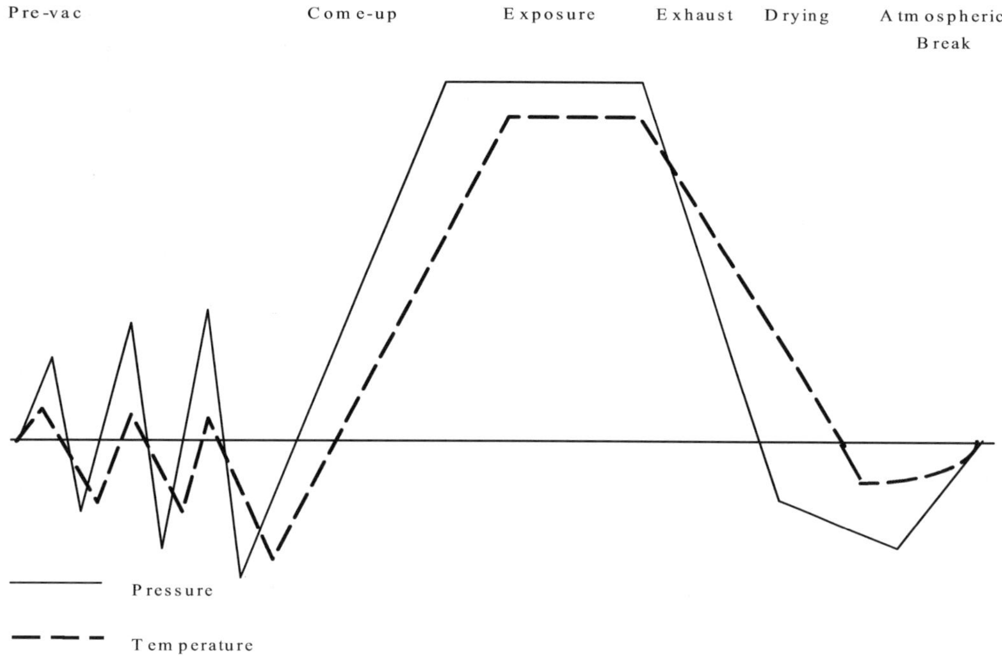

FIGURE 13.3 Typical Pulsed Vacuum Cycle

13.5 MICROBIOLOGICAL STERILITY ASSURANCE DEVELOPMENT

The designation of a sterilization cycle and development of that cycle is dependent upon product characteristics, specifically heat stability and bioburden. Cycles that have been classified previously based upon their "mechanical" modes may be further classified according to stability and microbiological characteristics as follows:

- Overkill cycle approach
- Biological Indicator/Bioburden cycle approach
- Bioburden approach

Bioburden and Biological Indicator/Bioburden cycles are predominantly utilized in the sterilization of liquid-filled containers, whether for final products (terminal sterilization) or for the sterilization of laboratory/production media and certain in-process liquids. The methods and practices associated with their sterilization is defined in the following chapter. The remainder of this chapter will only discuss the validation of items using the overkill method. Overkill sterilization can be utilized for certain stable liquids, in which case the methods utilized are a hybrid of that presented in this chapter and the succeeding one.

13.5.1 Overkill Cycles

When sterilizing heat-stable materials, an overkill approach may be adopted. Loads sterilizable by overkill include filters, container-closures, hoses, filling parts and other hard goods, and soft goods, such as gowns. Remember that the accepted criterion for sterility is the probability of survival of no greater than 10^{-6}. The objective of the overkill cycle is to assure that level of assurance, regardless of the number and the heat resistance of the organisms in the load. Extremely high F_0 values are generally used. Because items in the load are presumed to be heat stable, thermal degradation is of little concern and only the minimum F_0 in the load is considered. This may be chosen to provide at least a 12-log reduction of microorganisms with a D-value of 1 minute at 121°C [8]. A quick calculation will arrive at $F_0 = 12$ as a minimum overkill cycle.

Most microorganisms found in production environments have D-values ranging from 0.5 minutes or less. Thus, a population of up to 10^6 (one million) spores per unit of even the most heat-resistant strains of these environmental or material microbial isolates will be reduced to well below a PNSU of 10^{-6} by the minimum overkill cycle. Nonresistant organisms and the much smaller populations usually found in the clean room environments employed in preparing loads will be reduced to a much greater extent.

Despite the high degree of sterility assurance provided by the minimum overkill cycle, most validation teams will often choose to increase the cycle to ensure the deactivation of the bioindicators used to confirm sterilization during validation studies. These bioindicators are most often strips or suspensions containing from 10^4 to 10^7 spores of a highly heat-resistant organism (usually *G. stearothermophilus*). D-values for these organisms have been observed to range from 1.5 to as high as 4 minutes. Therefore, overkill cycles as high as $F_0 = 48$ (4 minutes × 12-log reduction) will be required to provide a PNSU of 10^{-6} for the bioindicator with 10^6 spores. Note that it is not required that the extent of overkill attain these F_0 levels. Nevertheless, because the failure to inactivate even one bioindicator may cast doubt over the validation study, it is common practice to adjust the cycle to effect total kill of whatever biological indicator has been selected. Unfortunately, the misguided use of such an approach may have unforeseen negative effects. An example of this might be the poor machinability of rubber stoppers that have been overprocessed in this manner. Another example would be the possible reduction in useable life for items that are subject to multiple sterilization cycles, especially those with elastomeric components.

Bioburden studies are not required, nor are they usually carried out when an overkill cycle is planned. It is important, however, that items to be sterilized be cleaned and handled in a manner that serves to keep the bioburden under control. Therefore, parts preparation areas are usually controlled environments (ISO 8 in most facilities). Even with this in mind, an occasional check on the bioburden would provide valuable confirmation.

With or without bioburden data, an assumption of the D-value (either $D = 1$ minute or the bioindicator D-value stated by the manufacturer) and a minimum target of a 12-log reduction are adopted. It would be extremely rare to find a spore-bearing organism with a D-value approaching 1 minute in a pharmaceutical clean room. Bioburden populations are also nearly certain to be considerably $<10^6$. Therefore, the assumptions adopted provide for a significant margin of safety and are used to calculate the F_0 to be delivered to the cold spot in the load. Cycle parameters of time and temperature, as well as the location of the cold spot, are determined during the validation studies.

13.6 STERILIZATION PROCESS PERFORMANCE QUALIFICATION

When viewed as a life-cycle activity, the most important key to validation is to understand the sterilization process as related to both the physical (thermal, mechanical) and microbiological parameters. This is most effectively done during the developmental program on both the R and D and production scales. These studies form an important foundation for the final testing and documentation that has been commonly referred to as "*validation*". Sometimes this is called *process confirmation* or *process validation* (PV). The author adheres strongly to the life-cycle approach; thus, references to *validation* in this section shall not be limited to those activities that are commonly included in this final confirmatory phase. For this phase, it is preferable to use the term *process performance qualification (PPQ)* as suggested by the US FDA Guidance for Industry *Process Validation: General Principles and Practices.* Even though this Guidance does not address sterilization, this terminology may become the more broadly accepted jargon in the future.

With this in mind, it is futile (or at least very risky) to initiate a PPQ study for a sterilization process that has not been shown to work. Whether this evidence is determined on an R and D scale, on a production scale, or both, is dependent upon the load and the sterilizer used. For example, hard goods and parts cycles in a pulsing vacuum-type sterilizer can be predicted fairly successfully from any scale of preliminary study. In fact, because sophisticated temperature measurement and monitoring systems can actually calculate lethality instantaneously, some of these cycles may forego developmental studies. (The author does not recommend this.)

In addition to an understanding of the process or cycle, the sterilizer and associated equipment must be qualified; that is, determined to be suitable for its intended purpose and able to meet all critical requirements, as defined by the sterilization process.

13.6.1 Measuring Temperature

In all qualification studies, the ability to accurately measure temperature is critical. Several items will be required to measure and record temperature effectively. The most versatile temperature-sensing devices for validation are thermocouples. These are constructed from wires of two dissimilar metals. They can be encased in a flexible sheath, Teflon being widely used. Type T (copper-constantan) thermocouples are most applicable in steam sterilizer qualification work. Their working temperature range is wide, and they are resistant to corrosion in moist environments. A high-grade thermocouple wire should be chosen. Standard grades have an inherent error as high as 1°C. This is very significant when calculating the experimental lethality. Premium grades of wire, accurate to as close as 0.1°C at 121°C, are recommended. These must then be calibrated against a temperature standard traceable to the National Institute of Standards and Technology (NIST; formerly, National Bureau of Standards), DIN, British Standard, or an acceptable national standard.

The temperature standard may be a mercury-in-glass thermometer or platinum RTD. The RTD is recommended because of its greater durability and accuracy. In fact, this is the same device most often specified to measure the chamber temperature used to control the sterilization cycle. The thermocouples to be calibrated are placed in a highly stable temperature source (controlled ice point device, hot reference device, or controlled temperature bath) along with the reference standard. The differences in the readings between the thermocouple and the reference device are recorded. The acceptable error should be no greater than the sum of the thermocouple wire accuracy (e.g., +0.1°C to −0.3°C) and the degree of traceability of the reference instrument (e.g., +0.1°C to −0.3°C). Thermocouples that do not meet this criterion should be replaced.

The calibration of thermocouples should be carried out at two temperatures. One of these is an ice-point reference at 0.0°C. The other should be a hot point slightly higher than the expected sterilization temperature (130°C is commonly chosen). If the thermocouple meets the accuracy criterion specified above, it is then permissible to apply a correction factor to bring the thermocouple reading to the same as that of the reference RTD. Data loggers and recording devices are often programmed to perform this correction automatically. Once correction factors are applied at both calibration temperatures, then the response of the thermocouple over the temperature range can be linearized. The corrected temperature measurements are used in lethality calculations. Calibration should be repeated after a series of validation runs. Experience has shown that weekly recalibration is perfectly adequate.

Thermocouple access should be considered in the design of the sterilizer. Most sterilizer manufacturers routinely include one or more unused ports in their pressure vessels. These can be tailored to the specific needs of the validation team. All penetrations must be made before the vessel's code compliance is stamped. To make modifications at a later date is troublesome. These must be made by a board-certified welder and are subject to reinspection and test by a code inspector. Special gland adapters are joined to the access ports that allow the thermocouples to pass into the chamber without developing pressure leaks. The adapter can be a pressure gland made of two mated flanges separated by two flexible gaskets. The thermocouples pass between the gaskets and the flanges and are bolted together tightly to prevent leaks. Another method is to use special-purpose fittings made for the express purpose of thermocouple access (e.g., Conax). This is a specially drilled rubber gland within a compression-type housing.

Qualification runs usually involve numerous temperature measurements. These can be recorded in a number of ways. Because of the frequency and number of recordings, a data logger is often employed. This is a digital output, multichannel device capable of frequent printouts or data transmission of many temperature measurements. These can be very sophisticated and can be preprogrammed to make thermocouple calibration corrections, store data, and even calculate F_0 and print or electronically record cumulative values. The best data loggers have large capacity (32–48 channels), precision to 0.1°C or better, fast scan rate of all thermocouples (one cycle per second or better), and the ability to interface with a computer, either by way of stored data or in real time.

13.6.2 Commissioning and Equipment Qualification

Prior to the initiation of PPQ studies, it is important that the sterilizer be suitably qualified to perform its function. This qualification ensures that the system meets critical requirements of the sterilization process as defined by the sterilization scientists (i.e., the "users"). These requirements should be clearly defined in terms of the sterilization process to be executed. The author recommends a specific User Requirements document for this purpose. These requirements always include the ability to achieve and maintain sterilizing conditions throughout the chamber. Typical critical requirements that are considered to affect the sterilization process (e.g., "quality-critical" requirements) are:

- Accurate temperature and pressure measurement.
- Chamber integrity (leak rate).
- Temperature distribution and uniformity throughout the chamber (Uniformity requirement may vary depending upon the cycle defined. For bioburden-based terminal sterilization, this is often ±0.5°C. For overkill cycles, this limit may be more flexible, although this is seldom greater than ±1.0°C).
- Precision of temperature control (usually the ability to maintain a control point temperature of ±0.5°C around the set point).
- Precise control of the sequence of operations and timing.
- Alarms to indicate out-of-specification conditions.

User Requirements should not be engineering specifications. The technical approach to achieving requirements is best left to engineers and sterilizer manufacturers. Therefore, design issues, such as the choice of instrumentation, line sizing, chamber configuration, etc. are not subject to formal equipment qualification (EQ) but to commissioning by the vendors and engineers who have developed the specifications. A commissioning study is a rigorous series of inspections and tests to ensure that the specifications are met. There is, however, an inherent flexibility within commissioning that allows for adjustments, corrections, and even modifications to bring the system to an acceptable state. These, of course, are thoroughly documented and reported. However, approval beyond the engineering and user groups is not required.

Once a sterilizer has been commissioned, it will be ready to be formally qualified. There is a school of thought emerging in the industry that qualification may be reduced to a paper exercise that merely confirms that the commissioning study has clearly proven that all quality-critical requirements have been satisfied. This probably achieves the level of assurance necessary to proceed with PPQ studies. However, in the current regulatory environment, this approach is not universally accepted, and the sterilizer qualification protocol may need to address some level of confirmatory inspection and testing. Because of the high degree of risk to patients presented by sterilization, the author does not recommend replacing equipment qualification of a sterilizer in its entirety by leveraging commissioning.

As a minimum, a sterilizer qualification protocol should confirm both the critical installation (IQ) and operational (OQ) requirements, as defined by the user. It is not, however, necessary to perform a detailed analysis against engineering details, as this has been completed in commissioning. Using the previous list of typical critical requirements for reference, the qualification of a sterilizer should include the following (note that installation and operational requirements may be covered in a common protocol or in separate protocols, as the investigators see fit):

- Calibration of temperature and pressure sensors (traceable to an accepted national or international standard)
- Air removal (where required; usually measured by vacuum level achieved vs. defined requirement)
- Demonstration of the sequence of operations, including cycle timing

- Confirmation of alarms and interlocks
- Precision of temperature control
- Temperature distribution and uniformity

13.6.3 Empty Chamber Temperature Distribution

This study has traditionally been considered a critical aspect of sterilizer qualification. The intent of this study is to demonstrate the temperature uniformity and stability of the sterilizing medium throughout the sterilizer. Temperature distribution studies should initially be conducted on the empty chamber. Temperature uniformity may be influenced by the type, size, design, and installation of the sterilizer. A satisfactory empty chamber temperature uniformity should be established by the User Requirements. A narrow range is required and is generally acceptable if the variation is less than ±1°C (±2°F) the mean chamber temperature.

With modern sterilizers, temperature deviations greater than +2.5°C (+4.5°F) may indicate equipment malfunction [8]. Stratified or entrapped air may also cause significant temperature variations within the sterilizer chamber. The investigator is cautioned to determine that the sterilizer has been specified to maintain the temperature uniformity profile required by the most-demanding sterilization cycle. It is of no value to assign arbitrary acceptance criteria to a sterilizer that has not been specified or built to meet them. It is important to note that the drain of a sterilizer is expected to be its coldest point and is outside the sterilization zone. For this reason, it is recommended that the drain not be included in calculations of the mean chamber temperature and not be subject to chamber distribution requirements. It should also be noted that uniformity across the chamber may be expected only at steady state. Uniformity measurements are meaningful only after the control point temperature has stabilized at the desired set point.

Initially, a temperature distribution profile should be established from studies conducted on the empty chamber. Confidence may be gained through repetition; therefore, empty-chamber studies are often conducted in triplicate to obtain satisfactory assurance of consistent results. After more than 40 years of sterilizer advancement, uniformity has become a virtual certainty. When qualifying sterilizers, the design of which a firm has previously qualified, it is usually permissible to conduct a single empty chamber distribution study. For proven sterilizer design from manufacturers known to the user, a single empty chamber run may also be suitable for qualification.

Subsequent to the empty-chamber studies, it has been a common practice to conduct maximum-load temperature distribution studies, ostensibly to determine if the load configuration influences the temperature distribution profile obtained from the empty-chamber studies. This is normally done during the PPQ trials. However, the value of this practice is doubtful and has been challenged by sterilization experts [5]. Consequently, mapping the chamber with a load is often ignored.

The thermocouples used in the empty chamber studies are distributed geometrically in representative horizontal and vertical planes throughout the sterilizer. The geometric center and comers of the sterilizer should be represented. An additional thermocouple should be placed in the exhaust drain adjacent to the sensor that controls vessel temperature, if possible. The number of thermocouples used in the study will be dependent on the sterilizer size. In a production-size sterilizer, 15–20 thermocouples should be adequate. The autoclave cart can be a useful support for the thermocouples as well as providing easily identifiable and reproducible positioning

13.6.4 Container and Object Mapping

For overkill cycles, certain non-product load items may pose concerns relative to consistent and effective heating throughout the item. Filter housings, hoses, containers filled with stoppers, and small filling assemblies can present both air removal and steam penetration problems. Developmental studies should be performed on these and similar objects with several thermocouples to determine the slowest-to-heat zone within an object. Once this has been determined, sensor placement within objects may be specified to probe the slowest-to-heat zone (i.e., the "cold spot"). Note that this type of study need not be repeated for every cycle, or even for every specific object, as long as classes of objects (e.g., hoses, filters, etc.) have been characterized adequately to determine the appropriate probe location. In performing this evaluation, attention should be paid to the proper orientation of the load items as this can have substantial impact on air and condensate removal.

13.6.5 Heat-Penetration Studies

Heat-penetration studies comprise the core of sterilization PPQ. The intent of these studies is to confirm that the slowest-to-heat objects within a specified load have achieved the requisite lethality. Cold spots originate because of the varied rate of heat transfer throughout the load.[9][11] Therefore, it is imperative that developmental heat-penetration studies be conducted to determine slow-to-heat items within a loading pattern and assure that those items are probed during confirmation studies to ensure that they are consistently exposed to sufficient heat lethality [8]. Penetration thermocouples are positioned at points within the process equipment suspected to be the most difficult for steam heat penetration. For homogeneous loads (e.g., a load of stoppers or packs of gowns), thermocouple placement in the load should cover the entire profile of the autoclave, including the geometric center, corners, and near the top and bottom of the chamber. Temperature data are obtained from representative maximum loads to establish temperature profiles defining load cold spots within the homogenous load pattern. Equipment load configurations may be designed to allow reasonable flexibility for the operating department by permitting the use of partial loads. For this, partial loads would be defined as a portion of the established maximum validated load. Thus, minimum load studies are not required provided that air removal is performed near identically despite load size variations. They are often run, however, to provide additional assurance.

Another question on heat-stable loads is the geometric configuration of heterogeneous or mixed loads. In these cases, studies have shown that the cold spot in the load is related to a specific object (the most difficult items for steam to penetrate), rather than to a specific location within the autoclave. Therefore, when performing heat-penetration studies in mixed loads, it is important to probe each type of component in the load. Triplicate studies will determine the hardest component to penetrate. For added assurance, it is a good practice to reconfigure the load between runs. This will add evidence to support the "hardest to penetrate" theory. Having determined this, future loads will not be subject to rigid configuration as long as the cycle chosen provides adequate lethality to sterilize the most difficult parts.

The most difficult to sterilize parts ("cold" items) established for a specified load or configuration will eventually be used to determine the exposure time in subsequent routine production runs. Ideally, the temperature sensor(s) that control sterilization-cycle exposure time at the process temperature may be positioned within the load adjacent to or within the previously detected cold item. Unfortunately, this is rarely practical or even possible. The most-used approach is to correlate the control probe temperature with the cold item. This correlation is established through the validation study, relating the coldest point and the control point. Then the cycle can be adjusted to provide adequate time for the coldest item to reach the desired value.

Lethal rates can be determined from the temperature data obtained from the heat penetration studies. The temperature data are converted by the following formula:

$$L = \log^{-1}\frac{T_0 - T_b}{z} = 10^{\frac{T_0 - T_b}{z}}$$

$$L = \log^{-1}\frac{T_0 - T_b}{z} = 10^{\frac{T_0 - T_b}{z}}$$

where:

L = instantaneous lethality

T_0 = temperature within the object or container

T_B = base process temperature (121°C)

z = temperature required to change the D-value by a factor of 10, where z = 10°C

The total lethality of the cycle is then determined by integrating over time as follows:

$$F_0 = \int 10^{(T-121)/10}\, dt$$

which approximates to:

$$F_0 = \sum 10^{(T-121)/10}\, \Delta t$$

where Δt is the chosen time interval and T is the average temperature over that interval. The smaller the interval chosen, the more accurate the calculation, so typically intervals of less than 1 minute are chosen. With the use this formula, it is possible to measure the theoretical lethality of the entire cycle despite the fact that that the temperature may deviate from the traditional reference sterilization level. It should be noted that the lethality calculation for steam sterilization by direct contact (i.e., parts, hard goods) should be terminated at the end of the sterilizer dwell cycle in order to avoid the presumptive accumulation of lethality during cooldown when saturated steam conditions may not prevail.

13.6.6 Microbiological Challenge Studies

Because heat penetration studies can only confirm the temperature and not the other conditions required for effective moist-heat sterilization, microbiological challenges are employed to provide the required necessary assurance that adequate lethality has been delivered to all parts of the load. These are most often conducted in parallel with heat penetration studies and are essential to a rigorous sterilization validation, despite some authorities downplaying the microbiological challenge in favor of physical data [6]. Calibrated biological indicators used for this purpose function as bioburden models, providing data that can be employed to calculate F_0 or to substantiate and supplement the physical temperature measurements obtained from thermocouples [8]. The microorganisms most frequently used to challenge moist heat sterilization cycles are *Geobacillus stearothermophilus*. These spore-forming bacteria are selected because of their relatively high heat resistance. For the bioburden cycles, in addition to the selection of an appropriate organism for use as a biological indicator, the concentration and resistance of the indigenous microbial population is established.

Modest reductions in sterilizing conditions used for the microbial challenge studies are a common practice in parts sterilization to afford an additional safety measure. Reduction of sterilizing dwell time by 1–3 minutes and /or reduction in sterilizer set point temperature of 1°C–2°C have been successfully utilized for years. The half-cycle approach originally developed for ethylene oxide (ETO) sterilization validation is an extreme example of this practice, and although effective, it extends processing times unnecessarily and is not recommended.[2]

When inoculating solid materials, the spores can be introduced onto the surface of the item. Subsequent to inoculation, the spore suspension is allowed to dry on the surface. Recovery counts should be conducted on selected inoculated components to verify the delivered concentration of spores. Commercially available spore strips may also be used when the confirmation loads are composed of devices and solid materials.

Microbiological challenge studies are typically conducted concurrently with the heat penetration studies. Similarly, when spore strips are used, they should be placed adjacent to a thermocouple probe. When inoculating items with a spore suspension, the resistance of the spore can be altered, and D-value determination is recommended. To expedite recovery and eliminate possible confusion, any directly inoculated items should be identified by markings or other suitable means.

After the sterilization cycle is complete, the inoculated items or spore strips are recovered and subjected to microbiological test procedures. Strips are immersed in a suitable growth medium (soybean casein digest medium is typical) and incubated for up to 7 days. The incubation temperature for *G. stearothermophilus* is 50°C–55°C. For overkill cycles, it is expected that all spore strips will be negative (not exhibit growth). To provide further assurance, both positive (unsterilized strips) and negative (growth medium with no spores) controls should be incubated along with the challenge samples.

13.6.7 Sterilizer Filter Evaluation

Microbially retentive filters are employed on most parts sterilizers to ensure that loads are not contaminated by the air used to vent the chamber as it cools or dries. Product loads are protected from such contamination by their primary containers (vials, bags) and many non-product loads are protected by wraps to provide a microbial barrier. Nevertheless, because of the possibility of pressure differentials between the chamber and the sterilized article during cooldown or vacuum-drying, filters are valuable.

For filters, two issues are of concern: sterility and integrity. Filters are typically sterilized during the load sterilization cycle. Filters should be probed with thermocouples upstream and downstream of the membrane. The most rigorous practice is to apply a microbial challenge in the form of a spore suspension to the filter itself. In practice, however, most firms consider that spore strips inserted upstream and downstream of the filter media to be an adequate challenge. Because filters are resistant to heat, this should be similar to the challenge applied in the load for overkill cycles.

The integrity of the filter must also be evaluated according to recommendations by the filter manufacturer. To ensure that filters will remain functional under all expected conditions, the integrity tests should be done following the maximum cycle time and temperature allowable under standard operating procedures. The filter life after repeated sterilization must also be considered, and their periodic replacement interval defined based upon the filter manufacturer's recommendations. Triplicate studies are recommended.

13.6.8 The Validation Report

Record keeping is a prime requirement of current GMPs. The records required for a validated steam sterilization cycle follow. They are usually stored in a secure central file, but they must be readily accessible. It is wise to assign the task of organization and retrieval of records to a single group. These records are as follows:

- Qualification reference documents (specifications, drawings, and calibration records)
- Operational qualification protocol and record
- Approved process confirmation (validation) protocol
- Raw calibration and validation data
- Approved validation report

The validation report is the guideline to maintenance of a validated sterilization process. It describes the cycle and the operating conditions that have been proved to give adequate assurance of sterility. It explains in detail how the manufacturing group can obtain results consistent with the validation study.

Several formats and degrees of complexity are used in report writing. However, all reports should contain some common elements, as follows:

- Identification of the task report by number.
- Reference to the protocol under which it was carried out.
- A brief summary of the range of operational conditions experienced and how they were controlled.
- A procedure for maintaining control within the approved range. This may be in the form of a standard operating procedure.
- A summary and analysis of the experimental results. This will include the range of lethality and degree of sterility assurance.
- A brief description of any deviation from expected results.

The range of lethality is calculated directly from the temperature data. It is important that a range be reported in the case of heat-labile products. The upper range of temperature exposure is critical to product stability. A sterilization cycle is also a product-processing step. Its effect on the product, as well as on the microbial population, must be considered. The description of such effect need not be included in the validation report itself. It should be the subject of adjunct analytical or stability studies. A discussion of the importance of this consideration is included later.

Cycle development reports are not usually a part of the validation report, although these are important to the life-cycle validation concept preferred by the authors. Some reference to how the cycle was chosen may be included in the validation report. This can be the title of the cycle development report or a brief summary of the results of that report. This should include the type of mechanical cycle recommended (high pre-vacuum, air-steam mixture, or other), heat resistance and bioburden data or assumptions thereof, and level of lethality required.

Bioindicator data is the ultimate proof that the sterilization cycle has been successful. As such, it should be highlighted in the validation report. The microbiology section of the report should include the methods used, a summary of the results, and conclusions. On completion, the final report is circulated for approval. This is generally by the same people who approved the protocol.

13.6.9 Maintenance of Validation

The last, and often overlooked, step in validating any process is the program to ensure that conditions established in cycle development and confirmed in process confirmation or validation studies are controlled and maintained. If this is done, and no major changes in equipment or process are made, periodic repetition of selected validation studies is not really required.

What is required is a periodic review of the system for adherence to the validation criteria. This may be very simple if a good program of *validation maintenance* is established. This term is used rather than "revalidation" to emphasize the continuity of the program. A validation maintenance review report may be issued to commit to record the attention being paid to this critical aspect of validation. Some key points of a good *validation maintenance* program are as follows:

- A routine calibration program for all instruments critical to the operation of the sterilizer and its support systems.
- A preventative maintenance program for other system components. This should include periodic operational rechecks and comparison to the Operational Qualification Record.
- Routine monitoring of the bioburden and (optionally) periodic bioindicator challenges.
- Well-maintained and accessible operating records and equipment logs.
- Process and equipment change-control procedures. These subject proposed changes to prior review to establish whether additional validation experiments are required.

Because of the critical nature of sterilization and to satisfy regulatory expectations, it is recommended that studies be performed on an annual basis to supplement the *validation maintenance* program. These should entail representative loaded chamber heat penetration and microbiological challenges. Many firms choose to perform only one such study on a selected "worst case" load, often that with the lowest delivered F_0.

The basis of continued validation maintenance is communication among the various operating groups (Manufacturing, Quality Assurance, Validation, or other). Sterilization processes were the first for which validation was emphasized. They continue to be the most heavily reviewed. It is important that the state of control of these processes be strongly maintained and the subject of concern to all these groups.

13.7 CONCLUSION

The proposed approach to validation of steam sterilization in autoclaves follows the basic life-cycle concepts applicable to all validation programs. Understand the function sterilization process, develop and understand the cycles to carry out the process, and define a suitable test or series of tests to confirm that the function of the process is suitably ensured by the structure provided.

Sterilization of product and of components and parts that come in direct contact with sterilized product is the most critical of pharmaceutical processes. Consequently, this process requires a most rigorous and detailed approach to validation. An understanding of the process requires a basic understanding of microbial death, the parameters that facilitate that death, the accepted definition of sterility, and the relation between that definition and sterilization parameters. Autoclaves and support systems need to be designed, installed, and qualified in a manner that ensures their continued reliability. Lastly, the test program must be complete and definitive.

Fortunately, steam sterilization in autoclaves is very effective. Failure of autoclave cycles in modem pharmaceutical manufacturing operations is rare. Nevertheless, the consequence of failure is so great that it is easy to justify the effort required to validate this critical operation.

BIBLIOGRAPHY

1. Agalloco J, Akers J, Madsen R. Revisiting the moist heat sterilization myths. *PDA Journal of Pharmaceutical Science and Technology* 2009, 63(2).
2. Agalloco J. Too much by half: Misapplication of the half-cycle approach to sterilization. *Pharmaceutical Technology*, November 2016.
3. Agalloco J. Maximum and minimum loads in steam sterilization. *Pharmaceutical Manufacturing*, February 21, 2018.
4. Agalloco J. Empty chamber studies (aka much ado about nothing). *Pharmaceutical Manufacturing*, July 19, 2018.
5. Agalloco J. Loaded chamber temperature distribution studies. *Pharmaceutical Manufacturing*, October 7, 2018.
6. Agalloco J. The bugs don't lie. *PDA Journal of Pharmaceutical Science and Technology* 2019, 73(6).
7. Ball CO, Olson F. *Sterilization in Food Technology*. McGraw-Hill, 1957.
8. Hougan O, Watson KM, Ragatz RA. *Chemical Process Principles, Part Two—Thermodynamics*. John Wiley and Sons, 1968.
9. Hughes K, Pavell A. A risk-based approach to variable load configuration in steam sterilization. *PDA Journal of Pharmaceutical Science and Technology* 2010, 62(2).
10. International Standards Organization. ISO standard 17665-1 sterilization of health care products: Moist heat, Part 1: Requirements for the development, validation and routine control of a sterilization process for medical devices, 2006.
11. Parenteral Drug Association. *Technical Monograph No. 1, Validation of Steam Sterilization Processes*, 1978.
12. Perkins JJ. *Principles and Methods of Sterilization in Health Sciences*. Charles C. Thomas, 1973.
13. Pflug IJ, ed. *Microbiology and Engineering of Sterilization Processes*. 3rd ed. Environmental Sterilization Services, 1979.

14. Pflug IJ. *Syllabus for an Introductory Course in the Microbiology and Engineering of Sterilization Processes*. 4th ed. Environmental Sterilization Services, 1980.
15. Phillips GB, Miller SW, ed. *Industrial Sterilization, International Symposium, Amsterdam*. Duke University Press, 1973.
16. Stumbo CR. *Thermobacteriology in Food Processing*. 2nd ed. Academic Press, 1973.
17. United States Food and Drug Administration. *Current Good Manufacturing Practices: Large Volume Parenterals* (proposed), 41 Federal Register 22031, 1976, June 1.
18. United States Food and Drug Administration. Current good manufacturing practice for finished pharmaceuticals. *Code of Federal Regulations*, Title 21, Part 211, 2020.
19. United States Pharmacopeia <1229.01> "Steam Sterilization by Direct Contact", *USP 43-NF 38*, November 2020.

14 Validation of Terminal Sterilization

Kevin D. Trupp and Thomas J. Berger PhD

CONTENTS

14.1 INTRODUCTION TO PARENTERAL PRODUCT STERILIZATION

The previous chapter discussed the steam sterilization approach for the processing of hard goods or porous loads. This chapter will discuss the sterilization validation approach that can be used in the processing of parenteral products by terminal sterilization using moist heat. The underlying principles of steam sterilization are applicable to both hard goods and terminal sterilization of parenteral products, but both have their unique characteristics.

An organized sequential flow of activities must occur as new parenteral formulations are developed and subsequently processed in the manufacturing facility. The moist heat sterilization of pharmaceutical solutions is established and verified through a series of activities that confirm the product has received a defined thermal exposure that renders the product free of living microorganisms while maintaining the desired chemical and physical attributes of the container/closure system. R and D activities can include sterilization developmental engineering studies consisting of sterilization cycle development; container thermal mapping; microbial closure validation, D- and z-value analysis; container-closure integrity validations as well as final formulation stability studies.

The subsequent production phase activities must include initial sterilization vessel qualification, which demonstrates that the

DOI: 10.1201/9781003163138-14

vessel will deliver the defined sterilization process in a consistent and reproducible manner. Also, solution and applicable container-closure microbial validation studies must be conducted at sub-process production sterilization conditions employing heat-resistant microorganisms. Equipment validation, filtration studies and assessment of the bioburden on component parts, as well as the environment, must also be ascertained.

The developmental and production phases of sterilization technology activities are then drafted into documents that are submitted as part of a New Drug Application for the particular parenteral formulation. These reports must follow applicable regulatory requirements for products that are terminally sterilized. Such studies allow one to establish, with a high level of sterilization assurance, the correct sterilization cycle (e.g., F_0, temperature, time, pressure, product time above 100°C, etc.) to be used for the steam sterilization of a specific parenteral formulation in a particular container-closure system.

14.2 STERILIZER DESIGN FOR TERMINAL STERILIZATION

Products in glass containers may be able to utilize the saturated steam processes as described in the previous chapter, but many products and containers require the use of air overpressure during the sterilization process. This section will discuss some of the key design considerations for terminal steam sterilizers and provide some specifics for the various types of steam sterilization processes utilizing air overpressure.

14.2.1 Typical Design Considerations for Steam Sterilizers

1. A pressure vessel constructed according to the American Society of Mechanical Engineers (ASME) or equivalent international code. This must withstand at least 50% in excess of the required internal pressures.
2. A safety door mechanism to prevent opening while the unit is under pressure. The locking device may be actuated directly by internal pressure or indirectly through an automatic switch. The door itself may be of the swing-out or sliding type.
3. Process control system (typically a Programmable Logic Controller (PLC) for controlling and monitoring the process).
4. Process Data Recorder and Data Collection system.
5. Product racks designed to hold/support the sealed product containers and to provide adequate heating/cooling media flow throughout the product zone.
6. Pressure safety relief valves for both the chamber and jacket (if equipped with a jacket).
7. The steam utilized for the terminal sterilization process can vary from basic industrial plant steam to Pure Steam (i.e., steam condensate meeting WFI requirements for chemistry and endotoxin). Industrial plant steam is typically used for jacket heating or indirect water heating or steam generation. Process steam (with defined feedwater and chemical requirements) can be utilized for most types of terminal sterilization processes that utilize direct steam injection. When using Pure Steam for the terminal sterilization process, the steam does not need to be tested for superheat, dryness and non-condensable gasses.
8. The water utilized for the terminal sterilization process must meet user-defined requirements. In some cases, city water can be utilized for the terminal sterilization process, but the chlorine content in the city water can cause premature stress corrosion cracking of the stainless-steel vessel. If using the direct injection of cooling water during the cooldown portion of the cycle, the microbial content of the water must be controlled to minimize the potential recontamination of the product. For most recirculating water processes, the water used to cool the product has been sterilized at the same time as the product. The water in contact with the product containers is typically cooled indirectly via a heat exchanger that can utilize any type of cooling water on the dirty side of the heat exchanger.

Note: A microbial retentive vent/air filter would not typically be required for processes used for terminal sterilization as there is no direct contact between the heating/cooling media and the contents of the containers.

14.2.2 Steam-Air Mixture Sterilization

A steam-air mixture process is typically utilized when the container closure system requires air overpressure and when the container system is not compatible with water (except as condensate), which can cause cosmetic issues with the container (e.g., moisture in the plunger section of a prefilled syringe). Steam-air mixture processes typically utilize large recirculating fans to prevent the formation of cold/hot spots within the sterilizer load. The steam-air mixture process typically uses an indirect cooling method such as cooling of the jacket or with cooling coils within the sterilizer. Because of this indirect cooling method, the cooling rate of the product is typically slower than direct exposure of the product containers to cooling water.

Some of the specific sterilizer design considerations for a steam-air mixture process include the following:

1. An optional jacket and insulation: If the sterilizer is equipped with a jacket, plant steam can be utilized in the jacket during the heating and exposure phases of the cycles to minimize condensate generation within the chamber, and cooling water can be introduced into the jacket during the cooling phase of the process. When cooling is provided by internal heat exchangers, a jacket is not required.
2. A thermostatic steam trap to efficiently remove condensate from the chamber: This is open when cool (in contact with the air or condensate) and closed when in contact with steam. As condensate collects, the trap opens owing to the slight temperature reduction and the condensate is discharged. There is also a steam trap to remove steam condensate from the jacket (if the sterilizer is equipped with a jacket).

3. Fan(s) to continuously recirculate the steam-air mixture during heat up and exposure and to recirculate the air during cooling.
4. Cooling provisions (e.g., cooling coils) to cool the air/product.

14.2.3 Recirculated Superheated Water Sterilization

Sterilization with recirculating superheated water (sometimes referred to as a water cascade or raining water process) is more efficient than a steam-air mixture and is therefore more common. There are many types of recirculating superheated water processes, the most common is a process where the bottom portion of the sterilizer (below the product zone) is filled with water and a recirculation pump is used to continuously recirculate water from the bottom of the sterilizer to spray nozzles or water distribution pans above the product zone. Another version of the recirculating superheated water process is to completely submerge the product in water, but this process is inefficient from a utilities consumption standpoint. All recirculating superheated water processes utilize air overpressure and the overpressure can be controlled during the sterilization process to minimize stress on the container-closure system. The maximum overpressure that could be utilized for the terminal sterilization process would be limited by the chamber pressure rating. The minimum overpressure will be driven by the temperature being used, the pressure needed to maintain the desired product characteristics and the required overpressure needed to prevent the recirculation pump from losing prime. These recirculating processes are typically heated and cooled indirectly with external heat exchangers located in the water recirculating water loop, but direct injection of steam (typically Process or Pure steam) and cooling water (w/ defined microbial limits) can also be used.

In addition to the typical sterilizer design considerations mentioned earlier, a superheated water sterilizer would also include a large recirculating water system (e.g., pump, pipes, heat exchangers, headers, spray nozzles) including specific water level control valves and monitoring devices.

14.2.3.1 Rotary and Shaker Sterilization

In some cases, certain products (i.e., suspensions and emulsions) require agitation during the sterilization process. For those types of products, it is typical to use a rotating rack within the sterilizer, but other agitation methods such as an internal shaking device are available. Refer to Figure 14.1 for the typical design of a rotary sterilizer and Figure 14.2 for the typical design of a sterilizer using a shaking mechanism. It is possible to use any of the sterilization processes listed with product agitation.

14.2.3.2 Continuous Sterilization

For this version of the superheated water process, the containers are terminally sterilized in a continuous sterilizer by a process in which the containers move through a constantly controlled environment in carriers with individual compartments. The time, temperature, and pressure requirements are set to predetermined values and are automatically and continuously controlled, monitored, and recorded. Refer to Figure 14.3, which depicts the pattern that containers (e.g., parenteral flexible product containers)

FIGURE 14.1 Photo of a sterilizer with a rotary mechanism.

Source: Fedegari Autoclavi SpA, Albuzzano, Italy.

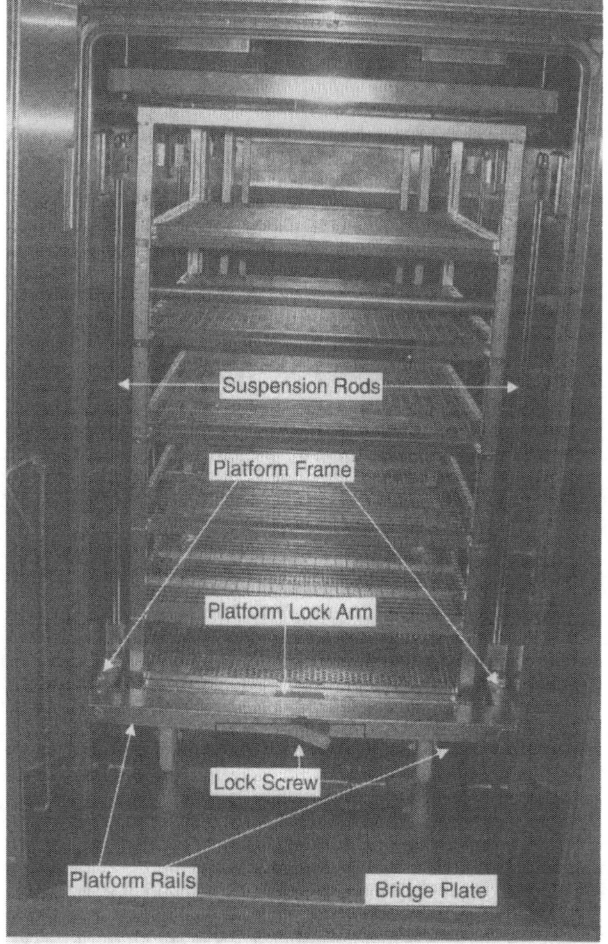

FIGURE 14.2 Photo of a sterilizer with a shaking mechanism.

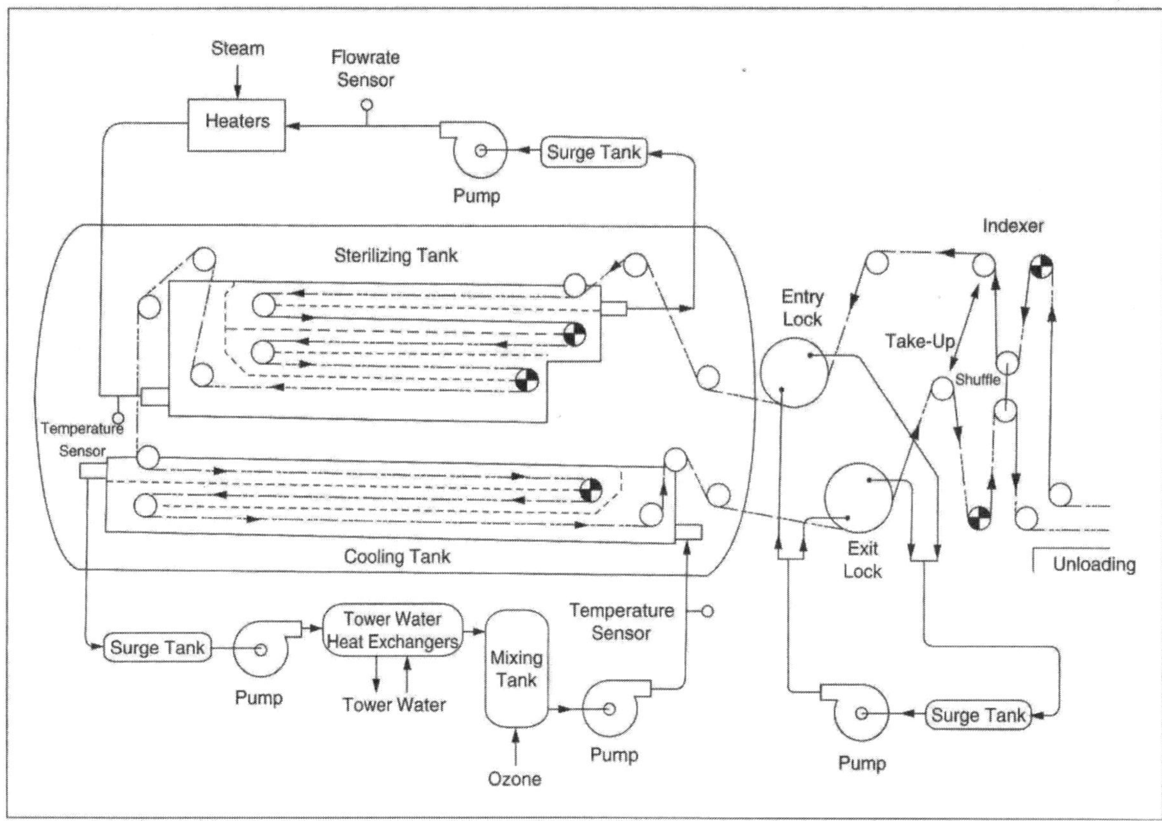

FIGURE 14.3 Schematic of a sterilizer for "continuous" processing of flexible containers.

follow as they move through the continuous sterilizer. The water locks of the pressure vessel are used to provide product entrance into and out of the overpressure environment. The overpressure environment is constantly maintained within predetermined limits.

The sterilizing phase begins as the product enters the hot water environment within the pressure vessel. The hot water environment may be a superheated water spray, which is circulated over the top of the continuously moving carriers. The residence time of the product within this sterilizing environment and the water temperature are controlled within predetermined limits to assure the required heat input.

Cooling begins as the product transfers from the sterilizing environment and enters the cooling environment, which is within the same pressure vessel. The cooling water environment is a cool water spray that is circulated over the top of the continuously moving carriers. The temperature of this water is controlled within predetermined limits to assure the required degree of cooling is achieved before the product leaves the cooling environment. A system of fixed temperature sensors located in the entering and exiting recirculating water for both the heating and cooling sections continuously monitors, records, and controls the temperature of the process water. Air overpressure is required to protect the container from stress while it is exposed to the high sterilizing temperatures.

14.3 STERILIZER CONTROL SYSTEMS

Sterilizers that maintain a specific water level (i.e., recirculated water processes) should be equipped with liquid level sensors. These sensors may be in the form of a single point level type probe or a continuous level sensor. Regardless of what type of sensor is used, a separate high-level sensor must also be provided. The separate high-level sensor provides greater assurance that the collected water at the bottom of the vessel remains below the product level.

Sterilizers that rely on recirculated water as part of the sterilization process typically include flow sensors. The flow sensor may be a direct measurement such as a flow meter (i.e., Coriolis, ultrasonic, magnetic, etc.) or an indirect measurement such as a differential pressure sensor across the recirculation pump.

The design and development of the sterilizer control system software should follow the principals of the International Society for Pharmaceutical Engineering (ISPE) Good Automated Manufacturing Practice (GAMP) 5: A Risk-Based Approach to Compliant GxP Computerized Systems (1). This guideline details a software life cycle from conception through decommissioning.

14.4 STERILIZATION CYCLES

The type of steam sterilization cycle to be utilized is dependent on product needs and equipment availability. As discussed in the previous chapter, the sterilization of hard goods or porous loads typically requires the use of a pulsed pre-vacuum cycle as it is preferable to remove the air from the porous materials being sterilized, whereas in the terminal sterilization of aqueous solutions in sealed containers, the major concern is to provide rapid heat transfer to the wall of the filled product containers, and air removal is not required nor even necessary as the hydrating moisture is contained within each container. Parenteral products may be filled into rigid or flexible containers. In either container type, there is typically air or nitrogen present in the headspace above the liquid. As the solution is heated, this gas expands and adds to the internal pressure increase resulting from the evolution of water vapor from the aqueous vehicle within the heated container. Thus, the pressure within the container will exceed the chamber pressure during steam process for sealed containers.

Glass vials can be sealed with special closures to withstand this pressure. As long as the pressure differential between the chamber and the containers does not become too great during the steam exhaust portion of the cycle, the vials will not burst or cause the closure systems to fail. If rapid cooling of the load is desired, the pressure differential might become significant enough to cause container-closure integrity to be lost.

Plastic bags, semi-rigid containers and syringes present a greater problem because they do not have the inherent strength of glass and may burst or deform as the pressure differential increases. To prevent this, air must be injected into the chamber to raise the pressure above the saturation pressure of the steam. This is particularly important during the cooling cycle, when the chamber pressure is reduced at a much faster rate than that within the container.

The following section provides a description of the various steam sterilization cycles used for parenteral products in sealed containers.

14.4.1 Saturated Steam—Pre-Vacuum Cycle

For a saturated steam process, the most common (and perhaps most effective) method to remove the entrapped air from the sterilizer is to remove it mechanically during the pre-exposure phase of the sterilization process. This is done by means of a mechanical vacuum pump or steam eductor. The saturated steam process can be utilized for stable products in small glass containers, but these types of sterilizers are not typically equipped with cooling provisions for liquid-filled products.

14.4.2 Steam-Air Mixture Cycle

It is important to understand the physical principle involved in a mixture of steam and air. The fixed relationship between temperature and pressure seen in the Chapter on Steam Sterilization in Autoclaves no longer applies. Dalton's law states that the pressure of an ideal mixture of gasses is equal to the sum of the partial pressure of the gasses, or:

$$P = P\text{A} + PB + Pc \ldots$$

Raoult's law further states that, for ideal mixtures, the partial pressure of the gas is equal to its vapor pressure multiplied by the mole fraction in the liquid. For steam in equilibrium with pure condensate, this reduces to:

$$P\text{A} = p^*\text{A}$$

where PA is the partial pressure of steam and p^* A is the vapor pressure of the condensate. The difference between the observed chamber pressure P and PA is the partial pressure of air.

The presence of air, although necessary to maintain container integrity, does reduce the heat transfer efficiency. The objective of the design in the "air overpressure" cycle is to maintain a well-mixed chamber. This assures that the heat transfer to the load will be uniform regardless of the presence of air. Steam-air sterilizers use fan(s) built into the top or end of the chamber, which circulates and mixes the air and the steam. Some steam-air sterilizers are capable of using water during the cooldown process to cool the containers more rapidly. This rapid cooling may also be necessary for product stability. Various methods (i.e., direct, injection, recirculation through a heat exchanger, etc.) for introducing the cooling media can be utilized.

14.4.3 Recirculating Superheated Water Cycle

The typical recirculating superheated water process (sometimes referred to as a water cascade or raining water process) begins by the addition of water to the sterilizer to a predefined level (below the product zone). Then a water recirculation pump is started to continuously recirculate water from the bottom of the sterilizer to spray nozzles or a water distribution pan above the product zone. The recirculation pump is kept on throughout the heat up, exposure and cooldown phases. During heat up, the water is heated at a predefined rate via a heat exchanger in the recirculation loop or with the direct injection of steam. Also, during heat-up, compressed air is added to the chamber to attain the desired overpressure levels. Once the temperature set point is achieved, the controller advances to the hold portion of the cycle and the temperature and pressures are maintained at the desired levels. For cooling, the steam supply is shut off and the recirculating water is cooled at a controlled rate by introducing cooling media to a heat exchanger installed in the water recirculation loop or by the direct injection of cooling water into the recirculating loop. This type of process does not require the use of a jacket, but does require specific water level controls.

The recirculating superheated water process is efficient and the temperatures and pressures can be tightly controlled during the entire process, thus minimizing container stresses.

14.5 STERILIZATION CYCLE DEVELOPMENT

This section will address sterilization and the associated microbiological activities that occur in Research and Development areas as well as the production environment when using the Biological Indicator/Bioburden approach in support of a parenteral product terminal sterilization. The table following depicts some of the sterilization engineering and microbiological activities as a parenteral product moves through development. These studies or similar ones are ordinarily conducted in developmental sterilizers or may occur as investigative engineering studies in a production sterilizer, as appropriate. The overkill method can be used for some of the more stable parenteral formulations, and its validation is accomplished as described in the previous chapter.

Sterilization development activity	Activity statement
Cycle development	Develop preliminary container sterilization specifications with engineering parameters such as temperature, pressure, time and F_0.
Container thermal mapping	Determine the cold spot within the filled and sealed container.
Formulation development	Perform analytical feasibility studies prior to product finalization with the validated methods.

Parenteral solution microbiological evaluations:

Moist heat D- and z-value analysis	Perform triplicate D-value analysis with the biological indicator to be utilized for sterilizer validation on each parenteral formulation at three temperatures, e.g., 112°C, 118°C and 121°C, and then calculate the z-value.
Antimicrobial preservative efficacy (APE)	Perform on the final product if it contains a preservative or if there is a multidose claim for the container.
In-process bioburden analysis	Perform studies with a panel of microorganisms to validate 70% recovery for the filtration process.
Spike hold time studies	Inoculate parenteral product with bioburden and growth promotion compendial microorganisms to evaluate the product's ability to support microbial growth.

Container closure evaluations:

Microbial closure inactivation	Perform kill curve kinetics using bioburden and biological indicator (spores) inoculated onto the worst-case closure site.
Container integrity	Perform dye ingress, microbial challenge or physical integrity tests following exposure to maximum sterilization conditions that stress the container.
Stability runs	Perform analytical chemistry and microbiological evaluations at various temperatures and times per ICH and or compendial requirements.
Bacterial endotoxin testing (BET)	Perform test method validation of API, excipients and final product per compendia requirements.

Abbreviations: APE = antimicrobial preservative efficacy; API = active pharmaceutical ingredient; BET = bacterial endotoxin testing; ICH = International Conference for Harmonisation.

Sterilization engineering personnel primarily focus their efforts in determining whether a parenteral formulation packaged in a particular container configuration can be sterilized in a current cycle or whether a new cycle must be developed. The referenced European Medicines Agency (EMA) (2) decision tree is followed when evaluating a new parenteral product in a large volume parenteral (LVP) or small volume parenteral (SVP) container. Sterilization feasibility studies are conducted in a sterilizer to ascertain the physical effects of the cycle on the product in question. Product attributes that can be affected by a cycle are closure integrity, product potency, product impurities, pH, color, shelf-life stability, visible and subvisible particulates as well as final product sterility. Once the basic engineering parameters (e.g., temperature, pressure, time and F_0) are established, then engineering thermal container mapping studies are performed (3, 4).

14.5.1 Sterilization Temperature Determination

Unlike porous and hard good load sterilization in which high temperatures (e.g., 121°C–124°C and high heat inputs (i.e., high F sub zeros) are typically utilized, the temperature and heat input of the terminal sterilization process can be much lower. Lower sterilization temperatures (e.g., 115°C) may improve the uniformity of the heat input across the load, can reduce stress on the container/closure system and can ensure sufficient microbial kill (at the same F_0) for the closure systems. The primary disadvantages for lower sterilization temperatures are longer sterilization/cycle times and, for some products, increased liquid product degradation (e.g., potency drop, impurity generation and color formation because of the longer exposure times required to obtain the desired product F_0 range.

14.5.2 Product F_0 Determination

When liquid products are exposed to sterilization temperatures, product and container/closure degradations are common. Example product degradations could include but are not limited to: potency loss, impurity generation, pH shift and color formation. Examples of container-closure degradations could include, but are not limited to: particulate generation, leachables from plastics and rubbers and loss of integrity. Thus from a finished product chemistry perspective, the lower the F_0 the better the product chemistry will be. Some container systems (e.g., ampoules) and products (e.g., saline) can tolerate high F_0 cycles but high F_0 cycles are not required to attain a sterility assurance level of 10^{-6}. Even very low F_0 (e.g., <4) cycles can be utilized to obtain sterility assurance levels of 10^{-6} for terminally sterilized products as long as the presterilization bioburden levels in the filled containers are low (e.g., < 100 cfu per container).

Thus the objective during process development is to develop the minimum F_0 requirement that is needed for sterility, and the maximum F_0 that can be tolerated by the product and the container/closure system. Lower F_0 cycles will result in more regulatory scrutiny (e.g., the EMA sterilization decision tree (2) does not allow for terminal sterilization with an F_0 <8), will require more routine bioburden monitoring and will have lower safety margins than a higher F_0 cycle.

14.5.2.1 Container Thermal Mapping Validation Studies

An R and D sterilizer is smaller than a production facility sterilizer but can simulate the sterilization cycles conducted in the larger production vessels. Container thermal mapping studies (when applicable) are typically performed in a laboratory sterilizer:

1. To locate the coldest zone or area inside a container
2. To determine the cold zone in the container and its relationship to the location monitored during validation studies
3. To generate data that may be used during the setting of production sterilization control parameters

When conducting thermal mapping studies, there are various factors to be considered, and these are dependent upon the:

1. Type of container (flexible or rigid)
2. Container orientation
3. Container size and fill volume (Note: Container mapping studies are not typically performed for container sizes of 100 mL or less as the solution temperature within the small container is very homogenous.)
4. Cycle type and temp
5. Viscosity
6. Sterilizer trays/design/surface contact
7. Sterilizer spray patterns/water flow

Typical container mapping data obtained for lipid emulsions contained within a 1000 mL glass container are shown as an example in Tables 14.1 and 14.2. The following summarizes the process for obtaining heat map data from the glass

TABLE 14.1

	1000 mL glass I.V. containers-heat mapping study (lipid emulsion)				
	Run CLHK00.049		Run CLHK01.050		
TC number	btl 1	btl 2	btl 1	btl 2	Average (SD)
1,12	7.91	C7.28	8.13	C7.36	7.67 (0.415)
2,13 (PC)	7.79	7.49	8.02	7.64	7.74 (0.226)
3,14	C7.46	7.40	C7.71	7.47	C7.51 (0.137)
4,15	7.64	7.80	7.87	7.96	7.82 (0.135)
5,16	12.66	12.90	12.95	12.91	12.86 (0.132)
6,17	12.73	12.46	12.77	12.68	12.66 (0.138)
7,18	12.78	12.69	12.95	12.91	12.83 (0.120)
8,19	13.32	13.33	13.42	13.78	13.46 (0.223)
9,20	14.21	14.33	14.03	14.56	14.28 (0.222)
10,21	H15.87	H17.24	H15.18	H16.09	H16.10 (0.856)
11,22	15.47	16.56	14.77	16.07	15.72 (0.773)
H–C	8.41	9.96	7.47	8.73	8.64 (1.028)
PC–C	0.33	0.21	0.31	0.28	0.28 (0.053)

Note: H denotes hottest TC location; C denotes coldest TC location; PC denotes approximate location of the production profile TC; Data from TC#9 used with a postcalibration variance of + 0.25°C at 100°C; All heat input values are calibration corrected.

TABLE 14.2

	1000 mL glass I.V. containers-heat mapping study (lipid emulsion)				
	Run CLHK00.49		Run CLHK01.050		
	btl 1	btl 2	btl 1	btl 2	Average (SD)
Coldest location					
Thermocouple number	3	12	3	12	–
Time to 100°C	19.0	19.0	19.0	19.0	19.00 (0.000)
Time ≥ 100°C	21.0	21.0	21.0	21.0	21.00 (0.577)
Time ≥ 120°C	4.0	3.0	4.0	3.0	3.50 (0.577)
Time ≥ 120–100°C	4.0	5.0	4.0	5.0	4.50 (0.577)
Maximum temperature (°C)	120.82	120.77	120.92	120.77	120.82 (0.071)
Heat input (F_0)	7.46	7.28	7.71	7.36	7.45 (0.187)
Production profile TC location					
Thermocouple number	2	13	2	13	–
Time to 100°C	19.0	19.0	19.0	19.0	19.00 (0.000)
Time ≥ 100°C	22.0	21.0	22.0	21.0	21.50 (0.577)
Time ≥ 120°C	4.0	3.0	4.0	3.0	3.50 (0.577)
Time ≥ 120–100°C	5.0	5.0	5.0	5.0	5.00 (0.000)
Maximum temperature (°C)	120.91	120.82	120.91	120.92	120.89 (0.047)
Heat input (F_0)	7.79	7.49	8.02	7.64	7.74 (0.226)

Note: H denotes hottest TC location; C denotes coldest TC location; PC denotes approximate location of the production profile TC; Data from TC#9 used with a postcalibration variance of +0.25°C at 100°C; All heat input values are calibration corrected.

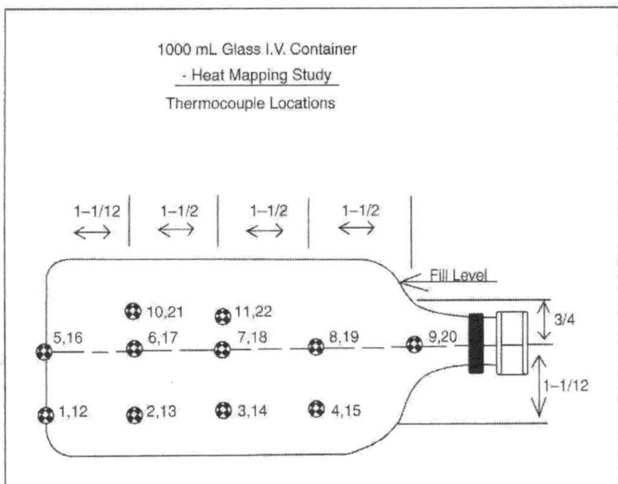

FIGURE 14.4 Heat mapping study using thermocouples at defined locations in a 1000 mL glass container.

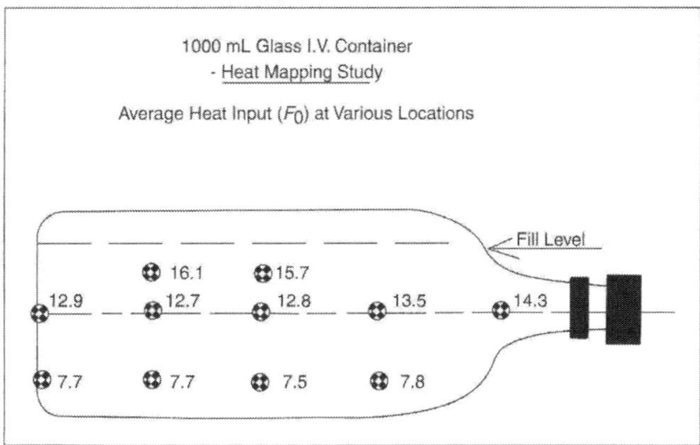

FIGURE 14.5 Heat mapping study with average heat input (F_0) at various locations in a 1000 mL glass container.

intravenous container filled with approximately 1000 mL of lipid emulsion:

Thermocouple probes (Copper Constantan, type T, 0.005 inch diameter) were used to monitor 11 locations within the 1000 mL container. The thermocouple probes were positioned at various distances as depicted (Figure 14.4). Each container was filled with approximately 1000 mL of the lipid emulsion, evacuated to 20 inches of mercury and sealed with aluminum overseal.

A flat perforated rack on a reciprocating shaker cart was used in the sterilizer. The cycle's target temperature was 123°C, recirculating water spray cycle with 70 rpm of axial agitation and 30 psig (pounds per square inch) of air overpressure.

When the sterilization cycle was controlled to give a heat input of approximately 7.5 F_0 in the coldest emulsion area, the average coldest emulsion area was found to be measured by thermocouple number (TC#) 3, 14. The average hottest emulsion area was measured by TC# 10, 21. The difference between the hottest and coldest emulsion areas ranged from 7.5 to 10.0 F_0 with an average of 8.6 F_0. Therefore, when the coldest emulsion area registered 7.5 F_0, the hottest emulsion area would average 16.1 F_0.

The emulsion area approximating the validation TC location was measured by TC # 2, 13 and averaged 7.7 F_0 when the coldest emulsion was approximately 7.5 F_0 (Figure 14.5).

14.5.3 Solution/Product Moist Heat Resistance D- and z-Value Analysis

A BIER vessel is an acronym for a biological indicator evaluator resistometer vessel that meets specific performance requirements for the assessment of biological indicators per American National Standards developed and published by AAMI (Association for the Advancement of Medical Instrumentation) (5). One important requirement for a BIER steam vessel is the capability of providing a square wave heating profile.

Refer to Figure 14.6 for a schematic of the steam BIER vessel used to generate the D and z-value data. D-value is the time in minutes required for a one log or 90% reduction in microbial population. The z-value is the number of degrees of temperature required for a 10-fold change in the D-value (Refer to Chapter 11 for additional details on F-, D- and z-values).

14.5.4 Master Solution/Product Concept

The family category of lipid emulsions and their respective $D_{121°C}$ and z-values as well as classification in terms of microbial resistance is shown in Table 14.3. A categorization of parenteral formulations with associated $D_{121°C}$ and z-values and their potential impact on microbial resistance using the biological indicator, *Clostridium sporogenes*, were previously reported (6). In addition, the methodologies used for D- and z-value analysis were likewise cited. The data in Table 14.3 indicate that the # 1 emulsion is at the top of the list, because it affords the most microbial

TABLE 14.3

List #	Solution	D121	z-value	Predicted spore log reduction
1	20% Emulsion	0.7	10.6	7.1
2	10% Emulsion w/increased linolenate	0.7	11.4	7.5
3	10% Emulsion w/100% soybean oil	0.6	10.1	8.0
4	20% Emulsion w/100% soybean oil	0.7	12.8	8.2
5	20% Emulsion w/increased linolenate	0.6	10.6	8.3
6	10% Emulsion w/50% safflower & 50% soybean oil	0.6	10.7	8.4
7	20% Emulsion w/50% safflower & 50% soybean oil	0.6	12.7	9.5
8	10% Emulsion	0.4	11.1	12.9

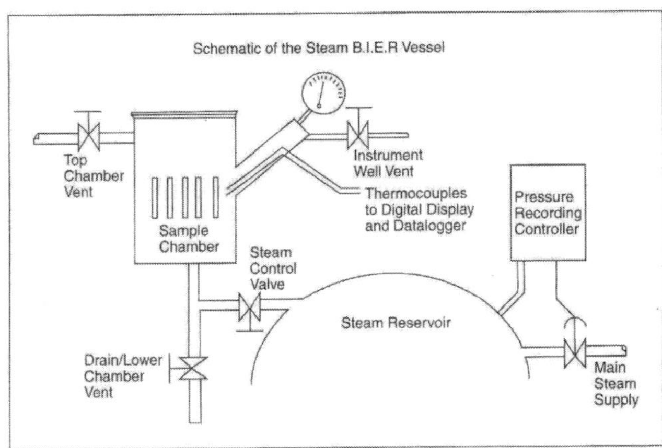

FIGURE 14.6 Schematic of a steam BIER vessel used for generating moist heat resistance D- and z-values for inoculated parenteral solutions or for biological indicators.

Note: BIER = biological indicator evaluator resistometer.

moist heat resistance. It is therefore the emulsion that should be microbiologically challenged (inoculated with spores) as part of the emulsion validation scheme. D- and z-value data have been reported for other biological indicators such as *Geobacillus stearothermophilus* (6, 7, 8) and *Bacillus subtilis* 5230 (9). There are many factors that can affect moist heat resistance including a biological indicator's age, the sporulation media used, as well as the particular spore strain employed (10).

14.5.5 Predicted Spore Logarithmic Reduction Values

Lipid emulsion moist heat resistance values ($D_{121°C}$ and z-values) were generated in the steam BIER vessel using the biological indicator *C. sporogenes* as shown in Table 14.3. The columns in the Table list the representative code or list number of the product, the emulsion or product name, its average $D_{121°C}$ value and z-value and finally the Predicted Spore Logarithmic Reduction (PSLR) value. Those parenteral formulations with the lowest PSLR value(s) are those that should be used for the microbial validation at sub-process conditions, because these provide the most microbial resistance (6).

14.5.6 Accumulated F (Bio) for Lipid Emulsions

Accumulated F_{bio} and z-values (Table 14.4) were used to construct the PSLR ranking for lipid emulsions as previously discussed for Table 14.3. The F_{bio} is the heat input for the biological solution based on the emulsion's moist heat D- and z-values. By inputting the product temperatures from the coldest thermocouple (TC) of an engineering run for a particular container/sterilization cycle, the emulsion can be ranked according to PSLR values. The combined $D_{121°C}$ and z-value allows comparison of moist heat rankings between emulsions.

The data in Table 14.4 demonstrate that the #1 emulsion has the lowest PSLR (7.105), thereby affording the highest moist heat resistance upon inoculation. Generation of this table allows prediction of which emulsion to microbiologically challenge as part of validation in the production sterilizer.

14.5.7 Microbial Closure Inactivation Validation in a Developmental Sterilizer

In lieu of using the large-type steam sterilizers in the production environment, microbial inactivation at the closure/bottle interface of an emulsion container can be assessed in a developmental

TABLE 14.4

			Solution							
Temperature (°C)	**Time (min)**	***F* (PHY) z=10.0**	**1 z=10.6**	**2 z=11.4**	**3 z=10.1**	**4 z=12.8**	**5 z=10.6**	**6 z=10.7**	**7 z=12.7**	**8 z=11.1**
105.4	1	0.0269	0.0330	0.0419	0.0278	0.0592	0.0330	0.0340	0.0579	0.0384
110.1	1	0.0793	0.0915	0.1082	0.0813	0.1380	0.0915	0.0935	0.1359	0.1019
114.1	1	0.1991	0.2181	0.2427	0.2023	0.2834	0.2181	0.2212	0.2806	0.2336
116.2	1	0.3228	0.3442	0.3709	0.3265	0.4134	0.3442	0.3476	0.4106	0.3611
118.1	1	0.5000	0.5200	0.5445	0.5035	0.5819	0.5200	0.5232	0.5794	0.5356
119.1	1	0.6295	0.6462	0.6663	0.6324	0.6966	0.6462	0.6489	0.6946	0.6591
119.4	1	0.6745	0.6897	0.7079	0.6772	0.7352	0.6897	0.6921	0.7334	0.7014
119.2	1	0.6442	0.6604	0.6799	0.6470	0.7092	0.6604	0.6630	0.7073	0.6729
118.5	1	0.5483	0.5672	0.5903	0.5515	0.6253	0.5672	0.5703	0.6230	0.5819
117.8	1	0.4667	0.4872	0.5124	0.4702	0.5513	0.4872	0.4905	0.5487	0.5033
116.2	1	0.3228	0.3442	0.3709	0.3265	0.4134	0.3442	0.3476	0.4106	0.3611
114.1	1	0.1991	0.2181	0.2427	0.2023	0.2834	0.2181	0.2212	0.2806	0.2336
110.6	1	0.0889	0.1020	0.1197	0.0911	0.1510	0.1020	0.1042	0.1487	0.1130
105.9	1	0.0301	0.0367	0.0463	0.0312	0.0648	0.0367	0.0379	0.0634	0.0426
101.7	1	0.0115	0.0148	0.0198	0.0120	0.0305	0.0148	0.0153	0.0296	0.0178
Total *F*		4.7436	4.9734	5.2646	4.7826	5.7366	4.9734	5.0107	5.7044	5.1573
D value			0.70	0.70	0.60	0.70	0.60	0.60	0.60	0.40
PSLR			7.105	7.521	7.971	8.195	8.289	8.351	9.507	12.893

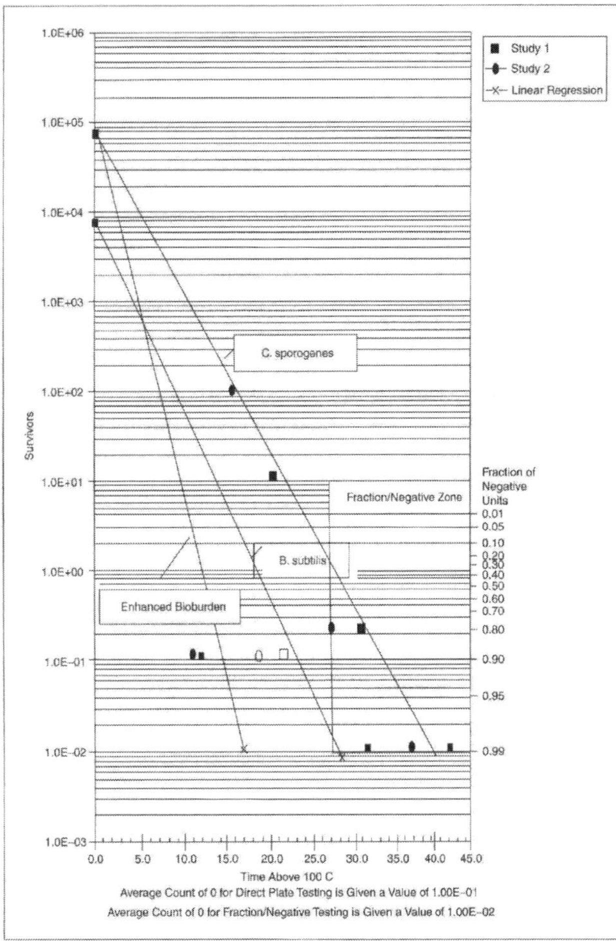

FIGURE 14.7 Microbial kinetic inactivation of bioburden as compared with biological indicators, *Bacillus atrophaeus* (formerly named *B. subtilis*) and *Clostridium sporogenes.*

sterilizer. The closure microbial inactivation (kinetic) studies can determine how the size of the container, type of closure compound used as well as the closure preparatory processes (e.g., leaching, washing, siliconizing, sterilizing, drying) influence microbial inactivation. Microbial closure kinetic studies are conducted at various time intervals in a given sterilization cycle. Direct Plate (DP) count or Fraction-Negative (F/N) methodologies are used to evaluate the surviving organisms. Test data were generated demonstrating the value of using both a moist heat organism (*C. sporogenes*) and a dry heat organism (*B. subtilis* now known as *B. atrophaeus*) as biological indicators for the sterilization validation of closure systems (11, 12). A typical graphic representation of the inactivation kinetics is illustrated in Figure 14.7. The previous studies may also be performed in a production sterilizer as engineering or feasibility studies.

14.5.8 Container Closure Integrity Validation

Container closure integrity or Maintenance of Sterility (MOS) validations are run on all moist heat terminally sterilized products with closure systems of a parenteral container. This validation is performed to demonstrate that the closure system of a container is capable of maintaining the liquid product and fluid path in a sterile condition throughout the shelf life of the product.

In a typical maintenance of sterility study, the product container is sterilized at a temperature that is higher than the upper temperature limit of the chosen sterilization cycle and for a time that is greater than the maximum time limit for the cycle or producing an F_0 level greater than the maximum F_0 level for the cycle. The rationale for the selection of the maximum temperature and heat input level for the pre-challenge sterilization is that rubber and plastic closures are subjected to thermal stresses during sterilization and those stresses are maximized at the highest temperature and the longest time allowed.

In some cases, the closures, e.g., administration or additive port are claimed to be sterile by a radiation process. In such cases, the closures are sterilized in bulk exceeding the maximum end of the radiation process e.g., 40 kGy, then fabricated to the flexible container and exposed to steam sterilization cycle conditions exceeding the maximum temperature end of the cycle. Thus, the closures are stressed by a joint process of radiation as well as steam prior to performance of the closure integrity test.

14.5.9 Product Validation for Endotoxin

Endotoxins are lipopolysaccharides from the outer cell membrane of gram-negative bacteria. Endotoxins can be detected by the manual gel-clot method known as the Limulus Amebocyte Lysate (LAL) test. There are also various quantitative methods (Turbidimetric and chromogenic) that use more rapid automated methodologies. All final product formulations have regulatory requirements to be tested for endotoxins and the method must be validated using three different lots of final product. LAL testing should be performed on final product formulations per FDA Guidelines and other regulatory compendia. Emulsion formulations, if colored or opaque, cannot be tested by the turbidimetric method and therefore may use a comparable test e.g., LAL, chromogenic or kinetic.

The LAL test is for products other than oral and topical products (e.g., parenteral solutions, some devices). Endotoxin testing is usually required at three different times in the cycle of the product. First, endotoxin testing should be performed on the lot of drug being used in clinical studies to ensure that the product is safe for the patients with respect to pyrogen levels. Second, in the developmental stages, endotoxin testing is usually required at the beginning and end of the stability studies. Finally, once the product is ready to be marketed, each lot of the product requires endotoxin testing prior to release.

To improve in-process control, a process should also be in place to decide if endotoxin testing should be performed on the active pharmaceutical ingredients (APIs) and/or excipients used in the product. In order to determine this, the International Conference for Harmonisation (ICH) guidelines for quality should be used; Q7A "Good Manufacturing Practice Guidance for Active Pharmaceutical Ingredients".

14.5.10 Production Facility Sterilization Development

The Production Table following depicts some of the sterilization engineering and microbiological activities associated with a parenteral product as it moves into the production environment. These studies occur in a production environment as appropriate.

Production facility activity	Activity statement
Heat penetration and temperature distribution (P and Ds)	Perform triplicate studies for minimum and maximum loading conditions using temperature sensors within the product containers and outside the containers to measure the sterilizer heating medium temperature.
Solution (master) microchallenge validation	Perform microbial validation of a parenteral solution or master solution at sub-process conditions in the production sterilizer.
Container closure microchallenge validation	Perform microbial validation of the container closure system (as applicable) at sub-process conditions in the production sterilizer.
Hold time studies	Microbial, chemical and endotoxin studies are performed to establish the longest time that a product can be held following manufacture but prior to filling and sterilization.

14.5.11 Heat Penetration and Temperature Distribution Validation

Perform triplicate studies in the production sterilizers with minimum and maximum loading configurations with temperature probes (typically Type T thermocouples or wireless sensors) penetrating the product containers as well as temperature probes/sensors distributed outside the product containers in a production sterilizer at nominal operating process parameters. Per, United States Pharmacopeia (USP) chapter <1229.2>, a fixed loading position within the sterilizer may be necessary for proper sterilization to ensure uniformity of heating and cooling in routine use. Once the load is positioned properly, its size can vary within a defined range. Load-mapping studies should be performed to determine the coldest and hottest locations within the load. These locations may not be specific individual containers but rather regions. This ensures that the containers are neither under- nor overprocessed in routine operation of the sterilizer. Validation of variable-size load patterns is accomplished using a bracketing approach for which success with maximum and minimum loads (avoiding both under- and overprocessing) establishes the acceptability of intermediate size loads. However, evaluation of intermediate load sizes may be beneficial. In product sterilization, only a single-size container with a single product lot is processed concurrently.

14.5.12 Microbial Solution Validation in a Production Sterilizer

Table 14.5 shows the microbial solution validation conducted at sub-process conditions in a fully loaded production sterilizer. For this example, the acceptance criterion of 6 spore logarithmic reduction (SLR) was set-up for the biological indicator *C. sporogenes* and a 3 SLR for the higher moist heat resistant biological indicator, *G. stearothermophilus*. Each emulsion (20 containers) is inoculated with the appropriate biological indicator spore suspension at a target level of 1.0×10^6 and 1.0×10^2 for *C. sporogenes* and *G. stearothermophilus*, respectively. The 20 inoculated containers are distributed throughout the production sterilizer for sterilization at sub-process conditions. The test containers are then returned to the lab for testing by the F/N test method.

TABLE 14.5

Fraction negative method

Organism	Code	Average no. spores/bottle	No. positive[a]/ no. positive controls	No. positive[a]/ no. negative controls	No. positive[a]/ no. test samples	Spore[b] logarithmic reduction
C. sporogenes	5C6	4.8×10^5	2/2	0/4	0/20	>7.0
C. sporogenes	15C6	6.4×10^5	2/2	0/4	0/20	>7.1
G. stearothermophilus	5B2	7.6×10^1	2/2	0/4	0/20	>3.2
G. stearothermophilus	15B2	7.7×10^1	2/2	0/4	0/20	>3.1

F_0 Range[c]: 5.8–7.6; Temperature Range: 120–125°C; Agitation: 67–73cpm

[a] Positive for the indicator microorganism.

[b] Spore logarithmic reduction = log a – log b; where a, initial population of spores; b, 2.303 log (N/q) = In (N/q); where N, total number of units tested; q, number of sterile units.

[c] F, integrated lethality or equivalent minutes at 121.1°C for hottest and coldest thermocoupled containers.

TABLE 14.6

Microorganism	Initial population/ stopper	No. positive[a]/ no. positive controls	No. positive[a]/ no. negative controls	No. positive[a]/ test samples	Spore[b] logarithmic reduction
C. sporogenes	8.4×10^3	2/2	0/4	0/20	>5.2
B. subtillis	3.0×10^4	2/2	0/4	0/20	>5.8

F_0 Range[c]: 5.8–7.6; Temperature Range: 120–125°C; Agitation: 67–73 cpm

Sterilization validation of 200 mL bottle inoculated closure surface coated with I.V. fat emulsion in cycle with agitation.

[a] Positive for the indicator microorganism.

[b] Spore logarithmic reduction = log *a* – *log b*; where *a*, initial population of spores; *b*, 2.303 log (*N*/*q*) = ln (*N*/*q*); where *N*, total number of units tested; *q*, number of sterile units.

[c] *F*, integrated lethality of equivalent minutes at 121.1°C for hottest and coldest thermocoupled containers.

14.5.13 Microbial Closure Validation in a Production Sterilizer

Table 14.6 shows the microbial closure validation at sub-process conditions in a fully loaded production sterilizer. The biological indicators used were *C. sporogenes* and *B. subtilis* (now referred to as *B. atrophaeus)*. Acceptance criteria of 3 SLR must be achieved for the moist heat (*C. sporogenes*) and dry heat (*B. subtilis*) indicators. The surface of the stopper that comes into direct contact with the sidewall of the bottle was inoculated with the appropriate biological indicator, dried and then a few drops of emulsion were placed over the inoculum to simulate manufacturing conditions. The inoculated closure was assembled onto the finished container, exposed to sub-process sterilization conditions in the production sterilizer and subsequently tested in the lab by the F/N test method. The data demonstrate that a >3.0 SLR was achieved at sub-minimal process conditions.

14.6 ANCILLARY SUPPORT PROCESS TESTING

14.6.1 Bioburden Analysis for Closures and Commodities

Determine the microbial load on closures and commodities as well as their moist heat resistance analysis.

As part of the microbiological quality control program, products and commodities are routinely sampled during the production process in order to assess the microbial load. This assessment is performed via the bioburden test for terminally sterilized products. The bioburden test method is developed during the product development stage prior to transfer to the production plant. This test assesses the microbial load of a solution prior to terminal sterilization (In-Process Bioburden Test). Micro R and D is responsible for the validation of the bioburden method prior to transfer to the production plant. The validation will demonstrate that recovery of microbial load at a relatively low level can be achieved.

The Microbial Limits Test is essentially a bioburden test of the raw materials used to make the final product. The test method and validation are conducted in much the same manner as the bioburden test. The limit for the Microbial Limits Test is calculated as follows: Final Product Action Level/Maximum Concentration of API in the Final Product. This limit is then "normalized" by dividing by the total amount of APIs in the final product.

In addition, the production bulk solution is monitored for total bioburden load including spore formers. Refer to Figure 14.8. The screening allows the plant quality lab to ascertain if there are any moist heat resistant microflora present in the bulk solution prior to the terminal sterilization of the parenteral solution in its finished container. This is accomplished using a boil or heat shock test.

14.6.2 Antimicrobial Preservative Efficacy

Performed on those formulations containing a preservative and those container configurations that have a multidose claim. This validation is performed per compendial requirements.

14.6.3 Sterility Testing (If Required)

Validated and compendial test methods must be utilized to Sterility Test terminally sterilized products. If the terminally sterilized product is approved by the applicable regulatory bodies for Parametric Release, then sterility testing is no longer required for routine product release testing nor can it be used as an alternative in case parametric release parameters are not met.

14.6.4 Biological Testing Support of R and D and Marketed Product Stability Programs

There are a number of analytical and microbiological tests performed over the shelf life of a product. A number of microbiological tests include bacterial endotoxin testing (BET),

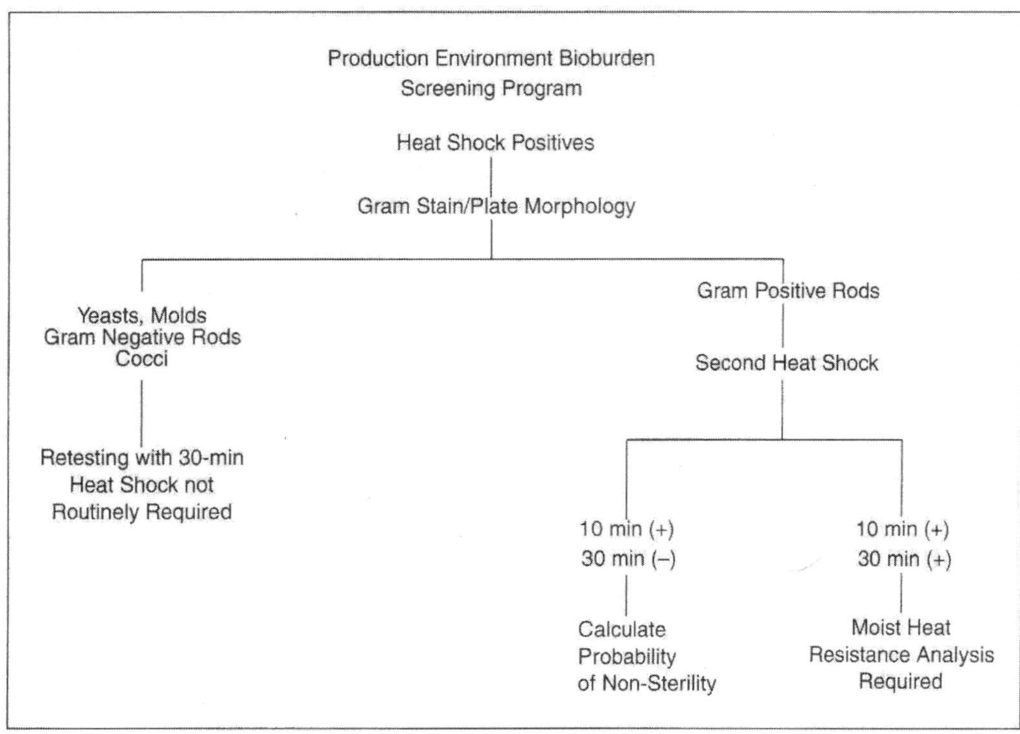

FIGURE 14.8 Representative production environment bioburden screening program.

Container Closure Integrity and Antimicrobial Preservative Efficacy (APE) if applicable.

14.7 CONCLUSION

As one reviews the final configuration that a terminally sterilized parenteral product is packaged in, it is not surprising that a similar evaluation occurred when contemplating how to present the new product as being both sterile and non-pyrogenic. The product development team focused on the various designs of sterilizers and the various manufacturing site locations for the support of currently marketed products. Once the team decided the appropriate facility for manufacture, then the various sterilization cycles discussed in this chapter were evaluated in order to select the appropriate one best suited for that parenteral product in its final container configuration. If a product is destined for the international market, then R and D personnel will follow the EMA decision tree to determine if the product can be sterilized at overkill conditions of 121°C for 15 minutes. If it cannot, then a justification is documented explaining the reason for selection of an alternate sterilization cycle with lower sterilization temperatures and/or lower product heat input. Personnel perform the applicable studies in a developmental sterilizer as detailed in this chapter, as well as feasibility studies in development or production sterilizers to monitor and test the physical attributes of the final designed container. Once the parenteral product's designs as well as the sterilization process parameters have been finalized, then the plant/site can perform their standard Heat Penetration, Temperature Distribution and microbiological studies in the production sterilizer.

REFERENCES

1. ISPE, GAMP 5 A Risk-Based Approach to Compliant GxP Computerized Systems, February 2008.
2. EMA, Guideline on the sterilisation of the medicinal product, active substance, excipient and primary container. (EMA/CHMP/CVMP/QWP/850374/2015)
3. Young JH. Sterilization with steam under pressure. In Morrissey RF, Phillips GB, eds. *Sterilization Technology*. New York: Van Nostrand Reinhold, 1993: 120–151.
4. Owens JE. Sterilization of LVP's and SVP's. In Morrissey RF, Phillips GB, eds. *Sterilization Technology*. New York: Van Nostrand Reinhold, 1993: 254–285.
5. AAMI. *BIER/Steam Vessels (ST45)*. Arlington, VA: American National Standard for the Advancement of Medical Instrumentation, 1992.
6. Berger TJ, Nelson PA. The effect of formulation of parenteral emulsions on microbial growth-measurement of D- and z-values. *PDA J. Pharm. Sci. Technology*. 1995, 49(1), 32–41.
7. Feldsine PT, Schechtman AJ, Korczynski MS. Survivor kinetics of bacterial spores in various steam heated parenteral solutions. *Developments. Ind. Microbial*. 1977, 18, 401–407.

8. Moldenhauer J. How does moist heat inactivate microorganisms? In Moldenhauer J, ed. *Steam Sterilization, a Practitioner's Guide*. Bethesda, MD: PDA and Godalming, Surrey, UK, 2003: 1–15.
9. Caputo RA, Odlaug TE, Wilkinson RL, et al. Biological validation of a sterilization process for a parenteral product-fractional exposure method. *J. Parent. Drug Assoc.* 1979, 33: 214–221.
10. Pflug IJ, Holcomb RG. Principles of thermal destruction of microorganisms. In Block SS, ed. *Disinfection, Sterilization and Preservation*. 3rd ed. Philadelphia, PA: Lea and Febiger, 1983: 759–766.
11. Berger TJ, May TB, Nelson PA, et al. The effect of closure processing on the microbial inactivation of biological indicators at the closure-container interface. *PDA J. of Pharm. Sci, and Technol.* 1998 March–April, 52(2), 70–74.
12. Berger TJ, Chavez C, Tew RD, et al. Biological comparative analysis in various product formulations and closure sites. *PDA J. of Pharm. Sci, and Technol.* 2000 March–April, 54(2), 101–109.
13. USP 40, Chapter <1229.2> Moist Heat Sterilization of Aqueous Liquids.
14. PDA Technical Report #1, Validation of Moist Heat Processes: Cycle Design, Development, Qualification and Ongoing Control, 2007.

15 Steam Sterilization-in-Place

James Agalloco

CONTENTS

15.1 INTRODUCTION

Sterilization in place (SIP) is a process that adapts sterilization methods used for items placed in fixed chambers to large pieces of equipment alone or assembled into process train. It has been defined as: "the sterilization of a system or piece of process equipment in situ."[1] Often the sheer size of the equipment precludes its placement inside a sterilizer. Process trains where several pieces of equipment are sterilized in situ together avoid the need for aseptic assembly of the equipment following sterilization of the items individually. The predominant means of SIP is by saturated steam, although other agents can be used. This chapter is focused on the use of steam SIP.

It is essential that the reader have a full understanding of SIP concepts in order to properly apply the technology and subsequently validate SIP. The available literature on SIP is extremely limited when compared with that for other sterilization processes. Articles that have been published on this subject have focused on process fundamentals, heat penetration, and filter sterilization.[2,3,4,5,6,7,8,9,10,11] Each of these articles has proven valuable in helping SIP become more fully understood and in advancing the industry's awareness of the subject. The criticality of system design to achieving sterility with SIP systems is such that it must be considered more closely than any other aspect of the SIP validation effort. Unfortunately, guidance on system design for effective SIP is sorely lacking.

The majority of the papers on steam sterilization have focused on products, materials and equipment positioned inside an autoclave. The autoclave provides the means for control of the sterilization process parameters. Correlation of the autoclave documentation to the process lethality delivered to the materials inside the chamber is achieved through the validation effort. There are large pieces of process equipment utilized in the production of parenterals whose size and configuration will not allow them to be placed inside an autoclave for sterilization. To ensure a higher degree of

DOI: 10.1201/9781003163138-15

sterility assurance for these items, they should be sterilized in situ rather than sanitized. Steam-in-place sterilization can enable an entire processing system to be sterilized as a single entity, eliminating or greatly reducing the need for aseptic connections. Manufacturing tanks, bioreactors, lyophilization chambers, processing equipment, filling lines and other large process systems are normally sterilized in this manner.

The subject of steam sterilization has been so ingrained in the minds of validation specialists that key design aspects of the sterilizer are generally overlooked. SIP, which employs the same moist heat mechanism as steam sterilization in autoclaves, forces an attention to detail in system design that far exceeds that of steam sterilizers. The reason for this increased emphasis is straightforward. When a firm applies SIP to its systems and equipment, it becomes the designer of the sterilizer itself, a role usually adopted by the autoclave manufacturer. Autoclave manufacturers have had many years of experience in designing their equipment, and consequently sterilizer designs are highly evolved and differences in sterilizers are relatively minor. Most of the important features of sterilizers that ensure their effectiveness are also required in the design of SIP systems. What may not be evident to practitioners of SIP is the application of sterilizer design concepts to the more flexible circumstances that arise in SIP systems. In order to better understand the design and qualification effort, a review of the nuances of SIP system engineering as it relates to the physical elements is required. A brief summary of SIP fundamentals is provided by way of introduction to the subject.

15.2 SATURATED STEAM AND SIP

Saturated steam is a steam–water mixture in which the vapor phase (steam or gas) is in equilibrium with the liquid phase (water or condensate). Saturated steam can exist at only one temperature and pressure along the saturation curve (see Figure 15.1). The addition of heat to saturated steam can result in its desaturation (or superheating). The loss of heat from saturated steam will result in additional condensation. Steam sterilization occurs most effectively when **saturated steam** contacts a surface or microorganism. The presence of

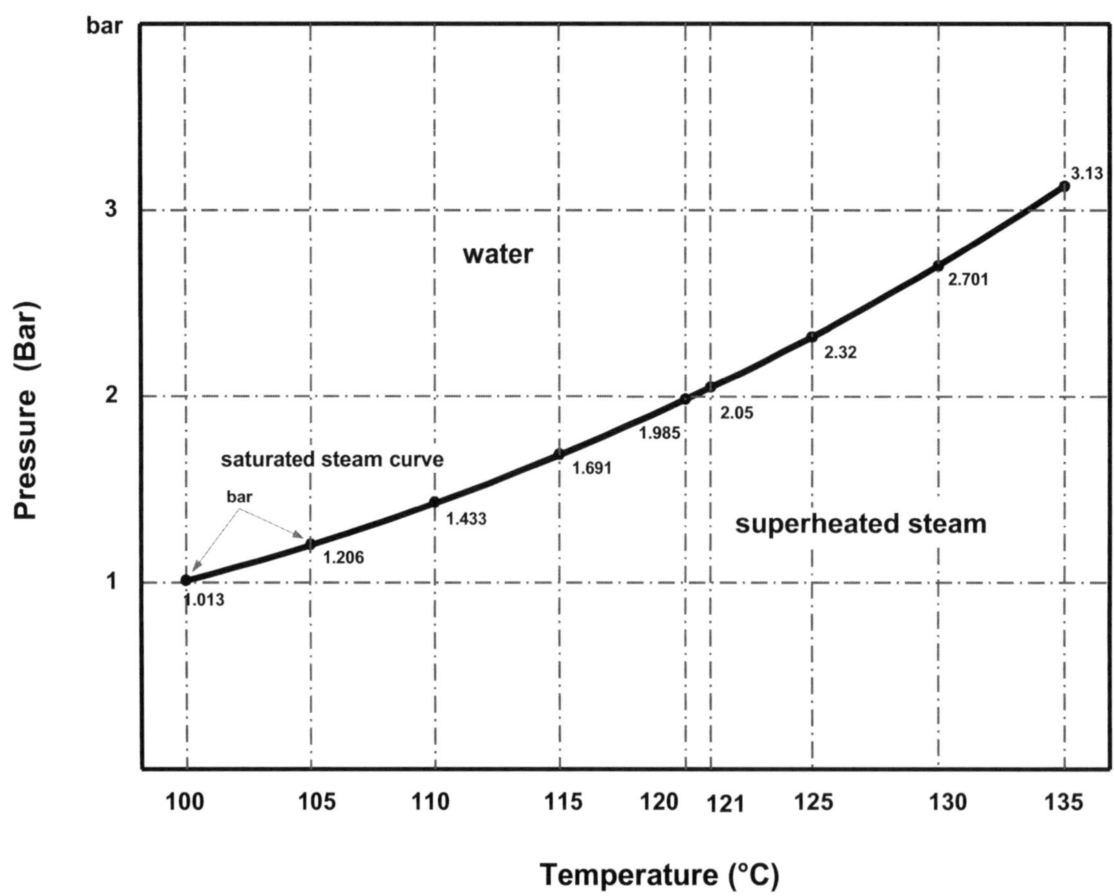

FIGURE 15.1 Saturated steam—temperature vs. pressure.

liquid water is required for the effective sterilization through denaturation of proteins in the cell wall at temperatures in the range of 121°C. Saturated steam is thus far more effective as a sterilizing medium than superheated steam in which liquid water is absent. In order to raise the temperature of an object with saturated steam, the steam must undergo a phase change to the liquid state at which time the heat of condensation is released. Of necessity, this produces a large amount of condensate especially at the beginning of the sterilization process when the process starts with the equipment at ambient temperature (see Figure 15.2). **Superheated steam** is steam that has been heated above its saturation temperature. The presence of this additional thermal energy converts any liquid condensate in equilibrium with the steam to the gas phase. The absence of a liquid phase in superheated steam markedly reduces its lethality to microorganisms. In effect, saturated steam behaves similarly to dry heat as a sterilizing vehicle. Dry heat is significantly less effective at the conventional temperatures (115°C–125°C) at which moist heat sterilization is employed. Caution must be exercised to assure that the steam utilized in a SIP process is saturated and not superheated.

Consider the difficulties in effecting the steam sterilization of a system or piece of equipment in situ. In order to be effective at the conventional temperatures for steam sterilization of approximately 121°C, the process must use saturated steam. The need to heat large masses of stainless steel from ambient temperature to 121°C and the loss of radiant heat to the surrounding room will result in the creation of large quantities of condensate, especially during the start of the process. Although the condensate will initially be in equilibrium (exist at the same temperature-pressure as the steam), it will continue to transfer heat to the surrounding cooler surfaces and continue to drop in temperature (and become less effective as a sterilizing agent at these lower temperatures). Supplying additional saturated steam to the system in an attempt to raise the temperature of this condensate will only result in the formation of additional condensate! The only solution to maintaining systems at the proper temperature for effective moist heat sterilization is through the removal of condensate from all parts of the system. This can only be accomplished by the positioning of condensate drains at every low point in the system. With this discussion as background, the emphasis placed on condensate removal in SIP processes found later in this chapter will be better understood.

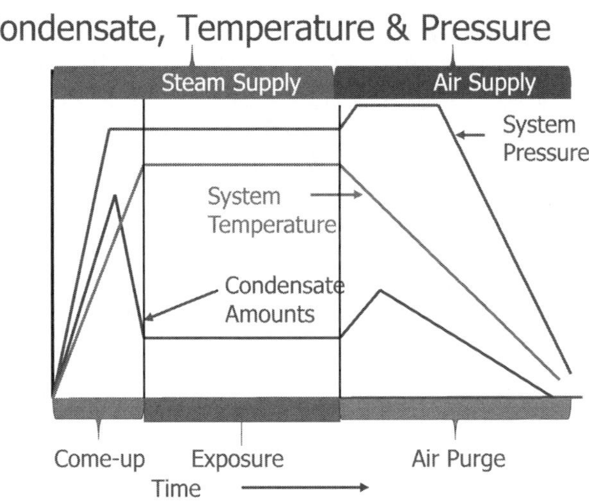

FIGURE 15.2 Temperature, pressure and condensation progression during SIP.

15.3 SIP FUNDAMENTALS

SIP differs only slightly from steam sterilization in autoclaves. The major difference is that for effective SIP, the sterilization scientist must ensure that the elements necessary for process effectiveness inherent in the autoclave design and operation are provided in the SIP system design. The following measures are of particular importance in SIP and they must be properly addressed in the design if the sterilization process (and its ultimate validation) is to be successful:

1. Complete displacement and elimination of air from the system.
2. Constant bleeds of steam at all low points to eliminate condensate retention.
3. Strict adherence to the sterilization sequence.
4. Proper maintenance of the system sterility after the process.

Each of these plays a major role in the design of the system and an expanded discussion of each is necessary to understand their importance. In order to better understand these concerns relative to the implementation of an SIP process, a parallel review of relevant steam autoclave technology is beneficial. Several essential features of steam autoclave design will be contrasted with the parallel aspects of SIP technology (Table 15.1).

The contemporary steam sterilizer is much more than a large pressure vessel fitted with a clean steam supply. It is a

TABLE 15.1
A Comparison of Steam Autoclave and SIP Design Features

Steam autoclave design feature	SIP system equivalent
Vacuum pump	Either not present or not utilized
Low drain	Multiple drains / bleeds required
Vent filter	Required (process filters may also be present)
Sequencing controller	At times (majority are manually operated and even automated systems include some manual steps)
Temperature controller/recorder	Temperature recorder Temperature control less well suited
Pressure recorder	Pressure Controller / recorder more common
Insulation	Required (where possible)
Steam heated jacket	Not recommended (or left empty)
Atmospheric break to drain	Atmospheric breaks on all bleed lines

sophisticated system designed to rigid specifications intended to achieve a set of restrictive conditions reproducibly. Modern steam sterilizers include numerous design features that make them far more reliable and effective when compared with their predecessors (see the earlier chapter in this text for additional details on steam sterilizer design).

15.3.1 Air Removal

The first step in many steam sterilization cycles is the removal of air from the load. (A notable exception are autoclaves designed for the terminal sterilization of filled containers where the presence of air is sometimes necessary to maintain container integrity.) If excessive air remains within the chamber, steam penetration is slowed and the development of cold spots within the load is more likely. Sterilizer manufacturers commonly utilize mechanical removal of air via multiple pre-vacuum cycles to improve air removal. Air removal serves to shorten come-up times (time to sterilizing temperature), improve temperature uniformity, increase steam penetration and consequently improve sterility assurance. The pre-vacuums alternate steam and vacuum pulses, with the vacuum removing air and condensate. Condensate removal is simple as the number of drain lines in a sterilizer are limited and are generally close to the vacuum pump.

In contrast, relatively few SIP systems employ pre-vacuums to assist in air removal largely because of the system scale and presence of multiple low points where condensate is potentially retained, which can slow the vacuum pulse. It is the responsibility of the SIP system designer to provide other means for the elimination of air and condensate. The most common method employed is the addition of bleeds to the system. Consider for a moment, a fixed tank with multiple inlet lines for vents, filter housings, rupture disks, process fluid, pressure gauges, etc. in the headspace of the vessel (Figure 15.3). Each of these offers a potential location for air entrapment and should be evaluated to determine if an air bleed is required. Bleeds should be added at the end of each line and at each low point in the system to facilitate air removal. Positioning outlets at the points in the system farthest from the steam supply facilitates steam penetration to those locations. The selection of bleed locations is facilitated by the need to remove condensate from the system at many of these same locations throughout the process (see the following discussion).

Caution must be exercised in displacing the air inside the equipment too rapidly, as it may result in entrapment of air in locations that might be removed under a slower depressurization of the system. Steam introduction is usually through filters and positioned in the upper portions of systems, using downward displacement of the colder (and denser) air. This step has many similarities to the gravity displacement of air in autoclaves that was universally used for many years in sterilizers.

It has been noted that pre-vacuums are not the norm in SIP systems. Lyophilizers are a possible exception, as their physical design and operational characteristics makes the use of pre-vacuums more attractive. Pre-vacuums have been successfully utilized in other SIP systems where the additional complexity of the vacuum system is deemed to be offset by the more rapid steam penetration achieved. Where vacuum is

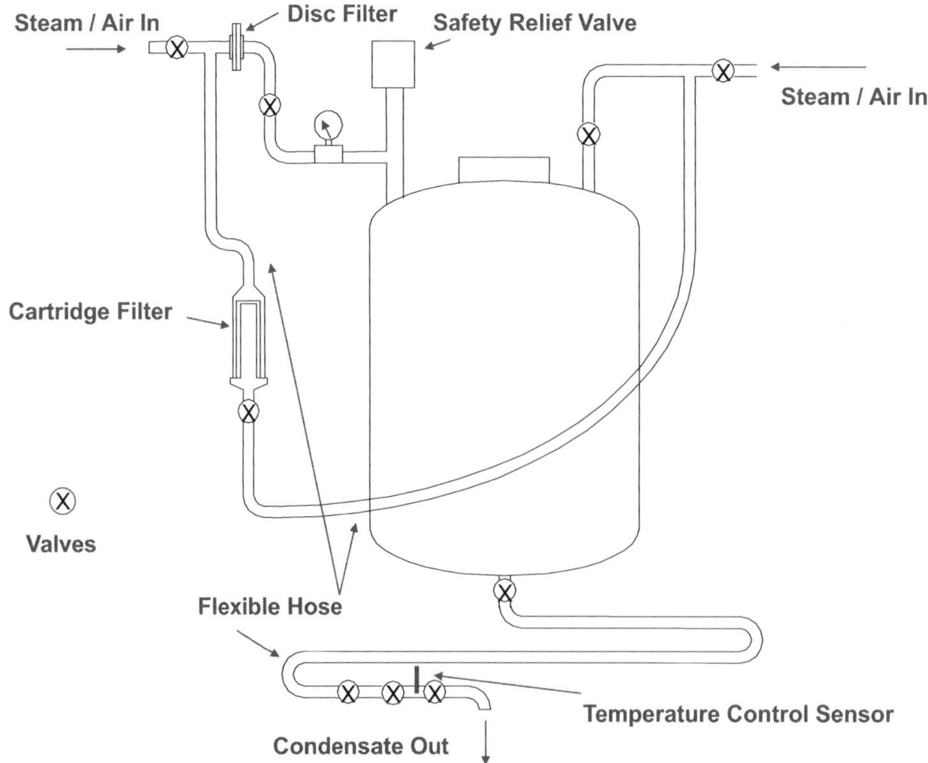

FIGURE 15.3 Portable tank as configured for SIP.

utilized in any SIP system, it would be drawn using a water-filled liquid ring pump, and not the more capable, but incompatible oil-filled vacuum pumps used for deep vacuum.

15.3.2 Condensate Removal

Another aspect of steam autoclave design that is of importance in SIP systems is the reduction and removal of condensate. In modern steam sterilizer design, this is accomplished through several design elements: the use of a surrounding jacket, the presence of a thermostatic steam trap and the application of insulation to the external surfaces of the sterilizer. These features serve to reduce the steam requirements for the chamber and to facilitate the removal of condensate formed in the chamber.

Steam autoclave design conventionally includes a steam jacket, operating at a temperature and pressure slightly lower than that of the sterilizing chamber. Steam sterilizers are also generally well insulated to avoid excessive heat loss. One of the purposes of the external jacket and insulation is to reduce the quantity of chamber steam needed to heat the body of the autoclave with a corresponding reduction in the amount of condensate formed (with the additional benefit of a slight reduction in overall cycle time). SIP systems are either un-jacketed (to reduce their complexity) or in cases where a jacket is already present, the jacket is maintained empty in the SIP process to avoid superheating of the internal steam. SIP systems are also less likely to have insulation on all of the exposed piping (a result of size, weight, location, needed flexibility and set-up time considerations). The absence of a jacket and insulation on large parts of the system means that the typical SIP system will produce considerably more condensate than an autoclave of similar internal volume.

Steam sterilizers have carefully sized and positioned thermostatic traps to maintain internal pressure while effecting rapid removal of air and condensate. Extra care is taken during the installation of sterilizers to level the unit, thus preventing the accumulation of condensate at other than drain locations. The condensate in SIP systems is removed via bleeds/drains at all low points in the system. These drains are most often thermostatic steam traps but may be adjustable (manual or automatic) valves or even fixed orifices. Condensate retention is always detrimental to SIP process effectiveness. The condensate will be at a temperature lower than that of the steam because of the loss of radiant heat to the uninsulated piping and unless removed can become cold enough to prevent adequate sterilization (recall that the rate of death for microorganisms is an exponential function of temperature and small differences in temperature can mean large differences in sterilization process lethality). There is documented evidence that the resistance of spores (especially those on spore strips) is increased when in the presence of water relative to a steam at the same temperature.[12] A well-designed SIP system will have bleeds/drains installed in each horizontal leg and at every low point in the system.

In the design of a tank and piping system, which is to be sterilized in place, the proper sloping of lines will assist in conveying condensate to the appropriate drain location. The size of each bleed in an SIP system, whether for air or condensate, is an important consideration in system design. Bleeds in SIP systems should vary in size in relation to the amount of condensate expected to collect at a particular location. An overly large bleed will result in the use of additional steam to maintain system pressure, whereas a bleed that is too small risks non-sterility because of condensate buildup and/or air retention. Air retention and condensate accumulation are so detrimental to the execution of an SIP process that system design should err on the side of caution, using a greater number and larger bleeds (and consequently more steam) rather than a lesser number and smaller ones.

15.3.3 Procedural Conformance

A critical part of contemporary steam autoclaves is the control system. The control system regulates the temperature within the sterilizing chamber, performs the sequencing of steps that brings the unit through its process cycle and documents the process during its execution. The control system of the autoclave ensures that even the most complex cycle can be carried out reliably and consistently. The proper positioning of valves, regulation of temperature and correct sequence of sterilizer operation is assured by the presence of a well-designed process control system.

In contrast, many SIP systems have no control system and rely on an operator's conformance to a detailed Standard Operating Procedure (SOP) and careful monitoring of process variables to achieve success. An operator is responsible for the execution of the process steps in the correct sequence at the appropriate time. This task is made more difficult by the large number of air and condensate bleeds whose manipulation at the appropriate time is essential for the proper completion of the sterilization process. In a very large system, this could require multiple operators working cooperatively.

Where an SIP system is automated, consideration of procedural conformance must be factored into software development. The correct sequence of operation must be established in the software, and changes to the software may be necessary during both SIP cycle development and after completion of the validation effort. Manufacturers of large pieces of process equipment such as lyophilizers, fermenters, etc. will often provide a microprocessor-based control system that can increase process reliability markedly, a major concern given the greater complexity of those systems. In these larger and more sophisticated systems, automation of the SIP procedure is more common although some manual tasks may be required. A control system makes the successful execution of SIP processes roughly comparable to the operation of an autoclave. However, it is still safe to say that the majority of SIP systems have less automation and are more dependent on the operator(s). For manual operations, a comprehensive SOP is essential to a successful sterilization procedure, as a mistake in timing or sequence could result in a compromise to sterility.

15.3.4 Post-Sterilization Integrity

A key element of steam sterilizer design that has relevance for SIP systems is post-sterilization integrity. In an autoclave, the exposure period is generally followed by a vacuum drying cycle and an eventual return of the chamber to atmospheric pressure just prior to unloading. Safeguards inherent in all new autoclave designs are leak tested chambers and vent filters to maintain the sterility of the load between the end of the steam exposure and the unloading of the sterilizer. The load items in the sterilizer are often wrapped, which protects their sterility post-cycle.

SIP systems must have similar capabilities. It is essential to maintain sterile conditions in the vessel from the start of cooldown until the system is ready for use. In SIP, the system internal surfaces must be protected from contamination post-sterilization. Maintenance of sterility is often accomplished by the introduction of a pressurized gas into the system through an appropriate filter at the end of the steaming step. In SIP processes, a high-pressure gas (air or nitrogen) is introduced into the system through a sterilizing filter while the system is still under positive steam pressure. The system must then be purged of the residual steam and condensate and maintained under positive gas pressure until ready for use. Additional fittings, valves and piping may have to be added to the system to protect the connection points and other components from microbial contamination prior to use. The introduction of a gas purge can dry the system, an issue of some importance if the product or material to be manufactured in the equipment is nonaqueous in nature. This is accomplished by keeping the low point bleeds open until no more condensate is exiting. If the compressed gas supply to the vessel is not maintained until the tank has completely cooled, there is a potential for the development of an internal vacuum within the system that could result in post-sterilization contamination of the internal surfaces.

SIP systems and steam sterilizers share similarities in their critical functions: air removal, condensate drainage, procedural conformance and post-sterilization integrity. The difference between a steam sterilizer and an SIP system is straightforward. The autoclave is exclusively designed to perform sterilization processes, whereas SIP systems are designed for other processes under aseptic conditions that necessitate their sterilization. The autoclave is not in direct product contact, it provides a means to sterilize wrapped parts that will be in product contact later in the process. Major parts of the SIP system surface will be in direct product contact and their sterility must be assured. Despite these physical differences, the functions necessary for sterilization effectiveness can be achieved in both; however, that result is attained in different ways. Success with SIP is largely based upon proper attention to design details. Focusing on air elimination, condensate removal, proper sequencing and post-sterilization integrity should lead to greater success with SIP design and implementation.

15.4 SIP SYSTEM DESIGN

The application of SIP system design concepts can be achieved through a review of the equipment elements that make up that system. As most SIP systems are composed of combinations of many smaller components, the reader is encouraged to review the relevant parts of the text for items present in their system. SIP system designs are largely based upon the field experience rather than rigorous engineering designs. There have been efforts to provide a more rigorous scientific basis for the empirical nature of the design, but unfortunately the gulf between theory and practice is still quite large.[13,14,15,16,17,18] The recommendations that follow are derived from experience with a range of SIP system designs, with consideration given to the underlying heat and mass transfer concerns found in the referenced materials.

A useful approach in SIP system design is to establish the "system boundary" that delineates the extent of the sterilized surfaces. The system boundary should extend somewhat beyond the equipment surfaces that will be in product contact during equipment use. Validation of the SIP must demonstrate that all points within the system boundary are properly sterilized and maintained sterile until use. In a very large system, it may be beneficial to sterilize it in sections (more on this later in this chapter). In this case, multiple overlapping system boundaries would be developed.

15.4.1 Pressure Vessels

The term "pressure vessel" includes equipment such as fixed and portable tanks, bioreactors, blenders, centrifuges, freeze dryers, crystallizers and other equipment that can be sterilized in place. For ease of discussion, the term "tank" is utilized in this section to represent all of these types of equipment. The majority of SIP systems consist of pressure and full vacuum rated tanks, with associated piping. A typical tank will have numerous nozzles that are used for manholes, sight glasses, lights, rupture disks, pop-off valves, pressure gauges, temperature wells, dip legs, etc. (Figure 15.3). These items are supplementary to any process piping, vent and process filters and associated valves required on the tank. Design concerns for vessels are numerous:

1. All inlet lines to the tank should be kept as short as possible to minimize air holdup and reduce condensate formation. Valves should be placed as close to the tank as possible. Lines entering the headspace of the tank should be vertical if at all possible. If "horizontal" lines are present, they should be pitched (typically 1/100) to assist condensate flow out of the system.
2. Traditionally, the "6D" rule would have been recommended for all piping connected to the tank.[19] With modern fitting fabrication and improved welding techniques, this can be improved to "2D" (for branch sizes less than 1") or less. These rules relate the length of the branch pipe to its diameter. They derive from the principles of fluid dynamics, more specifically the Reynolds' number, a dimensionless number describing the relationship of various forces within the fluid stream. The higher the Reynolds number, the greater the turbulence. The greater the turbulence, the less the likelihood of a dead leg causing a sterilization problem. A common error in applying the "6D" rule is to utilize the diameter of the larger pipe rather than the smaller pipe (Figure

15.4). Even when properly applied to SIP systems, adherence to the "6D" rule is not always sufficient to adequate sterilize the "dead leg" segment. Any portion of the system that extends from the body of the tank or out of the main process flow stream in a pipe must be minimized. Even with short piping lengths, however, the installation of a bleed or trap at the far end of the "dead leg" should be considered. Comprehensive discussion of the sterilization difficulties associated with "dead legs" has been discussed in the literature.[15,16,17,20]

3. Rupture discs are generally preferable to relief valves because of the reduced dead volume and cleaner interiors. In addition, rupture discs can be positioned closer to the tank proper, minimizing the distance that the steam has to travel from the tank proper. If a relief valve must be employed, it may be necessary to open it slightly during the sterilization to ensure steam penetration.
4. The use of pre-vacuums to assist in the removal of air prior to the introduction of steam can be employed to expedite the air removal process. Although this is an effective technique, it may not be beneficial. The use of pre-vacuums is most prevalent on lyophilizers where vacuum rating of the equipment is built in, and the ability of the system to maintain a tight vacuum can be used to advantage. The use of vacuums on tank SIP sterilization is less common, though those firms that employ pre-vacuums feel strongly about their utility in aiding steam penetration. Those that do not use vacuum are equally convinced in their choice. Regardless of whether pre-vacuums are used or not, any vessel subject to SIP must be vacuum rated.
5. Condensate bleeds should be placed at every low point in the system to facilitate condensate removal. Preference as to the use of traps, valves or fixed orifices as bleeds in a SIP system varies from firm to firm. Each can be successfully employed in a SIP system. The size of the bleed required at any location will vary with the amount of condensate expected at that location. As a general rule, the bleed should be somewhat oversized, as retained condensate is a far more serious problem than excessive steam consumption (An expanded discussion of the choice between valves, traps and orifices is provided later in this chapter).
6. The amount of steam required for the sterilization of a system is directly related to the size and mass of the system. Bleed locations and size should be selected to facilitate drainage accordingly. A 6-inch difference in the location of a bleed can make the difference between an acceptable and unacceptable system.
7. Caution should be exercised in the use of jacket steam in the sterilization of equipment. Although providing heat to the jacket may reduce the internal steam requirements to bring the system to sterilizing temperature, it may create problems with superheating of the steam in the vessel proper. Jacket steam is generally supplied at pressures considerably higher than that required for sterilization in order to facilitate the transfer of heat into a liquid-filled tank through a limited surface area. If jacket steam is utilized, the internal saturated steam may be heated to a temperature above saturation thus rendering the process ineffective as a means of sterilization.
8. A frequent question with regard to SIP systems concerns the use of superheated water in lieu of saturated steam as a sterilizing medium. This was addressed by J. Carlson in which the advantages of superheated water over saturated steam were outlined with regard to the energy savings possible through the use of superheated water.[21,22] The substitution of superheated water for steam is possible, provided one could ensure adequate flow throughout the system. If the objective is to sterilize a pipeline, then superheated water may be preferable. If the system includes complex piping such as found above a typical piece of parenteral manufacturing equipment, then steam systems are generally simpler and easier to control. In discussions with Mr. Carlson, there was general agreement with this distinction as to sterilizing medium preference.[23]
9. An atmospheric break should be provided between the discharge of each bleed and any collection point. Direct discharge of multiple condensate lines into a common sealed header risks potential competition among the lines for clear discharge, and the buildup of excessive back pressure that could inhibit condensate removal. Immersion of the bleed(s) in a collection vessel or drain sump is not recommended.
10. Insulation should be installed on tanks wherever feasible to reduce steam consumption and improve worker safety.

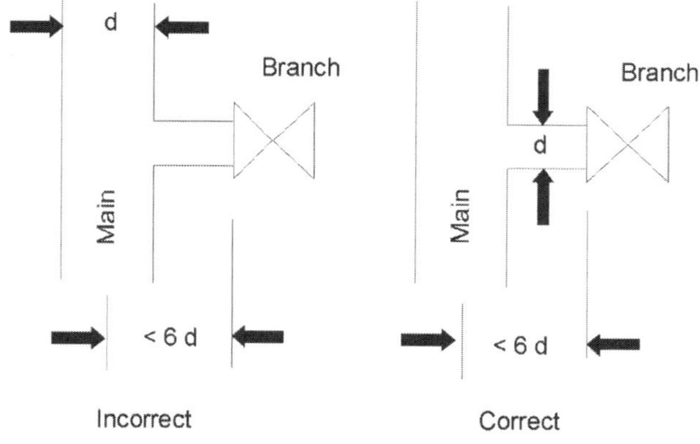

FIGURE 15.4 The '6D' rule visually.

15.4.2 Piping Systems

Piping is utilized to connect multiple pieces of equipment into a process train (common in larger systems such as sterile bulk and biotechnology facilities), or for utility systems as used for compressed gases where the piping constitutes virtually the entire system. Attention to detail in the design of the piping system is essential if SIP is planned:

1. Sanitary design principles should be adhered to including considerations such as: materials of construction, surface finish, cleanability and drainage.
2. Pipe runs should not be perfectly horizontal; the lines should be pitched (approximately 1/100) to provide for condensate drainage. Steam inlets to piping systems should be located at the highest points in the system. Bleeds must be placed at all low points in the system.
3. Valves should be placed in vertical runs of pipe to the extent possible, to minimize the potential for condensate retention during SIP. Where a valve must be closed during SIP, provision for air and condensate removal as close as possible to the sealing surface should be made to avoid creation of a dead leg. Where a valve must be open during SIP, the bleed should be placed beyond the valve extending the system boundary beyond the valve (Figures 15.5 and 15.6).
4. Sanitary diaphragm valves should be used wherever possible. Where tight shutoff is required, ball valves or other types may be required, notwithstanding the limitations in sterilizing the hidden surfaces of the vakve. The functionality of the intended valve should prevail over the desire to have the system and valves as sanitary as possible. A similar situation results with needle or other types of valves that cannot conform to sanitary design concepts. Sterility is essential, but if the equipment cannot be operated properly with a sanitary valve, then a non-sanitary valve that performs properly should be installed with whatever changes in materials of construction, design, etc. necessary to minimize its impact on the sterility of the system. The use of a diaphragm valve as a shut-off valve on a lyophilizer will result in an unacceptable leak rate; thus, a ball valve would be utilized despite its non-sanitary design.
5. Piping configurations shaped like the letters, "W", "M", "V", or "N" should be avoided as each of these will have a low point at which condensate can collect. Condensate removal (using valves, traps or orifices) must be provided for at any low points in the system.
6. Where pumps are utilized in the system, they should be capable of being sterilized in situ. Most centrifugal pumps are acceptable, whereas piston, vane and similar pumps with multiple chambers cannot be easily sterilized and should be avoided in SIP systems. There should be provision for condensate drainage from the low point of the pump housing.
7. Where flexibility in piping arrangement is required, permanent piping manifolds with appropriate valves or spool pieces to make temporary connections should be employed. The use of flexible hoses to make temporary connections may result in the inadvertent creation of low points that can retain condensate and reduce the effectiveness of the SIP process (Figures 15.7 and 15.8). Permanent welded systems are generally preferable to systems assembled from individual fittings, as control over system configuration is essential for proper SIP performance.
8. The use of insulation on the exterior of piping is recommended to reduce condensate formation because of radiant heat loss. Where insulation is not provided, the resultant loss of heat will increase condensate formation.

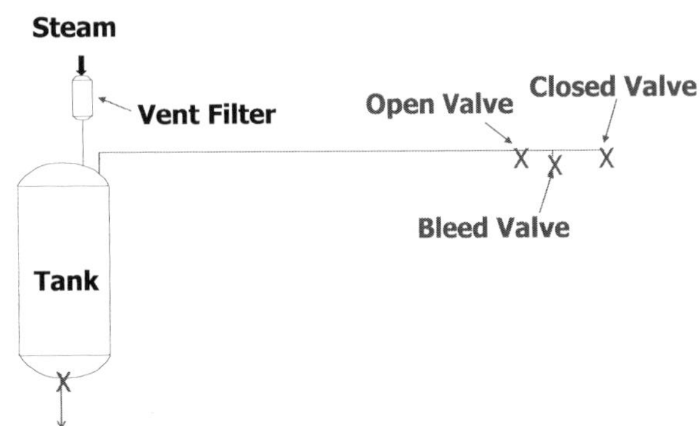

FIGURE 15.5 Tank with lengthy horizontal input line.

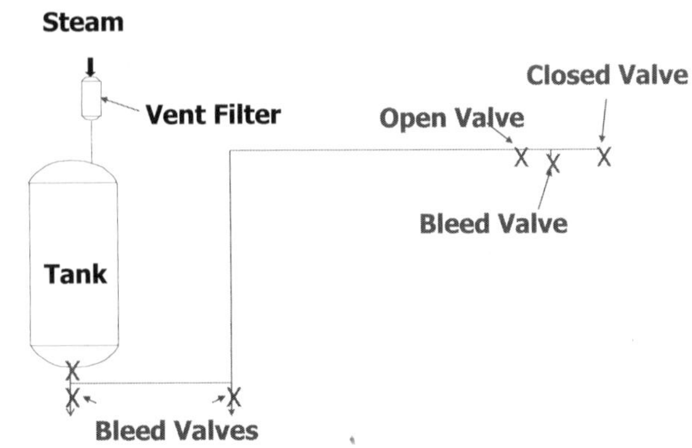

FIGURE 15.6 Tank with lengthy horizontal and rising output line.

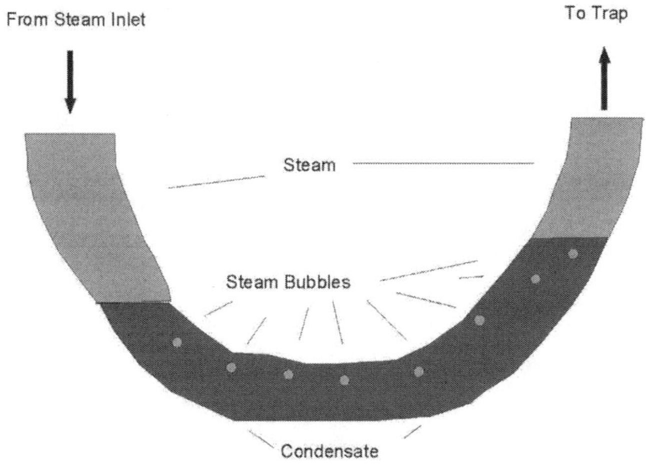

FIGURE 15.7 Condensate retention within hose low point.

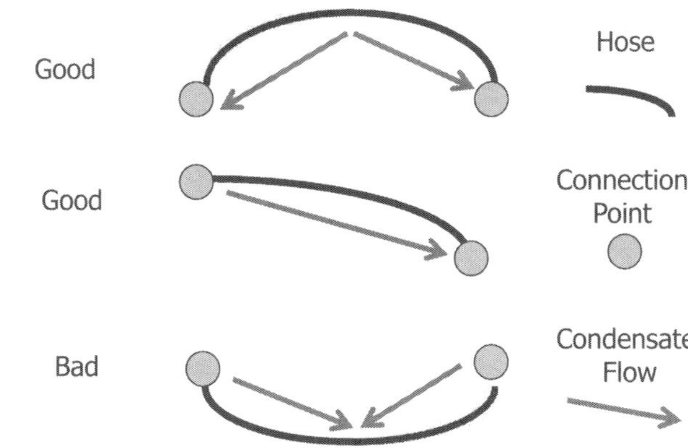

FIGURE 15.8 Proper flexible hose orientation.

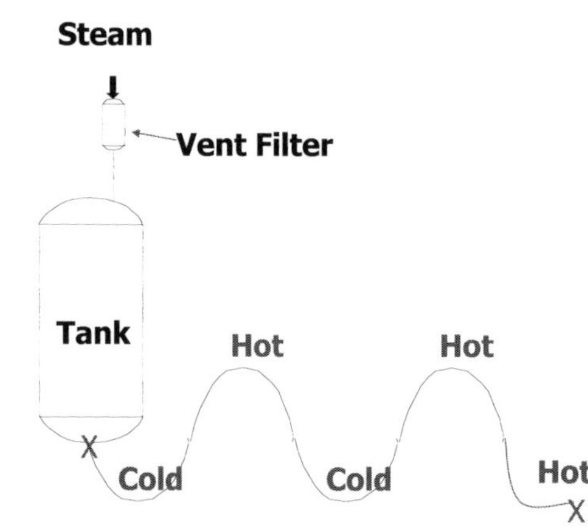

FIGURE 15.9 "Roller Coaster" discharge line.

9. Minimize dead legs in accordance to avoid inadequate sterilization of the divergent leg. (Note that the "6d" rule is no longer a benchmark and shorter branches are achievable.)
10. Where a low point in the system is not bled, condensate buildup can occur despite the presence of acceptable temperatures farther along in the system. Sufficient amounts of steam can flow along the top of the pipe to sterilize portions of the system farther from the steam source (Figure 15.9). Alternatively, "bubbles" or "bursts" of steam can intermittently pass-through condensate flooded areas to provide steam for downstream piping. The presence of sufficient temperature downstream of a potential collection point for condensate should not be interpreted as an indication that the intermediate point is adequately sterilized.
11. Gauges and instrumentation installed on the system should be of a sanitary design, with wetted parts having minimal surface area, easily cleaned surfaces, etc. The evolution of instrumentation design is such that virtually all major instrumentation types are available in sanitary designs (e.g., diaphragm type).
12. In liquid processing or fluid distribution systems, valves must be utilized to avoid fluid loss during use, even if traps or orifices are utilized during SIP to remove air and condensate. The choice between valves, traps and orifices is largely one of personal preference. Each has advantages and disadvantages in their application in an SIP system (Table 15.2). The size, location, amount of condensate, ease of operation, system usage, etc. dictate selection of the appropriate item. The case can also be made for a system that utilizes both a trap and a valve at each location where air or condensate must be removed. The use of both items in the same location affords the user the advantages of both, with the disadvantages of increased cost and complexity. The practitioner should view the choice between traps, valves or orifices as a choice among near equals, with the preference in each situation dictated by the specific circumstances involved.

One similarity that sterilizers and SIP systems should share is the presence of an atmospheric break at every drain location. This was recommended in the proposed United States Food and Drug Administration (FDA) large volume parenteral (LVP) regulation in 1976 to prevent potential backflow into the sterilizer from the drain during sterilizer cooldown.[19] Many installations with complex piping systems, especially in biotech facilities, lack this feature because of extreme concerns regarding environmental protection, which can compromise the ability to sterilize the product contact surfaces of the system.[24] The sharing of a single trap by multiple bleed lines

TABLE 15.2
Advantages and Disadvantages of Valves, Orifices and Traps

	Advantages	Disadvantages
Valves	No additional piping for liquid systems	Positioning can have an effect on condensate removal
	Frequently needed for fluid handling anyway	Waste steam compared with same size traps
	Can be easily automated	More expensive than comparable sized traps
	Some valves can be regulated between full open and full closed	Must be manually or automatically operated
	True sanitary design is possible	
	Can provide feedback to control system	
Orifices	No manipulation necessary for operation	Wastes more steam than valves or traps
	No moving parts	Nonadjustable
	Little maintenance required	Proper sizing is essential
	Least expensive	Requires valve for tight shutoff
	More sanitary than steam traps	Less sanitary than valves
Traps	No manipulation necessary for operation	Must be utilized with a valve in liquid systems
	Easier application in automated systems	Periodic maintenance required for proper operation
	Operate to remove air and condensate only when necessary	Can fail without obvious fault
	Conserve steam	Do not provide tight shutoff
	Usually less expensive than a valve	Either fully open or fully closed
	Smaller size than a valve	No true sanitary design available

is another common error. It can result in condensate retention / reentry.

The ideal bleed design would have a shutoff valve (which demarcates the system boundary) installed as close to the vessel as possible. The next component in line would be a trap or orifice plate. The system would open directly to the atmosphere after that. The bleed piping should be as short as possible and slope continuously to the discharge point.

15.4.3 Filters

In order to maintain the post-sterilization integrity of a SIP system, the presence of a microbially retentive vent filter is mandatory. In addition, processing systems often require numerous filters for process liquids, process gases, and vents. In considering how filters are to be integrated into a SIP system design, consideration of the design details described prior remains essential. There are some additional aspects that bear further explanation with regard to filters and filter housing sterilization:

1. Filter housings may require modification to provide for upstream and downstream bleeds to facilitate air and condensate removal. The bleeds that are positioned on the filter housing by the filter manufacturer are intended for use with the filtration and may not reflect the appropriate locations for a SIP process. The proper number and position of bleeds is shown in Figure 15.10.
2. Cartridge filters must be positioned so that the open end is down to facilitate condensate removal (Figure 15.11). It is useful to think of the two major parts of the filter housing as the base and the dome rather than the head and the bowl when positioning the housing (This terminology was changed by the filter industry in the late 1980s without any discussion or defined rationale).
3. The filter cartridges should be examined to see if they are suitable for use in an SIP process. Subtle differences in cartridge design that allow condensate accumulation against the membrane surface can inhibit effective sterilization. The dead volume where condensate can be retained in the cartridge must be kept to a minimum if the filter is to be successfully sterilized in situ.
4. Disc filters are often utilized as vent filters on smaller tanks. When employed for this purpose, the filter housing should be oriented in a vertical plane and bleeds may need to be added/relocated. (Figure 15.12).
5. The use of a "loop" to facilitate sterilization of the membrane filter in its housing may be necessary to reduce the pressure drop across the filter and allow the use of lower steam supply pressure (an expanded discussion of the "loop" method is provided later in this chapter) (Figure 15.13).
6. Depending on the size of the system, it may be advantageous to utilize more than one steam inlet. Attempting to introduce all of the required steam for an SIP system through the process and/or vent filter(s) may not be possible without damaging the filter(s) because of the higher pressure (and correspondingly higher temperatures) needed to deliver sufficient steam to a large system. For large tanks and systems, the use of a direct steam inlet is becoming commonplace, with secondary steam inlets for each filter (This is another application of the "loop" method).

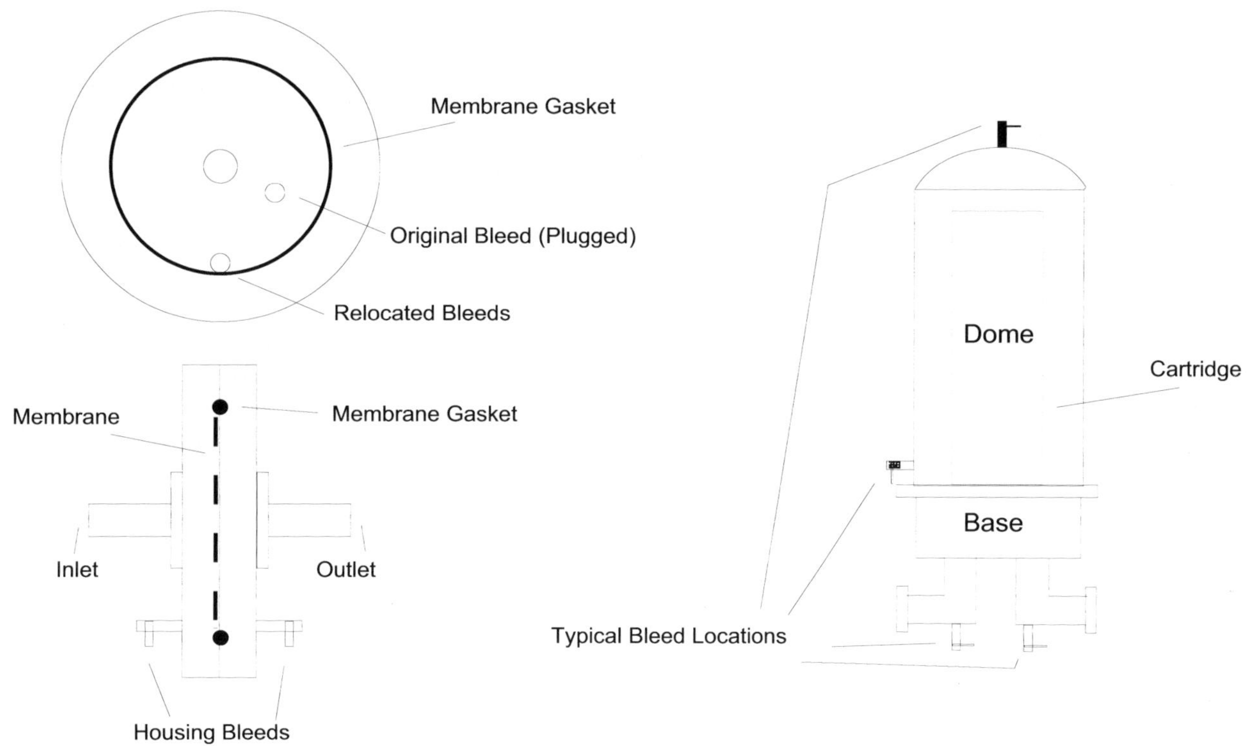

FIGURE 15.10 Typical filter housing modifications for SIP.

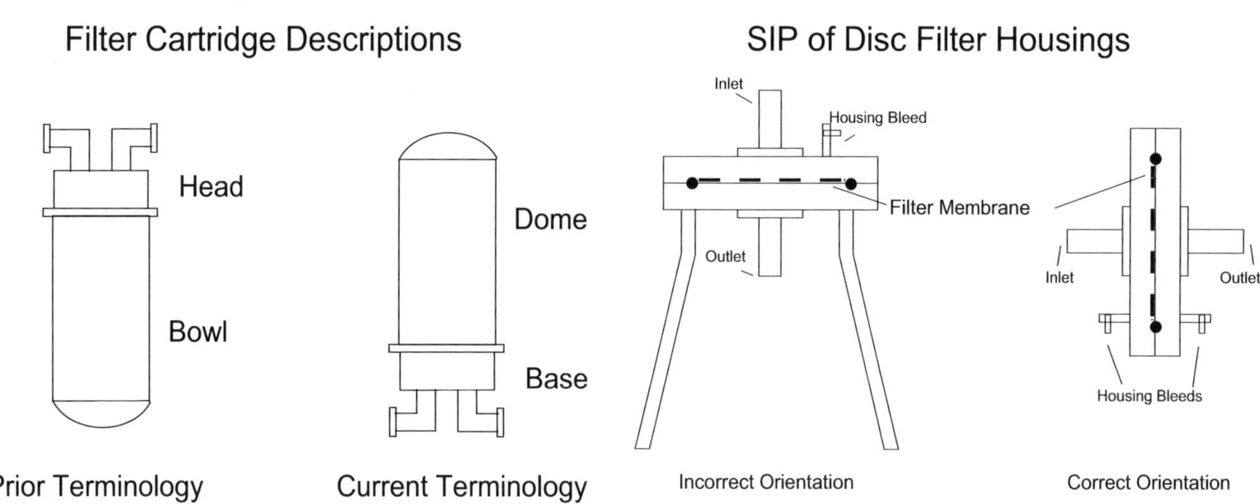

FIGURE 15.11 Cartridge housing orientation.

FIGURE 15.12 Orientation of disc filters for SIP.

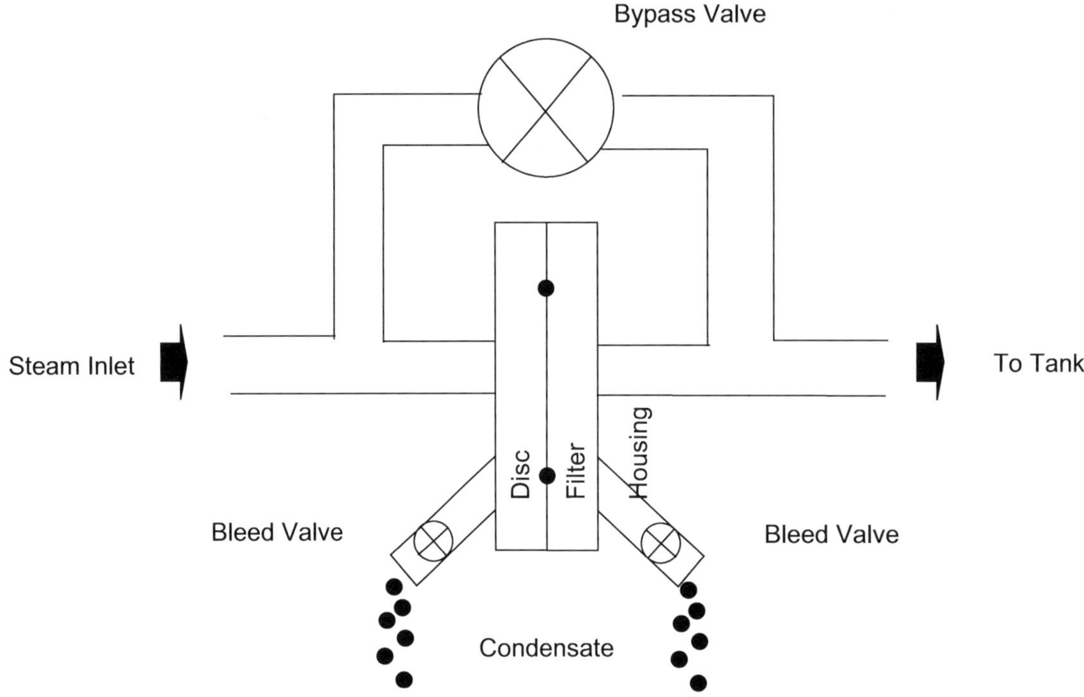

FIGURE 15.13 Typical SIP "Loop" configuration.

7. When using multiple steam inlets to a system, caution must be taken to avoid introducing steam at a higher pressure to the downstream side of the cartridge filter than to the upstream side. Most cartridge filters cannot withstand a differential pressure of more than 2–3 psi in the reverse direction and steaming a filter in the reverse direction may result in the loss of filter integrity. If disc filter membranes are employed, the use of back pressure support screens as a routine measure is recommended. Problems with excessive back pressure can be overcome through proper procedures in starting and completing the sterilization cycle. Similar problems can occur during drying or purge cycles if appropriate precautions are not taken.

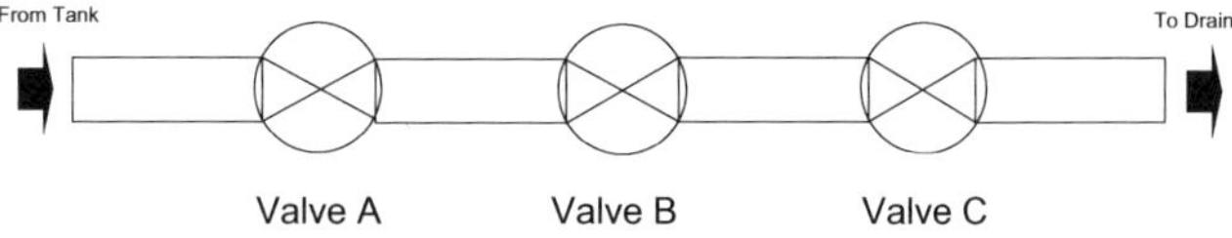

FIGURE 15.14 Triple valve arrangement for SIP and aseptic connection.

15.5 STERILIZATION OF SYSTEMS

Designing for SIP of larger systems follows the same principles established earlier, with the added complications associated with a more complex arrangement of tanks, lines, filters, valves, etc. Any arrangement of piping that results in a low point for condensate collection must be treated as described earlier. In large systems, particularly those in which the vessels are located on the same floor of a facility, there are numerous opportunities for this type of arrangement to occur. The system must be designed in a way that condensate can be readily removed. In order to achieve this objective, it may be necessary to sterilize the system in sections, in which each section process sterilizes a portion of the larger system. When using this type of approach, some portions of the system must be sterilized more than once to ensure that all portions of the system are fully covered.

An example of this is portable equipment that must be sterilized in one location and connected aseptically in another. Consider a portable holding tank with three valves installed in series so that the interior surfaces of the first two can be sterilized in the open position by regulating pressure within the entire system by adjustment of the last valve (Figure 15.14). After completion of the SIP process with a positive gas pressure internally, the second valve is closed and the piping after it is removed. While the system remains sealed prior to use, the closed second valve is utilized to maintain system integrity. When the system is ready for use, the second valve is removed, and an aseptic connection is made to the first valve (now the only one remaining on the line) and that valve is utilized to regulate flow through the line. In this instance, a single aseptic connection is required to connect the line. This arrangement might be utilized on a portable tank that is transported to the filling machine and aseptically connected to it. It is also possible to re-sterilize after making the aseptic connection.

A simple system that depicts how SIP can be utilized is depicted in Figure 15.15. In this system, any condensate formed in the holding tank (where the majority of the system condensate will be created) has no easy means of egress from the piping system. The condensate must be forced upwards from point B past point E before exiting the system at point D. A simple modification of the piping system incorporating an additional steam inlet and a condensate drain below the tank can be utilized to eliminate the condensate retention problem. In the new system (Figure 15.16), there are two separate sterilization patterns. The first sterilization pattern passes steam from the tank via points A, B and C, whereas the second pattern sterilizes the two lengths of piping via E, B, C and E, D. The use of a second pattern allows for the steam to enter the piping at point E in the system, with condensate being removed at the low points. A small portion of the piping near B will likely be sterilized twice in this installation. The addition of a second pattern to be sterilized causes no loss in cycle time, as the second pattern can be sterilized while the tank is receiving material through the liquid filters. The system can be further improved by modifying the piping delivering the solution to the filling machine. Note how, in Figure 15.16, a series of aseptic connections must still be made to add the polishing filter to the system and connect the piping to the filling machine. A modest refinement results in Figure 15.17, in which the polishing filter and its associated piping are sterilized-in-place and a single aseptic connection is made under the laminar flow hood. In an ideal installation (Figure 15.18), the filling machine itself could be sterilized in situ and there would be no aseptic connection required.[25,26]

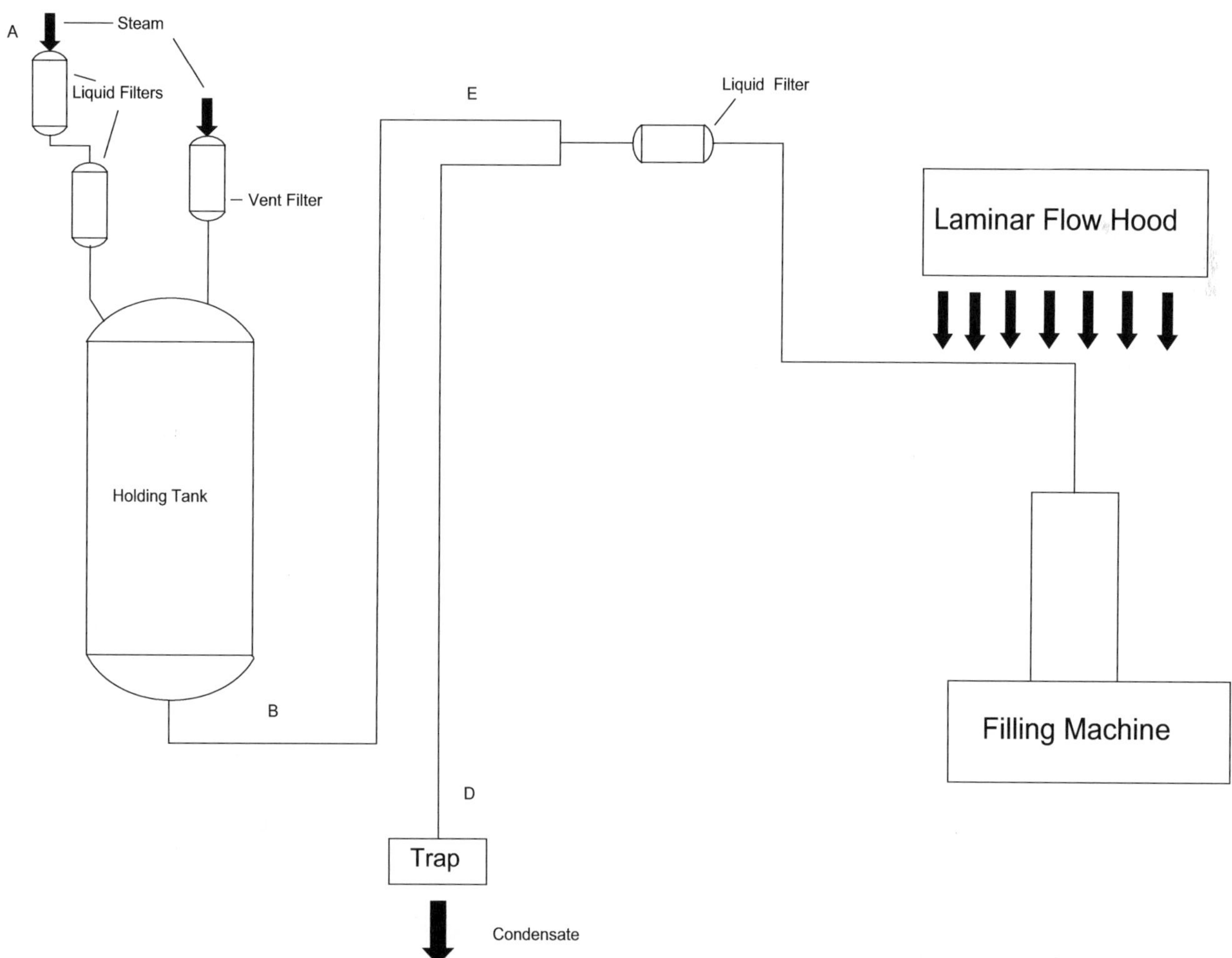

FIGURE 15.15 Tank system as originally configured.

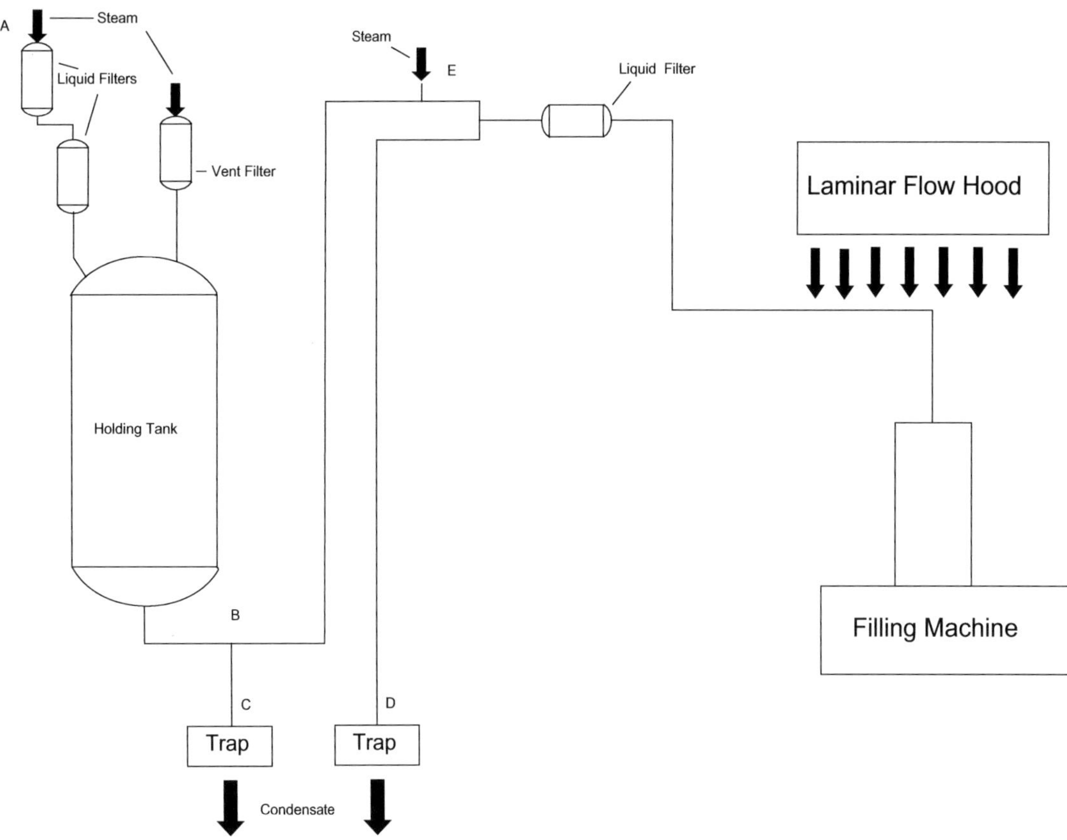

FIGURE 15.16 Initial improvement to system.

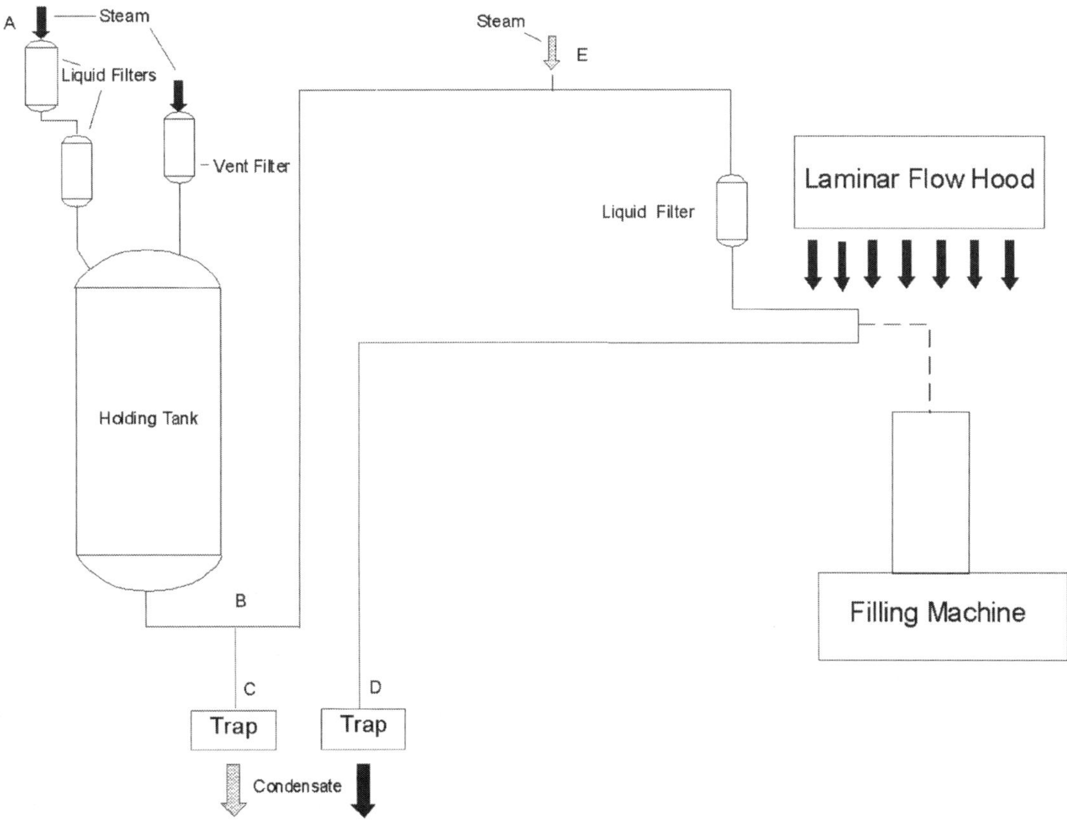

FIGURE 15.17 Second stage improvement to system.

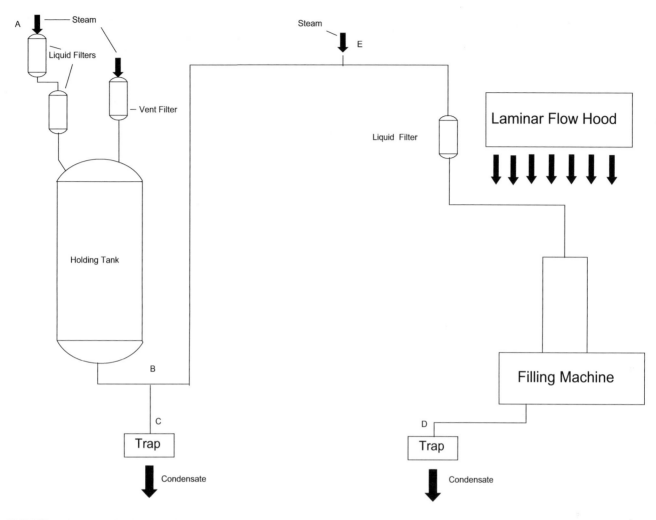

FIGURE 15.18 Possible further refinement to system.

A common arrangement of equipment is shown in Figure 15.19, in which multiple vessels are connected in parallel to a common line leading ultimately to a single outlet. This type of a piping arrangement can be found with lyophilizers where the chamber and condenser are piped to a common drain. The use of the single drain is intended to simplify the control of the sterilization process by allowing the process to be regulated by temperature control based upon conditions at a single point in the system. What has been created is an interactive system, in which the two vessels compete for the use of the drain, and what happens is that each vessel actually uses the drain intermittently and there may be long periods of time when one vessel or the other is operating without an effective condensate drain! This might seem to be of little consequence if the temperatures in the overall system are acceptable. In fact, this type of problem can have serious adverse consequences because of condensate retention and the resultant temperature reduction. This vessel and piping arrangement is quite common and is also present when tanks or filters are installed in parallel. If the vessels are different sizes (Figure 15.20), the larger vessel may have near exclusive use of the drain, with the smaller vessel potentially being significantly colder, containing large amounts of

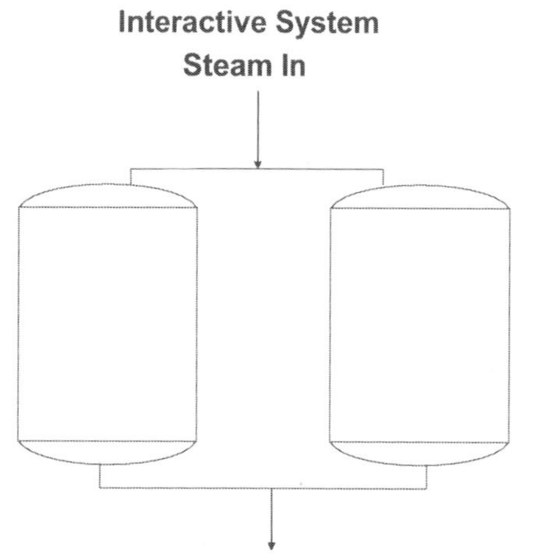

This type of design should be avoided

FIGURE 15.19 Interactive system with equal size vessels.

condensate and thus inadequately sterilized. A proper SIP design avoids this type of piping arrangement. A modification is shown in Figure 15.21 that eliminates the interaction

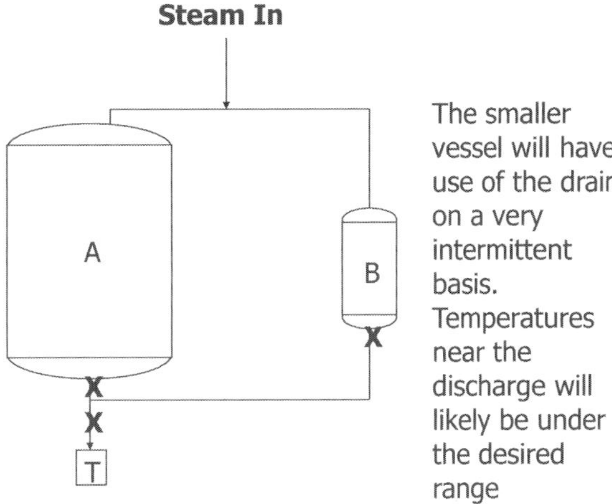
Interactive System
Steam In
A
B
T
The smaller vessel will have use of the drain on a very intermittent basis. Temperatures near the discharge will likely be under the desired range

FIGURE 15.20 Interactive system with different size vessels.

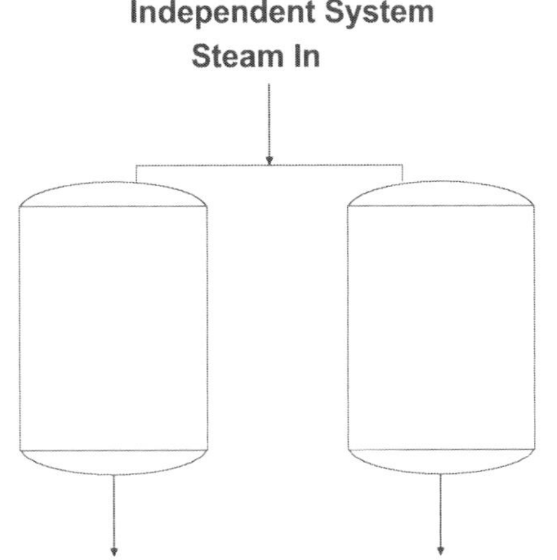
Independent System
Steam In
This design provides more reliable sterilization

FIGURE 15.21 Improved (no longer interactive) system.

between the vessels. A similar appearing arrangement is shown in Figure 15.22, in which two steam supplies on the upstream side of the filters are installed on the same tank (one filter for the process fluid, while the other filter serves as the vent). This type of a system does not operate as an interactive system because the tank acts as a large buffer between the steam supplies and both of them can successfully share a common drain.

A common question with SIP systems is whether they should be located in aseptic processing areas (APAs) at all. It might seem obvious that an SIP system should be in an APA, but if the basic concepts of SIP are adhered to and aseptic connections are completely eliminated then the enclosure of the SIP system in a controlled environment is an unnecessary expense.[23] By strict adherence to design principles and operating procedures, it is certainly possible to effect aseptic processing within sealed vessels outside an APA without risk of contamination. Consider lyophilizers, fermenters and even steam sterilizers, where most of the equipment exterior surface is not located within an aseptic processing area. The application of SIP to large process trains where large portions of the process train is outside of the APA is common in sterile bulk production. These systems have successfully produced sterile products without difficulty with only minimal portions of the system in an APA. Proper attention to the nuances of SIP system design is of greater importance than any additional sterility assurance provided by a controlled environment in the surrounding room to which the product is never exposed.[27]

15.6 SIP STERILIZATION OPERATING PROCEDURES

The details of any individual SIP procedure will vary according to the specific configuration. Despite the uniqueness of each SIP system, following the design concepts outlined previously with regard to air elimination, condensate removal, procedural compliance and system integrity will usually result in a fair degree of similarity among SIP sterilization procedures for different systems. The following section describes some of the more common aspects of SIP procedures. The user must of course develop an SIP procedure appropriate for their specific system. This discussion can serve as a general guide to the preparation of such a procedure.

Prior to the start of any SIP process, the starting position (open or closed) of each valve in the system should be specified. For convenience's sake, all valves should be in the same initial position, and the closed position is generally preferred. The procedure should then indicate exactly when each individual valve is manipulated.

Prior to the introduction of steam, the major drain valves, typically valves or traps below the major vessels and at other low points, should be opened. Steam is then introduced into the system usually at points high in the vessel or piping system. The steam can be fed directly into the system or it can enter through a membrane filter(s). After steam has been introduced, minor bleeds are opened, until steam or condensate is observed at their outlet. Eventually wisps of steam and condensate will be observed at all outlets of the system. The pressure (and therefore the temperature) in the system can be raised by increasing the steam supply or throttling back on the main drain valves. Throttling back on the drain valves is also necessary to build pressure in the system, as the amount of condensate formed will drop as the internal surface approaches the temperature of the steam. Adjustment of either the steam supply to the system or the main drains will be necessary to raise the internal temperature to a steady state where effective sterilization can be accomplished. Adjustment of the bleed valves may be necessary once steady state has been attained, as the amount of condensate formed will be reduced once the entire system is at sterilizing temperature (Figure 15.2). During the dwell portion of the sterilization process, the amount of condensate formed will be near constant and the system will come to a steady state condition where the need to adjust valves on the system will be minimal. Under this condition, a small amount of steam or condensate should be

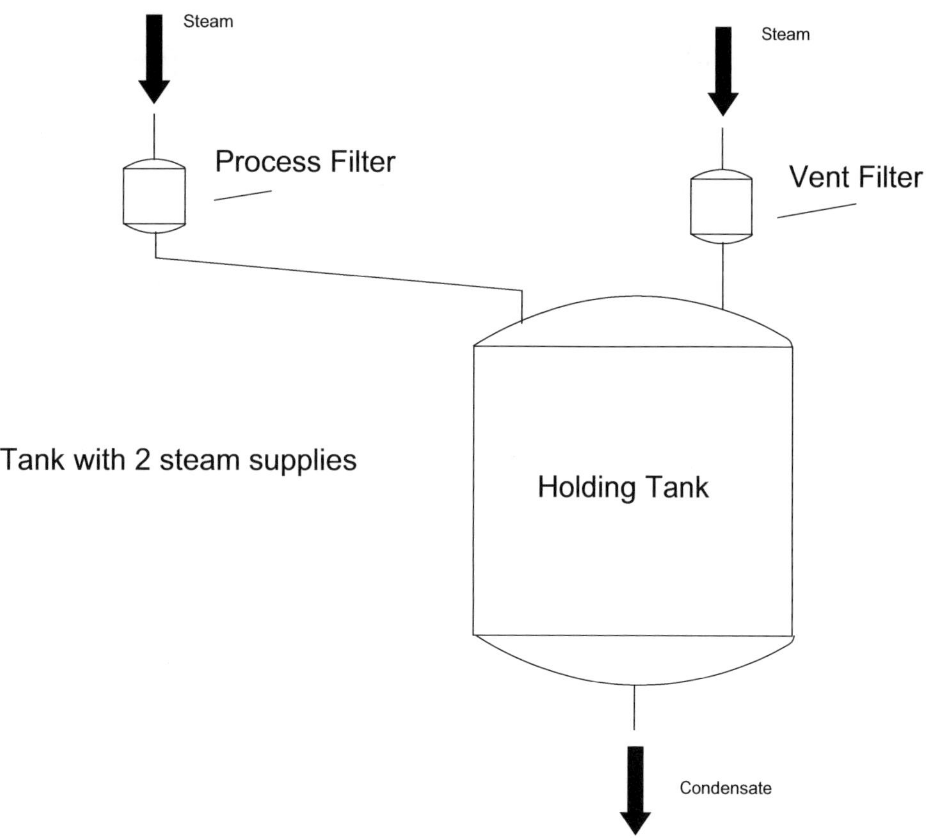

FIGURE 15.22 A non-interactive system.

observed being discharged from each low point in the system. In the sizing and positioning of the valves, the loss of steam from the system because of a valve that is open slightly more than necessary should be tolerated. The alternative situation in which the valve is open slightly less than is required will result in the retention of condensate and the potential compromise of the process. It is always preferable to waste some steam than to risk process failure.

A note of caution must be exercised in bringing the system to the desired sterilizing temperature/pressure. During the initial stages of the process, the bleed locations are required to permit both air and the large amount of condensate formed during the heating of the system. This may require a substantial period of time especially in larger systems or those with small openings. The temptation may be to open the steam input completely in an attempt to shorten the come-up time. This can result in inadequate air removal from the system and reduced cycle effectiveness, despite the apparent successful temperatures attained. What can happen is compression of air trapped in the system by the steam. As the pressure of this trapped air increases so will its temperature. This can provide the appearance of effective sterilization; however, hot air is an extremely poor sterilant. A simple means to confirm the effectiveness of air removal is to compare the temperature/pressure relationship with the desired saturation conditions immediately after attaining the set point temperature. A slower come up is generally preferable as it allows sufficient time for air removal.

Once the desired time at temperature has been achieved, the system must be shut down in an orderly manner. The simplest approach to shutting the system down involves the simultaneous closure of the steam supply valve(s) and introduction of a high-pressure supply of a sterile purge gas. The use of a gas supply pressure higher than that of the steam is necessary to insure the continued outflow of steam/condensate and eventually the purge gas itself from the system as the system cools. All portions of the system must be supplied with a sufficient volume of the purge gas to maintain positive pressure. Then as each point in the system drops in temperature, the purge gas pressure will prevent the recontamination of the system from the surrounding environment. As the steam is replaced with the purge gas, the temperature in the system will drop and the amount of condensate formed may increase slightly (Figure 15.2). The first bleeds closed are typically those in the upper portion of the system, allowing the condensate to continue to drain from bleeds in the lower portion. The last bleeds to be closed are those that served as the major condensate discharge locations, as these may continue to discharge condensate for some time. These bleeds can be left open for some period of time while the gas purge continues. The usually low dew point of the purge gas and the heat retained by the metallic components of the system will assist in drying the last quantities of condensate, as the hotter metal components will transfer heat back to the remaining condensate and facilitate its evaporation.

Once the system has been adequately dried, the last open bleeds can be closed while the purge gas remains on and a slight positive pressure is allowed to remain in the system. This pressure can be maintained in the system until it is ready for use thereby preventing the ingress of contaminants. The addition of the purge gas through a 0.2-micron filter allows for the maintenance of this positive pressure over an extended period of time. In some applications, a portable vessel can be successfully transferred from one APA to another while the pressure is maintained without contamination of the contents.

15.6.1 Process Control

To this point, the control methodology utilized for the SIP process has not been described in detail. Earlier the point was made that SIP systems are frequently controlled by pressure rather than by temperature. This is a considerable departure from the approach utilized in most steam sterilizers. Most steam sterilizers have a single outlet through which the air and condensate must pass in order to exit the chamber. A single temperature probe located in the drain line may be able to serve as a measurement of the coldest temperature in the entire sterilizer and be utilized to control and document the sterilization process. In contrast, the SIP system may have many locations where air and condensate are eliminated from the system. Recognizing that the pressure in the system will equilibrate throughout enables the use of pressure measurement in the system to control the steam supply for the SIP system. The singular relationship between temperature and pressure for saturated steam makes this possible. Although the ability to control pressure will not ensure effective sterilization across the entire system, it affords a workable parameter for steam regulation. Choosing a single temperature measurement location within the entire system to control the SIP process is possible, but that places a larger burden on the system designer to assure that the location chosen is correct. In a simple tank system, a single point may be both identifiable and usable in SIP process control. In more complex systems, control of the system by the use of internal pressure may be both simpler and more effective. In order to confirm sterilization effectiveness over a complex system, multipoint temperature measurement can be utilized as an adjunct to pressure control on the steam supply to the system.

15.6.2 Multipoint Temperature Measurement

In performing SIP procedures on large systems with multiple condensate discharge locations, the use of multipoint temperature recorders is necessary to confirm that the appropriate temperatures are realized throughout the system. This is especially important where the SIP process is manually operated and the manipulation of valves is left to process operators. What more effective way to establish that a large number of valves were operated properly during the SIP procedure than through the use of a multipoint temperature recorder? The appropriate locations to monitor would be determined during the Performance Qualification (PQ) portion of the validation program. It should be noted that the temperature probe inside a tank is almost never the coldest location in the system. The temperature in the discharge line of the tank only a few feet away is almost always lower and of greater importance in routine sterilization process confirmation. In the monitoring of temperature on SIP systems, the introduction of the temperature probe into the fluid stream may not be necessary. Temperature sensors can be attached to the outside of piping and insulated to minimize heat loss to the surrounding environment.[28] When temperature probes positioned outside the piping in this fashion achieve the required temperature, there is reasonable assurance that the interior of the piping is at least that hot.

In measuring the temperature in the system, the goal should be to measure the surface temperature and thus thermocouples should be in contact with the surface. Unlike an empty chamber study in an autoclave in which steam temperature uniformity is desired, the objective in SIP is effective sterilization of the product contact surfaces, and thus surface temperatures should be measured. There is no requirement to have a narrow temperature distribution in an SIP system; heat loss to the surrounding environment will generally preclude a tight range. The primary objective is attainment of lethal conditions at the equipment surface and provided that is accomplished a wider temperature range is of little consequence. Temperature ranges >10°C have been observed in the largest system.

15.7 MEMBRANE FILTERS AND SIP

All systems that are sterilized-in-place include at least one membrane filter. It is difficult to conceive of an SIP system that does not include at least one microbially retentive filter or how a system could be operated without one. The necessity to maintain a system sterile subsequent to SIP will ordinarily result in the introduction of a least one 0.2-micron filter that is utilized to introduce an air or nitrogen purge into the system after the sterilization dwell period. The filtered gas is initially utilized to purge the system of steam and then to establish and maintain a positive pressure on the system prior to use. This ensures that any leakage on the system will be in a direction away from the interior sterile surfaces. If a filter is required for this purpose, it seems logical that the filter should be sterilized with the system. After all, why bother to SIP at all, if an aseptic connection is needed just to render the system secure from microbial contamination prior to use?

15.7.1 Filter Housing Configuration and Orientation

With the presence of a microbially retentive filter in the system near certain, what special concerns relative to the filter presence must be addressed? The basic concerns cited earlier for tanks and other vessels are directly applicable to filter housings. Air and condensate removal from the filter must be considered, as well as the sterilization procedure and sterility post-sterilization. The removal of air and condensate at first appears to be a simple task, because filter housings are supplied with a number of bleed valves. It may be necessary to modify filter housings, whether for disc membranes or for cartridge filters to facilitate the removal of condensate[3].

An example based upon the simplest of all filter housings, the disc membrane filter housing, illustrates this point. Most disc housings available are designed with legs to support the housing that maintains the membrane in a horizontal plane for use on a bench top. If the filter is to be sterilized in place, how will the condensate formed on the upstream of the membrane surface be removed? There is no effective way to remove this condensate other than by passage through the membrane, which may not be easily accomplished. A workable solution is to position the membrane in the vertical plane (Figure 15.12). Once oriented vertically, relocation and/or addition of bleeds on both the upstream and downstream sides of the filter at locations just inside the O-ring at the lowest portion of the filter housing has been shown to be necessary to permit effective discharge of the condensate[3].

Having solved the condensate retention difficulties for disc membranes and housings, applying the same principles to pleated cartridge filters is straightforward. The primary recommendation for cartridge filters is to always position the cartridge with the open end down. This will allow any condensate formed on the interior of the cartridge to exit the housing without requiring it to pass through the membrane. Additional bleeds on the upstream side of the filter housing are usually required at the lowest points in the housing. This is accomplished in housings where the inlet and outlet are at the same end. The filter housing is most easily adapted to SIP when the housing has the appearance of an inverted "T". When the inlet and outlet are at opposite ends of the housing, sometimes identified as an in-line housing, the need for modification is usually greater.

A further consideration with filter sterilization is the direction of steam flow. Disc membranes should be provided with both upstream and downstream support screens. The presence of the support screen reduces the stress on the membrane during sterilization and should result in lessened problems with filter integrity post-sterilization. For pleated cartridge filters, the general rule is to sterilize them from the outside (upstream side) in. In this manner, the differential pressure developed by the introduction of steam into the filter serves to push the filter together rather than to blow it apart. The application of steam in this manner keeps the filter properly seated in the housing, where flow in the reverse direction might tend to lift it. Several filter manufacturers incorporate a locking arrangement that serves to hold the cartridge in the housing. Although this feature is helpful, it should not be relied upon to hold the filter in place when steam is traveling upwards through the membrane.

15.7.2 The "Loop" Method

No discussion of filter sterilization would be complete without some further mention of the so-called "loop" method (Figure 15.13). When the sterilization of large systems employing filters was first attempted, damage to the filter membrane occurred frequently. The majority of filters in use at that time were made of cellulose acetate, cellulose nitrate or a mixture of the two esters, which become extremely brittle and break easily when exposed to excessive heat (around 125°C). With larger systems, the steam pressure needed to bring the entire downstream system to sterilizing temperature would sometimes require the filter to be exposed to temperatures that would impair its integrity. To overcome this limitation, a "loop" or "bypass" would be added to allow the vessel to receive steam directly, without having to pass all of the required steam through the filter.[2, 3, 4] This "loop" would of course be closed prior to the end of the steaming process to preserve the integrity of the system. The use of a "loop" became common where these earlier types of membranes filter materials are sterilized in place. This practice is still necessary when using filter membranes with limited thermal stability. Many of the newer filter media on the market have greater thermal stability. For these types of filter media, the "loop" is not required and all of the steam necessary to sterilize large systems may be passed through the filter membrane without difficulty. The advent of ever larger systems has seen a return to the "loop method" as the increasing demands for steam and shortened process time have made even today's more heat-resistant filters rate limiting. Supplying steam directly to a vessel simultaneously with process / vent filters applies the same principle without any resemblance to a "loop".

15.7.3 Filter Sterilization

In reviewing many of the observations made previously regarding sterilization of filters, the obvious question to be raised is: How were these problems with condensate retention in filters found to be a problem? The basis for much of the information regarding difficulties with filter sterilization was a series of validation studies I personally directed during the early 1980s. When faced with the necessity to confirm the sterilization of filters that were to be steam sterilized, the decision was made to utilize a spore suspension and apply the spores directly to the membrane surface. Considering that only the downstream side of the filter was required to be sterile and understanding that 0.2-micron filters are microbially retentive, the biological challenges were placed on the downstream side of the membrane. It was soon discovered that when a resistant bioindicator such as *Geobacillus stearothermophilus* was utilized that sterilization was not as easily achieved as would be assumed from the time-temperature or F_0 values observed.[2,3,4,5,8] In resolving these sterilization problems, it became clear that accumulated condensate on or near the filter was the primary cause for the inability to sterilize the membrane surface. In subsequent efforts, the lessons learned from these original studies were confirmed repeatedly in SIP systems of varying size and complexity.

Spore-inoculated membrane filters were also found to be more reliable indicators of SIP system effectiveness than ordinary spore strips.[3,4] The validation of SIP systems utilizing filters was approached in the same manner as other sterilization

TABLE 15.3
Filter Challenge Results

	Challenge results	
No. of trials	**Filters**	**Spore strips**
88	Positive	Negative
110	Negative	Negative
0	Negative	Positive

validation studies. Once filter inoculation was found to be a superior indicator compared with spore strips in the filter housing, they were utilized together over a long series of validation studies. This practice was continued over a period of years with the following results (Table 15.3). There were nearly 90 individual validation trials in which the filters that were positive after sterilization while the spore strips were present during the cycle were negative for growth. The number of spore strips present in the system during these trials varied from 2 to 20 depending upon the size of the system. The spore strips were placed throughout the system, including in the housing with the inoculated filters. Here too, the validation trials comprised a range of system sizes ranging from individual tanks with a single vent filter to larger systems containing one or more tanks and multiple filter housings. Filters and filter housings used in these studies included all of the major manufacturers, different membrane materials, several housing designs and both hydrophilic and hydrophobic filters in a range of sizes from 47 mm disc membranes to multiple 10" cartridges. In the majority of these studies, the coldest location or minimum F_0 location was not the filter housing but some other location in the system. In fact, the filter housing was often the hottest location because of its proximity to the steam inlet. In every case, the organisms present on the spore strip were destroyed whereas those present on the filter surface demonstrated growth or no-growth depending on the local conditions present at the filter surface. Over this same period, there were >100 individual validation trials in which neither the filters nor the spore strips were found to be viable after sterilization. Note that although there were many trials in which the organisms present on the spore strips were destroyed while organisms survived on the filter surface, there was never an incident were the reverse was found. The inadequacy of the spore strip as a means for establishing the effective sterilization of a membrane filter surface became evident.

This is likely because of the spore strips (and thermocouples) were being placed in the housing remote from the filter surface, whereas the inoculation of the filter surface places the challenge directly upon the surface of critical interest. Clearly, **sterilization of the filter surface is the critical concern** for these types of systems, and the use of a microbial challenge on the filter surface is a "worst case" and thus a more representative challenge of a critical location in the system. The results summarized previously led to the conclusion that the filter was the "worst case" location in any SIP system, and the use of spore strips at all locations within SIP systems was subsequently discontinued.

Two examples of the enhanced sensitivity of inoculated filters relative to spore strips are described following:

1. In sterilizing a multi-cartridge vent filter on a steam sterilizer, positive filters were repeatedly found despite F_0 values in the filter housing that exceeded 300 minutes! When the condensate retention problems were resolved, the filters were readily sterilized.
2. In the sterilization validation of a large system containing numerous filters, one filter cartridge was found to be non-sterile after repeated trials despite F_0's >60 minutes. When the supplier of the cartridge was changed, no difficulty was encountered in sterilizing the new cartridge, which was able to drain more easily than the original cartridge. No other changes in the system or procedure were made.

Several publications have included references to the difficulties associated with the sterilization of filters.[29,30] Although these mentions of similar experiences have not been explicit with regard to the details of the studies performed, independent verification of the difficulties encountered in filter sterilization further substantiates the author's already strongly held beliefs in this area.

There is an indirect benefit to the use of filter inoculation rather than spore strip testing in SIP systems. Where spore strips are utilized in SIP validation, the system must be cooled, tank entry permits obtained, one set of strips recovered and a new set of biological indicators put in place. The accomplishment of these tasks can take several hours especially if the system is quite large. Where inoculated filters are utilized, tank entry is not required, and quite large systems can be readied for the next validation run in a relatively short period of time. In time-critical validation activities, the use of inoculated filters can thus save considerable time over the use of spore strips.

15.7.4 Filter Inoculation and Testing

In order to utilize inoculated filters in SIP, suitable methods must be available for the inoculation and testing of the challenged filters. Inoculation is straightforward, the filter to be challenged is wetted—water for hydrophilic membranes or water/alcohol mixture for hydrophobic membranes—using a pressure can (Figure 15.23). After complete wetting of the filter surface, a spore suspension of *Geobacillus stearothermophilus* in water/alcohol is added to the wetting fluid, to provide the equivalent of approximately 100 spores/cm^2 of filter area. The suspension is filtered through the membrane in the reverse direction leaving the spores on the downstream membrane surface. An additional quantity of wetting fluid is added to rinse the pressure can and complete the transfer of the spores to the filter. The filter is then purged with a gas stream to remove the residual wetting fluid from the housing and lines. Care is taken throughout this procedure not to

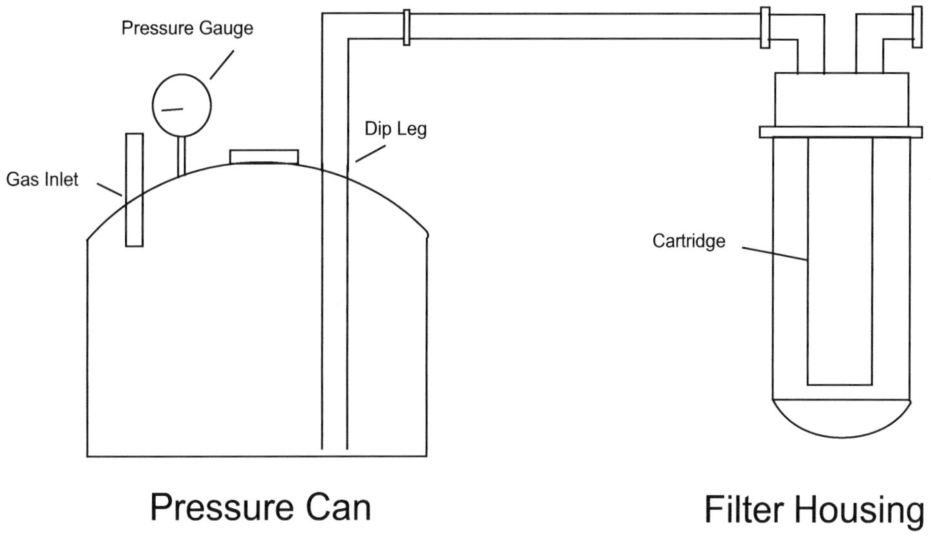

FIGURE 15.23 Configuration for filter inoculation.

exceed 2–3 psi on the filter to avoid causing physical damage to the membrane when filtering in the reverse direction (an important concern with pleated cartridges). The ends of the filter are covered with a sterile wrap and the filter is ready for use as a biological indicator. The sterile wrap is removed from the filter just prior to installing the filter on the equipment.

Testing of the filter is equally simple. After completion of the sterilization process, the filter is removed from the equipment and the end caps are re-covered with a sterile wrap. The filter is transported to the laboratory where it is removed from the filter housing. In the case of disc filters, a sterile knife is used to cut the membrane just inside of the O-ring that seals the filter membrane in the housing. The center portion of the membrane is placed into Soybean-Casein Digest Medium (SCDM) and incubated at 55°C–60°C for 7 days. Testing of cartridge filters requires a slightly different procedure. The cartridge housing is opened and the entire cartridge is removed from the housing using sterile gloves. The cartridge is placed into a container with the open end up, and SCDM is aseptically added to the container. The container is covered with a sterile lid and incubated as described previously. It may be necessary to fabricate special containers to accommodate 20" and 30" filters.

An often-asked question in regard to filter sterilization when the filters are inoculated on the downstream side is: "Could the absence of the spores after sterilization be attributed to physical action rather than microbial death?" The answer to this is simple: the number of times that inoculated filters were non-sterile after completion of the process indicates that sufficient viable spores remained on the surface of the filter. Further confirmation of this was established in a study in which the condensate from the downstream side of the filter was collected during the SIP process. Less than 10 viable colony forming units were found in the condensate after sterilization, and the challenged filter was confirmed to be sterile.

Extraordinary precautions are not generally required when conducting challenges with *Geobacillus stearothermophilus* in the manner described above. Adventitious contamination that might find its way into the media prior to the incubation is unlikely to grow at the 55°C–60°C incubation temperatures utilized. The ordinary mesophilic organisms present in the environment and on personnel are unable to grow at the incubation temperature.

15.7.5 Filter Sterilization in Autoclaves

One last point remains with regard to the sterilization of membrane filters. The description of this effort is somewhat anecdotal, but bears careful consideration given the broad implications. Several years ago, after preparing a filter cartridge as described earlier for an SIP validation study, the run was canceled and rather than discard the filter it was placed into a steam sterilizer and tested as part of the validation of a multi-vacuum sterilization cycle. The cycle was completed normally and achieved a minimum F_0 in excess of 30 minutes. Much to the investigator's surprise, the filter was found to be non-sterile when tested as described previously, whereas the spore strips elsewhere in the load were sterile. Puzzled by this unusual turn of events, the investigator made a series of additional trials using additional time, additional vacuums, alternative configurations, removal from the housing and higher temperatures all to no avail. Apparently, sterilization of filters in a sterilizer where a resistant indicator is located on the surface of the membrane is not straightforward.

The most plausible explanation for these results is condensate retention in the cartridge. In every SIP process in which filter sterilization was found to be a problem initially, the problem was resolved by eliminating condensate retention in the system. In a steam sterilizer with steam able to contact both sides of the filter membrane, the condensate formed may have nowhere to exit the filter. As steam penetrates the membrane surface to reach the spores, condensate forms and is retained in the filter matrix. As the steam is saturated, the addition of more steam will not return the condensate already formed back to steam. In SIP mode, the steam pressure on the upstream side forces condensate through the filter and out of contact with the filter. It appears that condensate retention in the filter matrix as a result of the absence of a pressure differential would explain why SIP works and sterilization in an autoclave (when tested in this more rigorous manner with spores on the membrane) would not. Testing of the filter in its housing with spore strips is common and given that condensate is unlikely to accumulate where the strip is located, the absence of growth on strips is not surprising. There are references to similar results having been observed by others.[25, 26]

The sterilization of filter cartridges in autoclaves in steam sterilizers is such a widespread practice it is surprising to find that it is not nearly as easy to accomplish as the industry would believe. That the failure to sterilize these filters in an autoclave is not the cause of significant sterility failures in the industry is probably because of the much lower resistance of ordinary organisms that might be on the filter surface to steam sterilization relative to *Geobacillus stearothermophilus*.

15.8 VALIDATION

15.8.1 Installation and Operational Qualification

The validation of SIP follows the conventional approach including Equipment Qualification (EQ) and PQ. The emphasis placed on system design and procedural conformance earlier in this chapter should reinforce the importance of the activities. During the EQ, the SIP system should be carefully scrutinized for conformance to the physical design details described earlier. As a general rule, if there is a concern that air or condensate may be retained at a given location then a bleed should be added to the system at that point. Another useful part of the EQ is to ensure that all valves in the system are clearly identified, which makes procedural conformance more certain. It is conventional during the EQ of systems and equipment to define the locations of all instruments on the system. This is sound advice but must be approached with some degree of flexibility with SIP systems. The cycle development may result in the need to relocate some of the permanent temperature probes to more critical locations.

The preparation of an extremely detailed procedure for the sterilization of the system is essential to success in validating the system and is often a part of the EQ. The procedure should include step-by-step directions for the manipulation of each valve in the system from the start of the procedure through to completion. The installation of any additional lines, hoses, fittings, temperature probes, etc. should all be detailed in the SOP. The use of diagrams to facilitate adherence to the procedure is certainly beneficial. The cycle development that follows the EQ utilizes thermocouples to confirm temperature prior to the introduction of biological indicators. Any procedure developed prior to the completion of the PQ should be considered a draft until its effectiveness has been confirmed.

15.8.2 Cycle Development

In conducting cycle development studies for SIP, the initial step is to thermally map the entire system with thermocouples. Thermocouple placement in the system should include the following locations (in relative order of importance to process control and SIP performance):

- Location(s) where permanent sensors are positioned that control the SIP process. These are usually near the lowest point(s) in the system and should be the coldest locations in the entire system.
- Locations elsewhere in the system that are being routinely recorded for other purposes. These are most often internal to the vessel and although not a cold spot, they do confirm consistency of both validation and routine SIP results.
- In proximity to all sterilizing filters (upstream and downstream if possible). These are always the hardest to sterilize locations in any system despite temperature results that might suggest otherwise. Failure to attain temperature in the filters is a major concern as noted earlier.
- At the primary discharge port on the bottom of any vessel. This is almost always the coldest point in the vessel.
- At the system boundaries where lines extend some distance from the vessels. These can retain air and/or condensate if their removal is not managed throughout.
- In vessel nozzles and other dead legs where instruments are located. These can also retain air if its removal is inadequate at the beginning of the process.
- Other locations in the vessel interior are typically noncritical. The system geometry and gravity should work together to move condensate towards the bottom outlet. It may be easy to gather data internally, but the data is never critical to performance.

The number of thermocouples to use varies with the scale of the system, it could be as few as 5–6 in a small system, or several hundred in a large, far more complex system (e.g., multiple vessels and connective piping). The cycle development objective is thermocouple placement to locate those portions of the system where condensate will be retained and the temperature will be reduced. In conducting studies on tanks and similar pieces of equipment, thermocouple placement should focus on the piping entering and leaving the vessel and any permanent temperature measurement location inside the vessel. The placement of multiple probes elsewhere in the vessel has little utility, those points are easily accessible to steam and probes in the bottom discharge of the vessel are nearly always the coldest (a result of the large amount of condensate created within that can only exit at the bottom outlet). Where temperature is being measured in piping systems, consideration can be given to placing the thermocouples outside the pipe rather than inside. This eases the task of thermocouple placement but is also a "worst case" measurement of the temperature inside the pipe. The placement of internal thermocouples must be accomplished with care. The effectiveness of the sterilization must be unaffected by the addition of these additional temperature probes in the system. The use of appropriate pressure fittings for thermocouple ingress is suggested to ensure that air and/or condensate removal is not altered. Equal care must be taken in placing the probes to make certain that condensate removal is not restricted because of the presence of the thermocouple in the system. In small diameter piping, the use of external or smaller gauge thermocouples should be considered if there is a possibility that the use of conventional probes would have an adverse effect.

The goal of the cycle development is to identify a configuration in which conditions within the system boundary are relatively uniform and sufficient for sterilization as well as identify the process sequences necessary to bring the system to that state from process start and to leave the system in a protected state after the process dwell period. The cycle development should also help establish the permanent temperature monitoring locations for routine process control and documentation. Once a suitable design and sterilizing procedure are identified that demonstrates sufficient temperature throughout the system, the PQ can commence.

15.8.3 Performance Qualification

In the author's experience, temperature measurement is not always a good predictor of microbial death, especially with filters, but the thermocouples can help to identify problem locations that if microbially challenged would surely result in sterilization difficulties. It is the author's recommendation to utilize biological challenges on the downstream side of all sterilizing filters, and not to use spore strips for the reasons outlined earlier. Despite this recommendation and publications documenting the limited utility of spore strips, these are more commonly used because of their convenience. When used, spore strips should be placed adjacent to each thermocouple in the system, as well as upstream and downstream of all filters. The placement of spore strips in SIP systems is not always an easy task. One of the simplest ways to ensure that strips are not lost is to place them in a permeable container that can be firmly affixed in the equipment. Good success has been achieved with tea infusers (perforated stainless-steel containers with a light chain attached) in which a spore strip is placed and the chain utilized to secure it to the equipment. This technique is only adaptable to locations large enough to accept a relatively large object without obstructing the fluid flow. Using tape to secure spore strips to the side of the vessel is also possible, but care must be taken to ensure that steam can access the entire surface of the strips. The use of spore strips in closed piping is of such difficulty that many practitioners employ strips only in tanks, filters and other easily accessible locations. There are special fittings / gaskets available for use in-line piping. The use of loose or poorly secured strips is not recommended, as they can be swept away by steam or condensate and lost. Wherever strips are placed, care must be taken to avoid obstructing steam entry or condensate removal because of their presence.

Possible alternatives to spore strips are inoculated stainless-steel coupons or wires that can be placed in the system just as spore strips would. These have the advantage of having sufficient mass that they would remain where they are placed despite the passage of steam or condensate over their surface and yet are small enough not to obstruct air, condensate and steam flow. Direct inoculation of other stainless-steel components in the system, i.e., cups placed over filling needles for clean in place (CIP)/SIP and coupons held by wire in the vessel is also possible. Where these alternates to spore strips are utilized, the end user must confirm the microbial population and resistance of the spores on the substrate.

In performing the PQ studies, consideration should be given to an experimental design that maximizes data utility while minimizing the number of required studies. Good success has been obtained with a bracketing approach that utilizes two different "worst case" assumptions (Figure 15.24). The first series of runs (typically three) are performed using time–temperature conditions that exceed that intended for use in operation, such as 127°C for 60 minutes. After completion of the full series of runs, all of the filters installed in the system, whether for liquid filtration or gas filtration (tank vent or compressed gases) are tested for integrity in their housing. In critical installations, the testing may be performed in situ, if that is the normal means for integrity testing. With the types of membranes in vogue today, there is little necessity to test the filters after each individual SIP run. If the filters are integral after three consecutive studies, there should be no concern that they would be damaged in a single run. Minimum studies with microbial challenges are performed next, with a possible test condition of 122°C for 30 minutes. Three complete sets of filters will be required if filter inoculation is utilized, otherwise a single set of filters can be used. If the conditions cited prior were representative of the "worst case" efforts, and proved successful in the PQ studies, then the routine sterilization SOP might provide for conditions of 124°C–126°C for 40–45 minutes. Thus both the time and temperature conditions utilized in the PQ studies would provide a considerable safety margin over that employed in routine sterilization.

Limits for the range of temperature across the system being validated should be extremely flexible. In large systems, temperatures may vary in excess of 10°C from the hottest to the coldest location. More appropriate is a requirement that all monitored locations exceed a minimum F_0 value or a

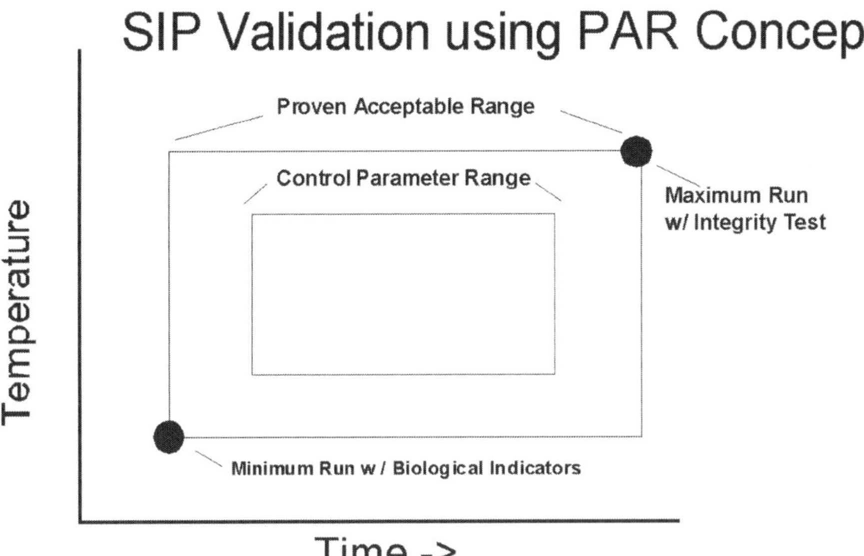

FIGURE 15.24 "Worst Case" matric for validation.

minimum time at temperature as determined during the cycle development effort.

15.9 AUTOMATION

The application of computer control systems in the pharmaceutical industry has been on the increase for several decades. The application of a computer control system to a system subject to SIP can facilitate procedural adherence and ultimately sterility assurance. To achieve this, the application engineer must delay the completion of the software (and possibly the hardware) necessary to automate the SIP process until the sterilization PQ has been completed. Only then can it be assured that the control system will utilize the proper sequence of valve actuation to effect sterilization. As the timely and correct manipulation of the valves in the system is of critical importance to success, the use of computer control provides considerable benefits over a manual operation. The use of a control system also eases the review of multipoint temperature measurements allowing for more precise timing of the SIP cycle. Control systems for SIP are commonly found on equipment that already has a control system present for other operations, e.g., freeze dryers and bioreactors. The application of automated SIP on an otherwise nonautomated piece of equipment is unusual because of the extra expense entailed.

There are obstacles to the expanded use of automation for SIP systems. In a large piping system, the sheer number of automated valves required can result in a considerable expenditure in software and hardware to properly execute the SIP procedure. The cost associated with this additional complexity can be a deterrent to automation. A second difficulty is the small bleeds necessary at some locations; automated valves of appropriate design are sometimes not available in the necessary sizes. Hybrid systems including both automated and manual manipulation of valves are most common.

15.10 BULK STERILIZATION

A subject often discussed in conjunction with SIP is the bulk sterilization of liquids. In bulk sterilization, an aqueous-based fluid that cannot be filter sterilized because of the presence of solids (either active materials or other excipients in the formulation) is sterilized in a closed vessel. The methods utilized to perform this sterilization are derived from those utilized for conventional SIP procedures:

1. Prior to the start of the bulk sterilization, the empty vessel and its associated piping should be sterilized-in-place in accordance with the methods described earlier.
2. The fluid material should be introduced into the vessel with a minimum degree of splashing against the upper portions of the tank. All valves in the system other than the vent and the line utilized to introduce the fluid should remain closed.
3. Agitation of the vessel contents should be started and heat applied to the jacket (In bulk sterilization elevated pressures of steam in the jacket are necessary in order to achieve rapid heat-up of the vessel contents, contrary to the advice provided earlier for SIP).
4. As the temperature of the fluid rises (as measured by a fluid product probe), small amounts of the aqueous phase will be lost. The agitation and increase in temperature will also assist in the expulsion of air from the liquid and the headspace above it.
5. When the temperature of the fluid approaches 98°C–100°C, the vent and fluid entry line valves should be closed. The temperature of the fluid will continue to rise, and some adjustment of the steam to the jacket may be necessary to control the

temperature in the desired range for the required time period. If the heating phase is overly lengthy, there can be measurable loss of water from the formulated material.

6. Upon completion of the sterilization portion of the cycle, the jacket of the vessel is emptied of steam, and cooling water is introduced to reduce the temperature prior to further processing. Filtered air or nitrogen is introduced to maintain a positive pressure in the vessel as the contents cool.

If the amount of liquid to be sterilized in the vessel does not reach to the upper portion of the vessel's jacket, thermocouples in the headspace may show unusually high temperatures compared with those immersed in the liquid. These temperatures can appear anomalous but may be correct and result from superheating of the steam and residual air in the headspace by the exposed jacket. This will be most often observed with the minimum batch size in the tank.

Validation of bulk sterilization is a hybrid of terminal sterilization of fluids in sealed containers and SIP. Considerations prevalent in terminal sterilization validation such as bioburden determination, D-value determination in product and stability of the formulation at elevated temperatures must all be considered. Similarities with SIP procedures are found in the placement of multiple temperature probes and strict adherence to sterilization procedures.

It is tempting to proceed directly to the bulk sterilization of the liquid in the vessel without performing an initial SIP of the empty tank. Steam created by heating the liquid is used to sterilize the tank headspace and the connecting piping installed on the top of the vessel. When this approach is used, there will be some added loss of liquid volume in the vessel that could affect the formulation. This approach does not allow for the complete sterilization of the bottom discharge of the tank and for this reason is not recommended.

15.11 NON-STEAM STERILIZATION IN PLACE

This chapter has focused on the use of steam for SIP because it is the primary agent used for that purpose. There are other less common agents used for SIP.[1] Hot air sterilization has been successfully used for aseptic spray dryers. The process takes much longer than a comparable steam SIP because air has only limited heat capacity and lower lethality compared with saturated steam. Recirculation of superheated water has been successfully used for SIP of water distribution systems.[21] Superheated water sterilization of a process piping system is possible in principle, but there are no documented applications. Vapor sterilization of process equipment has been accomplished using hydrogen peroxide or formaldehyde. Vapor processes are acknowledged as difficult to control, and this limits their utility in complex configurations. Gas sterilization of process equipment was more prevalent in the past when ethylene oxide was not considered such a major safety hazard. Despite the availability of numerous alternative sterilizing agents, saturated steam is used for nearly all SIP applications.

15.12 CONCLUSION

The development and validation of SIP procedures is among the most challenging of all validation activities. The reader is encouraged to remain current as new developments in SIP technology are published, for despite the interest in the subject, the existing base of published information is minimal and improvements in our knowledge basis are certain to occur.

NOTES

1. USP 39, Sterilization-in-Place, <1229.13>, 2016.
2. Myers T, Chrai S, Design Considerations for Development of SIP Sterilization Processes. *Journal of Parenteral Science and Technology* January 1981.
3. Myers T, Chrai S, Steam-in-Place Sterilization of Cartridge Filters in-Line with a Receiving Tank. *Journal of Parenteral Science and Technology* May 1982.
4. Kovary S, Agalloco J, Gordon B, Validation of Steam-in-Place Sterilization of Disc Filter Housings and Membranes. *Journal of Parenteral Science and Technology* March 1983.
5. Berman D, Myers T, Chrai S, Factors Involved in the Cycle Development of an SIP System. *Journal of Parenteral Science and Technology* July 1986.
6. Agalloco J, Shaw R, Steam-in-Place Sterilization. *Parenteral Drug Association Course Notes* June 1987.
7. Perkowski C, Controlling Contamination in Bioprocessing. *BioPharm* September 1987.
8. McClure H, How to Sterilize Liquid Filling Equipment at Point of Product Contact. *Pharmaceutical Engineering* January 1988.
9. Chrai S, Validation of Filtration Systems: Considerations for Selecting Filter Housings. *Pharmaceutical Technology* September 1989.
10. Agalloco J, Steam Sterilization-in-Place Technology. *Journal of Parenteral Science and Technology* September 1990.
11. Cappia JM, Principles of Steam-in-Place. *Pharmaceutical Technology*, Filtration Supplement, 2004: S40–S46.
12. Pflug IJ, *Microbiology and Engineering of Sterilization Processes*. Minneapolis, MN: Environmental Sterilization Laboratory, 1999.
13. Noble PT, Modeling Transport Processes in Sterilization-in-Place. *Biotechnology Progress* 1992;8(4): 275–284.
14. Noble PT, Transport Considerations for Microbiological Control in Piping. *Journal of Pharmaceutical Science & Technology* 1994;48(2): 76–85.
15. Young JH, Ferko BL, Temperature Profiles and Sterilization within a Dead End Tube. *Journal of Parenteral Science & Technology* 1992;46(4): 117–123.
16. Young JH, Sterilization of Various Diameter Dead-Ended Tubes. *Bioengineering and Biotechnology* 1993;42: 125–132.
17. Young JH, Ferko BL, Gaber RP, Parameters Governing Steam Sterilization of Deadlegs. *Journal of Pharmaceutical Science & Technology* 1994;48(3): 140–147.
18. Bacaoanu A, Steam Sterilization of Stationary Large-Scale Freeze-Dryers (Lyophilizers). *American Pharmaceutical Review* May–June 2004: 40–47.

19. Food and Drug Administration, 21 CFR 212 Proposed Current Good Manufacturing Practice in the Manufacture, Processing, Packing or Holding of Large Volume Parenterals for Human Use, June 1, 1976 (rescinded Dec 31, 1993).
20. ASME, Bioprocessing Equipment, BPE-2019, New York, 2019.
21. Carlson VR, Response to Question & Answer. *BioPharm* March 1990.
22. Haggstrom M, Sterilization-in-Place Using Steam or Superheated Water. Proceedings of PDA Basel Conference, PDA, Bethesda, MD, 2002.
23. Agalloco J, Carlson VR, private communication, 1990.
24. Agalloco J, Closed Systems and Environmental Control: Application of Risk Based Design. *American Pharmaceutical Review* 2004;7(4): 26–29, 139.
25. Yoshida S, Development of a New Concept for Time /Pressure Filling. *PDA Journal of Pharmaceutical Science and technology* 2003;57(2): 126–130 and 57(4): 316–317.
26. PDA, Process Simulation Testing for Sterile Bulk Pharmaceutical Chemicals. PDA Technical Report #28, *PDA Journal of Pharmaceutical Science and Technology* 1998;52(4), Supplement.
27. Odum JN, Closed Systems in BioProcessing: Impact on Facility Design. *Bioprocessing Journal* January–February 2005: 41–45.
28. Thorp G, Zwak J, Measuring Process temperature in Small Diameter Lines. *Pharmaceutical Engineering* 2004;24(5): 8–18.
29. Meltzer T, Steere W, Operational Considerations in the Steam Sterilization of Cartridge Filters. *Pharmaceutical Technology* September 1993.
30. Christopher M, Keating P, Messing J, Walker S, Inoculation and Sterilization of Hydrophilic Filters. Proceedings of Parenteral Drug Association International Symposium, Basel, Switzerland, Parenteral Drug Association, 1994.

16 Validation of Dry Heat Sterilization and Depyrogenation

George Sheaffer and Kishore Warrier

CONTENTS

DOI: 10.1201/9781003163138-16

16.1 INTRODUCTION

Advances in equipment design, process control, and process documentation, combined with robust Commissioning and Qualification (C and Q) life cycle approaches to validation and equipment maintenance, have evolved to solidify dry heat sterilization and depyrogenation as efficient and reliable manufacturing process solutions.

The advances also provide the capability to consistently control and verify Critical Process Parameters (CPPs) and ensure Critical Quality Attributes (CQAs) related to product and process requirements are consistently met.

The equipment utilized to provide the dry heat process conditions must be validated and maintained to ensure that the system is able to provide sterile and/or depyrogenated components on a reproducible basis. A traditional validation approach with an integrated C and Q approach will be presented, including the verification of applicable CPPs. An overview of the life cycle approach to validation and routine maintenance will be discussed.

Dry heat technology is one of the most commonly used methods to sterilize and/or depyrogenate pharmaceutical components and products. Most often, depyrogenation of parenteral containers is performed utilizing a dry heat oven or depyrogenation tunnel. The depyrogenation process is also utilized on certain heat-stable components, glass containers, metal equipment, etc. to render the item and final parenteral product free of pyrogens. Dry heat sterilization can also be used for heat stable oils, ointments, and powders.

The validation of a dry heat sterilization and depyrogenation process involves approaches and procedures that parallel those utilized for steam sterilization. The efficiency of any heat treatment is determined by the design and source of the heating medium. Hot air is substantially less efficient as a thermal transfer medium compared with steam.

A substance with a high thermal conductivity is a good heat conductor, whereas one with a lower thermal conductivity is a poor heat conductor. Although air is not considered a good heat conductor, the contact of hot air with good heat conductors, such as metal objects, will provide a fairly rapid heat transfer rate. Metal items such as stainless steel equipment will heat up faster than other materials (such as glass) because of greater thermal conductivity.

This chapter will detail the steps of a program that may be employed to properly specify equipment and validate a dry heat process.

16.2 TYPES OF DRY HEAT STERILIZERS

The most common types of dry heat sterilizers employed in the pharmaceutical industry today are forced-convection batch sterilizers and tunnel sterilizers. Continuous flame tunnel sterilizers are also used for ampoules. Microwave sterilizers are less commonly used.

16.2.1 Batch Ovens

Batch ovens are the most commonly used type of dry heat equipment in the industry because of their flexibility in unit and load size (Figure 16.1). The oven operates on the principles of convective or radiant heat transfer. Ovens can employ a range of cycles (by varying time and temperature settings) for processing utensils, glassware, stainless steel equipment, or pharmaceutical products. A typical cycle might employ temperatures in the range of 180°C–300°C (Degrees Celsius). The temperatures at the lower end of this range will sterilize, whereas the higher temperatures in the range are suitable for depyrogenation. Cycle effectiveness is dependent on both temperature and total cycle time. There is a cooling phase at the completion of the heating cycle, which serves to minimize thermal shock and increase handling safety.

Advances in equipment design and process control allow for a reasonably uniform controlled heat up, dwell time (at temperature), and controlled exhaust/cooling phases of the cycle. Sizing the heating elements to efficiently accommodate the intended mass to be processed and integration of cooling coils in the High-Efficiency Particulate Air (HEPA) filtered air will decrease overall cycle time and provide a more efficient process. Additionally, this optimization may allow for less heat stable products/commodities to be processed.

The load items are usually prepared in a classified non-aseptic area, with controls in place such as limited access, reduced particulate levels, known air quality, personnel gowning, etc. The preparation of glass containers consists of washing with high-quality water, steam, and/or filtered air prior to loading the containers in covered trays and placing them on racks or carts in the chamber. In most installations, a double-door oven is employed in which items are loaded from a non-aseptic area and the load is removed from the oven into an aseptic processing area. Packaging of the commodities to be processed in the oven must not only be heat-compatible but also be designed based on further use/processing requirements after the items are unloaded from the oven to maintain sterility, protection from contamination, etc.

16.2.2 Tunnels

The different types of tunnel sterilizers include forced convection and flame. Tunnels operate continuously and are typically kept hot for long periods of time with reduced temperatures at night or on weekends to conserve energy. Dry heat tunnels have the capability to process a larger quantity of glass vials and ampoules than batch ovens. The tunnels sterilize, depyrogenate, and cool glass containers. Figure 16.2 illustrates glass vials processed in a tunnel.

The continuous tunnel is capable of processing glass containers in a relatively short time frame relative to that of batch units. Belt speeds are designed and defined based on container size and desired throughput (required processed containers/minute) and typically target a container temperature above 250°C. The time for the forced-convection type tunnel to process containers is approximately 40 to 60 minutes from washing to delivery, dependent on the container size and throughput design requirements. This efficiency reduces packaging and additional handling, which improves sterility assurance levels by removing operators from the transfer of processed containers to the filling line.

The forced-convection tunnel is heated by electric elements and employs the same principles as forced-convection ovens. Bottles, vials, or ampoules are washed and loaded on the non-aseptic end of the tunnel. Most tunnels are integrated with the washer for continuous processing and to maintain a constant supply of components for filling operations (Figure 16.3).

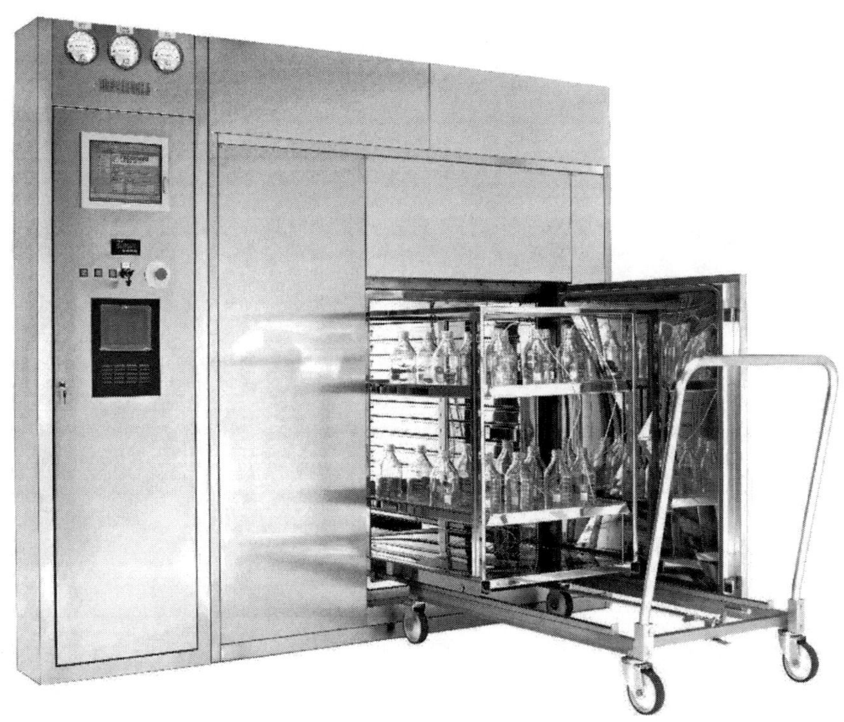

FIGURE 16.1 Batch oven.

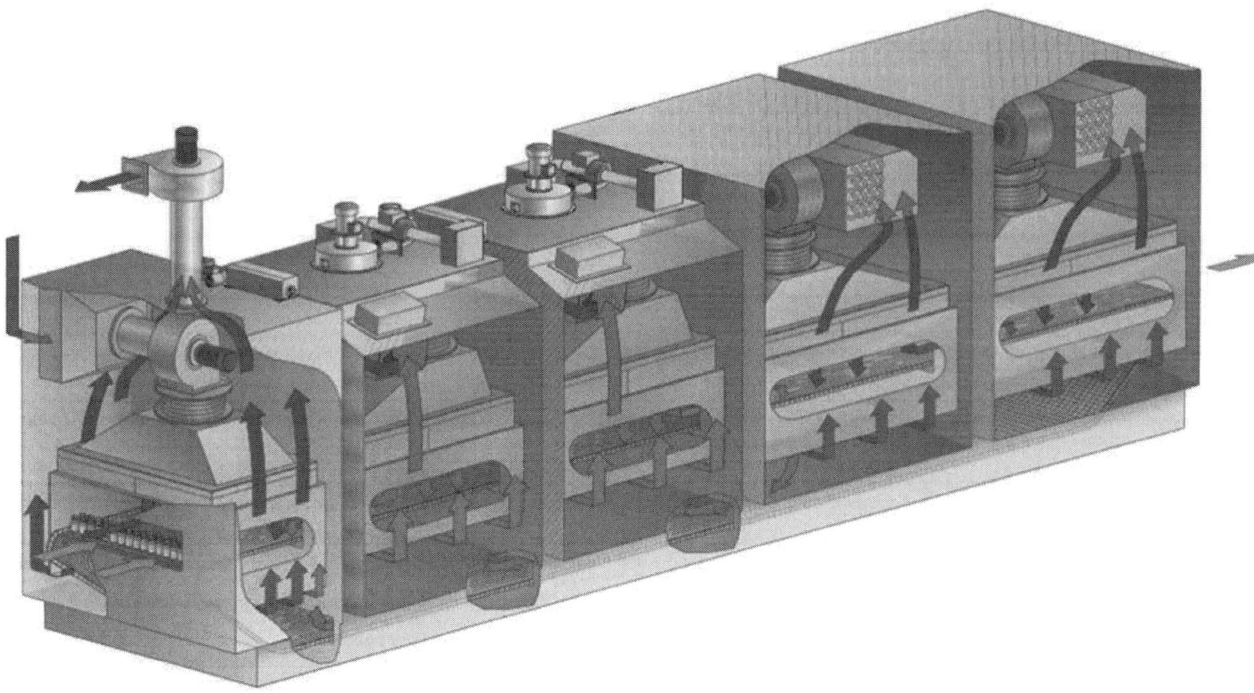

FIGURE 16.2 Dry heat tunnel schematic.

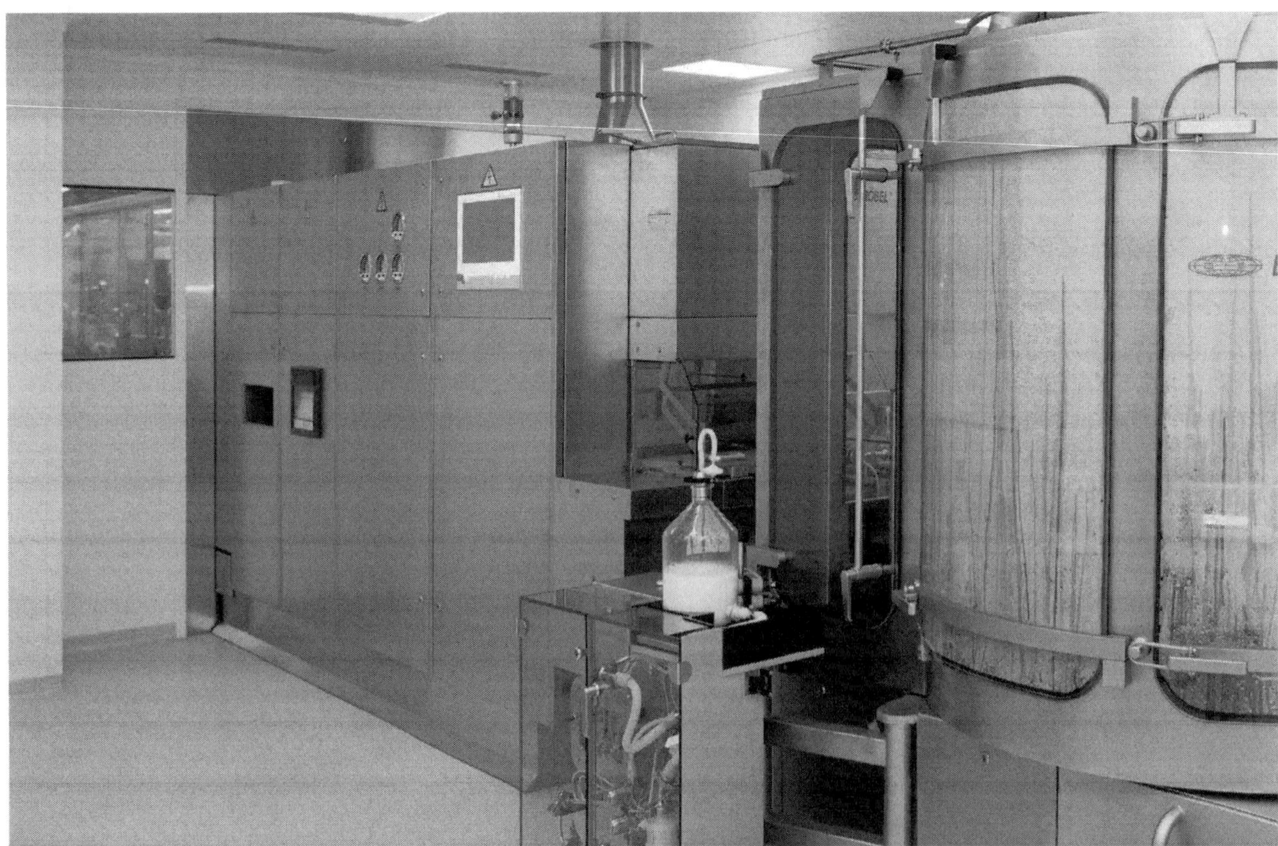

FIGURE 16.3 Dry heat tunnel with integrated vial washer.

A densely packed arrangement also provides container stability for smaller component sizes transported through the tunnel.

The containers are conveyed the length of the tunnel, the design of which is based on component size and required throughput (components per minute to be processed). The glassware is heated in the heating zone and typically maintained at 250°C–450°C (dependent on container size and throughput) and gradually "cooled-down" by HEPA-filtered air prior to leaving the tunnel at the aseptic end. The discharge temperature after cooling must be determined and incorporated into the design of the tunnel's cooling zone. A discharge container temperature of 30°C is a common design target. The integrated tunnel line can supply components to an isolator filling system, and include trayloading to a freeze dryer, if required, to further increase sterility assuance levels and manufacturing efficencies (Figure 16.4).

Tunnels, such as those manufactured by Bosch, IMA, Bausch+Ströbel and others, are now available with the capability to automatically sterilize the cool zone when the tunnel is not in use. This feature is considered essential and should be included as a design configuration.

Temperature sensors to monitor conditions are located within the various zones. Heat up and exposure time of processed components is affected by the geometry, color, surface, and composition of the item being treated, as well as conveyor speed (belt speed) and air temperature [1].

The controlled heating zone is supplied with HEPA-filtered air separately from the initial entry pre-heat zone and cooling zone. The heating zone is usually operated at a higher temperature or supplied with more heated air than the entry pre-heat zone. The cooling zone reduces the container temperature to avoid glass cracking from thermal shock upon exiting the tunnel.

The flame tunnel design utilizes conduction and convection heat transfer in the continuous processing of ampoules. It can process up to 10,000 ampoules per hour. Ampoules are placed on a conveyor belt, washed with water-for-injection, and channeled onto spokes of a rotating wheel. As the wheel rotates, the ampoules are heated to 425°C by natural gas heat for approximately 1 minute. The tunnel has a series of baffles in the chamber to increase the uniformity of heating. The ampoules then pass from the heating chamber into a cooling chamber, where they are gradually cooled by HEPA-filtered air. The cooled ampoules are then filled and flame sealed.

16.2.3 Microwave Tunnels

Microwave units, which are not commonly used, use electrical or electromagnetic energy to heat materials by conductive, near-field coupling, or radiative techniques [2]. With microwave technology, the microorganisms are heated without intense heating of the container. Studies have been performed on *B. subtilis* and *B. stearothermophilus* applied to glass vials and heated for various periods of time with microwaves. A 10^{12} reduction of *B. subtilis* was determined to occur within 3 minutes of treatment with microwaves [3].

Studies with dry *B. subtilis* spores treated in a dry heat oven and a microwave at the same temperature demonstrated the mechanism of sporicidal action of the microwaves was caused by thermal effects. A dwell time of 45 minutes at 137°C was required for both the dry heat oven and microwave oven; however, the microwave ramp-up time to sterilization temperature was shown to be four times faster than that of the dry heat oven [4]. A spore temperature above the vial temperature is achieved in a microwave oven and results have been reported of a spore reduction of 10^{12} in 4 minutes with the vial temperature of 160°C as compared with a dry heat oven cycle of 90 minutes at the same temperature [5].

The fluctuations of temperatures and power output, cost, and lack of production designed equipment has prevented microwave technology from becoming a viable alternative to dry heat processing.

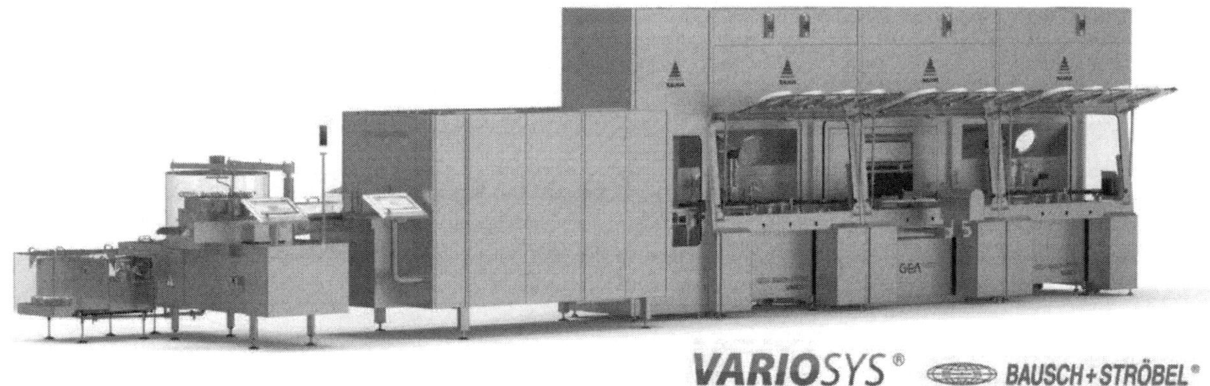

FIGURE 16.4 Integrated tunnel with vial washer and isolator filling systems.

16.3 PRINCIPLES OF HEAT TRANSFER AND CIRCULATION

The dry heat process must effectively heat the article, and the air surrounding the article, to achieve sterilization or depyrogenation. Although there are similarities between dry heat and moist heat validation approaches, there are several differences in the processes that direct the emphasis in validation. In moist heat, the condensation of the steam sterilizer releases large amounts of heat energy that serve to rapidly heat the items in the sterilizer. In dry heat processes, the hot air carries significantly less heat energy than an equivalent volume of saturated steam.

Because hot air has both a low specific heat and poor thermal conductivity properties, there is a need for longer sterilization cycles, at higher temperatures, than those required in steam sterilization. In dry heat processing, the penetration of heated air is not facilitated with the pre-sterilization cycle vacuum air removal commonly utilized in steam sterilization. With dry heat processing, the load is generally slower to both heat and cool and has a greater tendency for temperature stratification (causing temperature variations often >10°C during the cycle).

Despite these limitations, dry heat may be chosen as the preferred method to induce sterilization and/or depyrogenation over moist heat or other methods in certain instances. Some items are ideally prepared by dry heat methods. These include glass, stainless steel equipment having surfaces difficult to penetrate with steam, and items that may corrode with moisture. Some products are damaged or contaminated by the presence of water (e.g., petrolatum, oils, nonaqueous vehicles, fats, and powders). Additionally, it is not practical to use a continuous steam sterilizer, similar to those used in the food industry, in an aseptic facility, compared with a continuous tunnel sterilizer, especially for the supply of sterile/pyrogen free dry glass vials, ampoules, etc. to a filling line.

16.3.1. Convective Heat Transfer

Dry heat processes use convective and conductive heat methods to increase the temperature of the materials. Convection is a form of heat transfer whereby heat flows from one body to another because of the temperature difference between them. In dry heat processes, air is heated by convective methods by passing it across heating elements. Energy is transferred to the air from the heating elements. The heated air transfers energy to the items being treated, because those items are at a lower temperature than the air. The rate of heat transfer (how fast the items will heat up) is related to the specific heat and mass of the materials.

Air has the disadvantage of having a relatively low specific heat; therefore, it transfers energy at a slow rate. Saturated steam is an excellent material to use for heat transfer, as it has a relatively high specific heat (c_v = 1.0 Btu (British thermal unit)/lbm °F) as compared with air (c_v = 0.1715 Btu/lbm °F) [6]. However, as discussed previously, steam heat is not appropriate for all sterilization procedures. The low specific heat of air means it is also slower in cooling of materials back to safe handling temperatures after treatment.

16.3.2 Conductive Heat Transfer

Conduction of heat consists of the flow of thermal energy through a material from a higher to a lower temperature region. Conduction is another means of transferring energy, achieved by molecular interaction in which atoms at high temperatures vibrate at high energy levels and impart energy (as heat) to adjacent atoms at lower energy levels. Adjacent materials (such as air surrounding the product) will convectively transfer energy from the higher-temperature material to the colder material because electrons in an excited state collide with electrons of the lower-energy (colder-temperature) material. The excitation of the molecules of the object increases the level of molecular energy, which increases its temperature [7]. The rate at which heat conduction occurs is dependent on the materials involved.

16.3.3 Circulation

To aid the process, a system to increase the circulation of the air is employed. During the heat up cycle, air circulation helps by removing cool air from the chamber or heating zone and preventing temperature stratification. Higher temperature air replaces the cool air, and the load is heated more rapidly.

Air is also circulated at the completion of a cycle to cool the load. A depyrogenation tunnel utilizes recirculated air in the cooling zone to cool the containers post sterilization and prior to exit from the tunnel. The recirculated air protects the sterile load by maintaining a positive pressure between the sterile load inside the chamber and the non-aseptic load preparation area. Booster fans or blowers are used in the oven or tunnel in order to increase the circulation of the heat throughout the load (Figure 16.2).

Monitoring motor/fan speeds to ensure consistency of air flow rates within the oven and/or tunnel is essential, because it is an important factor in the rate of heat transfer. Currently most fan/motor systems are controlled by a Variable Frequency Drive (VFD) that facilitates balancing/tuning of the systems. The VFDs can be programmed to alarm if a fault occurs during operation. In addition to the VFDs, an anemometer is used to determine the flow rate of air in the heating and cooling zones of depyrogenation tunnels and may be used in batch ovens to monitor intake, exhaust, and circulating air.

To maintain consistent operation of each zone in a depyrogenation tunnel (preheating, if equipped, sterilizing and cooling zones), the differential pressure (DP) of each zone may be controlled, monitored, and alarmed to ensure consistent operation. The differential pressure across the HEPA filters is also monitored. This will ensure consistent airflow and system operation at operating parameters.

If the batch oven is installed in an aseptic environment between different classified environments, balancing of the air systems in the room is necessary. A slight positive pressure should exist from the higher classified aseptic area to the open batch oven to prevent contamination of the aseptic area. The batch oven must be slightly positive in pressure to the non-aseptic area (lower classification) to prevent the flow of dirty air into the oven. Pressurization air is filtered through a set of HEPA filters.

In a depyrogenation tunnel, pressurization is monitored and controlled by differential pressure between the component preparation room to the cooling zone, the filler room (tunnel discharge) and/or isolator to the cooling zone, or both the component prep and filler room to the cooling zone. Tunnel pressure is controlled to be positive to the component prep room and less positive than the filler room and/or isolator.

Air is usually supplied to the equipment by a central HVAC system or directly from the aseptic area (room). A filtered HVAC air supply is preferred because it has a lower particulate load and is usually both temperature and relative humidity controlled. Room air may be used but may have a higher particulate count and may vary in temperature and humidity levels. Because HEPA filters are incorporated in the equipment and associated air intake, the use of room air is an acceptable practice.

Relative humidity is an important factor in trying to heat air and maintain a consistent temperature and is an important factor in the qualification of open-ended tunnel sterilizers. Validation information should include monitoring relative humidity of the make-up and surrounding air.

Manufacturing facility and product process requirements must be considered in the installation and integration of dry heat equipment. The International Society for Pharmaceutical Engineering (ISPE) Baseline Guide, Sterile Product Manufacturing Facilities, provides a useful guideline for these design considerations [8].

As previously noted, HEPA filters must be used for both air supply and circulating air. The filters in the air circulation path must be designed to withstand high operating temperatures and should be monitored periodically for integrity. The HEPA filter may have a ceramic or heat-resistant housing with high temperature sealing gaskets.

HEPA filters that are subject to extremely high temperatures should not be integrity tested with aerosols, as they outgas and smoke when heated and may contaminate the contents of the oven or tunnel. In place of integrity testing, air nonviable particle testing should be performed.

The air introduced into the sterilizer and circulating within the unit should be tested for particulates at multiple locations while the fan is operating. It is preferable that the tests be run at the normal operating temperatures if the air particulate monitoring equipment is designed for operating temperatures. If conducting the test at operating temperatures, a water-cooled probe must be used and the operator needs to wear appropriate protective equipment suitable for high temperature work.

High particulate counts may be caused by improperly balanced air, improperly installed or damaged filters, vibration and/or shedding materials, or inadequate sanitization practices.

Circulating air must meet ISO (International Organization for Standardization) 5 or Grade A requirements. Air cleanliness classes and clean room/clean zone requirements are defined by the International Standard ISO 14644–1 with limits set for the maximum number of particles per cubic meter of air by micrometer size [9]. ISO 5 (Grade A) particulate counts may not be achieved during temperature ramp-up and/or cooldown phases of the cycle because of moisture evaporation, glass temperature gradients, the contraction and expansion of HEPA filters, and other system equipment/components.

The quality and reliability of HEPA filters used in the heated zones of tunnels and batch ovens has evolved from the original design of stainless steel frames, paper filter media, and ceramic sealing compounds to the current state of design consisting of stainless-steel frames and glass fiber media without sealing compounds (Figure 16.5).

The latter design has many advantages over the previous design including longer service life (may achieve >8 years of reliable operation), less heat sensitive materials, reduced risk of media to frame seal leaks (no sealing compound is used between the filter material and the frame, eliminating particle formation because of different thermal expansion properties), lack of air leaks as ceramic sealing compounds can break down and cause air leakage over time in older designs, and maintains air classification over a wide range of operating temperatures and changes in temperature (i.e., during pre-heating ramping to temperature and cooling down phases).

Some manufactures, such as Bausch + Ströbel, have offered glass fiber media filters as standard equipment since 2002, demonstrating long-term reliability. In most cases, sterilizing tunnels that are equipped with paper filters can be modified to use glass fiber filters.

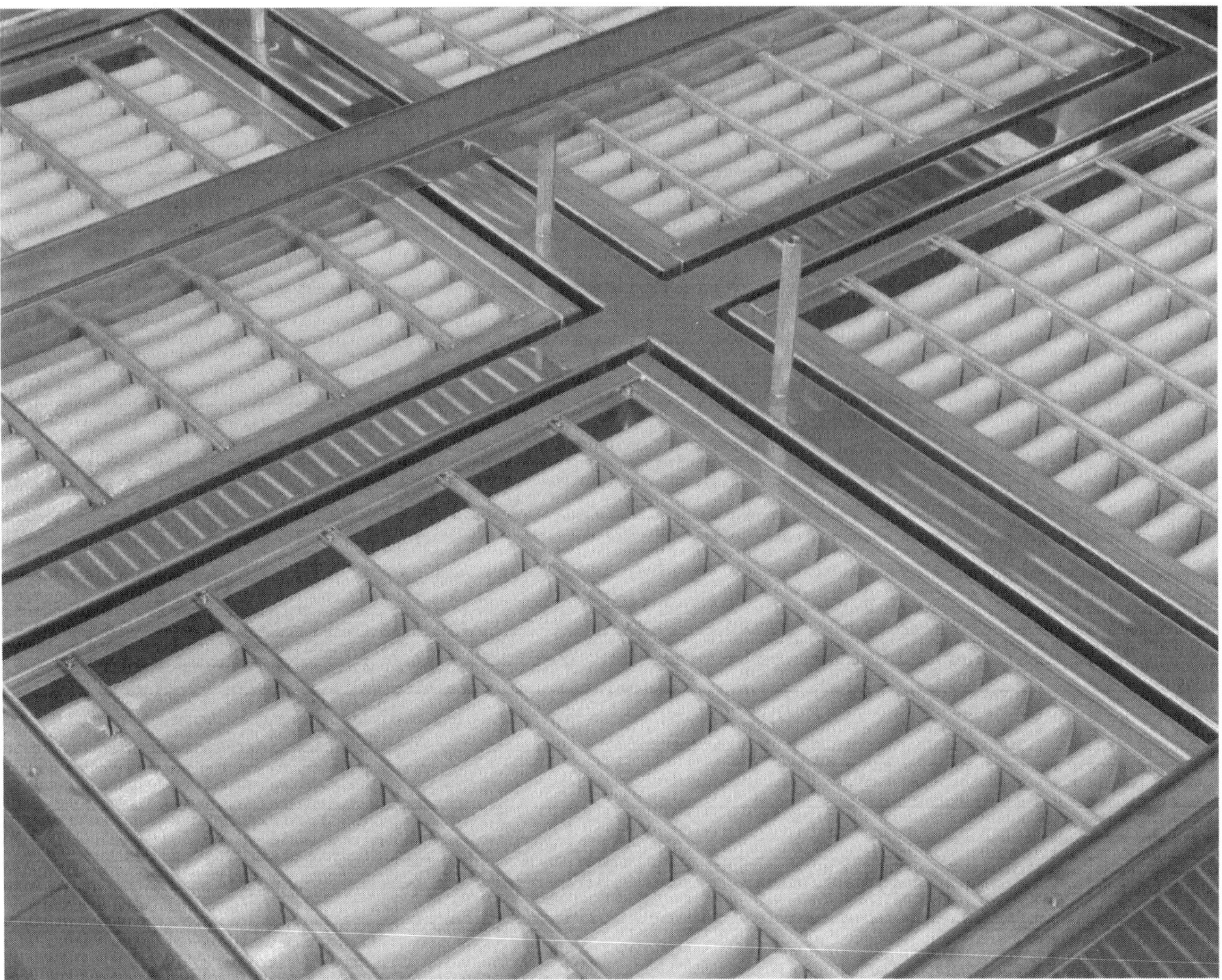

FIGURE 16.5 High temperature glass fiber HEPA filter.

16.4 LIFE CYCLE APPROACH

Experienced engineers, operations personnel, validation practitioners, and guidance organizations such as ISPE (International Society of Pharmaceutical Engineers), ASTM (American Society for Testing and Materials), etc., as well as regulatory authorities have long recognized the need to accurately define requirements to procure the correct equipment based on process needs. Validating and then properly maintaining the equipment in a validated state through effective maintenance, change control, and periodic revalidation is also required. This process not only supports Good Engineering Practices (GEP) to ensure a successful implementation of the equipment, it ensures that the equipment and sterilization/depyrogenation process remain in a state of control and compliance. The concept of the "cradle to the grave" life cycle approach has been represented in various illustrations over the years. Diagram 16.1 effectively depicts this concept.

The benefit of a defined life cycle process is to adequately integrate the dry heat equipment into the existing quality systems and to ensure the equipment is properly specified (meets users requirements and cGMP compliance standards); is verified (development and review of the design documents and design reviews) to meet process requirements; confirmed to be adequate and meet the intended purpose in the validation phase; and once released to production, adequately maintained and monitored to ensure validated state of compliance through service life until retirement.

Section 16.5 addresses the validation approach and application of this information. Section 16.6 discusses continued maintenance of the equipment until retirement.

16.4.1 User Requirements Specification

The user requirements specification (URS) defines the process requirements that will directly relate to the equipment suppliers' design criteria for sizing components of the dry heat equipment such as heaters, cooling systems, air

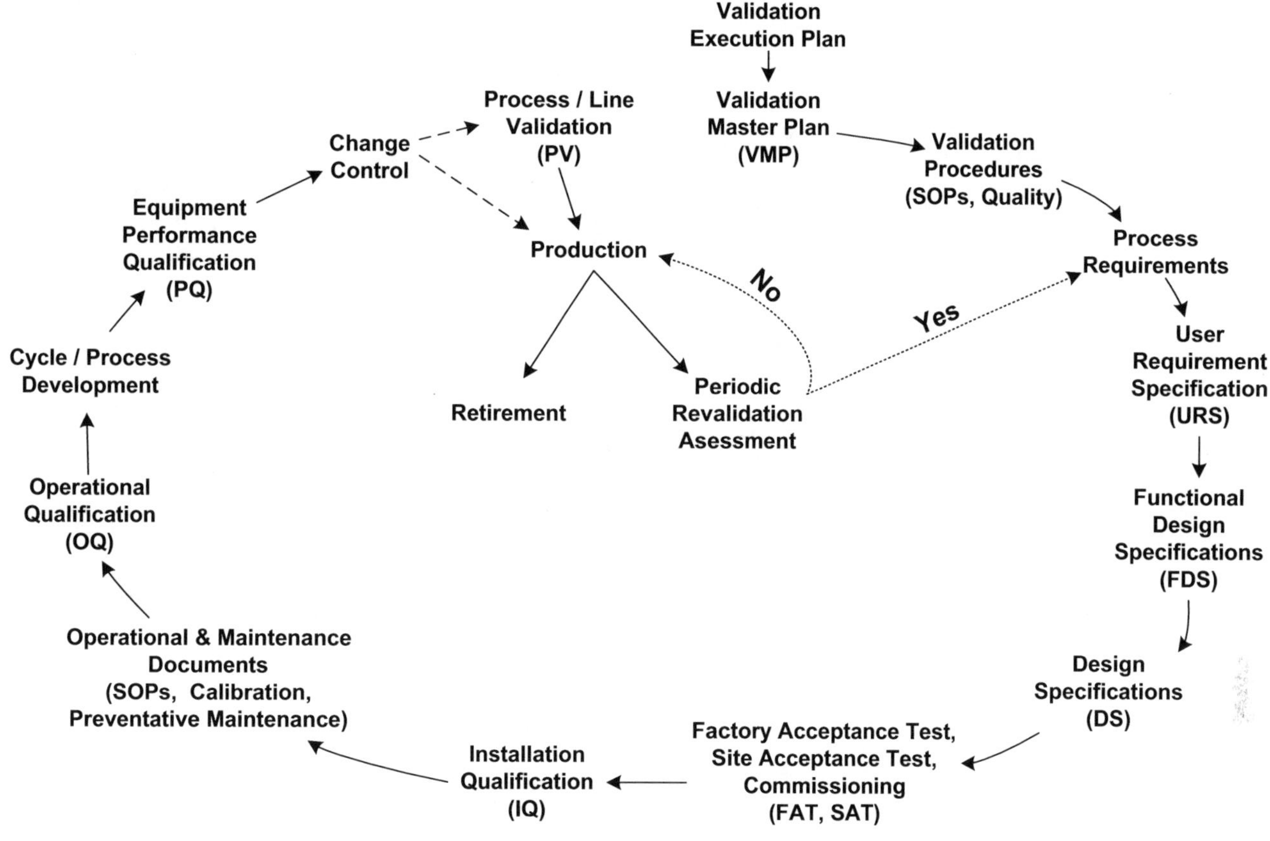

DIAGRAM 16.1 Validation life cycle process.

TABLE 16.1
Critical Design and Control Process Parameters

- Temperature of circulating air pre-heating zone (if incorporated in the equipment)
- Temperature of air circulating in sterilizing zone
- Temperature below sterilizing zone
- Temperature of air circulating in cooling zone
- Air flow in the sterilizing zone
- Air flow in the cooling zone
- Differential pressure between cooling zone and component preparation room
- Differential pressure between cooling zone and filler room and/or isolator
- Conveyor belt speed

circulation systems, capacity, etc., and providing the correct manufacturing solution. The URS criteria, in addition to supporting equipment design, defines the CPPs that need to be controlled, monitored, and verified. Once the system is validated, these parameters also serve as the routine CQAs used to confirm satisfactory operation and release of the processed components. Table 16.1 lists typical CPP parameters that are routinely controlled and recorded for a dry heat process.

16.4.2 Design Review

During the design review process, it is important to understand how the CPPs are being met, controlled, and monitored. The review process provides a road map to the equipment suppliers' approach to controlling these parameters and provides valuable information on associated equipment components and instrumentation that will need to be evaluated in the validation process. Additionally, equipment and process instrumentation need to be properly integrated into the routine maintenance and calibration programs.

16.4.3 Life Cycle Document Process Flow (V-Model)

The process and design documents illustrated and developed in the life cycle approach serve as the foundation for equipment design and fabrication and are used in the development of test plans and acceptance criteria of the commissioning/validation protocols. A commonly accepted illustration of this relationship is depicted in a V-model diagram (Diagram 16.2).

This V-model approach is applicable to integrate both equipment and software requirements into a combined design and validation program.

16.5 VALIDATION APPROACH

This section is intended to detail two common approaches to the validation of dry heat sterilization equipment. A traditional approach is to execute validation activities independently from any predecessor activities performed during factory testing through site installation and associated testing.

A more efficient approach utilizes an integrated C and Q approach that allows leveraging of predecessor activities from factory testing through site installation and associated testing, if corporate and/or site procedures define this approach.

16.5.1 Traditional Validation Approach

The traditional validation approach involves factory testing, start-up, and testing once the equipment is installed on site with subsequent Installation Qualification (IQ), Operational Qualification (OQ), Cycle Development (CD), and Performance Qualification (PQ) activities documented in their respective protocols (see Appendix I for details).

The testing initiated and performed at the factory and/or the manufacturing site may be repeated and documented in a different manner in the IQ, OQ, and PQ phases. A combined Installation and Operational Qualification protocol may be utilized based on corporate or site-specific procedures. Each protocol (IQ, OQ, IOQ, Cycle Development, and PQ) typically will have a unique final qualification report summarizing the results of the qualification (See Appendix II for details). If this traditional approach is utilized, details for typical activities for IQ, OQ, Cycle Development, and PQ are detailed in their respective sections (Sections 16.5.7, 16.5.8, 16.5.9, and 16.5.10, respectively, for details).

16.5.2 Integrated Commissioning and Qualification Approach

An integrated C and Q approach involves leveraging activities and testing beginning with Factory Acceptance Testing (FAT) through site activities such as Commissioning or Site Acceptance Testing (SAT) and start-up of the equipment or line. Typically, if this approach is utilized, a Validation

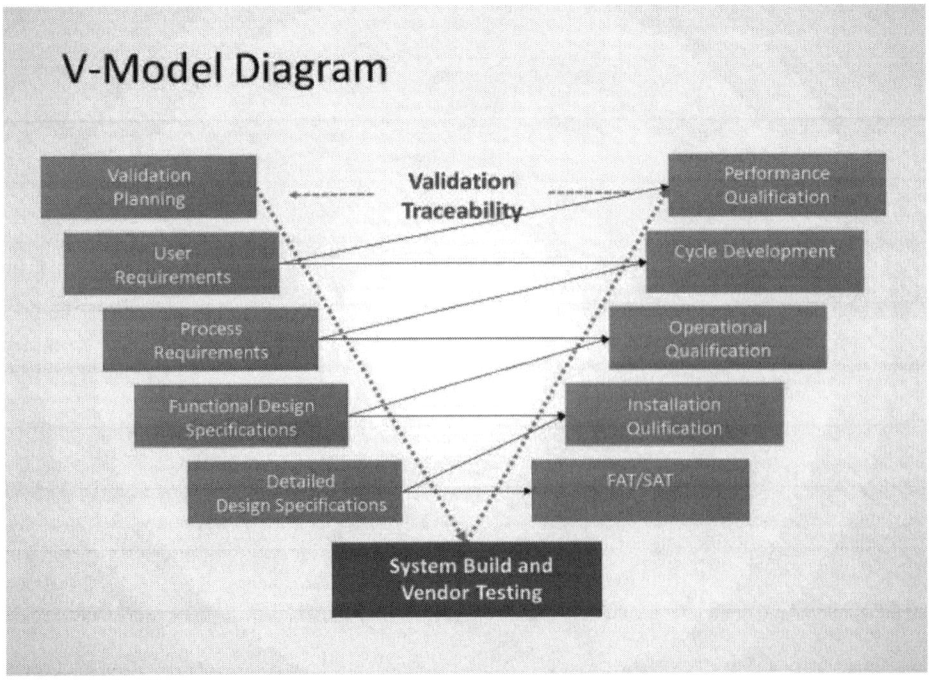

DIAGRAM 16.2 V-model diagram.

Note: FAT = factory acceptance test; SAT = site acceptance test.

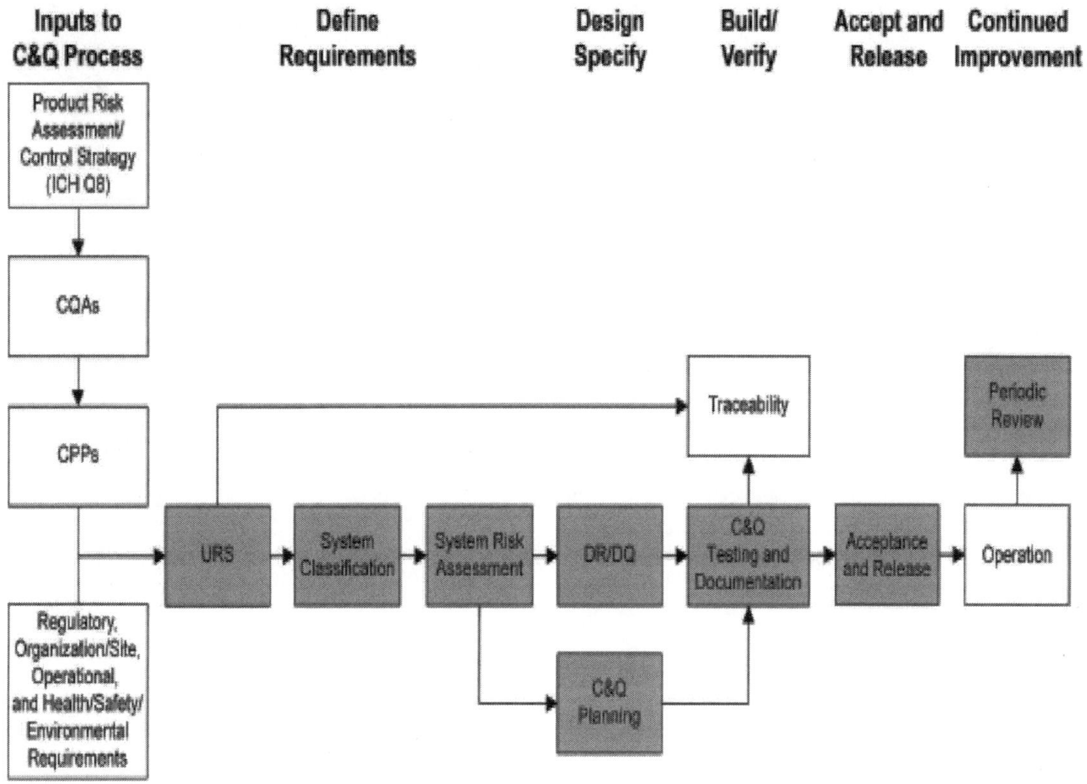

DIAGRAM 16.3 Integrated C and Q process diagram.

Note: C&Q = Commissioning and Qualification; CPPs = Critical Process Parameters; CQAs = Critical Quality Attributes; DR/DQ = Design Review/Design Qualification; ICH = International Conference for Harmonisation; URS = User Requirement Specification.

Master Plan (VMP) is generated and approved by company site stakeholders such as Engineering, Operations, Quality, and Validation (if separate departments as applicable) that aligns with corporate and site-specific guidance documents and procedures. It is important that this approach and any documented implementation plan is approved by stakeholders so that the team is aligned on what testing can be leveraged and how this testing can be utilized in a manner compliant with procedures. This plan should detail the testing to be performed at FAT, Commissioning/SAT that can be leveraged into subsequent IQ and OQ activities documented in their respective protocols, and define requirements, documentation, and conditions where testing can be utilized. It is beneficial to share this approved plan with Dry Heat Sterilization equipment vendors as well as any separate vendors that may provide automated controls, so that they understand and agree to the approach and support the documentation and testing to be leveraged. If this integrated approach is utilized, details for typical activities outlined in this section for IQ and OQ may be initially tested and verified at FAT and/or Commissioning/SAT so that they are not repeated in IQ and/or OQ.

The ISPE Commissioning and Qualification Baseline Guide, Volume 5, Second Edition, provides guidance on an integrated C and Q process. An illustration of this process via flow chart is in Diagram 16.3.

To utilize this integrated approach, a change management process to document changes that may occur in FAT, Installation, Commissioning, and/or SAT is documented and verified in subsequent validation steps. This process ensures that changes are assessed, and that the impact these changes have on previous life cycle documents is evaluated and updated as part of this process.

16.5.3 Automation/Controls Validation and Equipment

Equipment suppliers are providing more robust automation and process controls, frequently utilizing Programmable Logic Controller(s) (PLC) and Human Machine Interface (HMI) systems. These systems require qualification to 21 CFR Part 11 compliance standards (where applicable). An integrated approach combines the qualification of the equipment and controls systems to reduce the number of protocols generated and approved for execution as well as final reports. If using an integrated C and Q approach, the equipment vendor may have internal expertise to provide the required test scripts for PLC/HMI and the components of the system to be included in the validation approach. An alternative acceptable approach is to have separate IQ and OQ protocols for the PLC and HMI systems as part of automation or computer system validation, separate from the IQ and OQ equipment protocols. A firm should evaluate the need to utilize a computer system validation and/or controls subject matter expert (SME) to support the automation validation.

16.5.4 Validation Test Equipment

Whether the traditional or integrated C and Q approach is utilized, it is important that test equipment be documented in IQ, OQ, CD, and PQ protocols and reports. All test equipment employed in the validation studies must be traceable to an NIST (National Institute of Standards and Technology), ISO, or recognized standard as defined in protocols and other relevant guidance documents.

Some companies have specific guidance or procedures defining acceptable standards and tolerances required for calibrated test instruments or equipment. Use and calibration of the validation test equipment must be fully documented in a Standard Operating Procedures (SOPs).

The equipment used for validation testing of dry heat processes may include the following:

16.5.4.1 Voltmeter and/or Amp meter

The voltage, amps, and ohms are tested on the oven/tunnel if corporate or site procedures define this approach, as well as ancillary components during the Installation Verification activities at the site, usually during IQ (or if using the leveraging approach this can be verified in Commissioning/SAT). It is important that spikes or drops in the line voltage be investigated, as facilities age and additional production equipment may result in an electrical supply issue.

16.5.4.2 Signal Generator

Many equipment manufacturers utilize analog input and output devices for measuring and controlling parameters, often based on 4–20 milliamp (mA) or 0–10-volt output. Input/Output (I/O) and Loop Checks/Verifications for these analog devices are often verified as part of I/O testing, calibrations, and Loop Verifications during the validation lifecycle.

16.5.4.3 Resistance Temperature Detectors

The "primary" temperature measurement device is a Resistance Temperature Detector (RTD). RTD devices consist of platinum and copper wires and provide the most linear temperature measurement range between 0°C–400°C [10]. These RTD devices are often utilized to calibrate the thermocouples used in thermal mapping studies.

16.5.4.4 Thermocouples

Thermocouples are the most widely used devices for "portable" temperature measurements. Thermocouples are utilized for thermal mapping studies.

The choice of the thermocouple type and insulation surrounding the wires is dependent on the operating temperature and required temperature accuracy. For dry heat sterilization or depyrogenation processes, both type T (copper and constantan) and type J (iron and constantan) thermocouples can be used. The insulation most commonly chosen for high-temperature applications is Kapton-H by Dupont. This insulation is rated to 350°C, sufficient for depyrogenation use.

Thermocouples generally have a level of accuracy of 0.1°C.

16.5.4.5 Data Loggers

Multipoint recorders are commonly used during validation studies to record temperatures measured by the thermocouples or portable temperature devices. Data loggers are utilized as part of thermal mapping studies. The thermocouples must be calibrated against a NIST standard. Calibrations must be performed on the data logger/thermocouple system before and after the validation thermal mapping runs.

16.5.4.6 Wireless Temperature Logger

Units are available that combine the temperature sensor and the data logger in a wireless compact unit. These wireless loggers can be utilized as part of thermal mapping studies. The temperature loggers can be loaded into insulating canisters and placed in the dry heat oven or tunnel. At the end of the cycle, the data loggers are placed into a reader station, and the temperature data is downloaded to a computer for data analysis.

As the temperature loggers in the canisters can withstand a temperature of 170°C for 165 minutes, up to maximum temperature of 360°C for 45 minutes, the wireless temperature loggers are best suited for shorter sterilization and depyrogenation cycles, such as those in a tunnel. Similar to thermocouples, the temperature loggers must be pre-calibrated and post-calibrated when used as part of any thermal mapping studies.

16.5.4.7 Infrared Thermometer

Infrared thermometers are ideal for measuring vials exiting the cold zone of a tunnel.

16.5.5 Commissioning and Start-Up

The Commissioning and Start-Up phase, as part of the Validation approach for dry heat equipment, can either be employed as a precursor to the traditional Validation approach or an integrated C and Q approach as previously described. This phase begins with FAT and involves coordination through site activities such as Commissioning or SAT and start-up of the equipment or line.

Tests outlined in the IQ and OQ sections that follow may be used during this phase to either verify that the equipment or system is provided with components, process controls, and functionality per life cycle documents (URS, Functional Specifications) as part of the "debugging" exercise to mitigate risk for deviations during IQ and OQ or can be leveraged based on an integrated C and Q approach.

If the traditional validation approach is used, it is beneficial to document verifications that ensure that equipment is installed properly and that the vendor has met contractual and

User Requirements as part of FAT and site Commissioning/SAT testing aligned with Good Engineering Practices (GEP). It is highly recommended that the following verifications are documented in a formal turnover package, commonly termed as an Engineering Turn-Over Package (ETOP) and provided to the stakeholders prior to initiating validation activities. Approval of this ETOP documents the acceptance for validation activities to proceed.

Installation verifications such as walkdown of drawings and schematics and electrical wiring and I/O checks, mechanical and electrical component verifications, PLC and HMI hardware and software verifications, and verification of filter installation must be performed to verify that requirements have been met.

Functional verifications for the integrity of critical HEPA filters that service the ISO 5 sterilization zones, balancing of airflows to meet differential pressure requirements between equipment and the background clean room, initial temperature distribution verifications to meet specifications, and particle verifications for the ISO 5 zones to meet requirements according to ISO 5 limits (maximum number of particles per cubic meter of air by micrometer size) must be performed to demonstrate that requirements have been met.

16.5.6 Qualification Protocol

Validation protocol(s) must be written and approved prior to the start of actual validation work for IQ, OQ, CD, and PQ. Each of these protocols is designed to outline the program to be employed, the specific tests that will be conducted, and the acceptance criteria for each test. Once the protocol has been written, it must be approved by the designated responsible individuals.

A sample outline for a protocol is detailed in Appendix I.

16.5.7 Installation Qualification

The IQ is intended to compare the provided mechanical/electrical, and where applicable the automation system, against the URS and the manufacturer's specifications for proper compliance and installation. As previously noted, the IQ can either be a stand-alone document that may repeat any previous testing from commissioning/start-up, or the IQ may leverage testing from FAT through SAT/Commissioning per approved validation master plan.

Certain testing/verifications performed at the FAT that are impacted by final installation at the facility must be re-verified as part of the Commissioning/SAT or IQ phase of the Qualification effort.

All main and ancillary equipment, utilities, and connections must be checked against the URS and manufacturer's recommendations. Any automated components related to the PLC and HMI hardware and software for the system must also be verified.

Records of modifications made on the unit should also be checked against the equipment URS and the manufacturer's recommendations.

A schematic of all utilities supplying the equipment should be available to confirm that all connections are as specified and meet design criteria (per URS and functional specifications), local and state codes, and Current Good Manufacturing Practices.

The IQ documents must be reviewed and approved by designated responsible individuals prior to executing any tests or verifications. Procedures may require a copy of the original preapproved IQ protocol to be used as the execution copy. The execution of the IQ protocol should comply with site-specific procedures.

The IQ should include copies or references to the following information; identification of test instruments/equipment utilized and references to calibration file for these test equipment or copies of calibration records attached, manufacturer's specifications, user and functional requirement specifications, unit model number, serial number, corporate and/or department identification numbers, mechanical and electrical component verifications against specifications, PLC and HMI hardware components and documentation of initial software and associated software versions, wiring and I/O verifications, preventive maintenance (PM) programs, verification that calibration of critical instruments are in the current state of calibration, and the identification and location of all critical drawings pertinent to the unit.

If the IQ is leveraging previous commissioning verifications, it is beneficial to include a traceability matrix to identify which verifications or testing have been completed and in what FAT/Commissioning/SAT protocol and appropriate section. Any testing that may have been designated initially in the validation master plan to be leveraged from previous testing that could not be leveraged because of incomplete testing, incorrect documentation, and/or failures resulting in discrepancies must be repeated in the IQ as determined by stakeholders.

16.5.7.1 Structural

Verify dimensions, presence of identification plates (check for accuracy against records), correct leveling, proper insulation, presence of seals, integrity of gaskets and materials of construction, and inspect for structural damage.

16.5.7.2 Filters

All filters used within the system must be recorded, such as those used with air (supply, recirculating, and exhaust), or in other utilities (such as steam, water, or nitrogen) where applicable. Verifications on the filters should confirm that the filters have the proper identification (filter identification, serial number), type and rating, size, change frequency, air capacity, flow rate, and meet temperature limits and any integrity testing required, (with SOPs referenced).

Some HEPA filters may need to be checked periodically by performing an integrity test using polyalphaolephin (PAO) (Emery 3004), or in place of integrity testing, air nonviable particle testing to verify particulate levels. As previously noted, air cleanliness testing is recommended for filters that are heated.

To facilitate filter air cleanliness testing and scanning, Bausch + Ströbel has designed a linear, bidirectional twin

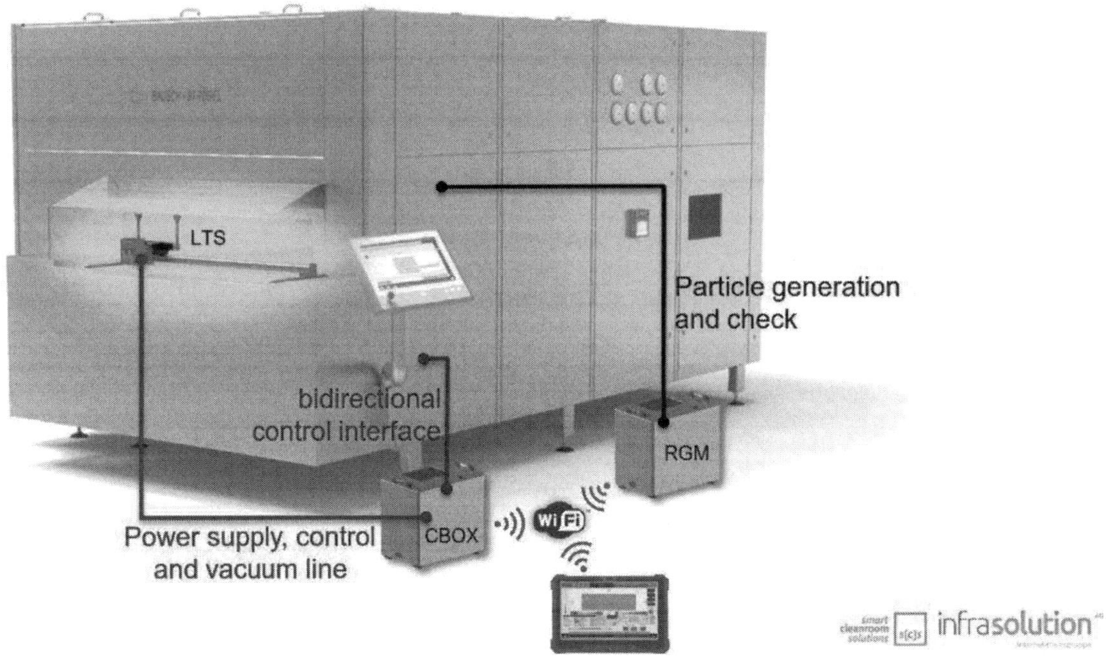

FIGURE 16.6 HEPA air cleanliness testing system.

particle counter system that automatically and precisely measures air cleanliness in the tunnel sterilizer (Figure 16.6).

If this linear automated sampling system is not available, a traditional isokinetic cone attached to an extension holder may be used to conduct the air cleanliness test at room temperature.

Filters used on ducts exhausting outside of the building require integrity testing and may require an environmental air quality permit.

16.5.7.3 Electrical Utility

Conformance to National Electrical Code standards, proper identification for source house panel and breakers supplying the equipment, and the presence of safety disconnects where applicable is required. A verification to document the measured voltage, amperage for each electrical feed to the equipment, PLC and/or HMI using calibrated equipment should be performed. It is beneficial to document the source panel and breaker and breaker trip rating when measuring voltage and amperage for equipment from the electrical distribution system as a baseline for maintenance and safety purposes.

Electrical and wiring verifications for wire size and type, terminations to control cabinets or PLC, communications between PLC and HMI, and I/O testing should be performed against manufacturer's and functional design specifications.

16.5.7.4 Air Supply and Circulation Components

Identify the source (direct from the HVAC system or room air), duct size, and duct material is suitable where exposed to the clean room environment, and the air classification (as per ISO Standard 14644–1) [11].

For systems that have baffles or louvers as part of the circulation system, the integrity of all baffles or louvers must be checked. Ensure that the baffles are not damaged, misaligned, or missing entirely.

Record each heating element manufacturer's model number, the number of heating elements, and the voltage, amperage, and wattage of the elements for the heaters. If these elements have separate electrical feed from the equipment, electrical voltage measurements should meet the requirements specified in the electrical IQ section.

16.5.7.5 Blowers

The blower must be mechanically sound and correctly balanced. Verify that the fan motor(s) rotate in the correct direction. Check for use of the correct fan belt (if belt driven technology is used), and that it is in good condition.

16.5.7.6 Ventilation

Check that the exhaust duct discharges to an appropriate area (not to an aseptic environment), duct material is suitable where exposed to the clean room environment, and verify that devices are present to prevent backflow.

16.5.7.7 Non-Electrical Utility (Cooling Medium)

Identify the type, source, pipe size, and pipe material of construction if exposed to the clean room, the presence of pipe insulation where applicable, type and size of cooling coil tubes, cooling water temperature, and flow rate.

16.5.7.8 Instruments

Identify all devices, controllers, and recorders both critical and noncritical to the operation of the unit. Critical instruments control or monitor critical process parameters and safety devices. These devices may include temperature, cycle timers, pressure, belt speed, air flow, differential pressure, and indicator lights. For each instrument, note the serial number, site identification number, instrument output range, and verification that applicable instruments requiring calibration are current and have calibration sticker and calibration reports available or copies attached.

Certain analog loop instruments must be loop checked/calibrated (temperature, differential pressure, belt speed) using a calibrated signal generator or other approved method and displayed values on the HMI verified as part of the loop check/calibration. It is highly recommended that critical devices and instruments be calibrated in situ where possible after final installation is completed.

16.5.7.9 Lubricants

All lubricants should be evaluated for suitability in the intended operating conditions and to prevent contamination of the material being sterilized or depyrogenated. Where available, sealed motors, bearings, and housings should be utilized to reduce the use of lubricants.

16.5.7.10 Electrical for Automated Systems (PLC and HMI)

Conformance to National Electrical Code standards, proper identification for the source house panel and breaker feeding the PLC and/or HMI, and the presence of safety disconnects must be verified. A verification to document the measured voltage, amperage for each electrical feed to the PLC cabinet and/or HMI using calibrated equipment is required. It is beneficial to document the source panel and breaker and breaker trip rating when measuring voltage and amperage for cabinet and HMI that requires electrical feed from the electrical distribution system as a baseline for maintenance and safety purposes.

Electrical and wiring verifications for wire size and type, terminations to control cabinets or PLC, network communications between PLC and HMI, and I/O testing should be verified against manufacturer's and functional design specifications.

16.5.7.11 Environment for Automated Systems (PLC and HMI)

The environment (temperature and relative humidity) where PLC cabinets/hardware and HMI is installed is typically verified against the manufacturer specifications using calibrated equipment.

16.5.8 Operational Qualification

For clarity, the IQ and the OQ have been described separately. It is becoming common practice in the industry to combine the IQ and the OQ into one protocol. If the OQ is separate from the IQ, the IQ final report must be approved prior to proceeding into OQ.

After the equipment has been checked for proper installation, as detailed in the IQ, it is necessary to determine that the oven and/or tunnel operates as designed for all programmed cycles. The components of the system must satisfy the operating ranges as determined by the specifications set by the pharmaceutical manufacturer. The oven and/or tunnel must be operated to confirm that it functions correctly on a repeated basis. The OQ document should be reviewed and signed by the required department representatives.

Each of the following process components must be identified and the operating performance and ranges determined.

16.5.8.1 Temperature Monitors

Temperature controllers, recorders, and sensors on the process equipment must be calibrated before the unit can be operated reliably. The units are generally calibrated at the time of installation by the manufacturer or user and should be recalibrated at set periodic intervals.

The calibrations should be performed by measuring actual temperatures in addition to electronic methods, such as checking voltage or resistance readings at various set points. It is also essential that actual temperatures of the unit be checked at the set points, as described under the component mapping studies in Section 16.5.9.2. The controller must prove reliable in maintaining the temperature within the specified set points.

16.5.8.2 Cycle Timer

The accuracy of the timer must be determined so that assurance is provided for the cycle length. The recorder must accurately display the cycle time.

16.5.8.3 Door Interlocks

If a unit is equipped with double doors, the interlocks must operate such that the door leading to the aseptic area cannot be opened if the door to the non-aseptic area is open, if the cycle has not been successfully completed, or if the temperature is too high for safe handling.

16.5.8.4 Heaters

All of the heating elements must be functional. It is preferable to have them monitored continuously with ammeters to immediately detect burned-out elements. A failed element could cause a substantial change in the operating performance.

16.5.8.5 Blowers

Properly adjusted blowers (fans) are very important to the effectiveness of the circulation in the oven and/or tunnel. The blower should deliver an air velocity consistent with manufacturer's specifications, which may be accomplished by adjusting the speed of the fan (the air velocity and motor speed should be noted in the OQ records). It is essential that the blades are rotating in the proper direction.

An airflow switch or other sensor present to detect and alarm a blower failure should be tested.

16.5.8.6 Cooling Coils

To enable a faster cooldown cycle, the air is often circulated across cooling coils. The effectiveness of the cooling coils can be verified by cooldown cycles in an oven and as part of mapping studies for cooling zone temperature within a tunnel, in both empty and loaded chamber mapping studies.

16.5.8.7 Tunnel Component Belt Systems

The component delivery belt speed is a critical operating parameter in both continuous hot-air tunnels and flame sterilizers. Belt speed is a critical process parameter and must be monitored and recorded. The belt speed and operating temperature are interrelated, in that a slower belt speed at a lower temperature will produce the same effect as a faster belt speed at a higher temperature. The belt drive motor speed setting and range should be verified in OQ testing.

A depyrogenation tunnel often incorporates belt speed as a critical parameter for recipes of different glass vial sizes processed within the tunnel. The belt speed, after calibration, should be verified as part of each glass vial configuration run performed in the OQ.

16.5.8.8 Chamber Leaks

The perimeter of the doors for batch oven, and tunnels where applicable, should be checked for air leakage while operating, especially when a decrease or fluctuations in chamber temperature are identified as part of any thermal mapping studies.

16.5.8.9 Particulate Counts

Particulate counts should be checked within the containers before and after the process to quantify the particle load contributed by the process.

16.5.8.10 Functional Tests for Automated Systems (PLC and HMI)

Verify that the PLC and HMI operates per functional design specifications (including hardware and software design specifications if they are not included within functional design specifications). Alarms must be verified across the operating range for each alarm (discrete or analog) point programmed for the system, with an emphasis on critical alarms that affect the operation of the unit or capability of the equipment to meet critical process parameters.

Equipment shutdown simulations for the control system should be verified, including start-up of the equipment after simulated shutdown, and all configurable parameters should restart correctly and not reset to default parameters.

Critical configurable cycles or recipe parameters must be verified, with an emphasis on confirming that these configurable parameters only accept modifications within the specified range from design documents.

HMI screen navigation for personnel operation should be verified and documented. Based on screen navigation testing, it is beneficial to have operational/use SOPs for the equipment that can be reviewed and redlined to align with the verified screens and associated navigation tree. These verifications may be performed starting at FAT.

Alarm verifications and HMI screen navigation confirmations may be leveraged from FAT and/or site Commissioning/SAT for the OQ, especially if there were no software updates that impact the functionality of these verifications between FAT/SAT and OQ.

16.5.8.11 Functional Tests for 21 CFR Part 11

Any 21 CFR Part 11 related compliance features that are part of the dry heat equipment must be verified as part of the automated controls testing in the OQ according to specifications in the functional design specifications (including hardware and software design specifications if they are not included within functional design specifications).

Access levels for administrators, operators, and users and the associated security verifications for what each of these groups are able to view, modify, and/or acknowledge must be tested and documented. Data integrity for any cycle run data and recipe data and the functionality to print data or export data to the house server must be tested and documented.

Audit trails for personnel who have access levels to modify recipe or run data, configure parameters, or acknowledge alarms must be verified. Audit trail reports that provide evidence of who made edits, changes, or acknowledgement of alarms with date and time stamps are often verified as part of this testing.

If electronic signatures are implemented, then signature functionality must be verified along with audit trail verifications. These verifications may be performed starting at FAT.

The 21 CFR Part 11 and security/access and data integrity verifications may be leveraged from FAT and/or site Commissioning/SAT for the OQ, especially if there were no software updates that impact the functionality of these verifications between FAT/SAT and OQ.

16.5.8.12 Pre-Calibration of Validation Test Equipment

Test equipment used for thermal mapping studies must be calibrated, refer to Section 16.5.4 for calibration requirements for the primary test equipment. This pre-calibration verification must be performed, documented, and verified to meet stated acceptance criteria prior to performing any empty chamber testing studies as described in the next section.

Calibration of the thermocouples and data logger against a traceable standard is a critical step to be performed before and after thermal mapping studies (empty and/or loaded chamber). Pre-calibration ensures that all the thermocouples are in working order and compares each temperature reading against a known standard. The post-calibration of thermocouples after runs is performed to ensure that the recorded temperature data is valid.

Thermocouples, RTD probe and monitor, data logger, and temperature bath are required to perform the calibration.

16.5.8.13 Empty-Chamber Testing

The initial testing is performed on an empty oven or tunnel to establish the uniformity of temperature distribution. The thermodynamic characteristics of the empty unit are depicted in a temperature distribution profile. The temperature profile will serve to locate hot or cold areas by mapping temperatures at various locations. The temperature profile is obtained by placing an appropriate number of thermocouples, preferably in a geometric pattern distribution in the empty tunnel or batch oven to determine heat profiles.

More thermocouples than the protocol requires should always be included in case there is a problem with individual thermocouples. In ovens, the eight (8) corners and center locations, at a minimum, and locations adjacent to any chamber controlling probe and chart recorder monitoring probe should be monitored as part of the mapping study. Other locations (center of shelves within the oven for example) may be added as locations for the empty mapping study. The thermocouple tips should be suspended to avoid contact with any solid surfaces (wall, ceiling, support rods, etc.). In the flame sterilizer, the thermocouples should be placed at the level of the ampoules.

For ovens, an acceptable profile should demonstrate uniform temperatures during the last 3–5 minutes of the cycle (typically ±15°C for an empty chamber). The temperature range must conform to the protocol requirements and user and manufacturer specifications.

For empty tunnel profiles, the temperature across the tunnel should be monitored. A single empty tunnel study is sufficient for each unique temperature set point and belt speed combination. The mapping of the empty tunnel is a bit easier than the loaded tunnel, as explained in the next section.

All environmental factors should closely represent the actual manufacturing conditions (e.g., relative humidity, room temperatures, static air pressure, and balance). All control settings must be recorded, including any variables that will affect the cycle. Key process variables such as temperature set points for a different zone, blower speeds, heating element settings, cycle-timer set point, belt speed, etc. must be recorded. The cycle timer (for batch ovens), belt speed (for tunnel or flame sterilizers), controller operating temperature, and production recorders can be verified by a multipoint temperature recorder with an internal timer. Air velocity profiles across the unit can be of significant benefit in identifying distribution temperature issues.

The empty chamber runs should confirm ISO 5/Grade A conditions at sterilizing temperatures. Sampling probes are positioned within the unit to measure total particulates. These studies can be conducted at both ambient and operating temperatures. The use of sample probe coolers or high temperature sampling cells of glass or metal is necessary to avoid contaminating or damaging the particle counter. Often ISO 5 conditions are met at ambient and elevated temperatures but are not achievable during heat up and cooldown because of expansion and contraction of the filters and equipment.

A thermocouple should be placed adjacent to any heat-controlling or monitoring temperature sensor to confirm that the operational controls are maintaining the desired heating specifications. It is important to document the ramp-up time (the time to reach the temperature set point) and the cooldown time (the time from the end of the cycle or dwell period to the time the components are cool enough to remove from the sterilizer), because data variances may indicate electrical or mechanical malfunctions in the batch oven. It is also critical that the tunnel and flame sterilizer be closely monitored within the sterilizing zone, because temperature variation is most critical at this location. The empty-chamber cycle can be one of maximum time with production operating temperatures or a shorter time period at a predetermined temperature, such as 250°C, or stated manufacturer's design and associated uniformity specifications. It is common to observe a greater temperature range within an empty chamber study because of the lack of mass that can affect the airflow within the chamber impacting temperature distribution and uniformity. This observation is also dependent on how the manufacturer configures the ramp-up temperature and time to allow the chamber to "pre-heat" until the set point temperature is met and then stabilized. Uniformity specifications should be determined based on load mapping studies and verified to be acceptable in PQ.

A detailed diagram of the location of the thermocouples should be included in the empty-chamber mapping study. This file will be extremely valuable when revalidation of the sterilizer is performed. The empty-chamber data file must include originals (or copies) of all graphs, temperature printouts, data calculations, and observations pertaining to the empty chamber mapping runs.

16.5.8.14 Post-Calibration Verification of Validation Test Equipment

After the empty chamber (and cycle development/loaded chamber mapping as described later in the PQ section) validation studies are completed, the thermocouples and data logger must be post-use calibrated, to verify that the thermocouples are accurately measuring the temperature during the entire period of use. This must be performed without any adjustment of the data logger.

This post-calibration verification may be made after any number of validation runs in OQ, (or CD and/or PQ as described later), but there is a risk that all of the validation tests may have to be repeated if the thermocouples fail to correctly post-calibrate.

A suitable interval for post calibration of the thermocouples as part of risk mitigation strategy should be defined and agreed upon by company stakeholders as part of the qualification effort.

16.5.9 Cycle Development

The dry heat cycle is utilized to inactivate or remove any contaminants that can cause deleterious effects on the patient recipient of the parenteral. Sterilization cycles are designed to inactivate the heat-resistant spores (e.g., *Bacillus atrophaeus*) as well as any vegetative cells that could potentially be present during processing. Depyrogenation cycles target the removal or inactivation of bacterial endotoxins (e.g., *Escherichia coli* lipopolysaccharide [LPS] or naturally occurring endotoxins [NOE]), which are fever-producing substances in the gram-negative cell outer membrane. The endotoxins, or pyrogens, retain their potency even when the cell is destroyed or lysed. The mechanism of inactivation of desiccation-resistant spores by dry heat has been suggested to be because of drying or moisture loss resulting in thermal oxidation and at higher temperatures, organism incineration. Typically, dry heat processes will concentrate on the more stringent cycle treatment of removing pyrogenic substances.

An appropriate sterilization or depyrogenation cycle must be developed before validation testing commences. Complete records and documentation of the cycle development must be referenced or included within the validation documentation. The cycle development must include:

16.5.9.1 Operating Parameters

All operating parameters must be defined during cycle development; including temperature settings, cycle time, penetration temperature profiles, and belt speed (for tunnel or flame sterilizers). The laboratory studies should imitate actual manufacturing conditions. Laboratory studies can identify the manufacturing operating parameters required to deliver an effective sterilization or depyrogenation cycle.

16.5.9.2 Component Mapping Studies

Before conducting the loaded-chamber heat penetration and distribution (load mapping) studies, component mapping should be conducted. This study determines the coolest point within a specific item. In subsequent loaded-chamber studies, penetration and distribution thermocouples should be positioned within the components at that location. For example, component mapping will determine if some areas of an item are heating at a slower rate than other areas. Mapping studies can be initially conducted in a laboratory scale oven and confirmed in the manufacturing equipment.

16.5.9.3 Load Heat Penetration and Distribution Temperature Mapping Studies

For cycle development purposes, the loads tested must be representative of standard items and quantities. Ideally, each size and type of material should be tested by heat penetration and heat distribution studies.

The component mapping results should designate the locations for the heat penetration thermocouples as part of the load mapping.

In determining the representative load for mapping, especially for loads that may include dissimilar items, the mass of the containers is an important factor to consider, rather than a bracketing approach of smallest and largest items to determine which items and associated load are the most difficult for heat to penetrate (i.e., because of dense mass or tight packing).

Detailed loading patterns should be developed; an exact detailed diagram of the thermocouple locations must accompany all temperature data. Photographs of the load patterns provide clarity in the final report. The diagram is necessary to identify where the hot and cold areas are within each specific load to be challenged for PQ studies. Hot areas in the load are more important for heat-labile items. Cold areas are important to monitor for sterility or depyrogenation assurance.

Load factors will be significant because air has poor conductive and convective properties. As a result, the hot and cold areas may vary for each type/arrangement of load. This is most likely to occur if each material heats at a different rate (because of size, mass, and packing configuration). In the batch oven, it is recommended that the penetration thermocouple locations be moved around after each run to obtain a broader view of heat penetration.

As in the empty-chamber testing, load mapping studies utilizing a partially or fully loaded chamber must include testing with thermocouples placed near the heat-controlling temperature sensors.

It is not uncommon for the presence of a load to improve the uniformity of heat distribution, or for the smaller load to be the "worst-case" load. The system should be tested with a "cold start" to reflect "worst-case" operating conditions and greater assurance of process effectiveness.

Penetration information is critical in a partially or fully loaded chamber because materials will heat at a rate different from the surrounding air. The rate of heat penetration will depend on the type of material in the load, mass, and how it is packed (loading configuration) and the temperature of the air supplied.

Heat penetration data is obtained by placing thermocouples inside the container, component, or item in such a way as to ensure contact with the surface (the thermocouple should read the surface temperature not the air temperature) at the locations previously determined.

It is important in the loaded chamber to document both ramp-up and cooldown rates of the air and product. The ramp-up time of the distribution thermocouples will describe the time required for the air to reach the temperature controller set point from ambient temperature. The ramp-up time of the penetration thermocouples will describe the time required for the load to reach the desired temperature. There is a heating lag as the components in the load reach the minimum required temperature after the air reaches that temperature (as measured by the distribution thermocouples). The heating lag is defined as the difference between the time required for the product to reach the minimum required temperature and the time required for the sterilizer air to reach the minimum required temperature. A heating lag will be magnified during the maximum load of product (i.e., maximum density or mass).

The total time the items are at or above the required temperature is documented in the loaded-chamber study, as well as the final F_H or F_D value.

The tunnel or flame sterilizer temperature data may have large variations between runs because the product is heated to high temperatures for a short period of time, as compared with batch ovens with long sterilization periods. As a consequence, establishing a correlation between different types of sterilizers/ovens/tunnels is difficult to achieve. In tunnels, the "worst-case" locations are generally the first and last rows of glass, and the middle of a dense pack of glass.

Loading and running thermocouples in a tunnel takes a great deal of time and coordination. The thermocouples are placed inside the glassware and spaced evenly across the width of the belt. A stainless steel bar may be used to hold the thermocouples in place. The thermocouples are fed through the tunnel as the belt is run at the maximum speed and lowest temperature, until they exit the tunnel. If portable temperature sensors are used, the canisters should not be placed adjacent to the glassware, as the canisters will act as a heat sink and the temperatures will appear lower than they actually are.

Care must be maintained to observe temperature ranges and fluctuations with awareness of any maximum component temperature restrictions. Distribution temperatures (empty sterilizer studies), penetration temperatures, come-up time, and F_H (or F_D) data can be evaluated for reproducibility between replicate runs using statistical methods. In the batch oven, it is often the load items closest to the bottom of the unit and nearest to the doors that are most likely to be the "cold spots".

Tunnel and flame sterilizers are particularly sensitive to changes in load configuration. Continuous runs (where bottles, vials, or ampoules are flush side to side) are generally the "worst-case" loads because of the rapid come-up time, short length of sterilization period, and variations in component packing and movement. The continuous sterilizer's "cold spots" tend to be found at the edges of the belt, and the first and last components to go through the unit. The presence of other components at the same temperature generally ensures the units in the middle of the load are heated more uniformly because they are better insulated from heat loss.

At the completion of the load mapping heat penetration and distribution studies, the monitoring and challenge locations within the components and within the loaded chamber have been identified for PQ studies. Cycle development studies have also established ramp-up time and temperatures, chamber distribution temperatures and cooldown rates and temperatures, cycle times, and belt speed (for tunnel or flame sterilizers), and F_H or F_D for each load configuration.

This load mapping cycle data can also be utilized for future requalification studies.

16.5.9.4 Microbial Lethality Requirements

In addition to the operating parameters specified in the cycle development study, the required F_H value and the bioburden/pyroburden levels of the components being treated must be determined. It is important to note that there is no accurate correlation between F_H value and successful Depyrogenation. Current best practice is to use F_D as an indicator to distinguish between sterilization and Depyrogenation. Any oversterilization concerns (for heat labile materials) should also be addressed.

If the cycle is required to render the container free of pyrogens as well as viable microbes, the cycle must demonstrate a 3-log cycle reduction of bacterial endotoxin (1/1000 of the original amount is inactivated). The pyrogen challenge can consist of inoculating an article with a minimum of 1×10^3 USP endotoxin units (EU) of bacterial endotoxin. A number of guidelines addressing cycle development and validation of sterilization cycles are referenced in guidance documents such as those from the Parenteral Drug Association (PDA) and PDA Technical Report No. 3 or United States Pharmacopeia (USP) chapters <1228> and <1229>.

Typically overkill cycles are utilized (based on biochallenges), and when utilized it is generally not necessary to evaluate the bioburden or pyroburden present on the incoming glass containers or equipment. The usual assumption considers that the relatively low numbers of organisms, or minute concentration of pyrogenic material present on the surface of glassware for parenteral use, are well within the challenge destruction of 10^6 spores of *Bacillus subtilis* or a 3-log reduction in endotoxin. The enumeration and identification of bioburden and pyroburden is considered of little value.

For sterilization processes only, the presence of high numbers of gram-negative organisms on a component before sterilization can raise concerns about the presence of endotoxin on the components prior to processing, because the overkill sterilization cycle will inactivate the vegetative cells and spores without reduction of the pyrogenic activity of the endotoxin. As glass containers are molded at high temperatures (1500°C) and shrink-wrapped by the supplier for shipping, they are unlikely to be contaminated with gram-negative organisms. Extensive studies performed on glass containers received in the shrink-wrap and before washing showed them to be free of endotoxin or have very low levels per container volume (<0.003 EU/mL). Washing of the articles will also decrease the endotoxin levels, if any are present [12].

The biological indicators of choice for validating and monitoring dry heat sterilization are commercial spore strips of *Bacillus atrophaeus* spores. When the strips are utilized, the manufacturer's D-value can be used. Spore strips can be purchased prepared with 10^4 to 10^9 spores. There are strips that can be utilized for temperatures between 250°C and 450°C. Any browning of the paper without crumbling does not interfere with usage or testing.

Alternately, a spore suspension can be inoculated on a container or piece of equipment to closely represent surface bio-challenge [13]. If the spores are dried on the surface of the article to be sterilized, the user must establish the D-value for each type of component to be tested. Heat-resistant organisms have D-values of only a few seconds at the temperatures typically used for endotoxin inactivation and for this reason, a spore challenge is not required in any process in which depyrogenation is demonstrated [14].

16.5.9.4.1 D- and Z-Values

The microorganism used as a biological indicator must have resistance characteristics (D- and Z-values) that are documented and appropriate for the sterilization cycle. The D-value is defined as the time required to reduce the microbial population by 90% (one logarithm). The relationship of lethality to temperature is expressed in the Z-value. Z-value studies will define the number of degrees that are required for a change in the D-value by a factor of 10. The microorganism used as a bioindicator must have resistant characteristics (D-value) that are documented and appropriate for the cycle [15]. The bioburden data and D- and Z-values are used to calculate the minimum F_H value required.

16.5.9.4.2 Sterilization—Bioburden Calculations F_H

The F_H value for *sterilization* is the integration of lethality at a reference temperature of 170°C. The F_D value for *depyrogenation* is the integration of lethality at a reference temperature of 250°C. A conservative approach to determining a minimum sterilization F_H would utilize the heat-resistant spores of *Bacillus atrophaeus* and assume a D_{170} value of 3 minutes (at a reference temperature of 170°C) and a Z-value of 20°C. The lethal rate determines the increment of lethal heat effect obtained over various temperatures using the Z-value (as compared with a reference temperature). The F_H value is derived by integration of the lethal rate with respect to time. The F_H value (equivalent time at the reference temperature) accumulates the total lethality.

When sterilization temperatures other than 170°C are used, the F_H value is reported as process equivalent time at the reference temperature of 170°C [16].

The equations used for equivalent time are as follows:

$$F_H = D_{170\,C}(\text{Log a} - \text{Log b})$$

where:

a = the bioburden per item
b = the probability of survival
D = the time at 170°C to reduce the population of most microorganisms in the product by 90%
F_H = the equivalent time in minutes at 170°C and a Z-value of 20°C

and

$$F_t^Z = \frac{F_{170}^Z}{L}$$

where:

F_t^Z = the equivalent time at temperature t delivered to a container for the purpose of sterilization with a specific Z-value
F_{170}^Z = the equivalent time at 170°C delivered to a container for the purpose of sterilization with a specific Z-value (when Z = 20°C, then $F_{170}^Z = F_H$)
L = lethal rate

$$L = \log^{-1}\frac{T_0 - T_b}{Z}$$

or

$$L = 10^{(T_0 - T_b)/Z}$$

where:

T_0 = temperature within the container or item
T_b = base temperature of 170°C

As a supplement to temperature data, spores of *Bacillus atrophaeus* are used to monitor the lethality of dry heat sterilization during the validation runs. Heat-labile products require strictly controlled sterilization cycles, because under-processing will result in a non-sterile material whereas over-processing may cause degradation of the material. Cycle development will determine the minimum amount of dry heat required to ensure that the probability of survival of the bioburden is less than 10^{-6}. The equivalent sterilization time and temperature can be described by the F value with a reference temperature of 170°C and assuming a Z-value of 20°C [17].

16.5.9.4.3 Depyrogenation—Pyroburden Calculations

Heat-stable materials, such as glassware and stainless steel equipment, can withstand temperatures well in excess of 250°C. Operating temperatures can be very high and the loading configuration may be less restrictive than that with heat-labile products.

The overkill method relieves the requirement for bioburden and bioburden resistance studies during cycle development and validation. Component preparation is still very important, because cleaning and handling procedures can serve to minimize the level of contamination of both viable and nonviable particulates, including endotoxins.

Calibrated *Escherichia coli* or naturally occurring endotoxin challenges are placed in the load. The coldest location in the loading pattern must be challenged. The endotoxin challenge should be based on the pyroburden of the components, taking into consideration the desired safety factor.

The presence of residual endotoxin can be detected by the Limulus Amebocyte Lysate (LAL) test.

The depyrogenation process is not as well understood as the sterilization processes, even the mechanism of endotoxin inactivation by dry heat is still being researched. The mathematical representation of thermal destruction of bacterial spores is well established for sterilization processes. Mathematical representations for LPS or NOE destruction are not completely established and continue to be explored. The initial studies published by Tsuji et al. [18] used a second-order model based on the sterilization process mathematical approach for spore inactivation to linearly describe the dry heat inactivation of LPS.

The studies demonstrated the inactivation curves for purified endotoxin could be made linear and that inactivation for a dry heat process can be predicted given the product heating curve. The isothermal kinetic changes were expressed by the following equation:

$$\log\frac{Y}{t}=\frac{1}{n}\log A+\left(1-\frac{1}{n}\right)\log Y$$

where:

Y = any parameter that changes with time t and temperature
A, n = constants at a specific temperature

The linear equation was represented as follows:

$$\text{Log } Y = A + B(10^{Cx})$$

where:

A, B, C = constants at a given temperature
x = minutes of heating time
Y = percentage of LPS remaining after heating and the same value as Y in the kinetic equation

Anderson and Kildsig [19] evaluated Tsuji's linear equation within a defined range of temperature values, because depyrogenation does not follow the semi-log standard time/temperature model developed for dry heat sterilization especially at lower temperatures. The model is a tool to determine the minimum inactivation temperature.

Akers et al. [20] continued use of the mathematical equation by developing F_D-value requirements for endotoxin inactivation.

Ludwig and Avis [21] have performed various studies to evaluate the minimum temperatures of depyrogenation, the mathematical application, and the type of LPS pyrochallenge. The inactivation of endotoxin and purified derivatives was viewed to be two linear biphasic slopes at temperatures of 225°C and 250°C and possibly a second-order inactivation that is temperature dependent. The calculated D-values by Ludwig and Avis at 225°C ranged from 0.15 to 1.20 minutes. The maximum calculated Z-value was 46.73°C [22]. The study explored the difficulty in using the standard F_D-value concept where an accurate Z-value needs to be derived from the D-value, and in calculating the D-value the inactivation rate must be first-order for the entire process [23]. Therefore, it is a requirement for demonstration of a 3-log reduction of endotoxin, which must be demonstrated to validate the dry heat cycle [24]. The inactivation rate of the endotoxin is dependent on the formulation, purification, and concentration of the challenge.

Nakata performed studies with 10,000 EU and determined D- and Z-values by linear regression analysis. Nakata views the inactivation curves as monophasic. He reported D-values of 43.8 and 1.7 minutes at 200°C and 250°C, respectively. The calculated Z-value was 30.9°C.

Nakata discusses the heat up primary phase inactivation as insignificant and the secondary phase inactivation at dwell temperatures as the stage in which almost all of the endotoxin inactivation occurs. The data demonstrated a 3-log reduction in 77 minutes at 200°C and an overkill cycle achieved in 30 minutes at 250°C [25].

Tsuji expressed the rate of endotoxin (LPS) destruction at 250°C using a Z-value of 46.4°C and a D_{250} value of 4.99 minutes [26].

The cycle should be designed utilizing a worst-case assumption, in which the required minimum time and temperature parameters are defined. The F_H can be calculated for a depyrogenation cycle using the general F_D-value equation with a reference temperature of 250°C.

16.5.9.4.4 *Examples of F_H and F_D Values*

Table 16.1 show various examples of F_H and F_D values for sterilization and depyrogenation studies; F_H (170°C) with a Z-value of 20°C, and F_D (250°C) with a Z-value of 46.4°C, at various temperatures (for 1 minute):

Temp. (°C)	$F_H{}^{20}{}_{170}$	$F_H{}^{46.4}{}_{170}$	$F_D{}^{46.4}{}_{250}$
170°C	1.0	1.0	0.02
210°C	100.0	7.3	0.14
250°C	10,000.0	53.0	1.0
270°C	100,000.0	142.0	2.7

Assuming a cycle of 250°C for 30 minutes, the minimum F_H values for the total cycle would be:

Temp. (°C)	$F_H{}^{20}{}_{170}$	$F_H{}^{46.4}{}_{170}$	$F_D{}^{46.4}{}_{250}$
250°C	300,000.0	1590.0	30.0

The F_D (250°C) that is determined can be used to calculate the amount of endotoxin that will be reduced. This can be calculated by integrating the heat penetration-lethality curves. The F_D for depyrogenation is used as a method to predict and quantify the endotoxin inactivation.

The endotoxin challenge will require inoculation of articles in the sterilizer load. It is much more difficult to verify the initial expected recovery of the endotoxin once placed directly on the component, because the endotoxin may tend to bind or adhere to the surface, decreasing initial recovery by 30%–80% before the component is depyrogenated by the dry heat. The presence of residual endotoxin is usually confirmed by the LAL test. The LAL test is very sensitive for reaction with endotoxin or LPS, which is a component of the gram-negative microorganism outer membrane.

16.5.9.4.5 *LAL Test*

The LAL test is an important monitoring procedure to test for the presence of endotoxin. The LAL test is based on the initiation by endotoxin of a blood-clotting cascade in the horseshoe crab. Clotting is measured and related to endotoxin concentration by one of three common in vitro methods: gel clot, turbidimetric, and chromogenic [27].

An acceptable endotoxin challenge should be based on the history of the pyroburden of the container and its contribution to the end product filled in the container on a per milliliter (mL) or milligram (mg) basis [28]. A meaningful challenge would evaluate the fill volume of the final container in view of the limit for the final product Bacterial Endotoxins Testing limit or sterile Water for Injection, USP, which has an endotoxin limit of 0.25 EU/mL or approximately 0.05 ng/mL. The threshold pyrogenic dose for man and rabbit is reported as 0.1 ng/mL (approximately 0.5 EU/mL) [29].

16.5.10 Performance Qualification

Upon completion of IQ and OQ efforts, CD (if applicable) and approval of the associated protocols, PQ testing may begin. The testing will be based on load mapping studies for heat penetration and distribution performed in the OQ and/or CD as a baseline to determine the rationale for PQ testing consisting of heat penetration and distribution studies containing Biological challenges.

16.5.10.1 PQ Heat Penetration and Distribution Studies

For validation purposes, the loads tested must be representative of standard items and quantities. Ideally, each size and type of material should be tested by penetration and associated distribution studies. For ovens, the time and temperature set points should be reduced. For tunnels, the temperature set point should be reduced and the belt speed increased if possible. To reduce the number of test combinations, a selection of representative items is made with consideration to size, number, mass, and geometry of the loading pattern. The representative loads could include the smallest and largest items as part of a bracket approach when items are similar in nature, e.g., glass vials.

In determining the representative load(s) for PQ, especially for loads that may include dissimilar items, the mass of the containers is an important factor to consider, rather than a bracketing approach of smallest and largest items to determine which items and associated load are the most difficult for heat to penetrate (i.e., because of dense mass or tight packing).

If CD studies were completed, the representative detailed load patterns for the PQ have been established.

Thermocouple placement should utilize the same procedures utilized in CD. CD parameters related to ramp-up and cooldown rates and times should be monitored in PQ. Particulate counts of the air must be checked in the equipment. The particulate counts are highest at the beginning of the cycle "ramp-up" and during cooldown. In some cases, higher counts are seen where particulate counters may be measuring the water vapor evaporating from the load articles if they are loaded wet. Generally particulate counts are lower when the system is in a steady state of operation (Hot or Cold).

The air particulate counts are to be within ISO 5/Grade A limits and the containers demonstrate results of 5 particles/mL ≥ 5 µm in size. Analysis of particulate matter within a container is performed by adding filtered particle-free water and shaking. The container contents are tested by electronic particle counter or microscopic analysis. These studies are typically performed independently of the thermal studies.

If the temperature profile is acceptable, three consecutive replicate runs in combination with bio-challenge/pyro-challenge studies are utilized to demonstrate loaded sterilizer and cycle reproducibility. The replicate runs must verify that the minimum required F_D value is being achieved within the coldest portion of the load.

16.5.10.2 Bio-Challenge/Pyro-Challenge Studies

The challenge should demonstrate the lethality delivered by the cycle with either microorganisms or endotoxin. The challenge can be accomplished using commercial strips or suspensions of *Bacillus atrophaeus* spores for sterilization or NOE for depyrogenation. The concentration of the challenge for overkill processes must demonstrate adequate sterility assurance.

For dry heat sterilization/depyrogenation to occur, there is little consensus on the minimum times and temperatures to achieve the desired result. A variety of time and temperature combinations can be utilized for dry heat sterilization/depyrogenation.

The bio-challenge will demonstrate the lethality delivered by challenging the cycle with either microorganisms or endotoxin. A suitable challenge must represent the pyroburden or bioburden for heat-labile materials or exceed it for overkill processes. The pyroburden and/or bioburden calculations were previously determined during CD (refer to Section 16.5.9). The concentration of the challenge for overkill processes must demonstrate adequate assurance in process efficacy.

Bio-challenge studies can be performed concurrently or separately from temperature penetration studies. If studies are performed concurrently, place the challenge items adjacent to items containing thermocouples. Studies can be performed by placing the bioindicators in the coldest areas (minimum F_H or F_D values) of each load. An alternative to adding the bio-challenge to each load involves determining the load with the absolute coldest area and minimum F_H or F_D value. This load is then considered to be the worst-case load. Successful bio-challenge of the worst-case load would eliminate the need to challenge all other previously tested loads (with higher F_H or F_D values).

The bio-challenge work is usually achieved by inoculating components with a known concentration of the challenge microorganism or endotoxin (i.e., *B. atrophaeus* suspension or *E. coli* endotoxin). In sterilization cycles, a challenge of 10^6 concentration of *B. atrophaeus* is common. In depyrogenation cycles, there appears to be no consensus on the challenge level used; concentrations must be recoverable and detectable to demonstrate a greater than 3 log reduction in endotoxin, typically 1×10^3 EU is used as a minimum.

The required number of challenged units should be predetermined during CD and cited in the validation protocol.

After the sterilization or depyrogenation cycle, the inoculated products are recovered along with unchallenged items (for negative controls) and tested for spore viability or endotoxin inactivation along with positive controls. If the challenge has spore survivors or residual endotoxin, the amount must be quantified and analyzed with respect to the achieved F_H or F_D value. The results of this study confirm that the sterilization or depyrogenation process is effective.

16.5.10.3 Qualified Operating Parameters

Final qualified critical and recipe parameters must be documented after successful runs in the PQ protocol, including temperature settings, cycle time, penetration temperature profiles, and belt speed (for tunnel or flame sterilizers).

It is also important to note that the parameters utilized under worst-case conditions as part of PQ runs (i.e., belt speed, temperatures in chamber or zones) may not be the final qualified operating parameters included as recipe parameters to be utilized by operations. These final qualified operating parameters must be included in the final SOPs that describe the use of the Dry Heat Sterilization equipment, and personnel must be trained on the SOP to make it effective.

16.5.11 Qualification Report

After each protocol (IQ, OQ, CD, and PQ) has been completed, the data must be analyzed to ascertain that all testing requirements have been achieved. The results of the tests, verifications, bio-challenge studies, and F-value computation must demonstrate the required degree of lethality (sterilization or depyrogenation) according to the protocol.

A sample outline for a Qualification Report is detailed in Appendix II.

16.6 LIFECYCLE SUPPORT SYSTEMS AFTER VALIDATION

Once the equipment has been validated for the sterilization or depyrogenation process, the unit must be monitored so that it remains in a state of control. This is achieved through various programs, including preventive maintenance, calibration, change control, training, and revalidation.

16.6.1 Preventive Maintenance

The preventive maintenance (PM) program provides maintenance tasks and frequencies by which the equipment is maintained. These PM tasks includes physical checking of the system, periodic inspection and replacement (as necessary) of door gaskets/seals, testing and/or changing of filters, testing of heater elements, calibration of controllers and recorders, and inspection and maintenance of fans, belts and motors.

The PM tasks and frequencies may be obtained from the oven/tunnel manufacturer and/or developed by the user based on the operating history of the unit. A proper PM program will help to prevent breakdowns during production and maximize the efficiency and operation of the equipment. The specific adjustments and maintenance that are made to the unit, scheduled or un-scheduled, must be recorded in the equipment logbook and PM program documentation (paper based or via computer maintenance management system).

An effective maintenance program ensures that the equipment will remain in the same operational state tested in the validation program. Routine maintenance, as documented in the preventive maintenance program, and like-for-like part changes can be performed without prior approval of the committee, if site quality procedures allow. These repairs would still be noted on a change control form to ensure that good equipment records are properly maintained.

16.6.2 Calibration

The calibration program provides calibration points and frequencies by which the critical instrumentation as part of the equipment is maintained in a current state of calibration. These instruments to be calibrated include belt speed, temperatures, and velocity.

The calibration points for each instrument and associated frequencies may be obtained from the manufacturer and/or developed by the user, based on the operating history of the unit. A proper calibration program will help to maximize the efficiency and operation of the equipment. The routine calibration of the equipment must be recorded in the equipment logbook and calibration program documentation (paper based or via computer maintenance management system).

An effective calibration program ensures that the equipment will remain in the same operational state tested in the validation program.

16.6.3 Change Control

Changes to the equipment that might compromise the validated state must be brought to the attention of the group or individual in charge of the change control program. A change control form may be completed by a trained person requesting the change, outlining the modification or repair required, the reason for the change, the potential impact to the validated state of the equipment, and the expected results. The request form should be reviewed by a committee consisting of delegated representatives from the Validation department, Quality Assurance, Engineering, and Manufacturing (or per site specific company procedures). The committee evaluates the modification to be made (or already completed if done on an emergency basis) and determines if it alters the validated status of the equipment.

If the change is deemed to cause an impact, representatives then agree to revalidation testing to return the equipment to approved use status. Any completed revalidation testing or other documentation as specified by the approved change control must be attached to the change control and accepted as part of a review process in order for the committee to then close out the change control.

16.6.4 Training

Depyrogenation processes rely heavily on scientific principles for the effective destruction or removal of endotoxins. Depyrogenation is an interdisciplinary activity in which the combined knowledge of a group of individuals is generally required for the establishment of a reliable process. Individuals responsible for the maintenance and operation of depyrogenation processes must be trained appropriately to ensure that they have an understanding of the process and equipment and can repeatably operate the unit or understand the requirements for the PM and calibration of the equipment. The operators are often the first to identify changes in process performance because of their intimate involvement with it. Effective training programs should emphasize depyrogenation principles, adherence to established processes and procedures, and the importance of documenting deviations from normal operations. Training programs often include training of personnel for specific SOPs developed for the operation and maintenance of the equipment, which includes documented evidence of training, and an evaluation of frequency for training when procedures and associated processes are modified via change control during the life cycle of the equipment.

16.6.5 Revalidation

Revalidation studies may be required after changes or repairs are made on the unit as determined by maintenance work orders or change control quality reviews, or at a predetermined periodic interval per company corporate or site-specific procedures. Revalidation usually does not include all the original validation studies but should include the worst-case load in terms of mass (load with the minimum F_H or F_D) for one or more container configurations that was previously validated.

16.6.6 Retirement

Retirement of Dry Heat Sterilization equipment must follow corporate or site-specific procedures. Retirement of equipment typically involves a documented retirement or decommissioning process that verifies that the equipment asset information is removed from life cycle and Quality Systems documentation.

This process ensures that PM and calibration records and associated frequencies are updated, applicable SOPs and associated training requirements are revised and/or retired, validation plans are updated, and site asset management records including financial tracking for assets are updated. In addition, final cleaning and/or decontamination is performed and documented as part of the retirement program prior to removing the equipment from the installed location or the site.

16.7 DOCUMENTATION OF QUALIFICATION ACTIVITIES

All qualification information should be easily identified and kept in a permanent controlled-access central file, where it can be readily retrieved.

A sample of Qualification documentation to be retained is depicted in Appendix III.

16.8 CONCLUSION

Sterilization and depyrogenation processes are an integral part of the requirement to provide sterile and safe products to patients. Dry heat is a commonly used method to sterilize and depyrogenate products, equipment, and/or components. This chapter provided an overview of life cycle practices to specify, procure, install, commission, qualify, operate, and maintain dry heat equipment. Additionally, the establishment and application of life cycle document specifications for equipment performance and operational critical process parameters to be confirmed in the qualification program, and routine monitoring of equipment performance, was presented.

An outline for a qualification program, including an efficient leveraged qualification approach, and maintenance and change control process was described. Following the outlined methodology will result in a compliant program documenting a reproducible dry heat process.

16.9 ACKNOWLEDGEMENT

We wish to acknowledge the authors of the previous edition of this chapter, Laurie A. Case and Gayle D. Heffernan, who provided the practical and scientific foundation related to applications and efficacy of dry heat processes. Their rationale is still relevant today.

REFERENCES

1. Validation of Dry Heat Processes Used for Sterilization and Depyrogenation, Parenteral Drug Association Technical Report No. 3, Revised 2013, p. 21.
2. Kirk-Othmer. Microwave technology. *Encyclopedia of Chemical Technology* 1981, 15, 494.
3. Lohmann S., Manique F. Microwave sterilization of vials. *Journal Parenteral Science and Technology* 1986, 40(1), 25–30.
4. Jeng DK, Kaczmarek KA, et al. Mechanism of microwave sterilization in the dry state. *Applied and Environmental Microbiology* 1987, 53(9), 2133–2137.
5. Perkins J. *Principles and Methods of Sterilization in Health Sciences*. Springfield, IL: Charles C. Thomas, 1973: 286.
6. Halliday D, Resnick R. *Physics Parts I and II*. New York, NY: John Wiley & Sons, 1967: 549.
7. Parker S. *McGraw Hill Concise Encyclopedia of Science and Technology*. 5th ed. McGraw-Hill Pub. Co., 1987: 455.
8. International Society of Pharmaceutical Engineers. *Sterile Product Manufacturing Facilities Baseline Guide*. Vol. 3. 2nd ed.: 61.
9. International Standard ISO 14644, Cleanrooms and Controlled Environments—Part 1: Classification of Air Cleanliness, 1999.
10. Green D. *Perry's Chemical Engineers' Handbook*. 6th ed. Vol. 22. McGraw-Hill Co., 1984: 32–33.
11. Wood RT. Parenteral Drug Association Short Course on Dry Heat Sterilization Validation and Monitoring on 17 June 1982, pp. 12–19.
12. Feldsine PT, Ferry EW, Gauthier RJ, Pisik JJ. A new concept in glassware depyrogenation process validation. I: System characteristics. *Journal of Parenteral Drug Association* 1979, 33, 125–131.
13. Guidance for Industry Sterile Drug Products Produced by Asptic Processing, U.S. Department of Health and Human Services Food and Drug Administration, 2004.

14. Validation of Dry Heat Processes Used for Sterilization and Depyrogenation, Parenteral Drug Association Technical Report No. 3, 1981, p. 43.
15. Validation of Dry Heat Processes Used for Sterilization and Depyrogenation, Parenteral Drug Association Technical Report No. 3, 1981, p. 28.
16. Validation of Dry Heat Processes Used for Sterilization and Depyrogenation, Parenteral Drug Association Technical Report No. 3, 1981, p. 24.
17. Validation of Dry Heat Processes Used for Sterilization and Depyrogenation, Parenteral Drug Association Technical Report No. 3, 1981, pp. 34–39.
18. Tsuji K, Lewis A. Dry heat destruction of lipopolysaccharide: Mathematical approach to process evaluation. *Applied and Environmental Microbiology* 1978, 36(5), 715–719.
19. Anderson NR, Kildsig DO. Alternate analysis of depyrogenation processes. *Journal of Parenteral Science and Technology* 1983, 37(3), 75–78.
20. Akers MJ, Ketron KM, Thompson BR. F value requirements for the destruction of endotoxin in the validation of dry-heat sterilization/depyrogenation cycles. *Journal of the Parenteral Science and Technology* 1982, 36, 23.
21. Ludwig JD, Avis KE. Dry heat inactivation of rough strain lipopolysaccharide and diphosphoryl lipid a on the surface of glass. *Journal of Parenteral Science and Technology* 1991, 45(1), 35–39.
22. Ludwig JD, Avis KE. Dry heat inactivation of endotoxin on the surface of glass. *Journal of Parenteral Science and Technology* 1990, 44(1), 4–11.
23. Akers MJ, et al. Parenteral fundamentals: Dynamics of microbial growth and death in parenteral products. *Journal of the Parenteral Drug Association* 1979, 33, 372.
24. Validation of Dry Heat Processes Used for Sterilization and Depyrogenation, Parenteral Drug Association Technical Report No. 3, Revised 2013, p. 27.
25. Nakata T. Destruction of challenged endotoxin in a dry heat oven. *Journal of Pharmaceutical Science and Technology* 1994, 48(2), 59–62.
26. Gould M. Microorganisms: Evaluation of microbial/endotoxin contamination using the LAL test. *Journal of Ultrapure Water* 1993 September, 44–46.
27. McCullough KZ. Process control: In-process and raw material testing using LAL. *Pharmaceutical Technology Journal* 1988 May, 46.
28. Groves F, Groves M. *Encyclopedia of Pharmaceutical Technology, Chapter on Dry Heat Sterilization and Depyrogenation*. New York, NY: Marcel Dekker, 1991: 468.
29. Akers MJ, Avis KE, Thompson B. Validation studies of the fostoria infrared tunnel sterilizer. *Journal of the Parenteral Drug Association* 1980, 34(5), 331–336.

BIBLIOGRAPHY

Akers MJ, Avis KE, Thompson B. Validation studies of the fostoria infrared tunnel sterilization. *Journal of the Parenteral Drug Association* 1980, 34(5), 330–347.

Akers, M. J., Ketron, K. M. and Thompson, B. R. F-value requirements for destruction of endotoxin in the validation of dry heat sterilization. *Journal of Parenteral Science and Technology* 1982, 36(1), 23–27.

Avis KE, Jewell R, Ludwig J. Studies on thermal destruction of E. Coli Endotoxin. *Journal of Parenteral Science and Technology* 1987, 41(2), 49–55.

Baird R. Validation of dry heat tunnels and ovens. *Pharmaceutical Engineering* 1988 March/April, 8(2), 31–33.

Ernst RR. *Sterilization by Heat in Disinfection, Sterilization, and Preservation*. Philadelphia: Lea & Febiger, 1977: 481–521.

Gould MJ. Microorganisms. *Journal of Ultrapure Water* 1993 September, 43–47.

Halliday D, Resnick R. *Physics*. New York: John Wiley & Sons, 1967.

Holman JP. *Thermodynamics*. New York: McGraw-Hill Book Company, 1969.

Kirk O. *Encyclopedia of Chemical Technology: Dry Heat Sterilization and Depyrogenation*. Vol. 4. New York, NY: Marcel Dekker Inc., 1991: 447–484.

Kirk O. *Encyclopedia of Chemical Technology: Microwave Technology*. Vol. 15. New York, NY: John Wiley and Sons, 1981: 494–517.

Kirk O. *Encyclopedia of Chemical Technology: Sterilization Techniques*. Vol. 21. New York, NY: John Wiley and Sons, 1983: 626–643.

LAL Users Group. Preparation and use of endotoxin indicators for depyrogenation studies. *Journal of Parenteral Science and Technology* 1989 May, 43(3), 109–112.

Ludwig J, Avis KE. Dry heat inactivation of endotoxin on the surface of glass. *Journal of Parenteral Science and Technology* 1990, 44(1), 4–12.

McBride T. Hot air sterilizer GMP compliance. *Pharmaceutical Engineering* 1984 January, 41–45.

Parenteral Drug Association. Depyrogenation, Technical Report No. 7, 1985, pp. 1–13, 101–107.

Parenteral Drug Association. Validation of Dry Heat Processes Used for Sterilization and Depyrogenation, Technical Report No. 3, 1981, pp. 5–26.

Perkins J. *Principles and Methods of Sterilization in Health Sciences*. Springfield IL: Charles C. Thomas, 1973: 286–310.

Pflug IJ. *Sterilization of Space Hardware, Environmental Biology and Medicine*. Vol. 1. Minneapolis: Gordon & Breach, 1971: 63–81.

Remington's Pharmaceutical Sciences. 15th Edition. Edited under the direction of Arthur Osol and John E. Hoover. Mack Publishing Co., Easton, PA 18042. First published June, 1976.

Reynolds WC, Perkins H. *Engineering Thermodynamics*. New York: McGraw-Hill Book Company, 1977: 536–581.

Simmons PL. The secret of successful sterilizer validation. *Pharmaceutical Engineering* 1981 May/July, 38–46.

Swarbrick J, Boylan JC. *Encyclopedia of Pharmaceutical Technology, Chapter on Dry Heat Sterilization and Depyrogenation*. Groves J, ed. New York, NY: Marcel Dekker, 1991: 447–483.

Tsuji K, Harrison S. Dry heat destruction of lipopolysaccharide: Dry heat destruction kinetics. *Appl. Environ. Microbiol.* 1978, 36, 710–714.

Tsuji K, Lewis A. Dry heat destruction of lipopolysaccharide: Mathematical approach to process evaluation. *Appl. Environ. Microbiol.* 1978, 36(5), 715–719.

United States Pharmacopeia, USP28-NF23 Mack Publishing Co., Easton, PA.

Weary M, Pearson F. A manufacturer's guide to depyrogenation. *Journal of Biopharmaceuticals* 1988 April, 22–29.

Whitiker G. Sterilization, validation, theory and principles, in Parenteral Manufacturers Association. Proceedings, Validation of Sterile Manufacturing Processes, Reston, VA, March 15, 1978.

Wood RT. *Parenteral Drug Association Short Course on Dry Heat Sterilization Validation and Monitoring*, 1982: 12–19.

Appendix I

Sample Qualification Protocol Outline

I1 PROTOCOL OUTLINE

The following format may be utilized in a qualification protocol:

I1.1 Objective Statement

A concise statement that defines the objective of the validation protocol.

I1.2 Scope Statement

A concise statement that defines the scope of the validation protocol.

I1.3 Responsibility

Identification of specific departments and their responsibilities in the validation process. This will ensure that each group understands the specific information or materials required to support the validation process.

I1.4 Test Program and Equipment

The test program should include a description of the tests that will be performed during the IQ, OQ, CD, and/or PQ protocols. The calibrated test equipment and software used to perform the studies must be described and any type and form of biological or endotoxin challenges to be used must be stated. All SOPs for each piece of equipment or testing process must be referenced.

I1.5 Rationale

The rationale of the tests that will be performed during the IQ, OQ, CD, and/or PQ protocols is defined.

The rationale on what will be verified in empty chamber, vial, or container configuration runs, and the location of temperature measuring devices and/or locations of biological or endotoxin should be described. In addition, biological or endotoxin challenge runs and loaded chamber runs should be described, especially if a bracketing or reduced run time is proposed in the protocol. Any guidance or corporate or site-specific documents that support the rationale being proposed should be referenced.

An important documentation requirement is to list the critical parameters of the oven/tunnel and commodities being sterilized or depyrogenated. It is recomended that a cycle run sheet is used to record this information. This form is filled out with the appropriate information at the time of the run. A sample of a run sheet is provided.

I1.6 Acceptance Criteria

Acceptance criteria must be listed for each test, with limits or ranges specifically identified. The limits or ranges chosen should be those commonly used by the firm, determined during commissioning, SAT, OQ, and/or URS and design documents.

I2 CHANGES IN SCOPE OF WORK

Changes in the scope of the work after any protocol has been approved for execution may be addressed in Protocol Deviations, Supplements, or Addenda according to company procedures. These supplements must be approved by all parties as designated by the validation master plan and corporate or site-specific procedures. The validation final report must refer to the issued protocol and all supplements.

Qualification Run Documentation Example

Note: Copy page as needed
Page:___ of ___

Equipment Description: List Equipment Name/ ID Number:	Protocol Number:
	Run Number:
Test/Run Date:	Recipe ID: ___________________________
Vial Size: _____________________	Vial Item/Part #: _______________

Parameter	Setting	Unit	Entered By/Date	Verified By/Date
Required Temperature of Sterilizing / Depyrogenation Zone		°C		
Min. Temperature of Sterilizing / Depyrogenation Zone		°C		
Max. Temperature of Sterilizing / Depyrogenation Zone		°C		
Min. Temperature Below Sterilizing / Depyrogenation Zone		°C		
Max. Temperature Below Sterilizing / Depyrogenation Zone		°C		
Cool Zone Temperature		°C		
Conveyor Belt Speed				
Entrance Baffle Plate Height				
Heating Zone(s) Baffle Plate Height				
Cooling Zone(s)/Exit Baffle Plate Height				

Comments:

Appendix II
Sample Qualification Report Outline

II.1. QUALIFICATION REPORT

The following information should be provided in each qualification summary report (IQ, OQ, CD, and PQ):

II.1.1 Scope

A statement reflecting the scope that the final report is intended to detail.

II.1.2 Summary of Data

A summary of the results of verifications and/or data collected in the validation protocol. Typically, each test or verification section included in the protocol is summarized in this section of the final report and a disposition on whether that section met specified acceptance criteria is included. Any discrepancies or deviations encountered during the execution of each section should be identified and resolved. Specific run data for mapping, bio or endotoxin challenge studies performed during the validation runs, including summary of thermocouple/data logger data with temperature results by location, minimum and maximum F values, and disposition of acceptance criteria for this data for each run should be included in the final report for each applicable protocol where this testing was performed. Raw data is generally not circulated within the report but should remain in a secured archive linked to the specific protocol.

II.1.3. Diagrams

Diagrams depicting the load and the placement of thermocouples, bioindicators, and/or inoculated parts should be included. Detailed diagrams of unusual items should be shown. Photographs can be extremely valuable in documenting studies as well. Other data may be included in the report as desired because of differences in protocol and equipment specifics.

II.1.4 Deviations

Results that did not meet acceptance criteria in any test section or tests that were not or could not be executed within a protocol should be documented in discrepancy or deviation form, according to protocol instructions and/or site-specific procedures. Each discrepancy or deviation should be included in the final report for each protocol, with either brief or detailed summary of the description of each deviation, investigation, impact to the protocol/study, and final disposition of the deviation. Exceptions to the acceptance criteria should be explained, including justifications if certain tests were not performed or are to be performed in the future under an addendum.

Appendix III
Sample Qualification Documentation Retention

The qualification documents to be retained, compliant with company retention policy, should include the following information:

III.1 QUALIFICATION PROTOCOLS

All executed IQ and OQ protocols, Cycle development protocols and/or PQ protocols including associated final reports for the equipment and/or process. All original data, results, and conclusions must be contained in this controlled document file and location. All reports should be dated, signed, and approved by the responsible individual(s).

III.2 CHAMBER STUDIES

All original data, results, calculations, and conclusions must be retained for empty and loaded chamber and bio-challenge studies performed in OQ, CD, and/or PQ protocols, as well as Revalidation protocols. One of the most important records from these studies is the run sheet, a form that is filled out with the appropriate information at the time of the run.

Load diagrams (depicting the actual placement of components, thermocouples, and bioindicators) are kept with the run sheets in the executed protocols to which the diagrams refer and may be included in the associated protocol final reports. The diagrams should include the empty chamber load, and the different loads used in the loaded chamber and bio-challenge studies. Other data to be identified with chamber studies would include calibrations, original temperature printouts, equipment temperature charts, bioindicator details and calibrations and test results, and calculation sheets (such as F_H or F_D values and temperature ranges).

III.3 ROUTINE MONITORING

All change control information and post-validation changes are recorded along with any revalidation work associated with the specific Dry Heat equipment asset. This will prevent the voiding of previous validation studies and provides a history of the validation of the equipment for traceability and future assessments. It is important to consider the validation effort as being protected by proper documentation and permanent files. The initial validation data is necessary for comparison with subsequent validations, and the overall validation program is only as reliable as the traceability of its documentation.

17 Depyrogenation by Inactivation and Removal

Karen Zink McCullough and Allen Burgenson

CONTENTS

17.1 INTRODUCTION

A pyrogen, from the Greek, "pyros" meaning fire, is anything that can elicit a fever in a mammal. Pyrogens can originate from microbes (bacteria, fungi, viruses), or they might be nonmicrobial in origin, such as some antibiotic and chemotherapeutic agents (Pearson, 1985). In some cases, pyrogenic materials are purposely used as adjuvants for vaccines (Chilton et al., 2013), but for the vast majority of parenteral pharmaceutical products, pyrogens are contaminants that may pose safety risks for patients. *Bacterial endotoxin*, a structural component of the outer cell leaflet of the outer cell membrane of most gram-negative bacteria, is the most prevalent and potent pyrogen commonly found in pharmaceutical manufacturing environments. Therefore, this chapter will discuss depyrogenation in terms of removal or reduction of bacterial endotoxins in parenteral drugs and manufacturing materials.

Because current analytics can't measure "0" pyrogens or "0" pyrogenic activity, the term, "depyrogenation" encompasses a variety of different processes designed to eliminate or reduce pyrogens in parenteral products and primary packaging components to clinically safe levels. There are two basic approaches to depyrogenation (Weary and Pearson, 1988). The first is depyrogenation by inactivation. Inactivation can be accomplished by incineration using dry heat (USP, 2020b) or by chemical alteration of the lipopolysaccharide (LPS) molecule such as exposure to acid and base (Niwa et al., 1969; McCullough and Novitsky, 1985). Please note that depyrogenation by dry heat is discussed at length in a separate chapter in this book.

DOI: 10.1201/9781003163138-17

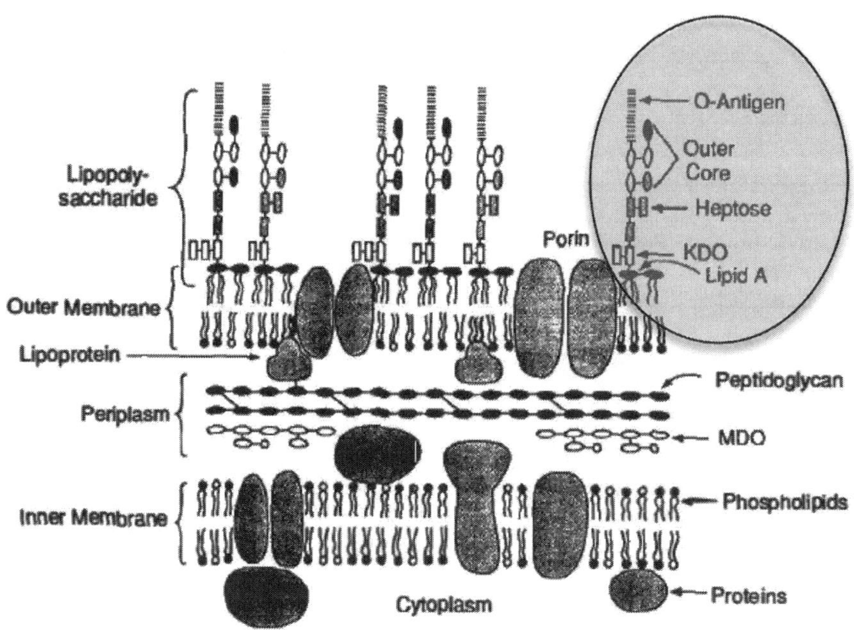

FIGURE 17.1 Structure of the gram-negative cell envelope.

Source: Raetz, 1990.

However, many materials, including biological product process streams and proteins, cannot withstand the high heat or harsh chemical conditions needed to inactivate endotoxins. For heat-labile materials, depyrogenation may be accomplished by removal using processes such as rinsing, filtration, or chromatography (Magalhaes et al., 2007; USP, 2020c, 2020d). These methods rely heavily on the chemistry of the endotoxin complex including, but not limited to, size, charge, and solubility. In this chapter, we explore various methods utilizing two different approaches to depyrogenation of bacterial endotoxins: depyrogenation by inactivation and depyrogenation by removal (USP, 2020a).

17.1.1 Endotoxins Chemistry

Endotoxins are found in the outer leaflet of the outer cell membrane of most gram-negative bacteria (Figure 17.1). Lipopolysaccharides (LPS) are the structural components that make up the majority of the outer leaflet of the outer cell membrane of gram-negative bacteria and are designated in Figure 17.1 by the shaded oval (Raetz, 1990; Beveridge, 1999; Silhavy et al., 2010).

LPS is an amphipathic molecule, meaning that it has a hydrophobic portion, Lipid A, that is embedded in the cell membrane, and a hydrophilic portion, called the oligosaccharide or O-antigen, that is exposed to the organism's external environment (Silhavy et al., 2010). The molecule is comprised of three distinct regions: The O antigen that confers immunological specificity on the organism, the inner/outer core oligosaccharides, and the Lipid-A moiety, which is the biologically active portion of the molecule. Lipid A is a di-phosphorylated di-glucosamine that may contain from four to seven acylations of varying length depending on the genus, species, and growth conditions of the microorganism (Ribi et al., 1962; Raetz, 1990; Trent et al., 2006). For example, the Lipid A from the gram-negative bacterium *Escherichia coli* has six acyl chains, each 12–14 carbons long, and is a powerful pyrogen (Figure 17.2). However, *E. coli*, the organism

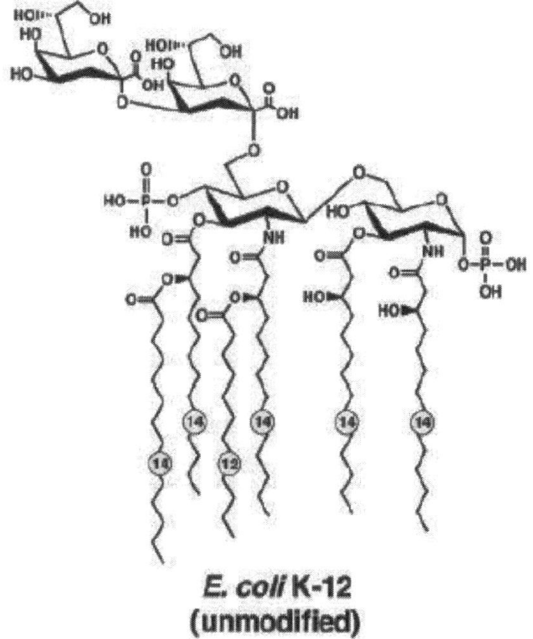

FIGURE 17.2 Chemical composition of *Escherichia coli* lipopolysaccharide (LPS).

Source: Trent et al., 2006.

from which the calibration standard is derived, is an enteric microorganism that is prevalent in livestock pens but not in pharmaceutical manufacturing facilities.

17.1.2 Endotoxins from Bacteria Autochthonous to Parenteral Manufacturing

The term, "autochthonous" is used in microbiology to describe microorganisms that are indigenous to a unique environment, in the case of parenteral processing, the manufacturing environment. Most of these autochthonous microorganisms are identified as "Non-Fermenting Gram-Negative Bacilli" (NFGNB) and as many as fifteen diverse "families" of NFGNB such as *Pseudomonas*, *Acinetobacter*, and *Flavobacterium* are commonly found in pharmaceutical waters (Deshnukh et al., 2013; Sandle, 2015; Reid, 2019). Unlike the *E. coli* Lipid A structure in Figure 17.1, these non-fermenters most often have seven acyl chains and are less potent (biologically active) than endotoxins from organisms whose Lipid A contains six chains (Figure 17.3).

In the parenteral industry, water systems, particularly pretreatment and deionized water upstream of stills, or inadvertent standing water can provide oligotrophic (low nutrient) conditions to which gram-negative bacteria can adapt and proliferate (Akers et al., 2020; Sandle, 2015). LPS from microorganisms autochthonous to pharmaceutical or medical device manufacturing facilities will exhibit materially and analytically distinctive LPS fingerprints because of adaptations to their unique environments to ensure survival (Morita, 1997). This adaptation often involves the modification of the organism's endotoxin structures through the bacterium's Two Component Signal Transduction System that allows for the "remodeling" of the endotoxins to stabilize the cell envelope and ensure the organism's survival (Nikaido, 2003; Raetz et al., 2007; Capra and Laub, 2012; Li et al., 2012; Bonnington and Kuehn, 2016; Sohlenkamp and Geiger, 2016; Bertani and Ruiz, 2018; Norris et al., 2018; Akers et al, 2020). These remodeled endotoxins are structurally very different from the standard calibration endotoxins that are extracted and purified from *Escherichia coli*. It is therefore important to understand that purified LPS from *E. coli*, which is the current analytical standard, is <u>not</u> representative of the adapted LPS generated by autochthonous NFGNB that are found in pharmaceutical manufacturing environments.

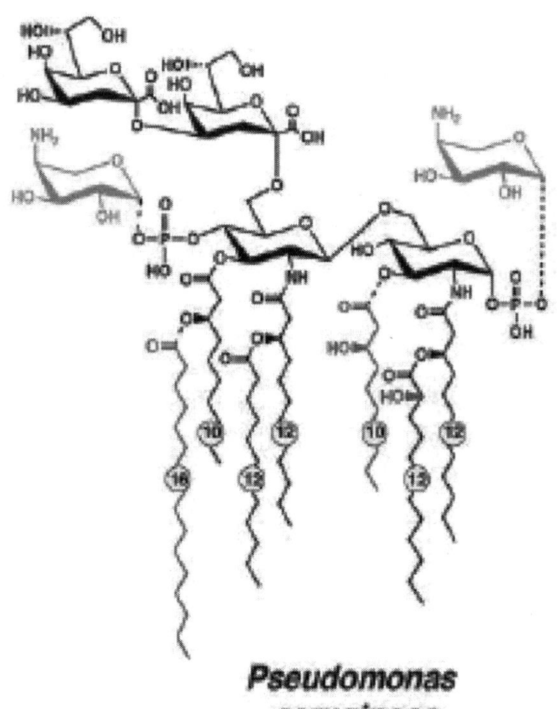

FIGURE 17.3 Chemical composition of *Pseudomonas aeruginosa* lipopolysaccharide (LPS).

Source: Trent et al., 2006.

As part of their life cycle, and especially in response to the harsh environments often found in pharmaceutical manufacturing facilities, gram-negative bacteria produce "buds" or microspheres of outer membrane constituents called Outer Membrane Vesicles, or OMV (Brogden and Phillips, 1988; Ellis et al., 2010; Bonnington and Kuehn, 2016.) These OMVs cannot replicate and therefore are not viable, but they can contain the same LPS constituents that were part of the cell membranes from their "parent" organisms or LPS that is displaced as the result of adaptation. The size of OMVs range from 20–250 nm, so they will easily pass through a 0.22 μm sterilizing filter (Schwechheimer and Kuehn, 2015; Figures 17.4A and 17.4B). The methods used to destroy or eliminate live microbes such as autoclaving or sterile filtration may kill or remove whole bacterial cells and autoclaving may destroy OMVs, but any cell wall fragments or OMVs that remain after these processes may still contain active endotoxins. Note, as well, that current research suggests that there is no evidence that LPS monomers from the outer leaflet exist in nature (Ellis et al., 2010).

Although water is the most ubiquitous source of endotoxins in parenteral manufacturing, other product and formulation-specific sources of endotoxins include manufacturing materials, especially those derived from natural plant or animal sources. Starting cells for cell and gene manufacturing, particularly manufacturing of autologous products where the patient is both the donor and the recipient of the therapy, may be contaminated with endotoxins present in their own blood, particularly if the patient carries a gram-negative infection. Vectors and plasmids used in cell and gene therapy manufacturing processes are often procured from university or research laboratories where they may not be prepared under standard Good Manufacturing Practice (GMP) conditions, meaning that they may contain endotoxins, possibly from the water used in their manufacture.

17.1.3 Detecting Endotoxins Activity

Lipid A is not only the portion of the LPS molecule that is toxic to humans; it is also the moiety that reacts with Factor C of the *Limulus* Amebocyte Lysate (LAL) reagent used to measure

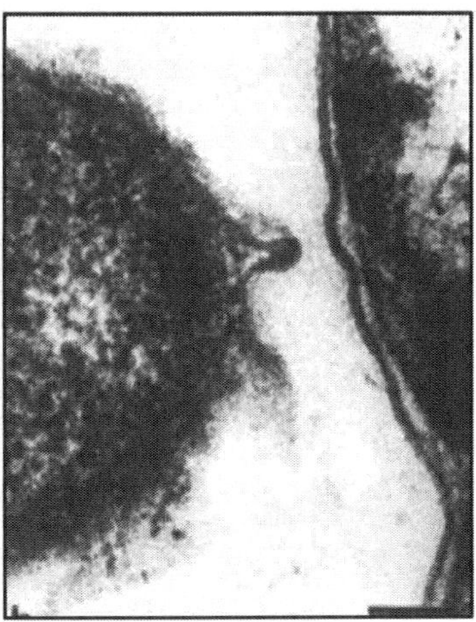

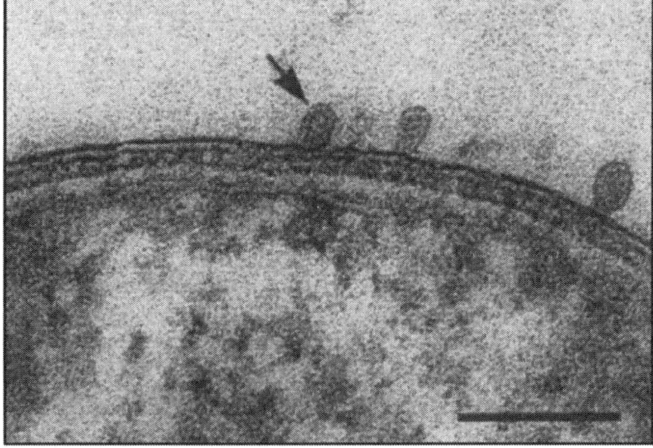

FIGURE 17.4A, B Electron micrograph: Formation of outer membrane vesicles.

Source: A) Brogden and Phillips (1988); B) Beveridge (1999).

endotoxins activity according to United States Pharmacopeia (USP) <85> "Bacterial Endotoxins Test" (BET). The BET is a compendial chapter that describes a number of assay methodologies for the detection and quantitation of bacterial endotoxins activity in pharmaceutical products (USP, 2020b). The content in USP <85> is harmonized with the European Pharmacopoeia and the Japan Pharmacopoeia and is the most commonly used method for the determination of the efficiency of depyrogenation studies that utilize bacterial endotoxins as the target analyte. Because of the variability in the fine chemistry of LPS across the spectrum of gram-negative bacteria, endotoxins are not measured by weight or mass but rather by their biological activity, called Endotoxin Units (EU).

17.1.4 Pyrogenicity and Sterility

Endotoxins are remarkably heat stable (Tsuji and Harrison, 1978; Tsuji and Lewis, 1978; Avis et al., 1987; Ludwig and Avis, 1990). Although depyrogenation by exposure to high heat for long periods of time will both kill the organisms and inactivate any remaining endotoxins by incineration, not all methods of depyrogenation are sterilizing and not all methods of sterilization are depyrogenating. For example, methods used to destroy or eliminate live cells such as autoclaving or passage through a 0.22 μm sterilizing filter may kill or remove whole bacterial cells and may incrementally reduce endotoxins activity in the solutions being sterilized, but any remaining cell wall fragments or OMVs may still contain active endotoxins. Therefore, sterility and "non pyrogenic" are not synonymous terms, as a sterile product may still contain assayable levels of endotoxins activity that can accumulate to clinically significant pyrogenic levels. Likewise, a product that is non-sterile because it is contaminated with a microorganism other than an LPS-containing gram-negative organism may be non-pyrogenic.

17.2 DEPYROGENATION BY INACTIVATION

Inactivation involves the chemical or physical treatment of endotoxins in a manner such that the process effectively reduces or eliminates the biological activity of Lipid A. The most prevalent methods of inactivation are detoxification by the use of acid or base, but other methods have been studied, validated, and used as appropriate. It must be noted that any depyrogenation procedure that utilizes chemicals must have well-defined, validated, and well-controlled operating parameters, including normality, exposure time, heat, and the neutralization or elimination of residual chemicals that may adversely affect the product.

17.2.1 Depyrogenation by Base Hydrolysis

Base hydrolysis, or saponification, is one of the most widely used techniques for depyrogenating process equipment that can withstand high pH. The process of treating an article with high levels of alkali for long periods of time cleaves the acyl groups from the Lipid A portion of the LPS molecule (Niwa et al., 1969) and in the process reduces or eliminates the toxicity of Lipid A as measured by the BET. The efficiency of the saponification process is dependent on both the time of exposure and the normality of the base used. Figure 17.5 is a graphic representation of the reduction of high levels of endotoxins activity when exposed to various normalities of Sodium Hydroxide.

17.2.1.1 Mechanism of Action

Figure 17.6A–C show the chemistry of the saponification reaction.

A typical application is the use of 0.1 N sodium hydroxide (NaOH) in the depyrogenation of non-autoclavable equipment such as plastics. NaOH is also used in cleaning of affinity chromatography columns for the purification of proteins

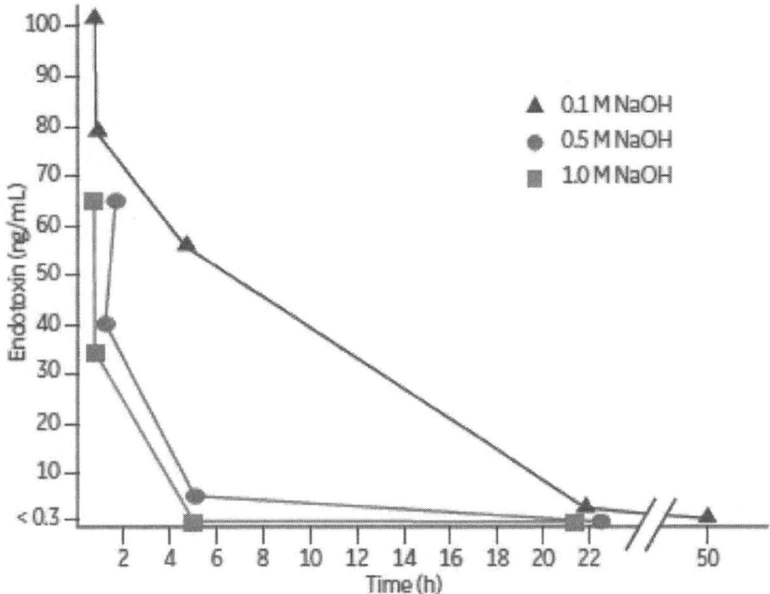

FIGURE 17.5 Detoxification of endotoxin by NaOH.

Source: Cytiva, 2020.

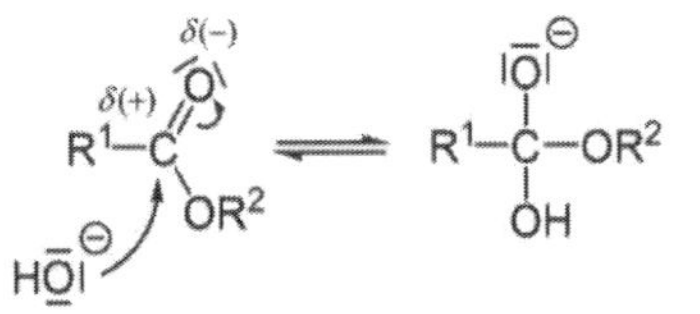

FIGURE 17.6A The hydroxide anion of the salt reacts with the carbonyl group of the ester binding the fatty acid to the diglucosamine. The immediate product is called an orthoester.

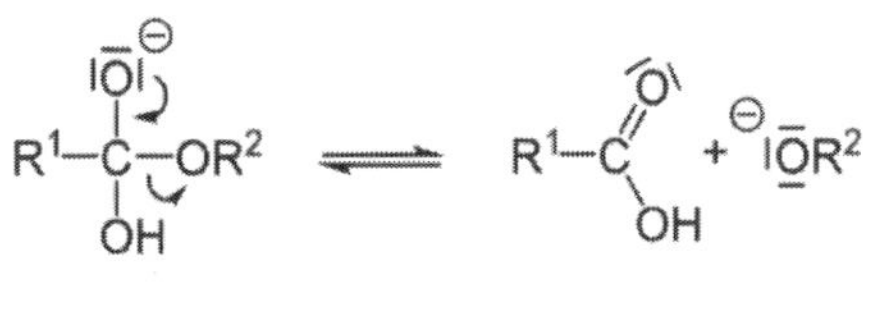

FIGURE 17.6B Removal of the alkoxide generates a carboxylic acid.

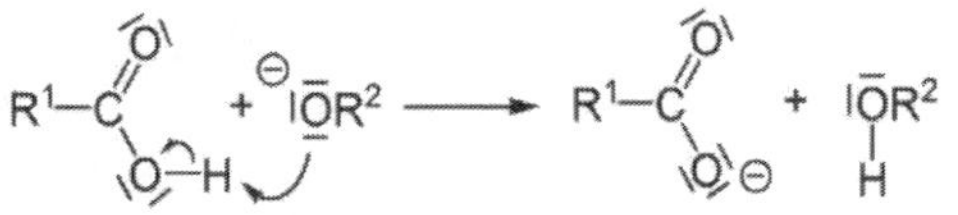

FIGURE 17.6C The alkoxide ion is a strong base so that the proton is transferred from the carboxylic acid to the alkoxide ion creating an alcohol.

and anion exchange columns for the removal of endotoxins (Hale et al., 1994; Cytiva, 2020). For example, Protein A is the preferred affinity column for monoclonal antibody production and may cost over $1 million per 1 m packed column. Therefore, to protect both the product and the company's investment in technology, it is essential that such columns be adequately cleaned, sanitized, depyrogenated, and neutralized between uses. Of course, rinsing of these materials with copious amounts of sterile Water for Injection (WFI) is needed to remove any residual alkali that might impact on the product.

17.2.2 Depyrogenation by Acid Hydrolysis

Treatment of endotoxins with acid cleaves the ketosidic linkage between the KDO and glucosamine residues of the Lipid A and results in two chemically distinct fractions: soluble polysaccharide and insoluble free Lipid A, which is not biologically active. But when hydrolysis is incomplete, there can be variability in the endotoxicity of the Lipid A fraction as reported by Haskins and co-workers (1961). As well, when the free Lipid A is complexed to a protein making the Lipid A soluble, it regains its activity (Haskins et al., 1961; Galanos et al., 1972). Acid hydrolysis may also act on the Lipid A itself by cleaving off fatty acid molecules, thereby changing the conformation of the molecule and simultaneously changing its solubility. Data collected to date suggest that endotoxicity is therefore related to solubility (McCullough and Novitsky, 1985; Weary and Pearson, 1988; Raetz, 1990).

Reports in the literature suggest that relatively low normalities of acid combined with high heat (100°C) for relatively long periods of time can depyrogenate (Ribi et al., 1961). Weary and Pearson (1988) report that boiling in 0.05N HCl for thirty (30) minutes or boiling in 1% glacial acetic acid for 2–3 hours can depyrogenate. Raetz (1990) adds that boiling in 0.1M HCl for fifteen (15) minutes can break the ketosidic linkage as well.

Acid depyrogenation is often used on surfaces that are not affected by low pH or heat. An example of such a process is glassware, where the effects of repeated base hydrolysis or heat depyrogenation may cause the glass to gradually etch. Glass is not subject to the effects of most acids and may respond better to depyrogenation by acid hydrolysis than to base hydrolysis. Phosphoric or nitric acid is often used for in-place passivation of stainless-steel vessels in the manufacturing area to remove surface rouging because of the presence of free iron ions on the surface of the vessel. Passivation also serves to reduce the endotoxin levels remaining after a process and cleaning. As with base hydrolysis, care must be taken to rinse surfaces with WFI in order to remove traces of acid that may impact the product.

17.2.3 Depyrogenation by Oxidation

Perhaps the earliest report of depyrogenation by oxidation was presented by Hort and Penfield (1912) who reported that a *Bacillus typhosus* (now known as *Salmonella typhi*) cell pellet treated with water yielded a supernatant with significant pyrogenic properties, but a pellet from the same organism treated with hydrogen peroxide did not. Oxidation works by peroxidation of the fatty acid in the Lipid A region of the LPS molecule. A number of authors have reported that combining the oxidative capabilities of hydrogen peroxide with heat will increase its depyrogenation proficiency (Campbell and Cherkin, 1945; Cherkin, 1974; Gould and Novitsky, 1985).

Oxidation by hydrogen peroxide is capable of depyrogenating surfaces where product contact may occur or for the depyrogenation of the surfaces of implantable medical devices. The chief disadvantage of using oxidation is that with the concentrations and heat needed to successfully depyrogenate, the treatment may adulterate the surface or solution that it is intended to depyrogenate, so careful attention must be paid to the product or surface-specific process parameters.

17.2.4 Depyrogenation by Moist Heat

Autoclaving is generally NOT considered to be a consistent depyrogenating process. However, there have been studies published that report a significant reduction in endotoxins activity post autoclaving. Decker and colleagues (2018) demonstrated that subjecting synthetic spider silk proteins to multiple autoclave cycles (121°C, 15 minutes) reduced the endotoxin activity by 10–20-fold (Decker et al., 2018). Miyamoto and colleagues (2009) demonstrated that subjecting water spiked with 2000 EU/mL LPS to a "soft thermal" treatment of the solution (140°C for 30 minutes) at 100% steam saturation reduced endotoxins activity by up to four (4) logs. Bamba et al. have published a number of studies on the concentration-dependent effects of autoclaving endotoxins in aqueous solutions under varying conditions of divalent salts and nonionic surfactants (Bamba et al., 1996a; Bamba et al., 1996b)

Tim Sandle (2011) compared three different depyrogenation treatment methods for glass vials to which a nominal standard endotoxins activity of 5000 EU were added: dry heat treatment at 250°C for 30 minutes, autoclaving at 121°C for 30 minutes, and treatment with 0.1N NaOH for one minute. His data on log reduction of these three methods (summarized following) demonstrate that although dry heat was the most efficient at reducing endotoxins activity, autoclaving and treatment of the glass with 0.1N NaOH also showed not insignificant reductions in endotoxins activity.

TABLE 17.1
Comparison of Endotoxin Destruction: Three Different Treatments

Treatment	Minimum log reduction	Maximum log reduction	Mean log reduction
Dry heat, 250°C for 30 minutes	5.0	5.4	5.2
Autoclaving 121°C 30minutes	1.0	2.5	1.7
0.1N NaOH, 1 minute	1.0	2.9	1.7

Source: (Sandle, 2013)

17.2.5 Depyrogenation by Irradiation

Irradiation kills microorganisms by creating highly reactive free radicals that break down DNA, and in the process can create a level of damage to the cytoplasmic membrane. Although standard methods of sterilization by irradiation might incrementally reduce levels of endotoxins activity under well-defined conditions including the type and dosage of radiation used and the materials under test (Guyomard et al., 1987, 1988), it cannot be assumed that irradiation will meet the current United States Food and Drug Administration (FDA) requirement of a three-log reduction in activity (see following).

To study the effects of irradiation on endotoxin contamination on medical devices, Guyomard and colleagues (1987) performed experiments to study the reduction of endotoxins activity of *E. coli* 055:B5 LPS that had been dried onto polystyrene plastic and after being exposed to increasing doses of ionizing radiation (Figure 17.7A and 17.7B).

17.3 DEPYROGENATION BY REMOVAL

Many materials including elastomeric closures, plastics, and biological product process streams cannot withstand the harsh conditions needed to inactivate endotoxins by dry heat, chemical, or physical treatments. For heat-labile materials, depyrogenation may be accomplished by removal using processes such as rinsing, chromatography, or filtration (Magalhaes et al., 2007; USP, 2020b, 2020c). These methods rely heavily on the chemistry of the endotoxin complex including, but not limited to, size, charge, and solubility. The firm's contamination control plan should discuss the risk of resident "pyroburden" on items, particularly on items that cannot be depyrogenated

7a. Effect of e-beam

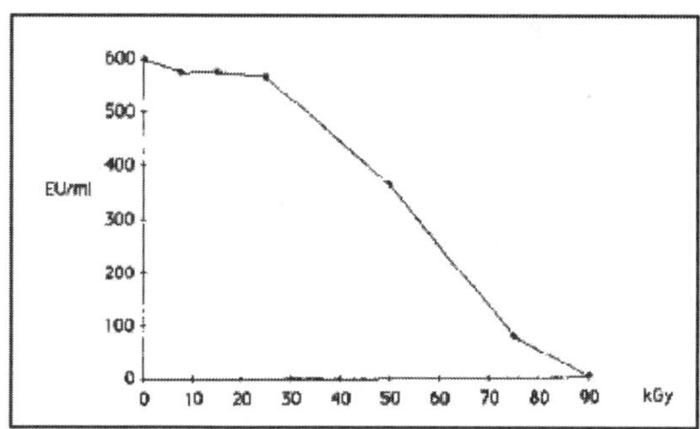

7b. Effect of Gamma Irradiation

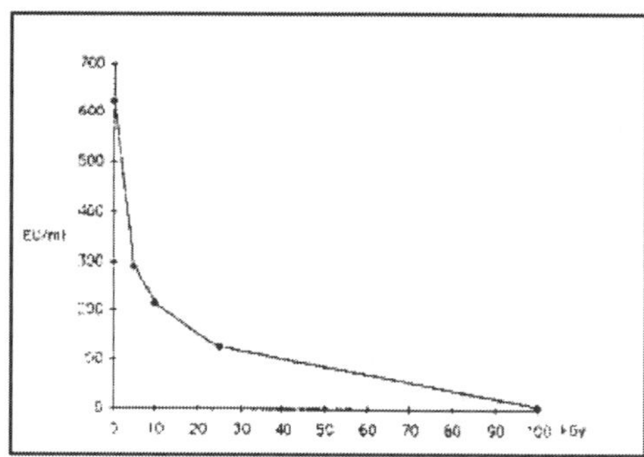

FIGURE 17.7A, B Effect of ionizing radiation on endotoxins activity.

Source: Guyomard et al., 1987.

by dry heat. This plan should contain an assessment that identifies the possible sources of endotoxins in/on these materials and appropriate mitigations.

17.3.1 Depyrogenation by Distillation

Distillation is a process involving the conversion of a liquid into a vapor that is subsequently condensed back to liquid form, allowing the separation of pure water from any volatile impurities, including gram-negative bacterial endotoxins. Pioneering work by Dr. Florence Seibert (1923) concluded that the pyrogens in contaminated IV solutions were filterable substances (i.e. able to pass through filters that retain whole bacteria) from gram-negative bacteria found in the water used to formulate these products (Seibert, 1923). Dr. Seibert also found that a single-pass distillation, using an apparatus containing a spray-catching baffle, would make pyrogen-free water. There are two types of distillation apparatus commonly used in the pharmaceutical operations: Vapor Compression distillation and Multiple Effect Stills. Both are capable of routinely producing water that meets USP standards for WFI, including the required endotoxin limits. The quality of the distilled water, however, will depend on (1) the quality of the feed water and upstream pretreatment maintenance, (2) the efficiency and maintenance of the still, and (3) storage and distribution conditions of the water downstream of the still. Ambient storage and distribution will require a validated sanitization protocol to assure that the risk of microbial access and growth, including the possibility of establishing a biofilm, is minimized. Hot loop systems, operated at 80°C or above, will not provide an environment for the proliferation of microorganisms. However, point-of-use valves, particularly downstream of heat exchangers, must be flushed and sanitized to assure that the risk of microbial growth and therefore endotoxins contamination in drawn water is mitigated.

17.3.2 Depyrogenation by Rinsing (Dilution)

Many materials used in the manufacture of parenteral products, for example elastomeric closures or plastics, are heat-labile and must be depyrogenated by a method that does not impact either their chemistry or integrity. One such method is depyrogenation by repeated rinsing of the material with a well-characterized solvent, usually sterile WFI (USP, 2020c). This method will physically remove endotoxins from surfaces and ultimately reduces the total endotoxins activity by the circulation of large volumes of solvent, and therefore dilution of the pyroburden, meaning any endotoxins that might be present on the materials.

Solvents other than WFI, for example dilute alkali or detergents, may be used for depyrogenation by rinsing. The concern with using nonaqueous solvents is the possibility of residual solvent in/on the material being depyrogenated and the adverse effects that might have on product stability and patient safety. In the case of a nonaqueous solvent, the validation must include a mechanism to rinse the depyrogenated material with sterile WFI until traces of the solvent are undetectable by appropriately sensitive testing.

Regardless of the solvent chosen, there are a number of critical factors that must be considered during and after validation (Berman et al., 1987):

- Solvents must be free of detectable endotoxins activity.
- A full description of the rinse solvent including normality or concentration and source, if appropriate.
- The optimum temperature of the rinse solvent must be defined. Very often hot WFI is drawn from a recirculating loop and is used as the solvent for rinsing. If the WFI is too hot for the material, it will need to be cooled to avoid any adverse impact. However, cooling the water under conditions that do not protect its sterility increases the risk for microbial contamination, microbial proliferation, and even the generation of endotoxins. Therefore, if hot water needs to be cooled, it should be drawn from the hot circulating loop and transferred using a sterile hose into a sterile vessel. The conditions of the transfer and storage of WFI used for depyrogenation by rinsing must be controlled in terms of maximum

hold time (generally 3–4 hours), hold temperature, and hold vessel.

- The optimum pressure of the rinse solvent, if it is injected or sprayed into a chamber or directly onto the articles being depyrogenated.
- Solvent flow through the system (liters/minute) must be controlled and monitored. Recirculation of solvents could be problematic, as endotoxins removed from materials may be redeposited. Therefore, a justification for and validation of recirculating solvents must be documented.
- For materials being rinsed inside of a chamber, for example, a stopper washer, the maximum load size per stopper type (chemical composition, size, shape) should be identified. Validation protocols must be written to include or bracket all stopper types and sizes.

17.3.3 Depyrogenation by Filtration

Depyrogenation by filtration or chromatography involves the removal of endotoxins from a liquid, either by adsorption or size exclusion, or sometimes both. These methods, although they can be effective means to depyrogenate, cannot be assumed to sterilize as well. Filtration media are manufactured from many different materials and in many different pore sizes and configurations to meet manufacturing needs. A cautionary note: Some filter media may contain cellulose or other materials that may be derived from cellulose and may leach a class of biomolecules originally called "LAL Reactive Material" later identified as glucans (Pearson et al., 1984; Roslansky and Novitsky, 1991). Glucans can enhance an endotoxin reaction with the LAL test, resulting in an over-quantitated result or even a "false positive." Although filter manufacturers have been attuned to this problem ever since the identification of "LAL Reactive Material" and have worked to minimize the use of these materials out of their media, it is still possible that the composition of some filter media may shed LAL-reactive glucans. Users are encouraged to discuss their needs with potential filter suppliers to understand the filter media composition and the possible leaching of glucans not only in water but also with the actual product or justifiable substitute material to be depyrogenated.

Microporous filtration works through a sieving action of small pores, with retention ratings of as small as 0.1 µm. Although these filters may retain endotoxins associated with whole bacteria, bacterial aggregates, or large fragments of bacterial cell walls, they are not as efficient at retaining endotoxins associated with free-floating small cell wall components and OMVs.

Until April 2017, the production of WFI in Europe had been limited to distillation only. At that time, the Ph. Eur. monograph for Water for Injections (0169) was revised to allow the production of WFI by a purification process proven to be equivalent to distillation, such as Reverse Osmosis (RO), which may be single-pass or double-pass, RO membranes in series, or which may be coupled with other appropriate techniques such as electro-deionization (the use of electric current and ion exchange membranes to deionize water), ultrafiltration, or nanofiltration (EMA, 2020). RO utilizes the tightest membranes that are currently in use in the pharmaceutical industry. They work by size exclusion and can separate dissolved salts and sugars from water, and they can also separate endotoxins from water. RO processes are performed at high pressures and under ambient conditions. Because they may retain gram-negative bacteria, they must be properly maintained to prevent microbial adaptation and possible grow-through.

Ultrafiltration is another membrane process whereby a fluid is passed, usually via a tangential flow, under pressure through membranes with pore sizes of between 1 and 100 nm. Although pore sizes have been measured, ultrafilters are actually not rated by pore size, but rather they are rated by their "Molecular Weight Cutoff." If the endotoxins and product have sufficiently different molecular weights (endotoxin being the larger and higher), the "retentate" will concentrate any endotoxin activity, and the filtrate will contain the product, free of endotoxins activity. This process has been used successfully to depyrogenate buffers and small-molecule drugs such as penicillins (Sweadner et al., 1977). A cautionary note: if the endotoxins and the product to be depyrogenated are not sufficiently different in molecular weight as with some biological products, the process could well co-concentrate the endotoxins and the product (Christy and Rubin, 2002).

17.3.4 Depyrogenation by Attraction, Adsorption, and Affinity

LPS is negatively charged at a pH >2. To leverage that attribute, many filter media are "charge modified" meaning that they are surface modified to retain a net positive charge that will attract the negatively charged LPS molecules (Gerba and Hou, 1985). These filters may be disc, cartridge, or depth filters. Charge-modified depth filters work to depyrogenate not only by their net positive charge, but they also provide a so-called "tortuous path" that may trap bacteria, OMVs, and larger cell wall fragments during filtration (Hou and Zaniewski, 1990).

Charged carbon filtration media, also called "activated carbon", contain very small particles of carbon prepared from organic materials and activated by steam or chemical treatment. The small size of the particles results in a very large surface area that can adsorb a range of organic molecules that may affect the color, odor, and endotoxins content of a process stream. Carbon filters are often used as part of the pretreatment train of WFI purification systems. However, care must be taken to validate the frequency and methods for regeneration of these media, as the very same bacteria that may be retained by the carbon filter may also proliferate there, creating biofilms and/or more endotoxins. Carbon filters may also take out desirable organic molecules in a drug product, and also may generate small, nonviable particles that could find their way into drug products.

When endotoxins and the target molecule, usually a protein, are of similar size and cannot be separated by size exclusion filtration methods, affinity chromatography may be a good choice. In this method, a ligand that contains cationic binding sites is attached to the column packing. The product is run through the column, and the endotoxins are removed by attraction (adsorption or binding) to the ligand. Sakata and co-workers (2005) describe the use of poly(ε-lysine) attached to the packing material, in this case beads, to remove endotoxins from protein solutions. Clearly, methodology for affinity chromatography must be developed to meet the needs of the solution being depyrogenated, keeping in mind that these methods can be affected by pH, temperature, electrolytes, and flow rate. To remove any endotoxins post treatment, the column needs to be cleaned and depyrogenated prior to its next use. Storage conditions in between runs must not provide conditions under which microorganisms can proliferate.

17.4 VALIDATION OF DEPYROGENATION

Historically, the pharmaceutical industry has thought of validation of depyrogenation processes as a stand-alone event or a series of repeating events that will demonstrate that the equipment and process can achieve what it purports to do. However, it is more appropriate to see depyrogenation as one of many process control parameters for process validation of the manufacture of the drug product.

Establishing control for any endotoxin removal or reduction activity requires that all depyrogenation requirements for a new drug product be identified as early in the development of the manufacturing process as possible so that product Critical Quality Attributes (CQAs) and critical process parameters (CPPs) that are supported by depyrogenation can be identified or anticipated (FDA, 2009, 2011). One CQA that is required for any parenteral drug product or medical device is that it is non-pyrogenic, meaning for endotoxins, that it may not have endotoxins activity detected at or above the calculated or assigned endotoxin limit. Assessment of the manufacturing process should identify all of those places where endotoxins are likely to enter the system (e.g. water systems, raw materials, primary packaging components) and identifying appropriate mitigation strategies for each of these sources.

17.4.1 User Requirement Specification

Establishing control for a depyrogenation process begins with the development of a User Requirements Specification (URS). Arguably the URS document is the most important step in the validation process because it is this list of required or desired functionalities that will ultimately drive the choice of equipment and vendor and provide a road map for equipment qualification, washing cycle development, column packing, solvents to be used, dwell times, etc. The URS is prepared, reviewed, and signed by equipment owners and stakeholders including engineering, operations, quality, maintenance, and IT and is reviewed/signed by the Quality Unit.

17.4.2 Preparation and Use of Endotoxin Indicators

To test a method's ability to inactivate or remove endotoxins activity from an item or product stream, materials undergoing depyrogenation are inoculated or "spiked" with a known level of endotoxins activity prior to the depyrogenation process. These "spiked" units are called "endotoxin indicators" or EIs, as they function in a manner that is analogous to biological indicators in sterilization studies (LAL Users' Group, 1989; USP, 2020e). Comparing the assayable endotoxins activity in the EIs before and after processing will provide a measure of the impact of the depyrogenation process on biological activity. EIs may be used to support process development activities, but they must also be used in the ultimate process validation study.

The current requirement per FDA's current Aseptic Processing Guidance is that the EI is prepared to contain at least 1000 assayable Endotoxin Units (EU) taken from a solution of purified LPS (FDA, 2004). Unlike endotoxins from natural sources that contain proteins and phospholipids in addition to LPS (Beveridge, 1999), purified LPS has a tendency to adsorb to surfaces or other molecules, and activity is often not recovered at 100% of the initial spike activity level (Bryans et al., 2004). Recovery rates can vary with:

- The genus/species of the organism from which the LPS originated.
- The level of activity in the initial inoculum.
- The type of surface onto which the LPS is inoculated.
- The method of purification and the purity of the LPS preparation, meaning not only the efficiency of the extraction and purification, but the impact of any excipients that might be in the analyte.
- The method chosen by the lab to inoculate the carrier and fix the endotoxins to the surface.
- Methods chosen for extraction and recovery of endotoxins activity from study surfaces.

It is not unusual to inoculate an indicator with 1000 EU of purified LPS activity and recover only 500 EU. This means that to meet the FDA suggested level of a starting value of 1000 recoverable EU, the initial spike may often be much higher than 1000 EU. Endotoxin indicators for dry heat depyrogenation of vials can be purchased from a vendor of the lysate reagent, but EIs for other types of processes must be prepared in-house. If EIs are prepared in-house, the laboratory must engage in its own method development activities to inoculate, extract, and measure endotoxins activity from an EI.

17.4.2.1 Preparation of Endotoxin Indicators for Solid Surfaces

To prepare an EI for a *surface* to be depyrogenated (e.g. a rubber stopper), place a small volume (<100 μL) of highly concentrated LPS onto the surface to be depyrogenated. For stoppers that rely on rinsing for depyrogenation, the inoculum is usually placed in the well on the underside of the stopper, as that site is considered to be the most difficult place on the stopper for the rinse solution to access. The inoculated EIs are dried generally under a vertical laminar flow hood. The exact method for drying is the laboratory's choice, but because the method for drying the LPS onto the EI surface can also affect the recovery of activity, the method for drying must be documented in a Standard Operating Procedure (SOP) to assure consistency from study to study (Novitsky et al., 1986; Jensch et al., 1987). Unlike most Control Standard Endotoxin preparations used as calibration analytes, purified LPS for EIs should not be formulated with fillers or stabilizers, as these excipients may affect the LPS inactivation profile (Ludwig and Avis, 1990). EIs must be differentiated from the other materials in the load for easy retrieval and extraction (see following). For example, in some cases in which stopper compositions are the same, EIs may be in pink stoppers rather than gray stopers, or EI vials may be clearly marked with a water- and chemical-resistant marker. In short, there is no one "correct" way to prepare EIs. However, because of the many sources of variability in preparing EIs, it is suggested that the laboratory choose a method, document it in an SOP, and stick with it.

To extract and recover endotoxin activity from the EI, a small volume of Water for BET (also known as LAL Reagent Water or WFI) is added to the indicator and the indicator is vortexed or sonicated to assist in the extraction of LPS from the carrier surface. As with the preparation of the EIs, there are many variations in methodology for the extraction of endotoxins activity from EIs and may include intermittent vortexing, sonication, or both for a prescribed period of time (usually an hour). Once the laboratory develops a method that gives consistent recoveries, they should document the procedure in an SOP and follow it for subsequent studies to assure comparability. The extracts from the unprocessed EI controls ("recoverable" endotoxins activity) and processed EIs ("residual" endotoxins activity) are tested for endotoxins activity using the compendial BET (LAL Users' Group, 1989; USP, 2020e).

17.4.2.2 Preparation of Endotoxin Indicators for Product Streams

In the case of the depyrogenation of a product stream by chromatography or filtration, the solution upstream of the column/filter is spiked with endotoxins to a concentration equal to 1000 EU of endotoxins activity. As with the solid endotoxin indicators, determination of the initial spike endotoxins activity may be affected by the product formulation (inhibition). This inoculated solution is the endotoxin indicator for the product stream. Endotoxins activity remaining in the solution downstream of the depyrogenation process step ("residual" endotoxins activity) will be compared with endotoxins activity in the untreated control samples ("recoverable" endotoxins activity) upstream of the depyrogenation step to determine the efficiency of the process.

17.4.3 Evaluation of Data

The United States Food and Drug Administration's (FDA's) 2004 Aseptic Processing Guidance states that a depyrogenation cycle is sufficient if there is a greater than 3 log reduction in activity between the recoverable endotoxin activity in the unprocessed EIs (controls) and residual endotoxin activity in the processed EIs (FDA, 2004). Log reduction is calculated as follows:

$$\text{Log}_{10}\text{ reduction} = \text{Log}_{10}\text{ of recoverable LPS activity} - \text{Log}_{10}\text{ of residual LPS activity}$$

For example, if the recoverable activity of an unprocessed product stream is 5925 EU/mL and the residual activity of the processed EI is 0.83 EU/mL, the log reduction is equal to:

$$\text{Log}_{10}\text{ reduction} = \text{Log}_{10}\ 5925 - \text{Log}_{10}\ 0.83$$
$$\text{Log}_{10}\text{ reduction} = 3.77 - (-0.08) = 3.77 + 0.08 = 3.85$$

In this example, there is a 3.85 log reduction in endotoxins activity, which meets the FDA suggested minimum inactivation level of three logs.

If the measurement for residual endotoxins falls below the bottom point on the referenced standard curve for quantitative BET assays, then the calculation looks like the example following. For this example, the lab is performing a kinetic chromogenic assay with a referenced standard curve of 5.0–0.05 EU/mL. The recoverable endotoxins activity is the same 5925 EU/mL as the previous example, and the test for residual endotoxins activity is below the referenced Limit of Quantitation (LOQ, or test method sensitivity) of 0.05 EU/mL. Because the activity of the residual endotoxin was BELOW the LOQ of the assay, the equal sign is replaced by a > sign to indicate that the inactivation was measured to be greater than the LOQ of the referenced assay, in this case greater than 5.26 logs.

$$\text{Log}_{10}\text{ reduction} > \text{Log}_{10}\text{ of recoverable LPS activity} - \text{Log}_{10}\text{ of residual LPS activity}$$
$$\text{Log}_{10}\text{ reduction} > \text{Log}_{10}\ 5925 - \text{Log}_{10}\ 0.05$$
$$\text{Log}_{10}\text{ reduction} > 3.77 - (-1.3) > 3.77 + 1.3 > 5.07$$

A cautionary note. All BET results are reported in endotoxin units/milliliter. If an EI, for example a stopper, is extracted with 1 mL of Water for BET, then the measured value can be interpolated directly from the standard curve, because all of the endotoxins activity will be contained in the 1 mL extraction volume, and the standards are diluted in endotoxin units/milliliter. However, if the EI is extracted with a different volume of Water for BET (e.g. 10 mL), then the result has to be corrected by a factor of 10 to account for the dilution of the initial inoculum by the extraction volume.

17.5 QUALIFICATION OF DEPYROGENATION PROCESSES

Depyrogenation Process Qualification (PQ) studies represent the confluence of a routine operating process: all equipment has been qualified, process parameters have been developed, EI preparation and extraction has been developed, SOPs have been developed and at least in their final draft, and operators have been trained. When possible, PQ studies should be conducted on real manufacturing materials to confirm and replicate process parameter development studies that define critical operating parameters (time/temperature/normality/flow rate) for a given product or material composition. PQ studies may bracket the critical operating parameters to anticipate and allow for any run-to-run and day-to-day variability in the depyrogenation process.

A successful PQ run requires that all of the acceptance criteria documented in the protocol are met: Process parameters have been met, endotoxin analyses are valid, and the final log reduction meets the prescribed criteria.

For any validation study, it is common practice in the pharmaceutical industry to follow the "rule of three" (three consecutive runs with acceptable data). However, a clear understanding of the incoming levels of resident endotoxin ("pyroburden") on the materials subject to the process may allow a firm to justify fewer than three consecutive successful runs. Conversely, if there is a highly variable pyroburden in the incoming materials, it may require a firm to perform more than three runs to validate the depyrogenation process parameters. The company must understand that a "validated" process must be capable of providing the needed level of safety assurance across the range of known variables. Just as with any process validation, the protocol for the study must be data driven, scientifically based, and risk based (FDA, 2011).

At the end of the PQ study, a final report is written to confirm that the process can consistently achieve what it purports to do and demonstrate that the proposed process-specific critical operating parameters produce consistent and reproducible results.

Once the validation studies are completed and signed off, the depyrogenation processes can be implemented for routine manufacturing, but the established level of control must be maintained. Control means that systems and processes are consistent, that all measuring equipment, controllers and sensors are properly and routinely calibrated against a traceable standard, and that preventive maintenance is provided on a regular basis. All of these control measures must be properly documented in SOPs, log books, and if possible, a validated equipment management software.

Front-line manufacturing operators are often the first to identify changes in the performance of depyrogenation processes because of their daily interaction with the process and equipment. Continued training of operators in identifying deviations to the process, properly documenting the process, and keeping up with any changes to the process are key to continuing control during routine operations.

Vendors of materials received as depyrogenated should be qualified and audited to assure that their processes are in a state of control and that they have systems in place to properly develop and validate performance parameters. However, they must also have procedures in place to reduce the risk of gram-negative contamination of the materials post depyrogenation. Quality Agreements, particularly for primary product containers or closures, should stipulate that the pharmaceutical manufacturer must be notified by the vendor in the event of a loss of control, a major change in manufacturing process, or a change in raw materials at the supplier facility (FDA, 2016).

17.5.1 Monitoring Depyrogenation Processes

Monitoring and testing are not control measures. Rather, they are tasks performed to assure that implemented control measures are successful. Key monitoring tasks to demonstrate a continued state of control include:

- Physical measurements of the required and measurable process parameters (time/temperature/normality/flow rate) that are recorded and reported by manufacturing personnel and confirmed by QA after each processing run to assure parameters were within the ranges specified by the PQ studies. Reviewed cycle parameter data must be included in the drug product batch records
- Regardless of the depyrogenation process employed, a surveillance program for incoming "pyroburden" levels on materials routinely depyrogenated in house should be established. A rise in the endotoxin activity levels on incoming materials may (1) impact the validated depyrogenation process and/or (2) suggest that something has changed at the vendor, and may require an additional assessment of the supplier.

17.5.2 Documentation and Change Management

Because of the criticality of depyrogenation processes to the safety of parenteral products, depyrogenation process qualification and supporting data as well as routine paperwork describing the process parameters for each load/run are GMP records and must be reviewed by QA and archived, preferably in the appropriate batch record. Of course, Good Documentation Practice must be followed.

Any change to a validated process requires that the change be submitted to the company's change management process. For depyrogenation, proposed changes to any critical operating parameter must be assessed by the firm's Change Board for any impact to the validated state prior to implementation. If revalidation is warranted, it must be documented via a protocol/report and must be traceable back to the original change request. If the Change Board decides that revalidation is not warranted, justification for the decision must be part of the change management record.

17.6 A PARADIGM SHIFT: NEW WAYS TO THINK ABOUT DEPYROGENATION

The United States Pharmacopeia (USP) has recently undertaken a project to look at depyrogenation of materials from a scientific as well as a pragmatic standpoint. These reviews and suggestions are provided in the <1228.x> series of informational chapters (USP, 2020a, 2020b, 2020c, 2020d, 2020e) and are summarized following:

1. There is a difference between sterilization and depyrogenation. For a very long time, depyrogenation was viewed as an adjunct to sterilization and was only briefly mentioned in USP informational chapter <1211>. USP now has separate series for sterilization (<1229.x series) and depyrogenation (<1228.x> series). Sterility is an absolute by definition. However, because of the constraints of sampling and the current analytics, sterility is really a probability. Sterility testing is binary, meaning that the test is either positive or negative. However, testing for endotoxins activity is quantitative and can be easily compared with calculated/assigned endotoxin limits or endotoxins activity reduction goals.
2. Depyrogenation requires a measured reduction in endotoxins activity (as a representative of the universe of pyrogens) to a predetermined level. Ideally, this should be accomplished using endotoxins from autochthonous microorganisms. Is the calibration standard (primary or secondary purified LPS) the most appropriate analyte for depyrogenation studies? A purified *E. coli* calibration standard is not representative of the universe of endotoxins that might be found in a pharmaceutical manufacturing plant.
3. Is the concept of a 3-log reduction obsolete? The 3-log rule was proposed in a 1984 version of USP informational chapter <1211> on Sterilization of Compendial Articles. It may have been chosen because vials were not shrink-wrapped in plastic back in 1984, and there were often cardboard particles that were loaded with endotoxins in the vials. OR, the 3-log rule could have come from constraints connected to the analytics at the time, as quantitative BET assays were in their infancy. OR, it could have been an application of the standard "rule of three." For example, rubber stoppers, because of the temperatures required to melt and mold the material and the use of modern packaging materials that eliminates the risk of additional particulates and microorganisms, generally come into the pharmaceutical facility with little or no detectable endotoxins on them. Does it make sense to load up stoppers with 1000 EU of a purified material that cannot possibly contaminate them in "real life" and look for a 3-log reduction?

   The new thinking is that depyrogenation is better defined as a reduction of measurable pyroburden activity to safe levels. Of course, this suggestion is much less prescriptive than the current one, because it requires that a) a company understand the levels of endotoxins activity that are in starting materials and b) the company must have a good understanding of the product and target patient populations to determine what "safe" levels are.
4. Historically, the performance of depyrogenation processes is verified once or twice a year. The new thinking is that once a depyrogenation process is initially qualified using EIs and the appropriate physical and chemical measurements, then there is little value in revalidation; particularly for a process such as stopper rinsing that is totally dependent on physical parameters (time, temperature, load), revalidation with EIs is a time consuming, expensive task that provides little value-added information when a firm already knows the load-specific critical operating parameters required for depyrogenation using EIs. Note: If there is a change to product formulation, major change to equipment, or a proposed change in time/temperature, EIs may be appropriate to support the revalidation effort. The impact of such changes should be examined and justified in the change management process.
5. Given the current compliance emphasis on Quality by Design, Risk Management, and Lifecycle Management, perhaps the focus in the industry should shift from *removing* endotoxins contamination to *preventing* contamination. How would a company do this?
   - Control of bioburden. If gram-negative organisms can't grow, they can't produce endotoxins. Therefore, there should be emphasis on the control of processes and conditions under which gram-negative organisms are likely to be found including water systems, pooling water in manufacturing areas, equipment that is not stored dry, raw materials, particularly from natural sources, and fermentation processes, particularly where a gram-negative organism is the host for the production of proteins.
   - There must be a robust vendor qualification/auditing for manufacturing materials at high risk for contributing endotoxins to assure control of pyroburden at the material source, including controls for endotoxins contamination and measurement of endotoxins activity.
   - Process mapping should be performed and reviewed periodically to identify steps in the manufacturing process at which endotoxins are likely to get into the system and what proactive **controls** are needed to keep them out.
   - Decisions must be data-driven, scientifically based, and risk based when considering the appropriateness of methods and critical operating parameters for depyrogenation processes.

17.7 CONCLUSION

We must engage in a paradigm shift and pivot from an emphasis in pharmaceutical manufacturing of *removing* endotoxins from materials and products to *preventing* the introduction of endotoxins in parenteral products. Evidence-based tools to accomplish this proposed shift exist: validation, risk assessment, process mapping, quality by design, auditing.

However, there are times when a depyrogenation process may be necessary. In those cases, it is important to understand the levels of endotoxins activity that are resident in or on the material and the target "safe" levels of endotoxins activity that are needed. Excessive validation requiring unrealistic levels of reduction in endotoxins activity of a calibration standard rather than the same endotoxins that might be found in the product potentially makes product more expensive, and in some cases may delay a product's entry into the market. A product can never be "pyrogen free" because current analytics can't measure and assure "0", but based on current understanding of threshold pyrogenic doses and our ability to measure endotoxins activity, we can measure and assure "safe."

REFERENCES

Akers J, Guilfoyle DE, Hussong D, McCullough K, Mello R, Singer D, Tidswell E, Tirumalai R. Functional challenges for alternative bacterial endotoxins tests part 3: The centrality of autochthonous endotoxins in the assurance of patient safety. *Am. Pharm. Rev.* 2020. www.americanpharmaceuticalreview.com/Featured-Articles/569998-Functional-Challenges-for-Alternative-Bacterial-Endotoxins-Tests-Part-3-The-Centrality-of-Autochthonous-Endotoxins-in-the-Assurance-of-Patient-Safety/

Avis KE, Jewell RC, Ludwig JD. Studies on the thermal destruction of *Escherichia coli* endotoxin. *J. Parent. Sci. Tech.* 1987, 41(2), 49–56.

Bamba T, Matsui R, Watabe K. Effect of steam-heat treatment with/without divalent cations on the inactivation of lipopolysaccharide from several bacterial species. *J. Parent. Sci Tech.* 1996a, 50(2), 129–135.

Bamba T, Matsui R, Watabe K. Enhancing effect of non-ionic surfactants on the inactivation of lipopolysaccharide by steam-heat treatment. *J. Parent. Sci. Tech.* 1996b, 50(6), 360–365.

Berman D, Kasica T, Myers T, Chrai S. Cycle development criteria for removal of endotoxin by dilution from glassware. *J. Parent. Sci. Tech.* 1987, 41(5), 158–163.

Bertani B, Ruiz N. Function and biogenesis of lipopolysaccharides. *EcoSal Plus* 2018, 8(1). doi: 10.1128/ecosalplus.ESP-0001-2018

Beveridge TJ. Structures of gram negative cell walls and their derived membrane vesicles. *J. Bacteriol.* 1999, 181(16), 4725–4733.

Bonnington KE, Kuehn MJ. Outer membrane vesicle production facilitates LPS remodeling and outer membrane maintenance in salmonella during environmental transitions. *mBio.* 2016, 7(5), 1–15. doi: 10.1128/mBio.01532-16

Brogden KA, Phillips M. The ultrastructural morphology of endotoxins and lipopolysaccharides. *Electron Microsc. Rev.* 1988, 1, 261–277.

Bryans TD, Braithwaite C, Broad J, Cooper JF, Darnell KR, Hitchins V, Karren AJ, Lee PS. Bacterial endotoxin testing: A report on the methods, background, data, and regulatory history of extraction recovery efficiency. *Biomed. Inst. Tech.* 2004, 73–78.

Campbell DH, Cherkin A. The destruction of pyrogens by hydrogen peroxide. *Science* 1945, 535–536.

Capra EJ, Laub MT. Evolution of two-component signal transduction systems. *Annu. Rev. Microbiol.* 2012, 66, 325–347. doi: 10.1146/annurev-micro-092611-150039

Cherkin A. Destruction of bacterial endotoxin pyrogenicity by hydrogen peroxide. *Immunochemistry* 1974, 33(6–7), 625–627.

Chilton PM, Hadel DM, To TT, Mitchell TC, Darveau RP. Adjuvant activity of naturally occurring monophosphoryl lipopolysaccharide preparations from mucosa-associated bacteria. *Infect. Immun.* 2013, 81(9), 3317–3325. www.ncbi.nlm.nih.gov/pmc/articles/PMC3754217/

Christy C, Rubin D. *Selecting the Right Ultrafiltration Membrane for Biopharmaceutical Applications,* 2002. www.pharmtech.com/view/selecting-right-ultrafiltration-membrane-biopharmaceutical-applications

Cytiva. *Use of Sodium Hydroxide for Cleaning and Sanitization of Chromatography Resins and Systems*, 2020. https://cdn.cytivalifesciences.com/dmm3bwsv3/AssetStream.aspx?mediaformatid=10061&destinationid=10016&assetid=20986\

Decker RE, Harris TI, Memmott DR, Peterson CJ, Lewis RV, Jones JA. Method for the destruction of endotoxin in synthetic spider silk proteins. *Nature Scientific Reports*, 2018. www.nature.com/articles/s41598-018-29719-6

Deshmukh DG, Zade AM, Ingole KV, Mathai JK. State of the globe: Non-fermenting gram-negative bacilli challenges and potential solutions. *J. Glob. Infect. Dis.* 2013, 5(4), 125–126.

Ellis TN, Leiman SA, Kuehn MJ. Naturally produced outer membrane vesicles from pseudomonas aeruginosa elicit a potent innate immune response via combined sensing of both lipopolysaccharide and protein components. *Infection and Immunity* 2010, 78(9), 3822–3831.

European Medicines Agency. *Guideline on the Quality of Water for Pharmaceutical Use*, 2020. www.ema.europa.eu/en/documents/scientific-guideline/guideline-quality-water-pharmaceutical-use_en.pdf

Galanos C, Rietschel E, Luderitz O, Westphal O. Biological activities of lipid a complexed with bovine-serum albumin. *Eur. J. Biochem.* 1972, 31, 230–233.

Gerba C, Hou K. Endotoxin removal by charge-modified filters. *App. Env. Microbiol.* 1985, 50(6), 1375–1377.

Gould MJ, Novitsky TJ. Depyrogenation by Hydrogen Peroxide. Parenteral Drug Association Technical Report Number 7, 1985.

Guyomard S, Goury V, Darbord JC. Effects of ionizing radiations on bacterial endotoxins: Comparison between gamma radiations and accelerated electrons. *Int. J. Radiation Appl. Inst.* 1988, 31(4–6), 679–684.

Guyomard S, Goury V, Laizier J, Darbord JC. Defining of the pyrogenic assurance level (PAL) of irradiated medical devices. *Int. J. Pharm.* 1987, 40, 173–174.

Hale G, Drumm A, Harrison P, Phillips J. Repeated cleaning of protein a affinity column with sodium hydroxide, 1994.

Haskins W, Landy M, Milner KC, Ribi E. Biological properties of parent endotoxins and lipoid fractions with a kinetic study of acid-hydrolyzed endotoxin. *J. Exptl. Med.* 1961, 114, 665–684.

Hort EC, Penfold WJ. Microorganisms and their relation to fever. *J. Hyg.* 1912, 12(3), 361–390.

Hou KC, Zaniewski R. Depyrogenation by endotoxin removal with positively charged depth filter cartridge. *J. Parent. Sci. Tech.* 1990, 44(4), 204–209.

Jensch UE, Gail L, Klavehn M. Fixing and removing of bacterial endotoxin from glass surfaces for validation of dry heat sterilization. In: *Detection of Bacterial Endotoxins with the Limulus Amebocyte Lysate Test.* New York: Alan R. Liss, 1987.

LAL Users' Group. Preparation and use of endotoxin indicators for depyrogenation process studies. *J. Parent Sci Tech.* 1989, 43(3), 109–112.

Li Y, Powell DA, Shaffer SA, Rasko DA, Pelletier MR, Leszyk JD, Scott AJ, Masoudi A, Goodlett DR, Wang X, Raetz CRH, Ernst RK. LPS remodeling is an evolved survival strategy for bacteria. *Proc. Natl. Acad. Sci.* 2012, 109(22), 8716–821.

Ludwig JD, Avis KE. Dry heat inactivation of endotoxin on the surface of glass. *J. Parent. Sci. Tech.* 1990, 44(1), 4–12.

Magalhaes PO, Lopes AM, Mazzola PG, Rangel-Yagui C, Penna TCV, Pessoa A, Jr. Methods of endotoxin removal from biological preparations: A review. *J. Pharm Pharmaceut Sci* 2007, 10(3), 388–404.

McCullough KZ, Novitsky TJ. Detoxification of Endotoxin by Acid and Base. Parenteral Drug Association Technical Report Number 7, 1985.

Miyamoto T, Okano S, Kasai N. Inactivation of Escherichia coli endotoxin by solft hydrothermal processing. *Appl. Env, Microbiol.* 2009, 75(15), 5058–5063.

Morita, RY. *Bacteria in Oligotrophic Environments.* New York: Chapman & Hall, 1997.

Nikaido H. Molecular basis of bacterial outer membrane permeability revisited. *Microbiol. Mol. Biol. Rev.* 2003, 67(4), 593–656. http://dx.doi.org/10.1128/MMBR.67.4.593-656.2003

Niwa M, Milner KC, Ribi E, Rudbach JA. Alteration of physical, chemical, and biological properties of endotoxin by treatment with mild alkali. *J. Bacteriol.* 1969, 97(3), 1069–1077.

Norris MH, Somprasong N, Schweizer HP, Tuanyok A. Lipid a Remodeling is a Pathoadaptive Mechanism That Impacts Lipopolysaccharide Recognition and Intracellular Survival of Burkholderia Pseudomallei, 2018.

Novitsky TJ, Schmide-Gengenback J, Remillard JF. Factors affecting recovery of endotoxin adsorbed to container surfaces. *J. Pharm. Sci. Tech.* 1986, 40(6), 284–286.

Pearson FC, III. *Pyrogens, Endotoxins, LAL Testing and Depyrogenation.* New York: Marcel Dekker, Inc., 1985.

Pearson FC, Bohon J, Lee W, Bruszer G, Sagona M, Jakubowshi G, Dawe R, Morrison D, Dinarello C. Characterization of limulus amoebocyte lysate-reactive material from hollow-fiber dialyzers. *Appl. Env. Microbiol.* 1984, 48(6), 1189–1196.

Raetz CRH. Biochemistry of endotoxins. *Ann Rev. Biochem.* 1990, 59, 129–170.

Raetz CRH, Reynolds CM, Trent MS, Bishop RE. Lipid a modification systems in gram-negative bacteria. *Annu. Rev. Biochem.* 2007, 76, 295–329. http://dx.doi.org/10.1146/annurev.biochem.76.010307.145803

Reid N. A global perspective for quantifying all endotoxins within pharmaceutical water systems. Presented at PharmaLab, Dusseldorf, Germany, 2019.

Ribi E, Haskins WT, Landy M, Milner KC. Symposium on bacterial endotoxins. I: Relationship of chemical composition to biological activity. *Bacteriol. Rev.* 1961, 25(4), 427–456.

Ribi E, Haskins WT, Milner KC, Anacker RL, Ritter DB, Goode G, Trapani R-J, Landy M. Physicochemical changes in endotoxin associated with loss of biological potency. *J. Bacteriology.* 1962, 84, 803–814.

Roslansky PF, Novitsky TJ. Sensitivity of Limulus amebocyte lysate (LAL) to LAL-reactive glucans. *J. Clin. Microbiol.* 1991, 29(11), 2477–2483.

Sakata M, Yamaguchi Y, Hirayama C, Bemberis I, Todokoro M, Kunitake M, Nakayama M. Affinity chromatography removes endotoxins. *BioPharm* 2005, 18(1). www.biopharminternational.com/view/affinity-chormatography-removes-endotoxins

Sandle T. A comparative study of different methods for endotoxin destruction. *Am. Pharm. Rev.* 2013. www.americanpharmaceuticalreview.com/Featured-Articles/148858-A-Comparative-Study-of-Different-Methods-for-Endotoxin-Destruction/

Sandle T. *Characterizing the Microbiota of a Pharmaceutical Water System: A Metadata Study*, 2015. https://symbiosisonlinepublishing.com/microbiology-infectiousdiseases/microbiology-infectiousdiseases33.php

Sandle T. A practical approach to depyrogenation studies using bacterial endotoxin. *J. GXP Compliance.* 2011, 15(4), 90–96.

Schwechheimer C, Kuehn MJ. Outer-membrane vesicles from Gram negative bacteria: Biogenesis and functions. *Nat. Rev. Microbiol.* 2015, 13(19), 605–619.

Siebert F. Fever-producing substance found in some distilled waters. *Am. J. Physiol.* 1923, 67(1), 90–104.

Silhavy TJ, Kahne D, Walker S. The bacterial cell envelope. *Cold Spring Harb Perspect Biol* 2010, 2, a000414.

Sohlenkamp C, Geiger O. Bacterial membrane lipids: Diversity in structures and pathways. *FEMS Microbiology Reviews* 2016, 40(1), 133–159. https://doi.org/10.1093/femsre/fuv008

Sweadner K, Forte Lita M, Nelsen L. Filtration removal of endotoxin (pyrogens) in solution in different states of aggregation. *Applied and Environmental Microbiology* 1977, 34(4), 382–385.

Trent MS, Stead CM, Tran AX, Hankins JV. Diversity of endotoxin and its impact on pathogenesis. *J. Endotoxin Research* 2006, 12(4).

Tsuji K, Harrison SJ. Dry-heat destruction of lipopolysaccharide: Dry-heat destruction kinetics. *Appl. Env. Microbiol.* 1978, 36(5), 710–714.

Tsuji K, Lewis AR. Dry-heat destruction of lipopolysaccharide: Mathematical approach to process validation. *Appl. Env. Microbiol.* 1978, 36(5), 715–919.

United States Food and Drug Administration. *Contract Manufacturing Arrangements for Drugs: Quality Agreements*, 2016. www.fda.gov/downloads/drugs/guidances/ucm353925.pdf

United States Food and Drug Administration. *Guidance for Industry: Process Validation: General Principles and Practices*, 2011. www.fda.gov/downloads/drugs/guidances/ucm070336.pdf

United States Food and Drug Administration. *Guidance for Industry: Q8(R2) Pharmaceutical Development*, 2009. www.fda.gov/downloads/drugs/guidances/ucm073507.pdf

United States Food and Drug Administration. *Guidance for Industry: Sterile Drug Products Produced by Aseptic Processing: Current Good Manufacturing Practice*, 2004. www.fda.gov/downloads/Drugs/Guidances/ucm070342.pdf

United States Pharmacopeia. 2020a. <1228>, "Depyrogenation".

United States Pharmacopeia. 2020b. <85>, "Bacterial Endotoxins Test".

United States Pharmacopeia. 2020c. <1228.4>, "Depyrogenation by Rinsing".

United States Pharmacopeia. 2020d. <1228.3>, "Depyrogenation by Filtration".

United States Pharmacopeia. 2020e. <1228.5>, "Endotoxin Indicators for Depyrogenation".

Weary ME, Pearson FC, III. A manufacturer's guide to depyrogenation. *BioPharm.* 1988, 1(4), 22–29.

18 Ethylene Oxide Sterilization

James Agalloco

CONTENTS

18.1 INTRODUCTION

The sterilization of pharmaceutical products and medical devices must preserve the materials' properties, which often eliminates methods of sterilization that rely on thermal energy or radiation. The simplicity and speed of heat (moist or dry) and radiation sterilization ordinarily makes those the sterilization methods preferable; however, the effects of these processes on materials can be detrimental to the essential material properties. Chemical methods, primarily Ethylene Oxide (ETO) gas sterilization, is frequently used in these situations. This chapter explores the principles and validation approaches used for ETO sterilization and delineates the necessary routine process control requirements. The laws of physics mandate that gases are uniform in the concentration of all components present. As a consequence, ETO processes are relatively simple to develop, validate, and operate.

The concentration of the chemical agent has the largest impact on the lethality of the sterilization process; however, there are other factors essential for effective sterilization by ETO. Moisture must be present as well for effective sterilization to assist in penetration of the agent through the spore coat.[1] In ETO sterilization, moisture is commonly provided by the humidity present in the gas. In ETO sterilization processes used for medical devices, there are protective layers surrounding the device and supportive packaging present requiring extensive pre-humidification of the load to ensure that adequate humidity permeates the load to the target location.

18.2 STERILIZATION BASICS

Sterilization is a process that completely destroys or removes microorganisms. In the context of this chapter, the emphasis is on the completeness of the treatment. In sterilization processes, a microbiological death curve can be graphically described by the logarithmic number of microorganisms remaining alive when plotted against time, resulting in a straight line, also termed the death curve[2]. This line can be extrapolated to estimate the number of possible survivors in a very large number of units or the potential for a surviving microorganism in a single unit (see Figure 18.1). This is termed the Probability of a Non-Sterile Unit (PNSU). An acceptable PNSU has been defined as not more than 1 positive unit in 1,000,000 units or one chance in a million that a single unit is non-sterile (this is a risk value originally developed for food safety).

The slope (the inverse of which defines the D-value) of the microbial death curve is a property of the microorganism and the conditions of the sterilization treatment itself. The slope of the curve is related to the time in minutes for the microbial population to be reduced by 90% (or 1 logarithm) and is commonly termed the D-value.[3] Accurate determination of the D-value requires precise measurement of the lethal conditions to which the microorganism is exposed and must be reported with the D-value. D-values for gases are easily developed as the agent concentration, relative humidity (RH), and temperature can be easily measured. Validating the destruction of microorganisms relies largely on differences in the relative resistance of a biological indicator (BI) and bioburden organisms (see Figure 18.2).[4]

DOI: 10.1201/9781003163138-18

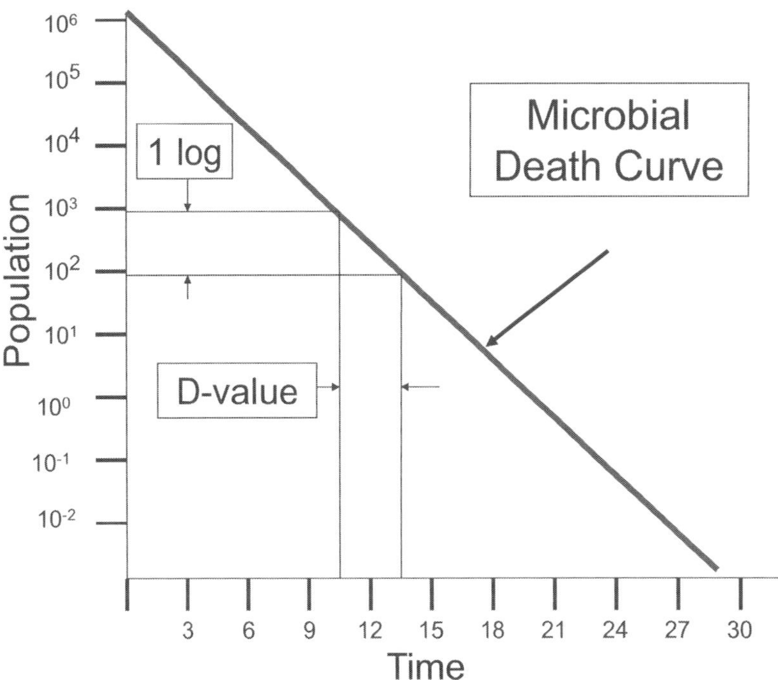

FIGURE 18.1 Microbial death curve.

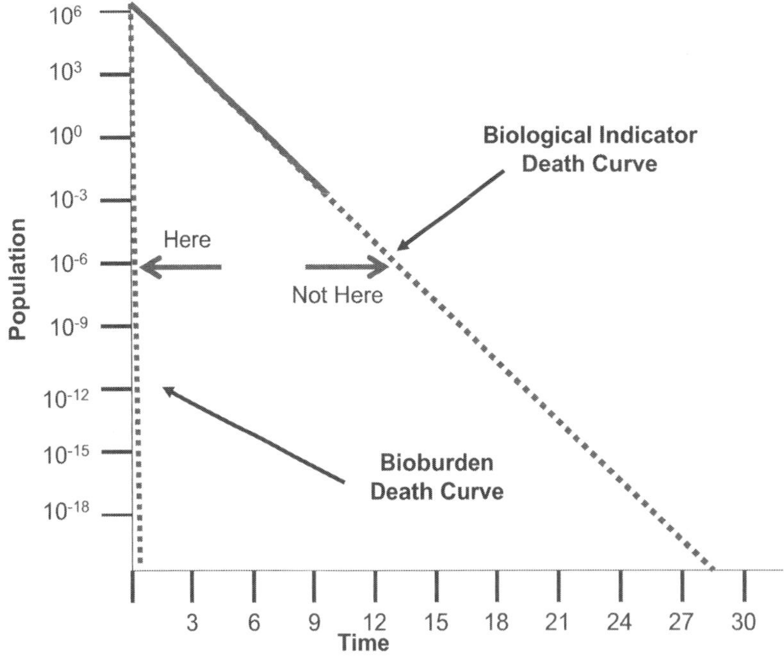

FIGURE 18.2 Relative resistance of bioburden and biological indicator.

The validation of ETO sterilization relies on these principles to establish reliable processes. There are three primary means to sterilization cycle development (and validation, which must follow the same approach) established in ISO 11135: the half-cycle approach, the overkill approach, and the biological indicator/bioburden approach.[5] The half-cycle method is the most conservative and widely used approach, especially for medical devices. The overkill approach mimics that used for other sterilization methods and can result in sterilization cycles that are somewhat shorter than those of the half-cycle approach. Both of these are discussed at greater length following. The biological indicator/bioburden approach relies on differences in their relative resistance to the sterilization process. Although technically possible with many sterilization processes, there are no publications describing its use with ETO.

18.3 ETO STERILIZATION FUNDAMENTALS

18.3.1 Process Overview

The typical ETO sterilization processes in contemporary use are defined by their extensive usage for medical devices and are minimally adapted for other applications. Delivering a lethal process requires that the important parameters be delivered consistently across the entire load:

- ETO concentration—the optimal range of concentrations for ETO falls between 300 and 1,200 mg/L. Higher concentrations may be increasingly lethal; however, the cost of materials and extended aeration times because of increased adsorption by materials can offset the benefits of more rapid kill.
- Temperature—these can range from ambient conditions (20°C–25°C) to as high as 60°C. The rate of kill is believed to double with each 10°C increase in process temperature.
- Humidity—the presence of moisture plays a vital role in ETO lethality, as moisture is believed essential to absorption of the gas by spores. Humidity levels of <30% should thus be avoided. A RH of 60% is considered optimal; however, to ensure adequate moisture throughout loads containing considerable corrugate and paper, sterilizing conditions typically target a higher level of moisture.
- Time—the duration of the dwell period is the easiest to control and is determined experimentally to ensure sufficient lethality is delivered once the other parameters have been defined.

The conventional process includes the following major steps:

- Preconditioning—The load items are exposed to elevated humidity (90%–95%) to increase the moisture levels on the device surface prior to introduction into sterilizer. The preconditioning makes the microorganisms more susceptible to ETO. In temperate climates, this step is essential to ensure a consistent process despite seasonal variations in humidity that would otherwise alter process lethality. This step is performed in an area specifically designed (and qualified) for this purpose. The preconditioning can also serve to heat the load items to the desired temperature when a cycle using higher temperatures is performed.
- Reconditioning—After transfer from the preconditioning area to the sterilizer, reconditioning in the sterilizer may be performed to re-humidify those peripheral portions of the load that may have dropped in humidity while in transit. This would be accomplished by low-pressure steam injection into the sterilizing chamber after loading. This step is optional as the loss in humidity may not be significant.
- Evacuation—To mitigate the explosive potential of ETO sterilization, air is removed from the chamber. Unlike steam sterilization, in which air removal is sometimes carried to excess, the removal of air must be tempered with the desire to maintain adequate humidity within the load. A single pre-vacuum is typical as desiccation of the load should be avoided. The vacuum is replaced with nitrogen as the maximum oxygen level must be maintained at <3% when ETO is present to minimize the explosion potential.
- Humidification—The maintenance of moisture levels within the load throughout the cycle may be necessary.
- Charge—100% ETO gas is introduced to attain the desired concentration. To minimize ETO emissions and increase safety, the internal pressure is maintained below ambient for the remainder of the process. The final pressure adjustment is typically made with nitrogen.
- Exposure—The lethal conditions are maintained for the preestablished process dwell period at the target concentration, RH, and temperature.
- Exhaust—The chamber is evacuated to reduce the ETO level prior to opening. The exhaust gases are treated to minimize ETO emission to the atmosphere. The exhaust cycle commonly includes multiple alternating pulses of nitrogen and vacuum to reduce the explosion risk while ETO is still present. The last vacuum is broken with air.
- Unloading—The sterilizer is unloaded and the items transferred to the post-conditioning area. Operating personnel performing this task may need to wear personal protective equipment to minimize their exposure to trace amounts of ETO diffusing from the materials.
- Post-Conditioning—The load is aerated so that levels of ETO (as well as the common process byproducts ethylene chlorohydrin and ethylene glycol) can be reduced to levels safe for the intended use. The United States Food and Drug Administration (FDA) established its expectations for these residuals in 1978.[6] This remains the only published expectation with respect to acceptable residual levels. The aeration area is typically supplied with 100% fresh air to aid in residue removal.

The in-chamber portion of a typical ETO sterilization cycle is depicted in Figure 18.3.

18.3.2 History

When first introduced in the 1950s as a sterilization method, ETO was largely supplied as a liquid under pressure in cylinders combined with either carbon dioxide or various hydrochlorofluorocarbons. Changes in cylinder weight were used to establish the amount of ETO introduced into the chamber. The mixtures of gases potentially result in variations in ETO concentration as the liquid phase evaporates. The concentration uncertainties and the growing global environmental

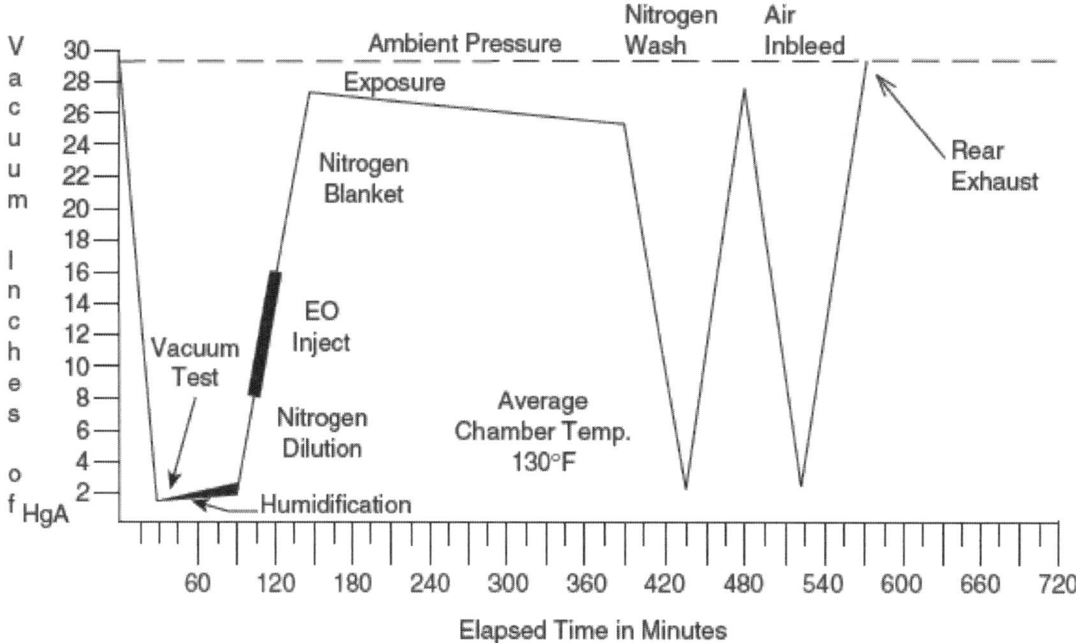

FIGURE 18.3 Typical ethylene oxide (ETO) sterilization cycle.

concerns regarding hydrochlorofluorocarbon (HCFC) emissions led to a gradual shift to the use of 100% ETO.

The growing awareness of ETO's safety and toxicity concerns in the 1980s resulted in changes of a different sort (see next section). Pharmaceutical manufacturers that once employed ETO in-house for equipment and other sterilization gradually changed their practices. Equipment sterilization was changed to steam-in-place, and other ETO sterilization processes were shifted to contractors.

18.3.3 Environmental and Patient Safety Considerations

ETO is a versatile and effective sterilant, which has enabled it to stay in widespread use despite the numerous hazards associated with its usage. It is a confirmed carcinogen and mutagen and restrictions on its residuals in pharmaceutical products were drafted by the FDA in the 1970s.[7,8] The safety concerns and residue 1978 limits prompted many manufacturers to outsource their ETO sterilization to contract sites who were more willing to implement suitable safety and environmental controls given their higher utilization. Although the 1978 limits were withdrawn with the implementation of ISO 10993–7, which focused on residuals in medical devices, they remained the primary guidance for drugs and biologicals as they are linked to patient dose as opposed to the mass-related limits in ISO.[9] Safety issues extend beyond sterilized product residuals as workers must also be protected from exposure. The United States Occupational Safety and Health Administration (OSHA) requirements define a maximum of 5 ppm ETO in air for 15 minutes and NMT 1 PPM of ETO over an 8 hour period.[10] In 2019, in response to problems with ETO emissions detected in residential areas, the US Environmental Protection Agency (EPA) and the FDA launched a joint initiative to reduce the usage of ETO sterilization within the United States.[11] This has resulted in the closure of ETO contract sterilization facilities and potential shortages of devices and drugs. The FDA has offered innovation grants to firms seeking to reduce emissions and implement effective alternatives to ETO.[12] The growing environmental and safety concerns will undoubtedly have an adverse effect on ETO sterilization use in the future; however, the barriers associated with its replacement are numerous such that continued usage can be anticipated for some years to come.

18.3.4 Material Effects

Sterilization processes are designed to kill microorganisms, and as such, they utilize conditions that may be destructive of essential material properties. ETO processes are not exempt from this phenomenon, and material evaluation is required customarily at conditions exceeding those utilized for routine sterilization. ETO and its degradation products Ethylene Glycol and Ethylene Chlorohydrin tend to remain on or within the materials post-processing, presenting a potentially toxic effect, and the amounts of residuals are closely regulated. Consideration of each of these possible adverse consequences must be an integral part of process selection, cycle development, and process validation.

18.3.5 Product Configuration/Packaging for ETO Sterilization

Unlike thermal and radiation processes, ETO must be able to reach the target surface, which can present difficulties with some materials. Sealed containers and installed syringe plungers are examples of items that preclude the use of ETO sterilization alone. Complex medical devices or drug-delivery systems may require extreme lengthy sterilization cycles. The items to be ETO sterilized should be designed to allow reasonable access by the gas and humidity to all target areas.

The primary package commonly used for ETO sterilization of medical devices is often a rigid plastic tray with a Tyvek lid. These are placed in individual folding cartons (IFCs) with package inserts. Multiple IFCs are placed in a corrugated shipper, with multiple shippers on a pallet. Penetration of the gas and humidity must reach target surfaces within all of that material. The package system components must permit penetration of the sterilizing gas and humidity during the process and allow the diffusion of residual gas, moisture, and breakdown residues from the package while still affording a microbial barrier over the shelf life of the component.

18.3.6 Preconditioning

The use of a separate preconditioning room (where sufficient product volume justifies it) is recommended to ensure that materials delivered into the sterilizer are at a constant condition of temperature and RH. Without a separate environmental room, the load will be subject to the normal fluctuations in conditions evidenced in the environment. This will include lower temperatures than are optimal for effective sterilization, and more importantly, lower humidity levels within the load (especially during the winter months). The proper conditions for sterilization can be more rapidly achieved in the sterilizer if the load has been subjected to prior treatment in a preconditioning room. These rooms will typically maintain 35–40°C and 90%–95% RH year-round. They are provided with recorders for temperature and RH. The necessary dwell time in the room can be determined by distribution and penetration mapping.

18.3.7 Process Equipment

ETO sterilization is ordinarily carried out in jacketed chambers much like those utilized for steam sterilization. The safety and toxicity concerns with ETO are such that contract sterilization firms provide the majority of the global capacity. There are instances where ETO has been used for closed vacuum/pressure rated systems such as lyophilizers, however this usage is waning. To ensure greater process reliability, external and/or internal mixing may be utilized to enhance the uniformity of the ETO, RH, and temperature throughout the sterilization chamber. This recirculation can be an aid in aeration of the chamber post-cycle. The jacket provides improved temperature control, whereas the pressure (and vacuum-rated) chamber serves to contain the ETO within the equipment during the process. The process is executed by a control system that provides sequencing, regulation of process parameters, and documentation. Contemporary control systems for sterilization systems are electronic, either programmable logic controllers (PLCs) or minicomputers. These systems incorporate various features including operator interface, recipe management, process execution and control capability, documentation, and interfaces with surrounding systems. The control system is vital to sterilization success. A well-designed control system facilitates operation of the system and is essential to maintaining a compliant sterilization process. Its importance cannot be overstated. It is critical for providing the control necessary to support and maintain a validated sterilization process.

ETO gas is used for sterilization of materials and equipment liable to damage by other sterilization methods. The polymeric materials commonly used in medical devices can be difficult to sterilize by other means. When final packaged for delivery into operating and other critical settings, the medical device's primary packaging must be sterile as well.[13,14] In the medical device industry, it is common for multiple pallets of shipping containers containing multiple peelable trays containing the medical device to be sterilized with ETO. ETO's excellent ability to penetrate corrugate and permeable materials enables these processes to be successful. None of the other commonly available sterilizing gases (O_3, ClO_2, and NO_2) can match ETO's ease of penetration. ETO behaves as a true gas and will not condense under the typical sterilizing conditions. Sterilization efficacy is commonly enhanced by pre-humidification of palleted load items prior to sterilization at 90%–95% RH. After transfer of the load from the pre-humidification area to the sterilizer chamber, re-humidification may be necessary to ensure that the outermost portions of the load are sufficiently moistened. The control systems on the sterilizer execute the process and regulate gas concentration, RH, and temperature during the cycle to provide consistent lethality, as changes can alter the process lethality. Humidification in the chamber (pre- and mid-cycle) is accomplished by clean steam injection into the sterilizing chamber. Single-point monitoring of the gas concentration, RH, and temperature can provide adequate process control over the sterilization process. Despite this seemingly minimal monitoring, regulatory approval for parametric release for ETO sterilization is widespread.

ETO is a powerful oxidizing gas that kills microbes by chemical reaction with various sites in microorganisms, primarily those having—NH_2, —SH, —COOH, and —CH_2OH groups.[15] Microbial kill with ETO approximates first-order kinetics and is related to the provided gas concentration, RH, and process temperature.[16] Sterilization methods for ETO for pharmaceutical process applications ordinarily follow medical devices practices because of the extensive experience with ETO for those materials. ETO sterilization is effective across a wide range of conditions: gas concentration (300–1,000 mg/L), RH (35%—85%), and temperature (20°C—65°C), although the usual processing ranges are somewhat narrower. ETO is an extremely potent material and has been identified as being mutagenic, carcinogenic, and neurotoxic, as well as being highly explosive.[17] Trace residuals from ETO sterilization are associated with human toxicity so careful design of the equipment, facility, and post-sterilization aeration are essential for safe use. For these reasons, internal usage within operating companies has decreased. There are a number of contract firms providing ETO sterilization that have invested in the necessary controls to ensure both worker and patient safety, and these provide most of the available industrial capacity for ETO sterilization.

As ETO processes are so extensively utilized for medical devices, the cycle development and validation approaches are largely tailored to the specific requirements of their sterilization.

The pre-process humidification ensures that adequate moisture is present on the surface of the materials for effective kill throughout the load. Post-processing aeration chambers are utilized with ETO to reduce residuals to safe levels after exposure. ETO sterilization processes introduce essentially all of the gas at the start of the process, and only minor adjustment is performed during the exposure to maintain constant pressure. Humidity is introduced using clean steam to the chamber pre-exposure for reconditioning after transfer, and adjustments may be required through the end of the exposure period.

ETO process control, like all sterilization processes, relies on a combination of physical measurements and biological assessments. BI kill in conjunction with data from the sterilizer instrumentation is utilized in evaluating process effectiveness. A lethality model has been proposed that mimics those utilized for steam and dry heat.[18] Its broader adoption by ETO practitioners is possible, as it simplifies lethality confirmation.

The extensive experience with ETO in medical devices has allowed firms to implement parametric release in lieu of sterility testing of ETO-sterilized materials. Parametric release replaces sterility testing with a defined set of requirements derived from the initial validation exercise that must be satisfied in conjunction with the execution of each subsequent sterilization cycle.[19,20] Submission to regulatory agencies is required prior to implementation and must be supported by comprehensive data derived from prior practice. Once implemented, the user is obligated to utilize parameter evaluation exclusively. In conjunction with parametric release, parallel testing with BIs is commonplace. The preferred BI is *Bacillus atrophaeus* (ATCC 6633) on paper strips. This microorganism was at one time considered *Bacillus subtilis* var. *niger*, and older references may still identify it as such.

18.3.8 Post-Process Aeration

It is generally necessary to fully quarantine the load after sterilization for a period of time to minimize worker exposure to residual sterilizing agent. Most loads will continue to off-gas quantities of ETO during this period, which can affect worker safety. To reduce the effect of this off-gassing, the load is commonly placed into an aeration chamber providing substantial fresh air changes, which serves to isolate the load until the residual levels are no longer toxic. The aeration room is temperature controlled, vented to an emission control system (if needed), and should be at a lower static pressure than the surrounding work environment.

18.3.9 Process Control Devices

The review and acceptance of individual sterilization process records is a compliance requirement to assure the process conforms to expectations. In ETO sterilization, the recorded process data is often that generated by instruments located within the sterilizing chamber, which cannot be directly related to sterilizing conditions at the target location. In the sterilization of medical devices, there can be numerous layers of packaging between the target surface and the chamber, making it difficult to confirm sterilization efficacy solely on chamber measurements. The Process Control Device (PCD) is used in the device industry for this purpose. BIs are one form of PCD; however, they require incubation before final results can be obtained. ISO 11140–1 "Sterilization of Health Care Products-Chemical Indicators-Part 1: General Requirements" describes the various types of PCDs and their application.[21] There are six classes of PCDs that provide varying amounts of information regarding the sterilization process. The selection of suitable PCDs for a specific application is an integral part of the cycle development process. The use of more than one type may be appropriate: simple PCDs on the exterior of a pallet would be used to confirm exposure, whereas more sophisticated PCDs would be used internally to confirm sufficient lethality has been delivered at hard to sterilize locations. The PCDs used for agent penetration are expected to emulate the ability of ETO to reach the least accessible parts of the medical device.

18.4 VALIDATION METHODS

The performance qualification or "validation" activity has been described as documentation that the process or product conforms to expectations as determined through independent parameter measurement and/or intensive sampling or challenge. It is the focus of regulatory attention for any sterilization process. It is common practice in performance qualification to utilize "worst-case" challenges in validation, and that is most prevalent with sterilization processes. Typical "worst-case" challenges for sterilization processes include reducing the process (set point) temperature, reducing the cycle dwell time, reducing both time and temperature, reducing the agent concentration, and using resistant biological challenges as bioburden surrogates. More detailed information on the expected practices can be found in the myriad of industry and regulatory publications on this subject.[22,23]

Historically, gas sterilization processes have been validated using the half-cycle approach, which uses conservative assumptions about the microbial resistance and number of bioburden microorganisms and was originally developed for use with ETO.[24] Early in the development of ETO sterilization, accurate information on gas concentration, RH, and temperature was largely unavailable, so the half-cycle method was utilized as a "worst-case" approach. D-value determination was not a concern given the difficulties in parameter measurement. A half-cycle approach was selected that avoided physical measurements and relied solely on the kill

TABLE 18.1
Half-Cycle, Full Cycle, and Materials Cycle Comparison

Cycle	Concentration	RH	Temperature	Time
	PPM	%	°C	Hours
"Half-cycle" process	600–650	70–75	30–35	4
"Full-cycle" process	600–650	70–75	30–35	8
Material evaluation cycle	600–650	70–75	30–35	>8

Abbreviation: RH = relative humidity.

of a high population of a resistant BI microorganism. Cycle development was extremely simple. The half-cycle approach mandates a minimum sterilization dwell period that destroys not less than 10^6 spores of a resistant BI. In routine operation, the process dwell period is doubled (thus the term half-cycle) and supports a minimum PNSU of 10^{-6}. The other process parameters are held constant. Table 18.1 includes data outlining how typical half-cycle validation cycle parameters relate to the full cycle condition. A materials evaluation cycle that is longer than the routine cycle is sometimes used to evaluate the effect of extended ETO exposure of the materials in the load.

The half-cycle method as utilized for gas sterilization is graphically depicted in Figure 18.4. The half-cycle method does not rely on the resistance of the BI (as a surrogate for the bioburden), because complete destruction of the indicator is required in the "half cycle". This method lacks the scientific rigor of those used for moist heat, which would accept the ETO "half cycle" as equivalent to "overkill". The half-cycle method is extremely conservative and given the limited analytics available when ETO sterilization was introduced provided a robust cycle albeit one that was not efficient. Determination of the biological indicators' D-value at the chosen parameters requires additional effort and was largely ignored for many years. Half-cycle approaches are inherently robust, and little effort was initially made to optimize the process dwell period when it will be arbitrarily doubled in routine use anyway. The half-cycle method evaluates only the impact of dwell time, assuming that the effect of lethality of variations in the other essential parameters such as gas concentration, RH, and temperature can be ignored. This is a severe limitation of the half-cycle method.

A refinement of the half cycle was D-value determination for the BI in order to estimate lethality more accurately (see the chapter on "Microbiology of Sterilization Processes"). This would allow more precise determination of the "half-cycle" dwell period necessary to inactivate the BI and thus provide some estimation of delivered lethality. As medical device complexity increased, this became increasingly important to ensure that the hardest to reach locations were being appropriately sterilized.

Another method suited for sterilization validation is a bracketing approach that better supports the extremes of the operating ranges for the critical process parameters.[25] In the bracketing approach, a cycle with lower concentration, lower RH, and a shorter dwell period is confirmed by microbial indicator destruction using what are less lethal conditions. This would parallel the cycle development effort needed to establish BI kill as in the "half cycle". Material effects are evaluated in a cycle employing a higher concentration, higher

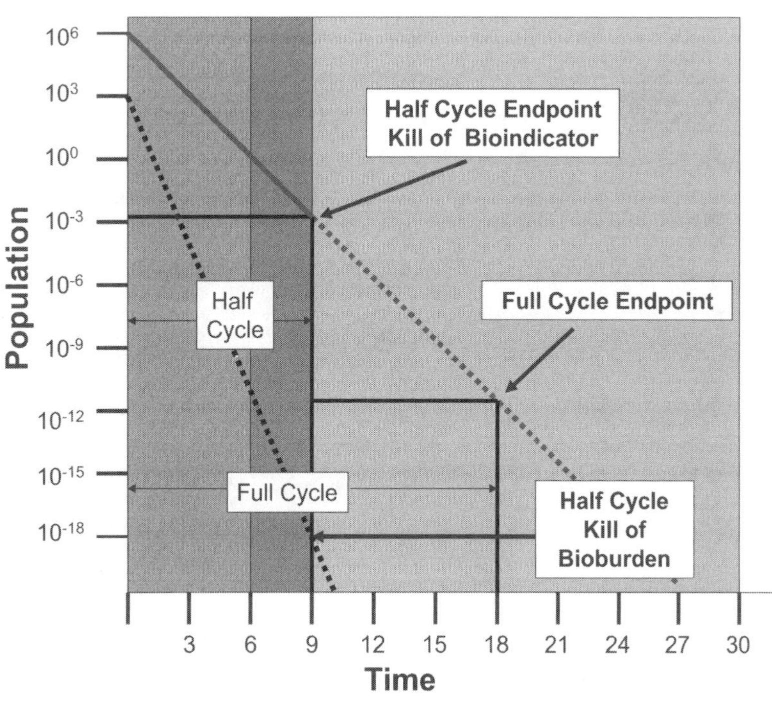

FIGURE 18.4 Half-cycle method.

TABLE 18.2
Bracketing Approach Cycle Parameters

	Concentration	RH	Temperature	Time
	PPM	%	°C	Hours
"Worst-case" sterilization	600–650	70–75	30–35	4
Routine process	650–700	75–80	35–40	6
"Worst-case" material effect	700–750	80–85	40–45	8

Abbreviation: RH = relative humidity.

RH, and a longer dwell period where the adverse impact is believed to be greater (Table 18.2).

Routine operation of the system utilizes conditions that fall between the process extremes that have been evaluated (Figure 18.5).

The half-cycle and bracketing approaches are fully compatible with ETO sterilization, giving the practitioner a choice of methods. The half-cycle method is by far the more widely used method, whereas the bracketing method will typically result in a shorter overall cycle time. The material effects should be evaluated at the extreme conditions of lethality, which result in exposure of the materials to the highest concentration of agent, higher humidity, and longer exposure time.

ETO sterilization and other non-moist heat sterilization processes are restricted in their ability to estimate delivered lethality. Aside from a few common BI microorganisms, the resistance of other microorganisms to the process is largely unknown and thus estimations of PNSU are indeterminate. Additionally, the multiple factors (gas concentration, RH, and temperature) influencing ETO kill make it difficult to determine how much lethality is actually delivered in a given cycle.

Regardless of the validation method utilized, there are common elements in all validation efforts:

- Equipment qualification—The equipment utilized for the sterilization process (pressure vessel) as well as any rooms utilized for pre- or post-cycle processing must be fully documented with respect to installation details and operational characteristics. Equipment qualification serves as the basis for change control for the physical equipment. This effort must of course include calibration of instrumentation and qualification of the control system.
- Empty chamber/vessel parameter distribution—Parameter measurement within the sterilization chamber/vessel is appropriate. This can be either single or multiple point, with the cost of measurement an important consideration. The goal in this exercise is to be able to correlate the values obtained with the routine monitoring location(s). Where the vessel is mixed during the process (as is almost universally desirable), this study confirms the effects of that mixing. Overmixing in these processes is not a consideration, as additional mixing can only improve the uniformity of the process parameters. BIs are not required in the evaluation of the empty chamber/vessel uniformity.
- Component and load mapping—These activities are not a part of ETO sterilization validation, because

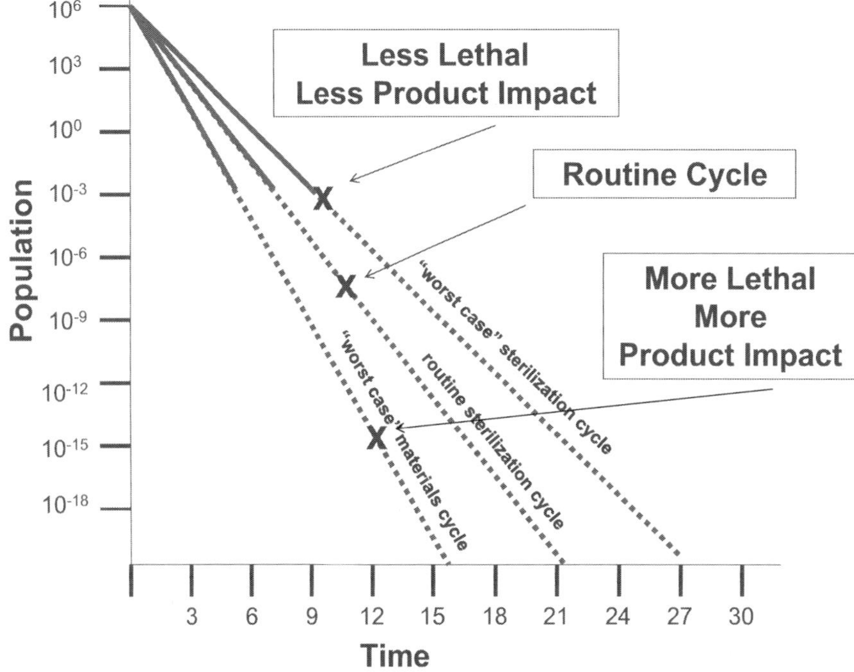

FIGURE 18.5 Bracketing approach visualization.

sampling systems placed within the load items would alter gas penetration. This evaluation is best provided by passive BIs placed within the load items. The use of physical/chemical indicators placed within the items can be used to support this effort, but as there are no available chemical integrators, this practice is of limited use.

- Biological indicators—The use of a BIs for initial validation and routine process control is an integral part of validation efforts for most ETO sterilization. There are a few exceptions as parametric release has been successfully accomplished by numerous practitioners. The BI serves as a "worst-case" surrogate for the bioburden present in routine operations. BIs are conventionally spores of a microorganism (most often a *Bacillus* species) chosen specifically for its greater resistance to the ETO sterilizing process than the expected bioburden. Inactivation of the BI during the validation establishes the lethality of the process across the items being sterilized. The measurement of physical conditions during the validation exercise and routine operation allows for some estimation of process lethality. Spore challenges may be either a spore strip positioned within the load or less commonly spores inoculated directly on a load item. Inoculated items should have their population determined by the end user, and when possible, their resistance to the sterilization process confirmed. Indicators are placed among the load items at the locations believed to be hardest for the penetration of ETO and RH.
- Process control devices—As described previously, PCDs can serve as useful and rapid indicators of process lethality. When used to merely confirm whether items have been exposed or not, placement on the exterior of the package or pallet is sufficient. When used for penetration confirmation, their locations and response should match that of BIs placed in nearby locations.
- Process confirmation/microbiological challenge—The core of the validation activity is the confirmation of acceptable process parameters and inactivation of the microbial challenge. Proof of cycle efficacy is provided in replicate studies in which the BIs are killed and physical measurements are taken as documentation. Differences in resistance between the BI and the expected bioburden present are exploited in the validation for ease of validation and routine process control.

18.5 ROUTINE PROCESS CONTROL

Sterilization processes must be subject to routine controls that support the efficacy of the cycle over time. Validation is not a one-time activity project, but an integral part of a Current Good Manufacturing Practices (CGMP)-compliant facility that must be sustained over the useful life of the facility and its products.[26] Control over sterilization processes is commonly achieved through practices defined specifically for that purpose including Calibration of Instruments, Physical Measurements of Process Parameters, Use of Physical Integrators/Indicators (and in some cases BIs), Change Control, Preventive Maintenance, and Periodic Reassessment. In the absence of approvals for parametric release, BIs are utilized for routine release of each sterilization load along with documentation from the control system.

18.6 CONCLUSION

This chapter provides an overview of ETO sterilization and validation. This chapter has broadly outlined the primary considerations with respect to this important sterilization process. The reader is encouraged to review the substantially larger body of knowledge available on ETO sterilization processes before implementation.

NOTES

1. Pflug IJ, *Microbiology & Engineering of Sterilization Processes*, 8th edition. Minneapolis, MN: University of Minnesota, 1995.
2. Bigelow WD, The Logarithmic Nature of Thermal Death Time Curves. *Journal of Infectious Diseases* 1921;29: 528–536.
3. Bigelow WD, The Logarithmic Nature of Thermal Death Time Curves. *Journal of Infectious Diseases* 1921;29: 528–536.
4. Agalloco J, Understanding Overkill Sterilization: Putting an End to the Confusion. *Pharmaceutical Technology* 2007;30(5): S18–60–76.
5. ISO 11135, Sterilization of health-care products—Ethylene oxide—Requirements for the development, validation and routine control of a sterilization process for medical devices, 2008.
6. FDA, Ethylene Oxide, Ethylene Chlorohydrin and Ethylene Glycol: Proposed Maximum Residue Limits and Maximum Levels of Exposure. *Federal Register* 1978;43(122): 27473–27483.
7. DHSS, *National Toxicology Program, Report on Carcinogens*, 14th edition. http://ntp.niehs.nih.gov/go/roc14, 2016.
8. 21 CFR 211.70, Ethylene Oxide, Ethylene Chlorohydrin, and Ethylene Glycol: Proposed Maximum Residue Units and Maximum Levels of Exposure. *CFR* 1978;43(122): 27474–27482.
9. ISO 10993–7, *Biological Evaluation of Medical Devices: Part 7: Ethylene Oxide Sterilization Residuals*. Arlington, VA: ANSI/AAMI/ISO, 1995.
10. OSHA, Ethylene Oxide Standard: 29 CFR 1910.1047.
11. Crotti N, *Federal Agencies Will Control the Fate of Medtech's Most-Used Sterilization Method*. www.medicaldesignand-outsourcing.com/federal-agencies-will-control-the-fate-of-medtechs-most-usedsterilization-method/ Medical Design and Outsourcing, November 15, 2019.
12. Sookne K, FDA Comments on Potential Device Shortages in the Face of ETO Facility Interruptions/Closures. *Healthcare Packaging*, October 25, 2019.
13. Agalloco J, *Course Notes, Sterilization: Processes & Validation*. Princeton, NJ, 2016.

14. Joyslyn L, Gaseous Chemical Sterilization. In *Disinfection, Sterilization and Preservation*, 5th edition. Block SS, ed. Philadelphia, PA: Lippincott Williams & Wilkins, 2001.
15. Sintin-Damao K, Other Gaseous Sterilization Methods. In *Sterilization Technology: A Practical Guide for Manufacturers and Users of Health Care Products*. Morrissey R, Phillips GB. New York: Van Nostrand Reinhold, 1993.
16. Gillis JR, Mosley G, Validation of Ethylene Oxide Sterilization Processes. In *Validation of Pharmaceutical Processes*. Agalloco JP, Carleton FJ, eds. New York: Informa USA, Inc., 2007.
17. Burgess D, Reich R, Industrial Ethylene Oxide Sterilization. In *Sterilization Technology: A Practical Guide for Manufacturers and Users of Health Care Products*. Morrissey R, Phillips GB, eds. New York: Van Nostrand Reinhold, 1993.
18. Mosley GA, Gillis J, Whitbourne J, Calculating Equivalent Time for Use in Determining the Lethality of EtO Sterilization Processes. *Medical Device and Diagnostic Industry* 2002;24(2): 54–63.
19. EMEA, Guideline on Parametric Release, EMEA/CVMP/QWP/339588/2005, 2006.
20. PI-005–2, Guidance on Parametric Release. *Pharmaceutical Inspection Cooperation Scheme*, 2004.
21. ISO 11140–1, Sterilization of Health Care Products-Chemical Indicators-Part 1: General Requirements: 2005 (R) 2010.
22. Hugo WB, Russell AD, Types of Antimicrobial Agents. In *Principles and Practices of Disinfection, Preservation and Sterilisation*. Russell AD, Hugo WB, Ayliffe GAJ, eds. London: Blackwell Scientific Publications, 1982.
23. Hugo WB, Russell AD, Types of Antimicrobial Agents. In *Principles and Practices of Disinfection, Preservation and Sterilisation*. Russell AD, Hugo WB, Ayliffe GAJ, eds. London: Blackwell Scientific Publications, 1982.
24. Krisher AS, Siebert OW, Materials of Construction. In *Perry's Chemical Engineers' Handbook*, 6th edition. Perry RH, Green D, eds. New York: McGraw Hill, 1984.
25. FDA, *Guidance for Industry for the Submission of Documentation for Sterilization Process Validation in Applications for Human and Veterinary Drug Products*. Rockville, MD: Center for Drug Evaluation and Research: Center for Veterinary Medicine, 1994.
26. Agalloco J, The Validation Life Cycle. *Journal of Parenteral Science and Technology* 1993;47(3): 142–147.

19 Validation of Chlorine Dioxide Sterilization

Mark A. Czarneski and Paul Lorcheim

CONTENTS

19.1 INTRODUCTION

Chlorine dioxide (CD) is a highly effective sterilizing agent that has many applications in the Life Science research, food facilities, health care, medical device, and pharmaceutical industries including the sterilization of components, environments, and medical devices. It has key benefits and characteristics that make it extremely effective and well suited for use in component and device sterilization, small chamber (isolator) decontamination, room/suite decontamination, and facility/building decontamination. It is a true gas at typical use temperatures and therefore can penetrate into hard-to-reach areas such as those found in needles, lumens, or devices with complex geometries. It is efficacious at ambient room temperatures, so it is excellent for temperature-sensitive materials and devices. It has a yellowish-green color, which allows its concentration to be precisely monitored and controlled by ultraviolet visible (UV-VIS) spectrophotometry. This ensures a repeatable and robust cycle by providing tight process control from beginning to end.

19.2 HISTORY/BACKGROUND

Chlorine dioxide was recognized for its disinfecting properties in the 1930s by Schaufler[1] (1933) and by Kovtunovitch and Chemaya[2] (1936). Chlorine dioxide was first prepared in 1802 by Chenevix[3] and later independently prepared and verified in 1811 by Sir Humphrey Davy.[4] Chlorine dioxide is primarily used in the pulp and paper industry, accounting for 95% of all chlorine dioxide produced. About 5% of large water-treatment facilities (serving more than 100,000 persons) in the United States use chlorine dioxide to treat drinking water. An estimated 12 million persons may be exposed in this way to chlorine dioxide and chlorite ions. It is also estimated that there were 743,015 pounds (337,026 kg) of chlorine dioxide released into the atmosphere from over 100 manufacturing, processing, and waste disposal facilities in 2000.[5]

The United States Food and Drug Administration (FDA) permits chlorine dioxide to be used to wash fruits and vegetables according to 21 CFR Part 173.300 "Chlorine Dioxide".

DOI: 10.1201/9781003163138-19

Additionally, chlorine dioxide has several applications approved for generally recognized as safe (GRAS) along with several approved food contact notices (FCNs) or food contact substances (FCSs). Section 409 of the FD and C Act defines an FCS as any substance that is intended for use as a component of materials used in manufacturing, packing, packaging, transporting, or holding food if such use of the substance is not intended to have any technical effect in such food. Chlorine dioxide is also approved for use in organic food and organic food processing according to 7 CFR part 205, Subchapter M—"Organic Foods Production Act Provisions". Chlorine dioxide gas has also been shown to be more effective than chlorine dioxide dissolved in aqueous solution.[6,7] Studies have shown that with equal concentrations (3 mg/L in gas and 3 mg/L in liquid), chlorine dioxide in the gas phase had a 7.4 log reduction of *L. monocytogenes* on peppers compared with 3.6 log reduction with 3 mg/L of gas dissolved in solution.

Chlorine dioxide is a single-electron-transfer oxidizing agent with a chlorine-like odor. This odor is the only similarity between chlorine dioxide and chlorine. Chlorine dioxide is not sufficiently stable to be bottled and shipped, so it must be generated at the point of use. Chlorine dioxide can be generated in a variety of methods as a gas dissolved in liquid or as a dry gas. The dry gas method generates chlorine dioxide gas by passing chlorine gas through solid sodium chlorite, which generates a pure ClO_2 gas (>99%)[8] with no byproducts:

$$Cl_{2(g)} + 2NaClO_{2(S)} \rightarrow 2ClO_{2(g)} + 2NaCl_{(S)}$$

Chlorine Dioxide chemical properties can be found in Table 19.1.

Other generation methods involve using an acid mixed with a sodium chlorite solution to generate chlorine dioxide and other byproducts that can cause corrosion.

Sodium Chlorite—Hydrochloric Acid method:

$$5NaClO_2 + 4HCl \rightarrow 5NaCl + 4ClO_2 + 2H_2O$$

Chlorite—Sulfuric Acid method:

$$8ClO_2- + 4H_2SO_4 \rightarrow 4ClO_2 + 2HClO_3 + 4SO4_2- + 2H_2O + 2HCl$$

Sodium Hypochlorite method:

$$2NaClO_2 + 2HCl + NaOCl \rightarrow 2ClO_2 + 3\ NaCl + H_2O$$

Chlorine dioxide is stable in concentrations of <10% in air at atmospheric pressure. The gas can become unstable at high concentrations (>10%) by contact with substances that catalyze its decomposition such as organic materials, phosphorus, potassium hydroxide, sulfur, mercury, and carbon monoxide, as well as exposure to heat.[10] When this explosive decomposition occurs, the volume increase is small, such that the decomposition is referred to as "puffing".[11] When using chlorine dioxide gas for disinfection/decontamination/sterilization the concentrations are much lower, typically 0.04%–1.1% (1–30 mg/L). The typical concentration for large-volume decontamination is 1mg/L with isolators having a concentration of 5 mg/L and vacuum sterilization runs typically occurring at 30 mg/L. In more recent studies in 2009, it was confirmed that the lower limit for decomposition was 9.5% (262 mg/L), confirming that there is no puffing or explosion hazard at these conditions.[12]

TABLE 19.1
Chlorine Dioxide Properties

Chemical Formula:	ClO_2
Molecular Weight:	67.45 g/mole
Melting Point (°C):	−59
Boiling Point (°C) at 100% concentrations:	+11
Boiling Point (°C) at isolator use concentrations:	−22
Boiling Point (°C) at room use concentrations:	−40
Density:	2.4 times that of air

Chlorine dioxide's method of microbial inactivation is different from that of chlorine (oxidation vs. chlorination), thus it is gentler on materials and provides a highly controllable and reproducible process. It does not react with organic materials to form chlorinated species or with ammonia to form chloramine. Additionally, chlorine dioxide is well suited for sterilizing components and medical devices because it is compatible with the many materials found in those components such as stainless steel, aluminum, glass, and most plastics.

Chlorine dioxide's method of kill is by oxidation and as such it can oxidize materials. It has a measured oxidation potential of 0.95V. This corrosion potential is lower than that of other common decontaminating/sterilizing agents such as hydrogen peroxide (1.78V), ozone (2.07V), sodium hypochlorite (1.49V), and peracetic acid (1.81V).

Stainless steel and most plastics (Teflon, KYNAR (PVDF), PVC, PE, PP) and gasket materials (silicone, EPDM, Buna, Viton, neoprene) that are commonly used in facilities and chambers have good material compatibility with chlorine dioxide gas. Aluminum will show signs of

$$O - Cl^{\bullet} = O$$

FIGURE 19.1 Structure of chlorine dioxide.[9]

oxidation over time, but if anodized, it does not. Chlorine dioxide gas will cause corrosion on ferrous metals if they are left unpainted, untreated, or uncoated. Electronics are acceptable as shown by Girouard (2016).[13] In other decontaminations of whole facilities, installed lab and manufacturing equipment were shown to be functional after being exposed.[14,15,16,17,18,19,20,21,22]

The rapid sterilizing activity of chlorine dioxide is present at relatively low gas concentrations of 1 to 30 mg/L compared with those of Ethylene Oxide (EtO), which requires higher gas concentration at ambient temperatures. Chlorine dioxide gas has lower ambient temperatures than steam, and it does not raise the temperature of the chamber.

19.3 EFFECTIVENESS

Chlorine dioxide is an oxidizing agent and reacts with several cellular constituents. Its most widely reported mechanism is the oxidation of the cell surface membrane protein[23,24] and free fatty acids.[25] Studies have shown efficacious results as a gas and in solution against a wide variety of microorganisms including viruses, bacteria, fungi, spores, and protozoa. Chlorine dioxide studies have shown it is capable of inactivating rotavirus, polioviruses, enteroviruses, and human immunodeficiency virus (HIV).[26,27,28,29,30,31,32,33] Chlorine dioxide was proven effective on major bacterial pathogens responsible for outbreaks in the food industry such as *Escherichia coli* O157:H7, *Listeria monocytogenes*, and *Salmonella enterica*[34] and also effective against various protozoal, fungal, and algal species, such as *Cryptosporidium parvum* oocysts, *Streptomyces griseus*, and yeasts[35,36,37,38,39]. Chlorine dioxide has also shown promising results against prions and prion-associated diseases.[40,41] The sterilant or sporicidal capabilities were initially demonstrated by Rosenblatt et al[42,43,44] and Jeng and Woodworth[45,46].

With *E. coli*, it was reported that chlorine dioxide causes surface damage and degradation followed by the damage of several inner cellular components.[47,48] Membrane damage has also been shown with bacterial spores including damage to the inner cell membrane, change in cell permeability, and interruption of germination of *B. subtilis* spores.[49] With *B. cereus*, similar damage was shown to cell walls noted as surface roughness and indentations.[50] With poliovirus, chlorine dioxide alters capsid proteins.[51] Chlorine dioxide does not act via chlorination and as such does not form trihalomethanes.[52] These reactions with organic matter have been studies by Gordon (1972),[53] Masschelein and Rice (1979)[54] and Aieta and Roberts (1985).[55]

Phenolic compounds are easily oxidized by chlorine dioxide[56], and it has been used to reduce the toxicity of chlorinated phenolic compounds.[57] Chlorine dioxide has also been shown to denature proteins[58,59] and has shown good results in inactivating beta-lactams from pharmaceutical production facilities.[60,61] This ability is significant because buildings that produce beta lactams cannot be used to produce other non-beta lactam products. The residues from beta lactams remain in a facility for long periods and can cause allergic reactions. Anaphylactic reactions to penicillin can be fatal. Beta lactam allergic reactions are the most common cause of adverse drug reactions mediated by specific immunological mechanisms.[62] The United States Centers for Disease Control and Prevention (CDC) states that 3%–10% of adults in the United States have had an allergic reaction to penicillin.[63] In other studies, chlorine dioxide has shown promising results in endotoxin reduction on various surfaces (304 and 316 stainless steel and titanium [anodized and un-anodized]).[64]

A series of square wave studies was performed in a two glove 23 ft^3 (0.65 m^3) flexible wall isolator to determine the effect of chlorine dioxide gas concentration on the inactivation rate of *Bacillus atrophaeus* spores. The D-value (the time at a specified chlorine dioxide gas concentration required to reduce the microbial population by 1 log or 90%) of *B. atrophaeus* spores on unwrapped paper carriers, when exposed to gas concentrations of 3 mg/L and 5mg/L, was determined using the Stumbo-Murphy-Cochran Method (Table 19.2 and Table 19.3). Each biological indicator (BI) was stored at 75% ±b2% relative humidity (RH) prior to entering the isolator and preconditioned in the isolator at 75% ± 2% RH for 30 ±1 minutes prior to exposure. The decline in %RH during the gas injection and exposure phases of the cycle was recorded for each of the D-value runs.

Data calculations for the 3 and 5 mg/L exposure concentrations utilizing the Stumbo-Murphy-Cochran are in Table 19.2 and 19.3. The results can be seen in Table 19.4.

TABLE 19.2
D-Value Determinations Using the Stumbo-Murphy-Cochran (SMC) Method (3 mg/L)

Gas Conc. (mg/L)	U	n	r	Nu	Log Nu	D-Value
3	21	10	0	N/D	N/D	N/D
3	24	10	1	2.30	0.362	3.92
3	27	10	1	2.30	0.362	4.41
3	30	10	2	1.61	0.207	4.78
3	33	10	6	0.51	–0.292	4.87
3	36	10	8	0.22	–0.651	5.05
3	39	10	9	0.11	–0.977	5.23
3	42	10	8	0.22	–0.651	5.89
3	45	10	10	N/D	N/D	N/D
					AVERAGE:	4.88 min

Note: When the number of sterile replicates (r) is 0 or 10 the D-value is not determined.

Stumbo-Murphy-Cochran formula: D-value = U/log No–log Nu

Key:

U = time in minutes

n = number of replicates tested

r = number of sterile replicates out of the number tested

Nu = natural log of *n/r* (ln *(n/r)*)

No = population of unexposed 81 (3.00 × 10^6 CFU/strip)

TABLE 19.3
D-Value Determinations Using the Stumbo-Murphy-Cochran (SMC) Method (5 mg/L)

Gas conc. (mg/L)	U	n	r	Nu	Log Nu	D-value
5	18	10	1	2.30	0.362	2.94
5	21	10	2	1.61	0.207	3.35
5	24	10	7	0.36	−0.448	3.46
5	27	10	9	0.11	−0.977	3.62
5	30	10	10	*N/D*	*N/D*	*N/D*
5	33	10	9	0.11	0.977	4.43
5	36	10	10	*N/D*	*N/D*	*N/D*
5	39	10	10	*N/D*	*N/D*	*N/D*
5	42	10	10	*N/D*	*N/D*	*N/D*
					AVERAGE:	3.56 min

Note: When the number of sterile replicates (r) is 0 or 10 the D-value is not determined.
Stumbo-Murphy-Cochran formula: D-value = U/log No–log Nu
Key:
U = time in minutes
n = number of replicates tested
r = number of sterile replicates out of the number tested
Nu = natural log of *n/r* (ln *(n/r)*)
No = population of unexposed 81 (3.00×10^6 CFU/strip)

TABLE 19.4
Results and Conclusion: D-Value vs. Chlorine Dioxide Concentration

Chlorine dioxide concentration (mg/L)	D-value (minutes)
3	4.88
5	3.56
10	0.75
20	0.27
30	0.12

Note: Spore strips were preconditioned for 30 minutes at 70% relative humidity and ambient temperatures.

19.4 SAFETY/TOXICITY

The Occupational Safety and Health Administration (OSHA) 8-hour Time-Weighted-Average (TWA) for chlorine dioxide is 0.1 ppm. The 15-minute Short-Time-Exposure-Limit (STEL) is 0.3 ppm. Chlorine dioxide is a respiratory/mucous membrane irritant. One of the important safety features of chlorine dioxide is that it has a 0.1 ppm odor threshold, which makes it self-alerting.[65] Most other sterilants are well over their STEL before they can be detected by the user's sense of smell. Because chlorine dioxide has such widespread usage in the water treatment and paper and pulp industries, there is a wide selection of environmental monitors and personnel safety badges available. Because of this broad usage, there have been numerous safety studies conducted both for environmental effects, inhalation, as well as ingestion. Because it is also used in the food industry for sanitization and disinfection, there are allowable limits from the US government for ingestion. The Environmental Protection Agency (EPA) has set the maximum concentration of chlorine dioxide in drinking water at 0.8 mg/L.[66]

Chlorine dioxide's special properties make it an ideal choice to meet the challenges of today's environmentally concerned world. Chlorine dioxide is an environmentally preferred alternative to EtO. The major concerns with EtO center on its flammability and high reactivity. Acute exposures to EtO gas may result in respiratory irritation and lung injury, headache, nausea, vomiting, diarrhea, shortness of breath, and cyanosis (OSHA Safety Fact Sheet).[67] Chronic exposure has been associated with the occurrence of cancer, reproductive effects, mutagenic changes, neurotoxicity, and sensitization (NIOSH 2019).[68]

Chlorine dioxide is used in large quantities in the water treatment and pulp and paper industries and can be used safely and effectively. Although chlorine dioxide is a sterilant and chemical disinfectant, it has not revealed clear evidence of other adverse health effects.[69] Chlorine dioxide is considered a mucous membrane irritant, and inhalation of excessive amounts can lead to pulmonary edema.[70]

Animal studies done with chlorine dioxide showed no significant association with any teratogenic potential effects of chlorine dioxide,[71,72,73] except for one study by Suh et al.[74] Data currently available support that chlorite, (the US-EPA considers chlorite and chlorine dioxide synonymous) is not considered carcinogenic,[75] and there is no published research related to chlorine dioxide carcinogenicity. Low concentration exposures (0.05 ppm and 0.1 ppm) of continuous exposure for 6 months showed no chlorine dioxide gas-related toxicity.[76] The rats showed no weight gain or changes in water or food consumption, and normal relative organ weight was observed. There were no changes in respiratory organs and no chlorine dioxide gas-related toxicity was observed.

Chlorine dioxide underwent a full review in 2006 in the Reregistration Eligibility Decision (RED) for Chlorine Dioxide and Sodium Chlorite (Case 4023). The US-EPA considers chlorine dioxide and sodium chlorite synonymous because both have the same toxicological end points. In the findings, the food quality protection safety factor was reduced from 10X to 1X. The food quality protection safety factor is for the protection of infants and children in relation to pesticide residues in food, drinking water, or residential exposures. This reduction was based upon a complete database for developmental and reproductive toxicity. Also in the RED, chlorine dioxide was designated with a toxic category of II for exposure by the oral route, a toxic category of III for skin toxicity, a toxic category of III for eye irritation, and a toxic category of II for inhalation toxicity. Toxicity Category I is considered DANGER, Toxicity Category II is WARNING, Toxicity Category III requires CAUTION, and Toxicity Category IV is safe.

19.5 KNOWN INCOMPATIBILITIES

Chlorine dioxide reacts with carbohydrates, such as glucose, to oxidize the primary hydroxyl groups first to aldehydes and then to carboxyl acids.[77] Ketones are also oxidized to carboxyl acids.[78] Although chlorine dioxide has "Chlorine" in its name, its chemistry is very different from that of chlorine. When reacting with other substances, it is weaker and more selective. Chlorine dioxide, as with other oxidizers as well as water, causes oxidation to uncoated ferrous materials as well as other materials subject to oxidation. Control of moisture during the decontamination process mitigates the oxidation potential. Chlorine dioxide gas is soluble in water and does not dissociate in water to form hydrochloric acid. It is easy to confuse chlorine dioxide and chlorine, but they are different and do not affect materials the same way.

19.6 STABILITY OF THE GAS

Chlorine dioxide is not a stable gas that can be produced, bottled, and shipped; it is typically produced by a system at the point of use. Chlorine dioxide exposure to light can lead to decomposition. The photochemical reaction is a homolytic fission of the Chlorine-Oxygen bond to form ClO• and O•.[79] When light catalyzes the reaction mechanism for decomposition of dry gaseous chlorine dioxide, this is postulated as:

$$ClO_2 + hv \rightarrow ClO\bullet + O\bullet$$
$$ClO_2 + O\bullet \rightarrow ClO_3$$
$$2ClO\bullet \rightarrow Cl_2 + O_2$$

When gaseous chlorine dioxide is exposed to light in the presence of moisture, a visible mist may form. This mist does not contain chlorine but rather a complex mixture of hypochlorous and other acids.[80] The following mechanism has been proposed for the photolytic decomposition of chlorine dioxide in the presence of moisture:[81]

$$ClO_2 + hv \rightarrow ClO\bullet + O\bullet$$
$$ClO_2 + O\bullet \rightarrow ClO_3$$
$$2ClO_3 \rightarrow Cl_2O_6$$
$$ClO\bullet + ClO_2 \rightarrow Cl_2O_3$$
$$CL_2O_6 + H_2O \rightarrow HClO_3 + HClO_4$$
$$Cl_2O_3 + H_2O \rightarrow 2HClO_2$$
$$2HClO_2 \rightarrow HClO + HClO_3$$

Because chlorine dioxide is a true gas and will not condense, the stability of chlorine dioxide as the sterilizing agent is enhanced over methods using vapors (mixtures of a gas with a suspended liquid). Because concentration can be easily monitored and controlled, the concentration is precisely maintained throughout the cycle.

19.7 CYCLE DESCRIPTION

All sterilization or decontamination cycles are similar and have the same general steps. There is a sterilant injection phase, then an exposure phase followed by a removal or aeration phase. For dry gasses such as chlorine dioxide, the cycle is similar to the EtO cycle such that an additional moisture conditioning phase is required. The typical chlorine dioxide cycle can be carried out at pressures from negative pressure (2 KPa) to slightly above atmospheric. Figure 19.2 shows an example cycle of an ambient pressure chlorine dioxide cycle.

Even though the general steps for sterilization are similar, each process has specific steps for sterilization. The chlorine dioxide gas cycle steps are as follows:

- Pre-Condition
- Condition
- Charge
- Exposure
- Aeration

19.7.1 PRE-CONDITION

Pre-condition is the first step of the chlorine dioxide cycle. At this point, the chamber should be leak tested. When using any sterilant, it is good practice to perform a chamber leak test prior to each decontamination cycle to ensure chamber integrity. For a vacuum chamber, vacuum is pulled down to a desired level, and then the chamber is held static for a period of time. The pressure difference from the beginning of the dwell time to the end is noted. If the pressure rise is not within acceptable parameters, the chamber must be properly sealed and retested before any sterilant is injected into the chamber. If ambient pressure chambers are used, the leak test pressurizes the chamber then monitors for pressure decay to be acceptable before continuing the cycle. Once the chamber has been leak tested, the chamber can be brought to the RH set point (60%—75%). Humidity can be generated by a variety of methods such as steam, atomization, etc. Steam offers the quickest, cleanest, and most efficient way to raise the humidity.

19.7.2 CONDITIONING

Once the humidity is at the proper level (60%–75%), the cycle can advance to the next step. Conditioning allows the load to pick up moisture. During the conditioning time, typically 30 minutes, the RH is monitored. If the RH drops by any significant amount (>5%), additional moisture is added to raise the RH. Once the conditioning time is completed, gas is then introduced into the chamber in the Charge phase. Product can be pre-conditioned with higher humidity prior to starting the cycle, which will shorten overall process time.

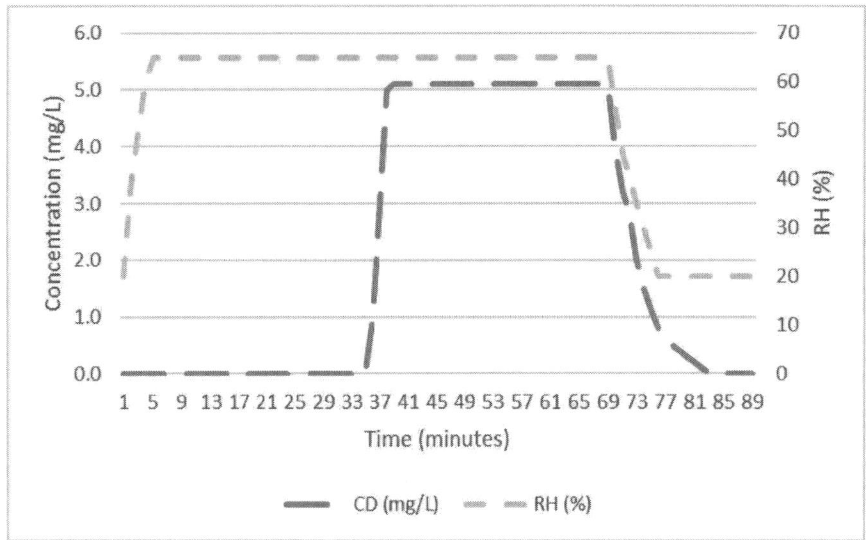

FIGURE 19.2 Cycle chart.

Note: CD = chlorine dioxide; RH = relative humidity.

19.7.3 Charge

During Charge or gas injection, chlorine dioxide gas is generated and introduced into the chamber to achieve the defined concentration. The target concentration is dependent on different factors: cycle time, cost, type of load, etc. If cycle time is extremely important, a higher concentration is selected to achieve a faster kill. At higher concentrations, the D-values are much shorter thereby shortening the overall cycle. If agent cost is the driving factor, then a lower concentration can be selected to preserve consumables, but the exposure time must be extended accordingly. Usually, a higher concentration is selected when using vacuum to ensure penetration into complex loads. Under vacuum conditions, the penetration of chlorine dioxide gas is quite rapid. Chlorine dioxide is a surface sterilant and does not have the penetrating abilities of EtQ. Chlorine dioxide does not penetrate into plastic polymers or through cardboard but it does reach confined areas (inside of syringes, bottles, tips and caps, lumens, stents, etc.). As chlorine dioxide does not penetrate into the polymers, there is an additional advantage of rapid aeration.

Because the chlorine dioxide concentration is easily measurable in real time, the target concentration can be repeatedly achieved, thus giving the assurance of a reproducible sterilization cycle. When gas concentration reaches the target concentration, the cycle proceeds to the Exposure step

19.7.4 Exposure

During Exposure, the concentration of chlorine dioxide gas is monitored and maintained to keep the concentration at the target concentration for the entire exposure time (typically 30–45 minutes). If the gas concentration drops during the cycle, additional gas is injected to ensure the gas concentration remains at the set point during the Exposure step.

19.7.5 Aeration

The Aeration step starts once the exposure step is completed. In this step, the chlorine dioxide gas is removed from the chamber. For vacuum chambers, this is accomplished by a series of vacuum pulls and filtered air or Nitrogen backfills.

Table 19.5 calculates the amount of chlorine dioxide used for a typical cycle.

Table 19.6 details the same cycle aeration curve. Aeration time is primarily dependent on the rate that the vacuum pump can evacuate the sterilization chamber. Aeration times of 15 minutes are usually attainable. This aeration brings the chamber environment to safe levels of 0.1 ppm or less.

Table 19.6: This table shows the aeration time from 5 mg/L down to 0.0002 mg/L. The aeration rate is one air exchange per minute. So, if the room is 1000 ft^3, the exhaust rate is 1000 CFM (cubic feet per minute).

19.8 CYCLE DEVELOPMENT

19.8.1 Moisture Conditioning

As mentioned previously, the presence of moisture in the load is critical to obtaining optimal lethal rates and effective sterilization with gaseous chlorine dioxide. Moisture is critical for the inactivation of spores regardless of the sterilizing agent.[82,83,84,85] Important points to consider when developing

TABLE 19.5
Quantity of Chlorine Dioxide for the Given Chamber

Chamber Volume (Cu Ft)	100 ≅ 28.32 cu meters
Target Concentration (mg/L)	5
Exhaust Rate (CFM)	100 ≅ 169.9 cu meters/hour
Amount of chlorine dioxide in the chamber	14.16 g

TABLE 19.6
Time and Amounts of Chlorine Dioxide during Aeration for above Chamber and Concentration. Aeration Is Down to OSHA Safe PEL Levels of 0.1 ppm

Air exchanges	mg/L	PPM	Time (min)
1	2.5000	905.00	1
2	1.2500	452.50	2
3	0.6250	226.25	3
4	0.3125	113.13	4
5	0.1563	56.56	5
6	0.0781	28.28	6
7	0.0391	14.14	7
8	0.0195	7.07	8
9	0.0098	3.54	9
10	0.0049	1.77	10
11	0.0024	0.88	11
12	0.0012	0.44	12
13	0.0006	0.22	13
14	0.0003	0.11	14
15	0.0002	0.06	15

Note: In each air exchange 1/2 of the chlorine dioxide present is removed

and optimizing the chlorine dioxide sterilization process are as follows:

- What moisture condition has the load been exposed to/stored at prior to sterilization?
- Can the moisture level be affected by seasonal RH variation?
- Are there components or packaging materials that may become desiccated during storage in a dry environment prior to sterilization?
- Could the density of the load or its physical geometry affect the penetration of moisture into the least accessible areas?

The choice of a moisture conditioning time in a traditional sterilizer-based application is a function of the prior concerns as well as the approach used to perform the moisture conditioning. Moisture conditioning can be accomplished either in an external chamber or within the sterilization chamber itself. Typical loads can attain the required moisture with 30 minutes of conditioning as part of the sterilization cycle. Appropriate validation studies are important to assure moisture penetration into the least accessible areas.

19.8.2 Exposure Time/Gas Concentration

Early studies with a traditional sterilizer-based application used a gas concentration of 30 mg/L. This concentration was chosen because of the density and composition of the sterilization load. Rapid inactivation of the BIs was observed with sterilization of 10^6 *B. atrophaeus* (formerly *subtilis* ATCC 9372) spores occurring in <15 minutes in almost all cases. In one application, testing was performed at a chlorine dioxide gas concentration of 3 mg/L with reproducible sterilization of 10^6 BIs. As would be expected, the required total gas exposure time was longer than that used at higher gas concentrations.

Based upon studies using a number of test systems, the following guidance can be given with respect to the choice of gas concentration and exposure time in process development studies:

In a chlorine dioxide sterilizer, the recommended gas concentrations for process development studies are 15–30 mg/L. A number of sterilization exposures have been performed using a chlorine dioxide gas concentration of 30 mg/L. In almost all cases, complete kill of 10^6 BIs was observed with 15 minutes of gas exposure. At a chlorine dioxide concentration of 5 mg/L, complete kill of 10^6 BIs should be observed with 30 minutes of exposure.[86] These results were obtained in sterilizers that were not densely loaded but did include a desiccated load. In the event of a dense, desiccated load where moisture penetration may be impeded, a longer conditioning and/or exposure time may be required for similar sterilization efficacy. Isolator studies have shown excellent penetration abilities with densely loaded transfer isolators at 5 mg/L for 30 minutes of exposure along with gas penetrating into half-suit armpits with sleeves down. These results were based upon a fixed concentration for a fixed time. In this work, the charge time was not considered and only the exposure concentration used. The cycle time can be shortened if both phase times (Charge and Exposure) are considered and combined to accrue a contact time or dosage. Dosage is the accumulation of concentration over time.

To calculate the dosage, the chlorine dioxide must be converted from mg/L to ppm. The calculations following can convert mg/L to ppm.

ppm calculation for 1 mg/L chlorine dioxide concentration
$$\text{ppm} = (\text{mg/m}^3)\,(24.45)/\text{molecular weight}$$
$$= (\text{mg/L})\,(1000)\,(24.45)/\text{molecular weight}$$
$$= (1\ \text{mg/L})\,(1000\ \text{L/m}^3)\,(24.45)/67.5$$

= 362.2

NOTE: The number 24.45 in the preceding equations is the volume (liters) of a mole (gram molecular weight) of a gas at 1 atmosphere and at 25°C.

So the dosage for a 2-hour exposure at 1 mg/L concentration has a dosage or contact time of 724 ppm-hrs (362 ppm × 2 h = 724 ppm-h).

Dosage or contact time (CT) starts accumulating when the monitoring system starts reading actual gas concentrations. This has the effect of combining the charge time and exposure time, thereby shortening the overall cycle time.

Other studies have shown 5-log reduction of spores at 400 ppm-hrs[87] in a large animal hospital. In isolators and processing vessels, a 6-log reduction of spores has been demonstrated in several publications.[88,89,90,91] Czarneski showed reductions at a dosage of 900ppm-hrs, whereas Eylath demonstrated 6-log reductions at dosages of 540 ppm-hrs, 600 ppm-hrs, 900 ppm-hrs, and up to 1800 ppm-hrs. Leo demonstrated a 4-log reduction at a low dosage of 180 ppm-hrs.[92] In these same studies, Leo documented a predicted 16.65-log reduction based upon four quartile results accrued. To follow up these studies, tests were performed to see if the results were based upon the concentration or dosage and Lorcheim demonstrated a 6-log reduction of spores at the same dose (720 ppm-hrs) with varying concentrations (0.3, 0.5, 1, 5, 10, and 20 mg/L).[93] In these studies, the dose was held constant and the concentration varied thus indicating that dose is the critical factor.

TABLE 19.7
Examples of Gaseous Chlorine Dioxide Process Development Studies Using *B. atrophaeus* Spores

Preconditioning		Chlorine dioxide exposure		# Nonsterile/ # Tested	Comments
Min	% RH	mg/L	Min		
30	75	10	5	10/10	Spores on unwrapped paper spore strips stored at 23% RH prior to use. Duplicate series of runs on different days.
30	75	10	10	5/10	
30	75	10	15	0/10	
30	75	10	5	8/10	
30	75	10	10	1/10	
30	75	10	15	0/10	
30	70	5	30	0/20	Spores on paper spore strips in Tyvek envelopes.
30	75	10	15	0/10	Spores on paper strips, unwrapped
30	75	10	30	0/10	
30	75	10	15	10/10	Spores on paper strips in blue glassine envelopes
30	75	10	30	0/10	
30	75	10	15	0/10	Spores on paper strips in Tyvek envelopes
30	75	10	30	0/10	
30	75	10	15	0/10	Spores on glass fiber discs in Tyvek envelopes
30	75	10	30	0/10	

Abbreviation: RH = relative humidity.

19.8.3 Examples of Chlorine Dioxide Process Development

Table 19.7 presents examples of gaseous chlorine dioxide process development studies in sterilizers using 10^6 *B. atrophaeus* (formerly *subtilis*) spores as the BI. This work evaluated different substrates.

19.8.4 Biological Indicators

Historical data has shown *Bacillus atrophaeus* (ATCC 9372) spores as the appropriate BI for chemical sterilants such as chlorine dioxide. To confirm the applicability for gaseous chlorine dioxide, tests were done with *B. atrophaeus* as well as other commonly used BIs.

Four spore-forming organisms were initially selected; *Geobacillus stearothermophilus* (ATCC 12980), traditionally used in steam sterilization activities, *Bacillus pumilus* (ATCC 27142), most often used in irradiation studies, *G. stearothermophilus* (ATCC 7953), used for hydrogen peroxide systems, as well as *B. atrophaeus* (ATCC 9372). A study was developed to expose each type of BI to a fixed chlorine dioxide cycle. In each of three runs, 15 BIs of each type were exposed to the chlorine dioxide cycle, removed from the chamber, and aseptically transferred to nutrient media. Microbial growth, as indicated by media turbidity, was recorded as a positive result. This testing was performed in triplicate.

The results are shown in Table 19.8. As Table 19.8 shows, *B. atrophaeus* spores were more resistant (highest number of BIs remaining non-sterile) than either of the G. *stearothermophilus* strains or *B. pumilus*. Based on this data, the use of spores of *B. atrophaeus* as the BI for gaseous chlorine dioxide is a good choice. Additionally, these results demonstrate that *G. stearothermophilus* is also a good choice. Either BI could be used because both *B. atrophaeus* and *G. stearothermophilus* show high levels of resistance to chlorine dioxide. G. *stearothermophilus* has the benefit of a higher incubation temperature, thereby reducing the potential for false positives because not many organisms grow at the elevated temperature of 55°C –60°C. When using chlorine dioxide gas and spore strips, the indictor strip should be wrapped in a Tyvek/Tyvek or Tyvek/mylar envelop. Tyvek does not impede the penetration abilities of chlorine dioxide gas. Glassine envelopes (blue envelopes) do impede the penetration of the gas and will require a slightly longer exposure time. It is also recommended to use spores inoculated on paper strips. Spores inoculated on rigid carriers such as stainless steel can have clumping issues depending upon the inoculation procedures. The size of the stainless-steel carrier can also have an effect on the spore loading. If the spores are inoculated on

TABLE 19.8
Biological Indicator Resistance Study Results

Biological Indicator	Run 1 # nonsterile/ total tested	Run 2 # nonsterile/ total tested	Run 3 # nonsterile/ total tested	Total # nonsterile/ total tested
B. atrophaeus (formerly *subtilis*) ATCC 9372	10/15	13/15	15/15	38/45
B. pumilus ATCC 27142	0/15	2/15	1/15	3/45
G. stearothermophilus ATCC 12980	1/15	2/15	2/15	5/45
G. stearothermophilus ATCC7953	9/15	9/15	8/15	26/45

Note: Cycle parameters for above studies: 30 mg/L gas concentration, 90% relative humidity pre-humidification, 6-minute exposure time

paper carriers, the same issue does not arise because of the paper creating a wicking effect causing the spores to spread throughout the carrier. Paper carriers also can be considered a porous load and this further tests the sterilization process for penetration into porous loads. Studies at ClorDiSys Solutions have been done to show chlorine dioxide does not absorb into the paper strip and cause residual kill.[94,95] These studies were performed using guidance from the EPA neutralization of disinfectants and used exposure at two different concentrations, 1 mg/L (final dosage of 729 ppm-hrs) and 5 mg/L (final dosage of 1960 ppm-hrs). The BIs were pulled from the isolator immediately after exposure (pre aeration), after aeration was completed, 30 minutes post aeration, and 60 minutes post aeration. Aeration was completed when concentrations in the isolator were at or below 0.1 ppm. Post-exposure/pre-aeration BIs should contain the most amount of chlorine dioxide absorbed into the spore strips. These samples were not capable of providing residual kill on the microorganisms added to the test tubes prior to incubation. All other samples removed after increasing lengths of aeration contained far less chlorine dioxide gas absorbed into the spore strips, and as a result, were also unable to provide any residual kill during incubation.

19.8.5 Validation

In determining sterilization parameters, the cycles utilized are based upon the "overkill concept" and are intended to provide a 12-log reduction of a resistant biological indicator. To provide an additional margin of safety, the sterilization parameters utilized during validation studies should represent "worst-case" conditions when compared with those provided for routine sterilization.

To determine final cycle parameters, reductions could be made in the following areas: Conditioning—the target humidity could be reduced (from the typical set point range of 60%–75% RH). The Charge/Exposure—the gas concentration and Exposure dwell time can be reduced (from the typical 45 min). They can be reduced by utilizing the contact time or dosage. In doing this, the charging phase and exposure phase can be combined to reduce the overall cycle time. Additionally, the condition time could be combined with the exposure phase to further reduce overall cycle time.

Additionally, load patterns can affect the efficacy of the sterilization. The maximum load type should be used because this will typically be the most difficult sterilization. Each load type needs to be tested in three consecutive microbial challenge studies to achieve a validated sterilization cycle.

19.8.6 Measurement/Quantification

A UV/VIS spectrophotometer should be integrated into the sterilization system. It precisely monitors and controls the chlorine dioxide concentration during Charge and Exposure, and the Aeration steps until it reaches approximately 0.1 mg/L. Chlorine dioxide gas is green colored and can be measured by photometric methods. In these methods, light is passed through a continuous sample of the gas brought to the photometer by a sample pump from a location within the chamber. The monitoring location is typically chosen by the user in the most difficult location. Typically, the most difficult location would be buried in the load, but using a sample pump in this location would actually improve penetration by drawing gas to this location, so the user should rely on the BI to confirm the decontamination. Typically, the gas sample tubing is placed in corners or behind objects in the chamber. The gas will absorb light, and this absorption can be measured and quantified. The adsorption is converted to concentration, thereby providing an accurate and, more importantly, repeatable concentration measurement. Additionally, as chlorine dioxide is a true gas it does not condense, making aeration quick and repeatable. This repeatability allows for aeration to be validated, ensuring that safe conditions are attained for repeat applications. There are also devices such as Draegar tubes or other sensors, which can verify that safe levels are attained prior to opening the chamber.

19.9 IN-PROCESS CONTROLS

Ease of process control is one of the greatest strengths of the chlorine dioxide technology and is superior to that of most other methods of sterilization. Chlorine dioxide is a gas at use concentrations (boiling point of −40°C at 0.04% and −22°C at 0.18%) that distributes rapidly and evenly throughout the chamber. Because it is a true gas, issues with temperature gradients, cold spots, heat sinks because of the materials of construction, and other issues that can affect the condensation of vapor decontaminating agents such as hydrogen peroxide and peracetic acid do not affect the decontamination effectiveness of chlorine dioxide. Also because of its gaseous properties, it can easily penetrate down long lumens and effectively sterilize complex components sealed in Tyvek bags.

An RH/Temperature probe monitors the RH and temperature conditions inside the chamber. A pressure transmitter monitors the chamber pressure.

The tight process control and accurate concentration monitoring, along with a detailed run record, can lead to parametric release when used for product sterilization as well as expedite validation efforts for all applications.

19.10 DELIVERY SYSTEMS

The ClorDiSys Solutions, Inc. Steridox-VP sterilizer and Cloridox-GMP Sterilization System (Figure 19.3) can be used for component or device sterilization. The Steridox-VP is a stand-alone sterilizer. The Cloridox-GMP is a portable chlorine dioxide gas generator system designed for interfacing with an existing steam or EtO sterilization chamber. Additionally, the Cloridox-GMP can be used to decontaminate isolators, clean rooms, processing vessels, pass-throughs, bio-safety cabinets, or other sealed chambers. Chlorine dioxide sterilizers provide a rapid and highly effective method for component or device sterilization. These systems feature a sophisticated sterilant concentration monitoring system to ensure a tightly controlled process. All instrumentation, including the photometer for concentration monitoring, is easily calibrated to traceable standards. The Human-Machine Interface (HMI) system features a password protected, recipe management system with historical and real-time trending. The process is easy to validate because of the repeatable cycle, tight process control, and highly accurate sterilant monitoring system. A run record is produced that contains the date, cycle time, cycle steps, as well as the critical operating parameters of relative humidity, temperature, pressure, and chlorine dioxide concentration.

19.11 MEDICAL DEVICE STERILIZATION

There have been few advancements in the medical device sterilization industry compared to most other industries. EtO has been facing increasing pressure due to potential environmental health hazards. Even with the issues, its market share has not decreased and has held steady since the mid-1990s. This is due to the industry resistance to change and as such innovation has taken a back seat. Recently, the first commercially available medical device has been sterilized with CD gas and has hit both the US and European markets.

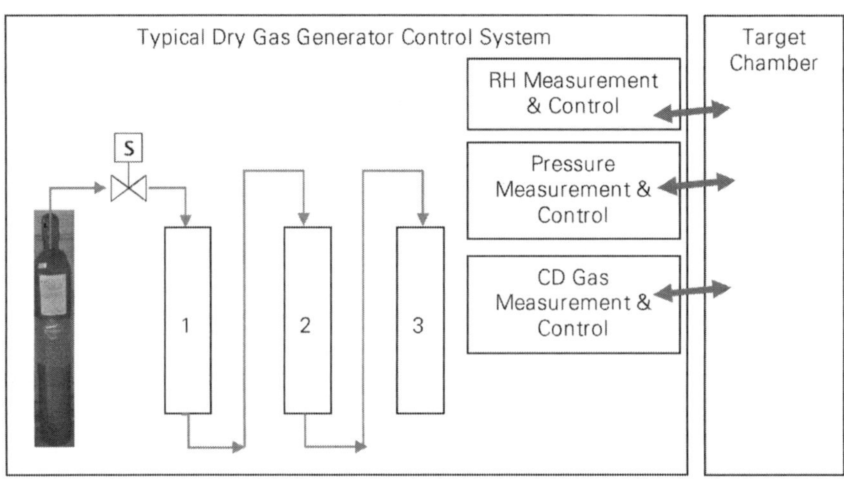

FIGURE 19.3 Cloridox-GMP.

In 2006, the EPA released a draft of its review of EtO and determined it is a human carcinogen, after reviewing studies by the National Institute of Occupational Safety and Health (NIOSH). In 2018, the EPA released its latest National Air Toxics Assessment based on industry supplied emissions data from 2014. The report showed that 109 of the 73,057 census tracts within the US faced cancer risks due to exceeding emissions levels on EPA guidelines. One of the areas with very high EtO levels was in Illinois near an EtO contract sterilization facility. In February 2019, the Governor decided to ban the use of EtO at the plant, but a legal settlement eventually allowed the company to resume operations after installing equipment to reduce EtO emissions.

In the late 2020, the FDA approved a contract sterilization facility for medical devices solely utilizing CD gas. This facility recently processed the first commercially sold medical device sterilized with CD gas that has both US and EU approval. A draft reference file for the use of CD gas has been created mimicking ISO and AAMI EtO guidelines to aid in submission packages to both the FDA and EU regulatory agencies. This provides a regulatory roadmap for validating the sterilization process outside of the typical EtO, gamma irradiation, E-beam / X-ray, steam, and dry-heat methods.

19.11.1 Recent Updates

CD gas medical device sterilization validation studies challenged the process with External Process Control Devices placed outside the load and Process Control Devices placed inside the product load. The studies were done at much lower concentrations (3 mg/L) compared to early sterilization studies (30 mg/L) with CD gas. CD gas was also found to penetrate the display box and did not require any external aeration. After exposures, the product was tested for endotoxin residuals. Endotoxins were tested using the LAL gel-clot test technique and <0.03 EU/mL or <l. 2 EU/device was found. This result is well under the maximum allowable limit of 20.0 EU/device for medical devices and under the 2.15 EU/device limit for cerebrospinal contact devices. CD gas residuals are typically not detected on surfaces, since they are aerated and removed as part of the sterilization process and does not condense onto materials like other methods can. Residual testing for chlorine dioxide gas examines its byproducts that can be chlorite, chlorate, and chloride. Chlorite has a maximum acceptable level of 6.83 mg/device, and testing was unable to detect any (minimum detection level 3.78 μg/device). Chlorate has no maximum acceptable level set, and testing showed an average of 216.5 μg/device over 3 runs. Chloride has a maximum acceptable level of 70.65 mg/device with none detected from the testing (minimum detection level 1.89 μg/device).

Other studies with elastomeric closures showed no clear detrimental effect of CD gas sterilization on the functional properties, even when a 20-fold overexposure was used. There was minor increase in the number of fragments and piercing force observed for styrene-butadiene compounds, although the difference was too small to be able to assign this as an effect of the sterilization process. The effect of CD gas sterilization on the chemical and functional properties of styrene-butadiene compounds and coated and uncoated bromobutyl compounds was studied. The coating was a fluorinated polymer coating. The compounds were investigated and tested according to the tests as described in the USP. In general, it was found that sterilization with CD gas can be competitive with the classical sterilization techniques. The stoppers perform equally well or even slightly better after being subjected to CD gas compared to gamma radiation or EtO.

19.12 OTHER NON-STERILIZATION APPLICATIONS

Chlorine dioxide has a long history of use, and it has been used longest in the potable water industry, and the pulp in paper industry is the largest user. Chlorine dioxide gas, as an oxidizing agent, has been shown to oxidize many things including chemical weapons. Gordon at Public Health Agency in Canada demonstrated chlorine dioxide gas as an effective method to inactivate anthrax toxins.[96] Chlorine dioxide gas has also been shown to be effective (>99%) against the chemical nerve agent VX, but has been shown to be ineffective or partially effective for soman and sarin.[97] Chlorine dioxide has been used for years for odor control in residential and industrial markets. As an example, NosGuard SG (Fort Lauderdale, FL) is used to eliminate odors caused by mold, mildew, pets, food, smoke, and more. There are many other companies with products and services based around odor control. Chlorine dioxide gas has also been shown to be effective again insects. Lowe showed the gas was able to kill bedbugs (*Cimex lectularius* and *C. hemipterus*) with a 100% mortality rate after exposure to various concentrations (1, 2, and 3 mg/L) at various dosages (519, 1029, 1132, and 3024 ppm-hrs).[98] Czarra has shown chlorine dioxide gas was effective again pinworm eggs.[99] Pinworm infections are considered nonpathogenic but with certain research mice they can have adverse effects on behavior, growth, intestinal physiology, and immunology that may affect research.

19.13 CONCLUSIONS

Chlorine dioxide is an oxidizing agent and as such it has shown good biocidal efficacy against viruses, bacteria, fungi, and spores. It is registered as a sterilant, which means it kills all forms of life. It has been shown effective at inactivating other substances such as proteins, endotoxins, beta lactams, chemical weapons, smoke odors, and odors of various kinds. It is used as a gas and a gas dissolved in solution as a sanitizing, broad-spectrum disinfectant and sterilant. It is used in many industries from food to life science research to the pharmaceutical, medical device, and biologics industries. Although it is an oxidizer, it has shown good material compatibility with most materials found in the pharmaceutical industries including electronics. This chapter documents many of the applications of chlorine dioxide gas in its true form, as a gas. As a gas it shows excellent penetration and distribution properties, which is critical in any decontamination. It is easily monitored with a photometric device, which makes for good repeatability, which makes for a good and thorough decontamination/sterilization.

NOTES

1. Schaufler C, Antiseptic Effect of Chlorine Solutions from Interactions of Potassium Chlorate and Hydrochloric Acid. *Zentralbl Chir* 1933;60: 2497–2500.
2. Kovtunovitch GP, Chemaya LA, Bactericidal Effects of Chlorine Solutions from Interaction of Potassium Chlorate and Hydrochloric Acid. *Sovetskaya Khirurgiya* 1936;2: 214–221.
3. Sidgwick NV, *Halogen Oxides: The Chemical Elements and Their Compounds*, vol. 2. Oxford: Oxford University Press, 1951: 1202–1218.
4. Davy H, On a Combination of Oxymuriatic Gas and Oxygen Gas. *Philos Trans* 1811;101: 155–161.
5. Agency for Toxic Substances and Disease Registry (ATSDR), Toxicological Profile for Chlorine Dioxide and Chlorite: U.S. Department of Health and Human Services, Public Health Service, Agency for Toxic Substances and Disease Registry, 2004: 2.
6. Han Y, et al., Reduction of Listeria Monocytogenes on Green Peppers (Capsicum Annuum L.) by Gaseous and Aqueoous Chlorine Dioxide and Water Washing and Its Growth at 7°C. *Journal of Food Protection* 2001;64(11): 1730–1738.
7. Gómez-López VM, Rajkovic A, Ragaert P, et al., Chlorine Dioxide for Minimally Processed Produce Preservation: A Review. *Trends Food Sci Technol* 2009;20: 17–26.
8. Gates DJ, *The Chlorine Dioxide Handbook*. Water Disinfection Series. Denver, CO: AWWA Publishing, 1998.
9. Battisti DL, *Development of a Chlorine Dioxide Sterilization System*. PhD dissertation, University of Pittsburgh, Pittsburgh, PA, 2000.
10. CDC—NIOSH Pocket Guide to Chemical Hazards—Chlorine Dioxide. www.cdc.gov/niosh/npg/npgd0116.html accessed 5/28/2019.
11. Cowley G, Safety in the Design of Chlorine Dioxide Plants. *Loss Prev* 1993;113: 1–12.
12. Jin R, Hu S, Zhang Y, et al., Concentration-Dependence of the Explosion Characteristics of Chlorine Dioxide Gas. *J Hazard Mater* 2009;166: 842–847.
13. Girouard D, Czarneski MA, Room, Suite Scale, Class III Biological Safety Cabinet, and Sensitive Equipment Decontamination and Validation Using Gaseous Chlorine Dioxide. *Applied Biosafety* 2016;21(1): 34–44.
14. Cole B, Czarneski MA, A Multiple-Staged Site Remediation of a Medical Research Animal Facility Affected by a Rodent Pinworm Infestation. Association of Biosafety for Australia & New Zealand 5th Annual Conference, Canberra, QT, November 9–13, 2015.
15. Takahashi E, Czarneski MA, Sugiura A, Japan's RIKEN BSI: Whole Facility Chlorine Dioxide Gas Decontamination Approach for a Barrier Facility: A Case Study. *Appl Biosaf* 2014;19: 201–210.
16. Czarneski MA, Microbial Decontamination of a 65-Room New Pharmaceutical Research Facility. *Appl Biosaf* 2009;14: 81–88.
17. Luftman HS, Regits MA, Lorcheim P, et al., Chlorine Dioxide Gas Decontamination of Large Animal Hospital Intensive and Neonatal Care Units. *Appl Biosaf* 2006;11: 144–154.
18. Leo F, Poisson P, Sinclair CS, et al., Design, Development and Qualification of a Microbiological Challenge Facility to Assess the Effectiveness of BFS Aseptic Processing. *PDA J Pharm Sci Technol* 2005;59: 33–48.
19. Sherman MB, Trujilloe J, Leahy I, et al., Construction and Organization of a BSL-3 Cryo-Electron Microscopy Laboratory at UTMB. *J Struct Biol* 2013;181: 223–233.
20. Lowe JJ, Gibbs SG, Iwen PC, et al., A Case Study on Decontamination of a Biosafety Level-3 Laboratory and Associated Ductwork within an Operational Building Using Gaseous Chlorine Dioxide. *J Occup Environ Hygiene* 2012;9: D196–D205.
21. Lowe JJ, Gibbs SG, Iwen PC, et al., Impact of Chlorine Dioxide Gas Sterilization on Nosocomial Organism Viability in a Hospital Room. *Int J. Environ Res Public Health* 2013a;10: 2596–2605.
22. Lowe JJ, Hewlett AL, Iwen PC, et al., Evaluation of Ambulance Decontamination Using Gaseous Chlorine Dioxide. *Prehosp Emerg Care* 2013b;17: 401–408.
23. Aieta EM, Berg JD, A Review of Chlorine Dioxide in Drinking Water Treatment. *J Am Water Works Assoc* 1986;78: 62–72.
24. Roller SD, Olivieri VP, Kawata K, Mode of Bacterial Inactivation by Chlorine Dioxide. *Water Research* 1980;14: 635–642.
25. Fukayama MY, Tan H, Wheeler WB, et al., Reactions of Aqueous Chlorine and Chlorine Dioxide with Model Food Compounds. *Environ Health Perspect* 1986;69: 267–274.
26. Alvarez ME, O'Brien RT, Mechanisms of Inactivation of Poliovirus by Chlorine Dioxide and Iodine. *Appl Environ Microbiol* 1982;44: 1064–1071.
27. Sansebastiano G, Mori G, Tanzi ML, et al., Chlorine Dioxide: Methods of Analysis and Kinetics of Inactivation of Poliovirus Type 1. *Igiene Moderna (Parma)* 1983;79: 61–91.
28. Berman D, Hoff JC, Inactivation of Simian Rotavirus SA 11 by Chlorine, Chlorine Dioxide and Monochloramine. *Appl Environ Microbiol* 1984;48: 317–323.
29. Harakeh MS, Butler M, Inactivation of Human Rotavirus SA 11 and Other Enteric Viruses in Effluent by Disinfectants. *J Hyg* 1984;93: 157–163.
30. Olivieri VP, Hauchman FS, Noss CI, et al., Mode of Action of Chlorine Dioxide on Selected Viruses. In *Water Chlorination Chemistry: Environmental Impact and Health Effects*, vol. 5. Jolley RL, ed. Chelsea, MI: Lewis Publishers, 1985: 619–634.
31. Sansebastiano G, Cesari C, Bellelli E, Further Investigations on Water Disinfection by Chlorine Dioxide. *Igiene Moderna (Parma)* 1986;85: 358–380.
32. Chen YS, Vaughn JM, Inactivation of Human and Simian Rotaviruses by Chlorine Dioxide. *Appl Environ Microbiol* 1990;56: 1363–1366.
33. Farr RW, Walton C, Inactivation of Human Immunodeficiency Virus by a Medical Waste Disposal Process Using Chlorine Dioxide. *Infect Control Hosp Epide* 1993;14: 527–529.
34. Gómez-López VM, Rajkovic A, Ragaert P, et al., Chlorine Dioxide for Minimally Processed Produce Preservation: A Review. *Trends Food Sci Technol* 2009;20: 17–26.
35. Korich DG, Mead JR, Madore MS, et al., Effects of Ozone, Chlorine Dioxide, Chlorine and Monochloramine on Cryptosporidium Parvum Oocyst Viability. *Appl Environ Microbiol* 1990;56: 1423–1428.

36. Whitmore TN, Denny S, The Effect of Disinfectants on a Geosmin-Producing Strain of Streptomyces Griseus. *J Appl Bacteriol* 1992;72: 160–165.
37. Wang Y, Jiang Z, Studies on Bactericidal and Algaecidal Ability of Chlorine Dioxide. *Water Treat* 1995;10: 347–352.
38. Bundgaard-Nielson K, Nielsen PV, Fungicidal Effect of 15 Disinfectants against 25 Fungal Contaminants Commonly Found in Bread and Cheese Manufacturing. *J Food Prot* 1996;59: 268–275.
39. Whitmore TN, Denny S, The Effect of Disinfectants on a Geosmin-Producing Strain of Streptomyces Griseus. *J Appl Bacteriol* 1992;72: 160–165.
40. Brown P, Rohrer RG, Moreau-Dubois MC, et al., Use of the Golden Syrian Hamster in the Study of Scrapie Virus. *Adv Exp Med Biol* 1981;134: 365–373.
41. Brown P, Rohrer RG, Green EM, et al., Effect of Chemicals, Heat and Histopathologic Processing on the High-Infectivity Hamster Adapted Scrapie Virus. *J Infect Dis* 1982a;145: 683–687.
42. Brown P, Gibbs CJ, Jr, Amyx HL, Chemical Disinfection of Creutzfeld-Jacob Disease Virus. *N Engl J Med* 1982b;306: 1279–1282.
43. Rosenblatt DH, Rosenblatt AA, Knapp JE, inventors. Use of Chlorine Dioxide as a Chemosterilizing Agent. *U.S. Patent* 1985;4: 504, 442.
44. Rosenblatt DH, Rosenblatt AA, Knapp JE, inventors. Use of Chlorine Dioxide as a Chemosterilizing Agent. *U.S. Patent* 1987;4: 681, 739.
45. Jeng DK, Woodworth AG, Chlorine Dioxide Sterilization under Square-Wave Conditions. *Appl Environ Microbiol* 1990a;56: 514–519.
46. Jeng DK, Woodworth AG, Chlorine Dioxide Gas Sterilization of Oxygenators in an Industrial Scale Sterilizer: A Successful Model. *Artif Organs* 1990;14: 361–368.
47. Berg JD, Roberts PV, Matin A, Effect of Chlorine Dioxide on Selected Membrane Functions of Escherichia Coli. *J Applied Microbiology* 1986;60: 213–220.
48. Cho M, Kim J, Kim JY, et al., Mechanisms of Escherichia Coli Inactivation by Several Disinfectants. *Water Res* 2010;44: 3410–3418.
49. Young SB, Setlow P. Mechanisms of Killing of Bacillus Subtilis Spores by Hypochlorite and Chlorine Dioxide. *J Appl Microbiol* 2003;95: 54–67.
50. Peta ME, Lindsay D, Brözel VS, et al. Susceptibility of Food Spoilage Bacillus Species to Chlorine Dioxide and Other Sanitizers. *S Afr J Sci* 2003;99: 375–380.
51. Alvarez ME, O'Brien RT, Mechanisms of Inactivation of Poliovirus by Chlorine Dioxide and Iodine. *Appl Environ Microbiol* 1982;44: 1064–1071.
52. Aieta EM, Berg JD, A Review of Chlorine Dioxide in Drinking Water Treatment. *J Am Water Works Assoc* 1986;78: 62–72.
53. Gordon G, Keiffer RG, Rosenblatt DH, The Chemistry of Chlorine Dioxide. In Progress in Inorganic Chemistry, vol. 15. Lippaer SJ, ed. New York: Wiley Interscience, 1972: 201–287.
54. Masschelein WJ, Rice RG. *Chlorine Dioxide Chemistry and Environmental Impact of Oxychlorine Compounds.* Ann Arbor, MI: Ann Arbor Science Publishers, 1979: 111–145.
55. Aieta EM, Roberts PV, The Chemistry of Oxy-Chlorine Compounds Relevant to Chlorine Dioxide Generation. In Water Chlorination: Environmental Impact and Health Effects, vol. 5. Jolley RL, ed. Chelsea, MI: Lewis Publishers, 1985: 783–794.
56. Grimley E, Gordon G, Kinetics and Mechanism of the Reaction between Chlorine Dioxide and Phenol in Aqueous Solution. *J Inorg Nucl Chem* 1973;35: 2383–2392.
57. Kaczur JJ, Cawfield DW, Chlorine Oxygen Acids and Salts. In *Kirk-Othmer Encyclopedia of Chemical Technology*, vol. 5, 4th ed. New York: John Wiley & Sons, 1992: 968–997.
58. Ogata N, Denaturation of Protein by Chlorine Dioxide: Oxidative Modification of Tryptophan and Tyrosine Residues. *Biochem* 2007;46: 4898–4911.
59. Ooi BG, Branning SA, Correlation of Conformational Changes and Protein Degradation with Loss of Lysozyme Activity Due to Chlorine Dioxide Treatment. *Appl Biochem Biotechnol* 2017;182: 782–791.
60. Lorcheim K, Chlorine Dioxide Gas Inactivation of Beta-Lactams. *Appl Biosaf* 2011;16: 34–43.
61. Cole B, Site Remediation of a Penicillin Production Facility Using Chlorine Dioxide Gas as a Sterilant. Poster Presentation at Association of BioSafety for Australia and New Zealand, 2014. ABSANZ Conference.
62. Torres MJ, Blanca M, Fernandez J, Romano A, de Weck A, Aberer W, et al. (for ENDA and the EAACI interest group on drug hypersensitivity). Diagnosis of Immediate Allergic Reactions to Beta-Lactam Antibiotics. *Allergy* 2003;58: 961–972.
63. Centers for Disease Control and Prevention (CDC), *Sexually Transmitted Diseases Treatment Guidelines*, 2006. www.cdc.gov/mmwr/preview/mmwrhtml/rr5511a1.htm accessed 5/2019.
64. Czarneski MA, Lorcheim P, Drummond S, Kremer T, Endotoxin Reduction by Chlorine Dioxide (ClO2) Gas. Kilmer Conference 2019, Dublin, Ireland, June 3–6, 2019.
65. May J, Solvent Odor Thresholds for the Evaluation of Solvent Odors in the Atmosphere. *Staub—Reinhalt* 1966;26(9): 385–389.
66. Agency for Toxic Substances and Disease Registry (ATSDR), Toxicological Profile for Chlorine Dioxide and Chlorite: U.S. Department of Health and Human Services, Public Health Service, Agency for Toxic Substances and Disease Registry, 2004, 3.
67. OSHA Ethelene Oxide Fact Sheet. www.osha.gov/OshDoc/data_General_Facts/ethylene-oxide-factsheet.pdf accessed 5/28/2019.
68. Ethylene Oxide (EtO): Evidence of Carcinogenicity May 1981, DHHS (NIOSH) Publication Number 81–130, Current Intelligence Bulletin 35. www.cdc.gov/niosh/docs/81-130/ accessed 5/28/2019.
69. Smith RP, Willhite CC, Chlorine Dioxide and Hemodialysis. *Regul Toxicol Pharmacol* 1990;11: 42–62.
70. White, GC, *Handbook of Chlorination.* New York: Van Nostrand Reinhold, 1972.
71. Couri D, Miller CH, Jr, Bull RJ, et al., Assessment of Maternal Toxicity, Embryotoxicity and Teratogenic Potential of Sodium Chlorite in Sprague-Dawley Rats. *Environ Health Perspect* 1982;46: 25–29.
72. Carlton BD, Basaran AH, Mezza LE, et al. Reproductive Effects in Long-Evans Rats Exposed to Chlorine Dioxide. *Environ Res* 1991;56: 170–177.
73. Skowronski GA, Abdel Rahman MS, Gerges SE, et al. Teratologic Evaluation of Alcide Liquid in Rats and Mice. *J Appl Toxicol* 1985;5: 97–103.

74. Suh DH, Abdel Rahman MS, Bull RJ, Effect of Chlorine Dioxide and Its Metabolites in Drinking Water on Fetal Development in Rats. *J Appl Toxicol* 1983;3: 75–79.
75. Ishidate M, Jr, Sofuni T, Yoshikawa K, et al., Primary Mutagenicity Screening of Food Additives Currently Used in Japan. *Food Chem Toxicol* 1984;22: 623–636.
76. Akamatsu A, Lee C, Morino H, et al., Six-Month Low Level Chlorine Dioxide Gas Inhalation Toxicity Study with Two-Week Recovery Period in Rats. *J Occup Med Toxicol* 2012;7: 2.
77. Masschelein WJ, Rice RG, eds., *Chlorine Dioxide Chemistry and Environmental Impact of Oxychlorine Compounds*. Ann Arbor, MI: Ann Arbor Science Publishers, Inc., 1979: P98, 111–145.
78. Kaczur JJ, Cawifield DW, Chlorine Oxygen Acids and Salts. In *Kirk-Othmer Encyclopedia of Chemical Technology*, vol. 5, 4th ed. New York, NY: John Wiley and Sons, Inc., 1992.
79. Masschelein WJ, Rice RG, *Chlorine Dioxide Chemistry and Environmental Impact of Oxychlorine Compounds*. Ann Arbor, MI: Ann Arbor Science Publishers, 1979: 111–145.
80. Spinks JWT, Porter JM, Photodecomposition of Chlorine Dioxide. *J Am Chem Soc* 1934;56: 264–270.
81. Nielson AH, Woltz PJH, Infrared Spectrum of Chlorine Dioxide. *J Phys Chem* 1952;20: 1879–1883.
82. Agalloco J, Carleton P, Frederick J, *Validation of Pharmaceutical Processes*, 3rd ed. New York, NY: Informa Healthcare USA Inc., 2008.
83. Jeng DK, Woodworth AG, Chlorine Dioxide Gas Sterilization under Square-Wave Conditions. *Appl Environ Microbiol* 1990;56: 514–519.
84. Westphal AJ, Price PB, Leighton TJ, Wheeler KE, Kinetics of Size Changes of Individual Bacillus Thuringiensis Spores in Response to Changes in Relative Humidity. *Proc Natl Acad Sci U S A* 2003;6: 3461–3466.
85. Whitney EAS, Beatty ME, Taylor TH, Jr, et al., Inactivation of Bacillus Anthracis Spores. *Emerg Infect Dis* 2003;6: 623–627.
86. Kowalski JB, Sterilization of Medical Devices, Pharmaceutical Components, and Barrier Isolator Systems with Gaseous Chlorine Dioxide. In *Sterilization of Medical Products*. Morrissey RF, Kowalski JB, eds. Champlain, NY: Polyscience Publications, 1998: 313–323.
87. Luftman HS, Regits MA, Lorcheim P, et al., Chlorine Dioxide Gas Decontamination of Large Animal Hospital Intensive and Neonatal Care Units. *Appl Biosaf* 2006;11: 144–154.
88. Czarneski MA, Lorcheim P, Isolator Decontamination Using Chlorine Dioxide Gas. *Pharma Technol* 2005;4: 124–133.
89. Eylath AS, Wilson D, Thatcher D, et al., Successful Sterilization Using Chlorine Dioxide Gas: Part One-Sanitizing an Aseptic Fill Isolator. *BioProcess Int* 2003a;7: 52–56.
90. Barbu N, Zwick R, Isolators Selection, Design, Decontamination, and Validation. *Pharm Eng* 2014: 6–14.
91. Eylath AS, Madhogarhia ER, Lorcheim P, et al., Successful Sterilization Using Chlorine Dioxide Gas: Part Two-Cleaning Process Vessels. *BioProcess Int* 2003b;8: 54–56.
92. Leo F, Poisson P, Sinclair CS, et al., Design, Development and Qualification of a Microbiological Challenge Facility to Assess the Effectiveness of BFS Aseptic Processing. *PDA J Pharm Sci Technol* 2005;59: 33–48.
93. Lorcheim K, Melgaard E, Linearity of the Relationship between Concentration and Contact Time for Sterilization with Chlorine Dioxide gas. American Biological Safety Association 58th Annual Biological Safety Conference, Providence, RI, October 9–14, 2015.
94. Czarneski M, Study Title: Chlorine Dioxide Gas Aeration, Neutralization Not Needed for 1mg/L Study, Internal Documentation, Study Completion Date: 11–25–2015.
95. Czarneski M, Study Title: Chlorine Dioxide Gas Aeration, Neutralization Not Needed for 5mg/L Study, Internal Documentation, Study Completion Date: 1–18–2016.
96. Gordon D, Krishnan J, Theriault S, Inactivation Studies of Lipopolysacchride and Anthrax Toxins Using Gaseous Decontamination Methods, Canadian Biosafety Symposium, Halifax, SK, May 31–June 2, 2009.
97. Snyder E, Systematic Decontamination of Chemical Warfare Agents and Toxic Industrial Chemicals. Report on the 2008 Workshop on Decontamination Research and Associated Issues for Sites Contaminated with Chemical, Biological, or Radiological Materials, EPA/600/R-09/035, 2009.
98. Gibbs SG, Lowe JJ, Smith PW, et al., Gaseous Chlorine Dioxide as an Alternative for Bedbug Control. *Infect Control Hosp Epidemiol* 2012;33: 495–499.
99. Czarra JA, Adams JK, Carter CL, et al., Exposure to Chlorine Dioxide Gas for 4 Hours Renders Syphacia Ova Nonviable. *J Am Assoc Lab Anim Sci* 2014;53: 364–367.

20 Liquid Phase Sterilization

James Agalloco

CONTENTS

20.1 INTRODUCTION

There are instances in the manufacture of pharmaceutical products and medical devices when an item must be sterilized, but the material's properties can eliminate methods of sterilization based upon thermal energy (from either moist or dry heat) or from radiation. The simplicity and speed of heat (moist or dry) and radiation sterilization ordinarily make those the methods of choice; however, the effects of these sterilization processes on many materials are detrimental to essential material properties. When faced with these circumstances, the practitioner often turns to other methods in which microorganisms are destroyed by exposure to chemical agents in gas, vapor, or liquid form. Although all of these processes rely on a chemical action against microorganisms, there are elements of liquid agent delivery for sterilization purposes that are substantially different from other sterilization methods. This chapter will outline liquid chemical sterilization processes, define cycle development practices, describe validation practice and outline routine process control requirements.

It should be recognized that as liquid phase sterilization has only recently come into vogue, the process is viewed with some skepticism. Liquid sterilization processes are viewed as merely sanitization or high-level decontamination treatments. There are at least four reasons why this is the case:

- First, the same liquid agents are often used for room and equipment sanitization where the outcome is non-absolute and thus, they may be considered less than fully effective sterilants.
- Second, liquid processes are quite simple and lack the sophistication that is often associated with sterilization processes.
- Third, when a liquid process is poorly conceived, the processed items can be contaminated during the agent removal/deactivation step, thus compromising the lethality of the process.
- Lastly, a liquid sterilization may be considered only a sanitization because that is all that the process is expected to achieve.

With the growing usage of liquid phase processes, these biases may eventually disappear, but they will likely persist for some time.

20.2 STERILIZATION BASICS

Sterilization is a process that kills or removes microorganisms. The use of an agent described in this chapter without adequate control measures should not be considered a sterilization process. In lethal sterilization processes, the microbiological death curve can be plotted with the logarithmic number of microorganisms remaining alive on the *y*-axis against time on the *x*-axis.[1] A best-fit of the points results in a straight line. The inverse slope of the line is the D-value (decimal reduction time).[3] The D-value is the time (ordinarily in minutes) to reduce the population by 90% (one logarithm). The death curve line can be extrapolated to estimate the number of possible survivors in a large number of units (Figure 20.1). This is termed the Probability of a Non-Sterile Unit (PNSU). An acceptable PNSU has been defined as a probability of one chance in a million that any individual unit is non-sterile (a risk value originally developed for food safety).

The D-value is an inherent property of the microorganism and the conditions of the sterilization treatment. For liquid chemical sterilization, the essential factors are the concentration of the agent and the temperature of the system. The slope of the curve is the time in minutes for the microbial population to be reduced by 90% (or 1 logarithm) and is commonly termed the D-value. D-value determination requires

DOI: 10.1201/9781003163138-20

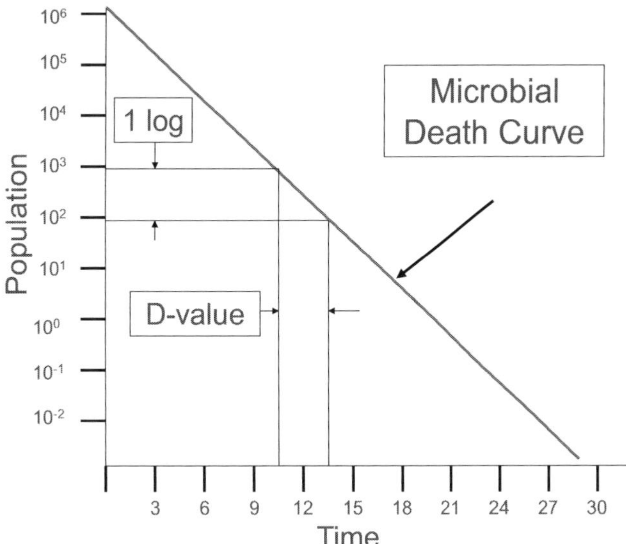

FIGURE 20.1 Microbial death curve.

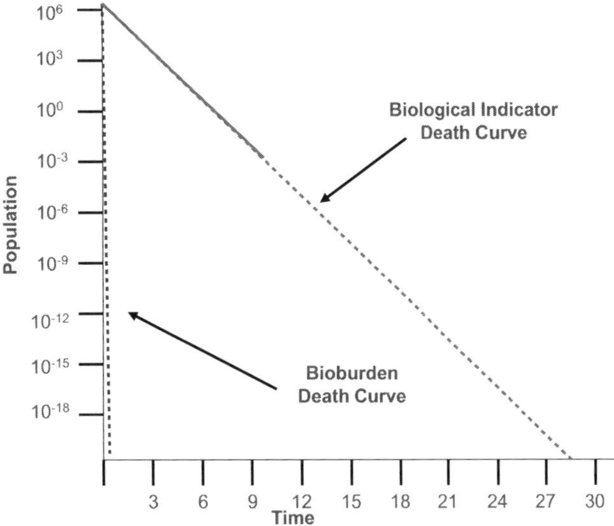

FIGURE 20.2 Relative resistance of bioburden and biological indicator organisms.

measurement of the lethal conditions to which the microorganism is exposed, and these should be stated with the D-value. Validating the destruction of microorganisms relies in part on differences in the relative resistance of a biological indicator and bioburden organisms (Figure 20.2).[2]

The concentration of the chemical agent has the largest impact on the effectiveness of the sterilization process. Substantially higher concentrations of a chemical agent are possible in liquids as compared with the gas phase There are other factors essential for effective sterilization of microorganisms by chemical agents. Moisture presence is assured as the diluents for all agents is water. Liquid water is needed for penetration of the agent through the spore coat.[3] The essential process parameters are comparatively easily determined for liquid sterilization: concentration of the agent (assuming the remainder is water), temperature and agitation (type/design of mixing system and speed of rotation).

20.3 LIQUID AGENTS

There are a variety of liquid chemicals capable of rapid kill of both vegetative cells and spores. Acids, bases, aldehydes, halides and strong oxidants in aqueous solution are all effective liquid sterilants.[4] Liquid chemicals in aqueous solution capable of sterilizing physical objects as described previously include:

- Aldehydes—e.g., Glutaraldehyde, formaldehyde
- Acids—e.g., Peracetic, nitric, sulfuric
- Bases—e.g., Sodium hydroxide, potassium hydroxide
- Oxygenating compounds—e.g., Hydrogen peroxide, ozone, chlorine dioxide
- Halides—e.g., Sodium hypochlorite, chlorine

Similar to gas sterilization, liquid sterilant lethality varies with both concentration and temperature (humidity is provided by the water present), provided sufficient mixing is present to ensure process consistency. Other factors potentially impacting lethality processes include pH, agitation, and the presence of soil or other contaminants that might protect microorganisms on the surface of the materials. The use of a validated cleaning process is strongly recommended prior to any liquid chemical sterilization process. Effective liquid chemical sterilization processes are otherwise straightforward because of their simplicity.

As with other forms of sterilization, the effect of the sterilization (including neutralization and rinsing) on the materials must be thoroughly evaluated. The chemical activity of the agents on the materials can be substantial. Extreme pH, significant oxidation and the reaction potential of many agents, all of which help make the agent effective against microorganisms, can also play havoc on materials (and processing equipment). Fortunately, information on the chemical compatibility of these agents with a variety of materials is widely available.[5]

20.4 FUNDAMENTALS OF LIQUID PHASE STERILIZATION

There are multiple applications for liquid phase sterilization. Although they adhere to the same core concepts for each application, the differences between the various uses of liquid agents are substantial.

Sterilization of Insoluble Materials—This process bears close resemblance to gas sterilization processes with which the reader is perhaps more familiar. Objects to be sterilized are carefully placed into a pre-sterilized closed container (which may be as simple as a sealed roller bottle to as complex as an agitated pressure-rated vessel). The sterilizing agent is added, agitated for the required time period, and then removed. Agent removal could be accomplished by chemical action (e.g., neutralization), physical methods (e.g., filtration) or their combination. The serial use of multiple agents is also possible. Concentration, temperature and agitation are controlled to be reproducible over time. The wetted surfaces of the container are sterilized during the process by exposure to the agent. Once sterilized, the materials are kept in the vessel until ready for use. This process is potentially suitable for a variety of items: i.e., complex medical devices; insoluble nanoparticles; and human, animal and synthetic bone and tissue. Alternatively, the container could be opened in an aseptic environment for further processing of the contents.

Terminal Sterilization of Filled Products—In a few instances, liquid products with inherent antimicrobial properties can be terminally sterilized in their final container. The formulated liquid is filled with appropriate bioburden control and sealed in the container. A defined agitation/rotation period with or without increased temperature provides for sterilization of the contents.

Sterilization of Process Piping/Equipment—The application of a liquid sterilant to equipment surfaces is a clear possibility mimicking the use of gas sterilizing agents. This is a specialized form of sterilization in place applying a chemical process rather than a thermal one and would require that the sterilizing agent contact the entirety of the equipment surface. Liquid agents that have a higher vapor pressure may provide lethality in the gas phase that would be in equilibrium with the liquid. In these processes, circulation and spraying are utilized to improve the uniformity of the process conditions throughout the system. Applications of this liquid process are rather limited; however, there is potential for expanded use.

20.4.1 Material Effects

Sterilization processes are designed to kill microorganisms and as such they utilize conditions that may be destructive of essential material properties. Liquid sterilization processes have potentially a greater effect than other methods because there is continuous contact with an active chemical agent in aqueous solution. The strong oxidative powers of chemical agents, pH extremes of acids and bases, and immersion of surfaces can cause significant changes in the materials being sterilized. The effect of the liquid agent(s) on the processing equipment must also be considered. The sterilizing chamber is composed of many different materials all of which must be tolerant of the sterilant and sterilizing conditions. The potential adverse consequences must be an integral part of process selection, equipment design and fabrication, cycle development and process validation.

20.4.2 Process Equipment

Large-scale liquid sterilization can be performed in closed agitated vessels or chambers much like those utilized for chemical reactions. A jacket can provide temperature control for fixed vessels. In large processes, a control system that provides sequencing, regulation of process parameters, and documentation can be beneficial. For small processes, the equipment might be as simple as a non-pressure rated container where the items to be sterilized are submerged. Agitation, temperature control and sequencing would be provided by the operator using laboratory apparatus and room/chamber environmental controls.

The item to be sterilized is directly exposed (without a protective wrap) to the liquid and allowed to remain in contact for the required time period. Following the dwell period, the item is either removed from the agent and treated to remove the agent or the chemical agent is chemically neutralized or removed by rinsing with a sterile fluid. The steps that follow the sterilization dwell proper must be performed in an aseptic manner that preserves the sterility of the materials. There are only two primary process alternatives: 1) remove the sterilized object from the sterilizing agent and then neutralize/wash the chemical agent before wrapping or 2) remove the agent and neutralize/wash the materials within the vessel.

- Removing the item from the chemical agent mimics the removal of a previously sterilized object from its protective wrap. Process design must consider how to protect the now sterile but unwrapped materials from contamination from this point onward. This may be accomplished on a small scale using aseptic practices in an ISO 5 unidirectional air environment.
- The chemical agent can be removed and neutralized in situ within a closed vessel containing the materials. The vessel affords protection of the sterile materials until the next process step.

Depending upon the end use of the items and the chemical activity of the materials used, the neutralization may also have potentially adverse material effects the impact of which must be considered. In validation of liquid chemical sterilization, agent removal (whether accomplished by physical or chemical means) is an important and integral part of the overall process.

20.5 VALIDATION OF STERILIZATION METHODS

The performance qualification or "Validation" activity has been described as documentation that the process or product conforms to expectations as determined through independent parameter measurement and/or intensive sampling or challenge. It is the focus of regulatory attention for any sterilization process. Liquid chemical sterilization is one of the least documented sterilization processes. Although there is information available on the effects of various chemical agents against spore-forming microorganisms, there is minimal information on their effect against non-spore formers.[6] For this reason, the most prevalent means of sterilization appears to be the half-cycle method in which kill of a high population of a resistant spore former establishes the "half cycle"

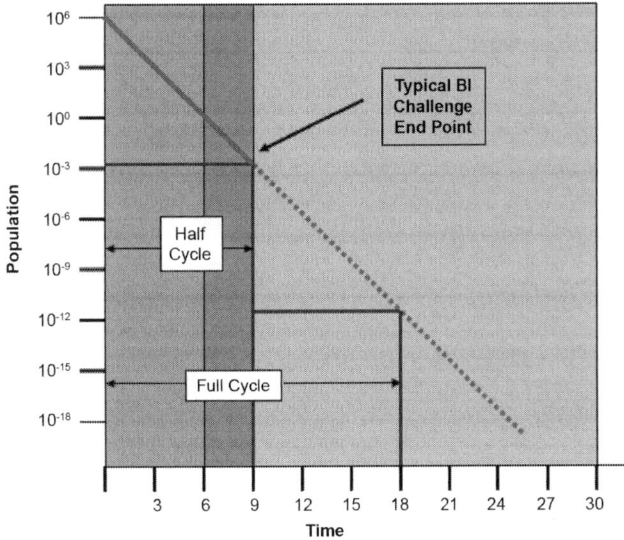

FIGURE 20.3 Half-cycle sterilization validation.

Note: BI = biological indicator.

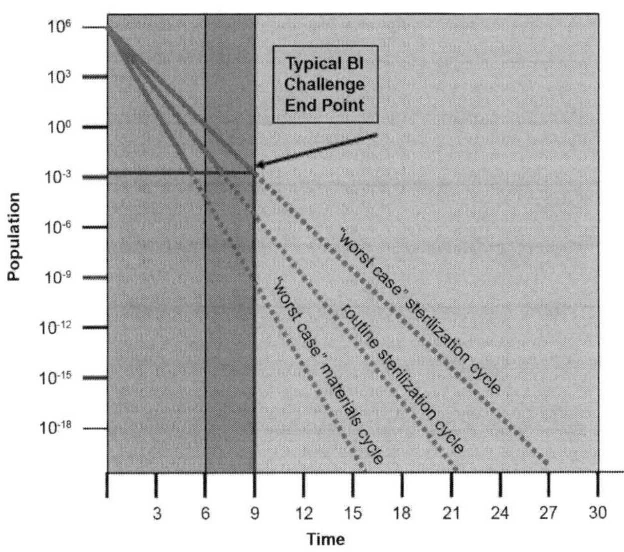

FIGURE 20.4 Sterilization validation using bracketing.

Note: BI = biological indicator.

and routine sterilization employs a "full cycle" with twice the dwell period. The half-cycle approach, which relies upon assumptions regarding the microbial resistance and population of the bioburden, was originally developed for use with ethylene oxide.[7] This approach mandates a sterilization dwell period that destroys not less than 10^6 spores of a resistant biological indicator. The assumption made with the half-cycle approach is that the bioburden present would be lower in both population and chemical resistance to the biological indicator. In routine operation, the process dwell period is arbitrarily doubled (thus the term half-cycle) and supports a PNSU of 10^{-6}.[4] The half cycle method is graphically depicted in Figure 20.3. The half-cycle method does not meaningfully consider the resistance of the biological indicator as a surrogate for the bioburden, because complete destruction of the indicator is required in the "half-cycle". Actual determination of the indicator D-value at the chosen sterilization parameters requires more effort and is largely ignored. Half-cycle approaches are inherently conservative, and little effort is made to optimize the process dwell period when it will be arbitrarily doubled in routine use anyway. The half-cycle method evaluates only the effect of time on lethality, assuming that the effect of variation of the other essential parameters, agent concentration and temperature, are ignored. This results in a severe limitation of the half-cycle method in the opinion of many practitioners.[8]

It is commonplace in performance qualification studies to utilize "worst-case" challenges, as is the prevalent practice with moist and dry heat sterilization processes. Typical "worst-case" challenges for liquid sterilization include reducing the process (set-point) temperature; reducing cycle dwell time; reduction of both time and temperature; reduction of the agent concentration and the use of resistant biological challenges as bioburden surrogates. More detailed information on the expected practices can be found in the myriad of industry and regulatory publications on this subject.[9,10]

Another method suited for sterilization validation is a bracketing approach that better supports the extremes of the operating ranges for the critical process parameters.[11] In the bracketing approach, a cycle with lower concentration, lower temperature and a shorter dwell period is confirmed by microbial indicator destruction using what are less lethal ("worst-case" sterilization) conditions. Material effects are evaluated in a cycle employing a higher concentration, higher temperature and a longer dwell period where the adverse impact on materials is believed to be greater ("worst-case" material effect). Routine operation of the system utilizes conditions that fall between the process extremes that have been evaluated (Figure 20.4).

Regardless of the validation method utilized, there are common elements in all validation efforts:

- Equipment Qualification—The equipment utilized for the sterilization process (pressure vessel or stirred tank) should be documented with respect to installation details and operational characteristics. Equipment qualification serves as the basis for change control for the physical equipment and the importance. This effort must of course include calibration of instrumentation and qualification of the control system.
- Empty Chamber/Vessel Parameter Distribution—Parameter measurement within the sterilization chamber/vessel may be appropriate. If the vessel is mixed during the process (as is almost universally desirable), this study confirms the effects of mixing. Overmixing in these processes is not a consideration, as additional mixing can only improve the uniformity of the process parameters. Biological indicators are not required in the evaluation of the empty chamber/vessel uniformity.

- Component and Load Mapping—These activities are not considered for liquid sterilization, because sampling systems placed on the load items could alter agent penetration. This is best established by evaluation by biological indicators placed within/on the load items. Presently, there are no physical/chemical indicators or integrators available for use with liquid sterilization.
- Biological Indicators—The use of a biological indicator for initial validation is required. The biological indicator serves as a "worst-case" surrogate for the bioburden present in routine operations. Biological indicators are conventionally spores of a microorganism (most often a *Bacillus* or *Geobacillus* species) chosen specifically for its greater resistance to the sterilizing process than the expected bioburden. Inactivation of the biological indicator during the validation establishes the lethality of the process across the items being sterilized. Spore challenges may be either a strip or coupon positioned within the load or spores inoculated on a load item. Inoculated items should have their population determined by the end user, and where possible their resistance to the sterilization process confirmed. Indicators are placed among the load items at locations believed to be hardest for the agent to penetrate. The use of biological challenges for liquid sterilization is limited to the validation of the process, as the materials must be in direct contact with the liquid agent, making placement and recovery of suitable biological indicators problematic in routine processing. Liquid sterilization processes are parametrically released (a typical situation with many sterilization processes that are utilized in-process).
- Process Confirmation/Microbiological Challenge—The core of the validation activity is confirmation of acceptable process parameters and inactivation of the microbial challenge. Proof of cycle efficacy is provided in replicate studies in which the biological indicators are killed, and physical measurements are taken as documentation. Differences in resistance are exploited for ease of validation and routine process control.

20.6 ROUTINE PROCESS CONTROL

Sterilization processes are subject to routine controls that support the efficacy of the cycle over time. Validation is not a one-time activity project, but an integral part of a Current Good Manufacturing Practices (CGMP) compliant facility that must be sustained over the useful life of the facility and its products.[12] Control over the sterilization processes is commonly achieved through practices defined specifically for that purpose including: Calibration of instruments, Training of Personnel, Physical Measurements of Process Parameters, Change Control, Preventive Maintenance and Periodic Re-Assessment.

20.7 CONCLUSION

This chapter provides an overview of the liquid sterilization methods and their validation. The relative novelty of these processes precludes a more detailed approach; however, as these methods come into increased use additional insight will be possible. The accompanying bibliography provides recommended information sources on this emerging process.

NOTES

1. Bigelow WD, The Logarithmic Nature of Thermal Death Time Curves. *J. Infect. Dis.* 1921;29: 528–536.
2. Agalloco J, Understanding Overkill Sterilization: Putting an End to the Confusion. *Pharmaceutical Technology* 2007;30(5): S18–60–76.
3. Pflug IJ, *Microbiology & Engineering of Sterilization Processes*, 8th ed. Minneapolis, MN: University of Minnesota, 1995.
4. Hugo WB, Russell AD, Types of Antimicrobial Agents. Chapter in *Principles and Practices of Disinfection, Preservation and Sterilisation*. Russell AD, Hugo WB, Ayliffe GAJ, eds. London: Blackwell Scientific Publications, 1982.
5. Krisher AS, Siebert OW, Materials of Construction. Chapter in *Perry's Chemical Engineers' Handbook*, 6th ed. Perry RH, Green D, eds. New York: McGraw Hill, 1984.
6. Block SS, *Disinfection, Sterilization and Preservation*, 5th ed. Philadelphia; Lippincott, Wilkins and Williams, 2001.
7. Gillis JR, Ethylene Oxide Sterilization. Chapter in *Validation of Aseptic Pharmaceutical Processes*. Carleton FJ, Agalloco JP, eds. New York: Marcel Dekker, 1986.
8. Agalloco J, Too Much by Half, Misapplication of the Half-Cycle Approach to Sterilization. *Pharmaceutical Manufacturing* 2016;15(11): S2.
9. FDA—Guidance for Industry for the Submission of Documentation for Sterilization Process Validation in Applications for Human and Veterinary Drug Products, 1994.
10. *Validation of Pharmaceutical Processes*. Agalloco JP, Carleton FJ, eds. New York: InformaUSA, 2007.
11. Agalloco J, Validation Protocols, 1980–2008.
12. Agalloco J, The Validation Life Cycle. *Journal of Parenteral Science and Technology* 1993;47(3): 142–147.

BIBLIOGRAPHY

CDC, Guideline for Disinfection and Sterilization in Healthcare Facilities, 2008.

FDA. FDA Cleared Sterilants and High-Level Disinfectants with General Claims for Processing Reusable Medical and Dental Devices, September 2015.

ISO 14160. Sterilization of Single-Use Medical Devices Incorporating Materials of Animal Origin—Validation and Routine Control of Sterilization by Liquid Sterilants, 1998.

McDonnell G, Russell AD, Antiseptics and disinfectants: Activity, action, and resistance. *Clinical Microbiology Reviews* 1999 January, 12(1), 147–179.

Sagripanti JL, Bonifacino A. Comparative sporicidal effects of liquid chemical agents. *Applied. Environmental Microbiology* 1996 February, 62(2), 545–551.

21 Vapor Phase Sterilization and Decontamination*

James Agalloco

CONTENTS

21.1 INTRODUCTION

The sterilization of pharmaceutical products, materials and medical devices must consider the potential impact of the process on the materials being processed, which can play an important role in the selection of an appropriate process. Gases and liquid treatments are among the possible alternatives to ambient temperature sterilization to avoid thermal impact, but gases are highly chemically active, and liquids are only just coming into use. Vapor sterilization can be a useful process for heat-labile materials and equipment. In less critical applications, such as rooms and enclosures, decontamination using similar systems can be useful to control microbial populations. This chapter will explore vapor sterilization and decontamination processes and describe their development, validation and operation.

Vapors are in widespread use for both sterilization and decontamination and there are numerous variations in the processes being employed. An important aspect of vapor treatments is the presence of two phases, gas and liquid. This duality presents unique challenges that must be appropriately considered for success. A brief review of the relevant physical chemistry reveals important distinctions regarding vapor treatments. Consider the following definitions of gases, liquids and vapors:

Gas—"*a substance possessing perfect molecular mobility, and the property of indefinite expansion, as opposed to a solid or liquid.*"[1]

Liquid—"*composed of molecules that move freely among themselves, but do not tend to separate like those of gases; neither gaseous or solid.*"[1]

Vapor—"visible exhalation, as fog, mist, steam, smoke, or noxious gas diffused through or suspended in the air."[1]

Vapors must be understood as being composed of two phases, gas and liquid present simultaneously and in equilibrium with each other. When there is only a small amount of liquid present, a vapor can appear to be a gas, whereas at higher concentrations the liquid phase may condense on surfaces and become increasingly visible. Vapors used for sterilization and decontamination are produced by heating aqueous solutions of the chemical agent above the boiling point to where the agent and water are both gases, dispersing that mixture into a hot air stream and introducing the heated gas combination to the target system. All liquids have a temperature-dependent vapor pressure created by the evaporated gas in equilibrium with the liquid phase. As the temperature increases, the vapor pressure also increases, resulting in a higher concentration of the material in the gas phase. The same phenomenon occurs in reverse when gases are cooled. At their dew point, gases begin to condense and return to the liquid phase. Chemical agents such as hydrogen peroxide and peracetic acid are utilized for sterilization in ways in which both the liquid and gas phases may be present simultaneously, and this is termed a vapor.

As noted earlier, when large amounts of the liquid are suspended in the gas, it can have the appearance of a fog or cloud (Figure 21.1A). When the system is heated, the liquid evaporates and appears as a gas (Figure 21.1B). These images were taken from the same vantage point in the morning (Figure 21.1A) and later in the afternoon (Figure 21.1B) when

DOI: 10.1201/9781003163138-21

Vapor & Gas Sterilization

FIGURE 21.1A and **B** The Mondsee and Schafberg with and without "vapor".

the temperature had increased and some of the suspended liquid (aka cloud) has reverted to the gas phase.

- Gases are more penetrating, more uniform in concentration and less subject to variations in temperature and relative humidity (RH).
- Vapors have different concentrations in each phase and unless well mixed are non-uniform. When a vapor has two possible condensable components, it is even more difficult to predict conditions.

The presence of two phases in vapor processes makes for numerous difficulties because the sterilizing agent and the humidity required for lethality vary in concentration depending upon the phase present. In addition, because of the temperature differences, the concentrations in each phase vary with location and time. The original premise behind vapor processes was that by rapidly boiling off the liquid, all components are rapidly converted into a gas and will maintain the same concentration despite the phase change. This is impossible because the heated gas (or vapor) must lose heat to its surroundings and the least volatile component will begin to condense. Variations in temperature will result in differing amounts of condensation at different locations. Locations near the inlet and where the temperature may be higher (near operating equipment) may not have any condensation. Because everything is introduced as a hot gas, the amount of material available to condense is small, and condensate may not be visible to the naked eye. All of this tends to make any processes relying upon heated vapor extremely complex and far more difficult to manage than either gas or liquid sterilization.

As with gas and liquid sterilization, the concentration of the sterilizing agent has the largest impact on process lethality regardless of the phase. Higher concentrations of the agent and humidity will always be present in the liquid phase relative to the gas phase because of its greater density. The agent concentration and humidity levels will also be different in each phase conforming to Raoult's Law. Thus, the process lethality will differ between the vapor and liquid phases and within each phase as well because of localized differences in concentration. Just as the amount of lethal agent varies with temperature, location and time in a vapor process, the amount of moisture present will also vary.

There are other factors necessary for effective kill of microorganisms. Moisture must be present to assist in penetration of the chemical agent through the spore coat.[2] In liquid sterilization, the presence of liquid water is assured. In gas sterilization, the moisture is provided by the humidity present in the gas phase. In vapor processes, the moisture necessary may be present as either a gas or liquid depending upon the phase present. The temperatures across the system are critically important in vapor processes, as they influence the agent concentration, RH level and rate of kill.

Vapor processes are further complicated by measurement uncertainty. In all vapor processes, the agent concentration, humidity levels and temperatures are rarely constant across the processing environment (and process cycle), and thus measurements taken in the gas phase do not allow for concentration determination in the liquid phase. As vapors are hot gases when introduced into ambient temperature systems, the process temperature is not constant over the duration of the process and concentration measurement is of little utility as it represents only the local conditions in one phase and is not representative of the remainder of the system.

21.2 STERILIZATION/DECONTAMINATION BASICS

Lethal sterilization processes completely kill microorganisms. Decontamination treatments have a less absolute objective and can be assessed by either their ability to meet a required end point as confirmed by post-process sampling, or by delivering a specified minimum log reduction in population. In sterilization processes, the microbiological death curve is defined by the logarithmic number of microorganisms present.[3] When plotting the log population against time, a straight line results. This line can be extended to estimate the number of possible survivors in a large number of units (Figure 21.2). This is termed the Probability of a Non-Sterile Unit (PNSU). A minimally acceptable PNSU is considered not more than 1 positive unit in 1,000,000 units (a risk value originally developed for food safety).

The slope of the microbial death curve is an inherent property of the microorganism and the conditions of the sterilization treatment itself. The inverse of the slope of the curve is the time in minutes for the microbial population to be reduced by 90% (or 1 logarithm) and is commonly termed the D-value.[3] Accurate determination of the D-value requires precise measurement of the lethal conditions to which the microorganism is exposed, and these must be reported with the D-value. D-values for gases and liquids are easily developed as the agent concentration, RH and temperature can be accurately measured.

None of this applies to vapors where the conditions at the point of kill are unknown. D-values for vapor sterilization processes are indeterminate, as the sterilizing conditions (phase, concentration, humidity level) to which the microorganism are exposed are unknown. When these are provided by biological indicator (BI) manufacturers, they typically include temperature and injection rate data and may be indicative of relative resistance only in their system. The manufacturers do this at customer request and understand it's not an actual D-value. The simultaneous presence of two distinct phases, gas and liquid, in vapor processes precludes D-value determination. Lacking an accurate D-value, the types of calculations possible with other sterilization methods are unattainable with vapor processes. Death curves and PNSU estimations are not possible for vapor processes, as the conditions of kill are both variable and indeterminate. Applying that information accurately to systems operating at different conditions is impossible. Thus, an empirical approach relying on BI kill is required. Nevertheless, some inferences can be made with respect to microbial death when using a vapor process:

1. The concentration of the lethal agent will be higher on a milligram/milliliter basis in the liquid phase than that in the vapor phase because of the significantly higher density of the liquid phase.
2. Microbial kill will be much faster in the liquid phase as a consequence of the higher concentration of the agent.
3. The D-value in the liquid phase can be accurately determined using inoculated coupons with a total immersion and rapid neutralization. Published data supports rapid kills at the expected concentrations in the liquid phase.[4] Comparable studies providing microbial kill rates in the gas phase are possible; however, an apparatus capable of assuring that only a gas phase is present has not been identified.

The absence of accurate concentrations in either phase at the point of kill in a vapor process makes the availability of liquid D-values less useful, as conditions across the system are variable with respect to phase, concentration and RH. At best, it could be theorized that a microbial destruction process using a vapor-delivered agent would resemble that shown in Figure 21.3.

This figure is a best-case representation of what might occur at a single location within a system being exposed to a lethal agent in the vapor phase process. As the conditions during vapor processes are neither constant nor precisely known, representing the kill rates as linear with respect to time is unlikely, and the kill over the process duration might be better represented by the colored bands in Figure 21.4 (A–C). The

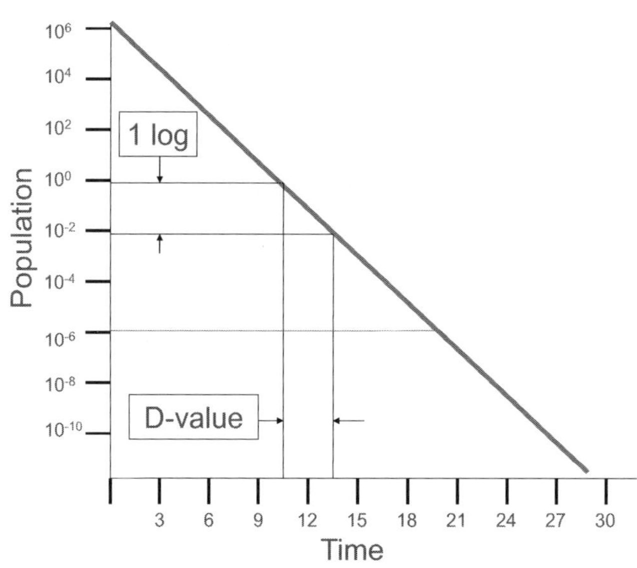

FIGURE 21.2 Microbial death curve.

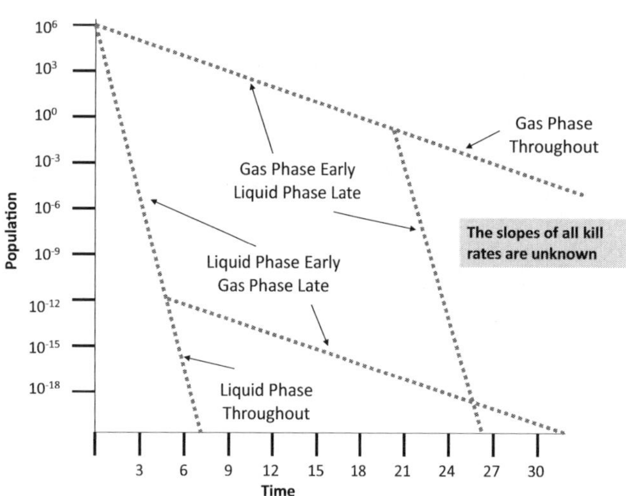

FIGURE 21.3 Estimation of biphasic kill rates.

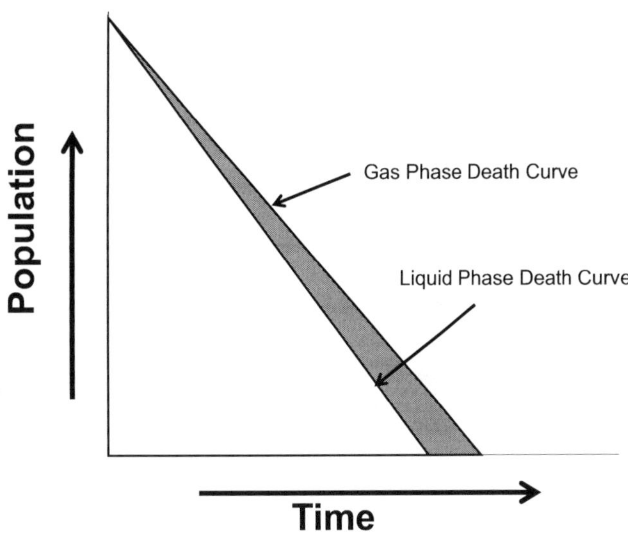

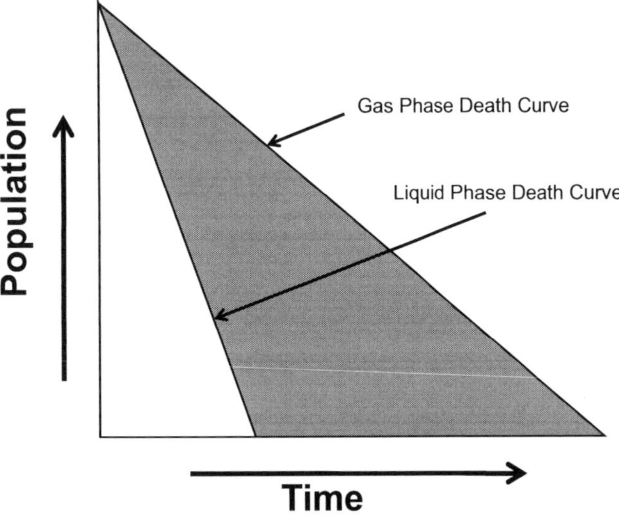

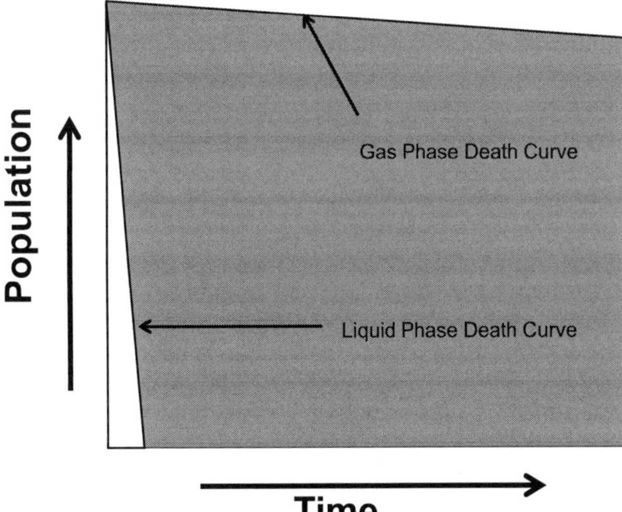

FIGURE 21.4 Vapor phase death curve (A, B, C).

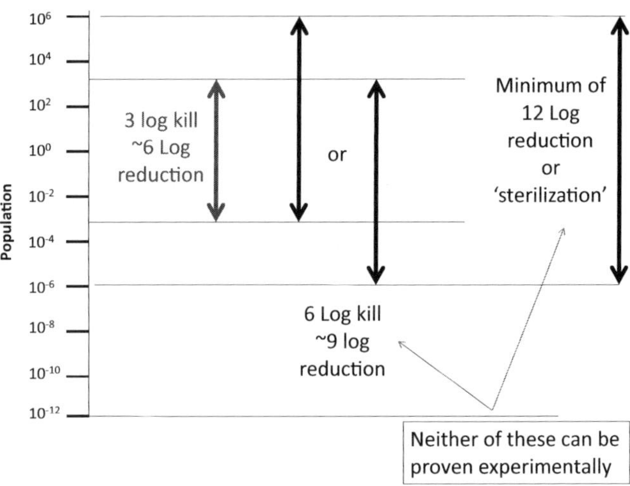

FIGURE 21.5 Log kill and sterilization.

real-world death curve at any location is of unknown shape and slope within the colored band, and given the variations in concentration and RH, a different death curve is possible for each location within the system!

There is consensus agreement that the death occurs substantially more rapidly in the liquid phase than in the gas phase, so condensation, at least in modest amounts, on the materials to be treated is desirable.[4] Thus the relative death curves in the different phases are believed to resemble those in Figure 21.4C rather than in Figure 21.4A.

Putting all the differences and difficulties with vapor processes aside, the objective of these processes is reliable and complete destruction of the microorganisms present. For vapor sterilization processes, this can only be demonstrated microbiologically by complete destruction of the BI. Details on validation approaches are included later in this chapter.

Decontamination using a vapor has the same operational limitations as sterilization but with less demanding acceptance criteria. The varied expectations are depicted in Figure 21.5.

These vary with the individual application, with isolators used for aseptic processing often unnecessarily subjected to sterilization-type treatments, whereas manned clean rooms and other targets are considered less rigorously. Rather than a minimum PNSU of 10^{-6}, the target is often expressed as a 3–4 minimum log kill, which is comparable to a 6–7 log reduction in the population of the BI.[5,6,7,8] The primary difference between sterilization and decontamination is in the microbial challenge population, and this alters the duration of the process dwell period. There are numerous vapor processes for aseptic filling systems in which product contact parts are sterilized with the simultaneous decontamination of the remainder of the system. This is managed by the use of 10^6 BIs on the product contact locations and 10^4 BIs elsewhere. The only difference being the BI population allows for near identical treatment of vapor process treatment and validation for both sterilization and decontamination.

21.3 VAPOR PROCESS FUNDAMENTALS

21.3.1 Material Effects

These processes are designed to kill microorganisms and as such they utilize conditions that may be destructive of essential material properties. Vapor processes are not exempt from this phenomenon, and material evaluation is required. The strong oxidative powers of the lethal agents and the presence of substantial moisture can cause significant changes in the materials being sterilized.[9] The lethal agents may remain on/in the materials post-processing, presenting a potential for an adverse effect. Low levels of residual have been shown to be inhibitory to microbial growth, and this can be an important consideration for microbial media handling. The gas portion of the vapor can enter Tyvek packaged materials, potentially altering their utility. The agent may also impact the material of construction used in the system. The system may be composed of many different materials all of which must be tolerant of the process conditions. Consideration of possible adverse consequences must be an integral part of process selection, equipment design, cycle development and process validation.

21.3.2 Process Application

Vapor processes are used in a variety of settings including multiple interconnected rooms, isolators and small pass-throughs. Occasionally they are used to sterilize or decontaminate fixed equipment such as a lyophilizer or other process unit. To assure greater uniformity of internal conditions, especially temperature, external and/or internal mixing is utilized to enhance the uniformity of the lethal agent and the RH throughout the system. In the treatment of room and enclosure systems, the amount of mixing provided by unidirectional air systems is limited, and supplemental means of mixing should be provided. Given the importance of uniform temperature in vapor processes, overmixing the air is generally preferable. Where the equipment has a jacket, it should be utilized to improve temperature control.

The process can be executed by a control system that provides sequencing, regulation of process parameters and documentation on the process. Contemporary control systems are electronic, either programmable logic controllers (PLCs) or mini-computers. These systems include various features including operator interface, recipe management, process execution and control capability, documentation and interfaces with surrounding systems. A well-designed control system facilitates operation of the system and is essential to maintaining a validated process. There are stand-alone systems that can be used to supply and, in some instances, exhaust simple systems. In these instances, the end user is responsible for interfacing their process equipment with the freestanding control system. Temperature regulation, pressure/vacuum capabilities and other operational features must be provided independent of the vendor-provided stand-alone controller.

21.4 PROCESS EXECUTION

Sterilization using vapors presents substantial difficulties because of the two-phase nature of the agent delivery. The most commonly utilized agent is hydrogen peroxide, although other materials such as peracetic acid or formaldehyde can also be utilized. A sequential vapor/gas process using hydrogen peroxide and then ozone gas received approval by the United States Food and Drug Administration (FDA) for use in hospitals in 2016.[10] Because H_2O_2 is the most commonly used vapor-delivered agent, it will be used to describe the typical vapor process. Hydrogen peroxide is typically supplied in aqueous solution (30%–50%) and introduced into the process with substantial amounts of water. The heated vapor is delivered to the target system as an elevated temperature gas (vapor). In either case, the agent injection will ordinarily result in temperature and RH variation across the chamber initially and throughout the process. Attaining a consistent uniform process with vapors requires constant and significant mixing.

The rapid heating of the aqueous solution converts it to the gas phase, which is mixed with hot air and introduced into the target system. Upon entry into the system, which is typically ambient temperature (colder than the inlet hot gas stream), some of the gaseous material will revert to the liquid phase as the heated gas stream cools. A small amount of condensate may be suspended in the gas phase, but most will eventually deposit on the cooler internal surface of the system/enclosure.

Vapor processes typically operate at or near room temperature and are thus appropriate for heat-sensitive materials. As the water vapor is also subject to condensation, it too can be in either phase. The concentration of agent and water condensed at each location will vary based upon the temperature at that location. The concentration in the gas phase will be uniform to the extent that the target system is well mixed. The concentration on the internal surfaces and materials will vary with the amount of condensation, which is in turn related to the local temperature of the surface.

Penetration of the gas phase through permeable materials is possible; however, it is unlikely to occur once condensed as a liquid. Because of the greater lethality of the condensed liquid, vapor processes are thus rarely utilized where penetration through multiple layers or wrapping is required. Penetration of H_2O_2 into materials in the gas phase is likely comparable to that of water vapor (steam), whereas liquid penetration through permeable barriers which are often hydrophobic is minimal.

Vapor sterilization and decontamination require appropriate agent concentration, RH and temperature. The difficulty created by the presence of two phases is that concentrations of the agent and RH will not be constant across the entire chamber. Concentration determinations in the gas phase (where concentration can typically be measured rather easily, if not inexpensively) do not correlate well with concentrations in the liquid phase. Dalton's and Raoult's Laws apply at each location within the system as temperature variations across the chamber and process dwell alter local dew points and vapor pressure. This substantially complicates precise control of the vapor process, as the target microorganisms are on the surfaces and presumably at a lower temperature than the vapor. Because vapor processes are typically too brief to attain steady-state conditions, the conditions are changing throughout the process. Kill rates of microorganisms in vapor systems vary with concentration and phase. The vapor process is further complicated by continuing temperature variations across the chamber that create localized concentration and RH differences. Nevertheless, provided the system maintains reasonable temperature control and is well mixed, process uncertainties can be minimized and effective treatments demonstrated.

BIs for vapor systems cannot have defined resistance in the form of D-values, as the effective concentration of the agent in contact with the microorganism cannot be determined with precision because of the condensation potential. Thus, although microbial destruction is certainly evidenced by vapor processes, the rate of kill is inexact. As the process parameters cannot be accurately determined, correct D-value determination is currently unavailable and vendor-reported values not transferrable to other systems. Therefore, PNSU estimation is not possible with vapor processes.

The vapor process may incorporate either an initial evacuation or a pre-dehumidification (drying) step to allow for increased H_2O_2 concentration without condensation. The value of these measures is debatable. If condensation of H_2O_2 on surfaces is desirable to increase kill, then condensation should be encouraged rather than avoided. In the most efficient (shortest) vapor processes, drying is rarely included. After exposure, the chamber is aerated/evacuated to remove H_2O_2 from the materials. This portion of the cycle is often the longest, as diffusion of adsorbed H_2O_2 is extremely slow and evaporation of condensed H_2O_2 typically requires more time than the other segments of the overall process.

21.4.1 Peracetic Acid

Peracetic acid, which is often supplied as a mixture with H_2O_2, is an effective sterilant because of its strong oxidizing potential.[4] It is explosive at temperatures above 110°C and thus is introduced into systems as a liquid mist at ambient temperature. A small amount of the liquid peracetic acid will evaporate into the gas phase. Surfaces to be sterilized must be exposed directly to the liquid because the concentration in the gas phase is extremely low. As a strong oxidizing agent, it is corrosive to many materials and also presents considerable handling and safety issues.

21.4.2 Formaldehyde

Formaldehyde (CH_2O) is another effective sporicidal agent. It was a commonly used agent for the fumigation of pharmaceutical facilities in the past and still sees use in some locales. Formaldehyde is a potent carcinogen, and its handling requires extreme caution. The means for introduction resembles that of H_2O_2 and peracetic acid. After processing, it may remain in its polymeric form, paraformaldehyde, as a white reside on surfaces. Its use as a sterilant is possible but extremely problematic.

21.4.3 Other Vapor Processes

There are a substantial number of aerosol delivery systems that utilize H_2O_2 for decontamination/sterilization that share the complications associated with the use of vapor introduction. These systems incorporate a variety of process elements intended to enhance performance. Some of the systems include multiple elements.

- Atomization—the introduction of the fluid as a fine liquid spray in order to increase surface contact and penetration.
- Ionization—addition of an electrical charge to increase free radicals and improve lethality.
- Additives—inclusion of additional antimicrobial agents to improve kill.
- Plasma creation—addition of substantial energy to create a plasma that can increase its antimicrobial activity.
- Combination—following an initial H_2O_2 process with a second lethal process using another agent such as ozone.

Regardless of the details, the various processes all result in a two-phase system with all of the attendant limitations described previously. They are intended for use in a variety of settings: room/facility decontamination, airlock/pass-through processing and sterilization chambers. The performance claims for these systems vary with the application.

21.5 VALIDATION

The performance qualification or "Validation" activity has been described as documentation that the process or product conforms to expectations as determined through independent parameter measurement and/or intensive sampling or challenge. It is the focus of regulatory attention for any sterilization/decontamination processes. It is common practice in performance qualification to utilize a "worst-case" challenge in sterilization/decontamination processes. Typical "worst-case" challenges for vapor processes include: reducing cycle dwell time, reduction of agent concentration and the use of resistant biological challenges as bioburden surrogates. The approaches used for sterilization and decontamination differ only slightly and both will be described.

Historically, gas sterilization processes were validated using the "half-cycle" approach, which uses conservative assumptions about the microbial resistance and number of bioburden microorganisms and was originally developed for use with ethylene oxide.[11] The "half-cycle" approach mandates a sterilization dwell period that destroys not less than 10^6 spores of a resistant BI. In routine operation, the process dwell period is doubled (thus the term "half-cycle") and supports a PNSU of 10^{-6}.[2] The "half-cycle" method as utilized for gas sterilization is graphically depicted in Figure 21.6. The half-cycle method relies on the resistance of the BI (as a surrogate for the bioburden) because complete destruction of the indicator is required in the "half-cycle". Use of the "half-cycle" method provides overkill: *"Overkill sterilization can be defined as a method in which the destruction of a high concentration of a resistant microorganism supports the destruction of reasonably anticipated bioburden present in routine processing."*[12]

The absence of a BI D-value with the vapor process is a non-issue that can thus be largely ignored. "Half-cycle" approaches are inherently conservative, and little effort is made to optimize the process dwell period when it will be arbitrarily doubled in routine use anyway. The "half-cycle" method evaluates only the effect of time, assuming that the effect of lethality of variations in the other essential parameters, agent concentration, RH and temperature can be ignored. This is generally considered a severe limitation of the "half-cycle" method by some practitioners; however, when applied to a vapor process in which the lethal parameters are indeterminate, its reliance on empirical results is not only convenient, but practically speaking the most conservative means possible.

Adaptation of the "half-cycle" method to vapor processes is easily accomplished. The straight line associated with gas sterilization is replaced with the sterilization band depicted in Figure 21.4A-C to provide for the expected kill variation in the different phases. This is depicted in Figure 21.7A and B, which suggest the extremes in delivered lethality between the gas and liquid phases.

Some "half-cycle" validation programs include evaluation of an extended cycle dwell to confirm that the full-cycle process will not result in adverse effects on the materials and equipment. These are also depicted in Figure 21.7A and B.

Another suitable method for validation of vapor processes is the bracketing approach, which better supports the extremes of the operating ranges for the critical process parameters.[13,14] In the bracketing approach, a minimum cycle with lower concentration, lower RH and a shorter dwell period is confirmed by BI destruction using what are understood as less lethal conditions. Material effects are evaluated in a cycle employing a higher concentration, higher RH and a longer dwell period where the adverse impact is believed to be greater. Routine operation of the system utilizes conditions that fall between the process extremes that have been evaluated (Figure 21.8). The routine cycle duration using the bracketing approach is typically not double the duration of the complete kill end point, and this allows for a shorter overall routine cycle duration. A "worst-case" materials cycle can be employed to support material quality.

Applying the bracketing approach to vapor processes allows the practitioner a choice of operating cycles within the validated range (Figure 21.9). The cycles differ primarily in the injection rate and length of exposure dwell period.

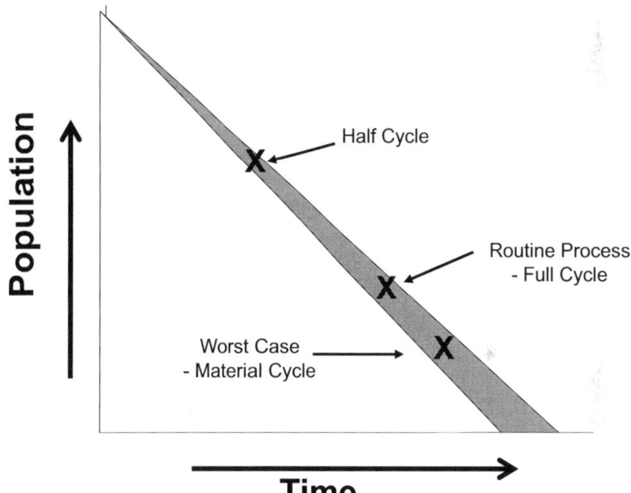

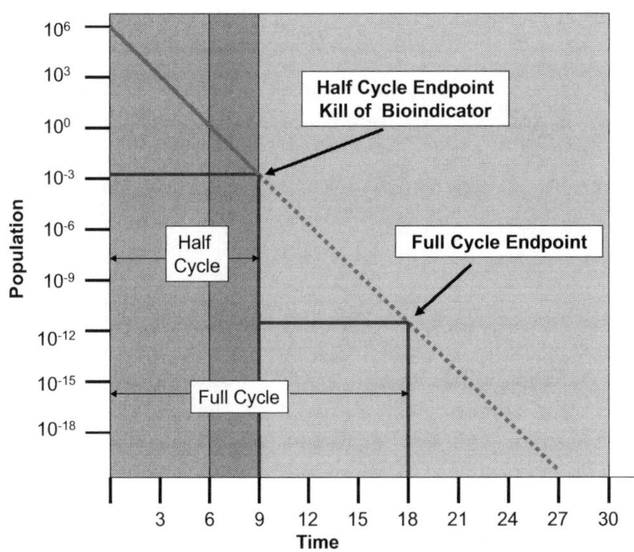

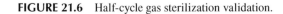

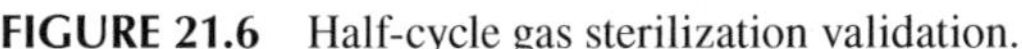
FIGURE 21.6 Half-cycle gas sterilization validation.

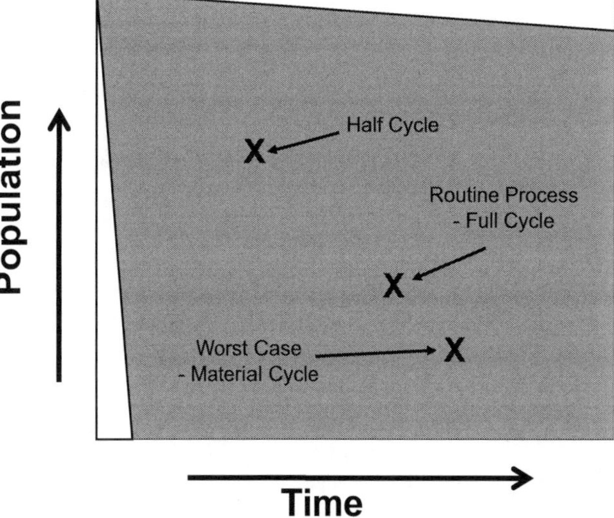

FIGURE 21.7B Half-cycle approach for vapor processes (A, B).

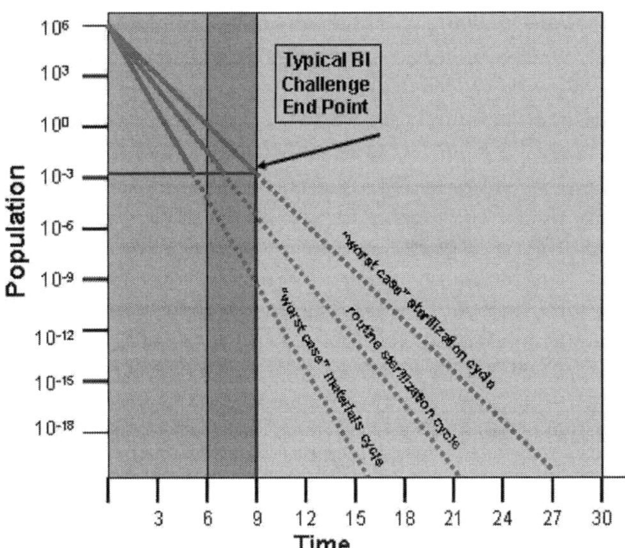

FIGURE 21.8 Gas sterilization validation using bracketing.

Note: BI = biological indicator.

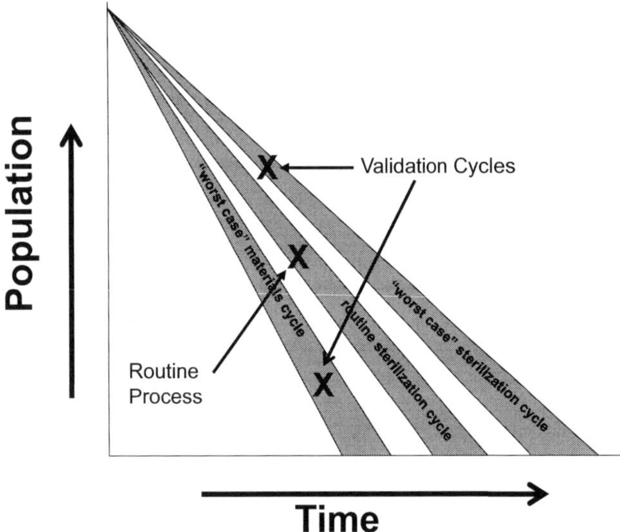

FIGURE 21.9 Bracketing approach applied to vapor processes.

The absence of actual D-values for vapor sterilization requires an empirical approach applied to the methods described previously.[15,16,17] Validating vapor sterilization relies upon demonstrating complete BI kill at either the half-cycle point or the "worst case" cycle and then increasing agent injection rates and process duration for the process to provide more lethal conditions. The material effects are evaluated at the extreme conditions of lethality that result in exposure of materials to the highest concentration of agent, higher humidity and longer exposure time.

Regardless of the validation approach, the use of multiple (3–5) BIs at all locations can add greater confidence in the process' robustness. Because the lethal conditions differ across the system, the locations overall are not replicates, which is only possible with multiple BIs at individual locations. In the validation of vapor sterilization, regardless of the approach utilized, all BIs should be fully inactivated. There is a false belief in "rogue" BIs whose survival is blamed on the BI rather than the use of an insufficiently lethal cycle.[18] The "rogue" BI argument is used to rationalize surviving BIs, which can be avoided when sufficiently lethal conditions are utilized. In the author's experience, process conditions that do not induce condensation are more likely to be determined inadequate.

A recent Medicines and Healthcare products Regulatory Agency (MHRA) blog post cited regulatory concerns regarding the use of vapor processes, stating that they should not be used for the sterilization of product contact parts and surfaces.[19] This regulatory position is believed excessive by the vast majority of vapor phase sterilization users, as many systems lack an easily implemented alternative and its acceptability has been demonstrated in numerous applications.[20]

Decontamination can be demonstrated in an identical manner albeit with multiple lower population BIs (10^3–10^4), which corresponds to a >6–7 log kill when fully inactivated.[5,6,7,8] The simultaneous decontamination of isolator non-product contact surfaces and sterilization of product surfaces using two different population BIs is widespread for aseptic processing systems.

Regardless of the validation method or process goal, there are common elements in the validation efforts:

- Equipment Qualification—The equipment target for the sterilization/decontamination process must be fully documented with respect to installation details and operational characteristics. Equipment qualification serves as the basis for change control for the physical equipment. This effort includes calibration of instrumentation and qualification of the control system.
- Empty Chamber/Vessel/System Temperature Distribution—Concentration and RH measurement within the target system is minimally useful with vapor processes because of the presence of multiple phases. Temperature mapping can be useful in identifying "hot" and "cold" locations that may represent the extremes in process conditions and need to be considered in the biological challenge studies. The system should be mixed during the process to minimize variations across it. Unidirectional air flow is insufficient for obtaining uniformity across a room or enclosure. Overmixing is not a consideration, as additional mixing can only improve the uniformity of the process parameters. BIs are not required in the evaluation of the chamber uniformity. Process consistency can be improved by maintaining the room in which the system is located with ± 2.5°C and ±5% RH.
- Component and Load Mapping—These activities are not a part of vapor sterilization or decontamination, because sampling systems placed within the load items would alter agent penetration. This evaluation is best provided by passive BIs or chemical indicators placed within the load items. This is appropriate

TABLE 21.1
Biological Indicators for Common Chemical Agents

Process / chemical agent	Preferred biological indicator
Hydrogen peroxide	*G. stearothermophilus* or *B. atrophaeus*
Peracetic acid	*G. stearothermophilus*
Formaldehyde	No established BI

only when a vapor process is used where penetration of load items is required.

- Biological Indicators—The use of a BI for initial validation and routine process control is an integral part of validation efforts for vapors. The BI serves as a "worst-case" surrogate for the bioburden present in routine operations. BIs are conventionally spores of a microorganism (most often a *Bacillus* or *Geobacillus* species) chosen specifically for their greater resistance to the sterilizing process than the expected bioburden. Inactivation of the BI during the validation establishes the lethality of the process across the items/surfaces being treated. The BIs of choice for the various sterilizing agents are listed in Table 21.1.

Spore challenges may be either a spore strip or stainless-steel coupon positioned within the chamber and load items.

Process Confirmation/Microbiological Challenge—The core of the validation activity is the confirmation of acceptable process parameters and inactivation of the microbial challenge. Proof of cycle efficacy is provided in replicate studies in which the BIs are completely killed, and chemical/physical measurements are taken as documentation. The differences in relative resistance between the BI and native bioburden are exploited in the validation of vapor sterilization/decontamination. Given the critical importance of BI destruction with vapor processes, this must be given more emphasis than anything else.

21.6 ROUTINE PROCESS CONTROL

Sterilization and decontamination processes must be subject to routine controls that support the efficacy of the cycles over time. Validation is not a one-time activity project, but an integral part of a Current Good Manufacturing Practices (CGMP) compliant facility that must be sustained over the useful life of the facility and its products.[12,21] Control over lethal processes is commonly achieved through practices defined specifically for that purpose including: Calibration of instruments, Physical Measurements of Process Parameters, Use of Physical Integrators/Indicators, Change Control, Preventive Maintenance and Periodic Reassessment.

21.7 CONCLUSION

This chapter provides an overview of the prevalent vapor sterilization and decontamination methods and their validation. This chapter has broadly outlined the primary considerations with respect to these processes. The reader is encouraged to review the larger body of knowledge available on these processes before their implementation.

NOTES

* This chapter was developed before information on a patented 'dry process' deep vacuum H_2O_2 sterilization was widely available and its use is not discussed.

1. http://dictionary.com
2. Pflug IJ, *Microbiology & Engineering of Sterilization Processes*, 8th ed. Minneapolis, MN: University of Minnesota, 1995.
3. Bigelow WD, The Logarithmic Nature of Thermal Death Time Curves. *J. Infect. Dis.* 1921;29: 528–536.
4. Block SS, Peroxygen Compounds. In *Disinfection, Sterilization and Preservation*, 5th ed. Block SS, ed. Philadelphia, USA: Lippincott Williams and Wilkins, 2001: 185–204.
5. USP 28, <1208> *Sterility Testing—Validation of Isolator Systems*, 2000.
6. PDA, TR #34, Design and Validation of Isolator Systems for the Manufacturing and Testing of Health Care Products, 2001.
7. FDA, Industry Guideline on Sterile Drug Products Produced by Aseptic Processing, September 2004.
8. Pharmaceutical Inspection Co-Operation Scheme, Recommendation on Isolators Used for Aseptic Processing and Sterility Testing, PI-014-2, July 2004.
9. Yim S, Detecting Low Levels of Vapor Phase Hydrogen Peroxide (VPHP) and Protecting Biotech Products. Presentation as ISPE Washington Conference, Tampa, FL, June 2010.
10. www.TSO3.com
11. Gillis JR, Ethylene Oxide Sterilization. Chapter in *Validation of Aseptic Pharmaceutical Processes*. Carleton FJ, Agalloco JP, eds. New York: Marcel Dekker, 1986.
12. USP 38, 2nd Supplement,<1229> *Sterilization*, 2013.
13. Agalloco J, Validation Protocols, 1980–2008.
14. USP 37, <1229.7>, Gaseous Sterilization, 2014.
15. USP 38 <1229.11> *Vapor Phase Sterilization*, 2015.
16. Agalloco J, Akers J, Hydrogen Peroxide: Highly Potent & Highly Problematic. *Pharmaceutical Technology* 2013;37(9): 46–56.
17. Agalloco J, Real World H_2O_2 Decontamination. Submitted for publication in *Pharmaceutical Technology*, 2019.
18. PDA TR #51, Biological Indicators for Gas and Vapor Phase Decontamination Processes: Specification, Manufacture, Control and Use, Bethesda, MD, 2010.
19. https://mhrainspectorate.blog.gov.uk/2018/04/20/vhp-vapour-hydrogen-peroxide-fragility/, 2018.
20. www.linkedin.com/pulse/open-letter-response-mhra-blog-vapour-hydrogen-jim-agalloco/, 2018.
21. Agalloco J, The Validation Life Cycle. *Journal of Parenteral Science and Technology* 1993;47(3).

22 Validation of the Radiation Sterilization of Pharmaceuticals

Geoffrey P. Jacobs

CONTENTS

DOI: 10.1201/9781003163138-22

22.1 INTRODUCTION

Gamma irradiation for the sterilization of pharmaceuticals has been a recognized method of sterilization for over 50 years (British Pharmacopoeia, 1963; United States Pharmacopeia, 1965). However, radiation sterilization may also be carried out using electron beam irradiation or the more innovative application of X-rays—uptake of the latter technology for pharmaceuticals has still to gain momentum.

Although high-energy gamma irradiation is predominantly used in the health care industries for the sterilization of disposable medical devices, there has been over the years a gradual increase in the number of pharmaceuticals being radiation sterilized. Today, drugs manufactured by leading pharmaceutical companies are radiation sterilized. These include ophthalmic preparations, topical ointments, parenterals and veterinary products. Unlike medical devices, which are clearly labeled that they are radiation sterilized, pharmaceuticals are not required to be labeled with the mode of sterilization and therefore information on whether a particular drug is radiation sterilized is often not readily available. The increased development of new drug delivery systems, including "combination products", has taken advantage of this sterilization technology.

22.2 RADIATION SOURCES

22.2.1 Gamma-Ray Sources

Gamma-ray photons are emitted by many radioactive isotopes or sources. The principal sources for industrial applications are Cobalt-60 and Cesium-137, with the former being by far the more common. Cobalt-60, with a half-life of 5.3 years, is produced by neutron bombardment of the inactive Cobalt-59. Each disintegrating ^{60}Co atom invokes emission of a beta particle of energy of up to 0.3 MeV, and two gamma photons of 1.17 and 1.33 MeV energy, with the production of an atom of Nickel-60.

The less widely used ^{137}Cs, with a half-life of 30 years, is a fission product of uranium. It emits a single gamma photon of 0.66 MeV energy. However, the use of ^{137}Cs has been limited to small, self-contained dry storage irradiators used primarily for the irradiation of blood and for insect sterilization.

22.2.2 Electron-Beam Production

Electron accelerators used in radiation processing are of three types:

- Direct current high-voltage accelerators where a constant beam is extracted (for example, Dynamitron [IBA, Belgium]).
- Multi-pass through a single cavity radiofrequency accelerator (for example, Rhodotron [IBA, Belgium]).
- Linear microwave pulsed-type accelerators.

They are often classified according to their electron energy as low-energy accelerators (400 to 700 keV); medium-energy (1 to 5 MeV); and-high energy (5 to 10 MeV). The latter provide the highest penetration and are commonly used for radiation sterilization.

The key parameters for electron accelerators are the beam energy (determining the thickness of the product that can be irradiated), beam current (determining the dose rate), scan width and uniformity, output window to container distance, and product speed control.

The direct high-voltage accelerators (or DC type) operate by accelerating electrons across a large drop in potential with a resultant high average beam power. The necessary DC voltage power supplies are usually based on high power, oil or gas filled HV transformers with a suitable rectifier circuit.

The single cavity radiofrequency accelerator (RF) is of pulse or continuous wave type, where lower radiofrequency (100–200 MHz) accelerates electrons with each amplitude. Continuous wave RF-type accelerators provide a DC-like beam current at higher energies.

The use of microwave energy in the electron accelerating process is the main feature of linear accelerators (linacs). Power supplies consist of pulsed microwave generators.

A more detailed review of the technology of electron beam accelerators has been undertaken by Chmielewski et al. (2008).

22.2.3 X-Ray Production

Like gamma rays, X-rays, or bremsstrahlung, are electromagnetic radiations with short wavelengths, very similar to gamma photons, and high photon energies that can stimulate chemical reactions by creating ions and free radicals in irradiated materials. The X-ray electromagnetic photons, emitted when high-energy electrons strike any material, are produced by an electron accelerator. The heavier the element, the greater is the X-rays' conversion efficiency. Therefore, very few X-rays are generated in materials consisting of elements with low atomic numbers (such as plastics—mostly C, H and O) whereas metals like Tantalum (Ta) or Tungsten (W) are very good X-ray generators. X-rays may penetrate deeper than Cobalt-60 photons, depending on the energy, and much deeper than particle-based e-beam units.

There are several types of accelerators that are considered suitable for X-ray production:

- L-band linacs (accelerating RF in the range of 1 GHz; single pass-through multiple cavities; e.g. Impela)

- DC accelerators (direct current; e.g. Dynamitron)
- Rhodotron (an RF-type accelerator; multi-pass through a single cavity)

Incidentally, processing materials and commercial products with high-energy X-rays can produce beneficial changes that are similar to those obtained by the use of gamma rays emitted by Cobalt-60 sources.

22.3 IRRADIATION FACILITIES

22.3.1 Gamma Irradiation Facilities

Industrial gamma-irradiation facilities are of both batch and continuous types. In the former, the product is loaded into tote boxes, which are manually placed on a conveyor within the irradiation chamber. The Cobalt-60 source is then raised from the storage pool and the product moved mechanically around the source rack. A typical source rack typically comprises six modules arranged, for example, in a rectangular plane, where each module contains up to 48 "source pencils". Slugs of Cobalt-60 are enclosed in an inner stainless-steel capsule (a "source element"), with two elements encapsulated in a stainless-steel rod to form the "source pencil" (Figures 22.1 and 22.2).

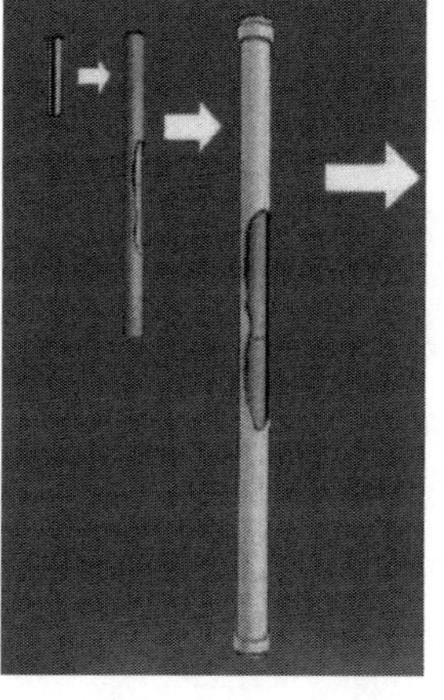

FIGURE 22.1 Slugs of cobalt-60 are enclosed in an inner stainless steel capsule (a "source element") with two elements encapsulated in a stainless steel rod forming a "source pencil". A source module contains up to 48 "source pencils".

Source: Nordion Inc., Canada.

On delivery of the desired radiation dose, the source rack is lowered back into the pool, and the product batch is removed. An alternative batch design employs aluminium carriers suspended from a monorail for holding the product. During irradiation, the carriers move around the source on the monorail system. The design lends itself for upgrading to a continuous or automatic design. Although batch designs are ideal for low throughput, they are particularly useful for small product batches requiring different radiation doses.

Continuous or automatic facility designs are essential for a large throughput. In these designs, the product is moved automatically into the irradiation chamber and conveyed around the radiation source in such a fashion that uniformity of dose and optimal radiation utilization are achieved. Movement within the irradiation chamber is usually in stages, with fixed but different radiation dose increments delivered at each stage or station in such a way that the total of the dose increments is equivalent to the total desired dose. The products are then automatically removed from the irradiation chamber to the unloading and storage areas. The product may be carried around the irradiation source in tote boxes, in carriers suspended from a monorail, or, for large volumes of product, in pallets often with capacities of up to 2 m^3, also suspended from a monorail.

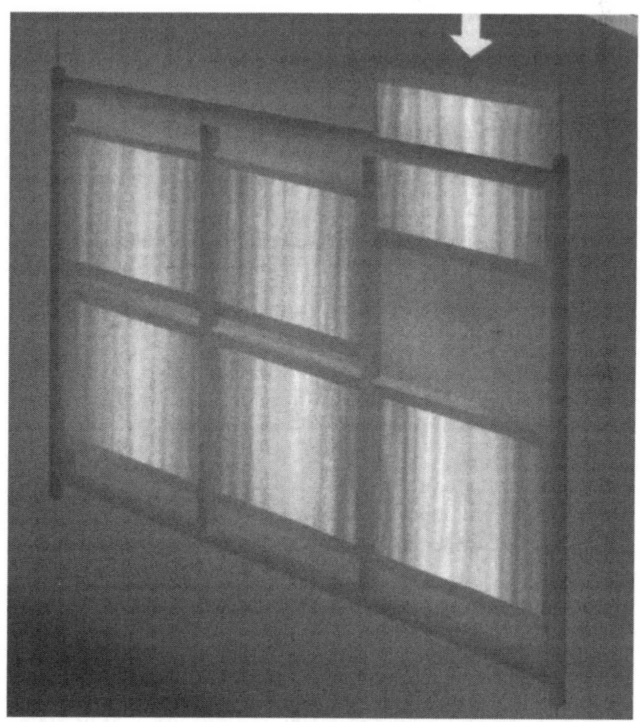

FIGURE 22.2 A source rack comprising six modules arranged in a rectangular plane. Modules may be in different configurations depending on the irradiator design.

Source: Nordion Inc., Canada.

The majority of irradiators used for health care product sterilization are referred to as Category IV irradiators (Mehta, 2008; Gamma Industry Processing Alliance and the International Irradiation Association, 2017).

Most users of gamma irradiation take advantage of existing contract sterilizers. Today, in excess of 200 large-scale commercial gamma irradiators are in operation in about 50 countries, utilizing some 400 million curies (Ci) of Cobalt-60 to irradiate more than 400 million cubic feet of product annually (Gamma Industry Processing Alliance and the International Irradiation Association, 2017). The International Atomic Energy Agency (2011) has published listings of gamma and electron beam irradiation facilities.

22.3.2 Electron-Beam Irradiation Facilities

The basic features of electron accelerators were discussed earlier. It is estimated that there are about 70 to 80 electron beam facilities for sterilization of health care products (Gamma

Industry Processing Alliance and the International Irradiation Association, 2017). A more recent development is the sterilization of small medical devices that are fed directly from a blister packaging machine to the electron beam sterilization unit, which operates at an equivalent speed to that of the packaging system. Other developments in electron beam irradiation were presented, for example, in 2011 at the 16th International Meeting on Radiation Processing held in Montreal (IMRP, 2012), in 2013 at the 17th Meeting in Shanghai (IMRP, 2014), in 2016 at the 18th Meeting in Vancouver (IMRP, 2018), and in 2019 at the 19th Meeting in Strasbourg (to be published). Useful sources of further information on the design and construction of electron accelerators used for radiation sterilization are Cleland and Parks (2003), Sarma (2004), Chmielewski et al. (2008) and Woolston (2009).

22.3.3 X-Ray Irradiation Facilities

The recent introduction of high-power, high-energy accelerators that achieve desirable dose uniformity ratios is changing the relatively slow uptake of X-ray sterilization technology. Modern industrial accelerators have increased throughput, making this mode of sterilization competitive with medium and large Cobalt-60 facilities. Today there are X-ray sterilization facilities in Europe, Japan and North America (IBA, commercial literature). The system in Europe, for example, is based on the IBA Rhodotron (Jongen et al., 2004), delivers accelerated electrons at 7 MeV onto a target, then converts the electron energy into X-rays. The X-rays are directed onto pallets of product undergoing the sterilization process. The system boasts a throughput potential in excess of 124,000 m^3 per year and a dose uniformity ratio (maximum to minimum dose spread) of better than 1.5 for product with a density of 0.3 g/cm^3. For further information the reader can refer to Meissner et al. (2000), Auslender (2002), Jongen et al. (2004), and Kroc-Pl et al. (2017).

22.4 MECHANISMS OF INTERACTION OF RADIATION WITH MATTER

As with all methods of sterilization, irradiation involves a compromise between inactivation of the contaminating microorganisms and damage to the substrate or product being sterilized. The imparted energy in the form of gamma photons, electrons or X-rays does not always distinguish between the two.

22.4.1 Gamma Irradiation

The usual mechanism for interaction between the high energy gamma radiation and matter is the formation of ion pairs caused by the ejection of an electron, leading to free radical formation, and excitation. The free radicals are extremely reactive as a result of the unpaired electron on one of the outer orbitals. Their reactions may involve gas liberation, formation and scission of double bonds, exchange reactions, migration of electrons and cross-linking. In fact, any chemical bond may be broken and any potential chemical reaction may take place. In crystalline materials, this may result in vacancies, interstitial atoms, collisions and thermal spurs, as well as ionizing effects. Polymerization is particularly common in unsaturated compounds. In microorganisms, radiation-induced damage may express itself in various biological changes that may lead to cell death. Although DNA is generally considered the major target for cellular damage, membrane damage may also play a significant role in reproductive cell death (Bacq and Alexander, 1961; Dertinger and Jung, 1970; Alper, 1979).

In solutions, a molecule may receive energy directly from the incident radiation (the "direct effect") or, for example in aqueous solutions, by transfer of energy from the radiolysis products of water (for example, hydrogen and hydroxyl radicals and the hydrated electron) to the solute molecule (the "indirect effect"). Although none of the radical species formed are stable, they may react with other components of the product whether the active ingredient, or the excipients, or both. The only resultant final products, however, in addition to H_20, are hydrogen peroxide (H_20_2) and hydrogen gas (H_2). In fact, the primary radiolysis products of water, hydrogen (·H) and hydroxyl radicals (·OH) and the hydrated electron (e^-_{aq}) disappear within 10^{-3} seconds following irradiation (Spinks and Woods, 1976), and therefore their relevance to aqueous pharmaceutical systems is confined to their being possible contributors to the chemical decomposition of added solutes or the packaging material. On the other hand, relatively stable species such hydrogen peroxide may not only contribute to the instability of the irradiated solutions, but may, although not necessarily, also introduce a problem of toxicity in the use of the aqueous pharmaceutical product. Any hydrogen formed will be at around half the concentration of hydrogen peroxide (Spinks and Woods, 1976), and its formation is not considered significant.

Although the radiation-induced short-lived radical species may not be a stability consideration, the *direct* and *indirect* radiation effects on the packaging material have to be considered. Although it is mandatory to use packaging materials made of radiation stable polymer(s) (see following), only long-term stability studies will show whether any deleterious effects have occurred.

For a better understanding of radiation chemistry and radiation biology relevant to cellular inactivation and molecular damage, the reader may refer, *inter alia*, to these references, Spinks and Woods (1976), Alper (1979), and Tallentire (1979).

22.4.2 Electron Beam Irradiation

The process of radiation-induced damage by electrons is similar to that for gamma photons. In electron irradiation, the high-energy electrons produced externally to the target molecule cause ionization of the molecular species as they pass through the medium and release their energy. The ionization process leads to the production of secondary electrons (known as delta rays), with a range of energies capable of bond breakage in the medium in the vicinity of the ionization event. The high-energy electrons are usually produced either by a direct current machine, by accelerating them across a large drop in potential, or by a linear or circular electron accelerator.

22.4.3 X-Ray Irradiation

X-rays are electromagnetic photons emitted when high-energy electrons strike any material and can therefore be produced by an electron accelerator. As stated previously, X-rays can initiate chemical reactions by creating ions and free radicals in irradiated materials, and their action is similar to that of gamma photons.

22.5 ADVANTAGES OF IRRADIATION FOR STERILIZATION

The advantages of irradiation for the sterilization of health care products are:

- Its high penetrability, thus allowing the product to be sterilized in its final container—even in its shipping container.
- The very low temperature rise (normally <5°C), therefore being compatible with heat sensitive products.
- Fewer process variables than other methods of sterilization—this improves process control with sterility rejections for radiation-sterilized products being the lowest reported.
- No remaining sterilant residuals.

Electron beam irradiation has the added advantages that the sterilization dose can be delivered in just a few seconds compared with several hours or even days with conventional gamma irradiation. This has an added advantage of easier control of the environmental conditions of the irradiation process, which may be important for radiation sensitive products (see the section "Materials Compatibility"). There is also the advantage of the flexibility of allowing individual product treatment when required.

X-ray sterilization is not as fast as electron beam irradiation. Because electron beam and X-ray machines are powered by electricity, there are no disadvantages of handling, shipping and disposal of radioisotopes.

A disadvantage of electron beam irradiation has been their low penetrating power, although the more modern machines have overcome this problem. X-ray machines may be even more penetrating than gamma rays.

For comprehensive reviews of radiation sterilization the reader is referred to, *inter alia*, Fairand (2001), European Commission (2011), and Jacobs (2013).

22.6 VALIDATION OF THE OF THE RADIATION STERILIZATION PROCESS

Validation of the radiation sterilization process, as an integral aspect of Good Manufacturing Practice, comprises the following components that relate either to the irradiation facility itself or the product being irradiated:

- Installation Qualification (IQ)
- Operational Qualification (OQ)
- Performance Qualification (PQ)
- Materials Compatibility
- Selection of Sterilization Dose
- Routine Process Control

It is common practice because of economic or feasibility considerations for a manufacturer of a radiation-sterilized product to use an outside contractor to provide irradiation services. The criteria used in choosing such a contractor must be the same as those used for choosing other outside contractors for pharmaceutical processing. It must be shown that the irradiation facility operates in a manner consistent with current Good Manufacturing Practices (cGMP), that it is registered with the appropriate regulatory authority such as the United States Food and Drug Administration (FDA), European Union (EU) or local health authority, and that it meets all national (or federal) and local regulations. Although many aspects of the validation of the process are usually undertaken by the contract sterilizer, the drug manufacturer bears overall responsibility for the sterility of the product. Essentially, the contract sterilizer is responsible for guaranteeing the delivered radiation dose.

For the more general aspects of validation of pharmaceutical processes, the reader is referred to other chapters in this book.

The authoritative guidelines for validation of radiation sterilization are (1) those published by the Association for the Advancement of Medical Instrumentation (AAMI), TIR 29: *Guide For Process Characterization and Control in Radiation Sterilization of Medical Devices* (AAMI, 2013), formerly *Process Control Guidelines for Radiation Sterilization of Medical Devices*, and (2) the standards published by the International Organization for Standardization, namely:

ISO 11137–1:2006 (reapproved in 2016), *Sterilization of health care products—Radiation—Part 1: Requirements for development, validation and routine control of a sterilization process for medical devices*), with Amendment 2, 2018 (ISO, 2006)

ISO 11137–2:2013 (reapproved in 2018), *Part 2—Establishing the sterilization dose for radiation sterilization* (ISO, 2013a),

ISO 11137–3:2017, *Part 3—Guidance on dosimetric aspects for radiation sterilization* (ISO, 2017a),

ISO/TS 11137–4:2020, *Part 4—Guidance on process control*, Technical Specification, (ISO, 2020).

The preceding AAMI document in its original form (AAMI, 1982) formed the basis for the International Organization for Standardization standard, ISO 11137, first published in 1984 (ISO, 1984), which in turn became the new AAMI/ANSI standard. ISO 11137 was originally published in one part (for example, ISO 11137, 1996a) until 2006, when it was republished in the several parts as listed previously.

22.6.1 Installation Qualification

Installation Qualification (IQ), or irradiator commissioning, is to ensure that the irradiator has been supplied and installed in accordance with its specifications and that the plant will perform consistently within predetermined limits when operated according to the process specification. IQ includes:

- Plant commissioning
- Defined and documented operating procedures and process specifications including:
  - Dosimeter placement (including frequency and rationale)
  - Product handling before, during and after irradiation, including means for segregation of nonirradiated and irradiated products
  - Process release
  - Instrument calibration and recalibration
- Description of the irradiator and associated conveyor systems including:
  - Dimensions
  - Premises and location of the irradiator
  - Materials and construction of the irradiation container
  - Radiation source configuration and characteristics
  - Correct functioning with design specifications of electromechanical systems and associated software
  - Documentation for any modifications
  - Cycle timer setting
  - Choice of dosimeters (see following)

For Gamma Irradiators—

- The activity of the source and dose rate
- The means of indicating the position of the gamma source
- The conveyor path and range of conveyor speeds
- The means of automatically returning the source to its storage position, and ceasing conveyor movement if process control timer or conveyor system fails
- The means of returning the source to its storage position and ceasing conveyor movement or identifying affected product if gamma source is not at its designated position

For Electron Beam and X-Ray Irradiators—

- The characteristics of the beam (electron energy, average beam current, scan width and scan uniformity)
- The means of ceasing irradiation if any failure of conveyor that affects dose occurs
- The means of ceasing conveyor movement or identifying affected product if any fault in the beam occurs
- The dimensions, materials and nature of construction of the X-ray converter

Some aspects of IQ may be considered as part of the operational or performance qualification.

22.6.2 Dosimetry

In radiation sterilization, measurement of the radiation dose is the essential parameter that has to be controlled during all stages of development, validation and routine monitoring, particularly when using gamma irradiation. This is achieved by using dosimeters—chemical or physical systems that respond quantitatively to the absorbed radiation dose. In irradiation practice, although not necessarily at the operational level, four types of dosimeters are used. Three types are used as standards, namely, primary, reference and transfer dosimeters, and a fourth group, routine dosimeters, are used for routine measurement. Understanding and correct use of well-established dosimetry systems and procedures, as well as the interpretation of dosimetry results from OQ and PQ studies, are essential for the validation, commissioning and control of the irradiation process.

22.6.2.1 Primary Dosimeters

Primary dosimeters are the highest quality dosimeters and are maintained by national standards laboratories. The two most commonly used primary standard dosimeters are ionization chambers and calorimeters (Spinks and Woods, 1976).

22.6.2.2 Reference and Transfer Dosimeters

Reference and transfer dosimeters (or secondary dosimeters) are used for calibration of radiation sources and routine dosimetry. The most commonly used reference standard dosimeters are the ferrous sulphate (Fricke) and dichromate dosimeters for gamma and X-ray use, and calorimetry for electron beam applications. In chemical dosimeters (ferrous sulphate and dichromate), the chemical change in a suitable substrate is measured. For example, the concentration of ferric ions formed from the radiation-induced oxidation of an aerated ferrous sulphate solution is determined spectrophotometrically.

Calorimetry, probably the most direct method of determining the amount of energy carried by a beam of radiation, is based on the increase in temperature of a block of material placed in the path of the beam. The material must be such that all the absorbed energy is converted to heat. Graphite or metals are used for this purpose. Other chemical reference standard dosimeters are the alanine, ceric-cerous and ethanol chlorobenzene dosimeters. Most of these reference standard dosimeters may also be used as transfer standard dosimeters.

Transfer standard dosimeters are usually sealed, packaged dosimeters that are sent to the irradiation facility for irradiation to nominal agreed-upon absorbed dose levels in a prescribed geometrical arrangement. The unopened packaged dosimeters are then returned to the national standardization institute (for example, the National Institute of Standards and

Technology [NIST]) to be read and evaluated, thus providing calibration of the client's irradiator.

For electron beam irradiation, the commonly used reference standard dosimeters are calorimeters: alanine, ceric-cerous, ethanol-chlorobenzene, ferrous sulphate and dichromate systems. However, they may be limited by the energy range being used.

22.6.2.3 Routine Dosimeters

Routine dosimeters are used at the irradiation plant level for monitoring and quality assurance in routine irradiation processing. Examples of routine dosimeters for gamma and X-ray use are dyed or clear polymethylmethacrylate, cellulose triacetate, ceric-cerous sulphate, radiochromic dye and ferrous-cupric systems. Most of these systems may also be used for electron beam irradiation.

In selecting a dosimetry system, consideration has to be given, *inter alia*, to the suitability of the dosimeter for the applicable absorbed dose range and for use with a specific product, stability and reproducibility, ease of calibration, ability to correct responses for temperature, humidity and dose-rate deviations, ease and simplicity of use, resistance to damage during routine handling, and inter- and intra-batch responses. It is a requirement that dose measurements are traceable to an appropriate national or international standard, and that their level of uncertainty is known. The period of validity of the calibration should be stated, justified and adhered to.

Practical information on radiation dosimetry, including calibration, selection, characterization, interpretation of results and use can be found in the various Standards published by the American Society for Testing and Materials—ASTM (for example, ASTM, 2009, 2019) as well as IAEA (2013) and ISO 11137–3 (ISO, 2017a) and ISO 11137–4 (ISO, 2020).

22.6.3 Operational Qualification and Performance Qualification

These have been included in one section as opinions may vary as to whether a particular operation is classified as OQ or PQ. The essential point is that all aspects of the validation are undertaken. OQ is to demonstrate that the installed irradiator can operate and deliver appropriate radiation doses within defined acceptance criteria. PQ is effectively dose mapping.

22.6.3.1 Dose Mapping

For the dose mapping procedure, the irradiator is filled to the upper limit of its design specification with containers packed with dummy products or a representative product of uniform density. Dose mapping is carried out on a sufficient number of irradiation containers to allow determination of the distribution and variability of the dose between containers. Dosimeters are placed throughout a minimum of three loaded irradiation containers, which are then passed through the irradiator while surrounded by similar containers or dummy products. The positioning of the dosimeters is dependent on the size of the irradiation container. The results of dosimeter readings obtained will give minimum and maximum absorbed doses in the product and on the container surfaces for a fixed set of plant parameters, product density and loading pattern. This will allow determination of the relationships between the minimum and maximum dose and the dose at the routine monitoring position. In a case of more than one conveyor path, dose mapping should be undertaken for each path.

According to ISO 11137–3 (ISO, 2017a), replicate dose mapping exercises should be carried out in order to obtain information on the variability of the measured doses caused by irradiator variation, product variation and dosimeter measurement reproducibility. A minimum of three dose mapping exercises, each carried out using a separate irradiation container, is recommended in order to obtain statistically valid data. For replicate dose mapping exercises, it could be sufficient to place dosimeters only in areas of dose extremes rather than carry out a full dose mapping exercise.

If partially filled irradiation containers are to be used during routine processing, the effect of partial filling on dose distribution within irradiation containers and the dose and dose distribution in other irradiation containers present in the irradiator has to be determined. For gamma irradiators, the relationship between timer setting, conveyor speed and dose has to be established. For electron beam and X-ray irradiators, variations in the characteristics of the beam during dose mapping have to be within the limits of the irradiator specification, and the relationship between the characteristics of the beam, the conveyor speed and the dose established.

For a dose mapping procedure for electron irradiation, dosimeters should be placed between layers of homogeneous absorber sheets making up a dummy product, or between layers of representative products of uniform density, such that at least ten measurements can be made within the maximum range of the electrons.

Consideration can be given to the use of surrogates for dose mapping exercises involving expensive products or when the availability of product is limited, for example for dilute solutions of aqueous products, water could be used as the surrogate.

22.6.3.2 Operational Qualification and Performance Qualification at the Operational Level

OQ and PQ at a practical level include:

- Information on the dimensions and density of the packaged product
- Orientation of the product within the package
- Product loading patterns, including the possibility of underdosing of dense product or their "shadowing" of other products
- The effect of process interruption
- Dose distribution mapping for assessment of radiation dose ranges within the product package
- Reproducibility within products

Information generated by IQ, OQ and PQ have to be reviewed and documented.

More specific details of dose mapping can be found in the appropriate ISO guidelines (for example, in section 9 of ISO 11137–1 (ISO, 2006)).

22.6.3.3 Process Specification

A process specification for each product should be prepared and documented. Details of such a process specification for gamma, electron beam and X-ray irradiation can also be found in ISO 11137–1 (section 9.4) (ISO, 2006) and include:

- A description of the packaged product, including the dimensions, density and orientation of the product within the package, and acceptable variations
- The loading pattern of the product within the irradiation container
- The conveyor path to be used
- The maximum acceptable dose
- The sterilization dose
- For products that support microbial growth, the maximal interval of time between manufacture and completion of irradiation
- The routine dosimeter monitoring position
- The relationships between the dose at the monitoring position and the minimum and maximum doses
- For products that have to be given multiple exposures to the radiation field, any required reorientation between exposures.

Specifically, for electron beam and X-ray irradiation, the process specification should also include the irradiator operating conditions and limits, that is, beam characteristics and conveyor speed.

Documentation should include reconciliation between the numbers of containers received, irradiated, and dispatched, and with the associated documentation, as well as certification of the range of doses received by each irradiated container within a batch or delivery.

22.6.4 Materials Compatibility

Any processing, such as sterilization, in the manufacture of a pharmaceutical product must cause little or no degradation. This also holds for radiation processing. In the first instance, data on the feasibility of irradiating a pharmaceutical, whether the final product, the active ingredient, or excipients, should be obtained from the scientific literature.

22.6.4.1 Literature Reviews

Reviews on the effects of gamma (and electron beam) irradiation are readily available (Association of the British Pharmaceutical Industry, 1960; Wills, 1963; Geue, 1973; Phillips, 1973; Trutnau et al., 1978; Schuttler et al., 1979a; Schuttler et al., 1979b; Schnell and Bogl, 1982; Schuttler and Bogl, 1982a, 1982b, 1984; Zalewski et al., 1988; Hasanain et al., 2014).

The last volume of the series published by Bogl and his colleagues (in German) have summaries of many of the earlier papers, contains a review of the previous seven volumes and reviews of the results of the 467 investigations of 311 substances published in 134 papers! Part of their review has been published elsewhere (Boess and Bogl, 1996).

An extensive literature review was published by Gopal (1978), which was updated by Jacobs (1985). These last two reviews cover most of the literature published during the period from 1955 to 1985 on radiation effects in pharmaceutical systems. A further extensive review has been published by Jacobs (1995), in addition to a review on the irradiation of blood components (Jacobs, 1997, 1998). More general overviews of the literature and earlier reviews, including analysis of the data, have been published by Razem (2008) and Marciniec and Dettlaff (2008)

A review on irradiating newer drug delivery systems has been published by Abuhanoglu and Ozer (2014). Other reviews, perhaps less extensive, as well as multiple drug studies, have also appeared in the scientific literature (Jacobs and Wills, 1988; Onori et al., 1996; Duroux et al., 1996; Olguner-Mercanoglu et al., 2004; Wilczyński, 2012).

Although many of the cited investigations report only superficial examination of the irradiated drug, the reported data give useful insights into overall radiation stability of these products and indicate whether more extensive testing of the product is worth undertaking.

Studies on the application of X-rays for the sterilization of pharmaceuticals are sparse (May et al., 2002).

22.6.4.2 Assessing Materials Compatibility

It is necessary to examine each new compound to assess its radiation stability, even though data may be available for closely related compounds. No single testing protocol is universal when setting up a feasibility study for the radiation sterilization of a specific group of compounds, and an appropriate testing strategy has to be developed for each group. A thorough knowledge of radiation chemistry would be necessary to infer the behavior of one compound from another. Furthermore, with a formulated medication, the stability of an individual component may change when irradiated as part of a product.

Although radiation sterilization doses are usually in the order of 25 kGy (see next section), the use of a higher dose such as 50 kGy is useful for feasibility studies as a means of identifying the type of radiolytic decomposition that may be expected at sterilization dose levels. Guidance on the establishment of the maximum acceptable dose for a product has been published by the UK Panel on Gamma and Electron Irradiation (2016).

A number of different analytical tools should be used to detect radiation-induced degradation. Each technique usually reveals a change in a specific moiety of the irradiated molecule as illustrated by our own data (Jacobs, 1983), and it is therefore essential to examine all generated data to obtain an indication of the extent of degradation. Wherever possible, stability-indicating assays should be undertaken. In high-pressure liquid chromatography analysis, for example, any discrepancy between the percent recovery value (or assay) and that for the total radiolysis can be attributed to the 1% to 2% inherent error in the methodology, and to the fact that, in general, molar detection sensitivity need not necessarily be the same for the parent compound and its products. Furthermore, there may be peaks that remain undetected because only a limited number of wavelengths are examined, and certain radiolysis products may not absorb in the ultraviolet (UV) region of the spectrum. The purity of the peaks should be determined from absorbance to be sure that other peaks with similar retention times are not being masked, and that each observed peak is representative of a single entity.

It is often useful to use the percentage change derived from chemical assays to estimate G-values. A G-value is defined as the number of molecules (or species) produced or changed for each 100 eV of radiation energy absorbed. On the assumption of a linear relationship between the radiation dose and the number of molecules decomposed, the G-values for the same molecule irradiated under similar conditions but with different radiation doses should theoretically be similar and independent of the dose.

Microbiological assays are still commonly used to analyze antibiotics, and such assays are still found in many compendia. However, large errors are inherent in these procedures. Ideally, these assays should only be used for initial screening for radiation-induced damage.

It has been suggested (Dziegielewski et al., 1973) that compounds in the form of salts or esters are generally less susceptible to radiolysis than the free acids, which, in turn, are more stable than their hydrates. However, other factors must play a significant role in affecting radiation susceptibility. For example, it is hardly credible that the radiolytic decomposition of cephradine monohydrate [G(-cephradine) of 141] is a result of water content, when G(-cefroxadine dehydrate) is 2, and G(-cefadroxil monohydrate) is 9 (Jacobs, 1983). No doubt, the partially saturated ring structure of cephradine contributes significantly to its radiation susceptibility. This is not to say that water of crystallization or loosely bound water plays no role in radiolysis. In fact, one study showed that reduction of the moisture content of a dried antibiotic by one order of magnitude was sufficient to reduce radiolysis by a few percent (Jacobs, unpublished data). This finding is compatible with the consensus that drugs in the solid state or in a nonaqueous milieu are more stable than those in aqueous solution. In this connection, one must be aware of long-term deleterious effects on the drug, such as oxidation by hydrogen peroxide as a radiolysis product of water.

As with all stability studies, assays should be carried out over an extended time period to indicate the long-term stability of the product. Accelerated aging, under conditions recommended by the appropriate regulatory authority such as the United States Food and Drug Administration (FDA), may be undertaken.

Even when radiolysis products are within acceptable compendial limits, it has to be conclusively established that any products formed are without any adverse effect at the concentration found. However, other studies (for example, Kane and Tsuji, 1983) showed that such radiolysis products are generally not unique to irradiation. It would often suffice to show that radiolysis products are the same, and at no greater concentration, than those found when the drug is subjected to other sterilization procedures. In this connection, guidance from the FDA/International Conference on Harmonisation (2006) is useful. It is noteworthy that the FAO-IAEA-WHO Expert Committee (WHO, 1981) has recommended that food items irradiated at doses of up to 10 kGy pose no danger to the consumer and can be unconditionally cleared. Appropriate inferences can be made to pharmaceuticals.

Prior to commencing any determination of the maximum tolerated dose for the product, it is essential to determine if any of the components of the product have or will have received prior radiation treatment. Because radiation effects are cumulative, any prior radiation treatment will affect the interpretation of dose-effect experiments.

22.6.5 Minimization of Radiation Damage

In cases in which radiolysis products are formed, these can sometimes be reduced by appropriate action. For example, irradiation may be undertaken *in anoxia* or at low temperatures, or by incorporation of suitable additives, providing that degradation pathways are known (Werner et al., 1990; Maquille et al., 2008).

Of course, such additives must not be toxic or interfere with the efficacy of the drug. This may be achieved, *inter alia*, by the use of energy transfer systems, thiol group-containing molecules, scavengers of radiolysis products of water, or reagents that convert radiolysis products to the parent compound. One example of such a radiation-tailored formulation is that of urea broth, used for identification of *Proteus* sp., and its differentiation from other gram-negative intestinal bacteria (Eisenberg and Jacobs, 1985).

In some cases, radiolysis may be reduced by use of electron beam irradiation rather than gamma irradiation (Slegers and Tilquin, 2006). Here, *dose rate* may be an important factor. Although there is no general rule, many drugs show less breakdown at the higher dose rate, that is, with electron beam irradiation. This may be because of consumption of all the oxygen (which generally increases radiation damage), with sterilization being completed before oxygen can be replenished; and possibly because of too short a time for production of long-lived free radicals that may increase radiation-induced damage. On the other hand, the high dose rate could in some cases cause increased damage because of the "high concentration" of gamma photons close to the substrate.

22.6.6 Radiation Effects on Packaging Materials

The packaging of a pharmaceutical is an integral part of the product, and therefore the radiation stability of packaging and container materials must never be overlooked when considering radiation compatibility. An irradiation process specification should include details of the packaging of the product. Lists of radiation-compatible packaging materials are readily available (for example, Gopal, 1978; HIMA, 1978; Jacobs, 2013; Shang et al., 1998; Massey, 2005, Haji-Saeid and Chmielewski, 2007; Croonenborghs et al., 2007; Nordion, 2017; AAMI, 2017; Meschini et al., 2018)

It should be emphasized that to ensure their stability, these materials are often formulated specifically for radiation processing by inclusion of, for example, aliphatic antioxidants rather than aromatic ones that are often responsible for yellowing following irradiation. "Regrind resins" must not be used. Very often excellent advice can be obtained from the polymer product manufacturer (the bag or pouch manufacturer) as to the ideal polymer and configuration for the pharmaceutical product at hand. Stability testing of the final preparation (drug plus packaging) will indicate any deleterious effect of the packaging on the drug.

Certification must be obtained that the specific polymer can be irradiated at sterilization doses. In addition, it must have FDA approval for medical use including compliance with biocompatibility and toxicity requirements, such as ISO 10993 Part 1 (2018a) and other parts of ISO 10993, as well as the pertinent monographs of the United States Pharmacopeia (2019) (for example, <87>, <88>, <1031> and <661>). Information on the use of the product in other FDA-approved pharmaceuticals or devices (for example, for combination products) is desirable. For a more detailed discussion of the radiation chemistry of polymers and the irradiation of polymer materials used for packaging, the reader is referred to the following references: Charlesby (1960), Ley (1976), Gopal (1978), Clough (1988), Hemmerich (2000), Massey (2005), Berejka and Kaluska (2008), Drobny (2012), and AAMI (2017).

22.6.7 Decontamination of Pharmaceutical Ingredients by Irradiation

Many powders used in the pharmaceutical industry are heavily contaminated with microorganisms because of their natural source, thus presenting a health hazard. Frequently they do not withstand heating processes to reduce the initial microbial load, so a low radiation dose (<10 kGy) may be sufficient to reduce the bioburden by several orders of magnitude. For example, a study on tragacanth, a natural carbohydrate thickening and suspending agent, has shown that gamma irradiation at 5 kGy greatly reduces the initial microbial load without unduly affecting rheological properties (Jacobs and Simes, 1979), a criterion not fulfilled when employing other methods of decontamination.

For products that can only be prepared aseptically or for those products that cannot withstand a terminal sterilization process to achieve maximally a sterility assurance level of 10^{-6}, consideration of the use of sublethal radiation doses (sometimes referred to as a "polishing dose") may be feasible. Approval of such action may be sought through appropriate regulatory bodies after consideration of a risk assessment of the process (Ley, 1969; Fairand and Fidopiastis, 2010; ISO TS 19930, 2017b).

22.7 SELECTION OF THE STERILIZATION DOSE

Selection of a radiation dose for sterilization is an integral part of validation of the sterilization process and is based on a knowledge of the number and/or resistance to radiation of the bioburden. Any deviation from the selected dose could result in either compromising the sterility of the product (in other words, the predetermined sterility assurance level may not be realized) or, in the case of a high radiation dose, chemical damage to the product. The radiation dose is dependent upon the activity and geometry of the source, the distance from the source to the product container, the duration of irradiation controlled by the timer setting or conveyor speed, the composition and density of the material, the path of the containers through a continuous irradiator or the loading pattern in a batch irradiator.

A radiation dose of 25 kGy (equivalent to 2.5 Mrad) has generally been accepted as suitable for sterilization purposes (for example, United States Pharmacopeia, 1989; British Pharmacopoeia, 1988, and more recent references following). The choice of a sterilization dose, initially for disposable medical products, was founded on the results of basic and applied research in microbiology as well as considerable practical experience (Ley, 1975). A series of investigations in the United States in 1956 by Chandler and his colleagues (Pepper et al., 1956; Koh et al., 1956; Bridges et al., 1956), were aimed to give supporting evidence for the choice of a suitable sterilization dose. Similar experiments in the U.K. (Darmady et al., 1961) confirmed the highest resistance in spores of *Bacillus pumilus* E601.

The minimum 25 kGy sterilizing dose claim originated from a study performed by Charles Artandi and Walton Van Winkle (1959). They determined the "minimum killing dose" for over 150 different species of microorganisms. As a conclusion of their study, they chose 25 kGy as the sterilizing dose, stating that "this [25 kGy] is 40% above the minimum to kill the most resistant microorganisms". Consequently, 25 kGy became established as a suitable minimum irradiation dose for sterilization. This historical sterilizing dose of 25 kGy can also be sufficient to eliminate viable bioburden and provide a high level of microbial control when a validated sterile claim is not required (Tallentire, 1980).

However, today the choice of a radiation dose is based on initial (pre-sterilization) microbial contamination, or bioburden, and the desired sterility assurance level (SAL) of the product[1]

Prerequisites for the establishment of the sterilization dose include (1) the availability of a competent microbiological laboratory to perform determinations of bioburden per product representative of that to be produced routinely, for example, by ISO 11737–1 (ISO, 2018b), (2) tests of sterility per ISO 11737–2 (ISO, 2019), and (3) an appropriate source of radiation capable of precisely and accurately delivering the required doses.

As part of a qualification program to demonstrate the effects of ionizing irradiation on the product, the determination of the product's *maximum tolerated dose* must be undertaken (see Materials Compatibility). In addition, the *maximum process dose* and the *minimum process dose* will also be set (see Dose Mapping). The *maximum tolerated dose* is that dose of radiation above which an unacceptable change in the analytical profile of the pharmaceutical is induced.

22.7.1 Selection of Dose by the AAMI/ANSI/ISO Standard

The most commonly accepted methods for dose selection are those published by the International Organization for Standardization, known as ISO 11137, and more specifically ISO 11137–2, (2013a) (see previous).

The basis of the dose setting methods described in the AAMI/IANSI/SO standards owe much to the ideas first presented by Tallentire and his colleagues, and are based, in part, on extensive studies of the effects of sub-sterilization doses on different microbial populations (Tallentire et al., 1971; Tallentire, 1973; Tallentire and Kahn, 1975, Tallentire and Kahn, 1978). Subsequently, standardized protocols were developed (Davis et al., 1981, 1984).

The ISO dose setting methods are detailed in ISO 11137–2 (2013a). In effect, there are three methods, Method 1 and Method 2, both product-specific sterilization dose methods, and VD_{max} Methods based on a substantiated sterilization dose. A method providing equivalent assurance to that of Methods 1 and 2 in achieving the specified requirements for sterility may also be used.

22.7.1.1 ISO Method 1

The first ISO method, designated Method 1 (and once known as the "first AAMI method" and designated method B1—method 1 detailed as Appendix B of the AAMI Guidelines [AAMI, 1984]), is certainly the most common method used for dose selection for sterilization of medical devices and those pharmaceuticals that are radiation sterilized. The method essentially requires determination of the average microbial contamination of representative samples of the product. Note that the radiation resistance of the microbial population is not determined, and dose setting is based on the resistance of microbial populations originally derived from data obtained from manufacturers (Whitby and Gelda, 1979). The assumption is made that the distribution of the resistance chosen represents a more severe challenge than that presented by the natural bioburden on the article to be sterilized. This assumption is verified experimentally by irradiating 100 samples at a given verification dose and accepted if there are no more than two contaminated samples. The sterilizing dose, appropriate for the average bioburden per sample and the desired sterility assurance level for the product, is then read from a table.

Prior to applying this method (or for that matter any of the other ISO methods) of dose determination, it is essential that the reader be familiar with all the details contained in the Standard. Such a detailed enumeration is beyond the scope of this chapter.

22.7.1.2 ISO Method 2

The second method (Method 2; once known as AAMI method B2) does not entail enumeration of the bioburden but relies on a protocol for a series of incremental-dose experiments to establish a dose at which approximately one in a hundred samples will be non-sterile. A sterilization dose is then established by extrapolation from this 10^{-2} sterility level, using a dose-resistance factor calculated from observations of the incremental-dose experiments that characterizes the remaining microbial resistance. This resistance is estimated from the lowest incremental dose at which at least one sample is sterile, and from the dose at which the surviving population is estimated to be "0.01 microorganisms" per sample. A disadvantage of this method is that at least 280 product items must be taken from *each* of three independent production batches.

22.7.1.3 Earlier ISO Methods

In the original AAMI Guidelines, other more elaborate procedures (originally known as AAMI Methods B3 and B4) were described for dose setting. These methods were not commonly used because of the extensive experimentation involved.

22.7.1.4 ISO VD_{max} Methods

Some years ago, a newer method (Method VD_{max}) specifically for substantiation of a 25 kGy dose was included in the AAMI/ISO guidelines (AAMI, 2005). In essence, it verifies that the bioburden on the product is less radiation resistant than a microbial population of maximal resistance consistent with the attainment of an SAL of 10^{-6} at 25 kGy.

This method was first officially introduced as ISO TR 13409 (1996b). Subsequently it was published as an AAMI Technical Information Report, TIR 27 (2001), and then as an ISO Technical Specification (ISO TS 13409, 2002), but only for small and infrequent production batches. It has now been incorporated in ISO 11137–2 (2013a, reapproved in 2018).

This approach to sterilization dose substantiation was first outlined by Kowalski and Tallentire (1999), who provide an overview of the method. From subsequent evaluations involving computational techniques (Kowalski et al., 2000), it was concluded that the method is safe, robust and yields unambiguous results. The method employs as its basis the standard distribution of resistances (SDR) on which Method 1 is also founded and embodies the following three principles:

1. The existence of a direct link between the outcome of the verification dose experiment and the attainment of an SAL of 10^{-6} at the sterilization dose.
2. Possession of a level of conservativeness at least equal to that of the SDR.
3. For a given bioburden, use of a maximum verification dose (VD_{max}) commensurate with substantiation of a selected sterilization dose.

In 2006, the "VD_{max} method" was expanded to include additionally a dose of 15 kGy (11137–2, 2006). Although the current ISO Standard (11137–2, 2013a) incorporates both the

VD_{max} 25 and 15, a Technical Specification (ISO TS 13004, 2013b, and reconfirmed in 2017) has been published to include VD_{max} 17.5, 20, 22.5, 27.5, 30, 32.5 and 35. This parallels a similar document published earlier by AAMI (TIR33:2005). Application of the VD_{max} approach to doses other than 25 kGy is discussed by Kowalski and Tallentire (2003).

VD_{max} 25, for the substantiation of 25 kGy as a sterilization dose, is for products with an average bioburden less than or equal to 1,000, and VD_{max} 15 for the substantiation of 15 kGy as a sterilization dose for products with low average bioburden (less than or equal to 1.5). Each of the other VD_{max} doses provide a methodology for the substantiation of a range of sterilization doses each of which is valid only for a specified unique range of average bioburden on product. The application of this method is no longer limited by batch size or production frequency, and the number of product units irradiated in the verification dose experiment remains constant.

Verification is undertaken at an SAL of 10^{-1} with ten items irradiated in the performance of the verification dose experiment. The dose corresponding to this SAL (verification dose, VD_{max}) reflects both the magnitude of the bioburden and the associated maximal resistance. If there is no more than one positive test in the ten tests of sterility, the chosen sterilization dose is substantiated. This method is applied with some modification to both single and multiple batches.

ISO also allows substantiating a 25 kGy dose using Methods 1 and 2. In addition, ISO 11137–2 allows dose setting by any other method that provides equivalent assurance to the previous methods in achieving the specified requirements for sterility.

22.7.1.5 Dose Audits

In accordance with the ISO 11137–2, the continued appropriateness of the established sterilization dose must be demonstrated through the conduct of periodic bioburden determinations and sterilization dose audits to monitor any change in the radiation resistance of the bioburden.

The frequency of bioburden determinations is dependent on the product average bioburden and the method that has been used for dose selection. If the average bioburden is equal to or greater than 1.5, the maximum interval of time between bioburden determinations is set at three months. For an average bioburden of less than 1.5 and the sterilization dose was either set using Method 2 or a sterilization dose of 25 kGy was selected, the maximum interval is also 3 months. However, if the sterilization dose was set using Method 1 or a sterilization dose of 15 kGy was selected, the maximum interval is one month. In cases in which the time interval between manufacture of batches is more than one or three months, determinations of bioburden are performed on each production batch.

An appropriate investigation must be undertaken if the outcome of bioburden determinations exceeds the specified limit. In such a case, the ISO guideline stipulates the appropriate action, which must then be followed by a dose audit. Depending on the outcome of this audit, whether successful or unsuccessful, procedures described in detail in ISO 11137–2 are to be undertaken.

The frequency of sterilization dose audits is either quarterly or based on an appropriate rationale that takes into account characterization of the bioburden and related microbiological considerations.

An increase in the interval of time between performance of sterilization dose audits is permitted if (1) at least four consecutive sterilization dose audits, whose outcomes have required neither dose augmentation[2] nor dose reestablishment, have been performed at the previously selected intervals, and (2) data are available that demonstrate the stability of the bioburden (based on bioburden determinations and characterization performed at least every three months) within the bioburden specification over these same intervals. However, the maximum permitted interval between performance of audits is twelve months.

Although published in conjunction with the 2006 edition of ISO 11137–2, a still useful guidance on the application of the ISO methods for dose selection is the U.K. Panel on Gamma and Electron Irradiation overview (2012).

22.7.1.6 European (EN) Standards

The original European Standard, EN 552:1994 (European Committee for Standardization, 1994), for validation of radiation sterilization, albeit for medical devices, has been withdrawn and, under the Vienna Agreement (ISO and CEN, 2016), is published as an ISO document.

22.7.2 Other Dose Selection Methods

22.7.2.1 United States Pharmacopeia

The United States Pharmacopeia 42 (2019), section <1229.10>, "Radiation Sterilization" states that the methods used to establish appropriate radiation doses to achieve the desired sterility assurance level are as defined in the various parts of ISO 11137.

This General Chapter continues that once the dose is selected, all materials exposed to radiation, especially the drug product and its primary container, should be evaluated for immediate and long-term effects, and that product stability, safety and functionality should be confirmed over the product's intended use period. Some materials may appear unchanged initially, and the effects may only become evident over time.

22.7.2.2 European Pharmacopoeia Procedures

The European Pharmacopoeia (2017), on Ionizing Radiation Sterilisation, in the section "Methods of preparation of sterile products", states:

> For this method of terminal sterilisation the reference absorbed dose is 25 kGy. Other doses may be used provided that it has satisfactorily been demonstrated that the dose chosen delivers an adequate and reproducible level of lethality

> when the process is operated routinely within the established tolerances. The procedures and precautions employed are such as to give an SAL of 10^{-6} or better.

The European Pharmacopoeia gives no guidance on how to estimate doses of <25 kGy.

22.7.2.3 European Medicines Agency

In the draft European Medicines Agency (EMA) guideline on the sterilization of medicinal products, active substances, excipients and primary containers (EMA, 2016), it is stated in the section "Ionization radiation sterilisation" that data as requested in Note for Guidance, "The Use of Ionization Radiation in the Manufacture for Medicinal Products" (EMEA[3], 1991) should be provided, supplemented as necessary by data requirements given in ISO 11137 and the European Pharmacopoeia chapter 5.1.1 (2017).

The preceding EMA guideline (EMA, 2016) includes three decision trees intended to assist in the selection of the optimal sterilization method. In the decision tree for dry powder products, nonaqueous liquid or semisolid products, is the option of whether the product can be sterilized with "an absorbed minimum dose of ≥ 25 kGy or using a validated lower irradiation dose" with a reference to the ISO 11137 guidelines (ISO, 2019).

A comment from the Parenteral Drug Association (PDA) on the then draft guidance (letter of October 13, 2016) is that the decision tree "allows the adoption of a 25 kGy sterilizing dose without the requirement for proper validation." The PDA recommends adding the requirement to validate all radiation doses per ISO 11137 and to amend the decision tree question to ask whether the product can be sterilized to an SAL of 10^{-6} per ISO 11137, without reference to the 25 kGy dose. This recommendation has not been incorporated in the final Guideline (EMA, 2019).

Irradiation has not been included in the decision tree for sterilization choices for aqueous products but has been included in the third decision tree for sterilization of containers.

22.7.2.4 Japanese Pharmacopoeia

The Japanese Pharmacopoeia (2016) states that radiation sterilization may be performed in accordance with ISO 11137–2 or the equivalent Japanese Specification, JIS T 0806–2.

22.7.2.5 Parenteral Drug Association Procedures

The PDA had made its own recommendations for dose setting procedures specifically for parenteral products (Parenteral Drug Association, 1988). These procedures, however, are similar to those already in use for other sterilization technologies.

One method is really a biological indicator (overkill) method in which the sterilization dose is at least double a radiation dose needed to achieve a six logarithmic inactivation of *Bacillus pumilus* spores on or in the product. In practice, the sterilization dose does not differ much from the classical "25kGy".

Another method involves determination of the maximum bioburden. The logarithm of this bioburden (with three standard deviations), plus a six-logarithm sterility assurance factor is multiplied by the decimal reduction factor (D_{10}) for *Bacillus pumilus* spores to estimate the sterilization dose. The decimal reduction factor is the radiation dose to reduce the number of viable microorganisms in the product by 90%.

22.7.2.6 International Atomic Energy Agency (IAEA) Procedure for Dose Setting

The International Atomic Energy Agency (IAEA), following an Advisory Group Meeting on the Code of Practice for Radiation Sterilization of Medical Supplies (Colombo, November 1986), adopted a pragmatic approach to the selection of a sterilization dose. The Guidelines (IAEA, 1990) developed at this Meeting state:

> It is a basic assumption that the product to be sterilized is manufactured under conditions that comply fully with the requirements of GMP. In the present context, it is particularly important that practices be implemented, and actions taken, which ensure that the number of micro-organisms on product items destined for radiation sterilization processing is consequently low.
>
> A dose of 25 kGy has been found to be an effective sterilizing dose. It is generally believed that this dose provides maximally a SAL of 10^{-6}. Where it is not feasible to generate data on the radiation resistance of the natural microbial population present on product items, a minimum sterilizing dose of 25 kGy can be used.
>
> It is more rational to base selection of a sterilizing dose on a knowledge of the resistance of the natural microbial population present in product items to be sterilized and on a reasoned selection of a maximal SAL. Methods of dose selection using this approach are Methods 1 and 2 in Appendix B of the AAMI Process and Control guidelines for gamma Radiation Sterilization of Medical Devices (corresponding to the current methods 1 and 2 of ISO 11137.

Although it is this author's belief that the methods of dose selection presented in ISO 11137 are the methods of choice, the IAEA approach some 30 years later is still rational particularly for less developed countries.

22.7.2.7 Pharmaceutical Inspection Convention

The Pharmaceutical Inspection Convention in its Guide to Good Manufacturing Practice for Medicinal Products (PICS, 2018), while discussing radiation sterilization, presents no specific guidance regarding dose selection other than the process must be validated.

22.7.2.8 World Health Organization

The World Health Organization (WHO, 2011), like the Pharmaceutical Inspection Convention, presents no specific guidance regarding dose selection other than stating that all sterilization processes should be validated.

22.7.2.9 Other Dose Setting Procedures

Other dose setting procedures have been proposed in the scientific literature, including those of van Asten and Dorpema

(1982), Davis et al. (1984), and Darbord and Laizier (1987) and the UK Panel on Gamma and Electron Irradiation (1989).

The preceding demonstrate the approaches to the choice of dose by different regulatory authorities. Close examination, however, shows the similarity of the various approaches.

22.8 ROUTINE PROCESS CONTROL

Routine Process Control includes process specification, pre-irradiation product handling, product irradiation, product loading and unloading, radiation dose monitoring during irradiation, processing records and documentation, process interruption and routine and preventive maintenance.

Irradiation containers should be packed in accordance with a specified loading pattern established during validation. Radiation indicators should be used as an aid in differentiating irradiated from non-irradiated products but should not be used as an indication of satisfactory processing. The design of the irradiation facility should ensure complete segregation of non-irradiated and irradiated products at all times.

During continuous gamma irradiation processing, dosimeters should be placed so that at least two are exposed during irradiation at all times. For batch modes, at least two dosimeters should be exposed in positions related to the minimum dose position (PICS/S, 2018; IAEA, 2008, 2013; EC, 2011). On the other hand, ISO 11137, Part 3, in section 10.2 (ISO 2017a) appears to be less demanding, or at least less specific.

During electron beam irradiation, a dosimeter should be placed on every container (PICS/S 2018; IAEA, 2008, 2013; EC, 2011). There should be continuous recording of average beam current, electron energy, scan-width and conveyor speed. These variables, other than conveyor speed, need to be controlled within the defined limits established during commissioning, because they are liable to instantaneous change.

ISO 11137 does not require that biological indicators be used for validation or monitoring of radiation sterilization, nor does it require that a pharmacopoeial test for sterility be carried out for product release.

The continued effectiveness of the established sterilization dose must be demonstrated through the conduct of bioburden determinations and sterilization dose audits as described earlier.

22.9 LEGISLATIVE CONSIDERATIONS

Although radiation sterilization has appeared in the United States Pharmacopeia since 1965, the FDA regards a radiation-sterilized drug as a "new-product" (that is, a requirement for the submission of a New Drug Application (NDA), albeit abbreviated, but see following) with the manufacturer responsible for proving its safety. The current United States Pharmacopeia (2019) makes the observations regarding radiation sterilization of drugs (1) that radiation sterilization is unique in that the basis of control is principally that of absorbed radiation dose, which can be precisely measured; (2) that evaluation of the irradiated "item" should consider all of the materials exposed to the radiation processing, especially the drug product and its primary container; and (3) that product stability, safety and functionality should be confirmed over the product's intended use period.

In the United Kingdom, for example, sterilization by exposure to ionizing radiation has been a recognized method since 1980, when the Ministry of Health agreed to accept materials exposed to a radiation dose of 25 kGy. Medicines controlled under the Medicines Act 1968 are subjected to individual assessment by the Committee on Safety of Medicines of the Medicines and Healthcare products Regulatory Agency. This committee requires, in addition to proof of sterility, proof that the potency of the drug is unaffected by the process and that any degradation products would not be harmful.

Similarly, although the British Pharmacopoeia (2019) recognizes gamma irradiation as a suitable sterilization process, it is the responsibility of the manufacturer to prove that no degradation of the product has taken place.

Most European countries allow pharmaceuticals to be radiation sterilized, provided that authorization has been obtained from the appropriate health authorities.

According to the European GMP guidance (European Commission, 2011), "Ionising radiation may be used during the manufacturing process for various purposes including the reduction of bioburden and the sterilisation of starting materials, packaging components or products and the treatment of blood products." A "Note" by the Committee for Proprietary Medicinal Products of the European Agency for the Evaluation of Medicinal Products (EMEA, 2001) states that parametric release can be applied to radiation sterilization, with the minimum dose "generally" 25 kGy, although "lower doses can be acceptable when justified." As stated earlier, this EMEA Committee has published decision trees for the selection of sterilization methods in which radiation sterilization is included for "dry powder products, non-aqueous liquid or semi-solid products".

In a recent draft Reflection Paper by the European Medicines Agency on GMP and Marketing Authorisation Holders (MAH) (EMA, 2020) in relation to the use of ionizing radiation in the manufacture of medicinal products, there are certain responsibilities for the MAH documented in Annex 12 of the GMP guide (EC, 2011). One is a responsibility for the MAH to agree on the design of irradiation cycles with the manufacturer. The guide states that "When the required radiation dose is by design given during more than one exposure or passage through the plant, this should be with the agreement of the holder of the marketing authorisation and occur within a predetermined time period. Unplanned interruptions during irradiation should be notified to the holder of the marketing authorisation if this extends the irradiation process beyond a previously agreed period."

The direct requirement for the MAH to work with the manufacturer with regard to the design of irradiation cycles is not considered a task that may be delegated by the MAH to the manufacturer (EMA, 2020).

The Japanese Pharmacopoeia (2016) requires that pharmaceuticals to which radiation sterilization can be applied must be sterilized so that a sterility assurance level of 10^{-6}

or less is generally obtained. This Pharmacopoeia states that because sterilization can take place at room temperature, both methods (gamma and electron beam irradiation) can be applied to heat-labile items, and items can be sterilized while packaged because the radiation rays will penetrate the packaging.

The Pharmaceutical Inspection Convention states that irradiation is used mainly for the sterilization of heat-sensitive materials and products (PICS, 2018), and continues (in Annex 12—Use of Ionising Radiation in the Manufacture of Medicinal Products) that this method is permissible only when the absence of deleterious effects on the product has been confirmed experimentally. The Annex states that ionizing radiation may be used during the manufacturing process for various purposes including the reduction of bioburden and the sterilization of starting materials, packaging components or products and the treatment of blood products.

Similarly, the World Health Organization states that sterilization can be achieved, *inter alia*, by irradiation with ionizing radiation, and that this mode of sterilization is mainly used for heat-sensitive materials and products (WHO, 2011). It continues that because many pharmaceutical products and some packaging materials are radiation-sensitive, this method is permissible only when the absence of deleterious effects on the product has been confirmed experimentally.

Of significance is the FDA guidance on applications for parametric release (FDA, 2010), which states: "The principles in the guidance may also be applicable to products sterilized by other terminal sterilization processes, such as radiation sterilization, which may be suitable for parametric release." The author is aware of FDA requests for companies to determine the feasibility of terminal sterilization with gamma irradiation to enhance sterility assurance.

Countries whose regulatory authorities are known to have granted approval for specific radiation-sterilized pharmaceuticals include Australia, Bangladesh, India, Indonesia, Israel, Mexico, Norway, Taiwan, U.K., U.S.A. and some other European countries.

Of particular interest is the recent Federal Register notice of a proposed rule by the FDA (2018) to repeal the regulation requiring an Approved New Drug Application (NDA) or an Abbreviated New Drug Application (ANDA) for any drug product that is sterilized by irradiation. According to the Notice,

> Repealing the irradiation regulation would mean that over-the-counter (OTC) drug products that are generally recognized as safe and effective, that are not misbranded, and that comply with all applicable regulatory requirements can be marketed legally without an NDA or ANDA, even if they are sterilized by irradiation. FDA is proposing to take this action because the irradiation regulation is out of date and unnecessary. The technology of controlled nuclear radiation for sterilization of drugs is now well understood, and our regulations require that OTC drugs be manufactured in compliance with current good manufacturing practices (CGMPs). Appropriate and effective sterilization of drugs, including by irradiation, is adequately addressed by the CGMP requirements.

The notice states: "This proposed rule would repeal the irradiation regulation, which provides that any drug sterilized by irradiation is a new drug."

NOTES

1 Sterility Assurance Level (SAL) is defined as the probability of a single viable microorganism surviving on a product after exposure to a given sterilization process. SAL can also be regarded as the probability of a single viable microorganism in a population of terminally sterilized product items. SAL is normally expressed as 10^{-n}. Although the majority of authorities give n a value of 6, the FDA does allow values of n of less than 6 (for example, 10^{-3}) for non-invasive products. However, it should be noted that SAL has a quantitative value and an SAL of 10^{-6} takes a lesser value than an SAL of 10^{-3}. Hence, there is a greater assurance of sterility associated with a lesser SAL.

2 The method for augmentation of the sterilization dose is based on a method proposed by Herring (1999) in cases where the number of audit positives exceeds the limits imposed by ISO 11137. It uses the information from the failed sterilization dose audit and the principles underlying Method 2 (outlined previously), together with a conservative estimate of the resistance of the most radiation-resistant component of the microbial population of the product.

3 The European Medicines Agency (EMA), was, prior to 2004, known as the European Agency for the Evaluation of Medicinal Products or the European Medicines Evaluation Agency (EMEA).

REFERENCES

Abuhanoglu G, Ozer AY. Radiation sterilization of new drug delivery systems. *Interv. Med. Appl. Sci.* 2014, 6, 51–60.

Alper T. *Cellular Radiobiology*. Cambridge: University Press, 1979.

Artandi C, Van Winkle W. Sterilization of pharmaceuticals and hospital supplies by ionizing radiation. *Int. J. Appl. Radiat. Isotop.* 1959, 7, 64–65.

AAMI. *Association for the Advancement of Medical Instrumentation, Process Control Guidelines for Radiation Sterilization of Medical Devices* (N. RS-P 10/82). Arlington, VA: AAMI, 1982.

AAMI. *Association for the Advancement of Medical Instrumentation, Process Control Guidelines for Radiation Sterilization of Medical Devices*. Arlington, VA: AAMI, 1984.

AAMI. *Association for the Advancement of Medical Instrumentation, Technical Information Report, TIR 27, Radiation Sterilization—Substantiation of 25 kGy as a Sterilization Dose—Method* VD_{max}. Arlington, VA: AAMI, 2001.

AAMI. *Association for the Advancement of Medical Instrumentation, Technical Information Report, TIR 33, Sterilization of Health Care Products—Radiation—Substantiation of a Selected Sterilization Dose—Method* VD_{max}. Arlington, VA: AAMI, 2005.

AAMI. *Association for the Advancement of Medical Instrumentation, Technical Information Report, TIR 29, Guide for Process Characterization and Control in Radiation Sterilization of Medical Devices*. Arlington, VA: AAMI, 2013.

AAMI. *Association for the Advancement of Medical Instrumentation, Technical Information Report, TIR 17, Compatibility of Materials Subject to Sterilization*. Arlington, VA: AAMI, 2017.

Association of the British Pharmaceutical Industry. *Use of Gamma Radiation Sources for the Sterilization of Pharmaceutical Products*. London: ABPI, 1960.

ASTM International. *Standard Practice for Dosimetry in Radiation Processing, ASTM E 2628–09*. West Conshohocken, PA: ASTM, 2009.

ASTM International. *Annual Book of ASTM Standards*, Vol. 12.02. West Conshohocken, PA: ASTM, 2019.

Auslender VL. Accelerators for e-beam and x-ray processing. *Radiat. Phys. Chem.* 2002, 63, 613–615.

Bacq ZM, Alexander P, *Fundamentals of Radiobiology*. Oxford: Pergamon Press, 1961.

Berejka AJ, Kaluska IM. Materials used in medical devices. In *Trends in Radiation Sterilization of Health Care Products*. Vienna: IAEA, 2008, 159.

Boess C, Bogl W. Influence of radiation treatment on pharmaceuticals—A review: Alkaloids, morphine derivatives, and antibiotics. *Drug. Dev. Ind. Pharm.* 1996, 22, 495–529.

Bridges AE, Olivo JP, Chandler VL. Relative resistances of microorganisms to cathode rays. II. Yeasts and molds. *Appl. Microbiol.* 1956, 4, 147–149.

British Pharmacopoeia 1963. London: Her Majesty's Stationary Office, 1962, A209.

British Pharmacopoeia 1988. London: Her Majesty's Stationary Office, 1987.

British Pharmacopoeia 2019. London: The Stationary Office, 2018.

Charlesby A. *Atomic Radiation and Polymers*. New York: Pergamon Press, 1960.

Chmielewski AG, Sadat T, Zimek Z. Electron accelerators for radiation sterilization. In *Trends in Radiation Sterilization of Health Care Products (STI/PUB/1313)*. Vienna: International Atomic Energy Agency, 2008, 27–47.

Cleland MR, Parks LA. Medium and high-energy electron beam radiation processing equipment for commercial applications. *Nucl. Instrum. Methods Phys. Res. Sect.* 2003, 208, 74–89.

Clough RL. Radiation resistant polymers. In: Kroschwitz JI, ed. *Encyclopedia of Polymer Science and Engineering*, Vol. 13, 2nd edn. New York: Wiley, 1988, 667–708.

Croonenborghs B, Smith MA, Strain P. X-ray versus gamma irradiation effects on polymers. *Radiat. Phys. Chem.* 2007, 76, 1676–1678.

Darmady EM, Hughes KEA, Burt MM, Freeman BM, Powell DB. Radiation sterilization. *J. Clin. Pathol.* 1961, 14, 55–58.

Darbord JC, Laizier J. A theoretical basis for choosing the dose in radiation sterilization of medical supplies. *Int. J. Pharmaceut.* 1987, 37, 1–10.

Davis KW, Strawderman WE, Masefield J, Whitby JL. Gamma radiation dose setting and auditing strategies for sterilizing medical devices. In Gaughran ERL, Morrissey RF, eds. *Sterilization of Medical Products*, Vol. 2. Montreal: Multiscience Publications Ltd., 1981, 34–102.

Davis KW, Strawderman WE, Whitby JL. The rationale and computer evaluation of a gamma sterilization dose determination method for medical devices using a substerilization incremental dose sterility test protocol. *J. appl. Bact.* 1984, 57, 31–50.

Dertinger H, Jung H. *Molecular Radiation Biology*. London: Longman; Berlin: Springer-Verlag, 1970.

Drobny JG, *Ionizing Radiation and Polymers—Principles, Technology, and Applications*. Amsterdam, Netherlands: Elsevier, 2012.

Dziegielewski J, Jezowska-Trzebiatowska B, Kalecinska E, Siemion IZ, Kalecinski J, Nawojska J. Gamma-radiolysis of 6-aminopenicillanic acid and its derivatives. *Nukleonika* 1973, 18, 513–523.

Duroux JL, Basly JP, Penicaut B, Bernard M. ESR Spectroscopy applied to the study of radiosterilization: Case of three nitroimidazoles. *Appl. Radiat. Isotop.* 1996, 47, 1565–1568.

Eisenberg E, Jacobs GP. The development of a formulation for radiation sterilizable urea broth. *J. appl. Bact.* 1985, 58, 21–25.

EMEA, European Agency for the Evaluation of Medicinal Products. *Note for Guidance—The Use of Ionization Radiation in the Manufacture for Medicinal Products (3AQ4A)*. London: EMEA, 1991.

EMEA, European Agency for the Evaluation of Medicinal Products. *Committee for Proprietary Medicinal Products, Note for Guidance on Parametric Release*. London: EMEA, 2001.

EC, European Commission. *Use of Ionizing Radiation in the Manufacture of Medicinal Products, Annex 12. EudraLex, Vol. 4, Good Manufacturing Practice (GMP) Guidelines*. Brussels: European Commission, 2011.

European Committee for Standardization. *EN 552:1994, Sterilization of Medical Devices—Validation and Routine Control of Sterilization by Irradiation*. Brussels: CEN, 1994.

EMA, European Medicines Agency. *Guideline on the Sterilisation of the Medicinal Product, Active Substance, Excipient and Primary Container, Draft (EMA/CHMP/CVMP/QWP/850374/2015)*. London: EMA, 2016.

EMA, European Medicines Agency. *Guideline on the Sterilisation of the Medicinal Product, Active Substance, Excipient and Primary Container, (EMA/CHMP/CVMP/QWP/850374/2015)*. London: EMA, 2019.

EMA, European Medicines Agency. *Reflection Paper on Good Manufacturing Practice and Marketing Authorisation Holders—Draft (EMA/457570/2019] /457570/2019)*. Amsterdam: EMA, 2020.

European Pharmacopoeia 2017. Edition 9.0, Section 5.1.1, Methods of Preparation of Sterile Products. Strasbourg, France: Council of Europe—European Directorate for the Quality of Medicines, 2017, 575.

Fairand BP. *Radiation Sterilization of Health Care Products*. Boca Raton: CRC Press, 2001.

Fairand BP, Fidopiastis N. Radiation sterilization of aseptically manufactured products. *PDA J. Pharm. Sci. Tech.* 2010, 64, 299–304.

FDA, Food and Drug Administration, Center for Drug Evaluation and Research. *Guidance for Industry, Submission of Documentation in Applications for Parametric Release of Human and Veterinary Drug Products Terminally Sterilized by Moist Heat Processes*. Silver Spring, MD: Department of Health and Human Services, 2010.

FDA, Food and Drug Administration. *Repeal of Regulation Requiring an Approved New Drug Application for Drugs Sterilized by Irradiation, Proposed Rule* [Docket No. FDA—2017—N—6924]. Silver Spring, MD: Department of Health and Human Services, Federal Register 2018, 83, 46121–46126.

FDA/International Conference on Harmonisation. *Guideline on Impurities in New Drug Products (Q3B(R)), Revision 2*. Silver Spring, MD: Department of Health and Human Services, 2006.

Gamma Industry Processing Alliance and the International Irradiation Association. *White Paper: A comparison of*

Gamma, E-beam, X-ray and Ethylene Oxide Technologies for the Industrial Sterilization of Medical Devices and Healthcare Products. Swindon, UK: International Irradiation Association, 2017.

Geue PJ. *Radiosterilization of Pharmaceuticals, A Bibliography, 1962–1972.* Lucas Heights, Australia: Australian Atomic Energy Commission, 1973.

Gopal NGS. Radiation sterilization of pharmaceuticals and polymers. *Radiat. Phys. Chem.* 1978, 12, 35–50.

Haji-Saeid M, Chmielewsk AG. Radiation treatment for sterilization of packaging materials. *Radiat. Phys. Chem.* 2007, 76, 1535–1541.

Hasanain F, Guenther K, Mullett WM, Craven E. Gamma sterilization of pharmaceuticals: A review of the Irradiation of excipients, active pharmaceutical ingredients, and final drug product formulations. *PDA J. Pharm. Sci. Tech.* 2014, 68, 113–137.

Hemmerich KJ. Polymer materials selection for radiation-sterilized products. *Med. Dev. Diag. Ind.* 2000, 22, 78–89.

HIMA, Health Industry Manufacturers Association. *Radiation Compatible Materials* (Report No. 78–4.9). Washington, DC: HIMA, 1978.

Herring C. Dose audit failures and dose augmentation. *Radiat. Phys. Chem.* 1999, 54, 77–81.

IAEA, International Atomic Energy Agency. *Guidelines for Industrial Radiation Sterilization of Disposable Medical Products (Cobalt-60 Gamma Irradiation), IAEA—TECDOC—539.* Vienna: IAEA, 1990.

IAEA, International Atomic Energy Agency. *Trends in Radiation Sterilization of Health Care Products.* Vienna: IAEA, 2008.

IAEA, International Atomic Energy Agency. *Database for Gamma and Electron Beam Irradiation Facilities in IAEA Member States.* Vienna: IAEA, 2011. Available from: www-nds.iaea.org/iacs_facilities/datasets/foreword_home.php

IAEA, International Atomic Energy Agency. *Radiation Technology Series No. 4, Guidelines for the Development, Validation and Routine Control of Industrial Radiation Processes.* Vienna: IAEA, 2013.

IMRP. Proceedings of the 16th International Meeting on Radiation Processing (Montreal 2011), *Radiat. Phys. Chem.* 2012, 81, 915–1282.

IMRP. Proceedings of the 17th International Meeting on Radiation Processing (Shanghai 2013), *Radiat. Phys. Chem.* 2014, 105, 1–108.

IMRP. Proceedings of the 18th International Meeting on Radiation Processing (Vancouver 2016), *Radiat. Phys. Chem.* 2018, 143, 1–94.

ISO, International Organization for Standardization, ISO 11137:1984. *Sterilization of health care products—Requirements for validation and routine control—Radiation sterilization.* Geneva, Switzerland: ISO, 1984.

ISO, International Organization for Standardization, ISO 11137:1996. *Sterilization of health care products—Requirements for validation and routine control—Radiation sterilization.* Geneva, Switzerland: ISO, 1996a.

ISO, International Organization for Standardization, ISO/TR 13409:1996. *Sterilization of health care products—Radiation sterilization—Substantiation of 25 kGy as a sterilization dose for small or infrequent production batches.* Geneva, Switzerland: ISO, 1996b.

ISO, International Organization for Standardization, ISO/TS 13409: 2002. *Sterilization of health care products—Radiation sterilization—Substantiation of 25 kGy as a sterilization dose for small or infrequent production batches.* Geneva, Switzerland: ISO, 2002.

ISO, International Organization for Standardization, ISO 11137–1:2006/AMD 2:2018 (reapproved in 2016). *Sterilization of health care products—Radiation—Part 1: Requirements for development, validation and routine control of a sterilization process for medical devices.* Geneva, Switzerland: ISO, 2006.

ISO, International Organization for Standardization, ISO 11137–2:2013 (reapproved in 2018). *Sterilization of health care products—Radiation—Part 2: Establishing the sterilization dose for radiation sterilization.* Geneva, Switzerland: ISO, 2013a.

ISO, International Organization for Standardization, ISO/TS 13004:2013. *Sterilization of health care products—Radiation—Substantiation of selected sterilization dose: Method VD_{max}.* Geneva, Switzerland: ISO, 2013b.

ISO, International Organization for Standardization, ISO 11137–3:2017. *Sterilization of health care products—Radiation—Part 3: Guidance on dosimetric aspects for radiation sterilization.* Geneva, Switzerland: ISO, 2017a.

ISO, International Organization for Standardization, ISO TS 19930:2017. *Guidance on aspects of a risk-based approach to assuring sterility of terminally sterilized, single-use health care product that is unable to withstand processing to achieve maximally a sterility assurance level of 10^{-6}.* Geneva, Switzerland: ISO, 2017b.

ISO, International Organization for Standardization, ISO 10993–1:2018. *Biological evaluation of medical devices—Part 1: Evaluation and testing within a risk management process.* Geneva, Switzerland: ISO, 2018a.

ISO, International Organization for Standardization, ISO 11737–1:2018. *Sterilization of health care products—Microbiological methods—Part 1: Determination of a population of microorganisms on products.* Geneva, Switzerland: ISO, 2018b.

ISO, International Organization for Standardization, ISO 11737–2:2019. *Sterilization of health care products—Microbiological methods—Part 2: Tests of sterility performed in the definition, validation and maintenance of a sterilization process.* Geneva, Switzerland: ISO, 2019.

ISO, International Organization for Standardization, ISO/TS11137–4. *Sterilization of health care products—Radiation—Part 4—Guidance on process control, Technical Specification.* Geneva, Switzerland: ISO, 2020.

ISO, International Organization for Standardization, and CEN. *Comité Européen de Normalisation, Guidelines for the implementation of the Agreement on Technical Cooperation between ISO and CEN (the Vienna Agreement),* 7th ed. Geneva, Switzerland: ISO, and Vienna, Austria: CEN, 2016.

Jacobs GP. Stability of cefazolin and other new cephalosporins following gamma irradiation. *Int. J. Pharm.* 1983, 17, 29–38.

Jacobs GP. A review: Radiation sterilization of pharmaceuticals. *Radiat. Phys. Chem.* 1985, 26, 133–142.

Jacobs GP. A review of the effects of gamma radiation on pharmaceutical materials. *J. Biomat.* 1995, 10, 59–96.

Jacobs GP. *Effects of Ionizing Radiation on Blood and Blood Components: A survey. IAEA—TECDOC-934.* Vienna: International Atomic Energy Agency, 1997.

Jacobs GP. A review on the effects of ionizing radiation on blood and blood components. *Radiat. Phys. Chem.* 1998, 53, 511–523.

Jacobs GP. Sterilization: Radiation. In Swarbrick J, ed. *Encyclopedia of Pharmaceutical Technology*, Vol. V, 4th ed. London: Taylor and Francis, 2013, 3412–3426.

Jacobs GP, Simes R. The gamma irradiation of tragacanth: Effect of microbial contamination and rheology. *J. Pharm. Pharmacol.* 1979, 31, 333–334.

Jacobs GP, Wills P. Recent developments in the radiation sterilization of pharmaceuticals. *Radiat. Phys. Chem.* 1988, 31, 685–691.

Japanese Pharmacopoeia 17th ed. Tokyo: The Ministry of Health, Labour and Welfare, 2016.

Jongen Y, Abs M, Bol J-L, Mullier B, Poncelet E, Rose G, Stichelbaut F. Advances in sterilization with X rays, using a very high power Rhodotron and a very low DUR pallet irradiator. In *Emerging Applications of Radiation Processing, IAEA-TECDOC-1386*. Vienna: International Atomic Energy Agency, 2004, 44–54.

Kane MP, Tsuji K. Radiolytic degradation schemes for Co-60 irradiated corticosteroids. *J. Pharm. Sci.* 1983, 72, 30–35.

Koh WY, Morehouse CT, Chandler VL. Relative resistances of micro-organisms to cathode rays. I. Nonsporeforming bacteria. *Appl. Microbiol.* 1956, 4, 143–146.

Kowalski JB, Tallentire A. Substantiation of 25 kGy as a sterilization dose: A rational approach to establishing verification dose. *Radiat. Phys. Chem.* 1999, 54, 55–64.

Kowalski JB, Tallentire A. Aspects of putting into practice VD max. *Radiat. Phys. Chem.* 2003, 67, 137–141.

Kowalski JB, Aoshuang Y, Tallentire A. Radiation sterilization: Evaluation of a new method for substantiation of 25 kGy. *Radiat. Phys. Chem.* 2000, 58, 77–86.

Kroc-Pl TK, Thangaraj JCT, Penning RT, Kephart RD. *Accelerator-Driven Medical Sterilization to Replace Co-60 Sources.* Illinois: Fermilab National Accelerator Section, Illinois Accelerator Research Center, 2017.

Ley F. Irradiated Products Ltd., U.K. personal communication, 1969.

Ley FJ. Radiation sterilization: An industrial process. In Nygaard OF, Adler HI, Sinclair WK, eds. *Radiation Research, Proceedings of the Fifth International Congress of Radiation Research 1974.* New York: Academic Press, 1975, 118–130.

Ley FJ. The effect of irradiation on packaging materials. *J. Soc. Cosmet. Chem.* 1976, 27, 483–489.

Maquille A, Habib Jiwan JL, Tilquin B. Cryo-irradiation as a terminal method for the sterilization of drug aqueous solutions. *Eur. J. Pharm. Biopharm.* 2008, 69, 358–363.

Marciniec B, Dettlaff K. Radiation sterilization of drugs. In *Trends in Radiation Sterilization of Health Care Products (STI/PUB/1313).* Vienna: International Atomic Energy Agency, 2008, 187–230.

Massey L. *The Effect of Sterilization Methods on Plastics and Elastomers*, 2nd ed. Norwich, NY: William Andrew, 2005.

May JC, Rey L, Lee C-J. Evaluation of some selected vaccines and other biological products irradiated by gamma rays, electron beams and X-rays. *Radiat. Phys. Chem.* 2002, 63, 709–711.

Mehta K. Gamma irradiators for radiation sterilization. In *Trends in Radiation Sterilization of Health Care Products.* Vienna: International Atomic Energy Agency, 2008, 5–25.

Meissner J, Abs M, Cleland MR, Herer AS, Jongen Y, Kuntz F, Strasser A. X-ray treatment at 5 MeV and above. *Radiat. Phys. Chem.* 2000, 57, 647–651.

Meschini K, Porto BG, Napolitano CM, Borrely SI. Gamma radiation effects in packaging for sterilization of health products and their constituents paper and plastic film. *Radiat. Phys. Chem.* 2018, 142, 23–28.

Nordion. *Gamma Compatible Materials.* Ottawa, Canada: MDS-Nordion, 2017. Available from: www.nordion.com

Olguner-Mercanoglu G, Ozer AY, Colak S, Korkmaz M, Kilic E, Ozalp M. Radiosterilization of sulfonamides: I: Determination of the effects of gamma irradiation on solid sulfonamides. *Radiat. Phys. Chem.* 2004, 69, 511–520.

Onori S, Pantaloni M, Fattibene P, Signoretti EC, Valvo L, Santucci M. ESR identification of irradiated antibiotics: Cephalosporins. *Appl. Radiat. Isotop.* 1996, 47, 1569–1572.

Panel on Gamma and Electron Irradiation, Microbiology Working Group, *Overview of Selected Aspects of ISO 11137–2:2006.* Redditch, UK: Panel on Gamma and Electron Irradiation, 2012.

Panel on Gamma and Electron Irradiation, *Guide on the Establishment of the Maximum Acceptable Dose ($D_{max,acc}$) for a Product.* Redditch, UK: Panel on Gamma and Electron Irradiation, 2016.

Parenteral Drug Association. *Sterilization of Parenterals by Gamma Irradiation, Technical Report No. 11.* Washington, DC: PDA, 1988 (published as a Supplement to J. Parent. Sci. Technol., 1988, 42).

Pepper RE, Buffa NT, Chandler VL. Relative resistances of micro-organisms to cathode rays. III. Bacterial spores. *Appl. Microbiol.* 1956, 4, 149–152.

Phillips GO. Medicines and pharmaceutical base materials. In *Manual on Radiation Sterilization of Medical and Biological Materials.* Vienna: International Atomic Energy Agency, 1973, 207–228.

PIC/S, Pharmaceutical Inspection Co-Operation Scheme, *Guide to Good Manufacturing Practice for Medicinal Products (PE 009–14), Annex 12, Use of Ionising Radiation in the Manufacture of Medicinal Products* (Geneva: Pharmaceutical Inspection Convention), 2018.

Razem D. Radiation sterilization of pharmaceuticals: An overview of the literature. In *Trends in Radiation Sterilization of Health Care Products (STI/PUB/1313).* Vienna: International Atomic Energy Agency, 2008, 175–185.

Sarma KSS. Development of a family of low, medium and high energy electron beam accelerators. In *Emerging Applications of Radiation Processing, IAEA—TECDOC—1386.* Vienna: International Atomic Energy Agency, 2004, 73–77.

Schnell R, Bogl W. *Der Einfluss der Strahlenbehandlung auf Artzneimittel und Hilfstoffe. Eine Literaturstudie, Teil VI, Bericht des Instituts fur Strahlenhygiene des Bundesgesundheitsamtes.* Berlin: Dietrich Reimer Verlag, 1982.

Schnell R, Bogl W. *Der Einfluss der Strahlenbehandlung auf Artzneimittel und Hilfstoffe. Eine Literaturstudie, Teil VII, Bericht des Instituts fur Strahlenhygiene des Bundesgesundheitsamtes.* Berlin: Dietrich Reimer Verlag, 1984.

Schuttler C, Bogl W, Stockhausen K. *Der Einfluss der Strahlenbehandlung auf Artzneimittel und Hilfstoffe. Eine Literaturstudie, Teil II, Bericht des Instituts fur Strahlenhygiene des Bundesgesundheitsamtes.* Berlin: Dietrich Reimer Verlag, 1979a.

Schuttler C, Bogl W, Stockhausen K. *Der Einfluss der Strahlenbehandlung auf Artzneimittel und Hilfstoffe. Eine Literaturstudie, Teil III, Bericht des Instituts fur Strahlenhygiene des Bundesgesundheitsamtes.* Berlin: Dietrich Reimer Verlag, 1979b.

Schuttler C, Bogl W. *Der Einfluss der Strahlenbehandlung auf Artzneitmittel und Hilfstoffe. Eine Literaturstudie, Teil IV, Bericht des Instituts fur Strahlenhygiene des Bundesgesundheitsamtes.* Berlin: Dietrich Reimer Verlag, 1982a.

Schuttler C, Bogl W. *Der Einfluss der Strahlenbehandlung auf Artzneitmittel und Hilfstoffe. Eine Literaturstudie, Teil V, Bericht des Instituts fur Strahlenhygiene des Bundesgesundheitsamtes.* Berlin: Dietrich Reimer Verlag, 1982b.

Shang S, Ling MTK, Westphal SP, Woo L. Radiation sterilization compatibility of medical packaging materials. *J. Vinyl Additive Technol.* 1998, 4, 60–64.

Slegers C, Tilquin B. Final product analysis in the e-beam and gamma radiolysis of aqueous solutions of metoprolol tartrate. *Radiat. Phys. Chem.* 2006, 75, 1006–1017.

Spinks JWT, Woods RJ. *An Introduction to Radiation Chemistry*, 2nd ed. New York: Wiley, 1976, 67–121.

Tallentire A. Aspects of microbiological control of radiation sterilization. *J. Radiat. Ster.* 1973, 1, 85–103.

Tallentire A. Radiation microbiology relevant to radiation processing. *Radiat. Phys. Chem.* 1979, 14, 225–234.

Tallentire A. The spectrum of microbial radiation sensitivity. *Radiat. Phys. Chem.* 1980, 15, 83–89

Tallentire A, Kahn AA. Tests for the validity of a model relating frequency of contaminated items and increasing radiation dose. In *Radiosterilization of Medical Products.* Vienna: International Atomic Energy Agency, 1975, 3–14.

Tallentire A, Dwyer J, Ley FJ. Microbiological control of sterilized products: Evaluation of model relating frequency of contaminated items with increasing radiation treatment. *J. appl. Bact.* 1971, 34, 521–534.

Tallentire A, Kahn AA. The sub-process dose in defining the degree of sterility assurance. In Gaughran ERL, Goudie AJ, eds. *Sterilization by Ionizing Radiation*, Vol. 2. Montreal: Multiscience Publications Ltd., 1978, 65–80.

Trutnau H, Bogl W, Stockhausen K. *Der Einfluss der Strahlenbehandlung auf Artzneitmittel und Hilfstoffe. Eine Literaturstudie, Teil I, Bericht des Instituts fur Strahlenhygiene des Bundesgesundheitsamtes.* Berlin: Dietrich Reimer Verlag, 1978.

UK Panel on Gamma and Electron Irradiation. Code of Practice for the Validation and Routine Monitoring of Sterilization by Ionizing Radiation. *Radiat. Phys. Chem.* 1989, 33, 245–249.

United States Pharmacopeia XVII. Rockville, MD: The United States Pharmacopeial Coonvention Inc., 1965.

United States Pharmacopeia XXII. Rockville, MD: The United States Pharmacopeial Convention Inc., 1989.

United States Pharmacopeia (USP 42), Radiation Sterilization <1229.10> *(Rockville, MD.: The United States Pharmacopeial Convention Inc.),* 2019.

United States Pharmacopeia (USP 42), *(Rockville, MD.: The United States Pharmacopeial Convention Inc.),* 2019.

van Asten J, Dorpema JW. A new approach to sterilization conditions: The IMO concept. *Pham. Weekblad, Sci. Ed.* 1982, 4, 49–56.

Werner IA, Altorfer H, Perlia X. The effectiveness of various scavengers on the γ-irradiated, methanolic solution of medazepam. *Int. J. Pharmaceut.* 1990, 63, 155–166.

Whitby JL, Gelda AK. Use of incremental doses of Cobalt 60 radiation as a means to determine radiation sterilization dose. *J. Parenter. Drug Assoc.* 1979, 33, 144–55.

WHO, World Health Organization. *Wholesomeness of Irradiated Foods, Report of a Joint FAO/IAEA/WHO Expert Committee, Technical Report Series, No. 659.* Geneva: WHO, 1981.

WHO, World Health Organization. *Good Manufacturing Practices for Sterile Pharmaceutical Products, Annex 6, Technical Report Series, No. 961.* Geneva: WHO, 2011.

Wilczyński S, Pilawa B, Koprowski R, Wróbel Z, Ptaszkiewicz M, Swakoń J, Olko P. EPR studies of free radicals decay and survival in gamma irradiated aminoglycoside antibiotics: sisomicin, tobramycin and paromomycin. *Eur. J. Pharm. Sci.* 2012, 45, 251–262.

Wills PA. Effects of ionizing radiation on pharmaceuticals. *Aust. J. Pharm.* 1963, 44, 550–557.

Woolston J. Radiation processing: Adapting to a changing world. *Eur. Med. Dev. Tech.*, January 1, 2009. Available from: www.emdt.co.uk/article/radiation-processing-adapting-changing-world

Zalewski Ch, Schuttler C, Bogl W. *Der Einfluss der Strahlenbehandlung auf Artzneitmittel und Hilfstoffe. Eine Literaturstudie, Teil VIII, Bericht des Instituts fur Strahlenhygiene des Bundesgesundheitsamtes.* Berlin: Dietrich Reimer Verlag, 1988

Contact the Author

Dr Geoffrey P Jacobs, Dr Geoffrey P Jacobs Associates (Consultants), P.O.B. 16352, Jerusalem 9116202, Israel. email: hida@zahav.net.i.

23 Validation of Sterilizing-Grade Filters

Suraj B. Baloda[1]

CONTENTS

23.1 INTRODUCTION

Sterile filtration is one of the most critical final steps in the pharmaceutical manufacturing process, and the challenges of filtration processes have also evolved over time because of constantly changing demands for removal of specific bioburden from the final product. Besides the filter manufacturers and biotechnology and the pharmaceutical industry's own pursuit for perfection in producing safe drugs, the regulatory authorities worldwide also mandate the need for adherence to aseptic processing guidelines. Since the publication of the third edition of this book, a series of guidelines and regulations have been revised or updated. The final draft of the United

DOI: 10.1201/9781003163138-23

States Food and Drug Administration's (FDA's) 1987 *Aseptic Processing Guideline* was published in September 2004 (*Sterile Drug Products Produced by Aseptic Processing–Current Good Manufacturing Practice*) (1). Earlier, the Parenteral Drug Association (PDA) published the authoritative summary of best practices in sterile filtration and validation of sterile filtration in its 1998 TR (*Sterilizing Filtration of Liquids–PDA Technical Report No. 26, Revised 2008*) (2). The PDA TR 26 is a comprehensive monograph that highlights the history of sterile filtration, provides valuable details on the criteria of selection and functioning of filters, and also explains validation considerations and integrity testing methods. Additionally, United States pharmacopeia (USP) Chapter <1299.4> ("Sterilizing Filtration of Liquids") provides a comprehensive overview of various factors that affect the filtration process and contribute to the effectiveness of any sterilizing filtration process (3).

Filter validation is an important aspect of the sterilization process that needs greater understanding to improve sterility assurance and efficiency. Unlike validation of terminal sterilization, which end users generally master with little difficulty, validation of aseptic filtration remains somewhat enigmatic. Thus, before designing a validation protocol, it is essential to understand the objectives of the desired process. This can be accomplished by generating a *filter requirements specification*. From this specification, validation becomes a matter of proving claims made for the filter, as outlined in this chapter.

The investment of initial cost as well as the effort in understanding the scientific rationale of the validation of aseptic filtration process by the end users goes a long way in assuring the quality of the product and reducing operating costs by a decrease in scrap or rework. Moreover, the validation of aseptic filtration, apart from being a regulatory requirement, also makes good business sense.

23.2 WHY VALIDATE?

There are two main reasons to validate processes. The first is obvious, whereas the second is often known but rarely quantified.

- Good manufacturing practice (regulatory requirements)—Various worldwide regulatory bodies require validation of manufacturing processes for LVPs and SVPs, ophthalmics, veterinary medicines, bulk chemicals, and in vitro diagnostics.
- Good business practice—It is essential to understand that an out-of-control process increases the amount of rework or scrap incurred, thereby increasing cost. A controlled process gives reproducibility and product consistency within known limits. A controlled process also aids regulatory compliance and, therefore, provides a license to do business.

The ultimate aim in both cases is patient safety.

23.3 WHY VALIDATE FILTERS?

As mentioned previously, filtration is an important final step in the manufacturing process and before we discuss the topic further, it is assumed that:

- The reader is familiar with filters and aseptic processing and has a basic understanding of microbiology.
- The reader is familiar with the general principles of validation.
- The filters have been chosen and are correctly sized for the required operation.
- The studies are designed to incorporate all prerequisite filter processing steps. For example, when performing a study to determine potential filter extractables, all normal prerequisite processing steps are first performed on the study filter, such as preflush and sterilization.

The very nature of aseptic processing presumes that the filter is one of the most critical components in the process and that it provides sterility assurance of the final product. It is, for example, as important as an autoclave in a terminal sterilization process. Therefore, a filter requires stringent controls and attention to ensure consistent and reproducible results. Again, both reasons for process validation come into play when talking about filters:

- Good manufacturing practice—Validation of filters is a regulatory requirement throughout the world (see References). As a quick reference, a snapshot of the relevant sections of the FDA's Aseptic Processing Guidelines (2004) are also incorporated and depicted as "FDA Guidelines" in the body of the chapter.
- Good business practice—Aseptic filters are used to ensure the sterility of a final product. Just before filtration, a drug product undergoes several value-added steps. After passage through the filter, the drug product is usually placed in its final container. The cost of rework at this stage can be extremely high. Indeed, many manufacturers will scrap product rather than attempt to rework it at this stage.

23.4 WHAT NEEDS TO BE VALIDATED? HOW IS IT DONE?

The purpose of filter validation is to ensure that the filter will *reproducibly* remove undesirable components (e.g., microbial bioburden), while allowing passage of desirable components.

Therefore, it is necessary to understand, and include in a validation study, the parameters such as how the filter is manufactured, expected filter operations (how the filter will be sterilized, how many times the filter will be used), the potential for toxic byproducts, and the obvious aspects of retention and inertness. Before any work is initiated to validate a filter, two prerequisites must be considered and satisfied:

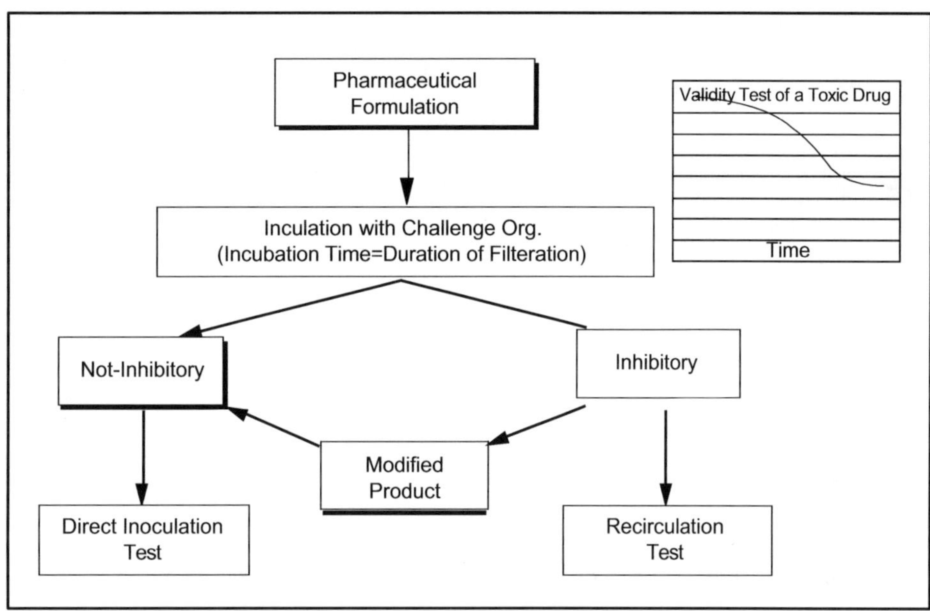

FIGURE 23.1 Flow chart for testing the inhibitory effect of a pharmaceutical formulation on bacterial viability.

1. The properties of the filter itself must be consistent and reproducible from lot-to-lot. This will be dealt with in some detail later in this chapter.
2. The drug product (properties) must be consistent and reproducible from lot-to-lot. (The subject of drug product reproducibility will not be dealt with in this chapter. Reproducibility is a basic underlying assumption on which the entire filter validation will be built.) Without drug product consistency and reproducibility, it is not possible to obtain consistent, reproducible results for flow rate, throughput, retention, filter inertness, and so on. If the validation process uses a consistent, reproducible drug product, then known processing times, flow rates, and throughputs can be set for routine manufacturing. Any changes in drug product (which may be a result of changes in raw materials, processing conditions, concentrations, or the like) may change the filter's performance. This final point cannot be stressed too highly because, not only can the filter performance be affected, but also changes to the drug product or processing conditions can *invalidate* the original study. Furthermore, it is also important to bear in mind and evaluate the impact of the drug product or the pharmaceutical formulation on the viability of the test organism (or bioburden), which would impact the retention test results.

FDA guidelines state:

> Direct inoculation into the drug formulation is the preferred method because it provides an assessment of the effect of drug product on the filter matrix and on the challenge organism. However, directly inoculating *Brevundimonas diminuta* into products with inherent bactericidal activity against this microbe, or into oilbased formulations, can lead to erroneous conclusions. When sufficiently justified, the effects of the product formulation on the membrane's integrity can be assessed using an appropriate alternate method. For example, a drug product could be filtered in a manner in which the worst-case combination of process specifications and conditions are simulated. This step could be followed by filtration of the challenge organism for a significant period of time, under the same conditions, using an appropriately modified product (e.g., lacking an antimicrobial preservative or other antimicrobial component) as the vehicle. Any divergence from a simulation using the actual product and conditions of processing should be justified.

Figure 23.1 outlines the procedure that can be used to test the viability of the test organism prior to retention testing.

To begin a validation study, it is prudent to first document the requirements of the filter. This should be done in sufficient detail to ensure the proper selection of a filter system that can be readily validated. The next step is to conduct those studies that can be done off-line or scaled-down (examples include bacterial retention and filter extractables). Then, full-scale tests are performed to verify correct, consecutive operation and reproducibility. Finally, validation must include ongoing evaluation and provisions that permit change of controls to be implemented over time.

There is considerable debate on whether to use scaled-down or full-scale filters for validation studies. The choice depends on the scientific rationale and the goal of the specific element of the validation study being considered. For example, if bacterial retention is under study, then it should be sufficient to prove that the membrane alone is capable of the required log

TABLE 23.1
Validation Study Elements

Physical	Chemical	Biological
Reproducibility	Inertness (i.e., compatibility, extractables, adsorption)	Endotoxins
Sterilization	Activity/stability	Toxicity
Integrity	Consistency and reliability	
	Operation	
Shedding		
Particulates		
Fibers		
Particle retention		Microbial retention

reduction. This assumption is clearly dependent on an accurate integrity test that can detect a manufacturing defect (such as incomplete filter seaming) in the full-scale filter.

If, however, the ultimate aim of the validation study is the verification of the scaled-down simulation or the validation of the filtration process itself, the validation should be run at full-scale, with a minimum of three consecutive, full-scale runs with three filter lots and three drug product lots. This, however, is neither the beginning nor the end of a filter validation.

Table 23.1 gives an overview of the elements of a validation study. The elements of interest for the validation depend on the requirements and specification set for the filter. It is not necessary to study each element to validate a filter, but the decision to study (or not study) a particular element must be based on the scientific and process rationale. In any case, the final validation data generated must be able to stand on its own merits (and scientific rationale) and be able to pass further regulatory scrutiny.

Table 23.2 lists some of the elements and tools available to demonstrate that the elements listed in Table 23.1 are true and fully understood. Some tools can be used to describe more than one goal. Other elements, such as integrity, can themselves be used as tools once they have been qualified for the application.

23.5 WHO IS RESPONSIBLE FOR VALIDATION?

FDA guidelines state:

> When the more complex filter validation tests go beyond the capabilities of the filter user, tests are often conducted by outside laboratories or by filter manufacturers. However, it is the responsibility of the filter user to review the validation data on the efficacy of the filter in producing a sterile effluent. The data should be applicable to the user's products and conditions of use because filter performance may differ significantly for various conditions and products.

Ultimately, the drug manufacturer is responsible for filter validation. Therefore, the drug manufacturer should select a filter manufacturer that not only provides sufficient information but also the services required to facilitate proper validation. Because certain validation requirements are in demand from all filter users, the filter manufacturer should be in a position to provide some of the required information. Thus, some of the burdens of responsibility can be alleviated by choosing a filter manufacturer that can provide this commonly required information.

Table 23.3 outlines the validation elements and also the subparts of these elements that could be reasonably expected to be either the responsibility of the filter manufacturer or the filter user. It must, however, be noted that the ultimate responsibility lies with the filter user.

TABLE 23.2
Elements and Tools of Filter Validation in the Manufacturing Process

Elements to be demonstrated	Tools in the manufacturing process
Filter performance reproducibility	Filter documentation available in filter validation guides and elsewhere: adherence to good manufacturing practice, change control in the filter manufacturing process, change notification, certificates of quality (C of A), lot release criteria, validation guides
Product reproducibility	Drug product characteristics: consistency, viscosity, particulates, bioburden, concentration, impurities
Sterilization	Heat distribution profile evaluation (thermocouples), heat penetration evaluation (spores)
Integrity test	Correlation to microbial retention, qualification of methods
Operation	Flow rate, throughput, temperature, pressure, integrity, visual
Particulates	Monitoring of effluent, on-line monitoring
Fibers	Filter manufacturer adherence to 21 Code of Federal Regulations (CFR), preuse flush
Inertness (compatibility)	Manufacturer's documentation (charts), flow rate, throughput, integrity, weight, visual, pH, conductivity
Inertness (extractables)	Weight change, gravimetric extractables, oxidizable substances, ultraviolet, high-performance liquid chromatography, total oxidizable carbon, Fourier Transform Infrared, NVR
Adsorption	Concentration analysis before and after use
Activity/stability	Concentration analysis, activity analysis, stability trials, conformational confirmation
Microbial retention	Challenge with microorganism in drug product
Bacterial endotoxins	Filter lot release, limulus amebocyte lysate testing of in-process samples
Toxicity	Materials of construction, class VI plastics test, cytotoxicity studies, clinical trial data

Abbreviation: NVR = nonvolatile residue.

TABLE 23.3
Validation Elements and Responsibilities of Filter Manufacturers and Filter Users

Element	Filter manufacturer	Filter user
Filter reproducibility	Validate filter claims and the filter manufacturing process	Review all data and audit manufacturer
Product reproducibility		Ensure consistency and reproducibility
Sterilization	Provide recommended procedures with limits for time, temperature, and number of cycles	Operate within manufacturer's limits Validate the procedure in use
Integrity test	Provide procedures and test specifications Provide test correlation with bacterial retention	Follow manufacturer's procedures Ensure correlation to bacterial retention exists Validate test method Perform integrity ratio work if wetting with product
Operation	Provide limits for operating, temperature, pressure	Design system to meet filter requirements and limits
Particulates	Provide data for removal	Verify required limits are achieved
Fibers	Meet non-fiber-releasing claim [21 Code of Federal Regulations 210.3b(6)]	Ensure NFR, preflush filters according to recommendations
Microbial retention	Provide retention claims, test methods, and service	Have the microbial retention test performed with the drug product and microorganism (bacteria, *Mycoplasma*, virus of interest)
Inertness	Provide charts for all materials of construction	Document compatibility
Compatibility		Perform studies
Extractables	Develop methods, identify components	Perform studies
Adsorption	Indicate known problems	Perform studies
Drug activity and stability	Indicate if known problems exist	Verify no conformational changes or activity losses Perform stability studies
Endotoxins	Perform analysis on a per lot basis	Verify low endotoxin levels from filters Ensure process operation does not contribute endotoxin (e.g., prolonged use with intermittent sterilization)
Toxicity	Perform testing and provide results (class VI plastics, cytotoxicity)	Obtain results and reports. Include filter with clinical trials or perform toxicological review

Abbreviation: NFR = nonfiber releasing.

23.6 ELEMENTS OF A VALIDATION STUDY

After a filtration process is properly validated for a given product, process, and filter, it is important to ensure that identical filters (e.g., of identical polymer construction and pore size rating) are used in production runs.

The following sections detail the purposes and the background of the various elements and tools listed in Tables 23.1–23.3. As mentioned earlier, not all of the elements need to be studied for any given filter. Elements and tools should be chosen as a direct result of the filter requirements, such as the purpose, function, or operating limitations of the filter. If a documented filter requirement specification is unavailable, progressing with the filter validation study may result in unnecessary testing, while the functional requirements of the filter remain untested.

23.7 PERFORMANCE REPRODUCIBILITY

Reproducibility applies to both the drug product and the filter. Success of the validation and the drug manufacturing process will depend on these two key elements.

FDA guidelines state:

> Factors that can affect filter performance generally include (*i*) viscosity and surface tension of the material to be filtered, (*ii*) pH, (*iii*) compatibility of the material or formulation components with the filter itself, (*iv*) pressures, (*v*) flow rates, (*vi*) maximum use time, (*vii*) temperature, (*viii*) osmolality, and (*ix*) the effects of hydraulic shock. When designing the validation protocol, it is important to address the effect of the extremes of processing factors on the filter capability to produce sterile effluent.

23.7.1 Drug Product

The chemical attributes of a drug product (such as formulation concentration, chemical composition and constituents, pH, viscosity, density, ionic strength, and osmolarity) should be known and controlled within defined limits. These limits, upper and lower, will determine important process characteristics, such as flow rate, processing time, throughput for a given filter surface area, and so on. It cannot be stressed too highly that, once set and validated, changing any of these chemical attribute limits may affect not only process characteristics but could also negate previous validation work.

Sometimes solution reproducibility is not possible. In this event, it is necessary to define limits and measure incoming raw materials for compliance. Operating with reproducible solutions and filters enables specifications to be set for flow rate, processing times, and throughput. Such specifications provide a very effective tool for in-process monitoring. A significant change in flow rate, throughput, or processing time could indicate a change in a solution's or a filter's consistency.

Defined limits for time, flow rate, and throughput are also necessary to conduct validation studies covering all anticipated operating limits. For example, the longest time is the worst-case for a retention study, because of the potential for bacterial movement through the filter (referred to as grow-through). Although there is some concern over low-flow conditions, it is generally accepted that high-flow rates are worst-case, because they lead to less residence time of a fluid within the filter. High-flow rates have the potential to cause greater differential pressures across the membrane, thereby creating a greater possibility of retention failures from membrane damage. Higher rates of flow frequently lead to higher differential pressures, which in nonsterilizing-grade filters has led to decreased microbial retention and longer times. Drug product reproducibility must be specified, monitored, and controlled by the drug manufacturer.

23.7.2 Filter Performance Reproducibility

FDA guidelines state:

> After a filtration process is properly validated for a given product, process, and filter, it is important to ensure that identical filters (e.g., of identical polymer construction and pore size rating) are used in production runs. Sterilizing filters should be routinely discarded after processing of a single lot. However, in those instances when repeated use can be justified, the sterile filter validation should incorporate the maximum number of lots to be processed.

There are several questions that must be answered relative to filters used for validation and those that will be supplied long-term by the filter manufacturer:

1. If a scale-down filter is used, is it the same membrane that is used in the full-scale filter?
   Filter manufacturers may manufacture filters of the same polymer by different processes. This could result in a scale-down filter and full-scale filter having significantly different characteristics. The membrane in both devices should be manufactured, tested for quality and performance, and released under the same conditions to ensure that the study is valid.
2. Is the filter being validated representative of the current catalogue item from the filter manufacturer and of all future filters?
   The filter manufacturer should be consulted to ensure that:
   a. All filter claims have been qualified and there is control over filter raw materials.
   b. The filter manufacturing process has been validated within specified operating windows.
   c. In-process and final release testing is performed on a per lot basis (particularly critical parameters such as retention and endotoxin).
   d. Certificates of quality assurance and validation guides are available.
   e. The filter manufacturing plant may be audited.
   f. There are policies for, and strict adherence to, change control.
   g. The filter is representative of the current catalogue item—several lots should be obtained from salable inventory.

Although these items are not regulatory requirements, possession of the information is useful during the validation process and in the preparation of marketing applications.

Assessment of a filter manufacturer's policy on implementation of change control during the manufacture of filters or change in filter product characteristics will help access the answer to question 2. It is highly likely, and desirable as technology advances, that changes will be made in a filter manufacturing process. It is necessary, therefore, to ensure that all changes are evaluated before implementation and that requalification takes place as required. Furthermore, because the filter manufacturer cannot judge the subtle effects of filter changes on the drug-manufacturing process, it is necessary that there be a change notification policy. That is, for significant changes, all end users should be notified in sufficient time to evaluate the change. Additionally, end users should be notified of all other changes on a periodic basis.

23.8 STERILIZATION

To validate use of a sterilizing-grade filter, it is not only necessary to prove that the filter is (and will continue to be) adequately sterilized, but also that the sterilization method does not damage the filter.

There are many sterilization methods available, but the preferred method of filter sterilization is with moist heat (steam) (4) because it is relatively easy to use and it minimizes potential sources of residual chemicals.

Important considerations (variables) for steam sterilization of filters are time, temperature, pressure, air and condensate removal, heat up, cooldown, and the total number of sterilization cycles. Any one of these variables, if uncontrolled, could lead to filter failure.

Current practice for autoclave validation and steam-in-place cycles are that both thermocouples and biological indicators (BIs; suspensions or spore strips) be used. The thermocouples verify that adequate temperatures are achieved, and the BI verify kill by moist heat. This chapter does not address steam sterilization validation, as the subject has been addressed in a guideline elsewhere (6). In practice, the validation of the steam sterilization of filters may be summarized as follows:

1. Obtain all relevant performance specifications of the filter and the filter housing from the manufacturer, such as the maximum recommended operating temperature, thermal resistance, the maximum number of sterilization cycles, the maximum allowable hydraulic pressure resistance, and so on.

2. Install the filters and filter housings to ensure that they self-drain of air and condensate (prevention of cold spots; Figure 23.2).
3. Perform cold spot mapping of the filter system (heat distribution studies), followed by thermocouple and BI analysis (heat penetration studies). At a minimum, thermocouples and BIs should be placed both upstream and downstream from the filter (Figure 23.3). These locations should ensure that both the upstream and downstream high points and low points are monitored (i.e., verification of low-point condensate removal and high-point air removal).
4. Perform ongoing monitoring of the sterilization for temperature and pressure (to ensure presence of saturated steam conditions) and differential pressure.
5. Ensure that ongoing operating conditions (normal and sterilization) are within the filter manufacturer's defined limits.

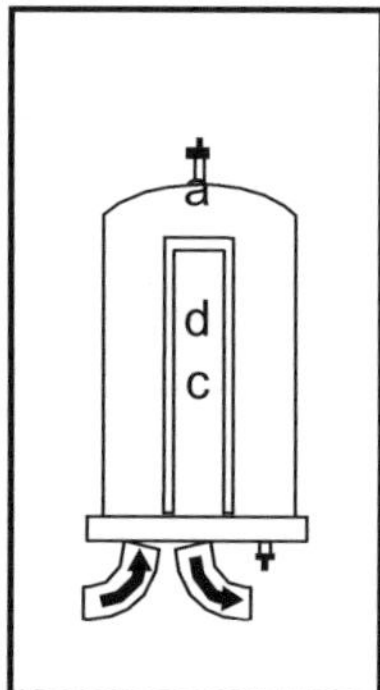

FIGURE 23.2 Correct filter installation to facilitate self-drain of air and condensate. a, vent; b, condensate bleed; c, inner core of filter; d, high point air; e, low point condensate.

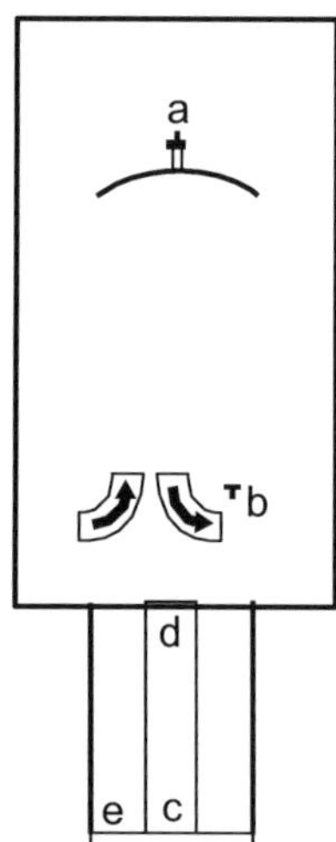

FIGURE 23.3 Suggested thermocouple/biological indicator placement. a, vent; b, condensate bleed; c, inner core of filter; d, high point air; e, low-point condensate.

23.9 INTEGRITY

An integrity test for routine manufacturing use should be nondestructive (bacterial retention tests are by nature destructive), provide an indication of "fitness for use" and, above all, be correlated with bacterial retention. This chapter assumes that the reader is knowledgeable about integrity test methods (such as bubble point and diffusion). This information will not be presented. The theory of these tests has been the subject of many publications (7, 8).

Provision of the test methodology and correlation is the responsibility of the filter manufacturer, and qualification of how the test is used is the responsibility of the end user; that is, it is not sufficient to merely put the procedure into use without adequate operator training and qualification of the test equipment. It is the responsibility of the filter manufacturer to demonstrate the correlation of the integrity test value with microbial retention. These correlations should take the form of an integrity test parameter versus retention. The data on the correlation between nondestructive filter integrity testing (e.g., diffusion test or bubble point test) and the retention testing (destructive test) by the filter manufacturers provides valuable information to the end user. As a result, the filter users should be able to assess the rigor with which a filter device has been qualified by reviewing the validation guide. In the filter validation guide, the filter manufacturer should provide an integrity test method that clearly cites wetting fluids, test gases, test pressures, test temperatures, test times, and pass/fail criteria.

All filter manufacturers should have established a relationship between a recommended physical integrity test and microbial retention. These data are used by the filter vendors to establish minimum integrity test specifications for their sterilizing grade filters. The pharmaceutical manufacturer is responsible for establishing minimum integrity test values for the filter when it is wetted with the pharmaceutical product to be filtered. These specifications will be used for release of the drug product, because microbially challenging each filter used in production is not practical.

23.10 QUALIFICATION OF INTEGRITY TEST USE

Many filter users struggle with the choice of test method. This is often confounded by conflicting expert opinions and by filter manufacturer claims. Regulatory authorities may allow the choice of any of the major tests or may suggest a method but in reality, it is the user's responsibility to document the rationale for their choice (1).

As mentioned previously, the user must ensure that the integrity test has been correlated with microbial retention. The manufacturer's recommended test must be strictly adhered to (time, temperature, wetting fluid, pressure, method). Test use must be qualified—in other words, it is necessary to qualify the test accuracy and reproducibility, regardless of whether the test is performed manually or with an automatic tester.

FDA guidelines state:

> Integrity testing of the filter(s) can be performed prior to processing, and should be routinely performed postuse. It is

important that integrity testing be conducted after filtration to detect any filter leaks or perforations that might have occurred during the filtration. Forward flow and bubble point tests, when appropriately employed, are two integrity tests that can be used. A production filter's integrity test specification should be consistent with data generated during bacterial retention validation studies.

Qualification of integrity test use, whether manual or automated, is not unlike qualification of any other method or instrument. For manual and automated tests, it is important that the accuracy and precision of the measurement instrument (usually pressure gauges or flow meters) is capable of discerning accurate values within the test time frame. For a manual test, it is important that operators are properly trained and qualified in use of the test. This may be performed through a documented training program, followed by examination of technique with filters of predetermined integrity test values (but unknown to the examinee). In addition, an incorrect filter (either a nonintegral filter or a larger-pore-sized filter) should be included to verify that operators are capable of detecting a filter failure. Qualification of an automated test instrument should also address instrument calibration, verification of test accuracy and reproducibility (usually vs. a manual test), and verification of alarm and security features. In addition, the software driving the unit (whether in disk, EPROM, or PLC format) should be qualified. The latter requirement, obviously, cannot be performed by the end user. Therefore, verification should be obtained from the instrument manufacturer that the qualification has been completed (through data or by an audit). Furthermore, besides the reliance on automatic integrity test instrumentation, it is important to be aware of the fact that the operator training is critical with regards to operation of automatic testers as it is possible to get incorrect results from an improperly operated instrument and incorrect filter installation during testing, e.g., when the filter installation set-up is not self-draining during testing (Figure 23.4).

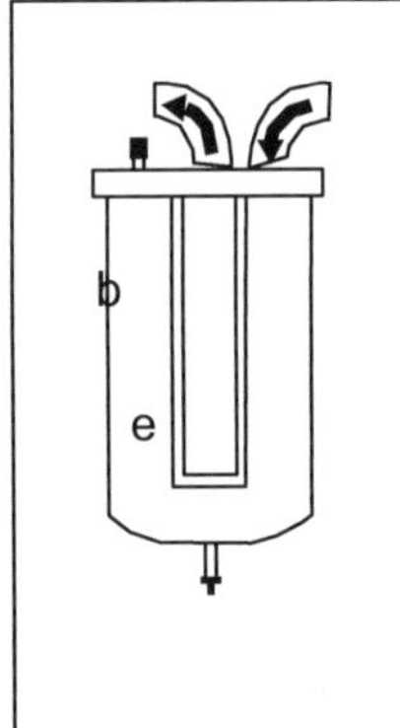

FIGURE 23.4 Incorrect filter installation demonstrating that the setup is not self-draining. a, vent; b, condensate bleed; c, inner core of filter; d, high point air; e, low-point condensate.

23.10.1 Drug Product-Based Integrity Tests

Use of the filter manufacturer's recommended wetting fluid as a control is critical to obtain a correct pass/fail value. However, in certain cases (particularly when performing a postfiltration integrity test), ensuring purity of the specified wetting fluid can be problematic owing to difficulty in flushing drug product from the filter.

Integrity tests are derived from physical laws and relations and, therefore, are dependent on specific variables. For instance, the bubble point test is dependent on vehicle surface tension and wetting contact angle with the particular membrane. Diffusion is dependent on the solubility and diffusivity of the gas in the liquid vehicle. Therefore, drug products (or any other nonspecified liquid) can cause enhancement or suppression of integrity test values compared with those that would be obtained with the manufacturer-specified liquid (e.g., water). Therefore, if the nonspecified liquid is not completely flushed before integrity testing, the resulting value may be greater than, or less than, the value that would be obtained when using the pure, specified liquid. The end user, therefore, is required to qualify the post-use test in the following manner.

Determine whether the drug product, compared with the specified liquid, suppresses or enhances the integrity test value. This may be performed on a scaled-down version of the filter, for example, a 47-mm disk (2, 5). The study should take account of drug product variability, usually by using at least three lots of drug product (assuming that raw materials are carefully controlled). Integrity tests are conducted on the filter membrane wetted with the filter manufacturer's specified wetting liquid. The filter membrane is then allowed to completely dry and is wetted with the drug product. To achieve accurate, reproducible results, a direct scale-down of drug volume to filter surface area should be used (i.e., if production batches are 10,000 L through 5000 cm2 of filter area, then the scale-down to 10 cm2 of filter area would use a wetting volume of 20 L). This will determine not only whether the product is enhances or suppresses, but also generate a preliminary integrity test ratio:

$$\text{Ratio} = \frac{\text{Value obtained with drug product}}{\text{Value obtained with filter manufacturer specified liquid}}$$

If the ratio obtained is 1 (e.g., the value obtained with the specified liquid is 50 psi and that obtained with the drug product is 50 psi), then the value obtained with product will be the same as that obtained with the specified liquid and there is no issue with performing the integrity test with product. If the ratio indicates that the product suppresses the integrity test value (that is less than 1.0; i.e., causes the bubble point to be lowered), then use of the drug product to wet the filter while still using the pass/fail for the specified liquid will be overly conservative. For example, the value obtained with the specified liquid is 50 psi, whereas that obtained with drug product is 45.5 psi. By using adequate validation data, it is possible to accept a drug product-based integrity test value of 45.5 psi. If the criteria are not met, the drug manufacturer would then flush the filter with water and retest (with 50 psi as the minimum acceptable value).

If the ratio indicates that the product is a bubble point enhancer (i.e., greater than 1.0; causes the bubble point to be increased), then use of the drug product to wet the filter while still using the pass/fail for the specified liquid will be invalid. For example, the value with specified liquid is 50 psi, whereas that with drug product is 54.5. In other words, if the minimum bubble point for the specified liquid is 50 psi, the product ratio is 1.1 (indicating that a value of 54.5 or higher with product would be an integral filter), but if the drug manufacturer decides to determine 50 psi as pass/fail for the product, then there is the potential for accepting an out-of-specification filter (e.g., one that when wet with drug product gives a value of 52 psi, which is below the product- compensated specification of 54.5 psi). Therefore, the pass/fail criteria should be 54.5 psi or higher. If the criteria are not met, that is, the value is 54.5 or higher, the drug manufacturer could then flush the filter with water and retest. However, although wetting the filter is an ideal, it is often impractical. Thus, a different approach can also be used. A scaled down test could be run with a set volume followed by confirmation of the values generated on a small scale by monitoring the actual test results in the full-scale.

However, the use of 50 psi as the minimum acceptable value for the test after a specified liquid flush may not be valid. The drug manufacturer must validate the flush volume necessary to ensure that only the specified liquid is present in the filter: all products have been removed from the filter. Validation of the flush volume can be achieved by measuring the product in the flush (concentration assay, spectrophotometric analysis, and so on) or by integrity testing after incremental flush volumes (e.g., flush 50 L then test, flush another 50 L and test—the procedure is repeated until a consistent, stable bubble point value is obtained).

If product integrity ratios are to be used, the scaled-down study is only the initial part of the qualification. The second part would be to obtain additional full-scale data and to monitor the ratio on an ongoing basis. This would be part of the ongoing manufacturing runs. The ongoing ratio of specified liquid preuse to product wet postuse testing should be trended to ensure that the ratio is not changing owing to raw material changes, lot-to-lot inconsistency, and so on.

Qualification of flush volume to ensure product removal should be performed at full scale, because it will be dependent on the filter configuration (size, support materials) and the installation (pipe size, valves, housing). For additional guidance, consult PDA TR No. 26 (1).

23.11 OPERATING CONDITIONS

The validation study must ensure that within the anticipated worst-case operating conditions the filter is not compromised. This is conducted by first obtaining data from the filter manufacturer on maximum recommended operating limits. It is then the responsibility of the end user to operate within these limits. The actual validation must then encompass the anticipated worst-case conditions that may be encountered during the process (i.e., those expected in the end user's process, rather than the filter manufacturer's maximum limits). In other words, it is only necessary to validate the filter for its proposed conditions, rather than validating for the manufacturer's stated maximum limits. Likewise, the manufacturer's maximum limits should not be exceeded. Lack of damage to the filter can be verified by performing several tests such as integrity, flow rate, throughput, retention, extractables, or others.

23.11.1 Time

Operating time is a concern for the following reasons:

1. Long processing times could allow bacteria, which have been trapped by the filter, to die, thereby resulting in increased endotoxin levels. This time criteria must be accounted for in endotoxin studies.
2. Long processing times may increase the probability for bacteria to penetrate the filter (9).

Time considerations during validation must ensure that the worst-case is covered. For example, if the drug manufacturing process results in intermittent filter blockage, then the effect of changing out a clogged prefilter during processing should be incorporated into the maximum time of all experiments. If the sterilizing grade filter is changed intermittently during the filtration process, this must also be considered. Time for the sterilizing grade filter would be the longest period of time that any sterilizing grade filter would be in place. It would not be the full amount of time to process the batch. The impact of long processing time potentially leading to the increased probability of bacteria penetrating the filter originates from the fact that microporous membrane filters have been demonstrated to allow, over extended periods of time, the penetration of bacteria through the filter and into the effluent (7, 9–11). However, filter penetration is simply a term used to describe the depth to which a target particle will travel through the filter before it is removed from the fluid stream. In fact, sterilizing filters, as a rule, allow some penetration into the depths of the filter. This is clearly consistent with both size exclusion and adsorptive removal mechanisms. The term "grow-through" describes a theoretical phenomenon in which organisms initially trapped by filter will multiply on and in the filter. In the process of multiplying, the organisms pass deeper and deeper into the filter matrix until they eventually emerge downstream from the filter. This "phenomenon," similar to the "blow-through," is a hypothesis that has not been rigorously studied and demonstrated to be a real event in the pharmaceutical process. It needs further detailed investigation because several factors, such as lack of proper system sterilization and improper aseptic manipulations (resulting in false positive) may contribute to such observations.

The reliability of sterile filtrations can nevertheless be increased by limiting processing time. Filter manufacturers can provide the data on the retention tests that have been conducted for a specific membrane or a device for extended time that generally suggests that filters should retain bacteria in excess of 48 hours. Accordingly, filter manufacturer's recommendations that aseptic processing is completed within a preestablished period of time is based on their experience. Thus, limiting the filtration time to a time less than that which is recommended by a filter manufacturer provides increased assurance against the phenomenon of grow through (or blow through) if it is a real phenomenon.

29.11.2 Temperature

The operating temperatures a filter will experience must be carefully reviewed to ensure that the manufacturer's recommended limits are not exceeded. These limits will probably be stated in terms of time at a specified temperature. For example, if a filter is required to process hot oil at 60°C for 12 hours, then the filter must be rated for that use. It must also be rated for the maximum anticipated sterilization temperature and time. Just because a filter is rated for 100°C for 100 hours does not mean it will withstand 121°C for any length of time.

Similarly, it is common practice for vent air filters to be left on water-for injection (WFI) tanks for extended periods. These filters should be rated for extended life at high temperatures (typically WFI is held at 70 °C –90°C) for the required times. Filter components may oxidize to varying degrees at elevated temperatures.

Operating temperatures will also have a significant effect on the filter's ability to withstand differential pressure.

For all these reasons, the validation must take into account normal operating temperatures as well as sterilization or sanitization temperatures.

29.11.3 Pressure

There are several aspects of pressure that must be considered. The inlet pressure to the filter must be monitored to ensure that there is no potential for structural damage. The differential pressure across the membrane must comply with the filter manufacturer's recommended limits. Differential pressure is normally stated as a function of temperature. For example, a filter may have a maximum recommended differential pressure of 80 psi at 25°C. This same filter may have a differential pressure recommendation of a maximum of 5 psi at 121°C.

Another aspect that must be accounted for is the direction of applied pressure. Filters are manufactured such that the allowable maximum pressure differential in the forward direction may be different from that in the reverse direction. For example, the aforementioned filter may also have a maximum limit of 50 psi differential in the reverse direction at 25°C and 2 psi reverse at 121°C.

All of these pressure factors must be considered when determining a filter's fitness for use. The anticipated maximum pressures should be incorporated, when possible, into the bacterial retention study and the sterilization study. They should also be considered for normal operating conditions. Fitness for use can be verified by retention, integrity, flow rate, and throughput.

When the limits for pressure are being decided, two areas tend to be overlooked. The first, sterilization pressures, has been discussed. The second, hydraulic stress, must also be considered. Hydraulic stress can occur, for example, when a valve is opened suddenly, causing a filter to receive full line pressure immediately, or by the action of in-line filling when surge tanks are not in use (causing rapid pressurization and depressurization of the filter). These forward and reverse pressures must be quantified, monitored, controlled, and validated.

29.11.4 Flow Rate and Throughput

This chapter assumes that initial flow decay studies for the given filter and drug product combination have been carried out to ensure that there is adequate surface area to obtain the required flow rate and throughput. These studies are ordinarily performed on a scaled-down filter, for economic reasons. Usually, there is sufficient safety margin added to defer verification until the full-scale validation batches are run.

Full-scale processing times, throughput, and flow rates should be established during the validation batches as a reference for all future batches. These flow rates and throughputs provide an effective tool for in-process monitoring of a validated process. In other words, if throughput or flow rate show intermittent problems, this is an indication of either variation in raw materials, inadequate filter sizing, or changes to the filter.

23.12 PARTICULATES

When discussing particulates in the context of filter validation, there are two questions that must be answered:

1. Is the filter contributing to the particulate load of the solution?
2. Is the filter specified as reducing the particulate load of the solution? That is, is the reported purpose of the filter to remove particulates of a specified size range? The USP defines *particulate matter* as follows:

   "Particulate matter consists of mobile, randomly sourced, extraneous substances, other than gas bubbles, that cannot be quantitated by chemical analysis due to the small amount of material that it represents and to its heterogeneous composition." (8)

Sources of particulates may be as varied in use as raw materials, process equipment, process environment, or the filters themselves. Indeed, the filter itself may contribute to the particulate load or may provide a particulate load reduction. There are specified regulatory guidelines for particulate limits in pharmaceutical solutions. For example, the USP limits (USP 43–NF38 <788>) for particulate matter in injections when tested by using the Light Obscuration Test Particle Count method (8), are as follows:

LVPs; i.e., single dose of more than 100 mL:

1. Not more than 25 particulates per mL ≥10 μm
2. Not more than 3 particulates per mL ≥25 μm

SVPs; i.e., single or multiple doses of 100 mL or less:

1. Not more than 6000 particulates per containers ≥10 μm
2. Not more than 600 particulates per containers ≥25 μm. Proof of compliance with these limits is normally performed by sample testing of the final solution by methods such as optical microscopy and light obscuration (8). Because of potential limitations with these techniques, other methods have also been investigated, such as light microscopic image analysis and scanning electron microscopy (12, 13).

All methods involve filtering a solution through a suitable analysis membrane and counting or sizing the resulting particulates. In analyzing particulates from a specific filter, it would be necessary to filter the pharmaceutical solution, when practical, through the test filter and the analysis filter. The counts could then be obtained from the analysis filter. It is critical that the background particulate level (i.e., the counts on the test system without a test filter in line) be accounted for in the analysis. It is for this reason that it may be more practical to use a clean liquid, rather than the pharmaceutical solution.

The validation must simulate all anticipated manufacturing operations. For example, if the filter is flushed before use, the initial particulate counts may be reduced. If the filter is then sterilized, the particulate counts may increase. Likewise, particulate counts on a time/volume throughput should be analyzed. This ensures lack of bias in count estimation during the course of a filtration. Examples are counts that start high and quickly drop to below detection limits (indicating that a preflush would be beneficial) or vice versa (indicating a maximum useful life or an incorrect filter type for the function).

If the stated or implied function of a filter is particulate removal, then the removal of a given size and number of particulates must be validated. This procedure should be performed, when possible, with the pharmaceutical solution. A standard test method (14) is used to challenge the filter with a known quantity and size distribution of particulates. The amount of particulate retention can then be verified. As with all particulate measurement methods, performance of adequate assay controls is critical to test result accuracy.

If the stated function of a filter is not removal of particulates (other than microorganisms), then the need to validate it for this function is obviated (see the section entitled Microbial Retention).

23.13 FIBERS

Fibers, just as particulates, are of concern for two reasons:

1. Is the filter shedding fibers into the solution?
2. Is it the filter's function to remove fibers?

Fiber is defined by 21 CFR, part 210.3 (5, 6) (i.e., U.S. GMPs; 1) as follows: "Fiber means any particulate contaminant with a length at least three times greater than its width." Section 211.72 of the same document further states:

> Filters for liquid filtration used in the manufacture, processing, or packing of injectable drug products intended for human use shall not release fibers into such products. Fiber-releasing filters may not be used in the manufacture, processing, or packing of these injectable drug products unless it is not possible to manufacture such drug products without the use of such filters. If use of a fiber-releasing filter is necessary, an additional NFR filter of 0.22 μm maximum mean porosity (0.45 μm if the manufacturing conditions so dictate) shall subsequently be used to reduce the content of particles in the injectable drug product. Use of an asbestos-containing filter, with or without subsequent use of a specific NFR filter, is permissible only upon submission of proof to the appropriate bureau of the FDA that use of a NFR filter will, or is likely to, compromise the safety or effectiveness of the injectable drug product.

The U.S. GMPs 21 CFR part 210.3, subpart 6 defines a NFR filter as follows:

> Nonfiber-releasing filter means any filter, which after any appropriate pretreatment such as washing or flushing, will not release fibers into the component or drug product that is being filtered. All filters composed of asbestos are deemed to be fiber-releasing filters.

The first requirement when more than one filter is used in a filtration train is to ensure that, at a minimum, the final filter in line is a NFR filter of 0.22 μm standard (sterilizing grade). It is preferable, where possible, to specify all filters in the filtration train as NFR. Filter manufacturers will certify that their products meet 21 CFR 210.3, subpart 6, based on lack of fibers in their manufacturing components. In addition to this, if fibers are a concern, the end user can test for fibers in the effluent using the particulate measurement systems discussed previously. This is not common practice unless the filter is specified for fiber-particulate reduction.

The German Federal Health Office has published standards for asbestos particulate limits in parenterals (14). The document states that the length of asbestos fibers is the decisive risk factor and has established three size categories:

1. Fibers exceeding 2.5 μm must be eliminated.
2. Fibers between 1 and 2.5 μm "may only be found in low concentrations."
3. Fibers not longer than 1 mm are of no concern.

The test method referenced is the U.S. EPA test, Analytical method for determination of asbestos fibers in water (16).

Considering these limitations for the final dosage form, it would be prudent to either obtain by certification from the filter manufacturer assurance that the filters (regardless of materials of construction) do not exceed these levels of contaminating asbestos fibers, or to perform the measurements. It is more logical to obtain information from the filter manufacturer who can verify filter raw materials at source. In addition, data from the filter manufacturer showing the ability of filters to prevent passage of these fibers would be beneficial for drug manufacturers.

23.14 MICROBIAL RETENTION

FDA guidelines state:

> Filtration is a common method of sterilizing drug product solutions. A sterilizing grade filter should be validated to reproducibly remove viable microorganisms from the process stream, producing a sterile effluent. Currently, such filters usually have a rated pore size of 0.2 mm or smaller (17). Use of redundant sterilizing filters should be considered in many cases. Whatever filter or combination of filters is used, validation should include microbiological challenges to simulate worst-case production conditions for the material to be filtered and integrity test results of the filters used for the study. Product bioburden should be evaluated when selecting a suitable challenge micro- organism to assess which microorganism represents the worst-case challenge to the filter. The microorganism *B. diminuta* (ATCC 19146) when properly grown, harvested and used, is a common challenge microorganism for 0.2 μm rated filters because of its small size (0.3 μm mean diameter).

It is necessary to demonstrate microbial retention with a given filter and drug product for the following main reasons:

1. To ensure that the filter is not undergoing degradation, deformation, or some other change under the conditions of use.
2. To ensure that the drug product is not causing the organism to shrink, thereby resulting in nonsterilizing conditions.

Initially, a filter is qualified as sterilizing grade by the filter manufacturer based on its ability to completely retain high levels of microorganisms as demonstrated by the results of a microbial retention test. By definition, a *sterilizing-grade filter* is one that, when challenged with 107 *B. diminuta* ATCC 19146 per square centimeter of filter area will produce a sterile effluent. In addition, each filter manufacturer has manufacturing lot-release criteria, which should include some representative microbial retention testing of each lot of filters.

Filter manufacturers generally base their testing method on the methods published by the American Society for Testing and Materials (ASTM Committee F-21, 1988) and on guidance provided in the FDA's *Guideline on sterile drug products produced by aseptic processing* (2, 18). Furthermore, ISO 13408–2:2018 also specifies requirements for sterilizing filtration as part of aseptic processing of health care products (19). It also offers guidance to filter users concerning general requirements for setup, validation, and routine operation of a sterilizing filtration process to be used for aseptic processing of health care products. ISO 13408–2:2018 is not applicable to removal of viruses. Sterilizing filtration is not applicable to fluids containing particles as effective ingredient larger than the pore size of a filter (e.g., bacterial whole-cell vaccines). ISO 13408–2:2018 tends to be more conservative and prescriptive than the PDA TR26. From a regulatory perspective, the pharmaceutical manufacturer is responsible for providing retention data that support the claim of filter validation in their manufacturing process. However, drug manufacturers have neither the experience nor the facilities to perform the test, nor the desire to introduce even small volumes of microbial suspensions into their production facility. Consequently, the pharmaceutical manufacturer has looked to filter vendors for advice on how to best challenge the filters. Indeed, many pharmaceutical manufacturers contract with the filter vendor to perform the test.

The drug product and the filter rating-specific validation of microbial retention, in contrast to those retention tests performed by the filter manufacturer, evaluates the influence of the drug product's physical and chemical attributes on the performance of the filter and efficiency of the filtration under simulated processing conditions. The relevant physical and chemical conditions are listed in Table 23.4.

It is possible (and often desirable) to perform this product-specific testing on a scaled-down version of the process filtration system. Generally, flow rate (mL/min) per unit surface area is the scaling factor, with the manufacturing process scaled down to 13.8 cm2 of effective filtration area (a 47-mm-disk filter). The processing temperatures, times, and pressure differentials modeled are those used for full-scale processing. A typical test schematic is shown in Figure 23.5.

The test's basic components are (1) a test filter, (2) a system to deliver the microbial (bacterial) challenge suspension to the test filter, and (3) a system for assaying the test filter effluent. The test filter can range from flat stock disks in filter

TABLE 23.4
Physical and Chemical Processing and Drug Product Attributes

Physical attributes	Chemical attributes
Pressure differential	pH
Flow rate	Viscosity
Duration (contact time)	Osmolarity
Temperature	Ionic strength
Batch size	Surface tension
Surface area	
Filter type and series	

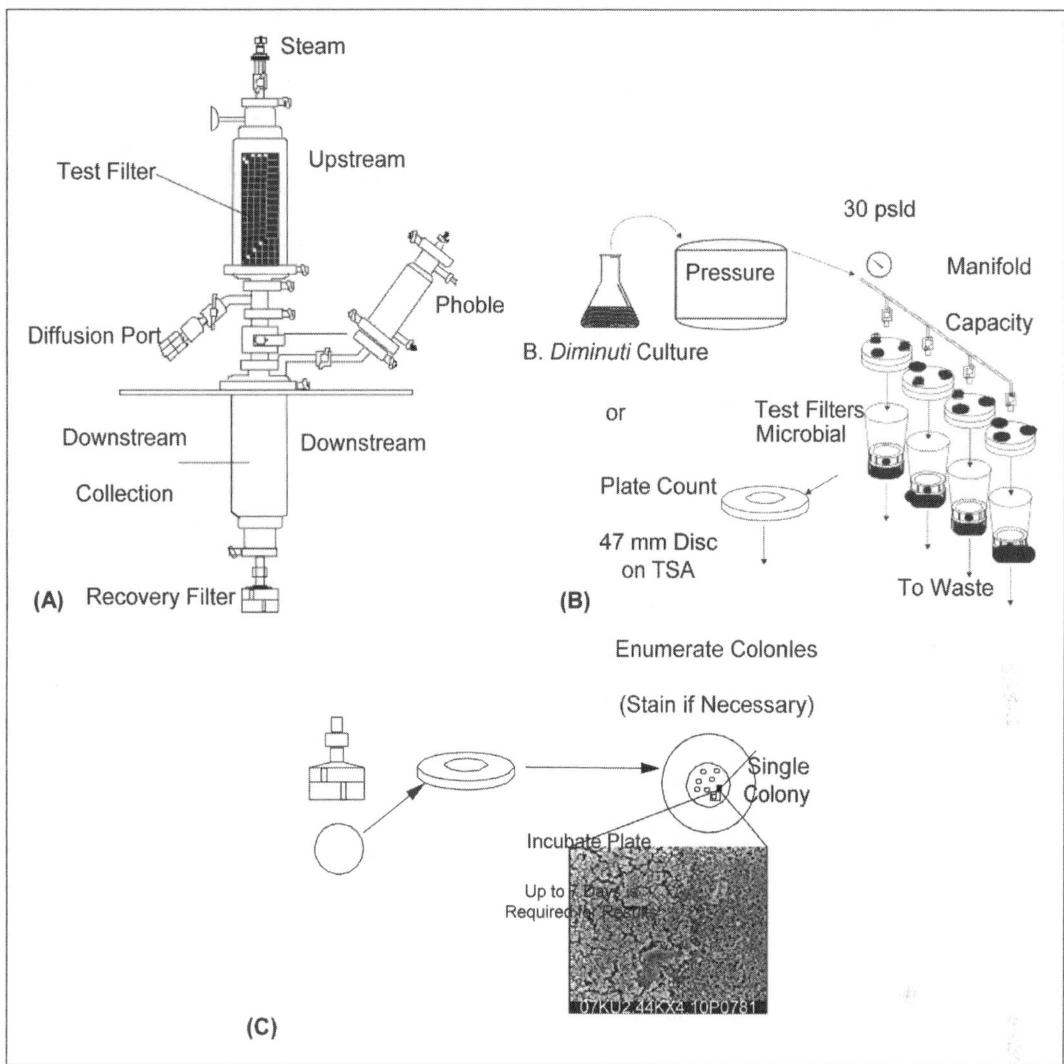

FIGURE 23.5 (A) Test stand showing filter cartridge setup for retention testing (full-scale version). (B) Retention test setup for flat-stock disks (scaled-down version). (C) Processing of assay (recovery) filters after retention testing.

holders to relatively large fabricated devices, such as stacked disk devices or pleated cartridges. When modeling fabricated device performance in retention testing, it is essential that the membrane used in the devices be the same as the flat stock membrane used in microbial retention studies. Specifically, if flat stock testing is to be used to validate a device, then the two membranes should be identical. Thus, the scale-up retention test data should be compared and be comparable.

Each retention test system should have the following basic features:

1. Valves should be included in appropriate positions throughout the system, and they should not be installed if they are not necessary.
2. Pressure gauges should be placed upstream and downstream from the test filter to accurately determine a pressure differential (D*P*) across the filter. This is especially important if the assay filter is used concurrently with the test filter. A D*P* across the assay filter must be accounted for when D*P* across the test filter is being measured.
3. Hydrophobic vent filters should be part of every reservoir that will require air exchange during emptying or filling of the reservoir. Aseptic connections should always be minimized.

Only those filters for which no test organism *(B. diminuta* or other relevant microorganism) passage is detected downstream are deemed sterilizing-grade. Although the colonies may "appear" on the assay filter as a result of false positive or extraneous contamination, the "passage" of even one bacterial colony on an assay filter renders the filter nonretentive. Thus, results of "zero" and "one" are considered significantly different, and size controls (more open filters are usually used) are required to ensure that one does not obtain either of these values illegitimately (17).

Areas of Concern for Product-Specific Bacterial Retention Testing

The following areas are of concern for product- specific bacterial retention testing:

1. The organism must be cultivated to ensure a small and consistent size.
2. The challenge concentration should not be less than 107/cm2 of filter area unless justified.
3. The drug product must not be toxic to the organism (i.e., it must not reduce the challenge to below 107/cm2 *B. diminuta* per square centimeter of filter area over the course of the test). This can be ensured by performing toxicity studies.
4. Viability controls should be run before testing.
5. When the drug product is shown to be toxic to the organism the following course can be followed:
   a. Expose the filter to the drug product (to determine if there are any filter–drug product interactions)
   b. Rinse the filter (to remove toxic residue of the drug)
   c. Challenge with the organism suspended in a vehicle that has properties as close to the product as possible (e.g., the product minus the toxic components)
   d. Or, alternatively, expose the filter to the toxic product for the actual period of processing. Then, suspend the challenge microorganism in the toxic product, and challenge the test filter for a shorter period of time (choosing a duration that ensures maximum microorganism survival)
6. The size of the collection or assay filter must be considered. Different testing laboratories will use either a 0.45-μm or a 0.22-μm filter as the assay filter. There are arguments for and against the use of either filter for organism recovery. Basically, it can be argued that the 0.45-μm filter will allow passage of organisms. The arguments for use of this filter are that it is the referenced method for sterility testing in the USP (20) and also that it is easier to cultivate organisms on a 0.45-μm filter than on a 0.22-μm filter (21). The arguments for and against a 0.22-μm filter are the opposite of those for the 0.45-μm.

The definition of *worst-case-processing* conditions is continuously under debate and it is recommended that the choice of such conditions be discussed with the regulatory agency before testing.

23.15 FILTER INERTNESS

The purpose of an aseptic filter is to remove bacteria and unwanted contaminants from a liquid or gas stream. Beyond this, the filter should be inert (i.e., neutral). It should neither add anything to the fluid or gas (extract) nor remove anything from it. Obviously, this is an ideal, but may be an unrealistic expectation. A small amount of extraction or adsorption is a possibility with any and all commercially available filters, depending on the nature of the drug product and the materials of construction of the filter. Therefore, it is unnecessary to categorically prove or disprove inertness, but rather to quantify its effects. This should be accomplished through empirical studies.

As an example, take a drug product filter *extraction* and *adsorption* study. The purpose of an extraction study is to detect trace amounts of filter extractables that may leach into a process on drug solvent stream. Several critical issues must be addressed:

- Characterization or identification of the extracted species—Quantification of extracted species (extractables)
- Sensitivity of analytical methods for detection and quantification—Potential effect of identified extractables on drug performance or stability
- Potential effect of identified extractables on filter performance—Potential toxicity of extractable species.

The purpose of an adsorption (binding) validation study is to determine whether a given filter adsorbs component(s) from a drug product. Clearly, the extent of any adsorption should be determined. The issues are:

1. The effect on the final drug product formulation. If the product is filtered into a bulk holding tank, then the effect on the final product may not be significant. However, if the product is directly filtered into its final package (such as ampoules), then some of the packages may not contain adequate amounts of critical ingredients (such as proteins or preservatives).
2. The effect on drug product stability.
3. The cost (in dollars) of adsorption.

The first step in testing filter inertness is to compare the drug product to be filtered with the membranes proposed for use, to determine if there are known incompatibilities. Historically, this comparison, coupled with integrity testing and product-specific bacterial retention testing, would have been considered sufficient for the purposes of a study on inertness. Today, however, both the drug industry and regulatory authorities are pursuing a course of action to qualify and quantify inertness caused by interaction between any drug product and filter components. This should not be considered surprising, because filter inertness has always been a concern. Historically, however, only gross measurement techniques, such as weight changes, integrity tests, visual inspections, and so on, could be relied on to warn of any potential problems. Today, more sophisticated and sensitive analytical measurement techniques are routinely used to measure inertness.

Table 23.5 lists various techniques that have been used to measure filter inertness. A brief description of each follows. Note carefully that these tests must incorporate all materials of construction.

23.16 CHEMICAL COMPATIBILITY CHARTS

Chemical compatibility charts are provided by filter manufacturers (22) and are typically included in product literature or validation guides. They should clearly show compatibility and

TABLE 23.5
Techniques for Determining Filter Inertness

Extractables	Adsorption
Compatibility	Compatibility
Oxidizable substances	Formulation analysis
pH	
Conductivity Gravimetric	
extractables Weight changes	
Advanced analytical techniques	

suitability information for all materials of construction of a filter unit across a broad spectrum of solvents. Although these charts are not all-encompassing, they do give an excellent general indication of the scope of potential chemical environments to which a particular filter unit may be exposed. Their real use, however, is not in what is listed as compatible, but rather what is listed as known to be incompatible. (With a properly defined filter requirements specification, this study would have been completed and documented before filter selection.) If a solvent stream of interest is not indicated, then the filter manufacturer should be consulted to determine (at least on paper) the compatibility of the primary solvent streams with the filter materials of construction. This study should be documented by a qualified individual with knowledge of the filter materials of construction (such as a filter manufacturer's developmental or analytical chemist). *The compatibility review should be considered a first pass only.* It should be supplemented with a combination of integrity tests, gravimetric extractables, oxidizable substance tests, and advanced analytical techniques.

23.16.1 pH and Conductivity

Either or both of these tests may be performed by taking measurements immediately before and immediately after filtration. This provides indication of gross changes to the filtered fluid. If changes occur, then further investigation is warranted. However, if no changes occur, then this indicates only that no gross change has occurred; it does not by itself indicate filter inertness.

23.16.2 Oxidizable Substances

The USP method (23) measures oxidizable substances levels in various flush volumes of water. Manufacturers will generally provide information on their product in the form of a statement such as: "Meets USP oxidizable limits after a flush of *X* liters of water." It is important for filter manufacturers to provide such flush data. It is equally important for drug manufacturers to perform flush operations before using filters. These flush operations may be carried out in conjunction with the preuse integrity test.

23.16.3 Gravimetric Extractables

A gravimetric extractables test is a USP test method (24) that determines the weight of a dry sample of the filter material before and after static soak in water or an appropriate solvent. Attention to detail while performing this apparently simple procedure is paramount in isolating measurement error. The test will not specify the source of extractables and, therefore, must be tightly controlled; nor will the test give an indication of what the extractables are.

23.16.4 Weight Change

A weight change test is a gross test to determine nonvolatile filter extractables. After determination of dry weight, the filter is soaked in the drug product and in water (as a control). The soak time is chosen to mimic manufacturing conditions. After soaking, the filter is again dried to a constant weight and checked against the original weight and the water control. Any discrepancies warrant further investigation, such as more detailed extractable analysis or adsorption studies. As with the previous tests, an indication of no change pre- and post-test does not indicate lack of filter inertness.

23.16.5 Advanced Analytical Techniques

Several more-advanced analytical techniques may be used to detect and quantify filter extractables in solvent streams (25). These methods require special expertise and are complicated by the fact that the solvent stream being analyzed must be clean initially. If, for example, a drug product with several constituents (and additional trace constituents) is analyzed, potential extractables may be masked by excessive background noise. For this reason, extractable studies are normally performed on the primary solvent stream(s) alone. However, it may be necessary for regulatory purposes to provide additional experimental data that clearly demonstrate that attempts to determine extractable levels in the presence of actual drug product yield meaningless results.

Some of the more commonly used analytical techniques are the following:

23.16.5.1 High-Performance Liquid Chromatography

Samples are analyzed before and after filtration. Additional peaks in the postfiltration chromatogram would indicate the presence of extractables. Extraction is normally performed in a pure solvent stream alone, thereby reducing background noise. It is usually performed under static soak conditions (this tends to concentrate extractables). If performed under static soak conditions, the level of extractables found must be considered in the light of the total process volume to obtain a "per volume of drug product extractable level." To date, the use of dynamic extraction techniques has infrequently been observed.

23.16.5.2 Fourier Transform Infrared Spectroscopy

This is a qualitative technique that is used to obtain chemical or chemical group information on solutes that are found by preparative techniques such as high-performance liquid chromatography (HPLC). Any extractables found are compared against a spectral library to determine the chemical makeup of the extractable and ensure that the source is a construction material. Positive identification of the extracted species then allows direct reference to previously documented toxicity studies performed by the filter manufacturer on the filtration device.

23.16.5.3 Total Oxidizable Carbon

This method is designed to detect the accumulated total of most organic carbon present in aqueous solutions and provides greater value. It is a quantitative technique that relies on the oxidation of carbon. There are two components of a total oxidizable carbon (TOC) analysis: (1) the chemical or physical oxidation of the organic material and (2) the measurement of the products of that oxidation. All current TOC analyzers measure the resultant carbon dioxide. Carbon dioxide is measured using non-dispersive infrared (NDIR) spectroscopy or by measuring the conductivity of the bicarbonate ion (HCO3-). USP 43 chapter <643> (23) provides detail on establishing test requirements for use in a pharmaceutical water TOC monitoring program. These test technologies can be applied to filters.

23.16.6 Adsorption

Adsorption can cause loss of drug product, conformational changes, as well as reduced activity or stability (26, 27). The validation must address all of these issues. Adsorption tests are best carried out by the end user. The issue of adsorption to filters has been covered in the literature (28) and, as such, the theory and mechanisms will not be dealt with in this text.

The method involves analyzing the drug product immediately before and after filtration for the compound(s) of interest. The intent is to detect differences in concentration of the active ingredient or other component of the formulation (such as stabilizers and preservatives). In addition, the drug product activity should be verified after filtration and over the course of a stability study. For proteins, it would also be desirable to ensure that conformational changes are not taking place (29).

If an adsorptive interaction or conformational change is discovered, then the drug manufacturer must determine if the interaction affects drug safety and efficacy. (If safety or efficacy is affected then use of the filter for processing the drug must be rejected.) If the adsorptive effect does not affect safety or efficacy, it should be determined if it is possible to compensate for the effect. For example, when directly filtering into vials, it may be discovered that the initial vials lack preservative, owing to adsorption. In such an event, it may be possible to pretreat the filter with preservative and fill all potential binding sites, ensuring that subsequent drug product filtration would not result in the first vials lacking preservative. As with any method, this would require stringent validation controls. If a costly ingredient, such as the drug active or proteins, is adsorbed, such a pretreatment may not be possible. For this, it would be necessary to obtain a more inert filter.

23.17 DRUG PRODUCT STABILITY

It is important to ensure that drug product stability should not be affected by filtration. The section on adsorption discussed the potential loss of activity, lowering of activity, or conformational changes as a result of filtration. These effects may not be readily apparent; therefore, the stability study should investigate any potential adsorptive effects.

23.18 ENDOTOXINS

A validation study must address that a filter does not add endotoxin to a drug product. Endotoxin can come from the filter itself, when new, or as a result of the way the filter is used.

The endotoxin content of a new filter will depend on the quality control processes of the filter manufacturer, the filter-manufacturing process itself, and the water used in filter manufacturing. These are all areas that are covered by choice of a filter vendor and through verification of endotoxin levels stated on a filter certificate of conformance or quality. Verification can be achieved by a drug manufacturer through testing incoming filters (30) or by auditing the filter manufacturer's methods and data.

The endotoxin levels of a filter can also be affected by filter use. The drug manufacturer must design and validate a manufacturing process to ensure low or no detectable levels of endotoxin. This is achieved by operating with low initial bioburden (following good aseptic techniques to minimize the source of endotoxins) and ensuring filter change-out frequency based on time, bioburden, and endotoxin analysis. Special caution should be exercised if filters are being used for extended periods or are to be re-sterilized for reuse. In these cases, bacteria that have been concentrated on the filter surface may begin to disintegrate, resulting in higher endotoxin levels downstream from the filter. Therefore, endotoxin determinations on a per-volume and per-time basis should be used to decide on appropriate filter change outs.

23.19 TOXICITY

A validation study should determine that passage of the drug product through a filter does not cause any toxicological effects. All filter materials of construction should be addressed—not just the membrane. Other components to be covered include any support layers, cage and core (the inner and outer hard plastic supports), end caps (the solid plastic pieces that hold the top and bottom of the filter together), and O-rings. The construction materials themselves should be listed as nontoxic. The filter manufacturer should provide relevant test data, such as a compendial plastics test similar to the USP class VI tests for plastics (31) and the USP mouse safety test (32), for all construction materials.

It is then the responsibility of the drug manufacturer to ensure that contact of the filter and drug product does not result in any toxic byproducts. This is achieved by ensuring that any extractables are materials of construction of the filter.

These extractables should be quantified for levels in the drug product dose and further qualified as to potential for toxicological effect. This review should be carried out by a qualified toxicologist.

23.20 PRE-USE/POST-STERILIZATION INTEGRITY TESTING

Sterilizing grade filtration has been reliably used for decades in the manufacturing of drugs by aseptic processing. Although these filters undergo a high degree of inspection by the filter manufacturers to assure filter integrity, it is incumbent upon the end user to follow applicable regulatory guidelines and demonstrate that the integrity of the sterilizing grade filter has further been maintained during the drug manufacturing process to assure sterility of the final product.

FDA guidelines state:

> Integrity testing of filter(s) can be performed before processing, and should be routinely performed post-use. It is important that integrity testing be conducted after filtration to detect any filter leaks or perforations that might have occurred during the filtration.

EU GMP Annex 1 guidelines state:

> The integrity of the sterilized filter assembly should be verified by integrity testing before use, to check for damage and loss of integrity caused by the filter preparation prior to use. A sterilizing grade filter that is used to sterilize a fluid should be subject to a non-destructive integrity test post-use prior to removal of the filter from its housing. Test results should correlate to the microbial retention capability of the filter established during validation.

The concept of Pre-use/Post-Sterilization Integrity Testing (PUPSIT) has been widely debated over the years. A primary rationale for the requirement to conduct an integrity test on a sterilized filter prior to use is that a potential damage to the filter that occurred during sterilization or subsequent handling or transportation could become "masked" during the filtration process and may not be detected by the post-use integrity testing (33).

The integrity test of a sterilizing grade filter must thus be ascertained to provide sterility assurance of the final product. Filter integrity is most commonly performed after the filtration process (post-use). Some filter users test the integrity before the filtration process and before the filter is sterilized (pre-use). PUPSIT has been a widely debated topic for the last several years because of the fact that such tests would require downstream, filtrate side manipulation generating a potential process risk (34). Therefore, most regulatory authorities request the test of the integrity of a sterilizing grade filter be performed after the filtration, post-use, but only recommend an integrity test pre-use, without specifying whether it should be performed pre- or post-sterilization (1, 19, 35, 36).

However, as stated previously, the requirement for verifying the integrity of a sterilizing grade filter before use and after its sterilization (PUPSIT) has been stated in EU GMP Annex 1 (37). The main argument for the need to perform a PUPSIT is the possibility that a flawed sterilizing grade filter pre-use could pass the post-use test, because the flaw could be masked by the separated contaminants from the fluid stream. The Annex 1 guideline also recognizes that PUPSIT may not always be possible after sterilization because of process constraints (e.g., the filtration of very small volumes of solution) and in these cases, an alternative approach may be taken providing that a thorough risk assessment has been performed and compliance is achieved by the implementation of appropriate controls to mitigate any risk of non-sterility.

A pre-use test before sterilization assures that a filter was installed properly, and it was not damaged during shipment or handling. Performing a pre-use test after sterilization detects the damage that may have occurred during the sterilization cycle and it potentially limits the associated risk. It must also be recognized that the PUPSIT requires costly consumables, documentation, and manpower to have a validated system that minimizes or does away with the risk of downstream contamination. The challenge of multiple risk factors associated with downstream exposure to the product as a result of performing PUPSIT must thus be recognized and taken into consideration for remediation. It is thus incumbent on the process owner to comply with the applicable regulatory expectations and concerns related to product sterility assurance and safety. Thus, in addition to achieving EU Annex 1 regulatory compliance, the use of PUPSIT is a practice also based on perceived risk assessed to reduce business loss risk especially in manufacturing products targeted for marketing in the European Union.

23.21 PREFILTRATION BIOBURDEN CONTROL

Microbial control during the drug substance and drug product manufacturing process is critical for ensuring product quality and safety. Thus, control of the microbial load at the sterile filtration step is an important component of the overall microbial control strategy for sterile biological drug products such as finished dosage forms that are typically manufactured by sterile filtration followed by aseptic processing. Because the presence of high bioburden levels in the drug product at the pre-filtration step increases the challenge to the sterilizing filter and may also lead to other quality issues, a pre-filtration bioburden limit should be established. Both FDA (1) and EMA regulatory guidelines (38) stipulate that a maximum acceptable bioburden level, which is referred to as a pre-filtration bioburden level, should be stated at the point immediately prior to the sterile filtration step.

FDA Guidelines state:

> Manufacturing process controls should be designed to minimize the bioburden of the unfiltered product (FDA Guidance for Industry, 2004).

EMA GMP guidelines state:

> For routine commercial manufacturing, bioburden testing should be performed on the bulk solution immediately before sterile filtration. In most situations, a limit of NMT

> 10 CFU/100 ml (TAMC) would be acceptable for bioburden testing. If a pre-filter is added as a precaution only and not because the unfiltered bulk solution has a higher bioburden, this limit is applicable also before the pre-filter and is strongly recommended from a GMP point of view. A bioburden limit of higher than 10 CFU/100 ml before pre-filtration may be acceptable if this is due to starting material known to have inherent microbial contamination. In such cases, it should be demonstrated that the first filter is capable of achieving a bioburden of NMT 10 CFU/100 ml prior to the last filtration (38).

Furthermore, PDA TR 26 also states that process bioburden should be evaluated in order to establish *B. diminuta* as a relevant challenge organism for filter validation and the evaluation should be based on bioburden identification and risk assessment (2).

Thus, qualitative and quantitative analysis of pre-filtration bioburden of raw material components and contact surface(s) for any given product during the formulation and manufacturing process should become part of bioburden identification and risk assessment. Although the FDA guidance (1) doesn't implicitly state that bioburden must be known, there has been increased scrutiny by the regulatory authorities on the impact of environmental monitoring on sterile processes. Additionally, increased regulatory expectations and the need to mitigate risk have also been a factor in the use of redundant filtration of bulk and final fill operations. As stated in the PDA TR 26 (1), redundant filtration is a type of serial filtration in which a second sterilizing-grade filter is used as a backup in the event of an integrity test failure of the primary sterilizing filter.

FDA guidelines state:

> Use of redundant sterilizing filters should be considered in many cases. Whatever filter or combination of filters is used, validation should include microbiological challenges to simulate worst-case production conditions for the material to be filtered and integrity test results of the filters used for the study.

EU GMP Annex I guidelines state:

> Due to the potential additional risks of the filtration method as compared with other sterilization processes, a second filtration via a further sterilized micro-organism retaining filter, immediately prior to filling, may be advisable. The final sterile filtration should be carried out as close as possible to the filling point.

As stated previously, the requirement is for bioburden understanding and control that is established in a robust and validated manner. The key aspect of the entire process related to pre-filtration bioburden control is the risk mitigation, which in individual cases is afforded by using two sterilizing filters rather than one. The need for redundant filtration is process dependent and should be based on a fully document risk assessment that includes bioburden, filter-retention validation, filter-integrity history, processing time, product value, and reprocessing potential, and thus, drug makers are free to use their own discretion about the need for this strategy, even as they are bound to prove that their processes ensure the microbial safety of their products.

23.22 CONCLUSION

Filter validation performed by the drug manufacturer is not just a regulatory requirement; it also makes good business sense. The validation process should start with a filter requirements specification. From this specification, it is simply a matter of choosing the methods that allow verification of the requirements.

Although the regulatory requirements may appear to be more specific in one geographic area than another, with the continuing efforts at ICH and future attention of specific pharmacopoeias to additional elements of processes validation globally aligned regulatory requirements for sterilizing-grade filters may not be far away.

With respect to PUPSIT and pre-filtration bioburden evaluation in the course of manufacturing, it is incumbent upon the drug manufacturer to meet regulatory requirements and market expectations by evaluating the risks associated with alternative approaches used during drug processing and final filtration.

NOTE

1 *Abbreviations used in this chapter*: ASTM, American Society for Testing Materials; ATCC, American type culture collection; BI, biological indicator; CFR, Code of Federal Regulations; EPA, Environmental Protection Agency; EPROM, Electronically Programmed Random Only Memory; FDA, Food and Drug Administration; GMP, good manufacturing practice; HPLC, high-performance liquid chromatography; ICH, International Conference on Harmonisation; ISO, International Organization for Standardization; LVP, large volume parenteral; NDIR, non-dispersive infrared; NFR, nonfiber releasing; NVR, nonvolatile residue; PDA, Parenteral Drug Association; PLC, programmable logic controller; SVP, small volume parenteral; TOC, total oxidizable carbon; TR, Technical Report; USP, *U.S. Pharmacopeia*; WFI, water for injection.

REFERENCES

1. FDA Guidance for Industry: Sterile Drug Products Produced by Aseptic Processing—Current Good Manufacturing Practice. Pharmaceutical CGMPs. September, 2004.
2. PDA Sterile Filtration Committee. PDA Technical Report No. 26: Sterilizing Filtration of Liquids. Bethesda, MD: Parenteral Drug Association, 1998 (52 pages), Revised 2008 (60 pages).
3. United States Pharmacopeia 43, <1229.4>. Sterilizing Filtration of Liquids. 2020.
4. Steere W, Meltzer T. Operational considerations in the steam sterilization of cartridges. *Pharm Technol* 1993, 17, 98–110.
5. Desaulniers C, Fey T. The product bubble-point ratio and its use in filter integrity testing. *Pharm Technol* 1990, 14, 42–52.
6. DeSantis P. Validation of steam sterilization. In Carleton FJ, Agalloco JP, eds. *Validation of Pharmaceutical Processes, Sterile Products*, 2nd ed. New York: Marcel Dekker, Inc., 1998, 413–449.

7. Peterson AJ, Hankner DO. Assessment of long-term bacterial retention. *Med Device Diagn Ind* 1985, 150–155.
8. United States Pharmacopeia 43-NF38, <788>. Particulate matter in injections, 2020.
9. Simonetti JA, Schroeder HG. Evaluation of bacterial grow-through. *J Environ Sci* 1984, 27, 27–32.
10. Wallha¨usser KH. Durhwarchsund durchblaseffekte bei langzeit-sterilifiltration-prozessen (English: Grow-through and blow-through effects in long-term sterile filtration processes). *Pharm Ind* 1983, 45, 527–531.
11. Leahy TJ, Gabler R. Sterile filtration of gases by membrane filters. *J Biotechnol Bioeng* 1984, 26, 836–843.
12. Barber TA, Lannis MD, Williams JG. Method evaluation: Automated microscopy as a compendial test for particulates in parenteral solutions. *J Parenter Sci Technol* 1989, 43, 27–47.
13. Bhattacharjee HR, Paley SJ, Pavlik TJ, Atterbury O. The use of scanning electron microscopy to quantify the burden of particles released from clean-room wiping materials. *Scanning* 1993, 15, 301–308.
14. ASTM. Standard practice for determining the performance of a filter medium employing a single-pass, constant-rate, liquid test. F95–88.
15. German Federal Health Office (BGA). Promulgation regarding measures to reduce the asbestos contamination of parenteral products. November 1993.
16. Chatfield EJ, Dillon MJ. *Analytical Method for Determination of Asbestos Fibers in Water.* Mississauga, Canada: Ontario Research Foulndation, U.S. Environmental Protection Agency, 1983.
17. Carter J. Evaluation of recovery filters for use in bacterial retention testing of sterilizing-grade filters. *PDA J Pharm Sci Technol* 1996, 50, 147–163.
18. American Society for Testing Materials (ASTM). *Standard Test Method for Determining Bacterial Retention of Membrane Filters Used for Liquid Filtration.* ASTM Standard F838–83. Philadelphia, PA: ASTM, 1983 (revised 1993).
19. ISO 13408–2:2018: Aseptic processing of health care products—Part 2: Filtration. International Organization for Standardization 2018.
20. United States Pharmacopeia 43, <71>. Sterility tests, 2020.
21. Levy RV. Letter to the editor. *Pharm Technol* 1991, 15(1), 18.
22. Filter Manufacturers Validation Guides. Millipore Corporation. Optiseal Durapore Cartridges, LAGL 0.22 (m TP).
23. United States Pharmacopoeia 43, <643> Oxidizable Subatances, 2020.
24. United States Pharmacopoeia 43. <661>. Gravimetric extractables, 2020.
25. Stone TE, Goel V, Leszczak J. Methodology for analysis of filter extractables: A model stream approach. *Pharm Technol* 1994, 18, 116–130.
26. Sarry C, Sucker H. Adsorption of proteins on microporous membrane filters: Part 1. *Pharm Technol Int* 1992, 16, 72–79.
27. Sarry C, Sucker H. Adsorption of proteins on microporous membrane filters: Part 2. *Pharm Technol Int* 1993, 17, 60–70.
28. Meltzer T. *Filtration in the Pharmaceutical Industry.* New York: Marcel Dekker, 1987, 174–175.
29. Truskey GA, Gabler R, Dileo A, Manter T. The effect of membrane filtration upon protein conformation. *J Parenter Sci Technol* 1987, 41, 180–191. Erratum in *J. Parenter Sci Technol* 1988, 42, 102.
30. United States Pharmacopeia 43, <85>. Bacterial Endotoxins Test. 2020. U.S. Department of Health and Human Services, Public Health Service, Food and Drug Administration. Guideline on validation of the limulus amebocyte lysate test as an end-product endotoxin test for human and animal parent- eral drugs, biological products, and medical devices. December 1987. (This guideline was withdrawn on 12 July 2011 and FDA refers to USP <85>).
31. United States Pharmacopeia 43, <88>. Biological reactivity tests—in vivo, class VI plastics test, 2020.
32. United States Pharmacopeia 43, <88>. Biological reactivity tests—in vivo, mouse safety test, 2020.
33. Thome B, Joseph B, Dassu D, Gaerke J, McBurnie L, Dixit, M, Stering M, Tomlinson S, Mills S, Ferrante S, Weitzmann C. Data mining to determine the influence of fluid properties on the integrity test value. *PDA Journal of Pharmaceutical Science and Technology* September 2020, 74(5), 524–562.
34. Ferrante S, McBurnie L, Dixit M, Joseph B, Jornitz M. Test process and results of potential masking of sterilizing-grade filters. *PDA J Pharm Sci and Tech* 2020, 74(5), 509–523.
35. Ministry of Health, Labour and Welfare (MHLW), Sterile Drug Products Produced by Aseptic Processing, Tokyo, 2011.
36. Pharmaceutical Inspection Convention, Pharmaceutical Inspection Co-operation Scheme (PIC/S), Recommendation on the Validation Of Aseptic Processes, Geneva, PI 007–6, 2011.
37. EU Guidelines to Good Manufacturing Practice Medicinal Products for Human and Veterinary Use. Annex 1. Manufacture of Sterile Medicinal Products, 2008.
38. EMA/CHMP/CVMP/QWP/850374/2015 European Medicines Agency; Guideline on the sterilisation of the medicinal product, active substance, excipient and primary container. 2019, 1–25.

24 Disinfecting Agents
The Art of Disinfection

Arthur Vellutato, Jr.

CONTENTS

24.1 INTRODUCTION

The art of disinfection has been successfully accomplished since the early 1800s in a multitude of capacities. Disinfectants, or rather, antimicrobial agents have long been in our world. They are found naturally in our environment and are created synthetically by humans. They range from mild, safely ingestible chemistries to toxic and corrosive agents that would most certainly produce a lethal effect in humans. But as with many things in the modern world, confusion surrounding where, when, and how antimicrobial agents are employed is commonplace. Complexities such as effectiveness, toxicity, corrosion, labeling, residuals, overuse, vapors, shipment, and suitability all cause confusion. Of more concern is the "lack of knowledge" factor that is further complicated by our continuation of past practices instead of current best science, our belief that what is said or required by regulatory agencies is correct, and our overall lack of complete understanding of the limitations of such antimicrobial agents. Every day such agents are most certainly inappropriately used, with consequences that are detrimental to humans, our final product, our facilities, and the ecosystem at large.

There are a great number of registered drug products, over-the-counter drug products, soaps, creams, gels, foams, cleaners, fabrics, building materials, plants, animal extractants, and even paints with antimicrobial claims. And more and more products are being added to the list claiming antimicrobial properties. In one sense, this progress is good, but it is also worrying that most people may never read the label on such products. We tend to read the big words: the product name or subtitle claim. We just pour some in a bucket with a "glug, glug" theory of measure and assume more is better. Directions for correct dilutions and solution and container disposal are ignored. This lack of detail leads us to misuse.

DOI: 10.1201/9781003163138-24

As an example, the most notable misinterpreted claim is a product that "kills 99% of household germs". Most assume that means that 99% of the organisms in the world are destroyed by the agent. But they are not. This statement means that 99% of the organisms that were tested, and at the population that they were tested, were proven to be destroyed by the agent to a level of 99%. This leaves 1% of each organism tested still alive.

The pharmaceutical, biotechnology, health care, and lab animal world are no different than the consumer world in this venue. We scramble to find the most efficacious chemical but ignore the real art of disinfection that would most certainly assure success. We happily follow past practices and ignore any past imperfection and do not consider potential future practices. As we cling tightly to the way it was always done, we potentially overuse our antimicrobial agents in both active ingredients and frequency. We fill our clean rooms with residues that are difficult to completely remove. We place our personnel using such products in uncomfortable and possibly adverse health situations. And we corrode our expensive equipment and surfaces because of overuse and lack of true cleaning. Often, we choose the quick answer and discard the reliable more involved method as we rush to complete our task to either get our operations back in motion or to meet the all-important CAPA (corrective and preventative actions) deadline. And in doing so, we sacrifice science for completion.

24.2 DISINFECTION PROCESS AT A GLANCE

Disinfection is a general term that is used to describe the destruction of a microorganism. But the term is also used in a singular, noun format of disinfectant to describe a classification of agents that prove effective in killing certain vegetative cells. The dually used term causes some confusion, but this connotation has been used since the beginning of antimicrobial agents and will most likely never be changed. The destruction of a cell on a hard surface or in liquid is based on the saturation and penetration of the cell wall or an organism with an efficacious chemical agent, for a specific contact time. Concern is vested in the potential soil load on the surface or the liquid to be disinfected that may either compromise the contact of the chemical agent with the organism or may soak up the actives in the chemical agent preventing it from contacting the cell. Most agents work on the theory of cytoplasmic disruption whereby the actives penetrate the cell and explode the nucleus. The penetration is dependent upon surface contact of the chemical agent with the exterior of the cell and a contact or wetted time sufficient to allow this penetration or saturation to occur.

Disinfection is effective in many places and in other places virtually impossible. On hard surfaces in a clean room, disinfection is normally successful but at different levels based on the surface type. A hard surface like stainless steel provides very little surface irregularity, but a more porous surface like epoxy, plastic, vinyl, Mipolam, Kidex, ceramic, and plexiglass presents a more difficult challenge. Microorganisms can lodge themselves into pores inherent in the surface, and thus contact with the disinfectant on a sufficient surface area also becomes more of a challenge. At the same time, dirtied surfaces may block the chemical agent from reaching the surface of an organism as the organism may be lodged in a pore or imperfection of the surface and surrounded by a soil load.

In the venue of disinfection, surfaces are categorized into several types. Metals are the easiest surfaces to disinfect. Glass and plexiglass can have pores and can present minimum problems. Epoxy, Terrazo, Mipolam, Kidex, vinyl, and like surfaces present further problems as pores and imperfections are more prevalent in the makeup of the surface. These surfaces trap more particulate foreign matter and residuals and can further complicate disinfection. Ceramic, tile, grout, polymers, sealers, and many plastics create an even more difficult challenge as they have further imperfection and pore issues that act as previously described to block contact of the agent or lodge contaminants in such imperfections. Cloth, foam, uncoated or poorly coated dry wall, concrete, cinderblock, paper, and other cellulose and noncellulose products that incorporate the ability to soak up water or contamination have an enormous potential to lodge organisms that may never be reached by chemical agents. These surfaces present the greatest challenge, especially with mold infestation. If contaminated with mold, such surfaces may never be disinfected as the mold reaches deep into the pores and the surface may become virtually impossible to disinfect. In this situation, removal of the surfaces or structures may be the only option.

Disinfection of air is virtually impossible. Many think that vapors or spray into the air is effective, but it is not. The ability of the organism to become saturated for any length of time suitable for destruction of the cell simply cannot occur. And thus, the contact time or dry time is too short. Disinfection of air is normally engineered from an air filtration process. However, filtration cannot assist when contamination is allowed to be continuously introduced into the clean room. As an example, imagine you are an organism on a particle that came in through a door from the outside. First you have to avoid all the obstacles in the clean room. Next, you have to navigate your way through the currents inherent and created in the clean room so that you can make your way to the return vent. But air keeps pushing you down and people keep blowing you left and right. Now, the entire time you are floating you have one thing in mind: finding food, water, and surviving. You find food in particulates and other organisms, and the search for food is your main direction to sustain life. So most likely you will land on a surface as the potential for available food is greater. However, if you do not, you next have to avoid the covering of the return vent and be taken on a road trip through the heating, ventilation, and air conditioning (HVAC) process to the filter in the ceiling where most likely you will be retained. As explained, the removal of airborne microorganisms is a difficult challenge. Gassing systems can assist but again, you have to create a toxic environment whereby the air is of a corrosive nature to the cell and because of this, over time, you will have either a chemical disruption on the wall of the cell or deplete the cell of oxygen. Processes like vaporized hydrogen peroxide (VHP) in this venue are not toxic enough, and if you were to be able to create such a toxic environment, you would most certainly simultaneously create an extremely toxic environment for humans. Such processes are effective for surface disinfection provided that shadowing and other criteria are met completely. And the surface is never cleaned by fogging or gassing. If one seeks to destroy an organism that can survive without oxygen, called an aerotolerant organism, or the more used slang term of "anaerobic organism", depletion of oxygen from the clean room is necessary to assure its demise. This is seen in processes with nitrogen blankets as a means of removal of oxygen from the area. Again, a practice that could not be employed in the entire facility but only in a localized area and, in the end, also does not clean the surface. The only truly assured methodology to destroy or rather prevent an airborne organism is control.

Although we focus in this chapter on antimicrobial agents, it is extremely important to realize that the agent alone cannot assure an acceptable environment. A wholistic approach therefore needs to be invoked to assure we can attain what we desire—acceptable environmental conditions to meet product and regulatory requirements. This means you are going to have to go further than spreading some antimicrobial agents onto the surface. In fact, the firm that controls more and disinfects less always wins the battle. The firm that decides to have minimal control and disinfects more to combat the invasion of contamination seems to always have excursions that riddle operations with expensive investigations, placing bandaids on unsound practices and routine regulatory comments. Truly, the wholistic approach is the desired means to the end.

24.3 TERMINOLOGY—SANITIZERS, DISINFECTANTS, SPORICIDES, AND STERILANTS

The varying terminology surrounding disinfection and disinfectants worldwide is one of the most confusing aspects in the scope of the subject matter. So much confusion is created initially by governmental agencies, industry, labeling, shipping regulations, safety data sheets (SDS), and past beliefs. One may ask, why is this so? The answer is very simple. Each country has its own set of rules for how disinfectants will be registered, tested, and labeled. To add to this confusion, past beliefs coupled with varying regulatory comments contradict each other and create terminology confusion. Later in this chapter, these regulatory discrepancies between agencies will be discussed in detail.

For the good manufacturing practice (GMP) professional, only four terminologies are required to be understood. These are: Sanitizer, Disinfectant, Sporicidal Product, and Sporicide/Cold Sterilant. All have different meanings and characteristics, however, the slang terminology used for all of these terms is "disinfectant". Definitions for each are presented following.

24.3.1 Sanitizer

A sanitizer is a compound that will reduce the number of bacteria to a safe level. Determined by the United States Environmental Protection Agency (EPA) as *Escherichia coli* and *Salmonella typhi*, or *Staphylococcus aureus* for food product contact surfaces. For example: 70% isopropyl alcohol (IPA) or EtOH.

24.3.2 Disinfectant

A disinfectant is a chemical or physical agent that destroys or removes vegetative forms of harmful microorganisms. Determined by the US EPA as a 106 reduction of *Salmonella choleraesuis, Staphylococcus aureus*, and *Pseudomonas aeruginosa* on stainless-steel penicylinders with the presence of a soil load. For example: phenol or quats.

24.3.3 Specific Sporicidal Product

A specific sporicidal product is a compound that when tested destroys a particular organism at an acceptable level as defined by registering bodies. An example of this would be "Sporicidal Against *Clostridium difficile* (*C. diff*)". Determined by the EPA as the 106 reduction of *Clostridium difficile* or the organism claimed on penicylinders with the presence of a soil load. For example: lower-level peroxides, bleaches, and IPA H202.

24.3.4 Sporicide/Cold Sterilant

A sporicide/cold sterilant is a compound that destroys all vegetative microorganism and bacterial and fungal spores. Determined by the EPA as the 106 reduction of *Bacillus subtilis* and *Clostridium sporogenes* on stainless-steel penicylinders with the presence of a soil load. For example: bleach or IPA H202

So, what's the real difference between these compounds, chemicals, or physical agents? A sanitizer is the weakest of the claims. In this classification are alcohols that can be used during manufacturing. Every other class, compound, or agent should not be used during manufacturing because of the possibility of residuals contaminating the final product, overspray of possibly toxic agents that may enter the product, harmful and toxic vapors, and most importantly the contamination of gloves with possibly toxic or harmful agents that can cause problems during aseptic manipulations or handling of aseptic connections.

A disinfectant, by its true definition, is a chemical that destroys vegetative cells and has little, if any, effect on spores. And such an agent is used routinely because the routine use of the specific sporicide or the sporicide/cold sterilant can cause significant corrosion of surfaces when used at that frequency. The use of the specific sporicide or the sporicide/cold sterilant daily to address contamination noticed in environmental monitoring signals a control problem and not an infestation of spores that linger in the room. The specific sporicide should be considered a dangerously claimed product if the end user is not paying attention as it is only valuable for the spore that is listed and not all spores. It does not have the broad efficacy performance required for the sporicide/cold sterilant. Many products like to claim a specific spore is killed, such as *Clostridium difficile* (*C. diff*), which in reality is very easy to destroy. The testing only needs to be done on carrier surfaces and not sutures like the sporicide/cold sterilant. End users should read the label to see what exactly the agent is capable of destroying prior to use. The assumption that it destroys all spores is incorrect. The sporicide/cold sterilant is the strongest label claim and to destroy *Bacillus subtilis* and *Clostridium sporogenes* it requires a minimum of 1.0 × 106 in a blood serum soil load. This claim requires surfaces and suture testing to be performed and is the strongest claim available.

24.4 CONTAMINATION CONTROL

Maintaining control of the environment is critical. It is the most important aspect surrounding a cleaning and disinfection program. We need to understand that control of the environment does not come from using a chemical agent that will destroy all microorganisms. Rather, control of the environment is the assurance level we establish to reduce or possibly eliminate the contamination from ever entering our controlled areas. Once a chemical agent has been applied to the surface and subsequently dries, its destructive capabilities against a microorganism are complete. Disinfection (or sporicidal) characteristics of a chemical agent require the organism to be wetted by the agent. Although certain residues from chemical agents may have some remaining antimicrobial properties, the destructive capabilities of the residue are minimal. Control over what enters the controlled area now takes over. Control of the environment has nothing to do with disinfection. Disinfection is complete when control takes over. Manufacturing operations will not disinfect surfaces while production is occurring. Disinfection is done prior to manufacturing, and the environment is released to produce a product. Our success or failure in control is measured each time as environmental monitoring is conducted during our filling operations.

Addressing contamination prior to its entry will assure that we will not have to contend with its presence. Although disinfection of surfaces always takes the lead in structuring our procedures, control over the environment should remain one of our main focuses.

The criteria for reducing the bioburden that enters the controlled area requires us to evaluate the entry of personnel, components, water, tanks, carts, and even disinfectants to name a few. This is done to assure that we do not undermine our disinfection efforts by introducing high levels of contamination after disinfection is complete. We must assure each item's appropriateness in the room classification to which it enters. In Grade A/B aseptic operations, we need to carefully evaluate each item as clean and sterile prior to entry. We need to assure the cleanliness, sterility, and appropriate fit of garments for our personnel. In Grade C/D, although our concern is less, we must assure that such environmental conditions do not adversely affect the Grade A/B area. In short, a room can be monitored as having very little if any bioburden at rest, however, in a static condition, we can easily corrupt our previously achieved results through poor control.

24.5 CLEANING VERSUS DISINFECTION

Too often cleaning is confused with disinfection. They are not the same. Cleaning characterizes the removal of particulates, microbes, and possibly existent residues from surfaces. Cleaning requires that a nondestructive mechanical action be applied that loosens and removes contaminants from the area. Procedurally, contaminants and residues are loosened and rinsed to the floor. Subsequently, the dirtied solution on the floor is collected and removed from the area (normally by a squeegee). By lessening the level of particulates, microbes, and residues on the surface, our disinfection efforts become simpler. First there are fewer organisms to destroy as most have been removed from the area. And second, as the bioburden and residues are lower, the possible obstructions blocking the chemical agent from contacting the organisms are minimized. In short, cleaning prepares the surface for disinfection.

Disinfection relates to the saturation and penetration of the cell wall of an organism by a chemical agent. It further requires that an organism remain wetted for a specified contact time with a chemical agent capable of killing the organism in question. Disinfection depends upon temperature, saturation and penetration of the cell wall, contact time, surface and bioburden of the surface, existent soil load, concentration of the chemical agent, and pH. Provided the appropriate chemical agent is utilized, the key to disinfection in the clean room is keeping the surface wetted for 5–10 minutes. This is difficult, as the movement of air via laminar flow tends to dry surfaces quickly.

As discussed, there is a significant difference between cleaning and disinfection. Cleaning tries to remove contamination from the surface whereas disinfection attempts to destroy what viable cells exist on the surface. We can use a toothbrush and mouthwash as real-life examples. If we were to discuss the options of not using a toothbrush anymore and only utilizing a daily mouthwash rinse with our dentist, he/she would inform us that we would soon have no teeth. Residues, particulates, and microbials will build up on the surface of our teeth, which would eventually deteriorate. This scenario depicts what occurs too often in the pharmaceutical and biotechnology industry. We forget to brush and just try to kill anything that exists on the surface. Eventually our surfaces become residue laden and more difficult to disinfect and eventually deteriorate. Within our note to technical brilliance, we sometimes forget simple common sense. And unfortunately, the phrase, "simple common sense" is not the title of any GMP, Code of Federal Regulations (CFR), or guidance document.

The effect of the buildup of residues, particulates, and possibly microbials is also aided by the surface itself. Clean room surfaces are irregular in nature as depicted in the scanning electron microscope (SEM) photos within this chapter. Such surfaces trap residues and other contaminants, which makes the surface more difficult to disinfect. (See SEM photos of residues and surfaces throughout this chapter.)

Sooner or later, we have to clean. The frequency of cleaning can vary from a daily function to a monthly function. It is usual to clean surfaces either biweekly or on a monthly basis. Some may say, "But that's an additional cleaning operation we need to do". The correct response is "It needs to be done".

Within the health care setting and most commonly reported in the hospital setting, test reports have shown the effect that cleaning the surface has on the microbial levels in controlled areas. Many publications purport this concept. In the pharmaceutical and biotechnology setting, a test report conducted and published in 1989 by A. Vellutato, Sr. and A. Vellutato, Jr. of Veltek Associates, Inc. demonstrated the effect that cleaning the surfaces has on the level of microbial contamination found on the surface. The study focused on the concept that cleaning alone would remove most of the existing microbial contamination. In the report, all surfaces were cleaned with a sodium lauryl sulfate detergent and mechanical cleaning action on a daily basis. Such cleaning was conducted in a Grade A/B, Grade A, and ISO 5 area. Environmental monitoring was routinely conducted at air, surface, and personnel sampling locations. The manufacturing operations filled an average of 4,000 units of 500 ml bags of USP water for injection. Manufacturing operations ran for four hours per day and upon completion of manufacturing, the clean room was completely cleaned. The cleaning mechanism utilized a mop and two-bucket system, a sprayer, and a squeegee. The procedure for cleaning was to apply the sodium lauryl sulfate detergent (DECON-Clean) to the surface utilizing a top to bottom approach. Upon completion of the mopping, the chemical was then sprayed onto the surface and all excess liquid on the surfaces pulled downward by squeegee to the floor. The remaining liquid was then collected on the floor and removed. The results for 30 days met industry limits for Grade A/B, Grade A, and ISO 5. At 45 days, control was lost, the results exceeded the limits, and a sporicidal agent was used once on day 45. The limits returned to acceptable levels for 31 days (day 76 of the study). The final conclusion was one to two uses of a sporicidal agent with the cleaning regime controlled the environment. Cleaning is based on a few physical factors: (1) the surface to be cleaned, (2) the contamination vested on the surface, (3) the chemical used to clean the surface, (4) the effect of the chemical agent on the surface to be cleaned, (5) the level of surfactants in the chemical agent, and (6) the effect of the chemical on the contamination, moreover residue, that exists on the surface.

Knowledge of the type of surface to be cleaned is very important. And, it is important to understand the concept of surface irregularity, surface tension, and residue buildup so that when antimicrobial effectiveness studies are performed, one can account for these inconsistencies. As depicted in Figures 24.1–24.6, surface irregularity varies as the composite materials change.

The surface irregularity may compromise cleaning and disinfection efforts for two basic reasons. First, cleaning

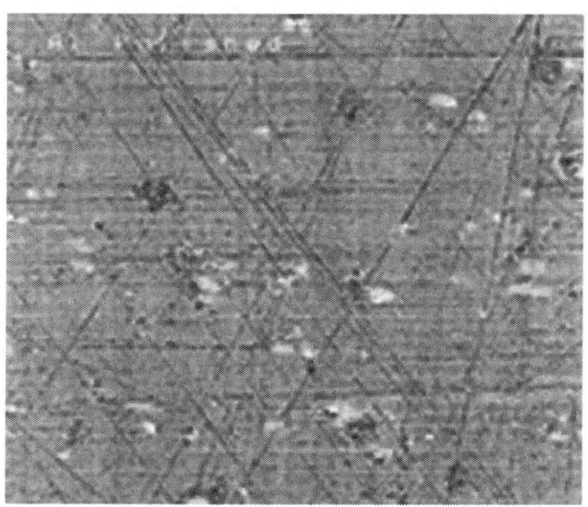

FIGURE 24.1 Aluminum surface.

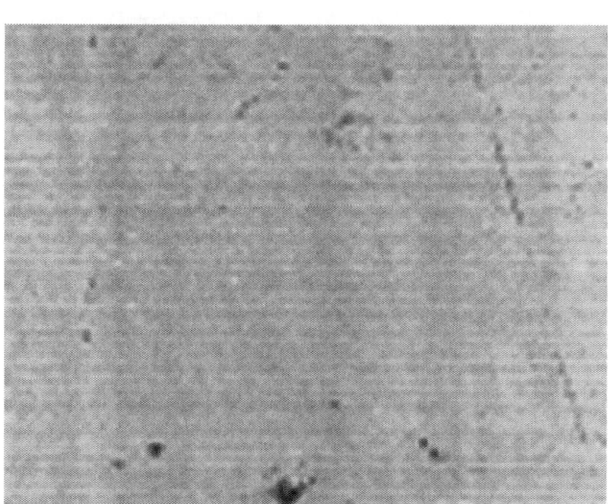

FIGURE 24.2 Stainless-steel surface.

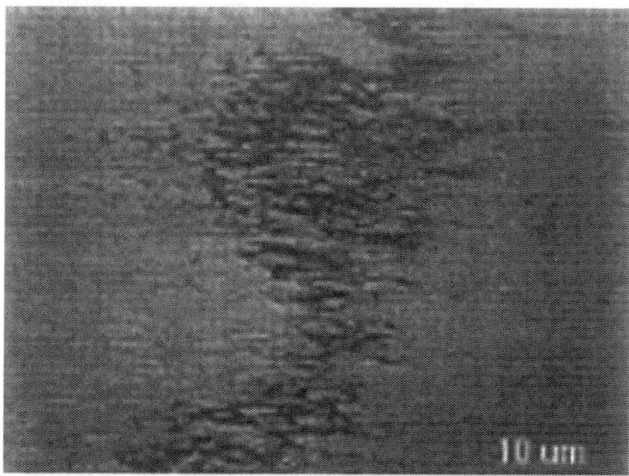

FIGURE 24.3 Epoxy-coated surface.

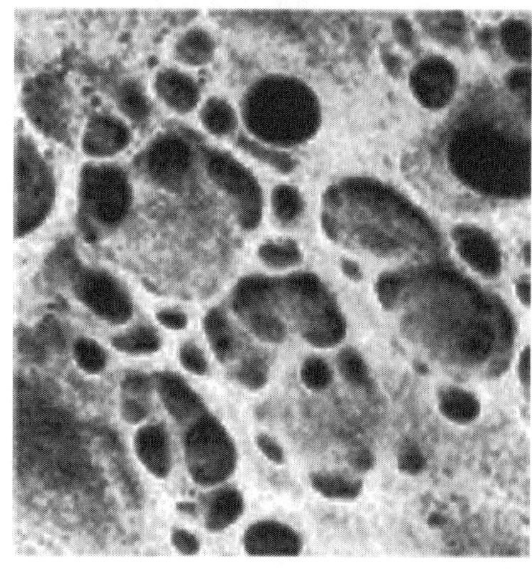

FIGURE 24.4 Clean room curtains #1.

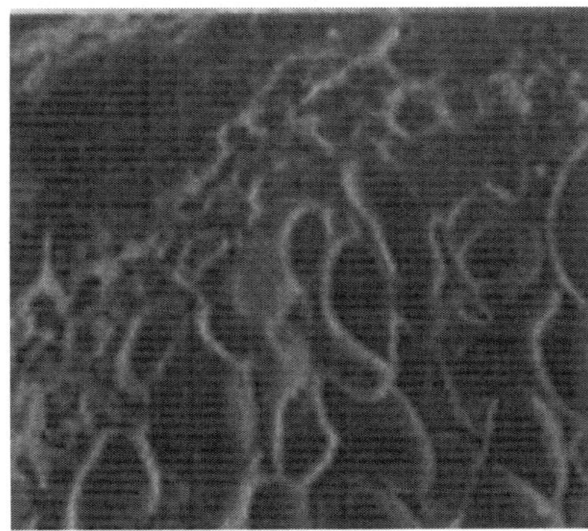

FIGURE 24.5 Clean room curtains #2.

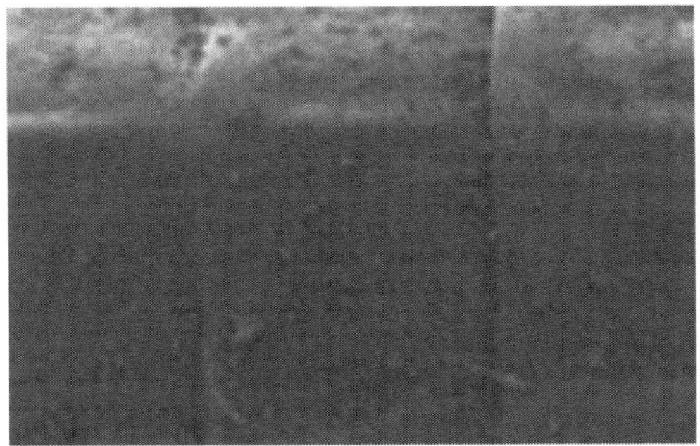

FIGURE 24.6 Vinyl surface.

becomes more difficult. And second, disinfection becomes harder as the chemical agent cannot contact all the surfaces for the required wetted time period (as obstructions may exist). Cleaning becomes more difficult as the surface's irregular nature may allow particulates, residues, and even microbial contamination to lodge itself within the rutted or porous areas of the surface. These "nooks and crannies" are very hard to clean with most clean room apparatus designed for cleaning. Most clean room mops and wipes are flat and do not allow the penetration of the fibers or surfaces of the cleaning mechanism to reach into the crevices. The lack of such an abrasive cleaning action bypasses the opportunity to loosen and remove these contaminants. Most irregular surfaces are commonly so across the span of the material, so a multitude of contamination sources exist on the surface to be cleaned.

Cleaning of an irregular surface requires one to use a cleaning device that can penetrate into the "nooks and crannies" and make contact with the existing contaminants and residues. As an example, we would not use a wiper or a pad to try to clean our teeth. We would use a toothbrush. Unfortunately, a clean room brush is not a commonly available item in the industry. Nor would requiring a controlled area to be cleaned with it be a reasonable request of production personnel. This is a subject that the cleaning apparatus manufacturers need to address more closely. The lack of available products forces the clean room professional to adapt a non-clean room product to address this specific need.

The second basic complication is that if particulates and residues exist in such "nooks and crannies", the possibility for a disinfecting agent to contact the surface within the "nooks and crannies" is improbable. Thus, we are disinfecting the surface of the particulates and residues and never disinfecting the actual surface. Without such assurance for cleaning and disinfecting the actual surface, the possibility for contamination to be lodged underneath the particulate and residue may be probable.

In the upcoming sections of this chapter, we will discuss antimicrobial effectiveness testing. In these sections, we will come to understand that during such validation testing, we need to look to inoculate a surface with a known enumeration of microorganisms and soak such surfaces in a disinfecting agent for a predetermined time period. Upon completion of the soaking, we will then test the surface to account for the possibly remaining viable contamination. When we attempt to use such data in the field, we will find our testing skewed for two basic reasons. First, our testing inoculates a variety of surfaces with a multitude of microorganisms. Some surfaces are more irregular than others. In our test, and after we have soaked the surfaces, we need to rinse clean the surfaces of any possible contamination to a filter that is then plated to a growth medium. However, we find the rinsing of the irregular surfaces more difficult as microorganisms may lodge and cling within the irregular areas of the surface. This means the ability to rinse free a microorganism from a smooth surface is easier than from an irregular surface. What we may learn from this testing is that it may seem that our disinfecting agent is more effective on the irregular or porous surface, but this is because of our inability to rinse free all of the microorganisms present. The smooth surfaces are more easily rinsed and the existing microorganisms are removed and can grow in our growth medium. However, the microorganisms from the irregular or porous surface may have never been removed from the surface itself. This means that they were never rinsed to the filter and plated for growth. Thus, we could conclude that smooth surfaces are harder to disinfect than irregular or porous surfaces in our manufacturing and testing areas. The effectiveness report may show higher remaining colony-forming units (CFUs) for the smooth surfaces than that for the irregular or porous surfaces. This assumption would be incorrect. In the manufacturing or testing areas, the opposite occurs, and we find it more difficult to disinfect the irregular or porous surfaces. Understanding this concept is critical to successful disinfection. We will have the same trouble, if not a more complicated problem, in disinfecting and rinsing the microorganisms from the irregular or porous surfaces in our manufacturing and testing areas. And because of this, we may leave viable contamination on such surfaces. Care needs to be taken when disinfecting irregular or porous surfaces, as they are harder to clean and disinfect than smoother surfaces.

In general, clean room material grades can be separated into six basic categories. Although a more specific list of material sub-categories could be rendered, for our discussion and understanding we can divide them into: aluminum, stainless steel, epoxy-coated finishes, plastics, vinyl, and glass.

In Figure 24.1, we see an aluminum surface. This surface is a metal grade that is soft and easily scratches and is deteriorated by chlorine solutions, glutaraldehyde and peroxide, and peracetic acid and hydrogen peroxide solutions (from a list of basic disinfecting agents). Although aluminum is a commonly used metal, deterioration of the surface from chemical exposure normally shows as a turquoise bluish-gray bubbling on the metal. Aluminum is also easily stained or discolored by the residues from phenol, quaternary ammonium, and iodine to name a few. However, among clean room surfaces, aluminum is normally easy to clean and disinfect.

In Figure 24.2, we see a 316 L stainless-steel surface. This surface is very smooth and contains few impurities in the metal that may be deteriorated by a disinfecting agent. Most impurities in the metal grade are those of the carbon family and are deteriorated by chlorine solutions, glutaraldehyde and peroxide, and peracetic acid and hydrogen peroxide solutions (from a list of basic disinfecting agents). Stainless steel comes in a grade based on the level of impurities in the metal. Normally, deterioration of the surface is in the form of a rusting that pits the surface or surface rust. Rusting of the stainless steel itself is from chemicals or water oxidizing and/or reacting with impurities in the metal. Surface rust is the rusting of airborne heavy metals that deposit atop the surface. Within the clean room operation, many metal grades exist from 302 to 402 stainless steel. Many in the industry demand the use of 316 L stainless steel; however, this metal grade cannot be used for every component because of its brittle nature. Some perfect examples of this would be a spring or a solenoid mechanism. Utilizing a metal that is too hard will cause the spring or moving part to routinely break as friction of movement on the metal will cause stress and fracture. Stainless steel is also easily stained or discolored from the residues from by phenol, quaternary ammonium, and iodine to name a few. However, among clean room surfaces, stainless steel is normally one of the easiest surfaces to clean and disinfect.

In Figure 24.3, we see an epoxy-coated surface. Epoxy-type surfaces are very numerous in types and materials. They are a coating that is applied in a liquid form, which then hardens. This is where disinfection becomes complicated. Its irregular surface appearance makes epoxy challenging for the disinfection professional. Most clean room walls and floors are made of an epoxy material or similar material so understanding cleaning and disinfection of this surface is critical. Common buildup of particulates and moreover residues in the crevices complicate cleaning and disinfection. Problematic situations can arise in the crevices of the surfaces as air pockets can form and, once disinfection is complete, be broken and possibly release existent contamination to the environment. The most difficult task with the epoxy-coated surface is to clean the contaminants from the surface so that the disinfectant can be applied and have the ability to address the surface itself. Normally epoxy-coated surfaces are deteriorated by chlorine solutions, peroxide, isopropyl alcohol, ethanol, and peracetic acid and hydrogen peroxide solutions (from a list of basic disinfecting agents). The normal deterioration occurs in the form of over drying and cracking of the material finish. The material becomes powder-like and orange to blue in color (dependent upon the material). Epoxy-coated surfaces need routine replacement or refinishing and as such remain continuously on a preventative maintenance schedule. The epoxy-coated finish is also easily stained or discolored from the residues from phenol, quaternary ammonium, and iodine to name a few.

The next basic clean room surface type is a collage or porous material such as plastic, vinyl, plexiglass, Delron, and other similar type products. In Figures 24.4–24.6, we can see what these surfaces may look like when magnified through an SEM microscope. Characteristically they have pore openings that are rather deep. Cleaning and disinfection are very difficult with these materials. Normally these types of surfaces need to be replaced over time. These materials average about two to three years before requiring replacement. Replacement is costly but required and should be viewed as a cost of doing business. Normally porous surfaces are deteriorated by chlorine solutions, peroxide, isopropyl alcohol, ethanol, and peracetic acid and hydrogen peroxide solutions (from a list of basic disinfecting agents). The normal deterioration occurs in the form of over drying and cracking of the material itself. A yellowing or changing of color to an orange or brown are common symptoms of the drying process. The porous material is very easily stained or discolored from the residues from phenol, quaternary ammonium, and iodine to name a few, and such residues and/or stains only increase the deterioration process. Most notably, these materials are used as curtains that separate process control areas. Items such as plastic curtains are one of the most widely used materials in the clean room environment. Cleaning of these surfaces is normally done frequently if not daily, as they represent the second closest non-product contact surface next to the filling line itself. This overcleaning shortens the life of the material.

The last category is glass. Glass is a relatively smooth surface with the characteristic of not deteriorating. Glass is very difficult to clean but remains a constant example of how difficult all surfaces are to clean. The reason glass seems harder to clean is that one can see through it. All other materials discussed to this point are not clear. If such materials were clear, as is glass, the cleanliness of these surfaces would look horrifically dirtied in comparison with that of glass. Glass is very easily dirtied. It is a very difficult material to keep clean. However, this surface is used more and more each day in clean room operations as it allows viewing of the operation by supervisors and visitors. In general, glass does not stain easily from disinfectants or sporicides. However, residues build up and it is discolored from the residues of phenol, quaternary ammonium, and iodine to name a few.

We have seen from this section the importance of cleaning. In Figures 24.7–24.10, we can see what these culprit residuals look like through SEM photos.

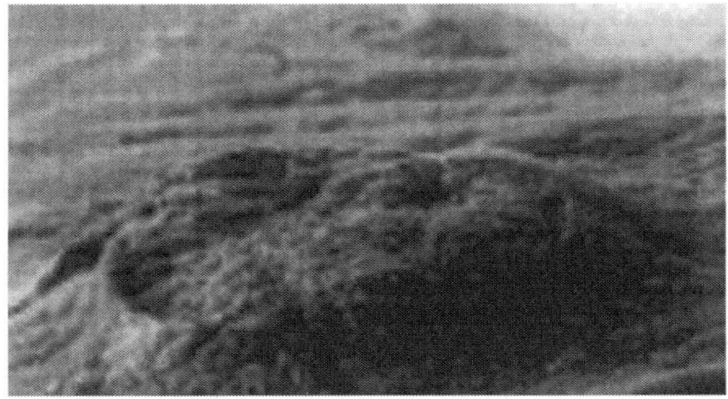

FIGURE 24.7 Phenolic residue.

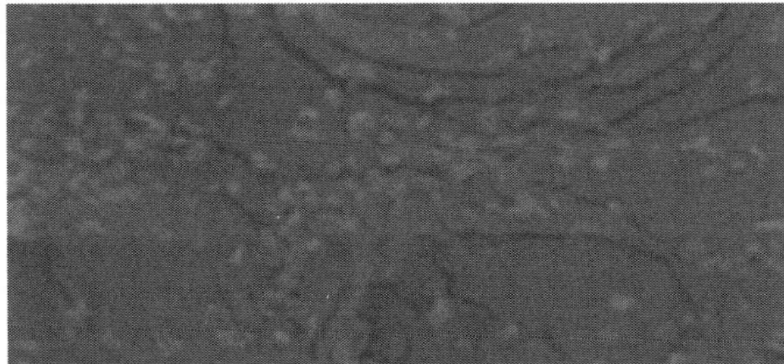

FIGURE 24.8 Sodium chloride residue.

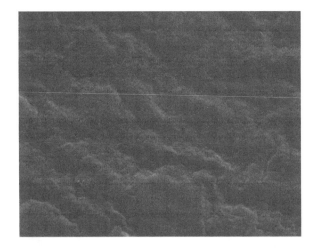

FIGURE 24.9 Quaternary ammonium residue.

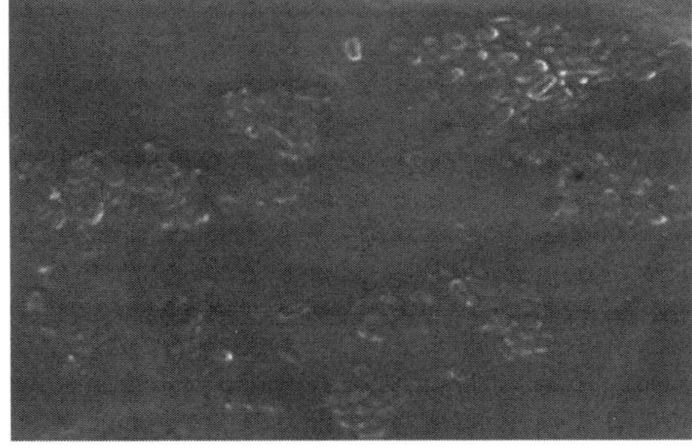

FIGURE 24.10 Peracetic acid and H202 residue.

Simply, the effect of cleaning surfaces assures the best possible opportunity to disinfect the surface as the microbial levels and the possibly existent residues and particulates will be lower. Later in this chapter, we will discuss the mechanisms to clean such surfaces. However, first we must understand the chemical agents that cause the main problem associated with a dirtied surface, the residue.

24.6 WORLDWIDE REGULATORY REQUIREMENTS FOR AGENTS AND GOOD MANUFACTURING PRACTICE

As a subject, worldwide regulatory requirements could run to several books. Prior to the year 2002, the main worldwide registering bodies were the United States Environmental Protection Agency (EPA), the Australian Therapeutics Goods Administration (TGA), the Canadian Health Canada, the Chinese Ministry of Health (MOH), and several registering bodies that were country-specific throughout Europe prior to the creation of the Biocidal Products Directive 98/8/EC (BPD). However, in recent years, this list and the requirements for such registration throughout the world has exploded to a point where nearly every country seems to now have an authority for the registration of antimicrobial agents. Although this controls chemical and process antimicrobials, it also increases costs dramatically, lengthens approval times, institutes varying types of labeling, and most importantly tries to marry one requirement from one country to another, which the world sees as virtually impossible.

Some of the most confusing scenarios in worldwide regulatory registration are labeling, varying claim requirements, and varying test procedures. Let us first discuss labeling. Well that is simple. Incorrect thinking would be to create the label and make it available in many languages and this should work for the world. Unfortunately, it is not that simple. Each country has a layout they desire. Most agree with the four-panel theory that places required labeling aspects into four distinct panels that each take up a side of the container. They are left, right, front, and back. Although four panels are available, in some instances this does not provide enough space. Thus, putting secondary labeling material inside the label in a peel and reveal label format is normally acceptable if the container is too small or the text required is lengthy. The inconsistent part of labeling worldwide is that each country requires differing information on each panel. Although the name of the brand, the subtitle claim, and the ingredients are standard on the front label, some desire first aid on the front panel and some on the side right or back panel. The remainder of the required label text is often different in each country. Label warnings like directions for use, precautionary statements, storage/disposal, general information, and efficacy performance are all located in varying sections on the label in every country. Unfortunately, there is not one approved standardized format for labels that can serve the world. This causes confusion for the end user and moreover management who will write internal standard operating procedures (SOPs) and safety documents. A firm with multiple locations throughout the world will have to create separate documentation for each country even though they may be using the same product at each location as the pertinent locations for vital information is different in each country. Two other problems are the requirements for each country referencing the claim made on a product and the varying test procedures required to be used to make each claim for a product. And moreover, the main confusion is that the testing and approved label claims for efficacy performance really do not serve pharmaceutical, biotechnology, lab animal, or the compounding pharmacy industries. There are many standard tests that are employed by each country that include but are not limited to AOAC (Association of Official Analytical Chemists), Annex IIB and VI of the BPDEU or EN Testing (EU Biocidal Directive), and many ASTM test protocols (per country/product/claim). A firm registering an antimicrobial agent will have to register the product in virtually every country in the world and know that the hundreds of thousands of dollars spent, the time and effort incurred, and the final product information will not meet the requirements of a GMP facility. And as such, GMP operations will have to reconduct antimicrobial effectiveness testing accordingly to meet governmental regulations for drug products.

24.7 ANTIMICROBIAL EFFECTIVENESS

Determining what chemical agents will destroy a known level of one's environmental isolates or ATCC cultures is the next step. Prior to conducting either a time contact kill study (tube dilution or on user surfaces), or an AOAC Protocol Study, one needs to review the available disinfecting agents and determine which is initially appropriate for their operations. Upon choosing one or two disinfecting agents and a sporicide, one can continue with the antimicrobial effectiveness studies.

Validating one's sanitizing, disinfecting, and sporicidal agents requires delineation of the organisms to be tested. One could use a list of ATCC cultures; however, utilization of one's environmental isolates nets a more exacting test for each unique manufacturing operation. Testing ATCC cultures such as *Bacillus subtilis*, *Aspergillus niger*, *Pseudomonas aeruginosa*, *Staphylococcus aureus*, *Candida albicans*, and *Escherichia coli* is acceptable, but not as exacting as testing a plane of one's known environmental isolates. The use of one's own environmental isolates is a preferred methodology by most regulatory agencies.

Antimicrobial effectiveness studies need to be based on realistic bioburdens that may be noticed in the controlled areas. It is usual is to test an enumeration greater than or equal to 1.0 × 104 CFUs. Our goal is to prove a three-log reduction. However, some guidelines like the British Standard BS EN 13697:2001 calls for a four-log reduction as proof.

In determining which test to conduct, one needs to review how one will address an organism in the clean room. As the organism will be on the surface, a time contact kill study that confirms the destruction of a known enumeration of cells on an end user's surface is more depictive than a time contact kill study done in a tube dilution (in suspension). The reasoning for this is organisms dried on a surface better depict the situation of disinfecting an organism in the clean room. The available surface area of the organism that can be contacted by the disinfectant that rests on the surface is 270o. The available surface area of the organism that can be contacted by the disinfectant in the tube dilution study is 360o. Obviously, the surface test presents a more realistic scenario. AOAC protocol testing is required by the EPA to register a claim for a disinfecting agent. It utilizes 60 carriers (a ceramic penicylinder) and requires a high enumeration value of equal to or greater than 1.0 × 106. Protocols use either AOAC use-dilution or AOAC sporicidal tests procedures. Although this is the method used for registration, it may be too involved and expensive for pharmaceutical and biotechnology firms to utilize as a method for testing antimicrobial effectiveness.

The EPA supports the use carrier methods for the evaluation of a disinfectant product's efficacy. This test requires that the microorganism is to be dried on a nonporous carrier. The rationales behind the choice of a carrier method are the beliefs that (1) microorganisms that are dried are more difficult to chemically inactivate than those microorganisms in suspension, and (2) that in the health care setting, microorganisms are more often found in the dried state than in suspension.

Within the framework of antimicrobial effectiveness testing, we also need to incorporate realistic contact times to depict representative dry times of the disinfectant on clean room surfaces. Because of the significant movement of air from laminar flow in the clean room, the normal dry time is five minutes at best. Although our floors may remain wetted longer (possibly up to ten minutes), the vertical surfaces will dry faster (three to five minutes). Thus, antimicrobial effectiveness testing should incorporate a worst-case scenario and utilize a three to five minutes dry time. An exception to this rule would be sanitizing agents such as isopropyl alcohol, ethyl alcohol or AAA ethanol at a concentration of 70%. These products dry faster, nearing one to two minutes, and testing should be altered accordingly. In recent years, many firms have begun to test three time frames to prove the disinfectant's activity over varying time periods. This author supports this practice and would suggest the time periods of three, five, and ten minutes be tested consecutively. Normally in three minutes, the disinfectant shows average activity, in five minutes good activity, and in ten minutes excellent activity. Testing of a variety of times assures one has data in the file to support a complete range of dry times (contact times) that may be noticed in their manufacturing or testing operations. Upon completion, this testing will provide the justification for utilizing the chemical agents to destroy the known and possibly existent contamination in the facility. As time progresses, we need to continually update our profile of organisms versus our chemical agents.

Although the type of test, the enumeration level of microorganisms, and the contact time are some of the most critical factors to assess, other critical factors also need to be determined. One of these is the expiration of the disinfectant. Expiration for effectiveness can be determined by incorporating a simple variable in our test called the aging of the disinfecting agent. Simply done, one would open a bottle of disinfectant and age it for the time period that they plan to use it, say 30-days (concentrate and ready-to-use). A ready-to-use solution is tested at the 30-day period. For a concentrate, the aged concentrate solution is diluted to the prescribed use-dilution. The use-dilution is then aged for the time period that it will be used (for example, seven days). At the expiration of the use-dilution, it is then tested for antimicrobial effectiveness. This system of expiration proves that a solution, in use for "X" time period, can destroy an acceptable level of microorganisms that we have determined to be present in our operations.

Expiration dating of disinfectant effectiveness can also be tested by conducting an antimicrobial effectiveness test on a newly open bottle of disinfectant (ready-to-use or concentrate-to-use dilution) and then subsequently aging the solution for the expiry period and testing the active ingredients. The correlation between the active ingredients at the time of opening and their satisfactory stability at the end of the expiry provides the needed data to support the use of the agent for the time period. However, in this test, there must be the understanding and explanation of the relationship between the tested solution and the expiry active data that followed.

In performing antimicrobial effectiveness testing, some other factors come to the foreground that warrant attention. The first is soil load. Normally clean room operations do not have a soil load present, and the most soil that exists would be that of disinfectant residues. In the case of an existing soil load, one should conduct their testing in a similar situation. Commonly used as a soil load, an increased level of protein (a fetal bovine serum) at 5% v/v is added to the organism challenge to test the ability of the disinfectant or sporicide under the circumstances of a soil load (dirtied) condition.

Another factor that may surface is the hardness and temperature of the water. All registrants of disinfecting agents must state in their labeling the ability or inability of the agent to achieve antimicrobial effectiveness in the presence of hard water (400 p.p.m. as CaCO3). At the same time, the temperature of the solution may cause its effect against microorganisms to vary, as elevated temperatures tend to increase the ability of the chemical agent's performance.

Although disinfectant validations are expensive and time consuming, they are foremost in most regulatory agency's minds, and thus are required to assure the effectiveness of a cleaning and disinfection system that will prove an acceptable environment during the manufacture of product.

24.8 ASSAY OF DISINFECTING AGENTS

Analytical validation of our disinfecting agents tests that the required percentage of the chemical agent is present to assure antimicrobial effectiveness. If the appropriate use instructions are followed, a ready-to-use product or a formulation from concentrate is normally easy to prove as having a sufficient amount of the active ingredients to reconfirm the required percentage that was validated in our antimicrobial effectiveness studies. The use of this product past the first use is where we start to see the possibility that the percentage of the active ingredients may begin to slowly become too low to warrant continued use. Different products have varying in-use time periods. Time periods range from seven to 30 days. This scenario needs to be validated for each chemical agent and each container type. As an example of the same chemical having varied in-use time periods, an isopropyl alcohol solution in an aerosol form can be used for a longer time period than one in a trigger spray bottle. The reasoning for this difference is the aerosol container is sealed in a pressure vessel. The reduction in percentage of the active ingredient because of evaporation or the question of sterility over time is not found in this container type. On the other hand, the trigger spray container may slowly aspirate room air to the master reservoir, which may compromise the level of active ingredient and the sterility of the container over time. The basic validation question is what time frame we can prove the chemical's active percentage and sterility to be valid for, once opened.

24.9 STERILITY OF DISINFECTING AGENTS

Through our antimicrobial effectiveness studies, we realize that disinfectants do not kill all organisms. As all chemical agents may have an inherent bioburden (normally spores), we must assure that such bioburden is removed prior to their entry into our controlled areas. The transfer of such organisms through our disinfectants to our controlled areas, especially our aseptic filling areas, should be viewed as a catastrophic event. To even further reinforce the issue, we spread the disinfecting agents all over our walls, ceiling, and floors. Controlling the contamination from ever entering is much easier than subsequently removing it from the controlled area. If we review regulatory expectations in this area, we will find the requirement to sterilize all disinfectants and sporicides prior to entry into the controlled environment. The FDA has stated in its "Sterile Drug Products Produced by Aseptic Processing Draft" that, "Upon preparation, disinfectants should be rendered sterile, and used for a limited time, as specified by written procedures". Likewise, the European Union in EU Annex 1 states, "Disinfectants and detergents should be monitored for microbial contamination". The Parenteral Drug Association's (PDA's) Technical Report #70 states, "To ensure sanitizers, disinfectants, and sporicides do not represent a source of contamination, they should be sterile-filtered or sterilized before use in Grade A (ISO 5) and adjacent Grade B (ISO 5/6) areas."

Purchasing a disinfectant or sporicidal product as sterile from an audited vendor requires one to review the following critical items as a quality control measure to assure what one is using is sterile prior to use:

1. Assessment of the bioburden of the solution.
2. Assessment of the bioburden of the container that the solution is to be filled into.
3. Pre-washing containers (cleanliness level).
4. Filter validation providing microorganisms are retained by the filter.
5. Assay of the solution (RTU or concentrate) to an acceptable active percentage.

6. Filtering the solution at 0.2 microns.
7. Aseptically filling the product into pre-sterilized containers or exposing the entire contents to a terminal sterilization process such as gamma irradiation.
8. A lot-by-lot sterility test per current USP or EP compendia.
9. Conducting sterility testing with the completion of *bacteriostasis* and *fungistasis* (B/F) testing. This proves that the sterility test is capable of growing organisms in the presence of the chemical agent.
10. Assessment for expiration dating of an unopened container.

If the disinfectant or sporicide is to be processed sterile in-house, then the review of the following critical items as a quality control measure is suggested to assure what one is using is sterile prior to use:

1. Assessment of the bioburden of the solution.
2. Assessment of the bioburden of the container that the solution is to be filled into.
3. Pre-washing the containers (cleanliness level).
4. Filter validation providing microorganisms are retained by the filter.
5. Assay of the solution (RTU or concentrate) to an acceptable active percentage.
6. Filtering the solution at 0.2 microns.
7. Aseptically filling the product into pre-sterilized containers or exposing the entire contents to a terminal sterilization process such as gamma irradiation.
8. If the product is aseptically filtered, then the assurance that everything that comes in contact with the product after the filter is rendered sterile.
9. Validation performed using a lot-by-lot sterility test per current USP or EP compendia or a bioburden analysis for at least three lots at the beginning and subsequently routinely tested as a quality control check. Sterility testing on a lot-by-lot basis need not be performed if sufficient validation testing is conducted.
10. Sterility testing with the completion of *bacteriostasis* and *fungistasis* (B/F) testing. This proves that the sterility test is capable of growing organisms in the presence of the chemical agent.
11. Assessment for expiration dating of an unopened container.

Validation of our sterility claim for disinfectants should be a focus of our validation efforts. Normally the testing of three processed lots for sterility provides sufficient data. The pre-sterilization of our chemical agents prior to entry into the controlled area is simple common sense and a common practice in the industry.

The container type is critical when using a sterile disinfectant. Container types include aerosol, trigger spray, squeeze bottle, and larger closed containers (one to five gallon and larger). An aerosol container is the most lucrative type, as the vessel does not aspirate the room air to the master reservoir. However, the container and its contents are required to be sterilized via gamma radiation or all the components pre-sterilized and subsequently filled via a validated aseptic filling operation. For obvious reasons, gamma sterilization of the entire contents is considered a far superior methodology for achieving a sterile product. In 1992, the first sterile disinfectant, DECON-AHOL, a sterile USP IPA, was marketed under US patent 6,123,900, Method of Sterilization. This patent and product showed the industry the effectiveness of this type of container. Thus, no viable or particulate contamination is returned to the solution contained. A pre-sterilized aerosol or pressurized vessel assures assay of the active ingredients (for the time period they remain stable) and sterility for the expiration period designated by the manufacturer.

Other smaller containers include the squeeze bottle and trigger spray bottle. Although acceptable containers for all disinfectants and sporicides, the container itself aspirates the room air back to the master reservoir. Thus, assay and sterility are compromised after the initial use. The same is true for containers of one to five gallons and larger. Once opened, sterility is compromised.

For varying disinfecting and sporicidal agents, a variety of containers need to be utilized. For ready-to-use mixtures we must decide how the product will be used. If in a smaller aerosol, trigger spray, or squeeze bottle, we may want to utilize a product that is pre-sterilized in this smaller form rather than attempting to pour or filter such solutions to an empty pre-sterilized container. We may also want to limit the capping of product "to be used later" as our assay and sterility may be compromised over time. For concentrated products that need to be diluted with a quality water grade, we may want to look to implement unit dose bottles that incorporate a premeasured dose and are sterile. This system is superior to that of pouring a concentrate disinfectant or sporicide into a measuring cup and capping the remainder for later use. Questions may arise as to the assay and sterility of this remaining solution over time.

Thus, no viable or particulate contamination is returned to the solution contained. A pre-sterilized aerosol or pressurized vessel assures assay of the active ingredients (for the time period they remain stable) and sterility for the expiration period designated by the manufacturer.

24.10 AVAILABLE SANITIZERS, DISINFECTANTS, AND SPORICIDES FOR GMP OPERATION

The following chemical agents are the options for GMP operations. The choice is delineated by assessment of one's environmental flora derived over time and the compatibility of the agents with the surface substrates where they will be used.

Sanitizers

- Isopropyl Alcohol 70%
- Ethanol at 70%
- Ethyl Alcohol at 70%
- Iodine (not normally used in clean room operations)

Disinfectants

- Phenols
- Quaternary Ammonium
- Hydrogen Peroxide at 3% or below
- Sodium Hypochlorite below 0.10%

Sporicides/Sterilants

- Sodium Hypochlorite above 0.25%
- Hydrogen Peroxide above 6% (sporicidal reduction at 6%)
- Peracetic Acid and Hydrogen Peroxide
- Glutaraldehyde
- Formaldehyde

In reviewing the basic types of chemical agents used in pharmaceutical and biotechnology operations, we can come to understand their basic differences and applicability in our operations. To follow is a brief description of each of the most used chemical agents in the industry. As a complete chapter could be written on each, the summaries are brief, so we can develop a basic understanding of each chemical.

24.10.1 Isopropyl Alcohol, Ethyl Alcohol, Ethanol Solutions

Alcohols have been used for years in pharmaceutical and biotechnology operations for three basic purposes: (1) as a sanitizing spray for gloves, surfaces, carts, etc.; (2) as a cleaning or wipe down agent to remove possible existent residues from critical non-product contact surfaces; and (3) as a product contact cleaning agent (ethyl alcohol only). In testing antimicrobial effectiveness of the products, a 70% solution demonstrates far superior efficacy performance than higher or lower concentrations. Alcohols come in a variety of forms. The most used forms in the clean room operation are isopropyl alcohol and ethyl alcohol (ethanol and alcohol). Over 90% of industry operations will utilize a 70% isopropyl alcohol solution to address clean room organisms, as it has been proven more efficacious than ethyl alcohol (190–200 proof diluted to 70%) or ethanol (a mixture with the base as ethyl alcohol [63%] and spiked with methyl [3%] and isopropyl [4%]) at a small percentage (rendering it not drinkable). Alcohols demonstrate rapid broad-spectrum antimicrobial activity against vegetative bacteria (including mycobacteria), viruses, and fungi but are not sporicidal. They are, however, known to inhibit sporulation and spore germination, but this effect is reversible. As alcohols are not sporicidal, they are not recommended for sterilization and are widely used for hard surface disinfection and skin antisepsis. As we have previously stated, of the existing alcohols, isopropyl alcohols demonstrate superior effectiveness against clean room organisms. However, for viruses, ethyl alcohol or ethanol seem to be slightly more effective and are used basically within the confines of the laboratory environment.

Published alcohol effectiveness or results of alcohol effectiveness are presented when sprayed to the surface and allowed to air dry. This should not be confused with the product's effectiveness if used in a saturated wipe. A saturated wipe contains a limited amount of the chemical agent. When used, its dry times are significantly less, and thus, its destructive power substandard to that of the liquid itself on the surface. Although the mechanism of wiping destroys cells in its action, the dry time of the alcohol is significantly faster.

Isopropyl alcohol wipes carry few if any claims. Usually they are used to wipe a surface in a cleaning operation (IPA wipe down) to remove existent disinfectant residues. Here also such products have a problem. As they are saturated with isopropyl alcohol, their ability to soak up residues and clean the surface is minimal. Most of the time these products just move the contamination around the surface. A superior methodology would be the use of an isopropyl alcohol and a dry wipe. Like cleaning a window, we would spray the solution onto the surface and wipe. This will remove most of the contaminants on the surface. This can be proven in a home experiment by using a saturated wipe to clean a window. When completed, the window will have streaks or swirls and look as though the dirt was just moved around. If one were to conduct the same experiment with a dry wipe and spray isopropyl alcohol, the surface would appear much cleaner. The soaking of the liquid with contaminants in suspension from the surface into a dry wipe is a superior methodology. During isopropyl alcohol wipe downs of non-product contact but critical surfaces, we need to employ the cleaner of the two methodologies. Rendering of the products as sterile is a must prior to use in Class 100 (Grade A and B, ISO 5) and adjacent Class 10,000 (Grade C, ISO 7) areas. Sterilization of disinfection agents is discussed in depth in this chapter in Section 24.9.

24.10.2 Phenols

Phenolics have been used for years as a disinfecting agent. Phenolics are effective against gram-positive and gram-negative organisms. However, phenols exhibit better antimicrobial effectiveness against gram-positive organisms than they do against gram-negative organisms. They have limited activity against fungi and certain virus strains such as HIV-1 (AIDS virus) and Herpes Simplex, Type 2. Phenolics normally are available in an alkaline and acidic base version. The theory surrounding the rotation of these two compounds is described later in the chapter. Overall, phenolics demonstrate superior antimicrobial effectiveness in an acidic base as opposed to an alkaline base. Some of the most common chemical compounds in phenolic germicidal detergents are in a low pH phenolic: an ortho-phenyl-phenol and ortho-benzyl-benzyl-para-chlorophenol and in a high pH phenolic: a sodium ortho-benzyl-para-chlorophenate, sodium ortho-phenylphenate, or sodium para-tertiary-amylphenate. Although phenols provide good broad-spectrum disinfection, they are not sporicidal and have major drawbacks in their use. One drawback is the horrific residues that are noticed from long-term use of the products. Phenolics are normally an amber or light tan color when manufactured. This color darkens with age and its exposure to light (especially fluorescent light). Residues start as a "dripping droplet" that is not easily removed. The use of 70% isopropyl alcohol or certain residue removers can remove such residues in their early existence on surfaces. Although somewhat effective, both residue-cleaning products eventually give way to the darkening phenolic that stains the surface with a dark-brown color. Transfer of such residues to an unwanted location is a concern, and precautionary measures need to be implemented to assure minimization of this scenario. Compatibility with most chemicals is normally very good; however, the effect of the anionic characteristic of the chemical in relation to applications in conjunction with a cationic surfactant such as a quaternary ammonium has been reported as problematic. Expiration of a formulated phenol also carries some concern. The formulation in a closed container should remain stable for a seven- to 30-day period (depending upon storage). Formulated solutions should be marked accordingly. The normality in the industry is seven days and less if the solution is in an open container (such as a bucket), in which case it should be discarded each day. Rendering these products as sterile is a must prior to use in Class 100 (Grade A and B, ISO 5) and adjacent Class 10,000 (Grade C, ISO 7) areas. Sterilization of disinfection agents is discussed in depth in Section 24.9.

24.10.3 Quaternary Ammonium Compounds

Quaternary ammonium products are used in more disinfection applications than phenolic germicidal detergents. Their spectrum of use is very broad and ranges throughout the industrial world, through hospital and institutional settings, and even home use. Quaternary ammoniums have excellent detergency. They are one of the best cleaners among the spectrum of disinfecting agents. These cationic solutions also have excellent deodorizing capabilities. Quaternary ammonium compounds are effective against gram-positive and gram-negative organisms. However, phenols exhibit better antimicrobial effectiveness against gram-positive organisms than they do against gram-negative organisms. They have limited activity against fungi and certain virus strains such as HIV-1 (AIDS virus) and Herpes Simplex, Type 2. In fact, and because of competition in this arena, quaternary ammonium compounds have the most organisms registered as label claims with the US EPA. Their mechanism of antimicrobial effectiveness is related to their positively charged molecule. Simply, the positively charged molecule is attracted to the microorganism's negatively charged cell wall. The cycle of kill is complete when the chemical agent is absorbed into the cell and spread throughout the organism. Quaternary ammoniums are available in both alkaline and acid-based compounds. Quaternary ammonium compounds normally utilize an alkyl dimethyl benzyl ammonium chloride or a dimethyl ethyl benzyl ammonium chloride. Although a multitude of formulations exist in the industry, these are two of the most popular components. As with phenols, quaternary ammoniums do leave sticky residues that become problematic over time. However, they are not of the scale of phenolic residues. Expiration of a formulated (from concentrate) quaternary ammonium also carries some concern. The formulation in a closed container is relatively stable; it should remain stable for a 30-day period in a closed container. Formulated solutions should be marked accordingly. The normality in the industry is seven to 30 days and less if the solution is in an open container (such as a bucket), in which case it should be discarded each day. However, in recent years, quaternary ammoniums have been made in ready-to-use formulas that are very stable. Rendering these products as sterile is a must prior to use in Class 100 (Grade A and B, ISO 5) and adjacent Class 10,000 (Grade C, ISO 7) areas. Sterilization of disinfection agents is discussed in depth in this chapter in Section 24.9.

24.10.4 Sodium Hypochlorite

Sodium hypochlorite solutions are one of the oldest known disinfectants and sporicidal agents. The product is available in many forms including ready-to-use 0.25% and 0,52% solution, to concentrate 5.25% and 10% solutions, to powders that are mixed with water to formulate a variety of solutions. Bleach as we know it is normally found in a 5.25% concentration; however, in recent years such formulations from the Clorox Corporation have been increased to a near 7% solution to assure continued stability of the active percentage.

One of the main problems with the use of sodium hypochlorite is that it is used at too strong of an active percentage. Sodium hypochlorite is used throughout the health care setting and normally diluted to concentrations of 0.25% or 0.52%. One of the problems with sodium hypochlorite formulations is the method of formulation designation that varies from firm to firm. Some formulate to a part per million (ppm), some to a percentage of a solution, and some to a dilution such as 1–10. At a use-dilution of 0.25% (or a 1–20 dilution or 250 ppm), sodium hypochlorite is effective against gram-positive and gram-negative organisms, viruses, fungi, and bacterial endospores. At a slightly increased use-dilution of 0.52% (or a 1–10 dilution or 500 ppm), sodium hypochlorite is effective against gram-positive and gram-negative organisms, viruses, fungi, and more effective against a wider range of bacterial endospores. Both formulations are used throughout the pharmaceutical and biotechnology industry.

In formulation of a sodium hypochlorite solution, many choose to acidify the solution, which makes it a more potent mixture when focusing on bacterial endospores. However, acidification causes rapid degradation of the active elements in the solution, and use of the product is limited to approximately two hours. Acidification may not be necessary as the product demonstrates excellent sporicidal characteristics in its neutral state, and bacterial endospore levels in clean rooms are not exuberant in numbers (above 1.0 × 106), and there is no soil load. A formulation of 250 or 500 ppm has excellent sporicidal activity. However, expiration of the solution can become a problem. A normal 5.25% concentrate sodium hypochlorite normally carries a one-year expiration for an unopened container. Some companies have validated and increased this expiration with applicable assay data over time to 18 months. Once opened, whether in a ready-to-use formula or a concentrate product, the product needs to be used within a 30-day period. Open containers (such as buckets and open bottles) have a shorter expiration as the chlorine in the solution begins to burn off leaving only the sodium chloride. Open containers should be formulated and used within the same day.

Some drawbacks with sodium hypochlorite are mainly focused on the residue and corrosiveness of the product. As previously stated, the chlorine in the solution begins to burn off leaving only the sodium chloride. This white crystal-like residue attacks the impurities in stainless steel (as an example) over a longer time frame. When using a sodium hypochlorite solution, it is imperative to remove the sodium chloride residues frequently to minimize this corrosive action. Rendering these products as sterile is a must prior to use in Class 100 (Grade A and B, ISO 5) and adjacent Class 10,000 (Grade C, ISO 7) areas. Sterilization of disinfection agents is discussed in depth in Section 24.9.

24.10.5 Hydrogen Peroxide

Hydrogen peroxide is one of the most common disinfectants in the industrial marketplace. The product is commonly used as an antiseptic in hospitals or for consumer use at a 3% solution. In the pharmaceutical and biotechnology industry, hydrogen peroxide is used at 35% as a sterilant in isolators and at 3%–10% for surface disinfection. Dependent upon the concentration used, hydrogen peroxide is effective against bacteria, yeasts, viruses, and bacterial spores. Destruction of spores is greatly increased both with a rise in temperature and increase in concentration. Hydrogen peroxide is a clear, colorless liquid that is environmentally friendly as it can rapidly degrade into water and oxygen. Although generally stable, most hydrogen peroxide formulations contain a preservative to prevent decomposition. At lower concentrations (3%–10%), the chemical is effective against gram-positive and gram-negative bacteria, viruses, and bacterial endospores in lower enumerations. However, at higher concentrations and longer contact times, the product exhibits superior sporicidal reduction of bacterial spores. Some of the positive features of the product are its mild, if any, odor and its low residue characteristics. However, the product also has some drawbacks in exceeding OSHA exposure limits if used in confinement in too large a quantity. Precautions should be taken for its use. Rendering of these products as sterile is a must prior to use in Class 100 (Grade A and B, ISO 5) and adjacent Class 10,000 (Grade C, ISO 7) areas. Sterilization of disinfection agents is discussed in depth in Section 24.9.

24.10.6 Peracetic Acid and Hydrogen Peroxide

Peracetic acid and hydrogen peroxide mixtures have received much attention in the pharmaceutical and biotechnology industry in recent years. The chemical is considered a more potent biocide than hydrogen peroxide as it is bactericidal, virucidal, fungicidal, and sporicidal at very low concentrations (<0.3%). The product was originally designed for the sterilization of medical devices. The chemical destroys the cell by destroying vital membrane lipids and DNA and denaturing proteins and enzymes. Peracetic acid and hydrogen peroxide mixtures decompose to acetic acid and oxygen. Active percentages of marketed products range from 0.3% to 1.3% as a sterilant in both ready-to-use and concentrate solutions. Ready-to-use solutions require a very high level of acetic acid as a stabilizer, nearing 5.2%, whereas concentrate products need smaller amounts of acetic acid in the formulation. Concentrate solutions incorporate approximately 8% acetic acid and upon formulation to a use dilution, this value drops to near 0.4%.

One of the main misconceptions with the use of peracetic acid and hydrogen peroxide mixtures as well as most registered sporicides is that the product needs to be used at the sterilant label claim active percentage. First, we must understand the requirements set forth by the EPA. The EPA requires all registrants making label claims to do so by following test methods outlined in the AOAC protocol. The sterilant claim on products/labels follows the AOAC sporicidal test. This test requires the complete reduction of 106 of *B. subtilis* and *C. sporogenes* in a 60-carrier test, at 20°C, in a soil load. In the clean room, the bioburden is significantly lower. Thus, the active percentage needed to destroy the flora normally seen is significantly less. Simply, registered sterilant label claims are too strong for what is noticed in a clean room. Coupled with the high enumeration of the AOAC test parameters is a soil load, which is not present in clean rooms. Thus, end users should look to validate a concentration of peracetic acid and hydrogen peroxide mixtures as well as other registered sporicides at realistic bioburden values as discussed later in this chapter in the section "Determining Antimicrobial Effectiveness". This will significantly reduce odors, deterioration of surfaces, problematic user situations, and residues.

Peracetic acid and hydrogen peroxide mixtures have a pungent vinegar smell that is offensive, if not intolerable, to many users. Because of the horrific smell and the characteristic drying of mucosal membranes, peracetic acid and hydrogen peroxide mixtures cause dissatisfaction among end users. Facilities that have used sodium hypochlorite (bleach) for years find the transition to peracetic acid and hydrogen peroxide very difficult in terms of worker satisfaction. The product, if used in a clean room environment that may have 15%–20% fresh air, may easily exceed required levels when industrial hygiene testing is performed. Safety precautions for end users should be assured prior to its use. Although reports of the product deem it noncorrosive to metals, industry reports have shown this product reacts adversely with most stainless steels, aluminums, plastics, epoxies, and most clean room surfaces. After application, a white cloudy residue is normally left that requires either an IPA wipe down or a water for injection (WFI) rinse to remove.

24.10.7 Glutaraldehyde

Glutaraldehyde has been used for some time as a disinfectant and sterilant for endoscopes and surgical equipment.

Glutaraldehyde is normally sold in a 2.0% solution. The product is usually supplied as an amber solution with an acid pH. Glutaraldehyde is a powerful biocidal agent having the advantage of continued activity in the presence of organic material. Glutaraldehyde has broad-spectrum activity against bacteria, bacterial spores, viruses, and fungi. The mechanism of action involves the destruction of the outer layers of the cell. Glutaraldehyde is the only aldehyde to exhibit excellent sporicidal activity. In recent years, glutaraldehyde's use has been focused mainly on the hospital environment. Many pharmaceutical and biotechnology organizations do not use a glutaraldehyde product in their operations. The product is very toxic, and specific handling precautions must be employed prior to its use. Especially noted are the gaseous fumes and the possible absorption through human tissue (skin).

24.10.8 Formaldehyde

Formaldehyde is widely known as a fumigant for rooms and buildings. It has been shown to be effective against bacteria and bacterial spores and vegetative bacteria. Acklund et al. (1980) showed that at 20°C and a relative humidity of approximately 100%, a six-log reduction of *B. subtilis* spores was obtained after one and a half hours exposure to 300 μg/L whereas at 250 μg/L only a four-log reduction was obtained after six hours of exposure. The mechanism of action of formaldehyde is assumed to be because of the reaction with cell protein and DNA or RNA (Russell, 1976).

Formaldehyde is normally used in the pharmaceutical and the biotechnology industry to bring back an area after shutdown or major maintenance. During its implementation, very stringent safety precautions are assured for personnel protection that include areas and building clearance and hold times of areas prior to release. Although formaldehyde is effective, this chapter has focused on routine methods of cleaning and disinfection. Formaldehyde does not fit appropriately as a choice in this venue and would be used as a mechanism in opening a new area or as a method when coming back from a shutdown period.

24.11 ANTIMICROBIAL EFFECTIVENESS TESTING CONDUCTED ON GMP ENVIRONMENTAL ISOLATES AND ATCC CULTURES AT VAI LABORATORIES

Most all GMP regulatory inspectors will look at the product label claims and say to the GMP firm, "Have you verified this efficacy using your isolates on your clean room substrates?" When first hearing this, many GMP firms will want to push back until they realize that what the inspector is asking is what the antimicrobial registering body does not include in their requirements to register the product. The first discrepancy is related to specific isolates recovered from the GMP facility to show that the agent being utilized can adequately destroy the organisms. The second discrepancy is the dry time or contact time that was used in the approval of the antimicrobial agent with the antimicrobial agent-registering body. Most probably for a disinfectant this was ten minutes and for a sporicide/sterilant it could have been hours. This difference is related to the population each body feels should be destroyed by the antimicrobial agent. A government agency like the US EPA would want to see minimally one million CFUs or a 1.0×10^6 population destroyed by the agent in the contact time prescribed. However, in no GMP firm would such bioburden exist and so although the GMP regulatory inspector wants to see a faster time for efficacy, they also agree the bioburden should be less, from 1,000 CFUs (1.0×10^4) to 10,000 CFUs (1.0×10^5). Nor would surfaces have a dry time or contact time for the agent on the surface for that length of time. This time period would be much shorter nearing 30–60 seconds for alcohol-based products and three to five minutes for water-based products. The third discrepancy relates to the carrier surface the test was performed on. For an AOAC test to meet many authorities such as the US EPA or Australian TGA, ceramic penicylinders are used and sporicidal sutures are also employed. There are very few firms that would have ceramic as a surface substrate or would need to know the efficacy on sutures. However, a hospital would desire this information, and this is where worldwide antimicrobial agent registration has its basis. Clean room substrates defined earlier in this chapter represent what we would expect to see in GMP operations, and a GMP firm needs to know the expected antimicrobial effectiveness of agents on such surfaces that inherently have such bioburden.

As testing commences, the GMP laboratory professional must also take into account realistic attributes of what he or she is doing. Your goal is to provide useful data that can be utilized in operations. In reality, you are performing a lab study on a coupon that measures anywhere from 1" × 0.5" to 2" × 2". On that coupon will be 10^4–10^5 of an organism. The surface size is not depictive of a clean room surface that will be disinfected. The enumeration of microorganisms is not representative of the flora that will be noticed on any surface in any area in your facility. In fact, the enumeration is outrageously high. On the surface you test there will be no residues and very little particulate that would characterize your areas. You are applying the agent perfectly, which might not be done day after day. You are not wiping or mopping, which will surely remove more contamination from the surface. In short, it is a very controlled experiment in the lab that does not take into account personnel training in cleaning and disinfection nor SOPs that will need to be written. It is important to understand that what you are doing is extreme overkill and what you assess in acceptance criteria should be realistic. Realistic acceptance criteria is > one log per PDA Technical report #70. In that first log, you will have destroyed 90% of the inoculate. A second and third log of reduction are miniscule compared with what you attained in the first log. And by requiring the second and/or third log, you most probably have invoked much higher residuals in your facility and placed cleaning personnel potentially into a higher vapor cleaning scenario. And, in doing so, have not really increased the performance efficacy to any value that matters as the extra strength was not required. All that you are doing needs to be scientifically assessed and assured that it can be repeated day-in and day-out in operations.

Many times, quality professionals possibly think too much about the testing scenario and forget about what will happen in operations. Imagine being part of the cleaning crew. For our example I will use paint as the chemical agent and brushes and rollers as our mops and wipers. Each day you come in and say to your manager "What are we going to do today?" The manager's response is "We are going to paint the room." So the manager gives you paint brushes, rollers, paint, and tape. And you do a great job painting the room making sure you remove outlet covers, light switches, tape windows, etc. The next day you come in you ask the same question, and the response is the same. So again, you do a fantastic job painting the room because the manager must want two coats. The next day and for the following 100 days, the manager responds to the question with the same answer, "We are going to paint the room". Now you are frustrated, bored, and the repetition is ridiculous. So each time you do not do as good a job because the next day you will have to do the same thing, paint the room. At day 300, you miss the left wall, you paint over top of the windows and your sense of perfection is definitely lacking. This is what it is like being part of the cleaning crew. To add to your joy, they make you wear gowns that cause you to sweat. Your goggles fog constantly, the vapors from the chemicals dry your nose, and you smell the smell even after you are home for the evening. This is what being part of the cleaning crew is like and no antimicrobial effectiveness testing (AET) study or validation study can prepare an operation for this type of scenario. So it is lost and forgotten but it is the thing that causes problematic environmental situations, as contamination that was let in has a flawed corrective action. So when we assess stringent acceptance criteria, we further complicate this scenario as corrosion occurs, disinfection is done more quickly, and residues are left remaining. It is food for thought to take yourself out of quality assurance or validation and imagine yourself in production services.

So, what will the GMP firm expect to see in antimicrobial effectiveness results? For over 25 years, VAI Laboratories have assisted GMP firms with antimicrobial effectiveness testing. The results presented represent results from lab studies for various agents against various organisms on various surfaces at various dry time or contact times.

Disclaimer: The following test report and the data contained therein is not intended to, and does not have any relationship to, nor does it amend or is intended to amend any information contained in any labels or registrations required or approved by the United States EPA or any state pursuant to any law, including but not limited to the Federal Insecticide, Fungicide, and Rodenticide Act (7 USC Sections 136 et seq), or any state analog, or any other foreign or international legal body or jurisdiction, for the antimicrobial agents utilized and referenced in the study. The data presented represents a private testing study and regimen ("Study") that investigated solely "whether microorganisms could be destroyed on carrier surfaces, as tested at lower populations which would depict bioburden in the clean room, and shorter contact or dry times, which is depictive of the GMP setting and an expectation by regulatory agencies such as the US Food and Drug Administration (FDA)." Prior to use of any product containing such chemical germicides or antimicrobial agents that are referenced and/or used in the study, all persons are advised by this Disclaimer to review all applicable labels and registrations associated with each such product.

Organism tested	Surface	Wetted period	Initial positive control inoculate	Results in Log reduction
Peracetic Acid and H_2O_2 in WFI				
Bacillus subtilis	Stainless steel	5	5.50×10^4	4.74
Aspergillus brasiliensis	Stainless steel	5	5.15×10^4	2.49
Staphylococcus aureus	Stainless steel	5	3.60×10^4	4.56
Pseudomonas aeruginosa	Stainless steel	5	2.74×10^4	4.44
Candida albicans	Stainless steel	5	1.48×10^4	5.17
Micrococcus luteus	Stainless steel	5	4.55×10^4	4.66
Penicillium rubens	Stainless steel	5	5.80×10^4	4.76
Bacillus cereus	Stainless steel	5	1.15×10^3	1.54
Cladosporium cladosporioides	Stainless steel	5	2.15×10^4	4.33
Penicillium glabrum	Stainless steel	5	1.39×10^4	4.14
Bacillus subtilis	Stainless steel	10	5.50×10^4	4.74
Aspergillus brasiliensis	Stainless steel	10	5.15×10^4	3.49
Staphylococcus aureus	Stainless steel	10	3.60×10^4	4.56
Pseudomonas aeruginosa	Stainless steel	10	2.74×10^4	4.44
Candida albicans	Stainless steel	10	1.48×10^4	5.17
Micrococcus luteus	Stainless steel	10	4.55×10^4	4.66
Penicillium rubens	Stainless steel	10	5.80×10^4	4.79

Organism tested	Surface	Wetted period	Initial positive control inoculate	Results in Log reduction
Bacillus cereus	Stainless steel	10	1.15×10^3	3.06
Cladosporium cladosporioides	Stainless steel	10	2.15×10^4	4.33
Penicillium glabrum	Stainless steel	10	1.39×10^4	4.14
Bacillus subtilis	Epoxy	5	7.70×10^4	4.89
Aspergillus brasiliensis	Epoxy	5	5.05×10^4	2.88
Staphylococcus aureus	Epoxy	5	3.90×10^4	4.59
Pseudomonas aeruginosa	Epoxy	5	2.55×10^4	4.41
Candida albicans	Epoxy	5	1.50×10^5	5.18
Micrococcus luteus	Epoxy	5	5.10×10^4	4.71
Penicillium rubens	Epoxy	5	6.50×10^4	4.85
Bacillus cereus	Epoxy	5	1.55×10^3	1.12
Cladosporium cladosporioides	Epoxy	5	1.54×10^4	4.19
Penicillium glabrum	Epoxy	5	9.55×10^3	3.98
Bacillus subtilis	Epoxy	10	7.70×10^4	4.89
Aspergillus brasiliensis	Epoxy	10	5.05×10^4	2.78
Staphylococcus aureus	Epoxy	10	3.90×10^4	4.59
Pseudomonas aeruginosa	Epoxy	10	2.55×10^4	4.41
Candida albicans	Epoxy	10	1.50×10^5	5.18
Micrococcus luteus	Epoxy	10	5.10×10^4	4.71
Penicillium rubens	Epoxy	10	6.50×10^4	4.85
Bacillus cereus	Epoxy	10	1.55×10^3	1.97
Cladosporium cladosporioides	Epoxy	10	1.54×10^4	4.19
Penicillium glabrum	Epoxy	10	9.55×10^3	3.98
70% Isopropyl Alcohol in WFI				
Micrococcus luteus	Stainless steel	30s	1.53×10^4	4.18
Staphylococcus aureus	Stainless steel	30s	1.13×10^5	3.83
Pseudomonas aeruginosa	Stainless steel	30s	3.10×10^5	4.27
Staphylococcus epidermidis	Stainless steel	30s	1.19×10^6	3.63
Staphylococcus warneri	Stainless steel	30s	1.02×10^5	3.49
Kocuria polaris	Stainless steel	30s	9.50×10^3	3.94
Staphylococcus hominis	Stainless steel	30s	4.25×10^4	3.11
Corynebacterium aurimucosum	Stainless steel	30s	1.08×10^5	5.03
Penicillium chrysogenum	Stainless steel	30s	1.39×10^6	3.33
Neurospora crassa	Stainless steel	30s	9.40×10^4	3.05
Micrococcus luteus	Stainless steel	60s	1.53×10^4	4.18
Staphylococcus aureus	Stainless steel	60s	1.13×10^5	4.56
Pseudomonas aeruginosa	Stainless steel	60s	3.10×10^5	5.49
Staphylococcus epidermidis	Stainless steel	60s	1.19×10^6	3.63
Staphylococcus warneri	Stainless steel	60s	1.02×10^5	3.19
Kocuria polaris	Stainless steel	60s	9.50×10^3	6.00
Staphylococcus hominis	Stainless steel	60s	4.25×10^4	4.63
Corynebacterium aurimucosum	Stainless steel	60s	1.08×10^5	5.03
Penicillium chrysogenum	Stainless steel	60s	1.39×10^6	3.60
Neurospora crassa	Stainless steel	60s	9.40×10^4	2.87
Micrococcus luteus	Nitrile glove	30s	1.56×10^5	1.95
Staphylococcus epidermidis	Nitrile glove	30s	2.06×10^4	1.88
Staphylococcus warneri	Nitrile glove	30s	2.21×10^5	2.05
Kocuria polaris	Nitrile glove	30s	1.02×10^6	5.09
Staphylococcus hominis	Nitrile glove	30s	4.15×10^4	4.62
Corynebacterium aurimucosum	Nitrile glove	30s	1.07×10^5	5.03
Penicillium chrysogenum	Nitrile glove	30s	1.50×10^5	2.54
Neurospora crassa	Nitrile glove	30s	1.13×10^5	2.29
Micrococcus luteus	Nitrile glove	60s	1.56×10^5	2.3
Staphylococcus epidermidis	Nitrile glove	60s	2.06×10^4	2.61

Organism tested	Surface	Wetted period	Initial positive control inoculate	Results in Log reduction
Staphylococcus warneri	Nitrile glove	60s	2.21×10^5	3.08
Kocuria polaris	Nitrile glove	60s	1.02×10^6	6.01
Staphylococcus hominis	Nitrile glove	60s	4.15×10^4	4.62
Corynebacterium aurimucosum	Nitrile glove	60s	1.07×10^5	5.03
Penicillium chrysogenum	Nitrile glove	60s	1.50×10^5	2.54
Neurospora crassa	Nitrile glove	60s	1.13×10^5	2.56
Staphylococcus aureus	Stainless steel	30s	6.60×10^4	4.82
Escherichia coli	Stainless steel	30s	2.17×10^4	4.34
Pseudomonas aeruginosa	Stainless steel	30s	9.20×10^4	4.96
Micrococcus luteus	Stainless steel	30s	1.17×10^4	4.07
Staphylococcus aureus	Stainless steel	60s	2.31×10^4	3.14
Escherichia coli	Stainless steel	60s	2.17×10^4	3.12
Pseudomonas aeruginosa	Stainless steel	60s	9.20×10^4	3.44
Micrococcus luteus	Stainless steel	60s	1.17×10^4	4.07
High pH Phenol in WFI				
Pseudomonas aeruginosa	Stainless steel	5 min	6.10×10^3	3.79
Staphylococcus aureus	Stainless steel	5 min	1.06×10^6	6.03
Bacillus subtilis	Stainless steel	5 min	2.29×10^4	1.74
Escherichia coli	Stainless steel	5 min	2.00×10^5	5.3
Aspergillus brasiliensis	Stainless steel	5 min	1.53×10^6	3.11
Candida albicans	Stainless steel	5 min	1.23×10^4	4.09
Clostridium sporogenes	Stainless steel	5 min	1.74×10^4	2.72
Micrococcus luteus	Stainless steel	5 min	1.35×10^5	5.13
Bacillus cereus	Stainless steel	5 min	6.00×10^5	6.56
Staphylococcus epidermidis	Stainless steel	5 min	3.65×10^6	3.54
Pseudomonas aeruginosa	Stainless steel	10 min	6.10×10^3	3.79
Staphylococcus aureus	Stainless steel	10 min	1.06×10^6	6.03
Bacillus subtilis	Stainless steel	10 min	2.29×10^4	1.76
Escherichia coli	Stainless steel	10 min	2.00×10^5	4.08
Aspergillus brasiliensis	Stainless steel	10 min	1.53×10^6	3.81
Candida albicans	Stainless steel	10 min	1.23×10^4	4.09
Clostridium sporogenes	Stainless steel	10 min	1.74×10^4	2.54
Micrococcus luteus	Stainless steel	10 min	1.35×10^5	5.13
Bacillus cereus	Stainless steel	10 min	6.00×10^5	6.56
Staphylococcus epidermidis	Stainless steel	10 min	3.65×10^6	3.65
Pseudomonas aeruginosa	Epoxy	5 min	6.20×10^3	3.79
Staphylococcus aureus	Epoxy	5 min	1.08×10^5	6.03
Bacillus subtilis	Epoxy	5 min	2.23×10^4	1.6
Escherichia coli	Epoxy	5 min	2.07×10^5	3.31
Aspergillus brasiliensis	Epoxy	5 min	1.72×10^6	2.57
Candida albicans	Epoxy	5 min	1.21×10^4	4.08
Clostridium sporogenes	Epoxy	5 min	1.83×10^4	1.89
Micrococcus luteus	Epoxy	5 min	1.35×10^5	5.13
Bacillus cereus	Epoxy	5 min	6.00×10^5	5.56
Staphylococcus epidermidis	Epoxy	5 min	3.65×10^6	3.57
Pseudomonas aeruginosa	Epoxy	10 min	6.20×10^3	2.57
Staphylococcus aureus	Epoxy	10 min	1.08×10^5	6.03
Bacillus subtilis	Epoxy	10 min	2.23×10^4	1.47
Escherichia coli	Epoxy	10 min	2.07×10^5	5.31
Aspergillus brasiliensis	Epoxy	10 min	1.72×10^6	3.24
Candida albicans	Epoxy	10 min	1.21×10^4	4.08
Clostridium sporogenes	Epoxy	10 min	1.83×10^4	1.68
Micrococcus luteus	Epoxy	10 min	1.35×10^5	5.13
Bacillus cereus	Epoxy	10 min	6.00×10^5	6.56
Staphylococcus epidermidis	Epoxy	10 min	3.65×10^6	3.62

Organism tested	Surface	Wetted period	Initial positive control inoculate	Results in Log reduction
Low pH Phenol in WFI				
Staphylococcus aureus	Stainless steel	5 min	2.70×10^6	4.73
Pseudomonas aeruginosa	Stainless steel	5 min	8.15×10^6	5.21
Candida albicans	Stainless steel	5 min	2.66×10^6	4.72
Aspergillus brasiliensis	Stainless steel	5 min	1.61×10^5	2.30
Penicillium citrinum	Stainless steel	5 min	1.83×10^5	3.56
Penicillium glabrum	Stainless steel	5 min	1.56×10^5	3.49
Bacillus licheniformis	Stainless steel	5 min	2.30×10^4	0.49
Micrococcus luteus	Stainless steel	5 min	3.40×10^4	4.53
Bacillus subtilis	Stainless steel	5 min	1.90×10^4	2.36
Brevibacillus borstelensis	Stainless steel	5 min	6.10×10^4	1.91
Staphylococcus aureus	Stainless steel	10 min	2.70×10^6	4.73
Pseudomonas aeruginosa	Stainless steel	10 min	8.15×10^6	5.09
Candida albicans	Stainless steel	10 min	2.66×10^6	4.72
Aspergillus brasiliensis	Stainless steel	10 min	1.61×10^5	2.73
Penicillium citrinum	Stainless steel	10 min	1.83×10^5	3.56
Penicillium glabrum	Stainless steel	10 min	1.56×10^5	3.49
Bacillus licheniformis	Stainless steel	10 min	2.30×10^4	0.62
Micrococcus luteus	Stainless steel	10 min	3.40×10^4	4.56
Bacillus subtilis	Stainless steel	10 min	1.90×10^4	2.28
Brevibacillus borstelensis	Stainless steel	10 min	6.10×10^4	2.27
Staphylococcus aureus	Stainless steel	5 min	3.75×10^6	4.50
Pseudomonas aeruginosa	Stainless steel	5 min	7.20×10^7	3.99
Candida albicans	Stainless steel	5 min	2.70×10^6	3.17
Aspergillus brasiliensis	Stainless steel	5 min	1.25×10^5	1.57
Penicillium citrinum	Stainless steel	5 min	1.48×10^5	1.78
Penicillium glabrum	Stainless steel	5 min	1.71×10^5	2.20
Bacillus licheniformis	Stainless steel	5 min	3.00×10^4	0.64
Micrococcus luteus	Stainless steel	5 min	4.00×10^4	4.60
Bacillus subtilis	Stainless steel	5 min	8.50×10^3	1.59
Brevibacillus borstelensis	Stainless steel	5 min	5.60×10^4	1.23
Staphylococcus aureus	Stainless steel	10 min	3.75×10^6	4.87
Pseudomonas aeruginosa	Stainless steel	10 min	7.20×10^7	3.78
Candida albicans	Stainless steel	10 min	2.70×10^6	3.34
Aspergillus brasiliensis	Stainless steel	10 min	1.25×10^5	1.87
Penicillium citrinum	Stainless steel	10 min	1.48×10^5	2.39
Penicillium glabrum	Stainless steel	10 min	1.71×10^5	3.16
Bacillus licheniformis	Stainless steel	10 min	3.00×10^4	0.76
Micrococcus luteus	Stainless steel	10 min	4.00×10^4	4.60
Bacillus subtilis	Stainless steel	10 min	8.50×10^3	1.33
Brevibacillus borstelensis	Stainless steel	10 min	5.60×10^4	1.17
Quaternary Ammonium in WFI				
Aspergillus brasiliensis	Stainless steel	5 min	4.60×10^4	4.66
Staphylococcus aureus	Stainless steel	5 min	8.20×10^4	4.91
Pseudomonas aeruginosa	Stainless steel	5 min	3.85×10^4	4.59
Candida albicans	Stainless steel	5 min	1.31×10^4	4.12
Micrococcus luteus	Stainless steel	5 min	1.10×10^4	4.04
Penicillium rubens	Stainless steel	5 min	9.05×10^4	4.96
Bacillus subtilis	Stainless steel	5 min	1.34×10^4	1.25
Staphylococcus epidermidis	Stainless steel	5 min	2.16×10^4	4.33
Escherichia coli	Stainless steel	5 min	2.00×10^5	5.36
Clostridium sporogenes	Stainless steel	5 min	1.74×10^4	2.72
Aspergillus brasiliensis	Stainless steel	10 min	4.60×10^4	4.66
Staphylococcus aureus	Stainless steel	10 min	8.20×10^4	4.91

Organism tested	Surface	Wetted period	Initial positive control inoculate	Results in Log reduction
Pseudomonas aeruginosa	Stainless steel	10 min	3.85×10^4	4.59
Candida albicans	Stainless steel	10 min	1.31×10^4	4.12
Micrococcus luteus	Stainless steel	10 min	1.10×10^4	4.04
Penicillium rubens	Stainless steel	10 min	9.05×10^4	4.96
Bacillus subtilis	Stainless steel	10 min	1.34×10^4	1.59
Staphylococcus epidermidis	Stainless steel	10 min	2.16×10^4	4.33
Escherichia coli	Stainless steel	10 min	2.00×10^5	4.08
Clostridium sporogenes	Stainless steel	10 min	1.74×10^4	2.54
Aspergillus brasiliensis	Epoxy	5 min	5.65×10^4	4.75
Staphylococcus aureus	Epoxy	5 min	6.80×10^4	4.83
Pseudomonas aeruginosa	Epoxy	5 min	5.15×10^4	4.71
Candida albicans	Epoxy	5 min	1.31×10^4	4.12
Micrococcus luteus	Epoxy	5 min	1.26×10^4	4.10
Penicillium rubens	Epoxy	5 min	7.60×10^4	4.88
Bacillus subtilis	Epoxy	5 min	1.61×10^4	1.20
Staphylococcus epidermidis	Epoxy	5 min	1.90×10^4	4.28
Escherichia coli	Epoxy	5 min	2.07×10^4	3.31
Clostridium sporogenes	Epoxy	5 min	1.83×10^4	1.89
Aspergillus brasiliensis	Epoxy	10 min	5.65×10^4	3.23
Staphylococcus aureus	Epoxy	10 min	6.80×10^4	4.83
Pseudomonas aeruginosa	Epoxy	10 min	5.15×10^4	4.71
Candida albicans	Epoxy	10 min	1.31×10^4	4.12
Micrococcus luteus	Epoxy	10 min	1.26×10^4	4.1
Penicillium rubens	Epoxy	10 min	7.60×10^4	4.88
Bacillus subtilis	Epoxy	10 min	1.61×10^4	1.33
Staphylococcus epidermidis	Epoxy	10 min	1.90×10^4	4.28
Escherichia coli	Epoxy	10 min	2.07×10^4	5.31
Clostridium sporogenes	Epoxy	10 min	1.83×10^4	1.68
Sodium Hypochlorite at 0.52% in WFI				
Candida albicans	Stainless steel	5 min	1.09×10^4	4.04
Aspergillus brasiliensis	Stainless steel	5 min	5.00×10^4	4.71
Staphylococcus aureus	Stainless steel	5 min	2.75×10^3	3.44
Escherichia coli	Stainless steel	5 min	4.30×10^3	3.63
Pseudomonas aeruginosa	Stainless steel	5 min	1.55×10^3	3.19
Bacillus subtilis	Stainless steel	5 min	6.65×10^3	3.82
Penicillium glabrum	Stainless steel	5 min	4.50×10^3	3.73
Streptomyces flavovirens	Stainless steel	5 min	2.20×10^4	4.34
Micrococcus luteus	Stainless steel	5 min	2.33×10^4	4.37
Bacillus infantis	Stainless steel	5 min	1.44×10^5	3.38
Candida albicans	Stainless steel	10 min	1.09×10^4	4.04
Aspergillus brasiliensis	Stainless steel	10 min	5.00×10^4	4.7
Staphylococcus aureus	Stainless steel	10 min	2.75×10^3	3.44
Escherichia coli	Stainless steel	10 min	4.30×10^3	3.63
Pseudomonas aeruginosa	Stainless steel	10 min	1.55×10^3	3.19
Bacillus subtilis	Stainless steel	10 min	5.30×10^3	3.72
Penicillium glabrum	Stainless steel	10 min	5.40×10^3	3.73
Streptomyces flavovirens	Stainless steel	10 min	2.20×10^4	4.34
Micrococcus luteus	Stainless steel	10 min	2.33×10^4	4.37
Bacillus infantis	Stainless steel	10 min	1.44×10^5	3.70
Candida albicans	Epoxy	5 min	1.61×10^4	4.21
Aspergillus brasiliensis	Epoxy	5 min	7.95×10^4	4.9
Staphylococcus aureus	Epoxy	5 min	1.10×10^3	3.04
Escherichia coli	Epoxy	5 min	6.20×10^3	3.79
Pseudomonas aeruginosa	Epoxy	5 min	1.65×10^3	3.22
Bacillus subtilis	Epoxy	5 min	6.45×10^3	3.81

Organism tested	Surface	Wetted period	Initial positive control inoculate	Results in Log reduction
Penicillium glabrum	Epoxy	5 min	8.40×10^3	3.92
Streptomyces flavovirens	Epoxy	5 min	2.28×10^4	4.36
Micrococcus luteus	Epoxy	5 min	7.30×10^4	4.86
Bacillus infantis	Epoxy	5 min	1.07×10^4	4.03
Candida albicans	Epoxy	10 min	1.61×10^4	4.21
Aspergillus brasiliensis	Epoxy	10 min	7.95×10^4	4.9
Staphylococcus aureus	Epoxy	10 min	1.10×10^3	3.04
Escherichia coli	Epoxy	10 min	6.20×10^3	3.79
Pseudomonas aeruginosa	Epoxy	10 min	1.65×10^3	3.22
Bacillus subtilis	Epoxy	10 min	6.45×10^3	3.81
Penicillium glabrum	Epoxy	10 min	8.40×10^3	3.92
Streptomyces flavovirens	Epoxy	10 min	2.28×10^4	4.36
Micrococcus luteus	Epoxy	10 min	7.30×10^4	4.86
Bacillus infantis	Epoxy	10 min	1.07×10^4	4.03
Sodium Hypochlorite at 0.25% in WFI				
Staphylococcus aureus	Epoxy floor	5 min	1.89×10^6	4.58
Pseudomonas aeruginosa	Epoxy floor	5 min	1.69×10^6	4.53
Candida albicans	Epoxy floor	5 min	2.20×10^6	4.64
Aspergillus brasiliensis	Epoxy floor	5 min	5.85×10^4	3.07
Penicillium citrinum	Epoxy floor	5 min	1.40×10^5	3.7
Penicillium glabrum	Epoxy floor	5 min	1.66×10^5	3.52
Bacillus licheniformis	Epoxy floor	5 min	4.30×10^4	1.58
Staphylococcus aureus	Epoxy floor	10 min	1.89×10^6	4.58
Pseudomonas aeruginosa	Epoxy floor	10 min	1.69×10^6	4.53
Candida albicans	Epoxy floor	10 min	2.20×10^6	4.64
Aspergillus brasiliensis	Epoxy floor	10 min	5.85×10^4	3.07
Penicillium citrinum	Epoxy floor	10 min	1.40×10^5	2.59
Penicillium glabrum	Epoxy floor	10 min	1.66×10^5	3.52
Bacillus licheniformis	Epoxy floor	10 min	4.30×10^4	2.37
Staphylococcus aureus	Vinyl floor	5 min	2.00×10^6	4.60
Pseudomonas aeruginosa	Vinyl floor	5 min	4.93×10^6	4.99
Candida albicans	Vinyl floor	5 min	1.87×10^6	4.57
Aspergillus brasiliensis	Vinyl floor	5 min	7.15×10^4	3.15
Penicillium citrinum	Vinyl floor	5 min	1.48×10^5	3.47
Penicillium glabrum	Vinyl floor	5 min	1.89×10^5	3.58
Bacillus licheniformis	Vinyl floor	5 min	5.10×10^4	2.07
Staphylococcus aureus	Vinyl floor	10 min	2.00×10^6	4.60
Pseudomonas aeruginosa	Vinyl floor	10 min	4.93×10^6	4.99
Candida albicans	Vinyl floor	10 min	1.87×10^6	4.57
Aspergillus brasiliensis	Vinyl floor	10 min	7.15×10^4	3.15
Penicillium citrinum	Vinyl floor	10 min	1.48×10^5	3.47
Penicillium glabrum	Vinyl floor	10 min	1.89×10^5	3.58
Bacillus licheniformis	Vinyl floor	10 min	5.10×10^4	3.19
Hydrogen Peroxide at 6% in WFI				
Staphylococcus aureus	Stainless steel	5 min	9.00×10^3	1.85
Pseudomonas aeruginosa	Stainless steel	5 min	3.50×10^3	2.14
Candida albicans	Stainless steel	5 min	2.00×10^5	0.91
Cladosporium cladosporioides	Stainless steel	5 min	1.70×10^4	0.53
Staphylococcus aureus	Glass	5 min	6.00×10^3	2.89
Pseudomonas aeruginosa	Glass	5 min	8.50×10^2	2.93
Candida albicans	Glass	5 min	2.15×10^5	1.36
Cladosporium cladosporioides	Glass	5 min	1.65×10^4	1.3

Abbreviation: WFI = Water for Injection.

24.12 RESISTANCE THEORIES TO GERMICIDAL AGENTS

In present times, resistance to antibiotics by certain organisms in the human body has become more than a theory. Recent publications have even shown the possible presence of "superbugs" that have become immune to certain antibiotics and require new interventions. The human body has a very complex internal system whereby mainly blood makeup and circulation, tissue, and absorption make it very difficult to assure that a drug product is adequately delivered to the specific location where an infection is noted in the concentration required to destroy the contaminate in question. At the same time, the level or dose of antimicrobial agent (an antibiotic for example) is minimal as reactions, toxicity, and adverse effects of too strong a drug product may cause a multitude of complications within the human body that can range from simple rashes to liver and kidney complications.

With the initial fever surrounding resistance in the human body, many scientists in the pharmaceutical and biotechnology world began to ponder a theory of whether such resistance could occur between biocidal agents and organisms present in the clean room environment. However, such complexity does not occur between biocidal agents and microorganisms. Within the clean room environment, hard surfaces that may or may not have porosity are the "patient to be treated". Only surfaces can be disinfected with biocidal agents. Air within the clean room cannot be disinfected to any feasible degree, only filtered. On a surface, very little blocks the disinfectant from contacting the contamination. To a small degree, organisms may lodge themselves in pores of the substrate and at the same time soil load or residual can block the biocidal agent from contacting the organism. Although the porosity of the surface and soil load are a concern, this concern is not equivalent to the concerns that exist with regard to the human body. Antimicrobial effectiveness of biocidal agents does not resemble at all that of human products. The simplicity of the surface to be disinfected and the ability to use biocidal agents of enormously high strength increase the disinfectant's ability to destroy microorganisms far beyond that of an antibiotic.

Studies performed and published by this author showed that organisms subjected to antimicrobial effectiveness testing and surviving organisms regrown and retested showed that resistance was unfounded (see "Assessing Resistance and Appropriate Acceptance Criteria of Biocidal Agents" in the References section). Simply said, after conducting studies for five years on over 70 organisms, resistance was not found to exist. This theory, supported by PDA Technical Report #70 is just that: an unproven theory.

24.13 CONCLUSION

Unfortunately, antimicrobial agents are not the single means to attain an acceptable environment. There is no magic chemical that when used suddenly creates the perfect environment. Many integral factors in a wholistic approach combine to create a successful system. Control of the environment is one of the keys to success. Antimicrobial agents are the corrective action for a poor control system. This being said, they are an imperfect solution and provide no preventative assurances after they have dried. Understanding the entire contamination control spectrum is the key to success. Often in the GMP world, we just invoke a few levels of a complete contamination control system and by doing so find ourselves with contamination excursions. The successful contamination control professional takes into account all aspects required for a complete system and assures that all aspects are implemented, tested, documented, and include the appropriate training and retraining of personnel. Only a complete system assures success and routine, repeatable success.

24.14 SPECIAL ACKNOWLEDGMENT

A special thank you to Kelly Rocco, Manager QA, Veltek Associates, Inc. for collecting and reporting VAI Laboratory Antimicrobial Efficacy Performance Testing Summaries.

BIBLIOGRAPHY

1. Block, SS (ed.), *Disinfection, Sterilization and Preservation*, 4th ed. Philadelphia, PA: Lea and Febiger, 1991
2. McDonnell, G, Denver Russell A. Antiseptics and disinfectants: Activity, action, and resistance. *Clinical Microbiological Reviews*, Jan 1999.
3. Vellutato, AL. Jr. Assessing resistance and appropriate acceptance criteria of biocidal agents. In Madsen R, Moldenhauer J, eds. *Contamination Control in Healthcare Products Manufacturing*, DHI, LLC, Chapter 9, 319–352, 2014.
4. Vellutato, AL. Jr., Implementing a cleaning and disinfection program in pharmaceutical and biotechnology clean room environments. In Moldenhauer J, ed. XX, DHI, LLC, Chapter 8, 179–230, 2003.
5. Aseptic Processing, Inc. (API) a division of Veltek Associates, Inc., Validation and Testing Studies Surrounding the Resistance of Biocidal Agents, 2009.
6. Center for Drugs and Biologics and Office of Regulatory Affairs, Food and Drug Administration, Guidelines on Sterile Drug Products produced by Aseptic Processing, p. 9, June 1987.
7. Center for Drugs and Biologics and Office of Regulatory Affairs, Food and Drug Administration, Sterile Drug Products Produced by Aseptic Processing Draft, Concept paper (Not for Implementation), September 27, 2002.
8. EU GMP Annex 1: Manufacture of Sterile Medicinal Products, November 25, 2008 (rev).
9. AOAC International. *Official Methods of Analysis*, 16th ed., 5th Revision. Gaithersburg, MD: AOAC, 1998.
10. Parenteral Drug Association (PDA). Technical Report #70 (TR70). Bethesda, MD: PDA, 2015.
11. European Committee for Standardization, Chemical Disinfectants and Antiseptics—Quantitative non-porous surface test for the evaluation of bactericidal and/or fungicidal activity of chemical disinfectants used in food, industrial, domestic and institutional areas—Test methods and requirements without mechanical action (phase 2/step2), EN 13697 E, ICS 11.080.20; 71.100.35. Brussels, 2001.

12. FB Engineering, Efficacy Testing of Biocidal Products—Overview of Available Tests. At the request of the Swedish Chemicals Agency. Table 1, pages 24–189, 2008.
13. Australian Government Department of Health and Aging, Therapeutics Goods Administration (TGA) Guidelines for the Evaluation of Sterilants and Disinfectants, February 1998.
14. Connor DE, Eckman MK. Rotation of Phenolic Disinfectants, *Pharmaceutical Technology*, September 1992, 148–158.
15. Vellutato, A, Sr., United States Patent 6,123,900. United States Patent Office, 1992.
16. Vellutato, A, Jr., Validation of the Core2Clean Spray Mop Fog Systems, Internal Validation Report, Veltek Associates, Inc. January 2000.

25 Cleaning and Disinfecting Laminar Flow Workstations, Bio Safety Cabinets and Fume Hoods

Arthur Vellutato, Jr.

CONTENTS

25.1 INTRODUCTION

Pharmaceutical, biotechnology and compounding pharmacy operations may use vertical laminar flow workstations, horizontal laminar flow workstations, biosafety cabinets, or fume hoods in a multitude of areas within their operations. Some users employ them for processing final product, processing ingredients, processing subcomponents, processing cultures, evaluating environmental monitoring samples, bioburden analysis, separations, sterile transfers, antimicrobial effectiveness studies (AET), compounding and a multitude of other areas of processing or testing. Within each type of system are an array of units similar in functionality; however, the material of construction of each can vary immensely. It is this disparity together with the vast array of processes that are performed within the unit that complicates the cleaning and disinfection of these systems.

The complication does not end at the type of process and construction types. Potential residues, proteins, product spills, microorganisms and viruses that may be exposed within and/or exist in the surface can add another layer of concern. Finally, variable classifications for the background areas are also present. Many organizations assume that they can spray or wipe down the units and attain a compliant environmental condition within which to process a future product or test only to detect either cross-contamination or the eventual deterioration of the unit itself.

For microbial efficacy, it is important to understand that viable contamination is destroyed by saturation and penetration of the cell wall over a specified contact (dry) time with a lethal agent. Destruction of the cell is complicated by the presence of irregular surfaces, soil and residue that may block coverage of the cell with the chemical agent. The effective means for destruction of the cell is normally cytoplasmic disruption or exploding of the nucleus of the cell. There are exceptions to cytoplasmic disruption, as some microorganisms called Prokaryotes do not have a nucleus. Most Prokaryotes are small, single-celled organisms that have a relatively simple structure and the basis for the demise of the

DOI: 10.1201/9781003163138-25

organism resides in saturation or the use of antibiotics. To destroy an organism with a disinfectant or sporicidal agent, it must be on the surface. Airborne microorganisms cannot be destroyed to any great degree by vapors or gasses of disinfecting agents unless we create a chamber in which the agent to oxygen ratio is weighted dramatically in the agents favor. We see this in fill lines with nitrogen tunnels that can deplete the oxygen level and destroy aerobic organisms. Anaerobic organisms can grow without oxygen so the mechanism for kill is normally an oxygenated environment and/or normal disinfecting agents. To assume our sporicidal actions in cleaning surfaces or fogging is destroying airborne microorganisms would be without scientific merit. Even vapor phase hydrogen peroxide (VHP) and paraformaldehyde eradication systems have controversial data regarding the ability to demonstrate airborne microorganism destruction.

Although this chapter is focused on the cleaning and disinfection of the various workstations and bioburden reduction for items and personnel entering the workstation space, it is critical to the design plan to include a review of the construction, airflow, potential for external contamination entering and the product that may be processed. It is essential to have a complete analysis of the application prior to choosing agents and developing the cleaning and disinfection plan of action. It is also critical to understand what regulatory agencies will require as validation data to be considered compliant. In the end, it is all about how well we do the tasks at hand and not that we just do them.

25.2 CONSTRUCTION

Assuring the construction materials, what the unit is made of, for vertical laminar flow workstations, horizontal laminar flow workstations, biosafety cabinets and fume hoods is a critical concern requiring attention to address the cleaning and disinfection of a unit. One of the biggest problems confronting Good Manufacturing Practices (GMP)/(Good Laboratory Practices (GLP) operations is that they try to use what they already have first. Sometimes their units are very old and at times, made of lower-grade stainless steel, carbon steel, aluminum, cast aluminum, alloy steel, lower-grade epoxy, plastics, coated particle board, laminates and many other materials that may deteriorate over time when exposed to specific chemically aggressive sanitizer, disinfectant and sporicidal agents. Not only do older units represent a potential problem, but newer, less-expensive units can also present a problem because of their materials of construction. In short, some units in the marketplace are made with materials that do not fit every type of operation and more importantly, every type of disinfecting chemical. The first thing that needs to be reviewed are the material types used in the construction of the units and the availability of replacement parts.

The second issue in construction is that return vents are present, and agent vapors and contamination can be circulated or emitted. In many units, perforated metal or aluminum grates are incorporated into the design and engineered to act as a return air vent. When material perforation is disinfected with chemical agents, the agent's residue will accumulate around and inside the vent. The exterior of the vent may be easily cleaned, but the interior perforations may build up residual from the disinfecting agents, or more concerning drug product, proteins, chemotherapy agents, blood and overall unwanted contaminates. Over time, the vents themselves can deteriorate. The vents recirculate air to High-Efficiency Particulate Air (HEPA) filters or are vented to external scrubbers or bioload/hazardous material collectors. The concern is how internal ducts or airflow channels will be cleaned and/or disinfected and at what frequency. The unit is essentially a machine with working parts such as fans or blowers, most of which are enclosed within the unit. It is not enough to clean and/or disinfect the exterior surfaces without the evaluation of what needs to be done internally. The issues extend externally to the unit to environmental neutralizing equipment. If residual and chemical buildup or corrosion occurs in these internal areas and external systems, then operation and safety may be compromised. After review of surfaces both inside and outside, one can determine what must be done based on the product(s) produced or products exposed in the unit, and the cleaning and disinfecting products and procedures that will be implemented.

25.3 PRODUCTS PRODUCED/TESTED IN VERTICAL LAMINAR FLOW WORKSTATIONS, HORIZONTAL LAMINAR FLOW WORKSTATIONS, BIOSAFETY CABINETS OR FUME HOODS

Workstations, sometimes termed hoods, laminar flow hoods, LFWs, BSCs and safety cabinets, are classified into three (3) categories. They are Level I, II and III. Level I provide ample protection to both personnel and the exterior environment but minimal protection to the product. Level II provide an aseptic environment and may have fume safety options. They are intended for Biological Safety Levels (BSL) 1, 2 and potentially 3. Level III is the highest classification and safety protection for personnel and product. They are suitable for BSL 4 dangerous human pathogens, infectious substances and fumes. The three different designs of workstations are presented following.

25.3.1 Laminar/Horizontal Flow Workstations/Hoods

Vertical laminar flow workstations and horizontal laminar flow workstations are commonly associated with aseptic manufacturing, manipulation, cells culture, or the testing/reading of microorganisms. They are also used in compounding operations whereby several sterile drug products are combined. These workstations are acceptable for working with BSL 2. Their protection is based on protecting the product, not the person. As such, the main concern in these areas would be on what did we bring into the hood, what goes airborne during use and what may have touched or exists on the surfaces. Most of the time, product spill, cell culture spill, testing reagents, microorganisms and particulates are the main surfaces or air contamination culprits. In this venue, the most difficult cleaning operation would be for product spills and residuals on the surfaces.

25.3.2 Biosafety Cabinets

Biosafety cabinets are a level of control upward from laminar/horizontal flow workstations and are used specifically for products or processes that require operator protection from pathogens or toxic chemicals. This type of system provides safety and options to vent toxins outward to purification systems. Toxic chemicals can be split into different categories that include hazardous chemicals and hazardous drug products. These workstations, with appropriate controls, are sufficient for working with BSL 1, 2, 3 and 4. BSL 4 will require additional personnel and safety precautions. BSL 4 is the highest level of infectious disease and/or highly dangerous or exotic microorganisms. Cleaning and disinfection are more complicated in the biosafety cabinets, because drug product or toxic chemicals may be evacuated from the unit. Thus, cleaning of internal and external ducting together with other operational systems may need to be addressed along with the cleaning and disinfection of internal surfaces. Normally these units are rated as Level III.

25.3.3 Fume Hood

A fume hood is normally used in the processing of toxic products or the combination of products that create toxic fumes or conditions to humans. These units operate as a mechanism that pulls air away from personnel working in the front of the unit for safety reasons relating to toxic exposure. Cleaning and potentially disinfection may require the internal cleaning of internal mechanisms in addition to the unit's work surfaces. Moreover, cleaning and disinfection of the work surfaces may be complicated by product spillage or residues that may adversely interact with the cleaning agents. Caution needs to be exhibited in the mixture of two or more chemicals and the potential reaction. The problem with all workstations is that most components, airflow ducts and electronics are incompatible with most cleaning and/or disinfecting agents. One needs to prepare for this activity rather than try to create a plan after signs of a problem. Normally these units are rated as a Level I depending upon the available functions.

25.4 EXPECTED BIOBURDEN IN LAMINAR/HORIZONTAL FLOW WORKSTATIONS, BIOSAFETY CABINETS, OR FUME HOODS

Vertical laminar flow workstations, horizontal laminar flow workstations, biosafety cabinets and fume hoods are normally monitored for microbials and particulates just like classified areas. The background to these systems can be anywhere from a Grade A to a Grade D environment depending upon the criticality of the product or testing being produced. Because of the obstructive nature of the unit and its effect on laminar flow, a Grade A background environment may be difficult to attain. Most backgrounds are defined and classified as Grades B–D. The concept of a vertical laminar flow workstation is exactly as it sounds. HEPA-filtered air at 99.9995% filtration efficiency at an airflow of 90 feet/minute is emitted from the filter mounted in the top of the unit. The air cascades downward to the surface and then moves out of the unit into the room. A horizontal laminar flow workstation is also exactly as it sounds. HEPA filtered air is directed from the filter located on the back wall of the unit. The air flows outward past the workspace, the operator and then out of the unit into the room. A biosafety cabinet works much like a laminar flow workstation with the exception that the unit incorporates a glass shield that pulls down in the front and both laminar flow air and room air are pulled into vents on the front edge of the unit, which is recirculated back to the HEPA filter. Some biosafety cabinets also have a fume hood feature so that air can be dispelled from inside the hood to the appropriate treatment before release to the environment. A fume hood pulls air from the workspace and room into the hood and expels it to an outside treatment location.

The vertical laminar flow workstations, horizontal laminar flow workstations and biosafety cabinets should all provide a clean, compliant airflow with little impact from the exterior unless contamination is brought into the unit by personnel or external items. Of these, biosafety cabinets are the superior design from a worker protection perspective. If microbiological cultures or cells are processed in the workstation, contamination in these units is more likely from what is occurring inside the hood rather than contamination coming in from the exterior of the hood. In this case, the internal contamination is known while the exterior potential contamination is unknown. To address introduction of item and personnel, the GMP/GLP operation needs to conduct microbial surface (RODAC) and microbial air testing and develop over time an environmental isolate list. These are the flora most identified in the units during operations. This testing should also be conducted in the background classified, which as described before should be from Grade A to D depending on the product requirements. The risk of microorganisms entering from the background into the unit is most probable if adequate control is not invoked. This list combined with surface (RODAC) testing of incoming items should provide a complete list for planning for cleaning and disinfecting the unit and assure the first step in regulatory compliance once validation studies have been completed.

Assuring aseptic conditions is more difficult with the fume hood. As the unit evacuates air from the hood, air from the background environment is pulled into the hood. This air will be of the quality of the background classification. A standard fume hood is normally used for hazardous chemical vapor removal in a non-sterile environment. If aseptic conditions are needed, then a biosafety cabinet is normally employed. Attempting to have Grade A conditions in a background area next to a fume hood is very difficult to manage.

Let us remember that cleanrooms and classified areas, if properly designed and maintained, will operate appropriately

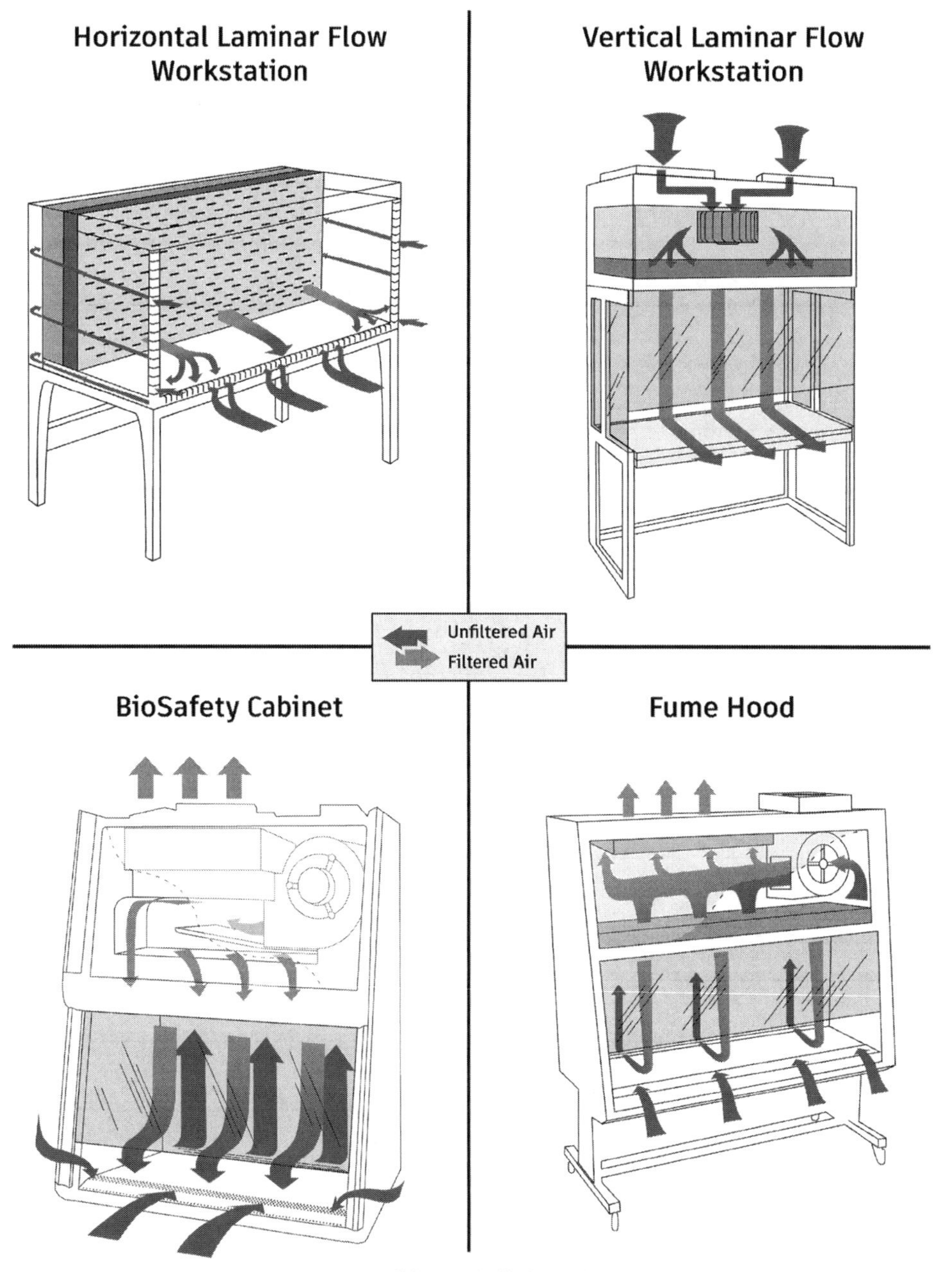

FIGURE 25.1

to desired classifications without personnel or component/items entering. As these contamination contributors enter, the system is stressed. Cleaning and disinfection efforts in the classified areas and the hoods are corrective actions to address the contamination. If we do not let the contamination in, we do not have to contend with its presence.

25.5 SELECTING CLEANING AND DISINFECTING AGENTS

Many firms choose their cleaning and disinfecting agents after validation so that they may review the effectiveness of the agents. This plan would not be prudent. Why? Some sanitizers, disinfectants and sporicides will be disqualified from the scope because of potential corrosion to work surfaces, corrosion to internal mechanisms or because they leave excess residuals that may adversely affect the product being produced. Another disqualifying factor would be chemicals that would cause a reaction with existent chemicals on the surface in a liquid form or the residuals from such chemicals. So, we must ask three questions before we choose and start validation.

The first question to ask is are the sanitizers, disinfectants and sporicides compatible with the surfaces of the workstation and with the interior mechanisms of the workstation. Deterioration of the workstation is a problem as corrosion or deterioration of surfaces may lead to such contamination affecting the product being produced. Internal deterioration may compromise appropriate operation of the workstation. Finally, if a surface is deteriorated in a workstation, a replacement part or piece may not be available as the turnover to new models in the industry occurs very quickly. This will prompt the end user to either buy a new unit or to attempt, at high costs, to custom manufacture a component or piece to fill the gap. In short, deterioration may become an enormous problem that should be avoided.

The second question is whether vapors from sanitizers, disinfectants and sporicides will harm the internal mechanisms over time. This would require replacement parts, and it is advisable to purchase such replacement parts when commissioning the unit as they may not be available years later. The concern in this venue is mostly vested in the airflow return cycle. Applying the agent differently can help reduce the internal deterioration and is explained later in this chapter. Internal mechanism and ducting deterioration are things to consider up front.

The third question would be is there a chemical product in liquid form on the surface or a residual of a chemical on the surface that is not compatible with the cleaning and/or sanitizers, disinfectants and sporicides we plan to use. This mostly relates to fume hoods. Potentially we could be working with a chemical on the surface such as ammonia. When chlorine bleach is mixed with an acid, chlorine gas is given off. Chlorine gas and water combine to make hydrochloric and hypochlorous acids. These are all extremely dangerous to humans.

These are questions to ask oneself before choosing disinfecting agents and conducting expensive and time-consuming AETs. The answer to these questions is normally unique to each operation, process or product being tested or manufactured. Careful evaluation should be done in advance of validation and commencement of use.

25.6 ANTIMICROBIAL EFFECTIVENESS TESTING VALIDATION DATA

The flora recovered from the surfaces and air from the workstations is combined with the bioburden from incoming items and personnel. The background bioburden is also evaluated. The GMP/GLP firm then looks at what microbial cells they may be working with and adds such potential flora to the isolate list. These isolates are then tested against the disinfecting agents in an AET. This study attempts to demonstrate that known bioburden microorganisms will be destroyed first by the incoming component/item disinfection procedure and agents and then by the routine cleaning and disinfection regime.

More than any other subject matter, the AET raw data associated with the studies and the final reports are among regulatory agencies most reviewed documentation surrounding cleaning and disinfection activities. If done in a scientific manner with appropriate and scientifically formulated acceptance criteria, they serve the GMP/GLP firm with valuable information that the firm is addressing things correctly and implementing an effective plan with procedures that will reduce contamination inside the workstations.

At the same time, product residual removal is also critical. Cross contamination and personnel safety sometimes go hand in hand. In some instances, such as with chemotherapy products (discussed later), just wiping with a chemical agent and a dry wipe or a saturated wipe is not enough. Appropriate chemicals together with wipes need to be utilized to not only clean the residual but also to inactivate the product to so that it is deemed harmless.

As defined in most every worldwide regulatory guideline including the FDA Sterile Drug Products Produced by Aseptic Processing—Current Good Manufacturing Practice and EU Annex 1 is the directive that AET be conducted and updated routinely. So how are studies done? Simply, each surface type within the workstation together with surfaces that will be disinfected for components/items that will be placed into the workstation are defined and 1" × 1/2" test surfaces are created from the material or like material (per Parenteral Drug Association (PDA) Technical Report #70). If one follows USP <1072>, such surfaces should be 2" × 2". But note, a 2" × 2" surface is difficult to handle during this testing, and the sterile saline and neutralizer containers must be wide mouthed to accept the surface. These are not commercially available sterile and so the end user will have to make them sterile. As depicted following, the test consists of five (5) separate tests using serial dilution. One can also conduct a filter membrane study, but the data from that is harder to assess as overlapping of organisms can and will occur. Porous surfaces, such as epoxy or plastic, are more difficult to disinfect than less porous surfaces such as stainless steel. The porous surface creates a less efficacious scenario as microorganisms may become trapped in the pores.

In the study, the first test is a negative control. This is to prove that no viable contamination existed on the coupons prior to testing. It is done by aseptically taking a coupon that has been autoclaved and placing it into a test tube containing a neutralizing agent that is selected decided upon based on the chemical agent being tested. Many times, it is Tween and Polysorbate 80. The test tube with the coupon is vortexed for 1 minute. At that point, 1 mL is pipetted out of the test tube and placed into a test tube containing a saline solution. The test tube is vortexed for 1 minute and then 1 mL is pipetted out and plated onto a 100 mL Trypticase Soy Agar plate. The plate is then incubated for 3–7 days, read and the results recorded.

The second coupon is considered the toxicity carrier. The coupon is inoculated with the test organism at minimally 1.0×10^4 (less if warranted) and allowed to air dry. However, the 1.0×10^4 inoculate is specified for open classified areas, not a workstation. A workstation is located inside the clean room. It has three confined walls and a filtration system with positive pressure. It does not have the traffic nor potential to become as contaminated as an open manufacturing room so lowering of the inoculated to 1.0×10^2 may be warranted. However, when working with microorganisms within the workstation, this value may need to be adjusted upward. The test procedure follows the same procedure as for the negative control. The purpose of this test is to prove that the neutralizer used to stop the killing power of the disinfectant is not toxic to the microorganisms being tested. This is proven if the viable recovery from the coupon is sufficient.

The third coupon is the neutralization confirmation test. The coupon is inoculated with the test organism at minimally

1.0×10^4. (less if warranted) and allowed to air dry. The test procedure follows the same procedure as is done for the negative control except a small amount of disinfectant is added to the test tube before vortex. This represents approximately ±0.5 mL (the quantity may vary for each agent). The purpose of this test is to prove that the neutralizer used is sufficient and capable of inactivating the disinfectant so that the microorganisms tested can survive and grow. This is proven if the viable recovery from the coupon is sufficient.

The fourth test is the positive control. The coupon is inoculated with the test organism at minimally 1.0×10^4 (less if warranted) and allowed to air dry. The test procedure follows the same procedure as is done for the negative control with the exception that the test is drawn out to a 6-phase serial dilution. This means that there are 6 test tubes (after the neutralizer) where 1 mL is extracted by a pipette after vortex and transferred to the next test tube in the serial dilution. The purpose of this test is to prove that the positive controls can grow through each phase of the serial dilution. This is proven if the viable recovery from the coupon is sufficient in each phase (test tube).

And the fifth test is the test carrier. The coupon is inoculated with the test organism at minimally 1.0×10^4. (less if warranted) and allowed to air dry. After drying, the coupon is exposed to the disinfectant by trickling disinfectant onto the coupon and covering all visible surface area. As soon as this is done, the test is timed for the desired contact (dry) time (example 5 minutes). After 5 minutes of exposure, the test follows the same procedures as for the positive control. The purpose of this test is to prove that the sanitizer, disinfectant or sporicide used can destroy the population of viable cells on the coupon. This is proven if the test tube for each phase shows no growth or shows reduced growth as compared with the positive control.

A successful AET would show no growth in the negative control. The toxicity carrier and the neutralization confirmation should show viable growth. The positive control should show viable growth and the test carrier should show no growth or reduced growth compared with the positive control carrier.

Knowing the results allows the implementation of a proven defensible cleaning and disinfection procedure for the items entering the workstation and for the workstation itself.

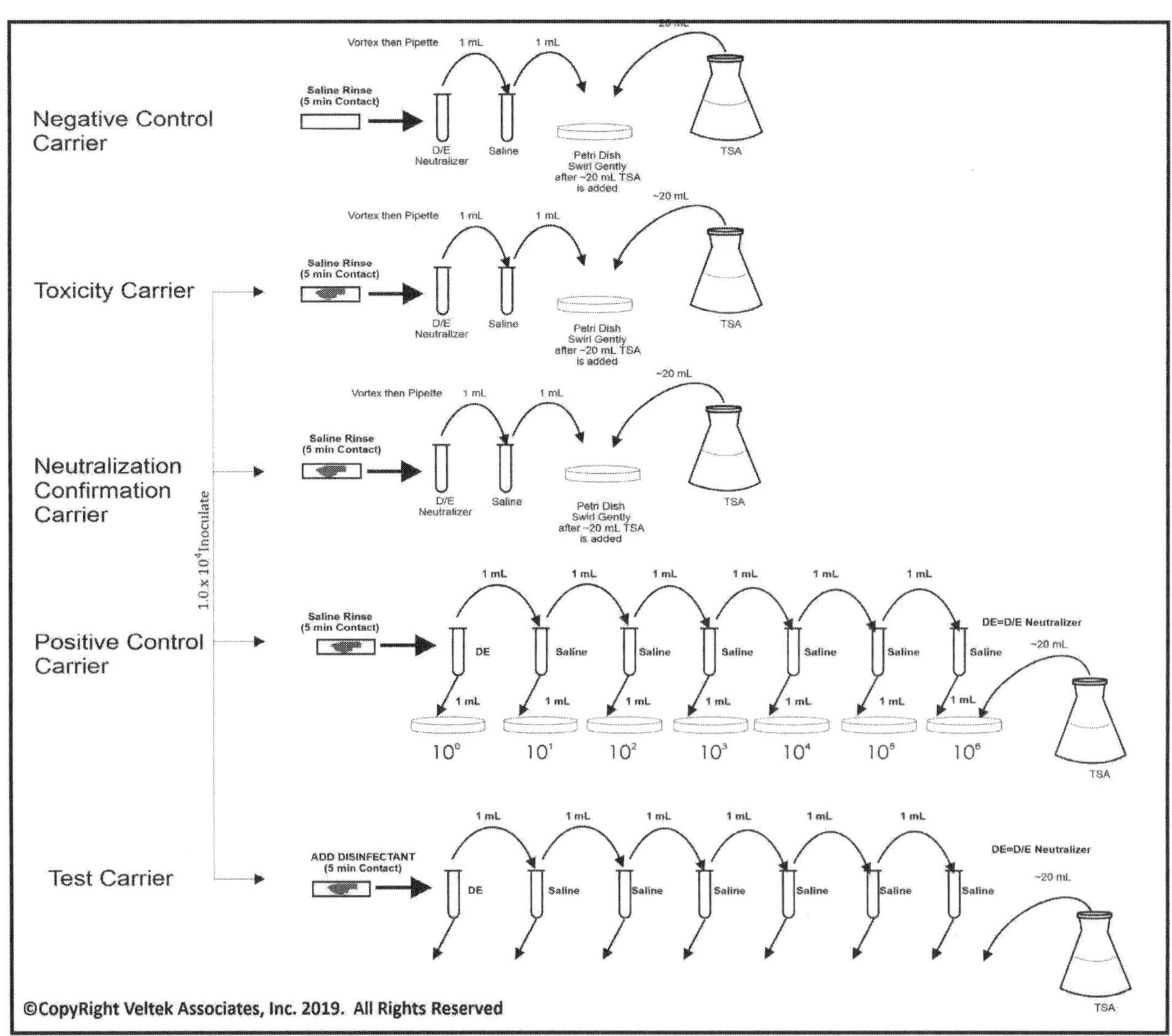

FIGURE 25.2

25.7 CLEANING AND DISINFECTION OF THE WORKSTATION

For this discussion, assume the background environment to the workstation has been appropriately cleaned to compliant classification requirements. In these areas, we would also assume that any items or components destined for the Grade A workstation have already been appropriately sterilized or disinfected using a transfer method that is based on the classification to where they originated. For a Grade A and B background environments, assume that personnel are gowned in a full sterile gown, mask, goggles, boots and double-gloved hands. This would be standard donning for these classifications. If personnel are not gowned accordingly, the area would not be compliant. In Grade C, most aseptic operations will employ gowning as in Grades A and B. However, in some Grade C--D operations, gowning may be reduced to coveralls, boots, hoods and one pair of gloves. A mask may or may not be invoked for the area and again is dependent on the desired environmental conditions and the product. In these areas, we would also assume that any items or components destined for the Grade A workstation have already been proficiently disinfected based on the classification of where they originated. Within the contents of this book, a chapter entitled, "Contamination Control for Incoming Components to Classified Areas—"War at the Door®"" details how to transfer items and components into various classifications. What we are trying to determine is what bioburden level do personnel and items/components destined for the Grade A workstation possess and how likely is such contamination able to breach the Grade A workstation protections.

25.7.1 Cleaning the Workstations

In all workstations, first remove all items from the workstation. Understand that we are going to clean the workstation from top to bottom and overlap our cleaning strokes so as not to miss any surface areas. It is better to over clean than to under clean. Cleaning is not disinfection. Cleaning is the removal of microbes, particulates, residuals and potentially product from the surfaces by a mechanical cleaning action and using a material wiper with excellent absorbent characteristics. In the literature for wipers this will be defined as "sorbancy". The chemical agent to use will be either sterile 70% isopropyl alcohol, sterile 70% ethanol or a sterile high surfactant cleaner like DECON-Clean. For a fume hood in which chemical residual or remaining wetted puddled surfaces may exist, one needs to ensure that the cleaning chemical is compatible with the residual on the surface prior to using the chemical agent. Always, a high surfactant cleaner is a better choice. A surfactant-based product allows particulates, microbes, residues and any residual product on the surface to be lifted and put into solution. Isopropanol and ethanol are low-level cleaners but leave minimal and insignificant residual on non-product contact surfaces. During the cleaning operation, it is critically important to understand that wetting of the filter is to be absolutely avoided under any circumstances. A filter wetted with a cleaning or disinfecting agent will evaporate the chemical agent first, leaving a water-wetted filter that can proliferate the growth of microorganisms, namely molds. From below the top filter area, a dry wipe can be sprayed with one of the cleaning agents previously mentioned. The surfaces of the filter grid and close to the filter can be wiped with this. The use of a pre-saturated wipe is acceptable if it is followed with a dry wipe. The idea is to wipe on wet and wipe off dry. That is cleaning. Cleaning is not spreading around the water wet cleaning agent. The next step is to wipe the side and back walls from top to bottom with overlapping horizontal strokes. Using one wiper for an entire hood is not cleaning. It is spreading the contamination around. The wiper should be folded in half, and two to three top-to-bottom horizontal strokes performed. The wiper should be flipped over and two to three more strokes performed. At this point, the wiper should be discarded and a new wiper used for the continuing cleaning operation. This procedure should be repeated for all three sides (vertical laminar flow and biosafety cabinet) or for the top and two sides of the horizontal laminar flow. The biosafety cabinet will also have a difficult to clean up and down sliding glass or plexiglass door. The door should be raised and lowered, and the same cleaning technique of overlapping horizontal strokes should be employed on the inside of the door and at the end of cleaning on the exterior of the door. In most biosafety cabinets, this surface is often dirtied from product spills and past disinfectant residuals as one must reach inside and upward to clean, and that operation is extremely awkward. When cleaning the fume hood, all sides and the top should be cleaned.

This leaves only the work surface to be cleaned. First all spills should be pre-cleaned. The operator then cleans from back to front with overlapping stokes using the same chemical agent used for cleaning the other surfaces and the same wiper lifespan. Caution should be taken to ensure the work surface is visually clean, as the worksurface will likely be more difficult to clean as spills or residual from operations may be dried on the surface. There is no reason why an operator could not initially scrub back and forth to remove stubborn contamination, but they should end the cleaning with a back to front wipe of the surface when complete. The work surface for a vertical or horizontal laminar workstation will often incorporate perforated return vents on the back or the side of the unit. In some units these are removable, in others they are not. Their purpose is to return as much of the HEPA-filtered air within the workstation back to the HEPA filters. This reduces the amount of makeup air the unit must pull from outside the workstation. The more air from outside the unit it must use as makeup air, the greater the chance for contamination and the increased frequency of servicing the filter. The biosafety cabinet will incorporate minimally two perforated return vents: one in the front by the glass shield and one at the rear of the unit. Some units may have them on the sides. It is critical to clean all perforated return vents thoroughly and to ensure disinfectant, product residual and corrosion is removed.

25.7.2 Disinfecting the Workstations

Now that the workstation has been cleaned, disinfect the surfaces. The key component of assured disinfection is exactly what was proven in the AET, wetted or dry time. The surface must remain wetted for the validated period. Spraying and allowing the surface to air dry is the best mechanism; however, overspray of the filter, as previously discussed, should be avoided. Controlled and light spraying to coat the surface should be performed from top to bottom and back to front, and leaving the workstation to air dry is one option. Over spraying and or overuse of the agent is not recommended, and puddling of the agent should be discouraged and dried with a dry wiper. If disinfection is done with care with the correct chemical agent, corrosion or residual buildup should be kept at a minimum. A better option would be to wet the dry wipe with enough chemical agent to coat the surface. This is more controlled, and the chance of overspray or overuse is less. However, the surface must be assured to be wet for the prescribed time period defined in the AET validation studies. A problematic situation with wiping the chemical agent on the surface is coverage. Care needs to be taken to assure the surface is wetted, streaking does not occur and the wipe remains wet through the wiping process. A saturated wiper can help in this regard, but remember that the liquid entrapped in the wipe will be spread to the surface and deplete quickly. This means the saturated wiper will need to be changed more often.

Prior to validating disinfectants and sporicides, the agents are reviewed for potential compatibility with the surfaces, corrosion to certain surfaces, buildup of residual and the effect when exposed to the product or agents used. Within the workstation, there is no need to rotate agents as we would in the clean room. Access to the workstation is limited, the unit is never turned off and personnel (except for their arms) should not be a source of contamination. Most of the time, the contaminating source is the product, cells, or microorganisms exposed within the hood. Registered disinfectants such as phenol or quaternary ammonium are not good options for cleaning the workstation. Both of these products leave residuals, only destroy vegetative cells and have little or no effect on bacterial or mold spores. Where bioburden reduction or bioburden elimination is needed in the workstation, the best potion is a sporicidal agent. However, sporicidal agents are not alike. Each agent type's mechanism of kill and limitations are different. Most leave residues and corrode in varying ways and over varying time periods. Following are the equipment surfaces that are predominant in workstations. The following lists attempt to identify most surfaces but are not all-inclusive.

External and Inside Workstation Construction Materials

- Aluminum
- Various grades of Stainless Steel (302, 304, 316, 316L)
- Glass
- Plexiglass
- Plastics
- Epoxy
- Epoxy-Coated and Painted Surfaces
- Gasket Materials such as Neoprene, Rubber, Vinyl
- Lighting
- Contactors

Internal Mechanism Materials

- Galvanized Steel (ducting)
- Plastic
- Electronics (circuit boards, coated wire, uncoated wire, LED, switches, lights)
- Steel (fan blades)
- Filter Medium
- Housing Material
- Meters (velocity, pressure, RH, temperature, pressure)

Concern needs to be invested in preserving and not damaging the internal mechanisms and in not corroding or chemically altering the external surfaces. On the worksurface of the workstation, corrosion may be visible and may progress over time. As stated before, perforated return vent deterioration may be seen on top but not underneath or within the return ducting. Such deterioration would likely be from chemical agents or corrosive chemical overspray or vapors that make their way into the return ducting. At the same time corrosion occurs, we must concern ourselves with microbiological and dangerous product residuals that accumulate over time that may require routine cleaning and disinfection. Many times, disinfection of a microbial contaminant inside will require first cleaning and then treatment with a system such as VHP. But it should be done sparingly as even VHP will have harmful effects on low grade metals, plastics, polycarbonates and gasketing. As stated previously, with each sanitizer, disinfectant and sporicide there are a unique set of characteristics that may make it either suitable or unsuitable for the workstation surfaces.

Now that the workstation interior surfaces and internal components have been cleaned and disinfected, the last concern is the exterior of the unit. This is a difficult challenge as most exterior surfaces such as the outside walls, rear, top, legs and underneath portions may or may not be made of materials that are compatible with disinfecting agents. At the same time, the top and rear of the equipment can be difficult to reach, clean and disinfect. This leaves the question of how to do this, with what and how often. The answers to those questions are based on the background grade established for the unit and how the exterior of the unit will affect the environmental conditions of that environment. Minimally we need to wipe all surfaces with either a sterile 70% isopropyl alcohol, sterile 70% ethanol or a sterile high surfactant cleaner and dry wipes. Most of the time this is sufficient. Routinely, we need to RODAC test these surfaces to find what bioburden is present. This can be used to scientifically address the exterior disinfection that may be required with the appropriate disinfectant or sporicide. One should not rush to disinfect the unit with a disinfectant or a sporicide frequently without this data. Such actions will typically result in residual buildup and corrosion to the exterior of the unit. Rather, find what contamination seems to be routinely present and implement a defined action plan from that point forward.

Characteristics of each type of well-known and routinely used sanitizers, disinfectants and sporicides are discussed following. Uniquely, each has a specific purpose and pros and cons for using such agents. The goal is to review the EM data, clean routinely, evaluate risk and then choose the appropriate agent and its frequency of use based upon the data and classification of the background.

25.7.3 70% Isopropyl Alcohol or 70% Denatured Ethanol

Alcohols are relatively safe for most surfaces but do have corrodibility to certain plastics and rubbers. They will also dull epoxy surfaces over time. The efficacy performance of alcohols used in clean room operations is minimal and limited to some vegetative cells at lower levels. Overall, these agents are used for cleaning, glove sanitization, sanitization of surfaces during manufacturing/processing/testing or as a final spray down to items entering the workstation. For fume hoods, cleaning with 70% isopropyl alcohol or 70% denatured ethanol or a high surfactant cleaner would be good choices provided the materials being used within the fume hood do not react and are compatible with the cleaning agents.

25.7.4 Phenolics

Phenolics are good disinfectants and have minimal corrodibility issues with surfaces at the active percentages used in germicidal detergents. However, they characteristically leave a residual that is difficult to clean over time. This phenolic residue darkens with age and exposure to light showing a tan or purplish mark on surfaces. Once phenolic residuals harden, they are almost impossible to remove without refinishing the surface. These agents also become sticky and tacky and may permanently stain certain surfaces. After use of these products in a workstation, and after the respective contact (dry) time, a 70% isopropyl alcohol or 70% denatured ethanol wipe down is required to remove the phenolic residue on the surface. These products are not recommended in workstations.

25.7.5 Quaternary Ammoniums

Quaternary ammoniums are good disinfectants and have minimal corrodibility issues with surfaces at the active percentage in the agents used. However, they characteristically leave a residue that is difficult to clean over time. Unlike phenolic residues, the quaternary ammonium residue does not darken with age. Although an excellent cleaner, the residual is a cause for some alarm as such residuals will build up over time and become difficult to remove. These agents also become sticky and tacky and may stain certain surface substrates. After use of these products in a workstation, and after the respective contact (dry) time, a 70% isopropyl alcohol or 70% denatured ethanol wipe down is required to remove the residual on the surface. These products are not recommended in workstations.

25.7.6 Hydrogen Peroxides

Hydrogen peroxide at ≥6% are efficacious products that do not leave residuals. Therefore, an application does not require a follow up 70% isopropyl alcohol or 70% denatured ethanol wipe down. Efficacy performance of hydrogen peroxide at ≥6% is good against vegetative cells and is also good against lower populations of bacterial and mold spores. This product is a sporicidal product that has associated registrations through the U.S. EPA and other worldwide agencies (STERI-PEROX). Some drying may occur to plastics and other porous items over time but overall, the product is relatively compatible with most surfaces. The product does not have equivalent sporicidal efficacy to peracetic acid and H_20_2 mixtures or 0.52% sodium hypochlorite when values reach 1.0×10^5 or greater. However, bioburden in the workstation does not warrant this level of efficacy as the bioload in the workstation is typically minimal. There may be some inhalation concern in continuous spray down in confined areas with low fresh air. However, this will not include the small space of a workstation, at the minimal time used and with the continuous airflow.

25.7.7 ≥0.25%–0.52% Sodium Hypochlorite

Sodium hypochlorite at >0.25% (optimally at 0.52%) has excellent efficacy against both vegetative and spore formers. At this level, most clean room flora can be eliminated at reasonable levels in a short contact time period. The product performs well at higher bioburden levels; however, it does leave a sodium chloride residue that can corrode all forms of metals if not removed. The product may be too strong for a workstation unless a BL3–BL4 condition is established or the process required working with high bacterial or mold spore levels. Chemotherapy drugs that may be processed within the area may require this agent as one step in an inactivation process. At that point, a triple step cleaning is required. This requires a first wipe step of 5.25% sodium hypochlorite for deactivation, a second step wipe of 2% USP sodium thiosulfate for decontamination and a third step wipe of 70% USP isopropyl alcohol for disinfecting/cleaning. The residual in this product can be cleaned so its use is acceptable in workstations. After use of these products in a workstation, and after the respective contact (dry) time, a 70% isopropyl alcohol or 70% denatured ethanol wipe down is required to remove the residual from the surface. Although effective in destruction of microbial contamination, the residue from the sodium hypochlorite needs to be removed. Without such residue removal, these products are not recommended in workstations.

25.7.8 ≥0.25%–0.52% pH Neutral Adjusted Sodium Hypochlorite

Sodium hypochlorite at >0.25% (optimally at 0.52%) with an altered pH to a neutral value has unmatched efficacy against vegetative and spore formers. At this level, most clean room flora can be eliminated in a short contact time period. The product performs well at higher bioburden levels; however, it does leave a sodium chloride and a slight pH adjustment chemical residue that will not corrode all forms of metals nor other substrates. The product may be too strong for a workstation unless a BL3–BL4 condition is established or the process required working with high bacterial or mold spore levels. The residual in this product can be cleaned so its use is acceptable in workstations with the condition of cleaning. After use of these products in a workstation, and after the respective contact (dry) time, a 70% isopropyl alcohol or 70% denatured ethanol wipe down is required to remove the residual on the surface. These products are not recommended in workstations.

25.7.9 Peracetic Acid and Hydrogen Peroxide Combinations

Peracetic acid and hydrogen peroxide mixtures have excellent efficacy against vegetative and spore formers. At this level, most clean room flora can be eliminated in a short contact time period. The product performs well at higher bioburden levels; however, it does leave a peroxyacetic acid and acetic acid residual that is very acidic in nature and can corrode metals if the residual is not removed. This product will also dry/ outgas plastics, vinyl plastics, elastomeric gaskets and other soft surfaces. The product may be too strong for a workstation unless a BL3–BL4 condition must be established or the process requires working with high bacterial or mold spore levels. The residual in this product can be removed so its use is acceptable. After use of these products, a 70% isopropyl alcohol or 70% denatured ethanol wipe down is required to remove the residual on the surface. These products are not recommended in open workstations.

Overall, the best cleaning agents one can use to clean workstations (provided chemistries and spilled product residues are compatible) are either sterile 70% isopropyl alcohol, sterile 70% ethanol or a sterile high surfactant cleaner. If achieving Grade A environmental conditions in the workstation is required, the use of a hydrogen peroxide at ≥6% would be the best choice because of its lack of residue and good efficacy. Once a residue is placed on a surface, it will never be completely removed. If extremely strong disinfection is required, then one of the other sporicides mentioned previously remain options with a post wipe down.

Prior to entry into any workstation, all components and items need to be disinfected and/or residual wiped from the package. An operator entering the workstation may be subject to an over gown procedure whereby either lab coats or protective sleeves are donned over room gowns. The operator must disinfect their hands with sterile 70% isopropyl alcohol (ex. DECON-AHOL WFI), don another pair of gloves and disinfect their hands again. The use of any chemical that would leave a residual on gloved hands is frowned upon. All items and components entering the workstation should have been prepared clean and either sterile or disinfected appropriately. This is completed after cleaning/disinfection and prior to working within the workstation as we do not want to introduce more bioburden than is already present in the unit. Normally in a Grade A/B background area only sterile 70% isopropyl alcohol or sterile 70% ethanol are used to disinfect items before entry to the workstation. Other chemical agents, with the exception of ≥6% hydrogen peroxide, are frowned upon during manufacturing or testing operations because of vapors and surface residual. Within the Grade A/B background environment, the use of the alcohols should be sufficient. In a Grade C/D background environment, this must change. There are many types of Grade C/D environments and one must specifically look at the unique area profile before the use of only a sterile alcohol could be deemed suitable.

25.8 CONCLUSION

Cleaning of vertical laminar flow workstations, horizontal laminar flow workstations, biosafety cabinets, or fume hoods is not merely a spray and wipe procedure. This chapter has defined the importance of the many critical elements such as workstation construction, knowing one's product, bioburden reduction prior to the workstation, AET, corrosion/compatibility, cleaning agents and disinfecting agents. Without the careful analysis of each area discussed, complications can arise that may adversely affect processes or final product and potentially prevent the workstation from serving as the clean and potentially sterile barrier that it was intended to be. The most important element of cleaning and disinfection of workstations is to know all the facts before acting. That will provide the best chance for success.

REFERENCES

1. Block, SS (ed.), *Disinfection, Sterilization and Preservation, Lea and Febiger*, 4th ed., Philadelphia, 1991.
2. Vellutato, AL. Jr., Assessing resistance and appropriate acceptance criteria of biocidal agents. In Madsen R, Moldenhauer J, eds. *Contamination Control in Healthare Products Manufacturing*, DHI, LLC, Chapter 9, 319–352, 2014.
3. Vellutato, AL. Jr., Implementing a cleaning and disinfection program in pharmaceutical and biotechnology clean room environments. In Moldenhauer J, DHI, LLC, Chapter 8, 179–230, 2003.
4. Center for Drugs and Biologics and Office of Regulatory Affairs, Food and Drug Administration, Sterile Drug Products Produced by Aseptic Processing Draft, Concept paper (Not for Implementation), September 27, 2002.
5. EU GMP Annex 1: Manufacture of Sterile Medicinal Products.
6. *Official Methods of Analysis* (1998) 16th Ed., 5th Revision, 1998, AOAC INTERNATIONAL, Gaithersburg, MD.
7. PDA Technical Report #70 (TR70), Parenteral Drug Association, Bethesda, MD, 2015.
8. PDA Technical Report #13 (TR13), Parenteral Drug Association, Bethesda, MD, 2015.
9. European Committee for Standardization, Chemical Disinfectants and Antisepics—Quanitative non-porous surface test for the evaluation of bactericidal and/or fungicidal activity of chemical disinfectants used in food, industrial, domestic and institutional areas—Test methods and requirements without mechanical action (phase 2/step2), EN 13697 (2001) E, ICS 11.080.20; 71.100.35 Management center: ue de Stassart 36, B-1050 Brussels
10. DECON-AHOL®, Veltek Associates, Inc., Untites States Patent Office, No. 6123900, 6333006, 6607695. www.sterile.com/patents
11. DECON-AHOL® is a Registered Trademark of Veltek Associates, Inc.
12. RODAC® is a Registered Trademark of Becton Dickinson Corporation.

26 Contamination Control for Incoming Components to Classified Areas

"War at the Door®"

Arthur Vellutato, Jr.

CONTENTS

26.1 INTRODUCTION

The art of controlling the ingress of the particulate and microbial contamination that may be conveyed through components, items, equipment, and personnel and enter a good manufacturing practice (GMP) facility is a monumental task. It requires organizations to develop not only appropriate cleaning, wiping, disinfection, sterilization, and transfer procedures, but also to develop a cultural understanding of where, when, and how contamination may enter. The difficult part of assessing these factors is that the where, when, and how is constantly changing based on a variety of factors. Some of these factors include varying operational requirements in varying classifications in the facility, personnel in varying classifications, varying levels of cleanliness and microbial flora on incoming components, and potential seasonal variations. It is a constantly moving target that needs routine reevaluation. To further complicate the task, the particulates and microbials that we are concerned about cannot be seen, only imagined. In a sense, we are protecting against an invisible enemy and effectively designing our plan based on past test data. Each day we may find different contaminants that are uncovered through our routine environmental monitoring program. However, the problem with particulate excursions is the origins are difficult to trace and microbial excursions and the associated results are only determined many days later. In short, this is an extremely difficult task to say the least.

In quest of a perfect system, many firms assume that exterior contamination should be destroyed by spraying a sporicide onto the surface. However, we hear very few conversations about the cleaning of particulates until they show up in our nonviable environmental monitoring (particle counts). Particulates and fibers become an enormous issue when we find them in the final product such as in vials, syringes, or liquids. And with fibers, there is no available environmental monitoring test specifically for their detection. The GMP world sometimes focuses too little on the particulate/fiber side of contamination and too much on the microbial side. This can prove detrimental. It seems we sometimes believe that it is acceptable to have sterile dirt. Many times, firms end up in total frustration when things suddenly go amuck when

DOI: 10.1201/9781003163138-26

excursions are related to particulate or fibers. If the excursion is microbial in nature, more available testing resources are available to find the root cause.

For microbial control, one cannot haphazardly spray every surface with a sporicide and claim victory. There is much more to the subject matter than imposing a strong corrective action procedure for the lack of control. There are many factors that need to be assessed beyond what chemical agent destroys a potential incoming microorganism. These factors include but are not limited to the chemical agents used, expected bioburden of the item, particulate levels of the item, procedure used, surface coverage, dry period (contact time), end user safety issues (mostly relating to toxicity, inhalation, dermal), end user comfort pertaining to vapors, packaging of items entering, surface types being disinfected, effectiveness of the disinfection procedure, potential sterilization cycles (autoclave, irradiation, ETO, X-ray, NO_2 or equivalent), grade or classification of final transfer, impact of cleaning/disinfection on final product, release time (VHP chambers), assurance for component/equipment to migrate to unwanted areas and the validation and routine documentations for such activities. Simply, assessing the multitude of critical criteria listed previously is essential to the successful implementation of procedures to introduce items to controlled areas.

For particulate and fiber control, the task is considerably harder, as we cannot test surfaces for their presence as easily as we can for microbials. Most testing is tendered via airborne particle counts. So many times, we are left blindly fighting an invisible enemy as particulate evaluation of surfaces in many aspects proceeds without a proven scientific testing methodology. And so the industry, without available science, is forced to accept what they assume is clean. Realization of these facts make selection and prior approval of control practices critical in relationship to particulate and fiber levels of items deemed suitable for clean room operations.

26.2 REGULATORY EXPECTATIONS AND INDUSTRY GUIDANCE

On the scale of factors that can disrupt and contaminate GMP classified areas, the entry of incoming items, equipment, components, and personnel with contamination ranks as one of the highest potential contributory scenarios challenging the GMP professional. The classified clean room without these items and personnel presence should remain stable over time in environmental data for particulates and microbials. As soon as materials and personnel enter the classified area, environmental conditions can degrade quickly and dangerously if appropriate precautions are not in place. However, regulatory guidance worldwide generally skips right past this subject. There are minimal guidance and/or requirements presented in governmental published writing to address these silent enemies. At the same time, despite the unwritten expectations of most regulatory agencies, the need for a GMP firm to address such contamination is often foremost in an inspector's mind. In the FDA "Sterile Drug Products Produced by Aseptic Processing—Current Good Manufacturing Practice" and EU "Annex 1", the notation is made in varying instances that the overall system must be in control and caution should be vested in control of incoming contamination. If you are a young professional trying to find guidance in either of these documents, you would find an almost total lack of specifics. When an inspector enters a facility, the review of items and personnel entering the classified areas is often a main attraction of the tour. Many of us have experienced this personally as an inspector watches for extended times incoming items and personnel to see what controls we have in place. Unfortunately, most governmental documents follow the same path in describing the goal but not specific requirements or expectations. In fact, if we wanted to create a slide deck of regulatory writings on this subject, our PowerPoint presentation would be extremely short. Noteworthy in their content are industry guidelines such as the Parenteral Drug Association's (PDA's) Technical Report #70 and #13. In PDA TR #70, pages of guidance discuss the subject matter in great length. However, actual regulatory expectations are not specifically noted to any degree that would help a GMP professional to implement a compliant plan.

26.3 CONSIDERATIONS IN DEVELOPMENT OF THE PLAN

Developing a plan for the control of contamination for incoming components requires a firm to understand that they are under siege each day. And that the army attacking could change or add more troops of a different flora. Each classification grade (D to C to B to A) is under siege from the lesser grade. Throughout the organization, all need to understand that the answer to this siege is accepting a "War at the Door®" offensive position and culture. The concept of War at the Door requires us to know as much as we can about our enemies, make plans to control their ingress, and constantly reevaluate our defensive plans to ensure that we are accounting for changes in the offensive strategy that is continually being waged upon our controlled environments.

Some in the industry assume that the contamination we fight comes from the exterior environment alone. Under such a belief, the War at the Door is only concerned about contamination from the exterior to the interior, as our controlled environments are always clean and can only be corrupted by external factors. This would be an incorrect assumption. Years ago, Veltek Associates, Inc. (VAI) conducted airborne microbial testing at the multiple locations within a facility to discern the location and levels of contamination. Testing areas were outside of the classified areas and the Controlled Not Classified (CNC) areas where the firm did not normally monitor. The goal of the testing was to determine what is coming from the outside to the inside in areas we did not routinely monitor. Airborne testing was conducted as airborne microorganisms were believed to be the main contaminating source to incoming components. It was not expected to be surfaces the operators would touch. The following are the results of the testing. In review of testing in Figure 26.1, the results may not be depictive of every facility. In fact, levels are expected to vary based on one's location in the world and specific facility characteristics. It is advisable to generate, in some form, this sort of data so that baseline exterior to interior controlled environment microbial levels can be ascertained.

In this example and through this testing, one can see that the exterior environment did not show the highest levels of contamination. Although contamination control is a must from the exterior to the interior, it is also critical that such control be vested internally from uncontrolled to CNC and then progressively from each classified area to the next higher classification. In the course of this testing, we learned that airborne molds are higher outside compared with bacteria even though counts are lower than areas inside. The molds can trickle in and find an acceptable environment to survive. This is often a darker (no UV light), cool, porous location with enough moisture that only gets dramatically reduced when located within a classified area. Bacteria is more prevalent within the facility, which would be expected as this represents the conjugation of personnel. It is this contamination, seen in every uncontrolled area, that has opportunity to insert itself onto components, carts, equipment and personnel that enter the classified areas.

Although identifying the level of bacteria and molds can serve as a basic analysis, speciation is key to completely evaluating what flora will be present and how we can establish a continued plan to stop its ingress into the classified areas. However, speciation of collections from the unclassified and outside areas is an expensive proposition. Knowing that it was a bacteria or a mold is sufficient from an outside and unclassified area perspective. It is not perfect and if we could speciate everything, we would have a complete analysis. Knowing whether the contaminant is a mold or bacteria will help us develop the control plan and speciation can be performed when EM is collected in the CNC and classified areas. At that point, one needs to know the microorganism as best they can as the plan that is to follow will address specific contamination expectations. As one cannot just collect samples from the unclassified areas once a year and call it a research project, it is suggested to conduct this testing four (4) times in a year for a year to attain a baseline. This will address potential seasonal variations. Although the example in Figure 26.1 shows that most of the contamination is inside, we can make plans to eliminate the exterior contamination and potentially reduce the interior contamination from the knowledge we gained. Seasonal variations are very real, and the astute GMP operation knows that this may be one source of contamination to the facility and to incoming items and personnel. In Southern Pennsylvania, USA, are located the largest mushroom-producing factories in the world. When they open the underground mushroom doors to conduct the harvest in April, spores flood the surrounding area. It is an occurrence that happens each year without failure. GMP operations need to prepare for this. In March–April, in the Southeast and Midwest USA, farming and the plowing of soil increases bacillus levels. And in Puerto Rico, in August and September, seasonal rains cause an increase in molds. Seasonal variations are noteworthy, and a good GMP operation knows their surrounding environment and plans appropriately for their periodic occurrence. Without the nonclassified and external data, one is blind to this issue and to what degree it occurs, where it occurs most and therefore does not plan effectively to prevent such contamination from potentially entering their operations.

As one creates the nonclassified research data and eventually the internal classified speciated isolate list, one becomes more knowledgeable of what the enemy looks like that is threatening their classified areas. The identification process in the CNC and through the classified areas would require minimally the use of the industry prominent DNA Sequencing of the 16sRNA (bacteria) and ITS2 rRNA (mold). The internal isolate list for an aseptic operation is defined as an expectation

Location	Test Volume (Cubic Feet)	Flow Rate (CFM)	Total Results in CFU's	Bacteria	Molds	Ranking in Total CFU's
Outside Air Personnel Entry Door #1	40	5	11	2	9	12
Outside Air Personnel Entry Door #2	40	5	14	4	10	10
Outside Air Loading Dock #1	40	5	16	3	13	9
Outside Air Loading Dock #4	40	5	18	3	15	8
Inside Warehouse Location #1	40	5	21	7	14	6
Inside Warehouse Location #2	40	5	19	8	11	7
Men's Locker Rooms	40	**5**	**27**	**21**	**6**	**5**
Women's Locker Rooms	40	**5**	**33**	**24**	**9**	**4**
Men's Restrooms	40	**5**	**41**	**33**	**8**	**3**
Women's Restrooms	40	**5**	**47**	**39**	**8**	**1**
Cafeteria	40	**5**	**42**	**31**	**11**	**2**
Office Location #1	40	5	12	8	4	11
Office Location #2	40	5	10	11	1	13
Hallway Location #1 to CNC	40	5	3	2	1	15
Hallway Location #2 to CNC	40	5	4	1	3	14

FIGURE 26.1 Unclassified airborne microbial air sampling test.

in the FDA "Sterile Drug Products Produced by Aseptic Processing—Current Good Manufacturing Practice" and EU "Annex 1". As an industry standard and an expectation, facility isolates should be used to validate disinfecting agents used in the facility. By RODAC or swab testing items entering the CNC to Grade A area, we have, in fact, provided a road map of what contamination we may see and then devised an appropriate methodology for destroying such bioburden. Too many times organizations utilize facility isolates recovered from facility surface substrates or from airborne testing alone. The surfaces of components and personnel entering often represent more of a danger to the classified area than EM recovered surface or air samples within the classified areas. The example shown in Figure 26.1 shows that where employees enter the facility, the mold levels are higher. Those entrances and exits will allow mold to enter the facility at a higher and more frequent level. This is important data and a warning that we would not have known had we not done some monitoring in the unclassified areas. At VAI, this prompted us to disinfect floors four times per day instead of only once in those areas. After doing so, mold counts on our floors were dramatically reduced. It also prompted us to create secondary airlocks to control the airborne molds. By doing so, the counts were virtually eliminated in a few days. Through the unclassified testing and addressing the situation properly, we reduced one source of future difficulties relating to high mold counts.

Evaluation of the existent flora outside the facility and within the uncontrolled areas of the facility is only one data point that warrants concern. These data points represent the level of contamination that our external environment and our operation can bestow on items entering the facility. The second set of important data that warrants concern is what flora are others sending us on the items we purchase. These items are commonly packed in a box or a drum that has a bioburden. They may be bagged or unbagged. The exterior bag or exterior of the unbagged item has its own bioburden. Simply, such item's external surfaces and internal contents may contain flora not realized in our local environment or internally in our operations. And so, in a sense, we have transported bioburden from another location into our facility. The products may be labeled sterile or non-sterile or not labeled at all, which would designate non-sterile. They may be labeled clean or cleaned or not be cleaned prior to delivery. All of these scenarios require either cleaning, disinfecting, and/or sterilization. As an example, a bouffant hat made in Mexico can come into our facility. Everyone will put it on and the flora from the non-sterile bouffant hat will disperse into our environment. As another example, an airborne microbial air sampler used in our facility will require the exterior be disinfected before use in a classified area. But because the inside of the sampler cannot be decontaminated, when turned on, the air flow from outside to inside and then to outside again releases whatever bioburden may exist inside the unit to our external environment. And as a final example, a product with the claim on the double bag package as sterile requires decontamination on the outside of the exterior bag prior to being transferred into the facility's classified areas. However, this may be done based on the product claim and not a review of the validation data from the vendor providing such product as sterile. One important note on receiving product labeled as sterile. A non-FDA registered manufacturer or a non-FDA registered product does not have to conduct any testing or validation to place the word clean and sterile on the label or literature. In fact, in most regions of the world, this is commonplace. What we are doing may be transporting contamination from a product that is not really clean or sterile into our facility. Simply put, we must carefully scrutinize each item that enters our operations.

It is critical that the Validation and the Quality Assurance (QA) departments investigate the data proving the product as clean and/or sterile and approve the product for use in the facility prior to use. Sterility is one issue that is well understood, but clean is completely different. What is a clean wiper, or paper, or cart, or pen, or mop, or goggle, or garment, or piece of equipment? The answer for this is virtually unknown to those working in GMP operations, and at times we allow what may be termed "Sterile Dirt" to enter and unfortunately plague our operations. If care is not taken with respect to incoming items, excursions can occur and end up in an investigation with the undesired root cause conclusion as, "We should have reviewed that data or set incoming acceptance criteria for the items before using the item."

Another important factor in the design of a well-made plan is that a compliant, designed, classified clean room should operate AT REST consistent with the expectations suggested in guidance documents such as the FDA "Sterile Drug Products Produced by Aseptic Processing—Current Good Manufacturing Practice", EU "Annex 1", ISO 14644, ISO 14698, and others. The design of any compliant clean room should assure classification compliance AT REST (or Static conditions). The clean room design should assure compliance and prepare for an unstated level of bioload that is removed or reduced through the system's air filtration system. This would be termed IN OPERATION (or Dynamic conditions). However, the nightmare of any well-designed clean room is the introduction of a bioload that stresses the system or that cannot be gathered effectively enough into the return vents of the HVAC system and then taken to the filters or is disbursed in high levels to areas where the air return is not as effective. This leads to the retention of particulates and microbes within the clean room that during operations may compromise the products. For many years, laminar airflow smoke studies, conducted by the GMP firm, would delineate airflow across the Grade A filling machine and associated component ingress and exit areas. This may show disturbances in the laminar air flow. However, these studies are not routinely conducted in Grades D, C, or potentially B. Thus, actual airflow in these areas is unknown. This may allow for the recirculation of air and thereby the retention of contaminants in pockets within these areas. A further goal of a GMP firm is to minimize the contaminated pockets by reducing the incoming bioburden from items that may enter the environment.

To address this concern, a GMP firm should invest time in developing assurance that items that enter the classified area are of the lowest particulate and microbial levels possible.

The classified clean room operator's nightmare is the potential overload of particulates and microbes from personnel, components, supplies, and equipment that enters the areas. It is important to consider all the issues prior in developing a plan. Without doing so, we are just haphazardly attempting to perform a function that is critical to ensuring acceptable environmental conditions in our classified areas.

26.4 DISCERNING WHAT BIOBURDEN EXISTS ON ITEMS ENTERING CLASSIFIED AREAS

Once a firm understands the flora that exists in their exterior environment and the uncontrolled areas, it is important to develop a comprehensive list of the potential bioburden that will be introduced from each and every item that will enter classified areas. This includes Grade A–D (ISO 5–ISO 8) areas. RODAC or swab testing for bioburden of all items, which may include bags, parts, assemblies, and other consumable products, should be performed as an initial assessment of not less than 10 items. If the item is a piece of equipment or used in limited numbers, the item should then be RODAC or swab tested for bioburden, but the numerical value may vary. Routine monitoring of these items should be employed to detect changes in bioburden over time. This would include Identification of microorganisms as previously discussed. The accumulation of data provides confidence in the existing procedures both internally and to meet regulatory expectations. If an item is stable in its bioburden level and flora, and there are no excursions that would relate to the item pending, then present procedures employed are satisfactory. If an item's bioburden is trending upward during routine testing, more attention should be paid to the disinfection procedures being employed. One needs to be able to pull up a trending graph showing the item's bioburden and assess if arrangements of incoming sterilization or disinfection are acceptable.

It is important to identify where each item will finally reside. The basis for a final decision on the required cleaning and disinfection procedure will depend on the process and the product. There is not a synonymous methodology for all items entering a type of grade. In short, one process and product may require a certain procedure for Grade C introduction, but a different process and product may require a more stringent procedure for the same Grade C classification. The basis for decisions can be ascertained by asking the following questions for each item:

1. What classification (Grade A, B, C, D, CNC) will the item be used in?
2. What region(s) of the world is the item manufactured, co-manufactured or packaged in?
3. In what classification grade (if any) is the item manufactured or packaged?
4. Is the final product produced a sterile product?
5. If the product is not a final product, is bioburden a concern to the product?
6. What is the expected bioburden of the item?
7. What classification is the final destination of the item?
8. Is the final area of destination using an open or closed process?
9. Is the item packaged in multiple bags? How many?
10. Is the incoming item supplied to the GMP firm sterile?
11. If not, can the item be autoclaved (steam), gamma irradiated, or ETO sterilized?
12. If the item can be sterilized, should it be exposed to sterilization processes.
13. Can the product withstand Vaporized Hydrogen Peroxide (VHP)?
14. If the product cannot be sterilized by one of the methods listed in #11 previously, can it be chemically disinfected without compatibility issues?
15. Is the product/equipment capable of being disinfected on 100% of all surfaces?
16. If the product/equipment cannot be 100% disinfected on all surfaces, what percentage can be disinfected?
17. What cleanliness is required for the product?
18. Does the item require cleaning prior to use and to what level?
19. Can the item be wetted?
20. Can the item be exposed to heat?
21. Is there a problematic situation with using a particular disinfectant that would adversely affect the final product?
22. Is there a final filtration of the product?

The preceding questions provide a framework for personnel to qualify where the item is coming from, how it is packaged, what may need to be done with it upon arrival, what processing will be required before use, what grade will be its use point, and whether the item will become part of a sterile or non-sterile product. Upon assessing the preceding, one can then decide, based on the criticality to the process and product, what will need to be done. Following are potential issues for each Grade/ISO classification for the item's final destination and some of the processes that may occur in these areas.

<u>Grade D (ISO 8)</u>—If an item's path leads it to a final destination of Grade D (ISO 8) then the cleaning and disinfection of such item may be, but is not guaranteed to be, less stringent. Most items in Grade D (ISO 8) are non-sterile except for certain liquids (excipients/ingredients). Personnel are normally dressed in facility uniforms, lab coats, shoe covers, beard covers and bouffant hats with the exception of aseptic filling operations whereby full coverall, hood, boots, masks, goggles, and gloves will be donned. Operations like blending, bioreactors, tank formulation, COP, vial martialing, glass preparation, component assembly and many other processes and assembly operations. Most all of the operations that occur in these areas surround preparation of something for use in a higher classification.

Grade C (ISO 7)—If an item's path leads it to a final destination of Grade C (ISO 7) then the cleaning and disinfection of such item can go in many directions. In older days, Grade C was divided into subcategories to delineate the different processes that would occur in each type of C (ISO 7) area. Grade C (ISO 7) areas can have no adjacent Grade A (ISO 5) or B (ISO 6) areas, or they may have an adjacent B (ISO 6) area or they may have an adjacent B (ISO 6) and subsequent A (ISO 5) area. Items in Grade C (ISO 7) treatments vary dependent upon whether it is part of an aseptic filling process. Personnel are normally dressed in full coverall, hood, boots, masks, goggles, and gloves if this area is adjacent to a Grade B (ISO 7) and subsequent Grade A (ISO 5) area. In some operations, the adjacent B (ISO 6) area with no adjacent A areas provides some reduction in requirements for gowning; however, if product may be exposed, the normality is full gowning. In some operations there are no adjacent grade B (ISO 6) or A (ISO 5) areas and no open product so gowning may be reduced to fit such operations where personnel are less intrusive. Potential activities vary widely in these areas and are too numerous to name and be comprehensive. It represents the largest range of processes in GMP areas. Besides the operation that considers Grade C (ISO 7) the final processing room for their product, the main function of the Grade C (ISO 7) is either manufacture or assembling a product or a part of the product that gravitates to the Grade B (ISO 6) area and subsequently to the Grade A (ISO 5) area.

Grade B (ISO 6)—If an item's path leads it to a final destination of Grade B (ISO 6) then the cleaning and disinfection of such item is treated as a Grade A (ISO 5) area as the Grade B (ISO 6) area is adjacent or the background operation to the Grade A (ISO 5). Personnel are dressed in facility uniforms as part of an aseptic filling operation whereby full coverall, hood, boots, masks, goggles, and two sets of gloves will be donned. Any disinfection of components into this area during manufacturing operations would be limited to filtered and sterile 70% USP isopropyl alcohol or 70% USP ethanol as vapors, over splash and residual during manufacturing could prove a contaminate to the final product. All operations in a Grade B (ISO 6) support the filling operations that would occur in the Grade A (ISO 5) area. The filling done in Grade A (ISO 5) and supported by Grade B (ISO 6) can be a multitude of processes that include vaccines in vials or syringes, pharmaceutical injectable, sterile liquids, bulk final product destined for vial or syringe in another operation, sterile compounds, sterile powders, and many others.

Grade A (ISO 5)—If an item's path leads it to a final destination of Grade A (ISO 5), then the cleaning and disinfection of such item is treated as the most critical area having potential to adversely affect final fill operations. Personnel are dressed in facility uniforms as part of an aseptic filling operation whereby full coverall, hood, boots, masks, goggles, and two sets of gloves will be donned. Any disinfection of components into this area during manufacturing operations would be limited to filtered and sterile 70% USP isopropyl alcohol or 70% USP ethanol as vapors, over splash and residual during manufacturing could prove a contaminate to final product. The filling done in Grade A (ISO 5) can also be a multitude of processes that include vaccines in vials or syringes, pharmaceutical injectables, sterile liquids, bulk final product destined for vial or syringe in another operation, sterile compounds, sterile powders, and many others.

26.5 INCOMING INTRODUCTION COMPONENT CLEANING AND DISINFECTION METHODOLOGIES

A multitude of methodologies exist for sterilization or disinfection of incoming component items that can be employed in GMP operations today. Very few methodologies are as generic as cleaning of incoming component items. Aside from wiping with a dry or saturated clean room wiper, sonication, and Clean Out of Place (COP), there are very few methodologies that can be offered as a cleaning methodology. Such cleaning methods are normally specific to the product and discussed in some detail in the options following. Consumable items or components that are used normally have related and specific ASTM or EN published standards or directives. With respect to the cleaning of product contact or near product contact, there are few guidance documents that define exactly what is required. In such guidances or federal regulations, the "how to" is not present and only the "it must be" exists. This means the guidance document will tell you it needs to be clean and sterile and validated, but it normally does not tell you how to accomplish the task. The FDA "Sterile Drug Products Produced by Aseptic Processing—Current Good Manufacturing Practice" or EU "Annex 1" are prefect examples of this situation. One may ask why. The reason is a government law or guidance document cannot support one method over another if the method may not be available to all. In short, favoritism is frowned upon. If a firm can attain the same final requirements by doing something in another way, that is acceptable to regulatory agencies. Overall, some methodologies are superior to others for both disinfecting and cleaning. Many operations may not be able to accommodate every method presented following. However, it is important to recognize the options and what to strive for in present or future operations.

26.5.1 Sterilization via Steam (Autoclave), Gamma Radiation, Ethylene Oxide (ETO), X-Ray, and Nitrogen Dioxide (NO_2)[1]

For aseptic manufacturing operations requiring the sterilization of items that may come in contact with product or be located close to exposed sterile product, sterilization via steam (autoclave) or gamma radiation is the best methodology. For steam sterilization, the items must be packed in a non-linting, non-shedding, low extractable material/bag or be processed through an autoclave that opens directly into the aseptic Grade B (ISO 6) classified area. For gamma radiation, items need to be packaged in minimally two outer bags and subjected and validated to a sufficient dose that is based on bioburden, density, and load configuration and is confirmed with dose mapping.

Other methodologies that also provide efficient and proven sterilization include X-ray and NO_2. X-ray is an effective sterilization source for high-density, large-volume polymer products but has limitations because of oxidation effects. Nitrogen Dioxide (NO_2) is a fast sterilization method that utilizes low temperatures and has no cytotoxic residuals. However, process operations for X-ray and NO_2 are minimal worldwide and in their infancy with respect to commercialization. Ethylene Oxide (ETO) has been around for decades. ETO is utilized widely in hospital operations and medical devices and may not be an acceptable sterilization source for pharmaceutical and biotechnology operations as the residual from the gas may be problematic to final product and personnel. ETO requires longer sterilization and aeration after sterilization time periods. And lastly in this category would be Electronic Beam (E-Beam). E-Beam is effective for low-density items but is not effective when the item density is too great, making it less desirable for the entire scope of items that may need to be processed by a firm.

These sterilization methods do not clean surfaces. That would require a separate and prior cleaning operation that would be specific to each item. From a microbiological perspective, if the destruction of spores has been proven in a validated cycle, we can be assured that the microbial content is nonexistent. It is critical to perform validation studies to confirm the effectiveness of each methodology.

26.5.2 Vaporized Hydrogen Peroxide Pass-throughs

VHP pass-throughs and chambers have existed for some time but are more prevalent in newer facilities normally because of space and venting constraints in older facilities. The concept is effective for large items entering controlled environments and less effective with a multitude of small items or packages. The design of the pass-through or chamber is simple in its' mechanics but more complicated in implementation. Simply, a sealed pass-through/chamber is created in a predetermined space whereby sealed locking entry and exit doors are established with sealed walls, ceilings, and floors. These units can be walk-in or smaller and resemble a simple manual pass-through. The entrance door is opened, and items are placed into the chamber or pass-through. Both doors are then locked and sealed. The items placed inside need to be placed in a configuration whereby 360°, or all surfaces, can easily have exposure to the VHP gas. The chamber, upon completion of the cycle, will require aeration through an air scrubber and then emitted externally per local regulations. As an example, if a stainless-steel can is placed on the floor of the large chamber, the VHP cannot contact the bottom area of the can. Likewise, with smaller pass-throughs, packaged or smaller items cannot be piled on top of one another as exposure to all surfaces by the VHP is hindered. In this case, items need to be placed with space between them or hung by clips to assure unobstructed exposure to the VHP. The materials inside are fully exposed to the VHP. This is more than what is normally done in large scale VHP procedures as the load is most probably much higher. VHP is introduced so that an equalized distribution is accomplished throughout the chamber or pass-through. Exposure and chamber aeration times will vary with each type of item to be sterilized. The limitations of VHP are time to expose to, assuring surface exposure, post-exposure aeration dwell time, and safety for personnel. Overall, this is one of the better methods and can be equivalent to autoclaving if done appropriately.

25.5.3 ABCD Component Introduction System

The ABCD Component Introduction System is a trademarked system by Veltek Associates, Inc. (VAI). The system is simple to implement and can dramatically reduce bioburden for incoming gamma-irradiated items. The system provides four (4) barrier bags for each item. For example, VAI's patented DECON-AHOL WFI Sterile is supplied in 11-ounce aerosol cans. Each can is packaged in a double bag and the case quantity of 24 cans are packaged in a double bulk barrier bag. Items that are now quadruple bagged are gamma irradiated with a validated dose and lot sterility tested. Upon reaching the end user, the fiberboard box is discarded in a nonclassified area. The outer bag containing the 24 cans is sprayed down prior to entry into the CNC area or Grade D area (dependent on facility design). This is necessary as the exterior of the outermost bag was exposed to the fiberboard carton and air in the noncontrolled area and should be treated as potentially having surface spores. The exterior bag should be sprayed with a sporicidal agent such as 0.52% sodium hypochlorite, peracetic acid/hydrogen peroxide, or 6%+ hydrogen peroxide and allowed to air dry for a validated contact time (normally 3–5 minutes). Once in the Grade D area, with the exterior bag disinfected, the Grade D operator sprays their hands with sterile 70% isopropyl alcohol and removes the outermost bag. The inner bag of 24 cans can be taken to the transfer area to the Grade C area. Upon arrival, the carrier of the 24 bagged cans sprays their hands with sterile 70% isopropyl alcohol and opens the inner bag. The Grade C person, in the Grade C pass-through, grasps the individually double-bagged sterile bagged cans and takes them into the Grade B transfer area. No spray down should be necessary as no contact other than with the sterile interior of the 24 can bag has occurred. At this point, the Grade C person sprays their hands with 70% sterile isopropyl alcohol and opens the third bag (one of the double bags on the can) and the Grade B person grasps the inner sterile bag in the pass-through and moves the can(s) to the Grade A pass-through or martialing location for introduction into the Grade A area. The Grade B person then sprays their hands with sterile 70% isopropyl alcohol and opens the final (fourth) bag and the Grade A person sprays their hands with sterile 70% isopropyl alcohol and grasps the sterile can. This bagging and transfer system, if done correctly, eliminates the need for spray down operations between classifications. The system is an aseptic handling exercise that eliminates many of the limitations prevalent in normal component spray down procedures as defined in the following paragraphs. This system can also be performed with autoclaved materials provided that the initial packaging provides the required number of bags per each classification. This system will not work for items that cannot be sterilized either by gamma radiation, steam, X-ray, ETO, or other similarly approved methods.

25.5.4 Incoming Component Spray Down Method

Many firms utilize a spray down methodology between different classifications. Many invoke procedures that simply spray items with a sporicide from one classification to another. Others invoke a multistep (most times a triple step) decontamination procedure where the item is first sprayed with a disinfecting agent like a phenol, or quaternary ammonium, or low active percentage bleach, hydrogen peroxide or peracetic acid. This is done to reduce bioburden. The second step is normally a validated sporicide, which should create the absence of life. The third step is normally some type of 70% isopropyl alcohol wipe down using either dry wipe and bottled isopropyl alcohol or saturated 70% isopropyl alcohol wipes on the item to remove the chemical residue from the first two steps.

However, it is the "if done correctly" in this procedure where problematic situations arise. Many times, operators do what could be termed the "Pharmaceutical Spritz". This means the operator haphazardly sprays the item or bag with some uncontrolled mist and really does not assure 100% wetting of all surfaces for a validated contact (dry) time of 3–5 minutes. They also do not assure that nooks and crannies or twist tied bag areas are saturated. The third step of removing residual from the two previous chemical agents is minimally effective if a dry wipe and 70% isopropyl alcohol is used and virtually ineffective if a saturated 70% isopropyl alcohol wipe is used as it just spreads around the residual. What is accomplished by a poorly executed triple step decontamination scenario is potentially not assuring each surface has been dosed with the appropriate agent, a potential lack of contact (dry) time, two residues on the surface with the inability to remove them, and the creation of a potential high vapor environment for employee exposure.

Many times, this methodology may end up with contamination detection that is noticed in environmental monitoring or traced to procedures during a product sterility failure. If this occurs, corrective actions become difficult as we have already used the strongest chemical we can in the procedure. And so, the question then that is often asked is "What do we use now?" Many then search the world for a stronger sporicidal agent, when it is most likely the decontamination procedure that is actually flawed. If they find one, they add more overkill and more problems in implementation relating to vapors, safety, and residues. Simply put, many ignore the prior suggestions for planning details and considerations presented earlier in this chapter by just saying, "Spray it with a sporicide before it enters". Many do not conduct testing, do not know the flora in nonclassified areas, nor do they know the bioburden level or the type of bioburden of what they are trying to disinfect. Most probably, at the end of the investigation, many may not be able to say with confidence that it was the sporicide that was the problem, and that is why we changed to a new one.

Imagine if you were having a party with twenty 8-year-old children and during your party you decided to throw a basket of candy into the air in the middle of the children as a fun event. The first error you made was the incorrect assumption that each child will get the same number of pieces of candy as you flung them into the air. It sounded like a good idea, but in the end, many calculations will be off. Some children are taller, some faster, and some slower, the candy projected in a trajectory that was not a uniform pattern, the girls in one group and the boys in another, some of the boys push the girls, some of the girls pushed the boys and some children were just not paying attention. In the end, the repercussion of the poorly made plan most probably showed a lack of success or victory. In fact, later in the week you might even hear from some parents that "We heard you had a candy throw, but my child did not get any candy and they were upset".

Spray down procedures and controlling contamination from entering a facility is much like the children and candy example prior. One does not just haphazardly spray a surface and claim success. One must consider a multitude of criteria, as defined in this chapter, assure that the plan addresses such criteria, implement the plan, and reevaluate the plan's success routinely.

In addressing the appropriate methodology for conducting a truly effective spray down procedure, we must add three issues. The first defines the appropriate chemical agents for incoming item disinfection. The second relates to vapors, operator comfort, operator health, and operator safety. And the third is the effect of agents other than 70% isopropyl alcohol or 70% ethanol in use while product is being manufactured.

A triple step decontamination procedure does not require a disinfectant, followed by a sporicide and an isopropyl alcohol wipe down in that order. Sporicides create the absence of life if allowed to reside on the surface in a wetted form for the prescribed or validated time period. So prior use of the disinfectant, which includes agents that are capable of destroying vegetative cells but not spores, would be without scientific merit. Examples would be phenolic, quaternary ammonium, or low percentage bleach, hydrogen peroxide or peracetic acid. We do not have to reduce the bioburden as the expected bioburden on incoming clean room items is lower than the population tested to assure regulatory claims as a sporicide. The item would have to exceed 1,000,000 CFU's per ½" × 1" carrier (per A.O.A.C. Sporicidal or Sterilant test protocol as done on ceramic pennicylinders and sutures). That is millions of CFUs per item based on surface area calculations. Items from outside the facility most probably have an inherent spore bioload, but not to the level required for registration or validated in internal AET conducted by the GMP firm. A sporicide is typically all one needs, but it is not the first step. The first step, and the rule for any disinfection process, is to pre-clean a surface before applying a disinfectant so as to remove soil, particulates, fibers, residues, microbes, and other elements that may soak up the active ingredients of the disinfecting agents or prevent their reaching the surface and make them less effective. So first one must clean the surfaces. This may be done with a dry wipe and a cleaner that incorporates a surfactant (like DECON-Clean, Simple Green, or equivalent) and a dry wipe. Cleaning with a dry wipe and a surfactant-based cleaner allows the chemical to loosen contaminates and residuals and put them into solution. The dry wipe can then soak them up and remove them from the surface. Cleaning is an extraction process. A saturated wipe, like the dry wipe, is made of a material that is either woven or nonwoven. When it becomes saturated, it can only spread the chemical agent. It cannot soak up any more liquid as it already is saturated. So, in this venue, this is a poor cleaning choice. Once the item is cleaned, it is prepared for disinfection.

The second step would be to accurately and systematically apply the sporicide. If applied via a wipe, the contact (dry) time is dramatically reduced and so is efficacy. The sporicide needs to be sprayed equally onto every surface of the item without failure and afforded a contact (dry) time period per AET that most probably is from 3–5 minutes. That is difficult to accomplish in a uniform manner, every day, through all our shifts, and with varying operators. Every time we miss a spot, we may potentially let in contamination. Equally as difficult is applying the sporicide to a multitude of items and placing them on a surface to air dry. Stacking them atop of one another leads to items taking longer to dry, which is good for efficacy but bad for handling of wetted packages. Training, supervision, and routine RODAC testing is critical to ensure this function is being executed appropriately.

The last step of a three-step decontamination procedure is the removal of the sporicidal residue. The reason for this is so the sporicide residual is not transferred to a critical site or product contact location. As stated before, the wipe down procedure minimally reduced the residual unless the wipe down is done with the utmost efficiency. This three-step decontamination is normally done when items are introduced from an unclassified area to the Grade D area or potentially from the Grade D area to the Grade C area. So, what is the answer to this seemingly unsolvable problem? If the item cannot be terminally sterilized via one of the methods discussed, or the use of a VHP chamber/pass-through cannot be accomplished and the utilization of the ABCD Component Introduction System is not applicable, then the answer is to review what you really have to accomplish from a scientific basis. If you know the bioburden of the item, and the bioburden of the area where it was or will be exposed before disinfection, and you know the specific isolates most prevalent on the item, and you have conducted AET studies, then you can decide what strength sporicide to use or what type of sporicidal product is an alternative. When the word sporicide is said, most jump to products on the market that are registered sporicides. That normally is Peracetic Acid, H_2O_2, 0.52% bleach, Glutaraldehyde, or similar agents. These agents are often too strong and would be an overkill for the task at hand. At the same time, one needs to realize that 70% isopropyl alcohol, 70% denatured ethyl alcohol (ethanol), and hydrogen peroxide leave virtually no residual unless one is concerned with direct application on product contact surfaces. At that juncture, we should review the options further which would require COP, sterilization, and/rinsing procedures. But for the incoming components, that will be handled by an operator who has sanitized their gloves with 70% isopropyl alcohol, the residual is nominal. Of the three options prior, ≥6% Hydrogen Peroxide is the only sporicide. In most laboratories, alcohols are used to preserve spores so they would only be used in a bioburden scenario where only vegetative cells were present. However, the problem is that people use not scientifically derived criteria in their AET studies. Some want a 3-log reduction. They also want an inoculate on the coupon (1/2" × 1" or per USP <1072> 2" × 2") that is minimally 1.0×10^5. Why? There is no scientific rational for this. No one would ever experience that high a level of spores on an incoming item. As an example, a 24" × 36" cart would have over 3,000 square inches in surface space. When you calculate bioburden understanding that every 1/2" × 1" at an inoculate of 1.0×10^5, you realize we are saying that the cart has the potential to have an inherent bioburden of greater than 450,000,000 CFUs. This would never be the case. As the Chair of the PDA Contamination Control Task Force writing PDA Technical Report #70, I suggested correction of this situation and received the unanimous consensus of the group. We addressed this misnomer and suggested guidance to state the value for reduction during AET studies at >l log with an inoculate of 1.0×10^4 and lower if scientifically justified. When we invoke a lower inoculate and the required reduction, we are preparing for potentially 30,000,000 CFUs. This is still extremely high, but more realistic as we do not expect the cart to have such high bioburden on all of its surface. Although this justification is dramatic as we will never see this bioburden, it helps to understand the reality of imagining either microbial population level on a 1/2" × 1-inch space on any item. That means on each 1/2 inch × 1 inch surface space we have either 100,000 CFUs (10^5) or 10,000 CFU's (10^4), respectively, in the examples prior. In the end, when we lower the value of the inoculate, a product like ≥ 6% hydrogen peroxide, or equivalent, can be readily validated. If we require the ≥ 6% hydrogen peroxide to be equal or better than a peracetic acid/H_2O_2 mixture or a 0.52% bleach, then we have unrealistic expectations that were not developed by scientific means. The ≥6% hydrogen peroxide leaves no residue and can be used as a spray down agent on incoming items. This approach eliminates the post wipe down after disinfection that was not effective.

The second area of concern as defined previously are vapors, operator comfort, operator health, and operator safety. Transfer areas are normally small environments and as we approach the Grade A area, they become even smaller. The buildup of vapors, especially from peracetic acid/hydrogen peroxide mixtures, presents a difficult work environment and potential severe health risks. Health risks extend further than the vapors but also into dermal chemical burns, eye contact, throat irritation, and nosebleeds. It also creates a slipping scenario whereby liquid from disinfection builds up on the floor.

The last area of concern is that during manufacturing only 70% isopropyl alcohol or 70% ethanol should be in use close to Grade A or in Grade A. Why? Potential residuals, vapors, and the toxicity from agents other than these alcohols can potentially adversely affect final product. Transfer area sprays after an initial sporicidal application in Grade D (or C in some operations) should only use 70% isopropyl alcohol or 70% ethanol. This should be sufficient if there is appropriate contamination control beforehand as our main concern in Grades B/A would be microorganisms shed by personnel. If we find bacterial/mold spores or microorganisms uncommon to humans, our preceding decontamination/sterilizations systems simply let them in. At that juncture, we need to return back to our control system to address the problem.

26.6 STATE-OF-THE-ART TECHNOLOGIES

When particulate and microbial contamination occur, it is our duty to investigate the source and find the root cause. Anyone who has tried to do this has most probably wished with all their heart that cameras that could see particulate and microbials and identify them as such were installed so that we could see where and when the contamination occurred. But alas, this is not the future. We live in a time period where technology increases daily; however, at present our investigative tools are minimal. Many times, and as the saying goes, "we are trying to find a needle in a haystack". Basically, the review of environmental monitoring tells us something is wrong and is our blueprint to solving the contamination breach root cause dilemma. Our environmental monitoring affords us results on surface, air, and potentially but generally minimally incoming component testing. Although firms have been required to dramatically increase testing since the 1980s, the testing we routinely conduct is minimally effective in helping us discern a root cause.

Presently, some systems can aid in helping us to do such a task. They do not answer the question but provide us more knowledge. The more knowledge we have the better. In one book chapter previous written by this author, the cost of the average investigation exceeded $45,000 when all involved testing and corrective actions were complete. Some investigations cost millions of dollars, especially if the product was released to the market and a contamination was noticed after therapy to humans or animals. So how can we improve our systems to help us in investigations?

Rapid Microbial Methods (RMM) is one technology that has afforded us closer to real-time information. Although the industry and regulatory have focused more on the use of RMM for release of product to the market related to sterility testing, there is an enormous advantage rarely spoken of for their use in contamination control. RMM has attempted and had success in shortening the delay from collection of a microorganism to the time of notification. This period was anywhere from 3–7 days in the past. This time frame reduction is enormous in assisting the contamination control program. Before RMM technology, EM would be conducted on say a Monday. Cleaning and disinfection would have continued per schedule and SOPs through to the next Monday when results would have been rendered for the past Monday. During the week of processing, cleaning, and disinfection, a variety of agents may have been used or not used based on predetermined schedules. If the results showed high spore levels, a week would have passed if a sporicide was not used in the time period. Obviously, a week without a corrective action is not desirable, but it is the best science allowed as time for growth of EM plates lags behind operational cleaning and disinfection.

Although RMM have shortened the time frame, we still are only afforded the knowledge that it was a bacteria or a mold, at best. Some systems only let us know viable life in the number of microorganisms detected. For anyone who has been part of an investigation into particulate, microbial, or pyrogen contamination, they know the grueling and lengthy process that most times does not end with an exact root cause that is without question. To the best of our ability, we attempt to trace the origin of particles, microbes, and pyrogens and how to stop their ingress, but this is an extremely difficult task.

Another and more powerful contamination control tool is Radio-Frequency Identification (RFID). Basically, RFID uses electromagnetic fields to automatically identify and track tags attached to objects. Most people know and use such technology more in their personal lives than in an industrial setting. Some common consumer RFID uses are with cell phones, luggage tags, and security access cards. Prior to the development of the Core2Scan system by VAI and contractual partners, RFID in the pharmaceutical and biotechnology world was basically limited to asset tracking. At the same time, a coated RFID tag, capable of being autoclaved, disinfected, and irradiated was unavailable. This made it difficult to implement RFID tags into classified aseptic processing areas as they could not be sterilized or disinfected as assured failure would occur.

The development of the coated RFID tag that could be autoclaved, gamma irradiated and disinfected solved the clean room concern for their entry to and use in classified areas. The innovations extended beyond tags and included accompanying software and middleware so one could now place presterilized tags on equipment, personnel, samplers, and incoming components and know not only where they are, but when items came together, and where they had been. This is critically important, especially to incoming items and components. Rarely are incoming items and components monitored in any great degree. Why? Monitoring would require swabbing and/or RODAC testing that would require growth media such as Trypticase Soy Agar. Exposing near product items or components to sampling interventions or media would disqualify the item for use. As such, minimal monitoring is done. However, with an RFID tag attached to the item or component or tray of items/components, we now can trace all the locations the item/component has been. If we also incorporate an RFID tag on personnel and equipment, we now have what is called Contact Tracing. Contract Tracing is something all in the world have become familiar with during the COVID pandemic. Mediocre success has been achieved worldwide with Contact Tracing with respect to COVID. In pharmaceutical and biotechnology operations, the task is much easier and can provide data for evaluation. The problem that exists is, are we smart enough to use it appropriately? As a simple example of tracing, suppose Operator's #1, #2, #3, and #4 all have an RFID tag on their wrist or in a necklace or incorporated into their gown. Operator #1 works on the vial filling. A package of saturated alcohol wipes is bought into the facility from Grade D to Operator #1 in the Grade A space. Operator #2 is responsible for external disinfection and transferring the alcohol wipes from the uncontrolled space to Grade D. Operator #3 is responsible for disinfection and transfer of the wipes from Grade D to Grade C. And Operator #2 is responsible for transfer of the alcohol wipers from Grade B to Operator #1 in Grade A. If there was a contamination noticed on RODAC testing of the external packages of the wipes, in any area where the wipes may have been exposed or one of the operators RODAC tested out of the area with high counts, the RFID system would show each and every location the four operators were throughout their shift. It would know if the right wipe was used. It would know the path they took past pieces of equipment with RFID tags on them. The RFID system also knows that Operator #1 was using the alcohol wipes on Equipment #1 including lot and expiration. The now available data is immense and invaluable to investigations. It can stop the ingress of unwanted items or personnel, including alarm capabilities, and can narrow the investigation path more scientifically.

In the future, state-of-the-art technologies will evolve and help to continuously improve our ability to introduce and trace items and components and reduce incoming contamination. However, innovation does not need to be a new design or a miraculous invention. Sometimes, changes in present practice can be a state-of-the-art method. One example of such change in our present practice is the bagging of items and components after autoclave sterilization or cleaning. Everyone has walked through controlled areas and seen racks of autoclaved items or components in autoclave paper or autoclave bags. The items/components were originally autoclaved in a classified area. They then could be transferred on carts and/or racks through a lesser grade area and the exterior sprayed down with a chemical agent required to introduce the item/component into a higher classification. Similar practice is done for larger (50 L–18 L) bulk glass or single-use bottles or single-use bags, respectively. This is seen in many vaccine operations that make a bulk product and then either move the bulk to another building or even another site. If immediately after sterilization or filtration into the desired bulk container, we were to package it in clean, sterile single or multiple bags for transfer of the items to the next operation, bioburden and particulate would most likely be much less. We would then open the bag, spray the inside bag or bottle, and probably would not need to use a sporicide as the item was well protected throughout. By simple easy changes, potential incoming contamination is dramatically reduced. The cost to gamma irradiate a case of bags is approximately twenty to thirty dollars plus transport. The time to bag the item is minimal and both costs are minimal compared with the cost of spray downs.

26.7 INCOMING COMPONENT INTRODUCTION METHODOLOGIES

There are a plethora of items, components, and equipment that make their way from unclassified areas to classified areas. One rule in contamination control is that such items be consumable items, not equipment, tools, replacement parts, or transport mechanisms. Equipment and parts should be area specific or be always located in classified areas. In that venue, we can then make categories of how these items should be introduced. Following is an analysis of the types of categories and the methods for transporting them to classified areas.

26.7.1 Transfer of Carts, Tanks, and Wheeled Vessels

As a precedent, GMP firms should, if possible, make area specific all items in this category. When a wheeled item enters from an unclassified area, there are many concerns that are difficult to address. First, the item must be cleaned from top to bottom prior to disinfection and in that order as we would clean that type of equipment. The tops of items are easier to clean as they likely are cleaner than the lower sections and wheels. Items coming from unclassified areas are more likely to have levels of molds and bacterial spores. This requires the item to be cleaned and have a sporicide applied to the surfaces, followed by a dry wipe/70% isopropyl alcohol wipe down to remove the sporicidal residual. These items may also be cleaned and then autoclaved or placed in a VHP chamber, if applicable. In cleaning wheeled vessels, one must take care for safety to assure personnel are not injured during cleaning or disinfection. To clean the wheels, the detergent is applied to the surface. Some type of abrasive non-abrasive friction-like clean action is then required to lift soil from the surface, followed by rinsing. At this juncture, the wheeled vessel needs to be rolled forward out of the soiled liquid on the floor to a clean dry floor surface that has been previously cleaned and disinfected. Once this is done and the wheel is completely dry, the sporicide can be applied. The selection of a sporicide is going to depend on the wheel type and chemical compatibility. After application and drying any, residual needs to be wiped free from the wheel or corrosion will occur over time. Another option for carts and smaller tank systems is an option that allows the top of the cart to be transferred off the wheel base and allows the top portion to move to a new, area-specific, cleaned base (Ex. Patented Cart2Core System by Veltek Associates, Inc.) This eliminates trying to clean wheels in place and allows the inverting of the bottom base and wheels for cleaning and disinfecting.

26.7.2 Tools

Tools are an item that is often stored and used in an unclassified area and then sprayed with only 70% isopropyl alcohol prior to entry because of fear of corrosion or deterioration. Obviously, this is an unacceptable practice. Tools need to be area specific and cleaned and sterilized/disinfected routinely. Loose tools should not be stored in the area. They will never be cleaned, disinfected, nor sterilized. Older fill machines have storage drawers and unsterilized or disinfected tools cannot be placed in such locations without cleaning and reprocessing. The best option for tools that can be autoclaved is to create or purchase autoclavable tool packs. A multitude of packs are purchased, bagged, and steam sterilized. When a mechanic needs a tool, the kit is opened, the tools used out of the kit, and then the entire tool kit is cleaned and re-sterilized.

26.7.3 Unbagged Items from Outside the Classified Areas

Unbagged items from outside classified areas may have *Bacillus* and mold spores from the fiberboard carton and the unclassified air and obviously need to be cleaned, a sporicide applied and then, if required based on residual, wiped with a dry wipe and 70% isopropyl alcohol. Upon completion of this operation, these items can be placed in a Grade D area. They should be immediately bagged accordingly to the classification where they will finally be used. After removal of the bag in Grade A, B, or C, only a sterile 70% isopropyl alcohol spray down is required. If the item is unbagged, then it is exposed to contamination for the entire time it resides in a classified area. Depending on the item's final destination, a potential cleaning and disinfection/sporicidal application to the item may be required. This can become difficult if the material is exposed and there is potential for vapors and overspray; careful analysis should be performed.

26.7.4 Bagged Items from Outside the Classified Areas

Bagged items from outside classified areas may have *Bacillus* and mold spores from cartons and unclassified air. The outer bag needs to be sprayed with a sporicide and allowed to air dry for the validated time period, which is normally 3–5 minutes. If the item has multiple bags, and such bagging allows the final bag to be removed in its final destination (as discussed earlier in this chapter), it does not need to be sprayed. The outer bag only needs to be removed. If the bag is a low-density polyethylene, it then can be wiped with a dry wipe and 70% isopropyl alcohol. If the bag is nylon or polyethylene on one side and incorporates a breather strip normally made of Tyvek, then it should not be wiped as wiping may cause the material to shed airborne fibers. The key with bagged items is to have enough bags to remove one in each classification. Per the procedure described previously, spray down should not be needed.

26.7.5 Documentation Systems: Papers/Forms/Tags/Labels

One of the most difficult items to introduce into classified areas is documentation. Documentation includes paper, forms, tags, and labels. The concern is two part: (1) shedding of the material from cellulose-based products, and (2) how to disinfect these paper items. It is important to define the industry standards and state-of-the-art methodologies that upgrade industry methodologies daily. The main problem with papers is that they are normally made of cellulose. Cellulose-based documentation or papers are an insoluble substance, made of polysaccharide consisting of chains of glucose monomers. In short, cellulose paper is made from small fibers that are pressed together with a slight binder and can be released very easily from abrasion or tearing. As they are released, they can go airborne and can ravish clean room operations if the correct product is not used. In this venue, selection is key prior to implementing such items to the clean room.

Electronic documentation is the future to eliminate much of the use of paper. However, it will never remove all uses of the item. An electronic batch record can be implemented, but the use of tags, labels, and notebooks will seemingly always be in our scope. The use of electronic records has been implemented by many, but this is not feasible for every operation and carries a very high price tag as data integrity, including CFR 21 Part 11 validation is a must do. RFID can bridge some of the gaps but has limitations that will still require some papers and labeling. Barcode, QR code, and any printed label are routinely used in some clean room operations. Forms and logbooks also hold stake in many operations. The first key to success in this venue is choosing the right product to use. The days of going to a paper supplier and buying reams of paper that are 500 sheets and packaged in paper sheathings are long past. But many firms still do this. They also take the paper, print it in an office setting, and then attempt to disinfect or autoclave it for entry into classified areas. These are all outdated technologies.

First, clean room operations should be using paper or paper substrates that are designed for clean rooms. Alternate substrates like Teslin (PPG) or CleanPrint 10 (VAI) are available. These alternatives are made of plastic and have no fibers in their construction. Thus, they do not shed cellulose fibers and are a much cleaner choice. Coated cellulose clean room paper and/or alternative substrates are available from many vendors. This includes Berkshire, DuPont, Kimberly Clark, and many more. This paper uses cellulose but adds binders to coat the paper, which reduces the shedding characteristics. So, making the correct choice is the first step to successfully implementing a safe documentation system for classified areas. The second critical item to consider is "where will one print on the paper?". The office or uncontrolled environment can only add viable and nonviable contaminants. Furthermore, what ink will be used? Toner in laser printers, or photocopier toner, or ink jet are bad choices as these inks will be released into the environment. For this venue, firms need to consider printing within the classified areas using process ink printers. They also need to consider clean room printing systems that seal the contamination from printing into a sealed HEPA-filtered cabinet and use a process printer such as the patented Core2Print (VAI). If clean room printing systems are not employed, how then will one clean and disinfect a printer located in or near the classified areas? This is a difficult task that most probably consider cannot be done. With either a clean room printing system or a process printer, the best that can be done is to gas sterilize the equipment into the classified areas and then develop or purchase a containment system. The combination of the appropriate paper, print mechanism, appropriate ink, and location of printing all add up to ensure a secure and cleaner product for use in classified areas.

The third critical item for documentation systems is "how will it be disinfected or autoclaved into the classified area?" This comes with two different scenarios. The first is a firm has a preprinted document or document system that is complete and will not change. And the second, primarily with batch records, is that the document was printed and a page or several pages need to be added during manufacturing and taken to the area. The first is considerably simpler than the second. In the first scenario, documentation items can be preprinted, bagged, exposed to a validated dose of gamma radiation or potentially X-ray. This can be done by the firm or performed externally. ETO, VHP, E-Beam, or any gas sterilization will not work as they cannot penetrate the density constraint of papers piled on top of each other or are incompatible with cellulose. The second scenario requires pages to be sent to the classified areas on an immediate basis. There is no time for bagging and processing through the preceding methods. The documentation will have to undergo immediate decontamination. This can be done by printing to the clean room printing system; by disinfection; or by autoclave. Disinfecting documentation is difficult as one has to go page by page and not use a product that contains alcohol of any type or the ink will run. The problem is not disinfecting one page. The problem is disinfecting many pages as they must be laid out or hung, in a sense, like a clothesline to dry. Autoclaving sometimes works, but the paper may come apart or wrinkle. Neither is desirable. This remains a near unresolvable problem at many firms.

26.7.6 Equipment with Electronics

Equipment with electronics and closed cabinetry is difficult to clean and disinfect as the inside of the unit cannot be adequately cleaned or disinfected. If the item has a fan (such as a particle counter or microbial air sampler, then the inside bioburden will be emitted to the environment once in use. The best procedure for items like this is to have them bagged and ETO sterilized into the area if the equipment will tolerate it. From that time on, the equipment should be area specific unless the expense does not warrant multiple units to be purchased. In this case and on a routine schedule the unit should be re-sterilized for each entry. The easiest and most compliant way to eliminate the re-sterilization is to perform validation studies to prove bioburden is not emitted from the unit.

26.8 CONCLUSION

The transfer of items and components from the exterior environment challenges firms each and every day with the elimination of what can be constantly changing bioburden. The firm that effectively evaluates what is in the exterior environment, what is in the classified areas, and what bioburden the item entering has may possess the critical data to develop a compliant plan. Without the up-front data and tools to investigate excursion situations, firms are blindly guessing as to what should be done to compliantly introduce items to classified areas. It is a simple practice that does not allow errors. The clean room system will provide compliant environmental conditions unless we allow the lack of control from items and personnel entering to corrupt it. Once corruption occurs, the price could be magnitudes higher than the cost of doing what needed to be done to provide a safe system for the product and the patient in the first place. Simply, firms need to understand that the "War at the Door®" relates not only to the external environment but also to each classification from the lower classification.

NOTE

1. Details on the validation of these and other sterilization methods are provided elsewhere in this handbook.

REFERENCES

1. Block, SS (ed.), *Disinfection, Sterilization and Preservation, Lea and Febiger*, 4th ed., Philadelphia, 1991
2. McDonnell G, Russel D. Antiseptics and disinfectants: Activity, action, and resistance. *Clinical Microbiological Reviews*, Jan 1999.
3. Vellutato AL, Jr., Assessing resistance and appropriate acceptance criteria of biocidal agents. In Madsen R, Moldenhauer J, eds. *Contamination Control in Healthare Products Manufacturing*, DHI, LLC, Chapter 9, 319–352, 2014.
4. Vellutato AL, Jr., Implementing a cleaning and disinfection program in pharmaceutical and biotechnology clean room environments. In Moldenhauer J, ed. DHI, LLC, Chapter 8, 179–230, 2003.
5. Center for Drugs and Biologics and Office of Regulatory Affairs, Food and Drug Administration, Sterile Drug Products Produced by Aseptic Processing Draft, Concept paper (Not for Implementation), September 27, 2002.
6. EU GMP Annex 1: Manufacture of Sterile Medicinal Products.
7. *Official Methods of Analysis* (1998) 16th Ed., 5th Revision, 1998, AOAC INTERNATIONAL, Gaithersburg, MD.
8. PDA Technical Report #70 (TR70), Parenteral Drug Association, Bethesda, MD, 2015.
9. PDA Technical Report #13 (TR13), Parenteral Drug Association, Bethesda, MD, 2015.
10. European Committee for Standardization, Chemical Disinfectants and Antisepics—Quanitative non-porous surface test for the evaluation of bactericidal and/or fungicidal activity of chemical disinfectants used in food, industrial, domestic and institutional areas—Test methods and requirements without mechanical action (phase 2/step2), EN 13697 (2001) E, ICS 11.080.20; 71.100.35 Management center: ue de Stassart 36, B-1050 Brussels.

11. DECON-AHOL®, Veltek Associates, Inc., Untites States Patent Office, No. 6123900, 6333006, 6607695. For complete patent detail see www.sterile.com/patents
12. Cart2Core®, Veltek Associates, Inc., Patents—US 1051879, 10518793,9994244; US Design D809730, D871708; Australia 2015369797; Canada 2965162; Europe Design 003533314–0001, 0002, 0003; Japan 6538851; Korea 10–1996740, 10–2070006; New Zealand 732617; Singapore 1120170315335, 11201811030V. Applications—US 16679948, 15637059, 29/630655, 29/569050, 29/718431; Australia 2017280954, 2018292136; Canada 3027512, 3028485, 3067845; China 201580069038.7, 201780037312.1, 202010902516.0, 201880044122.7; Europe 15874251.0, 18824388.5, 17815974.5; HK 18105297.9, 9128830.7; India 201717014496, 201817045836, 201917047848; Japan 2018–566218, 2019–106526, 2019–571745; Korea 10–2019–7001028, 10–2020–7000518; New Zealand 749502, 748864, 759381; Singapore 10202005805R, 11201910990U. www.sterile.com/patents
13. Core2Print®, Veltek Associates, Inc., Patents—US 10525750, 9566811, 9643439, Design D893602, D822102; Australia 2014321364; China 107683212B, ZL201730077431.2, ZL201480055669.9; Europe 3046771 (Austria, Belgium, Denmark, Finland, France, Germany, Greece, Ireland, Italy, Norway, Spain, Sweden, Switzerland, Netherlands, Turkey, Great Britain); Europe Design DM096039; HK 1225348; India Design 292744, 292688; Japan 6671455, 6424227; Singapore 11201710411, 11201602120Y Applications—US 16/709643; Australia 2016294408; Canada 2984143, Canada 2924904; China 201910218222.3; Europe 16825089.2, 19172950.8; HK 18112236.9, 42020005122.5; India 201717046687, 201617011992; Korea 10–2017–7032009, 10–2016–7010136; New Zealand 739021, 718756; Singapore 10201810107R. www.sterile.com/patents
14. PRINTED PLANAR RADIO FREQUENCY IDENTIFICATION ELEMENTS, Vanguard Identification Systems, Inc., Patent—US 7,909,955 B2.
15. Rodac® is a Registered Trademark of Becton Dickinson Corporation.
16. DECON-AHOL® is a Registered Trademark of Veltek Associates, Inc.
17. Cart2Core® is a Registered Trademark of Veltek Associates, Inc.
18. Core2Print® is a Registered Trademark of Veltek Associates, Inc.
19. War at the Door® is a registered trademark of Veltek Associates, Inc.

27 Aseptic Processing of Sterile Dosage Forms

James Agalloco and James Akers

CONTENTS

DOI: 10.1201/9781003163138-27

27.1 INTRODUCTION

There are essential drugs, biologics, and devices medicinal products that are produced in a manner to ensure that they are free of viable microorganisms, a condition commonly referred to as "sterile". Some products are sufficiently resistant to heat or ionizing radiation to permit the use of a validated sterilization process in their final product container. There are other products for which the use of a lethal process on the final dosage form is not possible. In those cases, the formulation and container-closure components are sterilized independently, filled, and sealed to create the finished product container. The processing steps subsequent to the sterilization of the various materials and prior to closure of the final container must be carried out in a manner that minimizes the likelihood of product contamination (viable and nonviable). This activity is termed aseptic processing and entails the proper execution of numerous design elements and operational practices to ensure success.[1] The recommended approach for sterile product manufacturing applies Quality by Design concepts to the preparation of sterile products in a manner aptly termed "Sterility by Design" (Figure 27.1).[2]

The importance of the control elements varies with the core process—aseptic or terminal. Terminally sterilized products rely on the lethal sterilization process with the other elements providing means for control of bioburden and particle levels. In aseptic processing, the importance of process control elements depends on facility, equipment, and process specifics, and there is no single solution that applies universally. This chapter includes content on those supportive elements.

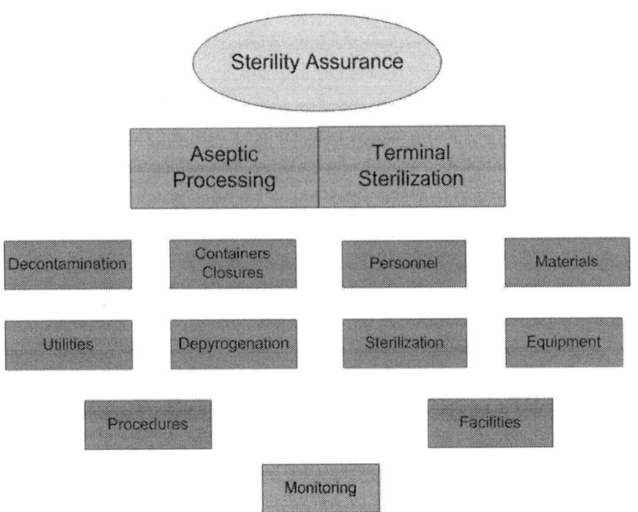

FIGURE 27.1 Sterility by design elements.

The inherent risks associated with the use of aseptic processing are widely recognized; regulators and practitioners across the globe acknowledge that aseptic processing should only be used where lethal treatments are impractical. A lethal process directed at final containers is in most cases preferable to one that relies on the uncertain exclusion of microorganism as a final product is assembled from a number of individually sterilized materials. However, they may be drug delivery systems, for example, which cannot be sterilized but are considered so advantageous in use that may be made aseptically, although the product contained in them may be heat and or radiation stable.

As aseptic processing ordinarily requires activities be performed by personnel, there is significant concern regarding risk from human borne microbial contamination. The regulatory preference for terminal sterilization has been formalized by both the United States Food and Drug Administration (FDA) and the European Medicines Evaluation Agency (EMEA).[3,4,5] Realistically, although the intent is that terminal sterilization will be utilized wherever possible, there is a plurality of products for which aseptic processing is currently the only possible means for preparation, i.e., sterile bulk antibiotics, freeze-dried formulations, most (but not all) biological products, and materials without sufficient moisture content (a minimum of approximately 5%). Aseptic processing is the only means available to produce "sterile" products for products that can't be sterilized in their final container.

Users and regulators must recognize that aseptic processing cannot be fully evaluated in-process against defined specifications to demonstrate "sterility". Commonly required and frequent evaluations such as microbial or particle monitoring cannot definitively establish the suitability of an aseptic process. This fundamental limitation is central in understanding both the meaning and value in the "validation" of aseptic processing.

Because there is no means to directly establish through process or final product analysis the achievement of sterility in any aseptic process, it cannot be "validated" in quantifiable probabilistic terms as an autoclave can. The literature may be ripe with references to "validation of aseptic processing"; however, all that has ever been demonstrated by a satisfactory media fill is that the potential for successful operation exists.

A media fill, as has long been recognized, is a snapshot in time. Over the course of decades, regulatory policy has resulted in media fills that are considerably more risk intensive, or in the common parlance more "worst-case" than are routine production operations. However, a media fill test with zero contaminated units cannot establish the sterility of a production lot made the day before (or the day after). Aseptic processing in clean rooms and Restricted Access Barrier Systems (RABS) is an uncertain method; subtle variations in operator practice can profoundly affect the result, and yet there are no ready means to detect or eliminate those variations. There are, however, means to reduce variability through sterility by design, but too often these principles are not effectively applied to existing aseptic systems and processes.

Nevertheless, firms endeavor to "validate" aseptic processing on a regular basis. That "validation" (or "revalidation") of the aseptic process is expected but often represents a misunderstanding of the information that a media fill (also called process simulation) or any of the data obtained from the process can provide.

27.2 SCOPE

This chapter reviews the supportive elements of sterile production using aseptic processing for finished pharmaceuticals. Other chapters in this volume address the similar (and interrelated) subjects of aseptic processing for sterile bulk pharmaceutical chemicals and manual aseptic processing. The focus will be on the unique considerations associated with aseptic processing for finished pharmaceuticals including process design and process simulation. That this chapter is located within a comprehensive text on validation is fortuitous, as the details of many of the supportive activities are described in close proximity. The reader should consult the more extensive treatment provided on these subjects, rather than the brief synopsis of each that appears following, the intent of which is to place these subjects in proper context. That context must include the rational application of analysis and measurement. Aseptic processing cannot provide the engineering and analytical surety possible where process conditions can be evaluated parametrically in real time or analyzed with statistically assured accuracy and precision in their final product.

27.3 BUILDINGS AND FACILITIES

The aseptic production of sterile products is performed in classified environments supplied with H13 or H14 medical grade high-efficiency particulate air (HEPA) filters. The classification of the environments used for aseptic processing varies with the criticality of the specific activity being performed. The aseptic manipulation of sterile materials should be conducted in ISO 5 environments that have been treated to essentially eliminate the background microbial count. Less critical activities such as washing and component preassembly are carried out in ISO 7 environments where microbial control expectations are more relaxed.

27.3.1 Critical Area

The most important activities are carried out under ISO 5 environments in which environmental control measures have been taken to effectively minimize the risk from microorganisms. This expectation encompasses every location where sterilized and/or depyrogenated components are exposed to the environment. In manned clean rooms, risk is minimized by disinfection of the room and equipment surfaces with antimicrobial agents. Aseptic processing isolators are decontaminated for the same reason but to different practical effect. Depending upon the specific design of the equipment/facility, unidirectional air may be provided within the critical area (or zone). Expectations for unidirectional airflow were established years ago when manned clean rooms were predominant. The transition to isolator-based filling has resulted in increasing questioning of the need for unidirectional air. The very first aseptic isolators successfully employed turbulent air systems, and although many newer designs incorporate unidirectional air, there is no definitive proof that it is necessary at all.

Microbial control measure in manned clean rooms and most RABS are accomplished by disinfection of room and equipment surfaces with antimicrobial agents. Aseptic processing isolators are decontaminated with sporicidal agents to eliminate contamination. These activities are thought to be identical in their outcome, but there are two crucial differences:

1. The isolator and some closed RABS environments are decontaminated using sporicidal vapor delivered automatically in a reproducible manner, whereas most clean rooms are manually decontaminated.
2. Personnel are restricted to the external environment by isolators and closed RABS which minimizes or effectively eliminates recontamination of the environment by personnel. In manned clean rooms, the environment is constantly re-contaminated by microorganisms released by the gowned human operators present.

Depending upon the specific design of the equipment and facility, unidirectional air may be provided within the critical area (or zone). Expectations for unidirectional airflow were established years ago when manned clean rooms were predominant. The transition to isolator-based aseptic processing left this unchanged; however, there is no evidence to support the use of unidirectional air. Certainly, decontamination is important, but it plays a secondary role to the proper control of personnel-released contamination, which experience has taught is the most prominent cause of microbial contamination.

The primary objective for advanced aseptic processing technologies is to reduce or eliminate the impact of personnel within the critical zone to the maximum extent possible.[6] The presence of personnel within or adjacent to the critical zone is recognized as the primary source of microbial contamination. Aseptic processing technologies have always endeavored to reduce the impact of personnel through a variety of means:

- Separation of Personnel—Separative approaches dominated industry practice for many years. Aseptic gowning and partial barrier systems were introduced in the 1950s when addition of HEPA filters resulted in the first clean rooms for aseptic processing. The most evolved separative technology is the aseptic processing isolator.

- Limiting Personnel Interaction—This entails technologies where machine automation and/or robots perform many of the activities customarily executed by gowned personnel. The first examples were Blow-Fill-Seal and Form-Fill-Seal for aseptic filling. Other examples can be found in lyophilizer loading/unloading and other conventionally manual tasks.
- Removal of Personnel—In these designs, the aseptic process is conducted without direct personnel exposure. The first of these to become widely available was the closed isolator for sterility testing. Newer examples include gloveless isolators and closed vial filling,
- Combinations of the previous—Many of the newer aseptic processing technologies being introduced incorporate multiple elements from the preceding categories.

27.3.2 Closed Systems

Closed systems offer a new approach to the execution of aseptic processing. Complete separation of personnel from the operating environment in either a single-use disposable system or fully closed environments provides a level of protection to sterile materials unattainable by other means.[7,8,9,10] Environmental monitoring in closed systems has no operational benefits and increases the risk of contamination by potential disruption of the closed system integrity. The role of the background or surrounding environment is markedly reduced, and its classification is not typically beneficial.

27.3.3 Supportive Clean Area

In manned clean rooms, the area immediately surrounding the critical zone is nearly as important. The personnel required for operation of the equipment are always present in this environment, and as they must occasionally access the critical zone, preventing contamination from them is essential. This extends to corridors and gown rooms used for access, and these are ordinarily classified as ISO 5 as well. Restricted access barrier systems require the same environmental background as manned clean rooms. The background environment surrounding an isolator is of substantially less importance because of the more certain separation between the critical zone and the environment where the personnel are located. Current regulatory expectations are for ISO 8 externally, but there is little evidence that external classification of any type is beneficial for aseptic processing.

27.3.3.1 Environments for Component Preparation

Activities prior to the sterilization/depyrogenation of materials are conducted in a variety of classifications. Once washed, items are protected by ISO 5 air until they are either wrapped or enter a sterilization/depyrogenation process. The preparations area proper is commonly defined as ISO 8. The background environment for compounding is almost universally ISO 7, with localized ISO 5 if necessary.

27.4 PERSONNEL

Personnel performance must be the focus of attention in aseptic processing. The operator is often required to perform precision activities (e.g., set-up of filling equipment from individual component parts) without introducing microorganisms to any of the product contact parts. That this can be accomplished on a consistent basis is a tribute to the skill of these employees as the gowned human operator is universally acknowledged to be ***the primary source*** of microbial contamination. The routine accomplishment of such actions without shedding bacteria is accomplished through careful attention to the precepts of aseptic technique. Aseptic technique in the pharmaceutical manufacturing environment is a loose assembly of practices originally conceived for microbiological laboratory manipulations; however, the principles are fully adaptable to aseptic processing of pharmaceuticals. Among the typical aseptic techniques intended to protect sterile materials from contamination by personnel are:

- Every surface of the gowned person is considered non-sterile
- Never touch a sterile object with a non-sterile object
- Never place a gloved hand over an open sterile container
- Use a tool for every activity wherever possible

Additional principles can be found in various microbiology texts and must be recognized as suggestions rather than hard rules, and of course must be adapted to the specific circumstances of the equipment and materials being handled.

The operators who work in aseptic processing should be cognizant of the basic principles of microbiology, sterilization, disinfection, aseptic technique, gowning practices, as well as the details of their assigned tasks. Training of personnel should include both classroom sessions and practical exercises in which their ability to perform the required aseptic procedures or other similar tasks can be evaluated. Training should be near continuous and the media fill (process simulation) is the ultimate evaluation of the operators' proficiency. As with almost any activity in our industry, the aseptic training program should be well documented.

Personnel should be monitored upon exit from the aseptic core, as sampling during the process itself risks residual media on gown or glove surfaces that can unnecessarily increase the risk in aseptic processing because it could enhance the survival of microorganisms shed by the sampled individual. Sampling at the end of the process addresses the potentially weakened integrity of the gowning system after a lengthy and perhaps rigorous work period within the aseptic core. Routine monitoring on exit typically focuses on the operator's gloves and forearms, as these are often closest to the sterile materials. Gowning certification is a usual prerequisite for entry into the aseptic core and ordinarily entails sampling of many more surfaces.[11] Monitoring requirements for personnel in isolators is generally restricted to isolator glove impressions taken at the conclusion of the process, as gown surfaces are not present.

27.5 MATERIALS, COMPONENTS, AND CONTAINER/CLOSURES

The sterilization and/or depyrogenation of components and materials used in the aseptic process can be performed using any of the many methods available: steam, dry heat, radiation, gas, or filtration. With respect to the aseptic process, it is essential that regardless of the sterilization method selected, it is validated to attain a minimum Probability of a Non-Sterile Unit (PNSU) of 1×10^{-6}. This PNSU determination is properly based upon the population and resistance of the bioburden, not the biological indicator.[12] An important consideration is the selection of a package configuration that allows for adequate sterilization yet affords adequate protection of the sterilized materials until ready for use in the process. The maximum time interval between sterilization and production use of the materials (other than the process solution itself) is established in conjunction with the execution of the process simulation by using materials that have been held for the maximum specified time period.

27.6 TIME LIMITS

Time limits are imposed to minimize the duration of the aseptic process in order to reduce the risk of microbial contamination. Time is an important consideration for a variety of reasons:

- Operator fatigue can result in poorer adherence to required technique.
- Extended and/or high work activity use of gowning materials can lead to their failure/compromise.
- Microbial proliferation in pre-filtration solution can occur during hold periods resulting in either filtration issues and/or pyrogen concerns.

Process duration related to operator fatigue is easily factored into the aseptic process simulation (see the following section), whereas that associated with microbial growth in the formulation is product specific and must be addressed in the validation of the formulation. There is no evidence to support an increase in microbial population within an aseptic environment provided care is taken to avoid the presence of nutrients (such as media residues) and moisture. From a risk perspective, the potential for microbial proliferation over time in these extremely clean locations should be considered minimal provided proper precautions are taken. Because aseptic processing HVAC systems are specifically designed to remove contamination shed from personnel and other sources, this is not surprising.[13]

27.7 ASEPTIC PROCESS SIMULATION

A widespread misunderstanding persists that validating an aseptic process allows the estimation of a "sterility assurance level". Actually, a media fill test only allows the estimation of a contamination rate. In cases in which human employees must conduct some portion of the aseptic process, variability in personnel performance limits the utility of the assessment. Process simulations establish that the methods and practices are capable of success; they cannot support that materials produced using identical methods at a different time and place achieve the same contamination control performance.

The oft cited sterility assurance level (SAL) of 1×10^{-3} for aseptic processing is nothing of the sort; it is merely the established maximum projected contamination rate associated with the successful filling of 3,000 media units.[14] Any SAL claim for aseptic processing based upon media filling is without basis; all that is known is the contamination rate of the units filled that day.

The only value that may be ascribed to a media fill is to establish that under the specific circumstances of an individual simulation that the facility is suitable for use in aseptic processing. The inability to consistently achieve the expected result—zero growth in any of the filled containers—is an indication that the process is not as capable as it should be.[15] It must be recognized that a process simulations represent "worst-case" challenges of the process in that the increase in interventional frequency associated with media fills and absence of process inherent characteristics (i.e., bacterial inhibition by the formulation) should increase the potential for microbial contamination in the media filled containers relative to a production batch. Industry surveys established that fewer than 5% of all media fills evidence some contamination.[16] Provided that the level of contamination remains at low levels within the expected acceptance criteria, it should not be considered problematic. The identification of a single positive in a media fill should rarely trigger a full-blown investigation into the source of the contamination as the incidence rate is within expectations and the assignment of a true assignable cause is generally impossible. The recent PDA survey reported that as many as 25% of respondents accept only media fills that are completely without microbial contamination.[17] Even perfect media fills do not allow for claims of absolute sterility, as no microbiological test can demonstrate the attainment of sterility. In cases in which no contamination is observed, the proper statement of the result should be "no growth", rather than "sterility". Because the media fill process simulation is a periodic capability assessment or "snapshot" of performance in finite terms, it is not an absolute proof of anything.

27.7.1 Study Rationale and Design

Before embarking on an initial media fill program (and periodic evaluation thereafter), the firm should prepare a study rationale outlining how its program supports the overall capabilities of its aseptic operations. For a single product facility, this can be quite easy to prepare as the permutations of lot size, fill volume, fill speed, container-closure, and other process details are likely to be rather limited. In facilities where each filling line produces a variety of products, the possibilities increase substantially. The study rationale should provide justification for each filling line, indicating how the chosen process simulation studies performed support the various product presentations filled on that line. The rationale should be reviewed periodically to ensure its appropriateness consistent with any changes in products, components, practices, personnel, or equipment that could alter the circumstances. Provided next are some of the more common considerations and choices to be made in developing this rationale.

27.7.1.1 Media Sterilization

Although it may appear counterintuitive, the sterilization of the media is not a meaningful concern. The media, because of its differing formulation from the product(s) being simulated, may be more confidently sterilized using a different means than the product(s). Provided that it is introduced into the sterilized system using identical procedures and equipment, it can be pre-sterilized by steam, alternative filters, and even radiation. The intent of the process simulation is to confirm the acceptability of the processing procedures with sterilized equipment not to validate the sterilization of the product by the filtration system. Validation of the sterilizing filtration must be carried out for each formulation and the ability of the filtration system to sterilize the media is irrelevant to that product-by-product validation. At the same time, the media utilized in the simulation must be sterile for a valid challenge of the aseptic process, but proof of that sterilization is relevant only to the media process and nothing more. The use of heat or radiation sterilization of media may be necessary to destroy mycoplasma that might penetrate sterilizing filters. The addition of pre-filters upstream of the sterilizing filter for the media can be beneficial in eliminating particles present in the media that might slow the filtration.

27.7.1.2 Frequency and Number of Runs

The initial media fills for a facility are defined in the study rationale and typically at least three (3) trials per filling line are performed. In facilities making a variety of product presentations, the number of initial studies required may increase commensurably. Once the baseline capability has been established, a minimum of two fills per line per year is considered compliant with Current Good Manufacturing Practices (CGMP).[18] The conduct of additional media fills may be useful for a variety of reasons, i.e., environmental contamination because of a unique event (power loss, water leakage, major breach of asepsis; substantial change in the equipment, processes, components, etc.; adverse environmental trend; or sterility test failure).[19]

Although it is certainly preferable to await definitive results from a 14-day incubation period (and satisfactory growth promotion), there is no obligation to do so.[20] There are firms that conduct their media fills just prior to periodic shutdowns as confirmation of capability at the end of a long operating period and take advantage of overlapping the incubation and shutdown periods. Other firms conduct their media fills immediately after their shutdown period as demonstration of renewed capability including any minor changes to the facility or equipment.

27.7.1.3 Duration of Runs

The seemingly best answer to the required minimum duration of a media fill is that it should exceed the duration of the longest routine filling process, and this is oft cited by regulators.[21] Although that approach may seem the soundest, it presents some substantial problems for those firms making very large lot sizes. The most comprehensive advice on this subject is provided in PDA's TR #22 in which recommendations for the complete range of process batch sizes are provided.[22] The PDA document is based upon 5,000-unit media fills, but the advice given could be easily adapted to a larger batch size.

27.7.1.4 Very Small Batches—N < 1,000

For batches of this size, which are common in certain clinical and radio-pharmaceutical operations, a process simulation test at the maximum batch size is recommended. Forcing the production of 5,000 or even 1,000 units may produce situations so different from the normal operation that the results may be meaningless. For simulation of these batch sizes, the process simulation test must evidence no growth in any of the filled containers to be acceptable.

27.7.1.5 Small Batches—5,000 > N > 1,000

For this batch size, which might be common for a clinical batch or other developmental situation, the minimum process simulation batch size should be equal to the standard maximum batch size. Although this does not afford the level of statistical confidence frequently associated with full process simulation tests, it is a reasonable compromise given the limitations of the small batch size.

27.7.1.6 Conventional Batch Sizes—100,000 > N > 5,000

For these common production scale processes, the number of units to be filled with medium can approach the size of the full production batch, especially with the trend toward larger and larger process simulation tests. Current practice is to produce larger and larger media fills to accommodate the required interventions into the simulation.[23]

27.7.1.7 Large Batch Sizes—N > 100,000

A number of possible approaches have been utilized for large production batches:

- Fill 5,000 units, switch to sterile water-for-injection (WFI) for an extended period, fill an additional 5,000 units.
- Fill 5,000 units, simulate filling for an extended period, fill an additional 5,000 units.

- At the completion of a regular production batch, disassemble/reassemble with sterilized equipment; fill 5,000 units.
- After an extended WFI fill, disassemble/reassemble with sterilized equipment, fill 5,000 units.
- Simulate filling for an extended period of time, disassemble/reassemble with sterilized equipment; fill 5,000 units.
- These practices can be easily adapted to accommodate batches such that the filling process extends for substantially more than a single shift.

What is almost universal in simulation design is that the fill is representative of the production process. With larger production fills, this forces the media fill to reflect a duration that is a realistic representation of the production process. A lengthy process can hardly be supported by a simulation that is over in less than an hour, nor is there any merit to a 4-hour minimum simulation duration for what might ordinarily be a 2-hour fill session.

27.7.1.8 In-Process or End of Batch Media Fills

The conduct of a simulation supporting the production of very large batches can be accomplished in part by the performance of a media fill immediately after the completion of the production fill. The filling line is cleared of the last containers of the production batch, the liquid line is flushed to remove any traces of the product, a vessel of sterile media is connected, and filling is restarted with media into the same components used for the production fill. Alternatively, the product contact parts used for the product can be replaced with a freshly sterilized set of parts. The other aspects of the simulation are essentially unchanged from the practices described in this chapter. The results of the media fill must be considered in the lot release decision for the production lot.

In-process media fills are particularly useful in the support of very large batches, as their successful execution at the end of a long production batch can support that even under the adverse environmental conditions expected after the end of production that successful aseptic filling is possible. The use of in-process media fills as the sole means of supporting aseptic processing is inadequate, as the impact of the ordinarily highly manipulative system assembly is deemphasized by the time period between the initial set-up and the in-process fill execution. The potential for flush out or inhibition of any set-up related microorganisms by the product being filled must be considered as well.

This practice is easily adapted to campaign production with an end of batch (actually an end of campaign) media fill being performed after the last batch in the campaign has been filled.

27.7.1.9 Line Speed

Supporting the full capabilities of a filling line utilized for different containers is easily accomplished. Filling lines will often operate at a variety of speeds with smaller fill volumes ordinarily associated with higher filling rates, a consequence of the smaller volume being dispensed into each container. One set of media fills for a line would utilize the smallest container operating at the highest speed, as this may present the greatest handling difficulty. As handling difficulty is associated with an expected greater need for human intervention in either inherent or corrective activity, then this is an obvious "worst case" selection for the process simulation program. The largest container filled at the slowest speed presents the greatest opportunity for airborne contamination to enter from above and is often selected as the other "worst case" extreme for filling systems.

It is advantageous to assess a processing systems process capability (Cpk) for key process step(s). This will enable a firm to judge the failure rate per units filled in statistical terms. This allows the optimization of the process through adjustment of equipment for best performance and also indicates where component specifications may need reconsideration.

27.7.1.10 Container/Component Selection

The largest and smallest containers are often chosen as they represent the extremes of either exposure duration or handling difficulty, but other selections may be appropriate. Consider small vials of similar diameter such as 1 mL, 2mL, and 3mL units. The 3 mL because of its higher center of gravity may present more of a handling difficulty than its smaller companions and thus be a more suitable choice for use in the process simulation. Similarly, the elastomeric closure chosen for use should be the one that presents the greatest handling concerns. Recognition that excessive handling represents higher contamination potential may result in a simulation regimen that includes more than the obvious choices of largest and smallest containers.

There have been regulatory recommendations to replace opaque containers with clear ones to aid in the detection of contaminants post-incubation.[24] That is an accepted practice provided the removal of the coloring agent or wrapping does not alter the material handling characteristics of the container, in which case the opaque container is preferable despite the added inspectional difficulties (see later section).

27.7.1.11 Media Fill Volume

In the execution of a media fill, the amount of media filled in the container can be modified from that ordinarily filled. The media amount is ordinarily reduced to extend the duration of the fill with a limited media quantity (media quantity is sometimes limited by sterilization constraints on the media that can restrict the maximum amount available for use). There is no minimum media volume that need be utilized provided: there is adequate media to contact the entire sealing surface; there is sufficient media to allow for detection of growth; and there is sufficient media to pass growth promotion. In a few instances, the volume filled in the container has been adjusted upwards by firms from that typically used in the container to address one or more of the concerns cited.

27.7.1.12 Media Selection

Selecting the test media to be utilized is at the core of the simulation process, and in the vast majority of cases is accomplished by the use of Soybean Casein Digest Medium, SCDM. This general-purpose media is the usual choice because of its ability to support the growth of a variety of aerobic environmental and human derived organisms. In only very limited instances is another media appropriate.

> A firm with persistent low-level microbial contamination in its inventoried products never detected microbial contamination in media fills that utilized SCDM. When media fills were performed with media that resembled its product substrate were conducted the contamination source was identified.[25]

27.7.1.13 Anaerobes/Inert Gassing

Expectations that media fills address anaerobic contamination are only appropriate in limited situations. True anaerobic conditions are not attainable in manned clean rooms even where inert gassing is utilized. Low percentages of oxygen (>0.3%) are toxic to true anaerobes and thus anaerobic media fills using Fluid Thioglycollate Media (FTM) are largely unnecessary. Anaerobic media fills have been used in inert gas filled closed isolators where extremely low oxygen levels are required for product stability, but that is a relatively rare circumstance.

In ordinary media fills, to facilitate microbial recovery, air is often substituted for the inert gas on the filling line. This practice hopes to eliminate the potential microbial inhibition that the inert gas might impact on aerobic organisms that might find their way into the gas distribution system during post-sterilization assembly.[26]

27.7.1.14 Manual Filling

In the preparation of small-scale lots, there is often a heavy reliance on personnel to perform many of the functions provided by a filling line, i.e., container movement, closure placement, seal administration, etc. The operator essentially replaces some or all the filling equipment required for the process. In this instance, each operator assigned to this process should perform triplicate initial and semi-annual repeat media fills to demonstrate their aseptic processing proficiency.[27] This subject is addressed in further detail in a separate chapter in this book.

27.7.1.15 Aseptic Assembly

The execution of an aseptic process will often necessitate some preparation steps to configure the equipment and materials. The most obvious task is the set-up/assembly of the fill line from individually sterilized components into a complete line ready for the fill. Adjustment/assembly of the conveyor, limit switches, vibratory feeders, and perhaps the fluid material pathway may all be a part of this activity. The aseptic process begins with these steps, and they must be performed and evaluated with the same care devoted to the process itself. These activities are an inherent part of the process simulation, as the equipment must undergo the same preparatory steps; nevertheless, observation of these activities and environmental monitoring must be incorporated. The last portions of this task include product introduction, weight adjustment, and clearing of all set-up components prior to commencement of the filling process. Environmental air and personnel monitoring should address these activities; however, surface sampling should be avoided.

27.7.1.16 Environmental Monitoring

The aseptic processing environment utilized for the aseptic process should be monitored in accordance with the routine program used for the operation of the facility. The temptation to increase the monitoring during the media fill should be resisted as it may have an adverse effect on the results by increasing the personnel present or what is sometimes called "bioload". Environmental monitoring, especially microbial sampling, must be recognized as an intervention in the aseptic core and increasing its intensity above normal monitoring may result in the introduction of microbial contamination that might not otherwise be introduced. The conduct of environmental sampling is an intervention, and there must be a balance between the desire to gather information about the conditions proximate to the sterile materials and the potential introduction of microbial contamination as a consequence of the human presence required to obtain that information. There is "no free lunch"; the gathering of environmental data must not compromise the validity of the test.

27.7.1.17 Microbiological Monitoring

The conduct of microbial monitoring during the simulation is performed according to the normal regimen for the process being simulated. The hope is that the monitoring results will identify microbial contamination that matches any detected in the filled containers. This is of course the ideal result, as effective corrective actions can then be simply identified and implemented.

The monitoring should embrace all of the conventional methods employed in aseptic environments (see the earlier chapter in this text for more detailed coverage of this subject) and include dynamic air sampling, settle plates (or bottles), surface samples, and personnel monitoring. The limits for the monitoring are the same as those used in routine production. If expanded sampling is performed at the conclusion of the fill, those same limits are ordinarily employed. As noted earlier, monitoring during the set-up of the equipment (something that can occur several hours or even the prior day) should be conducted as well.

27.7.1.18 Nonviable Particle Monitoring

Sampling for nonviable particles must be performed during the media fill for no other reason than to ensure that all relevant interventions are a part of the simulation. The relevance of the nonviable particle monitoring data is often rather minimal; however, excursions may provide insight into routes of contamination introduction.

27.7.1.19 Incubation Time and Temperature

At one time, this was among the more controversial and variable practices in effect at many firms.[28,29] The selected approaches included incubation at multiple temperatures with transfer from one condition to another after 7 days of incubation. There was even confusion as to whether it was preferable to begin at a higher temperature (30°C–35°C) followed by a lower temperature (20°C–25°C) or begin the incubation at the lower temperature and then move to the higher one. Recognition that growth promotion is required regardless of the actual conditions selected has led to a more broadly defined practice where the incubation temperature can range from 20°C–35°C including a single temperature for the entire 14-day period.[30] Provided the selected temperature is uniform (the usual range is ±2.5°C) the use of a single incubation temperature eases the execution of the media fill. This practice allows for flexibility of approach easily accommodating the greatest potential for microbial recovery. Despite the FDA's acceptance regulation of a single temperature for incubation, the use of multiple temperatures persists.[31]

27.7.1.20 New Facilities and Lines

The start of operations in an aseptic facility must be supported by initial simulation studies that establish the capability of the facility, equipment, procedures, and personnel to manufacture sterile products.[32,33] Depending upon the specific circumstances of the products being manufactured, the number of required media fills may be as few as three; as line complexity/capability increases, this may entail additional studies. These studies can ordinarily be matrixed to reduce the overall number; however, even if the facility has lines composed of identical components, each line must be evaluated independently of the others.

27.7.1.21 Suspensions and Aseptic Manufacturing

The process simulation should embrace all portions of the aseptic process from the point of sterilization through closure of the container. All the aseptic interventions (sampling, filter integrity testing, etc.) that are a portion of the formulation process must be included in the simulation. The vessel utilized for the media fill should be identical to that used for commercial operations—the use of a carboy for the media fill where the commercial product uses a stainless-steel vessel is inappropriate. This can prove more challenging where the formulation includes aseptic steps such as required in the preparation of suspensions, ointments, and other more complex products.[34] These processes may require extensive sterilization-in-place (SIP) and complex equipment, and thus present some unique issues in the design of the simulation. The practices originally designed for sterile bulk can be adapted for use in these instances.[35] In some instances, the simulation process may require the use of a sterile solid (generally a placebo material) in portions of the simulation (see the following section).

In the preparation of suspensions and in many of the less common aseptically produced sterile products and containers, overlapping simulations addressing the overall process may be appropriate with some portions of the process being largely conducted using a sterile powder and the remainder with a sterile liquid media. This is acceptable practice provided the entire process is covered by overlapping where one part of the simulation ends, and the subsequent section begins.

27.7.1.22 Sterile Powders

Sterile powder processing and filling presents a unique difficulty in the conduct of the simulation, as the equipment utilized for powder processing cannot easily accommodate the liquid media ordinarily utilized for the simulation. Usually, execution of the process simulation will require the addition of both sterile liquid media and a sterile powder placebo into the container. The order of the additions and the extent to which the powder filling process is adapted to accommodate the liquid fill can make this one of the more difficult simulations to execute. PDA's TR #22 provides a description of the processing options ranked in order of preference.[36] TR #28, also from PDA, provides considerations in the selection and preparation of the sterile placebo powder. As liquid filling on a powder line is an infrequent event, some firms chose to fill a number of liquid-only containers in conjunction with the powder fill to establish that this activity is not the cause of any detected contamination. As the sterilizing filtration for sterile powders may be conducted in a separate facility (or by a separate firm), simulation concerns at the filling site are generally restricted to the activities performed there including milling and blending as appropriate. The sterile bulk supplier is responsible for simulation activities of their aseptic process, and this subject is addressed elsewhere in this book.

27.7.1.23 Placebo Materials

When a placebo material is required in the process simulation as is necessary in suspension or bulk powder processes, a placebo is commonly used. The selection of the placebo is a compromise between a number of factors, i.e., ease of sterilization, handling properties similar to the production materials, ease of clean-up, lack of microbial inhibition, etc. PDA's TR #28 provides useful information on the selection of the placebo material.[37] Regardless of the placebo material chosen for the simulation, the placebo material should be packaged in an identical fashion as the materials they are substituting for. The sterilization of the placebo must be validated to ensure that it does not become a source of contamination in the simulation.

27.7.1.24 Other Aseptic Filled Dosage Forms and Formulations

Process simulation studies are required wherever aseptic processing is utilized for the manufacture of sterile drugs. The base case for all simulations is the solution fill, and adaptations to that configuration are added to accommodate the equipment and processes used for other products. Some of the modifications are quite simple, (the incorporation of a freeze dryer or filling into a plastic tube) whereas others may introduce substantial complications (a multi-chambered syringe with a lyophilized powder with liquid diluent, or a liposome formulation requiring extensive pre-filling processing). The added complexity of these more intricate processes may entail modifications to permit simulation and thus increase the potential for failure. Nevertheless, their association with simple solution filling precludes the use of looser acceptance criteria reflecting the difficulties associated with the simulation. There are instances where process simulation of the types described here for pharmaceuticals have been adapted for the aseptic preparation of medical devices albeit sometimes with even greater modification to accommodate their rather different processing requirements.

27.7.1.25 Campaign Production

The campaign filling of a series of batches without intervening cleaning/sanitization (and in rare instances sterilization of all product contact equipment such as stopper bowls) is a common practice for some large-volume sterile products. A media fill program can be developed to support campaign production in this fashion by applying the methods described earlier for large batches and/or in-process media fills.

27.7.1.26 Interventions

The production of sterile products in either manned clean rooms or isolators relies on the execution of any number of manual tasks by the operator. These interventions are either inherent (part of the process) or corrective (to remediate a fault) in nature. An inherent intervention is one that is either an integral part of the process (i.e., set-up of the equipment, initial supply/replenishment of components, etc.) or required procedurally (i.e., product sampling, environmental monitoring, fill weight adjustment, etc.) Corrective interventions occur at random intervals in response to container-closure jams or misfeeds, or other mechanical problems. The inclusion of inherent interventions in process simulations is relatively simple; they need only be included at the same frequency as they would occur during a production lot. Corrective interventions must be integrated into the media fill in the event that they do not occur as a natural consequence of the process, and their frequency should match their incidence in the production process.[37,38,39,40,41] As noted earlier, the extent of the interventions required whether inherent or corrective is an important consideration in the selection of the appropriate components/process to be simulated. Practices for all intervention should be carefully defined to ensure consistency between routine production and simulated operations.[42]

The most importance aspect of interventions is their proper design and execution. First and foremost is the awareness that the safest intervention of all is the one that is not performed. The aseptic process should be designed to eliminate interventions (inherent or corrective) of all kinds, or at the very least minimize the need for their execution. The premium paid for more uniform components, higher quality equipment, and preventive maintenance is well spent if it results in more reliable filling. Preassembly of components prior to sterilization, leave behind samples, and careful attention to equipment design can help eliminate interventions that can adversely impact asepsis.

27.7.1.27 Execution of the Process Simulation

The process simulation should be performed following a defined procedure outlining the various requirements beginning with the sterilization of the media. The use of a batch record as least as detailed as that used for production filling is recommended; however, it may be necessary to supplement this in order to adequately document the interventions included during the process. The time of execution for each intervention should be recorded and, if possible, correlated to a specific portion of the filled units for use in problem resolution. An observer positioned outside the critical area (and preferably outside the aseptic area) can provide a level of documentation well beyond that of the aseptic operator(s) without risking contamination. Firms have found the use of video recording beneficial in media fill execution as it can capture substantially more detail than an observer (the simultaneous use of video tape and an observer/supervisor has also been used); however, in some jurisdictions, labor laws may preclude the recording of operators at their jobs.

27.7.1.28 Initial Inspection of Filled Containers

It is customary to inspect the media-filled units immediately after sealing and prior to incubation to remove nonintegral containers from the test units to be incubated. Nonintegral containers should not be incubated and their removal prior to incubation avoids the unanswerable and inevitable later question when a nonintegral container is found to be contaminated post incubation.[43] The temptation to discount nonintegral contaminated units post-incubation must be resisted, as there is no means to establish whether the container was nonintegral prior to incubation. Once a container has passed initial inspection, any contamination detected must be counted against the simulation as these units are intended to represent materials that would be released for distribution. Integral containers that would otherwise be rejected for some type of cosmetic defects (i.e., particle in solution, fibers, marks on container, etc.) are not culled in this preliminary inspection, as their removal in a post-fill inspection is not certain, and thus they represent potentially marketed units. The number of units placed into the incubator should be accurately determined.

Media-filled units should be manipulated to ensure that there is contact between the media and the container-closure seal surfaces. Physical contact between the media and the seal surfaces ensures that those surfaces of the container that are more likely to have been contaminated during the process are accurately assessed. For syringes and ampoules, incubation of the media-filled containers can be performed in a random orientation to maximize the contact between the media and the sealing surfaces. Vials are generally inverted briefly prior to incubation (and midway through the incubation period if there is an intermediate 7-day inspection of the units). Some firms have chosen to invert vials during incubation, but that is not a universal practice.[44]

27.7.1.29 Mid-Incubation Inspection

There is no requirement to perform an inspection during the incubation period; however, many firms have included this practice in order to be aware of potential problems sooner. A preliminary inspection is sometimes employed as support for the commencement of aseptic filling at risk pending the final results.

27.7.1.30 Post-Incubation Examination of Media-Filled Units

After conclusion of the incubation (14-day incubation is universal), the containers are carefully inspected to detect microbial contamination. This inspection can be performed by trained personnel with a qualified microbiologist present to support the selection process. Microbiologists may be preferable for the entire inspection where the media must be removed from fully opaque containers, as might be the case with plastic tubes or other difficult to inspect items. Units suspected to be contaminated are identified, counted, and set aside for further evaluation. The total number of units inspected should be recorded.

27.7.1.31 Growth Promotion

Upon conclusion of the inspection, sterile units are selected randomly from the filled units and individual units are inoculated with <100 CFU/container of selected microorganisms. The usual choices for these microorganisms are those identified in USP/EP for the verification of media efficacy, plus some additional microorganism(s) of the firm's choice. Where microorganisms are added to the panel, they are usually selected from common environmental isolates not already represented in the compendial panel or those isolated in sterility test failures. The inoculated units are incubated at the same conditions as the test units and must demonstrate satisfactory growth in a limited timeframe, typically 2–3 days.[45]

Firms would prefer to select units at random immediately after filling and use those in a concurrent growth promotion test in an effort to shorten the timeframe to obtain definitive results; however, regulators frown on this practice as potentially obscuring contamination that might be in the units randomly selected for the growth promotion. Given the low incidence of contaminated units observed in contemporary media fills such caution hardly seems justified; nevertheless, growth promotion is generally performed post-incubation.[46,47]

27.7.1.32 Microbial Identification

Where positive units are detected in the post-incubation inspection, they should be identified to the extent necessary to assess their origin. Although most microbial contamination can be expected to be human derived, it may be possible for genetic identification to provide a more precise assignment of genus and species and enable the best assessment of the potential source. Correlation of a microbial identification in a media fill to the environmental monitoring during the simulation may be useful. However, common human organisms predominate, so the precise pinpointing of a source is unlikely. Regardless of what information is gathered about the microorganism, the objective in the identification is to determine its characteristics and gain insight into its potential source. However, in practice, environmental monitoring is a low-resolution activity and should be considered qualitative and informational rather than quantitative and definitive.

27.7.1.33 Accountability

Media fills endeavor to establish that the contamination rate for filled units is less than the targeted acceptance criterion. Counting the number of positives found post-incubation is generally easy given the generally successful results observed. Regulatory inspectors have raised concerns that accountability of filled units should be 100%, and that missing units must be considered as positives. That perspective seems excessively conservative, and accountability for the media fill that is comparable to that of a similar sized production fill should be considered acceptable.

27.7.1.34 Acceptance Criterion

Selection of an acceptance criterion for process simulation is the province of regulatory agencies. For many years, the standard of acceptance for media fill contamination rate was a criterion of not more than 0.1%.[48,49] When the first written guidance was published, no statistical treatment was provided.[50] Over the years, aspects of statistical confidence following a Poisson distribution were added.[51,52] Use of a Poisson distribution was considered appropriate as it was believed that microbial contamination in media fills was a random occurrence associated with a variety of possible causes. This expectation reached its zenith in publications that appeared at the end of the last century in which the statistical treatment included alert and action levels for the evaluation of aseptic processes.[53,54] This approach seemed inappropriate given the growing realization that environmental contamination recovered from aseptic clean rooms (and media fills as well) is predominantly derived from the human operator and thus is likely to be associated with operator activities rather than any random source. This perspective was first voiced in PDA's TR #22, in which the limitations of statistical treatment were addressed.[55] The use of statistics allows a number of contaminated units (9 in 15,710) that is less than 0.1% of the number of units filled. Approached in this manner, an aseptic process capable of slightly less than 0.1% contamination would be considered acceptable (under investigation certainly, but acceptable nevertheless). This realization led to changed expectations in newer regulatory guidance in which an expectation of zero contamination as the goal of every aseptic process as first defined by PDA are included.[56,57] The latest regulatory word on acceptance criteria for aseptic processing is that provided by the FDA in its "Guideline on Sterile Drug Products Produced by Aseptic Processing" that extends the most recent thinking and takes it a bit further.[58]

This guidance has a goal of zero contamination, but accepts no more than 1 contaminated unit in either 5,000 or 10,000 units (the document can be interpreted to require either). Perhaps most troubling of all is the absence of an acceptance criterion that can be applied for very large media fills (typically those associated with very large commercial batches). Current regulatory guidance suggests a maximum of one contaminated unit regardless of the number of filled units, an approach that may have the unintended consequence of smaller media fills as filling more units increases the risk of failure as the second positive in any fill is considered a failure.

27.8 IMPLICATIONS OF RESULTS FOR ASEPTIC FILLING

When the results of the media fill are available, there is certainly no issue when all the filled containers are free of microbial contamination. Fortunately, at the present time, media fills with any contamination are increasingly rare. Given that microbial contamination is almost always human derived, contaminated units may provide insight into potential sources if they are associated with an intervention. In the absence of linkage to an intervention, any investigative effort is likely to be inconclusive. Nevertheless, an investigation is mandated when any contaminated units are detected, as it is a regulatory expectation.[59,60]

In conjunction with the detection of contamination in the process simulation, the firm must make a determination on whether action should be taken relative to lots produced in proximity to the media fill. When the contamination rate is below the acceptance criterion (an unusual condition as limits are ever tighter), the contamination is generally considered as little more than a caution to the firm. In the event of an actual simulation test failure, the first action commonly taken is to place on hold the release of lots produced before and after the media fill. How far to extend this review is a matter of some discussion and can be based upon either a defined time period, the number of lots the firm still maintains control over, or a specific number of lots before and after the simulation.[61] In extreme cases, firms have considered the release implications for all lots produced back to the last successful media fill; however, this would be an unusual circumstance. To minimize the disruption to operations by potentially failed simulations, some firms have chosen to perform periodic media fills in conjunction with planned shutdown periods, scheduling them just before start or immediately after the shutdown period. There are cogent arguments for the use of either approach.

Any lots held pending the investigation are reviewed to determine whether the contamination detected in the simulation could have also contaminated the production materials. A thorough investigation of the media fill contamination and detailed batch records are essential to the release decision that must be made for each lot under quarantine. Fortunately, the reduced level of contamination evidenced in contemporary media fills has made instances of quarantine less frequent than in prior years and is likely to decrease further as aseptic processing performance continues to improve.

27.9 ISOLATION TECHNOLOGY

The use of isolators (or other advanced aseptic processing technology) alters little in the conduct of process simulations. All of the aspects and rationale considerations cited previously are applicable essentially unchanged. Regulators have generally indicated that isolators (and perhaps other advanced technologies) by virtue of their superior performance potential may be demonstrated capable using media fills of shorter duration than in conventional clean rooms.[62] However, in actual practice, this has not proven to be the case. In the early days of isolator validation, some media fills were impractically long and led to profound media supply issues. For this reason, process simulation tests are being performed incrementally and supplemented with mock or water fills to avoid unworkably large sample sizes.

The ability to process a lengthy campaign in an isolator is highly desirable. The validation of campaign length uses methods similar to those employed for very large batches provided earlier in this chapter. In-process media fills are another means for the establishment of campaign length in isolators. PDA's TR #28 provides some useful guidance in means for the establishment of isolator integrity over the campaign duration.[63]

27.10 CONCLUSION

Demonstration of aseptic processing proficiency as provided by media fills (process simulations) is an integrated exercise incorporating every aspect of the process. Success is only possible when each of the individual elements has been properly defined and controlled. Industry performance in aseptic process simulation has improved substantially over the past 20 years indicating continual improvement in the safety of sterile products produced aseptically.[64,65,66] This improvement is the result of the emphasis placed on aseptic processing by everyone concerned. Even when properly designed, executed, and controlled, aseptic processing is an activity requiring continual vigilance. The process simulation is just one element of the necessary controls used foraseptic processing.

NOTES

1. USP 41, <1211> *Sterility Assurance*, 2019.
2. Agalloco J, Akers J, The Myth Called Sterility. *Pharmaceutical Technology* 2010;34(3), Supplement: S44–45. Continued online at Pharmtech.com
3. FDA, Guidance to Industry for the Submission Documentation for Sterilization Process Validation in Applications for Human and Veterinary Drug Products, 1994.
4. FDA, Guideline on Sterile Drug Products Produced by Aseptic Processing, 2004.
5. EMEA, Decision Trees for the Selection of Sterilisation Methods (CPMP/QWP/054/98), 1998.

6. Akers J, Agalloco J, Madsen R, What Is Advanced Aseptic Processing? *Pharmaceutical Manufacturing* 2006;4(2): 25–27.
7. Agalloco J, Closed Systems and Environmental Control: Application of Risk Based Design. *American Pharmaceutical Review* 2004;7(4): 26–29, 139.
8. PDA TR# No. 22, revised. Process Simulation Testing for Aseptically Filled Products. *PDA J Pharm Sci Technol* 2011.
9. Agalloco J, Hussong D, Quick J, Mestrandrea L, Closed System Filling Technology: Introducing a New Paradigm for Sterile Manufacturing. *PDA Letter* 2015;51(11): 26–28. www.bioprocessonline.com/doc/closed-system-filling-technology-a-new-paradigm-0001
10. Agalloco J, Closed Systems: Changing the Aseptic Manufacturing Paradigm. Accepted for publication in *Pharmaceutical Technology*.
11. PDA TR# 36, Current Practices in the Validation of Aseptic Processing—2001. *PDA Journal of Pharmaceutical Science and Technology* 2002;56(3).
12. USP, <1229> *Sterilization*, 2013.
13. Cundell A, et al., Statistical Analysis of Environmental Monitoring Data: Does a Worst Case Time for Monitoring Cleanrooms Exist? *PDA J. Pharma. Sci. Technol* 1998;52(66): 326–330.
14. Agalloco J, Akers J, Letter to the Editor: Re: Apples, Oranges and Additive Assurance. *Journal of Parenteral Science and Technology* 1992;46(1).
15. PDA TR# No. 22, revised. Process Simulation Testing for Aseptically Filled Products. *PDA Journal of Pharmaceutical Science and Technology* 2011.
16. PDA, 2017 Aseptic Processing Survey, 2017.
17. PDA 2017 Aseptic Processing Survey, 2017.
18. EU GMP Regulations, Annex 1, Manufacture of Sterile Medicinal Products, 1998.
19. PDA TR# No. 22, Process Simulation Testing for Aseptically Filled Products. *PDA Journal of Pharmaceutical Science and Technology* 1996;50, Supplement S1.
20. PDA TR# No. 22, revised. Process Simulation Testing for Aseptically Filled Products. *PDA Journal of Pharmaceutical Science and Technology* 2011.
21. FDA, Guideline on Sterile Drug Products Produced by Aseptic Processing, 2004.
22. PDA TR# No. 22, revised. Process Simulation Testing for Aseptically Filled Products. *PDA Journal of Pharmaceutical Science and Technology*, 2011.
23. PDA TR# 36, Current Practices in the Validation of Aseptic Processing—2001. *PDA Journal of Pharmaceutical Science and Technology* 2002;56(3).
24. FDA, Guideline on Sterile Drug Products Produced by Aseptic Processing, 2004.
25. Agalloco J, personal communication, 1994.
26. PDA TR# 36, Current Practices in the Validation of Aseptic Processing—2001. *PDA Journal of Pharmaceutical Science and Technology* 2002;56(3).
27. PDA Technical Report No. 22, Process Simulation Testing for Aseptically Filled Products. *PDA Journal of Pharmaceutical Science and Technology* 1996;50, Supplement S1.
28. Agalloco J, Gordon B, Current Practices in the Use of Media Fills in the Validation of Aseptic Processing. *Journal of Parenteral Science and Technology* 1987;41(4): 128–141.
29. PDA TR# 36, Current Practices in the Validation of Aseptic Processing—2001. *PDA Journal of Pharmaceutical Science and Technology* 2002;56(3).
30. FDA, Guideline on Sterile Drug Products Produced by Aseptic Processing, 2004.
31. PDA, 2017 Aseptic Processing Survey.
32. PDA TR# No. 22, Process Simulation Testing for Aseptically Filled Products. *PDA Journal of Pharmaceutical Science and Technology* 1996;50, Supplement S1.
33. FDA, Guideline on Sterile Drug Products Produced by Aseptic Processing, 2004.
34. FDA, Guideline on Sterile Drug Products Produced by Aseptic Processing, 2004.
35. PDA TR# 28, Process Simulation Testing for Sterile Bulk Pharmaceutical Chemicals. PDA Technical Report #28, revised, *PDA Journal of Pharmaceutical Science and Technology* 2006;60, Supplement S-2.
36. PDA TR# No. 22, revised. Process Simulation Testing for Aseptically Filled Products. *PDA Journal of Pharmaceutical Science and Technology* 2011.
37. PDA TR# 28, Process Simulation Testing for Sterile Bulk Pharmaceutical Chemicals. *PDA Journal of Pharmaceutical Science and Technology* 1998;52(4), Supplement.
37. Agalloco J, Management of Aseptic Interventions. *Pharmaceutical Technology* 2005;29(3): 56–66.
38. Agalloco J, Akers J, The Truth about Interventions in Aseptic Processing. *Pharmaceutical Technology* 2007;31(5): S8–S11.
39. Agalloco J, Akers J, Revisiting Interventions in Aseptic Processing. *Pharmaceutical Technology* 201135(4): 69–72.
40. Agalloco J, Uncommon Sense in Execution of Process Simulations. *Pharmaceutical Manufacturing* 2013;10(3): 28–32.
41. Agalloco J, Complications in Process Simulation Execution. In *Pharmaceutical Technology Biologics and Sterile Drug Manufacturing eBook,* May 2020: 10–14, 43. www.pharmtech.com.
42. PDA TB# 2003–02, Incubation of Intervention Units in Aseptic Process Simulation Tests (Media Fills), 2003.
43. PDA TB# 2003–01, Damaged Containers in Aseptic Process Simulation Tests (Media Fills), 2003.
44. PDA TR# 36, Current Practices in the Validation of Aseptic Processing—2001. *PDA Journal of Pharmaceutical Science and Technology* 2002;56(3).
45. PDA TR# No. 22, revised. Process Simulation Testing for Aseptically Filled Products. *PDA Journal of Pharmaceutical Science and Technology* 2011.
46. PDA TR# 36, Current Practices in the Validation of Aseptic Processing—2001. *PDA Journal of Pharmaceutical Science and Technology* 2002;56(3).
47. PQRI Aseptic Processing Working Group—Final Report—2002. ww.pqri.org/aseptic/imagespdfs/finalreport.pdf
48. FDA, Guideline on Sterile Drug Products Produced by Aseptic Processing, 2004.
49. EU GMP Regulations, Annex 1, Manufacture of Sterile Medicinal Products, 2008.
50. PDA TR# No. 22, Process Simulation Testing for Aseptically Filled Products. *PDA Journal of Pharmaceutical Science and Technology* 1996;50, Supplement S1.
51. PDA TM# 2, Validation of Aseptic Filling for Solution Drug Products. PDA, 1980.
52. PDA TR# 6, Validation of Aseptic Drug Powder Filling Processes. PDA, 1984.
53. Technical Monograph #4, *The Use of Process Simulation Tests in the Evaluation of Processes for the Manufacture of Sterile Products.* The Parenteral Society, 1993.

54. ISO 13408–1, *Sterilization of Health Care Products—Aseptic Processing—Part 1: General Requirements.* International Standards Organization, 1998.
55. PDA TR# No. 22, Process Simulation Testing for Aseptically Filled Products. *PDA Journal of Pharmaceutical Science and Technology* 1996;50, Supplement S1.
56. CEN /TC 204WG 8 N 38—Recommendations on Validation of Aseptic Processes, 1998.
57. PIC/S PE002–1—Recommendations on Validation of Aseptic Processes, 1999.
58. FDA, Guideline on Sterile Drug Products Produced by Aseptic Processing, 2004.
59. FDA, Guideline on Sterile Drug Products Produced by Aseptic Processing, 2004.
60. EU GMP Regulations, Annex 1, Manufacture of Sterile Medicinal Products, 2008.
61. PDA TR# 36, Current Practices in the Validation of Aseptic Processing—2001. *PDA Journal of Pharmaceutical Science and Technology* 2002;56(3).
62. FDA, Guideline on Sterile Drug Products Produced by Aseptic Processing, 2004.
63. PDA TR# No. 22, revised. Process Simulation Testing for Aseptically Filled Products. *PDA Journal of Pharmaceutical Science and Technology* 2011.
64. PQRI Aseptic Processing Working Group—Final Report—2002. www.pqri.org/aseptic/imagespdfs/finalreport.pdf
65. Agalloco J, Madsen R, Akers J, Ph.D., Aseptic Processing: A Review of Current Industry Practice. *Pharmaceutical Technology* 2004;28(10): 126–150.
66. PDA TR# 36, Current Practices in the Validation of Aseptic Processing—2001. *PDA Journal of Pharmaceutical Science and Technology* 2002;56(3).

28 Manual Aseptic Processes

James Agalloco and James Akers

CONTENTS

28.1 INTRODUCTION

This chapter outlines the means for process control of aseptic processing for sterile products manufactured using predominantly manual procedures. Typical processes/products for which this type of guidance might prove beneficial include vaccine preparation, cell culture, gene therapy, Investigational New Drug manufacturing, clinical manufacturing, etc. where a substantial portion of the manufacturing process is aseptically performed by operators and thus susceptible to adventitious contamination. As the aseptic operations in these processes are unique to the specific process and are also substantially different from those employed for either final formulations involving conventional aseptic filling equipment or sterile bulk drugs, none of the existing regulatory or industry guidance is fully appropriate. This chapter addresses aseptic procedures in which manual activities constitute the majority of the process; those other manual steps associated with more conventional aseptic processing such as sampling or aseptic connection are readily integrated into the simulation of those processes and will not be addressed further here.

The evaluation of manual aseptic processes is adapted from the better-defined methods utilized for aseptic processing on a larger scale with automated or semi-automated equipment. It is expected that the reader is familiar enough with those practices to follow how they have been adapted for execution in a more manipulative setting.

28.2 BACKGROUND

Aseptic processes are vulnerable to contamination from a variety of sources; however, the greatest source of microbial contamination is the personnel who participate in its execution.[1,2] An aseptic process wherein personnel perform virtually all of the important processing steps must be carefully designed and executed to minimize the potential for microbial ingress. Like any other aseptic processing activity, a largely manual aseptic process requires an appropriate environment, effective sterilization of the materials, equipment, and components, and well-defined procedures/batch records.

28.3 BUILDINGS AND FACILITIES

The only meaningful facility concern for manual aseptic processing is the selection of an appropriate environment in which to perform the required activities. In general, the choice is an ISO 5 (EU Grade A, FS 209 Class 100) environment in which unidirectional air flow protection is provided to the materials during the procedure. In practice, this can mean many different locations are possible including: a portion of a larger environment of the same class, a localized unidirectional air flow hood protecting a specific portion/area within a lower classified environment; a table-mounted unidirectional flow hood (with either horizontal or vertical air flow), and an isolator (open or closed). Each of these systems is in current use for the conduct of manual aseptic processes. In some instances, the activity might be limited to a hose connection, sample removal, or other brief task, whereas in other cases it could include the entire aseptic process. Clearly, the more complex the process, the more sophisticated the operating environment that will be required. The manual processes considered in this chapter are ordinarily carried out in either a unidirectional flow hood or an isolator. These environs provide appropriate conditions for the execution of predominantly manual operations and are considered the critical zone in the parlance of regulators and industry.

DOI: 10.1201/9781003163138-28

Biosafety cabinets (BSCs) are not appropriate for manual aseptic processing as they are designed for the containment of microorganisms, and there is a net flow of air from the background environment into the BSC to provide that containment during their operation. BSC use is only appropriate for aseptic processes in which the need for worker protection from the product is significant. With the increased availability of single-use disposable systems for cell and gene therapy processing, in the future providing greatly improved separation from materials and workers ISO 5 environments may no longer be necessary. Additionally, the increased use of separative technologies (isolators) and automation will reduce human exposure to hazardous materials and product contamination risk to levels unattainable by the use of BSCs.

The supporting clean area outside the critical zone is typically either ISO 7 or the hybrid EU Grade B when a unidirectional air flow hood, or BSC is utilized for the manual process. The surrounding clean area is where the personnel performing the manual process are located, and appropriate gowning facilities are required that will vary substantially depending upon the background environmental requirements. The execution of the aseptic process is ordinarily supported by various sterilization processes for the materials and equipment required, and these may also be located in the support areas.

Closed isolators provide the greatest protection of product and operator and reduce the requirements for the surrounding area considerably. Evidence from decades of use indicate that an unclassified surrounding environment can be safely used with a closed isolator, but most new pharmaceutical and biopharmaceutical installations employ an ISO 8 background because of regulatory policy. Although closed isolators are the safest alternative, their cost and complexity can be greater. There is evidence that closed isolators when used in conjunction with automation can produce both lower levels of contamination risk and superior production outcomes in the manufacture of cytotherapy products.[3] For high-value products manufacturing in very low to moderately low volumes, closed isolators with some automation may be the best technical option. The drawbacks that exist owe mostly to excessive regulatory conservatism rather than technical limitations.

The overall design of the aseptic facility approximates that of large-scale environments utilized for equipment-based aseptic filling whether in a cleanroom or an isolator albeit on a much smaller scale. In some instances, the manual aseptic process is carried out in a portion of a larger aseptic processing facility.

28.4 PERSONNEL TRAINING AND QUALIFICATION

People are the single most important aspect of manned aseptic processing, and when the aseptic process relies on them almost exclusively for success, their proficiency at their assigned tasks must be beyond reproach. A manual aseptic process places such importance on the skill of the operators; their skills must be honed to near perfection.

28.4.1 Training of Personnel

The training requirements for these personnel typically include the usual elements of current Good Manufacturing Practices (CGMP), microbiological principles, sterility assurance, sterilization, gowning practices, and the other knowledge requirements of ordinary aseptic operating personnel. Theoretical knowledge of these areas is insufficient; the operators must be able to adapt the classroom discussions to the real-world environment. Where these individuals must excel is in the execution of those tasks that directly impact sterility assurance: aseptic gowning, aseptic assembly, and aseptic technique. They must be able to consistently perform precision tasks while minimizing the introduction of contamination onto the materials they are working with. It is important to understand that no training program regardless how well-designed or intensive can ensure that working humans will not release contamination into a working environment.

28.4.2 Gowning Qualification

Assessment of the operators' proficiency in their assigned tasks can be accomplished through the execution by these workers of practical exercises in which their skills are challenged and evaluated. The most basic of these, and ordinarily the first which the operator must succeed at, is aseptic gowning. This entails repetitive gowning in full aseptic garb under the observation of a fully qualified individual followed by monitoring of gown surfaces. The number of surfaces varies with the firm, but typically includes: gloves, forearms, and chest area (the locations closest to any manual activity the operator must perform).[4] The operator must successfully demonstrate their ability to meet the defined monitoring levels after each gowning exercise. Gowning certifications are conducted on a periodic basis to confirm that the operators maintain consistent gowning practices. Once the operator has passed initial gowning certification, they are granted access to the aseptic core for the continued instruction in aseptic processing.

28.4.3 Aseptic Handling Challenges

The conventional means for establishing personnel proficiency in aseptic processing is through participation in a media fill (also known as a process simulation).[5,6] These require the operator to perform aseptic interventions during the normal course of the simulation, and for those charged with aseptic assembly, to assemble the sterilized equipment prior to the fill. For individuals who perform manual aseptic processing, these activities may have little or no relevance for a variety of reasons:

- Automated filling practices are less susceptible to human contamination
- Interventions on automated filling systems are infrequent and facilitated by proper equipment design
- Automated filling does not require continuous human intervention
- Automated filling systems are often designed for SIP, requiring minimal set-up

Thus, an individual that has demonstrated proficiency at aseptic processing must still demonstrate their capabilities in the more rigorous requirements of manual aseptic processing. This is often accomplished by various forms of challenge tests in which the operator must directly handle sterile equipment and materials (usually with media) to affirm their aseptic technique (some firms use this approach for operators conducting automated filling). These tests may bear little resemblance to any specific manual aseptic process, but merely serve to evaluate personnel proficiency. The usual requirement is that the operators achieve perfect results in these evaluations.

28.5 EQUIPMENT, COMPONENTS, AND CONTAINER/CLOSURES

The equipment, raw material components, containers, closures, and other items required for manual aseptic processing vary with the requirements of the process. Perhaps the single most common factor with these items is their reduced size and number, which may allow them to be supplied to the processing environment in a sealed package after depyrogenation/sterilization. The preparation methods prior to depyrogenation/sterilization mimic those associated with automated aseptic processing although on a smaller scale of operation. The depyrogenation/sterilization methods for all items must be validated.

28.6 TIME LIMITATIONS

Limiting the amount of time an operator can engage manual aseptic processing is much more important than with more automated aseptic processes. After all the operator's skills and ability to focus can deteriorate given the concentration and attention to detail manual aseptic processing entails. In very short or non-repetitive processes such as aseptic connection and aseptic manufacture, time may be of little relevance (except as it may relate to material stability), and its impact lessened if not actually ignored. Where the operators must perform repetitive tasks such as container filling/stoppering, biological transfers, cell passage, or similar tasks, the effect of fatigue must be considered in both routine operation and process simulation. A "worst case" evaluation of fatigue would entail a process simulation equal or greater in time duration to the longest period an operator might perform the task without interruption, exit, and reenter the aseptic environment. Even after successful evaluation of the impact of fatigue on individuals, there may be situations in which the operator finds themselves unable to perform optimally. In such a case, the operator should relieve themselves and allow time for rest and recovery.

28.7 DESIGN OF MANUAL ASEPTIC PROCESSES

The heavy reliance on personnel capability and execution in manual aseptic processes makes the design of the process to minimize the impact of personnel critical to success. Although it is impossible to establish detailed design criteria, general process design principles can be followed that will increase the probability of success.

28.7.1 Manual Aseptic Process Design Principles

- Significant aseptic assembly should be avoided through the use of sterilized preassembled items. This will serve to minimize the extent of manual assembly required.
- Tools and utensils should be employed wherever possible rather than the direct contact with the operator's hands. Provide supports for the tool inside the ISO 5 environment to minimize contact between the tool and work surfaces within the workspace. Isolators, because they undergo very high-level decontamination, allow a great margin for error in conducting manipulations and handling tools or instruments.
- Perform as much of the process inside the ISO 5 environment as possible to minimize removal and subsequent reentry of sterile items in/out of the ISO 5 environment. This may require the placement of small equipment within the hood.
- Wherever possible, liquid transfers should be made using peristaltic pumps rather than through the use of automatic pipettes. Containers should be pre-marked to indicate the amount of material to transfer. Aseptic filling operations should be automated to the greatest extent possible; the less reliant a process is on human intervention, the safer it is.
- Materials being introduced into the process should be pre-measured prior to sterilization and addition.
- Utilize a 2nd (and 3rd, if required) person to supply/remove items to/from the ISO 5 environment. The 2nd person should wear sterilized gloves while holding the wrapped item and never contact the sterile item itself. The primary operator should never contact the wrapping materials.
- Electrical equipment and controls should be located outside the hood if possible. If that is not possible, a 2nd operator and not the primary operator should adjust them.
- When items must be temporarily removed from the ISO 5 environment, they should be sterile wrapped with minimal exposure of critical surfaces to the surrounding less clean environment. Upon return to the ISO 5 environment, the overwrap is removed and discarded.
- Sanitize the operating environment while empty and sanitize each item as it is first introduced. Do not introduce a large item into the environment mid-process.
- Plan the process so that samples can be taken with minimal risk. Take all samples from a container in a single step, and then subdivide that sample as required. If appropriate, leave material for samples

in containers from which the remaining materials have been transferred for further processing.

- The entire process should be documented in sufficient detail to ensure conformance to the desired practices. The 2nd (or 3rd) operator should complete the batch record.
- Environmental monitoring practices should be non-intrusive to avoid the potential for dissemination of contamination in the ISO 5 environment. The use of settling plates and post-process RODACs/swabs to monitor environmental conditions is preferable to active air sampling. Environmental monitoring in the close quarters of a unidirectional flow cabinet can be as risky as the processing itself. Human bioload should be kept to the lowest possible level at all times.
- The process should be rehearsed using all of the required items and placebo materials to refine the steps, location of items, etc.
- Any processing steps not required to be aseptic should be performed outside the ISO 5 environment by the 2nd (or 3rd) operator.
- The operators should work as a team. The primary operator should perform all tasks inside the ISO 5 environment. The 2nd operator assists in the introduction/removal of items from the ISO 5 environment and may assist the primary operator with some tasks inside that environment. A third operator may be necessary in some cases to support activities performed exclusively in the surrounding environment.
- The hands of the primary operator should remain in the ISO 5 environment at all times. The hands of operators should never reside in the airstream above open sterile materials.
- The 2nd operator should don fresh sterile gloves/sleeves prior to any activity by them inside the ISO 5 environment, or in transfers of items to/from the primary operator.
- The operators should decontaminate their gloves on a frequent basis. Because manual filling requires extensive use of gloves, they are subject to considerable potential for wear and tear. Ideally gloves should be changed every 90–120 minutes.
- Extra subassemblies and utensils should be sterilized and available for immediate use in the event a replacement is needed.

28.8 MANUAL ASEPTIC PROCESSING—PROCESS SIMULATION

Note: This section addresses only those elements of manual aseptic processing simulation that are unique to the extensive participation of personnel in the process. Details of study design provided in the chapters on validation of aseptic processing for either filling or sterile bulk production should be consulted for those activities that are essentially unchanged when manual procedures are employed.

Study Design—The development of a supportive rationale for the manual aseptic process simulation is essential. The rationale must define the adaptations to the production process necessary for the execution of the simulation. The smaller scale of the manual process lends itself rather easily to these adaptations, as in many instances, only minimal changes to the process are required. Because of the limited number of units filled in these processes, samples of the media taken during the process can be beneficial in determining at what point contamination was introduced (in the event of a failure investigation) and definition of these sample points should be included in the rationale. The study rationale should be kept current with changes to process, products, components, or equipment that could impact the acceptability of the process.

Manual aseptic processing can encompass a wide variety of activities but can be divided into 4 major categories: filling/subdivision activities; compositing/assembly activities; formulation/compounding activities; and manipulative steps performed in conjunction with other processes. The validation of each category is addressed in a different manner.

- Filling/subdivision processes involve repetitive actions in which sterile materials are transferred from a bulk container into smaller containers, closed, and sealed. This practice is common in IND and early clinical stage manufacturing of sterile products and in the manufacture of extremely small lot sizes. Validation of these processes mimics the practices defined for automated filling.[7,8]
- Compositing/assembly processes involve repetitive actions in which sterile materials in smaller amounts are pooled. Such practices are common in vaccine manufacturing in which the contents of incubated eggs are composited early in the formulation process. Adaptation of the validation methods for sterile filling and bulk materials may be appropriate.
- Formulation/compounding procedures in a manual setting might use laboratory glassware and utensils in which a sterile bulk formulation is produced. The smaller scale of the operation mandates changes in transfer methods. The methods utilized for sterile bulk materials are most appropriate in these processes.[9]
- Manual activities such as sampling, aseptic connection, etc. are often an integral part of other aseptic processes. As they are an inherent part of those processes, there is no need to address them independently.

Additional details on each of the various methods will be provided in conjunction with each of the elements addressed within the overall rationale.

Media Sterilization—Preparation of media for use in manual aseptic process simulation is rarely difficult. It may be more convenient to purchase pre-sterilized media; in that case, terminally sterilized media is the most reliable choice. The bulk container size is typically small enough that it can be sterilized in an autoclave prior to introduction into the process. For compositing/assembly processes, the availability of suitable sterile materials to use in the process simulation may be difficult and there may be little choice other than the production materials themselves suitably sterilized/adapted (if possible and necessary) for use in the simulation. For the simulation, each of the liquid containers containing sterile materials required for the process should have their contents replaced with media.

Frequency and Number of Runs—The only existing recommendation for frequency/number of simulation studies with manual processing is that each operator who performs the manual steps in the process should be qualified semiannually.

Duration of Runs—Simulation studies should slightly exceed the expected maximum duration of a single working session by a single operator. In compounding simulations, the length of the simulation should mimic that of the commercial manual process with the exception that process hold times without activity can be shortened dramatically.

Size of Runs—The size of the process simulation is largely dictated by the time period that a single operator would remain performing the same activity. The actual numbers of simulation units produced in that time period should meet or exceed the production quantity that the operator would normally handle in that time period.

Media Fill Volume—During the aseptic filling simulations, the amount of media transferred should be sufficient to wet the product contact surfaces of the container and be sufficient to detect growth. In compositing simulations, the amount transferred should be identical to that normally handled to mimic the process duration more accurately. In manufacturing simulations, the volumes of media and other fluids (which should all be replaced with media) should be identical to that in the process to be simulated.

Anaerobes/Inert Gassing—The methods utilized for machine-based aseptic processing are adopted without change. Air should be substituted for inert gases in all systems, except in those rare instances in which an isolator providing true anaerobic conditions is utilized for the production process. In those situations, the usual inerting gas would be utilized and the chosen media would most likely be Fluid Thioglycollate Media.

Environmental Monitoring—The conduct of environmental monitoring for manual aseptic processing uses methods identical to those for other highly controlled ISO 5 environments. The same cautions exercised with monitoring in other applications apply as well. Performing the sampling must neither introduce contamination into the environment, the simulated materials nor into the environmental sample. The smaller size of the environmental systems utilized for manual aseptic processing means that the monitoring methods must be chosen for their lack of impact on the environment. Given the small workspace, it is best to conduct surface sampling only at the beginning and end of an operation. Similarly, all personnel monitoring should be done at the end of a work period rather than during work.

Execution of the Simulation—The process simulation should be performed in a manner that properly documents the activities. A batch record designed specifically for the simulation is a common approach. As with other simulations, the presence of an observer who documents the simulation can be beneficial. The methods and principles defined for automated filling or sterile bulk chemical production can be utilized with relatively minor modifications.

Additional Samples—Testing of materials (e.g., sideparts, simulated chemistry samples, etc.) other than those directly representative of the sterile materials being simulated may be useful in the event of failure to determine when/where in the process the contamination may have been introduced. Should these samples be contaminated in the same simulation in which the simulated production materials are found sterile, the contamination may not be considered significant.

Pre-Incubation Inspection—The methods utilized for machine-based aseptic processing are adopted without change. As the containers in manual aseptic processing are more likely to be hand stoppered/sealed, this inspection must be performed with utmost care, as the potential for a deficient seal might be higher than a machine applied closure (The same caution should be applied to inspection of hand-sealed production units for the same reasons). Units with defective or suspected defective seals should be segregated from the materials sent for incubation.

Incubation Time/Temperature—The same considerations as machine-based aseptic processing apply. A single incubation temperature between 20°C–35°C ± 2.5°C for 14 days is appropriate.

Post-Incubation Inspection—The methods utilized for machine-based aseptic processing are adopted without change.

Growth Promotion—The methods and considerations relevant to machine-based aseptic processing are adopted without change.

Interpretation of Test Results—The smaller size of production lots produced by manual procedures is typically less than the current minimum simulation size of 5,000 units.[10] Thus simulations conducted in support of container filling must be devoid of contamination in any of the filled units. In compositing or formulation simulations, the simulated bulk material container(s) should be sterile.

28.9 CONCLUSION

This chapter may be the only document that addresses this subject in a comprehensive manner. As such, it includes recommendations that establish precedents in the absence of guidance for regulators or industry associations. Firms that perform these types of processes must respect the uncertainties associated with any process that is so heavily reliant on personnel to excel at all times. In new installations, we strongly encourage the use of isolation technology to minimize the microbial contamination potential from personnel, however well trained they might be.[11]

NOTES

1. Agalloco J, Conrad D, et al, Process Simulation Testing for Aseptically Filled Products. PDA Technical Report #22, *PDA Journal of Pharmaceutical Science and Technology* 1996;50(6), Supplement.
2. FDA, Guideline on Sterile Drug Products Produced by Aseptic Processing, 2004.
3. Kino-oka M, Mizutani M, Medcalf N, Cell Manufacturability. *Cell and Gene Therapy Insights* 2019;5(10): 1347–1359.
4. Akers J, Agalloco J, et al., Design and Validation of Isolator Systems for the Manufacturing and Testing of Health Care Products. PDA Technical Report #34, *PDA Journal of Pharmaceutical Science and Technology* 2001;55(5), Supplement.
5. FDA, Guideline on Sterile Drug Products Produced by Aseptic Processing, 2004.
6. Agalloco J, Conrad D, et al., Process Simulation Testing for Aseptically Filled Products. PDA Technical Report #22, *PDA Journal of Pharmaceutical Science and Technology* 1996;50(6), Supplement.
7. FDA, Guideline on Sterile Drug Products Produced by Aseptic Processing, 2004.
8. Agalloco J, Conrad D, et al., Process Simulation Testing for Aseptically Filled Products. PDA Technical Report #22, *PDA Journal of Pharmaceutical Science and Technology* 1996;50(6), Supplement.
9. Agalloco J, et al., Process Simulation Testing for Sterile Bulk Pharmaceutical Chemicals. PDA Technical Report #28, *PDA Journal of Pharmaceutical Science and Technology* 1998;52(4), Supplement.
10. PDA, Guideline on Sterile Drug Products Produced by Aseptic Processing, 2004.
11. Agalloco J, Akers J, The Future of Aseptic Processing. *Pharmaceutical Technology* 2005, Aseptic Processing Supplement: S16–S23.

29 Aseptic Processing for Sterile Bulk Pharmaceutical Chemicals

James Agalloco and Phil DeSantis

CONTENTS

29.1 INTRODUCTION

The majority of sterile dosage forms are formulations of solutions; however, there are a number of sterile products in which the active pharmaceutical ingredient (API) is not in solution but may be a suspension or sterile powder. In many of these products, the API is aseptically filtered while in solution in an appropriate solvent system and then crystallized. The sterile solid is then separated from the liquid phase, dried, size modified as needed, blended (if necessary), and then bulk packaged for shipment to a dosage form site for final formulation and/or filling. All of these operations are performed aseptically. A similar process may be appropriate for sterile additives that are used in formulations such as Arginine Hydrochloride and Sodium Carbonate. In this chapter, the sterile bulk pharmaceutical chemicals (BPCs) can be either an API or a sterile excipient. Support for the aseptic processing activities utilized in the production for all sterile BPCs is a regulatory requirement.[1] There are sterile bulk pharmaceutical chemicals that are prepared by other than sterilizing filtration. There validation is not considered within this chapter.

29.2 BACKGROUND

Establishing the sterility of aseptically processed sterile drugs is one of the more difficult, if not the most difficult, tasks in the industry. As there are no direct means by which the sterility of aseptically produced materials can be determined, firms utilize a variety of measures to support their practices. Solution drug products are supported by such practices as: sterilization validation for components and equipment, filter validation for the production solution, decontamination of environments, environmental monitoring of air, surfaces, and personnel, sterility testing and process simulations. The situation is no different with sterile BPC manufacturing. On one level, the only apparent difference may be the size of the container being filled. In actual practice, though, because the sterile bulk materials undergo a phase change after sterile filtration from liquid to solid, process confirmation is far more complex.

Besides the commonality that BPC process validation shares with all other processes and the additional elements raised for APIs (described elsewhere in this text), the sterile bulk processing must consider the challenge of maintaining sterility.

Process simulation used to challenge sterile solution manufacturing employs a liquid media as a direct replacement for the product formulation. Direct substitution of media into the BPC simulation is complicated by the presence of both liquid and solids handling equipment in the process train for the sterile BPC. The use of a single material (liquid or solid) from the point of sterilization (typically 0.2 µm filtration) to the end of the process is problematic. Solids cannot be easily handled through the beginning (liquid portion of the process), whereas liquids are difficult to process in the later (powder handling) part of the process.

The regulatory expectations for sterile BPC production have been the subject of a rare direct interchange of views between the FDA and industry, as described in the listed references,[2] These documents express the differences in perspective between industry and regulators. In their inspection

DOI: 10.1201/9781003163138-29

guide, the FDA expressed their desire that process simulations should be performed in support of every sterile bulk process. In marked contrast, industry felt that there were instances in which simulations were either unnecessary or inappropriate. This exchange was followed by a brief hiatus during which PDA and PhRMA developed an industry guidance document defining process simulation practices and methods.[3] This document enjoyed only limited success. Although it was warmly received by industry, the FDA found several elements objectionable. Concerns were raised regarding the definition of closed systems, requirements for simulation of closed systems, sampling of materials, and other matters. When these comments were reviewed by the task force that had developed the industry guidance, the technical report was revised and a meeting with the FDA was requested. Unfortunately, the meeting was never held, and the revision process to TR #28 was suspended for a period of almost five years. In 2004, because of continued issues of contention between industry and the FDA, a further revision of the technical report was developed.[4] This chapter draws upon relevant aspects of that document and interprets them for implementation, while incorporating other relevant facility and operating aspects that are not addressed in the guidance document.

29.3 BUILDINGS AND FACILITIES

The facility for sterile BPC production must accommodate the different scale of the process equipment utilized for these processes. The crystallizers, dryers, filter-dryers, mills, and other equipment commonly used can be substantially larger than the typical aseptic manufacturing and packaging equipment used in the preparation and filling of finished sterile dosage form containers. Consequently, ceiling heights are higher and processing rooms larger to accommodate the larger equipment. In some instances, the use of solvents in the process will mandate explosion-proof equipment. Those portions of the facility housing process train outside the aseptic core may have features more comparable to ordinary BPC processing areas, albeit with somewhat greater attention to microbial and particle control. This is possible because much of the process is contained within a closed system.[5,6]

Critical Areas—The aseptic processing core of the sterile BPC utilizes materials and finishes similar to those found in aseptic filling facilities (see other chapters in this book for information on the details of their design). The majority of the aseptic environment is designed to maintain ISO Class 7 conditions during operations that will realize EU Annex 1 expectations for Grade B under static conditions.[7,8] Within this aseptic background environment, localized ISO Class 5 environments (EU Annex 1, Grade A) are provided where aseptic operations such as: seed introduction, material sampling, and subdivision into bulk containers is performed. In older facilities, aseptically garbed personnel enter these environments to perform aseptic interventions to complete the process. Newer designs employ a variety of barrier and/or isolator designs to provide greater separation between the personnel and the sterile materials and components. Facilities using isolation technology throughout could have these systems located entirely in an ISO Class 8 environment.

29.3.1 Closed and Open Systems

The production of sterile BPCs relies on 'closed' systems to protect the materials throughout the process within the aseptic core. Technical Report #28 originally defined 'closed' systems in its initial release in 1998, and this was altered slightly in the 2005 revision to clarify the requirement. A closed system can be described as:[5]

- "Is constructed, installed and qualified in a manner which demonstrates integrity is maintained throughout the full range of operating conditions, and over a time period inclusive of the longest expected usage (i.e., manufacturing campaign). The qualification is done according to a formal protocol, following generally accepted engineering principles, and is documented.
- Is sterilized-in-place (SIP) or sterilized while closed prior to use using a validated procedure.
- Can be utilized for its intended purpose without compromising the integrity of the system.
- Can be adapted for fluid transfers in and/or out while maintaining asepsis.
- Is connectable to other closed systems while maintaining integrity of all closed systems (e.g., Rapid Transfer Port, steamed connection, etc.).
- Is safeguarded from any loss of integrity by scheduled preventive maintenance.
- Utilizes sterilizing filters that are integrity tested and traceable to each product lot for sterilization of process streams."

In the most advanced designs, gowned personnel are never located in the same environment as sterile materials and packaging materials, as the process utilizes 'closed' systems throughout. The use of 'closed' systems for the liquid handling portion of the train is fairly easy to accomplish. The tanks, sterilizing filters, and other liquid equipment are easy to 'close', as they ordinarily operate in that manner anyway. As the process transitions to powder handling and final subdivision, isolators can be used to enclose the powder handling equipment as pressure rated, SIP capable equipment is not generally available for the powder handling steps.

Regulatory comfort with 'closed' systems is not assured. In their review of the 1998 version of TR #28, the FDA explicitly noted, "that all of the aseptic processing difficulties with sterile bulk pharmaceuticals has been with systems that were 'closed'." Whether that experience preceded the TR #28 definition issuance is uncertain, but some degree of caution is nevertheless warranted. Some of the regulatory caution is associated with an earlier perspective (stated in the 1994 PhRMA position paper) that if a system was fully 'closed' then a supportive process simulation was not required. [3] This view was restated less firmly in the 1998 industry guidance and removed in the 2006 version.

Barriers and other designs in which all of the attributes associated with 'closed' systems are not present are considered 'open' and are acknowledged as less capable. Requirements for process simulation in 'open' systems have never been a point of contention because of the greater potential for contamination associated with the less certain separation between personnel and sterile materials. The methods utilized for 'open' systems resemble dose filling simulations albeit on a much larger scale.

29.3.2 Supportive Clean Areas

The aseptic operations are ordinarily supported into two separate environments that serve different purposes.

Dissolution Area—The production materials to be made sterile are dissolved in large vessels in which solvents and other items (amorphous carbon for impurity removal/de-colorization) may be added. Use of organic solvents mandates an explosion-proof environment. This area is often separate from that utilized for the other preparation processes. The processing environment in which the dissolution is carried out can be either ISO 7 or 8 depending on the firm's risk assessment.

Preparations Area—The rest of the items utilized in the aseptic core are processed in an ISO 8 environment (with localized ISO 5 to control particle counts over selected activities). In this area, items are readied for sterilization/depyrogenation through double door units that can be later unloaded from inside the aseptic core. Airlocks and/or pass-throughs are utilized to facilitate the transfer of sealed sterile bulk containers from the core. Final packaging and labeling may be performed in this part of the facility. Wash areas for utensils, wrapping stations, spare parts, and other support systems are located within the preparations area.

29.3.3 Co-Location of Bulk Manufacturing and Dose Filling

A few sterile BPC production facilities have been constructed as a single aseptic suite with both bulk manufacturing and final dose filling capability. This type of co-location within a combined aseptic suite can reduce the capital expenditure as many of the infrastructure and support systems can be shared. The single suite permits a degree of integration and cooperation between the two portions of the process that facilitates operations.

29.4 PERSONNEL TRAINING AND QUALIFICATION

The personnel who work in sterile BPC facilities must have comparable skills to those of other aseptic operators. They must be proficient in aseptic gowning and aseptic technique as well as all of the required process-related skills associated with chemical synthesis operations. The methods and equipment utilized in sterile BPC production are substantially different than those in small container filling, and operating job training must be adapted accordingly. Perhaps the best measure of their performance is their participation in a successful process simulation incorporating all of the relevant interventions required.

29.5 CONTAINER-CLOSURE SYSTEMS

Sterile bulk materials are unique in that nearly every finished container will exit the aseptic core, be warehoused, shipped (perhaps globally), and later be introduced into the critical zone at another aseptic facility. This can place extraordinary demands upon the packaging system, and the best package designs provide maximum protection to the product throughout its shelf life. Further consideration regards the practices and methods utilized for initial filling and sealing of the containers at the manufacturing site, as well as the dispensing of the sterile material at each of the filling sites. The container-closure system is inexorably tied to the equipment and procedures utilized to sterilize, transfer, and fill the containers/closures and thus has substantial impact on the facility design as well as the integrity of the aseptic process. This may include means for taking representative samples for sterility and other testing in a noninvasive fashion during the filling/sealing process. The preferred system from a shipping/storage perspective may not be as easily filled or discharged. The choice of a packaging system must consider all aspects of its use from preparation, through filling, sampling and ultimate discharge. Presented following are brief descriptions of the primary packaging systems utilized for sterile bulks materials along with the key advantages/disadvantages of each.

Glass—The first container employed for many sterile bulk materials may be a glass vial with conventional stopper, as the initial scale of production may preclude the use of an alternative package. Scaled-up versions of these become wide mouth screw capped jars, with tape providing an additional layer of security. These have the advantages of being easily sterilized and/or depyrogenated, impervious to moisture, relatively inert, and can be inspected while still closed. Negatives to their use include uncertainties of seal integrity, difficult protection post-sterilization, and susceptibility to physical damage. Heat or twist sealed sterilized LDPE over-wrapping can be added to simplify introduction into the aseptic filling suite. Custom designed shipping containers are often added to enhance the container integrity during shipping. This packaging system is more likely to be used in early development and replaced as the scale of operation increases and as a consequence is less common than the others in commercial operations.

Aluminum/Stainless Steel—The model for these containers is the milk can once so common in the dairy industry. These rely on an elastomeric closure to seat the lid to the body of the container. The lid can be secured with wire or a lever system to ensure its integrity. The canisters can be either single (aluminum) or multiple (stainless steel) use design. Advantages are ease of sterilization/depyrogenation, greater moisture and light protection, and physical strength. Disadvantages include the inability to inspect the contents, difficult cleaning (if reusable containers are employed), awkward protection post-sterilization/pre-filling, and large opening size. Stainless steel cans that were once almost universal have been largely supplanted by single-use aluminum cans. Additional external layers of sterilized LDPE are added to facilitate entry into the aseptic core at the filling site. Fiber drums or other protective containers are sometimes provided as an outermost layer during storage and shipping.

Plastic Bags—The use of ultra-clean sterilized LDPE (or other plastic) is another common alternative. The bag in contact with the product can be heat-sealed or twist sealed (with a surrounding heat-sealed bag). An internal vacuum or pressure can be utilized to provide a direct indication of container integrity to the end user. Additional bags with laminate layers to eliminate moisture or light penetration are commonplace. The completed multilayered package is shipped in corrugated boxes or fiber drums. Advantages include low cost, low weight, and ease of inspection. Among the disadvantages are the absence of structure, potential static-charge, and risk of product charring in the heat seal. Package integrity during air shipment can be assured through qualification of the heat sealer.

Sterility/Product Samples—The preparation of a sterile BPC commonly includes the collection of sample materials for sterility and other testing. These samples must be obtained from the batch in a representative fashion that assures the validity of the sample and yet maintains the sterility of the materials being processed. Some adaptation of the product package is often used to collect samples in a manner that provides this confidence. It should be recognized that opening/accessing the sterile bulk containers to collect samples represents an unnecessary risk and is to be avoided. The use of a single composite container inserted into the subdivision process for sample collection is only slightly less risky. The preferred approach is to collect samples at intervals during the subdivision process and composite them for the various tests in a well-controlled (isolator?) environment.

29.6 TIME LIMITATIONS

Aseptic processes often utilize time limits on process steps prior to sterilization to preclude excessive bioburden in the filtration process, as well as prevent endotoxin from becoming a problem. Sterile BPCs are no exception, and appropriate time limits should be individually placed on the dissolution and filtration steps to provide the required controls. Time limitations are also imposed on the length of production campaigns between sterilization of the process train.

29.7 PROCESS SIMULATIONS

Study Design and Rationale—The development of a supportive rationale for the process simulation study to be conducted should be developed that outlines the assumptions/choices made in the design of the simulation. The primary considerations in the development of the rationale are presented next:

Materials/Media Sterilization/Introduction—The materials used in the process simulation should be sterilized using a validated method. There should be no reason to doubt the effectiveness of the sterilization process. The process utilized for sterilization of the challenge materials need not be the same as that utilized for the commercial process. The container utilized for the simulation materials should be the same as that employed for the commercial materials so that the handling methods can be identical.

Frequency and Number of Runs—Three initial process simulation studies are conducted for new facilities or after major changes to the process or equipment. As a minimum, annual process simulation studies are used to support the ongoing acceptability of the practices/system for continued production usage.

Duration/Size of Runs—The simulations should include the same number of manipulations as the commercial process being simulated. The amount of material used in the simulation must be sufficient to contact the entirety of the internal surfaces of the process equipment. The acceptance criteria for these studies corrects for differences in the size of the simulation batch relative to the commercial batch size.

Test Methods—The conduct of a process simulation for sterile BPCs must encompass the full breadth of the process from sterilization through filling into the bulk container. In that scope, it mimics the conduct of simulations for dosage forms. The most challenging part of the simulation is that sterile BPCs ordinarily start as solutions that are converted to solids during the aseptic process. The transition in material characteristics is accomplished within the process train, which must contain liquids at the beginning of the process and then be capable of handling powders in the later steps. The development of a simulation approach embracing the entire process using a single material throughout introduces concerns unique to the sterile BPC process.

One possible approach is to conduct separate studies for the different steps in the process, allowing the use of a liquid material for the early steps and a powder material near the end. The simulation components must overlap, and the results of both are considered in the evaluation of the overall aseptic process.

Test Material Selection—The selection of the test material to be used in the simulation is a major concern. The choices for test material include: growth promotion (or growth supportive) media; an inert placebo material (which can be a liquid or powder depending upon the portion of the process to be evaluated); a production material, provided it does not inhibit microbial growth (Lactose and Sodium Carbonate are possibilities); and lastly, a phantom material (evaluation is performed using post-process environmental and surface monitoring). PDA's TR #28 identified the major considerations relative to the material selection:

> Inherent in the selection of a test material, and the decision to use a test material at all are considerations of potential adverse effects implicit in the use of the material. As a general rule, nothing should be introduced into the system, whether media or placebo, which may present a problem in subsequent processing. The material (if used) must be able to be easily removed from the equipment in order to prevent an increased potential for contamination of production materials that would later enter the system.[4]

The choice of test material has profound impact on the simulation design and one of the major considerations is the testing performed on the material after completion of the simulation study. Evaluation of the simulation requires testing of the entire amount of material produced, and this too may influence the material selection process.

The test materials need not be sterilized in the same manner as the production materials so long as they are introduced into the system prior to the sterilizing step. When introduced mid-process, as might be the case in a simulation requiring different materials in different portions of the system, the introduction should be made in a manner that minimizes exposure to the environment.

Interventions—One of the essential requirements in the execution of an aseptic process simulation is the inclusion of manual interventions at a comparable level of frequency as that encountered in routine process operation. The conduct of a process simulation for a bulk process should include the appropriate number of inherent and corrective interventions to support their execution during production operations. As bulk processes typically use lower levels of process automation compared with aseptic filling of finished containers, the number of different interventions required may be fewer. Because of the more manual nature of many bulk processes, however, they may be more invasive (risky). Therefore, it may be appropriate to include an identical number of interventions in the simulation as are required in the production process.

Testing—Once a process simulation has been completed, testing of the materials produced in the simulation is required. As the packaging for the simulation must be identical to that utilized for the product, there may be instances where direct observation of a media cannot be performed on the bulk containers. If the placebo material used for the simulation is not a growth media, sterility testing of the entire quantity produced is required. This performance of post simulation testing introduces environments, sterilization processes, and interventional activities not required in the preparation of the production materials. It is these activities that introduce additional risk to the process and are an inherent part of the decision to use a non-zero acceptance criterion. Although it is essential that this testing be performed in a well-controlled environment with appropriate controls, the testing must be recognized as an aseptic process in its own right with a potential for adventitious contamination.

The testing process may require the dissolution of the entire placebo in a large volume of a suitable diluent followed by membrane filtration of the entire diluent in an adaptation of the membrane sterility test.[9] The execution of such a test clearly introduces complexities not present in either the bulk production process or its routine sterility test.

Incubation/Inspection—Aside for the complications associated with obtaining the material to be incubated (either a large number of clear containers that can be inspected or a single membrane filter in a growth media), the incubation of these materials is easily managed. The items are placed in an incubator at 20°C–35°C (at a single temperature with all locations within ±2.5°C) and held for 14 days.[10] Alternative incubation approaches can be utilized provided growth promotion requirements are satisfied. Post-incubation inspection of the containers is performed using methods similar to those employed for final dosage containers, with allowances for the larger size of the container. Upon completion of the inspection, samples of the media utilized in the test should have its growth promotion properties confirmed.[11]

Interpretation of Test Results—The acceptance criterion86 for sterile bulk process simulation was first defined by PDA.[4] It was chosen to provide a level of confidence in sterile bulk production comparable to that afforded to finished dosage forms. The basis of the criterion was that the sterile bulk will ultimately be filled into final containers and was set at a maximum contamination level in the bulk simulation that projected to less than 1 CFU/ 10,000 filled final product container. In this instance, the smallest batch size with the largest fill volume represents the worst case. The additional complexities associated with the sterility testing of large quantities, typically multiple kilo amounts, precludes the adoption of no allowable contamination as the acceptance criterion.

Campaigns—The complexities of sterile bulk systems dictate lengthy cleaning and sterilization cycles, and for this reason many sterile bulk manufacturers have adopted a campaign production model. Each campaign is preceded by sterilization of the process train, followed by multiple production batches and an eventual cleaning of the system. Preventive maintenance and other activities are performed during the interval between campaigns. Campaign lengths are selected by the firm based upon their risk tolerance, as the loss of one batch in a campaign to either sterility or other key quality attribute failure may necessitate the rejection of the entire campaign. If the failure can be absolutely associated with a mechanical failure, action would only affect those lots produced subsequent to the failed lot.

The use of a campaign operational mode must be supported by appropriate process simulation studies. Process simulation studies for open systems (most often the final subdivision of the material) utilized in campaign manufacturing must be supported by full duration challenge studies in which the length of the campaign and the number of interventions is identical to the production campaign. End of campaign simulations for open systems may be possible, provided trace residuals can be satisfactorily inhibited. If that cannot be accomplished, then the process simulation should match the full campaign duration with an equal number of interventions.

For closed systems, a different approach can be utilized. The process simulation can be performed matching a single production batch duration and activity levels. Campaign length is supported by maintenance of the closed system and confirmed by reaffirmation of the system leak rate as matching that at the end of a full campaign. End of campaign simulations for either open or closed systems are not practical with many of these systems, as the production materials are inhibitory of microbial growth to an extent that this cannot be adequately compensated for in the process simulation. As the materials must be fully inactivated prior to the simulation (without compromise to the systems asepsis), this approach is impractical.

29.8 STERILIZATION

The process train for the production of the sterile BPC should be subjected to a validated sterilization process. The most widely used method is steam sterilization-in-place (SIP), and this has been acknowledged by regulators. SIP is best suited for the closed portions of the system that often coincide with the liquid portions of the train, as the equipment for those steps is ordinarily both vacuum and pressure tight. The caveats stated earlier for closed systems are just the beginning of the requirements for a steam sterilizable system. Detailed coverage of SIP is provided elsewhere in this volume. Other sterilization methods for closed systems have been utilized including formaldehyde and peracetic acid; however, these are more difficult to validate than steam and as a result have been viewed with skepticism by regulators.[1] Dry heat has been successfully used for sterile BPCs where the conversion from liquid to solid is accomplished by spray drying. Hydrogen peroxide has been successfully utilized for both process equipment and isolators that are increasingly popular in the later more open stages of the process train where the materials are in powder form.

Open systems represent more difficult circumstances. The equipment can be sterilized by any of a number of methods, usually combinations covering all of the equipment items including incidentals required for the process. In open processes, the surrounding environment may be a conventional clean room or an isolator.

In order to produce a sterile BPC, sterilization of many other items is required. Utensils, tools, sample containers, environmental sampling materials, and other Items must all be available for use in the process. Sterilization can be accomplished by any of the common methods including steam, dry heat, ethylene oxide or radiation. Most importantly, the container/closure system utilized for the sterile bulk must be sterilized. Provision should be made in all of these processes to ensure that the items can be introduced into the environment in a manner that minimizes the risk of microbial ingress. This may entail multiple layers of protective materials that protect the items from the sterilizer to the point of use.

The last element of the process is the sterilizing filtration used for the liquid product. The validation of filtration and all of the other sterilizations required for the process is essential. This book includes chapters describing the validation of many of these sterilization processes.

Depyrogenation of equipment is accomplished by a variety of means: environmental control, caustic washes, WFI rinses and other methods appropriate for the items and equipment being treated. Content on depyrogenation processes is provided in this volume.

29.9 LABORATORY CONTROLS

Environmental Monitoring—The conduct of an aseptic process is ordinarily supported by an environmental monitoring that assesses the conditions during the process. In the liquid portions of the process train, this monitoring is restricted to the surrounding environment, and thus the results are not indicative of material impact. In open systems, normally associated with the powder subdivision steps, the environmental conditions must be carefully controlled to protect the product quality attributes. The practices required are essentially identical to those defined for final dosage forms. The single biggest caveat is that the presence of fine powders of antibiotics can inhibit the growth of microorganisms in viable samples and interfere with nonviable particle counting as well. The monitoring systems must be adapted to ensure that the results of the monitoring are valid.

Details on the conduct of both viable and nonviable monitoring are provided elsewhere in this book.

Sterility Testing—The performance of sterility testing for sterile BPCs ordinarily presents no unusual hurdles. Aside from the need to inhibit antibiotic antimicrobial activity in order to conduct a valid test, the test is performed in accordance with standard practices. The obtaining of samples for the testing was discussed in conjunction with the selection of the bulk container and should be performed in a manner that does not compromise the sterility of the materials in either the bulk container or the sample itself.

Chemical Testing—Samples for chemical and other testing should be obtained in the same manner as those for sterility.

29.10 CONCLUSION

The production of sterile BPCs represents perhaps the most difficult of all aseptic processes. The complexity and size of the equipment train, the large openings of the bulk containers and the predominantly manual tasks associated with the more critical operations during the final subdivision of the material all serve to make this process a significant challenge to any firm. The methods described previously provide some rudimentary guidance to the practitioner, who must be knowledgeable in the unique aspects of bulk chemical processes, sterilization, and aseptic processing.

NOTES

1. FDA, Guide to Inspection of Sterile Drug Substance Manufacturers, 1994.
2. PhRMA, Sterile Bulk Pharmaceutical Chemicals: A PhRMA Position Paper. PhRMA QC Section Bulk Pharmaceuticals Committee and Sterile Bulk Pharmaceutical Chemicals Subcommittee, Pharmaceutical Technology, August 1995.
3. Technical Report No. 28, Process Simulation Testing for Sterile Bulk Pharmaceutical Chemicals. *PDA Journal of Pharmaceutical Science and Technology* 1998;53 (S3).
4. Technical Report No. 28 revised, Process Simulation Testing for Sterile Bulk Pharmaceutical Chemicals. *PDA Journal of Pharmaceutical Science and Technology* 2006;60, Supplement S-2.
5. Agalloco J, Closed Systems and Environmental Control: Application of Risk Based Design. *American Pharmaceutical Review* 2004;7(4): 26–29, 139.
6. Agalloco J, Closed Systems: Changing the Aseptic Manufacturing Paradigm. Accepted for publication in *Pharmaceutical Technology*.
7. ISO 14644–1, Cleanrooms and associated controlled environments, 2015.
8. EU Annex1, Sterile Medicinal Products, May 2004.
9. USP <71> Sterility Tests.
10. FDA, Guideline on Sterile Drug Products Produced by Aseptic Processing, 2004.
11. PDA, TR #22, Process Simulation Testing for Aseptically Filled Products. PDA Technical Report #22, *PDA Journal of Pharmaceutical Science and Technology* 1996;50(6), Supplement.

30 Qualification and Validation of Advanced Aseptic Processing Technologies

James Agalloco and James Akers

CONTENTS

30.1 INTRODUCTION

Aseptic processing is used for the manufacture of sterile products in which the materials, containers/closures and process equipment are separately sterilized and assembled into a sealed container in a contamination-controlled environment. It is widely utilized in the food, medical device, biotechnology and pharmaceutical industries for the manufacture of products whose essential properties would be damaged by terminal sterilization processes using steam, radiation or other means.

For many years, aseptic processing was performed in classified environments with little or no separation between gowned personnel and sterilized product, components and product contact equipment. A gradual shift in approach began in the 1970s with the recognition that the aseptically gowned personnel were the ***primary*** source of contamination, both viable and nonviable, in aseptic processing, increasing separation between personnel and the sterilized items was instituted as a means of limiting contamination risk. Initially the separation consisted merely of partial barriers surrounding the filling and stoppering operations. A major step forward was taken in the 1980s when the first aseptic isolators provided for near-absolute separation between the critical areas where sterile items are exposed and the surrounding environment where personnel were present. Difficulties at some firms during the 1990s with the implementation of isolation technology resulted in the development of a derivative technology—Restricted Access Barrier Systems (RABS) which sought to provide isolator-like performance using a less sophisticated approach. By the end of the century, there was a confusing myriad of isolators and RABS available. As the technologies differed so did their performance, and a means to distinguish them became increasingly necessary. Advanced aseptic processing technologies were defined as:

> An advanced aseptic process is one in which direct intervention with open product containers or exposed product contact surfaces by operators wearing conventional cleanroom garments is not required and never permitted.[1]

This definition delineates the real difference between technologies that truly separate personnel and sterile materials and other less capable systems. This chapter addresses the approaches for the design, qualification and validation of advanced systems whose implementation differs substantially from that required for less separative and thus less capable systems.

Although there is a global preference for the use of terminal sterilization because it provides superior assurance of sterility, the adverse effects of many sterilization processes are such that an estimated 85% of sterile pharmaceuticals are made using aseptic processing.[2,3,4] Sterile products are considered high risk, as they are customarily administered in hospital and clinical settings where the patient may be immunocompromised. Should the product contain microbial contamination, there is a potential for patient injury, and in a "worst case" situation even death. The criticality of concern

DOI: 10.1201/9781003163138-30

is such that aseptic processing is subject to intense scrutiny to assure that the aseptically produced health care product does not contribute to patient risk.

30.2 EVOLUTION OF CONTEMPORARY ASEPTIC PROCESSING TECHNOLOGIES

The origins of aseptic technique go back to the Joseph Lister who introduced them to the medical profession in the 1860s to better manage infection risk in patients undergoing surgery.[5] Lister's efforts revolutionized the way surgery was conducted, and similar aseptic techniques spread to the then nascent science of medical microbiology and ultimately to the manufacture of sterile products and medical devices. Early aseptically produced sterile products were crudely made, and a few simple practices were adopted to improve their safety with the limited technological capabilities available. As late as the mid-1950s, sterile products were made without the benefit of high-efficiency particulate air (HEPA)-filtered air supply and many other present-day practices. Prior to the 1950s, aseptically produced sterile products were manufactured using two similar methods:

- Manual/semiautomated assembly by gowned personnel in clean, but unclassified environments (the concept of room classification did not yet exist).
- Manual assembly by personnel using a glove box (a non-ventilated sealed unit accessed via gloves).

The inherent limitations in these processes are historical fact; there were only limited available means for improvements in safety during the entire period from Lister's first efforts until the middle of the 20th century, and largely these improvements focused on manual aseptic practices, a growing number of chemical disinfectants and perhaps most critically preservatives. The production of sterile products by aseptic technique, even without sophisticated means, employed the same principles of physical separation and gowning in evidence today, although the gowns were quite limited in capability. Although the technology was comparatively crude and rather primitive, products of the pre-clean room era were effective and safe. It is important to recognize that hospitals, as is the case today, were confronted with even more daunting contamination control challenges. Industry's ability to control microbiological contamination improved significantly when the HEPA filter (developed for the Manhattan Project in World War II) was declassified and became available for general usage. The HEPA filter allowed entire rooms to reach levels of particle and microbial cleanliness not previously attainable, and wholesale changes in facility designs and operating practices resulted. With large volumes of air effectively free of microbial contamination, it was possible to dilute and/or remove from the environment human borne contamination, which is always the greatest source of contamination. The advent of HEPA filtration made it possible to utilize equipment inside a clean room to perform most if not all the aseptic process, with gowned personnel in support. This allowed the introduction of automatic bottling and assembly equipment resulting in the production of much larger volumes at lower cost with substantially less human activity and therefore greatly reduced risk. Over the years, this concept evolved into the manned clean room that remains the most commonly used environment for aseptic processing operations. The use of gloveboxes continued at a much-reduced scale for non-sterile developmental activities of very potent compounds but only rarely for aseptic processing given the safety limitations inherent in the simple early designs. The manned clean room became the industry preference for aseptic processing because it was amenable to automated filling, inclusion of large-scale processing equipment and could be easy adapted for supportive activities such as product formulation and component preparation. By the 1980s, clean room designs incorporated flexible plastic and/or rigid barriers to provide greater separation between operators and sterilized materials, products or components. Throughout this period, as contamination control practices became more effective, there was a growing understanding that personnel were the primary source of contamination in aseptic processing, and there were improvements introduced in manned clean rooms that resulted in performance advances.[6]

Although the technological changes were significant, they were insufficient. Regulatory pressures for continued improvement in aseptic processing performance were substantial, and a noted FDA inspector defined the problem with manned clean rooms in a straightforward manner:

> It is useful to assume that the operator is always contaminated while operating in the aseptic area. If the procedures are viewed from this perspective, those practices which are exposing the product to contamination are more easily identified.[7]

Clearly something beyond the manned clean room was needed because of the inherent limitations the operator's presence creates if aseptic processing was to improve further. It had become obvious that the future in aseptic processing lay not with continued incremental improvements in manned clean room operations, but rather the elimination of direct human involvement altogether.

The early 1980s had witnessed the reemergence of the glovebox with substantial improvements that incorporated elements of clean room technology, novel transfer systems and a means for internal decontamination.[8] Some of the manipulative and transfer technologies used in these newer gloveboxes or isolators came from the nuclear fuel industry, and this high-level contamination and containment technology was married to improved antimicrobial decontamination systems. The first pharmaceutical isolators were utilized for sterility testing where human-derived contamination from the analyst was a nagging concern. By the early 1990s, there were frequent reports regarding remarkable reductions in sterility test positives and the absence of detectable microbiological contamination in sterility test isolators. Their adoption spread rapidly across the industry, to where they are essentially considered CGMP in all developed countries (Figure 30.1).[9]

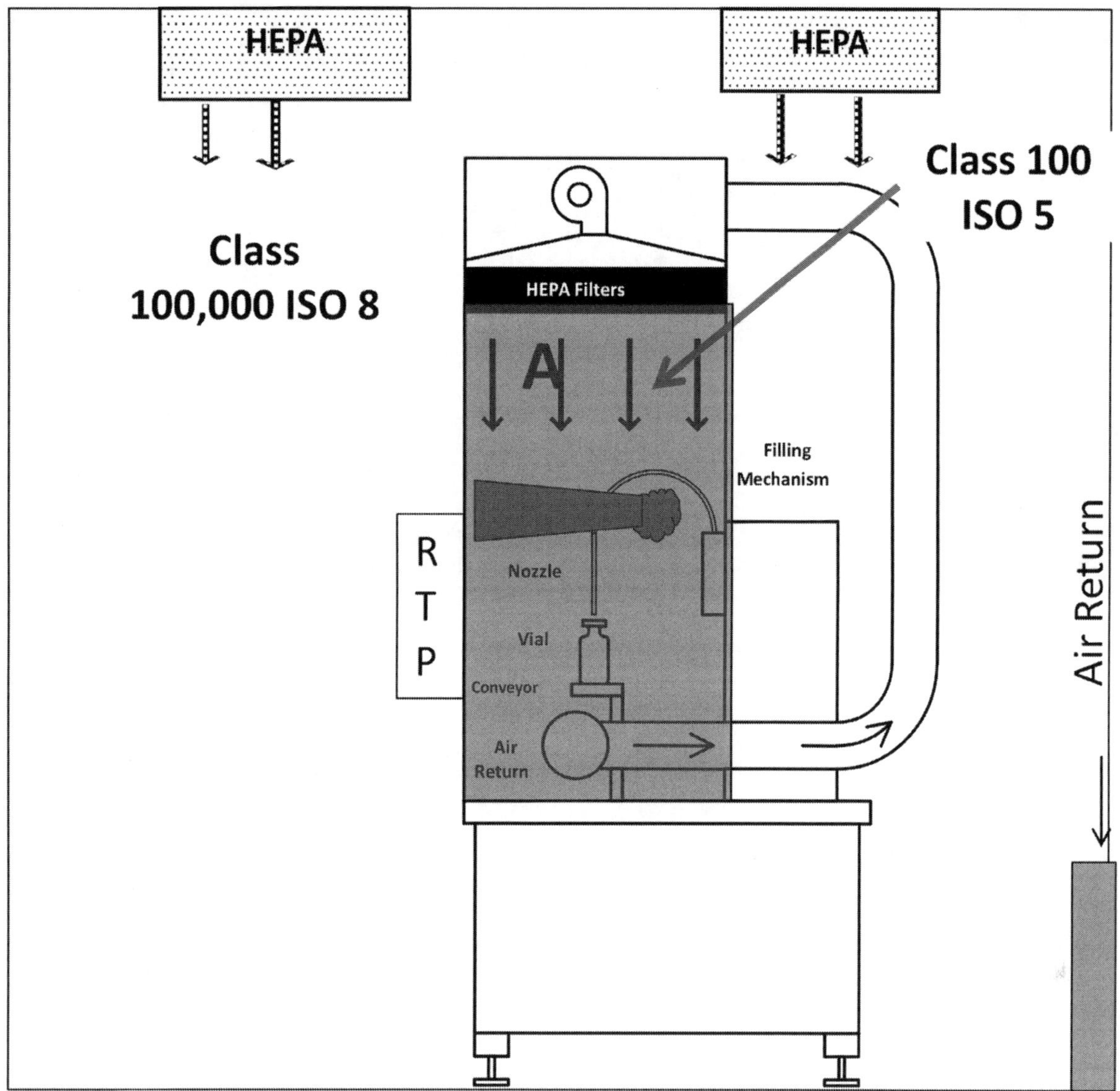

FIGURE 30.1 Simplified isolator configuration.

Note: HEPA = high-efficiency particulate air; RTP = rapid transfer port.

Success with isolation technology for sterility testing led to its adoption for aseptic filling, and the first of these facilities became operational in the mid-1980s.[10] Despite initial skepticism, it was evident that the isolator was potentially an enormous improvement in contamination control relative to even the most sophisticated clean room designs then available. The ability to completely exclude personnel from the operating environment where sterilized materials are exposed was recognized as an enormous advantage over manned clean room systems. The first production isolators for aseptic processing began to appear with expectations comparable to those observed for sterility testing. There are two basic types of isolators (closed and open):

> An isolator is sealed or is supplied with air through a microbially retentive filtration system (HEPA minimum) and may be reproducibly decontaminated. When closed it uses only decontaminated (where necessary) interfaces or Rapid Transfer Ports (RTPs) for materials transfer. When open it allows for the ingress and/or egress of materials through defined openings that have been designed and validated to preclude the transfer of contamination. It can be used for aseptic processing activities, or containment of potent compounds or simultaneously for both asepsis and containment.[11]

Over the last 20–25 years, isolators have become the technology of choice across much of the biopharmaceutical industry, especially in new facilities.[12] Industry experience has demonstrated both their superior performance and cost effectiveness relative to competing designs.[13,14] The latest innovation in isolation technology is the gloveless isolator where robotics and other automation eliminate any human involvement with sterile materials.[15]

Some were so enthusiastic as to claim that sterility assurance equivalent to terminal sterilization would be possible with

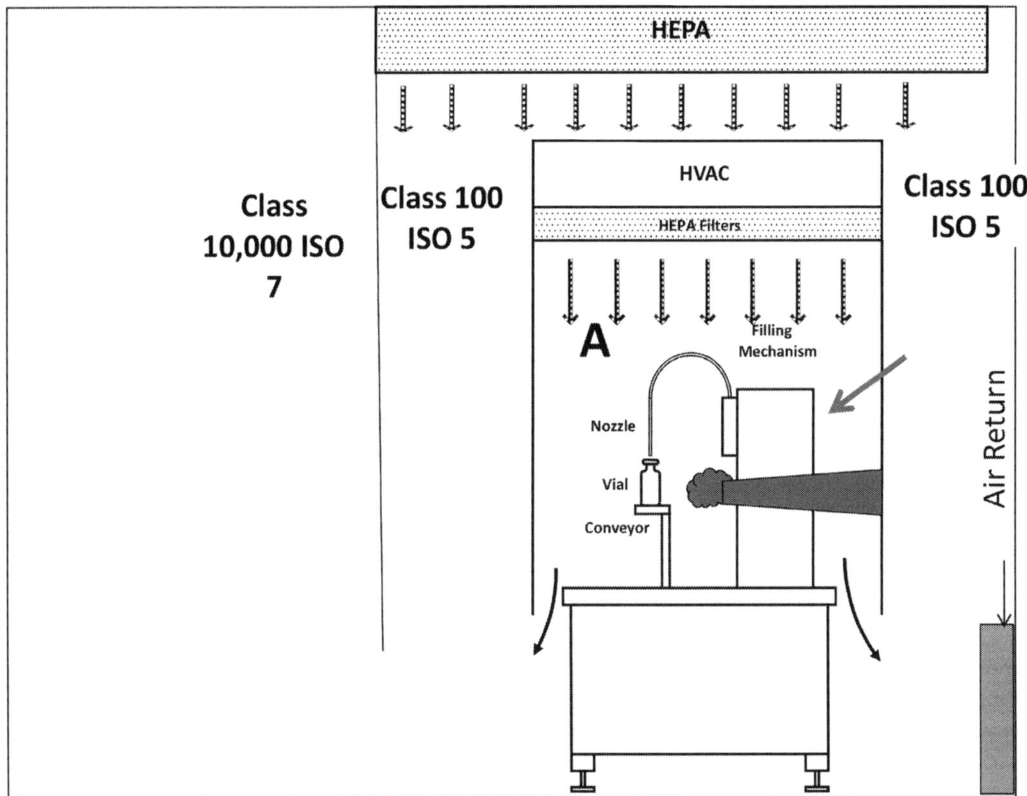

FIGURE 30.2 Simplified RABS configuration.

Note: HEPA = high-efficiency particulate air; HVAC = heating, ventilating, and air conditioning; RABS = restricted access barrier systems.

this new technology. This claim led to vigorous debate and had the unfortunate and unintended consequence of focusing attention away from the main advantage of human-free processing, which is reduced product contamination risk. As with many technologies, implementation took time, especially in the extremely conservative pharmaceutical industry, and mistakes were made. Although some firms implementing isolator technology experienced considerable and well-publicized problems, there were more successes than failures.[16,17] A number of firms, frustrated by the lack of progress they encountered with isolator implementation (complaints ranged from difficulty with decontamination efficacy and cycle time; persistent leaks; changed ergonomics, limitations in access for operation; and maintenance), sought other means to realize the operational performance improvements of isolators without the problems.

The result of these was the RABS concept that, although visually similar to isolators, was conceived as a means to overcome the perceived validation and operational hurdles[18,19] The principal issue RABS was intended to resolve involved vapor hydrogen peroxide decontamination of isolator enclosures. It was expected that RABS could employ an alternative means to decontamination using what some advocates defined as "high level disinfection". However, this came with a different, but still challenging, set of problems. RABS, assuming decontamination and full human separation under all operational conditions, are thought by some to provide the sterility assurance reliability of isolation technology with fewer technical complications. The RABS is a highly evolved barrier system that builds upon the separative approaches used in manned clean rooms through the use of material handling elements like those used with isolators. RABS may be derived from isolation technology, even though in some applications they may have more in common operationally with the manned clean room (Figure 30.2).

RABS systems are also further characterized as "closed" RABS and "open" RABS. The closed designation means that, like the isolator, direct human intervention is not allowed in this style of RABS operation. In other words, the "closed" RABS approach is from the separative perspective effectively identical to an isolator in operation. In contrast, "open" RABS as conceived contemplated operations in which the barriers were closed most of the time, but open-door interventions are allowed. As even "closed" RABS are sometimes opened between use for format changeover and decontamination, the distinction between RABS modalities is further blurred. The definitional difficulties for standard setting and compliance policy associated with this duality of operational control have proven to be difficult to resolve. The definition of advanced aseptic processing allows for a clear distinction to be drawn between "closed" and "open" RABS.

As interest increased in the implementation of RABS, ISPE developed an initial definition:

> A Restricted Access Barrier System (RABS) is an advanced aseptic processing system that can be utilized in many applications in a fill-finish area. RABS provides an enclosed environment to

> reduce the risk of contamination to product, containers, closures, and product contact surfaces compared to the risks associated with conventional cleanroom operations. RABS can operate as "doors closed" for processing with very low risk of contamination similar to isolators or permit rare "open door interventions" provided appropriate measures are taken.[19],

This was subsequently condensed by ISPE in 2015 and consequently lost some of the clarity provided in the original:

> An aseptic processing system that provides an enclosed, but not closed, environment meeting Grade 5 conditions utilizing a rigid-wall enclosure and air overspill to separate its interior from the surrounding environment.[20]

The PDA isolator and ISPE RABS definitions rely upon the operational characteristics of the system to define it appropriately, and each includes the option of operating in a "closed" or "open" manner. It might seem from the similarity of the definitions that these systems would have comparable capabilities. Nevertheless, there are important differences between them that bear further discussion. The nuances of each design will be briefly addressed in this chapter.

Isolators

Isolators can be either "open" or "closed" depending upon their operational state and may operate at positive, neutral, or negative pressures with respect to the surrounding environment. When "closed", isolators do not exchange unfiltered (HEPA or better) air with the surrounding environment. When "open", isolators should not exchange unfiltered air with the surrounding environment as air overpressure or overspill precludes the entrance of contamination. In fact, in many current isolator designs, the opening or "mouse hole" is used only for the exit of fully sealed product. Thus, the performance differences between "open" and "closed" isolators with proper design and operation are negligible.

Isolators are customarily decontaminated using automated equipment that injects vaporous or gaseous sporicidal agents that can reproducibly render the interior of the isolator free of microorganisms. The result is what is an effective "germ-free" environment for aseptic processing. These sporicidal processes are typically used to treat parts or component supply hoppers and feed mechanisms in situ. At one time, this engendered a great deal of discussion regarding differences in concept between sterilization and decontamination. However, experience has confirmed that in practice the antimicrobial effects achieved on these parts supply systems are not a source of risk. In fact, treatment in situ is less risky by far than sterilization using a validated autoclave cycle followed by aseptic assembly in the critical zone by gowned human operators.

Isolators can be operated so as to maintain a pressure differential between the enclosure and the surrounding environment that can be beneficial in ensuring operator/product safety. Isolators are intended to be operated without opening during use, the designation of "closed" or "open" with respect to an isolator relates solely to material infeed/discharge in a batch (closed) or continuous (open) mode.

Restricted Access Barrier Systems

RABS can provide many of the same operational advantages as isolators while eliminating what are to some the more challenging aspects of isolator design. RABS typically operate in a continuous mode just like an "open" isolator. A "closed" RABS is one that restricts all operator activities subsequent to decontamination in the same manner as an isolator. An "open" RABS allows for the enclosure to be opened during the processing to allow direct access by aseptically garbed personnel. Thus an "open" RABS is decidedly less capable, as its performance in the open condition is no different than that of a manned clean room. Additional details on "open" RABS will not be included herein as this technology does not meet the advanced aseptic processing technology definition.

RABS are prepared for use using high-level disinfection with sporicidal materials by aseptically gowned personnel. The treatment is preferably performed with the enclosure closed. A few "high-level" disinfection methods have been employed. In some designs, the RABS is sealed just like an isolator, decontaminated and then operated normally. Other users decontaminate both the RABS enclosure and the surrounding room simultaneously. The less separative and least automated approach is manual decontamination by aseptically garbed personnel, which may also be the most common methodology. A reliance on manual decontamination operations makes validation dependent on human performance, which increases the potential for variability.

As with all new technologies, the general descriptions of isolators and RABS presented previously cannot fully accommodate the nuances of difference within and between each category of equipment. This was recognized early on by both PDA and ISO, each of which developed a visual continuum to explain the capabilities of the various systems. Neither continuum includes RABS as a specific point of clarification, and ISO does not mention isolators using instead a more general term "separative enclosures" (Figures 30.3 and 30.4).[21,22]

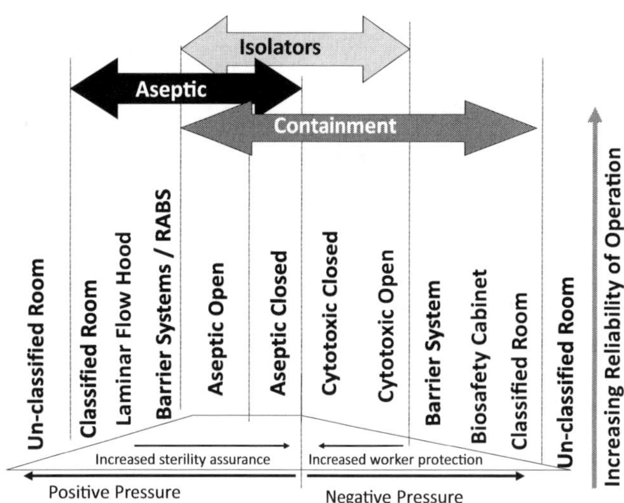

FIGURE 30.3 PDA's isolation continuum.

Note: PDA = Parenteral Drug Association; RABS = restricted access barrier systems.

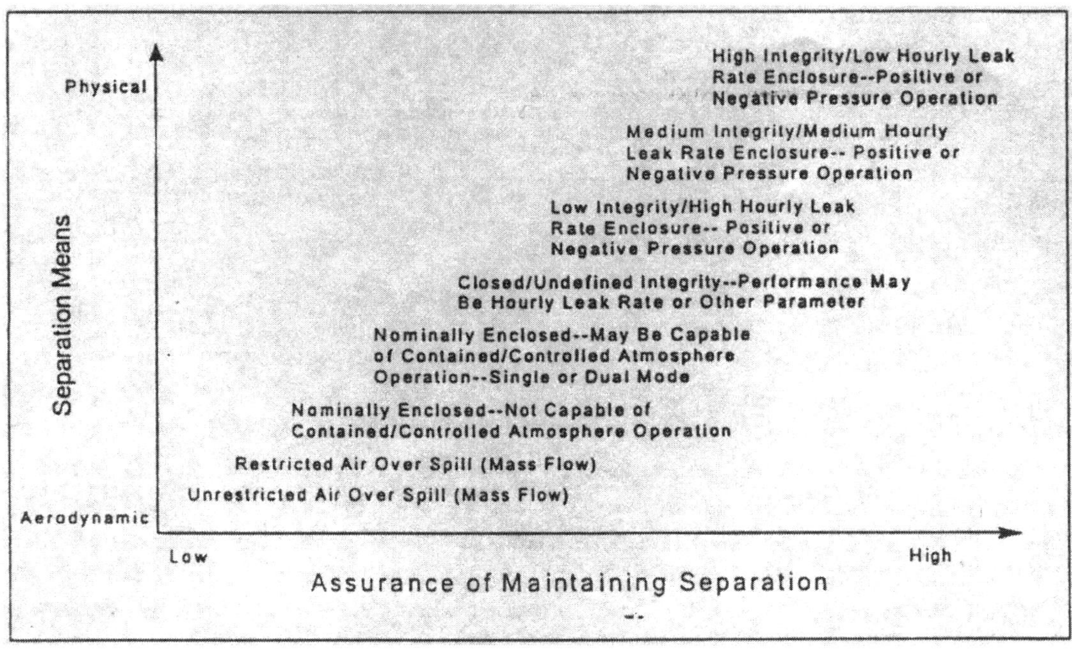

FIGURE 30.4 ISO 14644–7 continuum.

30.3 WHY ADVANCED ASEPTIC PROCESSING TECHNOLOGIES?

The driving forces behind the application of advanced aseptic processing technologies for sterile products relates foremost to improved separation between personnel and sterile materials.[23] When isolator technology is utilized, several added benefits are realized:[24]

- The ability to create a truly germ-free operating environment.
- A substantial reduction in operational cost for the facility.
- A shortened time period from conceptual design to full implementation compared with standard facilities intended for the same process.
- Containment of toxic materials to an extent previously unattainable.
- Substantial reduction in gowning requirements for personnel.
- Increased labor efficiency because of more flexible use of staff.
- Ability to produce in a campaign mode over extended periods.

When RABS technologies are utilized, a slightly different set of unique benefits are possible:[25]

- A means for evolutionary upgrade to existing facilities with reduced downtime.
- Lower capital cost when retrofitting an existing operation (this cost advantage for RABS over isolators generally disappears when new production lines are specified).
- Easier implementation because practices and equipment are more like predecessor designs.

The highly automated processing equipment available further blur the distinctions described previously. Aseptic processing environments essentially disappear in the midst of an automated machine. We are entering an era in which intervention-free systems will be the norm rather than the exception.

30.4 ADVANCED ASEPTIC PROCESSING SYSTEM QUALIFICATION AND VALIDATION

Qualification and validation are required activities for compliant and robust manufacturing across the global health care industry. Many firms have obsessed about these activities to the point where they have been blamed for extensive delays in project implementation. Much of the pain associated with the validation of advanced aseptic processing technology has been self-inflicted.[26] Validation has been defined by FDA as:

> Process validation is a documented program which provides a high degree of assurance that a specific process will consistently produce a product meeting its pre-determined specifications.[27]

Unfortunately, as the definitive statement on validation from the world's largest drug regulatory body, that definition leaves substantial room for interpretation. The one element of this definition that seems to create the most difficulty is establishing what actually is a "high degree of assurance". For established processes, such as steam sterilization or even cleaning validation. the criteria for acceptance have been widely discussed and are thus largely consistent across the industry. Isolators and RABS present unique challenges as they combine elements of sterilization, aseptic processing, environmental control, clean rooms and containment into a single system. Within the context of the multiple concerns that must be satisfied and the expected superior performance of these systems (as contrasted with ordinary human scale clean rooms for aseptic processing), well-intentioned, but unfortunately inexperienced (at least with isolation technology and decontamination) individuals and firms suggested criteria that mandated perfection in all aspects. Their stated objective was an aseptic filling isolator capable of providing finished product containers equivalent in sterility assurance to terminally sterilized products! This is a patently unattainable goal; no technology that relies on exclusion of microorganisms from sterile materials can ever realize the same degree of confidence as a terminal treatment that destroys their viability. Nevertheless, the damage was done. Isolators and to only a slightly lesser extent closed RABS were portrayed as systems that must be capable of operating contamination free for extended periods in order to be acceptable. The unfortunate consequences of this expectation were requirements for absence of leaks, especially in gloves; perfection in decontamination/sterilization; complete absence of internal microbial presence and other equally unattainable constraints.[28,29] Given unrealizable goals, laudable as they might be, it is no wonder that the validation of advanced aseptic systems can become an exercise in futility.

PDA addressed the subject of acceptance criteria by focusing on the definition of appropriate user requirements specifications. Justification for this approach can also be found in another definition of validation:

> Validation is a defined program which in combination with routine production methods and quality control techniques provides documented assurance that a system is performing as intended and/or that a product conforms to its predetermined specifications. When practiced in a "life cycle" model it incorporates design, development, evaluation, operational and maintenance considerations to provide both operational benefits and regulatory compliance.[30]

When the acceptance criteria are derived from the operational needs of the process rather than arbitrary expectations, they become more realistic, and thus easier to satisfy. If this model is followed, rigorous criteria can still be established; however, those criteria are defined by satisfaction of operating needs. An isolator (or other technology) need not be perfect in order to fully satisfy operational requirements; it need only be suitable for its intended use.[31]

The "life cycle" approach to validation provides for cradle to grave consideration of a system's compliance in a validated state[32,33]. Discussion of this model is best considered in an essentially chronological order. The various stages of the model include conceptual design and planning; detailed design and fabrication; equipment qualification; sterilization/decontamination cycle development (if required); performance qualification; operational use of the system and maintenance. Development of a well thought out user requirements specification is considered essential to success. The various elements of that specification form the essential criteria against which the design and performance of the system will be evaluated.

30.5 CONCEPTUAL DESIGN AND PLANNING

The origin of any system begins with its basic design, which must focus on what it is intended to do. Questions to be answered include the what, where, when and most importantly, the how. Implementation options are reviewed, discussed, refined, discarded and resurrected as the design process proceeds. Ultimately a system design is developed that satisfies the firm's operational requirements and capabilities. At this stage, the design incorporates decisions regarding isolator/RABS configuration, expected capacity, location, classification of the surrounding environments, adaptation to existing equipment and facilities, and preliminary process description (use, cleaning, and decontamination/sterilization). Once the design is completed and accepted, the project proceeds to detailed design and fabrication. The larger and more complex systems being developed often include a hybrid of RABS and isolators exploiting the unique operational advantages of each technology. RABS sections provide for protected infeed/discharge of materials. Isolators are employed for the more critical activities of filling and stoppering. Projects that are complex enough to require a validation plan should start its development at the completion of the conceptual design.

30.6 DETAILED DESIGN AND FABRICATION

Early in detailed design the system will be preliminarily sized for its intended use. For systems that are custom designed for a specific purpose, a mock-up is recommended to confirm that the intended design meets the end user needs. As these systems generally rely upon operators to perform a variety of tasks, ergonomics can play an important role in the suitability of the final design. The mock-up is used to confirm that the operators can readily perform all of the required functions. It is best to consider the full complement of operators who will use the system and the effect fatigue may have on their ability to perform the necessary tasks. The time and expense associated with a mock-up evaluation is usually well spent. A substantial amount of useful ergonomic information can be gleaned from a simple mock-up made of plywood. The extent of detailed design for "standard" isolators or RABS, if there are such things, is relatively minor. Many applications such as sterility testing or aseptic filling are performed in nearly identical fashion at multiple firms. In these instances, neither mock-ups nor an individual detailed design may be necessary.

Where detailed design is performed it expands upon the conceptual design defining the specifics of the overall system. Detailed design culminates in a set of drawings that will be used for construction and assembly. Where the system includes some measure of functionality (CIP, automatic purging of oxygen, system decontamination, etc.) these are defined in written specifications for the system's operation. The final drawings and specifications are ordinarily submitted to the customer for approval before fabrication is begun.

Fabrication against the approved specifications is performed at the system manufacturer's facilities and may include a factory acceptance testing or FAT. The FAT, which is often a part of the formal qualification, is ordinarily performed by the owner's staff prior to shipment. The original intent of the FAT was to confirm that the system was ready to be relocated from the vendor to the user, i.e., there were no required modifications that had to be made by the vendor at their site before it was relocated. In recent years, the FAT has tended to become a first step in the qualification of the equipment, while still providing the important acceptance of the completed system.

30.7 VENDOR INSTALLATION/COMMISSIONING

After acceptance of the system by the customer, the vendor will ship the system to the site and provide installation services to prepare it for on-site commissioning and completion of qualification. For larger systems, this may be the first time the entire system is assembled as an operational whole. The installation is ordinarily not considered a controlled activity in that it is generally not performed in accordance with a set of written procedures. The later portion of the installation and commissioning is often a convenient opportunity to check for conformance to specifications (IQ) as well as prepare standard operating procedures. The commissioning serves to ensure that the system is ready for operational qualification. Although this activity may seem redundant, it makes little sense to institute formal qualification activities on a system where the vendor may have made some minor error of omission or commission. The formal qualification activities should only commence after any installation miscues have been rectified. If there are changes made to the system that might affect its configuration performance, these should be documented and formally approved.

30.8 BACKGROUND ENVIRONMENTAL CONSIDERATIONS

Advanced aseptic processing systems are ordinarily installed in controlled and/or classified environments. Small isolators such as those used for sterility testing, low-volume clinical manufacturing and similar activities are rarely integrated into the room in which they are installed. In these instances, the isolator is essentially brought into a classified room, connected to the appropriate utilities and placed into operation. Its impact on the surrounding environment often isn't significant, as the background environment is typically no higher than ISO 8 (Class 100,000) and in some instances is merely provided with controlled temperature and humidity with restricted access. Larger isolator and RABS systems become an integral part of the room (and even the facility) and thus greater attention must be paid to maintaining proper conditions throughout the system and facility. Pressure differentials through mouse holes, tunnels and pass-throughs should be addressed. One of the important considerations for the facility is the need to maintain controlled temperature and humidity specifications across the environment to facilitate uniform decontamination/sterilization as required by the system specified. This is especially important with hydrogen peroxide and peracetic acid, which are two-phase systems and condensation has a major impact on their lethality. Other systems may require elevated humidity levels of 70% or more for optimal efficacy.

30.9 EQUIPMENT QUALIFICATION

Qualification of equipment is a preliminary step in the overall validation that has been well described in the literature (and elsewhere in this volume). Although it may be common in the industry to divide qualification into separate activities entitled installation qualification and operational qualification, there is in fact no reason to do so. FDA describes this activity as equipment qualification making no distinction between those activities that focus on the installation details relative to those that focus on the operation of the equipment. The separation is both artificial and cumbersome and execution of the qualification in a unified manner is recommended. Other than this consolidation, the qualification of advanced aseptic technologies as described herein should present no surprises to the experienced practitioner. The physical system and its performance are measured against the specified drawings and user requirements. Critical to success are such aspects as leak rates, air system performance, pressure differentials in various modes, and integration with other equipment, i.e., sterilizers, tunnels, decontamination equipment and environmental sampling apparatus. The more enclosed and separated the external and internal environment are the more the qualification effort mimics that of physical equipment. Where the degree of separation is less extensive as in closed RABS systems, then elements of clean room design and classification must be included. Thus, there are few truly unique aspects to the qualification of these advanced aseptic systems.

The qualification is little more than confirmation that the system conforms to the system specifications approved for its fabrication, installation and operation. This is accomplished through a protocol which confirms that the system is everything it is expected to be. The system specifications including drawings, schematics and performance requirements are compared to the final delivered system to establish its conformity to expectations. Deviations to the design are noted and are either corrected through modification of the design or by revision to the specifications if the system can be accepted although somewhat different from the intended design.

30.10 PRIMARY VALIDATION CONCERNS

The next activity typically required is the performance of performance qualification studies addressing the various critical processes that the system is required to perform. These include: decontamination validation of the isolator (and if required, sterilization/decontamination validation for any infeed systems), process validation for the system's intended use (formulation, aseptic filling, subdivision of potent compounds, medical devise assembly, etc.), and cleaning validation for the enclosure and any installed process equipment. The aseptic processing will require prior attention to all of the required decontamination/sterilization processes utilized, as well as environmental monitoring, and ultimately a process simulation study. Details on each of these activities are provided elsewhere in this volume and will not be repeated here.

Qualification of the advanced aseptic processing system's operation mimics its use as an environmental control measure and little else. It is the equipment installed within it that is of greater importance, and that equipment should be minimally impacted by the enclosure design details. If intended for aseptic processing, then validation activities include: decontamination of the surfaces, monitoring and control of the environment and aseptic process simulation, just as they would in a manned clean room. The only isolator-specific concern would be its leak rate to minimize product exposure. Note that in all design instances for advanced aseptic use, there is additional process equipment that is validated independent of its location. The methods used for these are largely unrelated to the separative technology employed and are essentially unchanged from those utilized for manned environments. Demonstration of ISO 5 is nearly identical whether in an isolator or a clean room. The relevant qualification issue for the enclosure is that it provides an environment suitable for the application performed within.

30.11 ROBOTICS AND MACHINE AUTOMATION

An increasingly important design element in advanced aseptic systems is the inclusion of robotic or specialized machine automation often for loading and unloading of batch operations systems such as lyophilizers or sterilizers. These are sometimes supplied with filling systems for the handling of components such that the filler only operates when the robotic devices perform as required. Separate qualification of this form of robotics is not necessary when they are fully integrated with the filler itself. Robotics may be used to replace human manipulations for purposes such as picking and placement or simply transferring materials from one section of an isolator network to another.

However, increasingly robots may be utilized to conduct critical operations. These include repetitive process critical interventions/activities such as pipetting and sample taking, and often allow a specific process operational procedure to be introduced by the owner. Therefore, security is critical to ensure that "ad hoc" modifications do not occur without proper change control, and where necessary validation testing.

30.12 USE OF THE SYSTEM

An advanced aseptic system is built for a purpose, and its qualification/validation must support that purpose. The PQ efforts confirm the acceptability of the design and procedures to provide the required functionality for which the system was built. Each of the primary standard operating procedures required for use: decontamination/sterilization, operation and cleaning are usually the direct result of a supportive PQ study that demonstrated how the system performs in each procedure. In conjunction with supportive procedures such as: instrument calibration, leak testing, environmental monitoring, etc., these procedures define how the system will perform. The application of change control to the procedures and physical system ensures that changes are reviewed for their potential impact on the validated state of the overall final system. The only other aspect of the systems usage which must be considered is training of operating personnel in the proper execution of these procedures.

30.13 SYSTEM MAINTENANCE

As with any other piece of mechanical equipment, an advanced aseptic enclosure should be supported by a continuing preventive maintenance program to ensure it operates reliably over time. Among the ongoing maintenance considerations are: calibration of instruments, leak testing/inspection/lubrication/replacement of seal surfaces, gloves and half-suits, periodic cleaning of internal and external surfaces, filter integrity testing, filter replacement, filter change, system leak testing and any required preventive maintenance for non-isolator equipment that are part of the overall installation.

30.14 CONCLUSION

Despite what may appear to some to be a near-impossible and never-ending task, the qualification/validation of advanced aseptic processing systems is not particularly difficult. Where realistic (the only kind which should be defined) requirements are established for the systems performance they can be readied for use rather simply. As noted, the earlier problems are largely associated with expectations no system could possibly attain. The classic qualification/validation methods used for other process equipment/facility systems are wholly adequate to bring these systems through validation. Experience with sterilization, process, and cleaning validation as well as some familiarity with ordinary clean rooms is certainly helpful.

NOTES

1. Akers J, Agalloco J, Madsen R, What Is Advanced Aseptic Processing? *Pharmaceutical Manufacturing* 4(2): 25–27.
2. FDA, Guideline on Sterile Drug Products Produced by Aseptic Processing, 2004.
3. EMEA, Annex 1, Sterile Medicinal Products, 2008.
4. PIC/S, Decision Tree for the Selection of Sterilization Methods, CPMP/QWP/054/098, 1999.

5. Lister J, Antiseptic Principle of the Practice of Surgery. *The Lancet* 1867.
6. Agalloco J, Madsen R, Akers J, Aseptic Processing: A Review of Current Industry Practice. *Pharmaceutical Technology* 2004;28(10): 126–150.
7. Avallone H, FDA Field Investigator Training curriculum, circa 1985.
8. Bristol-Myers France, circa 1982.
9. Wagner C, Raynor J, Industry Survey on Sterility Testing Isolators: Current Status and Trends. *Pharmaceutical Engineering* 2001;20(3): 134–140.
10. Martin P, Isolator Technology for Aseptic Filling of Anti-Cancer Drugs. In *Advanced Aseptic Processing Technology.* Agalloco J, Akers J, eds. London: InformaHealth, 2007.
11. PDA, Design and Validation of Isolator Systems for the Manufacturing and Testing of Health Care Products. PDA Technical Report #34, *PDA Journal of Pharmaceutical Science and Technology* 2001;55(5), Supplement.
12. ISPE, ISPE Barrier Survey 2020. Presented at ISPE Aseptic Conference, March 2020.
13. PDA, 2017 Aseptic Processing Survey, 2017.
14. Ferreira J, et al., A Comparison of Capital and Operating Costs for Aseptic Manufacturing Facilities. In *Advanced Aseptic Processing Technology.* Agalloco J, Akers J, eds. London: InformaHealth, 2007.
15. www.vanrx.com
16. Lysford J, Porter M, Barrier Isolation History and Trends. International Society for Pharmaceutical Engineering Washington Conference: Barrier Isolation Technology, ISPE, Tampa, June 2, 2004.
17. Agalloco J, Paradise Lost: Misdirection in the Implementation of Isolation Technology. *Pharmaceutical Manufacturing* 2016;15(4): 34.
18. Lysford J, The ISPE RABS Definition: An Introduction. *Pharm. Eng.* 2005;26(10): 116, 120.
19. ISPE, Definition of a RABS, August 2005.
20. ISPE, Definition of a RABS, August 2015.
21. PDA, Design and Validation of Isolator Systems for the Manufacturing and Testing of Health Care Products. PDA Technical Report #34, *PDA Journal of Pharmaceutical Science and Technology* 2001;55(5), Supplement.
22. ISO 14644–7, Cleanrooms and Associated Controlled Environments—Part 7—separative enclosures (clean hoods, glove boxes, isolators and mini-environments), 2001.
23. Akers J, Agalloco J, Madsen R, What Is Advanced Aseptic Processing? *Pharmaceutical Manufacturing* 2006;4(2): 25–27.
24. Agalloco J, Opportunities and Obstacles in the Implementation of Barrier Technology. *PDA Journal of Pharmaceutical Science and Technology* 1995;49(5): 244–248.
25. Agalloco J, Akers J, Madsen R, Choosing Technologies for Aseptic Filling: Back to the Future? *Pharmaceutical Engineering* 2007;27(1): 8–16.
26. Agalloco J, Paradise Lost: Misdirection in the Implementation of Isolation Technology. *Pharmaceutical Manufacturing* 2016;15(4): 34. Continued online at Pharmmanufacturing.com. Reprinted in Aseptic Processing Trends eBook, pp. 9–17, July 2017.
27. FDA, Guidance for Industry, Guideline on Principles of Process Validation, March 1984.
28. Akers J, Agalloco J, Isolators: Validation and Sound Scientific Judgment. *PDA Journal of Pharmaceutical Science and Technology* 2002;54(2).
29. Agalloco J, Paradise Lost: Misdirection in the Implementation of Isolation Technology. *Pharmaceutical Manufacturing* 2016;15(4): 34. Continued online at Pharmmanufacturing.com. Reprinted in Aseptic Processing Trends eBook, pp. 9–17, July 2017.
30. Agalloco J, Validation: Yesterday, Today and Tomorrow. Proceedings of Parenteral Drug Association International Symposium, Basel, Switzerland, Parenteral Drug Association, 1993.
31. Akers J, Agalloco J, Isolators: Validation and Sound Scientific Judgment. *PDA Journal of Pharmaceutical Science and Technology* 2000;54(2).
32. Agalloco J, The Validation Life Cycle. *Journal of Parenteral Science and Technology* 1993;47(3): 142–147.
33. FDA, Guidance for Industry, Process Validation: General Principles and Practices, 2011.

31 Total Particle Counts

Mark Hallworth

CONTENTS

31.1 INTRODUCTION

Manufacturing pharmaceutical products is a highly controlled process, whether the end product is aseptic, terminally sterilized, lyophilized, or even an originating bulk ingredient. Therefore, the environments in which the activities of manufacture are performed must be controlled and, through monitoring, proven to be in control.

The primary mechanisms for controlling particles in the manufacturing environments are filters and the clean rooms or local workstations they supply. High Efficiency Particle Arrestor (HEPA) filters are used to clean the

DOI: 10.1201/9781003163138-31

air provided to a clean room. There are four types of clean room clean environments used in pharmaceutical manufacturing:

1. Conventional, using turbulent flow air circulated at an optimum rate to dilute the concentration of particles in an area to an acceptable limit.
2. Unidirectional flow, using the velocity of air as a shower to wash particles from the critical areas.
3. A combination of these two technologies.
4. Isolated environments where the filtered internal air is separated from the surrounding area.

This chapter will review how monitoring total particle counts within controlled areas (clean rooms) is performed, the presence and interaction of particles in clean rooms, fundamental particle counter technologies and design principles, the legislative requirements for monitoring, how monitoring can be performed, and the considerations behind sampling techniques applied to monitoring.

31.2 RATIONALE FOR ENVIRONMENTAL MONITORING

Particle monitoring is primarily used to demonstrate the control of clean manufacturing environments used in the production of pharmaceutical product; regulatory agencies have issued guidance on what their expectations are regarding the control of those clean environments as part of their Good Manufacturing Practices (GMP), and it is expected that a company be able to demonstrate a site Contamination Control Strategy (CCS) that includes total particle counts.

31.2.1 Fundamental Requirements

Particle monitoring is required to prove contamination control of an environment. This includes the measurement of air in a clean room associated with personnel and process equipment activities, with regard to risk of finished product quality.

The United States Pharmacopeia General Chapters refer to testing a finished product to prove that the product is free from viable contaminants.[1] It is the presence of viable organisms in the production environment that is of the greatest concern for contamination of the product.

The average time taken to prove that an environment has been maintained at a suitable level of sterility is between 3 and 5 days. This time is needed for sampling, incubation, analysis, and reporting.

The USP <797> "Pharmaceutical Compounding—Sterile Preparations" refers to the intent of its fundamental requirements as being to "prevent harm and fatality to patients that could arise from microbial contamination, [or] excessive bacterial endotoxins."[2] Although compounding and manufacturing differ, the difference is primarily in the scale of manufacturing as opposed to the intent.

Although viable organisms cause the greatest concern in the clean room, control over the total particle count contaminants is required for several reasons:

1. Demonstration of control over particulate contaminants
2. Demonstration of control over viable contaminants
3. Proof of control over clean room activities, both personnel and processes

31.2.2 Proof of Control Over Particulate Contaminants

Particles present in a clean room are primarily because of personnel, process, or they might arise from the atmospheric abundance of particles brought into a clean room through the air handling/filtration system; these are listed in typical abundance within a clean room. The data in Figure 31.1 shows that a predominance of airborne particles in the atmosphere arises from fugitive dust and wind erosion sources. The particles from these activities are composed of aluminum, silica, and other oxides, are relatively large in size, and can be assumed to be inert from a viable activity perspective.

The Environmental Protection Agency (EPA) states that smaller particles (<2.5 μm) are formed by two primary mechanisms:

1. Heterogeneous nucleation of vapor phase material
2. Homogeneous nucleation of vapor phase material

These formation mechanisms combine to generate a set of particles that are different from larger sized particulate matter.

The U.S. EPA has classified particles into the four size categories shown in Figure 31.2: Ultrafine, Fine, Coarse, and Supercoarse.

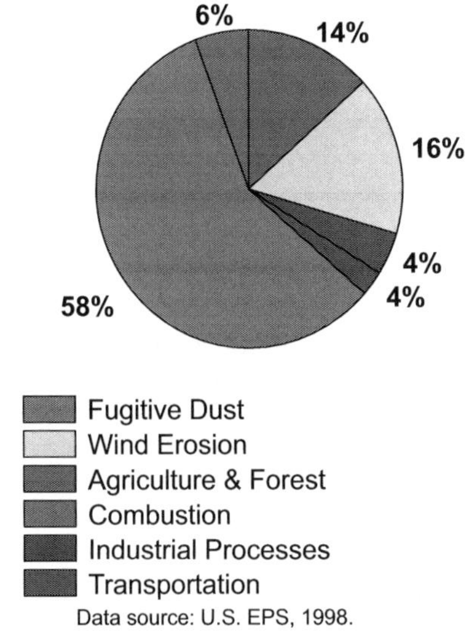

FIGURE 31.1 Major sources of particulate matter (PM10).

Source: U.S. EPA, 1998.

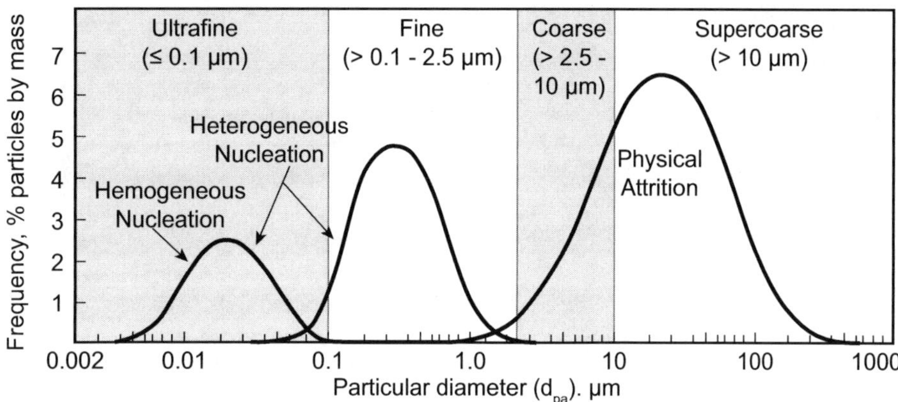

FIGURE 31.2 Tri-modal distribution of particle matter in the atmosphere.

The Most Penetrating Particle Size (MPPS) of most HEPA filters is between 0.1 μm and 0.2 μm with increasing efficiency of capture for smaller and larger particles. If it is assumed that all externally sourced particles in the clean room are essentially inert (i.e. they are oxides that have passed through the filter media), then impact on product is either chemical (i.e. react with the formulation or within the patient) or physical (i.e. trigger a physical response from a patient); in either case, quantities from external filtered sources should be considered minimal.

The control over total particle burden is important irrespective of source (operational conditions lead to process-related particles, at rest conditions lead to atmospheric borne particles), and international standards have been defined and applied to clean rooms since the early 1970s with the earliest applications of the Federal Standard FS209.[3] The Federal Standard FS209E (its last iteration) was superseded by the International Standard ISO14644–1 in 2002 and most recently with a revision of ISO 14644–1 (2015) (*Cleanrooms and associated controlled environments—Part 1: classification of air cleanliness by particle concentration*).

31.2.3 Proof of Control Over Viable Contaminants

The previous assumption, that all particles are chemically inert, does not completely hold true, as most particles within a clean room can be attributed to the personnel, or the activities of those personnel, within the clean room. Most particles brought into a clean room by personnel are associated with the risk of organic/viable contaminants.[4]

It is generally accepted by industry that a roughly proportional relationship exists between the viable component and the total particles within a clean room, and that this varies with the class of the clean room (the higher the classification of the clean room, the more likely there will be microbial activity). Therefore, having control over the total particle concentration within a clean room also offers a degree of control over the viable proportion of the total burden. It does not, however, provide useful quantitative information on the microbiological content of the environment.

To support the monitoring of all particles to prove environmental control, studies have been undertaken to verify that the control of total particle count has a direct influence on the viable contaminant levels. However, the data generated through such studies have been inconclusive and unable to prove that any direct correlation exists. Recent studies have also maintained that even through improvements in clean room design, the relationship of total counts to viable proportion is still undefined. [5]

This would be disconcerting as a formulated link is an essential element of room validation; however, William Whyte, during his studies, found that a free-floating viable contaminant existed at sizes between 10 and 15 μm and not smaller; this is because of the desiccating nature of the dry clean room environment potentially denaturing the microorganism.[6] Ljundqvist and Reinmueller found a strong correlation between larger particles and the viable risk to process operations,[7] and the USP <1116> "Microbial Evaluation of Clean Rooms" states that "while airborne micro-organisms are not free-floating or single cells, they frequently associate with particles of 10–20 μm."[8] Therefore proof of control over the macro-particles (ISO 14644–1 definition of particles >5 μm) would also offer a degree of control over the risk of exposure to viable particles.

The Current Good Manufacturing Practices (CGMP) guides support this expected relationship, as particle cleanliness is required to be demonstrated along with a requirement to prove beyond simple cleanliness. By controlling the large particles (>5 μm), it is believed that clean room sterility, or risk of losing sterility, can be monitored. Although controlling the total particulate burden offers control over the viable contamination risk, it does not alleviate the requirement for monitoring for the viable fraction using other technologies (dynamic monitoring, settle plates, etc.).

This application of the associated risk between viable particles and macro-particles is discussed later in the chapter when the limits imposed by various regulatory agencies are applied.

31.2.4 Proof of Control Over Clean Room Activities

A properly designed clean room in the "at rest" state, that is no personnel present and no operational machinery, should essentially be completely free from particles. It holds, therefore, that should this balance change, then so too will the level of particles generated. The correlation of particle activity change and associated changes to operational conditions is widely known and published, and therefore, mapping the changes in particle concentration to known activities within a room can improve knowledge of the process and contribute to process control.

The better informed we are, the better we can control particulate contamination. For example, studies have shown that a poorly gowned operator sheds more particles than one that is properly gowned, and that new clean room gowns shed fewer particles than those laundered multiple times.[7] Using this information can improve the control of the particles created by personnel within a clean space by emphasizing gowning techniques and the frequency at which gowns are replaced.

Another example is using data from monitoring to better manage situations that can occur. A broken vial or a filling line blockage because of machinery or system failure generates particles, and the ensuing intervention will almost certainly generate more particles that would not have occurred during "normal" operations. The impact and frequency of the interventions can be monitored using the particle data generated during the period of failure and following statistical process control (SPC) standards. An acceptable limit on anticipated interventions over unexpected interventions could assist in the risk assessment of the process.

31.2.5 Requirements to Satisfy Regulatory Standards

To meet the current requirements of total particle counting, a pharmaceutical manufacturer, as part of a CCS, must undertake two components of particle count information to prove regulated use of pharmaceutical grade clean rooms. The first step is the classification of the room: this is performed to international standards, primarily ISO 14644–1 (2015). Second, the demonstration that the room can maintain its particulate limit during operations by routine monitoring. The guidelines for monitoring are defined in the CGMP guides relevant to the region to which the product is released. Each of these topics is discussed later in this chapter.

31.3 CLEAN ROOM CERTIFICATION

There are two aspects of standards and specifications for non-viable particle counting that are important to pharmaceutical manufacturers: (1) the procedures within the standards used to characterize the particles must be accurately defined and (2) the test methods must be carefully determined. When the preceding are present in an international standard, it allows a "calibration" of the clean room to be performed in absolute terms. Once this baseline has been established, then determining the physical nature of the various mechanisms affecting particle generation, transport, and deposition becomes all-important in understanding why, when, how, and where to monitor particles in a pharmaceutical manufacturing environment. Knowledge of these mechanisms also assists in understanding the application limits of the monitoring instrumentation.

The original standard universally adopted for clean room certification was the Federal Standard FS209. The final version of this was revision FS209E, which was replaced in November 2002 by a new international standard, ISO 14644–1 (1999), and this was updated with a new release in 2015. References to ISO 14644–1 in the remainder of this text will only discuss the application of the standard ISO14644–1 (2015) version.

The certification state of the clean room must be determined in advance of testing; three states exist within the context of ISO14644–1:

1. <u>As Built</u>—a completed room with all services connected and functional, but without production equipment or personnel within the facility.
2. <u>At Rest</u>—a room where all the services are connected, all the equipment is installed and operating to an agreed manner, but no personnel are present.
3. <u>Operational</u>—all equipment is installed and is functioning to an agreed format, and a specified number of personnel are present working to an agreed procedure.

The limits for the clean room concentration of particles greater than a prescribed size are defined in Table 31.1.

TABLE 31.1
Airborne Particulate Cleanliness Classes for Clean Room and Clean Zones

ISO 14644-1:2015 classification number (N)	Maximum concentration limits (particles/m^3)					
	0.1 µm	0.2 µm	0.3 µm	0.5 µm	1.0 µm	5.0 µm
ISO Class 1	10					
ISO Class 2	100	24	10			
ISO Class 3	1000	237	102	35		
ISO Class 4	10000	2370	1020	352	83	
ISO Class 5	100000	23700	10200	3520	832	
ISO Class 6	1000000	237000	102000	35200	8320	298
ISO Class 7				352000	83200	298
ISO Class 8				3520000	832000	29300
ISO Class 9				35200000	8320000	293000

These limits have been defined in accordance with the calculation from the following standard:

$$C_n = 10^N \times \left(\frac{K}{D}\right)^{2,08}$$

where:

C_n = the maximum permitted concentration of airborne particles that are equal to and greater than the considered particle size. C_n is rounded to the nearest whole number, using no more than three significant figures.

N = the ISO classification number, which shall not exceed a value of 9 or be less than 1.

D = the considered particle size, in micrometers, that is not listed in Table 31.1.

K = a constant (0.1), expressed in micrometers.

The relationship of particle size to its abundance within a population is therefore a function of $1/D^{2.08}$. If the particle size is plotted against its concentration on a log/log scale, the slope of the curve for each class is 2.08 as shown in Figure 31.3.

The designation for clean room certification should include the following elements:

- The room classification number expressed as "ISO Class N."
- The occupancy state.
- The considered particle size. It is also possible to certify a clean room at multiple sizes; if this is the case, then the sample volume requirement for the largest particle size is used.

To demonstrate continued compliance to the standard room classification, testing needs to be repeated on a frequency defined by ISO14644–2 (2015). For all ISO classified clean rooms, the interval is defined as every 12 months, regardless of classification, although some internal risk assessments may choose to increase this frequency. This interval can also be extended providing that the pharmaceutical company shows that "no significant change" has occurred in the control of their clean room by evidence of continued compliance.

31.4 CLEAN ROOM MONITORING

Once a room has been certified as meeting a specific room classification in accordance with ISO14644–1, the room can be used for its intended purpose. However, the interval between recertification tests is insufficient to meet the current requirements for monitoring to CGMP. Both the US and the European regulatory agencies require a room to be regularly tested for compliance based upon the risk to finished product quality. The second phase of proving compliance with regulatory agencies requires the monitoring of the clean room environment.

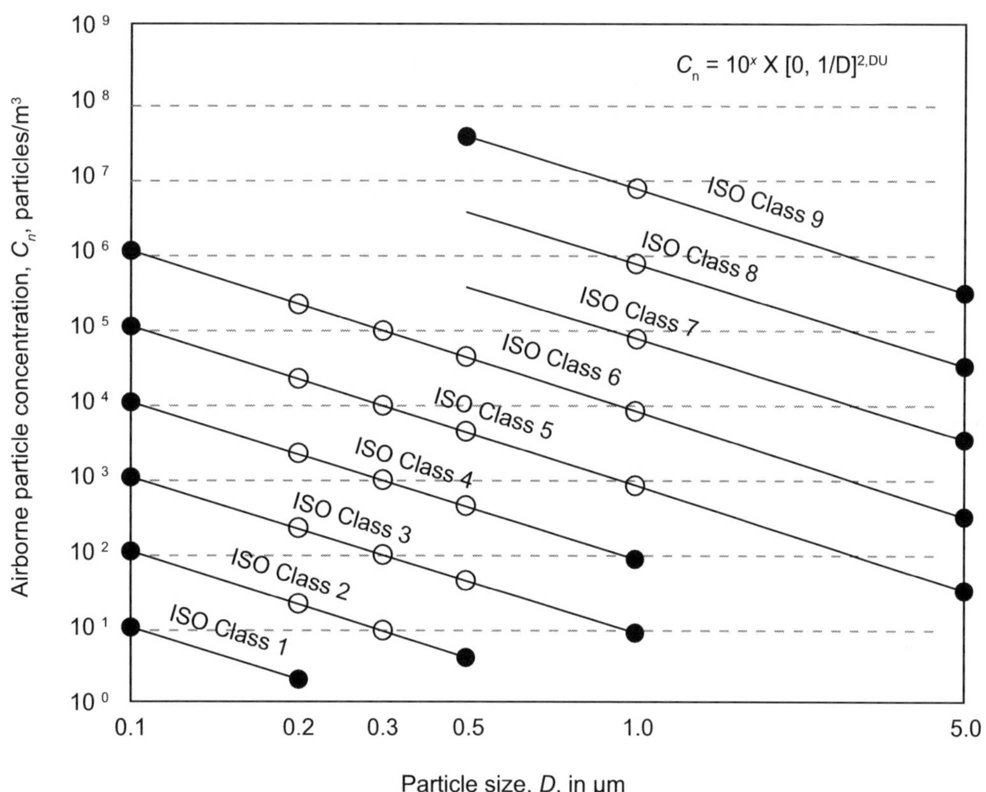

FIGURE 31.3 Graphical representation of ISO class concentrations limits for selected classes.

Note: ISO = International Organization for Standardization.

Clean room monitoring is defined as the observance of the condition within a production area during normal conditions. This means that monitoring is performed in a pharmaceutical production area for at least one of the following reasons:

1. To obtain continuous particle measurements following the operation or performance of a specific machine, or of the conditions at a particular critical location.
2. To obtain particle count results over a long period to establish control limits over a process or standard operational procedure (SOP).
3. To gather a record of monitoring information that may be required to verify operations in accordance with in-house and regulatory specifications.

The two primary regulatory bodies are the Food and Drug Administration (FDA) in the United States and the European Medicines Agency (EMEA). These agencies enforce their own requirements for CGMP. In addition, there are regulatory bodies responsible for individual nations. If export of product is intended to reach a locally regulated market, then an audit from one of these inspectors is also required. It is therefore important to be able to satisfy the more stringent requirements. Of these practices, those required for aseptic manufacture impose the greatest level of requirements, especially for proof over both nonviable and viable contaminants.

31.4.1 FDA Requirements

The FDA CGMP for drug products specifies the practices for manufacturing area operations, controls, validation, and documentation.[9] Adherence to the CGMP is verified by inspection, and the FDA has issued guidelines in the Code of Federal Regulations (CFR) on required practices and recommendations for compliance with these requirements.[10] The guidelines describe building and facility requirements, especially the critical and controlled areas.

21 CFR 211.42 (c) states,

> Operations shall be performed within specifically defined areas of adequate size. There shall be separate or defined areas or such other control systems for the firm's operations as are necessary to prevent contamination or mix ups during the course of the following procedures: Aseptic processing, which includes as appropriate: (i) Floors, walls, and ceilings of smooth, hard surfaces that are easily cleanable; (ii) Temperature and humidity controls; (iii) An air supply filtered through high-efficiency particulate air filters under positive pressure, regardless of whether flow is laminar or non-laminar; (iv) A system for monitoring environmental conditions [10]

Therefore, separate or defined areas in an aseptic processing facility should be controlled to achieve the required air quality that is dependent on the nature and risk of the process.

The FDA describes the area that poses the greatest risk to finished product quality as "critical." This is where sterilized product, glassware, and other associated components exposed to the general supply air are maintained in a sterile environment. All personnel activities conducted in these areas are monitored so that they do not compromise the efficacy of the environment. Table 31.3 shows the limits defined by the FDA 2004 *Guideline on Sterile Drug Products Produced by Aseptic Processing.*

This area is deemed critical because product is exposed to potential contamination and is not subsequently sterilized. It is therefore essential that the environment in which aseptic operations are conducted be maintained.

> Air in the immediate proximity of exposed sterilized containers/closures and filling/closing operations would be of appropriate particle quality when it has a per-cubic-meter particle count of no more than 3520 in a size range of 0.5 μm and larger when counted at representative locations normally not more than 1 foot away from the work site, within the airflow, and during filling/closing operations. This level of air cleanliness is also known as Class 100 (ISO 5). We recommend that measurements to confirm air cleanliness in critical areas be taken at sites where there is most potential risk to the exposed sterilized product, containers, and closures. The particle counting probe should be placed in an orientation demonstrated to obtain a meaningful sample.[11]

This ties the guide to the ISO14644–1 clean room standards document and to the level of cleanliness required to perform

TABLE 31.3
FDA Guidance on Room Classifications.

Clean area classification (0.5 μm particles/ft^3)	ISO designation	> 0.5 μm particles/m^3	Microbiological active air action levels (cfu/m^3)	Microbiological settling plates action levels (diam. 90 mm; cfu/4 hours)
100	5	3,520	1	1
1000	6	35,200	7	3
10,000	7	352,000	10	5
100,000	8	3,520,000	100	50

Abbreviation: FDA = Food and Drug Administration; ISO = International Organization for Standardization.

aseptic manufacturing, especially when personnel are present. This also gives insight into the requirements for sample point locations. Sample points should be placed where they are likely to witness any anomalous conditions, closest to where glassware and product are exposed. This adoption of a risk-based approach is echoed in the definition for the support areas and also to the required frequency of monitoring. The guide goes on to state that "regular monitoring should be performed during each production shift. It is recommended to conduct nonviable particle monitoring with a remote counting system. These systems are capable of collecting more comprehensive data and are generally less invasive than portable particle counters."[10]

Therefore, the greater the risk to the finished product quality, then the greater the degree of control required. This can be expressed though the number of allowable particles, the close proximity of the sample to the product, or the frequency with which samples should be taken.

In the case of powder filling, many of the particles are elements of the product and therefore do not pose a risk of being a contaminant. Erroneous results during the process or filling stage lead to false observances of high counts. It is recommended then that the process be monitored in one of two ways:

1. Prove compliance up to the time of batch start and again after a short cleanup period once the batch is finished. Supporting evidence of control will need to be supplied by way of other environmental parameters, viable counts, differential air pressure, and air flow velocity.
2. Select a sample point that reflects the quality of the air in the room or process without being impacted by it. This may lead to a sample point being placed close to the filter face of the area in question.

Also, any air monitoring samples from the critical areas should yield no microbiological contaminants. The 5.0 mm particle observations could be used as a means to identify if such exposure risks occur based upon the arguments prior.

The critical zone is the central core of the sterile manufacturing environment. This is, however, surrounded by a zone of varying classification possibilities.

The FDA describes the supporting clean areas as having various classifications and functions in which non-sterile components, products, equipment, and containers are prepared, held, or transferred. These environments should be designed to minimize the level of particle contaminants in the final product. The activity conducted in a supporting clean area determines its classification based on its risk to final product quality. The FDA guidance recommends "the area immediately adjacent to the aseptic processing line meets, at a minimum, Class 10,000 (ISO 7) standards (Table 31.4) under dynamic conditions. Manufacturers can also classify this area as Class 1,000 (ISO 6) or maintain the entire aseptic filling room at Class 100 (ISO 5). An area classified at a Class 100,000 (ISO 8) air cleanliness level is appropriate for less critical activities (e.g., equipment cleaning)."[10]

No information relating to how these areas are to be monitored is given in the guidance. However, as the whole program issued is one of risk assessment, the potential risk to product

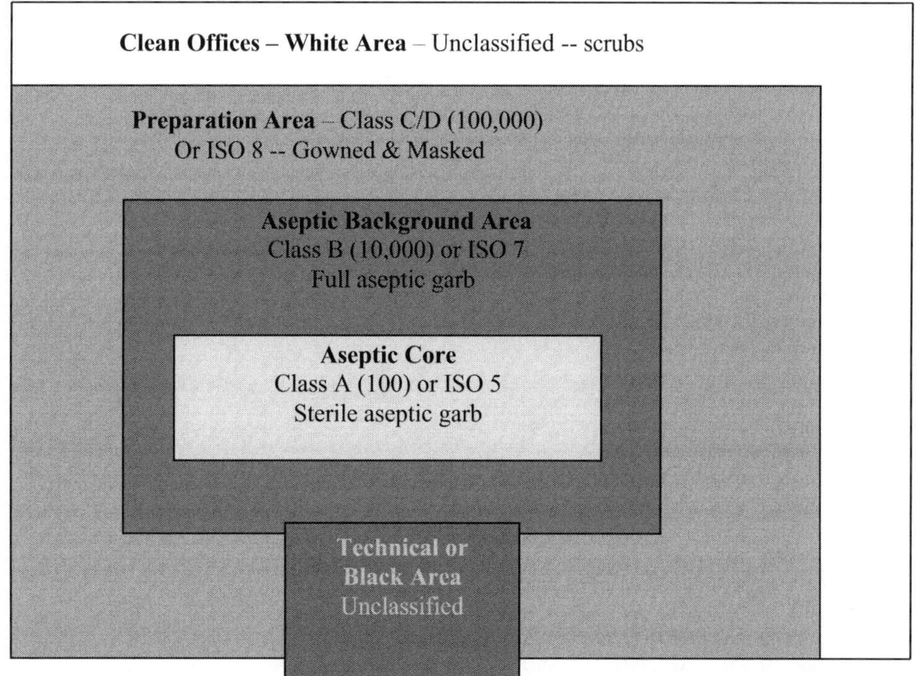

FIGURE 31.4 General clean room design.

TABLE 31.4[11]
Particle Monitoring Requirements for Sterile Support Areas

Air cleanliness classification	Type of operation	Frequency of sampling in operation
Grade A ISO Class 5	Critical aseptic preparation and filling areas.	At least once per shift.
Grade B ISO Class 7	Areas immediately surrounding the Grade A areas (filling suite). Includes sterile corridors and rooms; areas for sterile product and components storage; gown room exits, etc.	At least once per shift. (Immediately surrounding Class A) At least once per day. (All other grade B areas)
Grade C ISO Class 8	Non-sterile filling of terminally sterilized products; areas for equipment and component preparation.	At least once per week.
Grade D ISO unclassified	Equipment and component washing and handling; gowning rooms; general corridors.	At least once per month.

quality can be reviewed based upon the activities within these areas. The PDA proposed the following recommendations:

31.4.2 EMA Requirements

The EMA European Commission (EC) enforces its requirements on the manufacture of drug products using the "Good Manufacturing Practices—Medicinal Products for Human and Veterinary Use", Volume 4, 1998. The applicable annex for sterile manufacture is Annex 1, revised in May 2009. The rationale for the current revision was that the guidance had been reviewed following the release of the standard ISO 14644–1, and the revision was based upon the application of that standard over the preceding 5 years.[12]

The EC recognized that the manufacture of sterile products requires special consideration in order to minimize the risks of both microbiological and particle contamination. The manufacture of sterile products should be performed in clean areas where controls over access and gowning can be enforced. These areas should be maintained at an appropriate level of cleanliness based upon classic clean room design principles. The risk associated with the activities in each of the production areas should be classified and, based upon the findings, the room certified as a particular grade. Each of the assigned grades needs to be monitored in accordance with the assigned grade for particles and microbiological contaminants.

The two operation states are "Operational" and "At Rest." There are also four distinct grades:

Grade A—The local zone for high-risk operations, e.g. filling zone, stopper bowls, open ampoules, and vials, making aseptic connections. Such conditions are provided by laminar airflow. The maintenance of laminar conditions should be demonstrated and validated.

Grade B—For aseptic preparation and filling, the background environment for the grade A zone.

Grades C and D—Clean areas for carrying out less critical stages for the manufacture of sterile products.

Table 31.5 defines the limits set by the EC for maximum particle concentrations.

Again, the basic principle of a risk-based approach to environmental control is shown. Areas where the finished product is exposed to the environment must be maintained with a minimum particle exposure rate. Areas where activities offer a reduced risk of contamination by particle contact, or are subsequently cleaned and/or sterilized, are operated in an environment where the maximum permitted particle concentration is significantly higher.

The exposure to the particle risk needs to be monitored in accordance with the recommendations proposed by the regulation. The most critical area, Class A, needs to be monitored with the method that offers the greatest level of control (continuous system of sampling). The EMA recognizes this in the notes that follow the table of maximum permitted particle concentrations in Annex 1. It states that a continuous measurement system should be used for monitoring the concentration of particles in the Grade A zone and is recommended for the immediately surrounding Grade B areas.

The limit of 20 particle/m^3 at 5.0 µm is not in line with limits applied by the ISO14644–1 guidelines for ISO 5 room cleanliness and also the model of expected size distribution (1/ Particle diameter ^2.08, ISO146544–1) for a given population of particles within a clean room, they therefore recommended that this area be deemed to be an ISO class 4.8, which is allowable under the partial classes defined in ISO14644–1 (2015). The EC CGMP requires that these areas be expected to be essentially free from particles of size ≥5 µm. As it is impossible to demonstrate the absence of particles with any statistical significance, the limits are set to 20 particle/m^3 solely for certification. During the clean room qualification, it should be shown that the areas can be maintained within the defined limits."

This ties the limits on microbial monitoring (<1 cfu /m^3) to the number of total particles that are >5.0 µm and, as such, pose a risk of being a viable contaminant. The EMA is

TABLE 31.5
EC Annex 1 Particle Monitoring Classifications

Grade	Maximum permitted number of particles per m^3 equal to or greater than the tabulated size			
	At rest		In operation	
	0.5 µm	5.0 µm	0.5 µm	5.0 µm
A	3520	20	3520	20
B	3520	29	352,000	2900
C	352,000	2900	3,520,000	29,000
D	3,520,000	29,000	Not defined	Not defined

therefore looking to prove control over room cleanliness by using ISO guidelines and 0.5 µm particle concentrations and over room sterility by using the data for the larger particles.

If continuous monitoring is not performed in the critical areas, and routine certification or monitoring is executed using a portable particle counter (see Section 31.5), then the ties to ISO 14644–1 (the underlying reason for release of the revision) become apparent and a minimum volume is stipulated for these tests. The equations following offer reasoning as to why a minimum volume is required.

EC CGMP Sample Volumes Using ISO 14644–1 Calculations

In a Class A clean room, operational, the limits are 0.5 mm = 3500/m³ and 5.0 µm = 1/m³. If we use a particle counter with a flow rate of 28.3 liters/minute = 1 cubic foot per minute (1 CFM), the following times are established for testing each location:

$$0.5\text{mm. Vs} = \frac{20}{C_{n.m}} \times 100 = \frac{20}{3500} \times 1000 = 5.71\text{listers} =$$

seconds, however a one-minute sample minimum is required for ISO.

$$0.5\text{mm. Vs} = \frac{20}{C_{n.m}} \times 1000 = \frac{20}{20} \times 1000 = \text{ listers = minutes,}$$

using suitable instruments

ISO 14644–1 states that for room certification using multiple sizes, the maximum calculated sample period must be used and this typically reflects the largest particle size. Therefore, to follow ISO would prove impractical in a manufacturing environment, and a requirement of a 1 m³ sample should be taken during routine testing. It also follows that this routine testing is periodic. Therefore, confidence in control over an area comes not only from the number of samples taken but also from the volume of that sample.

No definition is set out for monitoring the support areas, whether it is by portable or continuous means. Neither is a limit set at a minimum volume requirement, as the ISO calculations for minimum volume yield an acceptable period of sampling.

31.4.3 Other International Requirements

There are two other international regulatory guidelines that document a need to prove compliance to particle counting limits. The World Health Organization (WHO) and the Pharmaceutical Inspection Co-operation Scheme (PIC/S) both state limits for maximum permitted concentrations of 0.5 µm and 5.0 µm particles.

31.4.3.1 World Health Organization

The WHO CGMP limits for particles follows the original limits imposed by the EC Annex 1, stating that no macroparticles should be allowed in the critical areas where product is exposed directly to the environment in which it is processed.

TABLE 31.6
WHO Airborne Particulate Limits for Sterile Manufacture

Grade	At rest		In operation	
	Max. permitted particles/m³		Max. permitted particles/m³	
	≥0.5µm	≥5.0µm	≥0.5µm	≥5.0µm
A	3520	20	3520	20
B	3520	29	352,000	2900
C	352,000	2,900	3,520,000	29,000
D	3,520,000	29,000	Not defined	Not defined

Abbreviation: WHO = World Health Organization

TABLE 31.7
WHO Comparative Table of Different Particle Standards

WHO (CGMP)	United States (209e)	United States (customary)	ISO/TC (209)	EEC (CGMP)
Grade A	M 3.5	Class 100	ISO 5	Grade A
Grade B	M 3.5	Class 100	ISO 5	Grade B
Grade C	M 3.5	Class 10,000	ISO 7	Grade C
Grade D	M 3.5	Class 100,000	ISO 8	Grade D

Abbreviations: CGMP = Current Good Manufacturing Practice; EEC = European Commission; ISO/TC = International Organization for Standardization Technical Committee; WHO = World Health Organization.

They do not offer guidance on the frequency of monitoring, only that monitoring to prove compliance must be performed.

They also show how the limits enforced are related to other limits imposed by the FDA and EMA regulations.

The direct relationship between the WHO classifications and the EMEA classifications is evident as is the combination of Class A areas with ISO 5, that which the FDA requires to be maintained in proving control over a critical area. Harmonization between the various standards points toward a common requirement for a risk-based monitoring program. The WHO is also an observer for the Pharmaceutical Inspection Cooperation Scheme (PIC/S).

31.4.3.2 Pharmaceutical Inspection Cooperation Scheme

The Pharmaceutical Inspection Convention, active within Europe and elsewhere, requires a common standard for inspections between member states. This is in an effort to remove trade barriers within a common market. They agreed to harmonize on the rules of a common CGMP[13]; the EU Guide to Good Manufacturing Practice for Medicinal Products and its Annexes was adopted. The particle counting limits for the PIC/S are therefore the same as those identified in the EMA CGMP Annex 1 and are shown in Table 31.5. The scheme is being adopted as a global standard by all major countries for pharmaceutical CGMP.

Regardless of governing body, particle counting certification and monitoring must be proven, and a system of monitoring needs to be employed. There are two philosophies for sampling that can be adopted:

1. Routine portable testing of an area
2. Automated monitoring of an area

The following portion of the chapter describes the implementation of a monitoring system and the selection of sample points that best reflect the activities within a production zone.

31.5 PARTICLE MONITORING PROCEDURES

The effective monitoring of a clean room involves the measurement of multiple environmental variables. Particle counting, in addition to the measurement of airflow patterns and air velocity, temperature, and relative humidity, and of differential air pressure between adjacent rooms, can be vital in ensuring operations are at optimum performance. The monitoring plan should include measurement of all these parameters where necessary.

31.5.1 Routine Portable Testing of an Environment

There are three primary reasons for the portable monitoring of a clean room:

1. Routine verification of performance
2. Diagnosing particle contamination from a specific machine or an operation used to clean the clean room (filter testing) or from a new contamination source (new machine or filter installation)
3. Providing data showing compliance with regulatory standards

The testing of a clean room will require an operational procedure, which will depend on operational requirements and form testing. All monitoring operational procedures should define:

- The parameters monitored (if other parameters are to be monitored, they should also be listed).
- Sampling locations.
- Sampling frequency.
- Sampling duration.
- Target cleanliness levels.
- Alert and alarm threshold levels (Alert or warning levels mean that operators are prepared to take remedial measures. Alarm levels will require that the clean room operation be halted, and steps taken to protect product in the area while remedial measures are implemented to control the contamination source.).
- Any actions taken against threshold limits, hardware errors, or clean room comments.

It is necessary to record contamination levels verifying that clean room cleanliness levels have been maintained. It is also necessary to maintain detailed records of monitoring device (particle counter, temperature probe, pressure sensor, etc.) validation as well as the monitoring activity results, proving the room is clean and that the instrumentation used is traceable.

Concerns in the use of a portable particle counter are:

- It introduces an additional person into the clean room, unless the manufacturing staff is responsible for testing.
- It does not always reflect the actual conditions of the process or the normal practices of personnel present in the clean room because of its intrusive nature.

Portable sampling is <5% of the process period compared with continuous data.[14] The automation of the monitoring addresses these issues described previously.

31.5.2 Automated Monitoring of an Environment

There are two primary means of automated sampling within a clean room: continuous or sequential (manifold) monitoring. The decision as to which method to use is based upon the nature of activities within the clean areas.

31.5.2.1 Continuous Monitoring

Particle sensors are located at each sample collection point and are operated continuously. Data is fed to a central data-processing system. The particle sensor consists of a small enclosure housing an optical system, a light source (laser diode), and signal generation electronics. The sensors often require an external vacuum source and signal communication cable to transmit data to the central monitoring computer. The advantages of such a system are:

- The sensors continuously monitor and report data to the system, therefore detecting short-lived particle burst situations.
- They are simple and have low-cost installation.
- The ease of relocation to alternative positions.
- Provides the highest level of confidence.

31.5.2.2 Manifold Monitoring

Manifold monitoring transfers sample air from each point through tubing to a sample manifold or sequencing valve, which transfers the sample to a single particle-measuring device in a programmed sequence. Manifold systems are very common and consist of a centrally located manifold and single particle counter with up to 32 sample tubes radiating from this central location. Each tube is capable of drawing a sample from a distance of up to 38 m (125 ft.) from the manifold. The advantages of such a system are:

- Low cost per sample point monitored.
- Low maintenance and calibration costs; only a single instrument per manifold to calibrate and service.

The factors that determine the selection of either of the systems include statistical validity of the data that is collected, large-particle sample line losses, differences in response time from individual sensors, requirements for frequency of data collection, system installation, and operating/maintenance costs.

Both system types have disadvantages to their selection.

- The multisensor system (Figure 31.5) requires that all sensors are calibrated for correct particle sizing and sample inlet flow rate. The problems of inter-sensor correlation require extremely careful calibration and frequent maintenance to reduce these errors to an acceptable level.
- The manifold system (Figure 31.6) uses multiple sample lines throughout the facility. Each of these lines draws a sample through from sample point to

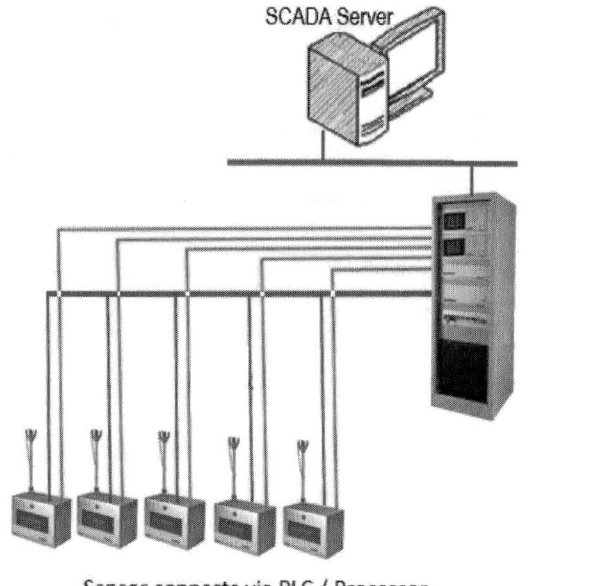

FIGURE 31.5 Multisensor monitoring system types.

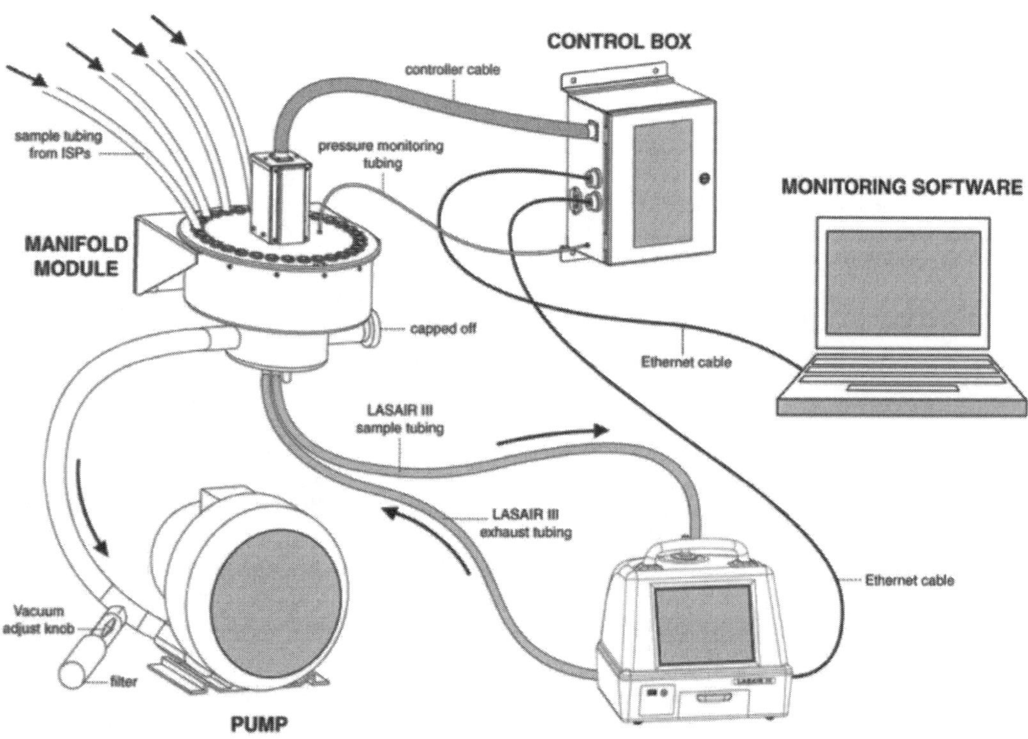

FIGURE 31.6 Multiplexed manifold monitoring system.

manifold. They require the transport of air through this tubing over distances up to 40 m. Particle losses caused by turbulent deposition and electrostatic effects may exceed 50% for particles of 5 μm in diameter. Losses of particles in the size range 0.1–1 μm are negligible and within experimental error. Any sample measurement must allow 10–15 seconds of purge time to clear a previous sample from a long line.

31.5.3 Combination System

The ideal solution is usually found in a combination of techniques to fully satisfy the monitoring requirements of a manufacturing facility. Various combinations are presented in Table 31.8.

TABLE 31.8
Matrix of Sampling Techniques

Sampling technique	Comments
Continuous	Capable of collecting most comprehensive data. May miss events because of the location selection. Cannot be used to certify the clean room. Expensive to give sufficient coverage for all clean rooms.
Manifold	May miss events because of either the location of the sample point or the frequency or sequential nature of the sampling. Particle losses in tubing a concern, especially in critical areas. Cannot be used to certify the clean room.
Portable	Very labor intensive for both sampling and reporting. Can be used to certify the clean room.
Continuous and Manifold	Continuous at critical zones and manifold in surrounding clean zones. Cannot be used to certify clean rooms.
Continuous and Portable	Continuous at critical zones and portable in all other areas. May miss events in surrounding rooms. Labor intensive. Can be used to certify clean rooms. Best solution for CGMP compliance.
Manifold and Portable	May miss events because of the location of the sample point or the frequency or sequential nature of sampling. Particle losses in tubing a concern, especially in critical areas. Frequency of monitoring using portables in critical activities. Labor intensive.
Continuous, Manifold and Portable	Full automation of all clean room activities. Continuous at critical sites, manifold in support areas, and complimented using portables for fault diagnosis and certification exercises. Total CGMP solution.

Abbreviation: CGMP = Current Good Manufacturing Practice.

31.6 FUNDAMENTALS OF OPTICAL PARTICLE COUNTERS

Optical particle counters (OPCs) have been used for counting and sizing particles in air since the mid-1950s. To understand the application of the use of a particle counter, it is important to understand the underlying principle behind its operation.

The original test method for determining the cleanliness of an environment meant employing some classic, empirical techniques. A known volume of air was drawn across a filter and submitted for subsequent analysis. The filter would be either paper or cellulose fiber. To control the volume of air, either a volume flow meter was used for a specified period of time or a total volume meter was used. The sample was taken in the clean room and prepared. The deposited particles were then counted by eye through a microscope. This technique required an extensive period of time and so was only adequate for annual measurements required for a specific area. However, it was unsuitable for routine monitoring.

The onset of instrumentation and optical particle counters was inevitable. Particle counters provide reliable, accurate, real-time, and repeatable measurements. This is ideal for evaluating the cleanliness of a clean room and provides the level of monitoring required to establish control limits over various processes.

31.6.1 Basic Operation

Airborne-particle counters work on a light scattering principle. They utilize a very bright light source to illuminate the particles. The current standard source of the illumination is a laser diode. Previous sources used were gas lasers (Helium-Neon, "HeNe") and "white light" bulbs. This bright light source is focused and shines through an optical block. Within the optical block are collection mirrors and one or more photodetectors. Sampled air is drawn through the laser beam, and entrained particles in the sample air pass through the beam. The laser light interacts with the particles and is scattered.

The resulting scattered light is collected using parabolic mirrors and focused on the photodetectors. These photodetectors convert the scattered light from each particle into a pulse of electricity. By measuring the signal height of the electrical pulse and referencing it to the calibration curve of the instrument, the size of the particle can be determined. By counting the number of electrical pulses in a sample volume or over a period of time, we can determine quantity. Particle counters are able to size and count the number of particles for a population of particles within a clean room, giving full quantitative information.

"Light scattering" is a general term and is composed of various different physical phenomena. Scattering is made up of:

- Reflected light—when a light hits a particle and is angularly deflected.
- Refracted light—when a light goes through the particle and its direction of travel is changed.

- Diffracted light—where light comes close to the particle and is bent around it.
- Phosphorescence—light is absorbed as one frequency and emitted as another.

The interaction of light and particles depends upon the particle composition, its refractive index, and the difference between the particle and the background medium. (For clean room particle counters this medium is air.)

During operation, the instrument compares the response from the particle signal to its calibration curve, which is previously generated using latex spheres of a known size, shape, and refractive index. The instrument is therefore not counting and sizing particles, it is counting and sizing pulses of light and mapping them to a similar electrical response from the latex spheres. A result of this activity that users should be aware of is that particles with different refractive indices and shapes create scattering responses either smaller or larger relative to the latex standard. For example, an alumina oxide particle, because of its high reflectivity, will scatter a great deal of light and so will appear larger than a respectively sized latex sphere. A carbon particle that absorbs light will size small relative to the latex standard. These sizing differences (deviations from the latex standard) will assign the particles into larger or smaller size channels, so the only absolute, and therefore the standard for calibration, is latex spheres in clean dry air.

The original technique of passing a volume of air through a membrane filter has not been cast aside. It still remains a valuable method of extracting a sample for subsequent analysis as to particle identification and is an approved method by the EPA for environmental monitoring.

31.6.2 Particle Counter Calibration

A particle counter is calibrated against size thresholds, not count values. This is because the instrument is considered volumetric: only at the extreme range of concentrations is a saturation limit reached or does coincidence, two particles resident in the laser beam simultaneously, occur. The particle counter size calibration procedure is carried out with monodispersed (very narrow distribution), spherical latex particles. The ASTM F 50, section 3.1.10 references particle size, which is defined clearly as "the equivalent diameter of a particle detected by an instrument using light scattering. The equivalent diameter is the diameter of a reference sphere having known refractive index and generating the same electrical response in the photo detector of the particle counter as the particle being measured."[15]

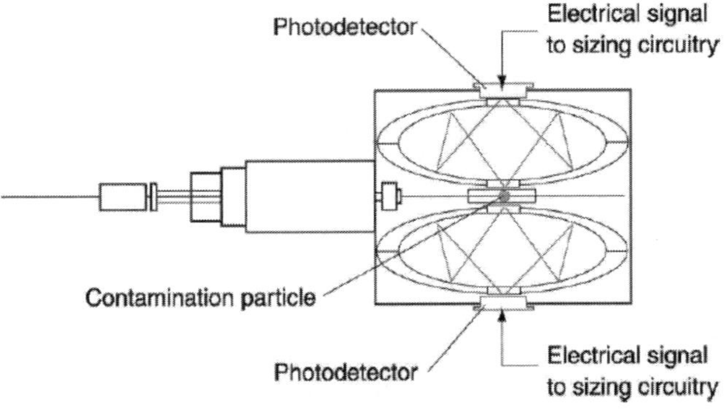

FIGURE 31.7 Laser particle counter fundamental principles.

The particle counter measures the amount of light scattered by a particle and places the relative size of the particle in a "size bin." A size bin is determined from the boundary thresholds. If a particle counter has several channels, such as a 0.1 μm sensitive instrument, the first channel on the display is shown as 0.1 microns. The second channel is shown as 0.2 microns. The third channel is shown as 0.3 microns. If a particle falls in the first channel, it is sized as being between 0.1 micron and 0.2 micron. It may be as small as 0.105 micron or as large as 0.195 micron but still falls in the first channel.

Another function of a particle counter is to normalize the counts per unit of flow (counts per cu. ft., counts per m^3). Therefore, the flow rate through the counter must be correct.

ASTM F 328 and 649 were written for aerosol particle counters. The F-328 procedure's basic principle is the use of monodispersed particles for primary calibration of the instrument and the comparison of the test instrument to a "referee" to determine the counting accuracy. The F-649 procedure's basic principle is the comparison of the test instrument to a reference instrument. This is called secondary calibration.

31.6.3 Error Control and Minimization

Once calibration has been performed, particle counters will still produce data that has inherent errors. To minimize errors, the particle counter performance specifications, beyond calibration, must also be acceptable. The following elements of performance have significant importance for particle counting and are discussed following.

31.6.3.1 Signal-to-Noise Ratio

The data produced by the particle counter should be produced solely by the detection of particles passing through the optical chamber. If the signal from these particles is not significantly greater than the electrical noise produced by the particle counter's electronic circuits, then some of the data will result from that background noise level rather than from actual particles. An increased signal-to-noise ratio will reduce electrical noise frequency and allow the detection of increasingly smaller particles; however, the added cost for an improved particle counter may not be required for the clean room when only two channels (ex: 0.5 and 5.0 mm) are required. Acceptable particle counter signal-to-noise ratio specification can be determined when more than two particles are recorded in the smallest size channel when sampling clean dry air; this is called the "zero count" capability of the instrument.

31.6.3.2 Particle-Sizing Accuracy

Particle-sizing accuracy is required because most count information is defined in terms of concentration (number per unit

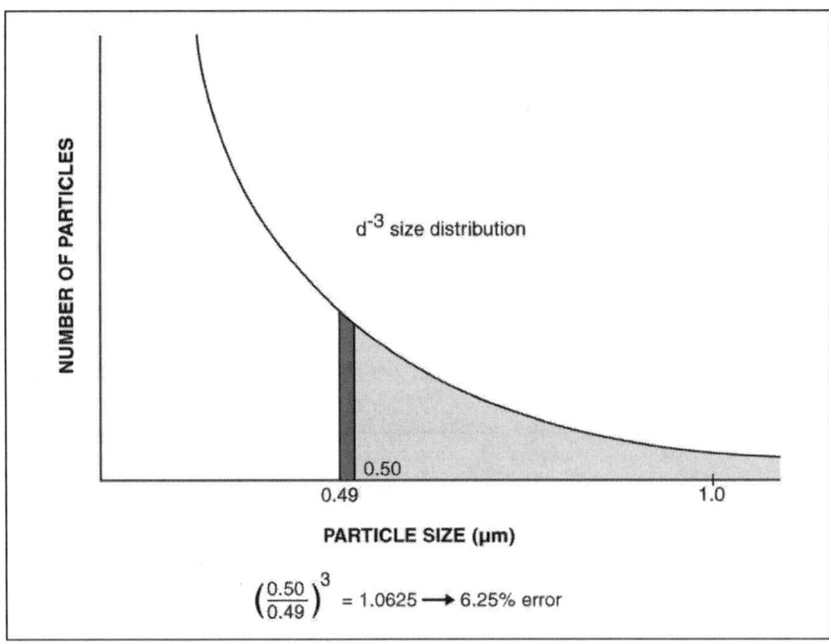

FIGURE 31.8 Laser particle counter fundamental principles.

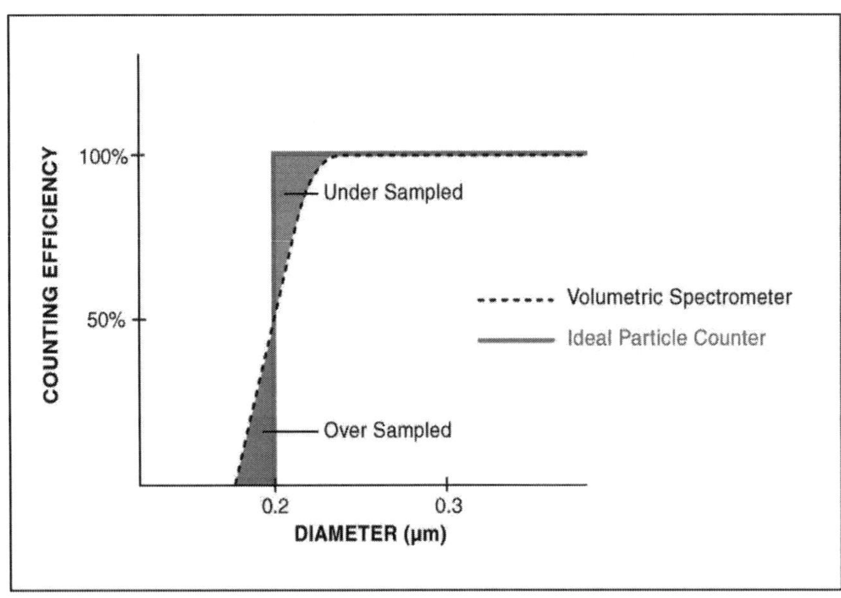

FIGURE 31.9 Counting efficiency vs. particle size.

volume) equal to or greater than a specified particle size. The particle size distribution usually encountered in clean rooms will cause an error in particle size measurement, which in turn creates a concentration error that varies with the associated particle size distribution function. Figure 31.8 shows how a sizing error of 2% can produce a concentration error of 6.25% for a third-power particle size distribution. Sizing accuracy can be converted to counting error based on the particle size distribution in the clean manufacturing area of interest. It is generally assumed to be 1/d2.08 for clean rooms based on the ISO 14644–1 standard.

31.6.3.3 Counting Efficiency

Counting efficiency is an expression of the probability that an OPC will sense, and therefore count, a particle passing through the particle counter's sample volume. This probability is a function of size up to a certain critical size above which all particles are normally sensed and counted. Figure 31.9 displays the plots of counting efficiency versus particle size. Note that although the signal produced by the particles is distributed symmetrically about the nominal most sensitive threshold, the exponential relationship between particle size and signal returned causes the counting efficiency curve to be asymmetrical.

31.6.3.4 Sensor Resolution

A particle counter's resolution is its ability to determine small differences in particle size. A number of factors combine to cause the resolution of a particle counter to be other than perfect. These include the uniformity of illumination of the sampling volume, the quality of the optical system, the quality of the electronics in the instrument, and the electrical background noise. If it were possible to introduce particles all exactly the same size to a real-world particle counter, the preceding factors would cause the reported distribution to be the familiar "bell curve" (normal or Gaussian) shape. Figure 31.10 shows the reported distributions that would result from introducing a group of particles, all exactly the same size, into particle counters with "excellent," "average," and "poor" resolution. Note that with excellent resolution, the OPC would always put each of the particles in the same size class regardless of the width of the size class. Note that the minimum possible width of the OPC size classes or "channels" tends to be determined by the fundamental resolution. Thus an OPC with average resolution can have more size classes across the range of the instrument than an OPC with poor resolution. The minimum possible width of the size classes is limited by the fundamental resolution.

31.6.3.5 Sample Flow Rate

For most critical pharmaceutical activities, very low particle levels are present, especially in the ≥5.0 μm size range, resulting in very little data produced by the particle counter. At these low concentrations, it can be expected that the random occurrence of particles could result in widely varying data if a series of small samples are taken. A particle counter with the largest available sample volume flow rate or sampling at the largest possible volumetric sample flow rate will give statistical confidence over a data set in the shortest period of time.

31.6.3.6 Particle Concentration Capability

For the majority of pharmaceutical clean room operations, a particle counter's ability to measure high particle concentrations is not necessary. However, when monitoring some clean room operations, such as powder filling, it is possible that a burst of particles may appear. If this occurs, the particle

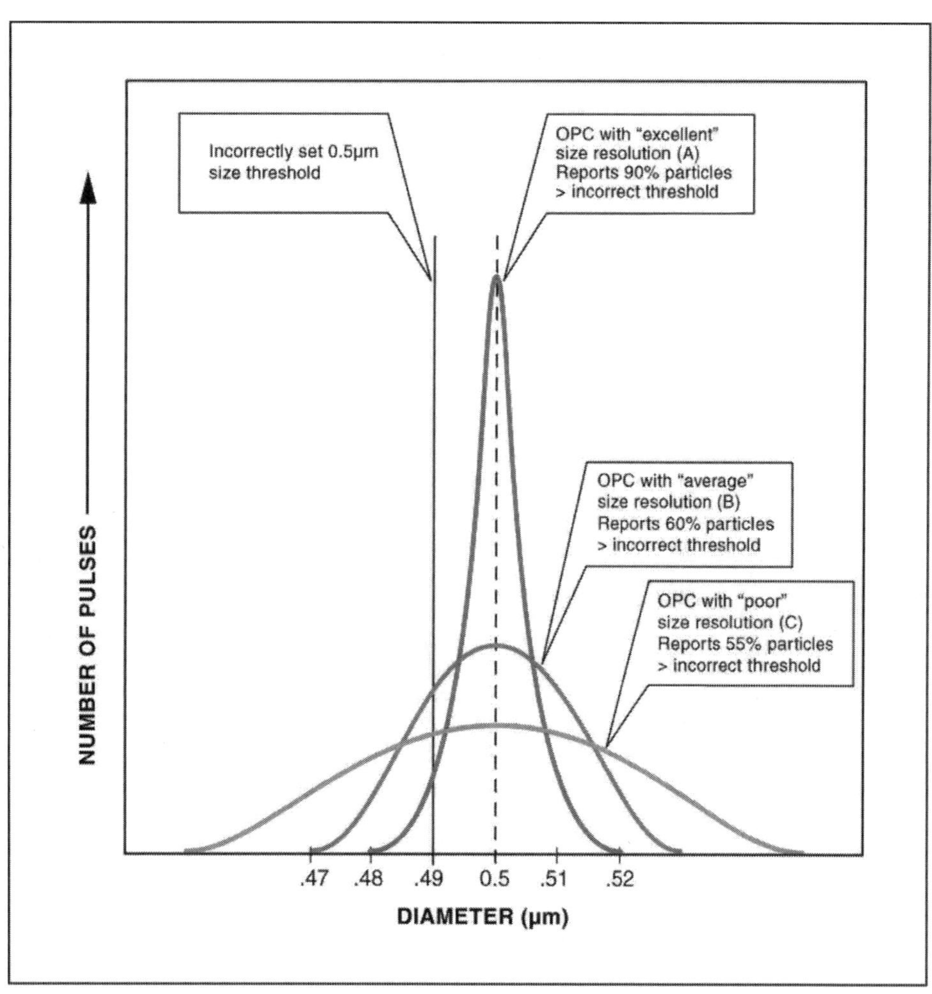

FIGURE 31 10 Variances in particle counter resolution.

Note: OPC = optical particle counter.

counter used to sample the area or the potential local contaminant source should be capable of detecting particle concentrations several orders of magnitude above the normal, with minimum error caused by coincidence. The selection the particle counter used in such applications may conflict with the previous one of maximum sample size capability.

31.7 AEROSOL MONITORING CONSIDERATIONS

The selection of sampling procedures and various rationales for monitoring are largely determined by what the system is designed to observe. This has been addressed in previous sections. There is still an underlying principle that governs the selection of sample point location. This involves the physics of how particles move within a body of media and the accuracy to which they can be sampled, irrespective of the particle counting method chosen.

When samples containing particles are taken, it should be considered that the distribution of those particles (even if size is maintained as a constant) within the media (air) is not uniform and may change over time. The concentration of particles within the sample will be low, especially the larger particles, and the direction of particles is not always matched by the direction of the air flow. Also, additional forces are acting on the particles that do not act on the air flow because of the differences in mass of the particles. These additional forces affect particle motion and are discussed following.

31.7.1 External Forces Affecting Particle Migration

Clean room monitoring activities can be defined as being performed to qualify and quantify the dynamics of a fluid; that fluid is the body of air within a confined space. This space may be either the air in the general clean room, a transport duct, or a laminar flow zone. The following terms describe the mechanisms of how particles behave in air and should assist in the understanding of sampling and management of samples, thus improving the efficiency of sampling.

- The *Stokes Number* is the ratio of a particle's radius to the dimension of an obstacle to fluid flow. This is an important factor in determining when a particle in motion will be collected by an obstacle or will pass around it. An obstacle could be a filter fiber, the sample inlet, or a component that should be kept clean, such as the opening to a vial.
- The *Drag Coefficient* is the ratio of the force of gravity to the inertial force on a particle. It indicates how a particle will resist any force that could cause a change in the particle velocity. Smaller particles have smaller drag coefficients because of their lesser mass.
- The *Relaxation Time* is the time for a particle initially in equilibrium with a moving fluid to match a change in fluid velocity. Large particles have a long relaxation time. When an aerosol stream moves through tubing that contains small-radius bends or elbows, the large particles will deposit on a tube wall because they cannot adapt easily to sudden velocity changes, though they continue in their original direction until they make contact with the tube wall. A related term to relaxation time is *stopping distance*, which is defined as the distance for a particle initially moving within a gas stream to come to a stop when the gas flow is halted, as by an obstacle.
- The *Deposition Velocity* or *Sedimentation Velocity* is the ratio of particle flux, distance per unit time for sedimentation to occur relative to the ambient particle concentration.

TABLE 31.9
Settling Velocities of Particles in Stagnant Air

Particle size (μm)	Settling velocity (cms^{-1})
0.00037	
0.01	6.95×10^{-6}
0.1	8.65×10^{-5}
1.0	3.50×10^{-3}
10	3.06×10^{-1}
100	2.62×10^{1}

There are also additional forces in effect on particles. These forces and the particles subsequent response to those forces control the particles migration through the air:

- Viscous Forces—The fluid dynamic force from a moving fluid stream. The viscous nature of an air stream will "pull" particles along that flow path.
  - If the flow is *laminar*, then additional forces act upon the larger particles causing settling and deposition. Smaller particles remain buoyant.
  - In *turbulent* flow, when large particles settle, they are re-entrained back into the airflow. Smaller particles are more prone to additional forces acting upon them, preventing them from being transmitted through a tube.
- Brownian Motion—As particles migrate through a body of air, random impacts from individual molecules will cause them to veer from course.
- Gravitational Force—Varies with particle mass and the difference between particle and air density; the larger the particle the greater the effect.
- Electrostatic Forces—Varies with the particle's electrical charge (surface area controlled) and the strength of the electrical field in which the particle is located. Electrostatic charge can develop as a particle "slips" through the air stream. It is important, therefore, to minimize

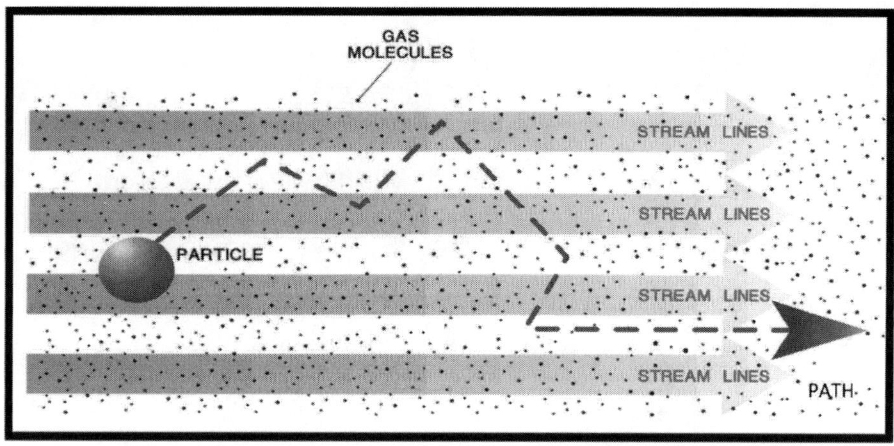

FIGURE 31.11 Migration of a particle along a mean free path because of Brownian motion.

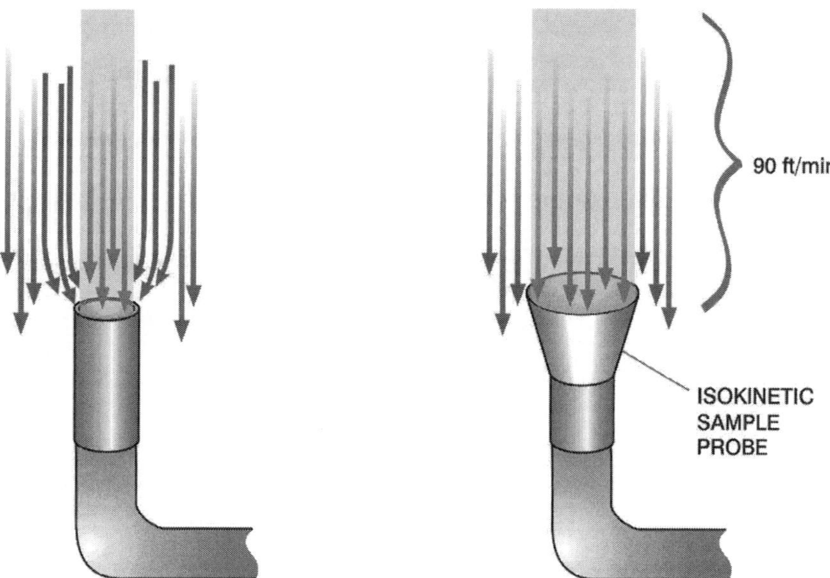

FIGURE 31.12 Isokinetic sampling.

these interactions to ensure all particles reach the final destination.

- Diffusion Force—Varies inversely with a particle's radius. Smaller particles are more prone to interactions because of diffusion.
- Thermophoretic Forces—(mainly for small particles) varies with the particle's surface area and temperature gradient.

The particle's response to these forces is controlled by the particle's size, mass, shape, and electrical charge. For essentially all these forces, the major particle parameter is size. The magnitude of the forces varies with particle size squared (or cubed).

31.7.2 Practical Considerations

To overcome the known forces on a particle, it is possible to design a system that minimizes the impact of the forces and the errors that may occur as a consequence of them.

31.7.2.1 Isokinetic Sampling

In laminar flow environments or in ducts leading to a filter, air flow is considered to be unidirectional, i.e. predominantly flowing in a single direction. The air must be neither over nor under sampled. This requirement is satisfied with isokinetic sampling, which ensures that the velocity in the supply air is the same as the velocity in the particle counter's sample-tubing inlet.

If the velocities differ, either a positive or negative sample collection error occurs. An isokinetic sample error increases with particle size. This is not of great concern for particles <1–2 μm. FS209E shows that isokinetic sampling errors >5% are not expected for small particles when using a sample probe with an inlet diameter of 2 mm or larger, even when sampling and sampled air velocities differ by an order of magnitude. However, when particles >5 μm are to be measured, isokinetic sampling is required. Anisokinetic correction can be calculated and incorporated into the monitoring strategy as required.

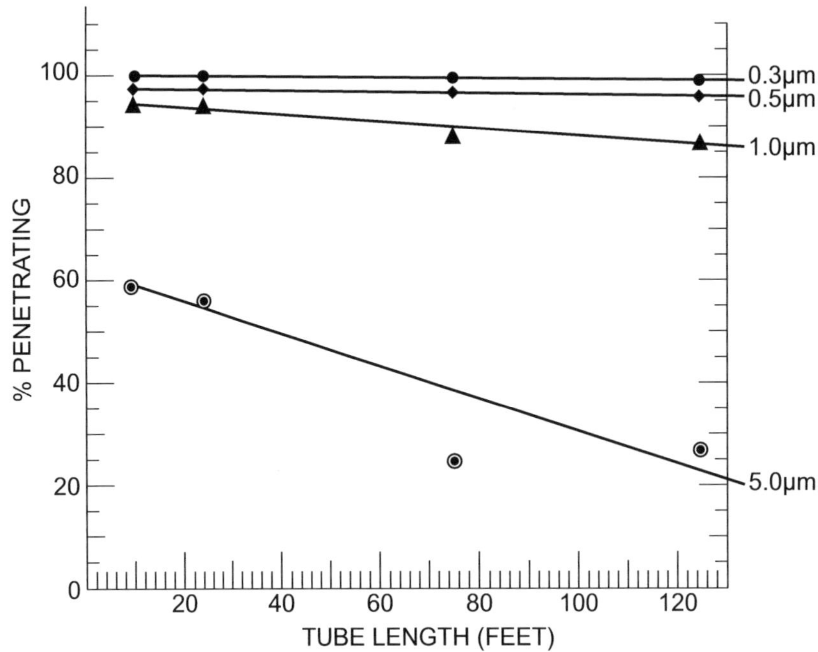

FIGURE 31.13 Particle loss in tubing (1/2" tubing at 100 liter/min).

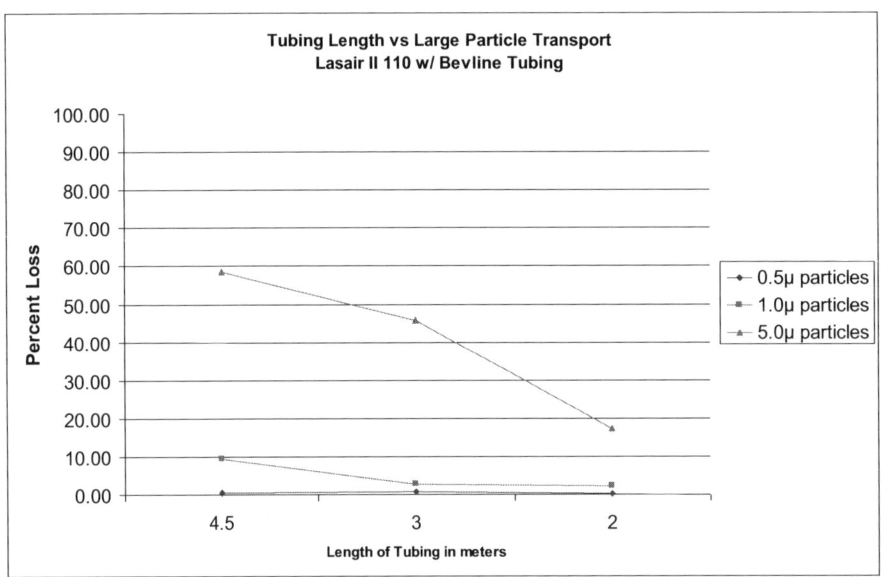

FIGURE 31.14 Particle loss in a portable particle counter (3/8" tubing at 28.3 liter/min).

31.7.2.2 Particle Loss in Transport Tubing

When a sample is taken, it is common that the sample probe head is in a separate location from the particle counter. Sample tubing is used to connect the probe head to the particle counter. If the sample is to be transported any significant distance in the tube, some particle loss will occur. These losses are dependent on tubing velocity, diameter, and distance. Large particles are lost by a combination of gravitational settling and inertial deposition on the walls of the tubing. Small particles are lost to the duct walls by Brownian motion and diffusion effects.

Figure 31.10 shows the penetration of different sized particles through an installed manifold system over distances up to 125 feet. Particles <1.0 μm in diameter show no significant losses, and the differences are essentially experimental error. Larger particles show a significant level of loss even over very short distances. This is one reason why aerosol manifold sampling is not a desired method for critical environments.

When portable particle counters are used, the flow rate in the tubing is significantly reduced, and so the maximum permissible distance is also reduced. Figure 31.11 shows a similar pattern to that of manifold sampling but over much shorter distances.

Electrostatic forces also account for a proportion of the losses in a sample. To reduce the effect of these additional forces, various types of material were tested to establish a

TABLE 31.10 Particle Transport Line Material Preference

Stainless Steel
Bev-A-Line (Hytrel-lined PVC tubing)
Polyester (as polyurethane)
Polyester-lined vinyl
Copper
High-density polyethylene
Glass
Teflon

Abbreviation: PVC = polyvinyl chloride.

suitable standard (Table 31.10). The order, most preferred first, is based on a combination of particle loss rate, electrical conductivity, and potential for oxide or sulfide formation when the tubing is exposed to urban air.

The diameter of the tubing should be selected to ensure the Reynolds number (defined in FS209E) is between 5000 and 25,000. The Reynolds number range is one for which no significant turbulent deposition occurs for particles <5–10 μm. Time in the tubing should be no more than 10–20 seconds to ensure the transmission of particles >0.1 μm before any significant losses occur.

31.7.2.3. Sample Point Selection

The selection of the sample point location is based upon a risk assessment performed for each process. There are various measures that will influence the final location including:

- Pointing the probe into the sample flow where the flow is unidirectional (laminar or iso-axial). In unidirectional flow, it is important to perform isokinetic sampling, especially for macro particles.
- Analyzing relative risks to product based upon activities within the room environment. Analyze workflow patterns to establish worst-case scenarios for background environments.
- Conducting airflow tests ("smoke tests").
  - Verify flows lead away from product and out of enclosure.
  - Verify minimal recirculation.
  - Identify particle traps and recirculation zones.

It is also possible to perform three-dimensional airflow studies of existing facilities combined with intensive particle monitoring to determine the operational characteristics of the clean room and identify any worst-case locations that may exist. The evaluation is best performed in Class B environments because of the number of available particles in the air and the variations in air flow patterns. (Establish critical contamination risk locations in environmental monitoring by means of three-dimensional airflow analysis and particulate evaluation.)[16]

31.8 EXAMPLE OF CLEAN ROOM CONTROL STRATEGY

Assume there is a clean room to be used as an aseptic filling area. The whole room needs to meet ISO Class 5 at 0.5 μm, operational, as the filling operations, although behind a hard curtain, are open to the environment. The following steps show how to determine the room's classification and monitoring requirements.

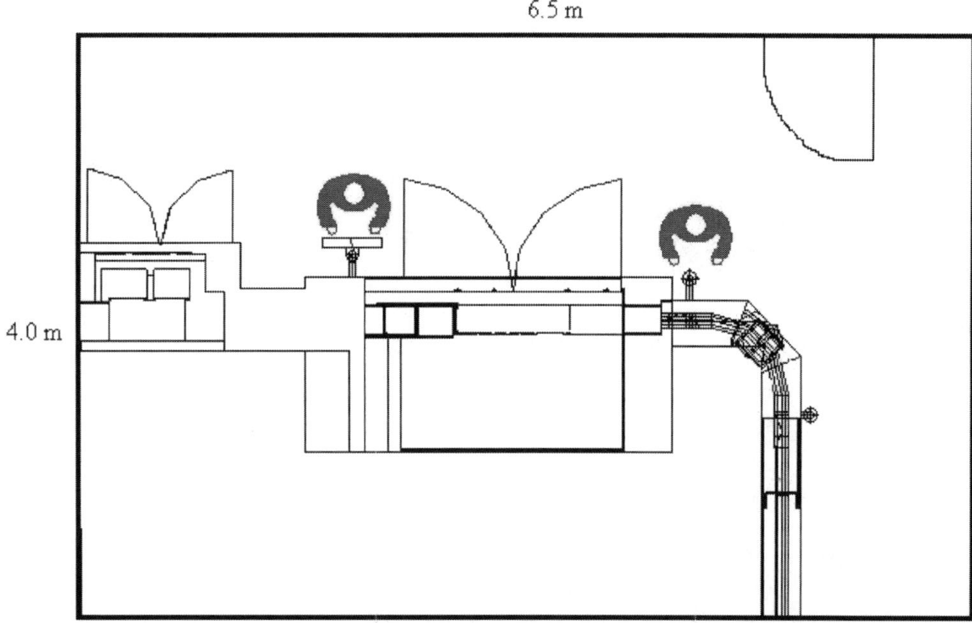

FIGURE 31.15 Sample layout of a clean room.

31.8.1 Example—Certification to ISO14644-1

The room is 6.5 m by 4 m and has an enclosed filling machine in the center.

1. Determine the maximum permitted particle concentration:

$$C_n = \frac{(0.1)}{(0.5)} \times 10^5 = 3517(\textit{rounded to } \textbf{\textit{3520}} \textit{ n / m}^3)$$

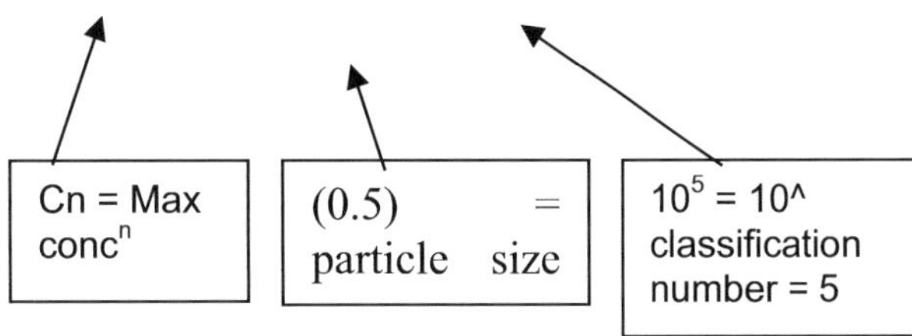

2. Determine the number of sample locations:

   a. Room Area = 4 × 6.5 m = 26 m²
   b. From a Table in ISO 14644–1 (Table 31.11)

The number of sample locations required to statistically test a room of 26 m² =

Rounded to the nearest, highest value = **7 locations**

These sample locations should be placed within an approximately 2 m² area within the clean room and at any location within each identified quadrant. If critical locations also need adding to the classification process more locations can be added at this time.

3. Determine the minimum sample volume:

$$V_s = \frac{20}{C_{n.m}} \times 1000$$

$$= \frac{20}{3517} \times 1000 = \textbf{\textit{5.69}} \text{ liters}$$

From ISO 14644–1 Section B.4.2.2, Minimum Volume = 2 liters and Sample period = 1 minute. Standard particle counters run at 28.3 L/min. Therefore, a 1-minute sample at each of the 7 locations is needed to meet the specification.

4. Determine the measurements for each location
5. Report

Each Location max = 3140 n/m³ <3,520 class limit = **PASS**

This room meets the specification for an ISO class 5 clean room at 0.5 mm, operational, and can now be used for the purpose for which it was designed.

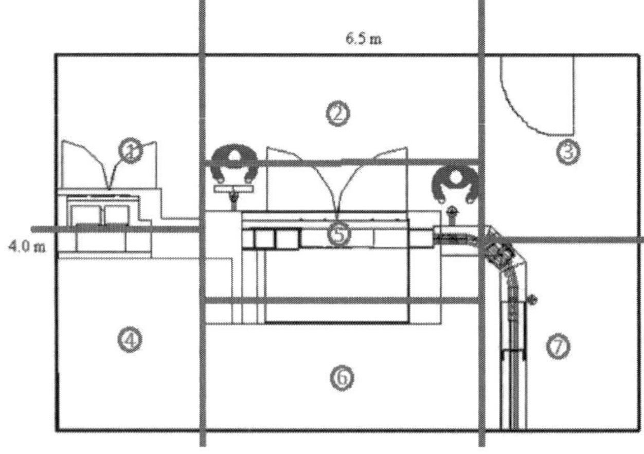

FIGURE 31.16 Sample point locations in a clean room.

TABLE 31.11
Sampling Locations Related to Clean Room Area

Area of clean room (m²) less than or equal to	Minimum number of sampling locations to be tested (N_L)
2	1
4	2
6	3
8	4
10	5
24	6
28	7
32	8
36	9
52	10
56	11
64	12
>1000	see Formula

Note: 1. If the considered area falls between two values in the table, the greater of the two should be selected. 2. In the case of unidirectional airflow, the area may be considered as the cross section of the moving air perpendicular to the direction of the airflow. In all other cases the area may be considered as the horizontal plan area of the clean room or clean zone.

Source: ISO 14644–1

TABLE 31.12
Number of Measurements Required

Location	Number/m³
1	2340
2	1467
3	3140
4	1509
5	1966
6	825
7	877

31.8.2 Example—Locations of Monitoring Sample Points

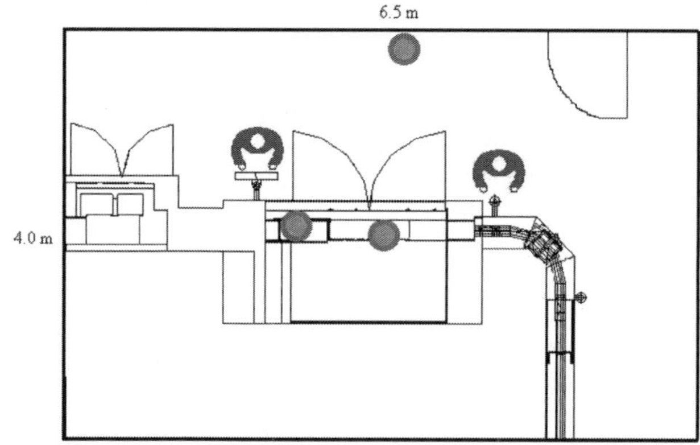

In this example, the sample locations are continuous—this best reflects the requirements of the EU CGMP that the entire process is monitored throughout all phases of operations (including set-up) and that no transient events are missed.

The locations on the filling process are determined based upon an individual risk assessment and the sample probes located considering the airflow patterns, nature of activities taking place, severity of contamination, and accessibility. The same rational can be used for the location of the dynamic monitoring for viable organisms. The single sample in the background reflects this is the highest traffic area and where operators remain during processing, unless line intervention may be required.

31.8.3 Example—Data Processing

As the room is generating continuous data at each sample locations each minute (sample duration based on risk, but ideally does not exceed 1 minute), the capture and storage of data becomes an important aspect of any monitoring system, and the data must be securely stored. Monitoring software solutions will need to demonstrate that the data can be archived and stored according to current requirements for integrity, access, and reporting. As the data is communicated directly from the sensor, via a time stamped packet, the integrity of origin is maintained, and a log of any and all out of tolerance readings will be displayed as an alarm result.

Reporting not will focus on individual results as the room demonstrates a continued state of control, unless an out of tolerance event occurs. These can be either single events of extraordinary magnitude, or an accumulation of several events above a defined alert level, as trended data becomes the primary tool in determining control over the room environment.

REFERENCES

1. The United States Pharmacopeia General Chapters (USP 28, 2005).
2. USP <797> Pharmaceutical Compounding—Sterile Preparations.
3. Federal Standard FS209 (Airborne Particulate Cleanliness Classes in Cleanrooms and Clean Zones, Federal Standard No. 209E. Washington, DC: General Services Administration, 1992.
4. Cleanroom garment system application and use, Berndt, Burnett & Spector, S2C2 monitor No.39, April 2000.
5. PDA Journal, Abreu, Pinto & Olivera, Volume 58, No.1. January 2004.
6. Cleanroom Design, 1999, John Wiley & Sons.
7. USP <1116> Microbial Evaluation of Clean Rooms.
8. Particle Measuring Systems Application Notes, 53, 47, 68 & 7.
9. Modern cleanroom clothing systems: People as a contamination source, Reinmueller and Ljundqvist, PDA Journal of Science Volume 57 No.2, March 2003.
10. Title 21, Code of Federal Regulations, Parts 210 and 211.
11. Guidance for Industry Sterile Drug Products Produced by Aseptic Processing, September 2004.
12. PDA Journal of Pharmaceutical Science and Technology, Volume 57, No2, Mar 2003, Supplement.
13. Medicinal Products for Human and Veterinary Use, Volume 4, 1998.
14. REVISION TO ANNEX 1, Manufacture of Sterile Medicinal Products, September 2009.
15. PE 009-1, 1 September 2003.
16. Process Analytical Technologies and Non-viable Particle Counting, Particle Measuring Systems Technical Paper, 2005, Mark Hallworth.
17. Katayama, Higo, Tokunaga, Hiyama and Morikawa, Fujisawa Pharmaceuticals Japan, PDA journal of Pharmaceutical Science and technology Vol. 59, 2005.

32 Environmental Monitoring

Jeanne Moldenhauer

CONTENTS

32.1 INTRODUCTION

Regardless of the type of pharmaceutical product manufactured, there are regulatory expectations that environmental monitoring be performed. This applies to both sterile and non-sterile pharmaceutical products, as well as products from compounding pharmacies, medical devices, and biological products. There are a variety of methods that can be used for monitoring of viable microorganisms and particulates in the air, on surfaces, or on personnel. Monitoring of water systems, steam systems, and compressed gases are usually included in the realm of environmental monitoring. These

DOI: 10.1201/9781003163138-32

monitoring methods may be the conventional methods using for hundreds of years in microbiology or the newer rapid or alternative microbiological monitoring methods available in recent years. The purpose of this monitoring is to ensure that the production areas utilized in the manufacture of these products is appropriately controlled to minimize the risk of microbiological contamination of the products being manufactured. This aids in ensuring products of the highest possible quality being manufactured.

Many methods used for pharmaceutical products are specified in the pharmacopeia, also known as compendial methods. Use of compendial methods assumes that the method is already qualified by being in the pharmacopeia. Unfortunately, the methods used for environmental monitoring are not compendial. As such, each laboratory is responsible for generating data to show that the method is scientifically valid and appropriate for use. Selecting a method for use is based upon the likelihood of contamination of the process, resource requirements, design and cost. Alternative or Rapid Microbiological Methods (RMMs) include some other considerations as well, for example, whether a regulatory filing is required prior to approval and implementation of the method, whether the numbers enumerated by the newer system will be significantly different from those generated by conventional methods, time to results, ease of use, and whether there are new concerns regarding whether the data is valid (e.g., does the system have the ability to distinguish between living and dead cells? This may also be called viable and non-viable cells).

When monitoring for viable microorganisms, some may grow aerobically (in the presence of air) or anaerobically (without the presence of air). Typical environmental monitoring programs are focused on aerobic monitoring, as most manufacturing processes have air in the manufacturing process. A limited discussion of anaerobic monitoring is provided. In most cases, periodic anaerobic monitoring is conducted to ensure that there is not a high level of contamination that should be of concern. Some anaerobic microorganisms are capable of also growing aerobically. However, there are some manufacturing processes that have tightly controlled oxygen levels, in some cases the manufacturing is under anaerobic conditions. Some products like those with sesame oil may be conducive to anaerobic contamination, which should be monitored in the product bioburden sampling. Product bioburden monitoring is considered outside the scope of this book chapter.

It is common to look at both viable and nonviable particulate monitoring as methods of assessing contamination control. This can be useful, but the methods do not monitor the same things. Particulates are affected by the activities taking place in the area. For example, if you have stoppering equipment or most robotic systems, they generate particulates when being properly operated. As such, you could have an increase in particulate because of proper operation. There is no direct correlation of viable microbial contamination to nonviable particulate particles. Viable microorganisms may come from the particulates present, using the particulate like a raft. They may also come from personnel present in the operating area, e.g., through improper gowning, improper handwashing, torn gowns, and the like. Microbiological monitoring is designed to assess the transfer, migration, and elimination of viable contaminants. (Whyte et al., 1982)

32.2 WHY PERFORM ENVIRONMENTAL MONITORING

Around 1990, regulators in Europe and the United States increased their focus on environmental monitoring, especially for aseptic manufacturing processes. At this time, there was a significant amount of literature available to indicate the many weaknesses of the compendial sterility test methodology. Although many believe that this test identifies whether any viable microbial contamination is present in the product, the conventional test method is not capable of doing this. The conventional test method lacks the necessary level of detection, is dependent upon the growth conditions in the method, and is not a statistically valid test. Contamination cannot get into the product unless it is present in the production area. This led to a conception that the monitoring is a *de facto* sterility test (Moldenhauer and Sutton, 2004; Moldenhauer, 2017a). However, environmental monitoring should be viewed as an evaluation of microbiological process control.

In 2004, the Food and Drug Administration (FDA) issued their *Aseptic Processing Guidance*. This document has a section devoted to environmental monitoring. It provides a high level of detail on what is expected in a program, allowable action levels, program design, and implementation. (FDA, 2004) This document is widely used by investigators for the environmental monitoring requirements of terminally sterilized products and non-sterile products as there are no specific regulatory requirements for these processes. Typically, the biggest differences between processes are the allowable alert and action levels and the frequency of monitoring. (Moldenhauer, 2017a)

The European Union has similar types of requirements specified in Annex 1 of Volume 4 of the European Good Manufacturing Practices (GMPs) (Eudralex, 2009). These requirements do not have the same types of details as the United States' requirements, but the action levels established are comparable. However, the intent for microbiological control is the same.

Another guidance frequently cited for environmental monitoring is ISO 14644–1. (ISO, 2015) The ISO document specified the requirements for clean room particulate levels, used for room or area classifications. Although often cited as requirements for microbiological monitoring, this document does not provide any information on the viable microorganism levels that should be met.

32.3 DESIGNING AN ENVIRONMENTAL MONITORING PROGRAM COMPLIANT WITH REGULATORY REQUIREMENTS

Environmental monitoring programs are dynamic. A program that was developed and compliant 20 years ago is not necessarily compliant today. The regulatory expectations for these programs change based upon updated guidance, new regulatory information, and most frequently because of investigatory citations. It is useful to routinely check regulatory compliance databases to see what types of observations have been made regarding environmental monitoring. Environmental monitoring programs also make good business sense. Keeping your environment within the stated process controls reduces the need for product holds, deviations, and investigations because of microbiological issues.

A comprehensive environmental monitoring program should be established. The intent of this program is to monitor the state of microbiological process control in the controlled and critical areas of the facility. The program goes far beyond the collection of media and room area samples. Some of the typical components of this type of program include the following (Moldenhauer, 2017a):

- A site or production area contamination control strategy including methodology for implementation.
- A microbiology master plan, either as a separate document or a detailed section of the site's master validation plan.
- Establishment and implementation of a system for room classification and air cleanliness levels.
- Completion of a program for cleaning and disinfection, including selection of cleaning/disinfectant solutions to be used and whether they are rotated or not. A sporicidal cleaning agent must be specified for use including the frequency of routine use.
- The heating, venting, and air conditioning (HVAC) system should be installed and qualified prior to establishing the environmental monitoring program.
- Qualification of environmental monitoring personnel on the gowning required for the area as well as the techniques to be used in monitoring.
- An environmental monitoring oversight committee. (Depending upon the company, this may be called a sterility assurance committee, which includes environmental monitoring.)
- Appropriately selected and qualified monitoring equipment (conventional methods or RMMs).
- Qualified/verified testing methods.
- Conduct of a risk assessment to determine the sample sites to utilize.
- Conduct of a validation protocol to justify the incubation and media conditions to be utilized, e.g., supporting data for single media and temperature incubation.
- Qualified or verified methods specifically for monitoring of water, compressed gases, clean steam, and surfaces (including personnel).
- Use of appropriately qualified production equipment for sampling activities (e.g., Installation Qualification and Operational Qualification (IQ/OQ) completed at minimum).
- Performance of the Environmental Monitoring Performance Qualification (EM PQ). Frequently this is performed concurrently with media or water fills. This is the time at which the sampling locations for routine monitoring are typically solidified.
- Establishment of the routine monitoring program and implementation of same.
- Implementation of procedures for data collection, analysis, and trending.
- Establishment of appropriate alert and action levels for monitoring.
- Establishment and implementation of a trending system; automated is preferred.
- A system for investigation of microbial data deviations (MDDs) and resolution of issues.
- Identification of objectionable organisms. (Note: In a sterile product, all organisms found are objectionable. Recent regulatory inspections indicated an expectation for "objectionable organisms" prior to sterilization. For example, *Ralstonia pickettii* may be a problem in an aseptic filtration process as it can go through a 0.22 μm filter.)
- Establishment and implementation of a system for identification of contaminants, using qualified methods.
- Establishment and maintenance of a library of environmental isolates for your facility.
- Development and implementation of a system to conduct product impact analyses for aberrant or unusual results obtained in environmental monitoring.
- Ensuring that a corrective and preventative action program (CAPA) is effective including microbiological deviations.
- A system for routine evaluation and update of the environmental monitoring program should be established.

32.4 CONTROLLED AND CRITICAL AREAS

The intent of the viable environmental monitoring program is to evaluate the microbiological quality of the controlled and critical areas in a sterile product manufacturing facility. With an understanding of the baseline (as is) microbiological quality, it is possible to design an effective control program, e.g., HEPA filtration, cleaning and disinfection procedures, personnel gowning, and aseptic behavior and practices. Anaerobic monitoring is frequently said to be "unimportant" in a sterile manufacturing program; however, one should assess the risk of anaerobic contamination to the product. Many individuals carry *Propionibacterium acnes,* which is a gram-positive microorganism found on human skin and is associated with acne. It prefers anaerobic conditions for growth. Taking anaerobes into consideration of risks provides for a better monitoring design.

Controlled areas refer to those rooms designated ISO 7 or 8, Grades C or D, and Class 10,000 to 100,000. In these areas, there is typically minimal product or component exposure. As such, they can have higher microbiological quality levels. These areas have HEPA-filtered air, which removes 99.999%

of the small particles from the air. Unidirectional airflow is not required and is infrequently used except in small localized areas. Controlled areas are used for activities like: preliminary product formulation in closed systems, storage of wrapped sterile components, bulk formulation areas, sterilizer loading rooms, corridors, air locks, and preparation or staging rooms. In the European GMPs, Annex 1, there are required room classifications for the different types of processing steps. (Eudralex, 2009)

Critical areas refer to those rooms designated ISO 4.8 or 5, Grade A or B, and Class 100. The air supplied in these areas is HEPA filtered. The airflow may be horizontal or vertical. Depending upon the processing needs, the entire room may have unidirectional airflow, or only specific areas, e.g., a module covering the exposed product or componentry. There are manufacturing processes that are performed entirely within an isolator rather than filling suites or rooms. The aseptic manufacturing isolator process typically includes: aseptic processing filling suites or aseptic bulk manufacturing rooms and a dry heat tunnel providing the sterilized and depyrogenated glassware with an outlet into the area. Other supplies are brought into the aseptic area using appropriate aseptic behaviors. Alternatively, one may find the sterilizer outlet or primary barrier modules within controlled rooms. For terminally sterilized products, the product flow may take place in lower air cleanliness classifications prior to sterilization. However, some companies use aseptic processing even for terminally sterilized products.

Because environmental monitoring is expected for controlled and critical areas, there are additional considerations that should be taken including:

- The frequency of monitoring.
- The locations or sites to be monitored.
- The direction of the airflow and the air-cleanliness designation of the room.
- The maximum number of people needed to run the process in that area.
- The training and expertise of the people conducting the monitoring and the production process.
- The level of activity during the monitoring.
- The time at which the monitoring is performed during the shift or the fill day.
- The duration of the monitoring when an air test is performed.
- The type of media to be used for monitoring that will maximize recovery of contaminants.
- The incubation time and temperature to be used for monitoring samples.
- The alert and action levels established for each controlled or classified areas.
- How to determine when corrective or preventative actions should be taken.
- Identification of whether an investigation should take place and whether it should be a laboratory or production area investigation.
- What types of documentation are required and necessary to meet the site's needs.
- Methods for data management and data recovery, trending methods, and how and when reports will be generated. Will this be an automated or manual process?
- How will contaminants be isolated? Will this be performed in-house or will a contract laboratory be utilized?
- How will an environmental isolates library be established? What form of isolate will be maintained, e.g., slants, lyophilized cultures, or Bioball? Who is responsible for maintaining the cultures?
- What procedures or work instructions will be required for the area, and who is responsible for generating them?

In recent years, there has been increased regulatory scrutiny on how senior management is involved in the environmental monitoring program. As such, there must be an established communication pathway to accomplish this. For companies using purchased trending software, many systems have an e-mail notification system to aid in accomplishing this task. Other companies utilize the Environmental Monitoring Committee as a method to have routine senior management communication on what is happening, how investigations are conducted, whether actions taken are corrective or not, and the like.

32.5 SELECTION OF RECOVERY MEDIA AND TEST CONDITIONS FOR ENVIRONMENTAL MONITORING

An activity frequently forgotten when establishing an environmental monitoring program is the generation of actual test data to support the media selected for growth and the ideal incubation conditions to ensure the recovery of organisms. There are many articles identifying issues with the correct media and incubation conditions to use. To address this, a protocol should be written to evaluate selected media and incubation conditions. In most cases, a single media and incubation condition can be identified. Details of these types of studies are included in Moldenhauer (2014).

It is important to utilize a microaerophilic organism, e.g., *Propionibacterium acnes*, in these studies. These organisms are slow growing in aerobic conditions and aid in the definition of the minimum incubation time. One company indicated that they could continue to use a 3-day incubation time, even though the *P. acnes* did not grow, as it never showed up in their environmental program. However, having only used a 3-day incubation, it would never be evident as it takes about 6 days to grow out under these conditions. (Moldenhauer, 2014)

32.6 VIABLE ENVIRONMENTAL MONITORING—GENERAL

An ideal monitoring system would identify all the microbiological contaminants present. However, no monitoring system is perfect. There are advantages and disadvantages for all the systems available. The accuracy of each method is also different. Although RMMs may yield results sooner and may have increased accuracy and better limits of detection, they also have disadvantages or weaknesses. As such, the results obtained from monitoring are considered an estimate of the microbial contaminants present. When a contaminant is found it should be considered "indicative" of other contaminants being present.

32.6.1 Written Program

Regulatory expectations necessitate having a written document that describes the entire environmental monitoring program. This document is also useful in explaining your program during regulatory investigations. This should be a detailed document.

32.6.2 Air Monitoring

Monitoring of the air in aseptic areas is conducted using at least three different methods: active air sampling, passive air sampling, and nonviable particulate monitoring. Understanding the contamination level in the air as well as the activities taking place in an area can provide important information of the contamination risks for the product being manufactured.

Active air monitoring utilizes equipment that can determine the number of microorganisms present in a specified volume of air. This is very useful because you can compare counts for specific samples sizes. Depending on the equipment used, either a cubic foot or a cubic meter of air is sampled.

Passive air monitoring uses media plates or broth samples that are exposed to the environment to assess the number of microorganisms that "fall-out" of the air and land on or in the media. Many believe that this sampling is more representative of the risks to the product being contaminated. Selection of where to place the media in the testing environment can play a significant role in the contamination that will be detected.

Nonviable particulate monitoring detects and enumerates the number of particulates present that are greater or less than a specified size. For most pharmaceutical applications, testing is conducted for particles ≥0.5 μm. There are some regulatory agencies that also require monitoring of particles ≥5 μm.

Air monitoring can be conducted continuously or at discrete sampling times. It is common to use continuous monitoring for nonviable particulates. The newer RMMs allow for continuous monitoring of viable microorganisms also. The benefit of the RMMs is that results can be obtained in real time.

It can be difficult to correlate the number of viable microorganisms found on an operator's gloves following a difficult manipulation. Positive trial-filled units can help substantiate risks that occurred or reinforce the appropriateness of action levels. As an example, a glove count of 8 colony forming units (cfu) after a line speed adjustment, where the same type of organism was found in a filled vial, would indicate that a risk of contamination exists for that action. However, a count of 200 cfu in the same situation where no contamination results were found for the product does not necessarily mean that a count of 200 cfu would be an appropriate action level. Additionally, during routine production, the collection of some portion of the product sterility test samples at the time period when routine monitoring occurs can be useful in support of failure investigations

Today, most action levels for aseptically filled products are specified in regulatory guidance.

32.6.3 Surface Monitoring

Another aspect of environmental monitoring is the sampling of facility and equipment surfaces to determine the number of viable microorganisms' present. Although RODAC plates are commonly used, there are also other applicable methods. The selection of an appropriate method is the number of samples to be taken, whether the sampling method interferes with the sample site, and the time it takes to process the sample.

Sample site selection for machinery surfaces may be more difficult than that for air sampling. The surfaces contact the media for monitoring. This raises a concern that a critical surface may become inadvertently contaminated in the sampling process. It is important to be able to sample those sites most likely to cause or indicate product contamination if they become contaminated because of the activities associated with the process. It is often wise to identify "indicator" sites that are near but not in contact with product contact surfaces. These indicator sites can be selected such that disinfection or operator interaction there is much the same as at the product exposure sites. The detection of contamination here "indicates" that critical sites are at risk and an investigation would be appropriate.

Surface sampling provides useful information on the effectiveness of the cleaning and disinfection process. Personnel are often cited for carrying and dispersing the viable contaminates in a clean room. This may be a result of touching surfaces or shedding.

Although there are many control systems in the clean room design, personnel contamination occurs at locations where the control systems do not remove them. During cleaning and disinfection, it is critical that the cleaning agents encounter the contaminated surface and that the procedures used have the appropriate contact time to eradicate these organisms.

Ideally, the gowning procedures used by personnel should prevent contamination of the area. However, use of poorly designed gowning or gowning with inappropriate materials can contribute to contamination.

Personnel monitoring is considered a subset of surface monitoring.

32.6.4 Frequency of Monitoring

There are a variety of considerations in establishing the frequency of monitoring, including the type of sterilization process, the applicable regulatory guidance, and the state of control of the facility. Prior to 1987, typically monitoring was conducted "at least daily." Since that time, it is commonly expected to perform monitoring for each batch during each shift of production. For some aspects of the process, additional sampling may be required, e.g., additional sampling may be required when manipulations are performed or special interventions are conducted. The frequency of sampling should be included in the written program.

32.6.5 When to Sample

In most sterile operations, surface sampling is conducted after the completion of the production. This is to ensure that the sampling does not compromise the surfaces in the clean room. Air sampling may be conducted at this time or during the entire process, e.g., continuous viable or nonviable monitoring and use of settle plates. How you selected the time to sample should be included in the written program.

The time of testing during a shift should be considered. Depending upon the type of clean room and the amount of activity, samples taken at different times during the shift may yield different results. The count comparison may reveal that there is greater accumulation in some areas as the day progresses. This may not be true of every operation or type of clean room but should be considered a factor that could influence the results. The monitoring should ideally be done to provide a sampling of various times.

32.6.6 Number of Samples to Take

Some guidance is provided in ISO 14644 on the number of samples to take (ISO, 2015). A risk assessment should be conducted to determine where to sample. This should provide guidance on how to determine the number of samples to take for monitoring. The number of sample sites during qualification may be higher than those used for routine sampling. For example, in qualification one may take all of the samples identified based upon risk, along with some samples identified from other methods like grid sampling. The benefit of this is potentially identifying other risks that were not previously identified.

32.6.7 Establishing the Frequency of Monitoring and Setting Final Limits (Levels)

Ideally, 6 to 12 months of test data should be collected daily or more often while varying the time and shift of testing. This aids in considering the following:

- Seasonality
- Shift relationships
- Cleaning frequency
- Day of the week
- Amount of activity
- Intervals between HEPA filter changes

Initial performance qualification studies may utilize 1 to 3 months of data. At that time, companies may advance into production activities. However, the protocol should state that "official" frequencies and "levels" will be established at the end of 6 to 12 months.

The activity and traffic patterns may account for dramatic differences in data from different locations within one room. The FDA Aseptic Processing Guideline (FDA, 2004) requires air testing within the "critical zone." These zones are the locations proximate to product exposure or closure operations. A critical zone could also be the outfeed of a tunnel sterilizer or any other location where sterile components or product is exposed even if this exposure occurs within a protective shield or under unidirectional air flow modules. In bulk or formulation areas, this could be the location of an aseptic transfer or aseptic addition. In controlled areas, there may not be an easily identifiable zone more critical than the rest of the room. In the absence of a critical zone, survey of the traffic pattern and the activity zones will indicate where the potential sources exist.

32.6.8 Qualified Personnel

It is very important that your site have subject matter experts (SMEs) for environmental monitoring. Many companies have implemented aseptic processing limits for non-sterile and terminally sterilized products, because the personnel at the site are not adequately informed to explain the rationale for your limits at the site. This can result in a program that is much more restrictive than necessary.

32.7 AIR MONITORING METHODS

There are numerous systems available for air monitoring. Traditional methods using culture media have been around for many years. There are also rapid or alternative microbiology methods that may be growth-based or viability-based methods. Each available system has advantages and disadvantages that must be considered in the selection of a system. It is also important to consider consistency in operations. Rarely would it be appropriate to have a single laboratory with five or ten different monitoring systems assuming that they could be used interchangeably. Each system has its own variability and limit of detection. They may measure different volumes of air. As such, your data may not be useful if you routinely change systems without keeping track of which data is based upon which instrument.

Conventional active air samplers draw air in and collect organisms from it. The organisms are subsequently incubated on or in media and enumerated. Each type of sampler has an "area of influence" and "efficiency." In this context, the area of influence is the space around a sampler from which the organisms are reliably collected. The efficiency refers to the number of organisms captured from a grouping entering the sampler. It is important to know the area of influence so that the location of the samples can be established and supported. In areas with a high air cleanliness rating, e.g., ISO 4.8 or 5, the efficiency should be very reliable as there are few organisms present (Figure 32.1).

FIGURE 32.1 An example of a SAS isolator air sampler for viable microorganisms.

Source: BioScience International.

There are several factors that affect the area of influence for an air sampler including:

- The design and configuration of the sampler, e.g., the size of the sampling port
- The sampling rate and volume of air taken, e.g., 1 cubic foot or 1 cubic meter
- The configuration of the air pathway to and around the sampler

In most cases, the area of influence covers several cubic inches of space around the sampler.

The efficiency can be affected by the nature of the collection medium and the number of capture opportunities within the sampler, the rate of sampling, and the method of capture (e.g., impact on a collection surface, trapping in a fluid or filter matrix) and the sampling time. The capture efficiency is influenced by whether the instrument can capture particles in a variety of sizes, measured in milli- or micrometers. Because there are differences in the efficiency and design among air samplers, data from one type of monitoring device cannot always be compared with the readings from another type of monitoring device.

It can be difficult to measure the efficiency as it assumes that every organism found downstream (escaping) a sampler under test can be captured by some other sampler. It also needs a method to dispense a known concentration of organisms in an air chamber that can be used to evaluate the sampler's detection. This problem is further complicated by the need to evaluate the efficiency of the device with very low levels of contamination. Very few commercial isolators are designed to use aerosols with low level contamination dispersed in them. Frequently this type of isolator is used for military or biological warfare types of evaluation. Assumptions must be made that the sampler serving as the "referee or standard" has a certain constant efficiency so the results of the sampler under test can be compared and expressed as 90% or 110% of the referee sampler's results. In most cases, the best we can evaluate is that one sampler may sample a higher or lower proportion of the organisms presented in a controlled test method. It should never be assumed that every sampling method can work with 100% efficiency.

Passive air samplers are the placement of exposed nutrient agar plates or broths, which depend upon the chance settling of an organism onto the surface of the media. In this case, you cannot determine the amount of air being sampled. It is difficult to determine the area of influence for passive air samplers. Another name for these samplers is settling plates or gravity exposure plates. They are generally considered qualitative indicators of air quality. Whyte and Niven (1986) presented one of the most complete discussions on settle plates. Their studies revealed that the fears of agar dehydration are not as significant as was suspected. Exposure for 24 hours in still air dehydrates agar by only 13%. This takes 6 hours in a laminar flow hood and 1 hour at a rate of 1 ft^3/min of concentrated air blowing directly onto the plate. This loss in water content correlates to only an 8% loss of viability in test organisms. However, they concluded that these plates must be exposed for extended periods to maximize the probability for them to accurately estimate airborne concentrations of organisms.

Whyte indicated when the airborne concentration falls below 0.5 cfu/ft^3 of air, no organisms can be detected during a 30-minute exposure because the probability of them settling is also influenced by the amount of space separating them. According to Whyte, given equal sampling time, a settling plate is 17 times less efficient than an active sampler.

32.8 RAPID MICROBIOLOGICAL METHODS FOR AIR MONITORING

Recently, several new technologies have been commercialized for the detection and enumeration of viable microorganisms in the air. These technologies are either growth-based or viability-based. Growth-based technologies are based upon the growth of the captured microorganisms. Viability-based technologies do not require growth of the microorganisms, but rather use some type of viability marker to determine if they are present. The following discussion describes some of these methods. A current, comprehensive matrix of rapid methods is provided at www.rapidmicromethods.com.

32.8.1 Viability-Based Methods

Viability-based methods typically eliminate the concern about whether the media and incubation conditions are appropriate to detect all microorganisms present. Because no growth is required, the organisms are enumerated regardless of their special growth needs. Not all the systems using viability technology capture the microorganism for further identification. This can potentially be a disadvantage when investigating data excursions. New systems are frequently developed, so this listing may not encompass all available systems.

32.8.1.1 Use of Mie Scattering and Riboflavin (NADH and Dipicolinic Acid) Detection

There are several similar instruments that draw air into the system. The particles are passed through a 405 nm diode laser and are sized (0.4 to ≥10 microns). Enumeration occurs using a Mie scattering particle counter. Particles that contain biological targets like riboflavin, NADH, and dipicolinic acid can auto-fluoresce when they pass through the laser. The instruments also include a fluorescence detector to enumerate these auto-fluorescing particles. Some systems only report the viable particles whereas others also include enumeration of the nonviable particles. In general, results are obtained in real time. There is no need for culturing and growing the microorganisms, eliminating the need for consumables, reagents, and media (Anonymous, 2018).

Biovigilant manufactures the IMD-A, which uses Mie Scattering and biological particles detection. For this system, results are provided in real time. This system is unique in that one can also generate a video of what was happening at the time of an excursion. This can be very useful when training operators in aseptic behaviors, as you can enumerate their counts and video their actions to see what happens when they perform specific operations. This system does not capture the counted biological particle, so microbial identifications cannot be performed.

TSI also manufactures an instrument, the Biotrak, for instantaneous microbial detection using Mie scattering and biological particle detection. This system passes air through a 685 nm laser for particle sizing. This system does not provide the video capability, but it can capture the biological particle detected and sends the particle to a gelatin-based nutrient agar plate. Use of the gelatin plate as media extends the length of time the plate can be utilized in the system, while maintaining its growth-promoting qualities. This allows for further identification and analysis of the biological particle counted. This system is depicted in Figure 32.2.

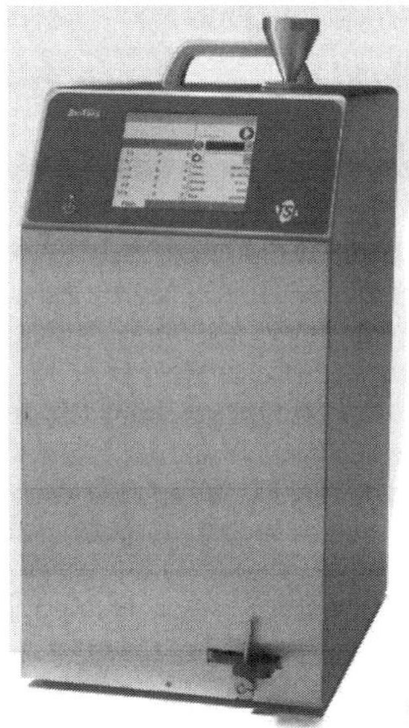

FIGURE 32.2 TSI's Biotrak system for air monitoring.

Source: TSI.

32.8.1.2 Other Viability-Based Systems Using NADH or Riboflavin Detection

Particle Measuring Systems, Inc. has the BioLaz Real-Time Microbial Monitor. In this technology, the air is pulled into the instrument sensing area using a stainless-steel sample probe. The drawn air is illuminated by a laser. As with other systems, the biological particles that contain NADH or riboflavin auto-fluoresce and are counted in one of two channels (Anonymous, 2018).

32.8.2 Growth-Based Technologies

These technologies incorporate steps that require the microorganisms to be cultured prior to enumeration. One of these systems is the Growth Direct by Rapid Micro Biosystems. This system is like a cross between conventional methods and rapid methods. Samples are prepared in the same way as the conventional method, but using special nutrient media plates provided by the vendor. Once collected and brought to the laboratory, the samples are processed through an automated system that controls the incubation time and conditions via preset parameters. At the completion of the testing, digital imaging automatically enumerates the micro-colonies in about one-half the time of traditional methods. Imaging is accomplished by passing the nutrient plate through a blue light that excites the micro-colonies to auto-fluoresce. Enumeration takes place using a CCD imaging system. Only growing colonies are counted. This test is nondestructive and if desired, the growing colonies can be sampled and subjected to additional identification and testing (Anonymous, 2018).

32.9 SURFACE MONITORING METHODS

It is important to select the appropriate monitoring method early in the development of the monitoring program. There is variability in the sensitivities and recoveries for different methods. They are not directly comparable to one another. Considerations in selecting a method include:

- Number of tests performed
- Ease of processing
- Configuration of the sampling site

RODAC plates are the most common surface sampling devices. They are typically 50 mm diameter agar contact plates where the agar has been poured in such a way that it forms a dome. The agar is pressed against a flat surface and organisms will stick to the agar medium. About 3.99 square inches are sampled by each plate and this can easily be converted to the number of microorganisms per square inch. The efficiency has been compared with that of quantitative swabs with between 10% and 50% recovery of what is present on a surface. RODAC plates are available commercially from a variety of sources, or they can be made in-house. The trick to preparing them is to produce a good convex surface on the plate or it will not make proper contact. Another product, the Hycon contact plate, is similar to the RODAC but square in design and contained in a peel open flexible plastic envelope. Hycon plates bend and offer this as an advantage over the rigid RODAC plate. Hycon also offers a smaller sampler that has the media on a slide. Following the use of any contact sampler, the surface contacted should be disinfected to remove any residual agar.

Cotton swabs or foam swabs wetted with isotonic solutions can also be used in a quantitative or qualitative manner. Qualitatively, they are excellent because they can be used to sample almost any accessible surface after which they can be placed into any liquid medium for culturing. Quantitatively, they must be relieved of the organisms collected by rinsing in buffer, which is then plated or membrane filtered so that the organisms can be counted. Cotton swabs rinse quite well. Foam swabs tend to resist efforts to rinse out organisms. One aspect of both types of swabs that must be considered is the risk of leaving cotton fibers or foam particles behind after sampling. This particulate matter may pose a risk to product. Quantitative versus qualitative is the choice of the user, although information about numbers of organisms cannot be generated using swabs qualitatively. There also may be a tendency for aggressively growing organisms to overgrow others in the liquid media, meaning that they may be completely hidden.

The recovery on swabs is like that of contact plates. The difference is the ability to use the swab on irregular surfaces or in small spaces. One type of swab that yields good quantitative results has calcium alginate swabs. It combines the quantitative recovery efficiency of a RODAC plate with the ease of a cotton swab. After moistening the swab and sampling a surface, it is placed into a small quantity of sterile Ringer's solution. Sodium hexametaphosphate solution is added and the calcium alginate fibers dissolve. The resulting liquid can then be pour plated either partially or entirely to yield a microbial count. With this method, the whole swab is plated and there is no presumed loss of organisms. A drawback is that some organisms may not grow in the presence of calcium alginate and/or sodium hexametaphosphate. In recent years, specialized swabs have been developed to increase recovery efficiencies, e.g., Copan flocked swabs.

Another type of sampling performed is rinse sampling. This is used most often in tanks or other areas where it is not feasible to routinely use other sampling methods. Much as it sounds, fluid is processed through the container. It should contact the entire surface (as much as it can be controlled). The solution exits the container, and it is captured. The solution captured is subjected to either membrane filtration or pour plate methods to enumerate the viable organisms present.

It may be valuable to perform sampling using different methods, depending upon the manufacturing process utilized.

32.9.1 Rapid Methods for Surface Monitoring

To date, surface monitoring methods have not been the focus of many RMMs. One system, the PallChek, was sold for surface monitoring, but this instrument is no longer commercialized. The Growth Direct described under air monitoring methods can also be used for surface sampling. This system utilizes a proprietary media rather than a RODAC plate media. The company has data to support use of the alternative media.

Some companies have adapted RMMs designed for sterility testing (liquid samples) to use for monitoring of surfaces. In these cases, the sample is collected using a rinse method or swab. With a swab, the organisms are released from the swab into the liquid. In both cases, the liquid is processed as a sterility test sample. For environmental monitoring, one would want to select a quantitative methodology. Instruments that have been evaluated for this purpose include: Scan*RDI* and Milliflex Rapid.

bioMerieux (Chemunex) has the Scan*RDI*. Using this technology, the test sample is filtered through a polyester membrane. Organisms retained on the filter are labeled with a nonfluorescent substrate. Inside the cytoplasm of metabolically active cells, the substrate is enzymatically cleaved using esterase to release a fluorochrome. Only cells with intact membranes will retain the fluorescent label. An argon laser scans the surface of the membrane (approximately 3 minutes). Viable cells are detected (this system uses 20 different discrimination parameters to assess viability). Auto-fluorescent particles, membrane fluorescence, and background noise are rejected in the discrimination process, and a total viable count is reported. Viable cells may be subsequently observed using a phase-contrast microscope and an automated stage (Anonymous, 2018). This test is considered nondestructive, but specialized procedures must be developed and implemented to recover any contamination present. As such, it can be difficult to identify contaminants.

Millipore manufactures the Milliflex Rapid. This system utilizes a membrane filter to capture individual cells. The sample is filtered, and the membrane is placed on an appropriate agar medium to allow the growth of micro-colonies. Micro-colonies are then treated with an ATP-releasing reagent, followed by the addition of luciferin and luciferase. Photons of light from each micro-colony are detected by a luminometer. The cell count is reported. It may be possible to continue incubation to form larger colonies for subsequent microbial identification (Anonymous, 2018).

Although rapid methods exist, companies have been slow to adopt these methods for implementation. Various fears exist among company management. Some of these include:

- With a viability method, if I cannot identify the microorganism detected, how will I conduct the investigation?
- Will the regulators accept these methods?
- Will the counts obtained be substantially different from those of conventional methods?
- When will the cost-savings be realized?

A risk-benefit analysis should be conducted to clearly identify the risks of the new method as compared with the benefits expected.

32.10 WATER MONITORING METHODS

In most companies, monitoring of the water system is considered part of the environmental monitoring program. Depending upon the company, the chemistry department may trend the total organic carbon (TOC) data or the microbiology department may track it with the environmental data.

Conventional water monitoring methods include collection of water samples from ports throughout the system or from use points. The sampling containers frequently have a two-part design, where the sample is collected in the top container. It filters (using gravity) and then the filter is removed aseptically when the samples get back to the laboratory. The membranes are place on a nutrient media, incubated, and enumerated.

32.10.1 Rapid Methods for Water Monitoring

One of the most recent advances in the use of rapid methods is the design and commercialization of in-line systems that can detect the viable microorganisms present in real time. Depending upon the system selected, the contaminants may be recovered for subsequent testing. One system can also identify the contaminant.

Two systems utilize Mie scattering and NADH/riboflavin technologies (as described for air monitoring) to detect water organisms. BioVigilant produces the IMD-W. This system is connected to the high-quality water system or can be configured to select discrete volume laboratory samples. Results are immediately available. Alarms are generated for aberrant results. Mettler-Toledo-Thornton also manufactures a system, the 7000RMS Microbial Detection Analyzer. This system is also installed on the high-quality water system. Water is drawn into the system at 30 mL/minute. The sample passes through the light of a laser diode that is subject to Mie scattering and fluorescence. The Mie scattering works to size the particles. The fluorescence identifies viable cells, based upon NADH and riboflavin. Discrimination parameters are used to analyze the particles as viable cells or not. This system is sensitive down to the single cell level.

Mettler-Toledo-Thornton manufactures the 7000RMS, which also is installed in-line. It uses a similar technology to the IMD-W. This system is depicted in Figure 32.3.

FIGURE 32.3 7000RMS.

Source: Mettler-Toledo-Thornton.

Battelle manufactures the REBS system. This technology utilizes RAMAN spectroscopy. The RMM, Resource Effective Biological identification System (REBS) provides rapid, automated analysis of liquid samples to detect, enumerate, and identify microorganisms within those samples. The REBS returns results within hours of sampling and eliminates the need for a growth stage prior to traditional detection, enumeration, and identification. The REBS can process liquid samples in a fully-automated, in-line fashion or can process liquid samples off-line. This system is depicted in Figure 32.4.

Other systems designed for testing of fluids can also be used for monitoring. In most cases, this would be a laboratory instrument and samples would be taken to the laboratory and tested.

Although many companies are evaluating rapid methods for water monitoring, the movement towards implementation has also been slow. There seems to be a lower level of concern in monitoring water systems than was present in air monitoring systems. Some concerns with these systems include:

- Will the counts be much higher using a viability-based method?
- How much change to the water system is required to incorporate the new technology? How much validation is required for the system?
- What are the risks concerning adding this system to the water system?
- What is the cost per sample versus the existing costs?
- How is this system affected by system sanitization or use of disinfectants/cleaning agents?
- For viability methods, is the counted biological particle recovered for subsequent identification?

It is important to have considered all of the risks prior to implementation of these systems.

FIGURE 32.4 REBS (Resource Effective Biological Identification System).

Source: Battelle.

32.11 USE OF GROWTH MEDIA

When performing growth-based culture methods, a growth-promoting media is required. Historically, it was common to use a general-purpose media like Soybean Casein Digest (SCD) or Trypticase Soy Agar (TSA) for bacterial recovery. A second media, Sabouraud's Dextrose Agar (SDA) was utilized for recovery of fungi (i.e., yeasts and molds). However, there are a variety of general-purpose media available including Standard Methods Agar and Brain Heart Infusion Agar. In recent years, some companies have adopted use of Schaedler's Modified Blood Agar.

Careful selection of the appropriate media and incubation conditions will provide the most thorough information. Performing a justification protocol to support the chosen incubation conditions (as described earlier in this chapter) can also aid in reducing labor and the cost of consumables. Inactivators or neutralizers (e.g., penicillinase, Tween, lecithin, and sodium pyruvate) may be needed in the media to neutralize the effects of bacteriostatic or fungistatic agents that may be picked up and transferred to the media in the process of the sampling. Some examples may be Tween and lecithin to inactivate the residues of disinfectants or preserved product aerosols, or penicillinase or cephalosporinase to inactivate/break down antibiotic powder. Sodium pyruvate can be utilized to inactivate residual vaporized hydrogen peroxide used in isolators. The need for inactivators must be researched within the context of the production operation, and the quantity of inactivator added should be in excess of what is needed but not to such excess that it becomes inhibitory to microbial growth. It is a potentially citable offense to fail to add inactivators to microbiological monitoring media where residues of a bacteriostatic agent are likely to enter the media.

It is a compendial requirement to test all media for the ability to support the growth of a wide range of microorganisms. Standard cultures to use in the testing are specified in the compendia. Additionally, environmental isolates should also be included. Some regulators also want a quantitative assessment of the level of recovery in this test. One method used to accomplish this is to take the form of inoculation of the newly produced batch and a previous acceptable batch, with requirements that the counts on the new batch must be within 70% to 130% of the previously acceptable batch. A count range should be near that expected during use. Therefore, the growth promotion test should be conducted with as low a count as possible. Current USP guidelines specify below 100 cfu. The numbers of organisms typically found on clean areas are small and the detection of low numbers during growth promotion tests confirms that the media have sensitivity to these numbers.

32.12 DATA INTEGRITY WITH ENVIRONMENTAL MONITORING SAMPLES

The reconciliation of the plates used for sampling and the plates incubated and read is an important activity although a largely administrative one. Plates must be 100% reconciled in both the sampling and incubation processes. All plates included in a monitoring session must be recovered and reported. Dropped plates must be clearly labeled and a process established for their handling prior to submitting for incubation.

In recent years, there has been some concern about how to verify that the counts obtained on plates are accurate. Some have proposed having at least two individuals reviewing and counting plates. Other companies have implemented systems using an automated colony counter or use of systems that automate the incubation and recovery steps to eliminate human vision used in the colony counting.

32.13 ESTABLISHING ALERT AND ACTION LEVELS

The Following Definitions Are Frequently Used to Describe These Levels

Alert Level

A count that exceeds normal operating levels but does not adversely affect product quality. It serves to alert appropriate officials to a potential adverse trend in control.

Action Level

A count or trend that exceeds normal operating levels that could adversely impact on product quality and requires action on the part of the firm.

Establishment of alert and action levels are key elements of an environmental monitoring program. Today, action levels are established in most regulatory documents for aseptic processing. These levels may be stringent if the process utilizes terminal sterilization. The reason is that terminal sterilization processes are designed to kill many logs of environmental isolates present in products. When we have a cycle validated to yield a probability of a non-sterile unit (PNSU) or sterility assurance level (SAL) of 10^{-6} using biological indicators, this same kill level is much higher based upon environmental bioburden, which has a much lower resistance to heat sterilization. Some companies have reported values of 10^{-30} or 10^{-50} for environmental isolates using terminal sterilization.

Alert levels are set below the action levels and are typically based upon historical data. For ISO 4.8 or 5 (Grade A) areas, where the action level is 1 in aseptic processes, it may not be feasible to have an alert level. When historical data is not available, e.g., a new facility, one may set the alert levels at one-half or on-third the action level while accumulating data to support the alert level. For example, in a Grade C (ISO 7) area, where the action level is 10,000 particles, and the alert level is set at 7,000 but all of the counts are routinely at 3,000–4,000, then the level of 7,000 would likely be deemed not appropriate as you would already be showing an increasing trend long before attaining a count of 7,000. Another method commonly used for setting alert levels is to take the mean and add one standard deviation and an action level at plus two standard deviations. For this assessment, one needs to assess whether the data is normally distributed. Some companies rank the data so that the alert level is just below the action level at the point where 95% of the counts are below it and only 5% exceed it. This approach minimizes the effect of a few high counts in the data. Whatever method you choose should be documented in your written environmental monitoring plan.

32.14 DOCUMENTING AND REPORTING OF ENVIRONMENTAL DATA

There must be written procedures that cover the gamut of the environmental monitoring program. These documents need to be updated and kept current. The data collected in the monitoring program should be documented. Typically, companies either have a manual system that incorporates forms for data collection or via entry into a computerized database system that maintains and tracks the data. Automated systems may be part of the company's LIMS system (Laboratory Information Management System) or a stand-alone automated system. Regardless of the system used, the company should ensure that all the data is integral. The data should link directly to a specific sample site, a specific test method, and should be fully identified and documented.

This data should be presented in several formats (FDA, 2004). For example, one should be able to identify all the locations where a specific organism has been found in the facility. It is also important to be able to identify any trends that may be occurring. Ideally, you want to find the trend before it goes out of limits. It is also useful to look at data for different seasons, different rooms, and the like.

A system should be evaluated periodically to assess the appropriateness of the data obtained and the action and alert levels. Many companies do this annually. It is also useful when the automated system provides a method to notify senior management of adverse trends.

32.15 OUT-OF-TREND AND UNUSUAL CIRCUMSTANCES

It can be useful to collect monitoring data when you have documented out-of-trend results, mechanical breakdowns, and power outages. This data can be invaluable in assessing the risks to the product being manufactured at the time. It is also useful in assessing the type of corrective action to implement. Failure to appropriately investigate and document an unusual circumstance may lead to many questions and observations in a regulatory inspection.

32.16 CHARACTERIZATION OF MICROORGANISMS ISOLATED

During a recent pharmaceutical microbiology conference, one European regulatory investigator indicated that there is no value in monitoring unless you identify the microorganisms recovered. For several years, European regulators were pushing for identification of all isolates found. Different strategies were proposed to address this issue. Some strategies involved use of simple clinical microbiology tests to assess the hazards of microorganisms isolated, e.g., use of oxidase or catalase tests to assess for clinical significance.

The FDA devotes an entire page in its Aseptic Guidance (FDA, 2004) to the characterization of isolates recovered. This data can be useful in determining the source of the contamination as well as provide data for comparison to product failure isolates.

Many different systems are available for microorganism identification. They range from simple analysis of biochemical reactions to genetic sequencing, with many other systems in between. Although genetic sequencing can provide the most detailed information, many of the systems available require molecular biologists to run the system and more importantly to understand the reports given. In recent years, many systems common to the chemistry laboratory have been used for microbial identification, e.g., FTIR, gas chromatography, NMR, RAMAN, and MALDI-TOF.

MALDI-TOF has been commercialized by two companies for microbial identifications. This has been an attractive method because of its low cost, i.e., less than $1/test, and its high throughput, i.e., 100 samples per hour. This technology has been presented to have a high level of accuracy also.

Several of the available technologies are dependent upon the performance of a Gram stain and entering a Gram stain result. This can be problematic as the accuracy of the Gram stain is dependent upon many factors, e.g.:

- Age of the culture being stained
- Appropriateness of the decolorization step
- Appropriateness of the counterstain step
- Whether controls were processed simultaneously
- Visual acuity

If this result is not accurate, then the corresponding identification is not likely to be correct. Gram staining is not required for use of MALDI-TOF.

An ideal identification system should have provisions for inclusion of organisms that cannot be identified, along with an identification number. This allows for sending the culture out for sequencing and updating of your system's database of known organisms. This is also important for analysis of adverse trends and product failure investigations.

It is important to maintain a database of the isolates found at the facility. It is expected in some compendia that environmental isolates be utilized in method validation studies. The rationale for selection of these organisms should be documented. This also necessitates that the site has a policy that describes how the library of environmental isolates is maintained. (Moldenhauer, 2017b)

At minimum, all isolates recovered should be maintained until the product is released. Failure to do this can make it difficult or impossible to accurately complete product investigations that may be needed.

32.17 ANAEROBIC MONITORING

The need for anaerobic monitoring is not well understood. Few pharmaceutical companies routinely operate under anaerobic conditions. The act of using a nitrogen overlay was interpreted by some as anaerobic conditions, but it is not a true anaerobic condition. Establishing a program to monitor for anaerobic growth is like the program for aerobic growth. The frequency of performing this monitoring is based upon the type of processes conducted. If all processes are aerobic, a simple plan to monitor once or twice a year may be enough. When conducting anaerobic monitoring, the procedures used need to ensure that the contaminants are quickly placed under anaerobic conditions. It is useful to verify that these procedures are effective with anaerobes prior to the actual monitoring.

Specialized transport and incubation systems are available for anaerobic monitoring. All these procedures should be verified prior to using them for routine monitoring. If this monitoring indicates a prevalence of anaerobic organisms, more frequent monitoring may be required to ensure product quality.

Special precautions may need to be used to process anaerobic organisms for identification. Gas chromatography identification methods may be useful in the identification process. If you are using contract laboratories for identification, you need to ensure that the anaerobic conditions are maintained during the shipping process.

32.18 MONITORING IN PRODUCTION ISOLATORS

When monitoring is conducted inside a production isolator, it is important that the act of monitoring does not contaminate the isolator. As such, systems that use probes (like particle monitors) are useful as they do not introduce media into the environment.

Dr. Akers (2010) identified a proposal for environmental monitoring in a production aseptic isolator. His proposal includes the following:

1. No more than two 1 m^3 air samples need be taken in a reasonably sized filling isolator over a 4-hour period.
2. Indications are that active air sampling is more sensitive than settle plates so there is no good reason to do settle plates at all.
3. Microbiological sampling of surfaces that have been decontaminated with *vaporized hydrogen peroxide* is unlikely to yield anything and should be minimized.
4. Physical measures of isolator performance may be of more value than microbiological measures. In other words, if an isolator air filtration system is working well, pressure differentials are properly maintained, gloves are integral then contamination risk is low.
5. Total particulate counts using electronic devices are likely to give more direct and more immediate indications of changes in isolator performance status than microbiological sampling.

Dr. Akers (2010) further concludes that we should try to minimize the difficulties associated with environmental monitoring samples being placed into and out of isolators. Several of the RMMs identified as viability-based methods under the air sampling section of this chapter can resolve this issue, as there is no media to be placed or removed with use of these systems.

32.19 MONITORING THE STERILITY TEST AREA

It is important to conduct environmental monitoring as part of your sterility testing program. As much as possible, the methods should be the same as those used in production monitoring. One wants to be able to compare data from this environment to the production environment. A critical concern in this area is the generation of false sterility test positives. It could result in the product being rejected although it is sterile, or the test may be identified as a false positive, the product released, and it is really contaminated.

In the case of a sterility test positive, a thorough and persuasive investigation must be conducted. There is guidance on the requirements of the investigation provided in the sterility test section of the FDA's Aseptic Processing Guidance (FDA, 2014). Assessing a sterility positive is false needs very clear and persuasive data. Ideally, genetic sequencing should be used for the identification of both the contaminant and the believed source of the laboratory contamination. Failure to have this type of data makes it difficult to meet current regulatory expectations.

To ensure that sufficient data is available to support these types of investigations, it is necessary to perform frequent monitoring and ideally associated with each testing session in the laboratory or isolator. With the increased use of isolator technology and closed filtration systems for sterility testing, it is very unlikely that a false positive has been generated.

32.19.1 Selection of Monitoring Sites

A risk assessment should be conducted to justify the sampling locations selected. Testing is more manual than automated, so there are different areas of risk. There are many opportunities for adventitious contamination. Analysts should be trained and qualified. Ongoing monitoring of their performance should also be conducted.

The design of the sterility test system should be based upon a process that is well known relative to how the test is conducted. Depending upon the production process, there may be many samples that need to be processed through the area. The data generated in the laboratory should be maintained and trended just like the data from the production area. This data should be reviewed to determine whether any adverse trends are occurring. This data is also useful in assessing the ongoing operator performance and aseptic technique.

Specific test sites in the laboratory may vary but should include air tests in the areas of sample receipt, testing, and incubation, surface monitoring of the analyst's gloves, sleeves, testing bench (or isolator); tests near sample receipt, floors, walls, incubators, gowning areas, and pass-throughs should be conducted. Isolators with a proved sterility history can be monitored less frequently after the microbiological monitoring that supports the maintenance of their sterility has been completed. Glove testing provides key information for isolators. Monitoring should be performed appropriate to the volume of work with a minimum of once each test day. If test volume is high, twice each test day may be more appropriate. Glove integrity has emerged as the more vulnerable and most likely failure point for isolation sterility testing. Changing to Hypalon gloves has significantly reduced the number of failures because of glove integrity (Akers, 2010).

32.19.2 Alert and Action Levels

These levels should be developed and assigned in the same manner as in the production area. When the levels are exceeded, the disinfection procedures should be reviewed for adequate frequency. One may also need to determine whether the disinfectant is effective against the isolated organisms.

32.19.3 Data Recording

The data should be collected, reviewed, and maintained in the same system as the production data. It is important to ensure that this data is reviewed as part of product testing failures.

32.19.4 Identification of Isolates

The identification of organisms from the sterility test area monitoring are as important as those in the production area. As mentioned for false positives, genetic sequencing may be required. Unfortunately, in many cases the same organisms are found in both the sterility and production areas, making it difficult to resolve microbial investigations.

32.20 CONCLUSION

To have meaningful data, it is important to have a program that has been carefully considered and is documented and implemented with the appropriate supporting qualification/justification data. This data can be invaluable in supporting product investigations and regulatory investigations.

ACKNOWLEDGEMENTS

This chapter is an update to a chapter previously written by Ms. Pamela Deschenes, currently with Veltek Associates, Inc. Her organization and content were heavily relied upon in the update of this chapter. Thanks so much for the good chapter you wrote previously.

Thanks also to the representatives from Mettler-Toledo-Thornton, Battelle, and TSI who provided information and photos of their technologies.

REFERENCES

Akers J. Risk and Scientific Considerations in the Environmental Monitoring of Isolators in Aseptic Processing. *American Pharmaceutical Review On-Line*, 2010. Downloaded from: www.americanpharmaceuticalreview.com/Featured-Articles/117497-Risk-and-Scientific-Considerations-in-the-Environmental-Monitoring-of-Isolators-in-Aseptic-Processing/ on September 24, 2018.

Anonymous. RMM Product Matrix. *Rapid Micro Methods*, 2018. Downloaded from: http://rapidmicromethods.com/files/matrix.php#quantitative on September 19, 2018.

Eudralex. EudraLex—Volume 4—Good Manufacturing Practice (GMP) guidelines. Annex 1 Manufacture of Sterile Pharmaceutical Products. Commission Européenne, B-1049 Bruxelles/Europese Commissie, B-1049 Brussel—Belgium, 2009. Downloaded from: https://ec.europa.eu/health/sites/health/files/files/eudralex/vol-4/2008_11_25_gmp-an1_en.pdf on September 9, 2018.

FDA. Guidance for Industry—Sterile Drug Products Produced by Aseptic Processing—current Good Manufacturing Practice, pharmaceutical cGMPs. Unites States Department of Health and Human Services, Center for Drug Evaluation and Research, Center for Biological Evaluation and Research, and the Office of Regulatory Affairs, 2004.

ISO. ISO 14644–1:2015. Cleanrooms and Associated Controlled Environments. International Standardisation Organization, 2015.

Moldenhauer J. Justification of Incubation Conditions. *American Pharmaceutical Review (Online)*, 2014. Downloaded from: www.americanpharmaceuticalreview.com/Featured-Articles/158825-Justification-of-Incubation-Conditions-Used-for-Environmental-Monitoring/ on September 18, 2018.

Moldenhauer J. Developing or Updating your Environmental Monitoring Program to Meet Current Regulatory Expectations. *IVT Network*, 2017a. Source URL: www.ivtnetwork.com/article/developing-or-updating-your-environmental-monitoring program-meet-current-regulatory-expecta on September 9, 2018.

Moldenhauer J. Establishing a Library of In-House Isolates. *IVT Network*, 2017b. Source URL: www.ivtnetwork.com/article/establishing-library-house-isolates on September 9, 2018.

Moldenhauer J. Objectionable Organisms. *IVT Network*, 2017c. Source URL: www.ivtnetwork.com/article/objectionable-organisms on September 9, 2018.

Moldenhauer J, Sutton VW. Towards an improved sterility test. *PDA J Pharm Sci Technol.* 2004, 58(6), 284–286.

Whyte W, Bailey PV, Tinkler J. An evaluation of the routes of bacterial contamination occurring during aseptic pharmaceutical manufacturing. *J Parenter Sci Technol.* 1982, 36(3), 102–107.

Whyte W, Niven L. Airborne bacteria sampling: The effect of dehydration and sampling time. *J Parenter Sci Technol.* 1986, 40(5), 182–188.

33 Validation of Container Preparation Processes

William G. Lindboe Jr.

CONTENTS

33.1 INTRODUCTION

Pharmaceutical products are more than formulations, the product must be supplied in a primary package system that protects its quality from the time of manufacture until ready for use, and the proper delivery of the product may be supported by the packaging system as well, with regard to dose, safety, and other attributes. This is true of all pharmaceutical dosage forms, and although it is of particular importance for sterile products, the proper preparation of containers is essential for all drugs. This chapter will review the validation concerns for primary packaging components, specifically containers and closures used in the packaging of sterile and non-sterile products prior to filling. The treatment of this topic in the first two editions of this volume[1] primarily addressed the developmental and selection of physical and chemical attributes of polymeric closures and the pre-sterilization processing of closures and sterile glass containers. This edition continues the approach in the third edition by expanding the scope to include plastics and explores non-sterile dosage forms, referencing sources for developmental and material selection information. The approach excludes traditional autoclaving, dry-heat depyrogenation, and on-line post-fill container integrity testing technologies. These topics are addressed elsewhere within this volume and/or in widely available references, and the reader confronted with Stage 1 process design, especially the selection of materials for containers and closures, is urged to consult these additional references. The chapter will suggest methods for performing Stage 2 process performance qualification of container preparation

DOI: 10.1201/9781003163138-33

processes from a practical viewpoint. Process validation nomenclature and sequence follows the 2011 FDA Guidance document ("the Guidance").[2]

The major trends in this area since the publication of the earlier editions has been the continued increase in the use of sequential computer-controlled operations, Process Automated Technology (PAT), and the continued expansion of the use of plastic containers as primary packages and delivery systems. Validation strategy has also changed in that a risk-based approach[3] and the Validation Life Cycle[4] has been widely accepted and incorporated into the Guidance as forming the basis of testing strategies. Specific strategies have emerged beyond these primary references to enable a uniform logical approach to the subject. Procedures by Akers and Agalloco[5] provide the basis for companies to perform their own risk analysis. Contract services also provide approaches to perform risk analysis, as described by Ricci and Fraiser.[6] In either case, the selection of what to validate and how to validate must be based upon sound logic and a determination of patient risk. The advantage of this approach is that very unlikely events and events with low risk assigned to them can be validated by proportionate measures.

Container preparation operations follow the major subdivision designations of sterile and non-sterile with the latter being historically trivial compared with the former. Recent advances in aerosol and inhalant administration have placed additional criticality on non-sterile packaging components. For sterile primary packages, current compendial requirements for sterility assurance,[7] particulates,[8] and pyrogens[9] must be addressed by container preparation operations. It should be noted that international specifications may be more stringent, specifically particulate specifications in Japan, and the reader must confirm all specifications prior to initiating any validation study. The "pyrogen free" requirement is now widely interpreted for preparation process as rendering a "three-log reduction." This practical quantifiable process result has the benefit of facilitating validation of washing and form-fill-seal polymer operations. Non-sterile primary packages typically address particulates, specifically dust from corrugated cardboard, within which the empty containers are typically shipped and stored. Functionality, both pre- and post-filling, is the major concern for metered-dose inhalers. The increased use of plastics as primary packaging material and the aseptic filling into these materials have brought about a resurgence in the nontraditional sterilization methods[10] that include the use of chemical and radiological sterilizing agents and their associated technology. The reader is urged to consult the USP chapter[11] covering containers to obtain a background on the particular process to be validated.

The assumption is made that the normal sequence of Stage 1, Process Design; Stage 2, Process Qualification Installation Qualification and Operational Qualification has been performed and documented. This chapter will address Process Qualification Process Performance Qualification. Stage 3 Continued Process Verification is not within scope; however, some discussion of the requirements will be presented. Process Performance Qualification (PPQ) protocols must verify personnel training, standard operating procedures, and in-place validated analytical test methods. Additionally, a vendor qualification program including periodic audits and bioburden/pyroburden studies for sterile containers needs to be in place. The details of sterilization and dry-heat depyrogenation process validation are covered elsewhere in this volume and will not be repeated in this chapter. Stage 1 Process Design programs for the various components using glass have the less rigorous "grandfathered" approach. Newer polymers and associated technology will require extensive development and product compatibility efforts, especially with vendor supplied "ready to use" components. The availability of diverse polymers and associated forming equipment make each validation unique to a certain extent. However, the same general requirements must be addressed and the reader will have to develop validation approaches to the specific technology being employed.

The major focus of this chapter will be on the validation of cleaning, nontraditional sterilization, and depyrogenation operations. In all cases, confirmation of function and physical properties must follow these preparatory operations. This testing generally provides a good opportunity for the utilization of quality statistics suggested by the Guidance such as Cpk and Cpm[12] on large amounts of test results. Typical testing levels are from three to six times in-process and/or release testing and protocols should include multiple container lots. Validated analytical procedures need to be in place for residual cleaning agents, lubricants, and the process challenges (spore suspension, sodium chloride, dye marker, and/or endotoxin) for washing validation. The approach is similar to cleaning validation and the reader is advised to consult the chapter addressing cleaning validation within this volume for additional background and considerations. Specifications for the components to be received in non-shedding materials coupled with procedurally controlled unpacking/unwrapping operations are essential to controlling particulates. For sterile products, this should extend back to the off-site manufacture, especially for plastics and web rolls for form-fill-seal operations.

The rule of three applies to all batch type processes. Continuous operations will have three different trials across shifts and personnel, as well. Although the Guidance permits less than three trials, this assumes extensive Stage 1 and early Stage 2 studies that may not always be in place. Every component size and variety should be challenged and bracketing (Family Groups) should specify performing at least one trial on each component entity. Validation is not the place to cut corners, especially when manufacturers are clamoring for end-product testing relief through parametric release. Additionally, in-place annual programs for Stage 3 Continued Process Verification (re-validation) for sterilization and depyrogenation operations are expected by auditors. This program typically involves a repeat of part of the PPQ and a thorough review of all production taking place within the review period. Packaging operations with multiple functions are usually divided into individual PPQ protocols for each operation. The system as a whole is typically documented by a summary report according to a validation plan or Process Qualification consisting of all PPQs including media fills for sterile products and extensive end-product testing for non-sterile products.

33.2 STERILE PRODUCTS

33.2.1 Glass Containers

Glass containers for sterile products have largely been relegated to small volume parenteral multidose, single dose syringes and ampoules. Historically, half-liter and liter-sized bottles were used for large volume parenterals, necessitating resource allocation to container pre-sterilization storage and preparation. These sizes still exist and legacy production operations will confront the validation team. Low thermal expansion borosilicate glass (USP designation Type I)[13] is used to accommodate thermal sterilization of the primary container for either aseptically filled or terminally sterilized product. Glass containers are produced at temperatures in excess of 1000°C and, at that point, are both sterile and pyrogen free. It is the subsequent handling that poses the potential for contamination, although this is generally minimal. This should be confirmed with bioburden and pyroburden studies on received glassware within an on-going program for each vendor. This will enable subsequent "bio" and "pyro" challenges during validation studies to be kept at a minimum in classified areas, especially for washing. Other non-parenteral glass includes treated soda-lime glass (type II), used for buffered aqueous solutions, soda-lime (type III), used for oils, and type IV glass, so called NP (non-perishable food designation), which has been largely replaced by plastic.

33.2.1.1 Glass Container Washing

Depyrogenation by washing, or dilution, has been effectively employed for glass primary packages for terminally sterilized large volume parenterals. Modern continuous process washing and dry heat depyrogenation in conveyer tunnels provides higher sterility and depyrogenation assurance. The batch-type dry-heat oven depyrogenation of the large one-liter bottles was so cumbersome that washing alone was an attractive alternative. The vials or bottles are washed, rinsed, and dried in a continuous conveyer operation. Companies with large capacity depyrogenation tunnels typically use it on all size vials and bottles, rather than perform the more rigorous washing validation and assume the greater risk of the non-thermal process. Validation protocols include depyrogenation studies and particulate evaluation of the washed glass containers. Recovery studies for endotoxin challenges need to be performed. Endotoxin standard, available from Associates of Cape Cod,[14] must be reconstituted and allowed to air dry on the challenged units. Studies using lyophilized endotoxin challenges directly can be criticized because this form of the challenge is more easily removed by washing. Challenged units must be marked and altered in a way, not to invalidate the study, but to absolutely assure that it is impossible to be used without detection in product for distribution. *Limulus* Amoebocyte Lysate (LAL) is the analytical testing method of choice. Recovery studies for the analytical methods must be performed, and it should be noted that reproducible recovery rates around 25% are common at the level of challenge utilized. Many firms have recovery requirements of 75% or higher, and these levels are impossible with this type of challenge. Company documentation must address this exception, and protocols need to have the results of specific recovery studies. There is always resistance to bringing endotoxin and/or a viable challenge organism into a production area, and that is why the challenges should be as low as practically possible to demonstrate the three-log reduction requirement, supported by bioburden, pyroburden, and recovery studies. It should be remembered that there was considerable resistance historically to bringing media into sterile areas for media fills for fear of encouraging contamination. However, history has proven that in well-maintained areas, there is no problem and today, media fill validation of aseptic processes is the accepted standard. Validation resource planning should be apprised of all component requirements and budgets should accommodate challenge studies for each size container to be validated. The temptation to reuse the non-challenge containers in subsequent runs to save money should be resisted because questions about the legitimacy of the study, especially for particulates, will be raised, even if they are artificially soiled. The reader is urged to consult the chapters on depyrogenation in this volume. In the author's experience, usually a minimum of ten endotoxin challenged containers should be used for any run. The challenges should be introduced to cover worst-case processing conditions of temperature and speed (cold and fast) and must include start-up, steady state, and shutdown challenges. Replicate challenges at any time point should be used in the event of lost or dropped samples; so many more challenges are usually required. A typical challenge number would be 50 to 100 containers.

Particulate testing consists of reconstitution with Quality Control-approved water for injection (WFI) and particle counting by visual observation followed by instrumental analysis. Microscopic examination of dried containers may also be employed. Sample containers are selected following washing with the quantity being approximately the same as the pyrogen-challenged containers. For those processes that have a thermal depyrogenation following washing, samples need to be taken following the thermal treatment to assure the particles are not generated during the processing. A check for broken heating elements should be made prior to each run for legacy oven and tunnel equipment to avoid failing for particulates.

The utilities servicing washing equipment must be qualified; this includes purified water, WFI, clean steam, filtered compressed air, vacuum, instrument air, and power. PPQ protocols should include washer point-of-use testing and verification of these utilities if not part of the system qualifications. If it is included, or has been performed previously, the studies must be referenced in the washing protocols. Washer early Stage 2 qualification protocols must verify temperatures, flow rates, capacities, speed, and confirm limits claimed by the equipment manufacturer and identified during these earlier studies. As a testimony to the longevity of washing equipment, the author has observed the reinstallation of legacy washers manufactured by Cozzolli and Metromatic from old facilities into brand new parenteral facilities. In this case, a full Stage 2 Process Qualification is needed to ensure proper reinstallation of the equipment.

33.2.1.1.1 Thermal Processing

It has been said that if glass were a new material there would be a hesitation in its use because it poses a safety concern if it is broken. That being said, the advantage of glass is its strength, inertness, barrier properties, and its ability to tolerate high temperature (Class I containers). As mentioned previously, the specifics of sterilization and depyrogenation are covered elsewhere in this volume. These techniques apply even in continuous combination washing and thermal processing equipment because the two operations can be addressed separately in targeted studies. Thermal depyrogenation of glass is the process of choice, when available. The rigidity of glass must be taken into account during processing operations. Glass does not flex under pressure. This means that the closure must flex and be put under additional strain during external heating and internal (cooling phase) pressure changes. In one instance, the author recalls the inner septum of the stoppers on 12 mL glass vials being burst, even though the stoppers remained crimped, because of the lack of pressure control during a cooling operation. The same conditions for a flexible plastic container might not result in a "popped" stopper.

33.2.1.1.2 Siliconization

Silicone (polydimethylsiloxane) aqueous suspensions are typically applied to glassware as an aid to draining the container, for providing a good meniscus for reading volume graduations on the container or syringe, and to improve appearance (provide a "polish") to the container and liquid content. The use of silicone must be fully supported in the clinical and scale-up phases of drug development. The silicone, typically Dow medical grade,[15] is applied as an emulsion in one of the washing stations immediately prior to drying. It can easily be evaluated by visible physical testing (such as the water brake test, distilled water meniscus, or aqueous dye adhesion) and must be inert to the product as shown in Stage 1 Design (development studies). The important thing is to ensure that it does not interfere with depyrogenation washing studies. The Parenteral Drug Association has published a Technical Report[16] on the use of Silicone on parenteral primary packaging components. An analytical method using infrared spectroscopy on silicone extracted using refrigerant 113 is described therein. The validation team is usually concerned with the qualitative presence of silicone on the vials beyond development and, for smaller containers, a composite sample must be used. Regardless of the analytical method or the type of component being washed, protocols must ensure that each nozzle position must be sampled with representation of start-up and shutdown. In the author's experience, a base sample of twenty-five vials from each nozzle for each of three validation trials of a full shift of washing operations is appropriate. Note that recharging reservoir levels, washer stoppages, and/or any maintenance activity must trigger additional samples that will increase the total number.

33.2.2 Vials

Irrespective of the material of manufacture, vials have a size variability that must be considered during preparation activity. Container washers and depyrogenation ovens and/or tunnels must be qualified for all anticipated sizes of containers. The small vials are generally the worst-case load in batch-type sterilizers because for a given volume of sterilizer there are more containers and they are more densely packed. Care must be taken to ensure that the placement of small vials is not overly dense in order to facilitate the penetration of clean steam or hot air into the loads. Small vials can be the worst case for washing as well because once full with liquid, later rinses can no longer enter. Draining must be possible for effective removal of contaminants. Washing equipment is generally a serial individual process; however, the nozzles on this equipment must be appropriate for all sizes washed. Additionally, flow rates must be adjusted for the smaller vials as they may actually be blown off cleaning nozzles. The proper change parts for this equipment must be confirmed for each vial type and size during commissioning, development, and qualification.

33.2.3 Ampoules

Ampoules differ from vials in that they are unit dose containers, flame sealed, and each container undergoes a container integrity test following sealing. Additionally, ampoules are not coated with silicone during the washing process. Ampoules require special washing equipment change parts because they are typically small and light. Dry-heat depyrogenation follows washing and is addressed elsewhere within this volume. All ampoules are considered contaminated by glass fragments to some extent when opened. Considerable development effort is needed to assess the effectiveness of the scoring equipment and glass quality to minimize this. There are suppliers of closed ampoules that are pre-washed and depyrogenated, placing the majority of the validation burden on the supplier (see following).

33.2.4 Syringes/Cartridges

Syringes, or rather syringe barrels, and cartridges are treated in the same matter as glass vials. Preprocessing will be addressed following; however, washing, siliconization, and depyrogenation of the glass and stoppers are required. Extrusion force and break-loose force are effective physical measures of the effectiveness of siliconization.

33.2.5 Tubes

Tubes (evacuated) are used largely for blood collection and may be considered a reverse syringe. Technically, they are medical devices; however, processing follows that of drug-containing syringes with the exception that they are not siliconized. The geometry of the tubes requires specialized washing equipment as for ampoules and syringes. Tubing is also used with intravenous kits associated with Large Volume

Parenterals. In these instances, the kits are typically irradiated in a batch-type process. Irradiation sterilization is covered elsewhere within this volume.

33.2.6 Closures

The closures for sterile containers usually have some secondary functionality in addition to providing a sterile seal for shipping and storage prior to use. This functionally could be the insertion of a syringe needle, connection to intravenous tubing, or in the case of ophthalmic solutions and syringe plungers, provide a means of administration. It is important that any validation of the preparation operations for closures contains testing to assure this secondary functionality is not diminished. As with containers, parenteral closures must be shown to be sterile, pass applicable particulate testing, and be pyrogen free. The latter quality is usually determined by testing a rinse solution after processing using the LAL test. It is assumed that a rigorous development process and vendor qualification program has occurred that has identified and confirmed all critical properties of the closures and that validated analytical methods, including recovery studies, are in place for the testing. Washing processes need to demonstrate a three-log reduction as in glassware washing.

33.2.6.1 Stoppers/Diaphragms

Stoppers/diaphragms are typically synthetic rubber or other synthetic polymers that are injection molded. Technical support should be a major criterion for selection of any vendor. Although in-house development documentation is the primary reference for validation protocols, often they are truncated summaries and detailed information can be obtained from the vendor.

As with glass containers, there is reluctance to bring organisms and endotoxin into classified areas. The Huber batch-type stopper washer is common in the industry (www.hubermachines.com last accessed 1/6/2019) (Figure 33.1).

FIGURE 33.1 Huber D540 stopper washer.

Source: Huber Systems

The machines can be equipped with CIP and SIP systems that, when validated, enable appropriate microbial and endotoxin challenges for component washing validation. The challenge should be kept as small as reproducible recovery studies will allow, while assuring the three-log reduction process requirement. As with glass, the endotoxin challenges or biological indicators (BIs) should be air-dried on to simulate a natural bioburden. Use of spore strips or spores on a model matrix, rather than dried on spore suspension on the actual unit, has been criticized as not representative of a realistic system. Well-marked units should be used with a minimum of ten challenged stoppers per validation run. It is best to test for residual detergent and silicone along with the BI runs. Stoppers and diaphragms are typically autoclaved after washing and evaluation after this stressed condition is necessary in extensive development studies. Newer technologies have emerged for closure treatments, for example as reported by Dublin and Wittler.[17] The process requirements are the same as those for the traditional batch-type unit operations, and considerable ingenuity may be required to demonstrate this in protocols. Closure component manufacturers are increasingly supplying this item as either ready-to-sterilize or ready-to-use, which transfers a portion or all of the validation responsibility to the vendor (see following).

33.2.6.1.1 Stopper/Diaphragm Washing—Particles

The reduction and/or more preferably the elimination of particles is the primary purpose of washing. Stoppers and diaphragms are typically rinsed vigorously with WFI and any other parenteral solvents used with the rinse solution being tested for particles. Additional testing for residual detergent is performed.

33.2.6.1.2 Stopper/Diaphragm Siliconization

Stoppers are siliconized to aid in the operation in stoppering equipment and/or movement within a syringe/cartridge barrel. Syringe stoppers also include extrusion and break-loose force reduction and reproducibility as necessary attributes. Physical testing is generally used with a confirmatory composite analysis to confirm siliconization. Sampling from each load of batch-type washers should sample from different locations. The stoppers and diaphragms are well mixed; however, this is necessary to ensure stoppers rolling around in the middle of the washer get the same treatment as those on the perimeter. Overloading is a common manufacturing error, and equipment manufacturers recommendations must be confirmed during PQ or Process Validation as well as in earlier qualifications. Use of different stopper lots of receipts is generally done.

33.2.6.2 Droppers

Droppers are drug delivery devices that consist of a barrel and functional closure. In some cases, the entire primary package can be considered a dropper, as in ophthalmic solutions. Components are typically washed to eliminate particulate and pyrogens. Dropper glass is rarely borosilicate high-temperature resistant and development studies must be in place to show that washing reduces particulates and depyrogenates effectively to a three-log reduction for ophthalmics and other solutions purporting to be sterile. Batch-type washers are generally used and validation sampling and testing generally follows that of vials and stoppers.

33.2.7 Plastics

Plastic is that class of polymers that has the ability to be formed (into beads) and reformed (into components) by the application of heat and pressure. The reader can obtain a good review of plastics used in pharmaceutical packaging in Remington.[18] Additional background can be found in the Parenteral Drug Association Technical Report[19] on Sterile Pharmaceutical Packaging. The trend toward plastic "bags" for the large volume parenteral primary packages can largely be summarized by the word convenience. Additionally, there is a reduced cost of manufacture and decrease in shipping costs. Ease in handling soft bags and IV additives, both in manufacturing and within the hospital, have been augmented by unique closure and dosage administration developments that would be difficult or impossible to achieve with glass containers. Flow in the flexible plastic bag of IV solution can be initiated by squeezing and collapses on emptying without air entry. The development of diverse polymers, co-polymers, and laminates with associated forming technologies has facilitated the transition to plastic containers. There have been studies[16] in the Food Industry showing a 50% packaging cost reduction of plastics over glass and metal cans without the higher weight, breaking, and denting. However, autoclavable food pouches consisting of PET, aluminum foil, and polypropylene laminates can be up to three times as expensive.[20] It must be remembered that molding and converting facilities are usually not current Good Manufacturing Practice (cGMP) and the big advantage of this technology is the ability to move the forming operation within controlled environments adjacent to the filling. With the convenience of plastics comes the loss of the inertness of glass and the impact on the environment when containers are not properly disposed of. Polymers can contain non-reacted monomers, plasticizers, preservatives, UV stabilizers, mold-release agents, and lubricants. The polyvinyl chloride (PVC) IV bag has largely been phased out because it contains a harmful leachable plasticizer, di-2-ethylhexyl-phthalate (DEHP) to make the bag soft. Incineration is a preferred method for disposing of medical waste, and it has been shown that PVC can produce dioxin and acids when burned. Additionally, the glass transition temperature of PVC is about 70°C[21] as compared with that of polypropylene (PP), which is −10°C.[22] This means that the latter material does not undergo the physical changes during thermal processing that are seen with PVC. This enables thermal processing at 121.1°C and *G. stearothermophilus* BIs as in glass, which is another reason to change from PVC to PP laminate. Validation teams face the qualification and validation of packaging systems and their associated machinery. Another quality of plastics is that physical properties of the unmolded resin can change with time. These factors require the vendor qualification for plastic container suppliers to be rigorous. Changes in formulation from vendor polymer manufacture throw drug and device manufactures into a panic and cause long work hours for validation teams. The author recalls one instance for a device manufacturer where a ten-year supply of a plastic was purchased upon notification of the discontinuation of a particular formula. Problems with storage of the unmolded resin "beads" in the latter years of this inventory caused manufacturing problems. This resulted in a panic vendor and polymer change that could have been avoided by an initial orderly transition to a second qualified vendor and/or formula. Plastics pose a difficult challenge to the validation teams regardless of the motivation for their use. An associated problem with plastics is that extruders are frequently cleaned by "burning" them out and relying on a purge of the initial moldings of the next batch to clean the equipment. This results in sporadic fine black particulate matter that can play havoc with particle specifications for the containers and the filled product. Validation teams need to verify cleaning procedures back to the extruder.

Non-sterile plastic packaging is typically purchased from external vendors, as is large web rolls of sheet polymers for form-fill-seal of non-sterile and aseptic fill operations. Static charge on plastic components is a magnet for all types of particulates, especially other plastics. Qualified filtered compressed air showers and/or washes are needed to ensure that the efforts to supply low particulate components by the vendors are not compromised by in-house unpacking. In any event, these operations should not take place in a warehouse environment. The typical approach is to have a staged unpacking moving progressively to cleaner areas using non-shedding, low particle-generating packing.

33.2.7.1 Sterilization of Plastic Containers

Although heat-resistant polymers can be autoclaved, it should be noted that the processes might be quite different from those used for glass containers. Terminally sterilized filled containers and autoclaved empty containers for aseptic filling are not as resistant to rough handling as glass, and special considerations for this are needed. Lower sterilizing temperatures, such as 116°C have been historically been used for PVC. New laminates of PP enable higher temperatures, as mentioned previously. Sealed ports for IV sets and injection of medications to be infused can be sterilized by the steam terminal sterilization of the filled bag. Use of a dry heat indicator such as *B. atrophaeus* is used to challenge the ports, independent of the fill challenge. Pflug et al[23] have shown that the maximum D-value for this organism occurs at about 30% equilibrium relative humidity, and challenges should be prepared in this environment to provide a worst case. Ports are also conveniently sterilized by E-beam following attachment to the bags, prior to filling. The autoclave jacket temperature may not be well controlled in older autoclaves, and care must be taken to ensure that plastics do not melt on surfaces overheated by jacket temperatures set arbitrarily higher for glass and stainless-steel filler parts. The nature of plastic usually means that siliconization is not required. Cleaning is required for particulate removal and depyrogenation for containers not molded in-line within classified GMP areas. Validation protocols should contain rigorous initial and ongoing pyroburden and particulate monitoring studies for plastics. These data are needed to justify using a minimum challenge to sterilizing operations. Traditional methods of sterilization include: dry heat, moist heat, ethylene oxide, radiation, and other chemical (hydrogen peroxide, chlorine dioxide). These methods historically have been applied in chambers (ovens or autoclaves) or in separate facilities. As with glass containers, the reader will find other chapters within this volume covering traditional methods of sterilization validation. Nontraditional methods of sterilization include: high intensity light, UV light, combined vapor and glass plasma, electronic beam, and injected low concentration ethylene oxide. These nontraditional methods are typically performed in-line on containers or form-fill-seal webs. Some physical assessment is needed for radiation methods, and plastic dosimeters are typically used. Sterilizer parameters, extruder temperature, pressures, and other parameters must be measured with National Institute of Standards and Technology (NIST) traceable devices and documented during validation. Critical parameters are subsequently monitored with appropriate alarms during routine production. Biological challenges need to consist of dried-on suspension at the lowest level supported by recovery studies and bioburden studies. The use of spore strips for studies can always be criticized as not being representative of an actual contamination, other than confirming uniform sterilizing medium. If high levels of spore suspension are used, post study equipment cleaning and sterilization must be specified in the validation protocol. The location of challenges should be well marked and challenge the entire width of webs. Provisions to capture the web immediately following sterilization can be made followed by cutting and incubation in media. Clear containers formed from the challenged web can be captured before sealing and filled with media within the laboratory or filled with media on-line followed by incubation.

33.3 FORM FILL AND SEAL

The prevalent trend in plastics and the primary advantage is form, fill and seal (FFS) or injection, blow mold, fill, and seal. FFS and blow, fill and seal (BFS) are technologies that were originally developed in the Food Industry. The first high volume application was the Brick-Pack process. These are the small rectangular containers of fruit juice or milk that are stored at room temperature. The concept was to apply a high temperature short time in-line thermal processing prior to using heat exchangers without the insulating effect of the containers. The container materials fed in continuous webs are chemically sterilized separately using hydrogen peroxide liquid baths. The hydrogen peroxide drained and the last residue destroyed by UV light and the containers are formed and aseptically filled within a classified modular isolator containment area. This simple process forms the basis of some pharmaceutical FFS and BFS processes. In BFS processes, there is hot melted plastic that is extruded into a tube, the tube is cut into parisons that are expanded out into molds by gas pressure to form the container. A review of the BFS process is provided by Ljungquist et al.[24] The plastic is extruded at temperatures around 200°C at which temperature sterilization and depyrogenation can occur. Containers must meet sterility, pyrogen, and particulate requirements. It has been shown that spore suspension dried onto plastic resin beads is nonviable after extrusion.[25] Additionally, it has been shown by Poisson et al.[26] that air dispersed organisms in environments of molding processes produce contaminated vials directly proportional to the challenge concentration. A similar challenge study with endotoxin coupled with pyroburden monitoring programs addresses the depyrogenation requirement. Empty containers from challenged resin are tested using LAL. Sampling of the parisons immediately following extrusion limits possible contamination of the line. Proprietary equipment can be designed and customized for a given validation sampling approach along with provisions for CIP and SIP of the line. Conventional media fills for sterilization process validation and cooled WFI fills for pyrogen and particulate testing can be used for process validation.

33.3.1 Inhalers

Inhalers consist of sterile plastic components that are molded in-line. Traditional batch-type sterilization or in-line radiation is used, as previously.

33.4 VENDOR-SUPPLIED COMPONENTS

Outsourcing has emerged as a cost-effective business option, and this practice has manifested itself within the parenteral drug industry as the receipt of prepared components. There are two major types, ready to sterilize and ready to use. This practice puts an additional burden on auditing groups, because the validation of the preparation operations has also been outsourced. Some in-house complimentary systems are needed so that the preparation is not lost through contamination from unloading and handling. Validation of storage, unpacking, and transfer mechanisms may be required.

33.4.1 Ready to Sterilize

These prewashed depyrogenated components are received in autoclavable packages that may be double wrapped. Hold time studies are required to demonstrate that stored components are the same as those components used in filled product stability studies. Extensive vendor qualification is required, including an audit of the entire validation effort. Although "qualified" audits by regulatory agencies have been used in vendor qualification programs, it is necessary that the entire validation package of the vendor be reviewed by the user.

33.4.2 Ready to Use

Ready-to-use components frequently require specialized complementary equipment (pass-throughs, etc.), are typically proprietary, and require specialized complementary filling equipment to use. Thilly et al.[27] describe a ready-to-use aseptic filling system using cyclo-olefin co-polymer (COC) vials within a Restricted Access Barrier System (RABS). GSK has partnered with Aseptic Technologies, with core technology licensed from Medical Instill Technologies, for this proprietary process. Electron beam surface sterilization is utilized at a 35 kGray intensity for gamma irradiated vials, which has the advantage of eliminating the "shadow effect" of UV and pulsed light systems. Filling is accomplished with needle perforation of the closure and laser resealing. Large media runs of 6000 vials support the validation efforts of a single head system. Polypropylene non-shedding packing materials are utilized for shipping the prepared components.

33.5 NON-STERILE PRODUCTS

The preparation requirement for non-sterile primary packaging is essentially cleanliness. This is accomplished by gravity by inverting the container and applying a filtered compressed air cleaning process. Both of these activities are typically performed in-line immediately prior to filling operations. Overhead feed systems may have helical tracks where the containers are swirled with the opening facing outward to use centrifugal effect to dislodge foreign material. These cleaning operations can be challenged within validation protocols by adding typical contaminant (cardboard dust) to the containers followed by inspection. The amount of dust added should be realistic. There is a tendency to place too much contaminant within the containers, which will result in a failed validation. The author recalls one validation of four-liter jugs that were to be filled with PEG 2000 and electrolytes for reconstitution. The challenge used was about 500 ml of ground cardboard in a series of jugs, which was about double the product fill. The challenge cardboard agglomerated together and would not fall out when the container was inverted. The subsequent ionized filtered air stream blew the cardboard all over with the result that containers, filling line, and validation personnel all failed to pass visual inspection for cleanliness. As with aseptic plastics, most of the innovation for non-sterile packaging is derived from the food industry. Systems proven in food industry but new to pharmaceuticals must be demonstrated to be effective. It must be remembered that the initial mandate for validation in the pharmaceutical industry was caused by the use of the twist-off bottle closure from the beverage industry on LVP bottles.[28] Ionized filter air showers, which are common to most non-sterile packaging preparation, can be validated by environmental particle counting equipment, borrowed from parenteral technology, or widely available through contract validation services.

33.5.1 Liquids/Semi-Solids/Solids

33.5.1.1 Glass

Types II and III glass may be encountered and some washing may be required. The term "sanitization" can be used to describe operations for some products. Bio challenges should be minimal to demonstrate a three-log reduction. Other pretreatments to assure the pH stability of filled products are validated with increased sampling and reconstitution under stressed conditions.

33.5.1.2 Plastic

Laminated tubes are now widely used and have largely replaced aluminum tubes for semi-solids. In-line inspection devices can be challenged with appropriately marked challenge containers with known physical defects. Surface treatments, such as UV light, to facilitate direct surface printing can be validated by monitoring treatment parameters and increased inspection of printed components. Ionized filter air showers can be validated by environmental particle counting equipment, borrowed from parenteral technology. PE bottles are widely used for solid dosage and require little more than filtered air cleaning.

33.5.1.3 Caps

The criticality of the cap function determines the level of validation. Beyond filter air cleaning, closure integrity for multidose containers is an attribute of concern. Increased sampling and use testing are a common validation strategy.

33.5.1.4 Delivery Systems

The award winning[29] GSK Advair Diskus typifies combined technologies providing the epitome of convenience to patients. Consisting of 14 components that are automatically assembled, the powder inhaler provides a display of remaining doses along with ease of use. The device combines blister packing with automatic assembly of components. Similar modern systems can be validated by monitoring and documenting machine parameters. Vision systems can be validated by challenging on-line reject mechanisms with edge of failure components that have been well marked to ensure they are not used for commercial production. A phase-wise validation consisting of protocols addressing each unit operation of the assembly and filling, followed by extensive sampling of centerline manufactured product can be used. Testing follows that specified in the USP[30]

33.5.1.5 Aerosol Cans

Coated aluminum cans are utilized in pressurized metered-dose inhalers (pMDIs). On-line inspection for can and coating defects by vision system can be validated as in Delivery Systems previously. The plastic actuator and valves have similar inspections and the filled primary packages, verified for content in-line can be challenged with well-marked edge of failure fills.

33.5.1.6 Blister Packaging

PVC films rolled on large webs, along with paperboard and other laminates are utilized. Machine operating parameters are documented during validation. Molding and sealing temperatures and pressures are monitored and combined with seal-strength testing for validation. The computer-controlled aspects of this machinery are covered by computer validation approaches outside the scope of this chapter.

33.5.1.7 Foil Laminates

Aluminum foil laminates are used typically with form-fill-seal equipment. The preparation processes may consist of filtered air cleaning and inspection of the web prior to forming. Validation of these types of processes follows that described in previous sections. Filled product testing for seal integrity follows worst-case filling and sealing conditions such as maximum and minimum line speed.

33.6 CONCLUSION

The validation of container preparation processes has evolved from traditional methods to new methods driven by technological change. Most of this innovation has come within the area of plastics and associated in-line container preparation processes. Validation approaches may be unique to each specific application. However, fundamental quality attributes must be confirmed and documented during validation. With new technology, considerable ingenuity to obtain the required testing and parameter measurements may be required. A joint approach, where pharmaceutical manufacturer partners with packaging equipment supplier and/or container vendor to accomplish validation, has proven to be effective for both new technologies and traditional processes.

NOTES

1. Ventura DA, Goodsir SW, Component Preparation Processes. In *Validation of Pharmaceutical Processes: Sterile Products*, 2nd ed. Agalloco JP, Carlton FJ, eds. New York: Marcel Dekker, Inc, 1998: 703–720.
2. Anonymous, *Guidance for Industry Process Validation: General Principles and Practices.* www.fda.gov/downloads/drugs/guidances/ucm070336.pdf accessed 8/26/2020.
3. US Food and Drug Administration, Department of Health, Education, and Welfare. Pharmaceutical cGMPs for the 21st Century-A Risk Based Approach. Final Report, Fall 2004. U. S. Government Printing Office.
4. Agalloco JP, The Validation Life Cycle. *Journal of Parenteral Science and Technology* 1993;47: 142.
5. Akers J, Agalloco J, A Revised Aseptic Risk Assessment and Mitigation Methodology. *Pharmaceutical Technology* 2017;41(11): 32–39.
6. Ricci MT, Fraiser HE, From Quality by Inspection to Quality by Design: A Roadmap for Success. *Pharm. Technol.* 2006;30(9s).
7. United States Pharmacopeia and National Formulary (USP 41-NF 36), Rockville, MD: United States Pharmacopeial Convention, 2016. https://online.uspnf.com/uspnf/document/GUID-AC788D41-90A2-4F36-A6E7-769954A9ED09_1_en-US accessed 1/18/2019.
8. The United States Pharmacopeia 41st Revision, <788> Particulate Matter in Injections, The United States Pharmacopial Convention, Inc. Rockville MD, 2006: 2722.
9. The United States Pharmacopeia 29th Revision, <151> Pyrogen Test, The United States Pharmacopial Convention, Inc. Rockville MD, 2006: 2546.
10. US Food and Drug Administration, Updated 510(k) Sterility Review Guidance K90–1. Final Guidance for Industry and FDA, November 16, 2001.
11. USP <661> Containers, USP 29 NF 24 the United States Pharmacopeia 29th Revision. The United States Pharmacopial Convention, Inc. Rockville MD, 2006: 2996.
12. www.winspc.com/what-is-spc/spc-tools/ accessed 8/26/2020.
13. USP <661> Containers. USP 29 NF 24 the United States Pharmacopeia 29th Revision. The United States Pharmacopial Convention, Inc. Rockville MD, 2006: 2656.
14. www.acciusa.com/ last accessed 8/26/2020
15. www.dupont.com/transportation-industrial/healthcare.html accessed 8/26/2020.
16. Anonymous, PDA Technical Report 12: Siliconization of Parenteral Drug Packaging Components. *Parenteral Drug Association* 1988;42(4S), Supplement.
17. Dublin M, Wittler H, A New Approach for Stopper/Closure Treatment System. *Pharmaceutical Engineering* May/June 1998; 18(3): 86.
18. Rainbow BE, Roseman TJ, *Plastic Packaging Materials: Remington, the Science and Practice of Pharmacy*, 22nd ed. Baltimore MD: Lippincott Williams and Wilkins, 2012: 1005.
19. Wang YJ, Chen YW, *Technical Report No. 5, Sterile Pharmaceutical Packaging: Compatibility and Stability.* Washington, DC: Parenteral Drug Association, Inc., 1983.
20. Halek GW, Yam KL, Lordi N, *Course: The Technology of Packaging II.* Rutgers the State University of New Jersey, 1984.
21. Reding FP, et al., Glass Transition and Melting Point of Poly(vinyl Chloride). *Journal of Polymer Science* January 1962;56(163): 225–231.
22. www.polymerprocessing.com/polymers/PP.html accessed 8/26/2020.
23. Pflug IJ, Holcomb RG, Gomez MM, Principles of the Thermal Destruction of Microorganisms. In *Disinfection, Sterilization and Preservation*, 5th ed. Block SS, ed. New York: Lippincott Williams & Wilkins: 97.
24. Ljungquist B, Reinmuller B, Lofgren A, Dewhurst E, Current Practice in the Operation and Validation of Aseptic Blow-Fill-Seal Processes. *PDA Journal of Pharmaceutical Science and Technology* 2006;60(6): 254.
25. Gustafsson A, The Bottle Pack System: Plastics and Their Pharmaceutical Applications. The Secretariat to the Convention for the Mutual Recognition of Inspectors in Respect to the Manufacture of Pharmaceutical Products, 1986: 94.
26. Poisson P, Colin SS, Tallentire A, Challenges to a Blow/Fill/Seal Process with Airborne Microorganisms Having Different Resistances to Dry Heat. *PDA Journal of Pharmaceutical Science and Technology* 2006;60(5): 323.
27. Thilly J, Conrad D, Vandecasserie C, Aseptic Filling of Closed, Ready to Fill Containers. *Pharmaceutical Engineering* 2006;26(2): 66–74.
28. Lindboe WG, Hayakawa K, Comparative Terminal Sterilization. *Journal of Parenteral Science.* Parenteral Drug Association, Washington DC 1993;47(3): 138.
29. Hartman LR, Winning with Plastic. In *Packaging Digest.* Oak Brook IL: Reed Business Information, 2003: 80.
30. USP <601> Containers. USP 29 NF 24 The United States Pharmacopeia 29th Revision, The United States Pharmacopial Convention, Inc. Rockville MD, 2006: 2617.

34 Validation of Lyophilization

Joseph P. Brower

CONTENTS

DOI: 10.1201/9781003163138-34

34.1 INTRODUCTION

In an ideal world, validation would begin with parallel product research and development activities. Validation for lyophilized products occurs more often during scale-up to manufacturing. Under growing regulatory pressure and the realization of the greater benefits, however, validation activities are being undertaken while the product is along the development pathway. There are also circumstances in which validation is required for existing commercial products, either because of changes requiring additional study or to meet current regulatory standards. This chapter will approach validation as an integral part of developing a new product. Appropriate application of the principles discussed may be applied for either a change control procedure or for revalidation, based upon specific needs.

34.2 PREPARING FOR VALIDATION

Components of a comprehensive validation program include equipment qualification (EQ), process engineering and process validation. The EQ portion focuses on the lyophilization equipment, including semiautomated or automated loading and unloading equipment, and is valid for processing the particular product or, for a multiproduct operation, any number of products. Conversely, the process for each product is unique and applies only to one product, and therefore the process validation is specific to that product.

The experience gained and data compiled during development are a significant part of the scientific rationale that forms the basis of validation studies. It also provides a critical reference for integrating a product into a production environment. This saves on adjustments to the process and further development studies at the time scale-up and validation is attempted in manufacturing. Completing process studies at both the ideal target parameters as well as boundaries of the process parameter range results in greater safety and efficiency in the parameters and a more robust process. Establishing a proven acceptable range (PAR), first introduced in 1984 by Chapman, is recognized as a "best practice" as part of lyophilization process engineering (1).

Activities for validation of a "legacy" product where development and initial validation may not meet current industry practices requires constructing a historically based file. Data such as pre-formulation, product and process development may not be available with historical manufacturing experience for the commercial product manufacturing. The most challenging task in this instance is justification for the product formulation and process design. This is particularly difficult in circumstances with commercial products that were developed prior to the awareness of the benefits of validation.

When a new product is in the development phase, a comprehensive report that includes technology transfer for Phase III clinical trial material should be assembled prior to scale-up and technology transfer to manufacturing. This report addresses the starting raw materials, including the drug substance, excipients and packaging components, along with formulation, compounding methods, dosage information and lyophilization process engineering activities. Each facet of product manufacturing needs to be included, beginning with testing of the product components through processing and final packaging requirements. The finished product qualities must also be defined. The report should clearly explain the scientific rationale and justification for the formulation and manufacturing procedures.

This development report is a crucial reference for integrating a new product into a manufacturing operation. The acceptance criteria for any validation study are based upon product and process requirements outlined in the development report. The report provides an invaluable reference for technology transfer, change control program management and troubleshooting during routine manufacturing.

EQ is best considered at the time of equipment specification and selection. The advantages include more effective project management, ease of completing the validation package and speed of bringing the equipment online. Equipment requirements and performance are based upon the processing parameters necessary for manufacturing the product, as identified during processing engineering studies completed during development.

As with the specification and purchase of any new piece of equipment, well-written equipment specifications include the intended validation activities for qualifying the equipment and assuring it meets the processing requirements. Defining testing and documentation expected during the Factory Acceptance Testing (FAT) at the vendor's facility is also a useful contractual agreement.

34.3 SOURCES OF INFORMATION

Sources of information include the Research and Development, Engineering, Clinical Supplies Manufacturing, Quality Control, and Regulatory Affairs groups. A development report is a key source of technical information regarding the characteristics of the active pharmaceutical ingredient (API), product and formulation design, and product processing requirements. The physiochemical character of the active substance, if appropriate, along with the functions of the excipients of the final product formulation and liquid and solid-state stability data are critical parts of such a report. Process development data and finished product characteristics should be available within the development report. Specific information on the equipment design and performance for the EQ is often archived by the Engineering and/or Validation department. Other engineering references include maintenance and calibration procedures. Operating procedures covering product handling and operation of the lyophilizer are available within manufacturing documentation. These would include the unique aspects for processing lyophilized fill volume frequencies and tolerances and operation. Finished product testing methods for the active ingredient, reconstitution and residual moisture should be available from the development scientists, Analytical Development group as standard testing methods for use within Quality Control. The Regulatory Affairs staff should be consulted for commitments made in regulatory filings and communications to regulatory agencies.

34.4 RECOMMENDATIONS FOR A VALIDATION PROTOCOL

The differing circumstances under which a validation study is prompted often dictate the best approach. Agreeably, prospective validation, where the validation studies are all completed and approved prior to shipment of any product, is preferred. There are, however, opportunities to complete certain validation studies during preparing material for Phase III clinical studies in which the product is to be administered to patients. Here, validation is concurrent with producing these materials. In addition, when implementing validation studies on an existing marketed product to bring the operation up to current regulatory expectations, concurrent validation would also be appropriate. Retrospective validation would be applied to a review of historical data of an existing process and product. Examples would be the review of the lyophilization processing data, finished product batch release test data and stability data from the commercial stability testing program.

The design of the validation testing and the composition of the protocol reflect the circumstance under which the study is conducted. For retrospective validation, the "test" may be statistical analysis of batch release data such as assay, pH, physical appearance, residual moisture, reconstitution time and constituted solution appearance. This retrospective process validation would be intended to show that the process is within an adequate level of control and the product is of consistent quality. A critical review of the processing conditions in a retrospective validation may consist of a test comparing actual processing conditions during lyophilization to ideal parameters, showing not only adherence to the defined processing conditions but also demonstrating process reproducibility.

Concurrent validation studies may be used during clinical manufacturing and scale-up activities. Additional testing or an increased number of samples, as in the case of transferring a product to another site or manufacturing material in a new production lyophilizer, may be conducted as a concurrent validation study. This would be reasonable if the parameters have not changed and the process has already been adequately validated. In addition to finished product testing, short-term accelerated stability may be appropriate prior to actually releasing the batch for distribution. Long-term stability studies should be done at the recommended storage conditions, up to and including at least six months beyond the desired expiry date.

Although there are circumstances in which retrospective or concurrent validation may be appropriate, prospective validation is preferred. This entails the testing, review of the data and approval of the completed validation studies prior to releasing product for distribution and use. Identifying the target process parameters and a proven acceptable parameter range, along with demonstrating consistent product quality and stability would be highly desirable prior to introducing the product into a manufacturing environment. It could also decrease the amount of time necessary for getting a new product to market.

Numerous studies to support process validation can be completed during the development phase. These studies correlate the product formulation, presentation and lyophilization processing parameters with finished product attributes and long-term stability. In addition, the reproducibility of the process would be demonstrated along with the consistency of finished product attributes. Uniformity studies during the first batches being integrated into manufacturing are often the last leg in the sequence of validation studies for bringing a product to market. Depending on the supporting data available from development studies, limited or short-term accelerated stability may be sufficient.

34.5 PREPARATION OF THE PROTOCOL AND SOPS

Each activity performed as part of the Installation Qualification (IQ) and Operational Qualification (OQ) should be organized into discrete functions and documents. During the IQ, the review and verification of utility connections, piping of the refrigeration and heat transfer system, reconnecting the vacuum system, rewiring of the control system, start-up and testing may be organized into a distinct document for each activity. This "modular" approach becomes more effective and efficient as the complexity of the procedures and equipment increases. Each aspect of bringing a lyophilizer online or integrating a new product into a manufacturing environment often involves a number of individuals or departments. Arranging the overall effort into smaller packages correlating with distinct activities makes the communication between individuals and departments more manageable. For example, the project engineer responsible for installation of a new lyophilizer may use a mechanical contractor to reconnect the piping and connect the utilities and an electrical contractor to connect the control system wiring. In such a case, a documentation package covering each activity may be issued and completed for each part of the project involving each contractor. A documentation package organized in such a manner is also a useful tool for project management.

Such an approach is also applicable for product and process validation. Considering the ranges of formulation aspects such as the acceptable pH range, a focused study to correlate the pH, phase transition temperature and finished product aspects upon processing would be well suited as a distinct protocol. This protocol may parallel studies already conducted during development. Another example is establishing the PAR for the processing parameters. Identifying such ranges is accomplished by processing at extreme shelf temperatures, chamber pressures and times, following the PAR approach as referenced earlier.

34.5.1 Establishing Acceptance Criteria

The selection of acceptance criteria is dependent upon the circumstances under which validation is being undertaken and requires judicious consideration. Challenges to the equipment, for example, may depend upon whether the lyophilizer is being first installed or whether validation is being completed for an existing unit currently in use. Where the lyophilizer is new, the acceptance criteria based upon the performance requirements that are identified within the equipment specifications would be warranted. The advantage of acceptance criteria based upon the stated equipment capabilities is that any process that is within the performance capabilities of the equipment could be utilized for lyophilizing product. For testing an existing unit in production, however, the most rigorous processing conditions would be a justifiable test challenge. The jeopardy of test challenges based upon the most current processing conditions is that if a process for a new product is outside of the parameters tested, then additional testing or qualification at the new parameters would be necessary. Details of constructing validation protocols, designing studies and establishing acceptance criteria will be presented in each respective section of this chapter. In approaching validation, it is more important to test and document what is critical for gaining a high degree of confidence that the process is well-defined and reproducible, the procedures are adequate and appropriate, the equipment is suitable for completing the process and the product is of consistent quality, purity and stability. In addition, it is a valuable opportunity to collect useful information for supporting a change control program. Validating for the sake of simply documenting information in a protocol, not having a clear understanding of what is significant, or creating a voluminous collection of data because more is better should be avoided. As a general rule, do what is necessary and do it well!

For some studies, as in the OQ, references will be made to known performance capabilities of the equipment. These are intended to be examples rather than standards. A few general notes are, however, appropriate. Most importantly, selecting acceptance criteria needs to be based upon a justifiable scientific rationale. This is applicable whether qualifying an existing piece of equipment for commercial product manufacture or validating a process during clinical manufacturing. Selecting appropriate processing ranges to be encompassed within the validation has a significant long-term effect in manufacturing. For example, when the range of residual moisture correlated with suitable stability is adequately determined during development, then any batch in manufacturing exhibiting a moisture within the boundaries of that range would be acceptable. If there were a batch where the residual moisture was beyond the boundary, then there would be reasonable questions whether that batch should be released. Adopting such a philosophy provides a clear and reasonable approach for successful routine manufacturing of high-quality product. There is also little question when a batch is found to be outside the PAR. This eliminates the scenario of placing a batch on stability or doing additional testing when there is a question of what a suitable envelope of processing conditions or product quality aspect would be for a batch to be released. Establishing a PAR, or PAR approach, becomes a valuable asset in a manufacturing environment.

34.5.2 Equipment Qualification

EQ can be conducted as part of and entail portions of an IQ and OQ, as well as a FAT and Site Acceptance Test (SAT). Activities encompassed within an EQ are a useful endeavor when begun as an integral part of the FAT and carried through installation of a new lyophilizer. These activities encompass verifying that the equipment is designed, constructed and performing as anticipated when compared with the equipment specifications. This assumes that the specifications are based upon current or anticipated needs for processing the products and agreed to between the vendor and purchaser. This would include verifying the engineering documentation and construction and assembly of the lyophilizer, along with demonstrating adequate performance of the system.

For the acquisition of a new lyophilizer, the EQ comprises a series of tests to assure that the lyophilizer meets the performance expectations necessary for its intended use and identified within the purchase specifications. These series of tests are useful as part of the FAT with the intent to measure and verify the performance capabilities of the lyophilizer prior to its shipment to the purchaser's site.

Incorporating the qualification requirements in the equipment specifications package to the vendor assures that proper attention is given by both the vendor and purchaser staff. These validation requirements include the EQ tests along with control system validation and extend into the IQ and OQ. Identifying the testing to be done at the factory to complete the FAT allows sufficient planning for both manpower resources and time at the vendors' facility. Validation of the automated system controlling the lyophilization process, along with the complementary processes such as sterilization-in-place (SIP), clean-in-place (CIP), in-process integrity, FIT, and data flow integrity needs to be started at the control system design and software development stage of the project. This follows the Life Cycle (2) approach that has become industry practice for validation of computer automation systems.

Part of the EQ that comes before any actual performance testing is the review and verification of the lyophilizer design. This is sometimes completed as a separate task and is often referred to as a DQ. This step, whether as a separate DQ or as part of the EQ, entails a review of the engineering documentation to verify that the equipment will meet the requirements of the specification prior to construction and assembly of the lyophilizer. Such a review includes the general layout of the equipment, piping arrangements for the CIP and SIP systems, refrigeration and heat transfer fluid system drawings, electrical elementary schematics, and Piping and Instrumentation Diagrams (P and IDs). This review of the engineering drawings should be documented and become part of the validation package.

Equipment performance tests, as a major part of the FAT, involve testing to demonstrate that the lyophilizer functions, performs and has the processing capacities as specified. These tests may mimic those planned as part of the OQ but do not negate the need for completing the OQ at the installation site. Often duplicating the testing for an OQ, tests encompass function, control capability and performance for freeze drying and support processes. The testing regime should include specific tests as listed in Table 34.1. Complementary functions such as CIP, SIP, and FIT should also be included when the lyophilizer has such capabilities. Testing of the loading and unloading would be appropriate as an integral part of the performance test, particularly if automated loading/unloading equipment is integrated with the lyophilizer(s).

This testing program is useful as part of the validation package, along with being part of the equipment acceptance. Circumventing the testing at the vendors' facility should be avoided, no matter how complex or unique the final installation. In addition, successfully completing the EQ does not negate the need to complete a SAT and comprehensive IQ/OQ at the final installation site. Factors such as assembly of the lyophilizer after being dismantled for shipping and differences in utility supplies warrant the need for testing prior to bringing the unit online for manufacturing product. The more complex and unique the lyophilizer design and final configuration, the more such efforts assure the success of the project. Some parts of the IQ could be completed at the factory and not repeated after installation. Such items may include instrumentation and hardware documents, testing of the control system and verification of as-built drawings, to cite a few examples.

34.5.3 Installation Qualification

The IQ is the first validation activity completed when the lyophilizer arrives at the final installation site.

Implementing the protocol may begin as the lyophilizer is being installed. Verification of the electrical wiring and piping may be accomplished as part of the assembly activities. The appropriate approach to completing the IQ is strongly dependent upon the specific circumstances of the project.

The IQ consists of a description of the lyophilization equipment, a system hardware and component list and the documentation of the installation procedures, equipment start-up and operator training. The IQ also includes references to the purchase specifications, engineering review and SOPs. The objective is to assure that the equipment design and construction are appropriate for the intended use, it is installed properly, the utilities are suitable and adequate and that procedures are in place for proper calibration, operation and maintenance

34.5.4 Equipment Description

The description of the lyophilization equipment provides a general overview of the lyophilizer, the installation site, functions and use in operation. The description also identifies the major components of the system. Complementing the list of the major components, a more specific description of each item provides greater detail. Such information is highlighted in Table 34.2. This data becomes an integral part of the change control system for the equipment hardware. The major components of the lyophilizer that should be included are the chamber and its components, condenser and its components, refrigeration units, heat transfer fluid, heat transfer circulation pump, heater elements, primary vacuum pumps, secondary vacuum pumps, utility piping including valves and filters and the control instrumentation.

34.5.5 Installation Activities

Documentation of the installation can also be included within the IQ section of the validation package. Part of this documentation may take the form of an installation checklist. This checklist would include each specific activity necessary for the installation of the lyophilizer, who completed and checked the work and the date the work was completed. These activities would include assembly of the various lyophilizer parts if dismantled at the factory for shipment, as well as the connection to utility supplies.

In addition to the early project activities of the engineering review and factory testing completed as part of the EQ, certain parts of the IQ should also be planned well in advance of receiving the equipment. These include the utility verification, specific installation location, start-up and training. The utility verification, identifying the quantity, quality and source

TABLE 34.1
Test Functions for Factory Acceptance Test, Site Acceptance Test and Operational Qualification Test

Test	Function in loaded (bulk water) and unloaded conditions
Shelf temperature	Range, rates, control, uniformity
Chamber pressure	Range, control
Condenser	Chilling rates, ultimate temperature
Vacuum system	Evacuation rates, ultimate pressure
Sublimation	Rate, ice capacity

TABLE 34.2
Major Lyophilizer Components Perspective Functions Documented in the Installation Qualification

Test	Function
Chamber/condenser	Pressure/vacuum vessels and components (shelves, doors, coils)
Shelf heat transfer system	Transfer of heat between a circulating fluid to product
Refrigeration system	Chilling heat transfer fluid and condenser
Heating system	Providing heat to shelf heat transfer fluid
Vacuum pumps	Removing non-condensable gases
Utility piping valves and filters	Control of heat transfer fluid, process vacuum/pressure steam
Control and automation	Equipment, process variables, sequencing steps and process data acquisition and storage

TABLE 34.3
Common Utility Supplies Documented in the Installation Qualification

Test	Function
Electrical power	Powers equipment
Chilled or tower water	Cools refrigeration units
Clean steam	Used for sterilization of critical areas
Water for injection	Used to clean chamber and condenser surfaces
Nitrogen gas	Used to maintain partial pressure and to backfill containers
Instrument air	Used to drive system pneumatics
Clean air	Used for system aeration and media fill backfill
Potable or city water	Used for water ring pump seal and pump cooling

of the utilities is best completed during the initial phase of the project and prior to operation of any of the lyophilizers systems. These encompass electricity, cooling water, process gases, sterilant and discharges for the lyophilizer. The listings in Table 34.3 are common utility supplies.

Physical installation includes the rigging into place and connection of the subsystems. With large sized units and those with external condensers, reconstruction at the installation site is an involved project in itself and includes mechanical, electrical and refrigeration mechanics. After the installation is complete, most vendors provide a service technician to start-up the system and provide training. Such activities should be documented and included within the IQ portion of the protocol.

34.5.6 Operational Qualification

The OQ focuses upon the equipment rather than a process for any specific product. Although not associated with any product or process, the OQ is a series of tests that measures performance capabilities and demonstrates the ability of the lyophilizer to complete critical processing steps. Functions of the lyophilizer, such as the shelf cooling rate and pressure control, are process related. They are, however, focused on measuring the performance capabilities of the equipment rather than demonstrating any processing capabilities relating to producing a particular product.

34.6 SCOPE AND OBJECTIVES

The OQ demonstrates the equipment performance for the range of processing functions at the installation site. The tests performed may be expanded as compared with those completed as part of the FAT at the vendor's facility. Additional activities such as CIP and SIP process development and validation as well as data flow and data integrity testing are also performed after the IQ has been successfully completed.

34.6.1 Measuring Equipment Performance

Although the testing at the factory would have demonstrated the performance capabilities, these tests also need to be performed at the final installation site. Different utility capacities such as cooling water and steam supply influence the equipment performance. These tests also verify that the utility supplies are adequate and meet the demands of the operating system. This testing is particularly valuable for large systems disassembled and shipped as smaller packages, where the unit is reconstructed at the installation site. Testing is necessary to demonstrate that installation was completed properly, and the equipment still meets the performance levels previously demonstrated during the FAT.

34.6.2 Verification of System Capabilities

The OQ evaluates each equipment function and the capacity to meet the performance standards. Reducing the lyophilization process into each function also has advantages for managing a change control program. For example, one test would focus on cooling rates used for the freezing step, whereas a separate test would be implemented to evaluate the heating function used during primary and secondary drying.

The advantage of having a separate and distinct testing protocol for each process step is that there is a specific and separate test function for each step. For example, when a significant change is made to the shelf cooling equipment or there is a question as to performance capabilities, a detailed and specific protocol could be implemented to demonstrate that there is no significant change to the system operating performance. Considering each function of the equipment for each step in the process allows segregation of each equipment function, with a respective test that demonstrates a specific performance capability. Table 34.4 summarizes the critical processing steps that should be tested during OQ.

34.7 EQUIPMENT PERFORMANCE TESTS

Performance capabilities and capacities can be evaluated using a separate test for each function of the lyophilizer, focusing on the operation of selected subsystems and the capacity for the specific functions during lyophilization. These subsystems include the heat transfer system, condenser and vacuum system. An overview for testing of each major subsystem is presented in the following sections. Also included are examples and illustrations for performance ranges.

These examples do not, however, reflect the capabilities of a specific lyophilizer nor are they intended to suggest any industry standard.

34.7.1 Heat Transfer System

The heat transfer system provides cooling required for freezing the product and the subsequent heat needed to establish rates of sublimation. Temperature control is required over the entire process temperature profile, from the time the product is loaded onto the lyophilizer shelves until it is removed after stoppering. Therefore, cooling and heating rates, along with control within a reasonable range at temperatures that embody the intended operating range, as well as shelf temperature uniformity, must be tested.

Maximum cooling and heating rate tests are intended to demonstrate the optimal performance of the equipment. The cooling rates, defined as an average change in temperature per unit time, are measured from room temperature to the ultimate achievable freezing temperature. Heating rates are measured from the lowest to the highest operating temperature for the lyophilizer. For a lyophilizer currently in use, the acceptance criteria may be the average rate across a temperature range that exceeds the current process requirements and should extend beyond to fully envelope the routine ranges of operation. Test results are expressed as an average rate of change as measured at the shelf inlet. Because the performance of the lyophilizer is dependent upon the specific design, acceptance criteria vary. Cooling and heating rates are heavily influenced by the heat capacity of the product; therefore, acceptance criteria should be established both for the empty chamber and for the loaded conditions. Bulk water can be used to simulate loaded conditions.

Shelf temperature uniformity across any one shelf and all of the shelves of the lyophilizer needs to be within an acceptable range to assure batch uniformity of the dried product. The temperature at any location is compared with either the mean of the measured values or the temperature indicated on the controlling instrument. The allowable range is dictated by the reference used, with tighter tolerances appropriate when comparing the actual to the mean of the measurements. Stated capabilities for shelf temperature uniformity by many of the lyophilizer vendors are 0.1°C at steady-state conditions. Appropriately completed under no load conditions, these functions may again be demonstrated under load conditions during the sublimation/condensation test.

TABLE 34.4
Critical Processing Step and Objectives During Lyophilization

Process step	Critical objectives
Loading	Shelf temperature control to provide uniform thermal history. Operation of the loading system, including mechanical operations, and data transfer and integrity between various equipment.
Freezing	Temperature change rates and ultimate temperatures to ensure solidification of the product prior to sublimation
Primary drying	Temperature and pressure control to ensure that sublimation of solvent is complete
Secondary drying	Temperature and pressure control to ensure that desorption of solvent achieves desired residual moisture level

34.7.2 Condenser

Measuring the cooling rate and ultimate lowest temperature of the condenser is useful in generating baseline data for future reference, such as monitoring the condition of the refrigeration system. Rates will vary based upon the size, type and number of refrigeration units on the system. The ultimate condenser temperature necessary is dependent upon the solvent system in the product formulation. For a completely aqueous solvent system, the maximum allowable temperature is commonly –50°C. For processing some organic solvents, the condenser temperature necessary is dependent upon the solvent being processed. For example, ethanol vapors must be chilled to below –115°C before condensation and solidification will occur. Pure tertiary butyl alcohol requires only slightly colder than room temperature.

In the sublimation/condensation test, the condensation rate and ice load capacity are demonstrated. In these tests, the actual performance is more critical than the baseline test of cooling rate and ultimate temperature. The rate of condensation, expressed as kilograms of ice per hour, or normalized to kilograms of ice per measure of shelf surface area, becomes a limit to the processing parameters that may be used in design of the lyophilization cycle. The results of the total ice capacity test become a limit to the product batch size.

34.7.3 Vacuum System

Like the cooling rate and ultimate temperature tests for the condenser, evacuation rates and lowest achievable pressure are baseline tests that indicate the performance of the vacuum pumping system. Typical evacuation rates allow for reaching 100 m within 20 to 30 minutes. The lowest achievable pressure is commonly 20 m or less.

Associated tests to include are the baseline leak rate and vacuum integrity test. Both tests are based upon the pressure rise of a sealed chamber and condenser that are isolated from the vacuum pumping system. Detailed presentations on the subject are covered in various classic technical publications on vacuum technology (3, 4). Each of these tests, briefly described following, is well-suited to stand-alone protocols.

The leak rate test is a baseline measurement that is intended to determine the presence of leaks in the freeze dryer chamber and condenser. This test is implemented with the chambers being clean, dry and with low levels of outgassing. Eliminating any vapors that may outgas and contribute to any pressure rise requires that the test should be done only after the system has been maintained at a low pressure for a number of hours. Common acceptance criteria used are the specifications agreed to by the equipment vendor and end user. The standard may be expressed as units of pressure per unit time and are best referred to as units of pressure and volume per unit time.

Different than the leak test, the vacuum integrity test is an in-process method used in manufacturing after the completion of sterilization and prior to loading product. First, a study to establish an acceptable value is necessary. Justifying an acceptable value is accomplished by determining a rate of pressure rise that includes any contribution of outgassing of water vapors remaining after the sterilization process. This requires that this development study be completed after the sterilization process has been validated, because the sterilization conditions may influence the amount of outgassing of residual moisture that remains after sterilization. The result of this study yields a value expressed as a pressure increase per unit time and does not need to be normalized, as in the leak rate test. Although there have been discussions on the topic published, there is no industry standard established that is based upon either empirical data or having a justifiable scientific rationale (5, 6).

34.7.4 Control Functions

Whether a control system is composed of distinct instruments for nominal control functions and process monitoring or an integrated control system, a nominal set of tests are appropriate. The tests described are intended to encompass both controller capability and equipment performance.

34.7.4.1 Shelf Temperature Control

Different than the achievable rates for the equipment alone, shelf temperature control tests combine the system capabilities in implementing a range of cooling and heating rates and control at a specific set-point across the operating range of the system. For cooling and heating, minimum, maximum, or specific controlled rates are challenged. These rates may be based upon either specific required processing conditions or the vendors stated equipment performance over the operating range of the system. Rates for both cooling and warming may range from a minimum of 0.1°C to a maximum such as 2.0°C per minute.

Shelf temperature control tests show the system's ability to maintain the steady-state shelf temperatures used for the freezing and drying process and should be within an acceptable range around a target set-point. If the acceptance criterion is other than the vendor's stated operating range, then control points used for the test must envelope the temperature ranges to be used for processing. Equipment capabilities range from ±1°C to ±5°C from the target set-point, as measured at the control point. Typical manufacturing units often achieve a range within ±3°C.

34.7.4.2 Pressure Control

The pressure control capability, critical as a process parameter, needs to show accuracy and precision of pressure control across the range anticipated for lyophilization cycles. This range can be a pressure as low as 20 μ to as high as 1600 μ. The results of the test are compared with the target values at a low, intermediate and high pressure. Acceptance criteria are specified as an acceptable range around the three-target set-points. An acceptance criterion of ±10 μ is readily achievable.

34.7.4.3 Process Monitoring

Defining the process as critical parameters of shelf temperature, chamber pressure and time dictates monitoring these conditions with suitable accuracy and precision. Product temperatures, being less critical because of intrinsic limitations, are also commonly monitored. The ability of the monitoring system to reflect the actual process status is assured by an appropriate calibration program. Although not normally a separate study within the OQ testing program, it may be appropriate to complete a comparison of values measured if multiple instruments are used for monitoring the same conditions, or if data is transferred from a recording instrument or programmable logic controller (PLC) to a separate computer workstation.

34.7.4.4 Sequencing Functions

With an automated control system, verifying the sequencing functions may be appropriate during the OQ testing. The first step is verifying the interfaces to the field devices such as pumps, motors and valves and their proper operation. This should also include operation of proportional control valves. This verification may have been completed separately as part of the control system qualification, conducted during the FAT, and would therefore not be necessary during the OQ studies. In verifying the control sequence functions, the hardware engaged for each step and the successful progression through the process is compared with that identified within the control system flow chart. Whether completed during the OQ or separately during control system validation is of little importance. Although it is preferred that the control system be qualified prior to implementing any of the OQ testing, it must be completed before testing any integrated control functions such as the lyophilization process tests described later in this chapter. Computer control system validation has unique requirements for validation and would best be accomplished as a separate validation study.

Data flow testing and data integrity qualification should be completed as part of the OQ. The United States Food and Drug Administration (FDA) requires electronic data records to meet the same standards as written records. The acronym ALCOA is a useful mnemonic; A-attributable, L-legible, C-contemporaneous, O-original, and A-accurate. The FDA Guidance, "Data Integrity and Compliance with Drug CGMP Questions and Answers Guidance for Industry" December 2018 provides a useful reference for control system validation (39).

34.7.5 Sublimation/Condensation Rate Tests

The capacity for sublimation and condensation, as well as the condenser ice load is demonstrated during this test. The performance of the heat transfer system, chamber and condenser configuration, refrigeration system, as well as the condenser are challenged with demands for accelerated rates and under stressing load conditions. The ability of the equipment to adequately sublime and condense the water vapor during processing is demonstrated in this study, challenging both the rate of sublimation and the condenser capacity.

Water is sublimed using some optimal conditions for achieving a maximum sublimation rate. The capability of the heat transfer system to provide a sufficient quantity of energy to promote sublimation of the ice part is one aspect of the lyophilizer capacity quantified during the sublimation/condensation test. The capacity of the lyophilizer as a system, including that of the condensing system that encompasses the condenser surface and the refrigeration units, is quantified during this test as well.

During such a test, the condenser temperature is monitored and is not necessarily part of the acceptance criteria. Once ice begins to collect on the condenser surface, the sensor is buried beneath an amount of ice and the surface exposed to the vapor stream may be measurably warmer than the indicated condenser temperature. Therefore, the condenser temperature is less significant than the critical parameter of chamber pressure.

34.7.6 New Technologies

Lyophilization manufacturers continually strive to introduce new technologies to better control and monitor the lyophilization process. A few examples will be discussed, but validation of any new technology must include not only the IQ, OQ, and PQ of the system, but also the impact on other critical requirements such as cleaning and sterilization.

34.7.6.1 Automatic Loading/Unloading Systems

Automatic Loading/Unloading Systems (ALUS) technology varies, but all involve automatic movements of both the lyophilizer and conveyance equipment to achieve loading and unloading of product containers without human intervention. ALUS equipment can also track rejected vials from the filling operation, through the loading and unloading process. With the advent of ALUS technology, the lyophilizer is no longer a stand-alone piece of process equipment but must be considered integrated with upstream and downstream equipment and processes. Qualification of ALUS (and other) equipment can be combined with the lyophilizer qualification or as a stand-alone system; however, if qualified as stand-alone, mechanical and communication interfaces between the equipment must be included in either the lyophilizer qualification or the ALUS qualification, including communication between ALUS equipment and upstream and downstream equipment.

34.7.6.2 Isolators

Lyophilization and/or ALUS equipment that are integrated into isolator systems must be qualified to ensure surface sterilization. Studies must be conducted to demonstrate whether the surfaces of the lyophilizer that encounter the isolator sterilant are able to achieve the required conditions during each phase of the lyophilization cycle. For instance, the portion of the lyophilizer door that is within the isolator may be too cold during the freezing step and cause condensation to form, inhibiting proper decontamination of the impacted surfaces.

Isolators may also impart challenges for proper clearance of the ALUS and lyophilizer post batch. It is common to have installed a door separate from the product loading door that serves as a maintenance entrance into the lyophilizer. Design and procedural considerations must be made to allow for thorough routine line clearance as well as maintenance activities.

34.7.6.3 Nucleation Technology

Several technologies are commercially available that allow for the nucleation of all product vials at the same shelf temperature. Nucleation is the point at which, following some degree of supercooling, the solvent starts to freeze. The degree of supercooling is proportional to the ice crystal size. The deeper the supercooling, the smaller the ice crystals can be. By controlling the temperature at which nucleation starts, it is possible to create larger ice crystals. This provides two primary benefits. One; the larger ice crystals provide better pathways for vapor transfer in the product cake, and therefore faster drying and faster reconstitution. Two; controlling nucleation provides a more homogeneous population of containers in the lyophilizer with respect to drying rates and residual moisture. All nucleation control technologies work in essentially the same way. They introduce small ice crystals at the air/liquid interface while the liquid is supercooled. The disturbance to the liquid layer starts the nucleation process. Figure 34.1 demonstrates the mechanism.

The various technologies all utilize extraneous equipment attached to the lyophilizer in one way or another. Validating the sterilization of both the equipment and the gas/liquid/vapor streams must accompany validation of the process itself.

34.7.6.4 Process Monitoring

The lyophilization process has long resisted attempts to holistically monitor the freeze-drying process. Especially at commercial scales, conditions within the freeze dryer are dynamic and subject to localized environmental factors that bias local measurements. For example, product thermocouples placed into 6–10 vials in a batch of tens of thousands fail to give accurate data representative of the batch. In fact, the mere presence of the thermocouple creates a bias within the probed

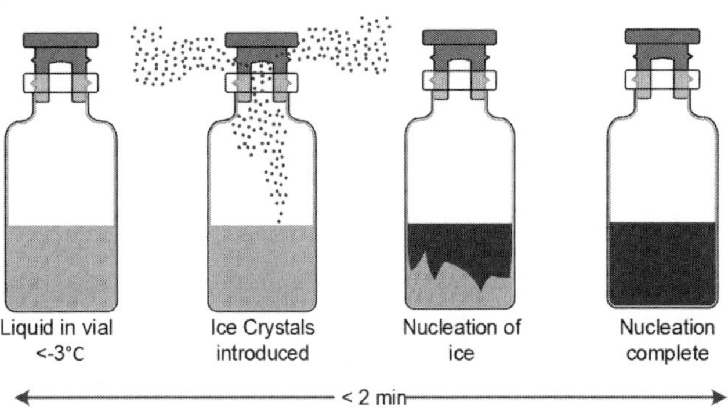

FIGURE 34.1 Controlled nucleation.

container. Several methods have been developed that measure batch outputs as opposed to individual containers. One relatively simple example is the comparison between the chamber pressure during the drying process using both a capacitance manometer and a Pirani gauge. The Pirani gauge is a resistance element that behaves differently when there is water vapor in the chamber than when the chamber is essentially dry. A capacitance manometer uses a deflecting diaphragm to measure pressure and does not change behavior regardless of the composition of the measured gasses. By observing the comparative curves of the two devices, the endpoint of drying can be detected.

More advanced detectors can quantitate residual moisture as well as other gaseous components of the chamber environment. The ability to detect and differentiate types of silicone provides the capability to detect internal silicone leaks and differentiate from silicone used on container components. As with other technologies, the impact of the technology on the cleanliness and sterility of the freeze dryer must be evaluated and qualified.

34.7.6.5 Process Testing

Process testing combines functions evaluated during the OQ studies, completing processing of a model product and presentation. Processing a surrogate product provides a challenge for the functions and capacities to demonstrate the equipment's performance under load conditions. Process parameters of shelf temperature, chamber pressure and time are compared with target values. The equipment capabilities with the integrated control functions can also be used to demonstrate batch uniformity capabilities. Process testing may be independent of any specific processing parameters associated with any particular product presentation. A series of well-controlled tests may be used to demonstrate the capability to reproducibly implement the critical lyophilization process parameters and yield consistent dried material qualities. Trial runs used to assess adequate process parameter control of shelf temperature and chamber pressure under load conditions can also be useful in demonstrating uniformity of processing conditions within the lyophilizer. The batch size and process parameters do not necessarily need to duplicate those for any actual product. Studies can be designed, with an appropriate model presentation as a surrogate, to challenge the equipment. Several surrogates have been proposed in presentations in the literature, ranging from a placebo of a specific product formulation to a combination of mannitol and arginine, in vial sizes from 10 to 100 mL (7). A model product may provide a sufficient challenge to demonstrate the equipment's performance capabilities under load conditions. During such studies, shelf temperature, chamber pressure and time, the critical process parameters, are compared with target values.

34.7.7 Uniformity Within a Lyophilizer

Demonstrating that the processing and environmental conditions are uniform within the lyophilizer can be included as part of the OQ. As with many batch drying processes, processing as well as local differences in environmental conditions may affect lyophilized product quality attributes. These attributes include the dried cake physical structure and appearance, as well as residual moisture content. Evaluation of these attributes can provide an assessment of the uniformity throughout the batch. The potential differences are also influenced by varying processing parameters relative to location in the lyophilizer. Studies by Greiff have shown that there is a measurable effect of location within the lyophilizer on both the amount of ice sublimed and the residual moisture when lyophilizing a 2% serum albumin solution (7). These studies also quantified the range of moisture that varied with the shelf temperature, the shelf position and elapsed processing time. Dried product attributes can be mapped at the same locations on the shelves used to demonstrate shelf temperature uniformity: each of the four corners and the geometric center. Product temperature and finished product attributes of samples are evaluated at each of these five locations on each shelf. Critical points and the objectives to be achieved at the completion of each significant step of the process are highlighted in Table 34.4.

The range of product temperatures for material at each of the locations can be compared at the completion of each major processing step: loading, freezing and primary and secondary drying. Comparison of temperatures and finished product attributes unique to lyophilized products at each of the locations can be evaluated statistically based on the processed material response at the conclusion of each step at the sampled locations. Temperature range and standard deviation for the monitored locations can be compared with the average temperature for each set of processing conditions. Evaluation of the data needs to accommodate a significant variation at times when the conditions of the process are changing and the process has not reached steady-state conditions. For example, a significant influence is the container type and location of the thermocouple placement in addition to differences in mass transfer of the water vapor through the dried product (10, 11). Figure 34.2 illustrates the variation in product temperature for a formulation containing 5% Human Serum Albumin in both tubing and molded vials and at the inlet and outlet of various shelves. Therefore, the significance placed upon any variation and the conclusions drawn from such data need to account for such inherent influences. Dried material attributes such as physical appearance, reconstitution time and residual moisture are finished product attributes affected by differences in process conditions. Although many of the attributes such as physical appearance are subjective evaluations, residual moisture is more quantifiable in reflecting the magnitude of variation because of location. A material or combination of excipients may be used as a surrogate for conducting these uniformity studies.

The surrogate selected needs to be measurably influenced by processing conditions and able to reflect significant differences in temperature during processing and dried product attributes. The surrogate formulation, concentration, and fill volume may influence the variation in physical structure and density and therefore affects the rate of mass transfer of

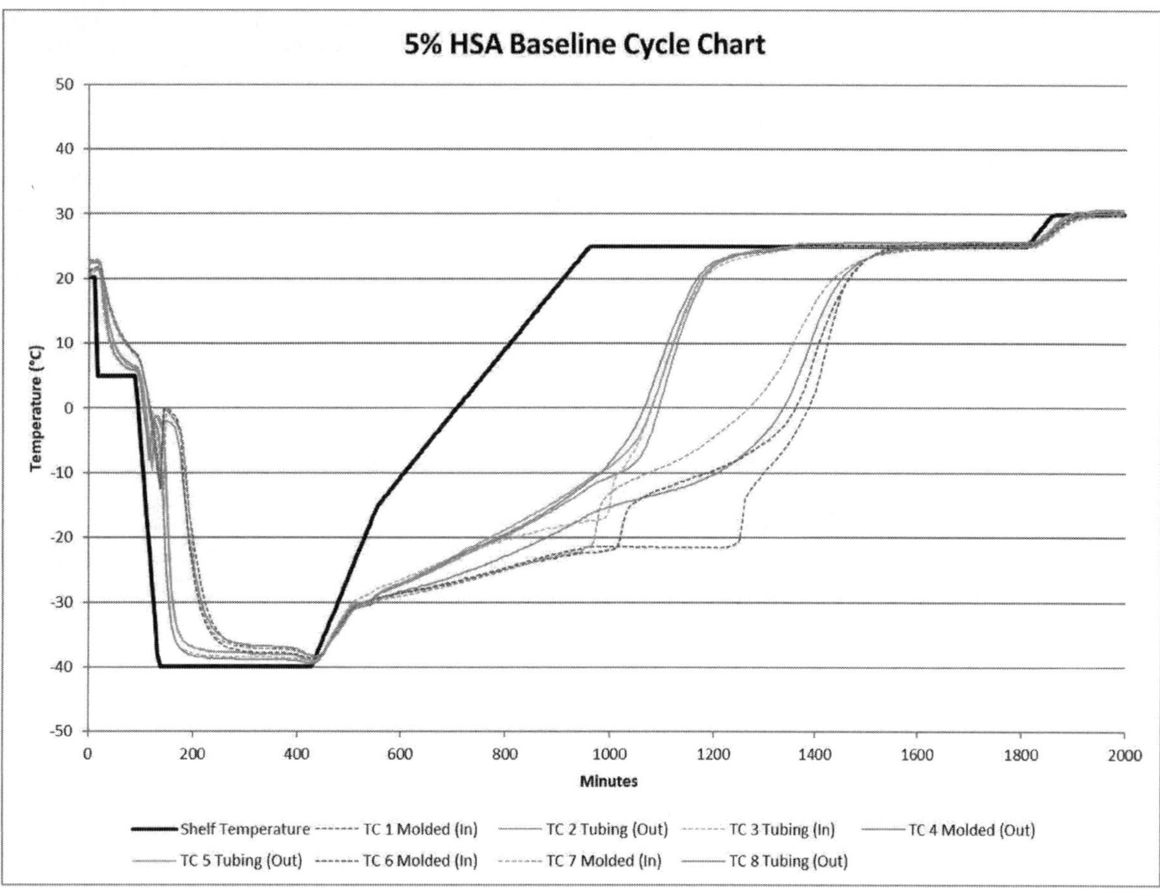

FIGURE 34.2 Temperature variation during drying for tubing and molded vials with 5% HSA formulation.

Note: HSA = human serum albumin; TC = thermocouple.

water vapor during sublimation (12). Surrogate preparations consisting of mannitol, polyvinylpyrrolidone and simple ionic salts such as potassium chloride, in the range of 5% to 12% w/v, normally solidify to form a dense, uniform structure, regardless of the rate at which the material is cooled during freezing. A significant difference in structure can be observed when polysaccharides such as dextran, sucrose and lactose, are solidified under different rates of cooling during freezing (13). Such readily measurable differences during processing and for the dried material provide useful product temperature and residual moisture results that are quantifiable data for evaluation.

Product temperature and dried material attributes that are influenced by differences in processing conditions can be correlated with location and product thermal history. Attributes of reconstitution time and residual moisture may be quantified. Results of statistical analysis of the temperature during processing and dried material data can be used to identify a location that is most representative as well as the most extreme throughout the entire process. These results can be used to demonstrate the extent of batch uniformity and identify appropriate locations for monitoring and product sampling during actual product validation studies. Combining extensive process monitoring with dried material testing, uniformity of the dried material can be evaluated and shown to be independent of location. Process testing combines functions tested separately in the preceding steps of the OQ studies and may utilize a surrogate preparation. This study challenges the integrated control functions as well as demonstrates batch uniformity. It is important to note that process testing is independent of any particular processing parameters and any specific product presentation. Rather, it is a series of well-controlled tests, designed to demonstrate the equipment capability to reproducibly implement the lyophilization process and yield consistent product qualities, independent of the location within the lyophilizer.

34.7.8 Integrated Control Functions

Integrated control functions encompass the lyophilization process itself, along with alarm functions and fail-safe responses to out-of-range process conditions. Critical parameters of shelf temperature, chamber pressure and time and the success in controlling these parameters within an acceptable range are demonstrated during the actual lyophilization of material. For ease of completing the testing and as a precursor to implementing a process with test material, the lyophilization cycle may be completed using an empty chamber, with alarm

TABLE 34.5
Process Fail Safe and Alarm Tests

Alarm	System response
High shelf temperature	Heater disabled; redundant refrigeration activated
Low shelf temperature	Refrigeration disabled
High chamber pressure	Gas bleed disabled; secondary vacuum pump activated; emergency cooling
Low chamber pressure	Gas bleed enabled

function tests and fail-safe responses challenged. During this test, alarm conditions such as the shelf temperature and chamber pressure may be altered by physically forcing such conditions. For example, directly engaging the heaters would cause the shelf temperature to warm above the allowable target set-point range. Engaging the refrigeration when the shelves are at the target set-point would cause the shelf temperature to fall below the range, also creating an alarm condition. Escalating fail-safe responses would also be tested in a similar manner. Table 34.5 highlights just a few of the critical parameters that would be appropriate to test during such a simulation.

34.8 LYOPHILIZATION PRODUCT QUALIFICATION AND PROCESS VALIDATION

Lyophilization is a method of preservation in which the conditions necessary for the process are dependent upon the characteristics of the material to be processed. The finished material is dependent upon the processing parameters used for freezing and freeze drying. This then requires that the physiochemical character of the material be well-defined and understood in order to develop a suitable process. For routine processing, the consistency of the starting material dictates the level of success for the outcome of the process. Such data is a prerequisite to designing an appropriate set of lyophilization conditions. There may be characteristics of the material that allow quantifying the level of success of processing. This requires that the characterization of the starting material be considered when undertaking a validation study and are discussed within the following sections.

Definitions for validation published in section 210.3 the Federal Register in May 1996 emphasize the distinctions between process validation and suitability (14). Process validation is defined as "establishing, through documented evidence, a high degree of assurance that a specific process will consistently produce a product that meets its predetermined specifications and quality characteristics." Process suitability is described as "established capacity of the manufacturing process to produce effective and reproducible results consistently." Section 211.220, describing process validation also includes demonstrating reproducibility of the process as a requirement.

The application of lyophilization is for the preservation of materials unstable in the presence of water, demonstrating that a process produces product of suitable quality characteristics at the time of release and extending over the shelf life of the product. Preservation is then an inherent objective of the process and requirement for the product, placing a greater emphasis upon correlating product attributes and stability to processing conditions. This emphasis is carried through the process development. Considering this approach to applying these validation concepts to lyophilized processes and products, the significance of development activities and the suitability of validation during development are apparent.

34.8.1 Pre-Formulation Data

As part of the pre-formulation activities, investigations include physiochemical character, purity, solubility, stability and optimal pH studies. Potential product formulations considering route of administration and solution stability are initially studied in preparation for producing material to be used in clinical studies, Unique to dosage form development studies for lyophilized products, thermal analysis of the drug substance and product formulations are also necessary. Data generated during this phase of product development is useful for future development activities, along with validation.

For lyophilized drug products, the active substance purity and morphology, formulation procedures, excipients used and initial concentration may affect the behavior during processing as well as the dried material stability, with a wide variety of examples in the literature. For example, certain beta-lactam antibiotics may solidify to an amorphous or crystalline morphology. Each different form exhibits different physiochemical properties such as solubility and stability (15). In addition, pH may be an influencing factor in the phase transition of the substance (16). The presence of certain excipients may also alter the morphology of the active substance (17). Degradation pathways involving hydrolysis, common for products that require lyophilization, are also significant. For biopharmaceuticals, numerous biochemical reactions such as hydrolysis, oxidation, deamidation, beta-elimination and racemization play an important role in the stability of the final product (18). It has also been reported that residual levels of an impurity in mannitol as low as 0.1% w/w was involved in the degradation of a polypeptide upon storage (19). There are often, therefore, critical behavioral characteristics that need to be considered in the manufacturing of lyophilized products and assessing the success of the formulation and processing methods developed and subsequently validated.

Well-documented studies, summarized within a development report on the physiochemical aspects, drug substance attributes and finished product characteristics becomes an important reference to support the design and the acceptance criteria for the validation studies. Such data is also valuable for future integration into a manufacturing operation. This includes the scientific rationale for formulating and bulk handling procedures, lyophilization processing parameters and finished product analysis.

34.8.2 Development Activities

Development activities encompass both the initial conditions for the drug substance or formulation and packaging considerations for a drug product, along with the lyophilization cycle. For a drug substance, the upstream processing and the condition of the starting material need to be quantified. This includes solvents present and, for multiple-solvents, the ratio of one solvent to another, impurities and related substances. As part of product development, compounding procedures, formulation components, active ingredient quality and selection of the container/closure in addition to process engineering of the lyophilization cycle are studied.

Acknowledging that development is a precursor and critical to designing the studies necessary for validation, considerations for each major phase of the development activities will be reviewed. This review starts with studies of the drug substance and progresses through finished product testing.

34.8.3 Drug Substance

The physiochemical character of the active ingredient steers the formulation design and selection of excipients for the finished product. If, for example, the drug substance has a propensity to form either an amorphous or crystalline phase, the method of freezing and the character of the material needs to be assessed during development. Material solidified during freezing as a crystalline form is more thermodynamically stable than an amorphous from. For example, studies have shown that the solid-state decomposition of cefoxitin sodium can occur at a significant rate. The amorphous form yields a 50% loss of the active ingredient within one week at accelerated storage conditions of 60°C. The crystalline form degrades to less than 10% loss in eight weeks (20). Investigating the characteristics of the active material therefore needs to be studied during development. As well, the influence of processing conditions on the morphological form is also critical, as discussed later in this chapter.

The specific physiochemical character of the material may be a useful means for verifying reproducibility during the validation studies. Materials that will form a crystalline morphology and have good bioavailability and stability may be formulated with mannitol as an excipient, where both the active ingredient and mannitol readily crystallize. However, some excipients will alter the morphology of other excipients or the drug substance. These differences may be quantified with analysis by X-ray diffraction. Peptides and globular proteins tend to inhibit the crystallization of some excipients. An example of this is the effect of human growth hormone (hGH) on the morphology of glycine and mannitol (21). In such circumstances, the physiochemical characteristics of the substance can be useful in qualifying the formulation design of the finished product. It may also be a useful tool in assessing process reproducibility and product consistency.

Other factors that need to be considered are the purity profile of the active substance. For example, a synthesized drug substance precipitated out of an organic solvent may contain trace amounts of the crystallizing solvent. Even residual levels of the solvent or other impurities can affect the measured phase transition of the material (22). Therefore, the amount of allowable trace solvents or impurities and their impact upon product behavior during processing need to be included in early development studies and may also be appropriate as a monitoring requirement during validation.

Upstream purification of peptides and proteins often uses varying combinations of organic solvents and acids to elute the substance from the chromatography column, depending on the substance. For a peptide that may orient itself in either an alpha-helical or beta-pleated sheet configuration depending upon the presence and concentration of an organic solvent, behavior in solution or during the freezing process may differ substantially for each conformation. Trace amounts of solvents and acids may affect the behavior of the substance in solution and during freezing. Such details of the requirements, sensitivities and behavior of the active substance need to be defined in the scheme of development and considered during validation activities. An appropriate purity profile should be established and monitored to show control for the starting raw material. Residual substances, including processing solvents, chemical intermediates, precursor fragments, along with microbiological quality are also necessary. Degradation products from upstream processing and bulk solution stability also need to be established during development and may be used during scale-up and manufacturing validation studies.

Based upon the active substance characteristics determined during development, acceptance criteria of the validation studies can be established. These criteria apply to demonstrating the consistency of the dried material processed within a PAR in the development phase and adequacy of the scale-up to manufacturing. To be comprehensive, numerous aspects, although not necessarily applicable to all products, are presented as illustrations.

In circumstances in which the active or any excipient may crystallize, monitoring of the morphology in evaluating the dried product may be warranted. If differences in solubility, reconstitution rate or stability are influenced by the morphology of either an excipient or the drug substance, then a quantitative method should be included for assessing the finished product attributes. Methods of analysis for dry powder include infrared spectroscopy, nuclear magnetic resonance, particle morphology and thermal analysis (23).

Degradation products because of hydrolysis, oxidation, or specific biochemical reactions should be monitored by an appropriate stability-indicating assay. Polymerization, aggregation and denaturation levels should also be included in the finished product and stability monitoring protocols, if appropriate.

34.8.4 Finished Product Formulation

The solubility and stability profile at different pH are important in identifying the acceptable pH range for the product formulation. In some instances, there is a compromise between solubility and stability, either for the bulk solution or dried product. For example, a 1 pH unit difference from pH 5 to pH 4 for penicillin increases the solubility along with opportunistic degradation reactions by one log (24).

Understanding the effect of bulking agents and their interactions should be studied during development. As with measuring the degree of crystallization discussed earlier, this may provide a quantitative measurement that may be useful for demonstrating process reproducibility and product consistency. Formulations containing excipients that tend to crystallize such as sodium chloride, phosphate buffer, mannitol or glycine are examples where this may be useful.

The effect of the variation in pH adjustment or the influence of any buffering system also needs to be studied. For the range of pH, any influence on the behavior of the active or excipients during freezing and the phase transition upon warming must be measured. As an example, in a biopharmaceutical formulation containing glycine with the pH adjusted using sodium hydroxide, sodium glycinate would be formed. The behavior of sodium glycinate in the formulation may be different than that expected of glycine in the free acid form. The difference in behavior and phase transition temperature has been evaluated by Akers (25).

Unless there is a specific and critical function of an excipient, an assay is not normally considered to be necessary during validation. There are, however, formulations in which an excipient is critical to the function of the active ingredient. For example, in some in vivo imaging agents, the reduction of stannous chloride is necessary in the coupling of a radiolabeled compound. For Amphotericin B, deoxycholate sodium is used as a solubilizing agent and needs to be at a minimum concentration to assure that the drug is completely soluble upon reconstitution. The concentration of the excipient in these two examples is critical and an assay would be appropriate.

34.8.5 Determining Thermal Characteristics

Establishing the temperature necessary to completely solidify the product during freezing and to maintain the product below during drying is imperative early in product development. If a component of formulation has a propensity to crystallize, it would be best to occur during freezing; processing parameters used for cooling the product may be crucial to induce such crystallization and need to be explored during development.

Using low temperature thermal analysis, the phase transition during cooling and warming is critical data necessary to identify appropriate lyophilization parameters and to justify the process. This is necessary for determining the temperature below which to cool the product during freezing and the maximum safe threshold temperature during primary drying. Results of the thermal analysis studies are used to support identifying a threshold temperature for processing. This threshold is the temperature during freezing the product is to be cooled below. In primary drying, it is critical for the product to complete drying with retention in the presence of ice and early in secondary drying. For example, the solid—liquid phase diagram for sucrose presented by MacKenzie indicates that there is a glass transition at –32°C to –34°C when the sucrose is in the presence of ice and prior to any amount of significant desorption (26). Commonly used methods for the low temperature thermal analysis needed for lyophilized products are highlighted in Table 34.6. There are a number of methods available that are commonly used for low temperature thermal analysis, each having particular advantages. Although the nature of the material dictates the most applicable method, confirming the analysis by a second method is a valuable tool for greater insight and understanding of the behavior of the material under freezing and freeze-drying conditions. Differences in measurements and observations and the impact upon the drying conditions designed for processing warrant the use of confirming methods.

TABLE 34.6
Methods of Low Temperature Thermal Analysis

Method	Principle	Indication
DSC (differential scanning calorimetry)	Change in molecular heat capacity	Glass transition and eutectic melt
ER (electrical resistance)	Change in electrical conductivity	Glass transition or eutectic melt
FDM (freeze drying microscopy)	Direct microscopic observation	Fluid flow and structural collapse

34.8.6 Assessing Bulk Solution Stability

With hydrolysis being the prominent reaction contributing to product degradation, stability over the length of time the product remains in the presence of liquid water as a bulk solution is critical. Controlled storage conditions for the bulk solution prior to filling and lyophilization are necessary as part of assuring finished product consistency and batch uniformity. Such an evaluation is important to verify that no appreciable degradation has occurred and justifies the time limits for bulk storage. This would include the time from when the product was formulated to the end of the filling operation, from the first container filled to the last. Because lyophilized formulations are unpreserved and do not contain a bacteriostatic or bactericidal agent, microbiological quality, including endotoxin levels, is important to monitor to assess the product microbiological quality.

Slight differences in the nature of the formulation because of aging may also influence the product phase transition. For example, absorption of oxygen or carbon dioxide from the air over an extended period of time may cause a pH shift, consume one component of a buffering system, or promote degradation. For a peptide or protein with both a hydrophilic and hydrophobic nature, alterations to desired secondary or tertiary structure may develop. As a result, polymerization, aggregation, or denaturation may occur. Any one of these may alter the behavior during processing and finished product characteristics. If such an opportunity would exist, any conformational changes evidenced by aggregation, perhaps leading to gelation, may occur. These need to be monitored using an appropriate analytical technique best suited for detecting such subtle and potentially significant changes. In addition, it is imperative to justify and validate the allowable bulk storage conditions such as temperature or atmospheric conditions, including a suitable time.

34.8.7 Justification of Processing Parameters

During the process development phase, ideal processing conditions should be devised to yield desired finished product qualities and acceptable stability. Target processing parameters of shelf temperature, chamber pressure and time that are safe, effective and efficient need to be established. As a matter of routine, target conditions are selected and studied as part of process engineering activities during the development phase. A temperature for completely solidifying the product during freezing is established based upon results of thermal analysis studies.

Thermal analysis data also dictate the maximum product temperature allowable during primary drying, as discussed previously. Shelf temperatures and chamber pressures are then selected to assure that the product remains below this critical threshold temperature during primary drying. Secondary drying conditions necessary to achieve suitably low residual moisture are also identified. Determining these processing parameters requires well-designed laboratory studies to define optimal conditions for a safe, effective and efficient process.

The result of process engineering studies would be definition of an ideal set of processing parameters for shelf temperature, chamber pressure and time as a target set of conditions for routine manufacturing. Control of these parameters and monitoring processing conditions begins when the product is first loaded onto the shelves of the lyophilizer until the product is stoppered and removed. In addition, the rate of change from one shelf temperature to another also needs to be predefined and controlled. These rates of changes, referred to as ramps, include cooling rates during freezing, warming of the shelf at the beginning of primary drying and the transition from primary to secondary drying.

As an example, the complete process description for methylphenidate hydrochloride, a product in which the active ingredient has a phase transition of −11.7°C and the formulation contains mannitol, may be described as outlined in Table 34.7 (27). Material processed according to the predetermined conditions would be expected to yield product of acceptable quality, purity and stability. Reproducibility of these parameters is demonstrated by comparing the actual processing parameters for any one batch with the ideal target parameters identified as a result of development studies. Evaluation of the finished product qualities and assessment of the stability over the desired shelf life demonstrates that the processing conditions are suitable and appropriate. Demonstrating the same process conditions and achieving the same finished product qualities and stability shows that the process is reproducible, and the product qualities are consistent. It is also appropriate that the range of acceptable conditions that produce product of acceptable quality are defined, scientifically justified and the impact upon finished product quality demonstrated. These include a range for the shelf temperature during freezing and drying, rates of change from one temperature to another during the process, the chamber pressure for drying and a minimum time at each condition. Selection of the suitable ranges for the processing conditions should be based upon empirical data rather than arbitrary selections.

TABLE 34.7
Definition of Target Process Parameters for a Methylphenidate HCl Preparation

Process step	Shelf temp.	Rate	Chamber pressure	Time
Product loading	5°C		Atm.	2
Cooling rate		0.5°C/min		
Freezing	−20°C		Atm.	4
Ramp to 1° drying		0.5°C/min	80 μ[a]	
1° drying	65°C		80 μ	
Ramp to 2° drying		0.5°C/min	80 μ	
2° drying	40°C		80 μ	8

[a] The pressure reported ranged from 210 μ to 15 μ; 80 μ was selected as a reasonable level for discussion.

Following an experimental design approach for developing a matrix of variables is undoubtedly a preferable method for conducting process engineering studies. With the numerous and complex influences on processing requirements, a complex matrix based upon numerous variables such as combinations of temperatures separately varied for each process step of freezing and primary and secondary drying is, at best, laborious, time consuming and often limited by the availability of API and equipment. Such an approach to process validation often requires an exhaustive number of studies. In the absence of the large number of studies to fulfill a complex matrix, a simpler matrix based upon the edges of a defined range would be reasonable and scientifically valid.

Process conditions that affect both the product temperature and rate of drying are shelf temperature and chamber pressure. For these process conditions, target parameters and suitable ranges around the parameters need to be defined during the process engineering studies. Validating the process therefore requires demonstrating that if conditions existed in which the process was completed at the extremes of these conditions, the finished product would have the same qualities and long-term stability as if the batch was processed at the target conditions. Because both the shelf temperature and chamber pressure are independent parameters, the various combinations of both conditions at the extremes and at the target would establish a PAR. This notion of a PAR was first introduced by Chapman in 1984 and is well-suited for lyophilization (1). The goal of the process validation studies is to verify that if the process was completed within any combination of the two variables, then the finished material would be of consistent quality and stability.

Designing a series of studies based upon the variables of shelf temperature and chamber pressure would encompass, as a minimum, permutations of high and low conditions for each. Demonstrating reproducibility is also an objective during validation, such that three batches processed at the target conditions would also be necessary. This therefore would require a minimum of seven batches: three at the target parameters to demonstrate reproducibility and four for the combinations of high and low conditions.

In addition to the shelf temperature and chamber pressure, the time to complete secondary drying will influence the residual moisture content of the dried material. Assuming that target residual moisture content has been identified, the validation studies should also encompass a range of time at the secondary drying conditions necessary to achieve the desired residual moisture. The range of time could be incorporated within the three batches at the target shelf temperature and chamber pressure. As an illustration and using the cycle defined for methylphenidate described in Table 34.7, the variations in shelf temperature, chamber pressure and time in secondary drying are presented in Table 34.8.

The parameters outlined in Table 34.8 consisting of high shelf temperature and high chamber pressure would provide the upper level of processing conditions. During freezing, the shelf would be controlled at the maximum or warmest temperature at which solidification would occur. During primary and secondary drying, the warmest shelf temperature and highest chamber pressure would result in the greatest amount of heat transfer. This increased heat transfer, as compared with that at target conditions, would result in the greatest rate of sublimation and warmest product temperature. The greatest amount of heat transfer would be expected to result in the warmest product temperature and possibly the shortest processing times. In considering the impact during secondary drying, the high levels would provide potentially higher rates of desorption and therefore the lowest residual moisture content. The end result should be the slowest freezing rate, fastest drying rate, warmest product temperatures during the process and lowest residual moisture.

The study encompassing the coldest shelf temperature and highest chamber pressure would be expected to yield a different rate and therefore time to complete each part of the lyophilization process. In this study, a decrease in the rate of sublimation as compared with the preceding cycle conditions is anticipated because the shelf temperature is lower and a resulting decrease in the amount of heat energy to support sublimation would occur. There would be, however, a contribution in heat transfer by the increased chamber pressure, as compared with the target processing conditions. The rate of sublimation and desorption would be expected to be lower than that for the set of processing conditions in the first study.

A higher chamber pressure would provide greater efficiencies in heat transfer from the shelf. Any increase in the overall amount of heat transfer relative to the parameters of lower shelf temperature and higher chamber pressure would depend upon the specific parameters selected. The greatest anticipated effect would be on the product temperature because of the increase in chamber pressure. This effect would be strongly dependent upon the specific processing pressure. For example, the impact of a 20 m increase is greater at a target pressure of at 80 μ as compared with when the chamber pressure is at a target pressure of 400 μ. For these sets of processing conditions, product temperatures during each process phase, rates of drying and residual moisture content would be intermediate as compared with the other studies.

Compared with a higher pressure and lower shelf temperature, drying rates with the reversed conditions of lower pressure and higher shelf temperature would be expected to be slower. Comparatively, freezing would be expected to require more time. Primary drying rates would also be reduced because heat transfer rates would be less, product temperatures lower and residual moisture higher.

The longest times for the product to reach completion for each cycle phase would result from reduced heat transfer because of a lower shelf temperature. Coupled with a lower chamber pressure, it may be expected that the processing rates would be the most significantly reduced. In this study, the principal objective is to demonstrate the times allocated for each portion of the process are adequate, even under the

TABLE 34.8
Varied Process Parameters for a Proven Acceptable Range

Process condition	Product loading	Cooling rate	Freezing	Ramp to 1° drying	1° drying	Ramp to 2° drying	2° drying
Shelf temperature							
High	10°C	0.5°C/min	−15°C	0.5°C/min	60°C	0.5°C/min	35°C
Target	5°C	0.5°C/min	−20°C	0.5°C/min	65°C	0.5°C/min	40°C
Low	0°C	0.5°C/min	−25°C	0.5°C/min	70°C	0.5°C/min	45°C
Chamber pressure							
High	Atmosphere	Atmosphere	Atmosphere	100 μ	100 μ	100 μ	100 μ
Target				80 μ	80 μ	80 μ	80 μ
Low				60 μ	60 μ	60 μ	60 μ
Time	2 hr		2.5 hr	2 hr	7 hr		6 hr 8 hr 10 hr

conditions in which the heat transfer was low and times were longest as compared with the target parameters; the heat transfer would be lowest, and therefore the freezing and drying require the longest time. The product temperature would also be expected to be the lowest as compared with the other processing conditions. Processing under these conditions of the least heat transfer demonstrates that there is sufficient time designed within the cycle parameters to accommodate such variations in rates of drying.

PARs of processing parameters during primary and secondary drying would be expected to yield some range of residual moisture. This range would result from different variables of shelf temperature and chamber pressure along with the conditions for desorption in secondary drying. The least significant impact is often variations in time. Depending upon the characteristics of the formulation and the association of residual moisture in the product, the allowable range of time in secondary drying needs to be correlated with a residual moisture content. This should be accomplished during the process engineering phase. Sequential stoppering or use of a sample extraction device to determine the change in residual moisture content over time is a convenient method for measuring the extent of moisture decrease. Another method used during development activities to justify the time necessary in secondary drying is generating a desorption isotherm. Examples, such as the sorption isotherms for polyvinylpyrrolidone have been presented by MacKenzie (28). Methods for conducting such studies have also been more recently described by Teng et al. (29).

34.9 FINISHED PRODUCT ATTRIBUTES

There are unique dried material quality attributes associated with lyophilized materials. The term dried material is used loosely here and meant to encompass both lyophilized drug substances and intermediates, as well as drug products intended for administration. Quality attributes are nearly identical for each type of material. Stringent microbiological quality is also a requirement for sterile products. In addition to chemical or biological assay and specific requirements for a finished product, such as those for parenteral administration, the condition of the dried cake also needs to be identified. These include the physical condition and appearance of the dried cake and the ease with which the dried material reconstitutes into a solution.

The result of a successful and effective freeze-drying process is the retention of the physiochemical attributes of the starting solution and preferably, retention of the structure established during freezing. Assay of the constituted solution assures the preservation of the desired activity present in starting material. Assay of multiple samples of dried material is used to demonstrate content uniformity.

34.9.1 Physical Appearances

The appearance of the dried material should be uniform in structure, color and texture. A material having ideal pharmaceutical elegance would be a dense white cake with fine, uniform structure as illustrated in Figure 34.3. As described earlier, successful freeze-drying results in the retention of the structure established during the freezing step. If the material forms a desired appearance upon freezing and that structure is retained throughout the drying, then the process should yield a finished product with an appealing appearance.

For some formulations, and typically particularly those with low solids content, the dried cake may shrink from the original volume upon drying, as evidenced in the sample in Figure 34.4A. Such shrinkage is dependent on the concentration of the starting solution, nature of the active ingredient and the amount and type of excipients used. However, the shrinkage is often uniform throughout the container as well as throughout the batch. Although requirements for stabilizing the drug substance and route of administration supersede pharmaceutical elegance, the design of an ideal formulation would yield a dense cake, uniform in color and texture, with good physical strength and friability (30).

A decrease in total volume or localized loss of structure can also be associated with a condition referred to as collapse. This condition, as described by MacKenzie (31), occurs when the frozen or partially dried material exceeds the phase transition where the material may soften and again become fluid. Samples of dried product in Figure 34.4B and C illustrate the different appearance of the cake structure because of extensive collapse.

With the material softening and becoming fluid, there is a loss of desired structure established during freezing. A loss of structure often coincides with entrapment of water as the material loses its structure and collapses. This entrapment of water into relatively larger masses may also prevent adequate

FIGURE 34.3 A cake that is uniform in appearance, texture and color, occupying the original volume of the liquid fill epitomizes a pharmaceutical elegance for a lyophilized product.

Source: Leonard Amico.

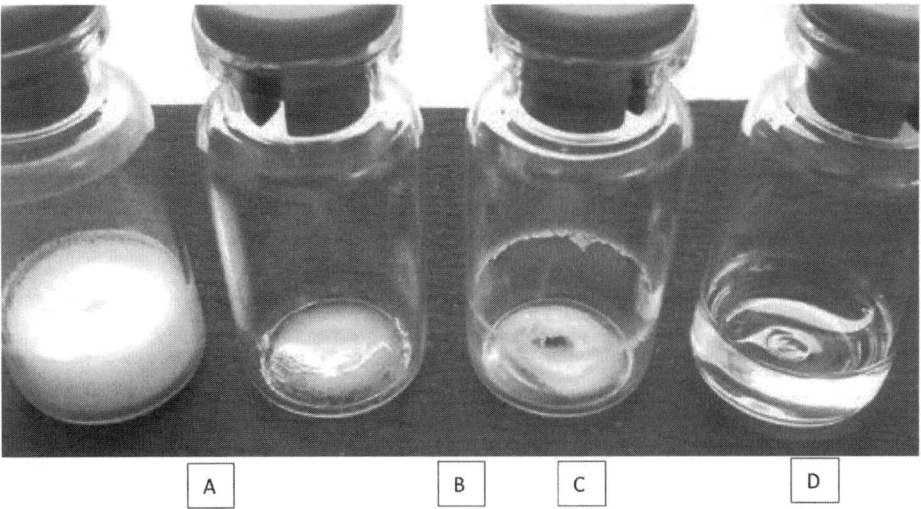

FIGURE 34.4 The slight gap between the dried cake and the side wall of the vial (A) exemplifies shrinkage that may occur with some formulations. This shrinkage may be attributed to either low concentrations or be characteristic of the materials in the formulation. Loss of the initial structure during drying because of the product being warmer than the phase transition temperature yields a varying amount of collapse: Extensive collapse with minimal similarity of color and a dimensional proportion to the original cake (B and C). Failure to dry altogether caused by misplaced closures during the filling operation (D).

Source: Leonard Amico.

desorption that normally occurs during secondary drying, resulting in a high residual moisture content. Reconstitution times may also be lengthened because of a "case hardening," making it more difficult for water to permeate the dried material upon reconstitution. In some cases, poor vapor transport because of seated closures can prevent drying altogether as illustrated in Figure 34.4D. Because the objective of this process is the preservation of the lyophilized material, the presence of collapsed material is suspect. As described previously, collapse may simply be considered a cosmetic defect. When the collapsed material exhibits an increased reconstitution time or poor dissolution, the presence of collapse is categorized as a quality concern. If the collapsed material retains a higher amount of residual water, where this water becomes involved in degradation of the product through hydrolysis, there is a more serious concern. The presence of a significant amount of residual water in localized regions of collapsed material may promote degradation of the product. There would also be a concern regarding the degradation products toxicity or any influence on the therapeutic effectiveness. Both potential results should be considered during product development.

Additional defects related to physical appearance may arise because of container filling and/or handling. The severity of these defects must be evaluated on a case-by-case basis. For instance, dried product in the shoulder of the vial may meet residual moisture and analytical requirements but may be indicative of product dripping from a filling nozzle. Residual dried product on the vial sealing surface would have long-term stability impact because of leakage of air into the vial during storage.

34.9.2 Residual Moisture

The primary objective for lyophilized materials is to minimize or eliminate water that would be chemically active during long-term storage of the product. Any water readily available may become involved in hydrolysis reactions, the common cause of degradation for lyophilized products. This therefore requires that a sufficiently low residual moisture content be achieved. An acceptable range of moisture content, identified during development, is a primary indication that the lyophilization process was successfully executed.

The established residual moisture suitable for acceptable long-term stability may approach the variability of the moisture determination method or may be as great as a few percent. For example, many lyophilized products within the USP have a finished product residual moisture specification of <2% of dry weight. Other products, such as Amphotericin B have a residual moisture limit of 8.0% (32). Whether the allowable residual moisture specification is small or large, a range of acceptable residual moistures should be identified and correlated with suitable long-term stability.

The analytical method for moisture determination must be validated prior to use during any process validation studies. There are numerous techniques for moisture analysis that range from physical methods such as loss on drying to chemical methods such as Karl Fisher. The most common and preferred method is extraction using anhydrous methanol titrated with a colorimetric Karl Fischer method. A comparative review of the conventional techniques is presented in an overview by May (33). The use of near-infrared methods has recently been introduced as an alternate method (34).

34.9.3 Residual Solvents

There is an increasing use of combinations of mixed solvents; i.e., aqueous and organic solvents (35–37). Residual moisture of the dried material assesses the successful completion of secondary drying. With the use of combinations of water and organic solvents, the moisture of the dried product would be assessed upon completion of lyophilization. Residual solvents should also be assessed as a measure of effective decrease to acceptable low levels. Various methods are available for quantifying residual solvent content, such as thermogravimetric analysis or headspace analysis by gas chromatography (GC), as outlined in the USP (31).

34.9.4 Reconstitution

The times required for reconstitution and the appearance of the constituted solution are also of importance. The nature of the dried material as a result of lyophilization yields a product that is highly hygroscopic. Reconstitution is often instantaneous upon adding the diluent. For ease of use in a clinical setting, reconstitution times are often less than two minutes. Whatever time is required to resolubilize that material, the constituted solution should be clear and free of any visible particulates or insoluble materials, meeting the compendial requirements such as those outlined within the USP (31).

The method of reconstitution is also important. For example, the package insert for lyophilized somatropin indicates that during aspiration, the diluent stream should be aimed against the side of the vial. In addition, the constituted solution should be gently swirled and not shaken (37). Vigorous motion could result in aggregation of the protein to form insoluble particles. For Amphotericin B, vigorous shaking is indicated until all of the crystalline material dissolves, forming a clear, yellow-colored colloidal dispersion (37). Whether the solution dissolves instantaneously or requires special handling, forming a colorless solution or a colored colloidal dispersion, the expected appearance of the constituted solution needs to be a quality attribute established and supported by development data.

34.10 ASSAY

Analysis of the active ingredient, whether by chemical or biological methods, would be the same for the constituted product as would be necessary for any ready-to-use preparation. Constituted solutions, however, have a limited shelf life after addition of the diluent. Depending on the solution stability, the package insert may indicate that the constituted solution be used immediately upon reconstitution or may be stored at selected conditions for a specified length of time. The stability of the constituted solution needs to be established during development and measured as part of the stability testing. The potency and purity must also be measured at the end of the indicated shelf life. This includes not only the solution after initial reconstitution but also after storage at the conditions indicated in the package insert. Analysis should also include assay of any degradation product.

34.11 CONCLUSION

Lyophilization is a complex unit operation, integrating multiple processing steps with varied conditions for long-term preservation of the pharmaceutical products. This same process is applied to processing drug substance as well as a compound formulation for a finished drug product.

Lyophilization processes consist of the manipulation of process parameters to create environmental conditions of sub-ambient temperatures and sub-atmospheric pressures. These extraordinary environmental conditions that promote the various conditions suitable for the respective processing mechanisms are created by the lyophilization equipment. The success of the process therefore relies heavily upon the operating performance of the lyophilizer. Confidence in the ability of the equipment to implement these processing parameters is achieved through the successful completion of a comprehensive IQ and OQ. Without the proper performance of the equipment, there is limited opportunity for successful processing of materials.

Throughout this chapter, emphasis is placed upon the need to develop an appropriate and adequate process. This includes challenging the process to develop a set of boundary conditions to create a PAR for the process. The result of such an approach is a rugged and robust process, yielding cycle conditions that are safe, effective and preferably, efficient. These processing conditions are demonstrated to be adequate and appropriate, ultimately through initial testing as well as after long-term storage of finished product. Of equal importance, this process is applied to preserve the quality of the material through processing and throughout the shelf life. Demonstrating the process suitability also requires correlating the process with product stability. The behavior of the material during the processes is strongly dependent upon the characteristics of the starting materials. The initial characteristics must also then be measured and quantified as well. This includes not only the quality of the starting raw material but also in the preparation and packaging prior to placing the product into the lyophilizer. Finally, how the characteristics and quality of the finished product is quantified is of equal importance. This includes the physical attributes of the dried material as well as the quality upon reconstitution. The level of quality must extend beyond the time of the initial testing for release of the batch to the final expiry date.

ACKNOWLEDGEMENT

This chapter is an update to a chapter previously written by Mr. Edward H. Trappler, of Lyophilization Technology. His organization and content were largely left unchanged in the update of this chapter. Many thanks to Ed for his excellent work here and his many contributions to the advancement of lyophilization technology over the years.

REFERENCES

1. Chapman K. The PAR approach to process validation. *Pharm Technol* 1984, 4(12), 47–54.
2. GAMP 5: GAMP Guide for Validation of Automated Systems (A Risk Based Approach to Compliant GxP Computerized Systems), ISPE March 2008.
3. Dushman S, Lafferty JM. *Scientific Foundations of Vacuum Technique*. New York: Wiley, 1962.
4. Ryans JL, Roper DL. *Process Vacuum System Design & Operation*. New York: McGraw-Hill, 1986.
5. Jennings TA. A model for the effect of real leaks on the transport of microorganisms into a vacuum freeze-dryer. *J Parenter Sci Technol* 1990, 44(1), 22–25.
6. The Parenteral Society. *Technical Monograph No. 7: Leak Testing of Freeze Dryers*. Wiltshire: Parenteral Society, 1995.
7. Greiff D. Factors affecting the statistical parameters and patterns of distribution of residual moisture in arrays of samples following lyophilization. *J Parenter Sci Technol* 1990, 44(3), 119–128.
8. Kobayashi M, Harasgunma K, Sunama R, Yao A. *Inter-vial variance of the sublimation rate in shelf freeze dryer*, Presented at the XVIII International Congress of Refrigeration, 1991.
9. Scheaffer G, Sum L, Trappler E. *Techniques in Demonstrating Batch Uniformity for Lyophilized Products*, Presented at the Annual Meeting of the PDA. Boston, MA, November, 1995.
10. Pikal MJ, Shah S, Senior D, Lang JE. Physical chemistry of freeze drying: measurement of sublimation rates by a microbalance technique. *J Pharm Sci* 1983, 72(6), 635–650.
11. Day L. Influence of Vial Construction and Material on Uniformity of Product Temperature during Freezing and Freeze Drying of Model Product Formulations. Presented at the Annual Meeting of the PDA. Boston, MA, 1995.
12. Pikal MJ, Roy ML, Shah S. Mass and heat transfer in vial freeze-drying of pharmaceuticals: Role of the vial. *J Pharm Sci* 1984, 77(9), 1224–1237.
13. Ryan R, Trappler E. Course notes, Fundamentals of Lyophilization, PDA.
14. FDA. Proposed rules. *Fed Regist* 1996, 61, 87.
15. Pikal MJ, Lukes AL, Lang LE, Gains K. Quantative crystallinity determinations for beta-lactam antibiotics by solution calorimetery: Correlation with stability. *J Pharm Sci* 1978, 67, 767.
16. Penner G, Trappler E. *Effect of pH on the Phase Transition Temperature of Model Excipients*, Presented at the Annual Meeting of the PDA. Philadelphia, PA, November, 1990.
17. Korey DJ, Schwartz JB. Effects of excipients on the crystallization of pharmaceutical compounds during lyophilization. *J Parenter Sci Technol* 1989, 43(2), 80–83.
18. Manning M, Patel K, Borchardt RT. Stability of protein pharmaceuticals. *Pharm Res* 1989, 6(11), 903–918.
19. Dubost D, Kaufman M, Zimmerman J, Bogusky MJ, Coddington AB, Pitzenberger SM. Characterization of a solid-state reaction product from a lyophilized formulation of a cyclic heptapeptide: A novel example of an excipient induced oxidation. *Pharm Res* 1996, 13(12), 1811–1814.
20. Orberholtzer ER, Brenner GS. Cefoxitin sodium: Solution and solid-state chemical stability studies. *J Pharm Sci* 1979, 68, 863–866.
21. Pikal MJ, Dillerman KM, Roy ML, Riggin RM. The effects of formulation variables on the stability of freeze-dried human growth hormone. *Pharm Res* 1991, 8(4), 427–436.
22. Her L, Deras M, Nail S. Electrolyte-induced changes in glass transition temperatures of freeze-concentrated solutes. *Pharm Res* 1995, 12(5), 768–772.
23. Brittain HG, Bogdanowhich SJ, Bugay DE, DeVincentis J, Lewen G, Mewman AW. Physical characterisation of pharmaceutical solids. *Pharm Res* 1991, 8(8), 963–973.
24. DeLuca PP, Boylan JC. Formulations of small volume parenterals. In Avis KE, Lachman L, Liberman HA, eds. *Pharmaceutical Dosage Forms: Parenterial Medications*. Vol. 1. New York: Marcel Dekker, Inc., 1984: 195.
25. Akers MJ, Milton N, Bryn SR, Nail SL. Glycine crystallization during freezing: The effects of salt form, pH, and ionic strength. *Pharm Res* 1995, 12(10), 1457–1461.
26. MacKenzie AP. Non-equilibrium freezing behavior of aqueous systems. *Philos Trans R Soc Lond B* 1977, 278, 167–189.
27. DeLuca PP, Lachman L. Lyophilization of pharmaceuticals I, effect of certain physical—chemical properties. *J Pharm Sci* 1965, 54(4), 617–624.
28. MacKenzie AP, Rasmussen DH. Interactions in the water-polyvinylpyrrolidone system at low temperatures. In Jellinek HHG, ed. *Water Structure at the Water- Polymer Interface*. New York: Plenum Publishing Corp., 1972, 146–172.
29. Teng CD, Zarrintan MH, Groves MJ. Water vapor adsorption and desorption isotherms of biologically active proteins. *Pharm Res* 1992, 8(2), 191.
30. Bashir JA, Avis KE. Evaluation of excipients in freeze-dried products for injection. *Bull Parenter Drug Assoc* 1973, 27(2), 68–83.
31. MacKenzie AP. Collapse during freeze drying-qualitative and quantative aspects. In Goldblith WA, Rey L, Rothmayer WW, eds. *Freeze Drying and Advanced Food Technology*. New York: Academic Press, 1975, 278–307.
32. USP 43/NF 38. Rockville, MD: United States Pharmacopieal Convention, 2019.
33. May JC, Grimm E, Wheeler RM, West J. Determination of residual moisture in freeze-dried viral vaccines: Karl Fischer, gravometric and thermogravimetric methodologies. *J Bio Stand* 1982, 10, 249–259.
34. Last IR, Prebble KA. Suitability of near-infrared methods for the determination of moisture in a freeze-dried injection product containing different amounts of the active ingredient. *J Pharm Biomed Anal* 1993, 11(11/12), 1071–1076.
35. Kasraian K, DeLuca PP. Thermal analysis of the tertiary butyl alcohol-water system and its implications on freeze drying. *Pharm Res* 1995, 12(4), 491–495.
36. Teagarden D, Baker D. Practical aspects of lyophilization using non-aqueous co-solvent systems. *Eur J Pharm Sci* 2002, 15, 115–133.
37. Den Brok M, Nuijen B, Kettenes-Van den Bosch J, et al. Pharmaceutical development of a parenteral lyophilised dosage form for the novel anticancer agent C1311. *J Pharm Sci Technol* 2005, 59(5), 285–297.
38. Trissel LA. *Handbook on Injectable Drugs*. 8th ed. Bethesda, MD: American Society of Hospital Pharmacists, Inc., 1994.
39. FDA. Data Integrity and Compliance with Drug CGMP Questions and Answers Guidance for Industry. Pharmaceutical Quality/Manufacturing Standards (CGMP). December 2018.

ABBREVIATIONS

Abbreviations used in this chapter:

ALUS, automated loading/unloading system

API, active pharmaceutical ingredient
CIP, clean in place
DQ, design qualification
EQ, equipment qualification
FAT, factory acceptance test
FDA, Food and Drug Administration
FIT, filter integrity test
GC, gas chromatography
IQ, installation qualification
OQ, operational qualification
P and ID, process and instrumentation diagrams
PAR, proven acceptable range
PDA, Parenteral Drug Association
PLC, programmable logic controller
PVP, process validation package
SAT, site acceptance test
SIP, sterilization in place
SOP, standard operating procedure
USP, *United States Pharmacopeia*.

35 Validation of Primary Packaging, Inspection and Secondary Packaging Processes

Charles Levine

CONTENTS

Packaging processes for sterile parenteral and ophthalmic products can be organized into the steps required to produce the primary package (sealed dosage unit) and the steps required to produce the secondary (labeled dosage unit) and tertiary package (final package unit, including leaflet and individual carton).

The primary packaging processes include the steps necessary to create a dosage unit with the intended critical quality attributes (CQA), assuming that the bulk drug product (suspension, emulsion, solution or powder) delivered to the packaging equipment is homogeneous. This chapter will focus on the CQAs controlled by the packaging processes for sterile products except for microbiological and biological CQAs, sterility and the absence of excessive levels of bacterial endotoxin, respectively. Some of the validation approaches presented in this chapter may be applicable to non-sterile dosage forms. Regardless of the sterile product dosage form, the CQAs are the same for all container types and packaging processes, with some slight variation depending on the patient indication.

A brief discussion of the process development and validation concepts for suspension packaging are also included. For the purposes of this chapter, no differentiation is made between single-dose and multidose products.

Each dosage unit (primary package) should possess the following CQAs:

- strength (labeled quantity of the active pharmaceutical ingredient).
- integral container, sufficient to maintain the CQAs (including microbiological) throughout the shelf life of the product.
- "essentially free" of foreign particulate matter.
- inert headspace, where required for product stability.
- functionality (prefilled syringes and cartridges).
- absence of cosmetic defects.[1]

The secondary and tertiary packaged units should possess all the following CQAs:

- labeled with the correct primary package label.
- inclusion of the correct patient insert.
- accurately coded with the lot number and expiration date.
- absence of cosmetic defects.
- application of serialization code or device to permit track and trace capabilities.

DOI: 10.1201/9781003163138-35

To consistently produce finished drug products that possess all the CQAs, packaging processes should operate within the Proven Acceptable Ranges (PAR)[2] for the established Critical Process Parameters (CPP), using equipment that has been appropriately qualified and maintained. Properly designed preventive maintenance programs are a critical aspect because of the many electromechanical systems used in packaging equipment. An assessment of product/component abuse is also a significant aspect for the qualification of automatic packaging equipment. Product/component abuse is considered any damage that results in units that may not be integral or units with cosmetic defects. This is typically evaluated over predefined time intervals during the factory acceptance test (FAT), site acceptance test (SAT) and operational qualification (OQ). The assessment of product/component abuse should be performed for each package size and configuration.

A detailed risk analysis is the basis for a well-designed validation and process control plan. Failure mode effects analysis (FMEA) assesses risk by ranking three characteristics of a defect: the severity of the defect from the patient's perspective (severity), the frequency of the defect occurring (probability) and the ability of the process to detect the defect (detectability). Understanding the modes of potential failure shapes the design of the qualification program. The overall effect of process validation is to reduce the probability that the defect may occur. This chapter provides an analysis of the risks and validation approach for each of the aforementioned packaging processes.

35.1 PRODUCT STRENGTH

Sterile products are available in a wide variety of package types and sizes, ranging from less than 1 mg of the active ingredient in a glass vial to 1 L bags of intravenous fluids. The production equipment and monitoring systems used to produce these products also exhibit widely different approaches. However, the approach to the manufacture and control of the product strength for sterile products relies on some basic concepts.

The labeled strength of the dosage unit is a CQA, which is well defined in USP General Chapter <697> Container Content for Injections:[3] "each container of an injection contains sufficient excess to allow withdrawal of the labeled quantity of drug. Such withdrawal shall be performed according to labeled directions, if provided." USP General Chapter Pharmaceutical Dosage Forms <1151>, Excessive Volume in Injections,[4] provides the recommended excess volume to ensure that the labeled quantity of a drug can be withdrawn from the container when using proper technique. See Table 35.1:

TABLE 35.1
Recommended Excess Volume

Labeled size (mL)	For mobile liquids (mL)	For viscous liquids (mL)
0.5	0.10	0.12
1.0	0.10	0.15
2.0	0.15	0.25
5.0	0.30	0.50
10.0	0.50	0.70
20.0	0.60	0.90
30.0	0.80	1.20
50.0 or more	2%	3%

In the FDA Guidance for Industry, Allowable Excess Volume and Labeled Vial Fill Size in Injectable Drug and Biological Products, June 2015,[5] the FDA states that excess volume should not be confused with an overage and cautions manufacturers that exceeding the recommendations for excess volume for single-dose containers may result in misuse of the product.

With all sterile products there are two process variables that must be controlled in order to comply with the labeled strength of the product: assay or concentration of the active pharmaceutical ingredient in the bulk drug product delivered to the filling equipment and the amount of the drug product filled into the dosage unit. This chapter will focus primarily on the control of the amount of drug product filled into the dosage unit. In addition, the maintenance of homogeneity when filling suspensions will be presented later in this section.

Liquid filling equipment systems for the manufacture of sterile products are diverse in their capacity, speed, precision and functionality. There are a number of concepts for delivering the proper amount of drug product into each container, and a partial list includes time pressure, rolling diaphragm pump, servo-controlled peristaltic pump, positive displacement pumps and gravimetric.

A popular design for powder filling equipment consists of a rotating dosing wheel at the bottom of a powder delivery hopper that utilizes vacuum and compressed air to accurately fill each cell within the dosing wheel and eject the contents of each cell into the individual vials. These systems use filters and vibratory mechanisms to maintain vacuum levels. The compressed air pressure, vacuum level and machine speed are process parameters to be considered during the validation.

Many equipment manufacturers continue to adapt the product delivery technology to the requirements for aseptic processing to create systems that can be cleaned and sterilized in place (CIP/SIP). Presterilized single-use disposable systems are another technology that eliminates all of the equipment preparation. Process development for the use of single-use systems requires specific capability studies. Peristaltic pump systems that are capable of precision filling at reasonable operating speeds, approximating 6,000 units per hour, are frequently used for large molecule active pharmaceutical ingredients.

Gravimetric systems are generally used for products in limited supply and high value, where one of the objectives is to minimize product loss. These systems may also be used for high potency single-dose products that require minimal dose variation. Gravimetric systems do not operate at the high speeds currently reached by the large-scale automated filling lines.

The gravimetric systems weigh each container before and after the filling station. Through computer control and servomotors, the systems can achieve great accuracies. The potential risks to be controlled for gravimetric systems include vibration, level (orientation of the weighing system) and airflow velocity, as each of these factors could affect the weighing accuracy.

The process parameters directly affecting these conditions include filling speed and air velocity uniformity. The acceptable range for the filling speed should be established during the FAT and confirmed during the SAT/OQ.

The air velocity uniformity must be established to provide unidirectional airflow, a requirement for Grade A/ISO 5 environmental classification. Once these velocities are established, the accuracy of the gravimetric product delivery system can be confirmed. Typically, the validation requires fill-weight verification performed using manual weighing techniques (not within the LAF).

Where in-process fill-weight confirmation must be performed manually, the design of the manual weighing procedure should consider the potential risk for contamination from the operator and the potential for error.

The weighing procedure representing the lowest risk for operator contamination is to remove the filled container after filling, weigh the filled container, completely empty the contents and weigh the empty container.

Because the complete removal of powders is difficult, the procedure for handling powder-filled containers requires weighing the container before and after filling to obtain an accurate fill weight.

Where tolerances permit averaging, the empty weight of the container could be used, so that only the filled container is weighed during the filling operation. The observed weight variation for some types of containers are more suitable to the averaging approach. Ampoules and tubing vials would be more likely to be candidates for this approach than molded vials.

While not the primary focus of this chapter, the selection of the proper filling system for each product type is necessary to ensure that the equipment is neither additive nor reactive. Biological drug products and large molecule drug products may be susceptible to sheer and are better suited to servo-controlled peristaltic pump systems or time-pressure filling systems where there are no moving parts in direct contact with the drug product solution.

The filling system must be designed to deliver the drug product with the required precision at the specified operating speeds, which requires proper sizing of the product delivery system and the product dosing system. All product contact materials must be compatible with the drug product solution, and in the case of servo-controlled peristaltic pump systems, the hardness of the tubing must be specified to ensure the precision of the system.

Many of these products are filled in small volumes, less than 1.0 mL per container, which places greater demands on the stability of the filling system to maintain accurate fill volumes. A paper in *Pharmaceutical Engineering* in 2007,[6] authored by Bosch Packaging and Technology, demonstrated a downward trend over a 100 minute test period when filling at a target of 0.7 mL. The test utilized silicone tubing with a wall thickness of 2.4 mm and an inner diameter of 0.5 mm. The average fill volume drifted from approximately 0.70 mL to less than 0.69 mL. The results of this paper emphasized the need to evaluate the physical stability of the tubing used in peristaltic systems.

The design of the filling needles may also affect the precision and the frequency of cosmetic defects if the surface tension and other physical characteristics of the drug product solution are not taken into consideration. If not properly designed, the drug product solution may drip from the needle after the valve is closed. Also, the filling of highly concentrated or viscous formulations, including suspensions, and increase the risk for product drying and needle clogging during extended stoppages of the filling operation. Product drying at the tip of the filling needle is also a function of the air velocity in Grade A environments, where unidirectional air at an average of 0.45 m/s is required.

A study by Y F Maa[7] evaluated the effect of filling needles of various materials of construction on formulation drying at the tip of the filling needles. This study evaluated an mAb formulation at 200 mg/mL with filling needles fabricated with stainless steel, glass, siliconized glass, polypropylene and Teflon coated stainless steel. The general conclusion of the study demonstrated that the propensity for clogging and drying of the product decreases with an increase in hydrophobicity of the materials of construction.

"Suck back" is another design feature that can minimize the presence of hanging drops on the end of the filling needle and thereby minimize the risk of product drying, clogging and the creation of cosmetic defects. "Suck back" is the ability of the dosing system to briefly create a vacuum on the tubing and filling needle downstream of the dosing system. This can be achieved by retraction of the piston or, in the case of rolling diaphragm pumps, application of vacuum on the non-product side of the diaphragm. "Suck back" can also be achieved in peristaltic pumps by reversing the rotation of the cam.

Another critical design factor is the withdrawal of the filling needle from the container during the filling phase to maintain an appropriate distance between the liquid level and the end of the filling needle. This design minimizes the potential for foam and prevents product contact with the exterior of the filling needle.

Generally speaking, the critical process parameters for a filling operation are limited to pressure of the drug product delivery system and filling equipment speed. However, in the case of time pressure filling systems, filtered nitrogen gas is used to enhance the pressure control on the supply vessel integrated into the filling system. The pressure of the nitrogen gas may also be a critical process parameter in this scenario.

Many filling systems are equipped with surge vessels positioned immediately prior to the dosing system. These surge vessels are normally equipped with a level control system to maintain an adequate supply of the bulk drug product to maintain control of the fill volume. The level control system consists of high and low level sensors that actuate the supply valve. The surge vessels may also be equipped with a second low level sensor that will automatically stop the filling operation. The second low level sensor is positioned below the low level sensor that opens the supply valve. Systems of this design do not require the development of end-of-fill procedures.

For less automated filling systems and those without 100% weight verification, end-of-fill procedures (when there is inadequate supply of the bulk drug product to maintain adequate fill level control) must be developed and qualified. The approach to the design of the end-of-fill procedure will depend greatly upon the level of automation. Where the quantity (weight or volume) of the remaining bulk drug product is continually measured by a load cell or level sensor, this process analytical technology (PAT) in conjunction with the filling speed can be used to establish the point at which the filling operation must stop.

These studies are typically designed to evaluate fill weights for individual machine cycles as the amount of available bulk drug product approaches the end point. This data can be used to establish a visual end point or electro-mechanical endpoint. These studies should be performed for each drug product and container size.

Suspensions represent a greater challenge to the design and validation of the filling process. The formulation of the suspension will greatly affect the design of the filling process. Drug products suspensions contain solid phase active pharmaceutical ingredients with a defined particle-size distribution in a solution that may contain thickening agents to maintain the particles in suspension. Generally, suspensions are not formulated to maintain uniformity throughout the shelf life of the product. Parenteral suspensions must be formulated with minimal thickening agents, which contributes to a greater propensity for solid phase settling. As the thickening agents reduce the surface tension, additional precautions must also be taken to minimize the entrainment of air.

Generally, a suspension filling process will include the following steps:

1. Bulk storage with continuous agitation to maintain homogeneity without entrapment of air. Continuous mixing for long periods of time may reduce viscosity and affect particle-size distribution.
2. Transfer to an intermediate recirculation vessel, in close proximity to the filling equipment, which may also be equipped with continuous agitation.
3. Recirculation system, equipped with a peristaltic pump to minimize shear forces, which could affect viscosity and particle-size distribution. The recirculation system must be appropriately sized to support the speed of the filling equipment.
4. Dosing system with a minimal number of mechanical parts to minimize the effect on viscosity and particle-size distribution.

For such a complex filling process, the critical process parameters could include mixing speeds, recirculation flow rates, holding times at various process points and filling-machine speed. In addition to sampling at the beginning, middle and end of the filling operation, the process development should critically evaluate three events; the start of the filling operation, any stoppage of the filling process and the end of the filling operation, when the agitation in the intermediate recirculation vessel must be stopped to avoid entrapment of air in the suspension.

The settling properties of the suspension will affect the stoppage times and the amount of product that can be filled at the end of the filling operation without continuous agitation. At each of the three events, sampling plans should be designed to collect units from each filling needle according to a defined number of cycles. The number of cycles should take into consideration the ratio between the labeled fill volume and the volume of suspension contained within the product delivery tubing between the recirculation vessel and the filling needles. Minimizing the volume of suspension present in the product delivery tubing will reduce the potential for product loss due to lack of homogeneity.

The identity of the filling cycle must be maintained for all units so they may properly be selected for analysis. To minimize the number of samples to be analyzed, one may consider the following approach. One should analyze the units from the last selected filling cycle. If these units comply with the acceptance criteria, one may analyze units from the midpoint between the first and last selected filling cycle. If these units comply with the acceptance criteria, one would analyze units midway between the last acceptable filling cycle and the first filling cycle. Once you have identified a nonconforming cycle, then analyze the cycles before and after the last acceptable cycle, identifying the first three consecutive cycles that meet the acceptance criteria.

Following a stoppage, the dead volume (suspension in the tubing between the recirculation line and the filling needles) may not be homogeneous, requiring the sampling plan to be designed to include the number of filling cycles to flush the dead volume. This exercise should be performed for the maximum stoppage time. The sampling plan could evaluate the beginning, middle and end of the dead volume as well as the first cycles from the suspension in circulation.

The end-of-batch scenario phase begins when the agitation in the recirculation vessel is discontinued. The sampling points during this phase can be established by the amount of suspension remaining in the recirculation vessel. The filling should continue until acceptable fill volumes can no longer be maintained. The analysis of samples should begin with the first cycle after the agitation has been discontinued and continue until nonconforming cycles have been identified. The determination of the end of batch should require multiple consecutive cycles with acceptable assays. Any stoppages during the end-of-batch filling will require additional studies.

Validation will mimic the process development, requiring sampling at the defined acceptable limits for the critical process parameters. Bracketing should be carefully considered. The smallest fill volume (lowest suspension flow rate) would likely be worst case. However, each drug product suspension should be evaluated separately, considering that the settling properties for the solid phase is likely to be different for each drug product suspension.

From the risk analysis perspective, units that do not contain the labeled quantity of the active pharmaceutical ingredient present the greatest potential for harm to the patient. The FMEA table for delivery of the labeled contents is provided in Table 35.2.

In each case, the probability of occurrence could be reduced if the user requirements specification for the filling equipment was properly developed and executed. The detectability is dependent on the in-process controls integrated into the process. Where economics and production scale support 100% automatic weight verification, the detectability is high, as the equipment responds to nonconforming results by machine stoppage or automatically rejecting the nonconforming unit. Some filling systems are equipped with robotic weighing systems to statistically sample and determine individual fill weights during the filling process. The electronic control systems normally have control chart capability.

Otherwise, firms develop periodic sampling programs to verify proper adjustment of the fill weight at the start of the operation and periodically thereafter throughout the filling operation. In many circumstances, the sampling at the start of the filling operation may include more than one unit from each filling needle. Subsequently, a single unit from each filling needle is collected at each sampling interval. The detectability for periodic sampling programs is ranked as medium, depending on the sampling frequency. The sampling frequency should be established based upon the equipment's ability to maintain the fill-weight variation over time. The use of historical data to calculate the process capability will be useful in this exercise.

TABLE 35.2
FMEA Labeled Contents

Defect	Severity	Probability	Detectability
Nonconforming fill weight/volume (clogged filling needles)	High	Low	*Medium* if periodically monitored *High* if 100% of the units are automatically weighed
Product/component abuse resulting in cosmetic defects	Medium	Low	*Medium* if periodically monitored *High* if 100% automatic inspection
Product/component abuse resulting in foreign particulate matter	High	Low	*Medium* if periodically monitored *High* if 100% automatic inspection
Dripping fill needles resulting in cosmetic defects	Low	Low	Medium
Suspension not homogeneous (content uniformity)	High	Low	Low content uniformity required for release testing

The results of the in process fill-weight determination are plotted in a control chart to allow operators to monitor the process and identify potentially adverse trends. There are a wide variety of systems available to capture the raw data from the fill-weight determination and present the data in the control chart form. Ancillary documentation should include indication of any equipment adjustment to maintain fill weights within the specified tolerances.

Statistically, fill-weight data is usually normally distributed, meeting one of the critical assumptions for the use of control charts. Generally, x – R (average – range) control charts are well suited for monitoring fill weights because the number of filling needles is usually less than 10. If more than 10, an x – S (average – standard deviation) control chart could be used.

Typically, fill-weight limits include a target value, limits beyond which adjustment is required (T1) and limits defining unacceptable product (T2). The lower control limit (LCL) for T2 should be equivalent to the label claim plus the USP recommended excess volume. The target fill volume (T) can then be described by the following equation:

$$T = 3\sigma + LCL$$

Where σ is the standard deviation for the normally distributed fill-weight data population.

Some firms may choose to set $T = 4\sigma + LCL$ to eliminate any possibility of nonconforming units as the normal distribution population described by $x - \pm 3\sigma$ accounts for 99.7% of the population, leaving a 0.15% probability of unit below label claim. For single-dose dosage forms, the use of $T = 4\sigma + LCL$ may result in excess drug product if the process variability is excessive.

The validation approach will be dependent on the in-process monitoring system envisioned for routine production. During qualification the accuracy of the 100% weighing systems or robotic sampling systems must be confirmed. This qualification is not a calibration, which must be performed, but a verification that environmental conditions, unidirectional airflow, electrostatic charge and equipment vibration do not create inaccuracies in the weighing system.

The in-process sampling is typically performed more frequently during validation and may also include multiple units per filling needle at each sampling interval. This approach provides confirmation of the variation of each filling needle as well as variation over time.

35.2 INTEGRAL CONTAINER

Because of the numerous container closure systems currently used for drug products, the objective of this section is to inform the reader about the general subject matter of container closure integrity to facilitate an informed application of the principles to ensure adequate control of container closure integrity during the product life cycle.

Container closure integrity is a topic where technological developments are increasingly defining the approaches that could be used during product development, process validation and commercial production and product stability. USP General Chapter <1207> Package Integrity Evaluation of Sterile Products[8] states that the product quality risks of a nonintegral package include entry of microorganisms, escape of products and entry of foreign materials and allowing changes in the gas headspace content.

The types of container closure systems currently used for sterile drug products are varied depending upon product stability and patient requirements. For each type of container closure system we will examine the necessary controls for the primary packaging components, the process parameters associated with the sealing process, process monitoring and, where appropriate, 100% inspection. These elements are considered the primary components of a well-designed quality control plan to ensure container closure integrity.

There are two general types of closure systems, physically mated and physiochemically bonded. The physically mated systems require the close physical mating of two dissimilar materials to create the seal, as is achieved by an elastomeric stopper and a glass serum vial. The physiochemically bonded system include glass ampoules and Blow Fill Seal containers (polyethylene) that are sealed by heating the open end.

Control of primary packaging components begins with a risk-based vendor selection process that results in a detailed specification for the primary packaging component and a quality agreement that defines the responsibilities of the vendor and customer. These agreements normally include a periodic audit of the manufacturing facilities, which can range from on-site visits to paper audits, depending upon the nature of the component and the vendor history.

The component preparation processes, including washing and sterilization, are increasingly performed by the primary packaging component vendors. Shifting processes which directly affect CQAs to the vendors increases the importance of periodic on-site audits.

Control of incoming shipments can include vendor certificates of analysis and, in some cases, statistical reports on critical dimensions. The automation utilized by many glass manufacturers includes the use of in-line and off-line dimension verification systems to generate these statistical reports. Performance history and risk analysis play an important role in the design of the sampling and inspection program for release of incoming shipments. Where vendor audit programs are in place, predelivery or ride-along samples collected by the vendor facilitate the collection of random samples.

A comprehensive specification for the primary packaging component should include the approved component drawing, the material of construction, the method of packaging used during transport to the customer and an acceptance sampling plan that categorizes the potential defects with an appropriate AQL. The level of detail present in the specification should reflect the processing of the component performed by the supplier

TABLE 35.3
Critical Defect List

Defect	Description*
Broken finish	A finish that actually has pieces of glass broken out of it
Chipped finish	Container with a small section or fragment of glass broken out or missing from the finish
Crack	Fracture that penetrates completely through the glass wall
Crizzle	A finish or neck that has a multitude of fine surface marks
Inside overpress	A rim of glass projecting upward from the inside edge of the finish that extends above the top of the sealing surface
Line over finish	A channel that can extend partially or completely across the sealing surface of the finish
Rough finish	A finish that has minute imperfections causing a rough surface
Split finish	An open vertical fracture that penetrates through the finish wall from the top of the finish downward toward the neck and from the internal diameter to the outer diameter
Malformed finish	Finish profile is incomplete
Warped finish	The top or sealing surface of the finish is not even

* Extracted from PDA Technical Report 43 Identification and Classification of Nonconformities in Molded and Tubular Glass Containers for Pharmaceutical Manufacturing:[9] covering ampoules, bottles, cartridges syringes and vials

Defects that may result in a nonintegral unit are classified as critical defects, which should have an AQL approaching 0.0%. Some of the critical defects for vials that may result in nonintegral units are described in the Table 35.3:

In some cases, firms also perform verification of critical dimensions in accordance with sampling procedures for inspection by variables, using the current version of ANSI/ASQ Z1.4.[10] The value of this verification is dependent upon the vendors' performance history and the scope and reliability of the statistical data that they provide with each shipment.

35.3 SEALING PROCESSES

35.3.1 Glass Vials with Elastomeric Stoppers and Aluminum Seals

The most widely used container closure system for parenteral products, serum glass vials (molded or tubing), form a seal with a plug elastomeric stopper, which is created when an aluminum seal compresses the stopper against the finish of the mouth of the vial. There are three sealing surfaces shown in Figure 35.1: (1) valve seal, (2) transition seal and (3) land seal.

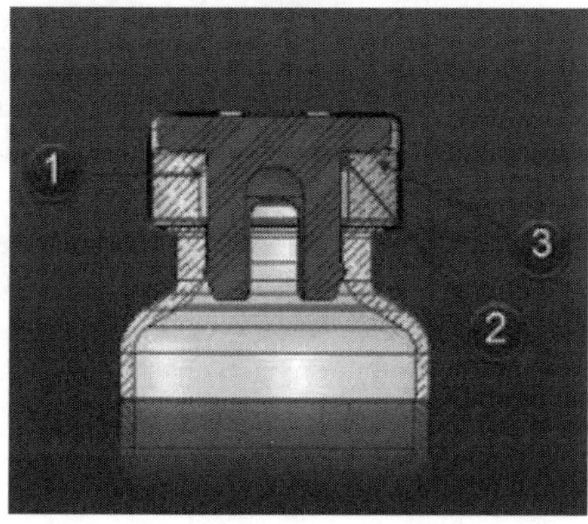

FIGURE 35.1 Plug Stopper Sealing Surfaces

TABLE 35.4
Mean Residual Seal Force Versus Capping Plate Plunger Distance

Vial	Capping plate plunger distance (mm)	Mean residual seal force (N)
6 mL	7.7	31
6 mL	8.1	49
6 mL	8.4	71
15 mL	7.7	52
15 mL	8.1	74

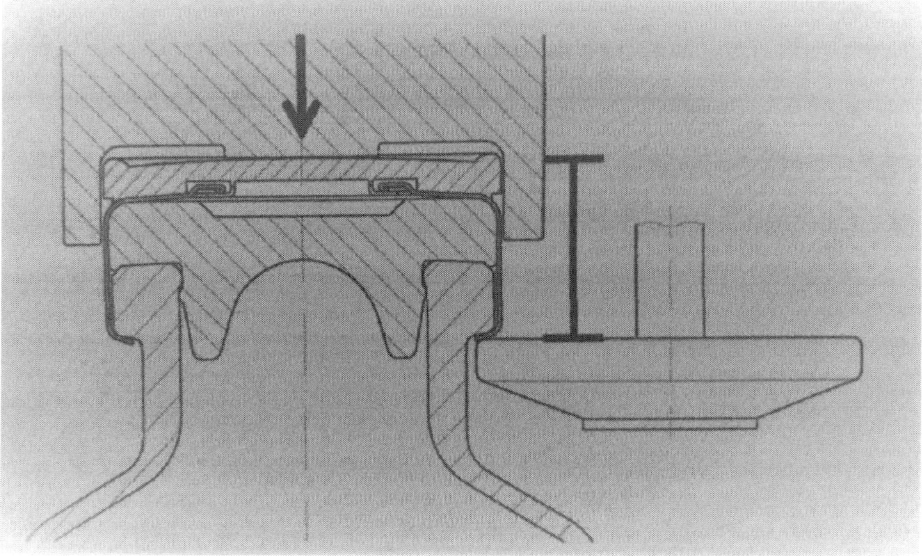

FIGURE 35.2 The vial capping process: Arrow = pre-compression force, caliper = distance capping plate-plunger.

Manufacturing capping equipment are designed to use one of the following options:

- spinning roller design, where the cap is compressed onto the stopper and spinning rollers rotate around the vial.
- sealing rail, where the cap is compressed onto the stopper and the vial is rotated against a fixed sealing rail.
- rotary capping plate, where the cap is compressed onto the stopper and a free turning rotary plate rolls the aluminum skirt below the lip of the vial finish.

Regardless of the machine technology used, when the aluminum skirt is properly folded under the lip of the vial, the flange of the stopper is compressed, creating the land seal. This compression force is referred to as the Residual Seal Force. An excellent study performed by Mathaes[11] et al. examined the effects of crimping equipment, machine speed and the capping plate plunger distance on the residual seal force. In Figure 35.2 the capping plate to plunger distance caliper is shown to the right of the vial, and the downward arrow represents the compression force.

While there were some differences among crimping equipment, the capping plate plunger distance was the only variable to have a significant effect on the residual seal force. Using test groups of 30 vials, the average residual seal force, measured in newtons, is presented in Table 35.4:

The study that Mathaes performed further evaluated the vials by performing a helium leak test on all of the vials. Regardless of the capping plate plunger distance, all vials demonstrated a helium leak rate equivalent to 10^{-8} mbar L/s.

The validation approach for a capping process depends upon the degree of PAT applied to the process. Systems with residual seal force monitors should require correlation of the minimum residual seal force to the appropriate container closure integrity test. The minimum residual seal force should be sufficient to provide a seal that meets all cosmetic requirements.

The validation of the capping process should confirm the proven acceptable ranges for the critical process parameters, machine speed and capping plate plunger distance. For capping systems equipped with residual seal force monitors, the values of the residual seal force can serve as the CQA, provided correlation to a validated leak test has been established. Further confirmation of acceptable units is achieved through the use of automated inspection systems that use vision system technology to verify that the aluminum seals are properly positioned and crimped on the vials and the cartridges.

Vision system (VS) technology at its most basic is the comparison of an electronic image of the container to an image of the standard container. The engineering to develop a VS that reliably rejects all defective containers without unacceptable levels of false rejects requires detailed definitions of an acceptable unit and all defect types. The electronic image can be considered a number of squares on a graph, where each square has a different of level on a gray scale. These squares are referred to as pixels.

The user requirement specification (URS) for the VS begins with the identification of the defects to be detected by the VS and the selection of limit samples. In doing so, one should recognize, as with particulate inspection, that VS is more reliable than humans.[12] If the inspection windows (the portion of the electronic image to be compared) are not properly created, false rejects and missed defects will be excessive.

The elements necessary to create a well-designed VS include the container handling system, the camera, the lighting approach and the computer system.

The handling system should be designed to ensure that every container is presented to the camera in the same position with minimal vibration. Minor variations in the position can be addressed by the computer system, as computer systems are normally programmed to adjust the position of the inspection windows or tools based on a reference point within the image.

The camera typically has four variables to be adjusted to ensure that the captured image maximizes the difference between acceptable and defective containers. These include f-stop, contrast, gain and shutter speed. The f-stop controls the amount of light that is received by the camera and therefore the amount of data used to generate the image.

The contrast affects the ability of the VS to define edges, and too much or too little contrast will adversely affect the ability to define the edges of containers. The gain is simply a brightness multiplier, which multiplies the level of brightness received by the camera as it is sent to the image processor. In a sense it improves on the image obtained by the actual lighting.

The shutter speed controls the amount of time that the camera's shutter is open and thus the amount of light. If the shutter is open too long the image could be blurred, and if the shutter speed is too short the image may be dark. Finding the proper adjustment is critical to obtaining an image that can reliably be compared to the standard.

Factory acceptance trials play a vital role in the development of the lighting configurations, camera settings and inspection tools. These settings can then be evaluated at a range of machine speeds to establish the optimal speed. However, factory acceptance trials (FAT) and site acceptance trials (SAT) and equipment qualification rarely consider all of the variables that arise during routine production activities.

One should be prepared to implement minor changes to the inspection tools, lighting and camera settings during initial activities. These changes should be managed through change control and qualified appropriately.

Container closure components provided by different suppliers may not be equivalent when inspected by a VS, which may require the creation of separate inspection programs in the computer system, as the inspection tools may require some adjustment.

The proposed EU Annex 1, published 20 December 2017,[13] makes the following statement about verification of container closure integrity during the manufacture of vials:

> Samples of other containers should be checked for integrity utilizing validated methods and in accordance with QRM, the frequency of testing should be based on the knowledge and experience of the container and closure systems being used. A statistically valid sampling plan should be utilized. It should be noted that visual inspection alone is not considered as an acceptable integrity test method.

This is a new requirement that will require the development of nondestructive tests that can be executed in an efficient manner or the validation of in-line systems that measure a parameter such as residual seal force. The residual seal force in this application could be considered to be a PAT. Other inline systems, such as vacuum decay, can be used for lyophilized products that are sealed under vacuum.

In 2015, the BioPhorum Operations Group[14] published an editorial to examine the requirement to perform 100% container closure integrity testing during routine manufacturing. The paper provides a summary of the manufacturing process controls similar to those previously described in this chapter and recommends that each manufacturer conduct a risk analysis for the existing manufacturing controls to determine if 100% container closure integrity testing is appropriate for their manufacturing process.

In theory, correlation of in-line PAT systems to a qualified and recognized container closure integrity test could satisfy this requirement.

The EU Annex 1, published 20 December 2017,[13] also clarifies the requirements regarding capping operations performed on stoppered vials outside of the aseptic area. In addition to the requirement to supply Grade A quality air to the area where the capping is performed, stopper height sensors must be installed and qualified to prevent capping for any vial whose stopper has become displaced. Stoppers can pop up if the silicone levels are not properly controlled and the stopper insertion station is not properly adjusted. However, the use of vials with the blowback feature in the neck will minimize the occurrence of stoppers popping up. The maximum allowable stopper height must be qualified using a container closure integrity test with a sensitivity equivalent to microbial immersion. This test would evaluate the uncapped vial with the stopper positioned at the maximum allowed elevation.

35.3.2 Ampoules

Most automatic ampoule filling/sealing equipment in the pharmaceutical industry utilizes the pull method for sealing the neck of the ampoule. In this technique the ampoule is rotated in front of a flame as the top of the neck is pulled away, which allows the molten glass to collapse and form a seal at the tip of the ampoule. The fuel source for the flame is natural gas and oxygen to ensure consistently high temperatures.

From a process validation viewpoint, the sealing process parameters include machine speed, container rotation speed (which may be integrated with the machine speed) and natural gas and oxygen flow rates. Considering that the integrity of the ampoule seal must be verified for 100% of the sealed units, the process parameters may not be classified as critical, considering that the detectability is high. In a sense, the 100% container closure integrity test is a form of PAT. However, these process parameters require validation to achieve acceptable reject rates for nonintegral units and cosmetic defects.

Bracketing is not recommended when qualifying the ampoule sealing process, as the process parameters must be adjusted for the diameter of the neck and the thickness of the glass in that portion of the ampoule.

Poorly designed or controlled filling processes can adversely affect the quality of the seal if product is present at the portion of the neck where the seal is formed.

Ampoules are supplied in two types, "open" and "closed." "Closed" ampoules reduce the capital expenditures required by the pharmaceutical manufacturer because they eliminate the need for ampoule washing and depyrogenation equipment. "Closed" ampoules, whose use is more prevalent in Europe than the United States, have a longer neck, which is sealed during the formation process. These ampoules are formed with a slight overpressure, and the depyrogenation capabilities (>3 log reduction of bacterial endotoxin) of the annealing process (beginning at 600°C) has been validated at several ampoule manufacturers.

The filling equipment for "closed" ampoules includes an opening station, where the neck of the ampoule is cut while the ampoule is in the inverted orientation. Localized suction draws away the particles created during the opening phase.

A study by Bernuzzi in 1993[15] compared the level of particulate matter rejects generated when filling "closed" and "open" ampoules. The study utilized two different automatic inspection systems to perform the particulate matter inspection and demonstrated a lower level of particulate matter rejects using the "open" ampoule process. Some variation among the automatic inspection systems was also observed. The rejection rate for the "open" ampoule process ranged from 0.154% to 1.248%, and the rejection rate for the "closed" ampoule process ranged from 1.434% to 3.86%. The use of "closed" ampoules requires the use of well-qualified automatic inspection systems and is primarily suited for the manufacture of low-cost drug products, where higher reject rates are economically acceptable.

High-voltage leak detection (HVLD) is the most prevalent technology used to perform the 100% leak detection required by the regulatory authorities for container closure systems that are physiochemically bonded. The HVLD equipment positions the ampoule between two electrodes at a high voltage differential. The presence of the sealed container presents a capacitance that minimizes the current flow between the electrodes. When there is a leak in the container and the drug product solution is conductive, the current flow between of the electrodes is significantly greater. The drawing below, originally published in the *PDA Journal of Pharmaceutical Science and Technology*[16] is perhaps the best example of the change in the electrical circuit when there is a leak present.

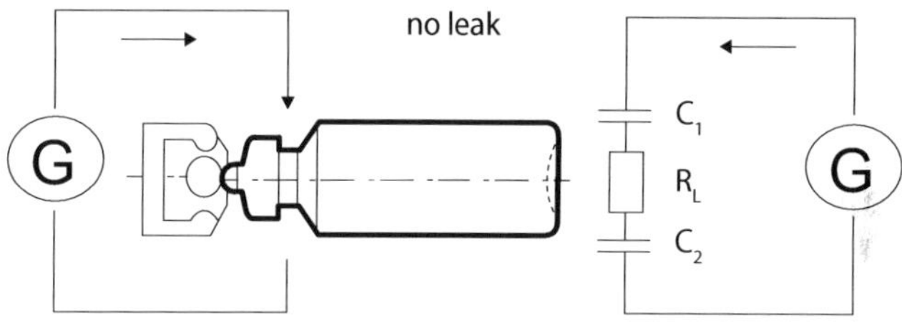

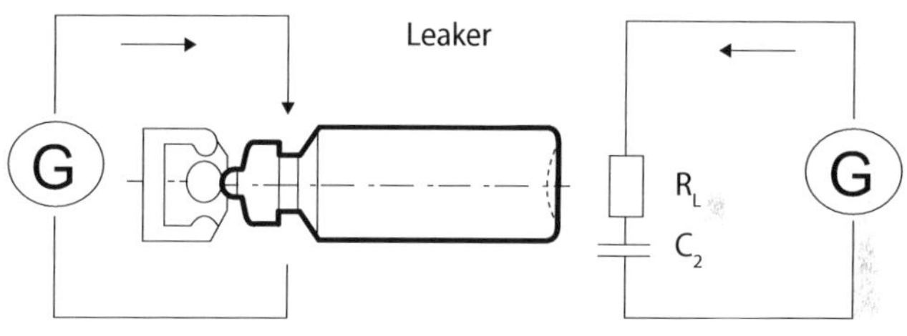

FIGURE 2—Schematic principle of high voltage leak detection G = Generator; C_1 and C_2 = Condenser; R_L = Resistor.

The high-voltage alternating current is applied across the package to ground, and the current is converted into a direct current signal, which is converted into voltage (sensitivity).

The development of the test method first requires identifying the optimum voltage setting, ensuring that the voltage is not excessive, which would create an electrical arc from the electrode to the ground, which bypasses the test package, resulting in a false reject. The method development also includes establishing the minimum voltage limit, to ensure that the test package is properly aligned and there is no system drift, and the establishment of the maximum voltage limit, which is indicative of a leaking unit. Within this range, studies should be performed with various defective units to establish the sensitivity above which a unit is rejected.

The determination of the sensitivity for rejected units should include trials with a sufficient number of units of multiple size leaks. In this manner, regression curves can be established to demonstrate correlation between the size of the leak and the rejection decision by the system.

HVLD systems can also be designed for plastic containers produced using blow fill seal technology, prefilled syringes and plastic laminate bags. The voltage differential for these systems is significantly lower than that for glass ampoules, as plastic represents much lower capacitance.

35.3.3 Cartridges

Cartridges are similar to syringes, as they are composed of a glass barrel that is closed on one end with a plunger/stopper and a serum finish on the other end that is sealed with a lined seal, consisting of an aluminum seal and an elastomeric disk. The end of the cartridge where the plunger is inserted may have a glaze finish to provide a dimensional fit to maintain the plunger/stopper within the cartridge. Cartridges are normally used with syringe devices for the delivery of insulin, biologicals or local anesthetics for use in dental procedures.

The cartridge barrels are prepared for filling by washing and siliconizing (interior surface only) prior to depyrogenation by dry heat. During the aseptic filling operation, the plunger is inserted into the bottom of the cartridge and filled with the drug product solution, and the seal is applied and crimped in a process similar to the application of aluminum seals for vials.

The plunger is usually designed with three rings to create the elastomeric glass seal. With this configuration, misalignment of the plunger may be detected by the presence of the drug product solution between the plunger rings. The rings reduce the contact surface area between the plunger and the glass barrel to minimize friction during sliding.

Some drug products supplied in cartridges are completely filled with drug product solution, which is sometimes referred to as a "no bubble fill." Generally, the "no bubble fill" is required to achieve drug product stability by eliminating headspace oxygen from the sealed container. Producing "no bubble fill" products requires that the plunger placement is properly controlled to ensure compliance with the volume of injection. The filling process for a "no bubble fill" requires that the cartridge is first overfilled, and then the liquid remaining above the glass finish is removed by vacuum immediately prior to the sealing process. If the vacuum process is not properly adjusted, drug product solution may remain on the surface of the cartridge finish, which may contribute to deformation of the elastomeric disk during the sealing process.

The sealing process is not controlled in the same fashion as the vial sealing process and relies more on the physical appearance. The height of the capper head is adjusted for the length of the cartridge, and the spinning roller or sealing rail are adjusted to form the aluminum under the cartridge finish. Because cartridges are sealed with an elastomeric disk and not a plugged stopper, inadequate adjustment and control of the crimping process may result in various critical defects.

Excessive downward pressure, created by setting the crimping head too low, may deform the disk to create a bulging disk. This defect could result in a non-integral unit and additional headspace, which could adversely affect product stability.

If the lined seal is not centered over the cartridge when applied, the seal will be off-center, which may result in a nonintegral unit and a misalignment with the delivery device.

Many cartridge manufacturing processes now include 100% automated inspection using vision system technology. The inspection windows can be designed to inspect the aluminum seal, glass cartridge barrel and the plunger position. The aluminum seal can be inspected for bulging disks, off center application and, where necessary, the color of the aluminum seal. The glass cartridge barrels can be inspected for cracks and cosmetic defects. The plungers can be inspected for position, alignment, color and the presence of drug product solution between rings.

Cartridges are also used to create dual-chamber drug products that use a central plunger and a bypass that permits the mixing of two substances just prior to dosing.

Functionality is an additional CQA for cartridge systems and is typically defined by the break loose force and the extrusion force. The break loose force is the force required to initiate movement of the plunger/stopper when the cartridge is activated by the syringe device containing the specified needle. The extrusion force is the maximum force observed during the travel of the plunger/stopper over the full length of the cartridge.

The greatest source of variation within the drug product manufacturing process affecting the functionality may be the application of silicone oil to the cartridge barrel and the plunger/stopper. The silicone oil is applied to the cartridge barrel as a silicone emulsion, using an umbrella-shaped needle that is fully inserted into the cartridge and is actuated (spraying the emulsion to the side) as the needle is withdrawn from the cartridge. This design ensures an even application and minimizes the risk of applying silicone to the finish and exterior of the cartridge. The silicone oil is cured during the cartridge depyrogenation process, resulting in a silicone oil layer thickness less than 80 nm.[17]

The critical process parameters affecting the application of silicone oil to the cartridge barrel include machine speed (washing/siliconization machine), concentration of silicone oil in the emulsion, volume of emulsion sprayed on each cartridge, depyrogenation temperature and the belt speed of the depyrogenation tunnel. Silicone emulsions are generally not stable and may separate over time, requiring mixing prior to filling the reservoir of the washer/siliconization machine.

The plunger/stopper is siliconized using silicone oil after washing and prior to sterilization. Most drug product manufacturers today purchase plunger/stoppers that are already washed and siliconized and packaged into sterilizable bags for steam sterilization. The manufacturers of elastomeric closures offer various levels of silicone oil. The appropriate level of silicone oil should be evaluated through product stability, machinability in the filling equipment and functionality through the shelf life of the product.

35.3.4 Syringes

The variety of designs for prefilled syringes and the associated filling processes is extensive, and each design presents different risks to be evaluated through validation. The level of preassembly defines the design of the filling line. Syringe barrels without any preassembly can be processed as a cartridge using in-line glassware washers and depyrogenation tunnels. As with the cartridges, the syringe barrels are washed, siliconized and depyrogenated as they are delivered to the filling equipment.

Preassembled syringes with staked needle or Luer lock with elastomeric or plastic components are washed, siliconized and packaged into nested containers sealed and sterilized by ethylene oxide by the glass manufacturer. The siliconization is achieved by spraying silicone oil, typically medical grade 360. Droplets of silicone oil have been reported as particulate matter in some biological products.

A study by Funke[18] in which an analytical method using white light interferometry was developed to measure the thickness of silicone oil (target level 0.5 mg per syringe) on 1 mL needle-staked glass syringe barrels (stored in the needle upright orientation) presented some interesting aspects regarding the distribution of the silicone oil. The study evaluated the distribution of the silicone oil from both a radial and longitudinal aspect. Radially, the thickness of the silicone oil ranged between 110 and 135 nm, and longitudinally, the two lots of syringe barrels respectively demonstrated thickness ranges of 100 to 190 nm and 90 to 190 nm. The empty break loose force (4–4.5 N) and the extrusion force (2–2.5 N) was considered acceptable for these syringes.

Migration of the silicone oil during storage of the empty syringe barrels has been documented in the literature,[19] demonstrating that the silicone will migrate towards the flange when stored needle upright and migrate radially to the bottom longitudinal axis when stored on its side. Migration of silicone oil is not a phenomenon associated with the curing of silicone emulsion on syringe barrels that have passed through a depyrogenation tunnel.

The CQAs of the syringe barrels to be considered during the process development and validation of the syringe filling process should include the target level of silicone oil and the age of the syringe barrels.

The preassembled syringes are filled in the inverted position and sealed by inserting the plunger/stopper into the syringe barrel. There are two methods for inserting the plunger/stopper: mechanical or by vacuum. The mechanical approach utilizes a mechanical insertion device and a vent tube, which allows the displaced air to exit the syringe barrel as the plunger is pushed into position. The vent tube recedes as the plunger/stopper is inserted. One disadvantage of this method is that the plunger/stopper may be compressed and the syringe barrel scratched.

The vacuum method creates a mini-vacuum chamber over the open end of the syringe while the plunger/stopper is moved into position. One disadvantage of this method is that as the plunger/stopper glides into position, it may create a ring of silicone.

When selecting the approach for insertion of the plunger/stopper, one should consider the size of the bubble in the sealed syringe and the final position of the plunger/stopper to ensure that an overpressure is not created.

From a validation perspective, the position of the plunger/stopper in the syringe barrel may be a CQA to prevent the internal pressure of the sealed syringe from being excessive.

35.4 CONTAINER CLOSURE INTEGRITY TESTING

Container closure integrity testing is performed as part of the product development process. It could also be performed as part of an in-process control and as a 100% verification. During product development, container closure integrity is an adjunct to product stability. For sterile products, regulatory submissions require the inclusion of a container closure integrity test to demonstrate that the primary packaging system maintains sterility throughout its shelf life and during normal usage. The controls utilized for the sealing process during drug product manufacturing should be directly correlated to the container closure system that demonstrated the required container closure integrity.

The USP now contains general Informational Chapters 1207, 1207.1, 1207.2 and 1207.3[20] that examine package integrity of sterile products, selection of integrity test methods during the product life cycle, leak test technologies and packaged seal quality test technologies. The General Chapter 1207 provides background information on the topic, including the risk associated with nonintegral units, a discussion of leaks and leakage rate, closure types and mechanics, and the package quality and maximum leak rate. The General Chapter 1207.1 examines the type of test methods to be used during the product lifecycle; development and validation, commercial manufacturing and ongoing product stability. It organizes the leak test methods into deterministic and probabilistic.

A "deterministic leak test method" is an objective test method where the leakage is measured using physicochemical technologies that are readily controlled and monitored. Some of the test methods include tracer gas, laser-based gas headspace analysis, pressure decay and vacuum decay. The technologies described in this USP chapter are not prescriptive methods but represent testing concepts that may be applied when leak testing sterile product packages. These test methods generally do not require sample preparation.

A "probabilistic leak test method" requires a larger sample size and test conditions to obtain meaningful results. Probabilistic test methods include microbial challenge and dye-leak test methods.

35.5 FREEDOM OF OBSERVABLE AND FOREIGN PARTICULATE MATTER

USP General Chapter <1> INJECTIONS AND IMPLANTED DRUG PRODUCTS (PARENTERALS)-PRODUCT QUALITY TESTS[21] requires

> Each final container of all parenteral preparations should be inspected to the extent possible for the presence of observable foreign and particulate matter (hereafter termed visible particulates) in its contents. The inspection process should be designed and qualified to ensure that every lot of all parenteral preparations is essentially free from visible particulates, as defined in USP general chapter <790>Visible Particulates in Injections.

The understanding of human visual inspection and the technologies used in automated inspection systems have advanced significantly in the past 20 years, and the approach for human visual inspection has become more standardized throughout the industry. A minimum standard for the conditions to be utilized during human visual inspection is established in USP General Chapter <790 > Visible Particles in Injections, where the minimum light intensity (2,000 and 3,750 lux), background (black and white), inspection time (approximately 5 seconds in front of each background) and container handling (gently swirled and/or inverted) are established.

Techniques for manipulation of the containers can vary depending on the drug product container. Suspensions, emulsions and viscous liquids may require a technique to roll the container in a horizontal orientation to examine the product as it coats the side of the container.

Advances have also been made in the design of automated inspection systems to accurately identify and reject containers with visible particulate matter. These include advances in camera technology, the use of LED lighting to provide brighter and more stable lighting levels and more powerful computing to permit the comparison of more images. The technological advances have also greatly improved the ability of these systems to accurately detect cosmetic defects, such as cracked containers, misaligned stoppers or plunger/stoppers, glass imperfections, the color of stoppers and/or seals.

Regardless of the level of automation incorporated into the inspection process, individual characteristics of the product will influence the design of the inspection process. Some of the factors to be considered include:

- container glass—molded versus tubing vial; flint versus amber glass.
- container type—vial; ampoule; syringe; cartridge.
- container volume.
- contents—clear colorless solution; opalescent solution; suspension; emulsion; viscosity; surface tension.

35.5.1 Manual Inspection

An effective manual inspection process relies on properly trained and qualified personnel, a suitable environment with the appropriate lighting and noise level, an inspection booth compliant with the requirements of USP General Chapter <790 >, ergonomic seating and product handling procedures to ensure that the inspectors periodically take breaks for stretching and eye rest.

Personnel training should be performed using containers with particulate matter that are representative of normal production. Where manufacturing operations are ongoing, most firms maintain a library of the type of particulate matter rejects found during routine production. As demonstrated by Knapp and Kushner in the numerous papers[12,22,23,24] they published, beginning in 1980, the detection of particulate matter by human inspectors is a probabilistic exercise. Since then numerous studies have been conducted to determine the size of a particle that can be detected by an inspector without the use of magnification. The graph[25] in Figure 35.3 demonstrates the variability observed by a number of investigators.

In the approach described by Knapp and Kushner, a test set should be created to contain units with varying Probabilities of Detection (POD). Approximately 68% of the units should have a POD between 0 and 0.30 (Accept zone as

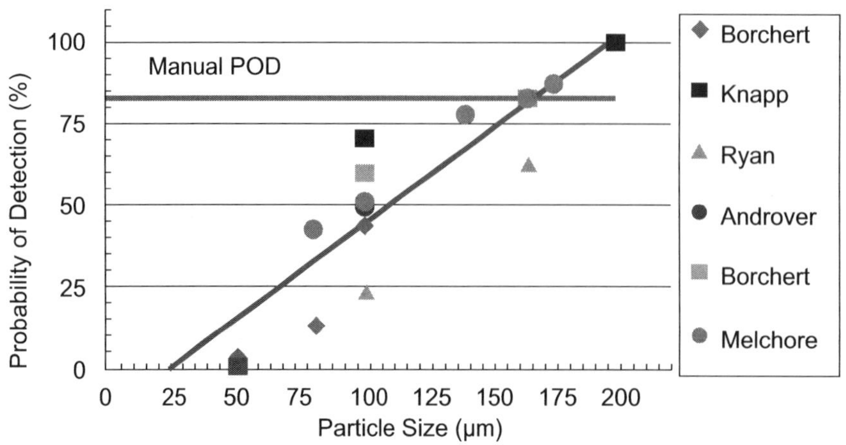

FIGURE 35.3 Probability of detecting particle sizes in small volume clear glass containers.

described by Knapp), approximately 16% should have a POD between 0.30 and 0.70 (Gray Zone as described by Knapp) and approximately 16% should have a POD between 0.70 and 1.00 (Reject Zone as described by Knapp). A 250-unit test set must be inspected multiple times by multiple inspectors to establish the probability of rejection for each unit. Knapp recommended 20 inspections to achieve statistically relevant results. Ideally, the best performing or most experienced inspectors should be selected to perform the multiple inspections of the test set.

Inspector training and qualification begins with a visual acuity examination, and the inspectors should have 20/20 corrected vision. Training should include detailed instructions to describe the handling of the container to ensure that particles can be put into motion without creating bubbles.

Where defective units are not available, they can be created either in-house, using particles certified for size and composition within local environmental protection to prevent particulate contamination during the seeding operation and appropriate systems to reseal the test unit. Certified particles are also commercially available, and external laboratories can prepare the test units.

Precautions should be taken during the creation of the test set to ensure that the unique coding of each unit does not interfere with the inspection process and the results of the inspection trials are not available to the inspectors to prevent bias. Test sets are not static and may change over time as a result of handling abuse, product degradation and in some cases adherence of the particles to the container closure system. Spare containers should be available to replace damaged units. This can be achieved by initially preparing a larger test set and reserving a portion of the units.

Additionally, data generated during trials with the test sets should be trended to evaluate the potential degradation of particles and their visibility.

The parameter created by Knapp to compare an automated inspection process to the human inspection is the Reject Zone Efficiency (RZE). The RZE is the average POD for all of the units in the Reject Zone. The RZE of the inspection process under evaluation must be greater than or equal to the RZE of the test set as created by the inspectors.

A similar comparison cannot be used to qualify new inspectors, considering that the RZE is an average of the most qualified inspectors, who participated in the development of the test set. Statistically, the RZE for a new inspector should be within the data population represented by the inspectors who developed the test set. For example, one could set the acceptance criteria at RZE–3σ, where σ is the standard deviation, derived from the individual RZEs collected during the 20 trials. Theoretically, an inspector meeting this criterion would be equivalent to the lowest-performing inspector who participated in the creation of the test set.

Initially, the new inspector may not be as effective as the average qualified inspector. To address this, it may be appropriate to establish a trial period for the newly qualified inspector to more closely monitor their performance for several months.

Acceptance sampling has been incorporated into USP General Chapter <1790> visual inspection of injections,[26] requiring a statistically valid sample collected from the units accepted by the visual inspection process. This sampling can also serve to monitor the performance of individual inspectors, if traceability to the inspector is maintained with the output from the inspection process.

Process trending is an important aspect of the product quality assessment and is expected for the 100% inspection process. The initial output from the process trending is the establishment of an action level for the percent rejected for particulate matter.

Typically, the data population for percent rejected is normally distributed, which provides the option of setting the action level using either graphical or statistical methods. Theoretically, the action level should be established to identify a batch where some aspect of the process control plan (process conditions, primary packaging component quality etc.) was not controlled within the approved process range.

When the action level is exceeded, the response may include, but is not limited to, root cause investigation, identification of the particles and reinspection where warranted.

Where processes are well controlled and exceeding the action level is rare, it may be appropriate to periodically identify the particulate matter from rejected units to maintain a current library and determine if there are new sources of particulate matter. This is especially useful when inspecting difficult-to-inspect products, where isolating the particles is the most effective method to obtain an accurate identification of the particles.

In some biological products and other large molecule products, product-related agglomerates represent an additional challenge to the inspection process. From a process control perspective, the inspection process should differentiate between product-related agglomerates, intrinsic particulate matter and extrinsic particulate matter. Intrinsic particulate matter is defined as particles that may be sourced from the primary packaging components or the bulk product manufacturing process. These particles could be sourced from materials such as glass, elastomeric closures, silicone tubing and stainless steel. Extrinsic particulate matter is defined as particulate matter generated from sources outside of the manufacturing process, such as human hair, insect parts etc.

35.5.2 Automated Inspection Systems

When inspecting liquid-filled containers, there are a variety of automatic inspection systems that utilize slightly different approaches for lighting and detection of particles in motion, but they all function according to several basic concepts.

1. Spin phase—Spin the container at a high rate of speed to create a vortex, which will rinse all surfaces inside the container and put the liquid in motion. The creation of bubbles will limit the spin rate.
2. Break phase—Stop the spinning of the container so that the liquid is in motion, the meniscus returns to

the original level and the container is stationary for the inspection phase.

3. Inspection phase—With the liquid rotating in the container, the detection system is making multiple comparisons of the light transmission or images within milliseconds. Where there is a difference in light transmission or images above a threshold value, the container is rejected.

The critical process parameters for an automated inspection system normally include the machine speed, the spin rate, the break time and the threshold value for container rejection. The break time, which is the time between the end of the spin phase and the start of the inspection phase, must be set to ensure that the meniscus returns to the original level, while ensuring that the liquid and any particles remain in motion at the inspection phase.

The investment in automatic inspection systems is significant, and most installations are designed to inspect multiple container sizes, container types and product types (solution, suspension, emulsion and lyophilized cakes). Bracketing products should be considered only with great caution. Each type of container has specific types of cosmetic defects, and the light transmission or camera image will be specific to molded vials and tubing vials. The physical characteristics (surface tension, viscosity and density) of the liquid will determine the spin rate and breaking parameters to put the liquid in motion without creating bubbles.

Generally, if the bracket is not created properly, a high false reject rate will be observed. Automated inspection systems are more reliable and have greater sensitivity than human inspectors. Depending upon economic factors such as direct material and labor costs, a higher false reject rate may be acceptable. In a paper presented by E Martinez at the Pharma Expo Conference in 2016, he estimated that it requires 82 inspectors to achieve an inspection rate of 300 units per minute, which is commonly achieved using the automated inspection systems available today.

The statistical approach, developed by Knapp and Kushner[23], created a mathematical approach to determine if automated inspection systems were equivalent to or better than human inspectors. The creation of a valid Knapp test set is critical to the validation of an automated inspection system.

From a quality perspective, the qualification of automated inspection systems requires that the RZE generated by the automated system (RZE_A) for the Reject Zone containers identified by the human inspection is greater than or equal to the human RZE (RZE_H). During the qualification of automated inspection systems, the POD for containers in the "accept" and "gray" zones provide some valuable information. The POD for "accept" zone containers is a direct measurement of the false reject rate. Typically, the POD for "gray" zone containers should be higher, as the automated inspection system is more sensitive than human inspectors.

A case study presented by Roy McClean of Hospira at the PDA Visual Inspection Forum in 2015 demonstrated that the POD_A is more consistent than POD_H over a range of particle sizes (100 to 2000µm), which appears to confirm that automated inspection systems are better at detecting particles near the boundary of human visual acuity.

35.5.3 Semiautomatic Inspection Systems (Presentation Devices)

To increase the productivity of the manual inspection process, many firms have introduced automated presentation devices. The overall inspection time is reduced because the handling is entirely automated. The inspector's function is to simply view the containers as they pass through the field of view and remove the defective units by pressing a reject button to automatically reject the unit. These presentation devices can operate at speeds up to 150 units/min.

The ergonomics of the operator's inspection area are well designed, and the units are oriented at a 60° angle to facilitate operator comfort and inspection capabilities. The inspection booth is also equipped with an ergonomically adjustable magnifying glass.

Similar to the automated inspection systems, the units are spun in order to put the liquid into motion, and when the container is passing in front of the inspector, it is rotating slowly to allow for the inspector to view the full circumference of the container. The presentation devices can also be equipped with mirrors above and below the container to allow for the inspector to conduct a full cosmetic inspection concurrently with the inspection for visible particulate matter.

The capabilities of the presentation systems should be confirmed by each manufacturer. Some firms have organized their inspection process to have two presentations of the systems in series, designating one for particulate matter inspection and the other for cosmetic inspection. In my experience, it is difficult to qualify inspectors at the stated maximum operating speed using the POD approach.

35.5.4 Difficult to Inspect Products

There are a variety of forms of parenteral products that are categorized as "difficult to inspect products" (DIP). DIP products include lyophilized products, products in amber glass, products in translucent containers, suspensions and emulsions. Since the effectiveness of the 100% inspection is severely limited by the product form, greater emphasis must be placed on eliminating the potential sources of particulate matter from the manufacturing process.

The holistic approach to control particulate matter will include verification that the primary packaging components, as they are delivered to the filling equipment, and the bulk product transfer system are free of particulate matter. The assessment of the bulk product transfer system should be focused on the portion of the system downstream of the final filtration step, where visible particles will be removed. Where the bulk product transfer system is aseptically assembled and not prepared using CIP/SIP, the components of the bulk transfer system must be packaged in hermetically sealed, non-fiber-releasing autoclave bags. Where the drug product form permits some level of inspection (solutions prior to lyophilization), evaluation of the first units filled will provide information regarding the cleanliness of the bulk product transfer system.

Generally, the majority of particulate matter present in the bulk product transfer system will be transferred into the first portion of the filled units. Where the filling equipment permits, the solution could be filled into a large container that has been prepared in an ISO 5 environment, after which the contents can be filtered and evaluated microscopically for the presence of visible (≥ 100 μm) particles. This approach could be Incorporated into the qualification of the filling process.

The assessment of the handling of primary packaging components may include more frequent assessment of the cleaning processes, whether performed by the component manufacturer or by the drug product manufacturer, and careful design and monitoring of the component handling procedures during the filling operation. The breakage of glass containers during handling is a major source of particulate matter, placing greater emphasis on the equipment design, recovery procedures in the event of glass breakage and monitoring of the glass handling process. Some manufacturers have integrated 100% automated inspection of the empty glass containers immediately prior to the filling operation. These systems can be qualified to automatically reject damaged containers, units with cosmetic defects and those containing glass particles.

Enhanced control of particulate matter sourced from elastomeric closures includes the selection of the appropriate level of silicone. The minimum level of silicone required to achieve acceptable machinability in the filling/sealing equipment and product functionality, in the case of cartridges and syringes, should be determined during product development. Reduced levels of silicone minimize the potential for silicone droplets in the drug product and the accumulation of particles in the component-handling equipment. Additionally, cleaning of component bowls and tracks to remove residual silicone oil must be implemented to maintain particulate control.

The 100% inspection of lyophilized products can only evaluate the appearance of the cake and the presence of particles on the surface of the cake. Automatic inspection systems ensure 360° inspection by cameras with various technologies to detect the difference in images. USPGgeneral Chapter <1790> recommends that a small sample of units be visually inspected after reconstitution in a clean environment, using appropriate particle control measures. Typically, S-3 or S-4 plans from ANSI/ASQ Z1.4 can be applied, which results in 20 units being sampled from batch sizes ranging from 3,201 to 150,000 units. The AQL for this inspection is set at 0.65% or less, which requires that no units with particulate matter be detected.

During the initial qualification of a filling process for the manufacture of a lyophilized drug product, consideration should be given to fill a placebo solution so that a 100% inspection for particulate matter can be performed. If easily inspected drug products are also filled on the same filling line, information from the 100% inspection of those products will demonstrate that the level of particulate matter in lyophilized products is well controlled.

35.6 INERT HEADSPACE

The number of drug products requiring inert headspace to achieve a useful shelf life has been increasing, considering the number of biological and large molecule products that continue to be approved by the regulatory authorities. Lyophilization and stoppering under vacuum is one approach that is used for products that are not stable in aqueous solutions. Liquid products utilize a variety of designs to achieve the inert headspace.

For most firms, nitrogen is the gas of choice because of cost and ease of handling. The dissolved oxygen level in the drug product formulation must be controlled during preparation and maintained during delivery to the filling system. Dissolved oxygen levels of less than 1 ppm can be achieved during formulation steps, prior to the addition of the oxygen labile active ingredient. This is easily achieved through the use of sparging tubes and standard agitation systems, especially when using Water for Injection (WFI) at 80°C where the dissolved oxygen is approximately 3 ppm.[27]

After reducing the dissolved oxygen to the required level, the WFI can be cooled under nitrogen atmosphere to the required temperature prior to addition of the oxygen labile active ingredient.

Depending upon the level of oxygen permitted in the container headspace, the filling approach can range from the use of an isolator with a full nitrogen atmosphere to a variety of devices to substitute the headspace of the individual containers with the inert gas during the high speed filling operation. These devices may include elements such as pre-flushing with the inert gas, concentric filling needles that deliver the drug product liquid and the inert gas simultaneously, post fill flushing with inert gas and a mini environment between the filling and stoppering station that is supplied with the inert gas.

Both in-line and off-line systems are available to measure headspace oxygen in the sealed containers. As a result of the detectability, as defined in FMEA risk analysis, the in-line systems (such as Frequency Modulated Spectroscopy[FMS]) that measure the oxygen levels in 100% of the containers will reduce the classification of the process parameters that control the oxygen levels in the container headspace. In this application, FMS is a (PAT). FMS is a laser absorption technique based on the absorption of near infrared (NIR) light centered at 762 nm.

The in-line FMS system is capable of accurate determinations regardless of the type of glass, including molded and tubing vials made of amber or clear glass. The in-line FMS systems will also function accurately with translucent plastic containers constructed of low-density polyethylene. A calibration curve for each container size and shape must be developed, as the absorption of light by the headspace oxygen in the container is dependent on the inner diameter of the container. The calibration curves are developed by filling containers with nitrogen/oxygen gas containing oxygen at 0,

1, 2, 4, 8 and 20%. Once a calibration curve is established, the acceptable oxygen level can be programmed into the control system to automatically reject any container whose headspace oxygen level does not conform to the acceptance criteria.

The in-line system should be integrated into the filling line downstream of the capping system. It may be necessary to develop a separate unit operation for those products that characteristically foam during filling, as the presence of product and the light path may result in a false reading through absorption or diffraction of light.

Validation of these systems will require the use of challenge containers, with oxygen levels above the acceptable limit. Units from that portion of the calibration curve can also be used for the purpose of validation.

The off-line systems, requiring a periodic sampling plan, will increase the classification of the process parameters to critical.

The critical process parameters to create an inert headspace are machine speed, nitrogen flow rate to the various components of the purging system and the position of the purging needles. Factors that could affect the bracketing approach should include container type, container size, ratio of the fill volume to headspace and product surface tension. However, if an inert atmosphere isolator is used, the importance of these parameters may be greatly reduced, with the most critical parameter being the oxygen level in the isolator.

35.7 SECONDARY PACKAGING PROCESSES

One aspect common to equipment used in primary and secondary packaging operations are detection systems that automatically reject nonconforming units or stop the equipment. Proper setup procedures, qualification and performance verification during commercial manufacturing will ensure that these detection systems consistently perform the intended task.

The user requirements specifications must clearly define the type of defect to be detected and automatically removed from the acceptable product output. Limit samples for crooked labels, damage to product codes and poorly printed variable information (lot number and expiration date) should be established through discussions between the drug product manufacturer and the equipment vendor. The limit samples could be characterized by establishing the probability of detection (POD), similar to the approach used for visible particulate matter.

The qualification of the detection systems should be performed using the established limit samples. The qualification must verify that each defective unit is both detected and rejected by the electromechanical system. The qualification should be performed at the maximum and minimum operating speeds. While the placement of the defective units within the test group should be random, it may be appropriate to place some defective units in sequential positions to better simulate a machine malfunction, such as poor printing of variable information.

Performing these qualifications three times for an individual container closure system/size is not justified by the equipment and process under examination. Variability of the equipment performance to properly identify and reject defective units is primarily a function of the complexity of the equipment setup procedure, dimensional variability of the primary container closure system, printing supplies and variability of the secondary packaging components. If the setup procedure is complex, one may want to consider multiple trials following a full equipment setup.

Performing a single qualification for each container closure system/size is a more rational approach. Most packaging equipment is designed to package multiple formats and various sizes of containers. In some circumstances, this requires that the positions of the sensors be changed for each format. To minimize the need for operator adjustment, pin systems should be used to fix the sensor position for each format. Where the equipment is not designed with a pin system, gauges or tools should be provided to equipment setup personnel to ensure that the sensor is properly positioned. Another aspect to be considered is the potential for sensors or rejection devices to become loose and malfunction during operations.

During routine production, at a minimum, verification of sensor functionality should be performed at the start and end of each batch, providing confirmation that the in-line sensors functioned properly during the packaging process. Sensor functionality is verified through the use of a challenge set of containers/packages containing defective units.

Many secondary packaging lines also include in-line check weighing systems. The functionality of the check weigher should be described in the user requirements specification and should be integrated into the secondary packaging process control plan. A check weighing system is utilized to detect a missing unit or component that cannot be detected through other means. The check weighing system must have adequate sensitivity to detect the missing unit, while taking into account the weight variation of the missing component and the weight variation of the other components of the package at that point in the packaging process. Here again, at a minimum, during routine manufacturing the functionality of the check weighing system should be verified at the start and end of each batch.

Another aspect of the qualification of secondary packaging equipment, which is similar to primary packaging equipment, is the evaluation for product/component abuse. It is likely that the risk for component abuse is greater for secondary packaging equipment, considering that the equipment is handling labels, inserts, individual cartons etc. High-speed packaging equipment utilizes a variety of techniques to place individual components into the final package.

35.8 SERIALIZATION

Serialization, sometimes referred to as track and trace, is currently being implemented by many countries throughout the world to address the proliferation of counterfeit drug products. The deadline for some countries has already passed, and some extend out to 2022. Each company should understand the deadlines and requirements of the markets that they serve, as there are no uniform requirements and defined machine-readable codes. This section will focus on the requirements established by the FDA in the United States.

The Drug Supply Chain Security Act, enacted by the US Congress in 2013, established the requirement for unique identifiers to permit the traceability of each individual salable unit through the drug supply chain in the United States. The original deadline for manufacturers to comply with the requirements of this act was delayed until November 26, 2018.

The definitions described within the act are critical to understanding the requirements. These include:

HOMOGENEOUS CASE—a sealed case containing only product that has a single National Drug Code number belonging to a single lot.

PACKAGE—the smallest individual saleable unit of product for distribution by a manufacturer or repackager that is intended by the manufacturer for ultimate sale to the dispenser of such product.

INDIVIDUAL SALEABLE UNIT—the smallest container of product introduced into commerce by the manufacturer or repackager that is intended by the manufacturer or repackager for individual sale to a dispenser.

PRODUCT IDENTIFIER—a standardized graphic that includes, in both human readable form and on a machine-readable data carrier that conforms to the standards developed by a widely recognized international standards development organization, the standardized numerical identifier, lot number, and expiration date of the product.

STANDARDIZED NUMERICAL IDENTIFIER (SNI)—a set of numbers or characters used to uniquely identify each package or homogenous case that is composed of the National Drug Code that corresponds to the specific product (including the particular package configuration) combined with a unique alphanumeric serial number of up to 20 characters.

The FDA guidance, Product Identifiers Under the Drug Supply Chain Security Act Questions and Answers, September 2018, describes the product identifiers as follows:

> The NDC and serial number are the two components of the Standardized Numerical Identifier (SNI) as defined in section 581(20) of the FD&C Act. The Product Identifier requires the SNI, lot number, and expiration date. The drug package label must include the product identifier information (i.e., the NDC, serial number, lot number, and expiration date) in both the human readable form and the machine-readable, 2D data matrix barcode format. FDA recognizes that variations may exist in how to abbreviate the human-readable portion of the label for the NDC, serial number, lot number, and expiration date. For example, "No." may be used instead of "number," or may not be listed at all.

The product identifying label on the homogeneous case can be a 1D or 2D barcode. Most manufacturers utilize systems that establish "inference" between the product identifier on the homogeneous case and the product identifiers of the individual salable units within the case.

The qualification of a serialization system comprises both the qualification of the packaging line equipment for printing or applying the labels and the qualification of the computer system that transfers the product identifying numbers to the packaging equipment and uploads the data to an electronic database for use by repackagers and distributors within the supply chain.

In general, the qualification of the serialization system will be organized into the packaging line equipment used for printing the standard numerical identifier and the computer system used for the generation of the SNI and capture and transfer of the information to the electronic database. The qualification of the packaging line equipment should include verification that the printed barcode is accurately read and that no duplicate numbers are printed.

The qualification of the computer system should be designed in accordance with GAMP[28] requirements and an in-depth risk analysis. The scope of the computer systems can vary widely, as with all validation exercises, the user requirement specification is the starting point for the validation. From the user requirement specification, the risk analysis can determine the functionalities with GMP impact.

In firms where the production floor is fully integrated with corporate IT systems, the SNIs for a batch, generated by the Material Resource Planning (MRP) system, are downloaded to the coding equipment on the packaging line. The positive verification of the SNI on the individual salable unit, performed by the packaging equipment, may then be communicated to the MRP system. The information from the MRP system is then uploaded to the supply chain's electronic database. From a computer validation perspective, the accurate transfer of information between IT systems and packaging equipment are major elements to be considered.

35.9 CONCLUSION

The packaging processes for sterile drug products represents a nexus for technology transfer where the selected primary packaging components and bulk drug product are brought together in a commercial setting. The primary packaging components are selected based on functionality and drug product stability. The bulk drug product is formulated to deliver a prescribed clinical outcome to the patient. The packaging processes must integrate these elements to reliably produce the drug product with all of the CQAs in a cost-effective manner.

The packaging processes are designed to maintain the prescribed CQAs for a sterile drug product; labeled potency, sterility, freedom from foreign particulate matter, absence of bacterial endotoxin, container closure integrity and, where applicable, inert headspace. The validation of the packaging processes should be constructed based on a risk analysis that identifies the potential modes of failure for each of the CQAs.

This chapter has provided some validation approaches for each of the CQAs, but other approaches may be appropriate because of the wide variety of primary container closure systems and manufacturing equipment.

NOTES

1. Individual dosage units with cosmetic defects may not be a CQA, but the occurrence of dosage units with cosmetic defects should be less than a predefined acceptable quality level (AQL).

2. Chapman KG, The PAR Approach to Process Validation. *Pharmaceutical Technology* 1984: 22–36.
3. General Chapter <697> Container Content For Injections (2015, May 1). Retrieved from USP: https://online.uspnf.com/uspnf/document/GUID-36AE7C6F-5983-4F42-9426-24574CEDB1A4_1_en-US
4. General Chapter <1151> Pharmaceutical Dosage Forms. (2018, August 1). Retrieved from USP: https://online.uspnf.com/uspnf/document/GUID-431F93A9-1FEC-42AE-8556-AA5B604B2E36_2_en-US
5. *Guidance for Industry*, June 2015. www.fda.gov/downloads/Drugs/GuidanceComplianceRegulatoryInformation/Guidances/UCM389069.pdf
6. Peterson A, Isberg E, Schlicht A, Capability of Filling Systems to Dispense Micro-Doses of Liquid Pharmaceutical Product. *Pharmaceutical Engineering* July/August 2007.
7. Maa Y-F, Filling of High Concentration Monoclonal Antibody Formulations into Prefilled Syringes: Investigating Formulation: Nozzle Interactions to Minimize Nozzle Clogging. *Journal of Parenteral Science and Technology* 2015: 417–426.
8. *USP General Chapter 1207 Package Integrity Evaluation of Sterile Products*, 2018.
9. PDA. 2013. "technical report 43."
10. ASQ. 2013. *Sampling Procedures and Tables for Inspection by Attributes.*
11. Mathaes R, Influence of Different Container Closure Systems and Capping Process Parameters on Product Quality and Container Closure Integrity in the GMP Drug Product Manufacturing. *Journal of Parenteral Science and Technology* 2016: 109–119.
12. Knapp JZ, Particulate Inspection of Parenteral Products: From Biophysics to Automation *PDA Journal of Pharmaceutical Science and Technology* 1982: 121–127.
13. European Commission, September 17, 2018. https://ec.europa.eu/health/sites/health/files/files/gmp/2017_12_pc_annex1_consultation_document.pdf
14. BPOG, White Paper: Container Closure Integrity Control versus Integrity Testing during Routine Manufacturing. *PDA Journal of Pharmaceutical Science and Technology* 2015: 461–465.
15. Bernuzzi M, Evaluation of the Influence of Open and Closed Ampoule Technologies on Particulate Matter in Small Volume Parenterals. *Journal of Parenteral Science & Technology* 1993: 265–269.
16. Moll F, Doyle D, Haerer M, Guazzo D, Validation of a High Voltage Leak Detector for Use with Pharmaceutical Blow Fill Seal Containers: A Practical Approach. *PDA Journal of Pharmaceutical Science and Technology* 1998: 215–227.
17. Friess W, et al., Analysis of Thin Baked on Silicone Layers by FT IR and 3-D: Laser Scanning Microscopy. *European Journal Pharmaceutical Biopharm* 2015: 304–313.
18. Funke S, Methods to Determine Set Silicone Oil Layer Thickness in the Sprayed on Syringe Barrels. *PDA Journal of Science and Technology* 2018: 278–297.
19. Wen ZQ, Distribution of Silicone Oil in Prefilled Glass Syringes Probe with Optical and Spectroscopic Methods. *Journal of Parenteral Science and Technology* 2009: 149–158.
20. *USP general chapter 1207 package integrity evaluation of sterile products*, 2018.
21. USP general chapter 1 injections and implants drug products (parenterals)—Product quality tests, 2019.
22. Knapp JZ, Kushner HK, Generalized Methodology for Evaluation of Parenteral Inspection Procedures. *PDA Journal of Parenteral Science and Technology* 1980: 14–61.
23. Knapp JZ, Kushner HK, Implementation and Automation of a Particulate Detection System for Parenteral Products. *PDA Journal of Parenteral Science and Technology* 1980: 369–393.
24. Knapp JZ, Kushner HK, Abramson LR, Particulate Inspection of Parenteral Products: An Assessment. *PDA Journal of Parenteral Science and Technology* 1981: 176–185.
25. Berdovich M, Considerations for Design and Use of Container Challenge Sets for Qualification and Validation of Visible Particulates Inspection. *PDA Journal of Pharmaceutical Science and Technology* 2012: 273–284.
26. 2019. USP General Chapter <1790> Visual Inspection of Injections.
27. Geng M, Duan Z, *Predicition of Oxygen Solubility in Pure Water and Brines Up to High Temperatures and Pressures.* Beijing, January 11, 2010.
28. 2008. GAMP 5 A Risk-Based Approach to Compliant GxP Computer Systems: ISPE.

36 Validation of Active Pharmaceutical Ingredients

James Agalloco and Phil DeSantis

CONTENTS

36.1 HISTORY

Validation was initially introduced in the 1970s to the pharmaceutical industry as a means for more firmly establishing the sterility of drug products where normal analytical methods are wholly inadequate for that purpose. In following years, its application was extended to numerous other aspects of pharmaceutical operations: water systems, environmental control, tablet and capsule formulations, analytical methods and computerized systems. Individuals working with bulk pharmaceutical chemicals (BPCs) were particularly reluctant to embrace validation as a necessary practice in their operations. Industry apologists explained this lack of enthusiasm in terms of differences in facilities, equipment, technology, hygienic requirements, cleaning methodologies, operational practice and numerous other aspects of disparity that seemingly justified the recalcitrance of this segment of the industry. This view was widespread in the bulk chemical industry through the end of the 1980s.

The extension of the concepts that have made validation such an integral part of practices across the health care industry to the production of BPCs seems obvious in retrospect. Yet for many years there existed a general reluctance to introduce validation into BPC activities. While there were some modest efforts, it was not until sometime after the biotechnology industry became technically and commercially viable that any significant effort was initiated. The production of biological products (e.g., vaccines, blood derivatives) for registration in the United States currently requires the approval of the FDA's Center for Biological Evaluation and Research (CBER). When it was initially charged with oversight over all biological products, CBER required extensive validation of fermentation, isolation and purification processes utilized in their preparation.[1] An objective comparison of BPC operations relative to those performed in the early stages of biologicals would reveal minimal differences. The production methodologies for many classical BPCs, for example, penicillin's, cephalosporin's and tetracycline's, are nearly indistinguishable from those utilized to prepare tPA, EPO and other biologicals. With this realization, the advent of validation for BPCs was apparent to all and was increasingly imposed upon the industry.

In 1990, the United States Pharmaceutical Manufacturers Association (now called PhRMA) formed a committee to define BPC validation concepts.[2] This committee's efforts culminated in 1995 when they issued their finished draft. This document served as a guide to the authors in the development of this chapter. Of necessity, considerable clarification and expansion of the material contained has been necessary to complete this effort.

In the late 1990s, a new term started to appear, first in Europe, but soon it spread across the entire industry—active pharmaceutical ingredients or APIs. Those who first used the new term suggested that it was synonymous with bulk pharmaceutical chemicals or BPCs. Since that time, it has become increasingly common in the industry speak only of APIs. A part of the rationale for this initiative has been voiced as a move toward harmonization. In publications since that time, API has largely supplanted BPC as the descriptive term for

DOI: 10.1201/9781003163138-36

these products. It will be used throughout the remainder of this chapter.

The official requirement for validation of API processes was formally established in Guidance for Industry, Q7A *Good Manufacturing Practice Guidance for Active Pharmaceutical Ingredients*.[3] This was the result of a multiyear effort by the International Conference on Harmonization (ICH), which resulted in this harmonized guidance document. This guidance document addresses the subject of validation briefly and employs the same definition FDA has adopted for other processes (see next paragraph). This chapter provides recommendations for validation consistent with the Q7A guidance.

In 2011, FDA published a revised guidance document on process validation and explicitly included active pharmaceutical ingredients as falling under the guidance.[4] This has created greater consistency of validation approaches across dosage form and API manufacturing. Although this chapter does not follow the FDA guidance precisely, it is in complete alignment with the concepts presented therein.

36.2 DEFINITION OF VALIDATION

There are numerous definitions of validation that have been written over the nearly 40 years since its appearance in the pharmaceutical industry. Rather than foster new definitions within the context of this chapter, the authors have chosen to draw upon some of the more widely quoted definitions. The FDA had long defined process validation as: "Process validation is establishing documented evidence which provides a high degree of assurance that a specific process will consistently produce a product meeting its predetermined specifications and quality characteristics".[5] This definition is referred to in FDA's subsequent guidance specific to APIs.[6] The more recent FDA guideline, however, defines validation as follows:

> For purposes of this guidance, *process validation* is defined as the collection and evaluation of data, from the process design stage through commercial production, which establishes scientific evidence that a process is capable of consistently delivering quality product. Process validation involves a series of activities taking place over the lifecycle of the product and process.

This is more aligned with the authors' long-held view of validation as a 'life cycle' concept.

36.3 REGULATIONS

Regulations specific to the control APIs are a relatively new concept; for many years FDA's policy was to apply a limited enforcement of the subpart 211 regulations for finished pharmaceuticals.[7] In recent years, FDA has endeavored to harmonize its approach to API regulation with the rest of the world and has issued a guidance document that draws heavily on subpart 211.[8] This effort followed the issuance of a Pharmaceutical Inspection Convention document that addressed the same subject in a different format.[9]

Some discussion of validation approaches utilized for APIs is essential to following this chapter. The approaches for APIs are essentially the same as those utilized for other processes and systems. This discussion serves to highlight the nuances of validation as they apply to APIs.

36.4 LIFE CYCLE MODEL

Contemporary approaches to the validation of virtually any type of process or system utilize the 'life cycle' concept.[10] The use of a life cycle concept was embraced by the FDA in its 2008 draft guidance on process validation, reaffirming the appropriateness of the approach that had been adopted previously by numerous practitioners. The life cycle concept entails consideration of process or system design, development, operation and maintenance at the onset. Use of the life cycle helps to provide a system that meets regulatory requirements but is also rapidly placed into service, operates reliably and is easily maintained. While the life cycle is best suited to new products, processes or systems, it certainly has applicability for existing systems as well. Existing systems that have never been previously validated can be reviewed against the same validation criteria that would be imposed for new systems. While these systems are likely to be deficient with regard to current requirements, the life cycle model provides a means for upgrading their programs to be on a par with newly developed systems. This is especially important for APIs, given that the validation of these processes has lagged behind many of the other areas of the industry where validation has already been instituted.

36.5 VALIDATION OF NEW PRODUCTS

The validation of a new API entails practices that parallel those utilized for the introduction of a new pharmaceutical formulation. Thus, a large part of the initial validation effort must be linked to the developmental activities that precede commercial scale operation. The similarity is such that aspects of reaction and purification methodologies should be as similar as possible, given of course the difference in the scale of the equipment utilized in the commercial facilities. Any differences between the API processes utilized for the formulation batches used to establish clinical efficacy and the commercial material must be closely evaluated, along with their impact on the API: chemistry, purity profile, stability, crystal morphology and other key attributes. The developmental laboratory has the responsibility for determining optimal reaction conditions including time, temperatures, raw material purity, molar ratios, solvent selection, crystallization method, wash volume, drying conditions, etc. Of primary concern is the identification of critical control parameters, that is to say those which impact quality, purity, safety and efficacy. The concerns to be addressed in any individual API validation program are of course unique to that process; the inclusion or exclusion of any single factor as a consideration in API validation should be determined by the manufacturer. Chemical reactions are among the more complex processes to be subjected to validation, and the number of critical factors in even a single reaction can be quite extensive. The amount of information that must be generated during development to support a validated API process is correspondingly extensive. The necessary information can be assembled into a technology transfer document that conveys the collected experience gained during development to those responsible for the commercial production of the API. The success of a developmental organization is better assessed by the quality of the information they convey to document their efforts than it is by the sophistication of the chemistry utilized to make the API. The technology transfer document is likely to be of central interest to FDA inspectors during the conduct of a Pre-Approval Inspection of the facility prior to approval.[11]

Alternative approaches may be acceptable. For example, the PPQ batches may be spread out over time, as would be necessary for an API that is infrequently made. Each individual batch could be released if it complies with release specifications while enough data is accumulated over time to establish the validated state of the process. It may even be possible for a legacy API process, in which data from previously produced batches wherein the critical process parameters have been recorded and correlated to API quality attributes could be used in lieu of a newly initiated PPQ study.

36.6 VALIDATION OF EXISTING PRODUCTS

Validation of active pharmaceutical ingredients should no longer be a new concept for the industry to address. It is true, however, that some legacy API products were introduced to the market without any significant validation in place. As a consequence, the first efforts to validate these products may have employed retrospective methods. The trending of results derived from in-process and release testing of these products and processes were initially deemed adequate to serve as the basis for these efforts. Given the FDA's general dissatisfaction with retrospective approaches, it is strongly recommended that these early efforts not remain the only approach utilized. Application of the current FDA process validation guidance[4] and its EMA counterpart[12] is the preferred approach.

In the absence of an adequate technology transfer document, it will be necessary to research the developmental records in order to ascertain the critical process parameters and their relationship to the critical quality attributes of the API. These attributes will generally include purity but depending on the use of the API in formulation of drug product, this list may be extensive. In the absence of developmental data, it may be necessary to revisit retrospective data, but with a broader perspective on both quality attributes, process parameters and their relationship. In this manner, the rationale for operating ranges to be validated is established. Having established the important correlation between process parameters and API quality, the control strategy is then defined. Once applied, this strategy is used to ensure that the process is validated.

For new products, guidances recommend a demonstration of control and consistency over a selected number of batches, referred to as Process Performance Qualification (PPQ) (historically termed 'prospective validation'.) Although three batches are common, the number of batches selected should be justified.

Where numerous APIs have not been adequately validated according to current guidance and industry practice, the establishment of priorities for validation generally follows economic concerns, with those products that provide the largest contribution to the firm's profitability being the initial focus of activity. Regardless of how the first validation efforts were completed, the adoption of the life cycle model for maintaining products in a validated state has become accepted industry practice and meets current regulatory expectations.

36.7 IMPLEMENTATION

The validation of any process or product relies upon several supportive activities. Validation in the absence of these activities has only minimal utility, as it is only through the integration of these other practices that meaningful validation can be accomplished. Several of these activities are defined in CGMP regulations while others are an integral part of a company's organization structure or are closely associated with 'validation' itself.[13]

Equipment Calibration—The process of confirming the accuracy and precision of all measurements, instruments, etc. to ensure that the measured variable is being accurately monitored. Calibration includes demonstration of conformance to applicable national standards such as NIST, DIN or BS for all key parameters. This is a universal CGMP requirement across the globe.

Commissioning—Maturation of facility qualification concepts and an awareness that a hierarchy exists for the systems installed in a facility has increasingly led to the adoption of an approach to system/facility start-up termed 'commissioning'. Commissioning is a Good Engineering Practice in which systems are confirmed as acceptable for use. All systems need commissioning in order to ensure their fitness for purpose and compliance with detailed engineering and vendor specifications. Also, commissioning affords the technical experts the opportunity to inspect, start up, correct and adjust systems without resorting to the more formal and rigorous expectations associated with equipment qualification (see the following paragraph).[14] Systems and equipment within the facility can be reviewed from a risk-based perspective, and those without significant risk to product quality may then be placed into use without further testing or review. This would be appropriate for such systems such as heating and cooling sources (e.g., steam, cooling water), non-product-contact gasses (e.g., instrument air) and HVAC systems serving closed processes (e.g., reactor halls).

Equipment Qualification—Where systems or equipment are assessed to have a significant risk related to product quality, most firms require additional oversight, including that of the quality unit. Equipment qualification is a precursor to process validation that focuses on physical and functional aspects of equipment that are determined to affect product quality, in this case the quality of the API. Equipment qualification expectations should be linked to User Requirements Specifications (URS), which are in turn derived from critical process requirements.[15] For example: if a synthesis reaction occurred best at pH 1, it would mandate that glass or Hastelloy be utilized, which would then become a qualification expectation for materials of construction on the vessel and process piping. As API facilities are commonly utilized for more than one reaction over the course of their operational life, flexibility should be the goal in performing equipment qualification. The equipment should be evaluated across the full breadth of its performance capabilities as future uses may require conditions quite different from those needed initially.

There is **no** requirement for a formal separation of the activity into distinct elements, such as installation and operational qualification. It has become increasingly common in recent years to combine these activities under a single effort. For the sake of those who still separate the activities, individual descriptions have been provided.

Installation Qualification—Documentation that the equipment was manufactured and installed in accordance with the intended design. This is essentially an audit of the installation against the critical equipment requirements affecting quality of the API.

Operational Qualification—Confirmation that the equipment performs as intended entails evaluation of performance capabilities. It incorporates measurements of speed, pressure and other critical process parameters.

Process Development—The development of products and processes, as well as the modification of existing processes, should be conducted to provide documented evidence of the suitability of all critical process parameters and operating ranges. This effort serves as a baseline for all product validation activities. The integration of development into commercial scale operations became a requirement with the advent of the FDA's Pre-Approval Inspection Program.[16] The importance of well-documented developmental activities to support subsequent commercial-scale production is essential in the validation of API. It is customary for many unit operations (reactions, separations, catalyst reuse, solvent reuse, etc.) to be initially confirmed on a laboratory or pilot scale prior to their eventual 'validation' on a commercial process scale.

Process Documentation—An oft-overlooked activity wherein the results of the development effort are delineated in sufficient detail in process documentation so that the variations in the process as a result of inadequately defined procedures are eliminated. While master batch records have long been a CGMP requirement, their adequacy is essential to the maintenance of a validated state.

Process Performance Qualification (Testing)—That portion of the overall 'validation' program that deals specifically with the evaluation of the process and the initial confirmation of control and consistency. It includes the protocol development, data acquisition, report preparation and requisite approvals. In the distant past this activity was considered 'process validation', but over the years the industry has come to realize that 'validation' encompasses a broader spectrum of activities, and continued use of the word 'process' is limiting.

Change Control—A CGMP requirement that mandates the formal evaluation of the consequences of change to products, processes or equipment. At least two distinct types of change control exist because of the different disciplines that are central to the evaluation of each.[17]

Process Change Control—A system whereby changes to the process are carefully planned, implemented, evaluated and documented to assure that product quality can be maintained during the change process. This type of change control is the province of the developmental scientist and production personnel. These changes include materials, process parameters and sequence of operations.

Equipment Change Control—A mechanism to monitor change to previously qualified and/or validated equipment to ensure that planned or unplanned repairs and modifications have no adverse impact on the equipment's ability to execute its intended task. This procedure usually entails close coordination with the maintenance and engineering departments.

36.8 ACTIVE PHARMACEUTICAL INGREDIENT VALIDATION—WHAT MAKES IT DIFFERENT?

The focus of this chapter is API validation. To this point, aside from the history section, the information presented would apply to almost any process validation. That commonality with other older validation efforts is deliberate. API validation is unique only to the extent that APIs are unique. The underlying maxims of success for validation, the knowledge and understanding of the scientific basis upon which the equipment or process is based, are universal. Mastery of the overall approach equips one to effectively employ those concepts in a variety of settings. On the other hand, knowledge of some unique concerns in the production of APIs is essential to understanding how the validation of their preparation should be carried out.

Life cycle validation concepts apply to all stages of API manufacture. This means that each step from the introduction of a starting material should be well characterized and understood based upon developmental data and production experience. ICH Q7A defines an API starting material as "a raw material, intermediate or an API that is used in the production of an API and that is incorporated as a significant structural fragment into the structure of the API".[18] Steps prior to the 'starting' step often take place outside of the API manufacturing site and are not considered in this discussion.

Although validation concepts of process understanding and control apply, many firms do not require formal validation documentation until some later process stage that is determined to have a direct effect on the quality of the final API. One approach suggested in the FDA's API guideline is to begin formal validation documentation at the so-called API step, the step where the molecule contains the chemical moiety determined to embody the pharmacological activity of the API.

In any case, regardless of when and where formalized documentation is required, the principles of this chapter apply across all API manufacturing steps. Adequate documentation should be provided to ensure a continuing understanding and control of even the early steps of the process and address the following, each of which is presented as it relates to APIs.

Unit Operations—APIs are the result of a series of chemical reactions in which materials called reactants are brought together under appropriate conditions whereby the reaction occurs and the reaction product is formed. Under even the most ideal circumstances, the desired product must be separated from unreacted raw materials, by-products, solvents and processing aids before it can be utilized in further processing. In the analysis of these processes, chemical engineers have found it convenient to divide the overall process into a series of unit operations, some of which rely on chemical processes, while others are physical in nature. The unit operations approach is beneficial because a complex many-step process can be separated and better understood as a series of simpler activities (unit operations) that are more easily understood and controlled.

Among the more common unit operations are mixing, heating, drying, absorption, distillation, condensation, extraction, precipitation, crystallization, filtration and dissolution. There are other less common unit operations, but the more important aspect is the subdivision of a lengthy process into smaller and more readily understood segments. The benefits to be gained from this approach are obvious, once the underlying principles are understood for a specific unit operation; those concepts can be reapplied in other steps or processes where that same unit operation is employed. In the validation of API processes, the ability to use standardized methods for each unit operation can make what would otherwise be an impossible task into a manageable one. The unit operation approach is of such utility that it has been applied in pharmaceutical dosage form manufacturing, where the same basic procedures are often encountered, that is, mixing, milling, filtration, etc.

Physical Parameters—A concern that has been sometimes neglected in the preparation of APIs relates to the control of physical parameters of the end product material. Often the focus of API development and processing is placed on chemical purity and yield, as those aspects tend to have the greatest economic significance. There is relative indifference to physical parameters such as size, shape and density compared to the seemingly more important concerns such as potency, impurity levels and process yield. The authors have observed numerous situations where this inattention has resulted in processing problems at the dosage form manufacturing stage. In each instance, it was often the case that the physical parameters of the end product had been virtually ignored in deference to concerns over chemical purity.[19] The FDA's Pre-Approval Inspection initiative indicated an awareness that these concerns have come to their attention during the course of NDA reviews and inspections.[20]

The most extreme circumstances where physical parameters are of critical importance are for those materials where different crystalline forms are possible. The different polymorphs may have decidedly different characteristics with regard to crystal shape, size and, most importantly, solution characteristics. Many important active ingredients exist in more than one crystalline form, and the manufacturer must ensure that only the desired form is being produced. One of the major concerns voiced by regulators is the potential hazard in using brokered active ingredients.[21] The ability to match the purity profile of an API is not sufficient if the crystallization is from a different solvent system or at different conditions. An entirely different material may result, with profoundly different pharmacological properties. The absence of detailed information on the isolation process used may cause difficulties should the real source of the material (the broker's supplier) change.

Chemical Purity—Central to the preparation of APIs are issues relative to the purity of the desired material. Until recently, the only concern was whether the material met the minimum potency requirements. A typical requirement would be a minimum potency specification of 98%. Any lot that had an assay higher than 98% would be acceptable. Awareness that the small amount of material that is not the desired molecule could cause adverse reactions led to the establishment of purity profiles for APIs. Using a purity profile approach mandates that the firm identify the impurities present. Current FDA expectations are that firms should characterize all impurities that comprise more than 0.1% of the final API and perform toxicity testing on any impurity that is at a concentration higher than 0.5%.[22] The establishment of a purity profile for a molecular entity assures that process changes that might result in a change in the byproducts and other materials isolated with the desired material do not impact the safety and efficacy of the final product.

Analytical Methods—As with other types of product validation activities, API validation cannot proceed without validated analytical methods. The most significant difference in the validation of APIs is the number of analytical methods that must be addressed. Analytical methods are needed for each intermediate stage, identifying and quantifying the major byproducts at each stage as well as the desired chemical moiety. Clearly, the scope of the analytical method validation for APIs represents a larger effort than is normally associated with process validation activities. A comprehensive review of analytical method validation can be found elsewhere in this volume.

Facilities—API facilities are vastly different from most other types of facilities in the pharmaceutical industry. The equipment is designed for specialized procedures and, as such, bears little resemblance to those that might be found in a dosage form facility. Most API equipment requires a broader range of utilities and a seeming maze of piping to perform properly. Chemical reactions are sometimes performed at temperatures well in excess of 120°C or less than 0°C and required specialized heat-transfer fluids to maintain those temperatures. Many reactions utilize solvents as reaction substrates or in the isolation of the materials. These solvents may be introduced via piping systems that supply the various pieces of equipment. Distribution systems for compressed gases used either in the reaction or to provide inert conditions within the equipment are also common. In many older API facilities, it is common to see multiple vessels at different elevations arranged around an open bay. In these facilities, several different chemical reactions might be underway in different vessels for different products at the same time. In a dosage form facility, this type of arrangement would be viewed with considerable skepticism. In API production, the reactions and vast majority of the unit operations take place within closed equipment, minimizing the potential for cross-contamination. The difference between API and dosage form facilities is most evident in warmer climates. In these areas, the API facility may be little more than a structural support for the equipment and staging areas for material, with no surrounding building. In effect, the equipment is outside, fully exposed to the environment. For certain API processes such as solvent recovery and hydrogenation vessels, the equipment is located outside even in northern climates, either because of sheer size or safety concerns. These types of arrangements are not typical for the later steps in the synthesis. Isolation of the completed API is usually performed in rooms specifically designed for that purpose and entails some level of environmental control to help ensure the purity of the API if it will be exposed. True clean rooms (i.e., rooms meeting the requirements of an international standard, such as ISO) are rarely employed.

Pure Rooms—In the preparation of APIs, it is common for the last step in the process to be completed in an environment far different from that in which the rest of the synthesis is performed. The term "pure room" is used loosely, as there are no regulatory requirements for these rooms, and the actual terminology varies considerably from firm to firm. Even without regulatory impetus, some firms have gone so far as to classify their 'pure' rooms at Class 100,000 (ISO 8) or better.[23,24] After the crystallization of the API, it is important to protect the product from airborne particulates and other foreign matter that might end up in the finished material. For this reason, it is common in many companies to perform a filtration of the active material while still in solution. The filtration removes particulates that may have accumulated in the material up to that point. After the filter, the solution is introduced into the crystallizer in the 'pure' room. The room itself is designed to minimize the opportunity for introduction of contaminants into the bulk material and may or may not be a classified environment. The crystallizer is often subjected to rigorous cleaning before the start of the process to ensure its suitability for the final bulk isolation. Following the crystallization, the API is centrifuged or filtered, washed, dried, milled and packaged in the 'pure' room. It should be noted that API processes which use 'pure' rooms are not intended to be sterile. The production of sterile APIs requires a much higher level of control over the environment, equipment and methodologies and is described elsewhere in this text.

Qualification of Equipment—The qualification of API process equipment—including reaction vessels, receivers, crystallizers, centrifuges, dryers, filters, distillation columns, solvent distribution systems, etc.—is a well-defined activity. While this equipment is somewhat different in design and operating features than the dosage form equipment that has been the subject of the majority of papers on the subject, the same general principles apply. Reaction vessels, receivers and crystallizers differ only minimally from formulation and Water for Injection tanks. Some API dryers are identical to those utilized in tablet departments. Solvent distribution systems are piping systems and may resemble pharmaceutical water distribution systems. Some pieces of equipment such as distillation columns and continuous reactors may not have counterparts in the dosage form side, but an understanding of the objectives of the equipment qualification should make the development of suitable protocols straightforward. In general, each unit operation including and subsequent to the API step must follow formal qualification procedures.

Configuration Confirmation—In multipurpose API facilities, the fixed equipment installed may be configured differently for different reactions. In these facilities, campaigns of one reaction may be followed by a reaction for a different product after a change in configuration. Configuration changes often include rerouting fluid flows and piping connections. Putting aside cleaning considerations for a later portion of the chapter, verification of the system's configuration should be performed. In effect, the reaction train must be requalified at the start of each campaign to ensure that the proper arrangement of valves, transfer lines, instruments and other items are established for the process to be introduced. Some firms run a water or solvent batch that simulates the process to verify that the proper connections are in place and that there are no leaks in the system. Following this trial batch, the system is then readied for use with the solvents that will be utilized in the process.

Environmental Control—The usual concerns relative to the environment in which the production activities are performed are not as significant in API manufacturing as they are for the preparation of pharmaceutical dosage forms. The introduction of microbial or particulate contaminants at early stages of the process is unlikely to be of significance. API reactions utilize high temperatures, extremes of pH and aggressive solvents that can minimize the impact of any microbial contamination. Filtration is a frequent part of API processing and may include activated charcoal treatments and other unit operations intended to remove unwanted by-products, reactants and solvents. In the course of these measures, incidental particulate contamination is also removed. The use of 'pure' rooms as outlined earlier serves to minimize contamination over the final steps.

Worker Safety—The safety of the personnel who work in the facility is always a major concern. Exposure to toxic substances is greatest when the operator is adding materials to or removing materials from the equipment. The use of air extraction equipment, isolation technology, automated handling and other means for minimizing human contact with toxic materials is nearly universal. The assessment of worker safety should also embrace exposure to vapor phase hazards, and leak testing of process trains should be performed where hazardous gases are present. Validation of the effectiveness of this equipment is not mandatory from a CGMP perspective but is certainly beneficial and should be tested in commissioning.

Process Water—The water used in API production is usually deionized water through the early process stages. If the product is isolated from a water solution in its last step, then a compendial grade of water, purified water or WFI may be utilized depending upon subsequent steps in dosage manufacture and the final use of the product. Cleaning of equipment can be performed with city water, provided the last rinse of the equipment is with the same quality of water utilized in the process step. The validation of water systems has been well documented in the literature.[25,26]

Process Gases—Some API reactions utilize gases as reactants or are performed under a gas blanket. The gas delivery system may start at either a large high-pressure bulk storage tank or a bank of gas cylinders. Attention should be paid during the installation of the system to assure that the materials of construction utilized in the system are compatible with the gas being handled. Distribution systems for these gases require qualification, but their similarity to gas distribution systems used in dosage form facilities means that the basic approach is well defined in the literature. For safety considerations, particular attention should be paid to proper identification of process gas lines throughout the facility (see following paragraph).

Compressed Air—Air which is classified as breathable should receive an intensive qualification effort, especially with regard to the verification of 'as built' drawings and confirmation of proper identification, as well as any safety and purity related issues. The emphasis given to breathable air is due to the number of unnecessary deaths that have occurred in the industry as a consequence of misidentified gas lines. Where air is utilized as a reactant in an API operation, it should be treated as described previously under "Process Gases". Instrument air requires the least intensive effort, as the adequacy of the installation can be often confirmed indirectly during the calibration and qualification of the process instrumentation. A single compressed air system could serve as the source for more than one of these air systems simultaneously. In this instance, the advice provided for the most critical application is appropriate throughout.

Jacket Services—It is common in API facilities, especially those that are reconfigured frequently to accommodate the production of different materials, to have each major vessel equipped with identical utilities, such as chilled water, plant steam and compressed air. The use of identical utility configurations on the vessels maximizes the flexibility of the facility, reduces the potential for operator error and simplifies the design of the facility. The control systems for these jacket services on the vessels would also be identical. Under these circumstances, the qualification effort is greatly simplified through the use of identical requirements.

Solvent Distribution—Many facilities use one or more solvents repetitively. In these instances, the installation of a dedicated distribution system for the solvent to the various use points can be justifiable. These systems may be lengthy lines from the bulk storage area (tank farm) to the various locations in the facility where the solvent is required. In some cases, a chilled solvent system may be present to provide chilled washes. Depending upon the solvent, specialized piping or gasket materials may be necessary to avoid leaks or corrosion of the system. Qualification of these distribution systems is easily accomplished.

Solvent Recovery and Reuse—The reuse of organic solvents in an API system is widespread, especially given the increased cost of these materials and the environmental difficulties associated with their proper disposal. This reuse is achieved through defined procedures for the recovery of the solvents from distillates, extractions and spent mother liquors. Where recovered solvents are utilized in the production of an API, the validation of the recovery process is strongly recommended. The validation of the recovery process would include all steps in the process and confirm the acceptability of the recovered solvent in the processes it will be utilized in. The validation of the use of recovered solvents could be a part of the development of the process. Repeated recycling of solvents could result in the concentration of trace impurities that could adversely affect reaction chemistry. At the very least, recovered solvents should be subjected to release testing and shown to be comparable in use to fresh solvent. The complexities associated with the validation and reuse of recovered solvents should not be overlooked.

Multiple Crops—In the crystallization of some APIs, multiple crystallization crops from the same mother liquor are sometimes utilized to maximize the amount of material isolated. Even where the cost of the materials being isolated is not high, the ability to increase the overall yield through the collection of multiple crops is frequently a routine part of the process. A related technique is to recycle the mother liquors without additional treatment from the crystallization back to the beginning of the process. Whether through multiple crops or recycling of the mother liquor, both of these processes may result in the concentration and/or retention of impurities. The validation of these practices must be a part of the development effort for the process and reconfirmed on the commercial scale.

Catalyst Reuse—Precious and semiprecious metals and other materials are often utilized as catalysts in the conduct of certain chemical reactions, for example, hydrogenation. While the quantity of catalyst required in any particular reaction is quite low, the cost of these metals is such that recovery is mandated. As the amount of catalyst required to support the reaction is generally supplied in excess, it is frequently possible to return the catalyst to the start of the process step without loss in effective yield. The reuse of the catalyst in this manner must be supported by appropriate development work.

Waste Treatment—The nature of the materials, by-products and solvents utilized in the preparation of APIs ultimately results in any number of waste treatment problems. The validation of these treatments is certainly **not** a CGMP required activity. Nevertheless, consideration should be given to those activities to insure their reliability. Such efforts can aid in attaining environmental approval for the facility.

36.9 IN-PROCESS CONTROLS

Bulk pharmaceutical chemicals resemble other types of products validated in the pharmaceutical industry in that they utilize various in-process controls to support and monitor the process through its execution. Typical controls that might be a part of an API process relate to materials include the following.

Material Specifications—The controls of reactants, solvents, intermediates and finished materials employ formal specifications for key parameters. The importance of these controls increases toward the end of the synthesis, and any of the controls that follow the API step are certainly important enough that the efficacy of limits set for these controls should be a major part of the developmental process. Foremost among the considerations in the latter process steps should be the purity profile of the key intermediates (see following paragraph). Physical parameters (size, shape, crystalline form, bulk density, static charge, etc.) of the finished API are sometimes considered less important them chemical purity. When the API is formulated in a solid or semisolid dosage form, these physical parameters may assume far greater significance.

Purity Profiles—Within the specification parameters, prominence is often given to the establishment of purity profiles for the key intermediates and finished goods. The FDA mandates the identification of all impurities with a concentration greater than 0.1% and generation of safety and other critical information for impurities at levels of 0.5% or higher.[27] The establishment of purity profiles for the final APIs provides for confirmation of the safety of the active material. It is often beneficial to establish purity profiles for intermediates earlier in the synthesis to prevent the carryover of impurities to the finished API. The maintenance of the purity profile mandates that a careful evaluation of process changes and potential alternate suppliers of solvents, raw materials, intermediates and APIs be made. The analytical method development and synthetic chemistry skills required to obtain the necessary data on impurities meeting the FDA's criteria are substantial. These efforts are well rewarded in an expanded knowledge of the process chemistry and analysis that can assure the quality of the desired active moiety.

Vendor Support to Validation—A frequent practice in API production is the sub-contracting of certain chemical steps to outside suppliers. As is the case with sub-contracted production for dosage forms, the owner of the NDA or DMF maintains responsibility for the validation of the process and must secure the cooperation of the subcontractor in the performance of any supportive qualification/validation activities. Agreement to this arrangement should be a precondition to the awarding of the contract to the supplier.

Supplier Quality Evaluation and Audits—Suppliers of intermediates, reactants, solvents and other materials should be subjected to the same types of evaluation utilized for other dosage forms. The extent of the assessment should vary with the importance of the material to the process. Precedence would be given to those materials whose purity would have the greatest impact on the finished API. Where the material being produced by the vendor has direct impact on the API's quality, as would be the case for chemical intermediates, a more intensive approach is required. Periodic audits of these key suppliers should be a part of the overall quality assurance program.

Sampling Plans—Obtaining samples of finished APIs or their intermediates presents the same difficulties encountered in the sampling of any similar material. When samples are taken of powder or crystalline materials, questions regarding the uniformity of the material being sampled must be addressed before the results of the sampling can be considered meaningful. APIs that are dried in rotary or fluidized bed dryers may be blended sufficiently as a result of the drying process. However, where tray dryers are utilized, a final blending of the dried material may be required before sampling for release to the next stage of processing. In certain instances, an intermediate or finished material will not be isolated as a dry powder but will be released as a solution in an appropriate solvent. Under these circumstances, concerns regarding the sampling of the material are minimized.

Particle Sizing—Milling and micronizing are common activities in the final stages of API manufacture. These procedures are utilized where the API producer has committed to providing a particular particle size for use in the formulation. Given the importance of particle size in many final dosage forms, where present these processes should be validated. Control of the final particle size for finished API should not rely on the milling/micronizing step alone. Control over the crystallization procedure is generally necessary to minimize the variation in the material that is to be sized in the mill. It should not be assumed that the milling/micronization procedure will be tolerant of a wide range of in-feed materials and still provide a consistently sized finished API product. The uniformity of materials is sometimes improved by passage through a particle sizing procedure or sifter, but this step alone should not be considered sufficient to achieve a uniform mix of the material prior to sampling (see previous paragraph).

Reprocessing—There is occasional need to reprocess an intermediate or finished API in order to alter its crystal size, reduce impurities, or otherwise recover off-specification material. Where these processes are utilized, their inclusion in the validation program is essential. FDA and other regulatory requirements on reprocessing and reworking of materials require the validation of any material reclaimed in this fashion. This is most readily accomplished as a part of the developmental process.

36.10 CLEANING VALIDATION

A comprehensive discussion of cleaning validation is beyond the scope of this chapter; the reader should refer to other sources on cleaning validation for details of this activity.[28,29,30] Within the context of this chapter, only those aspects of cleaning validation unique to API production will be presented. Additional guidance can be found in FDA's API Inspection Guide.[31] Cleaning effectiveness is an area of much greater concern in recent years, especially for multiuse equipment. These concerns increase for unit operations closer to the final isolation and purification of the API but are still significant in the earlier reaction steps. While debate continues over the acceptable level of carryover from one campaign to the next, most firms have established cleaning acceptance criteria that are difficult to satisfy with the traditional chemical manufacturing equipment used throughout most of the industry. The cleaning methods and practices described here often require multiple iterations to reach acceptable visual and chemical test limits. Newer facilities are beginning to incorporate design criteria to enhance cleanability. For example, final isolation, purification and drying steps may use equipment identical to drug product manufacture. Even reactors and piping, if not truly sanitary in design, are incorporating many of the cleanability concepts used for drug products. These include fewer fittings, elimination of screwed fittings and flanges, better surface finishes and more easily cleanable valves.

Boil-outs—Commonly used to clean API equipment, boil-outs entail the introduction of the solvent (it could be water) used in the just completed process and heating it to reflux. The expectation is that the evaporation/condensation will result in the dissolution of any residue on the equipment in the solvent. This will remove it from the internal surfaces that are ordinarily inaccessible for direct cleaning and thus clean them. Boil-outs are also utilized as one of the last steps in preparation of equipment for the start of a process or campaign.

Lot-to-Lot Cleaning—As the production of APIs often requires that solvents and materials with substantial toxicity must be employed, cleaning of the equipment after completion of the process has the potential for exposure of the worker to those materials. For this reason, it is common in API facilities to include some basic forms of waste treatment and equipment cleaning directly into the process in an effort to minimize worker exposure later on. In addition to these measures, many processes include the reuse of equipment and retention of materials in the equipment without cleaning. A typical instance would be leaving a heel in the centrifuge at the completion of the batch, thereby eliminating cleaning of the centrifuge after each batch. The retention of the heel must be validated because it represents a portion of the first batch, which may now become a part of subsequent batches. In fact, each batch in the entire campaign is potentially mixed with material from every prior batch! In this manner, the amount of cleaning required between batches of the same reaction step would be reduced. In those facilities where a process train is essentially dedicated to the same reaction step over a long period of time, the equipment and process are specially designed to minimize batch-to-batch cleaning of the equipment. There are of course instances where the presence of even trace quantities of finished material at the start of the reaction may create an undesirable outcome; in those circumstances, the equipment must be cleaned after the completion of each batch. Sparkler and other filters used to recapture catalysts, activated carbon used for decolorization and by-products may require cleaning after every batch.

Campaigns—The production of a number of batches of an identical synthesis in the same equipment is common in the manufacture of APIs. As mentioned earlier in relation to the qualification of equipment, production in a campaign mode may require the partial reconfiguration of the equipment train to allow for a new campaign. This may be a reaction leading to the same or a different API. To allow for campaign usage, the extent of cleaning required will generally be far greater than what is carried out between batches of the same process step. Cleaning limits for campaign cleaning are generally tighter than those applied for batch-to-batch cleaning. It is beneficial in campaign cleaning to follow a defined plan for changeover from one product to another.

Sampling for Residuals—In order to determine whether a piece of equipment has been appropriately cleaned, sampling is performed. Here again, the particular nature of the API materials makes for a more difficult situation. In dosage form manufacturing, relatively few of the materials likely to be retained on the surface of the equipment pose any substantial risk to the worker. In those dosage form processes where toxic or potent materials are handled, the design of the equipment with smooth surfaces, rounded corners, sanitary fittings, etc. reduces cleaning difficulty. The same equipment design principles make sampling of pharmaceutical equipment relatively simple due to provisions for access and inspection. The bulk of API equipment is designed to operate under more aggressive conditions and cannot always integrate the design features so commonly found in their pharmaceutical counterparts. Moreover, worker safety becomes a far greater concern, as the solvents and materials are not conducive to direct exposure to the employee. Sampling of API equipment may be restricted to fewer locations, and those locations may not be in the most difficult to clean or 'worst case' locations. For this reason, the residual limits for APIs may need to be far lower to accommodate the uncertainty of the sampling that can be performed.

36.11 COMPUTERIZED SYSTEMS

The application of computerized systems in the pharmaceutical industry is perhaps greater in API processing than in any other. Distributed Control Systems (DCS) have been utilized for many years in the control and regulation of chemical process plants. Their adaptation to API preparation is straightforward. The validation of computerized systems in the pharmaceutical industry has been extensively discussed, with the constant recognition that their broad usage in API production was a given.[32,33] Industry and regulatory guidance having always recognized this fact, this chapter could not hope to do justice to the subject, which has filled several textbooks on its own. The reader is encouraged to follow the recommended practices of PDA, PhRMA and GAMP.

Process Analytical Technologies are becoming of increasing importance in dosage form manufacturing as a consequence of FDA encouragement.[34] Those trained in process control see these as little more than adaptations of methods chemical engineers have utilized for years in continuous processing. Their application in API production is certainly possible and will likely be facilitated by the familiarity the professional staff has with both automation and process control.

36.12 PROCEDURES AND PERSONNEL

Where computerized systems are not utilized for the execution of the chemical synthesis, the chemical operator, following detailed batch records, is responsible for the operation of the equipment. The batch records must provide for sufficient detail to ensure that the worker can safely and properly perform the desired actions. In certain larger process trains, more than one operator will work simultaneously on the same batch. Provided that their activities are coordinated, there is little problem with this type of approach. The personnel must be trained in their jobs, and records of the training must be retained by the firm.

36.13 VALIDATION OF STERILE BULK PRODUCTION

The preparation of APIs that must be sterile upon completion of their synthesis and purification is a common activity in the pharmaceutical industry and increasingly common in biotech processes. The validation of sterile APIs represents one of the more difficult activities in the entire spectrum of validation. Not only must the final material meet all of the physical and chemical requirements associated with other APIs, it must also be free of microorganisms, endotoxins and particulates. In doing so, all of the considerations for validation of APIs outlined in this chapter must be addressed, with added concern for sterilization, environmental control, aseptic technique and other subjects associated with the production of sterile products. This subject has been addressed in a separate chapter in this volume.

36.14 CONCLUSION

This chapter has provided an outline of validation considerations relative to the production of bulk pharmaceutical chemicals. This is a subject that has only recently become of interest to the pharmaceutical community. The authors, while familiar with both validation and bulk pharmaceutical processing, have undoubtedly mentioned any number of issues that may not yet be embodied in validation protocols within operating companies. We have included these issues to ensure completeness in the presentation, not to suggest that they be included in every validation effort. As time passes, the industry will gain experience with the validation of APIs and will perhaps exclude some of these issues, while including other aspects we have not identified. Our intent in this effort has always been to integrate common validation practices with the unique aspects of bulk pharmaceutical manufacturing. By no means do we expect this to be the definitive effort on this complex subject. The reader is encouraged to monitor industry and regulatory developments relative to API validation as substantial changes in CGMP requirements for APIs appear likely.

NOTES

1. Food and Drug Administration, 21 CFR, Part 610.
2. PhRMA Quality Control Bulk Pharmaceuticals Working Group; PhRMA Guidelines for the Production, Packing, Repacking or Holding of Drug Substances, *Pharmaceutical Technology*, Part 1, December 1995, Part 2, January 1996.
3. Food and Drug Administration, Guidance for Industry, Q7A Good Manufacturing Practice Guidance for Active Pharmaceutical Ingredients, 2001.
4. Food and Drug Administration, Guidance for Industry, Process Validation: General Principles and Practices, 2011.
5. Food and Drug Administration, Guideline on General Principles on Validation, 1987.
6. Food and Drug Administration, Guide to Inspection of Bulk Pharmaceutical Chemicals. 1994.
7. Food and Drug Administration, 21 CFR, Part 211.
8. Food and Drug Administration, Guidance to Industry: Manufacturing, Processing or Holding Active Pharmaceutical Ingredients, March 1998.
9. Pharmaceutical Inspection Convention, Internationally Harmonized Guide for Active Pharmaceutical Ingredients—Good Manufacturing Practice, September, 1997.
10. Agalloco J, The Validation Life Cycle. *Journal of Parenteral Science and Technology* 1993;47(3).
11. Food and Drug Administration, Guide to Inspections of Oral Solid Dosage Forms Pre/Post Approval Issues for Development and Validation, January 1994.
12. European Medicines Agency, *Guideline on Process Validation for Finished Products-Information and Data to Be Provided in Regulatory Submissions*, 2014.
13. Agalloco J, Validation: Yesterday, Today and Tomorrow. Proceedings of Parenteral Drug Association International Symposium, Basel, Switzerland, Parenteral Drug Association, 1993.
14. ISPE, Baseline Guide: Volume 5—Commissioning and Qualification, 2001.
15. ASTM, E 2500–07, Standard Guide for Specification, Design and Verification of Pharmaceutical and Biopharmaceutical Manufacturing Systems and Equipment, 2007.
16. Food and Drug Administration, Guide to Inspections of Oral Solid Dosage Forms Pre/Post Approval Issues for Development and Validation, January 1994.
17. Agalloco J, Computer Systems Validation: Staying Current: Change Control. *Pharmaceutical Technology* 1990;14(1).
18. ICH, Q7A, Good Manufacturing Practice Guide for Active Pharmaceutical Ingredients, 2000.
19. Agalloco J, Personal Communications, 1972–1990.
20. Food and Drug Administration, Guide to Inspections of Oral Solid Dosage Forms Pre/Post Approval Issues for Development and Validation, January 1994.
21. Food and Drug Administration, Guide to Inspection of Bulk Pharmaceutical Chemicals, 1994.
22. Food and Drug Administration, Guide to Inspection of Bulk Pharmaceutical Chemicals, 1994.
23. Federal Standard 209E, Airborne Cleanliness Classes in Cleanrooms and Clean Zones, September 1992.
24. EU Guide to Good Manufacturing Practice for Medicinal Products, Annex 1—Manufacture of Sterile Medicinal Products.
25. Meltzer T, *Pharmaceutical Water Systems*. Littleton, CO: Tall Oaks Books, 1996.
26. Artiss DA, *Water Systems Validation@ in Validation of Aseptic Pharmaceutical Processes*. Carleton F, Agalloco J, eds. New York: Marcel-Dekker, 1986.
27. Food and Drug Administration, Guide to the Inspection of Bulk Pharmaceutical Chemicals, 1994.
28. Agalloco J, Points to Consider in the Validation of Equipment Cleaning Procedures. *Journal of Parenteral Science and Technology* 1992;46(5).

29. PDA, Points to Consider for Cleaning Validation. *PDA Journal of Pharmaceutical Science and Technology*, PDA Technical Report #29 1998;52(6), Supplement.
30. Voss J, et al., *Cleaning and Cleaning Validation: A Biotechnology Perspective*. PDA, Bethesda, MD 1995.
31. Food and Drug Administration, Guide to the Inspection of Bulk Pharmaceutical Chemicals, 1994.
32. Harris J, et al., Validation Concepts for Computer Systems Used in the Manufacture of Drug Products. Proceedings: Concepts and Principles for the Validation of Computer Systems in the Manufacture and Control of Drug Products, Pharmaceutical Manufacturers Association, 1986.
33. PDA, Validation of Computer-Related Systems. *PDA Journal of Pharmaceutical Science and Technology*, PDA Technical Report #18 1995;49(1), Supplement.
34. FDA, Guidance for Industry PAT—A Framework for Innovative Pharmaceutical Development, Manufacturing, and Quality Assurance, 2004.

37 Cell Culture Process Validation Including Cell Bank Qualification

Anne B. Tolstrup, Steven I. Max, Denis Drapeau and Timothy S. Charlebois

CONTENTS

37.1 INTRODUCTION

Fundamentally, the objective of any cell culture validation program with the purpose of producing biotherapeutics for human treatment is to demonstrate that both the production cell line and the upstream manufacturing process are suitable for their intended purpose and perform robustly within the established manufacturing ranges in such a manner that a consistent product of the desired quality will result. Manufacturing of biologics has traditionally been performed in a variety of microbial and mammalian host cells. Figure 37.1 (top panel) shows all the different living hosts that have been employed since the start of biologics manufacturing in the early 1980s. It is noted that *E. coli*, which in the beginning was the most frequently used host, was overtaken by the mammalian Chinese Hamster Ovary (CHO) cells in 2002. Use of mammalian expression systems has become necessary because the approved biopharmaceutical proteins became more complex in structure and size, often with the requirement of posttranslational modifications such as glycosylation. The change towards mammalian cells as the preferred host cell line has been driven by the huge success of monoclonal antibody therapeutics that have been approved over the past 30 years for treatment of cancer and autoimmune diseases. Figure 37.1 (bottom panel) shows the types of mammalian host cell used for manufacturing of all recombinant proteins including monoclonal antibodies approved for human therapy between 1984–2020. CHO cells are by far the most frequently used host cells, with an 80% prevalence among mammalian manufactured products and >50% among all biopharmaceutical products.

DOI: 10.1201/9781003163138-37

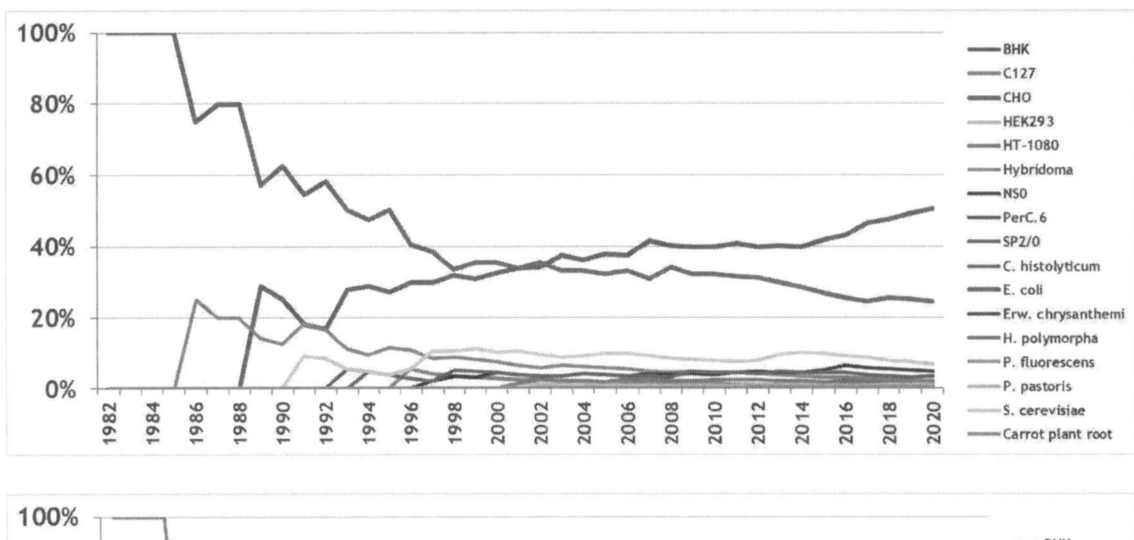

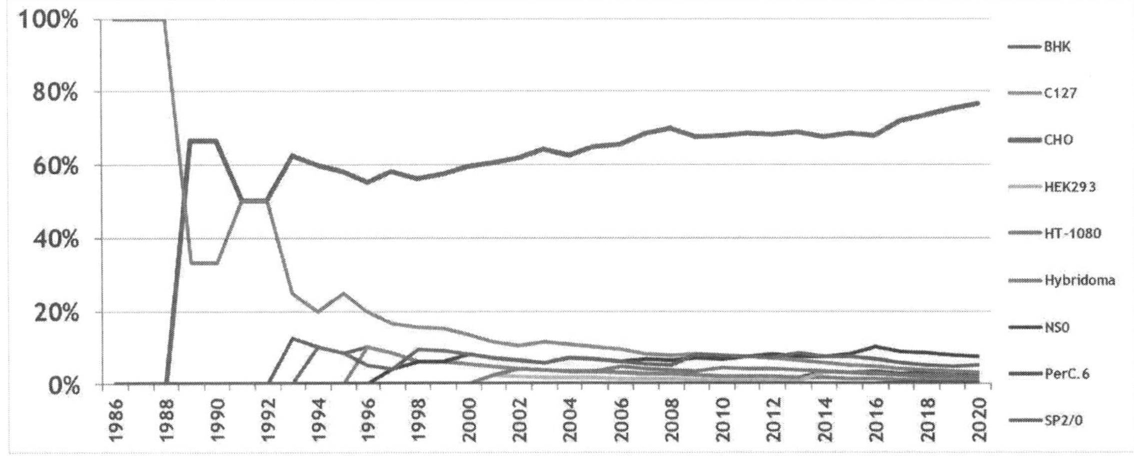

FIGURE 37.1 Top panel: All production cell lines used to manufacture approved recombinant proteins including monoclonal antibodies, shown as percentage of the total number of approved products from 1984–2020. Bottom panel: Mammalian production cell types used to manufacture approved biopharmaceuticals.

Source: Data and graphs are derived from BDO's BPTG bioTRAK® database: www.bdo.com/industries/life-sciences/bioprocess-technology by courtesy of Dawn Ecker.

In this chapter, the focus will be on mammalian cell culture processes used to produce monoclonal antibodies and other recombinant proteins for human therapy, since these present certain unique challenges from a process validation perspective. Case studies and examples are provided to illustrate how key validation principles and guidelines have been interpreted and applied in support of manufacturing processes currently used for commercial manufacturing. Over the last 20 years, mammalian biopharmaceuticals manufacturing has relied strongly on the use of CHO cells as a production system (Figure 37.1), and thus the examples are drawn substantially from this mammalian expression platform but are representative of the issues, thinking and approaches that must be applied to any mammalian cell culture process.

37.2 PROCESS VALIDATION GUIDING PRINCIPLES

As instructed by FDAs Guidance for Industry "Process Validation: General Principles and Practices" published in 2011, the process validation concept should be understood as a product life cycle approach comprising three stages:

- **Stage 1**: Process Design, which includes development and scale-up of the process followed by thorough process characterization;
- **Stage 2**: Process Performance Qualification (PPQ) to confirm that a robust and reproducible manufacturing process is available prior to commercialization; and
- **Stage 3**: Continued Process Verification (CPV), which is post-marketing, ongoing assurance that the process remains in control.

The EU guidance is in agreement with the FDA mindset around process validation, although they use slightly different language; for example, they speak about Ongoing Process Verification (OPV) rather than CPV in their Eudralex Chapter 4 GMP guidelines Annex 15 Qualification and Validation Guideline (2015).

The key areas of importance for cell culture with respect to process validation include evaluation of the production cell line suitability including cell line identity, purity and stability, upstream process design and development, process characterization and process performance qualification (PPQ) as well as CPV throughout the product life cycle. These points are addressed in turn with the main focus on Stage 1, which includes the cell line development, establishment of Master Cell Bank (MCB) and Working Cell Bank (WCB) and design and scale-up of the upstream manufacturing process. Stages 2 and 3 are described more briefly, outlining the general concepts for PPQ and CPV, respectively, with examples taken from the cell culture elements of the manufacturing process. Complementary information about process validation concepts can be found in Chapter 38.

37.3 CELL LINE DEVELOPMENT

Cell Line Development (CLD) is not done under GMP, and qualification or validation is therefore not a regulatory requirement. However, a very high level of documentation is required, since the CLD including the selected host cell line is the prerequisite for the most important raw material necessary for biotherapeutics manufacturing: the MCB. The history of the empty host cell line must be available with detailed descriptions of the raw materials used for cultivation as well as the tests that have been executed on the host cells in connection with the MCB qualification (see Sections 37.4 and 37.5). Increasingly after the mad cow disease occurrences in the 1990s, special attention has been given to animal-derived raw materials, especially to bovine materials and materials originating from other species susceptible to transmissible spongiform encephalopathies (TSE), and documentation of bovine spongiform encephalopathies (BSE)/TSE status of raw materials is a requirement.

Furthermore, during the past ten years FDA has had an intense focus on "clonality" documentation of the manufacturing cell line, and investigators are expected to provide detailed descriptions on how their production cell lines are generated and to demonstrate a high probability of "clonality" for any novel production cell line (see Rachel Novak, 2017). While biopharmaceutical industry professionals and academic experts have discussed and critically analyzed the rationale and feasibility behind this requirement intensely at conferences and in publications (Frye et al, 2016; Bandyopadhyay et al, 2017), FDA still expects such documentation to be submitted prior to market authorization at the latest. It should be noted that the European Medicines Agency (EMA) has not been following FDA regarding this request for documentation of clonality. The focus on clonality has led to development of new methodologies to support and enable cell cloning and documentation in a faster and more robust manner compared to the traditional (and still regulatory accepted) "two rounds of limited dilution cloning" approach. Examples of newer technologies and instruments which have gained acceptance as methods that can ensure a high probability of clonality include (1) single-cell sorting by flow cytometry instruments combined with imaging of the wells into which the single cells have been deposited; (2) ClonePix driven clone selection combined with one round of limited dilution cloning; and (3) the Beacon Optofluidic methodology, which uses optics/light to move individual cells into separate compartments (termed "pens") and—after expansion and productivity measurements in the pen—exports individual selected clones for further culture.

FDA does expect method validation studies to be conducted prior to performing the cloning activities, and they also expect a calculation of the "probability of clonality" with the regulatory filing prior to market approval.

In summary, CLD documentation requirements can be summarized in the following critical points:

1. Host cell line history documentation.
2. Cell Line Generation report including raw materials documentation and BSE/TSE risk assessments and/or certificates where animal-derived materials have been used.
3. Clonality method description and documentation.

The output from the CLD phase is selection of a production cell line (clone) with demonstrated robust and reproducible process performance and high productivity. A Research Cell Bank (RCB) of this production cell line is generated and cryo-preserved and will be used for establishment of the GMP manufacturing cell bank(s) (see Section 37.4). To reduce business risks at later development stages, although this is not a regulatory requirement, the RCB is often subjected to thorough testing, for expression of the correct recombinant primary sequence, for adventitious viruses, sterility, mycoplasma and for genetic stability over time. This is because a change of cell bank later in the upstream process development phase will lead to significant delays.

37.4 EVALUATION OF MANUFACTURING CELL BANK SUITABILITY

The establishment of a qualified cell banking system is critical to ensure that the starting point for GMP manufacturing remains consistent throughout the entire manufacturing lifecycle of a given biotherapeutic molecule. A two-tiered cell banking system is requested by regulatory agencies to secure unlimited supply of this essential starting material, and as such, this is now the adapted standard practice within the industry. Under this approach, a MCB is first generated from the RCB ultimately selected as the production cell line at completion of the CLD phase. The MCB is manufactured under GMP, and it is used to generate a practically endless number of WCB, thereby providing a continuous supply of well-characterized recombinant production cells for full-scale manufacturing. While it is well-accepted to initiate clinical trials (Phase I–II) based on manufacturing from the MCB, a WCB should be generated at the latest in connection with Process Characterization and PPQ of the final scaled-up manufacturing process to be used for commercial manufacturing.

Prior to start of clinical trials it is essential to confirm identity, purity and suitability of the recombinant cell used for production of the biopharmaceutical. This implies that both the host cell line used to establish the recombinant production cell line, the MCB and the WCB used for production of the biotherapeutic have to go through a number of tests demonstrating their identity (e.g., CHO cell origin), sterility, absence of adventitious virus contamination and mycoplasma and genetic stability over the defined duration of the upstream manufacturing process. Furthermore, as described in detail in Section 37.3, since 2011, FDA has requested documentation for "clonality" of the MCB.

The International Conference on Harmonization (ICH) has established guidelines containing recommendations for the testing and characterization of cell lines (ICH Topic Q5A, Q5B, Q5D and Q6B). However, the points addressed in the guidelines are not all-inclusive and may be subject to interpretation by the manufacturer. For example, it is expected that the host cell species origin (mouse, hamster, human) is considered when designing appropriate test panel for adventitious virus analysis. Also, awareness of emerging viruses such as Sars-Cov-2, including assessment of the need for specific testing, is required by the sponsor of biotherapeutics used in clinical trials. The following section will focus on efforts to establish and maintain a paradigm for the purity testing and genotypic characterization of CHO cell banks and production cell lines as it relates to industry standards and regulatory expectations. It is the intent of this paradigm to confirm the identity and purity of the cell bank as well as demonstrate the suitability of the cell line for its intended purpose.

37.5 DEMONSTRATION OF FREEDOM FROM ADVENTITIOUS AGENTS

The risk of contamination from adventitious agents is a feature common to all biological products derived from cell lines. A comprehensive program should include the use of a well-characterized host cell line, an MCB/WCB cell banking system manufactured under GMP, use of low-risk raw materials (non-animal derived), a rational testing scheme to detect a broad range of potential viral contaminants, and routine testing of production cultures. It is recognized that no cell bank testing regimen can guarantee the detection of all potential contaminants. As such, a rigorous evaluation of the ability of the downstream processes to remove and/or inactivate virus must be conducted and satisfactory results achieved. Regulatory expectations pertaining to these topics are addressed in detail in ICH Topic Q5A, and more detailed description of downstream process virus clearance concepts and studies can be found in Chapter 38.

Table 37.1 illustrates a typical cell bank testing and routine cell culture monitoring program for a recombinant CHO cell

TABLE 37.1
Cell Banks Qualification Testing

Test	Acceptance criteria	Host cell line	MCB	WCB	EOP (MCB)	EOP (WCB)
Microbial agents						
Mycoplasma	No mycoplasma detected	√	√	√	√	√
Sterility	No growth	√	√	√	√	√
Adventitious viruses						
In vitro assay (using 3 cell lines—e.g., Vero, MRC5, CHO for CHO host cells)	Negative	√ 28-Day	√ 28-Day	√ 14-Day	√ 28-Day	√ 14-Day
In vivo test for the presence of inapparent viruses	Negative	√	√		√	√
Hamster antibody production (HAP) test (for CHO)	Negative	√	√		√	
Mouse antibody production (MAP) test (for murine host cells)	Negative	√	√		√	
Mouse Minute Virus (MMV) DNA	Negative	√	√		√	
Bovine virus, 9 CFR	Negative	√	√		√	
Porcine virus, 9 CPR	Negative	√	√		√	
Retroviruses						
Transmission electron microscopic examination for A and C-type retroviruses	Report results	√	√		√	√
Mink S+L assays for retrovirus detection	Negative	√	√		√	√
Infectious retroviruses in Mus dunni Cells RT Assay	Negative	√	√		√	√
Identity						
Host cell line identity	Match to reference (e.g., CHO)	√	√		√	√

line. Testing of the MCB is extensive and serves to evaluate the purity of the cell line at the end of its development stage and to qualify the cells for introduction into the GMP manufacturing environment of production facilities.

To mitigate risks and avoid unnecessary surprises at a later stage, it is strongly recommended to evaluate the untransfected host cell line prior to transfection with the desired gene-of-interest (GOI) using a full identity and safety testing program equivalent to what is done for the MCB. At the point at which a WCB is created, a much less rigorous testing plan is typically employed, as the MCB cells are in culture for only a limited amount of additional time (~2 weeks) and are solely contained within a clean-room environment and cultivated under GMP during cell expansion and manufacturing of the WCB. Upon the successful completion of full-scale manufacturing, one-time testing of end-of-production (EOP) samples at the limit of in vitro cell age is performed to evaluate the cell line after it has undergone manufacturing conditions, where the opportunity may present itself for any latent virus not detected in the cell bank to be expressed. The unprocessed bulk (UPB) at time-of-harvest from all batches is routinely tested for mycoplasma, bioburden, retroviruses and (for CHO cells) also often MMV.

37.5.1 Microbial Agents

According to ICH Topic Q5D, cell banks used in the manufacturing of biologics for human use must be demonstrated to be free from microbial contamination, including bacteria, fungi and mycoplasma. The test for the presence of bacterial and fungal contamination is performed on 1% of the cell bank (minimum of two vials) and involves the direct inoculation of cell lysates into two different liquid media (Tryptic Soy Broth and Fluid Thioglycollate Media) intended to detect a broad range of aerobic and anaerobic microorganisms as well as fungi and yeast.

The mycoplasma assays performed according to *Points to Consider in the Characterization of Cell Lines to Produce Biologicals*, 1993, tests for the presence of any cultivable or non-cultivable mycoplasma species in cell lysates prepared from the MCB or other cell samples. Cultivable mycoplasmas are detected using semisolid broth and agar, which enhances the possibility of detecting both fastidious and easily cultivated strains. Non-cultivable mycoplasma species are detected through the use of the Hoechst staining procedure on Vero cells that have been inoculated with the test article.

37.5.2 Adventitious Viruses

The *in vitro* adventitious virus assay employs a panel of indicator cell lines capable of detecting a wide range of human and relevant animal viruses. Viral detection is based on the demonstration of cytopathic effects, hemadsorption or hemagglutination within any of the cell lines. According to ICH Topic Q5A, the choice of cell lines should include the species of origin of the cell bank as well as a human and/or a nonhuman primate cell line susceptible to infection by human viruses. Because of the susceptibility of CHO cells to infection by Murine Minute Virus (MMV) and the potential impact of such an event, an additional indicator cell line with increased sensitivity to MMV infection may be included in the panel, but many investigators also perform a specific MMV polymerase chain reaction (PCR) test.

Even though cell culture processes are increasingly free of animal-derived raw materials, as availability of chemically defined media without any components of animal-derived origin has become prevalent in the industry, early on in the host cell line development history, the cells will have been exposed to materials such as bovine serum and porcine trypsin. Therefore, the purity testing paradigm will most often include *in vitro* PCR tests specifically designed to detect the presence of bovine or porcine viruses.

Viruses that are not readily propagated or otherwise display cytopathic effects in the *in vitro* assay may be detected in the *in vivo* adventitious virus assay. Their presence is detected through the inoculation of adult and suckling mice, sometimes also guinea pigs and embryonated hens' eggs and the subsequent demonstration of symptoms associated with viral infection. Species-specific viruses may be detected via the mouse and hamster antibody production assays (MAP and HAP), which focus on the generation of antibodies in response to an *in vivo* viral infection. From time to time new viruses are discovered, and it is always necessary to perform individual risk assessments and design of MCB and other cell banks qualification test regimens to ensure that all risks are considered. A recent example of an emerging virus is the Sars-Cov-2. It is relevant to verify whether this would be detected by the standard virus tests or whether an additional specific assay such as a PCR test may be recommendable.

37.5.3 Retroviruses

Cocultivation of MCB or EOP cells with a cell line susceptible to retrovirus infection, such as *Mus dunni* or human rhabdomyosarcoma (RD), increases the possibility of detecting small amounts of any infectious retrovirus. Infectivity may be confirmed with a positive focus formation assay or reverse transcriptase assay. Furthermore, transmission electron microscopy (TEM) is used to evaluate the cells for the presence of virus-like particles.

37.5.4 Identity

Identity testing establishes the species identity of the cells and demonstrates that the cell bank is composed of a homogeneous population. The traditional assay is an isoenzyme analysis method that is based on the electrophoretic mobilities and banding patterns of various intracellular enzymes. Currently, genetic identity testing based on next generation sequencing (NGS) provides a more accurate assessment. Since the complete CHO cell genome sequence became available (Xu, 2011), CHO identity tests are now often performed by targeted sequencing utilizing NGS methodology.

37.5.5 UPB Testing

In addition to the testing of the host cell, MCB, WCB and EOP cell banks, UPB (cells and culture media at time of harvest) from each GMP production batch intended for processing into drug substance is evaluated for the absence of bacteria, fungi, mycoplasma and virus, including TEM and MVM tests when CHO cells are used. Testing at the cell culture stage prior to initiation of any downstream processing steps provides a suitable point for the detection of potential contaminating agents. This routine testing provides lot-to-lot coverage for adventitious agent contamination during the production process. ICH Topic Q5A provides guidance on the evaluation of viral safety of biotechnology products derived from human and animal cell lines. Even with satisfactory purity testing results from the cell banks, viral contamination may arise from introduction during the production process, and, as such, routine testing of unprocessed bulk is performed as recommended (ICH Q5A, 1997; FDA Points to Consider, 1993 and 1997). Contaminating viruses may be detected using a panel of indicator cell lines and subsequent observations of cytopathic effects, hemadsorption or hemagglutination. Most known viruses that may be detected by each of the indicator lines do not appear to represent a threat for CHO cell infection. Nonetheless, CHO cells have been shown to be susceptible to infection from a limited number of human viruses in the paramyxovirus and reovirus families (Wiebe et al., 1989). Furthermore, within the industry, contamination of CHO cells from both murine and bovine adventitious agents has been reported, including MMV, Blue tongue virus and Cache Valley virus, resulting in significant operational losses. Specific PCR tests for all these viruses as well as some others are routinely done during CHO cell MCB qualification by most investigators, while MMV tests have become prevalent at the UPB stage.

Unlike murine myeloma or hybridoma cells, which are known to contain endogenous retrovirus, CHO cells contain noninfectious retroviral-like particles (RVLP). However, because the RVLPs are biochemically and morphologically similar to infectious retrovirus, and because the possibility of infectivity cannot be completely excluded, there is a regulatory requirement to quantify these particles. This is accomplished through the evaluation of cell-free supernatants just prior to clarification from representative production batches via transmission electron microscopy (TEM). These data are then used in conjunction with downstream process clearance capability to assess the risk of RVLP exposure to patients, with the expectation that an extremely low risk profile should be achievable.

37.5.6 Summary Remarks on Sterility, Virus and Identity Testing

Overall, while the assays described here are routinely used by many investigators, they should in no way be considered an all-inclusive or definitive list. Instead, alternative assays and techniques for adventitious agent detection may be used when appropriate and demonstrated to have equivalent or better specificity, sensitivity and precision than the existing methods. As an example, quantitative PCR-based methods for RVLP quantitation have been cited as being highly comparable to the TEM method (de Wit, et al., 2000). Manufacturers are encouraged to discuss these types of alternatives with regulatory authorities and to establish appropriate lists of test methods for each individual product when mammalian manufacturing is used.

37.6 GENOTYPIC CHARACTERIZATION AND GENETIC STABILITY

Genotypic characterization of the recombinant production cells is considered a part of the overall evaluation of cell bank quality and is performed according to recommendations described in ICH guidance (ICH Topic Q5B). The objective of the characterization is to establish that the intact and correct coding sequence has been incorporated into the host cell genome and is stably maintained during culture from the MCB through to the end of full-scale production. This is addressed both through cDNA sequencing and via Southern Blot analysis of the integrated expression construct for rearrangements within the coding region, for number of independent sites of integration and for copy number. As DNA sequencing has now become a fast, cheap and very accessible methodology, primary sequence verification by DNA or cDNA sequencing is typically performed multiple times during expression vector construction and cell line development. Also, the availability of host cell genomes, including the full CHO and hamster genomes and transcriptional profiles (Xu, et al, 2011; Lewis, et al, 2013; Könitzer, et al, 2015), has made verification at the genomic level of the recombinant production cell lines much easier. It is, however, still a regulatory requirement to perform rigid sequence verification and copy number determination and conduct productivity and copy number stability studies over time for the MCB when applying for market authorization. In addition, mRNA evaluation is done to confirm transcript integrity. The integration site(s) may be verified by Fluorescent in situ Hybridization (FISH) analysis or by NGS sequencing; however, this is not a mandatory requirement. For cell lines that have progressed to the point of full-scale manufacturing, recombinant cells from the EOP are analyzed in the same manner to provide assurance that no significant changes have occurred in the recombinant genes over the prolonged duration associated with the scale-up of the manufacturing process. The data derived from these analyses are critical in helping to establish a limit of *in vitro* cell age (discussed later within this chapter). An example of a genetic test package to be submitted with a market authorization package is shown in Table 37.2.

TABLE 37.2
Example of a Genetic Characterization Package

ANALYSIS	PHASE I/II IND	COMMERCIAL REGISTRATION
Nucleotide Sequencing	cDNA sequencing (expression plasmid; research cell bank; MCB)	cDNA sequencing of MCB and EOP
Southern Analysis	Genomic DNA from MCB Cells	Genomic DNA from MCB and EOP Cells
Assessment	Integrity of Coding Region	Integrity of Coding Region
Assessment	Gene Copy Number	Gene Copy Number
Assessment		Consistency of Integration Site(s)
Northern Analysis	Total RNA from MCB Cells	Total RNA from MCB and EOP Cells
Complementary Analyses		FISH; NGS

37.6.1 Nucleotide Sequencing

To confirm that the mRNA transcript(s) produced by the recombinant cells are the expected primary nucleotide sequence, cDNA sequencing is performed of the coding region(s) a of the transcript(s). The nucleic acid sequence encoding the recombinant protein is often verified already at the expression vector stage as well as during cell line and production clone development. The guidelines indicate that the primary sequence should be identical to that of the expression plasmid, within the limits of the technique, and should translate to the expected protein sequence. However, cDNA sequencing is not intended to detect low levels of variant sequences; the method sensitivity is only capable of detecting variants present in 5–10% frequency. Sequencing at the genomic level is one of several complementary methods aimed at detecting mutations or variants. Intact mass analyses along with peptide mapping and MS-MS analysis at the protein level are also performed as part of the sequence verification.

37.6.2 Southern Analysis

The structure and integrity of the gene(s) integrated into the host cell genome is assessed by Southern blot analysis of genomic DNA digested with restriction enzymes that immediately flank the coding region. The inclusion of the expression plasmid DNA used to generate the recombinant cell line, digested in the same manner, provides a direct comparison and allows for confirmation of the presence of the appropriately sized restriction fragments. Furthermore, when diluted, the plasmid DNA allows for an estimation of the sensitivity of the Southern method to detect variant or aberrant sequences. Any rearrangements of the coding region may be revealed by the presence of hybridizing fragments that are larger or smaller than those predicted.

By choosing restriction enzymes that cleave only once in the expression plasmid, digestion of genomic DNA will generally yield genomic restriction fragments containing both plasmid sequences and host cell genomic sequences. These fragments are expected to be unique for each integration event. Southern blot analysis of these genomic end fragments can be used to assess the expression plasmid integrant structures as well as provide an indication of the number of independent sites of integration within the host cell genome. This strategy provides a unique genetic fingerprint of the production cell line and can facilitate detection of genetic changes in case these would occur over the duration of a manufacturing run. FISH or M-FISH is an alternative method for integration site analysis (Auer et al, 2018).

Analysis of the copy number of the expression construct is also enabled by the Southern blot method. This evaluation is performed to determine whether the copy number remains stable over the course of production cultures. It must be noted that the actual numerical value for copy number that is obtained by this analysis is only an approximation, as the inherent technical variability associated with this method contributes to a large variance in the final calculated value. As such, the method is only intended to look for gross changes in copy number. While more quantitative methods such as Q-PCR could be applied to this question, the relevance of small changes in copy number during cell culture has not been established. In specific cases where more information regarding the genetic profile of the MCB or the EOP may be desired, complementary analyses like FISH or NGS may be performed to enable a more in-depth understanding of the recombinant construct integrity and location in the transfected cell line.

37.6.3 Northern Analysis

Although not specifically addressed in ICH Topic Q5B, investigators often include integrity of the transcript encoding the recombinant protein in the genotypic characterization of the MCB and EOP to confirm that the expression vector transcribes the intended mRNA and that this remains qualitatively indistinguishable over the course of production.

37.6.4 Complementary Analyses

Next-generation sequencing of the entire genome of a given host cell has become an accessible technology, and some investigators are looking into whether it can bring value in the analysis and qualification of biotherapeutics production cell lines/MCBs. The same can be said for the FISH analysis, which when combined with whole-chromosome painting has been refined in recent years to become a quite sensitive tool to look for expression plasmid integration sites and examine integration stability over time.

The latter has become more relevant during the past ten years, ever since FDA started requesting documentation for "clonality" of the recombinant production cell line.

CASE STUDY

A recombinant CHO cell line expressing human bone morphogenetic protein (rhBMP-2) is used as an example to illustrate how some of the cell bank qualification methods are used in practice. rhBMP-2 is a member of the TGF-β superfamily and is expressed, in its mature form, as a homodimer with a mass of approximately 30,000 Da. A bi-cistronic expression plasmid containing the genes coding for rhBMP-2 and dihydrofolate reductase (DHFR, a selectable marker) was used to transfect CHO cells. Following selection and cloning, an individual clone secreting suitable levels of rhBMP-2 protein was identified. This cell line, designated EMC-G5, was selected as the production cell line.

An MCB and a WCB were sequentially established from the EMC-G5 cell line, the WCB being used as starting material for a full-scale (2,500 L cell culture) manufacturing process. Cells from the final harvest of the manufacturing process represented approximately 81 cumulative population doublings (CPD) from the MCB and are referred to as rhBMP-2 EOP.

Genotypic analysis of the expressed mRNA transcript was performed on nucleic acids isolated from rhBMP-2 MCB, WCB and EOP cells. The physical state of rhBMP-2 transcripts was assessed using Northern blot analysis (Figure 37.2). The results demonstrated a single bi-cistronic transcript (containing both rhBMP-2 and DHFR genes) of the expected size with no evidence of aberrant transcript greater than 1% of the total RNA population. Furthermore, the MCB transcript co-migrated and was qualitatively indistinguishable from the WCB and EOP transcripts, suggesting stability through cell culture scale-up and full-scale production.

The integrity of the rhBMP-2 expression plasmid incorporated into the genome of the recombinant CHO cell line was evaluated by restriction enzyme digestion and subsequent Southern blot analysis. Figure 37.3a shows the results of a Southern blot analysis of *Hin*d III digested genomic DNA isolated from rhBMP-2 MCB, WCB and EOP cells. *Hin*d III sites immediately flank the coding region; therefore, if this region of the expression plasmid is intact, the labeled rhBMP-2 probe should detect a single 1.6 kb fragment. The results showed a single band of 1.6 kb in rhBMP-2 MCB, WCB and EOP genomic DNA that co-migrated with the *Hin*d III-digested plasmid controls. No evidence of rearrangements within this region is observed even though the predicted fragment is detected in the plasmid control at a level of 25 pg (equivalent to ~10% load control).

Figure 37.3b shows a Southern blot analysis of *Bgl* II-digested genomic DNA hybridized to an rhBMP-2 probe.

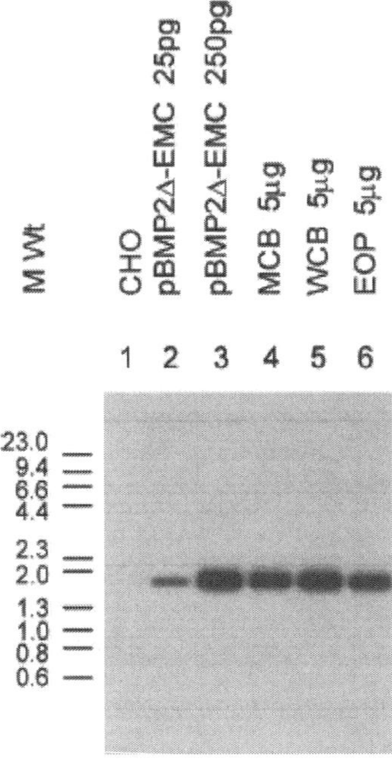

FIGURE 37.3A Southern blot analysis of *Hin*d III-digested rhBMP-2 MCB, WCB and EOP DNA.

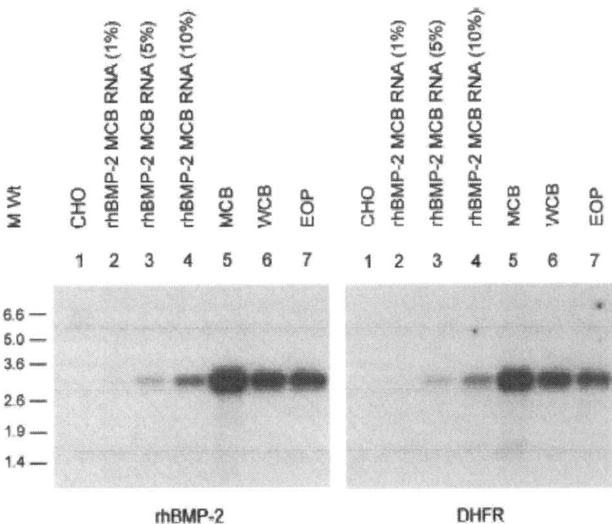

FIGURE 37.2 Northern blot analysis of total rhBMP-2 RNA isolated from MCB, WCB and EOP cells.

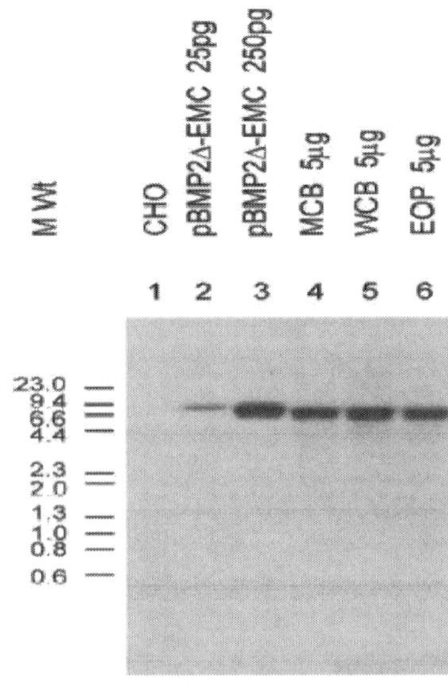

FIGURE 37.3B Southern blot analysis of *Bgl* II-digested rhBMP-2 MCB, WCB and EOP DNA.

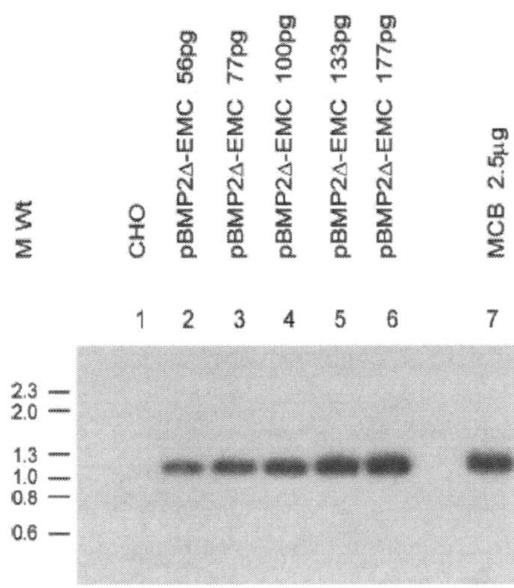

FIGURE 37.3C Comparative southern blot analysis of rhBMP-2 MCB and EOP DNA to assess rhBMP-2 gene copy number.

The expression plasmid contains a single *Bgl* II restriction site immediately upstream (5') of the rhBMP-2 gene and, as such, digestion of genomic DNA isolated from MCB, WCB and EOP cells would generate fragments that would be predicted to contain rhBMP-2 plasmid sequences across the site of integration to the first *Bgl* II site in the adjacent host cell DNA. The sizes of these fragments are dependent on the location of the flanking genomic *Bgl* II sites and expected to be unique for each integration event. The results demonstrate the presence of a single 3' genomic end fragment, suggesting that the expression plasmid integrated at a single chromosomal site within the host cell genome. Furthermore, identical fragments were detected in MCB, WCB and EOP cells, indicating the integrated plasmid is stable over the course of full-scale production.

Copy number estimates were obtained by comparative Southern blot analysis. MCB genomic DNA as well as varying amounts of plasmid DNA was digested with *Eco*R I, to excise the rhBMP-2 coding region, and compared by Southern blot analysis using an rhBMP-2 probe. As shown in Figure 37.3c, a single *Eco*R I fragment of the expected size is observed in DNA derived from MCB cells that co-migrates with the plasmid controls. Densitometry followed by quantitative analysis of multiple blots provided an approximate value of 70 rhBMP-2 gene copies per MCB cell, strongly suggesting that DHFR-based selection of the cell line in high levels of methotrexate resulted in amplification of the integrated plasmid.

In summary, the data indicated that a single copy of the expression plasmid had initially integrated at a single chromosomal site in the CHO genome. MTX-driven amplification resulted in approximately 70 rhBMP-2 gene copies per MCB cell. No evidence for gross rearrangement in any of the integrated rhBMP-2 plasmids or predicted transcripts was found. DNA sequence analysis confirmed the sequence predicted by the expression plasmid. Taken together with process data, the rhBMP-2 cell line was demonstrated to be stable and appropriate for full-scale manufacturing.

Genetic Characterization, Concluding Remarks

While the rhBMP-2 production cell line qualification example used here was performed more than 15 years ago, the regulatory requirements and the methodologies used for genetic characterization remains largely unchanged today, with the note that some novel methods have become available (FISH, M-FISH chromosome painting, NGS, to mention a few). In addition, FDA now requires distinct documentation of clonality probability with the market application. While modern methods are available for very detailed analysis of the genetic make-up of manufacturing cell lines, it is still not evident that such in-depth analysis would bring significant value to the cell bank qualification test package, and therefore, good old methods like Southern blot still prevail.

37.7 CELL LINE STABILITY

During the development of a recombinant production cell line and prior to establishment and characterization of cGMP MCB and WCB, it is common practice to evaluate how long the production cell line maintains phenotypic and genotypic stability during cultivation, relative to the cells at time zero (defined as the time of cell banking). Such stability studies are not mandatory regulatory requirements until the market application phase; however, it is recommended to perform at least preliminary studies already during the production clone selection phase, for example, for the Top3 RCB candidates, to avoid surprises at later development stages.

Typically, large scale mammalian cell-based manufacturing processes involve cell expansion and production periods for multiple weeks or even months, and as such it is incumbent on the manufacturer to ensure that the accumulation of cell population doublings does not lead to significant changes in cell characteristics that would affect their performance, or the quality of the product produced.

Changes in the expression phenotype of a production cell line can be brought about by several mechanisms, including chromosomal instability (loss of transgene copy number), gene silencing, genomic positional effects and population heterogeneity (ICH Topic Q5D). The genetic and phenotypic stability of a production cell line is defined within the limits of each individual cell culture manufacturing process. The link, if any, between cell line stability and the product quality profile of biopharmaceuticals will be influenced by the characteristics of the production cell line, the manufacturing process and the attributes of the protein biopharmaceutical itself. Accordingly, a stability profile needs to be established within the context of the specific manufacturing process employed for a given MCB or WCB.

The *limit of in vitro age* (LIVCA) for a production cell line is defined by phenotypic (growth rate and cellular productivity), genotypic (verification of intact integrated transgene sequence, transcript[s] of predicted size, verification of transgene coding sequence) and product quality attributes (verification that the cell line and the cell culture manufacturing process consistently produces the intended product quality [biochemical, physical and functional characterization]). Establishment of the LIVCA provides assurance of product consistency over the duration of the cell culture expansion and manufacturing process at full scale. During clone screening, a production cell line should be selected in which the productivity and the growth rate remain relatively constant, typically verified by demonstrated reproducibility in a number of laboratory scale manufacturing runs. Acceptance ranges for productivity changes over time are often set relatively wide (±20–30%) to accommodate known process and analytical variability. The goal is to secure a robust and reproducible manufacturing process resulting in an adequate product quality that meets the specification with respect to all CQAs (further described in Section 37.9) of the product.

37.8 ESTABLISHMENT OF THE LIMIT OF *IN VITRO* AGE

Establishing the LIVCA involves relating the cumulative population doublings (CPD) of a production cell line to phenotypic, genotypic and product quality attributes of the production cell line and cell culture process. This involves continuous passage of the production cells in a cell culture system mimicking the conditions used in the manufacturing process (i.e., the cell culture system should employ the same media formulations, seeding densities and other cell culture unit operations used in the cell culture manufacturing process envisioned at scale).

Currently, the most prevalent manufacturing mode by far for large biopharmaceuticals including monoclonal antibodies is fed-batch culture, although perfusion and continuous processing are becoming more popular, too, these years. In fed-batch, production cells are expanded through several seed train stages in vessels of increasing size prior to inoculation of the production bioreactor. The cells are further expanded here while the culture is supplied with nutrient feeds at specified time points. A temperature reduction to <37°C is often done after 4–6 days of exponential growth, after which the main recombinant protein production phase starts. The entire content of the production bioreactor is harvested as cell viability drops or productivity levels out, typically after about two weeks. Fed-batch processes are most often associated with a single thaw from a MCB or WCB according to a one-vial-one-run concept, although some companies do maintain prolonged seed bioreactor cultures for staggered inoculation of the production bioreactors. Perfusion cultures are also initiated by a cell expansion after cell bank vial thaw, but the production stage involves maintenance of a constant working volume in the bioreactor by continuous introduction of fresh culture medium and removal of spent medium through the use of a scalable cell retention system. Perfusion systems allow for extended culture durations (several weeks or months) and continuous harvest and purification operations. Variations to the perfusion mode, such as N-1 perfusion culture systems in which the cell culture inoculum is grown to high cell density in a seed bioreactor prior to inoculation of the production tank at high VCD, are also employed.

Since fed-batch manufacturing is still by far the most prevalent production mode used for large biologics such as mAbs, the next sections will focus on this manufacturing mode. However, if perfusion technology will be used, it is evidently critical to design LIVCA studies that mimic this mode of manufacturing as closely as possible.

For the purposes of determining the limit of *in vitro* age, the entire process from vial thaw until harvest, that is, all stages where population doublings accrue, must be considered. For a fed-batch process, most population doublings take place during the seed train expansion and the initial exponential growth phase in the production tank. Stability studies should closely mirror this scheme. This is done by cultivating the cells through regular passaging in small vessels while calculating the CPDs based on regular cell counts. Once these have outnumbered the theoretical CPDs needed to obtain a sufficient number of cells for inoculation of the intended commercial scale bioreactor (for example, a 15 kL tank), cells from the small-scale vessels are inoculated into small-scale production bioreactors and a scale-down fed-batch (or perfusion, if that is the intended manufacturing mode) production is executed using media, feeds and process parameters as close to the final upscaled production process as possible. Cell samples can be collected and frozen at different stages of the continuous passaging, for example, cells collected after 15, 30, 45 and 60 doublings. These cell samples of different age are subsequently thawed and expanded as needed in parallel with fresh MCB cells and compared in scaled-down manufacturing runs. If the fed-batch process performance, the titer and the product quality remain comparable for all five production runs in this example, the LIVCA has been verified to allow a seed train with up to 60 extra population doublings plus the fed-batch manufacturing process. Starting from an MCB or WCB with app 10×10^6 cells, the cell number after such an expansion would be more than sufficient to inoculate even very large bioreactors of 15 or 20 kL size. However, calculations should be done for the individual programs, where many factors, including the desired inoculum density, may vary. Also, as the stability test is often first performed during early clone selection, that is, at the RCB stage, CPDs sufficient for generation of the MCB and the WCB should be added to the calculation. These banks typically comprise 200–400 vials each with 10 to 20×10^6 or even higher cell numbers each.

37.9 ROLE OF PROCESS CHARACTERIZATION STUDIES

The goal of Process Validation Stage 1 is to establish a robust and reproducible scaled-up manufacturing process along with a (draft) Process Control strategy. Once the Process Design and Scale-up activities have been completed, including CLD and MCB/WCB generation and qualification, it is time for Process Characterization, the last element of Stage 1.

Full-scale Process Performance Qualification (PPQ) runs are performed during Process Validation Stage 2 to demonstrate that the process consistently generates drug substance of the desired product quality that meets the release specification and that all in-process controls (IPCs) are maintained within predefined acceptance ranges during manufacturing. To define the drug substance CQAs and establish their appropriate ranges and to identify Critical and Key Process Parameters (CPPs and KPPs, respectively), it is imperative to understand which input parameters have potential to affect drug substance quality (the CQAs) and/or process performance and, for each of these parameters, to confirm the range within which it must be maintained in order not to affect these in a negative way, thereby ensuring that the process is maintained in a state of control. This information about the process parameters must be derived from the process characterization studies. Thus, before moving to Process Validation Stage 2, that is, the PPQ, which implies full scale consistency runs in a GMP manufacturing facility, process characterization studies should be performed. Process Validation Guidance from FDA and EMA call for

1. defining the critical quality attributes (CQAs) of the active substance,
2. identifying process parameters that could affect these CQAs—most often categorized as CPPs (affecting product quality) and KPPs (affecting process performance), and
3. determining the appropriate operating ranges and action limits for these parameters.

These requirements are fulfilled by a combination of theoretical considerations, including risk assessment exercise(s), accumulated process knowledge and process characterization studies. For a cell culture-derived recombinant protein, the CQAs typically include specific activity, concentrations of product and non-product-related impurities, identity and sterility. The concentration of non-product-related impurities is generally influenced more by purification process than by cell culture process parameters. Therefore, cell culture process characterization studies generally focus more on specific activity and on relative amounts of product variants. These differ from each other with respect to characteristics such as composition of N-linked and O-linked carbohydrate, amino acid sequence at the N-terminus or C-terminus, or the presence of modified forms of certain amino acid residues.

It is well-established in the field of mammalian cell culture that each manufacturing process must be understood individually. This is due to the complexity of the living expression system employed where multiple components will influence the process performance. In short, a cell culture manufacturing process consists of four building blocks: the host cell, the expression vector, the media and the process (Figure 37.4). All four factors are strongly interlinked, and knowledge and experience with all these building blocks and with each specific manufacturing process resulting from this combination is key for a successful PPQ and for achieving a commercial manufacturing license.

Determination of CPPs and KPPs including appropriate operating ranges and action limits is done using scale-down model(s) (SDM) of the individual manufacturing steps, where the effect of varying critical input parameters is studied. For cell culture processes, examples of such parameters could be temperature and pH. Since effects of different parameters may be additive or may even show positive interactions, process characterization studies are ideally multifactorial and performed by means of Design of Experiment (DOE), combined with One Factor At a Time (OFAT) experiments where deemed appropriate. Cell culture process characterization studies may also evaluate the effects of short-duration excursions of an input parameter outside the acceptable range. While a parameter outside of its acceptable range for the full duration of a process phase (e.g., expansion phase or production phase) may lead to drug substance that doesn't meet its specifications, excursions outside the acceptable range for a shorter time period (e.g., 30 minutes) may not impact the drug substance quality. During routine manufacturing operations, an equipment failure will occasionally result in an excursion of this type, for example if a pH probe is defective but can be replaced within a short period of time. The impact on drug substance characteristics can be anticipated based on process characterization studies that move a parameter a known amount outside of the acceptable range for a known amount of time.

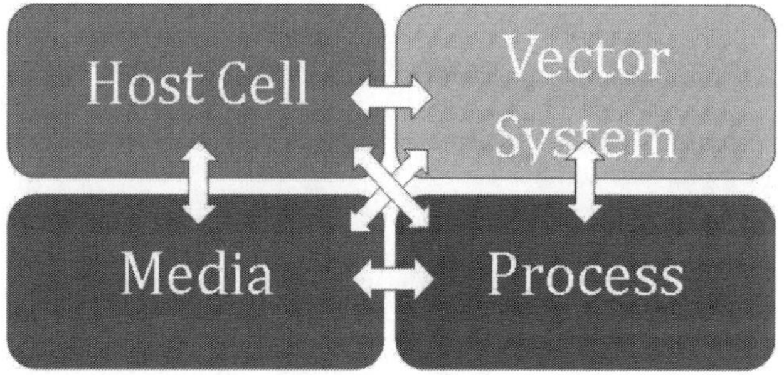

FIGURE 37.4 The four building blocks necessary for establishment of a cell culture manufacturing platform.

37.10 VALIDATION OF A SCALE-DOWN MODEL

Before cell culture process characterization studies can begin, the relevant scale-down models must be validated. A scale-down model (SDM) for the seed train expansion phase is relatively simple to establish. The cells are grown in batch mode in vessels of different size during the expansion phase, and knowledge about growth characteristics in these different vessels, for example, shake flasks versus wave bioreactors, have been collected during the development phase. Thus, expansion through small-scale shake flasks may often mimic the scaled-up seed train quite well. The more complicated SDM is the production bioreactor. This model is typically a laboratory bioreactor that has a working volume in the 1–5 L range, in combination with a scaled-down purification system. Recent availability of new and more automated equipment such as the ambr250 system provides further options for higher throughput SDMs. The SDM should accurately represent the full-scale process with respect to cell culture performance—growth rates and cellular productivities—and with respect to characteristics of purified drug substance.

It must be qualified before use with respect to all parameters judged to be critical, and the limits of the model should be understood before Process Characterization experiments are conducted. Multiple factors must be controlled to best mimic the process at scale, including but not limited to pH, temperature, gas sparging and stripping (O_2, CO_2, etc.), feed additions and agitation.

37.11 RISK ASSESSMENT

In parallel with establishment of the SDM, a risk assessment (RA) is performed to identify and rank the criticality of all process parameters and determine which of these need to be studied at small scale. The RA should follow a systematic approach, and tools such as Failure Modes and Effects Analysis (FMEA) are often used. Here, all process parameters affecting the process are listed, and the severity, occurrence and detectability of parameter excursions are evaluated and ranked. The parameters that are ranked highest are subsequently studied experimentally in the SDM. For a fed-batch cell culture process, examples of parameters that typically undergo SDM studies include pH range, temperature, inoculum density, feed regimens including glucose concentration, agitation rates, gassing strategy and time of harvest, while for the seed train critical parameters might be MCB/WCB thaw viability, inoculum density, stage duration and temperature. While biopharma companies and CDMOs often work with platform processes, it is important to emphasize that each process is unique and should be evaluated individually.

37.12 EXAMPLE OF PROCESS PARAMETER EFFECTS ON CQAS

As mentioned previously, several cell culture input process parameters are known to potentially affect quality characteristics of recombinant proteins. Bioreactor pH and temperature are almost always highlighted during risk assessment exercises as process parameters that may impact process performance as well as product quality with a high likelihood. It is well understood by experts in the field and from the literature that too low or too high pH may directly impact mammalian cell growth and thereby also the viable cell density and the viability. Similarly, temperature has a very direct influence on the cell culture growth rate, peak cell density and viability over the course of the production phase. A lower viability at time of harvest will generally result in higher levels of host cell DNA and host cell protein (HCP), some of the so-called non-product-related impurities that must be controlled within strict limits in the final product. Since pH and temperature have strong impact on the cell culture performance, for example, of their glucose uptake rate, these parameters may also indirectly affect product-related CQAs such as the glycosylation pattern (see Section 37.13).

37.13 EXAMPLE OF CQA AFFECTED BY SEVERAL PROCESS PARAMETERS

The N-linked glycosylation pattern of a biotherapeutic is very often a CQA, not least for monoclonal antibodies used for cancer treatment. Here, the glycopattern has been shown to influence potency, safety, immunogenicity and pharmacokinetics (Hossler, et al; 2009; Loebrich, et al 2019). Glycosylation is not a templated process; rather, each individual cell adds complex sugar structures to the expressed protein (for example to the Asn297 present in the backbone of most monoclonal antibodies), and these are subsequently trimmed during intracellular processing of the molecule. It is well known from the literature that multiple cell culture parameters can and will influence the final glycopattern outcome. In fact, all building blocks comprising a mammalian expression system may potentially influence the glycopattern (Figure 37.4), one example being the glucose level maintained in the culture supernatant during production, another the expression rate of the recombinant protein in the production cells. Recently, attempts to develop biosimilars have further emphasized this, since biosimilars need to have CQAs similar to those of originator molecules. This fact has driven the scientific experts in the field to better understand exactly which parameters can influence glycopatterns, and also how they may be modified. Since the CHO cell line and hamster genomes were published in 2011 and 2013, respectively, very significant progress has been obtained in understanding both the cellular and the extracellular (supernatant) aspects that impact glycosylation (Könitzer, et al, 2015; Fan, et al, 2015; Kildegaard, et al, 2016; Sumit, et al, 2019). Such knowledge is key when designing and conducting process characterization studies that can support establishment of relevant process parameter acceptance ranges.

37.14 ACCEPTABLE RANGE FOR A PROCESS PARAMETER

The results of cell culture process characterization studies are, along with all the process knowledge compiled during the entire process development stage, used to classify process parameters as CPPs or KPPs or to leave them non-classified and to establish acceptance ranges for the CPPs and KPPs. When an acceptable range has been identified for each cell culture parameter, this range is compared with the equipment capabilities, that is, the ranges that can be consistently maintained in the manufacturing bioreactor. The combined equipment capability and accepted process parameter range is often referred to as the "operating range". Some companies also use the terminologies Normal Operating Range (NOR) and Proven Acceptance Range (PAR). Ideally, the acceptable range is wider than the operating range, as this will lead to fewer deviations during manufacturing.

Along with classification of the input parameters, certain output parameters are typically established as In-Process Controls (IPC). This could, for example, be viability at time of harvest. IPCs should also have defined acceptance ranges. As described further by Riske and Levine in Chapter 38, it is strategically important to select only those parameters that have been shown through rigorous experimental testing and from process history to be CPPs in order to avoid unnecessary batch failure both during the PPQ runs and later during commercial manufacturing. This is because failure to meet a CPP acceptance range during GMP manufacturing will lead to a deviation and associated investigation that must be resolved and justified to not impact the product quality and process performance before the batch can be released. When all classified parameters are maintained within their acceptable ranges, drug substance can be expected to meet specifications if the principles for process design and validation have been followed, as described earlier. If this is the case, the goal of the process validation stage 1 has been met.

37.15 STAGE 2: PROCESS PERFORMANCE QUALIFICATION (PPQ)

Full scale PPQ runs serve **to confirm that a robust and reproducible manufacturing process is available prior to commercialization.** One purpose is to demonstrate that each input CPP, KPP and IPC is consistently maintained within the acceptable range that was established on the basis of the process development and process characterization studies. Another is to demonstrate that when each input parameter is maintained within the acceptable range, the drug substance consistently meets the release specification. A third is to verify the concentrations of non-product-related impurities that are generated by the cell culture process when each input parameter is maintained within the acceptable range. These impurities include HCP, host cell DNA, and endogenous virus-like particles that are known to be present in CHO cells, and they are typically tested during the PPQ runs as well as during commercial manufacturing depending on the nature of the impurity. For example, mycoplasma, retrovirus and *in vitro* virus tests of the unprocessed bulk harvest (UPB) are routinely done for each batch when manufacturing is done with CHO cells per ICH Q5A guidance, while host cell DNA and HCP in the UPB often is monitored only during the PPQ runs, as long as it can be verified that the downstream process has the capacity to efficiently remove those cell-derived impurities

The number of at-scale runs needed for a successful PPQ should be justified. In practice, three consecutive PPQ runs meeting all preestablished acceptance ranges and release criteria has historically been deemed sufficient, but regulatory agencies are requesting thorough justifications of the rationale for design and execution of the PPQ. A Master plan for the PPQ should be prepared along with specific protocols for each of the process stages subjected to PPQ. These protocols should contain (presumptive) process parameter classifications and predefined acceptance criteria. All equipment should be in a qualified state and appropriate raw material testing and control established. The PPQ Master plan should include the overall principles and descriptions of these aspects. It may also include description of employed bracketing approaches to the PPQ design, for example, when several identical or very similar production bioreactors are intended for use with the same process in large manufacturing plants. After completion of the PPQ campaign, all in-process and product quality data should be reported, with focus on whether the drug substance specification as well as the classified process parameter (CPPs, KPPs and IPCs) acceptance ranges were consistently met. Any deviations should be dealt with in detail in the reports, and it has to be shown that these did not impact the product quality, process performance or the validity of the PPQ runs in order for the PPQ campaign to be claimed successful.

37.16 STAGE 3: CONTINUED PROCESS VERIFICATION (CPV)

The final stage of process validation is the CPV, which is a post-marketing mandatory activity that should be undertaken to provide assurance that the process remains in control.

After completion of the PPQ campaign and the reporting, the Process Control strategy including the acceptance ranges should be reviewed and potentially modified or adapted based on the learnings from the campaign. For example, it may be decided that a KPP acceptance range should be tightened, or that IPCs like HCP should be monitored for some additional batches at scale (for example 10–20), after which it is evaluated again whether the added monitoring may be stopped. A Process Control strategy document should be established, (ideally, this was drafted prior to the PPQ campaign), and trending of all classified and controlled parameters should be implemented and done routinely according to a written CPV plan. It is recommended to conduct the CPV stage in two steps: First, the initial commercial batches (e.g., the first 30 batches) are conducted with a heightened monitoring level. Decisions of the exact parameters that should be monitored is often done based on a stage 3A risk assessment. After completing and trending data from these first commercial batches, a stage 3B risk assessment may be done to modify/reduce the number of parameters that needs to be monitored and, if needed, to adjust the acceptance criteria based on the accumulated data. Use of statistical tools are recommended both for the trending and the potential modifications of parameter classifications and/or acceptance ranges.

37.17 CONCLUSION

Validation of the cell culture manufacturing process, including qualification of the MCB and WCB used to initiate the manufacturing, is based on a thorough understanding of the host cell line used to generate the recombinant manufacturing cell line bank(s) along with the accumulated process development knowledge of the upstream manufacturing process. After scale-up of the process for clinical material supply manufacturing, a validated scale-down model is established and used during process characterization. The parameters studied during process characterization are identified using a structured risk-based ranking approach which starts with the identification of the critical quality attributes of the final product. After completion of the process characterization, critical process parameters and in-process controls are defined and verified during full-scale PPQ runs, described in the Process Control strategy and continuously verified within the CPV program to ensure robust and consistent commercial manufacturing.

REFERENCES

1. Guidance for Industry. *Process Validation: General Principles and Practices.* FDA, 2011.
2. EudraLex Volume 4 EU Guidelines for Good Manufacturing Practice for Medicinal Products for Human and Veterinary Use, Annex 15: Qualification and Validation (EU 2015).
3. Novak R. Regulatory perspective on the evaluation of clonality of mammalian cell banks, CDER/OPQ/OBP/DBRRI, January 23, 2017. FDA presentation.
4. Frye C, Deshpande R, Estes S, Francissen K, Joly J, Lubiniecki A, Munro T, Russell R, Wang T, Anderson K. Industry view on the relative importance of "clonality" of biopharmaceutical-producing cell lines. *Biologicals.* 2016 Mar, 44(2), 117–122.
5. Bandyopadhyay AA, O'Brien SA, Zhao L, Fu HY, Vishwanathan N, Hu WS. Recurring genomic structural variation leads to clonal instability and loss of productivity. *Biotechnol Bioeng.* 2019 Jan, 116(1), 41–53.
6. ICH Topic Q5A. *Quality of Biotechnological Products: Viral Safety Evaluation of Biotechnology Products Derived from Cell Lines of Human or Animal Origin.* International Conference on Harmonization, 23 September 1999.
7. ICH Topic Q5B. *Quality of Biotechnological Products: Analysis of the Expression Construct in Cell Lines Used for Production of r-DNA Derived Protein Products.* International Conference on Harmonization, 30 November 1995.
8. ICH Topic Q5D. *Quality of Biotechnological Products: Derivation and Characterisation of Cell Substrates Used for Production of Biotechnological/Biological Products.* International Conference on Harmonization, 16 July 1997.
9. ICH Topic Q6B. *Specifications: Test Procedures and Acceptance Criteria for Biotechnological/Biological Products.* International Conference on Harmonization, 10 March 1999.
10. Center for Biologics Evaluation and Research. Points to Consider in the Characterization of Cell Lines Used to Produce Biologicals. U.S. Department of Health and Human Services Food and Drug Administration, 1993.
11. Xu X, Nagarajan H, Lewis NE, Pan S, Cai Z, Liu X, Chen W, Xie M, Wang W, Hammond S, Andersen MR, Neff N, Passarelli B, Koh W, Fan HC, Wang J, Gui Y, Lee KH, Betenbaugh MJ, Quake SR, Famili I, Palsson BO, Wang J. The genomic sequence of the Chinese hamster ovary (CHO)-K1 cell line. *Nat Biotechnol.* 2011 Jul 31, 29(8), 735–741.
12. US Food and Drug Administration Center for Biologics Evaluation and Research Points to consider in the manufacture and testing of monoclonal antibody products for human use (1997).
13. Wiebe ME, Becker F, Lazar R, May L, Casto B, Semense M, Fautz C, Garnick R, Miller C, Masover G, Bergmann D, Lubiniecki AS. A multifaceted approach to assure that recombinant tPA is free of adventitious virus. *Dev Biol Stand* 1989, 70, 147.
14. de Wit C, Fautz C, Xu Y. Real-time quantitative PCR for retrovirus-like particle quantification in CHO cell culture. *Biologicals* 2000, 28, 137–148.
15. Lewis NE, Liu X, Li Y, Nagarajan H, Yerganian G, O'Brien E, Bordbar A, Roth AM, Rosenbloom J, Bian C, Xie M, Chen W, Li N, Baycin-Hizal D, Latif H, Forster J, Betenbaugh MJ, Famili I, Xu X, Wang J, Palsson BO. Genomic landscapes of Chinese hamster ovary cell lines as revealed by the Cricetulus griseus draft genome. *Nat Biotechnol.* 2013 Aug, 31(8), 759–765.
16. Könitzer JD, Müller MM, Leparc G, Pauers M, Bechmann J, Schulz P, Schaub J, Enenkel B, Hildebrandt T, Hampel M, Tolstrup AB. A global RNA-seqq-driven analysis of CHO host and production cell lines reveals distinct differential expression patterns of genes contributing to recombinant antibody glycosylation. *Biotechnol. J.* 2015, 10, 1412–1423.
17. Auer N, Hrdina A, Hiremath C, Vcelar S, Baumann M, Borth N, Jadhav V. ChromaWizard: An open source image analysis software for multicolor fluorescence in situ hybridization analysis. *Cytometry A.* 2018 Jul, 93(7), 749–754.
18. Hossler P, Khattak S, Li Z. Optimal and consistent protein glycosylation in mammalian cell culture. *Glycobiology* 2009, 19(9), 936–949.
19. Loebrich S, Clark E, Ladd K, Takahashi S, Brousseau A, Kitchener S, Herbst R, Ryll T. Comprehensive manipulation of glycosylation profiles across development scales. *MAbs.* 2019 Feb–Mar, 11(2), 335–349.
20. Fan Y, Del Val IJ, Müller C, Wagtberg Sen J, Rasmussen SK, Kontoravdi C, Weilguny D, Andersen MR. Amino acid and glucose metabolism in fed-batch CHO cell culture affects antibody production and glycosylation. *Biotechnol. Bioeng.* 2015, 112, 521–535.
21. Kildegaard HF, Fan Y, Sen JW, Larsen B, Andersen MR. Glycoprofiling effects of media additives on IgG produced by CHO cells in Fed-Batch Bioreactors. *Biotechnol. Bioeng* 2016, 113(2), 359–366.
22. Sumit M, Dolatshahi S, Chu A, Cote K, Scarcelli JJ, Marshall JK, Cornell RJ, Weiss R, Lauffenburger DA, Mulukutla BC, Figueroa B, Jr. Dissecting N-Glycosylation Dynamics in Chinese Hamster Ovary Cells Fed-batch Cultures using Time Course Omics Analyses. *iScience* 2019, 12, 102–120.

38 Validation of Recovery and Purification Processes

Frank Riske and Howard L. Levine

CONTENTS

DOI: 10.1201/9781003163138-38

38.1 INTRODUCTION

Pharmaceutical biotechnology combines microbiology, chemistry, traditional pharmaceutical technology, biochemistry, cell biology, and biochemical engineering with advances in genetic engineering technology to prepare proteins, vaccines, virus gene therapy, and other biologic products as therapeutic, ancillary, or diagnostic agents. Cell therapy will not be discussed in this chapter. These biologics are produced in a suitable host cell by fermentation or cell culture processes. Cells from a qualified master or working cell bank are expanded in culture until sufficient numbers are obtained for the desired production scale. During culture, the product of interest may be either retained intracellularly (perhaps in inclusion bodies) or secreted into the culture medium. If the product remains inside the cell, then the cells must be harvested and disrupted and the debris that is created removed to yield an extract for further purification. If the product is in an inclusion body, the inclusion bodies are collected and solubilized with denaturing agents such as GuHCL or urea, and the protein is refolded by dilution in the presence or absence of ox/redox agents. If the product is secreted, the cells must be separated from the conditioned culture medium prior to purification. These process steps, shown schematically in Figure 38.1, include initial product recovery followed by refolding (if necessary), isolation, purification, viral inactivation (if necessary), and finishing operations. Each of the process stages shown in Figure 38.1 may be achieved using a variety of unit operations, as will be discussed. The validation of a biomanufacturing process encompasses validation of each of these unit operations separately as well as in combination to demonstrate that the process can reliably and reproducibly produce the desired product.

The goal of process validation is to demonstrate that a process, when operated within established limits, produces a product of appropriate and consistent quality. The most efficient way to achieve adequate process validation is to first establish the Quality Target Product Profile (QTPP) and then determine the critical quality attributes (CQAs) of those outputs that effect product quality. Next, the goal is to establish the parameters and parameter ranges at each processing step that affect CQAs. This selection is based on best principles, process development, process history, and manufacturing and process characterization (PC). Critical quality attributes can be viewed as the measurable results of a particular process step, and the acceptable range for each constitutes the necessary specifications for that process step. The critical process

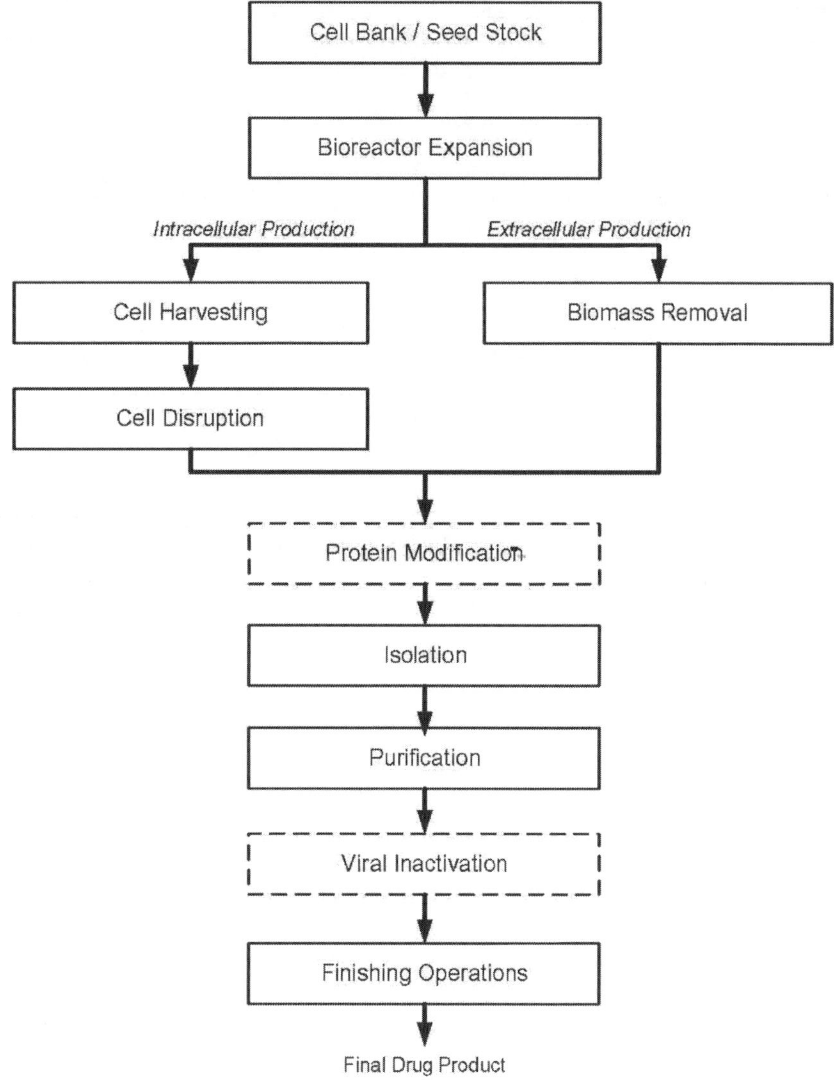

FIGURE 38.1 Major stages in the manufacture of a biopharmaceutical product.

parameters are a subset of all the independent variables related to a process step; the critical process parameters are just those independent variables that directly determine the values for the critical quality attributes. During early clinical studies, the QTPP, CQAs, and a preliminary classification of process parameters should be determined. Prior to process validation, the presumptive critical and key process parameters should be studied to demonstrate that the appropriate ranges for each can be met on a reliable and consistent basis and lead to an acceptable product.

Validation is a scientifically rigorous and well-documented study that demonstrates that a process or piece of equipment consistently does what it is intended to do. Due to the complex nature of proteins, viruses, cell therapies, and other therapies, it is often difficult to fully characterize a biologic product. Thus, final product testing alone is frequently insufficient to ensure consistent manufacture. Therefore, the processes used for the recovery and purification of biologics must be designed and validated to remove host- and product-related impurities, process additives, and potential contaminants (bioburden). These materials may arise from source material, equipment or reagents, and the environment and can include endotoxins, viruses, nucleic acids, and host cell proteins, as well as media constituents, process chemicals, ligands leached from chromatography media, and modifications or inactive forms of the protein itself.

Validation should be considered as early in the development of a process as is practical. In this way, data required for validation can be collected during development studies and the production of batches for clinical studies. For all new biotherapeutics, the evaluation of the product in humans under carefully monitored clinical trials provides the ultimate test of the safety and efficacy of the product.

Formal validation typically begins during the late phase of clinical development. This process includes activities at both small scale and production scale. The prerequisites for validation are that the biologic product should be defined in terms of its physical, chemical, and biological characteristics.

Specifications should be established to ensure product consistency, uniformity, and high purity. Assays used to determine product purity and composition and to release the product should be validated to meet standards of specificity, precision, accuracy, linearity, repeatability, and assay range with an LOD or LOQ. In addition, each of the facility/utility systems used to support the manufacturing process must be validated. Finally, all of the manufacturing process equipment must be validated through at least an operational qualification, as discussed in Section 38.7. Once all of these criteria are met, validation of the manufacturing process itself can begin.

Process validation of biopharmaceutical manufacturing will also include an assessment of the suitability of process raw materials; scaled-down model qualification and verification; viral clearance studies on virgin and end-of-lifetime resins; a determination of the process Design Space based on DOE and univariate experiments that includes an identification of CPP and KPP and their acceptable ranges;, process safety, which includes testing for adventitious agents in cell banks and crude harvest; column and resin lifetime and cleaning; tank and skid cleaning to appropriate levels; and the validation of the entire process as it is intended to be operated at commercial scale.

Process equipment qualifications are normally broken down into design qualifications (DQ), installation qualifications (IQ), operational qualifications (OQ) and performance qualifications (PQ). The DQ documents that the equipment ordered is consistent with a design already shown to meet the applicable user's requirements.[1] The IQ verifies that the equipment is complete, installed correctly, and calibrated and meets the design specifications. The OQ confirms each equipment component functions as specified and that all components work together correctly. The PQ establishes that the equipment functions as specified for the individual product(s) that it will be used to produce.

While much has been written in the past regarding process validation (c.f. reference),[2] advances in manufacturing and analytical technology, along with the increasingly complex nature and diversity of biotherapeutics products, has made validation more important than ever. Furthermore, process validation is not always a three and done exercise and post PQ continuous verification begins and extends for the lifetime of the product. The process critical and key parameters and outputs are measured, monitored and trended to understand how the process can be improved.[3] The lessons learned from continuous monitoring are added to the Process Control Strategy (PCS), a roadmap for conducting the process in the commercial facility. Process validation will establish the process operating range robustness and reproducibility and that it consistently meets all release specifications. In process validation, it is important that protocols clearly specify the procedures and tests to be conducted, the data to be collected, and the acceptance criteria. The protocol should also specify the number of sequential replicate process runs needed to demonstrate process reproducibility and process variability. Critical and key process parameters and outputs, acceptable and target ranges and intermediate and final specifications should be identified, monitored, and documented to establish the variability of process parameters for individual runs and will establish whether or not the equipment and process controls are adequate to assure that all of the product specifications will be met.

38.2 RECOVERY AND PURIFICATION PROCESS STEPS

38.2.1 Cell Harvesting

As shown in Figure 38.1, for those products produced intracellular entirely within the production host, the first recovery steps after bioreactor expansion are cell harvesting and cell disruption. In cell harvesting, the product-containing cells are isolated from the fermentation or cell culture broth, typically using one of the following unit operations: centrifugation or microfiltration.

In the case of viruses such as AAV or Lentivirus, used in viral gene therapy, the cells are often transfected or transformed with the gene of interest and genes responsible for the coat proteins and virus replication. The intact virus often buds from the cell, lyses the cell, or is found both intracellularly and extracellularly. Typically, the cells and cell debris are removed by depth filtration (virus extracellularly) or by cell lysis (detergent) followed by depth filtration.

38.2.2 Cell Disruption

The product-containing bacterial cells are typically broken apart using a homogenization unit operation step. Cells could also be lysed by chemical addition in which case a stirred-tank reaction unit operation would be used. Mammalian cells used to produce intracellular virus are often lysed with 1% Triton X-100 or a similar detergent.

38.2.3 Biomass Removal

With extracellular product production, the cells are simply removed from the product-containing fermentation or cell culture broth. Cell removal can be achieved using centrifugation, microfiltration, or depth filtration.

38.2.4 Protein Modification

Some products require modification after recovery. These modifications may, in some cases, occur immediately after isolation or following some of the purification steps. A number of potential protein modifications may be necessary for a particular product. This includes the denaturation and refolding of proteins trapped in bacterial inclusion bodies to allow it to achieve its fully active tertiary structure, chemical or enzymatic cleavage, fusion of a second protein to allow for efficient affinity purification, remodeling of glycoproteins,[4] or the covalent coupling of polyethylene glycol (PEG) to modify the pharmacological properties of proteins.[5] Protein modification is typically a stirred-tank reaction unit operation. Protein modification, as necessary, can be performed at any stage of the manufacturing process. PEGylation, for example, is most often performed on the purified protein obtained after most or all purification operations are complete. Regardless of where the modification step occurs in the process, complete process validation will include validation of this modification step as well.

38.2.5 Capture

Capture of the protein of interest from the clarified harvest broth refolded protein mixture (e.g., from inclusion bodies) or clarified cell lysate occurs during the capture step. Typically, this is a chromatography step such as Protein-A affinity resin for Mab and Mab-like molecules, affinity resins for several AAV subtypes, and traditional resins (cation and anion ion exchangers, hydrophobic interaction, mixed mode, and pseudo affinity resins for r-proteins. Other unit operations for this step include target precipitation liquid–liquid extraction or tangential flow filtration.

38.2.6 Polishing

There are typically at least two and sometimes several chromatography steps in a process to polish the molecule of interest. These steps remove/reduce product and process impurities. Additional process steps include viral clearance/inactivation technology (mammalian processes) and dead-end and tangential flow operations.

38.2.7 Viral Inactivation

Viral inactivation is required for products made from mammalian cells. Viral clearance may be achieved by chemical addition (detergent, alcohols, TNBP), lowering the pH to below 3.7, or heating the material. The first two methods inactivate only enveloped viruses, while the last (heat) can inactivate both enveloped and non-enveloped virus. For example, Adenovirus, a helper virus in some AAV processes, is activated at a lower T than AAV.[6] In all cases, the process will be held at a specified temperature long enough for the viruses to be inactivated. The relevant unit operation for viral inactivation is stirred-tank container, whether or not the process employs continuous mixing, as the solution is assumed to be homogenous throughout the hold time.

38.2.8 Finishing Operations

Finishing operations may include nanofiltration for additional virus removal (both enveloped and non-enveloped viruses), ultrafiltration for product concentration, and buffer exchange into the formulation buffer and sterile filtration (0.2 μm) of the formulated drug into the drug substance containers.

38.3 CONTINUOUS PROCESSING AND AFFINITY CHROMATOGRAPHY

ICH 8a[7] suggests that:

> definitions for both batch and lot are applicable to continuous manufacturing. A batch can be defined based on the production period, quantity of material processed, quantity of material produced or production variation (e.g., different lots of incoming raw material), and can be flexible in size to meet variable market demands by leveraging the advantage of operating continuously over different periods of time. A lot may also be considered a sub-batch. The actual batch or lot size should be established prior to the initiation of each production run. For batches that are defined based on time (e.g., a production period), a connection between material traceability and batch must be established to identify the specific quantity of the drug (21 CFR 210.3).

"A batch" should be defined early in development, as that definition will drive the process validation and subsequent continuous verification. Another consideration is the use of affinity capture ligands for non-monoclonal antibody (Mab) proteins. Affinity chromatography with Protein-A resin has revolutionized Mab bioprocessing. This capture step reduces the overall number of processing steps needed to produce a highly purified drug substance, hence increasing the overall process yield and reducing process time and costs. Affinity chromatography has been applied to non-Mab molecules.

Synthetic oligonucleotides with unique secondary structures, such as aptamers, have been used for bioprocess scale affinity chromatography to purify Vascular Endothelial Growth Factor (VEGF121)[8] and Factor VII, Factor H, and Factor IX from serum[9] The successful large scale purification of AAV9 using Capture Select AAV9 (camelid molecule) at the 10L and 50L scale, with >80% recovery, has been reported[10]

Affinity chromatography is a key[11] to enabling continuous and intensified processing of molecules due to the standardization of processes and facilities. Increasingly, processes are being developed in which the upstream is run in perfusion mode, or as an N-1 intensified batch, and clarified cell culture fluid (CCCF) is directly (continuously) loaded on the capture column. However, variability in CCCF composition can lead to variations in the composition of the eluate and challenges downstream of the capture step. An affinity-based capture step would act to normalize the process stream, as the elution pools from the capture step would likely have the same composition and concentration. Consequently, the remaining steps in the purification train will be standardized, which in turn will improve the overall downstream process robustness.

38.4 PROCESS VALIDATION

The fundamental basis of modern process validation starts with Quality by Design (QBD). QBD is a life cycle approach that equates to enhanced process quality and includes several key elements. These include that the product is defined to meet patient needs and performance requirements, that the process is designed to consistently meet product critical quality attributes, that the impact of starting raw materials and process parameters on product quality is well understood, and lastly, that the process is continually monitored, evaluated, and updated to allow for consistent quality throughout product life cycle[12,13,14]

The first step in a QBD approach is to identify critical sources of variability and control that variability through appropriate control strategies.[15,16] Process validation starts with an understanding of the process acquired during development and refined during process scale-up, manufacturing, and optimization. Solid process validation relies on extensive process characterization work performed using qualified scale-down models, the evaluation of manufacturing and PD data, additional viral clearance studies, and several other items. Once a process has been characterized, process parameter targets and ranges are finalized to ensure process and product robustness and product uniformity. All analytical methods required for detection and quantitation of the product and impurities are validated. At this point, validation of the manufacturing process can begin.

38.4.1 Critical Quality Attributes

For each process step there is a set of dependent results, properties of the intermediate process stream leaving that step, that are necessary for the successful production of the final drug product. These dependent results, outputs, or properties are often referred to as the critical quality attributes for the product that must be considered at each process step. The dependent results listed in the unit operation parameter tables (Tables 38.1–38.8) are examples of some quality attributes. Quality attributes include not only the intrinsic process variables necessary for the correct structure and activity of the protein of interest but also include host and process impurities such as host cell protein, DNA, and detergents. For each unit operation or process step, there is a subset of quality attributes that are critical to the successful outcome of the operation. These CQAs define the acceptable specifications for the intermediate process stream, and taken together, the CQAs define the ultimate product produced by a particular manufacturing process.

38.4.2 Key and Critical Process Parameters

For each process step, there are many independent variables (process parameters) that can be adjusted. The first step in classifying the process parameters is to conduct a process risk assessment to identify their putative importance. Both development and manufacturing data is gathered for each parameter, and personnel representing analytics, development, manufacturing, and other groups will gather to systematically go through each step and classify each parameter.

Typically, only a small subset of the independent process parameters are critical to the successful outcome of a particular process step. The critical process parameters (CPPs) are those that directly impact the CQAs for a given process step, parameters that must be controlled within an established acceptable range. Those parameters that do not affect product quality but do effect product recovery, are referred to as key process parameters (KPP). Strategically, it is important to select only those parameters that have been shown through rigorous experimental testing and from process history to be CPPs. Minimizing the number of CPPs controlling a process step will reduce the extent of validation required for that step.

To assist with establishing which process parameters are critical, it is important to have a clear understanding of the appropriate settings or ranges. The widest parameter range is the acceptable range, sometimes termed the proven acceptable range (PAR). The acceptable range is the widest range tested during process development or characterization or exercised during manufacturing that leads to an acceptable product. Outside the acceptable range, the process may fail a CQA, or the edges may represent the limit of product testing, which was stopped because the range is significantly wider than the target range. Imbedded within the acceptable range is the target range. This is the preferred operating range for the parameter of interest that will be validated during process characterization and will result in CQAs meeting their specifications. Typically, the acceptable and operating range and target are published in the master production record, which is the range for the independent process parameters (IPP) that will be specified in the master production record.[17]

Clearly, the operating range must lie within the validated range for the process to be compliant. If a validated range wider than the routine operating range is established, minor process variations beyond the operating range but within the validated range will not result in a failure of the CQA for the unit operation or overall process. In addition, establishing a validated range that is within the acceptable range ensures that validation studies are not performed near the edge of failure. Similarly, since many parameters may have a wide acceptable range, far outside the normal operating and acceptable ranges, these parameters are often called nKPP, or non-KPP. These parameters are not typically studied because they do not have an effect on quality or recovery. Conversely, CPP's often have a narrow acceptable range. In such a case, one should operate within a target range and validate at/near the targets.

38.5 PROCESS SCALING FOR DEVELOPMENT AND VALIDATION

38.5.1 Scale-Up/Engineering Runs

Scale-up from the initial laboratory-scale development studies to either pilot or full-scale manufacturing provides the first scale-related information. Some scale differences are intrinsic, such as heat and mass transfer differences, while others are equipment related. Hold times are frequently longer in larger scale systems, which may affect biopharmaceutical product quality. Scale-up of a manufacturing process increases the lot size to the scale that will be used for commercial manufacturing. The first lots produced at the larger scale should be engineering or shakedown runs that allow for the troubleshooting of the new large-scale equipment and any changes to process made between the early and late lots (process optimization). This is an important aspect of the process validation project timeline, namely allowing time for engineering runs, as all new manufacturing processes will require them to be operated a few times before the process steps will be predictable and reproducible.

38.5.2 Scale-Down

After scale-up to the commercial size and completing process optimization and early and mid-phase clinical studies, it is necessary to develop and qualify scale-down models for the major processing steps. The laboratory equipment should mimic the performance of the full-scale system. Scale-down model systems are used to conduct viral clearance studies, impurity removal studies and process characterization, to examine worst-case conditions for the step and the entire process, and to exercise column and membrane lifetime. The scaled-down process train can be used for evaluating changes in upstream process steps (such as the bioreactor), determining the robustness of a step when considering its impact on the performance of steps further downstream or troubleshooting process deviations in the manufacturing process. The approach to scale-down model construction and qualification is to accept the smallest model that accurately reflects the large-scale unit operation outputs for material purity, yield, and CQAs.

Feed streams from the full-scale manufacturing process or pilot scale material produced in the same manner can be used for developing and qualifying small-scale models. The unit operation inputs should be scaled-down based on the scale difference between the two systems. The same outputs should be measured in both systems, and both should be shown to be statistically comparable. Alternatively, the developer may find that the model is not statistically comparable for some parameters. If a consistent offset is present, this may be acceptable, or if not, the model would not be qualified for that particular parameter. All of the aspects of the full-scale system that can be replicated with the small-scale model should be. So, for example, if longer hold times exist in the full-scale process, then these should be replicated in the small-scale model (even if shorter hold times were possible in the original laboratory-scale development studies).

The unit operations that are most often validated in a scaled-down model process are chromatography steps, UFDF steps, and viral clearance steps. To accomplish the scale-down of a chromatography operation in an effective manner, the column media under test should be of the same type and, preferably, the same production lot as that used in the process-scale column. The column should be packed to meet the packing qualification outputs from manufacturing. Furthermore, all significant process parameters should be maintained as constant. In its guideline for viral safety, the International Conference on Harmonization (ICH) states that

> the level of purification of the scaled-down version should represent as closely as possible the production procedure. For chromatographic equipment, column bed-height, linear flow-rate, flow-rate to bed-volume ratio (i.e., contact time), buffer and gel types, pH, temperature, and concentration of protein, salt, and product should all be shown to be representative of commercial-scale manufacturing.[18]

The flow rate used in a validated small-scale model process should be scaled down by the ratio between the cross-sectional area of the production column and the scaled-down column so that the linear velocity remains constant. The column bed height should remain the same as that used in production so that the contact time of the feed solution with the media is not altered. For adsorption separations, gradient slope and volume should be scaled down by the ratio of the total volume of the production column to the volume of the scaled-down column. The ratio of product loaded to column volume should be kept constant, and the product should be present during the tests at the same relative concentrations that it is present during the actual manufacturing process. Finally, to be valid, the yield and purity of the product recovered from the scaled-down column should be comparable with that of the production column, and the chromatograms should be similar. The extent to which a given column-based separation is scaled down for validation will depend upon the actual production scale and the smallest scale that can reliably reproduce the production process.

In its guide to biotechnology inspections, FDA addresses the use of scale-down models for process validation and states that when

> scale-up is performed, allowances must be made for several differences when compared with the laboratory-scale operation. Longer processing times can affect product quality adversely, since the product is exposed to conditions of buffer and temperature for longer periods. Product stability, under purification conditions, must be carefully defined. Manufacturers should define the limitations and effectiveness of the particular step. Process validation on the production size batch will then compare the effect of scale-up. Manufacturers may sometimes use development data on the small scale for validation.[19]

Despite this, the guide continues, stating that "it is important that validation be performed on the production size batches,"[7] indicating that sequential full-scale qualification lots are still required as part of validation even if a firm uses scale-down models for most of its process validation.

38.5.3 Process Characterization

Process characterization studies are run under a process validation protocol and performed using scale-down models to quantify the relationship between parameters and CQAs for a process step, a series of process steps, or an entire process train. The setting of target values for the parameters during process characterization studies is based on their critical or non-critical designation. Typically IPPs that have been shown to be non-critical are set to their midpoints and treated as fixed, so that statistically based design of experiments (DOE) factorial designs can be executed to investigate presumptive CPPs and KPPs across a relatively wide range, wider than the range executed in manufacturing to date. These studies will determine the appropriate ranges for CPP and KPPs and parameter interactions so that appropriate operating values can be chosen that result in a satisfactory product.

38.5.4 Qualification Lots

The qualification or consistency lots are a demonstration under protocol that the entire process will produce a purified drug substance that passes all of the relevant release testing. Normally, additional testing beyond the planned routing testing for commercial lots is performed on the qualification lots, particularly on the intermediate process streams. The process validation protocol will normally state that the qualification lots will be "a minimum of three consecutive lots," although additional lots may be required if more than one reactor is used or the process is especially complex.

The parameters are set within their operating ranges for the execution of the qualification lots, typically at their target values, which are near the center of the operating range. The purpose of the consistency lots is not to challenge the parameter's proven acceptable range (PAR) or challenge outside that range (process characterization). Rather it is to show that at commercial scale and at the parameter targets, a consistent, acceptable product can be reproducibly manufactured.

38.6 UNIT OPERATIONS

This section presents the common recovery and purification unit operations encountered in the manufacture of biopharmaceuticals. A brief description of the unit operation is followed by a presentation of the typical input variables and the dependent output results generated for the unit operation. Scale parameters, the quantities to try to keep constant as the scale of the unit operation changes, are highlighted.

38.6.1 Centrifugation

38.6.1.1 Description

Centrifuges separate two different phases based on density difference. Typical solid–liquid separations include recovery or removal of cells (mammalian cell culture or bacterial), capture and removal of particulate solids, and recovery of product-containing intercellular solids after cell disruption. One popular configuration is the disk-stack continuous centrifuge, which can be envisioned as a cylinder spinning on its center axis with a sequence of disks encircling that axis at an angle. The feed stream flows continuously from the outside of the unit and through the spaces between the stacked disks. Centrifugal forces cause the denser solids to collect on the disk surfaces, while the lighter liquid phase continues toward the center of the spinning cylinder. The solids slowly move along the disk surface to the outside of the cylinder and are collected in the sediment collection space, from which they are either continuously or discontinuously discharged. The lighter liquid phase flows continuously to the center of the centrifuge and is discharged under pressure.[20]

38.6.1.2 Parameters

For each unit operation, relevant parameters will be presented in tabular form, as shown in Table 38.1, for centrifugation. The parameters presented include the key parameters that can be adjusted for the unit operation, the essential scale parameters held constant to assure comparable results as scale varies, key equations governing operation, and performance outputs that one typically measures. For centrifugation, the key parameters are typically revolutions per minute (RPM) (measured in the SI units of inverse seconds), the volumetric feed flow rate, the system pressure (measured in Pascals) and the temperature.

As the centrifugation scale changes, the ratio of the solid sedimentation rate (solids collected per time) to the solid feed rate is generally kept constant. Clearly, the solid sedimentation rate will rise with increasing RPM, while either a larger feed flow rate or an increase in the percent solids in the feed stream will yield a greater solid feed rate. Another centrifugation scale parameter to hold constant is the feed flow rate divided by Ambler's Sigma factor, which is the required surface area for the same amount of sedimentation in a gravity settling tank.[21] In scale-up or scale-down, sedimentation performance should be the same if the value of (Feed flow rate/Σ) is the same for two pieces of equipment. This is a widely used criterion for the comparison of centrifuges of similar geometry and liquid-flow patterns developing approximately the same centrifugal gravity; however, it should be used with caution when comparing centrifuges of different configurations.[22] Derivations of other equations of interest as well as diagrams of disk-stack centrifuges can be found in standard handbooks.[10]

Table 38.1 also presents the equation for centrifugal gravity, which is the product of the radius and the angular speed squared.

TABLE 38.1
Centrifugation Parameters

Item	Name	SI units
Key parameters	RPM	1/s
	Feed flow rate	L/s
	System pressure	Pa
	Temperature	°C
Scale parameters	(Solids sedimented)/(Solids fed)	(kg/s)/(kg/s)
	Feed flow rate / Σ	(L/s)/m^2
Equations	$G = \Omega^2 r$	–
	G = Centrifugal gravity	m/s^2
	Ω = Angular speed	1/s
	r = radius	M
Dependent results	Product yield	%
	Centrate clarity (turbidity)	NTU
	Particle size distribution	Number/size
	Solids sedimented	kg/s
	Cell lysis	%

38.6.2 Microfiltration and Nanofiltration

38.6.2.1 Description

Microfiltration is often used for clarifying perfusion bioreactor harvest and in other situations where the cell number in a material is an order of magnitude or more lower than that going into centrifugation. Microfiltration uses membranes with controlled pore sizes, typically in the range of 0.1 to 0.65 microns, to physically separate the biopharmaceutical product from cells. Proteins and all smaller molecules pass through the membrane, while cells and large particles are retained in the system. Nanofiltration uses much smaller, controlled pore sizes than microfiltration, typically in the range of 15 to 20 nanometers (0.015 to 0.02 microns), to physically separate the biopharmaceutical product from adventitious virus. In nanofiltration, proteins and all smaller molecules pass through the membrane, while viruses and large particles are retained at the membrane surface. For both microfiltration and nanofiltration, pressure is applied on the upstream side of the membrane, forcing the permeate stream through the membrane while retentate is retained upstream of it.

Both microfiltration and nanofiltration can be operated in either a once-through or a tangential flow mode. In once-through mode, the process stream flows perpendicular to the membrane, with most of the product containing liquid flowing directly through. Retained material builds up on the upstream side of the membrane and increases the resistance to flow. In tangential flow filtration, the feed stream flows parallel to the upstream side of the membrane, constantly sweeping away retained material.

38.6.2.2 Parameters

The key and critical parameters for nanofiltration and microfiltration are listed in Table 38.2. Nanofiltration is used specifically to enhance product safety, to dramatically reduce the likelihood of viral contamination in the drug substance if adventitious virus should enter into the process. The critical parameters for nanofiltration are input pressure, filter challenge (L/m^2 membrane) while maintaining a consistent temperature, and protein concentration. For microfiltration, there are usually no critical parameters, only key parameters, as this step is to remove cells and cell debris, and typically parameter changes only effect product yield. The key parameters are transmembrane pressure, feed flow rate, and solid content of the incoming material. For all microfiltration and nanofiltration systems, parameters to hold constant as the scale changes include the flow rate per filter surface area and the transmembrane pressure.

For tangential flow filtration systems, some additional scale parameters should be held constant with scale, including the individual channel height (or the diameter of the hollow fiber used) and the crossflow velocity. If shear-sensitive cells are present, the feed pumps used must provide sufficient crossflow to sweep the membrane surface while not damaging these cells; low-shear pumps are typically used, and the transmembrane pressure is routinely kept below 2 psi in these cases. Often the permeate flux is independently controlled with either pumps or valves to allow time for the tangential flow on the upstream side of the membrane to sweep away any accumulated debris.

During validation, the reproducibility and consistency of the flux within the statistical range determined during development and manufacturing will be established. Changes in flux below the acceptable range are the result of increased membrane fouling. For membranes that are reused, typically microfiltration membranes and not nanofiltration membranes, cleaning procedures will need to be established (and validated) that minimize fouling and maintain flux. The pressure profiles and product recovery should be shown to be consistent as well.

Product yield in microfiltration can be determined using HPLC or product-specific assays. When the process step is operated with relatively pure proteins such as nanofiltration, UV absorbance at 280 nm can be used. Product yield should be within the acceptable range over the lifetime of a membrane and with different lots of membrane. Yield changes may indicate a variation in product retention. Decreased permeate product yield may be due to a polarization layer impeding flow, aggregation, precipitation, or denaturation from operational changes or perhaps from inadequate cleaning procedures. The compatibility of the membranes as well as all of the wetted system components with each process stream must be determined. Membrane retention and selectivity should remain consistent with consistent operation and cleaning.

For all membrane unit operations, membrane manufacturers will ensure that the material of construction meet USP requirements for Class VI Biological Tests for Plastics and are nontoxic per the USP General Mouse Safety Test. Effluent from the filter must test negative for USP oxidizable substances after the appropriate flush volume. The membrane should not add leachate to the product. Removal of preservative, cleaning and storage solutions by the flushing procedure must be demonstrated and validated. While the manufacturer's procedures should be used as a guide, the actual flushing volumes and parameter settings should be determined for your product.

TABLE 38.2
Microfiltration (M) and Nanofiltration (N) Parameters

Item	Name	SI units
Key and critical parameters	Load volume (N)	L/m2
	Inlet pressure (M and N)	Psid
	TMP (M)	°C
	Chase volume (M and N)	L
	Protein concentration	mg/ml
Scale parameters	(Flow rate)/(Filter surface area) (M)	L/ m^2/h
	Transmembrane pressure (TMP)	Pa
	Channel height or hollow fiber diameter (M)	Mm
	Crossflow velocity[b]	m/s
Dependent results	Flux (M and N)	L/min
	Product yield (M and N)	%
	Permeate turbidity (M)	NTU
	Shear (M)	N/m^2
	Cell lysis (M)	%

38.6.3 Ultrafiltration/Diafiltration (UFDF)

38.6.3.1 Description

Ultrafiltration uses membranes with very small pores, defined by the nominal molecular weight cutoff (NMWCO) of spherical proteins that are retained by the membrane; typical ultrafiltration membranes used in bioprocessing have NMWCOs ranging from 1,000 to 1,000,000 Daltons. These membranes are designed to retain the biopharmaceutical product while allowing small molecules and buffer salts to pass through. Ultrafiltration is carried out in tangential flow mode. If the retentate volume is held constant by the addition of buffer as permeate is removed, ultrafiltration can be used for buffer exchange in a process known as diafiltration. If, on the other hand, the retentate volume is allowed to decrease as the permeate leaves, then concentration of the protein product will occur. Frequently, the product is first concentrated and then diafiltered at the lower process volume.

Scaled-down ultrafiltration setups often do not fully mimic the scale design and operation. However, they are often useful to examine protein recovery and integrity because operating parameters are varied. For example, the rotary lobe pumps used for larger-scale systems are often not available in scaled-down system. Rotary lobe pumps can operate at higher flow rates and pressures than peristaltic pumps and generate a different shear curve.

38.6.3.2 Parameters

UFDF is generally carried out using tangential flow systems where scale parameters such as the individual channel height (or the diameter of the hollow fiber used) and the crossflow velocity are kept constant and the transmembrane pressure is varied. The membranes can be multiuse or used only a single time and then disposed of.

As shown in Table 38.3, a simple equation links the final concentration of a solute after diafiltration to the initial concentration of the solute, the diafiltration volume of the new buffer, the system volume, and the retention of that component by the ultrafiltration membrane. For low molecular weight solutes such as buffer salts, R will be zero or very small because the solute is not retained by the ultrafiltration membrane. As the molecular weight of the solute approaches and exceeds the NMWCO of the membrane, R increases until it reaches a value of 1 for a fully retained large molecular weight solute.

TABLE 38.3
Ultrafiltration/Diafiltration Parameters

Item	Name	SI units
Key parameters	Protein load/membrane area	g/m^2
	Number of diafiltration volumes	system volumes
	Temperature	°C
	Permeate flux[a]	L/min
	Flush volume	L
Scale parameters	(Flow rate)/(Filter surface area)	L/ m^2/h
	TMP	Pa
	Channel height or hollow fiber diameter[b]	Mm
	Crossflow velocity[b]	m/s
Equations	$C_F/C_I = e[(V_D/V_S)(R - 1)]$	–
	C_F = Final concentration	kg/L
	C_I = Initial concentration	kg/L
	V_D = Diafiltration volume	L
	V_S = System volume	L
	R = Retention of component by the membrane	–
Dependent results	Flux	$L/(m^2\,s)$
	Product yield	%
	Retentate turbidity	NTU
	Shear	N/m^2

38.6.4 Depth Filtration

38.6.4.1 Description

Depth filtration removes cells and cell particulate debris from a liquid phase and is typically used for cell removal in mammalian systems. Depth filters, typically single use, remove materials by tortuous passage through asymmetric layers of materials. They can be made of polymers such as plastic and glass fibers and contain diatomaceous earth and a positive binder. The filters are available in lenticular or flat sheet format, and multiple units are secured in vertical or horizontal plate skids. Depth filters with positive binders are also used to reduce host cell proteins and DNA after the low pH hold step in protein purification.

38.6.4.2 Parameters

Depth filter choice is an empirical determination based on the harvest solid content, particulate distribution, and pore size of the membrane. Often two membranes, a coarser and a finer, are used for clarification. Normally, there are no critical parameters in depth filtration (Table 38.4). The key parameters include harvest percent solids and particle distribution, volume processed per membrane area, and flow rate.

TABLE 38.4
Depth Filtration Parameters

Item	Name	SI units
Key parameters	Load volume per membrane area	L/m^2
	Chase volume	L
	% solids in the feed	%
	Temperature	°C
	Flow rate	L/min
Scale parameters	Pressure versus flow rate	Psid
	(Solid load)/(Filter surface area)	kg/m^2
Dependent results	Product yield	%
	Turbidity	NTU

Normally the cell lysis during depth filtration is negligible, as this is typically a low-shear unit operation.

During validation, the robustness of the depth filters in producing a clarified harvest with an acceptable product recovery and clarity should be established over the multiple process runs. The depth filters themselves must be shown to be chemically compatible with the process streams and low leaching as determined by a risk assessment.

38.6.5 Homogenization

38.6.5.1 Description

Homogenization uses a large pressure drop over a short distance to disrupt and break open cells, typically bacterial cells. This pressure drop occurs inside a valve where the feed stream is forced through a small gap between the valve and the valve seat, accelerating the process stream rapidly to a high velocity. Cells are broken apart by the turbulent energy dissipating in the liquid going through the homogenizer valve. This energy generates intense turbulent eddies. In addition, the considerable pressure drop may lead to cavitation, generating further eddies. Homogenization can be carried out in either a single or two-stage valve configuration.

38.6.5.2 Parameters

Normally there are no critical, only key, parameters in homogenization (Table 38.5). This is because the extent of cell breakage normally affects only yield and the extent of disruption can be easily visualized by light microscope. The relevant scale parameters are the pressure drop and the ratio of the valve opening diameter to the volumetric flow rate.

During validation, the reproducibility and consistency of the cell disruption and the product yield for a given pressure profile should be established over multiple process runs. Changes in disruption and yield could result from variations in the properties of the feed stream, such as surfactant concentration, or from clogging within the homogenization valve. Validated cleaning procedures will need to be established that adequately clean all the wetted surfaces.

TABLE 38.5
Homogenization Parameters

Item	Name	SI units
Key parameters	Valve type	–
	Cell density	%
	Surfactant concentration in feed stream	kg/L
	Osmolality of the feed stream	mOsm
	Temperature	°C
Scale parameters	Pressure	Pa
	(Valve opening)/(Flow rate)	$m^2/L/s$
Dependent results	Cells disrupted (microscopy)	%
	Product yield	%

TABLE 38.6
Stirred-Tank Reaction Parameters

Item	Name	SI units
Key and critical parameters	Impeller type (K)	–
	Impeller speed (K)	1/s
	Baffles (K)	–
	Temperature (C)	°C
	Solubilization agent concentration, initial and diluted (C)	kg/L
Scale parameters	(Mixing power)/(Volume)	kw/L
	(Impeller diameter)/(Tank diameter)	m/m
	(Solubilization agent volume)/(Feed volume)	L/L
Dependent results	Product purity	%
	Product yield	%

38.6.6 Stirred-Tank Reactions

38.6.6.1 Description

Protein refolding is an example of a stirred-tank reaction. Overproduced proteins from foreign hosts in bacterial cells are often recovered as refractile or inclusion bodies (IB). The IB are typically 1–3 microns in size and contain (mainly) the protein of interest in a misfolded state. After cell disruption, these dense inclusion bodies are easily separated by centrifugation. Next, the inclusion bodies are washed and solubilized and the proteins refolded to obtain the biologically active product. Solubilization agents are chaotropes such as guanidine hydrochloride, urea, or sodium thiocyanate; surfactants such as sodium dodecyl sulfate or Triton X-100 are added in the presence of reducing agents. Refolding occurs when the concentration of the solubilization agent is reduced, typically be either dilution or diafiltration, sometimes in the presence of oxidation/reduction reactants. Aggregates may form if the protein concentration is too high. Finally, oxidation of the cysteine residues is needed for allow for correct disulfide bond formation in the native protein.

38.6.6.2 Parameters

For the stirred-tank reactor, usually only minimal mixing is required. This step has both key (K) and critical (C) parameters (Table 38.6).

During validation, mixing uniformity and product consistency at a given mixing power should be established over multiple process runs. Inadequate mixing could result from changes in the fluid properties of the feed stream, such as viscosity or percent suspended solids. High recovery in the solubilization/reduction/refolding process is critically dependent on solution composition and concentration, time, and temperature. All the internal components of the stirred-tank must be shown to be chemically compatible with the process streams and to be cleanable and maintainable in a low bioburden and endotoxin state.

38.6.7 Precipitation

38.6.7.1 Description

Fractional precipitation is used to remove broad classes of impurities, increase product purity, or concentrate proteins. The product may be precipitated or kept in solution. Typically, a precipitation step would be performed as the first purification process step, particularly following centrifugation, filtration, or homogenization steps. Precipitation is often carried out in two stages—the first to remove bulk impurities and the second to precipitate and concentrate the product protein. Precipitation is typically executed in standard cylindrical tanks using low-shear impellers. Typically, amorphous precipitates are formed owing to occlusion of salts or solvents or to the presence of impurities. As precipitates often have poor filterability, they are normally collected using centrifugation.

Precipitation can be caused by desalting, salt, nonionic polymers, or miscible organic solvent addition, by adjusting the pH to the isoelectric point of a given protein to minimize its solubility, or by increasing temperature.[23,24] Salts can precipitate proteins by "salting out" effects. The effectiveness of various salts is determined by the Hofmeister series, with anions being effective in the order citrate > $PO_4^{=}$ > $SO_4^{=}$ > CH_3COO^- > Cl^- > NO_3^-, and cations in the sequence NH_4^+ > K^+ > Na^+.[25] Ammonium sulfate is the most commonly used precipitant. The organic solvents commonly used for protein precipitation are acetone and ethanol; these are typically added with in-line mixers to minimize regions of high solvent concentration causing protein denaturation or local precipitation. Nonionic water-soluble polymers, such as polyethylene glycol, are also effective precipitants.

38.6.7.2 Parameters

Precipitation entails chemical addition, initial mixing, nucleation, precipitate growth, flocculation, and finally solid–liquid separation. While equilibrium solubility does not change with scale, the kinetics are scale dependent (Table 38.7). For two phase systems to have equivalent performance, a precipitate of a given size should form in the same time period. The rate-limiting step for precipitation varies, but typically, similar results at various scales can be achieved by keeping the ratio of mixing power to volume constant.[26]

TABLE 38.7
Precipitation Parameters

Item	Name	SI units
Key and critical parameters	Impeller type (K)	–
	Impeller speed (K)	1/s
	Baffles (K)	
	Temperature (C)	°C
	Precipitant concentration/addition rate (C)	kg/L
Scale parameters	(Mixing power)/(Volume)	Kw/L
	(Impeller diameter)/(Tank diameter)	m/m
	(Precipitant volume)/(Feed volume)	L/L
Dependent results	Product purity	%
	Product yield	%

During validation, product purity and yield at a given mixing power should be established over multiple process runs. Product variations, whether the product is precipitated or remains in solution, could result from changes in the precipitant purity or the feed stream. Use of a precipitant should be shown to be consistent within the acceptable range. The precipitant itself as well as all of the internal components of the precipitation vessel must be shown to be chemically compatible with the process streams and be cleanable and maintainable in a low bioburden and endotoxin state.

38.6.8 Liquid–Liquid Extraction

38.6.8.1 Description

Although liquid–liquid separations and precipitation of proteins has been studied for more than 60 years, they are rarely used today.[27] This is likely because about 70% of the projects in development are Mabs, which typically use Protein-A resin for capture. Precipitation and two-phase aqueous systems generally cannot complete with PA resin on recovery and purity.

Liquid–liquid extraction occurs with the partitioning of solutes between two immiscible phases. Because few proteins are soluble, let alone stable, in organic solvents, the systems of most interest for biotherapeutics are those created by the addition of certain pairs of hydrophilic polymers to aqueous solutions, causing a phase separation without the presence of any hydrophobic solvent.[28] One common system is created by the addition of both polyethylene glycol (PEG) and dextran, the PEG-rich phase being less dense than the dextran-rich phase. Aqueous two-phase systems can also be formed with PEG and various salts. Most proteins, as well as particulate matter and cellular debris, partition into the dextran-rich phase. A breakthrough in the usefulness of phase partitioning came with the attachment of ligands, mainly dyes, to PEG, which attracts specific proteins into the PEG-rich phase in a technique referred to as "affinity partitioning." By arranging the partitioning to occur in multiple stages, using classical countercurrent distribution, high product purity and yield can be achieved.[29]

38.6.8.2 Parameters

Protein partitioning in aqueous two-phase systems is strongly affected by pH and in polymer-polymer two-phase systems by the concentration of other ionic species as well (Table 38.8).

During validation, product purity and yield at a given pH, temperature profile, and equilibration time should be established over multiple process runs. Product variations could result from changes in the feed stream components or from variations in the quality of the phase forming raw materials. The phase forming materials as well as the internal components of the extraction vessel must be shown to be chemically compatible with the process streams and to be cleanable and maintainable in a low bioburden and endotoxin state. If the two phases are separated by use of centrifugation, then the section on that unit operation is applicable here.

TABLE 38.8
Liquid–Liquid Extraction Parameters

Item	Name	SI units
Critical and key parameters variables	pH (C)	–
	Ionic species (C)	kg/L
	Temperature (C)	°C
	Equilibration time (K)	s
Scale parameters	(Quantity of lighter phase material)/(Quantity of denser phase material)	kg/kg
	(Flow rate of lighter phase)/(Flow rate of denser phase)	L/s/L/s
	Number of extraction stages	–
Dependent results	Partition coefficient between phases	–
	Product purity	%
	Product yield	%

38.7 PROCESS EQUIPMENT QUALIFICATION

As an essential prerequisite for the validation of recovery and purification processes, the process equipment used for these steps should be qualified and validated. For recovery and purification processes considered here, this equipment will include systems for centrifugation, microfiltration, nanofiltration, ultrafiltration, depth filtration, homogenization, stirred-tank reactions, precipitation, liquid–liquid extraction, chromatography, and sterile filtration, as well as the equipment used for buffer preparation, process monitoring, and product mixing and storage. To initiate equipment qualification, all instruments must be properly calibrated to insure their correct and accurate operation,[30] and biosafety hoods should be certified to ensure the integrity of the HEPA filter and the proper circulation of air.[31] The validation of equipment used in biopharmaceutical processing involves installation qualification, operational qualification, and performance qualification.[32] The text includes chapters of commissioning and qualification, qualification of controlled environments, computerized and control system qualification and validation, validation of process utilities, and others that expand upon the content provided within this chapter.

38.7.1 Installation Qualification

The installation qualification (IQ) of biopharmaceutical recovery and purification process equipment verifies and documents that all aspects of the installation of the equipment adhere to the manufacturer's specifications, appropriate federal, state, and local safety, fire, and building codes, approved company specifications, and design intentions. The IQ demonstrates that the user of the equipment has purchased and installed the correct equipment for the specific task. This document demonstrates that the user has considered the relevant aspects of compatibility of the equipment with the process and that the user has standard operating procedures (SOPs) for keeping the equipment calibrated and in good operating condition through a calibration program, a preventative maintenance program, and a spare parts inventory. This document also demonstrates that the user has analyzed the operations of the equipment and determined the level of operator training required by preparing written SOPs covering these activities. Process equipment IQs should contain the following information:

38.7.1.1 System Application

This section should briefly describe what processes are to be performed and where the equipment is located. As an essential part of an overview of the system, a schematic diagram is included to support a complete understanding and description of the system.

38.7.1.2 Equipment Summary

A detailed description of the system including an equipment summary (manufacturer, model number, serial number) and a description of the components should be provided. Each component of the system should be listed and described separately, with sufficient information to clearly define the system. For example, in a chromatography system, the equipment summary might include feed tanks, tubing or piping, pumps, filters, pressure gauges, valves, detectors, and the column itself. For tangential flow systems, such as those used for cell harvesting or product concentration, the equipment summary should describe the pumps, piping, instrumentation and controllers, holding vessel, and the membrane (type, manufacturer, etc.).

Additionally, the equipment summary should include the design criteria for the equipment and a description of the review process used to ensure that this design is adequate for the equipment's intended use.

38.7.1.3 Supply Utility Descriptions

All utilities supporting the process equipment should be described and checked to ensure proper installation. For example, the electrical source (voltage, amperage, etc.) should be listed and checked against local codes and the electrical specifications of the system. If the system requires compressed gases or steam, these utilities should be validated, and their quality and source should be described and verified.

38.7.1.4 Standard Operating Procedures (SOPs), Manuals, and Drawings

The title and location of all appropriate manuals should be listed and a checklist prepared to ensure that all the manuals exist and have been referenced in the installation of a piece of equipment. All SOPs relating to the installation, operation, and maintenance of the equipment should be listed. These documents should also contain the piping and instrumentation drawings (P&ID) and schematics necessary for installing, maintaining, and repairing the system.

38.7.1.5 Spare Parts and Service Requirements

A detailed list of recommended spare parts and their location is usually included in the IQ. This spare parts list may either be a separate list or included in the manuals. The IQ should also list and review maintenance procedures to assure that prescribed maintenance can be performed without any detriment to either the process or product.

38.7.1.6 Operating Logs

A listing of the name and location of logbooks that document the use of process equipment is usually included in the IQ document.

38.7.1.7 Process Instrumentation

The type, manufacturer, part number, operating range, specific uses, and calibration schedule of all process instrumentation should be listed. This list should be divided into critical and noncritical instruments. A **critical instrument** is one whose failure would adversely affect the product's quality or safety. Depending on the system design and complexity, not all instruments are critical instruments. For example, if an ultrafiltration system is equipped with a flow meter but process performance is not a strong function of flow rate, then the flow meter may be considered a relevant but **noncritical instrument** in this system. The distinction is important because critical instruments will be calibrated and maintained on a more rigorous schedule than noncritical instruments. Also, change orders for a critical instrument will undergo a more extensive examination and failure of a critical instrument during the process will be reviewed more carefully than failure of a noncritical instrument. All instrumentation on the process system should be calibrated against standards traceable or comparable to the National Institute of Standards and Technology (NIST). The IQ should also list the SOPs that describe the calibration procedures for these instruments.

38.7.1.8 Materials of Construction

Those items that have product contact should be described and verified to be compatible with the product and process. All components of the system, including lubricants with the potential for contacting the product, filters, valves, tanks, etc. should be included. Equipment vendors can often provide appropriate compatibility data; however, the user may have to confirm such data with actual process fluids. If materials leach from the system into the product stream, then it should be demonstrated that subsequent process steps remove them.

38.7.2 Operational Qualification

The operational qualification is documented verification that a piece of equipment or process system, when assembled and used according to the standard operating procedures, does in fact perform its intended function. As with the IQ, the OQ is concerned with the equipment and not with the product or process per se. The OQ demonstrates that the user has tested the equipment and has found it to be free from mechanical defects or design defects before use in the production process.

Before starting the OQ for any process equipment system, the IQ on that system should be completed. Any required calibration for the system should also be completed, and in fact calibration may be part of the IQ. The IQ and OQ for supporting utilities such as water systems, lighting, heating/cooling, and electrical should also be completed prior to starting the OQ for the process equipment system. The OQ document should include the following information.

38.7.2.1 Training Verification

It should be verified that operators have received the proper training and are able to operate the equipment as intended by following the appropriate operating SOPs.

38.7.2.2 Check Automated Components

If the equipment is automated, the tests should verify that the equipment responds to the controller as designed. In addition, automated controllers will need to be validated to meet 21CFR Part 11 standards.

38.7.2.3 Check Manual Components

The manual elements of the system, such as hand-operated valves and traps, should be checked physically and/or visually to ensure proper operation.

38.7.2.4 System Integrity

The equipment should be tested to establish that it is capable of operating without leaks. The simplest means of detecting leaks is by visual inspection of the fluid path. Pressure hold tests on the components and piping can identify leaks before attempting operation. Leaks may also be detected in complex systems by demonstrating that the fluid output equals fluid input (fluid mass balance). Membrane manufacturers may be consulted to obtain recommended test procedures and specifications for verifying integrity of filters once installed in systems. If the system is designed to grow microorganisms or mammalian cells, the ability of the system to provide a sterile, contained environment must be shown prior to organism inoculation.

38.7.2.5 Flow/Pressure

Pumps should be tested to show that they deliver the required flow under normal operating conditions. Tolerances should be established for variations in flow rates.

38.7.2.6 Detectors/Recorders

If the data generated by detectors is used in process control, then the acceptable operating range, the limits of linearity response, the reaction time, and the response of the detectors and recorders with operating flow rates should be established.

38.7.2.7 Filters

Filter housings (with installed filters) should be examined to verify that they are appropriate for use with the flow rates and pressures likely to be encountered in the system. They should be suitable for their intended purpose, whether that be sterilization or particle removal. If filters are used for sterilization, they should be validated as such.

38.7.2.8 Computer Control

If computer control is to be used in the operation or cleaning of the equipment system, validation of the control software and hardware in the system must be addressed.[33] It should be shown that the software functions correctly and is protected from unauthorized alteration. The ability of the system hardware to perform its assigned task should also be shown. A schematic of the control logic, including "if-then loop paths," should be included.

38.7.2.9 Alarms

All alarms should be tested by simulation of "alarm conditions" either by actually challenging the system or by electronic simulation. For example, a pressure alarm may be tested by increasing the pressure in the system using pumps and valves; alternatively, the high pressure may be simulated by sending the appropriate voltage to the alarm mechanism.

38.7.2.10 Other Features/Components

Finally, each system may have unique features or components not found in conventional systems used for other applications. Appropriate tests to demonstrate the correct functioning of these features or components should be included in the OQ.

38.7.3 Performance Qualification

Performance qualification of process equipment will establish the reliability and reproducible performance of the equipment in the specific process used to make a product. The complexity of the performance qualification protocol varies widely—from fairly extensive in cases where the application of the equipment to a given process is unique or complex to not necessary at all for simple equipment where the specific process use is fully covered by the OQ conditions and ranges, cleaning and sterilization are covered by separate protocols, and any specific aspects of the equipment performance qualification may be incorporated into the process validation protocols for the entire process.

Two important aspects of equipment performance qualification that must be performed before overall process validation begins are equipment cleaning[34] and sterilization.[35]

38.7.3.1 Cleaning Validation

Cleaning validation is necessary for all product contact surfaces on reusable process equipment to prove that there is miniscule lot-to-lot carryover of process materials. Exceptions to this are disposable materials such as plastic liners, depth and sterile filters, nano-filters, single-use UFDF filters, and plastic process assemblies that are preassembled, dedicated for a single use, and then discarded. Additional detail on cleaning is found in a separate chapter within this volume.

A cleaning protocol must be developed, implemented, and validated. All equipment is cleaned prior to its initial use in processing harvest materials. The cleaning between production lots serves to remove any residual protein or chemical components from the previous lot. Cleaning protocols should be developed based on the users' experience and best practices and manufacturer recommendations on compatible cleaning solutions, contact times, circulation rates, rinsing cycles, and temperature of operation.

Sodium hydroxide (0.1 to 1.0 M) is a commonly used cleaning agent for biopharmaceutical recovery and purification process equipment. It can be removed with a water-for-injection (WFI) rinse using pH or the conductivity of the rinse water as the measure of cleaning agent removal. Cleanliness is often measured on the final system rinse by total organic carbon (TOC), a total protein assay, or product-specific assays (ELISA, HPLC). Product-specific assays are often not as useful because the protein is denatured and will not bind, or not bind in the expected fashion.

While it is tempting to set the specification for removal of an impurity or series of impurities to the limit of detection of the assay chosen to measure removal, this leads to a relative specification, in that as the assay technology improves, greater removal is required. A preferred approach is to set a specification based on the calculated potential carryover of the chemical into the dose that would be received by a patient and to show that this amount is at least three orders of magnitude below a level that would have any effect. If the concentration of the chemical that would have an effect is unknown, then an industry standard value such as "less than 1 ppm" may be acceptable.

38.7.3.1.1 Rinse Studies

A rinse of the process equipment includes the liquid-wetted product-contact surfaces. The final rinse is tested to show adequate cleaning and material removal. The rinse will include all surfaces, including the hard-to-clean ones, but not indicate from which surface the residues found originate. Swabs are used to determine this. In addition, not all remaining residues will necessarily be removed by a particular rinse solution or cleaning conditions—so one must be careful to verify that a given set of rinse conditions is sufficient to dissolve any residues present.

38.7.3.1.2 Swab Studies

Swabs are collected at defined "worst case" locations and are tested for residues. It is often not possible to swab all hard-to-reach locations, which are frequently some of the hardest to clean, so swabs by themselves do not constitute a sufficient procedure for a cleaning validation. Usually rinse and swab studies are combined to give a more complete assessment of cleaning. If a rinse shows no remaining residue above the acceptance criteria and a swab of some of the most difficult-to-clean locations show that the rinse has removed all residue at those locations, then the cleaning demonstration starts to be convincing.

38.7.3.1.3 Coupon Testing

Coupons are used to study cleaning agents and develop cleaning procedures. They are made of the same material as the process equipment and can be exposed to the process conditions inside (or outside) the process equipment, then tested for residuals. The coupon should receive cleaning that is no more vigorous than the process equipment. The coupon surface can be analyzed to understand residue formation and removal.

38.7.3.2 Sterilization

Sterilization validation, extensively covered in other chapters within this book, is necessary as part of the performance qualification for each aseptic process step as well as for superior bioburden control in other steps.

38.7.4 Lifetime Studies

Resin and UFDF membrane lifetime is an element of process characterization. Chromatography and UFDF membranes are either used once, for a campaign, or used multiple times over several campaigns. If used multiples times, the performance and cleanliness of the membranes and resins must be studied. Typically, a scale-down study is performed to a predetermined target number. The results of this study will be used to guide the cycle number at-scale. A scale lifetime protocol for a desired number of cycles will be opened and remain open until the requisite cycles are completed or a cycle does not meet the requirements. The outputs are that recovery and purity for the membrane and resin must be within the acceptable range and impurities must be present at a low level.

38.8 CLEARANCE STUDIES

Downstream processes contain unit operations designed to purify the protein of interest away from product and process impurities (host cell proteins, host DNA, etc.), inactivate or remove viruses, and make a safe product (low bioburden and endotoxin) while maintaining the potency of the therapeutic product. In addition to the impurities present in the bioreactor harvest, other impurities, such as detergents used during purification or ligands that may have leached from chromatography media into the product, are removed from the product during downstream processing. The elimination or many impurities and contaminants can be measured using assays (PCR enzyme immunoassays and protein blotting) specific for the material. However, not all contaminating materials can be specifically measured in the product milieu. Clearance studies with scale-down models are used to show that these materials are reduced by the process to low amounts. Typically, several logs of the material are spiked into the load buffer for those steps, with the potential to reduce the agent concentration. The clearance through the steps is calculated and compared to the amount added to the process. If the clearance is several logs greater than amount spiked, the material will be adequately cleared.[23]

Viral clearance studies are required for mammalian products.[36,37] These agents cannot be brought into a production facility due to the potential contamination of product, plant, and people. Instead, these studies are conducted in a separate facility using scale-down models where the process is accurately reproduced to assure that the data generated can be extrapolated to production-scale equipment.[38]

In spike/recovery experiments on process additives, a clearance factor can be calculated, as shown in Equation 38.1 by dividing the number of units introduced by the number of units recovered in the product after that step.

$$CF_i = I \div O \tag{38.1}$$

In Equation 38.1, CF_i is the clearance factor for the ith step in the process, where I represents the number of units introduced at the start of the process step and O represents the number of units recovered after the process step.

Each potential clearance step in a purification process should be challenged separately so that the clearance of a particular contaminant by each step of the process can be calculated. In general, the overall clearance factor for a manufacturing process (CF_t) is the product of the clearance factors for each step:

$$CF_t = (CF_1 \cdot CF_2 \cdot CF_3 \cdots CF_n) \tag{38.2}$$

The principal impurities which require clearance studies are media components (e.g., MTX, anti-foam, cell culture hydrolysates) and purification additives (e.g., detergents and solvents). Process impurities such as host cell proteins, host DNA, and Protein-A leachate are measured using specific assays. The primary process impurities are discussed in Sections 38.8.1–38.8.6.

38.8.1 Nucleic Acids

The concern of potential mutagenic and carcinogenic affects from the presence of nucleic acids in parenteral protein preparations led to the introduction of regulatory guidelines that limit the exposure of patients to DNA. Original guidelines from the World Health Organization and FDA placed these limits at 10 to 100 pg per dose per day, based on a perceived concern that potential oncogenes could be transferred to patients from mammalian production cell lines.[39,40,41] However, with further experience, the perceived risk associated with DNA has been downgraded. The initial guidelines have been relaxed such that a limit of 10 ng per dose per day is now acceptable under WHO and EU guidelines and may be accepted by the FDA as well.[42,43] Today, nucleic acids are measured by Q-PCR using probes specific for the host DNA.[44] During downstream process development, cGMP manufacture and PQ, the DNA amount at select process steps that typically remove this impurity are measured.

Historically, DNA[45] was measured by blotting and later by binding to DNA binding proteins (Threshold® System [Molecular Devices Corporation, Sunnyvale, CA]).[46] Most recently, quantitative polymerase chain reaction (Q-PCR) is used to amplify and detect specific sequences.[33] Q-PCR provides the most sensitive method for detecting residual DNA. Because of higher specificity and amplification potential, Q-PCR has a better signal-to-noise ratio and is therefore able to detect smaller quantities of DNA than total DNA methods.[47]

For the validation of Biogen-Idec's monoclonal antibody product Zevalin,[48] DNA removal throughout the process was shown using the Threshold Total DNA Assay. The data shows that the host cell DNA was reduced from approximately 3.1 × 10^7 pg DNA/mg antibody to <2.3 pg DNA/mg antibody, well within the acceptable limit for an injectable therapeutic product. By validating that the process consistently and effectively reduces DNA to an acceptable level, the manufacturer may be able to petition the regulatory body to eliminate residual DNA testing as a product release test.

In another example of a study to validate DNA clearance during purification of a recombinant protein, 21 consecutive purification cycles were performed using three different anion exchange chromatography media.[49] Radio-labeled DNA was spiked into the column load before each cycle and after every five cycles, and the clearance factor for removal of all DNA and DNA of greater than 50 base pairs was determined. For each chromatography media, the clearance factor was consistent throughout the validation study. The average clearance factor for two Sepharose Fast Flow anion exchangers (Q Sepharose and DEAE Sepharose) of approximately 1.5 million was obtained. For DE-52 Cellulose (Whatman, Clifton, NJ) the clearance factor was approximately half that of the Sepharose exchangers or 0.7 million.[35] Each of these validation studies demonstrated that a final concentration of DNA of less than 100 pg per dose of protein could be reproducibly achieved, which was the limit at the time of these studies.

38.8.2 Host Cell Proteins

Host cell proteins (HCP) have the potential to cause immunogenicity, or toxicity, of the intended product. This was considered a major issue, and many approaches to determine residual HCP levels in biopharmaceuticals have been developed.[50] To measure the clearance of HCP in a process, levels may be measured by using a direct immunoassay of these proteins in the actual process stream. For early-stage clinical trial material, some commercial kits are available for measurement of host cell protein from *E. coli*, CHO cells, and other commonly used production host systems.[51] The coverage of these kits has improved with time, and some, by mass spec or 2-D blots of 2-D gels, may show greater than 70% coverage of the total HCP in the starting clarified harvest. Although it may be possible to file for licensure using a commercial kit, if the coverage is high, in general, a cell line-specific host cell protein ELISA is used for product licensure. The production source proteins are prepared from the host organism or cell line, which contains a plasmid constructed to have all the DNA sequences except those for the gene encoding the protein product, or from the parent myeloma cell of a hybridoma. These proteins are isolated and polyclonal antibodies prepared against them.[52] The resulting antisera are used to develop a cell line specific immunoassay.

The overall clearance factor of HCP for a purification process will vary according to the types of resins, resin conditions, number of process steps, and the protein itself.

38.8.3 Endotoxin

The facility and each process steps should maintain a low bioburden and endotoxin environment. This is accomplished by following and maintaining CGMP requirements. Both are removed from product contact equipment by cleaning with 0.5–1.0 N sodium hydroxide.

Because of their high molecular weight and highly negative charge, endotoxin and other pyrogenic materials are commonly removed from proteins by either ion exchange chromatography, gel filtration, or ultrafiltration.[53,54,55] Among these methods, ion exchange chromatography is generally most useful in reducing bacterial endotoxin levels, provided that the selectivity of the media is such that co-purification of the bacterial endotoxin and product is avoided. Use of specific membranes to remove endotoxin is also increasing. Membranes specifically designed to remove endotoxin are available from Millipore, Sartorius, and Pall as well as other vendors. However, although endotoxin can be removed from a process through several means, its presence is an indicator of poor process hygiene (not for Gram-negative bacteria whose cell wall is composed of endotoxin) and indicative of the presence of Gram-negative bacteria in the system at some point. Therefore, an investigation should be conducted to try to identify the source of the contamination. The result may be that the material needs to be discarded.

The LAL assay for Gram-negative bacterial endotoxins is sensitive enough for detection at concentrations at least an order of magnitude below levels that will produce a pyrogenic reaction in the rabbit pyrogen test. The LAL test is a compendial test in the USP[56] and has been harmonized under the ICH. The possibility of inhibition or enhancement of the LAL assay by the protein product, however, must be ascertained through validation of the LAL test.[57]

Since a sensitive assay is available to detect the presence of pyrogens in in-process samples and final drug preparations, clearance studies demonstrating the removal of pyrogens are not necessary. Good process control and hygiene, high quality raw materials, appropriate facility environmental controls and overall cleanliness, the cleaning and sanitation of columns, tanks and product contact equipment, and the use of buffers and solutions of low endotoxin all contribute to maintaining a low bioburden and endotoxin environment.

38.8.4 Viruses

When mammalian cells are used as substrates to produce a protein product, there is concern that the cell lines may harbor viruses.[58,59] Endogenous retroviruses are widespread in animal populations and have been described in species as diverse as reptiles, birds, and many mammals. For example, murine hybridomas used in the production of monoclonal antibodies are known to harbor endogenous retroviruses that may have the potential to transform cells. Other rodent cell lines used in production of human therapeutics, such as Chinese Hamster Ovary (CHO), HEK-293, and Baby Hamster Kidney (BHK) have also been shown to contain these endogenous retroviruses, most of which are replication defective (Type A and C retrovirus-like particles). Production cell lines and cell banks are screened for viral contamination. In the absence of a specifically identified viral contaminant in the product cell line, the potential presence of retroviral particles is of greatest concern. The concerns regarding bovine viruses and prions from 10 to 15 years ago have decreased substantially with the removal of bovine serum from most cell culture processes.[60]

The most appropriate way to assure that viruses do not co-purify with the product is to test and select production cells and media components that are free from known adventitious viral contamination. Since most cell lines currently used are derived from sources that cannot be certified as free of endogenous viruses, and since adventitious agents may enter the production process and propagate in cells, viral clearance studies for products derived from cell culture are essential. Validation of the viral clearance capacity of a biopharmaceutical purification process is essential for determining the viral safety margin for the resulting product.[61,62,63,64] Viral clearance factors for select unit operations are determined and the overall clearance provided by the process demonstrated. Any process step for which viral clearance is claimed should clear at least 1 log of model virus, and steps that clear 4 or more logs are considered robust and independent of variability in processing parameters.[65]

Virus clearance is most readily measured by small-scale spiking experiments using scale-down models. Viral clearance should include both virus removal and inactivation, and clearance factors need to be at least 4 logs greater than the theoretical titer of RVLP (retroviral like particles) per dose of product.[66] A theoretical worst-case viral titer may be estimated from electron microscope (EM) pictures of the cell culture fluids from which the product is purified. This information, combined with the cell culture titer, the overall process yield, and the expected dose size and patient weight, is used to compute the RVLP/dose.[67]

In addition to characterizing the viruses contained in the cell line (generally only RVLP), it is important to characterize additional model viruses that cover physiochemical diversity (RNA and DNA and non-enveloped and enveloped). Typically, four viruses are used to support the pivotal viral study. It is desirable to perform spiking experiments with viruses that can be cultivated to a high titer, which have well-established detection assays and which do not present health hazards.

For proteins produced by recombinant DNA technology in mammalian or human cells (e.g., CHO, BHK, C127, HEK-293, and hybridomas), virus removal and inactivation validation include model viruses possessing a range of biophysical and structural features.[68] The viruses used almost always include Xenotropic Murine Leukemia virus (a model retrovirus),[69] Minute Virus of Mouse, and two or three additional viruses that are enveloped and non-enveloped and DNA or RNA viruses with different diameters and geometries. Examples of additional DNA viruses used in viral clearance studies include Herpes Simplex 1 (enveloped) and SV-40 (non-enveloped) and RNA viruses include Sabin Type I Polio (non-enveloped) and Influenza Type A (enveloped). Additional viruses commonly used in viral clearance studies include Infectious Bovine Rhinotracheitis Virus, Reovirus Type III, Epstein-bar Virus (hybridomas), and Pseudorabies Virus. When choosing an appropriate challenge virus, preference should be given to those viruses that display a significant resistance to physical and/or chemical agents.

Clearance studies are performed by spiking model viruses into step starting materials collected from CGMP or runs comparable to CGMP scale and measuring their removal on scaled-down models. Process steps known or probable to inactivate or remove virus should be studied. Typical steps to examine for Mab processes include Protein-A chromatography, low pH hold, anion exchange chromatography, and viral filtration (nano-filter). Although depth filters are often effective in reducing viruses, the viral reduction on these filters is not counted because they cannot be integrity tested.[70] The overall clearance factor can be determined by multiplying individual clearance factors. The steps must be orthogonal; for example, you cannot count two different anion exchange steps, but you can count a cation exchange and anion exchange step. Protein-A clearance is determined by Q-PCR because the low pH elution kills MuLV, and viral kill is measured at the next step, the low pH hold. The low pH, viral filter, and anion exchange clearance is measured by titer of the virus in the starting material and in the column eluate or pool. Only active viruses are measured; viral surface proteins or inactivated virus would not be quantitated.

Grun et al. provided a summary of virus removal by a variety of purification methods.[71] Average log clearance factors ranging from approximately 1.3 to 5.1 were noted for a variety of chromatography types. However, within each type of chromatography, the range of viral clearance varied widely and depended on the specific virus tested and the exact purification process used.

In 2003, Brorson published, on the robustness of the viral inactivation, that the results support the application of a modular viral clearance approach to the establishment of a design space for the low pH virus inactivation step[72] The A-Mab study expanded this modular approach to viral filtration and anion exchange chromatography.[73]

The A-Mab study was conducted on three licensed Mab. The study demonstrated that the low pH inactivation step has a wide design space and shows robust process performance. The study report states that, for all three Mab processes, virus inactivation kinetics were comparable and process performance was consistent and robust over a variety of conditions of feed streams, buffer composition, and protein loads. These results provide a high level of assurance for viral clearance in the A-Mab process when operated under the established design space conditions for the low pH inactivation step. The results of these studies demonstrate that consistent and reproducible inactivation of XMuLV is achieved at operating ranges of pH 3.2–4.0 for 60–180 minutes at protein concentrations of ≤ 35 g/L. The results of these studies for average log reduction factor (LRF) indicate that pH conditions between 3.2 and 4.0 for 60 minutes result in viral inactivation of greater than 6.6 LRF.

The A-Mab study and other reports show the robustness of viral filters for virus removal.[74,75] The results from small virus retentive filtration studies support modular claims for viral reduction. There was little or no dependence of virus removal on protein concentration or buffer characteristics, including buffer salt species, pH, and conductivity. The filtration load volume, chase volume, and filtration pressure were identified as process parameters that potentially impact the effectiveness of the virus removal.

The A-Mab authors said that anion exchange studies on multiple antibodies purified via generic template process (similar steps such as both AEX resin and AEX membrane, same sequence, run under similar conditions) or purified via a modular process (similar steps, different sequence, run under similar conditions) show similar LRF. A generic and modular viral clearance design space based on these results is proposed. Provided that pH and conductivity values are maintained at pH 5.7–8.5 and a conductivity of 0–15 mS/cm, LRF values of ≥ 5.5 for XMuLV and ≥ 4.0 for MVM are consistently achieved for four mAb. There was a minimal effect of viral load amount.

When conducting inactivation studies (low pH, detergent, solvent detergent), it is desirable to determine the kinetics of inactivation as well as the extent of inactivation because virus inactivation has been demonstrated in some cases to be a complex reaction with a "fast phase" and a "slow phase."[76] The inactivation study should be performed such that samples are taken at different times and an inactivation curve constructed. To do these studies, a high-titer virus stock is needed, as well as the appropriate infectivity assay.

A polymerase chain reaction using primers designed to detect very low levels of specific viral DNA or RNA sequences is frequently used to quickly determine the viral clearance potential of unit operations, both during process development and in viral clearance studies.[77] The FDA has recognized the potential of PCR to provide useful data for determination of the optimum process parameters for viral clearance, especially for retroviruses. When combined with infectivity assays, which measure active virus, PCR can also provide information about viral inactivation, since active and inactive viruses will be detected by PCR.

38.8.5 Process Related Components

Removal of potentially harmful or immunogenic components of the fermentation or cell culture media must be demonstrated through either residual testing or validation that the recovery and purification process adequately removes these components.[38]. In microbial production systems, antibiotics may be included in the fermenter to insure genetic stability of the production cell line. Frequently used antibiotics include kanamycin, tetracycline, and neomycin. Expression of the therapeutic protein in microbial systems is often repressed until sufficient biomass has been achieved, and then the expression is induced by a chemical reagent such as isopropyl-beta-D-thiogalactopyranoside (IPTG). Surfactants are often used to reduce foaming in the fermenter as well. Clearance of all these media components must be demonstrated by performing clearance validation studies on the individual unit operations and on the entire process.

Mammalian production systems also can contain media components whose removal must be validated prior to product licensure. Selective pressure is sometimes required to maintain genetic stability; components such as methotrexate, methionine sulfoximine, or gentamycin may be included in the production bioreactor. Antifoam is often used to reduce foaming and prevent clogging of the air filters, and Pluronic-F68 is added to media to provide shear protection to the cells.[78] Proteins such as recombinant insulin, transferrin, or albumin may be used in cell culture media to support high density cell growth, although their use has decreased dramatically over the last ten years due to the development of chemically defined media.

Recovery and early purification steps can also utilize components whose clearance through the later unit operations must be validated. These components include guanidine, urea, and dithiothreitol (*E. coli* refolding process), Benzonase® (viral gene therapy after cell lysis to degrade host DNA), and detergents and detergent/solvent (viral kill steps for mammalian processes). Sensitive assays are developed to measure low levels of the components that are introduced during the process. A combination of in-process testing and final product testing, or spike-recovery experiments at key steps, is used to determine an overall clearance factor.

38.8.6 Affinity Ligands

Increasingly, the initial capture and purification of recombinant proteins, especially monoclonal antibodies and antibody-based molecules (Fc proteins, bispecific antibodies, nanobodies, antibody fragments, and other antibody-like molecules), is accomplished using an affinity column such as immobilized Protein A or L. Viruses used in *in vivo* gene therapy such as AAV species can also be purified using resins with camelid-based affinity ligands specific for the AAV species[79] Any resin immobilized ligand, including Protein A, will leach during protein elution. Manufacturers of Protein A chromatography media and other affinity resins continue to improve their resin so they are less susceptible to leaching[80] and more stable to caustic cleaning agents[81] Removal of residual affinity ligands can be accomplished using several orthogonal chromatographic methods or tangential flow filtration. For example, ion exchange chromatography effectively reduces residual Protein A under conditions employed to reduce host cell proteins and host cell DNA.[82] Several commercial kits (e.g., Repligen Corp) are available for measurement of the Protein A in the range of 16–1000 pg/mL.[83] Typically, the rPA is measured at each process step during development and in the CGMP runs. Measurement of an affinity ligand in the final product might be removed if process validation demonstrates the consistent removal/reduction of the material by the purification process.

38.9 CONCLUSION

Validation of recovery and purification processes is based on process development knowledge for each of the process steps. Scale-up of the process for production of clinical trial material is often followed by construction of a validated scale-down model system, which is used for much of the process characterization work. The independent variables, relevant scale parameters, and dependent results of most interest depend on the unit operations employed for each of the process steps. Individual pieces of recovery and purification process equipment will be validated with equipment validation protocols prior to being used for process validation. The steps in process validation include identification of the critical quality attributes, determination of the critical process parameters, accurate scaling of the process, process characterization studies, and finally manufacture of full-scale qualification lots.

NOTES

1. ICH Q7A, Quality of Biotechnological Products: Good Manufacturing Practices Guide for Active Pharmaceutical Ingredients, November 10, 2000.
2. Levine HL, Castillo FJ, *Biotechnology: Quality Assurance and Validation.* Avis KE, Wagner CM, Wu V, eds. Buffalo Grove, IL: Interpharm Press, 1998: 51. Validation of Biopharmaceutical Processes.
3. International Conference on Harmonization, Technical and Regulatory Considerations for Pharmaceutical Product Lifecycle Management. Q12, November 2019.
4. Wrotnowski C, Neose Targets Complex Carbohydrate Products. *Genetic Engineering News* 2002;22.
5. Veronese F, Harris JM, Introduction and Overview of Peptide and Protein PEGylation. *Adv. Drug Delivery Review* 2002;54: 453–456.
6. Carter BJ, et al., Adeno-Associated Virus Autointerference. *Virology* 1997;92(2): 449–462.
7. Quality Considerations for Continuous Manufacturing. ICH 8a, HHS, FDA, CDER, February 2019.
8. Lönne M, Bolten S, Lavrentieva A, Stahl F, Scheper T, Walter J-G, *Biotechnology Reports* 2015;8: 16.
9. Forier C, Boschetti E, Ouhammouch M, Cibiel A, Ducongé F, Nogré M, Tellier M, Bataille D, Bihoreau N, Santambien P, Chtourou S, Perret G, *Journal of Chromatography A* 2017;1489: 39.
10. Hebben M, in *BioInnovation Leaders Summit*, 2015.
11. Lacki K, Riske FJ, Affinity Chromatography: An Enabling Technology for Large Scale Bioprocessing. *Biotechnol J* 2019;15: 1–11.
12. Rathore A, Winkle H, Quality by Design for Pharmaceuticals. *Nature Biotech* 2009;27: 26–34.
13. Winkle H, BPI Conference, 2010.
14. Castillo FC, et al., Biopharmaceutical Manufacturing Process Validation and Quality Risk Management, *Pharm. Eng.* 2016;36(3): 82–92.
15. Cooney B, et al., Quality by Design for Monoclonal Antibodies, Part 1: Establishing the Foundation for Process Development. *BioPharm. Int.* 2016;14(6): 28–35.
16. Cooney B, et al., Quality by Design for Monoclonal Antibodies: Establishing the Foundations for Process Development, Design Space, and Process Control Strategies. *BioPharm. Int.* 2018;16(6): E4, 5–22.
17. Chapman KG, The PAR Approach to Process Validation. *Pharmaceutical Technology* 1984;8(12): 22–36.
18. International Conference on Harmonization, Viral Safety Evaluation of Biotechnology Products Derived from Cell Lines of Human or Animal Origin. ICH Viral Safety Document: Step 2, ICH Secretariat, Geneva, Switzerland, December 1995.
19. U.S. Food and Drug Administration, Biotechnology Inspection Guide Reference Materials and Training Aids, 1991. www.fda.gov/ora/inspect_ref/igs/biotech.html.
20. Mannweiler K, Hoare M, The Scale-Down of an Industrial Disk-Stack Centrifuge. *Bioproc. Eng.* 1992;8: 19–25.
21. Ambler CM, The Theory of Scaling Up Laboratory Data for the Sedimentation Type Centrifuge. *J. Biochem. Microbiol. Tech. Engi.* 1: 185–205.
22. Leung W, Centrifuges. In *Perry's Chemical Engineers' Handbook*, 7th ed. Perry RH, Green DW, eds. New York: McGraw-Hill, 1997: 18–106 to 18–125.
23. Rothstein F, Differential Precipitation of Proteins: Science and Technology. In *Protein Purification Process Engineering.* Harrison RG, ed. New York: Marcel Dekker, 1994: 115–208.
24. Scopes RK, *Protein Purification: Principles and Practice*, 7nd ed. New York: Springer-Verlag, 1987: 41–71.
25. Belter PA, Cussler EL, Hu W-S, *Bioseparations: Downstream Processing for Biotechnology.* New York: John Wiley & Sons, 1988: 221–236.
26. Belter PA, Cussler EL, Hu W-S, *Bioseparations: Downstream Processing for Biotechnology.* New York: John Wiley & Sons, 1988: 229–235.
27. Cohn EJ, Strong LE, Hughes WL, Jr, Mulford DJ, Ashworth JN, Melin M, Taylor HL, Preparation and Properties of Serum and Plasma Proteins: IV: A System for the Separation into Fractions of the Protein and Lipoprotein Components of Biological Tissues and Fluids. *J Am Chem Soc* 1946; 68: 459–475.
28. Albertsson P-A, *Partition of Cell Particles and Macromolecules*, 3rd ed. New York: John Wiley & Sons, 1986.
29. Johansson G, Joelsson M, Akerlund H-E, An Affinity-Ligand Gradient Technique for Purification of Enzymes by Counter-Current Distribution. *J. Biotechnol.* 1985;2: 225–237.
30. Bremmer RE, Calibration and Certification. In *Validation of Aseptic Pharmaceutical Processes.* Carleton FJ, Agalloco JP, eds. New York: Marcel Dekker, Inc., 1986: 47–91.
31. Kruse RH, Puckett WH, Richardson JH, Biological Safety Cabinetry. *Clin. Microbiol. Rev.* 1991;4: 207–241.
32. Naglak TJ, Keith MG, Omstead DR, Validation of Fermentation Processes. *BioPharm* 1994;20: 28–36.
33. PMA Computer System Validation Committee, Validation Concepts for Computer Systems Used in the Manufacture of Drug Products. *Pharm Technol.* 1986;10: 24–34.
34. Brunkow R, Delucia D, Haft S, Hyde J, Lindsay J, McEntire J, Murphy R, Myers J, Nichols K, Terranova B, Voss J, White E, *Cleaning and Cleaning Validation: A Biotechnology Perspective.* Bethesda, MD: PDA, 1996.
35. Leathy TJ, Microbiology of Sterilization Processes. In *Validation of Aseptic Pharmaceutical Processes.* Carleton FJ, Agalloco JP, eds. New York: Marcel Dekker, 1986: 253–277.
36. ICH Q5A, Viral Safety Evaluation of Biotechnology Products Derived from Cell Lines of Human or Animal Origin. Q5A (R1), September 1999.
37. ICH Q5A, Viral Safety Evaluation of Biotechnology Products Derived from Cell Lines of Human or Animal Origin. Q5A (R2), June 2019.

38. PDA Biotechnology Task Force on Purification and Scale-up. Industry Perspective on the Validation of Column-Based Separation Processes for the Purification of Proteins. *J. Parenteral Sci. & Tech.* 1992;46: 87–97.
39. Office of Biologics Research and Review, Center for Drugs and Biologics, Points to Consider in the Production and Testing of New Drugs and Biologicals by Recombinant DNA Technology. Food and Drug Administration, Bethesda, MD, 1985.
40. World Health Organization Study Group, Acceptability of Cell Substrates for Production of Biologicals. *World Health Organization Technical Report Series* 1987;747: 1–29.
41. Office of Biologics Research and Review, Center for Drugs and Biologics, Points to Consider in the Characterization of Cell Lines Used to Produce Biologicals. Food and Drug Administration, Bethesda, MD, 1987.
42. Center for Biologics Evaluation and Research, Points to Consider in the Manufacture and Testing of Monoclonal Antibody Products for Human Use. Food and Drug Administration, Bethesda, MD, 1997.
43. *World Health Organization Weekly Epidemiological Record* 1997;72: 141–145.
44. Hussain MA, direct., qPCR Method for Residual DNA Quantification in Monoclonal Antibody Drugs Produced in CHO Cells. *J Pharm Biomed Analysis* 2015;115: 603–606.
45. Ostrove JM, Walsh W, Vacante D, Patel N, Molecular Hybridization Techniques and Polymerase Chain Reaction (PCR) as Methods for Safety Assessment of Animal Cells Used in Biopharmaceutical Production. In *Animal Cell Technology: Developments, Processes and Products*. Spier RE, Griffiths JB, MacDonald C, eds. Oxford, UK: Butterworth-Heinemann Ltd, 1992: 689–695.
46. Kung VT, Panfili PR, Sheldon EL, King RS, Nagainis PA, Gomez B, Jr., Ross DA, Briggs J, Zuk RF, Picogram Quantitation of Total DNA Using DNA-Binding Proteins in a Silicon Sensor-Based System. *Anal. Biochem.* 1990;187: 220–227.
47. Lovatt A, Applications of Quantitative PCR in the Biosafety and genetic Stability Assessment of Biotechnology Products. *Rev. Mol. Biotechnol.* 2002;82: 279–300.
48. Conley L, McPherson J, Thommes J, Validation of the ZEVALIN Purification Process: A Case Study. In *Process Validation in Manufacturing of Biopharmaceuticals*. Rathore AS, Sofer G, eds. Boca Raton, FL: CRC Press, Taylor and Francis Group, 2005.
49. Berthold W, Walter J, Protein Purification: Aspects of Processes for Pharmaceutical Products. *Biologicals* 1994;22: 135–150.
50. Wolter T, Richter A, Assays for Controlling Host-Cell Impurities in Biopharmaceuticals. *Bioprocess International* 2005;3: 40–46.
51. Cygnus Technologies, 2006.
52. Chang A, Sofer G, Dusing S, Expediting Compliance for Clinical Biotherapeutics. *Bioprocess International* 2004;2: 30–34.
53. Schindler R, Dinarello CA, Ultrafiltration to Remove Endotoxins and Other Cytokine-Inducing Materials from Tissue Culture Media and Parenteral Fluids. *Bio Techniques* 1990;4: 408–413.
54. Novitsky TJ, Gould MJ, Inactivation of Endotoxin by Polymixin B. *Depyrogenation, Parenteral Drug Association Technical Monograph No. 7*, 1985.
55. Nolan JG, McDevitt JJ, Goldman GS, Endotoxin Binding by Charged and Uncharged Resin. *Proc Soc Exp Biol Med* 1975;149: 766–770.
56. United States Pharmacopeia USP <85> Bacterial Endotoxins Test.
57. Center for Biologics Evaluation and Research and Center for Drugs Evaluation and Research, Guidelines on Validation of the Limulus Amebocyte Lysate Test as an End-Product Endotoxin Test for Human and Animal Parenteral Drugs, Biological Products, and Medical Devices. Food and Drug Administration, Bethesda, MD, 1987.
58. Quality of Biotechnological Products: Viral Safety Evaluation of Biotechnology Products Derived from Cell Lines of Human or Animal Origin, International Conference on Harmonization Topic Q5A, 1997.
59. Brorson K, Krejci S, Lee K, Hamilton E, Stein K, Xu Y, Bracketed Generic Inactivation of Rodent Retroviruses by Low pH Treatment for Monoclonal Antibodies and Recombinant Proteins. *Biotech. and Bioeng.* 2003;82: 321–329.
60. Report of a WHO Consultation on Medicinal and other Products in Relation to Human and Animal Transmissible Spongiform Encephalopathies. World Health Organization, Geneva, Switzerland, March 24–26, 1997.
61. Zhang M, et al., Quality by Design Approach for Viral Clearance by Protein a Chromatography. *Biotechnol. Bioeng.* 2014;111(1): 95–103.
62. Miesegaes G, et al., Analysis of Viral Clearance Unit Operations for Monoclonal Antibodies. *Biotechnol. Bioeng.* 2010:106(2): 238–246.
63. Chen D, Murphy M, Modular Retrovirus Clearance in Support of Clinical Development. Eli Lilli. CMC Strategy Forum Europe, 2018.
64. Burnham M, et al., Advanced Viral Clearance Study Design: Total Viral Challenge Approach to Virus. *Biopharm International* March 2018.
65. Wilkommen H, Schmidt I, Lower J, Safety Issues for Plasma Derivatives and Benefit from NAT Testing. *Biologicals* 1999;27(4): 325–331.
66. *Viral Safety and Evaluation of Viral Clearance from Biopharmaceutical Products*. Brown F, Lubiniecki A, eds. New York: Karger, 1996.
67. Lubiniecki AS, Wiebe ME, Builder SE, Process Validation for Cell Culture-Derived Pharmaceutical Proteins. In *Large Scale Mammalian Cell Culture Technology*. Lubiniecki AS, ed. New York: Marcel Dekker, 1990: 515–541.
68. Viral Safety Evaluation of Biotechnology Products Derived from Cell Lines of Human or Animal Origin. ICH Q5A(R1) Harmonized (1999).
69. PDA Biotechnology Task Force on Purification and Scale-up, Industry Perspective on the Validation of Column-Based Separation Processes for the Purification of Proteins. *J. Parenteral Sci. & Tech.* 1992;46: 87–97.
70. Venkiteshwaren A, et al., Mechanistic Evaluation of Virus Clearance by Depth Filtration. *Biotechnology Progress.* 2015;31: 431–437.
71. Grun JB, White EM, Sito AF, Viral Removal/Inactivation by Purification of Biopharmaceuticals. *BioPharm* 1992;5: 22–30.
72. Brorson K, et al., Bracketed Generic Inactivation of Rodent Retroviruses by Low pH Treatment for Monoclonal Antibodies and Recombinant Proteins. *Biotechnol Bioeng* 2003;82(3): 321–329.

73. Mab: A Case Study in Bioprocess Development. Product Development and Realization Case Study A-Mab The CMC Working Group. Version 2.1, 278 pages, (2009).
74. Sofer G, et al., PDA Technical Report No. 41: Virus Filtration. *PDA J. Pharm. Sci. Technol.* 2005;59(S-2): 1–42.
75. Gefroh E, et al., Use of MVM as a Single Worst Case Model Virus in Viral Filter Validation Studies. *PDA J Pharm.* 2014;68(3): 297–311.
76. AD Hoc Working Party on Biotechnology/Pharmacy, Commission of the European Communities, Validation of Virus Removal and Inactivation Procedures. Note for Guidance, Draft #7, January 28, 1991.
77. Farshid M, Taffs RE, Scott D, Asher DM, Brorson K, The Clearance of Viruses and Transmissible Spongiform Encephalopathy Agents from Biologicals. *Curr. Opin. Biotechnol.* 2005;5: 561–567.
78. Chang D, et al., Investigation of Interfacial Properties of Pure and Mixed Poloxamers for Surfactant-Mediated Shear Protection of Mammalian Cells. *Colloids and Surfaces B: Biointerfaces* 2017;156(1): 358–365.
79. Nass SA, et al., Universal Method for the Purification of Recombinant AAV Vectors of Differing Serotypes. *Mol Ther Methods Clin Dev.* 2017;9: 33–46.
80. GE Healthcare. rmp Protein A Sepharose Fast Flow Manual.
81. Rathmore A, et al., Re-Use of Protein A Resin: Fouling and Economics. *BioPharm* 2015;3(28).
82. Castro-Forero A, et al., Anion-Exchange Chromatographic Clarification: Bringing Simplification, Robustness, and Savings to MAb Purification. *Biopharm* 2015;13(6): 1–12.
83. For example, Protein-A EIA Kit TiterZyme® from Assay Designs, Inc.

39 Validation of Process Chromatography

Günter Jagschies

CONTENTS

39.1 INTRODUCTION

This chapter addresses validation of process chromatography, the key tool that provides today's highly pure therapeutic proteins. Process chromatography is also used to purify and thus enhance the safety of biologicals such as plasma-derived products and vaccines. For biopharmaceuticals, three to four chromatographic steps are typically employed to remove a large variety of impurities. Each chromatographic step can usually remove multiple impurities. Together with unit operations such as centrifugation and filtration, chromatography steps are an integral part of downstream processing, taking a product from crude feedstock to a purified form suitable for safe use as a biopharmaceutical, vaccine or health care agent.

Successful validation of process chromatography is equivalent to a confirmation of robustness. Robustness is what is really needed, not validation; things just need to be confirmed by validation. Creating robustness and confirming it become two sides of the same coin: the purification steps involved in processing the biopharmaceutical drug substance will consistently remove both product- and process-related impurities as well as any safety risk to the patient. Both aspects are discussed with similar focus in this chapter. Creating robustness begins with a solid understanding of what the process needs to be capable of delivering and of the tools applied to develop that and then selecting methodology and developing the process sequence accordingly.[1] Current practice in column-based chromatography validation follows a life cycle validation approach using risk-based tools.[2,3] The process life cycle begins early with structured, forward-looking development studies on impurity clearance capabilities using scale-down models of the final, intended manufacturing process. Manufacturing scale validation confirms the process performance and post-approval monitoring ensures that this performance is consistently repeated throughout the lifetime of the process in its original design. Process changes regulated via change-control programs start a new validation cycle, again from development to implementation at scale.

39.2 ROBUSTNESS—PURIFICATION CAPABILITY REQUIREMENTS

Biotherapeutics, vaccines and plasma proteins are derived from natural sources, typically either mammalian or microbial cells, and human plasma. The type of source determines

DOI: 10.1201/9781003163138-39

how the product is isolated from its production system. Mammalian cells may secrete the product into the culture medium, whereas yeast or bacteria may require applying specific cell disruption, extraction and isolation methods. In any case, insoluble cells and cell debris need to be removed as part of product preparation for the purification process. Dependent on the level of force applied to the cells in these initial steps, more or fewer cell-derived impurities end up in the feed stream for purification. While most of them are soluble, some may tend to precipitate when conditions in the process solution change or when the solution comes into contact with the surface of a chromatography resin. Some impurities, in particular certain cellular enzymes, may damage the target product through proteolytic or glycolytic activity. One distinguishes between process-related impurities (originating from the production cells, culture media or any materials added in or leaking from the process) and product-related impurities (generated through modification via imperfect biosynthesis during expression, chemical or enzymatic modifications during processing, or aggregation) (Table 39.1). The target product frequently displays a certain degree of heterogeneity.

Often, a limited and reproducible product heterogeneity (microheterogeneity) is acceptable as target purity in the final drug product given to the patient. In case any individual impurity exceeds a certain level, for example, 1% or more of total material, it needs to be separately characterized and a rationale provided that it is safe as an ingredient in the final drug product. Process-related impurities need to be removed to a level considered as safe for the patient, usually very low levels per dose and often below the detection limit.

TABLE 39.1
The Purification Challenge from Product- and Process-Related Impurities Describes the Capability for Impurity Clearance that Needs to be Confirmed in the Validation Project

Process-related impurities	Product-related impurities
Cell-culture nutrients, chemicals, fetal calf serum	Dimers, multimers, aggregates
Host cell proteins	Misfolded product and/or product with random disulphide bridge forms
Proteolytic enzymes, other enzymatic activity	Deamidated product variants
Endotoxins	Product with oxidation of methionine
Cellular DNA, other nucleic acids	Product with heterogeneity of post-translational modifications such as glycosylation, phosphorylation and acylation
Virus	Enzymatic degradation products
Cell debris, lipids	
Antifoams, antibiotics	
Leakage, e.g., from affinity columns	
Extractables, e.g., from plastic surfaces	
Water, buffers	

39.2.1 Robustness and Validation—Monitoring Process and Quantifying Quality

Analytical methods are your "eyes to look into the process". In order to follow purification progress during manufacturing and validate the capability of a step or a process sequence, analytical methods are required that are themselves qualified for the purpose (Figure 39.1). This means the capability of the analytical method of choice to measure, for example, product quality attributes or product and impurity concentrations in the actual concentration range and against the background of the material to be tested needs to be confirmed in an analytical method qualification study (guidance provided in ICH Q2, see Table 39.2, and in literature references).[4] As an essential guidance to process development, the stability window of the product under potential processing conditions needs to be defined very early on. Typical conditions to be tested for include ranges of protein concentration, redox potential, temperature, cosolvents, pH, proteases and cofactors. Without information about stability, there is increased risk that process development has to deal with unexpected yield losses and extra efforts to understand observations from experiments conducted. Process steps developed will almost by definition be hard to validate and not robust.

Monitoring methods often used in purification processes, such as UV, pH, conductivity, pressure and flow, are typically not able to quantify attributes of the product-containing solution during processing. However, an acceptable operating range for the signals these methods are providing should be defined and validated. Data points generated from monitoring should be trended in order to enable early detection of process variability. The robustness of performance of these methods relative to the decisions based on them should be verified at the typical ranges of solute concentration, and operating conditions such as pH and flow rate. The analytical tools used to check for the success of a step should offer higher resolution than the process step being controlled and, preferably, include a separation according to a mechanism that is different from the one used in the process step.

39.3 PRODUCT AND PROCESS UNDERSTANDING, DESIGNING FOR QUALITY

Process understanding is developed to ensure that the desired quality attributes of the target molecule are met and safeguarded by the manufacturing process. These attributes are in turn based upon the understanding of the molecule's mechanism of action in its medical indication and its safety when given to a patient. Quality is built in by design of the manufacturing process.

This concept is referred to as Quality by Design (QbD) and is described in the regulatory guideline on pharmaceutical

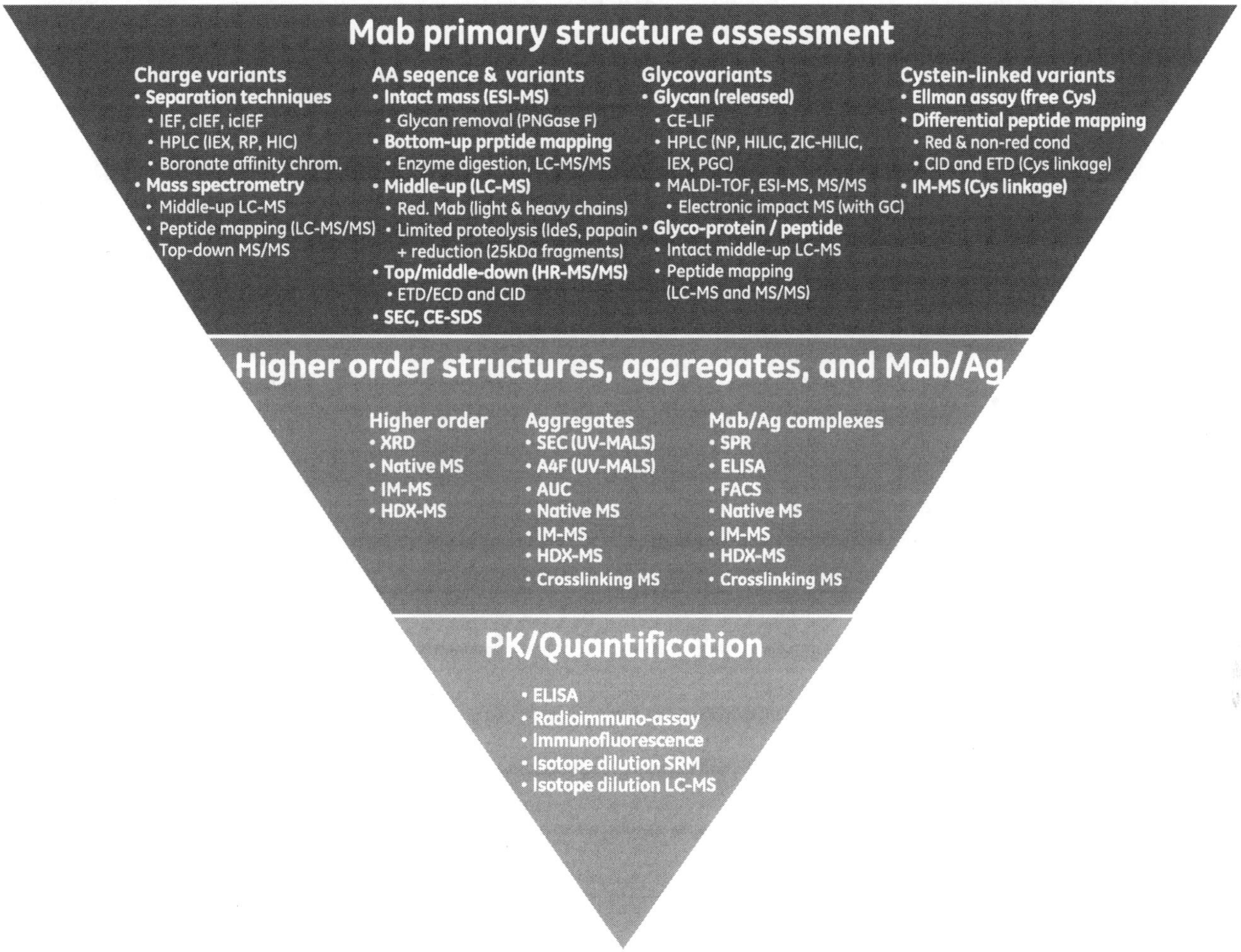

FIGURE 39.1 Analytical methods used to characterize therapeutic antibodies and related products.[5]

development ICH Q8, see Table 39.2, and in a dedicated textbook.[6] Essential activities involved include

- identification of critical quality attributes (CQAs) for the product and setting acceptable ranges for them.
- identification of critical operational parameters to be further investigated in scale-down and modelling studies using the Design of Experiments (DoE) approach. Accumulated process knowledge including from DoE studies and previous process development is used to eliminate noncritical parameters and reduce the study effort further down the path.
- use of DoE studies to identify interaction between variables and nonlinear responses
- identification of failure limits for the process and verification of scale-down models against the manufacturing scale process.

Figure 39.2 illustrates an experience-based concept of simplifying the approach to good process understanding and integration by dividing up and linking the tasks performed by the overall process to different design blocks in the three stages upstream processing (cell, expression and culture), product recovery (isolate, remove and prepare), and downstream processing (capture, removal and polishing). Close coordination and integration of technology choices and process designs contributes to robustness and is an absolute requirement in a successful validation program.

In addition, process understanding is built relative to and for judgment on scale and economic requirements on the final process including the implications they have for method selection. Detailed knowledge of the variety of purification methods and their design options that could be used to run different process steps (including associated consumables, materials and equipment), and of the impact different steps and method selections have on each other (integration) needs to be applied for an appropriate level of process understanding.

Activities and steps early in the process dictate what later steps need to be capable of, but the intrinsic capability of later steps may also put constraints on what can reasonably be done in earlier steps. A lack of consideration for this balance directly leads to a lack of process robustness and corresponding issues in the validation program.

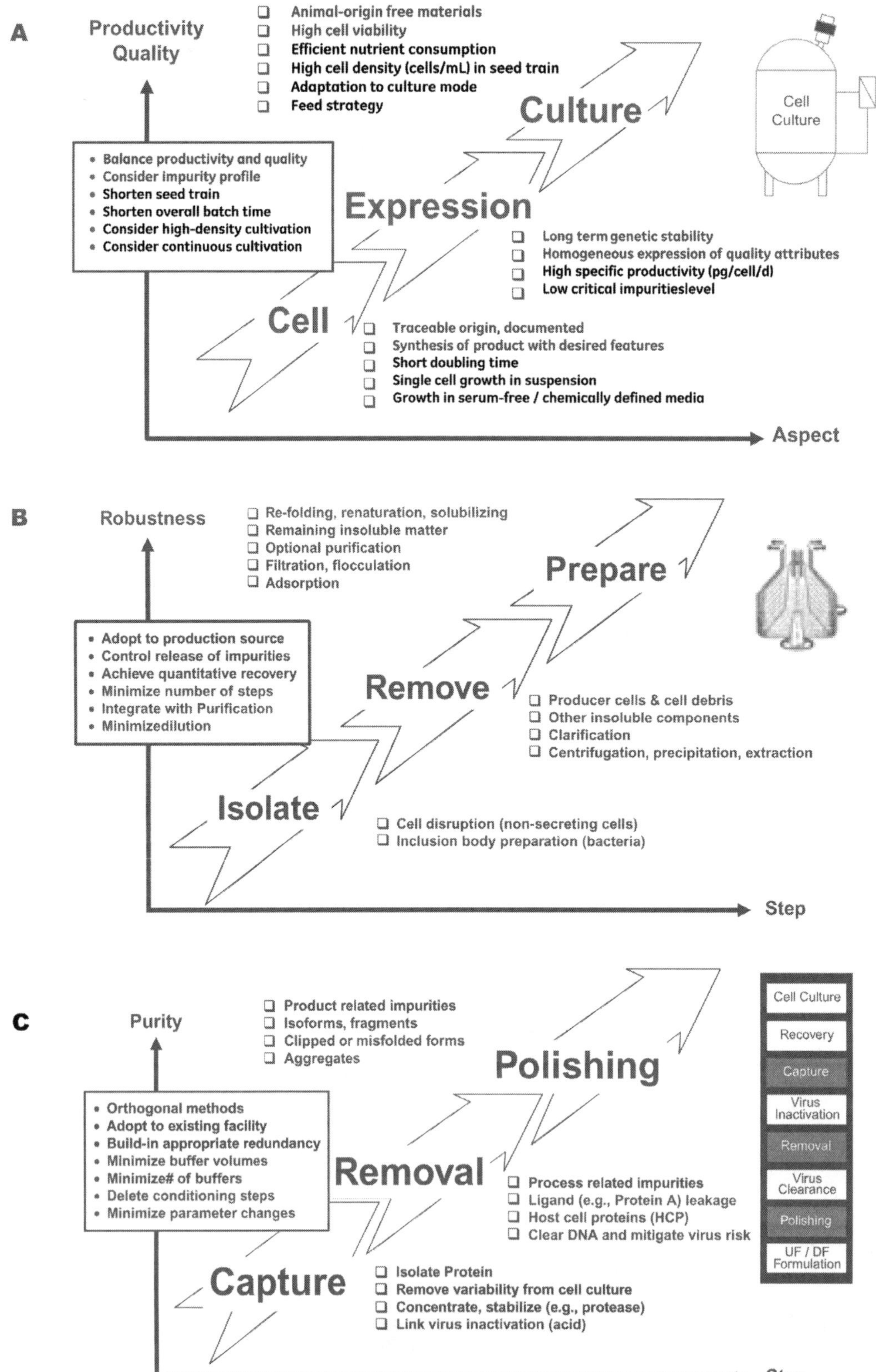

FIGURE 39.2 Experience-based schematic illustration of typical step requirements (red) and step adjustment topics (black) with a process for biotherapeutic proteins or vaccines (A = upstream, B = product recovery and C = purification).

Source: Reproduced with permission from Jagschies, G. et al. "Biopharmaceutical Processing: Development, Design and Implementation of Manufacturing Processes", Elsevier, Amsterdam 2018

Process development studies establish the basis for process validation through the operational conditions and their allowed variability (guideline provided in the ICH Q12 concept of "Established Conditions", see Table 39.2). This ensures robustness, equivalent to predictable delivery at the required quality level, and better definition of which changes need to be reported to authorities (previously, the concept of "Design Space" was used for this purpose). However, apart from quality, scalability of a process to intended commercial scale and economic efficiency of the process design also need to be well understood in order to avoid time delays prior to approval for clinical or commercial use of the process and to minimize wasting of development resources with a design that could turn out unacceptable under business considerations.

In order to be successful in all these aspects, teams need to consider the entire process from cell bank to bulk drug substance instead of just parts, such as the purification steps. It would be naïve to assume that every challenge to quality created in an earlier part of a process can be successfully addressed in purification, for example. Novel technologies that may be capable of overcoming any such design breaks do not tend to appear miraculously when needed; neither do workarounds to address the issue come without cost, and often repeatedly so, with each batch produced.

An illustrative example for such integrative approach to process development was provided by D.J. Cecchini:[7] his study looked at the interface between cell culture harvest and the first chromatography step (capture on a Protein A affinity resin) and revealed that lowering harvest pH to below 5.8 allowed the elimination of an expensive and environmentally unfavorable wash step with a high concentration chaotrope on the Protein A column. At the same time, impurity levels of DNA and HCPs were favorable, and issues with particulate matter getting to the Protein A step (increased pre-column filter area, potentially reduced resin lifetime, and need to remove high turbidity in the eluate) were reduced. All of these achievements support better robustness and validatability of the purification step. An extensive DoE study with four factors has been used to identify the best conditions.

39.4 REGULATORY FRAMEWORK AND GUIDANCE FOR PROCESS CHROMATOGRAPHY VALIDATION

General guidance and directives are published by national regulatory bodies such as the FDA in the United States, CDA (the China Drug Administration, formerly CFDA), or regional ones such as EMA. More specific guidance and recommendations are published, for example, by teams of members of the Parenteral Drug Association (PDA) in the United States.

39.4.1 ICH Guidelines and National Regulatory Bodies

Table 39.2 summarizes a selection of regulatory guidelines by the International Council for Harmonization of Technical

TABLE 39.2
Selected ICH Quality and Safety Guidelines

Code	Category	Chromatography-related key aspects covered
Q2 (A, B)	Analytical validation	Under revision! Purity analysis, use of spectroscopic or spectrometry data (e.g., NIR, Raman, NMR or MS); analytical procedure validation for appropriate test procedures and acceptance criteria setting (supporting Q6)
Q3 (A–D)	Impurities	Categorization of impurities, decision tree for action on impurities, reporting of impurities (Q3A)
Q5 (A–E)	Quality of biotechnological products	Viral contamination risks, clearance strategies, scale-down study design for clearance evaluation (Q5A), comparability and manufacturing changes (Q5E)
Q6 (A, B)	Specifications	Acceptance criteria determining appropriate specifications based on safety, process consistency, purity and analytical methodology considerations; specifications include in-process controls, bulk drug, final product and stability specifications (Q6B)
Q7	Good manufacturing practice	Update 2015! General rules for behavior in facilities and activity documentation; hygiene; containment; equipment cleaning, sanitization and maintenance; automation and calibration; sampling, testing and storage of production materials and consumables; production and in-process controls; training
Q8	Pharmaceutical development	QbD, design space, CQAs, CPPs, enhanced regulatory submissions with QbD
Q9	Quality risk management	Process, methodology and tools
Q10	Pharmaceutical quality system	Maintain state of control, management responsibilities, continuous improvement
Q11	Development and manufacturing of drug substances	Description and justification of the development and manufacturing process
Q12	Life cycle management	Management of post-approval Chemistry, Manufacturing and Controls (CMC) changes in a more predictable and efficient manner across the product life cycle
Q13	Continuous Manufacturing of Drug Substances and Drug Products	Concept paper November 2018
Q14	Analytical Procedure Development	New! Harmonize scientific approaches of analytical procedure development; facilitate more efficient, sound scientific and risk-based approval as well as post-approval change management of analytical procedures
M (1–10)	Multidisciplinary	Gene therapy, bioanalytical methods

Requirements for Pharmaceuticals for Human Use (ICH) for reference[8] that contain information relevant to chromatographic processes across aspects ranging from general GMP rules and quality systems, to strategies to develop products for quality and evaluate the risks to quality objectives, categorization and characterization of impurities, setting specifications and validating the analytical methods used to verify them, and also the aspect of changes over time (life cycle management). ICH guidelines are typically adopted into national law in the member nations of ICH. Thus, similar guidance is published by regional and local regulatory bodies such as the FDA, EMA and PMDA (Japan) and can be retrieved from their corresponding websites. These guidelines form the basis for inspections by regulatory bodies.

39.4.1.1 Outcome From FDA Inspections: Focus on Cleanliness and Documentation Issues

An analysis of the FDA Data Dashboard on Compliance/Inspections[9] reveals that the majority of findings during inspections of the biomanufacturing industry is related to nontechnical aspects such as absence or completeness of or lack of adherence with written procedures in general or SOPs in particular, observations on reporting, testing and release routines, and thoroughness of investigations and complaint handling as well as personnel training. The technical findings, where they come up, are mainly related to cleanliness, including cleaning and maintenance routines at both the building and the process level. This finding may support the view that process development following the guidance from regulatory bodies and expert associations (see Table 39.3, PDA) allows robust and safe biomanufacturing at scale. However, the human factor remains critical for successful implementation of what has been developed. Spotlessness, cleanliness and comprehensive documentation are not human strengths, and training, more training, and training again may be required to compensate for weaknesses.

With chromatography, such findings would point at nonoptimal or false handling of the technology (cleaning in place or CIP is normally an integral activity in a well-developed chromatography step) and thus have a direct impact on process robustness and validatability. Verification of the efficiency of cleaning regimes for chromatography resins and the corresponding columns with associated chromatography skids is essential unless a true (!) single-use operation is performed. In many scenarios labelled "single-use", a considerable degree of reuse remains present, however, and certain cleanliness precautions might still be required. In practice, the term "single-use" has no strict definition, and common production floor practice may favor reuse within limits due to economic considerations. A careful analysis of these scenarios is a key part of the validation of any single-use process set-up, including chromatography. Typical examples include but may not be limited to an extractables and leachables (E&L) study, the qualification of the single-use components supply chain, for example, with regard to supplier facility cleanliness and any sterilization methods applied to the products prior to delivery to the end user for bioprocessing, and the potential consequences of any reuse by the end-user of materials originally labelled "single-use" in the equipment documentation.

39.4.2 PDA Technical Reports

PDA Technical Reports listed in Table 39.3 offer guidance on validation topics relevant to process chromatography. These reports may be considered complimentary to regulatory guidance and directives, since they contain significantly more detailed descriptions on recommended strategies and procedures, mainly authored by biopharma and supply industry subject matter experts.

TABLE 39.3
Selection of Parenteral Drug Association (PDA) Technical Reports[10]

Code	PDA Technical Report title	Chromatography-related key aspects covered
TR 14 Rev. 2008	Validation of Column-based Chromatography Processes for the Purification of Proteins	Full detail on validation planning and activities for process chromatography, e.g., scale-down models, resin lifetime, viral clearance, cleaning and regeneration, storage; monitoring and maintenance
TR 42	Process Validation of Protein Manufacturing	Overall validation workflow, attribute setting, operational parameter setting
TR 47	Preparation of Virus Spikes used for Virus Clearance Studies	Viruses used in clearance studies, impact of virus spike on scaled-down model and on virus removal and inactivation
TR 49	Points to Consider for Biotechnology Cleaning Validation	Cleaning process design and development, cleaning agents, specifics to be considered with packed column cleaning
TR 56 Rev. 2016	Application of phase-appropriate quality system and cGMP to the development of therapeutic drug substance (API or Biological Active Substance)	What to do when during process development and clinical trials; gradual ramp-up of Quality System and cGMP standards while progressing from early to late stage development and through human trials; life cycle approach
TR 66	Application of Single-Use Systems (SUS) in Pharmaceutical Manufacturing	Qualification and verification of suppliers, materials, components and completed assemblies; commonalities and specifics in SUS vs. reuse
TR 74	Reprocessing of Biopharmaceuticals	Hypothetical cases study on an anion exchange step in a monoclonal antibody process (process details ref. "A-Mab: a case study")

39.4.2.1 Single-Use Technology and Cleanliness

With reference to the discussion of a cleanliness focus, the 2014 PDA Technical Report TR 66 provides extensive guidance on implementing single-use technology in biomanufacturing.

Regulatory body-authored guidance does not distinguish between a stainless-steel chromatography column and its piping or a column and its flow-path labelled as single-use when both are in fact being reused. In that case, such labelling would be irrelevant in a validation program and the same risk assessment for cleanliness would need to be applied, followed by the resulting precautions in the process.

The aforementioned PDA report states that "it is preferable to define a position from a risk-based perspective against reusing single-use components". The authors explain further:

> multiple caustic cycles over a single-use column may be used to process several pools from a batch or a series of harvests, but single-use tubing circuits in processing skids may not be able to withstand repeated caustic cycles. Therefore, while the single-use columns may be campaigned, the tubing circuits may need to be replaced for each cycle.

Dependent on the practice of single-use technology implementation in any given facility and the corresponding view on economics, a specific extension of the validation program may be required to cover limited reuse of components originally labelled as single-use.

39.5 PROCESS CHROMATOGRAPHY VALIDATION PROGRAM OUTLINE

The prerequisites of successful chromatography process validation include activities performed in the early process development phase under non-GMP conditions: characterization of the starting material containing the product, a production organism and pre-purification process specific impurity profile; selection of suitable purification steps and their characterization; designing a sequence of steps leading to the preliminary product purity targets; selection of meaningful analytical methods capable of characterizing the product, its impurities and the process impact on the product characteristics; and a comprehensive risk assessment program. The knowledge gained during these activities will be the foundation of robustness and of the validation program and should be carefully documented for later retrieval.

39.5.1 Robustness—Experience-Based Preselection of Materials and Methods

Certain preparation strategies prior to process development experimentation support the objective of achieving process robustness and facilitating validation. The selection of methods and in particular consumables and materials used with these methods must consider a number of supporting aspects in addition to fulfilling product purity objectives: scalability and robust supply at manufacturing scale are key among these considerations. A vendor and method selection activity with a corresponding supply chain audit helps with these decisions. Consulting the company's manufacturing unit, the QA function and the team of facility engineers is recommended to understand the existing experience and the requirements from these disciplines. Figure 39.3 illustrates the aspects to be studied. One objective is to reduce the vast amount of technology offerings to large-scale manufacturing practice prior to conducting large amounts of development experiments; another is to get an initial cost of ownership impression supporting business compatibility of the selections made at this early stage. Cost modeling software is increasingly used to consolidate an economics overview based on technical choices and confirm their validity for the intended process.

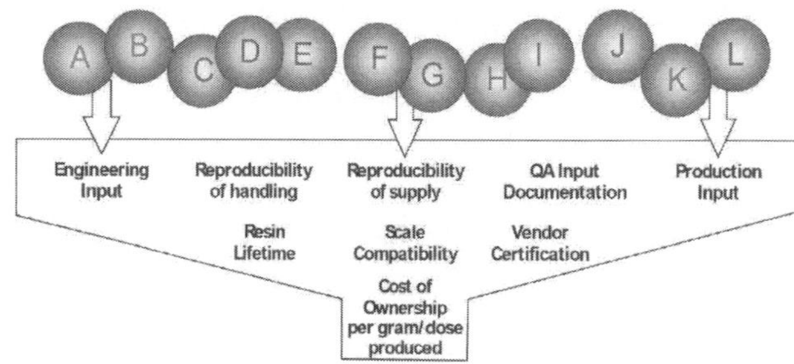

FIGURE 39.3 A filter of experience and studies supporting preselection of chromatography resins on the market (A–L) for their likelihood to enable robust and cost-efficient processing and to facilitate process validation.

Experience with key capabilities and features of chromatographic methods provides guidance for robust assignment of purification challenges to each method (see Figure 39.2 and Table 39.4) as well as advice on placement of the method in a well-integrated validatable process sequence.

In addition to the aspects summarized in Table 39.4, chromatographic methods have features that make them more robust and thus preferable at certain positions in the purification sequence.

- Stability of resins for caustic-based cleaning regimes is a general requirement for steps with risk for bioburden exposure as well as for steps with a risk for precipitation of solutes during the operation of the step.
- Resins, but also chromatography equipment, have pressure limitations that, if exceeded, can lead to catastrophic batch failures. Recommended flow rates and acceptable pressures provide guidance for setting these parameters both at lab and process scale.
- Resin features such as particle size distribution and ligand density can vary from batch to batch. They could impact load capacity and available resolution for the desired purity from the step. During development, the range of these resin variations should be tested for impact in order to avoid issues from setting narrower specifications than what is available from the resin vendor, which could result in delivery shortages.

TABLE 39.4
Planning for Robustness and Validatability—Purification Relevant Upstream Process Objectives and Challenges and Typical Downstream Process Steps with Assignment of Step Objectives and Considerations for Technology Selection

		Process step	Key objectives	Technology applied	Key challenges
Upstream		Mammalian cell culture Microbial fermentation	High product titer Product folding & post-translational modifications correct for intended function Product produced in native form, easy transfer to DSP	High producer cells: • Mammalian: CHO, NS • Microbial: *E. coli*, *S. cerevisiae*, *P. pastoris* Protein-free, chemically defined culture media	Achieve robust synthesis of target features at high titers Achieve shorter batch time Simplification of downstream processing • Maintain product solubility, minimize aggregate formation • Minimize damaging enzymatic activities: protease, glycosidase
Downstream	Recovery	Cell separation Product isolation	Isolate and prepare product for purification Remove producing cells and/or cell debris Protect purification steps from performance issues	Cell disruption methods Protein re-folding, renaturation Centrifugation Membranes: normal flow filtration (NFF), microfiltration (MF) Aqueous two-phase separation (ATPS), precipitation, flocculation	Achieve quantitative product recovery Maintain biological activity Control release of impurities from production cells Minimize generation of additional product-related impurities
	Purification	Capture	Secure process robustness & economy: • Transfer to stable environment • Reduce volume	Affinity chromatography Ion exchange chromatography (IEC) Multimodal chromatography (MMC)	High binding capacity for affinity resins Capacity in combination with fast handling of large volume Achieve very high product yield
		Purification	Removal of the bulk of process-related impurities High log reduction for virus	IEC, MMC, hydrophobic interaction chromatography (HIC)	Find best selectivity for broad range of impurities, e.g., HCP, DNA, aggregates Find step(s) that contribute high level of virus clearance
		Polishing	Removal of remaining traces of process-related impurities Removal of product-related impurities	IEC, HIC Reversed Phase (RPC), for small proteins/peptides Size exclusion (SEC), mainly for very large substances (e.g., plasmid, virus, possibly IgM)	Reproducibly remove wide variety of impurities at very low concentration Minimize generation of additional product-related impurities
		Virus clearance	Biosafety, control of risk for infection	Inactivation and removal methods (acid pH, detergents, filtration)	Flux and cost of virus filters
		Formulation	Transfer to formulation buffer	Ultrafiltration/diafiltration (UF/DF)	Long-term product stability

- High selectivity and capacity for the target product favors affinity chromatography at the beginning of the purification sequence via support to typical early downstream robustness challenges such as fast removal of product from exposure to stability/integrity risks and reduction of initially very large process volumes.
- Ion exchange chromatography (IEC) requires low ionic strength for product load; hydrophobic interaction chromatography (HIC) requires high ionic strength for loading. Placing these methods in a position where the product stream from the previous step matches the load requirements avoids extra conditioning. Load conditions for HIC may precipitate impurities, if used early in the purification train, and cause unwanted robustness issues from instability of the impurity profile as well as environmental issues from the salt load in solution.

39.5.2 Product and Process Characterization

While the product naturally has its own target characteristics identified as critical to ensure its safety and efficacy (Quality Target Product Profile: QTPP, and critical quality attributes: CQAs, see ICH Q8–R2 listed in Table 39.2), the process shall isolate and purify the product up to the final level of quality and shall, dependent on how they are defined, both create and preserve the critical attributes of the product while doing so. Product and process characterization are thus closely linked.

The interactive process commonly used to identify the critical quality attributes of a product is illustrated in Figure 39.4. Typically, any viral or bacterial contaminants as well as bacterial endotoxin, mycoplasma and any particles are highly likely CQAs. Due to the potential immunogenicity risks, any form of product aggregates, inactive or compromised product variants, and in general the higher structure order (secondary, tertiary and quaternary structural integrity) are also likely CQAs. Finally, host cell proteins and host cell DNA are usually considered critical quality attributes. The list of CQAs will vary with type of therapeutic modality being purified.

Examples of molecular attributes known to impact product quality through PK/efficacy include deamidation, oxidation, N-terminal sequence, mannose-6-phosphate (lysosomal targeting), fucosylation, gamma-carboxylation, higher order structure, sialic acids, high mannose and carbohydrate branching. Attributes known to impact safety and immunogenicity include galactose-alpha, 3-galactose (terminal sugars), fucosylation, free-sulfhydryl, aggregates and subvisible as well as visible particles. Details are provided in a dedicated book chapter[4].

39.5.3 Risk Assessment and Mitigation

Since safety and efficacy are the prime characteristics to be guaranteed in the product, all risks for achieving this objective need to be mitigated or removed by the process. A risk analysis is recommended for an appropriate judgment of process capability requirements (see Table 39.5). Apart from product quality and safety-related aspects, the risk assessment

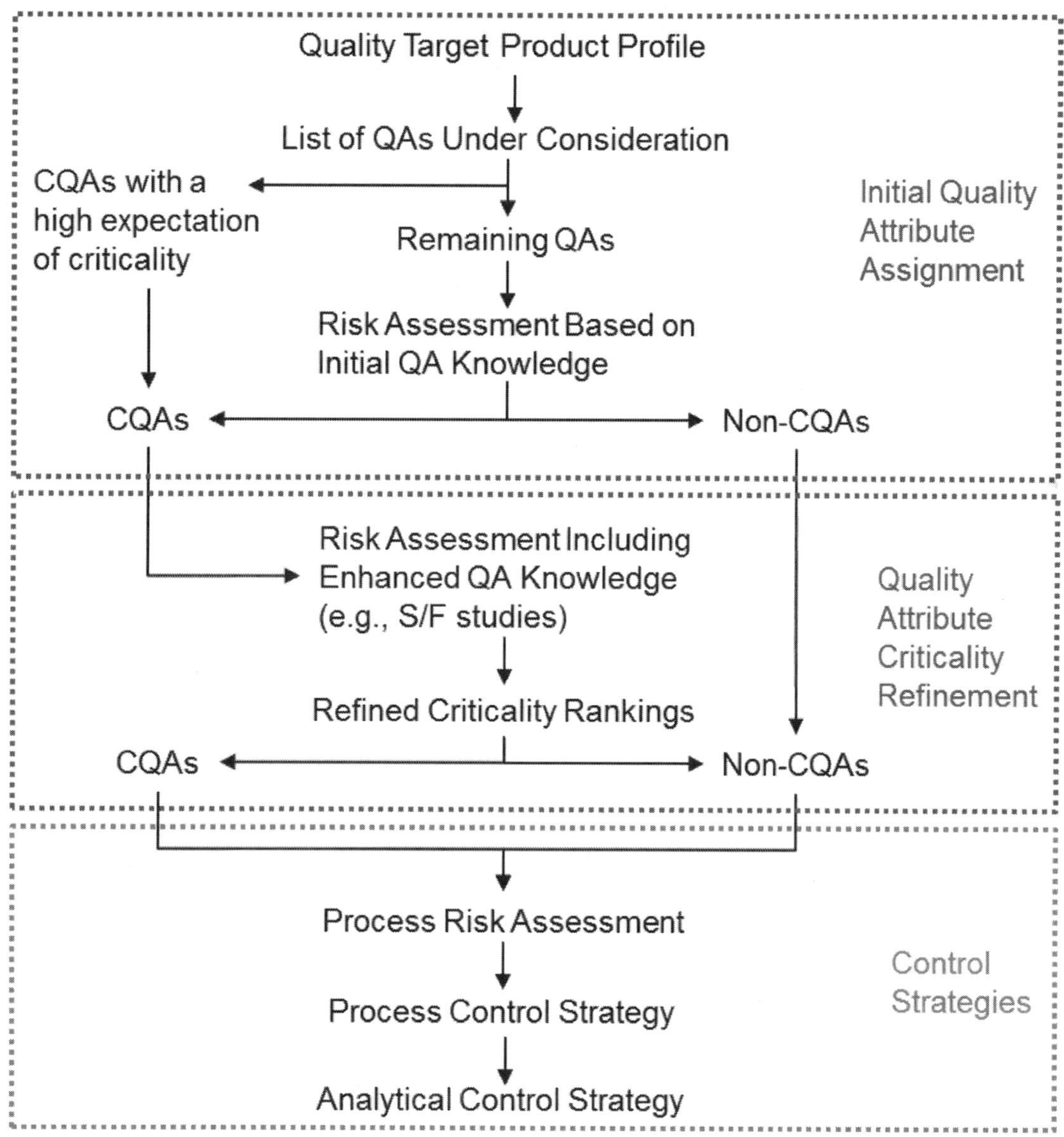

FIGURE 39.4 Overview of the interactive process for assessing quality attribute criticality, an input into the process and analytical control strategy (S/F = Structure/Function).

Source: Reproduced with kind permission from M. J. Traylor et al. "Analytical Methods" in G. Jagschies et al. "Biopharmaceutical Processing", Elsevier, Amsterdam 2018.

TABLE 39.5
Overview of Risks Associated with Typical Product- and Process-Related Impurities

Impurity	Risk characteristic	Mitigation strategy[11]	Priority
Any process variability	Safety risk for patient Supply risk	Design robust process based on design space and control strategy	Very high: all steps
Proteases	Damage/alteration of product	Minimize release, genetic engineering, pH-inactivation	Very high: 1st purification step
Virus	Safety risk for patient: infections	Cell line testing, virus clearance by two or more steps with orthogonal mechanisms	Very high: specific steps
Glycosidase, other enzymatic activity	Damage/alteration of product	Prevention of release from cells, fast inactivation/removal	High: 1st purification step
Lipids, cell debris	Performance change of resins or membranes	Recovery process, filters, centrifuges, flocculation	High: address prior to purification
Antifoams	Performance change of resins or membranes	Avoid use, filters	High: address prior to purification
Endotoxins	Safety risk for patient: fevers	Anion-exchange chromatography in FT, cation exchange chromatography in B/E	High: avoid introduction into cell culture by compliance with GMP
Host cell proteins	Safety risk for patient, immunogenicity	Orthogonal removal steps: IEC, HIC, MMIC	Medium to high: all steps
Product-related impurities	Safety risk for patient: altered potency, immune response	Orthogonal removal with polishing step: IEC, HIC, MMIC, RPC; SEC	Medium to high: all steps
Nucleic acids	Safety risks for patient: low risk of carcinogenic effect of nucleic acids	Anion exchange chromatography, cation exchange chromatography	Medium: polishing steps
TSE	Safety risk for patient: infections	Donor screening (human plasma), raw material selection (animal-free)	Medium: prior to processing
Mycoplasma	Safety risk for patient	Cell line selection, raw material testing	Low: prior to processing
Cell culture media	Safety risk for patient: antibiotics, immunogenicity of protein additives, potential adventitious agents	Cell culture without antibiotics and use of protein free cell culture media	Low: prior to processing

Source: Reproduced with permission from Jagschies, G. et al. "Biopharmaceutical Processing: Development, Design and Implementation of Manufacturing Processes", Elsevier, Amsterdam 2018.

will be extended onto process-related aspects as well. This includes any impact the facility environment, human interaction, materials such as buffers and consumables such as chromatography resins used in the process, processing equipment and not least the chosen process monitoring and control strategy could have on quality outcome.

A detailed review of risk assessment methodology relevant to biopharmaceutical manufacturing is available in dedicated publications.[12] From the risk assessment, criticality of identified product attributes and process parameters is derived, which leads to a categorization of these features and a corresponding prioritization of related validation and continuous monitoring efforts.

39.5.4 Chromatography Validation Program Timelines

The studies during chromatography validation will be performed both at small scale and at large scale: the tightly related cleaning-in-place and resin lifetime studies are conducted prospectively at small scale, for example, and are later confirmed at production scale. A similar approach is taken for the clearance of impurities, including product-related and process-derived impurities. Certain studies can reasonably only be performed at small scale, namely the investigation of viral clearance capabilities of the purification process. These experiments utilize spiking with virus samples that would be both impractical and too costly at large scale, ignoring the fact that one would not under any circumstances contaminate the production scale system with virus.

Once the parameters for operating the process are determined and experience from clinical material manufacturing has been gathered, batch records are created, and conformance batches are produced at scale to confirm that the process delivers as intended within the production environment. All aspects of running and controlling the process are recorded and compared with instructions in the batch records, and the quality profile of samples taken is compared with the intended product quality definition. This includes supporting processes such as column packing, buffer preparation, and maintenance activities for the purification line.

The sequence and timing of these activities follow the progress of process development from basic to fully matured and the progress in human clinical trials up to licensure.[13,14]

Typically, prior to entering first-in-human trials (Phase I), the "basic process" is characterized at unit operation level using qualified analytical methods and has its critical operational parameters identified. Characterization includes defining the critical control parameters, outputs, and working ranges for both inputs and outputs. Broad initial working ranges that are gradually narrowed with experience enable more rapid development and improve the ability to validate the process. Working in narrow ranges and then shifting out of those ranges can require repeating clearance studies, assay validation and even toxicity studies and clinical trials.

The number of purification steps that will be incorporated into the process and eventually require validation is determined by understanding the impurities each purification step is capable of consistently removing. One should also consider the ability to maintain product integrity, for example, by avoiding product modifications. It is important to keep in mind that clinical effects of purity may not be easy to verify. Since products entering clinical trials must be demonstrated to be safe, some validation of the downstream processing will be required even at the early stages. For the most part, the capability of the process to provide safety from viral infections is confirmed with an initial, limited series of viral clearance studies (with at least two model viruses), and clearance of any bioburden and harmful substances that may have entered the process or have come through from the upstream process is verified. Critical quality attributes are identified at this time and monitored using the assays developed for the process. A product impurity profile is required.

Around the time of Phase II, critical analytical assays should be validated, and detailed process characterization studies are initiated to establish the operating parameters with acceptable ranges and verify process robustness. Production equipment is being qualified by installation and operation qualification (IQ and OQ).

Toward Phase III, all analytical assays need to be validated and the virus clearance validation program completed. A specific assay for residual host cell protein is required, and major product-related impurities should be fully characterized. Process validation in downscale studies confirm resin lifetime and clearance for host cell proteins, DNA and—case dependent—specific additional impurities. Finally, prior to license application, full scale validation is performed with the manufacture conformance lots (at least three). Post-licensure, ongoing process monitoring is implemented, and change control and deviation management is established. Concurrent resin lifetime verification is performed at production scale. The "validated state" of the process is regularly audited throughout its life cycle.

39.5.5 Small Scale Studies for Validation

Small-scale systems are commonly used and can be a cost-effective approach in some process optimization and characterization/qualification tasks required for purification steps. Impurity clearance studies, resin life span studies, cleaning studies, process intermediate hold time studies and stability studies are often performed at small scale. When performed appropriately, these studies can be used to support establishment of protocols and acceptance criteria. In some cases, those small-scale studies provide supporting data, but typically reconfirmation is required during process validation at full scale. In other situations, they minimize the risks associated with hazardous materials. If the small-scale studies are to be used in validation, the small-scale system must be qualified.

Validation of a small-scale model requires that the chromatography system truly reflect the purification process that will be used for the licensed product. The feed stream for these studies should be taken from production. Materials such as chromatographic resins and buffers should be those that are approved by QA for use in manufacturing. Columns must be equilibrated in the way they are or will be in manufacturing. If measurements of pH, conductivity and UV are performed in manufacturing, they should also be performed for the small-scale column operation. Contact time is one of the most important factors for measuring comparability when chromatography is scaled up or down. Its importance is described in the ICH guideline on viral safety.[15] Oftentimes, column efficiency (height equivalent to a theoretical plate, HETP) and/or peak asymmetry (As) measurements are made to ensure the column packing is consistent at the two scales. Then product recoveries, product purity and the impurity profiles can be measured to demonstrate they are comparable to those observed in manufacturing.

There are some particular differences in chromatography systems of different scale that should be taken into account. The wetted materials may be different in both column and system at a small scale. Often stainless steel is used in large-scale chromatography, whereas plastic or glass is more common for smaller columns. Large-scale single-use chromatography equipment may use different plastic materials than lab-scale systems. Adsorption of both product and impurities may be greater with a particular material. Transport distances to monitors and collection vessels should be proportional at both scales. Column distribution systems are almost always different. Multiple inlet ports are typically used at large scale, while single ports are utilized for small columns. In spite of these differences, when properly designed, small-scale systems can be qualified and validated to correctly reflect manufacturing scale activities.

39.5.5.1 Spiking Studies for Impurity Clearance

Small-scale chromatography models are used for evaluating/validating clearance of viruses, transmissible spongiform encephalopathies (TSEs), DNA, host cell proteins and process impurities such as Protein A and cell culture media additives (e.g., BSA). To enhance sensitivity, spiking studies with a high concentration of the material that is to be removed by the chromatography step are performed. When spiking studies are performed in clearance studies, the impact of the spike on column performance has to be evaluated using multiple analytical methods. As analytical methods have become more sensitive, the need for scaled-down models for clearance studies has decreased.

Some clearance studies can now be performed at pilot or manufacturing scale. In particular, the use of PCR has decreased the need for small-scale models for DNA clearance. With the exception of viruses, TSEs and other potentially hazardous materials, clearance data acquired by small-scale studies can be confirmed at manufacturing scale by testing the final product or a selected process intermediate for the presence of the impurity. Once the clearance studies are performed and the product tested in validation runs, lot release testing can usually be eliminated. This can be a big cost saving. However, it is essential to maintain the assay capability in the event of a process change or Out-of-Specification (OOS) result.

Validation of virus and TSE clearance requires that the validation batches are run under the same conditions as those used in the small-scale clearance studies. This is sometimes overlooked when different departments are responsible for small-scale and manufacturing validation. When clearance studies are performed for potentially hazardous materials, such as virus and TSEs, sanitization studies are usually part of the study. This is done by demonstrating the effectiveness of the chosen sanitizing agent after a spike into a small-scale column. For equipment, coupons (i.e., cut-out pieces of equipment) may be used for the spiking study (for further details, see Section 39.7.3).

Viral clearance studies are required for mammalian cell culture derived products intended for both clinical trials and licensed product. Where there is potential risk, TSE clearance studies are also performed. Small-scale chromatography systems that are used to evaluate virus and TSE clearance are best designed and validated at the site where the analytical methods are most conveniently performed. For demonstration of removal and inactivation of hazardous materials such as virus and TSEs, spiking studies utilizing coupons can be performed in a safe environment away from the manufacturing facility.[16] Often, the actual spiking studies are performed in facilities where the safe handling of such agents can be assured. One problem that arises during early clinical studies is that subsequent process changes might invalidate the viral clearance. Planning for the changes, revalidating the small-scale system if needed and repeating the viral clearance study is necessary when changes are made to chromatography steps claimed to remove viruses.

39.5.6 Scale Up Confirmation/Engineering Runs

Once the purification process is optimized, typically prior to phase 3, it can be scaled up, and scale-up capabilities can be verified during confirmation (or "engineering") runs. Scale changes in chromatography may result in differences. Some of these differences are found in holding times, wetted materials, flow cells, distribution systems and process capability (e.g., pumping capacity). Although chromatography is one of the simpler unit operations to scale, some minor modifications may have to be made. The scale-up confirmation runs should be performed prior to starting the formal process validation. These preliminary runs can go a long way toward minimizing subsequent formal validation failures.

Columns must be packed to meet predetermined acceptance criteria. HETP and As determinations are often used to qualify columns.[17] In some cases, HETP and As measurements will not be the same in manufacturing scale columns as they were in smaller columns. This may not, in and of itself, be an issue. It can be the result of different column designs. The product purity profile, as measured by multiple orthogonal analytical techniques, and the impurity profile will be key determinants to ensure the scale up is acceptable and does not necessitate redesigning scale-down models and repeating clearance studies that were performed at small scale.

Once the scale-up capability is verified, it is essential to ensure that the conditions used for small-scale clearance studies are, in fact, those that are being used during validation. There are several FDA form 483s that note the conditions used in manufacturing do not reflect those that have been validated.

39.6 LARGE SCALE STUDIES FOR VALIDATION

Validation at manufacturing scale requires qualification of raw materials and equipment (see Table 39.6). All of the analytical methods used to evaluate the purification process effectiveness will be validated, with the exception of those used solely for characterization. Each purification unit operation must have predefined inputs and outputs that were established during development and that will be measured during the validation runs at manufacturing scale.

One issue that can arise for validation of chromatographic processes is the perceived need to run the process at upper and lower limits for each parameter. To do that might prevent a product from ever getting past the validation stage, since it may not be feasible to obtain product at both upper and lower protein loads or extremes of other critical parameters during the validation/conformance batches. However, development should have demonstrated which variables will need testing at the limits of the ranges. The use of designed experiments to establish process robustness has been addressed by Kelley.[18] Identification and establishment of operating ranges of critical process variables was presented by Gardner and Smith.[19] If small-scale studies have been validated and run

TABLE 39.6
Qualification Activities

Raw materials	Buffers, supply of water
	Chromatography resins
	Processing additives
Equipment	Columns
	Pumps
	Monitors
	Tanks
	Automated skids

under conditions representative of manufacturing, they may be used to support the outer limits. However, if these studies are to be used to support process limits for licensed product, it is recommended that QA sign off on the studies.

During process validation, all informative assays are typically used. This usually includes some assays that are not fully validated and that will not be used in routine manufacturing. Often included are highly sensitive assays for evaluating product purity and impurities after each chromatographic operation.

After validation, control charts and trending analysis are used to maintain the purification process in a validated state. Retrospective validation is seldom used for validation of chromatography processes. Its use is described briefly in a book chapter on protein purification issues.[20] However, now that the industry has a history of successfully producing biological products using chromatographic purification, some firms may have sufficient data to perform retrospective validation.

39.7 SPECIFIC POINTS TO CONSIDER IN CHROMATOGRAPHY PROCESS VALIDATION

Some issues deserve special attention in a discussion of validation for process chromatography. These include holding, processing and storage times; chromatography resin life span studies; and cleaning and sanitization.

39.7.1 Holding, Processing, and Storage Times

Holding times are defined in development and confirmed at manufacturing scale. For purification process intermediates, holding time studies include evaluation of product purity, stability and bioburden control. In small-scale studies, holding times can be extended to build in a safety margin. The smaller scales are first validated to represent manufacturing. The data, when appropriately documented and approved, can be an aid in releasing batches in the event of an unexpected hold time during manufacturing. Table 39.7 provides some FDA 483s and comments in approval letters related to holding times.

TABLE 39.7
Holding Time Issues Described in FDA Form 483s and Approval Letters

483s	Comments (approval letters)
Lack of data to support hold times	Institute bioburden monitoring of storage solution
Lack of container closure integrity studies	What is expected storage time based on validation studies for the regenerated column?
Hold times post sanitization/equilibration for column not validated	Submit results of hold period studies for in-process product intermediates that include container-closure integrity study and biochemical, bioburden, and endotoxin studies on a periodic basis

Comments in one post-approval inspection included "storage times in between runs for all of the purification columns have not been validated for the entire life cycle of the columns".

Processing time limits are established to ensure final product consistency and freedom from adventitious agents, such as bacteria and fungi. These time limits are also evaluated in development, but scale changes may necessitate some modifications that will be validated at full scale.

Storage times for chromatography columns are validated to demonstrate column integrity and control over bioburden. Storage time establishment is part of a chromatography resin life span study since the storage conditions can adversely impact column performance over time. Removal of column storage solutions is also validated. This is particularly relevant for small molecules, such as ethanol, whose presence might impact subsequent column performance. Contact time plays an essential role in cleaning (e.g., hydrolysis of residual contaminants on the resin surface is time dependent), and during storage further effects may occur in columns that appeared to be clean prior to storage. The capability of the start-up protocol to remove any residuals from the column should be validated. This is often done using one or more of the following assays: total organic carbon (TOC), conductivity, pH, UV.

39.7.2 Resin Lifetime

Chromatography columns are validated for consistent performance over their life span. An FDA Compliance Guide describes the need to have an estimated life span for each column type.[21]

As noted in the compliance guide, concurrent validation may be appropriate. For products derived from sources where there are no known viral risks, that is, those produced in bacteria or yeast, concurrent validation is often a good choice. Concurrent validation depends on the ability of in-process analytical tools to demonstrate performance consistency. Avoiding lengthy off-line assays can prevent holding time problems or further at-risk processing.

In most cases, a combination of both prospective and concurrent validation is used for life span studies. In one prospective study, up to 1,200 cycles were demonstrated.[22] In a concurrent study on the life span of an anion exchange column, 27 cycles or 4.5 years were validated. One of the parameters measured concurrently was consistency of DNA removal.[23]

One publication from the US FDA described factors that could predict degradation of Protein A column performance long before retrovirus clearance is decreased. Those factors include antibody step yield and breakthrough but not eluate impurity content. It was proposed that viral clearance studies of aged chromatography resins may not always be necessary.[24] While this approach may be accepted by some regulatory agencies, others may not find it suitable.

Parameters that can be used to evaluate continued performance are described in Table 39.8. There may be others, and each development project should decide which parameters provide them with the most relevant information.

TABLE 39.8
Parameters Indicative of Column Performance Deterioration

Indications of performance deterioration	Methods for determination
Decreasing product purity	HPLC, MS, CE, SDS-PAGE
Change in impurities profile	HPLC, CE, SDS-PAGE
Increase in specific impurities	Host cell protein, host cell DNA, Protein A
Decreasing product recovery	Product specific assays
Increasing breakthrough	Binding capacity, UV, HPLC
Changes in flow rate or pressure	Flow and pressure monitors
Changes in regeneration and/or requilibration profiles	pH, conductivity, UV monitors

The ability to clean and sanitize the column must remain constant over the life span. A decrease in product purity or increase in impurities often indicates contaminant build-up on resin surfaces that is not being removed by the cleaning process. Changes in regeneration volumes, column packing deterioration and increased backpressure are also indicative of decreased column performance.

39.7.3 Cleaning and Sanitization Validation

Cleaning and sanitization of chromatography resin and equipment are of paramount importance in the robust production of safe biotherapeutics. Chromatography resin and equipment suppliers provide recommendations for cleaning and sanitizing, and sodium hydroxide at concentrations between 0.1 and 1.0 M is the most frequently used agent for both cleaning and sanitizing columns. A recent book chapter details cleaning and sanitization strategies in purification processes.[25] It is important that the cleaning/sanitizing agents do not modify column performance. Each column and feed stream combination are evaluated to ensure the appropriate conditions for cleaning and sanitizing are selected. For example, in some cases, high salt is a good cleaning agent (e.g., NaCl at 1.0 M). If there are residual bound hydrophobic impurities, high salt could increase binding rather than provide a cleaning effect and use of organic solvents in the cleaning solution may become the only alternative. The latter case may lead to a reevaluation of alternative steps, as solvents would be considered both an environmental and a cost challenge. Some chromatography resins cannot tolerate harsh conditions, and cleaning problems can ensue after repeated use. Alternative cleaning regimes have been developed in such cases.[24]

Chromatography resins are dedicated to one product, but equipment, including column hardware, may be used for more than one product. Cross-contamination between process runs on multiuse equipment is prevented by using suitable cleaning routines and validating the absence of carryover from one product to the next. Although the acceptance of 10 ppm carryover from one drug product to another has been discussed in the pharmaceutical industry, this is not generally acceptable for highly potent biopharmaceuticals. Acceptance criteria are usually established by performing a risk assessment. Oftentimes, the acceptance limit is set at the detection limit of the assay. Unfortunately, in some cases, this is the only possible choice.

For dedicated equipment and packed columns, carryover of residual product and/or impurities is evaluated to ensure consistency of product. Carryover of degraded product or impurities can alter product immunogenicity as well as potency.

FDA 483s related to carryover include

- protein carryover from previous purifications not characterized.
- impact of carryover proteins not evaluated.
- cleaning validation not performed for removal of urea and cell culture media.
- no periodic monitoring of columns following cleaning.

Both small-scale studies and manufacturing scale runs are used for cleaning validation. The small-scale studies are really part of development and allow for higher concentrations and temperatures to be evaluated for removal of residuals and to build in a safety margin. The optimized conditions are then evaluated during conformance batches. Assays used to assess cleanliness include, among others, TOC, product-specific assays and total protein assays. Blank runs monitored by UV can indicate if there is protein carryover.

A problematic area for the biopharmaceutical/biologics industry has been control of bioburden in purification operations. Chromatography is not a sterile process. However, a combination of pretreatment/sanitization of the resin slurry used for column packing, slurry preparation in a closed tank system and closed packing of the pre-sanitized column body through a nozzle system supports appropriate microbial control. The use of vendor prepacked, sanitized columns and single-use pre-sterilized tubing and connectors is an accelerating trend in the industry. There are numerous FDA citations regarding bioburden control in downstream processing.

39.8 CONCLUSION

Chromatographic resins generally have a large surface area that provides high binding capacity. Proteins and other biological molecules, complex structures in themselves, have complex interactions with that surface area. Nonetheless, a great understanding of the variables involved in process chromatography has been accrued in the 40 years since the advent of modern recombinant biotechnology.[26] As a result, there is now greater acceptance of different validation approaches. For example, concurrent validation for chromatography resin life span, once considered unacceptable, is being applied to some processes. Generic validation is being considered in other cases when similar processes are applied to similar starting materials and products. In part, this progress is due to experience. It is also due to the availability of more sensitive analytical tools, such as PCR and advanced mass spectroscopy that provide a better understanding of virus, DNA and impurity clearance by chromatographic processes.

Process chromatography is an enabling technology for today's highly pure biologicals. Validation of those chromatography processes is a key part of ensuring consistent and safe biological products are being delivered to patients. It shall be emphasized once more, however, that the foundation for such consistency is robustness, whereas validation can "only" confirm the robustness of quality delivery from the manufacturing process. While validation is still based on "first principles of science", the approaches and certainly the terminology of this discipline has evolved several times over, not always with an outcome of simplification and reduction of effort. It is thus still a good advice to spend time with the preparations for a robust process design outlined in this chapter. These efforts will pay off when validation is performed.

NOTES

1 Jagschies G, Łącki KM, Process Capability Requirements. In *Biopharmaceutical Processing* Jagschies G, et al., eds. Amsterdam: Elsevier, 2018.

2 Parenteral Drug Association PDA Technical Report TR 14, revision 2008: "Validation of Column-based Chromatography Processes for the Purification of Proteins"

3 Parenteral Drug Association PDA Technical Report TR 56, revision 2016: "Application of phase-appropriate quality system and cGMP to the development of therapeutic drug substance (API or Biological Active Substance)"

4 Traylor MJ, et al., Analytical Methods. In *Biopharmaceutical Processing: Development, Design and Implementation of Manufacturing Processes.* Jagschies G, et al., eds. Amsterdam: Elsevier, 2018: 1001–1049.

5 Adopted from Beck A, et al., "Characterization of Therapeutic Antibodies and Related Products", in Special Issue: "Fundamental and Applied Reviews in Analytical Chemistry". *J. Anal. Chem.* 2013;85: 715–736.

6 *Quality by Design for Biopharmaceuticals: Principles and Case Studies.* Rathore AS, Mathre R, eds. Hoboken NJ, USA: Wiley & Sons, 2009.

7 Cecchini DJ, Applications of Design Space for Biopharmaceutical Purification Processes. In *Quality by Design for Biopharmaceuticals.* Rathore AS, Mathre R, eds. Hoboken, NJ, USA: Wiley & Sons, 2009: 127–142.

8 International Council for Harmonization of Technical Requirements for Pharmaceuticals for Human Use (ICH) guidelines: www.ich.org/products/guidelines/quality/article/quality-guidelines.html

9 https://datadashboard.fda.gov/ora/cd/inspections.htm

10 Available in print or as eBooklets from website www.pda.org/publications/pda-technical-reports

11 FT=flow-through mode operation; B/E=bind-elute mode operation; IEC=ion exchange chromatography; HIC=hydrophobic interaction chromatography; MMIC=multimodal interaction chromatography; RPC=reversed phase chromatography; SEC=size exclusion chromatography; TSE=transmissible spongiform encephalitis

12 Mollah AH, Long M, Baseman HS, eds., *Risk Management Applications in Pharmaceutical and Biopharmaceutical Manufacturing.* Hoboken NJ USA: Wiley, 2013.

13 Parenteral Drug Association PDA Technical Report TR 56, revision 2016: "Application of phase-appropriate quality system and cGMP to the development of therapeutic drug substance (API or Biological Active Substance)"

14 Hejnaes KR, Ransohoff TC, Chemistry, Manufacture and Control. In *Biopharmaceutical Processing.* Jagschies G, et al., eds. Amsterdam: Elsevier, 2018.

15 International Conference on Harmonization, ICH Q5A Viral Safety Evaluation, 1997.

16 Cleaning and Cleaning Validation: A Biotechnology Perspective, Chapter 5. Sampling Methods. PDA, Bethesda MD1996

17 Sofer G, Chromatography Media Lifetime. In *Biopharmaceutical Process Validation.* Sofer G, Zabriskie DW, eds. New York: Dekker, 2000: 197–211.

18 Kelley BD, Establishing Process Robustness Using Designed Experiments. In *Biopharmaceutical Process Validation.* Sofer G, Zabriskie DW, eds. New York: Dekker, 2000: 29–59.

19 Gardner AR, Smith TM, Identification and Establishment of Operating Ranges of Critical Process Variables. In *Biopharmaceutical Process Validation.* Sofer G, Zabriskie DW, eds. New York: Dekker, 2000: 61–76.

20 Winkler ME, Purification Issues. In *Biopharmaceutical Process Validation.* Sofer G, Zabriskie DW, eds. New York: Dekker, 2000: 143–155.

21 FDA Therapeutic Compliance Guide 7341.001: "There should be an estimated life span for each column type, i.e., number of cycles. Laboratory studies are useful even necessary to establish life span of columns. There are situations where concurrent validation at the manufacturing scale may be more appropriate. Continued use may be based upon routine monitoring against predetermined criteria."

22 Seely RJ, Wight HD, Fry HH, Rudge SR, Slaff GF, Validation of Chromatography Resin Useful Life. *BioPharm* 1994;7: 41–48.

23 Heitzmann M, Process Validation of the Basiliximab Purification Process. Oral Presentation. IBC Process Validation. February 2004, San Diego.

24 Brorson K, Identification of Chromatography Quality Attributes to Assure Retroviral Clearance Capacity of Multiply Cycled Resins. PREP, 2002, Poster presentation.

25 Grönberg A, Hjorth R, Cleaning-in-place and Sanitization. In *Biopharmaceutical Processing: Development, Design and Implementation of Manufacturing Processes.* Jagschies G, et al., eds. Amsterdam: Elsevier, 2018: 675–699

26 Łącki KM, Introduction to Preparative Protein Chromatography. In *Biopharmaceutical Processing.* Jagschies G, et al., eds. Amsterdam: Elsevier, 2018.

40 Single-Use Technologies and Systems

A. Mark Trotter and Derek Pendlebury

CONTENTS

DOI: 10.1201/9781003163138-40

40.1 INTRODUCTION TO SINGLE-USE TECHNOLOGY

We trace the beginnings of single-use technology to the 1970s and '80s, as applications were primarily small volume and scaled-down preparation of media and buffers and liquid sterile filtration of up to 100 L volumes. The use of carboys for storage and disposable filter assemblies for sterilizing filtration offered some economical and operational advantages. These initial drivers gradually created a demand for larger devices and products. The wide use in health care of intravenous (IV) bags and tubing sets for patient-side fluid delivery set the stage for transferring the disposability of filtration bag systems, single-use systems (SUS), to the biopharma markets. By the mid-1990s, larger volume bags, tubing and connector systems were often supplied with integrated 10-, 20- and/or 30-inch capsules filters and had become utilized by drug manufacturers. They in turn placed growing demand on single-use technology (SUT) suppliers to develop larger system approaches for pilot plant and production scale devices. As the biopharma industry entered the new "green" 21st century, the terminology of "disposables" was replaced by the more environmentally friendly term "single-use" devices. The scale-up trend accelerated, and SUT products and systems became widely accepted and adopted into biopharma processes, especially in larger protein molecule production uses such as cell culturing and fermentation. The mid- to late 2000 period show rapid development of more complex SUS and the introduction of large volume bags, upwards of 2,000 liters, that segued into scalable bioreactors and fermenters with similar large-scale capacities.

Single-use technology applications throughout the early 2000s to today have experienced explosive growth in replacing or substituting glass and stainless steel components, including the introduction of innovative, disruptive products and systems such as process controllers used for various filtration and purification processes, for example, tangential flow filtration, continuous chromatography platforms and fully integrated and automated downstream operations. These SUS combine multiple technologies into single skid-based systems that can be easily configured to undertake a wide range of operations. Many upstream process steps—from seed flasks, seed reactors and process scale reactors, to mid-stream cell harvesting, separations and clarification stages, to the downstream purification and final fill operations—have found SUT/SUS applications in all phases of biopharmaceutical therapeutic production.

There are key operations and economic drivers that accelerated the introduction and acceptance into this highly regulated with stringent industry standards that create high "barriers to entry" to the biopharma market.

The following list includes many of the key operations and economic drivers that will be examined.

- Time to market.
- Validation and qualification tests and studies.
- Increases in productivity.
- Decrease in microbial and product cross-contamination.
- Significant reductions in clean-in-place/steam-in-place.
- Reduction in chemical disposal costs.

These key factors concerning the SUT ability to reduce contamination (both product and microbial), the reduction or elimination of CIP/SIP and the resultant lowering of expensive pharmaceutical grade water consumption, all contribute to production efficiencies. The resultant benefits support the overall mandate of providing a secure drug supply while lowering drug costs to the patient and providing a return on investments.

40.2 SINGLE-USE COMPONENTS AND SYSTEMS

40.2.1 Filtration and Separation Technology and Application

Single-use technologies used for separation, clarification, purification and sterilizing processes are key operations to achieving production efficiencies and cost reductions. The following sections will discuss membrane filtration, depth filtration and viral clearance operations that utilize SUT devices. The separation and purification operations using tangential flow filtration and centrifugation of cell culture and fermentation products have seen innovative approaches to these costly and difficult applications. Single-use technologies can be incorporated into complex processed controlled systems for skid-mounted operations that can be automated, thereby achieving production efficiencies and lowering the cost of goods sold.

40.2.2 Filtration and Separation Technologies

40.2.2.1 Sterilizing of Liquids and Air/Gases

Sterilizing-grade and bacterial retentive filters are polymeric membranes typically having rigid pore structure through the cross-sectional matrix. These membrane filters are rated by filter manufacturers under designated pore sizes, such as 0.1 um, 0.2/0.22 um and 0.45 microns. However, to qualify as "sterilizing-grade," these membranes must meet stringent regulatory compliance and industry standards, that is, ASTM Standard Test F838-15a.

Disposable single-use filter cartridge and capsule configurations have been in use for decades. The first single-use capsules were introduced in the early 1970s as small area (100/200/500 cm^2 filtration area) devices. Standard size pleated cartridge cores were encapsulated in a polypropylene housing as an initial development. Larger cartridge capsules with filter surface area of 1000–4000 cm^2 were developed between 1990 and 2000. These larger-size disposable designs were in high demand, from pilot plant to production scale end users. These users required higher capacity disposable filters that were coupled to tubing and bags to make single-use systems (SUS). These bag/filter systems had applications in both upstream and downstream process operations; storage/transport/mixing/bioreactors/intermediate prep and final filling formulations. The demand for the large-scale separation/purification and sterile filtration continues explosively as the biopharma markets continue to increasingly adopt single-use technology products in the drug manufacturing processes.

Since steam-in-place (SIP) sterilization is not feasible with these polymeric connectors and devices due to melting and deformation, sterilization by gamma (γ) irradiation has permitted these pieces to come together, filters/tubing/connectors and bag, to introduce the first sterile (SUS) products.

As cited in the AAMI /ANSI/ISO 11137:2006, gamma irradiation sterilization cycles are designed to achieve a Sterility Assurance Level (SAL) of 10^6. Gamma irradiation is used, as other methods, autoclaving and Ethylene oxide (EtO) sterilization, are less suitable and include drawbacks such as water condensate and EtO residue toxicity problems in cell culture. Gamma irradiation sterilization can maintain the polymeric properties, for example, tensile strength, color, shape and fit and function.

Smaller capsulated filter devices were also needed by end users, and filter manufacturers responded by scaling down filtration trains in both upstream and downstream applications. Typically, 47 mm discs with a nominal 12–13 cm^2 surface area were used to demonstrate filtrate throughput and dirt holding capacities in filterability trials. These disc devices, either in stainless steel/plastic holders or capsule design, require less drug product or test solution to run multiple trials, typically using 1.0 liter or less. To verify the scale-up flowrate characteristic of larger high surface area (1000–4000cm^2) capsule filters, discs proved inefficient. This led to the development of small-scale pleated capsule devices with surface areas ranging in the 20–50 cm^2 range to simulate the flow path and pressure drops of the larger scaled-up capsules used in production.

40.2.3 Qualifying Industry Standard and Protocols

For use in liquid or gas sterilizing applications, membrane filters must meet or exceed requirements using validation and qualification protocols from ASTM Standard F838–15a, *Standard Test Method for Determining Bacterial Retention of Membrane Filters Utilized for Liquid Filtration.* It is also recommended to follow guidance from PDA Technical Report No. 26, *Sterilizing Filtration of Liquids.* The bacterial challenge testing (BCT) for these validation trials can require upwards of 5 liters of drug product to perform the set of three BCT trials. For disposable air/gas filter capsules, there is guidance in PDA Technical Report No. 40, *Sterilizing Filtration of Gases.* These air/gas filters feature the same single-use polymeric housing construction, size ranges in square centimeters (ft^2) and pressure ratings. Air/gas sterilizing filters must be validated using the same testing protocols as liquid filters as per protocols in Standard F838-15a.

Advances in membrane technology throughout the late 1990s to the mid-2000s were incorporated into sterilizing-grade filter cartridges and capsule configurations. Filter manufacturers marketed similar products for the sterilizing-grade 0.1/0.2/0.22/0.45 μm membranes. Filter manufacturers understood the commoditization of these filter products. Hence, to offer differentiation, many filter products were introduced that used unique pleating designs and newer membrane polymers. As of today, the membrane substrate material of choice, polyethersulfone (PES), is widely used by most filter membrane manufacturers. This choice is due to this polymer's robustness in temperature, chemical compatibility and pH extremes, while offering higher flow rates and dirt holding capacity than other marketed membrane polymers.

Analogous membrane substrates are still manufactured to meet the end user requirement for legacy drug manufacturing. These filter membrane polymers include cellulose acetate (CA), polyamide (PA—nylon), polyvinylidene difluoride (PVDF), polypropylene (PP) and polytetrafluoroethylene (PTFE—used primarily in air /gas/vent filtration), with 0.1/0.2/0.22 and 0.45μm pore size rating. These membranes are also used in special applications, for example, low protein binding or submicron particle removal. These effective and reliable membranes will continue to be used in new drug/protein product filtration due in part to their past performance and being "grandfathered" to validation compliance requirements in earlier testing and qualification.

40.2.4 Prefiltration/Depth Filtration

Used for clarification and prefilters, the materials of construction (MoC) used are typically pleatable sheets made from polypropylene, glass fibers and cellulose, ranging in pore size ratings from 1.0 to 100 um, most typically in the 1–10 um range. Membrane type prefilters with larger pore ratings (0.8–10 μm) are also used where compatibility and the same MoC are required. To increase or enhance throughput and capacities, other filter-aid materials such as diatomaceous earth (DE), activated carbon or silica-based materials may be added to the depth filter substrate or the filtrate itself. The general filtration mechanism is sieving throughout the depth of the filter media with smaller particle sizes retention augmented by adsorptive effect of the filter-aids. Many cellulosic depth filters are configured as lenticular (lens-shaped) devices using multiple stack-discs to increase surface area for larger-scale (500–2,000 liters) applications, for example, cell debris removal from bioreactor/fermenter harvest.

40.2.5 Mycoplasma Filtration

Although there are no standard practices like the ASTM F838-15a, *Standard Test Method for Determining Bacterial Retention of Membrane Filters Utilized for Liquid Filtration*, test laboratories can use *Acholeplasma laidlawii* as the test organism to qualify filters membranes for the 0.1 um removal ratings for mycoplasma retention. PDA has a task group developing growth media guidance and the current Technical Report No.50, *Alternative Methods for Mycoplasma Testing.* Each filter manufacturer has a differing test method and claims for the log reduction value (LRV) for their respective membranes. This leads to much confusion among end users, who must either use the filter manufacturer's test data, which may pose correlation risk, or perform actual testing using process parameters, drug product and the selected mycoplasma retentive 0.1 um membrane filter. These trials typical follow the ASTM F838-15a and PDA TR No.26 protocols for bacterial challenge testing. The membranes substrates are the same polymers used to produce 0.2/0.22 um filters, namely cellulose acetate, polyamide (nylon), polyvinylidene difluoride, polypropylene, polytetrafluoroethylene and polyether sulfone. These filters differ significantly in their LRV of *Acholeplasma laidlawii* under comparable testing conditions. Multiple factors affect the ability of 0.1 um filter membranes to retain mycoplasma. Due to the pleomorphic structure (lacking a cell wall) of mycoplasma and the small-size (0.15 to 0.3 um) high-differential pressures during filtration could channel the mycoplasma to larger pores in the flow path of the filter's pore structure. Additional factors that control retention efficiency are flow rate and viscosity. Higher flow rates can reduce residence time within the membrane, lowering adsorption and contact effects affecting the retention (LRV) potential. These factors and process parameters contribute to the overall capture ability of the filter.

A major application for 0.1 um filtration is cell culture media that may be contaminated with mycoplasma from animal sources. Some media suppliers and end users use redundant 0.1 um filters to reduce risk of mycoplasma penetration and subsequent bioreactor/fermenter contamination.

40.2.6 Virus Clearance—Removal Filtration

Viruses are retained by size exclusion and adsorptive mechanisms, typically defined as <0.05 µm or 50.0 nm. Nominally rated 50 nm filters developed in the late 1990 were used to retain larger sized viral particles (> 80–120 nm) while permitting large molecule weight biomolecules to flow through. A primary orthogonal method is nanofiltration. These nanofilters were considered the standard pore size rating until industry began recommending the use of 25 nm pore size rated filters for removal of mid-sized (30–40 nm) to small (>18 nm) viruses. The ratio of feed protein to effluent protein is the yield ratio or percent of protein transmission. Use of 25 nm or smaller (10 nm) will reduce protein transmission yield for larger biomolecules and is recommended for proteins < 160 kda (kilodalton). However, the use of nanofiltration is a misnomer, typically reserved for ultrafiltration with nominal molecular weight cutoff (NMWCO) of 100,000 MWCO or equivalent to a microfiltration of 0.1 um (PDA TR No.41 Virus Filtration 2008). As with bacterial filtration, process parameters such as time, temperature, pH and pressure will affect the robustness of filterative removals. Under normal operating conditions, nanofilters can reliably achieve greater than 4 logs of reduction of small non-enveloped viruses.

The basic concepts for disposable filter cartridge and capsule configurations are used for manufacturing single-use virus clearance product. There are significant differences between virus filter manufacturers products regarding removal capabilities and pore ratings. The selected virus filter needs to be robust, that is, maintain LRV of 4+ over the range of operating process variables, pH, process time, temperature, viral loading and buffering. Filter manufacturers qualify and validate their products differently, and using this information and data is subject to some variability. A general recommendation from regulatory guidance and industry practices is that end users should conduct virus spiking studies on scaled down capsules to simulate the removal claims. These single-use capsules can then be used to scale-up to the larger 10-, 20- or 30-inch (1,000/2,000/3,000 cm^2) production size capsules.

40.2.7 Tangential Flow Filtration and Hollow Fiber Modules

In the developing single-use market of the late 1990s to early 2000s, the filter manufacturers were aware that the market pressure for single-use technologies was extending into the downstream process operations. The use of tangent flow filtration (TFF), also known as crossflow filtration by some manufacturers, has traditionally been a clean and reuse product. The filtration parameters include extensive cleaning requirements (qualification and validation), the need for higher pressures (upwards of 45–50 psi (3–3.5 bar)), recirculating flow paths, pressure gauges, pumps, valves and sensors, all in contact with drug products. These multiuse cassettes and modules are typically cleaned by caustic or other digestive cleaning process between batches. When TFF systems are operated as clean and reuse systems, they typically involve multiple process steps; setup, clean-in-place (CIP), flush, water permeability, equilibration, processing, CIP, flush, and storage. Using a single-use TFF membranes/module can cut the process steps by more than half, leaving only the setup, equilibration, processing, and final clean-in-place steps. The result is lower labor and material costs that contribute to the economic justification. End users can identify these costs savings using an economic model based on these reductions in costs as well as other efficiencies. The complex qualification and validation of polymeric materials presented a major challenge for suppliers to develop a single-use system for TFF, as well as for end users to test and implement into their biopharma manufacturing operations. Materials of construction for cassette sheets and module fibers include polysulfone, polyether sulfone, regenerated cellulose and polypropylene, ranging in MWCO from 1–1000 kda.

The increasing concerns for product cross-contamination and potential for bacterial contamination along with greater regulatory scrutiny accelerated the drive to implement TFF as a single-use process. The cassette or hollow fiber designs required stainless steel hardware such as cassette holders, pumps and piping, valves and connectors. Through innovations in designing disposable tubing sets and containers for retentate recirculation, TFF manufacturers were able to deliver single-use systems that met cleanliness requirements. The major drawback, as with chromatography or membrane adsorber technologies, is that the base cost of the membranes is the same for single-use devices or reusable systems. The resultant cost of goods sold (COGS) per liter of drug product increases if the TFF use is not amortized over multiple batches. Elimination or lowering the cross-contamination risk can be achieved by dedicating drug product campaigns to specific cassettes and/or by implementing single-use programs. The negative aspect, as mentioned with chromatography, is these single-use material/device costs may be prohibitively costly to implement. The drug product value and the lowering or elimination of contamination risk (microbial or product) coupled with process efficiencies can justify these higher material costs and deliver an overall cost/liter reduction of drug product.

One major advantage of contamination risk mitigation to Contract Manufacturing Operations (CMO), for instance, permits the processing of multiple drug products in the same footprint and GMP facility with higher batch turn-around times. For product-wetted components used in conjunction with single-use TFF, such as peristaltic pumps, valves, sensors, bags and connectors/tubing, the availability of single-use technology alternatives that meet or exceed industry guidance and regulatory conformance and cost structures can provide end users a channel to a total single-use TFF system.

40.2.8 Centrifugation

Cell harvesting technologies such as conventional centrifugation or stacked disc depth filtration, which typically use stainless steel (SS) hardware, are difficult to implement as single-use systems. The very nature of SS hardware requires cleaning and sanitization processes between batch runs. The biopharma industry's growing dependency on disposables created new centrifugation techniques that are gaining acceptance. This trend is due in large part to well-known economic and efficiency advantages of single-use technologies, discussed later in this chapter.

One innovative technology utilizes a fully automated centrifuge system that employs a single-use polymeric container insert inside the centrifuge bowl. Another pioneering single-use practice uses fluidized bed centrifugation that balances centrifugal and fluid flow forces to capture cells in multiple single-use polymeric chambers. These techniques operate by retaining cells and cellular debris within the centrifuge, permitting the desired drug product concentrate to be continuously separated and discharged. Utilizing gamma (γ) irradiation sterilization for these SU polymeric designs eliminates the need for both CIP and steam-in-place (SIP) operations. All process contact surfaces are easy to install and are replaceable after each batch run.

The fluidized bed centrifugation presents features and benefits over conventional methods like cell harvesting and cell debris removal technologies that utilize conventional centrifuges such as disc-stacks. By eliminating the need for cleaning and sterilizing cycles, there is a lowered risk for cross-contamination between batches, as the product contact components are disposable. Lower shear stresses are produced by the fluid bed centrifugation during operation as compared with conventional centrifugation. Shear forces can damage mammalian cells, causing lysis, thereby increasing submicron particles counts that are not easily removed by the centrifuge method. Low-shear fluidized bed centrifugation systems and techniques coupled with depth filtration are preferred. The use of depth filtration can provide further clarification, removing smaller solid particulates.

40.2.9 Connectors, Valves, Tubing, Welders and Sealers

Single-use connectors are components of single-use systems that allow aseptic connections between either individual single-use technology components of a system or between single-use and stainless steel systems. Connectors always comprise two components, one on each system being connected, which are brought together to create the liquid flow path. Connectors can be classified as falling into four categories: open connectors, aseptic connectors, multi-use aseptic connectors and hybrid connectors.

Open connectors cannot make an aseptic connection between two systems unless the connection is made under aseptic conditions, typically ISO 5 (Grade A, Class 100) under a laminar flow hood. Aseptic connectors can be used to make a sterile connection between systems irrespective of the surrounding conditions. Multiuse aseptic connectors can be used to make several aseptic connections and disconnections to the fluid path without compromising the sterility of the system, while hybrid connectors are used to allow the sterile connection of a presterilized single-use system to a stainless steel vessel and can be steamed in situ once assembled. All connectors must be designed into a single-use system and installed during manufacturing, as they cannot be added to the system afterwards.

Open connectors such as the Medical Plastic Connector (MPC) were the first type of connectors adopted in single-use systems and, based on quantities used, are still the most widely used technology. Initially designed for medical applications, they are available in a range of sizes and formats. All are supplied capped to close the flow path when not in use and to maintain the sterility of the system the connector is attached to. To maintain the sterility of the fluid flow path during connection, the protective caps should only be removed and the two halves of the connector brought together under aseptic conditions. The two halves of open connectors are typically different design and are usually referenced as male and female connectors.

Aseptic connectors can be used to make sterile connections in non-aseptic environments. To achieve this, both halves of the connector are protected by a microporous membrane sealed across the fluid path of the connector that allows the passage of steam if an assembly is autoclaved but prevents bacterial ingress into the single-use assembly. The membranes attached to both halves of the connector are removed only after the two halves are connected. The membranes, sometimes called peel strips, can be made from a range of materials but are typically either a 0.2 um or high bacterial retention, medical grade polyethylene. Early examples of aseptic connectors were comprised of male and female halves, but advances in the design of aseptic connectors have led to the introduction of genderless designs, that is, where both connectors are the same. This enables single-use system manufacturers to simplify system design and reduce the number of products required to be stocked, with subsequent improvements in system lead times. Genderless systems also allow connections to be made between different tubing sizes allowing flow rates to be stepped up or stepped down as required.

Recently connectors designed to allow multiple aseptic connections and disconnections have appeared on the market. These systems have been validated by the manufacturers for a maximum number of cycles; however, at this time, these connectors have not made a significant impact on the industry, primarily because of their physical size and cost.

As single-use technology gained wider acceptance, the requirement to connect single-use systems to existing stainless steel equipment led to the design of hybrid connectors. These connectors are designed into the single-use assembly and are sterilized by gamma irradiation, but once the single-use system is connected to a stainless steel vessel, they are steamed in place. Some designs also allow the single-use system to be steamed off the vessel after use. This type of connector features a three-port design that allows steam to pass directly through two of the ports to "steam on" to the stainless equipment. Once the SIP cycle is completed, the connector valve is actuated, creating a sterile flow path between the stainless and single-use components. The valve can them be returned to the steam position once the sterile fluid transfer has been completed to allow the removal of the single-use component from the stainless equipment following a "steam off" cycle without compromising the sterility of the stainless component.

Fluid flow through flexible tubing in single-use systems is typically controlled at the macro level using pinch or roller clamps that provide a degree of control between fully opened and fully closed. Some of clamps must be installed when the system is being assembled, but they can be relocated along the tubing assembly to achieve the desired position. Other tubing clamp designs can be added to the tubing after assembly and can be positioned at any location on the tubing assembly; these usually operate by using a screw to tighten a plate across the tube, reducing or restricting fluid flow. Both pinch clamps and screw clamps provide a simple control process of tubing compression, and as such their efficient operation is dependent on the tubing durometer, wall thickness and the degree of compressibility of the tubing material.

Several companies have developed external tubing flow control valves that are positioned on the tubing assembly after the disposable system has been manufactured. They use a mechanically operated system to compress the tubing under very controlled and reproducible conditions. These systems offer the advantage of being outside of the fluid path, do not have to be gamma stable, can be positioned and repositioned on the tubing after assembly and provide a high degree of control and reproducibility, but at a higher cost.

Tubing used to manufacture disposable assemblies falls into two basic categories: tubing that can be heat welded and heat sealed or thermo plastic elastomers (TPE) and tubing that cannot. TPE tubing has the benefit of giving the end user the option of using either physical connectors or tube welders to make connections between assemblies. Tubing that cannot be heat welded, such as silicone, requires the use of connectors between assemblies. Both TPE and non-TPE types of tubing are available in a wide range of sizes, wall thicknesses, opacities and durometers to meet various applications.

40.2.10 Single-Use Bags and Films

Single-use bags used for storage of liquids were the first application in the move toward single-use technology in the biopharma market. The transition to them was a result of many factors. Rigid walled storage containers require storage space even when not in use. The containers, plus integral valves and pipework, must be cleaned and sterilized between use, which adds time and cost in the form of cleaning chemicals, WFI, labor, facility overheads, testing and validation, etc. These steps add considerable downtime to a process and require that larger numbers of containers be purchased and stored to ensure that enough are available to meet demand.

Single-use storage bags address many of these issues. They are flat packed before use and take up less storage space, require no cleaning, sterilization or validation before use, can be quickly installed in a support vessel and are immediately ready for processing. They are available in two basic designs: two dimensional (2D) bags that mimic either IV bags or large pillows and three-dimensional (3D) bags that are square or rectangular depending on the design. All bag styles and designs share several common features. All are made from a multilayer polymer film that is designed and manufactured to provide a low extractable and leachable profile. The film acts like a bio barrier, is extruded in a controlled environment to reduce particle loading and is formulated to maintain the efficacy of the material being stored inside. The bags themselves are typically manufactured in an ISO9001 facility in ISO Class 7 cleanrooms that are certified to ISO14644 and ISO14698 standards

Basic bags have one inlet, while most have multiple ports that allow the addition and removal of the material being stored. For 2D bags, with the same basic design as IV bags, the ports are usually positioned at one end of the bag, either as individual ports or in a molded assembly or boat. Due to the position of the port assembly, these bags are often referred to as end ported bags. For 2D pillow-style bags and for three-dimensional bags, tubing is connected to the bag through injection molded ports that are molded with a 360-degree flange to which the film is welded; this port design is commonly referred to as face ported. These ports can be either individual ports or a single injection molded assembly that has multiple connection ports incorporated into the design. The port flange is welded to the bag surface position in the required location and to provide a leak-free seal.

Two-dimensional bags are commercially available in sizes from 25 mL to 50 L. When being filled and emptied, these bags can be laid on a horizontal surface or hung vertically using the hanging port located at the end of the bag opposite the end port assembly. The ability to hang 2D bags vertically allows them to be completely emptied, using either a pump or gravity flow, which typically allows for greater than 99% recovery of the bag contents. Two-dimensional bags can also be stored frozen using metal or plastic support trays that are compartmented to support both the bag and the tubing/connector assemblies. The bags' stability under low temperature conditions is dependent on the type of film and tubing being used and must be validated for the application and temperature. These bags can also have face ports welded to the flat surface of the bag, which provides an additional port for the addition of liquids or allows a recirculation system to be set up to provide gentle mixing of the bag contents. Two-dimensional pillow bags are not commonly used for storage of liquids for several reasons. They take up a larger floor area but hold significantly less material than a 3D bag with the same footprint. A pillow-style bag requires a shallow support tray and shelving for storage of multiple trays. This design is not easily portable or movable and makes for an inflexible system, thus negating one of the main advantages of disposable systems for the customer. Two-dimensional pillow-style bags have been used successfully as plant cell bioreactors, where they are positioned on racks and are static for long periods of time. They are also the core technology behind rocking platform bioreactors supplied by many companies for small-scale production and for cell expansion in cell culture applications.

Three-dimensional storage bags are used for storage of buffers, media and intermediate compounds by almost all biopharmaceutical manufacturing facilities. They require the support of an external vessel that has cutouts or man-way access doors in the side of the vessel to allow for easy installation of the bag and positioning of sample and control ports on the bag, and easy access to these ports once installed in the vessel. Stainless steel vessels can also be jacketed to allow for storage under controlled temperature conditions or to allow heating or cooling of the bag contents. Vessels will also have a drain port in the bottom of the tank to allow for installation of the drain, tubing, clamps and valve from the bag and may also have a tray below the drain port where tubing can be coiled once the bag is installed to prevent it coming into contact with the floor.

Smaller bags and associated tanks are easily portable, and the tanks are usually mounted on wheels, which allows easy movement of the tank and contents within a facility. This allows buffers, media, intermediates, etc., to be brought to the point of use, in the quantities required, at the time required and in the sequence required to support operations within the facility. Larger tanks and bags can also be portable, but health and safety regulations within the facility may restrict the size of tanks that can be considered portable and may require the use of a motorized tug to move them. All sizes of tanks can be installed as static systems.

Three-dimensional bags have been supplied in sizes up to 10,000 L. They have multiple filling and venting ports on the top face of the bag. There will typically be a drain port in the bottom face of the bag, this being positioned in the center or off center depending on the design of the tank. The tank floor is usually sloped from the tank sides to the drain port to allow for better recovery of the contents. Three-dimensional bags can have multiple ports positioned along the lower side wall of the bag to allow for removal of samples for analysis or measurement of critical process parameters such as temperature, pH, CO_2, etc.

The film design choice is the critical component of the bag and must provide a unique combination of attributes, primarily flexibility, inertness and robustness. There are many varieties of film used in the manufacture of single-use bags, for applications that range from long-term storage of cells down to −80°C to aggressive mixing of highly viscous materials under ambient conditions. Most applications for flexible films in the biopharmaceutical market are for low pressure, physiological pH range applications that do not involve aggressive chemicals or operating conditions. Meeting the requirements of these applications has driven the development of the majority of films currently used.

The key requirements of the film are that it is physically and chemically stable, is resistant to puncture and tearing under normal operating conditions, exhibits a consistently low extractable and leachable profile and provides a barrier between the internal and the external environment of the bag that maintains the biological, chemical and physical integrity of the contents. Many different structures of films have been tested and evaluated over the last 20+ years, with the industry now standardizing animal-derived component-free (ADCF), multilayer, polyethylene (PE)-based films for 2D and 3D storage, mixing and cell culture applications. Ethyl vinyl alcohol (EVOH) is sometimes used in the manufacture of 2D bags

In both 2D and 3D bags, the inclusion of a layer of EVOH in the film sandwich acts as a barrier layer that significantly restricts gas diffusion across the film, measured as oxygen and carbon dioxide transmission per unit area over time. The inner product contact layer is PE, while the outer layer of the bag is a film material, such as a polyamide. This layer provides puncture and tear resistance to the film and gives a combination of excellent flexibility and physical stability to the complete structure. Films are tested by the manufacturer to ensure they meet specifications that provide assurances to the end user that the films are suitable for the applications intended. These include but are not limited to EP 3.2.2.1, USP<621>, USP<87> and USP<88>. Manufacturers also undertake and publish the results of extractable studies performed against standard solvents at specified temperatures and over predetermined times. The results of the complete physical and chemical testing program that a film undergoes is shared with customers in the form of a validation guide. This is typically done under a confidentiality disclosure agreement, as the validation guide contains information that is confidential and proprietary to the manufacturer.

More recently, industry organizations have become involved in the development of an industry-wide standardized means for determining the risk posed by leachables from single-use technologies through a more aggressive and wide-ranging extractable testing program using multiple solvents, temperatures and timelines. This has gone through several iterations and revisions, but the ultimate goal is to provide single-use technology users with a standard way of determining initial risk by the elimination of the variables inherent in the current vendor selected testing programs.

The availability of flexible film-based storage bags designed and manufactured with thousands of permutations of components and formats makes their adoption a very powerful addition to a manufacturing facilities inventory to drive increased production and manufacturing efficiencies, reduce installation and operating costs, improve facility flexibility and reduce down time between operations. New advances in film, design and monitoring technology can only further increase the adoption rate of disposable storage technology in the future.

40.2.11 Integrated Single-Use Systems

With the incorporation of suitable agitators, single-use bags can be used for the mixing of liquids, dissolution of powders into solution or the suspension of particles to produce an emulsion or suspension. With suitable gas inlet, gas outlet, sparging and operational monitoring technologies added to the bag, it can be used for cell culture applications.

As the complexity of the operation increases, so will the complexity of the bag design and the requirement to provide customized solutions. One of the primary benefits that single-use technologies bring to the biopharmaceutical manufacturing industry is the ability to be easily customized, by the addition of other single-use technologies, into an integrated, single-use solution for the customer.

The customization of single-use technologies brings new challenges to single-use manufacturers, as they are faced with the increased demand for complex, customized solutions while they are under considerable pressure from the industry to improve quality, control costs and reduce lead times for final assemblies. As a way of resolving these potentially contradictory requirements, many single-use manufacturers have adopted a multi-level tiered product design and availability approach to meet their customers' requirements.

40.2.12 Standard Single-Use Systems

Manufacturers produce a range of standard single-use system designs of storage, mixing and/or cell culture technologies that feature a range of integrated technologies to drive the operational efficiency of the solution. These standard products include filter capsules, tubing, connectors, probes, sensors, clamps, etc. Single-use manufacturers who also manufacture other single-use technologies, such as filters or connectors, will almost always utilize their own technologies as part of their standard single-use system product offerings. Single-use manufacturers who do not manufacture other single-use technologies will also supply standard product designs that incorporate single-use technologies through either preferred or contract supply agreements, usually with multiple vendors.

Single-use systems in this category are selected by manufacturers to allow them to rapidly and efficiently supply products to the market. They are either available from stock or have short lead times, as the single-use manufacturer will hold stock of the component parts to ensure a rapid response to demand. Component products held in stock by a single-use manufacturer are usually referred to as a component library. These products cannot be customized, as they are aggressively priced and are produced from standard, fully validated components. The decision on which single-use systems are identified as standard systems is typically based on market research, previous sales history, or customer input. Therefore, open and transparent communication with the users of the single-use systems in relation to projected demand is essential for the single-use manufacturer to be able to both plan production and manage their supply chain to meet delivery requirements.

40.2.13 Made to Order Single-Use Systems

Changing the design of a system to add single-use technologies that are different from those incorporated into the standard designs creates a custom product. The difference may be a small as selecting a nonstandard connector or selecting a different size filter capsule; however, the single-use manufacturer now must create new drawings, part numbers and bill of materials and change or amend manufacturing protocols and standard operating protocols. They may have to qualify a new vendor and establish inventory requirements and new quality processes, and additional supply chain and sterilization validation may also be required. As a way of expanding their ability to meet customer demand while simultaneously meeting delivery and cost targets, many single-use manufacturers have implemented a category of single-use systems called "Made to Order".

In this product group, while the basic design is fixed, for example bag dimensions, locations of critical elements such as agitators, spargers, sampling ports, etc. are different. The design has some custom capability in that additional ports and tubing sets may be added and filters, connectors or tubing type may be changed. The range of customization is limited to those options already selected and pre-validated by the single-use manufacturer, again typically selected based on prior experience with customers, but these options are already validated for use, using products from the single-use manufacturers standard component library, which significantly reduces lead time while keeping costs under control. This approach is highly flexible and offers many design options but is limited by the range of the single-use manufacturer's component library and by the customers' willingness to pay the higher costs associated with even partial customization.

40.2.14 Full Custom Single-Use Systems

Full customization, as the name implies, is the route to providing a fully dedicated, integrated solution for a customer that is driven by a very specific application that cannot be addressed by any standard or made to order system. Full customization typically means that the complete assembly, including bag dimensions, have not been manufactured before. Any single-use components can be specified into this type of single-use system, including ones never used before by the manufacturer, which will require full validation, new drawings, new part numbers, custom bill of materials, new supplier reviews, qualification and audit, etc. A full custom product will result in long lead times and significantly higher pricing than an equivalent-sized standard or made to order product.

In summary, when selecting the correct single-use system, the end user must balance the requirements against the ability of a standard product to support the application. If it is determined that a custom-designed single-use system is preferred, then the degree of customization must be determined, and the higher costs and longer time required to provide the custom solution must be weighed against the benefits provided by customization. There is no standard approach, and the requirement to support each application must be judged individually.

40.3 UPSTREAM AND DOWNSTREAM SINGLE-USE SYSTEMS AND APPLICATIONS

40.3.1 Mixing

Single-use mixing systems range in scale from a few liters to several thousand liters. The selection of mixing technology will depend on the format and miscibility of the materials being mixed. Mixing is achieved by one of three basic mechanisms, all of which have different levels of effectiveness depending on the mixer type, design and mixing power, bag format, fluid viscosities, temperature, particle solubility and particle loading.

As discussed earlier, all single-use mixing bags can be supplied in standard or customized formats by the bag manufacturers. Mixing bags can be installed into plastic or stainless steel vessels which can be either a single walled or jacketed to allow temperature control during mixing. Most systems can be supplied either as static or mobile systems and have the option of incorporation of load cells to allow gravimetric addition of materials.

40.3.2 Recirculation Mixing

In this method the bag is manufactured with a closed recirculation loop made of flexible tubing. Fluid is drawn out of the bag, passed through a peristaltic pump and pumped back into the bag.

The efficiency of mixing is determined by a combination of factors, including pump speed, fluid viscosity and miscibility, location of the outlet and inlet ports on the bag, and tubing dimensions. Irrespective of these factors, recirculation-based mixing is not a high efficiency mixing technology. It is used primarily for liquid to liquid mixing where the liquids being mixed have low and similar viscosities and are easily miscible, liquid to liquid mixing where the liquids have similar viscosities but are not miscible, that is, to manufacture low viscosity emulsions, or for mixing liquids with low amounts of easily dissolvable solids (salt and low concentration buffers).

The main advantage of recirculation mixing is that compared with other mixing technologies it requires low capital investment, the only equipment required being a vessel, the mixing bag and an appropriately sized pump. The major disadvantages are particle generation through spallation of the tubing when operated at high pump speeds and pulsed high shear conditions between the compressed tubing walls during the pumping process.

40.3.3 Bottom Mounted Impeller Mixing

Impellers mounted in the bottom of a bag can be driven either through a direct shaft drive through the bottom of the bag or more commonly through a magnetic coupling to an impeller mounted on either the bottom surface or the lower sidewall surface of the bag. There are many different types of impeller designs, sizes and locations—their description would take an entire chapter in itself. The most common designs of magnetic coupled mixers are (1) bottom mounted, with the impeller supported by either a non-particle-shedding surface or a non-particle-shedding ball race and (2) levitated magnetic impeller positioned on either the bottom or lower side wall of the mixing bag.

Impellers that are in direct contact with a support surface, such as the mixing bar supported on a Polyetheretherketone (PEEK) surface or a multi-blade impeller supported on a ball race, have the potential to generate particles within the system during mixing as a result of abrasion when moving surfaces are in direct contact. This can be minimized by the manufacturer's selection of appropriate materials, but particle shedding can never be totally eliminated. This type of mixer can perform any mixing application that a recirculation mixer can perform, but because they typically have a high mixing power they can also mix higher viscosity liquids, work with high powder loadings and are more efficient with harder-to-dissolve solids.

Levitated impeller systems use a superconducting cryo-magnet to both repel and rotate the magnetic impeller inside of the bag. This process allows the impeller to effectively float in free space and rotate without any frictional resistance or particle generation, as the impeller is not in contact with another surface. The tradeoffs are a lower rotation speed and lower mixing efficiency than the magnetic impellers. Due to the complexity of manufacturing cryo-magnetic systems, they are significantly more expensive than standard magnetic mixers and therefore find their main applications in mixing expensive APIs or finished products.

40.3.4 Top Mounted Mixers

In these systems, the SU bag contains either a rotating multi-bladed impeller or a rotating paddle, the shaft of which is sealed into the bag film through a welded bearing or a series of seals welded into the bag. The drive motor is supported above the tank, either directly above the midpoint or off center of the bag depending on the design of the system, and there is an easy-to-connect junction between the motor and the mixing shaft that enables rapid installation and removal. Because the motor is positioned above the tank, these types of systems are generally taller than other systems and require access to the tank from above, an important consideration where ceiling heights are low.

Multi-bladed impellers rotate around the center axis of the drive shaft to create turbulent mixing. If the drive shaft is angled through the fluid column, the mixing efficiency is improved, and the use of several impellers along the drive shaft create additional turbulence and improve efficiency. The downside of this design is that the tip speed of the impeller (the speed of movement of the tip of the impeller through the fluid) is always significantly higher that the rotational speed of the drive shaft, and this difference has to be taken into account when mixing shear-sensitive materials. In this type of rotational mixing, the impeller can only rotate in one direction and the entire liquid mass in the bag will quickly start to rotate in the same direction as the impeller, thus reducing mixing efficiency. In stainless steel vessels, baffles are built into the side wall of the tank to disrupt this rotation; in disposable systems the same disruption effect is obtained by using a square cross-sectional bag installed in a circular tank; the irregularities in the fit of the bag against the tank wall act as natural baffles to create turbulent mixing and disrupt fluid rotation.

Paddle-type mixers operate by rotating the paddle within the liquid column in the same way as a spoon is used to stir coffee. The paddle does not rotate but moves the liquid mass in front of it and creates mixing by liquid displacement behind the paddle. This allows effective mixing at very low speeds, and because the paddle itself does not rotate, very little shear is created at the edges of the paddle, making them a highly efficient solution for low-shear mixing applications. Paddle mixers have been shown to handle high solid loads and viscous liquids. Paddle mixers also create the same fluid mass rotation issues as impeller mixers; the use of square cross-sectional bags in circular tanks disrupts flow and prevents this issue. An alternative solution is to have the paddle reverse direction during operation, thus creating the disruptive flow required to prevent fluid column rotation.

The selection of the right mixer for an application is dependent on multiple factors but principally on the viscosity, miscibility, particle loading, solubility and shear sensitivity of the materials to be mixed. Temperature and time control considerations of the process and stability of the components being mixed, together with space and cost considerations at the facility, are also important factors to be considered in selecting the appropriate technology.

40.3.5 Single-Use Bioreactors and Sensors

Single-use bioreactors (SUB) and fermenters are replacing the traditional stainless steel hardware in the biopharma processing operation with a growing preference, and specifically in the cell expansion seed train area of production. With a strong growth curve and the resultant number of new manufacturers entering the biopharma market, there is a plethora of new technologies now available. This section will examine the traditional stirred tank, "wave" style rocker and paddle agitation reactors with newer innovative air lift, vertical wheel stir tank bioreactor designs.

The choice of bioreactor/fermenter design, form and function is dictated by the cell type selected to express the desired therapeutic, that is, the choice of bioreactor for cell culture applications (mammalian, human, insect) or fermenter for bacteria or yeast-based processes. Other key considerations are higher cell densities (10^8–10^9) and the resultant higher product yields, upward of 5–10 gm/l. The need for large 20,000-liter stainless steel vessels and supporting seed reactors is now difficult to find in biopharma processing. Multiple single-use bioreactors using the same or a smaller footprint of the larger hardware vessels can efficiently produce more products in less time. The impeller design, its placement in the tank or bag, oxygen consumption/demand, buffering and nutrient feeds along with temperature control and finally growth cycle/cell life curve coupled to induction of biomolecules expression—all of these influence the type, design and function of the single-use bioreactor. To facilitate the decision-making process, bioprocessors are increasingly turning to 3D computer simulation and modeling tools. Those approaches can optimize key process and operating parameters, for example, mixing, mass-transfer coefficients and foam prevention as well as the effects of hydrodynamic and mixing during scale-up.

The most commonly used single-use bioreactor is the stirred tank design, using either overhead or under-tank impellers. The transfer from stainless steel reactors to single-use bioreactors in both legacy and new therapeutic manufacture is facilitated by using the stirred tank design, which was already widely accepted, qualified and used by drug manufacturers prior to the introduction of single-use bioreactors. These impellers may be magnetic disposable, used below the bag, or shaft and blades of polymeric materials that pass through the top surface of the bag. The ability to scale-up single-use bioreactors from bench, to pilot plant, to full production is made significantly easier using predictive modeling supported by scale-up trial data. The adoption of stirred tank design also allowed existing controllers to be easily adapted to support single-use bioreactors, which in turn enabled single-use bioreactors to be easily and quickly integrated into plant-wide control platforms.

Rocking platform bioreactors were introduced in the 1990s and are based on a two-dimensional bag with the majority of ports and sensors located on the upper surface of the bag. The bags are paced on a movable rack that rocked back and forth, hence creating an induced "wave motion" down the longitudinal axis of the bag that provided simultaneous mixing and aeration. This method gained immediate popularity, especially for the R&D and bench scale-up purposes. The end user demand for larger "rocking" style bags and equipment created larger 2D bag designs together with sensor ports, heating/cooling blankets and process controllers. Two important considerations of this type of SUB are that (1) the working volume of these bags are 50% of the indicated bag volume, that is, a 100 L bag would have a 50 L working volume, and (2) the overall footprint is larger than a stirred tank design with the same working volume.

40.3.6 Paddle-Induced Bioreactors

Utilizing the same technology as paddle mixers, in paddle-induced bioreactors, the paddle does not rotate but moves the liquid mass in front of it and creates mixing by liquid displacement behind the paddle. This allows paddle-based bioreactors to create effective mixing at very low speeds, and because the paddle doesn't rotate, very little shear is created, making them a highly efficient solution for low-shear applications such as mammalian cell culture. When the sparger is positioned at the bottom of the paddle, the paddle combines efficient mixing with efficient sparging throughout the liquid column. Paddle-based bioreactors have been used in both circular and square tanks with equal efficiency, and 3D bioreactor bags have comparable functionality to classical cylindrical stirred tank bioreactors. The design of the system also results in a system that is shorter than a comparable top-mounted shaft drive bioreactor, which can be advantageous in facilities with low ceiling heights.

40.3.7 Single-Use Airlift Bioreactors/ Air-Driven Wheel Bioreactor

The pneumatically or magnetically driven vertical agitation design results in scalable homogenous mixing, reduced cell-damaging shear, improved mass transfer and uniform particle suspension in a smaller footprint than comparable volume single-use bioreactors. This mixing mechanism eliminates many of the problems associated with "rocker" and "stirred-tank" disposable designs due to their high maintenance cost and energy requirements. One significant concern is that the use of micro-bubbles can increase the height of the foam, which is undesirable due to fouling and impaired exhaust gas transfer.

A disposable airlift bioreactor as an alternative to the traditional airlift bioreactor would be especially attractive since it could provide considerable reduction in capital investment as well as operating costs. The replacement of the stainless steel bioreactor vessel with a pre-sterilized, pre-certified single-use bioreactor container or bag would eliminate sterilization (e.g., steam) and cleaning processes (CIP), thereby saving costs in time, labor and materials.

The distinguishing character of airlift bioreactors is the circulating fluid dynamics of the gas/liquid or gas/liquid/solid systems. These characteristics are denoted as gas holdup, liquid and solid velocities and mass transfer rate. Essential for the process development engineer is to identify whether to consider those variables collectively or separately for each of the sectors of the bioreactor. Either way, only an accurate understanding of the behavior and connection of the four key components of an airlift bioreactor (riser, separator, downcomer and bottom) will allow the reliable scale up from the laboratory to pilot or industrial size.

40.3.8 Bubble Column Bioreactors

Bubble columns (BCs) belong to the family of pneumatic bioreactors. These bioreactors do not have any mechanical or moving parts. The upper section of the BC is often widened to encourage gas separation. BCs require little maintenance or floor space and have low operating costs compared to other reactor types. The simple construction of bubble-column reactors makes them easy to maintain. In addition, it is possible to control the degree of shear uniformly within the reactor, which is critical to the growth of plant and animal cells in particular. The rising bubbles in the bioreactor are responsible for providing mixing and maintaining the cells in uniform suspension. Gas-liquid mass transfer behavior in BCs is closely tied to gas holdup through various flow regimes. Gas distributors used in BCs include (i) sintered, perforated or porous plates; (ii) membrane or ring-type distributors; (iii) arm spargers; or (iv) single-orifice nozzles.

40.3.9 Single-Use Sensors

Single-use (SU) sensors are built into single-use systems at time of manufacture and are sterilized with the system, usually by gamma irradiation. Their use eliminates the requirement to assemble and aseptically introduce a sensor into an already sterile system, along with the potential sterility breach that may occur as a result.

Single-use sensors can be integrated into a single-use system in three ways. They can be incorporated into the tubing feeding into or out of the system; typically, these are inline sensors and measure pressure, temperature, turbidity, UV absorbance, conductivity or liquid flow rates. Sensors can be incorporated into a loop connected to the main vessel (storage, mixer or cell culture vessel), and liquid is recirculated across these sensors. Sensors can be sealed into a plastic support that is welded directly to the bag in the same way that a tubing port is welded to the bag, so that the sensor is positioned inside the bag, while the connection to the measuring system is outside the bag. These optical sensors operate based on either fluorescence or luminescence; they are used to measure O_2, CO_2 and pH and are supplied by several companies. However, the biggest drawback to adoption at the time of writing is the limited number of operational parameters that can be measured with SU sensors.

Single-use sensors offer advantages over standard sensors in disposable assemblies because they are simpler to operate; they are noninvasive and therefore reduce the risk of contamination. Unlike invasive sensors, disposable sensors lend themselves to incorporation in a range of technologies such as disposable storage, mixing and cell culture bags, and potentially shaker and rocker flasks. Disposable sensors have not been as widely adopted in commercial manufacturing as have disposable systems, largely as a result of concerns over their long-term robustness, their ability to stand up to the challenges imposed by daily commercial manufacturing and concerns over long term accuracy as a result of drift, especially in continuous processing applications. There are also concerns around their ease of use, implementation into GMP manufacturing and integration into the control systems used to manage manufacturing.

40.3.10 Chromatography: Columns and Membrane Absorbers

Single-use technologies are expanding into downstream processes and are capable of replacing most hardware components, stainless steel vessels/piping/sanitary flanges/glass etc., with single-use polymeric devices.

Significant bottlenecks still exist in the single-use conversion process, specifically with chromatography purification systems with their diverse support equipment and instrumentation. Recognized as a significant downstream process bottleneck in scaling-up, increasing throughputs and improving flow rates, downstream chromatography has presented significant challenges to single-use system manufacturers. Recent developments have made it possible to undertake multiple load and elution cycles on smaller prepacked columns that reduces the required bed volume, eliminates column packing and reduces costs without sacrificing product yield or purity. The use of already validated resins in column format further reduces the barrier to adoption of single-use column technologies.

40.3.11 Disposable Chromatography Columns

Although small-scale prepacked columns and membrane filtration capsules and cartridges evolved concurrently, the scaling up of prepacked columns to large-scale process columns for production scale has seen limited adoption. This is primarily due to resin costs and the complexity of repacking columns. The development of single-use prepacked columns of 60 cm diameter and bed volumes in the 20–25 L range allows processing the outputs of large-scale single-use bioreactors, which makes a fully disposable process feasible. In the context of CMO operations, the investment in the column is fully chargeable to the client, so the column can easily be transferred back to a customer if a process needs to be established elsewhere.

There are several commercially available disposable chromatography column systems that meet regulatory cGMP requirements and industry standards. Besides the sanitary design, material of construction qualifications, IQ, OQ and PQ requirements, end users demand cost-reducing processes and systems that result in higher yields without sacrificing product purity. The critical features required to achieve these goals include reproducibility, scalability, speed, ease of use and operator safety. These features and benefits are summarized here.

Reproducibility: This can be a key consideration of risk during column set-up and packing. Reproducible performance is usually measured using peak symmetry, loading and elution profiles, and these should be near identical each time the column is packed when identical operational procedures are used. Reproducibility is more easily achieved using automated resin slurrying and column packing equipment; however, the costs and complexity of using such equipment on small to midscale columns puts it out of the reach of many customers.

Scalability: Optimization of the chromatography process from bench-top small volume column diameters to large process columns and vice versa must be scalable. The packing and condition steps should generate the same performance regardless of the column diameter when the same bed depth is used. One of the major issues is the transfer of processes from smaller particle chromatography (1.8 um to 5/8 um) typically used at analytical and prep scale to the larger particles (10 um to 100 um) used in prep-process to production scale operations.

Speed: The recent gain made in higher protein titers, upward of 10–20+ g/l, places greater focus on the downstream separation and chromatography purification processes as an important bottleneck. The ability of existing resins to handle high titer loads, which may require dilution of the loading solution, may negate the production advantages of high titers. Increasing flow rate to reduce processing time is usually not an option; however, reducing downtime by eliminating the time required to pack and unpack columns can reduce cycle time, thereby increasing production efficiencies. Using prepacked disposable column systems, the column maintenance time is eliminated, as are the nonproductive overhead costs.

Ease of Use: The use of either disposable chromatography system, prepack recyclable or single-use columns (campaigns) require these systems to be quickly connected and disconnected from other components in the process stream. The adoption of standard fittings and connectors will significantly help in the adoption of disposable chromatography technology.

Operational Safety: The benefits associated with single-use technologies used in biopharmaceutical applications carries over to single-use chromatography products. As with a single-use device, the use of aggressive sanitizing and cleaning chemicals can be minimized, and in most cases eliminated completely. As the columns are self-contained units and require no cleaning or sanitization, the risk of accidental chemical or drug exposure contact to the operator during takedown and setup and disposal may be greatly reduced or eliminated.

40.4 MEMBRANE ADSORBERS AND MONOLITHS

40.4.1 Traditional Columns and Membrane Chromatography Adsorbers

Traditional column chromatography relies on molecular diffusion to transport the target species through the microporous structure of the resin beads to the binding sites contained within the pore structure of the beads. With over 90% of the binding sites contained within the pore structure, the diffusion rate into and out of the porous matrix of the bead is the limiting factor governing capacity, flow rates and ultimately the overall processing time required. Column chromatography used in either flow-through or capture mode operate at flow rates measured in centimeters/hour. This relatively slow passage requires large diameter columns, up to 2.5 meters, to achieve reasonable flow rates. At these diameters, the column will require substantial volumes of resin, at substantial cost, to operate efficiently. The wider the diameter of the column, the more critical the uniform dispersion of the process solution to the upper surface of the column becomes; non-uniform dispersion will result in channeling of flow, typically through the center of the column, low binding efficiency near the column wall and peak tailing when the bound target molecule is eluted. There are also difficulties when purifying large molecular weight species, such as host cell proteins (HCP), DNA, viruses, endotoxins and large proteins. These species present their own diffusion challenges due to their size and inability to move freely into the micropores of resin beads.

The desire to overcome these drawbacks prompted the development of both membrane adsorber and monolith chromatography. The first applications were in ion exchange-based flow through applications for contaminant reduction/removal, for example, HCP, viruses, DNA, etc., and subsequently for use in protein purification applications. The contaminants are directly bound to the anionic or cationic ligands that themselves are directly bound to either the membrane or monolith surface. There is no diffusive flow as the liquid flows through the porous structure where the ligands are bound; the interaction between the ligand and the target molecule is immediate, and the binding capacity of the system is limited only by the density of the ligands within the porous structure.

The higher flow rates of membrane adsorbers over column systems are due to the open pore structure of the membrane, typically 5–8 μm, which offers little flow restriction to very high molecular weight species and is coupled with high density ligands bound directly to the membrane surface, which saturate the pore structure and lead to high binding capacities. These higher flow rates are usually measured in column volumes per minute (cv/m), significantly decreasing processing times and thus greatly improving dynamic throughput. Small syringe filter-style test systems have been shown to operate efficiently at >50 CV/min, while larger multi-liter bed-volume process scale systems can operate efficiently at between 5 and 10 CV/min. For a 5 L bed-volume process capsule, this means operating at 50 L/min, which translates to a 20-minute processing time for 1,000 L of material.

The 3–8 μm pore size membranes that most membrane adsorbers are based on have excellent capacity per unit volume (cm^3), but because the membrane thickness averages only 120 μm, this gives limited capacity per unit area, so multiple layers of membranes are typically used to increase the binding capacity of a device. This results in increased transmembrane pressure during operation, especially at high flow rates, and places great emphasis on the design of the membrane adsorber housing to minimize hold-up volume, reduce the amount of preconditioning, rinse and elution buffers required, and optimize the elution profile of bound target species. Membrane adsorbers have found a particularly important niche application in downstream processing in the polishing or removal of low levels of contaminants such as HCP, DNA, viruses, etc.

Comparing traditional steel or glass columns to membrane adsorbers in several cost models that consider relevant aspects of operation (capital equipment, materials, consumables, utilities and labor) showed that overall operating costs for membrane chromatography are reduced over traditional columns, despite an increase in consumable costs (ref). The comparable antibody yield for anion exchange membrane adsorber chromatography is comparable to traditional anion exchange column chromatography. The major advantages are processing speed and an overall reduction in processing time.

Monoliths, in chromatographic terms, are porous rod structures characterized by mesopores and macropores. These pores provide monoliths with high permeability, a large number of channels, and a high surface area available for reactivity. The unique structure of monoliths gives them several physio-mechanical properties that enable them to perform competitively against traditionally packed columns. Monoliths have very short diffusion distances while also providing multiple pathways for solute dispersion. Little of the surface area in a monolith is inaccessible to compounds in the mobile phase, and the high degree of interconnectivity in monoliths results in the highly desirable combination of low backpressure and high flow rates.

Disposable recyclable polymeric chromatography columns, membrane chromatography and monolith products continue to evolve as manufacturers bring innovative products to the downstream processes. Although membrane chromatography is not a new technology, the recent expansion of high salt tolerant robust absorber devices, anionic ligand technologies and support matrices that allow membrane adsorbers to be used upstream of Protein A columns in monoclonal antibody purification applications are fulfilling a wider range of end user requirements. Similarly, disposable column chromatography product offerings continue to flow into the downstream processes from simulated moving bed chromatography systems that increase productivity, reducing cycling down time to the pre-packed and dedicated "campaign" columns. On the horizon, monoclonal antibody purification by disposable Protein A column replacement may be the next downstream processing bottleneck to be tackled.

40.4.2 Single-Use Aseptic Filling System

A sterile, single-use fluid path is ideal for aseptic filling operations that use peristaltic pump-based filling systems. Peristaltic pumps achieve low shear and reduced spalling as a result of multiple rollers that create moderate tubing compression and significantly reduce fluid pulsing. The choice of the tubing materials is a critical consideration dependent on fluid characteristics such as viscosity, chemical compatibility, and adsorption as well as the key operating parameters such as time, pressure, temperature, etc. Platinum-cured silicone and thermoplastic elastomers are considered the most resistant to spalling and adsorption. Peristaltic pumps are employed to avoid direct contact of the pump with the sterile drug solution. Piston, rotary lobe and displacement pumps have seals, valves and moving parts in the fluid path that come into direct contact with drug solutions and require cleaning and sterilization between operations. Recent advances in the design and costs of disposable pump heads has allowed manufacturers of these types of pumps to compete with peristaltic pump manufacturers, especially in higher flow rate, process-skid-mounted operations.

Filling assemblies utilizing sterile fluid path assemblies with peristaltic pumps or disposable pump heads can produce cost of goods savings (COGS) by reducing change-out times, eliminating set-up errors and significantly lowering labor usage. In combination with platinum-cured silicone tubing or TPEs, single-use system filling assemblies can reduce or eliminate bacterial, particle and cross-contamination in these aseptic fills, which increases drug product availability and lower COGS. Another single-use technology benefit is the reduction or removal of potential operator contact with biohazard materials.

The disposable filling system comprises a supply connector, tubing, fittings, pump head, filling needles and, in some systems, a single-use container that acts as a buffer vessel and feeds the filling needles eliminating pulsed flow at the filling needle. The assembly is typically coupled to a multichannel filling machine with a disposable manifold that feeds all pumps and needles. This configuration is increasingly common in aseptic production-scale operations. The system can be assembled in a cleanroom and be gamma-irradiated, presterilized and double-bagged for cleanliness. The range of drug product container that can be filled include vials, cartridges, syringes and ampoules at both clinical and production scales.

In summary, single-use disposable sterile filling assemblies can reduce or eliminate microbial and product cross-contamination between campaigns and production runs, a critical consideration for contract manufacturers and multiple drug product facilities. Eliminating the cleaning and sterilization steps and their associated validation, which reduces the down time between operations, allows a facility to increase production efficiency, reduce fixed and variable overhead costs, eliminate cross-contamination and increase output while lowering cost of goods sold.

40.4.3 Key Operational and Economic Market Drivers

Main market driver to implement single-use systems: There are numerous journal articles, end-user surveys, conferences and meetings that address the economics and advantages of single-use systems over traditional stainless steel hardware and reusable components and equipment. The following overview of the main market drivers will focus on those internal operations that have the most significant impact on costs and ultimately the profitability of the drug manufacturing process.

40.4.4 Time to Market

40.4.4.1 Product and Process Development to Scale-Up—Timelines and Mileposts

"He who is first wins!"

Utilizing the significantly shorter lead times in qualification and commissioning processes and systems using prefabricated, ready-to-use polymeric devices and systems can provide a flexibility in process, device and facility design/construction not possible with stainless steel hardware. Reducing these timelines and costs to deliver a marketable drug product generally means millions of dollars of potential sales. Knowing or having a solid estimate of production scale-up timelines and costs can provide early feasibility projections.

Having a good estimate of the process development to production scale-up timelines and mileposts can offer insight into the probability of a successful program. Additionally, reduced change control documentation costs and timelines are achievable by using prequalified and acceptable polymeric materials with known risk evaluations. The timely qualification and commissioning of single-use systems is one of the major drivers for the CMOs who more readily implement single-use technologies. Other key factors for the implementation and acceptance of single-use technology are cost and time reductions achieved by using polymeric materials with known qualification, for example, leachables, and risk factor (toxicology) evaluations.

40.4.5 Increased Production Efficiency

While the installation of single-use systems requires operator-intensive hands-on assembly, multiple published studies have shown that the gains in more efficient plant utilization, elimination of expensive sterilization and cleaning processes and reduced utility and overhead costs along with significant reduction in the risks associated with cross-contamination result in a significant increase in production efficiency for facilities that utilize single-use technology appropriately. One unpublished study (private communication) demonstrated a 3.4-time increase in the maximum number of mixing operations that could be undertaken by converting from stainless steel magnetic mixers to a single-use magnetic mixing system. Not only did this saving directly translate into higher throughputs, reduced overheads and a lower cost of goods, it also postponed the requirement to implement a $25 million facility expansion by five years, thus providing the company with additional working capital to invest in other areas.

Single-use systems are supplied collapsed; therefore, the storage and inventory footprints are a fraction of similar volume sized stainless steel tanks. As the production volume of a drug increases through market demand and additional production capacity is required, single-use manufacturers can facilitate the continued use of their technologies while simultaneously simplifying the processes of production scale-up, change control and qualification by maintaining identical materials of construction, production equipment, methods of production, testing and quality procedures across their product ranges. Under these conditions, expanding or scaling up a process requires fewer SOP changes or modifications, carries a lower risk and may be considered a minor change rather than a major change, thus reducing the regulatory filling necessitated by the scale-up.

The "plug and play" advantages of single-use systems provide the flexibility and capability to rapidly reconfigure a manufacturing facility to process different sized batches or manufacture a different product entirely. These advantages apply equally to contract manufacturers or drug companies operating multi-product facilities. Single-use systems can be easily and quickly modified or adapted to provide different configurations, incorporate additional processes or expand capacity significantly more quickly than a traditional hard-piped stainless steel-based facility. The independent and discreet nature of single-use technologies means that the product flow path is discarded and replaced after each batch, and the risk of product cross-contamination between batches is virtually eliminated when using disposable consumables. Changeover times are significantly shorter than with standard systems; therefore, equipment and facility utilization are both significantly improved, which reduces facility overhead and total production costs.

40.4.6 Cost of Goods Sold—Process Production Costs

Single-use technologies offer many advantages that can directly impact their start-up costs and fixed and variable manufacturing costs and positively impact their costs of goods sold. The level of impact and importance of each of these advantages will vary depending on the facility, but the primary advantages and their impact on cost of goods sold are presented here.

- *Start-up Costs.* For new facilities, the time between breaking ground and supplying the first commercial batch can be several years and can cost hundreds of millions of dollars. Single-use systems can be installed as soon as the building is completed and be operational in a matter of months, not years. If movable, modular cleanrooms are used to house the single-use systems, and the time from breaking ground to commercial operation can be further reduced. Every month saved is money in the bank for the manufacturer.
- *Cleaning and Sterilization.* Facilities that utilize 100% single-use technologies do not require cleaning (CIP) or sterilization (SIP) capabilities. Eliminating the need for both CIP and SIP significantly reduces both initial capital investment, operating and disposal costs, with both having a positive impact on COGS
- *CIP/SIP Validation and Qualification.* The adoption of single-use technologies transfers the responsibility for cleaning and sterilization from the drug manufacturer to the single-use manufacturer, thereby reducing costs and time burdens on the end users to meet regulatory compliance.
- *Capital Investment.* The downtime between operations is significantly reduced when the product contact surface of tanks and vessels are replaced by single-use technologies. Stainless steel vessels and pipework are replaced by single-use systems and tubing, which provides for lower initial capital investments.
- *Process Execution Time.* It is widely acknowledged that the time required to remove and replace a single-use assembly is measured in minutes, and only in rare cases by an hour or more. The time saved per unit operation may not appear significant when that process step is viewed in isolation; however, when the complete manufacturing process dwell is considered, the time savings per step multiply very quickly to substantial cost savings for the drug manufacturer.
- *Footprint.* Many single-use technologies occupy less floor space than their stainless steel equivalents, and many skid systems are mobile and can be easily moved within the facility. The flexibility to be able to easily relocate equipment and reconfigure operations within the manufacturing facility can have a positive impact on fixed and variable overheads, which can result in significant reductions in operational costs
- *Maintenance.* Single-use system suppliers typically offer maintenance or service programs to support the hardware and control systems that they supply. The single-use components themselves require no maintenance, no spare parts are needed and the implementation of service contracts reduces the upkeep and maintenance cost overheads for the manufacturer, which reduces downtime and lowers COGS.
- *Waste Disposal.* Evaluation of different waste management streams can make it difficult to identify comparable costs and therefore cost savings. The disposal costs of the wash, clean and rinse solutions generated during the CIP process are a significant cost that can be many times more costly than the cost of the chemicals themselves. The quantities of purified water and water for injection are also significantly reduced. The elimination or significant reduction in these costs will have obvious benefits on the manufacturers COGS.

To determine cost estimates for a manufacturing suite or a complete biopharma facility can be difficult to quantify due to the variations associated with each process as well as local parameters. Capital equipment costs vary significantly based on selection of processes, systems and components. In general, facility capital costs are about 20% lower for manufacturing designs that fully utilize single-use technology in construction as compared to those plants that install stainless steel systems. The major cost reductions are associated with reduction in the number of vessels required and the CIP and SIP processes that have lower utilization or can even be eliminated. Media and buffer prep steps in single-use devices along with single-use bioreactors provide incremental cost savings. Advances in yields and purification capabilities combined with the overall trend to smaller production scales, means that the feasibility to implement single technologies throughout the process can be a reality.

40.4.7 Process Control—Risk Assessment

Compared to the higher risk associated with improper use or human error when performing clean-in place or steam-in-place operations, the clear advantages of disposable single-use devices become readily apparent. Every single-use device and system must meet stringent cleanliness and sterility as well as durability requirements. And as each is discarded after its single use, the risk of batch cross-contamination is largely eliminated. Batch cross-contamination through inadequate CIP procedures or failure to implement proper standard operating procedures can be reduced or eliminated by disposable devices. Since each assembled system is clean and sterile before use then discarded after use, there is little risk of cross-contamination.

Comparably, microbial contamination in either the upstream mixing or preparation of media and buffers along with the downstream product storage/aseptic filling can be reduced. The number and frequency of bacterial-contaminated lots has shown to be reduced using single-use systems, (*Media Fill Contamination PDA 2001 Aseptic Process Survey.*) These increases in productivity using single-use systems have reduced overall processing costs and increased drug supply.

The decrease or elimination of cleaning solutions and chemicals reduces the risk of residues or excessive exposure to operators. The concerns for biohazardous material containment and disposal may be lessened with the use of disposable technologies. Operator safety concerns and the needs for personal protection equipment (PPE), as required by OHSA regulations, may be met using disposable bags and containers systems. These systems minimize the concerns for biohazardous materials exposure to operators and for disposal.

These factors concerning the single-use systems reduction of contamination (both product and microbial), coupled with the elimination of CIP /SIP operations, result in lowering expensive pharmaceutical-grade water consumption, contribute to production efficiencies and meeting the overriding regulatory mandates for providing a secure drug supply and lower drug costs to the patient.

40.4.8 Waste Management

40.4.8.1 Single-Use Disposable Technology Environmental Impact

Single-use device manufacturers encourage the implementation of this technology to replace traditional hardware facilities with these disposable systems. Drug manufacturers' use of stainless steel and glass hardware currently prevails over the industry as the most common equipment used to produce biopharmaceutical drugs. Studies have compared the environmental impact of a traditional biopharmaceutical manufacturing facility to those implementing single-use disposable technologies. Cost reduction and greater design flexibility coupled with a lower environmental impact while improving manufacturing efficiencies are the perceived benefits. These studies show that disposables will result in an overall lower adverse environmental impact than traditional facilities. BPSA—*Guide to Disposal of Single-Use Bioprocess Systems.*

Traditional legacy biopharma processes generate a larger environmental impact owing to the requirements of steam sterilization (SIP) and chemical cleaning (CIP). As a result, there is an increasing interest in single-use disposable technologies to replace hardware. The resultant benefits are significant reductions in the amount of pharmaceutical-grade waters (PW/WFI) and cleaning chemicals used in production. Single-use technology implementation has also encouraged the growing interest in its environmental impact due in part to the significant amount of plastic waste generated.

The adoption of single-use technologies generally produces an overall waste stream reduction. For instance, manufacturing sites using disposables demonstrated a 25% reduction in carbon dioxide emissions compared to traditional stainless steel-equipped facilities. This emission reduction was primarily attributed to the lowered production volume of pharmaceutical-grade waters, which offsets the CO_2 emissions resulting from the production, shipping and disposal of single-use devices.

Recently published studies measured the following influential factors for the waste disposal of single-use devices:

- *Present Procedures*—company policy and practices applicable to disposable labware, cleaning supplies and cafeteria waste will already be established and will probably apply to single-use systems and components.
- *Regional Waste Management Firms and local EPA*—should applicable policy not already have been established, contact local environmental or waste management agencies for applicable guidelines and determine whether certified options are available.
- *Volume*—the volume of solid waste generated will affect the cost and practicality of disposal options.
- *Biohazard*—the level of biohazard in the waste may rule out some options.
- *Recyclability*—it may be possible to recycle some components of the system after disassembly.

The Bio-Process Systems Alliance (BPSA) Disposals subcommittee on single-use bioprocess systems developed the following table providing various disposal options and the advantages and disadvantages of each method. These options increase in cost from top of the list, "Landfill, untreated", down to "Pyrolysis". Although recycling is one of the most desirable options, most SU and disposable devices comprise a complex mixture of polymers and chemical additives. These components are not easily separated or feasible to isolate, thereby making recycling either too costly or impractical.

Comparison of Single-Use Bioprocess System Disposal Options:

Option	Advantages	Disadvantages
Landfill, untreated	Lowest operating cost, no capital cost	Not an option for hazardous waste; perceived as environmentally unfriendly
Landfill, treated	Inexpensive, no capital cost	Perceived as environmentally unfriendly
Grind, autoclave and landfill	Generally accepted as safe, reduces landfill volume	Significant capital cost, requires extra handling
Recycling	Environmentally appealing	Impractical for mixed materials
Incinerate	Generally accepted as safe	May be legally restricted and costly
Incinerate with generation of steam or electricity (cogeneration)	Most environmentally benign, some return on investment	May be legally restricted and presents the highest capital cost
Pyrolysis	Produces usable pure diesel fuel; fuel produced burns more cleanly than that produced from a refinery	New technology—few options available; subpar

> Appropriate disposal of single-use systems and components can be part of a process that is environmentally friendly by reducing overall energy consumption as well as chemical and cleaning demands associated with traditional stainless steel systems.
>
> **(BPSA 2007)**

40.4.9 Validation & Qualification/Industry Standards and Guidance

40.4.9.1 Validation and Qualification of Single-Use Technologies

The evolution of the qualification and validation of single-use Technologies devices began in the late 1990s with the FDA scrutiny of sterilizing-grade filters used in aseptic processing. The regulatory concern and later compliance guidance were eventually extended to all polymeric disposables used in biopharma applications. The highest level of concern was initially at aseptic filling. Soon, concerns extended to upstream operations, to disposable devices used in the separation and purification application. Eventually, even fermentation and cell culture buffers, media and nutrient feeds coming in contact with these products were brought under regulatory guidance. Industry professional associations and societies promulgated a myriad of technical guidance documents and standard testing protocols, which resulted in many conflicting interests between single-use device suppliers, drug manufacturer end users and global regulatory bodies. (references throughout this section)

Sterilizing filtration validation testing protocols and best practices were used as a platform to model other disposable devices' qualification and validation documentation. These filtration-support documents included validation and extractables guides.

Basic Validation Guides test protocols for sterilizing filtration products and polymeric based materials included:

- Viability and Bacteria Challenge Testing
- USP Particle Release Testing
- Extractables studies and product/process leachables testing
- Product Specific Integrity Test Values
- Chemical Compatibility Testing
- Adsorption of drug product and excipients

These standard qualificaton testing components led to similar testing of other disposable devices and components. Filtration product manufacturers were at the leading edge in developing qualification protocols for their filters. As disposable filteration systems developed using other polymeric single-use compoments (tubing, connectors, bags), it was logical to implement similar qualification platforms for these disposable products

40.4.10 Single-Use Systems and Component Validation and Qualification

Single-use suppliers and end users were required to further characterize and perform more thorough analysis and quality control/assurance to these devices. The increasingly number of industry specifications and standards grew from 2005 to now. Today, the vendor and/or disposable devise manufacturers is responsible for providing the proper documentation to demonstrate their products meet or exceed the complex regulatory requirements and industry standards.

The number and breath of these qualification tests and standard is beyond the scope of this chapter section, given the diversity of materials and components is use; however, herein are several major testing categories one should expect the device supplier/vendor to provide as validation/qualification guides and data sheets.

Single-Use Systems and Component Validation Testing Overview

1. Chemical Compatibility Bags /Containers:
    - Objective: To prove the mechanical resistance of container assemblies in contact with specific solutions.

- According to ASTM D543–06, *Method for Resistance of Plastic to Chemical Reagents.*
- Tested conditions: Customer's storage conditions (temperature and time) and customer's specific solutions.
  - Visual inspection (transparency, brittleness, physical integrity).
  - Drop test.
  - Physical properties: Tensile strength (film, seals, connections), and or integrity.
- Physio-chemical properties: Infrared analysis on films, weight test.

2. Extractables—Leachables Studies:
   - Objectives.
     - Identification and quantification of potential molecules migrating from the film material into the stored solution.
     - Allows a toxicological analysis in relation to the type of extractable and the quantity of extractable quantified.
     - USP/ASTM/ICH standards.
     - PDA /ISPE/BPOG/BPSA Guidance.
     - Toxicological assessment (end user).
   - Analytical Methods:
     - pH.
     - Conductivity.
     - Total Organic Carbon (TOC).
     - Non volatile residue (NVR).
     - Gas Chromatography/MS
     - Liquid Chromatography/MS
     - Metal ICP
3. Mechanical Integrity Testing:
   - Post-use integrity testing
   - Pressure decay for bags—leak test
   - Pressure and immersion testing
   - Ink penetration test on film sealing.
4. Microbial Integrity Test/Ingress Test;
   - To demonstrate impermeability to microorganisms of the bags and containers at seals and connector ports.
   - ISO 15747 "*Plastic container for IV injection*".
   - Test with *Bacillus atropheus* to meet the requirements of the ISO 15747.
   - Worst case simulation of process application.
   - Representative configuration of bags and containers.
5. Physico-Chemical Test;
   - To determine physical and chemical properties of plastics and their extracts based on the extraction of the plastic material.
   - USP<661> Physio-chemical tests on Plastic
   - Performed on irradiated at samples
   - Extraction in Water at 70°C for 24 hours—meet or exceed spec limits
     - Non-volatile Residue (NVR)
     - Residue on Ignition
     - Heavy metals
     - Buffering capacity
   - Single-use components similarly tested
6. Particulate Matter:
   - Objective**:** Determine the number of subvisible particulate inside the bags.
   - Method: The Particulate counting is performed according to the EP 2.9.19 & USP<788> Particulate Matter in Injections by light obscuration method.
   - Two sizes of particles are tested: 10µm and 25 µm.
   - Meets or exceeds the acceptance criteria EP & USP<788>.
7. Endotoxin Limits by LAL test:
   - Objective: determine the quantity of bacterial endotoxin (pyrogen toxin) inside bags /Chambers
   - Method: endotoxin test by LAL, chromogenic kinetic method according to USP<85> and EP 2.6.14 recommendation.
   - Kinetic enzymatic method with a very low sensitivity 0.005 EU/ml
   - Acceptance limit: < 0.125 EU/ml
8. Bioburden test:
   - Objective: Quantify viable microorganisms count.
   - Method: The product is filled with a qualified test volume of sterile NaCl 0.9%. Let the article set two hours (one hour for each side so all film surfaces will contact test fluid) at room temperature on a horizontal surface.
   - This extraction solution is filtered through a 0.45 µm membrane and cultured on a media for total flora (soybean casein digest agar at 30–35°C for five days.)
   - The acceptance limit is <1000 CFU in order to comply with industry-accepted practice of gamma sterilization (AAMI & ISO1137).

Table 40.1 reproduces ASTM Table A1.1, which lists the various testing and protocols for qualifying SUT devices and systems. The three (3) major parameters for testing are Testing Integrity, Particulates and Extractables. Each section details the type of test and references the method (ASTM, ISO, USP or proprietary), testing frequency and device/system in which these tests are applicable.

40.4.11 Qualification/Validation Responsibilities of Suppliers and End Users

At the end of this section is a list of the industry and regulatory documents and standards. These references to industry groups and regulatory bodies will require some investigation as to which entity, suppliers or end users or both, are responsible for what tests need be done and which documentation is required to fulfill regulatory compliance and/or meet or exceed these industry standards and guidelines. The following suggestions will assist in finding the appropriate path to achieve appropriate qualification and validation for these single-use devices.

TABLE 40.1
Test Method Matrices for SUS Validations

TABLE A1.1 Test Methods Matrices for SUS Validation

	Test	Method	Test Frequency	Applicability
Testing Integrity	Specific Gravity	ASTM D792	Qualification	Films and Containers
	Tensile Strength	ASTM D638	Qualification	Films and Containers
		ASTM D882	Qualification	Films and Containers
	Tensile Strength	ASTM D412	Qualification	Tubing
	Elongation at Break	ASTM D638	Qualification	Films and Containers
	Tear Resistance	ASTM D1004	Qualification	Films and Containers
	Low Temperature Brittleness	ASTM D1790	Qualification	Films and Containers
		ASTM D1709	Qualification	Films and Containers
		ASTM D746	Qualification	Films and Containers
	WVTR (Water Vapor Transmission Rate)	ASTM F1249	Qualification	Films and Containers
		ISO 15106	Qualification	Films and Containers
	Compression Set Constant Deflection	ASTM D395 – 02 (B method)	Qualification	Films and Containers
	Transmission Rate O2	ASTM D3985	Qualification	Films and Containers
		ASTM F1927	Qualification	Films and Containers
		ISO 15105	Qualification	Films and Containers
	Transmission Rate CO2	ASTM 1434	Qualification	Films and Containers
		ISO 15105	Qualification	Films and Containers
	Durometer	ASTM D2240	Qualification	Films and Containers
	Alternative Microbiological Method	USP<1223>	Validation	Packaging
	Compendial Procedures	USP<1225>	Validation	Packaging
	Integrity Evaluation	USP<1207>	Validation	Packaging
	Helium as the Tracer Gas	ASTM 2391	Validation	Packaging, Seal
	Accelerated Aging	ASTM F1980 – 07 (2011)	Qualification	Films and Containers
	Chamber Integrity - Pressure Decay	ASTM E2930	Qualification	Films and Containers
	Chamber Integrity - Helium Leak	ASTM F2391	Qualification	Films and Containers
	Irradiation Validation: Sterilization of healthcare products	ANSI/AAMI/ISO 11137 Method 1 Vdmax, Fluid Path	Qualification, followed by quarterly dose audits	Films, Containers, Filters, Tubing
		AAMI TIR27 Vdmax, Fluid Path	Qualification, followed by quarterly dose audits	Films, Containers, Filters, Tubing
	Burst Testing	ASTM F2054	Qualification	Seals
	Microbial Ingress Test (MIT)	Bacterial Challenge Test	Qualification	Films and Containers
	Integrity (Leak) Test	Pressure hold – Proprietary Method	Qualification	Films and Containers
		ASTM D499 – 94 (1999)	Qualification	Films and Containers
		ASTM E515	Qualification	Films and Containers
	Filter integrity testing with Microbial Retention assay	ASTM F838	Qualification	Filters
	Filter Sterilization	Proprietary methods	Routine	Sterilizing filters
	Visual Inspection	F1886/F1886M – 09	Routine	Seals
Particulates	Particulates Matter: Evaluate the presence of particulates in or a sample	USP<788> sub visible	Periodic	Films and Containers
		EP 2.9.19	Periodic	Films and Containers
		USP<788> sub visible	Qualification or lot release, or both	Connectors
		ANSI/AAMI BF7	Qualification or lot release, or both	Connectors
	Visual Inspection	USP<790>	Routine	
	Particulate Release	USP<788> Particulates in Large Volume Injection	Varies by filter supplier	Filters
Extractable	Biological Reactivity	USP<87>&<88>	Qualification	Films, Containers, Connectors, Tubing, Filters
	Transmissible Spongiform Encephalopathy (TSE) / Bovine Spongiform Encephalopathy (BSE) Tests	Proprietary Method	Qualification	Films, Containers, Connectors, Tubing, Filters
	Endotoxin	USP<85>	Qualification	Films, Containers, Connectors, Tubing, Filters
	Maceration	Proprietary Method	Qualification	Films, Containers, Tubing
	Reflux	Proprietary Method	Qualification	Films, Containers, Tubing
	Extractables and Extraction	Proprietary Method	Qualification	Films, Containers, Connectors, Tubing, Filters
	Gravimetric Extractable	USP<661>	Qualification	Films, Containers, Connectors, Tubing Filters
	Metal Analysis by ICP-MS	Proprietary Method	Qualification	Films, Containers, Tubing

ASTM International, Standard E3051–16: ASTM

Source: Reprinted, with permission, from ASTM E3051–16 Standard Guide for Specification, Design, Verification, and Application of Single-Use Systems in Pharmaceutical and Biopharmaceutical Manufacturing, copyright ASTM International, 100 Barr Harbor Drive, West Conshohocken, PA 19428. A copy of the complete standard may be obtained from ASTM International, www.astm.org.

Responsibility of Vendors

- Supply documented evidence regarding performance of system components under standard conditions based on tests commonly performed with model solutions as test fluids, for example, extractables studies and guidance for leachable testing.
- Quality control in standard operating procedures for selection of raw materials, production and product release.
- Qualification testing of finished products that meet or exceed internal specifications, industry standards and regulatory guidance and compliance.

Responsibility of Drug Manufacturers

- Perform the requisite DQ/IQ/OQ/PQ to document the proper utilization of these single-use devices in the drug manufacturing process.
- Perform the qualification testing and studies under actual process parameters using study drug product and the single-use device/system components. A typical validation requires three different runs using typical materials to be used in actual manufacturing processes.
- Qualification testing of finished product that meets or exceeds internal specifications, industry standards and regulatory guidance and compliance.

40.4.12 Industry Standards and Regulatory Requirements

Global guidance documents from the FDA, EMEA, ICH, ISO, etc., offer a myriad of differing regulatory and guidance documents for single-use device qualification and testing, for example, extractables and leachables, in addition to the various device manufacturers' quality testing; the end result is many are complex and either vague or contradictory.

The following references provide a convenient reference and are focused on the qualifications and validation of single-use polymeric devices used in biopharma manufacturing. These standards and guidance documents are referenced by global regulatory agencies and should be available from most suppliers and standard societies. Regulatory guidance requires that all new drug entities and processes have been examined and tested to meet these standards, for example, extractable profiles and leachable studies. Industry groups, professional societies and standards organization have all published white papers, standards and monographs about testing protocols, toxicology and testing limits and specifications, of which key references and guidance are listed following the concluding remarks.

Industry Standards and Guidance Weblinks

- ASTM International: (www.astm.org/) Standard Guide E3051–16
  - *Standard Guide for Specification, Design, Verification, and Application of Single-Use Systems in Pharmaceutical and Biopharmaceutical Manufacturing*
- BPOG: BioPhorum Operations Group (www.biophorum.com/)
- BPSA: Biopharma Product Suppliers Association (www.bpsalliance.org/)
- ISPE: International Society Pharmaceutical Engineering (www.ispe.org/)
  - Draft: *ISPE Good Practice Guide to Single Use Technology*
- PDA: Parenteral Drug Association (www.pda.org)
  - *Technical Report No. 66 Application of Single-Use Systems in Pharmaceutical Manufacturing*
- PQRI: Product Quality Research Institute (http://pqri.org/)
  - *Thresholds and Best Practices for Leachables and Extractables in Parenteral and Ophthalmic Drug Products*
- USP: United States Pharmacopeia (www.usp.org/)
  - < 661.3> Plastic Components and Systems Used in Pharmaceutical Manufacturing

40.5 CONCLUSION

The continued accelerating growth and penetration of single-use systems and technology in the biopharmaceutical development and manufacturing process will change the fundamental methods and means of producing 21st-century drugs and therapeutics. We fully expect future drug manufacturing will be accomplished on a single-use disposable continuous process platform. We know global regulatory authorities are keenly interested in promoting these new innovative technologies to achieve increased patient safety, lowering risks and drug costs. The industry looks to increasing productivity, creating a faster time to market and ultimately greater profitability and shareholder return. All are part of the major market drivers that promote further single-use product development and increasing acceptance in the biopharmaceutical manufacturing industry.

ACKNOWLEDGMENTS

In recognition of the substantial contribution from our early work published in *'Aseptic and Sterile Processing'* Chapter 19, *Single-Use (Disposable) Technology*, Trotter and Pendlebury (2017), Davis Health International Publishing, LLC, we gratefully accept their permission to extensively edit our original works for this book chapter. We recommend referencing our earlier work (see references) for an in-depth examination of this topic.

REFERENCES

AAMI/ANSI/ISO 11137:2006, "Sterilization of health care products—Radiation Part 1: Requirements for the development, validation and routine control of a sterilization process for medical products; Part 2: Establishing the sterilization dose; Part 3: Guidance on dosimetric aspects," (2006).

'Aseptic and Sterile Processing' Chapter 19, *Single-Use (Disposable) Technology*, Trotter and Pendlebury (2017), Davis Health International Publishing, LLC,

ASTM Standards

- D4169 Practice for Performance Testing of Shipping Containers and Systems
- E2363 Terminology Relating to Process Analytical Technology in the Pharmaceutical Industry
- E2500 Guide for Specification, Design, and Verification of Pharmaceutical and Biopharmaceutical manufacturing Systems and Equipment
- ASTM International Standard F838–15a, *Standard Test Method for Determining Bacterial Retention of Membrane Filters Utilized for Liquid Filtration.*

BPSA—*Guide to Disposal of Single-Use Bioprocess Systems*, by the Disposals Subcommittee of the Bio-Process Systems Alliance, November 2007

CFR 211.65 Equipment Construction
CFR 211.94 Drug Product Containers and Closures.

ICH Q7A Guidance for Industry: Good Manufacturing Practice Guidance for Active Pharmaceutical Ingredients (8–2001).

International Conference on Harmonization of Technical Requirements for Registration of pharmaceuticals for Human Use (ICH):4

- ICH Q7 Good Manufacturing Practice Guide for Active Pharmaceutical Ingredients
- ICH Q8 (R2) Pharmaceutical Development ICH Q9 Quality Risk Management
- ICH Q10 Pharmaceutical Quality System

ISO Standards

- ISO 13485:2003 Medical Devices: Quality Management Systems—Requirements for Regulatory Purposes
- ISO 14644 Cleanrooms and Associated Controlled Environments
- ISTA 3A General Simulation Performance tests

U.S. Food and Drug Administration (USFDA): *Guidance for Industry Process Validation: General Principles and Practices Pharmaceutical cGMPs for the 21st Century, A Risk-Based Approach*

Marketsandmarkets.com, (September 2020) Single-use Bioreactor Market by Product.

Parenteral Drug Association (PDA) (2008) PDA Technical Report No. 26, *Sterilizing Filtration of Liquids.*

Parenteral Drug Association (PDA) (2010) PDA Technical Report No. 50, *Alternative Methods for Mycoplasma Testing.*

PDA Technical Report No. 66 *Application of Single-Use Systems in Pharmaceutical Manufacturing. Consensus Quality Agreement Template for Single-Use Biopharmaceutical Manufacturing Products BioProcess Systems.*

Parenteral Drug Association (PDA) (revised 2008) PDA Technical Report No. 4, *Virus Filtration* Vol. 62: S-4.

Parenteral Drug Association (PDA) (2004) PDA Technical Report No. 40, *Sterilizing Filtration of Gases* Vol. 58, No. S-1.

USP Compendia Tests: USP <87, 88, 381, 661, 788,790, 1031>

- USP <87 & 88> Bioreactivity tests, acceptable predictors of toxicological activity but do not identify extractables or leachables.
- USP <381> Elastomeric Closures for Injectables: physicochemical tests are typically done in water, drug product or solvent vehicle. Test is gravimetric NVRs and is nonspecific.
- USP <661> Container Performance Testing, leaching polymers with PW, analyze for NVRs, residue on ignition, heavy metals. Do not identify specific leachables.
- USP <788> Particulate Matter in Injections/ USP<790> Visible Particulates in Injections
- USP <1031> Biocompatibility Materials in Drug Containers, Medical devices, and implants: extracted polymers do not alter stability of product or exhibit toxicity. (see <87, 88>)

References With a Focus on Container Closure Systems

- EMEA/205/04 Guideline on Plastic Primary Packaging Materials
- E.P 2.6.14 Bacterial Endotoxins
- E.P. 3.1.7. on EVA material and E.P. 3.1.5. on Polyethylene material
- E.P. 5.4 or ICH Q3C on Residual Solvents
- FDA CDER *Container Closure Systems for Packaging Human Drug and Biologics*
- ISO 15747 Plastic containers for IV
- ISO 10993 Biological evaluation of medical devices or
- ICH Topic Q3A, Q3B, Q3C (Impurities in drugs)

41 Considerations for Process Validation for Cell and Gene Therapies

Karen Zink McCullough, Anthony Thatcher and Merrick Endejann

CONTENTS

41.1 INTRODUCTION

The promise of cell and gene therapy (CGT) lies in a cure, sometimes as in autologous therapies a highly personalized cure, for illnesses that up until now were often managed by treating symptoms of the disease rather than rectifying its origins.

Cell and gene therapy products are a relatively new addition to the biologics universe. In 2010, FDA approved Provenge® (Dendreon), the first cellular immunotherapy, in this case an autologous cellular therapy for prostate cancer. Since then, FDA has approved 17 more cell therapy products (FDA, 2020). Deloitte reports that in 2019 there were over 930 companies developing and/or manufacturing cell and gene therapies and over 1,000 therapies in clinical trials (Deloitte, 2020).

Many of the standard process validation (PV) approaches utilized for biopharmaceutical aseptic manufacturing are applicable as well to CGT manufacturing. However, the industry has learned that standard approaches to process control and process validation for these products may not *always* fit hand-in-glove with the complexity and unique nature of CGT. In order to justify deviations from standard practice, CGT manufacturing requires a clear understanding and subsequent adaptation of the principles of Process Performance Qualification (PPQ) and Ongoing Process Verification (OPV) regulatory expectations. For example:

1. The most important issue for manufacturers of cellular products is that neither the formulated bulk product nor the finished drug product can be sterilized by conventional means, including aseptic (0.22 μm) filtration. Therefore, the need for strict adherence to asepsis in facility design, choice of equipment, process design, operator training and manufacturing operations becomes a product safety concern and may require justifiable deviations from or modifications to traditional thinking. Although gene therapies where the viral vector is the drug product can be terminally filtered, containment of viral particles and aerosols that may contain viral particles, especially

DOI: 10.1201/9781003163138-41

if more than one viral vector is used for manufacturing, can become a significant driver for facility design. These challenges will affect both process control and validation strategies.

2. CGT products often have very targeted patient populations with unmet medical needs. As such, clinical trial manufacturing, process validation and commercial manufacturing timelines may be compressed, with a goal of expedited regulatory review and approval of the product. In this case, early phase clinical trials (Phase I and II) often occur concurrently with late stage product and process development and the first two stages of PV (see Figure 41.1).
3. Variability in the starting materials (human cells for cell therapy) or in manufacturing materials that may or may not prepared under standard GMP conditions (e.g. viral vectors and plasmids) can pose validation and process control challenges to CGT manufacturers.
4. The nature and complexity of CGT products, compounded with a need to collect and organize data, particularly in support of expedited approval, means that the principles of Quality by Design (QBD) and risk management (RM) are essential tools for product and process knowledge integration, manufacturing control and validation strategy.
5. Very often, cell/gene therapies are combination products. For CGT, the designation of "Combination Product" often means that the therapy drug is filled into or adsorbed onto a regulated medical device (e.g. a sterile syringe or IV bag), or its proper use requires the utilization of a specialized regulated medical device (e.g. a special catheter).

Each if these challenges will be described in further detail in the following sections.

41.2 BACKGROUND

In the United States, somatic cell therapy belongs to a class of products called HCTP/S, or "Human Cells, Tissues and Cellular and Tissue-Based Products," regulated under 21 CFR 1271. These products may be autologous (same donor and recipient), allogeneic (different donor and recipient), or xenogeneic (cells from another species), which have been processed *ex vivo* for later transplantation (FDA, 1998).

With respect to the categorization and regulatory expectations, 21 CFR 1271 is explicit for cellular and tissue-based products. Products requiring minimal manipulation, such as peripheral or umbilical cord blood stem cells for autologous or first or second degree blood relative use, are considered to be "Public Health Service 361 products" and are regulated strictly under 21 CFR 1271, which focuses primarily on controlling the risk of transmission of communicable diseases through use of the products. More complex cellular therapies such as autologous and allogeneic products that exceed the minimal manufacturing requirements defined for 361 products are called "Public Health Service 351 products." These products are regulated in the United States by the drug and biologics regulations (21 CFR 210/211 and 21 CFR 600 series) and require an approved Biological License Application (BLA) for commercial distribution. Although PHS 351 products are the focus of this chapter, the concepts described can be applied to PHS 361 products as well.

Gene therapies are defined as the modification or manipulation of the expression of a gene or the alteration of one or more biological properties of living cells for therapeutic use. This transformation may be accomplished by using a viral vector or plasmid nucleic acid to manipulate somatic cells, either *ex vivo* for later transplantation (also considered to be a cell therapy) or *in vivo* as a stand-alone drug product (FDA, 1998).

Early attempts at Process Validation for CGT products were based upon a sequential and linear path for process validation first proposed by FDA in 1987 (FDA, 1987; retired) that included equipment qualification, utility qualification and facility qualification, culminating in a three-run process qualification. This linear model of process validation was more of a tactical than strategic philosophy that was independent of a product's unique nature and inherent variability, its life cycle status, the principles of Quality by Design and a robust Quality Management System. A number of initiatives proposed and implemented subsequent to the 1987 Guideline provided insight into Lifecycle Management (ICH, 2009), Quality Management Systems (FDA, 2006; ICH, 2008), Risk Management (ICH, 2006) and Quality by Design (ICH, 2009). These initiatives influenced the revision and subsequent interpretation of FDA's Process Validation Guidance document and prompted a redefinition of process validation (FDA, 2011). The generic process described in the 2011 Guidance and Figure 41.1 is meant to be a *continuum*, and the linear concept of process validation was replaced with process-specific *flexibility* based on risk and life cycle stage.

In the new model, process validation, although divided for convenience into three stages, is presented as a continuous, strategic initiative rather than a series of sequential "check box" tasks.

- Stage 1 is primarily a development activity where the firm accumulates knowledge of the product and process and assesses any variability that must be controlled for a consistently safe and effective therapy. At this stage, facilities for CGT are often designed to meet the needs of the product and process. Arguably, for unique new CGT products, Stage 1 is the most important stage of process validation, as what is learned about new systems for manufacture and associated variability will ultimately inform strategies for validation and control of manufacturing.
- Stage 2 is the qualification of the facility and equipment and the transfer of the process to the manufacturing facility. This transfer of process from development to manufacturing could involve scale-up of manufacturing and, once transferred, will set up a confirmation of the final processing in preparation for the Process Performance Qualification (PPQ) that generally happens at the end of Stage 2.

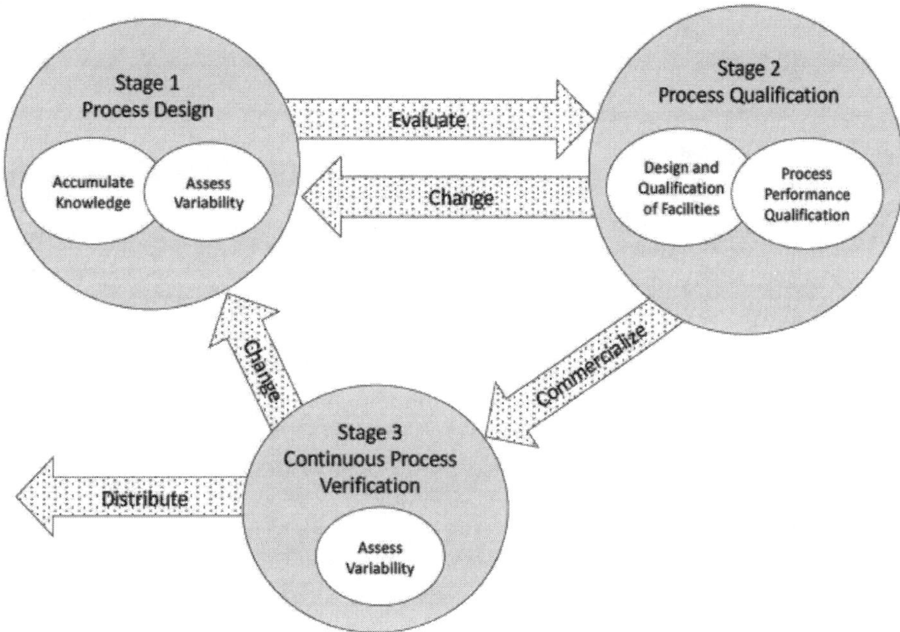

FIGURE 41.1 Process validation elements.

- Stage 3 is continuous process verification, meaning that for the rest of the product's commercial life, the manufacturer will engage in ongoing verification of the validated process by means of sampling and testing at identified critical control points in the process. Stage 3 is also a commitment to continuous improvement and possible revalidation that may be necessitated as knowledge, experience and ongoing process optimization necessitate.

As with all biopharma manufacturing, a Validation Master Plan (VMP) is prepared, reviewed and approved by a cross-functional team. The strategies and tasks described in the Validation Master Plan are critical to:

- Describing the scope and sequence of the validation effort, including any dependency relationships between activities and the regulatory requirements that the program intends to satisfy.
- Identifying and justifying risk-based decisions and any necessary controls, such as using development batches for early clinical trials
- Describing documentation requirements for protocols, reports and management of deviations in any qualification task.

41.3 CHALLENGES FACING CELL AND GENE THERAPY MANUFACTURERS

Starting at the time that a future CGT product is in research and development, the firm must anticipate, define and document the depth and breadth of the requisite activities that will make up the process validation continuum. Development scientists, in consultation with subject matter experts from engineering, facilities, manufacturing and quality, should work to consider the product design and intended use, manufacturing drivers, controls and logistics, facilities requirements and equipment needs (ICH, 2009).

41.3.1 Containment, Contamination and Cross-Contamination

Cellular and gene therapy products cannot be terminally sterilized, nor can cell-based therapies be subjected to aseptic (0.22 μm) filtration. While not "sterile" in the absolute sense, the manufacture of cellular therapies requires strict adherence to the principles of aseptic manufacturing in facility design, choice of equipment and manufacturing techniques to avoid contamination by unwanted cells (microbial contamination or product cross-contamination) in the drug product.

Understanding sources of microbial, endotoxin and particulate contamination are key to successful mitigation for any aseptic operation, particularly one with no terminal filtration. A Contamination Control Master Plan should be established for the proposed facility utilizing a product-, process- and facility-specific risk assessment to identify and prioritize major sources of potential microbial/particulate contamination and product or process cross-contamination. The master plan should cover the establishment of contamination control, the maintenance of control and a description of routine monitoring program including sampling sites, frequencies and the definition of action levels to confirm the effectiveness of the control. For autologous cellular products, the patient's cells must be considered as a potential source of contamination due to the possibility of existing patient infection or possible contamination at the collection site.

41.3.1.1 Facility

As a firm's development group accumulates experience with the product, they also accumulate knowledge of the processes and can design a facility to accommodate the identified Critical Process Parameters (CPP). A cross-functional group of subject matter experts can begin to design the facilities and identify specific equipment needed to accommodate and control the manufacture of the product in order to meet the Quality Target Product Profile (QTPP) requirements. See "Tools for Cell/Gene Therapy Manufacturing" (Section 41.3.4).

Unless the CGT manufacturing process is performed using closed-system technologies exclusively (and currently most cell and gene therapy processes are not), there likely will be some necessary open aseptic operations that will require the use of ISO 5 Critical zones (Koo et al., 2019). There are multiple methods that can be employed to reduce risk of contamination. Options, including EMA requirements, include:

- ISO 5/Grade A cleanroom with cascading room pressurizations and classifications to adjacent rooms of Grade B (ISO 5 at rest/ISO 7 in operation to meet European requirements).
- Biosafety cabinet (ISO 5) located within a background clean area of Grade B (ISO 5 at rest/ ISO 7 in operation to meet European requirements).
- Isolator or positive pressure isolator (ISO 5 internally) located within a background clean area of Grade D (ISO 8 at rest).

Note that whichever method is deployed, the manufacturing cleanroom must be thoughtfully designed, maintained and qualified to assure asepsis.

If more than one viral vector or plasmid will be used in the facility, separate manufacturing areas, each with its own personnel and material airlocks, air handler and process flows, should be considered. Due to the need to protect the product from contamination and cross-contamination, the facility design may incorporate a series of differential pressure sinks (spaces negative to surrounding rooms) and pressure bubbles (spaces positive to surrounding rooms), strategically placed as airlocks to lessen the risk of contamination and cross-contamination (Figure 41.2a–d.) This containment model is especially important in meeting European ATMP guidance for segregation of products using different viral vectors.

Figure 2a. "Standard" cascading pressures

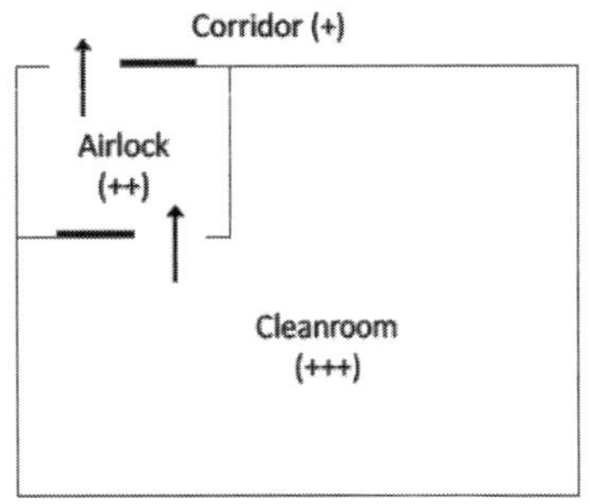

Figure 2b. Airlock "Bubble"

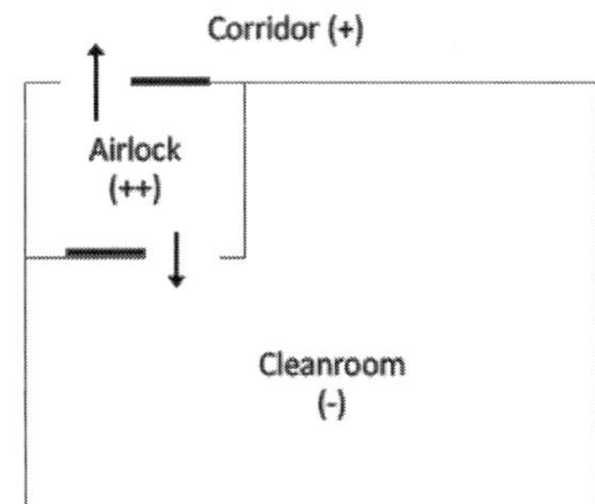

Figure 2c. Airlock "Sink"

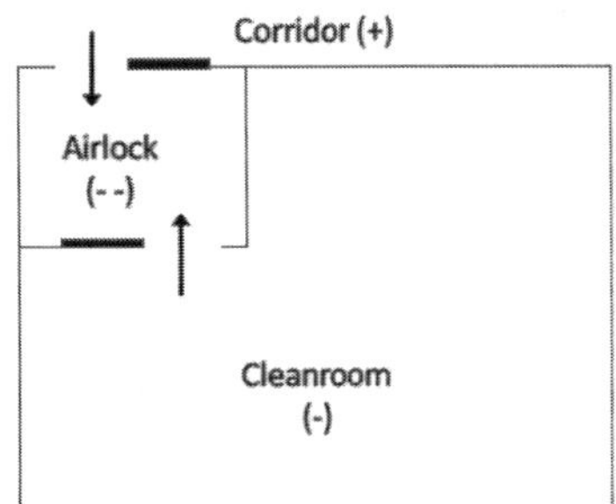

Figure 2d. "Dual Compartment" Airlock Sink

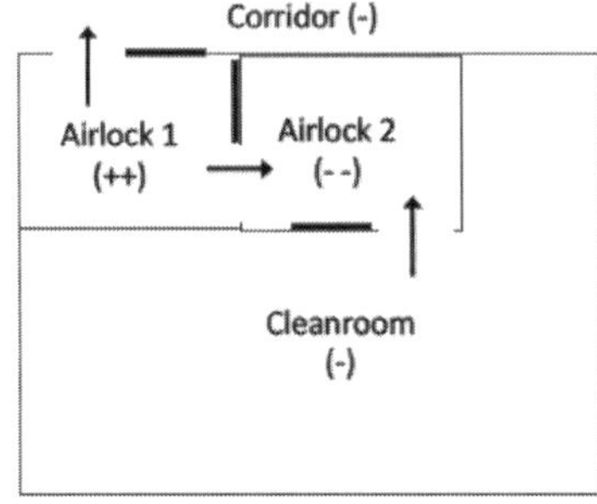

FIGURE 41.2A–D Generic configuration of pressure differential "sinks" and "bubbles" to meet containment requirements. In these figures, the arrows represent the airflow. The symbols (+) and (–) represent relative differential pressures for adjoining spaces.

For gene therapy manufacturing where the viral vector itself is the drug product, special attention must be paid to containment of aerosols generated by the production of the various vector products to guard against cross-contamination. This may be accomplished through the utilization of properly designed cleanrooms, suite-specific air handlers, single pass air (if needed) and pressure differential sinks at the material- and personnel-out airlocks. Note that when working with negative pressure spaces for manufacturing, it is critically important that the principles of asepsis be maintained despite the pressure sink.

Facility design must consider whether the product will be manufactured in sequence (campaigns) or by processing several lots in parallel. For autologous products, the so-called "arm to arm" chain of identity and batch tracking goes from the collection center, through the manufacturing and testing, and forward into the infusion site. Therefore, it is important to assure segregation for each lot as well as maintain the identity of batches to avoid mix-ups. Segregation of multiple batches within one manufacturing suite may be accomplished through the use of separate incubating spaces and procedural controls. An example of a procedural control would be not to allow more than one open aseptic manipulation at a time in any manufacturing suite or isolator chamber to protect against product or process cross-contamination.

Once built, cleanroom qualification activities are similar to what one would expect for any aseptic biopharmaceutical cleanroom qualification. Operational testing can provide focus on room pressurizations (including bubbles and sinks to contain contaminants, vectors and plasmids), air change rates, room light levels, smoke studies in Biosafety Cabinets (BSCs) and smoke studies to demonstrate the proper flow of air across pressure differentials. The Installation and Operational Qualification (IOQ) should provide key evidence necessary to support the next stage of qualification. The facility or room Performance Qualification (PQ) evaluates not only the facility's performance but also personnel cleanroom behavior, gowning, HVAC, and cleaning and disinfection routines as described in Standard Operating Procedures. The PQ should include the following key aspects:

- Environmental (viable air, nonviable air, viable surface) monitoring of the room under static (at rest) conditions following routine facility cleaning and sanitization.
- Environmental monitoring of the room and personnel under dynamic (operational) conditions with a maximum occupancy limit for personnel, particularly in transition spaces such as personnel air locks. In addition to personnel, the dynamic PQ should include the use of all equipment and movement around the room as it would be used during actual manufacture.
- For manufacturing areas that must meet EU requirements, a recovery assessment for nonviable particulates must be performed to demonstrate that the room, once operations are complete, can return to the requirements of its static state (EMA, 2008).

41.3.1.2 Flows (Materials, Personnel, Samples, Waste)

Identifying the proper flows, preferably unidirectional, to accommodate the anticipated manufacturing process is imperative. Very often, risk assessments regarding contamination in cleanrooms where cell therapies are manufactured identify two major risks: people and the transport of materials from the warehouse through to the clean zones. Appropriate cleanroom gowning and robust training on aseptic technique are ways to protect manufacturing processes from the operators. If a facility has a number of aseptic manufacturing suites, each with a different product, vector or plasmid, each suite should have separate gown in/out areas with procedural controls to mitigate the risk of cross-contamination

Materials for cell and gene therapies are often "kitted" using a batch-specific bill of materials (BOM). Product contact materials are generally sterile, disposable, single-use plastics. Cell growth media often are received as sterile, and supplements, if needed, are aseptically filtered and added. Many of these materials are unique to cell and gene manufacturing, and while the interior components of these packages are sterile, very often they are not presented with multiple wrappings that can be sequentially removed, so the immediate exterior packaging is not sterile. Given the risk of contamination and therefore the importance of assuring that exterior packaging is properly prepared, materials flow from warehouse receipt to the aseptic manufacturing space must be carefully mapped and controlled. The facility must be designed to accommodate sequential sanitization steps of exterior packaging as materials proceed from warehouse to the kitting room and ultimately as they enter the aseptic processing area. This procedurally controlled wipe-down process may require the use of appropriately sized active air pass-through units from areas of lower cleanliness classification to higher. Materials transport, including the effectiveness of sanitization, must be validated.

Frequent in-process sampling is often required at critical manufacturing steps to determine if the process can proceed to the next manufacturing step. These pauses for sample analysis are often time sensitive. To guard against mix-ups and assure timely delivery to the laboratory, sample flows must be carefully controlled, particularly when running multiple batches of autologous product in parallel. Where possible, electronic controls such as barcodes in both production and the laboratory should be used to assure sample identity and data integrity.

Waste flows must be segregated from product, ideally by unique unidirectional pathways from the production areas to the waste disposal area. Where possible, waste removal should be performed during production downtime and followed by appropriate sanitization of any shared pathways.

41.3.1.3 Utilities

The utilities needed for the manufacture of CGT will depend on the process. Because much of the media and other aqueous materials used in cell manufacturing are received as sterile and are pre-aliquoted into appropriate volumes, and cleaning agents are often received as "ready to use," there is little need for a large high purity water system. Although a small system may be installed for cleaning, small-scale reagent preparation or laboratory use, the costs of the design, installation, validation and ongoing maintenance and testing of such systems often outweigh the benefits. Therefore, sterile water needed for cleaning, small scale reagent preparation and laboratory use is often purchased and inventoried as a raw material.

Unlike standard biopharmaceutical manufacturing, some CGT manufacturers choose not to install autoclaves or depyrogenation ovens, as product contact and formulation materials are frequently purchased as plastic, single use, sterile and free of detectable endotoxins.

Process gasses may be important to the proper incubation, expansion and stability of cells and final drug product. Manufacturing gases may include liquid nitrogen for freezing cells and drug product and a compressed air system for isolator or equipment operation. Process gases such as nitrogen and/or carbon dioxide may be required for bioreactors and incubators. Gas distribution systems should include disposable sterile filters at the gas source, sanitary tubing runs to each piece of equipment and a point-of-use filter to assure that gas in contact with cells or product contact materials meets microbial and particulate quality of the cleanliness classification of the room in which it is utilized.

Electronic communication systems, including large software packages for materials and inventory management, scheduling, building management, laboratory management and quality systems management, must be identified, and the facility must be wired, including back-up generators, to accommodate all critical equipment.

41.3.1.4 Equipment

The average CGT manufacturing facility may contain incubators, biosafety cabinets, cell washers, cell separators (sometimes multiple types for a single process), wave bioreactors, large-scale bioreactors, centrifuges, tube welders, tube sealers, balances, microscopes, dialysis equipment and cell enumeration devices. Careful equipment placement is important, as air returns should not be obstructed. The qualification and validation (where necessary) of these systems is described elsewhere in this volume.

41.3.1.4.1 Isolators

Aseptic CGT manufacturing is often a manual process performed within a biosafety cabinet in a certified cleanroom. The use of isolators can provide some significant benefits by providing a barrier between the operator, who is the most likely source of microbial and particulate contamination, and the process. The use of isolators may impact the sequencing and logistics of unit operations development, as processes that were designed to go from station to station within a room may have to be redesigned to fit a linear isolator arrangement. For example, will any of the equipment be integrated with the isolator? Which equipment will be stand-alone within the chamber? Will any equipment be located outside of the chamber? If the product is manufactured in one isolator and filled in another, how will the bulk product be transferred to the filling isolator? If so, what additional process controls will need to be considered to accomplish this transfer in an aseptic manner? Each of these questions impacts the process development as well as facilities design, process control and process validation.

The placement of equipment within the isolator's chamber(s) must be carefully planned. Movement within the isolator chamber should be minimized, particularly when open manipulations are occurring. Process simulations with isolator mock-ups, using actual manufacturing personnel, can help ensure equipment positioning within the isolator is well thought out and appropriate for the product that will be manufactured.

In some cases, it may not be feasible or even necessary to place all equipment within an isolator chamber. Manufacturing activities that employ single-use technologies and are long in duration, such as closed bioreactions for cell incubation, may be located external to the isolator. When performing closed tasks outside of the isolator, assure that the classification of the background room meets appropriate regulatory requirements.

The FDA's Guidance for Industry on Sterile Drug Products—Produced by Aseptic Processing (FDA, 2004) provides the agency's thinking regarding the use of isolators. Appendix 1, for example, indicates that while the interior of the isolator should meet ISO 5 standards, the background clean area should be ISO 8, based on isolator design and manufacturing process.

The European Medicines Agency (EMA) has published a more recent Guideline on Good Manufacturing Practice specific to Advanced Therapy Medicinal Products (EMA, 2017) that provides the following general principles: a background clean area of Grade D or better is required for closed systems, isolators or positive pressure isolators. When a product is subject to open aseptic manipulation, a critical clean area of Grade A (laminar flow hood or biosafety cabinet) within a background clean area of Grade B is required.

The equipment qualification process for isolators must establish an adequate and repeatable decontamination process. Load patterns for transfer isolators, equipment placement and chamber configurations must be well characterized to consistently demonstrate decontamination. Proposed changes to manufacturing process, equipment placement and materials/components/samples/waste entering or exiting the isolators chamber must be evaluated to determine if validated decontamination processes are impacted. The decontamination process should establish:

- "Clean" hold time—the duration of time that exists between completion of a qualified decontamination process and the commencement of batching activities.
- Campaign length—the duration of time and/or number of consecutive batches that can be manufactured in succession prior to repeating the qualified decontamination process.

- Entrance of materials and components at the start of daily, batch or campaigned activities.
- Transition of materials, samples, components and waste out of the isolator.
- Setup, use and teardown of any installed equipment, such as tube welders and tube sealers.

Isolator qualification is similar to the PQ for any manufacturing room. It is necessary to qualify an isolator not only as a "static" system but also one that is dynamic and often in a state of change. Because of the relatively small scale of manufacture of CGT and the variety of activities that are necessary to produce a single batch across several days, the qualification can be a unique challenge. Qualification needs to focus not only on routine manufacturing activities but also the transitions between these same activities, particularly if there are transitions in/out of the isolator. Demonstrated and repeatable evidence should be generated during qualification efforts that the critical zone and surrounding areas are not impacted during routine manufacturing. Comprehensive discussion of advanced aseptic systems is provided elsewhere in this volume (see Chapter 30).

41.3.1.4.2 Incubators

CGT manufacturing incubation steps often represent the longest duration of a product's manufacturing process and can include one or more critical process parameters (CPP), such as temperature (for example: 37.0 ± 1.0° C), relative humidity (for example: ≥ 75% RH or 85 ± 5% RH) and gas concentration (for example: 5.0 ± 1.0 % CO_2). Humidity requirements may exist for cell incubation/expansion due to the delicate nature of cells, particularly during early static incubation steps. The moisture requirement may be presented as a single-sided CPP, such as ≥ 75% RH, or be bound by a range, such as 85 ± 5% RH. Many incubator manufacturers offer direct humidification systems using hot-plate generated, HEPA filtered steam or nebulization. When precise control is necessary, the use of a humidity pan to provide the incubator environment, as implemented in many development laboratories, is not sufficient to assure a narrow humidity range and may be an additional source of microbial contamination in the isolator or cleanroom. When incubation steps contain multiple CPPs, the qualification may be more akin to the qualification of a stability chamber than a standard laboratory incubator.

41.3.1.4.3 Cell and Gene Therapy Specific Equipment

Based on the relatively recent emergence of cell and gene therapies and the small-scale nature of the current manufacturing processes, there are limited options for commercial equipment for cell separation, expansion and washing systems. While custom manufacturing platforms can be one approach, the cost and time to develop and implement these automated platforms may be a deterrent. Firms may continue to use the smaller scale manufacturing platforms originally used to develop the product or may adapt their manufacturing processes to use these more robust custom manufacturing platforms for cell separation, expansion and washing steps as the product progresses through clinical trial manufacturing.

The advantages of using commercially available systems are standardized Installation Qualification (OQ) and Operational Qualification (OQ) packages that can and should be used. While these "vendor protocols" may not be designed specifically for an application or product, they can shorten the duration of qualification activities. They generally provide sufficient evidence of an equipment/instrument's proper installation and operation across a full range of conditions. Vendor protocols also typically provide basic and essential Computer System Validation (CSV) testing around the control and monitoring software and/or firmware associated with it.

Since vendor protocols are not designed specifically for a firm's product or process, it is often necessary to conclude the qualification effort for cell separation, expansion and washing with a Performance Qualification. After defining any risks, the equipment PQ may be partnered with a PPQ study concurrently, and that product can be used for additional validation studies, for example, laboratory method validation. Where vendor protocols must be supplemented, the company may include a "wrap around" to the vendor protocol describing and justifying any changes or additions.

41.3.2 Variability in Manufacturing Materials

Critical starting and manufacturing materials required for CGT can be highly variable and will affect strategies for process control and validation. To meet identified Critical Quality Attributes (CQAs), the sources and required quality of the starting cells, growth media, excipients, vectors, plasmids and sera as well as other product contact and final formulation materials must be determined early on to assure a safe and consistent supply for further development as well as for early clinical manufacturing. Collectively, these are called Critical Material Attributes, or CMA (Yu et al., 2014).

Human cells are the starting material for many cellular therapies. In the case of autologous therapies, the cells are diseased cells drawn directly from the patient. In the case of allogeneic products, the cells are normal human donor cells that are commercially available. In both cases, the nature of the cells is as varied as the individuals from which they were drawn, and this known variability must be taken into account when defining controls for critical manufacturing steps. Advances such as the use of induced pluripotent stem cells (iPSCs) to reduce this variability are on the horizon but are not yet proven substitutes for human-derived immune cells.

Some unique critical materials used in cell and gene therapies such as viral vectors and plasmids may be single sourced from university or hospital research laboratories and may not be manufactured under what we know as biopharmaceutical Good Manufacturing Practice. These materials and suppliers will require a risk assessment and an audit to understand the extent of the supplier's process controls and assure that materials can reliably meet their specific CMA. These data will inform the scope content of the PPQ studies, which will provide data on the efficacy of control of the variability often seen in critical materials.

Ancillary materials are defined by USP Chapter <1043> as a subset of raw materials that come in contact with the cells and are intended to activate, promote growth or have some other desirable effect on the cell line being promoted but are not intended to be in the final product. As such, their risk to the safety and efficacy of the product must be fully understood. Examples of these "transient" materials are viral vectors for CAR-T therapies, mRNAs for gene editing and human AB serum to promote cell growth.

41.3.3 Expedited Review and Approval

Cell and gene therapies often have very targeted patient populations with unmet medical needs. "Fast Track" and "Breakthrough" programs in the United States (FDA, 2019) and "PRIME" programs in Europe (EMA, 2016) provide for flexible and expedited review of manufacturing authorization applications for regenerative medicines based on early clinical data. This compression in standard pharmaceutical lifecycle may require that timelines and activities in Stages 2 and 3 of process validation described in Figure 41.1 often occur *concurrently* with product development, which is much earlier than in historical drug development and validation. Absent the ability to keep the processing completely closed, if early clinical material is prepared in the development laboratory or pilot plant, precautions must be taken to assure that all risks associated with product safety (sterility, endotoxin, mycoplasma, other adventitious agents) are properly identified and controlled. That assurance may include the use of closed systems where possible, ISO 5 biosafety cabinets or laminar flow hoods as needed, and the validation of filtration used for the filter sterilization for components that must be added to media or growing cell cultures.

The expedited approval process also forces a company to be organized, to be prudent about gaining as much knowledge of the product and process as early as possible and to use data from nonclinical and early phase clinicals along with principles of risk management to prioritize validation tasks (see "Integrating Product and Process Knowledge for Expedited Review and Approval," Section 41.3.4.4). All of these tools are components of a process called "Quality by Design," which ultimately not only will drive the process control and expedited review of the product but will also guide the process validation and continuous verification activities once the product is commercialized.

41.3.4 Tools for Cell/Gene Therapy Manufacturing

As the concept of process validation has evolved, so has the concept of a pharmaceutical Quality Management System (QMS), a series of interrelated systems, subsystems and processes that span the product lifecycle from development to product discontinuation (ICH, 2008). The QMS is not static. The QMS can be viewed as the glue that holds the overall process development and validation together. It should describe a number of tools that are used throughout the product's lifecycle, from research through to discontinuation. Two of those tools, Quality Risk Management and Quality by Design, are described in the following sections.

41.3.4.1 Quality Risk Management

Quality Risk Management (QRM) is a tool that is used proactively throughout the process validation continuum to make patient safety decisions based on scientific knowledge and process understanding (ICH, 2006). Table 41.1 describes the synergy between risk management and process validation.

To assure that all relevant concerns are heard, risk management is best accomplished using a cross-functional team of subject matter experts and stakeholders. To maintain its usefulness as a tool for continuous improvement, risk assessments

TABLE 41.1
Risk Management and Process Validation

Process validation stage	Risk management applications
1	• Development of Target Product Profiles and Quality Target Product Profiles. • Identification of critical quality attributes, critical material attributes and critical process parameters. • Identification and definition of critical operating ranges. • Identification and prioritization of critical quality attributes, critical material attributes and critical process parameters. • Development of a vendor management/materials management strategy for starting materials and manufacturing materials. • Description of stability profiles and storage requirements.
2	• Assessment of contamination and containment concerns. • Definition of the scope and content of qualification protocols (facility, equipment, test method) including appropriate acceptance criteria. • Development of preventive maintenance and calibration intervals.
3	• Definition of in-process sampling schemes. • Definition of routine monitoring of the environment. • Assessment of the revalidation requirements as the result of change management, CAPA management, continuous improvement, and laboratory management.

must be dynamic documents and should be revisited periodically to assure that they incorporate new data and changes so that they can remain current.

41.3.4.2 Quality By Design

The nature and complexity of CGT products compounded with a need to collect and organize data, particularly in support an expedited approval, means that the principles of Quality by Design and Risk Management are essential tools for product and process knowledge integration, manufacturing control and validation strategies. Because CGT is so new to biopharmaceutical manufacturing, Stage 1 of process validation may arguably be the most important stage of the PV continuum. Targeted diseases, less experience with new methods of transduction and transfection, and a high degree of variability in starting and manufacturing materials all require a solid understanding of the product and process.

Based on Juran's principles of Quality by Design (QBD), the pharmaceutical industry's QBD describes a systematic approach to development of new products that begins early in development with predefined objectives, emphasis on patient needs and product understanding, and consideration of necessary manufacturing controls. These considerations are based on the principles of risk management, which is a particularly important tool for assessing materials, products and processes with high inherent variability (Yu et al., 2014).

41.3.4.2.1 Target Product Profile

The Target Product Profile (TPP) provides a statement of the overall intent of the product as it will be reflected in the anticipated product labeling and provides information about the drug product at specific times of development (FDA, 2007; PDA, 2018). The TPP is a dynamic document and is expected to be revised as knowledge of the product and process are gained during development and clinical trials. An example of a TPP is provided in Table 41.2 for a fictitious product that is an implant for a knee meniscus tear. Note that each product attribute is defined by a target and points readers to supporting documentation.

41.3.4.2.2 Quality Target Product Profile

The Quality Target Product Profile (QTPP) provides another layer of specificity for the new product. It is the new CTG product's initial roadmap and eventually becomes the reference for its critical quality and material attributes (CQA, CMA), specifications, Critical Processing Parameters (CPP), Product Performance Qualification (PPQ) and ongoing verification studies (ICH, 2008). The depth and breadth of the QTPP is product-specific, and not all elements may be known or well understood at the time of the initial version (ICH, 2009; PDA, 2018). Some topics to be covered in the QTPP include but may not be limited those presented for a different fictitious product in Table 41.3.

From the QTPP, the company can create a list of Quality Attributes and Critical Quality Attributes (CQA) for the product (ICH, 2009). A CQA is a chemical, biochemical, physical, biological or microbiological property that poses the risk of an adverse effect on the patient should the product not meet the identified range for the attribute.

For example, the concept of product "potency" is difficult to define in cell and gene therapies because "potency" is process and product specific. Often, "potency" is defined as the number of viable, transduced (with viral vector) or transfected (with plasmid) cells and number of transgene copies per cell, or a combination thereof. Sometimes, particularly for allogeneic products, the cells must also be free from markers that are associated with potential rejection (e.g. % of viable, transduced, TCRαβ- cells). Therefore, depending on the definition, potency is not only a measure of efficacy but can also be a measure of purity. These attributes are therefore critical to the intended use of the product and are considered to be CQA. However, depending on the product, the color of the therapy may vary and is not likely to affect patient outcomes, so color might be considered a Quality Attribute rather than a Critical Quality Attribute. Once a CQA is identified, it becomes an important component of the overall validation effort. Some product Quality Attributes may not be identified initially as "critical", but as experience is gained and data are gathered, the distinction may become clearer and its designation may change.

TABLE 41.2
Example of a TPP for a Combination Allogeneic Cell Therapy Product

Product attribute	Target	Reference
Therapeutic indication	Orthopedic indication. Treatment of meniscus tear	Document PD2020-148
Dosage regimen	one implant	Document PD2020-149
Dosage form and potency	Allogeneic therapy prepared from healthy donor cells	Document PD2020-149
Route of administration	Implantation	Document PD2020-149
Delivery system	Cryopreserved cells on a scaffold	Document PD2020-149
Supply chain	Small lots manufactured due to short stability (60 days)	Document MM2020-398
Contraindications	Unknown at this time	Medical Risk Assessment PD2020-193
Warnings and precautions	Unknown at this time	Medical Risk Assessment PD2020-193
Drug interactions	Unknown at this time	Medical Risk Assessment PD2020-193
Use in specific populations	Unknown at this time	Medical Risk Assessment PD2020-193

TABLE 41.3
Example of a QTPP for a Combination Autologous Cell Therapy Product

Quality attribute	Target		
Summary of TPP	Indication	Cardiovascular	Document TPP2020-002
	Dosage form	Fresh product, IV Infusion	
	Dosing regimen	Three infusions, administered weekly	
	Volume/dose	100mL	
	Container/closure	Class II medical device sterile infusion bag (combination product)	
	Storage	2–8°C	
	Stability	48 hours post-manufacturing at 2–8°C	
Quality attributes		Quality criteria	
	Appearance	Colorless to slightly yellow, slightly opaque to opaque cell suspension, no visible particles	Document PD2020-412
	Potency	5–10^8 transduced, viable cells, ≥5.0% CAR positive viable cells	
		Safety criteria	
	Sterility	Sterile	Document PD2020-413
	Endotoxin	<2 EU/mL	
	Mycoplasma	Free of mycoplasma	

41.3.4.3 Process Capability

An understanding of process capability requires an understanding of process parameters and controls that will be required to meet the QTPP. A process parameter is considered to be a Critical Process Parameter (CPP) if a lack of control could adversely affect an identified CQA (ICH, 2009). For example:

- Transduction efficiencies for cellular products are variable and can be affected by the type of vector, the multiplicity of infection (MOI), the type, quality and number of target cells, the inoculum volume, the vector stability and the transduction exposure time (Zhang et al., 2004).
- Cell expansion will ultimately lead to cell harvest. While the standard time for cell expansion may be determined in the development laboratory to be, for example, seven days, lot-to-lot differences in donor cells, activation method, transduction efficiency and incubation conditions may justify a lot-specific expanded cell growth period to meet the requirements of total cells, viable cells and antigenic markers for harvest. Therefore, process operating ranges must be established for maximum flexibility while maintaining sufficient control to meet the requirements of the QTPP.
- Strict control of sources of microbial, viral and particulate contamination is particularly important for CGT. This requirement affects the process design (need to minimize open aseptic operations), equipment choice (use closed systems including isolators wherever possible) and facility design to focus on containment, contamination and cross-contamination.

41.3.4.4 Integrating Product and Process Knowledge for Expedited Review and Approval

Figure 41.3 and Table 41.4 depict an example of the complexity and interdependencies of product and process design during Stage 1 of PV for an autologous cellular product seeking expedited approval. This accelerated knowledge gathering process is common given the shortened or rolling clinical review process that many of these products receive. A stepwise approach is common for autologous and allogeneic cell therapies due to the patient-to-patient or batch-to-batch variability, respectively, during manufacturing. By revisiting the risk assessments of the process, clinical data and manufacturing analysis at discrete intervals defined by either patient cohorts or process improvements, companies can gain insight into evolving critical process parameters more quickly and accurately.

Using our example, this iterative process may be divided into three parts.

1. Part 1 contains steps 1–6 and is largely a knowledge and data-gathering exercise.
2. Although data gathering is a continuous process, Part 2 is given to data analysis to better understand product safety and efficacy as well as process variability.
3. If the data are sufficient to support the decision to apply for product approval prior to the completion of Phase III clinical trials, then Step 3 is to create a Product Characterization and Control Strategy documents that will identify Critical Processing Parameters (CPP), mitigate risks and provide for PPQ studies and ongoing verification. If sufficient data are not gathered to support this decision, then the process resets and steps 1–6 are repeated as needed until the needed data are collected

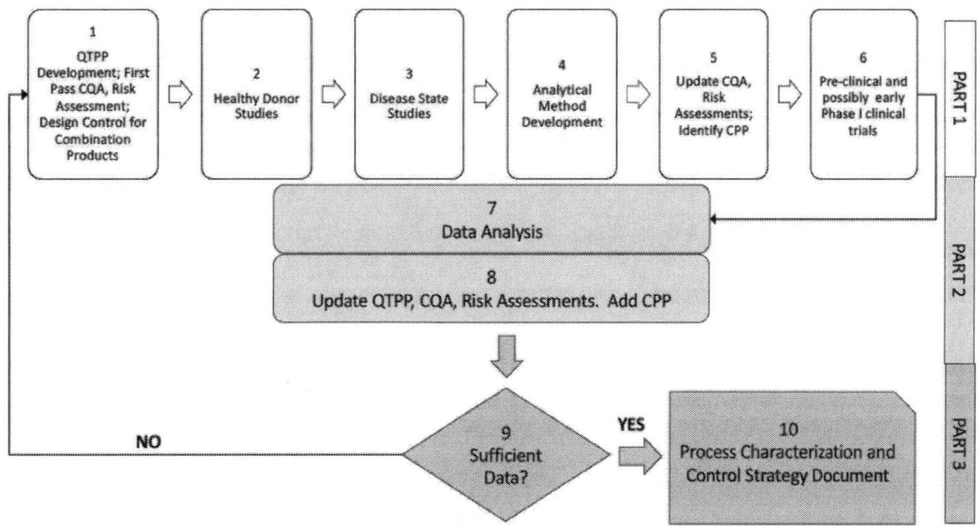

FIGURE 41.3 Example of the iterative nature of Stage 1 knowledge gathering for an autologous cellular therapy.

TABLE 41.4
Product, Process and Overall PV Considerations

Part	Step	Product design considerations	Process design and control considerations	Overall PV considerations
1	1	QTPP development. If the product is a combination product, begin design control	First pass identification of CQA, risk assessment.	Provides the foundation for overall PV.
	2	Healthy donor studies	Studies using applicable healthy donor collections for autologous cell therapies are highly important for gaining process knowledge. As mentioned, it is essential to establish where healthy donor material can and, more importantly, cannot stand in for disease state material. The use of disease state materials will follow once a process has been established.	Identify variability and flexibility. Initial work with the process will inform material, equipment purchases and facility design. Begin risk assessment for contamination control. Prepare basis of design for the facility.
	3	Disease state donor studies		
	4	Analytical method development	Analytical methods must initially address safety (sterility, endotoxin, mycoplasma) if late stage development lots are eventually to be used for Phase I clinical studies.	Analytical method development is ongoing as qualified methods are transferred from early development to support clinical trials, PPQ and process verification. Analytical methods are refined as needed, utilizing the experience gained through in-process and final product testing.
	5	Update CQA, risk assessment, identify CPP	Updating of basic documents based on accumulated knowledge and experience.	Using data from healthy and disease state runs, process development will be able to identify CPP as they relate to CQA using collected data and risk.
	6	Preclinical and possibly clinical studies	Process must be controlled to assure patient safety (contamination, cross contamination).	Experience with the process informs facility design and equipment choice in preparation for Stage 2. BoD documents prepared.
2	7	Data analysis	Robust assessment of process to date including product testing data, logistical concerns, building design, equipment list, CQA and CPP as they relate to the QTPP.	This interim analysis will result in a document that updates decision to move forward or go back to healthy and disease donor studies to further optimize the process. Facility plans for late-stage clinical and commercial manufacturing are finalized.
	8	Update documents	As a result of the analyses, the decision is made to move forward or not with continued Phase I/Phase II clinical studies.	
3	9	Decision to move ahead	Sufficient data may allow progress to step 10. Insufficient data requires a repeat of all or some of the studies in Part 1 to collect more data. The repeated initial phases logically conclude once enough data are accumulated to complete PV Stage 1.	Regardless of the decision, completion of the facility will allow for the commencement of Stage 2 validation tasks. These tasks and PPQ must be completed prior to commercialization. If product is to be commercialized prior to Phase III clinicals, the facility must be qualified, methods must be qualified and risks must be mitigated prior to commercialization.
	10	Decision made to move ahead	Create/finalize process characterization and initial process control documents.	

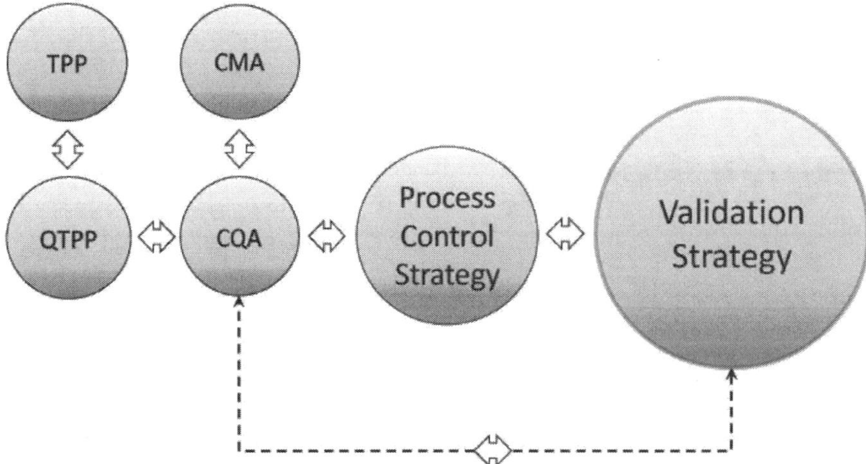

FIGURE 41.4 Relationship between QBD elements and validation strategy.

By the end of Stage 1 of process validation, particularly for those products for which expedited review is sought, the firm should be in a position to author a first-pass and sufficiently supported product characterization and process control strategy that provides justification for establishing CQA, CMA of incoming components and materials and CPP designed to reduce or mitigate variability to meet the requirements of the QTPP. Figure 41.4 summarizes the relationship between the essential elements of QBD, the resulting control strategy and the validation activities required to demonstrate a state of control.

41.3.4.5 The Laboratory

The laboratory is arguably the most important CPP in any manufacturing process, as data from the laboratory will provide information on the effectiveness of validation and qualification studies, the ongoing effectiveness of process controls and, ultimately, adherence to the elements of the QTPP. If not properly qualified (equipment, analyst training, test methods), the laboratory may generate inaccurate data, which would affect all phases of validation and clinical/commercial manufacturing. Analytical and microbiological method qualification and validation are addressed in separate chapters in this text (see Chapters 57 and 58).

As with any laboratory, laboratory equipment for CGT should be qualified for its intended use and all associated software should be properly validated. It is often advantageous to utilize vendor qualification packages (including Computer System Validation packages) to support the qualification effort of laboratory systems. As with manufacturing equipment, it may be justifiable to perform laboratory method or equipment PQ concurrently with the PPQ studies.

Wherever possible, it is advisable to use test methodology described in the enforceable chapters of the relevant pharmacopeias, because these methods are considered to be validated. When using a compendial method, the laboratory must demonstrate that the method is suitable for use with the material under test. Instructions for suitability tests are generally described within the compendial chapter. For example, the test for endotoxins activity, USP <85>, has a number of validated test methods described in the chapter and also has a method for demonstrating suitability called "Test for Interfering Factors."

If it is not possible to use a compendial method, then the test method will have to be validated in-house. There are a number of publications that describe requirements for test method validation, including USP <1225> and PDA Technical Report 57, "Analytical Method Validation and Transfer for Biotechnology Products" (PDA, 2012). If special tests need to be developed (e.g. for the inspecting and testing in unique incoming raw materials), it is important to have those methods validated and ready for implementation by PPQ or Phase II clinical trials.

Analytical methods for particulate detection, sizing and quantitation should be carefully developed as early in the process as possible. Nonviable particulate identification in a semiopaque or opaque cell suspension is extremely difficult. Demonstrated control over the particulate content of raw materials may justify the elimination of a final product specification for finished autologous products. However, given that allogeneic products are not patient specific, they may require a final inspection and method for particulate analyses. Because of the difficulty in developing these analytical methods, it may be helpful to discuss possible methods for detection with regulatory authorities during the development of the QTPP.

Throughput capacity of the laboratory is also critical, particularly for autologous products where delays in testing can result in unacceptable delays to treatment turn-around times. Appropriate space, equipment and personnel should be assessed as a part of Facility Capacity Qualification.

41.3.5 Process Performance Qualification and Beyond

CGT manufacturing requires an understanding and subsequent adaptation of the principles of Process Performance

Qualification (PPQ) and Ongoing/Continuous Process Verification. Short expiration times (as short as 18 hours for some "fresh" products) or extreme low-temperature storage/shipping conditions for some therapies may require product-specific definitions of stability and potency. Because of the specialized nature of CGT products, the lot or batch size is considerably smaller than is seen for standard biological products. This small batch size limits not only availability for commercial use but also limits availability of end product for testing. As a result, the quality of the product is often linked to the identification and management of Critical Processing Parameters that are linked to Critical Quality and Material Attributes.

41.3.5.1 Facility Capacity Qualification

For autologous products, where product is made on demand, a protocol should be put into place to assess the ramp-up of production capabilities to demonstrate the facility is capable of producing the full supply of products indicated on the license application. The protocol should:

- Definition of "full capacity" and the duration necessary to demonstrate "full capacity" of the facility.
- Identification and justification of any materials that may be used as surrogates during capacity qualification
- Assurance that personnel are mimicking actual production, testing and logistical tasks, with the quantity of personnel, for the same durations, of actual activities.

Capacity qualification may, depending on definitions of "full capacity," be demonstrated during a process simulation (media fill, see the following sections).

41.3.5.2 Process Performance Qualification

Although many manufacturing processes and supporting procedures for CGT products follow standard practices for aseptic manufacturing, there are some notable exceptions where consideration must be given to the unique nature of the products and processes. Some of those considerations are provided here. Where a deviation from standard practice is contemplated, it should be justified by a data-driven assessment, identifying the need, the risks and the benefits of the change.

As recent development and commercialization of autologous and allogeneic cell therapies can inform us, Stage 2 of process validation can occur much earlier than in historical drug development timelines. Many cell and gene therapy products have been approved based on Phase 2 clinical data due to accelerated health authority reviews because of targeted unmet needs in critical patient populations. When this occurs, it does not change the need to apply all of the process learning to date. However, by definition, the clinical data set collected prior to the commercialization of the product will be less robust. This should be a consideration when determining how many batches to manufacture under PPQ. Important as well is the nature of the product and starting materials, which may result in inconsistencies or broader operating ranges that are not generally seen in traditional biopharmaceutical manufacturing. Because of the "made to order" and small lot sizes, autologous therapies may require more PPQ runs than a typical biopharmaceutical product. This number needs to be defined based on the manufacturing data from pivotal clinical trials and how much process variability is observed in the patient populations. A case can be made for certain products for concurrent PPQ by including clinical manufactured batches in the PPQ data set. This is possible where there are demonstrated differences between healthy donor material and disease state material. If concurrent PPQ is to be considered, the criteria must be well defined in advance of manufacturing for selection of the clinical batches to be included in the analysis.

Regardless of the number of PPQ runs, transition from a relatively well-controlled clinical setting to a more complex commercial setting demands an absolute clear strategy from Stage 2 to a robust Stage 3 effort.

41.3.5.3 Aseptic Process Simulations

A component of any aseptic process validation is a process simulation, otherwise known as a "media fill" (FDA, 2004; PDA, 2011; WHO, 2011; PIC/S, 2011; EMA, 2017). Clearly, an autologous immunotherapy with a lot size of one unit isn't the average pharmaceutical or biopharmaceutical lot size. How do you execute a process simulation with a batch size of one?

The first thing is to understand the purpose of a process simulation. According to regulatory agencies, the goal of an aseptic process simulation is to *validate the aseptic process*. However, given that the simulation runs are performed at special times and under highly controlled conditions, they really are demonstrations of the *capability* of the process rather than a *validation* of asepsis for the process.

As an example of how to approach a process simulation with a much less conventional process, we use the example of an autologous cellular therapy. Some facts about this therapy: the lot size is one unit. Cells are isolated from the patient's leukapheresis collection. Cell expansion usually takes two weeks in a bag-based bioreactor, with multiple cell "feedings" and samplings to assure that cells are expanding properly and that the in-process samples meet the requirements of the sterility test. The cells are ultimately formulated (spun down and resuspended) in Lactated Ringers and are filled into a sterile infusion bag (Class II medical device), where it is sampled for the final time. How does a company adapt aseptic process simulations practices to this unique product?

1. *Who or what is the processing line in an autologous immunotherapy?*
   Assuming that the process is carried out in a biosafety cabinet by an operator, then one could argue that the operator *is* the processing line. The BSC is prequalified to meet ISO 5 requirements,

so in this scenario, the process simulation really becomes an assessment of the operator's (or operators') aseptic technique. Because one might consider operator aseptic technique qualification and process simulation as one and the same task, one set of simulations can qualify the operator and the process (for that operator) simultaneously.

2. *We have 12 separate processing rooms, each with a biosafety cabinet. Do we have to test all 12 rooms twice/year?*

   Some would argue that the BSC is what is being examined with the process simulation. But the BSC has been prequalified for its purpose separately and is not analogous to a filler or any specific aseptic processing equipment. Rather, the BSC is analogous to the room where aseptic processing equipment is held. While some companies prefer to perform process simulations on each BSC twice/year, that could get costly and not provide information about the real "variable" in the process, which is the operator.

3. *Since the cells expand for two weeks, do we have to run a two-week simulation?*

   No. You would simulate any "feeding" or sampling of the incubating cultures or centrifuging of the cells that are called for in the batch record and treat them as inherent interventions, compressing the entire timeline into a day or less.

4. *But we have only one drug product unit to incubate. Is that sufficient?*

   No, it is not. In addition to incubating the final filled "product", the firm would also incubate *any* samples taken for the lab, *any* waste contained in bags or syringes or tubes—in effect, anything that the media touches that is associated with the process should be incubated.

5. *Should we perform environmental monitoring (particulates, viable surface, viable air) during the process simulation?*

   Yes. You execute the same EM monitoring tasks that would be done during routine manufacturing.

6. *How do we document the process simulation?*

   Consistent with any process simulation, you must record all activities on a batch record. The batch record can be specially prepared (redlined) to show that lengthy incubation periods have been truncated and any sampling interventions have been added. However, the documentation is essentially the same as a normal batch record. If the process calls for two operators (an "operator" and a "verifier"), then both people need to participate in the simulation.

41.3.5.4 Continuous Process Verification

Continuous Process Verification or CPV (sometimes referred to as Ongoing Process Verification or OPV) is the ongoing analysis of batch data post PPQ to monitor, document and trend Critical Quality Attributes (CQAs) for each lot produced. Using statistical analysis tools and methodology such as Process Capability Indexing (Cpk), a company verifies the process in question remains within a qualified state, particularly given changes to the process prompted by routine/periodic process improvements, change controls, and Corrective/Preventive Actions (CAPAs). The goal of continuous process verification is to identify shifts in processed material quality early and address them appropriately before significant deviations or out-of-specification results become common. It is important to note that for cell and gene therapy products and autologous cell therapies in particular, with their inherent variability, common tools like Cpk may be inappropriate or inadequate. Regardless, it is an expectation of global health authorities that a tool for monitoring all CPPs that can impact CQAs be predefined and utilized. Even more important is establishing a review time frame to revisit the CPV tool set and confirm whether the applied analysis is still appropriate as post-commercial experience is gained. The shift from relatively controlled clinical populations to complex and varied commercial populations can cause significant process drift, which needs to be identified and compensated for.

41.3.6 Cell and Gene Therapies as Combination Products

Very often, cell/gene therapies are combination products. For CGT, this designation of "Combination Product" often means that the therapy drug is filled into a Class II regulated medical device (e.g. a sterile syringe or IV bag), or its proper use requires the utilization of a specialized regulated medical device (e.g. a special catheter).

Consider the example of an autologous immunotherapy that may be aseptically filled into a sterile Class II medical device infusion bag or a sterile syringe that will be used both as container/closure and for product administration. According to 21 CFR 3(e), "Jurisdiction." this therapy is a combination product. Or consider a cellular therapy that is filled into a standard container/closure, but product labeling requires a very specialized medical device, for example, a unique catheter, for administration. The catheter is a medical device, and the therapy, whether or not the cell/gene component is packaged with this device, is considered to be a combination product because of the labeling requirement to use this unique device.

Under 21 CFR Part 4, "Regulation of Combination Products," the component parts of a combination product must retain their individual regulatory status as drugs, biological drugs, medical devices or HCT/P products both before and after inclusion in the combination product. The manufacturer of a combination product must assure that each of the individual components is validated and manufactured in compliance with the appropriate GMP regulation and that when the individual components are used together as intended by the approved labeling, the combination product retains its safety, identity, strength, purity and overall quality.

In the application of multiple GMP requirements to the combination product, 21 CFR 4.4 provides for two options to guide manufacturers. The first is to essentially maintain two parallel product-applicable quality systems, one for the device component (whether or not it is manufactured in house) and one for the drug component. The second, 21 CFR 4.1(b), allows for a more streamlined method of assurance for cell therapy combination where the Primary Mode of Action (PMOA) is the formulated cell suspension that employs devices as the container/closure and delivery device.

a. 21 CFR 820.20 (Management Responsibility)
b. 21 CFR 820.30 (Design Controls)
c. 21 CFR 820.50 (Purchasing Controls)
d. 21 CFR 820.100 (Corrective and Preventive Action)
e. 21 CFR 820.170 (Installation, if required)
f. 21 CFR 820.200 (Servicing, if required)

When the combination product Target Product Profile is being created, the development team must be careful to take the device component of the combination product and document design inputs/outputs as required by 21 CFR 820. This exercise will help the team to understand the potential impact of the device on the process validation effort into account. For the example of a cellular therapy filled into an infusion bag, the development team must consider:

1. The size and shape of the container (bag(s), vial(s), syringes, etc.), including the proper number and configuration of ports and septa for the final product delivery.
2. Materials of construction of the container, including its tolerance for freezing (if the product is cryopreserved) and its tolerance for shipping at the temperature required for the therapy. In addition, the development team needs to understand the extractables and leachables of the container with the drug product formulation to assure that the combination of the specific product formulation and bag material does not adversely affect the efficacy or safety of the product. If products are cryopreserved, the ability of the bags to maintain their integrity and product purity against a freeze/thaw cycle is an important element to consider.
3. Materials of construction of any applied labels, particularly the adhesive of any labeling that will be applied directly to the container, are important considerations. Leaching of the adhesive into the drug product has happened, prompting companies to make a quick change to the label stock.
4. Container/closure integrity. Even though the bag is a delivery device, it's also a container/closure system, so it must be tested to assure the integrity of all of the seals in the product.
5. Expiration dating of the combination product must be set with the therapy in the actual specified bag to assure that long-term exposure does not impact the safety or efficacy of the drug product.
6. Critical Material Attributes for the bags must be determined. These may include but are not limited to requirements for sterility and limits on particulates and endotoxin content.
7. Potential suppliers of the bags, including their testing functions, must be screened for a written Quality Management System adherence to the device QSR.
8. Early consideration of appropriate transport containers for the combination product is important. Is the product cryopreserved? Will it be shipped in a stainless steel cassette inside of a dry shipper? Is the product fresh (no preservation)? Does it need to be packaged with cold packs? What about the inclusion of temperature monitoring? Sometimes the packaging has to be designed specifically for the combination product to meet temperature requirements during transit, so early attention to the ultimate configuration is critical.

41.4 CONCLUSION

The validation of cell and gene therapy manufacturing is no different, *in principle*, from validating any product manufactured using aseptic processing. However, due to the unique nature of these products, process validation requires the firm to think carefully about what the validation is asking them to demonstrate and the challenges that such a demonstration can pose.

Cell therapies cannot be terminally sterilized, nor can they be filter sterilized. The nature of the product requires the manufacturer to incorporate the most current thinking on asepsis, including the use of single-use closed systems wherever possible. Where closed systems are inappropriate or not possible, process design, facility design, choice of equipment and operator training are paramount in assuring patient safety.

The starting materials for cell and gene therapies are as varied as the individuals from which they are sourced. Manufacturing materials such as viral vectors and plasmids or even some specialized manufacturing equipment may be single sourced; it is possible that they were not made under what the biopharmaceutical industry considers to be "GMP." This variability in manufacturing materials inevitably leads to variability in processing, requiring the implementation of quality tools such as Quality by Design and Risk Management as early in product and process development as possible to identify and mitigate sources of variation that could adversely affect safety, potency and efficacy of the product.

Because of patients' unmet needs, cell and gene therapies may undergo an expedited review and approval process, which will condense the standard product life cycle timeline. The careful use and documentation of QBD and Risk Management will help to integrate product and process knowledge in support the early submissions, perhaps leading to an expedited approval.

Adapting the process validation philosophy to unique manufacturing processes requires that a manufacturer have a complete understanding of the purpose, depth and breadth of the validation continuum. Once that understanding is in place, it can be complemented with accumulating product and process knowledge to justify flexibility, adapt traditional thinking and define the science needed to properly validate these new products.

REFERENCES

Code of Federal Regulations. 21 CFR 3, "Product Jurisdiction".

Code of Federal Regulations. 21 CFR 4, "Regulation of Combination Products".

Code of Federal Regulations. 21 CFR 210, Current Good Manufacturing Practice in Manufacturing, Processing, Packing or Holding of Drugs, General.

Code of Federal Regulations. 21 CFR 211, "Current Good Manufacturing Practice for Finished Pharmaceuticals".

Code of Federal Regulations. 21 CFR 600, "Biological Products: General".

Code of Federal Regulations. 21 CFR 820, "Quality Systems Regulation".

Code of Federal Regulations. 21 CFR 1271, "Human Cells, Tissues, and Cellular and Tissue-Based Products".

Deloitte. *Cell and Gene Therapy: Opportunities and Challenges to Personalized Medicine*, 2020. www2.deloitte.com/us/en/pages/life-sciences-and-health-care/articles/challenges-in the-emerging-cell-therapy-industry.html

European Medicines Agency. EudraLex volume 4: Good manufacturing practice. *Guidelines on Good Manufacturing Practice, Medicinal Products for Human and Veterinary Use: Annex 1: Manufacture of Sterile Medicinal Products*, 2008. https://ec.europa.eu/health/sites/health/files/files/eudralex/vol-4/2008_11_25_gmp-an1_en.pdf

European Medicines Agency. EudraLex volume 4: Good manufacturing practice. *Guidelines on Good Manufacturing Practice Specific to Advanced Therapy Medicinal Products*, 2017. https://ec.europa.eu/health/sites/health/files/files/eudralex/vol-4/2017_11_22_guidelines_gmp_for_atmps.pdf

European Medicines Agency. *PRIME: Priority Medicines*, 2016. www.ema.europa.eu/en/human-regulatory/research-development/prime-priority-medicines

Food and Drug Administration. *Approved Cellular and Gene Therapy Products*, 2020. www.fda.gov/vaccines-blood-biologics/cellular-gene-therapy-products/approved-cellular-and-gene-therapy-products

Food and Drug Administration. *Expedited Programs for Regenerative Medicine Therapies for Serious Conditions*. Guidance for Industry, 2019. www.fda.gov/regulatory-information/search-fda-guidance-documents/expedited-programs-regenerative-medicine-therapies-serious-conditions

Food and Drug Administration. *Guidance for Industry and Review Staff: Target Product Profile, a Strategic Development Process Tool*, 2007. www.fda.gov/media/72566/download

Food and Drug Administration. *Guidance for Industry: Guidance for Human Somatic Cell Therapy and Gene Therapy*, 1998. www.fda.gov/regulatory-information/search-fda-guidance-documents/guidance-human-somatic-cell-therapy-and-gene-therapy

Food and Drug Administration. *Guidance for Industry Process Validation: General Principles and Practices*, 2011. www.fda.gov/files/drugs/published/Process-Validation-General-Principles-and-Practices.pdf

Food and Drug Administration. Guideline on General Principles of Process Validation (retired), 1987.

Food and Drug Administration. *Quality Systems Approach to Pharmaceutical Current Good Manufacturing Practice Regulations*, 2006. www.fda.gov/regulatory-information/search-fda-guidance-documents/quality-systems-approach-pharmaceutical-current-good-manufacturing-practice-regulations

Food and Drug Administration. *Sterile Drug Products Produced by Aseptic Processing: Current Good Manufacturing Practice*, 2004. www.fda.gov/regulatory-information/search-fda-guidance-documents/sterile-drug-products-produced-aseptic-processing-current-good-manufacturing-practice

International Conference on Harmonisation of Technical Requirements for Registration of Pharmaceuticals for Human Use. *ICH Q8 (R2), "Pharmaceutical Development"*, 2009. https://database.ich.org/sites/default/files/Q8%28R2%29%20Guideline.pdf

International Conference on Harmonisation of Technical Requirements for Registration of Pharmaceuticals for Human Use. *ICH Q9, "Quality Risk Management"*, 2006. https://database.ich.org/sites/default/files/Q9%20Guideline.pdf

International Conference on Harmonisation of Technical Requirements for Registration of Pharmaceuticals for Human Use. *ICH Q10, "Pharmaceutical Quality System"*, 2008. https://database.ich.org/sites/default/files/Q10%20Guideline.pdf

Koo L. *FDA Perspective on Commercial Facility Design for Cell and Gene Therapy Products*, 2019. https://cdn.ymaws.com/www.casss.org/resource/resmgr/cell&gene_therapy/cgtp_slides/2019_cmcs_koo_lily_slides.pdf

Parenteral Drug Association. Technical Report 81, "Cell-Based Therapy Control Strategy", 2018.

Parenteral Drug Association. TR 22. Process Simulation for Aseptically Filled Products, 2011.

Parenteral Drug Association. TR57. Analytical Method Validation and Transfer for Biotechnology Products, 2012.

PIC/S. *PI007-6, "Recommendation on the Validation of Aseptic Processes"*, 2011 https://picscheme.org/docview/3446

United States Pharmacopeia. 2020. <1043>, "Ancillary Materials for Cell, Gene and Tissue-Engineered Products."

United States Pharmacopeia. 2020. <1225>, "Validation of Compendial Methods."

World Health Organization. *Annex 6, WHO Good Manufacturing Practices for Sterile Pharmaceutical Products*, 2011. www.who.int/medicines/areas/quality_safety/quality_assurance/GMPSterilePharmaceuticalProductsTRS961Annex6.pdf?ua=1

Yu LX, Amidon G., Khan MA, Hong SW, Polli J, Ruju GK, Woodcock J. Understanding pharmaceutical quality by design. *The AAPS Journal* 2014, 16(4), 771–783.

Zhang G, Metharom P, Julie H, Ellem KAQ, Cleghorn G, West MJ, Wei MQ. The significance of controlled conditions in lentiviral vector titration and in the use of multiplicity of infection (MOI) for predicting gene transfer events. *Genet Vaccines Ther.* 2004, 2, 6.

42 Validation of Solid Dosage Finished Goods

William G. Lindboe Jr.

CONTENTS

DOI: 10.1201/9781003163138-42

42.1 INTRODUCTION

This chapter will present the techniques, procedures, data and documentation necessary to satisfy the current compliance requirement for validation of solid dosage finished goods, specifically tablet and capsule manufacturing operations. The emphasis will be on the practical inspectional requirement facing the validation team rather than a theoretical approach that does not reflect the practicalities (and problems) encountered when validating actual production operations. The major changes to the validation of solid dosage finished goods since the 3rd edition is the publication of the updated 2011 process validation guidance[1] ("the Guidance") by the FDA and the revision of the USP General Chapter <905> Uniformity of Dosage Units.[2] Background for the latter can be found in another guidance document[3] from the FDA. The reader must become familiar with these documents and change (or update) legacy nomenclature and methods contained in any preexisting standard operating procedures and/or documentation to these updated documents. The Guidance along with the updated USP <905> provides new nomenclature and consistency across other global regulatory publications. It provides a phased view of validation in addition to content and does lend itself to new production methods such as continuous processing and process automated technology. Additionally, some of the pillars of validation that had arisen from almost 30 years of industry practice as well as the Barr Decision[4] were largely ignored. This included, among other things, the requirement for three batches. The removal of this specificity presented a problem for validation teams who are continuously at odds with top management and production management about the time, materials and laboratory costs of validation. The best way to approach this problem is to have any costs (material and labor) of validation activity budgeted as a fixed cost in R&D and/or development accounts. If validation batches eventually become available for commerce, they become favorable to the Manufacturing account and Quality Control account, but only if the validation is successful. This built-in incentive is an approach that was used by James Agalloco and his coworkers at E.R. Squibb and Sons many years ago, which resulted in an efficient practice. The majority of this chapter will address the current Guidance term: process performance qualification ("PPQ"). It will convey some of the successes and failures that accompanied the development and application of validation protocols in what is now called PPQ to solid dosage forms, the validation of which began in the late seventies. Many of these errors are still being committed because of turnover in technical staff, lack of organizational learning and the narrow approach to validation that is a common by-product of outsourcing and contract validation service firms. Specifically, this chapter will discuss tablet and capsule manufacturing validation from batching through tertiary packaging and shipping. The ideas presented are an individual view based upon validation as practiced, from personal experience, published practice or discussions with others in industry and regulators. The specifics of the validation technique described herein may be debatable. However, in the author's experience, validation documentation based upon each approach has been audited at least one time by the FDA and/or EU with no adverse comment. The chapter includes commentary and opinion, which the reader may view as a "wish list" from someone who has worked continuously with direct validation responsibility for over three decades.

42.2 VALIDATION AND THE DOSAGE FORMING STEP

The initial thrust of validation was in the sterile manufacturing sector. That area has the single overriding concern of product microbial safety, specifically the probability of sterility of the product, and was disaster driven. The mathematical models utilized for the evaluation of this probability (equivalent time at temperature), initially developed by Charles Olin Ball and his coworkers, are well accepted.[5] There is no corresponding microbial safety concern or accepted general mathematical models applicable to solid dosage validation. The inspiration for this author's early protocols for non-sterile validation was Theodore E. Byers of the FDA. Byers referred to a recall incident with digoxin tablets as justification for solid dosage validation.[6] However, this example could be attributed to a general breakdown of cGMPs at the involved manufacturer. He also made the statement that validation was "not compendial testing," which implied that validation for solid dosage had to be something more than normal quality control in-process and release testing. There was no methodology or approach to solid dosage validation offered by regulators to satisfy this new requirement. The industry was told only what it was not.

It was apparent to those engaged in the development and/or manufacturing of tablets and capsules that dosage content uniformity was the major concern applicable to all products. In addition, the best point to evaluate content uniformity was during or shortly after the dosage forming (compression or encapsulation) manufacturing step. Consequently, it was natural for the validation protocols to emphasize the USP Content Uniformity Test. The justification for this was that content uniformity was an extensive property of the product, in that it depended on how much material was there in relation to the processing equipment, assuming a uniform blend. Uniformity would be the only "unknown" factor remaining to be confirmed from the development and scale-up (Stage 1) batches. The other properties of tablets and capsules were more intensive in nature, in that they were less dependent on the batch or equipment size. Additionally, in this author's experience and interaction in discussion groups, round tables, etc., the prevailing thought was emphasis should be on the final product (what is given to the patient). It was recognized that dissolution and other in-process physical tests were important also. However, at that time, dissolution was not applicable to all solid dosage products, and the specifications were sufficiently forgiving that the physical test of disintegration was considered indicative of dissolution. The physical in-process tests for solid dosage compressing and encapsulation had been performed in real time for many years (an early example of Statistical Process Control), and descriptive statistics were and still are widely utilized. The manufacturing steps prior to and subsequent to the dosage forming step were evaluated in terms of how they would affect testing results at the dosage forming step.

Currently, there is more of a balanced approach to solid dosage validation. Each unit operation is addressed, and there is no longer sole dependence on the dosage forming step, specifically relying on tablet and capsule testing results to validate the entire process. This includes raw material characterization, especially for direct compression formulations. It is not that validation testing of tablets and capsules has decreased but rather that validation testing in the other operations has greatly increased, especially mixing and blending operations. The operations of coating, polishing and imprinting are also considered. Packaging, once ignored, is now the subject of major validation activities, especially in automated blister pack lines, and is the subject of another chapter within this volume.

The fundamental premise of the U.S. Food and Drug Administration (FDA) and European Union regulatory agencies is that manufacturers are the experts in the technical aspects of their particular processes and products. Nevertheless, sound scientific practice and a risk-based approach are expected. The lack of specific procedures in the regulations in the area of solid dosage forms validation can give rise to differences in interpretation of requirements. Despite this lack of specificity, some unwritten rules have evolved for solid dosage form validation, based on inspectional observations from FDA FD 483s and warning letters. Over time, changes in practice have been necessitated in order to comply with judgments and consent decrees imposed on some firms because of failure to validate properly. Techniques will be presented that are current in industry and reviewed favorably by the FDA or EMA auditors. However, it should be noted that change is a constant factor, and as new techniques and equipment are developed, methodology will improve and newer approaches will be expected. Additionally, auditor's individual interpretation of the Guidance has not had time to become widely available, and, although not binding, the Guidance must be the basis for any validation master plans.

There may be some overlap with other chapters in this volume. This overlap is necessary in order to understand the basis for many practices in the validation of solid dosage manufacturing operations. When overlapping topics are discussed, the emphasis will be on their relationship with solid dosage validation, and the reader is urged to consult the other chapters within this volume.

42.3 2011 FDA GUIDANCE

The Guidance follows the Validation Life Cycle,[7] where validation starts with process design and continues through the product life cycle. It consists of process design and definition in Stage 1; process qualification in Stage 2; and continued process verification in Stage 3. The purpose of process validation in the first two stages is to characterize process variability and demonstrate and document control. Stage 3 demonstrates continued control and maintenance of the validated state throughout the product life. The emphasis is on characterization and control of process variability from design throughout the product life using statistics.

42.3.1 Stage 1 Process Design

The literature contains many detailed studies from which generalizations can be made that are valuable in Stage 1; additionally, processing experience with similar products can be used to support design when products are acquired and not developed in-house, typically over-the-counter drug (OTC) and generic products. The roll of validation personnel at this stage is to ensure that critical process parameters, independent variables (CPPs), and critical product attributes, dependent variables (CPAs), are identified and documented. This facilitates validation protocol development, which should begin at this stage.

42.3.2 Stage 2 Process Qualification

Qualification is defined in the FDA Guidance as "activities undertaken to demonstrate that utilities and equipment are suitable for their intended use and perform properly." While the FDA can use the terms "suitable" and "properly," it is a requirement of industry to define, demonstrate and document these terms for all utilities and each machine used in the manufacturing process. This activity has historically been divided into installation qualification and operational qualification. The former address the location, utilities, utilities and calibration of equipment, and the latter tests the operationally ranges of the equipment. Procedures for equipment operation and personnel training are usually prepared early within this stage.

42.3.2.1 Process Performance Qualification

This was referred to as "PQ" or "process validation batches" prior to the 2011 FDA Guidance. As defined in the Guidance: "The PPQ combines the actual facility, utilities, equipment (each now qualified), and the trained personnel with the commercial manufacturing process, control procedures, and components to produce commercial batches." Sampling based upon a statistical sampling plan and testing provides the assurance and confirmation of process design and that it performs as expected. Per the Guidance, "In most cases, PPQ will have a higher level of sampling, additional testing, and greater scrutiny of process performance than would be typical of routine commercial production." It is better to increase sampling for additional assurance rather than have separate validation acceptance limits because of the possibility of obtaining passing release limits and failing validation limits.

42.3.3 Stage 3 Continued Process Verification

This stage is usually addressed with a summary report of annual reports and change control. Statistical analyses such as control charts, cumulative difference charts (referenced below) are useful in this stage.[8]

42.4 VALIDATION NOMENCLATURE AND DOCUMENTATION

Undocumented activity can be equated to no activity, and verbal explanations in audits cannot be relied upon, especially given personnel turnover and outsourcing of validation activity. For solid dosage, nomenclature varies from company to company, and it is important to establish this in higher level policies and procedures and relate them to the Guidance. The reader may not agree with the use of a particular term for a described action used within this chapter or in what protocol it should be contained. It is important not to "get hung up" on terminology but to agree on the underlying descriptions and actions. This could be the case within large individual firms with multiple sites and operating divisions, as well. It is important to be consistent within the company and not to "reinvent" widely accepted terms. For example, the technical group performing solid dosage validation for a single site of a large company decided to call what was universally called a "deviation" elsewhere in the company an "observation." In another example, after a daylong meeting with a dozen experienced people hammering out a process validation testing plan (now, PPQ testing plan), one junior engineer proclaimed, "That is not process validation." In the latter case, the engineer was in full agreement with the testing plan and disagreed only with what it should be called.[9] The word "critical" is an often-used term that carries different meanings even within the same firm. A lesson can be learned from software development, where they use the Component Object Model.[10] In this approach, a specific function is coded once and only once, and this single subroutine or "object" is called upon wherever and whenever the function is needed. Similarly, all validation terminology should be defined in one place in higher reference procedures consistent with the Guidance and used only as defined there, both within documentation and at cross-functional team meetings.

Two concepts that are fully accepted and understood are that of the protocol and the deviation (the latter word may be different with the same basic intent, e.g., discrepancy, error; see previous paragraph). The protocol demonstrates the identification and control of variability within the previously established specifications and is fully approved with quality group oversight, both in approval and in execution. A basic rule of thumb for the validation protocol can be derived from an analogous situation in a jury trial: "Never ask a question you don't already know the answer to." The implication of this is that extensive trials must occur during Stage 1 development and qualification activity within Stage 2 to prevent unexpected or noncompliant results obtained under the formal protocol during PPQ. Deviations must be resolved with full investigation and corrective action addressing the underlying systemic cause to prevent a recurrence in subsequent protocol activity and/or in routine production. Beyond these, the term "executed protocol" refers to a protocol containing data capture spaces that has been annotated during execution to collect the data. In some instances, the executed protocol was routed for approval at completion with no accompanying report, especially where no deviations had occurred. However, the Guidance clearly expects a report, especially to discuss any unexpected events and/or results. The validation summary report is generally accepted to mean tabularized results and a statement of compliance, in addition to any deviation resolution, that is routed for approval to the same approvers of the protocol. The Guidance specifies that the report should include all unexpected events with full analysis. In some companies, failed studies have a report that is circulated to "close them out." This is dangerous in that the existence of an approved validation report implies that the validation itself was successful. Databases and validation status summaries must have the ability to differentiate between successful and unsuccessful efforts. Unsuccessful PPQ studies are expected to be nonexistent given adequate Stage 1 design and process characterization in addition to the facility and equipment qualification portions of Stage 2.

A recurring error in solid dosage validation is the phrase "validation not required." This phrase is used with some operations and compendial test methods, such as pH. One interpretation is that execution of the test method is permitted to be invalid. Conflicts of statement arise when prerequisite questions such as "all test methods validated" are answered "yes" and some methods are designated "validation not required." It is important to realize that typical solid dosage in-process test such as hardness (tablet breaking force), thickness and friability must be "valid," and while not validated as one would validate an HPLC assay, must be shown and documented as suitable under actual conditions of use.[11]

Applying process validation concepts originally developed for batch-type steam sterilizers to solid dosage forms had given rise to differing opinions on how process validation would be accomplished. Concepts such as "worst case" and "processing range validation" resulted in protocols that prescribe independent parameters to be set on "edge of failure." Running processes with all parameters set at a low level, or conversely set at a high level, proves nothing because it is not known if the effect of extremes of the various parameters can cancel each other. For example, running a tablet press at the minimum weight at minimum speed may be acceptable, where minimum weight at moderate or high speed may be unacceptable. The Guidance expects this type of limit justification to be performed with factorial experimentation (DOE) within R&D during Stage 1.[12] Formerly, process validation (now PPQ) was generally accepted to mean three batches run with parameters set at typical setpoint values. This may still be the case for a generic or OTC, where a process has been acquired with no accompanying development data. In this case, process history with similar products may support minimal development trials in determining independent process parameters for equipment settings. For new drugs, it is assumed that equipment-focused development and qualification studies for the specific are completed prior to PPQ. The results of these studies for facility and process are confirmed under formal protocol and are called design qualification (DQ), installation qualification (IQ) and operational qualification (OQ) and must include the parameter limits for speed, batch or component size, and any other process limits. The concept of Traceability Analysis (when consisting of a separate document, called Traceability Matrix),[13] borrowed from software development, has been frequently used to list every specification and/or limit and how and where it is supported as a prerequisite to PPQ.

Confusion between the terms "Performance Qualification," "Process Qualification," "Product Validation" and "Process Validation" was common in solid dosage validation and has been resolved with the Guidance, which defines process validation as a series of qualifications and an ongoing verification following initial product introduction. This approach, without the ongoing verification, was supported by the practice of engineering–construction firms that usually were involved in specific equipment and/or facility upgrades. Performing qualification on the equipment and facilities enabled them to complete the project with minimal interaction with the client firm's personnel and testing. Process performance qualification (PPQ) is the final test with everything in place, qualified and operating as specified prior to product introduction into interstate commerce. Deviations that occur during PPQ must be resolved to prevent the deviation from reoccurring. For example, a deviation occurred where an equipment temperature sensor malfunctioned during a (then-called) Process Validation batch. The initial deviation investigation found that the manufacture's recommended replacement interval had been exceeded and was not specified in the preventive maintenance activity for the sensor. Corrective action was initially limited to adding the recommended replacement interval to the Preventive Maintenance Schedule for the particular sensor. However, this same deviation had occurred two years earlier during another validation involving another sensor. Satisfactory resolution may not have occurred until the commitment was made to look at the firm's entire preventive maintenance program to ensure that sensor replacement interval is included and accurately specified.

In the context of solid dosage validation prior to the Guidance, qualification had meant equipment- and/or facility-focused studies. Today, the Guidance has generalized the term to include all validation related activities prior to product release into commerce. Process Qualification (PQ) is now defined within the Guidance to consist of facility and equipment qualification and PPQ. An evaluation of the original PPQ study is required when changes are made to the process. This evaluation, usually part of a change control program, specifies what additional PPQ is required. Typically, a subset or supplement to the initial protocol is required. Examples would be additional dosage strength for a given product or a revised individual unit operation of an existing manufacturing process, for example, drying of a wet granulation using a revised process or equipment. It is important to define terms and agree on the specifics for each product. A firm must have a formal change control program where such changes are evaluated and additional validation specified.

42.5 VALIDATION CONCERNS

The following concerns should be incorporated within solid dosage protocols in a defined format. They may serve as a resource when reviewing protocols for adequacy and completeness.

42.5.1 Prerequisites

In terms of solid dosage finished goods, PPQ requires the completion of a substantial number of prerequisites that must be documented and in place prior to initiation of the protocol. These generally consist of the following: development, equipment and facilities qualification (Stage 1 and initial part of PQ), analytical test method validation, documentation (manufacturing instructions, batch records and standard operating procedures) and training. Cleaning validation is frequently performed concurrent with PPQ, especially with limited development for OTC products and generics, and it is important to consider the effects on manufacturing facility schedules of hold times as well as potentially overloading the testing laboratories with validation samples. Other equipment qualification prerequisites, such as User Requirement Specifications, Functional Requirement Specifications, Factory Acceptance Testing, Site Acceptance Testing and Commissioning, may be required and are a practical necessity for new systems and equipment. An important deliverable from R&D is the identification of Critical Quality Attributes (CQAs) and Critical Process Parameters (CPPs). These may be available from the literature or firm experience with similar products, including OTC products and generic drugs. It is necessary to document these prior to PPQ protocol preparation, if these have not been explicitly documented in previous development reports. The personnel performing the validation should not be the ones designating what is critical, no matter how obvious, and documentation from supporting technical experts is essential. This can be in the form of a request for confirmation of the critical variables.

Many other activities can be considered validation in the general sense. However, those activities are rather more cGMPs than validation and are typically performed by functions other than those involved in process validation. In addition, these activities are general to all products and do not change meaningfully with solid dosage finished goods. Some of these activities are approved written procedures generation (document control program), preventive maintenance and calibration programs, specifications, test methods, analytical test method development, stability, personnel training and documentation, change control program and corrective and preventive action commitments tracking. It is important that there is close link between the change control program and the validation effort, since change can be the direct cause of new validation activity. The persons performing validation of solid dosage forms must ensure that applicable procedures have been published and that they are being followed. Once the protocols are approved, there may be a separate "training" on the protocols themselves including manufacturing personnel. This requirement has come about in recent years in an effort to avoid protocol deviations. Coordination with production planning and laboratory functions is essential. Budgeting in these support areas for validation is necessary, and communication during budget time can avoid scheduling problems later. The budgeting model for manufacturing described earlier can be extended to the laboratories to ensure compliance with protocol requirements. Checklists and separate procedures addressing prerequisites have been utilized in connection with validation master plans to allot resources and track this supportive activity.

The normal sequence for a solid dosage form is Stage 1 process design, Stage 2 process/equipment qualification and Stage 2 process performance qualification, followed by product release into commerce and stage 3 continued process verification. Most firms require the completion of one stage prior to the initiation of the next. However, it is acceptable to combine these stages within single protocols for OTC and generic products. A recurring problem is the requirement of completed "as built" drawings as part of facility qualification. These drawings are typically subcontracted out and may not be completed in time for PPQ to begin. It is important to put contingencies for this and other potential problems within the protocols to enable timely completion and to ensure that they are completed prior to product release into commerce.

42.5.2 Risk-Based Approach

There is no difference within the regulations as far as cGMPs requirements for filed or prescription drugs versus OTC solid dosage drugs. Historically, the auditing and enforcement approach has varied, and the experienced validation manager had considered this in prioritizing validation activity and resource allotment. Similarly, within products, those areas that were most problematic to the final product quality were given the most scrutiny during validation. The FDA has acknowledged this scientific approach and formalized it as being "risk based."[14] The amount of testing may vary, with the filed product typically having more testing performed as well as increased regulatory scrutiny. Examples would be a Class 1 dental device, denture cleaning tablet and a prescription tablet drug product. Although the sampling would be essential the same, the critical attributes of concern would be markedly different, as would the amount of additional validation testing performed. Equipment qualification would remain essentially the same in terms of parameters measured and documented. The extent of the documentation would be less for the Class 1 device than the ethical tablet.

The risk assessment must be documented. Techniques such as Failure Mode and Effects Analysis,[15] Ishikawa (herring bone) diagrams[16] and Potential Problem Analysis[17] can be used to formalize and provide structure to this process. Solid dosage forms typically have multiple dosage sizes for the same product. There are also multiple package sizes and packaging formats (bottles and blister packs). This diversity necessitates the use of family group bracketing or matrices during validation as a matter of practicality. It is important to describe the planned exclusion of some of the product/package presentations in a risk assessment along with rationale and justification.

Validation master plans and protocols should anticipate problems that may arise when products requiring validation are manufactured in the same plant location as those that do not require the same level or type of validation. For solid dosage products, these products are typically food supplements (i.e., vitamins). Many other products such as cosmetics and household products do not require validation and may be present in multi-product manufacturing facilities. Special training and awareness are required for personnel engaged in development and manufacturing activities for drug and devices to ensure overall compliance at these sites. Although the cGMPs are addressed by dedicated personnel with these products, supporting staff personnel with site-wide responsibility may be unaware of the additional documentation requirements and restrictions associated with drugs. Poor execution and/or documentation practice by support groups is especially detrimental during development. Although the Guidance acknowledges the gradual application of cGMPs beginning in early development, data generation may have to be repeated for lack of Good Documentation Practice (GDP) or noncompliant cGMPs. This usually results in missed deadlines, since PPQ, and specifically compressing and encapsulation PPQ, is typically the last activity to be completed prior to shipment into commerce of a new product.

42.5.3 Critical Attributes/Parameters

A critical attribute/parameter is a characteristic of the unit operation being validated that will adversely affect the identity, safety, efficacy and reliability of the final drug product. These, as stated earlier, have been subdivided into Critical Process Parameters (CPP) to represent independent variables and Critical Quality Attributes (CQA) to represent dependent variables. An independent variable is typically a machine setpoint or adjustment that is specified for the process. A dependent variable is a measured property that is an outcome of the process. As mentioned previously, for solid dosage the CQA of concern is often individual dosage content uniformity for the compressed tablets or filled capsules, followed closely by dissolution and assay. Other tests (for CQAs) may be important, depending on the particular drug substance(s) involved and the regulatory status of the product. The critical physical in-process testing, which is performed on-line at high frequency, primarily to preclude financial loss, is usually more than adequate process qualification for most of these parameters. Care should be taken that the level of in-process testing used for manually set-up and adjusted tablet presses is maintained for at least the validation batches of computer automated presses. This will contribute to the validation of these automated systems.

Often, overall testing can be reduced following validation by a clear description and justification of the critical attributes. This is where a development report can be essential to support process performance qualification sampling and testing and why process validation should be a concern at the earliest stages of development. Additional validation testing can concentrate on these attributes, while normal release testing may be reduced. Testing for information only should not be included in validation protocols. This research-type testing may wind up being required in routine release testing as an outcome of an audit.

42.5.4 Specifications

It is required to have the specifications finalized in Stage 1 and early Stage 2 before publication of the PPQ protocol. These are the prior approved product quality attributes required within the protocol. The concept of the Traceability Matrix, mentioned earlier, is a useful tool to ensure all specifications are addressed properly. Initially, limits are sometimes set differently than development because of the unknown variability of full-scale equipment relative to smaller pilot plant equipment for the particular formulation. This is often the case in tablet and capsule validation. Sometimes, the desire to adjust in-process specifications at the completion of full-scale batches to tighten ranges may be an objective of the protocol. This must be explicitly stated as such and the use of misleading phrases such as "established during validation" avoided. Machine operating ranges are often adjusted after the validation batches if they are the first full-scale batches (often the case with monograph OTC drugs). Most firms have a two-level approach to in-process limits by using internal warning limits, which can be adjusted to reflect a specific manufacturing equipment sequence. Tests include individual and/or average weight, disintegration (for both tablets and capsules), thickness (gauge), friability and hardness (tablet breaking force); the latter three are for tablets only. The internal warning limits should be tight enough to detect subtle change in equipment or material. Validation data is typically used to establish these internal limits or baselines. It is common for these limits to be adjusted with time to reflect raw material variation or sourcing changes, which also would require a documented change-driven revalidation evaluation.

Solid dosage specifications have less of an absolute character and certainly much less of a microbial safety concern than other dosage forms (e.g., sterile and non-sterile topical). The imposition of arbitrary microbial content specifications on solid dosage and associated cleaning validation based upon USP 34 <2>[18] must be made by medical groups only, and then with the full knowledge that solid dosage products and their components are manufactured and handled in unclassified areas and will fail any arbitrary specification sooner or later. Additionally, there are USP chapters (<61>, <62> and <1111>) that require the absence of certain organisms. This unrealistic expectation is discussed by Agalloco.[19] The moisture content of solid dosage products is typically less than 5%, and it is well known[20] that this is unfavorable to microbial growth. Personnel experienced with sterile manufacturing and less so with solid dosage that are involved in these decisions should be reminded that the funny taste in their mouth when they awake in the morning is caused by millions of bacteria. Any specification should not exceed that of potable water, which currently consists of a spot check for coliforms,[21] unless dictated by medical necessity.

Although acceptance criteria are fixed in absolute limits, the process generally allows for more variability within those limits. Tablet presses and capsule fillers are unique in that final dosage physical attributes are treated as independent variables to "set the machines in." On older machines, there are no numbers or units on weight, thickness and penetration adjustments, since the resulting product-measured parameter is indicated as the adjustment. Only the most modern computer controlled machines have anything resembling a true absolute independent parameter for these machine adjustments. Even in this case, there may not be any units associated with the adjustments. In addition, these adjustments are more susceptible to minor lot-to-lot variations in raw material parameters, especially in direct compression or direct encapsulation processes. These variations should be characterized in Stage 1 and Stage 2 process validation.

Machine speed specifications for tablet presses and encapsulation machines are often established during validation. It is important to establish a minimum and maximum speed for the dosage forming step. While a relative standard deviation for individual weight control of 1% may look nice in a process performance qualification report, it may also be indicative that the machines are running too slowly from an efficiency viewpoint. Compression and encapsulation are the only areas where machine speeds can be variable to any extent. It is important to consider this in the initial validation of a process so that manufacturing has the necessary range of speed to respond to the inherent variation of raw materials and environmental conditions. In one instance, the author recalled a maximum speed being established for OTC diphenhydramine tablets that was misinterpreted by plant personnel as an absolute requirement. When normal variation necessitated a speed reduction, the plant continued running at the maximum speed despite manufacturing unacceptable material.

42.5.5 Compressing and Encapsulation Facilities

Compressing and encapsulating rooms must have adequate separation from other products. It is permissible to have more than one press or filler in a single room; however, they must be running the same presentation of the same batch of the same product. If this cannot be accommodated, then individual smaller rooms are a better design choice.

Dust collection and proper room pressurization are major facility requirements for tablet presses and capsule fillers. Recirculated room air and dust collection must be adequately (HEPA) filtered. It is preferable to have these systems dedicated to individual rooms, although not usually practical. Elimination of cross-contamination via the HVAC system must be an essential design consideration. Additionally, dust collection air returned to the room, no matter how well filtered, usually creates a cross-contamination problem. Dust collection is better vented to the outside of the facility after separation of the dust.

Although it is usually not practical to move encapsulation machines around the facility, moving tablet presses is commonplace. Even in dedicated rooms, the presses are moved for press and/or room cleaning. Therefore, the doors must be of sufficient size and adequate design. The hospital-type swinging door has been successfully utilized for this purpose. They are more durable than sliding doors and are more easily replaced. They are also consistent with the sanitary nature of the room and surroundings. They close automatically using mechanical spring tension and provide an opening for air balancing purposes. Unlike sterile areas, it is preferable to have negative compressing and encapsulation rooms adjacent to service corridors. This is done for dust control, and the rooms themselves usually have HEPA filtration on exhaust vents. The use of positive pressure rooms is possible, but less common. In this case, the hall or service area would be negative to the process rooms and the rest of the facility. The pressure balancing design must be documented so that it can be verified during facility qualification and monitored during continued process verification.

42.5.6 Compressing and Encapsulation Equipment

Press tooling (upper punches, lower punches and dies) should be controlled items. It is essential to have cleaning and use logs on individual tooling sets. Security, mix-ups, usage ware and damage are all considerations in handling and maintaining tooling. Procedures for this control and maintenance must be in place prior to process performance qualification activity. This is true to a lesser extent with encapsulation machine change parts. Facility designs should include provisions for cleaning, maintaining and storing tooling and other change parts. Feed frames and filling rings or funnels are best purchased from the original machine manufacturer. The use of generic copies for these parts such as "plastic" feed frames and reconditioned parts is usually far more expensive in the long term. These generic parts are nonstandard and frequently behave differently than the original manufacturer's parts. It is important that consideration of "change parts" be included within the validation protocols. These parts can be changed with no additional validation required, provided the part meets original quality, usually determined by inspection. This simplifies the change control process and enables routine expendable parts (e.g., dust cups, dosators, cams, belts, hoses, etc.) to be replaced without a requalification interruption in production. Equipment manufacturers are the appropriate sources of information on the recommended "change parts," including their replacement frequency.

The modern rotary tablet press and rotary capsule filler are expensive and complex machines. Adequate documentation is generally available from the vendors or original manufacturer to ensure proper maintenance and training. A 21 CFR Part 11 assessment and validation is required for computer-controlled equipment and is performed independently of a product validation.

Accessory equipment such as conveyers, dedusters and polishers must be readily cleanable. It is often desirable to dedicate the accessory equipment to a particular product and dosage to simplify cleaning validation.

The required in-process testing equipment is often neglected in the early design phases. Machinery that increases output also suggests the application of corresponding automated individual tablet and capsule weighing equipment. Robotic machines in the lab can greatly reduce testing time for batches produced on high volume tablet presses. These are particularly helpful during the initial heavy additional testing required by validation.

42.5.7 Stage 2 PQ—Installation Qualification

Stage 2 PQ typically requires facility and equipment qualification, and both of these will entail installation qualification (IQ). This phase of qualification, as applied to solid dosage forms, will have checklists and data sheets as a minimum for preexisting (legacy) facilities and equipment and more extensive protocols for new facilities and equipment. It is important to realize that most compressing and packaging equipment is mobile, and this aspect must be addressed within the protocols. Generic IQ checklists designed to apply to all plant-wide equipment may not be appropriate for these operations. There is some overlap between this phase and operational qualification (OQ) in that the equipment is sometimes run without product and operational and maintenance procedures are drafted. It is important to demonstrate during solid dosage equipment qualification that calibrations of components on all movable equipment is unaffected by movement and the rigors of cleaning using organic solvents and /or water. This includes the strain gauges that control weight on automated tablet presses. Additional consideration for worst case testing must be made for the extremely dusty conditions in which this equipment operates. In most cases, vendors can provide written assurances that can be verified with minimal testing.

42.5.8 Stage 2—Operational Qualification

Material costs in solid dosage processes may be a limiting factor in Operational Qualification. Compressing batch sizes can consist of several million tablets, while equipment qualification can be achieved with smaller tests. It is possible to run a placebo batch (excipients only) during this phase to minimize expense. In this case, it is common to require three batches of actual product to be run during later process performance qualification. When purchasing used equipment, it is recommended to use placebo materials during initial Commissioning and OQ. During operational qualification of a used tablet press, it was discovered (much too late, unfortunately!) that the cams did not comply with IPT standard bevel angles. This resulted was destruction of a full set of tooling and lost product. Another operational qualification on a "remanufactured" press that used the equipment vendor's refinement to the tablet "kickoff" function resulted in tablets being kicked to the floor at moderate operating speed. This "refinement," which was a press fitted ring around the turret, had to be removed at additional expense. Purchasing groups should be reminded that used equipment must be fully qualified, and this expense must be considered and included in the initial purchase proposals.

Compressing and encapsulation equipment frequently have mechanisms that divert product into "reject" containers if monitored parameters fall out of specification. One use of this shut-off feature is when the supply of granulation to a press is interrupted. It is often sufficient for qualification to verify this logic with a single product or placebo. The procedure is to collect the last tablets or capsules not diverted to the rejection container and verify their individual weights manually with an independent calibrated balance.

42.5.9 Process Performance Qualification

The best way to validate a compressing and encapsulating process is to sample and test more extensively than the normal or proposed quality control release testing. The additional data facilitates variability characterization using statistics. As mentioned earlier, three such batches are typically required, or the equivalent for a continuous or semi-continuous operation. Centerline (or set-point) validation is used, meaning that the batches are not run on the edge of any limits ("edge of failure"). All operating limits should be supported by developmental runs or experiments in Stage 1 and/or early Stage 2. A possible exception to this is machine speed, as discussed previously. The assumption is conventionally made that, for a given machine, the product running at a slower speed is in a higher state of control and its quality need not be validated. This has been questioned, and some validation is generally required for the lower limit of speed on a given machine and certainly for a different, slower machine. Protocols must address machines that require different change parts for different speeds and weights such as fill cams, this being the equivalent of a different, slower machine. Other independent machine variables may not be critical, such as penetration (depth of compressing within the die). This parameter is adjusted as press tooling wears, proceeding from the lower part of the die to the upper part. All specifications for in-process physical testing and background environmental conditions (temperature and humidity) should be verified during development on small-scale batches. All independent operating parameters should be documented either on the manufacturing batch record or in supplemental validation records. These records are subject to FDA scrutiny and should be maintained and controlled the same as QC records or R&D notebooks.

42.5.10 Solid Dosage Validation Sampling Plans

For the dosage forming step, it is common practice in process performance qualification to divide batches into segments for testing purposes, the minimum being three (beginning, middle and end). The worst case points for sampling are often in the beginning (immediately after the machines are "set in" and the product is deemed acceptable) and end (immediately prior to shutdown). Machine shutdowns for lunch and breaks may also require sampling. In this author's experience, the "end" sample is particularly troublesome in terms of providing acceptable results. On one occasion for a legacy product, a particularly inexplicable uniformity problem resulted in the expensive practice of stopping batches with 30 kg left to compress in the feed hopper and discarding the remaining granulation. The cause was milling the active ingredient too fine relative to the excipients in a direct compression formulation. It is important to ensure that each compressing station or encapsulation dosator is representative of the sample. Protocols must have explicit language to that effect, since the relatively small amount of units tested in batch sizes numbering in the millions is predicated on the assumption of repeated use of identical tooling.

Since routine QC sampling involves composites, it is generally preferable to use point ("targeted") samples for validation, although composites of batches divided into six or more segments may also be acceptable for some tests, provided each compressing and encapsulation station is tested. The number of sample points and number of tablets and capsules sampled varies with the batch size and machinery type. At one firm, a cutoff of batch size of 1 million units has been used to increase from three to six segments. The sample size should ensure the equivalent of one complete rotation of the compressing machine (full filling ring or all filling funnels for capsule fillers) is obtained. Samples should be taken from all discharge points for dual- or quad-compression roller machines. This aspect should be included on the sample bottle label information. The segments may be identified in terms of filled bulk container number or in terms of time. A sufficient number of units should be obtained to perform within batch and between batch confidence interval testing for individual weight variation and USP content uniformity testing. Some parameters, such as disintegration, hardness, thickness and friability, are not readily compared in this way and vary within specifications both within batch and between batches. For example, the disintegration test is usually run with a timer set to some number of minutes. At the conclusion of the test, the basket is examined, and if the tablet or capsule has disintegrated, the timer setting is recorded, regardless of when the actual disintegration occurred. Resampling provisions within the protocol should explicitly specify sampling from the filled bulk container rather than directly from the machine discharge in case samples are missed or lost, because the discharge method cannot be repeated for a given batch once it is compressed or encapsulated and collected in a bulk container. Questions on resampling can be avoided by obtaining a contingent resample initially. There should be a sufficient number of tablets or capsules in each sample to perform all testing. In addition to routine QC testing, other sampling protocols typically in effect on initial production start-up are an R&D sampling protocol and a stability sampling protocol. These sampling protocols could be combined and use the same time points.

42.5.11 Blend Sampling

A blend (final mix with lubricant added) study executed by the author's coworkers across several products in the early 1980s resulted in blend content uniformity data that was largely unacceptable when the dosage forms (tablets or capsules) were entirely acceptable. After initial disbelief of this counterintuitive result, which occurred well before the three-dose maximum sample requirement (see below), the outcome was attributed to plastic scoops and containers that were used to obtain samples. With additional industry experience over the next several years, a PDA monograph was published to address this subject.[22] The situation was exacerbated by the Barr Decision,[11] which identified design deficiencies in a manufacturer's attempt to validate product. Two requirements emerged from that decision. They are (1) to sample a blend with a scale of scrutiny within one to three final dosage sizes and (2) to apply dosage content uniformity specifications to those samples. The latter requirement implies the taking of at least ten samples, since this is a requirement of the USP stage 1 Content Uniformity test. The FDA published and then withdrew a guidance[23] on the subject and never finalized another,[24] further attesting to the complexity of the problem. These attempts to address blend sampling did not fully address the mechanical action of sampling affecting results. Dr. James Bergum and his coworkers developed an evaluation scheme[25] for use with blend sampling as well as the dosage form. Although silent on the sampling difficulties, the acceptance criteria provide some level of assurance of meeting the dosage content uniformity specifications. The increased granulation testing requirement affects compressing and encapsulation validation in that there is a tendency to reduce the amount of tablet or capsule USP Content Uniformity testing performed, given limited laboratory resources. There is also a need to show that uniformity is not lost in storage and/or transfer to the tablet press. Liss et al.[26] surveys this phenomenon and demonstrates that segregation can occur within granulation while it is merely dropping through a pipe.

Theoretically, testing the blend makes good sense; the blend must meet dosage uniformity requirements, or it would be impossible to have tablets or capsules that meet these requirements without relying on serendipitous mixing in the tablet press feed frames or encapsulation granulation hoppers. Additionally, having demonstrated a uniform blend, performing individual tablet and capsule assays may be reduced and replaced by in-process individual weights and other physical properties of the dose, the sampling and testing of which represent a very small percentage of batches, which may number in the millions of doses. This testing plan, based primarily on individual weight, is much easier, faster and less expensive to perform. In practice, however, for blends, the process of sampling to simulate the dosing process and the subsequent analysis is nontrivial. The separating action of static electricity, movement caused by physical manipulation, fluidization, and surface affinity can occur in the simplest powder flows, such as filling a sample bottle or thief.[13] The forces may be different for a blend when a scoop (either at rest or flowing in mass) or thief sampler is used versus the "sample" taken by the mechanical dosing action of the tablet press or encapsulation machine. Local nonuniform samples can be "created" by the physical act of sampling when small samples are involved. This process can also occur in the laboratory unless the entire sample is analyzed.

It may be true that the same forces that make blend sampling invalid can give rise to nonuniformity in the tablets or capsules. In those cases where both blend and dosage test results are nonuniform, nothing short of major process and equipment upgrades and/or formulation changes will solve the problem.

A classic example is amoxicillin for oral suspension. Although not a tablet or capsule, it exemplifies the blend sampling problem. Uniformity testing of blend samples on the order of one to three doses can seldom achieve the assay (relative standard deviation) results of testing the filled bottles. The explanation is that the blend stays in its continuum until dosed; the chemist reconstitutes the entire bottle and uniformizes as instructed on the label, and withdraws a single dose aliquot. This process is usually a direct blend type (not wet granulation) formulation consisting of granular sugars and flavor and an active ingredient that is a fine powder.

Developments in compartmentalized sampling thieves designed to address this problem has brought needed help. These thieves are provided by Globepharma, New Brunswick, New Jersey (Figure 42.1). Sani Raju, a pioneer in solid dosage validation, developed these thieves to address the Warner-Lambert consent decree.[27] This type of thief enables sampling blenders using the same die filling mechanism of a tablet press or capsule filler ring. Such thieves can reproducibly sample "worst case" blender locations at the three unit dose level or less, which can be tested for active assay(s). The mixing action of blenders and "worst case" locations have been published[28,29] for each blender type in common use for tablet and capsule

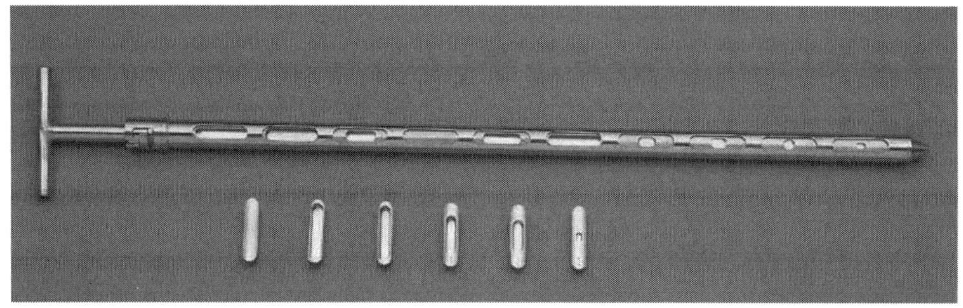

FIGURE 42.1

blends. These are generally located along the axis of rotation for the tumble types or in dead spots for mixing blade types. Demonstrating that these areas meet assay specifications in conjunction with extensive tablet or capsule content uniformity testing may be adequate validation for those products that have a discrepancy between blend and tablet (or capsule) uniformity results attributed to sampling. In these instances, theoretical explanations for well-documented inconsistency during development can be strengthened with extensive successful tablet or capsule validation testing and may justify exceeding the three-dose sample size limit for blends. Blend sampling should be validated in its own right during Stage 1 development, if possible, to avoid these problems. In practice, the author has found that delivering the entire filled thief to the QC lab and including the entire sample cavity in sample solutions removes most of the sampling effect on blend variability. It is important to remember that this is an analytical problem and not a pharmaceutical process problem. A commercial benefit from extensive successful blend sampling and testing for validation is that it may justify reduction of routine QC blend testing when validation is completed.

Blend sampling plans mirror USP Content Uniformity Test sampling, in that a minimum of ten samples are obtained from the blender and/or final blend container. The Bergum et al. sampling plan[19] should be followed for greater assurance. It is important to sample an equal number of "best case" areas to "worst case" areas to assess the maximum variability in the results.

42.5.12 Compressing and Encapsulating PPQ Validation Testing

The USP Content Uniformity test and Dissolution test are performed for all active ingredients on all additional validation sampling locations collected for compressed tablets, tablet cores, capsules and coated tablets. Multivitamins often have a reduced scheme. All other critical attributes of concern are also tested. Bergum has updated his procedure[30] for determining the probability of passing the USP Content Uniformity test based upon average weight and RSD. The tables[31] are straightforward and easy to use. His scheme for the Dissolution test[32] is entirely applicable and, similarly, easy to use. Cholayudth has provided this approach in MS Excel®[33] format. Normal or proposed routine quality control sampling should also be performed. It is better to use approved validated test procedures based upon those used during routine production rather than extra R&D type tests (e.g., lubricant surface area) that can be performed for information during Stage 1. Such informational sampling and testing can be part of an R&D testing protocol on the initial full-scale production batches. An example of this would be feed frame or machine hopper granulation samples. For many products, validation testing often delays product release. Schedulers must allow sufficient time to complete the validation testing, which can be several times the normal QC testing. A rule of thumb for the minimum multiple for normal QC testing is the "rule of three." This should be done for the smallest batch sizes, and the usual multiple, in this author's experience, is six, although other firms use from five to ten. Tablet batches with doses in the millions or continuous processing may require more. The limiting factor may well be the capacity of the laboratory. Groups responsible for validation must ensure that the extent and timing of validation activity is communicated as early as possible to scheduling and laboratory groups.

42.5.13 Blend, Tablet and Capsule Validation Acceptance Criteria

With the possible exception of USP Content Uniformity, the routine or proposed production limits are generally used for validation. This is done because, as mentioned earlier, it is almost impossible to explain passing release testing while failing validation testing. Using the segmental sampling plan described earlier, it has been the author's experience that any significant process problems will be detected over the course of the initial three batches. Using the Bergum acceptance criteria described in Section 42.5.11 for both Content Uniformity and Dissolution can provide much of the statistical analysis suggested in the Guidance. This type of acceptance criteria is ideal for development batches where processes may not be finalized and can serve as a means of evaluation to predict formulation performance at full-scale production.

Other important validation acceptance criteria are confirming that sufficient controls are in place to detect any deviations. This means that there is adequate in-process and physical testing of subsequent production batches. The advent of computer-controlled machines has enabled a reduction in in-process testing. Sufficient testing should be retained to detect machine malfunctions or granulation abnormality.

42.5.14 Standard Operating Procedures

Inherent to any validation is the assumption that approved procedures are in place covering all phases of the operation being validated and the validation activity itself. In essence, the procedures, as executed by trained personnel, define what is actually being validated. This is also true for solid dosage manufacturing operations. The enforcement of this aspect of cGMPs is usually performed by Quality Assurance. The people performing the validation will undoubtedly be involved in the generation of some of these procedures. A checklist is typically employed to ensure they are all in place and people are trained prior to the production and validation activities. The problem with checklists and "cookie cutter" form-type protocols is that once an incorrect box is checked or an applicable section is marked "not applicable," all the other entries can be questioned. All correctly checked boxes must now be reviewed, and all sections deleted must be reviewed for applicability. It is good documentation practice to have all results entered even if check boxes are used. It is preferred to list specific procedures and trained personnel within the protocol. Current practice is to have a database containing required training curricula and training dates for all employees.

It is important to ensure that procedures are followed and that they adequately describe the activity taking place,[34] much the same as the batch monitoring confirms adequacy of the process instructions. Although implicitly documented by the manufacturing batch record, a separate summary verification statement within the protocol provides an explicit statement of compliance.

Cleaning procedures are commonly validated under a separate protocol. Cleaning validation often runs concurrently with process validation for new solid dosage products and should be considered in scheduling activities. The cGMP requirement for clean equipment before and after a new product may need to be addressed and documented in equipment qualification and/or process validation protocols. Cleaning validation is addressed in another chapter within this volume.

42.5.15 Supportive Data From Other Areas

Compressing and encapsulating validation often requires supporting data from other operations. It is normal to validate a process from start to finish over an entire batch manufacturing process. Historically, there was a three-batch requirement, and for OTC and generic facilities, this remains a fallback position in the absence of Stage 1 development data. There are usually two opposing forces with respect to the number of batches or validation production. Manufacturing wants to do as little as possible, and technical support personnel want to do as much as possible. Upper level management approval of validation master plans can ensure that the validation effort is consistent with industry practice and FDA expectation. The accounting approach of Agalloco described above usually will placate Manufacturing and Quality Control financial concerns: however, the time element still has to be negotiated in most cases. It is important that the following supportive items from the prerequisites, discussed earlier, be in place: raw material specifications; raw material test methods; analytical test method validation; bulk specifications; finished product specifications; stability protocols and data; packaging component specifications; and test methods. The identification of these items may vary from firm to firm along with the source of the documents. Process Performance Qualification is typically the last thing that is accomplished prior to commercial release of product, and it cannot be finalized without the supporting studies. Validation master plans should include these deliverables from other groups.

42.5.16 Developmental Data

The need for development groups to provide documented data supportive of Stage 1 and Stage 2 process performance qualification cannot be understated. FDA compliance inspections for validation and the Pre-Approval Inspection Program have seen investigators poring over research in search of reports and raw data.[35] This scrutiny was partially the result of the "Generic Drug Scandal," where data were falsified and even the innovator's product was submitted for sample testing.[36] Publication of a development summary report (name may vary) directly supportive of validation should be a standard procedure. This summary should include identification of CPPs and CQAs, experimental designs used, testing results; rationale for conclusions; selection of specifications and justification of limits; bio-batch records; and testing results. It is far more convenient to stress limits of intensive parameters and manufacture under worst case conditions in development. These data must be available to justify "center-line" (as opposed to edge of failure) process performance qualification of full-scale batches intended for sale. "Worst case" as applied to "centerline/set-point" validation of tablets and capsules refers to scale of scrutiny, sample size and sample locations.

It is important to record machine parameters and characteristics during development even though the ultimate manufacturing equipment may be somewhat different. The rationale for equipment selection for a particular bio-batch must contain relevance to the final commercial process validation. For example, the use of a gravity feed frame press versus a rotary feed frame press may be considered an arbitrary choice for a limited production bio-batch. However, if the anticipated press for commercial production is a high speed rotary feed frame machine, the choice of the rotary feed frame for the bio-batches is more supportive of validation.

Validation testing criteria should be applied as early as possible in development so that commercial batch comparison to the bio-batches is facilitated. One question that may come up is the content uniformity of scored tablets when broken.[37] There is an FDA Guidance on this subject, and this, along with USP <905>, should be consulted for validation testing.[38] The author recalls one instance as a section head over a compressing and encapsulation operation when a scored tablet of a rarely manufactured product could not be broken in half. We needed to consult a senior experienced chemist in the QC laboratory who showed us the technique. The thumbnails were placed back to back in the score and pressure exerted by the index fingers on each end of the capsule-shaped tablet. Fortunately, these instructions were part of the insert for the product. The rational for tooling design must be included in supporting documentation. Similarly, the rationale for capsule size selection should be documented. The dosage uniformity approach of Bergum,[21] discussed earlier, is very useful to evaluate the merit of a particular tooling design or formulation during development batches.

42.5.17 Bio-Batch Equivalence

A key component of process validation is showing clinical or bio-batch equivalence.[18] It is a frequent subject of solid dosage form compliance inspections, since raw material variation can cause final product variation in uniformity, disintegration, hardness (now called tablet breaking force) and dissolution. It is desirable to use the same dosage form (tablet or capsule) for commercial batches as that used for the bio-batches. This facilitates the comparison of in-process testing results from validation batches to the bio-batches. Historically, early R&D batches have utilized dry filled hard gelatin capsules for convenience. Questions on the bio-batch equivalence to full-scale production can arise when the final commercial dosage form differs. Sufficient data should be collected during the bio-batches so a comparison can be made with the full-scale demonstration batches or validation batches. The statement of equivalence is usually made by taking into account the in-process testing results, raw material testing results and final product testing. The statement may be part of the validation report or in the conclusion of a development report that includes the full-scale data.

42.5.18 Raw Material Characterization

The raw material characterization must be appropriate for the type of process. Direct compression processes require meaningful particle size specifications for raw materials and usually prescreening of these raw materials during addition. Optional screening requires some type of documented in-process check after screening during validation and is a frequent focus of auditors. The trend toward global sourcing and commodity-type purchasing of raw materials makes vendor qualification essential. The cost savings in this type of purchasing may override the cost savings of a direct compression process over a wet granulation process, since the latter is more forgiving of raw material physical attribute variation. Process designers must keep this in mind, and purchasing departments need to be controlled by appropriate vendor qualification and change control programs following validation and bio-batch production of direct compression formulae. In one case in the author's experience, the formulators developed a "rescue process" along with the regular direct compression process, anticipating problems. The rescue process consisted of a roller compaction process (Chilsonator™) that changed the compression characteristics of the tablet. Two different validations had to be conducted and, given that this was projected to be and was a billion dollar product with over a 90% margin, the savings, if any, in manufacturing using a direct compression process were inconsequential.

42.5.19 Bulk In-Process Storage

The emphasis on hold times in sterile manufacturing has resulted in the validation requirement for bulk storage hold time of in-process solid dosage materials. While sterile manufacturing is concerned with microbial issues, solid dosage manufacturing is concerned with physical issues, such as granulation settling and compaction. In-process testing from the previous stage should be confirmed. It is good practice to include this in the process performance validation protocol, if not addressed in Stage 1 process design. More often, it is an addendum to a report or protocol or a separate R&D study. In-process materials should be stored in compliance with any restrictions placed on the environment during processing as specified in the manufacturing batch record. Areas of warehouses that will be utilized for the storage of in-process or final product must be qualified to maintain any labeled storage restrictions. Hard gelatin (or vegetable) capsule shells contained within fiber drums are susceptible to physical damage when stored in direct sunlight for extended periods. Some firms have minimum weight specifications on the incoming empty capsule shells, making them more expensive, but this reduces the variation and defects caused by overly brittle shells.

42.5.20 Encapsulation Machines

There are two main types of dry fill capsule fillers. They are the funnel dosator type and volumetric ring dosator type. The funnel type can be continuous or discontinuous. The ring type is described as being semiautomatic or fully automatic. The funnel variety can be treated like a tablet press, where each dosator is treated like a compressing station with a unique identity. The ring type has numerous cylindrical holes that are filled by force-feeding granulation into them, either by the use of an auger or tamping pins. The rings for these machines are often customized for a particular product, although standard depths for each capsule size are available, so they should be controlled with the same care as compression tooling to ensure the correct rings are used for a particular product. The author experienced a situation where unacceptably high weights were observed during the startup of the initial production batch of a new encapsulated product. This resulted in a panic change to the dosator-type filling machine from the ring-type and the accompanying massive change in documentation. Upon later investigation, it was learned the ring-type machines were erroneously set up with "deep fill" and "special depth" rings for the given capsule size that had been special ordered earlier for a deleted problem product. Encapsulation change parts do not wear as readily as press tooling, so records on the total amount of production on a given set are sometimes not maintained. Nevertheless, procedures should be in place to record total usage and control the handling of these encapsulation machine change parts, since problems can arise from age and mishandling.

Soft gelatin capsules are more appropriately addressed in a discussion on liquids and semi-solids validation. There is an additional microbiological concern for the gelatin (or vegetable) raw material. Otherwise, the segmental sampling scheme applicable to dry filled hard gelatin (or vegetable) capsules is appropriate. Appropriate attributes of concern specific to liquids should be identified and included in the additional validation testing.

42.5.21 Accessory Equipment

Dedusters and polishers are usually addressed solely in the installation and operational qualification and blanketed by their use during process performance qualification. The assumption is that they do not adversely affect the product and do not purport to do anything quantitatively. They are something put on as a contingency for additional assurance. Evaluation of the product as acceptable without going through these devices is one approach. The other approach is to provide some standard worst case challenge, but this can seldom cover all the possibilities and often gives rise to more questions. Salt polishing of dry filled capsules has been rendered obsolete by mechanical dedusters and the cleaner operation of automatic fillers. It is important that bio-batches and stability samples contain salt-polished product, if salt polishing will be part of the commercial process.

Metal detectors are similarly addressed during qualification. The approach is to check the machines against the manufacturer's claims in a "no load" situation. Following this, each product should be checked at maximum throughput. It is desirable to request upper size limits for metal fragments from an internal medical or toxicological group. Limits should cover ferrous (e.g., tool steel and iron) and nonferrous (e.g., stainless steel, aluminum, brass, copper) metal fragments for the purpose of metal detector qualification and validation. It may be assumed that the contamination would be an isolated production machine fragment. Also assumed is that in the event of a large increase of metal detector rejected material, the batch will be held and an investigation of the nature, source and size range of the fragments will be conducted. It is important to include these assumptions when requesting limits from the medical groups. Otherwise, the entirely impractical response of "zero" may result.

A different case occurs when product with a known contaminant has to be redressed or reworked, such as a broken compressing punch. In this case, specific challenges should be made up that emulate the contamination. These challenges must be performed with the strictest control, usually in the presence of Quality Assurance. This type of rework is an example of a "one-of-a-kind" effort, and concurrent validation is expected before release of the material.

Capsule classifiers are utilized on-line for the relatively rare events of empty or low-fill-weight capsules. It is more likely to find an empty capsule than a low-fill-weight capsule because of the close proximity of the empty shells to the filled capsules within production rooms. These devices are easily challenged and qualified with empty shells.

42.5.22 Rework or Redress Validation

Rework or redress that consists of 100% mechanical inspection—for example, sizing on engineering rollers and metal detection—may be concurrently released and consist of only the batch(es) affected. However, any more common rework process requiring milling of tablets or capsules (with subsequent sifting out the gelatin fragments); optional addition of active ingredient (spiking); reblending; etc., usually requires three batches for validation and a medical group opinion as to the equivalency to bio-batches. It is the author's opinion that one-of-a-kind reworks can be concurrently validated provided there is a clear assignable cause that does not invalidate the original process, and substantially more extensive product testing is ordinarily included. The repetition of a manufacturing step following a known error in that step, unless addressed in the original validation, requires extensive justification and developmental support.

The notion of reworking seems contrary to a validated process. However, machine and equipment maintenance events and human errors can result in material that must be reworked. Catching these mistakes confirms the validation of the original process and is good news in the sense that adequate controls are in place to catch errors. However, the product has been subjected to an unapproved process in most cases. This is where extensive documentation and validation-type testing on the bio-batches pays off. Sampling and testing designs performed on clinical or development batches may be used to validate the rework process. Rework entails validating a new full-scale process, and the level of scrutiny should be the same as in the original validation, and much more extensive if only one batch is involved. It is usually better to discard the material unless prohibitively expensive. Even then, recovery of the active in a validated API procedure may be preferred to rework of a dosage form process. Managerial accounting budgets should be created to encourage manufacturing to discard material in these instances.

Blending-off rework is forbidden for drug products; however, it is inherent in processes such as denture cleanser tablets (Class 1 dental devices—usually performed for appearance reasons) and has been used to recover material for food supplements. The added rework replaces excipients so there cannot be any upper limits, as in most vitamins or denture cleanser. The approach is to demonstrate an equivalent ability to meet final product specifications with the blended-off batches by additional scrutiny of the critical attributes of concern. These attributes are usually product efficacy and stability.

The redress of off-weight tablets and capsules requires the preparation of "edge of failure" challenges. Extreme caution must be used to ensure these challenge tablets or capsules do not get into commercial production. The disastrous effect of such an error may justify destroying the batch rather than risking a rework. The use of different color capsule shells of the same size makes this a less risky proposition for capsules than for tablets. Extensive sampling and testing is appropriate for these validations. As an example, in one such redress with a capsule classifier, over 5,000 individual weighings were performed. Modern capacitance- or weighing-based classifiers may require less testing for validation than a separation system relying solely on the flow of air. The systems that individually and reproducibly weigh each tablet or capsule are preferred, and their much higher expense may be justified by a single rework batch.

42.5.23 Evaluation of Data

The Guidance expects the use of statistics to identify CPPs and CQA in Stage 1 and to characterize variability in Stage 2. The use of extensive hypothesis testing as part of inferential statistical studies should be avoided for prospective validation, unless used with the product's own specifications and with the machines "set in" value for the dosage forming step. With large sample sizes it is easy to show that the arithmetic means of two batches are statistically different, even when they are both well within all specifications. Statistical studies should be limited to those normally performed for product release and the approach.[21] A better approach would be to use descriptive statistics or prove that the probability of getting a failure is below some standard level for the validation batches. Use of inferential statistics on granulation sample results is misleading, since a continuum is being sampled and not individual discrete doses. It is simpler to use established methods and limits that will be used after validation for normal in-process and final product testing once the process is validated. Over-complicated sampling plans are difficult to relate to routine operations and may lead to their imposition for routine use if problems occur later in time. Additionally, noncompliance with an untried sampling plan and/or technique is a typical deviation encountered resulting in missing data. In these instances, the runs should be repeated, if possible.

Care should be taken in the use of inferential statistics such as CpK, CpM, and multiples of the standard deviation (Six Sigma®) in the validation of blends, compressing and encapsulation. These statistics are useful tools borrowed from the automotive and heavy equipment industries and are based upon the process being the result of machines set in at the exact middle of process ranges. As stated earlier, tablet presses and encapsulation machines are adjusted until within the warning limits of in-process parameters and then run at the maximum validated speeds to produce compliant product. The statistics may not be optimized under these conditions and may not match or be as good as results achieved on smaller, low speed development equipment. It is important not to let these statistics become additional product specifications. Statistics are best applied at the very beginning during Stage 1 design and the final dosage form of Stage 2 PPQ.

Stage 3 Continued Process Verification can summarize annual review and stability program data that is already statistically based. The author has found that examining accounting records can confirm and can be a better indication of maintenance activity and undocumented changes than QA-based change control programs! Statistical tools such as cumulative difference[39] and the sequential probability ratio test[40] originally applied by Larry L. Simms[41] at Eli Lilly beginning 40-plus years ago are extremely useful.

42.5.24 Sustained Release

Sustained release dosage or modified release (also called extended release and other marketing names) typically has wide acceptance criteria for active ingredients at the end of the release period. This range is necessary to accommodate the inherent variability in raw materials and processing in this class of formulations. As stated earlier, especially in this case, limits should not be "tightened for validation." The dosage is intended to aid in patient compliance and/or for marketing. The dosage form typically consists of film-coated tablets. There is a usually a solid matrix in which the drug diffuses out. There are many different materials, usually polymers, to achieve the desired dissolution profile. There may be more than one dissolution test to include in validation testing, and separate core and coated tablet testing complicates batch subdivision into segments for valuation testing. In one type where hard gelatin capsules are used, nonpareils beads (sucrose spheres) are coated with solutions containing the active ingredient with different release rates. The beads are tested, and then different proportions are blended and then encapsulated to make the final dosage. In another case, the active ingredient is formed into granules in a fluid bed granulator, which are then coated with an inactive coating solution. The granulator consists of a vertical air-stream that suspends the solid particles. A granulating solution is sprayed into the stream to agglomerate the suspended particles to a desired size. Critical process parameters are air temperature, relative humidity, velocity, solution properties of temperature, viscosity and density. In another type of sustained release, a coated tablet is drilled with a laser beam to provide a zero order release (constant release rate with time). This proprietary process was marketed, so it may be encountered in multiple firms. Hole geometry is critical, and the tablets are cut in half and viewed microscopically and measured. The profile visually resembles a "golf tee," which makes a good in-process test for the drilling operation. Otherwise, the CPPS associated with the coating operation must be considered (see "Coating," Section 42.6.2). Formulations with two or more ingredients can be problematic with differing release rates. In one instance the author recalls that, in a tablet purporting to be effective for 24 hours, one component was problematic after eight hours. This firm went into Consent Decree for this and other problems. In another instance (Hydrochlorothiazide and Triamterene capsules), the dissolution profile could not be duplicated by generic manufacturers after the innovator's product came off patent.

42.6 PROCESSING AFTER COMPRESSING OR ENCAPSULATION

42.6.1 Coated Tablet Cores

The sampling and testing scheme for tablet core PPQ is the same as for compressed tablets. The problem is that most of the testing has to be repeated for the final coated tablets. This is a good idea for validation, since duplicate testing can be eliminated to some extent after the process is validated to save time and quality control testing costs. The on-line in-process testing of individual weights, thickness (gauge), hardness (tablet breaking force), average weight and friability are retained with as much precautionary chemical testing as risk will warrant for the compression or encapsulation manufacturing step.

42.6.2 Coating

Documentation of quantities of coating materials applied throughout the process is a validation requirement for manual sugar coating, which historically was an uncontrolled operation from the confectionary industry. Standardizing and recording drying air and bed temperatures reduce the variability in quantities used and has served to remove much of the "art" from this process. Film-coating operations and automated sugar coating have more precise and reproducible solution application equipment and are easier to validate. Validation schemes treat each coating pan as a batch, and care must be taken to maintain pan identity in the subsequent polishing, branding and inspection operations. Samples for chemical testing are usually taken at the end to ensure detecting any deleterious effects of these subsequent operations. Coating solution preparation is included in the validation protocols when the solution is dedicated to a single product. Validation for solutions prepared for multiple batches with extended hold times may have a separate protocol. Solutions containing active ingredients are validated following the sampling and testing scheme of liquids and semi-solid drug products. Press-coated tablets are treated the same as uncoated tablets or cores. Although it is possible to obtain a core sample from the "Drycota" presses, the testing should be performed on the coated tablet, unless the core and the coating have different active ingredients.

One big red flag for FDA investigators is seal coating with shellac. Older processes that use this material often do not specify the amount or the exact process in the filing. In these instances, the coating process should be optimized to eliminate any trace of a dissolution problem that seems to plague the use of this material.

42.6.3 Polishing

The application of wax in separate coating pans (often canvas lined) was associated with the high volume manual sugar coating operations of the past. The main concern is cleaning validation for these separate pan operations. The polishing of coated tablets should be considered as part of the coating operation and be performed in the same pans where possible. The quantity of wax applied should be monitored, either to the pans or directly on the rotating tablet bed. The application of print base at this stage should be controlled as in coating operations, in that quantities added must be documented.

Polishing dry filled capsules to remove surface granulation with salt, or salt and polysorbate mixtures, has largely been replaced with buffing machines and was discussed earlier in the "Accessory Equipment" section (42.5.21).

42.6.4 Printing

Etched imprinting rollers should be controlled similar to press tooling. Printing specifications are often subjective, and standards should be prepared for use during validation and subsequent production. Additional validation sampling for print evaluation is the usual validation approach. Samples for coating are normally obtained following printing since it is often the final processing operation for coated tablets prior to packaging. Inks should be treated as a raw material, with expiration date and quantities used recorded.

42.6.4 Inspection

The improvement in reproducibility and quality of film-coating operations coupled with product isolation within manufacturing areas has reduced or eliminated the need for manual visual inspection. However, the elimination of inspectors must be accompanied with procedures that ensure process isolation and minimize the risk of mix-up. Since empty hard dry filled capsules are often purchased or manufactured at a different facility, empty capsule inspection is still performed to ensure that no foreign capsules are present. For inspection operations, the so-called 200% inspection is employed. This is essentially two people watching the product go by on an inspection belt. The speed and density of the product are important parameters. The inspection belts with cavities to control the distribution of the product on the belt are best. Visual inspection is largely being replaced by computerized vision systems, which are reproducible and do not blink, become inattentive or fall asleep.

42.6.5 Packaging

Most packaging operations for tablets and capsules, other than label accountability, had been viewed as "grandfathered" in the respect that it is assumed they do not affect the product. However, packaging validation has become an area of focus for FDA field inspections because of labeling mix-ups. Roll labels with strict accountability are required. The current focus is beyond stability and labeling issues and consists of documentation of machine setup parameters, line speeds and operating procedures.

Blister packing machines may potentially stress tablets and capsules, since the blisters are formed by the application of heat and pressure. It can be easily demonstrated with stability studies and visual inspection that the tablets and capsules are not subjected to these stresses. Emphasis is also placed on seal integrity in terms of its effect on product shelf life and stability. Validation takes the form of increased in-process seal integrity testing in conjunction with stability testing. Line clearance during stoppages for product in heat-sealing stations must be addressed to ensure any potentially heat-stressed product is discarded. Additional detail can be found in the chapter on packaging validation within this volume.

42.6.6 Stability Data

Stability protocol preparation, sampling, and testing are usually performed by different groups than those who perform the qualification and validation studies. Often different laboratories actually perform the stability testing. Stability is considered part of validation in the broad sense of the definition. Initial product launches typically involve sample packaging. It is important to ensure that all package presentations are included in stability protocols. It is a good practice to put the process performance qualification batches (usually three) on stability. Stability data is supportive of packaging process qualification but is seldom a requirement for Stage 2 closure because of its continuous nature and is usually addressed in Stage 1 and Stage 3. Stage 3 reports should include a review of stability data.

42.6.7 Shipping Validation

Prepared packages containing temperature recording devices are staged in shipping areas for worst case periods and then shipped by typical carriers to worst case locations (e.g., Alaska during the winter and Arizona during the summer). The temperature (and sometimes humidity) histories for several shipments are combined to provide time–temperature profiles for stability chambers. The various product/primary container combinations are then placed on a shipping stability program. These studies are best considered as a qualification for a specific type of tertiary packaging so that study size and duration have some limit. Extensive use of matrixes (product family groups) and worst case environmentally vulnerable product provide the bases for initial validation.

42.7 CONCLUSION

The author has attempted in the preceding chapter to present the approach to validation of solid dosage forms and relate some areas of potential concern that should be addressed within protocols. Use of the new Guidance terminology and requirements was applied to preexisting techniques and approaches. A process validation consisting of a minimum of three batches with all prerequisites in place for all doses has been replaced with the "Lifetime" and risk-based approach.[3] Sampling and testing equivalent to three to six times routine in-process and final release testing is still usually employed along with supporting statistics. Use of the Bergum methods[21] for content uniformity and dissolution should satisfy most statistical requirements for PPQ Acceptance criteria are generally equivalent to tightened in-house limits.

NOTES

1 Guidance for Industry Process Validation: General Principles and Practices. www.fda.gov/downloads/Drugs/. . ./Guidances/UCM070336.pdf last accessed 1/7/2019.

2 The United States Pharmacopeia 34th Revision, (905) Uniformity of Dosage Units, The United States Pharmacopoeial Convention, Inc. Rockville MD, 2016: 736.

3 Guidance for Industry, Q4B Evaluation and Recommendation of Pharmacopoeial Texts for Use in the ICH Regions, Annex 6, Uniformity of Dosage Units General Chapter. www.fda.gov/downloads/Drugs/GuidanceComplianceRegulatoryInformation/Guidances/ucm085364.pdf Last accessed 1/7/2019.

4 United States District Court For the District of New Jersey, Wolin, District Judge. United States of America, Plaintiff, v. Barr Laboratories, Inc., et al., Defendants. Civil Action No. 92–1744 Opinion. 1993, www.gmp-compliance.org/guidemgr/files/1-23-1-WOLIN_JUDGEMENT.PDF Last accessed 1/9/2019.

5 Bigelow WD, Bohart GS, Richardson AC, Ball CO, Heat Penetration in Processing Canned Foods. *National Canners' Association*. Bulletin 16-L 1920.

6 Byers TE, GMPs and Design for Quality. *Journal of the Perenteral Drug Association* 1978; 32: 22–25.

7 Agalloco JP. The Validation Life Cycle. *Journal of Parenteral Science and Technology* 1993; 47: 142.

8 Box GEP, Hunter WG, Hunter JS, *Statistics for Experimenters: An Introduction to Design, Data Analysis, and Model Building*. New York: John Wiley &Sons, 1978, 556–563.

9 Lindboe Jr WG. Personal Observation, 2000

10 Microsoft Developers Network. *COM: Component Object Model Technologies*. https://docs.microsoft.com/en-us/windows/desktop/com/component-object-model-com-portal last accessed 1/7/2019.

11 The United States Pharmacopeia 43 Revision, (1225) Validation of Compendial Methods, The United States Pharmacopial Convention, Inc. Rockville MD, 2016.

12 Hicks CR, *Fundamental Concepts in the Design of Experiments*, 4th edition. New York: Saunders College Publishing, 1993.

13 *General Principles of Software Validation; Final Guidance for Industry and FDA Staff*. Rockville, MD: US Food and Drug Administration, 2002.

14 Pharmaceutical cGMPS for the 21st Century—A Risk-Based Approach: Final Report- Fall 2004. Department of Health and Human Resources, US Food and Drug Administration, 2004.

15 Palady P. FMEA—Author's Edition, Second Edition, PAL Publishing, 1997.

16 Anonymous, www.6sigma.us/etc/what-is-ishikawa-fishbone-diagram/ Last accessed 1/7/2019.

17 Kepner C, Tregoe B, *The New Rational Manager*, 1st edn. Princeton, NJ: Research Press, 1981.

18 The United States Pharmacopeia 43 Revision, (1225) Validation of Compendial Methods, The United States Pharmacopeial Convention, Inc. Rockville MD, 2019.

19 Agalloco JP, Non-Sterile Is Non-Sterile: A Reality Check on Microbial Control Expectations. *Pharmaceutical Technology* 2016; 2016 eBook(2): s31–s35.

20 Labuza TP, *Fundamentals of Water Activity Short Course*. Institute of Food Technologists, 2002.

21 Title 40 Code of Federal Regulations, Chapter 1, Subchapter D, part 141. 2016 www.ecfr.gov/cgi-bin/text-idx?SID=beca577f5dea458c00be8c9aa138cdfd&mc=true&node=se40.23.141_121&rgn=div8 Last accessed 1/7/2019.

22 Annonomous. Blend Uniformity Analysis: Validation and In-Process Testing. Technical Monograph No. 25, PDA, Bethesda, MD, 1997.

23 FDA Guidance for industry, Powder Blends and Finished Dosage Units—In-process Bend and Dosage Unit Inspection (Sampling and Evaluation) for Content Uniformity, Revised draft guidance, January 2004; http://pqri.org/wp-content/uploads/2015/12/FDADraftGuide.pdf last accessed 1/7/2019.

24 Guidance for Industry ANDAs: Blend Uniformity Analysis http://pqri.org/wp-content/uploads/2015/12/FDADraftGuide.pdf last accessed 1/7/2019.

25 Bergum J, Brown W, Clark J, Parks T, Garcia T, Prescott J, Hoiberg C, Patel S, Tejwani R, Content Uniformity Discussions: Current USP <905> Developments Regarding <905> and a Comparison of Two Relevant Statistical Approaches. *Pharmaceutical Engineering* July/August 2015; 35(4).

26 Liss E, Conway S, Zega J, Glasser B, Segregation of powders during Gravity Flow through Vertical Pipes. *Pharmaceutical Technology* February 2004: 78–96.

27 www.nytimes.com/1993/08/17/business/company-news-warner-lambert-to-halt-most-drug-production.html last accessed 9/5/2020.

28 Lieberman H, Lachman L, Schwartz J, *Pharmaceutical Dosage Forms: Tablets—Volume 2*, 2nd edn, 1990: 40.

29 Bajaj J, Brennan Jr. AK, Herzog S, et al. Tips for Improving Mixture Sampling. *Powder and Bulk Engineering*, Jan. 1993: 41.

30 Bergum JS, Li H, Acceptance Limits for the New ICH USP 29 Content-Uniformity Test. *Pharm. Tech.* 2007; 31: 90–100.

31 www.pharmtech.com/sample-content-uniformity-acceptance-limit-tables Last accessed 1/7/2019.

32 Bergum JS, Constructing Acceptance Limits for Multiple Stage Tests. *Drug Dev. Ind. Pharm.* 1990; 16 (14): 2153–2166.

33 Cholayudth P, Using the Bergum Method and MS Excel to Determine the Probability of Passing the USP Dissolution. *Test Pharmaceutical Technology* 2006;30(1): 86–98. www.pharmtech.com/print/238576?page=full&date=&id=&pageID=2&sk= accessed 1/7/2019.

34 Tetzlaff RF, Validation Issues for New Drug Development: Part I. *Review of Current FDA Policies, 1992, Pharmaceutical Technology* 16 (9): 44–56.

35 Avallone H, *Development/Scale-up of Pharmaceuticals.* Presented at FDA's NDA/ANDA Inspection Program, Dec. 12, 1990. College of Pharmacy, Division of Continuing Education, Rutgers, the State University of New Jersey.

36 www.sciencedirect.com/science/article/pii/S2211383513000762 Last accessed 9/5/2020.

37 Walker J, Abdulsalam A, Theobald AE, et al. Multiple-Scored Tablets: Weight and Content Uniformity of Subdivisions and the Distribution of Active Constituent within and between Tablets. *J. Pharm. Pharmac.* 1978; 30: 401–406.

38 www.fda.gov/media/81626/download last accessed 9/5/2020.

39 www.mathworks.com/matlabcentral/answers/46134-making-cumulative-difference-calculation Last accessed 1/7/2019.

40 Wald A, https://en.wikipedia.org/wiki/Sequential_probability_ratio_test Last accessed 1/7/2019.

41 Simms LL, *Process Control.* Arden House, 1980, "Specifications, Controls and Validation", Arnold and Marie Schwartz College of Pharmacy and Health Sciences Long Island University; The Industrial Pharmaceutical Technology Section, Academy of Pharmaceutical Sciences. 1980.

43 Validation of Oral/Topical Liquids and Semi-Solids

William G. Lindboe Jr.

CONTENTS

DOI: 10.1201/9781003163138-43

43.1 INTRODUCTION

The 2011 FDA guidance publication on process validation general principles and practices ("the Guidance") changed the validation nomenclature that had evolved in the industry over the last 30 years.[1] The approaches outlined assume an ethical pharmaceutical company and a drug developed by its innovator with deep research and development resources. Ignored was a large group in which generic and/or over-the-counter products are acquired with little or no development knowledge transfer and now must be validated in their changed venue. The good news was that the underlying activity required for validation had not changed. The scope of this chapter will be the compliance requirement for the validation of manufacturing processes for oral liquids and topical semi-solid pharmaceutical dosage forms. The FDA Guidance divides process validation into three stages: Stage 1, titled Process Design, Stage 2, titled Process Qualification (PQ), and Stage 3 titled Continued Process Verification. Stage 2 consists of equipment and facilities design and qualification (DQ, IQ and OQ) and Process Performance Qualification (PPQ). This latter term was formerly called performance qualification or process qualification and will be the focus of this chapter. PPQ begins with the batching of approved raw materials to the storage and shipping of packaged product. The chapter will be directed toward the individuals charged with the responsibility of preparing and executing validation protocols for these dosage forms. The variety of formulations that fall within the scope of this chapter will be surveyed, along with associated manufacturing equipment. The assumption is that Stage 1 design and development are essentially complete. However, a few common process problems will be discussed with the hope that the reader will be able to anticipate pitfalls and eliminate subsequent validation difficulties for those products with little or no development knowledge transfer.

Manufacturing validation of non-sterile liquids and topical semi-solids has been considered an afterthought following the implementation of parenteral and solid dosage validation. Although there is a concern for product bioburden that exceeds that of solids, the dosage forms within this category typically have adequate chemical preservative systems and are inherently low-risk products. That being stated, the validation of these products is anything but easy. The products typically consist of multiple components that are either hydrophilic or hydrophobic depending upon their route of administration (orally or topically) and often require numerous processing steps to achieve the desired final preparation. Additionally, the areas for manufacturing contain some basic equipment that is applicable to many different types of liquid and semi-solid products. Considerable development or similar product experience is ordinarily required to arrive at the optimal equipment parameter settings for each product. Many of the products are over-the-counter drugs that rarely receive the developmental attention of high value ethical pharmaceuticals. Additionally, these products are frequently exchanged between pharmaceutical manufactures, which can introduce subtle changes in equipment and facilities and renders the original developers and their accumulated knowledge unavailable to validation personnel. Another common practice is outsourcing of the manufacturing to a third party or private label manufacturer. This can further limit development efforts and alter manufacturing equipment options. As mentioned earlier, in these instances, the development, operating and validation personnel inherit a process and product with relatively limited information. Given these difficulties, the author has had considerable success in defending validation documentation of these dosage forms based upon the approaches that will be presented.

43.2 HISTORY OF ORAL/TOPICAL LIQUIDS AND SEMI-SOLIDS VALIDATION

The initial motivation for the Federal Food Drug and Cosmetic Act ("the Act") can be attributed to the dosage forms discussed in this chapter. The so-called "snake oil" and other sham preparations of the nineteenth-century charlatans were commonly oral or topical preparations.[2] Although in early incidents efficacy, rather than safety and manufacturing, was the primary concern, later refinements to the Act pertaining to safety were brought about by the sulfanilamide elixir adulteration, where the formulator used ethylene glycol as a solvent for the drug, with disastrous consequences.

Other than mislabeling, manufacturing of these dosage forms has not been a major concern of the regulators. These are the dosage forms that many retail pharmacists still prepare within the pharmacy. Many pediatricians prepare and dispense the final preparation of amoxicillin for oral suspension directly to their patients. There have not been any major manufacturing problems with these dosage forms that lead to public safety concerns. The development of versatile and highly effective manufacturing equipment, which will be discussed below, has aided this manufacturing experience. The product formulations often have diverse rheological properties, and in some cases difficult-to-achieve emulsions or uniform suspensions must be created. Originally, equipment advances related to the unfilled and bulk preparation and not the final packaged product. The development of sophisticated metered dose aerosols for oral inhalation and form-fill and seal-unit dose blister packages have added additional complexity and requirements to the filling and packaging. Packaging manufacturing personnel, who previously had only dealt with rugged and highly dependable Cozzoli and Arenco fillers, for liquids and semi-solids, respectively, now had to deal with novel sophisticated and unproven filling technologies. In some cases, manufacturing firms were not up to the task, as evidenced by a consent decree that was partially based upon improperly filled metered dose albuterol for oral inhalation.[3] The critical lifesaving application of this drug brought increased regulatory focus to non-sterile liquids and semi-solids and their validation. Although water-based inhalation drugs are sterile preparations and must meet the USP Sterility Test <71>, it was the non-sterile aspects of their manufacture that were problematic. These unit operations are covered within this chapter, while sterility is addressed in other chapters within this volume.

43.3 DOSAGE FORMS—SAMPLING, CPPS AND CQAS

Further discussion of validation terminology, other than simple definition, will be omitted because of overlap with other chapters. The reader is advised to consult these chapters for a more detailed discussion, along with the FDA Guidance. The acronym CPP (critical process parameter) pertains to machine- and product-independent variables (settings or properties) that affect end product quality. CQA (critical quality attribute) generally pertains to in-process and final product-dependent variables (test results) that affect product efficacy, purity, and/or safety. It is important that these parameters and attributes are developed with proper tolerances in Stage 1 (DQ) and within the equipment and facility qualification (IQ and OQ) prior to PPQ under a formal protocol. Design of experiments (DOE) can be applied at this stage to determine CPPs. As an example, evaluation of semi-solid rheology and characterization by an appropriate constitutive model using a Rheometrics Mechanical Spectrometer (Rheometrics Co., Piscataway, New Jersey) in a Stage 1 DQ is a dream instrument that is unavailable to most manufacturers and research groups within this dosage form category. However, the CPPs are usually easily identified, and experience with similar products can aid in their identification. Processing experience can fill the knowledge gaps within Stage 1 and is permitted in the Guidance. It is well known that rheology is affected by temperature, pressure and the manner in which the pressure is applied to the material. Not all semi-solids can be approximated by a Newtonian fluid. Many exhibit thixoplastic behavior, indicating a change in structure with the application of pressure. Pseudo-plastic behavior is also observed, and such properties may affect mixing, homogenization and uniformity when manufacturing processes are scaled up. It is not unusual for product batches to be heated by mechanical mixing. The author recalls one premix for a zinc oxide diaper rash cream that was heated solely by the action of a rotating mixing plate. All characteristics unique to these formulations must be considered in the preparation of validation sampling plans and documentation. A common manufacturing practice is to prepare a semi-solid "base," which is a placebo carrier to which a variety of active ingredients (alone or in combination) can be added for the purpose of minimizing development effort and providing a "family" of similar products. A large manufacturer may have several cream, lotion, and /or ointment bases that may differ only in viscosity, to which different amounts and types of API (active pharmaceutical ingredient) are added for the final preparation. When API percentages are low, extensive design of experiments (DOE) and its statistical analysis, as recommended by the Guidance, can be performed on the base without the API. The common and critical quality attribute of microbial content or bioburden can be assumed applicable to all liquids and semi-solids. As such, most topical and oral formulations include a chemical preservative system, unless the active ingredient itself happens to be bactericidal. The additional microbiological concerns of non-sterile liquids and semi-solids over solid dosage forms are based upon water content or water activity.[4] It is well known that microbial concerns increase with elevated water activity.[5] A typical aerobic plate count limit for these products is 100 colony-forming units per gram of material (CFU), corresponding roughly to the USP limit for purified water. Non-sterile liquids and semi-solids usually have a zero tolerance for pathogenic organisms, namely gram-negative bacteria. Preservative systems are intended to compensate for the natural variation in nonpathogenic bioburden and extended use of multi-dose primary packages by the consumer, not to cleanse the product of pathogens. Processes and raw material standards must be designed and have sufficient controls in place to preclude these dangerous organisms from being present, as discussed in USP <1115>.[6]

Non-sterile liquids and semi-solids require the use of USP Purified Water. The nature of the smaller OTC manufactures gives rise to water quality concerns in that the level of water system maintenance may be inadequate to address microbial considerations.[7] These firms may not have the resources to upgrade water systems, and many legacy systems contain some plastic piping. These types of systems require chemical sanitization because they cannot tolerate the heat typically necessary to sanitize the system. For older legacy systems, there is also a risk of periodic contamination caused by biofilms. Plastic pipe water systems should be equipped with modern chemical sterilization methods, such as ozone. Plastic system components should be replaced at specified intervals, and this cost and disruption should be considered when initially specifying these systems. Additionally, source water can be surface, well, or mixed, and chlorine content is seasonally variable; thus, bioburden is also variable, which adds to legacy system problems. Detailed content on water system design, operation, and validation is provided elsewhere in this volume.

Another common problem is the difficulty in obtaining representative samples of the bulk premix(es), as well as the final mixed bulk for suspension, emulsion, and highly viscous products. The difficulty increases with material viscosity and the manufacturing technique of "geometric dilution" that is widely utilized out of necessity when trying to disperse APIs in a standard, previously made base. Considerations for achievement of homogeneity range from the assumed, for aqueous liquids under agitation, to the extremely difficult, exceeding solid dosage in difficulty, for ointments, pastes, and adhesives. The act of sampling can bias results and adversely affect product quality when too much of a concentrate is removed for testing. Large numbers of samples for statistical analysis should be reserved for Stage 1 and the final package to be delivered to the patient. The following is an overview of the wide variant of dosage forms in this category, along with validation concerns. Additional description of the following discussion on these dosage forms may be found in the USP.[8]

43.3.1 Solutions—Oral, Aerosol for Inhalation

Solutions are drug active pharmaceutical ingredient(s) dissolved molecularly in a solvent system. Elixirs contain some alcohol in the solvent system and are for oral administration. Tinctures are alcohol based and are used for topical administration. Syrups, intended for oral administration, contain sugars or artificial sweeteners. The nature of properly mixed low viscosity solutions provides assurance they are uniform. Validation concerns focus on assurance of the adequacy of the mixing and bioburden. Manufacturing processes may involve premixes or "side pots" to facilitate dissolving solids. The premix is then diluted/added into a final bulk mixture that is tested and held prior to filling and packaging. Typical CPPs for the premix are evidence of dissolution (particle size) and refractive index. Excessive sampling of the premix can affect the final concentration of the bulk batch. The number and size of the samples must be considered in protocol preparation. The final mix may be sampled at any point, given the assumption of homogeneity. Syringe-based and bottle-based sampling thieves (Figure 43.1) have been developed by GlobePharma, New Brunswick, New Jersey (www.globepharma.com) to facilitate location-based sampling of low viscosity materials within final mix vessels and/or bulk holding tanks. Sample sizes for uniformity should follow the solid dosage blend edict of no more than a three-unit dose size, when possible. It is not uncommon to sample solutions from a sampling port intermittently while they are agitated. In this case, subdivision to the assay quantity must be performed in the laboratory. Primary packages are easily obtained during the course of filling and packaging operations. It is important to obtain the very first and very last containers filled to assure that these historic "problem areas" are included in the validation testing. All of the filling nozzles should be represented in sampling and testing. Bulk manufacturing CPPs include mixing speed, configuration of mixing blade, position of mixing blade, tank and /or kettle volume and geometry, pumping mechanisms and rates, temperature, pressures and pipe or hose diameters. All of these should have appropriate definition/specification and be properly measured and documented during validation. Filling should fully specify the equipment, especially the dispensing- and container-sealing systems, along with other mechanical settings. The presence of mixing and/or agitation of the bulk liquid within the filling system should be documented. Form-fill and seal-unit dose packages will require the documentation of the many parameters associated with the packaging equipment. Container crimping/capping parameters and container closure integrity should also be evaluated.

CQAs for the final mixed bulk and the filled product include drug assay(s), preservative efficacy, pH, viscosity, density, and/or specific gravity. Fill volume for multidose containers or dose uniformity for unit dose containers must be included. Filled solutions used as aerosols for inhalation require rigorous testing of valves, actuators, and containers prior to their receipt on the production floor. Filled aerosols and foams are two-phase systems consisting of the concentrate (bulk API containing solution) and the propellant. The propellant raw material testing and release must be documented. "Gas house" filling of propellants for aerosols must also be documented in detail. The parameters associated with the addition of the propellant and the pressure at which it is added must be included. The development of the pressurized metered dose inhaler ("pMDI") has gone through a transformation necessitated by the phase-out of CFC propellants for environmental reasons.[9] The accuracy and reproducibility of metered dose valves must be tested for the claimed number of doses within a container. A physical verification of the absence of unfilled and underfilled containers for critical drugs such as albuterol is expected given the availability of automated high-speed individual primary package weighing equipment

FIGURE 43.1 Liquid sampling thief.

Source: GlobePharma, Inc., New Brunswick, New Jersey.

from Mettler-Toledo (http://us.mt.com/us/en/home/products/Product-Inspection_1/checkweighing.html) and others.

43.3.2 Suspensions

Suspensions are liquids or semi-solids that have small solid particles dispersed within them. Some suspensions are prepared by the pharmacist and/or dispensing physician immediately prior to administration. These formulations are covered by the chapter on solid dosage validation within this volume. Suspensions require at least one premix—usually multiple premixes—where the solids are dispersed within the liquid using various dispersing and homogenization equipment that will be discussed below. CPPs for the premixed material are particle size of dispersed solids, density, temperature, and viscosity. Equipment CPPs include rotor and stator configuration, gap, rpm, pressure valve opening size, and location within vessels of the dispersing equipment. Final mix parameters include assay, viscosity, density and/or specific gravity, particle size, pH, and absence of entrained air. The greatest difficulty with suspension manufacture is typically the filling process, which often proves most problematic, especially for smaller fill volumes. The author has found it difficult to maintain suspension uniformity in filling hoppers. It is often necessary to require recycle lines from the filling machine back to the bulk holding vessel to prevent segregation during even brief interruptions in the filling process. The use of intermediate surge vessels outside the recycle loop was inadequate to assure content uniformity in the filled units. In one extreme case, recirculation was necessary directly above the filling needles to maintain uniformity. Consider that while the entire batch of suspension is certainly uniform if taken as a whole, the difficulty is assuring that all aliquots of that vessel (down to unit dose containers in some instances) are equally uniform with regard to ratio between the solid and liquid phases. Interruptions in filling caused by equipment failure and operator breaks must be considered in the sampling plan, as these are known problem areas. Testing of start and end of batch filling samples is required for the same reasons.

43.3.3 Emulsions

There are two types of emulsions, the oil-in-water and the water-in-oil. In both cases, the former is dispersed as small droplets in the latter or continuous phase. Usually, emulsifying agents are employed to keep the dispersed droplets from merging. This would lead to nonuniformity, since API presence is usually restricted to a single phase. Care is needed to ensure that the emulsion remains intact once achieved. Emulsions can be destroyed mechanically, thermally or chemically. Validation is concerned with equipment CPPs needed to create the emulsion in the first place and then preserve it through holding, filling, packaging, and shipping. These are temperature, disperser type and configuration, rotor–stator gap, homogenizer valve(s) opening, rpm, vessel configuration, pump speeds, pressure, hose diameter, and materials of construction. Product CPPs are viscosity, density/specific gravity, particle size, air entrainment, pH, and assay. CQAs are essentially the same as the previously discussed formulations.

43.3.4 Foams—Topical, Vaginal

Foams are similar to aerosols in that there is a concentrate and a propellant in the final primary package. The concentrate is typically a liquid or emulsion with a dissolved or suspended pharmaceutical active ingredient that combines with the propellant within the actuator to produce the foam. Garg et al. provides a comprehensive review of vaginal formulation excipients that will enable the selection of appropriate CPPs for inclusion in validation protocols, besides API assay(s).[10] Bioburden, preservative assay, and effectiveness, concentrate viscosity, pH, density, or specific gravity, and CPPs and CQAs related to the primary package delivery system are typical concerns.

43.3.5 Lotions—Topical

Lotions can be emulsions, suspensions or gels, and "lotion" is a general term associated with a low viscosity topical dosage form. Lotions frequently have the phrase "shake well" on their primary packaging and may have multiple active ingredients, such as a sunscreen. This combines to make lotions difficult to sample as both in-process and final mixed bulk. It is usually necessary to sample the material while under agitation or mixing during manufacture. The author recalls witnessing one lotion that immediately developed an intermittent "oil slick" on the top surface when the mixing was stopped. This product was to be filled into unit dose form-fill and seal sample packettes, and the delivered dose from this primary package was highly variable, despite having only a single active ingredient. Critical CQAs are assay, viscosity, and homogeneity.

43.3.6 Cream—Topical, Vaginal

Creams are emulsions consisting of oils dispersed in aqueous medium. The active can be in either the oil or the water phase. Alternatively, a non-active cream base may contain a suspended solid. In either case, the physical characteristics of the cream can govern the quality of the final preparation. The cream base CPPs and CQAs are generally restricted to the physical parameters of viscosity, specific gravity or density, and homogeneity. Final formulated bulk adds API assay(s), bioburden, preservative assay, and efficacy to the cream base CPPs and CQAs.

43.3.7 Ointment—Oral, Topical

Ointments can be an emulsion with an aqueous liquid dispersed in an oil phase. The oil phase can be mineral oil based and/or petrolatum based depending on the desired viscosity. Synthetic oils are also utilized. An ointment base is typically manufactured prior to the addition of active ingredient(s). They can also be single-phase nonaqueous suspensions, which would be just like an aqueous suspension except highly viscous. These are often heated to lower viscosity for filling.

The CPPs and CQAs are essential the same as creams. The high viscosity of ointments necessitates multistep formulation processes and exotic mixing and dispersing operations despite using a standard ointment base. It is important to specify the scale of scrutiny (how large a sample) when sampling and testing the intermediate manufacturing steps. Often, acceptance based on simple physical measurements of viscosity and dimensionless groups, arising from machine parameters and developed during process scale-up, is used.[11]

43.3.8 Pastes—Oral, Dental

Pastes are emulsions with suspended solids, often of very high viscosity, and usually do not have a "base" in which APIs are added. They are typically high volume–large batch size products that utilize large combination equipment (discussed below) to arrive at the final formulation. A common problem is the achievement of the emulsion, even though the API may be uniformly suspended. The inclusion of suspended insoluble abrasive solids in dental formulations increases the difficulty in determining uniformity prior to sampling the primary package. The APIs in dental formulations (fluorides) are typically bactericidal, and preservatives may or may not be present. Where they are absent, it is not unusual to have procedures specifying a 24-hour hold time prior to testing for bioburden to allow the active to reduce the microbial population in the final formulation. This is not a desirable practice and can be eliminated with proper raw material testing, process design, and vendor qualification.[12]

43.3.9 Gels

Gels are similar to ointments and pastes except they may exhibit solid-like behavior. They generally change to semi-solid or liquid with the application of heat and/ or pressure. Frequently, there is a hold time required for the gel to develop and gain the solid-like properties. Gels are suspensions with a higher viscosity caused by the interpenetration of the solids by the liquid. As in pastes, there is typically no standard base that is prepared prior to the addition of API(s). CPPs and CQAs include assay, specific gravity or density, and ability to be extruded from a multidose container through patient peristaltic manipulation (squeezing). An aqueous suspension of bentonite clay is widely used as a simulant for gels in development and equipment qualification where product cost is prohibitive to the required process development. A recent CQA of interest is gel strength.[13] Soft gelatin used for liquid filled capsules falls within this category.

43.3.10 Suppositories—Vaginal, Rectal

Suppositories are lipids, either natural or synthetic, with dissolved and/or suspended API(s). The lipid must have the characteristic of "melting" at body temperature so that the drug(s) can be delivered to the patient. The final bulk is heated and filled into chilled molds; today, typically, form-fill and seal-molding machinery is used. A common rectal suppository base is cocoa butter (theobroma oil). This is a natural product and a by-product of the confection industry, and it is common to have the heated liquid filtered through "cheese cloth" as the first manufacturing step. The CPPs of this operation, such as temperature and time of heating, are important, as is the ability to be molded. CQAs are assay, uniformity, and molded shape.

43.3.11 Adhesives—Transdermal, Ostomy, Denture

Adhesives are difficult to manufacture and substantially more difficult to clean. Equipment is usually dedicated. Cleaning must be considered, because a nonaqueous solvent such as mineral oil is typically used between batches of the same product, and there will typically be a permitted residue on equipment surfaces following multiple rinses. This residue should be estimated and included within the performance qualification protocol.

Transdermal adhesives may contain API, and the CPPs of drug diffusion and adhesion are of obvious concern. Tack, adhesion, release force, and cohesive strength are also critical.[14] There are various in vivo diffusion devices to measure drug flux.[15] CPPs of filling, extruding, and rolling equipment are critical, as the thickness of the adhesive is proportional to the delivered dose.

Ostomy and denture adhesives are medical devices included within this chapter because their manufacturing process utilizes typical pharmaceutical processes. With these products, machine variables and product physical characteristics are critical. As in transdermal adhesives, tack, adhesion, release force, and cohesive strength are CQAs. Extrudability is also included for denture adhesive filled in multidose tubes.

43.4 EQUIPMENT—PERFORMANCE QUALIFICATION AND CPPS

The equipment utilized for the manufacture of non-sterile liquids and semi-solids is just as diverse as the variety of dosage forms. Processes range from a simple dissolution of solids in an aqueous solution to multiple-step mixing, homogenization, filtration, milling, dispersion, and extrusion. Filling and packaging similarly range from a simple liquid fill, to difficult-to-maintain suspensions, to complex aerosol and form-fill and seal operations. Each piece of equipment will have its own unique set of parameters that must be controlled and documented for qualification/validation. Common CPPs are mixing speed (rpm), time, and temperature of any heating. It can be difficult to measure RPM using an optical tachometer on 316 SS surfaces. Special reflective tape can be used for these measurements. In many cases, for semi-solids, tachometers cannot be used because the mixing speed is too slow. In these instances, a calibrated stopwatch can be used to measure mixing speed. The following subsections describe the major equipment types currently in use for pharmaceutical manufacture of these products and most likely to confront the qualification/validation team.

43.4.1 Tanks

Tanks are primarily used to manufacture and hold liquids and low viscosity semi-solids. They are typically fabricated out of 316 SS or 316L SS, and this must be verified during equipment qualification. Companies that manufacture food and/or cosmetics in addition to pharmaceuticals may have a large variety of tanks, and it is important to maintain equipment history. In one instance, the author recalls the proposed use of portable tanks for a purchased diaper rash ointment. The tanks in question had been used previously for a lice treatment shampoo, the active of which is classified as an insecticide. After consideration of the implications, a different set of vessels were selected for the ointment. Tanks are usually constructed with a conical or curved bottom to facilitate gravity draining. They can be either fixed in place with permanent connection or portable, which may or may not be on wheels. In some cases, vessels will have a permanent in situ mixer with a defined agitator shape. For smaller and more portable vessels, the agitator motor and/or impeller is changeable, and it is important to document the specific motor and/or impeller used during the production and in the validation, as well. Each agitator motor and impeller should have its own identification markings so that performance can be related to equipment qualification studies. Depending upon the application, the tank may be equipped with a jacket for heating/cooling of the vessel contents.

Some extremely viscous intermediates and products may be stored in flat-bottomed vessels without a bottom outlet; product is removed using a mechanical compression system applied to the top of the material that descends as the product is dispensed.

43.4.2 Kettles

Kettles are essentially tanks intended for use with the more viscous and difficult-to-agitate materials. Kettles often have the ability to be heated, although this feature is not always utilized. Most have permanent mixers attached and range in size from essentially bench-top units to upward of 1,000 liters in capacity. The agitators on kettles will have an internal side-scraping feature that ensures viscous materials and solids are prevented from adhering to the sidewalls during ingredient addition and mixing. An inspection of these polymeric scrapers is necessary before and after each kettle use, as they have a tendency to wear. The kettles may also have covers to prevent excessive evaporation and allow vigorous mixing without product loss. Many kettles are designed with a primary and secondary mixing action, with independent agitators controlled by a dedicated motor drive. Lee Industries (www.leeind.com) manufactures a wide variety of kettles (and most other equipment categories discussed in this chapter) with different agitator designs/combinations (Figure 43.2) along with other processing equipment. In most cases, the manufacturer of the equipment can provide insight into equipment design and operation that will facilitate protocol preparation. Sampling of viscous materials is difficult and most easily performed during discharge. Worst-case location sampling can be performed with a sampling thief.

43.4.3 Mixers

Mixers are often associated with portable tanks and filling hoppers during primary package filling. They are generally used with liquids and the lower viscosity semi-solid products. The ubiquitous Lightnin Mixer (SPX Corporation, www.gowcb.com) will be found in virtually every liquid and semi-solid manufacturing facility. The impellers on these mixers are removable and tend to be handled roughly during cleaning. That, coupled with the longevity of this equipment, makes it essential to verify the impellers meet the original manufacturer's specifications. The location and angle of mixing must also be documented during validation. In some cases, the location of the impeller may affect mixing efficiency, especially of more viscous liquids, and affect dissolution times of solids. Most mixers, whether portable or fixed, have the ability to change the impeller or agitator configuration. These need to be identified and documented to prevent a change that might affect product quality. Mixing of high viscosity material requires slow agitation with the so-called "gate" impellers that gently move the material at low RPM, prevent the introduction of air, and facilitate air removal under vacuum. These mixers generally follow mixing with different equipment and/or impellers as a final mix.

43.4.4 Homogenizers—Dispersers

There are two principal types of homogenizers, the valve type and the rotor–stator type. The latter, while capable of achieving homogenization, is better classified as a disperser or colloid mill. The valve type usually has two stages, consisting of two valves in series, to prevent clustering of lipid globules after the first stage. The manufactures frequently encountered are the Cherry-Burrell and the older Gifford-Wood. The mechanism by which homogenization is achieved is not definitively known, although there are three prevailing theories, any one of which may dominate for a particular material and viscosity. These theories are based upon the generation of turbulent velocity arising from high pressure. The three mechanisms are as follows: shearing between globules, shattering of globules from impact with the valve surface, and the formation of pits or cavitations following passage through the valves resulting in the condensation of small vapor bubbles. There is an increase in viscosity following homogenization of oil-in-water formulations caused by the increase of surface area of the oil globules. This provides a convenient physical test to confirm the success of the operation. Microscopic examination is also often necessary to confirm homogeneity. Additional testing consists of particle sizing (Coulter Counter) and light scattering. Homogenization is a process where examination of the resultant product, especially upon standing, determines the adequacy of the machine settings. Manufacturing procedures may specify some variability in machine setting to accommodate for variations in input materials, especially natural products.

Lee R & D Tri-Mix

FIGURE 43.2 Versatile mixing kettle.

Source: Lee Industries, Inc., Philipsburg, Pennsylvania.

Typical first stage pressures are 2,000 psig, while second stage pressures are near 500 psig, with valve clearance around 500 microns. Positive displacement pumps are used and in many cases are an integral part of the homogenizers. These ranges are provided because validation personnel are often required to verify setting adequacy. In one instance, a homogenization process was transferred to another location after the sale of the particular product and equipment to another manufacturer. The original process consisted of eight homogenizers connected in parallel to accommodate the desired production volume. The new manufacturer hard-piped the homogenizers in series and then proceeded to apply the maximum pressure to both stages of all the homogenizers to try to achieve homogenization, which was never accomplished. When the author pointed out that the oil-water percentage was the same as ice cream mix and that appropriate homogenizer settings were widely available in the literature, the new manufacture still would not change the process and homogenizer configuration and attempted to validate the process, which included a 10% reject rate upon visual inspection.

The rotor–stator type has the critical CPPs of rotor speed and rotor–stator gap. Speeds are entirely variable, and gaps

FIGURE 43.3 Reversible homogenizer.

Source: Arde-Barenco, Inc., Norwood, New Jersey.

vary with the application. This type is used to disperse solids during premix operations and Arde-Barenco, Norwood, New Jersey, manufactures a versatile bidirectional unit (Figure 43.3) that is frequently encountered in manufacturing areas. The bidirectional feature serves to ensure that all the material passes through the rotor–stator, and the portability allows for use within tanks and/or kettles. This equipment can also be installed in-line, with recycle to the original vessel or with transfer to a second vessel.

These units are used with all types of nonsolution products to ensure greater dispersion of the solid phase in the liquid and can also be utilized to reduce the particle size of the solids (with the potential for heat generation).

Another form of dispersator is utilized to initially wet large quantities of poorly soluble solids for incorporation into a liquid base. These operate similar to a centrifugal pump but are termed "dispersers" by the manufacturer.

43.4.5 Combination Equipment

Some large capacity equipment can mix and homogenize in the same vessel. Koruma Kettles are an example of this type of equipment. In addition to large rotating mixing impellers, they have a high-speed disperser/homogenizer on the bottom of the kettle that can pump the material though a recirculation pipe to the top of the vessel. It is difficult to adjust the rotor–stator gap of the "Disho" dispersers on the bottom of the kettle, and a check of the effectiveness of premixing is usually necessary. These units are typically seen in high volume dental paste facilities.

43.4.6 Centrifuges

Centrifugation is a separatory process and is the opposite of homogenization. It is used as a preliminary operation to remove undesired components for further processing or discard. Products, including natural materials, frequently use centrifugation to reduce variability in physical properties. It has been well known that a mixture of materials of different densities will separate by the action of gravity. The disc–bowl type of centrifuge provides a centrifugal force to effect the separation, and its effectiveness increases with rotational speed. Another critical parameter is the separation between the discs, which is typically 0.020 inch to 0.050 inch (0.51 mm to 1.27 mm). Unlike the "art" of homogenization, centrifugation is modeled by the widely accepted Stokes equation, which provides the rate of separation as a function of material physical properties and centrifuge parameters. Frequently encountered production units are the Westphalia and Sharples (www.penwalt.com).[16] Forces in excess of 10,000 times the force of gravity are routinely used in production. A special type of centrifugal unit is the "Versator," which uses centrifugal force and vacuum to remove air introduced into both solutions and suspensions during earlier dissolution or mixing steps. The obvious CPP is rpm.

43.4.7 Pumps

Where gravity cannot be used, pumps are a necessity within a non-sterile liquid and semi-solid facility. Pumps can be positive displacement or not, with the former also used in metering and/or filling operations. Centrifugal pumps are used for lower viscosity materials, while lobe or peristaltic pumps are used for more viscous materials. Pumping rate is a CPP that is usually translated into some measurement of rotational velocity of the pump impeller or actuator. It is virtually impossible to directly measure the actual impeller speed. A careful analysis of the disassembled pump during qualification will enable the measurement of motor rpm to be translated into pump speed. Pumping rate can be estimated by pumping into a graduated vessel or a vessel that can be weighed. Materials of construction must be verified during qualification along with internal sealing. Waukesha pumps (Figure 43.4), SPX Corporation, are widely used in the chemical, food, cosmetic, and pharmaceutical industries, and it is important to ensure that the design of legacy equipment matches the current intended application. Specialized dispensing pumps have been successfully employed to precisely meter two or more solution streams and, when used with an in-line mixing system, can reduce tankage requirements by allowing mixing of a concentrate(s) with a separately stored diluent followed immediately by filling. More common in health and beauty

FIGURE 43.4 Waukesha pump.

Source: Waukesha Cherry-Burrell, Division of SPX Corporation, Delavan, Wisconsin.

aids, this process has been successfully utilized for large-volume OTC products.

43.4.8 Filters

Filters are used to remove undesired solids from a liquid or elevated temperature semi-solid. Cheesecloth is frequently used for OTC raw materials, and it is important that the type and quality of this material be specified and controlled. The cheesecloth and all filter materials, for that matter, should be tested and released as raw materials at least with a qualified vendor's Certificate of Analysis (COA). Ronnigen and US filter, along with sterile filtration suppliers Pall and Millipore, offer a variety of filters and filter media that can be used for clarification of liquids. In older processes, it is important to ensure that banned materials such as fiberglass and asbestos are not utilized. The inability to assure filter integrity by performing integrity testing on wider mesh semi-solid filters is reduced by the use of serial filters. The wide availability of filters for the food and parenteral manufacturing industries that are capable of integrity testing is covered in a separate chapter within this volume.

43.4.9 Fillers

Fillers are really an extension of the pump category, since they are a necessity to fill the primary package. Fill pumps are generally multiple small-scale positive displacement pumps similar to a syringe. Two manufacturers that can be found in many filling operations are the Cozzolli for liquids and the Arenco for semi-solids. These fillers are virtually indestructible, with the downside that old legacy equipment is frequently encountered in contemporary use. The simplicity of their operation and long history of success make qualification and validation relatively easy. Filling rates combined with extensive sampling of the filled tubes is generally sufficient for validation. Some materials may require a recirculation of the filling hopper during filling and/or mixing while filling (lotions) to maintain homogeneity. The rate and method of these operations must be documented. The author recalls one instance years ago, while observing a process during a validation trial, when in a moment of horrible comprehension, it was discovered that the unwritten practice of stirring the filling hopper with a wooden paddle was being routinely performed!

Peristaltic pumps can be used effectively for filling operations. It is important to document the type of tubing (usually USP medical) and the frequency of tubing change as part of the validation. The characteristics of the polymeric tubing will change with extended squeezing by the pumping mechanism and may affect the filling accuracy. To eliminate cleaning problems, all tubing should be single use. Dedication to a single product is not adequate to ensure the absence of cross-contamination or microbial build-up. An additional concern with reuse is the potential for the tubing material to degrade over time and slough particles into the product.

Sophisticated form-fill and seal equipment will still have the fundamental filling operation as one of its stages. Computerized control of the entire operation from web to final seal is standard. Separate and detailed qualification documents are necessary prior to process validation to ensure adequate validation. Additional detail on considerations for validation of filling/packaging operations can be found elsewhere in this volume.

43.5 ANALYTICAL TEST METHODS AND EQUIPMENT

The discussion of analytical test methods will be restricted to those that are frequently performed on the production floor as part of a validation trial or run. The reader is directed to the many widely available references on laboratory test methods and equipment pertaining to the quality control of pharmaceuticals as well as relevant chapters in this text.

43.5.1 pH Meters

The pH meter is usually a part of aqueous liquid manufacturing in-process quality checks. The meters are simple to use, and it is easy to make additional measurements for validation protocols—and therein lies the potential problem. Protocol designers often cannot resist placing specifications on statistics generated from pH data. It must be remembered that pH is the negative of the logarithm to the base 10 of the hydronium ion concentration. If concentration is normally distributed, then an exponential function of it, namely pH, will not be. The easy solution is to convert pH readings to concentration prior to generation of statistics and have the acceptance criteria reflect these statistics.

43.5.2 Hegman Gauge

One of the most versatile and useful in-process instruments is a device borrowed from the paint industry, the Hegman Gauge. This device is a graduated channel ground into a stainless steel plate where the material is spread along the channel by a stainless steel scraper. This device can detect nonhomogenized oil globules, non-dispersed solids, and air bubbles and enables one to estimate the size of these undesirables. It is a scientific replacement of the operator's spatula, where improperly written manufacturing instructions have the statement: "check if dissolved," or similar subjective evaluation.

43.5.3 Viscometers

Viscosity is an important property of semi-solids and their components. Viscosity is often highly variable with temperature, and a qualified temperature bath with specified hold times is required along with the qualified viscometer. Measurements on low viscosity liquids may be performed using the Ostwald viscometer, which times flow of a known quantity through a capillary tube. It is important to remember that the timer must be qualified prior to use, along with the viscometer. Stokes Fall is another timing measurement in which a ball of known size and mass falls through a given length of material. The most versatile and frequently used viscometer is the Brookfield type, which mimics Couette flow. This device measures the torque of a rotating cylinder in a cylindrical container of sample. This device can be used on more viscous materials and has the advantage of measuring force directly.

43.5.4 Refractometers

The measurement of refractive index is restricted to clear materials. However, it does provide an important physical constant when applicable. In many cases, the assay of a particular ingredient varies with refractive index, and custom instruments graduated in units of concentration of the ingredient of interest are used. These instruments are simple in design and can be handheld for convenient use within the manufacturing facility. Validation testing using these calibrated instruments is entirely acceptable as an alternative to the time-consuming sample transport to a laboratory for instrumented analysis. In many cases, an immediate reading pertaining to the process is needed.

43.5.5 Balances, Scales, Pycnometers

Balances and scales must be calibrated to NIST traceable standard weights. A validation concern for OTC liquids and semi-solids in "private label" manufactures is to ensure that they are actually used! It is not unheard of that supplier container weights are used in lieu of weighing for large additions of excipients. Another practice is to dump drums or bags of material into the batching vessel after a gross weight is obtained, and then weigh the empty container for the tare weight. If this practice is performed, steps must be taken to ensure all the correct amount of material gets into the batching vessel and that excess material is not added in error.

Besides the obvious measurement of mass in adding the ingredients to a formulation, balances and scales are used for the CPP of density or specific gravity. In this case a pycnometer, which is a glass vessel of precisely measured volume, is utilized. As in viscosity measurements, pycnometers must be used within qualified temperature baths to provide accurate measurements.

43.5.6 Thermometry

Temperature is an important CPP for liquids and semi-solids. The measurement of temperature follows that of parenteral manufacture and is well covered elsewhere in this volume. Sophisticated temperature measurement and calibration equipment, such as supplied by Kaye Instruments, are not always available at smaller manufacturers. Equipment is available for rent that will enable smaller manufactures to qualify their thermal sensors and validate their processes. It is important to calibrate temperature probes using the sensor and the transmitter using a temperature bath. In some cases, standard voltages are used to calibrate the transmitter and the gauge, or readout only, and this should be avoided (see the chapters on temperature measurement and calibration for additional information).

43.6 PROCESS PERFORMANCE QUALIFICATION PROTOCOL

Process Performance Qualification is ordinarily the last step with regard to startup and scale-up of non-sterile liquids and semi-solids. While this is true for all dosage forms, there is a tendency to combine development and validation for OTC products to save time and money. It has been demonstrated time and again that this often costs more in the long run and the proper sequence of qualification and process validation should be followed. Protocols should contain a documented check that specified activities required prior to the current protocol have been completed and are acceptable. The following is a discussion of important elements that are typically contained in a non-sterile liquids or semi-solids Process Performance Qualification protocol.

43.6.1 Prerequisites

As stated earlier, it is necessary to list all requirements and to have a positive verification of completion along with report or activity completion date. These activities include: development (Stage 1 Process Design); facilities design and qualification; equipment qualification (IQ, OQ); analytical test method validation; documentation (manufacturing instructions, batch records, standard operating procedures); and training. The reader is directed elsewhere within this volume for detailed discussion of these requirements.

Bulk raw material qualification is an important prerequisite. Bulk raw material issues can be characterized by one word: variability. This is often the fault of the buyer, where cost savings associated with commodity classified raw materials cause frequent vendor changes. Issues usually fall into the category of microbiological and /or physical. Natural products often have microbial problems when adequate vendor controls are not in place. The author recalls an instance when a lot of guar gum that was purchased at a considerable discount was found to fail because of microbiological contamination. An attempt was made to validate the ethylene oxide sterilization of the material at a bargain price processor. The unsuccessful result cost much more in time delay, wasted processing, and lost validation effort than the cost of purchasing the appropriate quality raw material from a qualified vendor. Mineral oil and petrolatum are major raw materials in this product segment and do not present microbial issues. They do have purity, physical, and contamination issues. Mineral oil is frequently purchased in large quantities, and that may involve transport and storage issues. The use of non-dedicated rail cars and motor freight tank trucks as well as outside storage tanks is a possible condition that must be considered during raw material and vendor qualification activity. Even the use of 55 gallon drums does not eliminate cross-contamination and storage issues unless these are dedicated to a single material. Vendor qualification includes selecting vendors with quality systems in place and who are audited by qualified auditors. Qualification includes sampling, testing, defined primary storage containers, shipping, and storage conditions.

Cleaning process development and cleaning validation are important prerequisites. Cleaning validation is frequently performed concurrent with other qualifications and/or validation. Cleaning of semi-solids and viscous liquids is difficult because of the oil and lipid content of these products. As previously indicated, the transport and storage of liquid and semi-solid raw materials prior to receipt and processing must be considered. It is important to develop cleaning processes prior to process validation because the delays and interruptions necessary for cleaning validation may alter the timing parameters of the manufacturing process. Globepharma, New Brunswick New Jersey (www.globepharma.com) has developed swabbing wands to facilitate micro and chemical swabbing of process equipment. Cleaning and contamination issues arise with the use of plastic tubing in manufacturing operations. Elastomeric tubing is widely used in liquid and semi-solid manufacturing and filling operations. Plastics may contain non-reacted monomers, plasticizers, UV stabilizers, preservatives, mold release agents, and lubricants. Chemical leaching studies need to be performed in conjunction with vendor qualification. High quality medical tubing is typically used. Tubing is usually discarded after a single use, as mentioned earlier with respect to peristaltic pumping. Dedicated use for a campaign of a fixed number of batches must be validated.

43.6.2 Sampling—Statistical Considerations

Validation sampling and testing typically is three to six times the usual QC sampling and testing. That being said, there needs to be a rationale and /or justification for the selected sampling plan, and the inherent process variability must be revealed. Wang and his coworkers have recently provided a scheme for screening CPPs.[17] The classic paper on this subject is by Plackett and Burman in 1946 and remains a standard.[18] Statistical sampling approaches should ensure that the samples are obtained from the entire container, filling run, etc., without emphasis on any one area, with problem areas included. For example, in-process samples are most easily obtained from the top of vessels and, as such, dominate sampling plans. These samples should be balanced by samples obtained from the bottom and middle areas. The "Square Root of N Plus One" approach has been highly criticized, and it is best to exceed this number of samples, especially when sampling relatively few bulk containers.[19] If standardized sampling plans are used for normal production, validation should specify some integral multiple of the plan to justify their routine use. Statistical acceptance criteria are best applied to packaged product, where the number of samples will enable meaningful results. In-process samples should be confirmatory and large statistical samples; both within batch and between batches should be reserved for Stage 1 and the final primary package of PPQ. Statistics such as CpK and CpM should be based upon historical results and manufacturing, since overfills are a common practice for multidose OTC liquid and semi-solid products. Hofer has provided a statistical means to determine the number of retests in the event of questionable results.[20] Note that results can be disqualified when root causes can be assigned to sampling or testing errors. This can occur with the high viscosity semi-solids and "lotion" products and can be anticipated within the protocol sampling and testing language. Composite samples should be avoided during validation sampling and testing, especially if composites will be used during routine production.

43.6.3 Sampling Plans—Bulk

Sampling from mixers is accomplished from recirculation sampling ports, where available. Location sampling should be based upon known "dead spots" or mixing voids identified for the given mixer geometry. Kukura et al. and Prodal et al. provide insight to these locations for typical mixer types and geometry.[21,22] A generally acceptable approach is to sample the top, middle, and bottom of a mixer of cylindrical or hemispherical geometry at the central axis, half-radius, and the side surface. If there is a central mixing shaft, then the samples are obtained next to the shaft. The side surface samples can be rotated in the geometric axis of rotation by 90° or 120° to the plane defined the center and mid-radius samples. This plan is best used if there are stratification concerns. The GlobePharma (New Brunswick, New Jersey) liquid- and syringe-type thieves have been successfully used in these applications.

For batches consisting of multiple drum storage containers, samples are most conveniently obtained from the filling hose while the drum is being filled. They are usually obtained from the top of the drum during routine production when separate sampling personnel are used. In order to justify this approach, validation samples should be obtained from additional levels of the drum. Drums may be skipped in these instances as long as the beginning and end drums are included.

43.6.4 Sampling Plans—Filled Product

Samples of the filled product are easily obtained, and this is fortuitous, as this is what the patient ultimately receives. As such, the preponderance of sampling and testing should occur at this final stage of manufacturing. It is important to include all manufacturing shifts for long filling and packaging runs in addition to all filling nozzles, as previously mentioned. Additionally, validation samples should be obtained to bracket filling line stoppages to provide worst-case samples and justify less rigorous sampling during routine production filling and packaging runs. Validation batches and/or filling trials are usually placed in stability programs. It is important to include all primary containers within stability programs, even though a bracketing approach may have been used in validation. For this reason, it is best to have a separate stability protocol and/or procedure that the validation protocol can reference.

43.6.5 Unit Dose and Metered Dose Containers

Large numbers of this type of primary package are typically sampled and tested. For life-saving products, it is difficult to justify any sampling plan unless some positive in-line check of each container is part of routine production. Validation protocols need to challenge fail-safe controls and testing with well-marked and monitored failing packages. These challenges are best conducted during qualification but must be fills of actual product at full production rates.

43.6.6 Multidose Containers

Testing should be conducted on all the doses delivered from a container and should exceed routine testing. The size of multidose containers may limit this practice because of limited laboratory resources. In this event, random samples from multidose containers throughout the filling run should be tested. Tubes are filled from the bottom and crimped with closures already in place. Cap torque testing needs to be obtained following filling to ensure that the filling and packaging operation have not detrimentally affected the closures.

43.7 PROCESS VALIDATION REPORT

A report is expected under the Guidance, especially if there are deviations to the protocol and/or unexpected results. Procedure-driven validation must contain forms analogous to protocols and reports. Documentation for non-sterile liquids and semi-solids follows the format of other dosage forms. PPQ reports for non-sterile liquids and semi-solids typically are approved by the same functional areas, as approved by the protocol. In many instances, the executed protocol containing annotated raw data and verifications are circulated for approval. Many firms circulate both in a combined document. An executed protocol alone may not be sufficient without some post-analysis or project "post-mortem" summary contained in a report.

43.8 VALIDATION LIFE CYCLE

The product should be monitored from inception, through validation, and during routine production in an approach called the Validation Life Cycle.[23] This approach has been acknowledged as extremely effective and has been incorporated in the Guidance as defined in its three stages. Change-driven and time-driven revalidation should be specified in validation master plans and/or in approved procedures. A Change Control program needs to be in place to trigger additional validation upon the implementation of significant changes. This must include raw material supplier changes, since many of the excipients for non-sterile liquids and semi-solids are commodities and are often purchased from the lowest cost vendor.

43.9 CONCLUSION

The author has attempted to provide an overview of non-sterile liquids and semi-solids validation. A survey of the types of formulations and associated equipment was provided along with anecdotal experience to give validation personnel an insight into this area of pharmaceutical manufacturing. Statistical sampling and testing plans emphasize Stage 1 process design and Stage 2 PPQ final primary package sampling and testing. Documentation parallels other areas of validation; however, the diversity of the materials and equipment within the scope of this chapter make each study unique.

NOTES

1 Guidance for Industry Process Validation: General Principles and Practices www.fda.gov/downloads/Drugs/. . ./Guidances/UCM070336.pdf last accessed 1/6/2019.
2 Immel BK. A Brief History of the GMPs for Pharmaceuticals. *Pharmaceutical Technology* 2001; 24: 44.
3 Martinez M, Schering-Plough Criminal Investigation May Center On Puerto Rico. *Puerto Rico Herald*. June 27, 2002.
4 Labuza TP. Fundamentals of Water Activity Short Course. *Institute of Food Technologists*, 2002.
5 www.drugfuture.com/pharmacopoeia/usp32/pub/data/v32270/usp32nf27s0_c1112.html <1112> Application of Water Activity Determination to Nonsterile Pharmaceutical Product.
6 The United States Pharmacopeia 43rd Revision, <1115> Bioburden Control of Non-Sterile Drug Substances and Products, The United Stated Pharmacopial Convention, Inc. Rockville MD, 2019.
7 The United States Pharmacopeia 43rd Revision, <1231> Water for Pharmaceutical Purposes, The United Stated Pharmacopial Convention, Inc. Rockville MD, 2019.
8 The United States Pharmacopeia 43rd Revision, <1151> Dosage Forms, The United States Pharmacopial Convention, Inc. Rockville MD, 2019.
9 Crowder TM, Louey MD, Sethuraman VV, Smyth HDC, Hickey AJ. 2001: An Odyssey in Inhaler Formulation and Design. *Pharmaceutical Technology* 2001; 21: 99.
10 Garg S, Tambwekar KR, Vermani K, Garg A, Kaul CL, Zaneveld JD. Compendium of Pharmaceutical Excipients for

Vaginal Formulations. *Pharmaceutical Technology—Drug Delivery* 2001, 14.
11 Block LH. Scale Up of Liquid and Semisolid Manufacturing Processes. *Pharmaceutical Technology Scaling Up Manufacturing—2005* March 2005, s26.
12 Kupp GD, Challenges, Considerations, and Benefits of Raw Material Testing. *Pharmaceutical Technology—Analytical Chemistry and Testing 2003*, 22.
13 Gupta P, Garg S. Semisolid Dosage Forms for Dermatological Application. *Pharmaceutical Technology* March 2002, 144.
14 Hopp M, Developing Custom Adhesive Systems for Transdermal Drug Delivery Products. *Pharmaceutical Technology* March 2002, 30.
15 Witt K, Bucks D. Studying in Vitro Skin Preparation and Drug Release to Optimize Dermatological Formulations. *Pharmaceutical Technology—Formulation, Fill & Finish* 2003, 22.
16 Pailhes M, Lambalot C, Barloga R. Integration of Centrifuges with Depth Filtration for Optimized Cell Culture Fluid Clarification Processes. *Bioprocessing Journal*, 2004, May / June 2004, page 56.
17 Wang K et al., Statistical Tools to Aid in the Assessment of Critical Process Parameters. *Pharmaceutical Technology* 2016; 40 (3).
18 Plackett RL, Burman JP, The Design of Optimum Multifactorial Experiments. *Biometrika* June 1946; 33 (4): 305.
19 Saranadasa H, The Square Root of N Plus One Sample Rule—How Much Confidence Do We Have? *Pharmaceutical Technology*, May 2003: 50.
20 Hofer JD, Routine Sample Size for a Retest Procedure. *Pharmaceutical Technology* November 2003: 60.
21 Kukura J, Arratia PC, Szalai ES, Bittorf KJ, Nzzio FJ. Understanding Pharmaceutical Flows. *Pharmaceutical Technology* October 2002: 48.
22 Prodal HS, Matice CJ, Fry TJ, Computational Fluid Dynamics in the Pharmaceutical Industry. *Pharmaceutical Technology* February 2002: 72.
23 Agalloco JP, The Validation Life Cycle. *Journal of Parenteral Science and Technology* 1993, 47: 142.

44 Validation of Non-Sterile Packaging Operations

William G. Lindboe Jr.

CONTENTS

44.1 INTRODUCTION

A seed can be viewed as a perfect package. Its function is the generation of new plants, and it has been shown to maintain this function for extended periods. Most vegetable seeds retain their functionality from one to six years when stored in dry conditions.[1] Seeds are easy opening, and their shape or exterior (fruit, flight-aid, etc.) helps with delivery. Pharmaceutical packaging aspires to be as efficient as the seed in that it is required to maintain full functionality through the end of a product shelf life. Packaging is a broad topic and is as critical as the item being packaged. The title of this chapter is "Validation of Non-Sterile Packaging Operations," which suggests a restriction to the final stage of manufacturing where finished non-sterile bulk material is mechanically placed within packaging components to produce the final product ready for shipment. Most of the information available on packaging validation concerns sterile or aseptic operations and sterile package integrity. This chapter will address aspects that concern the non-sterile manufacturing validation team and not with materials selection, which is an immense undertaking more appropriate for a volume on development. This chapter supplements the chapter on sterile packaging validation authored by Charles S. Levine within in this volume.[2] The current chapter includes non-sterile products, while excluding topics covered in other chapters such as container preparation; process automated technology, Environmental Quality Control, HVAC and others. The reader is advised to consult these chapters, as well as other sources, to get a full picture of facility-wide validation requirements. Despite this, there will be some unavoidable overlap with these chapters in an effort to ensure that all major topics are adequately covered. The reader may be new to non-sterile packaging, validation or both, and to address this, operational basics will be stated to introduce most topics. The experienced validation practitioner will excuse this but find the chapter useful as a handy reference and a resource for training.

Historically, the function of packaging has been preservation and extension of shelf life by protecting the product from changes caused during handling, shipping and storage. These

DOI: 10.1201/9781003163138-44

changes are caused by light, temperature, physical forces, chemicals, biological factors and the environment in general. Packaging also serves as a processing aid (terminal sterilization within the package), product identification and marketing. More recently, packaging has progressed from these legacy functions to being a critical element in dosing. This includes tablet and powder dispensers, prefilled syringes, metered dose inhalers, med ports and set ports on intravenous (IV) bags. This advance was enabled by the utilization of plastic in a majority of packaging components across all products. Over time, the use of plastic across many consumer product packaging has developed a substantial negative side effect. Plastic waste has found its way into the environment, with detrimental effects, through improper disposal.[3] Manufacturers may be required to provide specific disposal instructions if not providing the means for disposal, such as what is currently in place for printer toner cartridges. Firms should be proactive in this area to avoid mandated regulation and time unavailable mandatory suboptimal changes. Recall the Extra Strength Tylenol™ contamination event, where the industry had to implement tamper evident packaging very quickly.[4] Shortly after this event, all over-the-counter (OTC) products were required to have tamper-evident packaging. This change was relatively easy because of the preexisting technology and its partial implementation already under way.

44.2 ORGANIZATION OF NON-STERILE PACKAGING FACILITIES

Non-sterile packaging facilities are usually rooms that contain the machinery used to accomplish the packaging, although dedicated facilities used exclusively for non-sterile packaging operations are common. The packaging room is supplied from separate areas for bulk unpackaged product and container components. These components consist of containers, closures, stoppers, web rolls of plastic or laminate for form fill and seal, cotton, desiccants, tamper evident seals, etc. There is also a room or area for completed filled packages. Storage of completed packages may be in a separate facility. Public warehouses for this purpose should not be used until the product is released by the Quality Control organization for shipment.

The packaging room and component feed rooms may have controlled environments depending on the product being packaged. In some cases, the machinery itself is enclosed within barriers to maintain a controlled environment. These are referred to as restricted access barriers (RABS) or isolators. These are not used for non-sterile operations but could be used for toxic non-sterile materials.[5] Whenever there are controlled environments, the areas must be qualified under static and dynamic conditions as well as monitored during packaging operations. This is covered elsewhere in this volume and will be omitted from this chapter.

Product and components are transferred to the packaging room from the feed areas, either manually, on pallets, on conveyer belts or through pipes. In some cases, there is an overhead feed. Eventually, there may be mix-ups with these configurations when multiple feeds and machinery are involved. Along with measures to prevent this occurrence, there should be measures to detect it, if it occurs. Typically, it is the correct component or product but the incorrect lot number or batch number, or the material has not been released by the Quality Control organization. The latter can occur with any type of feed and is the basis for the public warehouse restriction for unreleased finished product mentioned earlier. It is important that only the Quality organization have the ability to change the release status of components.

44.2.1 General Considerations

Historically, packaging errors are the majority of product recalls.[6] As such, quality assurance controls must be put in place to address this. These controls include the use of rolled labels wherever possible, storage of labels in a secured area, label use reconciliation, line clearance between lots and between products by the Quality organization, etc. These controls, along with container stability studies, were implemented prior to the late 1970s. Because of this, non-sterile packaging has never been a validation priority. The development of additional packaging functionality, such as dosing and blister packaging, and new materials, primarily plastics, provided the impetus to apply validation to these manufacturing operations. In general, the outcome of non-sterile packaging validation is to demonstrate that the packaging equipment does not alter the critical quality attributes of the bulk product.

The sequence of topics in this chapter will follow the product movement through the packaging or fill-finish area. Process validation nomenclature and sequence follows the 2011 FDA Guidance document ("the Guidance").[7] Accordingly, non-sterile packaging validation consists of a series of qualifications. They are Stage 1, design, and Stage 2, installation qualification, operational qualification and process performance qualification. Stage 3 consists of continued process verification. The Guidance is written for new processes and assumes full development data are generated and available. In the author's experience, many non-sterile OTC processes or products are purchased or exchanged between companies and are packaged on existing legacy equipment of the receiving company. Some ingenuity is required to adapt the recommendations of the Guidance to these situations. Trial qualification runs may be used as development data to comply with design recommendations in the Guidance and to generate process knowledge. In general, machine parameter monitoring and product quality testing constitute process control for validation in packaging operations. All qualifications should be specified in a preapproved protocol. The protocol must describe the procedure to handle deviations in the requirements specified in the protocol and out of specification results (discrepancies) to the testing. The expectation of utility and equipment qualification is as follows: construction materials should not be reactive and/or particulate generating, such as stainless steel. The operation and performance characteristics are appropriate, meaning they are as specified by the equipment manufacturer or required by the product being manufactured. Utility systems and equipment are built and installed as specified and properly connected and calibrated. Installation is often subcontracted to firms with personnel not trained in current Good Manufacturing Processes (cGMP). Tradesmen have their own tools and parts inventory. Some of these unspecified components may be installed to save time. It is important that all specified component are staged for installation and installing personnel are supervised to ensure that the installation complies with specifications. The author has witnessed several instances where deviations had to be written and approved where unspecified piping had been installed. These situations can be anticipated with proper planning. Discovery by inspection after installation seldom results in identifying and correcting problems and is difficult to defend during a regulatory inspection. The anticipated operating ranges of all the equipment should be verified under load. Stoppages and restarts anticipated during normal operation should be simulated for the entire packaging line. Any logical or quality decision performed by the equipment should be challenged.

Packaging process performance qualification usually consists of running the validation batches or validation continuous production increment from the bulk production facility on the packaging line. Difficulty arises when validation batches are subdivided into different sized and/or type of primary containers ("put ups"). For example: a single validation bulk production batch may be subdivided into several different-sized containers and result in the same number of packaging validation batches ("packaging orders"). This production is usually placed on stability testing, so the entire process can be validated through the end of shelf life. Shipping ("cold chain") validation is usually simulated by placing temperature and humidity recorders within tertiary packages (corrugated cardboard shippers) and shipped to worst-case destinations in terms of time, temperature and humidity to generate an environmental history. Packaged product samples are placed in environmental chambers programmed to emulate the environmental history and tested as part of the stability program. This is usually a separate validation study.

44.3 PROCESS DESIGN

Packaging equipment differs from many bulk manufacturing operations in that there is more connectivity and interdependence between the equipment and less of a unit operation aspect. Stage 1 design is performed by engineering groups; however, there is usually a single principal engineer responsible. The design may be performed by internal, external or a combination of groups. The roll of validation personnel is to ensure that the design is properly documented and critical process parameters (CPP) and critical quality attributes (CQA) are identified. In some cases, equipment vendors supply a proprietary validation package. It is important that this package includes a design qualification that includes adjacent connected equipment, even though this equipment may be from another vendor. For this reason, validation personnel should interact with purchasing or procuring groups either directly or through the engineering principal. Validation master plans are useful to identify these requirements early in a project and obtain management approval at a high level. Design qualification documentation must identify CPPs and the resultant CQAs, or what this parameter does to the package and/or product. These are required for protocol preparation and should not be developed arbitrarily by the Validation group. It is essential that the selection of CPPs and CQAs come from Engineering or Development. The author has been in FDA inspections where parameters selected or omitted by the Validation group were challenged and brought to the attention of Development, which did not concur with the Validation group. This disagreement usually stems from the concept of the word "critical." A signed and approved protocol does not include higher level personnel, and there is always someone with a different opinion. This underlies the importance of the evolving master plan or sub-master plan, where these parameters can be identified and approved in advance by upper level personnel from the relevant disciplines.

As with other validation efforts, preapproved protocols are required. Protocol development is a large part of design qualification. Each parameter to be verified or challenged in qualification must be shown to be physically measurable and achievable in factory acceptance testing (FAT) or site acceptance testing (SAT). Parameters that cannot be measured once the equipment is installed will result in protocol deviations. This should be avoided, as the premise of the 2011 PV Guidelines is to have these issues worked out in advance along with designed in quality. The PV Guidance also expects a final validation report summarizing the validation effort. Usually, each individual qualification will have a report, and each stage will have a summary report. These reports serve to explain and justify all results and decisions, particularly any protocol deviations or unexpected results.

In many cases, product-line extensions and even new products are packaged on existing legacy equipment. This is possible because of the longevity and versatility of the equipment. The qualification documentation on this equipment may be lacking or even nonexistent. In these instances, trial qualification runs with the new components may be used as development. The major concern and goal of qualification is to ensure that the packaging equipment does not affect the product and/or primary package adversely. Factors such as mechanical stress, temperature, particulate generation and lubricants should be considered. Manufacturing experience with similar products and equipment can be used for decisions and justification for controls.

44.4 INSTALLATION QUALIFICATION

Non-sterile packaging installation qualification is concerned with the utilities and ranges of those utilities specified for equipment operation. Additionally, consideration must be given to connecting equipment such as conveyers and other feed systems to assure continuous operation without interference between equipment and other packaging lines. Utility ranges specified by the equipment manufacturer must be tested and documented to be in place with everything else running. Something as simple as boxes of components blocking access to another packaging line or function must be considered. Utility alarms and safeguards, such as uninterruptable power supply, should be challenged during qualification. Compressed gases such as air and nitrogen should have similar alarms. The entire packaging line may need to be shut down should one of these essential utilities be interrupted. Validation master plans should avoid requiring the completion of installation qualification (IQ) prior to starting operational qualification or any other phase. Frequently, drawings are completed late in the project and are part of IQ, thus delaying the final report and date of completion. Interim reports and the release from the Quality organization can suffice to proceed to the next stage of qualification.

44.5 OPERATIONAL QUALIFICATION

The approach of operational qualification (OQ) depends largely on the equipment being used. Equipment types widely used in non-sterile packaging will be reviewed below. New installations will be assumed; however, it is important that the concept of "change parts" be defined within operational qualification protocols. Frequently, during the life of equipment, maintenance events, upgrades and other changes require evaluation for revalidation. Mechanically equivalent parts such as belts, hoses, gears, cams, pins, rods, motors, fasteners and all expendable parts must be listed as not requiring additional qualification or have some qualification test specified in advance. Often, parts become obsolete or are replaced with an equivalent or upgraded part. In these instances, it is essential that a procedure is in place to define what further qualification is required (if any). Modern change control requires that each maintenance work order be reviewed for the requirement of additional qualification. There will less debate over decisions and/or bureaucratic delay if "change parts" and their qualification requirements are established, documented and approved ahead of time.

44.5.1 Scramblers

Scramblers or unscramblers (Figure 44.1) have the function of properly orienting the containers for transfer to the filling section of the packaging line. They may have additional container preparation functions such as dust removal and electronic charge removal. These functions need to be verified with reasonable challenges. A common error is to have too severe a challenge and have to deviate from the protocol in order to pass a less severe challenge. The full range of speeds and feeds need to be verified in the OQ. The empty containers are either mechanically or manually charged into a feed hopper, and operation at full and near-empty states need to be verified. It is impossible to anticipate all the container variations that marketing will request in the future. Therefore, a bracketing approach for size and shape should be specified in the OQ protocol.

44.5.2 Fillers

Container fillers are designed to fill solids, semi-solids and liquids, respectively. Solid fillers are further subdivided into tablets and capsules, granules (seeds), or powder. Historically, the filling station has personnel, or an "operator," manually running or observing an automatic or semiautomatic system. The tablet and capsule fillers have already-dosed product, whereas the other types assume a uniform continuous bulk product. This uniformity must be verified, and it is most convenient to

FIGURE 44.1 UNISORT 32 Unscrambler.

Source: NJM™.

do this with filled containers, rather than sampling the filling hopper. In some cases, there is mixing or some form of agitation present in liquid and semi-solid feed hoppers. This brings the additional difficulty of verifying mixing parameters and mixing blade configuration. Additional specifics for each type are described in the following subsections.

44.5.2.1 Slat Type Tablet Fillers

The Merrill™ type slat filler employs slats that accommodate a specified number of tablets or capsules per slat and the total fill being a multiple number of slats. Tablets can be problematic, since the thickness (gauge) is variable. When tablets get stuck in the slats from being "over gauge," the problem lies within the bulk compressing operation and not packaging. The cause is whole tablets and/or fragments bouncing back into the tablet press feed frames resulting from an improper kick-off setting on the table press. Filler qualification should verify an upper thickness limit by mechanical measurement and state that such an occurrence does not affect the qualification of the packaging equipment. The basic parameters are feeds and speeds. There may be additional controls on modern equipment that will require qualification.

44.5.2.2 Powder Fillers

Powder fillers commonly use screw-type augers rotating for a given time to fill primary containers. There is inherent variability in the flow characteristics of powders and granulations upon storage. Some type of fill-weight verification is required on every container. The wide availability and speed of in-line weighing equipment should be part of any modern packaging facility. This equipment may have additional vision inspection capability and can be employed after labeling, such as the CV35 Advanced Line Checkweigher from Mettler–Toledo (Figure 44.2). Examination of these quality attributes, along with equipment monitoring, constitutes control of powder filling operations.

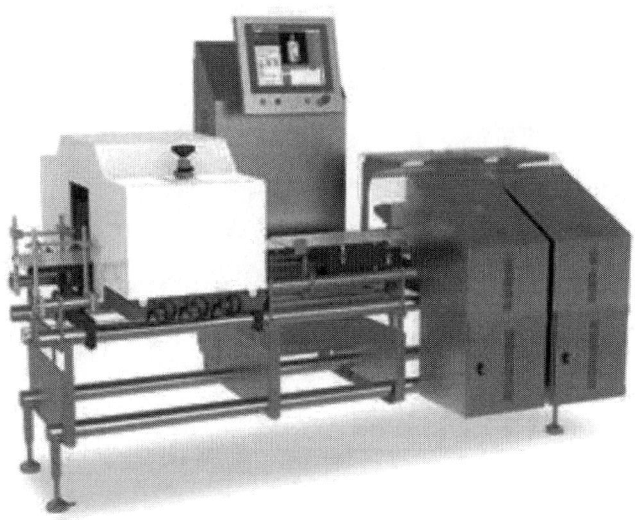

FIGURE 44.2 C35 Advanced Line Checkweigher.

Source: courtesy Mettler—Toledo.

44.5.2.3 Liquid and Semi-Solid Fillers

These fillers use pumps and valves to accomplish filling. One type uses peristaltic pumps with medical silicone tubing. This type requires determination of the effective life of the tubing. For product change, it is both quality and cost effective to replace the existing tubing with new tubing rather than attempt to validate the cleaning operation. Dedicated peristaltic pumps should also have a tube replacement schedule. This may be adjusted with filling experience after establishment of an initial time period. This approach should be documented in qualification protocols and validation plans. As stated earlier, the uniformity of the bulk after storage of any duration cannot be assumed without qualification. This qualification typically includes uniformity testing of the filled containers with the filling hopper at full and near-empty states. Any mixing or agitation of the filling hopper must be documented and qualified.

Semi-solids are often filled into preprinted laminated tube stock or formed tubes with the closure already in place. After filling, the tubes are rotated to an index mark and crimped closed. There is usually information on the crimp seal, such as expiration date and lot number. All setup parameters associated with the printing and sealing are critical.

44.5.3 Cottoners/Desiccant Addition

With solid dosage, there may be a station to add cotton and/or desiccants to the filled containers. These machines are qualified by determining feed and speed ranges. Protocols should contain documentation and acknowledgment of equipment "jam-ups," which inevitably occur. Inspection should be included to ensure that this operation does not adversely affect the filled product and/ or container.

44.5.4 Conveyors

This equipment is qualified by documenting feed and speed rates. Pauses in the operation occur where the containers are blocked, but the smooth belt continues to move beneath them. Inspection for quality should include material at each stage of a filling line for a maximum pause. There is a very low probability that any detrimental effect will occur; however, conveyor qualification should include consideration of this unlikely event.

44.5.5 Cappers/Stoppering Machines

Capper machines (Figure 44.3) apply a torque to a threaded closure ("cap") or press on a child-proof closure and/or tamper-evident seals and closures. Additionally, there may be a dosing function to the closure, such as a removable dropper or a dropper built into a closure cover, then covered by a cap.

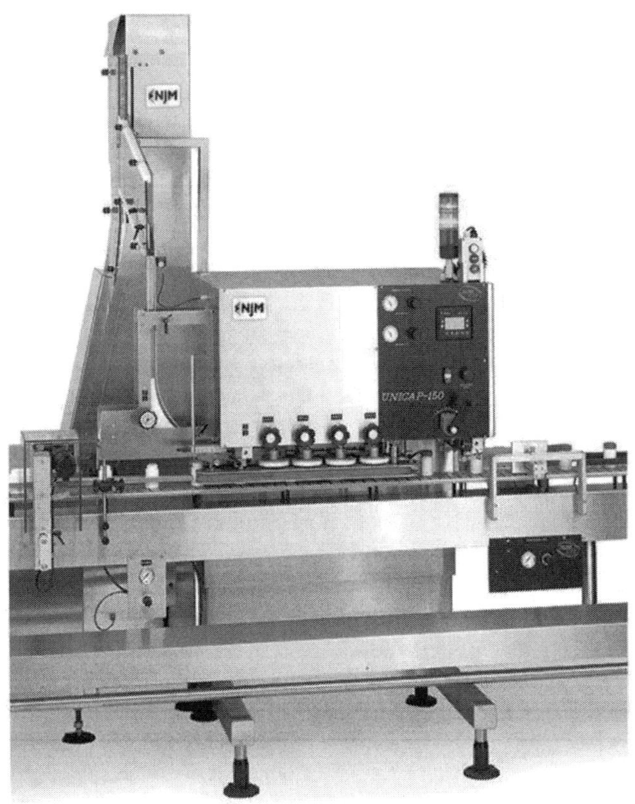

FIGURE 44.3 Unicap 150 Capper.

Source: Courtesy NJM.

Legacy equipment takes considerable adjustment to set the machines for effective operation. Speeds and feeds should be verified along with a means of detecting improper placement, seal and/or torque.

44.5.6 Labelers

These machines apply a preprinted label to the primary package. Alternatively, in-line printing equipment can be used. Speed and feed are critical process parameters. Controls that orient the containers for label application or printing need to be tested. Any reject mechanisms must be challenged. An example of labeling machines is shown in Figure 44.4.

44.5.7 Blister Pack Machines

Blister pack machines (the author has included strip packaging and aluminum foil packaging in this category) are the non-sterile equivalent of aseptic form-fill-seal equipment. The blister packaging is an instance where additional functionality beyond protection is present in the package. It provides easy product visibility for identification and aids in patient compliance by indicating the number of doses taken and the number of doses remaining. While the complication of sterility is absent, the variety of the various presentations and fills is extensive. An example of a modern machine is shown in Figure 44.5.

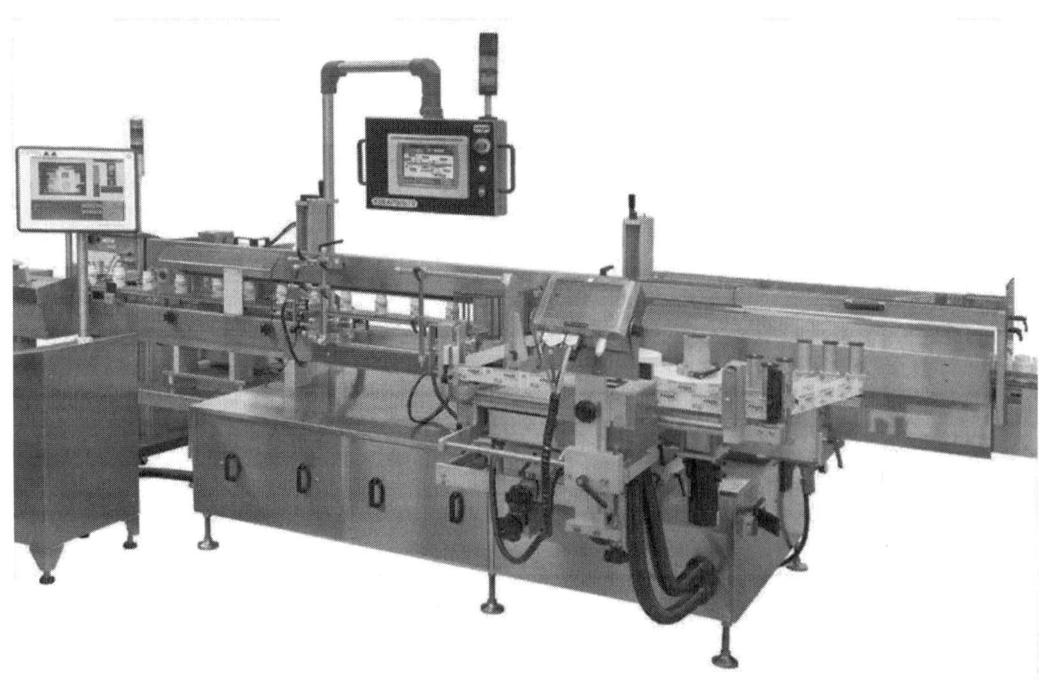

FIGURE 44.4 AutoColt IV Labeling Machine.

Source: Courtesy NJM.

FIGURE 44.5 A modern blister packaging machine.

Source: Courtesy Syntegon.

44.5.7.1 Operations of the Blister Pack Machine

There are five sequential operations common to blister pack machines. They are feeding, forming, filling, sealing and cutting. Some machines may have inspection and cartoning operations built in, as well. Tablets are the usual products packaged on these machines. Suppositories are often molded within a formed aluminum foil primary strip package. This introduces the complication of a warm product fill followed by cooling operation into the packaging sequence. Other fills include medicated or disinfecting gauze or towels and hard or soft gel capsules. There is a limitation to what can be packaged in that the force to remove the fill from the package should not result in damage to the fill. This is established in development, prior to qualification.

44.5.7.1.1 Feeding

This is a large roll web feed. This feed may be a polymer, laminate or aluminum foil laminate. Critical parameters are feed rate and physical dimensions of the web. The web tension may be critical, but this adjustment is usually optimized for a given speed.

44.5.7.1.2 Forming

This web feed is formed into the primary package by the application of heat and/or pressure by a die or die roller. Critical parameters are dwell time and forming temperature, if applicable.

44.5.7.1.3 Filling

The fill is accomplished with a bulk product feed that may be solid, semi-solid or liquid. The latter two have a "peel off" capability of the lid stock. This filling operation has the same critical parameters as a stand-alone filling machine, with the exception that the blister pack cavity is unit dose. For suppository manufacture, the formed strip remains in the forming die for filling. This die is cooled after filling. There may also be a card stock feed to the formed package. All feeds may be printed in advance by a third party. The latter introduces a concern in that they are seldom cGMP operations. Alternatively and preferably, printing may be done on-site.

44.5.7.1.4 Sealing

This is the operation that is most problematic during routine production. Use of design of experiments (DOE) is helpful to arrive at an optimal range of heat, pressure and dwell time to seal the various web types and thicknesses.[8] The machinery vendor and/or web vendor usually can provide information on these critical process parameters. Problems usually arise from dust or other particulates that interfere with the sealing process. Worst-case products should be identified in terms of particulate generation for inclusion in qualification.

44.5.7.1.5 Cutting

Feeds and speeds are the critical process parameters associated with the cutting operation. Blade maintenance is important, as dull blades may damage the package or fill or make an incomplete cut.

44.5.7.1.6 Computer Control

Modern blister pack machines are usually computer or programmable logic controller (PLC) controlled. This introduces the added difficulty of a computer-related system validation, which is best included and executed with the equipment qualification. Equipment suppliers may provide validation documentation and services. Some blister packaging machines have all the machines of a conventional packaging line built into a single machine. Accordingly, the qualification of these machines is extensive. Much of the programming is tailored and entered for the specific application, such as in the use of ladder logic. Source code may not be available. In these cases, all the functionally controlled by the computer must be qualified. All quality decisions made by the controller must be challenged.

44.5.8 Cartoners

These machines put the filled primary package into a secondary carton along with the package inserts. Other documents may also be inserted, such as coupons or promotional literature. Identification is the critical product attribute of concern. Accordingly, all parameters associated with the printing and labeling operation are critical. It is important that bar codes are of the appropriate size. The feed rate and speed should align with the adjacent equipment in the packaging line.

44.5.9 Case Packers

Case packing is an operation that is frequently manually performed, though modern machines are available for this operation. It consists of placing a number of filled secondary cartons into a tertiary corrugated cardboard shipping container (shipper). This operation is usually done as part of the packaging line; however, some medications are stored in a warehouse, and orders are "picked" into the shippers at a

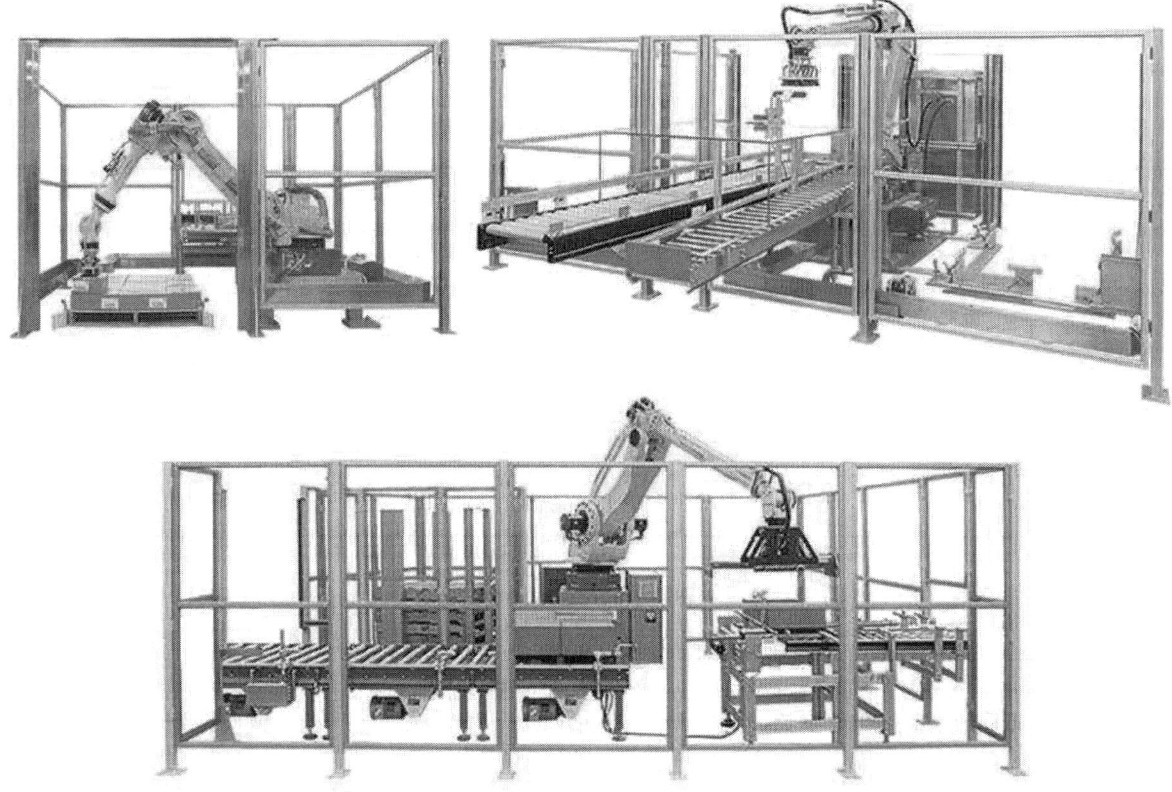

FIGURE 44.6 SERPA Palletizing Machine.

Source: Courtesy NJM.

later time. In the modern era of "just in time" shipping and receiving, this is a rare event; speeds and feeds are the critical process parameters.

44.5.10 Pallet Stackers

High volume products have entirely automated packaging lines, and this would include pallet stackers. These machines are essentially robot arms that fill pallets (Figure 44.6). Speeds and feeds are the critical process parameters.

44.6 PROCESS PERFORMANCE QUALIFICATION

Process performance qualification (PPQ) consists of bulk manufacturing validation product that is followed through the packaging operation. These lots, or continuous product, are usually placed on a stability storage and testing program. The product should be packaged in all of the container sizes. Shipping validation should run concurrent to the process performance qualification. A statistical sampling plan should be specified in the PPQ protocol.[9] Testing limits at time zero should not be tighter than release limits. Assurance should be obtained by additional sampling and testing and not by adjusting the limits; otherwise, the validation could fail an acceptable result for release. Stability limits are established separately in a stability protocol. The Guidance suggests preparing a report covering the PPQ and referencing the qualification studies. This report should discuss all results, both expected and unexpected.

44.7 CONTINUED PROCESS VERIFICATION

This requirement is generally covered by a combination of the stability program, the change control program and the annual report. This should be stated in the validation master plan. Statistical analysis of assays and uniformity results, along with other parameters that are included in annual reports, provide continued process verification.[10]

44.8 CONCLUSION

Non-sterile packaging validation was the last area to have validation studies because of the extensive cGMP and quality controls that were already in place. Modern computer controlled packaging equipment, along with the additional functionally of the package, such as dosing, has made packaging validation a current requirement. The interconnectivity of the various packaging machines and the variety of products and container sizes complicates qualification studies. Fully documented process validation should be in place for all non-sterile packaging operations.

NOTES

1. www.gardeningchannel.com/seed-life-cha5/rt-how-long-will-seeds-last/ last accessed 5/6/2020.
2. Levine CS, Validation of Packaging Operations. In *Validation of Pharmaceutical Processes*: Sterile Products, 3rd edition, Agalloco, JP, Carlton, FJ, eds, New York: Marcel Dekker, Inc., 1998, 703–720.
3. www.earthday.org/campaign/end-plastic-pollution/ last accessed 8/19/2020.
4. Haberman C, www.nytimes.com/2018/09/16/us/tylenol-acetaminophen-deaths.html last accessed 8/20/2020.
5. Dorn EJ, Frantz JJ, Valerio PF, ISPE Barrier Survey: Tracking the Journey of Barrier Technology. *Pharmaceutical Engineering* July/August 2020. https://ispe.org/pharmaceutical-engineering/2020-ispe-barrier-survey-tracking-journey-barrier-technology.
6. Halek GW, Yam KL, Lordi N, *Course: The Technology of Packaging II*, Rutgers the State University of New Jersey, 1984.
7. Anonymous. Guidance for Industry Process Validation: General Principles and Practices. www.fda.gov/downloads/drugs/guidances/ucm070336.pdf, last accessed 8/6/2020.
8. Hicks CR, *Fundamental Concepts in the Design of Experiments*, 4th edn. New York: Saunders College Publishing, 1993, 121–138.
9. https://stattrek.com/survey-research/sampling-methods.aspx last accessed 8/20/2020.
10. Box GEP, Hunter WG, Hunter JS, *Statistics for Experimenters An Introduction to Design, Data Analysis, and Model Building*. New York: John Wiley & Sons, 1978, 556–563.

45 Cleaning Validation for the Pharmaceutical, Biopharmaceutical, Cosmetic, Nutraceutical, Medical Device and Diagnostic Industries

Rebecca Brewer

CONTENTS

Cleaning validation has come a long way since the days of the Barr Laboratories Court Case[1] and since the first FDA guidelines referencing the subject of cleaning validation were published in 1991. At that time, the requirements for cleaning validation barely filled a single page of the Bulk Pharmaceutical Chemical and Biopharmaceutical guidance documents (now long obsoleted). Those early documents were then expanded to create the Guide to Inspection of Cleaning Validations by FDA (first published in 1992 as a Mid-Atlantic Inspection Guidance, then reissued as a national FDA guidance document in 1993).[2] Today, despite nearly 30 years of exposure to the requirements for cleaning validation, this topic remains one of the areas of validation that people frequently profess to know the least about.

Good manufacturing practice regulations have their basis in cleaning validation. Beginning in 1906 with Upton Sinclair's "The Jungle",[3] people demanded that the government improve cleanliness practices in the processing of food, giving rise to what is known today as the Current Good Manufacturing Practices for both food and drugs. While cleaning has always been part of the good manufacturing practice regulations, cleaning activities have not enjoyed the

DOI: 10.1201/9781003163138-45

limelight. The CGMPs that are followed today were predominantly written in 1978. References to cleaning and documentation associated with cleaning can be found throughout. As with many other areas of validation, however, there is no explicit reference to cleaning as a process to be validated. It is this very aspect of the CGMPs that was challenged in the Barr Laboratories court case. In that decision, Judge Wolin ruled that cleaning did require treatment as a process and therefore required validation. In 1996, proposed revisions to the CGMPs were drafted by FDA.[4] Although not adopted, these revisions proposed to redefine the manufacturing process as *beginning* with a cleaning operation.

Cross-contamination is a significant risk to patients. This is true whether through direct administration to a patient or, in the case of *in vitro* diagnostics, through the performance of a test on a patient sample. Cleaning and cleaning validation are two activities that can significantly reduce patient risk by assuring that cross-contamination does not occur. Cleaning validation is becoming more important with increasingly potent, increasingly complicated drug substances and increasingly complex biotechnology products. Products have greater risks of interaction with one another resulting in harmful effects to patients. To truly limit this risk, scientific approaches must be taken in all aspects of the cleaning and cleaning validation program.

When FDA published "Pharmaceutical cGMPs for the 21st Century: A Risk-Based Approach" in August of 2002 and reported on their progress in September 2004,[5] the continued importance of sound scientific rationales in pharmaceutical manufacturing and validation was reinforced. The pharmaceutical community as a whole renewed their efforts to ensure that sound quality principles were followed in the identification of critical to quality attributes for all measurements and analysis. Although risk-based decision-making in the establishment of scientific rationales was always a cornerstone of cleaning validation requirements, efforts have been renewed to ensure the incorporation of risk analysis documentation in cleaning programs.

The FDA Cleaning Validation guidance has not been updated in nearly 30 years;[2] nevertheless, this does not leave industry with a gap in guidance. It was highlighted in "Pharmaceutical Quality for the 21st Century: A Risk-Based Approach Progress Report" in May 2007[6] that FDA supports the use and development of voluntary consensus standards. In part, FDA highlighted that these standards are a way in which the industry and regulators can align on standards, adapt to new technologies and promote international harmonization. In the past 30 years, many such consensus and international regulatory standards on cleaning validation have been developed, driving what cleaning validation looks like today.

See Appendix A for some common terms and their definitions in support of cleaning validation that will be used throughout this chapter. In addition, Table 45.1 defines some of the key document types that are created in support of cleaning validation.

45.1 ORGANIZING FOR CLEANING VALIDATION

Due to the high number of risk-based rationales included in cleaning validation programs, strong policies are required to help drive the decision making. Some programs make use of cleaning validation master plans for cleaning validation in addition to cleaning validation policies. Whether a master plan or a policy, these guiding documents must include the decision-making framework appropriate to a plant site, manufacturing facility and/or dosage form.

Cleaning master planning could be the subject of an entire chapter unto itself, but suffice it to say that the cleaning validation master plan follows the same basic principles as any validation master plan. In fact, cleaning validation may be addressed as a section of an overall validation master plan (one that governs more than one type of validation) or as a stand-alone master plan. The cleaning master plan should:

- Provide an overview of the site/facility/area that is governed by the master plan.
- Provide an overview of the typical manufacturing process(es) that are to be performed in the area and the dosage forms that are produced.
- Provide an overview of the types of cleaning that are to be used (*e.g.*, automated Clean-In-Place (CIP) or Clean-Out-of-Place (COP), semi-automated cleaning or manual cleaning).
- Provide the responsibilities of the various departments having a role in cleaning validation activities.
- Provide the minimum requirements for the cleaning validation program, including:
    - Necessary scientific rationales in support of the program:
        - Residue selection.
        - Product grouping (if any).
        - Equipment characterization.
        - Equipment grouping (if any).
        - Process control strategy (including critical process parameters).
        - Product contact surface area calculation.
        - Limits calculation (and critical quality attributes).
        - Sample site selection.
    - Required studies in support of the program:
        - Cycle development for cleaning processes.
        - Analytical methods validation.
        - Sampling method recovery studies.
    - Essential programs that maintain the validated state and their required elements:
        - Cleaning and testing, if any, to be conducted upon the introduction of new or repaired equipment.
        - Monitoring of cleaning after validation completion.

- Routinely conducted compliance initiatives on site that maintain quality and will affect the company's ability to maintain the validated state:
  - Failure investigation.
  - Change control.
  - Preventive maintenance.
  - Calibration.
  - Revalidation.
- Important standard operating procedures (SOPs) governing cleaning and cleaning validation:
  - Development of cleaning standard operating procedures (especially for manual cleaning operations).
  - Equipment cleaning and use logs.
  - Visual inspection requirements for cleaned equipment.
  - Equipment quarantine and release.
  - Equipment sampling procedures for cleaning assessments (*e.g.*, swab, rinse, etc.).
- Provide the list of equipment and/or systems subject to cleaning validation.
- Provide a list of the products or product groupings that are produced on site.
- Provide a status summary of progress in the area of cleaning validation (for regulatory review—may also take the form of an annual summary report to the cleaning validation master plan).

Each of these topics will be addressed throughout the remainder of this chapter.

In some facilities, the cleaning validation policy or master plan serves as the most basic outline of the required elements for a successful cleaning validation program. In these facilities, scientific rationales are maintained as stand-alone documents. This approach is helpful in a facility or site where there are several dosage forms or product types and where the requirements for one dosage form or product type may be overly stringent for some or more flexible for others.

Where the rationales are maintained as separate documents, however, it becomes critical that the hierarchy in which these documents will reside be strictly maintained. Employees of many departments must be able to cross-reference the applicable documents to their area of interest with no ambiguity as to which document applies. Only through careful organization of the supporting documents can it be assured that consistent decision making is maintained over time.

Upon audit or review of older and/or existing programs, it is frequently discovered that prior rationales have been contradicted or forgotten and that the program has strayed from the previously established goals. Maintaining these documents over time becomes critical to ensure that no internal inconsistencies develop. Strong programs permit their policy and/or validation master plan to serve as an "index" to the risk-based rationales that will comprise the remaining portions of the program by cross-referencing their locations within the quality system. In this manner, the documents that comprise the program are always near at hand and are readily referenced when making decisions for new product introduction, new equipment introduction or changes in the factory. Requiring the periodic review of both master plans and their associated scientific rationale reference documents is recommended. A two to three-year review cycle is typically appropriate. Facilities that have an environment with frequent changes would require a highly frequent review and update of their rationales to ensure that cleaning validation strategies remain current.

See Table 45.1 for a review of common documentation types supporting cleaning validation initiatives. It should be noted that not all levels of policy/master plan will be available in all companies. In fact, the fewer levels of documents in the document hierarchy, the less opportunity for conflict in the requirements.

TABLE 45.1
Typical Document Hierarchy and Scope

Document name	Typical contents/requirements
Corporate Guidelines on Validation or Corporate Policy on Validation	• Policy document governing all sites • Multiple validation topics may be included (not just cleaning validation) • Scope/content is broad/general to apply to several sites and possibly even to diverse dosage forms • Corporate Guidelines or Corporate Policies are typical elements of a company's global Quality System and form an important starting point for cleaning program decisions • Common lexicon of cleaning validation (and other validation) terminology and how it applies to the corporate validation programs
Cleaning Validation Policy	• Typically established at the site level, a Cleaning Validation Policy is part of the Quality System and establishes the minimum required elements of the cleaning validation program • Cleaning validation may instead be a subset of instructions in a broader Validation Policy, but the intent remains the same • The Cleaning Validation Policy may only exist at the corporate level, and a Site Validation Master Plan, with sufficient detail, may serve to bridge between the corporate requirements and the site's actions

(Continued)

TABLE 45.1 (Continued)

Document name	Typical contents/requirements
Site Validation Master Plan	• Describes general principles of validation to be applied at specific site • May address multiple dosage forms or may be prepared for a single type of product or production area within the site • Outline of the details for specific types of validation—there may be several Master Plans dedicated to each type of validation or a single consolidated Master Plan addressing all types of validation, including cleaning • Site Validation Master Plans are not a regulatory requirement in all jurisdictions but are frequently the first document requested by regulators and auditors, since they provide a succinct view of the program elements
Cleaning Validation Master Plan (also may be called Cleaning Validation Approach Plan or similar variant)	• Contains details of philosophy and approach for a specific type of validation—in this case cleaning • May identify current program initiatives for improvement and/or ongoing initiatives associated with new facility introductions (either product or equipment) • Typically includes a product list and equipment list to help represent the scope of the cleaning validation initiative • Will remain in place, reflecting the current approach, although the content will change as activities progress and are completed • Typically reflects current program status and may have an annual summary prepared to define and defend both accomplishments and changes in priority • The need to generate this "topic-specific" validation plan is driven by the complexity and depth permitted in the Site Validation Master Plan
Scientific Rationales	• Documents that contain the details of the risk-based decisions reached for a specific product, group of products or group of equipment • Typically, scientific rationales may exist for, but are not limited to: • establishment of limits • identification of sampling sites • grouping or matrixing of equipment and/or products • residue selection • campaign strategies • worst-case dirty hold times to be challenged • Scientific Rationales follow the decision-making framework contained in the Site Validation Master Plan and Cleaning Validation Master Plan, but in this case the Scientific Rationales record the decisions made • Scientific Rationales are not always produced as stand-alone documents; they may be included in the body of a Master Plan or protocol—the higher the level of document within the hierarchy, the better for visibility and compliance
Project Plan	• Will be prepared for complex projects (*e.g.,* new product introductions, site transfers, facility renovations, etc.) and therefore may contain multiple types of qualification/validation activities to be performed • Will be developed as new projects warrant, to reflect project-specific needs • Will be completed and replaced with new projects—often receiving a summary report to the Project Plan that demonstrates that all activities were completed • Project Plans typically have a smaller scope than Master Plans • Project Plans are not always required as part of a compliant program; they are, instead, a convenience to help manage the complexities of a project without burdening the broader Site Validation Master Plan or Cleaning Validation Master Plan with all of the project-specific detail
Standard Operating Procedure (SOP)	• Specific directions for how to execute the various aspects of the validation program including process development, preparing documentation, executing the studies, collecting and testing samples, and preparing the summaries • SOPs are critical components of the Quality System • Compliance with contents of the SOPs is required, or exceptions/failure investigations must be generated • Demonstrated training in the contents of SOPs is also required
Protocol	• Defines the equipment, products and cleaning process to be confirmed (or that are covered by the validation through grouping) • Defines and justifies the number of cleaning runs that will be conducted and the way in which the equipment will be soiled (*e.g.,* single batch, campaign or simulated soiling) • Defines the sampling and testing that will be performed • Confirms, prior to the initiation of testing, the validated status of equipment, validated status of test methods and validated status of sampling methods, as well as the approved status of all procedures to be executed and the training of staff • Defines the acceptance criteria to be applied to the results, including all calculations • Contains data sheets, attachments or appropriate references to the documentation (*i.e.,* the documentation to be completed during execution of the validation) to provide the documented evidence that validation requires
Summary Report	• Directly responsive to the protocol, the summary reflects: • completed activities • data developed • deviations that may have occurred • conclusion of the studies

45.1.1 Multidisciplinary Nature of Cleaning Validation Activities

Cleaning validation, more so than any other kind of validation, is a multidisciplinary activity. To effectively clean, a deep understanding is required of equipment design, drug product formulation and chemical attributes, along with cleaning mechanics and chemistry. To validate cleaning requires precise analytical methods, as well as a clear understanding of how to sample and collect residues from surfaces reproducibly. All of these activities relate to the expertise of different disciplines, including operations, engineering, research and development, toxicology or medical personnel, validation, quality control and quality assurance.

Some of the elements that require so much interdisciplinary cooperation are the decisions involving:

- The residues to be assessed.
- Safe-levels of carryover.
- Grouping of products and/or equipment.
- Number and location of samples.
- Sampling and analytical method selection.
- Strategies to be employed for ongoing monitoring of cleaning activities.

Long before the risk-based CGMPs became a topic for discussion, cleaning validation had required the development of risk-based scientific rationales. These risk-based rationales form the basis for the cleaning validation protocol as well as forming the basis for the scientific design of the cleaning validation program in its entirety.

It is clear that an orderly approach to cleaning validation is required in order to ensure that all activities of the program are scientifically established. When embarking on cleaning validation for the first time, it is important to establish a cross-departmental team that will focus on all of the specialties required for the cleaning validation program. A flow chart of these activities has been provided in Figure 45.1.

Modern concepts for process validation, as established in the FDA's Guidance "Process Validation: General Principles

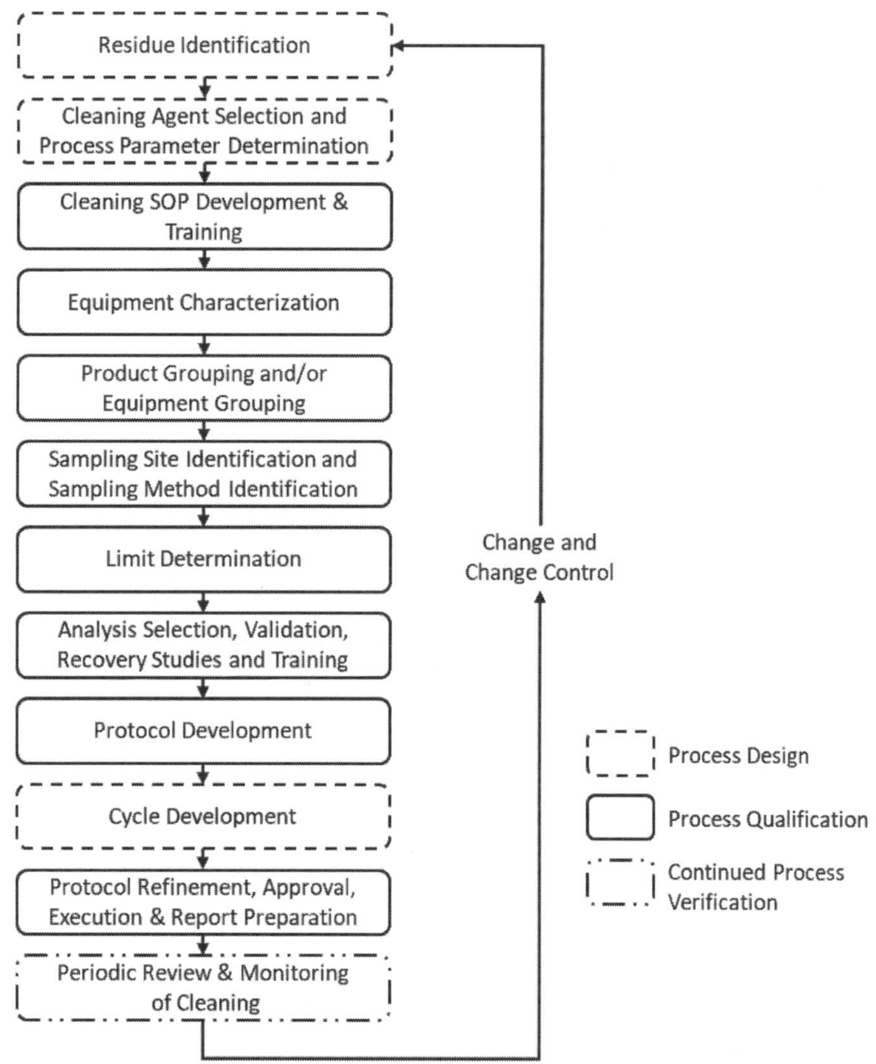

FIGURE 45.1 Activities of the cleaning validation program.

and Practices,"[7] require a three-phased approach to validation, including (i) process design, (ii) process qualification and (iii) ongoing monitoring, also known as continued process verification. These are reflected in Figure 45.1 with the different patterns of lines. The dashed lines are typical activities associated with design and development of the cleaning process. The solid lines are typical of those elements associated with the process qualification, and the dash-dotted line represents the continued process verification.

In a well-established cleaning validation program, a number of these elements may already be accomplished. For example, the equipment that is on-site may have already been fully characterized, and all grouping of that equipment may have been rationalized. The sampling sites for the equipment and the sampling methods associated with those locations may be well established. It is never a mistake, however, to evaluate each of the activities and ensure that all rationales for the program choices made are thoroughly documented and internally consistent.

Following through the flow chart in Figure 45.1 in a stepwise manner, each area of cleaning validation will be considered in the further sections of this chapter.

All compliance initiatives fall on a risk continuum. That is, there are options to make highly conservative choices or less conservative choices in the approach to compliance. Cleaning validation is no different. In evaluating the different activities in cleaning validation, one must consider the risk continuum and ensure that the position taken along that continuum is well defended (see Figure 45.2).

Validation is as much about what a firm will *choose to do* as it is about what a firm will *choose not to do.* By this, it is meant that every time an option presents itself, it is possible to create a scientific rationale both for and against that option. It is up to the personnel responsible for documenting the program to ensure that the selected options are defended not only for what was chosen but also for what was ignored or not selected. In this way, scientific rationales will be thoroughly defended.

45.1.2 Definitions Specific to Cleaning Validation

Due to the often nonspecific nature of the CGMPs and the diversity of products, processes and operating environments, it is critical to define terms. Each corporation should maintain a lexicon of terms or, at a minimum, provide definitions as part of controlled documents in order to ensure that regulators and participants in the various compliance programs will use terms consistently. While individual companies may establish terms that are different from those presented within Appendix A, the goal of these definitions is to provide a common understanding of the actions required to create a compliant cleaning validation program.

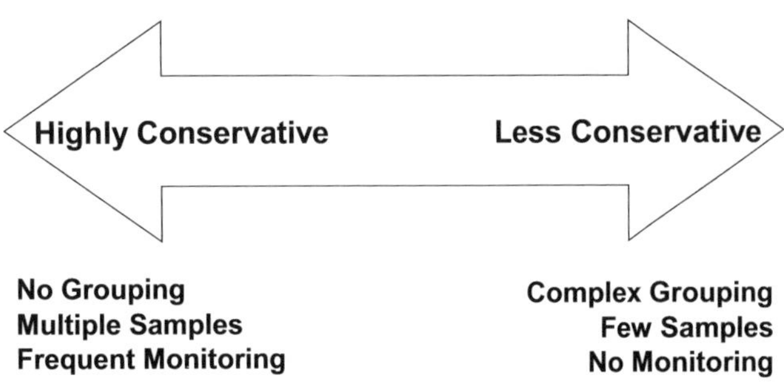

FIGURE 45.2

45.2 RESIDUE IDENTIFICATION

When performing cleaning validation there are a number of residues that must be considered for analysis, although the list may be tailored to the specific process in question:

- Active pharmaceutical ingredient(s).
- Constituents of the cleaning agent.
- Preservatives.
- Precursors or starting materials.
- Intermediates.
- Processing aids.
- Media.
- Buffer.
- Cellular debris or metabolites.
- Particulate.
- Bioburden.
- Endotoxin.
- Viral particles.
- Transmissible spongiform encephalopathies (TSE).
- Excipients.
- Colorants, dyes, flavors or fragrances.
- Water (particularly for water-free formulations/processes).

While all of these residues in the list above are *possible* residues to be considered, using a risk-based approach to determine which material(s) shall be considered as part of the cleaning validation program is an essential risk-rationale to develop. Some of the materials listed have serious health repercussions for the patient should they be carried over from one process to the next, for example, residual active pharmaceutical ingredients, bioburden and endotoxin, while other residues such as excipients or colorants might have no health consequences and might only create an abnormality in the product appearance or even no affect at all.

Therefore, it is essential that the risks associated with the selection be documented as part of the scientific rationale for the overall program, as well as for the specific product. The rationale typically includes a justification for which residues are:

- the most harmful (lowest limits or those with the greatest impact to safety, quality, identity, purity, and/or potency),
- the least soluble (hardest to clean) or
- that are present at the highest concentrations (greatest risk due to prevalence).

Active pharmaceutical ingredients (API) are those most commonly included in cleaning validation programs due to their potential harm to the next patient. Cleaning agents are the next most commonly selected materials, as they are not intended for consumption. They are ubiquitous to nearly all cleaning processes, so assessing issues such as removal and build-up becomes critical. Beyond the API and the cleaning agent, other materials such as preservatives, precursors or starting materials may also exhibit activity; the toxicities of these materials may require that they also be considered as cleaning validation targets. There may also be materials that need to be restricted from the subsequent process in order to ensure the efficacy of that process, to ensure the quality of the finished product, or to ensure the efficacy and safety of the subsequent drug product. These materials that may require limitation in the next process include precursors, starting materials, metabolites, cell debris, particulate, bioburden, endotoxin, viral particles or TSEs. Last but not least, a case can certainly be made to restrict colorants, dyes, fragrances or flavors in the subsequent product if it is going to affect customer product perception, such as appearance or functionality. Customer product perception can be harmful to product market share or could result in customer complaints.

Even after considering the selection criteria listed earlier, several residues may be identified as candidates. The next step, therefore, is to review the candidates, defend those that were not selected and then group the remaining elements, when possible. Grouping at this stage may include selecting a worst-case residue based on difficulty of removal as the representative of all other materials in the group and then testing for that residue to the lowest limit ascribed to any of the selected residues.

For products that have more than one active ingredient or more than one material that represents a risk to the patient, an element of the rationale will be the designation of one or more constituents as worst-case representatives of others that will not be tested. When using a specific test method, that is, a method that uniquely identifies each analyte, selecting all of the potentially harmful residues would be impractical, as the time to develop and validate the numerous methods required would be costly. Even if method development were rapid and inexpensive, the likelihood of having to collect samples under separate conditions to satisfy the requirements of diverse methods would result in inability to collect a meaningful sample set. This is one common reason that nonspecific methods, those methods that do not differentiate between materials but instead measure an attribute of the molecules, such as percent carbon (total organic carbon or TOC) or ionic properties (conductivity), are commonly selected when multiple residues are desired to be measured.

When considering residues in cleaning, it must be remembered that the cleaning process will not necessarily leave these materials unchanged. The alkaline or acid conditions associated with detergent cycles, contact with water or exposure to the air or heat can all promote physicochemical changes in the residues that are left on the surfaces. The likelihood of forming degradants can be assessed through laboratory forced degradation under those conditions/exposures that would typically occur during the cleaning operation.

It is appropriate to consider the safety of carryovers not only from the perspective of the native compound but from the modified forms such as degradation products or denatured materials. Nonspecific test methods can be an excellent choice when materials are known to be degraded easily. With some compounds, product specific assays can also detect some of the degraded forms of the product. The method validation will need to be reviewed to identify the capabilities of the specific assays in use.

Residue selection is an important first step in the cleaning validation program, as it will drive many of the other program decisions, including the establishment of analytical methods, the determination of limits and the identification of sampling techniques.

45.3 CLEANING AGENT SELECTION AND PROCESS PARAMETER DETERMINATION

All cleaning processes rely on the principle of TACT-WINS:

T ime
A ction
C oncentration/chemistry
T emperature

TACT represents the process parameters that are required to be controlled in any cleaning process, whether manual, semiautomated or automated. Changes in one TACT parameter may cause a commensurate increase or decrease in the other parameters. For example, for some soils, an increase in temperature can mean a possible decrease in chemistry or a decrease in the action applied. In all cases, however, the correct balancing of the TACT parameters requires proper knowledge and understanding of WINS:

W ater
I ndividual
N ature of the soil
S urface

WINS represents the parameters that affect the soil's removal from the surface. Each of the WINS attributes can affect the ability to apply TACT in a given situation. If the water quality is not appropriate, such as high hardness, it can affect the delivery of the surfactants (the "chemistry") to the surface. If the individual performing the training is improperly trained or does not have the correct tools to perform the job, the result can be inconsistent application of parameters such as the "action" on the surface. If the soil has been dried onto the surface or undergone other physicochemical changes, the soil removal might not be effective under the prescribed TACT conditions. If the surface differs from those tested in the laboratory either in terms of surface finish or the actual materials of construction, different TACT conditions might be required to get the residue off of the modified surface.

Cleaning chemistries fall into several broad categories:

- Water.
- Solvents.
- Commodity chemicals.
- Formulated cleaning agents.

Water is the universal solvent. If water alone will effectively clean the product without undue time or physical effort to remove residues, by all means employ water alone! For many cleaning processes, however, the use of water alone requires an unacceptable increase in time to get the cleaning accomplished. For these situations, one of the other approaches must be sought.

Solvents are typically applied in processes where solvent usage is already called for by the manufacturing process. For example, mother-liquors are commonly used as the solvents for cleaning of APIs. As the mother-liquor may be able to dissolve or effectively suspend the primary residue, it may offer a good choice for removal. Further, there is little risk in employing it for cleaning, as it is already associated with the process and therefore does not introduce any new residues to the surface. In addition, the facility is already equipped to handle the hazard and effluent issues associated with the solvent and may even have a solvent recovery process that enables them to reclaim the solvent after use. With today's increasing focus on environmentally friendly processes, however, companies are frequently trying to find ways to reduce their solvent usage, and eliminating solvents in cleaning is far simpler than removing solvents from synthetic pathways.

Commodity chemicals such as sodium hydroxide (NaOH) can be used for cleaning as well. Like their solvent counterparts, there may be hazard issues and effluent issues associated with these materials. Their typically high alkalinity or low acidity, however, often makes them effective at hydrolysis, oxidation and reduction of soils, which makes them helpful in inactivation processes. However, these chemicals lack the detergency of a formulated cleaning agent, and they may be inadequate to rinse degraded residues free from systems, taking larger volumes of water than would a formulated cleaning agent.

Formulated cleaning agents are by far the largest class of cleaners. This category includes solvent-based formulations and aqueous formulations. Typical formulated cleaning agents can include one or more alkalinity or acidity sources, surfactants, builders, sequestrants and chelants, as well as solvents and water. For industrial applications, unlike consumer-use products, these materials are formulated to be low-foaming and therefore are more readily rinsable and are appropriate for high impingement or high turbulence cleaning. Each formulated cleaning agent will have an optimal temperature range for use in which the surfactant will be most highly effective at helping to carry the soil away with rinse water.

To properly select a cleaning agent and establish cleaning procedures, one must understand:

- Soil (formulation, residue condition—dry, wet, baked on, layered on).
- Surface (materials of construction, surface finish, porosity, challenging geometries).
- Available cleaning methods (manual, semi-automatic, automatic).
- Available utilities (temperature, grade of water).
- Safety considerations (personnel protective equipment, likely aerosolization).
- Effluent considerations (temperature, pH, limited chemical constituents, limited volumes).

With this knowledge available, it is easy to screen detergents that meet facility and operational requirements for removing potential residues. Typical pharmaceutical cleaning agent suppliers often provide bench-top cleaning residue studies. These studies commonly make use of coupons (small samples of the materials of construction of the equipment) soiled with the residues in a manner that represents the process. The soiled coupons are then subjected to a variety of cleaning conditions including different cleaning chemistries at varying concentrations and temperatures for a variety of durations (time). In this manner the preliminary process parameters of time, chemistry/concentration and temperature (the T, C and T of TACT) that are optimal for the residue and equipment are established. While the "action" (the A in TACT) can be simulated easily at bench scale for some types of cleaning, a more accurate representation of action is often deferred to cycle development conducted at scale. This should not be viewed as a detriment, however. If a cleaning agent is successful in removing a soil with time, temperature and chemistry alone, the addition of action often improves the cleaning results, making the cleaning faster and more effective.

It is important to remember, when screening detergents, that for formulated products, the formulation may influence the ability to clean a soil more than the API. This is typically true because the excipients tend to be the majority constituents of dosage forms and recognition that all release modifying properties are typically provided through the excipients.

When selecting a cleaning solution (whether a solvent, commodity chemical or formulated detergent), it is important to understand the composition of those products in accordance with the Barr Labs court case.[1] In that case, Judge Wolin ruled that pharmaceutical manufacturers must know the composition of their detergents and must test for residuals from these detergents. As a result of this ruling, and due to demand from industry, pharmaceutical suppliers of formulated cleaning agents will typically reveal their formulations to their customers under confidentiality agreements. The composition information is provided for the purposes of understanding toxicity, solubility and markers for analytical detection. By revealing these formulations, it is possible for pharmaceutical companies to gain assurance that they have established appropriate scientific rationales for the removal of the cleaning agent.

Cleaning solutions, whether solvent, commodity chemical or formulated cleaning agent, should be treated as raw materials. That is, there should be assurance of control in the purchase, testing and specifications for the material. There should be a mechanism for complaints and investigations, along with the implementation of corrective and preventative actions, when necessary. Good manufacturing practice principles should be followed in terms of lot traceability and documentation. There should be a quality system in place that assures, for formulated cleaning agents, that no changes to the formulation will occur without prior notification. In cases where these changes are inevitable due to changes in environmental law or availability of specific chemistries, the pharmaceutical customer should be provided with studies, data and documentation to aid in bridging the gap between their original work and any new validation work.

Disinfectants are not effective cleaning agents, by and large. Although many of them contain surfactants, they are not designed for the heavy soil load associated with equipment immediately after processing. Disinfection and cleaning cannot occur in one step, as cleaning is required to remove the soil so that the disinfectant can be effective. Disinfectants are poor penetrants, and the disinfectant's active ingredients can be inactivated with excessive surface soil before having a chance to attack the bioburden that is their true target. Residues can provide protection for the microorganisms that are within and below the soil layer from the disinfectants that are applied to eliminate them.

In ideal situations, detergents are harmonized within the facility to ensure that there is no opportunity for mix-up, meaning that a single detergent is employed for the cleaning of multiple different product residues. However, in some cases the specialization of detergents may be unavoidable due to diverse residues, the materials of construction of a piece of equipment or the method of cleaning. In these cases, it is necessary to consider how best to ensure that the correct detergent is used each and every time the cleaning operation is performed. This may include a checklist as part of the cleaning documentation that is completed by the operator and may extend to double-checking the identity of the material that is being used as a part of the cleaning operation.

45.4 CLEANING SOP DEVELOPMENT AND TRAINING

Once the cleaning agent is known, the cleaning methods can be further refined. As stated previously, if a bench-top study was performed for the cleaning agent selection either in-house or by the detergent supplier, the end-user will typically have a starting point for time, concentration/chemistry and temperature for the cleaning process.

For manual cleaning processes, where an operator is wielding either a brush or a hose for performing cleaning, one of the biggest challenges associated with the cleaning is achieving reproducibility in the operators' actions. To ensure that reproducibility is possible, detailed procedures should be created, ideally with a corresponding checklist to be completed during the actual cleaning operation. Key elements in assuring operator to operator reproducibility are:

- Defining disassembly.
- Sequencing cleaning actions to prevent recontamination.
- Defining tool use and tool actions.
- Defining times for segments or activities in realistic durations.

Increasingly, industry has found it appropriate to create cleaning procedures in a form that is similar to batch records. By following a structured format with required periodic data entry, or acknowledgment signature, an effective record of the process control is established. In addition, a clear checklist of activities for the operator to follow is created. The challenge is always to determine how to define these procedures for the best effect. Interrupting the flow of the cleaning in order to create documentation for documentation's sake should be avoided, but capturing the critical steps is important. Preparing checklists, particularly for manual cleaning, can ensure that operators can be trained to a highly detailed standard operating procedure and can reliably execute that procedure using a checklist that highlights and standardizes the sequence and content of critical activities. These checklists commonly include the confirmation of achievement of the critical process parameters for cleaning, or TACT.

For semi-automated or automated cleaning, which might be performed with a parts washer or Clean-In-Place system, the TACT parameters are frequently fully instrumented and controlled by a microprocessor or programmable logic controller (PLC). In these cases, reproducibility of the cycle is not likely to be a problem if the recipe is secured from modification and selected correctly at the time of execution of the cleaning. What becomes of more concern are the human interfaces with the system prior to the initiation of the cycle such as making/breaking of piping connections, consistent disassembly and loading of a parts washer, and the like. Therefore, on these automated and semiautomated systems, the standard operating procedures focus on these critical human elements. In this manner, reproducibility concerns with semiautomated and automated washing are greatly reduced.

When reviewing standard operating procedures for manual cleaning in particular, asking a few simple questions with each step of the process can help ensure that the SOPs are consistent and sufficiently detailed:

- Are all appropriate personnel protective measures in place to protect from temperature, chemistries, aerosols, splashing, product residues?
- What is the duration or **time** of each step?
- What is the **action** to be performed at each step, and how is it defined?
- Are all tools associated with this action listed in the materials list and referenced consistently at this step?
- Are there directions as to when to discard a disposable tool when it has reached the end of its appropriate use period?

- What is the **chemistry/concentration** that should be applied to the surface during the wash step?
- How is the correct preparation of this cleaning chemistry/concentration assured?
- Are there instructions as to when to change the cleaning solution as it becomes increasingly dirty?
- What is the **temperature** for this step?
- How is the temperature controlled, if at all? Are there instructions for what to do when the temperature is out of range?
- What failures/discrepancies can occur in each step of this process, and what instructions are provided to the operator to deal with those failures?
- Are there directions for how to divide and document the cleaning activities that require more than one person to execute them?
- Are the steps performed in a sequence that prevents recontamination of surfaces that have already been cleaned?
- Are there instructions for cleaning, maintaining, inspecting and putting away all tools used in the cleaning?
- Is the drying time for the equipment defined?
- Is the location where equipment is to be dried defined?
- Are there instructions for the inspection of equipment upon completion of the cleaning?
- Are the tools that are required for inspection (*e.g.*, mirrors, flashlights) defined? Are there requirements for how these tools are to be cared for (*i.e.*, cleaned, stored, charged, handled, replaced)?
- Are there instructions for handling, protecting and storing equipment after completion of cleaning and drying?
- Are there instructions for the clean equipment hold time assignment (equipment expiration)? How are these communicated/documented?
- Are there instructions for the labeling/tagging of equipment and processing areas throughout the different stages of the process (*e.g.*, dirty, cleaning in progress, awaiting inspection, cleaned and ready for use, expired)?
- Is there a clear segregation between dirty and clean equipment, and is there an equipment and personnel flow that enforces that segregation?
- Are the methods of transport prescribed between production areas and the washroom to avoid contamination of the facility with product residues during transport?
- Are there time limits for how long equipment can sit dirty before the commencement of cleaning? How are these communicated/documented?

Training for cleaning and cleaning validation is of critical importance. In particular, operators must be made aware of the importance of cleaning and the importance of each step that they perform. Operators must understand the necessity to ensure that cleaning procedures are properly sequenced, that is, that activities are performed in the appropriate order to ensure they are not contaminating surfaces that have already been cleaned. General training in aseptic practices is worth its weight in gold for any facility; it provides a level of sensitivity to cross-contamination that is unparalleled.

Familiarization of employees with appropriate techniques to ensure that environmental contamination is not transferred to process and product contact surfaces is highly valuable. In particular, for manual cleaning of equipment such as automated conveyor systems or fillers, the equipment is traditionally cleaned with lint-free wipes and a bucket of a detergent solution. When cleaning this equipment, it is important to ensure that the operators understand a "top-down-center-out" approach to avoid contaminating already cleaned surfaces. It is not unusual to observe operators cleaning critical product-container contact surfaces and environmental surfaces with the same wipe and solution. It is important to help employees differentiate the surfaces and understand their role in keeping all surfaces clean.

In addition to training employees on the specifics of the cleaning procedures, it is important to train employees in the basics of cleaning validation, especially including their role in the validation. In particular, all operators and supervisors should be aware that during a cleaning validation trial they are not being judged, but rather the adequacy of the standard operating procedures (SOP) is being assessed—that the techniques the SOP describes are under challenge. Cleaning validation is also assessing elements such as the robustness and reproducibility of the training.

Another aspect of training is the education of inspectors (samplers). Inspectors are on the front line in helping to ensure that equipment is clean, both during the validation and after validation is complete. Visual inspection is important to every cleaning validation program because of the assurance it provides of the baseline cleanliness of all surfaces, not just those that are sampled. In addition, this visual assessment can also help to ensure that excipients and all other materials not subject to analytical testing are removed. Routine visual inspection after cleaning is the one common denominator between the validation and the routine operations that ensures that surfaces have met minimum cleanliness standards. Visual inspection also helps to ensure that the conditions that were achieved in the original testing remain consistent.

When training inspectors, it is necessary to ensure that they are made aware of appropriate inspection techniques and tools so that they do not contaminate the clean surfaces they are inspecting. Inspectors need to know where to look and how to identify residue on the surfaces. Because cleaning may be conducted in either a disassembled state or an assembled state, it is essential that inspectors understand whether equipment inspection should be conducted assembled or disassembled. Inspectors, like cleaning personnel, should always be cognizant of wearing appropriate protective equipment such as gloves and lab coats or other area-specific gowning to ensure they are not contaminating the equipment they inspect.

For routine inspection, standardized inspection tools should be used. Flashlights can be difficult to control unless they have rechargeable batteries and are placed on charge frequently. Intrinsically safe, electric lights may be a better option for inspection of deep vessels or other hard-to-illuminate areas. Other tools such as remotely operated digital cameras or borescopes may have great utility in enabling inspection of areas that need it most. This equipment has the added benefit of being able to capture images and save those images as part of the cleaning or cleaning validation record. However, when inserting a tool that might potentially come in contact with surfaces, the cleanliness of the tool itself must be considered, as it is necessary to safeguard the newly cleaned equipment surfaces.

Today's training can easily be augmented using digital images and video capture of proper techniques. Using such tools can help ensure that a standard curriculum is applied and that the same techniques are routinely taught.

45.5 EQUIPMENT CHARACTERIZATION

Cleaning validation involves not only the removal of residues but also the assurance that each and every piece of equipment associated with the process has been cleaned to acceptable levels. This is commonly referred to as a train-based approach. The "equipment train" is a series of equipment through which the product or products move as they progress through the manufacturing process.

Effective cleaning starts with effective equipment design. Contemporary standards such as the BioProcessing Equipment Standard from the American Society of Mechanical Engineers (ASME)[8] and similar design guidances that promote cleanability from the American Association of Medical Instrumentation (AAMI)[9] or the American Society of Testing and Materials (ASTM)[10] are available to help understand the design principles that must be observed to ensure cleanability. Some of the principles that have been recognized to be crucial to promote effective cleaning include:

- Limit or eliminate threaded connections—use clamp-type sanitary fittings or welded connections.
- Limit or eliminate dead-leg opportunities (Length: Diameter ratio <2:1 recommended).
- Limit or eliminate annular openings (*e.g.*, orifices).
- Orient instruments and connections to ensure limited possibility for entrainment of air, liquids or soils.
- Limit length of addition ports or instrument ports and place them in such a position that they may receive direct coverage with cleaning fluids or so that they may be used to introduce cleaning fluids.
- Consider agitator design or design of other obstructions within vessels or equipment carefully to ensure that they may be cleaned on all sides—consider alternate pathways to introduce cleaning chemistries to ensure all-over coverage.
- Ensure adequate slope for drainability (*e.g.*, 1/8 inch per foot).
- Employ sanitary valves and pumps to eliminate hold-up volumes and entrained product.
- Cove corners—no right angles.
- Ensure drain sizing is appropriate for hydraulic balance during cleaning.
- Employ vortex breakers in drains to ensure adequate drainage without blinding.
- Employ heat tracing to vent filters, where used, to prevent blinding of filter membranes with aerosols that may be created during the cleaning process.
- Ensure materials of construction are nonadditive, nonreactive, non-adsorptive.

While these are all appropriate goals, existing equipment that predates these standards or process requirements that limit the ability to comply with these requirements may be encountered. There may also be competing demands for plant safety or cost-reduction that may dictate less than desirable design choices, so that the overall facility can meet the design goals. In these cases, careful consideration must be made of the equipment in order to ensure that the equipment is cleanable. Any areas that do not meet ideal standards for cleanability should be addressed through additional steps taken during routine cleaning and validation to ensure that cleaning processes have overcome the design risks. Risk analyses may help to establish priorities for remediation of these challenging design features.

In order to assess that the equipment will be cleanable, characterize all of the equipment so that its design features are well known. Equipment characterization can assist cleaning validation initiatives in many ways:

- Promote more effective cleaning procedures by identifying cleaning challenges and ensuring that they are addressed in the cleaning methods employed.
- Identify hard-to-clean locations and high-risk locations in equipment for the purpose of sampling-site selection.
- Target materials of construction that will be included in sampling recovery studies and those that will not be included.
- Isolate materials that will be disposed of at the end of a production process and/or will be dedicated to a single product.
- Verify that all materials of construction are compatible with the selected cleaning agent and temperatures that will be used with the cleaning process.
- Collect product-contact and sample-site surface areas for the purpose of calculating limits and results.
- Confirm similar geometries, capacities and use of process equipment for the purposes of grouping that equipment.

When performed correctly, equipment characterization is the process whereby the features and attributes of equipment are catalogued, thereby ensuring that equipment can be cleaned reliably and reproducibly. Installation qualification is a likely opportunity to collect the equipment characterization information needed for new equipment. Furthermore, for new equipment, it is even more appropriate to make the documentation of the design attributes part of the bid specification prior to purchase so that the turnover package provided by the equipment vendor contains the necessary details.

In addition to pure design, the manner in which a piece of equipment functions is an essential part of characterization. Mechanical actions on the soils during or after processing could result in physicochemical changes to the soil (*e.g.*, heating, friction, drying). Since the mechanisms of soil deposition on the equipment are critical to that residue's cleanability from equipment surfaces, these mechanisms must also be considered in any characterization of the equipment.

Many factories maintain multiple pieces of equipment for the same function. This replication enables flexibility when scheduling production, scheduling cleaning activities and scheduling different products that make use of that same equipment. As a result, for cleaning validation, it is appropriate to group or bracket that equipment based on its nearest relatives. The equipment characterization—with its assessment of materials of construction, design dimension and features that affect how it is soiled, used and cleaned—makes an appropriate place to document decisions about grouping and bracketing of equipment.

For multiuse equipment, the equipment characterization is a document that can be shared between protocols. However, protocols often have a very narrow scope and are difficult to continually cross-reference for other studies. As a result, the equipment characterization activity may be one that is better to document as a stand-alone file that is then subject to change control for equipment modifications. In this fashion, all protocols that reference that same equipment can be assured that the identical information is available. This structure will also enable the ready grouping of equipment, as the similar members of the equipment family may be characterized and recorded in a single document to help highlight their equivalence and to help demonstrate why specific family members were selected to be studied for the validation.

Because equipment characterization assesses such things as how the equipment is soiled, how the equipment is cleaned, materials of construction, geometries, and surface areas, it becomes a logical place to also document the rationale for sampling sites. It is common during equipment characterization to photograph the equipment and sampling sites and to capture these images as part of the file. The sampling sites can be described in words and can be entered into routine sampling data sheets for the collection of data during protocol execution. In this manner, the equipment characterization becomes a living file that serves each cleaning validation protocol.

45.6 PRODUCT GROUPING AND EQUIPMENT GROUPING

Grouping, sometimes also called a family approach, is a method by which products or equipment are documented to be similar or equivalent for the purposes of cleaning validation. When considered similar, a worst-case member of the group or family is selected for demonstrating cleaning validation. When considered equivalent, any member of the family may be selected as representative of any other member.

Bracketing, a term that appears in EU GMP Annex 15 under Cleaning Validation,[11] has an equivalent meaning to grouping, although in many applications it may include an added burden for testing the extremes of a population (*e.g.*, smallest and largest equipment members of the equipment family, most soluble and least soluble members of the product group). In industry, this may also be referred to as matrixing. In this section, the term "grouping" has been used to represent all of the possible variants.

Grouping may be used to simply prioritize cleaning validation studies or may be used to eliminate some of the numerous possible combinations of product and equipment studies that might otherwise need to be performed.

When grouping products, all products must be:

- Manufactured on the same equipment group.
- Cleaned with the same cleaning agent.
- Cleaned with the same cleaning procedure.

Grouping considerations for products include:

- Similar patient risk levels (*e.g.*, therapeutic indication, patient population, route of administration, potency, toxicity for drugs/devices/nutraceuticals/cosmetics or, in the case of *in vitro* diagnostics, those products that have similar diagnostic uses).
- Similar formulations.
- Similar manufacturing processes.

Cleaning validation must always be carried out to meet the lowest limit of the entire product group by ensuring that the worst-case (or representative) product is tested to the lowest limit of the entire group. In this way, it can be assured that all products that are within the group are covered by the single validation.

When grouping equipment, all equipment must be:

- Used to produce products from the same product group.
- Cleaned with the same cleaning agent.
- Cleaned with the same cleaning method.

Grouping considerations for equipment include:

- Equivalence in terms of position or role in the manufacturing process.
- Similar functionality.
- Similar design (*e.g.*, geometry, materials of construction, capacity).

The surface area used in residue limit calculations must be the largest of all equipment included in the group to ensure the most conservative approach to setting limits.

TABLE 45.2
Different Grouping Decisions Employed for Different Points in a Cleaning Validation Program

Reason for grouping	Special considerations
Initial Validation	• All of the considerations listed previously are valid. • Conservative decisions to minimize grouping may be made if limited data exist for the factory on cleaning. • Conservative decisions to minimize grouping may be made if regulatory/customer review of the data is expected for a new product.
Revalidation	• Reason for the revalidation will be taken into consideration, for example: • Equipment change—the whole train may not be evaluated, only the affected equipment may be considered—existing groups may be redefined based on members affected by the change. • Product/formulation change—the individual product changed may be studied without affecting or considering the remainder of the group. • New cleaning agent introduction—the most challenging products based on prior data and established groups may be studied.
Monitoring	• More aggressive grouping than for the original validation may be employed, particularly if: • Some products were found to be well below predetermined levels of acceptability during the original validation. • Some equipment was found to be well below predetermined levels of acceptability during the original validation and was deemed not to be a challenge to the cleaning process.
Dirty Hold Time Studies	• Selected products or equipment for study may be based upon attributes that could make the equipment more difficult to clean after prolonged hold, such as: • Product—hygroscopicity, propensity to dry onto surfaces forming a hydrophobic film, degradation upon exposure to air or light, propensity to support microbial propagation (*e.g.*, specific constituents, unpreserved formulations, water activity). • Equipment—features that might retain excessive residual product, features that might promote drying or exposure to light, materials of construction that might become increasingly difficult to clean with time (*e.g.*, screens or membranes). These features may be different from those selected for the cleaning validation groups and may represent a different set of challenges.
Clean Hold Time Studies	• Cleaned equipment stored in the same environmental area with the same precautions of drying/covering/closing/sealing will store equivalently; therefore, more aggressive grouping will typically apply. • Product—typically not a consideration unless different products leave significantly different starting bioburden levels after cleaning is completed. • Equipment—typically equipment is stored based on broad "classes" that consider elements such as potential to retain moisture after cleaning (*e.g.*, complex geometries or polymers that may have a high relative humidity) or equipment that does not close or seal and therefore has to be stored with loose covers.

Grouping may be employed for the initial validation, for revalidation and for program changes, for monitoring, for clean and dirty hold time studies and the like. Different grouping decisions may be employed for these different studies based upon risk (see Table 45.2).

Any time that grouping is employed, recognize that an auditor can always ask the easy question of "Why didn't you study __________?" So be prepared to defend the grouping strategies utilized and to defend any worst-case members selected for testing. Remember the risk continuum in Figure 45.2 and ensure that the defined risks for grouping can be appropriately defended.

45.7 SAMPLING SITE IDENTIFICATION AND SAMPLING METHOD IDENTIFICATION

Sampling sites should be selected based on the most difficult-to-clean geometries of the equipment. If the difficult-to-clean locations are clean, it is expected that the easier-to-clean locations will be at least that clean or cleaner. These difficult-to-clean locations however, are frequently inaccessible—their very inaccessibility is what makes them difficult to clean! Therefore, when choosing sampling sites, one must always be cognizant of the difficulty of cleaning and the intended sampling methods for that location. Sampling methods have various advantages and disadvantages that make them suitable for various geometries and locations on the equipment.

Equipment may have both hot spots and critical sites. Hot spots are locations that are likely to become dirty during the manufacturing process and are difficult to clean. Critical sites are those locations that, if they were to remain dirty, would provide a disproportionate level of contamination to the next batch or portion of the next batch.

An example of a hot spot might be the bottom of an agitator or an instrument port inside a vessel that is likely to become soiled during the manufacturing process and might prove to be difficult to clean during the cleaning process. The use of an agitator in a mixing vessel means that any soil remaining on the sidewall of the vessel is likely to become homogeneously distributed within the next batch. Contrast this agitator to locations such as a filling needle, a tablet press table or a fraction collection valve on a chromatographic skid. Each of these locations has the opportunity to affect the next dose of the product or the next portion of the batch being produced. The residue that remains on these locations will not be homogeneously mixed throughout the batch but instead will disproportionately contaminate a small number of doses or, worst case, a single dose.

When selecting sample sites, a variety of locations must be evaluated, including hot spots and critical sites as well as some representative locations on the equipment. Remember to include those locations which might experience recirculation or redeposition of contaminants during the cleaning process. For example, in a vessel that might use fill-soak-and-agitate as the cleaning method, the agitation level might decrease as the liquid is drained from the vessel. This may mean that there are significant bathtub-ring risks on this type of equipment based upon the settling of suspended residues on surfaces as agitation levels are reduced. Although these locations are not "hard to clean" in the conventional sense, they become a deposition risk as a result of the mechanics of the cleaning process.

The number of sample locations selected for any individual piece of equipment should be based on the very same considerations that were addressed in sampling location selection:

- Difficult to clean geometries (hot spots).
- Locations that disproportionately contaminate a portion of the next batch (critical sites).
- Representative locations.

In addition, sampling sites and the number of locations selected may also be influenced by:

- Materials of construction (inasmuch as different materials might have different affinities to soil).
- Overall scale of the piece of equipment (to ensure that coverage issues are addressed top to bottom and side to side).

For example, in a fluid bed granulator which can be 6 meters in height or taller, there may be difficulties in coverage top to bottom as well as side to side. In order to ensure adequate cleaning, it may be necessary to sample several locations on the sidewall of this equipment despite the fact that the sidewall is all of the same material of construction and not a difficult-to-clean geometry.

In order to determine the sampling locations, several tools may be employed:

- Review of the equipment characterization for process attributes (*e.g.*, friction or heat), geometry, and materials of construction.
- Review and observation of manufacturing process and cleaning SOPs for potential areas of challenge or weakness or locations where any ill-controlled process parameters may result in variability.
- Interviews with and observation of operators to discuss their experience with difficult soil deposits, difficult-to-access areas or observed areas of visual failures.

Regardless of the tool selected, it is best practice to document the assessment in a sampling site rationale. Best practice is to be able to defend those areas that were selected and those that were not selected, and why.

When determining sample size for cleaning validation, target a sample size that will provide sufficient residue to the assay but not collect samples so large that residue recoveries may suffer. Most firms select a convenient surface area (*e.g.*, 100 cm^2 or 25 cm^2), but it should be noted that it is possible to vary sample sizes slightly from sample to sample, when the sample size is accounted for in the result. Sample size variation may occur naturally based on the geometry of the equipment. For example, if a valve has a product contact area of 105 cm^2, it is much more appropriate to sample the full surface and account for the small overage in the equation for the results (see "Limit Determination", Section 45.8) than it would be to instruct the sampling personnel to sample all of the surface except for 5 cm^2.

In general, the larger the sample, the better for the analytical method, as a higher limit of detection can be used because a larger representation of the soil on the surface is introduced to the test method. For this reason, 100 cm^2 is more appropriate than 25 cm^2, especially when working with low limits. However, care should be taken to confirm recovery levels with whichever sample size is selected, as larger areas may result in a swab that is too dry to effectively recover residues (see Recovery Studies section for additional detail).

There are a variety of sampling methods for cleaning validation. Any method can be used provided it can be demonstrated to be suitable to recover the soil reproducibly from the surface. This need for effective recovery is the reason methods validation is always coupled with an assessment of sampling method efficiency.

The most common sampling methods employed in cleaning validation are rinse sampling and swab sampling. Of the two, swab sampling is typically deemed by regulators to be preferable. There are clearly situations where both methods may apply, however, and it is important that any sampling method used should be selected based upon its strengths.

Swab sampling is the use of a material, usually absorptive, to physically wipe a surface and recover the analyte. Because of the need to physically wipe the surface, swab sampling is a preferred method in locations where the surfaces are readily accessible to a human hand or arm. (There are also commercially available extension tools that provide a telescoping arm that can be used to reach the bottom of a tank or a long distance into a pipe.) The swab is typically used with a diluent (water, solvent or a combination of the two), although the sampling may also be conducted dry if that is proven to be effective for recovery (although this is far less common). When used, the diluent for moistening the swab must be compatible with the analytical method or must be evaporated and replaced prior to analytical testing.

Rinse sampling, as compared to swab sampling, does not employ mechanical action on the surface other than that which is delivered by the fluid traversing the surface. Rinse sampling may be collected either as a portion of the final rinse of the cleaning process or as a rinse applied specifically for the purposes of collecting a validation sample. The advantages of rinsing specifically to collect a sample rather than as part of the final rinse include:

- The cleaning process is truly at its conclusion when the sample is collected rather than in mid-process.
- The quantity of rinse solution may be reduced for sampling, and therefore the dilution effect is minimized.
- The rinse sample, due to its limited size, can be made truly homogeneous before aliquoting the sample for the laboratory.
- The rinse sampling can be targeted to specific zones (depending on the method of application), which can result in sampling of critical spots and a potentially lower dilution factor.

As with swab sampling, the solvent employed for rinse sampling is selected due to its solubility and compatibility with the residue(s), although commonly rinse sampling is performed with the final rinse water. Rinsing with alcohol or other solvents may be possible with appropriate safety measures in place.

From the description of these two techniques, it is possible to see why swab sampling is typically preferred as opposed to rinse sampling. Regulators have long argued that the mechanical action provided by swab sampling provides benefits in determining whether the cleanliness of the surfaces has been achieved. In 1993, when the FDA drafted the cleaning validation guidance,[2] inspectors would frequently refer to a theory called the "baby in the bath water". They used the "baby in the bath water" theory to explain why they preferred swab sampling over rinse sampling. The theory was, if you are trying to determine whether or not a baby was clean, would you look at the baby? Or would you look at the bath water? The answer, of course, was that you would look at the baby!

In the question asked by the regulators, if the baby represented the equipment and the bath water represented the rinse sample, why did we believe that the rinse sample would represent the cleanliness of the baby? Their concern was that if the bath water (rinse) had not contacted all soiled areas or had not had sufficient contact time with the baby (equipment), if the residues on the baby (equipment) were poorly soluble in the rinse water, if the residues were not homogeneously distributed in the bath water (rinse) or if the residues became too dilute in the bath water (rinse), that examining the bath water (rinse) would not be an appropriate technique for establishing the baby's (equipment's) cleanliness. Therefore, they concluded that rinse would be an inappropriate technique for sampling the baby (equipment). In part, the FDA had answered its own question by asking all of those questions—those are the very details to be proved with properly executed recovery studies (see Recovery Studies section) and through appropriate sample site selection. If the recovery study and sample site demonstrate that the rinsing technique is adequate to demonstrate the cleanliness of the "baby" (equipment), there should be no further objections to the use of that rinse technique.

There is a fallacy that all piping can be rinse sampled effectively. While it is true that rinse sampling is often the best method for sampling long piping runs, if the piping is complex, having a lot of branches and tees, these locations may not get wetted appropriately during the cleaning and the subsequent rinsing operations. Air trapped in these tees, especially if they violate good engineering practices (*e.g.*, tees that are ≥ 2:1 length to diameter),[8] can resist getting wetted, even at very high flow rates. If cleaning fluids do not penetrate these locations, there is no reason to believe that rinse solutions, which are delivered in the exact same manner, would penetrate these locations and remove residues that remained intact for the entire cleaning cycle. These locations might need some limited disassembly or a reversal of the flow path to ensure that there is direct contact with these points.

Direct surface sampling is also possible in cleaning validation. For example, using mid-IR grazing angle spectroscopy (Fourier Transform Infrared Spectroscopy (FT-IR))[12] or photoelectron emission (optically stimulated electron emission (OSEE)) techniques,[13] a firm can both sample and measure the quantity of residue available on a surface. With these techniques, specific spectra may be obtained from residues remaining on the surface, thereby directly quantifying and identifying the designated residue. These techniques can be highly desirable, as they directly measure the quality of the surface or the "baby" (equipment). The ability to represent the sampling and the analysis in one step means that there is no real "loss" to the sampling system. As with swab sampling, however, the direct analysis of surfaces is limited to those areas that are accessible for inspection. These techniques often require the use of a moderately sized head that delivers the infrared signal to the surface and retrieves the reflected wavelengths. The detector processes the reflected spectra through an attached signal processor. Therefore, the sample sites are commonly limited to those that can accommodate the size of the head while still being within the distance that can be reached by the connectors to the external signal processor.

Visual techniques using the eyes or remote inspection cameras are another form of direct surface sampling. The various visual techniques are typically nonspecific, however, and are also nonquantitative—or, at best, semiquantitative—as results can only be stated to be "less than" a specific visual reference standard. With the advances in digital imaging, before and after images, or the comparison of appearance to prior cleaning events, may be accomplished. In some cases, a side-by-side comparison in the field against a stored image is possible.[14,15]

Coupon sampling involves the introduction of a soiled coupon of the materials of construction of the equipment into the process equipment for the purpose of later removal and analysis in the laboratory. In some facilities, small portions of the equipment, such as filling needles and spool pieces, may be similarly removed for the same purpose. The advantage is the opportunity to apply sampling techniques in the laboratory such as extended soaks, physical agitation or sonication and the application of more hazardous solvents than would be advisable on the plant floor. The technique of using coupons for sampling also lends itself to so-called "false soiling", if a worst-case soil condition is desired—for example, for sampling baked-on residue, or for quantitative soil removal using techniques such as gravimetric analysis.

Similar "coupon" approaches can also be used with either rinse or swab sampling with small swatches of fabric or materials where testing of the surface is very difficult. As an example, fluid bed dryers employ large bag filters. These bags have extraordinarily large surface areas. Because they are typically considered difficult to clean, they are frequently dedicated to an individual product. This dedication, however, does not exonerate the manufacturer from the responsibility to test for cleaning agents or materials used in the cleaning process. As a result, surfaces must be sampled for any residual cleaning agent. Swab sampling is typically deemed ineffectual for woven surfaces due to the complexity of the weave and the fact that residue may become trapped between the individual fibers. Rinse sampling can be very effective with woven surfaces because it provides a prolonged soak and will help to loosen or dissolve residues from the surfaces. Because of the immense size of some of these filter bags, however, it is impossible to provide an efficient sampling method. The equipment for soaking the part without using extremely large volumes of fluid is often not available. In these cases, a small swatch of material from a bag that is to be retired or a sample from a vendor is obtained and intentionally subjected to worst-case soiling. This coupon of fabric is subjected to the washing process with a routine filter bag, and the "coupon" is then returned to the laboratory for testing. Due to the small size of the swatch of fabric, the coupon can be fully immersed in a beaker for rinse recovery from that surface. Prolonged soaking of a coupon is possible to maximize recovery. Recovery studies would be necessary to confirm the recovery associated with this technique.

Concerns exist with the coupon sampling technique when manual cleaning is performed, inasmuch as the operator should not concentrate on cleaning the coupon. Although the use of this technique is fairly infrequent, it has great potential flexibility for facilities where insufficient product is available to soil the whole system, or where investigation into appropriate cleaning techniques is desired prior to completely soiling equipment. When using this technique to study cleaning processes, remember that if only the coupons are soiled, the recirculated soil load in cleaning solutions will be significantly lower than when the equipment is fully soiled. This may hamper the cleaning effectiveness on a fully soiled system, especially when considering soil redeposition issues.

An additional approach that can be taken for sampling includes the placebo approach. Placebo batches are recognized as both potential cleaning techniques and potential sampling techniques. In the former case, a placebo material produced using all typical excipients but no active ingredients would be passed through a process system for the purposes of scrubbing clean the system from the prior material. The principle is that the placebo would pass along the same pathways as the product and therefore would have an opportunity to scrub off residual product along those pathways.

In some cases, particularly in powder-based systems for safe product residues, such as those associated with topical consumer product powders, the placebo technique has been used to avoid wet-cleaning of surfaces and prevents potential stickiness of the surfaces due to residual moisture on the equipment that might wet the powders.

With placebo cleaning, it can be difficult to demonstrate the robustness and reproducibility of the process. The use of the manufacturing process to accomplish cleaning relies on the critical process parameters for the manufacturing process. These parameters may not be appropriate to ensure that there is adequate coverage of surfaces or sufficient mechanical action to remove the prior product.

Placebo sampling, on the other hand, is employing a placebo and passing it through the equipment for the purpose of measuring system cleanliness. The placebo sampling technique is very much like the "baby in the bath water" technique and depends upon:

- Excipients being fully "soluble" in the placebo.
- Sufficient contact time of the placebo to collect a representative sample.
- Placebo has adequate "coverage" of the process pathway to ensure removal of the placebo from all equipment locations.
- Quantity of the placebo and the residue being matched so that the residue is in a detectable range within the placebo (*i.e.*, not overly diluted).
- Residue being (somewhat) uniformly distributed within the placebo in order to ensure detection based on sampling any portion of the placebo.

As with rinse sampling, if a scientific case can be built for the use of the placebo method, it may be appropriate for either cleaning or sampling. Regulators have cautioned in the US cleaning validation inspection guidelines[2] that the placebo method may most appropriately be used in combination with other sampling methods.

The placebo sampling technique has been successfully employed in specific applications for a highly colored tablet granulation, where a pharmaceutical company used a white placebo to verify that all colored excipients had been removed from their tablet press hopper and feed frame. In this case, they use swab sampling to demonstrate API removal from the system but use the processing and inspection of a small quantity of placebo granulation and tablets for the verification and removal of tabletting excipients. They had demonstrated that the ability of an inspector to discern colored specks in a white tablet was far greater than the level of detection promised by a swab or rinse of the surface for that same-colored excipient.

Table 45.3 includes a summary of the major sampling techniques and their attributes. The clear message for all sampling techniques is that they all have advantages and disadvantages and should be applied based upon the selected sampling locations.

TABLE 45.3
Major Sampling Techniques and Their Attributes

Attributes	Swab	Rinse	Direct surface analysis	Coupon	Placebo
Physical sampling of surface	•	○	○	•	•
Robust technique *(low technique dependency)*	○	•	•	•	•
Non-invasive technique	○	•	○	•	•
Adaptable to hard-to-reach areas	○	•	○	○	•
Effective on flat surfaces	•	○	•	•	•
Effective on complex geometries	○	•	○	•	•
Controlled area sampling possible	•	○	•	•	○
Samples are homogeneous	•	○	○	•	○
Does not require prolonged contact time with surface	•	○	•	•	○
Adaptable to different solvents/materials for sample removal	•	•	N/A	•	•
Appropriate for on-line adaptation	○	•	•	○	•
Recovery study required	•	•	•	•	•
Frequency of use	High	High	Moderate	Low	Low

Key: • Effective or low risk; ○ Ineffective or high risk.

45.8 LIMIT DETERMINATION

Worldwide regulatory guidelines indicate that manufacturers must define their own limits for cleaning validation.[2,11,16,18,19,22,23,24] The large variety and type of products, as well as the number of processes and equipment, make it impossible for any regulatory agency to establish firm limits that would apply to every situation. What is clear in the regulatory guidelines, however, is that regulatory agencies have an expectation that cleaning will be pursued until the residues reach safe levels for patients who will receive the dosage form made in the equipment.[11,17,19,20,24]

The general strategy for how to construct a limit is well defined; however, it is up to the individual company to establish the basis for the selection of the individual terms that they use in their equation, including any conservative assumptions that they might make along the way.

While there are clear limits for drug products, the guidance can be less clear for products such as *in vitro diagnostics*, medical devices and processing aids that are removed as part of the process or for those materials that might be rendered inactive during the steps before or during cleaning. This section hopes to remedy that by establishing the different possible methods for setting limits in different segments of the industry.

The European Union introduced Health Based Exposure Limits[11] that were established on the basis of the "Permitted Daily Exposure" or PDE. According to that guidance, "[t]he PDE represents a substance-specific dose that is unlikely to cause an adverse effect if an individual is exposed at or below this dose every day for a lifetime." The principles of the PDE were already well established in worldwide industry guidance for residual solvents; however, this publication of PDE by the EMA[11] represented a shift in limit approaches in the pharmaceutical industry. The world's regulatory bodies (including the World Health Organization,[24] PIC/S,[19] Health Canada),[22] standard setting bodies (such as ASTM)[21] and industry associations (such as International Society of Pharmaceutical Engineers (ISPE),[25] Parenteral Drug Association (PDA)[26] and Active Pharmaceutical Ingredients Committee (APIC))[28] all followed suit to publish (or begin drafting) limit-setting approaches that aligned with the principles of PDE.

To calculate the PDE, a trained toxicologist is charged with the responsibility to identify the hazards represented by the substance for which the limit is being set. Animal and human data from safety and toxicology studies are reviewed to determine the critical effect and the level at which this material would have no observed adverse effect (NOAEL) on the patient.

The PDE is calculated as follows:

EQUATION 45.1: PERMITTED DAILY EXPOSURE

$$PDE = \frac{NOAEL \times Weight \times Adjustment}{F1 \times F2 \times F3 \times F4 \times F5}$$

Where: NOAEL = no observed adverse effect level

Weight adjustment = weight of the patient (or standard weight term)

F1 = A factor (values between 2 and 12) to account for extrapolation between species

F2 = A factor of 10 to account for variability between individuals

F3 = A factor 10 to account for repeat-dose toxicity studies of short duration, *i.e.*, less than four weeks

F4 = A factor (1–10) that may be applied in cases of severe toxicity, *e.g.*, non-genotoxic carcinogenicity, neurotoxicity or teratogenicity

F5 = A variable factor that may be applied if the no-effect level was not established. When only a lowest observed effect level (LOEL) is available, a factor of up to 10 could be used, depending on the severity of the toxicity

The toxicologist must prepare a technical report that justifies their selection of terms and identifies the studies referenced in the determination of the PDE and any uncertainty factors or safety factors that were applied. Non-standard safety factors can be applied if they are appropriately justified. This rationale should be retained by the pharmaceutical company to justify the limit employed in the cleaning validation program.

For developmental products, or those with limited safety studies, the "first human dose", also known as the "benchmark dose", may be selected as a conservative value commensurate with the uncertainty associated with the product at that stage of development.

The calculation of the PDE is only the first step in deriving an appropriate limit. The PDE represents the amount of the product to be cleaned that is safe for patient exposure. However, as stated earlier, the PDE is a substance-specific dose that is unlikely to cause an adverse effect if an individual is exposed at or below this dose every day for a lifetime. This material will be introduced to the patient as a carryover or contaminant within another product that the patient is prescribed. In order to determine the daily insult from the contaminating product, it is important to understand how much of the other product the patient will be receiving.

To calculate the amount of the carryover that will be administered to the patient, the size of the next batch, the maximum daily dose of that next product and the shared surface area that might carry the contaminant into the next batch, the equation for the maximum safe carryover is calculated as follows:

Equation 45.2: Maximum Safe Carryover (MSC)

$$\text{MSC} = \frac{\text{PDE} \times \text{MBS}}{\text{SSA} \times \text{MDD}}$$

Where PDE = permitted daily exposure (Product A)
MBS = minimum batch size (Product B)
SSA = maximum shared surface area (surface area shared between Products A and B)
MDD = maximum daily dose (Product B)

Some firms may still call this limit the maximum allowable carryover (MACO), which was the name for the limit when calculated using the therapeutic dose as the basis for the limit. However, other firms have switched terminology to MSC (or even maximum safe level (MSL)) to be unambiguous about the change in the basis of the limit to the NOAEL. Either name is fine, provided that the limit provides adequate safety to patients through careful selection of the remaining terms in the equation.

In a secondary manufacturing facility where formulated drug products are produced, the PDE term in the numerator represents the API (or residue of interest) weight only. The maximum daily dose term in the denominator represents the full dose weight of the subsequent product (inclusive of the API and all excipients). In manufacturing the next product, all ingredients, both excipients and actives, have the opportunity to pick up and carry with them the carryover from the prior active. Therefore, it is appropriate and conservative to consider the full dose weight as affecting the next patient. In some cases, the largest daily dose may even be substituted with the maximum dose administered during a longer term of administration.

The smallest possible next batch size is included in this equation in the numerator to represent a conservative assumption that the carryover is only minimally diluted by the next formulation entering the equipment train. This ensures the firm is considering the worst-case concentration that could affect the next patient.

The shared surface area represents the total of all product contact surface area that would be in common between the two products throughout the entire length of the equipment train at the manufacturing facility. Larger facilities commonly have more than one piece of equipment for each unit operation. As a result, there may be infinite possibilities of what could actually be shared between any two manufacturing events. It is common in these cases to select the largest equipment surface area that is in use for each unit operation. By selecting the largest area for the overall equipment train and ensuring that the quantity of residue would be safe for this larger area, any reduction in surface area during real operations would reduce the amount of residue-carrying surface area to which the patient would be exposed, making the actual condition safer for the patient.

As stated previously, other, more or less conservative, values may be substituted throughout the equation provided that the terms used are justified and ultimately provide safety to the patient receiving the next product. For equipment surface area in particular, it is possible to select the next product's equipment train or the worst-case equipment train as conservative values.

The full equipment train is considered because as the next batch traverses the surface of the equipment, all residues on the surface may be picked up and carried in the batch to the final filled doses of the product. The accumulated total contamination to the next batch, therefore, is represented by the accumulated total residue that remains on all surfaces.

When the permissible limit is divided by the total amount of surface area in the equipment train, it is mathematically assumed that there is uniform contamination of the surfaces—that every square unit of surface area would contribute the same amount of residue. This is known to not be true, as it is expected that some locations may be more highly contaminated than others. However, sampling sites are selected from the hardest-to-clean locations on the equipment. By collecting samples from the hardest-to-clean locations (those that would be the most likely to be dirty) and assuming that they are representative of all other surfaces of the equipment, a conservative assumption is made. Requiring that these hard-to-clean locations meet a limit that would be acceptable for all areas of the equipment is another conservative assumption, thereby ensuring that the limit and the residue quantities are safe for the next patient.

Another frequent concern when using the shared surface area approach is whether or not sampling of an individual piece of equipment can be accomplished if the limit is calculated based on the entire train. For example, if Tank #1 is soiled on Monday and cleaned the same day, but the product is final filled and packaged Friday, can Tank #1 be released before the filler is released? The answer is an emphatic "yes". Examining the units associated with the limit, observe the following:

Equation 45.3: Typical Units for the MSC

$$\text{MSC} = \frac{\text{mg/day}^{A} \times \text{mg or doses}^{B}}{\text{cm}^{2AB} \times \text{mg or doses/day}^{B}}$$

Where A = Product A
B = Product B
AB = shared surface area between Product A and B
Note: mg and cm^2 were selected as typical terms in this and subsequent examples but could also be represented by many other units, such as mL or µg, and in^2.

The result of Equation 3 is mg/cm^2 or, less specifically, mass per unit surface area. This means that every square unit of surface area of the equipment train may be measured independently and still assure that the total carryover will not exceed the limit for any individual piece of equipment. This is particularly helpful in facilities with prolonged processing or campaigns where different pieces of equipment may be freed for analysis at different times.

In general, if the MSC terms in the numerator are minimized by selecting conservative small values and the terms in the denominator are maximized, the overall limit will be reduced.

In multiproduct facilities, a firm can't always predict what the next product might be. In these cases, limits are calculated for all possible combinations of the next products on the equipment in order to determine the worst-case limit that must be applied to the product. In development facilities, where the characteristics of batch size, next daily dose and equipment train are not yet fully established for the next product, it is not unusual to make worst-case assumptions based on historical values or based on factors such as the minimum equipment processing capacity or maximum train typical of a particular dosage form. From a patient safety perspective, the actual conditions used for each batch should be calculated to ensure that none of the worst-case assumptions were violated in the production of the actual batch.

For APIs produced by synthetic routes, the equation would appear exactly as it does in Equation 45.2; however when considering the maximum daily dose, only the active quantity can be included in the limit because there are no excipients that would also carry over contaminants to the next batch. Although this will limit a term in the denominator and would typically provide a slightly larger limit, the extensive surface area in API factories compared to the final batch size typically provides a very conservative limit. It should be noted, that when considering the shared surface area between two APIs, only those pieces of equipment in the equipment train that see that residue need be included in the calculation of shared surface area. For example, in a four-reactor train with a centrifuge and a dryer, only the final reactor, centrifuge and dryer may actually be exposed to the final API. Earlier pieces of equipment in the equipment train would have been exposed to a different chemical precursor molecule that will be subject to its own limit. Therefore, only these last pieces of equipment would need to be considered when determining the shared equipment surface area with the next product.

This brings up an important differentiation for synthetic processes: cleaning validation for early stages of the process may include residues that may not have therapeutic indications. These precursor molecules may or may not be more toxic than the final API. If the precursor is not a known, studied substance, the toxicologist may be called upon to extrapolate likely effects from known analogs or might be asked to make a worst-case assumption of safe exposure levels.

For other materials that do not have a therapeutic index, such as cleaning agents, excipients, lubricants, polishing compounds or processing aids, a similar approach to that discussed for APIs is employed. Toxicologists should be engaged to research the compounds and establish a NOAEL, while applying all necessary safety and uncertainty factors.

For *in vitro* diagnostic products, the equation for the limit may be changed as follows:

Equation 45.4: Fraction of an Interfering/Enhancing Substance Approach for *In Vitro* Diagnostics

$$\text{MSC} = \frac{\text{MI/E} \times \text{MBS}}{\text{SSA} \times \text{MTV}} \times \text{SF}$$

Where MI/E = minimum quantity (Reagent A) that might cause an interfering or enhancing effect on the next material or test
MBS = minimum batch size (Reagent B)
SSA = shared surface area (surface area shared between Reagent A and B)
MTV = maximum test volume (Reagent B)
SF = safety factor such as 0.01, 0.001 or 0.0001 (optional term if there is uncertainty in the values selected)

The form for the mathematical formula for *in vitro* diagnostics is analogous to other products. However, instead of calculating the amount that would have an effect on the next patient, the amount that would have an effect on the next test is being calculated. Since there may be several reagents involved in the performance of a test, the limit for the interfering or enhancing quantity should be established for each reagent in use in combination with all subsequently produced reagents. Each reagent must be tested individually, or if grouping, the most critical reagent's limit may be applied to the hardest-to-clean reagent.

To determine the MI/E for a diagnostic product, some physical testing to determine the minimum inhibitory/enhancing concentrations that could affect the next reagent or its test is needed. The safety factor may be adjusted based upon the accuracy with which the interfering/enhancing quantities were determined and the degree of confidence in repeatable behavior.

For cosmetics and nutraceuticals, where an "active" substance may not be present or where that active might be Generally Recognized As Safe (GRAS), limits based on Equations 45.1 and 45.2 might be very generous. Even though the large quantity of residue has been calculated to be safe for the patient/consumer, it may be necessary to establish a quality threshold that represents a standard level at which equipment can be deemed clean. Quality threshold limits ensure that even when something is considered "safe" to carry over, the adulteration potential for that material is still limited in the next batch.

The most common quality threshold applied to pharmaceuticals is not more than 10 PPM appearing in the next batch. However, it is important to understand that this limit is clearly a convenience value when set at the number 10. Not more than 10 PPM in the next batch literally means not more than 10 parts of the prior product appearing in every 1,000,000 parts of the next product. This number is very clearly a concentration and is *not* related to patient safety. It can, however, be compared to an HBEL equation to demonstrate which limit is lower. If HBEL is lower, then it must always be pursued. If the 10 PPM limit is lower, some firms have opted to default to the 10 PPM limit as a standardized threshold for cleaning. Since this is a convenience value, it goes without saying that there are firms that pursue lower levels, at 1 or 5 PPM, and other firms that pursue higher levels, at 50 or 100 PPM, but they are all simply standardization of the permissible concentration appearing in the next batch and not safety related.

After the publication of the HBEL limit calculation approach, there was much discussion in the industry about doing away with the so-called 10 PPM limit, since it did not have a direct relationship to patient safety. However, many firms find it convenient to have a standard basis by which to set a cap on the permissible quantity of residue in the next batch when safety-based limits are higher. While the reduction in limits alone does nothing to increase patient safety, having a limit set below the safe threshold can provide firms with an alert limit that is closer to the actual results that they have been achieving, thereby providing an early warning system against adverse trends.

As with the PDE limit calculation, when calculating the 10 PPM limit, it must be understood how much of the next product will be produced and how much surface area will become contaminated with the carryover. This quality threshold limit, like HBELs, also includes an assumption that residues would be uniformly distributed in the next batch and uniformly distributed on the surfaces based upon the division of the limit with the surface area of the equipment train. As with the PDE equation, this assumption is offset by the careful selection of sampling sites.

Equation 45.5 contains the calculation for not more than 10 PPM in the next batch. For other concentrations, the terms 10 mg and 1 kg would be substituted, accordingly.

Equation 45.5: Not More Than 10 PPM in the Next Batch

$$\text{Quality Threshold} = \frac{10\ \text{mg}^{A} \times \text{MBS}}{\text{SSA} \times 1\ \text{kg}^{B}}$$

Where MBS = minimum batch size (Product B)
SSA = maximum shared surface area (Products A and B)

The units for Equation 45.5, like the HBELs, yield a result in mass per-unit surface area, as follows:

Equation 45.6: Typical Units for 10 PPM Limit

$$\text{Quality Threshold} = \frac{10\ \text{mg}^{A} \times \text{kg}^{B}}{\text{cm}^{2AB} \times 1\ \text{kg}^{B}}$$

Where A = Product A
B = Product B
AB = shared surface area between Products A and B

Because the units of measure on the HBEL and 10 PPM limits are the same, a direct comparison between the two values may be made. As the NMT 10 PPM limit includes the additional factors of batch size and equipment train, the limit rarely results in a nice round number like 10. Mistakes are frequently seen in cleaning validation programs when the number 10 is compared directly to the PDE equation. When working with limits, it is critical to ensure that equivalent units have been established prior to comparison.

All limits presented to this point have been expressed as mass per unit surface area. When reviewing a cleaning validation program, it might be identified that the limits are expressed in units that only include mass (without surface area) or are stated as mass per sample. Are these alternate limit formats incorrect? No, they are not incorrect; rather, they are simply a representation of the limit and the result as an equivalent expression with different units of measure.

The equation for the limit and the equation for the analytical test result are on either side of a comparator.

$$\text{Limit} > \text{Result}$$

When cleaning is successful, the result will be lower than the limit. Mathematical rearrangement of the terms in this equation can produce many equivalent, accurate expressions. When rearranging terms, it is important to remember to move them from the denominator to the numerator or from the numerator to the denominator when they cross from one side of the equation to the other.

Based upon Equations 45.2, 45.4 and 45.5, the terms for the analytical test result side of the equation (opposite from the limit) would typically be:

Equation 45.7: Terms for the Test Result

$$\text{Limit} > \frac{\text{SC} \times \text{SV}}{\text{SAS} \times \text{RF}}$$

Where SC = sample concentration (Product or Material A)
SV = sample volume (either swab diluent or rinse volume)
SAS = surface areas sampled (associated with either swab or rinse)
RF = recovery factor (for Product or Material A expressed as a decimal when in the denominator)

The recovery factor in this equation is intended to correct the amount of residue in the sample for any loss that occurred in sampling of the surface. After cleaning, the surface is sampled, but consider that the sampling method left some small amount of material on the surface of the equipment. It would be important to account for the material that was left behind, because when calculating the amount that a patient might be exposed to, the patient would be exposed to what was collected in the sample as well as any material that was left behind. Thus, sampling methods are corrected for any inefficiency in their collection. To correct for the recovery, the value from the sample is divided by the recovery factor, when it is expressed as a decimal. (If in the numerator it is a whole number percentage.) Mathematically, it does not matter whether this correction to 100% is performed from the initial analytical results (the concentration in the sample) or whether it is performed after it has already been converted to mass per unit surface area (the calculated result). The units for the results remain the same after the inclusion of the recovery factor. See the next section for more information on how to determine the recovery factor that should be applied to each sample.

Equation 45.8: Typical Units for the Result

$$\frac{\text{mg/mL} \times \text{mL}}{\text{cm}^2}$$

It can be seen in this case that the result side of the equation also yields mass per unit of surface area (*e.g.*, mg/cm^2) and therefore can be compared directly to the limits from Equations 45.2, 45.4 or 45.5.

If the reciprocals of the various terms are moved from one side of the equation to the other, the limits and results might be expressed as follows:

Equation 45.9: Examples of a Few Other Equivalent Limit Expressions

1. $\frac{\text{PDE} \times \text{MBS}}{\text{MDD}} > \frac{\text{SC} \times \text{SV} \times \text{SSA}}{\text{SAS} \times \text{RF}}$ Final Units: mg
2. $\frac{\text{PDE} \times \text{MBS} \times \text{SAS}}{\text{SSA} \times \text{MDD}} > \frac{\text{SC} \times \text{SV}}{\text{RF}}$ Final Units: mg
3. $\frac{\text{PDE} \times \text{MBS} \times \text{SAS}}{\text{SSA} \times \text{MDD} \times \text{SV}} > \frac{\text{SC}}{\text{RF}}$ Final Units: mg/mL

Where PDE = permitted daily exposure (Product A)
MBS = minimum batch size (Product B)
SSA = maximum shared surface area (Product A and B)
MDD = maximum daily dose (Product B)
SC = sample concentration (Product A)
SV = sample volume (either swab diluent or rinse volume)
SAS = surface areas sampled (associated with either swab or rinse)
RF = recovery factor (for Product A expressed as a decimal when in the denominator)

Other combinations are also possible beyond those shown in the various configurations of Equation 45.9. So why was the original division of terms shown in Equations 45.2, 45.4 and 45.5 established? The terms on the limit side of the equation represent the data about the product or its manufacturing process (*e.g.*, safety, batch size, dose, equipment used). All the terms on the result side of the equation have to do with how the residue is sampled or tested (*e.g.*, sample volume, sample surface area, sampling recovery factor). If terms associated with the sampling method are moved to the limit side of the equation, such as a sample surface area or diluent volume, it will create several limits: one for rinse and one for swab. If different sample sizes (different square areas) are swabbed or rinsed for different locations on the equipment, a limit will need to be calculated for each area that has a different surface area. This complexity might be better handled in the laboratory, where reportable results are routinely adjusted based on required factors, rather than in the limit, where the occurrence of multiple possible acceptance criteria might be confusing. More important than the potential complexity that this introduces to the limits is the fact that changes to some aspects of the sampling approach might be overlooked and the wrong limit might be applied.

Consider the example of a company that created a "per rinse sample limit". They moved their rinse volume term to the limit side of the equation, along with the surface area sampled. In this manner they had a limit that was expressed as mg/mL. They then took the worst-case limit in the factory and determined to apply it to all products they produced. They failed to consider, however, that different products would potentially use different solvents for rinse solutions, that they might choose to vary the rinse volumes depending on the product and its analysis and that the surface area sample might vary based upon the technique used in applying the rinse. By losing track of the terms that were included in the limit side of their "worst-case" equation, they had a limit that was not truly "worst-case" for all sampling scenarios. Converting the company's limits to mass per unit surface area enabled them to change their sampling approach moving forward without affecting their limits.

Finally, it should be noted that some facilities may have trouble achieving their HBEL limit (Equation 45.2). In these facilities, one of the options open to the manufacturer is to establish whether all equipment that is shared requires sharing, or whether batch sizes can be increased on the same equipment. Equipment dedication or use of disposable equipment are also possible options.

After a period of study of the cleaning process, it may be possible to establish a "process capability" limit as an alert limit that will serve as an early warning system for adverse trends. (This could be in addition to a quality threshold or in place of one.) The process capability limit represents what the cleaning process is routinely capable of rather than on theoretical calculation. As with other forms of alert limits, this limit will not be directly associated with product safety unless it is compared to an HBEL limit. The process capability limit should be lower than a traditional limit (*i.e.*, cleaning capability must exceed product safety needs). Process capability may serve as a better indicator of process consistency during the monitoring phase of the cleaning validation program than would the HBEL limit, especially when the cleaning process is highly capable. Be careful when employing process capability limits, however, as the temptation is to apply a single limit to the product or process. In reality, different pieces of equipment (and indeed different locations on a single piece of equipment) may exhibit very different process capabilities based on geometry, function or physical action on the soil. Be sure that process capability limits are set based on a realistic statistical review of a significant population of related results.

45.9 ANALYSIS SELECTION, METHOD VALIDATION, RECOVERY STUDIES AND TRAINING

After the identification of residues of interest, sampling methods and limits, it is appropriate to identify potential analytical methods. There are no restrictions on the analytical methods that can be applied for cleaning validation, provided that the methods are demonstrated to be sensitive and quantitative at the low levels required by most cleaning program limits.

Methods can be classified into two broad categories as either specific (direct) methods or nonspecific (indirect) methods. Specific methods uniquely identify the analyte of interest by composition or by comparison to a control sample. Typical specific methods applied for cleaning validation include High Performance Liquid Chromatography (HPLC), ion mobility spectrometry (IMS), atomic absorption (AA), and Fourier Transform Infrared (FTIR). Nonspecific methods measure something about the attribute of the residue such as ionic strength, acid-base character or carbon content. Four typical nonspecific methods that are applied are gravimetric analysis, pH, conductivity, and total organic carbon (TOC).

One advantage of using a nonspecific method for cleaning assessment is that the firm may be able to use a single analytical method to look for all (or most) types of residues.

In other instances, it is desirable to use a specific analytical method, which by definition requires that the residue(s) of interest to the cleaning validation be selected and be able to get separated from any other residues that might be present during the preparation of the sample or in the analysis itself.

Cleaning validation methods must be validated for their use. Methods validation for cleaning validation proceeds much like the validation associated with the potency assay. The method confirmation focuses on accuracy, precision, linearity, intermediate precision (ruggedness) and range. (Specificity may or may not be applied depending on whether a specific or a nonspecific method is used.) For cleaning validation, quantification limit (QL) and detection limit (DL) are highly critical. These two attributes are generally not confirmed by a potency assay but are essential for the trace analysis that is central to cleaning. The QL and DL are at the lower end of most assays' capability. The potency assay that is a desirable starting point for cleaning validation may not be sufficiently refined to reliably measure these parameters. Any method, whether employed on the surface of the equipment, in the laboratory, online or at-line, must be demonstrated to be suitable through appropriate methods validation.

As with all methods, other considerations included in the USP and ICH standards for methods validation should be observed, including:

- System suitability.
- Standards or controls to ensure that the assay is valid.
- Robustness to demonstrate that the assay is suitable under a potential known variability of the assay method and its parameters.
- Control over those materials and supplies, the consumable products that are used with the performance of the assay (*e.g.*, for HPLC—column manufacturer, mobile phase solvent grades, water quality, sample filters).

Total organic carbon (TOC) has been traditionally applied to cleaning validation. The method provides rapid results, and due to its nonspecific nature has been found to be useful in detecting all residues associated with complex processes such as biotechnology cell culture or fermentation operations. In these cases, the lack of specificity of the method is a benefit in that TOC will assess a broad spectrum of residues with a single test. The penalty for the lack of specificity is that all residues found must be attributed to the worst-case compound because the nonspecific result cannot be allocated to the potential contributing contaminants.

Some of the recent technologies that have been employed in cleaning validation include the use of ion mobility spectrometry (IMS), surface Fourier Transform Infrared (FTIR) detection and photoelectron emission. These methods are specific methods and offer specific advantages in the rapid identification of residues, or the ability to directly analyze residues on the surface of equipment, or the ability to measure trace residues after the filtration of large volumes of rinse, respectively.

For all assays, the method should be applied with some knowledge of the equipment, the cleaning process, and the intended sampling procedures. For example, when solvent sampling is required on polymeric materials of construction interference or leachables, this may conflate the analytical results. For other direct surface methods, limitations in terms of the physical geometry of the equipment will interfere with the ability to deploy the method because this method combines the sampling and analysis at the equipment site, which takes space. For equipment that is dried or disinfected with alcohol wipes, TOC may not be a good

TABLE 45.4
Comparison of Features of Typical Cleaning Validation Assay Methods

Attribute	pH	Conductivity	Total organic carbon	HPLC	Ion mobility spectrometry	Direct surface FTIR
Nonspecific	○	○	○	•	•	•
Does NOT detect in the presence of solvents	•	•	○	•	•	•
Requires a soluble/semi-soluble residue	○	○	○	○	•	•
Requires an ionizable residue	•	○	•	•	○	•
Is NOT typically rapid/real time	•	•	•	○	•	•
Does NOT typically have any On/At-Line Capability	•	•	•	○	•	•
Uses reagents/mobile phase/specialty gases	•	•	○	○	•	•
May require special sample preparation	•	•	•	○	•	•

Key: • No (advantage); ○ Yes (potential disadvantage).

choice for sampling due to the possible cross-contamination with volatile carbon.

Table 45.4 contains a quick comparison of some of the more common and innovative methods for cleaning validation sample analysis.

An adjunct requirement to the methods validation is the recovery study. Recovery studies are the evaluation of the performance of the sampling method to determine its recovery (or loss) of the analyte at the surface. Recovery factors are then applied, as described in the limits section (Section 45.8), to correct results to 100%.

During a recovery study, the residue of interest is spiked, at the limit concentration, onto the surface or coupon made of the same material of construction (and the same surface finish) as will be sampled in the field. All materials that are included in the selected equipment sample sites should be included in the recovery study. Ideally, there is a grouping rational that is prepared to show what materials of construction are represented by those that are tested in the laboratory—for example, hard metals, soft metals, high density polymers, low density elastomers and glass.

After spiking the residue onto the selected surface, the residue is typically allowed to dry, as would occur in post-cleaning prior to inspection and sampling. Drying in an oven may be employed if there is concern that the residual heat of the cleaning process will affect the residue on the surface.

Trained sampling personnel remove the residue from the surface in accordance with the standard operating procedure for the sampling method. It is important to assure that techniques used in the lab are the same as will be used in the field to ensure an accurate determination of recovery. In a case of rinse sampling, it may be difficult to simulate rinse sampling methods exactly in the laboratory. In these cases, worst-case assumptions about rinse contact time or reduced mechanical action in the laboratory should be applied and defended as part of the study.

After sample collection, the sample is analyzed using the validated analytical method and compared to the quantity obtained in a wet spike from the solution that was originally applied to the coupon. The percent recovery is calculated by comparing the wet spike to the amount recovered from the coupon samples.

Low recovery results may result in the need for the optimization of the technique, such as:

- Changing the sample container, lid, lid liner.
- Changing the swab type.
- Changing the solvent type/acidify solvent.
- Changing the swabbing method (*e.g.*, number of swabs, pattern of sampling).
- Changing the types and number of swabs applied (*e.g.*, one wet + one dry vs. two wet).
- Changing swab extraction (*e.g.*, duration of extraction, mechanical action applied).
- Observing personnel for differences in technique.
- Eliminating personnel as candidates for sampling.

For all recovery studies, it is a good idea to perform initial feasibility work with the sample kit (including the swabs, where appropriate) to ensure there is no inhibition or enhancement of the results.

When performing recovery studies, it is important to sample blank coupons (coupons not subject to spiking) to ensure that the recovery results that are obtained from the sampling methods demonstrate that there is no interference from residues on clean coupons themselves. Similarly, for swab sampling, blank swabs should be tested to ensure that they to do not provide any inhibition or enhancement of the result.

It is best practice to have replicate coupons tested for each sampling method to assure the validity of the result. Replicate coupons sampled by the same individual must show a relative standard deviation of less than 10–15% to demonstrate consistency of the sampling technique. Likewise, samples between operators sampling the same residue from the same materials of construction must show a similar relative standard deviation. Tighter relative standard deviations may be warranted for readily soluble substances. It is important to remember that all recovery studies are performed at trace levels and that the inherent variability, such as the spiking of coupons with trace materials, may include significant variation that will result in a larger standard deviation between replicates.

It is generally accepted that recovery should be greater than 50%. Usually, most companies will accept greater than 75% without any investigation. Some investigation and optimization of possible resolutions will be required when the recovery is between 50 and 75%. Results that demonstrate less than 50% recovery are generally subject to a mandatory investigation and optimization but may be accepted by quality assurance provided that due diligence in method optimization has shown that optimization is not increasing recovery. Inasmuch as all results are corrected for recovery (loss), the acceptance of a low recovery is more of a penalty to the manufacturer than it would be a risk to the patient or to the next product (provided that consistency in the recovery sample to sample has been demonstrated).

Recoveries that are greater than 100% are typically investigated to determine the source of the error; however, results are *not* corrected downwards. This is another conservative assumption in the determination of the actual amount on surfaces.

Swab sampling suffers in locations that are hard to reach. This includes many vessels in which manned tank entry is not possible, piping systems, small orifices like filling needles, or other locations that have a complex geometry. For samples that are difficult to reach, some companies have employed an extension tool to hold the swab to extend well beyond the reach of the human arm. In employing such a tool, it is important to also perform the recovery studies with that tool so that the full technique can be assessed.

It is important to remember that recovery factors are specific to the material of construction that was sampled. In accordance with recovery studies, the analyst must ensure that the correct material of construction correction factor is applied to each sample that is tested in the laboratory in the routine phase. Original analytical results and the corrected results must be available for inspection at the time of an audit. As such, it is sometimes helpful to explicitly show the mathematics performed to correct for rinse or diluent volume, surface area, and recovery. In cases where these mathematical or correction factors are part of the "reportable value" of the method, the investigator must be careful that the right correction factors are applied to each sample.

To facilitate the correct assignment of recovery factors to each sample, it is common to include a prominent table signifying the values that should be applied in a consistent location within the analytical method validation report.

When training operators to collect samples in the field, the supervisor must remind them that it is conservative to always collect more surface area and attribute the sample to less surface area. In this fashion, if they are responsible for estimating 100 cm^2 from a surface, it is safe for them to sample a slightly larger area than they are required to sample. Erring in this fashion will be safer for the patient, as more residue is collected and is attributed to the standard surface area. Similarly, operators should not worry about small losses of liquid on the surface during the sampling process; this will simply make the solution slightly more concentrated in the sample container. It is important, however, that these losses on the surface are not extreme, because if they are extreme, they may affect the ability to extract and recover the residue.

Appendix B contains additional tips on the techniques required to perform recovery studies.

45.10 PROTOCOL DEVELOPMENT

Now that much of the design, research and development for cleaning validation has been completed, the protocol may be prepared.

Cleaning validation protocols, like protocols in other areas, are formulaic. Typical sections include Purpose, Scope, (Background), Definitions, References, Responsibilities, Procedure/Testing, Acceptance Criteria, Deviations and Revalidation. What makes protocols in the specialized areas of validation different are the technical contents unique to that area of study.

For cleaning validation, the critical elements of the protocol (and the sections they affect) include:

- Identifying the other products and equipment that are included in the groups that are covered by this protocol—remember that successful completion of the worst-case or representative member typically means that all of the members of the group are now considered to be validated (Scope).
- Summarizing cycle development or bench work (such as testing performed by a cleaning agent vendor) that has resulted in the selection of this particular cleaning process (Testing).
- Cross-referencing scientific rationales that are external to the protocol for elements such as grouping of products and/or equipment, sampling site selection, limit determination and monitoring approaches (Scope, Testing, Acceptance Criteria, Revalidation).
- Incorporating rationales directly into the protocol for any topic that is not addressed in an external rationale document (Scope, Testing, Acceptance Criteria, Revalidation).
- Defining the conditions of the equipment prior to, during and after testing (Testing):

- Soiling method—normal processing, intentional false soiling, coupon use.
- Loading of worst-case soil—largest batch size, highest concentration of active, longest campaign, longest dirty hold time.
- Pre-cleaning activities conducted—minor clean operations during a campaign, rinsing prior to dirty hold time, or covering/closing during dirty hold time, maintaining equipment under nitrogen air flow during hold time.
- Sequencing of sample collection—noninvasive sampling methods first, followed by microbiological samples next, followed by invasive chemical sampling last (all samples must be coordinated to ensure that different locations were selected for each and that appropriate test methods get the representative hard-to-clean samples).
- Holding of equipment after cleaning (*i.e.*, equipment expiration or clean hold time)—preparation before storage, storage location, sampling frequency, routine interventions during storage, grouping of equipment that may be different from that which was used for the validation, sampling methods, sampling sites and analytical methods that may be different from those used for the validation. (Note: Because of these differences, equipment expiration studies are often subject to their own protocols so that the differences can be listed without confusing reviewers about the conditions of testing for the cleaning itself. This also serves to ensure that cleaning studies can be completed and summarized successfully, even when the equipment expiration studies are still ongoing.)

Of all the sections of protocols, the section that is most often least effectively prepared is the section on Revalidation. The goal of this section of all protocols, regardless of subject matter, is to define the conditions under which revalidation would be required. Instead of specifics, most firms choose to include a single sentence, such as "Revalidation will be required upon change." However, without identifying the scope and nature of the change or the impact of the change, this will not always be true. Table 45.6 (in Section 45.14) identifies some of the more typical potential effects of change on the validated state that might be mentioned in this section of the protocol. In addition, there may be ongoing monitoring of the process—through periodic review, visual inspection, instrumented measurements of the process result (*e.g.*, conductivity at the outlet of a CIP system) and/or periodic sampling for manual cleaning operations to demonstrate process consistency—that may signal a change that warrants revalidation.

45.11 CYCLE DEVELOPMENT

For automated and semiautomated cleaning, cycle development is a fairly rigorous process in which the results from individual trials are used to direct the process parameters to be used for the subsequent trial. Only after a successful or perhaps several successful cleaning outcomes will the final cleaning process for validation be determined. Optimization of the process parameters will take place to varying degrees at different firms depending upon their business needs for the cleaning process (in terms of time, water and/or chemical usage, for example). In other cases, once a successful process is reached, development halts, and no further refinement for the purpose of optimization is performed. Timelines and the frequent urgency to arrive at a validated state often limit the amount of optimization that is performed.

For manual cleaning, the usual process is to refine the SOP for cleaning until it is certain that the process parameters are reproducible and reflect adequate TACT (see the sections on SOPs, sections 45.3 and 45.4). After procedure definition, samples are collected to verify cleanliness. If they are successful, the cycle is determined to be adequate for use. The only optimization that is typically performed is to ensure that the cleaning practices are robust. It is recognized that manual cleaning processes require some overkill to ensure that day to day and operator to operator, the cleaning will be consistent.

When conducting cycle development (or perhaps they might be called engineering studies or pre-validation runs), a cycle development protocol is typically prepared and approved that outlines the intent and the process to be used. This document serves as a record of what was performed and can serve as the launching pad for additional process refinement. Whether for cycle development or for validation, protocols are required to identify deviations that occur in the process and address them. In cycle development, the deviation impact is minimal because the effect is on the immediate trial conducted, since processes are evolving during this time. For validation, on the other hand, the deviations have an impact on all cleaning events that have already been or are about to be run. Deviations mean that the consecutive successful completion requirement may be in jeopardy, and therefore the potential impact of the deviation must be evaluated within the context of all studies performed.

In all cases, whether cycle development or validation, the equipment must be demonstrated to be clean before being put back into use for clinical materials or marketed production. Remember that the effect of cleaning is on the next batch processed, and therefore, it is critical to ensure that equipment is successfully cleaned any time that the equipment will be put back into use. For existing equipment, it is not uncommon that cycle development activities are interspersed with production activities. It is therefore critical to observe a process of formal equipment quarantine and to release equipment only when results have been returned indicating that it is safe to use the equipment. The consideration for the next lot processed would also naturally lead us to understand that the level of testing in terms of the number of sample sites, the sampling methods and the analysis methods should be nearly as rigorous as, if not exactly like, the validation. Remember that the protocol is providing a "high level of assurance" and elements of "reproducibility" in the testing that it presents. It would be appropriate to ensure that a similar level of assurance is provided after cycle development before turning over the equipment to a subsequent product.

45.12 PROTOCOL REFINEMENT, PROTOCOL EXECUTION AND SUMMARY REPORTING

After defining the final cleaning process in cycle development, the elements of the protocol are revisited to ensure accuracy and to ensure that they reflect what was learned during cycle development. Next, it will be time to execute the protocol, making sure to capture any deviations that occur during the process.

There is typically a requirement to quarantine equipment after sampling before results from the study are returned. This quarantine will prevent the use of potentially unclean equipment and will also ensure that the equipment condition can be investigated should there be a failure.

Sampling is typically an invasive process where the surface of the equipment is touched with sampling materials (*e.g.*, swab samples) and non-process-related solvents (*e.g.*, swab diluent). As a result, there is typically a required recleaning, or at least a re-rinsing, after sampling has been completed before release of the equipment.

During validation, deviations may broadly be considered to be extrinsic to the validation—meaning those that are not process related (*e.g.*, power failure during a cleaning validation) or deviations may be considered intrinsic to the validation—meaning those that are directly process related (*e.g.*, failure to achieve a documented process parameter for the cleaning process). For the extrinsic deviations, the effect of the deviation is typically limited to the immediate trial. For the intrinsic deviations, the effect of the deviation could potentially influence all validation trials conducted up to that point in time. There will need to be an evaluation of whether the validation should be repeated in its entirety or whether only the immediate trial needs to be repeated.

Additional pointers for the execution of cleaning validation studies are included in Appendix C.

At the conclusion of all validation activities, a report is created that summarizes the testing performed and the results achieved. When summarizing results from any validation, the key is to ensure that all deviations that occurred during the process are defined and that any potential impact to the validation is identified and explained. Typical validation reports have sections that parallel the protocol sections so that the purpose, scope and testing requirements are reiterated along with the results summary. It is common to assemble paper copies of the validation packages so that the final report is located on top, with a copy of the signed protocol, executed protocol and raw data following the report.

45.13 PERIODIC REVIEW AND MONITORING OF CLEANING

Periodic reviews are a formal activity at a prescribed (and justified) frequency that assess all sources of intelligence from the quality system and evaluate whether there was any unintentional change or trend that remained undetected in our validated process. One important role of the periodic review program is to ensure that small changes or deviations that were individually judged to not affect the validated state, when viewed collectively, also indicate that there is no impact to the state of control.

Quality System programs such as change control, calibration, preventive maintenance, internal audits, deviations and CAPA (corrective and preventative actions) can all provide valuable data to help ensure that systems and procedures continue to be executed with the same control as was exhibited during the original validation. Statistics on programs such as visual inspection can also be an excellent source of intelligence about the success of cleaning.

For some manufacturers, their cleaning validation programs will require a monitoring program that conducts periodic resampling to confirm that the results of cleaning are continuing to meet the same analytical cleanliness levels that were fulfilled by the original validation. Monitoring was first described by FDA in its cleaning validation guidance[2] but continues to be echoed in world-wide guidance, especially as it relates to manual cleaning.[16,18,22,23] The reason that manual cleaning is singled out is the typical lack of instrumented monitoring during manual cleaning and the possibility of variability between operators performing the same procedure.

Commonly, manufacturers combine Monitoring and Periodic Review into a single program. This makes the best use of resources and enables data values collected during monitoring to be aligned with other events that could influence those results, such as change control, maintenance or calibration. Integrating the data reviews from these disparate areas adds strength to the review that is performed.

When establishing a monitoring program, a schedule is established that identifies the sampling points of interest based upon those products and equipment that represent the most risk during the validation. The risk assessment may consider, among other things:

- Highly potent or toxic products.
- Difficult-to-clean products.
- Products that demonstrated a high degree of variability during the original validation.

- Products for which the results were very close to the limits during the original validation.
- Products that are produced frequently for which build-up, migration or simple increase in the likelihood of cross-contamination exist.

For monitoring, risk-based decisions may be made (that are different from those associated with the original validation) with regard to grouping of products and equipment to be studied, location of samples, sampling and analytical methods, frequency of monitoring, and the like. Frequently, methods of sampling and testing are selected that are less time-consuming and less intrusive to the equipment in order to maximize equipment up-time. This is where methods such as IMS, TOC and FTIR can be highly valuable in saving equipment down-time. Over time, monitoring frequencies may be reduced or even eliminated when effective process control is demonstrated.

The key to effective monitoring is to establish control charts that help to track and trend monitoring results so that changes in the validated state can be detected early rather than waiting for a failing result. The concept of a cleaning process capability limit for cleaning validation was discussed in the limits section (Section 45.8). Whether through trending or establishing process capabilities, the company must ensure that only similar sample sites and similar products are included in the monitoring trends. It is inappropriate to assume that different equipment, different sampling sites or different product residues will clean to the exact same level. Including a diversity of samples in the process capability or monitoring trend assessment may provide a skewed view of inherent process variability based upon these differences.

While monitoring is discussed in every worldwide cleaning validation guidance, there are no specifics with regard to setting monitoring frequencies. Like limits, the inability for regulators to define a minimum standard is due to the high amount of variation in the industry with regard to product types, facilities, and the complexity of manufacturing. As was discussed in Figure 45.2, every risk-based compliance decision falls along a continuum, and it is up to companies to defend where they fall along that continuum. Some of the considerations that should be taken when determining monitoring frequency and approach are included in Table 45.5.

The most important aspect of any monitoring program is to remember to permit modification of the frequency, location of sampling, sampling approach and analytical approach as the monitoring program matures. The results may well show that it is not necessary to continue monitoring a specific piece of equipment or a specific cleaning process based upon its continued success in achieving results well below the level of interest. Creating a program that enables modifications over time will ensure that resources are being invested in the most meaningful samples. With the development of new analytical techniques and an increasing ability to apply Process Analytical Technology (PAT) to all aspects of the manufacturing process, it is important to leave room to replace the monitoring approach when true PAT becomes available in each factory. Every firm wants to ensure that the cleaning program in use will continue to permit the introduction of quality principles, thereby enabling the primary focus to be on those areas that have the greatest ability to affect quality!

TABLE 45.5
Monitoring Strategy Considerations

Low risk	High risk
Automated cleaning	Manual cleaning
In-process monitoring of TACT	No in-process monitoring of TACT
In-process cleanliness measurement (*e.g.*, PAT)	No in-process measurement
Ongoing cleaning validation projects	No new cleaning validation projects
Similar equipment types	Diverse equipment types
Ongoing retraining/certification	No retraining/certification
All surfaces visually available	Surfaces not visually available
Aqueous/readily soluble products	Insoluble/difficult-to-clean products
Low patient risk for carryover	High patient risk for carryover
Low "next process" risk for carryover	High "next process" risk for carryover
Few changes affecting cleaning	Many changes affecting cleaning that are not revalidated

45.14 CHANGE CONTROL

With the requirements for a comprehensive cleaning validation program established, the next significant event is change. In preparing for the cleaning validation, the products, processes, equipment and cleaning procedures were evaluated. Therefore, it should be evident that a change in any of these areas has the possibility of affecting the cleaning validation or the scientific rationales that formed the cleaning validation approach.

Table 45.6 includes a list of typical changes that might be made and a list of possible effects that these changes could have on cleaning validation.

Upon the identification of a change, the site change control process should evaluate the impact of the change and either document that there was no impact to the cleaning validation program and its rationales or document the affect and update the effected documents. The potential need for revalidation will have to be assessed based upon the impact of the change.

Remember that individual changes may each be assessed as requiring no revalidation. But when taken together, several insignificant changes might represent a significant change to a system. Effective change control takes these elements into consideration and establishes a mechanism to periodically review accumulated changes. Most commonly, this is conducted as a Periodic Review of accumulated changes (see section 45.13).

TABLE 45.6
Typical Change Control Impacts

Type of change *If change falls into more than one category, all possible affected systems should be evaluated.*	**Typical affected rationales / documents and typical studies required to support change** *Each to be performed as applicable to the change—any not considered should be defended in the scientific rationales provided in support of the change control; list may not be all inclusive.*
Product Formulation *(including quantities and types of excipients, quantities and types of actives, batch size, dose weight)*	• Product grouping • Worst-case product selection for monitoring studies • Limits rationale • Engineering trials may be required to demonstrate that the new formulation is/is not a harder-to-clean challenge • Repeat validation for this product and/or justify that new formulation does not represent a harder-to-clean challenge • Bioburden assessments
Production Process *(e.g., unit operations, processing parameters)*	• Equipment train definition • Justification of whether new process will result in different hard-to-clean locations or hard-to-clean residues • Demonstrate ability to clean equipment with worst-case products if the process change is likely to create a new worst-case; may consider worst-case product formulation only • Bioburden assessments
Production Equipment Modification	• Sampling site selection (new or different hot spot or critical site) • Materials of construction for recovery studies • Equipment surface area calculations • Limits rationale • Equipment train definition • Demonstrate ability to clean equipment with worst-case products if the equipment change is likely to have affect • Demonstrate ability to clean equipment to new lower limits if existing data do not support the new, lower limit • Bioburden assessments
Cleaning Agent *(change in manufacturer, type, or concentration)*	• MSDS if new cleaning agent • Health safety & environmental review if new cleaning agent or more concentrated than previously used (may require updates to personnel protective equipment in SOPs or changes in handling and cautions) • Cleaning agent methods validation and recovery studies • Limits rationale (for cleaning agents) • Demonstrate ability to clean equipment with worst-case products from each product grouping • Cleaning agent residual studies • Bioburden assessments • Standard operating procedure and equipment cleaning procedure updates
Cleaning Tools	• Health safety & environmental review if new tool could have more intimate personnel contact or (may require updates to personnel protective equipment in SOPs or changes in handling and cautions) • Demonstrate ability to clean equipment with worst-case products from each product grouping • Cleaning agent residual studies • Bioburden assessments • Standard operating procedure and equipment cleaning procedure updates
Cleaning Process Parameters *(including additional, eliminated and/or revised steps or set points)*	• Additional steps may not require testing with existing products/cleaning agents if a justification can be provided that the steps performed are additional to the existing validated process • Health safety & environmental review if new/revised parameters/steps could have more risk to users or (may require updates to personnel protective equipment in SOPs or changes in handling and cautions) • Demonstrate ability to clean equipment with worst-case products from each product grouping • Cleaning agent residual studies • Bioburden assessments • Standard operating procedure and equipment cleaning procedure updates

45.15 CONCLUSION

Cleaning validation is a highly risk-based endeavor. Risk-based decisions are made in terms of what residues to study, what parameters to include in the cleaning process, how to group equipment and products, how to select sample sites, which sampling methods to use, which analytical methods to use, what parameters should be included in the establishment of limits and what worst-case conditions are included in the validation protocol. This chapter has reviewed these elements, and more, in detail. It is essential to document the risk-based decisions that are made within each program so that the decision making can be evaluated periodically for accuracy and to ensure that the decisions can be reproduced consistently when new products or new equipment are introduced to the site.

Appendix A
Definitions Common to Cleaning Validation

CLEANING VALIDATION

Cleaning validation requires documented evidence to ensure that cleaning procedures are consistently removing residues to predetermined levels of acceptability, taking into consideration such elements as batch size, dosing, toxicology, equipment size and the like. A cleaning validation is typically completed by the accomplishment of replicate, consecutive successful cleaning operations. The number of cleaning operations to be performed is subject to justification.

CLEANING VERIFICATION

Cleaning verification requires documented evidence to ensure that cleaning procedures remove residues to a predetermined level of acceptability based upon the minimum of a single trial. A cleaning verification is performed to assure that the cleaning procedures used adequately clean the equipment when the manufacturing process or the cleaning procedures may be subject to change and therefore cannot be immediately subjected to validation. Cleaning verification includes but is not limited to the accomplishment of a minimum of a single cleaning trial for:

- Cleaning of equipment during development (*e.g.*, clinical trial production), when additional process, product or process scale changes might still occur.
- Initial cleaning operations upon relocation of equipment.
- Cleaning of new equipment.
- Cleaning operations when some cleaning process development is still required and the cleaning process is likely to change.

CLEANING CERTIFICATION

Cleaning certification requires documented evidence that a production area, including equipment and facility, are clean and ready for the next production use. This term typically implies that sampling for cleaning is performed and assessed against predetermined acceptance criteria, even after a cleaning validation is successfully completed. The primary reason for the certification is to ensure that production facilities and equipment associated with high-risk products, high-cost products and/or long-term and, therefore, high-cost production campaigns are ready for the subsequent operation. Certification frequently implies more than taking and testing cleaning samples; it often involves the performance of specific end-of-campaign activities such as the exchange of environmental filters, disposal of potentially absorptive materials of construction or other periodic environmental cleaning and sampling. Certification may be performed for products such as potent compounds, some biotechnology products or some active pharmaceutical ingredients.

CONTINUED PROCESS VERIFICATION FOR CLEANING, OR MONITORING

Monitoring requires periodic confirmation of previously validated cleaning procedures for the purposes of reconfirming the validated state. Monitoring for cleaning validation would fall under the category of Continued Process Verification (CPV for Cleaning) under the new FDA Process Validation guidance.[7] For monitoring, risk-based decisions may be made (that are different from those associated with the original validation) with regard to grouping of products and equipment to be studied, location of samples, sampling and analytical methods, frequency of monitoring and the like.

REVALIDATION

Revalidation is the repetition of all or part of the original validation. Revalidation is typically change-based and therefore is undertaken when a change has been made to the product, process, procedures for cleaning or equipment. The concept of time-based revalidation or the evaluation of the process at some interval to confirm that it continues to meet the validated state is less common, as Monitoring (CPV for Cleaning) can fulfill the role of time-based revalidation.

GROUPING OR BRACKETING

Grouping, sometimes also called a family approach, is a method by which products or equipment are considered to be similar or equivalent for the purposes of cleaning validation. Bracketing has a similar meaning to grouping, although it may include an added burden for testing the extremes of a population such as the smallest and largest equipment members in a matrixed approach or the most soluble and least soluble members of the product group.

CAMPAIGN PRODUCTION

Campaign production is the manufacture of batches of the same product in a consecutive fashion such that (a) no cleaning is performed between batches (typical of active pharmaceutical ingredient (API) manufacture, for example); (b) sufficient cleaning is performed to ensure mechanical functionality of the equipment but equipment does not reach a visibly clean level (also common in API manufacture); (c) cleaning is conducted to a visibly clean level with limited or no disassembly of the equipment (common to oral solid dose manufacture); or (d) full cleaning is conducted of product contact surfaces, but cleaning environmental surfaces and change-out of product-associated disposable parts (*e.g.*, gaskets, hoses) is not performed until end-of-campaign (common to high potency products and/or products associated with more stringent dosage forms). Such batch-to-batch cleaning within the same campaign (as in examples a, b and c) is typically not validated, as the risks of same product to same product carryover are considered minimal. In example d, the validation of the cleaning processes between batches may be validated in order to minimize the risk to the next batch.

Appendix B
Techniques for Conducting Recovery Studies

Recovery studies are governed by a protocol that defines the experimental design and the acceptance criteria. However, there are numerous best practice considerations in terms of handling and execution that should be included in local SOPs:

1. **Clean the Coupons**
   1.1. More coupons should be prepared than are required for the study. In this way, coupons with imperfect appearance or inaccurate spiking may be rejected without resulting in the need for a second preparation and spiking activity. (Note that the ideal coupon should be slightly larger than the intended sample size so that sampling does not have to take place at the edge of the coupon.)
   1.2. Clean and rinse coupons copiously and ensure that no residual cleaning agent remains, using purified water (or low TOC water). Solvents can also be used, but ensure that the solvents are fully removed before use.
   1.3. Passivate stainless steel coupons with hot nitric acid or through a room temperature overnight soak in nitric acid, if required (especially if new), then thoroughly rinse purified water (or low TOC water).
   1.4. Permit coupons to air dry for a short time in a protected area (*e.g.*, a clean rack like a CD-rack that will hold the coupons separated without allowing them to touch each other, and without a lot of contact to the face of the coupons). This will allow the water to run off the area of the coupon that will be tested.
2. **Clean Other Tools/Supplies**
   2.1. Assemble and clean other needed supplies (prepare extra of each for security). In particular the tools and supplies needed will include the implement that will be used to spike the coupons (reference point 3) and any volumetric containers or pipettes that will be used to create the stock solutions. Ensure that copious rinsing is employed to eliminate traces of any cleaning agents used in preparation. When using TOC, ensure final rinse is a low carbon water source.
   2.2. For glassware (and other impervious materials of construction) that are to be used for recovery that will be measured with TOC, acid pretreatment should be considered, followed by rinse using purified water (or low TOC water).
   2.3. Dry all equipment and supplies in a manner that prevents contamination from the environment or from other laboratory activities.
   2.4. If using disposables, precleaning may or may not be required. Be sure to evaluate any contributions from these supplies through the analysis of blank samples.
3. **Spike Delivery to the Coupons—Preparation**
   3.1. The lab should prepare the designated dilution of the media. Prepare the spiking solution so that the designated mass of residue can be delivered in a small quantity (*i.e.*, about 10–50 µL delivered—this can be as high as 0.5 mL (500µL); just be sure that the volume selected does not run across the plate when spiked).
   3.2. To perform the spiking, use any implement that will deliver the correct quantity of solution in a controlled manner, provided that the implement itself doesn't provide a lot of carbon (if using TOC), for example:
      3.2.1. Microbiological/cell culture automatic or repeating micropipettors.
      3.2.2. Gas-tight 50 µL or 100 µL syringe.
   3.3. Fill the delivery device with the spiking concentrate several times, discarding this solution each time (this in effect rinses and wets the interior with the spiking solution). Ensure that this process is performed slowing and deliberately to reduce the risk of entraining air bubbles or from incurring disconnected droplets within the reservoir.
   3.4. Fill delivery device with the appropriate quantity, per 3.1. For syringes, overdraw the solution and expel the excess to bring the device to volume while checking for air bubbles. Note that air bubbles may be specifically problematic with concentrated detergent solutions. For syringes, if air bubbles are drawn up, hold the delivery device upside down, viewing it against a dark background. Tap delivery device gently to dislodge air bubbles, then expel the air bubbles through the tip. For other devices, expel (to waste) and repeat the sample withdrawal or dispose of tip and retry.
   3.5. For gas tight syringes, these devices frequently have a triangulated tip—be sure to examine the configuration so that during the next step, the pierced side is pointed downwards towards the surface to ensure contact of the surface with the liquid.
4. **Spiking of Coupons**
   4.1. Perform the spiking of coupons either in a fume hood or biosafety cabinet or on a lab bench. The location is critical only in that it

needs to be appropriate for the handling and evaporation of the spiking solution and to reduce the opportunity for contamination from the surrounding areas during spiking and drying activities.

4.2. Place the coupons face up on a protective surface (*e.g.*, lint-free wipe, lab paper, autoclave wrap, tinfoil). Consider organizing the coupons in a logical layout based upon how they will be sampled in the next step of this procedure to minimize post-spiking handling (*e.g.*, group them by analyst, material of construction and spiking solution, with blanks last). Label the background on which the coupons are resting with the coupon number, the material of construction and the spiking solution so that the coupons can be identified easily in other steps of this procedure. This will also help with any notes needed during spiking activities, such as tracking which coupons have been spiked or which coupons should not receive spikes (blanks), but be aware that this background labeling will not be retained and that coupons could be rearranged on this background and become disassociated with the labels. Therefore, in addition to this, consider etching the coupons with a permanent identification number engraving on the face (outside of the area that would typically be sampled) to assist in tracking. Make equivalent notes in an appropriate laboratory notebook or worksheet that will be retained as part of the study. Avoid use of markers or other nonpermanent designation on the coupon itself that might contribute to contamination of the sample should the marking be inadvertently removed in the sampling process.

4.3. Spike an area that is well within the intended sampling boundary (*e.g.*, 10 cm × 10 cm for a 100 cm^2 sample) to ensure that all area swabbed is within the spiked region. (As previously noted, coupons should be just slightly larger than the area to be sampled. For example, when sampling 10 cm × 10 cm for 100 cm^2, the coupons would typically be approximately 12 cm × 12 cm, or even larger.)

4.4. Deposit the inoculum in small droplets evenly distributed across the entire area that will be sampled. Stay well within the potential sample collection area, but be sure that the droplets do not run together or run outside of the intended sampling area. The uniform distribution of the inoculum during spiking is to ensure the soil is present in a relatively even layer. The use of small droplets of the concentrated solution will ensure that surface tension holds each droplet in place after deposition on the surface. The small size of each droplet will also ensure rapid evaporation and drying on the surface. This approach will most closely simulate how soil is dispersed across the surface after cleaning and will prevent problems such as slow dissolution of a large crystalline mass of residue during sampling. Several techniques for dispensing are described as follows:

4.4.1. Without control over dispensing, spot the surface and then spread the inoculum around with the tip. This is a crude method and should work as long as excess liquid on the outside of the syringe or tip is avoided. Any losses in material will count against the result, so this is a conservative approach.

4.4.2. Dot the surface with small droplets within the sampling boundary (typically 20–30 drops required when spiking with µL quantities). This approach will work with many different kinds of dispensing systems and both soft and hard materials of construction of the coupons. This is good for residues with high surface tensions, too.

4.4.3. Make small chain-like circular motions with the tip of a syringe, moving the delivery device close to the plate and expelling the solution very slowly, making small circular motions from left to right. Repeat in additional rows until all inoculum is dispensed. This approach works well for devices where it can be dragged smoothly across the surface. It may not work well for soft materials of construction, such as polymers and elastomers, as the tip tends to snag on the surface.

4.5. Some coupon materials enable the visualization of the spiking solution as it is deposited on the surface. For white materials or surfaces without a lot of reflectance, look at the surface at an obtuse viewing angle to observe how the soil is being deposited and to check for uniformity.

4.6. Allow the coupons to dry undisturbed. Ensure that there are as few volatile organic carbons (VOCs) in the area as possible if planning to sample using total organic carbon (TOC). Often it is best to sample as soon as the coupons are dry rather than waiting for several days to sample. This will minimize the risk of adventitious contamination or loss of the spike. Note that 20–30 µL of most products on a coupon should be dry within an hour or so. Polymers and elastomers tend to dry more slowly, so spike them first before spiking the harder materials of construction.

4.7. Although experimental designs may differ, it is common to spike a minimum of three coupons for each sampler, for each material of construction.

4.8. Leave one blank clean coupon as a blank control for each person for each material of construction. Ensure that these coupons are subject to the same handling and drying environment. Some firms spike with the diluent alone (*i.e.*, no API or cleaning agent) in order to simulate contribution from the spiking process for the blank coupon.

4.9. Using the same dispensing apparatus used during the spiking of the coupons, dispense an equivalent full amount of the spiking solution directly into a test vial containing the sampling diluent. This will be used as the "theoretical quantity" for the determination of recovery.

   4.9.1. NOTE: In the case of residues with volatile constituents, including some detergents, spike this into an empty vial and place in the hood. Allow to dry along with the coupons, and only when it is fully dry, add the appropriate quantity of the sampling diluent to each vial for recovery. Agitate aggressively for recovery. The reason for this is that the low level of volatiles in the sample will evaporate upon drying. It must be ensured that the "theoretical" quantity (and all interferents) reflect the loss encountered during evaporation from the surface of the coupon but without the loss that might be incurred in sampling. This is a way to accomplish a direct spike but still allow the needed evaporation to occur.

5. Swab RECOVERY

5.1. Follow the internal SOP for the swabbing procedure during all recovery studies.

5.2. Ensure that the following are available for sampling portion of the recovery study:

   5.2.1. Swabs.

   5.2.2. Vials filled with the appropriate quantity of swabbing diluent, as specified by the test method.

   5.2.3. Wire cutters (ideal) or scissors (less ideal) for cutting swab heads (one pair per person, or have a non-contaminating method to clean them between people).

   5.2.4. Coupons according to the experimental design (typically, three (3) spiked coupons and one (1) blank coupon per person, per material of construction).

   5.2.5. Labels for the vials.

   5.2.6. Gloves (powder-free).

5.3. Set the coupons in the order to be swabbed (if not already laid out this way in step 4.2). Swab the blank coupon last in each case. It will help identify whether gloves, technique, cutters, environment or another element has contaminated the samples.

5.4. Label the sample vials.

5.5. Follow the site procedures for sampling by swab. In general, those steps will typically include the following:

   5.5.1. Remove the lid from the vial and make sure a clean location (*e.g.*, fresh lint-free wipe) is available to put them down, or use an aseptic technique to hold the lid while swabbing. Make sure the method will be practical to what will be done in the field.

   5.5.2. Take swab and moisten it by dipping it into the designated sample's vial. When wetting swabs in any diluent, hold the swab below the diluent level, swirl it and press it lightly against the sides of the vial in order to express any residual air bubbles trapped in the fabric of the swab. This will assist in wetting the swab fully.

   5.5.3. Pull the swab from the vial and press out the majority of the liquid by pressing all four sides at the top of the vial, permitting the excess to fall back into the vial. (Note: if using solvent-water or solvent as a diluent for swab sampling, do not press the liquid out so thoroughly that evaporation will remove all solvent before finishing swabbing; however, for water-only swabbing, it is important to not leave behind a lot of liquid.)

   5.5.4. Hold the swab so that the forefinger is 1–2 cm away from the fabric of the swab head but can still apply pressure to the swab head. Ensure that the swab head is flat against the coupon surface. This should yield a good bend in the swab handle, pressing the head flat but preventing fingers from dragging on the coupon surface. Another reason to keep fingers back from the swab head is to avoid including areas touched by the gloved hands in the sample when the swab head is cut off into the vial.

   5.5.5. Follow swabbing pattern from the local SOP. This pattern should require unidirectional, overlapping strokes performed perpendicularly to cover the full sampling area. It may be helpful to determine the number of rows/circuits it takes to cover the designated surface in each direction, if this isn't clear from the SOP. This number will serve as a good reference when sampling equipment surfaces during sampling on the plant floor.

5.6. Cut the swab heads, using clean cutters, over the vial that was used for wetting the swab, ensuring that no area that was in contact with gloved hands was included within the sample.

5.7. During feasibility for the specific analyte, the laboratory should have determined the number of swabs necessary for effective sample collection and whether more than one swab is recommended. That evaluation should have considered whether the swabs should both be wetted or whether the optimal combination is one wet and one dry swab. Note that only if there is a significant, meaningful difference in recovery with multiple swabs will that approach typically be implemented for routine sampling. If more than one swabs is used for each sample, both swab heads should be wetted and clipped into the same vial.

5.8. Close the cap on the vial and ensure that there is no opportunity for leaking.

5.9. Do not forget to sample the blank coupon when finished with all spiked replicates.

5.10. Store the samples in accordance with the test method requirements. Commonly, samples are stored at 2–8°C if they are to be held prior to testing.

5.11. Consider preserving the coupons after sampling to have them available for inspection and investigation, until test results are available.

6. RINSE RECOVERY

6.1. The method used to rinse coupons in the laboratory will need to be justified in the protocol, as rinse methods that are used on the plant floor may not be scalable or possible to perform in the same manner in the laboratory.

6.1.1. Justify the rinse technique as equivalent or worst-case as compared to the plant floor method (*e.g.*, shorter contact time, less volume per unit surface area)

6.2. Methods commonly involve pouring rinse solution across the spiked (or blank) coupons and recovering the solution in a sample container. For this reason, slightly different supplies might be needed for a rinse sample, for example:

6.2.1. Coupons might be curved or shaped to facilitate collection, or the bottom of a vessel such as a small beaker-shaped container might be used.

6.2.2. A funnel might be used to help direct rinse solution into the mouth of the sample container from the coupon (note: ensure that the funnel itself does not become a source of hold-up of the residue).

6.3. As with swabbing, label the vials, open the sample container and protect the lid.

6.4. Apply the rinse solution to the designated coupons, sampling the blank last. Ensure that gloves are not in contact with the rinse solution during rinsing or collection.

6.5. Close each sample container, mix and test according to analytical procedures.

6.6. Note that since rinse sampling is not typically considered to be subject to person-to-person variability in the field, it will not typically require multiple sampling personnel when performed in the laboratory.

7. INTERPRETATION OF RESULTS

7.1. The protocol should include the details of the sample extraction, sample preparation, analysis and calculation of the final, reportable result.

If using total organic carbon analysis, remember that carbon is all around us! Do not work on TOC sampling if there are solvents being used in the laboratory. Handle TOC swabs and equipment as though working in an aseptic environment. Ensure that sampling personnel are not the source of contamination. Do not touch things unnecessarily, and if there is any concern that a surface has been touched, require that gloves be changed.

When executing feasibility and recovery study qualification, always think about the logistics of sampling real equipment in the real world. Ensure that the techniques and practices used can be adapted to what will be practical in the production environment.

Appendix C
Additional Points to Consider in the Execution of Cleaning Validation

Have clear standards for "visibly clean" surfaces and ensuring that there are clear requirements for identifying a cleaning failure.

- Control for variables such as illumination levels, distance and angle of viewing.
- Characterize the surfaces in advance of initiating the inspection program, and keep that characterization up to date to be sure that surface condition is not inaccurately identified as a cleaning issue.
- Determine the amount of water that may still be present on a surface to classify as "dry" if the inspection takes place immediately after the completion of cleaning (*e.g.*, droplet v. puddle).
- Differentiate between hard water staining on surfaces and product residue—alcohol wipes have been brought to bear to help distinguish this at many firms.
- Decide how much discoloration may be present in small surface defects or scratches before it is assumed that it is either rouge or product residue—alcohol or wet wipes have been used to help distinguish this based on whether the stain is removable or not.
- Conclude whether or not silver or gray discoloration on swabs as a result of wiping the surface count as clean or unclean if no active residue is detected.
- Determine whether individual fine fibers from "lint-free" wipes should be considered as passing or failing (typically the observation of these fibers finds them less than 0.5 cm or 5,000 microns in length and narrower than 80 microns in width)—especially because they may be trapped on sharp edges or fittings on manually cleaned or manually dried equipment.
- Have a strategy to address visual failures based upon their location and magnitude including such elements as:
  - Location—environment/facility, non-product-contact equipment, potential product contact or direct product contact surfaces.
  - Residues characterization—quantity and location on the surface will dictate such elements as complete recleaning or part/spot cleaning.

Have a strategy to address dropped or failed samples, including:

- How are alternate sampling sites identified (especially if the sample site dropped was a critical site or hot spot that cannot be replicated)?
- Who is empowered to make decisions about the validity of the cleaning validation trial based upon a single (or multiple) missing samples?
- When samples are dropped, even if all remaining samples pass, may the equipment be used for the next process, or must it be recleaned under the assumption that the untested sample would have yielded a failing result?
- Who should determine the number of trials that are affected by the dropped sample in terms of consecutive successful testing (usually, only one trial would be deemed a failure, as this deviation is extrinsic to the process, but this decision must be rationalized and documented)?

Have a strategy for unknown peaks when using a specific method (note that many of these investigatory pathways have their roots in a complete and effective analytical method validation prior to initiating the validation; those method validation outcomes are then used to help investigate the unknown peak):

- Investigate the cleaning process and deviations that may have occurred including TACT parameters, tools used for cleaning, training of operators and the like.
- Investigate the activities prior to the cleaning in terms of failures or risks during the manufacture, calibration, maintenance or other unusual intervention before the validation and the like.
- Investigate primary degradation products of the actives and excipients (especially those formed during stresses of heat and pH associated with cleaning).
- Investigate cleaning agent residue and its impact on spectra or chromatographs.
- Investigate prior products processed, their actives, excipients and degradation products' appearance under the current chromatographic or spectrographic conditions.
- Identify additional test methods that may be used to investigate the molecular composition of the unknown material (especially as remaining samples may be limited).
- Determine the limits of the investigation and when recleaning and retesting will be performed.

BIBLIOGRAPHY

1. United States v. Barr Laboratories, Inc., 812 F. Supp. 458 (D.N.J. 1993). *Justia Law.* https://law.justia.com/cases/federal/district-courts/FSupp/812/458/1762275/ (accessed January 21, 2021).
2. Office of Regulatory Affairs. Validation of cleaning processes (7/93). *FDA*, July 1993. www.fda.gov/validation-cleaning-processes-793 (accessed January 21, 2021).
3. Sinclair, Upton. *Jungle. S.L.*, Ten Speed, 2019.
4. Federal Register, Volume 61 Issue 87, Friday, May 3, 1996. www.Govinfo.Gov, 3 May 1996, www.govinfo.gov/content/pkg/FR-1996-05-03/html/96-11094.htm (accessed January 21, 2021).
5. Office of Regulatory Affairs. Pharmaceutical cGMPs for the 21st century: A risk-based approach, final report. *FDA*, September 2004. www.fda.gov/media/77391/download (accessed January 21, 2021).
6. Center for Drug Evaluation and Research. Pharmaceutical quality for the 21st century a risk-based approach progress report. *FDA*, May 2007. www.fda.gov/about-fda/center-drug-evaluation-and-research-cder/pharmaceutical-quality-21st-century-risk-based-approach-progress-report (accessed January 21, 2021).
7. Center for Drug Evaluation and Research. Process validation: General principles and practices. *U.S. Food and Drug Administration*, January 2011. www.fda.gov/regulatory-information/search-fda-guidance-documents/process-validation-general-principles-and-practices (accessed January 21, 2021).
8. American Society of Mechanical Engineers. Bioprocessing equipment. 2019 ed., New York, NY, ASME, 2019. www.asme.org/codes-standards/find-codes-standards/bpe-bioprocessing-equipment-(1) (accessed January 21, 2021).
9. Association for the advancement of medical instrumentation. *AAMI.org*, 2019. www.aami.org/ (accessed January 21, 2021).
10. ASTM international: Standards worldwide. *ASTM.org*, 2019. www.astm.org/ (accessed January 21, 2021).
11. European Commission Directorate. *Guideline on Setting Health Based Exposure Limits for Use in Risk Identification in the Manufacture of Different Medicinal Products in Shared Facilities*, June 1, 2015. www.ema.europa.eu/en/documents/scientific-guideline/guideline-setting-health-based-exposure-limits-use-risk-identification-manufacture-different_en.pdf (accessed January 21, 2021).
12. Mehta N, et al. *Development of an in Situ Spectroscopic Method for Cleaning Validation Using Mid-IR Fiber-Optics.* Biopharm, May 2002. http://remspec.com/downloads/BP5875r.pdf (accessed January 21, 2021).
13. Mittal KL, Kohli R. *Developments in Surface Contamination and Cleaning, Volume 12: Methods for Assessment and Verification of Cleanliness of Surfaces and Characterization of Surface Contaminants.* Elsevier, 2019. www.sciencedirect.com/book/9780128160817/developments-in-surface-contamination-and-cleaning-volume-12 (accessed January 21, 2021).
14. El Azab W, Cousin S. *Visual Inspection Practices of Cleaned Equipment: Part I.* PDA Letter, April 14, 2020. www.pda.org/pda-letter-portal/home/full-article/visual-inspection-practices-of-cleaned-equipment-part-i (accessed January 21, 2021).
15. El Azab W, Cousin S. *Visual Inspection Practices of Cleaned Equipment: Part II.* PDA Letter, May 5, 2020. www.pda.org/pda-letter-portal/home/full-article/visual-inspection-practices-of-cleaned-equipment-part-ii (accessed January 21, 2021).
16. European Commission Directorate. *EU Guidelines for Good Manufacturing Practice for Medicinal Products for Human and Veterinary Use, Annex 15: Qualification and Validation*, October 1, 2015. https://ec.europa.eu/health/sites/health/files/files/eudralex/vol-4/2015-10_annex15.pdf (accessed January 21, 2021).
17. European Commission Directorate. *Questions and Answers on Implementation of Risk-Based Prevention of Cross-Contamination in Production and 'Guideline on Setting Health-Based Exposure Limits for Use in Risk Identification in the Manufacture of Different Medicinal Products in Shared Facilities'*, April 19, 2018. www.ema.europa.eu/en/documents/other/questions-answers-implementation-risk-based-prevention-cross-contamination-production-guideline_en.pdf (accessed January 21, 2021).
18. Pharmaceutical Inspection Convention Pharmaceutical Inspection Co-Operation Scheme. *Recommendations on Validation Master Plan Installation and Operational Qualification Non-Sterile Process Validation Cleaning Validation*, September 25, 2007. https://picscheme.org/docview/3447 (accessed January 21, 2021).
19. Pharmaceutical Inspection Convention Pharmaceutical Inspection Co-Operation Scheme. *Aide-Memoire Inspection of Health Based Exposure Limit (HBEL) Assessments and Use in Quality Risk Management*, June 1, 2020. https://picscheme.org/docview/1947 (accessed January 21, 2021).
20. Pharmaceutical Inspection Convention Pharmaceutical Inspection Co-Operation Scheme. *Questions and Answers on Implementation of Risk-Based Prevention of Cross-Contamination in Production and 'Guideline on Setting Health-Based Exposure Limits for Use in Risk Identification in the Manufacture of Different Medicinal Products in Shared Facilities'*, June 1, 2020. https://picscheme.org/docview/1948 (accessed January 21, 2021).
21. ASTM. *E3219–20 Standard Guide for Derivation of Health-Based Exposure Limits (HBELs)*, 2020, ASTM International. www.astm.org (accessed January 21, 2021).
22. Canada Health Products and Food Branch Inspectorate. *Guidance Document: Cleaning Validation Guidelines GUIDE-0028*, January 2008. www.canada.ca/en/health-canada/services/drugs-health-products/compliance-enforcement/good-manufacturing-practices/validation/cleaning-validation-guidelines-guide-0028.html. *Note: Draft update to include HBELs currently under final development as of Jan 2021* (accessed January 21, 2021).
23. World Health Organization. *WHO Technical Report Series, No 937, Annex 4: Supplementary Guidelines on Good Manufacturing Practices: Validation, Appendix 3: Cleaning Validation*, 2006. www.who.int/medicines/areas/quality_safety/quality_assurance/SupplementaryGMPValidationTRS937Annex4.pdf?ua=1 (accessed January 21, 2021).
24. World Health Organization. *Working Document 20.849 Points to Consider on the Different Approaches: Including HBEL: To Establish Carryover Limits in Cleaning Validation for Identification of Contamination Risks When Manufacturing in Shared Facilities*, Draft May 2020. www.who.int/medicines/areas/quality_safety/quality_assurance/QAS20_849_points_to_consider_on_cleaning_validation.pdf?ua=1 (accessed January 21, 2020).
25. International Society for Pharmaceutical Engineers (ISPE). *ISPE Guide: Cleaning Validation Lifecycle: Applications,*

Methods and Controls, Tampa, FL, 2020. www.ispe.org (accessed January 21, 2021).

26. Parenteral Drug Association (PDA). *PDA Technical Report No. 29, Points to Consider for Cleaning Validation*, Bethesda, MD, 2012. *Note: Draft under development to include HBELs as of 2020.* www.pda.org (accessed January 21, 2021).
27. Parenteral Drug Association (PDA). *PDA Technical Report No. 49, Points to Consider for Biotechnology Cleaning Validation*, Bethesda, MD, 2010. www.pda.org (accessed January 21, 2021).
28. Active Pharmaceutical Ingredients Committee (APIC). *Guidance on Aspects of Cleaning Validation in Active Pharmaceutical Ingredient Plants*, September 2016. https://apic.cefic.org/pub/APICCleaningValidationGuide-updateSeptember2016-final.pdf (accessed January 21, 2021).

46 Validation of Training

Chris Smalley

CONTENTS

46.1 TELL ME, SHOW ME, WATCH ME

In a previous edition of this book, a focus of the chapter on "Validation of Training" was on the testing that would be conducted to demonstrate that learning had occurred. Learning is indeed important, but our focus is now on understanding.

A didactic training environment will, for instance, require that multiplication tables be memorized. This is valuable learning, because this information will never change. Times tables might seem abstract; however, this information becomes useful when it is applied to a practical application, such as determining what portion of a salary needs to be turned over to the tax authorities. In this example, information becomes understanding.

In a GMP environment, it is not enough that training results in the trainees memorizing that Post-It ™ notes and Wite-Out ™ are not permitted to be used but rather the understanding of Data Integrity and its importance—understanding that nothing should be done to original data that will call its integrity into question.

A written test that would address this issue could be a problem statement. For example, "An operator is preparing to record a step they completed in the process. Unfortunately, their only pen ran out of ink. What should they do?"

The answer would indicate their understanding of documentation and Data Integrity. By the way, the best answer would be for the operator to prick their finger and write the entry in blood while protecting the product from biological contamination. Just kidding; but it is the discussion of the possible solutions that provides the understanding opportunity, such as the need to document promptly but also the risk of storing too much excess material in a process room.

In following the pathway of tell me, show me, watch me, a trainer is frequently capable of providing the tell me and show me elements but not always the watch me. Where training can be conducted in a workshop format rather than a classroom format, it is possible to provide all three. The safety departments at many companies use an outdoor workshop format, explaining the use of fire extinguishers, demonstrating the use of fire extinguishers, and then allowing the trainees to use a fire extinguisher. This is frequently the only way to train personnel correctly to aim the fire extinguisher at the base of the flames of a fire.

There are strong rationales for an organization to conduct training. Frequently mentioned are the requirements stemming from regulations, but also the reason nonregulated businesses conduct training is that it is good business. Ultimately, the reason not only to conduct training but to validate that the training is effective is when the trainees leave the classroom they are confident, competent and capable to implement the training. Manufacturing pharmaceutical and biopharmaceutical products with appropriately trained employees is no less important than training and testing people to drive cars—in both cases, lives are at stake.

It Is Good Business: An employee represents one of the largest investments that a business can make. Not only can an employee cost money by simply standing there, but if the employees don't perform the job properly, they can damage equipment and facilities, waste product, and consume an inordinate amount of time in correcting their errors. On the other hand, an employee qualified to perform their job will give the organization a competitive advantage in the marketplace. Well-trained employees can reduce waste, anticipate and prevent loss, have fewer on-the-job injuries and lost-time accidents, and perform with less direct supervision. The product that the customers receives, whether they are the next department to receive the component or the health care professional, will perceive a difference as well. After all, surveys have shown that when customers hear "It has been a pleasure to serve you" rather than "Have a nice day," their memories of the exchange are more positive. This can be achieved with a validated training program.

DOI: 10.1201/9781003163138-46

It Is Required by Regulation. Within the Code of Federal Regulations 21, Part 211, Current Good Manufacturing Practice for Finished Pharmaceuticals, there is a subpart for Organization and Personnel, which contains Section 211.25, Personnel Qualifications. In part, this section states that "Each person engaged in the manufacture, processing, packaging or holding of a drug product, shall have education, training and experience, or any combination thereof, to enable that person to perform the assigned functions."

Employees represent one of the greatest variables in pharmaceutical manufacturing. Their value over automation, in many instances, is their ability to use judgment in reaching conclusions and making decisions. It is this value that makes the case for validation of training so compelling.

In describing the validation of training, it is convenient to categorize the types of training that will be validated.

46.2 MOTIVATION OF A TRAINEE

What is the motivation for the trainee? It has been pointed out that it is good business and it is required by regulation, but those are motivations for the organization. Understanding and motivating the trainee should engender better, validatable training results.

Almost everyone wants to do a good job, and virtually everyone believes that they are performing at an above average level. A trainer/organization needs to capitalize on this belief, positioning the training as an opportunity to learn how to do their job even better. No one will believe that memorizing which citation within 21 CFR 211 will help them do their job better, so ensure that the training does not focus on such non-value-added content. All too often, a class would be conducted that drills into the trainees a 21 CFR citation, then crafts a valid question that provides multiple choices, and then trainers congratulate themselves that they have validated training. However, that violates the basic premise of validation, which is to prove that the User Requirements/Needs Assessment have been achieved.

User Requirements for training? Yes, a rationale for what training should achieve, and please don't make it "To Meet Regulatory Requirements." User Requirements will vary based on the objectives and the audience. Is the training for new employees? Is the training to upskill existing employees? Is the training for a new position for the employees or to use a new process or equipment? Let's delve into each of these.

New Employee Training. How does a new employee become educated in the skills needed to perform their job safely and effectively? Imagine for a moment that we are performing an Installation Qualification (IQ) similar to that for a new piece of equipment. Are your specifications adequate? That is, are the job description and other documentation that describes the job to be performed adequate? What are the minimum requirements for the employee being "installed"? Were those "Design Requirements" communicated to the Human Resource Department for recruiting and interviewing?

Can new employees read and understand the manufacturing batch record? Can they read and understand the Standard Operating Procedures (SOPs)? What education level are these documents written to? What grade level are new employees required to meet in reading comprehension? New employee training must consider:

- the requirements for the position being trained for,
- designing the training with the objective fulfilling those requirements,
- implementing the training, and then
- evaluating the training.

Evaluation is not possible unless the objectives (similar to Critical Process Parameters) are clearly set forth. Implementation is not possible unless a design has been made for the instructional material. That design, for new employee training, would usually consist of didactic training for the workplace and Standard Operating Procedures that would impact on that employee's tasks. For uniformity in presentation, commercially prepared audiovisual products such as those available from the PDA followed by a discussion period are useful, as would be a presentation from an experienced employee on Standard Operating Procedures.

For GMP overview issues, inviting a subject matter expert such as a professor from a local pharmacy college would make the presentation interesting to the new employees and afford the opportunity to audit the classroom to see who is paying attention or not mentally present.

Risk Management. The interest in Risk Management for validation activities has an impact on validation of training. Although some believe that Risk Management may be a rationale for reduced validation, in some areas the result of a Risk Assessment will highlight the need for additional mitigation. One of those areas frequently requiring additional mitigation is training.

The pharmaceutical industry frequently responds to problems by adding requirements to procedures. How often have gowning procedures been changed to add a statement requiring regowning when a person's knee touches the floor, or a calibration procedure been changed to require the technician to compare the accuracy of the calibration standard to the accuracy requirement of the instrument undergoing calibration? The real need is to train employees to think and understand. Every possible combination and permutation cannot be placed into procedures, and then training conducted on all of those possibilities. When examining the risks to the organization, failing to ensure that employees are capable of thinking and understanding independently are clearly among the greatest risks.

How, then, do we provide training that helps employees to think and understand? Training, in the content of procedures, focuses on the "what." Understanding relates to the "why." For training design, this entails not only presenting the words written in the procedures but the reasons why those words were written. One of the best approaches to training on this content is to use the Subject Matter Expert (SME) responsible for writing the procedures. A glaring fault in this approach, though, is assuming that the SME can properly present or train on the content. If the decision is made to use the SME in training to gain understanding, then a process for insuring that the training presentation is properly performed will need to be incorporated into the training design. This frequently takes one of two directions. One pathway would be to develop the training skills of the SME; the other pathway would be to have qualified trainers learn the material from the SME and then have them conduct the training.

Let's recap some of the topics raised in implementing the "IQ." They are training requirements, training design, training execution and evaluation of training. Embedded in these topics is the requirement to document. You may have what you assume to be an adequate documentation system, a system which records the date, the attendees, the presenters, and the topic. What those systems document is simply people sitting in seats, not proof that any content of the class was understood. Adequate documentation will record each of these four topics and together represent proof that training is under control and validated. And again, it is good business.

Program Documentation. It is a good idea to establish procedures for documenting training programs. Preparing needs objectives, design components, and requirements not only helps to clarify the planning phase but also increases the likelihood of a smoothly run program. Documentation allows the program to be repeated with little confusion and offers a chance to critique and tailor the design.

- An extensive outline of the program design, including a list of goals and objectives, notes for lectures, discussion questions, and so forth.
- A needs assessment that evaluates training needs. It is critical to clearly define the objectives of these needs. For instance, is the objective to learn to use different risk assessment tools such as HACCP or FEMA, or is the objective to understand the risk assessment tools and understand which one to select for a particular application?
- Additional design ideas, including use of game show formats or other interactive presentations that engage the attendees, as opposed to the classic PowerPoint presentations.
- References/readings on the content topic.
- Planning notes for the instructor and facilitators to use in the background of the lessons as reminders of content, without overloading the presentation or, even worse, appear to be reading the PowerPoint slides.
- Administrative information and forms.
- Evaluations with summaries of comments from program evaluation forms.
- Room arrangements/audiovisual needs.

The more explicit the documentation, the easier it is to update and repeat the program. Immediately after presenting a program, the trainer may make notes about what to do differently the next time. Before presenting the program again, the trainer can read the notes and evaluation comments from the previous presentation.[1] Program content that is critical for training success must be embedded in the training presentation materials so that different presentations do not vary in the depiction of that content.

Training Execution, or implementation, requires the development of a presentation and atmosphere that facilitate the transfer of information. The training design identified the target audience and the information to be presented; now comes the step involved in bringing that target audience together in such a way that they are both receptive to the idea that something can be learned and willing to participate in the training execution.

Avoid the training caricature of a group of people in rows of chairs nodding off in a darkened room while a speaker reads from the slides. Try to vary the format of the presentation. Use a table in the center of the room and, while having a discussion of the topics, move around the table to involve everyone. Give the students a break periodically to allow them to refresh themselves and recharge. Use a tour of an area that has relevance but perhaps one that the target audience does not actively work with. For instance, HVAC mechanics could tour the aseptic suite or the filling room operators tour the parts preparation and sterilization area.

Try to avoid the dry presentation of facts. Interactive formats can be developed using popular game shows such as "Wheel of Fortune" or "Jeopardy." Such formats require that the information be written in the language of the target audience. It is crucial that you understand your target audience and not necessarily just the goals identified in the design stage. Understand the mindset and motivation of your audience. If the group consists of empty nesters, do not use a rock music soundtrack for background fill-in. Conversely, do not use classical music background for a group consisting of young people. Do not use a game show format if it appears that the group may be shy. On the other hand, do not use a slide-tape presentation with an energetic audience.

Each format used to convey information additionally has advantages and disadvantages above being "right" for the audience. Slide-tape presentations can be made or purchased for a reasonable cost and customized to keep pace with procedure and facility changes. Motion pictures have a very high interest coefficient but are considerably more expensive and difficult to update. Videotapes have as high an interest and are as efficient as motion pictures at a lower cost; however, they, too, consume a great deal of resources to keep current. All of these formats should be incorporated into a total training presentation which includes tours, discussion, and games.

Evaluation of Training is both the most difficult and most important aspect of the validation effort.

46.3 SUBJECTIVE EVALUATION OF TRAINING

For the tell me, show me, watch me training method, the subjective evaluation of training is best. However, it is the more difficult. In most situations, the training will consist of tell me and show me, and it will fall to the trainee's supervisor to carry out the watch me portion. The organization must ensure that the supervisor has the "bandwidth" in order to perform this role, as it is very similar to that of managing an apprentice. However, assuring adequate resources is not the most difficult aspect: it is making the determination that training is complete and understanding has been achieved. Indeed, how has an organization demonstrated that the supervisors are qualified in their positions?

There will need to be a means to evaluate the success of the supervisors in accomplishing the subjective evaluation of training. For example, if the training was in documentation and Data Integrity, and the trainee committed some errors found on the Quality Unit review, and retraining is indicated for the trainee, it would be part of the evaluation of the success of that trainee's supervisor. Could aspects of the evaluation be delegated? Certainly, a SME could be tasked with performing aspects of the evaluation, especially those tasks that are new to the trainee or where a technique is crucial (for instance, aseptic gowning).

46.4 OBJECTIVE EVALUATION OF TRAINING

In objective evaluation, one of the tools that generates the most interest is testing. The goal of testing and evaluation are to evaluate the trainer and the program, not the trainees. Just as the Subject Matter Expert (SME) is an ideal source for training, the SME is the ideal source for preparation of the evaluation. The key to having an evaluation that is nonthreatening lies in planning the test effort during the training requirements and training design phases. In this manner, the testing can remain focused on how well the objective of changing the employee's knowledge or skill level was achieved, demonstrate areas for improvement in the program design, and provide the trainer with a performance monitoring tool. In keeping with the concept that training is good business, testing can provide management with information about the value and costs of training.

The test itself must be subjected to testing. A valid instrument measures what the trainer intends to test. Basically, there are four approaches to determining if an instrument is valid. These approaches, adopted by the American Psychological Association, are: (1) content validity, (2) construct validity, (3) concurrent validity, and (4) predictive validity. The actions taken to make the instrument valid are usually referred to as "defending" the validity of the instrument.

Content Validity—Content validity refers to the extent to which the instrument represents the content of the program. Content validity is probably the most important approach. Is it a representative sample of the skill, knowledge, or ability presented in the training program? To ensure content validity, no important items, behaviors, or information covered in the program should be omitted from the instrument.

Construct Validity—Construct validity refers to the extent to which an instrument represents the construct it purports to measure. A construct is an abstract variable such as the skill, attitude, or ability that the instrument is intended to measure. Examples of what a construct are include:

- Basic Construct: Ability to participate in an aseptic filling operation.
- Narrow Construct: Ability to read a volume having a meniscus.

Concurrent Validity—Concurrent validity refers to the extent to which an instrument agrees with the results of other instruments administered at approximately the same time to measure the same characteristics. Concurrent validity is determined by calculating the correlation coefficient between the results of the instrument in question and the results of a similar instrument.

Predictive Validity—Predictive validity refers to the extent to which an instrument can predict future behaviors or results.

If an instrument predicts a behavior, and a significant number of participants do exhibit that behavior, then the instrument possesses predictive validity. Predictive validity can be calculated and expressed as a correlation coefficient relating the instrument in question to the measure of the predicted results or behavior.[2] Certainly, adequate planning and definition of the job requirements are necessary to have a valid test. And the validity of the test can be supported by other evaluation methods that will be discussed later, including interviewing, group discussion, and on-the-job observation.

The type of test necessary to have a validated training program is the criterion-referenced test, where the employee's performance on the test is measured against the instructional objectives. Such a test need not be a written test, multiple-choice answer test or short answer fill-in-the-blanks test but rather practical application. The best tests relate to the job to be performed. Hence, if training were provided in gowning and contact plating, the order of events would be to design and execute the training and then, on three consecutive days, have the students gown, take the plates, and degown. Incubating the plates and reviewing the results would constitute the test and validate the training.

Let's take a moment to examine the role of the validity tests in the creation of the written test before further examining performance tests. Obviously, the training design included the training objectives, and these objectives are derived from the needs assessment that defined what competency is expected from the target audience, the conditions under which the competency will be displayed, and the accuracy with which this competency will be performed. The competency that the target audience is to display must be described in measurable and objective terms. Action verbs such as "sort and remove defects" are better than "inspect" and definable criteria such as "fill X ampules with media with NMT Y containers demonstrating contamination" as opposed to "aseptically fill" are necessary to prepare test items that can determine if the objectives have been achieved by the training program.

The objective attribute must be defined as well. Is the target audience to be the lead operators of the filling equipment or members of the team? Are the test subjects expected to refer to members of the team? Are the test subjects expected to refer to SOPs periodically or to perform competencies by rote? Are there time constraints on the performance of the competency? These conditions should be reflected in the construction of the test.

The accuracy with which the subject must perform the competency must be considered in the construction of the test. Is 100% accuracy necessary, or is the test audience subject to close supervision by a lead operator or supervisor, and some errors would be permissible? Is there expert judgment involved in evaluating and answering the situation presented in the training program? Are there elements within the training program that have no tolerance for any error, such as safety related items? In considering how well the learner must perform the indicated behavior, the issue of how much content may *not* be learned and still allows the learner to perform a GMP/GLP/GCP task needs to be answered.

With these clear objectives, a selection of written test design can be made that will evaluate the objectives. Remembering that the written test will only be one component in the evaluation of training; within the test there must be content validity and construct validity and other evaluation measures performed must demonstrate concurrent validity with the test, and finally, the supervisory feedback must subsequently demonstrate predictive validity.

Most instructional designers are aware of how important objectives are to the creation of instruction; many are less familiar with the role of objectives in testing.

Instructional objectives serve three fundamental purposes:

1. Objectives ensure that the test covers those outcomes important for the purposes that the training must serve. Remember that there are several different types of tests and that the content for these tests is derived by task analysis procedures that order objectives hierarchically. Matching test items to the appropriate course objectives within these hierarchies assures that the essential content is assessed.
2. Objectives increase the accuracy with which cognitive processes in particular can be assessed. A well-written objective is a roadmap for the creation of test items that will assess the specific competency described by the objective. Hence, objectives are essential to the construction of a validatable testing process.
3. The size of the domain covered by the objectives and the homogeneity of the objectives being assessed are important factors in determining how many items will need to be included in the test.

Of course, not all objectives are equally well written. Numerous authors have provided course developers with advice about how to write objectives. Most agree, however, that good objectives have four parts:

1. Who the learner is,
2. What behavior or competency the learner will perform,
3. Under what conditions the learner will perform the competency, and
4. To what standard of correctness the learner will perform the competency.

It is essential that the competency be described in observable, measurable terms; hence, the term "behavioral objective" used to describe the most useful statements of learner outcomes. When writing objectives, choose the most precise verb you can to state what the learner will be able to do. For example, the words "list," "categorize," "draw," and "evaluate" are better than "understand," "appreciate," or "know." The more descriptive the verb in an objective, the easier it will be to write test items that accurately assess the objective.

If well written, this part of the objective provides useful information to test writers, since the test essentially presents learners with a series of conditions under which they must demonstrate their achievement of the instructional objectives. Unfortunately, an aspect of the objective that is frequently omitted is the conditions element, by designers who do not realize how critical it is for clearly communicating the intent of the objective. Changing the conditions under which a behavior is to be performed can dramatically alter the difficulty and nature of the competency assessed.

For example, the behavior "assemble the tablet compression machine" is significantly altered depending upon whether the corresponding condition is "given the unassembled parts." The behavior with the latter condition can be expected to be significantly more difficult than with the former, and in fact the very nature of the intended competency specified by the objective changes depending upon which condition is used. Under the former condition, the objective describes skills in reading and using a repair manual, whereas under the latter condition the objective specifies mechanical skills.

Complete objectives include a statement of how well the learner must perform the indicated behavior. This component, however, is probably the most difficult component to write. It frequently takes the form "with 90% accuracy" or "correctly 80% of the time." It is helpful to realize that all standard statements need not be in the form of percentages. In fact, many competencies do not lend themselves to percentage standards at all. Other forms of standards are in terms of number of allowable errors, time limits, expert judgments, or negative consequences avoided—for example, "move pizza from oven to boxing counter *without burning the fingers*."

Good objectives are an essential precursor to sound testing systems. Translating objectives into rating scales for performance tests is usually easier than translating objectives into test items for paper-and-pencil tests. One strategy that can be helpful in this regard is to first classify the objectives according to the type of cognitive behavior each requires. Classifying objectives by cognitive skill assists item writers in choosing which item type—multiple-choice, essay, etc.—will most accurately and efficiently assess the objective and in deciding what the text of each item will be.

Several different classifications of cognitive behavior have been developed over the years. Bloom and his colleagues developed their system through an intensive content analysis of thousands of instructor-created test items. As a result,

Bloom's Taxonomy provides a particularly comfortable fit with and support to cognitive assessment.[3]

The classification scheme consists of six levels, with each given level building to the successive levels:

1. Evaluation
2. Synthesis
3. Analysis
4. Application
5. Comprehension
6. Knowledge

Understanding the nature of the cognitive performance to be assessed is a good first step to being able to write an appropriate test item. If a test writer can correctly identify the Bloom level of an instructional objective, a wealth of ideas about how to measure the objective become available.

Another important result of understanding Bloom's Taxonomy is an increased awareness of the cognitive behaviors beyond remembering, that is, beyond the knowledge level. Most of the tests taken in school at all grade levels and even at the college level are composed of knowledge levels questions. This circumstance is not difficult to explain, since knowledge level items are by far the easiest to write. However, developing tests that truly reflect on-the-job performance requires the ability to distinguish among different cognitive behaviors and skill in writing items at the higher cognitive levels, particularly the comprehension, application, and analysis levels.

There are six types of test items commonly used in paper-and-pencil or on-line tests:

1. True/false
2. Matching
3. Multiple-choice
4. Fill-in
5. Short answer
6. Essay

Of these six, multiple choice offers the most advantages for a paper and pencil test. Multiple choice questions present a question that may be identical to the question constructed for a short answer test and offers the recipient a number of choices consisting of a single correct answer and several distracters. Through the use of an "all of the above" or "none of the above" choices, not only is knowledge tested, but comprehension is evaluated as well. It is worth noting that if two of the multiple choice options appear to the participant to be correct, they can guess that "all of the above" is the correct answer. So this option should be limited. The probability of guessing the correct answer is lower that with true/false questions, and the process of scoring and providing feedback for the training program is much shorter that it would be for short answer or essay. It has the advantage of being able to assess most of Bloom's cognitive levels and yet can be easily scored by hand or by machine. In making comparisons, multiple choice will be the benchmark with a given item type. For each of these six item formats, a description of the item type and the kind of content for which the format is best suited, the Bloom levels assessable by the item type, the major advantages and disadvantages of using the item type, and a summary of the guidelines for writing each item type is presented.

1. True/False Items

   Description. The true/false item presents the test-taker with a statement that he or she must indicate is either true or false. This type of item is a sensible choice for "naturally dichotomous" content, that is, content that presents the learner with only two plausible choices. For example, assume our objective requires that, given an end-of-cycle sterilization report, learners will classify the cycle as pass or fail. You might construct a true/false question asserting that a given sterilization report is a "pass," to which the test-taker would respond "true" or "false." Content that is not naturally dichotomous is usually best assessed using the multiple-choice format, because true/false questions have some distinct limitations.

   Bloom Levels. True/false items can assess the knowledge, comprehension, and application levels. Unfortunately, however, they are most often used to assess only the knowledge level.

   Advantages. The primary advantage of true/false items is that they are typically easier to write than other types of closed-ended questions, that is, matching or multiple-choice. However, the reputed ease of construction is partly because most of these items are written at the knowledge level; it requires more thought to write true/false items at higher cognitive levels. The other advantages are that, like all closed-ended questions, they are easily and reliably scored, and test-taker responses can be submitted to statistical item analysis that can be used to improve the quality of the test.

   Disadvantages. The biggest disadvantage of true/false items is that test-takers have a fifty–fifty chance of getting the items correct simply by guessing. However, if the content that the true/false item covers is truly dichotomous, a multiple-choice item with more than two choices would be very difficult to write anyway. After all, multiple-choice items with only two choices also allow test-takers to guess correctly half of the time. Before writing true/false items, always examine the content and instructional objectives carefully to be sure that they are not more appropriately addressed by multiple-choice items. The key to using true/false items effectively is to use them only when the content is naturally dichotomous and to write true/false items that require more than mere memorization of content.

2. Matching Items

   Description. Matching items present test-takers with two lists of words or phrases and ask the test-taker to match each word or phrase on one list (hereafter referred to as the "A" list) to a word or phrase on the other (the "B" list). These items should be used only to assess understanding of homogeneous content, for example, types of sanitization agents, types of lubricants, types of switches, etc. Matching items most frequently take the form of a list of words to match with a list of definitions.

Bloom Levels. Matching items can assess the knowledge and comprehension levels. However, like true/false items, they are rarely written beyond the knowledge level.

Advantages. Matching items are relatively easy to write. Note, however, that one reason for this feature is that they do not assess beyond the comprehension level. Matching items can be scored quickly and objectively by hand and frequently also by machine. Responses to matching questions can be submitted to statistical item analysis procedures.

Disadvantages. Matching items are limited to the two lowest levels of Bloom's Taxonomy. Another disadvantage is that if these items are constructed using heterogeneous content, that is, if the words or phrases appearing on the "A" list are essentially unrelated to one another, matching items become extremely easy. For example, a list that contains a type of sanitizing agent, a type of lubricant, a type of switch, etc., will be easier to match to a corresponding "B" list than will a list that contains only names of different types of sanitizing agents. Another difficulty with matching items results from test writers including equal numbers of entries in both lists or allowing items from the "B" list to be used only once. Under these circumstances, test-takers can use the process of elimination to figure out cues to the correct matches.

3. Multiple-Choice Items

Description. The multiple-choice item presents test-takers with a question (technically called a "stem") and then asks them to choose from among a series of alternative answers (a single correct answer and several distracters). Sometimes the question takes the form of an incomplete sentence followed by a series of alternative completions from which the test-taker is to choose one. Sometimes the stem is a relatively complex scenario containing several pieces of information ending in a question. Dichotomous content can be assessed using multiple-choice questions with two optional answers; thus most true/false items can be converted to the multiple-choice format. In preparing a multiple-choice question, it should be remembered that the intention of the test is not to test the reading ability of the target audience: keep the language simple. To avoid misleading the subject, bold or underline negatives such as **no** and **not**. Use common errors that subjects make in developing distracters. Check the questions to ensure that the choices to one question do not indicate the correct answer in another question. The more questions that appear on a test, the greater the accuracy. For each training objective, approximately five questions should appear on the test. Obviously, the more critical the objective, the more questions, so that a GMP objective, for example, would be evaluated by more than six questions. Arrange the questions in order of difficulty, placing easier questions first to prevent frustration or discouragement. The number of choices does not need to remain constant but should not be less than three and only rarely be as many as six.

Bloom Levels. Multiple-choice questions can assess all Bloom levels except the two highest ones, synthesis and evaluation. The reason that these two levels are beyond the multiple-choice format is that they require totally original responses on the part of the test-taker. Since multiple-choice questions are closed-ended, that is, the correct answer appears before the test-taker who must recognize it, the test-taker's response is necessarily not original. However, multiple-choice allows assessment of more Bloom levels than any other closed-ended question format.

Advantages. Multiple-choice is the most flexible of all closed-ended item formats. Multiple-choice items can assess any kind of content at a variety of Bloom levels. Because the test-taker must choose among several optional answers, the probability of simply guessing the correct answer is lower than with true/false items. Furthermore, multiple-choice items are ideal for diagnostic testing. In other words, the distracters can target those learners who have specific problems; knowing the wrong answers test-takers chose can be important and useful information for instructors and course designers. In addition, multiple-choice questions are quickly and reliably scored either by hand or by machine and are ideally suited to statistical item analysis procedures that can lead to improved test quality.

Disadvantages. The major disadvantage of multiple-choice questions is that they are difficult and time consuming to write. Most testing authorities agree that well-written multiple-choice questions are usually worth the effort, especially if they can be used repeatedly with a large number of test-takers. (Re-ordering the choices after several uses may be appropriate to keep the test "fresh.") An additional weakness is that multiple-choice questions cannot assess objectives that require test-takers to recall information unassisted, since the correct answer does appear before the test-taker among the options, with the exception where the correct choice is "None of the above." Another disadvantage is their inability to assess directly the synthesis and evaluation cognitive levels. The principle disadvantage of multiple-choice question is the recognition factor. The typical performance setting will not present the test subject with information that may be recognized; rather, many situations will require the subject to recall information. Another disadvantage to multiple-choice question is the investment in time necessary to develop good distracters that are unambiguously wrong.

4. Fill-In Items

Description. Unlike the first three item formats discussed, fill-in items are open-ended, that is, the answer does not appear before the test-taker. Rather, the fill-in item is a question or an incomplete statement followed by a blank line upon which the test-taker writes the answer to the question or completes the sentence. Therefore, fill-in questions should be used when the instructional objective requires that the test-taker recall or create the correct answer rather than simply recognize it. Objectives that require the correct spelling of terms, for example, require fill-in items. Fill-in items are limited to those questions that can be answered in a word or short phrase; short answer and essay questions require much longer responses.

Bloom Levels. Fill-in items can assess the knowledge, comprehension, and application levels. They are written most often, however, at the knowledge level.

Advantages. Fill-in items are typically easy to write. They are essential for assessing recall as opposed to recognition of information.

Disadvantages. There are two major disadvantages of fill-in items. One is that they are suitable only for questions that can be answered with a word or short phrase. This characteristic typically limits the sophistication of the content that can be assessed with fill-in items. The second major disadvantage is that, like all open-ended questions, fill-in items present scoring problems. Because test-takers are free to write any answer they choose, sometimes there can be a debate over the correctness of a given answer. Test-takers are marvelously unpredictable when it comes to concocting an unanticipated answer to an open-ended question. Unlike the scoring of closed-ended questions, the scoring of all open-ended question requires judgment calls on the part of the scorer.

5. Short Answer Items

Description. These items are open-ended questions requiring responses from test-takers of one page or less in length. Short answer questions require responses longer than those for fill-in items and shorter than those for essay questions. Short answer questions are recommended when the objective to be assessed requires that the test-taker recall information unassisted (rather than recognize information) or create original responses of relatively short length.

Bloom Levels. Short answer question can be used to assess all Bloom levels except possibly the highest one, evaluation; most responses to evaluation questions would necessarily be somewhat longer.

Advantages. The major advantage of short answer questions is that they are able to elicit original responses from test-takers. For some objectives at the higher Bloom levels, only short answer and essay question are appropriate. Lower level short answer questions are typically easier to write than multiple-choice questions covering the same content. It is important to remember, however, that changing the format of a question can significantly alter the cognitive skills assessed. Short answer items are best reserved for those objectives that cannot be assessed using closed-ended questions.

Disadvantages. The disadvantages of short answer questions are, unfortunately, extremely serious ones. Most notably, short answer questions are very difficult to score reliably. The evaluation of short answer responses and essays are notoriously prone to error—resulting from halo effects, the placement of a given test in the scoring sequence, scorer fatigue, and especially, quality of handwriting. This disadvantage could be overcome by using an on-line test. In addition to being unreliable, the scoring of short answer responses is time consuming. Short answer questions also require far more time to answer than multiple-choice questions, thus sometimes limiting severely the content that can be covered by the test.

6. Essay Items

Description. Essay items are open-ended test questions requiring a response longer than a page in length. They are recommended for objectives that require original, lengthy responses from test-takers. Essay items are also recommended for the assessment of writing skills.

Bloom Levels. Essay questions can be used to assess all levels of Bloom's Taxonomy. They are the only item type with this capability, and writing is the only item type that can truly assess the evaluation level.

Advantages. The essay question's major advantage is its capacity to assess the highest cognitive levels. Of course, these levels, comprehension and knowledge, include the aspects of understanding and problem solving that are important objectives identified earlier. Essay questions that assess the lower levels are usually not difficult or time-consuming to construct. Those that assess the higher levels can be very difficult to write, requiring the provision of a great deal of stimulus material to which the test-taker responds in the essay.

Disadvantages. The disadvantages of the essay item are identical in nature to those of the short answer item; however, these problems are aggravated by the additional length of the responses. Essay questions are even more difficult to score reliably, take even more time to score, and use up even more testing time than do short answer questions. The principle disadvantage of short answer and essay questions lies in the debate that may ensue over the correctness of an answer. The number of unanticipated answers to open-ended questions has never ceased to be amazing. Secondarily, as mentioned earlier, the time to review and score the short answer or essay test will hinder the timeliness with which feedback can be provided to the training program. For these reasons, essay items are to be avoided if at all possible. Use essay questions only when the cognitive level of the objective requires it.

46.5 CONCLUSION

Training is an important aspect of the pharmaceutical process, crucial to every step and a factor in compliance with cGMPs, safety, productivity and customer satisfaction. Similar to process validation, cleaning validation, or sterilization validation, in training it is important to identify the requirements, the design of the means for testing against those requirements, the acceptance criteria, and documentation of the results.

Testing is a subcomponent of the validation of training, a component that has not been utilized for a variety of reasons but nevertheless a component central to the validation process. Testing must be conducted in a manner consistent with good science, as must any validation program, and like any validation program the testing instrument cannot stand on its own but must be supported and confirmed by other tools. Among these other tools are group discussion and observation by supervisors or other qualified individuals. The observations, whether performed by the supervisor or involving equipment assembly by a mechanic or gowning technique by a microbiologist, tend to be much stronger in cognitive assessment than written instruments. Although traditionally referred to as performance appraisal, these observation methods are in fact a non-written test method. The method is frequently criticized as being subjective; however, this fault can be minimized through the preparation of an evaluation period where a different supervisor other than the supervisor to whom the trainee normally reports performs the assessment. Where possible, utilize a subject matter expert (SME) to perform the assessment, such as a microbiologist for trainees participating in aseptic filling operations. Similarly, a SME would be necessary in determining the content validity of the written test. The most important aspect to remember in training and the validation of training, though, is not to fall into the trap of conducting and validating remedial training after disaster has struck but to make the investment in validation of personnel similar to the investment made in the validation of equipment and systems.

Much of this section has been devoted not only to the validation of training but also to make the case as to why validation should be performed on training and that training validation contains the same elements as other validation activities. Validation should be performed on training because it is good business as well as required by regulations. Training validation will have a Requirements phase identified using a Needs Assessment evaluation, a Design phase, an Execution phase, and an Evaluation phase. Like other validation activities, a matrix mapping the tests conducted during Evaluation back to Needs Assessment ensures the validity of the validation.

Appendix

Here are several guidelines for consideration in the construction of objective measures that can be used regardless of the type of training:

1. Arrange the items in order of difficulty; placing easier items first avoids discouraging participants unnecessarily.
2. Construct each item so that it is independent of other items. A series of items in which the correct answer to the first becomes the condition for the next item can prevent the measure from providing an accurate picture of the participant's knowledge or skills.
3. Avoid constructing items by taking quotes directly from a handout, overhead transparency, or book. Direct quotes tend to encourage memorization rather than understanding, and quotations taken out of context tend to be ambiguous.
4. Avoid trick questions. The intent of a performance measure is to determine the skills, knowledge, and attitudes of the participants, not to cause them to mark an item incorrectly.
5. As much as possible avoid negatives, especially double negatives. Such items take considerably longer to read and are often misinterpreted. If negative words must be used, underline or italicize the word or phrase to call attention to it.
6. Avoid providing clues to items in previous or subsequent items.
7. Use a variety of types of items in the performance measure rather than limiting the items to only one type. If the measure is lengthy, variety can add interest. When a variety of types of items are employed, group the items by type so participants do not have to constantly shift response patterns or reread instructions.

TRUE/FALSE ITEMS

Guidelines for construction of true/false items

1. Have participants circle the correct answer rather than write in T or F. Poor or careless writing can make the letters "T" or "F" or even the words "true" or "false" very difficult to read.
2. Avoid the use of "always," "all," "never," "only," and "none"; these words alert the participant to mark the item false. For example, the following is a faulty item:

   T F Effective managers always delegate.
3. Avoid the words "sometimes," "usually," "maybe," "often," and "may"; these words alert the participant to mark the item true. For example, the following is a faulty item:

   T F The tablet hardness is sometimes affected by the moisture content of the granulation.
4. If the statement is controversial, cite the authority whose judgment is referenced. For example, the following is a faulty item:

   T F The organization has a responsibility to provide assistance to employees who have drug problems.

   This statement as it is written appears to measure attitude rather than knowledge. To make it into a knowledge statement, it could be preceded by "The Federal Government has determined that . . ." or "The site Director has stated that . . ."
5. Do not include two concepts in one item. For example, the following is a faulty item:

   T F The trend toward quality circles began in the early 1980s and represents a big step forward in improving quality in the United States.

   Either of the two concepts (1) began in the early 1980s or (2) represents a step in improving quality in the United States could be false.
6. Each statement should be entirely true or entirely false without additional qualifiers such as "large," regularly," "sometimes," and "may." For example, the following is a faulty item:

   T F A media fill failure may indicate poor gowning technique.
7. Keep true and false statements approximately the same length.
8. Have approximately the same number of true and false items. True statements are easier to write, so there is a tendency to include more true than false statements.
9. Avoid making false statements by simple adding "not" to true statements. For example, the following is a faulty item:

   T F The cGMPs are not the guideposts of pharmaceutical manufacturing.
10. Avoid using trivial details to make a statement false.
11. Avoid a pattern of answers such as TTFFTTFF.
12. Place the central point of each statement in a prominent position or highlight it in some manner.
13. Avoid long and complicated statements that may test reading ability rather than the content. For example, the following is a faulty item:

    T F If data are recorded to the nearest .001 inch (for example), then the class width should be an integer multiple of .001 inch so that each interval will contain the same number of possible data values.

14. Avoid negative statements and eliminate double negatives. For example, the following is a faulty item:
    T F If a person has not had access to a patent, he or she cannot infringe the patent.

MATCHING ITEMS

Guidelines for construction matching items:

1. Place the list of descriptions (the longer list) on the left side of the page so the participant need read it only once. Place the options (the shorter list) on the right to be scanned as often as needed.
2. Make both the descriptions and portions lists homogeneous.
3. Provide at least three more options than descriptions or permit the use of each option more than once to reduce guessing.
4. Specify in the instructions the basis for matching and how to mark the answer.
5. Make each option plausible to the uninformed.
6. Arrange the options in some logical order, such as chronological, numerical, or alphabetical, to save reading time.
7. Allow between five and 15 options.
8. Number the descriptions and letter the options.
9. Specify in the instructions whether options can be used more than once.
10. Avoid having more than one correct option for each descriptor; item 1 in Column A should have only one option in column B that is correct.

MULTIPLE-CHOICE ITEMS

Guidelines for constructing multiple-choice items:

1. The stem (the part that precedes the responses) should clearly state the premise; the response options should be kept as short as possible. For example, the following is considered a faulty item:
   An income statement
   a. Reflects the firm's financial position.
   b. Is more important than the firm's balance sheet.
   c. Is a key financial statement.
   d. Is always performed on a cash basis.
   The stem, "An income statement," fails to state the basis or context for the response.
2. The response options should contain only one defensible answer. If more than one item is correct, the effect of guessing is increased.
3. Distracters (incorrect responses) should be plausible; use common mistakes and misconceptions to create distracters. For example, the following is a faulty item:
   Which element has been most influential in recent pharmaceutical development?
   a. Scientific research.
   b. Psychological change.
   c. Convention.
   d. Advertising promotion.
4. All response options should be grammatically and logically consistent with the stem; for instance, watch the uses of "an" and "the."
5. The length of correct responses should be approximately the same as the incorrect responses; there is a tendency to make the correct answer longer. For example, the following is a faulty item:
   Laminar flow
   a. never increases.
   b. gradually increases.
   c. provides a unidirectional flow of HEPA filtered air and conforms to the ISO standard.
   d. Is a means of covering surfaces.
6. It is more appropriate to ask what an item is rather than what it is not; knowing what is incorrect does not indicate whether the participant knows what is correct. For example, the following is a faulty item:
   The filtration rate is not affected by
   a. temperature.
   b. bioburden.
   c. pressure.
   d. container size.
7. Include from three to five responses options for each item. All items do not have to provide exactly the same number of response options.
8. "All of the above" is usually the correct answer and therefore makes the item too easy. The participant can guess "all of the above" is correct if two of the other options appear to be correct. For example, the following is a faulty item:
   Which of the following factors are involved in achieving a defect-free product?
   a. Quality.
   b. Maintenance.
   c. Personnel.
   d. Supplier.
   e. All of the above.
9. Rotate the position of the correct response from item to item. Instructors have a tendency to use "b" as the correct response more often than other response options.
10. Place any words that the response options have in common in the stem. For example, the following is a faulty item:
    Moisture content is an important factor to consider in tablet compression because
    a. Amount of moisture affects content uniformity.
    b. Amount of moisture affects "capping"
    c. Amount of moisture affects potency.
    d. Amount of moisture affects tablet thickness.
    The phrase "Amount of moisture" should be placed in the stem.

11. All options should be homogeneous in content. For example, the following is a faulty item:

    The misery index

    a. Should be calculated for each project under consideration.
    b. Is calculated quarterly by the Chamber of Commerce.
    c. Looks at our balance of trade position.
    d. May affect project viability in an indirect manner.

    The first two responses refer to time, the third appears to be a definition, and the last refers to a consequence.
12. It is acceptable to use either a direct question or an incomplete statement as the stem. The preceding item used an incomplete statement. The following item uses a direct question:

    Which type of data is used to express the number of defects found in a product at final test audit?

    a. Categorical.
    b. Numerical—discrete.
    c. Numerical—continuous.
13. Use "none of the above" sparingly. It tends to test only the participant's ability to identify incorrect answers. Recognizing that items are wrong does not mean that the participant knows the correct answers.

COMPLETION ITEMS

Guidelines for constructing completion items:

1. Use only one blank per item. For example, the following is a faulty item:

   The _______________ of cGMP compliance is measured by ______________.
2. Require single-word answers rather than phrases.
3. Use either direct questions or fill-in-the blank statements, such as "Write the formula used for determining standard deviation."
4. Place the blank near the end of the sentence rather than near the beginning. For example, the following is a faulty item:

   _______________ is likely to increase when a test is lengthened.
5. Word items so they have only one correct answer. For example, the following is a faulty item:

   Laminar Flow does not protect _______________.

   The answer the instruction wanted in this item is "covered items." However, a great many words could be placed in the blank without being incorrect.
6. Make sure that the word deleted from the sentence is a significant one. For example, the following is a faulty item:

   A customer can be _____________ as anyone who is impacted by the development of the product.
7. Use "a(n)" before a blank to avoid grammatical cues.
8. Structure all answer blanks to be the same length, regardless of the length of the word to be supplied. Blanks that correspond to the length of the word provide an additional clue to the answer.

NOTES

1. *Inside Training and Development*, Susan Warshauer, pp. 61–62.
2. *Handbook of Training Evaluation and Measurement Methods*, Jack J. Phillips, pp. 82–85.
3. *Development and Validation of Minicourses in the Telecommunication Industry*, Richard R. Reilly and Edmond W. Israelski, pp. 721–726.

BIBLIOGRAPHY

Bloom BS. *Taxonomy of Educational Objectives: Handbook 1. Cognitive Domain*. New York: McKay, 1974.

Bramley P. *Evaluating Training Effectiveness*. London: McGraw-Hill Book Co., 1991.

Caffarella RS. *Program Development and Evaluation Resource Book for Trainers*. New York: John Wiley & Sons, 1988.

Cocheau T. Training from the start. *Training and Development Journal* 1990, 44(10), 22–27.

Dick W, Carey L, Carey JO. *The Systemic Design of Instruction*. New York: Addison-Wesley, Longman, 2000.

Goad TW. *Delivering Effective Training*. San Diego: University Associates, 1982.

Phillips JJ. *Handbook of Training Evaluation and Measurement Methods*. Houston: Gulf Publishing Co., 1990.

Reilly RR, Israelski EW. Development and validation of minicourses in the telecommunication industry. *Journal of Applied Psychology* 1988, 73(4), 721–726.

Smith JE, Merchant S. Using competency exams for evaluating training. *Training and Development Journal* 1990, 44(8), 65–71.

Vesper JL. Defining your GMP training program with a training procedure. *BioPharm*, November 2000, 28–32.

Warshauer S. *Inside Training and Development: Creating Effective Programs*. San Diego: University Associates, 1988.

Wigging GP. *Assessing Student Performance: Exploring the Purpose and Limits of Testing*. San Francisco: Jossey-Bass, 1999.

47 Vendor Qualification and Validation

Maik Jornitz

CONTENTS

47.1 INTRODUCTION

Supplying the highly regulated biopharmaceutical industry with equipment and services generally means that vendors must adopt similar quality standards pertinent to the processes of the biopharmaceutical industry applied to the vendors' processes. These standards start with the qualification of the equipment during the development phase, the validation of the production process, in-process controls and documentation during the production process, release criteria, specifications and tolerance settings and complete traceability of the finalized product and product components. Once the equipment is ordered or supplied to the end-user, most commonly, the vendor will submit qualification documentation, support qualification and acceptance testing and, in some instances, offer product- or process-related validation services (5). The vendor's production processes often mirror the production processes of the relevant industry the vendor supplies to. Additionally, the vendors establish appropriate technical support structures to be able to react rapidly to support needs of the industry. This is of importance because the end-user must be able to answer to regulatory enquiries and also in case equipment requires maintenance, calibration or repair. Production interruptions cannot be tolerated, as it might result in multimillion-dollar losses in revenue and put drug product batches at risk.

An important factor for the end-user is the determination of the criticality of the supplied goods to the processes and the finished product of the end-user. As the end-user has an input control of goods received, the vendor has an output control or release criteria. However, the criticality of the products supplied to the end-user will also determine the evaluation scrutiny the product will undergo before it is shipped or transferred into the end-user's production process. For example, a drug product component or excipient will have a higher criticality than a sterilizing grade filter than a filter housing. The component's quality is essential, as it influences the end product's quality directly. A sterilizing grade filter has an indirect influence; however, it is still more critical than a filter housing due to the fact that the filter has an influence on the sterility of the end product. Therefore, vendor validation and qualification processes most definitely differ in stringency and scope due to the fact that some components have a lesser criticality than others, which does not mean that products of lesser critically do not need to meet the required specification; the control parameters are just different.

This chapter describes some of the qualification/validation work vendors undergo and establish to meet the biopharmaceutical end-user requirements.

47.2 VALIDATION WITHIN VENDOR DEVELOPMENT

Vendors strive to improve their products and processes to be able to supply the industry with state-of-the-art materials, components, equipment and improvements within the industry's processes. For this reason, vendors typically invest 3–8% of their revenue in the development of new products or improvement projects for existing products. However, every time a product is newly developed or revised, a similar documentation trail to that of pharmaceutical R&D has to be established. It begins with the choice of a qualified sub-supplier and ends with a fully qualified product and validated production process. Vendors development groups are multifunctional teams that work together with sales and marketing, supply chain and production to have an appropriate idea of what is required within the industry, whether the raw material needed is readily obtainable at the quality specification set and whether the production capacities as well as machineries are available. Once these cornerstones have been investigated and verified, the development of the product will start. Any effort to develop a piece of equipment without the knowledge of market needs, supply assurance and production feasibility is a wasted effort. As logical as it sounds, these cornerstones are the first milestones that are documented within a development process and will eventually mean audits by the vendor's quality assurance and supply

DOI: 10.1201/9781003163138-47

chain departments of the raw or sub-material supplier. These audits have to be well documented and are commonly applicable to a minimum of two suppliers. Supply assurance for a vendor is as important as for the end-user, as any supply change will result in a change notification and comparability studies, and in some instances a possible revalidation of the equipment at the end-user level and notification of the regulatory authorities. Therefore, changes within the vendor's processes, raw materials or specifications are to be avoided. Any change within the vendor's processes will ultimately influence the user's processes and regulatory filing requirements. Critical raw materials and components fall under long-term supply assurance contracts and might have multiple-year inventory levels within the sub-supplier or vendor level. Additionally, vendor development will involve quality assurance to analyze whether the sub-supplier's quality certification, systems and assurance meet the specification given to the vendor's development by the industry. As the end-user audits the vendor's processes, the vendors will do the same at the sub-supplier level. The more thorough the vendor's internal and external quality and supply investigations, the better the supply quality to the end-user.

Once quality sub-suppliers have been established, the vendor's development group will create prototypes and different versions of the product, which will be first tested in-house and at a later stage at a beta-site, which is commonly an end-user's process development or small-scale site. However, all stages of the pilot scale production and/or assembly have to be thoroughly documented to assure consistency and improvement. In instances of source code development for equipment control, any development of such code or any change within the code has to be documented (7, 9). The entire source code establishment requires audits and needs to be well documented. It is import to check the source code development on a frequent basis to avoid any late stage surprises. As Figure 47.1 shows, the costs of software revision become exponentially higher toward the later stages of software preparation.

Prototypes will be tested, and if these do not meet the specified requirements, the development must return to the drawing board and improve the equipment to the users' specifications. When done, the product parts production, assembly and packaging specifications must be

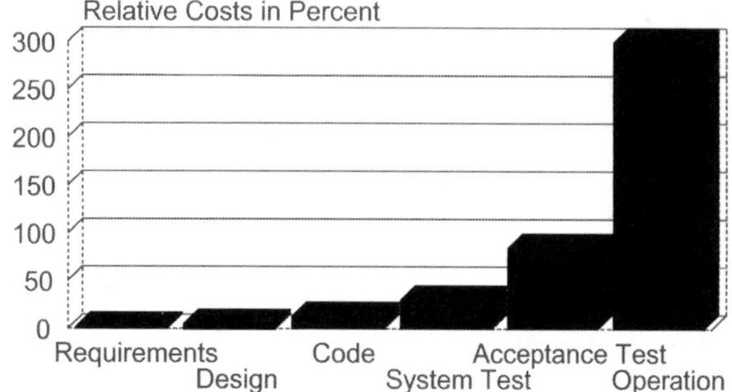

FIGURE 47.1 Costs of software development.

locked. This means that production parameters and tolerance are recognized and set, most commonly by repeated production batches, as required within the industry itself. At this point validation protocols and standard operating parameters (SOP) have been instituted. Validation tests are commonly set by publicly available international standards, for example, sterilizing grade filters have to meet current pharmacopeial requirements and will be tested accordingly. Nevertheless, the vendors have their own sets of tests, specifying which components and equipment will undergo to verify performance criteria set by the vendor and user alike. The test results of these tests can be found within the vendors' qualification documentation (validation guides), which are supplied to the end-user (see Figure 47.2).

These documents, however, will not replace process validation or performance qualification at the end-user's site. These documented tests establish the basis requirements for the equipment to (a) be able to work within the biopharmaceutical environment and (b) verify that the equipment meets regulatory requirements. If this scientific basis is not met by the developed product, the product will be scrapped.

Furthermore, these tests will also set the standards for which tests quality assurance will use to determine product consistency and reliability. Most commonly, vendors have already standard quality assurance tests defined by other production processes or equipment specifications. These can be utilized to a large degree; however, it could be that a specific piece of equipment requires additional tests or release criteria. For example, the in-process controls and release criteria for a sterilizing grade filter will differ for a membrane chromatography device or filter housing. The main release criteria for a sterilizing grade filter is its integrity, whereby the release criteria for a membrane chromatography device would be adsorptive capacity and for a filter housing surface roughness. These are only examples of release criteria. Most often, every product category has a multitude of release tests to fulfill before the product is shipped. In any case, all the product categories have to have appropriate controls and release criteria established to meet quality and consistency standards.

A vendor's development departments have to work in close conjunction with multiple departments. The development group not only creates new or improved products but has to assure sub-suppliers, determine appropriate production specifications and tolerances and the validation as well as a smooth hand-over into full-scale production. Additionally, with the validation of the vendor's production process, close collaboration with quality assurance is required to create appropriate SOPs, validation and qualification documentation, and development documents, especially for source code. Finally, the vendor's product management and technical service departments are supplied with performance data and specifications of the new product as established by the development department. These data have to match or exceed the criteria set by product management and, respectively, the end-user.

Table of Contents

FIGURE 47.2 Example of a validation guide supplied by a vendor.

47.3 VALIDATION OF VENDOR PRODUCTION PROCESSES

Depending on the complexity of the vendors' products, the production processes require process specific validation. Most commonly, vendor production processes are multistep processes meaning every step requires validation, appropriate operation procedures and qualification, training and certification of the personnel involved. For example, membrane casting for a sterilizing grade filter is one step within the production of a sterilizing grade filter. This casting process requires very specific environmental and process conditions and machine settings. During the casting process, the machine parameters are constantly monitored and samples of the casted product are taken and tested frequently (see Figure 47.3).

All parameters and test results are documented within the batch records of this particular cast and can be reviewed by auditors. The documented results also serve as a historical database to perform statistical evaluations, evaluate process performance or support development efforts. The casting process parameters determine the pore size of a membrane but also its pore size distribution; that is, the process has to be closely adjusted and monitored to achieve a narrow or desired distribution (see Figure 47.4).

Once the membrane batch is casted, it will be pleated, sealed, end-capped, welded, integrity tested, bagged and autoclaved. However, the procedures or individual production steps line-up does not matter: every single production step has defined process parameters within which the n-step has to be run. The timeframe between every step requires as much monitoring as the step itself. The process has to be validated as an individual step and in its entirety.

This means that production parameters and release criteria are defined for each step and described within SOPs. All parameter, specifications and tolerance levels are documented within a validation master protocol and cannot be changed without approval by a multitude of departments, most importantly, quality assurance. Release criteria are established by in-process tests. Once the criteria are met, the product can move to the next step of the process. However, if the product does not meet the criteria, an investigation will be initiated to analyze why the product is out of specification. Most commonly, such investigations occur during the development of full-scale production, when the products are moved from development's pilot scale to full scale. Scaling within vendors process can be as difficult as in the end-user's industry. This might be seen as a negative but should be seen positively, as it is better to amend root causes of undesired product quality early within the scale-up of the production process instead of within the final established production processes.

Once the production processes are established, maintenance protocols will assure that the production equipment will fulfill the criteria set. Maintenance protocols are written during the validation phase, as wear and tear can vary equipment performance and specific tools must be inspected during the validation phase. Any automated equipment utilized in the production process has to undergo installation, operation and performance qualification, especially in regard to the process and systems controls. For example, a CNC (computer(ized) numerical(ly) control(led)) cutting machine, which is utilized to cut a specific part for a medical device, requires as much qualification work and documentation as an autoclave. The product component delivered by the CNC has to be of consistent quality, complying with set tolerances and specifications. Another example are membranes used for sterilizing grade filters, which have to meet thickness, porosity, integrity, extractable, thermal and mechanical stability and particulate test parameters (5, 6). If any of these specified parameters are not met, the batch will not be released, and an out of specification investigation will be performed. These tests are described and used as release parameters for validation batches and later for commercially marketed batches. However, consistency in set quality parameters is the most important aspect in any stage of the production.

In certain production processes, the process cannot be automated and the production step is performed by personnel, for example, welding. The settings within welding can be described,

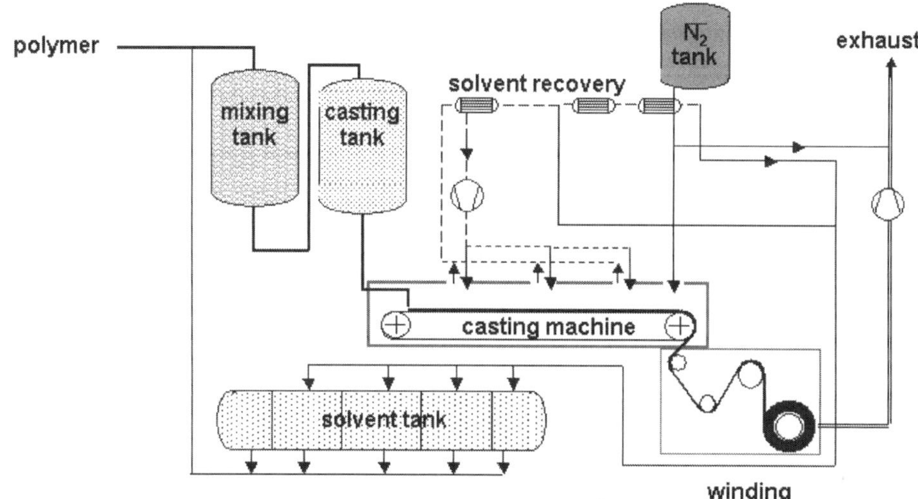

FIGURE 47.3 Casting process.

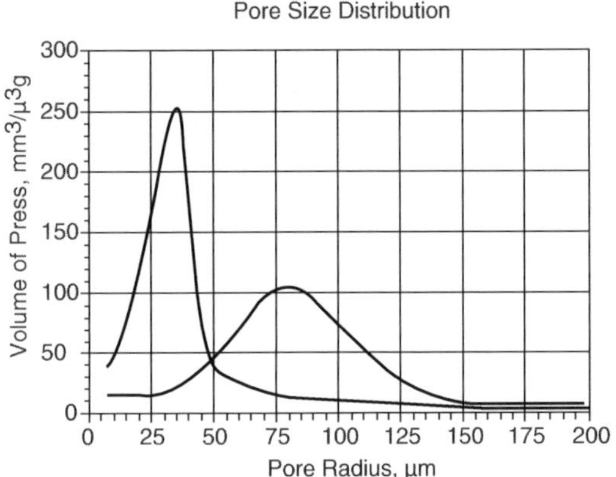

FIGURE 47.4 Pore size distribution example.

but only as indicator specifications, for example, the tube volume, material thickness that will determine the protective gas pressure and welding energy settings. However, due to the complexity and individuality of some equipment, for example, bioreactors or cross-flow systems, most welding might be done by the pure experience of the welder. The welders require specific certification and most often have many years of experience. The welding itself will be analyzed and inspected before release, but this does not minimize the skill level required for such welding tasks. Similar skills are required for cutting, honing, bending, polishing, etc. The validation within these processes is the certification and routine training of the personnel and use of log books as well as the quality of the raw materials used. Any raw material entering the facility will be sampled, inspected and documented and requires specific certification and log numbers. The raw material has to be traceable and of specified quality.

An additional piece of validation work on the vendors' part is packaging validations and tests. The goods will be packed in specifically designed packaging that assure robustness during transport. The vendors will test the packaging design using specific standards, for example, ASTM D 4169 and D 4728–95 (1, 2). These tests are drop and vibration tests. DIN ISO 12048 is a compression test, which will verify the stability of the packaging (3). As soon as the goods leave the factory, the vendor loses control over the handling of the goods. Therefore, packaging plays a major role to maintain the quality and integrity of the goods shipped. Moreover, robustness is not only attached to mechanical stability but also to thermal and chemical stability. Temperature changes during transport are not unusual, especially during overseas shipments. The packaging must be flexible enough to overcome any thermal expansion or shrinkage. It also should repel any condensation occurring due to temperature changes or changes in humidity. Oxidation due to sunlight is probably the most common photochemical attack to polymeric packaging. The packing has to be stabile under these circumstances; otherwise, polymeric degradation would result in weakening the packaging or particulate shedding of the packaging.

The ultimate tests for packaging are multiple shipments into the different regions supplied to and supplied by different carriers. At the end, these tests will create a grid of test data of different means of transportation at different environmental conditions, which will result in a tolerance band for the designed packaging. Only such tests can create practical data verifying the experimental lab data. Pure lab data would not support the assurance of structural integrity and safety.

47.4 VENDORS IN-PROCESS CONTROLS AND RELEASE CRITERIA

Depending on the vendors' products, the in-process controls and release criteria vary from narrowly defined step-by-step controls within the production process or as an end-result control and release (5). Most commonly individually produced components are tested when produced and again when the individual components are assembled. As previously described, the control and release criteria and tests are established within the development process and depend also on the criticality of the product supplied to the end-user. In some instances, control and release criteria are fairly simple and encompass only a single test criteria; most of the time, though, products distributed to the pharmaceutical industry undergo multiple tests within the parts and final product production process.

Raw materials supplied to the vendor are first checked to see whether the quality documentation is complete. Again, depending on the criticality of the component, the material might undergo specific tests to verify that the quality standards described are met. For example, polymer granulates undergo thermal profiles to check that the quality and type is the same as specified by the vendor to the sub-vendor. In other instances, the raw material is visually inspected, for example, stainless steel tubing in regard to surface finished and material stamps. If the raw material does not meet the one of the specifications, the material will not be released into production. All raw material batch records are kept with the batch records of the resulting product. The product has to be completely traceable to allow appropriate investigation, if necessary. Raw material suppliers are generally audited once a year, depending on the significance of the raw material supplied. However, if there has been an incident, the supplier will be audited immediately thereafter and corrective action verified.

For example filter cartridges, whether pre- or membrane filter, are tested for extractables (see Figure 47.5) to check whether there is any change within the profile that might not meet release criteria (5, 6). Similar tests are flow, throughput, mechanical and thermal robustness. Membrane filter are commonly individual integrity tested before release.

Stainless steel products also have specific definitions that need to meet the biopharmaceutical requirements (5). These are individual stamping of the steel goods, welding certification, material qualification and certification. The steel source can determine the quality of the steel. The steel components are required to be right, as these determine welding quality, corrosive robustness and the electrolytic behavior within a system. Nowadays stainless standards are set by the industry, which define for example the ferrite content or surface smoothness.

Depending on the application, the stainless steel equipment used differs greatly in the surface treatment. The smoother the surface, the greater the treatment steps and the costs involved.

Identified Extractables of Different Membrane Filter Cartridges From Several Filter Manufacturers							
Cartridge A	**Cartridge B**	**Cartridge C**	**Cartridge D**	**Cartridge E***	**Cartridge F***	**Cartridge G***	**Cartridge H***
Diethylphthalate	Cyclohexan	Propionic acid	Diethylphthalate	Acrylic acid	Dimethylbenzen	Etherthioether	Caprolactame
Stearic acid	Ethoxybenzoic acid	Diphenylether	12 oligo. aliphates	2 phenolic oligo	Etherthioether	Propionic acid	Butyrolactone
2, 6-Di-tert.-butyl-cresol	2,6-Di-tert.-butyl cresol	2, 6-Di-tert.-butyl-cresol	Hydroxybenzoic acid	2, 6-Di-tert.-butyl-cresol	2,6-Di-tert.- butyl cresol	2,6-Di-tert-butyl-cresol	Laurinlactame
2,2-Methylene-bis 4-ethyl-6-tert. Buty phenol	Cyclohexadiene 1,4-dion	4-Methyl-2,5-cyclohexadiene-1-on	Tert.-butyl-methyl-2, 5-cyclohexadiene-1-on	3 oligo. Benzyl-di-phenylmethan	3,5-Di-tertbutyl-4-hydroxyphenyl propionate	3,5-Di-tert.-butyl-methyl-2,5-cyclo-hexadiene-1-on	Laurinlactame derivate
Hydroxybenzoic acid	Phenylisocyans	Hydroxybenzoic acid	2,4-Bis(1,1-di-methoxy-1-ethyl)-phenol	Triphenylphosphite	2 N-containing high MW compounds	4,4-Dichloro-diphenylsulfone	4-Methoxy-4-chlor-diphenylsulfone
7 oligo.Siloxanes	Palmitic acid	Palmitic acid	Succinic acid	Stearic acid	Acetamide	Benzothiazolone	Adipinic acid
Bis-(2-ethylhexyl)-phthalate	Stearic acid	Dimethoxydiphenyl sulfone	3 oligo, siloxanes	Bis-(2-ethylhexyl)-phthalate	N-cont.aromatic high MW comp.	4-Hydroxypropyl-benzoate	Dibutylphthalate
12 oligo.Aliphates	12 oligo. aliphates	11 oligo. aliphates	Polyether	Polyacrylate	3 oligo. amides	6 oligo. aliphates	Ethylhexylphthalate
4-Methyl-2,5-cyclohexadiene-1-on	11 oligo. siloxanes	Methoxy-4-chloro-diphenylsulfone		Ethylacrylate	Bis-(2-ethylhexyl)-phthalate	Hydroxyphenyl acetamide	Dihydroethyl-phthalate
	Methyl-4-hydroxybenzoate	Bis-(2-ethylhexyl)-phthalate		Diphenylphthalate	10 oligo.siloxanes	Methoxy-4-chloro-Diphenylsulfone	2, 6-Di-tert.-butyl-cresole
	Etherthioether	Polyether		9 oligo. Siloxanes	2 oligo.aliphates	7 oligo.siloxanes	Disobutylphthalate
		7 oligo. siloxanes		6 oligo. aliphates		Cyclohexanone	Diacetylbenzene
		2,4-Bis(1, 1-di-methoxy-1-ethyl)-phenol					Cyclotridecanone
							4-4-Dichlorodi-phenyl-sulfone
							Propionic acid
							4 oligo, siloxanes
							3 oligo, aliphates

* Identification of the RP-HPLC peaks by FTIR is still in progress–extractables list of marked cartridges may be incomplete.

FIGURE 47.5 Extractable table of eight different sterilizing grade filters.

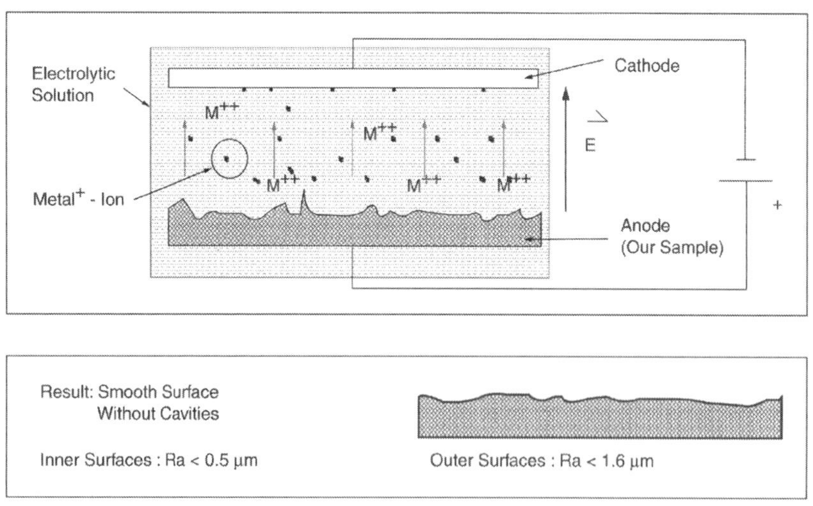

FIGURE 47.6 Schematic of an electropolishing process.

In some instances, surface treatments are not needed or are even undesirable. A glass beaded surface is sufficient. However, since cleaning is a major factor within the biopharmaceutical industry, the surfaces are required to be smooth and with a minimum groove rate. Any groove would allow pockets of microbial growth, which could result in a biofilm formation. Electropolishing after high grid polishing is utilized to cut any high peaks of material and avoid pockets (see Figures 47.6 and 47.7).

Most commonly, when automated equipment is supplied, appropriate qualification documentation is required before the equipment is released and shipped to the client. Without such documentation, the equipment would be of no use and the shipment might be rejected. It is essential that these documents are send to the client for preapproval. Once the approval is received, only then can the vendor ship the equipment to the client. Appropriate qualification documentation is an essential release criterion nowadays.

No.	Designation of procedural step	Remarks	Recommended abrasive	Grit	Peripheral speed in m/min
1a	Preliminary rough polish ("fettling")	Preliminary step for rough welds;only for very coarse work; recommended follow-up step: 1b, with 60-grain abrasive	Preferably grinding wheel with hard rubber or plastic bond	24/36	1,200-1,800
1b	Rough polish	First step for thick sheets, hot-rolled sheets or smooth welds	a) Grinding wheel with hard rubber or plastic bond b) Set-up wheel c) Grinding belt, if the shape of the piece permits	if 36 is necessary, follow up with 60	1,200-1,800
2	Finish grind	Standard step for cold- rolled sheet or coil	a) Set-up or rubber wheel b) Grinding belt, if the shape of the piece permits	80 / 100	1,500-2,400
3a	Precision grind	The surface finish corresponds to that of roll material in accordance with "Procedure o (IV)"	a) Set-up wheel b) Grinding belt, if the shape of the piece permits	120 / 150	1,500-2,400
3b	Precision grind	Preparatory step in producing a normal polish following step 3a.	a) Set-up wheel b) Grinding belt, if the shape of the piece permits	180	1,500-2,400
3c	Precision grind	Intermediate step in producing a normal polish following step 3b.	a) Polishing wheel b) Grinding belt, if the shape of the piece permits	240 abrasive paste for set-up wheel, or 240 grinding belt	2,400 - 3,000 Grinding belt: approx.1,500
4	Brushing	To produce a smooth, matte, silk luster. This step, following one of the "o (IV)" procedures,produces a surface finish that corresponds to the designation "burnished." Brushing finer (e.g., high-gloss polished) Surfaces Produces a very attractive efffect. The surface finish will depend on the brush speed and the abrasive used.	Tampico	Abrasive paste made of pounce or quartz powder. Other abrasives may also be used,depending on the desired surface finish.	600- 1,500
5	Polishing or lapping	Final step for producing a normal polish following step 3c (Note lappingleaves fine chatter marks)	Polishing wheel	Burnishing compound for stainless steels in stick or cake form	
6a	Polishing	a) Preparatory step for producing a high-gloss polished surface following step 3c	Polishing wheel	320-400 finish polishing compound in stick or cake form	2,400-3,000
		b) Preparatory step for producing high-gloss polished coil.	Polishing belt	Burnishing compound for stainless steels in stick or cake form.	approx. 1,500
7	Blasting	Final step for producing a matte, non directional surface structure	Glass beads Stainless steel grit Nonferrous quartz sand	various	

FIGURE 47.7 Table of different polishing methods and the end result.

Higher	Risk or Impact	Lower
API	LIMS	Cardboard Box
	Sterile Filter	
WFI		Integrity Tester
Isolator	Bioreactor	Cleaning Agent
Autoclave	Balance	
Lyophilizer		Manufacturing Robot

FIGURE 47.8 Possible example of risk and impact assessments.

From an end-user standpoint, the release criteria of the vendor have to meet the risk assessment criteria set by the end-user (and more often the regulatory authorities). That is, depending on the quality impact of a specific component or equipment supplied, the release criteria on both sides, vendor and end-user, will differ in stringency. The quality of supplied water for injection (if not produced within the facility) has a higher risk attached than a condensate valve on a tank. Different risk or impact classifications have to be defined for product and equipment supplies (see Figure 47.8). Some products have a direct impact on the quality of the end product, some have only a minor influence and some have no influence but are used to check on a component with a quality influence. For example, an integrity test system does not have a direct influence but is used to check the integrity, that is, the quality of a sterilizing grade filter, which has an influence on the quality. The release and test criteria for these products will differ and be defined in a way that will meet the necessary quality purpose. It would make no sense to use similar evaluation conditions for noncritical items. It would just raise costs and cause possible process delays. Therefore, these risk assessments have to be performed before release criteria are defined.

47.5 QUALIFICATION OF EQUIPMENT

The probably most descriptive and utilized guidance on qualification mechanisms is the GAMP (Good Automated Manufacturing Practice) guidance published by the ISPE (International Society of Pharmaceutical Engineering) (4). It describes thoroughly the necessary individual steps required to fulfill the quality expectations of automated systems. This guidance is used for a multitude of equipment utilized within the biopharmaceutical industry, for example, autoclaves, lyophilizers, filling machines, integrity test systems, bioreactors and others.

Within the GAMP documentation, specification steps are described but also three main qualification requirements; Installation Qualification (IQ), Operational Qualification (OQ) and Performance Qualification (PQ). There are other qualification tests that are quoted randomly, for example, Design Qualification (DQ) and System Qualification (SQ). However, the three major qualification segments are IQ, OQ and PQ and are applicable to every automated piece of equipment supplied.

A system design and the qualification steps all start with the URS (User Requirements Specifications). This is the foundation of any system that will be designed, and if it is defined inappropriately, the entire project might be prone to fail or at least will require rework, with additional costs involved. The URS can be seen as the foundation of a building: the better the foundation, the better the construction on it. Any of the previously mentioned qualification steps are the verification of the User Requirements Specification (URS), Functional Specification (FS) and Design Specification (DS).

Installation qualification—Documented verification that all important aspects of hardware and software installation adhere to the system specification.

Within this qualification, the entire system is checked whether all components are correctly installed and whether the entire documentation for the individual components is available. Most often, the IQ step runs through a thorough checklist to evaluate that everything meets the requirements set within the design or hardware specifications (see Figure 47.9) (5, 6, 7).

The IQ documentation is supplied by the vendor but checked by the end-user. Most typical and most practical would be to perform the IQ and OQ part during the Factory Acceptance Test (FAT), which verifies that the system is working.

Operational qualification—Documented verification that the system operates in accordance with the system specification throughout all anticipated operating ranges (see Figure 47.10).

These tests verify that the functional specifications are met by empirically checking and testing against the manufacturer's recommended test sequences all the critical operational

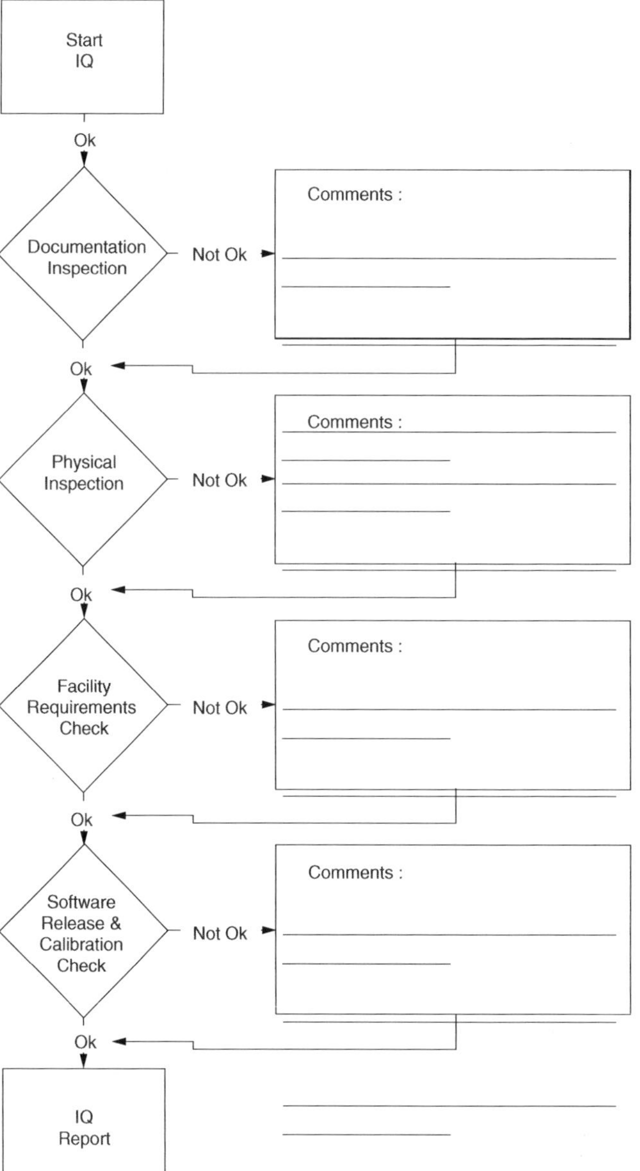

FIGURE 47.9 Typical flow diagram of the first layer of an IQ protocol.

and functional features and performance specifications of the machinery. These test sequences are performed within the vendor's facility, again most commonly during the FAT. Within this qualification phase, the system will run at the specifications given by the user. Therefore, vendors are required to have all supplies necessary to run the system, for example, water and steam supplies. The OQ can be performed within a few hours or weeks, depending on the complexity of the system build. Most commonly the OQ documentation is already established within the process of the functional specification, as every single function described requires being tested during OQ. If an FS happened and the documentation is not established, at this point the workload will be tremendous and the precision will suffer.

Once the system run through the FAT and OQ and all documentation is established, the system can be shipped to the

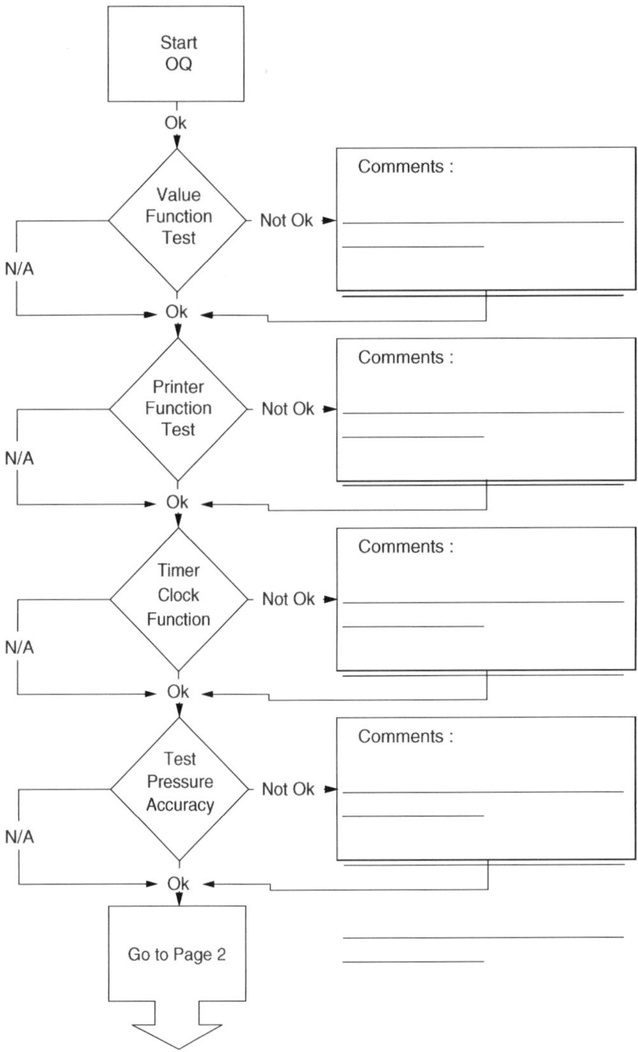

FIGURE 47.10 Typical flow diagram of the first layer of an OQ protocol.

vendor. At this point, the PQ is performed as the final part. The PQ is often also part of the SAT (Site Acceptance Test) or vice versa, depending on individual user procedures.

Performance qualification—Documented verification that the system operates in accordance with the User Requirements Specification while operating in its normal environment and performing the function required by the process to be validated.

These records include batch records, routine calibration and performance checks, which are commonly defined by the equipment used. Every piece of equipment has different requirements of compliance, with specifications defined within the specification phase. Moreover, the environment within the end-user's facilities varies. For this reason, performance qualifications check whether the equipment works within such an environment. Additionally, during the performance qualification phase, the equipment may be pushed to the limits to verify that it still performs and does not spiral out of control. In some instances, automated equipment might malfunction when, for example, the software is pushed to a limit. It could well be that the system shuts down or that certain controls and adjustments elevate themselves out of control or set tolerances. These stringent tests belong to a risk assessment program, which determines the functionality of the system. Will it still work in as robust way as requested, or will it perform in a way detrimental to the entire manufacturing process? The environment certainly has an influence on such functionality, as well as the process control system and its source code. It has been experienced that systems are not validatable due to commercially available software that is adjusted to the purpose but not fully compatible. Such software might not be able to cope with the stringency and demands of a production process and therefore show insufficient performance.

These three fundamental qualification processes are repeated during each phase of the validation process. In the qualification phase, a baseline level of performance information is obtained from the component manufacturers' data and test results, structural testing of the software and the associated vendor documentation (see Figure 47.11).

Equipment validation packages must be prepared and available for the user's own validation efforts and tests to verify proper functioning of the equipment. These validation packages are commonly very comprehensive and cover every function of the equipment. For example, the documentation for a complex fermentation system can result in close to 1,000 ring binders. In some instances, regulations applicable to the particular equipment will be quoted for the user to support other necessary validation or qualification processes within the facility. As previously described, the equipment supplier can and commonly does support the end-user with installation and operational qualification documentation; however, any process validation or performance qualification has to be performed within the facility and process environment. This will assure that the equipment is functioning properly aside the laboratory settings within the manufacturer's facilities.

Finally, maintenance and continued testing and verification are the responsibility of the end-user, who may seek assistance from the equipment manufacturer or its own maintenance department. Service manual establishment is required before equipment is supplied to assure appropriate maintenance possibilities. Such service manuals list spares required within specific frequencies. Commonly, the vendor has experience at which interval certain parts of the system need to be exchanged or replaced. These essential spares need to be defined and listed within the service manual as well as maintenance intervals. These tasks can also be performed by outside service organizations; however, the qualification of these organizations has to be verified. Most often, service contracts are established between the vendor service side and the user maintenance department.

Another important aspect should not be forgotten—training. All qualification and acceptance steps are good but useless if the staff utilizing the equipment is not trained effectively. Training protocols and standard operating procedures (SOP) need to be described before the equipment is used. Both training manuals and SOPs should be reviewed to assure correctness.

1.1	**Contents**	
1.	**Contents**	**4**
1.1	Contents	4
1.2	FDC	6
1.3	Declaration of Conformity	7
2	**Overview and Proper Use**	**8**
2.1	Overview	8
2.2	Proper Use	8
2.3	Technical Specifications	9
2.3.1	Dimensions and Weight	9
2.3.2	Connections Data	9
2.3.3	Pneumatics	9
2.3.4	Test Media	9
2.3.5	Product-Specific Data	10
2.3.6	Test Methods	10
2.3.7	General Data	10
3	**Safety**	**11**
3.1	Notes Explanations	11
3.1.1	Manufacturer's, to Label	11
3.2	Installed Safety Equipment	12
3.3	Safety Precautions	12
3.4	Operators Obligations	12
3.5	Safety Tests and Inspections by the Manufacturer at the Facilitites	12
4	**General Warnings**	**13**
4.1	Hazards	13
4.2	Operating Personnel and Service and Maintenance Personnel	13
4.3	Installation of Replacement Parts	13
4.4	Shutdown Procedure	13
5	**Installation**	**14**
5.1	Equipment Supplied	14
5.2	Receipt of Delivery	14
5.3	Installation Instruction	15
5.3.1	Ambient Conditions and Conditions at the Place of Installation	15
5.3.2	Transport and Unpacking the Equipment	15
5.4	Connections	15
5.4.1	Front Panel and Back Panel	15
5.4.2	Left and Right Panels	16
5.5	Connections with Internal Pressure Sensor	17
5.6	Connections with External Pressure Sensor	18
5.6.1	Inserting Paper Roll and Ribbon Cartridge	19
6	**Function**	**20**
6.1	General Functional Description	20
6.2	Test Programs	20
6.3	Managing the Test Results	20
6.4	Sarlocheck 4 Communication - PLC	20
6.5	Function Test	20
6.6	Cleaning	21
6.7	User Interrupt	21
6.8	Self Test for Broken Conductions and Short Circuiting	21
6.9	Calibration	21
6.10	Internal Pressure Sensor	21
6.11	Pressure Guage Function	21
6.12	Cleaning/Drying	21
6.13	Safety Functions	21
6.14	Software Update	21
6.15	Serial Number	21
7	**Operation**	**22**
7.1	Turning on the Unit	22
7.1.1	Operating Elements	23
7.2	Switching the Unit ON and OFF	24
7.2.1	Test Routine	25
7.2.2	Switching off the Unit	25
7.2.3	Switching off the Unit in the Event of a Power Failure	25
7.3	Tests	25
7.3.1	Test parameters	25
7.3.2	Test methods	25
7.3.3	Programming tests (F1– Main Menu)	26
7.4	Bubble Point Test	31
7.5	Diffusion Test and Bubble Point Test (Complete Test)	31
7.6	Water Intrusion Test Water Water Flow Test	31
7.7	Pressure Drop Test	31
7.8	Multipoint Diffusion Test	31
7.9.	Programming Tests (F2 – Main Menu)	32
7.9.1	Programmer's Input	32
7.10	Program Database (F4 – Main Menu)	34
7.10.1	Load Test Program from Internal Memory [F1]	34
7.10.2	Load Test Program from a Diskette [F2]	35
7.10.3	Backing up all Test Programs to a Diskette [F3]	36
7.10.4	Load Backup of all Test Programs from a Diskette [F4]	37
7.11	Visualization [F3– Main Menu] (Only with Multiunit Mode Activated)	38

FIGURE 47.11 Example of a validation documentation for an automated system.

47.6 STAGES OF EQUIPMENT SUPPLIES AND QUALIFICATIONS

Stages of the individual specification and qualification segments are mainly visualized within the V-model of the GAMP guidance (4). The V-model shows the different responsibilities but also interactions of specifications versus qualifications (see Figure 47.12). It is often modified to meet different requirements of different equipment suppliers.

Within the V-model, the individual tasks or steps are described but also responsibilities defined. In parts of the process the user is solely responsible, in other parts the supplier, and specifically in the qualification phase, the user and supplier share responsibilities, as most often these tasks are performed jointly. Every single step is of utmost importance and has to be viewed with stringency and thoroughness, as every step following depends on the quality of the previous task. The entire system can only be as good as the starting quality; therefore, multiple other process control and approval steps are involved that are not shown within the V-model. However, before a system is built, each function, software and hardware design has to undergo critical review to verify that the User Requirements Specifications are met. In some instances, specifications given might not be feasible to design or produce or subparts are not available or too costly. Sometimes, a cost focus might be not desirable, as cutting corners might result in a system that is not fulfilling the needs of the process defined. Examples have shown that shortcuts in respect to equipment or design qualities have resulted in higher adjustment costs at a later stage. In some instances, inadequate attention to the design of the system has resulted in yield losses or dysfunctions. The costs resulting from such failures are tremendous. The recommendation has to be that the user and vendor work closely together to find an optimal solution for the particular need. Costs have to reasonable but should not be the main focus.

The most important aspect is the User Requirements Specification, in which attention to detail is essential. Any rough idea given as URS will end in a back and forth between the user and vendor in the functional specification stage. Valuable man hours are wasted, which is undesirable for both parties. Often forgotten but always present is that the user is the specialist of the application and the vendor the specialist of the equipment. Utilizing both sets of experience will result in the best possible option. However, controls and measurements should be utilized during the milestones to assure that the system will function once built and implemented within the facility.

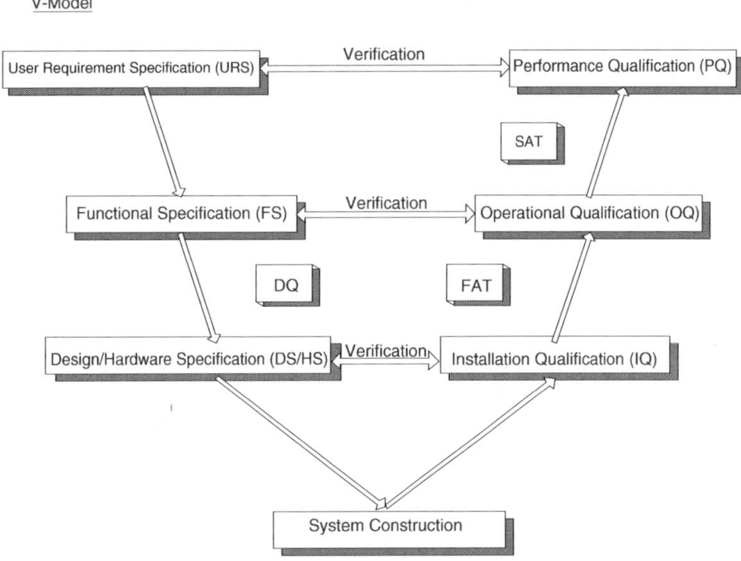

FIGURE 47.12 V-model.

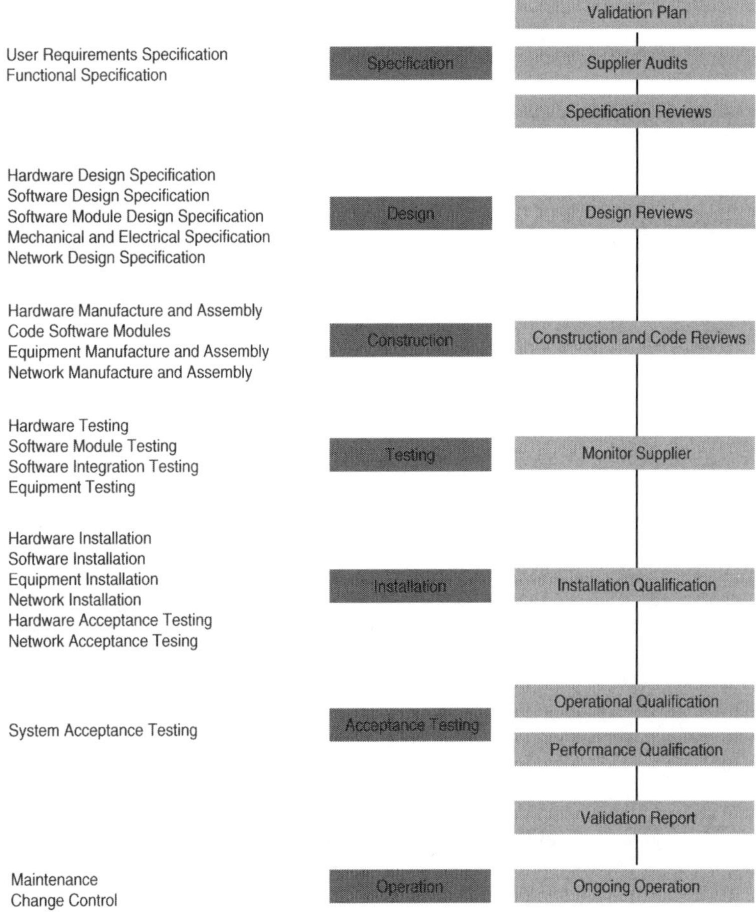

FIGURE 47.13 Validation schematic.

As previously described, the V-model creates an overview; however, project flows and detailed activity description require other tools, for example, specific project management software (see Figure 47.13). These tools will define activities in detail and also control points for parts of a system, the entire system or just the raw materials (7, 8, 9). The time frames will also be reviewed on a frequent basis, as time pressures commonly will result in human error. Every vendor has experiences with their equipment supplies and knows what quality system requirements need to be established within a detailed project plan. These control points also help the vendor to avoid any errors, which would create additional costs.

47.7 CONCLUSION

Validation and qualification of equipment within the end-user facility under the process conditions is an essential need and regulatory requirement. However, vendors of equipment, whether consumables or capital investments, perform a multitude of qualifications programs within their own facility. Such qualifications programs start during the development phase and commonly include not only the vendor's own processes but also sub-vendor sites, processes and product qualities. A vendor cannot just rely on the sub-suppliers, but has to assure their quality, just as any end-user needs to do. Furthermore, the development team receives quality milestones by the end-user. These specifications have to be kept, which means within the development phase, control mechanisms are defined that are used to verify that the specifications are met and that are also used as release criteria at full-scale production. Similarly, capital equipment receives User Requirements Specifications that are converted into functional specifications followed by software/hardware design specifications. Again, the fulfillment of the User Requirements Specifications have to be controlled at every stage to avoid any surprises and noncompliance. The capital equipment runs through different specification stages like a consumable product runs through a development phase. At the end of the day, both product groups require compliance to the user specifications.

Once the goods have been developed or built, the performance has to be qualified within the user's environment. Does the equipment perform under these circumstances? For example, sterilizing grade filters undergo process validation utilizing the actual or close at drug product and the process conditions. Evidence has to be given and documented to show the filter is performing to the set requirements under the environmental circumstances. The Performance Qualification stage does so for capital or automated equipment. Again, the equipment might be pushed to its limit to assure that it functions under worst-case conditions reliably. The tests are performed on-site to guarantee that any environmental condition does not have an adverse effect on the performance of the equipment. Lab tests at the vendor or pure certification cannot be accepted and will not meet regulatory requirements.

Vendors, nowadays, do not just produce and supply goods but make sure that these goods meet the requirements of the biopharmaceutical industry and its regulatory authorities. Moreover, once an item is sold, the vendors' efforts do not stop; they support the end-user with services to support any subsequent user qualification and validation effort. Both vendors' experiences and end-users' know-how will optimize the process reliability and, in combination, assure that the specification of the equipment will meet the needs of the process. The vendor has to be more than half way toward the end-user by supplying qualification data of the goods supplied, which can be utilized to either make a choice of equipment or be utilized within the filing documentation.

REFERENCES

1. ASTM D 4169. Standard Practice for Performance Testing of Shipping Containers and Systems, 2003.
2. ASTM D 4728–95 Random Vibration Testing of Shipping Containers.
3. DIN ISO 12048. Packaging—Complete, Filled Transport Packages—Compression and Stacking Tests Using a Compression Tester, 1994.
4. GAMP. *Guide for Validation of Automated Systems in Pharmaceutical Manufacture, Volume 1, Part 1: User Guide*. Brussels, Belgium: ISPE European Office, 1998.
5. Jornitz MW, Meltzer TH. *Sterile Filtration: A Practical Approach*. Chapter 7, "Filter Integrity Testing". New York, NY: Marcel Dekker, 2001.
6. Meltzer TH, Jornitz MW. *Filtration in the Biopharmaceutical Industry*. New York: Marcel Dekker, 1998.
7. Spanier HJ. Qualification of a filter integrity test device. *Pharm. Tech. Europe* 2001, 13(12), 45–52.
8. Technical Report No. 15. Industrial perspective on validation of tangential flow filtration in biopharmaceutical applications. *Journal of Parenteral Science and Technology* 1992, 46(Supp).
9. Wolber WP, Beech D, McAllister M. Validation of computerized membrane integrity test systems. *J. Parenter. Sci. Technol. 1988*, 106–110.

48 Validation for Clinical Manufacturing

Warren Charlton and Tom Ingallinera

CONTENTS

48.1 INTRODUCTION

The focus of this chapter is Validation of Clinical Trial Manufacturing (CTM) for Pharmaceutical Products. A review of validation activities has been complied for a range of drug products (e.g., non-sterile, sterile, biotech, blood/plasma, and nasal inhalants). Basic to all CTM products for human testing is the need for validation of analytical methods, equipment, utilities and other unique drug-specific factors such as environmental monitoring of the aseptic processing area used in the sterile filling. The selected products have very distinct differences in complexity, ranging from simple non-sterile powders for reconstitution to complex biotech product that is ultimately aseptically filled and freeze dried.

The intent of this chapter is to provide an understanding of "what" is needed and "when" it is needed but not "how" these are validated or developed. The drug development process is among the most complex, costly and regulated of human pursuits, and the statistical chances of success are horribly low. The chances of success for a drug compound entering Phase 1 trial has remained at slightly under 10% for the last two decades. When we couple this trend in patient disease knowledge with a disciplined phase-appropriate process for drug development, a more cost-effective, success-oriented strategy should consider phase-appropriate process validation.

In support of developing an overall Validation Master Plan, this chapter provides validation activities, a flowchart of phase-appropriate validation/critical activities by phase (e.g., pre-IND to Phase 3) and product-specific validation/supporting information. Appendix A is a clinical batch verification protocol (Phase 1). Appendix B is a Contract Manufacturing Requested Product Information.

It is understood within the pharmaceutical industry that major equipment and utilities directly involved in processing pharmaceuticals for human use, including clinical trial material, must be qualified (1). This involves, at a minimum, installation and operational qualification (IQ/OQ), and if appropriate, performance qualification (PQ). An excellent guide for the planning and execution of IQ/OQ/PQ is the ISPE Baseline Guide for Commissioning and Qualification (2). Not only should major process equipment such as granulators, mills, tablet presses, capsule fillers, etc. be qualified, but utilities directly impacting the process should be qualified as well. This includes but is not limited to HVAC, compressed air and pharmaceutical water systems. Since many, if not most, clinical batches are unique with regard to batch size, formulation or process, the approach for many firms has been to create a protocol for each unique batch and to let that batch data stand on its own.

48.2 PROCESS VALIDATION

Validation of pharmaceutical processes began in the late 1970s, based on comments by field inspectors and followed up by the establishment of the principles of validation for sterile injectables. The Updated FDA Guideline on General Principles of Process Validation was issued in 2011 (3) and is still in effect. A major focus of validation of solids (although certainly not the only one) has been content uniformity of blends and uniformity of dosage USP<905>, but it does not address (nor was it intended to address) clinical batch validation, for which process attributes are ill defined. In early clinical manufacturing, it may require 100% weight checks and have a high reject rate and low yields. The pharmaceutical equipment used may be manual or semiautomatic fills and compression. But in some situations, especially in Phase 3 clinical testing, a formulation and process may be well established, and a number of identical batches may be required over some months or years. In these cases, the process can be validated much like a commercial product/process is validated, as

DOI: 10.1201/9781003163138-48

long as the formulation, process and batch size do not change. Change control becomes vitally important here to ensure that changes are tracked in relation to validation documentation.

A healthy debate could be easily initiated among industry experts regarding at what stage in the continuum of dosage form development a clinical batch validation is expected. Most industry experts will agree that Phase 3 requires a level of validation or "verification" resembling commercial process validation. Many will perform at least a limited level of "verification" in Phase 2 to ensure that clinical batches used at this stage are uniform and have the quality and purity that is expected. The greatest debate would no doubt be generated at Phase 1. Many firms will do no more testing than that required to show that clinical batches in Phase 1 meet the minimum requirements of USP and the limited specifications established at this early stage. Others will go further and test blends for uniformity and test more than USP requires for tablet or capsule dissolution and uniformity of dosage units.

The FDA has an expectation of some track record of validation or "verification" testing of clinical batches during all phases of clinical batch manufacture (3, 6). See the protocol outline in Appendix A that has been presented for consideration in "verifying" the adequacy of the manufacturing process of clinical batches in the early stage of development (Phase 1) whose formulation and process are not finalized. For Phase 3 clinical batches, a validation protocol may closely resemble that of a commercial product.

The following is a checklist for phase-appropriate validation:

Validation activities		Supporting activities
• Initial validation of testing equipment • Initial viral clearance studies (biologicals/blood products) • Start human trials (safety) • Ensure product produced in a qualified facility (i.e., trained people and validated processes & equipment). • EMEA requires an "Authorized Person" who is responsible for ensuring that there are systems in place that meet the requirements of Annex 13 and should have a broad knowledge of the pharmaceutical development and clinical trial process. • Major process equipment & utilities impacting the process should be qualified. • Process validation as needed to ensure product safety, quality and uniformity of IND products. • In-process testing (PAT)	PHASE 1	• API-clear understanding of process • Defining properties (e.g., solubility hygroscopicity, melting points) • Initial analytical methods qualified • Initiate Reference Standards • Tentative Specifications (acceptance criteria) • Biologicals: initiate studies on viral clearance & protein characterization • Process & assay development as needed (e.g., biologicals) • Initiate clinical manufacturing • Analytical methods (8) • Suitability qualified • Acceptance criteria tabulated for impurities/isomers • USP/ICH guidelines apply • Change control in place • Initiate stability studies • Submit IND application • Flowcharting to demonstrate knowledge of process • Certificates of analysis for all drugs/chemicals, excipients, components and batch results need to be available • Batch sizing for eventual scale up • Batch acceptance limits including impurities • Initiate process validation plan for unique and complex processes (e.g., all sterile manufacturing processes; sterilizing operations; those effecting potentially adverse microbial growth or removal of endotoxins) • Biologicals/other products if appropriate: conduct viral clearance studies • Specifications in place (drug, chemicals, component, etc.)

Validation activities

- Validation of analytical procedures
- Validation Master Plan developed
- Validation protocols established/ approved
- Validation of development batches (small scale
- Validation Master Plan Document approved. Change control in place.
- Production scale validation (IQ, OQ and PQ).
- Media fills for sterile operations
- Manufacture validation batches (PQ/conformance batches)—normally 3.
- Validation summaries
- Validation team reviews the Master Plan and signs off verifying all requirements have been met with the last signature by QA (note: at least 3 team members from different areas for the company).

PHASE 2

PHASE 3

Supporting activities

- Quality Systems should be reviewed to ensure all systems are in place.
- Viral clearance studies are operational and continue to be expanded (biological processes and as needed to ensure safety).
- Preliminary specifications, tests, acceptance criteria and limits are operational and continue to be expanded (tabulated and validation in process).
- Complete understanding of manufacturing process, site of production, and batch release requirements.
- Critical manufacturing steps should be identified (e.g., training, equipment, process, storage, environment, sterilization) and verified that controlled documentation and/or validation are completed.
- Qualify and validate assays
- Tabulation of assay results available.
- Justification of impurity specification
- Drug Product Filter Validation completed
- Specifications are in place and full validation report available (8)
- Conduct pivotal trials
- Submit regulatory license application
- Pass preapproval inspection—note: normally expect FDA at one or more of these batches to view operation
- Obtain regulatory approval in writing
- Product can be shipped

48.2.1 Sterile Clinical Trial Manufacturing Discussion

Parenteral clinical supplies present numerous challenges. During preclinical and Phase 1 very little is known about the chemistry, dose, toxicity and pharmaceutical properties of the drug. One must take a systematic approach to get the drug into humans a quickly as possible. This is critical in order to establish a go or no go for a specific compound. Many development dollars and years can be wasted on elaborate formulation and analytical methodology only to find that the drug is ineffective or not safe. Thus, an injectable formulation may be the first dosage form to be utilized in clinical settings.

Developing acceptance criteria (preliminary specifications) is a priority during preclinical and Phase 1. Early in the development process, keep the drug product simple, either a simple solution or a lyophilized powder for injection. Start by evaluating the active compound for its pre-formulation attributes (e.g., solubility, salt selection, pH solubility profile, osmolality and density) and physicochemical properties (e.g., color, odor, melting point, hygroscopicity and optical activity). The effects of heat, light, pH, oxygen and ionic strength on the stability of the compound need to be confirmed. If the solubility of the drug is limited, cosolvents and other formulation changes may be necessary.

Initial container/closure can be a glass ampoule or a vial with a coated closure when limited compatibility data is available. Clear glass is preferred over amber, to allow inspection of the solution before injection. Alternate container/closure systems should be evaluated in development. Initial storage of clinical supplies can be refrigerated or frozen until more data is available. Note: always attempt to limit potential issues that have not yet been determined (e.g., protect the protect from light unless sure of no impact; store actives and finished product under conditions that are certain not to be any issue until data is available to demonstrate storing under normal room temperature is adequate).

The safety necessary to start Phase 1 clinical studies is determined by the preclinical/toxicology studies. The materials and formulation used in the preclinical/toxicology studies should be similar in purity. One of the challenges in using a single formulation is the fact that it may be necessary to dose small animals with 5 to 100 times the dose per kilogram of body weight to see toxicity and allow for a safety margin. The lyophilized formulation allows for flexibility in reconstitution, which lets one prepare a more concentrated solution for dosing animals.

Manufacture of early-stage clinical supplies needs to be made in a facility that adheres to cGMPs with Quality Systems in place and excellent past regulatory history. High speed filling equipment is rarely warranted during early-stage manufacturing. Terminal sterilization should be the first choice if possible. Data needs to be available if terminal sterilization is not a viable alternative.

Use of disposable technologies (e.g., plastic tanks, bags, etc.) fits well with potential potent and hazardous compounds that are not highly defined in the early stages of development. Product containment using disposables reduces the cleaning validation requirements, and the need for extractable studies can be based on similar product. For Phase 1, vehicle-based bubble point can be used for sterile filter integrity. Full sterile drug filter validation studies should use drug product from the conformance (registration) batches (final formulation drug product submitted in the new drug application (NDA), and the conformance batches must be the same concentration and formulation). Batch records can be difficult. Each batch may change in size, concentration, container/closure and manufacturing process. A complete history documenting these changes is critical for preparing documentation for filing.

48.3 VALIDATION REQUIREMENTS UNIQUE TO STERILE CTM DRUGS

The intent of this section is to provide *regulatory expectations of processes utilized in the manufacture of sterile drugs that must be validated* (i.e., IQ, OQ and PQ, unless indicated differently) and systems/controls in place prior to producing clinical trials for human use.

1. Processes and associated equipment that sterilize and/or depyrogenate drug, components, product contact equipment and product must be validated (IQ, OQ and PQ) prior to CTM Batches. Some examples are:
   a. Steam sterilization
      1. Product path components
      2. SIP cycles
      3. Freeze dryers
   b. Terminal sterilization of the product
      1. Steam autoclave
      2. Gamma radiation
      3. E-Beam
   c. Depyrogenation of components
      1. Dry heat tunnels/ovens
      2. Stopper washing (3 log reduction)
   d. Sterilization of all equipment and processes associated with product and product contact surfaces (e.g., tanks, fillers, stoppering)
   e. Utilities (e.g., WFI, pure steam, air/nitrogen and clean rooms)
   f. Special equipment and sterilizing systems
      1. Isolators
      2. Hydrogen peroxide sterilization
      3. Sterilization
      4. Others
2. Drugs, chemicals and excipients tested per endotoxin specifications. Bioburden limits should also be conducted during in development.
3. Filter validation Phase 1–2.
   a. Water for Injection (WFI) soluble products can initially use WFI bubble points/forward flow results prior to and after sterile filtration. Prior to post-filtration integrity testing, the filter should be purged (i.e., cleaned using WFI). Double filtration provides an additional assurance.
   b. Recommend that Phase 3 product use validated product bubble points/forward flow results (i.e., final drug product bubble point to be filed for NDA).
   c. Validation PQ batches should not begin until filter validation data is available.
4. Analytical test methods are required prior to first CTM batch to verify cleanliness.
5. Media fills:
   a. Media fill simulations used to test the container-closure system (not product specific).
   b. If the container closure is new to the filling line (e.g., different size or closure) the container closure requires three media fills unless equivalence or bracketing can be justifying less.
6. Hold times—at all stages within the process
   a. Validation justification as compared to actual allowance and product records
      i. Maximum period between start of bulk product compounding and sterilization (filtration).

   ii. Maximum permitted holding time of bulk if held after filtration prior to filling.
   iii. Product exposure on processing line.
   iv. Storage of sterilized containers/components.
   v. Total time for product filtration to prevent organisms from penetrating filter.
   vi. Maximum time for upstream filters used for clarification or particle removal.
   b. Process hold times and ramps (e.g., freeze drying)
7. Environmental monitoring room validation: Environmental validation of the room and associated filling/processing equipment in static and dynamic conditions is expected. Data should include smoke videos that can be used as training aids for aseptic processing.
8. EM training and qualification of operators/staff: Verification of personnel training to not adversely affect the sterility of the batch during normal run conditions.

48.4 VALIDATION REQUIREMENTS UNIQUE TO PLASMA/BLOOD PRODUCTS AND BIOLOGICALS

Blood/plasma and biological products, while having higher levels of regulatory requirements associated with sterile drugs, require additional verifications and controls to ensure the public is protected. The following are examples typical of these products. Note: with the changing new drug environment, including the combination of solids with sterile products (e.g., reservoir implants), these same issues may need to be considered. An example is whether the product needs a Comparability Exercise when modifications are made to the formulation.

Plasma/Blood Products Special Areas of Attention and or Validation

1. Process steps used for viral inactivation/removal (e.g., filtration, pH adjustment, chromatography, etc.)
   a. Log reduction of clearance
   b. Specific viral testing performed
2. If from blood or blood components:
   a. Certification that donors have met FDA blood donor requirements
   b. Materials have been screened per FDA blood product material requirements
3. For contract manufacturing/filling. Lots received should include COA that states the following have been tested:
   a. Adventitious viral agents
   b. Mycoplasma (culturable and non-culturable)

Biotechnology-Protein Manufacturing

1. Changes in the manufacturing process should include a Comparability Exercise (CE)
   a. An effect on efficacy and/or safety might be expected or cannot be ruled out.
   b. Need to justify that the change in the manufacturing process will not affect efficacy and/or safety.
   c. If a modification of the product is detected during the CE, it may indicate the need for further preclinical and/or additional clinical data.
   d. Potential for altering the profile and ratio of the impurities. The biological impact of changes should be considered prior to administration in humans.
2. Storage and shipment (e.g., shipment, receipt, cold room storage, cryogenic/frozen product storage, thawing, and product delivery system/filling process temperature control)—Validation of Storage and Handling
   a. Require significant controls/validation to insure the proper in-process control.
   b. When batches are scaled up, this also requires attention in the CE.
3. Process validation should include a Risk Assessment to ensure all parameters/controls have been considered in the overall Validation Master Plan.
   a. A quality risk assessment would review all unit operations/processes (flowchart of total operation from materials received through product shipment) that may affect the product's identity, strength, quality and purity. Included in this risk assessment should be an understanding of the impact of process variables (e.g., temperature, mixing speed, process time, flow rates, column wash volume, reagent concentrations and buffer pH) and developing alert and action limits.
   b. A thorough knowledge of the processing steps of all ingredients when supplied by others is critical. Changes to their processes/equipment/formulation, etc. should have impact under change control.

48.5 VALIDATION REQUIREMENTS UNIQUE TO NASAL PRODUCTS

Nasal products provide a less invasive method of administration than injectables, and patients are more willing to use these products when compared to self-injection. While it is an easy alternative for the patient, it is a very complex and demanding process requiring significant understanding of the associated development, processing and controls. This is a developing area with expectations of continuing new advances in products, drug delivery systems, processing operations, environmental requirements and of course regulation with delivery systems to the lungs or to the nasal mucosa including:

- Aqueous-based oral inhalation
- Pressurized metered inhalers
- Dry power inhalers

- Products for nebulization
- Metered dose nebulizers

1. Inhalation sprays are intended for delivery to the lungs and contain therapeutically active ingredients and excipients. The use of preservatives or stabilizing agents are discouraged. Aqueous-based oral inhalation-based drug products are required to be manufactured sterile (21 CFR 200.51). Manufacturers must also comply with 211.113(b), which requires them to establish and follow written procedures designed to prevent microbiological contamination, including validation of any sterilization processes.
2. Nasal spray products and pressurized metered-dose inhalers are designed to apply sprays to the nasal cavity and therefore are not subject to this rule for sterilization (13).
3. Unique validation/testing
   a. Particle size test and limits
      - Validated multipoint particle sizing method (e.g., laser diffraction).
      - Acceptance criteria to assure consistent size distribution in terms of total particles in a given size range.
      - Acceptance criteria are set based on the observed range of variation and should take into account the particle distribution of batches that showed acceptable performance in vivo, as well as the intended use of the product.
      - If alternate sources of drug substance are proposed, evidence of equivalence should include appropriate physical characterization and in vitro performance studies.

   b. Development tests conducted
      - Minimum fill justification
      - Extractable/leachables
      - Delivered dose uniformity/particle mass through container life
      - Shaking requirements
      - Actuator disposition
      - Low temperature performance
      - Cleaning requirements
      - Performance and temperature cycling
      - Physical characteristic
      - Robustness
      - Preservative efficacy

Table for Other Tests for Inhalation vs. Nasal products

Inhalation products	Nasal products
• Single dose fine particle mass	• Demonstrated deposition is localized in the nasal cavity
• Individual stage particle distribution	• Droplet formation—particle size distribution and full characterization of the product
• Droplet size distribution and drug output	
• Initial priming of container	
• Re-priming of the container	
• Compatibility	
• Effect of moisture	
• Safeguards to prevent multiple dose metering of dry powder inhalers	
• Breath-activated devices-data provided to demonstrate all target groups capable of triggering the device	
• Dry powder inhaler reservoir systems need a count indication for when the number of actuations indicated have been delivered	

Note: This table describes the regulatory differences in expectations for each dosage form.

Note: This information is a brief sampling of validation and operational needs of each type of nasal type product. Please refer to FDA and EU guidelines for more specific information. These products are often unique with the product and corresponding delivery system defining what needs to be validated and controlled during the manufacturing, filling, storage, testing and distribution of these products.

48.6 CONCLUSION

This intent of this chapter on clinical manufacturing validation is to provide an understanding of what is needed and a logical sequence of events required to bring a product through the regulatory process. In support of these objectives, various validation needs and a road map are provided.

This will become increasingly important as new drugs and drug delivery systems become more integrated, making it harder to understand the differences between different drugs as new technologies become reality.

Note: Significant regulatory activity is in progress as this chapter was being developed. So expect further changes in regulatory expectations and enhancements!

Appendix A
Clinical Batch Verification Protocol (Phase 1)

PURPOSE

The purpose of this Process Verification Protocol is to prescribe the testing to be performed, as it will apply to a batch intended for clinical trial use.

PROCEDURE

I. Protocol Content

The protocol may contain the following sections:

A. Product information (name and strength).
B. Protocol approval signatures.
C. Reference documents (a listing of documents to be cross-referenced such as—but not limited to—analytical standards, development or scale-up reports, SOPs and engineering studies).
D. A statement of purpose for the testing described in the protocol.
E. Rationale for the sampling plan and the acceptance criteria for the tests.
F. Description of the equipment and the process.
G. Critical process steps to be verified by the testing.
  - Solids: includes blend time, mill speeds, and screen sizes. Critical process steps to be verified by the testing.
  - Sterile: includes temperatures, dissolution time, mix speed, dissolved oxygen, pH, density and osmolality, and for lyophilized products, residual moisture after cycle. Other attributes to be tested in process are pre-filtration bioburden and endotoxins.

H. Test Functions: Based on knowledge from development and scale-up, a list of process steps that are to be tested should be listed. Each test function should show the acceptance criteria based on development work along with a rationale for that criteria.
I. Test Criteria: A description of sampling including sample size, location and number of samples, along with a rationale for these.
Note: For blend samples, sample size will be a weight equivalent to one to three dosage units.
J. The verification protocol will specify sampling methods and tests. Test data will be attached, and the document will be approved, certifying that all acceptance criteria were met. Unless specified by the client, the *minimum* testing is listed as follows:
This verification will be performed on a batch-by-batch basis. The data will be summarized and included in a final package with the approved protocol.

Process step	Sampling *(minimum)*	Tests
Non-sterile: solid dosage		
Initial blend	**Top, middle, bottom of container**	**Assay (for blend uniformity)**
Final blend	**Top, middle, bottom of container**	**Assay (for blend uniformity)**
Compressing or capsule filling	**Beginning, middle and end of run**	**Content uniformity, assay, dissolution, dosage unit weights**
Sterile products		
Initial compounding	Top, middle, bottom of mixing vessel	Assays, density, pH, osmolality
Filling	Beginning, middle and end of fill	Assays, oxygen in headspace, fill weights
Lyophilizing or terminal sterilization	Beginning, middle, end of load	Assay, moisture (shelf mapping), reconstitution time, pH, oxygen in headspace

L. Results: Assemble the results, write an analysis and conclusion and assemble the final package. The analysis and conclusion will state whether or not the test results met the acceptance criteria. If the acceptance criteria are not met, a full explanation is required.
M. Routing for Approval: The finished package is attached to the batch record and routed for approval to the same individuals who approved the protocol initially.

This protocol is presented as an approach to satisfying the expectations of FDA that an individual batch meets its specifications (5).

Appendix B
Contract Manufacturing Requested Product Information

(Information needed to support production and stability)

1. Product classification
   Examples: diluent, small molecule pharmaceutical, fermentation derived, cell culture derived, cytotoxic, anti-infective
2. Status (Phase 1, Phase 2, Phase 3, Commercial)
3. Registered by FDA (CDER, CBER, CDRH, CVM), EMEA or Others
4. Timetable by phase
5. Type of service needed
   a. Aseptic manufacturing
   b. Aseptic filling
   c. Lyophilization
   d. Validation
      i. Analytical
      ii. Equipment
      iii. Process
   e. Stability studies
   f. Labeling and packaging
   g. Regulatory
6. Analytical: types of methods needed or supplied for example:
   a. HPLC
   b. GC
   c. Karl Fisher Moisture Analysis
   d. Oxygen Headspace Analysis
   e. TLC
   f. Optical Rotation
   g. UV/ Visual
   h. pH (provide range)
   i. Specific Gravity
   j. Infrared Spectrophotometer
   k. Lowry Protein Determination
   l. Protein Nitrogen Units (PNU)
   m. SDS-Page
   n. Radial Immunodiffusion (RID)
   o. Polyacrylamide Gel Isoelectric focusing (PAG)
   p. Enzyme Linked Immunosorbent Assays (ELISA)
   q. Others
7. Microbiological
   a. Bioburden determination
   b. USP Particulate testing
   c. Bacteriostasis/fungi stasis validation
   d. Bacterial and fungal identification
   e. Endotoxin (Gel Clot Method)
   f. Sterility testing
   g. USP Antimicrobial Preservative Effectiveness
   h. Container Closure Integrity test (Microbial Ingress)
   i. Others . . .
8. Raw material specifications
9. Storage conditions for bulk, in-process, on-test, shipping and associated validation
10. Formulation requirements
    a. Aseptic additions required?
    b. Batch documentation
    c. Special equipment and need for dedication
       i. Disposable processing
       ii. Tanks
       iii. Mixing/recirculation
       iv. Filling requirements
       v. Temperature
       vi. Light protection
       vii. Inert gas blanketing
       viii. Product density and viscosity
       ix. Special process or techniques
       x. Product incompatibility
    d. Handling precautions
    e. Product and API Material Safety Data Sheets
    f. Disposal requirements
11. Components
    a. Presterilized and cleaned?
    b. Validation requirements?
    c. Specifications
    d. Surface treatment
    e. Manufacturer
    f. Stopper
       i. Type
       ii. Silicone treatment/limits
    g. Closure
12. Filtration and filling
    a. Aseptically filled or terminally sterilized
    b. Filter
       i. Single or redundant
       ii. Validation of integrity test
       iii. Sterilization validation
    c. Filler
       i. Type of filling pumps required
          1. Positive displacement
             a. Rolling diaphragm
             b. Ceramic pumps
             c. Stainless steel pumps
             d. Glass
             e. Other
          2. Time-pressure fill

3. Sterilization validation of unique items
4. Fill weight testing/validation
5. Container abuse studies
6. Other

ii. Nitrogen overlay required?
iii. Headspace analysis required?
iv. Protection from light?
v. Temperature requirements

d. Inspection and labeling
1. 100% visual inspection
2. Automatic inspection for:
i. Headspace volume (fill level)
ii. Particulate
iii. Container integrity
iv. Bulk vial identity
3. Labels

13. Process validation
a. Microbial filter retention study
b. Media fills
c. Terminal sterilization cycle development
d. Cleaning verification
e. Mixing verification
f. Fill homogeneity
g. Others

14. Lyophilization
a. Cycle development
i. Time/temperature limits between filling and loading
ii. Nitrogen or inert gas purge during cycle
iii. Ramp rate during cooling
iv. Ramp rate during heating
v. Product loading temperature and cycle ramps
vi. Product pressure limits during cycle based on critical points
vii. Eutectic and/or collapse points
viii. End product moisture requirement
ix. End point pressure (full vacuum vs. partial vacuum)
x. Have you seen issues with powder on shelves after development cycles?
xi. Stopper tested to insure it does not stick to shelf after stoppering?

REFERENCES

1. Spataro ME, Bernstein DF. Managing Clinical Trial Materials: Investigational Product Change Control. *Pharm Tech* 1999, 23(10), 144–150.
2. Baseline Guide Volume 5: *Commissioning and Qualification*. 2nd ed. Tampa, FL: ISPE, June 2019.
3. Current Good Manufacturing Practice for Phase 1 Investigational Drugs Guidance for Industry, July 2008.
4. Process Validation: General Principles and Practices. Guidance for Industry Food and Drug Administration, January 2011.
5. Hartman BW, Motise PJ. *Human Drug CGMP Notes*. Washington, DC: Center for Drug Evaluation and Research (CDER), Food and Drug Administration, September 1999.
6. INDs for Phase 2 and Phase 3 Studies Chemistry, Manufacturing, and Controls Information Food and Drug Administration, May 2003.
7. Annex13: Manufacture of Investigational Medicinal Products, EMEA Document Dated, February 3, 2010.
8. EMEA-Draft-Guideline on the Requirements to the Clinical and Pharmaceutical Quality Documentation concerning Investigational Medicinal Products in Clinical Trials, Deadline for Comments Dated, June 2005.
9. FDA-Q8 (R2) Pharmaceutical Development, November 2009.
10. EMEA-Draft Guideline on the Pharmaceutical Quality of Inhalation and Nasal Products, dated July 30, 2005 for End of consultation.
11. FDA-Guidance for Industry-Sterility Requirements for Aqueous Based Drug Products for Oral Inhalation.
12. *PDA Technical Report 42*. Process Validation of Protein Manufacturing. September/October 2005, 59(S-4).
13. CDER Guidance Document-Nasal Spray and Inhalation Solution, Suspension and Spray Drug Products-Chemistry, Manufacturing and Controls Documentation, July 2002.
14. Contract Manufacturing Requested Product Information. Provided by Pharmaceutics International, Hunt Valley, Maryland.

49 Validation of New Products

Norman Elder

CONTENTS

There are global expectations that require that before a new product can be released for commercial distribution, it must undergo successful validation. Unlike an Equipment Qualification, or EQ, for a piece of equipment, Product Validation examines the process under which a product is made for robustness and reproducibility. Product Validation is the documented evidence that a process will, with a high degree of scientific assurance, consistently produce product that meets its predetermined specifications and critical quality attributes, as required by the Current Good Manufacturing Practice regulations for finished pharmaceuticals, 21 CFR 211.100 and 211.110 (1) and the EC Guide to Good Manufacturing Practice (2). The validation phase for a new product should be viewed as a part of the entire life cycle of the product: as a journey, not a destination (3). In 2011, FDA issued an updated guidance document on Process Validation which embraced the life cycle concept as the desired approach for validation of systems and processes (4). EMA issued a document echoing FDA's perspective (5). Stage 1 of the life cycle describes the development process and its purpose, which is to establish the operating parameters, material/component choices and quality attributes of the product. The validation of products/processes is at the core of Stage 2 in the life cycle, where the commercial scale process is confirmed acceptable. That effort represents the core content of this chapter. The last segment of the life cycle is continuing validation (Stage 3), which is described elsewhere in this text.

The Stage 1 expectations of the new regulatory guidance are clearly the ones most impactful on the development of a new product. As the overwhelming majority of developmental efforts never result in a commercial product, the means by which validation data is gathered are not always in lockstep with the development. Expending extraordinary efforts to develop a validatable process from the onset is both time-consuming and wasteful of resources. It is more appropriate to document the development well enough in its early phases (pre-IND, Phase 1 and early Phase 2 clinical) such that confirmatory studies can be performed later to substantiate process details and establish critical control parameters. The early phase development materials would rely more heavily on biological, chemical, and physical analysis to support their use. Developmental projects that do not advance to later stage clinicals are documented but never actually validated. This is consistent with FDA's IND and Phase 1 developmental guidance (6). Those candidate projects that reach the late Phase 2 should have the core validation requirements identified even if not yet completed. By the time the product completes Phase 3 clinical, the process validation should be largely completed, except for those activities exclusive to the commercial site. To operate in this manner, it may be necessary to perform experimental studies that support the commercial scale and ensure its direct linkage to the core studies supporting product efficacy. This may entail establishment of Proven Acceptable Ranges (PARs) for important process parameters and specification limits based upon actual process performance (7). In some instances, this may entail performing scale-down experiments on the anticipated commercial process to support process parameters. While the execution sequence differs from the development

The FDA's "Compliance Policy Guide" on Process Validation Requirements (CPG7132c.08) (6) explains the enforcement policy for the Center for Drug Evaluation (CDER), the Center for Biologics Evaluation (CBER), and the Center for Veterinary Medicine (CVM) regarding the timing and completion of validation activities for certain products, including sterile and non-sterile processes.

Validation is required when a new product is introduced into a facility, when the facility is new, or when there is a change in an existing process that may affect the safety, quality, identity, potency, purity, or security of a product. A prospective validation approach is always preferable; however, there may be occasions when a concurrent validation approach may be applied—for instance, in the case of a low volume production demand where replicate batches are not readily available, or a modification to a well-defined, previously validated process. The validation process must go through several formal steps of implementation to ensure proper validation. These steps must be properly documented and executed in the prescribed order.

The validation of a new product, as with any validation project, should begin with validation planning. This plan is best developed when a developmental project is considered for Phase 3 clinical evaluation. The initial change control documentation should explain the background of the new product introduction or the scope of the change being requested and any impact to existing systems or processes. From that background and scope, a Validation Project Plan

DOI: 10.1201/9781003163138-49

can be created outlining all aspects of the validation activity, including a description of the product and overview of the process, equipment, facility, utilities, and components. Reference should be made to drawings, flow diagrams, specifications, test methods, in-process controls, procedures, supporting validation protocols, and any development work performed. This is a high-level document that creates the structure on which the protocol is built. The scope of the validation and rationale for the approach can be clearly stated at this point for consensus before proceeding with the actual validation work.

The next step in the validation of a product is protocol development. The protocol is the mechanism to capture evidence that the validation activities are performed in a controlled environment, by trained individuals, following approved procedures, and using appropriate materials and components. When developing the protocol, it is important to have a clear understanding of the process. Sufficient development work should be done in advance of validation such that the critical parameters and key process control points have been established and characterized with limits and end points. Through the utilization of Design of Experiments in the development work, those key parameters can be scientifically identified and established. The validation work should reflect the development process but not be part of it. The validation batches should follow the manufacturing record without necessarily challenging the limits of the established process. The validation exercise should be verification of an established process, not discovery or experimentation. The development report should be referenced in the validation work as a document to defend the validity of how parameters and limits were established.

The protocol should follow the scope outlined in the Validation Project Plan. In many firms, this is used as Stage 2 confirmation of the commercial process. The following elements should be included in a typical protocol or be verified during the execution or prior to execution of the validation:

- All impacted systems such as utilities, equipment, and computer systems should be qualified and/or validated and verified to be acceptable within the proposed operating limits of the product to be validated.
- Calibration of critical instruments, that is, any instruments used to collect data or control the process, must be current at the time of execution.
- New or revised Standard Operating Procedures (SOPs) directly involved with the process to manufacture the product must be in effective status and personnel trained.
- Batch Records must be in final approved state.
- Analytical Standards and Test Methods must be in place prior to conducting testing of validation samples and must be traceable and validated, respectively.
- Raw materials and components are tested, passed, and released for use.
- Cleaning procedures, test methods, and rationale for cleaning limits must be established.
- Specifications for raw materials, in-process checks, and finished product must be preestablished.
- Filter validation specific to the new product should be complete, including microbial challenge and bubble point information.
- Container/closure challenge and media fill simulation for new components/processes should be complete.

The procedures that provide specific detail relative to executing the validation testing are described in individual test functions within a protocol. The rationale used to establish the acceptance criteria may be inserted in each test function if not previously stated in the body of the protocol. Acceptance criteria contained within protocols are based on approved specifications and must be traceable to approved design specifications, requirements, or procedures. The acceptance criteria should clearly state what is required to pass each test function. It should be explicitly stated, unambiguous, and verifiable. Include a clear and concise description of the steps to be used when obtaining data to ensure that they are reliable and appropriate to evaluate against the test function acceptance criteria. The description should include what data and/or samples to collect and during what point of the process the data/samples are to be collected. The sampling plan should be clearly stated, including the method for collecting the samples, any special precautions to be considered, the number of samples, location, size of sample, and labeling information. Diagrams may be included to facilitate protocol execution and data collection. A prepared data collection sheet is a valuable tool to include in the test function to clearly map testing, sampling, and results. Some types of information that should be included are start and stop times for process steps or sample collection, signature of the person conducting testing and collecting data, and verification signatures of the person checking calculations. The data collected and its organization should be sufficient to allow a reviewer of the completed test function to make a full determination of the acceptability of the data.

Test functions should support and validate proposed in-process specifications and be consistent with drug product final specifications, as indicated in 21 CFR Part 211, subpart F, *Production and Process Controls*, section 211.110, *Sampling and testing of in-process materials and drug products.* In part, this section states that

> to assure batch uniformity and integrity of drug products, written procedures shall be established and followed that describe the in-process controls, and tests, or examinations to be conducted on appropriate samples of in-process materials of each batch. Such control procedures shall be established to monitor the output and to validate the performance of those manufacturing processes that may be responsible for causing variability in the characteristics of in-process material and the drug product. Such control procedures shall include, but are not limited to, the following, where appropriate:
>
> (1) Tablet or Capsule weight variation;
> (2) Disintegration time;

(3) Adequacy of mixing to assure uniformity and homogeneity;
(4) Dissolution time and rate;
(5) Clarity, completeness, or pH of solutions.

Additional testing or sampling may be stipulated in the protocol above and beyond the stated in-process and final product specifications, based upon product characteristics or unusual or extensive process steps.

The validation protocol should define the critical parameters, ranges, process steps and hold times for the new product. Typical parameters to be addressed for sterile dosage forms during manufacturing, filling, lyophilization, and sterilization are time, temperature, pressure, mixing speed, homogenizer speed, recirculation time, and speed. For true solutions, development studies should first define the solubility characteristics through a solubility study, which should then be confirmed as a test function during the validation at the established minimum time during the manufacturing process. Documentation of mixing volumes must exist and be justified with process tank configurations and equivalency of tanks used in manufacturing and filling. Testing should include content/uniformity across the batch, with a larger sample size than routine production requirements to provide additional statistical validity to the results. For products which are not solutions but instead suspensions that are recirculated, testing should be included to confirm that, during the maximum allowable down time, the content/uniformity of the product delivered to the filler is not affected; and similarly, after the specified down time is exceeded, that the minimum allowable flush time is adequate to reestablish appropriate content/uniformity. A test function should also be included for lyophilized products that examines content/uniformity across all shelves through a predetermined sampling plan. The protocol should also include a test function for fill volume verification at the maximum expected speed that the product will be filled. An appropriate sample size should be selected so that statistical methods can be applied to demonstrate a process capability (CpK) at or above 1.0.

Similarly, for solid dose products, protocol testing should reflect in-process and final product release testing. For tablets, samples should be included such as Loss on Drying from throughout the fluid bed dryer or tray dryer and following final blending, as well as uniformity of the final blended granules throughout the tote; they should also include tablet weight variation during compression as well as content/uniformity, appearance, hardness, thickness friability, disintegration, and any other in-process tests from the beginning, middle, and end of the batch. Also, dissolution testing should be performed as an in-process verification across the batch, and in the event of a film coating operation, final dissolution. In the case of capsules, testing should be included to evaluate variation in weight of individual capsules or individual components within a capsule as well as content/uniformity and dissolution.

Sampling plans should be outlined in the protocol, with supporting rationale or reference, to demonstrate suitability or statistical validity. A commonly accepted approach was developed by a collaborative effort between the FDA, industry, and academia, the Product Quality Research Institute (PQRI). The approach was accepted by the FDA and is incorporated into the draft guidance published by the FDA in October 2003 titled "Powder Blends and Finished Dosage Units- Stratified In-Process Dosage Unit Sampling and Assessment" (8), which can be used to substantiate sampling plans and acceptance criteria. The acceptance criteria may specify a minimum Relative Standard Deviation (RSD) for a location or process step, across a batch or between multiple batches. This broaches the topic of the number of required batches.

The March 2004 revision of the previously referenced FDA "Compliance Policy Guide," CPG7132c.08, deletes a reference to a specific number of required validation or "conformance" batches (6). The long-accepted industry practice has been to manufacture three batches to demonstrate reproducibility; however, in light of the emerging focus on a risk-based approach to validation, this may no longer be the case. As the CPG states,

> Advanced pharmaceutical science and engineering principles and manufacturing control technologies can provide a high level of process understanding and control capability. Use of these advanced principles and control technologies can provide a high assurance of quality by continuously monitoring, evaluating and adjusting every batch using validated in-process measurements, tests, controls, and process endpoints. For manufacturing processes developed and controlled in such a manner, it may not be necessary for a firm to manufacture multiple conformance batches prior to initial distribution.

This points to the importance of understanding and controlling the process prior to commencing validation and provides an avenue to develop a justification eliminating costly manufacture of validation batches for which there may be no opportunity to market. A firm using a well-established approach to "Continuous Quality Verification" as a process beginning with development may be able to reduce the level of validation based on ongoing assurance and demonstration of product quality. The EU Guidance on Manufacture, Annex 15, "*Qualification and Validation*," section 25, presents a slightly different stance when it states,

> In theory the number of process runs carried out and observations made should be sufficient to allow the normal extent of variation and trends to be established and to provide sufficient data for evaluation. It is generally considered acceptable that three consecutive batches/runs within the finally agreed parameters, would constitute a validation of the process.
>
> **(9)**

The number of required validation batches should be specifically stated in the protocol, with a rationale for the approach, and should be manufactured at full scale of the intended commercial batch. The number of batches may also depend on the complexity of the equipment or process (10). For instance, in a case of ten identical tablet presses, there may not be a need to validate a product on each press; however, in the case of three identical lyophilizers, an approach might be adopted to run three batches in the first one and one in each of the remaining two to demonstrate that the process in each identical unit produces identical results. Again, the approach and rationale should be clearly stated in the protocol.

Once a well-structured protocol has been developed, it must be routed to the appropriate functional areas for approval. Generally, these include validation, development, manufacturing, laboratories, and quality assurance. The author of the protocol could be any trained individual, not necessarily from the validation discipline. But in any case, a separate validation approval should be obtained so that the author is not approving his or her own work. The validation approver is responsible for assuring that all of the critical elements have been included in the validation approach; that sufficient rationale is included to justify the approach; that appropriate cGMP practices are followed; that the protocol follows and supports the Validation Project Plan; that critical parameters and process variables are clearly outlined; and that the test functions and acceptance criteria adequately challenge the functionality of the process. The development signature signifies agreement that the process steps and acceptance criteria are in agreement with data collected during development of the project, that test functions are technically feasible, and that manufacturing work orders and specifications are current and correct. The user, generally a manufacturing representative for new products, is responsible for ensuring that the objectives, acceptance criteria, and expected results adequately reflect the intended process, facility, and equipment and that process and system descriptions are accurate and complete. The approval by a laboratory representative assures that all laboratory testing requirements are feasible, that methods are adequate, appropriate and in a validated state, that sample size is adequate, and that acceptance criteria are appropriate. This approach also gives the laboratories advance notice of the upcoming validation work to allow proper resource planning to provide timely results. The quality assurance group is responsible for providing quality oversight to the validation process. The Quality review assures that the protocol and validation approach conform to internal procedures and cGMPs, that the test functions and acceptance criteria are supported by the specifications and manufacturing work orders, and that an appropriate batch disposition approach is identified and in place. When all approvals have been obtained, this constitutes approval to proceed with validation execution.

As noted previously, there are some activities that must be completed and documented prior to execution of the protocol or testing of the samples, such as current calibration of instruments being used to collect validation data; documentation of personnel training and a signature log of those involved in execution; validated test methods, SOPs, and work orders in an approved state; and equipment and facilities fully validated. When these precursors have been completed and documented, the protocol execution may begin. The execution must adhere to the approved protocol. Personnel executing the protocol must follow area SOPs and manufacturing work orders as they apply. Sampling and testing must be documented by the individual performing the sampling/testing. The validation representative is responsible for ensuring that the sampling is executed according to the approved protocol, in a controlled manner, recorded on the data collection form, and submitted appropriately to the laboratory for testing per the applicable test method. In some cases, a signature may be required for collecting data or performing a calculation, with a second person signing as verification. It may be worth noting that in these cases, the verification signature must be someone who is authorized to review the data and cannot be the same as the person collecting the data. Care should be taken that good documentation practices are followed at all times. During the course of execution, should a failure occur or an event which precludes adherence to the protocol exactly as stated, a discrepancy should be noted. A predetermined procedure should be established that discusses how to handle discrepancies, but it should be performed at the time of the discovery of the discrepancy. If all discrepancies are reported and discussed at the time of the summary, there may be undue pressure to yield to a resolution that may not have been accepted in real time during execution. Instead, discrepancies should be documented at the time they occur, with a resolution of the assignable cause, its impact on the process, and consensus, including quality assurance, of forward action. Depending on the nature and extent of the discrepancy, it may or may not impact the validation effort. In the case of a discrepancy related to a non-process-related event, such as sample handling or equipment malfunction, it may be necessary to repeat only a portion of the validation, for example, only the batch in question. If a discrepancy results from acceptance criteria not being met due to a critical parameter, then a process change needs to be implemented and the validation repeated. If a portion of the process that creates a discrepancy can be isolated from other steps, a case may be made to repeat only those portions that are suspect. For instance, if a tablet coating process has a significant discrepancy, the portion of the validation that deals with coating may need to be repeated in whole or in part, but the compression portion of the validation may not be impacted. In some cases, a discrepancy during validation may also result in a manufacturing deviation that will follow the batch and go through the QA disposition process.

At the completion of all validation activity, evaluation of data, and review of laboratory test results, a Summary Report must be written to capture the results and report a conclusion. The summary should reference the original protocol and any supporting documents, such as the initiating change control document or the Validation Project Plan. The original scope and objective of the validation effort should be restated as an introduction to discussion regarding the actual results of test functions. Each test function should be discussed in sufficient detail to describe the intended result, how and when the execution was performed, and a comparison between the acceptance criteria and actual results. The data should be presented in a format that allows the reviewer to clearly discern if acceptance criteria have been met without the tedious review of all of the raw data or further calculations. All original data collected during the testing, such as laboratory results, data collection sheets, and temperature data, should be clearly identified and retained as back-up documentation for easy reference but not necessarily with each test function in the summary. This approach provides the reviewer with a more concise and presentable document. Discrepancies encountered during execution should be summarized with the resolution and any impact on the validation. Although each test function may address the individual disposition based on acceptance criteria, a final statement should be made regarding the overall validation effort for the product. The summary statement should provide an overall analysis of data with a conclusion and final disposition of the acceptability of the validation effort and subsequent commercial manufacturing.

The final Summary Report should be routed for approval to the initial reviewers. The manufacturing representative is responsible for ensuring that operations adhered to appropriate SOPs and manufacturing work orders, that they were carried out according to cGMP practices, and that any manufacturing discrepancies are accurate in scope and disposition. The validation approver is responsible for the accuracy of data and analysis and review of acceptance criteria against test results, correct format, and clear documentation. The development approver should assure that the process was executed as prescribed, that any discrepancies are accurately described, that any resulting remediation is appropriate, and that any data submitted by development is accurate and follows cGMP practice. Finally, the Quality Assurance representative should review for completeness, conformance to established standards, adherence to cGMPs, attainment of acceptance criteria, and acceptability of overall conclusion. Upon approval of the Summary Report, the validation activities can be reported in the Validation Project Plan Summary along with any other requirements and can then be closed.

Approvals for New Drug Applications (NDAs) may be received in advance of completing the validation effort for the new product. During the Pre-Approval Inspection (PAI) for the product, if the validation activity has not been completed, the inspector may review similar products, processes, or equipment validation to gain a level of assurance that the firm's approach is sound. It is a good practice to have the non-executed protocols in an approved state for review with the investigator during the PAI. If the existing validation work is suspect or the firm has a history of noncompliance, the approval may be held until the validation can be completed and a Post-Approval Inspection conducted. If approval is granted in advance of the completion of validation activities for the new product, the successful validation must still be completed and approved before commercial batches may be distributed. The FDA may request the final validation report and, based on acceptable review and good validation history with the firm, waive a Post-Approval Inspection.

With approval by regulators, the firm is now allowed to distribute the product commercially having successfully completed Phase 2 of the validation life cycle. FDA recommends the firm continue to sample subsequent batches as defined in the validation protocol until the firm is confident in the reliability of the process, The extent of this post-approval expanded sampling and testing is not defined and presumably is based upon the robustness of the underlying process and evaluation of the results. Reduction in sampling to the firm's normal release sampling levels completes Phase 2. From this point onward, the product transitions into Phase 3, where it is subject to continuous (or continuing) process validation. In that phase, change control and statistical assessments can be used to affirm product quality.

49.1 CONCLUSION

The validation of a new product is an essential requirement in today's industry. Putting aside the regulatory expectation for its execution, it establishes confidence in the commercial scale process at the production site, as it will be made in routine manufacturing. This provides the firm with confidence in the process/product combination, something that can only be presumed from success on a smaller scale in the development setting. The Phase 2 activities formally confirm the acceptability of the processes used to establish the product's quality.

REFERENCES

1. US Food and Drug Administration. "Code of Federal Regulations". *Part 210: Current Good Manufacturing Practice in Manufacturing, Processing, Packaging, or Holding of Drugs: General* and *Part 211: Current Good Manufacturing Practices for Finished Pharmaceuticals*, Revised 2003.
2. Medicines Control Agency. *Rules and Guidance for Pharmaceutical Manufacturers and Distributors 2002*. 6th ed. Fifth Impression, 2005.
3. Agalloco J. The Validation Life Cycle. *Journal of Parenteral Science and Technology* 1993, 47(3), 142–147.
4. FDA. *Guidance for Industry: Process Validation: General Principles and Practices*, 2011.
5. EMA, Concept Paper on the Revision of the Guideline on Process Validation, EMA/CHMP/CVMP/QWP/809114/2009, 2010.
6. US Food and Drug Administration. "Compliance Policy Guide" on *Process Validation Requirements for Drug Products and Active Pharmaceutical Ingredients Subject to Pre-Market Approval* (CPG7132c.08), Revised March 2004.
7. Chapman K, The PAR Approach to Process Validation. *Pharmaceutical Technology* 1984 December, 8(12), 22–36.
8. US Food and Drug Administration. Powder Blends and Finished Dosage Units-Stratified In-Process Dosage Unit Sampling and Assessment, October 2003.
9. EMA, Annex 15, Qualification and Validation, March 2015.
10. Agalloco J. Risk-Based Thinking in Process Validation: Finding the Appropriate Number of Tests. *Pharmaceutical Technology* 2011, 35(2), 68–76.

50 Retrospective Validation

Kevin M. Jenkins

CONTENTS

50.1 INTRODUCTION

Retrospective validation is the validation of older/legacy products, processes or equipment. Retrospective validation establishes documented evidence that a system does what it is supposed to do based on a review and analysis of historic information. It is normally conducted on a product already being commercially distributed and is based on accumulated production, testing and control data. Often these products or systems have not been validated to contemporary standards due to their age. It should be realized that there is a diminishing point of return at which retrospective validation is no longer justified. Current processes, products and equipment are expected to meet contemporary regulatory standards by the relevant regulatory agencies. However, all is not lost—there is a rational approach to determine if retrospective validation is appropriate, and there is recent regulatory guidance on what is required. In these cases, the approach needs to be evaluated and completed in a timely manner, with significant data to support the product/process/system.

A key consideration in contemplating the use of retrospective validation is the regulatory agency point of view. It may be useful to discuss with your district, local or market region regulatory agency any concerns with the use of a retrospective approach. The recent FDA Guidance for Industry—Process Validation: General Principles and Practices provides guidance on the use of an approach for legacy products in stage 3 of the document. As always, an approach to retrospective validation should be documented in advance in the form of a validation plan. This chapter serves as a guide to developing a supportable retrospective validation approach in terms of prerequisites, recent regulatory guidance, an overall approach and examples. Finally, this chapter will explore retrospective validation in a process analytical technology environment where the traditional three-lot validation is replaced by a continuous stream of data. It will examine the continuous manufacturing process and how that can be accommodated with retrospective validation.

50.1.1 The Foundation for Retrospective Validation

Any retrospective validation project should start with a strong foundation in regard to the overall systems that support operations. As will be shown in Figure 50.1, there are five layers of systems leading up to retrospective product validation that need to be in place for success. These five systems are as follows:

- Site Culture/Commitment

The first layer is the overall culture of the site or operation. This includes the change control culture—are all changes captured and handled adequately in a system? Part of this culture includes the support of leadership for change control, quality and validation. These key quality systems need to be at the foundation. Education of the overall workforce is also critical at this level. Quality culture is a critical element, since a robust quality culture assures a right-first-time approach.

- Life Cycle and Change Control

DOI: 10.1201/9781003163138-50

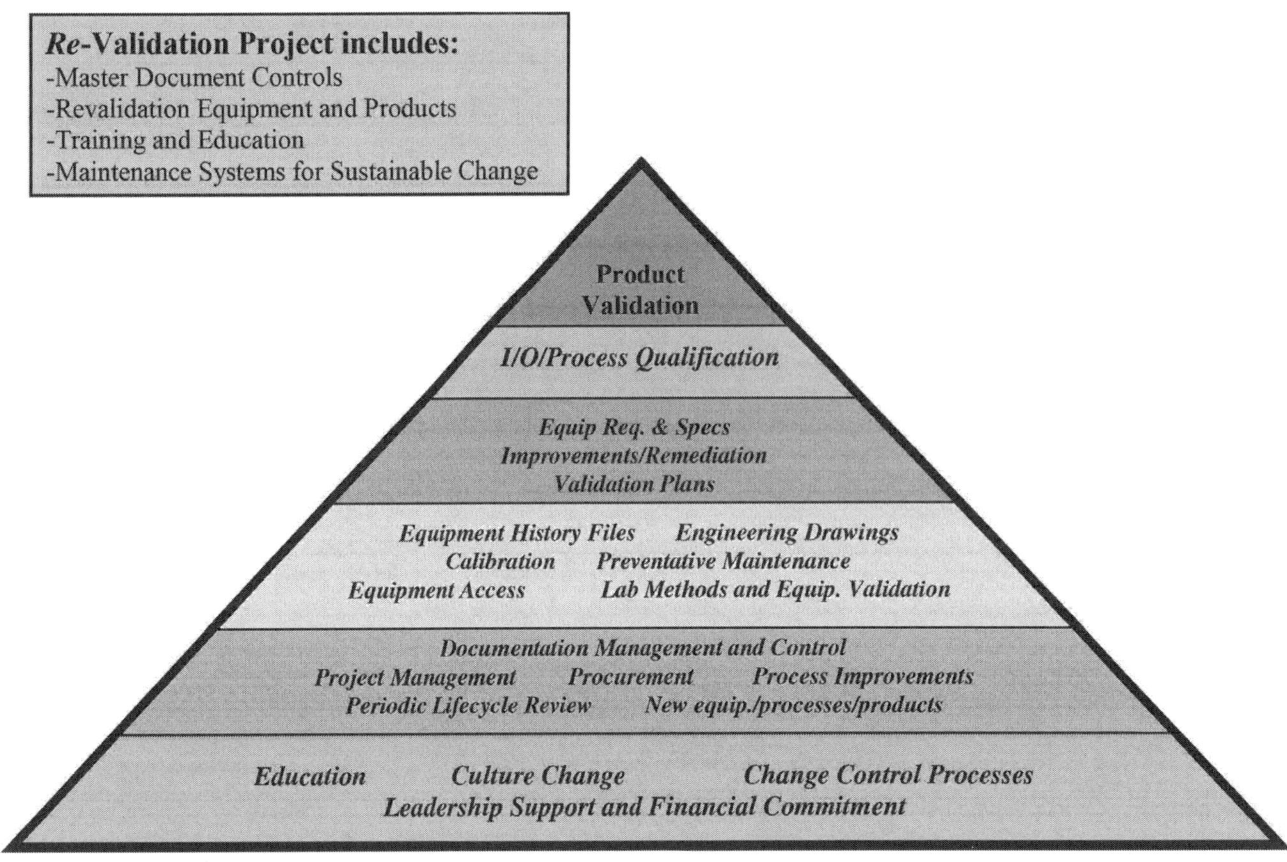

FIGURE 50.1 Validation foundation.

The next level includes functions such as project management, procurement and life cycle review. Are these well-controlled processes? Can it be assured that through these areas the correct components, replacement parts and processes are maintained?

- Equipment/Instruments History, Maintenance and Calibration

The third level includes equipment history, drawings, calibration and preventive maintenance. These are crucial to maintain equipment and processes in correct working order and under a state of control. Included as well in this level are laboratory methods and equipment—are they validated and maintained? Are these methods contemporary and compliant with the most updated compendia and regulations? This is critical to the foundation of a working quality system in the laboratory where the data is generated.

- Equipment Requirements and Specifications

The next level is the equipment requirements and specifications—do they exist? Are they maintained through a life cycle approach? These requirements and specifications are critical, since they are the foundation for the next level—qualification of equipment. They also contribute to the overall validation plan that describes the system as a whole, how it functions together and the approach to validation.

- Installation and Operational Qualification

The final level in advance of product validation is equipment installation and operational and process qualification. It is critical to have these in place to assure control of the process and the products produced as part of that process.

Each layer builds on the previous one. If elements are missing, then there needs to be some form of remediation at each stage to assure a solid foundation before undertaking any product or process retrospective validation. These elements are summarized in Table 50.1, with the relevant considerations for each item listed. This table should be consulted before undertaking a retrospective validation project or approach. Missing elements erode the foundation on which a case could be built for retrospective validation.

50.2 RETROSPECTIVE PRODUCT VALIDATION—ACTIVE PHARMACEUTICAL INGREDIENTS (APIS)

The first approach to retrospective validation explored is for active pharmaceutical ingredients (APIs). The elements discussed previously must be in place from a system perspective—now it will be approached from a product standpoint. The first requirement for retrospective validation applied to a particular API is that "it must be a well understood product/process with few deviations."(1) If this initial requirement

TABLE 50.1
Foundation Elements and Considerations for Retrospective Validation

Element	Considerations
Product validation	Need a strong product validation approach for prospective validation before approaching a retrospective project.
IQ/OQ and Process Qualification	The underlying IQ and OQ need to be robust for the related equipment/facility. In addition, any PQ for processes such as sterilization or aseptic processes must be in place.
Equipment requirements and specifications	Need to be in place and act as foundations for the related IQ and OQ. Should be current and under revision control.
Improvements and remediation	Any improvements or remediation to equipment, facility and processes should be captured in updated requirements, specifications and related IQ, OQ and PQ.
Validation plan	An overall comprehensive validation plan should be in place. Retrospective validation, where applied, should be included and justified in this plan.
Equipment history files	Equipment history files should exist which capture changes, updates and overall maintenance of the equipment.
Engineering drawings	Engineering drawings for the equipment should be up to date and linked to change management system.
Calibration	Calibration for the equipment and related instruments should be maintained and current.
Preventative maintenance	Preventative maintenance should be documented and current.
Lab method and equipment validation	The lab methods that will be used to support retrospective validation need to be validated to contemporary standards. All relevant instruments should have contemporary qualification.
Documentation management	Documentation should be current for site and under the change management system control.
Project management	There should be a strong project management group that assures systems are maintained.
Procurement	Procurement should control purchase of new equipment, spare parts and materials. Changes should be under the change management system.
Process improvements	Process improvements should be captured in the change management system in regards to drawing updates, revalidation and instructions.
Periodic life cycle review	A program should be in place to periodically review the status of all systems and recommend actions such as revalidation. This should occur on at least a three-year cycle.
New equipment, processes and products	Should be evaluated against current systems and included in validation plan and life cycle.
Education	Assure colleagues in operation are educated on requirements for validation, change control and validation planning at a minimum.
Culture change	Assure there is a culture shift if change control and validation were not robust in the past.
Change control processes	Assure robust change control process is in place for facilities, utilities, equipment, lab methods, processes and products. Part of the life cycle approach to assure processes continually in a validated state of control.
Leadership support and financial commitment	All of the items listed require leadership support and financial commitment, or they will not be sustainable and erode the overall foundation of a robust system. This also includes overall project management from beginning to end.

cannot be met, then a concurrent validation approach may have to be considered. If this initial criterion can be met, then there are a series of prerequisites that must be further considered as follows:

- The process is well understood and documented throughout the full-scale manufacturing process.
- Critical process parameters and critical attributes are identified, justified, documented and understood.
- Reliable test data was/or will be generated using a pharmacopoeia (USP/EP/JP) method or internally validated test methods with justified and established specification limits.
- There are no significant process or product failures, and any failures must be attributed to operator error or equipment failure or a "one-off" well understood and documented deviation investigation.
- Impurity profiles are well established for the API.
- Rationale for in-process and final API limits have a sound scientific rationale that is well documented.

Exploring each of these prerequisites in more detail:

1. *The process is well understood and documented throughout the full scale manufacturing process.* Processes that are not well understood are not acceptable candidates for retrospective validation. The execution of retrospective validation is not an acceptable time to gain knowledge and understanding of the process.
   There must also be sufficient documentation of the process to qualify for retrospective validation. The firm must be able to produce all relevant documentation such that demonstrate the conditions under which all batches included in the retrospective validation were produced. This task may be complicated by low volume production over time and changes during this time period. As discussed previously—part of the foundation is the data, documentation and change history.
2. *Critical process parameters and critical attributes are identified and understood.* Critical process

parameters contribute to critical attributes. For instance, the critical process parameter of drying time has a direct impact on the attribute of moisture content through physical measurement via loss on drying (LOD) testing. This attribute is thereby the measure of the process parameter. These process parameters need to be identified and the attributes measured and demonstrated to be in control. Any attribute not in control needs to be critically evaluated to determine if the batches under retrospective validation allow this approach to be utilized.

3. *Reliable test data generated using a pharmacopoeia (USP/EP/JP) method or internally validated test methods.* The data utilized as part of retrospective validation are critical to proving the process is well defined and controlled process. Data integrity is therefore crucial to the retrospective validation platform being built. All data utilized must be either from pharmacopoeia or an internally validated method. If it is a pharmacopeial method, assure that the current version of the method was used at the time these data were generated. In the case of an internally validated method, assure that it meets contemporary validation standards at the time generated. Failure to meet these requirements puts these data, and thereby the retrospective validation, in jeopardy.
4. *There are no significant process or product failures. Any failures must be attributed to operator error or equipment failure.* Often the significance is difficult to judge. Remember that any product failures should have been thoroughly investigated via a deviation investigation procedure. If the failure investigation determines the root cause was due to operator error or equipment failure and not the process itself, then this does not necessarily implicate the process. An example would be equipment malfunction—such as a centrifuge in the process that stopped mid-batch due to a power failure and resulted in an aborted or failed batch. If, however, the batch completed without attributed operator or equipment error and was out of specification—this would cast doubt on control of the process. This must be critically evaluated through a comprehensive deviation investigation for each batch included in the retrospective validation.
5. *Impurity profiles are well established for the API.* Since the firm will typically have experience with the API considered for retrospective validation, there should be substantial data on the impurity profile of the API. These data must be reviewed to assure there were no adverse trends or issues with impurities. Once again, the methods utilized for the impurities evaluation should be validated.
6. *Rationale for API limits has a sound scientific rationale and is well documented.*
The limits that were established at the initial time of validation should have a scientific rationale for how they were set. This is usually based on development data along with safety data and any clinical or medical studies. It is important that this rationale be well documented such that it can be explained and defended if necessary. It should also be reviewed as part of the retrospective validation to assure the limits meet contemporary standards and the compendia currently in place in the markets where the API will be utilized for final product.

50.3 RETROSPECTIVE PRODUCT VALIDATION PLANNING FOR API's

Current ICH Q7A[1] guidelines recommend between 10–30 consecutive batches be examined as part of retrospective validation. Fewer batches may be used to justify retrospective validation providing a documented sound scientific rationale is provided. These batches need to be statistically examined closely for any trends, deviations and/or out of specification results. Such data may call into question the applicability of a retrospective approach to validation unless there is significant evidence to indicate these trends or results are not indicative of the process.

The validation plan for retrospective validation needs to define the number of batches to include with a scientific rationale for the number. The frequency of production and age of the process (years produced) should be considered. For instance, an API that is produced infrequently, such as a once a year, but only produced for the past five years provides limited data over a short time span. Production of 10 batches per year over two years provides both a wealth of data and a shorter time span. The time span is an issue, since the longer the time span, the greater the chance of significant changes to the process, equipment and methods. Once again, this should be evaluated and factored into the rationale included in the plan.

The contents of validation plans and protocols were discussed previously in Chapter 64. The content of the plan should include the consideration of the prerequisites, number of batches to be included, critical process parameters, critical attributes, acceptance criteria and data analysis approach to be utilized.

50.3.1 Case Study—A Retrospective Validation Approach for APIs

This section explores retrospective validation for a fictional API called Cure-all-hydrochloride (Cure-all-HCl). First, examine some background on the API. Cure-all-HCl starts as a precursor, cure-all salt, that processes through a reaction process with hydrochloric acid to form the Cure-all-HCl form. This is accomplished by charging the cure-all salt into a 50 kg reaction vessel, V-1. The hydrochloric acid is transferred from a holding tank into the reaction vessel V-1. Next, a wash process with acetone is conducted to remove impurities and then an aqueous wash to remove water-soluble impurities. This is accomplished by transfer of the slurry into vessel centrifuge C-1 and charging the acetone from a separate transfer tank, T-1. After centrifugation, water is added. The next step is the drying process through centrifugation to remove excess moisture to a target of 3.0% in the centrifuge. The final product is then packaged into polyethylene lined drums.

The initial Cure-all-HCl process was developed and implemented 20 years ago. Each year, about 10 batches are produced. The equipment was not initially validated at installation 20 years ago, but the centrifuge was replaced five years ago and was completely validated (IQ/OQ) to contemporary standards. In addition a new control system for the vessels V-1, transfer tank T-1 and centrifuge C-1 were installed three years ago and full computer validation completed. The test methods for Cure-all-HCl consist of the following assays:

Potency by HPLC
Impurities by HPLC
Moisture by LOD
pH measurement
Residual solvents by GC

All methods were validated to contemporary standards five years ago. In addition, a retrospective validation project was conducted at the same time to qualify all instruments in the lab.

Before starting, determine if the API is appropriate for retrospective validation. Using criteria previously established by Trubinski, it is determined from Figure 50.2 that there are over 20 batches in the product history, and it is a product that the company intends to continue manufacturing.[2]

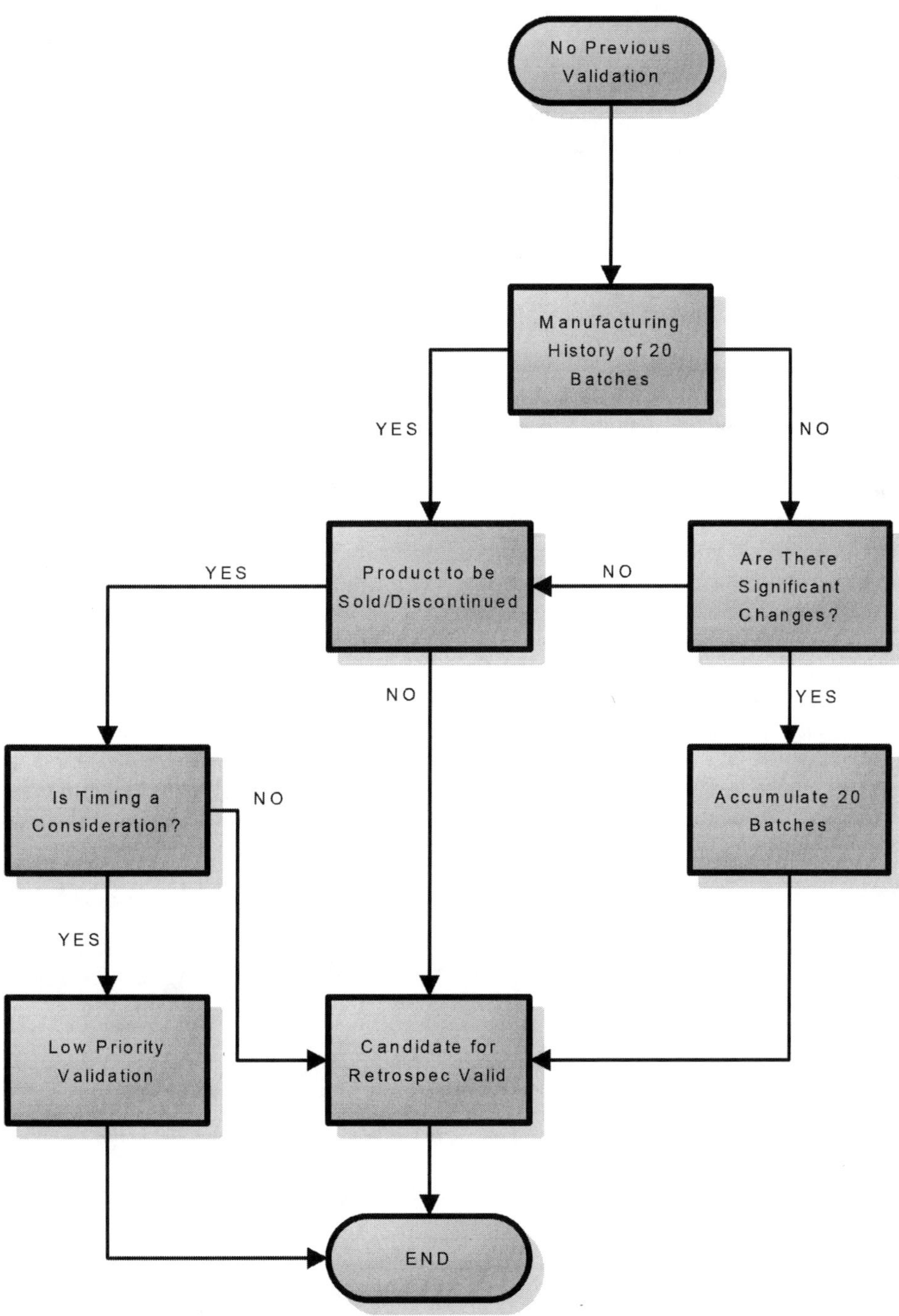

FIGURE 50.2 Retrospective validation decision tree.

Examination of the systems in place, as described previously in Table 50.1 and Figure 50.1, is conducted. All identified systems are robust, and therefore it appears to be a candidate for retrospective validation. At this point, a protocol to examine the 20 batches against predetermined criteria should be written. This approach will utilize each of the previously mentioned test methods and assure all data meet defined specifications. In addition, statistical analysis of the data will be performed to assure no values are out of trend over time. In this instance, there was one out-of-specification result for batch number 17. An investigation determined that this was due to adding a low volume of acetone, which resulted from a deviation to procedure. The batch deviation was determined post-completion and analysis. The batch was rejected since it was outside the registered specification for acetone addition quantity. Therefore, it is not considered to have an impact on the retrospective validation. The investigation is included as part of the validation package and was well documented along with a risk analysis performed. The conclusion will need to defend the use of the 19 remaining batches for retrospective validation.

50.4 DRUG PRODUCT

Retrospective validation for drug products has some different considerations than APIs. Routinely, the FDA expects prospective validation for any new products coming to market. Even for existing products, the window for retrospective validation "is closing if not already closed,"[3] as noted by the FDA. The guidelines for prospective validation have been around since 1983, and investigators tend to view retrospective validation in a negative framework, since it is expected that all products on the market at this point comply with those guidelines. As a result, retrospective validation is not as commonly used in drug product validation as it once was. Retrospective validation was applied when validation requirements were first put in place. Today, it is an expectation that firms have contemporary validation data.[4] Once again, as stated previously, retrospective validation applicability is dictated by the change history for the process. Therefore, one needs to define up front what will be evaluated as part of retrospective validation for drug products.

This was the view of the FDA and continues to be so, but there is recent guidance that may help support a retrospective validation approach with some new requirements. The recent (2011) FDA Guidance for Industry on Process Validation provides guidance for validation of legacy products. This guidance is what the FDA term as stage 3—Continued Process Validation. This section will explore a number of considerations and explore this new guidance in more detail for applicability to legacy product/process validation.

In the new guidance the FDA states in regards to legacy products:

> Manufacturers of legacy products can take advantage of the knowledge gained from the original process development and qualification work as well as manufacturing experience to continually improve their processes. Implementation of the recommendations in this guidance for legacy products and processes would likely begin with activities described in Stage 3.

The key elements to stage 3 include the following:

1. An ongoing program to collect and analyze product and process data that relate to product quality. This data includes key process parameters such that trending can occur to recognize quality issues. It should include the incoming materials/components, in-process material and the final product. Statistical tools should be used for analysis with oversight and review of the data and conclusion from a quality perspective. It needs to demonstrate a state of control for the process and product. This includes the evaluation of the capability index Cpk for the product. Generally, if the Cpk is greater than 2.0, the process is in control but should follow the stage 3 process for retrospective validation, as noted in the guidance. Table 50.2 shows possible Cpk values and what they indicate about the product robustness.

Cpk value	Action/result
>2.0	Robust process, but still utilize stage 3 revalidation
1.34 – 2.0	Good process; still follow stage 3, and may require some process improvement
1.00 – 1.34	Minimum—some type of process improvement warranted before stage 3
<1	Poor process—not reproducible; consider patient safety if on the market

2. This data collection should follow a preestablished data collection plan. The plan should document the statistical methods and procedures that will be used for collection and analysis. This includes the methods for trending the data. The quality group should review the data against a set of established criteria and take appropriate action when needed in regard to variability in the products and possible process improvements or CAPAs. The key here is to detect variation, characterize it and determine root cause and the corrective actions. The guidance recommends use of quantitative statistical methods and the examination of both intra-batch and inter-batch variation.
3. Continued monitoring and sampling at the level established during process qualification. This should take place until enough data is generated to create variability estimates. This will then provide a basis for establishing levels and frequency of routine sampling and monitoring. Clearly, this would be greater at first than would have been conducted in routine production.
4. Additional sources of variation need to be examined routinely. This would include defect complaints,

out-of-specification (OOS) findings, process/product deviations, yield variations, batch record review, incoming material nonconformance or trends and adverse event reports.
5. Maintenance of facility and equipment. This is another aspect of assuring control through maintenance records, calibration and change control. These are key indicators that can impact process/product variability.

As a result of these five key elements, there may be a need to improve or optimize the process. This is a life cycle approach, where change controls are evaluated and action may need to occur for larger-scale overall process change and subsequent revalidation.

This new guidance results in multiple paths that can be taken for existing products and new products. As noted previously, the Cpk values can be a guide, along with review of the product history for robustness.

TABLE 50.2
Multiple Paths Based on Guidance

Existing products	New products
Legacy products/process should be screened for capability	Those developed following the guidance—quality by design and stage 1 and 2 of guidance
Acceptable products should follow stage 3 approach of guidance	Transition through stage 2.5
Less capable products/process should be redeveloped and proceed through stage 1 and 2 of the guidance	Followed by stage 3 as defined in guidance for life cycle

This approach differs from traditional retrospective validation in a number of key ways.

Traditional Retrospective Validation vs Real-Time (stage 3)

Traditional retrospective	Real-time stage 3
Established documented evidence that a process does what it proports to do based on review/analysis of historical data	Confirming the acceptability of the production materials (in-process and final product) using data collected and evaluated as developed
Limited data set based on sampling frequency	Robust set of data based on statistical tools and data collection plan
Never well accepted by regulators	Recommended guidance in 2011 FDA publication (stage 3 section)
The results were essentially known before the data is gathered and analyzed	The results are in real-time against prior performance and established criteria; continual improvement process

Just as was the case for APIs, in the case of drug products, the product must be a well-understood product/process with few deviations. If this is not true, then retrospective validation for the particular product may not be the best choice, and a prospective approach should be considered. If a retrospective approach is followed, the validation requirements may need to be tightened, to add further batches and to provide a strong rationale for why this approach is still valid. As always, consultation in advance with regulatory agency or industry consultants should be considered as a logical choice.

Just as with retrospective validation for APIs, if this initial criterion is met, then there are a series of prerequisites that must be further considered, as follows:

- The process is well understood and documented throughout the full scale manufacturing process.
- Critical process parameters and critical attributes are identified and well understood.
- Reliable test data can be generated using pharmacopeias (USP/EP/JP) methods or internally validated test methods.
- *In-process controls and acceptance criteria are established and in use throughout the critical portion of the process.*
- There are no significant process or product failures. Any failures must be attributed to operator error or equipment failure and not the process itself.
- *Impurity profiles are well established for the API used in the drug product.*
- *Change control is in place and followed for the process. This includes planned and emergency changes. A retrospective review indicates no major changes which required revalidation that did not occur.*
- Maintenance and calibration program maintained and under a state of control.

It should be noted that generally these prerequisites are the same as they were for APIs—except in three cases. Those three cases are highlighted in italics and will be explored in further detail.

1. *In-process controls and acceptance criteria are established and in use*
   Generally, a drug product process involves a number of defined stages. In tablet production processes, for example, there are blending, drying, compression and coating stages, to mention a few. At each stage, there are critical in-process controls that assure the product is acceptable to progress to the next stage. As part of retrospective validation, these in-process controls and the relevant data should be examined as part of the validation. There should be acceptance criteria in place from a production standpoint, and retrospective validation criteria that are wider than the operating range but tighter than development ranges must be applied. Any deviations need to be investigated and explained as part of the validation.
2. *Impurity profiles are well established for the API used in the drug product*
   In the previous section on APIs, this was discussed in relation to the API itself. Now it needs to be

assured that the API or APIs used in the drug product have a well-defined profile such that there is analytical, impurity and stability data that characterizes the API. The data must be further reviewed to assure that there are no adverse trends or issues with degradation, by-products or impurities that would place the final drug product at risk.

3. *Change control is in place and followed for the process*
In the drug product/process, change control is a critical consideration. If the intention is to retrospectively validate a product for which there was an underlying lack of change control, how can the firm assure the process/product is consistent? Remember, the key definition of validation is proven evidence that the product/process has been consistent over the retrospective validation evaluation period. Change control assures the process is consistent with processes that are in control. Adequate change control is a premise to validation. Lack of adequate change control requires a detailed review of the process and some retrospective review of the change history. If the change control is inadequate, retrospective product validation may not be appropriate and a prospective or concurrent product validation approach should be strongly considered. This review should be documented as part of the validation. It is important that this review be comprehensive such that changes to product/processes, equipment, excipients, procedures and systems are reviewed.

50.4.1 Retrospective Product Validation Planning for Drug Products

The FDA's Codes of Federal Regulations (CFR) does not specifically identify retrospective validation. It only states that process validation is a requirement for the manufacture of pharmaceuticals and medical devices.[5] There is, however, a section on retrospective validation in the FDA guidelines on validation. These guidelines state that

> in some cases a product may have been on the market without sufficient pre-market process validation. In these cases it may be possible to validate, in some measure, the adequacy of the process by examination of accumulated test data on the product and records of the manufacturing procedures used.[6]

However, the new Process Validation Guidance by the FDA from January 2011 does provide clear guidance on retrospective validation under the section titled "Stage 3—Continued Process Verification," as described previously. Usually the greater concern for patient safety and compliance is with those currently marketed products which were validated long ago.

The question to ask is how much data is required for retrospective product validation? This chapter will explore some examples as part of case studies later following the new guidance for stage 3. For now it is sufficient to say that this depends on the process and its history. There is one other critical statement in this test from the guideline: "records of the manufacturing process used." It was stated in the opening of this section that for the process to be a candidate for retrospective validation, it must be well documented, understood and under change control. These are critical to allow this retrospective validation approach.

The second paragraph of the guideline specific to retrospective validation suggests that "retrospective validation can also be useful to augment initial pre-market prospective validation for new products or changed processes." Typically, these are prequalification batches that can be examined as part of developing a final prospective validation approach. However, in many cases these prequalification batches have inherent differences from the final batch process to be qualified. This must be taken into account as part of the approach. Statistical analysis of the prequalification batches as compared to retrospective data is a useful tool that will be discussed later as one strategy.

The last paragraph of the guideline states in regards to retrospective validation that "Test data may be useful only if the methods and results are adequately specific." The section goes on to state that "Specific results, on the other hand, can be statistically analyzed and a determination can be made of what variance in data can be expected." This statement suggests that statistical evaluation be performed on the method itself using tools such as Gauge repeatability and reproducibility (R&R) or Cpk calculations to determine variation of the methods and process robustness. If the variation is too extreme, it may not be possible to use the method to justify the previous set of acceptance criteria in the retrospection validation plan. In those cases, other methods may be necessary or tightening of the method variance required. That is an acceptable approach in a prospective validation plan where such adjustments can be made. In a retrospective validation approach, these data have already been generated. It may, however, allow a change to the approach or target specific methods as key to validation based upon the statistical analysis of the data. For instance, this analysis may determine one method is more critical than another and narrow the scope of data required for the retrospective validation.

The last sentence of the guideline section states that "Whenever test data are used to demonstrate conformance to specifications, it is important that the test methodology be qualified to assure that test results are objective and accurate." This was also one of the principles required for retrospective validation, that reliable test data is generated using a pharmacopeias (USP/EP/JP) method or internally validated test methods.

- EMEA view

The EMEA guidance for manufacturers has a section on retrospective validation in annex 15.31–15.35.[7] There are five major points (31–35) which support what has already been stated as requirements and prerequisites. The first point is "retrospective validation is only acceptable for well-established processes and will be inappropriate where there have been recent changes in the composition of the product, operating procedures or equipment." This supports the initial criteria that the process/product must be well understood. This guidance also points out the issue of change control, as is noted in the previous discussion on prerequisites. It brings up another point about consistency of the process, that changes to the process/product over time have not included major changes that may require significant revalidation. If that was the case, then concurrent validation should have been performed at that point, not retrospective validation at a later point.

The next point in the EU guide is that "validation of . . . processes should be based on historical data. The steps involved require the preparation of a specific protocol . . . leading to a conclusion and recommendation." It was indicated in planning for retrospective validation that a predetermined validation plan or protocol must document the approach and acceptance criteria. This guidance further reenforces the point and highlights the importance of the final conclusion and recommendation. Data are another key aspect of the retrospective validation. There is further guidance on the source of data.

> The source of data . . . should include . . . batch processing and packaging records, process control charts, maintenance log books, records of personnel changes, process capability studies, finished product data, including trend cards and storage stability results.

The point here is there should be a comprehensive data review from all of these documented sources. Every "stone" must be overturned, so to speak, as part of this review. This is a significant task in some cases that requires a thorough approach to assure all data is collected, analyzed and included in the retrospective validation package. These data must then be tested against the preestablished acceptance criteria. In some cases, this is such a significant undertaking that a concurrent validation approach will require less investment.

The next section deals with the batches selected as part of the validation. The EU guide states that "Batches selected for retrospective validation should be representative of all batches made during the review period, including any batches that failed to meet specifications." Previously, it was stated as a prerequisite that there may be no significant process or product failures, and any failures must be attributable to operator error or equipment failure. This guidance appears to allow inclusion of failed batches and notes that it must be a comprehensive set of batches over the review period. A word of caution is appropriate here—failed batches will have a great deal of scrutiny by a regulatory agency. This means that the firm must exert an even greater level of scrutiny and conclude if a consistent and thereby validated process/product exists. The investigations of these failed batches need to have identified root cause(s). That determined root cause(s) should not indicate a process/product failure—if it does, concurrent revalidation after the issue is resolved must be considered. This may also require development work to understand the source of the failure.

Another key point is made in the final sentence of this reference: "Additional testing of retained samples may be needed to obtain the necessary amount or type of data to retrospectively validate the process." During the development of the validation plan and protocol, the firm may find that the methods and data are not sufficient to support the validation. This could be due to assay variability, as mentioned previously, or that the method was not in place at the time the batch was initially analyzed. The firm may also encounter during data review that there are missing data. One additional word of caution—in the case of failed batches, do not use retained samples to try and retest and eliminate failed results unless investigation(s) can invalidate the initial result. If these issues do not apply, then available retained samples may be useful to supplement overall data and support the retrospective validation.

The final point in the EU guide provides guidance for the number of batches to include in the retrospective validation. The recommendation is "ten to thirty consecutive batches . . . but fewer may be examined if justified." This is consistent with previous ICH guidance for APIs. One important note is the word "consecutive" batches—the firm cannot pick and choose the batches it wants to use. The number of batches must be defined in the preapproved validation plan and protocol. As stated clearly in the guidance, any choice outside of 10–30 requires a rationale for that choice. In the examples section, this chapter will explore how to arrive at this number. It should be based on the specific attributes, history and characterization of the product.

Remember that a retrospective validation approach is not generally accepted unless plenty of data are included to make a case. The data are key to justification of the retrospective validation and should utilize statistical methods for evaluation.

50.5 STATISTICAL ANALYSIS OF BATCHES

In the case of measuring variation of the methods used to generate retrospective data, Gauge R&R and Cpk values are useful to assess each method. The Gauge R&R is an ANOVA gauge of repeatability and reproducibility that uses an analysis of variance random effects model. The Cpk is a statistical capability index that can be used to evaluate process or method robustness. This evaluation should be used to determine if the methods are suitable to provide data for a retrospective validation strategy.

Once the data are generated, there needs to be some evaluation against either previous development data or predetermined acceptance criteria. Significant testing, where there is a null hypothesis of no difference between the observed and known or previous obtained values, is useful.[8] Useful analysis methods include but are not limited to comparison of mean, paired t-tests, and F-tests for comparison of standard deviation. The overall lesson is that use of statistics provides a scientific basis for comparison to make a case that the process or product is comparable to previous experience or data and that the method is well under control.

50.5.1 An Approach to Retrospective Validation for Drug Product

50.5.1.1 Case Study—Examples

In this section, a retrospective validation approach for a fictional sterile product Steri-Cure will be provided. It is an aseptically filled liquid composed of the drug substance Cure-all Hydrochloride. Cure-all is dissolved in WFI using a 500 liter mixing tank, aseptically filtered and transferred to a holding tank for aseptic filling. Table 50.2 lists the critical quality attributes and parameters which impact these attributes.

Critical quality attributes	Critical process parameters affecting quality attributes
• Potency	• Formulation
• Degradation products	• Temperature
• pH	• API dissolution mix speed
• Color of solution	• API dissolution mix time
• Appearance of solution	• Final mix time
	• Final mix speed
	• Final mix pH
	• Hold times

Steri-Cure has been aseptically filled at this particular facility for 15 years. There is no contemporary validation for this product. Previous validation was performed on other aseptic products that utilize the same filling equipment and filling suite. This validation included all aseptic process validation, and media fills are conducted on a semiannual basis for this product and aseptic process.

Retrospective validation is proposed to validate the manufacturing process, since Steri-Cure, an older product, is only produced once every two years based on demand. We will first look at the process robustness using the capability index Cpk. We find that the Cpk for the process is 2.1, which is good, but we will still follow the FDA guidance stage 3 for continued process verification. A validation plan will be written that considered the change control history, development data, methods validation and drug substance characterization.

In this case, the change control on the tanks, pH meter and mixer used in the process are current, since they are shared with other products. There is cleaning validation for this equipment as well. All analytical methods for the product were validated two years ago as part of an overall program in the laboratory. A review of batch records indicated that dating back ten years there were 25 batches produced. The drug substance is well characterized, and method development data, while old, is still available.

Evaluation of this data for a retrospective approach indicates that while there are an adequate number of batches for evaluation, only two batches were manufactured since the analytical methods were fully validated. This adds a degree of risk to a retrospective validation approach. It is not necessary to include the filling equipment or filling suite, since adequate, contemporary validation exists for the filling of other comparable liquid aseptic products in the same equipment and line configuration. Although the same argument could be applied for the manufacturing process, it is not recommended.

The development data is older, and at the time, the documentation did not cover all of the critical quality attributes. It is discovered that the mixing time was changed along with a pH step addition about five years ago due to an incident where the drug substance was not completely dissolved after the normal mixing time, temperature and drug substance addition. Investigation at the time determined that mixing speed was not recorded in the development or in routine operation. It was determined that due to changes in the operators that previously performed this manufacturing process, a change was made in mixing. A study was conducted at that time to determine optimal mixing time, and a pH adjustment step was added.

Due to these significant process changes and the lack of previous batch data without adequate method validation, retrospective validation is not recommended.

A concurrent validation approach is therefore recommended that will follow the new guidance as follows:

First, we need to determine how much sampling data is necessary. The guidance calls for this to be greater than the original validation and statistically based. Since we have 25 batches to date but not a great deal of initial validation data, we will follow a statistical approach. It is determined that we should sample beginning, middle and end of each batch and take enough samples to meet or exceed the AQL used for current product release. We decide to make this about 10% of the batch to have an adequate set of data. We further determine that rather than three lots in this validation, we will include four lots and then reduce the sampling after the fourth batch based upon our review of the data for all four batches. As a result, we therefore decide to not release the four batches until all the data is evaluated, which is a prospective validation approach rather than concurrent.

Next, this is all documented in a preapproved sampling plan that is part of the validation plan and includes the following elements: the statistical evaluation for sampling plan, methods we will use to trend and evaluate the data, the criteria for acceptance and a recalculation of the Cpk against the previous 2.1 value. The plan also establishes that we are looking for intra-batch and inter-batch variation. We will include the four prospective validation batches and the 25 previous batches for comparison and evaluation.

We will utilize all of this data to determine what level of sampling will continue for routine production of batches after the fourth validation batch. We also are doing a retrospective review of CAPAs—were they closed and deemed effective after the corrective actions? This same type of retrospective review will take place looking at any previous adverse events in the field, out-of-specification results, recalled/rejected batches, yield variations and incoming material variations. A formal and comprehensive review of previous manufacturing deviations will look to determine if CAPAs were assigned and completed and whether there were any repeat issues without resolution. This same exercise will be performed for previous change controls. Were they completed, and are there any open change controls? If open, is there a plan to monitor or control the process or a short-term corrective action in place? For completed change controls, were they effective, and are there any repeat issues?

The last step will be to review the maintenance and calibration program to determine if they are in a state of control and look at any deviations there to assure they were corrected and not repeated.

In summary, our review indicated this product was a good candidate for prospective revalidation using the stage 3 Continued Process Verification guidance. This requires enhanced sampling via an approved plan and a comprehensive review of all the associated data to determine process/product robustness. We are looking for sources of variation. Factors that contributed to this decision are the change history, deviations in process, analytical method validation to contemporary standards, CAPA review, change control review, review of adverse events, OOS, recalls/rejected batches and maintenance/calibration program. As mentioned previously, all of these must be critically evaluated to look for variation and determine if in a state of control. The sampling data will determine to what extent and timing for continued enhanced sampling and any trends using statistical tools. Often, age of the product/process can work against a retrospective validation approach.

50.6 RETROSPECTIVE PROCESS VALIDATION

Retrospective process validation often has some of the same concerns and considerations as prospective validation, as mentioned previously. Overall, there are systems where it can be applied and those where it cannot. For instance, a system that has routine in-process data for evaluation and is controlled based on the data could be a candidate. In the case of a water system—perhaps there was some initial validation conducted long ago, or the system was only qualified. Since it is a requirement to continually monitor and control the system, that data can be used to retrospectively validate the system. The FDA guide to inspection of water systems requires phase I, II and III validation.[9] In the final phase, the data for chemical and microbiological analysis is required on a frequent basis under protocol. Clearly, there are USP criteria preestablished for these tests. Therefore, if the data history is available, then these USP criteria can be utilized for evaluation and retrospective validation of the system.

In other systems such as sterilization processes, it is clear these must have not only prospective validation but an annual or periodic requirement for revalidation that is prospective as well.[10] These sterilization processes are part of a larger overall aseptic process, and often there is not the level of in-process data generated as part of control for the water system example.

In summary, if the system has in-process controls and pre-established criteria, these data could be used to evaluate and validate the system.

50.7 RETROSPECTIVE EQUIPMENT QUALIFICATION/VALIDATION

Retrospective equipment qualification/validation applies in cases where contemporary IQ, OQ and PQ do not exist for equipment. In these cases, as mentioned previously, retrospective product validation is not recommended, since the underlying foundation is not sound.

In the case of retrospective equipment qualification/validation, many of the criteria mentioned in Table 50.1 and Figure 50.1 still apply. Criteria such as improvements and remediation to the equipment, equipment history/use, calibration records, preventative maintenance records and change control should all be evaluated. Leadership support and financial commitment is important, since retrospective qualification of some equipment may be more costly and at greater risk than to purchase and prospectively qualify new equipment.

If a retrospective approach can be justified, the first activity is to develop an accurate set of combined requirements and specifications (R/S). Since this equipment is already installed, these combined R/S should represent the "as found" condition unless there is a justification to change. For instance, if the mixer is rotating in the opposite direction from initial design, it should be left that way and captured as such in the R/S, since all batches previously produced were under these conditions. This assumes that all specifications were met.

Once this combined as found R/S is created, it can be used for the foundation of the Retrospective IQ/OQ and PQ, if applicable. Critical attributes and process parameters should be defined. The IQ will be minimal, since the equipment is already in place, but will confirm that installation was per original manufacturer recommendations and that critical documentation, spare parts listed, procedures and operating instructions are in place. The operational qualification will assure that the system operates as designed and all critical parameters are met.

50.8 RETROSPECTIVE VALIDATION IN A PAT ENVIRONMENT

Process analytical technology (PAT) is a system for designing, analyzing and controlling manufacturing through timely measurements (i.e., during processing) of critical quality and performance attributes of raw materials and/or in-process materials and processes with the goal of ensuring final product quality.[11] The specifics of PAT are covered in detail under Chapter 56. In this section we will only cover some considerations in a PAT environment related to retrospective validation.

In a process utilizing PAT, traditional validation principles may not apply. Three batches are insignificant where thousands of data points will be examined for the critical parameter(s) measured. It is important to be aware that PAT may show much more about the process than was previously known or understood.

As an example, consider a batch API process where moisture level after centrifugation and drying is critical. The capability to measure moisture using PAT via an in-line NIR (near infrared sensor) is now available. Figure 50.3 is a set of data in five-minute intervals for this process.

In order to make a change to the drying or centrifugation time to this process in the traditional validation model, it would be necessary to analyze three consecutive batches. In a PAT environment, continuous data is available that can be analyzed. The traditional three-batch approach would not apply since continuous data is present. Advantages of this approach include not only the additional data under real-time conditions but also the process understanding, quality of data and reduced validation time.

In fact, there is a preponderance of data—actually substantially more data than normally available in traditional validation. It is important to note when the change occurred, set preestablished criteria for the moisture level and measure or trend data after the change. If this data indicates a state of control within predefined specifications or limits, then it may meet validation criteria. In essence, it is possible to concurrently validate the process through use of on-line data generated as part of a PAT measurement process.

If this PAT process measurement is new, it would be necessary to compare variability of PAT method to an established, validated, analytical lab-based method as a reference. The PAT Guidance from the FDA in fact indicates that a test-to-test comparison may be required when implementing a new on-line process analyzer.

PAT opens up a new approach to validation—it provides real-time data to examine and validate process changes within the process itself.

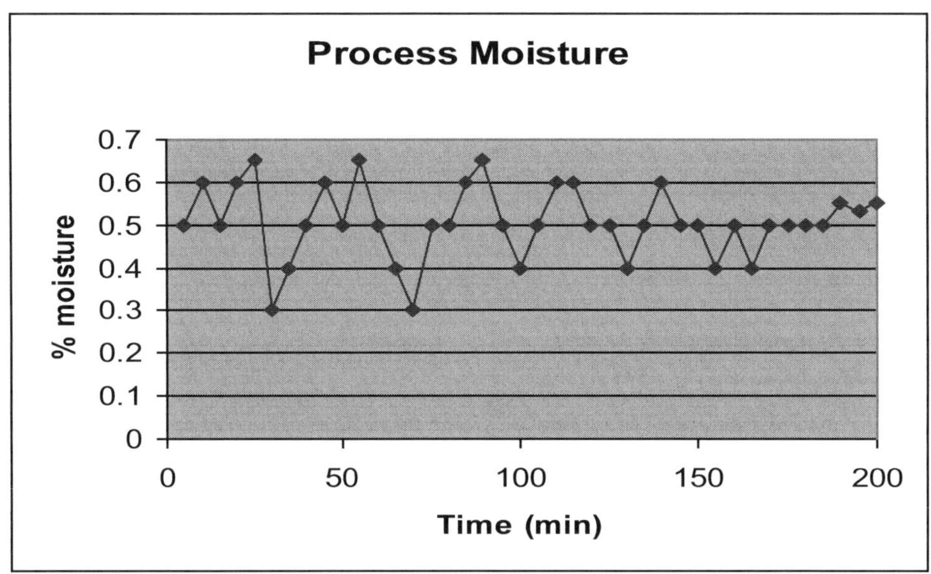

FIGURE 50.3 In-process data for moisture.

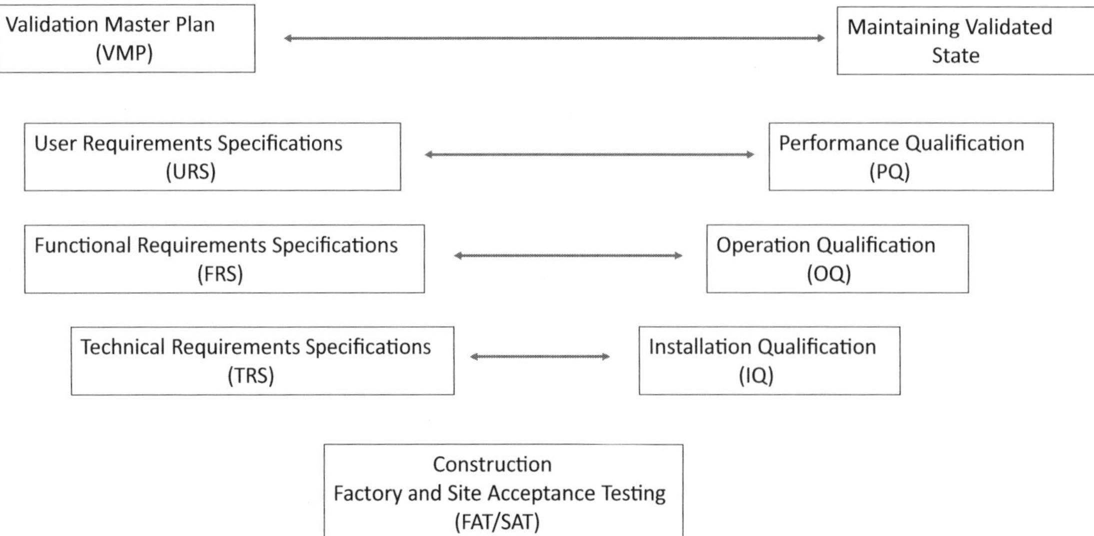

FIGURE 50.4

50.8.1 Continuous Manufacturing

Continuous manufacturing is where larger batches are produced or the process of production in continual in the sense of running much longer. As a result, these types of processes often are monitoring in real time and generate large amounts of process/product data. This data can be used to establish acceptability of the process and generates more data than the traditional three-batch validation. This also support the new FDA Guidance on Validation in regard to stage 3 with continual process verification. This robust data set allows for statistical methods to evaluate process stability and capability. This use of what is sometimes called "big data" has demanded new algorithms and computer analysis to recognize trends and be somewhat predictive in nature. This is a whole new area that goes beyond the traditional validation and allows much more process understanding.

This new technology and approach fits very well with the FDA Continued Process Verification guidance.

50.9 CONCLUSION

This chapter has discussed retrospective validation in relation to both drug substance and drug product. There are a number of prerequisites which must be in place for a process to be a candidate for retrospective validation. In addition to these prerequisites, there is an underlying set of elements that must be placed as a foundation to assure a solid ground for retrospective validation. The recent FDA guidance on validation and stage 3 continuous process verification were also discussed and lend themselves to a retrospective approach with defined criteria for what is required.

It also includes examples of test cases for both drug substance and product retrospective validation. These test cases utilized the prerequisites and elements as criteria for acceptance. Unless there are solid data sets, foundation and rationale, retrospective validation is not always justified as an approach. In many cases, prospective or concurrent validation is a better approach. All of these considerations must be taken into account to pursue this avenue for validation.

A brief examination of process and equipment validation has been provided. The same considerations apply in these instances. Finally, the chapter has explored retrospective validation in regard to a PAT environment. This area provides so much data that the traditional approach does not apply.

Overall, retrospective validation has a number of risks that must be carefully considered for the particular product, history and prerequisites described previously. If a solid foundation does not exist, it may not be the best approach. This pre-evaluation will allow the risks to be determined and make the validation approach decision based on all the data.

NOTES

1. International Conference on Harmonization. Good Manufacturing Practice Guide for Active Pharmaceutical Ingredients ICH Q7A. (www.ich.org/cache/compo/363-272-1.html).
2. Berry IR, Nash RA, eds. *Pharmaceutical Process Validation*. New York: Dekker Publishing, 1993: 252, (chapter 8).
3. Ask the FDA: Questions and Answers with Current and Former Agency Personnel. Institute of Validation Technology, website, April 20, 2001.
4. Barr D, Celeste A, Fish R, Schwemer W, *Application of Pharmaceutical cGMP's*. Washington, DC: Food and Drug Law Institute, 2003.
5. Current Good Manufacturing Practice Regulations for Finished Pharmaceuticals, 21 CFR Parts 210 and 211 and Good Manufacturing Practice Regulations for Medical Devices, 21 CFR Part 820.
6. U.S. Food and Drug Administration. Guidelines on General Principles of Process Validation, May 1987.
7. Medicines Control Agency, MCA Rules and Guidance for Pharmaceutical Manufacturers and Distributors 2002. Published for the MCA under license from the Controller of Her Majesty's Stationary Office. St. Clements House, 2–16 Colegate, Norwich NR3 1BQ, sixth edition, 2002.
8. Miller JC, Miller JN, *Statistics for Analytical Chemists*, 2nd ed. Chicheser, UK: Ellis Horwood Ltd, 1988, 53–77.
9. FDA Guide to Inspection of High Purity Water Systems. (www.fda.gov/ora/inspect_ref/igs/high.html).
10. FDA Guidance for Industry: Sterile Drug Products Produced by Aseptic Processing—Current Good Manufacturing Practice. U.S. Department of Health and Human Services, Food and Drug Administration Center for Biologics Evaluation Research (CDER), Office of Regulatory Affairs (ORA). (www.fda.gov/cder/guidance/5882fnl.htm).

    U.S. Food and Drug Administration. FDA Guidance for Industry: Process Validation: General Principles and Practices, January 2011, Revision 1.
11. PAT Guidance for Industry—A framework for Innovative Pharmaceutical Development, Manufacturing and Quality Assurance. U.S. Department of Health and Human Services, Food and Drug Administration (FDA). Center for Veterinary Medicine (CVM), Office of Regulatory Affairs (ORA), Pharmaceutical cGMP's; September 2004.

    Trubinski CJ, Retrospective Process Validation. In Wachter A, Nash R, eds. *Pharmaceutical Process Validation*, 3rd ed., 2003.

REFERENCES

Agalloco J. Compliance Risk Management: Using a Top Down Validation Approach. *Pharmaceutical Technology* 2008, 32(7), 70–78.

Agalloco J. FDA's Draft Guidance for Process Validation: General Practices & Principles: Can It Be Applied Universally? *Pharmaceutical Technology* 2009, 33(5), S22–S27.

Agalloco J. The Importance of Equivalence in the Execution & Maintenance of Validation Activities. *Pharmaceutical Technology* 2010, 34(12), 43–46.

Agalloco J. Process Validation Course Notes, 1985 to Present.

Agalloco J. Risk-Based Thinking in Process Validation: Finding the Appropriate Number of Tests. *Pharmaceutical Technology* 2011, 35(2), 68–76.

Agalloco J. The Validation Life Cycle. *Journal of Parenteral Science and Technology* 1993, 47(3), 142–147.

51 Validation and Six Sigma

Robert Bottome

CONTENTS

51.1 INTRODUCTION

Not too long ago, Six Sigma and Lean manufacturing programs were greeted with considerable skepticism by most pharmaceutical engineers and quality professionals, who questioned their relevance and validity. Process improvement professionals attempting to use these tools and methods to reduce variability and improve process capability sometimes encountered resistance centered on outmoded notions of what it means to have a "validated" pharmaceutical process. Sometimes, the very fact that the process has been validated using a traditional "single point" approach was used to justify resistance to process improvement proposals.

The perceived value of "Design for Manufacturability," "Design for Six Sigma" and "Quality by Design" has grown considerably in the last two decades—to the point where it is now codified in extensive industry guidance (1—ICH Q8 Pharmaceutical Development, Q9 Quality Risk Management and Q10 Pharmaceutical Quality System). As a direct consequence, we are learning how to effectively apply these tools and techniques from early development to initial approval and then throughout the product life cycle.

The objective of process validation "remains that a process design yields a product meeting its pre-defined quality criteria"[2]. However, the emerging consensus is that optimum outcomes result from the integrated use of several tools, including design of experiments to reveal which variables are robust in the face of variation across specification limits and which variables need to be carefully controlled. This so called "enhanced pharmaceutical development" affords us a deeper understanding of our quality target profile and the associated risks to product quality if certain parameters are permitted to vary beyond certain ranges, especially when they interact with each other. This deeper understanding and the resultant "design space" can then inform the "initial" validation applied to the first few batches produced at scale. Beyond initial approval, however, this enhanced approach to validation can also be used to focus control strategies and (when extended under an approved protocol across multiple commercial batches once the design space has been "verified") to simplify and accelerate tech transfers and process optimization.

51.2 DEFINITIONS

The term Six Sigma has come to mean many things, but in this chapter it refers to a methodology and an associated set of tools for reducing process variability. In fact, Six Sigma is merely the latest iteration in an evolving science of operations management that traces its roots to Taylor[3] and his initial effort to define labor standards, Deming's application of Shewhart's tools for distinguishing systemic variability from special causes[4], Juran's application of the Pareto principle to data segregation[5] and Crosby's arguments against end of pipe remedies[6]. As currently understood, Six Sigma methods incorporate the battle-tested fundamentals of the quality movement with a set of updated statistical tools and concepts. These include the idea that a process that operates at a Six Sigma level of reliability only produces 3.4 defects per million opportunities[7]. Essentially, Six Sigma methods take a practical problem and translate it into a statistical problem so that a set of statistical solutions can be derived and then translated back into a set of practical solutions.

More importantly, Six Sigma as a management philosophy has helped to solidify most of the key concepts that have characterized the quality movement across the twentieth century. For example, at its essential core, Six Sigma is about applying rigorous analysis to distinguish the critical few problems from the trivial many, so that the most talented people can analyze them using the best available tools and techniques before solutions are defined.

Lean manufacturing overlaps with Six Sigma to the degree that greater throughput and process efficiency is achieved by reducing variability (a shared goal of the two methodologies).

DOI: 10.1201/9781003163138-51

The Toyota production method, for example, consists in large part of an interrelated set of tools and techniques for reducing variability through standardization, smoothing and simplification.

Consequently, contemporary "Lean Sigma" exponents are likely to be passionate advocates of:

- Employee empowerment and driving decision-making close to the problem.
- Continuous incremental improvements with occasional breakthrough reengineering.
- Reducing variability: Convert bell-shaped curves into needles!
- Using financial analysis to prioritize and measure forecasted vs. actual value added.
- A focus on cycle time compression as a means to reduce variability, drive efficiency, enable responsiveness to shifting demand and provide tighter feedback loops that increase first-time-through conformance to requirements.
- Emphasis on Voice of the Customer, converting customer preferences and needs into critical-to-quality attributes (recognizing that the customer may be internal to the organization).
- A belief that inductive statistics and end of pipe monitoring is wasteful; a preference for 100% inspection of single-part production flows (continuous process monitoring) where handoffs are predicated on acceptance and verification of conformance to quality.
- Regular and routine application of a variety of statistical tools to understand and distinguish inherent or common cause variability from special or assignable causes (including the verification that measurement systems and instruments can provide us with the data we need).

Initially, the dedicated process improvement professional working within the pharmaceutical industry sought to optimize the process within validated ranges and the terms of the license. They also looked for gains to be had in the transactional environment outside the validated envelope—streamlining process, eliminating or modifying error-prone steps or reducing hand-offs. Rarely were the tools and methods applied to streamline and enhance the validation process itself—but all that began to shift as regulatory agencies in the United States, Europe and Japan began to take an interest in "Quality by Design."

51.3 THE EMERGING CONSENSUS

Early efforts to apply lean and Six Sigma methods to pharmaceutical processes almost immediately ran into objections predicated on a misunderstanding of the true role of validation. A process could not be improved, since improvement implied a change that would take the process out of its validated state. At the same time, it was not unusual in the industry to accept anomalous results or process variability in a validated process as long as the output met all of the specifications and did not drift outside the boundaries established by the license. One of the more interesting side-effects of the traditional approach was to encourage the validation of broad ranges or windows where possible and then seek to optimize yield or other attributes within these ranges (often through expensive and time-consuming investigations of quality events). Process owners struggled to establish alert and alarm limits within broad ranges established in the license. There was even (at least initially) a misplaced concern that application of modern process improvement methods to pharmaceutical processes (like control charts and Paretos) would earn the disapproving attention of the Food and Drug Administration. Inspection management teams would take down these essential charts before an inspector arrived in a well-intentioned but ultimately misguided attempt to eliminate a pretext for questions about process performance.

The traditional approach to validation (without the application of Six Sigma tools) could conceivably result in the acceptance of variability in process inputs within specification limits on raw materials and utilities. This variability could end up locked in a process where the output may not be capable of consistently meeting release criteria (for example, a validated process may only yield 70% acceptable material and require rejection of the remaining 30%, an approximate one sigma level of reliability).

The application of "Lean Sigma" tools and techniques immediately enhanced even traditional process validation efforts, where the focus is on a limited number of commercial scale batches at discrete time points using the proposed control strategy with an increased level and frequency of sampling. Examples of this include the effective use of Failure Modes and Effects Analysis in support of installation qualification. Basic mistake proofing and challenge testing are also often characteristic of this project phase. The use of these three tools early in the process typically translate into successful project delivery and lower overall costs of ownership post-approval. Operational qualification efforts are now routinely supplemented with the validation of the measurement system using "gauge R&R" on the instrumentation used. The "rolled throughput yield" so central to Six Sigma-centered process improvement efforts could be profitably applied to operational qualification where there was an opportunity to optimize the flow path, filter sizing and line losses with a consequent positive impact on process yields. The obvious challenge remained with Performance Qualification: how do you verify the capability of outputs relative to product specifications and relate limited testing at commercial scale to pilot or lab bench work?

The US FDA and the EMA (among others) have been quite vocal and explicit in their support of Six Sigma tools and methods. Consider these excerpts from US FDA guidance, issued in 2003, on Process Analytical Technology (PAT)[7]:

> Gains in quality, safety and/or efficiency will . . . likely come from reducing cycle times . . . preventing rejects, scrap, and re-processing . . . and manage variability. . . . [T]hese concepts are applicable to all manufacturing situations.[1]

> In a PAT framework, process validation can be enhanced and possibly consist of continuous quality assurance where a process is continually monitored, evaluated, and adjusted using validated in-process measurements, tests, controls, and process endpoints. . . .
>
> Continuous learning through data collection and analysis over the life cycle of a product is important. . . .
>
> A process is generally considered well understood when
>
> 1. All critical sources of variability are identified and explained;
> 2. Variability is managed by the process; and
> 3. Product quality attributes can be accurately and reliably predicted over the ranges of acceptable criteria. . . .
>
> The ability to predict reflects a high degree of process understanding.

The intent of PAT remains to

1. Accept and measure input variability (know which variables are worth monitoring);
2. Use the feed-forward indicators derived from these measures to make control system adjustments at critical control points (temperature, humidity, pH, etc.); and
3) Measure process variability and use the feed-forward indicators derived from these measures to inform control system adjustments downstream,
4) Such that the output itself is fixed and controlled, thereby achieving levels of reliability that closely approximate Six Sigma quality.

This fundamental shift in control philosophy can enable parametric, concurrent lot release (by exception), which would increase our industry's ability to turn inventory (currently at a sclerotic 1.2 turns per year) and so should translate into an ability to reduce cost of production.

51.4 WHEN VALIDATION IS AN OBSTACLE

As mentioned previously, the delusion that a "validated" system has had its variability reduced or its parameters optimized can be a real obstacle to necessary improvement. In addition, the way we approach validation can sometimes be the problem. Two scenarios seem to be the most common instances of this: over-inclusive scope and the use of constrained technical resources who engage only at the end of a project.

The scope and approach to validation has sometimes expanded from what began as an attempt to establish control over critical product safety and quality issues only into a validated envelope that now extends into unexpected areas. For example, modifications to a waste-water handling system can trigger revalidation of process equipment. Instead of asking "What is the duly diligent standard of care?" validation engineers are often left implementing a work plan that reflects the maximum that can be done given the constraints of time and resources.

Finally, the most common manifestation of validation as an obstacle comprises the bottlenecks that are created by constraints on resources. Although protocol development and testing can be outsourced to third parties or equipment vendors, getting these approved and signed off can be an ordeal. The review and approval of draft protocols and testing results ends up generally limited to a handful of over-utilized individuals in most management systems. As a result, there is virtually no early or iterative review of projects as they take shape, so few opportunities to guide the design in the most robust direction can be seized.

51.5 THE VALIDATION STEP VS. THE VALIDATION PROCESS

When the validation "step" gets crammed into a dwindling slice of time toward the end of a project after mechanical completion, the business value of the resulting outcome can be significantly diminished. Given the enormous constraints, most teams are content to take three runs at a "single point" target as part of the initial process validation. A great example from industry of missed opportunities to build robust specifications comes from a filling validation effort that used three runs from the same bulk material when diversifying the bulks used to cover a range of parameters would have defined a broader operating range that covers the expected range of lot allocation process parameters (Figure 51.1).

Engineers are characteristically driven primarily by the desire to deliver conforming qualifying lots and the pressure to hit a narrow target. There is little incentive and less luxury to try to broaden the acceptable range for critical parameters when every week of delay translates into lost market share. Since validation is often the last step, it often pays the price for upstream delays with compression pressure.

As a result, the process as transferred is not robust and can suffer from serious operability issues. Each significant excursion triggers a time-consuming and resource-intensive investigation with an array of associated experimentation

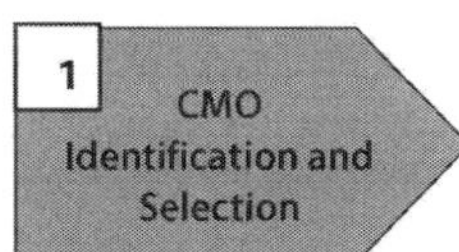

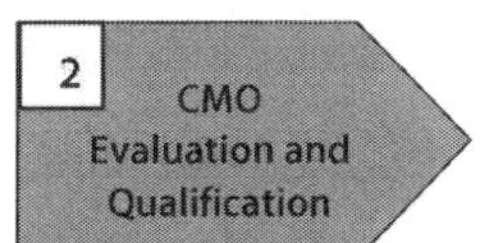

FIGURE 51.1 Generic process development timeline showing the relationship between the validation process and the validation step.

and testing to verify that there is no product impact. As this data accumulates, it may be possible to build robustness into a process after initial market approval, but no one would argue that this is the efficient, effective or desirable approach.

In recent years, most validation professionals have come to embrace the idea of a validation process that begins early in the life cycle of a product. This enhanced development process is characterized by systematic and extensive experimentation to define main effects. The design of these experiments is informed by the establishment of a quality target profile that spells out the design criteria for the product. Scientific rationale and the quality risk management process (once again using the Failure Modes and Effects Analysis) are applied to identify the critical to quality attributes and the critical process parameter candidates that will be the focus of the experimentation. Application of statistically valid design of experiments is the best way to explore two- and three-factor interactions and to fully illuminate cause-and-effect relationships.

Guidance issued in 2010 by the International Conference on Harmonisation of Technical Requirements for Registration of Pharmaceuticals for Human Use include this recommendation:

> For Design of Experiments involving single- or multiple-unit operations that are used to establish CPPs and/or to define a Design Space (DS), the inclusion of the following information in the submission will greatly facilitate assessment by the regulators:
>
> - Rationale for selection of DoE variables (including ranges) that would be chosen by risk assessment (e.g., consideration of the potential interactions with other variables);
> - Any evidence of variability in raw materials (e.g., drug substance and/or excipients) that would have an impact on predictions made from DoE studies.
> - Listing of the parameters that would be kept constant during the DoEs and their respective values, including comments on the impact of scale on these parameters.

The initial, single-target validation process that benefits fully from a normal operating range that was defined using an enhanced development effort is one characterized by the skillful use of "Design for Six Sigma" tools and concepts. By fully leveraging every opportunity for upstream testing and experimentation on the process, we can design an efficient, culminating "demonstration of robustness"—or "design space verification"—in which the scale-dependent parameters operate in a manner that is consistent with the bench and pilot scale experiments. If the verification is included along with a protocol for continuous process verification in the submittal, this can significantly accelerate and simplify post-market transfers and process improvements. In this updated scheme, the initial validation "step" itself becomes a wafer-thin element of a life cycle approach that comprises continuous monitoring and ongoing evaluation of manufacturing process performance.

51.6 HARNESSING THE FULL POWER OF THE VALIDATION PROCESS

Enhanced development as a precursor to initial validation can be achieved by applying the statistical principles first applied by Fisher (8) and then adapted to industrial uses by Box, Hunter and Hunter (9) and others, commonly referred to as "Design of Experiments" or full factorial experiments. This Design for Six Sigma (or Quality by Design) is an approach that is based largely on full application of the Good Automated Manufacturing Practices (GAMP) model: clear requirements, specifications and life cycle documents created BEFORE validation. More specifically, the process assumes that the developer will:

1. Establish critical to quality attributes (CTQs), using Define-Measure-Analyze-Design-Verify (DMADV) techniques, including FMEA and the rigorous voice of the customer (broadly defined includes operations) assessments.
2. Use design of experiments (DOE) methodology (e.g., factorial design) to translate these CTQs into specifications and control parameters by illuminating the cause and effect relationships between input variability and process control settings.
3. "Design for Robustness" implies that the experiments are designed to create a performance space/surface (design space or "sweet spot") that can be used to choose the proper input settings to achieve desired targets with minimum variability (to minimize the effects of noise on the main effects) and as the basis for ongoing statistical process control on the floor, not just to obtain licensure.

Traditional "one-factor-at-a-time" experiments are ineffective and inefficient, even when conducted early in the development process. One-factor-at-a-time repeated three times will not enable PAT or process optimization. An example of a traditional set of three experiments is provided in Figure 51.2

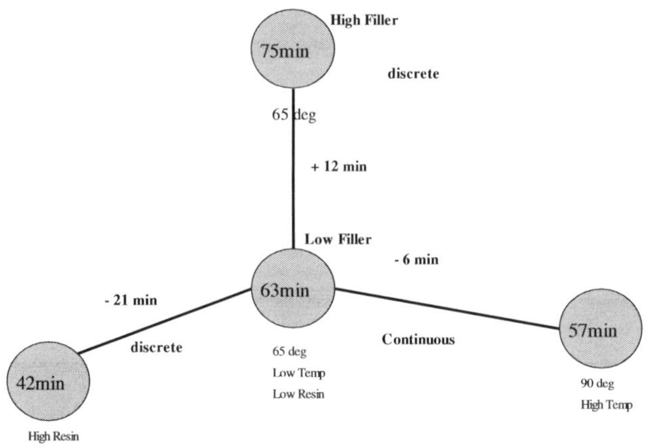

FIGURE 51.2 Traditional, one-factor-at-a-time approach to validation.

(*Torbeck*[7]). The process under development comprises intermingling of a filler and a resin at temperature. The outcome or desired effect is curing time. As depicted in Figure 51.2, a set of three experiments where only one factor varies at a time, will ensure that the results only illuminate effects along the axis. None of them reach "out in the cube."

As the FDA points out in their PAT Guidance document, "Traditional one-factor-at-a-time experiments do not effectively address interactions between product and process variables." In theory, since a design of experiments approach would entail using the same time and resources required for a one-factor-at-time approach, the change in approach should be practicable during bench and pilot scale production. Furthermore, the ICH guidance explains that "inclusion of a full statistical evaluation of the DoEs performed at early development stages (e.g., screening) is not expected. A summary table of the factors and ranges studied and the conclusions reached will be helpful."

Quoting once more from the FDA's PAT guidance document:

> Methodological experiments (e.g., factorial experiments) based on statistical principles of orthogonality, reference distribution, and randomization provide effective means for identifying and studying the effect and interaction of product and process variables.

A DoE allows us to estimate the performance of a system within its normal operating space (the cube) without having to measure system response at every point in that space. Using the curing time system described earlier and assuming that the effect is approximately linear (nonlinear systems can be tested using different techniques), we can create an equation that describes all of the effects of all the factors on curing time (the response): the effects of the factors by themselves (the three main effects), the three two-way interactions and one three-way interaction. Since we need to include the intercept of the linear equation, we end up with eight coefficients that need to be estimated with eight experimental runs. Since the system is approximately linear, we can assign high and low values to our three factors (−1 for low and +1 for high) and derive the following design (Table 51.1):

TABLE 51.1
A Three-Factor, Two-Level Experiment

Run (Yates' Name)	Temperature	Resin Type	Filler Type
(1)	−1	−1	−1
a	+1	−1	−1
b	−1	+1	−1
ab	+1	+1	−1
c	−1	−1	+1
ac	+1	−1	+1
bc	−1	+1	+1
abc	+1	+1	+1

This experimental design would be referred to as a "full factorial" since it contains all the possible combinations of the three factors ($2 \times 2 \times 2 = 2^3 = 8$ unique combinations, corresponding to the eight corners of our cube).

Having verified the capability of the measurement system, the engineer is ready to run the eight experiments. However, it is often frugal and wise to start with a "fractional factorial," especially if we have evidence that some effects do not need testing (three-way interactions, in non-biological systems at least, are rare). In the curing time system, four experiments would suffice if, as depicted in Figure 51.3, we substitute the High Filler/High Resin/High Temp experiment (out in the cube) for the Low Filler/Low Temp/Low Resin experiment (which was the origin of the cube we had already explored).

As it happens, High Filler/High Resin/High Temp results in a significant (28 minutes) impact on curing time. By running only four of the eight runs, we achieve a 2^{3-1} fractional factorial design (the "−1" indicates that we are only doing half of $2 \times 2 \times 2 = 8$).

An example of a 2^3 full factorial design for developing a tablet for ingestion is shown in Figure 51.4. In this case, the desired effect is a target % of active ingredient bound after a certain time after simulated ingestion, and there is no obvious opportunity to be frugal and get away with a fractional design. The three variables in this case (*Torbeck*[10]) are plasticizer (A or B), total spray time, and dry time (one discrete and two continuous variables).

The very model of the modern validation engineer applies their Six Sigma training by entering the design and the results into a software package (like Minitab) that can rapidly run an Analysis of Variation (ANOVA) to see if the differences in the effects and their coefficients are significant and to generate main effects plots along with interaction plots.

As shown in Figure 51.5, total spray time is the critical process parameter that needs to be carefully controlled. The other variables can vary across the specification ranges, and the response is still within its acceptance criteria (the system can be said to be robust relative to those variables).

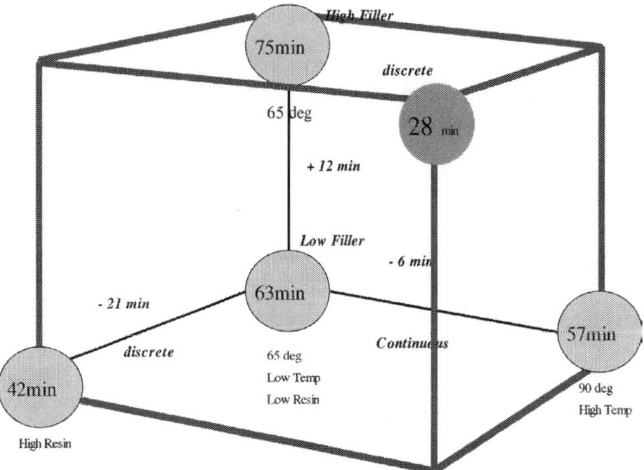

FIGURE 51.3 A Fractional Factorial Design with Results "out in the cube".

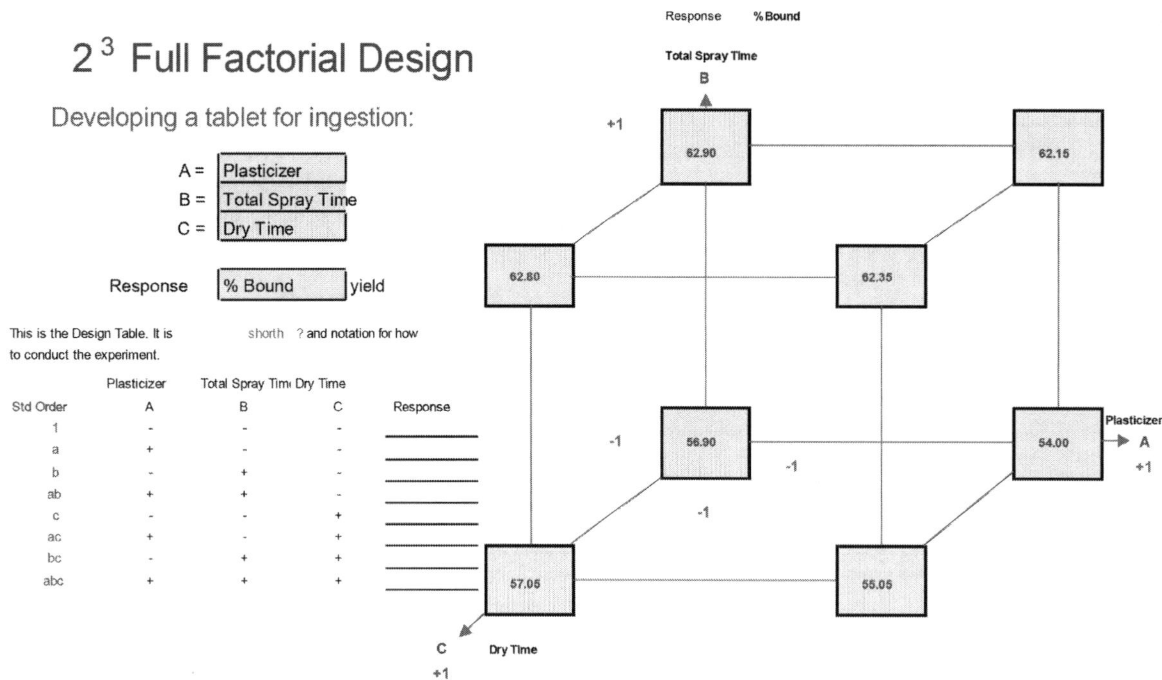

FIGURE 51.4 A full factorial design.

always to low to high

Std Order	Plasticizer A	Total Spray Tir B	Dry Time C	AB	AC	BC	ABC
1	-56.9	**-56.9**	-56.9	56.9	56.9	56.9	-56.9
a	54	**-54**	-54	-54	-54	54	54
b	-62.9	**62.9**	-62.9	-62.9	62.9	-62.9	62.9
ab	62.15	**62.15**	-62.15	62.15	-62.15	-62.15	-62.15
c	-57.05	**-57.05**	57.05	57.05	-57.05	-57.05	57.05
ac	55.05	**-55.05**	55.05	-55.05	55.05	-55.05	-55.05
bc	-62.8	**62.8**	62.8	-62.8	-62.8	62.8	-62.8
abc	62.35	**62.35**	62.35	62.35	62.35	62.35	62.35
Sum	-6.1	**27.2**	1.3	3.7	1.2	-1.1	-0.6
Effect	-1.525	6.8	0.325	0.925	0.3	-0.275	-0.15
% Effect	-2.578191	**11.496196**	0.5494505	1.5638208	0.5071851	-0.46492	-0.253593

Note the % effect is the effect divided by the average times 100.

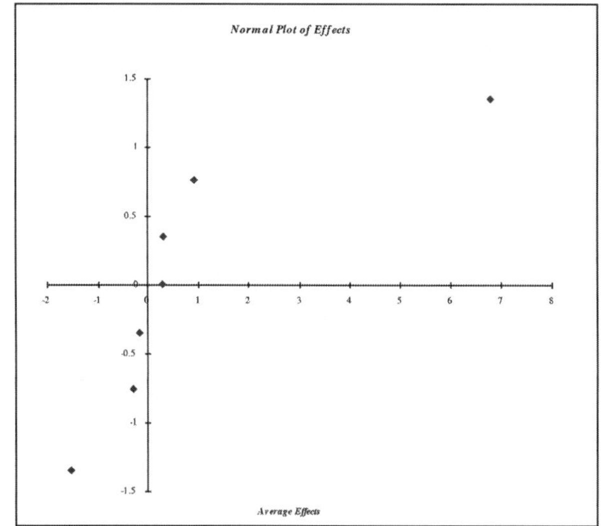

FIGURE 51.5 Results of a 2^3 full factorial design showing that total spray time is the critical process parameter.

51.7 CONCLUSION

By designing the experiments performed during the development of the process in this way, the opportunity to transfer a process that is truly robust can be seized. The parameters can be set at the optimum values during the initial validation at scale, and the "demonstration of robustness" can be used to verify the design space and lay the groundwork for continuous monitoring.

The primary benefit of the approach lies in the ability it confers to distinguish "signal"—which should be carefully validated and controlled—from "noise"—which can be cut from the scope of validation. By control charting the signal, or critical, parameters, wasteful investigations that conclude with a finding of "no product impact" can be avoided, freeing up constrained technical resources to concentrate on optimizing yield and stability.

As summarized by FDA in its PAT guidance, "When used appropriately, the [factorial experiments and other] tools described above can help identify and evaluate product and process variables that may be critical to product quality and performance." Once these variables have been identified and their interactions with each other understood as necessary, we can create a clear performance space for reliable and routine operations.

ACKNOWLEDGMENTS

Steve Sharon
Dr W. Rob Arathoon
Ken Kornfield
Ron Branning
Harry Lam
Torbeck & Associates
Mr. Lynn D. Torbeck, Statistician

REFERENCES

1. Committee for Medicinal Products for Human Use (CHMP) ICH Guideline Q8, Q9 and Q10, December 2010 EMA/CHMP/ICH/265145/2009.
2. *Ibid.*
3. Taylor Frederick W. *Scientific Management: Comprising Shop Management: The Principles of Scientific Management and Testimony before the Special House Committee.* New York: Harper and Row, 1964.
4. Deming WE. *Out of the Crisis: Quality, Productivity and Competitive Position.* Cambridge, MA: Cambridge University Press, 1982.
5. Juran JM. *Juran's Quality Handbook.* 5th ed. New York: McGraw-Hill, 1998.
6. Crosby P. *Quality Is Free.* Nairobi, Kenya: Mentor Publishing, 1979.
7. Breyfogle Forrest W. 1999. Implementing Six Sigma.
8. FDA. *Guidance for Industry: PAT: A Framework for Innovative Pharmaceutical Manufacturing and Quality Control.* August 2003.
9. Torbeck and Assoc. Training Course. "Validation by Design®".
10. Fisher RA. *The Design of Experiments.* Edinburgh: Oliver and Boyd Ltd, 1951.
11. Box G, Hunter W, Hunter S. *Statistics for Experiments.* New York, NY: John Wiley and Sons, 1978.
12. Torbeck L, Branning R. Designed Experiments: A Vital Role in Validation. *Pharmaceutical Technology,* 1996 June, 20, 108–114.
13. Christine MV, Moore, PhD. Quality by Design: FDA Lessons Learned and Challenges for International Harmonization. International Conference on Drug Development, Austin, TX, 2012.

52 Validation and Contract Manufacturing

Michael C. Beckloff

CONTENTS

52.1 INTRODUCTION

Since writing this chapter some many years ago, as is true for many of us in the pharmaceutical industry, I have been through several career transitions. I "retired" from big corporate "life" and co-founded a small start-up company focused solely on developing and commercializing pediatric friendly medications. The goal of the start-up was to provide commercially available, high quality CGMP complaint, clinically relevant products to this underserved patient population. We were focused on changing the way pediatric medicine is delivered in the United States. Approval of pediatric drugs in the United States is a very complex proposition. The United States, as the world's global leader in pharmaceuticals, really does need to do a better job with pediatric medicine.'

Our start-up merged with another small pharmaceutical company focused on the geriatric space. We are now a single company serving two important and underserved patient populations—pediatrics and geriatrics.

As previously mentioned, serving these patient populations created a unique set of circumstances with respect to Contract Manufacturing. In our initial start-up, we had a very specific strategy for product development. All products and primary analytical methods were developed in-house in our formulation and analytical development laboratory. All scale-up, manufacturing, and process validation activities were outsourced to various Contract Manufacturing Organizations (CMOs). This strategy allowed us to gain FDA approval and commercial launch of five pediatric products, all 505(b)(2) NDAs, in just under seven years.

Some of the unique challenges we faced with outsourcing included the following:

- The small batch sizes and low annual forecast volumes associated with pediatric products was unattractive to many of the larger upper-tier contract manufacturers.
- There is typically a steep drop-off in capabilities that occurs between the upper-tier and many of the smaller contract manufacturers. This includes:
  - General facilities, construction, equipment, upkeep, etc.
  - Availability of in-house expertise and skill sets
  - Capacity and throughput constraints
  - Ability to leverage specialized expertise across the organization
  - Actual commercial product experience, including FDA validation expectations
  - Level of FDA interactions and understanding of documentation requirements.
- With the larger organizations, the most capable performing staff ("A Team") is typically allocated to large pharma projects.

For a large percentage of pharmaceutical and biotechnology industry professionals, outsourcing of contract manufacturing services has become a common occurrence—essentially, a "way of life". One would be hard pressed to find a seasoned industry veteran who has not been involved in outsourcing of contract manufacturing services at some point in their career. The war stories are varied and span the full spectrum of results, from complete success to complete failure, with most falling somewhere in between.

Outsourcing of contract manufacturing services continues to be an integral component of strategic drug development, allowing the industry to effectively deal with technical capabilities, capacity, and timing issues it faces while struggling to find pathways to introduce new drug products to the market by more efficient, effective, and cost-competitive means. Contract manufacturing is utilized by all size companies. From big pharma and biotech, to generic houses, to tiny virtual companies—every size and shape company imaginable uses contract manufacturing as a means to bring products to market. Outsourcing of contract manufacturing

DOI: 10.1201/9781003163138-52

services occurs everywhere and for many different reasons. Domestically and internationally, contract manufacturing and process validation of the drug substances and drug products that they manufacture continues to be at the forefront of the pharmaceutical industry. Industry professionals continue to gain more experience with people, processes, and technologies to leverage contract manufacturing effectively and efficiently.

Successful validation of pharmaceutical processes within the contract manufacturing environment has a set of unique and complex challenges that intra-company programs do not face. As is true for all successful programs, validation of pharmaceutical processes via a CMO requires considerable planning and execution. Unfortunately, there are very few, if any, contract manufacturing "turn-key" operations that allow one to return when the product and process are fully validated, in the "box," launched, and ready to ship to the pharmacy. Successful contract manufacturing programs require full involvement and full participation on the part of the sponsor. After all, a contract manufacturer is just that—a contract manufacturer. The contract manufacturing industry is much more adept and efficient at exploiting a well-defined transferable process as compared to innovating a product and process that is poorly defined and characterized.

The critical success factors required when validating a pharmaceutical process using a contract manufacturing strategy are numerous and complex. Without exception, successful process validation begins with excellent product and process development. There is no substitute for a rugged, well-characterized, and scientifically sound product and process. If the goal is to "get it done" with a contractor as soon as possible, a properly developed process will allow for successful transfer, scale-up, and process validation with the least amount of starts, stops, and cost overruns. Sounds simple, and conceptually it is, but time and time again drug development programs are hindered by inadequate scientific design, lack of sufficient data, poorly executed technology transfer, and ultimately delayed or failed process validation. As the cliché goes, it seems there is never time to do what is necessary to get it right the first time, but there is always time to repeat it.

Other critical success factors include selection and qualification of the contractor, selection and management of the development team, planning and program management, and effective and efficient management of the regulatory requirements that drive and control the entire pharmaceutical and biopharmaceutical industries.

52.2 PRODUCT AND PROCESS DEVELOPMENT

The benefit-to-cost ratio must always be considered, and as obvious as it would seem, it is important to recognize that validation of pharmaceutical processes begins at the beginning. Successful validation hinges on a complete and thorough understanding of the product and the production process. Preformulation work, analytical methods development, production of Phase 1–3 Clinical Trial Material (CTM), scale-up, production of engineering batches, production of registration batches, and finally production of process validation batches must all "add up." A large body of meaningful data is generated during the product development process. Continuity, coupled with recognition and analysis of existing data, is critical to successful process validation. These data must be managed and properly communicated to the contract manufacturer. Conversely, data generated by the CMO must be properly communicated to and analyzed by the sponsor. It is the job of the sponsor to effectively manage and guide technology transfer from the development laboratory to the CMO.

Successful validation is constructed as part of a cumulative progression coordinated by the sponsor based upon what is learned and understood about the product as it is developed. Process validation cannot be completely "delegated" to a CMO; it must be effectively managed by the sponsor. It is the sponsor's job, not the CMO's, to ensure that all of the necessary data have been generated to support technical transfer, scale-up, and ultimately process validation. Without proper the proper science and planning, nothing "good" happens in process validation.

A challenge that often occurs in sponsor companies is frequent "mid-stream" change in development team personnel. When such personnel changes occur, the new sponsor team can become frustrated with a particular contractor and may move to another, believing this will solve all of the problems. In reality, this results in loss of continuity, loss of time, and cost overruns. This change of personnel occurs most frequently in small and virtual companies. If a proper process validation plan is developed early in the program, loss of continuity can be prevented.

The following tables, derived from numerous FDA guidance documents[1–16], etc., summarize the scientific and regulatory expectations for new chemical entities, both active pharmaceutical ingredients and finished dosage forms, from early Phase 1 through approval of the registration application. Of course, each product is unique, with a different set of specific requirements; however, these tables provide a point of reference and a place to begin to plan and develop a well-defined and characterized product and process. To a large extent, efficient product and process development is about understanding minimum requirements both from scientific and regulatory points of view. Minimum requirements should not be confused with minimum effort. Successful process validation in the CMO environment is more about understanding what is "mission critical" with respect to the product and process—what studies need to be designed and executed and how the data will be positioned, from a regulatory point of view, as the product proceeds from the pre-IND stage through clinical development, process validation, and finally to the commercial marketplace. The process is complex, intertwined, and very much interdependent. Successful development teams understand this interdependence and consider the product and process development, analytical methods development, and regulatory strategy in parallel, resulting in an efficient and effective development and process validation plan.

TABLE 52.1
Scientific and Regulatory Expectations, Drug Substance

Drug Substance

	Phase 1	Phase 2	Phase 3	NDA
General:				
Code Number (CADS Registry Number)	•	•	•	•
United States Adopted Name (USAN)	•	•	•	•
Chemical Name(s)	•	•	•	•
Compendial Name	—	—	—	—
Common (or other) Name(s)	—	—	•	•
Molecular Formula	•	•	•	•
Molecular Weight	•	•	•	•
Chemical Structure, Including Stereochemistry	•	•	•	•
Appearance, Color, and Physical State	•	•	•	•
For Proteins Add:				
AA Sequencing	•	•	•	•
Description (Disulfide bonds, shape, subunits, number of AA residues)	•	•	•	•
Biological Activity	•	•	•	•
Physicochemical Properties:				
Solubility (e.g., water, ethanol, ether)	•	•	•	•
pH Solubility Profile	—	•	•	•
pKa	•	•	•	•
Dissociation Constant	•	•	•	•
Bioactivity	•	•	•	•
Partition Coefficient	—	—	•	•
Hygroscopicity	—	•	•	•
Melting Point/Boiling Point	•	•	•	•
X-Ray Diffraction/Single Crystal	—	•	•	•
Chirality/Optical Rotation	•	•	•	•
Refractive Index	—	—	•	•
Polymorph Screen/Solvate/Hydrate	•	•	•	•
Particle Size Distribution	•	•	•	•
pH of Aqueous Solution	•	•	•	•
For Proteins Add:				
Isoelectric Point	•	•	•	•
Extinction Coefficient/Unique Spectra	•	•	•	•
Biological Activity	•	•	•	•
Structure Elucidation:				
Elemental Analysis	—	•	•	•
UV Spectroscopy	•	•	•	•
IR Spectroscopy	•	•	•	•
^{1}H NMR Spectroscopy	•	•	•	•
^{13}C NMR Spectroscopy	—*a*	•	•	•
Mass Spectrometry	—	•	•	•
Physicochemical Characteristics (TGA, DSC, DTA, X-ray, Raman, etc.)	—	—	•	•
Impurities	•	•	•	•
For Proteins Add:				
ELP, CEP, IEP, SEC-HPLC	•	•	•	•
Western Blot	•	•	•	•
Method of Manufacture:				
Name and Address of Manufacturer, Drug Establishment Registration Number(s)	•	•	•	•
List of Critical Equipment	—	—	•	•
Starting Material(s)	—	•	•	•
Starting Material(s) Specifications	—	•	•	•
Reagents and Solvents	•	•	•	•
Reagents and Solvents Specifications	—	•	•	•

(Continued)

TABLE 52.1
(Continued)

Drug Substance

	Phase 1	Phase 2	Phase 3	NDA
Synthesis Scheme	•	•	•	•
Flow Diagram	•	•	•	•
Description of Process/Process Controls	—	•	•	•
In-Process Controls/In Process tests (e.g., HPLC)	—	•	•	•
Key and Final Intermediate	—	•	•	•
Reprocessing/Reworking/Recovery/Regeneration	—	—	•	•
For Biologics or semisynthetics add:				
Storage and Transportation of Intermediates	•	•	•	•
Preparation Procedures (e.g., cleaning, drying)	•	•	•	•
Isolation Processes	•	•	•	•
Holding times/Storage Conditions	•	•	•	•
Traceability Procedures	•	•	•	•
Specifications (Includes Analytical Procedures and Acceptance Criteria):				
Appearance	•	•	•	•
Identification (UV, IR, HPLC-Chiral)	•	•	•	•
Counter Ion	—	•	•	•
Melting Point	—	•	•	•
pH of Aqueous Solution (optional)	—	—	—	•
Heavy Metals	•	•	•	•
Residue on Ignition	•	•	•	•
Residual Solvents	•	•	•	•
Water Content	•	•	•	•
Microbial Limits	—[b]	—[b]	•	•
Bacterial Endotoxins	•	•	•	•
Assay	•	•	•	•
Related Substances	•	•	•	•
Primary Degradation Product and Degradation Pathway	—	—	•	•
Justification of Specifications	—	—	—	•
For Proteins Add:				
Specific Biological Activity	•	•	•	•
Purity (dimers, oxidized forms, electrophoretic)	•	•	•	•
Chiral Drug Substance:				
Chiral Identity	•	•	•	•
Chiral Assay	—	•	•	•
Enantiomeric Impurity	—	•	•	•
Reference Standard:				
Working (Preliminary)	•	•	—	—
Primary	—	—	•	•
Analytical Methods:				
Summary	•	•	•	•
Complete Description	—	•	•	•
Sample Chromatograms	•	•	•	•
Method Validation:				
Linearity	•	•	•	•
Specificity	•	•	•	•
Forced Degradation:				
Acid pH	—	•	•	•
Basic pH	—	•	•	•
Heat	—	•	•	•
H_2O_2	—	•	•	•
UV Light	—	•	•	•
Accuracy	•	•	•	•
Repeatability	•	•	•	•
Intermediate Precision	—	—	•	•

Drug Substance

	Phase 1	Phase 2	Phase 3	NDA
Reproducibility (if needed)	—	—	•	•
Robustness:				
Mobile Phase pH	—	—	•	•
Mobile Phase Composition	—	—	•	•
Detector Wavelength	—	—	•	•
Column	—	—	•	•
Solution Stability	—	—	•	•
Limit of Detection[c]	•	•	•	•
Limit of Quantitation[c]	•	•	•	•
Method Validation Package	—	—	—	•
Related Substances:				
Method Validation	•	•	•	•
Identification	—	—	•	•
Qualification	—	—	•	•
Batch Analysis Data (Certificates of Analysis):				
Toxicology Study Lots	•	•	•	•
Clinical Study Lots	•	•	•	•
Impurity Profile Comparison	•	•	•	•
Container Closure System:				
Description	•	•	•	•
Label	•	•	•	•
Specifications	—	•	•	•
Drug Master Files	—	•	•	•
Stability:				
Summary, Shelf Life, Statistical Analysis	—	—	—	•
Post-Approval Commitment	—	—	—	•
Results of Stress Stability to Support Analytical Validation	—	—	—	•
Normal (long-term) (months)	1	2–6	12	12
Accelerated (months)	1	2–6	6	6
Photostability	•	•	•	•
Development Report	—	—	—	—
Process Validation Protocol	—	—	—	—
Process Validation Report	—	—	—	—

FDA Expectation (• = Expected;—= Not Required)

[a] = May be required at Phase 1 for peptides.

[b] = Reported for sterile drug substance or drug substance intended for use in sterile drug product.

[c] = As applicable to type of method used.

TABLE 52.2
Scientific and Regulatory Expectations, Drug Product

Drug Product

	Phase 1	Phase 2	Phase 3	NDA
Formulation Summary:				
Product Name	•	•	•	•
Dosage Form	•	•	•	•
Brand Name	–	–		•
Strength(s)	•	•	•	•
Excipients	•	•	•	•
Manufacturing Process	•	•	•	•
Composition Statement (include ALL components used to manufacture, regardless if not present in final product)	•	•	•	•

(Continued)

TABLE 52.2
(Continued)

Drug Product

	Phase 1	Phase 2	Phase 3	NDA
Package Container Closure System	•	•	•	•
Storage Conditions	•	•	•	•
Development Pharmaceutics:				
Excipient Compatibility[a]	•	–	–	–
Prototype Formula Evaluation	–	–	•	•
Formula Optimization	–	–	–	•
Process Optimization	–	–	–	•
Container Closure Evaluation	–	–	•	•
Selection of Commercial Formula, Process, and Container Closure System	–	–	•	•
Container Closure Integrity (for sterile products)	•	•	•	•
Process Scale-Up	–	–	•	•
Development Pharmaceutics Summary	–	–	–	•
Product Development Report	–	–	–	–
Specification for Drug Substances:	•	•	•	•
Specifications for Excipients:				
Compendial Status	•	•	•	•
DMFs for Noncompendial	•	•	•	•
Formula:				
Unit Formula	•	•	•	•
Batch Formula	•	•	•	•
Preservative Effectiveness Testing	•	•	•	•
Manufacturer (Including Packager, Labeler, Testing Laboratories):				
Name	•	•	•	•
Address	•	•	•	•
Establishment Registration Numbers	–	•	•	•
Identification of Processing Rooms and Filling lines (for sterile products)	•	•	•	•
Manufacturing and Packaging:				
Manufacturing Process Summary	•	•	•	•
Manufacturing Process Flowchart	•	•	•	•
Manufacturing Process Description	–	•	•	•
Process controls	–	•	•	•
Environmental Controls	–	–	–	•
Movement of Raw Material, Personnel, Waste, and Intermediates in and out of Manufacturing Areas (for protein and sterile products)	–	–	–	•
Potential Contamination with Adventitious Agents (for protein products only)	•	•	•	•
In-Process Controls/Critical Processing Variables/Justification	–	•	•	•
Type of Equipment for Each Unit Operation	–	–	•	•
Process Validation	–	–	–	–
Hold-Time Qualification	–	–	•	•
Reprocessing/Reworking	•	•	•	•
Executed Batch Production and Control Records	–	–	–	•
Master Production and Control Records	–	–	–	•
Specification (Including Analytical Procedures and Acceptance Criteria) and Justification:				
Appearance	•	•	•	•
Identification	•	•	•	•
Content Uniformity/Fill Weight	•	•	•	•
Related Substances	•	•	•	•
Primary Degradation Product	–	–	•	•
Loss on Drying	•	•	•	•
pH	•	•	•	•
Particulate Matter (for parenterals)	•	•	•	•
Volume in Container (for parenterals)	•	•	•	•
Viscosity	•	•	•	•
Osmolarity/Osmolality	•	•	•	•

Drug Product

	Phase 1	Phase 2	Phase 3	NDA
Microbial Limits	–	–	•	•
Sterility	•	•	•	•
Bacterial Endotoxins	•	•	•	•
Dissolution	•	•	•	•
Moisture Content	–	•	•	•
Polymorph (if needed)	–	–	•	•
Other Dosage-Form-Specific Tests as Needed	–	–	•	•
Analytical Methods:				
Summary	•	•	•	•
Complete Description	–	•	•	•
Sample Chromatograms	•	•	•	•
Method Validation:				
Linearity	•	•	•	•
Specificity	•	•	•	•
Forced Degradation:				
Acidic pH	–	–	•	•
Basic pH	–	–	•	•
Heat	–	–	•	•
H_2O_2	–	–	•	•
UV Light	–	–	•	•
Accuracy	•	•	•	•
Repeatability	•	•	•	•
Intermediate Precision	–	–	•	•
Reproducibility (if needed)	–	–	•	•
Robustness:				
Mobile Phase pH	–	–	•	•
Mobile Phase Composition	–	–	•	•
Detector Wavelength	–	–	•	•
Column	–	–	•	•
Solution Stability	–	–	•	•
Limit of Detection[b]	•	•	•	•
Limit of Quantitation[b]	•	•	•	•
Dissolution Sink Conditions	•	•	•	•
Method Validation Package	–	–	–	•
Batch Analysis Data (Certificates of Analysis):				
Experimental Batches	•	–	–	–
Clinical Study Batches	•	•	•	•
Bioavailability Batch	–	–	•	•
NDA Registration Batches	–	–	–	•
Container Closure System:				
Description	•	•	•	•
Specifications	–	–	•	•
DMF Letters of Authorization to Reference	–	•	•	•
Stability:				
Normal (long-term) (months)	1	2–6	12	12
Intermediate (months)	–	–	–	6
Accelerated (months)	1	2–6	6	6
Photostability	–	–	•	•
Thermocycling				•
Stability Summary, Shelf Life, Statistical Analysis	–	–	–	•
Long-term Stability Commitment	–	–	–	•
Label and Labeling	•	•	•	•
Environmental Assessment	–	–	–	•

FDA Expectation (• = Expected;—= Not Required)

[a] = Should be performed as early as possible.

[b] = As applicable to type of method used.

52.3 THE "RIGHT" PEOPLE IN THE RIGHT JOBS AND THE "RIGHT" TEAM EXPERTISE

At the larger pharmaceutical and biotech companies, the knowledge, experience, and the know-how of managing a process validation program at a CMO are usually in place at the upper levels of the management pyramid. The actual implementation of the program is left to lower level and sometimes inexperienced personnel. The personnel at these levels frequently struggle with the CMO process validation programs. In part, this is because their assignment to the project serves as a training mechanism to allow these personnel to obtain necessary experience and expertise, and also because CMO validation programs can be extremely challenging and consume significant amounts of time and effort—commodities that the upper level managers generally do not have available, given the other programs and projects they must direct. With proper oversight from experienced supervisory management, the job of validating a process at a CMO is usually accomplished more effectively and efficiently by large companies as compared to the small, less experienced companies, despite the assignment of junior level personnel for many, if not all, of the required tasks.

An important and distinguishing feature between the large and small companies is that the large-company personnel, although somewhat inexperienced, usually have the appropriate educational and scientific expertise available within their own organization to assist, which may allow them to better understand the validation requirements and principles. The smaller companies may not have sufficient internal technical expertise to properly direct the CMO process validation programs. This is further complicated by the fact that the upper levels of management in some small companies may also be lacking in these areas, resulting in very limited guidance. The rule, rather than exception, for small and virtual companies seems to be to expect people to run external CMO programs for which they have very little or no experience and/or training. For example, the staff toxicologist is asked to oversee the development, manufacturing, and process validation programs; the director of clinical development is asked to handle manufacturing; and the staff project manager is asked to oversee the product development and manufacturing plan. Although it might seem counterproductive, this occurs frequently and forces small companies into the difficult position of being more dependent on the CMO for technical support and guidance. For these reasons, smaller pharmaceutical and biotech companies generally have more difficulty managing and directing a process validation program at a CMO. Small companies scramble to conserve cash and raise funding; the clinical work is the glamorous "big bucks" Wall Street stuff; manufacturing and validation is usually an afterthought, and after all . . . "How hard can it be"? The answer is, of course, hard enough!

Having the "right people" in the "right jobs" during a CMO process validation effort is critical to the overall success and efficiency of the program. Establishing a program director with the experience and training to specifically "direct" the project as compared to using a less-experienced project manager to "manage tasks" is critical to designing and executing an efficient and effective process validation program. There is a significant value in having someone who understands product development and process validation directing the program as compared to using a project manager who may or may not have the necessary experience. Many projects end up being driven by inexperienced project managers managing task lists and timelines via Gantt charts rather than understanding the underlying science and regulatory requirements that should be driving the project. This practice leads to inefficiencies and cost overruns. The best program director is one who understands the specific production process, the critical parameters that must be evaluated during process validation, and the associated regulatory requirements that can ensure product registration and approval.

Finely tuned project teams are critical to successful product development. In the contract manufacturing environment, teams are divided into two distinct groups: first the sponsor's development team and second the contract manufacturer's project team. The secret to success is to get both teams to integrate and function as a single unit with the same goals and objectives. It is important for the sponsor and CMO to develop an attitude of cooperative partnership focused on the 'sponsor's mission and the benefits to the patients rather than one of only a "vendor" and a sponsor. Sponsors must understand that the contract manufacturing business is a very competitive and high-overhead business model. To survive, CMOs must operate on the volume of multiple products and multiple customers. The sponsor must appreciate this and understand that they are not the only customer the CMO must satisfy. Conversely, the CMO must understand that the sponsor has put a huge amount of trust and expense into the experience, expertise, and integrity of the CMO—in effect putting their "lives and jobs" in the hands of the CMO. It is a very challenging dynamic, but once again, having the goal of developing a cooperative, mission-critical partnership rather than one of only a "vendor" and a sponsor should be considered as a critical success factor.

Clearly, an "A-Team" is needed for all projects. What does an "A-Team" look like? On the sponsor's side, it is likely that the project will be high profile within the company; it will likely be a very high priority; and as in the case of the small and virtual companies, it may be the ONLY project on which the company is working and the ONLY way the company will survive. On the contractor's side, the sponsor's product will likely be only one of many projects on which the CMO is working. The sponsor "ponders" its only project as the CMO "ponders" how to get the next customer's product and project in and out the door. Ideally, the "A-Team" for a project would consist of the sponsor's project director along with a key decision maker from the CMO's side. In addition, it is helpful to have representation from both the sponsor and CMO in the areas of project management, synthesis and/or pharmaceutics, manufacturing, analytics, and regulatory. It is imperative to have the correct and complementary expertise on the sponsor's side as well as the CMO's side. Successful process validation is driven by proper and effective science. There must be proper technical expertise within the team on both sides. Finally, it is critical that the sponsor establish a person-in-the-plant for the engineering registration and process validation runs and probably the first several commercial production runs. Murphy's Law was apparently designed for process validation activities in the CMO environment. Having a person-in-the plant will help to raise the awareness of the CMO as process validation is initiated and will allow for more precise and faster response times should problems arise during the validation runs.

Regular meetings are critical to monitor progress, assign actions, and troubleshoot problems. Teleconferences and videoconferences are useful and cost-effective tools, but regular face-to-face meetings are extremely valuable and should not be overlooked or avoided. Face-to-face meetings also allow for overall team building and aids in helping to develop the cooperative partnership necessary for a successful sponsor/CMO relationship.

Because small and virtual companies may not have the expertise and resources and because the large companies may need to outsource lower priority projects, it may be necessary to use consultants or consulting groups to help manage the development and process validation programs. In hiring and managing consultants, it is important to determine whether they will be able to give the project the time and attention the project will demand. Like the CMO business model, it is important to also understand that the consulting business model is a time-based business model that requires the consultant to work on multiple projects for multiple clients. There are three primary considerations that need to be addressed when identifying the "right" consulting expertise:

- Does the consultant have the proper training to assist with the process validation program?
- Does the consultant have adequate time to devote to the process validation effort, and are they willing to guarantee this via a consulting contract?
- Has the consultant actually participated in this type of work previously, and how many times? Have they actually done this specific work, and if so, are they willing to provide references?

Assuming these three primary considerations are met, a fourth and extremely critical factor to consider is whether the consultant has the type of personality that will allow for effective integration into the project team. The consultant must be integrated into the overall cooperative partnership that is necessary for successful process validation. The sponsor must recognize that the consultant may, in fact, be spending more one-on-one time with the CMO than the sponsor. The sponsor must be certain that the consultant has the necessary interpersonal skills to build and maintain the proper sponsor/CMO relationship.

52.4 SELECTION AND QUALIFICATION OF THE CONTRACT MANUFACTURER

Thorough development of the product and the process can be the single most important critical success factor associated with technology transfer to the CMO and process validation. The second critical success factor that must be considered is the selection and qualification of the CMO. Selection, qualification, and management of the CMO can be divided into four primary phases, as illustrated in Figure 52.1:

CMO identification and selection is largely a matter of searching databases, reviewing trade journals, and attending trade shows, obtaining recommendations from consultants, industry colleagues, and contacts or direct personal experience. In some cases, prior provider relationships may exist, which may also serve as a source to identify an appropriate CMO. The most significant of these is personal experience, followed by recommendations from industry colleagues—have you worked with CMO X, and what was the outcome? What are their strengths, and what are their weaknesses? During this preliminary screening process, it will be necessary to evaluate multiple CMOs to ensure that a truly viable candidate will be identified. The practice of preparing a detailed Request for Proposal (RFP) and asking the top three to five CMO candidates to participate has proven to be a valuable tool.

Once potential CMOs have been identified, a detailed evaluation and qualification process must follow. Due diligence is often given inadequate consideration; however, the importance of this "homework" cannot be understated. Although thorough due diligence can require a significant amount of time and expense, this activity must be considered as an important investment toward successful technology transfer, scale-up, and process validation. Due diligence will pay great dividends to help ensure smooth technology transfer, scale-up, and process validation. Unless the team has personal experiences, each potential CMO should be visited by the CMO selection team as part of the due diligence effort. Some of the preliminary due diligence process can be handled using

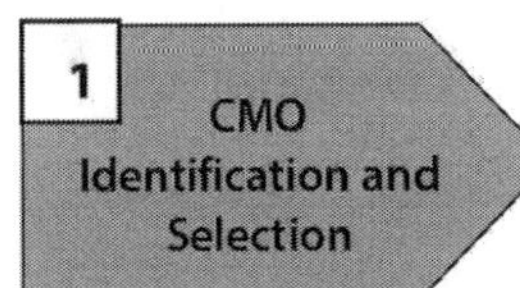

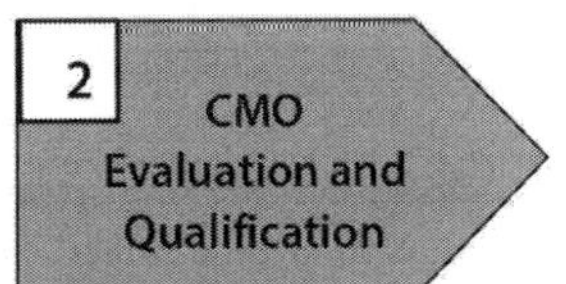

FIGURE 52.1 Primary phases of CMO selection, qualification, and management.

initial telephone interviews and questionnaires provided to the CMO for completion and return. The due diligence process must include a site visit by the CMO selection team. The site visit to the CMO gives the sponsor a "real-time" chance to evaluate the CMO and begins the process of developing a cooperative partnership.

Once a Confidentiality Agreement has been executed and appropriate technical information has been exchanged, a number of questions must be asked, answered, and considered as the selection process moves forward. A representative list of the types of questions that should be asked is provided next. Every product is unique, with an individual set of requirements; therefore, the following list should not be considered all-inclusive but rather should be used as a guide to help the CMO selection team to think through the specific requirements for the product to undergo technology transfer, scale-up, and validation.

Due Diligence Checkpoints

- Does the CMO currently manufacture products for the commercial market?
- How many and what types of products are produced by the CMO, and what are the commercial requirements for the products produced?
- What is the current capacity of the CMO for development projects?
- What is the current capacity of the CMO for commercial production?
- What is the projected capacity of the CMO for a development project at the time it will be ready to go to commercial production?
- How many customers does the CMO currently service?
- What is the procedure by which the CMO handles and manages technology transfer and process scale-up?
- What is the CMO's track record regarding regulatory inspections?
- What issues were identified in the last three years of regulatory inspections, and what Form FDA 483 observations have been cited?
- What are the typical lead times required by the CMO?
- Does the CMO track on-time completion of projects? If so, what is the ratio or percent?
- What are the typical costing mechanisms used by the CMO?
- Has the CMO been involved in recall situations, and if so, what were the details of the recalls?
- Does the CMO have the capacity to schedule the project in the time frame needed by the sponsor?
- By what procedure does the CMO typically prepare, review, and approve process validation protocols?
- With how many successful preapproval inspections (PAIs) has the CMO been involved?
- What is the sponsor's general impression of the CMO leadership team? Do they appear to be cooperative and knowledgeable, and do they appear to have integrity?
- What is the financial stability of the CMO?
- By what processes does the CMO handle project-scope changes and cost overruns?
- How does the CMO typically handle intellectual property aspects of a contract?
- Does the contract manufacturer have the appropriate equipment and experience to allow the process and technology to be transferred?
- Does the CMO have adequately designed production facilities to accommodate the proposed manufacturing process?
- Does the CMO have adequate facilities to accommodate the packaging requirements of the product?
- Does the CMO have adequate analytical laboratory facilities, equipment, and personnel to perform the required analytical testing?
- Does the CMO have appropriately qualified scientific personnel to transfer, scale-up, and validate the process?
- What is the typical or standard practice of the CMO regarding project teams?
- Do the project teams include in-house project managers?
- How many projects do the in-house project managers typically handle at once?
- How much actual industry experience do the project managers typically have?
- How does the CMO handle project management? What is the typical process for project management?
- Do the project teams include all crucial disciplines needed for successful project planning and execution?
- What does the CMO project and development team look like, and how is it established?
- Does the CMO have adequate numbers of trained staff in the production, laboratory, packaging, project management, quality assurance, and regulatory units?
- How does the CMO handle the preparation, maintenance, and submission of regulatory documentation? Are examples available for review?
- Does the CMO produce penicillin, cephalosporin, cytotoxic, or hormone products in the facility or anywhere on the manufacturing campus?
- Are investigational and/or early development products produced by the CMO, and are these products produced on common production equipment?
- How are investigational products evaluated, campaigned, and controlled within the facility?
- How is the change control process managed by the CMO?
- How are Quality Agreements handled and managed by the CMO? Is an example Quality Agreement available for review?
- Does the Quality Agreement cover the items and issues required, and meet the sponsor's needs?

- Is the CMO open to negotiation regarding the Quality Agreement?
- Is the CMO willing to issue a Certificate of Compliance for each batch produced?
- Does the CMO have experience writing development, technology transfer, scale-up, and process validation reports? Are examples available for review?
- Is the CMO willing to provide three to five customer references?
- How does the CMO manage and qualify vendors for active pharmaceutical ingredients (APIs), excipients, container/closures, packaging materials, and labeling components?

Initial due diligence should evaluate these items and provide the basis of a targeted matrix from which the most appropriate CMO can be selected. Once the CMO is initially qualified, the final component of the due diligence effort requires that a detailed and thorough Quality Systems audit be conducted by a qualified Current Good Manufacturing Practice (CGMP) auditor. Assuming the audit yields satisfactory results, contract negotiation can proceed.

Contract negotiation can be a long and arduous task. The legal challenges are many and complex, requiring significant time to complete. For this reason, assuming the initial CMO qualification is favorable, it may be beneficial to proceed with contract negotiation in parallel with some of the more detailed due diligence activities. The larger and more capable CMOs may have specific contractual requirements that may prove difficult. Identifying these issues as early as possible may save significant time should a legal impasse be reached, allowing the sponsor to "fall back" to a second qualified CMO candidate. Two critical considerations in the contractual negotiations are the willingness of the CMO to manage intellectual property issues and whether the CMO is willing to accept penalty clauses in the contract for project delays, quality issues, execution flaws, and potential regulatory deficiencies. Conversely, it is also advantageous to consider building incentives into the contract to reward the CMO for meeting or exceeding timelines that meet development, validation, and commercialization goals. Building such details into the contract will help the sponsor establish accountability and allow for better performance management of the CMO.

Managing the performance of the CMO and subsequent process validation effort is largely the responsibility of the sponsor's project team, and in particular the program director. The program director must drive the project and make the necessary decisions. The program director must monitor contractual obligations, quality requirements, and regulatory compliance continually during the course of the project. It is critically important that the program director work to ensure that a thorough project plan with detailed timelines, metrics, regular project meetings, and detailed action assignments is paramount to successful technology transfer, scale-up, validation, and commercialization of the sponsor's product.

52.5 REGULATORY CONSIDERATIONS AND MANAGEMENT

In addition to the CGMP requirements and regulatory expectations described earlier, there are several key regulatory issues that must be considered. Successful product approval ultimately depends on a successful FDA Pre-Approval Inspection (PAI). During this inspection, in addition to the general CGMP regulatory requirements, the FDA investigator will conduct a detailed data integrity review of the information submitted in the registration application, will evaluate the product development process and report, and will thoroughly evaluate the process validation data and final approved report. A successful PAI depends on the integrated efforts of the entire project team including sponsor and CMO members. It is the program director's role to ensure that all regulatory and data requirements have been properly addressed during the course of the project. The development and process validation reports will help guide the FDA investigator during the PAI. In this regard, it is extremely important that the final reports be prepared as the development program is underway and not left for the end of the project. Regular and detailed audits of raw data as well as final reports must be conducted to ensure that nothing is overlooked. When the team is confident that all required information is in order, a PAI readiness program and audit should be considered. This activity will serve as a dress rehearsal and will give the team confidence that nothing has been overlooked.

The unfortunate reality is that successful registration and product approval is simply the beginning of a continuing road of regulatory and scientific assessment, strategy, and compliance. Following launch, it is often necessary to make changes in the process to improve efficiency or allow for additional production scale. With Annual Reports, Changes Being Effected (CBE)-0, CBE-30, Prior Approval Supplements, and site changes or additions, batch failures, stability failures, and analytical methods changes, the CMO and process validation requirements must be constantly monitored and managed by the sponsor. In effect, changes in the process are managed using many of the same product and process development and program management techniques previously described.

52.6 TECHNICAL AND REGULATORY DOCUMENTATION

The amount of technical and regulatory documentation required for new drug product approval is extremely comprehensive. A new drug product often requires 8–12 years for development and regulatory approval. The data generated over the course of development and commercialization are enormous. Evaluating and managing these vast amounts of data can be a challenging proposition even in the context of a product developed using internal resources. The management and control of the associated documentation becomes even more complex when a CMO is entered as a variable in the drug development equation. The experience, document control systems, change control systems, and quality systems of each CMO largely dictate what the CMOs capabilities are with respect to technical and regulatory documentation. The smaller CMOs may be somewhat limited and inexperienced in handling documentation, which requires much more oversight and guidance, while the larger CMOs can accommodate more complex documentation issues via more mature technical and regulatory management systems.

Critical validation documentation can be divided into the following broad categories:

- Facilities Qualification and Validation
- Production, Packaging, and Laboratory Equipment Qualification and Validation
- Computer Systems Validation
- Cleaning Validation
- Production Process Validation
- Packaging Process Validation
- Analytical Methods Validation
- Sterile Product Sterility Validation

The adequacy of regulatory compliance and supporting validation documentation for facilities, equipment, and computer systems should be handled by thorough CGMP audits of the CMO. Many times, this is assessed during the initial due diligence phase of CMO evaluation. Following the initial CGMP evaluation, ongoing evaluation of these systems can be managed through a well-designed CMO change control system.

One of the more complicated CMO technical and regulatory documentation challenges deals with customer confidentiality issues. In the vast majority of cases, the CMO is bound by confidentiality agreements that prevent disclosure of technical scientific and validation data to third parties. This confidentiality restriction is most apparent in the area of cleaning validation and extends to the chemical identity of compounds, specific processes, and operating parameters, as well as the specifics of the analytical methods used to evaluate the effectiveness of cleaning procedures. Assuming that confidentiality must be maintained, and multiple drug products for multiple CMO customers are manufactured using common production equipment; it is practically impossible to unequivocally confirm that the CMO's cleaning validation program and data are acceptable. The product-specific cleaning validation protocols and reports simply cannot be provided for review by the CMO due to confidentiality. In this situation, the only option is to rely on the CMO's experience and regulatory track record. To assess the CMO's experience and understanding of the regulatory requirements, there must be a careful review of the cleaning validation policy and the Standard Operating Procedures (SOPs) used to drive cleaning validation for all products and equipment. In effect, the specific procedures and systems used to support cleaning validation for a specific product must be extrapolated to the other products produced on the common production lines with common equipment. The assumption is that proper cleaning validation procedures and regulatory documentation will be used for all products, based on what has been done for a specific product.

Production process, packaging process, and analytical methods validation, as well as many of the components of sterility validation, are product specific. There should be no confidentiality issues with respect to full disclosure of this information by the CMO. The CMO is often faced with data and documentation requests that are "custom" requirements based on what the sponsor believes will be needed to gain and maintain product approval or requirements that meet internal sponsor quality systems. These "customized" requirements vary greatly from sponsor to sponsor. Smaller, less experienced sponsors may make requests that are unrealistic and problematic for the CMO. Larger sponsors may make requests based on financial muscle and, again, may be unrealistic and problematic for the CMO. The CMO is faced with the struggle to find a flexible systematic approach to accommodate the many different sponsor-driven requirements while still allowing the CMO to comply with its internal quality systems. In most cases, the content and format of the technical and regulatory documentation are a negotiation point between the sponsor and the CMO and must meet the needs and requirements of both the sponsor and the CMO. If the needs of both groups cannot be met, the project will likely be destined for failure. Therefore, it is important for the sponsor to be fully engaged and proactive very early in the development process.

The product-specific technical data and regulatory documentation system should be designed to accommodate reviews and approvals by both the sponsor and the CMO. This will aid in ensuring that the sponsor has a complete and thorough understanding of what and how the CMO intends to meet the pertinent scientific and regulatory requirements. Meetings and discussions should be held in advance of development design and document preparation to make certain that both groups agree to the overall approach. The CMO should then draft the relevant documentation and circulate it to the sponsor for review and comment. This cycle should be repeated until agreement is reached. The documentation can then be distributed for final review and approval by both parties. This process will minimize unexpected surprises and possible delays that may adversely affect revenues and earnings for both parties.

It is most important that the sponsor minimize changes imposed on the CMO. Specifically, scientific content is critical to the sponsor but not necessarily the CMO's format. If the scientific and regulatory requirements are met, changes for the sake of changes should be discouraged. Minimizing such insignificant changes will assist the CMO in meeting project timelines and regulatory requirements. In all likelihood, some changes will be required; however, if the CMO is pushed too far outside of their normal practices, the sponsor will likely be introducing additional complexity into the project and, in the worst case, setting up the CMO for possible failure and subsequently delaying product approval. A well-designed Quality Agreement between the sponsor and the CMO can be used to drive the specifics regarding the preparation, review, and approval of the required scientific and regulatory documentation.

In most cases, the Chemistry, Manufacturing, and Controls (CMC) section of a regulatory application will be prepared under the strict oversight of the sponsor. The CMC section will be written to a level of detail with which the sponsor is comfortable, based on experience and the overall submission strategy. Technical data and documentation prepared by the CMO will be used to support the overall submission. A general rule for the regulatory submission is to provide a level of detail that will give the reviewing regulatory authorities sufficient information to conduct a thorough regulatory review upon which to base product approval yet is general enough to allow for maximum flexibility for both the sponsor and the CMO. For example, where possible ranges should be used to describe operating parameters such as times, temperatures, etc., excessive detail (e.g., a specific temperature, a specific time, a specific piece of equipment) in the regulatory application may inadvertently result in situations where a supplemental application may result. Such a submission requires approval by regulatory authorities and time for both pre-preparation and regulatory review. In extreme cases, excessive detail can result in compliance or production deviations which can be difficult to manage during CGMP inspections. It is critical that any ranges described in the regulatory application be based on actual data obtained during product development and, if necessary, validation. This is a delicate balance. Arbitrary ranges and overly general descriptions are rarely successful and usually result in comprehensive questions from reviewing chemists and, generally, delay approval of the regulatory application.

Tables 52.1 and 52.2 in this chapter listed the technical and regulatory requirements for each phase of development. Many of these data will be generated by the CMO or other outside laboratories or contractors. It is essential that the sponsor obtain complete, comprehensive, and approved reports for all development studies conducted. The sponsor should conduct audits of each of these studies to ensure data integrity between the raw data (e.g., laboratory notebooks) and those in the final reports. The sponsor should confirm that all information and data have been properly reported and evaluated. Because 8–12 years may be required before a product is developed and approved and because project personnel, CMOs, contract laboratories, etc., will very likely change over this period of time, it is imperative that all data and reports be finalized and approved in a form that will allow project continuity as changes in the development team occur. A successful regulatory submission will consider and address each component outlined in the FDA Guidance documents and should be considered in conjunction with the technical requirements provided in Tables 52.1 and 52.2 in this chapter.

52.7 CONCLUSION

Outsourcing of contract manufacturing services will continue to lead industry to effective new pathways to introduce new drug products to the market by more efficient, effective, and cost-competitive means.

Successful validation hinges on a complete and thorough understanding of the product and the production process that will be developed by or transferred to the CMO. Preformulation work, through scale-up and validation, must all fit together in a clear, scientific, and contiguous manner. Anything less complicates the process and can delay approval. After all, it is the sponsor's job, not the CMO's, to ensure that the process is properly developed and validated.

Selection, qualification, and management of the CMO can be divided into four primary phases: CMO identification and selection, CMO evaluation and qualification, contract negotiation, and CMO management. It is critical for the sponsor to establish a development team that includes an appropriate number of staff, as well as appropriate experience and expertise. A team with less than ideal experience will struggle to successfully complete the task on time and on budget.

Successful product approval ultimately depends on implementation of successful FDA regulatory strategies. Product development, production of CTM, and registration and validation batches, as well as GMP inspections by the FDA, require significant attention to the regulatory details and cannot be underestimated. The sponsor must work closely with the CMO to ensure all the regulatory "bases" are covered. Successful product approval may be the "easy" part of the entire drug development process. Following approval and commercialization, the product takes on a life of its own. Product approval is simply the beginning of a long road of scientific and regulatory complications, all of which must be effectively managed by the sponsor and CMO team.

One final thought: it is important for industry professionals to remember the fundamental overriding objective—the drugs and biologics that are under development will ultimately end up treating and benefiting patients. The industry's obligation is to get these products approved as quickly and efficiently as possible. Approved products allow for more funding of research and development of new and innovative products or, in the case of generic products, more affordable medicines. Industry is not doing its job if it bungles the responsibility of product development and validation. In the heat of the battle, it is easy to forget the patients; however, it is quite likely one of our loved ones will need one of the products that we helped to get approved and to the market. What is done day-in and day-out in the pharmaceutical industry is important work that can have a significant impact on a significant number of people and the quality of their lives.

ACKNOWLEDGMENTS

The author gratefully acknowledges the assistance of the following colleagues in preparation of this chapter:

Norma J. VanBunnen
Jonathan T. Beckloff PharmD MBA
Kristopher M. Beckloff MBA
Miguel de Soto-Perera, Ph.D.
Gary D. Hindman, PhD.

REFERENCES

1. U.S. Food and Drug Administration. *Content and Format of Investigational New Drug Applications (INDs) for Phase 1 Studies of Drugs, Including Well-Characterized, Therapeutic, Biotechnology-Derived Products*. Center for Drug Evaluation and Research (CDER): Center for Biologics Evaluation and Research (CBER), November 1995.
2. U.S. Food and Drug Administration. *INDs for Phase 2 and Phase 3 Studies: Chemistry, Manufacturing, and Controls Information*. Center for Drug Evaluation and Research (CDER), May 2003.
3. International Conference on Harmonization. Guidance on Q6A Specifications: Test Procedures and Acceptance Criteria for New Drug Substances and New Drug Products: Chemical Substances. Published in *Federal Register* 2000 December, 65(251), Friday.
4. U.S. Food and Drug Administration. ICH Harmonized Tripartite Guideline. *Q1A(R2) Stability Testing of New Drug Substances and Products*, November 2003.
5. ICH Harmonized Tripartite Guideline. *Text on Validation of Analytical Procedures Q2A*, March 1995.
6. ICH Harmonized Tripartite Guideline. *Q2B Validation of Analytical Procedures: Methodology*, November 6, 1996.
7. International Conference on Harmonization (ICH). *Q1B Photostability Testing of New Drug Substances and Products*, November 1996.
8. U.S. Food and Drug Administration. *Q1A Stability Testing of Drug Substances and Drug Products*. Center for Drug Evaluation and Research (CDER): Center for Biologics Evaluation and Research (CBER), August 2001.
9. CDER Manual of Policies and Procedures. *Chemistry Reviews of DMFs for Drug Substances/Intermediates*, August 1998.
10. U.S. Food and Drug Administration. *Drug Substance: Chemistry, Manufacturing and Controls Information*. Center for Drug Evaluation and Research (CDER), Center for Biologics Evaluation and Research (CBER), January 2004.
11. U.S. Food and Drug Administration. *Drug Product: Chemistry, Manufacturing and Controls Information*. Center for Drug Evaluation and Research (CDER), Center for Biologics Evaluation and Research (CBER), Draft Guidance, January 2003.
12. International Conference on Harmonization (ICH). *Guidance for Industry Q7A Good Manufacturing Practice Guidance for Active Pharmaceutical Ingredients*. Center for Drug Evaluation and Research (CDER), Center for Biologics Evaluation and Research (CBER), August 2001.
13. U.S. Food and Drug Administration. *Completeness Assessments for Type II API DMFs Under GDUFA*. Center for Drug Evaluation and Research (CDER), Center for Biologics Evaluation and Research (CBER), Guidance, October 2017.
14. U.S. Food and Drug Administration. *Analytical Procedures and Methods Validation for Drugs and Biologics*. Center for Drug Evaluation and Research (CDER), Center for Biologics Evaluation and Research (CBER), Guidance, July 2015.
15. U.S. Food and Drug Administration. *Applying ICH Q8(R2), Q9, and Q10 Principles to CMC Review*. Office of Pharmaceutical Quality, May 2016.
16. U.S. Food and Drug Administration. *Chemistry Review of Question-Based Review (QbR) Submissions*. Office of Pharmaceutical Quality, November 2014.

53 Computerized Systems Validation

Saeed Tafreshi

CONTENTS

DOI: 10.1201/9781003163138-53

The concept of validation was developed in the 1970s and is widely credited to Ted Byers, who at that time was the Associate Director of Compliance at the U.S. FDA. The concept was focused on:

> Establishing documented evidence which provides a high degree of assurance that a specific process will consistently produce a product meeting its predetermined specifications and quality attributes.

This concept continues to be followed, with some modifications, by the various authorities regulating GMP around the world. This definition also has been adopted for the validation business, manufacturing and laboratory computer systems.

53.1 COMPUTER VALIDATION HISTORY

The need to validate computer systems formally began in 1979 when the United States introduced GMP regulatory legislation that specifically referred to automation equipment. GMP is enforced by national regulatory authorities who can prevent the sale of a product in their respective country if they consider its manufacture not to be GMP compliant. Validation for GMP is a license-to-operate issue.

Over the last three decades, the manufacturing industry has increasingly used computer systems to control manufacturing processes for improved performance and product quality. This policy is often embedded in corporate strategy. Computer systems, however, by the nature of their complexity are susceptible to development and operational deficiencies that can adversely affect their control ability and effect product safety, quality, and efficacy. Common examples of such deficiencies include poor specification capture, design errors, poor testing, and poor maintenance practice. The potentially devastating outcome of GMP noncompliance of computer systems was demonstrated in 1988 when deficient software in data management system controlling a blood bank could have led to the issue of AIDS-infected blood. Additionally, computer systems can endanger public health through the manufacture and release of drug products with deficient quality attributes.

The first widely publicized FDA citation for computer validation noncompliance occurred in 1985; however, as early as 1982, the FDA was publicly stating that it was "nervous" if computer systems were used without being validated. In 1983, the FDA issued the Guide to Inspection of Computerized Systems in Drug Processing, Technical Report, Reference Materials, and Training Aids for Investigators that became known as the "Blue Book." This publication guided inspectors on what to accept as validation evidence for computer systems. The Blue Book formally introduced the anticipation of a life cycle approach to validation. The aim was to build in quality (QA) rather than rely on testing in quality (quality control).

Responding to the FDA's proactive position on computer systems validation, the PMA formed a Computer Systems Validation Committee to represent and coordinate the industry's viewpoint. The results were a joint FDA/PMA Conference in 1984 discussing computer systems validation and in the following year the publication of an industry perspective. The publication presented an approach for validation for both new and existing computer systems. GMP legislation is unusual in that it is equally applied to new production facilities and to production facilities built entirely or partially before the legislation (including amendments) was enforced.

Throughout the 1980s, computer systems validation was debated primarily in the United States Ken Chapman published a paper covering this period during which the FDA gave advice on the following GMP issues:

- Input/output checking
- Batch records
- Applying GMP to hardware and software
- Supplier responsibility
- Application software inspection
- FDA investigation of computer systems
- Software development activities

In addition, since the end of 1980s, the FDA and the pharmaceutical industry have debated the GMP requirements and the practicalities of electronic signatures. A resolution was achieved that became the FDA's proposed regulation.

Complementing the U.S. GMP guidance, the European Commission and authorities in Australia both issued GMP codes of practice, in 1989 and 1990, respectively. The European code, known as the "Orange Guide," was later issued in 1991 as a Directive superseding member state GMP legislation and included an annex covering computerized systems.

In most countries, GMP has been interpretive, and to prosecute a pharmaceutical manufacturing a court must be convinced that the charges reflect the intent to flout governing legislation. In the United States, however, a court declaratory judgment determined supplementary GMP information to be substantive. The net effect was that the FDA's advisory opinions became binding on the agency. In August 1990, the FDA announced that it no longer considered advisory opinions binding on the grounds that counsel considered such restrictions unconstitutional. Hence, the FDA interpretation of the regulations in Compliance Policy Guides, Guide to Investigators and other publications by FDA authors became nonbinding.

Computer systems validation also became a high-profile industry issue in Europe in 1991 when several European manufacturers and products were temporarily banned from the United States for computer systems noncompliance. The computer systems in question included autoclave Programmable Logic Controllers (PLCs) and Supervisory Control and Data Acquisition (SCADA) systems. The position of the FDA was clear; the manufacturer had failed to satisfy their "concerns" that computer systems should:

- Perform accurately and reliably
- Be secure from unauthorized or inadvertent changes
- Provide for adequate documentation of the process

The manufacturers thought they had satisfied the requirements of the existing GMP legislations, but they had not satisfied the FDA's expectations of GMP. Hence the adoption of current Good Manufacturing Practices (cGMP) to signify the latest understanding of the validation practices and standards expected by the regulatory authorities began.

In 1991, the U.K. Pharmaceutical Industry Computer Systems Validation Forum (known as the U.K. FORUM) was established to facilitate the exchange of validation knowledge and the development of a standard industry guide for computer systems validations. Suppliers were struggling to understand and implement the various interpretations and requirements of GMP presented by the manufacturers. ISO 9000 and TickIT accreditation for quality management provided a good basis for validation, but this does not fully satisfy GMP requirements. Then, the U.K. FORUM's guide came to fruition and was launched as a first draft within the United Kingdom. The guide is often referred to as the GAMP guide.

Meanwhile two experienced GMP regulatory inspectors, Ronald Tetzlaff and Tony Trill, published papers, respectively, presenting the FDA's and United Kingdom's MCA inspection practice for computer systems. These papers presented a comprehensive perspective on the current validation expectations of GMP regulatory authorities. Topics covered included:

- Life cycle approach
- Quality management
- Procedures
- Training
- Validation protocols
- Qualification evidence
- Change control
- Audit trail
- Ongoing evaluation

The pharmaceutical industry, in search of a common approach to computer systems validation, began incorporating these topics. Nevertheless, the FDA and MCA continue to encounter instances of noncompliance practice based on:

- Incomplete documents
- Insufficient detail in documents
- Missing documentary evidence

There was a clear need for guidance and standards on computer systems validation, and early in 1995 there were four milestones of significance to practitioners:

- The United States proposed new GMP legislation affecting electronic records and electronic signatures.
- After 16 years, the United States amended its legislation affecting computer validation, making a minor concession concerning the degree of input/output validation required for reliable computer systems.
- PDA presented a manufacturer's guide (Technical Report 18, since retired) to complement the PMA life cycle.
- The U.K. FORUM issued a revised draft of their supplier guide for European comment.

These initiatives helped the manufacturers and suppliers meet the challenge to validate computer systems effectively and efficiently. The initiatives which further clarified the requirements of validation included:

- The U.K. FORUM's investigation into the benefits of supplier audits shared by a number of participant manufacturers.
- The German APV (Information Technology Section) guide to Annex 11 of the European United GMP Directive regarding computerized systems.
- The German GMA Committee 5.8 and NAMUR Committee 1.9 joint working group's recommendations for computer systems validation.
- The coordination of the German initiatives with the U.K. FORUM supplier guide, and possibly the PDA TR 18, as announced at the ISPE computer validation seminar in Amsterdam in March of 1995.

What is clear to date is the mutual benefit of regulators, manufacturers, and suppliers working together towards a common GMP goal. GMP, while facilitating improvements to manufacturing performance, also is integral to the continuing high standing of the pharmaceutical industry.

53.1.1 Regulatory and System Development Requirements

In order for the industry to follow a common path in complying with the cGMP guidelines related to computer control systems, there is a need to understand the basics of proper system development and consider the overall cost into building a true business case. In doing so, it is necessary to follow the stages in sequence for the validation of a computerized control system to FDA requirements and their relationship to the development and implementation stages of an automation project. Typical System Development Life Cycle steps are:

- Definition
- Design
- Development
- Factory Acceptance Test (FAT)—Formal Management of Change (MOC) is required, going forward
- Installation
- Site Acceptance Test (SAT)
- Validation Testing
- Operation and Support
- Ongoing Maintenance

The very purpose of using computerized systems is to increase quality and efficiency. These can only be realized with forward-thinking plans. Success on this path in a regulated industry is to combine the regulatory and technology requirements from the start (planning stage). Ignoring this point has and will continue to result in overspending of time and resources that will drastically reduce the desired benefits. Consider the following in your development plans:

- DO NOT proceed without your User Requirements Specification (URS)
- Include your validation team early on
- Draft URS with all users
- Link your system purchase to URS
- Study the vendor-proposed changes within an approval process
- Hold your vendor responsible for System Specification Document (up to SAT completion)
- Approve your vendor Change Control during development
- Participate in FAT with real approval power
- Divide FAT and SAT economically
- Participate in SAT with real approval power
- Finalize your URS and System Specification Documents
- Turn on your Change Control
- Validate according to System Specification
- Accept ONLY minor discrepancies
- Link your Change Control to System Specification Document

Management Checklist for Computerized Systems

- Purchase the system you need—know your requirements
- Involve all stakeholders in drafting your system requirements
- Know your regulatory requirements
- Plan periodic team reviews with your vendor and system configurator
- Include your validation team prior to acceptance stage
- Review and approve your quality system plan in a life cycle approach that includes design, development, installation, acceptance, qualification, operation and maintenance

The Quality System regulation requires that "when computers or automated data processing systems are used as part of production or the quality system, the manufacturer shall validate computer software for its intended use according to an established protocol." This has been a regulatory requirement for GMP since 1978.

In addition to the aforementioned validation requirement, computer systems that implement part of a regulated manufacturer's production processes or quality system (or that are used to create and maintain records required by any other FDA regulation) are subject to the Electronic Records, Electronic Signatures regulation. This regulation establishes additional security, data integrity and validation requirements when records are created or maintained electronically. These additional Part 11 requirements should be carefully considered and included in system requirements and software requirements for any automated record keeping systems. System validation and software validation should demonstrate that all Part 11 requirements have been met.

53.2 COMPUTERIZED SYSTEMS AND APPLICATIONS

Computers and automated equipment are used extensively throughout the pharmaceutical, biotech, medical device, and medical gas industries in areas such as design, laboratory testing and analysis, product inspection and acceptance, production and process control, environmental controls, packaging, labeling, traceability, document control, complaint management, and many other aspects of the quality system.

What are the "systems" in "computer or related systems" in § 211.68? The American National Standards Institute (ANSI) defines systems as people, machines and methods organized to accomplish a set of specific functions. Computer or related systems can refer to computer hardware, software, peripheral devices, networks, cloud infrastructure, operators, and associated documents (e.g., user manuals and standard operating procedures).

Increasingly, automated plant floor operations have involved extensive use of embedded systems in

& PLCs
& digital function controllers
& statistical process control
& supervisory control and data acquisition
& robotics
& human–machine interfaces
& input/output devices
& computer operating systems

Computerized operations are now common in FDA-regulated industries. Small "minicomputer" systems are being used, sometimes in conjunction with larger computers, to control batching operations, maintain formula files and inventories, monitor process equipment, check equipment calibration, etc. The medical device industry is presently utilizing automatic test sets controlled by computers. In this application, the computer is relied upon as to whether a particular test parameter is within a specific tolerance. The operator does not see the values of the parameters measured but merely receives a green or red light indicating a go/no go situation. Products are accepted or rejected on this basis. In order to evaluate and/or report the adequacy of any computer-controlled processes or tests, the basics of computer construction and operation must be understood. The entire computer control system has been simplified as follows.

A computer is a machine and, like all other machines, is normally used because it performs specific tasks with greater accuracy and more efficiency than people. Computers accomplish this by having the capacity to receive, retain, and give up large volumes of data and process it in a very short time.

An understanding of computer operation and the ability to use a computer does not require a detailed knowledge of either electronics or the physical hardware construction. An overall view of the computer organization with emphasis on function is sufficient.

There are basically two types of computers, analog and digital. The analog computer does not perform computing directly with numbers. It accepts electrical signals of varying magnitude (analog signals) that in practical use are analogous to or represent some continuous physical magnitude such as pressure, temperature, etc. Analog computers are sometimes used for scientific, engineering, and process-control purposes. In the majority of industry applications used today, analog values are converted to digital form by an analog to digital converter and processed by digital computers.

The digital computer is the general-use computer used for manipulating symbolic information. In most applications the symbols manipulated are numbers and the operations performed on the symbols are the standard arithmetical operations. Complex problem solving is achieved by basic operations of addition, subtraction, multiplication, and division.

A digital computer is designed to accept and store instructions (program), accept information (data), and process the data as specified in the program and display the results of the processing in a selected manner. Instructions and data are in coded form the computer is designed to accept. The computer performs automatically and in sequence according to the program.

The computer is a collection of interconnected electromechanical devices (hardware) directed by a central control unit. The central control unit is the controlling device that supervises the sequence of activities that take place in all parts of the computer. Classically, the hardware consists of the mainframe (computer) for computation, storage, and control, and peripheral devices (input–output devices) for entering raw data and printing or displaying the output. Input data historically have been entered into the computer by teletypewriters, magnetic tape, punched tape, card readers, and now by field sensors signals through related input modules. Output may be displayed in the form of a hardcopy printout, magnetic tape, CRT, etc. The two units of input and output are often joined and referred to as input/output or simply I/O. A computer terminal with a CRT display is an example of a combined I/O device information display center.

Equally important as hardware in the effective use of the digital computer is the software. The numerous written programs and/or routines that dictate the process sequence that the computer will follow are called software. A computer can be programmed to do almost any problem that can be "defined." Defined means that the solution of a problem must be reduced to a series of steps that can be written as a series of computer instructions. In other words, the individual steps of the problem must be set up, including the desired level of accuracy, prior to the computer processing and solving the problem. The computer must be directed or commanded by a precisely stated set of commands or program. Until a program is prepared and stored in the computer memory, the computer knows absolutely nothing, not even how to receive input data. Therefore, the accuracy and validation of the program is one of the most important aspects of computer control.

Physical quantities are especially adaptive to binary digital techniques because most physical quantities can be expressed as two states: switches are on or off, a quantity level is above or below a set value, holes in cards are punched or not punched, electrical voltage or current is positive or negative or above or below a preset value. For such applications as process control, the digital computer executes predefined decisions by comparing input data to a predetermined value. The computer takes a course of action dependent on whether the input data is greater than, equal to, or less than the predetermined value.

The predetermined value and course of action the computer follows is in the form of a program stored in the computer memory. So actually, the computer *does not make decisions independently* but merely follows written program instructions. A printout or display of the actual values measured may be included as a part of the program. Verification of proper computer operation may be accomplished in this example by applying known inputs that are greater, equal to, and less than the predetermined value and subsequently reviewing the results.

Computers provide the platform to store and execute instructions in the desired sequence. Control cycle consist of:

- Logics written as programs (software)
- Outputs generated based on the program instructions
- State of field devices monitored and reported via inputs
- Reported new conditions compared with stored logic
- Continuous new output completes the control loop
- Process information is used to provide live visuals (HMI)
- Control access possible locally or from connected terminals
- Generated data captured, shared, and stored

In general, control systems used in the regulated industries can be listed as follows:

Programmable Logic Controller (PLC)—serves as a local device with limited capacity and capability. It is mainly used for local control and as an execution tool linked to more complex, remote configuration systems.

Distributed Control System (DCS)— employs computer processing power and memory to store and execute complex configurations in a network of multiple unit operations and local controllers. An example of more complex control configuration embedded within the main control system is Advanced Predictive Controller (APC) software. APC provides the capability of predicting the process behavior by use of mathematical models and process data. The component framework of DCS are defined by the variety of combination of software component, layer architecture, client servers, and software architecture selected.

Supervisory Control and Data Acquisition (SCADA)—very similar to DCS, with more expansion capability across networks. DCS is process oriented, as it focuses more on the processes in each step of the operation. SCADA focuses more on the acquisition and collation of data and its link to business information systems.

Integrated Control Systems (ICS)—provides much more capability and central control. This system can include a number of control systems, processors, enterprise systems, and more complex network of computers.

53.3 CONTROL SYSTEMS EVOLUTION AND THEIR CONCEPTS

The automation evolution has come a long way from mechanical and simple electromechanical devices to digital processors and now computer program driven systems (pre-artificial intelligence). These systems were created to automate manufacturing processes, and they all provided the needed process automation of their time. As process advancements required more capable control systems for automation, these systems evolved. The evolution from mechanical control systems (one way/output) to program-based control systems (two way/input and output) shares a basic concept of initiating a work command (output). In the program-based control systems, the work command is created from preconditions defined in the program. The selection of work command can now be in relation to the information received from input signals. This advantage changed the concept from a simple output only to an input, logic, and output concept that opened the control system range to include all conditions that can be defined in the program.

Simplifying computerized control systems to these three basic concepts facilitates the understanding of the combined system functionality and allows us to divide the control system components to achieve more efficient planning, design, development, qualification, and management of automation projects.

The general perception of computer systems is that they are complex. I could not disagree more! Computers are very simple tools that do exactly as we instruct them to do. The complications are mostly related to our methods of creating the instructions and, in some cases, our incomplete knowledge of configured instructions.

Automation projects, like any other project, require an orderly execution of many segments led by the expertise of multiple subject matter experts (SMEs). For successful management of such projects, it is essential to coordinate the business needs and the project requirements no later than the planning phase. Success and failure trends gathered by almost three decades of experience in automation projects clearly indicates the importance of a structured approach to acquiring, operating, and maintaining these systems.

Lack of proper documentation of system requirements and specifications that is intended to serve as the foundation of automated systems contributes to many of the challenges within manufacturing departments and ultimately the business goals. A brief review of the issued 483s (FDA findings reports) during inspections validates this shortcoming in the industry. Deviations from procedures and data integrity issues make up the majority of these findings.

A good understanding of the concepts can drive the needed focus for division of responsibilities and an orderly execution of automation project tasks.

Later in this chapter, we will review the required quality system for proper design, development, validation, and maintenance of computerized systems.

Following the proper quality system that is based on a practical approach and considers full understanding of the concepts will provide the path in which these systems can be managed efficiently through their life cycle.

In reviewing the evolution of control systems, it is important to realize the relationship between the system input and output, which provides the control capacity for that system. Based on this realization, we can categorize the system and identify the related components to build the requirements for their functionality and compliance.

Figure 53.1 shows simple work being accomplished that is directly linked to a cam rotation speed without feedback (a one-way control system).

Early digital processors were also a one-way control system, performing work via actuators, and did not offer feedback but, compared to previous control systems, provided only an easier setup and more channels to drive outputs. This was not a huge advancement compared with capabilities of previous mechanical and electromechanical control systems. The outputs in this system are similar to the previous mechanical and electromechanical devices that used the high point of cams to trigger work action. In digital processors, the work (outputs) is triggered based on the time defined for each output from the time slot(s) offered by the system timer cycle. This timed cycle repeats itself upon completion of each cycle.

Figure 53.2 shows a digital processor that could be programmed to generate output signals for performing a sequence of operation in a time-based order only.

One-way control without a feedback was a limiting factor that could not support the more advanced sequence of

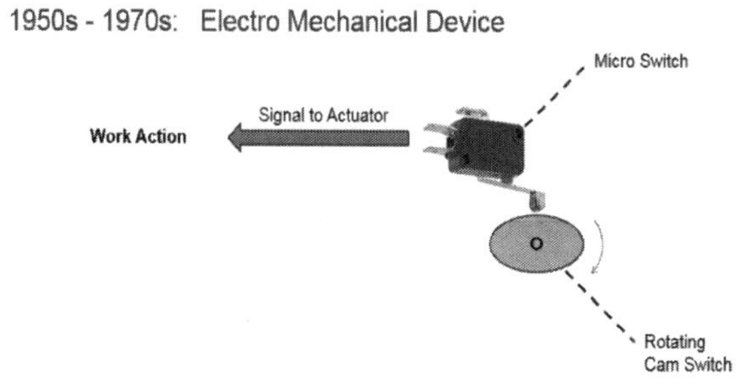

FIGURE 53.1

operations. This limitation meant that the work action had to be repeatable regardless of the field conditions—in other words, limiting the automation to those cycled actions that could be repeated without interruptions.

Introduction of the Programmable Logic Controller (PLC) was a major step that offered capabilities that are still the main part of our automated control systems. The advantage offered by PLC is the ability to capture and report actual field device position and process conditions in the form of input that is then compared with the desired process conditions stored in the program. Relative to this comparison, an output command is generated to manipulate the process. This action was made possible by introduction of "if" and "then" programming statements. A field process condition as an input to the program is measured (less than, equal to, or greater than) against a desired value to generate an output that contributes to stability and control of the process in the desired direction. The use of "if" and "then" programming statements introduced the two-way control capability where feedback to a control loop was made possible. With this advantage, more complex process conditions can be programed and controlled.

The ability to generate outputs based on this comparison in a continuous manner remains the back bone of our automated processes today. A typical PLC architecture layout is shown in Figure 53.3.

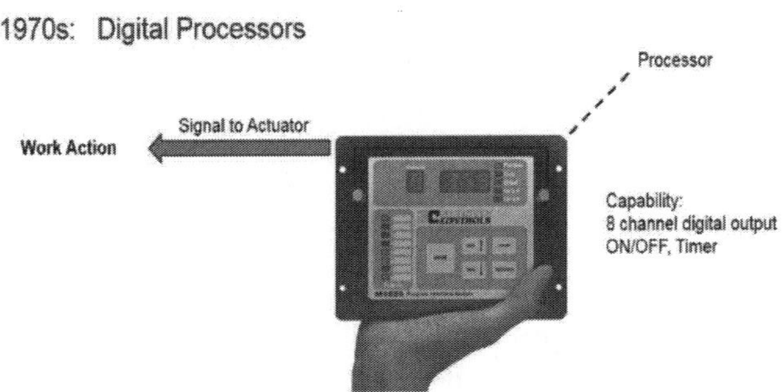

FIGURE 53.2

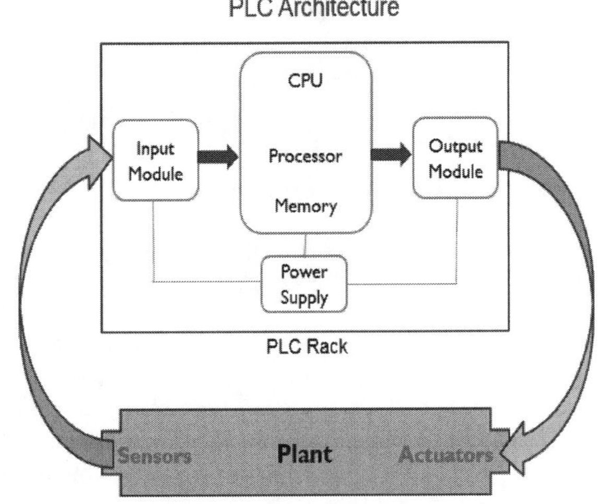

FIGURE 53.3

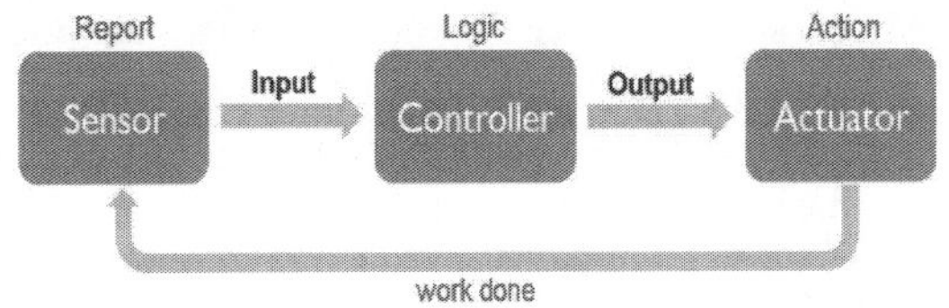

FIGURE 53.4

The key to successful automation project execution is understanding three simplified segments (applicable to all programmable control systems), by system designers, developers, quality units, and validators:

1. Input
2. Logic
3. Output

No matter how complex a control system is configured, they all follow the same basic principal, as shown in Figure 53.4.

Input—sensors are field devices that transmit process information to the controller as input data, such as temperature transmitters, pressure transmitters, level transmitters, speed indicators, position indicators (for valves and interlocks), humidity sensors, on/off switches (digital indicators).

Logic—our desired outcome provided to the computer as written programs is the logic that enables the computer to execute all predefined conditions. Programs are written in many formats based on the configured system, but they all serve the same purpose of providing instructions to the controller.

Output—actuators are field devices that perform the sequence of operations based on the output signals of the controller. Output signals to field devices can be identified as control valves %open or closed (for flow, heating, cooling, piston operations), motor start/stop, motor/pump rotation speed, etc.

53.4 COMPUTERIZED SYSTEMS VALIDATION

As noted earlier, automated systems (applicable to all control systems) can be divided into three parts: input, logic, and output. Components of these three parts are to be identified and documented in the early design phase. This information is collected to create the system specification document. This document is used in forming and in support of your validation testing protocols. In preparing and performing validation, consider the following:

1. All input signals received must be accurate in representing the field device conditions. In order to accomplish this task, identification of each field device is to be verified against the approved system drawings. Then perform and document I/O test to demonstrate proper connection from the field to the screen for input devices. Calibration of each field device for its range of operation is to be performed and documented. These tasks can be accomplished simultaneously in the order noted and include the output devices that require the same verification.

2. All output signals to the field devices (actuators) must accurately manipulate the field devices to the commanded positions. In order to accomplish this task, identification of each field device is to be verified against the approved system drawings. Performing and documenting the I/O test will demonstrate proper connection from control screen to the field for output devices. Calibration of each field device receiving the output signals within their range of operation is to be performed and documented.
3. The control logic (software) must be validated to demonstrate proper system functionality for intended use within all defined ranges of operation, as specified in the approved system specification document. This task begins by verification of the computer system hardware identification against the approved system drawings. All hardware verification of system I/O (part 1 and part 2) and computer system is performed by execution of the approved installation qualification (IQ) protocol. Execution of the approved system operation qualification (OQ) test protocol begins by verification of approved system operating procedures and training documentation for the personnel performing the test operations. Functionality tests consist of the following, with some variations, due to application of the computerized systems being considered such as manufacturing systems, IT infrastructure, data acquisition systems, enterprise systems, etc.:
    - System security test—verification of access and denial of access for every category of user specified.
    - HMI verification—operator interface and control screen functionality to perform as specified.
    - Interlock test—verification of safety and preconditions for operations as specified.
    - Alarm test—ensuring proper alarming notifications for the ranges of operation conditions specified.
    - Control loop test—verification of proper functionality of control loops in achieving the operational conditions specified.
    - Sequential control test—verification of orderly execution of the sequence of operation as specified.
    - Historical and real-time trend test—verification of proper capturing, displaying, and storing of the operation parameters as specified.
    - Loss of power and communication test—verification of proper performance of the system in disaster conditions and system recovery as specified.
    - Program installation and backup—verification of proper functionality of system software installation and backup based on the approved procedure as specified.

It is important to realize that a successful execution of the three parts listed previously relies on availability of complete and accurate system information. This information is created during the early phases of automation projects and may encounter revisions prior to validation phase. However, all system changes must be reflected in the system specification document as an essential part of any automation project.

When validating a computer control system, particular attention must be made to following established procedures and the documentation required during each stage to ensure that proper and sufficient documented evidence is provided to support validation inspection by the FDA.

The FDA has issued two validation definitions, which state the following:

1. "Establishing documented evidence that a system does what it is designed to do."
2. "Establishing documented evidence which provides a high degree of assurance that a specific process will consistently produce a product meeting its predetermined specifications and quality attributes."

The FDA audits against compliance with cGMP requirements. Rigid procedures are required to be followed, and those procedures must generate sufficient documentation to ensure that traceability and accountability of information (an audit trail) is maintained.

The FDA does not provide certification for a company and its procedures, nor does it approve what documentation should be produced. The company is responsible for demonstrating that procedures are followed and associated documentation generated to support the manufacture of the company's products.

The FDA's position was made clear in a statement made by Ronald Tetzlaff (when he was employed by the FDA) in *Pharmaceutical Technology*, April 1992. He stated that "Unless firms have documented evidence to ensure the proper performance of a vendor's software, the FDA cannot consider the automated system to be validated."

Therefore, it is important that companies have approved Quality Systems in place to ensure that procedures are followed and an audit trail is maintained.

53.5 COMPUTERIZED SYSTEM QUALITY STRUCTURE

Validation of a computerized control system to FDA requirements can be broken down into a number of phases that are interlinked with the overall project program. A typical validation program for a control system also includes the parallel design and development of control and monitoring instrumentation. This approach requires an integrated quality system to efficiently manage the multidiscipline automation projects. A typical Quality System includes the following phases:

53.5.1 Definition Phase

Validation starts at the definition (conceptual design) phase because the FDA expects to see documentary evidence that the chosen system vendor and the software proposed meet the customer's predefined selection criteria.

Vendor acceptance criteria, which must be defined by the customer, should typically include the following practices:

53.5.2 The Vendor's Business Practices

& Vendor certification to an approved QA standard. Certification may be a consideration when selecting a systems vendor. Initiative that promotes the use of international standards to improve the quality management of software development shall be considered.
& Vendor Audit by the customer to ensure company standards and practices are known and are being followed.
& Vendor end user support agreements.
& Vendor financial stability.
& Biography for the vendor's proposed project personnel (interviews also should be considered).
& Checking customer references and visiting their sites should be considered.

53.5.3 The Vendor's Software Practices

& Software development methodology.
& Vendor's experience in using the project software, including operating system software; application software; "off-the-shelf" and support software package (e.g., archiving, networking, batch software).
& Software performance and development history.
& Software updates.
& The vendor must make provision for source code to be accessible to the end user (e.g., have an escrow or similar agreement) and should provide a statement to this effect. Escrow is the name given to a legally binding agreement between a supplier and a customer that permits the customer access to source code, which is stored by a third party organization. The agreement also permits the customer access to the source code should the supplier become bankrupt.
Vendor acceptance can be divided into these areas:
& Vendor prequalification (to select suitable vendors to receive the Tender enquiry package).
& Review of the returned Tenders.
& Audit of the most suitable vendor(s).

Other documentation produced during the definition phase includes the URS, standard specifications and Tender support documentation.

The Tender enquiry package must be reviewed by the customer prior to issue to selected vendors. This review, called SQ, is carried out to ensure that the customer's technical and quality requirements are fully addressed.

53.5.4 Vendor Relationship Practices

- Technology purchases require system knowledge that are often provided by the vendor. Your vendor selection process plays an important role in achieving your desired goals of automation.
- Co-development projects are selected when a purchased technology is to be configured and installed by your firm or a third party.
- In all cases, the most critical phase is proper identification of your system requirements. Vendors and configurators generally draft their system specifications to meet your requirements.
- Audit and approval of your vendor and configurator quality system is your responsibility.
- Your system requirement and the vendor/configurator system specifications (that require your approval) will serve as your base document to support your development, installation, acceptance, qualification, operation, and maintenance tasks.
- The ownership of your system begins at the completion of the acceptance phase. So plan to work closely with your vendor and configurator in all design, development, installation, and acceptance phases with approval power.
- System specification document generated by your vendor and configurator must be accurate and complete. The efficiency of every phase following acceptance phase will depend on the quality of your system specification document. Work closely with your vendor and configurator on this task.
- From the regulatory view, the system owner (you) will remain responsible during the life cycle of the system.
- Plan for your needs of technical support in all phases of the system life cycle.

53.5.5 System Development Phase

The system development phase is the period from Tender award to delivery of the control system to site. It can be subdivided into four subphases:

& Design agreement
& Design and development
& Development testing
& Predelivery or FAT

The design agreement phase comprises the development and approval of the system vendor's Functional Design Specification, its associated FAT, Specification, and the Quality Plan for the project. These form the basis of the contractual agreement between the system vendor and the customer.

The design and development phase involves the development and approval of the detailed system (hardware and software) design and testing specifications. The software specifications comprise the Software Design Specification and its associated Software Module Coding. The hardware specifications comprise the Computer Hardware Design Specification and its associated Hardware Test Specification and Computer Hardware Production.

The development testing phase comprises the structured testing of the hardware and software against the detailed design specifications, starting from the lowest level and working up to a fully integrated system. The systems vendor must follow a rigorous and fully documented testing regime to ensure that each item of hardware and software module developed or modified performs the function(s) required without degrading other modules or the systems as a whole.

The predelivery acceptance phase comprises the FAT, which is witnessed by the customer, and the DQ review by the customer to ensure the system design meets technical (system functionality and operability) and quality (auditable, structured documentation) objectives.

Throughout the system development phase, the systems vendor should be subject to a number of quality audits by the customer or their nominated agents to ensure that the Quality Plan for the project is being complied with and that all documentation is being completed correctly. In addition, the vendor should conduct internal audits, and the reports should be available for inspection by the customer. The systems vendor also must enforce a strict change control procedure to enable all mediations and changes to the system to be thoroughly designed, tested, and documented. Change control is a formal system by which qualified representatives of appropriate disciplines review proposed or actual changes that might affect a validated status. The intent is to determine the need for action that would ensure and document that the component or system is maintained in a validated state.

The audit trail documentation introduced and maintained by the Quality Plan and the test documentation can be used as evidence by the customer during the FDA's inspections that the system meets the functionality required. In particular, the test and change control documentation will demonstrate a positive, thorough, and professional approach to validation.

53.5.6 Commissioning and In-Place Qualification Phase

The commissioning and qualification phase encompasses the System Commissioning on site, Site Acceptance Testing, IQ, OQ, and, where applicable, PQ activities for the project. The most important part of this phase must be identified as qualification based on system specification documentation. The system installation and operation must be confirmed against its documents. All system adjustments and changes occur in this phase must result in updating of the corresponding specification document. It is an assurance when building a reliable system base document in support of a life cycle approach during a phase that most last-minute changes are discovered. No benefit of any life cycle approach can be obtained when the system and its documentation do not match after completion of this phase.

53.5.7 Ongoing Maintenance Phase

The term maintenance does not mean the same when applied to hardware and software. The operational maintenance of hardware and software are different because their failure/error mechanisms are different. Hardware maintenance typically includes preventive hardware maintenance actions, component replacement, and corrective changes. Software maintenance includes corrective, perfective, and adaptive maintenance but does not include preventive maintenance actions or software component replacement.

Changes made to correct errors and faults in the software are corrective maintenance. Changes made to the software to improve the performance, maintainability, or other attributes of the software system are perfective maintenance. Software changes to make the software system usable in a changed environment are adaptive maintenance.

When changes are made to a software system, sufficient regression analysis and testing should be conducted to demonstrate that portions of the software not involved in the change were not adversely impacted. This is in addition to testing that evaluates the correctness of the implemented change(s).

The specific validation effort necessary for each change is determined by the type of change, the development products affected, and the impact of those products on the operation of the system. All proposed modifications, enhancements, or additions to the system should be assessed to determine the effect each change would have on the entire system. This information should determine the extent to which verification and/or validation tasks need to be iterated.

Documentation should be carefully reviewed to determine which documents have been impacted by a change. All approved documents (e.g., specifications, user manuals, drawings, etc.) that have been affected should be updated in accordance with the applicable site or corporate change management procedures. Specifications should be updated before any change is implanted.

53.5.8 Management of Change (MOC)

An essential component of a company's validation program, "change control" is a formal methodology for instituting changes, from their inception (a change request) to their incorporation, with the ultimate goal of keeping specifications current. Once incorporated, change control becomes the mechanism by which a state of control is maintained.

An overview of related components of a change is shown in Figure 53.5.

FDA Position on Change:

> The integrity of the data and the integrity of the protocols should be maintained when making changes to the computerized system, such as software upgrades, including security

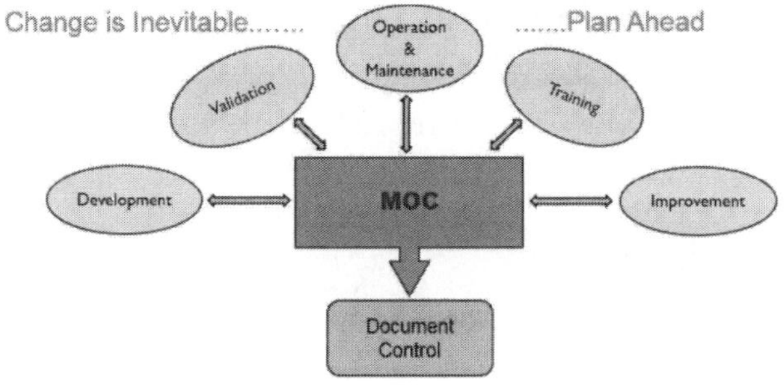

FIGURE 53.5

MOC Life Cycle:

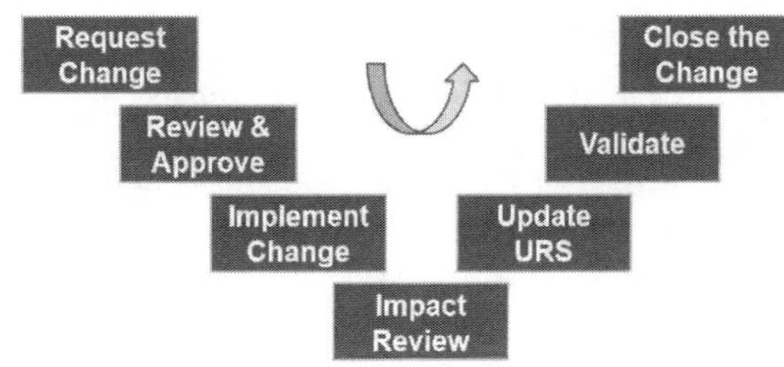

FIGURE 53.6

> and performance patches, equipment, or component replacement, or new instrumentation. The effects of any changes to the system should be evaluated and some should be validated depending on risk. Changes that exceed previously established operational limits or design specifications should be validated. Finally, all changes to the system should be documented.

Why is MOC critical in GMP?

To demonstrate that the automated systems and affiliated processes are:

1. In a state of control
2. Suitable for their intended purpose

Establishing an awareness of the importance of validation by all personnel, as well as gaining their commitment to MOC, provides a solid foundation for managing change.

A typical MOC life cycle view is shown in Figure 53.6.

53.6 SOFTWARE QUALITY SYSTEM AND QUALIFICATION

The Quality System regulation treats "verification" and "validation" as separate and distinct terms. On the other hand, many software engineering journal articles and textbooks use the terms verification and validation interchangeably, or in some cases refer to software "verification, validation, and testing (VV&T)" as if it is a single concept, with no distinction among the three terms.

Software verification provides objective evidence that the design outputs of a particular phase of the software development life cycle meet all of the specified requirements for that phase. Software verification looks for consistency, completeness, and correctness of the software and its supporting documentation as it is being developed and provides support for a subsequent conclusion that software is validated. Software testing is one of many verification activities intended to confirm that software development output meets its input requirements. Other verification activities include various static and dynamic analyses, code and document inspections, walkthroughs, and other techniques.

Software validation is a part of the design validation for the project but is not separately defined in the Quality System regulation. FDA considers software validation to be "confirmation by examination and provision of objective evidence that software specifications conform to user needs and intended uses, and that the particular requirements implemented through software can be consistently fulfilled." In practice, software validation activities may occur both during as well as at the end of the software development life cycle to ensure that all requirements have been fulfilled. Since software is usually part of a larger hardware system, the validation of software typically includes evidence that all software requirements have been implemented correctly and completely and are traceable to system requirements. A conclusion that software is validated is highly dependent upon comprehensive software testing, inspections, analyses, and other verification tasks performed at each stage of the software development life cycle.

Software verification and validation are difficult in nature because a developer cannot test forever, and it is hard to know how much evidence is enough. In large measure, software validation is a matter of developing a "level of confidence" that the application meets all requirements and user expectations for the software automated functions. Measures such as defects found in specifications documents, estimates of defects remaining, testing coverage, and other techniques are all used to develop an acceptable level of confidence before shipping the product. The level of confidence, and therefore the level of software validation, verification, and testing effort needed, will vary depending upon the application.

Many firms have asked for specific guidance on what the FDA expects them to do to ensure compliance with the Quality System regulation with regard to software validation. Validation of software has been conducted in many segments of the software industry for almost three decades. Due to the great variety of pharmaceuticals, medical devices, processes, and manufacturing facilities, it is not possible to state in one document all of the specific validation elements that are applicable. However, a general application of several broad concepts can be used successfully as guidance for software validation. These broad concepts provide an acceptable framework for building a comprehensive approach to software validation.

53.6.1 Requirements Specification

While the Quality System regulation states that design input requirements must be documented and that specified requirements must be verified, the regulation does not further clarify the distinction between the terms "requirement" and "specification." A requirement can be any need or expectation for a system or for its software. Requirements reflect the stated or implied needs of the customer and may be market-based, contractual, or statutory, as well as an organization's internal requirements. There can be many different kinds of requirements (e.g., design, functional, implementation, interface, performance, or physical requirements). Software requirements are typically derived from the system requirements for those aspects of system functionality that have been allocated to software. Software requirements are typically stated in functional terms and are defined, refined, and updated as a development project progresses. Success in accurately and completely documenting software requirements is a crucial factor in successful validation of the resulting software.

A specification is defined as "a document that states requirements." It may refer to or include drawings, patterns, or other relevant documents and usually indicates the means and the criteria whereby conformity with the requirement can be checked. There are many different kinds of written specifications, for example, system requirements specification, software requirements specification, software design specification, software test specification, software integration specification, etc. All of these documents establish "specified requirements" and are design outputs for which various forms of verification are necessary.

A documented software requirements specification provides a baseline for both validation and verification. The software validation process cannot be completed without an established software requirements specification.

53.6.2 Defect Prevention

Software quality assurance needs to focus on preventing the introduction of defects into the software development process and not on trying to "test quality into" the software code after it is written. Software testing is very limited in its ability to surface all latent defects in software code. For example, the complexity of most software prevents it from being exhaustively tested. Software testing is a necessary activity. However, in most cases software testing by itself is not sufficient to establish confidence that the software is fit for its intended use. In order to establish that confidence, software developers should use a mixture of methods and techniques to prevent software errors and to detect software errors that do occur. The "best mix" of methods depends on many factors including the development environment, application, size of project, language, and risk.

53.6.3 Time and Effort

To build a case that the software is validated requires time and effort. Preparation for software validation should begin early, that is, during design and development planning and design input. The final conclusion that the software is validated should be based on evidence collected from planned efforts conducted throughout the software life cycle.

53.6.4 Software Life Cycle

Software validation takes place within the environment of an established software development life cycle (SDLC). The software life cycle contains software engineering tasks and documentation necessary to support the software validation effort.

In addition, the software life cycle contains specific verification and validation tasks that are appropriate for the intended use of the software. No one life cycle model can be recommended for all software development and validation projects, but an appropriate and practical software life cycle should be selected and used for a software development project.

53.6.5 Plans

The software validation process is defined and controlled through the use of a plan. The software validation plan defines "what" is to be accomplished through the software validation effort. Software validation plans are a significant quality system tool. Software validation plans specify areas such as scope, approach, resources, schedules, and the types and extent of activities, tasks, and work items.

53.6.6 Procedures

The software validation process is executed through the use of procedures. These procedures establish "how" to conduct the software validation effort. The procedures should identify the specific actions or sequence of actions that must be taken to complete individual validation activities, tasks, and work items.

53.6.7 Software Validation after a Change

Due to the complexity of software, a seemingly small local change may have a significant global system impact. When any change (even a small change) is made to the software, the validation status of the software needs to be re-established. Whenever software is changed, a validation analysis should be conducted not just for validation of the individual change but also to determine the extent and impact of that change on the entire software system. Based on this analysis, the software developer should then conduct an appropriate level of software regression testing to show that unchanged but vulnerable portions of the system have not been adversely affected. Design controls and appropriate regression testing provide the confidence that the software is validated after a software change.

53.6.8 Validation Coverage

Validation coverage should be based on the software's complexity and safety risk and not on firm size or resource constraints. The selection of validation activities, tasks, and work items should be commensurate with the complexity of the software design and the risk associated with the use of the software for the specified intended use. For lower risk applications, only baseline validation activities may be conducted. As the risk increases, additional validation activities should be added to cover the additional risk. Validation documentation should be sufficient to demonstrate that all software validation plans and procedures have been completed successfully.

53.6.9 Flexibility and Responsibility

Specific implementation of these software validation principles may be quite different from one application to another. The manufacturer has flexibility in choosing how to apply these validation principles but retains ultimate responsibility for demonstrating that the software has been validated.

Software is designed, developed, validated, and regulated in a wide spectrum of environments and for a wide variety of applications with varying levels of risk. In each environment, software components from many sources may be used to create the software (e.g., in-house developed software, off-the-shelf software, contract software, shareware). In addition, software components come in many different forms (e.g., application software, operating systems, compilers, debuggers, configuration management tools, and many more). The validation of software in these environments can be a complex undertaking; therefore, it is appropriate that all of these software validation principles be considered when designing the software validation process. The resultant software validation process should be commensurate with the safety risk associated with the system, device, or process.

Software validation activities and tasks may be dispersed, occurring at different locations and being conducted by different organizations. However, regardless of the distribution of tasks, contractual relations, source of components, or the development environment, the manufacturer retains ultimate responsibility for ensuring that the software is validated.

Software validation is accomplished through a series of activities and tasks that are planned and executed at various stages of the software development life cycle. These tasks may be one-time occurrences or may be iterated many times, depending on the life cycle model used and the scope of changes made as the software project progresses.

53.7 SOFTWARE LIFE CYCLE ACTIVITIES

Software developers should establish a software life cycle model (SDLC) that is appropriate for their product and organization. The software life cycle model that is selected should cover the software from its birth to its retirement. Activities in a typical software life cycle model include the following:

- Quality Planning
- System Requirements Definition
- Detailed Software Requirements Specification
- Software Design Specification
- Construction or coding
- Testing
- Installation
- Operation and support
- Maintenance
- Retirement

Verification, testing, and other tasks that support software validation occur during each of these activities. A life cycle model organizes these software development activities in various ways and provides a framework for monitoring and controlling the software development project.

For each of the software life cycle activities, there are certain "typical" tasks that support a conclusion that the software is validated. However, the specific tasks to be performed, their order of performance, and the iteration and timing of their performance will be dictated by the specific software life cycle model that is selected and the safety risk associated with the software application. For very low-risk applications, certain tasks may not be needed at all. However, the software developer should at least consider each of these tasks and should define and document which tasks are or are not appropriate for their specific application.

53.7.1 Quality Planning

Design and development planning should culminate in a plan that identifies necessary tasks, procedures for anomaly reporting and resolution, necessary resources, and management review requirements, including formal design reviews. A software life cycle model and associated activities should be identified, as well as those tasks necessary for each software life cycle activity. The plan should include:

- Specific tasks for each life cycle activity
- Enumeration of important quality factors
- Methods and procedures for each task
- Task acceptance criteria
- Criteria for defining and documenting outputs in terms that will allow evaluation of their conformance to input requirements
- Inputs for each task
- Outputs from each task
- Roles, resources, and responsibilities for each task
- Risks and assumptions
- Documentation of user needs

Management must identify and provide the appropriate software development environment and resources. Typically, each task requires personnel as well as physical resources. The plan should identify the personnel, the facility and equipment resources for each task, and the role that risk (hazard) management will play. A configuration management plan should be developed that will guide and control multiple parallel development activities and ensure proper communications and documentation. Controls are necessary to ensure positive and correct correspondence among all approved versions of the specifications documents, source code, object code, and test suites that comprise a software system. The controls also should ensure accurate identification of and access to the currently approved versions.

Procedures should be created for reporting and resolving software anomalies found through validation or other activities. Management should identify the reports and specify the contents, format, and responsible organizational elements for each report. Procedures also are necessary for the review and approval of software development results, including the responsible organizational elements for such reviews and approvals.

53.7.2 Requirements

Requirement development includes the identification, analysis, and documentation of information about the application and its intended use. Areas of special importance include allocation of system functions to hardware/software, operating conditions, user characteristics, potential hazards, and anticipated tasks. In addition, the requirements should state clearly the intended use of the software.

The software requirements specification document should contain a written definition of the software functions. It is not possible to validate software without predetermined and documented software requirements. Typical software requirements specify the following:

- All software system inputs
- All software system outputs
- All functions that the software system will perform
- All performance requirements that the software will meet
- The definition of all external and user interfaces, as well as any internal software-to-system interfaces
- How users will interact with the system
- What constitutes an error and how errors should be handled
- Required response times
- The intended operating environment
- All ranges, limits, defaults, and specific values that the software will accept
- All safety-related requirements, specifications, features, or functions that will be implemented in software

Software safety requirements are derived from a technical risk management process that is closely integrated with the system requirements development process. Software requirement specifications should identify clearly the potential hazards that can result from a software failure in the system as well as any safety requirements to be implemented in software. The consequences of software failure should be evaluated, along with means of mitigating such failures (e.g., hardware mitigation, defensive programming, etc.). From this analysis, it should be possible to identify the most appropriate measures necessary to prevent harm.

A software requirements traceability analysis should be conducted to trace software requirements to (and from) system requirements and to risk analysis results. In addition to any other analyses and documentation used to verify software requirements, a formal design review is recommended to confirm that requirements are fully specified and appropriate before extensive software design efforts begin. Requirements can be approved and released incrementally, but care should be taken that interactions and interfaces among software (and hardware) requirements are properly reviewed, analyzed, and controlled.

53.7.3 Design

The decision to implement system functionality using software is one that is typically made during system design. Software requirements are typically derived from the overall system requirements and design for those aspects in the system that are to be implemented using software. There are user needs and intended uses for a finished product, but users typically do not specify whether those requirements are to be met by hardware, software, or some combination of both. Therefore, software validation must be considered within the context of the overall design validation for the system.

A documented requirements specification represents the user's needs and intended uses from which the product is developed. A primary goal of software validation is to then demonstrate that all completed software products comply with all documented software and system requirements. The correctness and completeness of both the system requirements and the software requirements should be addressed as part of the design validation process for that application. Software validation includes confirmation of conformance to all software specifications and confirmation that all software requirements are traceable to the system specifications. Confirmation is an important part of the overall design validation to ensure that all aspects of the design conform to user needs and intended uses.

In the design process, the software requirements specification is translated into a logical and physical representation of the software to be implemented. The software design specification is a description of what the software should do and how it should do it. Due to complexity of the project or to enable persons with varying levels of technical responsibilities to clearly understand design information, the design specification may contain both a high-level summary of the design and detailed design information. The completed software design specification constrains the programmer/coder to stay within the intent of the agreed-upon requirements and design. A complete software design specification will relieve the programmer from the need to make ad hoc design decisions.

The software design needs to address human factors. Use error caused by designs that are either overly complex or contrary to users' intuitive expectations for operation is one of the most persistent and critical problems encountered by the FDA. Frequently, the design of the software is a factor in such use errors. Human factor engineering should be woven into the entire design and development process, including the design requirements, analysis, and tests. Safety and usability issues should be considered when developing flowcharts, state diagrams, prototyping tools, and test plans. Also, task and function analysis, risk analysis, prototype tests and reviews, and full usability tests should be performed. Participants from the user population should be included when applying these methodologies.

The software design specification should include:

- & Software requirements specification, including predetermined criteria for acceptance of the software
- & Software risk analysis
- & Development procedures and coding guidelines (or other programming procedures)
- & Systems documentation (e.g., a narrative or a context diagram) that describes the systems context in which the program is intended to function, including the relationship of hardware, software, and the physical environment
- & Hardware to be used
- & Parameters to be measured or recorded
- & Logical structure (including control logic) and logical processing steps (e.g., algorithms)
- & Data structures and data flow diagrams
- & Definitions of variables (control and data) and description of where they are used
- & Error, alarm, and warning messages
- & Supporting software (e.g., operating systems, drivers, other application software)
- & Communication links (links among internal modules of the software, links with the supporting software, links with the hardware, and links with the user)
- & Security measures (both physical and logical security)

The activities that occur during software design have several purposes. Software design evaluations are conducted to determine if the design is complete, correct, consistent, unambiguous, feasible, and maintainable. Appropriate consideration of software architecture (e.g., modular structure) during design can reduce the magnitude of future validation efforts when software changes are needed. Software design evaluations may include analysis of control flow, data flow, complexity, timing, sizing, memory allocation, criticality analysis, and many other aspects of the design. A traceability analysis should be conducted to verify that the software design implements all of the software requirements. As a technique for identifying where requirements are not sufficient, the traceability analysis should also verify that all aspects of the design are traceable to software requirements. An analysis of communication links should be conducted to evaluate the proposed design with respect to hardware, user, and related software requirements. The software risk analysis should be reexamined to determine whether any additional hazards have been identified and whether any new hazards have been introduced by the design.

At the end of the software design activity, a Formal Design Review should be conducted to verify that the design is correct, consistent, complete, accurate, and testable before moving to implement the design. Portions of the design can be approved and released incrementally for implementation, but care should be taken that interactions and communication links among various elements are properly reviewed, analyzed, and controlled.

Most software development models will be iterative. This is likely to result in several versions of both the software requirements specification and the software design specification. All approved versions should be archived and controlled in accordance with established configuration management procedures.

53.7.4 Construction or Coding

Software may be constructed either by coding (i.e., programming) or by assembling together previously coded software components (e.g., from code libraries, off the-shelf software, etc.) for use in a new application. Coding is the software activity where the detailed design specification is implemented as source code. Coding is the lowest level of abstraction for the software development process. It is the last stage in decomposition of the software requirements where module specifications are translated into a programming language.

Coding usually involves the use of a high-level programming language but may also entail the use of assembly language (or microcode) for time-critical operations. The source code may be either compiled or interpreted for use on a target hardware platform. Decisions on the selection of programming languages and software build tools (assemblers, linkers, and compilers) should include consideration of the impact on subsequent quality evaluation tasks (e.g., availability of debugging and testing tools for the chosen language).

Some compilers offer optional levels and commands for error checking to assist in debugging the code. Different levels of error checking may be used throughout the coding process, and warnings or other messages from the compiler may or may not be recorded. However, at the end of the coding and debugging process, the most rigorous level of error checking is normally used to document what compilation errors still remain in the software. If the most rigorous level of error checking is not used for final translation of the source code, then justification for use of the less rigorous translation error checking should be documented. Also, for the final compilation, there should be documentation of the compilation process and its outcome, including any warnings or other messages from the compiler and their resolution or justification for the decision to leave issues unresolved.

Firms frequently adopt specific coding guidelines that establish quality policies and procedures related to the software coding process. Source code should be evaluated to verify its compliance with specified coding guidelines. Such guidelines should include coding conventions regarding clarity, style, complexity management, and commenting. Code comments should provide useful and descriptive information for a module, including expected inputs and outputs, variables referenced, expected data types, and operations to be performed. Source code should also be evaluated to verify its compliance with the corresponding detailed design specification. Modules ready for integration and test should have documentation of compliance with coding guidelines and any other applicable quality policies and procedures.

Source code evaluations are often implemented as code inspections and code walkthroughs. Such static analyses provide a very effective means to detect errors before execution of the code. They allow for examination of each error in isolation and can also help in focusing later dynamic testing of the software. Firms may use manual (desk) checking with appropriate controls to ensure consistency and independence. Source code evaluations should be extended to verification of internal linkages between modules and layers (horizontal and vertical interfaces) and compliance with their design specifications. Documentation of the procedures used and the results of source code evaluations should be maintained as part of design verification.

53.7.5 Testing by the Software Developer

Software testing entails running software products under known conditions with defined inputs and documented outcomes that can be compared to their predefined expectations. It is a time-consuming, difficult, and imperfect activity. As such, it requires early planning in order to be effective and efficient.

Test plans and test cases should be created as early in the software development process as feasible. They should identify the schedules, environments, resources (personnel, tools, etc.), methodologies, cases (inputs, procedures, outputs, and expected results), documentation, and reporting criteria. The magnitude of effort to be applied throughout the testing process can be linked to complexity, criticality, reliability, and/or safety issues.

Software test plans should identify the particular tasks to be conducted at each stage of development and include justification of the level of effort represented by their corresponding completion criteria.

An essential element of a software test case is the expected result. It is the key detail that permits objective evaluation of the actual test result. This necessary testing information is obtained from the corresponding predefined definition or specification. A software specification document must identify what, when, how, why, etc., is to be achieved with an engineering (i.e., measurable or objectively verifiable) level of detail in order for it to be confirmed through testing. The real effort of effective software testing lies in the definition of what is to be tested rather than in the performance of the test.

Once the prerequisite tasks (e.g., code inspection) have been successfully completed, software testing begins. It starts with unit level testing and concludes with system level testing. There may be a distinct integration level of testing. A software product should be challenged with test cases based on its internal structure and with test cases based on its external specification. These tests should provide a thorough and rigorous examination of the software product's compliance with its functional, performance, and interface definitions and requirements.

53.7.6 User Site Testing

Testing at the user site is an essential part of software validation. The Quality System regulation requires installation and inspection procedures (including testing where appropriate) as well as documentation of inspection and testing to demonstrate proper installation. Likewise, manufacturing equipment must meet specified requirements, and automated systems must be validated for their intended use.

Terminology regarding user site testing can be confusing. Terms such as beta test, site validation, user acceptance test, installation verification, and installation testing have all been used to describe user site testing. The term "user site testing" encompasses all of these and any other testing that takes place outside of the developer's controlled environment. This testing should take place at a user's site with the actual hardware and software that will be part of the installed system configuration. The testing is accomplished through either actual or simulated use of the software being tested within the context in which it is intended to function.

User site testing should follow a predefined written plan with a formal summary of testing and a record of formal acceptance. Documented evidence of all testing procedures, test input data, and test results should be retained.

There should be evidence that hardware and software are installed and configured as specified. Measures should ensure that all system components are exercised during the testing and that the versions of these components are those specified. The testing plan should specify testing throughout the full range of operating conditions and should specify continuation for a sufficient time to allow the system to encounter a wide spectrum of conditions and events in an effort to detect any latent faults that are not apparent during more normal activities.

During user site testing, records should be maintained of both proper system performance and any system failures that are encountered. The revision of the system to compensate for faults detected during this user site testing should follow the same procedures and controls as for any other software change.

The developers of the software may or may not be involved in the user site testing. If the developers are involved, they may seamlessly carry over to the user's site the last portions of design-level systems testing. If the developers are not involved, it is all the more important that the user have persons who understand the importance of careful test planning, the definition of expected test results, and the recording of all test outputs.

53.7.7 Maintenance and Software Changes

In addition to software verification and validation tasks that are part of the standard software development process, the following additional maintenance tasks should be addressed:

53.7.7.1 Software Validation Plan Revision

For software that was previously validated, the existing software validation plan should be revised to support the validation of the revised software. If no previous software validation plan exists, such a plan should be established to support the validation of the revised software.

53.7.7.2 Anomaly Evaluation

Software organizations frequently maintain documentation, such as software problem reports that describe software anomalies discovered and the specific corrective action taken to fix each anomaly. However, too often, mistakes are repeated because software developers do not take the next step to determine the root causes of problems and make the necessary process and procedural changes to avoid recurrence of the problem. Software anomalies should be evaluated in terms of their severity and their effects on system operation and safety, but they should also be treated as symptoms of process deficiencies in the quality system. A root-cause analysis of anomalies can identify specific quality system deficiencies. Where trends are identified (e.g., recurrence of similar software anomalies), appropriate corrective and preventive actions must be implemented and documented to avoid further recurrence of similar quality problems.

53.7.7.3 Problem Identification and Resolution Tracking

All problems discovered during maintenance of the software should be documented. The resolution of each problem should be tracked to ensure it is fixed, for historical reference, and for trending.

53.7.7.4 Task Iteration

For approved software changes, all necessary verification and validation tasks should be performed to ensure that planned changes are implemented correctly, all documentation is complete and up to date, and no unacceptable changes have occurred in software performance.

53.7.7.4.1 Benefits of Qualification

Software validation is a critical tool used to assure the quality of software and software automated operations. Software validation can increase the usability and reliability of the application, resulting in decreased failure rates, fewer recalls and corrective actions, less risk to patients and users, and reduced liability to manufacturers. Software validation can also reduce long-term costs by making it easier and less costly to reliably modify software and revalidate software changes. Software maintenance can represent a very large percentage of the total cost of software over its entire life cycle. An established comprehensive software validation process helps to reduce the long-term cost of software by reducing the cost of validation for each subsequent release of the software. The level of validation effort should be commensurate with the risk posed by the automated operation. In addition to other risk factors, such as the complexity of the process software and the degree to which the manufacturer is dependent upon that automated process to produce a safe and effective product, determine the nature and extent of testing needed as part of the validation effort. Documented requirements and risk analysis of the automated process help to define the scope of the evidence needed to show that the software is validated for its intended use.

53.8 EMERGING TECHNOLOGY

The current digital technology is evolving at such a rapid rate that at this time we are not fully able to comprehend how it will impact our world. Artificial intelligence (AI) has a big role to play in this evolution. It is the fastest growing technology with the greatest power for disruption. At its most basic, it eliminates mundane, repetitive tasks. At the other extreme, it will begin to replace higher-level cognitive human abilities with a far greater capacity for volume, analysis, accuracy, and consistency than can currently be performed by humans.

For now, it is defined thus: "Artificial Intelligence is the Intelligence of Machines and the branch of Computer Science which aims to create it."

An overview of AI is shown in Figure 53.7.

The following is a brief overview of AI development stages:

1. **Gathering of Data**: Quality and quantity of data will directly determine how good predictive model can be
2. **Data Preparation**: Loading the data into a suitable structure to prepare it for machine learning training
3. **Choosing a Model**: Identifying data relationships towards desired output
4. **Training the Model** (bulk of machine learning): Using the data to incrementally improving the model
5. **Evaluation**: Testing of model with fresh data (not used in model training)
6. **Parameter Tuning**: Additional improvement of model training
7. **Prediction**: Realization of the value of machine learning

The IEEE Standard has recently been published for the purpose of promoting clarity and consistency in the use of Software Based Intelligent Process Automation (SBIPA)

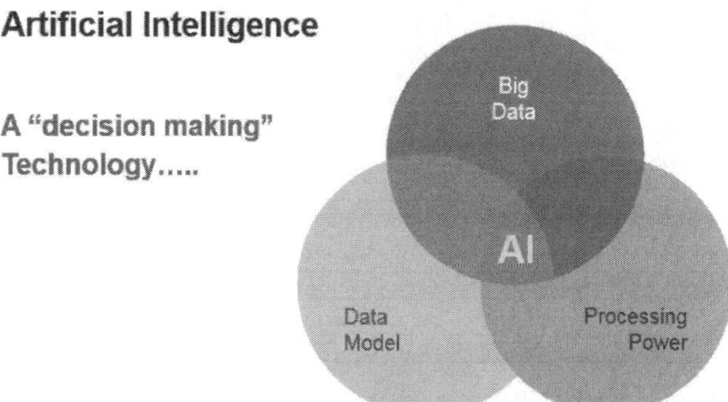

FIGURE 53.7

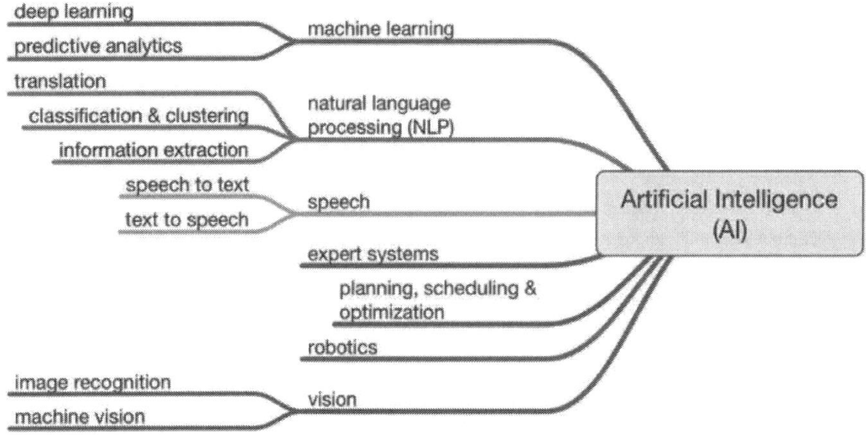

FIGURE 53.8

terminology—Electronic (ISBN: 978-1-5044-4354-8, Date of Publish: 28 Sept. 2017).

A single definition for AI that is universally accepted by practitioners has still not been adopted. Some define AI as a computerized system that exhibits behavior that is commonly thought of as requiring intelligence. Others define AI as a system capable of rationally solving complex problems or taking appropriate actions to achieve its goals in whatever real world circumstances it encounters.

Diversity of AI problems and solutions and the foundation of AI in human evaluation of the performance and accuracy of algorithms make it difficult to clearly define a bright-line distinction between what constitutes AI and what does not. For example, many techniques used to analyze large volumes of data were developed by AI researchers and are now identified as "Big Data" algorithms and systems. In some cases, opinion may shift, meaning that a problem is considered as requiring AI before it has been solved, but once a solution is well known, it is considered routine data processing. Although the boundaries of AI can be uncertain and have tended to shift over time, what is important is that a core objective of AI research and applications over the years has been to automate or replicate intelligent behavior. An early blueprint of AI applications is shown in Figure 53.8.

The following statement was published on October 2016 by the Executive Office of the President National Science and Technology Council, Committee on Technology:

> Advances in Artificial Intelligence (AI) technology have opened up new markets and new opportunities for progress in critical areas such as health, education, energy, and the environment. In recent years, machines have surpassed humans in the performance of certain specific tasks, such as some aspects of image recognition. Experts forecast that rapid progress in the field of specialized artificial intelligence will continue. Although it is very unlikely that machines will exhibit broadly-applicable intelligence comparable to or exceeding that of humans in the next 20 years, it is to be expected that machines will reach and exceed human performance on more and more tasks.

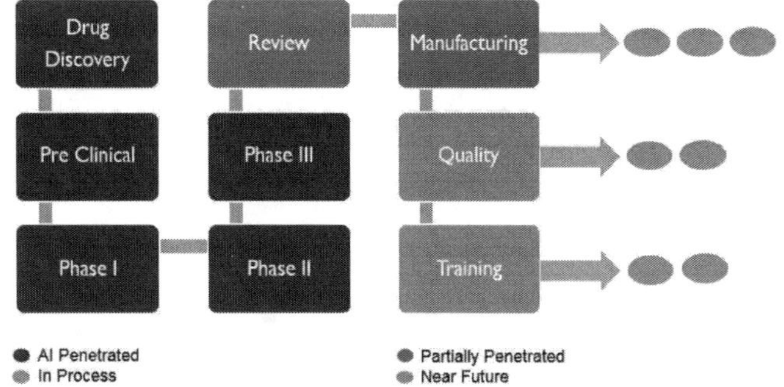

FIGURE 53.9

In the past decade, the use of AI is finding application across the automation world in almost all industries. Transition to this powerful and disruptive technology will continue to grow exponentially in the coming decade. The use of AI in the regulated health care industry is in its early stages, focusing on R&D areas today, but it will continue down the chain to impact the manufacturing and management of this industry. Using program-based computer systems, we have made significant improvements in our manufacturing machines and processes. AI's advantage is the ability to connect these systems by way of relating their operation data and therefore improving processes. Employing this new technology will result in reducing the number of humans performing knowledge tasks, offer greater control of processes, and introduce much wider capabilities, and at the same time it will drastically reduce our current processing time.

See Figure 53.9 for current AI penetration level in the health care industry.

53.9 AI REGULATION AND QUALIFICATION CHALLENGES

The technology0driven rate of change is now greater than we have ever experienced and it will continue to grow exponentially.

Clear understanding of the rate of this change is essential in our plans for the future. Artificial intelligence (AI) and machine learning (ML) based technologies have the potential to transform health care by deriving new and important insights from the vast amount of data generated during the delivery of health care every day. An example of potential high-value applications include earlier disease detection, more accurate diagnosis, identification of new observations or patterns on human physiology, and development of personalized diagnostics and therapeutics.

The health care industry is beginning to realize the potential benefits of employing AI in some capacity in its operations. The rate of this transition will increase when the full advantages of AI are realized. At this time, the use of AI is focused mainly on research and clinical trial areas. The regulators are already making plans and drafting approaches to guide the use of this technology in the areas of focus for today and will continue to expand it to greater capacity beyond research and clinical in the near future.

On April 2019, FDA Commissioner Scott Gottlieb, MD, released the following statement on steps toward a new, tailored review framework for artificial intelligence-based medical devices:

> We're taking the first step toward developing a novel and tailored approach to help developers bring artificial intelligence devices to market by releasing a discussion paper. Other steps in the future will include issuing draft guidance that will be informed by the input we receive. Our approach will focus on the continually evolving nature of these promising technologies. We plan to apply our current authorities in new ways to keep up with the rapid pace of innovation and ensure the safety of these devices.
>
> The artificial intelligence technologies granted marketing authorization and cleared by the agency so far are generally called "locked" algorithms that do not continually adapt or learn every time the algorithm is used. These locked algorithms are modified by the manufacturer at intervals, which includes "training" of the algorithm using new data, followed by manual verification and validation of the updated algorithm. But there is a great deal of promise beyond locked algorithms that is ripe for application in the health care space, and which requires careful oversight to ensure the benefits of these advanced technologies outweigh the risks to patients. These machine learning algorithms that continually evolve, often called "adaptive" or "continuously learning" algorithms, do not need manual modification to incorporate learning or updates. Adaptive algorithms can learn from new user data presented to the algorithm through real-world use. For example, an algorithm that detects breast cancer lesions on mammograms could learn to improve the confidence with which it identifies lesions as cancerous or may learn to identify specific subtypes of breast cancer by continually learning from real-world use and feedback.
>
> We are exploring a framework that would allow for modifications to algorithms to be made from real-world learning and adaptation, while still ensuring safety and effectiveness of the software as a medical device is maintained. A new approach to these technologies would address the need for the algorithms to learn and adapt when used in the real world. It would be a more tailored fit than our existing regulatory paradigm for software as a medical device. For traditional software as a medical device, when modifications are made that could significantly affect the safety or effectiveness of the device, a sponsor must make a submission demonstrating the safety and effectiveness of the modifications. With artificial intelligence, because the device evolves based on what it learns while it's in real world use, we're working to develop an appropriate framework that allows the software to evolve in ways to improve its performance while ensuring that changes meet our gold standard for safety and effectiveness throughout the product's lifecycle—from premarket design throughout the device's use on the market. Our ideas are the foundational first step to developing a total product lifecycle approach to regulating these algorithms that use real-world data to adapt and improve.
>
> We're considering how an approach that enables the evaluation and monitoring of a software product from its premarket development to post-market performance could provide reasonable assurance of safety and effectiveness and allow the FDA's regulatory oversight to embrace the iterative nature of these artificial intelligence products while ensuring that our standards for safety and effectiveness are maintained. This first step in developing our approach outlines information specific to devices that include artificial intelligence algorithms that make real-world modifications that the agency might require for premarket review. They include the algorithm's performance, the manufacturer's plan for modifications and the ability of the manufacturer to manage and control risks of the modifications.
>
> The agency may also intend to review what's referred to as software's predetermined change control plan. The predetermined change control plan would provide detailed information to the agency about the types of anticipated modifications based on the algorithm's re-training and update strategy, and the associated methodology being used to implement those changes in a controlled manner that manages risks to patients. Consistent with our existing quality systems regulation, the agency expects software developers to have an established quality system that is geared towards developing, delivering and maintaining high-quality products throughout the lifecycle that conforms to the agency's standards and regulations.
>
> The goal of the framework is to assure that ongoing algorithm changes follow pre-specified performance objectives and change control plans, use a validation process that ensures improvements to the performance, safety and effectiveness of the artificial intelligence software, and includes real-world monitoring of performance once the device is on the market to ensure safety and effectiveness are maintained. We're exploring this approach because we believe that it will enable beneficial and innovative artificial intelligence software to come to market while still ensuring the device's benefits continue to outweigh it risks.
>
> We have more work to do to build out this initial set of ideas and we'll rely on comments and feedback from experts and stakeholders in this space to help inform the agency as we continue to think about how we'll regulate artificial intelligence technologies to improve patient care. We anticipate several more steps in the future, including issuing draft guidance that'll be informed by the feedback on today's discussion paper.

> As with all of our efforts in digital health, collaboration will be key to developing this appropriate framework. We encourage feedback and welcome a diversity of opinions and thoughtful discourse, which will contribute to building the foundation of this regulatory paradigm. As algorithms evolve, the FDA must also modernize our approach to regulating these products. We must ensure that we can continue to provide a gold standard of safety and effectiveness. We believe that guidance from the agency will help advance the development of these innovative products.
>
> We've taken similar steps to advance other novel oversight frameworks for new technologies. Our Digital Health Innovation Action Plan laid the groundwork for new approaches to foster innovation in digital health. We're building our Digital Health Center of Excellence to develop more efficient ways to ensure the safety and effectiveness of technologies like smart watches with medical apps. Our Software Precertification Pilot Program is allowing us to test a new approach for product review. While I know there are more steps to take in our regulation of artificial intelligence algorithms, the first step taken today will help promote ideas on the development of safe, beneficial and innovative medical products.

As of now, the International Medical Device Regulators Forum (IMDRF) defines "Software as a Medical Device (SaMD)" as software intended to be used for one or more medical purposes that perform these purposes without being part of a hardware medical device. FDA, under the Federal Food, Drug, and Cosmetic Act (FD&C Act) considers medical purpose as "those purposes that are intended to treat, diagnose, cure, mitigate, or prevent disease or other conditions."

One of the greatest benefits of AI/ML in software resides in its ability to learn from real-world use and experience and its capability to improve its performance. The ability for AI/ML software to learn from real-world feedback (training) and improve its performance (adaptation) makes these technologies uniquely situated among SaMD and a rapidly expanding area of research and development. Regulators' vision is that with appropriately tailored regulatory oversight, AI/ML-based SaMD will deliver safe and effective software functionality that improves the quality of care that patients receive.

To date, FDA has cleared or approved several AI/ML-based SaMD. Typically, these have only included algorithms that are "locked" prior to marketing, where algorithm changes likely require FDA premarket review for changes beyond the original market authorization. However, not all AI/ML-based SaMD are locked; some algorithms can adapt over time. The power of these AI/ML-based SaMD lies within the ability to continuously learn, where the adaptation or change to the algorithm is realized after the SaMD is distributed for use and has "learned" from real-world experience. Following distribution, these types of continuously learning and adaptive AI/ML algorithms may provide a different output in comparison to the output initially cleared for a given set of inputs.

Constant adjustment of outputs based on new inputs is continuous and, unlike program-based logic, does not follow predefined boundaries. This is the main difference (advantage) of AI/ML compared to program-based control systems of today, where we solve the problems (make the decisions) and program the system to execute it. AI/ML is the first decision-making technology that has the capacity to provide the solutions even beyond our real-time comprehension. As is indicated in the current regulatory approach, the qualification of this technology requires validation of the model. It is important to realize that this validation will cover the model up to the first change. Any learning of the model that result in a change will require validation. Clearly this approach will be limiting since the continuous "learning" is the main advantage offered by this new technology unless a boundary for learning could be defined in relation with inputs and outputs. The difficulty will remain as how to limit a solution that is not known. The current limiting approach in qualifying this new technology will remain until the regulators draft and develop their own use of AI in inspecting and approving the use of this new technology.

Historically, technologies used by the regulators lagged years behind their industry, but they were still able to perform their compliance inspections because the tasks of word processing and data compiling did not present critical tool limitations. The new technology (AI/ML), due to its huge rate of change compared to the past technologies, requires a new approach by the regulators in allowing the industry to benefit from the greatest advantage ever provided by new technology. This change eventually will have to place the regulators on the path of planning the development and use of this technology ahead of their industry. Here are some reveling examples of the past (Figure 53.10):

For now, we continue on the path that our use of the new technology within a desired boundary requires validation of the selected model up to each change. Continued challenges introduced by the volume of change (within the data models) as a function of ML will pave the way to new realization that a practical approach requires technology solutions to challenges presented by technology.

Future validation methods must evolve to include the model variation as the new path is learned by relating inputs and outputs in the system. However, the qualification of such a system requires the following:

- **Quality of Input**
- **Validity of Model/Data Relationship**
- **Verification of Output**

Control concepts of Input, Logic, and Output (noted earlier) can also help us in qualification of AI/ML systems that use a neural network as its base. A simple neural network layout is shown in Figure 53.11.

There is no limitation on inputs in this system, and the mathematical model will serve as the logic. It is important

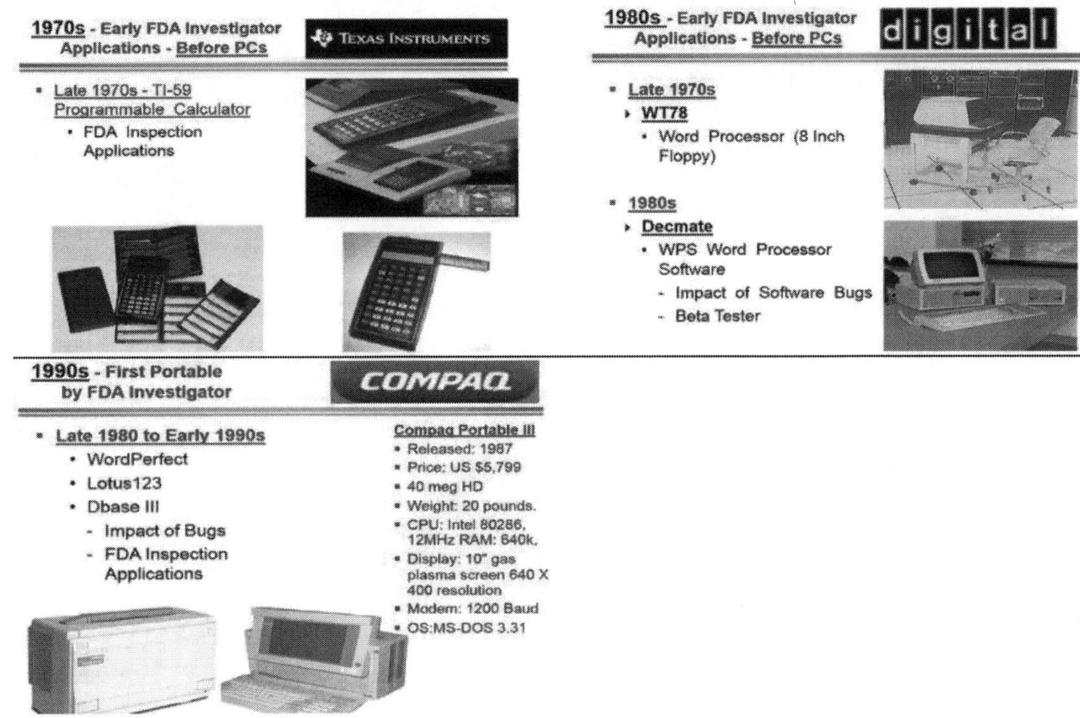

FIGURE 53.10

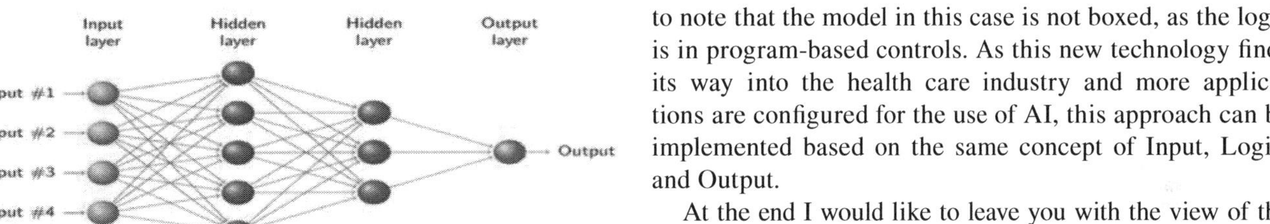

FIGURE 53.11

to note that the model in this case is not boxed, as the logic is in program-based controls. As this new technology finds its way into the health care industry and more applications are configured for the use of AI, this approach can be implemented based on the same concept of Input, Logic, and Output.

At the end I would like to leave you with the view of the technology evolution in Figure 53.12.

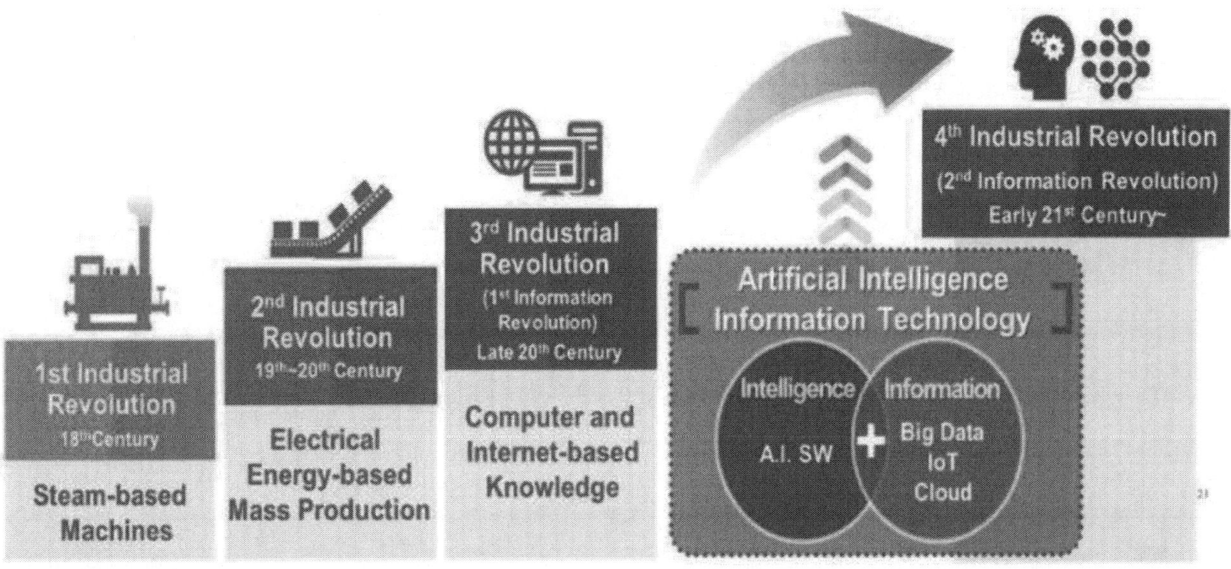

FIGURE 53.12

53.9.1 An Abbreviated Computer Validation History

- 1978—Validation for GMP concept developed by FDA
- 1979—The United States issues Federal Regulations for GMP including validation of automation equipment
- 1983—FDA Blue Book for computer system validation
- 1985—U.S. PMA published guideline for validating new and existing computer systems
- 1987—FDA technical report on developing computer systems
- 1988—FDA conference paper on inspecting computer systems
- 1989—EU Code for GMP including Annex 11 on computerized systems
- 1991—EU Directive for GMP based on EU Code for GMP
- 1994—U.K. FORUM draft guidelines to suppliers
- 1994—The United States proposes new electronic record and electronic signatures GMP regulations
- 1994—GAMP first draft distributed to U.K. for comments
- 1995—U.S. PDA publish validation guideline for manufacturers
- 1995—The United States amends GMP regulations affecting automation
- 1995—U.K. FORUM revise draft guidelines to suppliers
- March of 1997, FDA issues final part 11 regulations
- First Draft July 2000 (GAMP Europe)
- Final Draft March 2001 (GAMP Americas)
- Version 1 Quarter 2, 2001 (Co-Publication with PDA)
- GAMP4, December 2001, major revision and new content in line with regulatory and technological development
- February 4, 2003, FDA withdraws the draft guidance for industry, 21 CFR Part 11
- GAMP 5, 2008, A Risk Based Approach to Compliant GxP Computerized Systems

Abbreviations Used

AI, artificial intelligence
APC, advanced predictive controller
cGMP, current good manufacturing practice
CRT, cathode ray tube
DCS, distributed control system
DQ, design qualification
EU, European Union
FAT, factory acceptance test
FDA, Food and Drug Administration
GAMP, good automated manufacturing practice
GMP, good manufacturing practice
IAPT, International association for Pharmaceutical Technology
ICS, integrated control system
I/O, input and output
IQ, installation qualification
ISPE, International Society of Pharmaceutical Engineering
IT, information technology
MCA, medicines control agency
ML, machine learning
MOC, management of change
OQ, operational qualification
PDA, parenteral drug association
PLC, programmable logic controller
PMA, pharmaceutical manufacturers association
PQ, performance qualification
QA, quality assurance
SAT, site acceptance test
SaMD, software as a medical device
SCADA, supervisory control and data acquisition
SDLC, software development life cycle
SQ, system qualification or specification qualification
URS, user requirements specification
VV&T, verification, validation and testing

FOOD AND DRUG ADMINISTRATION REFERENCES

Deciding When to Submit a 510(k) for a Software Change to an Existing Device. www.fda.gov/downloads/medicaldevices/deviceregulationandguidance/guidancedocuments/ucm514737.pdf. 21 CFR 807.81(a)(3).

Glossary of Computerized System and Software Development Terminology, Division of Field Investigations, Office of Regional Operations, Office of Regulatory Affairs. Food and Drug Administration, August 1995.

Guideline on General Principles of Process Validation, Center for Drugs and Biologics, and Center for Devices and Radiological Health. Food and Drug Administration, May 1987.

C:\Users\Anitha\AppData\Roaming\Microsoft\Word\15064-2030-Ref Mismatch Report.docx - LStERROR_594Modifications to a Device Approved through a PMA Are Governed by the Criteria in 21 CFR 814.39(a).

Modifications to Devices Subject to Premarket Approval (PMA): The PMA Supplement Decision Making Process. www.fda.gov/downloads/MedicalDevices/DeviceRegulationandGuidance/GuidanceDocuments/UCM089360.pdf. 21 CFR 807.81(a)(3).

Pre-Cert Program Version 1.0 Working Model. www.fda.gov/downloads/MedicalDevices/DigitalHealth/DigitalHealthPreCertProgram/UCM629276.pdf.

Software as a Medical Device (SaMD): Clinical Evaluation. www.fda.gov/downloads/medicaldevices/deviceregulationandguidance/guidancedocuments/ucm524904.pdf.

Technical Report, Software Development Activities, Division of Field Investigations, Office of Regional Operations, Office of Regulatory Affairs. Food and Drug Administration, July 1987.

OTHER GOVERNMENT REFERENCES

Adrion WR, Branstad MA, Cherniavsky JC. NBS Special Publication 500-75, Validation, Verification, and Testing of Computer Software, Center for Programming Science and

Technology, Institute for Computer Sciences and Technology, National Bureau of Standards, U.S. Department of Commerce, February 1981.

Powell PB, ed. NBS Special Publication 500-98, Planning for Software Validation, Verification, and Testing, Center for Programming Science and Technology, Institute for Computer Sciences and Technology, National Bureau of Standards, U.S. Department of Commerce, November 1982.

Preparing for the Future of Artificial Intelligence, Executive Office of the President National Science and Technology Council Committee on Technology, October 2016.

Software as a Medical Device (SaMD): Key Definitions. www.imdrf.org/docs/imdrf/final/technical/imdrf-tech-131209-samd-key-definitions-140901.pdf.

Wallace DR, ed. NIST Special Publication 500–235, Structured Testing: A Testing Methodology Using the Cyclematic Complexity Metric. Computer Systems Laboratory, National Institute of Standards and Technology, U.S. Department of Commerce, August 1996.

INTERNATIONAL AND NATIONAL CONSENSUS STANDARDS

IEC 61506:1997. Industrial Process Measurement and Control: Documentation of Application Software. International Electro-Technical Commission, 1997.

IEEE 1012–1986. Software Verification and Validation Plans, Institute for Electrical and Electronics Engineers, 1986.

C:\Users\Anitha\AppData\Roaming\Microsoft\Word\15064-2030-Ref Mismatch Report.docx - LStERROR_606IEEE Software Based Intelligent Process Automation (SBIPA) terminology, Electronic ISBN: 978-1-5044-4354-8, September 28, 2017.

IEEE Standards Collection. Software Engineering, Institute of Electrical and Electronics Engineers, Inc., 1994. ISBN 1-55937-442-X.

ISO 9000-3:1997. Quality Management and Quality Assurance Standards-Part 3: Guidelines for the Application of ISO 9001:1994 to the Development, Supply, Installation and Maintenance of Computer Software. International Organization for Standardization, 1997.

ISO/IEC 12207:1995. Information Technology: Software Life Cycle Processes, Joint Technical Committee.

ISO/IEC JTC 1. Subcommittee SC 7, International Organization for Standardization and International Electro-Technical Commission, 1995.

PRODUCTION PROCESS SOFTWARE AND TECHNOLOGY REFERENCES

Grigonis GJ Jr, Subak EJ Jr, Michael W. Validation Key Practices for Computer Systems Used in Regulated Operations. *Pharm Technol* 1997.

Guide to Inspection of Computerized Systems in Drug Processing, Reference Materials and Training Aids for Investigators, Division of Drug Quality Compliance, Associate Director for Compliance, Office of Drugs, National Center for Drugs and Biologics, and Division of Field Investigations, Associate Director for Field Support, Executive Director of Regional Operations. Food and Drug Administration, February 1983.

Technical Report No. 18. Validation of Computer-Related Systems: PDA Committee on Validation of Computer Related Systems. *PDA Journal Pharmaceutical Technology* 1995, 49(Suppl. 1).

Technology Impacts: Dr. Michio Kaku on "the Future in the Next 5–20 Years". www.youtube.com/watch?v=59EjxpltRss.

C:\Users\Anitha\AppData\Roaming\Microsoft\Word\15064-2030-Ref Mismatch Report.docx - LStERROR_615Use of AI in Research at Benevolent. "Kenneth Mulvany Interview". http://benevolent.ai/video/hot-topics-interviews-ken-mulvany-founder-of-benevolentai/.

What Is AI?. www.youtube.com/watch?v=kWmX3pd1f10.

GENERAL SOFTWARE QUALITY REFERENCES

C:\Users\Anitha\AppData\Roaming\Microsoft\Word\15064-2030-Ref Mismatch Report.docx - LStERROR_617Dustin E, Rashka J, Paul J. *Automated Software Testing: Introduction, Management and Performance.* Addison Wesley Longman, Inc., 1999. ISBN 0-201-43287-0.

Ebenau RG, Strauss SH. *Software Inspection Process.* McGraw Hill, 1994. ISBN 0-07-062166-7.

Fairley RE. *Software Engineering Concepts.* McGraw-Hill Publishing Company, 1985. ISBN 0-07-019902-7.

Halvorsen JV. A Software Requirements Specification Document Model for the Medical Device Industry. Proceedings IEEE SOUTHEASTCON'93, Banking on Technology, Charlotte, NC, April 4–7, 1993.

Kaner C, Falk J, Nguyen HQ. *Testing Computer Software.* 2nd ed. Vsn Nostrand Reinhold, 1993. ISBN 0-442-01361-2.

Mallory SR. *Software Development and Quality Assurance for the Healthcare Manufacturing Industries.* Inter Pharm Press, Inc., 1994. ISBN 0-935184-58-9.

Perry WE, Rice RW. *Surviving the Top Ten Challenges of Software Testing.* Dorset House Publishing, 1997. ISBN 0-932633-38-2.

Wiegers KE. *Software Requirements.* Microsoft Press, 1999. ISBN 0-7356-0631-5.

54 Risk Based Validation of a Laboratory Information Management System (LIMS)

Roger D. McDowall and Jeff Eshelman

CONTENTS

DOI: 10.1201/9781003163138-54

54.1 INTRODUCTION

Pharmaceutical Quality Control laboratories must work electronically if they are to survive.

This statement is not made because of regulatory requirements but simply because of business pressures facing the pharmaceutical industry today: profit margins are under pressure from government pricing; also, delays in accepting or rejecting raw materials, active ingredients or finished products costs time and money. As analytical laboratories are at the end of the production chain, any delay is visible and can magnify the cost of other delays elsewhere in production. Therefore, any implementation of a Laboratory Information Management System (LIMS) in a QC laboratory has to provide tangible business benefits through the elimination of paper records and the use of electronic signatures with associated electronic workflows.

Validation of computerised systems has also been undergoing considerable change following the FDA's GMPs (Good Manufacturing Practices) for the 21st Century [1]. This has been followed by the GAMP Forum's publication on GAMP 5 [2], which takes a risk based approach to computer validation. There is also a GAMP Forum Good Practice Guide on Validation of Laboratory Computerised Systems, second edition [3]. With the current focus on data integrity, a number of regulatory agency publications have been published that strongly suggest implementing technical controls to ensure working practices are enforced and also protect electronic records that are generated during analysis [4–8]. In addition, there have been a GAMP Guide on Records and Data Integrity [9] and two good practice guides that can be applied to a regulated laboratory [10, 11]. Guidance documents have also been issued by the PDA [12] and APIC [13]. The former provides useful compliance guidance for chromatographic analysis and microbiology, and the latter contains a comprehensive checklist and practical examples for data process analysis for data integrity, which we will discuss in more detail in this chapter.

LIMS validation must be cost-effective and risk-based to help deliver the benefits from a process-driven implementation within a relatively short period of time, or there is little benefit to an organisation. Recent regulatory trends have given significant consideration to data integrity, with multiple citations being issued by various regulators globally. Data integrity concerns must be holistically integrated throughout the installation and validation of any electronic system such as a LIMS employed in a pharmaceutical Analytical Development or Quality Control laboratory.

To appreciate and understand the rationale for this new approach to implementing a LIMS, it is important to understand the problems that face current installations. These can be summarised as follows:

- **Poor LIMS Implementation**: It is difficult to perform an effective LIMS implementation in a laboratory if the implementation strategy is to simply automate the current process. This results in a very expensive typewriter being implemented. A better alternative is to understand and streamline the process ahead of implementing a system.
- **No Interfacing to Analytical Instruments**: Failure to interface analytical instrument computer systems that generate the bulk of the data in QC laboratories to a LIMS results in manual entry of data. Manual data entry is a slow task and requires transcription error checking to ensure accuracy and integrity. It is still surprising to find the number of LIMS implementations that are stand-alone and fail to consider interfacing instruments within the laboratory or applications outside of it.
- **Calculations are Performed Outside of LIMS and Instrument Data Systems**: Many calculations are typically performed in spreadsheets or by calculators outside of the LIMS. The reason for this is mainly the data system is unable to provide the calculation or spreadsheets are widely available and easy to use. Alternatively, laboratory staff cannot be bothered to read the data system manual to implement the calculations.
- **No Interfacing to Production Systems**: Information and specifications contained in production systems are not transferred electronically to the LIMS; these data have to be input manually into the system and manually checked to ensure accuracy. Commonly available enterprise resource planning systems and LIMS can be readily interfaced. However, resource availability and time constraints often lead to this aspect of LIMS implementation being bypassed.
- **Extensive Customisation of a Commercial System**: Instead of using the standard workflows within a system, many laboratories implement LIMS by changing the system functions to fit the laboratory's current ways of working. This is inefficient and assumes that a laboratory's processes are efficient and effective. This assumption is usually wrong and creates additional cost and time delays for a LIMS project.

Thus it is unsurprising that many LIMS implementations are inefficient, not cost effective and take a long time to validate.

This chapter on risk-based validation of LIMS describes how to deliver substantial and tangible business benefits required of a LIMS in a GMP laboratory by redesigning the process before coupling this with an effective risk-based computer validation to comply with regulations. This chapter is structured in the following sections:

- Understand and improve the current ways of working
- Design the LIMS environment
- Specify, implement and validate the LIMS

The approach outlined here is based on the use of a commercial LIMS with configuration only and avoiding customisation of the application as appropriate for a specific laboratory. It is better to change the business process rather than be extensively customised. This section is written primarily for the implementation and risk-based validation of a LIMS in a single laboratory or site. The modifications of the approach required for a multi-site or global LIMS validation are to define the core requirements that all laboratories will use within the system and the initial validation of the core system that must not be modified. Local additions to the core system may be permitted, but these need to be specified and validated locally. A GAMP Good Practice Guide on global information systems control and compliance may be useful in this context [14].

54.1.1 Qualification Terminology

The IQ, OQ, PQ terminology is obsolete, as discussed in section 3 of the FDA guidance on General Principles of Software Validation [15]. However, this terminology is still used widely by the pharmaceutical industry and consequently by LIMS suppliers. Qualification phases for the installation and checkout of the components and by a supplier (IQ and OQ) will still be used in this chapter; however, user acceptance testing (UAT) of the configured system with instruments interfaced to it will replace PQ.

54.1.2 Computerised System Validation or Computer System Assurance?

FDA are developing a different approach for demonstrating that a computerised system is fit for its intended use. This is called Computer System Assurance (CSA) as opposed to Computerised System Validation (CSV). A draft guidance for industry has been promised since 2018 but to date has not been issued by the agency. The key points are less about the generation of paper but:

1. Critical thinking to identify key functions to test and manage risk
2. Leveraging the work of the supplier during software development to reduce the amount of customer testing
3. Using undocumented testing

As the draft guidance has not been issued yet, this chapter will include discussion only on items 1 and 2 and ignore item 3.

54.2 Understand and Optimise the Business Process

For successful use of electronic signatures within a new or upgraded LIMS, an electronic workflow is required. Therefore, a QC laboratory has to migrate from paper-driven or hybrid processes to an electronic ones. This is the key to a cost-beneficial validation of any LIMS: map, analyse, understand the data vulnerabilities and then optimise the business process to work electronically, to ensure data integrity and to use electronic signatures effectively.

This understanding and redesign work is achieved through two process mapping workshops; using the process mapping terminology, these are the "As Is" (current) process and the "To Be" (future) process. These two workshops need to be 2–4 weeks apart, as they are relatively intellectually intense; time is needed for reflection between each workshop so that the resulting material can be reviewed critically. When undertaking this work, it is important to realise that the process starts and finishes outside of the laboratory, and therefore staff working in areas that interface with the laboratory need to be involved as well as QC staff.

54.2.1 Understand the Current (As Is) Process

The purpose of this workshop is to understand the way the laboratory currently operates and how computerised systems are utilised inside and outside the laboratory. This workshop establishes a baseline and allows the participants to critically analyse their ways of working.

The "As Is" workshop should cover the following topics:

- What is process mapping? There are a number of techniques, but either cross-functional process mapping or IDEF (Integrated DEFinition) are considered by the author to be the optimal approach (www.idef.com) to documenting and optimising a business process.
- Map the current process used in the laboratories.
- Map the boundaries of the current data systems and LIMS (if used).
- Identify spreadsheet and laboratory notebook use.
- Identify differences in working practices between laboratories.
- Identify SOPs/test methods used in the process.
- Identify process bottlenecks where delays occur and the reasons for them.
- Identify data vulnerabilities so that the new process incorporates technical means to ensure data integrity.
- Identify process steps where and why signatures and initials are used. This aspect is critical for identifying the proper alignment of electronic signatures and electronic records.
- Obtain process metrics: for example, how many, how much, how long and how often.
- Identify process improvement ideas.

It will soon become apparent from this workshop that processes are inefficient, as they are paper and hybrid driven, and that computerised systems are not being used to their full potential.

An example of an "As Is" process flow for a QC laboratory is shown in Figure 54.1; this shows that the process

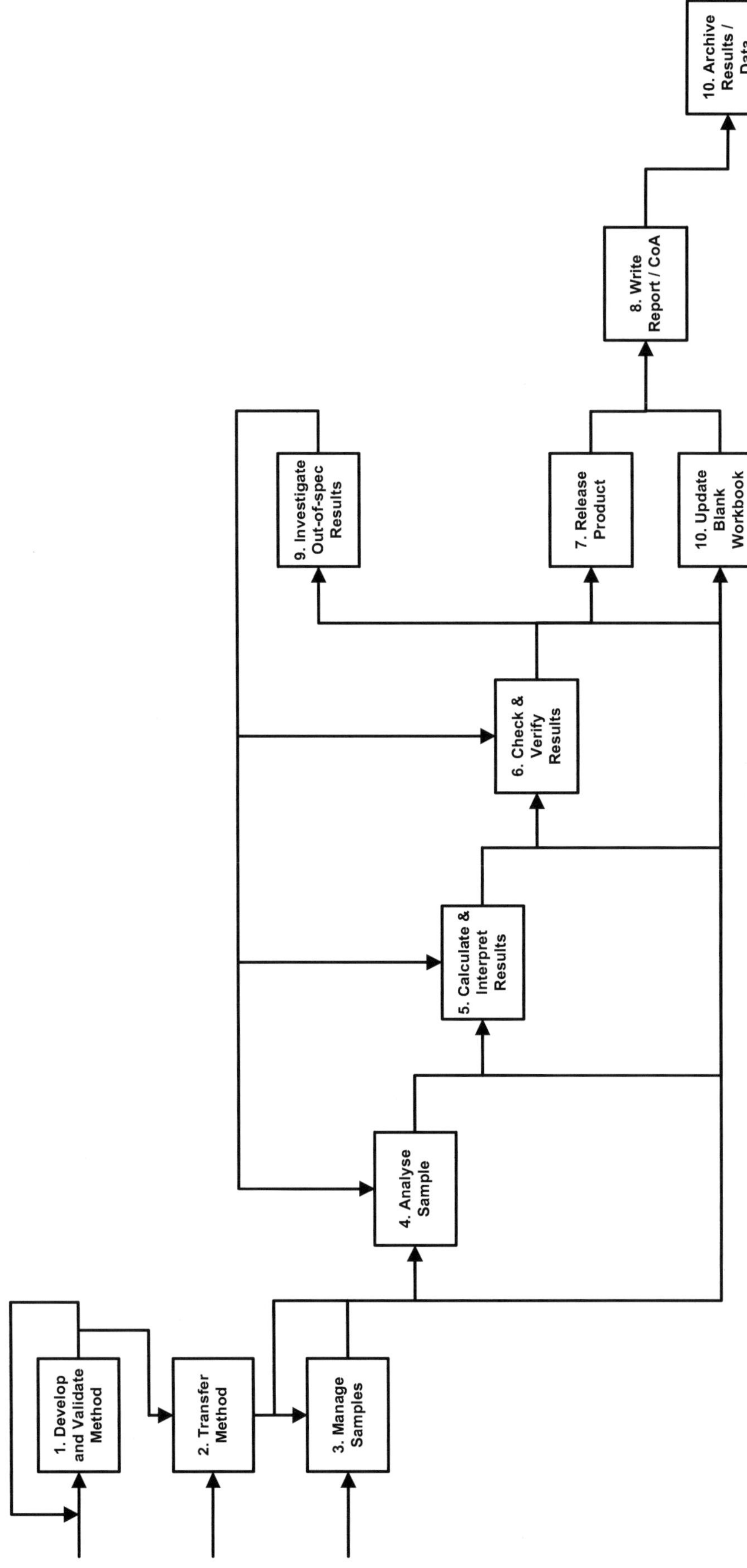

FIGURE 54.1 "As Is" process map in a QC laboratory

is paper driven, as blank and uncontrolled worksheets are maintained outside of the LIMS in addition to information stored within the instrument data systems and LIMS. Also, instruments are not connected to the LIMS, and calculations are performed with calculators and spreadsheets. The LIMS has all data manually entered into the system and is an expensive typewriter. When faced with a typical "As Is" process map, it is obvious that to implement and validate a LIMS in a QC environment will be a huge waste of resource with little if any payback for the organisation. If a global LIMS is required in an organisation, process mapping is invaluable, as it highlights where the differences are between laboratories and identifies these as areas for harmonisation in the new process.

54.2.2 Optimise the Process for Electronic Working

To improve and optimise the process, a second workshop is carried out after the draft report from workshop 1 was circulated for review. Note the careful phraseology: we are optimising the process, not re-engineering it; the reason is that much can be achieved with a short optimisation workshop rather than a full-scale process re-engineering project. The underlying assumption is that the basic operation of a regulated QC laboratory is sound; it is only the details that need to be improved or redesigned, not re-engineered.

The three basic operating principles of the electronic laboratory, according to Jenkins [16] and discussed by McDowall [17], are:

1. **Capture Data at the Point of Origin**: If you are going to work electronically, then data must be electronic from first principles. However, there is a wide range of data types that include observational data (e.g. odour, colour, size), instrument data (e.g. pH, LC, UV, NMR etc) and computer data (e.g. manipulation or calculation of previously acquired raw data). The principle of interfacing must be balanced with the business reality of cost-effective interfacing: what are the data volumes and numbers of samples coupled with the frequency of the instrument use?
2. **Eliminate Transcription Error Checks**: The principles for design are as follows: never re-enter data and design simple electronic workflows to transfer data and information seamlessly between systems. This requires automatic checks to ensure that data are transferred and manipulated correctly. Where appropriate, implement security and audit trails for data integrity and only have networked systems for effective data and information sharing.
3. **Know Where the Data Will Go**: Design data locations before implementing any part of the LIMS and the LIMS environment. The fundamental information required is what volumes of data are generated by the instrumentation and where the data will be stored: in an archive system, with the individual data systems or on a networked drive? The corollary is that security of the data and backup are of paramount importance in this electronic environment. In addition file naming conventions are essential to ensure that all data are uniquely numbered, either manually or automatically. If required, any archive and restore processes must be designed and tested so that they are reliable and robust.

These principles should be used to optimise the "As Is" process maps to define the new or "To Be" process in the second workshop, which typically will cover the following items:

- Review of the "As Is" process maps with modification where necessary to reflect the current working practices.
- Optimise and harmonise (especially between laboratories) the process and generate the "To Be" process using the following inputs:
  - Improvement ideas generated in workshop 1.
  - Eliminating unnecessary process steps.
  - Identifying any manual steps to be automated by the new LIMS or other computer systems.
- Defining the new boundaries of the LIMS and other computer systems inside and outside the laboratory.
- Identify data transfers between these systems.
- Estimate potential time and calculation of time savings from the new process.
- Identify any "Quick Wins" for rapid implementation (these are defined as improvement ideas that are cheap to implement but provide high benefit and give credibility to the overall approach).

The new process map is shown in Figure 54.2; the process has been made electronic and data are transferred to the LIMS electronically from the data systems to eliminate manual data entry. Electronic signatures have been implemented within the LIMS to eliminate much of the current paper records. Note that paper will not be eliminated entirely, but the majority of records will be electronic. The work list is no longer required, as the information will be maintained electronically. Although the main tasks in the process still remain, the time taken when working electronically between steps 5, 6 and 7 has been cut by approximately 50–60% because the systems are set up to work electronically. If records of sample preparation are to be eliminated, then a laboratory execution system (LES) or the LES functions of some LIMS can be used to automate this portion of the process.

Any record vulnerabilities identified in the As Is workshop need to be addressed during the To Be discussions on the process and throughout the implementation of the new LIMS and interfaced instruments and systems.

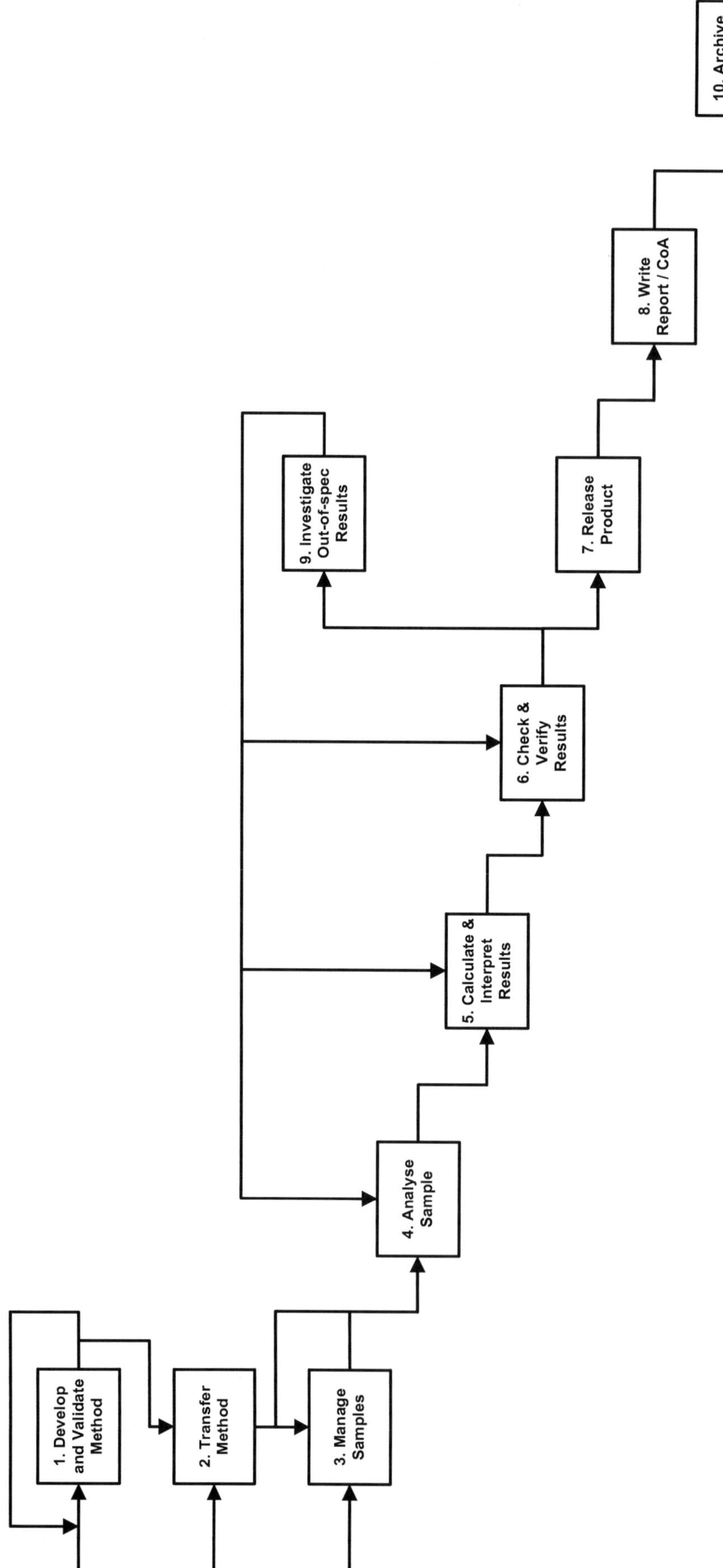

FIGURE 54.2 Optimised "To Be" electronic process for a laboratory.

54.2.3 Estimated Benefits of Working Electronically

When the new process has been defined and mapped, the new timings of the process can be estimated and compared with the current process. Based on the differences between the two, an estimate in overall time and resource saving can be calculated. Savings should be large enough to cost-justify the system on tangible business benefits alone, including faster product release, quicker acceptance and rejection of raw materials and holding less stock. While intangible benefits such as quality are important, the organisation needs to know if there will be a payback from the investment.

However, do not assume that better processes will result in head count reduction in a laboratory. What it means is that the overall laboratory process will work more efficiently and faster; however, the LIMS will mean changes in the laboratory staff roles. LIMS application administration will be needed where none existed before, for example, power users within the laboratory will be the first line of help for users, and staff will be needed for the inputting specifications (if not done automatically) and methods into the LIMS. Do not underestimate the amount of work that this will entail.

54.3 DESIGNING THE LIMS ENVIRONMENT

As stated in the introduction, QC laboratories must work electronically if they are to survive; therefore, the LIMS environment needs to be defined based on the optimised process. The first stage in designing this is to look at a LIMS as an interface between the laboratory and production. This will be followed by interfacing the LIMS and other computerised systems within the QC laboratory to produce an electronic LIMS environment to support the newly designed process.

54.3.1 Positioning a LIMS: Hitting Two Targets

It is important to realise that a LIMS should automate both the laboratory where it is implemented and the production facilities that the laboratory serves. To be effective a system should deliver benefit to both the laboratory and production. How should this be achieved? A LIMS is unlike any other piece of laboratory automation equipment available to the analytical chemist. It can provide benefits both within the laboratory and outside of it. Thus a LIMS has two targets:

- The laboratory: the information generator
- The organisation: the sample provider and the information user

The problem is how to implement a system so that it hits both targets effectively. Figure 54.3 shows an outline of the functions that a LIMS should undertake in a simplistic way. The diagram shows a LIMS sited at the interface between a laboratory and an organisation. Samples are generated in the organisation and received in the LIMS, and then the samples are analysed within the laboratory. The data produced during analysis are reduced within the LIMS environment to information which is transmitted back into the organisation. Figure 54.3 represents the ideal positioning of a LIMS: the organisation and the laboratory both benefit from the system.

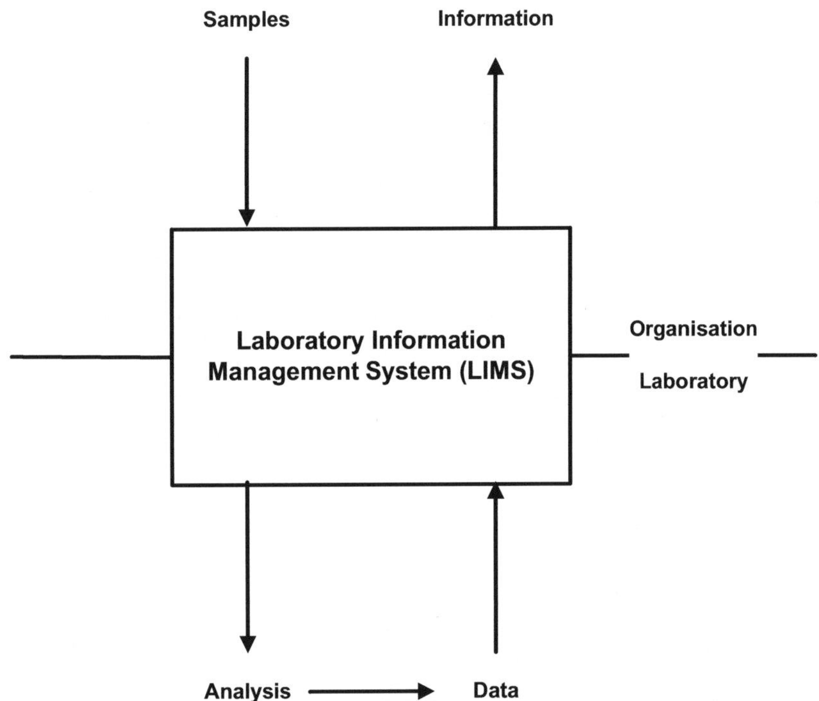

FIGURE 54.3 A LIMS delivering benefit to the laboratory and the organisation.

Quality Assurance and IT participation in the process is critically important, along with input from the LIMS supplier and any external service providers used for the implementation. Efforts must be made to ensure data integrity throughout the design and configuration process for LIMS, as with any system in a regulated environment.

54.3.2 The LIMS Environment

A successful LIMS implementation builds a LIMS environment to serve both the organisation and the laboratory. The key to success is that the LIMS must integrate the processes and the computerised systems in these two areas where analytical information is generated and used.

Some of the applications outside of a laboratory that a LIMS could be interfaced to design the LIMS environment are listed here and in the top half of Figure 54.4:

- E-mail systems for transmission of reports to customers or keeping them aware of progress with their analysis (care needs to be taken if e-mail is being used to transfer GMP relevant data or information).
- Web servers for laboratory customers to view approved results and also for contract laboratories to input data into the QC LIMS.
- Enterprise Resource Planning (ERP) systems for linking the laboratory with production.
- Applications maintaining product specifications.
- Data warehouses.
- Electronic Document Management Systems.
- Failure Investigation Systems.
- Electronic Submission Systems (for GMP laboratories in pharmaceutical R&D).

These are just a few of the possible applications that a LIMS could be interfaced to; the list of potential candidates will be based on the nature of the analytical laboratory and the production organisation it serves.

Some ERP vendors can claim LIMS functionality and that there is no need to implement a LIMS; however, the problem with this approach is that the ERP's concept of a QC laboratory often does not match the reality. For example, the sample process flows within an ERP tend to be high level and very simplistic and cannot automate all laboratory

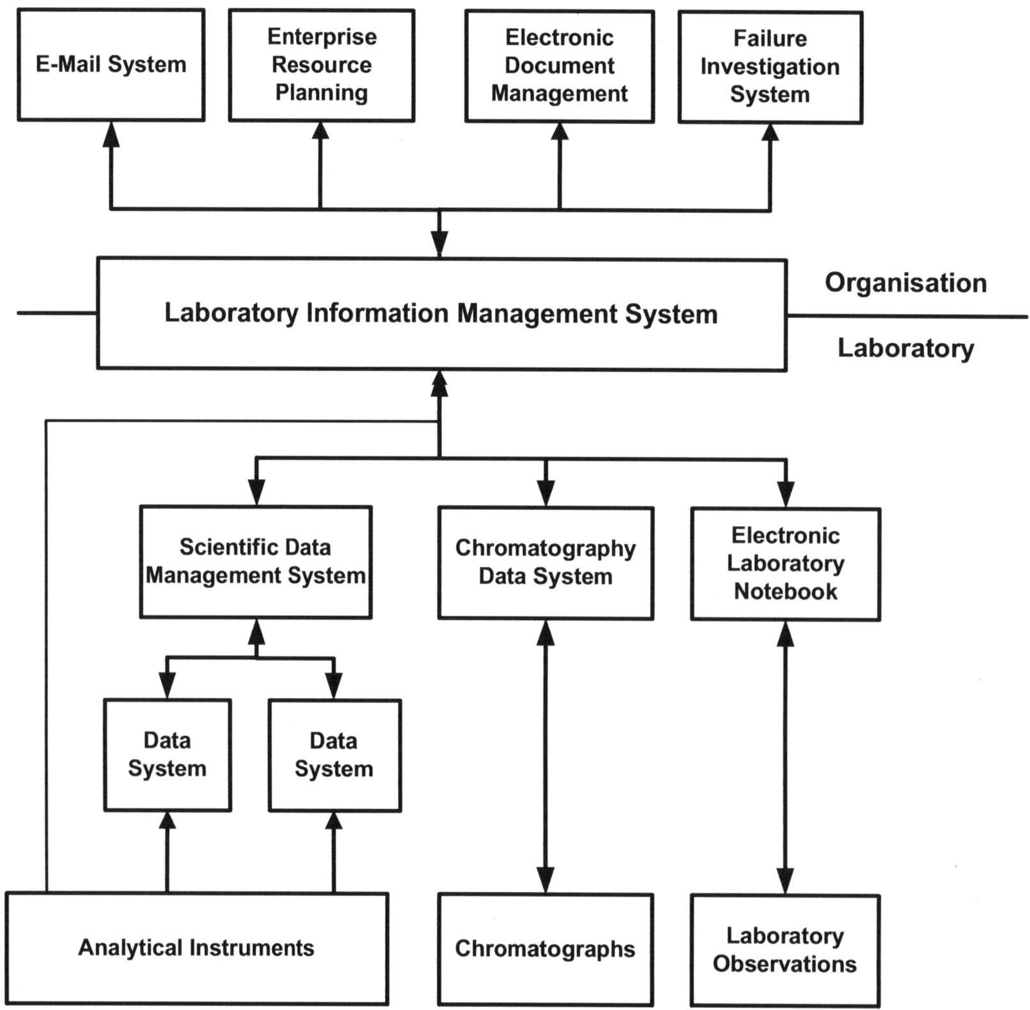

FIGURE 54.4 Options for a LIMS environment to integrate the laboratory with the organisation.

functions, for example, out of specification (OOS) investigations, without extensive writing of custom software. Once the organisational side of the LIMS environment has been designed, the LIMS environment within the laboratory needs to be designed.

Designing the LIMS environment means that you need to consider the other systems in the lab that must interface with the LIMS. This includes other laboratory applications such as scientific data management systems, chromatography data system (CDS) and electronic lab notebooks, as well as various data systems that may be attached to those or run independently. It also includes analytical instruments, chromatographs, and laboratory observations, as shown in the lower half of Figure 54.4. Data can be transferred to the LIMS by a variety of means:

- Direct data capture by the LIMS.
- Data capture by an instrument data system with analysis and interpretation, and only a reportable result is transferred to the LIMS.
- As above, but the results or electronic records are transferred to the LIMS via a Scientific Data Management System.
- Laboratory observations can be written into a notebook then entered manually into the LIMS or captured electronically via an Electronic Laboratory Notebook (ELN) and transferred electronically to LIMS.
- Bar codes (or Radio Frequency Identity—RFID) can be used to label samples and enter data rapidly into the LIMS.

Before implementing a LIMS, it may be appropriate in some laboratories to standardise and implement instrument data systems, for example, the chromatography data system. The rationale for this approach is that a data system can be quicker to implement than a LIMS and it will provide a firm foundation to build the LIMS above it. If it is done the other way around, the data system may need to be updated later, with a consequent change in working practices and more revalidation.

54.4 LIMS IMPLEMENTATION AND RISK-BASED VALIDATION

This section of the chapter deals with the life cycle of a category 4 computerised system as outlined by GAMP 5 [2] and modified by McDowall [17]. The aim is to realise and deliver through the LIMS those business benefits identified in the process redesign in a cost-effective manner.

54.4.1 Electronic Records Generated by a LIMS in a GMP Environment

A change of a system from a records-based approach to the validation of computerised systems was suggested by the FDA in the Guidance for Industry on Part 11 Scope and Application [18]. The GAMP Good Practice Guide for Compliant Part 11 Records and Signatures [19] took this and developed a risk-based approach to validation of a computerised system based on the impact of the records generated and managed by an application. This publication is now out of print and has been replaced by the GAMP Guide for Records and Data Integrity [9]. However, this bottom-up and record-focused approach to validation does not generate business process efficiencies that the process mapping and redesign approach will do for the implementation of any computerised system. However, with the emergence of both data falsification and poor data management practices after the Able Laboratories and Cetero Research fraud cases [20–22] and the focus by the FDA on data integrity via the Compliance Program Guide 7346.832, which was updated in 2010 [23] and again in 2019 [24], means that electronic records and the associated data and metadata are in the limelight again.

The approach for a LIMS implementation should be to focus on the business efficiencies, but during the prototyping phase of the project, identify the electronic records created by the system and assess their vulnerability. Implement technical controls to protect these records, as once validated they can be used time and time again. This is in preference to using procedural controls and training, as these are error prone and will need increased review scrutiny by a second person over an electronic process.

The records generated and managed by a LIMS in a GMP environment are high impact, as they are used in product release and/or product submission. Some examples of electronic records that could be contained within a system are listed here; this list should not be considered exhaustive, as it depends on how a specific LIMS has been implemented and used:

- Specifications of products, intermediates and raw materials
- Stability protocols
- Sampling methods
- Analytical methods
- Worklists
- Observations and results captured directly by the LIMS, e.g. pH and balance measurements
- Results transferred from analytical systems, e.g. spectrometers, chromatography data systems
- Comparison of results versus specification and identification of Out of Specification (OOS) or Out of Trend/Expectation (OOT/OOE) results
- OOS investigations and, where appropriate, additional results
- Electronic signatures
- Certificates of Analysis
- Audit trail entries
- Instrument qualification and calibration status

The electronic records need to be identified and documented [18]. It is important to understand that this is not a static process; as the LIMS is updated, the new version may contain new functions that may create new electronic records in addition to those listed. If new functions are added using the scripting language, then these may also create new electronic records. Therefore it is important to review this list on a regular basis; when the system is upgraded and during a periodic review are the obvious times.

These are high impact records, as defined by GAMP Guide on Records and Data Integrity [9], as they can impact product quality and/or patient safety. Therefore, a more rigorous approach should be adopted which includes:

- **Hazard Identification**: the hazards that the LIMS could face should be identified, along with the consequences of each one. However, although a hazard and its associated consequences may have been identified, we do not know if a specific one poses a risk to the system. To identify the potential risks to the system, a risk assessment needs to be undertaken.
- **Risk Assessment**: For each hazard identified, the severity of the consequence and probability of occurrence both need to be estimated; this is achieved by allocating high, medium or low levels. There are different classes of hazard such as human, software, hardware, IT support, physical and environmental. The system to detect each of the hazards identified the effect then estimated it as high, medium or low [3]. Risks will be classified as either as class 1, 2 or 3 (high, medium or low risks) to identify which risks are important enough to implement mitigation controls.
- **Control Selection**: Controls for electronic records and electronic signatures generated by systems can be implemented at a number of levels:
  - Organisation via policies and standards, e.g. validation policy and passwords.
  - Procedural (and implicitly training) via SOPs, e.g. user manual and change control.
  - Application and network via technical controls such as audit trail, application and/or network security and checks.
  - IT Infrastructure via network security, backup and recovery, hardware and network redundancy.
  - Computer system validation.

Owing to the nature of a LIMS in a Quality Control Laboratory in a GMP environment, the system will require validation plus other controls to mitigate risk and protect the electronic records such as application security and access control and one or more audit trails for working electronically. In addition, the server needs to have redundant components such as dual processors, disk controllers and resilient storage to ensure that data are protected and not lost due to hardware failure. This section will concentrate on the risk-based validation of a LIMS; it is intended to build upon the process redesign and design of the LIMS environment outlined earlier to ensure a successful and cost-effective LIMS implementation.

54.4.2 System Implementation Life Cycle Activities

For the purposes of simplicity, the implementation life cycle will begin with either the implementation of a new system or an upgrade of an existing LIMS. This means the writing of the initial user requirements specification (URS) used to generate the request for proposal (RFP) used in the system selection process and the system selection will be omitted from this chapter. Readers who want to understand this part of the life cycle process should read the appropriate chapters from McDowall [17]. However, to leverage the work of the LIMS supplier's software development into their LIMS validation, this will be covered in the next section.

Therefore, the start of the implementation life cycle here will be where either:

- A new LIMS will be configured and installed in a laboratory but with an outline URS used to select the specific system.
- An existing LIMS installation will be upgraded to the latest application version.

There are three main work streams to consider that are outlined here and presented in Figure 54.5:

- Specification, installation and qualification of the computer hardware
- Validation of the LIMS application
- Writing procedures and training the users

Tasks for the three streams are shown in Figure 54.5, and this will help to put the remaining tasks in this section into context.

54.4.3 LIMS Supplier Assessment

EU GMP Annex 11 [25] notes that assessment of suppliers should be risk-based. In this section we look at two aspects of supplier assessment:

1. Software development and the QMS
2. IT support if a cloud deployment of the LIMS is being considered

Ideally, supplier assessment should be performed and completed before the purchase order and the contractual terms and conditions are agreed. This is best in the time between final system selection and placing of the purchase order. It enables payment terms and any outstanding issues to be resolved.

Remember that the LIMS supplier will be an extension of your laboratory for the software and your IT partner for a cloud deployment. Therefore, it is essential to ensure you have the right partner.

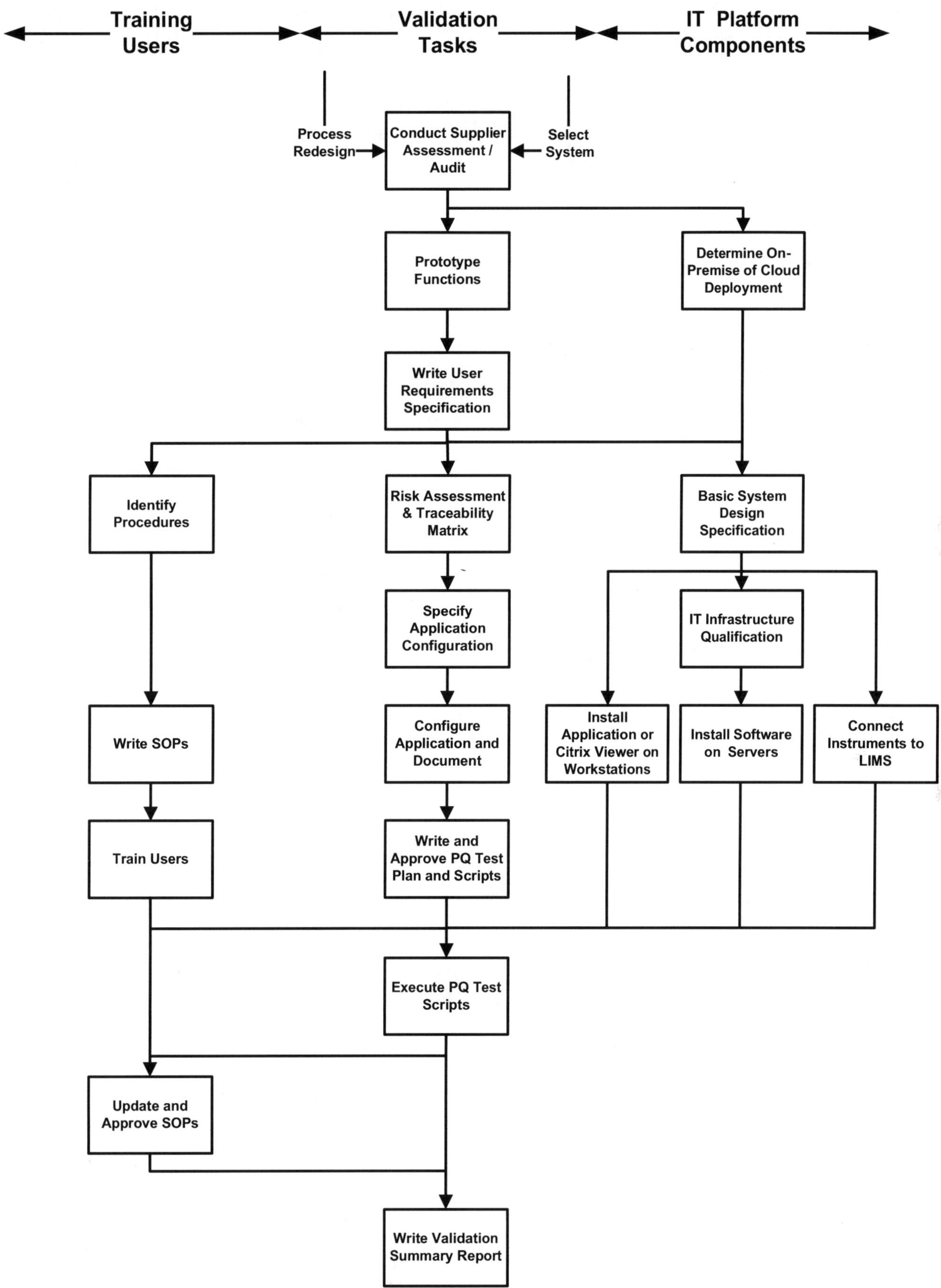

FIGURE 54.5 Outline of tasks involved in a LIMS risk-based validation.

54.4.3.1 Leveraging Supplier Software Development to Reduce CSV Effort

As part of a risk-based approach, it important to assess the supplier's software development process. In many organisations, this is performed by sending the LIMS supplier a questionnaire, checking the answers and filing the document with the rest of the validation material. This is not the way to leverage the supplier's work to reduce the overall validation effort.

A proactive approach is to gather information on the supplier's quality management system (QMS) and procedures for software development and support. This can be achieved by a questionnaire. However, there needs to be an assessment either on site or online that focuses on the following:

- There should be a short time to confirm the material in the questionnaire is correct for the QMS and associated procedures.
- The main focus of the assessment must be on how software functions are specified, defined further and programmed; how the code is peer reviewed, how any corrections are made and how it is formally tested; how errors are identified, resolved and integrated into the code base; how the software is release tested and the criteria for release.
- Support processes for documenting, assessing and resolving errors found by customers are also important.

The aim of this assessment is to see how well the software development process documented and the extent of specification and testing to answer the question of whether you can rely on the work performed by the supplier. If the company uses software engineering tools that support Agile development, these can be a great help to determine if the laboratory can accept supplier testing.

A LIMS application is a GAMP software category 4 application [2] in that the software can be configured to change the business process automated. However, within the overall category 4 application, there are category 3 functions that cannot be modified to change the business process. The functions may be parameterised, but they are essentially the same as tested by the supplier. Therefore, if the supplier's software development process is acceptable, all category 3 functions can be accepted as validated by the supplier. This must be documented, and functions in the URS should be identified as either category 3 or 4 to help target the laboratory's testing effort.

Validation focus can then be placed on the category 4 functions; however, there is a range of these in a LIMS:

- Simplest configuration: functions that can be turned on or turned off within the software, some of which may be linked with parameter, e.g. password length, password complexity and expiry. Further risk assessment can be applied, such as why would password expiry need to be tested when it is a timing function, and as long as there is a trusted time source, this can be assumed to work.
- Development of functionality using a language provided by the LIMS supplier. GAMP 5, in Appendix M4, notes that software developed using this approach is on the border between category 4 and 5 and should be treated as custom development [2]. Often, this is controlled after initial validation of the system by an SOP that defines the process along with expected evidence in order to demonstrate that the procedure was followed.

Using a supplier assessment in this way can reduce the overall amount of testing of the LIMS and interfaced instruments and systems.

54.4.3.2 Software as a Service Assessment

The traditional approach to LIMS deployment is changing slowly within the conservative pharmaceutical industry: instead of purchasing and operating physical infrastructure together with purchased applications installed on-site, companies are turning to the cloud and virtual infrastructure and leasing applications as Software as a Service (SaaS) including LIMS. Early in the project there should be a decision on the type of LIMS deployment either on-site or SaaS. It is important to be proactive in assessment of a LIMS supplier's IT service, as your data and its long term retention depend on this.

If an SaaS deployment of the LIMS is contemplated, then the assessment must be extended to include the IT support. Although technically achievable, a SaaS LIMS deployment should not be contemplated without understanding and mitigating both business and regulatory risks that this approach entails, as the LIMS supplier now becomes your IT provider.

In most regulated organisations, the IT function reports via the Finance Department to the Board. As such, expenditure on IT infrastructure and applications is open and visible to the group responsible for controlling costs. The SaaS approach has several advantages from the financial perspective:

- A move from capital to revenue spend. Revenue spend comes straight off the bottom line rather than a capital depreciation over several years.
- No hardware to install; this is managed by the hosting company or companies.
- Fewer IT staff required in-house to maintain the application; however, the technical contract and service level agreement (SLA) still need to be monitored.
- There is no need to order hardware, as a SaaS application can be designed to expand or contract to meet current needs (on-demand service), if required.

As we are discussing a cloud solution for a regulated laboratory, let us look briefly at what the regulations are for an application from an IT perspective.

The main regulations for defining implementation, validation, and operation of lab informatics software in a GxP regulated environment are:

- EU GMP Annex 11 regulations [25]
- OECD GLP No 17 Application of GLP Principles to Computerised Systems [26]
- 21 CFR 11, Electronic Records and Electronic Signatures [27], with interpretation by the underlying predicate regulation 21 CFR 211 [28]

Annex 11 provides the best overview of IT requirements for a computerised system in a regulated environment: system owner, staff training, physical and logical security, backup, user account management, change control, problem management and disaster recovery. In addition, both Annex 11 [25] and OECD No 17 [26] permit the use of SaaS applications and provide a requirement for agreements to cover the roles and responsibilities of both parties and define the scope of the agreement.

To appreciate the business and regulatory risks that must be managed, let us look at the situation with a traditional in-house approach. Traditionally, the landscape of IT in a regulated GxP company is:

- Regulated laboratory, e.g. analytical development, bioanalysis or quality control where the process owner resides
- IT department where the system owner is responsible for:
  - Physical and virtual infrastructure and infrastructure applications
  - Regulated application installation and monitoring
- Support services for the application, such as application administration and supplier liaison in addition to those already listed
- Quality oversight, which can be a single group overseeing both the business and IT or separate quality functions—one in the business and a separate IT quality group

All these functions are within the organisation and can be controlled within that organisation.

When moving to a SaaS application for your LIMS solution, most of the IT operations are outsourced to at least one if not more independent entities outside the direct control of the outsourcing company. The only control that the regulated laboratory has is via a contract, technical and quality agreement or service level agreement (SLA) that defines the roles and responsibilities of each party and the service levels provided. But:

- Do you know all parties involved?
- Are the providers compliant with procedures, and have staff been trained in GxP awareness?
- Are there records of backup, change control, etc.?
- Do you know where your data are?
- Are you sharing your server with a dating site or similar business with less stringent security needs?

Remember, all you have is a contract and service level agreement (SLA) with the service provider and the accountability for the LIMS and the data it contains remain with you.

To manage business risk, you need to understand what is happening with your data. SaaS can mean one of the following options:

- Single GxP entity: a SaaS company that operates the physical infrastructure, generates and monitors your virtual infrastructure, and then installs and operates an application on your behalf; i.e., the supplier is both an Infrastructure as a Service (IaaS) and SaaS provider.
- Dual GxP entity: a SaaS company installs an application on a GxP IaaS provider.
- Dual partial GxP entity: a SaaS provider uses a major IaaS provider such as Amazon Web Services (AWS), Google Cloud or Microsoft Azure.

We will discuss the third option, as it carries the greatest potential risk, and focus on the IaaS provider first. It will be highly unlikely that the laboratory will be allowed to audit the data centre or the organisation providing IaaS. Instead, they will rely on a combination of ISO 27001 certification [29] or equivalent plus a white paper, typically written by a third party, which justifies this stance. The problem with this approach is the one major omission: there is no mention of staff GxP awareness training. This is a risk that must be managed.

The question you have to ask is how does the SaaS supplier manage this risk, and is that management adequate? Remember the SaaS provider is now your IT department for this application, and you are responsible and accountable for compliance and risks to your data.

Assessing a SaaS supplier involves more than sending a questionnaire—it requires a multilevel approach to determine the adequacy of the provider's approach to the following considerations:

- Software life cycle development and support of the application (as discussed earlier)
- IT support services, including qualification of virtual IT infrastructure and GxP awareness training of the supplier's staff
- Qualification of the application by the supplier and how the application is configured prior to validation by the laboratory
- Software updates, revalidation (discussed in Section 54.4.8.3) and change management
- Risk management in dealing with the IaaS supplier
- Agreements with the supplier defining work to be performed during implementation as well as the operations already listed

We will discuss the key topic of software updates and SaaS. The traditional pharmaceutical industry approach to computerised system validation (CSV) is that once the project is finished, there should be no changes to the system, as validation is seen as expensive. This is wrong, and this attitude must change, because control of the application is lost when moving to the cloud. Often, software suppliers will use an Agile software development model, and this can result in new version releases every three to four months. This is obviously good news for CSV personnel who want a job for life but a nightmare for companies who will be on a validation treadmill.

A solution is to only use SaaS providers that batch software releases into a single annual upgrade. The upgrade should be accompanied by release notes to allow each laboratory to understand the changes, update and execute validation documents and write change requests. The supplier will require you to accept those annual updates, as it is easier for them to maintain the current and a previous version rather than a multitude of versions that only gets larger over time. It is far simpler to upgrade on an annual basis, as changes are likely to be relatively small compared to the big bang when a Dear Esteemed Customer letter arrives on your desk informing you that the current on-site application goes out of support in six months. This is when your choices get very expensive.

A key aspect of hosting in the cloud is planning for what to do at the end of the contract or when you want to change to a different application and provider in the future. Can you get your data back from the cloud, or do you have to keep paying support fees forever? The possibility of de-clouding is a mandatory topic in the planning process, to be discussed both internally and with the SaaS provider. Options for de-clouding should be included in the contract.

54.4.4 Updating the User Requirements Specification

For a new system, there will be an initial LIMS URS available for system selection; however it will need updating for the purchased new version of the application. The reason for this is that the URS used for system selection is general in character and is unlikely to be sufficiently specific to design tests for the validation of the selected LIMS. Therefore, an application and version specific URS needs to be written that defines the intended use of the system and contains the functions and the capabilities required by the system. It is this document that the user acceptance tests (UAT) tests will be based on using the risk assessment and traceability matrix documents.

For the LIMS upgrade, the existing version of the URS will not cover all the new functions available and this document needs to be reviewed and updated where appropriate. This should be the easier job as users will be trained on the current version of the system. The release notes provided by the application vendor will be useful to highlight areas where the URS will need to be updated if the laboratory intends to use them (see below). Each requirement in the URS should be prioritised as either mandatory (must have requirement) or desirable (nice to have but if not available the LIMS functionality is not impaired) [17].

54.4.4.1 Read the Application Release Notes

To focus on the changes that have occurred in a new release of the LIMS software, read the software release notes to understand the nature of the new features that have been added as well as the software errors that have been fixed. Although this sounds simple and straight forward, in reality this is more complex due to the way that the pharmaceutical industry handles with software in a regulated environment. If version 2 of a LIMS application has been installed and validated, typically the laboratory will miss the next release, version 3, unless there is a good reason for change. Version 4 of the LIMS will be implemented instead; thus, the laboratory implements every other version of the software rather than keeping current due to the perceived cost and effort of validation. Therefore, reading the release notes of the last two (or more) versions of the software and understanding the impact is the norm rather than the exception. This means that new features need to be understood and prototyped to understand their value and potential impact on the laboratory's ways of working.

However, the pharmaceutical industry's usual approach is to leave a system unchanged as revalidation is considered too expensive. This course of inaction can be dangerous, as a laboratory can find that the software goes out of support, resulting in a far bigger project than simply updating the software regularly with much less revalidation effort.

54.4.4.2 Using Process Maps to Define System Requirements

An additional advantage of the "To Be" process maps is their use in facilitating the requirements for the LIMS and the other systems used in the LIMS environment. The traditional problem with writing requirements for any computerised system is that obtaining requirements can be akin to extracting teeth from the users. The terminology "gathering requirements" implies that they are freely available to be written down but often nothing is further from the truth. The advantage of the process maps are that they provide an effective medium for obtaining requirements. Each activity in the process map has inputs and outputs defined from the workshops. A facilitator then needs to ask the laboratory users what happens in each process activity. This means that the requirements can be more precisely defined providing a greater certainty in system specification, selection and validation as the users are focussed on a specific task.

54.4.4.3 Better Definition of User Requirements: The Role of Prototyping

Prototyping is an important tool to help understand how a new LIMS application or the new features of an upgrade can work within a laboratory. The corollary is that users must have been trained on the new version of the software rather than reading the on-line help files. Features can be evaluated in an unqualified installation to identify if they are useful and then to refine how each one may be used to best advantage and business benefit. Although it is valuable, prototyping has to be handled with care, only two rounds of prototyping should be undertaken, e.g. high level to determine which functions should be in the implementation and which should be excluded followed by a second round for further in depth evaluation of the selected options and finalising the details of operation.

There is a danger that prototyping can be unstructured with little documentation from the exercise. From experience, the best way to tackle this is to have as defined outputs from each phase of the prototyping an update of the URS plus outline testing documents. If the LIMS scripting language is being used during this work, then documentation of the functions being modified needs to be generated and maintained.

All of these documents should be uncontrolled but unless they are available for review outside of the project team, the second phase of the prototyping work cannot proceed. This approach is intended to instil the discipline to ensure the work is documented as it goes on but also is an investment in time to reduce the amount of effort needed later to write the UAT test scripts.

54.4.4.4 Addressing Data Vulnerabilities

As the prototyping process draws to a close, it is important to identify the electronic records (both the data and associated contextual metadata) that are generated in the new process, including any interfaced systems. The technical controls in the software that have been activated to protect these records should be assessed to ensure that any vulnerabilities have been addressed and risks have been adequately mitigated. These controls should be documented in the system specification documents.

54.4.4.5 Defining Electronic Signature Use

During the prototyping phase, electronic signature use should be evaluated to support the electronic workflows that were designed in the process redesign phase. It is important to understand the need to differentiate between identification of actions and signing of records. The former is akin to the correction of an error in a laboratory notebook, where an entry is struck through without obliterating the original and corrected, and then the initials and date of the person making the entry are appended. The latter is the formal signing of the page in the laboratory notebook by the owner to state they accept responsibility for the correct data on the page.

For many companies, it is unfortunate that compliance has overridden the regulations and records are signed by custom and practice more than they need to be. 21 CFR 211 regulations simply state that only two signers are needed per test (§194(a) sub-clauses 7 and 8) [28], the first to state that the results generated are correct and a second person to say they have been checked for accuracy and the correct procedures have been followed. Therefore, the LIMS needs to reflect the regulation rather than electronically sign everything.

54.4.4.6 Writing Testable or Verifiable User Requirements

Writing user requirements that are either testable (a functional test based on the exact wording of the requirement) or verifiable (traced to an action, e.g. IQ of the application or writing of a procedure) is a skill that needs knowledge and expertise to perform. All too frequently, URS requirements can be all encompassing (the system will be 21 CFR 11 compliant) but untestable or simply too vague (the system will be user-friendly). A better approach is to follow IEEE Standard 1233 for Software Requirements Specifications [30].

A requirement must have the two major requirements, capability and condition, and may have a third of constraint. For example:

- **Capability**: Access to the system requires a user to have a password.
  - As written, the only test that can be designed is whether a user has a password or not.
- **Condition**: With a minimum length of 12 alphanumeric characters.
  - This now provides the ability to test with 11 characters, 12 numbers, 12 letters etc.
- **Constraint**: Passwords will be aged every 90 days.
  - Constraints can be requirements in their own right.

From this bullet list, we can have one requirement, as follows:

- Access to the system requires a user to have a password with a minimum length of 12 alphanumeric characters aged every 90 days.

The problem with this requirement is that there is too much detail for effective traceability and risk assessment. For example, the password aging is mixed with the complexity and length requirements. Are you going to wait 90 days until the password expires? A better approach, as suggested previously, is to have two requirements:

- Access to the system requires a user to have a password with a minimum length of 12 alphanumeric characters.
- Password aging will be every 90 days.

Shorter and more focused requirements enable more effective risk assessment. For example, why test password expiry when it is time-bound? If time is synchronised with a trusted time source, there is no need to test this requirement. Similarly, do you want to test password complexity when it is enforced by the application or operating system? These approaches must be documented and justified to succeed.

54.4.4.7 Requirements Traceability

There is a regulatory requirement from EU GMP Annex 11 clause 4.4 that requirements must be traceable throughout the life cycle [25]. Therefore, requirements must be uniquely numbered.

If requirements are further broken down into configuration or functional specifications, then the numbering method should allow traceability back to the URS as well as forward tracing to validation documents written later in the validation process. This can be achieved by a numbering convention within a word document or by the use of an automatic requirements management tool.

54.4.5 Write the Validation Plan

For a new system implementation, a validation plan is required to control the work of the validation. As a minimum, it should define the roles and responsibilities of all individuals working on the project, the life cycle to be followed and the documentation to be written at each stage of the project. In addition, for global or site projects, the overall validation strategy should be presented: how the development and validation of a core application for all sites will be achieved, along with the documentation to support it, then how it will be installed at each site and under which conditions an instance can be modified by a local site.

For an upgrade of an existing LIMS, a change control request could suffice to control the work, but inevitably a validation plan is written, as the work will usually involve replacement of the server and modifications of the current ways of working.

54.4.6 Combined Risk Assessment and Traceability Matrix

Risk assessment is now a key validation requirement following the FDA's Part 11 Scope and Application guidance [18]. After the URS has been written, both the system and the individual functions need to be assessed for regulatory and business risk using the Functional Risk Assessment methodology [17]. Here, individual system functions are assessed as either critical or not critical (C or N respectively) from a regulatory and/or business risk perspective. Coupled with the prioritisation in the user URS, each requirement is graded as either mandatory or desirable as well as either critical or non-critical; these can be plotted in a 2 × 2 Boston grid to determine overall risk. Only functions that were both mandatory and critical were considered for PQ testing; all other combinations are not considered any further. The rationale for this is based on the vendor's testing of the application.

Mandatory and critical functions were then evaluated further to see if they need to be:

- Explicitly tested and then assigned a specific test script number where similar functions are tested together.
- Assumed to work, as there was no access to the algorithm.
- Implicitly tested, such as the windows and some display functions.
- Verified during the qualification of the system.
- Traced to a procedure or an SOP.

This is a simpler process for a commercial system than the modified Failure Mode Effect Analysis (FMEA) outlined in the GAMP guide [2].

54.4.7 System Architecture

The vendor's recommendations should be used to size and specify the database and application servers for the system. Storage locations and failover provisions also need to be specified, as will the other LIMS instances used for evaluation, training, validation and production. Diagrams of the overall system architecture will help understand the approach taken and should be encouraged to be drawn for inclusion here. Increasingly, rather than have a server for each instance, virtual servers are used running within an environment such as VMware; this is useful to reduce hardware costs and maintain individual instances of the LIMS. Data can either be stored on the production server or on a Storage Area Network (SAN). All details concerning the system architecture should be documented in a System Design Specification (SDS) or equivalent, as this is an input into the configuration records for the overall system.

As data integrity is a key concern of regulatory authorities globally, time synchronisation and a trusted time source should be used to ensure that date and time stamps are accurate and dependable. There are a number of options for this, use of a national observatory, network time protocol server or the GPS system.

54.4.8 System Installation and Check-Out (IQ and OQ)

54.4.8.1 IT Infrastructure Specification and Qualification

IT infrastructure used for the LIMS, whether physical and/or virtual, with on-premises or SaaS deployment, needs to be adequately specified. Typically, this will use supplier specifications as much as possible to reduce the effort of writing but include sufficient detail (e.g. IP addresses, active equipment names etc). If a SaaS option is used, the LIMS supplier should have these specifications available.

Installation plans for all the servers (database and application instances as well as any Citrix servers used for the application) should be written by the IT supplier. These plans should include the installation of the hardware and documenting its configuration as well as installing and configuring the operating system and any utilities for each server, e.g. agents for backup, network management software etc. The installation of hardware, operating system and any utilities for all servers must follow these plans and record the actual details of each server installed such as serial number and configuration (memory, processor type and speed and IP address etc).

54.4.8.2 LIMS Database, LIMS Application and Instrument Interface IQ

The activities that are involved in this task should be:

- An evaluation of the vendor's installation qualification documentation to check that it is acceptable. The documents should be approved before and reviewed after execution in all cases. Both USP <1058> [31] and EU GMP Annex 15 clause 2.5 [32] permit the merging of IQ and OQ protocols to make execution of protocols easier, if practicable.

- Installing the LIMS database and software on the respective servers for each instance by either a member of organisation's IT staff or the vendor's service personnel. The application IQ is completed and followed by a review of the documents at the same time to ensure that the appropriate testing has been performed with acceptable results.
- The analytical equipment and instruments to be interfaced to the LIMS in the initial phase of the LIMS implementation will be interfaced now and checked to make sure the connections work. Again, this will be planned, and there will be documentation available to demonstrate the activities undertaken.

54.4.8.3 Establish Change Control and Configuration Management

Once the servers and application has been installed, the system needs to be placed under change control. Some organisations write a specific change control SOP for each system; however, the smarter ones will have a single procedure that is applicable to all regulated systems. Allied with change control is configuration management, which is just as important but often neglected. Configuration management is the definition of the configuration items (CIs) that constitute the whole system. CIs consist of:

- Hardware
- Software
- Documentation (ranging from vendor supplied material including electronic manuals to company-specific documents)

The level of detail required should be sufficient to provide business benefit from the information; for example, server information will require make, model, processor size and speed, memory, disk size and configuration, operating system and service pack, and network information such as IP address etc. Less information would be required of a workstation; for example, typically this would be the minimum specification available to run the application within the company, as many organisations change workstations every 3–4 years. When a change is made, the configuration management records before and after the change should document what CIs have been modified, added or removed.

54.4.8.4 Do I Need a LIMS Application OQ?

Traditionally, there is now an operational qualification to demonstrate that the LIMS application software works as the vendor intended it to. Here is where we can take advantage of a risk-based approach. Some OQ packages offered by LIMS vendors are their internal test suites, used either as is or modified for external sale for demonstrating that the unconfigured application works. Look at the process; the vendor produces and tests the base application and manufactures the media from which you install the same software. Do you need to execute essentially the same tests that the vendor has? No. Furthermore, the application that is installed will be configured by the laboratory to their own ways of working away from the base package, making the execution of a comprehensive OQ a further waste of time. Therefore, review the OQ document to see what is covered. All that is required is a demonstration that the application works from the supplier's perspective so that a qualified and unconfigured LIMS application is handed over to their customer. Anything more extensive should be omitted from a LIMS validation as it adds little, if any, value to the overall project and is relatively expensive and time consuming to perform. If the user acceptance testing is executed correctly against the user requirements, then the application functions as intended.

54.4.9 Configuring the System

LIMS do not have "ON" buttons, and therefore each installation will need to be configured to the laboratory's working practices, as noted earlier in this section under user requirements (Section 54.4.4). Depending on the vendor and application chosen, this can be achieved in a variety of ways, either alone or in combination:

- Configuration by selecting one of a series of options offered by the vendor. For example, selecting the access privileges for a specific user type.
- Configuration within the boundaries of the LIMS application by using the scripting language supplied by the vendor.
- Customisation by writing new functionality to extend the LIMS.

This section will look at how this work needs to be undertaken, tested and documented. It should be noted that customisation of software should be avoided wherever possible, as this raises the GAMP software category to 5, which requires a much more extensive system validation [3].

54.4.9.1 Do I Need a Functional Specification?

Not necessarily, as it depends on how any additional LIMS functionality is implemented and how extensive the work will be. What is important is that the configuration of the LIMS is recorded rather than what a document is called. The URS can contain the majority of requirements of the system, but the detail needs to be recorded in one or more configuration documents. However, if the LIMS will be extensively configured, then an overall functional specification is advised, and this will have traceability back to the URS. Note that a single URS requirement may generate more than one requirement within the functional specification.

54.4.9.2 Using the LIMS Scripting Language

Before starting any work with the scripting language, developers will need to be trained and understand the implications of use. Alternatively, this is an area in which the vendor and their staff could be engaged to develop on the behalf of the laboratory. If prototyping has been used earlier to generate requirements and workflows, the resulting scripts can used again here. If the vendor publishes any standards for using the language, these should be followed as good practice. Where possible, the scripts should be reviewed by a second person before being implemented. Copies of the scripts used to modify the LIMS functions should be maintained outside of the system in case of disaster; do not rely solely on recovery from magnetic backup tape to preserve them.

The functionality of the configuration should be tested against requirements or other specifications to ensure that they are correct. Correctly performing configurations are copied into the validation environment prior to the PQ.

54.4.9.3 Input of Methods and Specifications

Populating the database with methods and the corresponding specifications will take time and should not be overlooked when planning the project. Although this process will start during the configuration, the process will be ongoing throughout the operational lifetime of the LIMS as new products and specifications are added to the system. The ideal for specifications is to download the information from another system where it is maintained electronically; however, specifications often are maintained on paper, and this requires the laboratory to input then and check them manually before transferring them to the operational instance. Similarly, methods will need to be inputted to the LIMS and controlled; inevitably this will be a manual process, although, once entered, methods can be copied from one product and adapted to another one.

54.4.10 Write SOPs and Train Users

Users and IT operations staff will need to write or modify the SOPs identified in the URS for the various operations of the LIMS; this typically ranges from basic user operations through application and system support to database maintenance. Where applicable, the SOPs must provide direction on regular audit trail review and other data integrity topics required by regulatory guidance [1,2,16]. Either these SOPs need to be available in final draft form when the PQ is executed to enable any changes required to be incorporated before the documents are approved and released, or they are approved before the PQ. If any changes are required after the PQ, these will be identified as they are checked out during the PQ. All users of the system, including the IT support staff, need to be trained as appropriate to their tasks, and records maintained of these activities.

Initially, only the staff involved with the PQ will need to be trained on the SOPs; however, before other users are allowed to use the operational system, they will need to be trained. Care must be taken in training, as LIMS may have long learning curves, and it is neither fair nor reasonable to expect laboratory personnel to work at the same level immediately after LIMS training, as they will still be getting to grips with a new application. This is less of an issue with an upgrade; however, it depends how many versions have been skipped. End user training should be scheduled as closely with the LIMS implementation as possible, as extended time gaps will decrease the effectiveness of said training.

54.4.11 LIMS Performance Qualification or User Acceptance Testing

The purpose of the user acceptance testing or PQ is to demonstrate that the functions specified in the URS and as configured work as intended to meet both business and regulatory requirements for the LIMS. Terminology here can be confusing within the analytical laboratory, as IQ, OQ and PQ are used for both analytical equipment qualification and computerised system validation but mean different things [17]. Also, the terminology used to describe the documents generated in this phase of validation work can differ greatly; it is more important to remember that the work is done and documented rather than what a particular document is called.

54.4.11.1 Test Plan for Controlling the UAT

A test plan for the user acceptance testing adapted from IEEE software engineering standard 829 is used for overall control of the work [33], and the main elements are shown in Table 54.1.

This is achieved by defining the system to be tested and its scope; it can list the test scripts to be written and link these back to the URS requirements to be tested under each test script and the features not to be tested [17]. Testing cannot be exhaustive, so there is also a section on the assumptions, exclusions and limitations to the testing; this

TABLE 54.1
Outline of a LIMS Test Plan (Modified from IEEE Standard 829)

Original sections for a test plan from IEEE 829	Sections adapted for a LIMS UAT test plan
1. Introduction	1. Introduction
2. Test system/item	2. Test environment
3. Features to be tested	3. Features to be tested
4. Features not to be tested	Confirmation of application configuration
5. Test approach	Overview of the test suite
6. Pass/fail acceptance criteria	Test script design and traceability
7. Suspension criteria & resumption requirements	Implementation strategy
8. Test deliverables	4. Features not to be tested
9. Testing tasks	5. Test approach
10. Environmental needs	6. Outline pass/fail acceptance criteria
11. Responsibilities	7. Suspension criteria & resumption requirements
12. Staffing and training needs	
13. Schedule (test order)	

is a very useful way of recording contemporaneous notes of why testing was conducted in a particular way. Specifically, it documents why some requirements are either assumed to work or are excluded from testing, along with the justification for this.

54.4.11.2 UAT Test Scripts

The test scripts or protocols are written in sufficient detail to test the requirements in the URS; traceability back to the requirements is important from two perspectives. The first is to check coverage of testing versus requirements, and the second is to check that the requirements are testable or verifiable. Occasionally the URS may need to be updated at this stage to modify some requirements that cannot be tested or verified adequately, or it is realised that a requirement has been written incorrectly. This is normal, and to help reviewers of the updated specification, a table detailing the changes in any new version should point to those requirements that are new, modified or superseded.

Testing of the system should cover its main functions, including instruments and systems that have been interfaced with the LIMS plus the overall capacity of the system. If the LIMS is interfaced with an ERP system, then many of the test scripts will start in the ERP by generating a work order that is downloaded to the LIMS; at the end the analytical release will be sent to the ERP. It is important to realise that although the basic operations of the LIMS must be validated before operational release, many of the database population activities will be controlled by procedure and do not require validation per se.

Approved UAT test scripts are executed, and documented evidence in both paper and electronic form are collected; test results should be compared with explicitly stated acceptance criteria. It is important to use screenshots sparingly, evaluating where the system does not record information within the database or audit trail and where they add value. Similarly, witness testing is not an FDA requirement but a validation custom and practice; however, a second person review is mandatory. The results of this were documented in the respective test scripts and summarised in the validation summary report for the LIMS, and a specific UAT report need not be written.

54.4.11.3 UAT Testing Types

The aim of UAT testing is two-fold: primarily to demonstrate that the LIMS meets its intended purpose (as defined in the URS) and secondarily to discover and resolve software errors. As the basic software application is supplied by a vendor, the emphasis for the second aim should be focused where the scripting language has been used to add functionality. Some of the types of testing that can be incorporated into the PQ test scripts are:

- **Security Testing**: Testing carried out here will include security of the network, including remote (via Internet) and local access to the network running the LIMS application. In addition, the access allowed by defined user roles and groups within the application should be checked using representative testing. Access control testing should include both the business users and the IT support staff such as network and database administrators.
- **Functional Testing**: Testing of this nature shows that the LIMS works as intended and typically works within the boundaries of the system—that is, unless another application is interfaced to the LIMS, when the process flow may indicate that testing should start in the interfaced application and not the LIMS. Remember, if a supplier assessment has leveraged supplier development into the project, software functions that are GAMP category 3 do not need to be explicitly tested.
- **Boundary Testing**: Testing at the specified boundaries of fields within the configured portions of the LIMS.
- **Capacity Testing**: This is a load test that shows that the overall system is capable of operating correctly when under heavy load, e.g. large number of users or multiple instrument data capture etc. In addition to users performing the testing, the IT department should be monitoring the operation of the server hardware, e.g. processor and memory usage and disc access and usage.
- **Performance Testing**: In essence, this tests that the application response times are acceptable. As such, it can be combined with capacity testing.
- **Compliance Testing**: In addition to demonstrating that the system can perform its intended functions, the compliance aspects, such as the audit trail, electronic record and electronic signature integrity functions, are also adequately tested.
- **Disaster Recovery Testing**: This testing is intended to give confidence that the early stages of disaster recovery and resilience work as intended. Examples of this type of testing could include uninterruptible power supply (the system still operates when power is turned off), hot disk swap (a RAID 5 disk can be replaced while the test or validation instance is operational without impacting data integrity) and a full recovery of the system from a tape backup. If the IT department has tested some of these tests on a site-wide basis, they can be excluded from testing if justified.

Note that two or more types of testing can exist in a single test script; for example, functional testing could check access control of a user to a specific function or include modifying data for which an audit trail entry is created.

Any deviation from the acceptance criteria must be evaluated. Should the deviation require a modification to the existing configuration, the modification should be made and the functionality retested. Consideration should be given to the impact of the modification on previously tested functionality, and the rationale for or against extensive regression testing must be documented in the deviation report.

54.4.12 Write System Description and Definition of E-Records

A system description should be written and approved for the LIMS. This is an explicit requirement from EU GMP Annex 11 clause 4.3 [25] which states:

> For critical systems an up to date system description detailing the physical and logical arrangements, data flows and interfaces with other systems or processes, any hardware and software pre-requisites, and security measures should be available.

Much of this information may exist in validation documents; if so, the relevant information can be cross-referenced by the system description. In addition, the system description should also contain the definition of electronic records for the system and the fact that 21 CFR 11 applied to the application as required by the Part 11 Scope and Application guidance [18].

54.4.13 Reporting the Validation

Before writing any validation summary report, the first activity is to read the applicable validation plan and understand what the original intent of the validation was. This will identify whether deviations have occurred that have not been explained previously.

54.4.13.1 Write Validation Summary Report and Release the Core System

This validation report contained the summary of the validation of the core system and was issued after the validation of the core system (the first rollout). A statement in the validation summary report released the system for operational use, including electronic signatures. The report was reviewed and approved by the system owners and QA prior to releasing the system for operational use.

54.4.13.2 Write Validation Summary Report for Each Rollout of the System

Each additional phase of the system roll-out had a validation summary report written to describe the work that has been undertaken in that phase to maintain the original validation status of the system. These tasks included a summary of the evidence for:

- Any additional servers installed and qualified
- Interfacing of any new instruments or systems to the LIMS
- Updated configuration logs
- Any further or repeat UAT test scripts executed under the UAT test plan
- User training performed and an updated list of authorised users for the system

54.4.14 Conclusion

Risk-based validation of a LIMS opens up the opportunity for organisations to streamline the amount of work to be undertaken and focus the effort on the areas of highest business and regulatory risk. This requires careful thinking and judgement coupled with written justifications for the approaches taken.

REFERENCES

1. *FDA Pharmaceutical cGMPs for the 21st Century: A Risk-Based Approach.* Food and Drug Administration, Rockville, MD, 2002.
2. *Good Automated Manufacturing Practice (GAMP) Guide Version 5.* International Society for Pharmaceutical Engineering, Tampa, FL, 2008.
3. *GAMP Good Practice Guide a Risk Based Approach to GXP Compliant Laboratory Computerised Systems.* 2nd ed. Tampa, FL: International Society for Pharmaceutical Engineering, 2012.
4. *MHRA GMP Data Integrity Definitions and Guidance for Industry.* 2nd ed. Medicines and Healthcare Products Regulatory Agency, London, 2015.
5. *MHRA GXP Data Integrity Guidance and Definitions.* Medicines and Healthcare Products Regulatory Agency, London, 2018.
6. *WHO Technical Report Series No.996 Annex 5 Guidance on Good Data and Records Management Practices.* World Health Organisation, Geneva, 2016.
7. *PIC/S PI-041–3 Good Practices for Data Management and Integrity in Regulated GMP/GDP Environments Draft.* Pharmaceutical Inspection Convention/Pharmaceutical Inspection Cooperation Scheme, Geneva, 2018.
8. *FDA Guidance for Industry Data Integrity and Compliance with Drug CGMP Questions and Answers.* Food and Drug Administration: Silver Spring, MD, 2018.
9. *GAMP Guide Records and Data Integrity.* International Society for Pharmaceutical Engineering, Tampa, FL, 2017.
10. *GAMP Good Practice Guide: Data Integrity-Key Concepts.* International Society for Pharmaceutical Engineering, Tampa, FL, 2018.
11. *GAMP Good Practice Guide: Data Integrity by Design.* International Society for Pharmaceutical Engineering, Tampa, FL, 2020.
12. *Technical Report 80: Data Integrity Management System for Pharmaceutical Laboratories.* Parenteral Drug Association (PDA), Bethesda, MD, 2018.
13. *Practical Risk-Based Guide for Managing data Integrity, Version 1.* https://apic.cefic.org/pub/Data_Integrity_Best_Practices_Guide_for_API_FINAL_March-2019.pdf.
14. *GAMP Good Practice Guide Global Information Systems Control and Compliance.* Tampa, FL: International Society for Pharmaceutical Engineering, 2005.
15. *FDA Guidance for Industry General Principles of Software Validation.* Food and Drug Administration, Rockville, MD, 2002.
16. Jenkins S Presentation at a Seminar on the Paperless Laboratory. *Pittsburgh Conference on Analytical Chemistry and Applied Spectroscopy*, Chicago, 2004.

17. McDowall RD. *Validation of Chromatography Data Systems: Ensuring Data Integrity, Meeting Business and Regulatory Requirements*. 2nd ed. Cambridge: Royal Society of Chemistry, 2017.
18. *FDA Guidance for Industry, Part 11 Scope and Application*. Food and Drug Administration, Rockville, MD, 2003.
19. *GAMP Good Practice Guide a Risk-Based Approach to Compliant Electronic Records and Signatures*. Tampa, FL: International Society for Pharmaceutical Engineering, 2005.
20. *Able Laboratories Form 483 Observations*, December 23, 2019, 2005. www.fda.gov/media/70711/download
21. *Cetero Research Untitled Letter (11-HFD-45–07–02)*. Food and Drug Administration: Silver Spring, MD, 2011.
22. *FDA Letter to ANDA Sponsors Conducting Bioequivalence Studies at Cetero Research*, 2011. www.fda.gov/downloads/drugs/drugsafety/ucm267907.pdf.
23. *FDA Compliance Program Guide CPG 7346.832 Pre-Approval Inspections*. Food and Drug Administration: Silver Springs MD, 2010.
24. *FDA Compliance Program Guide CPG 7346.832 Pre-Approval Inspections*. Food and Drug Administration: Silver Spring, MD, 2019.
25. *EudraLex: Volume 4 Good Manufacturing Practice (GMP) Guidelines, Annex 11 Computerised Systems*. European Commission, Brussels, 2011.
26. *OECD Series on Principles of Good Laboratory Practice and Compliance Monitoring Number 17 on Good Laboratory Practice Application of GLP Principles to Computerised Systems*. Paris: Organisation for Economic Co-Operation and Development, 2016.
27. *21 CFR 11 Electronic Records: Electronic Signatures, Final Rule*, in *Title 21* 1997, Food and Drug Administration, Washington, DC.
28. *21 CFR 211 Current Good Manufacturing Practice for Finished Pharmaceutical Products*. Food and Drug Administration, Silver Spring, MD, 2008.
29. *ISO 27001–2018 Information Security Management*. International Standards Organisation, Geneva, 2018.
30. *IEEE Software Engineering Standard 1233: IEEE Guide for Developing System Requirements Specifications*. Institute of Electronic and Electrical Engineers, Piscataway, NJ, 1998.
31. *USP 41 General Chapter <1058> Analytical Instrument Qualification*. United States Pharmacopoeia Convention, Rockville, MD, 2018.
32. *EudraLex: Volume 4 Good Manufacturing Practice (GMP) Guidelines, Annex 15 Qualification and Validation*. European Commission, Brussels, 2015.
33. *IEEE Software Engineering Standard 829-2008 Software Test Documentation*. Institute of Electronic and Electrical Engineers, Piscataway, NJ, 2008.

55 Control Systems Validation

Phil DeSantis and Steven Ostrove

CONTENTS

55.1 INTRODUCTION

As discussed in the chapter on computerized systems (Chapter 53), the Food and Drug Administration (FDA) and other regulatory authorities consider computer systems that perform functions mandated by the applicable regulations (sometimes referred to as "predicate" rules) as requiring validation.[8]

Control systems that are "computerized" are a special class that may be described as equipment as well as computers. The general approach to qualifying equipment can be found in Chapter 4, "Commissioning and Qualification". All types of equipment used for the purpose of producing a pharmaceutical product must be qualified; control systems are no exception (ref. 21CFR211.68). The use of computerized control for manufacturing and quality control has grown substantially for several decades and continues to grow. The introduction of computerized control was the first impetus to identify the need for computer system validation (CSV).

The aforementioned computerized systems chapter fully discusses the history of computer validation as well as the development and validation of software. This chapter will focus on control systems that function within manufacturing processes. In addition to being considered "computerized" by their software functionality, they are also seen as equipment by virtue of their hardware interface to mechanical manufacturing equipment. For this reason, the terms "validation" and "qualification" are sometimes used interchangeably when addressing control systems. This chapter will preferentially use the widely used acronym CSV (for control systems validation) but will occasionally use "qualification" when it is considered appropriate.

Computerized control devices and systems are usually specified by a Process Engineering, Process Automation or Information Technology (IT) department. These areas are expert in designing, installing and maintaining the systems and providing the necessary service and training to allow the end user (Operations) the ability to employ their benefits. However, validation of the control components and their software must be performed by qualified personnel who are also familiar with CGMP (Current Good Manufacturing Practices) regulations and company quality system requirements related to validation.

55.2 TYPES OF CONTROL SYSTEMS

This chapter is a general guide to what is required to qualify/validate the control systems used in pharmaceutical, biotechnology or the medical device industries. The intent is to

DOI: 10.1201/9781003163138-55

provide the reader with an appreciation of the complexity and the similarities of all types of computer or automated system qualifications. It will also, however, indicate the variability in approach to CSV depending on the complexity of the device or system

This chapter will cover the various types of computer systems, which includes automated devices used in the control of pharmaceutical/medical devices. While the chapter on computerized systems focused primarily on software, this chapter will deal with the qualification of the various types of computer or automated control systems inclusive of their hardware. For the purpose of this discussion, we will expand the definition of computerized control system to include any device or system that performs a manufacturing function and has a programmable digital structure. Software development and qualification are discussed in detail in the computerized systems chapter and will be discussed in this chapter only as it dictates the function of the associated hardware.

As stated in the introduction, all control systems require validation; the level of qualification is dependent upon both function and structure. To help determine validation requirements, much of the industry has adopted the GAMP levels of software systems. There are four levels of systems according to the guide [4]; these are:

- Level 1—Firmware—This is the microchip type of system, programmed by the manufacturer. These are inaccessible to and unchangeable by the end user.
- Level 3—Standard Software Package—Non-configurable, also called "Off the Shelf" (NOTE: Level 2 has been deleted in the GAMP guide); programmed by the vendor or developer and not changeable by the user. Some specific user input is allowable, such as operator log-in, set point or alarm limit.
- Level 4—Configurable Software Package—Standardized packages that the owner can configure to fit their specific needs or operations. Often called COTS or "Configurable Off The Shelf", these systems consist of preprogrammed modules that offer variables that may be set by the user and can be ordered into any number of structures or sequences to perform a specialized function.
- Level 5—Custom Software—Prepared specifically for the operation (usually prepared by specialty firms or in-house programmers).

Each level requires its own level of validation, increasing as the level goes up with the greatest effort required by custom software systems. Note that the levels are related to the software and not hardware. This is because variability in validation requirements relates more to software than to hardware. Of course, the interaction of the software and hardware needs to be qualified (functional testing). It is not possible to fully qualify one without the other.

Several types of computerized controllers are used in the pharmaceutical industry. Each one has its own purpose in production control. Starting with the simplest to the most complex, there are (1) microprocessor-based devices, (2) single-loop controllers, (3) programmable logic controllers (PLCs), (4) personal computers (PCs), (5) supervisory control and data acquisition (SCADA) and (6) distributed control systems (DCS). Each of the control system types usually fits into one of or another of the GAMP levels, although there may be some overlap. Computerized systems that interface with these but do not actually perform control functions are dealt with in the computerized systems chapter. Examples of interfacing or networked systems include data historians, manufacturing execution systems (MES), inventory management systems and computerized maintenance management systems (CMMS).

Microprocessor-based devices are found in almost every type of unit used in manufacturing. Examples are digital balances and barcode readers. These devices are generally single-function, are programmed on chips and are read-only. They fit into GAMP Level 1. They may measure, monitor, compare and/or output information. The computer program is essentially a "black box", and testing is entirely focused on functionality. This type of device is usually a component of a larger process system and is most often qualified along with its parent system. Examples are a bar code reader on a packaging line or a level sensor on a vessel.

Only slightly more complex are single-loop controllers, which control a single variable by comparing a measurement to a set point and sending an output signal to a controlling mechanism such as a control valve. These are usually microprocessor-based and may consist of several components. Components include various combinations of sensors, transmitters, receivers, controllers and field devices (e.g., valves), any or all of which may be computerized. These are usually GAMP Level 3 in that user input is limited and their function and sequence of operation cannot be changed. These are generally preprogrammed in firmware that cannot be altered, with the exception of the few user inputs that are allowed, such as set points. Despite being more complex than Level 1 devices, single-loop controllers are relatively simple to validate. They are most often addressed within the qualification of the larger system to which they belong.

PLCs are more complex, multifunctional devices that can be programmed to control several variables as well as manage a sequence of operations with multiple options. They are found in packaging lines, filling machines, blenders, autoclaves or just about any other type of process equipment. PLCs usually fall with GAMP Level 4 because they are configurable by the user. The base software structure is provided by the vendor or developer and is fixed, usually in the form of modular code that is unalterable by the end user.

PCs are rarely used in control except when interfaced to control systems for the purpose of providing a human interface or data acquisition. PCs can be stand-alone (that is, operate independently) or linked together in a network with other components. Even if linked into a system, they may still perform functions independently of the others in the system, or they may call on another unit to complete a process. Because they perform an important function in process control, PCs are included in the discussion of control system validation. In this application, they most often employ software that is configurable by the user, placing them in GAMP Level 4.

SCADA and DCS systems are used to control a larger processes or multiple processes simultaneously. Although they differ in structure, SCADA and DCS function in a similar manner. Both can control, monitor and document the full operation of a process, from the initial raw material handling to the final package. Both operate within a "distributed" architecture where they interface with one or more control devices. SCADA systems usually interface with PLCs. They manage inputs to the PLCs that change their configuration from simple set point changes to complete recipes including multiple variables and operational sequences. They also obtain and store data from the process. DCS are structured and function similarly to SCADA except that they interface with specially designed digital controllers as well as PLCs. Once programmed, the PLCs and digital controllers in these distributed systems can operate independently, even if the central device shuts down. Both SCADA and DCS may also network to other functions such as inventory control, warehousing, labs or other systems needed for total plant operation.

Level 5 systems were more prevalent in the early pharmaceutical applications. These are custom designed software systems developed in house or by a software developer or systems integrator. Rather than configuring preprogrammed (and fully validated) modules, the code is developed in its entirety for a specific application. Obviously, such systems require a great deal more validation effort in order to test the code prior to its actual application. Such systems have been largely supplanted by GAMP 4 COTS system for process control but still occasionally appear in the industry.

The more unique and complex the individual control system and the greater degree to which it is networked to other systems will dictate the extent of the validation program. In general, computer system validation (CSV) is more complicated than mechanical equipment qualification because, rather than being focused on hardware alone, it must also take into account the software and the interaction between the hardware and the software.

55.2.1 Hybrid Control Systems—Utilities Management

In the computerized world, complex computerized systems have many functions, some of which may have a direct effect on product quality and some which may not. An example of this may be an enterprise management system that controls the release of production materials based on data received from a laboratory system. Obviously, this corporate or sitewide system may have many functions that do not affect quality.

This type of hybrid system is less common among control systems, but it does exist. The most common examples are systems that control utilities that directly affect product quality, such as environmental control and water purification. Utility systems have some functions and components that affect quality and some that do not. A Building Management System, for example, may monitor and control all of these. In these cases, risk assessment will determine which functions and components are subject to CSV. The complete design, installation, testing and maintenance of these systems are subject to good information technology (IT) practice and handled similarly to equipment as described in the chapter on commissioning and qualification.

55.3 LIFE-CYCLE APPROACH TO CONTROL SYSTEM VALIDATION

Control systems and devices that are used in pharmaceutical manufacturing are nearly always determined to have a direct effect on product quality and therefore need validation. This is determined through a risk assessment of the full system (hardware and software and all related mechanical equipment), just as would be done with any other process equipment or instrumentation (see "Commissioning and Qualification", Chapter 4). The main difference between general equipment qualification and CSV is that computerized systems must consider software. In a rigorous CSV program, the software is tested in two distinct phases: structural and functional. Structural qualification of software is carried out during its development and consists of inspection, quality control checks and simulated testing. This is done by the developer, whether internal or external to the user company. Except for custom software systems (GAMP Level 5), most of this testing is not included in the CSV program of the user. This is because the overwhelming percentage of control systems and devices do not use custom software.

With most control systems, CSV deals with the fully integrated system of computer hardware and software. In simpler systems (and sometimes even with complex ones) the total mechanical process may also be included in the testing.

In all cases, however, CSV follows the same basis qualification sequence as equipment:

- Define
- Develop
- Design
- Test
- Operate and maintain

What follows is a life cycle approach to CSV that has proven effective and acceptable over decades in the pharmaceutical industry. It follows the recommendations of GAMP 5 [4] and is widely accepted by regulatory authorities. What the life cycle does is to implement quality-by-design, wherein the design and development of the system provide the foundation for validation. There may be other approaches to CSV, but we will stick with this proven approach.

55.3.1 Define—User Requirements

Definition starts with the user group compiling requirements, most often in the form of a User Requirements Specification (URS). The URS may cover the mechanical system along with its control system and associated devices. This is usually the case when the control system is dedicated to the equipment system. It may interface with a higher-level system, but as long as the measurement and control function is limited to one mechanical system, this is an acceptable format.

For more complex systems such as DCS and SCADA, which may interface with multiple controllers, it is usually best to prepare a separate URS. In either case, the URS will list one or more requirements for the computerized system or device; the simpler the system, the fewer the number of requirements. These requirements should normally be limited to the user needs without providing a technical solution. An example might be the requirement to control temperature within a specified range. This does not indicate how temperature should be measured or controlled, which will be determined by the engineers charged with the next step in the sequence, design.

A most effective URS not only lists requirements but provides an initial assessment of the requirement category. Typical categories are: business essential, quality critical, safety and others. Defining quality critical requirements is extremely helpful in validating control systems by focusing on quality critical requirements that need to be satisfied in operation.

A segment of an example URS follows:

URS TABLE

Part 2—General Requirements for Manufacturing Equipment

Designation	Description	Type
URS-31	The control system must have logical and physical security to prevent unauthorized access to the system and its data.	QCA
URS-32	The control system must have controls to ensure electronic record integrity and security.	QCA
URS-33	The control system must have an audit trail to provide a record of user actions.	QCA
URS-34	Critical and noncritical instrumentation must be easily seen and accessible for calibration.	BEA
URS-35	The system must be automated with an industrial programmable controller for executing real-time control and monitoring of the equipment.	BEA
URS-36	The system must have a means of notifying users of abnormal equipment and/or process conditions.	BEA
URS-37	The system must be capable of operating with utilities available at the specified site.	BEA
URS-38	The system must be capable of being installed and operated within the designated area at the specified site.	BEA
URS-39	The control system should have a local operator interface.	BEA
URS-40	The control system should be capable of generating accurate and complete copies of GxP records in electronic form suitable for inspection and review.	QCA

URS TABLE

Part 2—General Requirements for Manufacturing Equipment

Designation	Description	Type
URS-41	The control system should be able to operate in either a manual (operator steps through each function) or recipe (system steps through each function, with operator intervention, when necessary) mode.	BEA
URS-42	The system should be designed to be energy efficient.	BEA
URS-46	The system must meet all local code electrical requirements.	Safety

In this example, BEA refers to Business Essential Attribute and QCA refers to Quality Critical Attribute. (This is an example only and is not intended to provide a specific assessment of quality impact.)

As will be evident, single-function microprocessor-based devices may be derived from as few as one requirement. The more requirements, the more complex the system is likely to be. In summary, the URS establishes *what the system needs to do.*

55.3.2 Develop—Functional Specification

The Functional or Technical Specification is sometimes called the FRS for Functional Requirements Specification, a term the author does not prefer because it causes confusion with the URS. The Functional Specification (FS) is a description of *how the system does its job*, describing each function in detail. It is developed by engineers, systems integrators or equivalent technical experts. Some, but not all, functions are derived from the URS. The design engineer or software developer will also include all of the system functions that are needed to have the system perform reliably and as expected. In essence, it provides a technical solution to the User Requirements plus everything else that the system needs to do.

The Functional Specification will describe all of the components of the mechanical and control system, both in text as well as diagrammatic format. Schematic diagrams that show the interface of the control system to the process equipment are essential. The most commonly used are Piping and Instrumentation Diagrams (P&IDs) that indicate these interfaces. For more complex control systems, a separate diagram of system architecture should be prepared. This will show all the computer hardware components, peripherals (printers, operator interface panels, input/output panels, etc.), as well as network connections to other systems.

Very simple single-function devices are usually included in the Functional Specification for the equipment system. For more complex multifunction controllers like PLCs, the Functional Specification is usually separated into a section of the equipment FS or contained in a separate document. Highly complex systems like SCADA and DCS, and definitely custom software systems, commonly require a standalone Functional Specification.

The importance of the FS and its relationship to the URS cannot be overemphasized. It is the pivotal document that defines the control system design. Each requirement in the URS must be addressed by the FS. From a CSV perspective, satisfying all the requirements identified as critical to quality must be confirmed. Many firms employ a traceability matrix at this stage in which quality-critical requirements are mapped to the part of the FS that address each.

55.3.3 Design and Develop

Using the FS as a basis, technical experts prepare specifications and drawings that are used in order to actually build the system. Of course, many components of control systems are already designed and merely need to be interconnected and programmed or configured to perform the functions outlined in the FS. Detailed design varies with system complexity, with simple manuals for microprocessor-based single-function devices. Design of these is usually integrated with the design of the mechanical system.

As systems increase in complexity, hardware and software require more specific documentation and specification in order to build the system. Designs include:

- Wiring diagrams
- Input/output (I/O) lists
- Control-loop diagrams
- Panel drawings
- Human machine interface (HMI) drawings and specifications
- Specifications for all system components (field instruments, control valves, power supplies, printers, alarm panels, network connections, etc.)
- Software:
  - Level 1—Built into chips for the simplest devices—usually not open to review by the user.
  - Level 3 (PLCs)—Programmed in proprietary language, usually by the equipment vendor. Source code may be proprietary and not reviewable by the user.
  - Level 4 (DCS/SCADA)—Operating system and software modules proprietary to the developer; source code not reviewable to user. Configuration (set points, sequences, durations, recipes, etc.) documented and open to user.
  - Level 5 (Custom Software)—Complete source code documentation, plus record of structural testing by developer.

Some firms attempt to extend the traceability matrix to detailed design. This can be very cumbersome and is of questionable value.

Where software is custom or configured, its development is often contiguous with design. As each module of code is planned and built, the programmer or systems integrator will perform various forms of structural review and inspection, followed by simulated functional testing to prove out the code.

Once the complete system is built, the testing phase will provide the final assurance that the overall system has been built correctly.

55.3.4 Testing

Many users look to testing as the core of CSV. In the life-cycle of control systems, however, it is only one in a series of phases that is required for reliable system performance. Testing confirms the required relationship between requirements and functionality, ultimately providing documented evidence in actual real-time use.

Because control system CSV is so closely tied to equipment, it is convenient to look at control systems in the same sequence, with some additional perspectives to account for software. These steps vary with the complexity of the control system and will be discussed further on in the chapter. The basic sequence is:

55.3.4.1 Design Qualification (DQ)

Inspection of the design to ensure that it is properly documented, follows specifications and will meet user requirements.

55.3.4.2 Commissioning

A series of engineering activities to inspect, start up and operate a system to confirm that it meets design and functional specifications. This phase allows for corrections of any issues or discrepancies encountered as long as they are documented and then rechecked to confirm conformance. Commissioning usually deals with system structure and functionality without regard to impact on product quality. It includes both installation and operation of the system. Commissioning is guided by good engineering practice (GEP) [6] principles that are applicable to controls as well as equipment. For complex control systems that interface with multiple process systems (e.g., SCADA, DCS) and are treated as stand-alone for testing purposes, this phase may be referred to as Structural and Functional Testing.

55.3.4.3 Installation Qualification (IQ)

IQ includes documentation of all system components, including computer hardware and software and confirmation that these meet the design and Functional Specification as well as applicable User Requirements.

55.3.4.4 Operational Qualification (OQ)

OQ comprises operational testing of the system to ensure that it satisfies the quality critical Functional Specification, especially as it applies to quality critical requirements. This testing may occur simultaneously with the operational qualification of the process system.

55.3.4.5 Performance Qualification (PQ)

PQ is the final User Acceptance Test (UAT) performed in real-time operation of the integrated control and equipment system to perform its full function, including product manufacturing operations where these are involved. PQ may be simultaneous with Process Performance Qualification runs (for production processes) or PQ runs of critical utilities and packaging operations (different firms use different nomenclature.) These runs should be focused on documenting acceptance criteria derived from quality critical requirements. Full system functionality should have been confirmed during commissioning.

55.3.5 Operation and Maintenance

Is it important that once a system has been accepted and placed into operation that it be maintained in a validated state. This means that an effective change control program be instituted for both computer hardware and software, as well as mechanical equipment. Proposed changes must be evaluated on two levels: (1) overall operation of the system and (2) effect on product quality. Evaluation or risk assessment will determine the level of documentation, inspection and testing required to accept the change. As with equipment, some control system changes may be accepted through routine maintenance procedures and documentation. Changes that present a significant risk to product quality, however, must be managed in the site's formal change control system.

55.4 CSV TESTING GUIDELINES

Control systems validation must be a rigorous exercise because requirements tend to be largely quality critical. This means that a large proportion of the structural details and operational functionality of these systems are subject to formal validation procedures as defined by a site's quality systems. This section will review what is normally included in CSV of control systems. This is not meant to replace complete and rigorous system commissioning, however. For clarity, this chapter will describe the test elements of control systems validation using the life cycle terminology in Sec. 3, that is, IQ, OQ, and PQ. The structural and functional testing applied during development of stand-alone software are more completely described in the chapter on computerized systems (Chapter 53).

The general requirements for qualification of manufacturing and critical utility systems are described in the chapter on commissioning and qualification (Chapter 4). The following guidelines are specific to the control systems associated with these mechanical systems, and more specifically to those mechanical systems that are assessed to have a direct quality impact. To clarify, if a mechanical system is assessed not to have a direct impact on quality (e.g., a boiler), then its control system would be categorized similarly and not require CSV (i.e., commissioned only).

As described for User Requirements, when a control system is dedicated to a single equipment system, the qualification of the controls is most often combined into a common protocol with the equipment. For more complex controls interfaced to more than one equipment systems, like a DCS or SCADA, stand-alone protocols are more prevalent. In addition, the choice to separate or combine IQ, OQ and PQ protocols is a matter of choice, often depending on system complexity, scheduling or other factors. Design Qualification (DQ) is nearly always a separate exercise. The following focuses on the control elements only. More detail for each qualification activity may be found in Chapter 4.

55.4.1 Design Qualification (DQ)

This stage is often waived for vendor-supplied standard control systems that have been established through wide industry use. This applies to single-function microprocessor-based devices as well as vendor-supplied controls dedicated to a standard equipment system, such as a PLC for an autoclave or a blender. The latter are usually programmed by the vendor and have limited configurability by the user, such as set points, durations and limited modifications to sequences of operation. For these basic systems, the user may choose to conduct an audit of the vendor's software development practices. The reputation of the vendor and extent of industry use may be taken into account when deciding on the type (virtual, in person) and depth of the audit.

The design of more complex control system is reviewed at the completion of coding and/or configuration and after the systems integrators have completed their structural testing. This phase is usually documented by a traceability matrix, wherein the User Requirements are compared to the Functional Specification to ensure that the requirements designated as critical to quality have been addressed.

DQ may be taken to the detailed design level by examining the designed inputs and outputs linking controls to equipment. Also, sequences of operation can be examined to ensure they are correct. The more custom a system is, the more valuable and necessary this phase will be.

55.4.2 Installation Qualification (IQ)

IQ ensures that the control system as installed meets the User Requirements and that critical installation requirements have been verified. The URS and FS guide the acceptance criteria for control system IQ. There is often confusion among users as to what constitutes IQ as opposed to OQ. This is really not a relevant discussion as long as the overall system requirements are met. To avoid meaningless discussion, recognize that it is often possible to combine IQ and OQ into a single combined protocol. For this discussion, IQ will be limited to static inspection absent functionality considerations.

Minimum IQ requirements for a control system include:

- A system schematic indicating interfaces between the control system and the equipment. This is most often in the form of an enhanced Piping and Instrumentation Diagram (P&ID).

- A control system architecture diagram showing input/output devices, network connectivity to other systems and peripheral devices (e.g., alarm panels, printers, operator panels/displays, etc.). This may not be required for simple systems fully described by the P&ID.
- Hardware listing, including vendor and model number. Includes control devices, interface devices, field devices (e.g., sensors, valves, transducers), power modules, data storage, etc.
- Hardware switch settings (e.g., dip switches of PLC).
- Input/output (I/O) registers.
- Instrument loop diagrams—Individual control loops including sensors, controller and controlled device. Loop continuity checks may be included in IQ, but many users include these in commissioning (because they are later confirmed in OQ and PQ).
- Instrument calibration—For control systems, this requires "full-loop" calibration, i.e., including all system components from the sensor to the display interface and electronic data record. (Note: Merely calibrating the sensor is not adequate.)
- Operating system identified with version number/date.
- Application software specifically identified (version number, etc.).
- Back-up/recovery software (controlled location).
- System documentation (URS, FS, design documents, operator's manual, maintenance manual).
- Parts list. (Not *spare parts.* It is a business decision to maintain spare parts.)
- Completed commissioning/functional test documentation—These are "engineering/IT" documents, but their completion should be considered a requirement for control systems qualification. The functional sections of these may be deferred to the start of OQ. See the chapter on Commissioning and Qualification (Chapter 4) for further discussion.

55.4.3 Operational Qualification (OQ)

OQ confirms that operational/functional requirements that are necessary to ensure product quality have been satisfied. Functional OQ tests may be performed in real time simultaneously with mechanical system qualification or employing a simulation in place of actual equipment, or some combination of the two. If simulated (off-line) testing is employed, then the software used in that testing must be secured and verified to be identical to the software downloaded onto the actual controlling device.

OQ requirements listed here are typical and applicable to most systems from GAMP Levels 3 to 5. Single-function microprocessor systems (GAMP Level 1) do not embody all of the described functionality and may often be limited to calibration and functional testing with the equipment system OQ.

- System security—Verification of software against pre-tested version (as required). This may be performed using a built-in or external system diagnostic.
- System access—Access should be tested at the designated permission levels (e.g., operator, supervisor, system manager) to ensure that permissions are not exceeded at any level. This should be performed using both acceptable and rejected attempts at log-in.
- Interference challenge—Ensures that the system functionality is not affected by external environmental changes. Typical interference challenge includes radio frequency interference as may be experienced from an external communications device or nearby AC motor.
- Audit trail—All human entries and electronic recording of data must be traceable and not erasable. Attempts to disrupt this must be unsuccessful (and should also be recorded).
- SOPs—For OQ, these may be in draft form, to be finalized prior to PQ. For control systems, typical SOPs include:
  - System Setup/Installation
  - Data Collection and Handling
  - System Maintenance/Calibration
  - Backup
  - Recovery
  - Contingency Plans (Emergencies)
  - Security
  - Change Control
  - Storage
- Sequence of operations—This is usually tested in real time using the actual process equipment, although OQ testing may be done in part in simulation mode. (Note that simulated process operations may be performed on a computer or using an electromechanical simulator, although in recent times this latter form of testing is rarely employed.) The sequence should include all operations from log-in to completion. Not every recipe variation needs to be tested in OQ, but a representative sequence should be chosen (see PQ for other recipes or sequences). Where sequences are significantly different (e.g., process vs. cleaning vs. sterilization), each type of sequence should be tested in OQ. Sequence of operations is usually tested under normal process conditions, without unexpected events.
- Control of process parameters—Process parameters determined to have a direct effect on product quality are best tested in real time using the actual process equipment. This is usually simultaneous with a sequence of operations run. Precision of control around set points and within limits must be ascertained to satisfy user requirements.
- Alarms and notifications—Routine notifications are verified during regular OQ runs. It is not expected that alarms conditions be forced, however. Alarms may be challenged by simulation (e.g., transmitting a false out-of-limits signal from the field to the control system). Complete alarm tests may be completed during commissioning or factory testing and leveraged into OQ as long as they are documented properly (see Chapter 4 on

commissioning and qualification). Any alarm condition that occurs during an actual OQ run must be properly recorded and any relevant issues resolved.

- Reporting—System reports must be verified. These may be on a batch basis, in which case an OQ run will generate a report, on a periodic basis or on demand. All report forms built into the system should be verified in OQ.
- Interfaces to peripherals and networked systems—Communications to all connected devices and systems must be verified.
- Recovery and restoration—Verification that system responds as specified after a power failure or other operational interruption.

55.4.4 Performance Qualification (PQ)

PQ is sometimes termed the User Acceptance Test (UAT) for a computerized system. For a control system, this test must be integral to the actual performance of a manufacturing (or supporting utility) process. As such, it must actually perform its complete function in real time with real product or production material (e.g., packaging, critical utility). Until it does this properly and reliably, it is not validated.

For manufacturing operations, many firms choose to include controls systems PQ with the Process Performance Qualification (PPQ) runs for a process. These are performed as normal production runs or batches and may result in commercial (or clinical) product. Combining CSV and PPQ provides the opportunity to qualify every recipe or sequence of a control system designated for use in manufacturing.

55.4.5 Protocols

This chapter does not provide guidance on the form or format of CSV protocols. There is significant information on protocols to be found elsewhere in this book and in the general validation literature. It is important to note that the accepted core protocol concepts apply equally to CSV as to process and other validation activities. Typical protocol requirements include:

- Purpose
- Experimental design
- Acceptance criteria

This chapter's advice on protocols is to focus on the user-defined requirements that have been assessed to be essential to product quality. Design tests and acceptance criteria around these as simply and clearly as possible. Do not become enmeshed in what has often become a ritualistic approach, wherein acceptance criteria are so narrowly established that the real intent is lost. Examples:

- Set the acceptance criterion reasonably. For alarm, use "alarm initiated and recorded at X^{0}C". There is no need to spell out the exact text of the alarm as an acceptance criterion. This avoids a needless deviation if the actual alarm printout sentence reads slightly differently from the protocol text.
- When confirming a step, record "confirmed" rather than rewriting the entire step description in the observed/actual entry. This avoids transcription errors and needless correction.
- Record numerical data when appropriate, not when there is no quality impact. Only critical process parameters need to be recorded.

55.5 DATA INTEGRITY

No discussion of control systems or any CSV can ignore the issue of data integrity. In December 2018, the FDA published a guideline for data integrity [10]. According to this guideline the FDA "expects that all data be reliable and accurate". This means that the data recorded electronically (or by any other means) can be retrieved and used for investigations, or other purpose. The FDA expects that all data is consistent with its original form. That is, an entry is stored as received and is unchanged when retrieved.

Data integrity is based on the principle summarized by the acronym ALCOA (attributable, legible, contemporaneously recorded, original/true copy and accurate). Its purpose is to prevent inaccurate information from being part of the batch record or other GMP documentation. All control systems must comply with these principles through design and confirmed through qualification testing. Appropriate challenges must be included in the test program before systems may be placed into use. Challenges range from log-in access, operator inputs, data review and reporting to data archiving/transmission to networked systems. Any data changes must be traced, documented and annotated. Original data may not be erased, similar to the single-line cross-out practice employed for manual batch records.

55.6 NETWORKS

As discussed previously, control systems may be networked either with other control systems or with other computerized systems that perform related functions. Networks are valued in that information may be shared across several platforms. Each of these connected systems needs to be evaluated as to the impact on product quality and their functionality under the applicable CGMP regulations in order to determine their validation requirements. A network is a group of individual units (PCs or PLCs) linked together so that information can be easily shared. Networks may be open or closed; that is, they may or may not interface with systems outside of the site (i.e., over the internet). In the pharmaceutical industry, the closed network is the preferred type, as the internet represents an open system and thus the greater possibility of data corruption.

If a control system transmits data used on batch records compiled on another system (e.g., a manufacturing execution system or MES), then this data is the active record, regardless of any printouts of this data or other forms. Data security needs to be tied to the record, and the record maintained in a validated and secured data base. In this case, if users were transmitting this data over the network, then the network itself needs some level of validation. The extent of validation is subject to risk assessment, as in every other related case.

Many firms have assessed networks as being somewhat similar to mechanical plant utilities in that they have an indirect impact on product quality (see Chapter 4, "Commissioning and Qualification"). This means that while they may not require a full CSV life cycle as described in this chapter, they do require the application of good IT design, installation, testing, and maintenance. These practices are analogous to the good engineering practices (GEP) used for equipment design, installation and commissioning. Network use is validated during the CSV studies in which data transmission across the network is confirmed to be secure and accurate. This testing should encompass all avenues of data transmission from the control system to related systems.

55.7 21 CFR PART 11

No discussion of computer or control system qualification will be complete without at least an overview of Part 11 (21 CFR Part 11). This part of the Code of Federal Regulations has caused the pharmaceutical industry great concern in recent years due to its perceived complexity. Part 11 has been around since 1997 and is now mirrored in the EU and PIC/s GMP Guidelines as Annex 11. Regulatory bodies have allowed considerable time for the industry to comply by updating their control systems, updating their operating procedures, training etc. before strict enforcement would be implemented. Although systems placed in operation prior to 1997 in the United States may technically be considered exempt from the Part 11 rules, the time for compliance to these regulations has long passed.

When one looks closely at the requirements, they are really quite understandable; however, their implementation requires specific attention. The FDA has issued two sets of guidelines for this part of the CFR. The first set of guidelines has been withdrawn, and a new "draft" guideline has been issued. [11] The current guidelines have made compliance to Part 11 regulations clearer to the industry. The regulations have not changed, only the industry perception.

Part 11 has three major sections, as follows:

Subpart A—General Provisions

11.1 Scope
11.2 Implementation
11.3 Definitions

Subpart B—Electronic Records

11.10 Controls for closed systems
11.30 Controls for open systems
11.50 Signature manifestations
11.70 Signature/record linking

Subpart C—Electronic Signatures

11.100 General Requirements
11.200 Electronic signatures components and controls
11.300 Controls for identification codes/passwords

Subparts B and C represent the main body of the requirements. Only an overview of the requirements will be presented; further study will be required to fully understand this section of the CFR.

Subpart B is concerned with any computerized system (of any size or type) or the personnel who use these systems. Both open and closed systems are included (21 CFR 11.10 and 21 CFR 11.30). In this part of the CFR the FDA specifies that any system used to "create, modify, maintain, or transmit electronic records shall employ procedures and controls designed to ensure the authenticity, integrity . . . and ensure that the signer cannot readily repudiate the signed record as not genuine". This means that the system(s) need to be validated/qualified and that, as with written records, there needs to be traceability of all data. Access to the systems and the data or records (electronic) needs to be limited and authorized.

Records that are maintained in paper format as the final official copy are not included in this section of the regulations. The predicate rule states that all of the regulations set forth in maintaining CGMP compliance take precedent over Part 11. Part 11 needs to fit into the compliance requirements of the system. This applies to all regulated items covered in Title 21. Thus, if a device is being manufactured, then Part 820 will be applied first, and then Part 11 requirements.

Subpart C deals with the actual control and requirements for electronic signatures. It describes the levels for security and access, the need for verification of the person signing. There are two types of identification discussed; these are biometric and non-biometric. The non-biometric form is most familiar to everyone. These include items such as identification badges (picture ID), sign in logs and password. If this type of identification is used, then two forms must accompany the signature (i.e., user identification and a password). On the other hand, a biometric identification would include items such as fingerprint identity, retinal scans of the eye or voice recognition. Biometric identification is becoming easier and less expensive and is available on some systems now.

As can be seen from this short discussion of Part 11, the regulations are not difficult; however, some aspects of the rules may be harder to implement. All control systems have or should have limited access to both the system and the various levels of data (e.g., operator, supervisor and administrator). Any change in the data needs to have an "audit trail" indicating "who" made the change and why the change was made (similar to changes in paper records). Thus, compliance to Part 11 has become achievable and, with the new Guidelines from the FDA, it has become more understandable. However, care needs to be taken with all computerized systems to be sure that all of the Part 11 regulations are implemented.

REFERENCES

1. DeSantis P. A Framework for Quality Risk Management of Facilities and Equipment. *Pharmaceutical Online*, January 2018. www.pharmaceuticalonline.com/doc/a-framework-for-quality-risk-management-of-facilities-and-equipment-0001.
2. DeSantis P. Facilities and Equipment Risk Management: A Quality Systems Approach. *Pharmaceutical Online*, December 2017. www.pharmaceuticalonline.com/doc/facilities-and-equipment-risk-management-a-quality-systems-approach-0001.
3. Eudralex, Volume 4, Good Manufacturing Practices (GMP Guidelines), Annex 11, *Computerized Systems*, 2011.
4. International Society for Pharmaceutical Engineering (ISPE), GAMP 5. *A Risk-Based Approach to Compliant GxP Computerized Systems*, 2008.
5. ISPE Baseline® Guide Volume 5. *Commissioning and Qualification*, 2nd ed., 2019.
6. ISPE Good Practice Guide. *Good Engineering Practices*, 2011.
7. United States Code of Regulations, 21 CFR part 210. *Current Good Manufacturing Practice in Manufacturing, Processing, Packing, or Holding of Drugs: General*, 2020.
8. United States Code of Regulations, 21 CFR part 211. *Current Good Manufacturing Practice for Finished Pharmaceuticals*, 2020.
9. United States Code of Regulations, 21 CFR Part 11. *Electronic Records: Electronic Signatures*, 2020.
10. United States Food and Drug Administration, Guidance for Industry. *Data Integrity and Compliance with Drug CGMP*, December 2018.
11. United States Food and Drug Administration, Guidance for Industry. *Part 11, Electronic Records: Electronic Signatures-Scope and Application*, August 2003.

56 Process Analytical Technology (PAT)

Understanding Validity of Pharmaceutical Quality Control and Assurance

Ajaz S. Hussain

CONTENTS

56.1 INTRODUCTION

The efforts to advance science, engineering, and technology, and mature their utility in practices to design and develop therapeutics and vaccines, reduce regulatory uncertainty, and make manufacturing of legacy and new products more reliable and efficient have been ongoing for decades. These efforts remain on a perpetual critical path because laws and regulations precede advancements in science and technology, and a guidance document is often insufficient to change crystallized habits [1, 2]. A crisis beyond chronic drug shortages is

DOI: 10.1201/9781003163138-56

necessary to call the attention of lawmakers to the importance of reliable pharmaceutical manufacturing [3].

In the United States, amid polarized politics, the preferred route to resilient manufacturing is for the US FDA to facilitate industry application of Emerging Technologies—Industry 4.0 [4]. A pilot program is in progress to learn how FDA can evaluate the "maturity" of corporate quality management systems (QMS) for traditional pharmaceutical manufacturing [5].

Once again, we seek urgent solutions for manufacturing therapeutics and vaccines at an unprecedented scale and yearn to secure a dependable supply of life science products. The US FDA is informing lawmakers to strengthen the nation's public health infrastructure and seeking public-private partnerships and inter-departmental collaborations such as with the National Institute of Standards and Technology (NIST) of the US Department of Commerce to improve management of measurement uncertainty on the technological path it has chosen to facilitate—Industry 4.0 [4, 6]. Furthermore, US FDA published an approach to operational *Resiliency* [7] to overcome the long-standing concerns on its capability to oversee the vast and variable global pharmaceutical supply chain and find ways to fill the backlog of deferred Current Good Manufacturing Practices (CGMP) inspections amid the COVID-19 pandemic [8]. Are these expectations reasonable? In polarized politics, can those responsible for the oversight of FDA operations, the public representatives, and lawmakers, ask questions that matter to the public?

The mismatch in maturity across a global pharmaceutical supply chain is in multiple domains. Money, raw materials, knowledge, understanding, and other resources needed to ensure adequate Pharmaceutical Quality control vary, as does the evidence required to consider a pharmaceutical process valid. In these circumstances, does it makes sense to write a chapter on the Process Analytical Technology or PAT initiative launched two decades ago?

56.1.1 Why Write a Chapter on Process Analytical Technology or PAT in 2021?

A familiar pattern is repeating. Some lessons in the trials and tribulations experienced in the PAT initiative are documented—for example, in [1–3] and [9–11]. Other lessons experienced in search of words that can explain. This chapter seeks to explain objectively lessons learned experientially.

As professionals yearning to be good practitioners, we acknowledge and guard against conflict of interests and disciplinary dogmas and biases limiting our ability to be self-correcting. Yet, we struggle to self-correct. We are generally aware of market failure, asymmetric information, weak signal-to-noise ratio, and poor feedback mechanisms in our systems. Do we also recognize the importance of maintaining awareness of residual uncertainty in training graduate scientists and engineers to be vigilant and educate the next generation to build a workforce that can control and deliver quality assurance with integrity? Not adequately, our practice in these aspects is highly variable [9–11]. Emphasizing *experiential learning* can reduce this variance is the premise of this chapter.

This chapter shares insights and outlines an approach to make sense and leverage *experiential learning* stipulated in the US Federal Food, Drug, and Cosmetic Act (FDC Act) [12] and Code of Federal Regulations about CGMP [13]. By definition, the verb *experience* is to feel (emotions). As professionals, we tend to discount our feelings because, in part, we lack a good process to integrate subjective feelings in our objective thinking.

The COVID-19 pandemic is an unprecedented crisis across multiple systems, and it has placed the world on a path to a "new normal," akin to redesigning systems [3]. We all are sensing uncertainty, information asymmetry, and apprehension that patients experience routinely, some more or less than others. In our lifetime, this pandemic is unprecedented, yet the experiences it is generating are familiar. It has made most professionals public and patient. As pharmaceutical professionals, we can learn from this experience to empathize with patients and the public.

We expect the COVID-19 pandemic will continue to inject pandemonium in systems that directly and indirectly relate to pharmaceutical supply, quality, and regulatory oversight. As a community of scientists, engineers, and technologists, we seek to advance solutions needed for dependable and sustainable manufacturing. We live and work in socio-technical systems, each a whole defined by its function in a larger system. Our professional development and maturity are causal, and the outcome is how we think and behave, the technologies we use, and the systems we build to guide our practice to be good. Understanding and making sense of lessons in the PAT initiative can contribute to the development and maturing of pharmaceutical professionals and enhance their capability and suitability to solve problems, prevent mistakes, and confront uncertainty with integrity.

In this broad context, the chapter shares insights on two processes (a) *understanding*—which the PAT Guidance described [14] and (b) *sensemaking* [15] to illustrate a PAT approach to make a meaningful contribution to the ongoing efforts of communities of knowledge and practice. It seeks to offer recommendations to reduce dependency on external regulators to detect errors and mistakes and facilitate self-assurance via self-authorship of good practices within communities of pharmaceutical practices responsible for giving others the assurance they need to experience therapeutic outcomes as claimed on the product label—better than placebo in the real world.

56.1.2 How Is the Narrative Generated?

In pharmaceutical operations, self-correcting without harming others and not depending on an FDA Warning Letter to notice deficiencies and deviations is a measure of good practices and maturity of QMS. By definition, science is a self-correcting process. In practice, however, we struggle to self-correct. History rhymes to alert us that amendments to laws direct the evolution of regulatory and corporate systems and practices. For instance, in 1962, the Keyfauver Harris Amendment to the Food Drug and Cosmetic Act of 1938 (FDC Act) [12].

Today the Keyfauver Harris Amendment is still a useful reference point in many ways. One of its provisions—"*substantial evidence of effectiveness*" is currently raising concern in FDA's ability to make independent decisions in the interest of patients and the public [16]. Although not noticed in mainstream news reports, what underpins this concern (root-cause) is the stipulated "*scientific training and experience*" in section 505(d) of the FDC Act to "*fairly and responsibly*" assess evidence from an appropriate combination of investigations, not just the customary one or two "pivotal" randomized controlled clinical trial. One can reasonably opine that this struggle is also evident more broadly, for instance, the public health system's response to the pandemic generally and particularly to new and repurposed therapeutics.

The notion of "pivotal" clinical trial for new drugs or bioequivalence study for generics exemplify dependence on "univariate" evidence and discounting of other parts of the evidence, that is, struggling to be holistic in considering "*totality of the evidence*." Sectors habituated on univariate measures struggle to achieve consensus on multivariate evidence of cause-and-effect relationships. The utility of evidence from an appropriate combination of investigations speaks to a *multivariate* aspect of assessing quality, a foundational pillar of the PAT initiative [1, 14].

Typical disciplinary education and training emphasizes analytical capabilities and builds expertise narrowly in a specific discipline. Analytical capabilities allow us to observe and understand aspects of a thin slice of reality we pay attention to and experience. Certainly, such expertise and analytical capabilities are necessary. Necessary but not sufficient to be holistic, to be suitable for assessing multivariate cause and effect relationships "*fairly and responsibly*," and to mature systems such as QMS.

Our analytical expertise is the foundation of our capability to remain within prespecified limits of (professional) tolerance; however, our suitability to respond to unaccounted sources of variance and disturbances is in our learning from experience. Among other things, to learn from *experience*, we need to be suitable and capable of making non-theory laden observations or good observations.

Buried within the volumes of amendments and legal precedence are the clear stipulations in section 505(d) of the FDC Act—we need "scientific training and experience" to make decisions "fairly and responsibly" [12]. And in the 21 CFR § 211.25—Personnel qualifications [13] match duties and responsibilities with an appropriate combination of "*education, training, and experience*" to ensure compliance with corporate policies and routines (SOPs) in adherence to CGMP. Yet, as a community of knowledge and practice, we chose not to define and regulate qualifying standards for pharmaceutical practitioners in the industry. Why? Perhaps because we think we maintain currency in our practice by learning from the experience of self and others, for example, Warning Letters. Or maybe the mesmerizing power of "*FDA Approved*" and "Validated" eases rationalizing—*if I don't look, there is no problem*, and we get comfortable in the "*2–3 sigma*" constraints [9].

The PAT Guidance described "*process understanding*" [14]. Typically, we "make sense" intuitively, and this process is unconscious. *Understanding and sensemaking* are two processes necessary to learn objectively from the experience of good observations. How to consciously make sense is essential to learn from experience. Serendipitously before the FDA Science Board endorses the PAT initiative on 16 November 2001 [1], a practical approach of making sense knowingly [15] is associated with the launch of the PAT initiative at the time of another unprecedented crisis-precipitated on 11 September 2001.

In learning from this experience [1, 15], the ensuing narrative provides a practical description of an approach to evidence lessons in making sense of PAT trials and tribulations. Some of the challenges are in a preserved source of PAT prior knowledge—the chapter entitled "*Process Analytical Technology and Validation*" by Bradley S. Scott and Anne Wilcock, published [17]. Utilizing this previous chapter as prior knowledge, the ensuing narrative reflects, highlights observations that remind of the struggle to self-correct incorrect understanding and discusses the concept referred to as "*new prior knowledge*," introduced elsewhere [10, 18]. The latter [18] calling the attention of regulators and industry to unaccounted or unappreciated physicochemical failure modes in currently marketed first-line therapy Narrow Therapeutic Index (NTI) drugs, and the former [10] seeks assured supply of generic medicines via integrated multivariate assessment, as opposed to a single "pivotal" biostudy mindset, of the four parts that complete assessment of *Therapeutic Equivalence—pharmaceutical equivalence, bioequivalence* when necessary, *CGMP* and product *label*.

Furthermore, it also places the QMS maturity pilot [5] in context considering the lessons in the continued challenges in manufacturing (shortages, quality failures, and recalls) and increasing inspectional observations concerning breaches in data integrity assurance or BAD-I, that compelled the FDA n 2015 to seek One Quality Voice [22] and re-establish a focus on Emerging Technologies [23] to, among other things, promote continuous pharmaceutical manufacturing which the PAT Guidance anticipated and facilitates.

Finally, lessons in these experiences are translated into points to consider to prepare ourselves, inform lawmakers, and participate in the FDA's public-private partnerships with a perspective on the relevance of collaborations with NIST [5,6] to improve how we manage measurement uncertainty is discussed broadly in the public's interest and trust the people, need and expect.

56.2 Understanding and Making Sense of PAT 2001 in 2021

In practice, we move between physical, information/digital, and cognitive domains, as in cells of Table 56.1 below, in observing, for instance, the outcome of an action complying with as SOP (bottom row), recording data with integrity, tabulating, and plotting data to inform ourselves and others to be aware of a situation. In the cognitive domain, we intuitively integrate data, information, perceptions, expectations, and emotions with our prior knowledge to think. Our education and training, disciplinary dogmas, assumptions, paradigms, and evolving worldviews influence how we feel, think, and act.

TABLE 56.1
Knowledge, Understanding, and Making Sense of CGMP and QMS

Cognitive domain	Knowledge – new and prior and ways of thinking →	Understanding individual and shared →	Making sense individually and collectively →	Intent/decision → ↓
	⍏ Awareness	⤭ ⤱ Emotions, perceptions, expectations, assumptions, beliefs ⤭ ⤱		Order ↓
Information and digital domain	⍏ Information			Policy ↓
	⍏ Data			SOP ↓
The world outside or physical domain	⍏ Outcomes	← Compliance with routines (SOPs)		← ↓Training

Subjectivity interacts with what we observe, think, and do to control quality. Hence, the subjective influencers are placed in Table 56.1 with interacting arrows. Arrows in cells on the right-hand side of Table 56.1 are bold to suggest presumed confidence or certainty—in a QMS, established quality policy, SOPs, and practices. These and other arrows in Table 56.1 offer a sense of direction for moving around and interacting to make meaning of the text. On the left-hand side, arrows cut with a bar or a dash signifies imperfections (e.g., Out of Specification outcomes), risk of BAD-I, or breaches in the assurance of data integrity, incomplete and asymmetric information, inadequate situational awareness, blind spots, and cognitive biases. We integrate information into prior knowledge to understand (e.g., the risk posed or opportunity offered) and decide the next course of action. In doing so, we make informed decisions. However, recurring errors and mistakes—and the excuse "root cause unknown" for OOS observations that repeatedly occur across the sector signify ignorance of common causes.

A bit harsh reminder, for instance, emotional outbursts are referred to HR for Anger Management. More routinely, to be objective in professional practice, we prefer to put aside our beliefs to acknowledge scientific evidence and corporate policies and professional norms, discount or disown the subjectivity of our emotions and expectations, at least temporarily. Still, we also often forget that our nonconscious assumptions and feelings remain influential and a source of our biases and blind spots.

Like assurance, quality is also subjective, and we must acknowledge and account for personal aspects and make these objectives of learning from experience. Information becomes knowledge when experienced in its application and self-authorship (as opposed to copy and paste). Therefore, instead of discounting or disowning the subjectivity of emotions, we acknowledge and account for actively learning from our experience to gain insight, acquire knowledge from information, understand and make sense in our routines, and when thighs are not routine. It relates to validation in that we seek validation of our peers or are self-assured, internally validated in our suitability and capability for a task at hand. Making contemporaneous notes of our experiences and tracking how we feel in complex and uncertain situations, that is, perhaps beyond our expertise, or when we feel anxious or threatened, opens the door to learning from experience.

We understand within the domain of our knowledge and expertise. Beyond it, outside the realm of our knowledge and expertise, we learn to make sense to arrive at a decision—Intent/Decision cell of Table 56.1.

An organization that intends to take action on something would follow a formal process of an order per a policy to assign a task carried out by operators using a proper procedure or SOP. Operators qualified via training for specific SOPs deliver expected outcomes and record their observations with integrity via supervised compliance to SOP. So then, why would they not achieve the predicted outcome? Perhaps the process was prematurely or erroneously deemed validated without understanding.

56.2.1 Process Understanding

In the context of Table 56.1, we can explore the relevance of our observation, data, and information and seek to achieve a shared understanding while avoiding falling prey to "groupthink." A multidisciplinary group or team can help reduce groupthink assuming all members are engaged to gain shared understanding. In an interdisciplinary group, the same words can mean different things, or their relevance is weighted differently. For instance, an engineer will use the word "control" in the context of a dynamic process with feedback and feed-forward information to achieve target attributes. A chemist or an industrial pharmacist thinks of control for testing before, during, and after production on a fixed in time process.

A shared understanding between a chemist and engineers would be based on evidence needed for real-time release in the PAT context. To aid in achieving this consensus, the PAT Guidance emphasized process understanding and offered the following characteristics of a well-understood process "(a) all critical sources of variability are identified and explained; (b) process design manages the variability, and (c) product quality attributes predicted accurately and reliably over the design space established for materials used, process parameters, manufacturing, environmental, and other conditions." The ability to predict and confirm reflects a high degree of process understanding and the means to end debate. The questions we will explore in the following section include, among others—was this understanding of the PAT Guidance understood and captured as prior knowledge [17]? If not, what should be our *new prior knowledge*?

56.2.2 Sensemaking, First Principles, and Knowledge Pyramid

Common causes underpinning lingering challenges posited in this chapter are our inability to account for differences in meaning we make and nonconscious biases; we must be a good observer—that is, setting aside preferences that often are theory-laden biases, and consciously learn from experience. These steps are necessary but not sufficient. We must go beyond our prejudices in a complex and chaotic system and learn to organize the unknowns—residual uncertainty and know-how to act in prevailing tension. It is an adaptive response to challenges that require a response outside our existing expertise. It is a different way of thinking than the analytical thinking discussed previously—sensemaking literature in organizational studies is rich and mature [20]. Still, this chapter's discussion on sensemaking offers a practical approach per Table 56.1 in the context of PAT.

In the PAT initiative, a knowledge pyramid with *first principles* at its apex, followed by mechanistic understanding, causal links that predict performance, univariate approach to decision making, and at the bottom data derived from trial-n-error experiments was used extensively in presentations to industry. Within FDA, for instance, Training for the FDA Pharmaceutical Inspectorate in 2004 [19], the knowledge pyramid is shown on slide 21 of this presentation available on the internet.

Making sense from *first principles* is getting to fundamentals underpinning disciplinary jargon, asking questions to know sincerely (i.e., not to put down). For instance—how do you know what you know about data, information, knowledge, and understanding of critical attributes and risk to quality (consequence and probability of occurrence) coupled with questioning the basis to claim validity—adequate calibration, reproducibility, and repeatability of SOPs and processes, and measurement systems in routine production operations can open the door to making sense even when a topic is outside one's area of expertise or when we confront a paradox.

Two paradoxes can be observed in the application of the US FDA PAT Guidance. Many readers may recall that the application of PAT Guidance was curtailed or never realized soon after it was published in September 2004 [20]. Yet, the guidance remains on the FDA internet—why? The answer to this question follows. Although the standalone PAT Guidance, in its regulatory utility, is now a historical icon, PAT principles and vocabulary are now part of other guidance documents such as the one Process Validation issued in 2011 [21]. They are also embedded broadly in the pharmaceutical lexicon and ways of thinking or paradigm, the reason to write this chapter.

56.3 PAT Prior Knowledge to New Prior Knowledge

The chapter on Process Analytical Technology and Validation in the 3rd edition of this book [17]—referenced as prior knowledge is placed at the end of this discussion in its original content and format. The authors Bradley S. Scott (McNeil Consumer Healthcare, Guelph, Ontario, Canada) and Anne E. Wilcock (Marketing and Consumer Studies, University of Guelph, Guelph, Ontario, Canada) provide a unique perspective, very different from this author's viewpoint. They wrote the chapter following an extensive literature search, including FDA Advisory Committee meetings related to PAT, the guidance development and principles, and challenges to PAT implementation.

Reading this chapter provides a means to introspect to gage what could have been done differently by the author and the FDA to communicate more effectively about the PAT. For instance, the first reference in this chapter returns to a white paper by the FDA's PAT team and Manufacturing Science Working Group summarizing their learning, contributions, and proposed next steps for moving toward the "desired state" of pharmaceutical manufacturing in the 21st century [1]. Including this white paper in the prior knowledge would have answered many questions about challenges in conventional manufacturing. For example, the following two short paragraphs from the FDA white paper summarize the key challenge.

> Pharmaceutical manufacturing operations are inefficient and costly. The cost of low efficiency is generally not understood or appreciated (e.g., manufacturing costs far exceed those for research and development operations). Low efficiency is predominantly due to "self-imposed" constraints in the system (e.g., static manufacturing processes, focus on testing as opposed to quality by design, approach to specifications based on discrete or the so-called "zero-tolerance" criteria, a less than an optimal understanding of variability, etc.). These constraints keep the system in a corrective action mode.
>
> Continuous improvement is an essential element in a modern quality system, and it aims at improving efficiency by optimizing a process and eliminating wasted efforts in production. In the current system, continuous improvement is difficult, if not impossible. Reducing variability provides a "win-win" opportunity from both public health and industry perspectives. Therefore, continuous improvement needs to be facilitated.

56.3.1 Journey to Industry 4.0

Given this context, measurement uncertainty is at the core of the challenge. The collaboration FDA announced with the NIST of the US Department of Commerce is, in part, to improve management of measurement uncertainty on the technological path it has chosen to facilitate—Industry 4.0 [4, 6]. The role of compendial or "market standards" vs. standards for real-time control and product release needs to be reconciled along with reference standards, calibration methods, and measurement uncertainty and benchmarks—particularly for physical, biological, and microbiological attributes generally and specifically nanotechnology. Not to mention the need for standards for artificial intelligence, algorithms for multivariate process controls, the Internet of Things (IoT), and real-world surveillance to capture the "voice of patients" and improve pharmacovigilance and feedback. Real-time regulatory review and inspection standards are also likely to evolve in parallel.

In the short term, we need a clear definition of "continuous improvement." For example, during the PAT efforts at FDA, the definition of continuous improvement as in QS-9000, Third Edition element 4.2.5, was an option [25]. For this option to apply how product quality is specified in regulatory applications and the USP Monograph would need to be revisited—that is, move away from "attribute" or "zero-tolerance" features (e.g., not units outside X-Y% limits to a continuous variable, target or mean ± variance format) as the collaboration with NIST progresses or a bridge between market standards and real-time release standards more firmly endorsed by the vested stakeholders. For some time, such as bridge has existed—*ASTM E2709–10 Standard Practice for Demonstrating Capability to Comply with a Lot Acceptance Procedure*, referenced in the FDA's Guidance on Process Validation [21]. However, much hesitance still persists due to regulatory uncertainty.

Another opportunity to harmonize analytical methods in the context of the PAT toolkit, such as spectroscopic tools like NIR or Raman that utilize multivariate methods, is the agreement to revise the ICH Q2(R1) Guideline on Validation of Analytical Procedures. The ICH's Management Committee endorsed the concept paper for ICH Q14: Analytical Procedure Development and Revision of Q2(R1) Analytical Validation—on 15 November 2018 [26]. Furthermore, ICH Q13: Continuous Manufacturing of Drug Substances and Drug Products [27] is progressing, and the first draft will likely be available shortly. Before it is final, additional clarification can be considered to remove regulatory uncertainty to facilitate the journey not just for continuous manufacturing but also to support continual improvement, clearing the path to industry 4.0.

56.3.2 One Quality Voice

The PAT initiative opted to reduce regulatory uncertainty via education, training, and certification of a PAT team of CMC Reviewers, CGMP Investigators, and Compliance Officers. "In a sense, we now speak the same language," noted George Pyramides in 2005, a PAT team member representing the Office of Regulatory Affairs. "This language 'bridge' allows for communication between the cadre members on a level that I have not seen among other Agency teams. I now feel part of a larger group that is beyond the scope of my field laboratory" [28]. Also, the PAT Guidance is in the "Pharmaceutical CGMPs" and not "Pharmaceutical Quality/CMC" classification to signify a "team approach," which George Pyramides aptly described as "In a sense, we now speak the same language."

The continued challenges in manufacturing (shortages, quality failures, and recalls) and increasing inspectional observations concerning breaches in data integrity assurance or BAD-I, compelled the FDA in 2015 to seek a structural change, Office of Product Quality, and One Quality Voice [22], and re-establish a focus on Emerging Technologies [23] to, among other things, promote continuous pharmaceutical manufacturing, which the PAT Guidance anticipated and facilitates. A lesson to carry forward from the PAT initiative and achieve One Quality Voice demands a true team effort, as summarized above. The struggle within the FDA to gain a "common understanding" of CGMP remains palatable [29]. Procedural or technical understanding is achievable among peers or experts. Others seek to make (common) sense by asking questions without prejudice from the first principles. This nuanced consideration is an important lesson as we advance.

56.3.3 QMS Maturity

The ability of the US FDA to maintain the needed rigor of its inspections is limited. The pressure on the FDA to approve applications keeps increasing. With the emergent dominance of public-private partnerships, there is a need for a higher level of transparency to maintain trust in public institutions and systems. QMS is a socio-technical system. Public-private partnerships often focus on advancing technology in the context of economic growth. Human development is assumed will follow. The maturity of a QMS is the maturity of people and their processes. Progressing the pilot QMS maturity project with external consultants poses a risk of not achieving the primary objective—One Quality Voice. Beyond the ongoing pilot, reaching One Quality Voice within FDA CMC Review, CGMP Investigation, and Compliance is essential.

In the context of statistical process control or the current FDA Guidance on Process Validation, the state of control of legacy products and their operations is largely unknown. A concerted effort is needed, and proposals such as [30] should be encouraged after proper vetting. Drugs, particularly generic drugs, are made by various companies distributed across the globe in developed and emerging economies. Assurance of lot-to-lot therapeutic equivalence demands knowing failure modes and controlling critical quality attributes in the post-approval phase and over a product's life cycle. The current system needs balancing to pay adequate attention, specifically to pharmaceutical equivalence and CGMP compared to bioequivalence assessment, typically a one-time assessment pre-approval and occasionally post-approval for certain SUPAC Level 3 changes. The journey ahead is best described as "chaos to continual Improvement" by averting the 2–3 sigma barrier imposed on new drug applications and breaking the 2–3 sigma barrier for products currently on the market [9, 31].

56.4 SUMMARY AND KEY CONSIDERATIONS

The technical know-how of achieving a low error rate (e.g., 3.4 dpm or six sigma), real-time controls, and continuous manufacturing has existed for many years. The steps to correct and prevent errors are known [31]:

- Training, qualification, certification, and mentoring support to ensure flawless execution
- A focus on supply chain controls and confidence

- Analytical characterization of raw materials, manufacturing processes, and products—not just in the development phase but also in the commercial setting (as needed)
- Continual monitoring to ensuring robust analytical methods, manufacturing processes, and products (e.g., using industry benchmark for analytical variability and decreasing assay variability)
- Management involvement and engagement in identifying, tracking, and controlling variation via process capability assessment

Outside the pharma sector, problems that we continue to struggle with have been solved using these tools. So what is the gap between what we know and what we can implement? Historically the pharmaceutical regulatory system evolved as a reaction to one crisis after another. Over the years, each new crisis adds a layer of regulatory requirements—which has become the "extensive regulatory oversight," utilizing management standards to manage a sector that "is unable to manage itself." The system has become so complex that we often find ourselves in over our heads and confused between common and special cause variability.

The PAT initiative did shift the paradigm, and it offers a mirror to reflect and introspect. "The pharmaceutical industry has a little secret: Even as it invents futuristic new drugs, its manufacturing techniques lag far behind those of potato-chip and laundry-soap makers" [34]. There are valuable lessons in this pharmaceutical experience amid and beyond the COVID-19 pandemic.

Before the COVID-19 pandemic declaration, Industry 4.0 was already a major force with significant organizational implications. At the core of several "Make in My Country" efforts to secure a competitive advantage in the geopolitical chaos are the weakening interrelationships that hold together the global supply chains. Integrity of the supply chain plus enabling intelligent manufacturing is the goal of Industry 4.0 via technologies such as cloud computing, cyber-physical systems, big data analytics, and artificial intelligence, and the IoT.

The pharma journey—*Chaos to Continual Improvement*, anticipated in 2019, before the pandemic, called the pharma community to recognize that chaos is a system—an unpredictable system exhibiting extreme sensitivity to "initial" or starting conditions—self, raw materials, and regulatory submissions [32]. Acknowledging we are in chaos—partly due to self-imposed constraints—offers us a way to remove these constraints to optimally develop products and processes that are resilient to environmental disturbances and are non-polluting and energy-efficient. The journey to Industry 4.0 not only demands a reset of every aspect of how we define and qualify starting conditions—raw materials, reference standards, and equipment, we also need to revisit the qualification of individuals capable of multivariate decision making, engineering solutions, and collaborating with artificial intelligence and learning at a rate to remain relevant [33]. To honorably journey to Industry 4.0, we must first break the self-imposed *"2–3 sigma" barrier*, recognize the legacy challenges, acknowledge that pharmaceutical science is not self-correcting to the extent it needs to be and that we must establish a process for *"new prior knowledge,"* focus on exquisite control of starting materials—discard the notion of "inactive ingredients," educate and train practitioners and do so before we attempt system-wide implementation of "continuous manufacturing."

The regulatory emphasis on "Good" practices, Process Validation, quality by design, and systems approach to quality management within the pharma sector holds valuable lessons within and without, that is, for the world. We have debated and sometimes struggled to comply; this struggle has contributed to the pharma's suitability and capability to advance emergency use therapeutics and vaccines. Paradoxically, the COVID-19 pandemic is showing us the best and worst of human nature. We are amazed at the teamwork and collaboration to take care of patients and develop therapeutics and vaccines. We are also horrified by the disregard for human life and profiteering from human vulnerabilities. Most of the solutions needed to journey ahead demand public-private partnerships, and maintaining trust in our institution is low and declining. This trend is worrisome, and we must act now to reverse it quickly.

To give others assurance, we first must be self-assured. Self-authorship is a step to self-assurance; it can be objective and measurable. A Continual Improvement Plan that incorporates elements of Continuous Professional Development based on self-authorship can be objective and practical. Continuous development and maturity relate to the capability and suitability of processes we develop and control. Using process understanding and validation, we can make sense of how to go beyond our traditional education and training, to be self-authored to learn from the experience of self and others, to know what we know and how we know it. Then we can be self-transforming in filling gaps between what we know and what we can implement.

56.5 PRIOR KNOWLEDGE: INTRODUCTION

Conventional pharmaceutical manufacturing practices have low manufacturing efficiencies and capacity utilization, high scrap and reject levels, and therefore a high cost of quality. Process Analytical Technology (PAT) is a toolkit used to increase operational efficiencies, capacity utilization and process understanding, while decreasing operating expenses and ensuring that quality is built into the product. Multiple benefits associated with PAT implementation have been identified, but these benefits are not without drawbacks such as limited employee technical knowledge and suboptimal return on investment. The benefits and challenges of PAT are reviewed in this chapter, together with a discussion of the conventional pharmaceutical manufacturing paradigm, PAT principles and tools, PAT guidance development to date, and PAT validation approaches.

Process Analytical Technology has been defined by the United States Food and Drug Administration (FDA) as a system to design, analyze, and control pharmaceutical manufacturing processes through the measurement of critical process parameters and quality attributes. Through the measurement of raw material and in-process material attributes and the control of critical process parameters, finished goods quality will increase. The FDA anticipates three main benefits to accrue from implementation of PAT in the pharmaceutical industry: increase in the understanding of process and product, improvement in the control of pharmaceutical manufacturing processes, and incorporation of quality into the product from the design stage (35).

Process and product understanding is an ongoing process; it is logical that it be enhanced by implementation of elements of PAT such as chemical/physical/microbiological process analyzers, mathematical and statistical analysis, and risk analysis. As a result of increased process and product understanding, critical sources of variability are identified and controlled, and product quality attributes can be accurately and reliably predicted, thus increasing the quality of the final product. Process and product understanding help to identify parameters that are critical to the process. The monitoring of process and product attributes during the manufacturing process will allow quality to be built into the product, an approach that is far superior to reliance on final inspection.

The PAT initiative in the United States has been led by the FDA's Center for Drug Evaluation and Research (CDER), which released the revised PAT Guidance for Industry in September 2004. This document is a result of the combined efforts of industry, regulatory agencies, and academic institutions; it provides a framework for PAT implementation in the pharmaceutical industry. It has two main components: (1) scientific principles and tools to support innovation in pharmaceutical development, manufacturing and quality assurance, and (2) regulatory strategy to support innovation (1).

The FDA's PAT Guidance for Industry, which includes the PAT Framework and PAT dialogue between the FDA and industry, has been used by both the FDA and CDER in an attempt to reduce barriers perceived by industry. Industry representatives have identified FDA regulatory uncertainty as a barrier, which has stifled change (2, 3, 4, 5). The perception that the FDA regulatory body is rigid, resisting change, and stifling innovation is a perception that the FDA and CDER would like to change with respect to PAT.

This chapter reviews PAT developments within the pharmaceutical industry with particular emphasis on the conventional pharmaceutical manufacturing paradigm, benefits of PAT, PAT Guidance Development, principles and tools of PAT, validation requirements, and challenges associated with PAT implementation.

56.6 THE CONVENTIONAL PHARMACEUTICAL MANUFACTURING PARADIGM

The conventional manufacturing paradigm in the pharmaceutical industry involves batch processing, with laboratory analysis of samples taken at predetermined intervals and processing steps. Sample collection and subsequent testing are primarily in-process and at the end of batch processing (1). The development of a new process/product, followed by subsequent transfer to full-scale manufacturing, involves five main steps: design of the product/process, development of analytical methods and controls, accumulation of process knowledge, transfer of technology, and production of batches on a commercial scale (3). This approach requires a system to ensure that the final product meets predetermined specifications. Current quality systems to ensure that the finished product meets specifications include process/method/equipment validation, process control by standard operating procedures (SOPs) and process instructions/master recipes, and off-line testing of samples at the end of each batch (6). The compliance infrastructure to support these quality systems is believed to be difficult to sustain from an economic perspective (7). This compliance burden has ensured that quality products are released to the patient, but at the same time, has resulted in cost of quality increases, which have impacted organizations' financial performance and the cost of products to patients and society.

The average cost of quality for the pharmaceutical industry has been reported to exceed 20% (7), with a 3-sigma level (66,807 defects per million opportunities). Impacting the cost of quality are variables such as time-based end points, process variability, raw material variability, time associated with sampling, variability as a result of powder/blend sampling errors, and sample preparation. Further contributing to the cost of quality are the multiple test methods that are needed to assess different attributes of raw materials/in-process materials/finished drug substances (1).

Several manufacturing metrics have been identified (7, 43) that show that the current pharmaceutical manufacturing paradigm is not functioning optimally. These metrics include utilization levels of 15% or less, scrap/rework of 5%–10%, average cycle time of 95 days, and exception/nonconformance reports that can increase the cycle time by more than 50%. Furthermore, validation requirements increase the time to market for new products while at the same time-consuming valuable resources. A sample validation time frame reported for Pfizer extended to 117 weeks to prepare validation documents (15 weeks for hardware, 1 week for software, and 101 weeks for system validation protocol) (38).

The issues listed in Table 56.2 illustrate the desirability of an alternative paradigm for pharmaceutical manufacturing. These results, which originate from a systematic review of the pharmaceutical literature, highlight an important point about quality. In the absence of technological monitoring and feedback controls, the current paradigm demands strict adherence to SOPs and process instructions/master recipes. This system is not conducive to change or process improvement initiatives. The previously identified deficiencies, along with current societal and economic issues, are the predominant factors that are driving the pharmaceutical industry to change.

An additional factor that is impacting the pharmaceutical sector is the downward spiral in the number of new chemical entities (NCEs) launched per year. This is of particular concern because pharmaceutical organizations are increasing their spending on research and discovery and not receiving adequate returns on their investments. One suggestion to counteract this downward trend was made by Bai, Nayar, Carpenter et al. (93),

TABLE 56.2
Issues/Problems Associated with the Conventional Pharmaceutical Manufacturing Paradigm

Issue/problem	References
Low process capability.	Hussain (44) Wechsler (5)
Elevated levels of scrap, rework, reject, recall.	Hussain (44) Wechsler (5)
Low capacity utilization.	Hussain (44)
Recurring problems that do not seem to get resolved.	Hussain (44)
Resolution of issues/investigations is slow.	Hussain (44)
High cost of compliance.	Dean (7) Hussain (44) Wechsler (5)
Risk of drug shortages.	Hussain (44)
Risk of releasing poor quality drugs.	Hussain (44)
Delay in approval of new drugs because of lack of data/process understanding.	Hussain (44)
Quality problems confounding clinical trials.	Hussain (44)
System not conducive to process improvements.	Hussain (44)
Strict adherence to SOPs and master recipes/process instructions required.	Hussain (44)
Predetermined testing is completed at specific time intervals or process stages.	Hussain (44)
Little learning after the validation phase.	Hussain (46)
High proportion of FDA resources needed to ensure adequate product quality.	Hussain (44)
Continued debates between FDA–industry, few permanent resolutions with respect to quality issues.	Hussain (44)
Silos of information. Time consuming process required to extract and analyze the data in these silos.	Neway (45)
Low efficiencies. Conventional analytical methods are time consuming and labor intensive (i.e. Karl Fisher titration for drug substance water content).	Neway (45) Zhou et al. (98)
Labor intensive and inefficient.	Lai et al. (99) Wechsler (5)

Abbreviations: FDA = FOOD AND DRUG ADMINISTRATION; SOP = STANDARD OPERATING PROCEDURES.

who proposed that alternative noninvasive analysis methods such as NIR would assist in the design and optimization of protein formulations. Such optimization would decrease research and development efforts by reducing cost and cycle time for new product development and product launch.

Other factors impacting the pharmaceutical sector include limited marketing exclusivity periods, increased competition with generic products, and price scrutiny. These factors have all contributed to decreased returns for shareholders. This has forced the industry to carefully evaluate sources of potential waste and cost reduction.

Pressures currently impacting the pharmaceutical industry, along with challenges associated with the conventional manufacturing process, are driving companies within the sector to seek an alternative manufacturing paradigm. The future may lie with the incorporation of PAT into the manufacturing process.

56.7 PAT BENEFITS

There are three main benefits to implementing PAT in the pharmaceutical industry: increase in process/product understanding, increase in manufacturing process control, and incorporation of quality into the product from the design stage (1). Other benefits include reduced operating costs, quality improvements, positive regulatory impact, improved occupational safety, positive research and discovery impact, and reduced environmental impact (Table 56.3).

From an industry perspective, reduced operating costs and quality improvements are probably the most attractive because of their direct impact on profits. Contributing to the reduced operating costs are increased capacity utilization and increased operational (processing/packaging) efficiencies. Current capacity utilization levels have been estimated to be <15% (7). This suggests that there is tremendous potential for increased capacity utilization. Decreases in operating costs can also be achieved through continuous process monitoring and parametric release that will translate into improved cycle times (i.e. the time from raw material receipt through the value-added steps of processing and packaging). Hussain (8) reported the average cycle time to manufacture and package a pharmaceutical product to be 95 days. Of the 95 days, Dean (7) reported that only three days consist of value-added activities of dispensing, granulation, compression, and coating. If the average cycle time could be reduced from 95 days to the industry's best practice of six days through the implementation of PAT, both inventory and warehousing costs would be reduced and additional capacity made available.

Capacity constraints in pharmaceutical quality control laboratories can be eliminated through the use of PAT. The conventional manufacturing paradigm requires samples to be transferred to the laboratory for analysis using techniques such as wet chemistry, Fourier Transform Infra-Red (FT-IR), Ultraviolet (UV) Spectroscopy, High-Performance Liquid Chromatography (HPLC), and Gas Chromatography (GC). The use of PAT analyzers can reduce sample collection and analysis time and thus cycle times. Reduction in analysis time has been demonstrated (10) by implementing an NIR reflectance model for the identification of blister packaging film. The reference method, infrared spectroscopy (European Pharmacopoeia 3.1.11) requires approximately 2 hours whereas analysis of the same film using the NIR reflectance method requires less than 2 minutes. Han and Faulkner (11) reported similar reductions in analysis time by use of NIR reflectance for analysis of moisture content, identification of active ingredient, and assay of granulation, tablet cores, coated tablets, and blisters. Traditional methods require 15 minutes to determine moisture content and 30 minutes for the identification and assay of the active ingredient (UV and HPLC methods), whereas the NIR reflectance protocol requires less than 1 minute for moisture content, identification, and assay analysis. Similar time savings have been demonstrated using Raman spectroscopy for the analysis of aspirin. Wang *et al.* (115) reported that Raman analysis of aspirin tablets for assay results required 15 minutes, whereas HPLC analysis required 90 minutes per sample.

TABLE 56.3
Benefits Associated with Implementing PAT in the Pharmaceutical Industry

Benefits category	Specific PAT benefits	Reference
Reduced operating costs	Increased operating efficiencies.	Balboni (2) Hammond (19) Hussain (17) Lin et al. (102) Neway (45) Rudd (46)
	Improved cycle time (reduced release times, parametric release, reduced sample preparation time, minimize reliance on end product testing, faster analysis times).	Balboni (2) Clarke (4) Dubois *et al.* (48) Dyrby *et al.* (41) Gold *et al.* (49) Gupta *et al.* (50) Han and Faulkner (11) Harris and Walker (51) Hussain (21) Hussain (17) Kamat *et al.* (52) Laasonen *et al.* (53) Laasonen *et al.* (10) Laitinen *et al.* (96) Lin *et al.* (102) Lonardi *et al.* (54) Neway (45) O'Neil *et al.* (55) Otsuka (56) Sanchez *et al.* (57) Tumuluri *et al.* (92) Wang *et al.* (115) Wechsler (5)
	Decreased operating costs.	Balboni (2) Clarke (4) Hammond (19) Hussain (47) Laasonen *et al.* (53) Laasonen *et al.* (10) McCormick (43) Sanchez *et al.* (57) Shah (13) Tumuluri *et al.* (92) Wechsler (5)
	Possible continuous processing.	Balboni (2)
	Real-time monitoring, feedback controls, and results.	Balboni (2) Harris and Walker (51) Lai *et al.* (99) Lin *et al.* (102) Lonardi *et al.* (54) McCormick (43)
	Inventory reduction (through parametric release and improved cycle times).	Neway (45)
	Increased capacity utilization.	Hussain (21) Workman (58)
	Attain production schedule.	Workman (58)
	Reduced reprocessing expenses.	McCormick (43) Watson *et al.* (20)
Quality improvements	Increased quality (decreased product variability, decreased number of rejections, scrap, batch failure, and systems failures, and increased product reliability).	Avallone (18) Balboni (2) Clarke (4) Hammond (19) Herkert *et al.* (59) Hussain (21) Hussain (17) Hussain (47) Frake *et al.* (39) Lin *et al.* (102) McCormick (43) Morisseau and Rhodes (60) Neway (45) Shah (13) Wechsler (5)
	Increased regulatory compliance.	Hussain (47) Neway (45) Shah (13)
	Increased product uniformity (ensure batch-to-batch consistency, decrease variation).	Balboni (2) Hammond (19) McCormick (43) Neway (45) Watson *et al.* (20) Wechsler (5)
	Process fingerprinting.	Balboni (2)
	Increased process understanding.	Balboni (2) Fevotte *et al. (101)* Hammond (19) Hussain (21) Hussain (17) Rudd (46)
	Quality designed into the process.	Herkert *et al.* (59) Lonardi *et al.* (54) Wechsler (5)
	Use of scientific, risk-based approach in decision making.	Hussain (17)
	Recall prevention/ avoidance.	Shah (13)
	Minimized patient risk including security of supply.	Hussain (21) Lonardi *et al.* (54)
	No sampling required or reduced sampling requirements (reduces/ eliminates sampling error).	Dyrby *et al.* (41) Hammond (19) Herkert *et al.* (59) Kamat *et al.* (52) McCormick (43) Skibsted *et al.* (103) Workman (58)
	Critical process control provided.	Herkert *et al.* (59) Rudd (46)
	Rapid identification of counterfeit drug substances.	Cui *et al.* (61)

TABLE 56.3 (Continued)

Benefits category	Specific PAT benefits	Reference
Positive regulatory impact	Moderate regulatory burden on FDA.	Wechsler (5)
	Improved scientific basis for regulatory functions.	Hussain (17)
Increase occupational safety	Decreased occupational exposure to toxic substances.	Avallone (18) Hammond (19) McCormick (43) Watson *et al.* (20) Woo *et al.* (62)
Positive research and discovery impact	Reduced product development life cycle/ time to market.	Bai *et al.* (93) Gupta *et al.* (50) Hussain (21) Hussain (47)
Minimize environmental impact	Reduced environmental impact (assurance that process and plant environments are maintained within environmental regulations).	Workman (58)
	Minimize waste (i.e. solvent waste) generation during manufacturing.	Wang *et al.* (115) Workman (58)

Abbreviations: FDA = FOOD AND DRUG ADMINISTRATION; PAT = PROCESS ANALYTICAL TECHNOLOGY.

The implementation of PAT in the pharmaceutical industry can lead to quality improvements as a direct result of continuous monitoring and the use of process control tools. Real-time monitoring of batch processing steps decreases product variability; the number of batch failures and amount of scrap material are reduced, and the consistency between batches is increased. Reductions in batch failures and scrap material would decrease the current scrap and rework levels of 5%–10% (7). This would translate into reduced cost of quality from the current levels that exceed 20% of the cost of goods sold (7). Reduced analysis times would improve the efficiency within the quality department, further reducing the cost of quality. Andre (12) decreased the work expenditure of the traditional identification method, HPLC, for 7-aminocephalosporanic acid by 98% by use of NIR reflectance spectroscopy. Reduced cost of quality equates with reduced operating expenses and thus an increase in shareholder value.

PAT can be used to reduce consumer risk and for recall prevention and avoidance purposes (13). Through the use of process analyzers such as NIR, it is theoretically possible to perform identification, assay, and dose uniformity on each tablet produced. This would reduce the risk of product cross-contamination, specifically in non-dedicated manufacturing facilities where multiple actives and dosage concentrations are produced. Process Analyzers, in combination with multivariate analysis and process control tools, can also be used to ensure batch-to-batch consistency in real time. Batch-to-batch consistency would improve uniformity within the process, ensuring that the product could be consistently produced within specifications. Through real-time monitoring and control, the consumer's risk of exposure to adulterated products or products that do not meet quality specifications is reduced. PAT can also be implemented for the rapid screening of raw materials (12–16) to identify counterfeit materials, further increasing consumer safety.

Wechsler (5) and Hussain (17) suggest that a positive regulatory impact would result from PAT implementation within the pharmaceutical sector. They hypothesize that PAT would improve the scientific basis for regulations, which would moderate the regulatory burden on the FDA.

Health, safety, and environmental benefits in the form of increased occupational safety and decreased environmental impact of manufacturing operations are expected outcomes of the use of PAT in this sector. Process analyzers can be used to monitor production facilities to ensure the environment is maintained within specific conditions. Avallone (18) and Hammond (19) emphasized decreases in employee exposure to toxic substances. Watson, Dowdy, DePue et al. (20) used an in-line FT-IR process analyzer to monitor a synthesis process to ensure reaction conditions did not generate an unstable hydroperoxide. PAT can also result in reduced waste/scrap from processing and packaging.

56.8 PAT GUIDANCE

To facilitate a paradigm shift in the manufacturing philosophy within the pharmaceutical sector, the FDA has been actively involved in dialogue and the development of guidance for the industry. To provide guidance for this shift, the FDA and CDER formed the Advisory Committee for Pharmaceutical Science (ACPS).

The objective of the ACPS was to identify the current status and future trends of PAT in the pharmaceutical industry and to develop a collaborative industry/academic/regulatory approach. ACPS, through industry and academic involvement, investigated regulatory challenges with respect to PAT implementation as well as method validation and specification requirements and the feasibility of parametric release (5). Parametric release is defined as the assessment of product attributes and process controls to ensure that the in-process or finished pharmaceutical product is of acceptable quality (1). These product attributes and process controls are based on scientific understanding of the process and product. Using the information gathered from this collaborative approach, ACPS recommendations were incorporated into the PAT Guidance for Industry, which was introduced in September 2004.

56.9 PAT PRINCIPLES AND TOOLS

The PAT Guidance for Industry identified the following principles and tools as being suitable for implementation in development and manufacturing activities in the pharmaceutical industry: (1) multivariate tools for design, data acquisition, and analysis, (2) process analyzers, (3) process control tools, and (4) continuous improvement/knowledge management.

56.9.1 Multivariate Tools for Design, Data Acquisition, and Analysis

Pharmaceutical dosage forms are complex systems in which chemically and physically relevant data are measured and analyzed. The conversion of this data to knowledge and the identification of multifactorial relationships can be achieved through multivariate analysis. Common multivariate tools that are housed within the PAT framework include library construction for NIR, design of experiments, principal component analysis, principal component regression, multiple linear regression, partial least squares, neural networks, and statistical process control charts (21).

Multivariate tools for quantification, in particular NIR calibration, have been documented previously (22–28) and only a description of common multicomponent analysis tools will be presented here (Table 56.4).

Library construction for NIR is an alternative to traditional qualitative analysis methods for pharmaceutical raw materials. Blanco and Romero (14) stress the following points when developing an NIR library for qualitative analysis: 1. Establish baseline NIR spectra using samples of known identity. 2. Samples of known identity should be composed of various batches so that they represent the physical-chemical variability of the raw material. 3. During construction of the library, the NIR pattern recognition method and construction parameters must be determined. 4. Internally validate the method to determine if spectra within the library are incorrectly identified or remain unidentified. 5. Construct sub-cascading libraries composed of related substances, degradation products, and

TABLE 56.4
Applications of Process Analyzers in the Pharmaceutical Sector

Reference	PAT	Process	Attribute analyzed	On-line/ In-line/ At-line/ Not Stated
Abrahamsson *et al.* (63)	NIR spectroscopy—transmission	Compression	Quantification of active ingredient	Off-line
Airaksinen *et al.* (40)	NIR spectroscopy—reflectance Ramon spectroscopy	Granulation	Identification of theophylline monohydrate	Off-line
Andre (12)	NIR spectroscopy—reflectance	Raw material	Quantification of 7-aminocephalosporanic acid (7-ACA)	Off-line
Betz *et al.* (64)	Temperature sensor—temperature increase power consumption of mixer motor	Granulation	Granulation end point (temperature and power consumption ratio—TPR)	In-line
Blanco *et al.* (65)	NIR spectroscopy—reflectance	Compression and coating—tablet	Identification and quantification of gemfibrozil	Off-line
Blanco and Romero (14)	NIR spectroscopy—reflectance	Raw material	Identification	Off-line
Blanco and Villar (66)	NIR spectroscopy—reflectance	Compression and coating—tablet	Quantification of miokamycin	Off-line
Yoon *et al.* (67)	NIR spectroscopy—reflectance	Tablets	Site of manufacturing identification	Off-line
Blanco *et al.* (68)	NIR spectroscopy—reflectance	Granulation	Quantification of nimesulide	Off-line
Clarke (33)	NIR microscopy	Granulation and compression—tablets	Spatial distribution and cluster size of ingredients.	Off-line
Cui *et al.* (61)	NIR spectroscopy—reflectance	Powder samples	Identification of sulfaguanidine	Off-line
Davis *et al.* (42)	X-ray powder diffraction	Granulation	Monitor the transformation of metastable polymorph to stable polymorph during granulation process	On-line
Dyrby *et al.* (41)	NIR spectroscopy—transmittance Raman spectroscopy	Compression	Quantification of active ingredient in Escitalopram tablets	Off-line; At-line
Fountain *et al.* (69)	NIR spectroscopy	Mucoadhesive thin-film composites	Quantification of testosterone	Off-line
Gupta *et al.* (50)	NIR spectroscopy	Milling of roller-compacted powders	Particle size and compact strength	On-line

Reference	PAT	Process	Attribute analyzed	On-line/ In-line/ At-line/ Not Stated
Harris and Walker (51)	NIR spectroscopy -transmittance	Technique can be applied to the drying of cakes, pastes, and slurries.	Solvent evaporation	On-line
Herkert *et al.* (59)	NIR spectroscopy—reflectance	Packaging line (blister)	Identification	On-line
Jorgensen *et al.* (70)	NIR spectroscopy—reflectance	Granulation	Wet granulation end point	On-line
Laasonen *et al.* (71)	NIR spectroscopy—reflectance	Packaging component identification	Identification of blister PVC films	Off-line
Laasonen *et al.* (53)	NIR spectroscopy—reflectance	Compression—tablets	Quantification of caffeine	Off-line
Laasonen *et al.* (10)	NIR spectroscopy—reflectance	Packaging component identification	Identification of blister PVC films and film thickness	Off-line
Otsuka (56)	NIR spectroscopy	Granulation	Particle size	Off-line
Ritchie *et al.* (72)	NIR spectroscopy—transmittance	Compression—tablets and capsules	Content uniformity and assay	Off-line
Watson *et al.* (20)	FT-IR	Active pharmaceutical ingredient	Monitor synthesis	In-line
Woo *et al.* (62)	NIR spectroscopy	Liquid manufacturing	Hydrogen peroxide formation	Off-line
Blanco *et al.* (73)	NIR spectroscopy—reflectance	Granule and tablets	Quantitation of ascorbic acid	Off-line
Blanco *et al.* (74)	NIR spectroscopy—reflectance	Capsule	Identification and quantification of pirisudanol dimaleate	Off-line
Blanco *et al.* (35)	NIR spectroscopy—reflectance	Powder	Moisture Content	Off-line
Gottfries *et al.* (75)	NIR Spectroscopy—Reflectance and Transmittance	Compression tablets	Quantification of metoprolol succinate	Off-line
Han and Faulkner (11)	NIR spectroscopy—reflectance	Granulation, compression, coating, and blistering of tablets	Moisture content, identification, quantification	Off-line
Last and Prebble (36)	NIR spectroscopy—reflectance	Freeze-dried injection	Moisture content	Off-line
Lonardi *et al*, (15)	NIR spectroscopy—reflectance	Powders	Moisture content and quantification	Off-line
Plugge and van der Vlies. (16)	NIR Spectroscopy	Powders	Identification, moisture content, and quantification	Off-line
Higgins *et al.* (76)	NIR spectroscopy—reflectance	Liquid manufacturing	Particle size	On-line
Rantanen *et al.* (38)	NIR spectroscopy—reflectance	Granulation—fluid bed	Granule moisture content	In-line
Watano *et al.* (32)	Image probe (CCD camera and high-energy xenon (XE) lighting system)	High shear granulation	Particle size	In-line
Watano *et al.* (31)	Image probe (CCD camera and high-energy xenon (XE) lighting system)	High shear granulation	Particle size Particle shape	In-line
Andersson *et al.* (77)	NIR spectroscopy—reflectance	Granulation—fluid bed	Particle size	In-line
Broad *et al.* (78)	NIR spectroscopy—transmittance	Aqueous suspension	Quantification of ethanol, propylene glycol, water	Off-line
Rantanen *et al.* (37)	NIR spectroscopy	Granulation	Granule moisture content	In-line
Sanchez *et al.* (57)	NIR spectroscopy—reflectance	Raw material	Moisture content	Off-line

(Continued)

TABLE 56.4
(Continued)

Reference	PAT	Process	Attribute analyzed	On-line/ In-line/ At-line/ Not Stated
Kirsch and Drennen (79)	NIR spectroscopy—reflectance	Compression	Tablet hardness	Off-line
Eustaquio *et al.* (80)	NIR spectroscopy—transmittance	Compression	Quantification of paracetamol	Off-line
Blanco *et al.* (81)	NIR spectroscopy	Granulation, compressed and coated cores	Quantification of gemfibrozil	Off-line
O'Neil *et* al. (55)	NIR spectroscopy—reflectance	Raw material	Particle size	Off-line
Frake *et al.* (39)	NIR spectroscopy	Granulation	Granule moisture content Particle size	In-line
Gold *et al.* (49)	NIR spectroscopy—reflectance	Capsule	Dissolution	Off-line
Morisseau and Rhodes (60)	NIR Spectroscopy—Reflectance	Compression	Tablet hardness	Off-line
Kamat *et al.* (52)	NIR spectroscopy—reflectance	Lyophilized	Moisture content	Off-line
Dubois *et al.* (48)	NIR spectroscopy—reflectance	Aqueous Suspension	Quantification of phenazone, glycerol, ethanol, lidocaine hydrochloride, sodium thiosulphate	Off-line
Blanco *et al.* (84)	NIR spectroscopy—reflectance	Blending	Identification and quantification of ferrous lactate dehydrate	Off-line
Blanco *et al.* (85)	NIR spectroscopy—reflectance	Blending, compression, coated tablets	Identification and quantification of otilonium bromide 400 mg/g	Off-line
Gupta *et al.* (86)	NIR spectroscopy—reflectance	Roller compaction	Acetaminophen content uniformity, moisture content, relative density, tensile strength, Young's modulus	In-line
Lai and Cooney (87)	Light-induced fluorescence technology (LIF)	Blending	Homogeneity end point and blend stability of triamterene (2,4,7-triamino-6-phenylpterine) powder.	On-line
Moffat *et al.* (88)	NIR spectroscopy—reflectance	Compression	Identification and quantification of paracetamol in intact tablets.	Off-line
El-Hagrasy *et al.* (89) El-Hagrasy *et al.* (90) El-Hagrasy and Drennen III (91)	NIR spectroscopy	Blending	Homogeneity of salicylic acid powder	On-line
Tumuluri *et al.* (92)	NIR spectroscopy—reflectance	Hot-melt extruded film	Quantification of clotrimazole	Off-line
Bai *et al.* (93)	NIR spectroscopy—reflectance	Lyophilized formations (vials)	Protein confirmation in lyophilized protein formations.	Off-line
Gupta *et al.* (94)	NIR spectroscopy	Roller compaction	Density, moisture content, tensile strength (TS), and Young's modulus	Off-line
Bai *et al.* (95)	NIR spectroscopy—reflectance	Lyophilized formations (vials)	Quantification of glycine crystallinity.	Off-line
Laitinen *et al.* (96)	Monochrome CCD camera	Granulation	Particle size analysis and end point determination of the granulation process	At-line
Seyer *et al.* (97)	NIR spectroscopy—reflectance	Lyophilized formations (vials)	Degree of crystallization	Off-line
Zhou *et al.* (98)	NIR spectroscopy—reflectance	Drying of drug substance	Differentiate between surface and bound water. Determination of water content in drug substance.	In-line
Lai *et al.* (99)	Light-induced fluorescence technology (LIF)	Compression	Total tablet content of active in tablet	On-line

Reference	PAT	Process	Attribute analyzed	On-line/ In-line/ At-line/ Not Stated
Lai *et al.* (100)	Light-induced fluorescence technology (LIF)	Blending	Blend homogeneity	On-line
Fevotte *et al.* *(101)*	NIR Spectroscopy—Reflectance	Crystallization	Qualitative and quantitative analysis of SaC (active pharmaceutical ingredient) during crystallization.	In-line
Lin *et al.* (102)	FT-IR with an attenuated total reference (ATR) probe	Pharmaceutical salt formation process	Real-time end point monitoring and determination for a pharmaceutical salt formation process. Pharmaceutical salt (4-*/*1-methyl-2-piperidin-4-yl-4-[3-(trifluorometryl)phenyl]-1H-imidazol-5-yl*/*-*N*-[(1S)-1-phenylethyl]pyridine-2-amine (freebase), an active pharmaceutical ingredient as a P38 MAP kinase inhibitor)	In-line
Skibsted *et al.* (103)	NIR spectroscopy—reflectance	Blending	Qualitative and quantitative analysis of API to assess blend homogeneity	In-line (Note—probe not installed in mixer, but inserted at specific time intervals)
Johansson *et al.* (104)	Ramon spectroscopy	Compression	Quantitative analysis of API in tablets	Off-line
Cogdill *et al.* (105, 106, 107)	NIR spectroscopy—reflectance	Compression	Quantification of API in tablet and physical parameters including hardness.	On-line
Hausman *et al.* (108)	Ramon spectroscopy	Granulation	Risedronate sodium solid-state form was continuously monitored using on-line Raman spectroscopy during drying.	On-line
Islam *et al.* (109)	Ramon spectroscopy	Topical gels and emulsions	Raman spectroscopy demonstrated as a process analytical technique for quality control of topical gel and cream formulations	Off-line
Johansson *et al.* (110)	Ramon spectroscopy	Compression	Laser-induced heating of compressed tablets during Raman spectroscopy analysis.	Off-line
Kontoyannis (111)	Ramon spectroscopy	Compression	Quantitative analysis of $CaCO_3$ and glycine in antacid tablets.	Off-line
Langkilde *et al.* (112)	Ramon spectroscopy	Pharmaceutical active ingredient	Quantitative analysis of two crystal forms of a pharmaceutical active ingredient.	Off-line
Szostak and Mazurek (113)	Ramon spectroscopy	Powder and tablets	Quantitative of acetylsalicylic acid and acetaminophen in tablets.	Off-line
Taylor and Zografi (114)	Ramon spectroscopy	Powder	Quantitative analysis of indomethacin crystallinity.	Off-line
Wang *et al.* (115)	Ramon spectroscopy	Tablets	Aspirin tablet assay.	Off-line

Abbreviations: ATR = attenuated total reference; CCD = charge-coupled device; FT-IR = Fourier-transform infrared; LIF = light-induced fluorescence; NIR = near-infrared; PAT = process analytical technology; PVC = polyvinyl chloride; TS = tensile strength.

enantiomers to ensure NIR pattern recognition method and construction parameters are optimized for substance identification. 6. Use an external validation set of unequivocally identified substances to perform external validation. Using the discriminating power designed into NIR libraries, document a qualitative method for raw material identification.

Design of experiments (DOE) uses factorial design to evaluate the effects of several factors on a process. When using DOE, varying the factors simultaneously instead of one at a time is an effective approach to identifying interactions among the factors. In complex multifactorial pharmaceutical dosage forms, these interactions affect the quality of the end product.

Pharmaceutical products and processes are composed of multiple components. To reduce the number of variables being analyzed, principal components analysis (PCA) may be used. PCA identifies combinations of variables or factors that contain the maximum variability within the sample set, and then uses these variables or factors as the principal components; a large data set is then reduced to a smaller number of components. The principal components can then be analyzed using further multivariate tools (2, 25). One application of PCA in PAT is to determine the correlation between a process analyzer result (e.g. NIR) and a final product/process attribute (e.g. moisture content, assay results, etc.).

Multiple linear regression (MLR) is the analysis of two or more variables as a linear combination of the dependent variables (28). The combination of MLR and PCA results in principal components regression (PCR), which has been used for multivariate analysis in PAT. PCR uses the principal components identified through PCA and performs regression on the resultant sample property to be predicted (25). Variance of the measured variables is used to determine the model of best fit. PCR incorporates most sources of variability and is more efficient than MLR.

Partial least squares (PLS) is a multivariate analysis tool that combines PCA and multiple regression (28). With respect to process analyzers, PLS can be used to identify variability between the spectral data and the product property. In contrast to MLR, the model of best fit for PLS is based on the product property of the measured variables. PLS is a common multivariate analysis tool for NIR calibration models used to quantify the active ingredient in granulation, compression, coated, and packaged pharmaceutical products (Table 56.4). If the response is nonlinear, an artificial neural network (ANN) could be used to identify relationships. ANN contains computational algorithms that process experimental data and transform these data using nonlinear logarithmic/exponential/quadratic functions to determine the response (25).

Multivariate tools allow scientists to identify the correlation between raw data and their impact on the process. With these multivariate models, continuous recalibration is essential if there are even minor changes to the production process, raw materials, or critical process parameters (12).

Trends in product quality attributes can be followed using statistical process control charts (SPCCs). SPCCs plot data on an ongoing basis in relation to upper and lower product specifications. These data can be utilized to determine process capability, i.e. the ability of the process to meet specifications consistently.

56.9.2 Process Analyzers

Process analyzers measure the physical, chemical, and biological properties of materials. They collect both quantitative data (moisture content, particle size, active ingredient quantification, microbial counts) and qualitative data (microbial identification, active ingredient identification). Data collection can be nondestructive, require minimal sample preparation, and have rapid or real-time response when compared with traditional methods (29). Data integrity is necessary to ensure compliance with the United States FDA 21 Code of Federal Regulations Part 11 (21 CFR Part 11), which requires specific controls with respect to electronic signatures, security, and audit trail functionality.

PAT Guidance for Industry (1) categorizes process analyzers into four categories that are differentiated from one another based on the stage of the process at which sample measurement occurs: at-line (the sample is removed, isolated from, and analyzed in close proximity to the process stream), on-line (the sample is diverted from the manufacturing process and may be returned to the process stream), in-line (the sample is not removed from the process stream and analysis can be invasive or noninvasive), and off-line (the sample is removed, isolated from, and analyzed away from the process stream as in a laboratory environment). Of these four categories, on-line and in-line process analyzers have the greatest potential to reduce operating costs and improve quality; both minimize sample requirements and sample handling compared with their at-line and off-line counterparts. Clevett (30) indicated that 80%–90% of errors associated with analysis were associated with sample handling, either directly or indirectly. On-line and in-line process analyzers reduce sample handling and sample preparation errors, thereby reducing retest and cycle times.

Process analyzers are used primarily to determine the following attributes of raw materials, in-process materials, and finished goods: particle size and shape, moisture content, active ingredient quantification, dissolution profiles, and tablet hardness (Table 56.4).

56.9.2.1 Near Infrared

Particle size of a granulation, powdered blend, or powdered pharmaceutical raw material is important in that it impacts physical properties such as powder flow, dissolution rate, compressibility, and tablet hardness. Monitoring particle size and control of the manufacturing process prevents overprocessing of the product. According to the literature, the most common process analyzer to be used in the determination of particle size of milled roller compacted powders, granulations, liquids, and raw materials is NIR (Table 56.4). NIR process analyzers have been evaluated on-line, in-line, and off-line; results of these evaluations compare favorably with those of traditional methods such as sieve analysis, digital microscopy, and particle size instrumentation.

The shape and spatial distribution of particles influence physical properties such as powder flow and filterability (28). Clarke (33) used NIR microscopy off-line to determine the spatial distribution and cluster size of ingredients in granulation and compressed pharmaceutical products. He concluded that NIR microscopy was a useful tool in the determination of particle shape, particle distribution, and cluster size of the chemical components of the sample.

Moisture content has commonly been measured using techniques such as NIR spectroscopy. In 1968, Sinsheimer and Poswalk reported the use of NIR in the determination of moisture content in pharmaceuticals (34). The use of NIR spectroscopy for the determination of moisture content in raw materials, pharmaceutical powders, and freeze-dried injectables is summarized in Table 56.4. Blanco, Coello, Iturriaga et al. (35) determined the moisture content in the raw material ferrous lactate dehydrate using off-line NIR reflectance spectroscopy. Multivariate analysis (PLS and MLR) methods provided similar results, with prediction errors of <1.5%. Lonardi, Viviani, Mosconi et al. (15) determined that a calibration model (MLR) containing a 50/50 mixture of laboratory samples/production samples provided the best predictive power and the lowest error. Plugge and van der Vlies (16) determined the moisture content of an antibiotic powder containing ampicillin trihydrate using off-line NIR spectroscopy and reported that the method was accepted by the FDA in 1992. NIR reflectance spectroscopy was also used by Last and Prebble (36) to determine the moisture content of a freeze-dried experimental injectable drug. Accuracy and precision of their off-line analysis were identified as limitations of the NIR models assessed; they recommended incorporating more samples and optimizing the NIR wavelength regions.

Process analyzers have been utilized in the determination of wet granulation end points. Wet granulation consists of three processes: mixing, spraying, and drying, with moisture content being a critical end point parameter in the final phase. Traditional methods of determining moisture content require sampling of the granulation and then analysis by Karl Fischer or 'loss of drying' moisture analyzers. These techniques are performed off-line/at-line and hence require operator intervention to collect the samples. To obtain real-time results, NIR spectroscopy has been evaluated as an off-line and in-line process analyzer (Table 56.4). Rantanen, Rasanen, Tenhunen et al. (37), Rantanen, Rasanen, Antikainen et al. (38), and Frake, Greenhalgh, Grierson et al. (39) used in-line NIR sensors to collect moisture content data in fluid bed granulators. The in-line calibration models provided enough predictive power for the determination of moisture content in the sample set.

NIR spectroscopy methods have been used to identify and quantify active ingredients and excipients in granulation (fluid bed and wet granulations), compressed bulk, coated bulk, aqueous products, and active ingredients in blistered product. NIR methods have also been used for identification of raw materials, solvent evaporation, packaging component identification, safety applications such as monitoring of hydrogen peroxide formation, and hardness determination. Use of on-line, at-line, and off-line NIR spectroscopy for these applications is summarized in Table 56.4.

56.9.2.2 Raman Spectroscopy

Raman spectroscopy is suitable for quantitative analysis of pharmaceutical products because of the relationship between signal intensity and active pharmaceutical ingredient concentration. Raman spectroscopy has been evaluated for identification and quantification of active ingredients in granulation, compression, drug pellet, and solid mixture; samples have been evaluated for both off-line and at-line use (40, 41,117, 118). Raman spectroscopy has also been used to monitor hydration states of active pharmaceutical ingredients as a method of process control (119, 120, 121).

When implementing Raman spectroscopy in pharmaceutical processes, it is important to consider the sampling area. Dyrby, Engelsen, Norgaard et al. (41) and Johansson, Petterson, and Taylor (104) concluded that the increased predictive error associated with the Raman model was a result of the tablet area sampled. Compared with the NIR transmittance model, Raman spectroscopy using surface sampling collects data on a smaller volume of the tablet, thus explaining the higher predictive error. Increasing the irradiated surface area by use of rotating sample holders has been shown to decrease the predictive error of the Raman model (41, 104).

56.9.2.3 CCD Camera

Watano, Numa, Miyanami et al. (31) and Watano, Numa, Koizumi et al. assessed particle size in a high-shear granulator in-line through the use of an image probe (CCD camera and high-energy xenon lighting system). The image probe was combined with a fuzzy logic control system to control granulation growth in the high-shear granulator, preventing excessive granule growth. The system was capable of accurately and reliably producing granules that met specifications, independent of the starting materials and operating conditions. Laitinen, Antikainen, Rantanen et al. (96) assessed particle size growth in a fluidized-bed granulation process using a monochromatic CCD camera. At-line analysis of granulation samples via the CCD camera successfully monitored granule growth and granulation end point for the fluidized-bed granulation process. The conclusion was that the imaging approach used provided rapid evaluation of granule particle size.

56.9.2.4 X-Ray Diffraction

On-line application of X-ray powder diffraction was evaluated by Davis, Morris, Huang et al. (42) for use in monitoring the transformation of metastable polymorph to stable polymorph during wet granulation of flufenamic acid. The on-line process analyzer was successful in monitoring the polymorphic transformation of the flufenamic acid. The results of this evaluation suggest that X-ray powder diffraction may be used as an on-line process analyzer to monitor granulation process and process parameters such as granulation end time.

56.9.2.5 FT-IR Process Analyzer

Process analyzers have been evaluated for active pharmaceutical ingredient (API) synthesis. Watson, Dowdy, DePue et al. (20) evaluated an in-line FT-IR process analyzer for the conversion of buspirone hydroxylation to 6-hydroxybuspirone. They recommended the use of the in-line FT-IR process analyzer to monitor and control the synthesis process because this process ensures active pharmaceutical ingredient quality and predicted the need for batch reprocessing.

Lin, Zhou, Mahajan et al. (102) demonstrated the ability to real-time monitor a pharmaceutical salt formation process with FT-IR coupled with an attenuated total reference (ATR) probe, a task which cannot be accomplished with traditional analytical instrumentation such as titration and HPLC. FT-IR ATR permitted differentiation between mono and bis-salts, allowing for real-time determination of the synthesis end point. Other benefits were improved quality monitoring, higher yields, and ease of method transfer between laboratories and FT-IR instruments, all of which contribute to improved efficiency.

56.9.2.6 Light-Induced Fluorescence

Light-Induced Fluorescence (LIF) Technology is selective for fluorescent materials (usually the active ingredient) within a drug formulation. LIF measures the emission wavelength as a result of wavelength excitation. LIF Technology is a nondestructive PAT tool for the analysis of powder mixing kinetics and blend homogeneity and tablet active ingredient content (87, 99, 100). Lai and Cooney (87) proposed that LIF would be especially useful within the pharmaceutical industry because 60% of the 200 main active ingredients fluoresce. Benefits of on-line LIF analysis in blending include real-time blend kinetic results and reductions in errors because of thief sampling (87).

56.9.3 Process Control Tools

Process control tools monitor and actively manipulate a process to ensure control. Process analyzers can be integrated into a process control application to measure critical process parameters and product attributes in order to achieve desired in-process and finished quality specifications. Shah (13) summarized those critical process parameters (moisture content, particle size, blend uniformity, content uniformity, tablet hardness, and viscosity), which could be monitored and controlled to ensure that in-process and finished quality specifications are achieved.

Watano, Numa, Miyanami et al. (31) and Watano, Numa, Koizumi et al. (32) evaluated a process control tool for monitoring and controlling a high shear granulation process. An image probe (CCD camera and high-energy xenon lighting system) and model-based system (fuzzy logic) were utilized to collect particle size images throughout the high-shear granulation phase. The processing conditions were varied to simulate normal manufacturing variation. The system accurately monitored and provided feedback during granulation, preventing excessive granule growth.

56.9.4 Continuous Improvement/ Knowledge Management/ Information Technology Systems

The integration of PAT (process analyzer, multivariate analysis, and process control tools) results in the generation of a large volume of data that must be converted from data to knowledge. Knowledge management tools provide a way of storing data as well as using models, process simulation tools, and pattern recognition tools to develop process knowledge and understanding. Knowledge management systems should be designed to expand as a product moves through design into the product development phase and subsequent manufacturing. The knowledge base includes information on critical process parameters; management of the data can be used as a basis for process improvements.

To maximize the benefits of PAT, the data must be collected, analyzed, and presented in a manner that demonstrates that the product meets all release criteria. This information can be summarized in an electronic batch record or external repository that centralizes data and process instructions from a variety of sources.

Quality Assurance can verify that the data within an electronic batch record pertaining to raw materials, in-process materials, and finished goods meet release specifications. It is also possible for computerized controls to be built into the electronic batch records to prevent raw materials, in-process materials, and finished goods from being released accidentally when their attributes do not meet specifications. Such computerized control ensures that quality is not compromised and reduces the risk of product recalls.

56.10 VALIDATION REQUIREMENTS

Process analytical technology consists of the integration of process analyzers, multivariate analysis tools, process control tools, and continuous improvement/knowledge management/information technology systems. The integration of such a complex system (Figure 56.1) requires that the following five validation/calibration activities be considered: analytical method validation, sensor calibration, computer system validation, equipment qualification, and process validation. Validation is the ability to demonstrate that a procedure, process, equipment, material, or system can consistently produce results that meet specifications. It involves examining and understanding those parameters/conditions of steps that are critical to the process (and ultimately the product) as well as establishing specifications.

To secure the benefit from qualification, it must be seen as an integral part of project. The first stage of the overall process involves assessing the Good Manufacturing Practice (GMP) criticality and defining the key validation requirements of the PAT application (including PAT analyzer, analytical method, processing equipment, and computerized system). The initial phase of the validation life cycle (Figure 56.2) includes pre-qualification, which consists of defining the system documentation: the user requirements specifications (URS), the request for proposal (RFP), the functional specifications (FS) and the design specifications (DS).

Real-time monitoring of manufacturing processes with PAT analyzers generates data that can be used to control manufacturing processes and/or develop a broader understanding of the manufacturing processes. During design of a PAT application, specific consideration of the desired outcome of the PAT should be contemplated because this will influence the validation requirements. Key stakeholders who should be consulted at this stage are representatives of Information Management, Quality, Operations, Engineering, and Research and Development.

Once the specific objective of PAT has been determined and a vendor selected, validation activities must be considered. Qualification is designed to ensure that the installation and operation of PAT analyzers and processing equipment are according to specification. As with most PAT analyzers and modern processing equipment, the qualification should include computerized system validation.

Qualification begins with Installation Qualification (IQ) and is followed by Operational Qualification (OQ) and Performance Qualification (PQ). Once qualification activities have been completed, the analytical method can be validated. Validation of the analytical method is critical to validating the manufacturing process. Upon completion of the validation phase, the PAT application

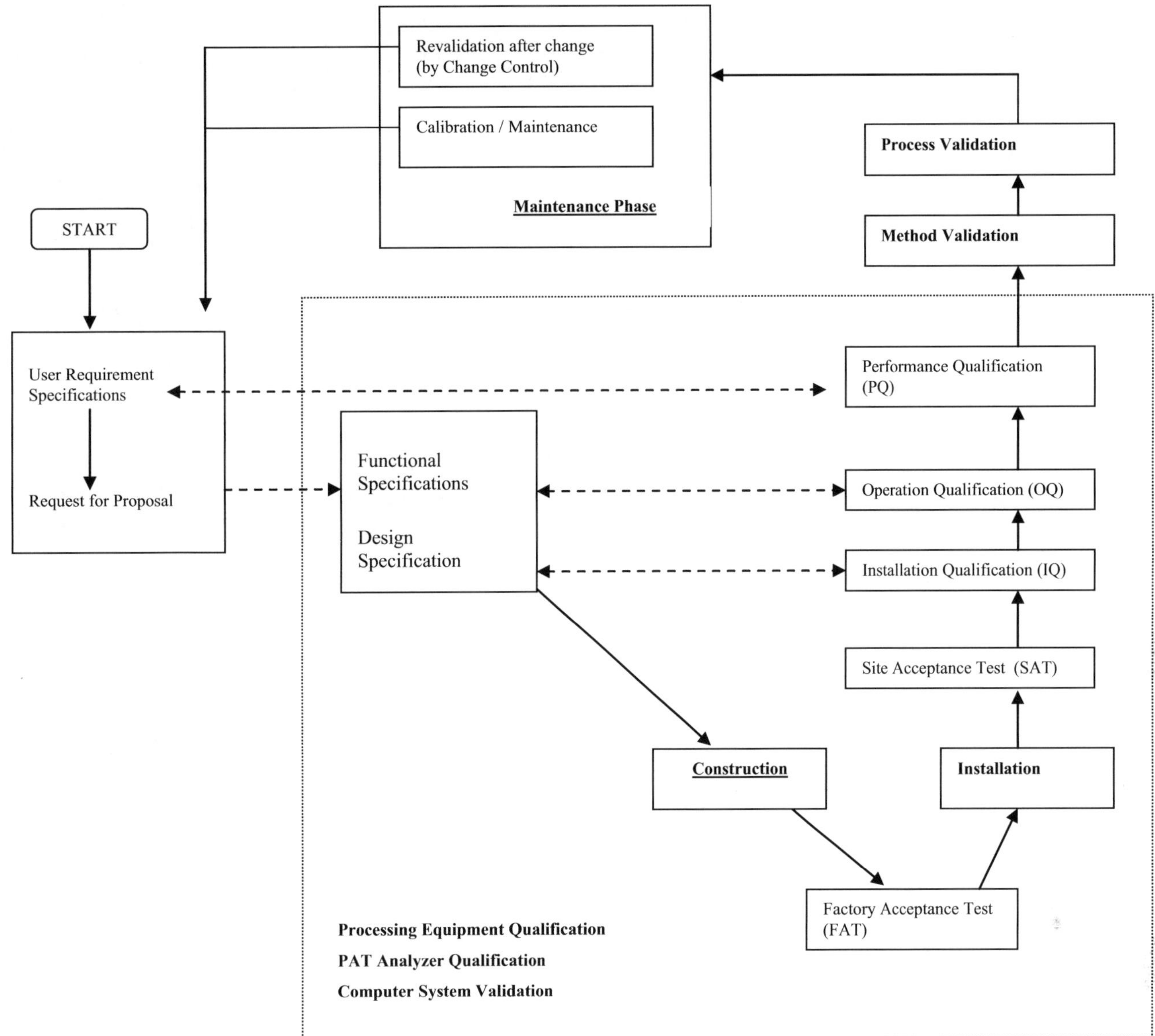

FIGURE 56.1 Validation life cycle for the implementation of a pharmaceutical PAT application.

Note: PAT = Process Analytical Technology.

can be transferred to the maintenance phase, which consists of calibration, preventative maintenance, and change control.

56.10.1 Analytical Method Validation

The main objective of analytical method validation is to demonstrate that the analytical procedure is suitable for its intended purpose. Most of the guidance documents published to date emphasize separation techniques such as HPLC rather than direct spectroscopic techniques such as NIR spectroscopy.

Elements of chromatography method validation include specificity, selectivity, linearity, range, accuracy, precision, repeatability, intermediate precision, reproducibility, ruggedness, detection limit, quantification limit, robustness, and system suitability. To develop an analytical method validation recommendation for PAT applications such as NIR, scientists validated analytical methods using the traditional method validation approach.

As a result of extensive scientific evaluation, key elements required to validate NIR spectroscopy for PAT application have been identified (Table 56.8). These elements were incorporated into the USP's General Chapter on Near-Infrared Spectrophotometry <1119> (116), which are summarized in Table 56.6. It should be noted that these are recommendations; an alternative validation approach may be required for on-line, off-line, at-line, and in-line PAT applications.

56.10.2 Computerized System Validation

All computerized systems, scientific instruments, and processing equipment governed by any GxP regulation (Good

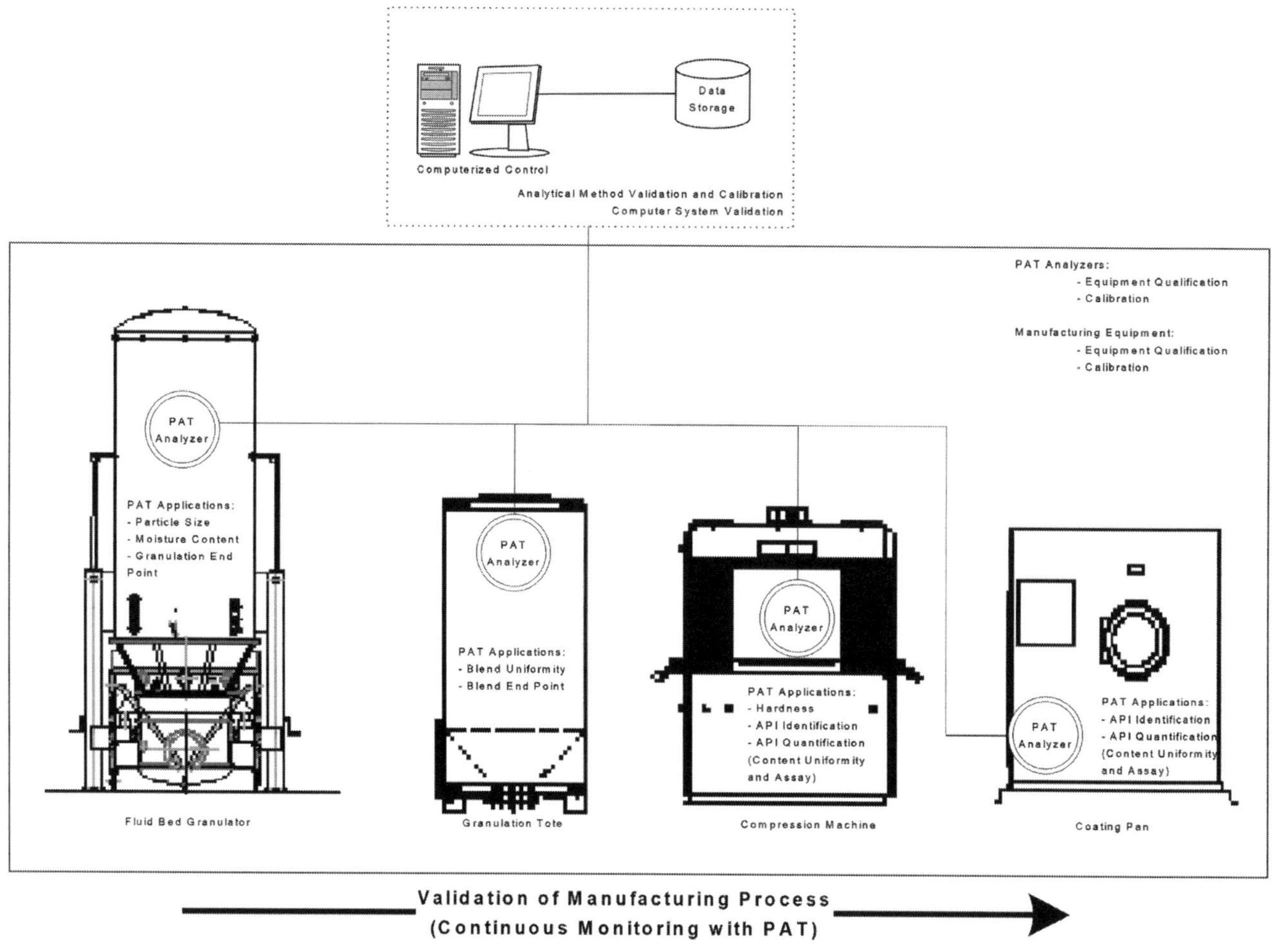

FIGURE 56.2 Application of Process Analytical Technology and validation considerations (solid dose oral manufacturing process). *Note*: API = active pharmaceutical ingredient; PAT = Process Analytical Technology.

Manufacturing Practices, Good Laboratory Practices, and Good Clinical Practices) must be validated. This includes computer hardware, software, network infrastructure, equipment, instruments, as well as procedures that create, modify, maintain, archive, retrieve, and/or distribute data used during development, testing, manufacturing, and distribution.

Data generated by a computerized system can be categorized as either an electronic record or an electronic signature. An electronic record is any combination of text, graphics, data, audio, pictorial, or other information represented in digital form that is created, modified, maintained, archived, retrieved, and/or distributed by a computer system. An electronic signature is a computer data compilation of any symbol(s) executed, adopted, or authorized by an individual to be the legal equivalent of that individual's handwritten signature.

A Validation Plan describes what activities will be performed in order to validate the GxP computerized system. The activities that may be addressed in a Validation Plan are listed in Table 56.10.

The IQ establishes confidence that process equipment (both hardware and software) and ancillary systems comply with appropriate codes and approved design intentions, and that the manufacturer's recommendations are considered. An IQ Protocol should define tests to be conducted during installation of the GxP computerized system and include acceptance criteria. An IQ Report should document the results of the execution of the IQ Protocol and state whether or not acceptance criteria were satisfied. All software should be archived and placed in version control upon installation. Elements of the Installation Qualification Protocol relate directly to the documentation and performance of a design specification.

The OQ establishes confidence that process equipment (both hardware and software) and subsystems are capable of consistently operating within established limits. An OQ Protocol should be written and approved by a designated cross-functional team and should define tests to be conducted during the operational qualification of the system and the acceptance criteria. An OQ Report should document the results of tests conducted following the OQ Protocol and state whether or not acceptance criteria were satisfied. Elements of the Operational Qualification Protocol relate documentation and performance of a functional specification.

The PQ establishes confidence that the computerized system is effective and reproducible. It should be prepared and approved by a cross-functional team and should define tests to be conducted during PQ and the acceptance criteria. A PQ Report should document the results of tests conducted following the PQ and state whether or not acceptance criteria were satisfied. Any policies or procedures that need to be modified or created for implementation of the new or enhanced GxP computerized system should be identified prior to PQ. Elements of the Process Performance Qualification Protocol relate directly to documentation and performance of a user requirement's specification.

56.10.3 Equipment Qualification

Equipment Qualification ensures that laboratory and manufacturing facilities and systems that are directly involved in the manufacture, testing, control, packaging, holding, and distribution of marketed products comply with Good Manufacturing Practices (GMP). It demonstrates the suitability of PAT analyzers and manufacturing equipment for their intended use.

Equipment Qualification consists of requirement and specification documents, installation qualification, operation qualification, and performance qualification (Figure 56.2). IQ is the documented verification that all key aspects of the installation adhere to the manufacturer's/engineering recommendations, design intentions, relevant electrical/building codes, and safety specifications. OQ is documented verification that the equipment can operate as intended and is capable of satisfactory operation over the entire range of operating parameters. OQ includes verification of operation to ensure that the equipment meets certified standards. Finally, PQ is documented evidence that the integrated equipment can perform as intended throughout anticipated operating ranges in the production environment, and that it satisfies user requirements.

Validation test criteria that should be evaluated when validating an NIR instrument for PAT application are summarized in Table 56.6. It should be noted that, although the USP General Chapter <1119> identifies minimum validation requirements, additional validation test cases may be recommended by the manufacturer of the equipment.

56.10.4 Process Validation

Process validation is the demonstrated ability of a process, including equipment, raw materials, environmental controls, and master recipe to produce finished goods within specifications consistently. The validation requirements and approaches for sterile dosage forms, liquid oral dosage forms, solid oral dosage forms, powders, ointments, and creams vary; various guidance documents have been developed by regulatory agencies to address these dosage forms. The traditional pharmaceutical validation paradigm requires that the manufacturing process be repeatable. This has been typically demonstrated by testing three consecutive batches using traditional analytical methods. In the proposed PAT framework, this traditional three-batch approach would be replaced with continuous monitoring of the process to ensure finished product quality. This monitoring would result in continuous updating of data in calibration and validation models. Through the accumulation of additional data, increased process knowledge and understanding would occur, thus ensuring that quality is built into the process instead of being inspected into the finished product. This increased understanding would also allow pharmaceutical manufacturers to adjust to variation in inputs (raw materials, process conditions) while ensuring critical outputs are achieved. This would result in consistent product quality attributes and fewer rejected batches.

Before using PAT for process validation, the PAT analyzer and equipment must be qualified and calibrated, along with the supporting information technology infrastructure. A validated analytical method for the PAT application must also be completed to ensure the validity of the data generated. Upon completion of these activities, PAT can be used to monitor the manufacturing process.

56.10.5 Calibration

Calibration systems and procedures are established for scientific instruments and processing equipment that are critical to quality in the manufacturing and testing of a product. The calibration program ensures that all equipment is performing accurately and reliably on a continual basis, according to inhouse requirements. Calibration is essential to maintaining a validated state. Two calibration issues that must be addressed in PAT are methods (i.e. NIR calibration models and qualitative libraries) and hardware (i.e. NIR instrument calibration).

Evaluation of the ongoing performance of a validated analytical method is critical to successful implementation of process analytical technology. The collection of samples must represent the range of product attributes. Numerous factors that should be evaluated when selecting samples for calibration methods are presented in Table 56.11.

After the calibration method or qualitative library has been established, methods should be updated on a regular basis with new samples. As indicated in USP <1119>, accuracy, precision, and critical validation elements should also be reevaluated at predetermined intervals in order to assess the performance of the analytical method. If a deviation occurs (i.e. the accuracy or precision decreases), the root cause should be investigated immediately (107). The findings of that investigation may dictate corrective action(s) that involve modification of the preventative maintenance procedure or equipment calibration process before putting the NIR method back into service. Changes should be documented through the site's change control process and may be subject to revalidation.

NIR sensor calibration is critical when sensors have been used in the identification and quantification of active ingredients, excipients, physical properties of drug substances, and other critical in-process parameters. One mechanism of calibration is internal performance tests (equipment calibration tests). The frequency of internal performance tests has been demonstrated as a daily requirement for NIR instruments (105, 106). Sensor calibration ensures that process analyzers in PAT applications are functioning as they were designed to function and ensures the accuracy and precision of the data collected from these sensors. Completion of sensor calibration can be accomplished by re-executing critical components of the equipment qualification at predefined intervals. Frequencies recommended by the manufacturer should be followed unless scientific justification supports alternative frequencies. Test cases to consider when performing routine sensor calibration include wavelength uncertainty, tolerances, photometric linearity, and spectrophotometric noise. Specific information on test case and acceptance criteria for these calibration elements can be found in the Equipment Qualification section.

56.10.6 Revalidation

When changes occur to systems, processes, methods, materials, or computerized systems, the effect of those changes is assessed through the change control process. If it is determined that the changes impact the validity of the previous studies, then the modified system/method/process is revalidated. All revalidation activity should be conducted in accordance with current regulatory requirements and site SOPs.

Specific revalidation criteria for NIR methods have been identified in USP <1119>, which recommends revalidation for qualitative NIR analytical methods when the following criteria arise: addition of a new material to the reference library, physical changes to material, supplier changes, and an expanded range of material characteristics. Quantitative methods may be revalidated if any of the following criteria arise: raw material grade changes (changes in purity, polymorphic changes), changes in manufacturing process (processing steps added, removed, or altered), changes in finished product composition (coating solution changes or reformulation), reference method changes, or major instrument maintenance or alteration (specifically, changes that may impact optics).

Table 56.6 identifies additional changes that may impact the validated systems/methods used in PAT. This table also identifies manufacturing process changes that may impact PAT applications. A site change control is required to assess the impact and identify validation activities.

56.10.7 Validation Summary

The level of validation required when implementing a PAT solution depends upon several variables. Through the demonstration of process understanding, control of critical parameters, and monitoring of physical/chemical/biological properties and environmental conditions, process validation activities can be reduced by implementation of PAT. In the absence of process understanding, the FDA has indicated that a test-to-test comparison may be required when implementing a new on-line process analyzer. This test-to-test comparison includes comparison of the data from the on-line process analyzer with that of conventional test methods (1).

56.11 PAT CHALLENGES

Challenges facing PAT implementation within the pharmaceutical industry have been discussed frequently (Table 56.5). A major challenge appears to be numerous perceived regulatory barriers. Regulatory challenges that have been identified include unclear PAT validation requirements and 21 CFR Part 11 requirements, as electronic data related to batch release parameters will be stored. Therefore, retention and management of these data must comply with 21 CFR Part 11.

TABLE 56.5
Challenges Associated with Implementing PAT in the Pharmaceutical Industry

Challenges category	Specific PAT challenges	Reference
Current infrastructure does not facilitate PAT Implementation	Information Technology (IT) infrastructure requirements may not exist in current facilities.	Balboni (2) GlaxoSmithKline (82) Neway (45)
	Lack of senior management support.	Balboni (2)
	Current resource constraints.	Balboni (2) Hussain (47)
	Difficulty in applying PAT when manufacturing Phase I and II drugs (drug formulations have not been finalized).	Bush (83)
	Large volumes of continuous data are produced (system constraints need to be considered during system design).	Dyrby *et al.* (41) Neway (45)
	System needs to handle real-time access for multiple users (system constraints).	Neway (45)
	Limited employee knowledge base.	Blumenstein (3) GlaxoSmithKline (82) Hussain (8) McCormick (43) Neway (45)
	Complex mathematical models can result in the introduction of misinterpretation.	Workman (58)
	24/7 instrument and software support required.	GlaxoSmithKline (82)

Challenges category	Specific PAT challenges	Reference
Regulatory challenges	21 CFR Part 11 requirements.	Balboni (2) GlaxoSmithKline (82)
	Validation requirements unclear.	Balboni (2) Shah (13)
	No perceived regulatory incentive.	Hussain (47)
Cost of PAT implementation	Return on investment.	Balboni (2) GlaxoSmithKline (82) Neway (45) Shah (13)
	Regulatory uncertainty, including regulatory approval delays.	Balboni (2) Blumenstein (3) Clarke (4) Wechsler (5)
	For calibration, need a wide range of samples, which are within and outside specifications, which will increase cost.	GlaxoSmithKline (82)
	Proving equivalency between PAT and traditional methods.	Shah (13)
Industry mindset and concerns	Attitude within Pharmaceutical Industry—"no reason to change—status quo is fine".	Hussain (21) Neway (45)
	Implementation of PAT into current manufacturing process may expose deficiencies in the manufacturing process.	Avallone (18) Clarke (4) Neway (45)
	Accumulation of data, which may show inadequacies in process which produce product, which are acceptable based on traditional testing methods.	Avallone (18) Clarke (4) Neway (45)
	Implementation of PAT could result in increased recalls.	Neway (45)
	No perceived benefit.	Hussain (47)
Technology Challenges	Process analyzers (sensors) prone to drift.	Lavine and Workman (26)
	Calibration model requires frequent updates to include product and process variation.	Dyrby *et al.* (41) Lavine and Workman (26)
	Processes are susceptible to unmodelled events.	Lavine and Workman (26)

TABLE 56.6
USP General Chapter <1119> Recommended Validation Criteria for NIR Equipment Qualification

Validation test case	Test details	Recommended specification
Maximum nominal bandwidth	Instrument bandwidth, based on the analyte/product matrix/process to be measured, should be assessed during the user requirements phase.	USP General Chapter <1119> indicates that a maximum nominal bandwidth of 10 nm at 2500 nm (NIR reflectance) or 16 cm^{-1} at 4000 cm^{-1} (NIR transmittance) is applicable for most applications.
Photometric linearity	Photometric linearity is typically expressed as a percent reflectance or percent transmittance. Traceable carbon-doped polymer standards are used for NIR reflectance. Typically a set of four standards is used to calibrate the NIR instrument, over the range of absorbances required.	A_{obs} versus A_{ref} at 1200, 1600, and 2000 nm. Slope = 1.0 ± 0.05 Intercept = 0.0 ± 0.05
Tolerances	Wavelength tolerances.	± 1 nm at 1200 nm (± 8 cm^{-1} at 8300 cm^{-1}) ± 1 nm at 1600 nm (± 4 cm^{-1} at 6250 cm^{-1}) ± 1 nm at 2000 nm (± 4 cm^{-1} at 5000 cm^{-1})
Wavelength uncertainty	One spectrum is collected and a minimum of three peaks is measured over the range.	Reflectance Mode USP Near-Infrared Calibrator RS_{USP29} Peaks occur at 1261, 1681 and 1935 nm. Transflectance mode NIST SRM 2035 Transmittance NIST SRM 2035 rare earth oxide in standard glass or NIST SRM 2036 $_{USP29}$

TABLE 56.7
Changes to Validated Systems/Methods That May Require Revalidation

Category	Change that may require revalidation
Process	• Process moved to a new location. • A significant change in the production volume (batch size, for example from single batch to double batch or double batch to triple batch) relative to the capacity of the processing equipment. • A significant change in the manufacturing facility and/or environmental controls under which products are manufactured. • Change from manual to automated manufacturing; or vice versa, when significant changes to the processing parameters are made. • Introduction of significantly different manufacturing equipment for use in a manufacturing process. • A significant change in the source/composition of manufacturing materials (i.e. drug substance and or excipient) used to manufacture a product. • Any changes to the formulation, packaging, equipment or process, which could impact, on the effectiveness or product characteristics (safety, purity, identity or strength of the product). • Whenever there are changes in product characteristics. • When changes are made to the raw material supplier, consideration should be given to subtle, potentially adverse differences in the raw material characteristics, which may have a significant impact downstream in the process. • Variations revealed by trend analysis through Annual Product Reviews (e.g. process drifts)
Equipment Qualification	• Changes to the equipment. • Move equipment to a different location. • Changes to the operation of the equipment. • Changes to PLC. • Changes to the operating parameters. • Changes to equipment optics.
Packaging	• New or modified products. • New or modified packaging materials. • New or modified equipment. • New material suppliers. • New or modified change parts. • New, modified, or relocated packaging lines. • New operation parameters.
Analytical Method	• Changes or modifications are made to equipment, manufacturing process, analytical procedure or the composition of the drug product has changed (i.e. new source or synthesis of drug substance or new impurity present).
Computer	• Changes to the hardware and/or software operating systems. • Install new software version.

Abbreviation: NIR = NEAR INFRARED.

TABLE 56.8

Reference	PAT method	Type of PAT procedure	Validation parameter	Conclusions
Blanco *et al.* (84)	NIR spectroscopy—Reflectance	Identification	• Specificity	Blanco *et al.* (84) identified selectivity as a critical validation element for identification methods.
		Quantification	• Precision • Repeatability • Intermediate precision • Accuracy • Linearity • Robustness	Repeatability demonstrated through 12 NIR reflectance determinations from a production batch by the same operator on the same day (Coefficient of Variation (CV) = 0.3%, acceptance criteria <1%). Intermediate Precision demonstrated by two different production batches analyzed by two operators on three different days (CV for batch 1 < 0.7%; CV for batch 2 < 0.7%, acceptance criteria <2%). Accuracy results between the reference method and the NIR reflectance method were not statistically different. NIR and reference standards were compared for linearity. NIR concentration was plotted versus reference concentration (r=0.994).

Reference	PAT method	Type of PAT procedure	Validation parameter	Conclusions
Blanco *et al.* (65)	NIR spectroscopy—Reflectance	Identification	• Specificity	Blanco *et al.* (65) identified selectivity as a critical validation element for identification methods.
		Quantification	• Precision • Repeatability • Intermediate precision • Accuracy • Linearity • Robustness	Repeatability demonstrated by the same operator on the same day, with 6 NIR reflectance determinations (CV within acceptance criteria <1%). Intermediate Precision was evaluated by varying operators and completing analysis between days. CV and ANOVA analysis indicate no systematic errors. Accuracy results between NIR and reference method (UV method). Methods were compared by a *t*-test. No significant difference between NIR and the reference method. NIR and reference standards were compared for linearity. NIR concentration was plotted versus reference concentration (r=0.988).
Blanco *et al.* (85)	NIR spectroscopy—Reflectance	Identification	• Specificity • Repeatability • Robustness	Validation of the identification method was demonstrated through repeatability and robustness. To demonstrate repeatability, each sample was analyzed 12 times (correlation coefficient ranged between 0.992 and 0.999. To demonstrate robustness, 10 samples produced over a 3-month period were analyzed (correlation coefficient ranged between 0.990 and 1.00)
		Quantification	• Precision • Repeatability • Intermediate precision • Accuracy • Linearity • Robustness	Precision was demonstrated by analyzing 1 sample, 12 times by the same operator on the same day (CV < 0.6%). Intermediate precision was demonstrated by analyzing one batch from each processing step on 3 different days by 2 analysts (CV 0.5%–0.8%; ANOVA no systematic error identified). Accuracy and linearity demonstrated by comparing NIR and reference method results. Robustness of the calibration model was evaluated over a 1-month time period (10 samples from new production batches analyzed). Calibration model was deemed stable.
Laasonen *et al.* (53)	NIR spectroscopy—Reflectance	Identification	• Specificity	To demonstrate specificity, caffeine and excipients were analyzed. Caffeine peaks did not interfere with excipient peaks (mean spectral residual + 3 SD). One sample *t*-tests confirmed that the mean residual of the excipient batch was significantly different from that of the production batch (including caffeine).
		Quantification	• Precision • Repeatability • Intermediate Precision • Accuracy • Linearity • Range • Robustness	Method repeatability was demonstrated by collecting NIR spectral information 6 times for a single dosage form (from two batches). A Design of Experiment (DOE) was performed for intermediate precision. The DOE evaluated operator, date of analysis, and batches (%RSD < 2%). Accuracy was demonstrated by comparing NIR and reference method (HPLC) results. Results were compared by a Student's *t*-test, and were determined to be not statistically different. Linearity was determined during the calibration of the NIR method. The NIR method was utilized to predict different batches of 60%–130% caffeine concentration. The results were compared with HPLC results by linear regression (method of least squares). The confidence interval slope included 1, and the *y* intercept did not statistically differ from 0 (*t*-test). Range was demonstrated through linearity, accuracy, and precision. Robustness of a NIR method can be influenced by: • Environmental conditions—environmental conditions (temperature/humidity/direction of sunlight/dust/vibrations) were controlled. • Changes in sample preparation—tablet holder was moved affecting spectrophotometer beam, which affected the results. • NIR source was replaced, which did not affect the results. This was demonstrated through a paired Student's *t*-test.

(*Continued*)

TABLE 56.8
(Continued)

Reference	PAT method	Type of PAT procedure	Validation parameter	Conclusions
Moffat *et al.* (88)	NIR spectroscopy—Reflectance	Identification	• Specificity	Interfering components such as excipients, degradation products, water, residual solvents, and impurities should be considered when developing a NIR method for identification. Zero-order spectra are reviewed to determine if interfering peaks are present. If interfering peaks are present, the spectra can be mathematically/chemometrically treated to remove spectral interference (i.e. second derivative).
		Quantification	• Precision • Repeatability • Intermediate Precision • Accuracy • Linearity • Range • Robustness	To demonstrate repeatability, nine determinations (3 replicates of 3 different concentrations) were analyzed. Precision was demonstrated by scanning the same sample multiple times. When evaluating precision, Moffat *et al.* (88) identified the following parameters that should be considered: • Thermal degradation should be considered for heat-sensitive active pharmaceutical ingredients (APIs) • Surface inhomogeneity • Low dosage API forms—use NIR transmittance instead of reflectance as NIR transmittance analyzes larger surface volume Intermediate precision was demonstrated by analyzing the same sample by different analysts on different days. Traditional intermediate precision involves the assessment using different NIR instruments. Moffat *et al.* (88) indicated there are few NIR instruments in the pharmaceutical sector, making it difficult to assess this parameter on intermediate precision. Accuracy was demonstrated by statistical comparison of NIR values and reference method values. NIR predicted assay values versus reference method values over a specific range were analyzed by linear regression (method of least squares) to demonstrate linearity of the NIR method. Moffat *et al.* (88) verified the range of the NIR method, but identified limitations that included difficulties obtaining samples with a wide range, as the samples need to be produced from the production process. Moffat *et al.* (88) identified factors to be considered when evaluating robustness: • Temperature/humidity should be controlled. If these factors are not controlled, they should be considered during robustness evaluation. • Variation in tablet compaction. • Sample containers if used need to be considered, as they impact NIR analysis. • Changes in sample presentation/orientation. • Tablet morphology (shape/scored/embossed/printed/coating). • Probes depth and probe installation.

TABLE 56.9
NIR Analytical Method Validation Requirements

Validation Parameter	Type of NIR Procedure	
	Qualitative	Quantitative
Specificity	+	+
Linearity	–	+
Range	+	+
Accuracy	–	+
Precision—repeatability	+	+
Precision—intermediate precision	–	+
Robustness	+	+

Abbreviation: NIR = NEAR INFRARED.

TABLE 56.10
Potential Elements of a Validation Plan

System description/configuration
Applicable policies, procedures, and guidelines
Responsible departments and/or individuals
Validation strategy
Risk assessment
Supplier assessment
Categorization of components
Vendor evaluation/audit
Assumptions/exclusions/limitations
Documentation—system, technical, and operational
Testing procedures
Acceptance criteria
Deviations/error reporting/resolution
Change control process standard operating procedures
Security
Backup/archive/disaster recovery
Training
Qualification protocols and reports—IQ, OQ and PQ*

*Installation qualification, Operational qualification, Performance qualification.

TABLE 56.11
Factors to Evaluate When Selecting Samples for Calibration Methods

Different processing conditions
Multiple batches of raw material (representative of the physical-chemical variation of the raw material).
Different API/excipient suppliers
Different API/excipient concentrations in drug formulation
Related substances, degradation products, and enantiomers
Variation in tablet compaction (tablet compaction force)
Sample containers (if used)
Changes in sample presentation/orientation
Granulation characteristics (moisture content, particle size)
Blending characteristics (blend uniformity)
Tablet morphology (shape/scored/embossed/printed/coating)

Abbreviation: API = ACTIVE PHARMACEUTICAL INGREDIENT.

To encourage the ongoing pursuit of PAT implementation, the FDA has released the Guidance for Industry on PAT and has sponsored multiple symposiums to generate dialogue between the FDA and industry.

Additional challenges include the lack of infrastructure within current manufacturing facilities. Information technology, including the ability to network manufacturing equipment with local area networks, may be a challenge in existing facilities. The installation of routers, switches, servers, and the network to an existing facility adds incremental costs to a PAT implementation project. Related challenges include the ability of a network to manage large volumes of data on a continuous basis and real-time access by multiple users. These factors impose constraints on the system, which may result in downtime or loss of data. To minimize downtime and the risk of data loss, continuous instrument, software, and local area network support is required, increasing the costs.

The cost of PAT implementation has resulted in limited senior management support in some organizations (Table 56.4). Questions have arisen about the return on investment as well as the accumulation of data, which may highlight inadequacies in a process. These inadequacies may not interfere with the production of finished products, which are acceptable based on traditional testing methods, but have a low process capability. In this scenario, the concern focuses on the potential for increased product recalls.

Overcoming the limited knowledge about PAT in the pharmaceutical industry is a major challenge. In McCormick's study (43) of a small cross section of pharmaceutical companies, only one-half of the organizations surveyed were aware of the FDA's PAT guidance. Only 14% of the organizations surveyed were currently implementing PAT.

The challenges discussed in this chapter, combined with current resource constraints faced by the pharmaceutical industry and the limited knowledge about PAT process analyzers appear to be the main factors limiting PAT implementation.

56.12 CONCLUSION

The conventional manufacturing paradigm in the pharmaceutical industry is based on batch processing, with laboratory analysis of samples collected at predetermined time intervals and processing steps. This manufacturing approach tests quality into final products, with resulting suboptimal efficiencies, high levels of rework and scrap, high cost of compliance, and low levels of continuous improvement. One major advantage of a shift from the current paradigm to PAT would be that quality would be built into products. Building quality into the process may translate into increased product quality per se, increased regulatory compliance, increased capacity and efficiencies, and/or decreased manufacturing and quality costs.

The PAT approach requires the integrated implementation of process analyzers, multivariate analysis tools, process control tools, and continuous improvement/knowledge management/information technology systems. The complexity of the PAT system has resulted in uncertainty with respect to both regulatory approach and validation. The FDA's PAT Guidance for Industry (1) was an attempt to reduce the uncertainty and perceived barriers. In addition to the guidance document, there has been a series of PAT conferences chaired by the ACPS and CDER. Although regulatory and validation uncertainty have been identified as barriers hindering the adoption of PAT implementation in the pharmaceutical industry, the largest barrier appears to be the return on investment, especially in the short term. PAT elements such as information technology infrastructure, process analyzers (i.e. NIR), process controls, and knowledgeable staff require a substantial financial investment. This barrier may perhaps be the limiting factor, especially n times of decreasing shareholder returns and market exclusivity, increasing generic competition, decreasing research and discovery productivity, and increasing research and discovery costs.

REFERENCES

1. US FDA white paper: Innovation and continuous improvement in pharmaceutical manufacturing: Pharmaceutical CGMPs for the 21st Century. The PAT Team and Manufacturing Science Working Group Report: A Summary of Learning, Contributions, and Proposed Next Steps for Moving towards the "Desired State" of Pharmaceutical Manufacturing in the 21st Century, September 2004. https://wayback.archive-it.org/7993/20170405121836/www.fda.gov/ohrms/dockets/ac/04/briefing/2004-4080b1_01_manufSciWP.pdf (accessed July 9, 2021).
2. Hussain AS. Viewpoint: The nation needs a comprehensive pharmaceutical engineering education and research system. *Pharm Technol* 2005. https://ispe.org/sites/default/files/membership/students/2021/ViewpointPE%20(Hussain)%20(2005).pdf (accessed July 9, 2021).
3. Gurvich VJ, Hussain AS. In, and beyond COVID-19: US academic pharmaceutical science and engineering community must engage to meet critical national needs. *AAPS PharmSciTech* 2020, 21, 153. https://doi.org/10.1208/s12249-020-01718-9 (accessed July 9, 2021).
4. FDA Voices: Accelerating the Adoption of Advanced Manufacturing Technologies to Strengthen Our Public Health Infrastructure. *Content Current as of 01/15/2021.* www.fda.gov/news-events/fda-voices/accelerating-adoption-advanced-manufacturing-technologies-strengthen-our-public-health (accessed July 9, 2021).
5. Quality Management Maturity for Finished Dosage Forms Pilot Program for Domestic Drug Product Manufacturers. *Program Announcement.* 85 FR 65824: 65824–65825, October 16, 2020. www.federalregister.gov/documents/2020/10/16/2020-22976/quality-management-maturity-for-finished-dosage-forms-pilot-program-for-domestic-drug-product (accessed July 9, 2021).
6. Memorandum of Understanding between the National Institute of Standards and Technology (NIST) of the United States Department of Commerce and the Food and Drug Administration (FDA). MOU 225-21-006. www.fda.gov/about-fda/domestic-mous/mou-225-21-006 (accessed July 9, 2021).
7. US FDA. *Resiliency Roadmap for FDA Inspectional Oversight.* www.fda.gov/media/148197/download (accessed July 9, 2021).
8. United States Government Accountability Office. FDA's future inspection plans need to address issues presented by COVID-19 backlog. *Statement of Mary Denigan-Macauley, Director, Health Care*, March 4, 2021. www.gao.gov/assets/gao-21-409t.pdf (accessed July 9, 2021).
9. Hussain AS, Morris K, Gurvich VJ. Pharmaceutical quality, team science, and education themes: Observations and commentary on a remarkable AAPS PharmSciTech theme issue. AAPS *PharmSciTech* 2021, 22, 88. https://doi.org/10.1208/s12249-021-01970-7 (accessed July 9, 2021).
10. Hussain AS, Gurvich VJ, Morris K. Pharmaceutical "new prior knowledge": Twenty-first-century assurance of therapeutic equivalence. *AAPS PharmSciTech* 2019, 20, 140. https://doi.org/10.1208/s12249-019-1347-6 (accessed July 9, 2021).
11. Hussain AS. Pharmaceuticals beyond 2020: Professionals and artificial intelligence: Indian pharmaceutical association. *Pharma Times* June 2020, 52, 14–19. https://ipapharma.org/3d-flip-book/pharma-times-june-2020/ (accessed July 9, 2021).
12. Public Law 87–781-, October 10, 1962. https://uscode.house.gov/statutes/pl/87/781.pdf (accessed July 9, 2021).
13. Code of Federal Regulations. Title 21: Food and drugs, Chapter I: Food and drug administration: Department of health and human services, Subchapter C: Drugs: General Part 211: Current good manufacturing practice for finished pharmaceuticals, Subpart B: Organization and personnel. Sec. 211.25 Personnel Qualifications.
14. FDA Guidance for Industry. PAT: *A Framework for Innovative Pharmaceutical Development, Manufacturing, and Quality Assurance*, September 2004. www.fda.gov/media/71012/download (accessed July 8, 2021).
15. Leedom DK. Final report: Sensemaking symposium. The US Command and Control Research Program: Office of the Assistant Secretary of Defense for Command, Control, Communications, and Intelligence, October 23–25, 2001. www.dodccrp.org/events/2001_sensemaking_symposium/docs/FinalReport/Sensemaking_Final_Report.htm (accessed October 17, 2020).
16. FDA head calls for inquiry into Alzheimer's drug review. *The Associated Press*, July 9, 2021. www.webmd.com/alzheimers/news/20210709/fda-head-calls-for-inquiry-into-alzheimers-drug-approval (accessed July 9, 2021).
17. Scott BA, Wilcock A. Process analytical technology and validation. In Agalloco JP, Carleton FJ, eds. *Validation of Pharmaceutical Processes*, 3rd ed. CRC Press, September 25, 2007.
18. Shah HS, Chaturvedi K, Hussain AS, Morris K. Physicochemical failure modes for first-line therapy Narrow Therapeutic Index (NTI) drugs: A call for attention NTI risk classification and New Prior Knowledge. *Europen Pharmaceutical Review*, July 2, 2021. www.europeanpharmaceuticalreview.com/article/157918/physicochemical-failure-modes-for-first-line-therapy-narrow-therapeutic-index-nti-drugs-a-call-for-attention-nti-risk-classification-and-new-prior-knowledge/ (accessed July 8, 2021).
19. Hussain AS. *Pharmaceutical Quality by Design: Improving Emphasis on Manufacturing Science in the 21st Century.* Rockville, MD: Training for the FDA Pharmaceutical Inspectorate, August 5, 2004. www.pharmamanufacturing.com/assets/Media/MediaManager/Hussein_Pharma_Quality_By_Design.pdf (accessed July 8, 2021).
20. Weick KE. Organizing and the process of sensemaking. In Weick KE, ed. *Making Sense of the Organization Volume 2: The Impermanent Organization.* John Wiley & Sons Ltd, 2009: 129–151.
21. FDA Guidance for Industry. *Process Validation: General Principles and Practices*, January 2011. www.fda.gov/media/71021/download (accessed July 8, 2021).
22. FDA Pharmaceutical Quality Oversight: One Quality Voice. *Office of Pharmaceutical Quality, CDER, FDA.* www.fda.gov/media/91721/download (accessed July 8, 2021).
23. FDA Guidance for Industry. *Advancement of Emerging Technology Applications for Pharmaceutical Innovation and Modernization*, September 2017. www.fda.gov/media/95444/download (accessed July 8, 2021).
24. www.accessdata.fda.gov/scripts/cdrh/cfdocs/cfCFR/CFRSearch.cfm?fr=211.25 (accessed July 8, 2021).
25. Hussain AS. Pharmaceutical 6 sigma and quality by design. The 28th Annual Midwest Biopharmaceutical Statistical Workshop, Ball State University, Muncie, IN, May 23–25, 2005. https://www2.slideshare.net/a2zpharmsci/pharmaceutical-

6-sigma-and-qbd-may-2005-ball-state-university (accessed July 8, 2021).
26. Final Concept Paper, ICH Q14: Analytical Procedure Development and Revision of Q2(R1) Analytical Validation, November 14, 2018. https://database.ich.org/sites/default/files/Q2R2-Q14_EWG_Concept_Paper.pdf (accessed July 8, 2021).
27. Final Concept Paper, ICH Q13: Continuous Manufacturing of Drug Substances and Drug Products, November 14, 2018. https://database.ich.org/sites/default/files/Q13_EWG_Concept_Paper.pdf (accessed July 8, 2021).
28. FDA's PAT Team: Shall We Dance? FDA's PAT team aims to take compliance and manufacturing from art to science. *Pharmaceutical Manufacturing*, April 7, 2005. www.pharmamanufacturing.com/articles/2005/216/ (accessed July 8, 2021).
29. CDER's Janet Woodcock: Nobody Can Really Tell Me If FDA Inspections Are Effective. *Pharmaceutical Online*. From the Editor, April 23, 2013. www.pharmaceuticalonline.com/doc/cder-s-janet-woodcock-nobody-can-really-tell-me-if-fda-inspections-are-effective-0001 (accessed July 8, 2021).
30. Pazhayattil A, Sharma S, Galande A, Ingram M, Rhoades R. Assessing legacy drug quality. *Pharmaceutical Technology* July 2021, 45(7), 54–62. www.pharmtech.com/view/assessing-legacy-drug-quality (accessed July 8, 2021).
31. Hussain AS. How to break the pharmaceutical 2–3 sigma barrier (Like amgen). *Pharmaceutical Online*. Guest Column, September 18, 2017. www.pharmaceuticalonline.com/doc/how-to-break-the-pharmaceutical-sigma-barrier-like-amgen-0001 (accessed July 8, 2021).
32. Hussain AS. Chaos to continual improvement: Path to harmonization. *CPhI Industry Report*, 2019. www.cphi.com/content/dam/Informa/cphi/en/cphi-insights/HLN19CPhI%20Insights-2019-Industry-Report.pdf (accessed July 8, 2021).
33. Hussain AS. Digitization in pharma and digital therapeutics: A migratory birds eye view for charting a path forward. *Pharma Times* December 2020, 52(12), 11–15. https://ipapharma.org/3d-flip-book/pharma-times-december-2020/ (accessed July 8, 2021).
34. New Prescription for Drug Makers: Update the Plants. After years of neglect, industry focuses on manufacturing: FDA acts as a catalyst. *The Wall Street Journal* September 3, 2003. www.wsj.com/articles/SB10625358403931000 (accessed July 8, 2021).
35. PAT Guidance for Industry: A Framework for Innovative Pharmaceutical Development, Manufacturing, and Quality Assurance. U.S. Department of Health and Human Services, Food and Drug Administration, Center for Drug Evaluation and Research (CDER), Center for Veterinary Medicine (CVM), Office of Regulatory Affairs (ORA), Pharmaceutical cGMPs, September 2004.
36. Balboni ML. Process Analytical Technology: Concepts and principles. *Pharm Technol* 2003, 27, 54.
37. Blumenstein J. Pfizer. Regulatory Challenges: PAT Application in NDAS. FDA's Advisory Committee for Pharmaceutical Sciences, Subcommittee on Process Analytical Technologies (PAT), Gaithersburg, MD, June 12–13, 2002. www.fda.gov/ohrms/dockets/ac/02/slides/3869S1_02_Blumenstein.ppt (Accessed May 2006).
38. Clarke PE. CDER Aims to Improve Drug Manufacturing. *News along the Pike: Center for Drug Evaluation and Research* 2002 July, 8, 1.
39. Wechsler J. Modernizing Pharmaceutical Manufacturing. *Pharmaceutical Technology North America* 2002, 26, 16–24.
40. Chrisholm RS. AstraZeneca. Perspective on Process and Analytical Validation. FDA's Advisory Committee for Pharmaceutical Sciences, Subcommittee on Process Analytical Technologies (PAT), Gaithersburg, MD, February 25, 2002. www.fda.gov/ohrms/dockets/ac/02/slides/3841s1_06_chisholm/index.htm (Accessed May 2006).
41. Dean D. PriceWaterhouseCoopers. Pharma Manufacturing: Why There Is a Need to Improve: The Role of PAT. FDA's Advisory Committee for Pharmaceutical Sciences, Subcommittee on Process Analytical Technologies (PAT), Gaithersburg, MD February 25, 2002. www.fda.gov/ohrms/dockets/ac/02/slides/3841s1_03_Dean/index.htm (Accessed May 2006).
42. Hussain AS. FDA's Advisory Committee for Pharmaceutical Sciences, The ACPS's Process Analytical Technology Subcommittee, Rockville, MD, November 28, 2001. www.fda.gov/ohrms/dockets/ac/01/slides/3804s1_02_hussain/sld001.htm (Accessed May 2006).
43. Hammond S. Pfizer. Validation Perceptions That May Slow PAT Development and Implementation. Subcommittee for Pharmaceutical Sciences, Subcommittee on Process Analytical Technologies (PAT), Rockville, MD, October 23, 2002. www.fda.gov/ohrms/dockets/ac/02/slides/3901S1_07_Hammond_files/frame.htm (Accessed May 2006).
44. Laasonen M, Harmia-Pulkkinen T, Simard C, et al. Determination of the Thickness of Plastic Sheets Used in Blister Packaging by Near Infrared Spectroscopy: Development and Validation of the Method. *Eur J Pharm Sci* 2004, 21, 493–500.
45. Han SM, Faulkner PG. Determination of SB 216469-S during Tablet Production Using Near-Infrared Reflectance Spectroscopy. *J Pharm Biomed Anal* 1996, 14, 1681–1689.
46. Andre M. Multivariate Analysis and Classification of the Chemical Quality of 7-Aminocephalosporanic Acid Using Near-Infrared Reflectance Spectroscopy. *Anal Chem* 2003, 75, 3460–3467.
47. Shah DN. Aventis Pharmaceuticals. Regulatory Challenges: Post-Approval PAT Applications. FDA's Advisory Committee for Pharmaceutical Sciences, Subcommittee on Process Analytical Technologies (PAT), Gaithersburg, MD, June 12–13, 2002. www.fda.gov/ohrms/dockets/ac/02/slides/3869S1_03_Shah/index.htm (Accessed May 2006).
48. Blanco M, Romero MA. Near-Infrared Libraries in the Pharmaceutical Industry: A Solution for Identity Confirmation. *Analyst* 2001, 126, 2212–2217.
49. Lonardi S, Viviani R, Mosconi L, et al. Drug Analysis by Near-Infra-Red Reflectance Spectroscopy: Determination of the Active Ingredient and Water Content in Antibiotic Powders. *J Pharm Biomed Anal* 1989, 7, 303–308.
50. Plugge W, van der Vlies C. Near-Infrared Spectroscopy as an Alternative to Assess Compliance of Ampicillin Trihydrate with Compendial Specifications. *J Pharm Biomed Anal* 1993, 11, 435–442.
51. Hussain AS. Opening Remarks. FDA's Advisory Committee for Pharmaceutical Sciences, Subcommittee on Process Analytical Technologies (PAT), Gaithersburg, MD, June 12, 2002. www.fda.gov/ohrms/dockets/ac/02/slides/3869S1_01_Hussan-%20Opening%20Remarks/index.htm (Accessed May 2006).
52. Avallone H. Johnson & Johnson. Development/Compliance Issues (tablets). FDA's Advisory Committee for Pharmaceutical

Sciences, Subcommittee on Process Analytical Technologies (PAT), Gaithersburg, MD, June 12–13, 2002. www.fda.gov/ohrms/dockets/ac/02/slides/3869S1_04_Avallone.ppt (Accessed May 2006).
53. Hammond S. Pfizer. Applications and Benefits of PAT. FDA's Advisory Committee for Pharmaceutical Sciences, Subcommittee on Process Analytical Technologies (PAT), Gaithersburg, MD, February 25, 2002. www.fda.gov/ohrms/dockets/ac/02/slides/3841s1_02_hammond/index.htm (Accessed May 2006).
54. Watson DJ, Dowdy ED, DePue JS, et al. Development of a Safe and Scalable Oxidation Process for the Preparation of 6-Hydroxybuspirone: Application of In-Line Monitoring for Process Ruggedness and Product Quality. *Org Process Res Dev* 2004, 8, 616–623.
55. Hussain AS. The Subcommittee on Process Analytical Technology (PAT): Overview and Objectives. FDA's Advisory Committee for Pharmaceutical Sciences, Subcommittee on Process Analytical Technologies (PAT), Gaithersburg, MD, February 25, 2002. www.fda.gov/ohrms/dockets/ac/02/slides/3841s1_01_hussain/index.htm (Accessed May 2006).
56. Blanco M, Coello J, Iturriaga H, et al. Critical Review: Near-infrared Spectroscopy in the Pharmaceutical Industry. *Analyst* 1998, 123, 135R–150R.
57. Workman J Jr, Veltkamp DJ, Doherty S, et al. Process Analytical Chemistry. *Anal Chem* 1999, 71, 121R–180R.
58. Workman J Jr, Creasy KE, Doherty S, et al. Process Analytical Chemistry. *Anal Chem* 2001, 73, 2705–2718.
59. Blanco M, Villarroya I. NIR Spectroscopy: A Rapid-Response Analytical Tool. *Trends in Analytical Chemistry* 2002, 21, 240–250.
60. Lavine BK, Workman J, Jr. Chemometrics. *Anal Chem* 2002, 74, 2763–2770.
61. Workman J Jr, Koch M, Veltkamp DJ. Process Analytical Chemistry. *Anal Chem* 2003, 75, 2859–2876.
62. Yu LX, Lionberger RA, Raw AS, et al. Applications of Process Analytical Technology to Crystallization Processes. *Adv Drug Del Rev* 2004, 56, 349–369.
63. Shabushnig JG. Pharmacia Corporation. Process Analytical Technology: An Industry Perspective. FDA's Advisory Committee for Pharmaceutical Sciences, Subcommittee on Process Analytical Technologies (PAT), Gaithersburg, MD, February 25, 2002. www.fda.gov/ohrms/dockets/ac/02/slides/3841s1_04_Shabushnig_files/frame.htm (Accessed May 2006).
64. Clevett KJ. Process Analytical Chemistry: Industry Perspectives: Trends in Applications and Technology. *Process Control Qual* 1994, 6, 81–90.
65. Watano S, Numa T, Miyanami K, et al. A Fuzzy Control System of High Shear Granulation Using Image Processing. *Powder Technol* 2001, 115, 124–130.
66. Watano S, Numa T, Koizumi I, et al. Feedback Control in High Shear Granulation of Pharmaceutical Powders. *Eur J Pharm Biopharm* 2001, 52, 337–345.
67. Clarke F. Extracting Process-Related Information from Pharmaceutical Dosage Forms Using Near Infrared Microscopy. *Vib Spectrosc* 2004, 34, 25–35.
68. Sinsheimer JE, Poswalk NM. Pharmaceutical Applications of the Near Infrared Determination of Water. *J Pharm Sci* 1968, 57, 2007–2010.
69. Blanco M, Coello J, Iturriaga H, et al. Determination of Water in Ferrous Lactate by Near Infrared Reflectance Spectroscopy with a Fibre-Optic Probe. *J Pharm Biomed Anal* 1997, 16, 255–262.
70. Last IR, Prebble KA. Suitability of Near-Infrared Methods for the Determination of Moisture in a Freeze-Dried Injection Product Containing Different Amounts of the Active Ingredient. *J Pharm Biomed Anal* 1993, 11, 1071–1076.
71. Rantanen J, Rasanen E, Tenhunen J, et al. In-Line Moisture Measurement during Granulation with a Four-Wavelength Near Infrared Sensor: An Evaluation of Particle Size and Binder Effects. *Eur J Pharm Biopharm* 2000, 50, 271–276.
72. Rantanen J, Rasanen E, Antikainen O, et al. In-Line Moisture Measurement during Granulation with a Four-Wavelength Near-Infrared Sensor: An Evaluation of Process-Related Variables and a Development of Non-Linear Calibration Model. *Chemom Intell Lab Syst* 2001, 56, 51–58.
73. Frake P, Greenhalgh D, Grierson SM, et al. Process Control and End-Point Determination of a Fluid Bed Granulation by Application of Near Infra-Red Spectroscopy. *Int J Pharm* 1997, 151, 75–80.
74. Airaksinen S, Luukkonen P, Jorgensen A, et al. Effects of Excipients on Hydrate Formation in Wet Masses Containing Theophylline. *J Pharm Sci* 2003, 92, 516–528.
75. Dyrby M, Engelsen SB, Norgaard L, et al. Chemometric Quantitation of the Active Substance (Containing c≡n) in a Pharmaceutical Tablet Using Near-Infrared (NIR) Transmittance and NIR FT-Raman Spectra. *Appl Spectrosc* 2002, 56, 579–585.
76. Davis TD, Morris KR, Huang H, et al. In Situ Monitoring of Wet Granulation Using Online X-Ray Powder Diffraction. *Pharm Res* 2003, 20, 1851–1857.
77. McCormick D. PAT Survey Reflects Optimism, Uncertainty. *Pharm Technol* 2005, 29, 24.
78. Hussain AS. Second Meeting of FDA/ACPS Process Analytical Technology: Closing Remarks. FDA's Advisory Committee for Pharmaceutical Sciences, Subcommittee on Process Analytical Technologies (PAT), Gaithersburg, MD, June 12–13, 2002. www.fda.gov/ohrms/dockets/ac/02/slides/3869S1_09_Hussan-%20Summary/index.htm (Accessed May 2006).
79. Neway JO. Filling the Void: PAT in a Connected Manufacturing Environment. *Pharm Technol* 2003, 27, 46–52.
80. Rudd D. GlaxoSmithKline. Product and Process Development: An Industry Perspective. FDA's Advisory Committee for Pharmaceutical Sciences, Subcommittee on Process Analytical Technologies (PAT), Gaithersburg, MD, February 25, 2002. www.fda.gov/ohrms/dockets/ac/02/slides/3841s1_05_Rudd/index.htm (Accessed May 2006).
81. Hussain AS. Office of Pharmaceutical Sciences, CDER, FDA. ACPS Process Analytical Technology (PAT) Subcommittee Meeting #3 Opening Remarks. Subcommittee for Pharmaceutical Sciences, Subcommittee on Process Analytical Technologies (PAT), Rockville, MD, October 23, 2002. www.fda.gov/ohrms/dockets/ac/02/slides/3901S1_01_Hussain_files/frame.htm (Accessed May 2006).
82. Dubois P, Martinez JR, Levillain P. Determination of Five Components in a Pharmaceutical Formulation Using Near Infrared Reflectance Spectrophotometry. *Analyst* 1987, 112, 1675–1679.
83. Gold TB, Buice RG Jr, Lodder RA, et al. Determination of Extent of Formaldehyde-induced Crosslinking in Hard Gelatin Capsules by Near-Infrared Spectrophotometry. *Pharm Res* 1997, 14, 1046–1050.
84. Gupta A, Peck GE, Miller RW, et al. Nondestructive Measurements of the Compact Strength and the Particle-Size Distribution after Milling of Roller Compacted Powders by Near-Infrared Spectroscopy. *J Pharm Sci* 2004, 93, 1047–1053.

85. Harris SC, Walker DS. Quantitative Real-Time Monitoring of Dryer Effluent Using Fiber Optic Near-Infrared Spectroscopy. *J Pharm Sci* 2000, 89, 1180–1186.
86. Kamat MS, Lodder RA, DeLuca PP. Near-Infrared Spectroscopic Determination of Residual Moisture in Lyophilized Sucrose through Intact Glass Vials. *Pharm Res* 1989, 6, 961–965.
87. Laasonen M, Harmia-Pulkkinen T, Simard C, et al. Development and Validation of a Near-Infrared Method for the Quantitation of Caffeine in Intact Single Tablets. *Anal Chem* 2003, 75, 754–760.
88. Lonardi S, Newby PJ, Ribeiro D, et al. GlaxoSmithKline: Why does PAT Need Rapid Microbiology Methods Subcommittee for Pharmaceutical Sciences, Subcommittee on Process Analytical Technologies (PAT), Rockville, MD, October 23, 2002. www.fda.gov/ohrms/dockets/ac/02/slides/3901S1_12_Lonardi_files/frame.htm (Accessed May 2006).
89. O'Neil AJ, Jee RD, Moffat AC. Measurement of the Cumulative Particle Size Distribution of Microcrystalline Cellulose Using Near Infrared Reflectance Spectroscopy. *Analyst* 1999, 124, 33–36.
90. Otsuka M. Comparative Particle Size Determination of Phenacetin Bulk Powder by Using Kubelka-Munk Theory and Principal Component Regression Analysis Based on Near-Infrared Spectroscopy. *Powder Technol* 2004,141, 244–250.
91. Sanchez MS, Bertran E, Sarabia LA, et al. Quality Control Decisions with Near Infrared Data. *Chemom Intell Lab Syst* 2000, 53, 69–80.
92. Workman J. Kimberly-Clark Corp. Chemometrics and PAT: An Opinion on Current Status and Recommendations for the Future. FDA's Advisory Committee for Pharmaceutical Sciences, Subcommittee on Process Analytical Technologies (PAT), Gaithersburg, MD, February 25, 2002. www.fda.gov/ohrms/dockets/ac/02/slides/3841s1_08_workman/index.htm (Accessed May 2006).
93. Herkert T, Prinz H, Kovar KA. One Hundred Percent Online Identity Check of Pharmaceutical Products by Near-Infrared Spectroscopy on the Packaging Line. *Eur J Pharm Biopharm* 2001, 51, 9–16.
94. Morisseau KM, Rhodes CT. Near Infrared Spectroscopy as a Nondestructive Alternative to Conventional Tablet Hardness Testing. *Pharm Res* 1997, 14, 108–111.
95. Cui X, Zhang Z, Ren Y, et al. Quality Control of the Powder Pharmaceutical Samples of Sulfaguanidine by Using NIR Reflectance Spectrometry and Temperature-Constrained Cascade Correlation Networks. *Talanta* 2004, 64, 943–948.
96. Woo YA, Lim HR, Kim HJ, et al. Determination of Hydrogen Peroxide Concentration in Antiseptic Solutions Using Portable Near-Infrared System. *J Pharm Biomed Anal* 2003, 33, 1049–1057.
97. Abrahamsson C, Johansson J, Sparen A, et al. Comparison of Different Variable Selection Methods Conducted on NIR Transmission Measurements on Intact Tablets. *Chemom Intell Lab Syst* 2003, 69, 3–12.
98. Betz G, Bürgin PJ, Leuenberger H. Power Consumption Measurement and Temperature Recording during Granulation. *Int J Pharm* 2004, 272, 137–149.
99. Blanco M, Eustaquio A, Gonzalez JM, et al. Identification and Quantitation Assays for Intact Tablets of Two Related Pharmaceutical Preparations by Reflectance Near-Infrared Spectroscopy: Validation of the Procedure. *J Pharm Biomed Anal* 2000, 22, 139–148.
100. Blanco M, Villar A. Development and Validation of a Method for the Polymorphic Analysis of Pharmaceutical Preparations Using Near Infrared Spectroscopy. *J Pharm Sci* 2003, 92, 823–820.
101. Yoon WL, Jee RD, Charvill A, et al. Application of Near-Infrared Spectroscopy to the Determination of the Sites of Manufacture of Proprietary Products. *J Pharm Biomed Anal* 2004, 34, 933–944.
102. Blanco M, Romero MA, Alcalà M. Strategies for Constructing the Calibration Set for a Near Infrared Spectroscopic Quantitation Method. *Talanta* 2004, 64, 597–602.
103. Fountain W, Dumstorf K, Lowell AE, et al. Near-Infrared Spectroscopy for the Determination of Testosterone in Thin-Film Composites. *J Pharm Biomed Anal* 2003, 33, 181–189.
104. Jorgensen AN, Luukkonen P, Rantanen J, et al. Comparison of Torque Measurements and Near-Infrared Spectroscopy in Characterization of a Wet Granulation Process. *J Pharm Sci* 2004, 93, 2232–2243.
105. Laasonen M, Rantanen J, Harmia-Pulkkinen T, et al. Near Infrared Reflectance Spectroscopy for the Fast Identification of PVC-Based Films. *Analyst* 2001, 126, 1122–1128.
106. Ritchie GE, Roller RW, Ciurczak EW, et al. Validation of a Near-Infrared Transmission Spectroscopic Procedure. Part B: Application to Alternate Content Uniformity and Release Assay Methods for Pharmaceutical Solid Dosage Forms. *J Pharm Biomed Anal* 2002, 29, 159–171.
107. Blanco M, Coello J, Iturriaga H, et al. Determination of Absorbic Acid in Pharmaceutical Preparations by Near Infrared Reflectance Spectroscopy. *Talanta* 1993, 40, 1671–1676.
108. Blanco M, Coello J, Iturriaga H, et al. Control Analysis of a Pharmaceutical Preparation by Near-Infrared Reflectance Spectroscopy: A Comparative Study of a Spinning Module and Fibre Optic Probe. *Anal Chim Acta* 1994, 298, 183–191.
109. Gottfries J, Depui H, Fransson M, et al. Vibrational Spectrometry for the Assessment of Active Substance in Metoprolol Tablets: A Comparison between Transmission and Diffuse Reflectance Near-Infrared Spectrometry. *J Pharm Biomed Anal* 1996, 14, 1495–1503.
110. Higgins JP, Arrivo SM, Thurau G, et al. Spectroscopic Approach for On-Line Monitoring of Particle Size during the Processing of Pharmaceutical Nanoparticles. *Anal Chem* 2003, 75, 1777–1785.
111. Andersson M, Folestad S, Gottfries J, et al. Quantitative Analysis of Film Coating in a Fluidized Bed Process by In-Line NIR Spectrometry and Multivariate Batch Calibration. *Anal Chem* 2000, 72, 2099–2108.
112. Broad NW, Jee RD, Moffat AC, et al. Non-Invasive Determination of Ethanol, Propylene glycol and Water in a Multi-Component Pharmaceutical Oral Liquid by Direct Measurement through Amber Plastic Bottles Using Fourier Transform Near-Infrared Spectroscopy. *Analyst* 2000, 125, 2054–2058.
113. Kirsch JD, Drennen JK. Nondestructive Tablet Hardness Testing by Near-Infrared Spectroscopy: A New and Robust Spectral Best-Fit Algorithm. *J Pharm Biomed Anal* 1999, 19, 351–362.
114. Eustaquio A, Blanco M, Jee RD, et al. Determination of Paracetamol in Intact Tablets by Use of Near Infrared Transmittance Spectroscopy. *Anal Chim Acta* 1999, 383, 283–290.
115. Blanco M, Coello J, Eustaquio A, et al. Analytical Control of Pharmaceutical Production Steps by Near Infrared Reflectance Spectroscopy. *Anal Chim Acta* 1999, 392, 237–246.
116. GlaxoSmithKline. Perspectives in Chemometrics. Experience from GlaxoSmithKline. FDA's Advisory Committee for Pharmaceutical Sciences, Subcommittee on Process Analytical

Technologies (PAT), Gaithersburg, MD, February 25, 2002. www.fda.gov/ohrms/dockets/ac/02/slides/3841s1_09_walker/sld001.htm (Accessed May 2006).
117. Bush L. New CGMP Plant at Purdue University Offers Multiple Benefits. *Pharm Technol* 2003, 27, 22.
118. Blanco M, Coello J, Eustaquio A, et al. Development and Validation of a Method for the Analysis of a Pharmaceutical Preparation by Near-Infrared Diffuse Reflectance Spectroscopy. *J Pharml Sci* 1999, 88, 551–556.
119. Blanco M, Coello J, Iturriaga H, et al. Development and Validation of a Near Infrared Method for the Analytical Control of a Pharmaceutical Preparation in Three Steps of the Manufacturing Process. *Fresenius J Anal Chem* 2000, 368, 534–539.
120. Gupta A, Peck GE, Miller RW, et al. Real-Time Near-Infrared Monitoring of Content Uniformity, Moisture Content, Compact Density, Tensile Strength, and Young's Modulus of Roller Compacted Powder Blends. *J Pharm Sci* 2005, 94, 1589–1597.
121. Lai CK, Cooney CC. Application of a Fluorescence Sensor for Miniscale On-Line Monitoring of Powder Mixing Kinetics. *J Pharm Sci* 2004, 93, 60–70.
122. Moffat AC, Trafford AD, Jee RD, et al. Meeting of the International Conference on Harmonisation's Guidelines on Validation of Analytical Procedures: Quantification as Exemplified by a Near-Infrared Reflectance Assay of Paracetamol in Intact Tablets. *Analyst* 2000, 125, 1341–1351.
123. El-Hagrasy AS, D'Amico F, Drennen III JK. A Process Analytical Technology Approach to Near-Infrared Process Control of Pharmaceutical Powder Blending: Part I: D-Optimal Design for Characterization of Powder Mixing and Preliminary Spectral Data Evaluation. *J Pharm Sci* 2006, 95, 392–406.
124. El-Hagrasy AS, Delgado-Lopez M, Drennen III JK. A Process Analytical Technology Approach to Near-Infrared Process Control of Pharmaceutical Powder Blending: Part II: Qualitative near-Infrared Models for Prediction of Blend Homogeneity. *J Pharm Sci* 2006, 95, 407–421.
125. El-Hagrasy AS, Drennen III JK. A Process Analytical Technology Approach to Near-Infrared Process Control of Pharmaceutical Powder Blending: Part III: Quantitative Near-Infrared Calibration for Prediction of Blend Homogeneity and Characterization of Powder Mixing Kinetics. *J Pharm Sci* 2006, 95, 422–434.
126. Tumuluri SVS, Prodduturi S, Crowley MM, et al. The Use of Near-Infrared Spectroscopy for the Quantitation of a Drug in Hot-Melt Extruded Films. *Drug Dev Ind Pharm* 2004, 30, 505–511.
127. Bai S, Nayar R, Carpenter JF, et al. Noninvasive Determination of Protein Conformation in the Solid State Using Near Infrared (NIR) Spectroscopy. *J Pharm Sci* 2005, 94, 2030–2038.
128. Gupta A, Peck GE, Miller RW, et al. Influence of Ambient Moisture on the Compaction Behavior of Microcrystalline Cellulose Powder Undergoing Uni-axial Compression and Roller-Compaction: A Comparative Study Using Near-Infrared Spectroscopy. *J Pharm Sci* 2005, 94, 2301–2313.
129. Bai SJ, Rani M, Suryanarayanan R, et al. Quantification of Glycine Crystallinity by Near-Infrared (NIR) Spectroscopy. *J Pharm Sci* 2004, 93, 2439–2447.
130. Laitinen N, Antikainen O, Rantanen J, et al. New Perspectives for Visual Characterization of Pharmaceutical Solids. *J Pharm Sci* 2004, 93, 165–176.
131. Seyer JJ, Luner PE, Kemper MS. Application of Diffuse Reflectance Near-Infrared Spectroscopy for Determination of Crystallinity. *J Pharm Sci* 2004, 89, 1305–1316.
132. Zhou GX, Ge Z, Dorwart J, et al. Determination and Differentiation of Surface and Bound Water in Drug Substances by Near Infrared Spectroscopy. *J Pharm Sci* 2003, 92, 1058–1065.
133. Lai CK, Zahari A, Miller B, et al. Nondestructive and On-Line Monitoring of Tablets Using Light-Induced Fluorescence Technology. *AAPS Pharm Sci Tech* 2004, 5, 1–10.
134. Lai CK, Holt D, Leung JC, et al. Real-Time and Non-Invasive Monitoring of Dry Powder Blend Homogeneity. *AIChE J* 2001, 47, 2618–2622.
135. Fevotte G, Calas J, Puel F, et al. Applications of NIR Spectroscopy to Monitoring and Analyzing the Solid State during Industrial Crystallization Process. *Int J Pharm* 2004, 273, 159–169.
136. Lin Z, Zhou L, Mahajan A, et al. Real-Time Endpoint Monitoring and Determination for a Pharmaceutical Salt Formation Process with In-Line FT-IR Spectroscopy. *J Pharm Biomed Anal* 2006, 41, 99–104.
137. Skibsted ETS, Boelens HFM, Westerhuis JA, et al. Simple Assessment of Homogeneity in Pharmaceutical Mixing Processes Using a Near-Infrared Reflectance Probe and Control Charts. *J Pharm Biomed Anal* 2006, 41, 26–35.
138. Johansson J, Pettersson S, Folestad S. Characterization of Different Laser Irradiation Method for Quantitative Raman Tablet Assessment. *J Pharm Biomed Anal* 2005, 39, 510–516.
139. Cogdill RP, Anderson CA, Delgado-Lopez M, et al. Process Analytical Technology Case Study Part I: Feasibility Studies for Quantitative Near-Infrared Method Development. *AAPS Pharm Sci Tech* 2005, 6, E262–E272.
140. Cogdill RP, Anderson CA, Delgado M, et al. Process Analytical Technology Case Study: Part II: Development and Validation of Near-Infrared Calibrations in Support of a Process Analytical Technology Application for Real-Time Release. *AAPS Pharm Sci Tech* 2005, 6, E273–E283.
141. Cogdill RP, Anderson CA, Delgado M, et al. Process Analytical Technology Case Study, Part III: Calibration Monitoring and Transfer. *AAPS Pharm Sci Tech* 2005, 6, E284–E297.
142. Hausman DS, Cambron RT, Sakr A. Application of On-Line Raman Spectroscopy for Characterizing Relationships between Drug Hydration State and Tablet Physical Stability. *Int J Pharm* 2005, 299, 19–33.
143. Islam MT, Rodrıguez-Hornedo N, Ciotti S, et al. The Potential of Raman Spectroscopy as a Process Analytical Technique during Formulations of Topical Gels and Emulsions. *Pharm Res* 2004, 21, 1844–1851.
144. Johansson J, Pettersson S, Taylor LS. Infrared Imaging of Laser-Induced Heating during Raman Spectroscopy of Pharmaceutical Solids. *J Pharm Biomed Anal* 2002, 30, 1223–1231.
145. Kontoyannis CG. Quantitative Determination of $CaCO_3$ and Glycine in Antacid Tablets by Laser Raman Spectroscopy. *J Pharm Biomed Anal* 1995, 13, 73–76.
146. Langkilde FW, Sjoblom J, Tekenbergs-Hjelte L, et al. Quantitative FT-Raman Analysis of Two Crystal Forms of a Pharmaceutical Compound. *J Pharm Biomed Anal* 1997, 15, 687–696.
147. Szostak R, Mazurek S. Quantitative Determination of Acetylsalicylic Acid and Acetaminophen in Tablets by FT-Raman Spectroscopy. *Analyst* 2002,127, 144–148.
148. Taylor LS, Zografi G. The Quantitative Analysis of Crystallinity Using FT-Raman Spectroscopy. *Pharm Res* 1998, 15, 755–761.
149. Wang C, Vickers TJ, Mann CK. Direct Assay and Shelf-Life Monitoring of Aspirin Tablets Using Raman Spectroscopy. *J Pharm Biomed Anal* 1997,16, 87–94.

150. 2006 USPC, Inc. Official 4/1/06-7/31/06 General Chapters: <1119> Near-Infrared Spectrophotometry.
151. Ryder AG, O'Conner, GM, Glynn TJ. Quantitative Analysis of Cocaine in Solid Mixtures Using Raman Spectroscopy and Chemometric Methods. *J Raman Spectrosc* 2000, 31, 221–227.
152. Clarke FC, Jamieson MJ, Clark DA, et al. Chemical Image Fusion: The Synergy of FTNIR and Raman Mapping Microscopy to Enable a More Complete Visualization of Pharmaceutical Formulations. *Anal Chem* 2001, 73, 2213–2220.
153. Chang H, Huang P. Thermo-Raman Spectroscopy. *Rev Anal Chem* 2001, 20, 207–238.
154. Ghule A, Baskaran N, Murugan R, Chang H. Phase Transformation Studies of Na_3PO_4 by Thermo-Raman and Conductivity Measurements. *Solid State Ionics* 2003, 161, 291–299.
155. de Jager H-J, Prinsloo L. The Dehydration of Phosphates Monitored by DSC/TGA and in Situ Raman Spectroscopy. *Thermochim Acta* 2001, 376, 187–196.

57 Validation of Analytical Procedures and Physical Methods

Francis E. Beideman

CONTENTS

57.1 INTRODUCTION

Method validation is currently defined by ICH and international regulatory agencies as the formal process of providing documentation that an analytical procedure is suitable for its intended purpose. Methods requiring validation may include those used for testing of Drug Substance/API (potency, impurities, identity, and other critical attributes such as moisture, particle size, etc.), excipients, and Drug Product/finished dosage (potency, impurities, degradation products, identity, and other critical product characteristics such as dissolution, moisture, hardness, viscosity, etc.). Method validation should follow the life cycle of the product, beginning during early product development through the various development stages including clinical testing, submission, manufacturing, and marketing/post-marketing. This process is depicted in Figure 57.1. Formal, protocol driven, method validation typically begins with the need to test Phase 1 clinical supplies. The method and validation thereof will then be updated, as needed, throughout product development until the validation for regulatory filing. The validation requirements will generally become more inclusive and restrictive as the intended purpose of the method is refined during the development of the product. Post-approval, the method validation status should be evaluated by monitoring method performance, change control, and regulatory requirements and then by performing revalidation, as necessary.

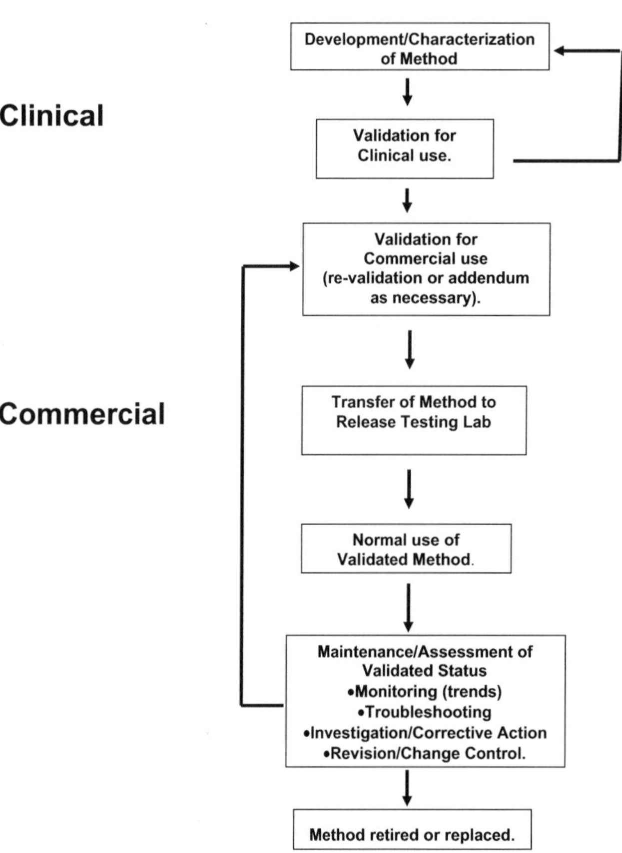

FIGURE 57.1 Schematic of test method life cycle.

57.2 REQUIREMENTS FOR TEST METHOD VALIDATION

It is important to recognize that method development and method validation are separate activities. That is, the method parameters and requirements must be understood and documented prior to validation. It is also important that all formal validation activities are carried out according to cGMPs, with appropriate documentation and QA review and

DOI: 10.1201/9781003163138-57

approval. Qualified instrumentation must be used and appropriate SOPs must be in place.

57.3 THE TEST METHOD VALIDATION PROCESS

Table 57.1 describes the method validation attributes prescribed in the "Guideline for the Validation of Analytical procedures" ICH Q2(R1). Note that requirements for Assay methods differ from requirements for limit tests and tests for dissolution and content uniformity.

Early-Stage Validation: The guideline also states that although "analytical methods performed to evaluate a batch of API for clinical trials may not yet be validated" per ICH Q2(R1), "they should be scientifically sound". Test procedures should be reliable to support clinical studies. Appropriate parameters and sound scientific judgment should be used. The testing required at this early stage of product development should be based on the Quality Target Product Profile (QTTP) and should determine those product characteristics that have a critical impact on product quality. For a solid finished dosage, those characteristics are often potency and dissolution. For the purposes of this discussion, both characteristics will be tested using an HPLC method. Accuracy, linearity, and repeatability are three analytical attributes that should be validated to assure that the method is fit for analysis of the clinical batches. Validation of specificity may also be required to show that the assay method is suitable for testing the identity of the batch and is stability indicating so that the product can be shown to be stable over the dosing schedule for the clinical study. The need for a protocol should be determined via discussion with the quality unit.

Late-Stage Validation: When product development is complete and preparation of pivotal submission batches are being planned, method development must be complete; the method must now be formally validated.

57.3.1 The Validation Protocol

The validation protocol is a document that describes the testing necessary to prove that an analytical method is suitable for its intended use. The protocol is reviewed and approved by the quality unit and should be suitable for regulatory submission. The protocol should include sufficient detail to allow the completion of the testing and may contain the following parts:

- Title (including version number)
- Approval Page
- Introduction
- Description of Testing (including acceptance criteria)
- A section to receive/document the results
- Conclusion
- Deviations including a log, investigations, and discussion of the impact on the validity of the method
- The Draft Method

Acceptance criteria for each method attribute must be developed based on knowledge of the QTTP and required specifications. For example, an acceptance criterion for % RSD = 3.00 may be appropriate if the potency specifications are 90.0%–110.0%. However, a more appropriate criterion might be 1.50% if the specification is 95.0%–105.0% or 0.5% if the specification is 98.0%–102.0%. Similarly, specifications for low-level impurities present at levels near the Quantitation Limit may have %RSD requirements such as ±20%.

TABLE 57.1
Validation Parameters Required for Different Types of Methods per ICH Q2(R1)

Method type	Identification	Impurity test		Assay Dissolution (measurement only) Content uniformity
Parameters		**Quantitative**	**Limit**	
Specificity [(1)]	+	+	+	+
Linearity	–	+	–	+
Range	–	+	–	+
Accuracy	–	+	–	+
Precision				
Repeatability	–	+	–	+
Intermediate Precision	–	+ [(2)]	–	+ [(2)]
Detection Limit	–	– [(3)]	+	–
Quantitation Limit	–	+	–	–

Note: "+" parameter is normally evaluated; "–"parameter is not normally evaluated.

(1) Lack of specificity of one analytical procedure could be compensated for by other supporting analytical procedures.

(2) In cases where reproducibility has been performed, intermediate precision is not needed.

(3) May be needed in some cases.

57.3.2 Validation Attributes

Linearity—The linearity of the method is validated by making 5–10 solutions with concentrations spanning the necessary range of the method. This range may include the low levels necessary for determination of impurities and degradation products. Linear regression should be performed and the R^2, slope, and y-intercept reported. $R^2 \geq 0.99$ is often used as an acceptance criterion. A requirement for the y-intercept to be ≤5% of the standard concentration is often used to justify use of a single-level standard. The application of nonlinear calibrations must be described and justified.

Specificity—The method must be shown to respond only to the substance being measured. It should not respond to excipients, impurities, or degradation products (If the method is to be used to monitor stability studies). Validation of specificity of chromatographic methods may be performed by individually injecting API solution, placebo (or individual solutions of excipients), known impurities, and degradation products. The method is considered to be specific if there are no peaks from the placebo/impurities eluting at the same retention time as the retention time of the API. An acceptance criterion for any interference might be set at 0.1% depending on the product attributes.

If a method is to be used for testing stability, additional studies are necessary. Forced degradation studies using a diode array detector and/or a mass spectrometric detector are often performed for this purpose. These studies are often carried out by exposing API solution, finished dosage solution, placebo, and blank to conditions likely to cause decomposition of the API. These conditions may include temperature, oxidation, acid hydrolysis, basic hydrolysis, photolysis, and reduction. The conditions must be designed to cause degradation at about 5%–20%. The degraded samples and related controls are then analyzed using the method. It is important that the conditions not be so extreme as to completely degrade the substance. The peak eluting at the retention time of the API is then examined by recording either the UV or mass spectra at various points during elution of the peak. This evaluation provides evidence that the peak used to calculate the assay only includes the substance of interest without interference from impurities or degradation products. At least three points are necessary, but additional spectra can be helpful. The spectra are then compared. The spectra recorded for the areas on the upslope and downslope of the peak should be comparable to the spectrum at the peak apex (Software available from the vendor of the chromatographic data system can be used for this process). The chromatograms should also be evaluated holistically, evaluating the peak shapes, any fronting or tailing, and the presence of small peaks on the front or tail of the main peak. The mass balance observed after degradation should be calculated and discussed.

Accuracy—The accuracy of the method is often determined by spiking known amounts of the API into the placebo and carrying out the complete method. The levels tested should cover the range of the product specifications. Therefore, a finished dosage having a specification of 90.0%–110.0% would require spike levels at 90%, 100%, and 110%. These should be done in triplicate. (Alternatively, finished dosages made at various levels are analyzed in triplicate). Acceptance criteria should be based on the product specifications but difference from nominal and %RSD criteria often range from 1%–3%.

Precision—The precision of the method is validated by performing studies to evaluate Repeatability, Method Precision, and Intermediate Precision.

Repeatability—This is often validated by making ten replicate injections of a sample and/or sample solution and calculating the average and RSD. Acceptance criteria need to be determined based on the application but are often ±1%–2%.

Method Precision—This can be validated by carrying out the complete analysis a number of times; 3–6 replicates are often used. The average and %RSD are calculated. Acceptance criteria for agreement and %RSD need to be determined based on the application but are ±2%–3%.

Intermediate Precision—Intermediate precision is often validated by executing the analysis using a second analyst utilizing a second instrument and a second column. Three lots are often used for this testing. Two of the lots are typically run in triplicate and the third run with N = 6 replicates. Execution of the Intermediate Precision experiment in a second laboratory can be used to provide documentation of the qualification of the second laboratory to perform the testing (Method Transfer). Acceptance criteria for within-laboratory and between-laboratory testing must be developed based on the product specifications but a value of ±3% is often used.

Range—The interval between the upper and lower amounts (concentrations) of analyte that have been shown to have an acceptable level of accuracy, precision, and linearity.

Robustness—These experiments will show that the method being validated is unaffected by small changes in method conditions and laboratory environment. Detector wavelength, column temperature, autoinjector temperature, mobile phase composition, and sample preparation conditions (stirring, temperature, extraction time, extracting solvent composition, filtration, etc.). The robustness attributes should be chosen based on knowledge of the method and the product. The stability of the standard and sample solutions always needs to be evaluated as part of robustness testing. Typical storage times are 24 hours (Overnight), 3 Days (Over the weekend) and 5 days. Acceptance criteria must be set based on the product specifications.

Detection Limit/Quantitation Limit—The validation of DL/QL will likely be necessary, particularly if impurities and degradation products are of interest. These attributes can be determined by injecting increasingly dilute solutions of API and known impurities and degradation products when available. The DL is the lowest concentration that can be distinguished from the blank and is the level of the analyte that can be detected but cannot be reliably quantitated. It is often defined as the concentration with a response having a Signal to Noise Ratio (S/N) in the range of 3–5. The QL is that level at which a reliable amount of the analyte can be reported. It is often defined as the concentration having a response with a S/N ratio of about ten. The DL/QL can also be calculated from the linearity study with the DL being defined as DL = 3.3σ/Slope, where σ can be estimated from the standard deviation of the blank responses, the residual standard deviation of the regression line, or the standard deviation of the *y*-intercepts of the regression lines.

$$QL = 10\ \sigma/\text{Slope}$$

57.3.3 Secondary Methods

Methods desired for the support of validation batches may need to be validated by comparing method attributes with those of the primary method. For example, the accuracy and precision of a Near Infrared method to be used for blend and content uniformity will need to be compared with the attributes of the primary HPLC method used for that purpose.

57.3.4 The Validation Report

Upon completion of the experiments described in the validation protocol, a comprehensive validation report will be written. The report will include all pertinent data with references to the original data. Any deviations to the protocol and data that do not meet the protocol acceptance criteria will be included with laboratory investigations into the impact, if any, on the validity of the method. A conclusion describing the outcome of the validation should be included. The report should also include a final version of the validated method. Typical sections of a method validation report are as follows:

- Title
- Approval Page
- Introduction
- Sections for data pertaining to each validation attribute
- Deviations and investigations
- Conclusion
- Final, validated method.

57.4 CONCLUSION

The formal validation of analytical procedures assures that the methods are appropriate for the testing of pharmaceutical ingredients and products and can be used in the process of assuring the safety and effectiveness of the pharmaceutical product.

REFERENCES

1. Gazzano-Santoro H, Broughton C, Schmalzing D. *Validation of Analytical Procedures and Physical Methods*. Validation of Pharmaceutical Processes, 3rd ed. CRC Press, 2007.
2. International Conference on Harmonization. ICH Topic Q2(R1). *Validation of Analytical Methods: Tests and Methodology*, 2005.
3. FDA Guidance for Industry. Analytical Procedures and Methods Validation for Drugs and Biologics, July 2015.
4. International Conference on Harmonization. ICH Topic Q7A Good Manufacturing Practice Guide for Active Pharmaceutical Ingredients. Draft Consensus Guideline. Geneva, Switzerland: ICH, 2000.
5. International Conference on Harmonization. ICH Topic Q6B. Specifications: Test Procedures and Acceptance Criteria for Biotechnological/ Biological Products ICH Harmonized Tripartite Guideline. Geneva, Switzerland: ICH, 1999, 64, 44928.
6. FDA Guidance for Industry. Analytical Procedures and Methods Validation for Drugs and Biologics, 2015.
7. Hermans A, et.al. Approaches for Establishing Clinically Relevant Dissolution Specifications for Immediate Release Solid Oral Dosage Forms. *The AAPS Journal* 2020, 22(34).
8. Jrado JM, et al. Some Practical Considerations for Linearity Assessment of Calibration Curves as Function of Concentration Levels According to the Fitness for Purpose Approach. *Talanta* 2017, 172, 221–229.

ABBREVIATIONS

Abbreviations used in this chapter:

API:	Active Pharmaceutical Ingredient
cGMP:	Current Good Manufacturing Practice
DL:	Detection Limit
FDA:	United States Food and Drug Administration
HPLC:	High-Performance Liquid Chromatography
ICH:	International Conference for Harmonisation
QA:	Quality Assurance
QL:	Quantitation Limit
QTTP:	Quality Target Product Profile
QTTP:	Quality Target Product Profile
RSD:	Relative Standard Deviation
S/N:	Signal to Noise Ratio
SOP:	Standard Operating Procedure
UV:	Ultraviolet

58 Validation of Microbiological Methods

Anthony Grilli

CONTENTS

58.1 INTRODUCTION

The word "validation" is used liberally in pharmaceutical microbiology. Media are validated with growth promotion tests, microbial methods are validated as suitable for specific drug products, and disinfectants are validated for efficacy against microbes on various surfaces. Each of these validation activities have different aims and may be readily replaced with other terms such as suitability, verification, or qualification, but they all strive to assure that a specific process will consistently meet specific quality parameters utilizing the microbiology laboratory as its tool. Although this chapter will review the requirements behind microbiological method validation, it is worth reviewing what critical control points in the microbiology laboratory process itself must be validated and held constant to achieve a valid test result in the first place.

58.2 MICROBIOLOGICAL METHODS AND VALIDATION

Classical microbiological methods based on broth and agar paradigms are inherently variable. Consistent results are dependent upon tightly controlled processes, from the manufacture of the media to its eventual application by the microbiologist. The challenge of microbiological method validation is apparent as soon as we look at the units of measurement—colony-forming units (cfu) and most probable number (MPN) for example. Because culture methods are dependent upon amplifying microscopic organisms to sizes that can be detected by the human eye (from 1 micron to 1 mm for example)—one cannot be clear on exactly how many actual discrete original cells that colony actually represents. Perhaps it was a single cell, perhaps it was a cluster, perhaps it was a piece of biofilm or a whole community of microorganisms drifting on human dander—all we know is what we can see—a single colony-forming unit. So right away, the very unit used to measure microbiological method validity is a source of potential variability. How does one manage this inherent variability to demonstrate that microbiological methods are accurate, precise, and repeatable? By tightly controlling, through qualification or validation, those laboratory processes involved in delivering that laboratory's "colony-forming units" test results.

Microbiological Laboratory Validations Required to deliver Reliable Microbiology Data:

- Water Purification System: used for preparing media, reagents, and labware.
- Incubators: controlled temperature distributed properly throughout
- Freezers: controlled temperature and alarms for storing cultures
- Autoclaves: pressure, temperature, time adequate for sterilization
- Primary Engineering Controls: control of viable and nonviable particles
- Clean Rooms: pressure differentials, particle counts, prevent external contamination
- Cleaning Processes: labware free of compounds inhibitory to microbes
- Microbial Identification Systems: consistent identification of microbes of concern

Consider the critical control points in the evaluation of a simple swab microbial enumeration test, often used in a drug manufacturing cleaning validation. Equipment surfaces are swabbed with polyester or Dacron swabs, and these are sent to the microbiology lab for enumeration and perhaps identification of surviving microbes. But for the results of the microbiology swab test to be meaningful, consider the validations that occur in the lab before cleaning validation, so that it might provide a meaningful result:

- Why a swab?
    - Are there other better mechanisms for removing microbes from the surface? Sponges, contact plates, rinses? Validate adequate recovery.
- What swab?
    - Have swab recoveries been performed to show microbes are adequately removed from the surface sampled? Is the swab of the proper size and

DOI: 10.1201/9781003163138-58

  - makeup to pick up and release contaminants over the desired surface area and surface? Validate adequate recovery.
- What neutralizer?
  - What cleaning agent was used in the cleaning study and has the swab transport medium been shown to neutralize any biocide that might still be present? Validate neutralization.
- What time to test?
  - Have the time and temperature of storage between sampling and testing been shown to be valid for the swab and neutralizing medium? Will microbes die over time? Will microbes increase over time? Validate transport and hold time.
- What recovery mechanism?
  - Will the swab be sonicated, vortexed, stomached? For how long? Validate recovery from swab.
- What medium?
  - Is the medium appropriate for recovery of organisms that would be considered objectionable to that dosage form? Validate media suitability.
  - Is the medium appropriate for recovery of all organisms that are likely to be present in that dosage form? Validate media suitability.
- Has the medium been validated for growth promotion?
  - Will the medium support the growth of low levels of contaminants that might be present? Validate growth promotion.

This last point alone consumes much of the microbiology laboratory's internal validation efforts if media are still manufactured in the laboratory. Consider the equipment validation or qualification for the growth promotion test alone. Has the autoclave used to prepare the media been validated for adequate kill cycles? Have the incubators been validated for proper temperature distribution over time? Has the water bath been validated for proper tempering temperature? Has the refrigerator used to store media and cultures been properly qualified? Have the temperature devices used to monitor and alarm these devices been qualified? Has the water system used to hydrate the media been validated to be free of potential microbial inhibitors? Has the water system used been validated to have low background contaminants? Has the cleaning process for glassware used in the manufacture of the media been validated to ensure no potential inhibitory residues? Are the test species used in the growth promotion tests shown to be no more than 5 transfers removed from original ATCC receipt, as required by USP? Have the Biological Indicators used to validate the autoclave been confirmed to have the proper count and D-value? And on and on and on.

It has been cited that roughly 15% of a pharmaceutical microbiology laboratory's hour is spent on internal quality assurance, validating equipment, facility, and supplies, and demonstrating that critical control points are operating within valid limits.[1]

In the swab example, instrumentation and supply qualification necessary to demonstrate a valid test result have been emphasized. And for the most part, equipment validation and supplies qualification are easily performed and readily demonstrated to be robust and repeatable once the process is defined and set. Of course, a much less consistent control point in the classic microbiological process is the biological specimen on the other side of the microscope. Pasteurian microbiology is extremely technique dependent. Rigorous emphasis and drilling on analyst technique are critical to ensure consistent results. Aseptic technique, accurate dilutions, attention to detail, discernment of colonial morphology, and even demonstration of color acuity all work towards reducing that variability.

Validation of a microbiologist's aseptic technique is obvious and is often demonstrated though media fill studies, gowning qualifications, transfer studies, and other means. But aseptic technique is not the only physical skill set crucial to meaningful validation of data that should be demonstrated during analyst qualification.

Consider serial dilutions. In a classical microbiology laboratory, the bioburden of any sample is quantified through a series of 1:9 mL dilutions of that sample in a buffered medium, and then aliquots of that buffer dilution are delivered to agar plates, incubated, and counted. The dilution that gives 25–250 cfu of bacteria or yeast or 8–80 cfu of filamentous fungi on that plate is the dilution that counts, and cfu count is multiplied by that dilution to give cfu/mL or gm.[2] This cfu range varies by reference and industry; these ranges are cited in USP <1227> *Validation of Microbial Recovery from Pharmacopeial Articles* dilution ranges and the FDA's *Bacteriological Analytical Manual*[3], the *Standard Methods for Examination of Water and Wastewater*[4], and the EPA recommended 30–300 cfu. Regardless of the range chosen, proper serial dilution is required to get there. Serial dilution is a basic laboratory technique but one with huge sources of variability. Starting with a 10^8 cfu/mL suspension of test species, with the goal of delivering 100 cfu in a 0.1 mL inoculum, 5 sequential 1:9 mL dilutions are made. And unlike in dilutions of chemical solutions, microbiological solutions are composed of discrete insoluble particles suspended in buffer, possibly bound together on a pellicle, which must be suspended uniformly throughout the matrix prior to each subsequent dilution step if accurate enumeration of "colony-forming units" is to be achieved. All while maintaining aseptic technique. Things to watch for in training new analysts in the skill of aseptic serial dilution are proper culture mixing or vortexing without overspilling, utilizing fresh sterile pipettes between dilutions, accurate delivery of dilution volumes between plates and between tubes, understanding of pipette volumes "to deliver" and "to contain", use of a fresh pipette when working down a dilution scale, use of a single pipette when working up a dilutions scale. And on and on.

Microbiologists themselves should be validated for proper enumeration techniques. *Standard Method for the Examination of Water and Wastewater* offers a useful procedure to validate the precision of the microbiologist in delivering plate counts.[5] The microbiologist performs duplicate plate counts on 15 water samples. Water samples have a more homogeneous distribution of planktonic bacteria than most other samples, removing this variable from the results and focusing the outcome on the microbiologist's technique. The plate counts are converted to Log10, and the range, or the difference between the two replicate counts is calculated. The average of the 15 analyses is calculated and multiplied by 3.27. If the range is greater than 3.27 times the average range, variability is deemed excessive and the analyst should be "recalibrated" to deliver more precise results. Even after serial dilution, tempered growth promoted media, and properly timed and temperature-controlled incubation, there is some acceptable variability in some compendia on what seems to most the simple act of counting colony-forming units on the plate. *Standard Methods* allows a microbiologist a 5% error rate when counting the same plate several times, and 10% disagreement is deemed acceptable between two microbiologists counting the same plate.[6] This type of personnel validation is critical to delivering critical validation results.

Or consider the very first rapid microbiological method, the Gram stain. In 1883, Dr. Gram was using iodine to remove gentian violet from his kidney cell slides. By way of happy accident, he noticed that some bacteria on his slides were not washing clear, even after alcohol was applied. Serendipity meets application and the Gram stain was born. The microbiology laboratory still uses his 140-year-old technique to narrow the field of suspected microbes in half, on its way to eventual genus species identification. The fork in the identification road that is Gram status is critical to proper microbial identification, particularly for biochemical identification systems. A bad stain or wrong interpretation can have dire consequences. The ID system may give you a genus species, but that does not make it the correct name. Validation of a microbiologist's ability to perform and interpret a Gram stain on a consistent basis is critical in obtaining meaningful results.

The composition of Gram statin reagents varies, particularly the decolorizing reagent that is available as ethanol or acetone alcohol mixtures. JW Bartholomew gave a thorough review of the importance of stain reagent formulation control in 1962.[7] The age of the bacterium is also important, as Gram variable microbes may stain pink or purple depending on which stage of the growth phase one catches them in. But the largest variable to control and validate is the microbiologist themself. Do they gently heat fix the culture on the slide with a heat source or do they burn it to oblivion? Do they gently decolorize the Gram stain or wash it right off the slide? The details of time and washing technique also vary from laboratory to laboratory. Some laboratory Standard Operating Procedures (SOPs) direct to rinse the slide with water between stains, others do not rinse with water after the iodine step but go right to decolorize. The time to stain, decolorize, and counterstain between laboratories varies. Is the microbiologist color blind? This is an actual problem in microbiology laboratories, where stains and biochemical reactions are color dependent. Have they been trained to use other morphological characteristics to inform their decision, such as the presence of spores or cell size and shape?

Now that the laboratory has validated its materials, processes, equipment, and staff, the use of these tools to validate a microbiological method can begin. USP offers guidance in its informational chapters on validating method recovery from pharmacopeial articles and validation of alternative microbiological methods.

USP <1227> "Validation of Microbial Recovery from Pharmaceutical Articles" focuses on the importance of neutralizing inhibitory properties of the product that may prevent microbes in the agar and broth system from amplifying to the point where they can be seen with the unaided eye. Multidose drug products are formulated with antimicrobial preservatives, and viable contaminants in these products may remain inhibited in the agar used to enumerate them, thereby not forming colony-forming units. Consider antibiotic formulations, clearly and antimicrobial drug product, and antiseptic washes, specifically designed to sanitize skin—yet these products are still vulnerable to microbial contamination and recalls of purported antimicrobial products occur with ever-increasing frequency. Various additives are recommended for media depending on the active ingredient the lab wishes to neutralize—thiosulfate for halogens, 'tween and lecithin for sorbates, etc.[8] For drug products that are aqueous, membrane filtration is a great means of physical separation of inhibitor from microbe and broth—but validation is still required as various filter membranes retain antimicrobials and continue to inhibit contaminants in broth or agar systems. These validations may be qualitative or quantitative, depending on whether the criteria are pass-fail or a stated maximum population.

USP <1227> terms this pursuit of effective antimicrobial neutralization "validation", USP <1226> "Verification of Compendial Procedures" calls the work "verification" when referring to USP <61>, <62>, <51>, and <71>, and USP <61>, <62>, <51>, and <71> themselves refer to these pursuits as "method suitability". One can forgive the pharmaceutical professional for being confused by these interchangeable terms, but in the end its all the same thing—can one recover tests species from a drug product microbial evaluation test system if that test system is inoculated with 10–100 cfu of test species. Microbiological methods are either quantitative or qualitative, and the validation required by either differs. The requirements to validate a quantitative method would be more rigorous as considerations of limit of quantification and linearity come into play. A classical quantitative microbiological method is an agar or broth that enumerates bioburden through colony-forming units or most probable number culture techniques. Examples include USP <51> "Antimicrobial Preservatives Effectiveness" and USP <61> "Microbiological Examination of Nonsterile Products: Enumeration Tests". A qualitative microbiological method gives a pass/fail or presence/absence result. Examples include USP <71> "Sterility Tests" and the USP <62> "Microbiological Examination of Nonsterile Products: Tests for Specified Microorganisms". These compendia are accepted as valid, although as stated, they must be "verified" to be suitable for various formulations, using the strategy outlined in the methods themselves and in USP <1227> "Validation of Microbial Recovery from Pharmacopeial Articles".

The European Pharmacopoeia gives direction on validating unique drug products with inhibitory properties that are not entirely chemical in nature: live biotherapeutic products (LBP) acting as drug substance and product. Microbiome research continues to demonstrate the importance of indigenous bacteria to digestive, dermal, and even emotional health, and many drug products are in development in which the active pharmaceutical ingredient (API) is a patented bacterium. These drug products and substances must be tested for microbial contaminants just as any other API or formulation would. But the challenges of detecting a single *Clostridium* cell in the presence of 10^8 cfu of anaerobic enteric mono-septic drug product are self-evident. Although polymerase chain reaction (PCR) and gene probes may seem to be potential tools to help find the proverbial needle in the haystack, they are by design very specific, targeting specific gene sequences for specific bacteria. USP pathogens are indicator species and do not represent the universe of potential objectionable organisms. Knowledge of microbial nutritional and incubation requirements are used to select out targeted pathogens bioburden while "neutralizing" or inhibiting the "active ingredient".

Ph. Eur. 2.6.36 "Microbial Examination of Live Biotherapeutic Products: Tests for Enumeration of Microbial Contaminants" and 2.6.38 "Microbiological Examination of Live Biotherapeutic Products: Test for Specified Micro-Organisms" follow the sequence of the harmonized USP Chapters <61> and <62> but offer flow charts recommending strategies such as increasing diluents, addition of LBT growth inhibitors, varying media formulations to enumerate potential contaminating bioburden, and characterizing specified microbes. For example, lactic acid bacteria growing aerobically on tryptic soy agar will manifest as small pinpoint colonies, easily differentiated from larger *Bacillus subtilis* colonies that may be used to validate recovery of spore-forming aerobes. Similarly, the addition of cycloheximide to tryptic soy agar would suppress growth of a *Saccharomyces cerevisiae* LBP, allowing bacteria in product to be cultured.

Validation of microbial recovery takes a different slant when assessing pharmaceutical container closures whose bioburden must be determined prior to sterilization. The dose of gamma irradiation or the length of time at 121°C will depend on how much and what kind of bioburden, on average, resides on a particular container closure system, syringe, patch, or other delivery mechanism. These methods are outlined in ANSI/AAMI/ISO 11737-1 *Sterilization of Medical Devices—Microbiological Methods Part 1—Estimation of Population of Microorganisms on Products.* In this case, microbial recovery takes a literal meaning, where the bioburden is washed off or removed from the surface of the device and plated and incubated. This removal step can be any of a variety of means including but not limited to diluting, rinsing, shaking, stomaching, sonicating, blending, grinding, or vortexing. Once the organisms have been rinsed off, the task of culturing them for enumeration remains, and then the questions of type of media, plate count or membrane filtration, incubation time and temperature must all be addressed. Given such freedom, the bioburden recovery method must be validated to confirm the result obtained reflects actual conditions, as sterilization parameters depend on it.

Validation of bioburden recovery from devices occurs in one of two ways. First is a product inoculation method where artificial bioburden is inoculated in known quantities and put through the recovery test. The microorganism used are desiccant resistant and easily differentiated from background bioburden by color or morphology. *Bacillus atrophies* spores work well for this test. The count of spores recovered by the culture technique is divided by the count of spores applied to the device, and a percent recovery is achieved that can be applied to similar devices processed in the same way. The second means to validate bioburden recovery is to perform an exhaustive recovery test on the device. In an exhaustive recovery test, a single device is put through a series of five sequential bioburden recovery tests. The counts from the first recovery are divided by the total count on all five recoveries, and a percent recovery efficiency is obtained and applied to other counts of that device.[9]

As mentioned at the onset in the swab cleaning validation thought experiment, validation of surface contaminants recovery is an important topic to consider in pharmaceutical cleaning validations and environmental monitoring qualifications. Many firms set their success or failure on this colony-forming units concept—and award finite number status to a datapoint that may be equal to one, or it may be equal to dozens of cells or spores linked, aggregated, or floating together on a dust particle. Prior to 2012, USP <1116> "Microbiological Control and Monitoring of Aseptic Processing Environments" offered finite numerical colony-forming unit criteria to meet to "pass" a clean room qualification.[10] The FDA aseptic Guidance for Industry on sterile drug manufacturing[11] and the EU in its "Manufacture of Sterile Medicinal Products"[12] offer similar finite pass/fail criteria for different ISO classified clean rooms and environments. In 2012, USP <1116> "Microbiological Control and Monitoring of Aseptic Processing Environments"[13] took a more rational approach stressing environmental monitoring trends versus a single finite viable unit number. In the current informational chapter, clean room performance and qualification are based on "percent of time" microbes are recovered, rather than a finite colony-forming unit number, giving homage to the imprecision of the colony-forming unit. USP <1116> is an informational chapter, it is not binding. People love finite limits and immediate numbers and are not eager to base clean room performance on this long-term trending technique. In my own experience performing microbial incubation and enumeration of agar plates taken from hundreds of clean rooms, I have yet to see a certifier, pharmacy, or manufacturing plant give up the finite action level criteria for a long-term trending process.

Yet even accepting that the colony-forming unit does not represent a single cell, how effective is the sampler's tool at collecting these contaminants. The contact plate is the industry's standard for recovering surface microbes in pharmaceutical processes. Yet an exhaustive recovery study of contact plates showed a 30% recovery efficiency of *Bacillus* spores dried on a stainless-steel surface.[14] A 30% recovery of viable surface contaminants where the action level is ≥1 cfu, which may be dozens of microbes, does highlight the differences between chemistry and microbiology validations.

Microbial variability has significant implications in performance of out of specification investigations. FDA Guidance *Investigating Out of Specification Test Results for Pharmaceutical Production* makes this distinction between chemistry and microbiology clear in its section on averaging test results. Averaging is appropriate for chemistry analysis, where the analyte is homogenously distributed, to get a more accurate result.[15] It also notes that USP prefers averaging for microbial products, given the inherent variability of microbial assays.

But the USP is referring to the average of multiple agar plates from a single sample prep., not averaging multiple sample preps or multiple product lots to lower bioburden average into specification. It is common.

The guidance gives a chemistry example where an HPLC assay results in a product potency result of 89.5% and the specification is 90%–110%. Phase 1 Investigation could not uncover an assignable cause. The same sample was retested seven times, and all results were in the high 98%–99% range with tight agreement. The guidance says in such a scenario these retest results could invalidate the original result.

The microbiology lab does not have it that easy. A high count in a sample of lactose monohydrate used in tablet formulation cannot disappear with multiple retests. Content uniformity is a chemistry test that is more analogous to the microbiology lab's retesting conundrum. A high potency result from 1 in 10 samples of a powder blend would show that the powder was not thoroughly blended. A high bioburden in a powder blend would show a hot spot of contamination that the sampler was fortunate to catch. In neither example would you discard the original result and retest.

Yet a look at the individual numbers behind a microbiologists OOS result may uncover an assignable cause. Do the individual dilution replicates that gave the average agree with one another by 50%–200%? Do the colony-forming units from serial dilutions follow the dilution breaks? If the answer is no, examine the technique of the analyst and whether they are properly calibrating, qualified, and validated to perform test data that are accurate, precise, and reliable.

For example, a microbiologist testing an excipient for bioburden generates an OOS result. The excipient has a specification of <1000 cfu/gm and generated a result of 23,000 cfu/gm. A second microbiologist, Analyst B, had tested the same sample of excipient the day before with a passing result. The variation in replicate plate counts and the lack of consistent serial dilution breaks in the failing analyst incriminates A and not the sample.

Analyst A	Analyst B
10^{-1}: 224, 50	10^{-1}: 90, 85
10^{-2}: 25, 60	10^{-2}: 9,8
10^{-3}: 9, 0	10^{-3}: 0,0
Cfu/g: 2840	Cfu/gm: 880 cfu/gm
Bioburden is OOS	Bioburden is in specification

At one time, the USP had a more liberal view on retesting. Before 2009, USP <61> "Microbiological Examination of Nonsterile Products: Microbial Enumeration Tests" was called USP <61> "Microbial Limits Tests". Simpler times, general test chapters had simpler names and laboratories had simpler procedures for managing OOS results. If a result looked odd, USP allowed a retest with a 2.5 larger sample size.

> For the purpose of confirming a doubtful result by any of the procedures outlined in the foregoing tests following their application to a 10.0-g specimen, a retest on a 25-g specimen of the product may be conducted. Proceed as directed for Procedure, but make allowance for the larger specimen size.[16]

Doubtful is a subjective word, most production managers under a deadline would be tempted to consider failing results doubtful. A retest with a larger samples size might appear to yield a more accurate result but may also serve to dilute bioburden that resides in a hotspot. The Barr Decision changed the thinking on this retesting process, requiring an assignable cause for error in the laboratory before retest, emphasizing the lab's need to have validated internal processes.

An OOS sterility test has much more serious ramifications. Ingestion of a tablet with high bioburden may have little or no consequence, depending on the identification of the microbe and the health status of the patient. Obviously, injection of any microbial load, or even a dead gram-negative microbial load can have dire consequences. The USP <71> "Sterility Tests" sampling scheme is arbitrary and notoriously ineffective at capturing low-level contaminants likely to be present in parenteral drugs. In a batch that has 0.001 frequency of contaminated units, the lab has at best a 2% chance of capturing it in their test.[17] A failing test result is rare because of small sample size, lack of homogenous microbial distribution, and ineffectiveness of two broth and incubation conditions to culture systems to capture every potential microbial contaminant. Yet a genuine contaminant has deadly impact and cannot be retested away. USP <71> "Sterility Tests" offers four potential reasons to retest:

- The data of the microbiological monitoring of the sterility testing facility show a fault.
- A review of the testing procedure used during the test in question reveals a fault.
- Microbial growth is found in the negative controls.
- After determination of the identity of the microorganisms isolated from the test, the growth of this species

(or these species) may be ascribed unequivocally to faults with respect to the material and/or the technique used in conducting the sterility test procedure.

Each of these assignable causes points to a microbiology validation defect. Bullet 1 indicates unvalidated primary engineering controls and sterile material transfer processes, bullet 2 aims at a faulty microbiological method validation, bullet 3 points to a faulty autoclave or media storage validation study, and bullet 4 connects the contaminant to flawed gowning qualification or media fill validation. These are all reasonable assignable causes. However, except for the second bullet where growth in a negative control is obvious (and should never happen), connecting the sterility failure to these validation failures is risky. Environmental monitoring may show a hit on a fingertip plate. But the likelihood of matching the genus species or 5s ribosomal RNA segment of a sterility test failure to a microbe found on the technician's gloved fingertip plate is magical thinking. A contact plate with a 30% recovery effectiveness picking up the same *Corynebacterium* strain that caused the test failure from the dozens of strains that may have escaped the clean room garb or material transfer process is not likely. Better to validate the test, the conditions of the test, and the test environment, and keep these systems in a state of tight control. Better to be confident of test results generated, then try to see an assignable cause in every failure.

Compendial methods are validated or more properly defined, determined to be suitable for various samples, but if an alternative to the compendia is sought, full method validation is required.

USP <1223> "Validation of Alternative Microbiological Methods" gives guidance on selection, evaluation, and use of noncompendial methods. Here again we are faced with the challenge of validating anything against a referee method whose basic unit of measurement is as imprecise as the colony-forming unit. There are different criteria for validating a qualitative versus quantitative test.[18]

Validation parameter	Qualitative test	Quantitative test
Accuracy	No	Yes
Precision	No	Yes
Specificity	Yes	Yes
Limit of detection	Yes	Yes
Limit of quantification	No	Yes
Linearity	No	Yes
Range	No	Yes
Robustness	Yes	Yes
Repeatability	Yes	Yes
Ruggedness	Yes	Yes
Equivalence	Yes	Yes

Again, these terms are easily understood for discrete analytes accurately measured at parts per million and parts per billion levels, but how do they apply in the world of the colony-forming unit? How does one demonstrate accuracy using as its standard a test, for reasons previously outlined, that is notoriously inaccurate? Although the validation parameters are the same as those for a chemistry test, the application is quite different.

<u>Accuracy</u>—Are the counts achieved in the alternative match those in the compendial method? Equivalent counts in the microbial world are 50%–200%

<u>Specificity</u>—Does the alternative method detect a range of challenge organisms expected? Range might mean phyla, order, or it may be based on patient risk, the manufacturing environment, or some other criteria.

<u>Limit of Detection (LOD)</u>—What is the lowest number of microbes that can be detected? The ability to test with a single cell is a challenge here. It is impossible to precisely spike a sample with 1, 2, or 3 colony-forming units to determine the LOD, never mind a single cell. Application of statistics and MPN inoculations are required to demonstrate LOD.

<u>Robustness</u>—Will minor variations in temperature, media preparation, and incubation time impact the outcome of the test?

<u>Precision</u>—Is there statistical agreement of multiple samplings of a homogenous material? As seen in the example of pour plate qualification, the precision is as much based on the technician's skill set as on the analysis itself.

<u>Ruggedness</u>—Will the same test used on the same sample give the same results with different analysts, on different days, with different media lots, etc.? Again, how well trained are those analysts? How well validated are the systems surrounding the test?

Whether validating a qualitative or quantitative method, establishing LOD is an essential parameter. The most probable number method is useful for selecting low limits where plate counts and colony-forming units fail. In MPN Methodology, series of broth (although this can just as readily be applied to agar) are inoculated with increasing dilutions of culture. The number of tubes that show growth in each dilution are statistically converted to a most probably number. Consider the ten-tube series offered by FDA BAM:[19]

TABLE 58.1
For 10 tubes at 10 mL Inocula, the MPN per 100 mL and 95 Percent Confidence Intervals

Positive tubes our of 10	MPN/100 mL	Confidence level low	Confidence level high
0	0	—	3.3
1	1.1	0.05	5.9
2	2.2	0.37	8.1
3	3.6	0.91	9.7
4	5.1	1.6	13
5	6.9	2.5	15
6	9.2	3.3	19
7	12	4.8	24
8	16	5.9	33
9	23	8.1	53
10	>23	12	-

Abbreviation: MPN = most probable number.

This is a more useful tool than the plate count to get low and accurate inocula. Consider the example following:

Objective: Determine the limit of detection of *E. coli* in an alternative method.

Procedure

1. Remove a culture of *E. coli* ATCC 8739 from –70°C culture collection.
2. Place frozen bead containing culture to Tryptic Soy Broth and incubate 24 hours.
3. Transfer a loopful of broth to a Tryptic Soy Agar plate and streak for isolation.
4. Incubate the plate at 30°C–35°C for 2 days.
5. Confirm purity of *E. coli* plate by colony morphology and Gram stain.
6. Prepare a 0.5 MacFarland Standard suspension of *E. coli* in phosphate-buffered saline as follows:
    a. Touch a single colony with a Dacron swab.
    b. Transfer the growth collected on the swab to 10 mL phosphate-buffered saline by squeezing it against the test tube side wall and swirling it.
    c. Vortex the culture well to adequately suspend the culture.
    d. Add phosphate-buffered saline or culture media as needed to achieve 0.5 MacFarland Standard.
7. Perform a series of 1:9 dilutions of this culture until 10^{-7} dilution is reached.
8. Dilute the 10—^{7} dilution 1:1 two times, resulting in $10^{-7.5}$ and 10^{-8}.
9. Use the three dilutions referenced previously to inoculate the alternative method. Placing 100 µL in product of 10^{-7}, $10^{-7.5}$ and 10^{-8} culture is assumed to provide 10, 5, and 0 cfu
10. Enumerate each inoculation to a series of ten test tubes containing 10 mL Tryptic Soy Broth
11. Incubate the test tubes for 48 hours at 30°C–35°C.
12. Record how many of the tubes are turbid, and confirm growth is *E. coli* by streaking to MacConkey Agar and incubating at 30°C–35°C for 48 hours.

Interpretation

The following results were obtained for the following inocula:

10^{-8} culture had 1 of 10 tube positive = 1.1 MPN
$10^{-7.5}$ culture had 3 of 10 tubes positive = 3.3 MPN
10^{-7} culture had 10 of 10 tubes positive = >23 MPN

Ten replicates of inocula delivered approximately 1.1 MPN of *E. coli*, providing a statistically valid method for delivering nearly single cells to determine limit of detection of alternative methods.

Note in this validation example alone, how many processes will have been validated or qualified to properly execute the validation itself. Equipment, supplies, media, staff, and technique qualification, verification, or validation are all essential to generate valid microbiological result.

58.3 CONCLUSION

Validation of microbiological methods is a critical component of safe drug delivery. Although testing end product with valid methods has its place, true quality assurance comes from controlling critical control points in the manufacturing process, and those control points often rely on microbiological data to demonstrate the process is in control. The bioburden of raw materials, efficacy of cleaning and disinfection, and control of viable air particles are just some of the control points that rely on valid microbiological methods to demonstrate they are operating within limits. If the method is not valid, equally important if the laboratory generating the data is not qualified and validated, the manufacturing process itself may be out of control, blindly generating adulterated product for public consumption

NOTES

1. Standard Methods for the Examination of Water and Wastewater, American Public Health Association, 9020A "Laboratory Quality Assurance".
2. USP 42-NF38 <1227> Validation of Microbial Recovery from Pharmacopeial l Articles.
3. FDA BAM Chapter 3: Aerobic Plate Count.
4. Standard Methods for the Examination of Water and Wastewater, American Public Health Association, 9215A.
5. Standard Methods for the Examination of Water and Wastewater, American Public Health Association, 9020A "Laboratory Quality Assurance" Section 4 Analytical Quality Control Procedures.
6. Standard Methods for the Examination of Water and Wastewater, American Public Health Association, 9215.10 Heterotrophic Plate Count Analytical Bias.
7. Bartholomew JW, Variables Influencing Results, and the Precise Definition of Steps in Gram Staining as a Means of Standardizing the Results Obtained. *Stain Technology* 1962; 37: 139–155.
8. USP 43-NF38 <1227> Validation of Microbial Recovery from Pharmacopeial Articles.
9. ANSI/AAMI/ISO 11737–1 Sterilization of Medical Devices—Microbiological Methods—Part 1: Estimation of Population of Microorganisms on Products.
10. USP 34<1116> Microbial Control and Monitoring Environments Used for the Manufacture of Healthcare Products.
11. Guidance for Industry—Sterile Drug Products Produced by Aseptic Processing—Current Good Manufacturing Practice, US Food and Drug Administration, 2004.
12. "Manufacture of Sterile Medicinal Products" In EudraLex—The Rules Governing Medicinal Products in the European Union, Volume 4 EU Guidelines to Good Manufacturing Practice—Medicinal Products for Human and Veterinary Use—Annex 1: Manufacture of Sterile Medicinal Products, European Commission, 2008.
13. USP35 <1116> Microbiological Control and Monitoring of Aseptic Processing Environments.

14. Grilli A, Sutton S, Agar Transfer Devices for Environmental Monitoring in the Compounding Pharmacy: Science and Compliance. *International Journal of Compounding Pharmacies* 2015; 19(1) Jan/Feb.
15. Guidance for Industry—Sterile Drug Products Produced by Aseptic Processing—Current Good Manufacturing Practices, Food and Drug Administration, 2004.
16. USP 26 <61> Microbial Limits Test.
17. Sutton S.
18. USP 43-NF38 <1223> Validation of Alternative Microbiological Methods Guidance.
19. FDA Bacteriological Analytical Manual Appendix 2 Most Probable Number from Serial Dilutions, Robert Blodgett, October 9, 2020.

59 Rapid Methods for Pharmaceutical Processing and Their Validation

Jeanne Moldenhauer

CONTENTS

DOI: 10.1201/9781003163138-59

59.1 INTRODUCTION

Microbiological testing is key to the operation of pharmaceutical facilities. This testing is used to assess the microbiological quality and attributes of the product, components, ingredients, environment, and the utilities. The conventional microbiological methods used are limited by the time it takes to grow the microorganism under the specified test conditions. These methods have traditionally required days until the results are obtained. Sterility testing and mycoplasma testing traditionally have been cited as the most serious offenders, with a minimum 14-day or 28-day release time, respectively. Significant costs are associated with holding the product during this time. As such, there has been an increasing interest in the use of rapid microbiological methods (RMMs) (Moldenhauer, 2013). Today, companies have successfully validated and implemented rapid sterility and mycoplasma detection methods. Different approaches were used to validate and implement these methods. Companies worked individually to select and validate methods for sterility testing. For mycoplasma testing, there was a large group of companies that worked together through the Parenteral Drug Association (PDA) to develop and validate a rapid method that was superior to the conventional method of testing.

These new methods can offer many advantages, from the ability to gain results in real time to those where the results are obtained in a much shorter time period. Depending upon the method, they may also improve the accuracy of the method, the limit of detection, or other key attributes associated with the method. One of the difficulties associated with these methods has been defining what is necessary to validate these methods. This chapter provides an overview of some of these methods and a description of the method validation.

59.2 THE HISTORY OF RAPID MICROBIOLOGICAL METHODS

RMMs were introduced to the pharmaceutical industry about 25 years ago. Vendors of these methods believed that the benefits of the new technologies would be immediately adopted in the pharmaceutical industry. But pharmaceutical companies were fearful of the new methods and the regulatory acceptance of these methods. The anticipated widespread implementation of the systems did not occur. Approximately 15 years ago, the first submission of an RMM for regulatory approval occurred in the United States when Glaxo submitted the PallChek technology using adenosine-triphosphate bioluminescence as a product release test for the microbial limits test. This was significant as it was replacing a compendial product release test for non-sterile product (Moldenhauer, 2013). During this process, Glaxo worked closely with the Food and Drug Administration (FDA) to validate and implement the method. Validation of the microbiological method was a subset of the company's plan to implement Process Analytical Technologies (PAT) across the entire facility.

Later, Glaxo submitted a proposal to perform bioburden monitoring of water using the Scan*RDI* system (Note: In some countries this same system is denoted the ChemScan.). This technology utilizes solid-phase laser-scanning cytometry to detect viable microorganisms in a few hours. For this method, there is no need for the organisms to be cultured or grown as part of the test method. This methodology replaced a compendial product release test for a sterile product (Moldenhauer, 2013).

Gressett et al. (2008) published a paper indicating that a rapid method for sterility testing using the Scan*RDI* had been approved by the FDA. Not much later, Jennifer Gray of Novartis gave presentations that a rapid sterility test method using the Milliflex Rapid was approved by the FDA (Gray et al., 2008). Many other companies have developed rapid sterility methods using instruments that use the same general adenosine-triphosphate (ATP) methods as the Milliflex Rapid (e.g., Celsis) (Moldenhauer, 2013).

BioVigilant introduced the IMD-A. The IMD-A is an instrument that has a probe placed into the environment to monitor the air for viable biological particles (viable cells). This instrument introduced companies to the concept of the potential for real-time results for viable microorganism monitoring in their environmental monitoring program. Because this technology did not require the use of culture media in addition to providing real-time results, companies saw the potential for cost savings associated with the method (Moldenhauer, 2013). As such, there was an increased interest in these new methods. The IMD-A had a unique feature of being able to video what was happening in the area when a "hit" or count was detected. TSI also introduced a system using similar detection methods for real-time air monitoring, the BioTrak. This instrument was unique in that the viable "hit" detected could be captured on a gelatin agar plate, allowing for the potential of subsequent identification of the organism.

More recently, a group of scientists formed a collaborative workgroup to guide the development and implementation of an online water bioburden analyzer (OWBA). This group included Hans-Joachim Anders (Novartis), Fred Ayers (Eli Lilly), Brian Fitch (Biogen), Ren-Yo Forng (Amgen), Scott Hooper (Merck & Co.), Michelle Luebke (Baxter), Jeanne Mateffy (Amgen), Peter Noverini (Baxter), Brian Termine (GSK), Lisa Yan (Shire), and Jeffrey Weber (Pfizer Inc.) on the Online Water Bioburden Analyzer Workgroup. This group has been analyzing the effectiveness of current online water bioburden analyzers for viable contamination in purified water systems. The systems available today utilize intrinsic biochemical autofluorescence of microorganisms that have been laser excited. Some of the molecules used include NADH and picolinic acid, which excite with 405 nm lasers. This is the same type of detection method used to detect microorganisms present in the real-time air monitoring systems. The results reported are autofluorescent units not colony-forming units (Anders et al., 2017). The work of this group is still in progress.

Another system that can be widely used is the Endosafe PTS. This unit is a handheld spectrophotometer that uses an FDA-licensed disposable cartridge to accurately perform endotoxin testing. This system provides the test results in just 15 minutes, which is considered "near real time" results (Anonymous, 2019).

The combination of today's total organic carbon (TOC), conductivity, real-time monitoring of bioburden, and the PTS method of endotoxin testing combined provides near real-time results for all aspects of water monitoring.

59.3 OVERVIEW OF MICROBIOLOGICAL METHODS

There are several different types of microbiological methods. This section provides an overview of these methods.

59.3.1 Types of Conventional Test Methods

There are three basic types of microbiological evaluations conducted: determination of whether an organism is present (presence-absence tests), if organisms are present determination of how much is present (enumeration tests), and if an organism is present what organism is it (identification tests) (PDA TR33, 2000).

59.3.1.1 Presence/Absence Tests

This type of technology simply determines whether a microorganism is present in a sample. Sterility testing is a typical test using this type of assessment (Moldenhauer, 2003, 2013).

59.3.1.2 Enumeration Tests

These tests are used to determine the number of microorganisms present. The results obtained are affected by many characteristics including: the test conditions, culture medium, incubation conditions, whether the organism is or is not stressed or shocked, and whether it is a stock culture or an environmental isolate (Moldenhauer, 2003, 2013). The variability with this type of test is high when using conventional methods. It is common to assume that values within 30% are equivalent (PDA Technical Report Number 21, 1990). This number was established based upon real "equivalency" data that compared growth rates of organisms to determine the approximate number for equivalency. A Poisson distribution was used and averaged to 30%. This type of test is used for determining counts in environmental monitoring and other tests where specific counts are necessary.

59.3.1.3 Identification Methods

Numerous systems exist for the characterization and identification of microorganisms. Many of them have their origins in the clinical setting. Some technologies are manual and use classical methods of identification whereas others are automated and utilize other technologies (Moldenhauer, 2013). "Rapid" identification systems have been readily recognized as appropriate for use in pharmaceutical applications. Both phenotypic and genotypic methods are routinely utilized.

59.3.2 Types of New Technologies—Rapid Microbiological Methods

Many of the new technologies available to replace microbiological methods have their foundation in other sciences like chemistry, molecular biology, optics, and so forth. These technologies have been categorized in the PDA's Technical Report Number 33 (2000) and its subsequent update revision (PDA, Technical Report Number 33(Revised), 2013) as growth-based methods, viability-based methods, cellular component or artifact-based technologies, and nucleic acid-based technologies. Since the publication of the PDA document, systems have been released using various types of spectroscopy. As more methods have been developed, there are also methods that incorporate more than one type of technology into the system, e.g., may be able to detect, enumerate, and identify microorganisms or addition of a viability assessment to a system that uses another technology like spectroscopy to detect microorganisms (Moldenhauer, 2013). An example of this type of system is Battelle's Resource Effective Biological identification System (REBS), which uses Raman spectroscopy to detect, enumerate, and identify microorganisms. Some publications refer to the systems that use more than one type of technology as combination systems, whereas others focus on one aspect of the system for categorization.

59.3.2.1 Growth-Based Technologies

Systems in this category assume that organism detection is dependent upon measurement of attributes that require growth of the microorganisms. Some of the types of systems that are included in this category are: ATP Bioluminescence (and its sub-set adenosine kinase), Colorimetric Detection of Carbon Dioxide Production, Measurement of Changes in Headspace Pressure, Impedance, and Biochemical Assays (Moldenhauer, 2003, 2013).

59.3.2.2 Viability-Based Technologies

This category of technologies allows the user to detect and/or enumerate viable microorganisms present without requiring growth of the microorganism. As such, it is possible to have different results using a viability-based technology for enumeration versus a growth-based technology. The viability-based technology is likely to include those organisms that are not able to be cultured under the conditions used by the growth-based technology. Some of the systems included in this category are: Solid Phase Laser-Scanning Cytometry and Flow Fluorescence Cytometry (Moldenhauer, 2003, 2013).

59.3.2.3 Artifact-Based or Component-Based Technologies

Technologies based upon artifacts or components utilize the presence of specific cellular components. For example, the bacterial endotoxin test method tests for the presence of endotoxin. Some of the systems using this technology are fatty acid profiles, enzyme-linked immunosorbent assays (ELISA), and fluorescent probe detection (Moldenhauer, 2003, 2013).

59.3.2.4 Nucleic Acid-Based Technologies

Technologies that are based upon some type of nucleic acid, e.g., ribonucleic acid (RNA) or deoxyribonucleic acid (DNA), fall into this category. Some of the examples of systems using this technology are DNA probes, ribotyping or molecular typing, and polymerase chain reaction (PCR) (Moldenhauer, 2003, 2013).

59.3.2.5 Spectroscopy Systems

Many of the instruments from the chemistry laboratory are now being used for microbiological applications. Some of these instruments include: MALDI-TOF (Shelep, 2011), Fourier Transform Infrared Spectroscopy (FTIR) (Wenning et al., 2010), gas chromatography (fatty acid identification systems), and Raman Spectroscopy (Ronningen and Bartko, 2009). The MALDI-TOF is used for microbial identifications, whereas Raman (Battelle's REBS) is used for detection, enumeration, and identification (Moldenhauer, 2013).

Optical spectroscopy measures the interactions between light and the subject of interest. The scattering of light that occurs when light is disturbed by its interactions with particles. This scattering can be used to determine whether particles are present in the air. It can be combined with laser technologies to detect fluorescence. This type of technology has been used for real-time environmental monitoring of air. Coupled with an appropriate viability marker, both viable and nonviable particles can be detected by some of these systems. Two of the systems available include the IMD-A and the BioTrak (Anonymous, 2011a; Moldenhauer, 2013).

More recently, these optical spectroscopy systems have also been used for monitoring of the viable counts in high-quality waters. Systems available include the IMD-W and the Mettler 7000.

59.4 REGULATORY GUIDANCE ON VALIDATION OF RMMS

The first information provided as guidance for the validation of RMMs came from the PDA's Technical Report Number 33 (2000). This document provided useful information; however, it was written before most companies had completed a successful validation of these systems. This document was updated and revised in 2013 (PDA, 2013).

The *United States Pharmacopeia (USP)* published an informational chapter <1223> "Validation of Alternative Microbiological Methods". It was formally completed in 2006 (USP, 2006). This chapter modified the criteria for Validation of Compendial Methods in USP <1225>, which described the requirements for validation (usually chemistry methods), and made them applicable to microbiological methods. Although this chapter addresses alternative methods, it does not have specific details relative to the validation of identification methods (Anonymous, 2011b). This document was updated and revised in 2015 (USP, 2015).

In recent years there has been a trend to separate "rapid" identification methods from other rapid or alternative methods. This is not a function of implying that they are not rapid methods, but rather that the validation is quite different from the other methods for qualitative and quantitative analysis. The USP has published a draft monograph <1113>, "Microbial Identification". It is currently being retitled to "Microbial Characterization, Identification, and Strain Typing". Included in this monograph is a description of how to verify these methods (Anonymous, 2011c).

The European compendia, *Pharm Europa,* also published a chapter 5.1.6, "Alternative Methods for Control of Microbiological Quality", which provides an overview of alternative microbiological methods (qualitative, quantitative, and identification methods) as well as guidance for the validation of these methods (*Pharm Europa*, 2006). This chapter was also updated in 2015 (Ph. Eur., 5.1.6).

The International Conference for Harmonisation (ICH) has published information on the validation of analytical methods in their document Q2 (R1), *Validation of Analytical Procedures: Text and Methodology* (ICH, 1995).

59.4.1 Why Were All the Documents Being Revised?

The USP Microbiology Expert Committee expressed an interest in updating USP <1223> in order to provide flexibility to accommodate future microbiological methods. Some of the concerns were related to specialized products like cellular therapy and compounded medications. Additionally, the qualification requirements were updated explaining how to show non-inferiority to the compendial methods (Miller, 2015).

An initiative was also undertaken by the European Directorate for the Quality of Medicines in updating the requirements in *Ph. Eur.* 5.1.6. "Alternative Methods for Control of Microbiological Quality". This chapter was rewritten to include real examples of how to validate methods. It also includes primary validation and validation for the intended use for qualitative, quantitative, and identification methods (Miller, 2015).

59.5 THE VALIDATION PROCESS FOR RMMS

Validation of RMMs is like the validation processes traditionally used for equipment and method validation. Agalloco (1993) defines validation as, "Validation is a defined program which in combination with routine production methods and quality control techniques provides documented assurance that a system is performing as intended and/or that a product conforms to its pre-determined specifications. When practiced in a "life cycle" model it incorporates design, development, evaluation, operational and maintenance considerations to provide both operational benefits and regulatory compliance." Method validation has been described as the program to confirm that an analytical procedure used for a specific test is reliable, reproducible, and suitable for its intended purpose. Most RMMs include equipment, software, consumables, reagents, and human interactions that all must be considered in the validation process (Anonymous, 2011b; Moldenhauer, 2013).

Because using RMMs in a pharmaceutical environment requires that they be validated, it is important to consider the needs and requirements for validation even before purchasing the system. It is useful to develop a validation strategy, a validation master plan, a User's Requirement Specification (URS), a test plan, and the appropriate validation protocols (Moldenhauer, 2013).

59.5.1 Purchasing a Rapid Microbiological System

A significant investment is required for many of the new rapid microbiological technologies on the market. As such, it is important to understand the requirements that must be met prior to purchase of the system. There are several different areas from which requirements may originate, for example validation needs and expectations, regulatory expectations, scientific expectations, and so forth. For some systems, even the size of the system can be critical to determine whether the instrument will fit into the laboratory (Moldenhauer, 2013).

59.5.1.1 What Are the Technical Capabilities of the System?

Prior to purchase, one should understand exactly what the equipment or system can do. For example, what is the method sensitivity? What categories of products can be tested using this technology? What types of testing can be performed using this technology? Are there product characteristics that might interfere with the system's detection capability? How many samples can be processed in what time period? What is the level of automation in the system? One should also assess whether the system will be able to meet the various requirements that must be met (Moldenhauer, 2013).

It is also important to perform feasibility or proof-of-concept testing. This may be performed by the end user site, a contract testing laboratory, or at the vendor's site. During this evaluation, it is important to assess whether the rapid technology being considered will work with the items to be tested with the system. For example, if the system requires filtration through a specified filter, it is important to determine whether your test articles can be filtered through the specific filter. For systems using fluorescence, you must assess whether your test articles will interfere with the test methods (Moldenhauer, 2013).

59.5.1.2 What Regulatory Requirements Must be Met?

Prior to purchase of the system, it is useful to determine what requirements must be met for implementation of the system. Additionally, one should take into consideration what regulatory requirements must be met to implement the system. Some systems may easily be shown to be equivalent to existing methods and no additional regulatory concerns need to be met. Others may require prior approval of a regulatory body before they can be implemented (e.g., most product release tests). For those requiring regulatory submission, it may be necessary to have substantial support from the vendor to be able to gain regulatory approval. In the United States, this may be aided by the vendor having a Type V Drug Master File (DMF) on file with the FDA. Some vendors choose to have a formal issued validation testing package that they provide for users to include in their submission. In Europe, one may need the vendor to provide data confidentially to the regulator for review as part of the product application. It is beneficial to work with the vendor to assess their willingness to support these activities prior to purchase. Some companies tie the necessary supporting requirements into their purchase agreement (Moldenhauer, 2013).

59.5.1.3 What Is the Cost?

It is important to understand the cost associated with the rapid technology selected. In some cases, the bulk of the expense is in the initial purchase of the equipment. For some systems, the bulk of the cost is associated with the ongoing costs of operating the system. Many companies require formal documented reviews of both costs (initial and ongoing) and how the money spent for the system will generate a return on investment (ROI). It is important to understand the impact of the spending on the company's bottom line. Additional costs are associated with the validation, and in some cases additional equipment that must be purchased in support of the new technology. For example, a new method for sterility testing may require that one purchase specialized isolators that can accommodate the needs of the new system that will be used for the new technology (Moldenhauer, 2013).

59.5.1.4 What Are the Capabilities and Sustainability of the Vendor?

For most of the rapid technologies currently available, there is a single source of the system. This can limit the end user's ability to obtain supplies for the system or to gain technical information on the system. As such, the vendor selected can have a significant impact on your ability to successfully validate the system and continue to use the system on an ongoing basis. In most cases, a significant amount of support from the vendor will be needed during the validation process and to support any issues that arise after implementation (Moldenhauer, 2013).

There are several considerations for selection of a vendor of a rapid technology including (Anonymous, 2011b):

- The systems used by the vendor to manufacture the system.
- The documentation associated with the manufacture of the system.
- The systems established for change control and the associated notifications for users of significant changes.
- The regulatory status of the company, e.g., regulatory inspections, deficiency letters
- Whether there are other users of the system that may be used as references.
- Do other companies offer the supplies, parts, or equipment for commercial sale?
- What type of training is provided?
- What support services are provided, e.g., calibration, preventative maintenance, troubleshooting, regulatory services?
- Are support services available locally, or if not, what are the associated costs with obtaining support services?
- Has the vendor validated the system and is it in a reference document like a Drug Master File?
- Are validation protocols or guides available to support validation activities?
- How often is the software updated and how is this handled with the end user? This may lead to the question of how much downtime is associated with software changes.

59.5.2 Risk Assessment

Quality Risk Management (QRM) is a key concept in pharmaceutical systems today. This concept involves a systematic approach to understand the risks associated with a system. It involves identifying potential risks, determining the likelihood of the risk occurring, whether you can detect the risk, and how severe the risk is to the product. It also should consider what level is acceptable for the process. Some risks can be mitigated by taking other actions. When possible, the "high" risks should be mitigated to lower the risks.

The risk assessment should be conducted prior to any validation activity. More details are provided in Annex 15 (EMEA, 2019).

A key concept of Annex 15 is provided following (EMEA, 2019):

> A quality risk management approach should be applied throughout the lifecycle of a medicinal product. As part of a quality risk management system, decisions on the scope and extent of qualification and validation should be based on a justified and documented risk assessment of the facilities, equipment, utilities and processes. Retrospective validation is no longer considered an acceptable approach. Data supporting qualification and/or validation studies which were obtained from sources outside of the manufacturers own programmes may be used provided that this approach has been justified and that there is adequate assurance that controls were in place throughout the acquisition of such data.
>
> **(EMEA, 2019)**

59.5.3 The Validation Approach

Rapid technologies vary in their complexity and their level of automation. The scientific basis for how the system works also varies. It is important to develop a validation approach that includes validation of the hardware, software, and method. Depending upon the system selected, this may be performed concurrently or separately. It is important to determine the steps that will be incorporated into the validation plan for the technology selected. There are significant differences between the technologies available. The level of complexity of the system and the intended use of the system may impact the considerations that are applicable in the validation strategy. You may also need to assess the technologies available to do the testing, e.g., some process control systems are currently testing in aerosol chambers where there are limitations to the microorganisms and concentrations that may be used (Moldenhauer, 2013).

Some of the considerations in developing a validation approach include (Anonymous, 2011b):

- User's Requirement Specification (URS)
- Functional Design Specification (FDS)
- Supplier Assessment
- Risk Assessment
- Validation Master Plan (VMP)
- Design Qualification
- Factory Acceptance Testing (FAT)
- Site Acceptance Testing (SAT)
- Installation Qualification (IQ)
- Operational Qualification (OQ)
- Performance Qualification (PQ)
- Method Validation (MV)
- Personnel Training
- Documentation Requirements, e.g., SOPs, Protocols
- Requirements Traceability Matrix (RTM)
- Summary Report

59.3.4 Validation Master Plan

A VMP may be established for the company site, microbiology, and/or the specific process or equipment to be validated. Regardless of the scope of the master plan, it provides guidance on all the required validation activities that must be conducted for the acceptance of the system. Some of the common considerations in a VMP include: description of the system(s) covered, scope of the document, the types of documentation that must be generated, required supporting systems, testing expectations, and in some cases, the acceptance criteria that must be met, training, and the departments or individuals responsible for the approval of protocols, execution of the studies, and final reports. Typically, there is also guidance on how one should address issues (deviations or exceptions) that occur during execution of testing and how the testing documents can be amended to make changes (Moldenhauer, 2013).

It is useful to show in the VMP how the risk assessment was utilized to determine validation requirements.

59.3.5 User's Requirement Specification

In order to purchase the right rapid technology, it is important to understand what your expectations for the system are. Some expectations will be specific to the applications of the system you want to buy, whereas others may be general requirements to meet your company's requirements. You should have a clear understanding of what you want from the system before you purchase it. These expectations are incorporated into a formal document called the User's Requirement Specifications (URS). The URS serves as the foundation for several other documents including the requirements traceability matrix and the various tests and acceptance criteria incorporated in the validation documents (Moldenhauer, 2013).

Using the wording Good Manufacturing Practices (GMP) equipment does not mean anything. There is no one piece of equipment that meets this requirement, nor is there a specific description of GMP equipment. There are many different requirements in the regulations for equipment. To meet all the requirements, there are items that must be met by the end user. For more information, see Moldenhauer (2019)

Some of the considerations in developing a URS are (Anonymous, 2011b):

- What is the intended purpose of the system, e.g., this system is intended to replace the conventional product release sterility test?
- What are the specific performance requirements that must be in the system?
- A description of how the system will be used, e.g., in the laboratory, on-line, in a manufacturing environment.
- Characteristics of the hardware that must be met, e.g., size requirements, power requirements.
- Characteristics of the software that must be met, e.g., stored on a chip, PC compatible software.
- Expectations for reports to be generated.
- Compatibility and/or communication with other systems, e.g., networks, testing equipment, utilities.
- Descriptions of how the data will or should be managed.
- Necessary requirements for safety.
- Requirements for support services or needs, e.g., preventative maintenance, calibration.
- Concerns for engineering or physical requirements, e.g., must fit in a specified space or area, must be clean room compatible.
- Training that should be provided.
- Minimum requirements for an acceptable supplier, e.g., audit support, regulatory expectations.
- Financial considerations that must be met, e.g., cannot exceed a price of X.

The complexity of the technology and its intended use will determine the complexity of this document and whether all the previously listed considerations are necessary considerations for the document.

59.3.6 Functional Design Specification

All the functional requirements for the system are described in the functional design specification (FDS). Some companies choose to include these requirements in the URS and skip the development of this document. The FDS requirements are intended to ensure that the URS requirements for performance will be met. This document is specific and detailed, which leads to development of a lengthy document (Moldenhauer, 2013).

The FDS should describe the system functionality, configuration, the necessary inputs and outputs, the environment in which the system must operate, the utilities required and/or used, system architecture, system interfaces, the types of data, and the system security. Some FDS documents also identify where and how the expectations will be evaluated to ensure compliance with the criteria for the requirement. The testing is typically conducted within the qualification protocols or depending upon the requirement in the factory acceptance testing (Anonymous, 2011b; Moldenhauer, 2013).

Some examples of the typical information included in the FDS are (Anonymous, 2011b):

- System Description
- Purpose
- Scope of the FDS requirements for system documentation, e.g., user manuals, procedures, technical documentation
- Physical specifications for the system, e.g., size, power, operating environment, utilities
- Specifications for the computer system, e.g., type of computer, system operating system (including version numbers in some cases), supporting software, requirements for the computer hardware and accessories, networking requirements, printer requirements, databases, and so forth.
- Requirements for system security, e.g., multilevel password system, methods for record retention, necessity of an audit trail, compatibility with requirements for 21CFR Part 11, and so forth
- Validation attributes that must be met, e.g., the compendial validation expectations like accuracy
- Customization required for the system

- How the system responds to deviations and errors, e.g., alarms
- Whether there are parts of the system that will not be utilized or tested

The complexity of the system and the intended use will determine the level of detail required in this type of document.

59.3.7 Supplier Assessment

One of the first concerns is whether the company is financially stable. For many new technologies, the companies have underestimated how much time is necessary to gain adoption of their system into the pharmaceutical market. This has resulted in some going out of business and others going through several different acquisitions. If the company is not solvent and is the sole supplier of your technology and supplies, this can be a problem.

Verifying that the supplier can provide the services and systems that are necessary for your operation is critical when you are dealing with a single-source supplier. This includes assessment of the quality systems utilized by the supplier, the testing used to release the products manufactured by the supplier, where the products are manufactured, and the financial stability of the company. Additional considerations should be assessed if any or all of the validation is conducted by the supplier, e.g., whether the supplier understands and follows the GMPs applicable to your industry. One might also want to know whether the regulator's have reviewed this documentation in the past (Moldenhauer, 2013).

These assessments may be accomplished in a variety of ways like use of questionnaires, soliciting written information from the supplier, and/or conducting an assessment at a supplier's site (Moldenhauer, 2013). During this assessment, you might want to review the quality of the validation documentation.

59.3.8 Design Qualification

Design qualification (DQ) is conducted by many facilities, though not all, to ensure that the proposed design of the system or equipment is appropriate for its intended purpose. There are a variety of methods available for how to conduct this qualification. One way to accomplish this is to determine if the vendor's design is appropriate to meet the various requirements you have established in the URS (Moldenhauer, 2013).

59.3.9 Factory Acceptance Testing

In some cases, testing is performed at the vendor site prior to shipping the equipment to the end user. This testing may include functions that require special equipment that the end user does not have. Some choose to perform testing at the factory site that would cause the system to be returned if found at the end user site. Depending on the company, this testing may be formally documented in protocols or in laboratory notebooks. In some cases, this testing is referenced in subsequent validation protocols without repeating the testing (Moldenhauer, 2013).

59.3.10 Site Acceptance Testing

Some companies perform testing once the system is obtained at the end user site prior to the official validation studies. These studies may be a repeat of the some of the tests conducted in the FAT in order to ensure that the system received operates the same way it did when at the vendor site (Moldenhauer, 2013).

The level of documentation generated and required varies depending upon the protocol method used (Moldenhauer, 2013).

59.3.11 System Integration

This is the step in which all the component subsystems are combined into a single operating system. At this stage, evaluation should verify that all the components are working properly. Included in this evaluation are things like: communication with other laboratory equipment and/or laboratory information systems (Anonymous, 2019).

59.3.12 Installation Qualification

The Installation Qualification (IQ) is a formal protocol that is designed to ensure that the system as received and installed at the end user site meets the specified requirements. In some cases, companies will contract with the vendor of the equipment to perform this testing. It is good practice for end user personnel to participate in this activity, even when the vendor conducts the studies, to gain knowledge of the system. The documentation generated becomes a technical handbook for the system as supplied to your site. This information can be invaluable when determining later whether replacement parts are or are not identical to those installed. It is at the end user's discretion on whether the hardware and software are tested concurrently or as separate documents (Moldenhauer, 2003, 2013).

Some of the considerations included in this evaluation include (Moldenhauer, 2003, 2013):

- Verification that the items received and installed agree with those specified in the purchase order.
- The items received were not damaged in shipping or during the installation process.
- All the required supporting documentation for the system was received.
- The required documentation is complete.
- Specified utilities are available and properly connected.
- Verification that the system was installed correctly.
- Wiring of the system is as specified, if applicable.
- Version numbers for software or EPROMs used are documented.

- Models, serial numbers, and operating ranges are documented for key components.
- Applicable hardware and software are documented to be present.
- Peripheral equipment and accessories are present and properly connected and configured for use, if applicable.
- Drawings are available for the system and/or installation, as applicable.
- Backup and/or recovery copies of all software have been made and are available.
- Dip switch settings are documented.
- Cabling connections are documented and as specified.
- Configurations are documented.
- Verification that logbooks are established and maintained for the system.
- Verification that the system has been incorporated into a change control system.

59.3.13 Operational Qualification

The Operational Qualification (OQ) is designed to document that the system operates as expected. Depending upon the complexity of the system, the hardware and software may be tested concurrently or separately. When developing these testing requirements, it is useful to establish testing requirements that will be repeated or reevaluated for system updates and revisions, for example, one might establish a standard test set of data. There are many ways to accomplish these expectations (Moldenhauer, 2003, 2013).

Some approaches to conducting this testing involve including actual product testing during the evaluation of the system (PDA, 2000). This is not the only way to conduct testing, however. For example, some companies have successfully used testing plans that conduct testing with a product or solution that serves as a standard, and then product testing is evaluated as part of the method validation. This type of model is like the approach used with equipment used for testing for bacterial endotoxins. One of the reasons this type of model is attractive is that when a failure occurs, one can attribute it to the system, rather than wondering whether the product or the system is at fault (Moldenhauer, 2003, 2013).

The amount of testing performed varies based upon the complexity of the system, the type of testing performed, and how the data is used, e.g., process control test or product release test (Moldenhauer, 2013).

Typical considerations in an operational qualification protocol include (Moldenhauer, 2003, 2013):

- Verification that the installation qualification protocol has been completed.
- All critical equipment requiring calibration have been calibrated.
- The required system SOPs have been issued and implemented.
- Certifications have been completed for any items requiring certification.
- Alarms are functional and operate as designed.
- Error messaging is functional and operates as designed. For some systems it is not possible to identify and/or test all error messages in the qualification. One may be able to use a risk assessment procedure to determine the number and type of error messages that should be tested in the qualification process.
- Instrument operation, e.g., generation of expected results, standard curves, assays, expected results with specified organisms, and so forth.
- Inputs and outputs operate correctly.
- Interfaces and connections operate correctly.
- Software structure and documentation is complete.
- The system sequences in the specified order, and timing, if applicable.
- Potential sources of interference have been evaluated and mitigated, if necessary. For example: radio frequency interference (RFI), electromechanical interference (EMI), shielding of wiring, light interference and so forth.
- Ancillary equipment is functional and interfaces with the system correctly.
- Backup and recovery procedures are effective.
- Security procedures are effective.
- Stress testing, e.g., the system operates correctly when the maximum number of systems are operating in the laboratory or under stresses of environmental conditions.
- Data management capabilities are operational as specified.
- Operator training or qualification is completed.
- System suitability or standard curves are completed and acceptable.
- Preventative maintenance programs have been established and implemented for the system.
- Safety programs have been established and implemented for the system.
- Data archiving occurs as specified.
- Audit trails are accurate.
- Reports are accurate.
- Electronic signatures comply with 21CFR Part 11, if applicable.
- Verification that validation attributes or criteria are met.

59.3.14 Establishing Validation Attributes or Criteria

Recommendations for validation criteria have been established in several different documents including the compendia and industry guidance. There are some minor differences in the criteria across documents. As such, it is important to clearly identify the requirements that must be met at your site. Within these documents, there are differences in the criteria to be met depending upon whether it is a qualitative, quantitative, or identification test method (PDA, 2000; USP, 2006; Ph. E., 2006; Moldenhauer, 2013).

The validation criteria may be described as follows; however, as all the guidance documents are currently under revision, one should check the most recent guidance when writing validation protocols (Anonymous, 2011b; Moldenhauer, 2013):

- Accuracy is used to describe the closeness of the test results obtained using the alternative method to those obtained using the conventional or reference method. It should be evaluated across the practical range for the test method. It is frequently expressed as the percentage of recovery of microorganisms using the method.
- Equivalence or Comparative Testing describes the testing and the results obtained from the alternative method compared with the same testing performed with the compendial or reference method. This assumes that the test samples utilize equivalent standardized microbial cultures. This is usually tested as part of the performance qualification. It also evaluates the equivalence in the accuracy, precision, specificity, limit of detection, limit of quantification, linearity, and/or range for both the traditional and the new methods. It is useful to use standardized cultures and separately look at different products/sample matrices that will be routinely tested (Anonymous, 2019).
- Note: The newest revision of USP <1223> (USP, 2015) has statements that imply the product does not need to be tested as part of these tests. There is a statement in the chapter that implies at least one product type is assessed in the equivalence evaluation. *Ph. Eur.* indicates that the equivalence is shown by performing the validation parameters using standardized organism preparations and alternatively, in addition, a panel of microorganisms that show parallel testing of a set number of test samples should be conducted to show equivalence for a period of time. A risk assessment should be conducted to justify the test conditions (Anonymous, 2019).
- Limit of Detection describes the smallest number of microorganisms that can be detected using the specified test conditions. This is the level at which the method can detect the presence or absence of microbes (e.g., in a qualitative test).
- Limit of Quantification describes the smallest number of microorganisms that can be accurately and precisely enumerated in a test sample using the method.
- Linearity describes the ability of the system to produce results proportional to the organism concentration in a test sample within a specified range of microorganisms.
- Precision describes the level of agreement among the individual test results when the method is used repeatedly for multiple samplings of the same test organisms across the range of the test. This parameter is associated with the use of the method in the same laboratory over a short period of time, using the same analyst and the same equipment. Some use the term repeatability for this parameter. It is frequently expressed as the standard deviation or the coefficient of variation of a series of measurements (Anonymous, 2019).
- Range describes the interval between the upper and lower concentrations of microorganisms that have been shown to be determined with accuracy, linearity, and precision.
- Robustness describes the ability of the method capacity to be unaffected by small but deliberate variations in the parameters used for the method. Usually this testing is performed by the supplier of the equipment and either included in the DMF or provided in a report to the end user.
- Ruggedness describes the degree of precision or reproducibility of test results that are obtained when the same test samples are processed using a variety of normal test conditions. This might include different analysts conducting the tests, testing the samples on different pieces of equipment, using different lots of reagents, testing in different laboratories, and so forth (Anonymous, 2019). This test is typically best performed by the equipment manufacturer.
- Specificity describes the ability to detect a range of microorganisms for the method that are appropriate with the intended use of the system. One should also evaluate the potential for false positives and false negatives with the system. False results may be a result of interference from other items used in the testing, inability to distinguish between viable and nonviable cells, background noise, and so forth. Specificity testing may sometimes be conducted concurrently with accuracy testing (Anonymous, 2019). For ANVISA (Brazil's regulatory agency), they have asked for a much larger subset of organisms to be used for this type of testing, e.g., 20–30 strains.

Table 59.1 describes the expected tests for different guidance documents and test types.

59.3.15 Statistical Evaluations

Statistical evaluations are expected in the compendial validation requirements for RMMs. Additionally, they are discussed in the PDA's Technical Report Number 33. However, many have struggled with the statistical methods described in these chapters. Numerous publications and presentations have been given that provide alternative methods for conducting these evaluations. Van den Heuval et al. (2011) and Schwedock (2011) provide alternative methods and explanations for the statistics frequently used.

The detailed statistical requirements should be evaluated from the compendial chapters. They are not specifically discussed in this chapter.

TABLE 59.1
Validation Criteria Expected for Different Test Types

Validation criteria	Type of test	USP <1223>	Pharm Eur. 5.1.6	PDA Technical Report Number 33
Accuracy	Qualitative		May be used instead of LOD	
	Quantitative	✓	✓	✓
	Identification	✓	✓	✓
Limit of Detection (LOD)	Qualitative	✓	✓	✓
	Quantitative	✓	✓	✓
	Identification			
Limit of Quantification	Qualitative			
	Quantitative	✓	✓	✓
	Identification			
Linearity	Qualitative			
	Quantitative	✓	✓	✓
	Identification			
Precision	Qualitative	✓		✓
	Quantitative	✓	✓	✓
	Identification		✓	
Range	Qualitative			
	Quantitative	✓	✓	✓
	Identification			
Robustness	Qualitative	✓	✓	✓
	Quantitative	✓	✓	✓
	Identification		✓	
Ruggedness	Qualitative		✓	
	Quantitative	✓	Assessed as intermediate precision	✓
	Identification			
Specificity	Qualitative	✓	✓	✓
	Quantitative	✓	✓	✓
	Identification		✓	
Equivalence or Comparative Testing	Qualitative	✓	✓	✓
	Quantitative	✓	✓	✓
	Identification		✓	

It is useful to involve your own statisticians early in the development of validation requirements and testing (Moldenhauer, 2013).

59.3.16 Performance Qualification

The Performance Qualification (PQ) should be conducted as an integrated unit, i.e., hardware and software. Additionally, items like printers or ancillary equipment required in the process should be concurrently evaluated during the qualification. This testing provides documented evidence that the system performs as expected operating in the user environment and utilizing the user's systems and procedures. Some companies validate the equipment and the method concurrently, whereas others separate the method validation out (Moldenhauer, 2003). It is useful to separate the method validation into a separate evaluation, especially when the item being tested has a potential to interfere with the test (Moldenhauer, 2013).

There are different ways to perform this testing. Some compare the conventional method with the alternative method for a period of time or across a specified number of batches to show equivalence. During the testing, a minimum of three separate evaluations should be performed. This may be three different lots of material or three unique study evaluations (Anonymous, 2011b; Moldenhauer, 2013).

The data generated in these studies should be evaluated for the equivalence of the alternative method to the compendial or reference method. Depending upon the rapid technology selected, many are superior to the existing method (Anonymous, 2011b). It is important to select a statistical method that is appropriate for the type of results that should be obtained, the number of samples being tested, and the expected level of microorganisms that should be detected using the method (Moldenhauer, 2013).

Some of the typical considerations in this type of protocol include (Moldenhauer, 2003, 2013):

- Testing to show that all of the compendial validation criteria are met under the conditions of testing, e.g., with the standard products tested and the test samples (some companies only evaluate these parameters during the OQ).
- Documentation of the system reliability, accuracy, reproducibility, and consistency over time.
- Verification that the system operates and performs as expected in the user's environment, using user procedures, and operated by users.
- Reports generated should be accurate.
- Changes to information that should be incorporated into audit trails are accurate.

Other examples of testing plans are included in the PDA's Technical Report Number 33 (PDA, 2000).

For some systems, like those used for air monitoring, one may want to run the conventional test method side by side with the alternative method for a period of time to show equivalence (Moldenhauer, 2013).

59.3.17 Method Validation or Method Suitability

There are a variety of different ways to conduct method validations. For example, alternative sterility test methods may be validated using the sterility test validation procedures specified in the compendia. Other test methods may repeat evaluation of the validation criteria when the test is conducted using specific products (Moldenhauer, 2013).

The testing should be conducted as part of a formal protocol (Moldenhauer, 2013).

One of the considerations in this testing is whether each test material is evaluated to interfere with the method and/or yield abnormal results, e.g., false positives. Another concern is how to evaluate results when there is cellular debris or cell culture materials. When you cannot resolve false positive results or false negatives, the new method may not be suitable for your product (Anonymous, 2019).

59.3.18 Validation of Microbial Identification Systems

Identification systems usually have testing for accuracy and precision. The end user should define acceptance criteria for each parameter. Some typical concerns are the isolates not in the database, the accuracy of the system to properly identify the organism, and the like (Anonymous, 2019).

59.3.19 Personnel Training

Different rapid technologies have different levels of complexity of operation. In some cases, minimal operator experience or training are required, whereas other systems require a specialized education and prolonged training before the system can be operated correctly. One should identify who is responsible for the training, both initially and on an ongoing basis. Many vendors are quick to provide training at the time of sale. The problem can arise in handling operator turnover. Frequently, only a few personnel are trained to use these new technologies, and the site may not have other individuals that can train new employees that may need to use the system (Moldenhauer, 2013).

There are some contract laboratories and/or consultants that may be able to provide the training required on an ongoing basis (Moldenhauer, 2013).

Systems should be established to verify the competency of the personnel that will be operating these systems. Additionally, requirements should be established describing at what phase in the project operator training should be conducted (Moldenhauer, 2013).

59.3.20 Documentation Requirements, e.g., SOPs, Protocols

The minimum expected documentation associated with the system should be defined. Additionally, one should define at what stage in the project the documentation should be issued (Moldenhauer, 2013).

Some of the typical types of procedures that should be generated include (depending upon the complexity of the system, these may be individual procedures, or the requirements incorporated into a small number of procedures) (Moldenhauer, 2013):

- Operation of the system
- Exceptional operation of the system, i.e., what to do when something goes wrong
- Preventative maintenance for the system
- Calibration of the system
- Cleaning of the system
- Method validation requirements, if applicable
- Methods for assessing whether products are compatible for use with the system, if applicable
- Receipt, testing, and acceptance or rejection of supplies for the system
- Qualification of analysts that operate the system
- Change control for the system (hardware, software, methods, and procedures)
- Security procedures, e.g., assignment of system administrator's, levels of passwords, control of passwords, and so forth
- Control of the system software
- How to handle system updates
- Procedures for data management and evaluation
- Backup, recovery, and archival procedures
- Disaster recovery and contingency plans, e.g., what to so if the single source system fails
- Specific procedures for different methods used

59.3.21 Requirements Traceability Matrix

Once the various requirements for the system have been established in the URS and/or FDS, it is important to assess whether all of the requirements have been successfully met by the system purchased. The requirements traceability matrix (RTM) is a document that provides a linkage between the requirements established and the location of the testing that has been established to ensure that the requirement is met. This document is useful in assessing whether the validation documents generated are appropriate to ensure that the system meets the established requirements (Moldenhauer, 2013).

This document is considered a living document that is updated and revised throughout the validation process (Anonymous, 2011b, 2013).

59.3.22 Summary Report

Final reports should be generated and approved at the completion of each phase of the validation testing. This report should be accurate, complete, and approved by those organizations that approved the protocol (Moldenhauer, 2013).

59.4 OTHER TYPES OF TESTING, E.G., AIR MONITORING

Some of the alternative methods available today are used for testing samples that are not liquid-based. Most of the original requirements specified in the compendia and the PDA's Technical Report Number 33 were written for liquid-based testing. As such, some of the companies for air monitoring equipment have struggled with application of these methods to air samplers (Moldenhauer, 2013).

In other cases, vendors out of necessity have performed much of the validation for validation criteria and included the testing in a DMF because many pharmaceutical companies do not have the necessary equipment to perform the testing, e.g., challenging an air sample with a single microorganism (Moldenhauer, 2013).

59.5 IMPLEMENTATION OF THE SYSTEM

Following completion of the initial validation, the system is released for implementation and use at the site. This release for site use may be dependent upon other factors like whether the method must be submitted to a regulatory agency for approval prior to release (Moldenhauer, 2013).

59.6 QUALIFICATION OF ADDITIONAL SITES OR ADDITIONAL EQUIPMENT

Depending upon the rapid technology selected and the intended use, more than one system may be needed at your site. In some cases, the methods are initially validated at a corporate site and then transferred to another site where routine testing is performed. With the substantial amount of work required to conduct the validation, there are frequently concerns on whether validation of subsequent systems can be reduced (Moldenhauer, 2013).

When the new technology is submitted to a regulatory agency as a comparability protocol or a request for scientific advice, it may be useful to describe the reduced testing plan proposed for reduced testing of subsequent systems. If no submission is made, one can also discuss these plans with the local regulatory inspectors for the site. The following is a description of how some companies have chosen to deal with testing of additional systems (Moldenhauer, 2013).

When considering a reduced testing plan, it is important to understand whether the new system is "identical" versus a system that is similar and uses the same basic technology. Identical systems may provide more opportunities for reduced testing (Moldenhauer, 2013).

59.6.1 Equipment Evaluations

There are many possibilities to create differences between systems based upon how they are installed. As such, it is common to perform a complete IQ on each system purchased. The contents of the IQ may be an exact duplicate of the initial IQ for the first system, other than data entries made (Moldenhauer, 2013).

It is appropriate to repeat sufficient testing from the OQ to ensure that the system is operating as intended. The total amount of testing conducted will be dependent upon the system and how it will be used (Moldenhauer, 2013).

If the repeated IQ tests and OQ tests show that the system functions the same way as the original piece of equipment, the PQ may be omitted or significantly reduced. A risk assessment should be conducted to assess the tests that need to be repeated (Moldenhauer, 2013).

59.6.2 Software Evaluations

When the same software is installed on more than one unit, the requirements for testing may vary. Some companies treat it the same way as equipment with an IQ and OQ for each system. Other companies have chosen to verify the software identification, including the version number and any other updates, to show that it is equivalent and only perform some minimal tests to ensure that all of the applicable sections of the software have been installed. The rationale for the test plan should be described along with assessments of the associated risks (Moldenhauer, 2013).

59.6.3 Method Evaluations

Typically, it is not necessary to repeat the method validations for additional pieces of equipment added. Rather, this type of situation is addressed by the method transfer programs within the facility (Moldenhauer, 2013).

59.7 ONGOING ACTIVITIES

Following the initial validation of the equipment, one should take care to maintain the equipment and methods in a validated status. As such, it is important to define the requirements and methods to be used for subsequent re-validation of the equipment. Additionally, one should ensure compliance with procedures for change control, preventative maintenance, and so forth (Moldenhauer, 2013).

It is important to ensure that you can document all software changes, e.g., made during service calls, and that the appropriate testing has been conducted to show that the system remains in a validated state.

The performance of the system over time should also be assessed to ensure that it is operating as shown in the validation. If there are a few or many deviations or exceptions, one should assess whether there are critical parameters that should be added to the validation criteria for the system. If the deviations or changes necessitate changes or modifications to the system, this should trigger an assessment of whether validation is necessary for the system (Moldenhauer, 2013).

59.8 CONCLUSION

Although the validation of alternative methods can be daunting, the benefits of completing these activities can offer improvements in the time to results and the quality of the data recovered or cost savings that make it worth the work to validate and implement the systems (Moldenhauer, 2013).

AUTHOR BIO

Jeanne Moldenhauer, Excellent Pharma Consulting, Inc., has nearly 30 years of experience in the pharmaceutical industry. She chairs the PDA Environmental Monitoring/Microbiology Interest Group, serves on the PDA Scientific Advisory Board, founded the Rapid Microbiology User's Group, and is a member of ASQ and RAPS. She is a certified quality engineer and a certified quality manager. She is the author of many publications and books.

REFERENCES

Agalloco J. The Validation Life Cycle. *Journal of Parenteral Science and Technology* 1993, 47(3), 142–147.

Anders H-J, Ayers F, Fitch B, Forng R-Y Hooper S, Luebke M, Mateffy J, Noverini P, Termine B, Yan L, Weber J. Practical Applications of Online Water Monitoring Bioburden Analyzers in Pharmaceutical Manufacturing. *Pharmaceutical Online*, 2017. www.pharmaceuticalonline.com/doc/practical-application-of-online-water-bioburden-analyzers-in-pharmaceutical-manufacturing-0001 (Accessed December 3, 2019).

Anonymous. *Optical Spectroscopy Technology*, 2011a. http://rapidmicromethods.com/files/overview.html (Accessed September 9, 2011).

Anonymous. *Validating Rapid Micro Methods*, 2011b. http://rapidmicromethods.com/files/validation.html (Accessed September 8, 2011).

Anonymous. *Validating Rapid Microbiological Methods*, 2019 http://rapidmicromethods.com/files/validation.php (Accessed December 3, 2019).

Anonymous. USP Microbial Identification: In Process Revision Including a Title Change, 2011c. www.psc-asia.com/index.php?view=article&catid=6%3Alatest-news&id=86%3Ausp-microbial-identification-in-process-revision-including-a-title-change&option=com_content&Itemid=8&lang=en (Accessed September 9, 2011).

Anonymous. Endosafe PTS from Charles Rivers Laboratories. *Select Science*, 2019. www.selectscience.net/products/Endosafe%c2%ae-PTS%e2%84%a2/?prodID=106732#tab-2 (Accessed December 3, 2019).

EMEA. Eudralex Volume 4: Guidelines for Good Manufacturing Practice for Medicinal Products for Human and Veterinary Use Annex 15: Qualification and Validation. Legal Basis for Publishing the Detailed Guidance: Article 47 of Directive. *European Commission Directorate-General for Health and Food Safety*, 2015. https://ec.europa.eu/health//sites/health/files/files/eudralex/vol-4/2015-10_annex15.pdf (Accessed December 3, 2019).

Gray J, Stärk A, Berchtold M. New Rapid Sterility Test. Presentation given at the ECA Rapid Methods Conference on December 9, 2008.

Gressett G, Vanhaecke E, Moldenhauer J. Why and How to Implement a Rapid Sterility Test. *PDA Journal of Pharmaceutical Science and Technology* 2008, 62(6), 429–444. Bethesda, MD.

ICH. (1995). ICH Harmonised Tripartite Guideline Validation of Analytical Procedures: Text and Methodology Q2(R1). INTERNATIONAL CONFERENCE ON HARMONISATION OF TECHNICAL REQUIREMENTS FOR REGISTRATION OF PHARMACEUTICALS FOR HUMAN USE. Current Step 4 version Parent Guideline dated 27 October 1994 (Complementary Guideline on Methodology dated 6 November 1996 incorporated in November 2005). Downloaded from: https://database.ich.org/sites/default/files/Q2_R1__Guideline.pdf on June 23, 2021

ICH. *Validation of Analytical Procedures: Text and Methodology*. London, UK: European Medicines Agency: 1–15.

Miller M. Rapid Methods Update: Revisions to a United States Pharmacopeia Chapter. *European Pharmaceutical Review*, 2015. www.europeanpharmaceuticalreview.com/article/34483/rapid-methods-update-revisions-to-a-united-states-pharmacopeia-chapter/ (Accessed December 3, 2019).

Moldenhauer J. Validation of Rapid Methods and Systems. In Moldenhauer J, ed. *Laboratory Validation: A Practicioner's Guide*. Bethesda, MD: PDA/DHI, 2003: 921–938.

Moldenhauer J. Validation of Rapid Microbiological Methods (Chapter 20). In Kolhe P, Shah M, Rathore N, eds. *Sterile Product Development: AAPS Advances in the Pharmaceutical Sciences Series 6* New York, Heidelberg, Cordrecht, London: American Association of Pharmaceutical Scientists and Springer, 2013: 513–534.

Moldenhauer J. GMP Compliant Equipment. *GXP Journal*. IVT Network, 2019. www.ivtnetwork.com/article/gmp-compliant-equipment.

PDA Technical Report 21. Bioburden Recovery Validation. *PDA Journal of Pharmaceutical Science and Technology* 1990, 44(3), 1–8. Bethesda, MD.

PDA Technical Report Number 33. Evaluation, Validation, and Implementation of New Microbiological Testing Methods. *Journal of Parenteral Science and Technology* 2000, 54(3). Bethesda, MD.

PDA Technical Report Number 33 (Revised). Evaluation, Validation, and Implementation of New Microbiological Testing Methods. *Journal of Parenteral Science and Technology* 2013. Bethesda, MD.

Ph E. Alternative Methods for Control of Microbiological Quality. *European Pharmacopeia (aka Pharm Europa)* 2006, 50106, 4131–4142.

RonningenTJ, Bartko AP. Microbial Detection, Identification, and Enumeration Based Upon Raman Spectroscopy. In Moldenhauer J, ed. *Environmental Monitoring Volume 4: A Comprehensive Handbook*. PDA/DHI, 2009: 183–198.

Schwedock J. Statistics of Validating and Alternative Sterility Test: Limits of Detection and Other Problems. In Moldenhauer J, ed. *Rapid Sterility Testing*. Bethesda, MD: PDA/DHI, 2011: 245–266.

Shelep D. Maldi-TOF Method for Identification of Microbial Isolates. In Moldenhauer J, ed. *Environmental Monitoring Volume V: A Comprehensive Guide*. Bethesda, MD: PDA/DHI, 2011: 67–80.

Van den Heuval E, Verdonk GPHT, Ijzerman-Boon PC. Statistical Methods for Detection of Organisms with Sterility Tests. In Moldenhauer J, ed. *Rapid Sterility Testing*. Bethesda, MD: PDA/DHI, 2011: 201–244.

Verdonk G. Personal Communication with Author of the Chapter.

60 Extractables and Leachables in Drug Products

An Overview

William Parker and Don DeCou

CONTENTS

60.1 INTRODUCTION

Regulatory agencies, including the United States Food and Drug Administration (FDA), have become increasingly concerned with regard to the presence of leachable elements and compounds in pharmaceuticals, drug delivery systems, and medical devices since the 1980s. This increased concern is the result of several well-documented incidents of contaminants leaching from containers and packaging, resulting in a potential or real risk to humans.

In the 1980s, the presence of volatile N-nitrosamines (nitrosamines) was observed in rubber baby bottle nipples and pacifiers. (1) The FDA's investigation showed that the nitrosamines could migrate into foods, such as milk and infant formula. Nitrosamines are carcinogenic compounds, and the exposure of infants to these compounds prompted the FDA to call upon the manufacturers to reduce nitrosamine levels in the products and assign specifications of low parts per billion levels to these components. (2)

In the 1990s, the FDA became aware of several carcinogenic classes of compounds that were leaching from rubber seals in metered-dose inhalers (MDIs) used for the delivery of asthma medication. (1) These included N-nitrosamines, polynuclear aromatic hydrocarbons (PNAs), and 2-mercaptobenzothiazole. The leaching of these compounds was recognized as a potential risk to human health by the FDA. As a result, the FDA introduced regulations prompting manufacturers to monitor and control these compounds in rubber components and drug products.

More recently (2008), McNeil Consumer Healthcare voluntarily recalled more than 500 lots of Tylenol, Motrin, Benadryl, and Zyrtec[1] products. (3) This was in response to consumer complaints about a "musty" odor in the products. The cause of the odor was determined to originate from 2,4,6-tribromoanisole. In this case, 2,4,6-tribromophenol was used to treat wooden shipping pallets. The compound migrated from the pallets into the products and was converted to 2,4,6-tribromoanisole, which has a very low odor detection threshold. The recall cost McNeil an estimated $665 million in sales. (4)

These events show the importance of addressing extractable and leachable compounds in the development of pharmaceutical and biopharmaceutical drug products, in regulatory filings, and also in assuring patient safety.

Careful consideration must be employed when choosing packaging and delivery systems as part of pharmaceutical and biopharmaceutical drug product development. It is important that packaging and delivery systems meet the chemical criteria needed for the successful storage and delivery of a drug in addition to meeting the physical and/or performance criteria. Chemical compatibility is a critical component of these systems, as leaching of organic and inorganic chemicals can have many adverse effects that may include direct toxicity to patients as well as reduced efficacy and quality of a drug product. There are no therapeutic benefits of packaging leachables, and they should be mitigated or eliminated entirely.

Exact prescription of how to conduct an extractables and leachables (E and L) program is not found in go-to references like the U.S. Pharmacopeia (USP), but the onus of understanding the E and L of packaging and delivery systems resides with the drug developer. Because of this, efforts have been made over the last 20 years by industry groups, composed of drug developers, regulatory experts, scientists, and manufacturers, to educate the industry and harmonize the processes for a variety of applications. The goal of this chapter is to provide a broad overview of E and L, including definitions, examples, and approaches to study design.

60.2 EXTRACTABLES AND LEACHABLES

In 1999, the FDA released their container closure guidance, which formed the backbone of E and L risk management (5).

DOI: 10.1201/9781003163138-60

A central part of that guidance is the classification of dosage forms by leaching risk and degree of concern based on route of administration, a recreation of which is shown here as Table 60.1. Because of this, the focus for many years has been on understanding the interactions between drug and packaging and the associated risks of those interactions, primarily for orally-inhaled and nasal drug products (OINDPs), parenteral drug products (PDPs), and ophthalmic drug products (ODPs). At the other side of that spectrum sits oral tablets and capsules where, historically, packaging is rarely assessed for leaching risks, outside of mere theoretical exercises.

E and L related documents that have been published recently include USP <1663> (6), "Assessment of Extractables Associated with Pharmaceutical Packaging/Delivery Systems", USP <1664> (7), "Assessment of Drug Product Leachables Associated with Pharmaceutical Packaging/ Delivery Systems", and USP <661.2> (8), "Plastic Packaging Systems for Pharmaceutical Use". These documents describe critical dimensions of E and L assessments.

USP <661.2> describes a "Chemical Safety Assessment" as it applies to extractable compounds. This chapter also considers the route of administration. It says:

> With regard to the testing of the packaging system (and/or its components of construction as appropriate) and the packaged drug product, an appropriate and rigorous chemical safety assessment would include extractables testing of the packaging system and leachables testing of the packaged drug product. It is expected that the design of the extractables and leachables study would be based on sound and justifiable scientific principles, and that the studies themselves would be consistent with 1) the nature of both the packaging system and packaged drug product, 2) the clinical use of the packaged drug product, and 3) the perceived safety risk associated with the packaging system and dosage form. Although no dosage form is excluded from this testing requirement, it is anticipated that the nature and degree of testing would be dosage form-dependent and consistent with a risk-based approach. In view of the considerable diversity of packaging systems, dosage forms, and packaged drug products, it is not possible to provide specific test conditions for performing extractables and leachables studies.

This suggests that dosage forms once considered as not requiring E and L assessment will now require attention in regulatory filings. A drug developer that does not include an E and L program during product development will face delays in approval of the product at best, or non-approval of the product.

Table 60.2 lists terms and abbreviations that will be used in this chapter and their definitions.

Extractables originate with the drug product container closure system. Leachables are a subset of extractables. Extractables may or may not be present as leachables and leachables may or may not originate with the container

TABLE 60.1
Examples of Packaging Concerns for Common Classes of Drug Products

Degree of concern associated with the route of administration	Likelihood of packaging component–dosage form interaction		
	High	**Medium**	**Low**
Highest	Inhalation aerosols and solutions; injections and injectable suspensions	Sterile powders and powders for injection; inhalation powders	
High	Ophthalmic solutions and suspensions; transdermal ointments and patches; nasal aerosols and sprays		
Low	Topical solutions and suspensions; topical and lingual aerosols; oral solutions and suspensions	Topical powders; oral powders	Oral tablets and oral (hard and soft gelatin) capsules

Source: Adapted from (5) *Guidance for Industry: Container Closure Systems for Packaging Human Drugs and Biologics*. United States Department of Health and Human Services Food and Drug Administration, May 1999. (www.fda.gov/downloads/drugs/guidances/ucm070551.pdf)

TABLE 60.2
Glossary

Extractable (as defined in USP <1663>)	Organic and inorganic chemical entities that can be released from a pharmaceutical packaging/delivery system, packaging component, or packaging material of construction and into an extraction solvent under laboratory conditions.
Leachable (as defined in USP <1663>)	Foreign organic and inorganic chemical entities that are present in a packaged drug product because they have leached into the packaged drug product from a packaging/delivery system, packaging component, or packaging material of construction under normal conditions of storage and use or during accelerated drug product stability studies.
Primary packaging	The components together that make up the container in which the drug is stored for its shelf life, often also referred to as the container closure system.
Secondary packaging	The components together that make up the system that contains the primary packaging system, and which do not directly contact the packaged drug.
LC/MS	Liquid chromatography/mass spectrometry
GC/MS	Gas chromatography/mass spectrometry
ICP/MS	Inductively coupled plasma/mass spectrometry
ICP/OES	Inductively coupled plasma/optical emission spectroscopy
IC/ECD	Ion chromatography/electrochemical detection
NVR	Non-volatile residue
FTIR	Fourier-transform infrared spectroscopy

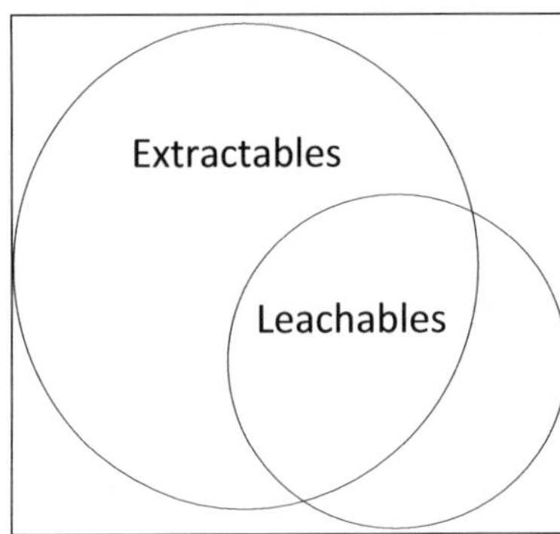

FIGURE 60.1 Venn diagram showing extractables and leachables.

closure system. This is illustrated in Figure 60.1 by way of a Venn diagram.

A variety of chemical-specific and non-chemical-specific analytical techniques can be used to assess the E and L of a material or packaging or delivery system. Non-chemical-specific techniques, such as NVR and FTIR, were used more prominently before mass spectrometric instrumentation became more common in analytical laboratories and before understanding of leaching risks became more refined. These techniques can still be used when a simple scouting for gross changes or differences in materials is useful or convenient; however, these techniques do not give the user data that is robust enough to identify specific chemicals, which is where the focus of an E and L program should be. The identification of E and L is key when assessing the risk to the drug product and patient that those individual chemicals may pose.

Standard techniques include LC/MS, GC/MS, IC/ECD, ICP/OES, and/or ICP/MS. Over the past 10 years, with high-resolution-accurate-mass (HRAM) spectrometers becoming more accessible, many E and L laboratories have been using HRAM with their LC/MS non-volatile organic compounds (NVOCs) methodology. The advantage over unit-mass-resolution mass spectrometers (i.e., single-quadrupole mass spectrometers) is the ability to use HRAM data as a starting point for the identification of unknown compounds. Although it can be at times impractical in time and cost to elucidate the structure of unknown compounds (depending on the material and extraction conditions), the goal should be to have no unknown extractables before progressing to leachables method development and validation.

GC/MS is typically used to analyze for semi-volatile organic compounds (SVOCs) and volatile organic compounds (VOCs). The former is performed through direct analysis of liquid extracts, whereas the latter is on the headspace gases generated by heating an amount of neat material in a sealed vial. Although these techniques have not changed much in recent years, HRAM mass spectrometers that can be used as extractables screening tools are becoming more accessible for GC/MS systems, particularly TOF (time-of-flight) mass spectrometers.

IC/ECD is a screening technique that can be used to analyze for a specific list of anions. This can be useful for determining, for example, if bromide and chloride anions can be extracted from halogenated butyl rubbers. Other anions that may be of interest, depending on the product and packaging, also should be targeted with this screening technique. It is not necessary to identify the source compound from which the anions originate (speciation), which generally would require more analytical work than the simple screening.

ICP/OES and ICP/MS are the techniques for analyzing extracts for a range of extractable elements quickly and simply. With the release of USP <232> "Elemental Impurities—Limits" (9) and <233> "Elemental Impurities—Procedures" (10), a greater focus has been made on modernizing analyses for elemental impurities in drug products. As in IC/ECD, it generally is not necessary to speciate the extracted elements (in solution as cations), but unlike extractable elements, there is not much in the way of guidance documentation in industry nor USP chapters dedicated to this class of extractables.

Many scientists working on the development of new pharmaceutical and biopharmaceutical drug products might look to the USP for guidance, and this should be the main reference used when working through an E and L program when seeking regulatory approval in the United States. The general chapters <1663> and <1664> are useful references, but a statement made in the first paragraph of <1663> is most helpful (emphasis added):

> Although intended to be helpful and generally applicable, the chapter is for informational purposes and does not establish specific extraction conditions, analytical procedures, or mandatory extractables specifications and acceptance criteria for particular packaging and delivery systems or dosage forms; nor does it delineate every situation in which an extractables assessment is required.

Simply put, the USP does not prescribe a specific path to follow, nor analytical techniques needed, to perform the assessment of extractables and leachables, even with the mainstay techniques mentioned previously. The technical aspects are left to the drug developer, hopefully with guidance from experienced scientists and toxicologists. There are a variety of resources beyond the USP to help the drug developer, including publications from the Parenteral Drug Association (PDA), the Product Quality Research Institute (PQRI) Leachables and Extractables Working Group, the BioPhorum Operations Group (BPOG), and guidelines from the International Conference on Harmonisation (ICH). Each of these publications are valuable resources and should be consulted when initiating an E and L program. The most important thing to remember throughout the process is that the main goal should be to identify leachables to monitor over the shelf life of the product.

An E and L assessment should be underpinned by at least a basic knowledge of factors that lead to leaching,

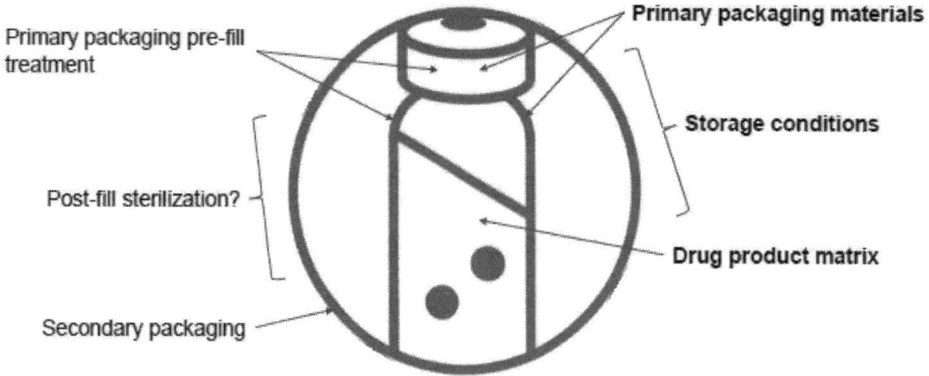

FIGURE 60.2 A simple representation of leaching factors.

which are visualized in Figure 60.2. The greatest factors to leaching are:

1. Primary packaging materials (because these are the main source of leaching over time).
2. Storage conditions because of the fact that the speed at which leaching occurs is directly related to the ambient temperature.
3. Drug product matrix, because of the basic chemical principle that chemicals dissolve more easily in similar chemicals.

Depending on the application and type of packaging, secondary packaging materials could also be a factor in leaching. If there is a post-fill sterilization, often performed at elevated temperatures through an autoclaving process, leaching could occur even then. How the primary packaging components are washed and/or sterilized prior to filling could affect their extractables profiles, thus leading to a change in potential leaching over time.

The process of determining the leachables of highest risk in a drug product starts with an assessment of the materials of construction of the primary packaging and/or delivery system and continues through a number of steps as shown in Figure 60.3. The flow chart is not meant to be representative of *every* E and L program, but instead shows a generic view of the typical process. In an ideal scenario, suppliers will provide robust chemical characterization data of each individual component that is intended for the system or manufacturing process. This data would include not just the specifications of the material of construction that typically can be found in a "data sheet" but would also include biocompatibility data of the material. For example, USP Class VI plastic is tested via USP<87> "Biological Reactivity Tests, In Vitro" (11) and <88> "Biological Reactivity Tests, In Vivo" (12). This should be looked at as the bare minimum of data needed before even considering a material for use in a medical application.

Beyond biocompatibility, again in an ideal scenario, extractables data (theoretical or experimental) would be provided by the manufacturer or supplier of packaging/delivery systems. Lack of this information would require a drug

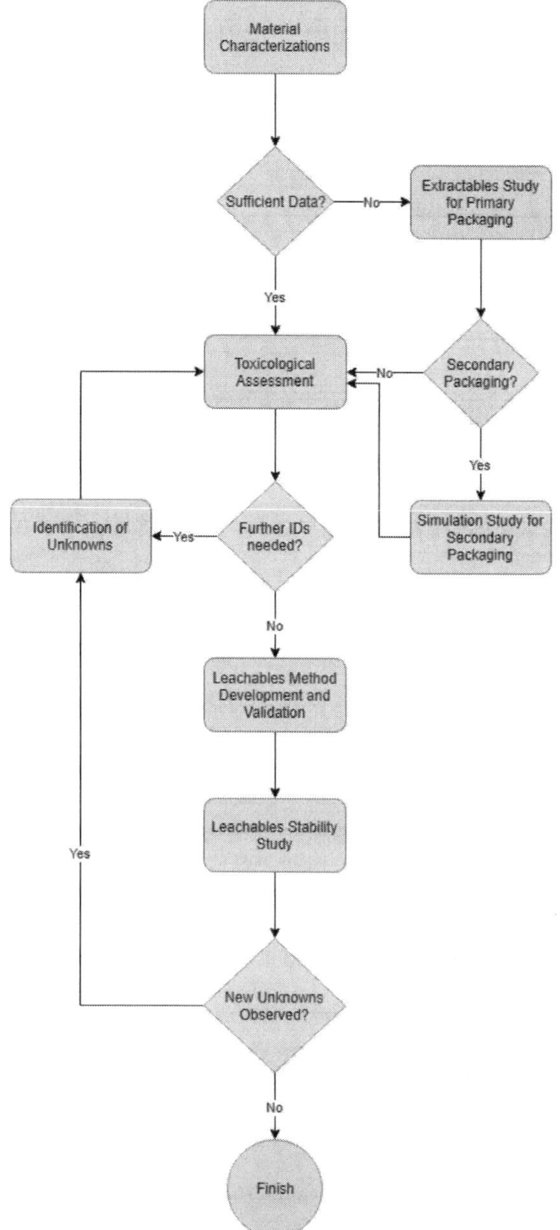

FIGURE 60.3 Typical E and L program flow chart.

developer to obtain the data experimentally. This data should include organic extractables covering NVOCs, SVOCs, and VOCs. Although this makes it appear that the organic compounds observed might be strictly classified as such, they typically are not, and there will always be overlap. For example, 2,6-di-tert-butyl-4-hydroxytoluene (BHT), a common antioxidant in rubbers, might be observed in a GC/MS analysis for SVOCs and an LC/MS analysis for NVOCs (and, less commonly, in a GC/MS headspace analysis for VOCs). For organic extractables, there should also be data that covers a range of solubilities and polarities of extraction solvents. Organization by solvent is typically how organic extractables are presented, so a drug developer can compare more easily with their own product and start to assess the likelihood of leaching of any individual extractable.

Inorganic extractables data are primarily focused on extractable elements. Although intended as a guideline for examining elemental impurities in drug products, ICH Q3D, "Guidelines for Elemental Impurities" (13) is a reliable reference that can be used to evaluate packaging components for extractable elements. If available, vendor-supplied data should include the Class 1 and 2A lists of elements at a minimum and preferably Class 2B and Class 3 as well. The Class 1 and 2A lists contain As, Cd, Hg, Pb, Co, V, and Ni—the absence of which must be shown in drug products per Q3D, regardless of route of administration and intentional addition. Therefore, it makes sense to confirm their absence in the packaging components that will contact the drug during its shelf life. Primary packaging as a source of elemental impurities, along with manufacturing processes, are specifically called out in the guideline, both of which can be addressed with appropriate vendor-supplied data or novel extractables studies. If Class 1 and 2A extractable elements are observed in an extractables study, then the developer must decide if they can be mitigated in the manufacturing process and/or during packaging and storage, or if it would be simpler to select an alternative component. It is notable, also, that following ICH Q3D can also fulfill the requirements stated in USP <232> and <233>.

To determine whether provided extractable data are sufficient, an assessment should be made with respect to the drug product. Drug product properties are listed in Table 60.3, along with the importance of each with regards to extractables data. These are the main factors in determining the sufficiency of any extractables data provided by suppliers.

Often, material characterization data from vendors are insufficient or not provided. In those cases, the drug developer must perform an extractables study. As the responsibility of obtaining this data resides with the developer, it is incumbent upon them to fill any gaps or perform a complete extractables study. Keeping in mind the factors shown in Table 60.3, the design of an extractables study for primary packaging should be based primarily on the specific drug product and its properties.

Secondary packaging should not be ignored. At minimum, it should be addressed as either being a risk or not. That determination is mainly dependent upon the materials of construction of the primary packaging system. Leaching from secondary packaging may not be considered a risk in many scenarios. Nonpermeable primary packaging containers such as glass vials, glass syringes, or aluminum canisters pose low risk for migration from secondary packaging. Many ODPs are packaged in semipermeable polymer primary packaging systems. Common polymers include polypropylene (PP), high-density polyethylene (HDPE), low-density polyethylene (LDPE), and linear low-density polyethylene (LLDPE). There are others, but in ODPs these are common primary packaging materials that are also used to deliver the drug to the eye. These primary packaging materials also account for the largest surface area through which migration may occur. Their material of construction is the primary factor in determining whether the drug product's secondary packaging should be examined as a source of leachables over the shelf life of the drug product.

TABLE 60.3
Drug Product Properties and Factors

Drug product property	Importance
Ingredients	The matrix of the drug product is the most important factor in solvent selection. Without extractables data generated with a solvent that resembles the drug product matrix, the data should be considered insufficient.
Storage conditions	The extraction conditions at which the extractables data were generated should be compared with the real storage conditions of the drug product. If the extraction conditions are not equivalent, the data should be considered insufficient.
Dosing	Dosing information is used to work backwards from an appropriate safety concern threshold (generally selected based on route of administration) to determine the analytical evaluation threshold (AET). If extractables data were collected above that theoretical AET, the data should be considered insufficient.

Manufacturing contact materials should also be considered as potential sources for leachable compounds. The current trend in biopharmaceutical processes is moving away from traditional stainless-steel reactors in favor of single-use systems. These systems contain multiple components of various polymers, all of which may contribute unwanted chemicals that may be present as leachables in the drug product. Many of the principles applied to extractables studies of primary packaging systems can be applied to manufacturing components that come into contact with the drug and/or its precursors. An examination of the manufacturing and filling processes should be done in order to identify the greatest leaching risks. Particular attention should be paid to tubing, filters, and bag systems. Typically, these have the highest chance of imparting leachables in the manufacturing process.

Up to this point, the majority of the discussion has been focused on extractables. In an E and L program, most of the effort in understanding the chemical properties of the packaging and delivery system components and how they may affect the drug is centered around extractables. The end goal of an extractables study is to distill the information into a list of chemicals that have been deemed likely to leach over the shelf life of the drug product. This has been mentioned before, and at this point in the chapter the reader most likely can see that much of the time spent on an E and L program is for the purpose of defining specific leachables, validating methods for those specific leachables, writing a leachables stability protocol, and then implementing the protocol created.

Leachable compounds and elements have no therapeutic benefits and may cause significant harm to patients or reduce the efficacy of the therapy. They should be limited where possible but cannot be completely removed. Selection of compounds and elements to be monitored as leachables is based on a variety of factors. These may include the propensity to migrate into the product (including solubility in the product), the effect on the active pharmaceutical ingredient (API) or other ingredients of the formulation, and the compound's toxicity to a patient.

For example, a large antioxidant such as pentaerythritol tetrakis (3,3,5-di-tert-butyl hydroxyphenyl propionate) more commonly known as Irganox 1010 may be extracted via organic solvent. It would not be soluble in a highly aqueous formulation and might not be selected as a target leachable. If the formulation contains a surfactant or an organic modifier it may increase the compounds solubility in the product formulation and hence, might be considered as a target leachable.

Simulation studies are often used to determine which extractable entities will migrate into drug formulations and thus become leachables. A typical approach to simulation studies was shown by Jenke et al. (14) using a model container closure system and model solvents. Each drug product and container closure combination will need to be assessed to select appropriate simulating solvents and accelerated storage conditions. If following Jenke et al. (14), the entire container closure system should be filled with the correct volume of the simulating solvent, sealed as intended, and then stored as intended with secondary packaging. The simulating solvent should be an exaggeration of the drug product matrix, and more than one can be selected, depending on the information that is desired from the study. It is recommended that the pH of the drug product solution be bracketed by multiple solvents (± 2.5 pH units), including the use of a solvent to mimic the organic contact of the formulation (e.g., 10%–20% v/v IPA/water). Common secondary packaging components include adhesive labels (with associated inks and lacquers), cardstock packaging, plastic wells (for prefilled syringes), and foil pouches. These should all be included in the simulation study. In order to "force" the migration of secondary packaging leachables toward the simulating solvents, the entire package should be stored for the duration of the study inside a sealed container. All of these parameters—the exaggerated simulating solvents, accelerated storage conditions, and storage in a sealed container—serve to exaggerate the real conditions and lend confidence in the data over time that the simulation does indeed represent the long-term storage, but over a shorter time period.

For greater confidence in being able to detect that migration indeed occurs through the primary packaging, positive controls should be used. Often the main source of secondary packaging migrants is from a label adhered directly to the primary packaging container. A known amount of a label extractable can be spiked directly onto the label before being adhered to the container, and this would serve as the positive control. This would not only confirm that migration occurs, but with early time points selected for testing, this could also show the rate at which the migration may occur.

In addition to a compound or element's ability to migrate into a drug formulation as a factor in selecting leachable targets, the compound's toxicity and its potential harm to a patient must be considered. The determination of an E and L potential harm to a patient requires further investigations. At this stage, the services of a trained toxicologist are required.

All of the extractables data collected through studies on the primary and secondary packaging should be assessed from a toxicological standpoint through the use of a toxicological risk assessment (TRA). A TRA in the context of E and L studies is a process that attempts to identify and estimate risks to a patient who may be exposed to leachable compounds while taking a drug product. TRAs typically consist of four primary steps: hazard identification, hazard characterization, exposure assessment, and risk characterization.

Hazard identification includes the identification of adverse health effects associated with exposure to a specific chemical. Hazard characterization includes the determination of the quantitative potency of any adverse effect of a chemical. Exposure assessment includes the measurement or prediction of the intake of a chemical in terms of magnitude, duration, and frequency of exposure. Risk characterization integrates the three to determine the probability of occurrence and severity of risk to human health from the leachables.

Often very little toxicological information can be obtained for specific extractable compounds. The Threshold of Toxicological Concern (TTC) is therefore a useful concept in risk assessments. The TTC defines a generic exposure value below which no appreciable risk to human health exists. The TTC was developed as a substitute for substance-specific information. The concept proposes that such a value can be identified for many chemicals, including those of unknown toxicity, when considering their chemical structures. (15)

The results of a fully executed extractable assessment, simulation studies, any preliminary leachable screening studies, and a TRA will aid the drug developer to select compounds and/or elements that should be monitored as leachables in stability studies.

Armed with a list of likely leachables (organic and inorganic), the first step is to employ a Good Manufacturing Practices (GMP)-compliant laboratory to develop and validate methods that are specific to the drug product and the individual leachables. How the leachables will be analyzed involves a balance among cost (number of methods to develop and validate), timeliness, and quality. If extractables are not selected as a target leachable, sufficient justification must be made with documentation in a TRA.

From a cost perspective, it may be desirable to develop a single method that includes all leachables of interest. This may not be practical as individual leachables may differ chemically and may not achieve the required quantitation limits that are needed for each leachable when bundled into a single method. This would, however, be a way to save on cost and time if sensitivity of one or more leachable can be sacrificed. On the other hand, developing a method for each individual leachable in order to optimize the specificity and sensitivity would prove to be quite costly. These two examples are the extremes, and the reality for any given leachables program will fall in between.

A drug developer must employ as many methods as necessary to quantify target leachables in the drug product. What is typically developed is one or two GC/FID and HPLC/DAD (or HPLC/UV) methods to cover organic leachables and one ICP/OES or ICP/MS method to cover elemental impurities, because these are generally available and provide sufficient quantitative data for any subsequent TRAs.

Validation of the quantitative leachables method should be accomplished according to industry accepted practices, criteria, and standards, such as <1225> "Validation of Compendial Procedures" (16) and <1664> "Assessment of Drug Product Leachables Associated with Pharmaceutical Packaging/Delivery Systems". A well-accepted validation guideline is ICH's *Validation of Analytical Procedures: Text and Methodology* Q2(R1) (1817). If this guideline is followed, a quantitative method for leachables should follow the scheme of "testing for impurities", which includes accuracy, precision, specificity, detection limit, quantitation limit, linearity, and range. A contracted GMP laboratory should write a protocol (or protocols) for method development and validation, for each method, and the protocol(s) should clearly state how each of these criteria is to be tested, as well as their pass/fail specifications. The protocol(s) should also include language about how failures will be addressed (i.e., laboratory investigations) and that the final result will be a method that is specific to the drug product, leachable(s), and client.

Once the methods are validated, a leachables stability protocol should be written that will define the storage conditions of the drug product and at which time points testing will be performed. If timed well, leachables stability can be performed at the same time as other stability testing (e.g., assay, impurities, and residual solvents). ICH's *Stability Testing of New Drug Substances and Products* Q1A(R2) (18) is a useful resource for defining stability conditions and, relevant to this chapter, states outright, "Stability testing should be conducted on the dosage form packaged in the container closure system proposed for marketing (including, as appropriate, any secondary packaging and container label)." Also described in this guideline are accelerated storage conditions, which are dependent upon the long-term (shelf life) storage conditions. For example, if the long-term storage condition is 25°C ± 2°C/60% relative humidity (RH) ± 5% RH, the appropriate accelerated storage condition would be 40°C ± 2°C/75% RH ± 5% RH. It is advantageous to have accelerated testing points for leachables stability. Theoretically, leachables are a subset of extractables and the rate of leaching is mainly dependent on the ambient temperature, described mathematically through the Arrhenius equation. Therefore, it makes sense that leachables results at an accelerated testing point could theoretically be representative of a later long-term testing point. This principle was described earlier when talking about simulation studies.

It is possible that a new "unknown" organic leachable may be observed at some point over the course of a stability study in either the long-term or accelerated conditions (or both). If encountered, it should be identified and assessed for toxicological impact, as described earlier. In order to mitigate the logistical complications and delays involved when this occurs, an increasing number of laboratories are offering unknown leachables screenings during leachables stability studies, to be performed alongside the validated, targeted leachables methods. The main advantage to this strategy is that if the screening is performed by the same laboratory that performed the original extractables study, the methodology will be similar, leading to a simpler overall process, and possibly simpler correlation with extractables data if a previous gap in data was identified. The main disadvantage to this strategy is the increased cost of running a leachables stability study. The additional screening portion could be performed at reduced time points defined in the stability study protocol.

At the completion of the leachable stability studies, the developer will have a complete extractable and leachable data set. The data should be compiled, summarized, and included in any regulatory submissions.

Overall, E and L studies are time consuming, resource intensive, and costly. They are often overlooked or delayed in favor of other development activities. In today's regulatory environment, some form of E and L studies will be required for most submissions. Developers that do not address E and L will face delays in approval.

By means of example, two case studies are presented following. These studies are meant to depict a common mistake of a developer regarding E and L along with a solution to the mistake, and an example of a developer who is committed to proactively address E and L in their submissions.

Case Study 1 involves the development of a sterile liquid drug product. The drug formulation was contained in a glass container with a rubber stopper and flip-top seal. In this example, the developer did not perform any E and L studies. The product was not approved, and the regulating body cited the failure to provide E and L data as one reason for this.

The developer consulted an outside laboratory to address the finding. The action that was implemented included both short-term and longer-term actions. The first action involved a short-term study to provide the regulating body with a timely response to the finding. This included using screening methods to evaluate multiple lots of end of shelf life drug products for volatile, semi-volatile, and non-volatile leachable compounds, as well as inorganic leachables. The drug product samples were prepared for analysis. In this case, the actual stoppers of the container closure system were prepared (extracted in a suitable solvent) and analyzed as positive controls alongside the drug product samples. Concentrations of compounds (and elements) that were observed in both the drug product samples and stopper extracts were determined and reported. The developer provided this data to the regulating body along with a commitment to perform a full E and L study.

The developer then performed a full controlled extraction study on both the rubber stopper and glass vial. Compounds were targeted for development of leachable methods that were then validated. A stability study was conducted and leachables were then monitored to the end of the product's shelf life.

The developer in this example encountered a lengthy delay in product approval. This type of situation happens all too often and is one that could have been prevented by the implementation of a timely extractable and leachable program.

Case Study 2 involves a developer who is committed to proactively address extractables and leachables in their submissions. This developer routinely used five commercially available rubber stoppers to manufacture drug products. It had performed extraction studies for each their drug products spanning several years. After many product-specific development projects in which similar studies were performed on a repeated basis, the developer implemented an alternate strategy.

The new strategy involved performing an extraction study on each of the five stoppers using a battery of common solvents spanning a range of aqueous pHs and strengths of organic solvents. Once completed, the developer was armed with data that would allow extractables data to be used in the selection of a primary closure system. Once the primary closure was selected, the developer would perform simulation studies using the specific drug product formulations. Armed with both the nonspecific data as well as the drug product-specific data, selection of leachable targets was simplified and allowed for the development and validation of leachable methods earlier in the product development process.

60.3 CONCLUSION

E and L studies are a vital aspect of the drug development process. As stated previously, all dosage forms will require some form of assessment ranging from OINDP and parenteral products requiring full assessments, to solid oral products in which a paper-only assessment may suffice. E and L studies are time consuming, resource intensive, and costly; however, the cost of not performing them may be delayed or failed product approval.

E and L studies can be intricate and complex, the details of which cannot be fully addressed in a single chapter. An excellent resource is *Compatibility of Pharmaceutical Products and Contact Materials, Safety Considerations Associated with Extractables and Leachables* by Dennis Jenke (19).

NOTE

1. Tylenol®, Motrin®, Benadryl® and Zyrtec® are registered trademarks of Johnson & Johnson.

REFERENCES

1. Schroeder A. Leachables and Extractables in OINDP: An FDA Perspective. Presented at the PQRI Leachables and Extractables Workshop, December 5–6, 2005, Bethesda, MD.
2. FDA Compliance Policy Guide 500.450 (CPG 7117.11). www.fda.gov/ICECI/ComplianceManuals/CompliancePolicyGuidanceManual/ucm074418.htm.
3. American Academy of Family Physicians. McNeil Recalling HundredsofLotsofContaminatedOTCMedications,January18, 2010. www.aafp.org/news/health-of-the-public/20100118mcneilrecall.html.
4. www.cbsnews.com/news/magic-number-tylenol-recall-cost-jj-665m-in-lost-sales/.
5. Guidance for Industry: Container Closure Systems for Packaging Human Drugs and Biologics. United States Department of Health and Human Services Food and Drug Administration, May 1999.
6. *United States Pharmacopeia and National Formulary* (USP 40-NF 35). Vol. 1. Rockville, MD: United States Pharmacopeial Convention, 2017: 2020–2035.
7. *United States Pharmacopeia and National Formulary* (USP 40-NF 35). Vol. 1. Rockville, MD: United States Pharmacopeial Convention, 2017: 2035–2047.
8. *United States Pharmacopeia and National Formulary* (USP 40-NF 35). Vol. 1. Rockville, MD: United States Pharmacopeial Convention, 2017: 554–558.
9. *United States Pharmacopeia and National Formulary* (USP 40-NF 35). Vol. 1. Rockville, MD: United States Pharmacopeial Convention, 2017: 295–297.
10. *United States Pharmacopeia and National Formulary* (USP 40-NF 35). Vol. 1. Rockville, MD: United States Pharmacopeial Convention, 2017: 298–301.
11. *United States Pharmacopeia and National Formulary* (USP 40-NF 35). Vol. 1. Rockville, MD: United States Pharmacopeial Convention, 2017: 169–172.
12. *United States Pharmacopeia and National Formulary* (USP 40-NF 35). Vol. 1. Rockville, MD: United States Pharmacopeial Convention, 2017: 172–177.
13. *ICH Harmonised Guideline: Guideline for Elemental Impurities Q3D*. Current Step 4 version, December 16, 2014. www.ich.org/fileadmin/Public_Web_Site/ICH_ Products/Guidelines/Quality/Q3D/Q3D_Step_4.pdf (Accessed January 14, 2019).
14. Jenke D, Egert T, Hendricker A, et al. Simulated Leaching (Migration) Study for a Model Container-Closure System Applicable to Parenteral and Ophthalmic Drug Products (PODPs). *PDA Journal of Pharmaceutical Science and Technology* 2016. https://journal.pda.org (Accessed January 16, 2019).
15. Stone T. An Overview of Risk Assessment Strategies for Extractables and Leachables. *BioPharm International* 25(1).
16. *United States Pharmacopeia and National Formulary* (USP 40-NF 35). Vol. 1. Rockville, MD: United States Pharmacopeial Convention, 2017: 1780–1785.

17. *ICH Harmonised Tripartite Guideline: Validation of Analytical Procedures: Text and Methodology Q2(R1)*. Current Step 4 version, October 27 1994. www.ich.org/fileadmin/Public_Web_Site/ICH_Products/Guidelines/Quality/Q2_R1/Step4/Q2_R1_Guideline.pdf (accessed January 15, 2019).
18. *ICH Harmonised Tripartite Guideline: Stability Testing of New Drug Substances and Products Q1A(R2)*. Current Step 4 version, February 6, 2003. www.ich.org/fileadmin/Public_Web_Site/ICH_Products/Guidelines/Quality/Q1A_R2/Step4/Q1A_R2_Guideline.pdf (accessed January 15, 2019).
19. Jenke D. *Compatibility of Pharmaceutical Products and Contact Materials Safety Considerations Associated with Extractables and Leachables*. Wiley, 2009.

61 Evolution and Implementation of Validation in the United States

James Agalloco and Phil DeSantis

CONTENTS

61.1 INTRODUCTION

Validation has been a widely discussed subject in the health care industry since the mid-1970s. Since that time, it has been applied in a variety of areas such that its scope can appear endless. Validation was once an exercise that was associated solely with sterilization processes but is now discussed in relation to virtually every aspect of health care manufacturing and distribution. To better understand how it is practiced today, it is useful to understand its evolution and how that shapes how an individual firm addresses its execution. This chapter will review the history of validation from its origins to provide greater insight into present day practice.

The dates utilized in this chronology are approximate. The introduction of practices described in this chapter varies with some firms leading the way, whereas implementation at others lagged significantly.

61.2 1972–1980—BEGINNINGS

The origin of validation in the United States can be traced to sterility failures with terminally sterilized large volume parenterals (LVPs) in the early 1970s. At roughly the same time, unrelated failures in the United Kingdom with similar products led to their first validation activities.[1,2] Food and Drug Administration (FDA) investigators and industry experts agreed on the need for independent verification of the sterilization process as a means of establishing unequivocal process reliability. This activity was termed 'validation' and became a required activity for all terminal sterilization processes. The small volume parenteral (SVP) industry, which manufactured the majority of their products by aseptic processing, recognized that despite their use of more robust overkill cycles for containers, closures, and filling parts, that they too would be expected to validate their sterilization processes. This recognition by the more numerous SVP manufacturers largely resulted in their mimicking the more evolved and necessarily more conservative practices of the LVP firms that had pioneered sterilization validation. In 1976, the FDA issued its proposed Good Manufacturing Practices for Large Volume Parenterals (the never approved 21 CFR 212 regulation), which defined their overall expectations. By that time, sterilization validation was being performed in a near identical manner regardless of whether the sterilization process was used for LVP finished products or SVPs in-process materials.[3] Rather than develop validation practices specific for overkill sterilization, nearly all early SVP validation protocols were identical to those used for LVPs, with the simple logic that if it was sufficient for terminal sterilization, it would be more than sufficient for overkill sterilization of parts and other materials. The implications of this expediency have continued to handicap steam sterilization validation to this day.[4]

These initial efforts to validate sterilization processes utilized monitoring equipment and approaches that are certainly crude by today's standards. Precise measurement of temperature was difficult, and resolution was limited to what was discernible on a multipoint chart recorder. F_0 values could only be obtained by hand calculation using logarithmic tables (calculators, data loggers, and personal computers were not yet available). Protocols, procedures, reports, and related documents were produced with rudimentary word processors or ordinary typewriters. The inclusion of drawings and tabulated data was both limited and crudely executed.

The sterilizers being utilized for the process were those that had been utilized for years. Confirmation of lethality via biological destruction and attainment of proper temperatures was considered sufficient proof of its acceptability. There was no consideration given to equipment qualification as it is understood today. The validation of other sterilization processes (e.g., ethylene oxide) was understood to require extensions and adaptations of what was learned performing steam sterilization validation.

Having its roots in sterilization, the early definition provided by FDA spokesmen, Ted Byers and Bud Loftus was, "Process validation is establishing documented evidence that a process

DOI: 10.1201/9781003163138-61

does what it purports to do."[5] This original emphasis FDA placed on validation as being an event or activity persists to this day and has been a focus of regulatory compliance ever since.

At its onset, validation of sterilization processes adhered closely to the principles of the scientific method as taught in high school. Given a premise, an experimental design is established for which obtaining a predefined expected result would support the original premise. After completion of the experiment, the data is reviewed to establish whether the initial premise is supportable based upon the experimental evidence. In the practice of validation, the protocol establishes the premise including the predefined acceptance criteria, and the report documents the results of the evaluation.

As time went on and sterilization validation became established, focus shifted to other processes. As the 1980s approached, validation of product formulation processes emerged, and content uniformity was recognized as a major consideration for product efficacy. Product validation practices varied from firm to firm as there were no absolute requirements as had been readily established for sterilization. Many initial efforts included some aspect of retrospective validation or control charting, given the absence of any meaningful prospective validation for all but the very newest products. These too were hampered by the limited availability of rapid statistical tools.

Validation maintenance or re-validation was given little consideration during this early period; as there was such a backlog of required validations, consideration of repeat efforts was largely deferred. Rudimentary change control was instituted; however, this too was constrained by the absence of software tools for information management.

In 1978, the Parenteral Drug Association (PDA) published the first of its numerous technical monographs that have helped shape sterilization validation.[6] These documents provided consensus practices for a generation of validation specialists, albeit still largely tied to the original terminal sterilization methods. Nevertheless, these have proved valuable as a guide to firms worldwide, especially in validation efforts for parenterals.

The future of validation in the industry remained unclear. Initially, many firms assembled task forces from existing staff members with varying backgrounds and experience to perform their initial validation studies. It was widely believed that validation was a one-time exercise and that once completed it could be used to support operations over an extended period. The expectation was that task forces could then be disbanded, and the members would return to their prior roles. One of the authors was asked in early 1980, "Why are you accepting a position in validation, in a few years you'll have it all done and you'll need to find a new job!" Around the same time, a few visionary firms understood that there would be a need to maintain validation efforts over time, and the first permanent validation departments were formed.

61.3 1980–1990—EVOLUTION

In 1980, one of the authors (then with E.R. Squibb and Sons) with others formed a validation discussion group to discuss common concerns. As the heads of validation at our respective sites, we wanted confirmation that our internal ideas and approaches for validation were consistent with those practiced elsewhere. We were fearful of overlooking some important nuance that others might adopt and being somehow deficient in our efforts. We were all enthusiastic about validation and expected that our employers would realize tangible benefits (Table 61.1). Through our discussions, we hoped to improve our understanding, streamline our approaches, and identify demonstrable advantages that would facilitate internal support for our efforts.[7] In actuality, we were making it up as we went along, there being no precedent to serve as a model!

TABLE 61.1

Benefits of Validation

- Increased Throughput\
- Reduction in Rejections, Reworks, Retests, Resamples
- Reduction in Utility Costs
- Avoidance of capital expenditures
- Fewer Complaints about process related failures
- Reduced testing - in-process and finished goods
- More rapid and accurate investigations into process upsets
- More rapid and reliable start-up of new equipment
- Easier scale-up from development work
- Easier maintenance of the equipment
- Improved employee awareness of processes
- More rapid automation

The first facilities designed and constructed with concern for validation requirements began to appear around 1980. The firms constructing these plants endeavored to provide facilities, equipment, and utility systems that were 'validatable'. A direct consequence of this was the emergence of equipment qualification. Because undocumented change could alter performance, design and construction activities were closely documented to ensure that the baseline configuration was identified. Documenting and maintaining equipment and systems in a known state prior to validation facilitates greater confidence in continued performance in conformance with the validation exercise. For existing plants, qualification at that time was predominantly documenting what was already in place, because nearly all of the equipment and systems in use with validated processes had not been subjected to a formalized qualification effort.

A shift in thinking occurred in parallel with the implementation of equipment qualification. Facilities, equipment, and systems were qualified as capable for use in the production of a variety of products/processes, so their performance needed to be established independent of any individual use. In contrast, products and the processes used to make them must be considered in combination.[8] The separation of overall activities into 'qualification' and 'validation' eliminated the artificial redundancy created when equipment is tied to a specific product/process. The realization that products and processes are inexorably linked facilitated the separation of product specific and product non-specific activities and the execution advantages that afforded.

Around the same time, change control was increasingly recognized as vital to ensuring that systems, processes, and products once validated would remain so over time. Applying change control more broadly led directly to the first considerations of a 'life cycle' for validation.[9,10] First acknowledged as useful with computerized systems, the utility of the life-cycle concept for other systems became evident.[11] This concept provided for consideration of validation requirements at the start of the process, product, or system design, confirmation of the desired characteristics during its initial validation, and formal review of the impact of changes over the life of the system. Best understood as a 'state' function, 'validation' was not something merely to be attained once, but a 'state' to be maintained over the product's time in the marketplace or the equipment/system's operational life.

As validation evolved beyond sterilization, it was no longer the exclusive province of the microbiologists and engineers. Formulation and process development scientists, analytical experts, maintenance, production, and quality managers were drawn in as necessary contributors to a firm's validation activities. Simply put, validation became a small part of everyone's job. Development was increasingly charged with defining a process that would prove reliable on a commercial scale. Anything less was unacceptable. The number of samples for analysis taken from development, scale-up, and commercial lots increased substantially in efforts to better define and confirm the important quality attributes of the product. Engineering and maintenance departments were charged with designing, fabricating, and maintaining facilities, utility systems, and equipment that could be more readily maintained in the initially qualified state. Production was responsible for collection of samples, quality control for their analysis, and quality assurance units provided compliance oversight.

The 1980s also witnessed the emergence of the validation service companies. These began as independent organizations providing a variety of validation services to firms lacking sufficient resources or experience to execute qualification and validation efforts without added support. Manufacturing firms faced with the qualification/validation of new facilities were among the first users of outside assistance, as they sought to manage the heavy workload of projects associated with a new facility start-up.

Other innovations aimed at facilitating validation began to emerge. A useful tool for thermal sterilization validation at that time was the Kaye Digistrip, which quickly became preferred for temperature recording, and, equally important, F_0 determination in real time. The personal computer (PC) became available early in the decade, offering word processing to virtually everyone. In addition, spreadsheets, databases, and enhanced graphics were now available for improved documentation. This had a profound impact on validation as protocols and reports were not only more easily produced, they also became far more detailed (and perhaps regrettably, substantially larger than before). Although the Digistrip was without question a vast improvement over the pen and ink recorders of the early 1970s, the PC could be considered a mixed blessing. Documentation expectations increased several-fold, and although the size and attractiveness of reports certainly increased, it is unclear whether the scientific basis for validation activities improved commensurably, if at all.

In parallel with the emergence of the PC in the company offices, its cousin—the microprocessor—began to appear across the factory floor. Where processing equipment had once been controlled by either the operator or relatively simple electromechanical devices, programmable logic controllers (PLCs) and other software driven systems were experiencing expanded use. This advance came with two big negatives: technologies beyond the understanding of many end users were placed into service, joined by a perhaps excessive fear of the possible negative consequences of using a computer for Current Good Manufacturing Practices (CGMPs) activities. Regardless, industry users of industrial controllers quickly recognized that these systems required both qualification and validation. The Pharmaceutical Manufacturers Association (PMA; now the Pharmaceutical Research and Manufacturers of America or PhRMA) formed a Computer Systems Validation Committee in 1983 that delivered an industry perspective on the subject in 1986.[12,13] A core aspect of this effort was the adaptation of software development life cycle practices to computerized systems in the pharmaceutical industry. Computerized systems validation and the now prevalent Good Automated Manufacturing Practices (GAMP) can be traced back to these initial efforts.[14]

In 1983, the FDA published it first draft guidance on Process Validation. This early draft included a comprehensive update of definitions and regulatory expectations and replaced a patchwork of earlier documents and inferences defining validation that had loosely evolved from FDA presentations, inspections, and other sources. The initial draft of this guidance mandated triplicate validation studies in support of validation. An industry suggestion to alter this to a "statistically significant number of batches" to help those firms that validated only in duplicate was quickly withdrawn, when it was recognized that this would dramatically increase the workload of already beleaguered validation departments. The final guidance was issued in 1987 and included a requirement that process validation be performed in triplicate.[15]

The decade of the 1980s also witnessed the beginning of the biotechnology industry as something distinct and seemingly different from traditional pharmaceuticals. With this came consideration of validation requirements in some unique new areas. How to validate fermentation, chromatography, ultrafiltration, and other biotech processes were mainstream subjects among the biotechnology industry. The technical concerns may have been different, but the means to accomplish them were derived from the practices used for pharmaceutical dosage forms. Concurrently, there was increasing understanding that validation was a useful tool in establishing the effectiveness of other dosage form-related processes for both biotech and traditional products: inspection, packaging, and aseptic processing.

Positive experiences with validation at the site level led several US firms to address it proactively globally before local regulatory authorities considered its utility. The author participated in several major validation projects outside the US, where the goal was full adherence to US practices to ensure product quality met corporate standards. Efforts in this regard were greatly accelerated in some multinational firms after Union Carbide's Bhopal incident.[16] Corporate validation standards were established to ensure product quality on a global basis. In this environment, it was common to speak of "a single standard of high quality worldwide". Success with the life cycle approach with computer systems led to its successful adaptation for process validation even where computers played only a minor role in their production (see Appendix).[17]

Merger mania struck the global pharmaceutical industry late in the 1980s. Prior to that time, there were few dominant firms, the industry being populated with many firms of roughly comparable size and relatively few firms having more than a 2%–3% market share worldwide. This changed rapidly as firms combined their resources in an effort to achieve critical mass in research, greater presence in overseas markets, and increased profitability. One by-product of this was the realization that there was substantial excess manufacturing capacity in the newly combined firms, which led to plant closures, divestitures, outsourcing, product transfers, and a vastly different operating climate than previously. Unfortunately, this also led to a major displacement of experienced personnel as firms offered separation packages in efforts to reduce their head count. The availability of numerous contract providers led many firms to reduce the size of their validation departments, coincident with the need to relocate products to other facilities. The use of contract validation services increased consequently with their ranks often filled by the very individuals who had recently been laid off.

61.4 1990–2000—ADOLESCENCE

Late in the previous decade, significant problems with the quality of generic drugs emerged and brought forth a change in FDA inspection practices that had a major impact on process validation activities. The FDA Pre-Approval Inspection program was instituted to confirm that sites possessed the ability to manufacture their products in accordance with their New Drug Application (NDA), Abbreviated New Drug Application (ANDA), or Biologic License Application (BLA) submissions.[18] For most formulation types, this required the completion of full-scale validation batches. Because of their shorter dating, sterile products were exempted from the at-scale validation requirement but still had to demonstrate adequacy of the sterilization validation prior to approval of the site for their production.

In the post-merger climate, the combination of staff reductions and product relocations resulted in an increase in contractor-supported validation activities, many of which were performed by ex-employees of the same or other firms. Whether this was cost- or time-effective was not a major consideration provided head counts at the pharmaceutical firm were reduced. This had unfortunate downsides, as the core competency for validation shifted largely to outside of the organizations that actually required it. Suppliers of validation services varied substantially in size and sophistication, and the heavy reliance on their expertise had a profound impact. In an effort to streamline project execution, standardized protocol templates became the rule rather than the exception. Protocols originally written for one design, process, or application were modified slightly and applied in different situations. The net result was protocols of excessive size, full of 'boilerplate' requirements, poorly focused and lacking in clarity. That these egregious practices were deemed acceptable was commented upon by numerous experts.[19,20] Validation services for facility, equipment, and utility systems were increasingly provided by organizations affiliated with the large engineering companies. Their goal was to integrate validation support, predominantly in the installation/operation qualification stages of a project, with the engineering and design effort. The intent was certainly positive but also had some unintended adverse consequences. In some instances, firms believed that their internal efforts should mimic the focus of the service providers. Massive equipment qualification documents were developed, documenting virtually every nuance of the equipment or system. Although these may or may not have been excessive, the larger concern was the loss of attention on the core process or product for which the equipment was intended. One smaller service provider was so bold as to state, "Our efforts will generate enough paper to bury the average inspector!" All of this led to increasingly bloated qualification efforts with little real support for what should always have been the critical concern: the quality of the end product. The emphasis on the qualification of facilities, equipment, and systems was further exacerbated as some firms neglected to maintain adequate internal capabilities for validation of their own production processes.

To ensure that firms provided sufficient and consistent information about the sterility assurance practices in new product submissions, the FDA defined its expectations for sterilization validation in a comprehensive guidance document related to CMC submission requirements for sterile products.[21] Based predominantly upon terminally sterilized products, this guidance presented the industry with a level of clarity that had not been previously available. This had the distinct advantage of defining what was specifically required as validation activities in support of sterilization and sterility assurance. It is noteworthy that the excesses observed in so many industry activities (see the preceding paragraph) were not a part of the FDA's guidance. It became evident that the industry had been somewhat misguided in its growing emphasis on equipment-related concerns, as opposed to far more critical processes, like sterilization.

There were other missteps. Early in our validation discussion group sessions we had talked about validation of cleaning processes. It had been apparent to us that cleaning of equipment was a process of equal importance to product manufacturing. Any hint of cross contamination of one product with another would result in rejection of the possibly contaminated materials. The challenge was, of course, to define an acceptable level of contamination. In the discussion group, we had discussed our strategies and concerns, but we would never openly discuss our cleaning procedures, cleaning limits, or anything substantive about our efforts. The validation of cleaning and discussion of it were essentially a 'taboo' subject across the industry. The dam burst in late 1991, and shortly thereafter one of the authors was invited to speak at an industry meeting on what had heretofore been an untouchable subject of cleaning validation. PDA began work on cleaning validation soon afterwards and later delivered the first industry consensus documents on the subject.[22,23]

The success of the biotechnology industry in validating their processes impacted the traditional small molecule pharmaceutical sector as well. If a 20-step biological process beginning at the seed bank could be validated, surely a small molecule synthesis process could be addressed in similar fashion. The active pharmaceutical ingredient (API) segment of the industry, which had long subscribed to the notion that they were 'too different' from dosage form manufacturing, began to feel increasing pressure to implement validation as a routine requirement for their operations. The Q7A initiative undertaken by the International Conference on Harmonization ultimately laid to rest any further objections to validation in small molecule synthesis operations.[24]

61.5 2000–2010 MATURITY

The dominant theme of the first decade of the 21st century within the global industry became outsourcing. Copying the business model of the electronics industry, pharmaceutical firms pursued outsourcing as a sustainable business model either as a user or provider of services, and sometimes as both. Although contract manufacturers had always existed, often providing specialty services such as BFS (blow-fill-seal), pre-filled syringes, and soft gelatin capsules; the mergers and divestitures of the prior decade created a number of firms whose primary business model is to produce drugs on a contract basis. Embraced initially by biotechnology firms seeking to focus on their core technologies, contract manufacturers enable the establishment of virtual pharmaceutical firms in which all of the production, testing, and distribution is performed on a contract basis. The continued rapid growth of outsourcing has adversely impacted validation practices. There is little doubt that it stresses the communication channels of the involved organizations, and that alone may have a negative effect on validation practice. This is further complicated by confidentiality concerns.

In the same time period, the expanding implementation of computerized systems in the pharmaceutical industry followed trends in other industries. A significant shock to this evolution was the publication by the FDA of 21 CFR 11 *Electronic Record, Electronic Signature* which threatened to paralyze further progress.[25] The focus of concern was whether electronic records, which computerized systems produce in abundance, could be considered sufficiently secure to avoid falsification, and thus endanger patients. The downside potential of this was perceived to be so extreme that projects involving process automation and computerization of CGMP data were stopped until full understanding of compliance concerns and their resolution was accomplished. The data integrity concerns associated with this persist to the current day, and computer systems validation has an added burden of proving system security to an unprecedented level.

Perhaps the decade's most profound impact on validation was that brought about by the FDA's changing perspective on product quality. Ajaz Hussain, who was then the FDA's Deputy Director, Office of Pharmaceutical Science, CDER, was perhaps the first to point out the lack of underlying science in many pharmaceutical processes. Dr. Hussain identified what he believed was a lack of real process understanding on the part of numerous firms, i.e., an apparent overreliance on end-product testing (and thus even validation itself) was supporting inadequately understood processes. He advocated a reemphasis on robust product development founded on sound science, with appropriately defined specifications and process parameters. The appropriateness of this recommendation cannot be faulted. His concerns prompted Dr. Hussain to propose Process Analytical Technology (PAT) as real-time affirmation of process acceptability, reducing the need for in-process and end-product testing.[26] The utility of PAT as a universal practice is uncertain. There are instances where it can offer clear advantages over classical drug manufacturing approaches; yet there are counter arguments that suggest it may be of little benefit in other situations. The passage of time will ultimately reveal the utility of PAT within the pharmaceutical industry. A more important and perhaps unanticipated outcome was the growing awareness that extensive process knowledge was central to assuring a consistent process and thus the reliable delivery of a quality product.

In 2004, the FDA issued its intentions to adopt a risk-based compliance initiative to streamline the development, approval, review, and preparation of pharmaceutical products.[27] This initiative remains in place, but it is unclear how well it has translated into a better regulatory process and eventually more robust production methods. Numerous attempts have been made to implement risk-based thinking to aid firms in becoming more compliant while simultaneously increasing their efficiency. Early applications of risk-based compliance have included equipment qualification, cleaning, environmental monitoring, inspection, and others. If these efforts are effective, some of the poorly defined and egregious qualification and validation efforts touted in the 1990s could be rendered obsolete.

61.6 2010–2020—REINVENTION

In 2008, the FDA issued its first draft of a radically revised guidance document on process validation.[28] This document, which was finalized in 2011, and a similar EMA version reflect contemporary regulatory thinking on process validation. The salient element of both documents is affirmation of the utility of the "Life Cycle Concept for Process Validation" as the preferred practice.[13,29] Implementation of the life cycle concept requires a dual approach: near term implementation of a continuous validation assessment for already marketed products (Phase 3) and longer-term transition to rigorous development, scale-up, and validation of new products (Phases 1 and 2). Starting with Phase 3 makes perfect sense as commercial products must be of the required quality. As the development of new products/processes takes many years, the effects of Phase 1 and 2 effort is delayed. To those familiar with the best practices of validation as early as the late 1980s, this is nothing new, but merely an acknowledgement of what many practiced more than two decades earlier.

The outsourcing trend witnessed explosive growth in this decade. Contract services of all types, but especially development and manufacturing, grew dramatically. Firms, large and small, sought the help of contractors to develop and supply pharmaceuticals. The service providers varied substantially in scale as well as aiding their customers in a myriad of ways. This made the execution of validation an exercise in effective communication more than ever before. Validation activities that were once carried out within a single organization now required simultaneous interaction across several. The effort was often further complicated by language and distance, increasing the potential for misinterpretation and error.

Later in the decade, compliance problems at various firms were linked to discrepancies in analytical and production data and resulted in global regulatory initiatives focusing on data integrity.[30,31] Although outwardly similar to the earlier concern for the veracity of electronic records under 21 CFR 211, it was actually a more egregious problem—whether the data as initially recorded reflected reality. Considered more broadly, the problem may not be so much data integrity, but the personal integrity of the individuals responsible for the collection, recording, and preservation of required records.

61.7 CONCLUSION

If there is one constant in the history of validation, it would have to be the continual evolution of perspective, practice, approach, and emphasis. In many instances, the evolution of practice has improved the certainty of our knowledge and thus the quality of our products/processes. The advantages of the datalogger for use in thermal studies compared with chart recorders, and logarithmic calculation of lethality are obvious. Whether the same can be said about 85-page installation qualification protocols for a laboratory incubator is certainly highly questionable. The advent of validation practices derived from a risk analysis perspective offers the possibility to revisit the entire subject. That it will result in another wave of change in validation is perhaps certain and, considering the history of validation, that is perhaps the only constant.

In the authors' opinion, far too much of present-day validation activities in the US have been little more than rote adherence to ever-increasing expectations. The sense that if a little validation is good then more must be better has gotten out of control. A return to the demonstrated need for scientific evidence prior to imposition of a new requirement is essential. If the FDA's risk analysis initiative fosters rethinking of validation expectations as well as the fundamental quality goals for a product/process, then future improvement in validation practice will be possible. An approach to validation that falls somewhere between the perhaps overly simplistic, yet effective protocols of the 1980s and the contemporary bloated validation efforts seems appropriate.

NOTES

1. Chapman K, A History of Validation in the United States—Part I, *Pharmaceutical Technology* 1991; 15(10).
2. Matthews B, The Davenport Incident, The Clothier Report, and Related Matters—30 Years On. *Journal of Parenteral Science & Technology* 2002; 56(3): 137–149, May-June.
3. FDA, Proposed Current Good Manufacturing Practices in the Manufacture, Processing, Packing or Holding of Large Volume Parenterals, Federal Register 22202–22219, June 1, 1976, Rescinded—December 31, 1993.
4. Agalloco J, A Tale of Two Sterilizers. *PDA Journal of Pharmaceutical Science and Technology* 2020; 74(1): 162–169.
5. PMA, Process Validation Concepts for Drug Products, PMA Validation Advisory Committee, 1985.
6. PDA, Technical Report #1, Validation of Steam Sterilization Cycles, Philadelphia, 1978.
7. The group is still in existence today serving as an informal network to exchange ideas more than 40 years later.
8. FDA, Guidance for Industry, Quality Systems Approach to Pharmaceutical CGMP Regulations, 2006.
9. Agalloco J, Computer Systems Validation—Staying Current: Change Control. *Pharmaceutical Technology* 1990; 14(1): 45–50.
10. Harris J, Agalloco J, et al., Validation Concepts for Computer Systems used in the Manufacture of Drug Products. In *Proceedings: Concepts and Principles for the Validation of Computer Systems in the Manufacture and Control of Drug Products*, Pharmaceutical Manufacturers Association, 1986.
11. Agalloco J, Process Validation and Computer Systems Validation—Similarities and Differences. In *Proceedings: Concepts and Principles for the Validation of Computer Systems in the Manufacture and Control of Drug Products*, Pharmaceutical Manufacturers Association, 1986.
12. Chapman K, A History of Validation in the United States—Part II, In *Pharmaceutical Technology*, Validation of Computer- Related Systems 1991; 15(11).
13. PMA Computer Systems Validation Committee, Validation Concepts for Computer Systems used in the Manufacture of Drug Products. *Pharmaceutical Technology* 1986; 10(5): 24–34.
14. Smith P, 20th Anniversary Special Feature: Validation and Qualification. *Pharmaceutical Technology Europe* 2008; 20 (2).
15. FDA, *Guideline on General Principles of Process Validation*, 1987.
16. Eckerman I, *The Bhopal Saga—Causes and Consequences of the World's Largest Industrial Disaster.*India: Universities Press, 2005.
17. Agalloco J, The Validation Life Cycle. *Journal of Parenteral Science and Technology* 1993; 47(3): 142–147.
18. FDA, CPG Sec. 490.100 Process Validation Requirements for Drug Products and Active Pharmaceutical Ingredients Subject to Pre-Market Approval, 1993.
19. Sharp J, Letter to the Editor. *J. Parenteral Science & Technology* 1993; 53(1).
20. Agalloco J, Validation: A New Perspective. Chapter in *Compliance Handbook for Pharmaceuticals, Medical Devices and Biologics*. Medina C, ed. New York: Marcel-Dekker, 2004.
21. FDA, *Guidance for Industry* for the Submission Documentation for Sterilization Process Validation in Applications for Human and Veterinary Drug Products, November 1994.
22. PDA, *Cleaning and Cleaning Validation: A Biotechnology Perspective*. Bethesda, MD, 1996.
23. PDA, Points to Consider for Cleaning Validation, PDA Technical Report #29, *PDA Journal of Pharmaceutical Science and Technology* 1999; 53(1, supplement).
24. ICH, *Guidance for Industry* Q7A Good Manufacturing Practice Guidance for Active Pharmaceutical Ingredients, 2001.
25. FDA, *Guidance for Industry* Part 11, Electronic Records; Electronic Signatures—Scope and Application, 1997.
26. FDA, *Guidance for Industry*: PAT—A Framework for Innovative Pharmaceutical Development, Manufacturing, and Quality Assurance, 2004.
27. FDA, Pharmaceutical CGMPs for the 21st Century—A Risk-Based Approach—Final Report, 2004.
28. FDA, *Guidance for Industry*—Process Validation: General Principles and Practices, 2011.
29. EMA, Concept Paper on the Revision of the Guideline on Process Validation, EMA/CHMP/CVMP/QWP/809114/2009, 2010.
30. FDA, *Guidance for Industry* Data Integrity and Compliance with Drug CGMP: Questions and Answers, 2018.
31. Eban K, *Bottle of Lies: The Inside Story of the Generic Drug Boom*. New York: Ecco/HarperCollins, 2019.

Appendix

Evolution of the Life Cycle Concept

The body of this chapter describes the evolution of validation as an essential activity within the global health care industry. Validation as first conceived had little structure and there was no consideration of how it was to be implemented. As early practitioners, there were no historical parallels or comparable activities to guide us. We were literally making it up as we went along. Considering the number and scope of unvalidated processes in every organization at that time, attention was focused on getting things validated rather than developing an overall strategy for execution.

Working in a newly constituted validation department in 1980, we were charged with the validation of all activities within the manufacturing environment including sterilization, products/processes, and the first extensively automated systems on site. As several of us engineers had prior exposure to computers and computerized systems, we were familiar with the debugging of software and hardware systems that was often necessary to get them to operate as expected. Our facility intended to use programmable logic controllers for the manufacturing of various injectable products in a to-be-constructed facility. Expecting the need to debug these systems before committing to automated production, we planned a series of development and testing stages leading towards full implementation (Figure 61.1).

Our implementation plan was not linear, as we recognized the need for software/hardware revision and further testing, so we included several recycle loops. The planned approach was followed throughout the project largely without alteration. As the project proceeded several of us became involved with the PhRMA (then PMA) Computerized Systems Validation Committee (CSVC) to share our experience and glean insights from others. At the first PMA Symposium on Computer Validation in 1985, a presentation was given on the Software Development Life Cycle. Next week at our lunch table, the 'aha' moment occurred. We recognized that the life cycle approach should be extended to the hardware components of a system as well. At the next CSVC meeting, we shared our insight, which ultimately led to the adoption of the concept for computerized systems by the CSVC (Figure 61.2).

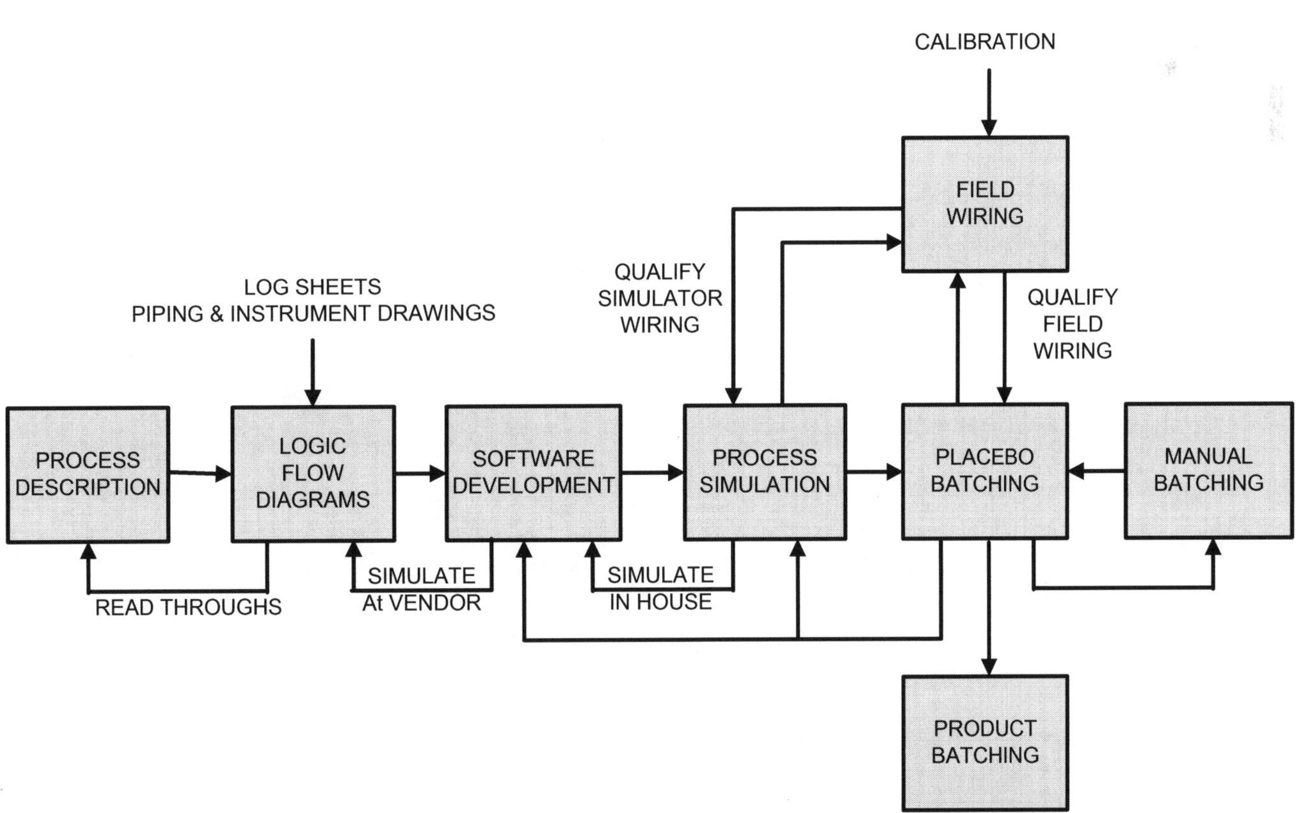

FIGURE 61.1 On the cusp of the life cycle.

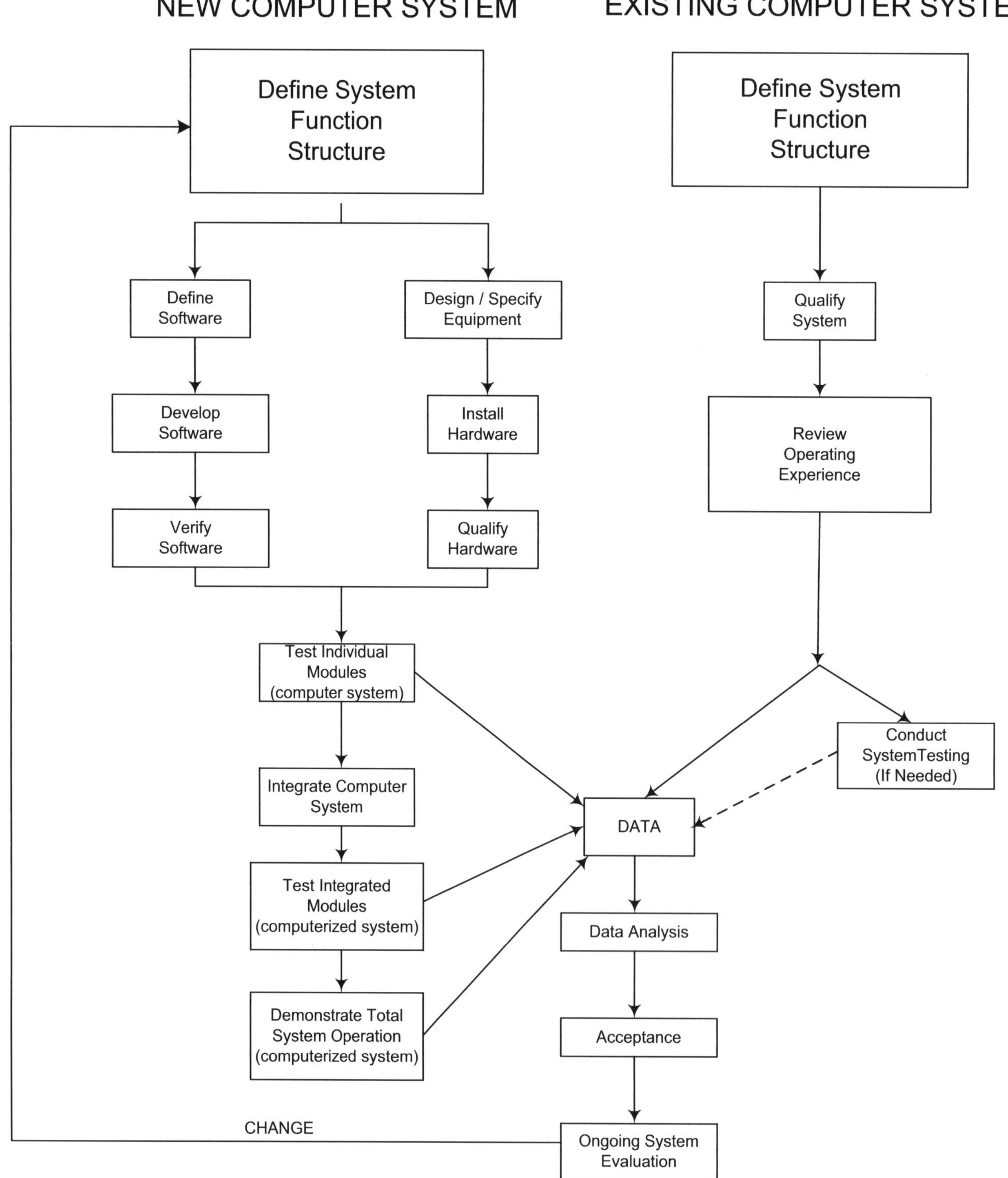

FIGURE 61.2 Computerized Systems Validation Committee life cycle.

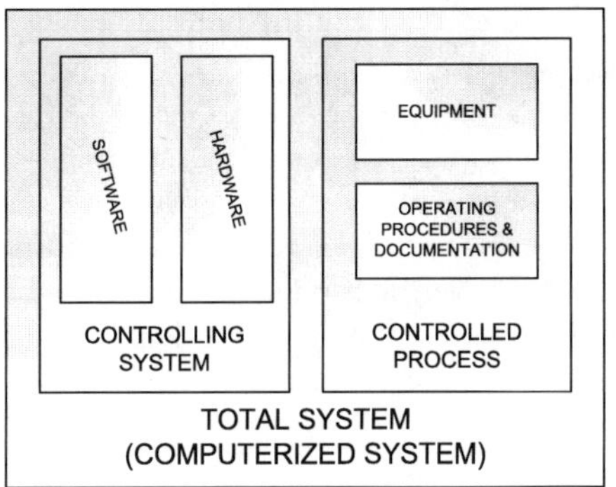

FIGURE 63.3 Computer and process systems.

One of the supportive elements in the CSVC concept paper was a diagram that depicted a parallel structure where process and computer/control systems were aligned (Figure 61.3).

We had pressed for this as our firm's project objective was the validated production of injectable products, and validation of the PLC systems was only a necessary prerequisite. We recognized that the software must deliver the process sequence and conditions needed as defined in our batch records. Similarly, the control hardware was merely an extension of the process equipment in which the product is to be made. The computerized system life cycle represented only a part of what was necessary to validate our products and their processes. A parallel and more essential life cycle for the product and processes themselves was also necessary. We realized that changes in product/process or process equipment would necessitate changes in the control system software and hardware. The process life cycle was acknowledged to be the more essential concern and fit well with product life cycle which was becoming increasing prevalent in marketing. We also understood that, although all of our production processes were not automated, the validation life cycle offered us a superior means to assure that all of our processes were in a validated state. We developed the validation life cycle for internal use and adhered to its principles across all of our validation programs whether automated or not by the mid-1980s (Figure 61.4). We adopted the life cycle at that time but did not publicly disclose until several years later.

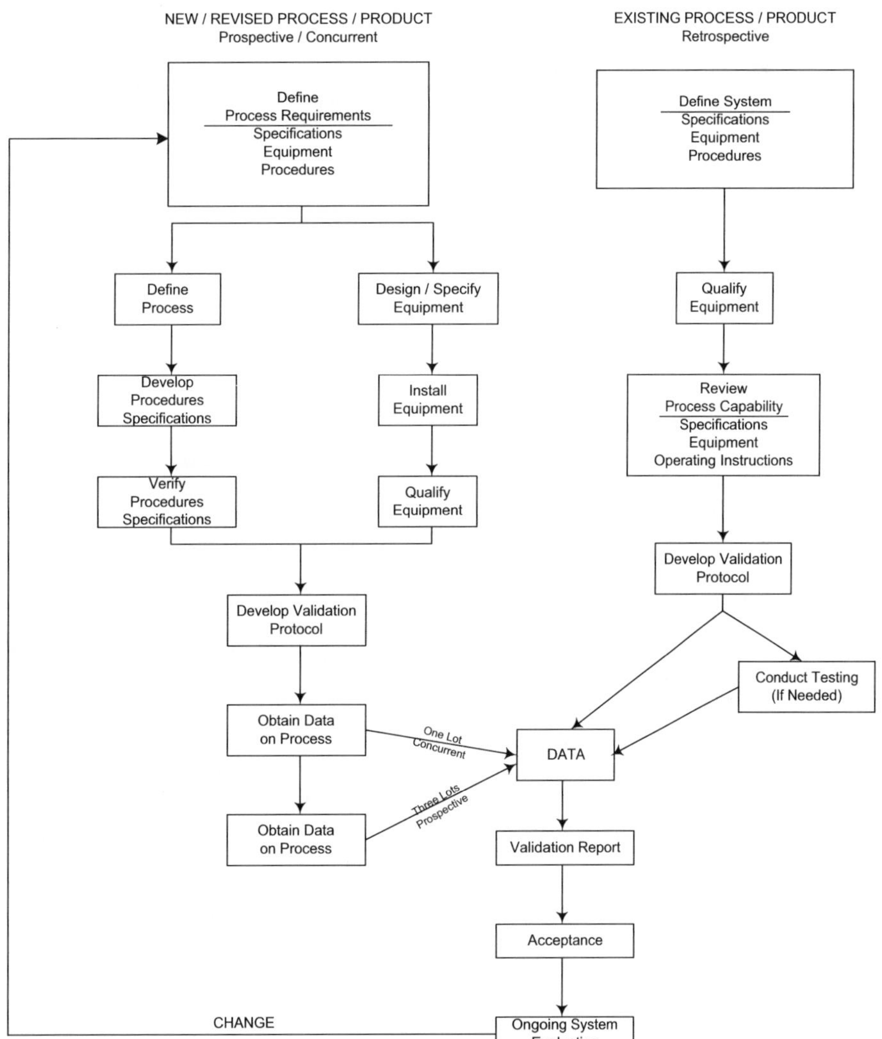

FIGURE 61.4 Process life cycle.

The parallels to the CSVC life cycle were both intentional and appropriate. Managing the validation of processes and products relies on activities of design and refinement and evaluation of the sequences, conditions, and measurements needed. Software and hardware are merely means used to realize that intent. As it was outside the scope of the CSVC's activities, the product/process life cycle for validation was not mentioned in their efforts.

As we used life cycle thinking in our daily activities, we soon recognized the need for an important improvement to the life cycle. In each step of the process, identification of a need for revision to any aspect of the system, whether it be procedural (e.g., SOP, software, etc.) or physical (e.g., process equipment, hardware, etc.) should be executed at that step, refining the outcome immediately (Figure 61.5). This would expedite project execution as corrections/adjustments to the process/product design and equipment would be implemented more rapidly. Depending upon the extent of the changes made, the revision might require returning to earlier stages in the life cycle and repeating a portion of that earlier work. The central

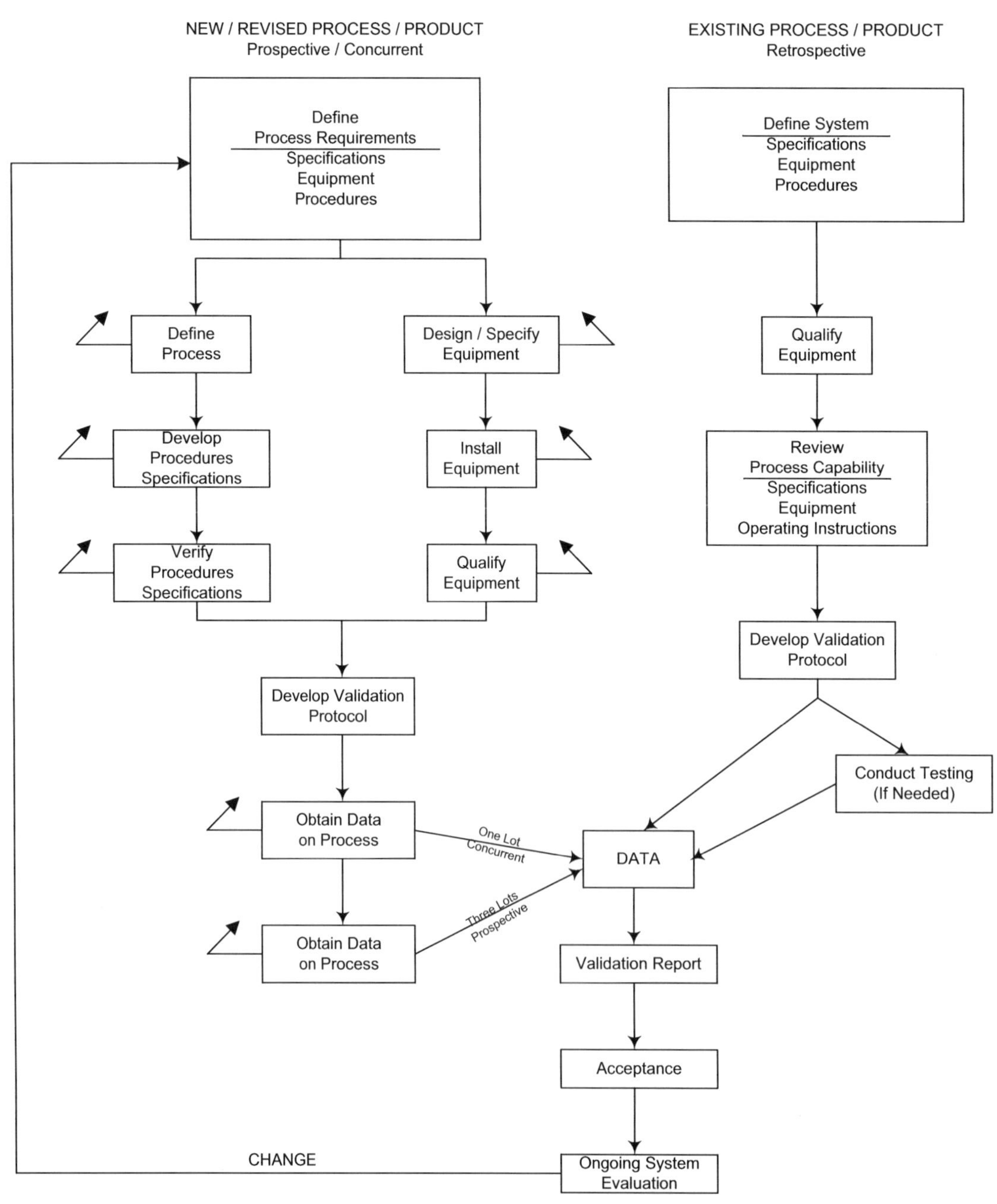

FIGURE 61.5 Real-world life cycle.

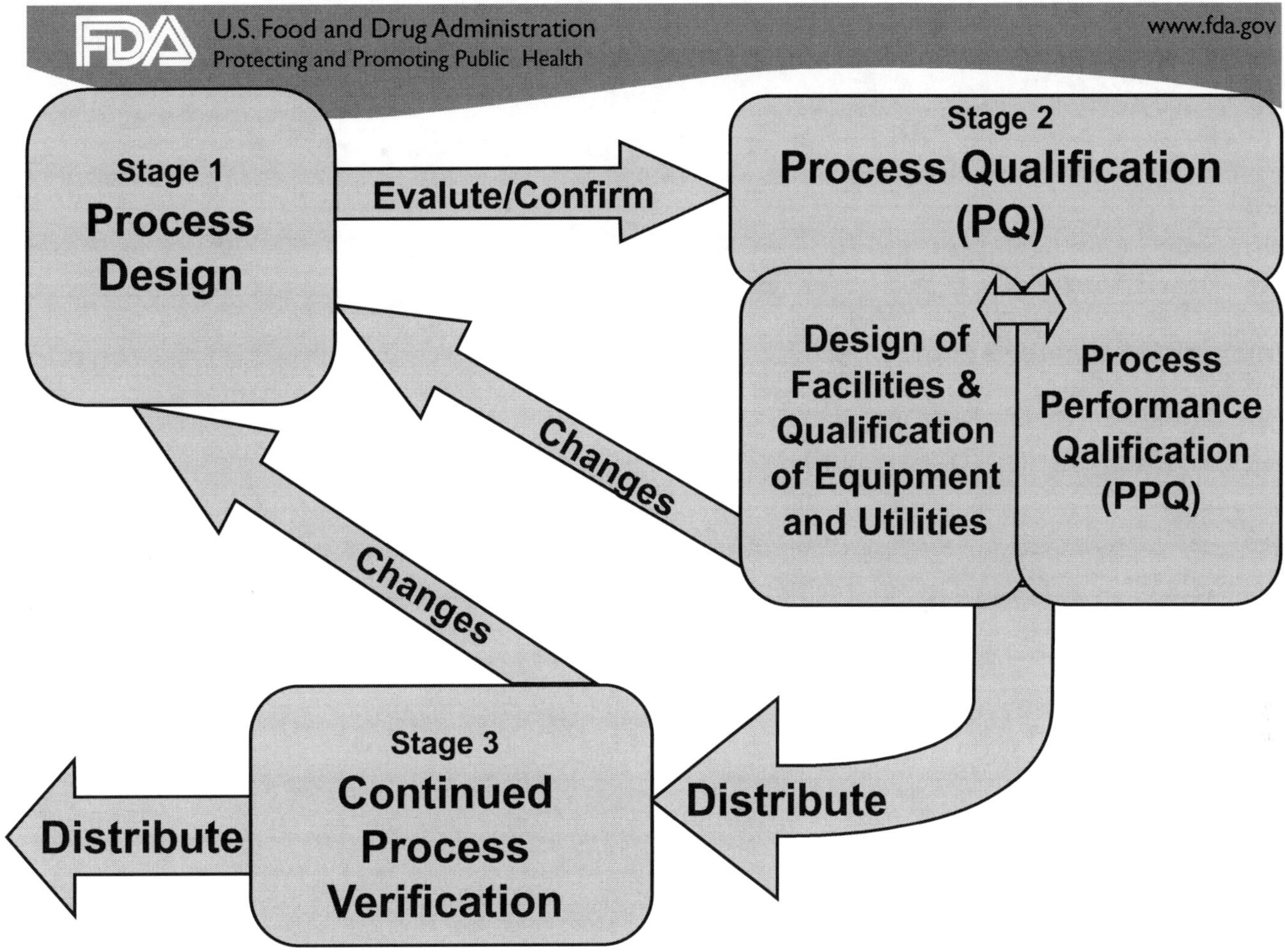

FIGURE 61.6 United States Food and Drug Administration—2011.

importance of change control programs to maintaining the validated state life was also recognized.

We adopted this as its value to our practices was clear and immediate. We offered this as a suggestion to the CSVC because it seemed a significant measure to embrace. They acknowledged its utility but withheld adoption for fear that the added complexity might overwhelm those only just becoming comfortable with computer systems validation. Nevertheless, we adhered to this practice internally as early as 1986 and subsequently included it in our training and consulting activities.

In 2009, the FDA issued its revised guidance on Process Validation and incorporated a life cycle concept (Figure 61.6). We were delighted to see it embraced by the FDA, and at the same time chagrined by the confusion its inclusion fostered in so many. The life cycle concept represents a superior means to manage validation activities for products/processes and all other systems/practices where validation is expected.

Confusion notwithstanding, it is important to credit the FDA (and subsequently the EU) with incorporating the concept into their process validation guidelines and, consequently, into their expectations. It should be clear, however, that the life-cycle concept in process validation is not new. It has been practiced and taught by the authors and other validation experts since the early 1980s. We are happy to see it be accepted as a foundational concept in process validation.

62 Validation in Europe—What Are the Differences?

Trevor Deeks

CONTENTS

62.1 INTRODUCTION

In order to understand the nuances of the European approach to validation, it is important to understand the history of the European Regulations and the history of validation in the pharmaceutical industry. In 1975, when the United Kingdom voted to enter the European Common Market, as it was then called, the UK "Orange Guide" (1) already existed and Good Manufacturing Practice (GMP) in the UK was already established. Furthermore, it was seen as the most advanced GMP guidance outside of the USA—and so it was! Therefore, almost by default, this was adopted as the baseline for the current Eudralex Volume 4 (2), now often referred to as the "European GMP Guidelines". Although the Eudralex document is referred to as "Guidelines", it is a statutory instrument of European law (EC Directives) and as such it is therefore enforceable by law in Europe.

By the late 1970s, driven by hospital sterilizer engineers, responsible for autoclaves used for sterile dressings packs, and by UK hospital pharmacists, the original HTM10 document (3), the forerunner of HTM2010 (4), was published. This document alone drove many of the current European practices associated with autoclave validation, including steam quality testing and the "Bowie Dick" test, both of which were really intended for autoclaves used for sterilizing porous dressing packs, not for aseptic processing equipment and for terminally sterilized products. This document has now been superseded by EN285 (5), but the testing requirements and methodology remain the same. When so broadly applied, these tests are as inappropriate now as they were then. They do not apply to the types of autoclave loads that are generally used in pharmaceutical processing. This issue is discussed in greater detail later in this chapter.

The other important historical fact is that in the 1970s, the word "validation" was not common parlance in the pharmaceutical manufacturing industry. The term started to appear in the lexicon in the late 1970s. Life was much simpler then. The unwritten expectation (based on HTM10) was that autoclaves were temperature mapped annually with a specified minimum number of thermocouples. There was also an expectation that clean rooms would be <u>tested</u> every 6 months. These were the only rules of validation in Europe at that time.

It is interesting to reflect that, in the UK, the expectations and frequency of autoclave temperature mapping, and the frequency of clean room <u>testing</u> have both been reduced to become annual events. But, as an adjunct to these "qualification" activities, other requirements crept in, and both the reduction in frequency, and the appearance of other requirements, has gradually evolved into a combination of industry norms, European and international standards, and the issuance of regulatory standards and guidance.

Engineering principles related to the technology of sterilization play a significant role in the science of validation. On the other hand, the determination of the appropriate frequency of requalification or revalidation are totally empirical. "Pick a number and write a justification for it" seems to

DOI: 10.1201/9781003163138-62

have been the approach to defining frequencies. Many other industries have established the frequency of maintenance, calibration and requalification of equipment based on reliability rather than on pure guesswork. Reliability centered maintenance (RCM) is practiced in the automobile industry, the power industry and in particular in the nuclear industry. There have been initiatives to introduce it to the pharmaceutical industry, but the bar to introduction is regulatory "expectation" and assumptions regarding current "best practice", which are often driven by historical industry norms rather than by hard data. One illustration of how abandoning historical norms for a more data-driven approach is seen in the experience of one major multinational pharmaceutical manufacturer in the UK. The Head of Engineering on the site had newly arrived from a nuclear reprocessing facility in the UK and set out plans to introduce RCM. The engineering department started to collect and trend performance data on every process and utility system on site. One of the first systems to evaluated was the purified water system, for which there was continuous on-line monitoring of conductivity, total organic carbon (TOC), flow rates and usage data for the distribution loop, as well as the monitoring of minimum and maximum levels in the storage tank. The temperature of the loop was also controlled and monitored. Furthermore, the quality of the water going into the storage tank was controlled via a valve connected to a conductivity meter, which switched the water to recycling through the purification system if the conductivity alarm level was exceeded. No water that exceeded 1.1 μS/cm at 25°C ever found its way into the tank. The system was almost fail-safe, and this new Head of Engineering knew all about "fail-safe" systems. It was discovered that each time that preventative maintenance was performed on the system, the on-line conductivity and TOC readings increased by about 50%. It was determined that the interventions involved in the system maintenance caused a reduction in water quality, and that the maintenance frequency could be reduced by half, without any observable impact on chemical or microbial water quality. The maintenance frequency was reduced, and the water quality was more closely monitored for 3 days after each maintenance period to ensure that it continued to meet compendial requirements. The water quality remained consistently high for longer periods and money was saved because of less frequent maintenance. This change was driven by data rather than by industry norms or manufacturer's recommendations and a "win/win" situation for the company.

62.2 REGULATORY AND HISTORICAL DRIVERS FOR VALIDATION IN EUROPE

In order to better understand the requirements and expectations for validation in the European Union (EU), it is important to appreciate the regulations and guidances, and some of the key events, that have driven them. It is also important to appreciate that these same regulations and guidances are also used by the countries that follow the PIC/S guidelines (6). The EU GMP Guidelines and all of the annexes that are associated with or provide guidance on qualification and validation (7) are exactly the same regulations and guidelines that are applied in the PIC/S countries, with the exception that the PIC/S guidelines do not reference the European Directives (European Law) from which they are derived. Annex 16 (8), the Qualified Person (QP) Batch Release guideline, is either not applicable in these PIC/S territories, or it is interpreted slightly differently and the QP is given a slightly different name.

The countries that follow PIC/S guidelines include Australia, New Zealand, South Africa and many smaller countries. The EU also has mutual recognition agreements (MRAs), which cover both GMP inspections and QP release, in place with a number of other territories. These countries include Canada and the European Free Trade Association (EFTA), which includes Switzerland. The EU also has MRAs in place with Japan and the USA, which cover some types of inspections. The influence of the EU and its approach to GMP is therefore very wide, covering a large portion of the industrial world. This influence has shaped the way that multinational pharmaceutical companies apply the principles of GMP and of validation in their manufacturing facilities. It has also been very influential in shaping the global approach to sterilization and the validation of aseptic manufacturing processes. These are discussed in more detail later in this chapter.

The history of validation in Europe starts in the 1970s following a number of disastrous incidents in the 1960s, which include the thalidomide tragedy. There were also a number of incidents involving Evans Medical Ltd., in Liverpool, England, between 1966 and 1972. This last series of incidents led to the deaths of at least five people because of microbially contaminated infusion fluids. The contamination was found to be because of problems with the autoclaves used to process terminally sterilized parenterals (9). It led to the publication of the Clothier Report and subsequently, in 1971, to publication of the first GMP rules in the UK, the aforementioned Orange Guide.

The Orange Guide continued to be published at various intervals (1977, 1983, 1993, 2002 and 2007) since then, but over the years it has become the "Rules and Guidance for Pharmaceutical Manufacturers and Distributors", and in 1993, it contained European GMP "Rules and Guidance" rather than British GMPs, which until "Brexit" in 2019, were one and the same. It still has an orange cover and it still represents GMP regulations in both the UK and the EU, although it remains to be seen whether this will continue to be the case as a result of "Brexit". It is important to note that it was the UK Guide to Good Manufacturing Practice that was the primary driver for the text of the EU GMP Guidelines, and it is reasonable to say that the UK has been one of the dominant voices in the evolution of the EU GMP Guidelines and of the practical interpretations of them by the inspectors of the member states. The UK Medicines and Healthcare Products Regulatory Agency (MHRA) is one of the larger European member state agencies and, until recently, it has enjoyed the benefit of close geographical proximity to the European Medicines Agency (EMA) in London.

The original Orange Guide made no mention of "validation" because at that time the word had not yet been coined with respect to pharmaceutical manufacturing or indeed to any pharmaceutical activities such as analytics (GMP or Good Laboratory Practice [GLP]) and clinical studies. The terms "commissioning and testing" were used in common parlance throughout the 1970s, and there was a particular expectation that these activities should apply to autoclaves, sterilizing ovens and clean rooms. At that time, Federal Standard 209E (10) was followed in Europe, and it was not until sometime later that grades A, B, C and D were defined for pharmaceutical clean rooms.

The terms "commissioning and testing" were used by the people responsible for these activities, now considered as qualification. These responsibilities fell to engineers and to the Quality Control unit, respectively. The activities performed at that time were rudimentary in comparison with what is performed today. For autoclaves, there was no mention of biological indicators (BIs) in HTM2010. Although there were some very strictly defined rules in the UK with regard to the number of independent temperature probes to be placed in the autoclave, the calibration tolerances of the temperature probes and of the autoclave controller probe(s) and with regard to the cycle parameters. There were also defined requirements for air removal, because it was the failure to remove the air that led to the Evans Medical incident.

Many things have changed since that time, and the changes have been driven primarily by improvements in technology and by a better understanding of the processes and the risks to product. However, the fundamental principles have not really changed. The approach to validation has become slightly more sophisticated, and although the volumes of paper generated have dramatically increased, the primary objective of validation was then, as it is now, to demonstrate that the process works as intended and that the data generated can be relied upon.

62.3 TERMINOLOGY

At this point, it is worth establishing some definitions of the terminology used in this chapter. The science of validation utilizes a number of terms that are often misused or wrongly interpreted. Therefore, is it appropriate to include some European definitions such that the reader is aligned with European terminology.

The first point to make in this respect is that the EU draws a clear distinction between "qualification", which is a term that is strictly applied to equipment, utilities and systems, and "validation", which is applied to processes (including sterilization and cleaning) and analytical methods. Strangely, it is also applied to computer systems, which are "validated" but the validation takes the form of a set of "qualifications" (DQ, IQ, OQ and PQ—see following for definitions).

Following are some of the more common terms that are used in this chapter. Most of these are shown as they are defined in the Annex 15 of EU GMP Guidelines (7), and others are included for clarity of meaning. The definitions are sometimes enlightening and have certain implications; for example, the definition of operational qualification (OQ) references the "anticipated operating ranges", thus implying that the operating ranges that are required by the process for which the equipment or system will be used need to be covered during the execution of the OQ. Although this would seem to be common sense, it is surprising how often companies performing such qualifications can lose sight of this requirement. Similarly, the definition of installation qualification (IQ) refers to the "approved design and the manufacturer's recommendations", indicating the importance of these criteria.

Aseptic Process Simulation (APS)—the process by which aseptic manufacturing processes are validated using growth media in place of product to detect any microbial contamination events.

Best Practice—a practice that is considered optimal for a particular activity and to which companies aspire when following continuous improvement programs.

Concurrent Validation—validation carried out during routine production of products intended for sale.

Design Qualification (DQ)—the documented verification that the proposed design of the facilities, systems and equipment is suitable for the intended purpose.

Expectations—regulatory requirements that are generally not specified in any regulation or guidance document but which are common interpretations of the guidance and are often cited in regulatory inspections if not met.

Installation Qualification (IQ)—the documented verification that the facilities, systems and equipment, as installed or modified, comply with the approved design and the manufacturer's recommendations.

Mutual Recognition Agreement (MRA)—an agreement between different regulatory agencies that recognizes the equivalence of regulations between the agencies and results in relaxations, such as relaxations of the need for inspections by a second agency, or of the need for a second release of a batch of drug product on import.

Operational Qualification (OQ)—the documented evidence that the facilities, systems and equipment, as installed or modified, perform as intended throughout the anticipated operating ranges.

Performance Qualification (PQ)—the documented verification that the facilities, systems and equipment, as connected together, can perform effectively and reproducibly based on the approved process method and product specification.

Process Validation—the documented evidence that the process, operated within established parameters, can perform effectively and reproducibly to produce a medicinal product meeting its predetermined specifications and quality attributes.

Prospective Validation—validation carried out before routine production of products intended for sale.

Requirement—a practice or standard that is specified within a regulation or a regulatory guidance document.

Retrospective Validation—validation of a process for a product that has already been marketed based upon accumulated manufacturing, testing and control batch data.

Revalidation—a repeat of the process validation to provide assurance that changes in the process/equipment introduced in accordance with change control procedures do not adversely affect process characteristics and product quality.

System—a group of equipment with a common purpose.

Worst case—a condition or set of conditions encompassing upper and lower processing limits and circumstances within standard operating procedures (SOPs) that pose the greatest chance of product or process failure when compared with ideal conditions. Such conditions do not necessarily induce product or process failure.

62.4 CURRENT EU REGULATORY GUIDANCE

Annex 15 of the EU GMP Guidelines is the annex that describes "Qualification and Validation". It defines the EU expectations for these activities. It first appeared in 2001, prior to the first publication of the United States Food and Drug Administration's (US FDA's) "Risk-Based Approach" document (11) in 2004 and, even then, it described a risk-based approach to validation. It was possibly the first regulatory reference to a risk-based approach in pharmaceutical manufacturing. The concept that there are risks associated with pharmaceutical manufacturing, and that these risks must be evaluated and managed has been recognized in Europe for a long time, and there have been many influential leaders in Europe who have shaped the industry norms for evaluating and managing risk. Many of these industry norms are now finding their way into the EU GMP Guidelines via revisions. The proposed revisions (12) to the EU GMP Annex 1 (7), the "Sterile Annex", are heavily influenced by industry norms. Many of the changes come as no great surprise to Europe-based manufacturers, because many EMA inspectors were already inspecting against these proposed new requirements. They were already "expectations" and the observations being cited, for example, included citations for such things as failure to perform clean steam testing, failure to design aseptic filling suites with unidirectional movement of personnel, failure to measure nonviable particles at both the 0.5 μm and the 5.0 μm levels during routine environmental monitoring, and of course the most contentious of all the proposed new requirements, pre-use, post-sterilization integrity testing (PUPSIT) of sterilizing grade filters. It should be noted that the wording in the Annex 1 revision has largely resolved the contention, but there still remain pockets of resistance to this proposal. The recent publication of the PDA Technical Report (13) on the implementation of PUPSIT provides more definitive, science-based guidance on this topic and provides probably the best review on the subject together with recommendations for a risk-based approach. This subject is also covered further, later in this chapter.

Annex 15 was the first regulatory guidance document to define DQ, well before the GAMP (Good Automated Manufacturing Practices) guidelines (14), which describes an approach to the validation of computer systems and process automation and which gives its own interpretation of DQ. Annex 15 was also the first regulatory guidance document to describe the need for periodic review of qualified and validated systems and processes and to introduce the concept of maintaining validated status:

> Facilities, systems, equipment and processes, including cleaning, should be periodically evaluated to confirm that they remain valid. Where no significant changes have been made to the validated status, a review with evidence that facilities, systems, equipment and processes meet the prescribed requirements fulfils the need for revalidation.

However, this statement does not completely define the expectations. For example, there are expectations for formal requalification of certain systems that are cited in other guidance documents and standards. These include the expectation of a periodic formal requalification of clean room environments. This is now more formally stated in the proposed revision of Annex 1 and the requirement again simply formalizes what was previously an EMA expectation that grade A and B environments should be requalified every 6 months and grade C and D environments should be requalified every 12 months (section 5.29). What is helpful in the proposed revision of Annex 1 is the reference to the ISO 14644 (15), and there is the clear implication that clean rooms should be qualified according to the methodologies described in this series of standards. The revised Annex 1 also clearly states that "For classification, the airborne particles equal to or greater than 0.5 μm should be measured." It also provides a table showing the ISO classifications and the equivalent EU grades (A, B, C and D). This helps to clear up much of the confusion that has existed within the industry. It is probably no coincidence that this table appears as it does, because there has been some confusion over the equivalence of the EU and the ISO classifications, particularly because of the dual status required in the EU classifications (at rest and in operation) and because of the confusion as to whether laminar flow is required for grade B or whether turbulent flow is acceptable. These are issues that have perplexed manufacturers in the USA in particular.

Annex 15 describes some validation requirements that are slightly different from the expectations of the FDA. These include:

1. A description of the contents of the Validation Master Plan (VMP) as a "validation programme" for all facilities, systems, equipment and processes to be validated, which implies that the VMP covers the entire facility. It recognizes, however, that large projects may require separate VMPs, which is more akin to the US FDA concept of the VMP.
2. A statement regarding the acceptability of a "worst-case" approach for cleaning validation.
3. Guidance regarding the qualification of established (in-use) facilities, systems and equipment.

Annex 15 was significantly revised in October 2015, and there were some subtle but significant changes in this revision. Perhaps most significantly, it makes clear statements about prospective, concurrent and retrospective validation. Concurrent validation should only be performed "in exceptional circumstances". It also makes an ambiguous and noncommittal statement related to retrospective validation that "processes in use for some time should also be validated (retrospective validation)". This statement is open to interpretation, but the common interpretation by EU regulators is that if a process has been in use for some time, then it should have been validated "some time ago". Retrospective validation is not acceptable for new processes. The expectations for process validation have been known for some time and as far as the EMA is concerned all registered processes should have been validated by now.

The annexes referred to all relate to Part I of the EU GMP Guidelines. This is the part that covers the "Basic Requirements for Medicinal Products". Part II of the EU GMP Guidelines covers "Basic Requirements for Active Substances used as Starting Materials", and it is essentially the same as ICH Q7 (16), the internationally harmonized guideline on active pharmaceutical ingredients (APIs). Part II has its own set of definitions and guidance on validation. This includes many of the more recent concepts related to Quality by Design (QbD) as described in the current FDA guidance on process validation (17), and although the language is different, in practice the differences are not so significant. The EMA has adopted all of the ICH guidelines including ICH Q8(R2), *Pharmaceutical Development* (18) and ICH Q9, *Quality Risk Management* (19) and accepts the principles of QbD. It describes approaches to process validation and the concepts of defining critical process parameters and critical quality attributes (CQAs). Therefore, it is realistic to say that a prospective process validation exercise that meets the expectations of the FDA will also meet the expectations of the EMA. It only needs to be done once! This is discussed in more detail later in this chapter.

In the EU, the Pharmaceutical Quality System (PQS), as described in ICH Q10 (20), needs to possess the same cornerstones that are required by the FDA. At the highest level, the company needs to declare its Validation Policy, and at the next level, a VMP should exist covering the entire site. Below this there should be a hierarchy of procedures, protocols and templates to cover the execution of qualification and validation activities and the recording of data. It is also quite normal for systems to be formally released by quality assurance (QA) following qualification and prior to being taken into use. Consequently, the system must cover the quarantine of systems under change control or when formal requalification is required. Again, a single PQS covering these requirements will meet global expectations.

What can be concluded from reading the current Annex 15 in conjunction with the proposed revision of Annex 1 is that these two documents provide the regulatory framework for the expectations related to most of the qualification and validation activities within the pharmaceutical manufacturing environment. However, Annex 15, just like many other regulatory guidance documents, is open to much interpretation. Most of this interpretation has historical origins, which have become "industry norms" or "best practice". Many of the European best practices are covered throughout the rest of this chapter.

62.5 PHASE APPROPRIATE QUALIFICATION AND VALIDATION

It has always been well understood in the EU that process validation, as it is defined in Annex 15, applies only to "products intended for sale". However, there is a further annex, related to the manufacture of clinical supplies materials (Investigational Medicinal Products or IMPs). This is Annex 13 (7), and it is also an annex of Part I of the EU GMP Guidelines (Basic Requirements). There has always been considerable discussion about which aspects of the Basic Requirements apply to IMPs. However, it has always been understood and frequently stated that the validation of sterilization processes and aseptic processes is a requirement for all IMPs (from Phase 1 onwards).

Phase appropriate requirements in the EU are largely driven by the requirements of the CMC regulatory dossier i.e. the Investigational Medicinal Product Dossier (IMPD). This in turn has been influenced by the existence of the European Clinical Trials Directive (21) and the GMP guidance related to that Directive, which is Annex 13. There is one very significant difference here between the EU and the USA and that is the level of enforcement. It is rare for any site that manufactures Phase 1 and Phase 2 clinical supplies to be inspected in the USA. The FDA have clearly and explicitly stated that they do not have the resources, and they rely on companies to police their own operations to a very great extent. On the other hand, sites located in the EU engaged in the manufacture of IMPs at any phase of development must have a current GMP Certificate, issued by the inspecting authority of a member state. IMPs include placebos, comparator products and any other medicinal product that is for human consumption in a clinical trial, such as a companion medication. The only exemption to this is if the product is already a licensed medicinal product in the EU, in which case the manufacturer will possess a Manufacturing License. It is important to remember, however, that if a licensed product is manipulated in any way and is presented in any form or alternative packaging from that specified on the Product License, then it is deemed to have been further processed and, as such, it is a different product, under the responsibility of the sponsor of the clinical trial. A good example of this further processing is when a comparator product is removed from its marketed pack and is over-encapsulated in order to blind it. In such cases, the sponsor must generate stability data to support its expiration date in the clinical study.

For sites outside of the EU that are manufacturing IMPs for EU clinical trials, the QP performing the Technical Release of the IMP under European Directive (22) must certify that the site complies with GMP "at least equivalent" to that referred to in European Directive 2001/83/EC (see section 2.2.ii of Annex 16), and this is done by means of a site audit. This puts the onus on the QP to establish European GMP compliance.

The regulation does allow for the audit to be performed by a competent auditor other than the QP, but in practice the QP must be confident that the auditor has a good understanding of European GMP requirements and expectations.

These EU requirements for clinical manufacturing have had a significant impact on GMP standards for Europe-based companies. Many US-based contract manufacturing organizations (CMOs) have also been cited during QP audits for not meeting EU GMP expectations. This is particularly the case for aseptically filled injectable products and biological drug substance manufacturers, whose premises and validation procedures often do not meet EU expectations, and because there is little or no enforcement from the FDA at Phase 1 and 2, it is likely that they do not meet FDA expectations either!

Process validation and full analytical method validation are not requirements for IMPs being used in EU clinical trials, but "premises and equipment are expected to be qualified" and "for sterile products the validation of sterilizing processes should be of the same standard as for product authorized for marketing" (see Section 17 of Annex 13) even at Phase 1. The EU regulation also makes specific reference to the "special problems" for validating aseptic processes when the batch size is small. The regulation makes a specific statement that the maximum number of media filled units may be the maximum number filled in production. However, it also states that "if practicable, and otherwise consistent with simulating the process, a larger number of units should be filled with media to provide greater confidence in the results obtained". This is one of the few definitive statements that the EU Annex 13 makes, and it is surprisingly rare to see that very few companies have adopted this approach. It is equally rare to find a risk assessment that states that such an approach is not "practicable". The challenges of aseptic process simulation (APS) are well understood by the authorities, and the European perspective on APS is covered in greater depth later in this chapter.

Another definitive statement in Annex 13 is the requirement for a demonstration of viral inactivation/removal and of the removal of "other impurities of biological origin" when required. It is presumed that this statement also refers to host cell DNA and host cell proteins as the other impurities of biological origin, because these are the ones that can be most toxic. Again, such a demonstration is required even at Phase 1.

There is also an expectation that analytical methods will be "fit for purpose" and that cleaning procedures will be "verified". Cleaning verification is covered in more detail in the Process Validation and Cleaning Validation sections following.

There is some debate about the need to perform *Brevundimonas diminuta* challenges on filter membranes at Phases 1 and 2, as the formulation may change at least once between Phase 1 and commercial production. It is advisable to at least perform a risk assessment to determine whether the formulated drug product is significantly different in terms of pH, ionic strength and viscosity, so that it could impact the membrane's ability to perform as intended. Regardless of whether a *Brevundimonas diminuta* challenge is performed, as a minimum requirement a bubble point integrity test must be performed following product filtration. This expectation is no different from that of the FDA.

In order for a manufacturing facility to be compliant with EU requirements for the manufacture of IMPs, and in order to qualify for an EU GMP Certificate, all clean utilities, clean rooms and equipment must be qualified in exactly the same way as required for commercial products. There are no relaxations in this respect. In addition, data integrity is important at all phases of development. Automated systems that generate, store, retrieve, or manipulate critical data must be fully validated. Annex 11 (7) states that "The extent of validation necessary will depend on a number of factors including the use to which the system is to be put" There are no relaxations for data generated at Phase 1 and 2. In fact any data that is to be included in a regulatory dossier (including the IMPD) is critical data. This calls into question the need for the validation of equipment and systems used in process development (PD). If the data that are used to support an IMPD dossier are generated in the PD laboratory, then the company should carefully consider how data integrity can be assured and maintained. This may result in a decision to use qualified analytical equipment in the PD laboratory and will certainly require the validation of data handling systems such as Laboratory Information Management Systems (LIMS) and data archives. In Europe, many companies are now considering the qualification of their PD laboratories and many companies have already done it.

In the EU, all IMPs must be assigned an expiration date, which must appear on the label. The expiration date must be supported by stability data that has been generated ahead of the expiration of the clinical batch. There are some rules that allow limited extrapolation of the data (23). The data must be generated on batches that are representative of the clinical batch and the expectation is that these data will be generated on qualified equipment using analytical methods that are "fit for purpose".

A similar consideration applies to PD activities in the later stages of development. These activities involve process understanding studies involving Design of Experiments (DOE). No matter how thorough the statistical model is for the DOE, the data are only as good as the equipment and methods used to generate, store, retrieve and manipulate them. Analytical method validation is therefore an important aspect for the product life cycle.

One term that has come into common usage in both Europe and the USA is the term "Analytical Method Qualification". This term is not cited in any GMP regulation or in any harmonized guideline. It has its origins in GLP when applied to methods used for nonclinical and preclinical safety testing. In the GLP laboratory, the term is used to imply a demonstration that the method will provide a level of specificity and accuracy when run under a given set of parameters but with no predetermined acceptance criteria. In the GMP arena, the term is commonly used to imply a partial validation of an analytical method. The absence of predetermined acceptance criteria seems to contradict any notion of a level of specificity and accuracy, because the level is defined by the acceptance criteria. It seems to be scientific nonsense! The EU GMP expectation is that both qualification and validation require predetermined acceptance criteria, but some CMOs and contract GMP laboratories have been known to offer Analytical Method Qualification without them. It is highly advisable to steer clear of such companies! The EMA will not accept a method qualification or partial validation without predetermined acceptance, and because the IND dossier and the globally harmonized common technical dossier (CTD) includes a section entitled "Validation of Analytical Procedures" (24), it is unlikely that any respectable regulatory authority would accept it either.

The EU GMP expectation is that analytical methods should be validated appropriately for the phase of development. What is meant by "appropriately"? At Phase 1 and 2, this typically involves a partial validation using the normal ICH validation parameters (25) but with wider acceptance criteria than would be expected at later phases to allow for the fact that the methods will not have been optimized at early phases. There will also typically be relaxations for method robustness and for intermediate precision. Typically, at Phase 1 and 2, the method will only be run in one laboratory, by a limited number of analysts, on a limited number of instruments and with a limited variety of sample matrices. These considerations have a major influence on what might be considered as "fit for purpose". During PD, analytical methods will inform the PD activities and will drive the process controls at Phase 1. The process controls will grow and change as the process goes from Phase 1 to commercial manufacturing, and the acceptance criteria for analytical method validation will in turn be driven by the need to control the process more tightly as process understanding increases. The general rule for the acceptance criteria is "start wide and tighten later", as with product specifications. This is certainly the approach in Europe, and if the company does not tighten them later, the regulatory authorities will force it to do so during evaluation of the Marketing Authorisation Application (MAA).

62.6 DESIGN QUALIFICATION

Turning again to the EU requirements for DQ, Annex 15 states that "during DQ the compliance of the design with GMP should be demonstrated and documented". It should be noted that, other than the use of the specific term "Design Qualification", the descriptions regarding design review, approval and confirmation are applicable to capital projects investigated by the US FDA and other non-European authorities.

Annex 15 also states that "the requirements of the user requirements specification should be verified during the design qualification". This statement has been interpreted in many ways and the greatest unanswered questions are:

1. What does DQ involve for different systems?
2. When is DQ required?
3. How should it be executed?

In answer to question 1. prior, the expectations for DQ are based on the complexity of the system. For computer systems, the DQ is well-defined by the current GAMP guidelines. This is a universally accepted guidance document, even though it has no formal regulatory status. The general view of this guideline is that if a system has been validated according to its recommendations, then it will generally withstand regulatory scrutiny. However, if the automated system is also part of a processing system or a laboratory instrument, the DQ also needs to consider the process(es) or analytical method(s) for which the equipment is being used including the design of the functional hardware. This should be included in the user requirements specification (URS) and should be verified during PQ. The URS therefore needs significant input from the user, not just the IT Department!

For a manufacturing facility, the DQ will also require a URS, but the design of a manufacturing facility will include multiple systems that will collectively deliver a manufacturing solution. Typically, the DQ of a manufacturing facility will be performed at defined stages during its design and construction. Typically, there will be a documented review at the end of the conceptual design. This is a set of very high-level design documents that capture the key requirements of the facility such as the footprint, the clean room classifications, the way that people, equipment and materials will move around the facility, any specific containment and disinfection requirements, and the heating, ventilation, and air conditioning (HVAC) and clean utilities required to support the design concept. At this stage, the specifications will be no more than broad definitions of what is required but without any level of detail of how these will be met. This detail comes later in the design phase of the project. Once the conceptual design has been reviewed and approved, a level of engineering change control is normally initiated for the project, such that anything that is likely to change the concept must be reviewed and approved.

The next phase of design, alternatively called basic engineering, preliminary engineering or basis of design (BOD), defines key systems schematically and develops specifications. It is at this stage that the design describes how the requirements will be achieved. For example, how large will the air handling units (AHUs) be and how much air will they need to supply? Which AHUs will supply which zones and how will that provide containment? Will there be recycling of the air or will it be 100% pass through? How many and what ratings will the filters need to be? How will the clean utilities be fed into the clean rooms? There is a more in-depth documented review at the end of this phase, often in the form of a formal Current Good Manufacturing Practice (CGMP) review or umbrella DQ.

Next, the detailed design provides, as the term suggests, a set of detailed drawings and other documents suitable for actual construction and installation. During this stage, designs of specific systems (e.g. sterilizers, critical utilities) may be reviewed and those designs are qualified.

Critical utilities (e.g. Water for Injection [WFI] systems) are often sufficiently complex to be subject to a separate DQ, which during the BOD stage will normally take the form of a review of the piping and instrumentation diagrams (P and IDs) and other relevant engineering drawings to evaluate whether the intent of the URS is being met. It will typically also include a review of the product contact surfaces. This is particularly relevant for pharmaceutical water systems. Piping details are then reviewed during detailed design, for example to check for dead-legs and design flow rates.

Through the construction/installation stage, it is the responsibility of the engineering team to ensure that approved designs and specifications are being met. Any required design changes during construction are subject to formal review and approval.

During this stage of the project, many of the major items of equipment will have been delivered or will be undergoing Factory Acceptance Testing (FAT). This is also the stage when commissioning plans, qualification protocols, SOPs and planned maintenance programs are being written for these major items. It is therefore a good time to take stock and determine if the facility, as designed, is meeting the user's expectations, as defined in the URS. For the engineering company, it is also an opportunity to iron out any major issues that have been encountered during construction.

The design reviews that are described prior are performed in any good capital project managed by a firm's engineering group, often in conjunction with a reputable engineering and/or construction company. However, the regulatory importance of these reviews is not always appreciated, and sometimes they are not properly documented. Design review must be considered essential to facility qualification and specific reviews should be identified expressly for that purpose. These reviews should be planned and formally documented, approved by the Quality unit, and should be treated as controlled documents. Of course, not every discussion and design decision is a DQ activity; however, at key points in the design, as described previously, DQ must be performed.

It is not unusual to find, during an audit of a new facility, that key documents are either missing, are not readily available or are not approved and controlled as they should be. In this respect, it is not uncommon to find that the process, personnel, material and equipment flow diagrams do not exist or are not available as controlled documents. These documents are important because they inform and support the procedures for movement around the facility. This is of particular importance for aseptic processing activities, because there are some specific requirements in Annex 1 for unidirectional movement of personnel in and out of the aseptic core. These documents are critical documents, as are the pressurization diagrams, the AHU zoning diagrams and the engineering documents that define the air handling capacities, the air filtration regimes and the utility supplies. In fact, any document to which the IQ and OQ refer should be considered as a controlled document, because these documents are the data that support the IQ and OQ and therefore should meet data integrity expectations (26, 27).

So far, this chapter has dealt with the DQ of facilities, clean utilities, complex systems and automated systems. However, there are a large number of smaller, simpler systems. These systems include smaller items of laboratory equipment such as pH meters, balances, etc. They also include some processing equipment, such as tanks, stand-alone laminar flow hoods, incubators, etc. It is unusual to see a formal documented DQ for such systems, even in European manufacturing facilities. However, there is almost always a specification of some kind for these items, and the specification should be associated with a purchase order. It is the purchase order that defines what was the user requirement. If you consider how decisions are made to purchase specific small items of equipment, typically, the user looks in a catalogue or looks at a manufacturer's specifications for the equipment and decides if that meets their needs. This is a conscious procurement decision, and it needs to be documented. In these cases, the manufacturer's specifications become the URS and the documentation is the user's approval of the purchase order.

Question 2. prior has already been partially addressed, at least for facility qualification. In these cases, the DQ is an ongoing activity. It needs to be performed at defined stages, but it should be completed before the IQ and OQ begin. If it later turns out that the design intent has not been met because the IQ, OQ or PQ testing has failed, then a change may be needed, hence the need for project change control. This is an essential part of the DQ documentation. Project change control does not need to be as rigorous as the operational change control that is in place during normal production, because it will not have any direct product or regulatory impact. At this stage of the project, no product has been made. Project change control needs to be "fit for purpose", which means that it needs to capture the change, evaluate it, determine what additional qualification or validation is required and document all of these steps in a concise but transparent manner.

Likewise, for computer systems, the DQ will be an ongoing part of the project and will not be totally complete until the PQ is completed and the system is ready to be implemented within the GMP environment.

For smaller systems, the DQ needs to be performed on delivery of the system, both for GMP considerations and for business considerations. If the system is accepted and there is a long delay between delivery and qualification, the warranty on the system may expire or may be voided. In some cases, the warranty may not start until the system is installed, in which case the timing is less critical. However, there may be a manufacturer's requirement that the system is installed within a defined time period, and users should be aware of such specifics.

The final question posed at the beginning of this section (question 3. prior) is "how should the DQ be executed?" For facility and major utility construction, this has for the most part been answered previously. However, the proof of the design is in the testing. In this respect, the EU definitions of IQ and OQ provide some insight as to how the testing should address the design. The IQ should document that the system has been installed in compliance with the design and the OQ should document that the system performs as intended throughout the anticipated operating ranges. If it does not, then perhaps the design was not suitable. However, it is the PQ that is the real "proof of the pudding" because it should document that the system performs effectively and reproducibly based on the approved process method and product specification. If the facility was built for a specific process or even for a generic process, then the PQ should finally confirm that the design was appropriate, and it can then claim to have been qualified. For computer systems, the GAMP guideline provides a more detailed interpretation of how the testing should address the URS in order to qualify the design. It describes a specific set of documents that can be used for this purpose and at the same time provide a documentation package that is easy to follow and can act as a guide for the regulatory inspector or auditor to review the validation of the system. The Traceability Matrix describes how the functional specifications address the URS and how these specifications are tested to demonstrate this. The Traceability Matrix is an invaluable document, and most companies will generate one. For those who do not, perhaps it is time to start.

62.7 CLEAN ROOMS AND LAMINAR FLOW UNITS

Annex 1 has some specific requirements for the qualification and requalification of clean rooms and clean air devices such as laminar flow hoods (LFHs) and biosafety cabinets (BSCs). There is, for example, a requirement for clean environments to reach specified air-cleanliness levels both in the "at rest" state and in the "in operation" state. The "at rest" state is defined as "the condition when the installation is installed and operating, complete with production equipment but with no operating personnel present". The "in operation" state is "the condition when the installation is functioning in the defined operating mode with the specified number of personnel working". These conditions must be defined for each clean room or suite of clean rooms.

The proposed revision of Annex 1 (11), referred to as a "consultation document" provides even more granular detail for the expectations for clean environments, and although these requirements have not yet been formally published, it is unlikely that much will change from the current consultation document, because the new details are, for the most part, current "expectations". They are the unwritten regulations based on industry norms and based on the interpretation of EMA inspectors. The consultation document has simply put into writing that which is already required. The details of these requirements, if not met, will already incur inspectional observations. Despite numerous comments from highly respected organizations (28), very little of the current wording is likely to change. It is therefore strongly advised that companies evaluate how they measure up right now and not wait until the final document is published.

The proposed Annex 1 revision contains some interesting specifics that help enormously to clarify many of the questions that have been asked regarding the current Annex and the current expectations. As previously stated, the relationship between the EU grades A, B, C and D and the ISO 14644 room classifications are well explained, and there are clear statements about when ISO applies and when the EU grade A, B, C and D classifications apply.

Clean rooms should be qualified and requalified according to the ISO 14644 standard Part 1, which gives the requirements for the numbers and locations of the sampling points for airborne particulate measurements, but it also states that a higher number of sampling points is required for aseptic processing rooms. The consultation document also makes some statements about the measurement of air speeds, air visualization studies and the "clean-up period", which should be determined during initial classification (= initial qualification). The clean-up period is a concept that is quite different from anything that is required by the FDA, and many US-based manufacturers have had difficulties in understanding this requirement. It is the time taken for the clean room to return to the "at-rest" state following the completion of operations, and it is specifically intended to be measured using a nonviable particle counter, although this is not specifically stated. It does not normally include the use of viable air counts. The simplest way to measure this value during initial qualification or requalification is to leave the particle counter running, stop operations, evacuate the clean room and let the particle counter run for a further 20–30 minutes. The "at-rest" particle levels should be achieved within 15–20 minutes following the end of operations. This is the guidance value provided in the current Annex 1, and it has been removed from the revision. The likely reason for this is that typical clean up periods for modern clean rooms are of the order of 5–10 minutes. The consultation document simply states that the clean-up period should be determined during the initial classification. The consultation document also gives specific guidance regarding the frequency of requalification, which is 6-monthly for grades A and B and 12-monthly for grades C and D. The difference in frequency is historical. It is based on a risk-based approach, and it is intended to mitigate the risk of damage to HEPA filters or of deterioration of the filter seals, which was a particular concern for LFHs, BSCs and depyrogenation tunnels. It was subsequently expanded to cover all grade A and B areas, although it is doubtful that HEPA filters in grade A and B areas are at greater risk than those that supply other grades of clean rooms. The consultation document also states that clean rooms should be requalified "following changes to equipment, facility or processes based on the principles of QRM" (QRM = Quality Risk Management).

The grade A, B, C and D classifications are much more than clean room classifications. The simplest way to understand the concepts described in the consultation document is to think of the ISO 14644 standard as a set of design specifications or engineering standards for clean rooms and perform the IQ, OQ and PQ to establish that these standards have been met. The EU grades A, B, C and D on the other hand are a set of operational requirements, which also need to be included in the URS and considered within the DQ but which define the activities that can be carried out, the gowning required, the way in which people, equipment and materials should move in and out, and the type, frequency and acceptance criteria for the monitoring in each environment. In short, Annex 1 describes a total operating philosophy for pharmaceutical clean room environments. The FDA Aseptic Processing Guidance (30), is the only other regulatory document that comes anywhere near to describing such a detailed operating philosophy. Much of this philosophy has not changed between the current Annex 1 and the consultation document. There are some extra details, but the most significant change is that it has been expanded to include processes other than those for manufacturing sterile products. The scope now includes "other products that are not intended to be sterile (such as certain liquids, creams, ointments and low bioburden biological intermediates)". This expansion in scope has far-reaching consequences for manufacturers that wish to comply with EU GMP expectations, but, as with the other changes to the document, these are already expectations of the inspecting authorities.

In the manufacture of low bioburden biologicals, and in the absence of any formal, globally harmonized regulatory guidance on the appropriate environmental levels for different activities associated with such manufacture, there have been some unique approaches. One of these was the creation of in-house room classifications. The adoption of an additional grade of room is common to many biological manufacturers. Typically, this is a classification that is below the grade D classification. In many companies, this is referred to as "controlled unclassified", and one company has created a grade "E" classification. In light of the proposed revisions to Annex 1, it is anticipated that these in-house approaches will gradually disappear. Already some companies are reexamining the need for additional classes of environment.

One important clarification in the consultation document, although it is implied in the existing Annex 1, is that clean room monitoring should not be based on the sampling locations described in ISO 14644, but rather the "locations, frequency of monitoring and incubation conditions" should be based on "appropriate risk assessments. . . . based on detailed knowledge of process inputs, the facility, equipment, specific processes, operations involved and knowledge of the typical microbial flora found, consideration of other aspects such as air visualization studies should be included." In this respect the trend analysis is implicated for consideration. (see Annex 1 Consultation Document section 9.4).

One further point that should be mentioned is that the consultation document provides some statements that align it more closely with FDA requirements in that it specifies the need for gowning qualification and requalification and it provides greater detail related to APS. It makes some statements about qualification of the reuse of clean room garments, and it includes statements about the validation of the cleaning of the clean rooms.

Any manufacturer who wishes to comply with EU GMP expectations for working in pharmaceutical clean rooms, now or in the future, is well advised to read Annex 1 and the consultation document. There is a significant amount of detailed guidance, and it helps to clarify the current expectations.

62.8 STERILIZATION (OR STERILISATION?) VALIDATION

The validation of methods of sterilization is largely harmonized today. Twenty-five years ago, the European agencies were very resistant to the idea of introducing live, highly resistant bacterial spores (BIs) into the clean room environment and especially into production equipment. Even now European inspectors may scrutinize the use of BIs. In particular, the way in which BIs are introduced into, and retrieved from, the production environment may be scrutinized during EMA inspections. This scrutiny may include ensuring that all BIs are accounted for on retrieval. The expectation is documented reconciliation of all the BIs used and procedures for ensuring that the production environment and the equipment are decontaminated in the event of a survivor during a validation run. The use of protective Tyvek envelopes for holding the BIs during use has gone a long way to allaying the concerns of EMA inspectors. The use of BIs in actual production runs is not an acceptable practice in Europe. However, some US companies are known to follow such practices.

There is an argument that, as a result of the improvements in sterilization technology, such as air detectors and improved design of autoclave cycles, involving multiple vacuum pulses and automation systems that will detect small discrepancies in sterilization cycles, that BIs are no longer necessary. Some work performed many years ago (29) on steam-in-place (SIP) systems indicated that the physical parameters may be more valuable in detecting sterilization failures in overkill cycles. A number of suboptimal cycles were run in which the acceptance criteria for physical parameters were not met, but in all cases, all of the BIs were killed and all of the positive and negative controls met their acceptance criteria. However, the use of BIs during the validation of sterilization processes has gained wide acceptance, and it is unlikely that this will change in the foreseeable future, either in Europe or anywhere else.

A much greater controversy is the EU GMP requirement for the testing of clean steam. The methodology for this requirement is also historical and is currently specified in EN285 (5). The current Annex 1 contains a simple statement that "care should be taken to ensure that the steam used for sterilization is of suitable quality and does not contain additives at a level which could cause contamination of product or equipment". On face value this might be interpreted as ensuring that the steam quality is that of clean steam that meets the chemical standards for WFI of the USP, Ph. Eur. and JP, but European inspectors have been citing companies for many years for failure to perform adequate steam quality testing that includes the testing described in HTM2010. These tests are the tests for non-condensable gases, superheat and dryness factor. These tests are historically derived from the previously mentioned HTM10, the forerunner of HTM2010. Both documents were developed by engineers and pharmacists working for the National Health Service in the United Kingdom. The original document was intended primarily as a guidance document for autoclaves used in sterile supply departments within the hospital service, whose primary function was to sterilize dressings and instruments for use in operating theatres. The incorporation of pharmaceutical production autoclaves by hospital pharmacists came later, but the incorporation of these tests was intended for the steam sterilization of porous loads such as dressings. It has been inappropriately applied to pharmaceutical production sterilizers. Although some loads, such as sterilizing grade filters and Tyvek bags, for wrapping sterile equipment, may be considered as "porous". They are not anywhere near as resistant to steam penetration as dressings, and the value of these tests must be challenged. The FDA has never seen any value in them, and the methodologies for these tests is imprecise, difficult to replicate and highly operator dependent. It is necessary to question the value of any test that is more of an evaluation of the operator's ability to perform the test than a test that provides any meaningful data! It is also important to evaluate the risks of not performing these tests when there are far more precise and established methods of validation of autoclave cycles. Nevertheless, the proposed revision of Annex 1 specifically names these tests for steam used for porous loads and SIP systems, in addition to tests for steam condensate quality. These tests are still under debate, but it seems unlikely that the EMA will back down on this requirement.

With the exceptions discussed previously, the initial qualification and requalification of autoclaves is performed in a similar manner in Europe as it is in the USA. The proposed revision of Annex 1 provides a number of helpful statements regarding acceptance criteria, the importance of the drain temperature, air removal, equilibration times and the correlation of temperature and pressure. These considerations are no different than those required by the FDA. The Annex 1 revision does stress the importance of loads being dry on removal from the autoclave, frequent leak tests on the chamber and of filtering the inlet air. It also specifies that the revalidation of steam sterilization cycles should be performed annually. The importance of establishing fixed loading configurations is stated. However, there is no guidance as to how to identify "worst-case" loads, perhaps implying that all loading configurations should be validated. In practice, the principles of QRM also apply to the validation of steam sterilization cycles and the identification of worst-case loading configurations is an accepted practice, although the way in which the "worst case" is determined may be closely scrutinized. The "worst-case" load might be the one that gives the lowest F_0 value for any single location within the load, rather than the lowest average F_0 value. The basis for this consideration is that it only takes one item in the load to be non-sterile for the entire process to be contaminated. Sterility is not a concept of averages. It is an absolute value.

There are no major differences between the EU and other territories with respect to the validation of other sterilization processes. Dry heat sterilization, sterilization by irradiation and chemical sterilization are all more globally harmonized than steam sterilization. These standards are generally described in EN and ISO standards, which are harmonized. For dry heat sterilization, it has long been accepted in Europe that a 3-log reduction in endotoxins is a far greater challenge than killing BI spores, and there is no expectation that BIs are needed for the validation of dry heat sterilization and depyrogenation cycles. However, when endotoxin-spiked containers are used, the Annex 1 consultation document specifically states that they need to be carefully managed with a full reconciliation performed, which has always been an expectation.

Isolator technology has been used for aseptic processing more widely and for a longer period in Europe than in other regions. During the 1990s, the FDA was apparently reluctant to accept this technology, but it was becoming increasingly popular in the EU and was demonstrated as a very effective technology during this period. There is probably greater expertise in Europe in the validation of isolators and especially in the validation of decontamination cycles that provide surface sterilization using vapor-phase hydrogen peroxide (VHP) or mixtures of VHP and peracetic acid. The EU has driven this technology, and the development of monolayers of BIs has been influenced by European isolator manufacturers and users. The validation of isolator cycles is only one aspect of the validation of the isolator. The detection of leaks, the removal of residues of gaseous sterilants, the air flow patterns and the background environment in the clean room housing the isolator are also important criteria for the validation. These are recognized in the current Annex 1, and there is very little change in the consultation document. However, there is more discussion pertaining to pressurization regimes. The increased use of negative pressure isolators and open isolator systems has led to more definition of how these systems should be housed, used, sanitized and validated. It is reasonable to assume that the validation of an isolator will involve the same requirements in Europe as it does in the USA. It is the technology, and the experts in that technology, that drive the validation requirements, not the regulatory authorities.

62.9 ASEPTIC PROCESS VALIDATION

The EU has always seen aseptic processing as the least desirable option for manufacturing sterile products. In the past, the UK authorities published a decision tree to help manufacturers to decide whether aseptic processing is the best option for their sterile products. This document has since been superseded by the concept of the risk-based approach. Nevertheless, both Annex 1 and the consultation document include references to the concept that consideration should be given to using "terminal bioburden reduction steps, such as heat treatments (pasteurization), combined with aseptic processing, to give improved sterility assurance". For most aseptically processed products, even a mild heat treatment will inactivate the product, as is the case for biological products. Consequently, this concept has never gained popularity, and there are very few examples. Some blood products, some ophthalmic products and some products in large volume containers may utilize such methods, but these do not represent a significant proportion of sterile pharmaceutical products in Europe.

Because of the huge increase in the number of biological products both in development and in commercial production, there has been a corresponding large increase in aseptic processing over the past 20 years. Yet it remains the least harmonized aspect of pharmaceutical manufacturing. There are a number of key differences in expectations between the USA and the EU, and these differences impact their validation.

There are a number of EU requirements for environmental cleanliness that are different from the FDA expectations. The first of these is the EU requirement for continuous environmental monitoring, of both viable organisms and nonviable air particles, within the critical processing zone (the ISO 5 or grade A area) for the full duration of critical processing, including equipment assembly. This is not currently specified in Annex 1, but it is in the concept document and, as previously stated, many of the requirements in the consultation document are current expectations, and failure to meet these expectations will result in inspectional observation. Additionally, there is a requirement for continuous viable monitoring in the grade B (background) environment for the full duration of critical processing, including equipment assembly. The FDA Aseptic Processing Guidance (30) on the other hand states that the monitoring program should "cover all production shifts", but nowhere is the word "continuous" used. This has led to some differences in the way that environmental monitoring programs and APS have been designed and performed. Most large-scale, automated, aseptic filling lines are equipped with permanently located viable and nonviable air sampling ports within the grade A or the ISO 5 zone and are therefore capable of achieving both EU and FDA requirements.

This differences in requirements are also coupled with two other differences. One is the EU requirement for particles at both ≥0.5 µm and at ≥5.0 µm to be continuously monitored. Room classification does not require particles at ≥5.0 µm to be monitored, but there has always been a requirement for these particles to be measured during routine environmental monitoring programs within the EU. This should not be a problem, because modern particle counters will count both particle sizes at the same time. However, the particle counter does need to be calibrated to cover both ranges. Another difference is the requirement for the crimping of vials within the grade A environment. The actual wording of the Annex 1 revision is that the environment must be "supplied with Grade A air". This statement recognizes that the nonviable particle levels may not achieve Grade A limits during crimping operations, but it is expected that the viable samples will meet Grade A limits. The background environment for crimping is not specified, and a Grade C environment is sometimes used. According to the current Annex 1, the vial is not considered "fully integral until the aluminum cap has been crimped into place". This crimping requirement is therefore implied, but it is further reinforced in the consultation document, which clearly states that "the stoppered vials should be protected with the grade A supply until the cap has been crimped".

These differences in requirements place additional demands on companies to demonstrate that the grade A environment is maintained at multiple locations within the critical processing zone, and these need to be considered during equipment design, qualification, APS and routine monitoring of batch production.

The EU has additional requirements for filtration through a sterilization grade filter. The use of a second sterilizing grade filter is an established expectation. In both Annex 1 and the consultation document, it is considered "advisable". It is further reinforced by reviewers of CMC dossiers, who will require a strong justification to permit any exemption from this expectation. This is not too difficult a challenge to overcome by manufacturers, and it normally involves minor changes to the process. However, such minor changes must be validated, and the validation may include the need to integrity test and establish process parameters for two filters instead of one, the need to ensure that the product stability is not adversely impacted by the double filtration and the need to establish leachables profiles for two filters in line. It may also be necessary to perform PUPSIT on two filters rather than one.

The concerns and controversy over PUPSIT are not new and are well documented (13,31,32). The current Annex 1 states that "the integrity of the sterilized filter should be verified before use". However, the interpretation of this requirement is that the integrity of the sterilized filter is performed "in-place", and this is now stated in the consultation document. The greatest concern related to PUPSIT is the risk of impacting the sterility of the filtrate during the performance of the integrity test. Another concern is the risk of other contaminants. Many companies have previously justified not performing PUPSIT based on a risk analysis. In a 2017 survey (33), 44% of the companies that responded were not performing PUPSIT. There is a large body of opinion that for filters that are gamma irradiated, the risk is lower than for filters that are steam sterilized.

The need for PUPSIT is probably the greatest concern related to the requirements of Annex 1 and the cause of the greatest objections to the consultation document. Most of the resistance to PUPSIT is stated to come from older facilities that do not have the capability to perform integrity testing in-line. Another argument presented is that, if the post-use integrity test passes, then there is a low probability that the pre-use test would have failed. If the post-use test fails, then the batch will be rejected. The argument presented is that there is very little risk to product quality or patient safety and that the risks of performing PUPSIT outweigh the risks of not performing it. Nevertheless, the requirement is being actively enforced and the "jury is out" on whether opposition to this requirement will be successful.

The validation of aseptic processes via process simulations using sterile media, generally referred to as "media fills", has in the past been a source of differences between Europe and the USA. Traditionally these simulations were, as the term implies, purely media fills i.e. the filling of sterile media on the aseptic filling line. However, in Europe, it has long been recognized that the filling operation is only one aspect of the process that is designed to maintain asepsis of the drug product, and there has long been an expectation that the APS will simulate the entire process and not just the filling of the sterile product into sterile containers. This includes simulating the sterile filtration step, the assembly of the filling line, any holding steps, all interventions and all critical manufacturing steps including aseptic operations "subsequent to the sterilization of materials". The European requirement also includes assessing "additional aseptic steps for non-filterable formulations". The EU requirement is not really so different from those stated in the FDA guidance. However, the interpretation has been quite different in the past. For example, companies were asked to prepare their media and sterilize it via the filtration method used for the product, in order to simulate the filtration step. It has since been argued (mostly successfully) that the validation of the filtration step is a separate activity requiring the establishment of filtration parameters and a demonstration of removal of *Brevundimonas diminuta*, generally referred to as "bacterial retention testing". The properties of the media are often very different from those of the product, and this will impact the ability of the filter to remove microorganisms. The practice of sterilizing the media by filtration is now falling into disrepute, and this is reflected by the wording of the consultation document, which states that the APS should use sterile nutrient media and/or placebo.

There are still some differences in perception and interpretation with respect to products that cannot be filtered through a sterilizing grade filter. In such cases, the drug substance is also treated as a sterile product. Examples of such products include live viral vaccines and gene and cell therapy products (ATMPs). For these products, it is necessary to perform the entire process as an aseptic process, starting with the creation of the Master Cell Bank, or in the case of cell therapy products, the handling of the donor cells. The European expectation is that all steps in the process will be simulated using media, and this can be done via a modular approach, simulating each step individually on three occasions for initial validation.

The FDA are more likely to accept a risk-based approach. Many of the steps in the production of the drug substance involve closed processes, and the risk-based approach may be to simulate only the open handling steps. If any of these are generic steps that are repeated a number of times during the process, then the simulation would not necessarily simulate each instance of open handling but might simulate the generic step once only for each run, but the initial validation would normally be performed three times.

There are some relaxations for gene and cell therapy products that recognize the challenges of these processes (34,35). However, sterility should not be compromised under any circumstances. Some of these products have a very short shelf-life once the product has been filled into its primary container. By the time a sterility test has been performed, the patient has had the dose. In such circumstances, there is a heavy reliance on both aseptic process simulation and on in-process testing for sterility, using rapid methods, at multiple steps. For such products, there is not a general consensus between the EMA and the FDA on how these processes might be simulated, but it is early days for these products and it is highly likely that a more harmonized approach will start to appear in regulatory guidance documents.

62.10 EQUIPMENT AND FACILITY QUALIFICATION

As has already been stated, there is a European requirement that premises and equipment are qualified even for the manufacture of products intended for Phase 1 clinical studies. This is a global requirement. Europe is no different in this respect. The differences are seen in the interpretation and enforcement of these requirements. Enforcement has already been discussed in this chapter. Without any enforcement at Phase 1–2 in the USA, facility qualification is less uniformly implemented. On the other hand, the European interpretation during early phase is very evident from the numerous inspections of IMP manufacturing facilities in the EU.

The more recent regulations on the application of the EU Clinical Trials Directive (22) and of the need for GMP compliance is not limited to the manufacturing and testing sites, but it is expanded to all sites that have any involvement in the supply chain, including virtual companies that outsource all regulated activities. Even though all GMP activities may be outsourced, there are still some GMP activities that rest with the study sponsor (the outsourcing company, or the "Contract Giver"). Batches of IMPs are not released by the CMO directly to the clinical site. Even the clinical packaging site, which often performs the QP technical release of the batch, is releasing the batch to the sponsor. The sponsor has the final responsibility to release the IMP for use in a defined study at a defined study site or sites. This is normally considered as a GCP activity. However, the sponsor site must also provide oversight of the CMO, as part of their obligations under the Quality/Technical Agreement, the sponsor is responsible for effecting any recalls and dealing with technical complaints, the sponsor is responsible for maintaining the IMPD, the sponsor is normally responsible for evaluating stability data and assigning expiration dates, and in some cases, the sponsor may take responsibility for performing QP release, in which case the sponsor site must possess a GMP certificate and is subject to inspections. The sponsor must therefore have a GMP compliant Pharmaceutical Quality System (PQS). This system is needed, as a minimum, to support the sponsor's responsibilities under the Quality/Technical Agreement and must include the activities listed previously. The most significant impact on equipment and facility qualification is the need for GMP-compliant data handling systems. These systems will process, store and retrieve all of the data associated with the IMPD. They are often the same systems that store and retrieve the protocols and reports associated with the clinical and preclinical studies. They must therefore be fully compliant with the GxP requirements for computer systems, which are the requirements of 21CFR Part 11 and EU Annex 11 (6). For these systems, there are no relaxations of requirements. The data they handle is absolutely critical, not just for compliance, but for the business. It is not uncommon for the retrieval of data to support the regulatory dossiers to be the most challenging and time-consuming part of writing the dossier.

For manufacturing and testing sites involved in GMP manufacturing at any phase of development, the EU expectation is that equipment, utilities and automated systems will be qualified to the same extent as they would be if they were being used for commercial manufacturing. The only differences come from the type of equipment that is likely to be in place, which is smaller scale and usually not fully automated, and the use to which the equipment is to be put. For many processes it may also be single-use, which eliminates a lot of cleaning validation and verification (see following). The probability is that equipment and control systems will be required to produce a wider range of products and control a wider range of processes. In Phase 1 and 2, rarely is the same product and the same process run more than once or twice before it is scaled up, changed, or discarded because the product failed to show clinical safety and/or efficacy. In practice, this means that the operational range for the equipment may be greater, with respect to temperature, speed, or component combinations, e.g. for aseptic filling lines. The OQ will therefore need to test a wider operational range and, in the case of aseptic filling, the APS program will need to bracket a wider range of component combinations and batch sizes.

The qualification of storage facilities and controlled-temperature units (freezers, refrigerators and incubators) is not well defined by either the EU or the FDA guidance documents. Nevertheless, there is a considerable degree of harmony between Europe and other territories regarding frequency and numbers of qualification probes and permanent monitoring probes, and regarding the temperature ranges and tolerances required. The temperature ranges are driven by the compendial (USP, Ph. Eur. and JP) requirements for product storage or for microbial incubation conditions, and the tolerances are generally driven by the compendial and by ICH guidance, but the frequencies and numbers of probes are often driven by vendor recommendations. There is one global guidance document that covers such equipment, and this is the World Health Organization (WHO) GMP guidance document (36). There seems to be little awareness of this guidance outside of developing countries, whose aim is to meet WHO GMP compliance. Very few companies in Europe or in the USA purport to follow this guidance, and yet many companies inadvertently meet it!

For equipment and utility qualification at any phase of development or commercial manufacture, maintaining qualified status is extremely important and EU inspectors put a lot of emphasis on this. The inspection of logbooks is a very popular activity during EU GMP inspections. Again, this probably stems from the UK tradition. UK inspectors regard the equipment logbook as a "potted history" of the equipment. They are looking to see that all activities that were performed on the equipment are logged in, even if it is only a one-line entry with a reference to another document. This will include usage, cleaning, calibration. maintenance and qualification. If there is a one-line entry every time the equipment is used for anything, this makes it easier for the user to look back when there is a problem. It also makes it easier for the inspector to find problems! Modern laboratory systems will often include an equipment log within the operating software of the system. When this is the case, it pays to ensure that it captures all of the activities listed previously.

Many companies have adopted the approach of keeping "Equipment History Files", which is an alternative approach whereby all of the documents related to an item of equipment are kept in a single file. These documents will include the initial qualification binders, any change control documents and calibration and maintenance records. However, the records documenting usage and cleaning records are normally part of the batch records and without an equipment log (book or electronic log), the Equipment History Files are not a complete "potted history". Historically, some companies were issued with inspectional observations for not keeping all necessary records in the equipment logbook. The level of sophistication and traceability of records has moved on since those times and provided the documentation system in operation can be demonstrated to work, then alternatives are now more generally accepted.

62.11 FACILITY READINESS

The readiness of the facility to be capable of running the process must be assessed at specific stages. This is generally done under change control to assess the impact of the process and the product on other activities in the facility. Initially when a new product/process is introduced into the facility, a more detailed assessment is required to establish that the operating ranges required of the equipment have been qualified to establish the risks for cross-contamination and to evaluate the need for changes to cleaning procedures, or to validate the cleaning of the new residue. The facility readiness needs to be reassessed when there are significant changes to the process or the process equipment, for example when new equipment is introduced into the aseptic filling room, the airflow patterns and the impact on environmental cleanliness must be reassessed.

If the change control is managed effectively and the assessments performed are thorough, there should be no surprises. Nevertheless, it is always advisable to perform a documented risk assessment of the capability of the facility to run the process at commercial scale. This should be performed prior to the performance of the PPQ and should include a demonstration that the critical process parameters (CPPs) can be met and controlled with a proven acceptable range (PAR) for the process. This is no different from the FDA expectation, but a European facility that has received frequent regulatory inspections (every 2 years) is more likely to be in compliance in this respect than a US facility that has never seen a regulatory inspection.

62.12 PROCESS VALIDATION

In Europe, as in the USA, the approach to process validation has changed significantly in the last 20 years. This change has been brought about partly because companies have been regularly challenged to justify the way in which they control their process and to justify the parameters they consider important and the validity of the outputs from the processes, the CQAs. How were these arrived at and how are they justified? It was often the case that the company simply did not have good answers to these questions. "We have always done it this way"! Sometimes the answer was that "it worked for other products so it should work for this one".

The resulting changes that brought about the concepts of establishing process understanding during the development phases, of establishing a PAR or a "Design Space", and of gaining a better understanding of process robustness and more meaningful CQAs, happened in Europe at the same time as it happened in the USA. The publication of the ICH harmonized guideline on Pharmaceutical Development (ICH Q8 (R2)) happened in all territories at the same time, and the FDA revised its process validation guidance shortly afterwards. The EU never published a separate detailed guidance document on process validation. It has always been contained within Annex 15. In 2015, Annex 15 was revised to bring it in line with both ICH Q8 and the FDA guidance document. This is a true testimony to the benefits of harmonization. It also means that companies do not need to validate their processes twice. If it is done correctly, once is enough for licensure in all territories. There are some nuanced differences in the expectations for maintaining validated status via continuous process monitoring. Annex 15 refers to three approaches, "traditional, continuous and hybrid". The "hybrid' approach is described as "a hybrid of the traditional approach and continuous process verification" and can be used "where there is a substantial amount of product and process knowledge and understanding which has been gained from manufacturing experience and historical batch data". This presumes perhaps an older process that has been running at the commercial scale for some time. Process knowledge at the commercial scale is probably the best knowledge of all. Annex 15 goes on to suggest that the hybrid approach may be used after changes or during ongoing process verification even though the product was initially validated using the traditional approach. The traditional approach is of course the manufacture of a number (normally 3) of batches under routine conditions.

Annex 15 states that in some cases it may be appropriate to perform PQ in conjunction with process validation. However, it does not elaborate any further. The PQ of autoclave loading patterns might be an example that was being considered when this statement was written. Annex 15 expands on the performance of PQ, using "production materials, qualified substitutes or simulated product" and tests should cover "the operating range of the intended process". Where such materials and tests are used, for example when performing APS, the inference is that this is representative of process validation for that process or part of the process.

As previously stated, process validation as described in Annex 15 is intended to apply to all pharmaceutical dosage forms, and there is a separate Part II of the EU GMP Guidelines that applies to APIs in which process validation of APIs is covered. However, this is harmonized because Part II is essentially the reproduction of ICH Q7 (16).

Continuous process verification has been adopted by Annex 15 following the revision of the FDA guidance on process validation (17). Annex 15 perhaps goes a bit further than the FDA guidance in one respect, as it states that if a product has been developed following a QbD approach, and the control strategy provides a "high degree of assurance of product quality", then continuous process verification can be used as an alternative to traditional process validation, rather than purely as an adjunct for ongoing maintenance of the validated status, which is the primary interpretation of the FDA guidance. In fact, Annex 15 distinguishes continuous process verification from ongoing process verification. The latter being regarded as the approach to validation maintenance in the EU. The success of this approach will depend on the degree to which the control strategy is accepted as providing a "high degree of assurance of product quality". Annex 15 also makes reference to the need for a regular evaluation of the control strategy and suggests the use of Process Analytical Technology and of multivariate statistical process control.

The main difference that Annex 15 draws between continuous process verification and ongoing process verification seems to be that the former approach involves a QbD approach, whereas ongoing process verification requires a documented periodic data review. The Annex implies that the documented review is the Product Quality Review required by the EU authorities. It also recognizes that there may be a need for additional actions, such as extra sampling, because of incremental changes to the process over time.

62.13 CLEANING VALIDATION

The approach to cleaning validation in Europe has changed significantly in recent years as a result of the publication by the EMA of a recent guideline on setting exposure limits (37). This is one of the few single-topic guidelines published by the EMA. It is being strongly enforced and it applies to different products manufactured in shared facilities. It requires that companies determine health-based exposure limits for all active substances manufactured in their facilities. These limits are referred to as Permitted Daily Exposure (PDE) Limits, from which the Maximum Carryover (MACO), more commonly associated with cleaning validation, is calculated. The PDE for a commercial product will normally be calculated from the no-observed-adverse-effect level (NOAEL) and the guideline provides the equation for performing this calculation, which also includes five factors (F1 to F5). These factors are intended to mitigate for pharmacological and toxicological variations between species, individuals and for repeated dosing. The calculation also considers the route of administration of the drug. Although the calculation of the PDE appears to be simple, the interpretation of the factors requires some expertise and the guideline requires that the calculated PDE value is approved by a scientific expert, which is normally a toxicologist. Therefore, the current inspectional expectation has rapidly developed, whereby the calculation and approval are performed by a qualified toxicologist. The guideline also describes an alternative approach for IMPs where there is very little toxicological data, such as data on reproductive or developmental toxicity, or where the NOAEL is not known.

Annex 15 still allows for the use of a "worst-case" approach for cleaning validation, but since the publication of the new guideline, the worst case must be considered based on the PDE as well as on formulation and other factors.

The verification of cleaning procedures during early-phase development is a relaxation from the requirements for a commercial product. This is based on practical considerations. At Phases 1 and 2, it is highly unlikely that a company will manufacture three batches at the same scale and in the same equipment, which is necessary to perform cleaning validation on three production runs. However, if the manufacturer uses reusable equipment, it is expected that the equipment will be effectively cleaned in readiness for the next batch or the next product. This requires a demonstration that the equipment has been effectively cleaned of residues from the previous batch to an acceptable level. If the new product can be shown to be bracketed by other products being manufactured in the same equipment, then a full verification is not needed, but in-process sampling will be needed (e.g. rinse water samples). If the product cannot be bracketed then "cleaning verification" is needed. Either way, this will involve determining acceptable carry-over levels and hence acceptable residue levels. It will also involve developing sufficiently sensitive and specific detection methods, and the ability to show adequate recovery from swab and rinse water samples.

62.14 VALIDATION OF STORAGE AREAS AND TRANSPORT

Good practices for the storage and transport of pharmaceutical products are described in detail in the EU Good Distribution Practice (GDP) Guidelines, which are contained within a European Directive (38). This Directive covers all aspects of receiving, warehousing, returns and destruction. However, details of the qualification/validation of storage areas and controlled-temperature storage units are not included. In practice, the qualification/validation of storage areas is no different from the practices followed in other regions, although, as stated previously, the WHO guidance seems to represent the "gold standard" for the qualification of storage areas, and there is very little knowledge of this standard either in Europe or in the USA, although most companies inadvertently seem to meet these requirements

Annex 15 contains some text covering the "verification" of transport. This includes the transport conditions defined in the marketing authorization (as described in the dossier) or "as justified by the manufacturer". This statement seems to imply that changes can be made without the need for preapproval, as long as they are justified. In other words, they must be validated. The Annex recognizes the variability of transport, but it requires that transport routes are clearly defined, and that seasonal and other variations are considered. This requires a risk assessment to consider the variables "other than those that are controlled and monitored" and gives as examples, delays, failure of monitoring devices and topping up of liquid nitrogen. The "verification" requires continuous monitoring and recording of critical environmental conditions "unless otherwise justified". It does not give any recommendations as to how this should be done or how many times, but the norm is, of course, three live shipments, the same as the industry norm. In addition, there is an expectation that simulated "worst-case" shipments will be performed, as a type of PQ, to verify that the shippers used will provide a suitable stable environment for the product when challenged with extreme environmental conditions. There is no mention of agitation of the product during transport, which is a common concern of FDA inspectors, unless it is implied in the statement concerning variables "other than those that are controlled and monitored".

Temperature-controlled trucks are also storage units albeit temporary ones. These should be qualified to the same standards as cold rooms and controlled-temperature units. The PQ for trucks, in particular, will need to consider the environmental extremes of the journey. The first thought is that in Europe these extremes may not be quite so variable as those between Alaska in the winter and Arizona in the summer. However, Europe includes parts of Scandinavia that are within the arctic circle and parts of the Mediterranean that can get above 50°C in the summer. When choosing a courier in Europe, it is important to be aware of the temperature ranges that can be experienced and to ensure that their trucks are suitably qualified. Surprisingly, this is an issue that often gets forgotten.

62.15 CONCLUSION

The title of this chapter asks the question "what are the differences?" This question was intended as "what are the differences between validation in Europe and validation in other industrialized countries". However, it has largely been interpreted as what are the differences between Europe and the USA, because, with the notable exceptions of Japan and China, most other major industrial countries have adopted the Pharmaceutical Inspection Convention/Cooperation Scheme (PIC/S) GMP Guidelines, which are the same as the European guidelines. The answer to the question, as this chapter suggests, is that the differences are sometimes in the written guidance but more often in the detailed interpretation of the guidance. In many cases, harmonization has been effective, but in other cases, there are still deeply held opinions on how validation should be performed, and perhaps the most significant of those is the way in which sterile products should be validated, as witnessed in the proposed revision to EU Annex 1, which is anything but an attempt to harmonize. Rather, it provides more granularity regarding the expectations that the EMA has been promoting and enforcing for many years.

There are many aspects of validation that are harmonized. Facility, equipment and clean utility qualification, and computer system and process validation are for the most part harmonized, and it is not too difficult for companies to plan their qualification of new facilities and their process validation such that they meet all global expectations. These are major undertakings, and it is reassuring to know that they only need to be done once to satisfy the expectations of most inspecting authorities. However, in addition to sterilization validation and APS, it is important for pharmaceutical manufacturing companies to be aware of the other differences in validation requirements in Europe. In particular, it is important for companies to be aware of the recent changes in expectations for cleaning validation, and it is particularly important for companies manufacturing clinical trials materials to appreciate the differences in expectations for "phase-appropriate" validation and the level of enforcement of those expectations by EMA inspectors and by QPs. Despite these differences, it is possible, with good planning and a good understanding of the differences, to write validation protocols for initial validation of sterilization, APS and cleaning that meet all global requirements, resulting in the need to perform these activities only once. It is also possible, with forward planning, to design a requalification and revalidation program that meets all global requirements despite some significant differences in expectations.

REFERENCES

1. Guide to Good Pharmaceutical Manufacturing Practice. The Stationery Office, London 1971.
2. Eudralex (The rules governing medicinal products in the European Union Volume 4. EU Guidelines for Good Manufacturing Practices for Medicinal Products for Human and Veterinary Use), Part 1, European Commission, 2013.
3. Health Technical Memorandum 10: Now obsoleted and Superseded by HTM2010 in 1994.
4. Health Technical Memorandum 2010, Sterilization, Part 3 (Including Amendment 1): Validation and Verification, Her Majesty's Station Office, London, Reprinted 1998, ISBN 0-11-321746-3.
5. BS EN 285, Sterilization-Steam Sterilizers-Large sterilizers, BSI 1997.
6. Pharmaceutical Inspection Convention, Pharmaceutical Inspection Co-operation Scheme, PE 009-13 (Part I and Annexes), Guide to Good Manufacturing Practices for Medicinal Products, Part I and Annexes, PIC/S, January 2017.
7. Eudralex (The rules governing medicinal products in the European Union Volume 4. EU Guidelines for Good Manufacturing Practices for Medicinal Products for Human and Veterinary Use), Part 1, Annexes 1, 11, 13, & 15, European Commission.
8. The European GMP Guidelines (The rules governing medicinal products in the EC. Vol. IV. Good Manufacturing Practices for medicinal products, Luxembourg: Office of Official Publications of the EC, Part 1, Annex 16: Certification by a Qualified person and Batch Release, 2015.
9. Matthews BR. The Devonport Incident: The Clothier Report, and Related Matters-30 Years On. *PDA J. Pharm. Sci. Technol.* 2002, **56**(3), 137–149.
10. Federal Standard 209E, Airborne Particulate Cleanliness Classes in Cleanrooms and Clean Zones, Cancelled November 29, 2001.
11. Pharmaceutical CGMPs for the 21st Century: A Risk-Based Approach, Final Report, US Food and Drug Administration, September 2004.
12. Annex 1 Revision Consultation Document. https://ec.europa.eu/health/sites/health/files/files/gmp/2020_annex1ps_sterile_medicinal_products_en.pdf, 2020.
13. PDA Technical Report: Points to Consider for Implementation of Pre-Use Post-Sterilization Integrity Testing (PUPSIT) and references therein.
14. GAMP 5, A Risk-Based Approach to Compliant GXP Computerized Systems, ISPE, 2008.
15. ISO 14644–1:2015 Cleanrooms and associated controlled environments-Part 1: Classification of air cleanliness by Particle concentration. www.iso.org/standard/53394.html, 2015.
16. ICH Harmonized Tripartite Guideline, Good Manufacturing Practice Guide for Active Pharmaceutical Ingredients, Q7, Step 4, November 2010.
17. Guidance for Industry, Process Validation: General Principles and Practices, Revision 1, FDA, January 2011.
18. ICH Harmonized Tripartite Guideline, Pharmaceutical Development, Q8 (R2), Step 4, August 2009.
19. ICH Harmonized Tripartite Guideline, Quality Risk Management, Q9, Step 4, November 2005.
20. ICH Harmonized Tripartite Guideline, Pharmaceutical Quality System, Q10, Step 4, June 2008.
21. Directive 2001/20/EC of the European Parliament and of the Council of 4 April 2001 on the Approximation of the Law, Regulations and Administrative Provisions of the Member States Relating to the Implementation of Good Clinical Practice in the Conduct of Clinical Trials on Medicinal Products for Human Use. "The Clinical Trials Directive" 2001.
22. Regulation (EU) No. 536/2014 of the European Parliament and of the Council of 16 April 2014 on Clinical Trials on Medicinal Products for Human Use, and Repealing Directive 2001/20/EC, Article 61, Authorisation of Manufacturing and Import, 2014.
23. MHRA, Points to Consider When Preparing an IMP Dossier. https://assets.publishing.service.gov.uk/government/uploads/system/uploads/attachment_data/file/317987/Points_to_consider_when_preparing_the_IMP_dossier.pdf.
24. ICH Harmonized Tripartite Guideline, The Common Technical Document for the Registration of Pharmaceuticals for Human Use: Quality-M4Q (R1), Quality Overall Summary of Module 2, Module 3: Quality, section 3.2.P. 5.3, Step 4, September 2002.
25. ICH Harmonized Tripartite Guideline, Validation of Analytical Procedures: Text and Methodology, Q2 (R1), Step 4, Step 4, November 2005.
26. Medicines & Healthcare Products Regulatory Agency (MHRA) 'GXP' Data Integrity Guidance and Definitions: Revision 1, March 2018.
27. Data Integrity and Compliance With Drug CGMP, FDA Guidance for Industry, U.S. Department of Health and Human Services, December 2018.
28. PDA Releases Comments on Annex 1: Manufacture of Sterile Medicinal Products, PDA, Bethesda, MD, March 2018.
29. Deeks T. Unpublished Data Presented to MHRA, 1999.
30. Guidance for Industry, Sterile Drug Products Produced by Aseptic Processing: Current Good Manufacturing Practice, FDA, September 2004.
31. Morris T, Jornitz M, Gori G, Baseman H. PUPSIT and the Annex 1 Revision. PDA Letter, August 29, 2019.
32. Mok Y, Besnard L, Love T, Lesage G, Pattnaik P. Best Practices for Critical Sterile Filter Operation: A Case Study. *Bioprocess International* 2016 May.

33. PDA PUPSIT Survey, ISBN:978-1-945584-01-5, Parenteral Drug Association, Inc., 2017.
34. Eudralex (The rules governing medicinal products in the European Union. Volume 4, Good Manufacturing Practice) Guidelines on Good Manufacturing Practice specific to Advanced Therapy Medicinal Products, European Commission, November 2017.
35. Guidance for Industry, Compliance with 21CFR Part 1271.150(c)(1): Manufacturing Arrangements (Compliance with CGTP Requirements), FDA, September 2006.
36. WHO Guidance Temperature Mapping of Storage Areas, Technical Supplement to WHO Technical Report Series, No. 961, 2011, Annex 9, January 2014.
37. Guideline on Setting Health Based Exposure Limits for Use in Risk Identification in the Manufacture of Different Medicinal Products in Shared Facilities. EMA/CHMP/CVMP/SWP/169430/2012. European Medicines Agency, November 2014.
38. European Guidelines on Good Distribution Practice of Medicinal Products for Human Use, 2013/C 343/01. European Commission, November 2013.

63 Japanese Approach to Validation

Satoshi Sugimoto, Mitsuo Mori, Kiyoshi Mochizuki, Keisuke Nishikawa, Takuji Ikeda, Yusuke Matsuda, Hiroaki Nakamura, and Yasuhito Ikematsu

CONTENTS

63.1 INTRODUCTION

Japan's Pharmaceutical Inspection Convention/Cooperation Scheme (PIC/S) membership was approved at the PIC/S general meeting in May 2014, and Japan became the 45th PIC/S member on July 1, 2014. With the participation in PIC/S, Good Manufacturing Practice (GMP) for the manufacture of sterile pharmaceutical products in Japan was expected to be equivalent to PIC/S GMP. There has been "Guidance on the Manufacture of Sterile Pharmaceutical Products Produced by Aseptic Processing (revised edition)"[1] and "Guidance on the Manufacture of Sterile Pharmaceutical Products Produced by Terminal Sterilization (revised edition)"[2] published by the Scientific Task Force of the Ministry of Health, Labour and Welfare of Japan (hereinafter abbreviated as MHLW) on the manufacture of sterile pharmaceutical products, in the form of supplementing PIC/S GMP even after joining PIC/S. Until the 16th revision of the Japanese Pharmacopoeia (hereinafter abbreviated as JP), "Parametric Release of Terminally Sterilized Pharmaceutical Products", "Media Fill Test (Process Simulation)", and "Microbiological Evaluation of Processing Areas for Sterile Pharmaceutical Products" were listed as general information in the JP.[3] These would be removed to prevent duplication of requirements when PIC/S-GMP Annex 1 is formally revised. Many sterile pharmaceutical products manufacturers in Japan have implemented their practices based on PIC/S GMP and other supplementary guidelines and notices. This chapter summarizes the actual operation in Japan, focusing on these. In particular, facility design, water system, environmental monitoring, bioburden

DOI: 10.1201/9781003163138-63

management, process simulation, and terminal sterilization are described, and pest control, which is especially important for operation and management in Japan, is described in a new section. In addition, regarding microorganism control, which is indispensable for manufacture of sterile pharmaceutical products, this section describes rapid microbiological methods combining the latest technology and knowledge. "Rapid Microbiological Method" was newly included in the general information of the 17th revision of the JP,[4] and this section explains mainly validation of the method.

The basic concept of the validation of aseptic processing in the manufacture of sterile pharmaceutical products is to combine hardware such as the manufacturing facility with software such as the operating methods and management. Sterile pharmaceutical products manufactured through the aseptic process contain various factors that cannot be verified during the development and design stages. Therefore, it is necessary to plan and implement qualification and validation as a system for the entire manufacturing site. By scientifically verifying aseptic operations such as sterilization and filling, maintenance of the manufacturing environment, facilities and equipment at the manufacturing site, and the risk of contamination in the manufacturing process, it is ensured that contamination is prevented. There is also a basic requirement to control the manufacturing process using validated operating procedures and manufacturing control parameters. Regarding these requirements, this chapter focuses on the important points and explain them.

63.2 FACILITY DESIGN

63.2.1 Classification of Manufacturing Areas by Air Cleanliness

Facilities for processing sterile pharmaceutical products comprise clean areas controlled based on predefined airborne particle and microbiological standards. The areas are classified as critical, direct support, and indirect support areas depending on the nature of the operation to be conducted.

Generally, the cleanliness in processing areas is defined by the number of airborne particles ≥0.5 μm in diameter per unit volume of air. The number of particles ≥5 μm in diameter may serve as a reliable parameter for early detection of environmental deterioration if regularly monitored and evaluated by trend analysis. Table 63.1 shows the air cleanliness requirements for classified areas.[1]

The critical area (Grade A) is a processing area where sterilized products and materials as well as their surfaces are directly exposed to the environment. The environmental conditions should be specified to be suitable for the elimination of contamination risks and preservation of the sterility of products. The following processes are conducted in this area: sterilization activities (e.g. sterile connections, addition of sterile materials) prior to filling, sterile filling, and sterile closure.

The direct support area (Grade B) is defined as a background area of the critical area when aseptic processing is conducted using an open clean booth or restricted access barrier system (RABS). The direct support area is a working area for personnel who operate machines installed in the critical area and for those who supervise the operation of the machines. The direct support area also serves as a route for the transfer of sterilized products, materials, and equipment to the critical area or for moving sterilized products from the critical area. In the latter case, appropriate measures need to be implemented to protect the sterilized products or materials from direct exposure to the environment.

The indirect support area (Grade C and Grade D) is an area used for processing materials and products prior to sterilization processes and hence materials and products are directly exposed to the environment. Examples of indirect support areas include an area for preparing drug solution prior to sterilization and an area for washing and cleaning sterilization equipment and apparatuses. The cleanliness of the indirect support area needs to be controlled by establishing specifications for acceptable airborne particle count with consideration of the required level of contamination control and type of works performed in the area.

TABLE 63.1
Categories of Clean Areas

Area		Air cleanliness[1]	Maximum allowable number of airborne particles(/m³)			
			Count under nonoperating conditions		Count under operating conditions	
			≥0.5 μm	≥5.0 μm	≥0.5 μm	≥5.0 μm
Aseptic processing area	Critical area	Grade A (ISO 5)	3,520	20	3,520	20
	Direct support area	Grade B (ISO 7)	3,520	29	352,000	2,900
Indirect support area		Grade C (ISO 8)	352,000	2,900	3,520,000	29,000
		Grade D	3,520,000	29,000	Dependent on process attributes[2]	

[1] The ISO class designation in parentheses refers to the count during operation.
[2] There are cases where a maximum allowable number may not be specified.

63.2.2 Considerations on Facility Design

Clean areas should be designed by taking into account the following as general requirements. 1) Clean areas should be clearly separated from rest rooms and eating areas. 2) Clean areas should be well-separated for intended purposes from other processing operations within a facility and should have sufficient space to allow proper conduct of all manufacturing operations that are to be done within them. 3) Clean areas should be designed to achieve efficient flow and control of materials, products, and personnel within the areas. The location of equipment in the areas should also be carefully planned to minimize crossing of personnel, products, and materials flows. 4) Material handling procedures or fixed depots should be effective in preventing a mix-up between clean and dirty or sterilized and non-sterilized apparatuses and utensils. 5) Facilities should be designed to facilitate ease of cleaning, maintenance, and operations and periodically inspected to verify that the facilities are maintained as originally designed. Particular consideration should be given to seals and packing of interior materials such as doors, walls, and ceilings in order to keep processing rooms tightly closed. Insulation materials to prevent moisture condensation should be maintained. 6) Use of non-smooth surface material and horizontal surfaces around windows and doors should be avoided to reduce dust and microorganism accumulation and to keep the airflow smooth. If such designs are unavoidable, these parts should be suitable for easy cleaning. Sliding doors may be undesirable for this reason. 7) Transparent (e.g. glass) windows or video cameras should be installed in the Aseptic Processing Area (hereinafter abbreviated as APA) to facilitate observation and supervision from non-aseptic areas. 8) When parenteral and other sterile pharmaceutical products are manufactured simultaneously in the same room, manufacturing equipment for preparation, filling, and sealing of drug products should be dedicated for each product and should be closed to avoid cross contamination. If any part of the equipment system is open, appropriate measures and activities should be implemented to prevent contamination. 9) The working areas for preparation, filling, and sealing of sterile pharmaceutical products and sterile API should be separated from the areas for processing non-sterile drug and non-sterile API. The separation is not necessary if there is no risk of contamination of products processed in the working areas. 10) Facilities should be structurally designed to be efficient in preventing or minimizing risks of cross contamination if used for processing highly pharmacologically active substances, pathogenic substances, highly toxic substances, radioactive substances, live viruses, or bacteria. 11) Walls, floors, and ceilings should be easily cleanable and durable against cleaning agents and disinfectants. 12) Drains and sinks should be prohibited in the APA. If drains are placed in Grade C areas in indirect support areas, drains should be fitted with traps or water seal parts that are easy to clean and disinfect to prevent contamination by backflow. If floor trenches are located, they should be shallow to facilitate cleaning. 13) Clean areas should be supplied with air filtered through an appropriate filter, e.g. a high-efficiency particulate air (HEPA) filter, to maintain an acceptable level of air quality and pressure difference between areas. The pressure difference should be monitored so that it is maintained as specified. 14) Temperature and relative humidity in clean areas should be controlled within ranges compatible with the properties of the materials and products being handled in the areas and also set at levels suitable for microbiological control. 15) Environmental temperature and relative humidity should be controlled within specified limits and, wherever feasible, monitored continuously. 16) Air pressure in clean areas should be maintained higher relative to adjacent lower cleanliness areas through doors, except for containment philosophy facilities for handling potent substances. 17) Direct support areas should be separated from adjacent areas by airlocks. Between direct support areas and adjacent areas there should be pass-through rooms and/or pass-through boxes that are used to transfer sterilized materials or used to disinfect materials. 18) The gowning room should be equipped with an airlock system and gowning areas and degowning areas should be physically separated from each other. Air particulate cleanliness in the gowning room should be maintained at the same grade as the area (at rest) into which it leads. In order to reduce rapidly the numbers of particles and microorganisms associated with gowning activity, the volume and/or air change rate of the room should be adequately considered. Supply air at a relatively high level and exhaust air at a lower level in the room are desirable. The air cleanliness of the pass-through box should be specified according to the intended use. 19) As for changing rooms to the direct supporting areas, it is desirable to have a separate entry changing room and exit changing room. As an alternative measure, it is acceptable to stagger the time of entry and exit at one changing room.

63.3 HEATING, VENTILATING, AND AIR CONDITIONING SYSTEM

Air in clean areas needs to be maintained at appropriate levels by designing, instituting, and managing a suitable heating, ventilation, and air conditioning (hereinafter abbreviated as HVAC) system. The integrity of the system should be ensured with respect to not only temporal variations because of operational activities, such as door opening and closing and facility equipment operation, but also long-term variations because of nonoperational activities, such as seasonal changes in outdoor conditions or aging of equipment and apparatuses. The HVAC system and its management program are composed of the following basic elements: temperature, relative humidity, airflow volume, air exchange rate, unidirectional airflow, pressure difference relative to adjacent rooms, integrity of the HEPA filter, airborne particle count, and microbial count.

63.3.1 Temperature and Relative Humidity

Temperature and relative humidity have a direct impact on the comfort of personnel and the possibility of microbial contamination in processing areas; therefore, these environmental parameters should be appropriately defined, controlled, monitored, and maintained at appropriate levels.

63.3.2 Air

It is critical to secure constant airflow from an area of higher cleanliness level to an area of lower cleanliness level in order to maintain the required environmental conditions of clean areas. 1) The pressure difference between the APA and indirect support areas should be adequately defined, monitored, and controlled. 2) Airlocks should be established between the APA and indirect support areas, and the pressure difference between these areas should be maintained at a level sufficient to prevent the reversal of defined pressure difference or airflow. For example, a desired pressure difference between areas, when both doors are closed, should be at least 10 to 15 Pa. Likewise, an appropriate pressure difference should be established and maintained between indirect support areas of different cleanliness levels. 3) As the pressure difference is an essential part of sterility assurance, the pressure difference between areas should be continuously monitored, and it is recommended to have an alarm system to enable the prompt detection of abnormal pressure differences. 4) Airflow in the critical area (Grade A) should be unidirectional and supplied at a velocity and uniformity sufficient to swiftly remove airborne particles away from the critical area. Airflow should also be supplied with sufficient care so as not to create reverse currents (eddies) from adjacent areas (direct support areas, Grade B) into the critical area to prevent contamination. When conventional clean booth and RABS are used, the recommended average flow rate is 0.45 m/sec ± 20%. Lower flow rates may be appropriate depending on isolator systems and other cases. 5) The airflow requirements stated previously should be verified by validation by a smoke test or other qualification tests at the installation of airflow equipment. Similar validation is also necessary when airflow patterns are changed or suspected of being changed. 6) Changes in flow velocity may alter flow direction where airflow is specified to be unidirectional. The velocity should be confirmed to be constant at a predetermined level by monitoring the velocity of airflow from HEPA filters at time intervals specified in the program. 7) An appropriate air change rate should be established by evaluating the potential for product contamination of individual processing areas and gowning rooms in the APA to maintain air cleanliness at specified levels. The generally recommended air change rate is 30 times per hour in the direct support area and 20 times per hour in Grade C work rooms among indirect support areas. These change rates should be monitored at regular intervals to verify that the rates are continuously maintained as specified. Airflow should be controlled so as not to stir up dust and bacteria from the floor and to prevent the deterioration of the work environment. The most common method of securing downward current is to install supply vents close to the ceiling and exhaust vents close to the floor. Similar considerations on ventilation are recommended for indirect support areas, depending on the potential risks of contamination with microorganisms and foreign matter. 8) The cleanliness of the work room must be promptly returned to the non-operating level after the manufacturing operation is completed and the operators leave the room. In the direct support area, the airborne particle count should preferably be returned to the counts at non-operating conditions in 15 to 20 minutes. 9) Intended differential pressure and airflow patterns during processing should be specified and documented and then validated. The impact of the turbulence created by the movement of personnel on the cleanliness of the manufacturing environment should be evaluated, and evaluation results should be reflected in relevant Standard Operating Procedures (SOPs).

63.3.3 Integrity of HEPA Filters

HEPA filters should be accompanied by a supplier's certificate of quality verifying that the filter is capable of eliminating at least 99.97% of particles ≥0.3 μm in diameter. A leak test of HEPA filters used in critical areas (Grade A) and direct supporting areas (Grade B) should be performed by using appropriate leak testing aerosols, e.g. poly-alpha-olefin (PAO). When alternate aerosols are used, such aerosols should be used after confirming that they do not promote microbial growth.

HEPA filters should be tested for leaks at installation and thereafter at suitable time intervals. The procedure and frequency of testing should be tailored to the environment, where the filters are installed, and their intended purpose of use. The integrity of HEPA filters in the critical area and direct support area should be confirmed at least once a year. The integrity check is recommended to be more than twice a year in the case that conditions of use in the critical area are severe or special considerations are required for the prevention of microbial product contamination. HEPA filters installed in the critical area (Grade A) should be tested for uniformity of air velocity across the filter at installation and thereafter at suitable time intervals. The frequency of integrity checks should be determined as stipulated previously. The pressure difference (pressure loss) between the HEPA filter's upper side pressure and lower side pressure should be tested at installation and thereafter at suitable time intervals. If filter clogging risk is high, it is recommended to include pressure difference monitoring in routine control procedures.

63.4 WATER SYSTEMS

In Japan, "Guidance on the Manufacture of Sterile Pharmaceutical Products Produced by Aseptic Processing (revised edition)"[1] describes in detail the concept of water system design and management. The basic design of the equipment and other subsystems applicable to pharmaceutical water production should be developed after establishing the procedures necessary for the efficient manufacturing and quality control of pharmaceutical water so as to maintain a constant supply of pharmaceutical water in the required quality. Critical points to consider on designing the water systems should include, but not be limited to, the following: 1) All of the grades, specifications, quantities, and control methods for pharmaceutical water(s) should clearly be defined. 2) The variant quality of source water including seasonal changes should thoroughly be ascertained. 3) The water system design should be predetermined on maximum momentary water flow rate, application time and frequency of water to be used, and such conditions demanded at the points of use as temperature, number of ports, and piping specifications including branches and the pipe's diameters. 4) Pharmaceutical water equipment should have such a reliable sterilization or sanitization system as to ensure microbial control is provided. 5) The location of water sampling ports for water quality control should be evaluated so well as to ensure a stable supply of pharmaceutical water that fulfils required quality specifications. Water samples should be collected from the locations not only of points of use but also other critical points for the pharmaceutical water process. Locations necessary for water quality assessment should be provided with structural features that facilitate the sampling for quality analysis.

Pretreatment equipment should be selected in consideration of the capacity suitable for maintaining invariable water quality within the specifications required and for maximizing water treatment efficiency and system life on the basis of elaborate investigation of the amounts of heavy metals, free chlorine, organic matter, microorganisms, and colloidal particles, *etc.* present in the source water.

The manufacturing equipment for water for injection (WFI) should be designed so as to facilitate periodic sterilization. If steam sterilization is inapplicable to the equipment because of its low heat resistance, an alternative system should be allowed to perform sterilization or sanitization using hot water or chemical agents. For WFI processing, distillation, reverse osmosis (RO), ultrafiltration (UF), and any combination of them are recommended.

WFI should preferably be used immediately after production to avoid any intermixture contamination with microorganisms and other chemical substances. The following factors should be taken into account in storing WFI and other high-purity pharmaceutical waters in tanks. 1) Storage tanks should be a closed-type with smooth inner surfaces. The nozzle of a level indicator attached to the tank should be minimum in number and as short as possible. 2) Storage tanks should be structured to prevent water stagnation and allow easy cleaning of the inner surfaces and to facilitate complete drainage. 3) The appropriate capacity of a storage tank should be determined to provide a water turnover at the highest possible rate. However, long-term storage of pharmaceutical water in the tank should be avoided. The maximum storage time should be established by validation. 4) The storage tank should be provided with a hydrophobic vent filter with micropores of 0.2 or 0.22 μm to prevent the invasion of microorganisms and foreign matter into the tank. The integrity of the vent filter should be ensured prior to installation and at regular intervals thereafter. 5) Where hot water is supplied into a storage tank, a heater should be installed around the vent filter to prevent obstruction of the filter because of condensation of the hot water. 6) When the tank is disinfected with hot water, the tank should be equipped with additional mechanisms to have heat spread over the whole inner surface of the tank including its upper part.

Pharmaceutical water stored in a tank is transported to the points of use through piping with relatively small diameters and structure as a closed system so that the interior of the piping, once installed, is difficult to examine and inspect. Therefore, thorough review of control methods and preventive measures in the piping system should be made at the design phase. Key points to consider in the piping systems are described following: 1) It is preferable that the piping systems should not be provided with bypasses or branches through which water is not constantly running. 2) WFI should preferably be circulated constantly at a temperature not lower than 80°C and at a turbulent flow rate to prevent microorganisms and organic matter from anchoring on the surfaces. Where no water circulation is provided, the unused water should be drained and refilled with new water. 3) Any loop circulation at ambient temperature should be provided with preventive measures for microbial growth. One example is the employment of UV lamps (Germicidal Ultraviolet Lamp) placed at an appropriate location along the piping. 4) The piping system designed as a closed loop should be provided with preventive measures in the loop to maintain a positive inner pressure against any backflow from points of use. 5) Where a no circulation system is adopted, the system should be provided with such preventive measures as hot water flushing or steam sterilization for microbial contamination prior to water use. 6) Every horizontally arranged piping should have an inclination of at least 1/100 given to prevent water from stagnation in the piping in drainage and steam or hot water sterilization. 7) As water will readily stagnate at "dead legs" occurring in T-shaped branches from the main piping leading thence to a closure mechanism such as valves, the distance between the diametrical axial center of the main piping and the closure mechanism in use should not be longer than six times the inner diameter of the branch, but preferably not longer than three times if possible. The water for pharmaceutical purposes in the holding tank is distributed to the points of use through the piping in the water system.

For heat exchangers, any contamination of feed water because of the leakage of the heat medium contained in the heat exchanger should be prevented. A double tube type or double tube-sheet type (shell-and-tube type) of heat exchanger is generally used. When a heat exchanger other than these types is used, a heat exchanger allowing no contamination of the feed water because of the heat medium should be selected. If any potential risk of contamination of the feed water is supposed, a positive pressure on the feed water side should be maintained at a level higher than that on the heat medium side, and an appropriate monitoring system and alarm for the pressure differential should be attached to the heat exchanger.

Adequate design and control of the points of use and sampling points should be made in consideration of the following: 1) No sterilization filters should be placed in the loop or at any points of use, because the filters may hamper adequate monitoring of microbiological contamination in the water system and endotoxins could be released from microorganisms retained by the filters or from dead microorganisms in the filters. If the inclusion of sterilization filters is unavoidable, the frequency of their disinfection/sterilization and replacement should be determined based on validation. 2) When water samples cannot be collected at the points of use, sampling ports should be installed as close to the points of use as possible, except for the case where the sampling and/or the installation of such sampling ports are regarded to be obviously disadvantageous.

Valves, gauges, and detectors installed in the water system need to be of a sanitary structure such as a diaphragm and have no static area of fluid or dead volume. In order to timely monitor the chemical quality of the water, installation of a TOC gauge and conductivity meter in the line is desirable. It is recommended to select a TOC gauge location where the water quality would be expected to be the poorest in the piping system.

Pumps should be a sanitary-type design with sealed casing protected against contamination and capable of withstanding hot water sanitization and/or pure steam sterilization.

In addition, "the JP, General information of Quality Control of Water for Pharmaceutical Use"[5] describes Microbiological Monitoring in detail. It is useful for microbiological monitoring to use the R2A Agar Medium, which is excellent for growing bacteria of oligotrophic type and can detect a wide range of bacteria including slow-growing microorganisms. In the media growth promotion test with the R2A Agar Medium, use the type strains of *Methylobacterium extorqus* and *Pseudomonas fluorescens* or equivalent to these strains. Prior to the test, inoculate these strains into sterile purified water and starve them at 20°C–25°C for 3 days.

63.5 ENVIRONMENTAL MONITORING/ PRODUCT BIOBURDEN[1,2]

63.5.1 Objective

The primary objective of environmental monitoring is to keep manufacturing environments for sterile pharmaceutical products clean by evaluating the levels of microorganisms and airborne particles within individual APAs and indirect support areas, Environmental monitoring tracks conditions to prevent environmental deterioration and product contamination and determine the efficiency of cleaning, disinfection, and decontamination procedures. Environmental monitoring may be classified into two categories: microbiological and particle control. Microbiological control is not intended to identify and characterize all microorganisms present in the environment but to scientifically estimate the bioburden of the environment, ensure that the manufacture of sterile pharmaceutical products is conducted in an appropriately controlled environment, and implement measures (e.g. disinfection) necessary for maintaining the environment at the required cleanliness level.

63.5.2 Environmental Monitoring Program

The risk of contamination to sterile pharmaceutical products varies depending on the type and quantity of the pharmaceutical product to be manufactured and the equipment for environmental control such as the HVAC system. Therefore, a monitoring program should be developed and monitoring of microorganisms and particulate matter should be routinely performed. In addition, SOPs for implementing the program should be developed, and the outcome should be adequately recorded.

Table 63.1 and Table 63.2[4] show the air cleanliness and the acceptance criteria for environmental microorganisms required for each processing area. These tables can be referred to, and the following items should be considered when developing the monitoring program.

1) Properties of substances to be monitored, frequency of monitoring, sampling locations, and action levels should be assessed and examined in order to estimate environmental contamination risks. 2) Because the risk of contamination varies depending on the formulation and size or volume of the pharmaceutical products, structure/function of manufacturing equipment, automation level, time of retention of closures, and availability and performance of equipment for environmental control such as the HVAC system, the appropriate environmental monitoring program should be developed as needed. 3) The environmental monitoring program should include all necessary items to verify whether air-cleanliness in each production area is constantly maintained. 4) The program should include periodic investigation on characterization of environmental flora and isolates to assess contamination risks to pharmaceutical products.

63.5.3 Monitoring Targets

Monitoring targets are microorganisms and airborne particles.

1) Target airborne particles are those ≥0.5 μm in diameter. Particles of other diameter (e.g. ≥5 μm) should be measured on an as-needed basis. 2) Target microorganisms are bacteria and fungi. 3) Target microorganisms are airborne bacteria and microorganisms on the surface of walls, floors, fixtures, equipment, gowns, *etc.* 4) The monitoring of airborne microorganisms on settle plates is performed as needed.

63.5.4 Monitoring Points

Environmental monitoring should be conducted in critical areas (Grade A) which are APAs, direct support areas (Grade B), and indirect support areas (Grade C or D) adjacent to APAs. The following items should be considered when determining monitoring points.

1) Monitoring points should include air in working areas, manufacturing equipment (and process control equipment, where appropriate), and aseptic environments; air for keeping the aseptic environment clean; and compressed air or gas that comes in contact with the environment and equipment. 2) Regular monitoring points in the manufacturing area should

be determined based on the risk assessment and the data obtained in monitoring for cleanliness classification, e.g. the near vicinity (e.g. within 30 cm) of a site where sterile materials are exposed to the surrounding environment, a site that is prone to potential sources of contamination because of frequent human interventions and traffic or because of susceptibility to lower cleanliness levels, or a site regarded as "worst case" based on the airflow analysis. 3) The points where critical operations are performed, where a contamination risk is considered high, and points that represent the cleanliness levels of the manufacturing area should be included. 4) Sampling of surfaces that come in contact with pharmaceutical products or other materials in critical areas should be performed immediately after the completion of filling or other aseptic processing operations. 5) Gases that may directly contact pharmaceutical products, primary containers, and surfaces that come into direct contact with pharmaceutical products should be periodically inspected and controlled to ensure the absence of microorganisms.

63.5.5 Monitoring Methods

Appropriate methods should be selected according to the items to be monitored. Consideration should be given to the potential contamination risks increased by interventions of personnel who are involved in sampling and the disturbance of airflow during sampling. The following methods should be considered.

1) For measurement of particulates, particle counters that can detect particulates of different sizes are used. 2) Measurement methods of microorganisms for environmental monitoring include active microbial sampling methods, measurement methods for microorganisms on surfaces, and settling plates. Appropriate samplers and measuring methodology should be selected according to the purpose of the monitoring and the items to be monitored. 3) For cultivation of microorganisms, growth promotion testing should be performed on all lots of prepared media. Media and extraction liquids should be sterilized in an appropriate manner. 4) Identification of microorganisms detected in Grade A and B areas to the species level is recommended.

63.5.6 Frequency of Monitoring

Recommended frequencies of monitoring during operation are given in Table 63.3.[4] The frequencies are set for general and conventional aseptic processing. In individual cases, appropriate monitoring frequency should be determined based on the results of risk assessment. The following items should be considered when determining monitoring frequency.

1) Sampling frequency should be determined in accordance with the air cleanliness level required for individual working areas under both operating and nonoperating conditions. 2) The sampling procedures should include specifications on the frequency of sample collection from gowns and other surfaces. 3) Sampling of particulate matter in Grade A and B areas should preferably be conducted via continuous monitoring beginning with equipment assembly until completion of critical operations. 4) The monitoring frequency for Grade C and D areas should be determined by the types of pharmaceutical products, processes, operations, *etc.* to be performed in the areas for appropriate quality control and risk management.

63.5.7 Alert and Action Level Specifications

In environmental monitoring, it is important to evaluate whether the specified cleanliness level for a monitored object is constantly maintained. Alert and action levels should be established for individual target substances and locations to be monitored.

Action level specifications may be established by referring to data contained in Table 63.2. The following items should be considered when establishing level specifications.

1) A risk assessment should be performed for each manufacturing facility and acceptance criteria should be established based on the identified risks. 2) Alert level specifications should be established based on results of PQ tests. 3) The monitoring program should include actions and measures to be taken (i.e. investigation of causes of noncompliance, suspension of manufacturing) when alert or action level specifications are met.

The following items should be considered in the evaluation.

1) The following items should be included in the evaluation: (i) Changes in numbers of microorganisms and airborne particulates over a period of time, (ii) Changes in the detected species of microorganisms, (iii) Changes in monitoring points, (iv) Review of the validity of alert and action levels, (v) Review of the frequency of positive results from each operator, and (vi) Changes that may impact the monitoring results during the monitoring periods. 2) In the event of deviations found in the environmental monitoring data, actions to be taken for the products manufactured and measures to be taken to recover the required cleanliness of the environment should be determined with consideration of the nature of activities

TABLE 63.2
Acceptance Criteria for Environmental Microorganism Count (During Operations)[1]

Cleanliness grade	Airborne microorganisms: Air (CFU/m³)	Airborne microorganisms: Settle plate[2] (CFU/plate)	Surface microorganisms: Contact plate (CFU/24–30 cm²)	Surface microorganisms: Gloves (CFU/5 fingers)
A	<1	<1	<1	<1
B	10	5	5	5
C	100	50	25	–
D	200	100	50	–

[1] Acceptance criteria are expressed as mean values.

[2] Measurement time per plate is 4 hours at maximum and the measurement is performed during processing operation.

TABLE 63.3
Frequency of Environmental Monitoring for Microbacterial Control

Cleanliness grade		Airborne particulate matter	Airborne microorganisms	Surface microorganisms	
				Equipment and walls	Gloves and gowns
A		During processing	Every working shift	After completion of processing	After completion of processing
B		During processing	Every working shift	After completion of processing	After completion of processing
C, D[1]	Area in which products and containers are exposed to the environment	Once a month	Twice a week	Twice a week	-
	Other areas	Once a month	Once a week	Once a week	-

[1] When a contamination risk is low, for example, where products are not exposed to the surrounding environment, monitoring frequency may be reduced accordingly.

performed at the time, distance between the product and the site where the deviation was found, and the severity of the deviation. 3) Caution should be exercised not to underestimate the contamination risk by averaging particle or microbial counts. 4) If microorganisms are detected in a Grade A area, the effect of such microorganisms on product quality should be evaluated even if the count meets acceptable criteria. 5) Surfaces and personnel should be monitored after the completion of critical operations. 6) For the adequate maintenance of the manufacturing environment, data obtained from routine monitoring should be analyzed to detect any trend changes in the environment and establish monitoring limits for trend analysis.

63.6 PROCESS SIMULATIONS

Sterile drug products may be manufactured through a single or several sterilization processes or a combination of sterilized components; an aseptic filling process is one of the manufacturing processes of drug products purporting to be sterile. In order to evaluate the sterility assurance of the drug products, the whole aseptic process should undergo process validation. Process simulation using media or other microbiological growth materials instead of actual products is one of the validation methods to evaluate not only the filling process but the whole aseptic manufacturing process. Included within the scope are the manufacturing process of sterile API and/or sterile in-process products and the overall manufacturing processes of drug products purporting to be sterile.

The operating personnel, operating environment, and processing operation should also reflect the actual manufacturing process, including "worst case" conditions. The necessary information for conducting the tests should be referred to the guideline that incorporates the General Information of JP XIV[4] as the basic concept.

Regarding the number of units filled during a process simulation test, a sufficient number of units should be used to simulate the aseptic manufacturing process.

In consideration of process simulation testing, it is recommended that the whole aseptic processing be simulated, because there may be a risk of contamination in the processes other than filling. When a process simulation test is conducted, it should be so planned that all contamination factors assumed in normal operations are included based on the identification of potential contamination factors. "Guidance on the Manufacture of Sterile Pharmaceutical Products Produced by Aseptic Processing (revised edition)"[1] recommends the conduct of process simulation tests considering the following eight points:

1) Cleaning of facilities and equipment and cleaning and disinfection of manufacturing equipment, containers, closures, and trays should be conducted in accordance with SOPs. 2) Process simulation should be performed for all routine activities at different manufacturing stages and temporal processing interventions. 3) Process simulation for temporal processing interventions which are known to occur on a routine basis (e.g. weight adjustment and supply of sterile materials, containers, closures, environmental monitoring) and anticipated but non-programmed interventions (e.g. modification of manufacturing line, adjustment of equipment conditions, repair or replacement of equipment parts) should be performed under practical operating conditions that simulate the worst possible intervention conditions. 4) Process simulation should be performed under equipment operating conditions (e.g. lines speed) that would most likely cause contamination. 5) Process simulation should be performed over the time period determined by taking the longest possible time of actual operations into account. 6) All personnel engaged in aseptic processing are required to participate in process simulation. The simulation test should be designed by simulating the largest possible number of participating personnel and working shifts. 7) Enough medium should be filled in the container to allow the medium to contact the entire inside surface of the container on rotation or inversion, thereby rendering a reliable judgment of bacterial growth. 8) Process simulation should be performed by replacing inert gas with air as the simulation test is not intended for detection of anaerobic growth.

The target of the acceptance criteria of a process simulation should be zero growth regardless of the number of units filled per simulation. Where contaminated units are found, actions shown in Tables 63.4 and 63.5 should be taken.[4]

During the initial validation when a process is newly established, it should be confirmed that the results of a minimum of three consecutive runs on each separate day comply with the criteria. In periodic validation (every half year as a principle except for the line for multiple use with partially different processing, in which case a process simulation run for each product should be required), the result of one process simulation run is employed. All personnel working in the critical processing area should be trained in aseptic processing and participate in a media fill run at least once a year. When filling lines have not been used for over six months, conduct appropriate numbers of media fill runs in the same way as for the initial performance qualification prior to the use of the filling lines. In cases of facility and equipment modification (interchanging standard parts may not require requalification), major changes in personnel working in critical aseptic processing, anomalies in environmental monitoring results, or a product sterility test showing contaminated products, conduct appropriate numbers of media fill runs in the same manner as for the initial performance qualification prior to the scheduled media fills.

TABLE 63.4
Initial Performance Qualification

Minimum number of simulations	Number of units filled per simulation	Contaminated units in any of the three simulations	Action
3	<5,000	≧1	Investigation, corrective measures, restart validation
3	5,000~10,000	1	Investigation, consideration of repeat of one media fill
		>1	Investigation, corrective measures, restart validation
3	>10,000	1	Investigation
		>1	Investigation, corrective measures, restart validation

TABLE 63.5
Periodic Performance Requalification

Minimum number of simulations	Number of units filled per simulation	Contaminated units	Action
Every half year	<5,000	1	Investigation, revalidation
	5,000~10,000	1	Investigation, consideration of repeat media fill
		>1	Investigation, corrective measures, revalidation
	>10,000	1	Investigation
		>1	Investigation, corrective measures, revalidation

63.7 TERMINAL STERILIZATION

The chapter of "Parametric Release of Terminally Sterilized Pharmaceutical Products"[4] in the General Information of JP XVII describes that the pharmaceutical products to which terminal sterilization can be applied generally must be sterilized so that a sterility assurance level of 10^{-6} or less (SAL ≦ 10^{-6}) is obtained. The SAL of 10^{-6} or less can be proven by using a sterilization process validation based on physical and microbial methods but cannot be proven by sterility testing. General requirements are the validation requirement of the sterilization process, microbial control program, sterilization indicators, change control, *etc.* Important control points that may affect the selected sterilization method are described in "Sterilization and Sterilization Indicators"[4] in the General Information of JP XVII.

Sterilization cited in the General Information of JP XVII "Sterilization and Sterilization Indicators"[4] includes moist-heat sterilization and dry-heat sterilization as heat method, ethylene oxide (EO) gas sterilization and hydrogen peroxide sterilization as vapor method, and γ-ray radiation and electron beam radiation as radiation sterilization and filtration method. And "Guidance on the Manufacture of Sterile Pharmaceutical Products Produced by Terminal Sterilization"[2] and "The chapter of Parametric Release of Terminally Sterilized Pharmaceutical Products in the General Information of JP XVII"[4] describes the terminal sterilization in detail, in which the moist-heat sterilization and γ-ray radiation method and electron beam radiation method as radiation sterilization are described. In addition, other sterilization methods can also be used as long as they can control the critical sterilization parameters and can consistently ensure a sterility assurance level of 10^{-6} or less.

The steam sterilization method of the heat methods is the most widely employed as terminal sterilization of drug products in Japan. This results from its greater capability of maintaining stability of the API. In terms of safety assurance of the drug product, the gas method and irradiation method both pose the possibility of complex degradation products other than heat decomposition that will likely call for an enormous amount of testing for identification of the degradation products as well as justification of safety qualification; thus, these have not been popular among pharmaceutical development companies.

In principle, moist-heat sterilization should be performed at 121.1°C for 15 minutes or, if not feasible, at F_0 value for no less than 8 minutes. If sterilization of pharmaceutical products at F_0 8 minutes is not feasible, alternative sterilization conditions should be established by identifying process parameters to ensure a SAL of $<10^{-6}$ by scientific validity. The commercial supply of large volume parenteral solutions requires that the manufacturer assesses the bioburden of raw materials of pharmaceutical products and manufacturing environments and establish moist-heat sterilization conditions for the sterilization of the solutions. The basic concept for the manufacture of large volume parenteral solutions is to secure a manufacturing environment that is free from spore-forming bacteria that have higher heat resistance than the indicator microorganism employed for establishing sterilization conditions and to ensure the sterility of pharmaceutical products by the routine monitoring and control of bioburden. The targeted value of bioburden for pharmaceutical products before sterilization should be <1 CFU/product. If bioburden testing reveals the presence of microorganisms exceeding the target value, the presence or absence of heat-resistant (spore-forming) microorganisms should be determined and, if present, the Sterilization Resistance Test should be performed with samples of detected microorganisms to verify that the sterility assurance level (SAL) is $<10^{-6}$. Critical parameters of moist-heat sterilization include temperature profile (usually indicated by F_0 value), temperature, steam pressure, exposure time, and loading pattern.

For the electron beam radiation method, the dose should be the dose required to achieve the SAL of 10^{-6}. In this method, irradiation time, absorbed dose, loading pattern, and density are recommended as critical parameters.

It is recommended that the propriety of sterilization by terminal sterilization methods should be judged by employing an appropriate sterilization process control and using a sterilization indicator suitable for the selected sterilization method. In moist-heat sterilization and hydrogen peroxide sterilization, *Geobacillus stearothermophilus* (strain name: ATCC7953, NBRC13737) is recommended and in dry-heat method and/or ethylene oxide (EO) gas sterilization, *Bacillus subtilis* (strain name: ATCC9372, IFO13721) is recommended as a sterilization indicator.

63.8 PEST CONTROL[6]

Pest control is important to maintain the cleanliness of the manufacturing environment. The cleanliness of the manufacturing environment might be influenced (affected) by insect migration, because they can carry microorganisms, such as spores, bacteria, and fungi, especially in APAs and indirect support areas. Thus, it is important to identify the species of the insects and their entry pathway and understand the barrier properties. Fungi tend to grow easily in the dead spaces, such as under the roof (ceiling) and within double-walled interiors in Japan because of high humidity. For that reason, it is a regional specificity and should be acknowledged that insects eating fungi are likely to occur in the facility, and that it is high risk that particularly small insects can enter the structural facility through narrow gaps that cannot be visually confirmed.

It is necessary to control small animals, because they might inhabit and nest in the dead space as described previously, and they might disrupt the facility and impede facility operation.

63.8.1. General

Captured insects inside pharmaceutical manufacturing facilities generally enter from outside and proliferate in a suitable inhabitable environment. It is important to not only make barriers depending on the needs among APAs, indirect support areas, and outside of the facilities, but also to avoid providing a suitable environment for target insects to inhabit in case of their entry. In addition, the environment should be maintained and managed to inhibit the insect proliferation in each segment of the facility.

It is believed that the risk of insects attached on carry-in materials is low. However, they might enter with carry-in materials and proliferate in the preceding areas. Therefore, not only control procedures for carry-in material operations, but also a control program for insects and small animals should be established (hereinafter called "Pest Control Program").

Pest control should be planned and conducted based on the results of risk assessment considering risks to products and the manufacturing environment. Suitable frequency and control criteria should be set after specifying the risk insects and small animals and evaluating the influence of these insects/small animals on products and the manufactural environment. Furthermore, confirm the influence of the sampling itself.

In the establishment of a Pest Control Program, the following items should be conducted (performed) as risk assessment. 1) Specifying the species of insects/small animals considering their ecology and entry pathway. 2) Analysis and evaluation of the specified insects/small animals risk. 3) Control criteria set (permitted risk level). 4) Treatment methods for risk reduction upon deviation from the control criteria value

63.8.2. Pest Control Program

Possess a documented Pest Control Program that is suitable for APAs and indirect support areas and create and maintain these records. It is recommended that the Pest Control Program include the following items. 1) A procedure for corrective and preventive actions (including monitoring). 2) Set the control criteria values and the treatment procedure for deviation from control criteria values. 3) A follow-up procedure after deviation from control criteria values. 4) A review of the cleaning plan based on the risk. 5) Facility inspection and training for engaged persons. 6) Validation and improvement of the program

The monitoring should be control (performed) in a wide range of controlled areas as well as immediate general areas, and it is recommended that evaluation of APAs and influence on products should be done according to the need of monitoring results. It is also recommended to investigate the target area after completion of facility construction. When the sampling procedure and sample size are set, the following items should be considered. 1) Monitoring timing of APAs and indirect support areas is determined by risk assessment. 2) Apparatus used for monitoring in APAs and indirect support areas should not contaminate the products and the facility environment. 3) The sampling procedure and sample size should be determined considering the ecology of the insects and small animals that may inhabit the manufacturing area and conducted (performed) in a valid way. 4) The instrument used for capturing insects during the monitoring procedure should be highly effective and should be suitable for the facility environment. 5) The equipment for capturing small animals (including rodents) during monitoring is not recommended because it may contaminate the facility environment. For the same reason, using bait for monitoring the eating trace by small animals should be prohibited.

When setting the control criteria values, consider the following items. It is known that the insects tend to show clumped distribution in the Pest Control Program. 1) Reflect (indicate) monitoring results on control criteria is recommended. 2) Evaluate separately the insect emergence and entry from outside. 3) The population and inhabit status need to be evaluated. 4) Evaluation should be performed by each area and each species. 5) Set level of action and alert according to the risk of product and also the content of the preventive treatment.

In case of unexpected deviation in risk assessment of the Pest Control Program, corrective action should be conducted (performed), at the same time, make an improvement of the preventive action plan according to the cause of the deviation. In order to maintain the appropriate environment without any deviation problem, the corrective action plan should be set considering cleaning, cleansing, maintenance of the building and facilities, and based on the past results of monitoring.

63.8.3 Measures to Be Taken in a Deviation from a Program

Appropriate and effective countermeasure should be performed based on the type of insects and small animals. When setting the countermeasure, consider the ecology of insects besides the insect emergence and entry sites. Most of the detection cases are found to be mainly insects in APAs and indirect support area; however, they came from outside of control areas mostly. Because the inhabitable state of insects is influenced by the niche space (microclimate) constructed by the building structure and the placement of facilities, inspection and inhabitable control countermeasures should be conducted with consideration of the preceding factors.

Aside from the preceding, it is recommended to reconfirm the prerequisite of pest control (including variable environment in other than control areas) when there is an entry of insects from outside or abnormal insect emergence. It is desirable to reconfirm the barrier function of structures and facility considering their aging degradation. Additionally, because changes that influence niche is considered as the prerequisite change of the Pest Control Program, risk assessment should be conducted again and reconfirmed, then the Pest Control Program should be considered to review if it is necessary.

63.8.4 Summary

Some insects can enter through small cracks and gaps less than a millimeter. Japan is a disaster-prone area, such as earthquakes and typhoons, physical forces such as vibrations caused by these disasters generate cracks and gaps in the joints of walls, ceilings, and floors. It is impossible to evaluate insect entry because of the detection limitation of these small cracks and gaps by monitoring of differential pressure, airborne microparticle, and microorganisms. Needless to say, continuous monitoring of the Pest Control Program is important to guarantee quality in manufacturing.

63.9 RAPID MICROBIAL METHODS[4]

63.9.1 "2. Validation", "Rapid Microbial Methods", General Information, JPXVII

The followings are all the contents described at "2. Validation", "Rapid Microbial Methods", General Information, JP XVII.

"To qualify introduced equipment, a standard component or strain, which represents the target of each method, should be utilized. That is, in direct measurement, type strains should be used, while in indirect measurement, standard components, *etc*., of the target bacteria are used.

To validate a protocol/procedure, it is required to demonstrate that the detection target is a suitable index/indicator for bacterial number or quantity. It is also important to state whether any special precautions are necessary in applying the protocol/procedure. When using a type strain, the result of validation should be equivalent to or better than that of the conventional method. However, because the detection principles of new methods are usually different from that of conventional methods, the correlation between them is not always required. For detection of environmental bacteria, it is important that the physiological state of the type strain should be maintained as close as possible to that of environmental bacteria, in order to obtain reliable results."

63.9.2 Explanation of the Main 4 Points About "2. Validation", "Rapid Microbial Methods", General Information, JPXVII

1. "To qualify introduced equipment, a standard component or strain, which represents the target of each method, should be utilized. That is, in direct measurement, type strains should be used, while in indirect measurement, standard components, *etc*., of the target bacteria are used."

Because a wide variety of measuring methods are used in the equipment for rapid microbial methods, there are a wide variety of methods for confirming the existence of microorganisms and counting units in each equipment. There are also methods that do not need cultivation of microorganisms. In particular, because there are many methods to confirm the existence of microorganisms by their metabolic activity and its components, it is necessary to pay attention to the fact that what is observed (indicator) differs depending on each measuring method. When conducting qualification based on this premise, direct methods, for example solid phase cytometry and flow cytometry, are evaluated using type strains available from ATCC (American Type Culture Collection) and NBRC (NITE Biological Resource Center), *etc.*, whereas indirect methods, for example, ATP method, use ATP standard reagents. Previously, both direct and indirect methods were qualified using the type strains. However, indirect methods were changed to be qualified using standard reagents such as components that are the targets of detection by each method.

2. "To validate a protocol/procedure, it is required to demonstrate that the detection target is a suitable index/indicator for bacterial number or quantity. It is also important to state whether any special precautions are necessary in applying the protocol/procedure. When using a type strain, the result of validation should be equivalent to or better than that of the conventional method."

In the text, "to demonstrate that the detection target is a suitable index/indicator for bacterial number or quantity "means, for example, that ATP methods are the methods to detect ATP, and that ATP can be detected from bacteria. For this purpose, it is possible to obtain calibration curve data by measuring the ATP amount derived from bacteria using type strains, *etc.*, and to clarify the scientific reason. In addition, "equivalent to or better than that of the conventional method" means, for example, that it takes 3 days to judge the result by the culture method, but it is possible to judge the result by the ATP method in 3 hours. Therefore, it would be better to show the advantages by each measuring method. With regard to "any special precautions", refer to Table 63.6.[7]

TABLE 63.6
Special Precautions in Applying the Rapid Microbial Methods

No.	Special precautions	Items to be considered
1	Clarify the purpose of introducing.	Clarify the microbial methods desired to be performed in Rapid Microbial Methods (RMM)
2	Consider the suitability (matching) of the microbial method.	Because RMM is diverse, the suitability of the desired microbial methods is judged.
3	Consider how to use the data obtained.	The management method in the new unit is considered, not in the absolute value by the culture methods.
4	Rebuild SOPs by adding or changing test methods	It is necessary to consider the operation procedure and establish a procedure based on it.

Abbreviation: SOPs = Standard Operating Procedures.

For the premise of the content of "2. Validation" it is important that "even if type strains exist for method validation, it is not easy to standardize the physiological activity" as stated in the preamble shown in the general information. It means that because obtaining absolute values for both the culture methods and the rapid microbial methods is difficult, it is necessary to show that bacteria can be detected for each measurement method and the object to be measured.

It is fundamental to validate the analytical method by the measurement method (equipment) to be used in the validation, and it is important for the validation of each applicable method to evaluate actual on-site measurement data of applicable samples to each method based on the validation protocol.

3. "When using a type strain, the result of validation should be equivalent to or better than that of the conventional method. However, because the detection principles of new methods are usually different from that of conventional methods, the correlation between them is not always required."

An important point in this description is that "the correlation between them is not always required". Previously, it has been considered that it is necessary to obtain a correlation between the conventional method (culture method) and rapid microbial method when considering rapid microbial methods. However, the culture method itself is dependent on culture medium, culture temperature, culture time, and technique for culture manipulation; in addition, other factors such as colonization, growth, death, and stationary phase cause changes in the growth condition (the number and the shape of colonies). What is called "variation" occurs. Moreover, because the measurement method in which physiological activity and components, *etc.* are observed depends on the metabolic activity of bacteria, *etc.*, measurement results of the same microbial species also vary. Because the results of both the culture method and the rapid microbial method vary depending on the conditions, the requirement for the correlation between the two methods itself could be considered as contradictory. Therefore, it could be said that finding a correlation between the measurement results from the two methods is not always necessary. However, to show a correlation between the two methods as an evaluation index is not denied.

4. "For detection of environmental bacteria, it is important that the physiological state of the type strain should be maintained as close as possible to that of environmental bacteria, in order to obtain reliable results."

As described in the explanation of 3. prior, it is generally known that microorganisms have different physiological activities and microbial components depending on the phase of growth, death, stationary phase, *etc.* With this understanding, it is clear that there is not necessarily correlation between conventional and rapid microbial methods. On the other hand, it is important to verify and understand the effects of these variables when validating rapid microbial methods. Taking pharmaceutical water as an example, microorganisms that live in water are often affected by various physical (heat, UV, pressure) and chemical (disinfectants, acids, alkalis) influences. Many of these damaged microorganisms and those suitable for nutrient-poor environments have unusual nutritional requirements that cannot be detected under normal culture conditions. General information of the JP XVII, G8 Water, "Quality Control of Water for Pharmaceutical Use", "4.4.2. Media Growth Promotion Test" can be referred to because this section describes "Prior to the test, inoculate these strains (type strains) into sterile purified water and starve them at 20°C –25°C for 3 days".

63.9.3 Summary Recommendation

The key points of validation when introducing rapid microbial methods are shown in Table 63.7.[7]

There are various methods of rapid microbial methods, and it is desirable to perform qualification and validation of a method suitable for its characteristics. For this, the cooperation of equipment suppliers is indispensable, and it is recommended that suppliers and users carefully consider the validation.

This topic was explained in "2. Validation, Rapid Microbial Methods, General Information, JPXVII". In addition, in the manufacture of regenerative medicine products, "Guidelines for aseptic manufacturing of regenerative medicine products", and "Research and Q&A on the international harmonization of GMP, QMS, and GCTP guidelines" issued as a government announcement on November 28, 2019, registered rapid microbial methods in the main chapter and recommended to introduce the latest scientific technology for microorganism control. From this background, it is expected that the rapid microbial methods will become widespread.

TABLE 63.7
Key Points of Validation of Rapid Microbial Methods

No.	Items	Recommendation
1	Equipment qualification	Direct measurement: Evaluate using type strains. Indirect measurement: Evaluate using standard reagents such as components to be detected.
2	Analytical method validation	Provide the scientific basis for the measurement of the number and amount of microorganisms. Show advantages over conventional methods. Indicate points to consider when using. Show that microorganisms can be detected for the analytical method and the measuring object.
3	Evaluation using type strains	Though it should be equal to or better than the conventional methods, it is not always required to show the correlation because of the difference in detection principle.
4	Type strains used in a test	It should be noted that the test results may vary depending on the conditions (physiological activity, *etc.*) of the microorganisms used in the test.

63.10 CONCLUSION

In this chapter, the validation method for manufacture of sterile pharmaceutical products in Japan was described. Since Japan also became a PIC/S member in 2014 as shown in the introduction, GMP for manufacture of sterile pharmaceutical products in Japan was expected to be equivalent to the PIC/S GMP. On the other hand, because there are various notices and guidelines that complement PIC/S GMP in Japan, this chapter summarized the actual operation in Japan focusing on these Japan-specific guidelines. In addition, as a topic, this chapter explained pest control, which is emphasized in Japan, and rapid microbial methods, which are worth noting as the latest microbiological management methods. It is necessary to appropriately validate various processes for sterility assurance of the sterile drug products. In terms of the PIC/S GMP Annex 1 draft, it is important to establish and implement a "Contamination control strategy". From such viewpoint, it is expected that this chapter will be useful for the understanding of the validation method and the advanced sterility assurance in Japan.

NOTES

1. The Scientific Task Force of the Ministry of Health, Labour and Welfare of Japan, Guidance on the Manufacture of Sterile Pharmaceutical Products Produced by Aseptic Processing (revised edition), April 2011.
2. The Scientific Task Force of the Ministry of Health, Labour and Welfare of Japan, Guidance on the Manufacture of Sterile Pharmaceutical Products Produced by Terminal Sterilization (revised edition), November 2012.
3. JP 16, G4 Microorganisms 2011.
4. JP 17, G4 Microorganisms 2016.
5. JP 17, G8 Water 2016.
6. The Scientific Task Force of the Ministry of Health, Labour and Welfare of Japan, A Guidance for the Aseptic Manufacturing Process of Regenerative Medicines, A10. Pest Control, November 2019.
7. Ikematsu Y, et al. *Design, Maintenance and Training of Operators for the Biological Clean Room. Chapter 5 Section 6*. Japan: Technical Information Institute Co. Ltd., 2018.

64 Organization of Validation in a Multinational Pharmaceutical Company

Kevin M. Jenkins

CONTENTS

64.1 MULTINATIONAL COMPANY

64.1.1 Introduction

This chapter explores validation in multinational companies. The organization dynamics required to support validation, the systems that need to be in place and the coordination of the activities for a successful outcome.

A multinational pharmaceutical company generally consists of a corporate headquarters with multiple worldwide manufacturing sites. This type of structure operates across multiple regulatory agencies such as the Food and Drug Administration (FDA) in the United States, European Medicines Agency (EMA) in parts of Europe and others. It can also include countries in other parts of the world with their own specific regulations (Pharmaceutical and Medical Devices Agency (PMDA) for Japan, Angelica Nacional de Vigilancia Sanitaria (ANVISA) for Latin America and the Irish Medicines Board (IMB) for Ireland as examples). This is important in determining the structure of a validation organization such that regulations across areas where products are manufactured and marketed are duly taken into account. Sites within the multinational company network may specialize in one type of formulation such as aseptic or combinations such as aseptic and solid oral dosage, which adds complexity and

DOI: 10.1201/9781003163138-64

perhaps different regulations depending on the markets they supply.

64.2 THE BUSINESS CASE FOR VALIDATION

Validation is often conducted at firms because of regulatory requirements; however, there is a strong business case for validation [Agalloco]. A strong validation program has benefits beyond regulatory compliance. Such a program can assure robust processes and likewise a world class qualification program assures equipment is performing properly. This can lead to less deviations, less downtime, effciency in operation and reduced costs. The robustness of a process/product also determines the quality of the end product, which leads to less issues in the field, less complaints and less out of specification results. All of this increases patient satisfaction, decreases costs and improves the company's overall reputation for reliability, quality and supply of critical pharmaceutical products to meet patient's needs. That is the primary objective that must be kept in mind and why a strong validation program has a business case.

It has also been noted that validation can be viewed as an extension of TQM (Total Quality Management). Agalloco noted that "Through appropriate validation methodologies a firm can increase the quality of its operations and realize benefits." Validation is sometimes used as an excuse not to make improvement, rather considered in combination with life cycle management, it is recognized as a tool to assure processes are designed appropriately, continuously improved and kept in a state of control as a result. All of this assures quality, controls cost and provides consistent delivery of product flow.

64.3 VALIDATION STRUCTURE

Jeater (et al.) list three possible structures for validation support, the consultant model, the task force model and the dedicated group. Snyder et al. describes the holistic validation approach, which is a function of multiple groups coming together in the organization to execute and document validation. There are also possible combinations of these three models to support validation at a large company, which we will call a hybrid model. It is useful to explore each of these models in more detail and look at the advantages and disadvantages of each.

64.4 THE CONSULTANT

These are generally resources hired on a contractual or consultant basis that are usually of a limited duration. However, a validation department could be staffed by a rotating group of contractors, although this type of resource is usually reserved for specific validation projects or to deal with ups and downs in workload. There is effciency because it provides flexibility and these contracts can be scoped for a specific duration or for validation milestone completion stages. These types of resources offer the advantage of viewpoint across the industry because they have worked at other companies in the industry. This provides best practices that may not be realized within a company-dedicated resource. As a short-term resource, the costs can be manageable but over a longer-term contract the resources can be quite expensive, especially if travel time, travel cost, housing and allowances are factored into the cost. External resources may not be familiar with local procedures and corporate standards and therefore may require training on these procedures and standards. There is also a loss of the subject matter experts (SME) and the consultant's knowledge once they leave the project as opposed to maintaining a long-term internal resource for validation. In addition, the validation work that was completed generally needs to be approved by a dedicated company resource, especially for the quality authority. Table 64.1 lists advantages and disadvantages of the consultant model as discussed.

TABLE 64.1
Advantage/Disadvantages of the Consultant Model for Validation Support

Advantages	Disadvantages
Provides flexibility	Expensive over long term
Brings industry best practices	Lack of company procedure/policy knowledge
Diverse skills	Lack of SME/loss of internal knowledge
Expertise across multiple companies	Approval of work by permanent employee

Abbreviation: SME = subject matter expert.

64.5 THE TASK FORCE

A task force approach to validation includes multiple disciplines, usually from quality (QA and QC), operations/production and engineering at a minimum. For process/product validation, research and development would be included as well on the team. Regulatory could be a support group for any filing or registrations required for submission to the appropriate agencies.

This type of structure works best for a specific project with a limited duration or time line and the correct expertise and skills would be required. For instance, a sterile injectable validation product would require different skills from a solid oral dose project. These colleagues on the team are usually from a different part of the organization and may need to return upon completion. Therefore, this is not the best long-term strategy for managing validation on a continual basis. It may also require a backfill in the department the task force colleague was on loan from. This strategy can be combined with others but will be discussed in the hybrid approach. Table 64.2 lists advantages and disadvantages of the Task Force model.

TABLE 64.2
Advantages/Disadvantages of the Task Force Model for Validation Support

Advantages	Disadvantages
Brings experienced colleagues from organization	Not a sustainable long-term solution
Internal and know process/procedures	Backfill of previous role for some duration
Good for limited duration project	
Cuts across multiple disciplines	

64.6 THE DEDICATED GROUP

A dedicated group for validation consists of individuals hired with specific validation skills to staff a full-time validation department. This is typically at the site level but can have some shared disciplines in a corporate center. One such example may be a global engineering, global validation or global microbiology group that supports the sites on a consultant-like basis. At the site level, there are usually validation professionals or engineering professionals with specialization in disciplines such as facility/utility qualification, equipment qualification and process/product validation. There may be specialty skills in computer/software validation or that could reside in a corporate IT (information technology) group. As mentioned previously, the type of product produced would require different skills for sterile injectables, solid oral dose, medical devices and APIs (active pharmaceutical ingredients). The role of this central group is to support the sites and they often have expertise that is shared across multiple sites. These individuals can be based at sites or a central function but provide the advantage of supporting projects across many sites and are there when needed. This concentrates the expertise in a central function and allows for these groups to share their collective expertise from projects across the global organization.

The site group may have links into the research and development group and other support groups such as regulatory for submissions and laboratories for analysis of validation samples.

Advantages of this approach are the presence of embedded expertise that is permanent and retention of these SMEs. This group becomes familiar with the site/company products, processes and equipment. The dedicated resources are generally less costly than contracted resources and there is an ownership culture with dedicated colleagues. The disadvantages are less flexibility in resources and lack of external benchmarking. Table 64.3 lists advantages and disadvantages of the Dedicated Group.

64.7 HYBRID STRUCTURE

The hybrid structure combines all three—consultant or contracted resources, dedicated resources and task forces as needed. This approach provides more flexibility because the dedicated resources can be supplemented with contractors or consultants as validation requirements increase or decrease. Generally, validation projects are dependent on expansion/growth, new product and processes. There is a baseline of re-qualification on an annual or periodic basis because of Good Manufacturing Practice (GMP) requirements and change control. That can be more predictable. The hybrid model is used extensively because it has more flexibility and can be easily adjusted based on changing needs in the validation life cycle and with temporary additions for one-time larger projects. Table 64.4 lists both the advantages and disadvantages versus the other models presented so far in the discussion.

TABLE 64.3
Advantages/Disadvantages of the Dedicated Model for Validation Support

Advantages	Disadvantages
Internal knowledge of equipment/processes retained	Less flexibility
Less costly than contractors	Lack of expertise outside the company
Can specialize in certain areas	
Ownership culture	

TABLE 64.4
Advantages/Disadvantages of the Hybrid Model for Validation Support

Advantages	Disadvantages
Combination of dedicated and contract resources provides flexibility	More complex to manage
Can balance/manage costs	Less accountability
Effciency	
Benchmarking and external expertise from contractors/consultants	

64.8 GLOBAL VALIDATION STANDARDS

The GMPs and Validation requirements are described in the 21 CFR parts 210/211 and Validation Guidelines such as the recent Process Validation: General Principles and Practices; Guidance for Industry. For the EU, Annex 15 describes the requirements for qualification and validation. Outside of these guidelines and regulations there are PIC/S (Pharmaceutical Inspection Co-operative Scheme) guidelines and ISO standards such as ISO 9001 Design Verification and Validation.

There are also specific guidelines/regulations in various other regions/markets/countries, although most follow the US CFR/guidance to validation, Annex I or ISO standards. The point is this all becomes very complex and for a multinational pharmaceutical company that produces product in multiple locations and distributes to multiple worldwide markets, it is critical to address all of the requirements.

To deal with this, it is best practice to have a set of global validation standards that embraces all regulations. There may also be separate targeted standards for aseptic products, medical devices, APIs and solid oral dosage forms. Table 64.5 lists possible validation standards and their main topic. Table 64.6 is a list of standards specific to aseptic products. There should be standards for the other dosage forms, such as APIs, solid oral dosage and medical devices.

TABLE 64.5
Proposed Validation Global Standards

Proposed standard title	Topics included
Equipment Cleaning Validation	Requirements for equipment cleaning validation, swabs/rinse, setting residual limits
Systems Validation	Overall process of setting specification, design review, factory acceptance testing, IQ/OQ and PQ
Validation of Analytical Methods	Guidelines for validation of analytical methods, acceptance criteria and methodology
Validation of Cleaning Assay methods	Similar to analytical methods validation but specific to cleaning validation—recovery studies, minimum recovery requirements
General Requirements for Validation	Role and responsibility for validation, documentation requirements, VMP/VPP, data verification, concurrent and prospective validation definitions
Packaging Validation	Specific to validation of packaging lines, equipment, container closure and labeling considerations for validation
Microbiological Methods Validation	Requirements for microbiological methods validation. Endotoxin recovery, microorganism recovery, sterility testing methods, bioburden and environmental monitoring
Lab Equipment Validation	Qualification requirements for laboratory instrumentation. Data integrity, computer system validation, precision and linearity of data.
Equipment Qualification	Specific requirements for installation and operational qualification of production equipment
Facility/Utility Qualification	Qualification of utilities and guidelines for high purity water, air, nitrogen. Building control systems, HVAC
Process/Product Qualification	Process and product qualification PQ/PV requirements including setting acceptance criteria and number of batches required for validation, statistical tools for determining and stability and analytical testing requirements
Computer System Validation	Requirements for computer validation including design verification, URS, and ALCOA data integrity considerations for computer and automated systems and computerized laboratory instrumentation

Abbreviations: ALCOA = attributable, legible, contemporaneous, original and accurate; HVAC = heating, ventilation and air conditioning; IQ = Installation Qualification; OQ = Operation Qualification; PQ = Performance Qualification; PV = Process Validation; URS = User Requirements Specification; VMP = Validation Master Plan; VPP = Validation Project Plan.

TABLE 64.6
Proposed Validation Standards Specific to Sterile Injectables

Proposed Aseptic Standards for Validation	Topics Included
Dry heat sterilization validation	Parameters and load patterns
Moist heat terminal sterilization validation	Parameters and load patterns
Sterilizing filters validation	Testing of filters
Depyrogenation validation	Validation of depyrogenation tunnels and processes
Visual inspection qualification (manual/automated)	Qualification of manual and automated inspection, defect types, categorization
Steam in place qualification	Use of steam in place for sterilization/ sanitization

64.9 QUALITY SYSTEM FOR VALIDATION AND CULTURE

The validation life cycle for validation is a key quality system for any pharmaceutical operation as supported by ICH Q10 Quality Systems. The validation life cycle assures that post through development of requirements/specification to Installation Qualification/Operation Qualification (IQ/OQ) and Performance Qualification/Performance Validation (PQ/PV) there is a defined mapping as shown in Figure 64.1 for the V model of validation.

Post validation completion, it is critical to have a robust change control system that assures changes are assessed, re-qualification takes place and based on the severity of the changes some periodic revalidation is performed. For aseptic products/equipment, annual re-qualification is a requirement for sterilization process, media fills and critical aseptic control process (ref aseptic guidelines).

The common standards provide consistency across the multinational sites, but the sites also need to stay in compliance through a periodic review process. This review should be triggered by (1) regulatory changes/guidance documents, (2) internal feedback and benchmarking, and (3) trends in inspection observations such as current Good Manufacturing Practices (cGMPs). This review requires a review and approval process and even if there is no trigger to update should have some periodic frequency for review. There also needs to be a process to implement the standards at the multinational sites and assure they are translated to local language and local operating procedures. Post implementation at the sites, it is useful to audit against the standards to assure they are in use as written.

64.10 CULTURE

Quality standards for validation are part of the culture of a multinational company. They set a standard but equally important they go to the heart of the culture for the organization. If

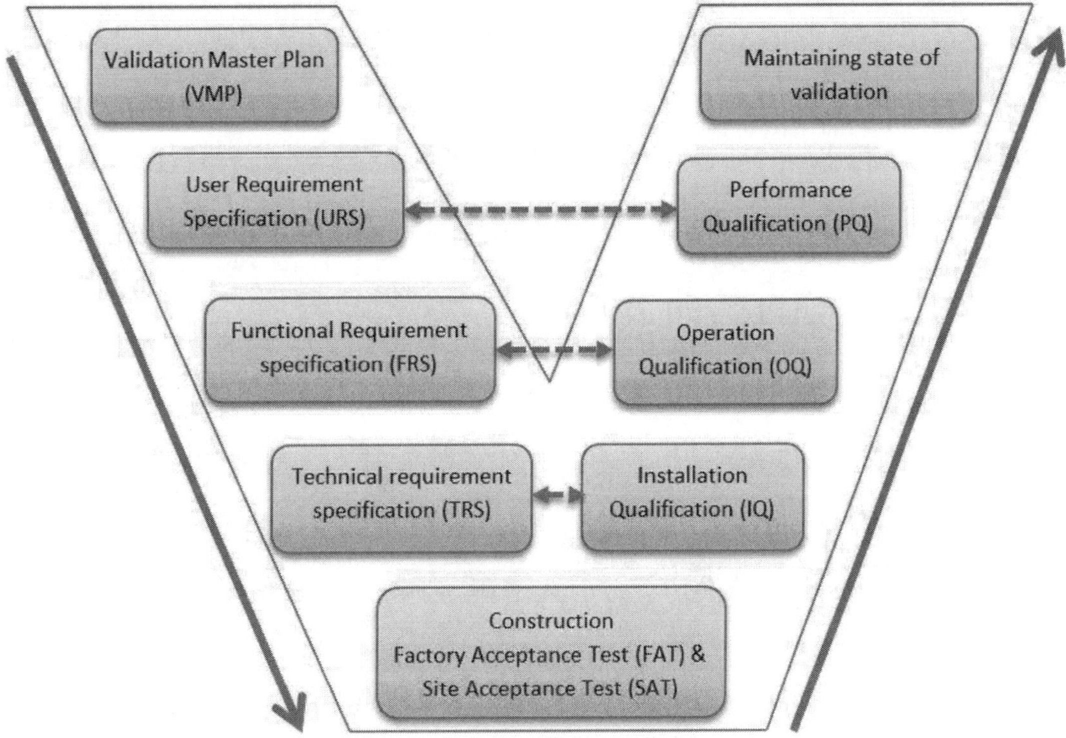

FIGURE 64.1 V model of validation.

standards are followed, the sites should be in compliance with worldwide regulations. There is more to quality culture, but this is a key element.

There is also a need to look not only at each individual change to equipment and process but also the cumulative changes as part of life cycle review. This continuous improvement loop is part of the overall quality system to assure we are in a state of control.

64.11 RESPONSIBLE DEPARTMENTS AND ROLE IN VALIDATION

Multiple departments have a role in validation. Direct supporting departments are the validation group, operations/production group and engineering group. There are other indirect/direct support groups such as technical services/R and D, laboratories, corporate quality and regulatory. It is useful to explore the role and responsibilities for each of these groups. Now that the structure is defined, it is important to determine who does what, their role and responsibilities.

64.12 VALIDATION REPORTING AND RELATIONSHIPS

64.12.1 Quality

The quality authority at the site or in the corporate function has the responsibility for approval of most if not all validation. They are assuring that standards were following, the validation meets applicable regulation and that the acceptance criteria were met. If the acceptance criteria were not met, there needs to be a deviation with appropriate root cause and corrective action. These actions may result in repeating portions of the validation to demonstrate a state of control. Certainly, there are other approvers as noted in a matrix of responsibilities, but quality is the ultimate approver. It is expected that those in quality releasing product know that the equipment/processes are in a validated state. Just as Operations is responsible for only using equipment/processes that are in a validated state of control.

The actual validation group may reside within the Quality organization but could also be part of a Technical Services, Operations or Engineering group. It is common for facility and equipment qualification to be in the Engineering group.

64.12.2 Technical Services

Technical Services is a very common group within large corporations. They are generally responsible for managing and executing validation at the sites or on a corporate level. This group can have a development aspect that bridges between R and D and the sites. They may participate in late-stage development, process changes, method implementation and final execution of new product/process changes and new product development/implementation. They are typically focused on the process and product. This could be the formulation and manufacturing process in many cases. Usually, they heavily rely on Engineering for the equipment installation and qualification but may be heavily engaged in the equipment specification to support the process/product. Technical services would be a key approver on the content of the validation and assure that the acceptance criteria were met to the design specification to assure the end process/product performs as designed.

64.12.3 Engineering

As noted previously, Engineering is relied upon for facility and equipment design/specification, installation and qualification. That portion of the validation group for facility/utilities and equipment may reside in this group. In either case, they need to be engaged in the execution phase of the project. They also have a responsibility for the technical content of the qualification. Usually within the site Engineering group there are critical support services such as calibration and maintenance. As part of the system qualification, calibration needs to assure the equipment is calibrated and that there is a schedule for routine calibration to assure a state of control. The maintenance department as part of qualification implements a routine preventative maintenance schedule and work instructions so that the equipment is properly maintained.

64.12.4 Operations

Operations essentially owns the equipment and facilities and ultimately the product that it produces. They must assure with the support of other departments that the equipment, facility and utilities are maintained, calibrated, validated and operated as they were designed to perform. They should approve qualification and validation as the owner of the final product in this process train. As with other functions, the validation group could certainly report into operations for both equipment/facilities and process/product. This does create a bit of a one-stop shop for validation, because the bulk of the activities occur in Operations and this allows full accountability.

64.12.5 Automation/IT

Today, most if not all equipment and utilities have some form of automation and generate streams of data. This makes automation and Information Technology (IT) critical to the qualification and validation process. As these systems produce more process data the management of the information, retention, analysis, and integrity becomes even more important. Automation has to assure the validation protocols have the correct tests for the computerized/programmable logic controller (PLC) system. There has to be connectivity and performance of the integrated systems. Many of the modern production systems, building control systems and some utilities are interconnected and therefore have a higher level of complexity. Automation assures that the systems perform as intended; the code is verified and maintained through the life cycle and change control.

64.12.6 Safety

Large companies generally have a Safety group at the site and at the corporate level. As part of a safety culture, it is imperative that Safety be aware of new processes/equipment to assure colleague/employee safety. They are consulted in regards to safety aspects of the equipment and processes at the very early stages to assure unsafe conditions are not put in place.

64.12.7 Corporate

As mentioned previously, in a multinational organization Corporate groups can exist for many of these functions if not all, but usually quality, engineering, and technical services. Corporate can take on a role of oversight and of consultant. They assure that global standards are followed, may provide or procure resources on a certain time line and manage the network of sites to assure corporate objectives are met. This can include creating corporate guidelines/procedures for consistency, assisting in writing validation documentation, training, approval of large validation projects and related protocols/reports. The advantage here is consistency and a global resource that can be utilized when needed by various sites and regions of the company.

64.12.8 Regulatory

Regulatory can be a central group with satellite resources at larger sites in the network. Their function is to be the interface with regulatory agencies on submissions, responses to submissions and changes to existing processes. For the US FDA this can be in the form of prior approval submissions (PAS) or Changes Be Effective (CBE-0, CBE-30). They may also be engaged during Prior approval Inspections (PAIs) to assure that the submission is clear, answer key questions or direct the inspector to the correct SME or supporting data. It is important to engage regulatory early in the process for new products/processes and equipment because they can determine what is needed for submission in the impacted markets. This is of critical importance for long lead time items such as stability data, for example, where studies will take time.

64.12.9 Responsibilities of All Functions

One method to capture these responsibilities is a Responsibility Assignment Matrix or RACI matrix. Table 64.7 lists for each function who is responsible (R), who approves (A), is consulted (C), is informed (I) and executes in regards to who writes the document or performs the task (E).

TABLE 64.7
Responsibility Matrix Example

Process/deliverables	Operations	Project management	Engineering	Automation/IT	Quality	Safety
Functional specifications	C	I		R A E		C
Design specifications	C	I		R A E		C
Validation master plan	C	R A E	C	A E	A	C
Design review	C	R E	I	E	C	C
Design review summary report	A	R A E	I	A	A	C
Verification testing protocol IQ/OQ	C	R A E	I	A	A	I
Verification testing report IQ/OQ	A	R A E	I	A	A	I
Product validation	R A E				R A E	

Abbreviations: IQ = Installation Qualification; IT = Information Technology; OQ = Operation Qualification; RACI diagram: R = Responsible; A = Accountable; C = Consulted; I = Informed; E = Enabled.

64.13 ORGANIZATION/PRIORITIZATION VIA THE VALIDATION MASTER PLAN AND DIFFERENCES WITH SITE MASTER PLAN

A Validation Master Plan is a key document for the site and at a corporate level. The Validation Master Plan should contain the following:

- Validation strategy and use of a life cycle approach.
- Summary of facilities, equipment, systems and processes included in the validation program at site or globally and their validation status.
- Organizational structure of validation activities and associated roles and responsibilities.
- Change management process.
- Deviation management process—specific to validation.
- Reference to existing documentation (i.e. Individual validation plans for cleaning and production processes, project verification plans, relevant SOPs).

This document needs to be a living document relative to the list of systems and status. It should be on an annual refresh cycle to maintain its utility. Some of the content should reside in the corporate standards discussed earlier in this chapter.

64.14 VALIDATION PROJECT PLAN

A validation project plan (VPP) is used to document validation methodology and includes the commissioning and qualification plan, including validation, change control and associated protocols. These are usually used for larger complex projects such as a new facility or production line where there is integration with multiple systems and equipment.

These documents introduce validation as part of project management. Project managers generally use tools such as gang charts and spreadsheets to track critical steps and interdependencies between these steps. Additionally, resources need to be planned regarding what specific skills are needed and when in the process. One of the significant reasons for large projects to become delayed is not planning correctly for the resource needs. Having the correct skill at the right time to avoid costly delays.

Planning is even more crucial when multiple sites in the network are involved. In a large multinational company, often a new product has key elements from different sites. For instance, the API may be from a site in Germany, and this API is used to formulate the product at the finished dosage site in the US then the final packaging occurs at multiple sites in Asia, Europe and the US to support those markets. All of this adds complexity and necessitates planning across the entire network of sites. Global standards, a validation master plan for each site and validation project plan assures that there is agreement across these sites on the plan and how it will be conducted. Timing and resource planning is also key to success across multiple sites, languages and cultures when there is one agreed upon plan.

64.15 RESOURCE PLANNING

As mentioned previously, resource planning for validation is critical to the cost and schedule. Most large-scale projects are a function of the resources, costs and time line. If not tracked appropriately then something will have to be adjusted and usually that extension of the time line results in added cost—the old adage that time is money. In large corporations, these projects are usually tracked at the site level but there is some corporate oversight. This oversight is generally in the form of approval processes with appropriate escalation for capital and funding approval. There is additional reporting for key milestones to assure that the project does not overspend and a process to request or deny additional funding and perhaps even cancel the project based on results at the key milestones.

The budgeting process is a key element in these controls. A budget plan is submitted and as the project is approved and funded latest estimates of the spend rate are provided for review.

Time line and tracking have a direct influence on the overall spend and become critical where there are interdependencies, such as the procurement of equipment and the long lead time that must be considered in the time line. There may be parallel tasks that can be completed such as completion of the facility where the equipment will be installed and late-stage development work to prepare for scale-up in the final installed equipment. All of these need to be tracked and accounted for in the time line and in the budget.

64.16 DOCUMENTATION

Earlier, the importance of corporate validation standards, the Validation Master Plan and Validation Project Plan was introduced. Underlying all of this are the requirements documents, specification documents, associated protocols and reports for IQ/OQ and PQ. There are usually supporting documentation including technical reports, engineering studies and analytical data that support the validation. Table 64.8 demonstrated how all of this comes together to comprise the validation documentation. This type of documentation may be centrally stored at the site or at the corporate level. The greater the consistency in format across the organization, the more transferable this documentation is for technical transfer and perhaps replication at multiple sites.

64.17 PROCESS AND PRODUCT VALIDATION

The goal of PQ is to prove a process is rugged and reproducible. The best approach is to understand the process through knowledge of process parameters and the limits that assure the desired product characteristics. For new processes and products, this information is usually from the research and development groups or technical transfer groups. Generally, this is documented in technical reports from development of the product/process.

Reproducibility of the process typically occurs with a three-batch qualification. This occurs with data generated at large scale, and statistical evaluation should occur to assure the robustness of the data. This also means that validation does not end at the three qualification batches but continues through the life cycle of the product and based on periodic review of the batch data. This task may fall to operations or in conjunction with a technical support group. The data analysis may warrant improvements to the process over time and may be repeated in terms of the qualification based on new equipment or process changes or both.

TABLE 64.8
Validation Documentation Organization

Validation documentation			
Corporate Standards			
Validation Master Plan (VMP)			
Validation Project Plan (VPP)			
Requirements documentation	Specifications	Protocols/reports	Supporting documentation

There are key requirements that must be provided prior to conducting process validation. The following is a checklist of knowledge that must be provided by R and D or Technical support groups:

- Critical product characteristics
- Critical process parameters, associated limits, recommended operating ranges, worst-case range, edge of failure
- Rationale for noncritical process parameters
- Process description
- Equipment, facility and materials required (excipients, APIs)
- Proposed manufacturing procedures (draft)
- Proposed cleaning methods and limits
- Recommended in-process and final product testing
- Validated analytical methods
- References to all development studies
- References to any previous validation studies

The types of studies usually conducted during PQ include process capability, uniformity, holding and reproducibility. During these studies, it is useful to operate on at least draft procedures and validated methods, with trained personnel and qualified/calibrated equipment. For most validation studies, three consecutive prospective batches will be utilized. More batches may be needed, and there should be a defined rationale for the number of batches chosen and perhaps a statistical analysis to confirm the choice. Data collection is critical during these runs to assure in-process ranges can be defined.

The PQ protocol should at minimum contain the following elements:

- The applicable batch record (may be a draft)
- List of procedures for qualification
- Analytical tests to be performed via validated methods
- Data collection plan
- Proposed acceptance criteria with test listed and rationale for criteria
- Number of batches to be included
- Critical product characteristics
- Critical process parameters
- Process for evaluation of results, which may include a statistical analysis
- Roles and responsibilities
- Process description
- Reference to supporting documentation

This protocol and report form the documentation to support the reproducibility of the process and the final product characteristics. It has to stand the test of time as the initial qualification work. It should be retained for the life of the product on the market and be a reference point for all future changes and re-validation in the future. Therefore, this type of documentation is generally secured in a central location with off-site backup to assure no loss of information over time.

It is important that critical parameters are captured in the PQ protocol along with acceptance criteria. These can be obtained from some of the following sources:

- Engineering studies
- Technical reports
- Product development reports
- Method development and transfer
- Research and Development reports

64.18 GETTING IT RIGHT THE FIRST TIME

Setting stringent criteria for key quality attributes and process limits is critical. A common mistake is to set criteria and limits so loose as to allow "passing" validation. Rather validation should challenge the limits of the process and through appropriate development work assure a tight process that produces a consistent high-quality process. This will lead to less out of specification results during routine production and less failures and lost batches. All of this contributes to success for the product history. We have talked about continuous improvement, but at this stage early in the process the goal is getting it right the first time. This includes assuring the critical process parameters are captured as mentioned before and included in the validation plans with defendable acceptance criteria. The sources of these parameters and acceptance criteria should be captured/documented and with sufficient rationale to defend the selection of the parameters and the ranges for acceptance.

64.19 THE FUTURE

Until recently, almost all pharmaceutical production was batch processing. However, new technology allows for continuous processing where there is no longer a unique batch but a continuous flow of materials to produce doses of tablets or filled vials on a continual basis. These continuous processes are also coupled with in-line analytical capabilities that can produce real-time analytical data on each and every dose. Where does validation fit in this new technology? How will our regulations evolve in this situation? These are questions that many just are starting to explore.

On the other side of these developments is the small "custom" pharmaceutical production in a small footprint, the so-called pharmaceutical plant in a box. This employs the use of digital printing to custom produce a single dose of specific drugs almost at the patient level. How will these systems be validated, and what data will be needed to assure quality and safety?

64.20 CONCLUSION

The implementation and maintenance of a process, facility and equipment in a validated state may seem like a daunting task but good science, sound engineering and a logical approach can assure success. Resources, time line and budget are three critical elements along with use of consistent standards based on experience and global regulations. Finally, validation is not only a regulatory requirement but good business, because it assures a consistent process that delivers quality product at the best cost.

REFERENCES

1. Jeater JP, Cullen LF, Papariello GJ. Organizing for Validation. In Berry IR, Nash, RA, eds. Chapter 2, *Pharmaceutical Process Validation*. 2nd ed. Vol. 57. New York: Marcel Decker.
2. Agalloco J. Validation: An Unconventional Review and Reinvention. *PDA Journal of Pharmaceutical Science & Technology* 1995 July–August, 49(4), 175–179.
3. Current Good Manufacturing Practice Regulations for Finished Pharmaceuticals, 21 CFR Parts 210 and 211: Good Manufacturing Practice Regulations for Medical Devices, 21 CFR Part 820.
4. FDA Guide to Inspection of Sterile Drug Manufacturers, 1994, FDA website reference.
5. FDA Guideline on Sterile Drug Products Produced by Aseptic Processing, 1987, FDA website reference.
6. Pharmaceutical and Medical Devices Agency PMDA for Japan, GMP's, www.pmda.go.jp.
7. Angelica Nacional de Vigilancia Sanitaria ANVISA for Latin America, portal.anvisa.gov.br.z.
8. Irish Medicines Board IMB (Now Health Products Regulatory Authority (HPRA)), www.hpra.ie.
9. Annex I to European Compendia "Manufacture of Sterile Medicinal Products".
10. ICH, Q7, Good Manufacturing Practice Guide for Active Pharmaceutical Ingredients, Sections 12.1–12.6, Validation.
11. U.S. Food and Drug Administration. Guidelines on General Principles of Process Validation, May 1987.
12. Guidance for Industry, Process Validation: General Principles and Practices, Food and Drug Administration, January 2011.
13. ICH Q10 Pharmaceutical Quality Systems, European Medicines Agency. https://ema.eurpoa.eu.
14. PIC/S, (The Pharmaceutical Inspection Convention and Pharmaceutical Inspection Co-operation Scheme), Guidelines for GMP in Pharmaceuticals, PI 007–6, "Validation of Aseptic Processes". www.pharmaguideline.com.
15. PIC/S, (The Pharmaceutical Inspection Convention and Pharmaceutical Inspection Co-operation Scheme), Guidelines for GMP in Pharmaceuticals, PI 006–3 "Validation Master Plan Installation and Operational Qualification Non-Sterile Process Validation, Cleaning Validation". www.pharmaguideline.com.
16. PMDA, Pharmaceuticals and Medical Devices Agency, Japan. www.pmda.go.jp.
17. ANVISA, Brazilian Health Regulatory Agency. www.gov.br/anvisa.
18. Barry IR, Nash RA. *Pharmaceutical Process Validation*. 2nd ed. New York/Basel/Hong Kong: Marcel Dekker, 2003.
19. Agalloco J, Carleton FJ. *Validation of Pharmaceutical Processes*. 3rd ed.
20. Mollan MJ, Lodaya M. Continuous Processing in Pharmaceutical Manufacturing. www.pharmamanufacturing.com/assets/Media/MediaManager/ContinuousProcessing PharmManufacturing.doc.

65 Validation in a Small Pharmaceutical Company

Stephen C. Tarallo

CONTENTS

65.1 INTRODUCTION

This chapter provides a perspective on a general approach to validation activities in a small pharmaceutical company. As a small company with a focus on contract manufacturing, Lyne Laboratories has successfully completed validation on numerous ANDAs, NDAs and OTC products. Our approach to validation meets the highest industry and regulatory standards and has consistently and effectively been used with small, large and virtual pharmaceutical companies. For Lyne Laboratories and other small companies, validation is both challenging and rewarding. Although it often taxes resources and demands intense and broad-based management involvement, it can also stimulate peak performance from the team and individuals within the company. As a crucial component in pharmaceutical manufacturing, managing the validation process requires leadership skills in addition to technical and scientific competency.

Validation principles date back approximately 30 years, and yet, even today these principles remain a standard for all new manufacturing processes. With advanced technologies, the scientist has been afforded more accurate means to accomplish these activities. This has greatly improved the quality process and thereby provided better scientific data.

Pharmaceutical companies of all sizes typically dedicate considerable resources, in terms of time, money and specialized personnel, to validate a current Good Manufacturing Practice (cGMP) facility or process. The regulatory agencies, appropriately, do not distinguish or make exceptions in terms of validation for small companies versus large companies. For a small pharmaceutical company, technical and financial resources will undoubtedly be challenged. In many cases, resources outside the company may be called upon to complement existing skills. Balancing internal and external resources is essential in order to maintain ultimate control and responsibility for the overall process. This can be overwhelming to a small company or plant with limited resources, so it is important to structure the validation team carefully. Leaders of small pharmaceutical companies must realize that process validation is critical not only to meet regulatory requirements but as a tool for evaluating the entire process from the supply of active pharmaceutical ingredient (API) to ensuring that the drug product meets its intended stability parameters. Validation in a small company is also an excellent management tool for developing the knowledge and skills of key personnel.

The design, construction, commissioning and validation of pharmaceutical facilities and processes pose significant challenges for project managers, engineers and quality professionals. Constantly caught in the dilemma of budget and schedule constraints, they have to deliver an end product that complies with all building, environmental, health and safety governing codes, laws, and regulations. The process must also comply with one very important criterion; it must be validated to meet current Good Manufacturing Practices (cGMP) regulations.

The cost of validation is determined by time spent on documentation, development of protocols and Standard Operating Procedures (SOPs), and time spent on actual fieldwork, data collection and analysis. Often, varying validation practices and methodologies result in inefficient implementation and costly delays. Too often, the validation process reveals a large burden of unfinished commissioning business, resulting in a delay in start-up.

In some cases, validation is carried out but involves a limited number of personnel within the organization. This lack of information sharing increases the misunderstanding of a manufacturing process by the most important people within the company—manufacturing and quality personnel.

It is easy to lose sight of overall objectives during the validation cycle. Companies can get very focused on the scientific aspect of pharmaceutical manufacturing and forget that it is a business. A company must run efficiently, produce quality products and meet the demands of the marketplace. Validation data should provide the baseline information that will become

DOI: 10.1201/9781003163138-65

the reference data and parameters for a given product during the product's life cycle. The emphasis for the validation process should be to develop as much information before, during and after validation because the process is not likely to change during the product's life cycle. A product process is evaluated annually to assess any changes or annual trends that may force the process out of control. This is part of the cGMP annual product review and ISO 9000 annual product review.

With potential limitations on technical, financial and staffing resources placing pressure on the organization and process, successful validation at a small pharmaceutical manufacturing company requires great planning, organization and vision. When the entire company is aware of a new process start-up and all participants are trained in their respective areas of expertise, validation can proceed smoothly and add valuable information, knowledge and processes to an organization.

65.2 VALIDATION PLANNING

The scope of validation work needs to be developed early in the project to help facilitate the writing of a Validation Master Plan. Validation planning allows the project and validation managers to prepare resource and scheduling requirements and ensures that design engineer specifications and detailed design are suitable for validation.

The Validation Master Plan should be designed to encompass all facets of validation activity that the company expects to employ at present and for future validation activities. The Plan should be a structured, detailed record defining all the testing, acceptance criteria and documentation required to satisfy the regulatory authorities and support the validation process. Based on an impact assessment, the plan will also clearly define the scope and extent of the qualification or validation process by listing the matrix of products, processes, equipment, or systems affected.

The Validation Master Plan applies to all facilities, equipment and processes that are subject to requirements of cGMPs. This includes but is not limited to facilities, process utility systems, manufacturing and finishing equipment, analytical equipment, calibration, test equipment and computer-related systems.

The Validation Master Plan assigns responsibilities for developing and executing validation program activities and gives a first look at an anticipated testing execution schedule. There are many variables that must be taken into consideration during the planning process. For example, a small pharmaceutical manufacturing company must determine whether outside analytical testing laboratories will be used because that will usually add significant time to the schedule.

At the inception of a project, it is necessary, and in fact essential, that the project team and project sponsor approve the Validation Master Plan to enable the release of sufficient financial and staffing resources to support the entire project.

The Validation Master Plan should include the various technical support personnel within the company who will have direct responsibility for facets of the Validation Master Plan. By means of a GMP audit, for example, early involvement by Quality Assurance (QA) should provide clear communication of regulatory requirements, ensuring that effective procedures and practices are established up front for incorporation into the project. Because validation activities assess the critical aspects of a given manufacturing process, the development department, from bench, pilot and scale-up, should be focused on a successful process transfer. At the start of process development, the focus should be on the commercial scale-up process. This will minimize the potential for problems during the technology transfer and manufacturing of scaled-up engineering batches.

A Validation Master Plan could include some of the items listed following:

Building	design and construction
HVAC	design and installation qualifications
Process Water	design, installation qualification
Utilities	electricity, gases, steam, refrigeration, design, installation qualifications
Process Equipment	design, construction, installation, operational qualification
Laboratory	analytical and microbial validation methods
Product Process	validation

The key to successful project implementation is a well-defined project scope that enables the validation team to determine the degree of effort and level of resources required, enabling them to focus on its defined responsibilities. It is the function of the facility, equipment, or utility that determines what level of commissioning and qualification are needed. Developing the project commissioning and validation scope is normally accomplished by conducting a risk analysis or impact assessment, whereby the impact of a system on product quality is evaluated, and the critical components within those systems are identified.

These are some of the critical areas that need to be considered when writing a Validation Plan:

<u>Installation Qualification</u> (IQ)—The documented verification that an equipment/system installation adheres to approved specifications and achieves design criteria. IQ documentation and protocols are developed from process and instrumentation diagrams, electrical drawings, piping drawings, purchase specifications, purchase orders, instrument lists, engineering specifications, equipment operating manuals and other necessary documentation.

<u>Operation Qualification</u> (OQ)—The documented verification that the equipment/system performs per design criteria over all defined operating ranges. Systems and equipment must function reliably under conditions approximating those of normal use. Draft SOPs must be prepared for the operation of each system and piece of equipment, if applicable. Those

procedures are to be finished and formally approved after completion of the Performance Qualification (PQ) evaluation of each system.

Process Performance Qualification—The purpose of performance qualification is to provide testing to demonstrate the effectiveness and reproducibility of the equipment, system or process. In entering the performance qualification phase, it is understood that the equipment has been judged acceptable on the basis of suitable installation and operational studies. Critical operating parameters must be independently measured and documented in each trial. Equipment, systems or processes should perform as intended, with expected yields, volumes and flow rates as described in appropriate SOPs. Components, materials and products processed by each system or piece of equipment should conform to appropriate specifications.

Product Performance Qualification—Establishing confidence through appropriate testing that the finished product produced by a specified process meets all release requirements for functionality and safety.

Prospective Validation—Validation conducted prior to the distribution of either a new product or a product made under a revised manufacturing process, where the revisions may affect the product's characteristics.

Retrospective Validation—Validation of a process for a product already in distribution based upon accumulated production, testing and control data. Technically, there is no such thing as re-validation because it always involves a current process. Retrospective validation provides an opportunity to verify that the process remains in control and on target.

Validation—Establishing documented evidence that provides a high degree of assurance that a specific process will consistently produce a product meeting its predetermined specifications and quality attributes.

Validation Protocol/Plan—A written plan stating how validation will be conducted, including test parameters, product characteristics, production equipment and decision points on what constitutes acceptable test results.

Worst Case—A set of conditions encompassing upper and lower processing limits and circumstances, including those within SOPs, which pose the greatest chance of process or product failure when compared with ideal conditions. Such conditions do not necessarily induce product or process failure.

Process Validation—Establishing documented evidence that provides a high degree of assurance that a specific process will consistently produce a product meeting its predetermined specifications and quality characteristics. Process Validation will include acceptable release testing of not less than three batches that meet the processing limits for all critical parameters.

Analytical Method Development Validation—Demonstrating that the analytical procedure is suitable for its intended purpose. A tabular summation of the characteristics applicable to identification, control of impurities and assay procedures is included.

Cleaning Validation—Ensuring cleaning effectiveness through a cleaning validation program that includes initial cleaning of new equipment and post-batch cleaning. Cleaning methods are developed and qualified to show that residuals or by-products from manufacturing and cleaning have been removed. Swab and rinse samples are collected from points identified in the cleaning validation protocols and analyzed using a qualified method. Validation of post-batch cleaning procedures includes acceptable results from not less than three batches.

65.3 VALIDATION TEAM

Management of the validation process is key to controlling the cost and time of validation. Pharmaceutical companies typically require considerable resources in terms of time, money and personnel to validate. In a small pharmaceutical company, a critical part of managing a validation project is the selection of personnel from within the organization to participate in preparing and executing the Validation Master Plan. Therefore, fundamental project management principles should be considered, with the primary objective to identify a project manager. It is essential that this individual have strong leadership skills and be capable of directing and motivating others. This individual must have a good understanding of cGMPs, pharmaceutical manufacturing processes and good communication skills in order to interact with the various team members and departments within the organization. The project manager will constantly be challenged by monitoring performance, meeting deadlines, costs, scheduling and rescheduling various activities and will need to outline the project activities with anticipated time lines in order for the project to proceed efficiently. Delays, communication problems and poor coordination of activities are just some of the problems that may be encountered.

The Project Team should be structured appropriately, and the roles and responsibilities clearly defined. Team members should be knowledgeable about validation with particular emphasis on the areas that they represent. The educational backgrounds of personnel involved with validation work are varied and may range from pharmacists, chemists and microbiologists, to chemical engineers, process engineers and others. The need for employees with diversified backgrounds is understandable. However, the validation group's responsibilities require a complete understanding of technical equipment, equipment controls, electronics, laboratory instrumentation and testing and product sampling and testing. Team members will have to balance daily activities with new added validation responsibilities.

Some of the departments involved in validation and their responsibilities are as follows:

Research and Development—Responsible for formulation development activities that include formulation ingredients listing and concentrations; process optimization, equipment types and facility requirements; and raw material and packaging component specifications, as well as product specifications.

Regulatory—Responsible for assessing the regulatory requirements to implement a new process. Typically, validation activities are required because of a new product under regulatory review by a regulatory organization. It will be necessary to interpret global regulations and standards to obtain global marketing authorization.

Quality Control/Analytical Laboratory—Responsible for preparation of SOPs related to testing of raw materials, in process samples, bulk drug product, finished drug product, cleaning validation samples, product process validation samples and stability studies.

Engineering—Responsible for participating in the design and installation of a new facility and/or equipment; preparation of SOPs for maintenance and set-up of equipment once the equipment is qualified and the process validated; and providing technical support for post validation activities.

Logistics/Material Control—Responsible for ordering materials used for the manufacturing of pre-validation and validation batches. Preparation of SOPs for purchase specifications, identifying and maintaining supplies, profiles and evaluating their performance during the product life cycle.

Manufacturing—Responsible for the design of the facility and equipment required for the manufacturing of the product to be validated; works closely with R and D during the development and optimization of the manufacturing process; and is responsible for all SOPs related to the manufacturing process.

Quality Assurance—Responsible for review and approval of all SOPs required for all activities from installation qualification through process validation, as well as cGMP auditing of all activities related to the entire project including facilities, equipment, analytical, manufacturing and validation; approves the validation report, ensuring that the validation process meets its intended criteria.

At times, it may be necessary for a small pharmaceutical company to seek outside resources because of technical expertise limitations and/or financial reasons. Finding the correct resources outside the company will in some instances prevent problems and delays. It may not be as simple as identifying and engaging the services of a single consultant, but rather engaging a consulting firm with varied staff that can support manufacturing, cGMP auditing, documentation and validation writing. In other circumstances, it may be more prudent to bring in a validation expert consultant to direct the team and delegate responsibilities within the staff. In either case, the key is for management to maintain control of the entire process and ensure that outside resources are complementary and accountable to the head of the project team.

Identifying and selecting consultants adds value as well as time to the validation process. If possible, a small manufacturing company should strive, as a regular course of business, to regularly network and become knowledgeable about the available expertise in the marketplace. Developing and maintaining industry contacts can save significant amounts of time when an outside resource is needed to supplement those already within the company.

65.4 DOCUMENTATION

The documentation required during validation organization is paramount to the success of the validation plan. The types of documents required range from qualification to process validation and include analytical testing documents and standard manufacturing and packaging documents. All of this information is requisite to the execution of the validation plan; any void will result in delays with poor integration of data. Typical documentation for the qualification (IQ/OQ) of a facility or building might include protocols that define the test procedures, documentation, references and acceptance criteria that will establish that the facility has met intended qualification.

In order to streamline the validation process, the validation team will need to perform a gap analysis to determine the required documents. Technical information should become available to the team as detailed design proceeds. This enables the team to begin developing a schedule of activities, staffing schedules, validation protocols, sampling plans, test plans and training materials.

Approaches to streamline the amount of paperwork required to give sufficient documented evidence of validation could include:

- Standardizing protocols and report templates wherever possible, so that reviewers become used to protocol formats and contents.
- Structuring executed protocols as reports to prevent the need for writing a separate report.
- Combining IQ and OQ documents, resulting in fewer documents to develop, review and approve.
- Establishing validation acceptance criteria based upon process capabilities and thereby meeting product quality standards. (Establishing unrealistic acceptance criteria will often lead to increased work loads and cost overruns.)
- Documenting all deviations. Attempt to determine assignable cause with a well-defined plan for corrective action.

- Ensuring that commissioning documentation for process systems are planned, structured, organized and implemented so that they may become an integral part of the qualification support documentation.

Examples of qualification (IQ/OQ) documents required:

- Building Installation
- Building Utilities—electrical, plumbing
- HVAC
- Compressed Air
- Utility Piping
- Process Piping
- Filling Equipment
- Packaging Equipment
- Process Equipment
- Analytical Instrumentation

Examples of process validation documents required:

- Standard Compounding Instructions
- Standard Packaging Instructions
- In-Process Testing Documents
- Finished Product Testing Documents
- Cleaning Procedures
- Cleaning Validation Protocols
- Analytical Testing Documents
- Sampling Protocols

65.5 VALIDATION IMPLEMENTATION AND EXECUTION

In order to meet the intended objectives of a successful validation plan, scheduling for validation is critical and offers a significant challenge to the project manager. Because many departments of a small pharmaceutical company are involved with the validation plan, the project manager must prepare and organize the activities well in advance so that adequate time is allocated to meet milestone targets. The project manager will need to develop integrated schedules with direct input from team members to ensure everyone remains committed to meeting the overall time line. Leadership becomes a very important aspect of project management during the implementation period. Clear, effective and unwavering direction is required for successful validation.

There is constant change during the project life cycle, especially if it involves construction and/or new equipment purchases. The project manager will need to identify, track and coordinate the changes. It may be necessary to establish a strategic meeting schedule to discuss such changes with the validation team. This will undoubtedly lead to changes in the master time line and possible delays, if the project manager has not added extra time to the schedule in anticipation of such delays. Of course, it is impossible to predict where the delays might occur, but good planning before initiating activities should minimize the downtimes.

It is recommended that all systems go through a shakedown or debugging phase before beginning qualification activities. This should improve the efficiency of transitioning from IQ to OQ activities and will help to reduce the number of changes required during the qualification phase. Typically the last phase of qualification, a performance qualification is usually the part of the validation program where product is produced on a large scale before engineering and commercial validation batches are produced. Because of the importance to the overall plan, the project manager should allocate sufficient time to the qualification activities which include, but are not limited to, equipment setup and maintenance, equipment outputs, equipment cleaning and personnel training.

As previously mentioned, it is important to assess internal resources at the beginning of a project to schedule activities appropriately, especially in terms of analytical testing. A small firm may be overwhelmed by the amount of test samples and commitment time required to analyze samples. Contracting with an outside analytical testing laboratory to back up the primary analytical laboratory will reduce delays with respect to validation testing and overall time lines, but can add up to three to four months to the schedule, for the most part related to methods transfer. This will only pose a problem if it is not accounted for early in the planning and scheduling process. At the same time, validation costs will increase because it will be necessary to transfer the methods to an outside laboratory before validation test samples can be analyzed. The transfer procedure can be performed early in the project cycle after the methods have been validated, so that the contract laboratory will be ready when the validation project begins. If the manufacturer does not have an analytical department, it may be cost effective to utilize the laboratory that developed and validated the methods to reduce redundant activities and added delays by searching for a new laboratory. In either case, the project manager will need to ensure that time lines accommodate a need for external resources.

65.6 CONCLUSION

As stated in the introduction, validation principles are a tool that, if applied properly, will result in a significant amount of scientific data for a given manufacturing process. A validation program should be a baseline for the industrial pharmaceutical scientist to use in tracking the process output throughout a product's life cycle. For small pharmaceutical companies without the resources available at large companies, a well-organized validation plan is essential to a smooth, cost-effective process. The focus of the validation project manager should be to:

1. Prepare and define the overall validation activities for both management and the validation team members.
2. Structure the activities in order to integrate them into the overall organization without disrupting daily operations.

3. Identify the most competent and team-oriented individuals within the organization and make them part of the validation team.
4. Complete the project on time and within budget.

These four steps will ensure the validation program not only is successful but becomes part of the company's standard routine. At the same time, it is critical that the resources and responsibilities for implementing the program be committed to an individual who can oversee, manage, schedule, coordinate, communicate and interact with a group of professionals from both within the company and without. The skills associated with this are not necessarily technical, but rather business savvy and leadership skills, allowing oversight and management of both the financial and technical resources for a given project. The ultimate responsibility as far as the regulatory agencies are concerned remains with the company—whether large or small—so it is essential that control of the business is maintained at all times.

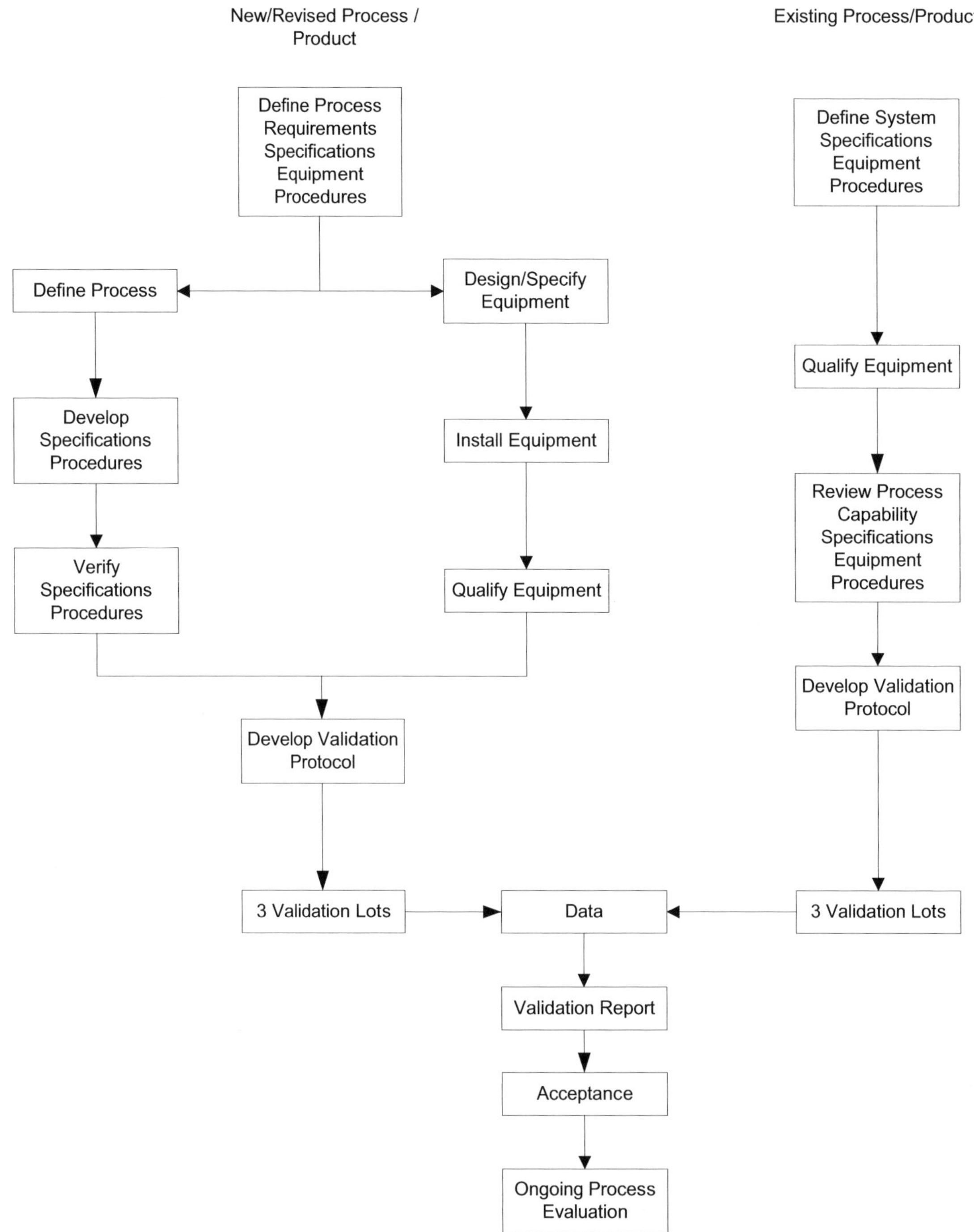

FIGURE 65.1 Validation process schematic

REFERENCES

1. *Journal of Validation Technology* 2004 August, 10(4).
2. *Journal of Validation Technology* 2004 February, 10(2).
3. Draft Guidance for Industry. *Analytical Procedures and Methods Validation.* US Department of Health and Human Services Food and Drug Administration, Center for Drug Evaluation and Research, Chemistry, Manufacturing and Controls Documentation, August 2000.
4. Guidance for Industry. *PAT: A Framework for Innovative Pharmaceutical Development, Manufacturing and Quality Assurance.* US Department of Health and Human Services Food and Drug Administration, Center for Drug Evaluation and Research, Pharmaceutical CGMPs, September 2004.
5. Draft Guidance for Industry. *Quality Systems Approach to Pharmaceutical Current Good Manufacturing Practice Regulations.* US Department of Health and Human Services Food and Drug Administration, Center for Drug Evaluation and Research, Pharmaceutical CGMPs, September 2004.
6. Guidance for Industry. *General Principles of Software Validation: Final Guidance for Industry and FDA Staff.* US Department of Health and Human Services Food and Drug Administration, Center for Drug Evaluation and Research, January 11, 2002.
7. Guideline on General Principles of Process Validation. US Department of Health and Human Services Center for Drug Evaluation and Research, May 1987.
8. Guide to Inspections of Validation of Cleaning Processes. US Food & Drug Administration, Office of Regulatory Affairs, Office of Regional Operation, The Division of Field Investigations, July 1993.
9. Zaret E PhD. The Value of Validation. *Pharmaceutical Formulation & Quality* 2002 June/July, 47–48.
10. Kropp M MD. Pharma's Continuing Validation Challenge: Remediation for 21 CFR Part 11 Compliance. *American Pharmaceutical Review* 2004, 7(3), 10–12.
11. Wrigley G, du Preez J PhD. Facility Validation: A Case Study for Integrating and Streamlining the Validation Approach to Reduce Project Resources. *Journal of Validation Technology* 2002, 8(2), 1–22.
12. ISPE. *GAMP Good Practice Guide: Validation of Process Control Systems*, 2003.
13. Carleton F, Agallaco J, eds. *Validation of Pharmaceutical Processes Sterile Products.* 2nd ed. New York, NY: Marcel Dekker, Inc., 1999.
14. Loftus B, Nash R, eds. *Pharmaceutical Process Validation.* Vol. 23. New York, NY: Marcel Dekker, Inc., 1984.

66 Regulatory Aspects of Process Validation in the United States

James Agalloco and Robert Mello

CONTENTS

66.1 INTRODUCTION

The third edition of this book chapter on Regulatory Aspects of Validation was published in 2007. Although many of the regulatory documents and concepts provided in that chapter remain valid today, there have been some substantial changes to the regulatory perspective on process validation. This chapter updates the explicit and implied regulatory concepts of pharmaceutical process validation within the United States in 2021.

First, before exploring what has changed, it is important to understand those concepts that have not been altered since 2007. This requires a look at preexisting laws, regulations and industry guidance documents issued prior to 2007 that are still relevant today.

Second, the most significant change has been the shift away from the initial practice of executing traditional Installation Qualification (IQ)/Operational Qualification (OQ)/Performance Qualification (PQ) data collection sets for equipment and processes in a linear manner. This has largely been superseded by a life cycle approach following that of the typical product as defined in the United States Food and Drug Administration's (FDA's) 2011 revision to the Guidance on Process Validation and mirrored in European Medicines Agency's (EMA's) largely similar document issued in 2012.

Another important change, seemingly minor to some, was made in 2008 with a change to 21 CFR 211.113(b), which included an explicit expectation that validation of aseptic processing could be fully realized. An increasing emphasis has been evident in the last decade, which has formalized expectations for the evaluation of risk requiring assessment and mitigation. These areas will be explored in some detail.

Lastly, not everything associated with the regulation of validation appears in formal documents. Practices across the industry are swayed by podium presentations, webinars, non-regulatory papers, interviews and blogs. In an industry in which compliance concerns are considered paramount, scientific uncertainties can be introduced that induce fear of delay, or outright rejection of a product or practice. These informal positions have impacted validation execution in a variety of ways.

66.2 WHAT HAS NOT CHANGED

What has not changed is the need for manufacturers to demonstrate to regulators that a product is fit for its intended use. Process validation accomplishes this. What has not changed are the core expectations of the global regulatory agencies: FDA, EMA, ANVISA, TGA, Health Canada, etc. The US Current Good Manufacturing Practices Regulations (CGMPs) are relevant to pharmaceutical, biopharmaceutical, medical device and human cell/tissue products. The EMA is gradually assuming overall regulatory control over medical devices, which had previously been more loosely coordinated under prior regulations.

The US CGMP requirement for process validation is stated in various sections of the Code of Federal Regulations (CFRs) and one of the more succinct statements is shown here:

> **§ 1271.230 Process validation**
>
> **(a) General.** Where the results of processing described in § 1271.220 cannot be fully verified by subsequent inspection and tests, you must validate and approve the process according to established procedures. The validation activities and results must be documented, including the date and signature of the individual(s) approving the validation.[1]

Although this citation makes the overall expectations for validation explicit, it does not provide any guidance to explain the specifics of which "validation" should consist. When first introduced in the US in the mid-1970s, a common, albeit informal, definition of validation was:

> Validation is the attaining and documentation of sufficient evidence to give reasonable assurance, given the state of science, that the process under consideration does, and/or will do, what it purports to do.[2]

DOI: 10.1201/9781003163138-66

The FDA issued its first guidance on process validation in 1987, and this outlined their core expectations at that time and included the first truly official regulatory definition:

> Process validation is a documented program which provides a high degree of assurance that a specific process will consistently produce a product meeting its pre-determined specifications and quality attributes.[3]

Some interpretation of the definition is necessary, expanding the concept of process to include aspects and items that, although not products, are precursors or directly supportive activities used for the manufacture of products such as: utilities, procedures, systems, analytical methods, etc. With this adaptation, the principles and scope of effort were clarified.

In Europe, where initial efforts with respect to "validation" began at approximately the same time period as those in the US, the terms "commissioning and qualification" where more commonly used (see the chapter on European Perspectives on Validation for expanded discussion on this subject). Multinational firms led the way in order to rationalize their global validation efforts. Thus, there was growing commonality in practice across the globe. An early definition of validation from Europe paralleled much of what the FDA described in 1987.

> Process Validation—the documented evidence that the process, operated within established parameters, can perform effectively and reproducibly to produce a medicinal product meeting its predetermined specifications and quality attributes.[4]

In the early 1990s, it was increasingly evident that improvements in drug quality and availability could be realized by improved coordination among the major regulatory agencies and pharmacopeia. The International Conference on Harmonization (ICH) composed of FDA, EMA, and the Japanese Ministry of Health, Labour and Welfare along with the United States Pharmacopeia, European Pharmacopoeia and Japanese Pharmacopoeia began the development of guidance documents that could simplify drug development, manufacturing and control practices globally. One of the many definitions found among ICH guidances states the following:

> Validation: A documented program that provides a high degree of assurance that a specific process, method, or system will consistently produce a result meeting predetermined acceptance criteria.[5]

There are many other definitions of validation in the literature, all of which are derived from the scientific principles that many of us were introduced to in elementary chemistry and/or physics courses. The scientific principles follow a simple sequential process:

- A stated premise (*e.g.*, the process is effective for microbial destruction).
- An experimental plan (*e.g.*, assess microbial destruction with measurements of temperature and destruction of resistant biological indicators).
- A written conclusion (*e.g.*, the temperatures exceeded the time-temperature required and all biological indicators were killed).

The validation protocol, its execution and summary report follow the scientific method and provide documentation confirming the premise. Activities surrounding these core activities constitute what was largely understood until 2011 as "validation". When validation activities commenced in the 1970s, the means for its execution varied substantially and practices were more *ad hoc* than what might be commonplace in a CGMP environment. The backlog of non-validated products and systems precluded any attempt to structure validation practice for the long-term. Activities and practices were singular, often unique to a particular firm and process. Beyond general adherence to the overall scientific method, there was little consensus on validation methodology. Although in the late 1980s, some practitioners in the US were prescient in their thinking and adopted a "life cycle" model for their firm's validation practices. That practice, however, was more the exception than standard practice (see the chapter on Evolution of Validation for additional content on this subject).

66.3 WHAT HAS CHANGED

In late 2007, the FDA proposed numerous revisions to 21 CFR 211 including one that exposed a significant disconnect between regulatory and industry perspectives on validation. Section 21 CFR 211.113(b) was revised to state:

> Appropriate written procedures, designed to prevent microbiological contamination of drug products purporting to be sterile, shall be established and followed. Such procedures shall include validation of all aseptic and sterilization processes.[6]

The objections raised by industry were related to the extent to which aseptic processing can be validated. The reliance on multiple variables, absence of accurate means for real-time monitoring and periodic (i.e., snapshot) nature of process simulations were all cited as making expectations of "validation" unfounded. Nevertheless, the proposed changes became official in 2008, but the opposition revealed the first schism between the FDA and the regulated industry on the means by which "validation" could be realized.

The ability to demonstrate that production processes, equipment and assays are performing as expected in order to produce a quality product is a universal regulatory expectation not only at the time of a submitted marketing application but over the entire life cycle of the product. The scientific method, however, was never intended to address maintenance of a validation condition over time. Despite this having been recognized by some practitioners as early as the late 1980s, there was no articulated regulatory position on the utility of the life cycle approach for validation. This changed in 2011, when the FDA revised its 1987 Process Validation Guideline. The 2011 Guidance for Industry—Process Validation: General Principles and Practices proposed a 3-stage life cycle method (Figure 66.1).[7]

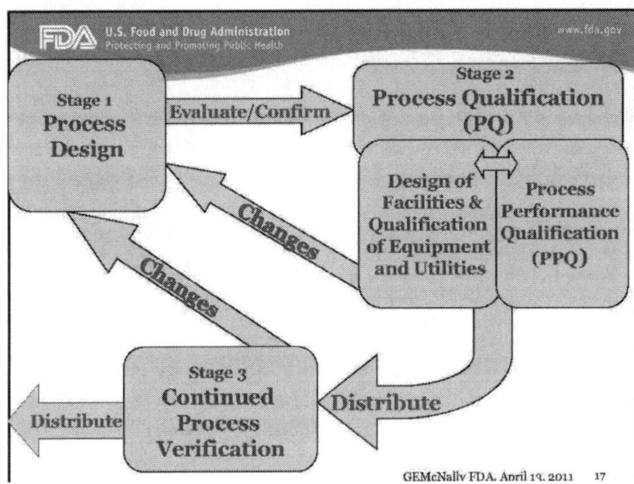

The multi-stage life cycle approach divides validation activities into distinct elements, each with a differing focus:

- Stage 1—Process Design—The development of a robust process typically requires significant experimentation to research and select the required controls necessary for reproducible results. This effort is the province of developmental scientists and engineers who are charged with gathering information on the relationship between material and process parameters and the desired quality attributes of the finished product. Key to this effort is organization of the experimental data gathered to provide a knowledge base on the product/process. The developmental effort may be performed at different scales over a considerable period of time.
- Stage 2—Process Qualification—This activity embodies the core aspects of the scientific method in which reproducible performance of the commercial scale process/product is confirmed through rigorous scientific assessment. This confirmation can be derived from independent assessment of parameters, expanded sampling regimens and other measures. Statistical tools can prove useful in establishing process robustness. This activity is often used as a part of technology transfer from the developmental organization to those responsible for commercial scale production. Many firms execute this stage in conjunction with product launch.
- Stage 3—Continued Process Verification—The longest period in the life cycle is Stage 3, which encompasses the entire production life of the product from introduction to removal. As such, it may be the most important stage as it directly supports the quality of products in the marketplace. It confirms that the output quality is unchanged over time. Few processes/products remain constant over their life. In nearly all instances, there are inevitable changes in materials, equipment, vendors, test methods, personnel and more. The passage of time and deterioration in equipment performance as a consequence of use can adversely influence performance. Even in the absence of identifiable changes whose impact can be prepared for and assessed prospectively, there may be unknown alterations that can impact the output. Continued process verification is typically composed of two distinct elements: comprehensive change control and trending of performance over time.

Counterintuitively, when the FDA Guidance was first introduced in 2008, it was not always recognized that the initial emphasis should be placed on Stage 3 as this assessed marketed product and potentially impacted current patients. Implementation of Stage 1 and 2 practices has little immediate effect on the patient, and the impact of these efforts can only be recognized later in the product's life cycle. The guidance has limitations as a universal practice when applied in health care settings that fall outside the product-centric guidance.[9] Areas where the guidance becomes a sub-optimal fit include:

- Utility and environmental systems whose design/development differs markedly from pharmaceutical products and lacks development's experimentally based evidence.
- Processes that are heavily reliant on personnel expertise/manipulation (such as manual cleaning/inspection) and that lack the precise controls associated with pharmaceutical, chemical and biological unit operations.
- Commonly used processes such as cleaning and sterilization where cycle development practices are poorly documented, or only indirectly define the operating conditions.
- Low-volume products and processes that lack extensive development and whose infrequent production reduces the utility of process trending.
- Aseptic processes where in-process and end-product testing is unable to provide ***definitive*** evidence of performance.

Despite these exceptions, the guidance has been well received as it supports an overall program for maintaining products in a validated state.

Once the FDA guidance was finalized, the EMA issued a similar guidance document of its own.[10] The differences between these are relatively minor allowing firms producing products for both markets to adopt a singular approach capable of satisfying both regulatory agencies.

A growing concern globally is the regulatory agencies increased usage of less definitive means of communicating their expectation using guidances and annexes. Both FDA (Guidances for Industry) and EMA (Annexes to CGMPs) employ these as a means of defining their expectations. These documents are binding on the regulator in that they agree to accept practices that conform to their recommendations but require those who chose alternative practices to seek regulatory approval before acceptance. These documents do not always adhere to the regulatory "what to" expectations, but frequently delve into the "how to" that has conventionally been the province of the manufacturer.[11] By defining regulatory acceptable "how to" they also, perhaps unknowingly, result in the restriction of innovation by making variance from them subject to added review and potential delay. The tension created as a consequence of added caution on the part of regulators places added burdens on validation personnel to develop sufficient supportive rationale.

Regulatory guidances on aseptic processing from the FDA[12] and EMA[13] include numerous criteria and expectations that challenge the underlying science and reason. These documents, although intended to define suitable practices, at times mandate results and conditions that are contrary to established science. Extreme examples include statements such as:

- A single anomalous result requires investigation and/or remediation.
- Air monitoring samples of critical areas should normally yield no microbiological contaminants (these are not, after all, completely "sterile" environments, especially if people have access to the area).
- It should be noted that for Grade A, the expected result should be no growth (again, not a sterile environment).
- Monitoring should include sampling of personnel at periodic intervals during the process (such actions create additional risk for microbial contamination, not less).

Most importantly, the existing and draft standards are derived from and designed for practices associated with manned clean rooms in which personnel intervention within the critical aseptic environment is required. In lieu of adaptation or any formal consideration of advanced technologies employing isolators, closed systems and single use technologies, regulators have remained static, expecting such systems, designed in the present-day environment, to match the vastly different technological norms derived from the 1970s. It should be evident that "how to" GUIDANCES don't always have positive outcomes for product quality and patient safety.

66.4 NON-REGULATORY GUIDANCE ACTIVITIES

Although regulations establish the overall requirements for validation, its contemporary execution is described more fully in different forms: regulatory guidances, pharmacopeia, trade organizations and publications from individual authors. As these incorporate a substantial amount of technical detail, they provide meaningful insight into contemporary validation practice in the United States.

For example, the United States Pharmacopeia began a major transition to its content in 2005, when the Pharmacopeial Convention leadership charged its expert committees with developing content that was grounded in scientific principles and more instructional than the existing content. This process is continuing more than a decade later and USP's changing content challenges regulators and stakeholders in a variety of ways.

The most significant USP changes relating to validation are those addressing microbial contamination in sterile and non-sterile products. USP's 1980 era content on sterilization has been dramatically revised and greatly expanded. The new chapter <1229> *Sterilization* included a significant change in sterility assurance expectations by identifying the pre-sterilization bioburden as the real target of all sterilization processes and thus affording greater flexibility in processing.[14] The subchapters included under this chapter provide content on sterilization methods and supportive practices. Discussions on the utility of environmental control practices for non-sterile products resulted in an innovative chapter <1115> *Bioburden Control of Nonsterile Drug Substances and Products.* This chapter asserted the futility of environmental monitoring as an effective means of control over microbial contamination in non-sterile materials and highlighted the importance of facility and equipment design, process and cleaning water quality, cleaning and operating procedures. The many limitations of microbial sampling as a means of assuring microbial absence are increasingly acknowledged.[15] This has prompted potential revisions to USP <1111> *Microbiological Examination of Nonsterile Products: Acceptance Criteria for Pharmaceutical Preparations and Substances for Pharmaceutical Use,* which would potentially remove expectations for absence of certain microbial strains.[16] The notion that the complete absence of any microbial strain could be possible in a non-sterile product is simply illogical.[17]

At USP, the changing perspectives on microbial control for non-sterile products prompted more consequential changes relating to sterile products given the near identical limitations of microbial sampling and testing. The increasing inadequacy of environmental monitoring, process simulations and sterility testing for the assessment of contamination control performance within newer, more advanced technologies created a need to reassess USP's relevant sterility assurance content. Revised chapter <1211> *Sterility Assurance* drew upon the rationale behind the <1115> revision with a similar approach that identified ten elements (Figure 66.2) that support the control over microorganisms and reduced the relative importance of microbial sampling.[18]

The combination of supportive practices and corresponding reduction in the longstanding emphasis on microbial sampling has been broadly termed "Sterility by Design".[19] This represents a dramatic shift in thinking and has the potential to shift the means for validation for the manufacture of sterile products.

Additional difficulties to the validation, compliance, production and quality control of pharmaceuticals are frequently

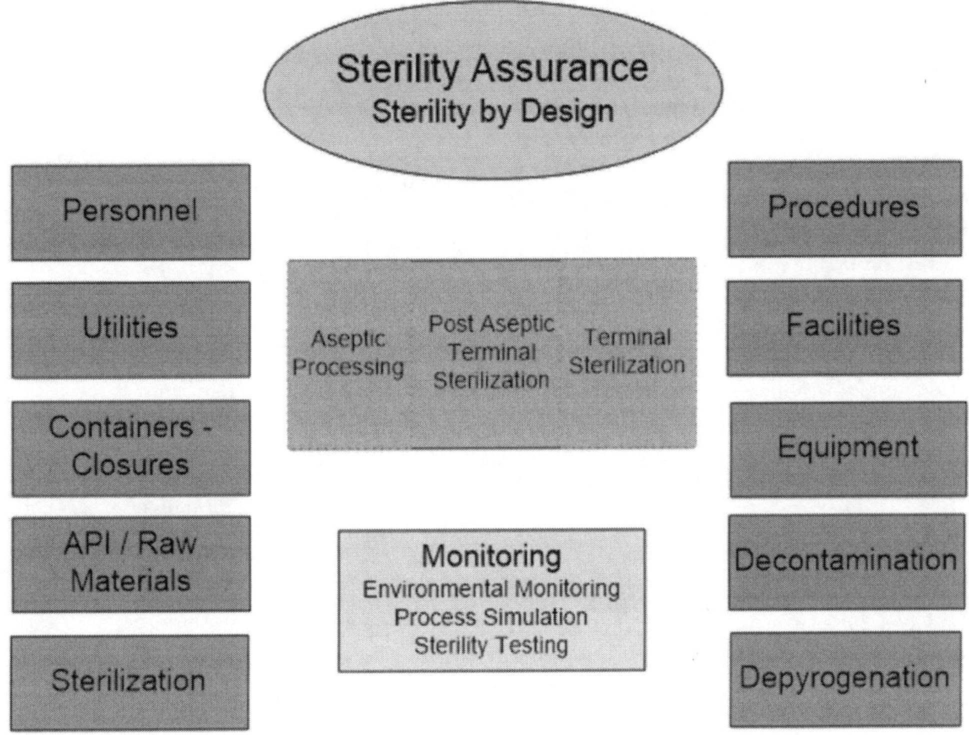

FIGURE 66.2 USP <1211> *Sterility Assurance* supportive elements.

Note: API = active pharmaceutical ingredient; USP = United States Pharmacopeia.

introduced by what could be called "pseudo-regulatory" positions on pharmaceutical operations. Examples include:

- Inspection observations—content from FDA inspections is often made available under Freedom of Information and can be taken out of context. Observations are the considered opinion of the inspector and may exceed regulatory requirements. Some observations are wholly subjective (e.g., adequacy of smoke studies) and none should be used as other than suggestions to follow.
- Podium positions—regulatory representatives make frequent presentations at conferences and meetings. Unless the statements made cite specific regulations, they should not be considered as anything beyond opinion. The recent difficulties with Low Endotoxin Recovery (perhaps more correctly termed Low Endotoxin Standard Spike Recovery) are but one example.
- Blogs, e-mails and other communications—these have no official standing and represent the author's opinion and should not replace regulations. For example, a MHRA blog suggesting inadequacies in Hydrogen Peroxide sterilization practices caused unwarranted consternation across the industry in 2019.
- Informal comments—these are hardest to document and the least consequential of all. A prominent example, "you can't validate a bad practice", implies that even robust validation data based on sound scientific principles can be considered questionable if it is contrary to a preconceived regulatory concept.

66.5 CONCLUSION

We have attempted to point out those validation concepts that have remained constant since the 2007 edition of this book. Because time and technology progress forward, we have also attempted to update the understanding of various process validation guidance documents as they exist in today's medical product manufacturing environment. We should always be aware that change is constant in the global regulatory and pharmacopeial world.

However, we have also attempted to provide distinction and, hopefully, some clarity, as to the manner in which some interpretations of the "how to" of process validation has drifted away from the technical, data-driven industry experts and more towards the theoretical aspects of some regulators embodied in concepts based on those well-worn phrases such as "it depends", "it could be . . . .", or "you can't validate a bad practice". Let's allow good science and solid data to prevail in the support of robust process validation practices.

In the last decade, validation activities of all types have become increasingly sophisticated. Although the scientific method has always been at the core of validation efforts, it could be argued that in the early years of validation that the acceptance criteria used in many instances lacked objectivity. Temperature ranges, minimum F0, number of runs and more were based upon arbitrary values, much of which were opinion based. In the initial rush to first implement validation, the science may have been less than robust. Over time, the rigor included in protocols and test methods gradually increased. This was aided by improved analytical methods, better process controls, improved instrumentation, automated analysis and perhaps most importantly a recognition that improved process knowledge was necessary to ensure process reliability. Present-day validation practice is substantially more scientific than previously. There can be little doubt that this trend will continue, and future validation activities will provide a continuing improvement in confidence derived from them. There's no going back, and this bodes well for us all.

NOTES

1. 21 CFR 1271.230
2. Byers, T, Presentation at PDA Annual Meeting, Philadelphia, PA, 1980.
3. FDA, Guidance for Industry, Guideline on General Principles of Process Validation, 1987.
4. EMA, Guideline on process validation for finished products—information and data to be provided in regulatory submissions, EMA/CHMP/CVMP/QWP/BWP/70278/2012-Rev1, Corr.1, November 2016.
5. ICH, Q7A Good Manufacturing Practice Guidance for Active Pharmaceutical Ingredients Guidance for Industry, September 2016.
6. Federal Register Vol. 72, No. 232 / Tuesday, December 4, 2007, pp. 68113–68116
7. FDA, *Guidance for Industry—Process Validation: General Principles and Practices*, January 2011.
8. McNally G, *Process Validation: Lifecycle Approach*, Presentation at PDA Annual Meeting, April 2011.
9. Agalloco J, FDA's Draft Guidance for Process Validation: Can It Be Applied Universally? *Pharmaceutical Technology*, Sterile Manufacturing Supplement, 2009.
10. EMA, Guideline on Process Validation, EMA/CHMP/CVMP/QWP/70278/2012-Rev1, March 2012.
11. Madsen R, Agalloco J, What, Not How, published on-line at LinkedIn.com, April 2020.
12. FDA, Guidance on Sterile Drug Products produced by Aseptic Processing, 2004.
13. EMA, Annex 1, Manufacture of Sterile Medicinal Products, draft revision, 2019.
14. USP 36, S1, <1229> *Sterilization*, 2013.
15. Madsen R, Agalloco J, Unknown and Unknowable, published in *Pharmaceutical Technology Advancing Development and Manufacturing* e-book, 4–9, May 2019.
16. Agalloco J, Singer D, Refining Microbiological Control for Non-Sterile Products. published on-line at pharmtech.com, October 2020 (www.pharmtech.com/view/refining-microbiological-control-for-non-sterile-products).
17. Agalloco J, Non-Sterile Is Non-Sterile: A Reality Check on Microbial Control Expectations. publication in *Bioprocessing and Sterile Manufacturing 2016*, a *Pharmaceutical Technology* eBook, pp 31–35, 2016.
18. USP 41, S2, <1211> *Sterility Assurance* 2017.
19. Agalloco J, Akers J, The Myth Called Sterility. *Pharmaceutical Technology* 2010; 34(3, Supplement): S44–45. Continued online at Pharmtech.com.

67 The Future of Validation

James Agalloco and Phil DeSantis

CONTENTS

67.1 INTRODUCTION

The first mentions of validation appear in the mid-1970s in response to difficulties with terminal sterilization in the US and UK.[1] It was poorly understood at first, and the implications for the health care industry were obscure as a consequence. It was gradually understood to be the acquisition of independent confirming information on process/product quality that went beyond that obtainable from routine release testing. Although the original driver was compliance related, some practitioners came to understand that the rigor provided by its use could provide tangible benefits to product quality, equipment/process/system reliability, and project execution.[2] Originally viewed as an extension of Quality Assurance activities, it has become a discipline of its own, requiring a variety of technical, organizational, and communication skills for success. Although regulators played an important role in its origin, end users and others have largely shaped the practices that are now commonplace. The regulatory preferences for use of a life cycle approach came some 20+ years after it was adopted by industry.[3,4] So what will the practice of validation be like in the future?

As new technologies (i.e., computerized systems, isolators, etc.) were introduced into the industry, validation concepts and practices adapted to address them. Innovation brings a degree of uncertainty that can only be assuaged by the acquisition of supportive evidence that the outcomes are as expected. The exact means to perform these qualification and validation activities cannot be predicted as they must be tailored to the innovation itself. Nevertheless, it is good business and good manufacturing practice to validate technologies as they are introduced. The past successes applying validation to technology advancement bode well for the future.

There are emerging technologies whose implementation is already underway, and the means for their validation will not be universal until a sufficient body of experience has been gathered. Areas where validation concepts are emerging include robotics, single-use systems, closed systems, continuous processes, combination products, personalized medicine, and artificial intelligence. Some of these have already been successfully validated, but there is no consensus on whether the approaches taken will become standard practice. Others present greater challenges and validation methods will take longer to evolve. Nevertheless, experience has shown that over time even the most daunting of tasks succumb to the efforts of science and engineering.

There are other trends emerging in validation practice that can be expected to endure. A prominent example is Quality by Design concepts that are increasingly common for product, process, and facility design. This has near-perfect synergy with the life cycle approach to validation and allows for the incorporation of validation issues at project onset as another component of the Quality by Design exercise. Beginning the project and its associated validation activities at the same time provides a stronger connection throughout the life cycle of the product and facilitates both. Linking the development activities with the ultimate commercial expectation is accomplished more readily when validation considerations are introduced at project onset.

A gradual shift in validation efforts towards the use of external resources has been underway since the late 1980s. The initial providers were often associated with engineering firms and independent providers. More recently, vendors of all types, but predominantly components and equipment manufacturers, have become more involved in validation efforts. Many component suppliers now provide pre-washed and pre-sterilized components, allowing manufacturers to focus on their product and avoid investment in facilities, equipment, and validation activities. The vendors undertake full responsibility for the quality of their now preprocessed components, including the necessary validation efforts. Equipment vendors have pursued a similar strategy and many offer comprehensive qualification and validation services for their products. These efforts shift non-product-related validation efforts from the end user to the supplier. This trend is expected to continue, with the vendors increasing proving validation services that were once the sole province of the drug manufacturer.

The expanding contract development and manufacturing organization (CDMO) business segment has taken on a major role in the validation of processes and products that they develop and manufacture. The transition to this business model requires a shift in practice that can prove challenging. The CDMO and its client must establish a level of communication that ensures complete validation while maintaining both organizations' intellectual property. As CDMO often file with regulatory bodies Type V Drug Master Files that are not accessible to the client, there are potential gaps within filing

DOI: 10.1201/9781003163138-67

and registration documents that can be problematic. As multiple clients typically engage the same CDMO, standardization of qualification and validation methods can become further complicated. The best practice might be the adoption of the CDMO's practices across all clients, although some clients will be hesitant to accept that. As CDMOs will be a large part of the foreseeable future, refinement/standardization of practices across both CDMOs and client firms, and eventually the entire industry, can be anticipated.

Equipment suppliers long exploited their familiarity with their own equipment to provide equipment qualification services to their customers. Their expertise with the processes and products typically manufactured on that equipment has led them to offer performance qualification support as well. Earlier concerns that vendors might be less rigorous in their efforts have dissipated as customers realize that the vendors would not risk their hard-earned reputations if they accepted substandard installations and performance.

The rate of technological change in the health care industry has steadily increased, paralleling what has occurred in the general economy. This evolution has stressed regulators and industry alike, as regulations have always lagged technology. Rules developed for technologies introduced in the 1960s constrain the ability to introduce improvements derived from innovative systems. Validation of novel approaches forces greater emphasis on the underlying science and engineering and less on regulatory dogma.

The growing consideration of risk in health care has altered and been altered by validation. The overblown qualification efforts of the 1980s and 1990s are now recognized as excessive given the limited impact many systems have on the product quality and process robustness. Scientifically sound approaches have enabled simplification of validation efforts facilitating rapid implementation while leveraging scientific and engineering experience. The "one size fits all" approaches that were once so prevalent are now encountered less frequently. Current thinking with respect to validation and risk places emphasis on assuring the product meets its intended quality attributes, which requires comparable attention on the process used to make it.[5] In non-product/process applications, this converts to fixation on achieving the correct result whether it be analytical, numerical, or other metric. With the goal of the effort properly identified, it is easier to establish relative importance (or not) of the systems and actions surrounding it. The priority that should be given in quality confirmation and the systems that support it for products and processes is depicted in Figure 67.1.

Validation has evolved considerably since its inception. Once considered solely a compliance driven activity, it has gradually transitioned into a core activity in all aspects of product/process development and implementation. The notion that it was a burdensome process that served only to assuage inspectors has largely disappeared. For those experienced in its application, it is difficult to imagine alternative means for

FIGURE 67.1 Top-down approach to validation.

Note: API = active pharmaceutical ingredient; CGMP = Current Good Manufacturing Practice

major project implementation that do not adhere to the core practices associated with validation.

67.2 LOOKING BACK TO LOOK FORWARD

At its origin, "validation" (with a small "v") emerged as an activity or an exercise to be completed. At that time even at its loftiest level, it was a project with a beginning (a plan or protocol), a middle (a series of tests), and an end (a report). Shortly thereafter, change control was applied and the need to "re-validate" was introduced. This was still, however, an activity to perform.

Gradually, with various influences and the pioneering of a "life cycle" concept (initially by the authors), the concept of Validation (with a capital "V") gained a foothold. This more scientific approach recognized that Validation must be more than "proof that a process does what it purports to do" (to paraphrase an early definition). It was not only essential to be able to confirm that a process worked but to fully understand "why" it worked. Process development came to be an important factor in validation, wherein critical quality attributes and process parameters were defined. The "life cycle" was completed with the application of change control and, later, continuous process monitoring (under various titles that mean essentially the same thing). In the future, validation will maintain a unique identity and increasingly a foundational role in the achievement of product quality.

67.3 CONCLUSION

There are some clear observations derived from 40+ years of validation experience that can help us anticipate its future:

- It is here to stay—considered a one-time activity initially, it will be a fixture across the health care industry in the foreseeable future.
- It provides tangible benefits—the fixation on compliance that shaped its origins has largely dissipated and there is general acknowledgement that validation is fiscally responsible.
- It relies on a scientific foundation—with origins in the scientific principle, it remains evidence and documentation based.
- It can be a career choice—there are individuals who've spent their entire working career in the validation field, finding it a challenging and rewarding profession.
- It can become a corporate lifestyle—the life cycle approach to validation management has proven effective and enduring.

The last edition of this text ended with a chapter much like this one that endeavored to forecast what validation would be in the future. It was remarkedly prescient given what our industry has witnessed and at the same time is equally suited as a vision of the future from this point forward:

> Over 30 years later, validation practices have evolved to suit the changing environment in which it operates, and it should be evident that it will be able to accommodate future changes as well. That new factors influencing validation will continue to emerge is near certain. What is equally certain is that validation will adapt to work within that new environment as it has in years past. "The only constant is change".[6]

NOTES

1. Chapman K, A History of Validation in the United States—Part I. In *Pharmaceutical Technology*, Vol. 15, No.10, 1991.
2. Agalloco J, The Other Side of Process Validation. *Journal of Parenteral Science and Technology* 1986;40(6): 251–252.
3. FDA, *Guidance for Industry*—Process Validation: General Principles and Practices, 2011.
4. EMA, Concept Paper on the Revision of the Guideline on Process Validation, EMA/CHMP/CVMP/QWP/809114/2009, 2010.
5. Agalloco J, Compliance Risk Management: Using a Top Down Validation Approach. *Pharmaceutical Technology* 2008;32(7): 70–78.
6. Herodotus, *The Histories*, Circa. 430 BC.

Index

Note: Page numbers in italics indicate figures; page numbers in bold indicate tables.

D

E

F

G

H

I

N

Q

T

W

X

Z